MODERN
ELECTRONIC
CIRCUITS
REFERENCE
MANUAL

BOOKS by JOHN MARKUS

ELECTRONIC CIRCUITS MANUAL

ELECTRONICS DICTIONARY

ELECTRONICS FOR COMMUNICATION ENGINEERS

ELECTRONICS FOR ENGINEERS

ELECTRONICS MANUAL FOR RADIO ENGINEERS

ELECTRONICS AND NUCLEONICS DICTIONARY

GUIDEBOOK OF ELECTRONIC CIRCUITS

HANDBOOK OF ELECTRONIC CONTROL CIRCUITS

HANDBOOK OF INDUSTRIAL ELECTRONIC CIRCUITS

HANDBOOK OF INDUSTRIAL ELECTRONIC CONTROL CIRCUITS

HOW TO GET AHEAD IN THE TELEVISION AND
RADIO SERVICING BUSINESS

MODERN ELECTRONIC CIRCUITS REFERENCE MANUAL

SOURCEBOOK OF ELECTRONIC CIRCUITS

TELEVISION AND RADIO REPAIRING

WHAT ELECTRONICS DOES

MODERN ELECTRONIC CIRCUITS REFERENCE MANUAL

Over 3,630 modern electronic circuits, each complete with values of all parts and performance details, organized in 103 logical chapters for quick reference and convenient browsing

JOHN MARKUS

Consultant, McGraw-Hill Book Company
Senior Member, Institute of Electrical and Electronics Engineers

McGRAW-HILL BOOK COMPANY

New York St. Louis San Francisco Auckland Bogotá Hamburg
Johannesburg London Madrid Mexico Montreal New Delhi
Panama Paris São Paulo Singapore Sydney Tokyo Toronto

Library of Congress Cataloging in Publication Data

Markus, John, date.
   Modern electronic circuits reference manual.

   "Over 3630 modern electronic circuits, each complete
with values of all parts and performance details,
organized in 103 logical chapters for quick reference
and convenient browsing."
   Includes bibliographical references and indexes.
   1. Electronic circuits—Handbooks, manuals, etc.
2. Integrated circuits—Handbooks, manuals, etc.
I. Title.
TK7867.M345   621.3815′3   79-22096
ISBN 0-07-040446-1

   67890 KPKP 898765

The editors for this book were Tyler G. Hicks and Joseph Williams
and the production supervisor was Sally Fliess. It was set in Univers
65 by University Graphics, Inc.

Printed and bound by The Kingsport Press.

Contents

Preface

Over 3,630 practical modern electronic circuits are arranged here in 103 logical chapters for convenient browsing and reference by electronics engineers, technicians, students, microprocessor enthusiasts, amateur radio fans, and experimenters. Each circuit has type numbers or values of all significant components, an identifying title, a concise description, performance data, and suggestions for other applications. At the end of each description is a citation giving the title of the original article or book, its author, and the exact location of the circuit in the original source.

This fourth in a series of state-of-the-art reference volumes illustrates dramatically the accelerated trend to integrated circuits that has taken place since publication of "Guidebook of Electronic Circuits" in 1974. About half of the applications now use ICs, and tube circuits have become a distinct rarity. This trend becomes even more evident when comparing circuits with those in the first and second books of the series, "Sourcebook of Electronic Circuits" and "Electronic Circuits Manual." The four books supplement each other and together provide a total of over 13,300 different practical circuits at a cost of only about 1 cent per circuit. The collection can serve as a basic desktop reference library that will match retrieval speeds of computer-based indexing systems while providing in addition the actual circuit diagrams.

The circuits for this new book were located by cover-to-cover searching of back issues of U.S. and foreign electronics periodicals, the published literature of electronics manufacturers, and recent electronics books, together filling well over 100 feet of shelving. This same search would take weeks or even months at a large engineering library, plus the time required to write for manufacturer literature and locate elusive sources.

Engineering libraries, particularly in foreign countries, have found these circuit abstracts to be a welcome substitute for the original sources when facing limitations on budgets, shelving, or search manpower. As further evidence of their usefulness in other countries, some of the books have been translated into Greek, Spanish, or Japanese.

Entirely new chapters in this book, further emphasizing evolution of the industry in recent years, include Clock Signal, Fiber-Optic, Game, Keyboard, Logic Probe, Microprocessor, Programmable, Switching Regulator, and Touch-Switch Circuits. Significant new circuits appear in chapters found also in previous books, particularly for Automotive, Burglar Alarm, Digital Clock, Fire Alarm, Flasher, Frequency Counter, Frequency Synthesizer, Instrumentation, Intercom, Lamp Control, Medical, Memory, Motor Control, Music, Power Control, Protection, Siren, Stereo, and Telephone Circuits.

To find a desired circuit quickly, start with the alphabetically arranged table of contents at the front of the book. Note the chapters most likely to contain the desired type of circuit, and look in these first. Remember that most applications use combinations of basic circuits, so a desired circuit could be in any of several different chapters. Scope notes following chapter titles define the basic circuits covered and sometimes suggest other chapters for browsing.

If a quick scan does not locate the exact circuit desired, use the index at the back of the book. Here the circuits are indexed in depth under the different names by which they may be known. Hundreds of cross-references in the index aid searching. The author index will often help find related circuits after one potentially useful circuit is found, because authors tend to specialize in certain circuits.

Values of important components are given for every circuit because these help in reading the circuit and redesigning it for other requirements. The development of a circuit for a new application is speeded when design work can be started with a working circuit, instead of starting from scratch. Research and experimentation are thereby cut to a minimum, so even a single use of this circuit-retrieval book could pay for its initial cost many times over. Drafting errors on diagrams are minimized because any corrections pointed out in subsequently published errata notices have been made; this alone can save many frustrating hours of troubleshooting.

This book is organized to provide a maximum of circuit information per page, with minimum repetition. The chapter title at the top of each right-hand page and the original title in the citation should therefore be considered along with the abstract when evaluating a circuit.

Abbreviations are used extensively to conserve space. Their meanings are given after this preface. Abbreviations on diagrams and in original article titles were unchanged and may differ slightly, but their meanings can be deduced by context.

Mailing addresses of all cited original sources are given at the front of the book, for convenience in writing for back issues or copies of articles when the source is not available at a local library. These sources will often prove useful for construction details, performance graphs, and calibration procedures.

To Joan Fife, student at the University of Santa Clara, goes credit for typing the complete manuscript directly from dictation while correcting this author's grammar and punctuation practices of yesteryear and even catching technical oversights. Handling of hyphenation, abbreviations, and citations was entirely her responsibility, along with final editing, markup for the printer, and production of the index.

To the original publications cited and their engineering authors and editors should go major credit for making possible this fourth encyclopedic contribution to electronic circuit design. The diagrams have been reproduced directly from the original source articles, by permission of the publisher in each case.

John Markus

Abbreviations Used

A	ampere	CRO	cathode-ray oscilloscope	F	farad
AC	alternating current			°F	degree Fahrenheit
AC/DC	AC or DC	CROM	control and read-only memory	FET	field-effect transistor
A/D	analog-to-digital			FIFO	first-in first-out
ADC	analog-to-digital converter	CRT	cathode-ray tube	FM	frequency modulation
		CT	center tap		
A/D, D/A	analog-to-digital, or digital-to-analog	CW	continuous wave	4PDT	four-pole double-throw
		D/A	digital-to-analog		
ADP	automatic data processing	DAC	digital-to-analog converter	4PST	four-pole single-throw
AF	audio frequency	dB	decibel	FS	full scale
AFC	automatic frequency control	dBC	C-scale sound level in decibels	FSK	frequency-shift keying
AFSK	audio frequency-shift keying	dBm	decibels above 1 mW	ft	foot
		dBV	decibels above 1 V	ft/min	foot per minute
AFT	automatic fine tuning	DC	direct current	ft/s	foot per second
		DC/DC	DC to DC	ft²	square foot
AGC	automatic gain control	DCTL	direct-coupled transistor logic	F/V	frequency-to-voltage
Ah	ampere-hour	diac	diode AC switch	F/V, V/F	frequency-to-voltage, or voltage-to-frequency
ALU	arithmetic-logic unit	DIP	dual in-line package		
AM	amplitude modulation	DMA	direct memory access	G	giga- (10⁹)
AM/FM	AM or FM	DMM	digital multimeter	GHz	gigahertz
AND	type of logic circuit	DPDT	double-pole double-throw	G-M tube	Geiger-Mueller tube
AVC	automatic volume control			h	hour
		DPM	digital panel meter	H	henry
b	bit	DPST	double-pole single-throw	HF	high frequency
BCD	binary-coded decimal			HFO	high-frequency oscillator
BFO	beat-frequency oscillator	DSB	double sideband	hp	horsepower
		DTL	diode-transistor logic	Hz	hertz
b/s	bit per second	DTL/TTL	DTL or TTL	IC	integrated circuit
C	capacitance; capacitor	DUT	device under test	IF	intermediate frequency
		DVM	digital voltmeter		
°C	degree Celsius; degree Centigrade	DX	distance reception; distant	IGFET	insulated-gate FET
				IMD	intermodulation distortion
CATV	cable television	EAROM	electrically alterable ROM	IMPATT	impact avalanche transit time
CB	citizens band	EBCDIC	extended binary-coded decimal interchange code	in	inch
CCD	charge-coupled device			in/s	inch per second
CCTV	closed-circuit television	ECG	electrocardiograph	in²	square inch
		ECL	emitter-coupled logic	I/O	input/output
cm	centimeter	EDP	electronic data processing	IR	infrared
CML	current-mode logic			JFET	junction FET
CMOS	complementary MOS	EKG	electrocardiograph	k	kilo- (10³)
CMR	common-mode rejection	EMF	electromotive force	K	kilohm (,000 ohms); kelvin
CMRR	common-mode rejection ratio	EMI	electromagnetic interference		
cm²	square centimeter	EPROM	erasable PROM	kA	kiloampere
coax	coaxial cable	ERP	effective radiated power	kb	kilobit
COHO	coherent oscillator			keV	kiloelectronvolt
COR	carrier-operated relay	ETV	educational television	kH	kilohenry
				kHz	kilohertz
COS/MOS	complementary-symmetry MOS (same as CMOS)	eV	electronvolt	km	kilometer
		EVR	electronic video recording	kV	kilovolt
				kVA	kilovoltampere
CPU	central processing unit	EXCLUSIVE-OR	type of logic circuit	kW	kilowatt
		EXCLUSIVE-NOR	type of logic circuit	kWh	kilowatthour
CR	cathode ray			L	inductance; inductor
				LASCR	light-activated SCR

LASCS	light-activated SCS	NMOS	N-channel MOS	QRP	low-power amateur radio
LC	inductance-capacitance	NOR	type of logic circuit	R	resistance; resistor
LCD	liquid crystal display	NPN	negative-positive-negative	RAM	random-access memory
LDR	light-dependent resistor	NPNP	negative-positive-negative-positive	RC	resistance-capacitance
LED	light-emitting diode	NRZ	nonreturn-to-zero	RF	radio frequency
LF	low frequency	NRZI	nonreturn-to-zero-inverted	RFI	radio-frequency interference
LIFO	last-in first-out	ns	nanosecond	RGB	red/green/blue
lm	lumen	NTSC	National Television System Committee	RIAA	Recording Industry Association of America
LO	local oscillator				
logamp	logarithmic amplifier			RLC	resistance-inductance-capacitance
LP	long play	nV	nanovolt		
LSB	least significant bit	nW	nanowatt		
LSI	large-scale integration	OEM	original equipment manufacturer	RMS	root-mean-square
m	meter; milli- (10^{-3})	opamp	operational amplifier	ROM	read-only memory
M	mega- (10^6); meter (instrument); motor	OR	type of logic circuit	rpm	revolution per minute
		p	pico- (10^{-12})	RTL	resistor-transistor logic
mA	milliampere	P	peak; positive		
Mb	megabit	pA	picoampere	RTTY	radioteletype
MF	medium frequency	PA	public address	RZ	return-to-zero
mH	millihenry	PAL	phase-alternation line	s	second
MHD	magnetohydro-dynamics	PAM	pulse-amplitude modulation	SAR	successive-approximation register
MHz	megahertz	PC	printed circuit		
mi	mile	PCM	pulse-code modulation	SAW	surface acoustic wave
mike	microphone			SCA	Subsidiary Communications Authorization
min	minute	PDM	pulse-duration modulation		
mm	millimeter				
modem	modulator-demodulator	PEP	peak envelope power	scope	oscilloscope
		pF	picofarad	SCR	silicon controlled rectifier
mono	monostable	PF	power factor		
MOS	metal-oxide semiconductor	phono	phonograph	SCS	silicon controlled switch
MOSFET	metal-oxide semiconductor FET	PIN	positive-intrinsic-negative		
		PIV	peak inverse voltage	S-meter	signal-strength meter
MOST	metal-oxide semiconductor transistor	PLL	phase-locked loop	S/N	signal-to-noise
		PM	permanent magnet; phase modulation	SNR	signal-to-noise ratio
MPU	microprocessing unit	PMOS	P-channel MOS	SPDT	single-pole double-throw
ms	millisecond	PN	positive-negative		
MSB	most significant bit	PNP	positive-negative-positive	SPST	single-pole single-throw
MSI	medium-scale integration	PNPN	positive-negative-positive-negative		
				SSB	single sideband
m^2	square meter	pot	potentiometer	SSI	small-scale integration
μ	micro- (10^{-6})	P-P	peak-to-peak		
μA	microampere	PPI	plan-position indicator	SSTV	slow-scan television
μF	microfarad			SW	shortwave
μH	microhenry	PPM	parts per million; pulse-position modulation	SWL	shortwave listener
μm	micrometer			SWR	standing-wave ratio
μP	microprocessor			sync	synchronizing
μs	microsecond	preamp	preamplifier	T	tera- (10^{12})
μV	microvolt	PRF	pulse repetition frequency	TC	temperature coefficient
μW	microwatt				
mV	millivolt	PROM	programmable ROM	THD	total harmonic distortion
MVBR	multivibrator	PRR	pulse repetition rate		
mW	milliwatt	ps	picosecond	TR	transmit-receive
n	nano- (10^{-9})	PSK	phase-shift keying	TRF	tuned radio frequency
N	negative	PTT	push to talk		
nA	nanoampere	PUT	programmable UJT	triac	triode AC semiconductor switch
NAB	National Association of Broadcasters	pW	picowatt		
		PWM	pulse-width modulation		
NAND	type of logic circuit			TTL	transistor-transistor logic
nF	nanofarad	Q	quality factor		
nH	nanohenry				

TTY	teletypewriter	V	volt	VSWR	voltage standing-wave ratio
TV	television	VA	voltampere	VTR	videotape recording
TVI	television interference	VAC	volts AC	VTVM	vacuum-tube voltmeter
TVT	television typewriter	VCO	voltage-controlled oscillator	VU	volume unit
TWX	teletypewriter exchange service	VDC	volts DC	VVC	voltage-variable capacitor
UART	universal asynchronous receiver-transmitter	V/F	voltage-to-frequency	VXO	variable-frequency crystal oscillator
		VFO	variable-frequency oscillator	W	watt
		VHF	very high frequency	Wh	watthour
UHF	ultrahigh frequency	VLF	very low frequency	WPM	words per minute
UJT	unijunction transistor	VMOS	vertical metal-oxide semiconductor	WRMS	watts RMS
UPC	universal product code	VOM	volt-ohm-milliammeter	Ws	wattsecond
		VOX	voice-operated transmission	Z	impedance
UPS	uninterruptible power system	VRMS	volts RMS		

Abbreviations on Diagrams. Some foreign publications, including *Wireless World,* shorten the abbreviations for units of measure on diagrams. Thus, μ after a capacitor value represents μF, n is nF, and p is pF. With resistor values, k is thousand ohms, M is megohms, and absence of a unit of measure is ohms. For a decimal value, the letter for the unit of measure is sometimes placed at the location of the decimal point. Thus, 3k3 is 3.3 kilohms or 3,300 ohms, 2M2 is 2.2 megohms, 4μ7 is 4.7 μF, 0μ1 is 0.1 μF, and 4n7 is 4.7 nF.

Semiconductor Symbols Used

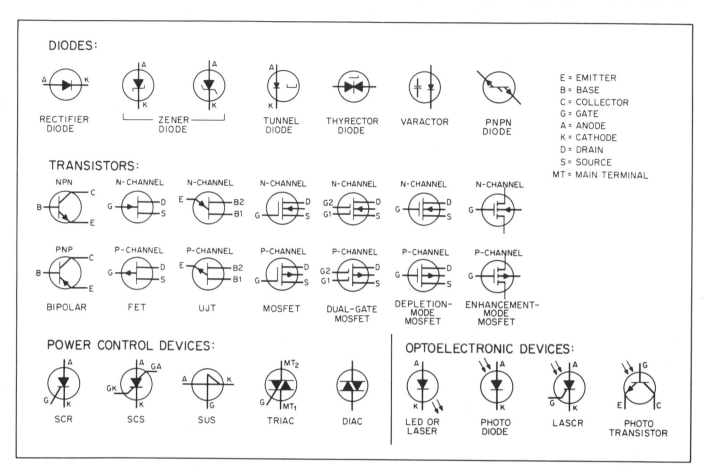

DIODES:

RECTIFIER DIODE   ZENER DIODE   TUNNEL DIODE   THYRECTOR DIODE   VARACTOR   PNPN DIODE

E = EMITTER
B = BASE
C = COLLECTOR
G = GATE
A = ANODE
K = CATHODE
D = DRAIN
S = SOURCE
MT = MAIN TERMINAL

TRANSISTORS:

NPN   N-CHANNEL   N-CHANNEL   N-CHANNEL   N-CHANNEL   N-CHANNEL   N-CHANNEL

PNP   P-CHANNEL   P-CHANNEL   P-CHANNEL   P-CHANNEL   P-CHANNEL   P-CHANNEL

BIPOLAR   FET   UJT   MOSFET   DUAL-GATE MOSFET   DEPLETION-MODE MOSFET   ENHANCEMENT-MODE MOSFET

POWER CONTROL DEVICES:

SCR   SCS   SUS   TRIAC   DIAC

OPTOELECTRONIC DEVICES:

LED OR LASER   PHOTO DIODE   LASCR   PHOTO TRANSISTOR

The commonest forms of the basic semiconductor symbols are shown here. Leads are identified where appropriate, for convenient reference. Minor variations in symbols, particularly those from foreign sources, can be recognized by comparing with these symbols while noting positions and directions of solid arrows with respect to other symbol elements.

Omission of the circle around a symbol has no significance. Arrows are sometimes drawn open instead of solid. Thicker lines and open rectangles in some symbols on diagrams have no significance. Orientation of symbols is unimportant; artists choose the position that is most convenient for making connections to other parts of the circuit. Arrow lines outside optoelectronic symbols indicate the direction of light rays.

On some European diagrams, the position of the letter k gives the location of the decimal point for a resistor value in kilohms. Thus, 2k2 is 2.2K or 2,200 ohms. Similarly, a resistance of 1R5 is 1.5 ohms, 1M2 is 1.2 megohms, and 3n3 is 3.3 nanofarads.

Substitutions can often be made for semiconductor and IC types specified on diagrams. Newer components, not available when the original source article was published, may actually improve the performance of a particular circuit. Electrical char-

acteristics, terminal connections, and such critical ratings as voltage, current, frequency, and duty cycle, must of course be taken into account if experimenting without referring to substitution guides.

Semiconductor, integrated-circuit, and tube substitution guides can usually be purchased at electronic parts supply stores.

Not all circuits give power connections and pin locations for ICs, but this information can be obtained from manufacturer data sheets. Alternatively, browsing through other circuits may turn up another circuit on which the desired connections are shown for the same IC.

When looking down at the top of an actual IC, numbering normally starts with 1 for the first pin *counterclockwise* from the notched or otherwise marked end and continues sequentially. The highest number is therefore next to the notch on the other side of the IC, as illustrated in the sketches below. (*Actual positions* of pins are rarely shown on schematic diagrams.)

Addresses of Sources Used

In the citation at the end of each abstract, the title of a magazine is set in italics. The title of a book or report is placed in quotes. Each source title is followed by the name of the publisher of the original material, plus city and state. Complete mailing addresses of all sources are given below, for the convenience of readers who want to write to the original publisher of a particular circuit. When writing, give the complete citation, exactly as in the abstract.

Books can be ordered from their publishers, after first writing for prices of the books desired. Some electronics manufacturers also publish books and large reports for which charges are made. Many of the books cited as sources in this volume are also sold by bookstores and by electronics supply firms. Locations of these firms can be found in the YELLOW PAGES of telephone directories under headings such as "Electronic Equipment and Supplies" or "Television and Radio Supplies and Parts."

Only a few magazines have back issues on hand for sale, but most magazines will make copies of a specific article at a fixed charge per page or per article. When you write to a magazine publisher for prices of back issues or copies, give the *complete* citation, *exactly* as in the abstract. Include a stamped self-addressed envelope to make a reply more convenient.

If certain magazines consistently publish the types of circuits in which you are interested, use the addresses below to write for subscription rates.

American Microsystems, Inc., 3800 Homestead Rd., Santa Clara, CA 95051

Audio, 401 North Broad St., Philadelphia, PA 19108

BYTE, 70 Main St., Peterborough, NH 03458

Computer Design, 11 Goldsmith St., Littleton, MA 01460

CQ, 14 Vanderventer Ave., Port Washington, L.I., NY 11050

Delco Electronics, 700 East Firmin, Kokomo, IN 46901

Dialight Corp., 203 Harrison Place, Brooklyn, NY 11237

EDN, 221 Columbus Ave., Boston, MA 02116

Electronics, 1221 Avenue of the Americas, New York, NY 10020

Electronic Servicing, 9221 Quivira Rd., P.O. Box 12901, Overland Park, KS 66212

Exar Integrated Systems, Inc., 750 Palomar Ave., Sunnyvale, CA 94086

Ham Radio, Greenville, NH 03048

Harris Semiconductor, Department 53-35, P.O. Box 883, Melbourne, FL 32901

Hewlett-Packard, 1501 Page Mill Rd., Palo Alto, CA 94304

Howard W. Sams & Co. Inc., 4300 West 62nd St., Indianapolis, IN 46206

IEEE Publications, 345 East 47th St., New York, NY 10017

Instruments & Control Systems, Chilton Way, Radnor, PA 19089

Kilobaud, Peterborough, NH 03458

McGraw-Hill Book Co., 1221 Avenue of the Americas, New York, NY 10020

Modern Electronics, 14 Vanderventer Ave., Port Washington, NY 11050

Motorola Semiconductor Products Inc., Box 20912, Phoenix, AZ 85036

Mullard Limited, Mullard House, Torrington Place, London WC1E 7HD, England

National Semiconductor Corp., 2900 Semiconductor Dr., Santa Clara, CA 95051

Optical Electronics Inc., P.O. Box 11140, Tucson, AZ 85734

Popular Science, 380 Madison Ave., New York, NY 10017

Precision Monolithics Inc., 1500 Space Park Dr., Santa Clara, CA 95050

QST, American Radio Relay League, 225 Main St., Newington, CT 06111

Radio Shack, 1100 One Tandy Center, Fort Worth, TX 76102

Raytheon Semiconductor, 350 Ellis St., Mountain View, CA 94042

RCA Solid State Division, Box 3200, Somerville, NJ 08876

Howard W. Sams & Co. Inc., 4300 West 62nd St., Indianapolis, IN 46206

73 Magazine, Peterborough, NH 03458

Siemens Corp., Components Group, 186 Wood Ave. South, Iselin, NJ 08830

Signetics Corp., 811 East Arques Ave., Sunnyvale, CA 94086

Siliconix Inc., 2201 Laurelwood Rd., Santa Clara, CA 95054

Sprague Electric Co., 479 Marshall St., North Adams, MA 01247

Teledyne Philbrick, Allied Drive at Route 128, Dedham, MA 02026

Teledyne Semiconductor, 1300 Terra Bella Ave., Mountain View, CA 94040

Texas Instruments Inc., P.O. Box 5012, Dallas, TX 75222

TRW Power Semiconductors, 14520 Aviation Blvd., Lawndale, CA 90260

Unitrode Corp., 580 Pleasant St., Watertown, MA 02172

Wireless World, Dorset House, Stamford St., London SE1 9LU, England

MODERN
ELECTRONIC
CIRCUITS
REFERENCE
MANUAL

CHAPTER 1

Amplifier Circuits

Includes general-purpose RF amplifiers covering various portions of spectrum from DC to 2.3 GHz at outputs up to 230 W, some with voltage-controlled gain, for pulses as well as video and other RF signals. See other chapters in book for RF amplifiers having specific applications.

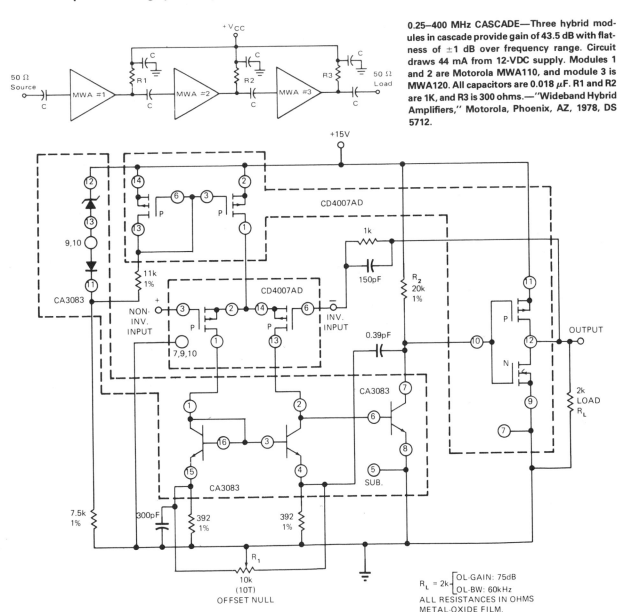

0.25–400 MHz CASCADE—Three hybrid modules in cascade provide gain of 43.5 dB with flatness of ±1 dB over frequency range. Circuit draws 44 mA from 12-VDC supply. Modules 1 and 2 are Motorola MWA110, and module 3 is MWA120. All capacitors are 0.018 µF. R1 and R2 are 1K, and R3 is 300 ohms.—"Wideband Hybrid Amplifiers," Motorola, Phoenix, AZ, 1978, DS 5712.

$$R_L = 2k \begin{cases} \text{OL-GAIN: 75dB} \\ \text{OL-BW: 60kHz} \end{cases}$$

ALL RESISTANCES IN OHMS
METAL-OXIDE FILM.

CMOS/BIPOLAR VOLTAGE FOLLOWER—Combination of two 4007 CMOS gate packages and one CA3083 transistor package provides gain of about 75 dB as voltage-follower amplifier and bandwidth of 50 kHz. Slew rate is about 30 V/µs, and settling time is 2 µs. Requires only single +15 V supply. Can be driven to within 1 mV of ground. Interfaces well with single-supply D/A converters.—B. Furlow, CMOS Gates in Linear Applications: The Results Are Surprisingly Good, *EDN Magazine*, March 5, 1973, p 42–48.

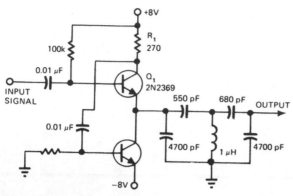

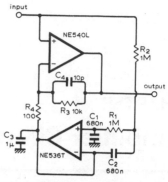

CAPACITIVE-LOAD EMITTER-FOLLOWER— Overcomes problem that develops with trailing edges of pulses when emitter-follower using NPN transistor is driving heavily capacitive load. Extra transistor is used to dump capacitor charge when emitter-follower stops conducting at trailing edge of input waveform. Pulse trailing edge thus tends to cut off Q_1 and saturate lower transistor so it discharges capacitor. Circuit works equally well with pulses, square waves, and sine waves. Transistors need not be matched. Reverse polarity of supplies to use PNP transistors. Useful for driving long coax lines or logic from high-impedance source, without inversion.—H. L. Morgan, Emitter Follower's Fall Time Is Independent of Load, *EDN Magazine*, Feb. 5, 1977, p 105.

AC WITH IMMUNITY TO LARGE DC OFFSET— Designed to amplify from about 250 kHz down to low frequencies in presence of large DC input offsets. Main NE540L amplifier has gain of 101, while NE536T has DC gain of unity and forms part of low-pass network that applies DC input offset as common-mode voltage to inverting input of main amplifier.—A. Royston, Low Frequency A.C. Amplifier, *Wireless World*, May 1976, p 80.

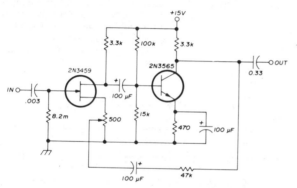

BASIC FEEDBACK AMPLIFIER— Combination of unipolar and bipolar transistors gives desirable amplifying features of each solid-state device. Circuit can be optimized for RF or AF by adjusting coupling, feedback, and emitter bypass capacitor values. Changes in feedback affect distortion, frequency response, and gain stability. To optimize for RF, reduce capacitor sizes. For both AF and RF response, capacitors shown can be paralleled by small ceramic or Mylar units. If FET and bipolar are selected for high transconductance and high gain-bandwidth product, overall voltage gain can be 20 or more for frequencies up to several megahertz.—I. M. Gottlieb, A New Look at Solid-State Amplifiers, *Ham Radio*, Feb. 1976, p 16—19.

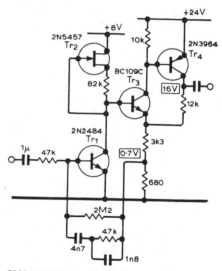

SMALL-SIGNAL AMPLIFIER— Combines features of virtual-earth and high-input-impedance amplifiers economically for such applications as a record amplifier, and provides several times the gain of a virtual-earth amplifier alone.—D. Rawson-Harris, Small Signal Amplifier, *Wireless World*, Feb. 1977, p 45.

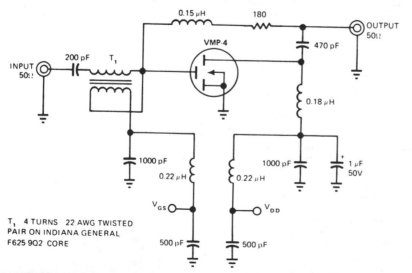

T₁ 4 TURNS 22 AWG TWISTED PAIR ON INDIANA GENERAL F625 9Q2 CORE

13 W at 160 MHz— Circuit uses Siliconix VMP-4 power MOSFET to provide 11-dB gain with 26-V supply, or 14 dB with 36-V supply. Broadband design permits operation over wide range of frequencies up to as high as 600 MHz.—RF Power MOSFET Outputs 13 W at 160 MHz with High Gain, No Breakdown, *EDN Magazine*, June 20, 1976, p 144—145.

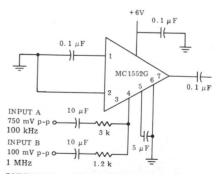

SUMMING/SCALING VIDEO— With Motorola MC1552G video amplifier connected as shown, summation of input signal currents is accomplished at pin 4 through input resistors whose values are chosen to give desired scale factor.— "A Wide Band Monolithic Video Amplifier," Motorola, Phoenix, AZ, 1973, AN-404, p 9.

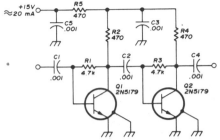

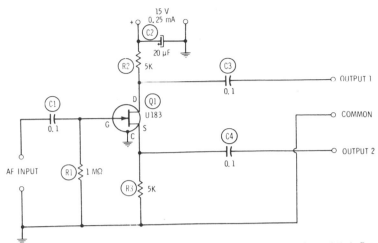

500 kHz TO 500 MHz—Two-stage general-purpose wideband small-signal amplifier provides nearly 14-dB gain at 150 MHz when inserted in 50-ohm transmission line with no tuned circuits at input or output. Noise figure with optimum source resistance is about 3 dB at 150 MHz. Amplifier is capacitively coupled common-emitter cascade. Capacitors make low-frequency gain begin dropping off below about 2 MHz. Increasing all capacitors to 0.0l μF will lower frequency response to about 200 kHz.—R. Rhea, General Purpose Wideband RF Amplifier, *Ham Radio*, April 1975, p 58–61.

PARAPHASE PHASE INVERTER—Uses 180° phase difference between source and drain outputs of Siliconix U183 FET to convert AF input to push-pull output without transformer. Voltage gain in each half of circuit is about 0.8. Frequency response referred to 1 kHz is flat within 3 dB from 50 Hz to 50 kHz, when using 1-megohm output load.—R. P. Turner, "FET Circuits," Howard W. Sams, Indianapolis, IN, 1977, 2nd Ed., p 29–30.

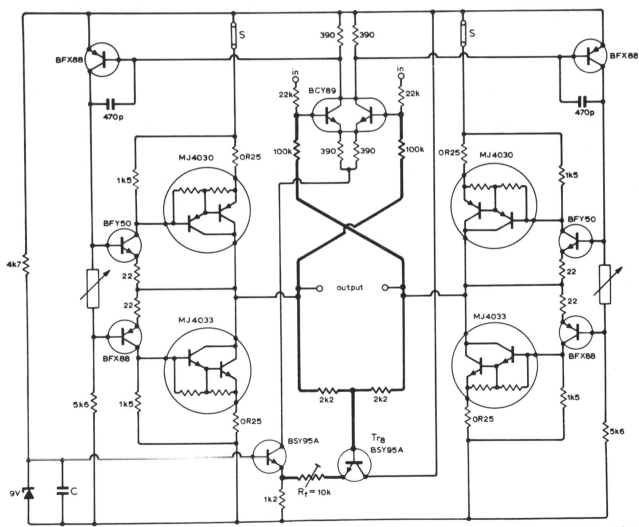

230-W WATER-COOLED—Used to excite magnetic specimens in frequency range of 0 to 110 kHz at outputs up to 12 A. Output stage uses two complementary pairs of emitter-followers connected so each pair forms half of bridge, using MJ4030 and MJ4033 Darlingtons mounted on liquid-cooled heatsinks. Article describes cooling arrangement and circuit operation in detail and gives suitable preamp circuit for driving inputs of BCY89 dual transistor. Designed for 32-VDC supply, which connects to top and bottom horizontal buses on diagram. Feedback circuits are drawn in heavy lines. Resistors in series with Darlingtons (0R25, representing 0.25 ohm) are wound from resistance wire since they must carry large currents. Output impedance of circuit is less than 0.5 ohm, for matching to low-resistance load.—I. L. Stefani and R. Perryman, Liquid-Cooled Power Amplifier, *Wireless World*, Dec. 1974, p 505–507.

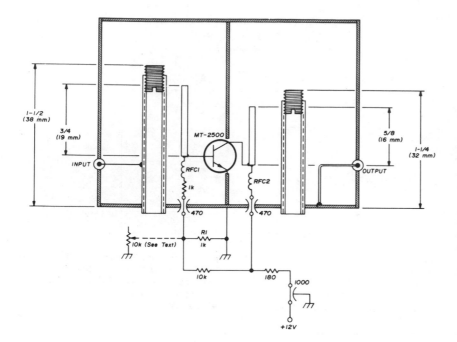

2304-MHz PREAMP—Narrow-band stage using Fairchild MT-2500 bipolar transistor gives gain of 6 to 9 dB and noise figure of 2.5 to 4.5 dB. Cavity resonators at both input and output give excellent frequency selectivity. Similar circuit can be used with Fairchild MT-4500, FMT-4005, or equivalent newer stripline-type transistors. RFC1 is 3 turns and RFC2 is 5 turns, air-wound with No. 26 enamel by using No. 52 drill as mandrel. Coupling strips on base and collector of transistor are 0.25-mm brass shim stock. Article gives construction and tune-up details, along with alternate design for HP-35821E and HP-35862E transistors using coupling loops. 10K pot is used only during tune-up.—N. J. Foot, Narrow-Band Solid-State 2304-MHz Preamplifiers, *Ham Radio*, July 1974, p 6–11.

MULTIPURPOSE MODULE—Flexible circuit using FET to drive bipolar transistor has −3 dB points at 100 Hz and 0.6 MHz. Components are noncritical and can be changed considerably in value to optimize gain, frequency response, power output, or power consumption. Load presented to FET is primarily input resistance of bipolar transistor, about 1000 ohms, which gives voltage gain of 4 for FET.—I. M. Gottlieb, A New Look at Solid-State Amplifiers, *Ham Radio*, Feb. 1976, p 16–19.

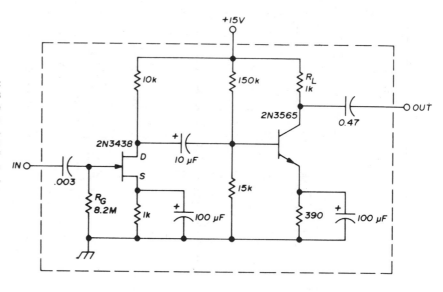

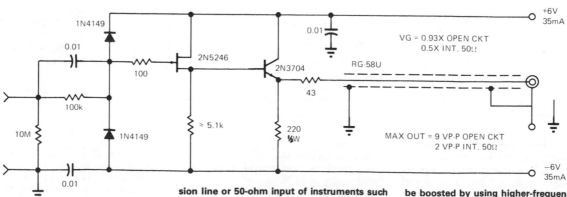

IMPEDANCE CONVERTER—Used to match 10-megohm input impedance to 50-ohm transmission line or 50-ohm input of instruments such as spectrum analyzer, video amplifier, or frequency counter. Voltage gain is exactly 0.5. Frequency response is from DC to 20 MHz and can be boosted by using higher-frequency transistor.—M. J. Salvati, FET Probe Drives 50-Ohm Load, *EDN Magazine*, March 5, 1973, p 87 and 89.

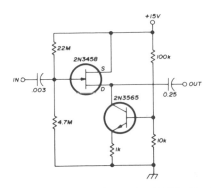

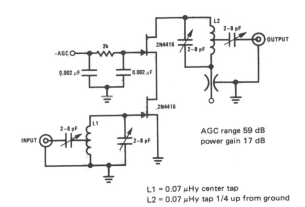

L1 = 0.07 µHy center tap
L2 = 0.07 µHy tap 1/4 up from ground

FET-BIPOLAR SOURCE FOLLOWER—Used where source follower with high output-voltage swing and voltage gain close to unity is required. Circuit has constant-current bias supply. Combination of unipolar and bipolar transistors gives desirable amplifying features of each solid-state device.—I. M. Gottlieb, A New Look at Solid-State Amplifiers, *Ham Radio,* Feb. 1976, p 16–19.

200-MHz CASCODE—JFETs give low cross-modulation, large signal-handling ability, and AGC action controlled by biasing upper cascode JFET. Neutralization is not needed.—"FET Databook," National Semiconductor, Santa Clara, CA, 1977, p 6-26–6-36.

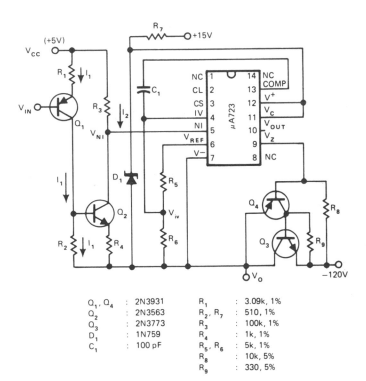

Q_1, Q_4	: 2N3931	R_1	: 3.09k, 1%	
Q_2	: 2N3563	R_2, R_7	: 510, 1%	
Q_3	: 2N3773	R_3	: 100k, 1%	
D_1	: 1N759	R_4	: 1k, 1%	
C_1	: 100 pF	R_5, R_6	: 5k, 1%	
		R_8	: 10k, 5%	
		R_9	: 330, 5%	

HIGH-VOLTAGE BUFFER—Circuit shown for µA723 voltage regulator permits use as high-voltage and high-current buffer in linear applications. Power dissipation of output transistor Q_3 is only limiting factor. I_1 is proportional to V_{IN}, I_2 is proportional to I_1, and output voltage V_0 is proportional to I_2 and V_{IN}.—G. Niu, Single Op Amp Implements High-Voltage/Current Buffer, *EDN Magazine,* Oct. 5, 1977, p 96 and 98.

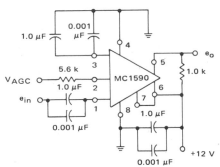

VIDEO AMPLIFIER—AGC capability of Motorola MC1590G makes it highly suitable for wideband video amplifier applications. Voltage gain is about 25 dB up to 50 MHz for 100-ohm load and 45 dB up to 10 MHz for 1K load. Several circuits can be cascaded to increase gain, using capacitive coupling.—B. Trout, "A High Gain Integrated Circuit RF-IF Amplifier with Wide Range AGC," Motorola, Phoenix, AZ, 1975, AN-513, p 9.

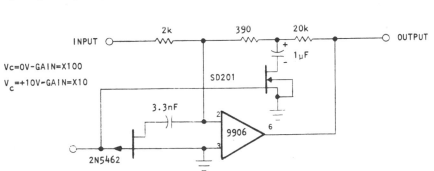

$V_c = 0V - GAIN = X100$
$V_c = +10V - GAIN = X10$

WIDEBAND VARIABLE GAIN—FET serves as gain-controlled device in feedback loop of Optical Electronics 9906 opamp. Resistive T network has SD201 MOS transistor as ground leg, with resistor values chosen so transistor is electrically close to summing junction, automatically limiting total signal voltage. Resulting arrangement of voltage-controlled feedback and compensation gives variable-gain amplifier with good linearity and constant wideband width for all gain levels.—"Wideband Variable Gain Amplifier," Optical Electronics, Tucson, AZ, Application Tip 10277.

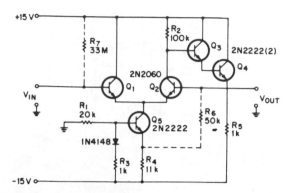

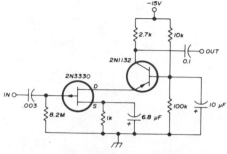

UNITY-GAIN VOLTAGE FOLLOWER—Measured gain is 0.9997 V/V with an error of ±0.1% over ±1.5 V swing. Circuit has infinite input impedance and zero bias current. Addition of dashed components to simple voltage-follower design gives near-perfect performance.—C. Andren, The Ideal Voltage Follower, *EEE Magazine*, Jan. 1971, p 63–64.

CASCODE—Combination of unipolar and bipolar transistors gives desirable amplifying features of each solid-state device. Ideal for use with tuned circuits in audio, video, IF, and RF applications.—I. M. Gottlieb, A New Look at Solid-State Amplifiers, *Ham Radio*, Feb. 1976, p 16–19.

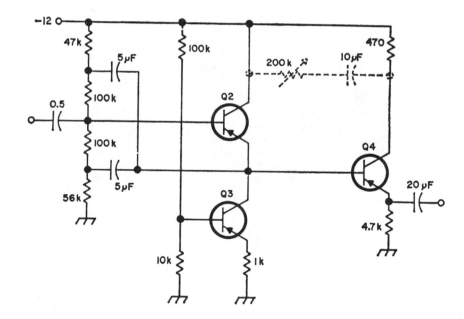

HIGH-Z PREAMP—Provides up to 20 megohms of input impedance and has essentially flat response from 10 Hz to 220 MHz. Q3 serves as emitter resistor for Q2, and emitter-follower Q4 reduces loading. Input impedance is further increased by adding optional components shown in dashed lines. Transistors are 2N2188, SK3005, GE-9, or HEP-2.—Circuits, *73 Magazine*, Feb. 1974, p 102.

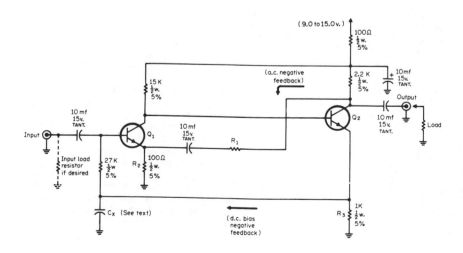

10-dB GAIN AT 0.01–100 MHz—High-gain wideband untuned general-purpose amplifier uses Fairchild 2N5126 or equivalent transistors in direct-coupled circuit. Design is stable for both power supply and temperature variations. Gain is adjusted with R_1, with maximum of 38 to 44 dB and maximum output of about 1 V P-P. Will drive low-level transistor circuits having load of about 1000 ohms. If several amplifiers are used in series for higher gain, shielding is required. Applications include amplification of pulsed light signals detected by photodiode. C_x can be 100-pF mica.—A. B. Hutchison, Jr., General Purpose Wide Band Amplifier, *CQ*, May 1972, p 22–23.

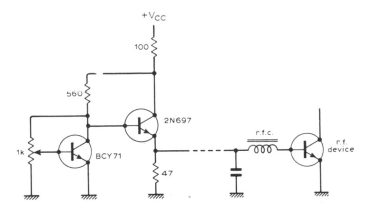

BIAS SUPPLY FOR CLASS AB—Two-transistor supply using PNP silicon transistor as amplified diode variable-voltage source gives improved-performance transistor RF power amplifier operating in class AB linear mode. Transistor types are not critical. Output impedance of bias source is about 1 ohm, and output voltage changes only up to 3½% for ±2.5 V change in input voltage V_{CC}.—C. P. Bartram, Bias Supply for R.F. Power Amplifiers, *Wireless World,* April 1976, p 61.

1–36 MHz DISTRIBUTED—Provides 18-dB gain over entire frequency range without use of special ferrite transformers. Gain contribution of each transistor, in phase with amplified wave as it passes down artificial transmission line, adds to that of other transistors. Capacitors marked with asterisks are low-inductance ceramic types such as Erie Redcap. Delay-line inductors L are 12 turns No. 24 closewound on ⅛-inch diameter Lucite rod, and L/2 units are 7 turns. Can be used as preamp for frequency counter and as auxiliary for other test equipment. Article covers construction, heatsinking of transistor, and testing.—H. Olson, Wide-Range Broadband Amplifier, *Ham Radio,* April 1974, p 40–44.

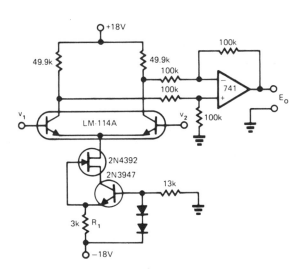

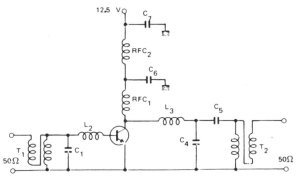

C_1 = 300 pF (chip)
L_2 = 4.2 nH (adjust)
L_3 = 8 nH (adjust)
C_4 = 130 pF (chip)
C_5 = 750 pF (chip)
C_6 = 2.2 μF

T_1 T_2
RFC_1 = 7 turns
    6.3 mm coil diameter
    0.8 mm wire diameter
RFC_2 = 3 turns on ferrite bead
C_7 = 0.68 μF

118-136 MHz BROADBAND—Designed for low-level amplitude modulation system. 50-ohm line transformers are wound with copper ribbon on ferrite core to give 4:1 ratio. Design and construction procedures are covered. Transistor is Motorola 2N6083, rated 30 W for 4-W input.—B. Becciolini, "Impedance Matching Networks Applied to R-F Power Transistors," Motorola, Phoenix, AZ, 1974, AN-721, p 17.

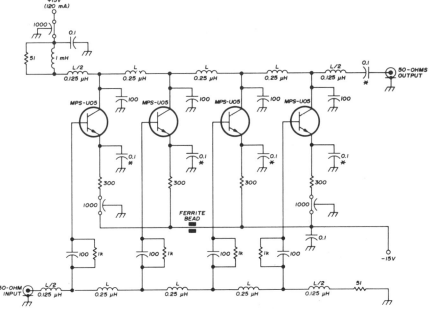

DIFFERENTIAL PAIR—Conventional differential amplifier circuit provides differential-mode gain of 96, common-mode input resistance of 500 megohms, CMRR of 106 dB, and current-source output resistance greater than 1 gigohm. Article gives design equations.—R. C. Jaeger and G. A. Hellwarth, Differential Cascode Amplifier Offers Unique Advantages, *EDN Magazine,* June 5, 1974, p 78 and 80.

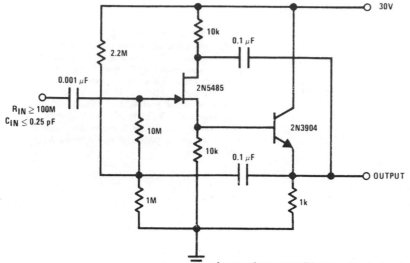

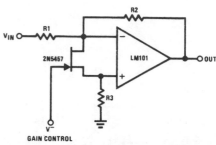

HIGH INPUT IMPEDANCE—Simple JFET input circuit is operated as source follower with bootstrapped gate bias resistor and drain to give maximum possible reduction in input capacitance. Used as unity-gain AC amplifier.—"FET Databook," National Semiconductor, Santa Clara, CA, 1977, p 6-26–6-36.

VOLTAGE-CONTROLLED GAIN—2N5457 FET acts as voltage-variable resistor between differential input terminals of opamp. Resistance variation is linear with voltage over several decades of resistance, to give excellent electronic gain control. Values of resistors depend on opamp used.—"FET Databook," National Semiconductor, Santa Clara, CA, 1977, p 6-26–6-36.

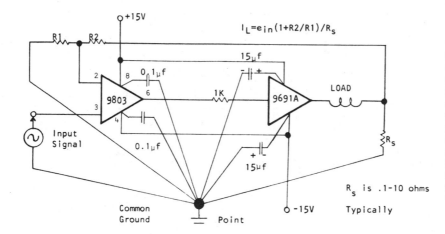

$$I_L = e_{in}(1 + R2/R1)/R_s$$

R_s is .1-10 ohms Typically

100 W FOR DC TO 500 kHz—Circuit using Optical Electronics opamps has high input impedance, high gain capability, and 100-W output capacity without use of transformers, for high-fidelity audio circuits, cathode-ray deflection circuits, and servosystems. Output currents up to 10 A require heavy output wiring, large power-supply bypass capacity, and heavy common ground point. Load is in feedback loop of opamp. Constant-current drive for load makes impedance matching to loudspeaker unnecessary.—"A High Gain 100 Watt Amplifier," Optical Electronics, Tucson, AZ, Application Tip 10205.

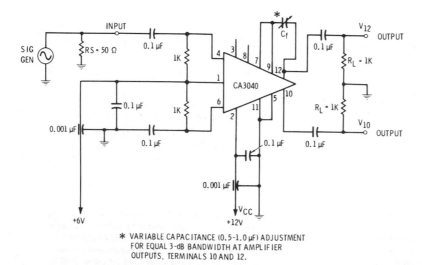

* VARIABLE CAPACITANCE (0.5-1.0 μF) ADJUSTMENT FOR EQUAL 3-dB BANDWIDTH AT AMPLIFIER OUTPUTS, TERMINALS 10 AND 12.

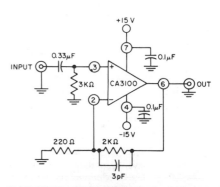

20-MHz WIDEBAND—RCA CA3040 IC is connected for single-ended input and balanced output, with no resonant circuits. Gain is above 30 dB over wide frequency range.—E. M. Noll, "Linear IC Principles, Experiments, and Projects," Howard W. Sams, Indianapolis, IN, 1974, p 162–163 and 168.

20-dB VIDEO—Simple circuit having gain of 20 dB provides 3-dB bandwidth of 20 MHz for CA3100 bipolar MOS opamp. Total noise referred to input is only 35 μVRMS.—"Circuit Ideas for RCA Linear ICs," RCA Solid State Division, Somerville, NJ, 1977, p 12.

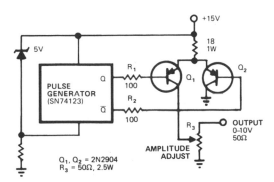

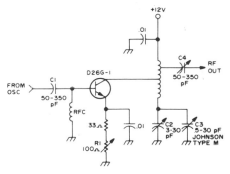

CONSTANT OUTPUT IMPEDANCE—Pulse output stage gives output range from millivolts to 10 V P-P across 50 ohms while optimizing waveform characteristics of output pulse. With 2N2904 output transistors, circuit delivers 200 mA with 20-ms rise and fall times.—W. A. Palm, Pulse Amplifier Varies Amplitude, *EDN Magazine*, Aug. 5, 1978, p 76.

50-MHz POWER—Developed for use with 50-MHz microtransistor crystal oscillator, using additional GE microtransistor for boosting RF output to about 75 mW. Article covers construction with microcomponents and gives other microtransistor circuits for low-power amateur radio use and possible bugging applications.—B. Hoisington, Introduction to "Microtransistors," *73 Magazine*, Oct. 1974, p 24–30.

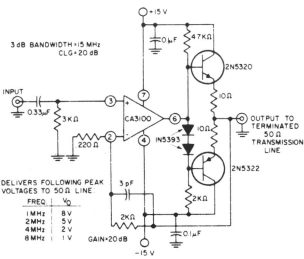

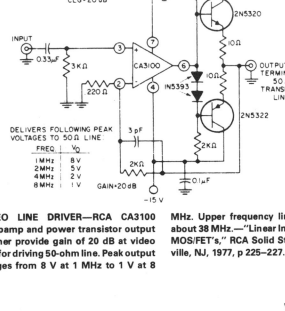

20-dB VIDEO LINE DRIVER—RCA CA3100 wideband opamp and power transistor output stage together provide gain of 20 dB at video frequencies for driving 50-ohm line. Peak output voltage ranges from 8 V at 1 MHz to 1 V at 8 MHz. Upper frequency limit for unity gain is about 38 MHz.—"Linear Integrated Circuits and MOS/FET's," RCA Solid State Division, Somerville, NJ, 1977, p 225–227.

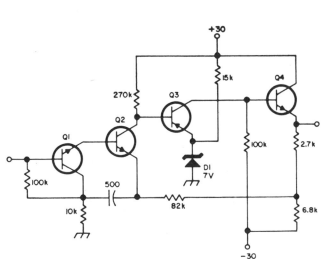

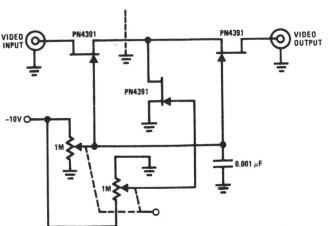

PREAMP FOR 0.5 Hz TO 2 MHz—Provides 11-dB gain over entire frequency range, with input impedance of 32 megohms. Q3 is GE-2 or HEP-52, and other transistors are SK3020 or HEP-53.—Circuits, *73 Magazine*, Jan. 1974, p 125.

VIDEO ATTENUATOR—FETs in T attenuator provide optimum dynamic linear range for attenuation of video signals with ganged 1-megohm pots. If complete turnoff is desired, attenuation greater than 100 dB can be obtained at 10 MHz by using appropriate RF construction to minimize leakage. ON resistance of transistor (between drain and source) is less than 30 ohms.—"FET Databook," National Semiconductor, Santa Clara, CA, 1977, p 6-26–6-36.

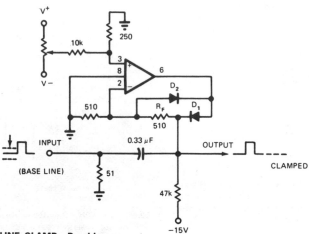

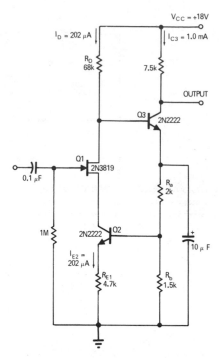

PULSE BASELINE CLAMP—Provides accurate clamping of baseline level of fast positive-going digital pulses with constantly changing duty cycle when capacitively coupled into level-sensitive circuits. Uses HA-2535 opamp. Diode D_2 clamps negative output swing of opamp to about 0.3 V, preserving amplifier recovery time in preparation for clamping next input transi-

tion. Pulse widths at input are less than 100 ns, with transition time under 15 ns and duty cycle ranging from 2 to 50%. Diodes are HP 2800 series. For clamping sine or triangle AF waves, opamp can be 741.—D. L. Quick, Clamp Speeds Restoration of AC-Coupled Base Lines, *EDN Magazine*, Sept. 5, 1975, p 76.

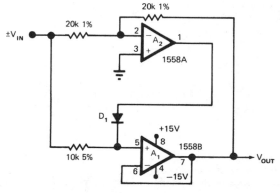

STABILIZED BIAS—Simple voltage-feedback loop stabilizes bias on direct-coupled FET and bipolar transistor stages. Arrangement uses constant-current source Q2 to maintain stable bias voltage on base of Q3. By choosing proper resistor values, DC voltage feedback from emitter of bipolar is made to control constant-current value. Any change in drain current produces opposite change in constant-current value, for stabilizing bipolar. Article gives design equation.—H. T. Russell, DC Feedback Stabilizes Bias on FET/Bipolar Pair, *EDN Magazine*, Nov. 15, 1970, p 51.

POLARITY-IGNORING VOLTAGE FOLLOWER— Absolute-value circuit is basically voltage follower A_1 whose input is positive regardless of polarity of V_{IN}. With positive input, inverting amplifier A_2 is disconnected by D_1. With negative input, inverting amplifier applies positive

input to voltage follower through D_1. Output voltage is thus absolute value of input voltage.—R. J. Wincentsen, Absolute Value Circuit Uses Only Five Parts, *EDN Magazine*, Nov. 1, 1972, p 44.

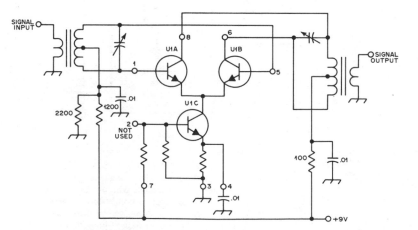

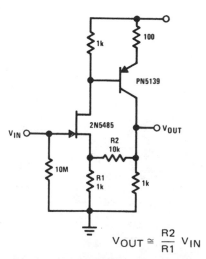

RF DIFFERENTIAL AMPLIFIER—Uses RCA CA3028A linear IC to provide power gain of about 32 dB at frequencies up to about 120 MHz. Values of tuned circuits depend on frequency

used. Unmarked resistors are on IC. —D. DeMaw, Understanding Linear ICs, *QST*, Feb. 1977, p 19–23.

HIGH-IMPEDANCE VIDEO—Compound series-feedback circuit using FET at input provides high input impedance and stable wideband gain for general-purpose video amplifier applications.—"FET Databook," National Semiconductor, Santa Clara, CA, 1977, p 6-26–6-36.

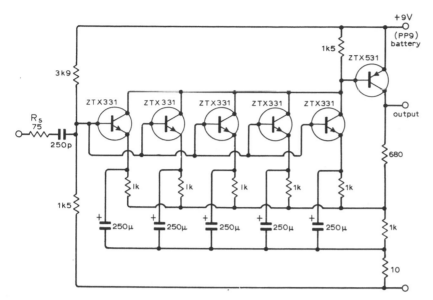

2.5-MHz BANDWIDTH LOW-NOISE—Paralleled transistors meet noise requirements of applications such as hot-wire anemometry for measuring gas flow and temperature, wherein typical signal voltages are as small as 1 μV peak over frequency range of 100 Hz to 200 kHz. Two feedback paths are used, one to provide low DC gain and stabilize bias voltages and the other for independent adjustment of AC gain. Design bandwidth is 7 Hz to 2.5 MHz.—J. A. Grocock, Low-Noise Wideband Amplifier, *Wireless World,* March 1975, p 117–118.

50-dB BROADBAND VIDEO—RCA CA3018 four-transistor array is connected as two pairs of common-emitter emitter-follower combinations, with two feedback loops providing high DC stability. One path goes from emitter of Q3 back to input, and other goes from collector of Q4 to collector of Q1. Values of C1, C2, and C3 give low-frequency cutoff (3 dB down) of 800 Hz. Upper cutoff is 32 MHz.—E. M. Noll, "Linear IC Principles, Experiments, and Projects," Howard W. Sams, Indianapolis, IN, 1974, p 165–168 and 174.

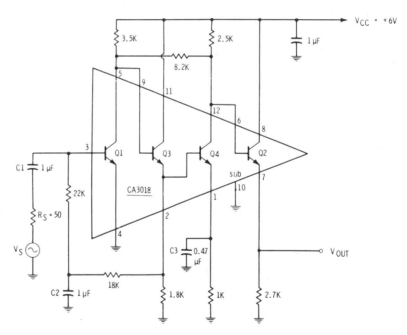

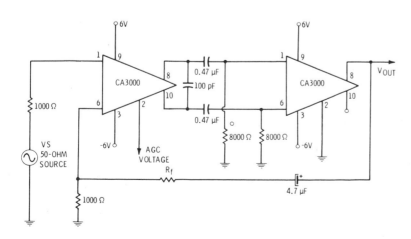

RC-COUPLED DIFFERENTIAL—Input signal is applied to base of first differential amplifier and push-pull output is obtained from pins 8 and 10 for transfer to inputs of second IC. Feedback is transferred through RC combination back to pin 6 of first IC. Gain is varied with AGC voltage applied to pin 2 of first IC. Gain is over 60 dB with flat response from 100 Hz to 100 kHz.—E. M. Noll, "Linear IC Principles, Experiments, and Projects," Howard W. Sams, Indianapolis, IN, 1974, p 89–91.

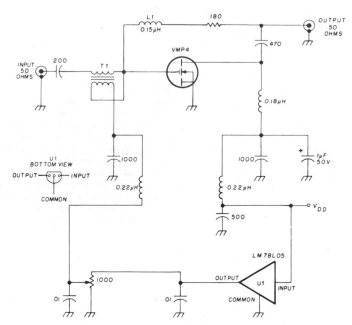

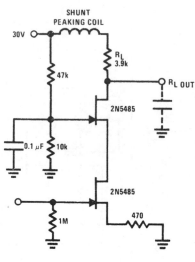

FET CASCODE VIDEO—Use of 2N5485 FETs gives very low input loading, with feedback reduced almost to zero. Bandwidth of amplifier is limited only by load resistance and capacitance.—"FET Databook," National Semiconductor, Santa Clara, CA, 1977, p 6-26—6-36.

40-265 MHz VMOS—Wideband power amplifier using Siliconix Mospower FET in negative-feedback circuit has flat gain within 0.5 dB over entire operational range of 40 to 265 MHz. Use 6 to 8 turns of No. 30 on ½-W 1-megohm resistor for L1 (not commercially molded choke). T1 is 4 turns No. 22 twisted-pair on Indiana General F625-9Q2 toroid core. Avoid static charges until transistor is soldered into circuit.—E. Oxner, Mospower FET as a Broadband Amplifier, *Ham Radio*, Dec. 1976, p 32–35.

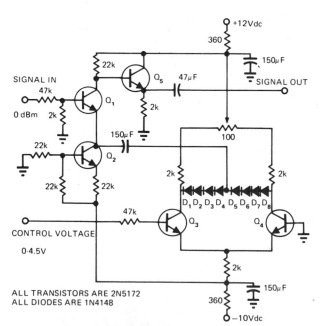

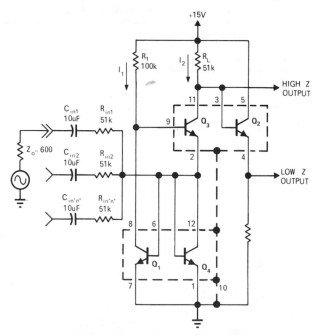

70-dB VOLTAGE-CONTROLLED GAIN—Amplifier Q_1 uses current source Q_2 as emitter resistor to provide correct current bias for class A operation. Coupling through 150-μF capacitor to silicon diode string D_1-D_8 provides variable resistance needed to achieve variable gain. Simple differential amplifier Q_3-Q_4 adjusts forward bias of diodes to change their forward resistance. Increasing positive control voltage from 0 to 4.5 V changes voltage gain from −74 dBm to about −4 dBm with respect to 0-dBm input signal.—N. A. Steiner, Voltage-Controlled Amplifier Covers 70 dB Range, *EDN Magazine*, March 5, 1975, p 72 and 74.

SUMMING AMPLIFIER—Uses RCA CA3018 four-transistor array as current-mirror triad with low-impedance buffered output, to serve as high-performance summing amplifier. Measured harmonic distortion is less than 1% at voltage gains up to 50 and with output swing of 10 V P-P. High output impedance of 51 kilohms can be buffered by Q_2 connected as emitter-follower.—W. G. Jung, Monolithic-Triad Current Summer, *EDN/EEE Magazine*, July 1, 1971, p 52.

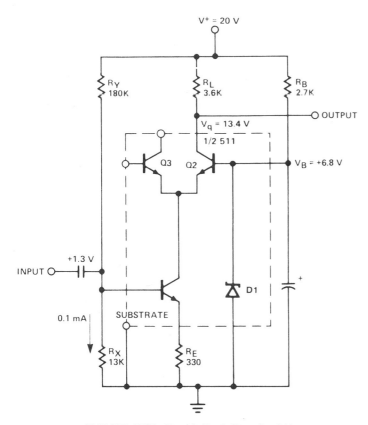

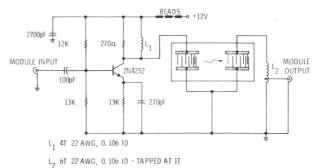

168-MHz BANDPASS—Gain stage provides gain of 6 dB from 162 to 174 MHz. Skirt slope immediately outside passband decreases at 80 db/MHz. Uses dispersive-design surface-acoustic-wave bandpass filter with 168-MHz center frequency, 7% bandwidth, and extremely steep skirt response. Parallel inductor at collector terminal matches capacitance of acoustic-wave device, and tapped inductor matches output terminal of filter to 50 ohms. Used in spread-spectrum communication receiver. Article covers design and construction of filter on quartz substrate.—T. F. Cheek, Jr., R. M. Hays, Jr., and C. S. Hartmann, A Wide-Band Low-Shape-Factor Amplifier Module Using an Acoustic Surface-Wave Bandpass Filter, *IEEE Journal of Solid-State Circuits,* Feb. 1973, p 66–70.

CASCODE RF/IF—Uses half of Signetics 511 transistor array to provide voltage gain of about 10 over bandwidth of 2 MHz with output voltage swing of 12 V P-P. Design procedure is given. Circuit provides excellent isolation between input and output.—"Signetics Analog Data Manual," Signetics, Sunnyvale, CA, 1977, p 746–747.

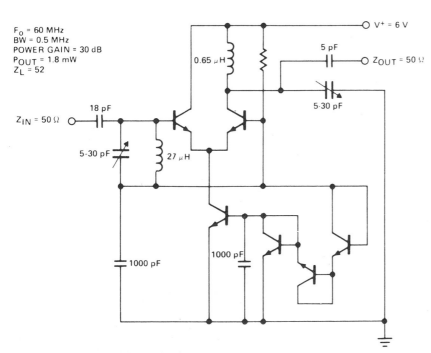

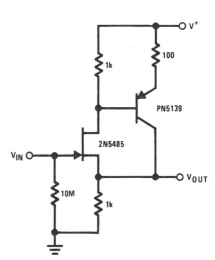

60-MHz NARROW-BAND—Signetics NE510/511 transistor array provides bandwidth of 0.5 MHz for 3 dB down and noise figure of 7 dB for power gain of 30 dB. Maximum output swing across 50 ohms is 300 mVRMS. Circuit is easily tuned.—"Signetics Analog Data Manual," Signetics, Sunnyvale, CA, 1977, p 749.

WIDEBAND BUFFER—Low input capacitance of 2N5485 FET makes compound series-feedback buffer serve as wideband unity-gain amplifier having high input impedance.—"FET Databook," National Semiconductor, Santa Clara, CA, 1977, p 6-26–6-36.

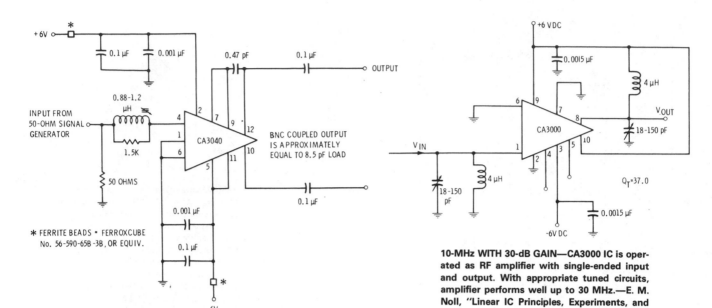

80 MHz WITH INPUT PEAKING—Response of CA3040 video IC is extended beyond 80 MHz in simple circuit that includes adjustable input peaking coil. Response is flat within 3 dB to well below 1 MHz, for gain of about 32 dB.—E. M. Noll, "Linear IC Principles, Experiments, and Projects," Howard W. Sams, Indianapolis, IN, 1974, p 163 and 169.

10-MHz WITH 30-dB GAIN—CA3000 IC is operated as RF amplifier with single-ended input and output. With appropriate tuned circuits, amplifier performs well up to 30 MHz.—E. M. Noll, "Linear IC Principles, Experiments, and Projects," Howard W. Sams, Indianapolis, IN, 1974, p 91–92.

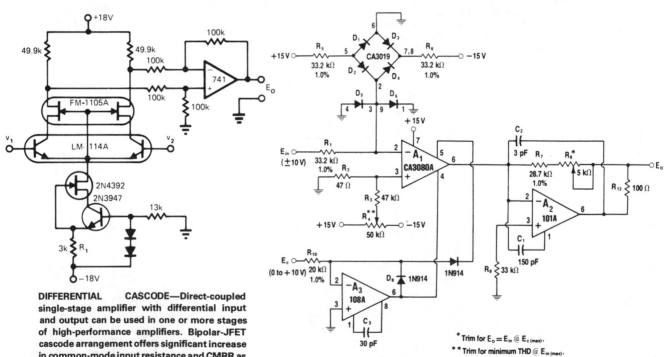

DIFFERENTIAL CASCODE—Direct-coupled single-stage amplifier with differential input and output can be used in one or more stages of high-performance amplifiers. Bipolar-JFET cascode arrangement offers significant increase in common-mode input resistance and CMRR as compared to conventional differential pair, with little or no degradation of other performance parameters. Differential-mode gain is 116, common-mode input resistance is greater than 100 gigohms, CMRR is greater than 160 dB, and current-source output resistance is greater than 1 gigohm. Article gives design equations.—R. C. Jaeger and G. A. Hellwarth, Differential Cascode Amplifier Offers Unique Advantages, *EDN Magazine,* June 5, 1974, p 78 and 80.

VOLTAGE-CONTROLLED OPAMP—CA3080A operational transconductance amplifier uses bridge to provide automatic temperature compensation of gain that is controlled by voltage between 0 and +10 V applied to opamp A3. With values shown, input and output signal-handling range is ±10 V. Once balanced, circuit provides linear gain control up to four decades.—W. G. Jung, "IC Op-Amp Cookbook," Howard W. Sams, Indianapolis, IN, 1974, p 455–456.

CHAPTER 2

Antenna Circuits

Includes circuits for measuring and adjusting VSWR, field strength, earth conductivity for grounds, and impedance, as well as antenna motor controls, radio direction finders, sferics receiver, active antennas, RF attenuators, remote antenna switching systems, RF magnetometer, and far-field signal sources for tuning beam antennas. See also Receiver, Transceiver, and Transmitter chapters.

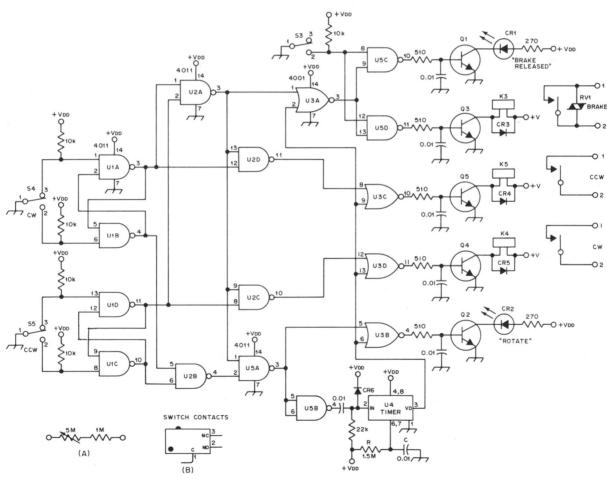

CR1, CR2 — Light-emitting diode, Motorola type MLED600 or equiv.
CR3-CR6, incl. — Silicon signal diode, 1N914 or equiv.
K3-K5, incl. — Switching relay, 12 V dc, 1200 ohms, 10 mA; contact rating 1 A; 125 V ac; Radio Shack 275-003 or equiv.
Q1-Q5, incl. — Silicon npn transistor, 2N3904 or equiv.
RV1 — Varistor, GE 750 or equiv.
U1, U2, U5 — CMOS quad NAND-gate IC, RCA CD-4011A or equiv.
U3 — CMOS quad NOR-gate IC, RCA CD-4001A, or equiv.
U4 — Timer IC, 555 or equiv.

DELAYED BRAKE—Protects antenna rotator on high tower from damage by delaying brake action automatically after rotation and by disabling direction-selector switches so antenna system coasts to stop before rotation can begin in other direction. For about 3-s delay in timer U4, use 2.2 megohms for R and 1 μF for C instead of values shown. RV1 is commonly listed as V150LA20A by GE. S3-S5 are original brake release and direction switches in CDE Ham-II rotor system. Article covers construction and installation, including modifications needed in control unit.—A. B. White, A Delayed Brake Release for the Ham-II, *QST*, Aug. 1977, p 14–16.

LADDER ATTENUATOR—Inserted in series with receiving antenna to provide 5 steps of attenuation for comparing performance of antennas or preamps. Resistors are ¼-W composition with 5% tolerance.—D. DeMaw, What Does My S-Meter Tell Me?, *QST*, June 1977, p 40–42.

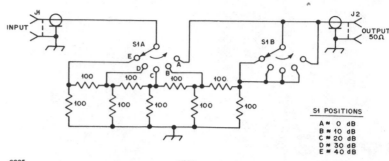

TWO-ROTATOR CONTROL—Low-cost Alliance C-225 TV antenna rotator and Alliance K22A rotator with control box are used with single transistorized-bridge control circuit. Rotators operate in tandem on same shaft to provide double torque for handling medium-size 20-meter amateur radio antennas. One arm of bridge is 520-ohm wirewound pot in which wiper position is proportional to heading. Article covers wiring and bench-testing of rotators.—F. E. Gehrke, Antenna Rotator for Medium-Sized Beams, *Ham Radio*, May 1976, p 48–51.

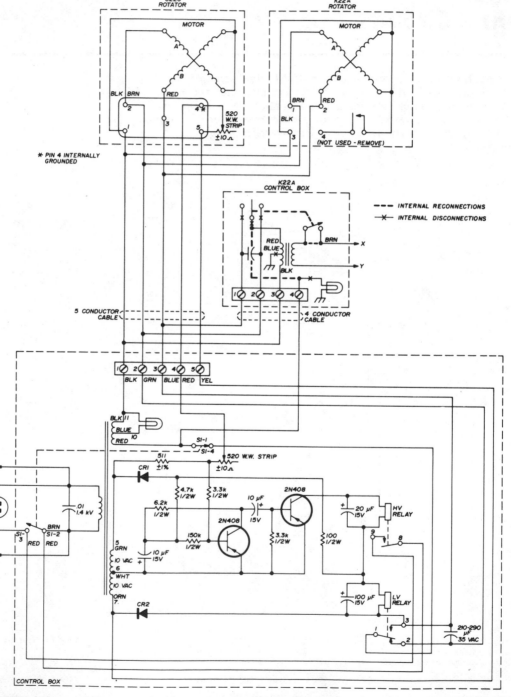

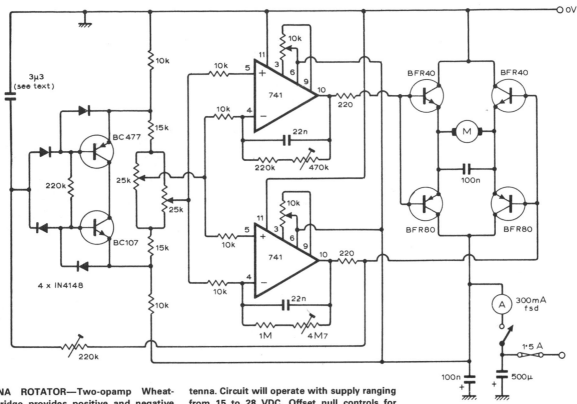

ANTENNA ROTATOR—Two-opamp Wheatstone bridge provides positive and negative error signals to give proportional control for 24-VDC motor used for remote positioning of antenna. Circuit will operate with supply ranging from 15 to 28 VDC. Offset null controls for opamps use 10K pots. Article describes operation and adjustment of circuit in detail. —D. J. Telfer, An Aerial Rotator Servo, *Wireless World*, April 1975, p 177–181.

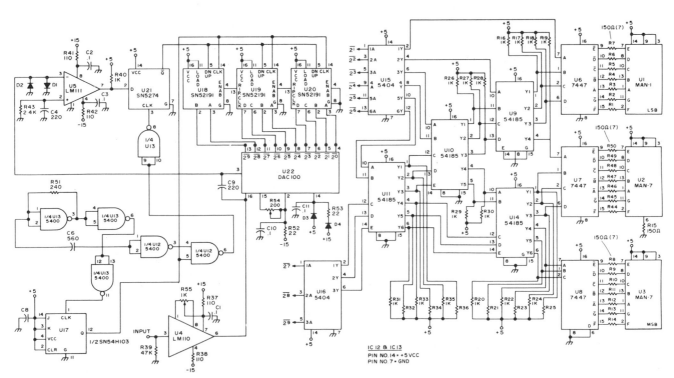

DVM FOR SWR—Converts voltage output from analog computer to drive for 3-digit LED display of standing-wave ratio. Circuit uses Precision Monolithics D/A converter A1MDAC-100CC-Q1. Requires regulated 5-VDC logic supply at 1 A for digital display, along with ±15 V supplies for logic. Article gives alignment procedure. Accuracy of digital reading is better than 0.1% over 0-8 V range.—T. Mayhugh, The Automatic SWR Computer, *73 Magazine*, Dec. 1974, p 86–87.

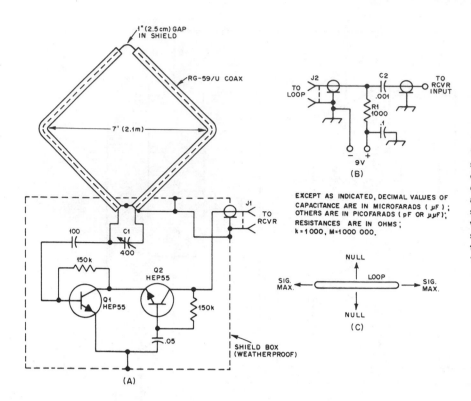

160-METER LOOP-PREAMP—Shielded 5-foot square loop and single preamp pull signals out of noise when propagation conditions make other antennas unsatisfactory. Operating voltage is supplied through coax feeder. R1 isolates signal energy from ground, and C2 keeps DC voltage out of receiver input. Nulls are off broad side of loop.—B. Boothe, Weak-Signal Reception on 160—Some Antenna Notes, *QST,* June 1977, p 35–39.

FAR-FIELD TRANSMITTER—Provides far-field signal source for tuning Yagi and other beam antennas used on amateur radio frequencies. Q1 is Pierce oscillator operating in fundamental mode of 7.06-MHz crystal to permit field-strength measurements at 14.12, 21.18, and 28.24 MHz for 20-, 15-, and 10-meter bands. Antenna uses two 5-foot lengths of wire connected as dipole. T1 is Amidon core T50-2 with 22 turns on primary and 20 turns center-tapped on secondary. T2 is same core with 22-turn primary and 5-turn secondary.—G. Hinkle, Closed Loop Antenna Tuning, *73 Magazine,* May 1976, p 32–33.

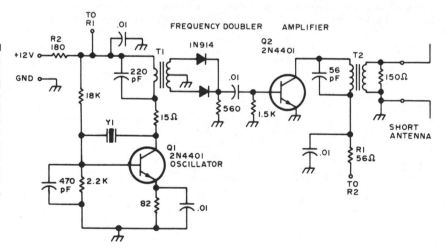

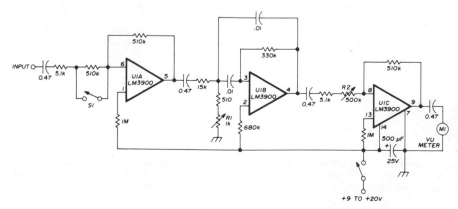

VSWR METER—Simple, easily transported VSWR meter consists of high-gain amplifier, narrow-bandwidth (100-Hz) selective amplifier tuned to 1000 Hz, and variable-gain output amplifier driving low-cost VU meter. Ideal for nulling-type VSWR measurements. Draws only about 6 mA from 9-V transistor battery. Closing S1 increases gain about 100 times for low-level readings. R1 sets U1B to 1000 Hz, while R2 sets reference on VU meter.—J. Reisert, Matching Techniques for VHF/UHF Antennas, *Ham Radio,* July 1976, p 50–56.

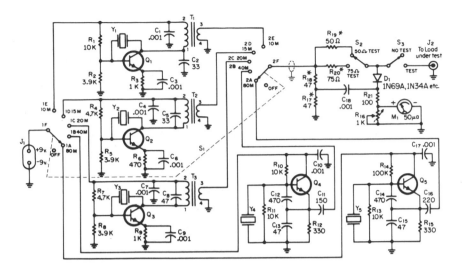

SELF-EXCITED SWR BRIDGE—Portable bridge has built-in signal sources for each band from 80 through 10 meters, for tuning antenna on tower before transmission line is connected. Oscillators are crystal controlled at desired antenna tune-up frequencies. Separate oscillators for each band simplify switching problems, so only supply voltage from J_1 and oscillator outputs to meter circuit need be switched. Current drain from 9-V battery is maximum of 12 mA. R_{17} and R_{18} should be closely matched, while R_{19} and R_{20} should have 5% tolerance.—T. P. Hulick, An S.W.R. Bridge with a Built-In 80 Through 10 Meter Signal Source, *CQ*, June 1971, p 64—66, 68, and 99.

Q_1-Q_5—RCA 40245.
S_1—2 pole 6 position subminiature rotary switch. (Centerlab PA-2005).
S_2—S.p.d.t. slide switch.
S_3—S.p.s.t. slide switch.
T_1—Pri.: 11 t. #36 e. Sec.: 3 t. #36 e. on Indiana General CF-101 Q2 toroid.
T_2—Pri.: 16 t. #36 e. Sec.: 4 t. #36 e. Same core as T_1.

T_3—Pri.: 20 t. #36 e. Sec.: 5 t. #36 e. Same core as T_1.
Y_1, Y_2, Y_3—Overtone crystals for 10, 15 and 20 meter bands respectively. HC-6U holders.
Y_4, Y_5—40 and 80 meter crystals respectively in HC-6U holders.

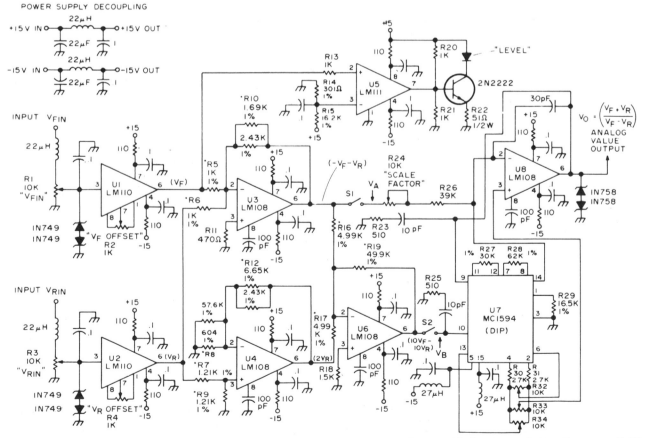

SWR COMPUTER—Automatically computes standing-wave ratio in 50-ohm coax feeding antenna and delivers analog voltage for driving meter or digital display. Inputs are forward (V_{FIN}) and reverse (V_{RIN}) voltages as conventionally measured for SWR checks. Requires regulated ±15 VDC supply at 40 mA. Article gives construction details and covers adjustment of critical resistors during alignment.—T. Mayhugh, A Digital SWR Computer!, *73 Magazine*, Nov. 1974, p 80—82, 84, and 86.

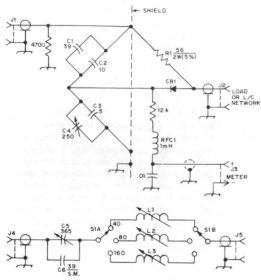

C1, C2 — 39- and 10-pF silver micas in parallel.
C3 — 5-pF silver mica.
C4 — 250-pF straight-line-wavelength variable (Hammarlund MC-250M).
C5 — 365-pF miniature variable (Archer-Allied 695-1000).
CR1 — Germanium diode.
J1, J2, J4, J5 — Coaxial receptacle.
J3 — Phono jack.
L1 — 15 turns No. 24 enamel close-wound on Miller 66A022-6 form (purple slug).
L2 — 30 turns like L1.
L3 — 63 turns like L1, but scramble-wound.
S1 — 2-pole 3-position wafer switch.

RF BRIDGE FOR COAX—Simplifies adjustment of vertical antenna for 40, 80, and 160 meters. S1 in add-on LC unit switches coil for desired band. Values of C1-C4 and standard resistor R1 give range of 10 to 150 ohms for measurement of radiation resistance. Meter can be from 50 to 200 μA full scale if 500 mW of power is available as signal source. For shorter-wavelength bands, change resistance in parallel with J1 to 5600 ohms and omit C6. L1 for 10 meters should then have 3½ turns No. 18 spaced to occupy ¼ inch on Miller 4200 coil form. L2 (15 meters) is 6 turns No. 16 enamel closewound on similar form. L3 (20 meters) is 11 turns No. 14 enamel on Miller 66A022-6 form.—J. Sevick, Simple RF Bridges, *QST*, April 1975, p 11–16 and 41.

5-STEP ATTENUATOR—Applications include comparing performance of various receiving antennas and measuring gain of preamp used ahead of receiver. Dashed lines represent required shield partitions. All resistors are ¼-W composition with 5% tolerance.—D. DeMaw, **What Does My S-Meter Tell Me?**, *QST*, June 1977, p 40–42.

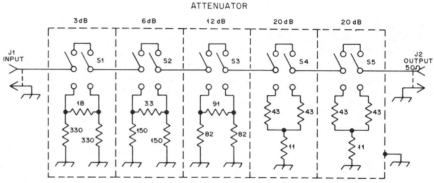

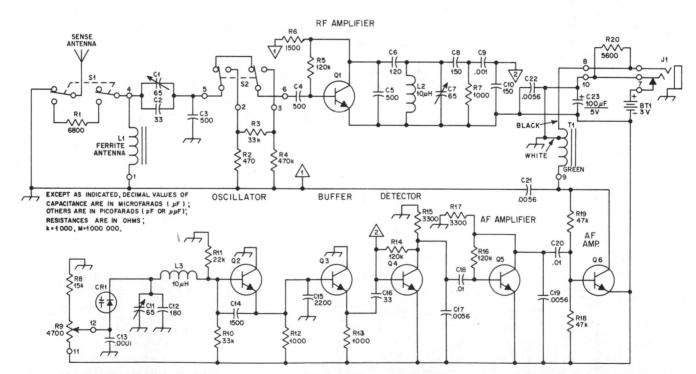

80-METER DIRECT-CONVERSION—Portable receiver with directional ferrod antenna and vertical sense antenna was developed for radio foxhunting at 1975 Boy Scout World Jamboree in Norway, in competitions for locating four low-power crystal-controlled transmitters hidden along 4-km course. Varactor-tuned oscillator provides 20-kHz tuning range with R9, adequate for the frequency used—3.566, 3.585, 3.635, or 3.680 MHz. T1 is subminiature autotransformer with 8-ohm and 2000-ohm sections, for 8-ohm headphones. For high-impedance headphones, connect headphone jack J1 to lug 9 of T1. ON/OFF switch is not needed. L1 is 22 turns No. 28 enamel wound over two 10 × 95 mm ferrite rods taped together. Q1-Q6 are NPN high-frequency small-signal transistors.—N. K. Holter, Radio Foxhunting in Europe, *QST*, Nov. 1976, p 43–46.

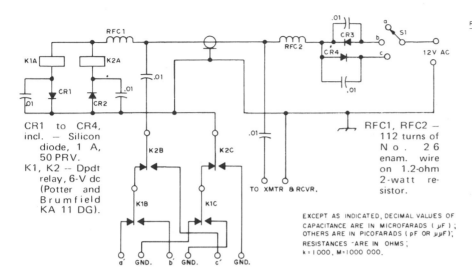

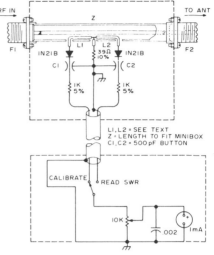

CR1 to CR4, incl. — Silicon diode, 1 A, 50 PRV.
K1, K2 -- Dpdt relay, 6-V dc (Potter and Brumfield KA 11 DG).

RFC1, RFC2 — 112 turns of No. 26 enam. wire on 1.2-ohm 2-watt resistor.

EXCEPT AS INDICATED, DECIMAL VALUES OF CAPACITANCE ARE IN MICROFARADS (μF); OTHERS ARE IN PICOFARADS (pF OR μμF); RESISTANCES ARE IN OHMS ; k = 1000, M=1000 000.

3-ANTENNA REMOTE SWITCHING—Single RF feed line serves for feeding transmitter power to tower and selecting desired one of three antennas. With S1 at a, neither K1 nor K2 is energized. RF energy then passes through cable to antenna terminals a' and GND. In position b, positive half-waves from 12-VAC supply operate relay K1 through CR1 and CR3, so antenna b' is energized. With S1 at c, K2 is energized through CR4 and CR2 for feeding c'.—U. H. Lammers, A Remote Antenna Switch, *QST*, Aug. 1974, p 41–43.

L1,L2 = SEE TEXT
Z = LENGTH TO FIT MINIBOX
C1,C2 = 500 pF BUTTON

SWR TO 500 MHz—Permits measuring standing-wave ratio well above limits of many inexpensive indicators. For transmitters up to 2 W, coupling loop L1-L2 can be about 1 inch long. For high-power transmitters, loop length can be reduced to about ⅛ inch.—W. E. Parker, UHF SWR Indicator, *73 Magazine*, June 1977, p 68–70.

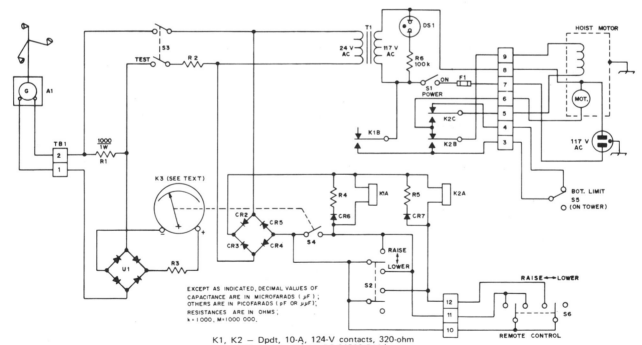

EXCEPT AS INDICATED, DECIMAL VALUES OF CAPACITANCE ARE IN MICROFARADS (μF); OTHERS ARE IN PICOFARADS (pF OR μμF); RESISTANCES ARE IN OHMS ; k = 1000, M=1000 000.

A1 — Three-cup anemometer (Taylor Instrument Corp. No. 14077Q).
CR2-CR5, incl. — Silicon diode 100 PRV, 1 A.
CR6, CR7 — 1N69.
F1 — 3.2 A, Slo-Blo.
DS1 — Neon indicator.

K1, K2 — Dpdt, 10-A, 124-V contacts, 320-ohm coil (Automatic Electric PG 24809-B11.)
K3 — Meter relay, 100 μA (Weston No. S46707).
R2, R3 — Approximately 12,000 ohm, 1-watt; see text.
S1 — Spst (JBT No. MS-35058-22).
S2, S6 — Dpdt center off (JBT. No. 35059-27).
S3 — Dpdt normally off (JBT No. MS-35059-30).

S4 — Part of K3.
S5 —Spdt, bottom limit switch (Microswitch No. BZE6-2RN).
T1 — 117-V primary, 24-v secondary, 300 mA
U1 — Bridge rectifier assembly (Bradley Labs. No. CO14E4F).

WIND-ACTIVATED CONTROL—Anemometer feeding meter relay energizes control relay for antenna-tower hoist motor, to lower tower automatically when wind exceeds preset safe speed and raise it again when wind drops well below danger level. When only K1 is energized, motor rotates in tower-lowering direction. When K1 and motor-reversing relay K2 are both energized, motor reverses and raises tower.—J. Bernstein, The Tower-Guard System, *QST*, Dec. 1974, p 25–28.

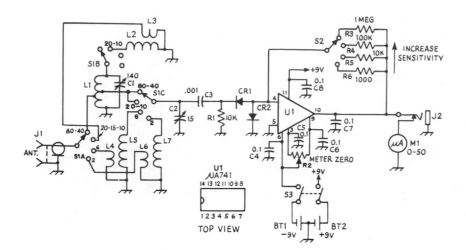

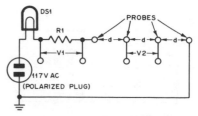

DS1 — 100-watt light bulb.
R1 — 14.6 ohms (5 watt).
Probes — 5/8-inch dia (iron or copper);
  spacing, d = 18 inches; penetration depth,
  D = 12 inches.

Earth conductivity = $(21) \times \dfrac{V1}{V2}$

  (millimhos/meter)..

C1 — 140-pF variable.
C2 — 15-pF variable.
CR1, CR2 — 1N914 or equiv.
L1 — 34 turns No. 24 enam. wound on an Amidon T-68-2 core, tapped 4 turns from ground end.
L2 — 12 turns No. 24 enam. wound on T-68-2 core.
L3 — 2 turns No. 24 wound at ground end of L2.
L4 — 1 turn No. 26 enam. wound at ground end of L5.

L5 — 12 turns No. 26 enam. wound on T-25-12 core.
L6 — 1 turn No. 26 enam.
L7 — 1 turn No. 18 enam. wound on T-25-12 core.
M1 — 50 or 100 μA dc.
R2 — 10,000-ohm control, linear taper.
S1 — Rotary switch, 3 poles, 5 positions, 3 sections.
S2 — Rotary switch, 1 pole, 4 positions.
S3 — Dpst Toggle.
U1 — μA741.

LINEAR FIELD-STRENGTH METER—Has sufficient sensitivity for checking antenna patterns and gain while positioned many wavelengths from antenna. Can be used remotely by connecting external meter at J2. L1 is tuned by C1 for 80 or 40 meters. For 20, 15, or 10 meters, L2 is switched in parallel with L1. L5 and C2 cover about 40 to 60 MHz, while L7 and C2 cover 130 to 180 MHz. Band-switched circuits avoid use of plug-in inductors. At most sensitive setting of S2, M1 will detect signals from pickup antenna as weak as 100 μV.—L. McCoy, A Linear Field-Strength Meter, *QST*, Jan. 1973, p 18—20 and 35.

EARTH CONDUCTIVITY—Simple AC measurement technique gives 25% accuracy, adequate for siting amateur radio antennas and designing radial ground systems. Measured values will range from 1 to 5 millimhos per meter for poor soil, 10—15 for average soil or fresh water, 100 for very good soil, and 5000 for salt water.—J. Sevick, Short Ground-Radial Systems for Short Verticals, *QST*, April 1978, p 30—33.

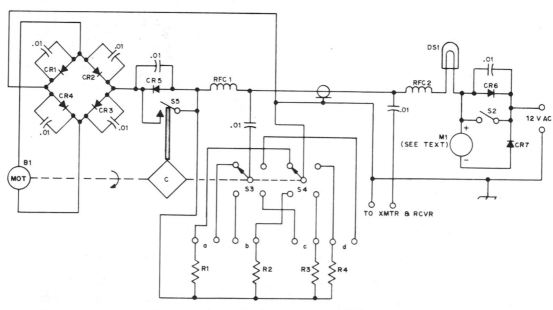

FOUR-POSITION MOTOR SWITCH—Single RF feed line also carries DC for 3-V permanent magnet DC motor B1 atop antenna tower, driving S3 and S4 for remote switching to antennas a, b, c, and d. Diagram shows switches set for feed to antenna a, with no drive applied to B1 since cam C has opened microswitch S5. CR5 and CR6 are now connected in series with opposite polarity, so neither positive nor negative half-waves from 12-VAC supply can drive motor. If S2 is closed, positive half-waves start B1. Once started, motor runs until cam opens S5; if S2 has not yet been released, motor continues running on positive and negative half-waves. Diode bridge CR1-CR4 makes motor rotate in only one direction for either drive polarity. If S2 is released, before S5 opens, motor stops. 6-V 1-A lamp DS1 comes on dimly when S2 is closed and brightens when S5 closes. If S2 is released now, B1 drives to next position and stops. If S2 is held down, switching continues. Meter M1 and CR7 identify position of switch. R1-R4, in range of 1K to 10K, are chosen to give ¼, ½, ¾, and full deflection of meter. Motor drives switch through 2860:1 reduction gears taken from alarm clock. All diodes are 50-PIV 1-A silicon such as 1N4001.—U. H. Lammers, A Remote Antenna Switch, *QST*, Aug. 1974, p 41—43.

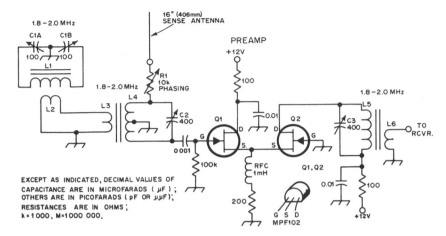

160-METER PREAMP WITH FERRITE LOOP—

Ferrite-rod antenna is combined with 16-inch wire rod to give cardioid radiation pattern for low-noise 160-meter antenna system. Preamp using MPF102 FETs has gain of 25 dB. L1 is 48 turns No. 14 enamel spread to 4.5 inches on 0.5-inch Amidon ferrite rod 7 inches long. L2 is 6-turn link wound over center of L1. L4 and L5 are each 50 turns No. 26 enamel on T80-2 powdered-iron toroid cores, with 6 turns for links L3 and L6.—D. DeMaw, Low-Noise Receiving Antennas, *QST*, Dec. 1977, p 36–39.

EXCEPT AS INDICATED, DECIMAL VALUES OF
CAPACITANCE ARE IN MICROFARADS (μF) ;
OTHERS ARE IN PICOFARADS (pF OR μμF) ;
RESISTANCES ARE IN OHMS ;
k =1 000, M=1 000 000.

CAPACITIVE ANTENNA—

Combination of short whip antenna and broadband amplifier gives antenna covering entire range of 3 to 30 MHz without frequency selectivity. Q1 is 2N3819 FET source follower driving three-transistor amplifier using 2N918, 2N6008, or other 200-MHz 20-V NPN transistors to provide 30-dB gain. Circuit rolloff starts at 3 and 35 MHz. High gain of amplifier makes combination simulate quarter-wave whip over entire frequency range.—R. C. Wilson, The Incredible 18″ All-Band Antenna, *73 Magazine*, March 1975, p 49–50.

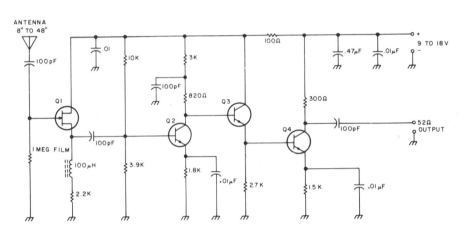

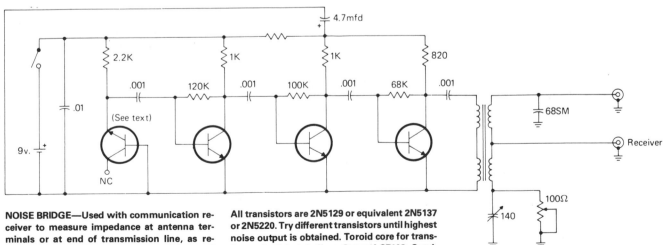

NOISE BRIDGE—

Used with communication receiver to measure impedance at antenna terminals or at end of transmission line, as required for adjusting antenna matching and loading devices for desired impedance at specific frequency. Consists of diode-connected transistor broadband-noise generator, 3-stage noise amplifier, and toroid transformer bridge. All transistors are 2N5129 or equivalent 2N5137 or 2N5220. Try different transistors until highest noise output is obtained. Toroid core for transformer is 3/8-inch Indiana General CF102. Quadrifilar winding has 4½ turns of four No. 28 enamel wires twisted together, wound on core and connected as on diagram. Noise bridge can also serve as wideband noise source for signal injection during troubleshooting in AF or RF circuits, and as noise source for aligning RF circuits.—J. J. Schultz, An Improved Antenna Noise Bridge, *CQ*, Sept. 1976, p 27–29 and 75.

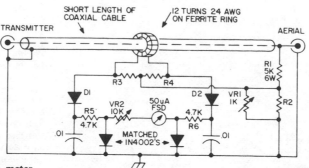

LOGARITHMIC WATTMETER—Single meter scale covers 1–1000 W, with equally spaced divisions for 1, 10, 100, and 1000. This log scale makes it possible to measure very low reflected powers and very high forward powers simultaneously with same percentage accuracy. Basis of operation is that voltage dropped across forward-biased 1N4002 silicon PN junction diode is proportional to logarithm of current through it. For 50-ohm line, use 220 for R2 and 27 for R3 and R4. For 75-ohm line, corresponding values are 180 and 33. Detector diodes are point-contact germanium rated at 80 PIV. Article gives construction details. Ground coax braid at one end only. Ferrite ring is 0.5-inch Mullard FX1596 or equivalent.—P. G. Martin, Some Directional Wattmeters and a Novel SWR Meter, *73 Magazine,* Aug. 1974, p 17, 19–21, 23–24, and 26.

1-kHz MODULATOR FOR VHF SOURCE—Used with 144-MHz signal generator driving VSWR bridge, for measuring and matching VHF antennas. R1 adjusts frequency of NE555 timer used in place of customary MVBR. Series-pass transistor increases output of MVBR about 2 dB.—J. Reisert, Matching Techniques for VHF/UHF Antennas, *Ham Radio,* July 1976, p 50–56.

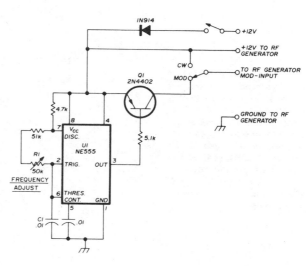

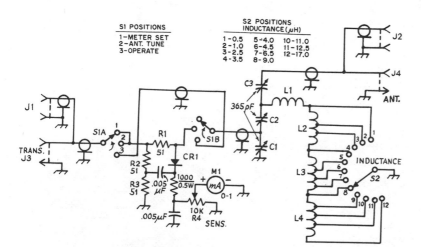

C1-C3, incl. — Miniature 365-pF variable (Archer/Radio Shack No. 272-1341 or equiv.).

CR1 — IN34A or equiv.

L1 — 15 turns No. 24 enam. wire, close-wound on 1/4-inch ID form. Remove form after winding.

L2 — 28 turns No. 24 enam. wire on Amidon T-50-6 toroid core. Tap 7 turns from each end. (Amidon Associates, 12033 Otsego St., N. Hollywood, CA 91607.)

L3 — 28 turns No. 24 enam. wire on Amidon T-50-2 toroid core. Tap at 5, 10 and 15 turns from L2 end.

L4 — 36 turns No. 24 enam. wire on Amidon T-68-2 toroid core. Tap at 6, 12 and 18 turns from L3 end.

M1 — 0 to 1-mA dc meter, 1-1/2 inches square. See text.

R1-R3, incl. — 51-ohm, 2-watt, 5-percent tolerance.

R4 — Miniature 10,000-ohm control, audio or linear taper suitable.

TRANSMATCH—Tapped variable inductance and three broadcast tuning capacitors are easily preadjusted to match low-power (QRP) transmitter to antenna for SWR of 1 in commonly used amateur bands. Resistance bridge is used only for initial determination of correct settings for C1, C2, C3, and S2 at each band to be used. Set S1 at 1, feed peak output of transmitter to J1 (5 W maximum), and adjust R4 for full-scale reading of M1. Next, connect 50-ohm resistive load between CR1-R1 junction and ground. Meter reading should now drop to zero, indicating null at 50 ohms. Move 50-ohm dummy load to J2, set S1 at 2, and adjust settings of C1, C2, and C3 for zero deflection of meter. Note settings, then repeat for each other transmitter frequency to be used. Repeat procedure with antenna or feed line in place of dummy load, using smallest inductance that gives SWR of 1. After completing adjustments, set S1 to 3 to bypass bridge for normal transmitter operation.—D. DeMaw, A Poor Ham's QRP Transmatch, *QST,* Oct. 1973, p 11–13.

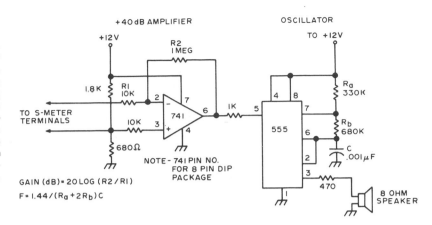

V/F CONVERTER—Voltage developed across S-meter is amplified by 741 opamp having gain of 40 dB, so full-scale voltage of 100 mV becomes 10 V at opamp output. This drives modulation input of 555 timer connected as free-running oscillator. Nominal 1-kHz output increases in frequency as drive current is reduced; conversely, drop in frequency corresponds to stronger signal at S-meter. Developed for use as audible guide when tuning Yagi and other beam antennas for amateur radio operation.—G. Hinkle, Closed Loop Antenna Tuning, *73 Magazine*, May 1976, p 32–33.

$$GAIN\ (dB) = 20\ LOG\ (R2/R1)$$
$$F = 1.44/(R_a + 2R_b)C$$

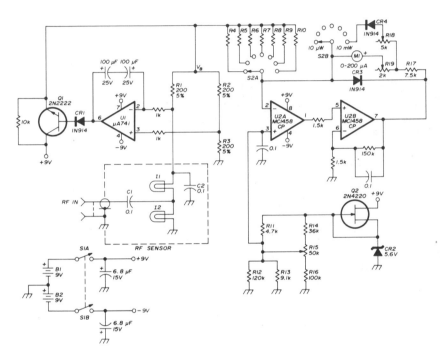

0.2 μW TO 10 mW—Accurate low-power wattmeter uses small lamps as barretters for measuring RF power up to 10 mW from 1 to 500 MHz. Applications include measurements of antenna gain, local oscillator frequency, VSWR, and filter response. Subminiature T-3/4 RF sensor lamps operate in bridge circuit with R1, R2, and R3. Voltage difference between bridge legs is amplified by opamp U1. Bridge current driver Q1 supplies current for balancing bridge. Equilibrium voltage of 3.5 V at V_B is fed to metering circuit including U2. Article covers calculation of values for calibration resistors R4-R10, which range from 5.715 to 7192 ohms.—J. H. Bowen, Accurate Low Power RF Wattmeter for High Frequency and VHF Measurements, *Ham Radio*, Dec. 1977, p 38–43.

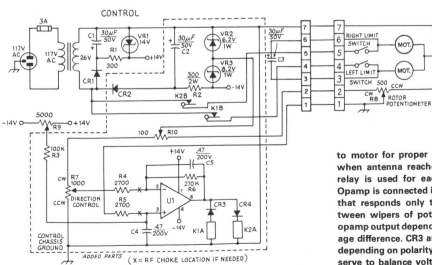

ANTENNA ROTATOR—Developed for use with CDE TR-44 antenna control using low-voltage AC motor having pot for bearing indication. Circuit eliminates need for holding control handle in position until antenna reaches desired bearing. Uses 12-VDC 1000-ohm 1-A relays, TI SN7274IL opamp U1, and wirewound 360° rotation command pot R7 operating from 14-V regulated supply of original control. When R7 is set to desired new heading, relay applies power to motor for proper direction, and drops out when antenna reaches desired heading. One relay is used for each direction of rotation. Opamp is connected in differential-input mode that responds only to difference voltage between wipers of pots R7 and R8. Polarity of opamp output depends on polarity of input voltage difference. CR3 and CR4 energize K1 or K2 depending on polarity of error signal. R9 and R3 serve to balance voltage difference remaining when R7 and R8 are at travel limits. R9 also nulls offset present when there is no input to U1. Accuracy is about 5°. Diodes are 100-PIV 0.5-A silicon.—K. H. Sueker, Automating the TR-44 Antenna Rotor, *QST*, June 1973, p 28–30.

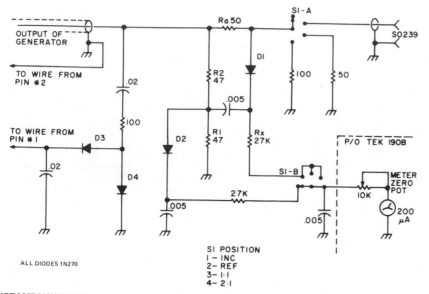

FIELD STRENGTH AT 7 MHz—Operates from single dry cell. Meter can be calibrated in decibels with Hewlett-Packard 606A or equivalent signal generator. Jack permits remote metering. L1 is 5 turns, and L2 is 30 turns wound on Amidon T68-2 core.—R. W. Jones, A 7-MHz Vertical Parasitic Array, *QST*, Nov. 1973, p 39–43 and 52.

INSTANT VSWR BRIDGE—Modified 190B Tektronix constant-amplitude signal generator is combined with 50-ohm resistance bridge to give stable high-accuracy instrument for measuring voltage standing-wave ratio as guide for tuning antennas. Range is 160 meters through 10 meters. Trim Rx so incident or forward voltage at position 1 of S1 equals reflected voltage at position 2. Article gives chart for finding VSWR. Step-by-step procedure for modifying signal generator is given.—D. Sander, Make Antenna Tuning a Joy, *73 Magazine,* May 1978, p 134–136.

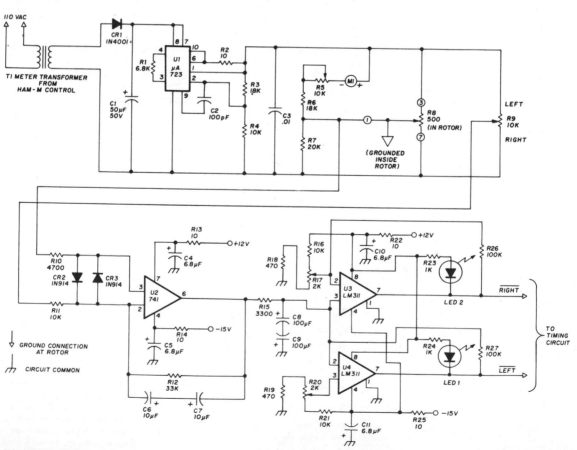

ANTENNA POSITION CONTROL—IC logic provides automatic brake release and positive position control for commercial Ham-M antenna rotator. Regulated power supply drives bridge having position-sensing pot R8 in rotator and R9 in control box. When antenna is in desired position, wiper voltages of pots are equal. When R9 is set to new position, voltage difference is amplified by error amplifier U2. Comparators U3 and U4 determine rotation direction needed for rebalance and deliver logic circuits to timing circuit (also given in article) that drives motor and brake release relays. Timer prevents jamming of circuit by operator error.—P. Zander, Automatic Position Control for the HAM-M Rotator, *Ham Radio,* May 1977, p 42–45.

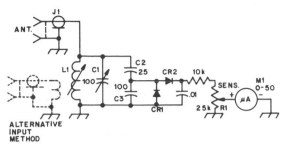

ALTERNATIVE INPUT METHOD

Band	160 M	80 M	40 M	20 M	15 M	10 M
L1 (μH)	100 (Nom.)	25 (Nom.)	10 (Nom.)	2.2 (Nom.)	1.3 (Nom.)	0.5 (Nom.)
C2 (pF)	25	25	15	15	10	10
C3 (pF)	100	100	68	68	47	47
Miller Coil	4409	4407	4406	4404	4403	4303

FIELD-STRENGTH METER—Useful for antenna experiments and adjustments in amateur bands from 160 to 10 meters. Increasing size of pickup antenna increases sensitivity. Far-field measurements are made with alternate input circuit, in which reference dipole or quarter-wave wire cut for frequency of interest is connected to input link. Diodes are 1N34A germanium or equivalent. M1 is 50 μA. Table gives values of tuned-circuit components for six amateur bands.—D. DeMaw, A Simple Field-Strength Meter and How to Calibrate It, *QST*, Aug. 1975, p 21–23.

MODULATION MONITOR—Provides off-the-air monitoring of RF signals up to 200 MHz by rectified detection of AM signals and by slope detection of FM signals. Can also be used as signal tracer, audio amplifier, or hidden-transmitter locator. High-gain audio amplifier has low-noise cascode input stage and output stage driving headphone or loudspeaker. S₁ selects RF signals detected by D₁ or AF applied to J₂. L₁ is 4 turns No. 18 for monitoring 75-150 MHz. Will also monitor VHF transmissions from pilot to ground stations while in commercial aircraft, using 24-inch wire antenna near window and earphone. Passive-type receiver is safe in aircraft because it has no oscillators that could interfere with navigation equipment.—W. F. Splichal, Jr., Sensitive Modulation Monitor, *CQ*, Jan. 1973, p 59–61.

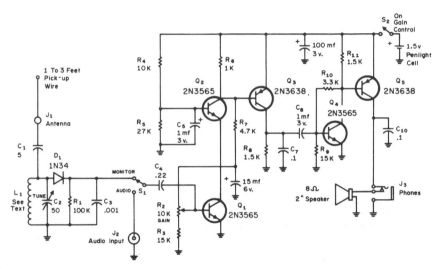

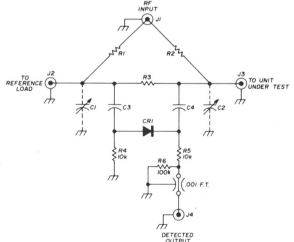

C1, C2	small capacitive tab required for balance
C3, C4	0.001 μF (small disc ceramic or chip capacitor)
CR1	1N82A or equivalent germanium diode
J1, J4	UG-290A/U BNC connector
J2, J3	UG-58/U type-N connector
R1, R2	47 to 55 ohms, matched
R3	51 ohms, ¼-watt carbon composition
R4, R5	10k ohms, ¼-watt carbon composition
R6	100k ohms, ¼-watt carbon composition

VSWR BRIDGE—Works well through 450 MHz for measuring and matching VHF and UHF antennas. If identical load impedances are placed at J2 and J3, signals at opposite ends of R3 are equal and in phase and there is no output at J4. If impedances are different, output proportional to difference appears at J4. Impedance values can be from 25 to 100 ohms, although circuit is designed for optimum performance at 50 ohms.—J. Reisert, Matching Techniques for VHF/UHF Antennas, *Ham Radio*, July 1976, p 50–56.

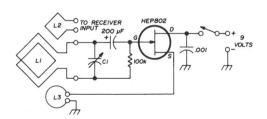

Q MULTIPLIER FOR LOOP—Improves performance of loop antenna on 40, 80, and 160 meters. Feedback control is obtained with adjustable single-turn loop L3 coupled to L1, and receiver input is taken from L2. L3 is rotated within field of L1 to adjust amount of regeneration, optimize circuit Q, and make directional null more pronounced. Article gives loop construction details. Ground lower end of 100K resistor to provide ground return for FET.—K. Cornell, Loop Antenna Receiving Aid, *Ham Radio*, May 1975, p 66–70.

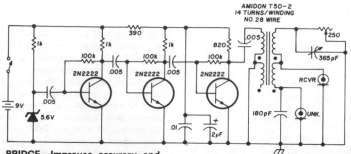

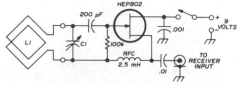

RF NOISE BRIDGE—Improves accuracy and measurement range for impedance measurements from 3.5 to 30 MHz, particularly resistive and reactive components of high-frequency antennas. Accuracy is 3 ohms RMS. Wideband noise, generated in zener followed by three-transistor amplifier, is injected into two legs of bridge in equal amounts by toroidal transformer having quadrifilar windings. With unknown impedance connected and detector (any communication receiver) set to desired frequency, reference impedances (250-ohm non-inductive pot and 360-pF variable capacitor) are adjusted for deepest possible null. Value of unknown impedance is then equal to parallel combination of references. Article covers construction and calibration.—R. A. Hubbs and A. F. Doting, Improvements to the RX Noise Bridge, *Ham Radio*, Feb. 1977, p 10–20.

LOOP PREAMP—Loop for lower-frequency amateur bands is connected to gate of HEP802 FET and output to receiver is taken from FET source. C1 is two-gang variable capacitor from old broadcast radio, with stators in parallel to give 600 pF. Article gives loop data for 40, 80, and 160 meters and for high end of broadcast band. For 40 and 80 meters, use 18-inch square loop with 2 turns spaced ¼ inch. Ground lower end of 100K resistor to provide ground return for FET.—K. Cornell, Loop Antenna Receiving Aid, *Ham Radio*, May 1975, p 66–70.

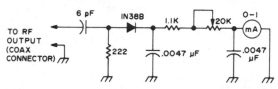

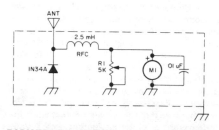

RF OUTPUT INDICATOR—Designed for use with amateur radio transmitters. Pot is adjusted for maximum desired indication on band used. For 20–10 meters, 6-pF capacitor is adequate. On lower bands (80-40 meters), use 7 or 12 pF instead.—Novice Q & A, *73 Magazine*, Holiday issue 1976, p 20.

RADIATED-FIELD METER—Gives quick check of overall transmitter performance, including antenna system. Meter can be 1 mA, but 0–200 μA or 0–50 μA will be more sensitive. The longer the reference antenna used, the greater will be the sensitivity of the meter. Keep lead lengths short. If measurements for various transmitter inputs are recorded when transmitter is working properly, they can serve as guide for later troubleshooting.—E. Hartz, Is My Rig Working or Not?, *73 Magazine*, Oct. 1976, p 56–57.

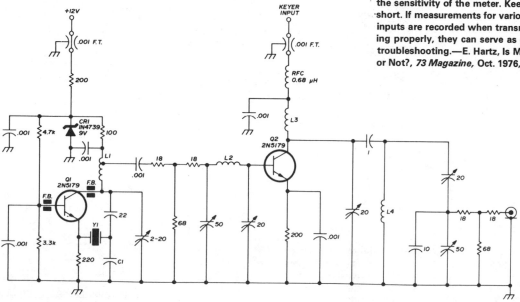

C1	as large as possible, consistent with good oscillator starting (100 pF typical)
FB	ferrite bead
L1	9 turns no. 24 (0.5mm) on Amidon T-37-12 toroid core; tapped 3 turns from cold end
L2	15 turns no. 28 (0.3mm) on Amidon T-25-12 toroid core
L3,L4	4 turns no. 24 (0.5mm), ¼" (6.5mm) inside diameter, ¼" (6.5mm) long
Y1	72-MHz, 5th-overtone, series-mode crystal

144 MHz FOR VSWR BRIDGE—Modulated signal source provides 10-mW CW output and 5-mW modulated output at modulation frequency of 1000 Hz. Spurious and harmonic outputs are 40 dB below desired output. 72-MHz crystal oscillator is followed by doubler stage. Oscillator runs continuously while doubler is keyed with simple ON/OFF square-wave keying. Freedom from load variations is obtained with double-tuned output filter providing up to 6-dB attenuation between generator and load. Use regulated power supply or batteries.—J. Reisert, Matching Techniques for VHF/UHF Antennas, *Ham Radio*, July 1976, p 50–56.

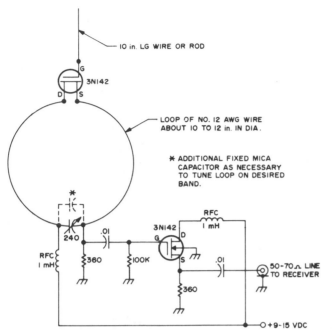

- 10 in. LG WIRE OR ROD
- 3N142
- LOOP OF NO. 12 AWG WIRE ABOUT 10 TO 12 in. IN DIA.
- ✱ ADDITIONAL FIXED MICA CAPACITOR AS NECESSARY TO TUNE LOOP ON DESIRED BAND.
- 240
- RFC 1 mH
- 360
- 100K
- 3N142
- RFC 1 mH
- .01
- .01
- 360
- 50-70 Ω LINE TO RECEIVER
- ○ +9-15 VDC

ACTIVE ANTENNA—Uses tuned loop with relatively low Q for broadband operation over one amateur band, phase-coupled by FET to 10-inch vertical sensing antenna to give unidirectional reception pattern. Loop is tuned to either 80 or 40 meters by trimmer capacitor at its base. Output of loop is coupled to another 3N142 FET used as source follower, to isolate output of loop from heavy loading effect of 50-ohm transmission line going to receiver. Performance is comparable to that of full-size quarter-wave vertical antenna on 40 meters. Battery source can be used because drain is only about 2 mA.— J. J. Schultz, An Experimental Miniature Antenna for 40 to 80 m, *73 Magazine,* June 1973, p 29–32.

PROTECTION FOR QRP TUNING—Simple resistive SWR bridge provides dummy load, relative power output indicator, and safe method of tuning transmitter without destroying transistors because of mismatched load. Input divider R1-R4 has total resistance of 50 ohms, using ½-W composition resistors, for dissipating transmitter output when S1 is in TUNE position. M1 indicates relative power applied to this load. Antenna is connected through Transmatch, and antenna tuner is adjusted for minimum deflection or lowest SWR. R5 isolates transmitter from antenna. With S1 in OPERATE position, M1 indicates relative power output into antenna.— A. S. Woodhull, Simplified Output Metering Protects QRP Transmitters, *QST,* April 1977, p 57.

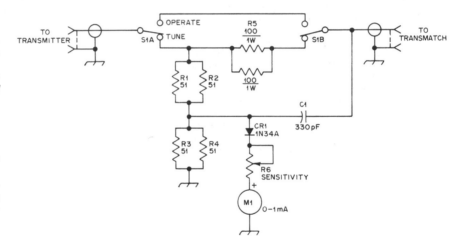

- TO TRANSMITTER
- OPERATE
- TUNE
- S1A
- R5 100 1W
- 100 1W
- S1B
- TO TRANSMATCH
- R1 51
- R2 51
- C1 330 pF
- R3 51
- R4 51
- CR1 1N34A
- R6 SENSITIVITY
- M1 0–1 mA

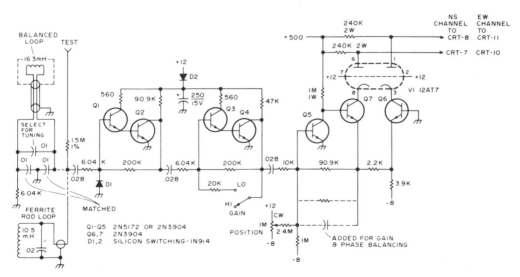

- BALANCED LOOP
- 16.3 mH
- TEST
- SELECT FOR TUNING
- .01 .01 .01
- 1.5M 1%
- 6.04K
- 6.04K
- FERRITE ROD LOOP
- 10.5 mH
- .02
- +12
- D2
- 560
- Q1
- 90.9K
- Q2
- 250 15V
- 560
- Q3 Q4
- 47K
- 6.04 K
- 200K
- .028
- D1
- MATCHED
- 6.04 K
- .028
- 200K
- 20K
- LO
- HI
- GAIN POSITION
- .028
- 10K
- Q5
- 90.9K
- 2.2K
- +12
- CW
- 1M
- 2.4M
- 1M
- −8
- Q1-Q5 2N5172 OR 2N3904
- Q6,7 2N3904
- D1,2 SILICON SWITCHING-1N914
- 240K 2W
- +500
- 240K 2W
- NS CHANNEL TO CRT-8
- EW CHANNEL TO CRT-11
- CRT-7 CRT-10
- 6
- +12
- 7
- +12
- 2
- 8
- Q7 Q6
- V1 12AT7
- 1M 1W
- 3.9K
- −8
- ADDED FOR GAIN & PHASE BALANCING
- −8

10-kHz SFERICS RECEIVER—Developed to measure direction and strength of atmospheric electromagnetic radiation (sferics) associated with severe weather conditions, for detection and tracking of tornadoes. Signals from crossed loop antennas feed deflection amplifier of CRO. Article covers problems involved and gives circuit for sense amplifier that resolves 180° ambiguity in oscilloscope pattern.—R. W. Fergus, A Ham Radio Severe Weather Warning Net, *73 Magazine,* Sept. 1974, p 27–30, 32, 34–36, and 38–39.

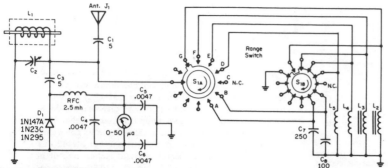

C₁, C₃—5 mmf tubular or disc ceramic.
C₂—5-120 mmf variable capacitor. Millen #20100.
C₄, C₅, C₆—.0047 mf tubular or disc ceramic capacitor.
C₇—1.8-2.2 mc—250 mmf tubular ceramic or mica capacitor.
C₈—2.2-3.2 mc—100 mmf tubular ceramic or mica capacitor.
L₁—3.2-8.5 mc—22 t. #22 on 4½" × 1/4" d. ferrite rod—see text.

L₂—6.3-17 mc—16 t. #22 on 1" l. × 1/4" d. ferrite rod.
L₃—14-38 mc—6 t. #22 on 5/8" l. × 1/4" d. ferrite rod.
L₄—26-70 mc—7 t. #22 air-wound, self-supporting, 1/4" i.d.
L₅—60-150 mc—2 t. #22 air-wound, self-supporting, 1/4" i.d. with 3/16" spacing between turns, spread or squeezed as needed to cover v.h.f. range.

RF MAGNETOMETER—Measures RF radiation and current distribution for antennas, transmission lines, ground leads, building wiring, and shields. Can also be used as sensitive portable field-strength meter. Will indicate orientation of field. High-Q circuit is tunable from 1.8 to 150 MHz for indicating frequency of fields produced by RF harmonic. Applications include detecting reradiation from rain gutters, metal fencing, towers, and guy wires that are distorting antenna field patterns, and detecting radiation from ground leads, appliance power cords, and hidden building wiring. When used as probe, will accurately pinpoint leakage of RF energy from joints, holes, or slots in shielded enclosures. Operation is similar to absorption-type wavemeter, except that pickup coil is electrostatically shielded by slotted aluminum IF transformer can to eliminate capacitive coupling. Inductor is wound on ferrite coil to give very high Q as pickup element.—W. M. Scherer, An R.F. Magnetometer and Field Strength Meter, *CQ*, April 1971, p 16–20.

DIRECTIONAL WATTMETER—Gives 10% accuracy between about 100 kHz and 70 MHz. Full-scale values of ranges are 1, 10, 100, and 1000 W. Low resistance in secondary circuit of current transformer is split into two equal parts, so sum and difference voltages are available at ends of secondary. Meters then indicate forward and reflected power values. For 50-ohm line, use 220 for R2 and 27 for R3 and R4. For 75-ohm line, corresponding values are 180 and 33. Detector diodes are point-contact germanium rated at 80 PIV. Article gives construction details. Ground coax braid at one end only. Ferrite 4ecring is 0.5-inch Mullard FX1596 or equivalent.—P. G. Martin, Some Directional Wattmeters and a Novel SWR Meter, *73 Magazine*, Aug. 1974, p 17, 19–21, 23–24, and 26.

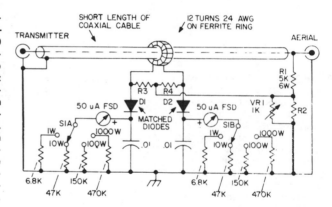

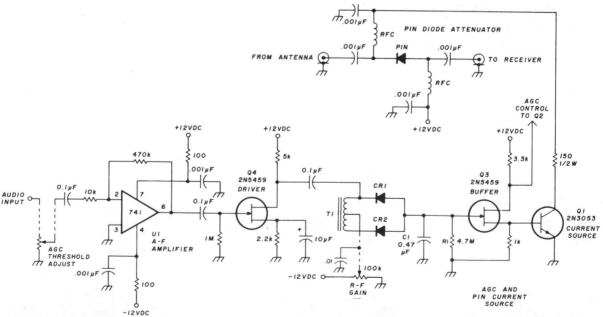

PIN DIODE ATTENUATOR—Designed for insertion between antenna and input of any HF receiver to improve adjacent-channel selectivity by providing attenuation ahead of mixer for entire tuning range. Hewlett-Packard 5082-3379 PIN diode has very low impedance when conducting and very high impedance when bias current is small. NPN transistor Q1 provides over 100 mA as current source to PIN diode. Q1 is driven by AGC circuit through JFET buffer Q3. AGC voltage is derived from top of audio gain control in receiver for rectification, with 200 mVRMS at input of opamp U1 giving maximum attenuation. Center tap of T1 (any small AF transformer) can be grounded. CR1 and CR2 are germanium diodes. Article also gives circuit of IF system using cascaded 9-MHz crystal filters to improve selectivity further and provide overall AGC control range of 70 dB.—M. Goldstein, Improved Receiver Selectivity and Gain Control, *Ham Radio*, Nov. 1977, p 71–73.

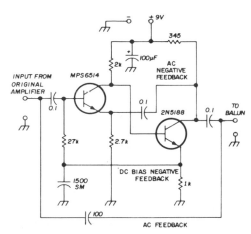

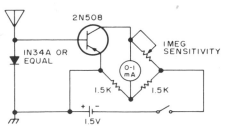

FIELD-STRENGTH METER—Easily assembled for checking performance of amateur radio transmitter and its antenna system.—Circuits, *73 Magazine,* Jan. 1974, p 128.

BROADBAND NOISE AMPLIFIER—Developed for use with antenna noise bridges for measurements at 20 meters. Provides 35 to 50 dB additional gain, not entirely constant over useful range of 1.8 to 30 MHz. Three strong feedback loops are introduced between driver and final amplifier. Use transistors specified, because substitutions may cut overall gain by 10 to 20 dB.—A. Weiss, Noise Bridge, *Ham Radio,* May 1974, p 71–72.

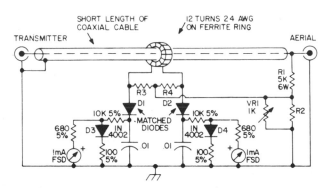

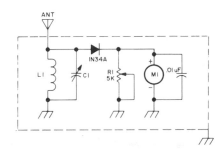

TUNED RADIATED-FIELD METER—Provides quick check of transmitter performance and approximate check of frequency. Values for L1 and C1 are chosen to cover desired frequency range. Meter can be 1 mA or 200 µA. Keep lead lengths short.—E. Hartz, Is My Rig Working or Not?, *73 Magazine,* Oct. 1976, p 56–57.

SWR METER—Gives direct measurement of standing-wave ratio on transmission line, independent of absolute power levels and of frequency. Voltages of 1N4002 silicon diodes are proportional to logarithms of their currents, which in turn are proportional to forward and reflected voltages. Meter scale is nonlinear, with maximum sensitivity as SWR approaches 1:1. For 50-ohm line, use 220 for R2 and 27 for R3 and R4. For 75-ohm line, corresponding values are 180 and 33. Detector diodes are point-contact germanium rated at 80 PIV. Article gives construction details and calibration curve. Ferrite ring is Mullard FX1596 or equivalent, with 0.5-inch outside diameter. Ground coax braid at one end only.—P. G. Martin, Some Directional Wattmeters and a Novel SWR Meter, *73 Magazine,* Aug. 1974, p 17, 19–21, 23–24, and 26.

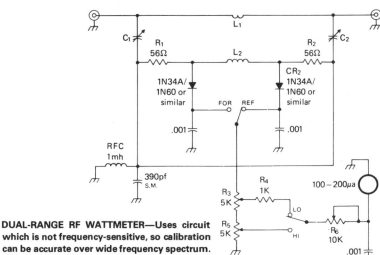

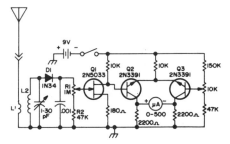

FIELD-STRENGTH METER—Developed for tuning all types of antennas, from mobile whips to four-element quads. Avoids shielding and other problems of switched T pads for calibrated attenuator by first detecting RF, then attenuating DC output. Technique has added advantage that circuit is no longer frequency-sensitive. To cover 13–24 MHz, L2 is 11 turns spaced out to about 1 inch, with 2 turns over top for L1. D1 can be any diode such as 1N34. R1 serves as calibrated attenuator, with R2 in series giving 0-dB point at junction. Article covers construction and operation.—J. L. Iliffe, An Amplified, Calibrated, Signal Strength Meter, *73 Magazine,* June 1973, p 85–86.

DUAL-RANGE RF WATTMETER—Uses circuit which is not frequency-sensitive, so calibration can be accurate over wide frequency spectrum. Ranges are 0–1 and 0–10 W. L2 is T-50-2 toroid wound almost full with No. 28 enamel, leaving only room for 2-turn link L1. C1 and C2 are 3–20 pF trimmers. Article covers calibration and use and gives table for reading SWR by comparing watts FORWARD with watts REFLECTED.—A. Weiss, The Silk-Purse In-Line Wattmeter, *CQ,* May 1977, p 50–52 and 74–75.

CHAPTER **3**
Audio Amplifier Circuits

Includes preamps for all types of inputs and AF amplifiers with power outputs up to 90 W and bandwidths up to 50 kHz, most using transistors with or without opamps and ICs. Circuits include variety of methods for reducing distortion and eliminating switching transients. See also Audio Control, Receiver, and Stereo chapters.

500-OHM INPUT—Simple audio amplifier having high gain, low noise, and excellent temperature stability can be used as mike booster, first AF amplifier stage in receiver, or for other preamp applications. With values shown, circuit will amplify down to about 10 Hz. To increase low-frequency cutoff for speech amplifier, reduce C2 to 1 μF or less.—E. Dusina, Build a General Purpose Preamp, *73 Magazine,* Nov. 1977, p 98.

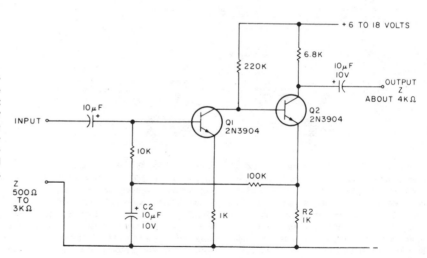

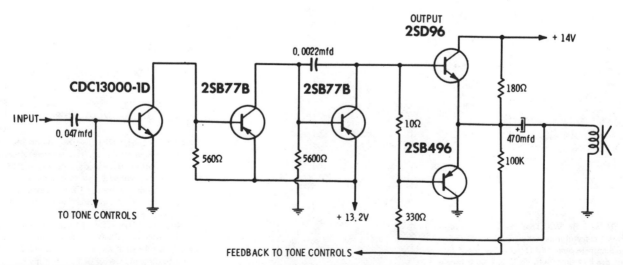

COMPLEMENTARY-SYMMETRY AMPLIFIER—Simplified version of circuit takes advantage of fact that PNP and NPN transistors require signals of opposite polarity to perform same function. Bases of output transistors are fed in parallel, with loudspeaker connected to common terminal of transistors. Drawback is difficulty of locating matched PNP and NPN transistors.—J. J. Carr, Solid-State Audio: A Review of the Latest Circuitry and General Troubleshooting Procedures, *Electronic Servicing,* Aug. 1971, p 38–43.

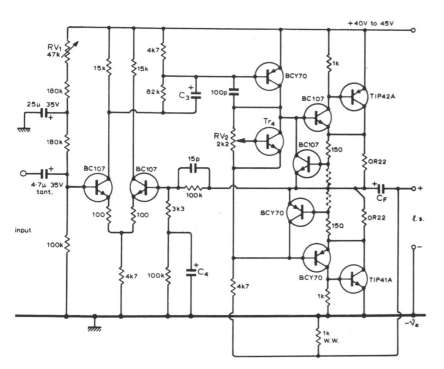

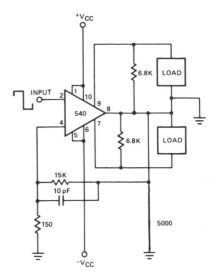

30-W—Designed for use with active filter crossover networks for three loudspeakers. For lowest-frequency channel, C_3 is 150 μF and C_4 is 50 μF. For middle channel, C_3 and C_4 are 25 μF. For high-frequency channel, C_3 and C_4 are 10 μF. Article includes circuit for active filter network.—D. C. Read, Active Filter Crossover Networks, *Wireless World*, Dec. 1973, p 574–576.

HAMMER DRIVER—Signetics 540 power driver handles either push-pull or single-ended inductive loads such as relays, solenoids, motors, and electric hammers. In push-pull connection shown, load is driven in either positive, negative, or both arms of output. Either output can be selected by appropriate choice of input pulse polarity. Supply can be ±5 to ±25 V.—"Signetics Analog Data Manual," Signetics, Sunnyvale, CA, 1977, p 764.

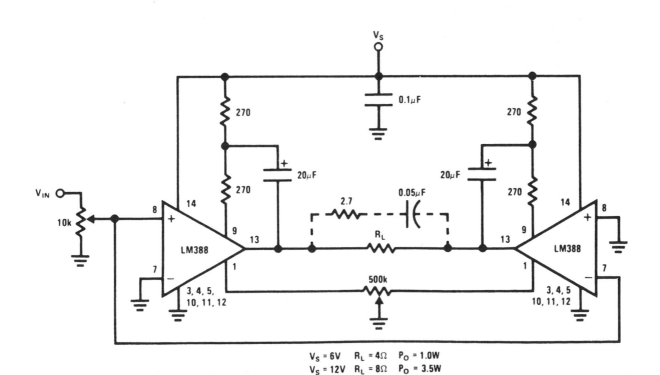

$$V_S = 6V \quad R_L = 4\Omega \quad P_O = 1.0W$$
$$V_S = 12V \quad R_L = 8\Omega \quad P_O = 3.5W$$

3.5-W BRIDGE AMPLIFIER—Bridge connection of National LM388 power opamps provides 3.5 W to 8-ohm loudspeaker when using 12-V supply. With 6-V supply and 4-ohm load, maximum power is 1 W. Coupling capacitors are not required since output DC levels are within several tenths of a volt of each other.—"Audio Handbook," National Semiconductor, Santa Clara, CA, 1977, p 4-37–4-41.

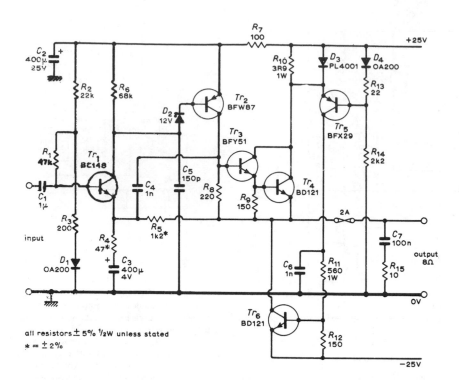

LOW-COST 30-W—Economical compromise gives 30 W into 8-ohm load at 0.1% distortion (mainly second harmonic) and hum level of −50 dBW. Article covers design and operation of circuit in detail.—P. L. Taylor, Audio Power Amplifier, *Wireless World,* June 1973, p 301–302.

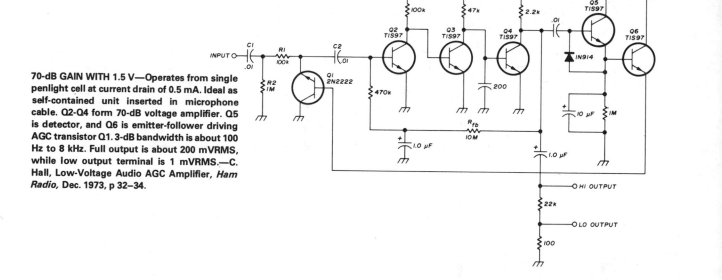

70-dB GAIN WITH 1.5 V—Operates from single penlight cell at current drain of 0.5 mA. Ideal as self-contained unit inserted in microphone cable. Q2-Q4 form 70-dB voltage amplifier. Q5 is detector, and Q6 is emitter-follower driving AGC transistor Q1. 3-dB bandwidth is about 100 Hz to 8 kHz. Full output is about 200 mVRMS, while low output terminal is 1 mVRMS.—C. Hall, Low-Voltage Audio AGC Amplifier, *Ham Radio,* Dec. 1973, p 32–34.

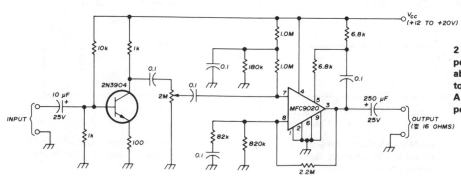

2 W WITH IC—Uses Motorola MFC9020 audio power amplifier to give maximum output of about 2 W for 16-ohm loudspeaker. Used in autopatch system for FM repeater.—R. B. Shreve, A Versatile Autopatch System for VHF FM Repeaters, *Ham Radio,* July 1974, p 32–38.

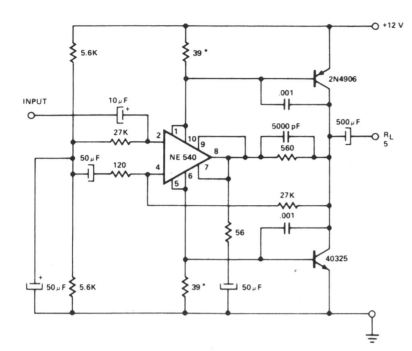

AUTO RADIO AMPLIFIER—Circuit shown permits operation of Signetics NE540 power driver from single-polarity 12-V supply of auto. Bipolar supplies for differential inputs of 540 are achieved by returning inputs to half of available supply or 6 V. Load is AC coupled because amplifier has DC gain of 1, and amplifier output is therefore 6 VDC. 39-ohm supply resistors are selected for minimum crossover distortion.—"Signetics Analog Data Manual," Signetics, Sunnyvale, CA, 1977, p 764.

HIGH-GAIN IC WITH TRANSISTOR—High input impedance of RS741C opamp permits use with any general-purpose crystal microphone. R2 is volume control, and R3 controls gain and frequency response of IC. Power transistor stage drives loudspeaker directly, without output transformer.—F. M. Mims, "Integrated Circuit Projects, Vol. 6," Radio Shack, Fort Worth, TX, 1977, p 79–88.

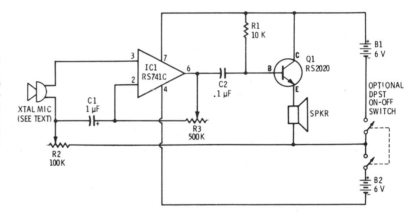

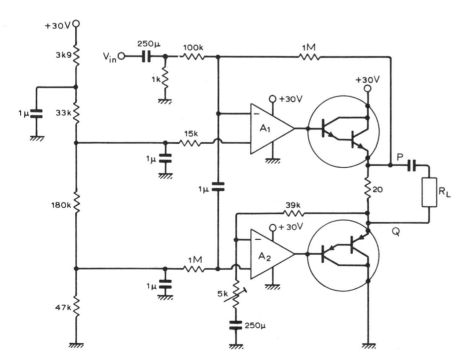

POWER AMPLIFIER WITH ERROR TAKEOFF—Voltage proportional to distortion is amplified for use in reducing nonlinear distortion at output, in circuit developed for use as single-ended power amplifier. Power Darlingtons are MJ4000 and MJ4010, and both opamps are 741. Preset 5K pot is adjusted initially for minimum distortion. Article gives theory of operation and design equations.—A. M. Sandman, Reducing Amplifier Distortion, *Wireless World,* Oct. 1974, p 367–371.

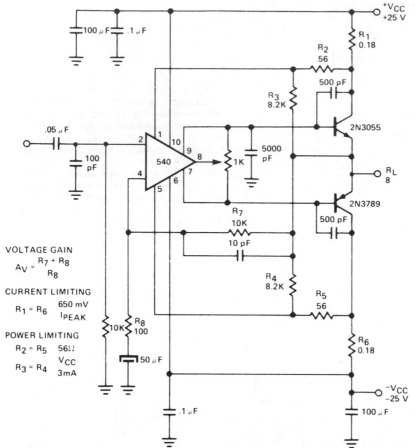

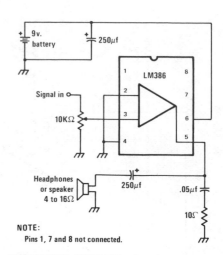

VOLTAGE GAIN

$$A_V = \frac{R_7 + R_8}{R_8}$$

CURRENT LIMITING

$$R_1 = R_6 \quad \frac{650 \text{ mV}}{I_{PEAK}}$$

POWER LIMITING

$$R_2 = R_5 \quad 56\Omega$$

$$R_3 = R_4 \quad \frac{V_{CC}}{3mA}$$

NOTE: Pins 1, 7 and 8 not connected.

HEADPHONE AMPLIFIER—Can be used with FM tuner in place of more expensive audio amplifier. For stereo, use one LM386 circuit for each channel. Can be mounted directly on headphones if weight of battery is not objectionable.—J. A. Sandler, 11 Projects under $11, *Modern Electronics,* June 1978, p 54–58.

35 W—Signetics 540 drives complementary output transistors to give high output current for driving 8-ohm loudspeaker. Feedback is adjusted to give AC gain of 40 dB. Gain rolls off to unity at DC to prevent DC offset voltages from being amplified to level that might damage loudspeaker circuit. Power limiting is provided by placing resistor network around output stage.—"Signetics Analog Data Manual," Signetics, Sunnyvale, CA, 1977, p 762–763.

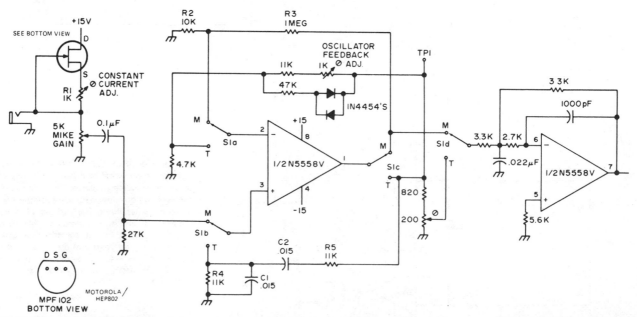

PREAMP WITH TEST TONE—Built around Signetics N5558V dual opamp or equivalent Motorola MC1458CP2, National LM1458, or Texas Instruments SN72558P. First half of opamp is used either as gain stage for increasing voltage level of carbon microphone or as AF Wien-bridge tone oscillator, depending on position of S1. Frequency is determined by values of C1, C2, R4, and R5. Silicon signal diodes form nonlinear control element. Adjust R6 until oscillator output at TP1 is 10 V P-P. FET provides constant current through variable resistance of carbon microphone, to give audio input voltage. Second opamp is active low-pass filter with 3.3-kHz cutoff, rolloff of 12 dB per octave, and voltage gain of 10.—H. Olson, An IC Mike Preamp That Doubles as a Tone Generator, *73 Magazine,* March 1974, p 45 and 47–48.

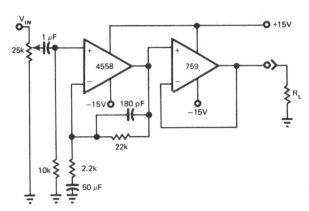

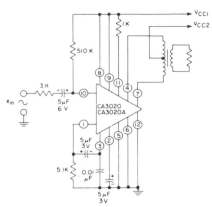

HEADPHONE OPAMPS—Dual low-noise 4558 opamp provides gain and reduces system noise and distortion, while 759 power opamp supplies output power of 0.7 W into 16-ohm load with less than 0.1% total harmonic distortion.—

R. J. Apfel, Power Op Amps—Their Innovative Circuits and Packaging Provide Designers with More Options, *EDN Magazine*, Sept. 5, 1977, p 141–144.

1-W CLASS B—Audio application of CA3020A wideband power amplifier provides 1-W output to loudspeaker load through AF output transformer with 10% total harmonic distortion. V_{CC1} is 9 V, and V_{CC2} is 12 V. With CA3020, both supply voltages are 9 V and maximum power output is 550 mW. Sensitivity is 35-45 mV.—"Linear Integrated Circuits and MOS/FET's," RCA Solid State Division, Somerville, NJ, 1977, p 105.

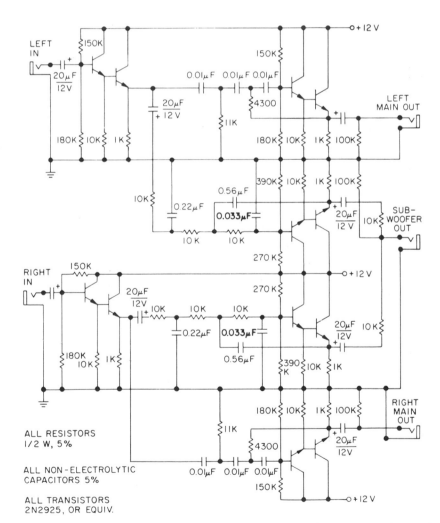

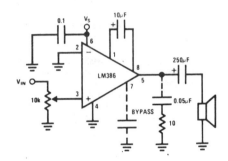

46-dB GAIN—Single National LM386 power amplifier provides gain of 200 V/V at maximum output power of 250 mW for 12-V supply. Optional 0.05-μF capacitor and 10-ohm resistor suppress bottom-side oscillation occurring during negative swing into load drawing high current.—"Audio Handbook," National Semiconductor, Santa Clara, CA, 1977, p 4-30–4-33.

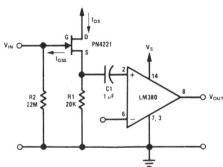

CROSSOVER FOR 20-Hz SUBWOOFER—Used at channel outputs of stereo system when reproducing music down to 20 Hz as synthesized by electronic function generators. Active crossover network drives subwoofer (low-bass loudspeaker) connected in bridged-center configuration, for handling sounds below range of normal woofer. Crossover consists of third-order Butterworth (18 dB per octave) networks providing 20-Hz cutoff along with 100-Hz crossover. Response of subwoofer should extend one octave above crossover. One advantage of active crossover is freedom from transient intermodulation distortion.—W. J. J. Hoge, Switched-On Bass, *Audio*, Aug. 1976, p 34-36, 38, and 40.

HIGH INPUT IMPEDANCE—Use of JFET as isolator boosts input impedance of opamp to 22 megohms for low-frequency input signals. Impedance drops to 3.9 megohms as frequency increases to about 20 kHz. Overall gain of circuit is about 45 dB when using 18-V supply.—"Audio Handbook," National Semiconductor, Santa Clara, CA, 1977, p 4-21–4-28.

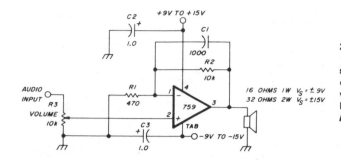

2-W MONITOR—Fairchild 759 opamp provides 1-W AF output when supply is ±9 V and loudspeaker is 16 ohms, and 2 W with ±15 V and 32-ohm loudspeaker. Use heatsink. Gain is 20 for values shown, with response rolled off at 15 kHz by C1.—W. Jung, An IC Op Amp Update, *Ham Radio,* March 1978, p 62–69.

60 W WITH DC-COUPLED OUTPUT—Q6 is Motorola MJE6044 complementary Darlington output transistor, and Q7 is MJE6041. Q1 and Q2 are MD8002 dual transistor, Q3 is MPS-A56, Q4 is MPS-A13, and Q5 is MPS-A06. For 8-ohm loudspeaker, supply is ±36 V with 6.2K for R4, 430 ohms for R5, and 33K for R7. Output center voltage must be maintained at 0 VDC to ensure maximum signal swing and prevent DC voltage from acting on loudspeaker. Frequency response is 10 Hz to 50 kHz for −1 dB points. Same circuit is used with different components for other powers down to 15 W and for 4-ohm loudspeaker.—R. G. Ruehs, "15 to 60 Watt Audio Amplifiers Using Complementary Darlington Output Transistors," Motorola, Phoenix, AZ, 1974, AN-483B, p 5.

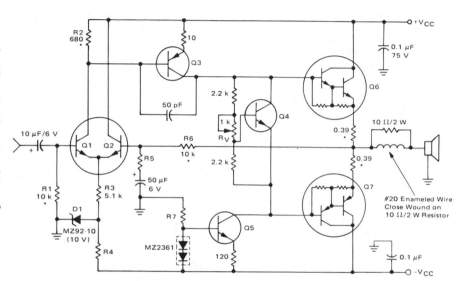

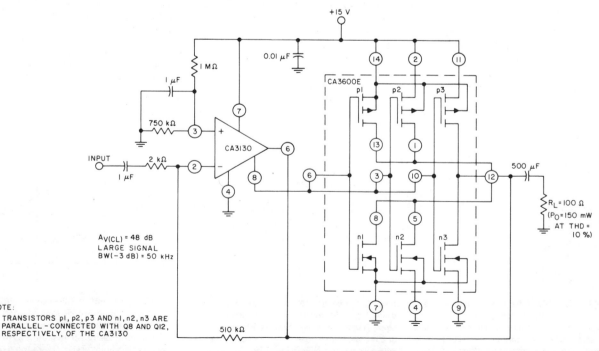

50-kHz BANDWIDTH—Three transistor pairs in CA3600E array are parallel-connected with output stage of CA3130 bipolar MOS opamp to boost current-handling capability about 2.5 times. Use of feedback gives closed-loop gain of 48 dB. Typical large-signal bandwidth is 50 kHz for 3 dB down.—"Circuit Ideas for RCA Linear ICs," RCA Solid State Division, Somerville, NJ, 1977, p 12.

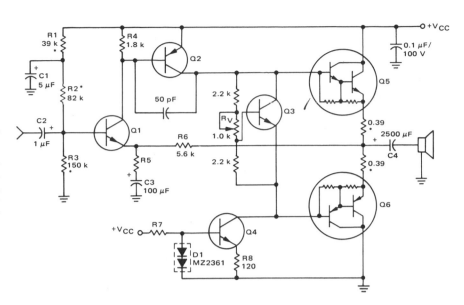

60 W WITH AC-COUPLED OUTPUT—Uses Motorola complementary Darlington output transistors, with MJE6044 for Q5 and MJE6041 for Q6. For 8-ohm loudspeaker, Q1 is MPS-A06, Q2 is MPS-A56, Q3 is MPS-A13, and Q4 is MPS-A06. Supply is 72 V. R5 is 220 ohms, and R7 is 68K. Same circuit is used with different components for other output powers down to 15 W and for 4-ohm loudspeaker. Frequency response is 20 Hz to 50 kHz for −1 dB points.—R. G. Ruehs, "15 to 60 Watt Audio Amplifiers Using Complementary Darlington Output Transistors," Motorola, Phoenix, AZ, 1974, AN-483B, p 3.

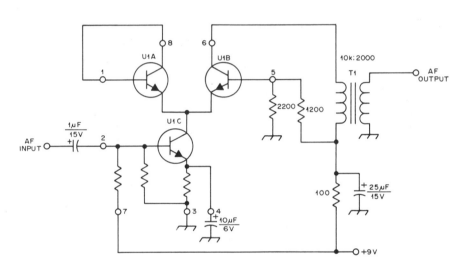

CASCODE AMPLIFIER—Uses two sections of RCA CA3028A linear IC (U1A is not used). Provides power gain of about 40 dB. Unmarked resistors are on IC.—D. DeMaw, Understanding Linear ICs, *QST*, Feb. 1977, p 19–23.

HIGH OUTPUT CURRENT—Uses CA3094A programmable opamp as driver stage for two parallel-connected transistors of CA3183AE array to develop 100-mA average AF current (peaks up to 300 mA) through 75-ohm load. Diode-connected transistors D_1-D_3 in array provide temperature compensation for output transistors.—"Circuit Ideas for RCA Linear ICs," RCA Solid State Division, Somerville, NJ, 1977, p 11.

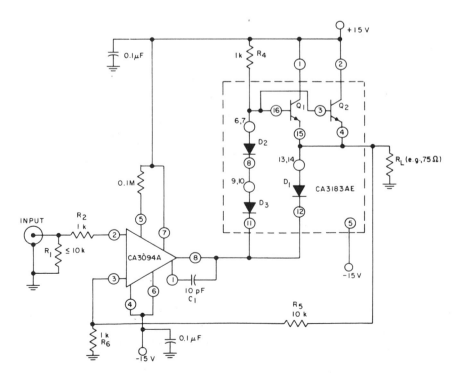

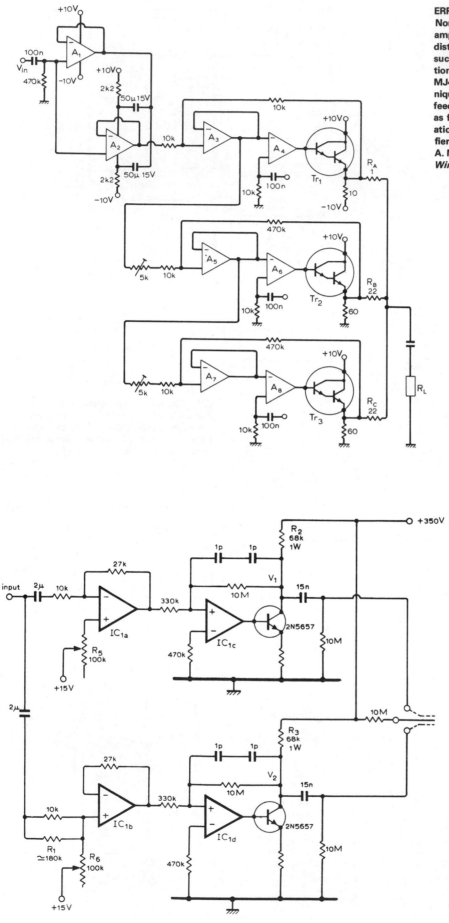

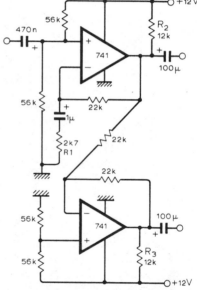

ERROR TAKEOFF REDUCES DISTORTION—Nonlinear distortion is reduced in single-ended amplifier by producing voltage proportional to distortion and amplifying this error voltage in such a way that it can be used to reduce distortion at output. Circuit uses 741 opamps and MJ4000 power Darlington transistors. Technique overcomes basic limitation of negative feedback wherein feedback loop gain decreases as frequency increases. Article also gives variation of circuit more suitable for power amplifier, and describes circuit operation in detail.—A. M. Sandman, Reducing Amplifier Distortion, *Wireless World*, Oct. 1974, p 367–371.

BALANCED OUTPUT WITH OPAMPS—Low-cost amplifier provides low-impedance balanced output from unbalanced signal output of preamp. Response is flat from 10 to 20,000 Hz, and distortion less than 0.1% at 800 Hz into 600-ohm load. Gain is 20 dB. Other opamps, such as LM307 or 747 (dual 741) can also be used.—K. D. James, Balanced Output Amplifier, *Wireless World*, Dec. 1975, p 576.

ELECTROSTATIC HEADPHONES—Uses LM3900N four-opamp IC and two transistors to step up headphone output signal of AF power amplifier sufficiently to drive pair of electrostatic headphones without introducing excessive distortion. Total harmonic distortion at 1 kHz is 1% at 300-V peak-to-peak output, and drops to 0.1% at 50-V output.—N. Pollock, Electrostatic Headphone Amplifier, *Wireless World*, July 1976, p 35.

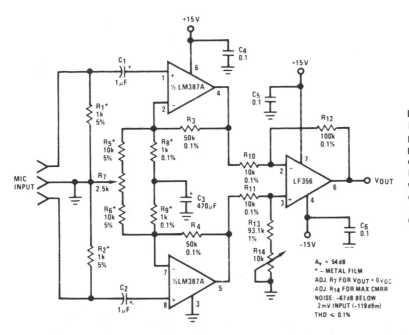

LOW-NOISE TRANSFORMERLESS PREAMP— Noise performance of balanced microphone preamp is improved with instrumentation amplifier configuration of three opamps. Each half of LM387A is wired as noninverting amplifier. LM387A serves to amplify low-level signals while adding as little noise as possible, leaving common-mode rejection for LF356.—"Audio Handbook," National Semiconductor, Santa Clara, CA, 1977, p 2-37–2-40.

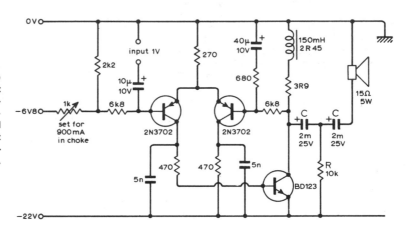

5-W CLASS A—Three-transistor feedback loop gives excellent DC stability, while arrangement of two capacitors and resistor feeding loudspeaker keeps these capacitors properly polarized as AF output voltage swings above and below zero level.—R. H. Pearson, Novel 5-Watt Class A Amplifier Uses Three-Transistor Feedback Circuit, *Wireless World,* March 1974, p 18.

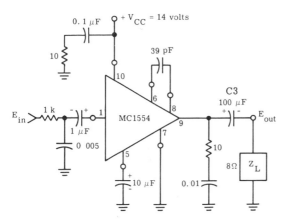

1-W NONINVERTING—Motorola MC1554 IC operates from single supply and uses capacitive coupling to both source and load, for voltage gain of 9 with frequency response (−3 dB) from 200 to 22,000 Hz. Input impedance is 10K, and total harmonic distortion is less than 0.75%. Use external heatsink.—"The MC1554 One-Watt Monolithic Integrated Circuit Power Amplifier," Motorola, Phoenix, AZ, 1972, AN-401, p 2.

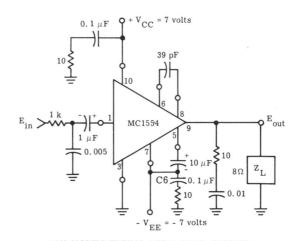

1-W NONINVERTING WITH SPLIT SUPPLY— Motorola MC1554 IC is connected for operation from ±7 V to provide voltage gain of 9 over frequency range (−3 dB) of 40 to 22,000 Hz. Input impedance is 10K, and total harmonic distortion is less than 0.75%. Use external heatsink.— "The MC1554 One-Watt Monolithic Integrated Circuit Power Amplifier," Motorola, Phoenix, AZ, 1972, AN-401, p 2.

BATTERYLESS MICROPHONE PREAMP—Transistor and two components can be mounted in microphone housing. R_2 is added to existing amplifier to provide operating voltage for preamp, while C_2 feeds preamp signal to input of amplifier.—W. H. Jarvis, Line-Powered Microphone Pre-Amp, *Wireless World*, Dec. 1976, p 43.

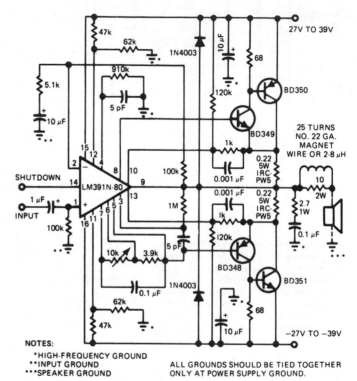

60 W—Combination of National LM391 audio driver IC and discrete power transistors provides 60-W output for loudspeaker at very low distortion. IC output can swing ±40 V. Total harmonic distortion of circuit is under 0.05%.—P. Franson, Consumer-Product IC's—New Offerings Trigger an Explosion in Markets Old and New, *EDN Magazine*, Nov. 5, 1977, p 54–65.

NOTES:
*HIGH-FREQUENCY GROUND
**INPUT GROUND
***SPEAKER GROUND

ALL GROUNDS SHOULD BE TIED TOGETHER ONLY AT POWER SUPPLY GROUND.

50°C/W HEAT SINK ON BD348 AND BD349
3.9°C/W HEAT SINK ON BD350 AND BD351

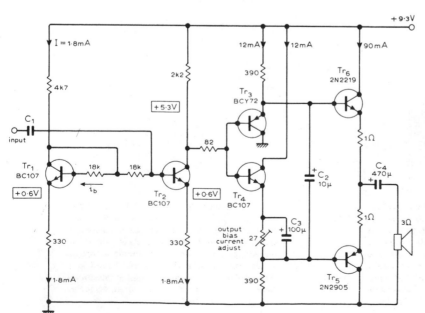

DIRECT-COUPLED PUSH-PULL—Provides high-quality sound at ample volume for car radio, operating from 9.3-V regulated supply. Push-pull emitter-follower stages are connected to give symmetrical low output impedance on both positive and negative portions of audio waveform. Input transistor Tr_1 provides temperature compensation, while driver Tr_2 provides temperature-compensated bias and maximum symmetrical voltage swings to output stages.—G. Kalanit, Low Voltage Audio Amplifier, *Wireless World*, Oct. 1976, p 74.

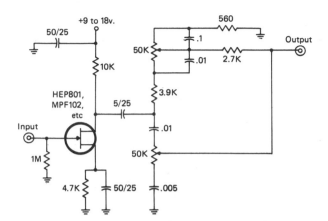

FET PREAMP WITH TONE CONTROLS—Developed for use with simple 2-W audio amplifier when testing very low-level output circuits and microphones. Will not load circuit to which input is connected. Optional bass/treble tone controls are included.—J. Schultz, An Audio Circuit Breadboarder's Delight, *CQ*, Jan. 1978, p 42 and 75.

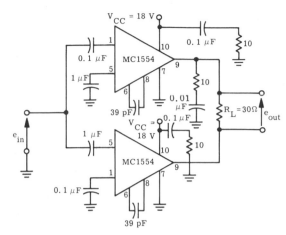

3-W DIFFERENTIAL—Upper Motorola MC1554 power amplifier is connected in standard configuration for noninverting gain of 9, while lower IC has inverting gain of 9 to give effective overall voltage gain of 18. Input impedance of upper amplifier is 10K while that of lower amplifier is 1K, with unequal input coupling capacitors providing required match of frequency responses. Differential output connection allows output voltage swing to exceed power-supply voltage.—"The MC1554 One-Watt Monolithic Integrated Circuit Power Amplifier," Motorola, Phoenix, AZ, 1972, AN-401, p 4.

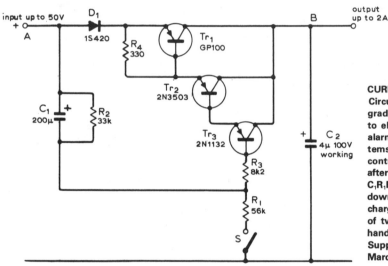

CURRENT CONTROL FOR POWER SWITCH—Circuit makes power supply current increase gradually from zero when supply is turned on, to eliminate transients that sometimes cause alarming loudspeaker thumps in audio systems. Current through silicon power diode D_1 is controlled by voltage on C_1, which charges up after closing of switch with time constant $C_1 R_1 R_2 / (R_1 + R_2)$. When switch is opened, rundown of supply current is controlled by discharge of C_1 through R_2. Article also covers use of two current control circuits in tandem for handling higher loads.—P. J. Briody, Power Supply Delayed Switching, *Wireless World*, March 1975, p 139–141.

10-W CLASS A—Highly stable circuit uses easily obtainable components. Transistor types are not critical. Short-circuit protection is provided by constant-current source D_1-Tr_2-Tr_5-Tr_4. Output transistors Tr_5 and Tr_6 require heatsinks capable of dissipating at least four times rated output power. D_1 and Tr_2 should be in thermal contact.—A. H. Calvert, Class A Power Amplifier, *Wireless World*, June 1976, p 71.

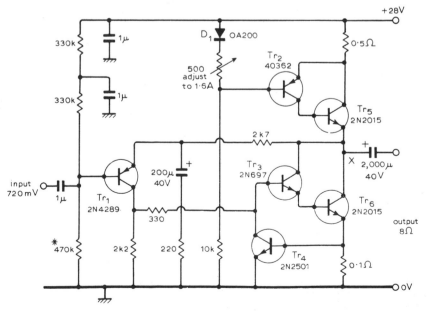

＊ adjust to make point X half the supply voltage (if necessary)

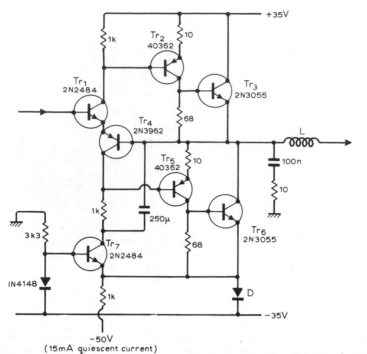

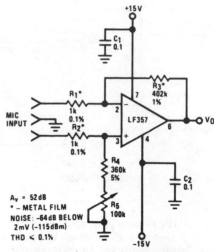

CLASS B WITHOUT TAKEOVER DISTORTION—
Circuit uses two transistor triplets in output stage, with quiescent current set at 15 mA by Tr₇ which serves as constant-current source. For small signals, Tr₁ and Tr₄ can be regarded as long-tailed pair without tail. For positive signals, Tr₁-Tr₄ become active and behave as super emitter-follower, while Tr₁ with Tr₄-Tr₆ serve for negative signals. One advantage of circuit is low output impedance.—N. M. Visch, A Novel Class B Output, *Wireless World,* April 1975, p 166.

TRANSFORMERLESS BALANCED-INPUT MI-CROPHONE PREAMP—Uses FET-input opamp to amplify differential signals while rejecting common-mode signals. Gain is set at 52 dB by ratio of R₃ to R₁. Input resistors R₁ and R₂ are made large compared to source impedance while being kept as small as possible, for optimum balance between input loading effects and low noise. Good compromise value is 10 times source impedance for R₁ + R₂.—"Audio Handbook," National Semiconductor, Santa Clara, CA, 1977, p 2-37—2-40.

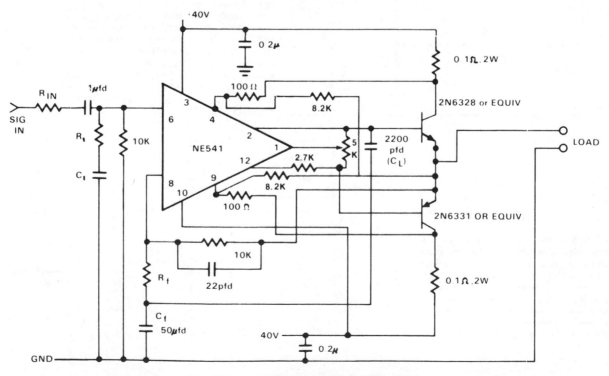

*LI—24 turns, single layer #36 on 10Ω, 1W res.

75 W WITH CURRENT LIMITING—Signetics NE541 high-voltage power amplifier provides current gain of 90 dB from 20 Hz to 20 kHz and output levels up to 20 VRMS from 300-mVRMS input. IC includes built-in short-circuit protection, with additional protection provided by external current limiting. Transistors in output stage boost power to 75 W for driving loudspeaker load.—"Signetics Analog Data Manual," Signetics, Sunnyvale, CA, 1977, p 765.

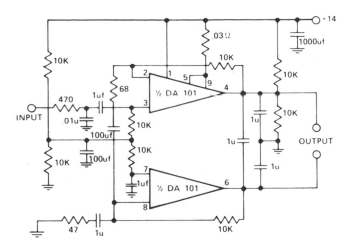

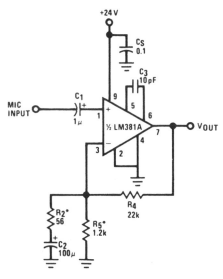

POWER OPAMPS IN BRIDGE—Bridge configuration is theoretically capable of 4 times power output of conventional quasi-complementary or complementary-symmetry amplifier. Use of bridge circuit in automotive AM/FM stereo receiver requires suitable protection of modules.

Article covers incorporation of protective controls in single module with dual opamps.—E. R. Buehler and B. D. Schertz, Fault Protection of Monolithic Audio Power Amplifiers in Severe Environments, *IEEE Transactions on Consumer Electronics,* Aug. 1977, p 418–423.

SINGLE-ENDED MICROPHONE PREAMP—Noise performance is −69 dB below 2-mV input reference point. Use metal-film resistors for R_2 and R_5. Total harmonic distortion is less than 0.1%. Gain is set by ratio of R_4 to R_2 and is 52 dB.—"Audio Handbook," National Semiconductor, Santa Clara, CA, 1977, p 2-37–2-40.

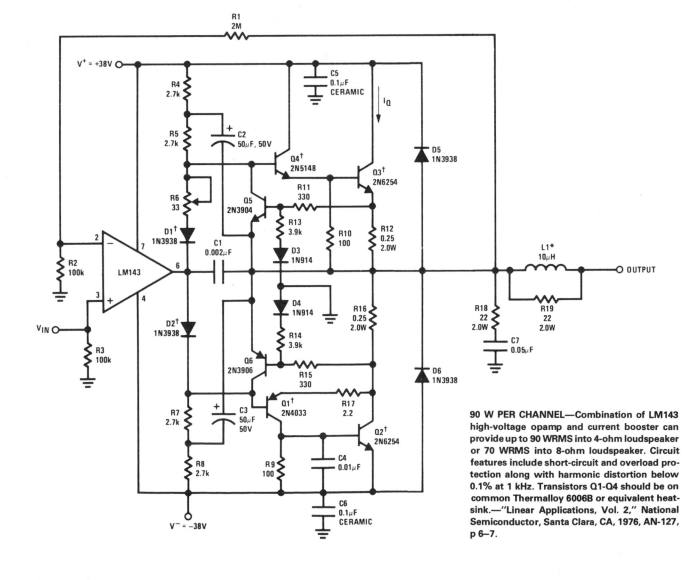

90 W PER CHANNEL—Combination of LM143 high-voltage opamp and current booster can provide up to 90 WRMS into 4-ohm loudspeaker or 70 WRMS into 8-ohm loudspeaker. Circuit features include short-circuit and overload protection along with harmonic distortion below 0.1% at 1 kHz. Transistors Q1-Q4 should be on common Thermalloy 6006B or equivalent heatsink.—"Linear Applications, Vol. 2," National Semiconductor, Santa Clara, CA, 1976, AN-127, p 6–7.

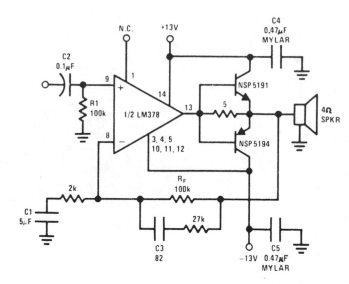

12 W WITH BOOSTER TRANSISTORS—At signal input levels below 20 mW, National LM378 opamp supplies load directly through 5-ohm resistor up to current peaks of about 100 mA. Above this level, booster transistors are biased on by load current through same resistor to increase output power. Transistors and opamp must have adequate heatsinks.—"Audio Handbook," National Semiconductor, Santa Clara, CA, 1977, p 4-42–4-43.

HIGH-GAIN JFET—Simple two-JFET circuit provides gain of 500 at low power. Reducing drain current increases gain at sacrifice of input dynamic range.—"FET Databook," National Semiconductor, Santa Clara, CA, 1977, p 6-26–6-36.

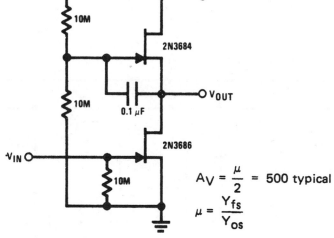

$$A_V = \frac{\mu}{2} = 500 \text{ typical}$$

$$\mu = \frac{Y_{fs}}{Y_{os}}$$

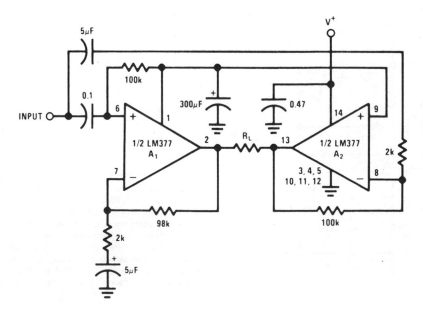

BRIDGE AMPLIFIER—National LM377 dual power amplifier is used in bridge configuration to drive floating load that can be loudspeaker, servomotor, or other device having impedance of 8 or 16 ohms. Maximum output is 4 W at gain of 40 dB, with 3-dB response extending from about 20 Hz to above 100 kHz.—"Audio Handbook," National Semiconductor, Santa Clara, CA, 1977, p 4-8–4-20.

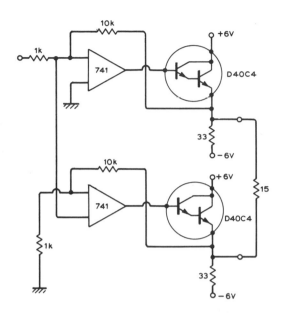

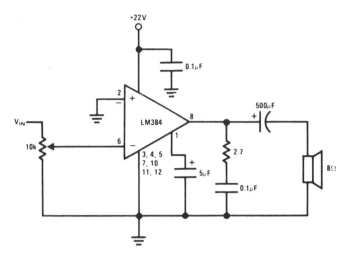

ERROR ADD-ON REDUCES DISTORTION— Based on fact that error at output of upper opamp also appears at input of this opamp. Error signal is taken from this input for lower opamp, where it is amplified by opamp and Darlington for addition to output of upper Darlington. Article gives design equations and intimates that open-loop gain improves at 12 dB per octave as compared to conventional 6 dB. Applications include reduction of loudspeaker distortion which cannot be handled by negative feedback.—A. Sandman, Reducing Distortion by 'Error Add-On,' *Wireless World,* Jan. 1973, p 32.

5 W WITH OPAMP—Simple circuit including volume control has low harmonic distortion. Higher allowable operating voltage of LM384 opamp gives higher output power, but heatsink is required.—"Audio Handbook," National Semiconductor, Santa Clara, CA, 1977, p 4-28– 4-29.

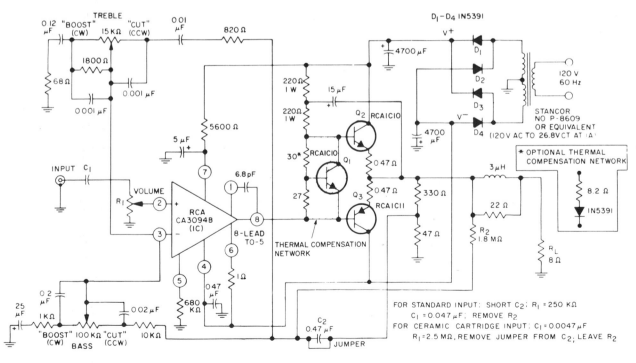

12-W OUTPUT—Uses CA3094B programmable opamp to drive complementary-symmetry power-output transistors. Intermodulation distortion is only 0.2% when 60-Hz and 2-kHz signals are mixed in 4:1 ratio. Location of tone controls in feedback network improves signal-to-noise ratio. Hum and noise are typically 700 μV (83 dB down) at output. Transistor Q_1 provides thermal compensation.—"Circuit Ideas for RCA Linear ICs," RCA Solid State Division, Somerville, NJ, 1977, p 11.

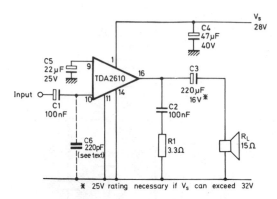

4.5-W CLASS B—Mullard TDA2610 drives 15-ohm loudspeaker with total harmonic distortion of less than 1%. Supply is 28 V ± 10%. Network C2-R1 ensures stability with inductive load.—"Audio Power Amplifier TDA2610," Mullard, London, 1976, Technical Note 35, TP1541.

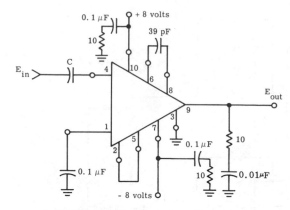

1-W INVERTING—Uses Motorola MC1554 power amplifier to provide voltage gain of 35 with components shown. Output voltage swing is 12 V P-P into 12-ohm load. For response down to low audio frequencies (below 100 Hz), large value of C is required, such as 1 μF. Input can be direct-coupled at sacrifice in output offset, but this can be corrected by properly biasing pin 1 or terminating it in about 250 ohms. Upper frequency limit for −3 dB is about 22,000 Hz.—"The MC1554 One-Watt Monolithic Integrated Circuit Power Amplifier," Motorola, Phoenix, AZ, 1972, AN-401, p 3.

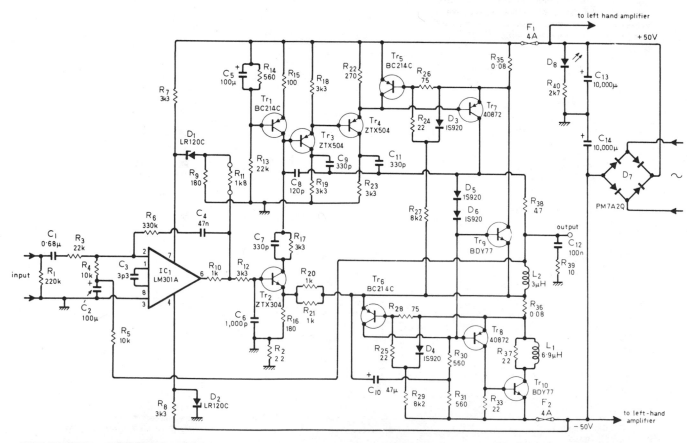

FEED-FORWARD CORRECTION—Circuit reduces distortion caused by nonlinearity of output power transistors by deriving error component that bypasses these transistors. Technique used is known as current dumping. Article describes operating principle in detail. Circuit shown gives application to commercial amplifier (Quad 405), in which midfrequency distortion is only about 0.005%. Features include elimination of adjustments, alignment procedures, and thermal problems during entire life of amplifier.—P. J. Walker, Current Dumping Audio Amplifier, *Wireless World*, Dec. 1975, p 560–562.

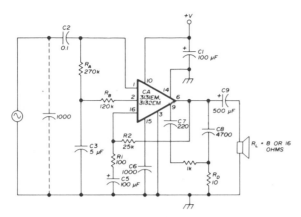

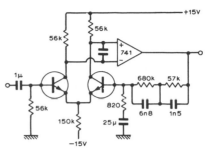

5-W IC—RCA IC includes preamps, power amplifier, and integral heatsink. CA3131 has internal feedback network that maintains 48-dB gain, while CA3132 requires external feedback network including R1 and R2 connected between pins 6 and 16. Input 1000-pF capacitor is required if input has open circuit. Electrolytic C1 should be placed as close as possible to pin 10. C6 sets 46-dB closed-loop gain point at 200 kHz. C7 equalizes gain for positive and negative signal swings. C9 sets low-frequency response of amplifier. Recommended supply voltage is 24 VDC.—E. Noll, Audio-Power Integrated Circuits, *Ham Radio,* Jan. 1976, p 64–66.

LOW-NOISE OPAMP PREAMP—Circuit combines noise features of discrete design with simplicity and high open-loop of IC opamp such as 741. Transistors can be 2N3708, BC109, or equivalent. Output impedance is low enough to drive headphones directly.—D. R. Hedgeland, Op-Amp Pre-Amp, *Wireless World,* Dec. 1972, p 575.

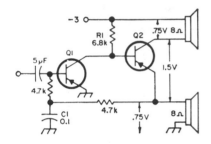

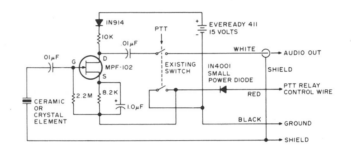

50 mW FLAT TO 30 kHz—Power amplifier achieves push-pull output with single transistor. Both transistors should be germanium such as 2N404, SK3004, or HEP-253.—Circuits, *73 Magazine,* Feb. 1974, p 100.

PREAMP IN MIKE—Common-source FET preamp and 15-V battery fit into Turner 350C hand mike for boosting output of ceramic element 20 dB. Frequency response of preamp is flat from 200 Hz to over 100 kHz. Drain is 200 μA only when push-to-talk is pressed, giving long battery life.—G. Hinkle, Self-Powered Mike Preamp, *73 Magazine,* Nov. 1976, p 65.

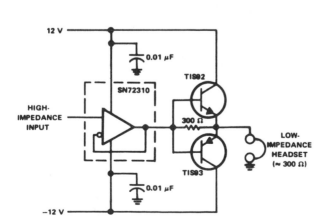

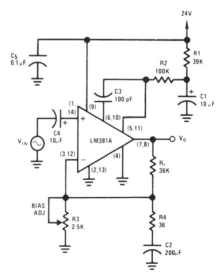

IMPEDANCE BUFFER—Complementary-transistor output stage provides required low impedance for driving headphones from SN72310 voltage-follower opamp. Supply is ±12 V.—"The Linear and Interface Circuits Data Book for Design Engineers," Texas Instruments, Dallas, TX, 1973, p 4–41.

ULTRALOW-NOISE PREAMP—Provides gain of 1000 over bandwidth of 20 Hz to 10 kHz, operating from 24-V supply and 600-ohm source impedance. Design procedure is given. Total wideband noise voltage is 43.7 μV, and wideband noise figure is 2.83 dB.—"Audio Handbook," National Semiconductor, Santa Clara, CA, 1977, p 2-15–2-19.

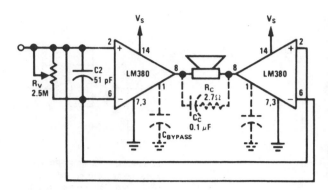

BRIDGE AMPLIFIER—Two opamps are used in bridge configuration to provide twice the voltage swing across loudspeaker load for given 18-V supply, increasing power capability to about twice that of single amplifier. To eliminate excessive quiescent DC voltage across load, non-polarized capacitor can be used in series with load or 1-megohm pot can be connected between pins 1 of opamps with position of moving arm adjusted to balance offset voltage. Components shown with dashed lines are added for stability with high-current load.—"Audio Handbook," National Semiconductor, Santa Clara, CA, 1977, p 4-21–4-28.

BASS BOOST—Compensates for poor bass response of loudspeaker by use of external series RC circuit between pins 1 and 5, paralleling internal 15K resistor of opamp. 6-dB effective bass boost is obtained if resistor is 15K, and lowest value for stable operation is 10K if pin 8 is left open.—"Audio Handbook," National Semiconductor, Santa Clara, CA, 1977, p 4-30–4-33.

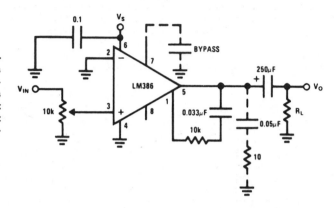

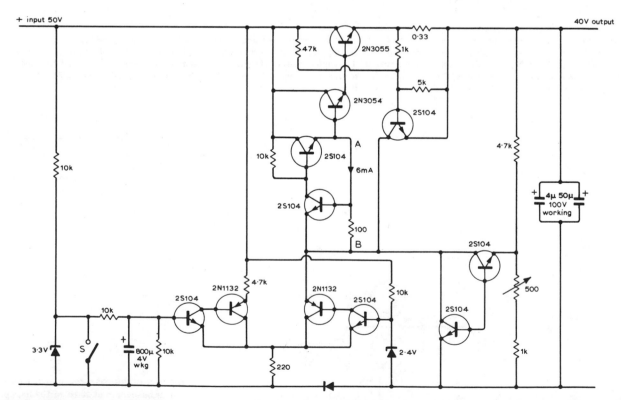

CURRENT-CONTROLLED SWITCHING—Addition of current control to 40-V regulated power supply for audio amplifier eliminates switch-on transients that sometimes cause alarming loudspeaker thumps. Switch S, which can be either relay or third pole on standard ON/OFF switch of amplifier, is opened to initiate charging of 800-μF capacitor that allows gradual buildup of output current. Similar transient suppression occurs when switch is closed to initiate current run-down as set is turned off. Run-up and run-down times are a few seconds each. Article also gives simpler current control circuit suitable for use with unregulated supplies.—P. J. Briody, Power Supply Delayed Switching, *Wireless World*, March 1975, p 139–141.

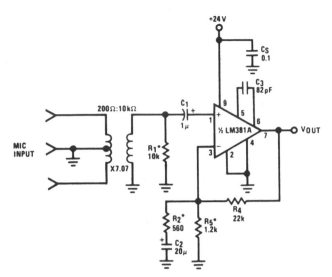

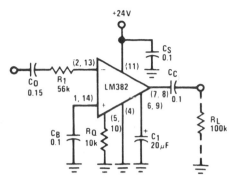

INVERTING AC AMPLIFIER—Provides gain of 40 dB when using 24-V supply and input impedance greater than 10K. Low-frequency performance is flat to 20 Hz. Design procedure is given.—"Audio Handbook," National Semiconductor, Santa Clara, CA, 1977, p 2-20–2-24.

BALANCED-INPUT MICROPHONE PREAMP—Use of two wires for microphone signal and separate wire for ground keeps hum and noise at minimum. Signal wires are twisted together in shield acting as ground. Net gain is 52 dB, giving 0-dBm output for nominal 2-mV input. Noise performance is −86 dB below 2-mV input level, and rejection of common-mode signals is 60 dB.—"Audio Handbook," National Semiconductor, Santa Clara, CA, 1977, p 2-37–2-40.

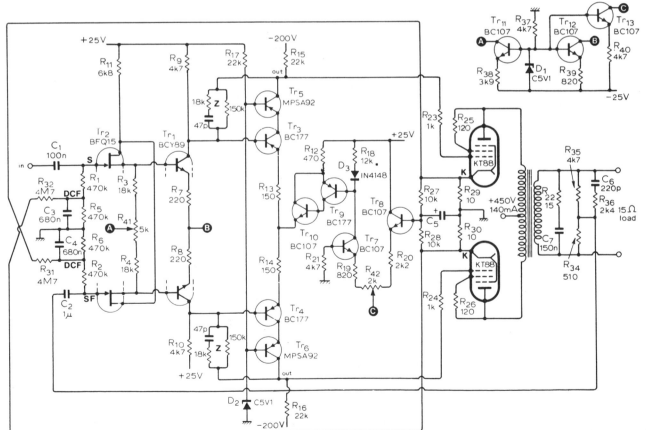

TWO-TUBE WILLIAMSON OUTPUT—Response is flat from 10 Hz to over 50 kHz for outputs up to 15 W, and total harmonic distortion for full power output at 15 kHz is only 0.25%. Output stage uses tubes connected as triodes in push-pull. Input transistor can be any in Philips BFQ10-16 family or equivalent replacement such as Siliconix E401. Article gives many suitable replacements for other transistors as well, and describes design of required feedback circuits in detail.—S. Berglund, Transistor Driver for Valve Amplifiers, *Wireless World,* April 1976, p 36–40.

CHAPTER **4**
Audio Control Circuits

Includes variety of tone controls, compressor, expander, compandor, clipper, mixer, clamp, automatic level control, click suppressor, attenuator, equalizer, speech filter, noise squelch, logic-controlled gain, voice- or tone-controlled relays, active crossover, and switching gate circuits for audio signals. See also Audio Amplifier, Filter, Receiver, and Stereo chapters.

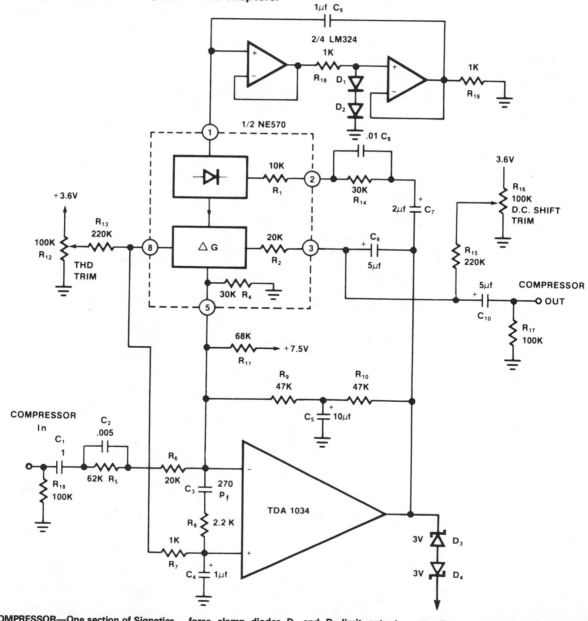

HI-FI COMPRESSOR—One section of Signetics NE570 dual compandor is used with external opamp for compression of large input signals in high-fidelity audio system. To prevent overload by sudden loud signal when compressor is operating at high gain for small signals, brute-force clamp diodes D_3 and D_4 limit output swings to about 7 V P-P. Limiting action prevents overloading of succeeding circuit such as tape recorder. Circuit includes input compensation network required for stability. Corresponding expander used for playback of recorded material should have same value for rectifier capacitor C_9 as is used in compressor.—"Signetics Analog Data Manual," Signetics, Sunnyvale, CA, 1977, p 804.

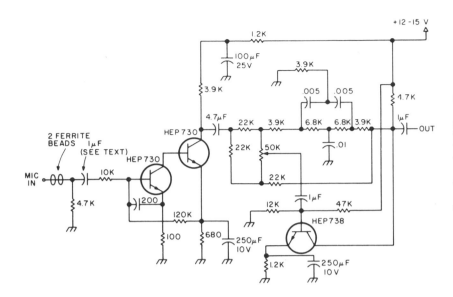

PREEMPHASIS AT 1500 Hz—Single-transistor peaking filter is combined with low-noise RF-protected preamp stage to improve speech intelligibility for any type of modulation. Effectiveness is most noticeable with deep bass voice, where soft peak around 1500–2000 Hz improves speech intelligibility. Can also serve as audio-type CW or SSB filter.—You Can Sound Better with Speech Pre-Emphasis, *73 Magazine*, Feb. 1977, p 42–43.

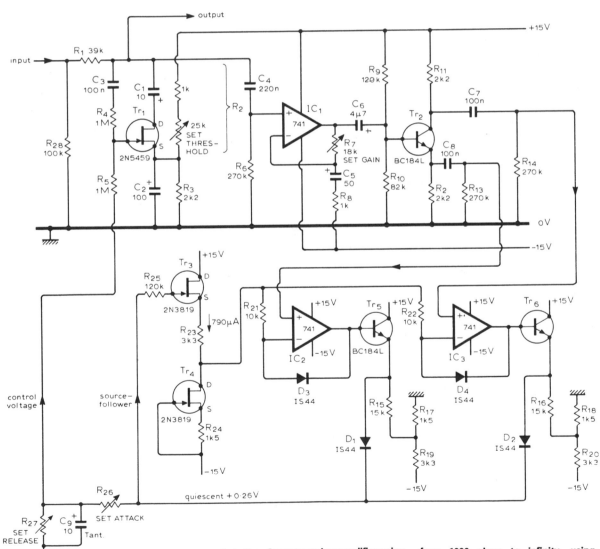

COMPRESSOR/LIMITER—High-fidelity circuit uses voltage-controlled attenuator to increase attenuation of input signal in response to voltage of control loop. Designed for use in modern sound studios. Output-sensing amplifier using IC_1 has gain of 19 over audio band. Tr_2 stage is phase-splitter driving precision rectifiers IC_2 and IC_3. Final part of circuit defines attenuation time constants; R_{26} sets attack time and R_{27} decay time. R_{26} can range from 0 to 1 megohm and R_{27} from 1000 ohms to infinity, using either switched or variable components. Article describes circuit operation and adjustment in detail. Tr_6 is BC184L or equivalent.—D. R. Self, High-Quality Compressor/Limiter, *Wireless World*, Dec. 1975, p 587–590.

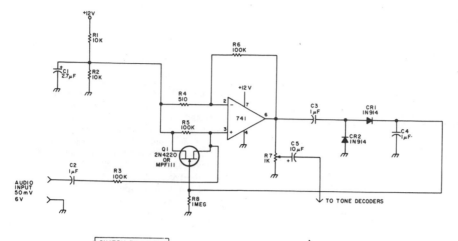

CONSTANT AF LEVEL—Provides constant output level even though input may vary between 50 mV and 6 V, for distribution to tone decoders of autocall system used to monitor simplex or repeater channel to which amateur radio receiver is tuned. Positive terminal of electrolytic C3 must go to pin 6 of 741.—C. W. Andreasen, Autocall '76, *73 Magazine*, June 1976, p 52–54.

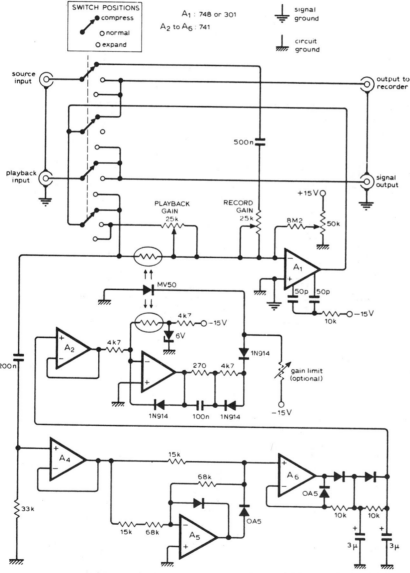

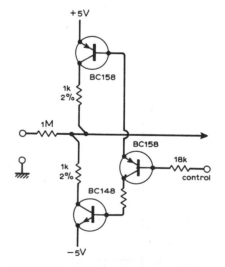

PANNING MIXER POT—Circuit gives best possible approach to sine law so $A^2 = B^2$ is constant for all positions of wiper. Calculated error is less than 1 dB over full range of wiper. Use wirewound pot to minimize crosstalk.—J. Dawson, Single Gang Pan-Pot, *Wireless World*, Feb. 1976, p 78.

COMPANDER WITH 100-dB RANGE—Simple square-law circuit preserves dynamic range of virtually any input signal when recorded by ordinary tape recorder. Suitable for speech signals as well as for recording or playback in noisy environments. Opamp A₁ should have separately decoupled supply. Switching provides compression during recording and expansion during playback. Tracking of photocells is es-sential for accurate power-law compansion. LED can be glued with clear epoxy to matched photocells. Use silicon signal diodes such as 1N914, 1N4148, or 1S44. Inexpensive photocells such as Vactec VT-833 gave suitably low distortion. Article gives performance characteristics and operating details.—J. Vanderkooy, Wideband Compander Design, *Wireless World*, July 1976, p 45–49.

AUDIO SWITCHING GATE—Can be used with programmed channel selectors, as required in music synthesis for controlling audio signals by means of TTL levels. DC offset at output is negligible when gate is off, simplifying design of subsequent stages. Use logic 1 (+5 V) to open gate, and logic 0 (0 V) to close it.—L. Cook, Analogue Gate with No Offset, *Wireless World*, Feb. 1975, p 93.

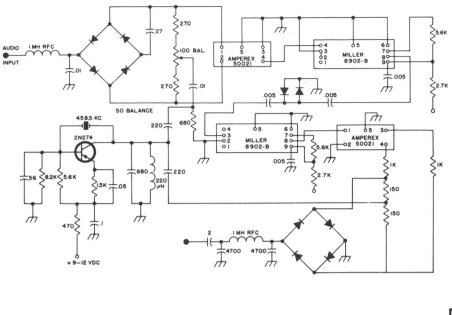

CLIPPER—Provides speech clipping at RF level for SSB transmitters. Based on use of two Amperex 455-kHz IF crystal filters having 4-kHz rated bandwidth, in place of costly conventional sideband filters. Miller 8902-B IF amplifier module simplifies wiring. Diode balanced modulator at upper left can use individual 1N63 or 1N914 diodes, or RCA CA-3019 hex diode array that also provides diodes for RF clipper. Audio input can be taken from speech compressor of receiver, or separate audio amplifier can be added to boost AF level. Same diode types are used in product detector for final SSB signal, lower right, which is diode ring demodulator taking injection voltage from carrier oscillator. Article tells how pin 8 on IF module (not connected internally) is bridged to solder line between link output of last IF transformer and AM detector diode, and gives other construction details.—J. J. Schultz, Inexpensive RF Speech Clipper, *73 Magazine,* Sept. 1974, p 61–64 and 66.

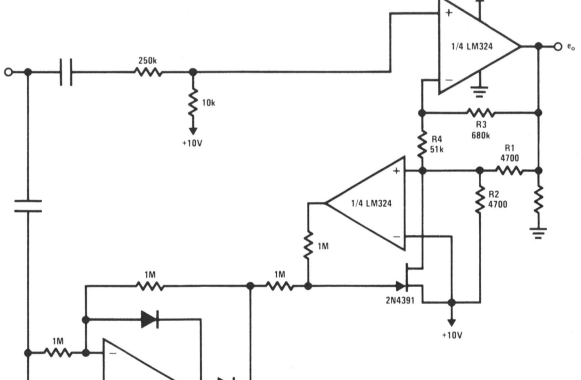

LINEAR 1:4 VOLUME EXPANDER—Gain varies automatically with strength of input signal, being lowest for weak signals. Resistors R3 and R4 modify linear control curve to give log curve desired for audio systems. Overall gain of circuit is about 0 dB at midrange of expansion. Lower pair of LM324 quad opamp sections function as full-wave precision linear peak detector. For stereo, only upper part of circuit needs to be duplicated for second channel.—"Linear Applications, Vol. 2," National Semiconductor, Santa Clara, CA, 1976, AN-129, p 7.

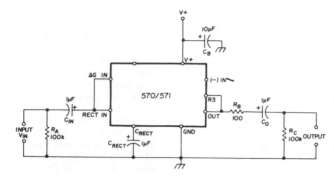

EXPANDER—Uses Signetics dual-channel compandor IC; 571 has lower inherent distortion and higher supply voltage range (6–24 V) than 571 (6–18 V). Values shown are for 15-V supply with either IC. Gain through expander is 1.43 V_{IN}, where V_{IN} is average input voltage. Unity gain occurs at RMS input level of 0.775 V, or 0 dBm in 600-ohm systems.—W. G. Jung, Gain Control IC for Audio Signal Processing, *Ham Radio*, July 1977, p 47–53.

SWITCHING-CLICK SUPPRESSOR—Correction network shown can be inserted in audio channel of mixing console without producing transients or level changes. Although Baxandall network is shown, switching technique is applicable to other filters. With S_1 normally open and S_2 normally closed, circuit operation is normal. If switch positions are simultaneously reversed, response remains flat regardless of positions of bass and treble pots, center-frequency gain remains unchanged, and phase shift is unchanged. There is then no transient interruption of AF signals. Switching clicks cannot occur because there is no direct current in the network.—J. S. Wilson, Click-Free Switching for Audio Filters, *Wireless World*, Jan. 1975, p 12.

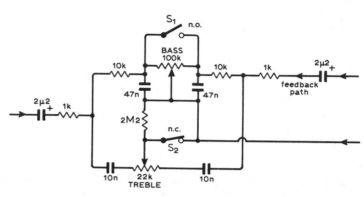

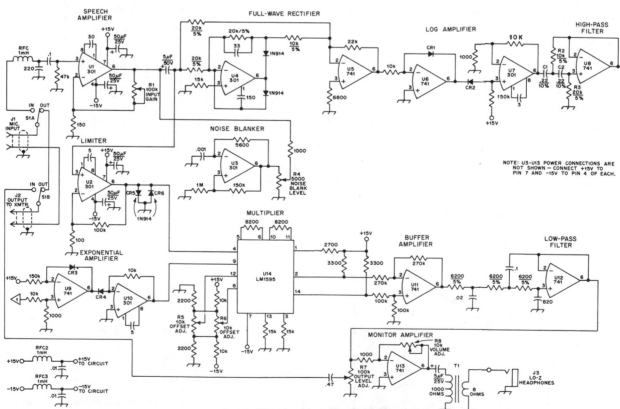

NOTE: U3–U13 POWER CONNECTIONS ARE NOT SHOWN — CONNECT +15V TO PIN 7 AND −15V TO PIN 4 OF EACH.

SPEECH PROCESSOR—Can improve signal strength and intelligibility of voice signals up to 6 dB without unpleasant changes in fidelity. Used between microphone and input of AM or SSB transmitter. Based on separation of signal envelope from constant-amplitude carrier that together make up voice signal. After logamp U6 separates components of speech waveform, envelope is filtered out by active RC high-pass filter U8 having 50-Hz cutoff, with exactly unity gain above cutoff. Filtered signal goes to exponential amplifier U9-U10 and is then multiplied by correct sign information in U14. Sign information is obtained by hard-limiting input voice signal with diode clipper CR5-CR6. Resulting square-wave output is multiplied by signal from UJT in U14. Processed signal goes to transmitter input through low-pass filter U12 having sharp cutoff above 3 kHz to eliminate unwanted high-frequency energy. CR1-CR4 are 1N914 or other matched silicon diodes. T1 is 250-mW audio transformer. Article gives construction and adjustment details.—J. E. Kaufmann and G. E. Kopec, A Homomorphic Speech Compressor, *QST*, March 1976, p 33–37.

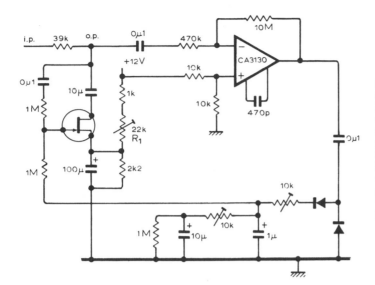

COMPRESSOR/LIMITER—Developed for use with microphone in public-address systems. Bandwidth is 15 Hz to 25 kHz. R_1 sets threshold voltage and compression law. Output of CA3130 inverting opamp is made as large as possible before being applied to rectifier and low-pass filter, to minimize effects of diode non-linearities and capacitor leakage. Low-pass filter gives required fast attack time of about 500 μs and long decay time of about 1 min.—M. B. Taylor, Speech Compressor/Limiter, *Wireless World,* May 1977, p 80.

THREE-INPUT MIXER—Motorola MC3401P or National LM3900 quad opamp serves for three input amplifiers each having adjustable gain range of 1 to about 11 and input impedance above about 100,000 ohms. Common outputs feed fourth opamp section connected as high-impedance amplifier. Maximum overall gain for mixer-amplifier is about 300. Use well-filtered 9–15 V supply or battery capable of supplying 25 mA.—C. D. Rakes, "Integrated Circuit Projects," Howard W. Sams, Indianapolis, IN, 1975, p 21–22.

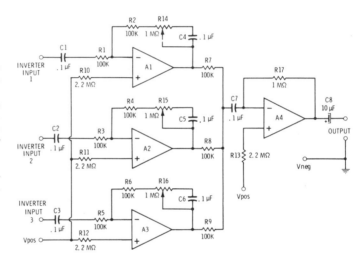

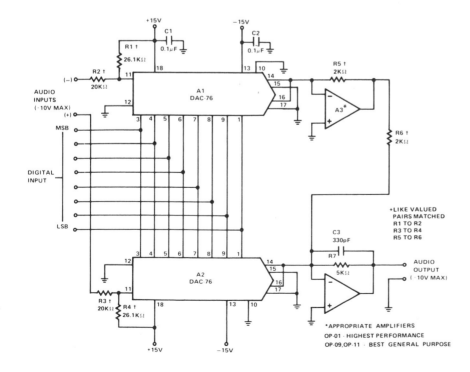

CLICKLESS LEVEL-CONTROL ATTENUATOR—Uses two Precision Monolithics DAC-76 D/A converters to eliminate gain-change transients while providing exponential control of audio signal level. Maximum (all 1s) gain is unity from either input to output, while differential input to output gain is +6 dB. Control range is 78 dB.—W. Jung and W. Ritmanich, "Audio Applications for the DAC-76 Companding D/A Converter," Precision Monolithics, Santa Clara, CA, 1977, AN-28, p 4.

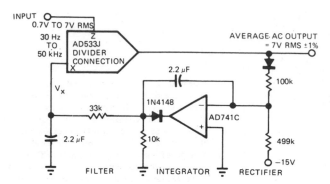

AF LEVEL CONTROL—AD533J analog multiplier used in its divide mode provides measure of automatic level control to compensate for variations in loudness occurring from microphone to microphone in public-address system. Divider output is first rectified and compared with −15 V reference. Difference is then integrated and fed into denominator of divider-connected AD533J as control signal V_x. Average AC output is held within 1% of 7 V.—R. Frantz, Analog Multipliers—New IC Versions Manipulate Real-World Phenomena with Ease, *EDN Magazine*, Sept. 5, 1977, p 125–129.

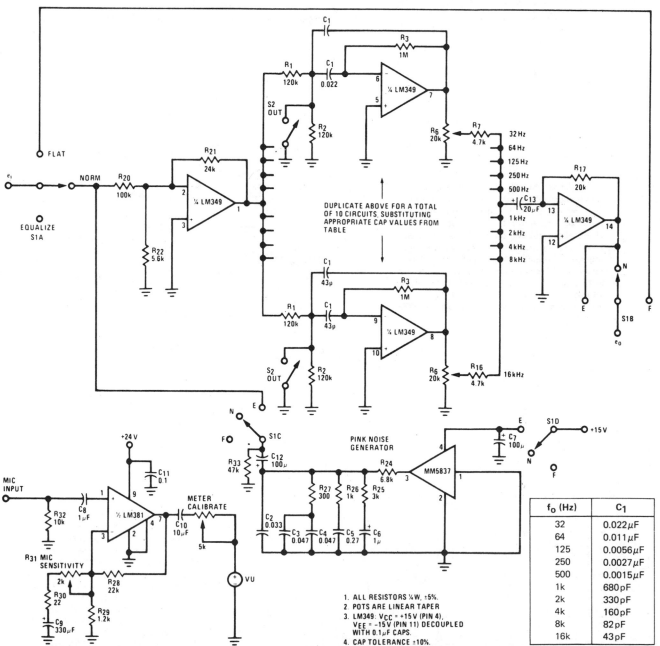

f_o (Hz)	C_1
32	0.022 μF
64	0.011 μF
125	0.0056 μF
250	0.0027 μF
500	0.0015 μF
1k	680 pF
2k	330 pF
4k	160 pF
8k	82 pF
16k	43 pF

1. ALL RESISTORS ¼W, ±5%.
2. POTS ARE LINEAR TAPER
3. LM349: V_{CC} = +15V (PIN 4), V_{EE} = −15V (PIN 11) DECOUPLED WITH 0.1μF CAPS.
4. CAP TOLERANCE ±10%.

ROOM EQUALIZER—Ten-octave equalizer is combined with pink-noise generator in such a way that all but one octave band can be switched out, with pink noise passed through remaining filter to power amplifier and loudspeaker. Microphone with flat frequency response over audio band is used to pick up resulting noise at some center listening point in room being equalized. Amplified output of microphone drives VU meter where arbitrary level is established for one filter section. Other filter sections are then switched in one at a time and adjusted to give same VU reading. Equalizer settings then give flat room response for all ten octaves. High end can then be rolled off or low end boosted to suit personal preference. Adjustments are readily repeated when furniture is changed in room. Table gives values of C_1 for each octave.—"Audio Handbook," National Semiconductor, Santa Clara, CA, 1977, p 2-53–2-59.

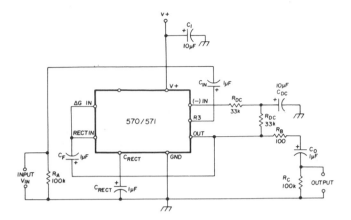

COMPRESSOR—Circuit has unity gain at 0.775 VRMS input and complementary input/output characteristic. Voltage gain through compressor is square root of $0.7/V_{IN}$, where V_{IN} is average input voltage. Uses Signetics dual-channel compandor IC; 571 has lower inherent distortion and higher supply voltage range (6–24 V) than 571 (6–18 V).—W. G. Jung, Gain Control IC for Audio Signal Processing, *Ham Radio*, July 1977, p 47–53.

CONSTANT 1.8-V AF FOR SSB—Uses Motorola MFC6040 voltage-controlled amplifier IC having 13-dB gain and maximum of 90-dB gain reduction. Q1-Q2 form microphone preamp, Q4 is AGC detector/amplifier for IC, and Q3 is output buffer. With 500-ohm dynamic microphone, output remains constant at 1.8 VRMS.—L. Novotny, Speech Compressor, *Ham Radio*, Feb. 1976, p 70–71.

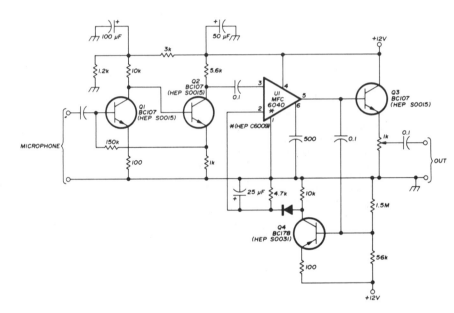

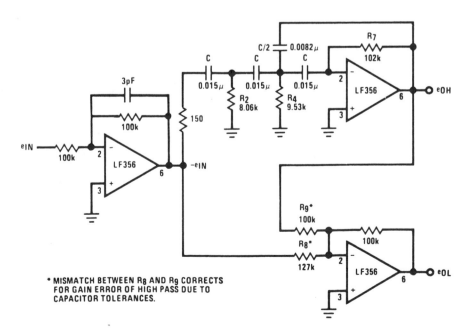

ASYMMETRICAL ACTIVE CROSSOVER—High-pass and low-pass active filters using National LF356 opamps are asymmetrical about 500-Hz crossover point. Sum of filter output voltages is always constant and equal to unity. Rolloff of low-pass filter is only −6 dB per octave, as compared to −18 dB per octave for high-pass filter.—"Audio Handbook," National Semiconductor, Santa Clara, CA, 1977, p 5-1–5-7.

* MISMATCH BETWEEN R₈ AND R₉ CORRECTS FOR GAIN ERROR OF HIGH PASS DUE TO CAPACITOR TOLERANCES.

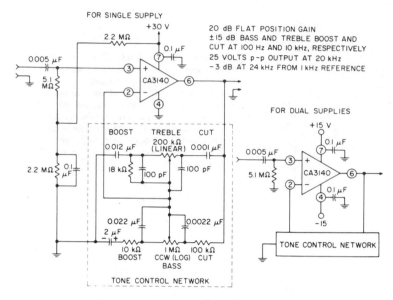

BASS/TREBLE BOOST/CUT—Using linear and log pots with tone control network in feedback path of CA3140 bipolar MOS opamp, circuit provides 20-dB gain in flat position and ±15 dB bass and treble boost and cut at 100 and 10,000 Hz. Output is 25 V P-P at 20 kHz and is −3 dB at 24 kHz from 1-kHz reference. Optional connection for ±15 V supply is also shown.—"Circuit Ideas for RCA Linear ICs," RCA Solid State Division, Somerville, NJ, 1977, p 10.

HIGH-ON LOGIC CONTROL—Uses Signetics NE571 or NE570 analog compandor. When control input is high, CR1 is off and current developed by R_{GAIN} flows into rectifier input, allowing audio to be amplified. Gain is unity (or other nominal value chosen by changing value of R3) for control inputs greater than 3 V. Switching is abrupt, with full attenuation below 1.5 V. Narrow transition width and nominal DC center of 1.8 V allow direct control from CMOS, TTL, DTL, or other positive logic. Supply voltage should be stable.—W. G. Jung, Gain Control IC for Audio Signal Processing, *Ham Radio*, July 1977, p 47–53.

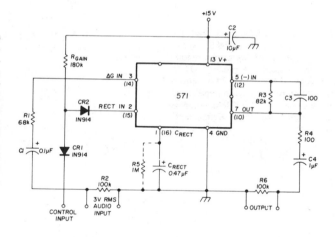

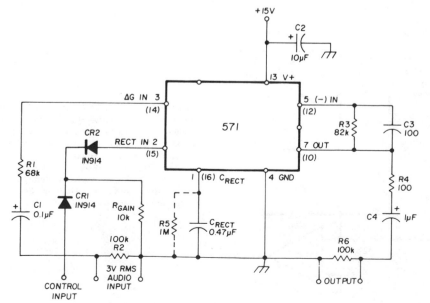

LOW-ON LOGIC CONTROL—Uses Signetics NE571 or NE570 analog compandor. Gain is determined by current developed through R_{GAIN} in conjunction with internal 1.8-V voltage reference. When control input is low, normal current flows through R_{GAIN}. When control signal is high, CR1 is forward-biased, interrupting current flow, and output is attenuated.—W. G. Jung, Gain Control IC for Audio Signal Processing, *Ham Radio*, July 1977, p 47–53.

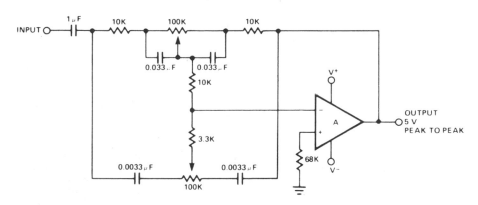

TONE CONTROL FOR OPAMP—Provides up to 20 dB of bass boost or cut at 20 Hz and up to 19 dB of treble boost or cut at 20 kHz. Turnover frequency is 1 kHz. Opamp can be 531 or 301.— "Signetics Analog Data Manual," Signetics, Sunnyvale, CA, 1977, p 638–640.

SPEECH FILTER—High-pass and low-pass filters in cascade provide corner frequencies of 300 and 3000 Hz for limiting audio bandwidth to speech frequencies. Rolloff beyond corners is −40 dB per decade. Input-to-output gain is 1.— "Audio Handbook," National Semiconductor, Santa Clara, CA, 1977, p 2-49–2-52.

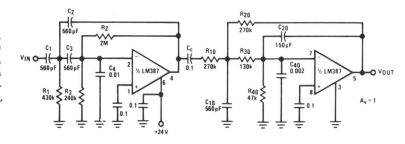

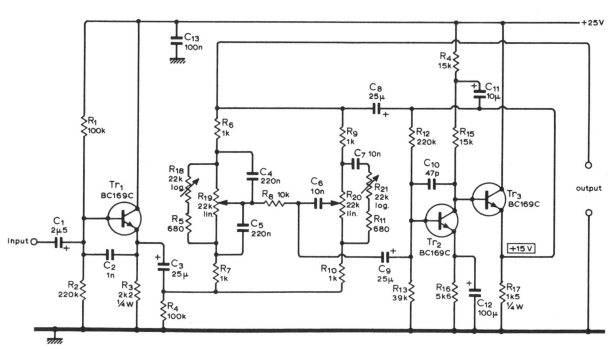

IMPROVED BAXANDALL CONTROL—Uses separate "effect" controls for bass and treble to limit maximum degree of boost and cut obtainable from bass and treble controls. R_{18} controls effect for bass and R_{21} for treble. Circuit has unity gain with controls set flat. Article gives response curves and describes operation of circuit in detail.—M. V. Thomas, Baxandall Tone Control Revisited, *Wireless World,* Sept. 1974, p 341–343.

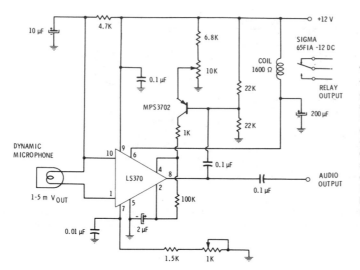

VOX WITH SPEECH COMPRESSION—Turns on transmitter automatically when operator begins speaking into microphone. Circuit switches back to receiving condition automatically at end of message. IC can be LS370 or equivalent such as LM370 or SC370. Amount of compression is adjusted with 10K pot, for reducing gain of IC automatically to maintain reasonably constant audio output at pin 8 despite different voice levels at microphone.—E. M. Noll, "Linear IC Principles, Experiments, and Projects," Howard W. Sams, Indianapolis, IN, 1974, p 344–347.

COMPRESSOR—Keeps output voltage constant as long as input signal is kept above AF threshold level. Opamp is MC3340P. —*Circuits, 73 Magazine,* Holiday issue 1976, p 170.

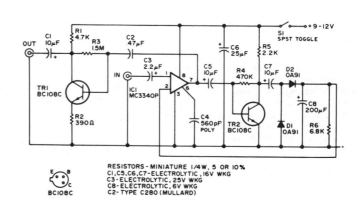

RESISTORS - MINIATURE 1/4W, 5 OR 10%
C1,C5,C6,C7-ELECTROLYTIC, 16V WKG
C3- ELECTROLYTIC, 25V WKG
C8- ELECTROLYTIC, 6V WKG
C2- TYPE C280 (MULLARD)

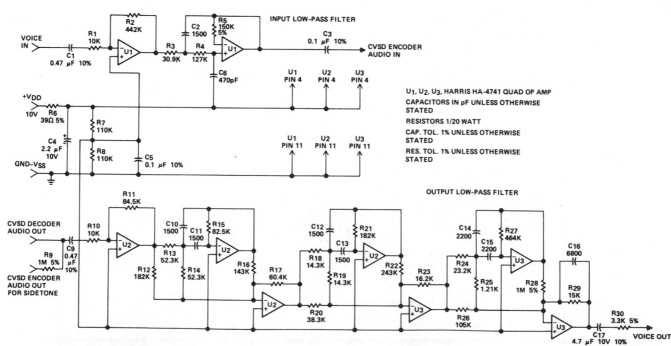

U₁, U₂, U₃, HARRIS HA-4741 QUAD OP AMP
CAPACITORS IN pF UNLESS OTHERWISE STATED
RESISTORS 1/20 WATT
CAP. TOL. 1% UNLESS OTHERWISE STATED
RES. TOL. 1% UNLESS OTHERWISE STATED

AUDIO FILTERS FOR SNR MEASUREMENT—Used in checking performance of Harris HC-55516/55532 half-duplex modulator-demodulator systems for converting voice signals into serial NRZ digital data and reconverting that data back to voice. Supply required for opamp sections is ±15 V. Response of input filter is down 3 dB at 3 kHz and is down 20 dB at 9 kHz. Response of output filter is flat up to 3 kHz and down more than 45 dB from 3.8 kHz to 100 kHz.—"Linear & Data Acquisition Products," Harris Semiconductor, Melbourne, FL, Vol. 1, 1977, p 5–10.

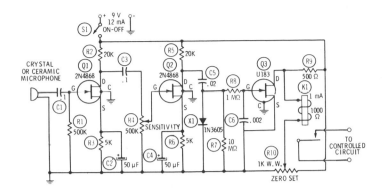

SOUND-OPERATED RELAY—Output of about 1.8 mVRMS from crystal or ceramic microphone will energize relay when sensitivity control R4 is at maximum. First two stages form high-gain RC-coupled AF amplifier, output of which is rectified by silicon diode X1. DC voltage developed across diode is applied to gate of Siliconix U183 FET which acts as DC amplifier driving Sigma 5F or equivalent relay. To adjust, short microphone terminals, set R4 for maximum sensitivity, then adjust R10 until relay opens.—R. P. Turner, "FET Circuits," Howard W. Sams, Indianapolis, IN, 1977, 2nd Ed., p 111–113.

SYMMETRICAL ACTIVE CROSSOVER—Provides −18 dB per octave rolloff (third order) and maximally flat (Butterworth) characteristics for crossover frequency of 500 Hz. Uses National LF356 opamps in high-pass and low-pass filters and same opamp as buffer having low driving impedance required by active filters. Power supplies are ±15 V. Design equations are given.—"Audio Handbook," National Semiconductor, Santa Clara, CA, 1977, p 5-1–5-7.

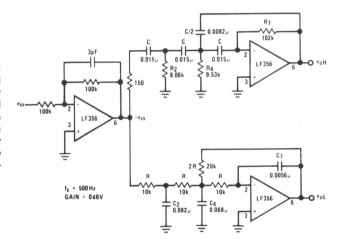

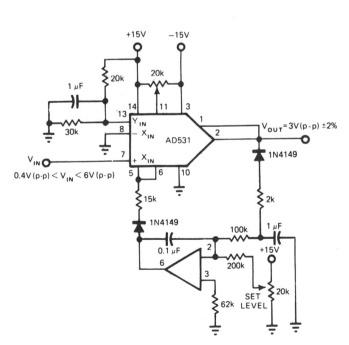

WIDEBAND AF LEVEL CONTROL—AD531 analog multiplier can hold output to 3 V P-P ± 2% for inputs from 0.4 to 6 V P-P, at frequencies from 30 Hz to 400 kHz. Opamp type is not critical.—R. Frantz, Analog Multipliers—New IC Versions Manipulate Real-World Phenomena with Ease, *EDN Magazine*, Sept. 5, 1977, p 125–129.

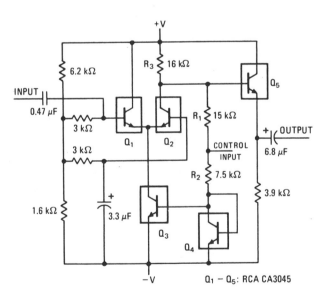

THUMPLESS CONTROL—Five-transistor circuit for audio amplifier applications eliminates thumping sounds that can sometimes be heard when level of input signal changes suddenly. Differential amplifier Q_1-Q_2, with R_1 in emitter-current control circuit, eliminates thump. Control input acts on identical transistors Q_3-Q_4 which make transconductance of differential pair Q_1-Q_2 vary in direct proportion to control voltage. Fifth transistor in array, Q_5, is used as output signal buffer. Amplifier gain is 30 for control voltage of 15 V.—P. Brokaw, Automatic Gain Control Quells Amplifier Thump, *Electronics*, Jan. 10, 1974, p 131–132; reprinted in "Circuits for Electronics Engineers," *Electronics*, 1977, p 46–47.

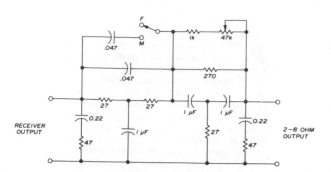

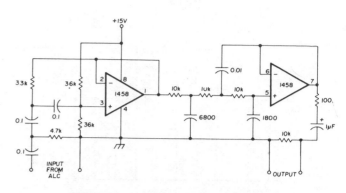

MALE-FEMALE VOICE SWITCH—Circuit developed by NASA engineers to improve intelligibility of voice communication during Apollo moon shots passes only the three portions of the speech spectrum required for clear speech: 300–400 Hz and 2500–3000 Hz for both sexes and 900–1700 Hz for males or 1100–1900 Hz for females. Pot adjusts null to about 600 Hz. Circuit improves readability of weak DX voice signals in noise.—J. Fisk, Circuits and Techniques, *Ham Radio,* June 1976, p 48–52.

SPEECH FILTER—Pair of Bessel-type high-pass filters removes undesired components created by peak clipping during audio signal processing. Developed for use with automatic level control applications of NE571 analog compandor.—W. G. Jung, Gain Control IC for Audio Signal Processing, *Ham Radio,* July 1977, p 47–53.

TWO-CHANNEL PANNING—Provides smooth and accurate panoramic control of apparent microphone position between two output channels, as often required in mixing consoles at recording studios. Requires only single linear pot. At each extreme of pot, gain is unity for one channel and zero for other. With pot centered, gains for both channels are −3 dB. R_2 depends on supply voltage used, which can be from 9 to 30 V.—"Audio Handbook," National Semiconductor, Santa Clara, CA, 1977, p 2-59–2-61.

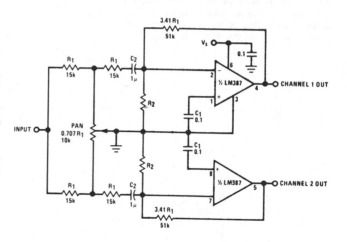

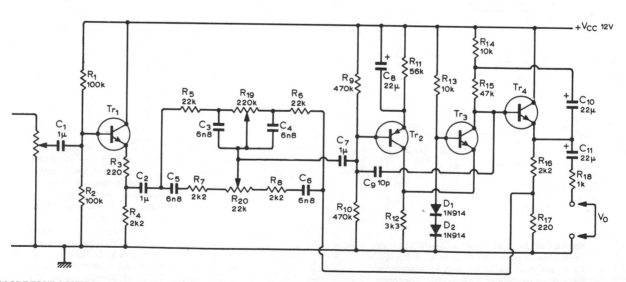

CASCODE TONE CONTROL—Circuit raises 100-mV input signal level to 1 V for driving power amplifier and uses cascode arrangement to improve S/N ratio of tone control network. Values shown give maximum bass boost or cut at 50 Hz with R_{19} and maximum treble boost or cut at 10 kHz with R_{20}. Tr_2 can be BC15, BC214, BC309, or equivalent. Other transistors can be BC109, BC114, BC184, or equivalent.—J. N. Ellis, High Quality Tone Control, *Wireless World,* Aug. 1973, p 378.

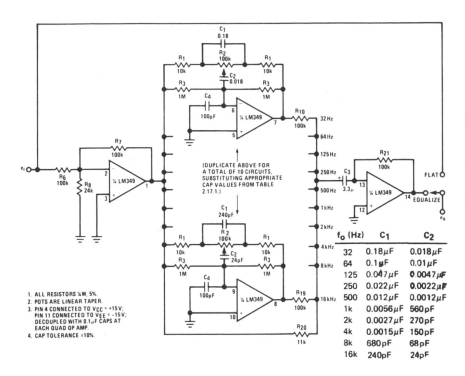

OCTAVE EQUALIZER—Provides ten bands of tone control, separated by one octave in frequency, with independent adjustment for each. Used to compensate for unwanted amplitude-frequency or phase-frequency characteristics of audio systems. Values of C_1 and C_2 for each circuit are given in table. With control R_2 in flat position, circuit becomes all-pass with unity gain. Moving R_2 to full boost gives bandpass characteristic, and moving in other direction to full cut gives band-reject or notch filter. For stereo, identical equalizer is needed for other channel.—"Audio Handbook," National Semiconductor, Santa Clara, CA, 1977, p 2-53–2-59.

f_0 (Hz)	C_1	C_2
32	0.18 µF	0.018 µF
64	0.1 µF	0.01 µF
125	0.047 µF	0.0047 µF
250	0.022 µF	0.0022 µF
500	0.012 µF	0.0012 µF
1k	0.0056 µF	560 pF
2k	0.0027 µF	270 pF
4k	0.0015 µF	150 pF
8k	680 pF	68 pF
16k	240 pF	24 pF

AUTOMATIC LEVEL CONTROL—Uses Signetics NE570 or NE571 analog compandor to provide automatic level control for audio signal processing, to give constant high percentage modulation despite varying input levels. Optional resistor R_X varies threshold of level regulation. Widest range of gain control is obtained with R_X open. When resistor value is lowered, larger input signal is required for full output. Peak-level clipping with pair of reverse paralleled LEDs controls overshoots on speech by limiting RMS output to 2.2 V P-P. R_Y regulates clipped amplitude.—W. G. Jung, Gain Control IC for Audio Signal Processing, *Ham Radio*, July 1977, p 47–53.

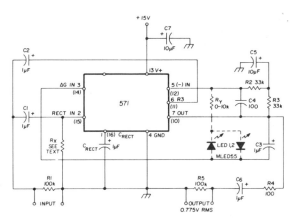

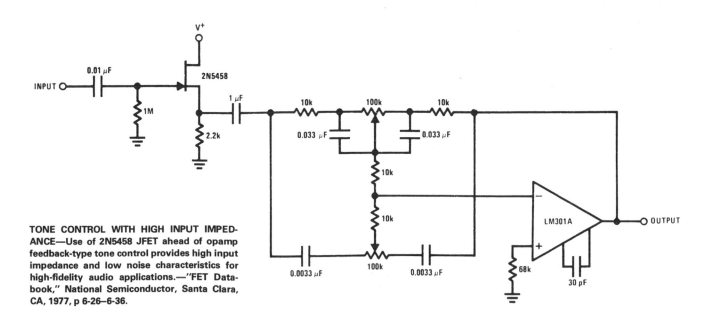

TONE CONTROL WITH HIGH INPUT IMPEDANCE—Use of 2N5458 JFET ahead of opamp feedback-type tone control provides high input impedance and low noise characteristics for high-fidelity audio applications.—"FET Databook," National Semiconductor, Santa Clara, CA, 1977, p 6-26–6-36.

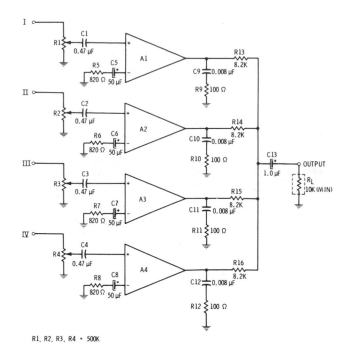

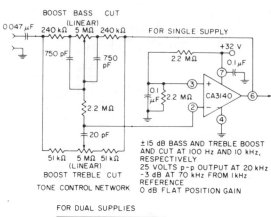

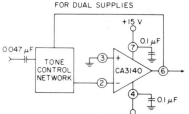

R1, R2, R3, R4 • 500K

FOUR-CHANNEL MIXER—All four sections of RCA CA3048 quad differential amplifier are utilized in linear mixer providing gain of 20 dB for each channel. Designed for use with load of 10K or larger. All inputs are high impedance.—E. M. Noll, "Linear IC Principles, Experiments, and Projects," Howard W. Sams, Indianapolis, IN, 1974, p 173 and 179.

BAXANDALL TONE CONTROL—Utilizes high slew rate, high output-voltage capability, and high input impedance of CA3140 bipolar MOS opamp to provide unity gain at midband along with bass and treble boost and cut of ±15 dB at 100 and 10,000 Hz. Optional connection for ±15 V supply is shown below.—"Circuit Ideas for RCA Linear ICs," RCA Solid State Division, Somerville, NJ, 1977, p 10.

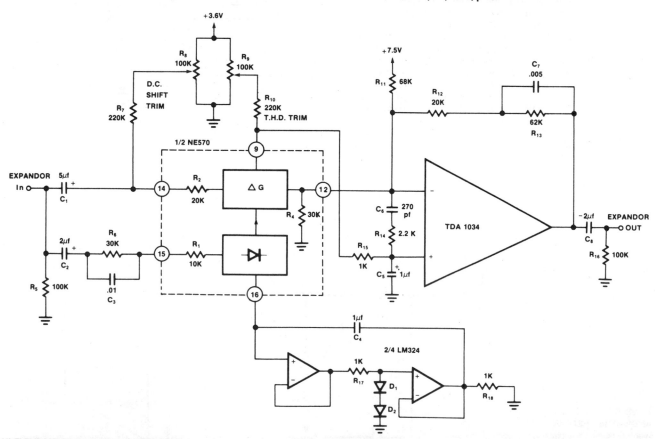

HI-FI EXPANDER—Used in playback of material that has been recorded with overload-preventing compressor. External opamp is used for high slew rate. Adjust distortion trimpot R_9 for minimum total harmonic distortion when using input of 0 dBm at 10 kHz. Adjust DC shift pot R_8 after this, for minimum envelope bounce with tone-burst input.—"Signetics Analog Data Manual," Signetics, Sunnyvale, CA, 1977, p 804–805.

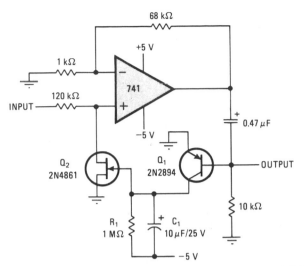

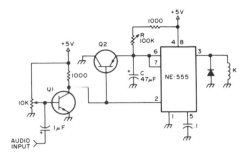

60-dB RANGE FOR AUDIO—JFET acts as voltage-controlled resistor in peak-detecting control loop of 741 opamp. Input range is 20 mV to 20 V, with response time of 1–2 ms and delay of 0.4 s. Output is about 1.4 V P-P over entire 60-dB range.—N. Heckt, Automatic Gain Control Has 60-Decibel Range, *Electronics*, March 31, 1977, p 107.

AUDIO-OPERATED RELAY—Addition of two general-purpose transistors to 555 timer gives audio-triggered relay that can be used for automatic recording of output of channel-monitoring radio receiver or data from any audio link. Adjustable time delay R keeps control circuit actuated up to 5 s (determined by R and C) to avoid cycling relay during pauses in speech or dropouts in data. Q1 is NPN, and Q2 is PNP. Attack time equals very short pull-in time of 5-V reed relay K. Adjust 10K input pot just below point at which K pulls in when there is no audio input.—R. Taggart, Sound Operated Relay, *73 Magazine*, Oct. 1977, p 114–115.

DIFFERENTIAL-AMPLIFIER CLIPPER—Provides gain as well as precise symmetrical clipping for improving intelligibility of speech fed into radio transmitter. Circuit reduces dynamic range of energy peaks to bring them closer to average energy level. When inserted in series with microphone, use of clipper gives at least 6-dB increase in effective power. Signals are passed up to certain amplitude but limited above this level.—B. Kirkwood, Principles of Speech Processing, *Ham Radio*, Feb. 1975, p 28–34.

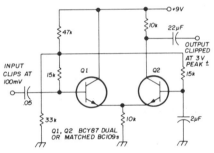

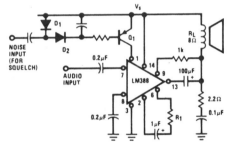

SQUELCHABLE AMPLIFIER—Circuit designed for portable FM scanners and two-way walkie-talkie radios can be turned off by noise or by control signal to minimize battery drain. When squelched, LM388 opamp-transistor-diode array draws only 0.8 mA from 7.5-V supply. Diodes rectify noise from limiter or discriminator of receiver, producing direct current that turns on Q_1 and thereby clamps opamp off. Voltage gain is 20 to 200, depending on value used for R_1. Power output without squelch is about 0.5 W for 8-ohm loudspeaker.—"Audio Handbook," National Semiconductor, Santa Clara, CA, 1977, p 4-37–4-41.

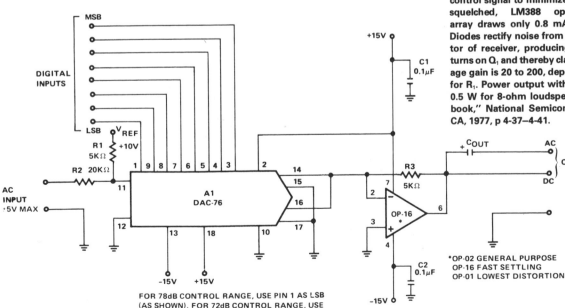

FOR 78dB CONTROL RANGE, USE PIN 1 AS LSB (AS SHOWN). FOR 72dB CONTROL RANGE, USE PIN 9 AS LSB, GROUND PIN 1.

TWO-QUADRANT EXPONENTIAL CONTROL—Decibel-weighted control characteristic of Precision Monolithics DAC-76 D/A converter matches natural loudness sensitivity of human ear, to provide much greater useful dynamic range for controlling audio level. Control range can be either 72 or 78 dB, depending on pin connections used. 8-bit word control input can be interfaced with standard TTL-compatible microprocessor outputs. To avoid annoying output transients during large or rapid gain changes, use clickless attenuator/amplifier (also given in application note).—W. Jung and W. Ritmanich, "Audio Applications for the DAC-76 Companding D/A Converter," Precision Monolithics, Santa Clara, CA, 1977, AN-28, p 2.

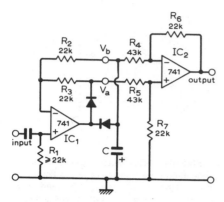

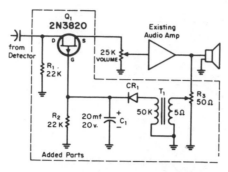

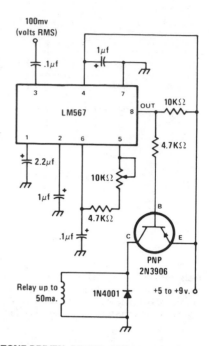

CLAMPING WITH OPAMPS—Circuit is used after stage of AC amplification to clamp minimum level of signal voltage to 0 V for signals having amplitudes between 10 mV and 10 V. With 250-μF electrolytic for C, sinusoidal waveforms between 3 and 10,000 Hz are clamped with little distortion. Overall gain is unity.—C. B. Mussell, D.C. Level Clamp, *Wireless World*, Feb. 1975, p 93.

AF COMPRESSOR—Developed for use in communication receiver where signals vary so greatly that even modern AVC systems cannot level all signals. Circuit is AVC that sets maximum audio level which will not be exceeded. Uses one FET as series attenuator controlled by DC voltage derived from audio output. R_3 permits adjustment of compression level.—C. E. Richmond, A Receiver Audio Compressor, *CQ*, June 1970, p 35 and 86.

TONE-DRIVEN RELAY—LM567 tone decoder will respond to frequency between 700 and 1500 Hz, determined by setting of 10K pot. When input of 100 mVRMS at preset frequency arrives, output of IC goes low and energizes relay through transistor. Tone can be obtained from audio oscillator or telephone Touch-Tone pad. Relay contacts can be used to turn desired device on or off.—J. A. Sandler, 9 Easy to Build Projects under $9, *Modern Electronics*, July 1978, p 53–56.

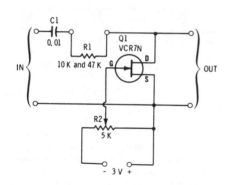

VOLTAGE-CONTROLLED ATTENUATOR—Used to control low-level audio signals with variable DC voltage of ±3 V. Control pot can be remotely located. Highest possible output is equal to input level, occurring when gate bias is set close to pinchoff value. Output is minimum when gate bias is zero.—E. M. Noll, "FET Principles, Experiments, and Projects," Howard W. Sams, Indianapolis, IN, 2nd Ed., 1975, p 258–260.

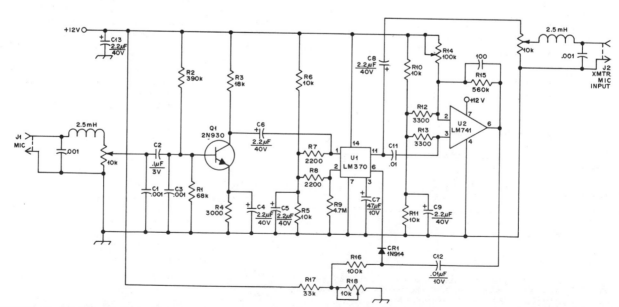

EQUALIZER—Designed for use between input jack and microphone of amateur transmitter, to keep bandpass response between limits of about 200 and 3100 Hz. Circuit also provides measure of volume compression, improving transmitter efficiency. Construction and adjustment details stress importance of eliminating ground loops and RF feedback. U1 is voltage-controlled amplifier in feedback loop, with 741 opamp U2 as compression detector. U2 is biased so output is almost at ground, and no feedback voltage is applied until input to U2 exceeds 0.9 V. U1 thus operates in linear mode at maximum gain until output voltage exceeds 0.9 V, when voltage is applied to U1 and gain of IC is reduced.—R. Tauber, The Equalizer, *QST*, March 1977, p 18–20.

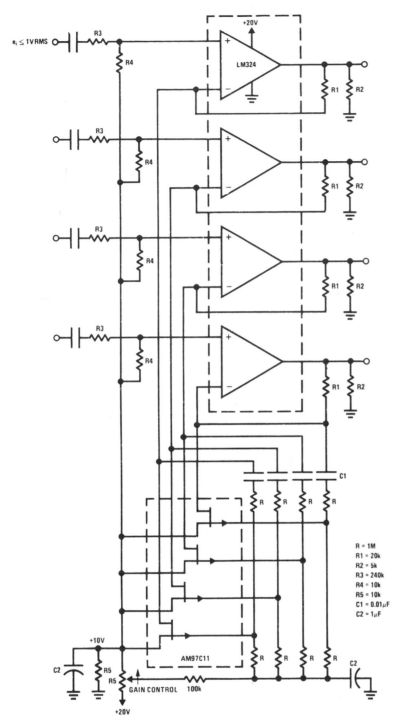

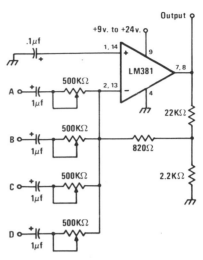

FOUR-CHANNEL MIXER—Combines AF signals from one to four sources into single audio signal for input of LM381 opamp that serves also as preamp. Shield mixer circuit and use shielded cable for all input leads to avoid pickup of 60-Hz field by high-gain opamp. Increasing supply voltage from minimum of 9 V boosts output signal voltage.—J. A. Sandler, 9 Easy to Build Projects under $9, *Modern Electronics,* July 1978, p 53–56.

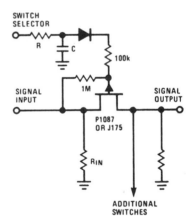

NOISELESS AUDIO SWITCH—Deglitched current-mode switch using JFET can be placed directly on printed-circuit board instead of front panel, to minimize hum pickup and crosstalk. JFET allows transition time of drive to be adjusted with series resistor R and shunt capacitor C to provide noiseless switching of AF signals. Diode type is not critical. Any number of switches can be ganged.—"Audio Handbook," National Semiconductor, Santa Clara, CA, 1977, p 2-62.

QUAD GAIN CONTROL—Combination of National AM97C11 quad FET and LM324 quad opamp gives tracking gain control having 40-dB range. Bandwidth is 10 kHz minimum, and S/N ratio is better than 70 dB for 4.3-VRMS maxi-mum output. Temperature sensitivity of FET can be reduced by using silicon resistor for opamp feedback resistor R1.—"FET Databook," National Semiconductor, Santa Clara, CA, 1977, p 6-39–6-46.

CHAPTER **5**
Audio Measuring Circuits

Includes S-meters and VU meters, along with circuits for measuring AF
distortion and flutter, peak program meter tester, and clipping-point indicator.
See also Frequency Measuring and Frequency Multiplier chapters.

FLUTTER METER—Signetics 561N PLL detects
frequency variations in 3-kHz tone recorded on
magnetic tape for test purposes. Frequency of
VCO in 561N is set to nominal 3 kHz with 5K pot.
Demodulated output is AC coupled to amplifier
having high input impedance. Either CRO or
true RMS voltmeter can be used to make RMS
flutter readings. To calibrate circuit, feed in 3-
kHz tone from oscillator and measure output
level shift when frequency is offset 1%.—"Signetics Analog Data Manual," Signetics, Sunnyvale, CA, 1977, p 860.

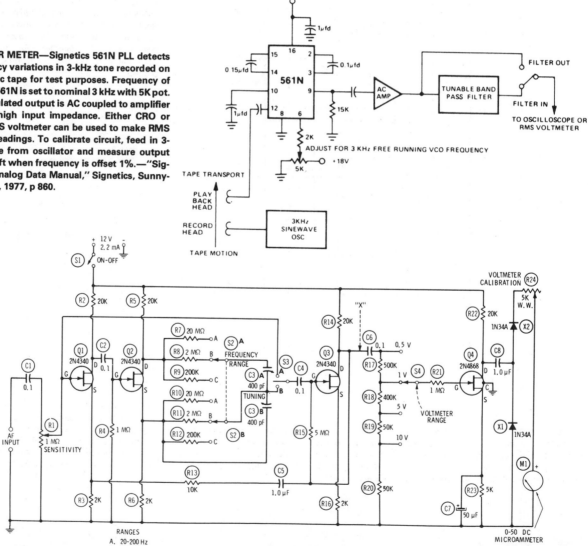

HARMONIC-DISTORTION METER—Used to
measure total harmonic distortion of audio amplifier, component, or network. Pure sine-wave
signal is applied to device under test, and output of device is fed to AF input of distortion
meter. After setting S2 to appropriate frequency range, close S1, set S3 at A, set S4 at
appropriate voltage range, and adjust R1 for
full-scale meter deflection. Record this voltage
as E_1. Set S3 to B, tune C3 for null, then set S4
to successively lower ranges for accurate reading of voltage at null. Record residual null volt-
age as E_2. Percentage distortion is then $100E_2/E_1$.—R. P. Turner, "FET Circuits," Howard W.
Sams, Indianapolis, IN, 1977, 2nd Ed., p 147–
150.

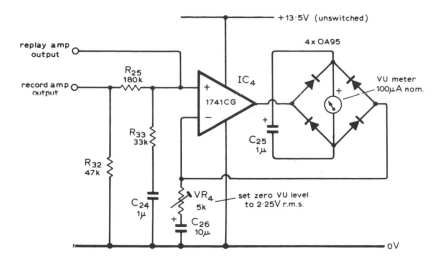

IC DRIVE FOR VU METER—Used in high-quality stereo cassette deck operating from AC line or battery. Meter rectifier bridge is in feedback loop of opamp, to give highly linear AC/DC conversion with flat frequency/amplitude response and short voltage rise time at low cost. Article gives all other circuits of cassette deck and describes operation in detail.—J. L. Linsley Hood, Low-Noise, Low-Cost Cassette Deck, *Wireless World,* Part 1—May 1976, p 36–40 (Part 2—June 1976, p 62–66; Part 3—Aug. 1976, p 55–56).

ADD-ON S-METER—Although designed for use with Clegg FM-27B 2-meter FM receiver, circuit can be readily adapted to other receivers. Amplifier brings low-level 455-MHz IF signal up to level suitable for driving meter. For other IF, such as 10.7 or 11.7 MHz, capacitor values should be changed accordingly. Any NPN transistor with beta of 30 or more at IF value can be used. Diodes can be any type. Supply should be regulated but can be 7–14 V. Output of diode detector will vary from 0 to 1 V at nominal impedance of 20K; for best result, meter with 20- to 50-μA movement can be used.—M. Stern, FM-27B S-Meter, *QST,* Dec. 1976, p 35.

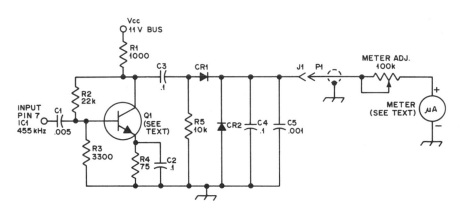

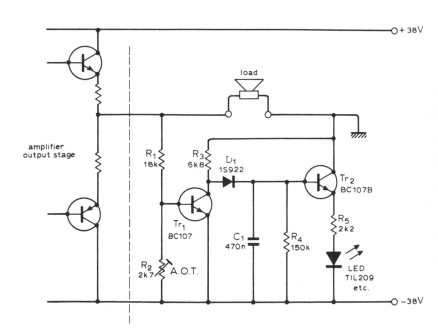

CLIPPING-POINT INDICATOR—Uses LED to indicate when clipping distortion begins in 50-W power amplifier. Display circuit is referenced to negative supply, making detection level independent of supply variation; circuit thus works equally well for instantaneous, music, or continuous overloads. Tr_1 is normally turned hard on and Tr_2 is off. When overload makes collector-emitter voltage of lower amplifier output transistor approach saturation, Tr_1 begins to turn off and C_1 charges through D_1 so LED turns on. Attack time is chosen to make single 3-ms overload transient visible.—J. Dawson and K. Northover, L.E.D. Clip Indicator, *Wireless World,* Jan. 1976, p 60.

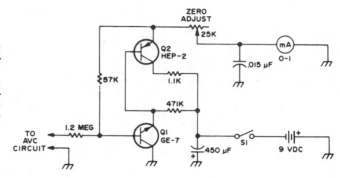

ADDING S-METER—Circuit works well with most all-band receivers. Q1 may also be SK-3011, NR5, TR-10, or DS75. Q2 may also be HE-1, SK-3005, or TR-06. Value of 1.2-megohm input resistor may need to be adjusted depending on AVC voltage, to prevent strong signals from overloading meter.—Novice Q & A, *73 Magazine*, Feb. 1977, p 127.

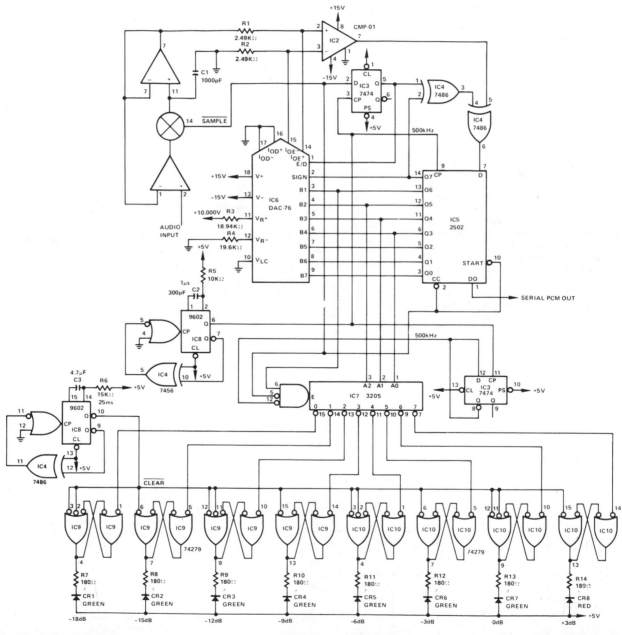

LEDs DISPLAY VU PEAKS—Exponential coding of Precision Monolithics DAC-76 D/A converter is used to good advantage in peak-reading VU indicator with logarithmic weighting, driving LED display. Input audio is converted by DAC-76, CMP-01 comparator, and 2502 successive-approximation A/D converter after being sampled by sample-and-hold input circuit. A/D converter is clocked at 500 kHz and completes conversion every 18 μs, which is fast enough to track audio signals. 4 most significant magnitude bits drive 3205 1-of-8 decoder which is enabled by most significant bit. Resulting eight output levels, separated by 3-dB increments, drive 8-bit RS latch using 74279 chips, updated every 25 ms by 40-Hz display multiplex clock.—W. Jung and W. Ritmanich, "Audio Applications for the DAC-76 Companding D/A Converter," Precision Monolithics, Santa Clara, CA, 1977, AN-28, p 6.

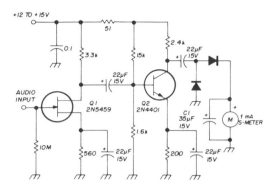

1-mA S-METER—Amplifier designed for 1-mA meter movement consists of two-stage voltage amplifier driving meter rectifier. FET input provides high impedance to detected audio and minimizes loading and distortion problems. Q2 is common-emitter voltage amplifier with simple positive-pulse rectifier for meter. C1 filters rectified audio signal. AF input for S-9 reading is 25–30 mV P-P and for full scale is 50–60 mV P-P. Frequency response is 500 Hz to 10 kHz.—M. A. Chapman, Solid-State S-Meters, *Ham Radio*, March 1975, p 20–23.

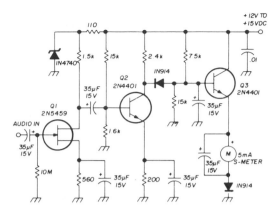

5-mA S-METER—Circuit designed for 5-mA meter movement uses two-stage voltage amplifier Q1-Q2 with emitter-follower output Q3 serving as impedance-matching stage. AF input for S-9 reading is 25–30 mV P-P and for full scale is 50–60 mV P-P. Frequency response is 500 Hz to 10 kHz.—M. A. Chapman, Solid-State S-Meters, *Ham Radio*, March 1975, p 20–23.

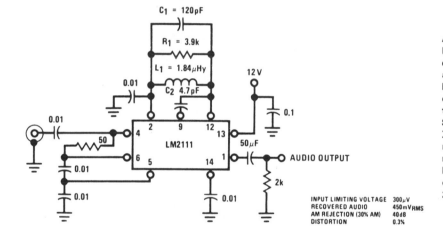

AUDIO-FREQUENCY METER—Covers 0–100 kHz in four ranges. Meter reading is independent of signal amplitude from 1.7 VRMS upward and independent of waveform over wide range. Linear response means only one point need be calibrated in each frequency range. Circuit uses two overdriven FET amplifier stages in cascade. Square-wave output of last stage is rectified by X1 and X2. Deflection of meter depends only on number of pulses per second passing through meter so is proportional to pulse frequency. Battery drain is 1.4 mA.—R. P. Turner, "FET Circuits," Howard W. Sams, Indianapolis, IN, 1977, 2nd Ed., p 129–131.

LEVEL DETECTOR—Circuit lights green LED if signal at output of audio preamp exceeds 1 V peak for predetermined period. Red LED comes on when tone-control stage at output of preamp is on verge of clipping. VU meter driver circuit is also provided. Entire circuit must be duplicated for other stereo channel. Article describes circuit operation in detail and gives all associated circuits used in high-performance audio preamp. D_1 is 1N914; red LED is TIL209 or equivalent; green LED is TIL211.—D. Self, Advanced Preamplifier Design, *Wireless World*, Nov. 1976, p 41–46.

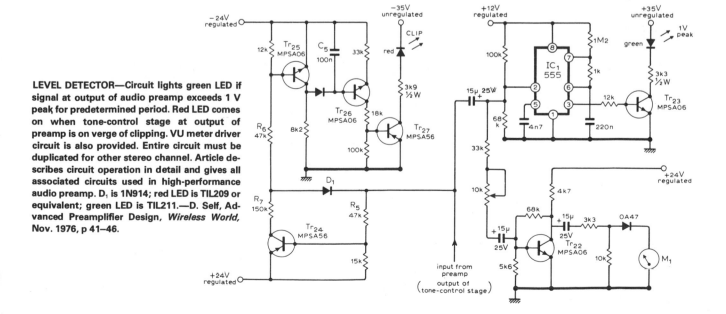

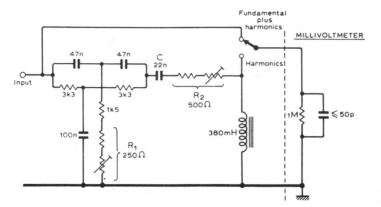

MEASURING AF DISTORTION—Passive high-pass 1-kHz filter is used with audio millivoltmeter to improve accuracy of distortion measurements for low-impedance sources at 1 kHz. Filter removes low-frequency noise from input signal and compensates for loss of harmonic frequency. Applications include setting bias and recording levels of tape recorder. Adjust R_1 for best null, then adjust R_2 and value of C to equalize responses at harmonics.—J. B. Cole, Passive Network to Measure Distortion, *Wireless World,* Jan. 1978, p 60.

AF VOLTMETER—Although not calibrated on absolute basis, either 3 dB or 10 dB of attenuation can be switched in with S1 for measuring purposes. Internal adjustments are made easily by tacking 51-ohm resistor temporarily across input, then driving input with step attenuator fed with audio power at −10 dB by generator having 50-ohm pad in its output. CR1-CR4 are 1N914.—W. Hayward, Defining and Measuring Receiver Dynamic Range, *QST,* July 1975, p 15–21 and 43.

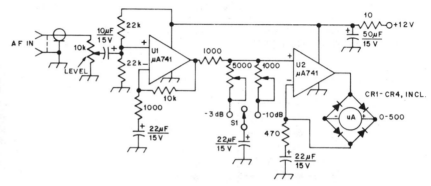

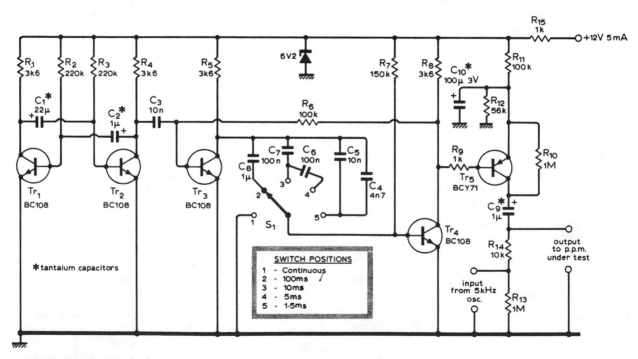

PEAK PROGRAM METER TESTER—Used with 5-kHz audio oscillator to produce tone bursts of 1.5, 5, 10, and 100 ms, as required for checking response of program meter to tone bursts. Transistors Tr_3 and Tr_4 form mono with switched timing capacitors. Article covers calibration and use.—E. T. Garthwaite, Tone Burst Generator for Testing P.P.Ms, *Wireless World,* Aug. 1976, p 53.

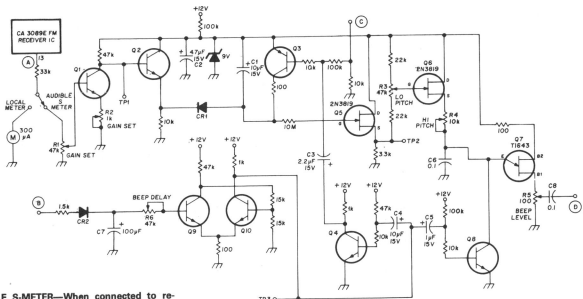

AUDIBLE S-METER—When connected to repeater, circuit generates tone burst 3 s after input signal has dropped out, with duration of 60 ms. Pitch of tone varies inversely with signal strength; highest pitch of 3500 Hz thus represents weak signal, and 350-Hz pitch corresponds to strongest input signal. Can be used to check performance of transmitters and antennas using that repeater. Repeater receiver must have S-meter, as in RCA CA3089E receiver, output of which can be fed to terminal A of circuit. Switch changes output from S-meter to audible encoder. Input B goes to squelch, C goes to +12 V source that is on when receiver is on, and D provides tone output for feed to audio amplifier and loudspeaker. Unlabeled transistors can be any medium-gain small-signal NPN and PNP silicon, comparable to European BC107 and BC177.—F. Johnson, Audible S-Meter for Repeaters, *Ham Radio*, March 1977, p 49–51.

CHAPTER 6
Automatic Gain Control Circuits

Includes circuits providing automatic control of gain for one or more stages in AF, RF, IF, video, or balanced modulator sections of receivers. See also Amplifier, IF Amplifier, and Receiver chapters.

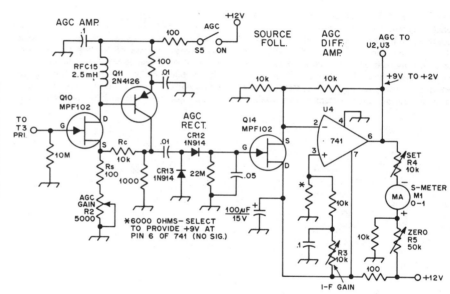

AGC WITH MANUAL CONTROL—Used in 1.8-2 MHz communication receiver having wide dynamic range. R3 serves as manual IF gain control. R2 provides gain variation from 6 to 40 dB for AGC amplifier. Delay is about 1 s. Input is taken from primary of transformer that drives product detector of receiver, and AGC output goes to CA3028A IF opamp. Two-part article gives all other circuits of receiver.—D. DeMaw, His Eminence—the Receiver, *QST*, Part 2—July 1976, p 14–17 (Part 1—June 1976, p 27–30).

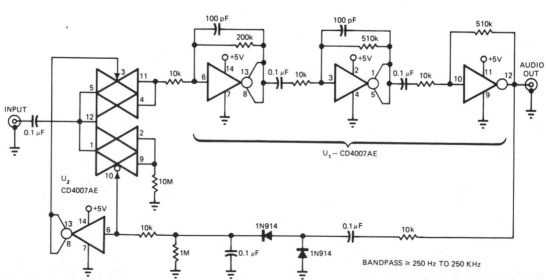

AGC WITH CMOS LOGIC—First stage U$_2$, using CD4007AE, is wired as two-line demultiplexer with only one output acting as transmission gate. Gain is lower in first stage to reduce noise. U$_1$ is used as three-stage high-gain audio amplifier in which first two stages have low-pass filtering for stability. AGC voltage, developed from audio output, is fed back to U$_2$ to turn transmission gate off when gain must be reduced. Audio output is about 2.5 V P-P for inputs of 2 mV and greater.—K. H. Fleischer, Turn Digital CMOS IC's into a Low-Level AGC Amplifier, *EDN Magazine,* Oct. 5, 1977, p 99.

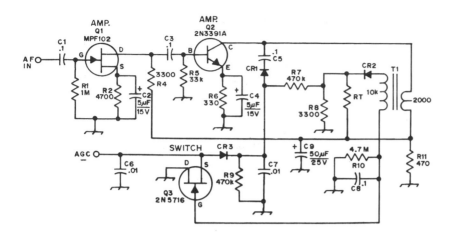

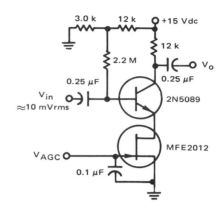

HANG AGC—Provides very fast attack time with no AGC pop. Diodes are 1N914. When voltage across R10-C8 decays below that across R9-C7, Q3 conducts and clamps AGC bus to ground. AGC threshold is determined by value of R_T, between 100K and 470K. AGC line must have high impedance, as with FET IF system. With IC or bipolar IF amplifier, use low-impedance driver. T1 is audio transformer with 10K primary and 2K secondary (Radio Shack 273-1378).—D. Stevens, Solid-State Hang AGC, *QST*, July 1975, p 44.

FET GIVES 30-dB GAIN RANGE—Only 1-V change in gate-source voltage of FET changes voltage gain over full range. Possible drawback is harmonic distortion due to unbypassed emitter degeneration.—"Low Frequency Applications of Field-Effect Transistors," Motorola, Phoenix, AZ, 1976, AN-511A, p 8.

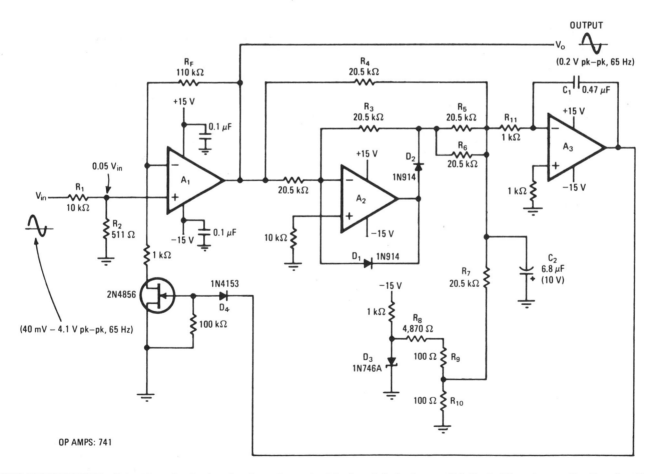

TWO-DECADE RANGE—Output is maintained at 0.2 V for inputs from 40 mV to 4.1 V. Voltage-controlled JFET serves as variable control element. Comparator A_3 produces error voltage that determines gain of A_1. A_2 and diodes form full-wave rectifier. Developed for use in radar seeker device to prevent overload of amplifier as target gets closer.—C. Marco, Automatic Gain Control Operates over Two Decades, *Electronics*, Aug. 16, 1973, p 99–100; reprinted in "Circuits for Electronics Engineers," *Electronics*, 1977, p 44–46.

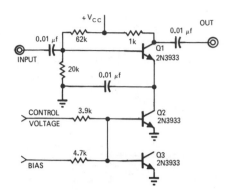

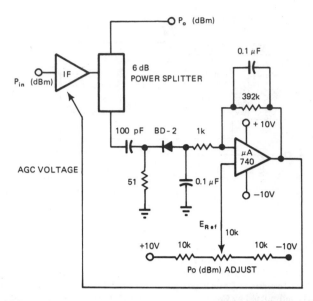

LOW PHASE SHIFT—Voltage-controlled amplifier has less than 3° phase shift over gain-control range of 40 dB at frequencies up to 10 MHz, as required for AGC circuits. Current generator Q2 controls gain of wideband resistance-coupled amplifier Q1. Gain of Q1 increases linearly with amplitude of positive control voltage on base of Q2.—A. H. Hargrove, Simple Circuits Control Phase-Shift, *EDN Magazine*, Jan. 1, 1971, p 39.

30-MHz AGC LOOP—Low-pass filter serves as loop giving closed-loop bandwidth of at least 5 kHz. Loop operates in square-law region of detector diode. Inputs to IF amplifier are in range from −60 dBm to −10 dBm, and AGC action provides 30-MHz IF output of −15 dBm. Power splitter ensures that detector also operates at −15 dBm. Article gives design equations and performance curves.—R. S. Hughes, Design Automatic Gain Control Loops the Easy Way, *EDN Magazine*, Oct. 5, 1978, p 123–128.

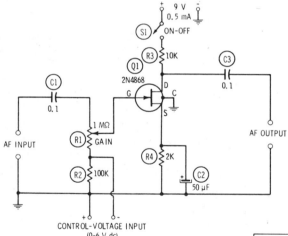

AUDIO AGC USING FET—DC control voltage obtained from key signal point in audio amplifier is applied to gate of FET to vary bias. Gain of stage varies inversely with gate bias voltage. When control voltage is 0 V, voltage gain of stage is 10 and maximum undistorted output signal is 1 VRMS. When control voltage is 6 VDC, output is reduced to 0.5 mVRMS, giving better than 90-dB range for AGC control.—R. P. Turner, "FET Circuits," Howard W. Sams, Indianapolis, IN, 1977, 2nd Ed., p 39–40.

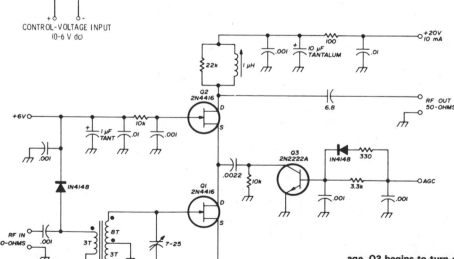

50–54 MHz RF-AGC AMPLIFIER—Developed for 6-meter SSB transceiver to give minimum of 15-dB power gain, low noise figure, and good signal-handling capability when AGC is applied.

AGC control resembles bipolar cascode circuit using differential pair with current source, although operation does not involve changing amplifier bias level. With increasing AGC volt-

age, Q3 begins to turn on, shunting more and more signal current away from Q2 and thereby decreasing stage gain. Input transformer is wound on small toroid core. Range of AGC voltage is 0–1.2 V.—A. Borsa, High-Performance RF-AGC Amplifier, *Ham Radio*, Sept. 1978, p 64–66.

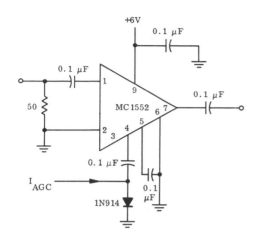

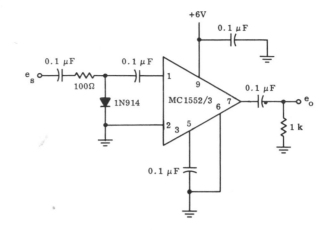

CONTROL WITH EXTERNAL DIODE—External resistances normally used with Motorola MC1552 video amplifier are replaced by 1N914 or equivalent diode so gain of amplifier is determined by AGC current through diode. Arrangement gives wide range of gain control, but lowest obtainable level of gain is normal unmodified gain of amplifier. Same circuit can be used with MC1553 high-gain video amplifier.—"A Wide Band Monolithic Video Amplifier," Motorola, Phoenix, AZ, 1973, AN-404, p 10.

CONTROL AT LOW GAIN LEVELS—Diode is used as variable impedance in voltage-divider network at input of video amplifier to provide AGC at lower gain levels than could be handled with more conventional external-diode circuits. Voltage gain for Motorola MC1552 decreases from about 50 for 1-mA AGC control current to about 20 for 8 mA. For MC1553 high-gain video amplifier, gain drops from 400 at 1 mA to 25 at 8 mA.—"A Wide Band Monolithic Video Amplifier," Motorola, Phoenix, AZ, 1973, AN-404, p 11.

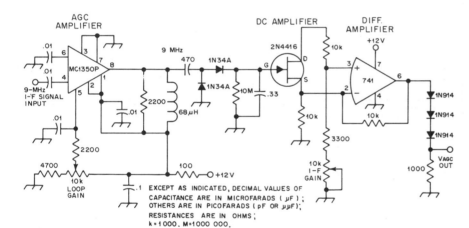

AGC LOOP FOR MOSFETs—Used at output of 9-MHz IF amplifier in commercial receiver to divide desired control-voltage magnitude and swing for FT0601 MOSFETs in IF strip. MOSFETs are biased by 2.1-V zeners in source leads in FETs, to drive gate-2 voltage sufficiently negative for full AGC action.—G. Ricaud, Modifying the W1CER/W1FB AGC Loop for Use with MOS-FET I-F Amplifiers, *QST*, June 1977, p 47.

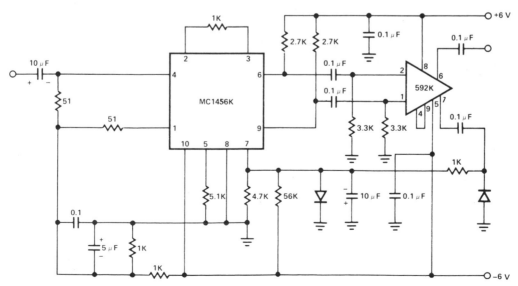

WIDEBAND AGC AMPLIFIER—Combination of 592K opamp and MC1496K balanced modulator gives DC output signal proportional to amplitude of AC input signal, for varying gain of balanced modulator. Unbalancing carrier input of modulator makes signal pass through without attenuation.—"Signetics Analog Data Manual," Signetics, Sunnyvale, CA, 1977, p 709–710.

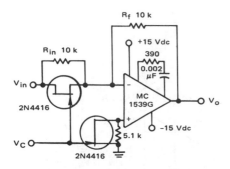

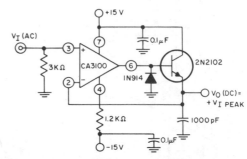

AGC AMPLIFIER—FET used in conjunction with opamp permits varying of gain by changing ratio of R_f to R_{in}. Offset voltage in output due to input bias currents is minimized by placing FET in parallel with 5.1K resistor between noninverting leg of opamp and ground, so resistance varies with changes of R_{in}.—"Low Frequency Applications of Field-Effect Transistors," Motorola, Phoenix, AZ, 1976, AN-511A, p 9.

POSITIVE PEAK DETECTOR—CA3100 bipolar MOS opamp is connected as wideband noninverting amplifier to provide essentially constant gain for wide range of input frequencies. Diode clips negative half-cycles, so output of transistor is proportional only to positive input peaks.—"Circuit Ideas for RCA Linear ICs," RCA Solid State Division, Somerville, NJ, 1977, p 16.

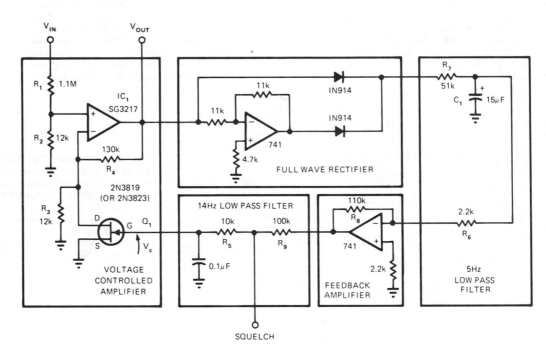

AGC WITH FET—FET serves as nonlinear element in fast-acting instrumentation circuit handling wide range of signals. R_1 and R_2 attenuate input signal so FET input is less than 25 mV for inputs up to 2 VRMS. Article covers design and performance. Gain is almost linear with gate voltage of FET.—R. D. Pogge, Designers' Guide to: Basic AGC Amplifier Design, *EDN Magazine*, Jan. 20, 1974, p 72–76.

CHAPTER 7
Automotive Circuits

Includes capacitor-discharge, optoelectronic, and other types of electronic ignition, tachometers, dwell meters, idiot-light buzzer, audible turn signals, headlight reminders, mileage computer, cold-weather starting aids, wiper controls, oil-pressure and oil-level gages, solid-state regulators for alternators, overspeed warnings, battery-voltage monitor, and trailer-light interface. For auto theft devices, see Burglar Alarm chapter.

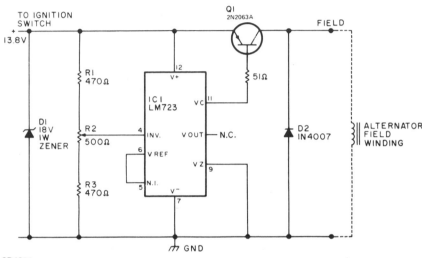

SOLID-STATE AUTO REGULATOR—Replaces and outperforms electromechanical charging-voltage regulator in autos using alternator systems. Prolongs battery life by preventing undercharging or overcharging of 12-V lead-acid battery. Uses LM723 connected as switching regulator for controlling alternator field current. R2 is adjusted to maintain 13.8-V fully charged voltage for standard auto battery. Article gives construction details and tells how to use external relay to maintain alternator charge-indicator function in cars having idiot light rather than charge-discharge ammeter. Q1 is 2N2063A (SK3009) 10-A PNP transistor.—W. J. Prudhomme, Build Your Own Car Regulator, *73 Magazine*, March 1977, p 160–162.

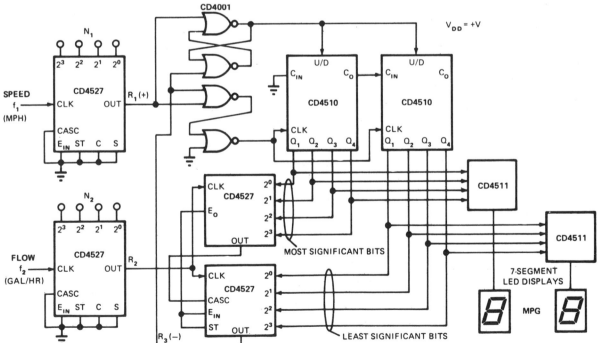

MILEAGE COMPUTER—Fuel consumption in miles per gallon is continuously updated on 2-digit LED display. Entire system using CMOS ICs can be built for less than $25 including gas-flow sensor and speed sensor, sources for which are given in article along with operational details.

Circuit uses rate multiplier to produce output pulse train whose frequency is proportional to product of the two inputs. Output rate is time-averaged. Speed sensor, mounted in series with speedometer cable, feeds speed data to CD4527 rate multiplier as clock input. Gas-flow sensor, mounted in series with fuel line, feeds clock input of other rate multiplier.—G. J. Summers, Miles/Gallon Measurement Made Easy with CMOS Rate Multipliers, *EDN Magazine*, Jan. 20, 1976, p 61–63.

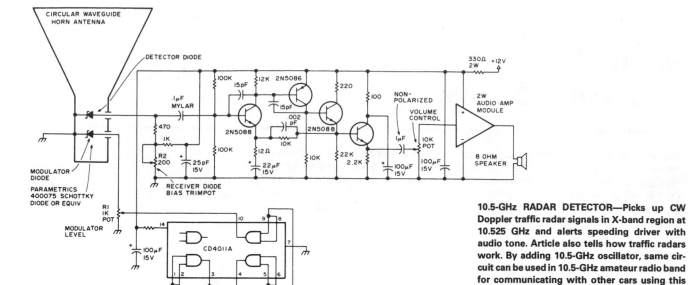

10.5-GHz RADAR DETECTOR—Picks up CW Doppler traffic radar signals in X-band region at 10.525 GHz and alerts speeding driver with audio tone. Article also tells how traffic radars work. By adding 10.5-GHz oscillator, same circuit can be used in 10.5-GHz amateur radio band for communicating with other cars using this band. Dimensioned diagram of horn is given.— S. M. Olberg, Mobile Smokey Detector, *73 Magazine*, Holiday issue 1976, p 32–35.

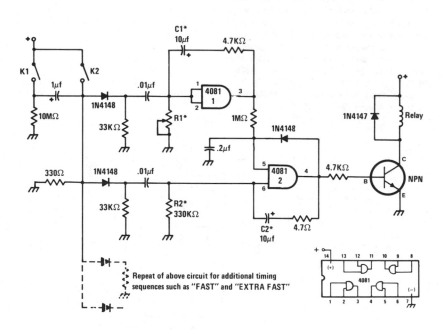

SPEED TRAP—Time required for auto to activate sensors placed measured distance apart on driveway or road is used to energize relay or alarm circuit when auto exceeds predetermined speed. If speed limit chosen is 15 mph, set detectors 22 feet apart for travel time of 1 s. Sensors can be photocells or air-actuated solenoids. For most applications, R1 can be 1-megohm pot. Transistor type is not critical. Values of R2 and C2 determine how long alarm sounds.—J. Sandler, 9 Projects under $9, *Modern Electronics*, Sept. 1978, p 35–39.

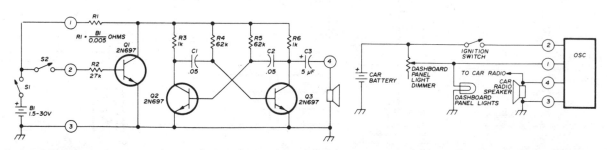

HEADLIGHT REMINDER—Uses basic oscillator consisting of Q2 and Q3 arranged as collector-coupled astable MVBR. Power is taken from collector of Q1 which acts as switch for Q2 and Q3. With S1 closed and S2 open, oscillator operates. Closing S2 saturates Q1 and stops oscillator. When used as headlight reminder for negative-ground car, B1 is omitted and power for oscillator is taken from dashboard panel lights since they come on simultaneously with either parking lights or headlights. If ignition key is turned on, Q1 saturates and disables Q2-Q3. With ignition off but lights on, Q1 is cut off and oscillator receives power. Audio output may be connected directly to high side of voice coil of car radio loudspeaker without affecting operation of radio. Almost any NPN transistors can be used. Changing values of R4 and R5 changes frequency of reminder tone.—H. F. Batie, Versatile Audio Oscillator, *Ham Radio*, Jan. 1976, p 72–74.

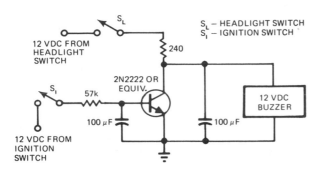

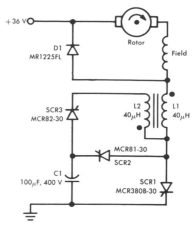

HEADLIGHTS-ON ALARM—Designed for cars in which headlight switch is nongrounding type, providing 12 V when closed. When both light and ignition switches are closed, transistor is saturated and there is no voltage drop across it to drive buzzer. If ignition switch is open while lights are on, transistor bias is removed so transistor is effectively open and full 12 V is applied to buzzer through 240-ohm resistor until lights are turned off.—R. E. Hartzell, Jr., Detector Warns You When Headlights Are Left On, *EDN Magazine*, Nov. 20, 1975, p 160.

ELECTRIC-VEHICLE CONTROL—SCR1 is used in combination with Jones chopper to provide smooth acceleration of golf cart or other electric vehicle operating from 36-V on-board storage battery. Normal running current of 2-hp 36-V series-wound DC motor is 60 A, with up to 300 A required for starting vehicle up hill. Chopper and its control maintain high average motor current while limiting peak current by increasing chopping frequency from normal 125 Hz to as high as 500 Hz when high torque is required.—T. Malarkey, You Need Precision SCR Chopper Control, *New Motorola Semiconductors for Industry*, Motorola, Phoenix, AZ, Vol. 2, No. 1, 1975.

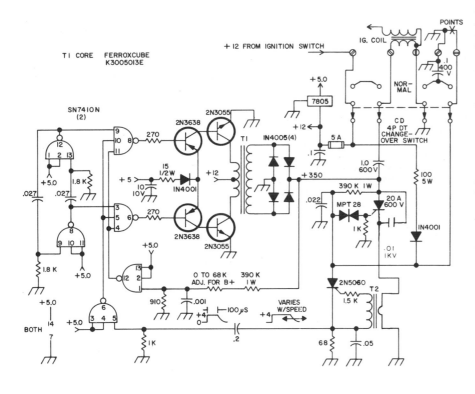

CD IGNITION—Uses master oscillator—power amplifier type of DC/DC converter in which two sections of triple 3-input NAND gate serve as 10-kHz square-wave MVBR feeding class B PNP/NPN power amplifier through two-gate driver. Remaining two gates are used as logic inverters. Secondary of T1 has 15.24 meters of No. 26 in six bank windings, with 20 turns No. 14 added and center-tapped for primary. T2 is unshielded iron-core RF choke, 30–100 μH, with several turns wound over it for secondary. When main 20-A SCR fires, T2 develops oscillation burst for firing sensitive gate-latching SCR. Storage capacitor energy is then dumped into ignition coil primary through power SCR.—K. W. Robbins, CD Ignition System, *73 Magazine*, May 1974, p 17 and 19.

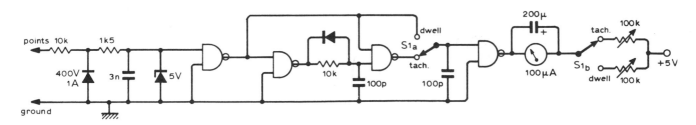

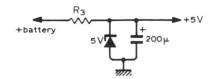

TACH/DWELL METER—Built around SN7402 NOR-gate IC. Requires no internal battery; required 5 V is obtained by using 50 ohms for R3 in zener circuit shown if car battery is 6 V, and 300 ohms if 12 V. Article gives calibration procedure for engines having 4, 6, and 8 cylinders; select maximum rpm to be indicated, multiply by number of cylinders, then divide by 120 to get frequency in Hz.—N. Parron, Tach-Dwell Meter, *Wireless World*, Sept. 1975, p 413.

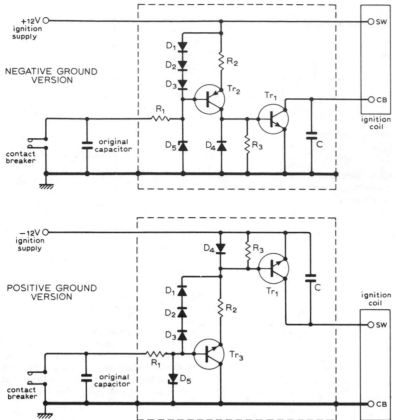

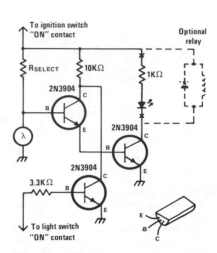

HEADLIGHT REMINDER—Photocell energizes circuit at twilight to remind motorist that lights should be turned on. Indicator can be LED connected as shown or relay turning on buzzer for more positive signal. Circuit can be made automatic by connecting relay contacts in parallel with light switch, provided delay circuit is added to prevent oncoming headlights from killing circuit. Mount photocell in location where it is unaffected by other lights inside or outside car.—J. Sandler, 9 Projects under $9, *Modern Electronics*, Sept. 1978, p 35–39.

TRANSISTORIZED BREAKER POINTS—Uses Texas Instruments BUY23/23A high-voltage transistors that can easily withstand voltages up to about 300 V existing across breaker points of distributor in modern car. Circuit serves as electronic switch that isolates points from heavy interrupt current and high-voltage backswing of ignition coil, thereby almost completely eliminating wear on points. Values are: Tr_2 2N3789; Tr_3 (for positive ground version) 2N3055; D_1-D_4 1N4001; D_5 18-V 400-mW zener; R_1 56 ohms; R_2 1.2 ohms; R_3 10 ohms; C 600 VDC same size as points capacitor. Article covers installation procedure.—G. F. Nudd, Transistor-Aided Ignition, *Wireless World*, April 1975, p 191.

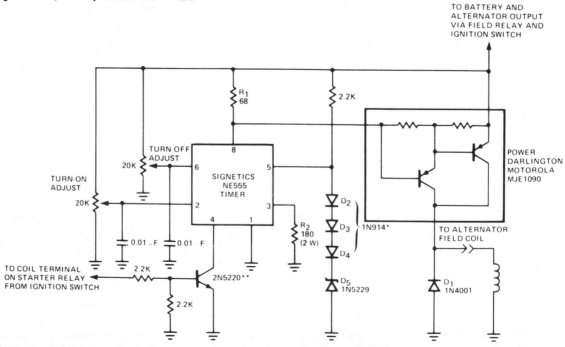

VOLTAGE REGULATOR—Timer and power Darlington form simple automobile voltage regulator. When battery voltage drops below 14.4 V, timer is turned on and Darlington pair conducts. Separate adjustments are provided for preset turn-on and turnoff voltages.—"Signetics Analog Data Manual," Signetics, Sunnyvale, CA, 1977, p 731.

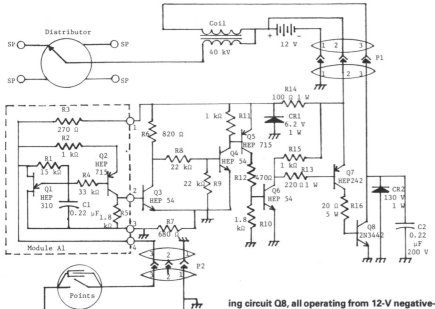

BATTERY MONITOR—Basic circuit energizes LED when battery voltage drops to level set by 10K pot. Any number of additional circuits can be added, for reading battery voltage in 1-V steps or even steps as small as 0.1 V. Circuit supplements idiot light that replaces ammeter in most modern cars. LED type is not critical.—J. Sandler, 9 Projects under $9, *Modern Electronics,* Sept. 1978, p 35–39.

COLD-WEATHER IGNITION—Multispark electronic ignition improves cold-weather starting ability of engines in arctic environment by providing more than one spark per combustion cycle. Circuit uses UJT triangle-wave generator Q1, emitter-follower isolator Q2, wave-shaping Schmitt trigger Q3-Q4, three stages of square-wave amplification Q5-Q7, and output switching circuit Q8, all operating from 12-V negative-ground supply. 6.2-V zener provides regulated voltage for UJT and Schmitt trigger. Initial 20,-000- to 40,000-V ignition spark produced by opening of breaker points is followed by continuous series of sparks at rate of about 200 per second as long as points stay open.—D. E. Stinchcomb, Multi-Spark Electronic Ignition for Engine Starting in Arctic Environment, *Proceedings of the IEEE 1975 Region Six Conference,* May 1975, p 224–225.

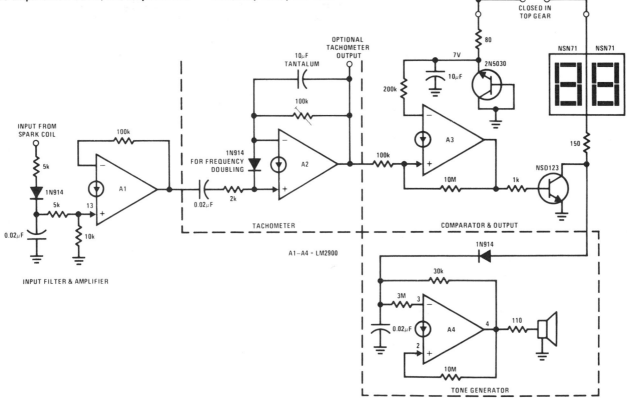

HIGH-SPEED WARNING—Audible alarm tone generator drives warning loudspeaker to supplement 2-digit speed display that can be set to trip when vehicle speed exceeds 55-mph legal limit. Engine speed signal is taken from primary of spark coil. Switch in transmission activates circuit only when car is in high gear. All functions are performed by sections of LM2900 quad Norton opamp. A1 amplifies and regulates spark-coil signal. A2 converts signal frequency to voltage proportional to engine speed. A3 compares speed voltage with reference voltage and turns on output transistor at set speed. A4 generates audible tone. Circuit components must be adjusted for number of cylinders, gear and axle ratios, tire size, etc. 10-μF capacitor connected to A3 can be increased to prevent triggering of alarm when increasing speed momentarily while passing another car.—"Linear Applications, Vol. 2," National Semiconductor, Santa Clara, CA, 1976, LB-33.

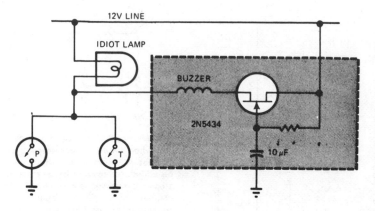

BUZZER FOR IDIOT LIGHT—Provides audible supplement to engine-monitoring indicator lamps that are often difficult to see in daylight. Uses 2N5434 JFET to provide delay of about 7 s each time ignition switch is turned on, to allow for peaceful starting of car and normal buildup of oil pressure when lamp is monitoring oil-pressure and engine-temperature sensors. Entire circuit can be mounted inside plastic housing of unused or disconnected dashboard warning buzzer in late-model car.—P. Clower, Audio Assist Gives "Idiot Lights" the "Buzz," *EDN Magazine,* June 20, 1976, p 126.

OIL-PRESSURE DISPLAY—Red, yellow, and green LEDs give positive indication of oil pressure level on electronic gage console developed for motorcycle. Transducer converts oil pressure to variable resistance R_T which in turn varies bias on transistors. LEDs have different forward voltages at which they light, so proper selection of bias resistors ensures that only one LED is on at a time to give desired indication of oil pressure.—J. D. Wiley, Instrument Console Features Digital Displays and Built-In Combo Lock, *EDN Magazine,* Aug. 5, 1975, p 38–43.

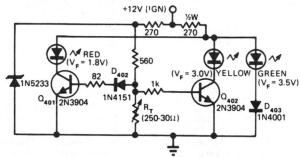

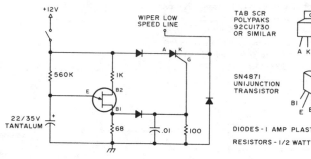

WIPER CONTROL—Operates wipers automatically at intervals, as required for very light rain or mist. Changing 560K resistor to 500K pot in series with 100K fixed resistor gives variable control of interval.—*Circuits, 73 Magazine,* July 1977, p 34.

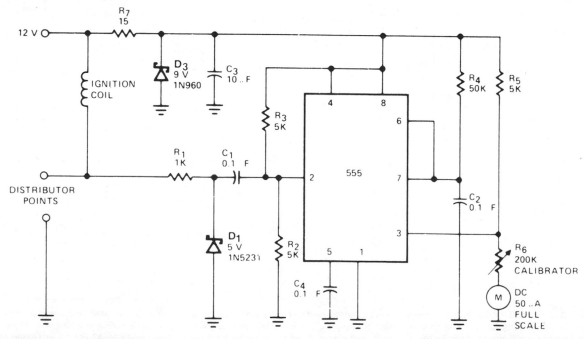

DISTRIBUTOR-POINT TACHOMETER—555 timer receives its input pulses from distributor points of car. When timer output (pin 3) is high, meter receives calibrated current through R_6. When IC times out, meter current stops for remainder of duty cycle. Integration of variable duty cycle by meter movement serves to provide visible indication of engine speed.—"Signetics Analog Data Manual," Signetics, Sunnyvale, CA, 1977, p 724–725.

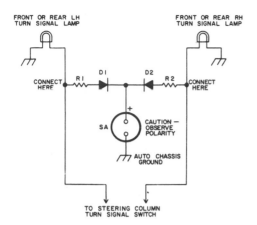

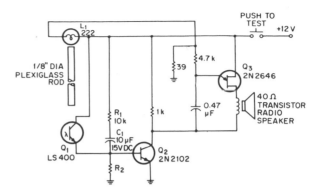

AUDIBLE TURN SIGNAL—Gives 3500-Hz audible tone each time turn-signal light flashes on, to warn driver that signal has not been turned off when making less than right-angle turns. Schematic shown is for 12-V negative-ground systems. For 6-V negative-ground systems, cut values of R1 and R2 about in half. For positive-ground systems, reverse connections to diodes and Sonalert. R1 and R2 are 2.7K 0.5 W. D1 and D2 can be any general-purpose small-current silicon diode. SA is Mallory SC1.5 Sonalert.—A. Goodwin, Turn Signal Reminder, *73 Magazine*, Holiday issue 1976, p 166.

OIL-LEVEL GAGE—Permits checking crankcase oil level from driver's seat. Sensor consists of light-conducting Plexiglas rod attached to dipstick, with lamp L_1 at top of rod and phototransistor Q_1 mounted at add-oil mark on dipstick, about ½ inch below bottom of rod. At normal oil level, oil attenuates light between Q_1 and bottom end of rod, making phototransistor resistance high. Pushing test switch makes C_1 charge and saturate Q_2 long enough to activate UJT AF oscillator Q_3 and give short tone verifying that lamp is not burned out and gage is working. When oil is low, enough light reaches Q_1 to keep Q_2 saturated after C_1 charges, giving continuous tone as long as switch is pushed.— L. Svelund, Electronic Dipstick, *EEE Magazine*, Nov. 1970, p 101.

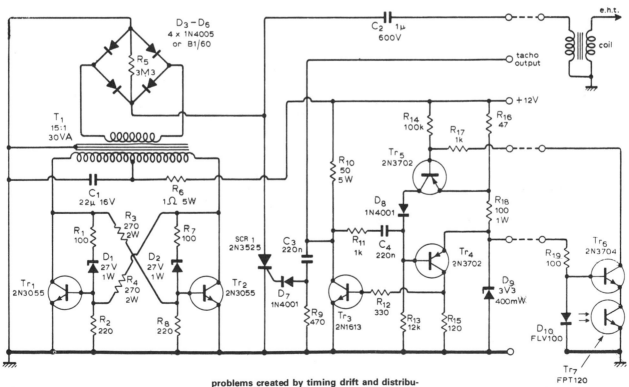

OPTOELECTRONIC IGNITION—Combination of low-cost point-source LED and high-sensitivity phototransistor forms optical sensor for position of cam in distributor. Technique eliminates problems created by timing drift and distributor-shaft play. Sensor head is small enough to fit most distributors. Article gives dimensioned drawings for shutter design and sensor mounting, and describes operation of associated capacitor-discharge electronic ignition circuit in detail. Leads to sensor do not require shielding.—H. Maidment, Optical Sensor Ignition System, *Wireless World*, Nov. 1975, p 533–537.

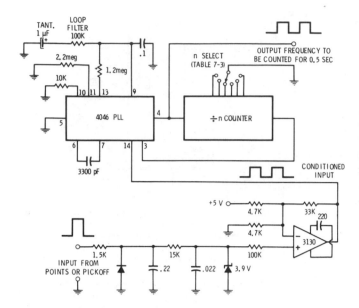

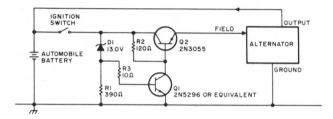

DIGITAL TACHOMETER—Pulses from auto engine points or other pickoff are filtered before feed to 3130 CMOS opamp used as comparator to complete conditioning of input. Pulses are then fed through 4046 PLL to divide-by-N counter that is set for number of cylinders in engine (60 for four cylinders, 45 for six, and 30 for eight). Output frequency is then counted for 0.5 s to get engine or shaft speed in rpm.—D. Lancaster, "CMOS Cookbook," Howard W. Sams, Indianapolis, IN, 1977, p 366–367.

REGULATOR FOR ALTERNATOR—Simple and effective solid-state replacement for auto voltage regulator can be used with alternator in almost any negative-ground system. Circuit acts as switch supplying either full or no voltage to field winding of alternator. When battery is below 13 V, zener D1 does not conduct, Q1 is off, Q2 is on, and full battery voltage is applied to alternator field so it puts out full voltage to battery for charging. When battery reaches 13.6 V, Q1 turns on, Q2 turns off, alternator output is reduced to zero, and battery gets no charging current. Circuit can also be used with wind-driven alternator systems.—P. S. Smith, $22 for a Regulator? Never!, *73 Magazine*, Holiday issue 1976, p 103.

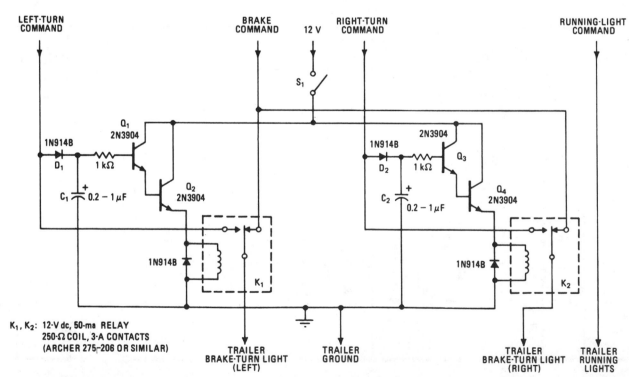

K_1, K_2: 12-V dc, 50-ma RELAY
250-Ω COIL, 3-A CONTACTS
(ARCHER 275-206 OR SIMILAR)

AUTO-TRAILER INTERFACE FOR LIGHTS— Low-cost transistors and two relays combine brake-light and turn-indicator signals on common bus to ensure that trailer lights respond to both commands. C_1 and C_2 charge to peak amplitude of turn signal, which flashes about 2 times per second. Values are selected to hold relay closed between flash intervals; if capacitance is too large, brake signal cannot immediately activate trailer lights after turn signal is canceled. Developed for new cars in which separate turn and brake signals are required for safety.—M. E. Gilmore and C. W. Snipes, Darlington-Switched Relays Link Car and Trailer Signal Lights, *Electronics*, Aug. 18, 1977, p 116.

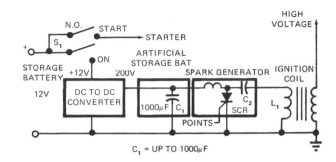

CAPACITOR SERVES AS IGNITION BATTERY— Developed for use with capacitor-discharge ignition systems to provide independent voltage source for ignition when starting car in very cold weather. Before attempting to start car, S_1 is set to ON position for energizing DC-to-DC converter for charging C_1 with DC voltage between 200 and 400 V. Starter is now engaged. If voltage of storage battery drops as starter slowly turns engine over, C_1 still represents equivalent of fully charged 12-V storage battery that is capable of driving ignition system for almost a minute.—W. Stalzer, Capacitor Provides Artificial Battery for Ignition Systems, *EDN Magazine*, Nov. 15, 1972, p 48.

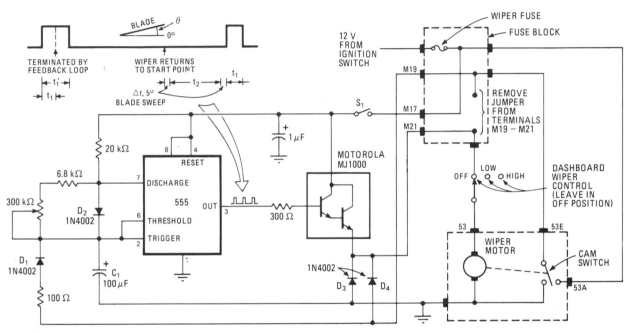

WIPER-DELAY CONTROL— 555 timer provides selectable delay time between sweeps of wiper blades driven by motor in negative-ground system. Article also gives circuit modification for positive-ground autos. Delay time can be varied between 0 and 22 s. Timer uses feedback signal from cam-operated switch of motor to synchronize delay time with position of wiper blades.—J. Okolowicz, Synchronous Timing Loop Controls Windshield Wiper Delay, *Electronics*, Nov. 24, 1977, p 115 and 117.

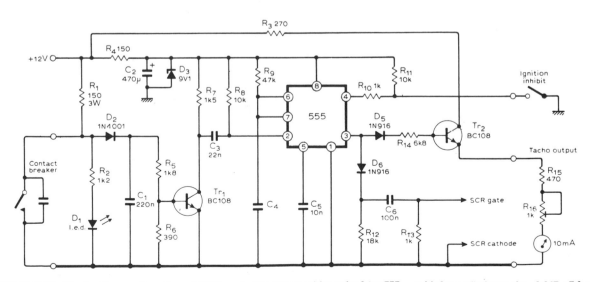

RPM-LIMIT ALARM— Used with capacitor-discharge ignition system to provide tachometer output along with engine speed control signal. When breaker contacts open, C_1 charges and turns Tr_1 on, triggering 555 timer used in mono MVBR mode. Resulting positive pulse from 555 fires control SCR through D_6 and C_6. When contacts close, D_2 isolates C_1 to reduce effect of contact bounce. With values shown, for speed limit between 8000 and 9000 rpm, use 0.068 μF for C_4 with four-cylinder engine, 0.047 μF for six cylinders, and 0.033 μF for eight cylinders. LED across breaker contacts can be used for setting static timing.—K. Wevill, Trigger Circuit for C.D.I. Systems, *Wireless World*, Jan. 1978, p 58.

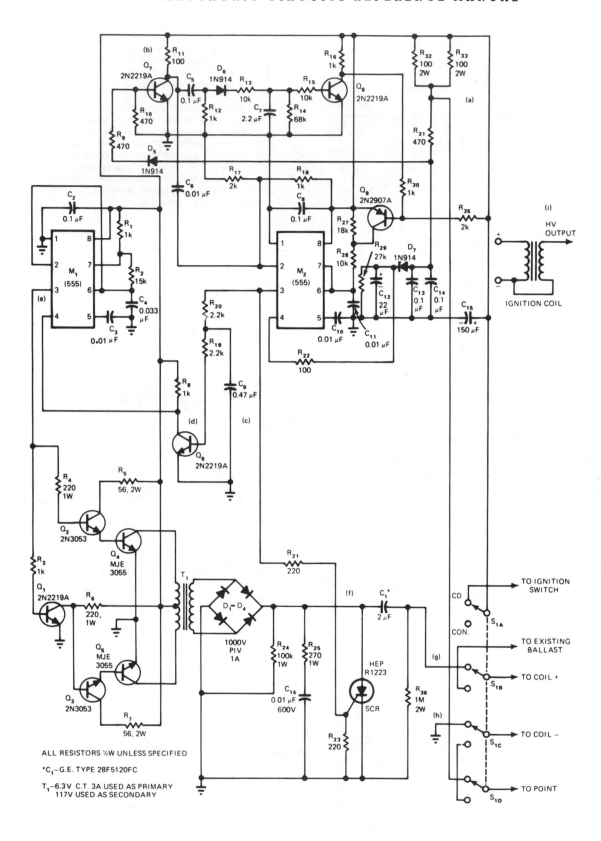

LOW-EMISSION CD—Solid-state capacitor-discharge ignition system improves combustion efficiency by increasing spark duration. For 8-cylinder engine, normal CD system range of 180 to 300 μs is increased to 600 μs below 4000 rpm. Oscillation discharge across ignition coil primary lasts for two cycles here, but above 4000 rpm the discharge lasts for one cycle or 300 μs because at higher speeds the power cycle has shorter times. Circuit uses 555 timer M_1 as 2-kHz oscillator, with Q_1-Q_3 providing drive to Q_4-Q_5 and T_1 for converting battery voltage to about 400 VDC at output of bridge rectifier. When distributor points open, Q_7 turns on and triggers M_2 connected as mono that provides gate drive pulses for SCR. Article describes operation of circuit in detail and gives waveforms at points a-i.—C. C. Lo, CD Ignition System Produces Low Engine Emissions, *EDN Magazine*, May 20, 1976, p 94, 96, and 98.

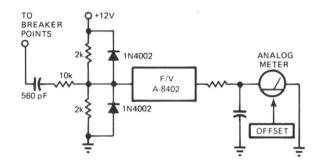

TACHOMETER—Intech/Function Modules A-8402 operating in frequency-to-voltage converter mode serves as automotive tachometer having inherent linearity and ease of calibration. Converter operates asynchronously, which does not affect accuracy when driving analog meter.—P. Pinter and D. Timm, Voltage-to-Frequency Converters—IC Versions Perform Accurate Data Conversion (and Much More) at Low Cost, *EDN Magazine,* Sept. 5, 1977, p 153–157.

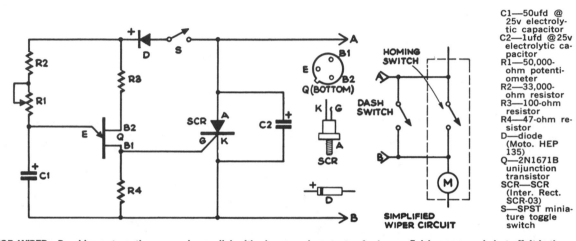

C1—50ufd @ 25v electrolytic capacitor
C2—1ufd @25v electrolytic capacitor
R1—50,000-ohm potentiometer
R2—33,000-ohm resistor
R3—100-ohm resistor
R4—47-ohm resistor
D—diode (Moto. HEP 135)
Q—2N1671B unijunction transistor
SCR—SCR (Inter. Rect. SCR-03)
S—SPST miniature toggle switch

TIMER FOR WIPER—Provides automatic one-shot swipes at preselected intervals from 2 to 30 s for handling mist, drizzle, or splash from wet road. Circuit shorts out homing switch inside windshield-wiper motor, which is usually in parallel with slow-speed contacts of wiper dashboard switch. With wiper switch off and ignition on, short two switch terminals at a time to find pins that start wiper. When blades begin moving, remove jumper; blades should then finish sweep and shut off. It is these terminals of switch that are connected to points A and B of control circuit.—V. Mele, Mist Switch—It's for Your Windshield Wipers, *Popular Science,* Aug. 1973, p 110.

CHAPTER **8**
Battery-Charging Circuits

Includes constant-voltage, constant-current, and trickle chargers operating from AC line, solar cells, or auto battery. Some circuits have automatic charge-rate control, automatic start-up, automatic shutoff, and low-charge indicator.

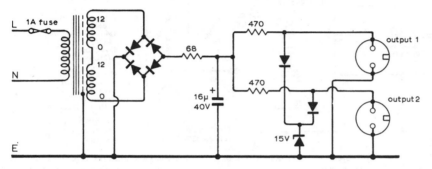

9.6 V AT 20 mA—Developed to charge 200-mAh nickel-cadmium batteries for two transceivers simultaneously. Batteries will be fully charged in 14 hours, using correct 20-mA charging rate. Zener diode ensures that voltage cannot exceed safe value if battery is accidentally disconnected while under charge. Diode types are not critical.—D. A. Tong, A Pocket V.H.F. Transceiver, *Wireless World*, Aug. 1974, p 293–298.

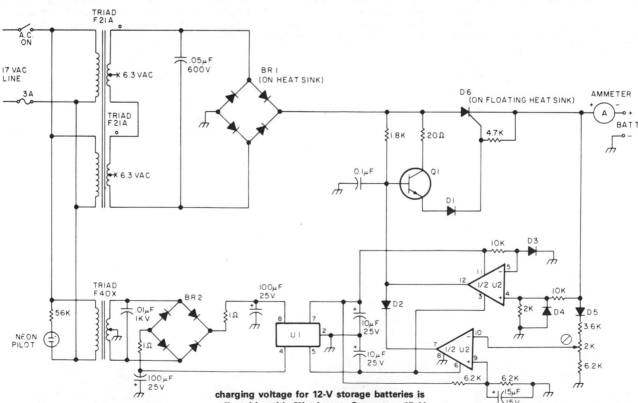

ADJUSTABLE FINISH-CHARGE—Uses National LM319D dual comparator U2 to sense end-of-charge battery voltage and provide protection against shorted or reversed charger leads. Final charging voltage for 12-V storage batteries is adjustable with 2K trimpot. Separate ±15 V supply using Raytheon RC4195NB regulator U1 is provided for U2. D1-D5 are 1N4002 or HEP-R0051. D6 is 2N682 or HEP-R1471. BR1 is Motorola MDA980-2 or HEP-R0876 12-A bridge. BR2 is Varo VE27 1-A bridge. Q1 is 2N3641 or HEP-S0015.—H. Olson, Battery Chargers Exposed, *73 Magazine*, Nov. 1976, p 98–100 and 102–104.

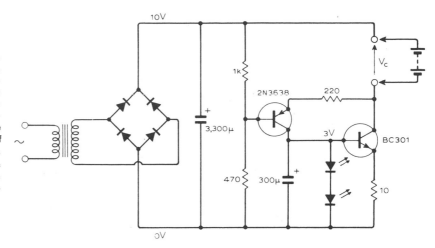

NICKEL-CADMIUM CELL CHARGER—Charges four size D cells in series at constant current, with automatic voltage limiting. BC301 transistor acts as current source, with base voltage stabilized at about 3 V by two LEDs that also serve to indicate charge condition. Other transistor provides voltage limiting when voltage across cells approaches that of 1K branch of voltage divider. Values shown give 260-mA charge initially, dropping to 200 mA when V_c reaches 5 V and decreasing almost to 0 when V_c reaches 6.5 V.—N. H. Sabah, Battery Charger, *Wireless World*, Nov. 1975, p 520.

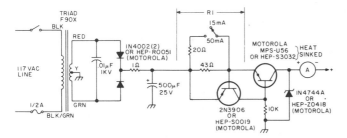

12-V FOR NICADS—Produces constant current with simple transistor circuit, adjustable to 15 or 50 mA with switch and R1. Zener limits voltage at end of charge. Developed for charging 10-cell pack having nominal 12.5 V, as used in many transceivers.—H. Olson, Battery Chargers Exposed, *73 Magazine*, Nov. 1976, p 98–100 and 102–104.

CHARGING SILVER-ZINC CELLS—Used for initial charging and subsequent rechargings of sealed dry-charged lightweight cells developed for use in missiles, torpedoes, and space applications. Article covers procedure for filling cell with potassium hydroxide electrolyte before placing in use (cells are dry-charged at factory and have shelf life of 5 or more years in that condition). Charge current should be 7 to 10% of rated cell discharge capacity; thus, for Yardney HR-5 cell with rated discharge of 5 A, charge at 350 to 500 mA. Stop charging when cell voltage

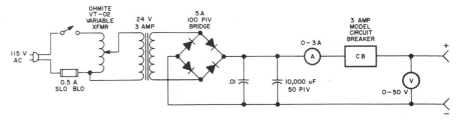

reaches 2.05 V. If used only for battery charging, large filter capacitor can be omitted.—S. Kelly,

Will Silver-Zinc Replace the Nicad?, *73 Magazine*, Holiday issue 1976, p 204–205.

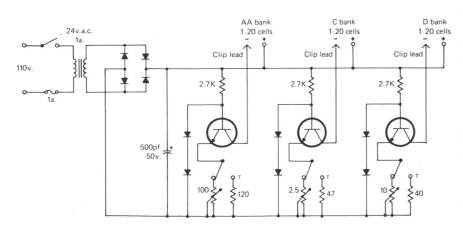

BULK NICAD CHARGER—Can handle up to 20 AA cells, 20 C cells, and 20 D cells simultaneously, with charging rate determined separately for each type. Single transformer and full-wave rectifier feed about 24 VDC to three separate regulators. AA-cell regulator uses 100-ohm resistor to vary charge rate from 6 mA to above 45 mA. C-cell charge-rate range is 24 to 125 mA, and D-cell range is 60 to 150 mA. Batteries of each type should be about same state of discharge. Batteries are recharged in series to avoid need for separate regulator with each cell. Trickle-charge switches cut charge rates to about 2% of rated normal charge (5 mA for 500-mAh AA cells). Transistors are 2N4896 or equivalent. Use heatsinks. All diodes are 1N4002.—J. J. Schultz, A Bulk Ni-Cad Recharger, *CQ*, Dec. 1977, p 35–36 and 111.

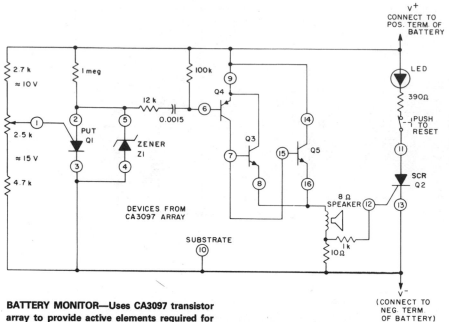

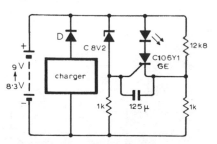

BATTERY MONITOR—Uses CA3097 transistor array to provide active elements required for driving indicators serving as aural and visual warnings of low charge on nicad battery. LED remains on until circuit is reset with pushbutton switch.—"Circuit Ideas for RCA Linear ICs," RCA Solid State Division, Somerville, NJ, 1977, p 9.

SOLAR-POWER BACKUP—If solar-cell voltage drops 0.2 V below battery voltage, circuit is powered by storage cell feeding through forward-biased OA90 or equivalent germanium diode. When solar-cell voltage exceeds that of battery, battery is charged by approximately constant reverse leakage current through diode. Battery can be manganese-alkaline type or zinc-silver oxide watch-type cell.—M. Hadley, Automatic Micropower Battery Charger, *Wireless World*, Dec. 1977, p 80.

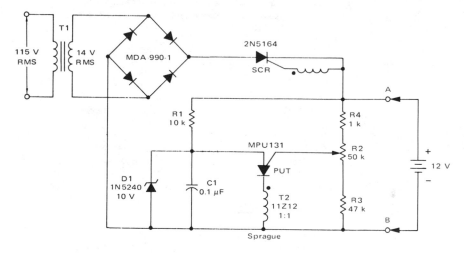

12 V AT 8 A—Charging circuit for lead-acid storage batteries is not damaged by short-circuits or by connecting with wrong battery polarity. Battery provides current for charging C1 in PUT relaxation oscillator. When PUT is fired by C1, SCR is turned on and applies charging current to battery. Battery voltage increases slightly during charge, increasing peak point voltage of PUT and making C1 charge to slightly higher voltage. When C1 voltage reaches that of zener D1, oscillator stops and charging ceases. R2 sets maximum battery voltage between 10 and 14 V during charge.—R. J. Haver and B. C. Shiner, "Theory, Characteristics and Applications of the Programmable Unijunction Transistor," Motorola, Phoenix, AZ, 1974, AN-527, p 10.

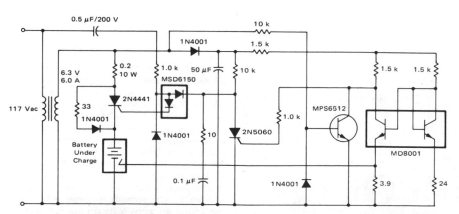

THIRD ELECTRODE SENSES FULL CHARGE—Circuit is suitable only for special nickel-cadmium batteries in which third electrode has been incorporated for use as end-of-charge indicator. Voltage change at third electrode is sufficient to provide reliable shutoff signal for charger under all conditions of temperature and cell variations.—D. A. Zinder, "Fast Charging Systems for Ni-Cd Batteries," Motorola, Phoenix, AZ, 1974, AN-447, p 7.

LED VOLTAGE INDICATOR—Circuit shown uses LED to indicate, by lighting up, that battery has been charged to desired level of 9 V. Circuit can be modified for other charging voltages. Silicon switching transistor can be used in place of more costly thyristor.—P. R. Chetty, Low Battery Voltage Indication, *Wireless World*, April 1975, p 175.

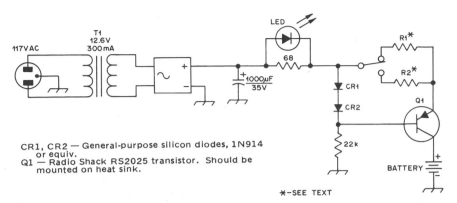

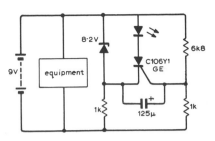

CR1, CR2 — General-purpose silicon diodes, 1N914 or equiv.
Q1 — Radio Shack RS2025 transistor. Should be mounted on heat sink.

∗—SEE TEXT

NICAD CHARGER—Switch gives choice of two constant-current charge rates. With 10 ohms for R1, rate is 60 mA, while 200 ohms for R2 gives 3 mA. Silicon diodes CR1 and CR2 have combined voltage drop of 1.2 V and emitter-base junction of Q1 has 0.6-V drop, for net drop of 0.6 V across R1 or R2. Dividing 0.6 by desired charge rate in amperes gives resistance value.—M. Alterman, A Constant-Current Charger for Nicad Batteries, *QST*, March 1977, p 49.

LED INDICATES LOW VOLTAGE—LED lights when output of 9-V rechargeable battery drops below minimum acceptable value of 8.3 V, to indicate need for recharging. Can also be used with transistor radio battery to indicate need for replacement. Zener is BZY85 C8V2 rated at 400 mW, with avalanche point at 7.7 V because of low current drawn by circuit. LED can be Hewlett-Packard 5082-4440.—P. C. Parsonage, Low-Battery Voltage Indicator, *Wireless World*, Jan. 1973, p 31.

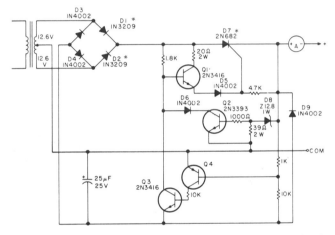

12-V AUTOMATIC—Circuit of Heathkit GP-21 automatic charger is self-controlling (Q1 and Q2) and provides protection against shorted or reversed battery leads (Q3 and Q4). Zener D8 is not standard value, so may be obtainable only in Heathkits. D1, D2, and D7 should all be on one heatsink.—H. Olson, Battery Chargers Exposed, *73 Magazine*, Nov. 1976, p 98—100 and 102—104.

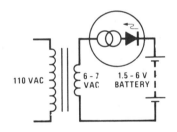

LED TRICKLE CHARGER—Constant-current characteristic of National NSL4944 LED is used to advantage in simple half-wave charger for batteries up to 6 V.—"Linear Applications, Vol. 2," National Semiconductor, Santa Clara, CA, 1976, AN-153, p 2.

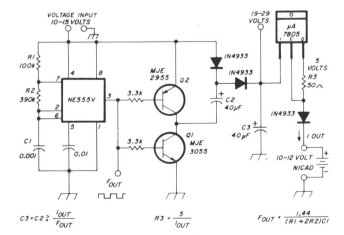

$$C3 = C2 \geq \frac{I_{OUT}}{F_{OUT}}$$

$$R3 = \frac{5}{I_{OUT}}$$

$$F_{OUT} = \frac{1.44}{(R1 + 2R2)C1}$$

NICAD CHARGER FOR AUTO—Voltage doubler provides at least 20 V from 12-V auto battery, for constant-current charging of 12-V nicads, using NE555 timer and two power transistors. Doubled voltage drives source current into three-terminal current regulator. Switching frequency of NE555 as MVBR is 1.4 kHz. Charging current is set at 50 mA for charging ten 500-mAh nicads.—G. Hinkle, Constant-Current Battery Charger for Portable Operation, *Ham Radio*, April 1978, p 34—36.

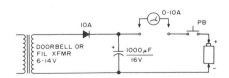

NICAD ZAPPER—Simple circuit often restores dead or defective nicad battery by applying DC overvoltage at current up to 10 A for about 3 s. Longer treatment may overheat battery and make it explode.—Circuits, *73 Magazine*, July 1977, p 35.

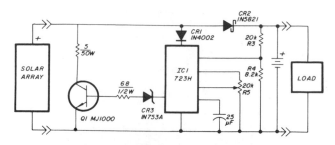

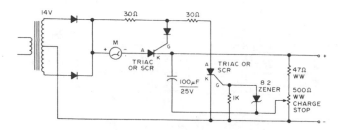

SOLAR-POWER OVERCHARGE PROTECTION— Voltage regulator is connected across solar-cell array as shown to prevent damage to storage battery by overcharging. Series diode prevents array from discharging battery during hours of darkness. Regulator does not draw power from battery, except for very low current used for voltage sampling. Battery can be lead-calcium, gelled-electrolyte, or telephone-type wet cells. For repeater application described, two Globe Union GC12200 40-Ah gelled-electrolyte batteries were used to provide transmit current of 1.07 A and idle current of 12 mA.—T. Handel and P. Beauchamp, Solar-Powered Repeater Design, *Ham Radio,* Dec. 1978, p 28–33.

AUTOMATIC SHUTOFF—Prevents overcharging and dryout of battery under charge by shutting off automatically when battery reaches full-charge voltage. Accepts wide range of batteries. Choose rectifying diodes and triacs or SCRs to handle maximum charging current desired. For initial adjustment, connect fully charged battery and adjust charge-stop pot until ammeter just drops to zero.—Circuits, *73 Magazine,* July 1977, p 34.

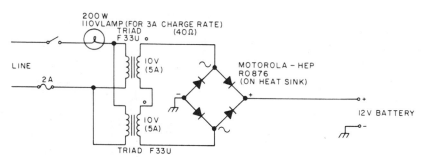

BASIC 12-V CHARGER—Uses 200-W lamp as current-limiting resistor in transformer primary circuit. Serves in place of older types of chargers using copper-oxide or tungar-bulb rectifiers.— H. Olson, Battery Chargers Exposed, *73 Magazine,* Nov. 1976, p 98–100 and 102–104.

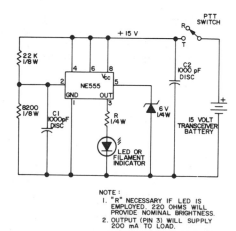

NOTE:
1. "R" NECESSARY IF LED IS EMPLOYED. 220 OHMS WILL PROVIDE NOMINAL BRIGHTNESS.
2. OUTPUT (PIN 3) WILL SUPPLY 200 mA TO LOAD.

NICAD MONITOR—Uses two comparators, flip-flop, and power stage all in single NE555 IC. When battery voltage drops below 12-V threshold set by R1 and R2 for 15-V transceiver battery, one comparator sets flip-flop and makes output at pin 3 go high. IC then supplies up to 200 mA to LED or other indicator. For other battery voltage value, set firing point to about three-fourths of fully charged voltage. Since battery voltage will show biggest drop when transmitting, connect monitor across transmit supply only so as to minimize battery drain.— A. Woerner, Ni-Cad Lifesaver, *73 Magazine,* Nov. 1973, p 35–36.

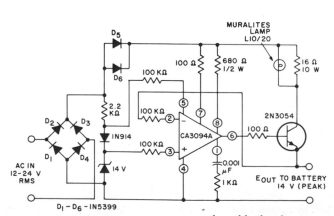

14-V MAXIMUM—Circuit accurately limits peak output voltage to 14 V, as established by zener connected between terminals 3 and 4 of CA3094A programmable opamp. Lamp brightness varies with charging current. Reference voltage supply does not drain battery when power supply is disconnected.—"Circuit Ideas for RCA Linear ICs," RCA Solid State Division, Somerville, NJ, 1977, p 19.

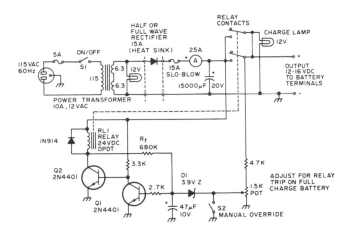

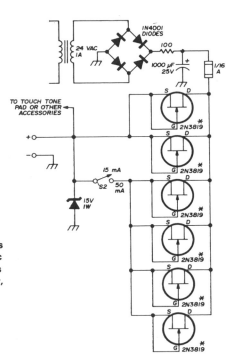

AUTOMATIC SHUTOFF—Charger automatically turns itself off when 12-V auto storage battery is fully charged. Setting of 1.5K pot determines battery voltage at which zener D1 conducts, turning on Q2 and pulling in relay that disconnects charger. If battery voltage drops below threshold, relay automatically connects charger again. S2 is closed to bypass automatic control when charger itself is to be used as power supply.—G. Hinkle, The Smart Charger, *73 Magazine,* Holiday issue 1976, p 110–111.

NICAD CHARGER—Pot is adjusted to provide 10% above rated voltage (normal full-charge voltage) while keeping charging current below 25% of maximum. For 10-V 1-Ah battery, set voltage at 11 V and current below 250 mA.—G. E. Zook, F.M., *CQ,* Feb. 1973, p 35–37.

NICAD CHARGER—Developed for recharging small nickel-cadmium batteries used in handheld FM transceivers. Field-effect transistors serve as constant-current sources when gate is shorted to source. Practically any N-channel JFET having drain-to-source current of 8–15 mA will work. FETs shown were measured individually and grouped to give desired choice of 15- or 50-mA charging currents.—G. K. Shubert, FET-Controlled Charger for Small Nicad Batteries, *Ham Radio,* Aug. 1975, p 46–47.

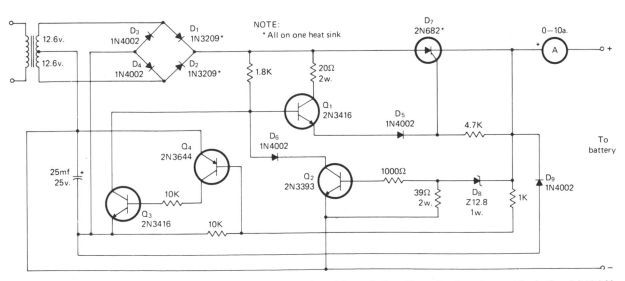

12-V CHARGER—Heath GP-21 charger uses SCR as switch to connect and disconnect battery at 120-Hz rate. Voltage at anode of SCR D_7 goes positive each half-cycle, putting forward bias on base of Q_1 through 1.8K resistor so Q_1 passes current through D_5 to gate of D_7 to turn it on for part of half-cycle and charge battery. D_7 stays on until voltage across it drops to zero. When battery has charged to 13.4 V, charging stops automatically. Rest of circuit protects against battery polarity reversal and accidental shorting of output leads. Special 12.8-V zener can be replaced by selected 1N4742 and forward-biased 1N4002.—H. Olson, We Don't Charge Nothin' but Batteries!, *CQ,* Feb. 1976, p 25–28 and 69.

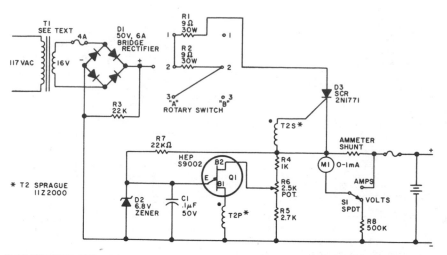

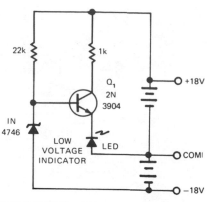

UJT CHARGER FOR 12 V—Keeps 12-V auto storage battery fully charged, for immediate standby use when AC power fails. Power transformer secondary can be 14 to 24 V, rated at about 3 A. Two-gang rotary switch gives choice of three charging rates. Pulse transformer T2 is small audio transformer rewound to have 1:1 turns ratio and about 20 ohms resistance, or can be regular SCR trigger transformer. UJT relaxation oscillator stops when upper voltage limit for battery is reached, as set by pot R6. If oscillator fails to start, reverse one of pulse transformer windings.—F. J. Piraino, Failsafe Super Charger, *73 Magazine*, Holiday issue 1976, p 49.

18-V MONITOR—Circuit turns on LED when ±18 V battery pack discharges to predetermined low level, while drawing less than 1 mA when LED is off. Zener is reverse-biased for normal operating range of battery. When lower limit is reached, zener loses control and Q_1 becomes forward-biased, turning on LED or other signal device to indicate need for replacement or recharging.—W. Denison and Y. Rich, Battery Monitor Is Efficient, yet Simple, *EDN Magazine*, Oct. 5, 1974, p 76.

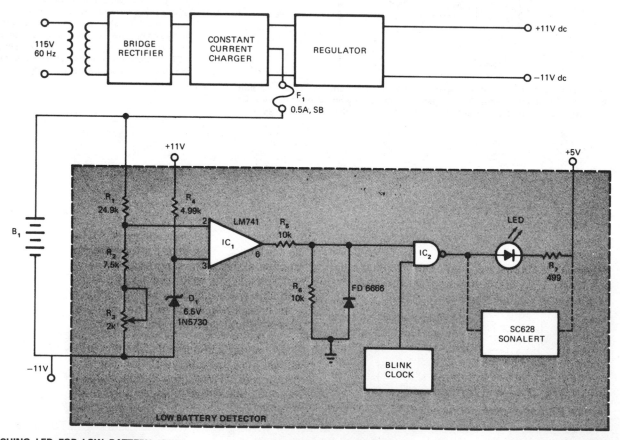

FLASHING LED FOR LOW BATTERY—Developed for use in portable battery-operated test instrument to provide visual indication that depletion level has been reached for series arrangement of 24 nickel-cadmium cells providing 32.5 VDC for regulator of bipolar 11-V supply. Instrument must then be plugged into AC line for recharging of batteries. Voltage across B_1 (nominally 32.5 V) is sensed by R_1-R_4 and D_1. When level drops 24.1 V, opamp comparator output goes positive and enables gate IC_2, so blink clock (such as low-frequency TTL-level oscillator) makes LED flash. Audible alarm is optional.—R. T. Warner, Monitor NiCad's with This Low-Battery Detector, *EDN Magazine*, April 20, 1976, p 112 and 114.

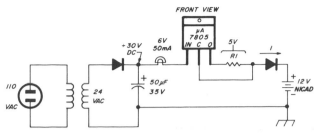

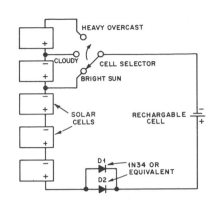

CONSTANT-CURRENT NICAD CHARGER— Constant current is obtained from voltage regulator by floating common line and connecting R1 from output to common terminal. Regulator then tries to furnish fixed voltage across R1. Input voltage must be greater than full-charge battery voltage plus 5 V (for 5-V regulator) plus 2 V (overhead voltage). Changing R1 varies charging current. If R1 is 50 ohms and V is 5 V, constant current is 50 mA through nicad being charged.—G. Hinkle, Constant-Current Battery Charger for Portable Operation, *Ham Radio*, April 1978, p 34–36.

SOLAR-ENERGY CHARGER—Single solar cell on bright day delivers 0.5 V at 50 mA, so three cells are used in bright sun to recharge secondary cell. Switch permits use of additional solar cells on cloudy days. Solar cells can be Radio Shack 276-128.—J. Rice, Charging Batteries with Solar Energy, *QST*, Sept. 1978, p 37.

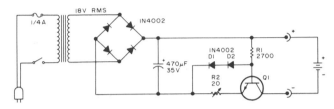

NICAD CHARGER—Regulated charger circuit will handle variable load from 1 to 18 nicad cells. Current-limiting action holds charging current within 1 to 2 mA of optimum value (about one-tenth of rated ampere-hour capacity) from 0 to 24 V. Q1 should have power rating equal to twice supply voltage multiplied by current-limit value. If charging 450-mAh penlight cells, charge current is 45 mA and transistor should be 2 W.—A. G. Evans, Regulated Nicad Charger, *73 Magazine*, June 1977, p 117.

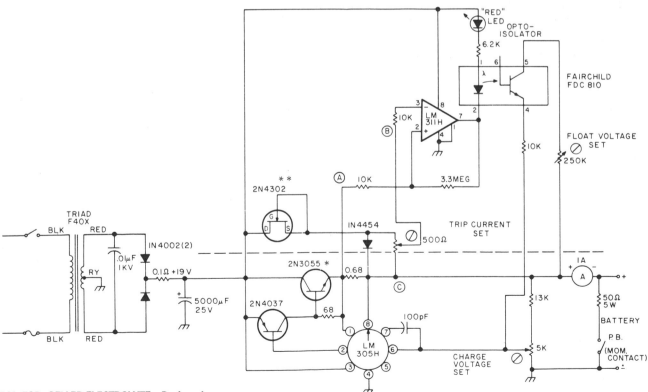

12-V FOR GELLED-ELECTROLYTE—Designed to charge 12-V 3-Ah gelled-electrolyte battery such as Elpower EP1230A at maximum of 0.45 A until battery reaches 14 V, then at constant voltage until charge current drops to 0.04 A. Charger is then automatically switched to float status that maintains 2.2 V per cell or 13.2 V for battery. Circuit is constant-voltage regulator with current limiting as designed around National LM305H, with PNP/NPN transistor pair to increase current capability. Circuit above dashed line is added to standard regulator to meet special charging requirement. Article covers operation and use of circuit in detail.—H. Olson, Battery Chargers Exposed, *73 Magazine*, Nov. 1976, p 98–100 and 102–104.

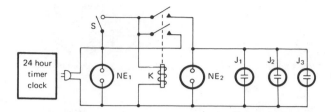

NICAD CHARGE CONTROL—Prevents double-charging if someone forgets to turn off 24-h time clock after recommended 16-h charge period. Nicad devices with built-in chargers are plugged into jacks J_1-J_3, and timer dial is advanced until clock switch is triggered. Neon lamp NE_1 should now come on. Momentary pushbutton switch S is pushed to energize relay K and start charge. When timer goes off, K releases to end charge.—M. Katz, Battery Charge Monitor, *CQ*, July 1976, p 27.

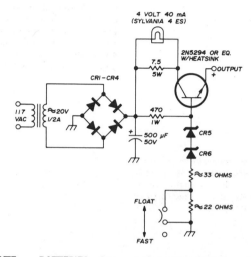

GELLED-ELECTROLYTE BATTERIES—Constant-voltage charger for Globe-Union 12-V gelled-electrolyte storage batteries can provide either fast or float charging. Constant voltage is maintained by series power transistor and series-connected zeners. Output voltage is 13.8 V for float charging and 14.4 V for fast charging.—E. Noll, Storage-Battery QRP Power, *Ham Radio*, Oct. 1974, p 56–61.

CHAPTER 9
Burglar Alarm Circuits

For auto, home, office, and factory installations. Sensors include contact-making, contact-breaking, photoelectric, infrared, Doppler, and sound-actuated devices that trigger circuit immediately or after adjustable delay for driving alarm horn, siren, tone generator, pager, or silent transmitter. Some circuits have automatic shutoff of alarm after fixed operating time as required for auto alarms in some states. See also Protection (for electronic door locks) and Siren chapters.

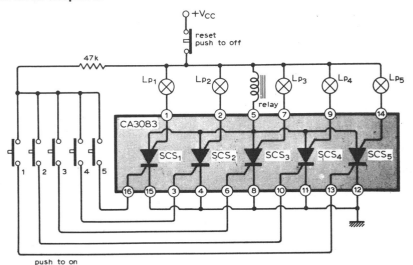

FIVE-INDICATOR ALARM—Single five-transistor IC uses NPN structures on P-type substrate as PNPN silicon controlled switches having common connection for anode (substrate). Relay serving as anode load is energized for actuating alarm if any of the SCS pushbutton switches is closed. Corresponding lamp is energized to identify door or window at which sensor switch has been closed by act of intruder. Alarm remains on until reset by interrupting power supply. Power drain on standby is negligible because SCSs act as open circuits until triggered, permitting use of batteries for supply. Two or more ICs may be added to get more channels.—H. S. Kothari, Alarm System with Position Indication, *Wireless World*, Feb. 1976, p 77.

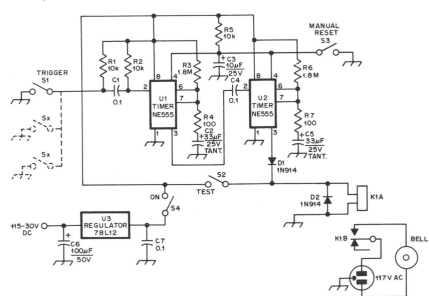

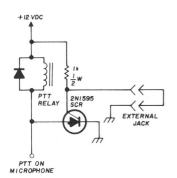

ENTRY-DELAY ALARM—First 555 timer provides delay of about 20 s after triggering by sensor before alarm bell is energized, to allow thief to be caught inside house or give owner time to enter and shut off alarm. Alarm then rings for about 60 s under control of timer U2. Alarm period was set short to attract attention without unduly annoying neighbors.—J. D. Arnold, A Low-Cost Burglar Alarm for Home or Car, *QST*, June 1978, p 35–36.

SCR LATCH—Turns on mobile transceiver or other mobile equipment when power is applied, if external circuit is broken when equipment is stolen. Transmitter will then put unmodulated carrier on air even with PTT switch disconnected or off, for tracing with radio direction finder. If added components are carefully concealed in equipment and new external wiring is worked into existing wiring harness, few thieves will be able to locate trouble. External wires are run under dash so thief must cut them to get out equipment. PTT relay should have protective diode. SCR is 100 PIV, 1 A, but HEP R1003 or R1217 can also be used.—E. Noll, Circuits and Techniques, *Ham Radio*, April 1976, p 40–43.

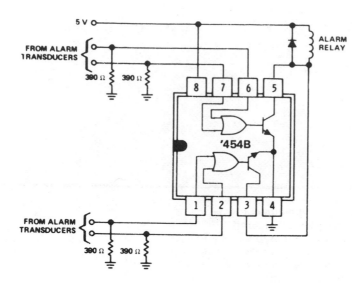

ALARM-SIGNAL DETECTOR—Texas Instruments SN75454B dual peripheral positive-NOR driver energizes alarm relay when alarm signal is received from any one of four different alarm transducers.—"The Linear and Interface Circuits Data Book for Design Engineers," Texas Instruments, Dallas, TX, 1973, p 10-66.

AURAL INDICATOR—Provides attention-getting chirp sound, warble, or continuous tone when turned on by high input from burglar-alarm sensor circuit. Second section of 556 timer provides optional frequency modulation of basic tone to give warbling effect. Chirp is achieved by gating tone oscillator on only during high states of warble oscillator. Aural sensitivity is maximum in range of 1–2 kHz, set by value of R_{t2}.—W. G. Jung, "IC Timer Cookbook," Howard W. Sams, Indianapolis, IN, 1977, p 232–235.

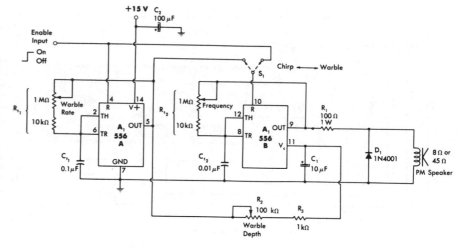

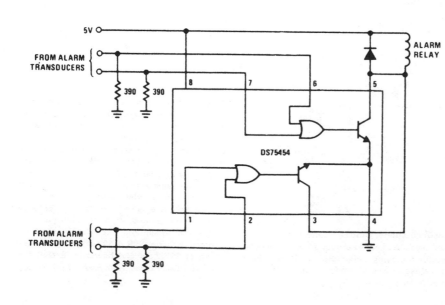

ALARM DETECTOR—National DS75454 dual peripheral NOR driver operating from single 5-V supply energizes alarm relay when one of alarm transducers for either section delivers logic signal as result of intruder action.—"Interface Databook," National Semiconductor, Santa Clara, CA, 1978, p 3-20—3-30.

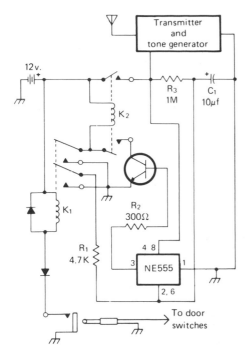

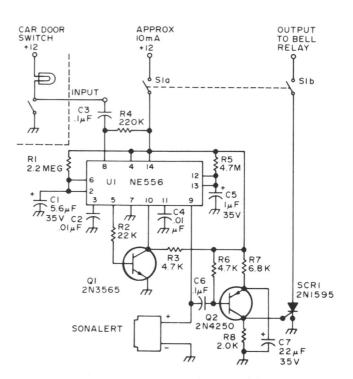

SILENT ALARM—When thief opens car door, relays K_1 and K_2 activate tone-modulated transmitter, which can be any legal combination of power, frequency, and antenna. A few milliwatts of power should be adequate. Thief hears nothing, but owner is alerted via portable receiver tuned to transmitter frequency. Transmitter remains on about 15 s (determined by R_3 and C_1) after door is closed until NE555 times out and removes power from transistor. Use any NPN transistor having adequate current rating for relay. If alarm is provided with its own battery and whip antenna, it cannot be disabled from outside of car.—A. Day, Soundless Mobile Alarm, *CQ,* April 1977, p 11.

CAR-THEFT ALARM—Alarm remains on even if signal from car door switch or other sensor is only momentary, so relay is wired to be self-latching until keyswitch S1 is turned off. Use hood locks or hood-opening sensors to prevent thief from disabling alarm by cutting battery cable. Circuit includes time delay of 6 s for entering car and shutting off alarm, to avoid need for external keyswitch. Sonalert makes loud tone during 6-s delay period to remind driver that alarm needs to be turned off. At end of 6 s, Sonalert stops and much louder bell is energized to further discourage intruder.—J. Pawlan, The Smart Alarm, *73 Magazine,* June 1975, p 37–41.

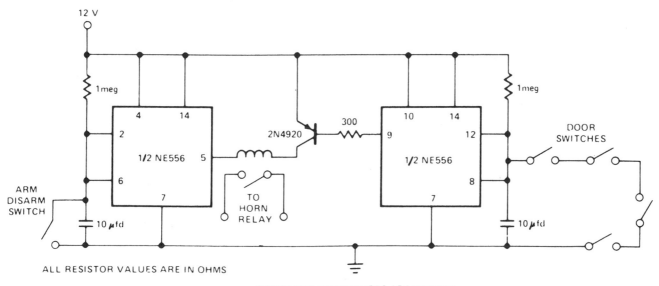

SHORT DURATION TIMERS ARE NEEDED TO ALLOW ENTRY AND EXIT

DELAYED ALARM—When normally closed arm/disarm switch is opened, first section of NE556 dual timer starts its timing cycle. After delay to allow for entry or exit, pin 5 goes low to ener- gize alarm circuit. Now, as long as all door switches are closed, PNP transistor is kept off because pin 9 is high. When any door switch is opened, transistor turns on after normal delay for owner to enter car, and horn is sounded un- less owner closes arm/disarm switch within delay time.—"Signetics Analog Data Manual," Signetics, Sunnyvale, CA, 1977, p 724–725.

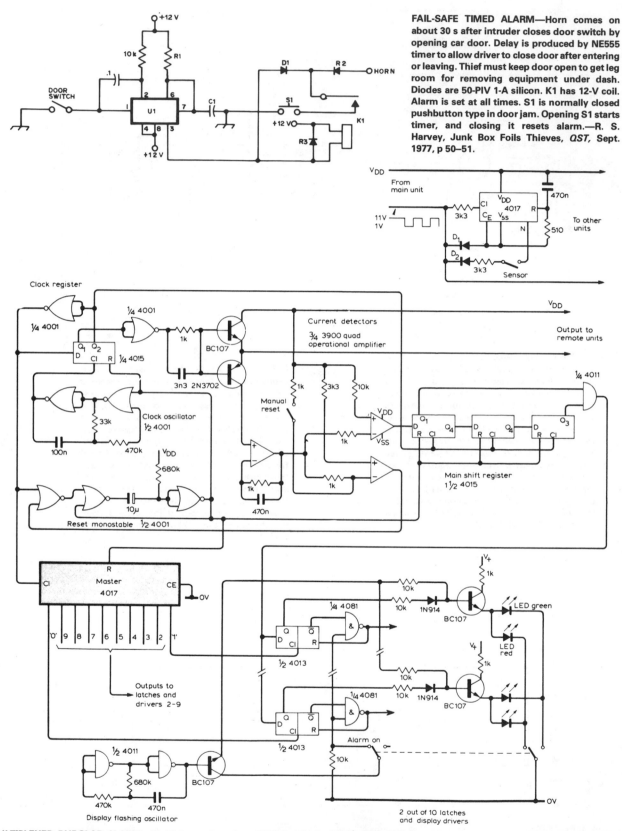

FAIL-SAFE TIMED ALARM—Horn comes on about 30 s after intruder closes door switch by opening car door. Delay is produced by NE555 timer to allow driver to close door after entering or leaving. Thief must keep door open to get leg room for removing equipment under dash. Diodes are 50-PIV 1-A silicon. K1 has 12-V coil. Alarm is set at all times. S1 is normally closed pushbutton type in door jam. Opening S1 starts timer, and closing it resets alarm.—R. S. Harvey, Junk Box Foils Thieves, *QST*, Sept. 1977, p 50–51.

MULTIPLEXED BURGLAR ALARM—Multiplexing technique provides for detection of state of up to 10 sensors, with immediate identification and location of activated sensor. Only one pair of wires runs from control unit to paralleled remote sensor circuits, one of which is shown at upper right. Each sensor location uses different output from one to zero. Multiplexer circuit is based on 4017 decade counter having 10 individual outputs, to give signals in 10 time slots. Power supply rail is used to reset counter. Clock line is eliminated by switching supply line as square wave. Sensor indication line is eliminated by detecting power supply current drain. Control unit uses oscillator and shift register to generate clocking waveforms. 3900 quad opamp converts sensor line current to logic levels for clocking by master 4017 to control 10 output latches and display driver. Two consecutive sensor-open signals are required to activate alarm, minimizing false alarms by interference pulses.—R. J. Chance, Multiplexed Alarm, *Wireless World*, Nov. 1978, p 73–74.

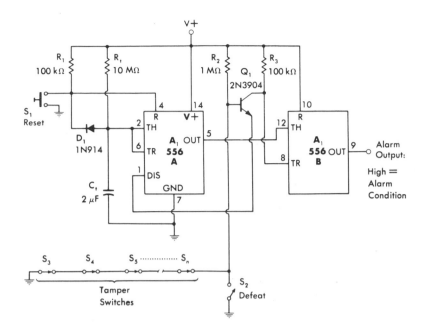

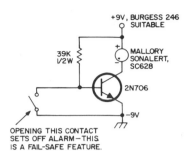

CIRCUIT-BREAKING ALARM—Operates from small 9-V battery, making it independent of AC power failure. Opening of switch or equivalent breaking of foil conductor removes ground from base of transistor, to energize alarm.—Circuits, *73 Magazine,* April 1973, p 132.

WINDOW-FOIL ALARM—Combination of power-up mono MVBR and latch, using both sections of 556 timer, drives output line high when sensor circuit is opened at door or window switch or by breaking foil on glass. Once alarm is triggered, reclosing of sensor has no effect; S₁ must be closed momentarily after restoring sensor circuit to turn alarm off. Circuit includes 22-s power-up delay that prevents triggering of alarm when it is first turned on.—W. G. Jung, "IC Timer Cookbook," Howard W. Sams, Indianapolis, IN, 1977, p 231–232.

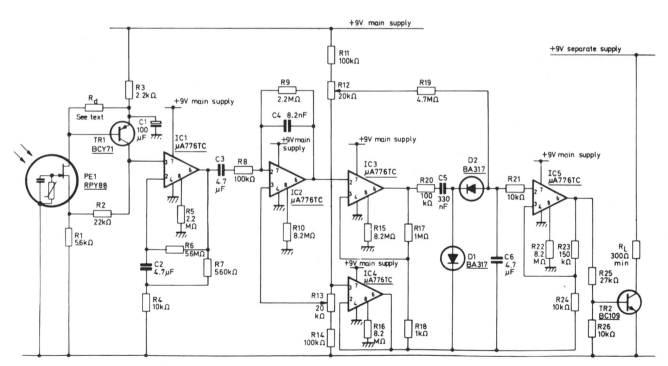

LOW-CURRENT INTRUDER ALARM—Use of programmable μA776 opamps reduces standby current of infrared alarm to 300 μA, permitting operation from small rechargeable cells. Detector is Mullard RPY86 that responds only to wavelengths above 6 μm, making it immune to sunlight and backgrounds intermittently illuminated by sun. Low-cost mirror is used instead of lens to concentrate infrared radiation on detector. R_d is chosen to make input to first opamp between 2 and 6 V. Circuit energizes alarm relay R_L only when incident radiation is changed by movement of intruder in monitored space.—"Ceramic Pyroelectric Infrared Detectors," Mullard, London, 1978, Technical Note 79, TP1664, p 8.

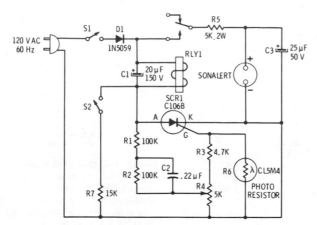

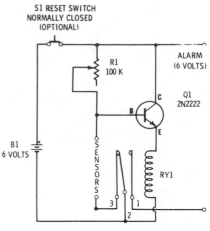

LIGHT-INTERRUPTION DETECTOR—Use of SCR as regenerative amplifier rather than as switch gives extremely high sensitivity to very slight reductions in light reaching photoresistor. Requires no light source or accurately aligned light-beam optics. In typical application as burglar alarm, light shining through window from streetlight provides sufficient ambient illumination so any movement of intruder within 10 feet of unit will energize Sonalert alarm. Sensitivity control R4 is adjusted so SCR receives positive pulses from AC line, but their amplitude is not quite enough to start regenerative action of SCR. Reduction in light then increases resistance of photoresistor enough to raise level of gate pulses for SCR, starting regenerative amplification that energizes relay. Use Mallory SC-628P Sonalert which produces pulsed 2500-Hz sound. With S2 open, alarm stops when changes in light cease. With S2 closed, alarm is latched on and S1 must be opened to stop sound.—R. F. Graf and G. J. Whalen, "The Build-It Book of Safety Electronics," Howard W. Sams, Indianapolis, IN, 1976, p 7–12.

LATCHING ALARM—Closed-circuit alarm drawing only 130 μA of standby current from battery is turned on by opening sensor switch or cutting wire. Automatic latching contacts on relay prevent burglar or intruder from deactivating alarm by resetting sensor switch. Relay is Radio Shack 275-004. Sensor can be foil strip around window subject to breakage.—F. M. Mims, "Transistor Projects, Vol. 3," Radio Shack, Fort Worth, TX, 1975, p 75–86.

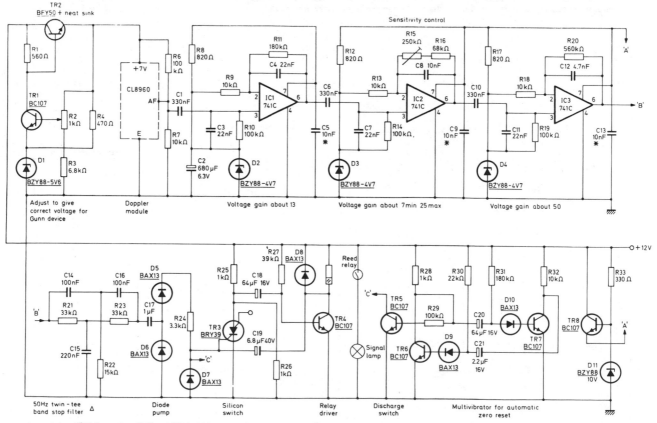

MICROWAVE DOPPLER INTRUSION ALARM—Mullard CL8960 X-band Doppler radar module detects movement of remote target by monitoring Doppler shift in microwave radiation reflected from target. Module consists of Gunn oscillator cavity producing energy to be radiated, mounted alongside mixer cavity that combines reflected energy with sample of oscillator signal. Transmitted frequency is 10.7 GHz. Doppler change is about 31 Hz for relative velocity of 0.45 m/s (1 mph) of relative velocity between object and module, giving AF output for velocities up to 400 mph. Filtered AF is applied through diode pump to trigger of silicon controlled switch TR3 that makes contacts of reed relay open for about 1 s. Relay action is repeated as long as intruder is in monitored area. Report covers circuit operation in detail.—J. E. Saw, "Microwave Doppler Intruder Alarms," Mullard, London, 1976, Technical Information 36, TP1570, p 6.

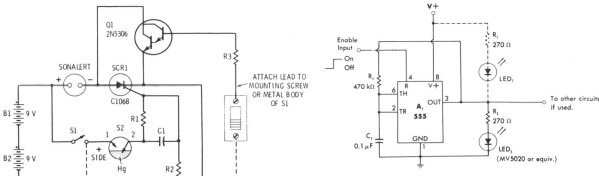

NOTE: TOUCHING MOUNTING SCREW OF S1 AND CASE OF S2
COMMUTATES SCR BY MOMENTARILY ENERGIZING Q1

HOTEL-ROOM ALARM—Alarm mounted in flashlight-shaped cylinder is positioned on floor inside hotel room in such a way that it is knocked over by intruder opening door. Mercury switch S2 then triggers SCR and activates Mallory SC-628P pulsed Sonalert alarm. Circuit latches on and can be turned off only by use of Darlington-amplifier touch switch. Connection from base of Darlington to positive terminal of battery must be made through fingertips as shown by dashed line in order to silence alarm. Once silenced, S1 can be opened to disconnect latch so alarm can be moved. Other applications include protection of unattended luggage. C1 is 0.1 μF, R1 is 1 megohm, R2 is 1K, R3 is 39K, and S2 is mercury element removed from GE mercury toggle switch.—R. F. Graf and G. J. Whalen, "The Build-It Book of Safety Electronics," Howard W. Sams, Indianapolis, IN, 1976, p 19–24.

VISUAL INDICATOR—When circuit is activated by high output of burglar alarm circuit, 555 timer operating as very low frequency MVBR makes LED₁ flash on and off during alarm condition. Alternate connection of LED₁ to V+ holds LED₁ on for standby while flashing it during alarm. Oscillator output is also available for other uses if desired. Indicator can be located remotely from alarm.—W. G. Jung, "IC Timer Cookbook," Howard W. Sams, Indianapolis, IN, 1977, p 232–235.

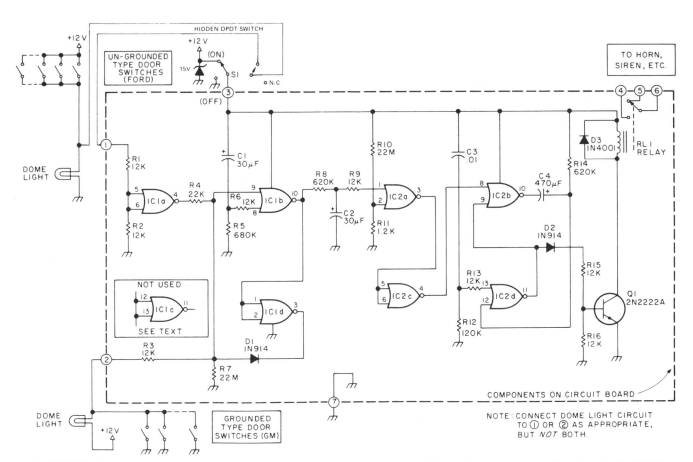

NOTE: CONNECT DOME LIGHT CIRCUIT TO ① OR ② AS APPROPRIATE, BUT *NOT* BOTH.

5-min SHUTOFF—Vehicle intrusion alarm shuts off automatically in about 5 min after being triggered, as required by law in some states. Drain on battery is negligible until alarm is set off by intruder. Once triggered, operation sequence is not affected by subsequent opening or closing of doors. System uses two CMOS CD4001AE quad two-input NOR gates for switching logic. IC1 provides sensor interface, latch, and entry/exit time delays. IC2 provides output through Q1 and relay, as well as automatic shutoff delay. Article gives construction details and layout for printed-circuit board.—W. J. Prudhomme, Vehicle Security Systems, *73 Magazine*, Oct. 1977, p 122–125.

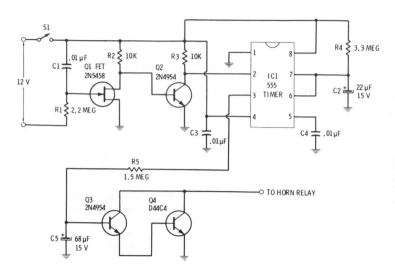

CURRENT-DRAIN SENSOR—Current drawn by dome light when door is open or by ignition when turned on triggers current-sensing stages Q1 and Q2 to start 555 timer and apply power to horn relay. Initial 15-s delay in sounding horn allows owner to enter car and open hidden switch S1 to deactivate alarm. If S1 is not opened during delay interval, horn sounds for about 90 s, then circuit automatically resets itself. C5 and R5 control duration of initial 15-s delay. C2 and R4 control total time that horn sounds.—R. F. Graf and G. J. Whalen, "The Build-It Book of Safety Electronics," Howard W. Sams, Indianapolis, IN, 1976, p 57–62.

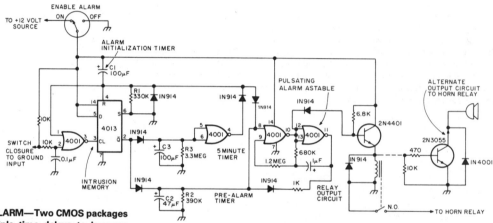

PULSED-HORN ALARM—Two CMOS packages incorporate multiple time delays to improve convenience and effectiveness of auto intrusion alarm. R1C1 gives 30-s delay for arming alarm after it is turned on by switch concealed inside car, to let driver get out of car. R2C2 gives 15-s delay before alarm sounds after door is opened, to allow driver to get back in car again and disable alarm. R3C3 turns off alarm in 300 s and resets alarm system for next intrusion. Car horn is pulsed 60 times per minute, so alarm would not be confused with stuck horn. Article tells how circuit works and gives detailed instructions for installation and connection to door and trunk switches.—G. Hinkle, Give the Hamburglar Heart Failure, *73 Magazine*, Feb. 1977, p 36–37.

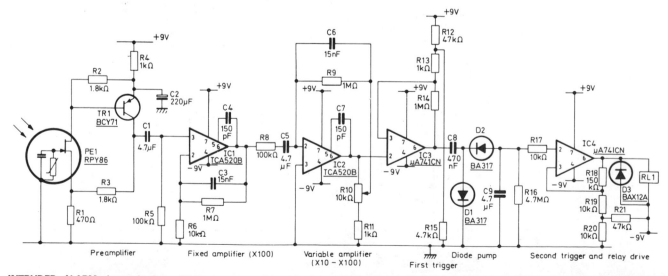

INTRUDER ALARM—Input is from Mullard RPY86 infrared detector responding to wavelengths above 6 μm, making it immune to sunlight and backgrounds intermittently illuminated by sun. Output signal is produced only when incident radiation is changed by movement of intruder in monitored space. Mirrors rather than lenses concentrate incident radiation on detector because mirrors do not require high-quality surface finish. Preamp is followed by two amplifier stages, with R10 varying gain of second stage between 10 and 100. Bandwidth is 0.3–10 Hz. First trigger, having threshold of about 1 V, drives second trigger through diode pump to energize alarm relay when intruder is present.—"Ceramic Pyroelectric Infrared Detectors," Mullard, London, 1978, Technical Note 79, TP1664, p 8.

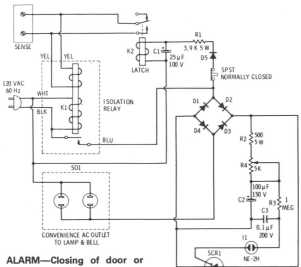

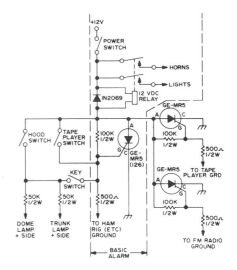

OPEN-CIRCUIT ALARM—Closing of door or window switch sensor or closing of normally open panic-button switch at bedside and other strategic locations in home trips alarm that sounds loud bell and flashes bright light on and off. Sensor shorts control winding of K1, allowing K1 to drop out and apply line voltage to alarm circuit. One AC path is through D5 which rectifies AC for energizing DC latch relay K2 to short sensor lines even though initiating sensor has opened. Simultaneously, AC is applied to diode bridge having SCR between DC legs. C2 starts charging through R2 and R4, and C3 charges through R3. When voltage across C3 reaches about 90 VDC, it fires neon and C3 discharges into gate of SCR. Full line voltage is then applied to lamp and bell plugged into load outlets. When C2 drops below holding current, SCR turns off during next AC cycle and load goes off until neon fires again. Setting of 5K pot R4 gives range of 15-80 flashes and horn pulses per second. To stop alarm, open SPST switch momentarily.—R. F. Graf and G. J. Whalen, "The Build-It Book of Safety Electronics," Howard W. Sams, Indianapolis, IN, 1976, p 75–80.

WIRE-CUTTING ALARM—SCR normally acts as open circuit in series with 12-VDC alarm relay because grid is made negative by voltage divider consisting of 100K in series with 500 ohms. If ground on 500-ohm resistor is removed, as by removal of tape player or CB set from car by thief, gate becomes more positive and SCR conducts, to energize relay, sound horn, and make headlights shine brightly. Additional triggering SCRs or alarm switches can be added as shown outside of dashed area for basic alarm.—A. Szablak, Another Burglar Alarm, *73 Magazine*, May 1974, p 45–46.

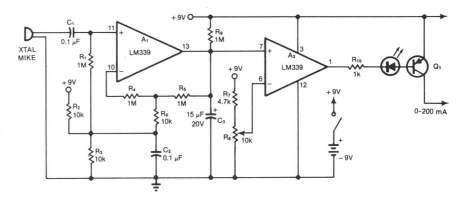

SOUND-ACTIVATED SWITCH—Can be used as sensor for burglar alarm or for turning on surveillance tape recorder to monitor conversations. R_8 is adjusted to give desired sensitivity at which A_2 triggers switch Q_1 to provide 200-mA load current and turn on indicator LED. First section of LM339 quad comparator serves as amplifier and detector providing gain of 100. Second comparator compares DC output of first with reference level selected by R_8.—D. R. Morgan, Sound Turns Switch On, *EDN Magazine*, Aug. 5, 1978, p 82 and 84.

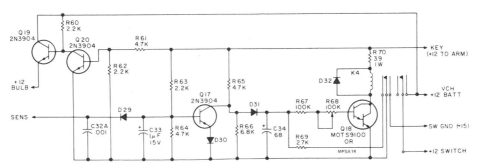

ALARM DRIVES PAGING BEEPER—Complete protection of vehicle is provided by multiplicity of door-switch, mat-switch, vibration, motion, and other sensors connected to common sensor input of alarm switching circuit that controls radio pager, 1-W GE Voice Command II transmitter operating around 147 MHz, 100-W electronic siren, and power horns. Closing of contacts in any sensor grounds common input (assuming keylock switch has been closed to arm circuit by applying +12 V), applying power to siren and pager system. Range is about 1 mi for Motorola Pageboy II cigarette-pack-size pager receiver. Article describes construction, operation, and installation in detail and gives complete circuit of pager.—J. Crawford, Build a Beeper Alarm, *73 Magazine*, Oct. 1977, p 68–77.

DOPPLER BURGLAR ALARM—Small radar transmitter operating at 10.687 GHz fills protected area with radio waves. Waves reflected from stationary objects are ignored by receiver, while waves undergoing Doppler shift in frequency by reflection from moving object such as intruder are selectively amplified for triggering of alarm. Single waveguide section is divided into two cavities, each having Gunn diode; transmitter cavity feeds points A and B of transmitter TR7-IC$_3$, and other cavity feeds points C and D of amplifier that drives alarm relay. Article covers construction and operation of circuit and gives sources (British) for parts and construction kits. Opamps are SN72748 or equivalent, IC$_3$ is μA723 or equivalent, Tr$_1$-Tr$_3$ are ZTX500 or equivalent, Tr$_4$-Tr$_6$ are ZTX302 or equivalent, Tr$_7$ is 3055, D$_1$-D$_8$ are 1N4001 or equivalent, D$_9$-D$_{10}$ are 1N914, SCR$_1$ is TIC44 or equivalent, Z$_1$-Z$_2$ are BZY88-C8V2, relay is 18-V with 1K coil, Doppler module is Mullard CL8960 or equivalent, and self-oscillating mixer for receiver is Mullard CL8630S or equivalent. Alarm stays on until reset by appropriate switch.—M. W. Hosking, Microwave Intruder Alarm, *Wireless World*, July 1977, p. 36–39.

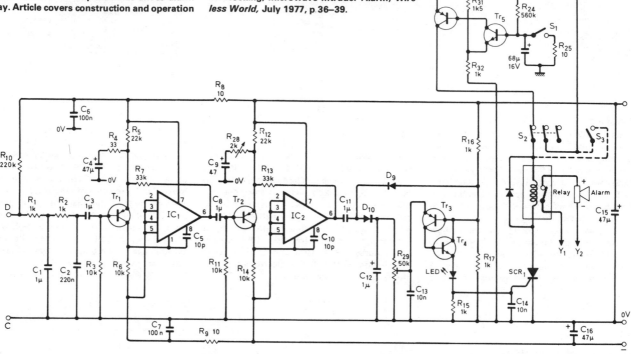

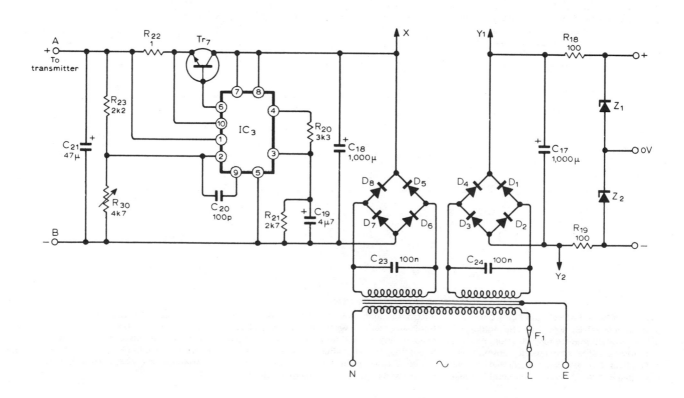

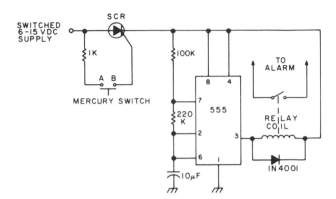

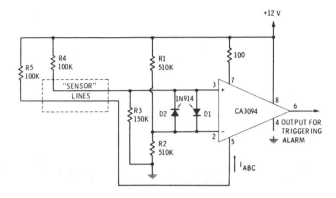

BEEPER—Intermittent alarm using 555 timer can be set to energize horn, lights, or other signaling device at any desired interval when tripped. When used on auto, sound cannot be mistaken for stuck horn. Choose SCR rating to handle current drawn by relay and timer. If alarm draws less than 200 mA, relay is not needed.—W. Pinner, Alarm! Alarm! Alarm!, *73 Magazine,* Feb. 1976, p 138–139.

OPEN/SHORT/GROUND ALARM—Pin 6 of CA3094 IC is high for no-alarm condition. When any one sensor line is open, is shorted to other line, or is shorted to ground, output of IC goes low and resulting output current serves for activating alarm system.—E. M. Noll, "Linear IC Principles, Experiments, and Projects," Howard W. Sams, Indianapolis, IN, 1974, p 316–317.

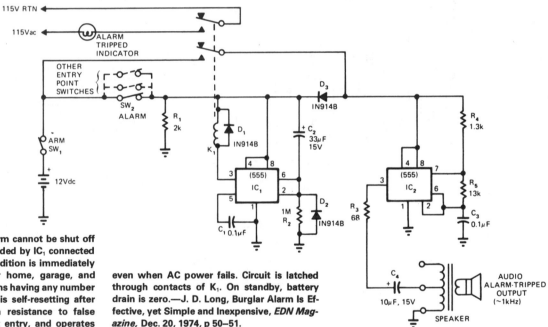

LATCH-ON ALARM—Alarm cannot be shut off for 12 s, with delay provided by IC_1 connected as mono, even if trip condition is immediately removed. Developed for home, garage, and auto burglar alarm systems having any number of trip switches. Circuit is self-resetting after delay interval, has high resistance to false alarms other than direct entry, and operates even when AC power fails. Circuit is latched through contacts of K_1. On standby, battery drain is zero.—J. D. Long, Burglar Alarm Is Effective, yet Simple and Inexpensive, *EDN Magazine,* Dec. 20, 1974, p 50–51.

CHAPTER 10
Capacitance Measuring Circuits

Timers, bridges, dip meters, counters, phase-locked loops, and microprocessors drive meters, digital displays, or audible indicators giving values of capacitors.

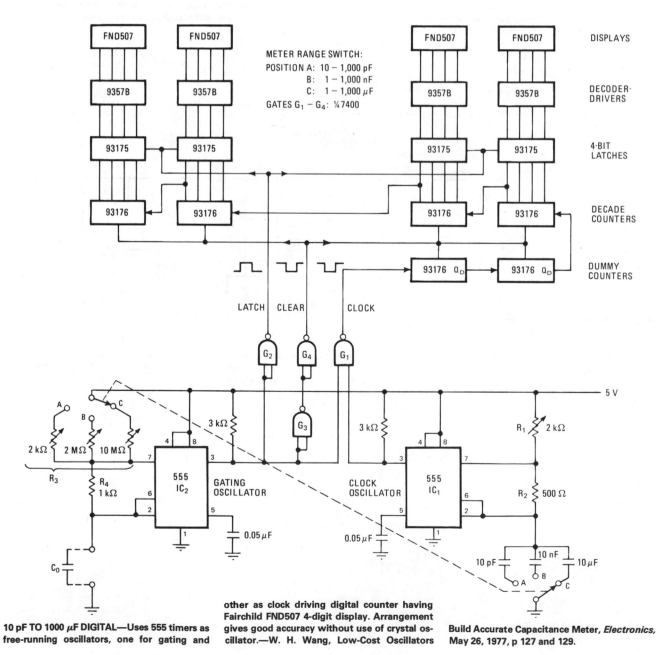

10 pF TO 1000 μF DIGITAL—Uses 555 timers as free-running oscillators, one for gating and other as clock driving digital counter having Fairchild FND507 4-digit display. Arrangement gives good accuracy without use of crystal oscillator.—W. H. Wang, Low-Cost Oscillators Build Accurate Capacitance Meter, *Electronics*, May 26, 1977, p 127 and 129.

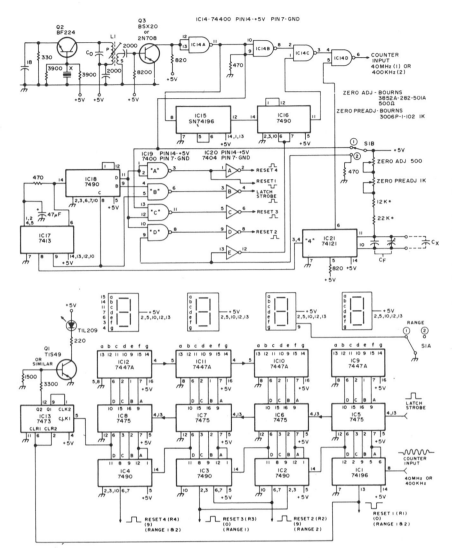

1 pF TO 1 μF—Presents instantly in digital form the value of unknown capacitor, in ranges of 1–9999 pF and 1–999.9 nF. Four digits are displayed, with leading-zero suppression and overflow indicator. Accuracy is better than 0.1% of full range ± 1 digit for higher values in both ranges. Mono MVBR IC21 produces pulse whose length is directly proportional to value of C_X plus about 980-pF total in C_F. This pulse enables gate IC14D whose output goes to counter. Oscillator Q2, buffer Q3, dividers IC15 and IC16, and gates IC14 together give 40-MHz (range 1) or 400-kHz (range 2) pulses that are counted while IC21 holds IC14D open. Article covers construction in detail.—I. M. Chladek, **Build This Digital Capacity Meter,** *73 Magazine,* Jan. 1976, p 70–78.

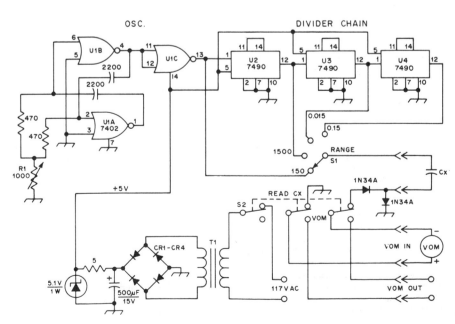

C WITH VOM—TTL-derived square-wave generator U1 charges unknown capacitor C_X to about 3.5 V at 285 kHz when using 150-μA scale of Heath MM-1 volt-ohm-milliammeter, to give 150-pF full-scale range. Larger values of capacitance are read by decreasing frequency with 7490 decade dividers. Use Mallory PTC401 for CR1-CR4. T1 is 6.3-VAC filament transformer. S2 restores normal VOM functions. Article gives design equations.—K. H. Cavcey, **Read Capacitance with Your VOM,** *QST,* Dec. 1975, p 36–37.

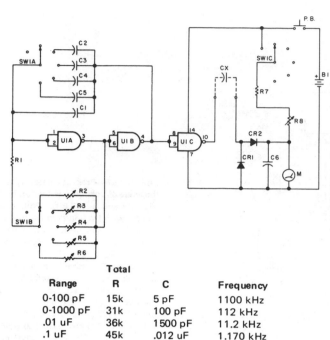

5 pF TO 1 μF—Consists of an oscillator using two gates from CD4011 quad NAND gate, separated from diode rectifier by another NAND gate. Increasing oscillator frequency gives more pulses per second and higher integrated meter reading. Each meter range is linear, so value of 5-pF capacitor can be read on lowest range. Diodes are 1N34 or equivalent. R1 is 12K, and R2-R6 are 50K trimpots set to values shown in table. R7 is 5K, and R8 is 10K trimpot. B1 is 9-V transistor battery. Article covers construction and calibration with known capacitors.—E. Landefeld, Build a Simple Capacitance Meter, *73 Magazine*, Jan. 1978, p 164–165.

Range	Total R	C	Frequency
0-100 pF	15k	5 pF	1100 kHz
0-1000 pF	31k	100 pF	112 kHz
.01 uF	36k	1500 pF	11.2 kHz
.1 uF	45k	.012 uF	1.170 kHz
1 uF	45k	.1 uF	109 Hz

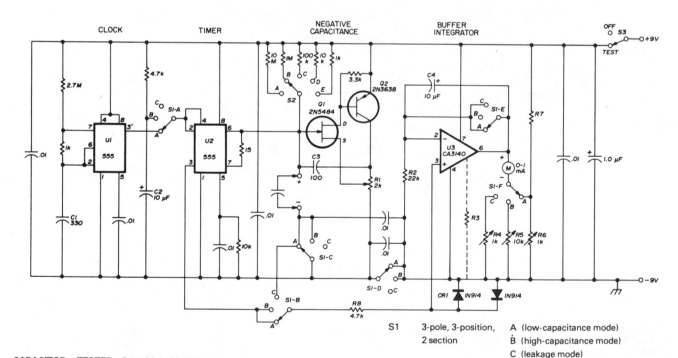

S1	3-pole, 3-position, 2 section	A (low-capacitance mode)
		B (high-capacitance mode)
		C (leakage mode)

S2	1-pole, 5-position	mode A μF	mode B μF	mode C
	A	0.0001	0.25	
	B	0.001	2.5	
	C	0.01	25.0	
	D	0.1	250.0	
	E	1.0	2500.0	leakage

S3	SPST (test)	

CAPACITOR TESTER—Portable instrument measures capacitance values to 2500 μF and leakage current with up to 8 V applied. Timer U1 operates as clock providing about 350 negative-going pulses per second to trigger timer U2 and unclamp test capacitor so it charges through switch-selected resistor to half of supply voltage. U2 then resets, discharging capacitor through pin 7. During charge, pin 3 of U2 is high (about 8 V) and duration of high state is directly proportional to capacitance. Resulting rectangular waveform is applied to unity-gain buffer opamp U3 that feeds meter through calibrating trimpot R6. Meter deflection is proportional to average value of rectangular output waveform and is therefore proportional to capacitance. Table gives switch functions. Mode B uses larger clock timing capacitor to permit measuring larger capacitance values, for total of 10 ranges. Article covers construction, calibration, and use.—P. H. Mathieson, Wide-Range Capacitance Meter, *Ham Radio*, Feb. 1978, p 51–53.

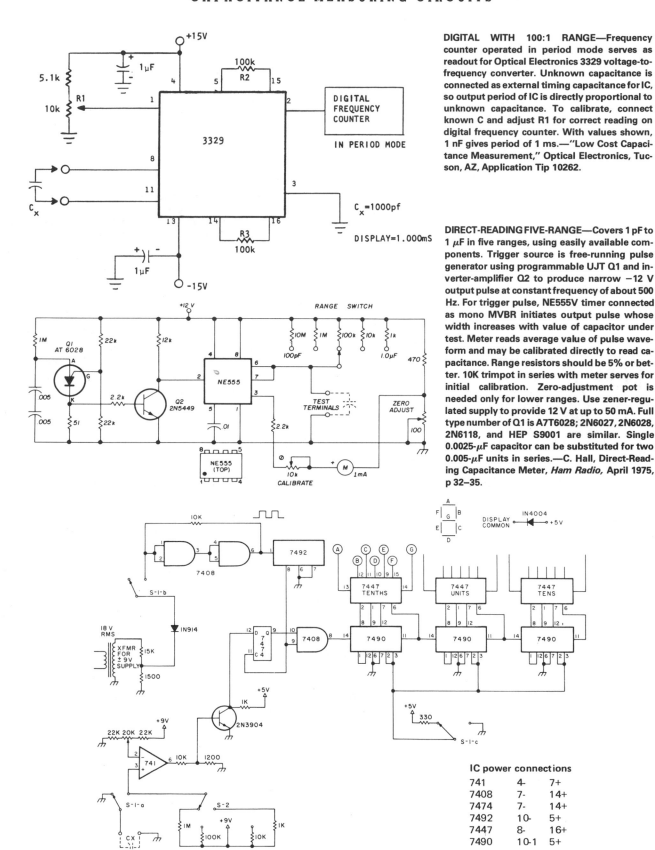

DIGITAL WITH 100:1 RANGE—Frequency counter operated in period mode serves as readout for Optical Electronics 3329 voltage-to-frequency converter. Unknown capacitance is connected as external timing capacitance for IC, so output period of IC is directly proportional to unknown capacitance. To calibrate, connect known C and adjust R1 for correct reading on digital frequency counter. With values shown, 1 nF gives period of 1 ms.—"Low Cost Capacitance Measurement," Optical Electronics, Tucson, AZ, Application Tip 10262.

DIRECT-READING FIVE-RANGE—Covers 1 pF to 1 μF in five ranges, using easily available components. Trigger source is free-running pulse generator using programmable UJT Q1 and inverter-amplifier Q2 to produce narrow −12 V output pulse at constant frequency of about 500 Hz. For trigger pulse, NE555V timer connected as mono MVBR initiates output pulse whose width increases with value of capacitor under test. Meter reads average value of pulse waveform and may be calibrated directly to read capacitance. Range resistors should be 5% or better. 10K trimpot in series with meter serves for initial calibration. Zero-adjustment pot is needed only for lower ranges. Use zener-regulated supply to provide 12 V at up to 50 mA. Full type number of Q1 is A7T6028; 2N6027, 2N6028, 2N6118, and HEP S9001 are similar. Single 0.0025-μF capacitor can be substituted for two 0.005-μF units in series.—C. Hall, Direct-Reading Capacitance Meter, *Ham Radio*, April 1975, p 32–35.

IC power connections		
741	4-	7+
7408	7-	14+
7474	7-	14+
7492	10-	5+
7447	8-	16+
7490	10-1	5+

1-99,900 μF—Circuit converts charging time of unknown capacitor to capacitance value shown on 3-digit display. S-1 is shown in OFF position, with unknown capacitor shorted. When S-1 is changed to other position for start of test, C_x is connected to measuring circuit through range switch S-2 and 741 opamp used as comparator. 60-Hz timing waveform is now applied to sine-wave squaring circuit using two sections of 7408 AND gate. This starts 7490 counters. Zener-regulated +9 V is applied to C_x through selected range resistor. When charging voltage of capacitor exceeds reference voltage on inverting input, 741 output goes positive and stops counter. Article describes circuit operation in detail. Range switch gives scaling factors of 1, 10, 100, and 1000.—A. S. Joffe, Now—a Digital Capacity Meter!, *73 Magazine*, May 1978, p 58–60.

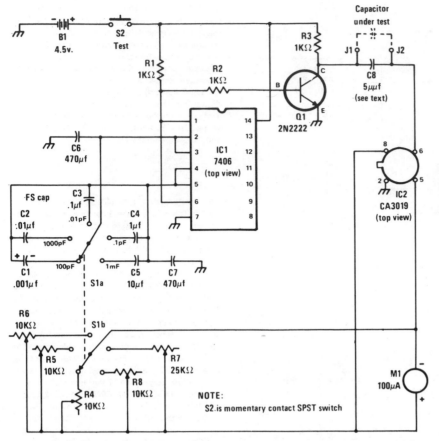

FIVE RANGES UP TO 1 μF—Direct-reading meter gives capacitance values in five ranges, all using same 0–100 scale on 100-μA meter. Operates from three penlight cells. To calibrate, connect known capacitor to jack, close S2, and adjust trimmer pot for each range in turn to give correct indication of capacitor value on meter.—C. Green, Build This Easy Capacitor Meter, *Modern Electronics,* Aug. 1978, p 78–79.

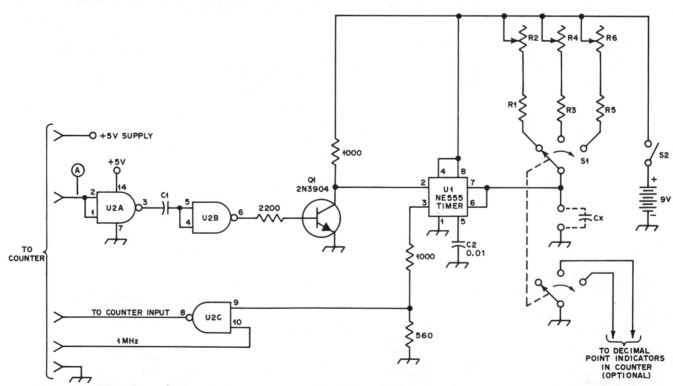

ADAPTER FOR COUNTER—Converts counter into digital capacitance meter for measuring values down to around 5 pF with better than 1% accuracy. Three ranges give full-scale values of 99,999 pF, 0.99999 μF, and 9.9999 μF. Positive-going count-enable command from frequency counter, applied to point A of gate U2A, re-

moves short-circuit from unknown capacitor C_x and enables gate U2C. Capacitor charges exponentially through R1 and R2 (range 1) to voltage at which threshold comparator at U1 makes flip-flop change state, shorting C_x and disabling gate U2C. During charge time, 1-MHz pulses are applied to counter input. Counter reading then

corresponds to capacitor value. C1 is 18 pF, R1 is 860K, R2 is 100K, R3 is 86K, R4 is 10K, R5 is 8.6K, R6 is 1K, and U2 is 7400 quad NAND gate.—R. F. Kramer, Using a Frequency Counter as a Capacitance Meter, *QST,* Aug. 1977, p 19–22.

CHECKING BY SUBSTITUTION—Uses 1-MHz crystal oscillator with fixed-tuned tank circuit L1-C2 link-coupled to resonant measuring circuit consisting of L4, C4, C5, and unknown capacitance. Simple RF voltmeter is connected across measuring circuit as resonance indicator. C4 and C5 have calibrated dials reading directly in picofarads. L1 is Miller 20A224RBI slug-tuned unit adjusted to 250 μH. L4 is Miller 41A685CBI adjusted to 60 μH. Links L2 and L3 are 2 turns each. To use, close S1, set C5 to maximum, and adjust C4 for peak deflection of M1. Connect unknown capacitance to XX with shortest possible leads, retune C5 to resonance, then subtract this capacitance reading of C5

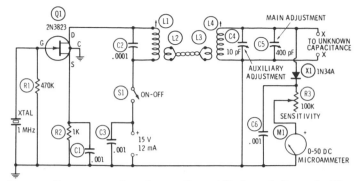

from maximum reading to get value of unknown capacitor.—R. P. Turner, "FET Circuits,"

Howard W. Sams, Indianapolis, IN, 1977, 2nd Ed., p 140–142.

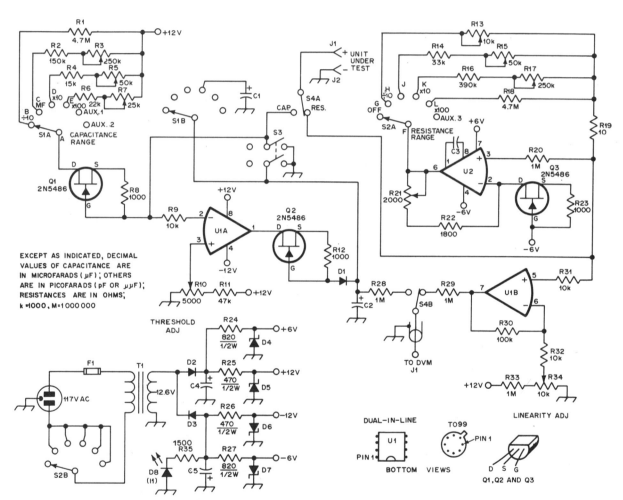

C1, C4, C5 — 220 μF, 16 V, Sprague 227G016CG or equiv.
C2 — 22 μF, 16 V, Sprague 226G016AS or equiv.
C3 — 130-pF disk, Sprague 1CC0G131X0100C4 or equiv.
D1 — Silicon small-signal diode, 1N914 or equiv.
D2, D3 — Silicon rectifier diode, 200 V, 1 A; 1N4003 or equiv.
D4, D7 — Zener diode, 6.2 V, 400 mW, 1N753 or equiv.
D5, D6 — Zener diode, 12.0 V, 400 mW,

1N759 or equiv.
D8 — 3/16-in. red LED, Motorola MLED50 or equiv. (I1 on pc board).
F1 — 1/2-A pigtail fuse, Buss MDV 1/2 A, 250 V.
J1, J2 — 5-way binding post. (Radio Shack package no. 274-661 includes red and black posts.)
Q1, Q2, Q3 — N-channel JFET, 2N5486 or equiv.
S1, S2 — 2-pole, 6-position rotary switch, CTS no. T206 or equiv.

S3 — Dpdt momentary toggle switch, Alco no. MTA206T or equiv.
S4 — Dpdt toggle switch, Alco no. MTA206P or equiv.
T1 — 12.6-V, 100-mA power transformer, Mouser no. 81PG120. Mounting centers 1-13/16 inch.
U1 — Dual operational amplifier, National Semiconductor type LM1458. Interchangeable with IC type 5558.
U2 — Linear IC operational amplifier, RCA type CA3130.

R AND C ADAPTER FOR DVM—Self-contained circuit provides four ranges of capacitance (0–1, 10, 100, and 1000 μF) and four ranges of resistance (0–1, 10, 100, and 1000 kilohms) when used with *QST* combination digital voltmeter and frequency counter. Auxiliary range positions on switches are provided for special measuring requirements such as temperature sensing, antenna elevation indication, and raingage measurements. For capacitors, constant-current source Q1 charges capacitor linearly. When charging voltage makes U1A switch from positive to negative, C2 stops charging. Voltage across C2, proportional to value of unknown C, is then fed to DVM. Article covers construction and calibration.—R. Shriner, New Tasks for the Digital Voltmeter, *QST*, March 1978, p 19–22.

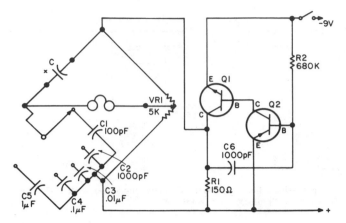

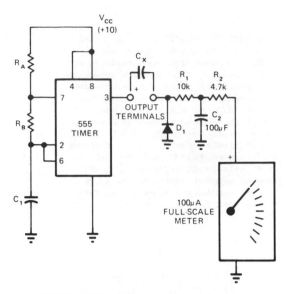

BRIDGE FOR 25 pF TO 10 μF—Uses five reference capacitors, one for each range. Linear pot VR1 serves for balancing. High-resistance headphones indicate null, and capacitor value is then read from setting of VR1. Scale of VR1 is marked for 100 to 10,000 pF for C2 range and 0.01 to 1 μF for C4 range. Scale values are multiplied or divided for other ranges. Calibration is carried out on C2 range, using known capacitor values. Tone oscillator can use almost any pair of transistors, one NPN and the other PNP.—F. G. Rayer, Adrift over Your C's?, *73 Magazine*, March 1976, p 106–107.

LINEAR SCALE—Wide frequency range and high output current of 555 timer contribute to linearity of operation as capacitance meter. Timer is connected as astable MVBR with frequency determined by values used for R_A, R_B, and C_1. When timer output is high, unknown capacitance C_x is charged almost to V_{CC}. When timer goes low, C_x discharges through D_1. Use 100 kHz for 100-pF full-scale reading, 10 kHz for 1 nF, 1 kHz for 10 nF, and down to 1 Hz for 10 μF. Use regulated supply.—R. Horton, 555 Timer Makes Simple Capacitance Meter, *EDN Magazine,* Nov. 5, 1973, p 81.

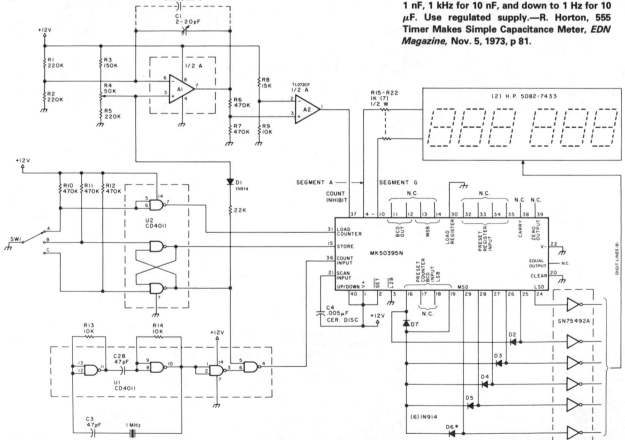

6-DIGIT VALUES—Digital capacitance meter provides display of capacitance values from 1 pF to 999999 pF (1.0 μF). Start-measurement switch drains charge from capacitor under measurement and diverts constant-current source to ground. Capacitor begins charging, and counter accumulates 1-μs pulses from crystal clock. When capacitor charge voltage reaches threshold of count-inhibit line for counter, contents of counter are displayed as capacitance value. Circuit uses Mostek MK50395N six-decade counter that provides 7-segment output data for display. Position A of SW1 is starting point, B stores data in counter display after capacitor measurement, and C initiates measurement. Display includes leading-zero suppression.—J. Garrett, What's Your μF?, *73 Magazine,* Dec. 1978, p 234–235.

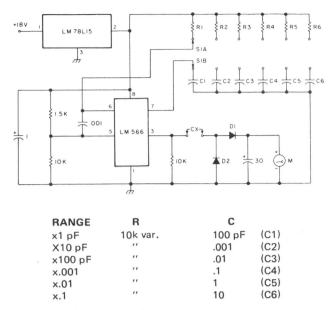

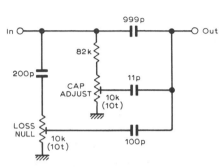

RANGE	R	C	
x1 pF	10k var.	100 pF	(C1)
X10 pF	"	.001	(C2)
x100 pF	"	.01	(C3)
x.001	"	.1	(C4)
x.01	"	1	(C5)
x.1	"	10	(C6)

PLL CAPACITANCE METER—Based on fact that alternating current floating through capacitor depends on applied voltage, frequency, and capacitance value. Circuit uses square wave for charging capacitor to full voltage, then measures current flow as linear function of capacitance. LM78L15 provides regulated 15 V for LM566 PLL VCO. Frequency of VCO depends on values of R and C selected by rotary switch S1, to give six linear scales: 0–10 pF, 10–100 pF, 100–1000 pF, 1000 pF to 0.01 μF, 0.01–0.1 μF, and 0.1–1 μF. Accuracy is about ±5%. Meter is 100 μA. Use small signal diodes.—S. Shields, How Many pF is That Capacitor, Really?, *73 Magazine*, March 1978, p 48–50.

PERFECT CAPACITOR—Simple circuit shown provides equivalent of perfect no-loss 1000-pF capacitor at frequencies below about 100 kHz. Principle can be used to construct fixed-frequency capacitance standards for use in high-accuracy capacitor bridge. All capacitors are silver mica. If mounted in oven, stability can be 1 PPM and residual phase-angle difference from pure capacitance only 1 microradian.—B. J. Frost, "No Loss" Capacitor, *Wireless World*, Dec. 1977, p 80.

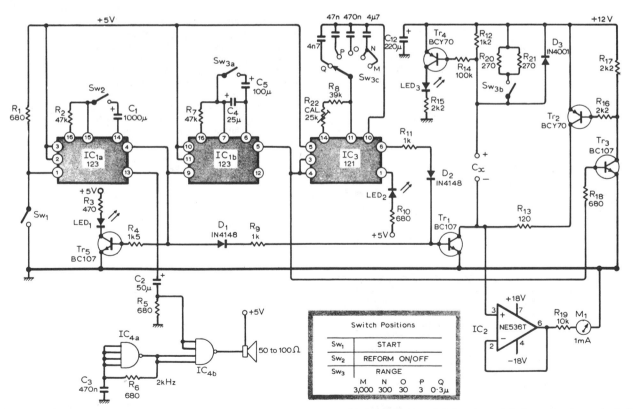

Switch Positions				
Sw₁	START			
Sw₂	REFORM ON/OFF			
Sw₃	RANGE			
	M	N	O	P	Q
	3,000	300	30	3	0·3μ

ELECTROLYTICS WITH REFORMING—Automatic tester for electrolytics applies voltage for about 15 s to repolarize dielectric before measurement is made. This provides sufficient reforming for test purposes, using 12 V through 1200 ohms, but test should be repeated if leakage current is high because of incomplete reforming. Tone from loudspeaker indicates end of 15-s reforming period. Green LED₁ indicates reforming process is ready to start. Red LED₂ indicates excessive current is flowing during reforming. LED₃ flashes to indicate test capacitor is being charged during measuring cycle. Article covers construction and calibration in detail. IC₁ is SN74123N, IC₃ is SN74121N, and IC₄ is SN7413N.—A. Drummond-Murray, Electrolytic Capacitor Tester, *Wireless World*, May 1977, p 47–49.

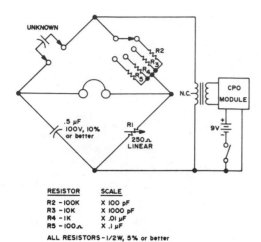

RESISTOR	SCALE
R2 – 100K	X 100 pF
R3 – 10K	X 1000 pF
R4 – 1K	X .01 µF
R5 – 100Ω	X .1 µF

ALL RESISTORS – 1/2W, 5% or better

0.0001 TO 1 µF IN FOUR RANGES—Simple bridge uses AF voltage from Cordover CPO-4 code practice oscillator module, fed through transistor output transformer connected in reverse for impedance matching. Earphones serve as null detector, but amplifier can be added for greater sensitivity or CRO used. Only one scale need be calibrated, using known values of capacitors.—W. P. Turner, Build a Basic Bridge, *73 Magazine,* Nov. 1974, p 95.

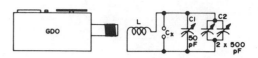

C BY GRID DIP—Values of unknown capacitances up to about 1000 pF can be measured with simple circuit used with grid-dip oscillator. Coil L can be 6 turns of stiff wire. To calibrate, close variable capacitors C1 and C2 fully, tune for dip, and note dip frequency at pointer position of C2. Now connect known capacitors up to 1000 pF one by one to CX, retune C2 for dip, and mark capacitor value on C2 dial. Close C2, then repeat calibration for C1 while using smaller capacitors up to 50 pF.—F. G. Rayer, GDO to Find C, *73 Magazine,* Aug. 1974, p 35.

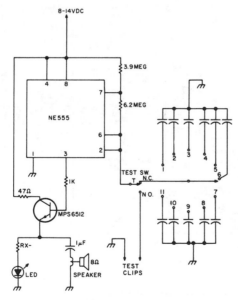

0.5-pF TO 0.001-µF COMPARATOR—Provides audio-tone comparison of built-in reference capacitor to unknown capacitor connected between test clips. Frequency of tone is about 8 kHz for 0.5 pF, dropping to 100 Hz as capacitor value goes up to 0.001 µF. Larger capacitor values merely turn LED on and off; 0.1 µF gives flashing at about 5 Hz. Any NPN audio or switching transistor can be used in place of MPS6512. Suggested reference values for capacitor bank are 0.7, 3, 5, 10, 25, 50, 100, 330, 470, 680, and 820 pF.—W. Pinner, The Capacitor Comparator, *73 Magazine,* March 1977, p 49.

COMPUTERIZED METER—With 4.7 megohms for R, simple 555 timer circuit used in conjunction with computer measures capacitors in five ranges from below 100 pF to 0.1 µF. For larger range, resistor value can be changed. Article includes BASIC software suitable for 8080-based systems, including calibration program based on known values of capacitance. 555 mono MVBR is triggered under control of computer output bit, with count being made while mono is timing out. Count is averaged over ten triggerings, then multiplied in computer by calibration factor to give capacitance value. Any desired type of output indicator can be used.— J. Eccleston, Computerized Capacity Meter, *73 Magazine,* July 1978, p 88–89.

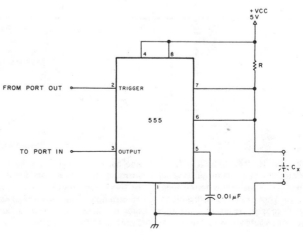

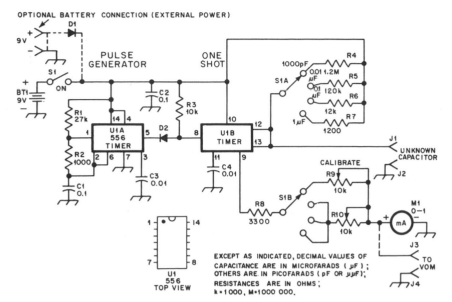

DUAL TIMER MEASURES C—One section of U1 (two 555s in single package) is connected as oscillator that serves as trigger for other section (U1B). Ratio of R1 and R2 determines length of pulse generated during each oscillation cycle, while C1 and same resistors set frequency at about 500 Hz. U1B produces predetermined-duration output pulse for each start pulse regardless of starting pulse length. Pulse duration is set by R4-R7 and external capacitor being measured. Smaller capacitor in given range produces shorter output pulse from U1B mono MVBR. Average pulse power increases with pulse length and increases meter reading linearly so capacitance value is indicated directly. Values shown give ranges of 1000 pF, 0.01 μF, 0.1 μF, and 1 μF full-scale. R10 serves as calibration resistor for all three higher scales. D1 is 50-PIV or higher silicon power-type diode, and D2 is 1N914 or equivalent.—D. A. Blakeslee, An Inexpensive Capacitance Meter, *QST*, Sept. 1978, p 11–14 and 37.

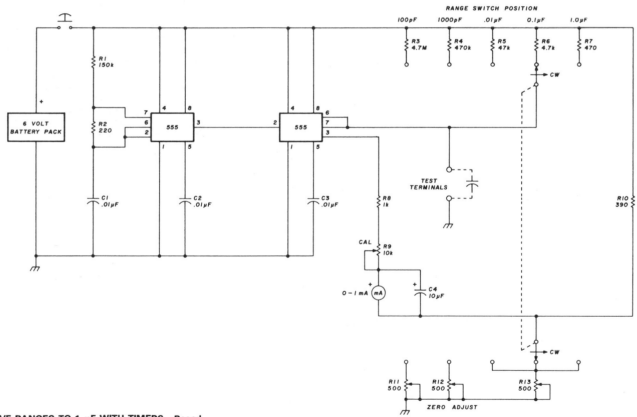

FIVE RANGES TO 1 μF WITH TIMERS—Based on fact that output pulse width of 555 timer varies linearly with value of timing capacitance used. If timer is triggered with constant frequency, average DC value of resulting pulse train is linear function of pulse width. DC meter then reads capacitance values linearly. Decade capacitance ranges are obtained by switching value of timing resistor. Trimpot for each range is adjusted for zero meter reading when pushbutton is pressed, without test capacitor.—C. Hall, Simplified Capacitance Meter, *Ham Radio*, Nov. 1978, p 78–79.

CHAPTER 11
Cathode-Ray Circuits

Includes probe circuits, preamps, deflection amplifiers, 2-channel and 4-channel trace multipliers, triggered sweep, dynamic focus correction, B-H and Lissajous pattern generators, time-mark generator, and TV typewriter circuits. See also Game, Power Supply, Sweep, and Television chapters.

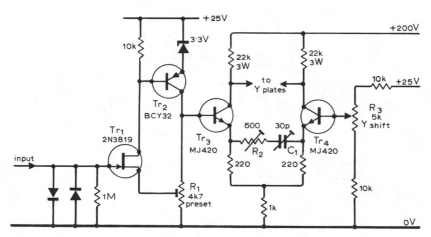

Y AMPLIFIER FOR CRO—Combines advantages of differential output stage and high-impedance JFET input stage. Silicon input diodes provide crude overload protection for input, while Tr_2 acts with Tr_1 for level-shifting as well as amplifying. R_1 is used to set quiescent output voltage of Tr_2 at about 15 V; this setting is critical, and may require multiturn pot. Article gives setup procedures.—G. A. Johnston, Deflection Amplifier for Oscilloscopes, *Wireless World*, April 1975, p 175.

CURRENT AMPLIFIER—Used in FET curve tracer to amplify drain current passing through R_{14} sufficiently to give required Y output for oscilloscope. Uses SN72741P opamp as difference amplifier. Article gives other circuits of curve tracer and calibration procedure.—L. G. Cuthbert, An F.E.T. Curve Tracer, *Wireless World*, April 1974, p 101–103.

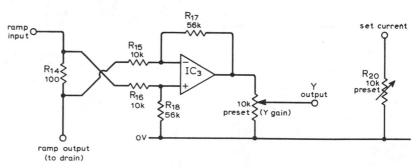

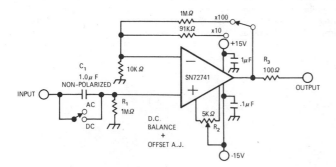

RANGE	FREQUENCY
X 10	D.C. 100 KHZ
X 100	D.C. 10 KHZ

Output Voltage P.P.	
D.C.	28V
10 KHZ	27V
100 KHZ	3V

SCOPE PREAMP—Extends vertical sensitivity range of scope or VOM at minimum cost. Voltage at output is in phase with input. Switch across C gives choice of AC or DC operation. Table gives frequency and output voltage limits. Input impedance is about 500 kilohms.—G. Coers, High-Gain AC/DC Oscilloscope Amplifier, *EDN/EEE Magazine*, Feb. 1, 1972, p 56.

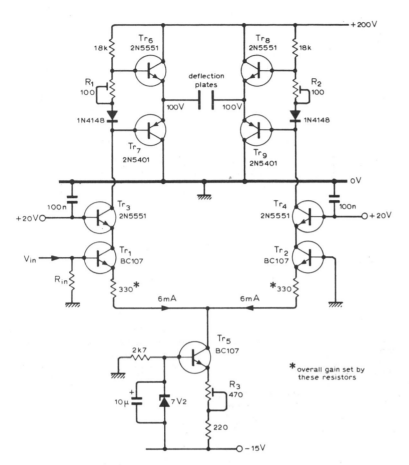

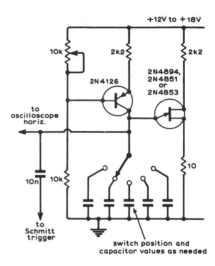

TRACE QUADRUPLER—Designed for use with DC oscilloscopes. Constant-current UJT oscillator produces linear sawtooth for triggering Schmitt trigger and serving as horizontal sweep voltage. Frequency is varied by switching capacitors, and can be up to about 100 kHz. Emitter-follower may have to be added to UJT output to prevent loading of timing capacitors by low impedance of Schmitt trigger. Used to quadruple maximum time-base frequency of oscilloscope.—J. A. Titus, Trace Quadrupler, *Wireless World,* Oct. 1972, p 479.

Y AMPLIFIER WITH 10-MHz BANDWIDTH—Rise time is 40 ns. Tr_1-Tr_4 form constant-current tail, with Tr_5 improving linearity. Tr_6 and Tr_7 are complementary emitter-followers, as also are Tr_8 and Tr_9, for feeding deflection plates. Input should be from 50-ohm source to achieve full bandwidth. Other complementary small-signal transistors rated above 200 V can be used.—B. J. Frost, Wideband Y Amplifier for Oscilloscope, *Wireless World,* June 1976, p 71.

DYNAMIC FOCUS CORRECTION—Provides sum of squares of vertical and horizontal position voltages, as required for focus correction in high-resolution flat-face magnetically deflected CRT. Circuit uses Optical Electronics 5898 four-quadrant analog multipliers to give required squared outputs, along with 9831 opamp having comparable bandwidth. Input summing resistors for horizontal and vertical deflection amplifiers are chosen for compatibility with amplifiers being used. Select current-sampling resistors to generate 10-V peak signal.—"Dynamic Focus Correction with Analog Function Modules," Optical Electronics, Tucson, AZ, Application Tip 10127.

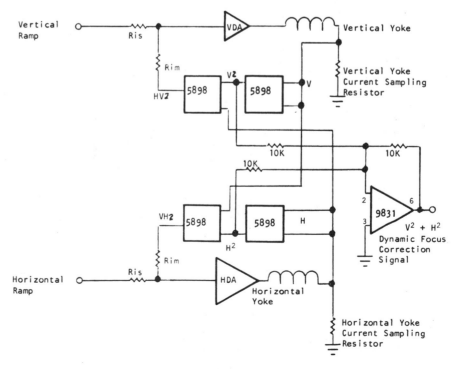

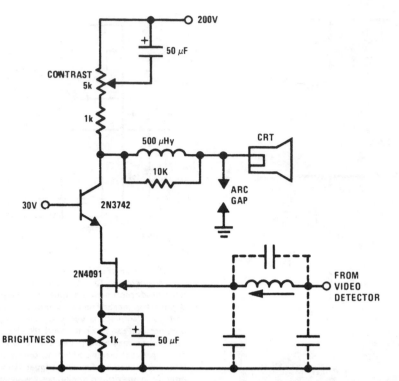

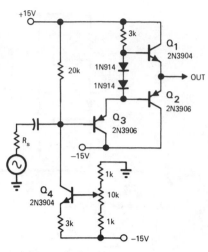

±7 VDC OFFSET—High-impedance current source Q_4 provides desired level shift for AC signals in video circuit whose DC level controls intensity of CRT. Input and offset signals are fed to base of Q_3 which drives complementary-symmetry emitter-follower Q_1-Q_2. For values shown, level can be shifted about ±7 VDC.—P. B. Uhlenhopp, Variable DC Offset Using a Current Source, *EDN/EEE Magazine,* Aug. 15, 1971, p 46.

CATHODE DRIVE FOR CRT—Cascode connection of 2N4091 JFET and 2N3742 bipolar transistor provide full video output for cathode. Gain is about 90. M-derived filter using stray capacitances and variable inductor blocks 4.5-MHz sound frequency from video amplifier. Cascode configuration eliminates Miller capacitance problems of JFET, allowing direct drive from video detector.—"FET Databook," National Semiconductor, Santa Clara, CA, 1977, p 6-26–6-36.

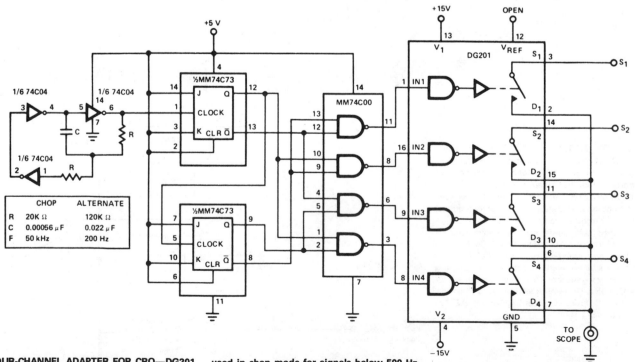

FOUR-CHANNEL ADAPTER FOR CRO—DG201 CMOS analog switch controlled by 50-kHz clock allows display of four input signals simultaneously on single-trace oscilloscope. Adapter is used in chop mode for signals below 500 Hz. Frequencies above 500 Hz are best viewed in alternate mode with clock frequency of 200 Hz. One of inputs is used to trigger horizontal trace of CRO.—"Analog Switches and Their Applications," Siliconix, Santa Clara, CA, 1976, p 7-63–7-66.

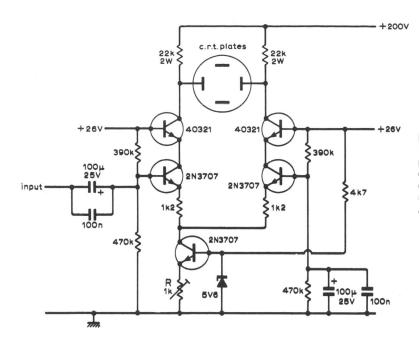

ELECTROSTATIC-DEFLECTION AMPLIFIER— Combines frequency response of cascode amplifier with linearity of long-tailed pair fed by constant current. Adjust R for 3 mA through each load resistor. Output transistors require small heatsinks.—G. A. Johnston, Deflection Amplifier, *Wireless World,* Nov. 1973, p 560.

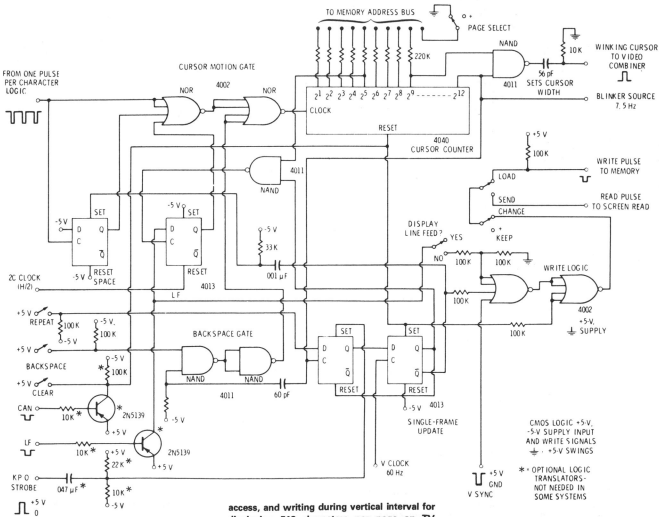

CURSOR FOR TV TYPEWRITER—Complete CMOS logic cursor and update system is shown for system using RAM memory, direct memory access, and writing during vertical interval for displaying 512 characters per page on TV screen. External 7.5-Hz source is required to make underline cursor flash to indicate position at which next character will be entered on screen.—D. Lancaster, "TV Typewriter Cookbook," Howard W. Sams, Indianapolis, IN, 1976, p 128–129.

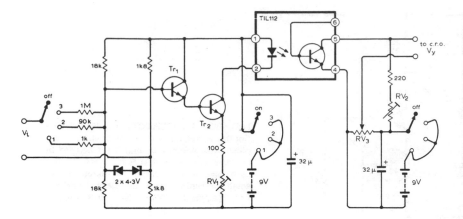

OPTOISOLATOR FOR PROBE—Offsets need for potentially dangerous practice of floating oscilloscope with respect to ground. Also permits simultaneous display of two voltages with correct polarity on double-beam oscilloscope when one of them is floating. Texas Instruments TIL112 optocoupler used has bandwidth of about 30 kHz. Three ranges give choice of 1:1, 10:1, and 100:1 input attenuation. Set RV_2 to bias phototransistor of optocoupler to center of its linear range (about 4.5 V between pins 4 and 5), then set RV_1 to give unity input/output ratio on range 1. RV_3 is set to give zero DC output when input terminals are shorted, but can be omitted if zeroing of output level is unnecessary.—A. F. Sargent, Simple C.R.O. Input Isolating Probe, *Wireless World*, Feb. 1976, p 76.

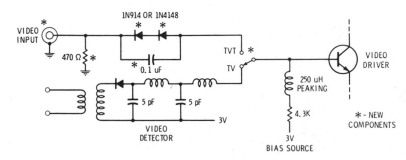

TV INTERFACE FOR TYPEWRITER—Video input circuit for black and white transistor TV receiver permits feeding video output of TV typewriter to video driver in set, for producing character or game display on TV screen. Use of direct coupling eliminates shading effect or changes in background level as characters are added. Diodes provide 1.2-V offset in positive direction so in absence of video the video driver is biased to blacker-than-black sync level of 1.2 V. With white video input of 2 V, driver is biased to usual 3.2 V of white level. Hot-chassis TV sets can present shock hazard.—D. Lancaster, "TV Typewriter Cookbook," Howard W. Sams, Indianapolis, IN, 1976, p 190.

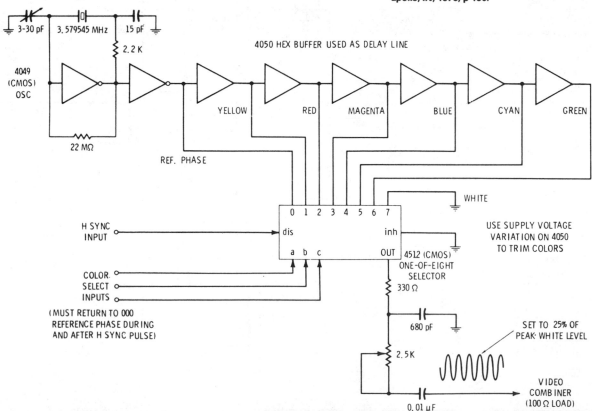

COLOR FOR TV TYPEWRITER—Uses 3.579545-MHz crystal oscillator to drive string of CMOS buffers forming digital delay line. Output delays caused by propagation times in each buffer can be used directly or can be trimmed to specific colors by varying supply voltage. Reference phase and delayed color outputs go to 1-of-8 data selector whose output is determined by code presented digitally to its three color select lines. Selector drive logic must return to 000 (reference phase) immediately before, during, and for at least several microseconds after each horizontal sync pulse so set can lock and hold on reference color burst. Sine-wave output chrominance signal is cut down to about one-fourth of maximum video white level.—D. Lancaster, "TV Typewriter Cookbook," Howard W. Sams, Indianapolis, IN, 1976, p 205–206.

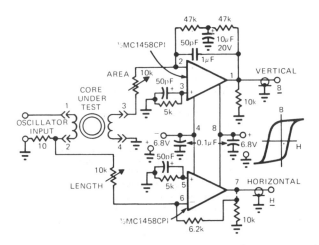

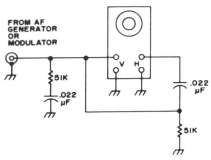

B-H LOOP DISPLAY—Low-cost dual opamp circuit allows display of hysteresis loop on calibrated XY oscilloscope. Two windings are placed on core to be tested, with opamp of flux-measuring system connected to secondary for deriving vertical deflection input representing flux B. Article gives design equations and details of circuit operation and use.—D. A. Zinder, X-Y Oscilloscope Displays Hysteresis Loop of Any Core, *EDN Magazine,* Feb. 5, 1975, p 54–55.

ELLIPTICAL PATTERN—Connection shown gives Lissajous-type elliptical pattern on CRO from ordinary AF signal generator. Modulation can be added to either vertical or horizontal feed for CRO.—Novice Q & A, *73 Magazine,* March 1977, p 187.

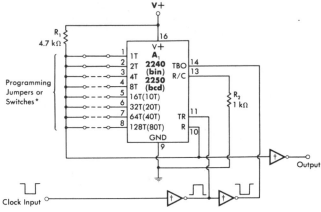

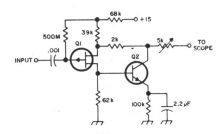

TIME-MARK GENERATOR—Produces precisely spaced output pulses suitable for calibrating CRO time bases. Can be programmed in binary by using 2240 for A_1 or in BCD by using 2250. Use crystal oscillator or other high-accuracy source for external clock. Time interval of output pulse is equal to clock width multiplied by $n + 1$, where n is number programmed into A_1 (1 to 255 for 2240 and 1 to 99 for 2250). Circuit can be programmed electronically by microprocessor if desired.—W. G. Jung, "IC Timer Cookbook," Howard W. Sams, Indianapolis, IN, 1977, p 218–220.

HIGH-Z PROBE—Provides about 1200-megohm input impedance to CRO, with unity gain. Pot adjusts equalization at higher frequencies. Q1 can be U112, 2N2607, 2N4360, or TIM12. Q2 can be 2N706, 2N708, 2N2926, 2N3394, or HEP 50.—Circuits, *73 Magazine,* March 1974, p 89.

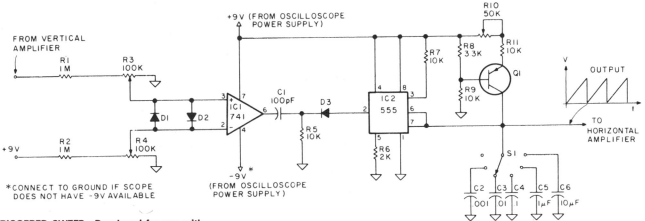

TRIGGERED SWEEP—Developed for use with general-purpose CRO in troubleshooting digital circuits, to provide one horizontal sweep of cathode-ray beam each time circuit is triggered by input signal pulse. Noninverting input of 741 opamp is connected to vertical amplifier of CRO, and inverting input is used to control trigger level. When input signal rises above trigger level, output of opamp swings to −V and makes output of 555 timer go high, allowing output capacitor to charge at constant current through transistor in series with resistor. Result is nearly perfect ramp voltage. All diodes are 1N914. Q1 is any PNP switching transistor.—W. J. Prudhomme, Trigger Your Oscilloscope, *Kilobaud,* Aug. 1977, p 34–38.

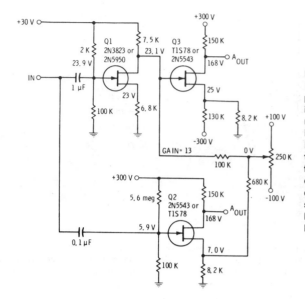

ELECTROSTATIC-DEFLECTION AMPLIFIER— Circuit develops equal-amplitude but opposite-polarity sawtooth outputs when sawtooth input is applied to gates of Q1 and Q2. Q1 is connected as common-source amplifier for applying opposite-polarity sawtooth to gate of Q3. Polarity at output of Q3 then becomes same as that of input, increased to amplitude suitable for deflection plates of CRT. Sawtooth at output of Q2 has opposite polarity. Circuit values are chosen to balance gain so both outputs have same magnitude.—E. M. Noll, "FET Principles, Experiments, and Projects," Howard W. Sams, Indianapolis, IN, 2nd Ed., 1975, p 229–230.

DEFLECTION-PLATE AMPLIFIER—Resistive collector network of symmetrical differential amplifier is replaced by constant-current source to improve slew rate of deflection amplifier driving capacitive load such as deflection plates of electrostatic cathode-ray tube. Q_1 and Q_2 are identical current sources. Network Q_1-CR_1-Q_7-R_6-R_8 forms current source for Q_1. Q_7 is used as 6.2-V zener diode.—W. Peterson, Current Sources Improve Amplifier Slew Rate, *EEE Magazine,* Nov. 1970, p 102.

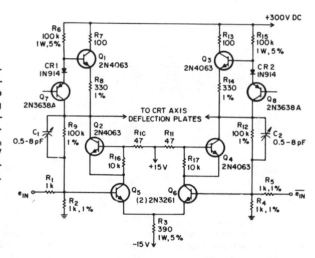

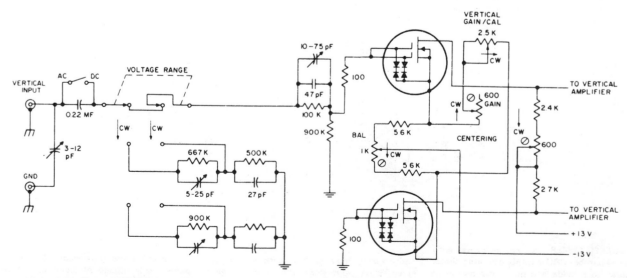

DIFFERENTIAL VERTICAL AMPLIFIER—Uses two RCA 40841 dual-gate FETs in vertical input stage of solid-state oscilloscope, with gates of each connected in single-gate configuration. Circuit is designed for frequencies up to 500 MHz. Wide dynamic range permits handling of large signals without overloading.—"Linear Integrated Circuits and MOS/FET's," RCA Solid State Division, Somerville, NJ, 1977, p 435–436.

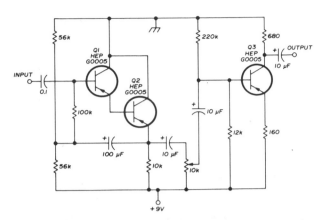

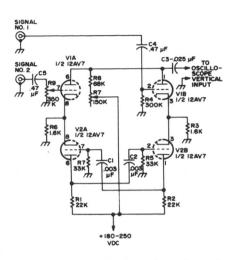

HIGH-Z CRO PREAMP—Darlington circuit provides extremely high input impedance (over 2.2 megohms). With input shorted, noise level is 78 dB down as read at output with VTVM. Linearity is within 1.5% for inputs from 100 μV to 1 mV, and frequency response is $\pm$2 dB from 100 Hz to 350 kHz. Originally designed to boost input to CRO, but can be adapted to many other applications requiring high gain, low noise, and high input impedance.—J. Fisk, Circuits and Techniques, *Ham Radio*, June 1976, p 48–52.

SIGNAL SWITCHER—Two-tube electronic switch serves in effect to provide simultaneous presentation of two different signals on CRO screen by switching signals alternately to vertical input at rate fast enough so both displays are seen.—Novice Q & A, *73 Magazine*, March 1977, p 187.

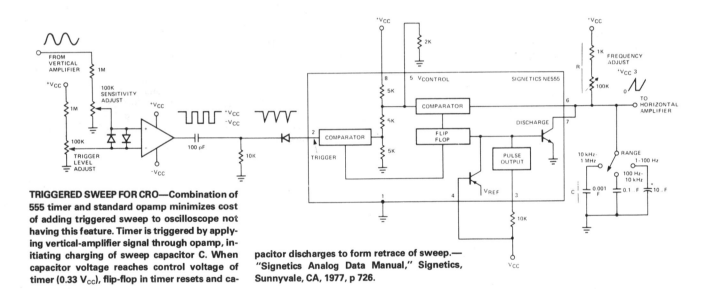

TRIGGERED SWEEP FOR CRO—Combination of 555 timer and standard opamp minimizes cost of adding triggered sweep to oscilloscope not having this feature. Timer is triggered by applying vertical-amplifier signal through opamp, initiating charging of sweep capacitor C. When capacitor voltage reaches control voltage of timer (0.33 V_{CC}), flip-flop in timer resets and ca-pacitor discharges to form retrace of sweep.—"Signetics Analog Data Manual," Signetics, Sunnyvale, CA, 1977, p 726.

CHAPTER **12**
Clock Signal Circuits

Covers circuits for generating clock pulses at frequencies ranging from 1 Hz to well above 30 MHz for use in digital circuits of multiplexers, memories, counters, shift registers, microprocessors, videotape recorders, and digital cassette recorders.

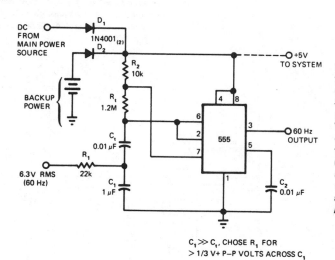

C₁ ≫ Cₜ, CHOSE R₁ FOR
> 1/3 V+ P–P VOLTS ACROSS C₁

60-Hz CLOCK OUTPUT FROM 555—Basic 555 timer IC produces constant 60-Hz rectangular output for use as noninterruptible free-wheeling clock source. C_1 introduces filtered 60-Hz power-line reference component across C_t at 2 V P-P. This signal overrides normal timing ramp of 555, causing it to act as amplifier or Schmitt trigger. When AC line power fails, C_t resumes normal function as timing capacitor for 60-Hz astable MVBR. Circuit can easily be adjusted for other reference frequencies.—W. G. Jung, Take a Fresh Look at New IC Timer Applications, *EDN Magazine,* March 20, 1977, p 127–135.

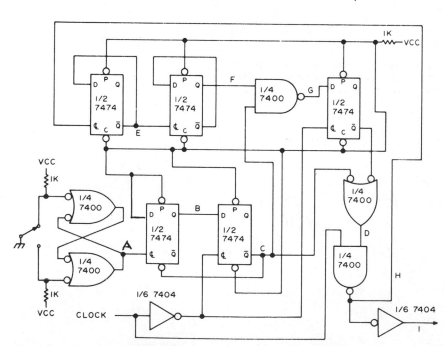

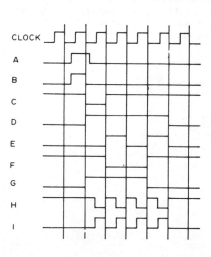

FOUR-PULSE BURST—Generates burst of four clock pulses each time switch is pressed. Modifications can produce any desired number of pulses in burst. Reliability of pulse count is ensured by use of debouncing latch using pair of 7400 gates. VCC is +5 V.—E. E. Hrivnak, House Cleaning the Logical Way, *73 Magazine,* Aug. 1974, p 85–90.

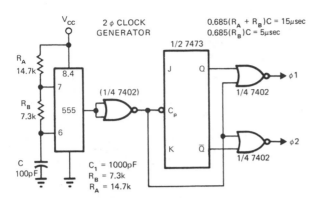

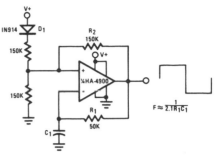

TWO-PHASE CLOCK TO 1 MHz—Signetics 555 timer is used as oscillator to generate nonoverlapping clock pulses as required for most two-phase dynamic MOS memories and shift registers. Duty cycle is determined by values of external resistors R_A and R_B which, together with timing capacitor C, determine frequency of oscillation. 7473 flip-flop controls phase that is switched on through 7402 NOR gates. Article gives timing waveforms and equations. Maximum operating frequency is 1 MHz.—G. Schlitt, Monolithic Timer Generates 2-Phase Clock Pulses, *EDN Magazine*, Aug. 1, 1972, p 57.

SQUARE-WAVE CLOCK—One section of Harris HA-4900/4905 precision quad comparator gives excellent frequency stability as self-starting fixed-frequency square-wave generator for clock applications. R_1 and C_1 determine frequency, and R_2 provides regenerative feedback. For higher precision at frequencies up to 100 kHz, crystal may be used in place of C_1.—"Linear & Data Acquisition Products," Harris Semiconductor, Melbourne, FL, Vol. 1, 1977, p 2-96.

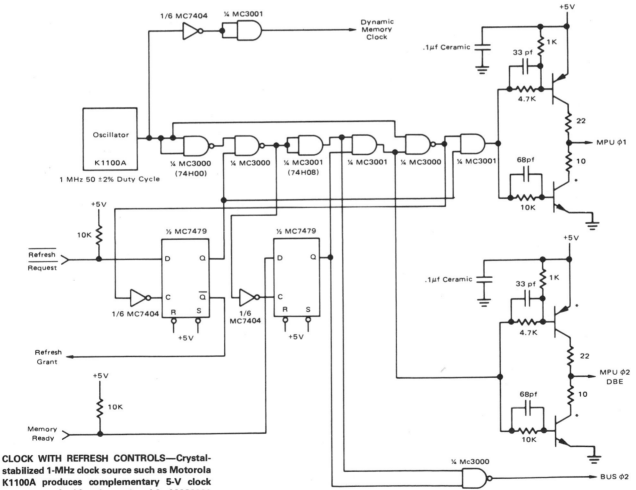

CLOCK WITH REFRESH CONTROLS—Crystal-stabilized 1-MHz clock source such as Motorola K1100A produces complementary 5-V clock outputs required for phases 1 and 2 of MC6800 MPU and also provides interface signals required for dynamic (refresh request and refresh grant) and slow (memory ready) memories. Refresh control circuit uses MC7479 dual latch, MC7404 hex inverter, and pair of 10K pull-up resistors. If refresh request state is low when sampled during leading edge of phase 1, phase 1 is held high and phase 2 low for at least one full clock cycle. Refresh grant signal is high to indicate to dynamic memory system that refresh cycle exists. If memory ready line is low when sampled on leading edge of phase 2, phase 1 is held low and phase 2 high until memory ready line is brought high by slow memory controller. All transistors are MPQ6842.—"Microprocessor Applications Manual" (Motorola Series In Solid-State Electronics), McGraw-Hill, New York, NY, 1975, p 4-57—4-58.

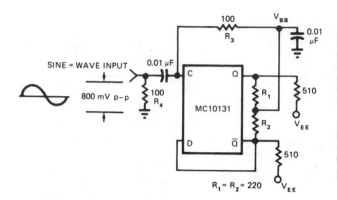

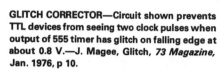

CENTERING CLOCK SIGNAL—Circuit generates DC bias across complementary outputs of Motorola MC10131 flip-flop for optimum operation with emitter-coupled logic (10,000 series). Bias is independent of state of flip-flop, which uses toggle frequency of about 150 MHz. Article covers applications for other flip-flops and counters requiring maintenance of best toggle frequency over wide temperature range.—T. Balph and H. Gnauden, Build a Clock Bias Circuit for ECL Flip-Flops, *EDN Magazine*, May 5, 1976, p 116.

GLITCH CORRECTOR—Circuit shown prevents TTL devices from seeing two clock pulses when output of 555 timer has glitch on falling edge at about 0.8 V.—J. Magee, Glitch, *73 Magazine*, Jan. 1976, p 10.

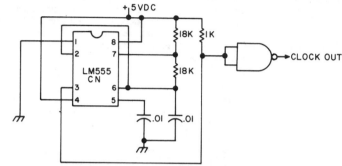

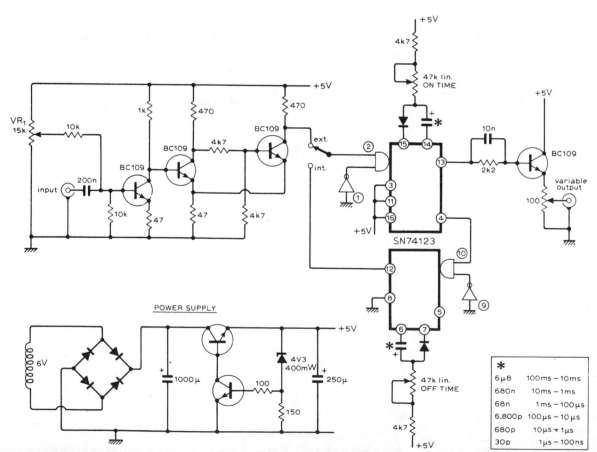

VARIABLE WIDTH AND PRF—Low-cost pulse generator uses versatile dual monostable IC to provide clock pulses that can be varied in width over wide range by changing sizes of two external capacitors and adjusting 47K linear pots. Switched bank of six capacitors can be used instead, to give on or off times ranging between 100 ms and 100 ns, as given in table. With switch in external position, on-time mono is driven by three transistors connected as Schmitt trigger giving pulse having same frequency as that of input signal. VR₁ sets trigger level. Suitable regulated 5-V supply circuit is also shown.—J. Garrett, Pulse Generator, *Wireless World*, Feb. 1976, p 78.

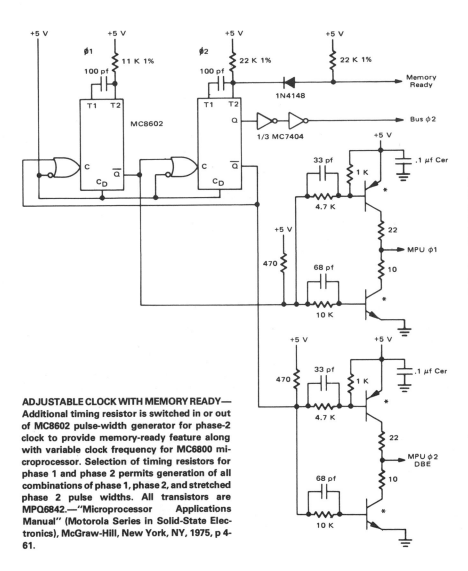

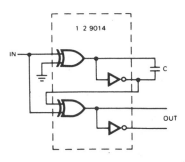

EDGE DETECTOR—Half of 9014 quad EXCLU-SIVE-OR gate serves for generating output pulse for both low-to-high and high-to-low transitions of input signal. Used for regenerating clock in self-clocking pulse-width modulation transmission system. Circuit acts as frequency doubler for square-wave input. With 1000 pF for C, output pulse width is 70 ns; for 200 pF, width is 30 ns; and when C is 0, width is 10 ns.—Circuits, *73 Magazine,* Aug. 1974, p 99.

ADJUSTABLE CLOCK WITH MEMORY READY—Additional timing resistor is switched in or out of MC8602 pulse-width generator for phase-2 clock to provide memory-ready feature along with variable clock frequency for MC6800 microprocessor. Selection of timing resistors for phase 1 and phase 2 permits generation of all combinations of phase 1, phase 2, and stretched phase 2 pulse widths. All transistors are MPQ6842.—"Microprocessor Applications Manual" (Motorola Series in Solid-State Electronics), McGraw-Hill, New York, NY, 1975, p 4-61.

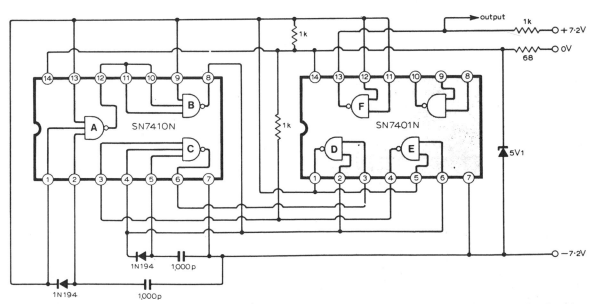

320 kHz FOR CALCULATOR—Two low-cost TTL ICs generate 320-kHz clock signals for electronic desk calculator. Output swings between +7.2 V and −7.2 V. NAND gates of ICs are connected to form free-running multivibrator, with self-starting gate C ensuring that clock waveform is available as soon as supply voltage is applied.—T. J. Terrell, Clock Generator for Electronic Calculators, *Wireless World,* Dec. 1975, p 575.

1-Hz CLOCK WITH BATTERY BACKUP—Circuit normally produces output pulses at 1-s intervals with basic accuracy corresponding to that of power-line frequency. Programmable 8260 timer operates as divide-by-60 counter producing output swing compatible with TTL or 5-V CMOS loads. With backup power applied to QR gate D_1-D_2, circuit operates reliably at 1 s over supply range of 5–15 V. Power drain is minimized at ±5 V.—W. G. Jung, "IC Timer Cookbook," Howard W. Sams, Indianapolis, IN, 1977, p 214–215.

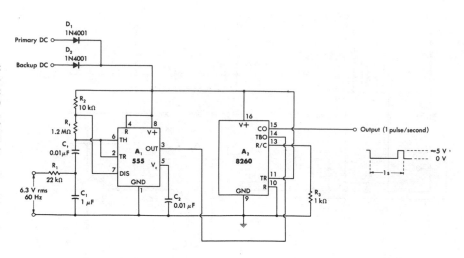

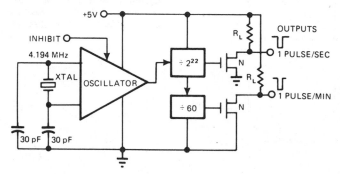

WATCH-CRYSTAL TIMER—When used with standard 4.194-MHz watch crystal, Intersil 7213 crystal-controlled timer generates outputs of 1 pulse per second and 1 pulse per minute, using internal divider chain. CMOS dynamic and static dividers keep power dissipation under 1 mW with 5-V supply.—B. O'Neil, IC Timers—the "Old Reliable" 555 Has Company, *EDN Magazine,* Sept. 5, 1977, p 89–93.

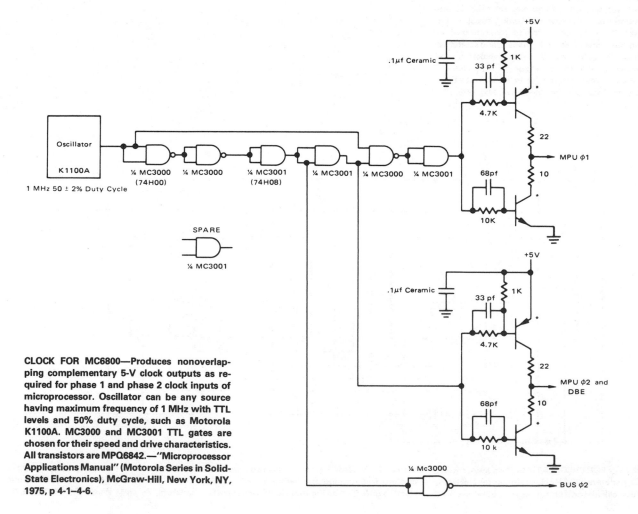

CLOCK FOR MC6800—Produces nonoverlapping complementary 5-V clock outputs as required for phase 1 and phase 2 clock inputs of microprocessor. Oscillator can be any source having maximum frequency of 1 MHz with TTL levels and 50% duty cycle, such as Motorola K1100A. MC3000 and MC3001 TTL gates are chosen for their speed and drive characteristics. All transistors are MPQ6842.—"Microprocessor Applications Manual" (Motorola Series in Solid-State Electronics), McGraw-Hill, New York, NY, 1975, p 4-1–4-6.

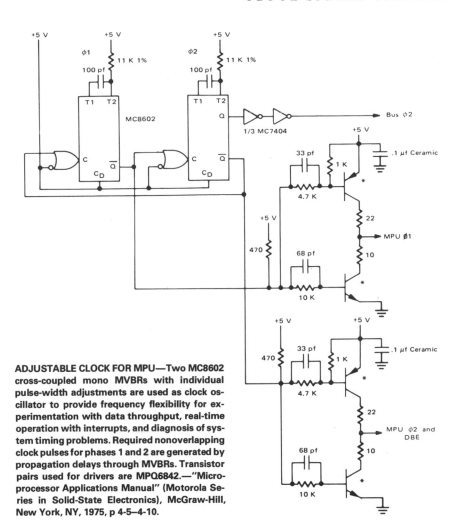

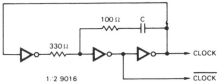

RC CLOCK—Simple TTL clock generator is suitable for most TTL systems. Requires only half of hex inverter package and three passive components. Clock frequency depends on value of C: 200 pF gives 5 MHz; 1600 pF gives 1 MHz; 0.018 μF gives 100 kHz; and 0.18 μF gives 10 kHz.—Circuits, *73 Magazine,* Aug. 1974, p 99.

ADJUSTABLE CLOCK FOR MPU—Two MC8602 cross-coupled mono MVBRs with individual pulse-width adjustments are used as clock oscillator to provide frequency flexibility for experimentation with data throughput, real-time operation with interrupts, and diagnosis of system timing problems. Required nonoverlapping clock pulses for phases 1 and 2 are generated by propagation delays through MVBRs. Transistor pairs used for drivers are MPQ6842.—"Microprocessor Applications Manual" (Motorola Series in Solid-State Electronics), McGraw-Hill, New York, NY, 1975, p 4-5–4-10.

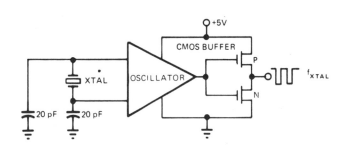

CRYSTAL-CONTROLLED TIMER—Intersil 7209 crystal oscillator provides buffered CMOS output capable of driving over five TTL loads at any crystal frequency up to 10 MHz. Used in applications requiring high-accuracy buffered timing signals for system clocks.—B. O'Neil, IC Timers—the "Old Reliable" 555 Has Company, *EDN Magazine,* Sept. 5, 1977, p 89–93.

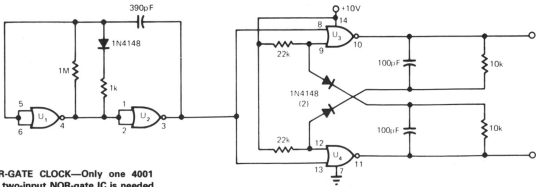

CMOS FOUR-GATE CLOCK—Only one 4001 CMOS quad two-input NOR-gate IC is needed for symmetrical complementary-output clock having good temperature and supply stability. Gates U_1 and U_2 form astable oscillator producing positive-going pulses used to trigger divide-by-2 flip-flop U_3-U_4. Circuit will operate over wide range of supply voltages and temperatures.—M. Eaton, Symmetrical CMOS Clock Is Inexpensive, *EDN Magazine,* March 20, 1974, p 80 and 83.

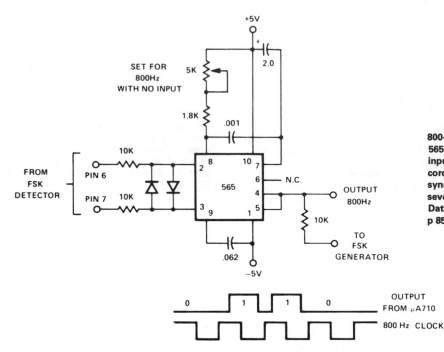

800-Hz CLOCK FOR CASSETTE RECORDER— 565 PLL is set for free-running at 800 Hz with no input. When data pulses extracted from FSK recorded data on cassette tape are fed in, clock is synchronized to data and stays in sync for up to seven 0s in succession.—"Signetics Analog Data Manual," Signetics, Sunnyvale, CA, 1977, p 859–860.

60-Hz GLITCH-FREE CLOCK—Circuit generates complementary gated 60-Hz clock pulses that are always wider than 2 μs, without glitches even if gate is enabled or disabled during clock pulse. Accuracy depends on stability of power-line frequency.—R. I. White, Gated 60 Hz Clock Avoids Glitches, *EDN/EE Magazine*, Nov. 1, 1971, p 52.

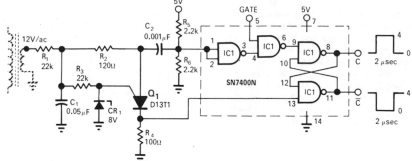

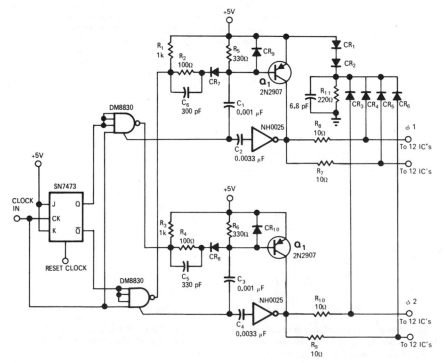

DRIVER FOR 24 MOS REGISTERS—With input of clock pulses preshaped to width of about 150 ns, circuit shown will generate 17-V 1.5-A clock signal required for driving 1024-bit serial MOS memories or shift registers. Article traces operation of circuit. All diodes are 1N3064.—R. D. Hoose and G. L. Anderson, Clock Driver for MOS Shift Registers, *EDN/EE Magazine*, Dec. 15, 1971, p 56–57.

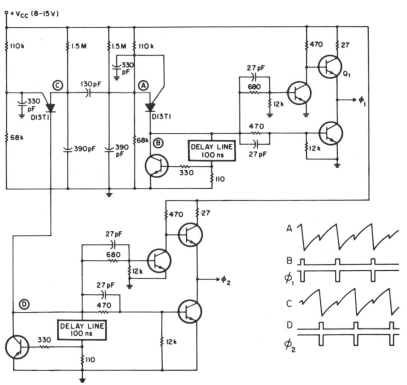

2.5-kHz CLOCK—Requires only 3-mW power to generate fast 15-V two-phase rectangular pulses with 100-ns widths and 20-ns rise and fall times. Suitable for many MOSFET shift registers. Uses two D13T1 programmable UJT oscillators, each having four-transistor driver stage using 2N2369 transistors. Oscillators are cross-synchronized by 130-pF capacitor. Timing and bias networks of UJTs set pulse repetition frequency.—G. A. Altemose, Low-Power Two-Phase Clock, *EDN/EEE Magazine*, May 15, 1971, p 49.

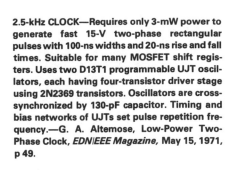

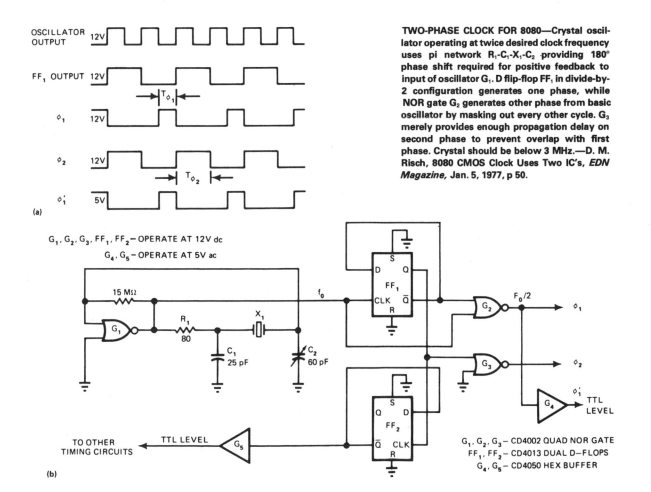

TWO-PHASE CLOCK FOR 8080—Crystal oscillator operating at twice desired clock frequency uses pi network R_1-C_1-X_1-C_2 providing 180° phase shift required for positive feedback to input of oscillator G_1. D flip-flop FF_1 in divide-by-2 configuration generates one phase, while NOR gate G_2 generates other phase from basic oscillator by masking out every other cycle. G_3 merely provides enough propagation delay on second phase to prevent overlap with first phase. Crystal should be below 3 MHz.—D. M. Risch, 8080 CMOS Clock Uses Two IC's, *EDN Magazine*, Jan. 5, 1977, p 50.

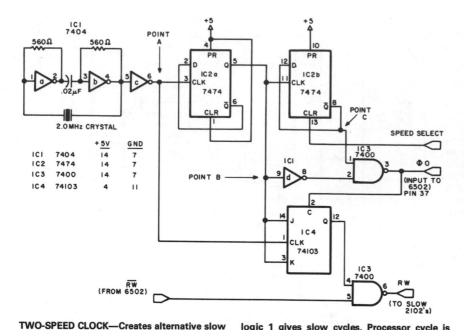

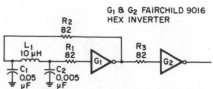

TWO-INVERTER COLPITTS—Simple low-cost clock uses two inverters in LC circuit. G₁ operates as Colpitts oscillator, with C₁ setting feedback level and L₁-C₂ setting frequency. Low DC resistance path through R_1, R_2, and L_1 provides high negative DC feedback around G₁ and biases it into linear region. G₂ squares output of G₁ to appropriate TTL levels. For values shown, output period is 1.2 μs and rise and fall times are under 20 ns.—C. A. Herbst, TTL Inverter Makes Stable Colpitts Oscillator, *EDN/EEE Magazine*, May 15, 1971, p 50.

TWO-SPEED CLOCK—Creates alternative slow clock cycle for MOS Technology 6502 processor on KIM-1 microcomputer card under control of SPEED SELECT line generated by slow memories. Control logic of 0 gives fast cycles, while logic 1 gives slow cycles. Processor cycle is maintained for 1 μs for fast memory access, but cycle is automatically stretched to 2 μs for slower 2102s.—Y. M. Gupta, True Confessions: How I Relate to KIM, *BYTE*, Aug. 1976, p 44–48.

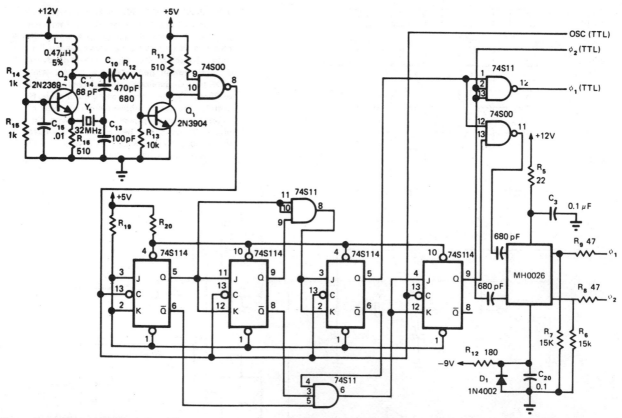

TWO-PHASE CLOCK FOR 8080—Typical clock design for Intel 8080 microprocessor system uses carefully designed logic sequence and level translation in combination with 32-MHz crystal to generate two-phase 12-V clock waveform.—S. G. Brannan, μP Clock Generators—Buy or Build? *EDN Magazine*, Sept. 20, 1975, p 53–55.

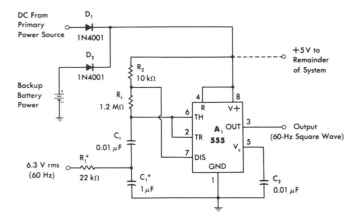

*$C_1 \gg C_t$. Choose R_1 for $> 1/3$ V+ p-p across C_1.

BATTERY BACKUP POWER—Provides 60-Hz square-wave output for driving electronic timekeeping circuits from battery during AC power failure. 555 timer is connected as square-wave astable MVBR normally locked to incoming 60-Hz power frequency. R_t should be trimmed for zero beat with 60-Hz source. When primary power fails, reference voltage disappears and effect of C_t on frequency is minimized. 555 now oscillates at frequency determined by R_t and C_t, which is 60 Hz. Use 6-V battery.—W. G. Jung, "IC Timer Cookbook," Howard W. Sams, Indianapolis, IN, 1977, p 201–203.

SYNCHRONIZER—For each switch closure, circuit produces one output pulse that is one clock period wide, synchronized with clock. When switch is closed, debouncing latch using 7400 gates goes high and makes flip-flop B high. Next clock pulse makes flip-flop C high and resets flip-flop B. At next clock pulse, flip-flop C goes low to complete cycle of operation.—E. E. Hrivnak, House Cleaning the Logical Way, *73 Magazine*, Aug. 1974, p 85–90.

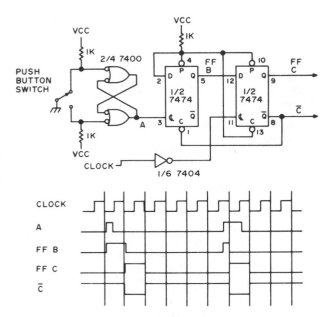

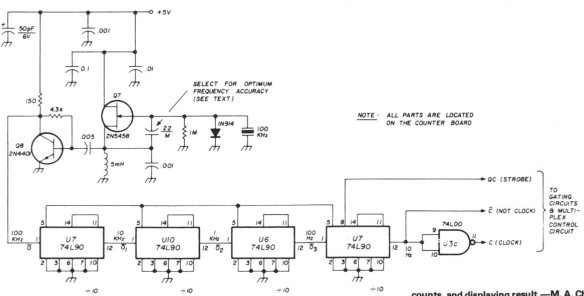

10-Hz CLOCK—Stable and accurate clock is generated by high-precision 100-kHz crystal oscillator and decade divider chain. Used in 20-meter receiver as part of digital display system that shows frequency of received signal after counting HFO, LO, and BFO outputs, summing counts, and displaying result.—M. A. Chapman, High Performance 20-Meter Receiver with Digital Frequency Readout, *Ham Radio*, Nov. 1977, p 56–65.

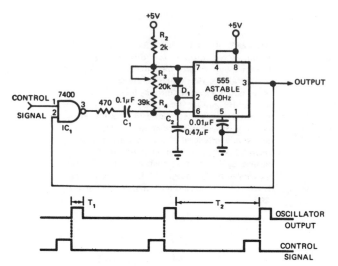

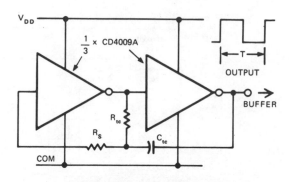

CLOCK FOR REGULATED SUPPLY—Single RCA CD4009A serves as clock generator and mono for driving regulated power supply having foldback current-limiting protection. R_{tc} and C_{tc} are major frequency-determining components. R_s should be made equal to or greater than $2R_{tc}$. Article gives equation for period T of oscillator, which ranges from about 2.2 to 2.5 times $R_{tc}C_{tc}$.—J. L. Bohan, Clocking Scheme Improves Power Supply Short-Circuit Protection, *EDN Magazine*, March 5, 1974, p 49–52.

VTR CLOCK—Locked oscillator using only two-input NAND gate and 555 timer provides logic clock signal for videotape recorder. Vertical sync signal, stripped from video information recorded on tape, is used as control signal. C_1 controls locking range for free-running frequency of 555. When C_2 is charging (555 output is high), R_2 and D_1 determine time constant T_1. During discharge of C_2, D_1 is reverse-biased and discharge time constant T_2 is determined by R_3 and R_4.—L. Saunders, Locked Oscillator Uses a 555 Timer, *EDN Magazine*, June 20, 1975, p 114.

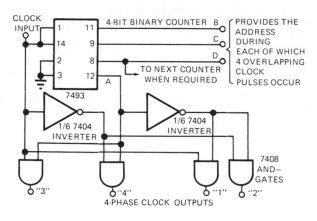

FOUR-PHASE CLOCK—Provides expandable 3-bit binary output and four overlapping clock pulses for each unique binary output. A-output of 7493 binary counter is used along with clock input to form four-phase overlapping clock function. Article includes timing diagram that shows sequence of output pulses. Developed for use in addressing multiplexers, ROMs, and other digital units.—B. Brandstedt, Clock Pulse Generator Has Addressable Output, *EDN Magazine*, Dec. 15, 1972, p 42.

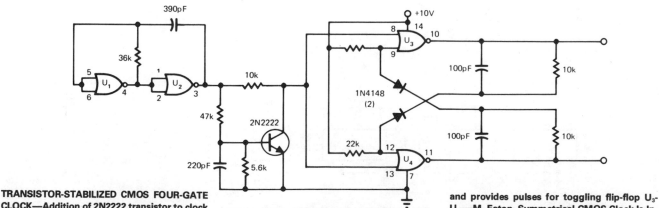

TRANSISTOR-STABILIZED CMOS FOUR-GATE CLOCK—Addition of 2N2222 transistor to clock using 4001 CMOS quad two-input NOR-gate IC boosts temperature stability to 0.05%/°C and supply stability to 0.05%/V. Transistor circuit differentiates output signal of oscillator U_1-U_2 and provides pulses for toggling flip-flop U_3-U_4.—M. Eaton, Symmetrical CMOS Clock Is Inexpensive, *EDN Magazine*, March 20, 1974, p 80 and 83.

CHAPTER **13**
Code Circuits

Covers Morse-code circuits as used in amateur, maritime, and other CW communication applications for keyers, monitors, code generators and regenerators, decoders, practice oscillators, CW filters, and call-letter generators. For circuits capable of handling CW along with other types of modulation and for circuits handling other types of codes, see also Filter, IF Amplifier, Keyboard, Memory, Microprocessor, Single-Sideband, Receiver, Transceiver, and Transmitter chapters.

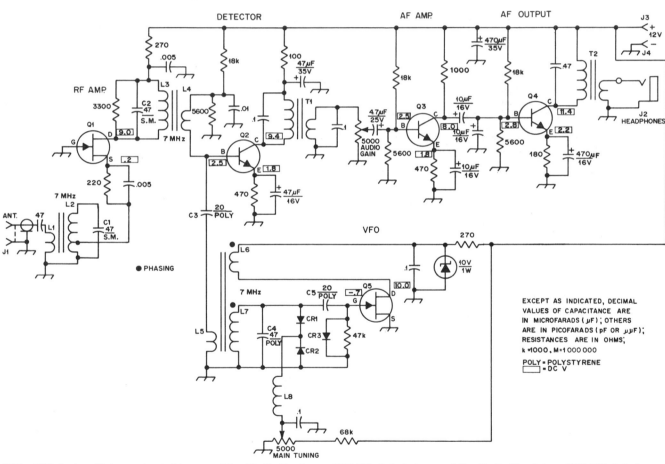

CR1 - CR3, incl. — High-speed switching diode (Radio Shack type 276-1620).
J1 — RCA-type phono jack.
J2 — 1/4-inch phone jack.
J3, J4 — Binding post.
L1 — 3 turns insulated hookup wire wound over (ground) end of L2.
L2 — Radio Shack type 273-101 rf choke. Tap at 4 turns above ground end.

L3 — Radio Shack type 273-101 rf choke.
L4 — 4 turns insulated hookup wire wound over cold end of L3.
L5 — 5 turns insulated hookup wire wound over ground end of L7.
L6 — 4 turns insulated hookup wire wound adjacent to high end of L7.
L7 — Radio Shack type 273-101 rf choke with six of the original turns removed.

L8 — Radio Shack type 273-102 rf choke.
Q1, Q5 — JFET (Radio Shack type RS-2035).
Q2 - Q4, incl. — Transistor (Radio Shack type 276-1617).
T1 — Audio transformer (Radio Shack type 273-1378).
T2 — Audio transformer (Radio Shack type 273-1380).

40-METER DIRECT-CONVERSION—Simple, foolproof circuit design uses discrete components mounted on printed-circuit board shaped to fit in oval herring can. Single 7-MHz RF stage and voltage-tuned VFO feed product detector

Q2 that drives 2-stage AF amplifier having peak response at about 650 Hz for most comfortable CW listening. VFO uses Armstrong or tickler-feedback circuit, with CR1 and CR2 connected as voltage-variable-capacitance diodes. Zener

regulator powers VFO circuit for good frequency stability. Receiver will tune any 100-kHz segment of 40-meter band.—J. Rusgrove, The Herring-Aid Five, *QST*, July 1976, p 20–23.

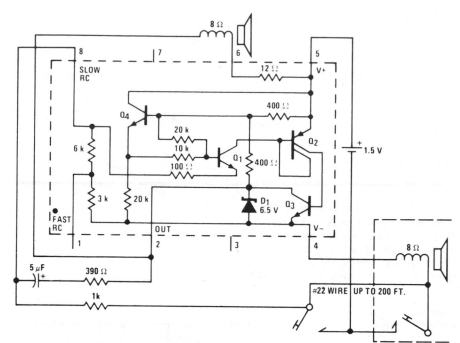

MORSE-CODE SET—National LM3909 flasher IC is connected as tone oscillator that simultaneously drives loudspeakers at both sending and receiving ends of wire line used for Morse-code communication system. Single alkaline penlight cell lasts 3 months to 1 year depending on usage. Three-wire system using parallel telegraph keys eliminates need for send-receive switch. Tone frequency is about 400 Hz.—"Linear Applications, Vol. 2," National Semiconductor, Santa Clara, CA, 1976, AN-154, p 5–6.

CQ ON TAPE—Frequently used code message such as amateur radio CQ call is recorded by keying audio oscillator with desired message and picking up oscillator output with microphone of endless-loop cassette or other tape recorder. Rewound recording is played back through single-transistor stage connected as shown for driving keying relay of transmitter. Circuit requires shielding.—Circuits, *73 Magazine*, July 1977, p 34.

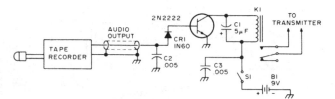

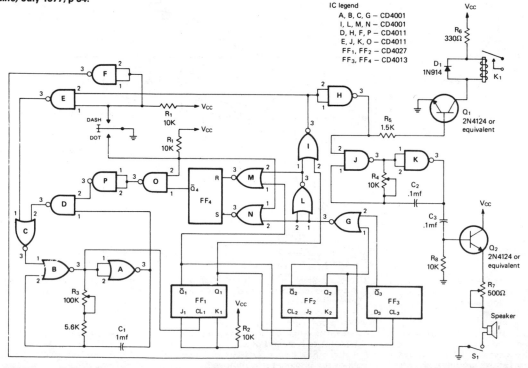

KEYER WITH MEMORY—Includes sidetone oscillator and dash-dot memory along with variable speed, automatic spacing, and self-completing dots and dashes. If dot paddle is pressed and released while keyer is generating dash, dot is generated with correct spacing after dash is completed. Gates A, B, and C form gated MVBR. Gates D, E, O, and P serve to complete characters. JK flip-flops FF_1 and FF_2, D flip-flop FF_3, and gates F, G, and L provide character-shaping required for dash-dot memory using gates M, N, and RS flip-flop FF_4. Gates J and K generate audio sidetone. K_1 is B & F Enterprises ERA-21061 SPST reed relay. Supply can be 9-V battery.—T. R. Crawford, A Low-Power Cosmos Electronic Keyer in Two Versions, *CQ*, Nov. 1975, p 17–24.

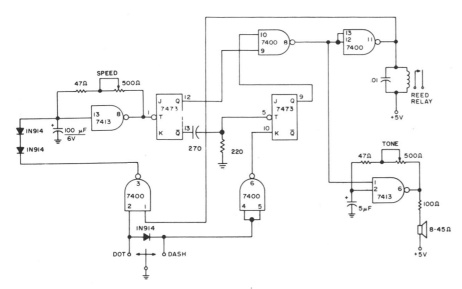

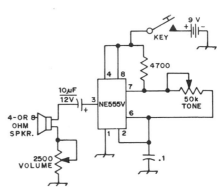

KEYER—Uses gating and flip-flop functions to generate dots and dashes under control of gated clock. SN7413 Schmitt trigger is connected as relaxation oscillator. Circuit provides minimum spacing between dots and dashes regardless of paddle movements.—A. D. Helfrick, A Simple IC Keyer, *73 Magazine*, Dec. 1973, p 37–38.

TIMER FOR CODE PRACTICE—Signetics NE555V timer operating on 9-V supply serves as AF oscillator providing adequate volume for classroom instruction. Output tone can be varied from several hundred to several thousand hertz.—J. Burney, Code Practice Oscillator, *QST*, July 1974, p 37.

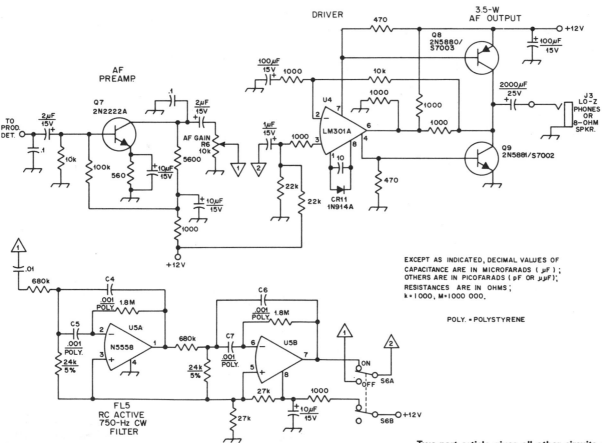

EXCEPT AS INDICATED, DECIMAL VALUES OF CAPACITANCE ARE IN MICROFARADS (μF); OTHERS ARE IN PICOFARADS (pF OR μμF); RESISTANCES ARE IN OHMS; k = 1 000, M = 1 000 000.

POLY. = POLYSTYRENE

3.5 W FOR CW—Discrete devices minimize distortion and eliminate fuzziness while listening to low-level CW signals in communication receiver covering 1.8–2 MHz. RC active bandpass filter peaked at 800 Hz improves S/N ratio for weak signals. Adjust BFO of receiver to 800 Hz. Two-part article gives all other circuits of receiver.—D. DeMaw, His Eminence—the Receiver, *QST*, Part 2—July 1976, p 14–17 (Part 1—June 1976, p 27–30).

SENSOR KEYER—Skin resistance of about 10K creates dashes when finger touches grid pattern on left side of paddle and dots when other finger touches pattern on other side. Transistors act as solid-state switches. Developed for use with Heathkit CW keyer HD-10. Supply is 10 V, obtained from 10-V zener connected through appropriate dropping resistor to higher-voltage source. Article covers construction of paddle by etching printed-wiring board.—T. Urbizu, Try a Sensor Keyer, *73 Magazine*, Jan. 1978, p 184–185.

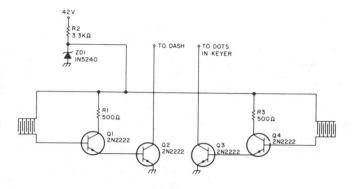

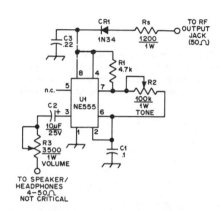

AF OSCILLATOR MONITORS CW—Can be added to any transceiver not already having built-in sidetone oscillator, to hear keying of transmitter. RF input from transmitter is rectified by CR1 to provide about 6-VDC supply. Keying of carrier on and off turns NE555 AF oscillator on and off correspondingly.—J. Arnold, A CW Monitor for the Swan 270, *QST*, Aug. 1976, p 44.

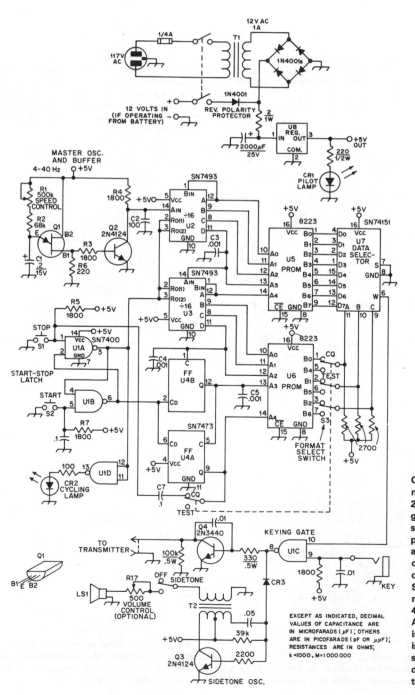

CQ CALL SYNTHESIZER—Uses only two Signetics 8223 256-bit PROMs for storing up to 2048 bits of code information, for automatic generation of Morse-code CQ calls, test messages, and other frequently used messages. Repeated words are stored in only one location and selected as needed, to quadruple capacity of memory. PROMs can be programmed in field or custom-programmed by manufacturer. Speed and timing of code characters are determined by UJT oscillator Q1, variable from about 4 to 40 Hz or 5 to 50 WPM. CR1 and CR2 are Archer (Radio Shack) 276-042 or equivalent. CR3 is 1N34A, 1N270, or equivalent germanium. Q1 is Motorola MU4891 or equivalent. Article describes circuit operation and programming in detail.—J. Pollock, A Digital Morse Code Synthesizer, *QST*, Feb. 1976, p 37–41.

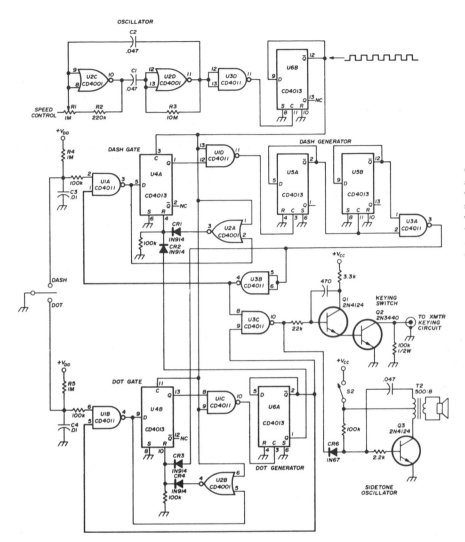

CMOS KEYER—Draws only 0.4 mA on standby and 2 mA with key down if supply is 10 V. Will work properly with 4 to 15 V. Features include self-completing dots, dashes, and spaces, along with sidetone generator and built-in transmitter keying circuit. Ratio of dashes to dots is 3:1, and space has same duration as dot. Time base of keyer is generated by NOR gates U2C and U2D connected as class A MVBR. Frequency of oscillator is inversely linear with setting of R1. Inverter U3D buffers oscillator and squares its output. Flip-flop U6B divides frequency by 2 and provides clock source with perfect 50% duty cycle. Once enabled, gates ensure completion along with following space. Article gives power supply circuit operating from AC line and 12-V battery.—J. W. Pollock, COSMOS IC Electronic Keyer, *Ham Radio*, June 1974, p 6–10.

SIDETONE MONITOR—Mostek MK5086N IC is used with crystal in range from 2 to 3.5 MHz as signal generator driving FET audio amplifier. Switch S1 gives choice of four AF tones, determined by dividing crystal frequency in hertz by 5120 for T1, 4672 for T2, 4234 for T3, and 3776 for T4. Can also be used as code practice set and as audio signal generator.—J. Garrett, A Sidetone Monitor-Oscillator-Audio Generator, *QST*, June 1978, p 43.

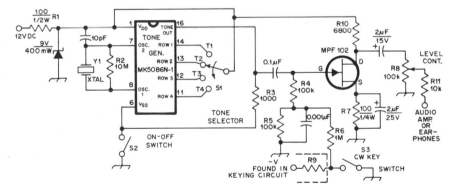

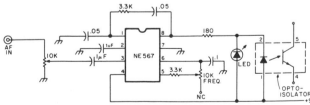

REGENERATED CW—Audio oscillator whose frequency can be varied is keyed in accordance with incoming CW signal, to give clean locally generated audio signal without background noise and interference. NE567 phase-locked loop serves as tunable audio filter and LED switch driver for activating NE555 variable-frequency tone oscillator. LED serves as visual tuning aid to indicate that PLL is locked on to incoming signal.—Regenerated CW, *73 Magazine*, Dec. 1977, p 152–153.

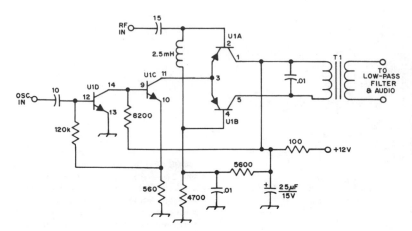

PRODUCT DETECTOR—Designed for use in 40-meter CW direct-conversion receiver, in which oscillator input is from 3.5—4 MHz VFO. U1 is RCA CA3046 transistor quad. Circuit provides bias stabilization for constant-current transistor and some amplification of AF output. T1 is audio transformer.—A. Phares, The CA3046 IC in a Direct-Conversion Receiver, *QST*, Nov. 1973, p 45.

TONE DECODER—Decodes audio output of amateur radio receiver. Resulting audio tone burst corresponds to CW signal being received, with tone frequency varying with receiver tuning. Center frequency of NE567 phase-locked loop is adjusted with R1. Audio is translated into digital format of 1s and 0s, with tones for 0s. Output can be fed into computer for automatic translation of Morse code and printout as text.—W. A. Hickey, The Computer Versus Hand Sent Morse Code, *BYTE*, Oct. 1976, p 12—14 and 106.

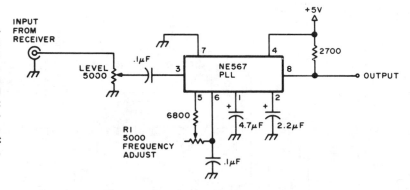

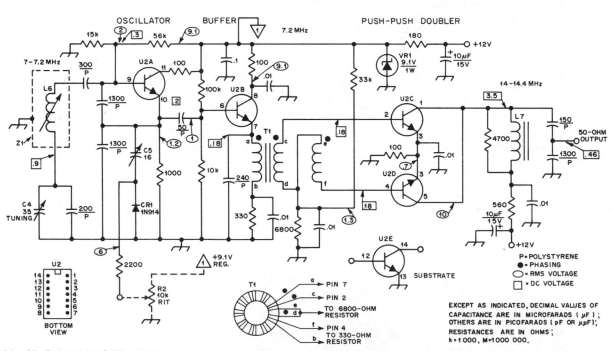

C4 — 35-pF air variable (Millen 26035 or Hammarlund HF-35).
C5 — 16-pF air trimmer, pc-board mount Johnson 187-0109-005.
CR1 — Silicon high-speed switching diode, 1N914 or equiv.
L6 — Slug-tuned inductor, 3.6 to 8.5 µH (Mil-

ler 42A686CBI or equivalent). Use shield can (35-mm film canister or Miller S-33).
L7 — Toroidal inductor, 0.9 µH. Use 12 turns of No. 24 enam. wire on a T50-6 toroid core. (See *QST* ads for toroid suppliers — Amidon, G. R. Whitehouse and Palomar Engrs.).

R2 — Optional circuit (see text). 10,000-ohm linear-taper composition control.
T1 — Trifilar-wound trans. 2 µH, 20 turns, twisted six turns per inch. No. 28 enam. wire on a T50-2 toroid core (see text).
U2 — RCA CA3045 array IC.
VR1 — Zener diode, 9.1 V, 1 W.

BFO FOR 20 METERS—Uses CA3045 transistor array, with U2A as series-tuned Clapp oscillator covering 7–7.2 MHz. Tuned emitter-follower U2B provides push-pull drive at 7 MHz to bases of push-push doubler U2C-U2D. Output of BFO is applied to product detector rather than to mixer of receiver. Audio signal from detector is frequency difference between BFO and incoming signal, typically 700 Hz for CW reception. Article covers construction and adjustment.—D. DeMaw, Understanding Linear ICs, *QST*, Feb. 1977, p 19–23.

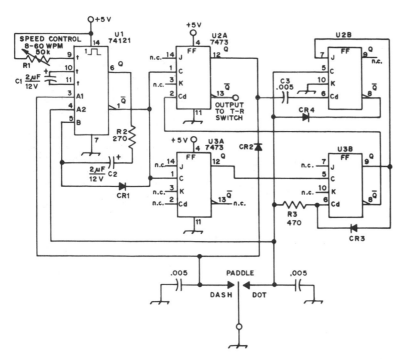

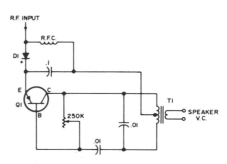

KEYER WITH DOT MEMORY—Features include self-completing characters, exact timing of characters, and dot memory. Timing circuit uses 74121 mono MVBR U1, serving dot generator and output stage U2A, dot memory U2B, and dash generator U3A-U3B. U2 and U3 are 7473 dual JK flip-flops. Length of timing pulse is determined by R1-C1, with R1 controlling speed of keyer. Pulse-width stability at all speeds is better than 5% between first and all following pulses. Dot memory U2B allows keying of dot at any time, even if dash has not yet been completed. Dot is held in memory and keyed out automatically after dash. Diodes are 1N914.—J. H. Fox, An Integrated Keyer/TR Switch, *QST*, Jan. 1975, p 15–20.

CODE MONITOR—Works with any transmitter, regardless of type of keying. Use any good PNP transistor. With NPN transistor, reverse connections to diode. Frequency of tone gets higher as resistance of 250K pot is reduced. Monitor is turned off at minimum resistance. Enough RF to operate monitor can be obtained simply by connecting it to chassis of receiver or transmitter.—J. Smith, Yet Another Code Monitor, *73 Magazine*, Sept. 1971, p 58.

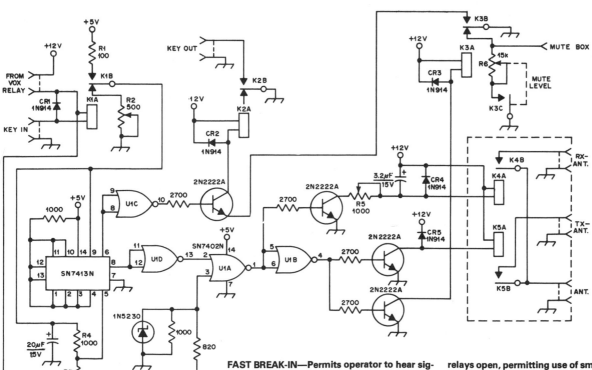

FAST BREAK-IN—Permits operator to hear signals even between dots while calling DX station, so call can be stopped if DX station answers someone else. Timing circuit ensures that transmitter is not producing power when relays open, permitting use of small high-speed relays. K1-K4 are common reed relays. K5 should have contacts rated for 300 VAC at 500 mA.—A. Pluess, A Fast QSK System Using Reed Relays, *QST*, Dec. 1976, p 11–12.

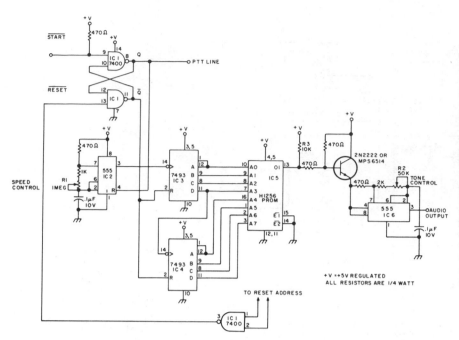

CW CALL GENERATOR—Basic CW identifier uses two gates of 7400 to form starting flip-flop of automatic message generator. Provides adjustable speed and tone, with up to 256 bits of storage in Harris H1256 PROM. IC2 is 555 astable MVBR providing clock signal for driving two 7493 4-bit binary counters that address PROM. When counters reach maximum address of 255, next clock count makes PROM restart at address 0. Each address turns on tone oscillator for one clock period, producing one dit. Three addresses in row turn on tone for three clock periods, producing one dah. Space between dits and dahs of same letter is equal to one dit, letter space is three dits, and word space is six dits (2 dahs). Thus, W takes nine addresses. Article describes operation in detail and tells how to modify circuit for use as RTTY message generator.—R. B. Joerger, PROM Message Generator for RTTY, *73 Magazine,* March 1977, p 94–98.

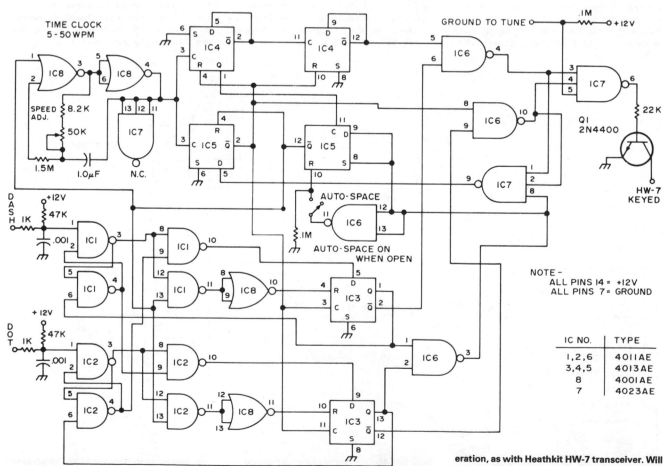

CMOS KEYER—Features include self-completing dots and dashes, dot and dash memories, iambic operation, dot and dash insertion, and automatic character spacing, all achieved with low-power CMOS digital devices that are compatible with low-power (QRP) transceiver operation, as with Heathkit HW-7 transceiver. Will operate directly from 3–15 V batteries of QRP transceiver, without regulation.—G. Hinkle, The QRP Accu-Keyer, *73 Magazine,* Aug. 1975, p 58–60.

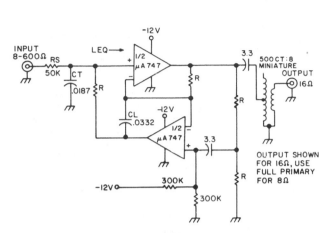

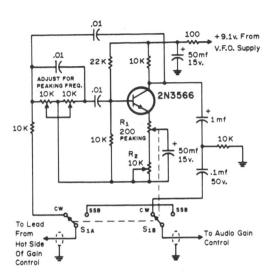

CW FILTER—Simple single-section parallel-tuned active filter uses negative-impedance converter or gyrator to replace hard-to-get inductor of passive code filter. Capacitor CL is gyrated from 0.0332 μF to effective inductance of 1.87 H. Filter has 6-dB gain at resonance and essentially zero output impedance. Bandpass is 85 Hz centered at about 865 Hz. Uses single −12 V supply. Resistors R are 7.5K, matched to about 2%.—N. Sipkes, Build This CW Filter, *73 Magazine*, June 1977, p 55.

REGENERATIVE CW FILTER—Can be added between product detector and volume control of SSB receiver or transceiver that does not have CW filter. Just before oscillation occurs, gain becomes extremely high with very narrow bandpass. Regeneration and bandpass can be adjusted as required. Filter typically has 40-Hz bandwidth centered on 800 Hz.—R. A. Yoemans, Further Enhancing the Yaesu FTDX-560 Transceiver, *CQ*, July 1972, p 16–18 and 20–22.

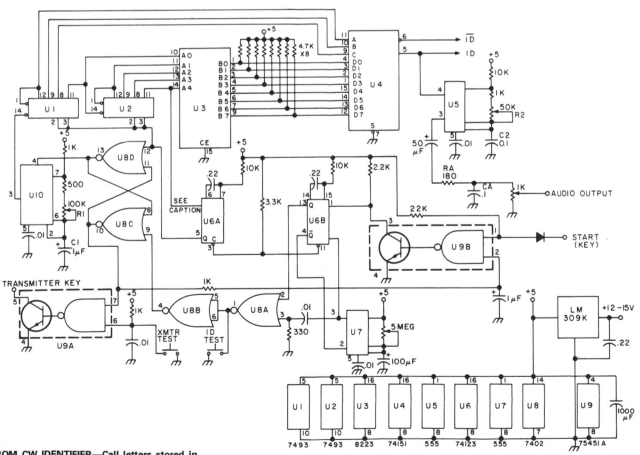

PROM CW IDENTIFIER—Call letters stored in 8223 PROM U3 drive 74151 multiplexer/data selector U4 for keying NE555 audio oscillator U5 which feeds transmitter mike input through RA. Timed holdoff keeps identifier from being rekeyed within specified time period, with reidentifying at end of period. Article covers operation of circuit and gives construction and programming details.—W. Hosking, ID with a PROM, *73 Magazine*, Nov. 1976, p 90–92.

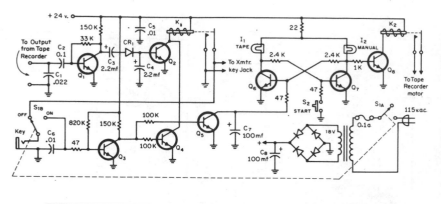

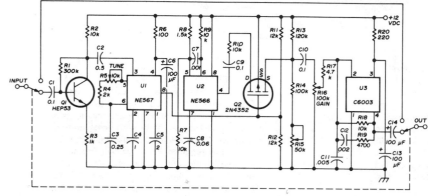

RECORDED-CODE KEYER—CW message recorded on cassette tape with keyed tone generator is played back for keying transmitter with frequently repeated messages. Transmitted signal is essentially perfect reproduction of recording. Automatic transmission can be stopped at any point by tapping hand key; this deactivates tape recorder for manual keying of transmitter. All transistors are 2N2222 or equivalent BCY58. Use 24-V reed relays with 1K coil resistance. Pilot lamps are 24 V rated up to 3 W. CR_1 is silicon diode. Whenever recorder emits beep, positive signal appears at base of Q_2, making it conduct and activate K_1 whose contacts go to key jack of transmitter.—A. Day, An Audio Tape-Controlled CW Keyer, *CQ*, Nov. 1971, p 31–32.

PLL CODE REGENERATOR—Permits comfortable listening to CW signals deeply embedded in noise, hash, and interference, by detecting one particular CW transmission and keying independent oscillator with its signal. Consists of signal amplifier Q1, narrow-band PLL frequency detector and trigger U1, PLL function generator U2, gate Q2, and AF output amplifier U3. In absence of triggering signal, output pin 8 of U1 presents high impedance to ground. With triggering frequency, output presents low impedance to ground. Oscillator U2 is gated by U1 through Q2.—C. R. Lewart and R. S. Libenschek, CW Regenerator for Interference-Free Communications, *Ham Radio*, April 1974, p 54–56.

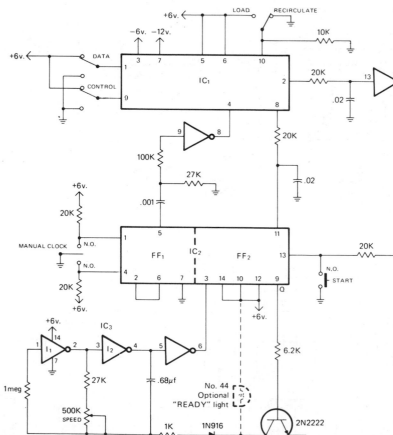

DATA and CONTROL switches to ground and **RECIRCULATE** switch to hold, then pressing **START** button (with speed control set fast) until register is full of zeros. Release START button and switch control to +6 V for about 10 s. Now set CONTROL to ground and start programming. To enter dah, switch manual clock through three complete up-center/down-center cycles, switch data line to ground, and cycle manual clock through one cycle to insert space into memory. To enter dit, switch manual clock through only one up/down cycle. Continue until entire message is entered in register, switch to RECIRCULATE, and push START switch. Message will now be sent in perfect code at any desired speed.—B. P. Vandenberg, An Inexpensive Memory Keyer for Contests, *CQ*, May 1976, p 50–51.

200-BIT MEMORY—Drawback of volatile code-storing memory is offset by low cost and comparative ease of programming. IC_1 is Signetics 2511D MOS dual 200-bit shift register whose digital levels are shifted one position for each clock pulse. Inverters I_1 and I_2 of CMOS hex inverter IC_3 form variable-frequency gated square-wave clock oscillator controlled by state of flip-flop FF_2 in 74C74 CMOS dual D flip-flop. Shift register must first be cleared by setting

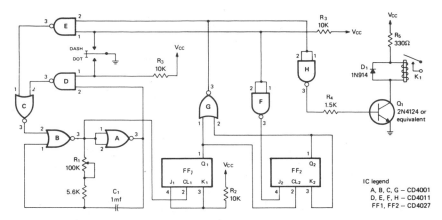

CMOS KEYER—Gives variable speed, automatic spacing, and self-completing dots and dashes. Gates C, B, and A form gated MVBR. Dot paddle initiates dot by making output of gate D go low, while gate E goes low for dash. Character-shaping section (gates G, H, and F with JK flip-flops FF₁ and FF₂) takes square pulses from pin 3 of MVBR gate B and gives perfectly spaced dots and dashes. FF₁ divides clock pulses by 2, making dot equal to one complete period of MVBR. FF₂ divides output of FF₁ by 2, and outputs of both flip-flops are logically ORed to provide dashes. K₁ is SPST reed relay. Supply can be 9-V battery.—T. R. Crawford, A Low-Power Cosmos Electronic Keyer in Two Versions, *CQ*, Nov. 1975, p 17–24.

IC legend
A, B, C, G — CD4001
D, E, F, H — CD4011
FF1, FF2 — CD4027

KEYER OSCILLATOR—Oscillator or clock starts when key is closed and can be held until dot, dash, or space is completed. U1 is SN7400, and U2 is SN74L04. Diodes at input of U2A form OR gate that controls oscillator. These inputs can be used to keep oscillator running for self-completing action. Time constant, set by C1 and R1, is 4 ms which is width of clock pulse. Values for C2, R2, and R3 give pulse repetition rate of 50 to 95 ms, corresponding to about 12 to 24 WPM. For higher speeds, reduce values of C2 and R2. CR1 and CR2 are signal diodes that prevent first pulse from being different; 250-pF capacitor on output prevents noise spikes from triggering keyer circuits falsely.—J. T. Miller, Integrated-Circuit Oscillator, *Ham Radio*, Feb. 1978, p 77.

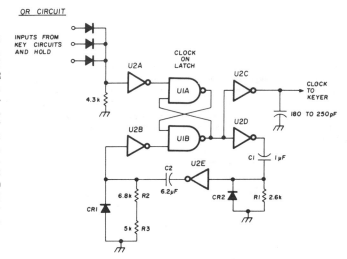

OR CIRCUIT

INPUTS FROM KEY CIRCUITS AND HOLD

CLOCK ON LATCH

CLOCK TO KEYER

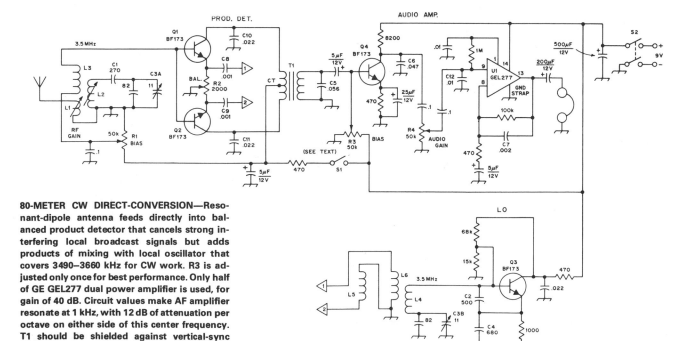

PROD. DET.

AUDIO AMP.

LO

80-METER CW DIRECT-CONVERSION—Resonant-dipole antenna feeds directly into balanced product detector that cancels strong interfering local broadcast signals but adds products of mixing with local oscillator that covers 3490–3660 kHz for CW work. R3 is adjusted only once for best performance. Only half of GE GEL277 dual power amplifier is used, for gain of 40 dB. Circuit values make AF amplifier resonate at 1 kHz, with 12 dB of attenuation per octave on either side of this center frequency. T1 should be shielded against vertical-sync magnetic fields of TV sets up to 60 feet away. T1 is transistor push-pull output transformer. Use twisted pair for connecting 1-1 and 2-2, to prevent imbalance effects. Article gives construction details for L1-L4.—B. Pasaric, A New Front End for Direct-Conversion Receivers, *QST*, Oct. 1974, p 11–14.

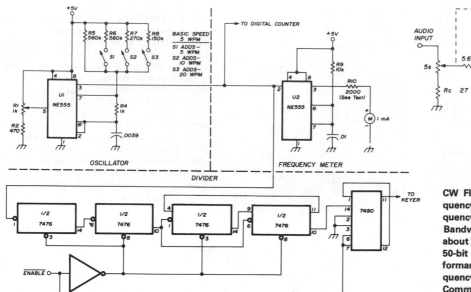

CW FILTER—Variable-bandwidth variable-frequency audio filter can be tuned to center frequency anywhere in range from 300 to 3000 Hz. Bandwidth can be as narrow as 50 Hz, which is about 3 times theoretical minimum of 15 Hz for 50-bit words and Morse code at 25 WPM. Performance can be improved over entire frequency range by using 741 opamp.—R. Skelton, Comments, *Ham Radio*, June 1975, p 56–57.

KEYER SPEED CONTROL—Electronic time base provides direct readout of keyer speed for 5 to 40 WPM in increments of 5 WPM by noting positions of three speed-control switches. Vernier adjustment pot R1 can be used for continuous speed adjustment if desired. Analog frequency meter provides alternate direct indication of keyer speed on milliammeter that can be calibrated in words per minute. Frequency of NE555 oscillator is 100 times keyer speed. Keyer clock is obtained by dividing oscillator speed by 120; thus, for 24 WPM, oscillator runs at 2400 Hz which can be read easily on digital counter. Time-base divider would supply 20-Hz clock frequency for 24-WPM keying.—G. Jones, Calibrated Electronic Keyer Time Base, *Ham Radio*, Aug. 1975, p 39–41.

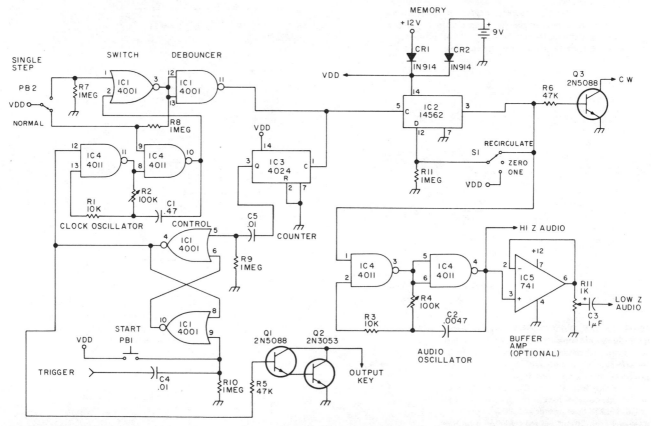

128-BIT CODE STORE—Draws almost no power, can be used on RTTY as well as CW, and can be reprogrammed in less than 1 min. Built around Motorola MC14562CP 128-bit shift register. Combination of 1 (high level) and 0 (low level) bits forming identifier message is fed into memory by placing S1 in program position zero and pressing start switch PB1 to dump contents of shift register and leave only 0s. Now, set S1 to desired first 1 or 0, push switch PB2 once, and repeat for rest of coded identifier. Set S1 to RECIRCULATE, push PB1 to cycle back to starting point, and unit is ready for use. If debouncer is not effective, move its wire from IC1-11 to IC4-10. Easier-to-get alternate values for C1 and R2 are 0.05 μF and 1 megohm.—C. W. Andreasen, Programmable CW ID Unit, *73 Magazine*, Oct. 1976, p 52–53.

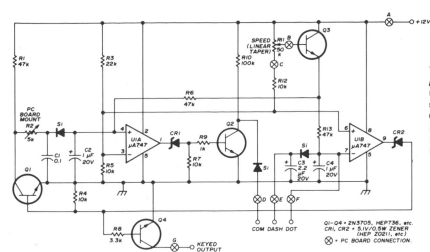

10–60 WPM KEYER—Uses µA747 (pair of µA741Cs in 10-pin TO-5 package). R2 adjusts relative length of first two dits to provide even spacing. Dot-dash ratio is set by C3 and C4, with C4 for dot and both in parallel for dash. Collector of C4 provides for keying positive voltage (20 V or less) to ground. Keying transistor will handle up to 50 mA without heatsink. Characters are self-completing. Used with low-power transceiver.—H. F. Batie, Introducing the Argomate, *Ham Radio*, April 1974, p 26–33.

KEYER WITH TR CONTROL—Provides automatic control of TR relay for break-in operation. 74221 TTL retriggerable mono MVBR forms dots or dashes, with paddle selecting side of IC that puts out pulse. Half of A2 (74123 dual retriggerable mono MVBR with clear) makes spaces between dots or dashes. Remaining half of A2 acts with A3b and A3c as TR switch.—B. Voight, The TTL One Shot, *73 Magazine*, Feb. 1977, p 56–58.

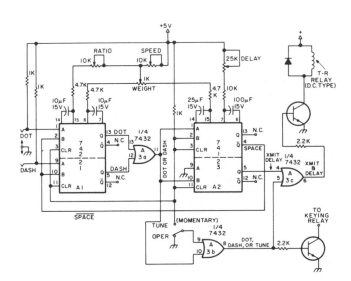

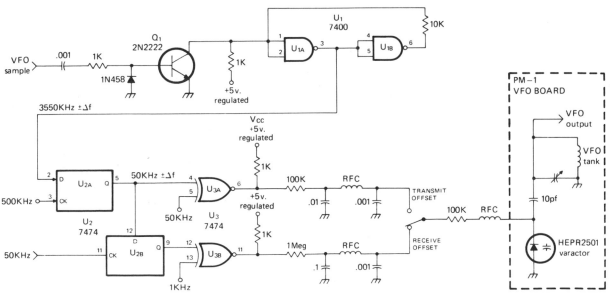

VFO CONTROL TO ±1 Hz—Used in coherent CW radio station to hold frequency of variable-frequency oscillator constant at 3550 kHz within 1 Hz so 12-WPM signal can be handled in bandwidth of only 9 Hz for greatly improved signal-to-noise ratio. Sample of VFO output, squared by Q_1 and U_1, goes to U_2 for mixing with 3500.000-kHz harmonic signal from 500-kHz frequency standard, to produce 50-kHz signal ± undesired drift for mixing in U_{3A} with 50.000-kHz signal from standard. If there is difference in frequency, U_{3A} generates control voltage proportional to amount of difference, applied to varactor tuning diode to pull VFO back to 3550.000 kHz. Same process occurs in receive offset chain, except that standard frequency in U_{3B} is such that receiver will be 1 kHz away from desired 3550.000 kHz and produce desired 1-kHz audio output.—A. Weiss, Coherent C.W.—the C.W. of the Future, *CQ*, June 1977, p 24–30.

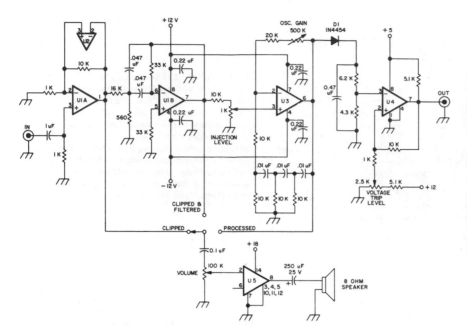

CODE REGENERATOR—Converts noisy CW output of receiver into TTL 1s and 0s for driving automatic Morse-code printer. Clipper U1A and 800-Hz active bandpass filter U1B feed injection-locked 800-Hz phase-shift oscillator U3 (National LM7401CN or Motorola HEP-C6052P). U1 is LM1458N or MC1458P. U2 uses part of LM709CN or HEP-C6103P as matched pair of 7-V back-to-back zener diodes. Diode detector D1 has time constant of about 5 ms, which is short enough for highest Morse-code speeds. Detector feeds LM311N or MLM311P1 voltage comparator U4 which can drive TTL or DTL directly. National LM380N audio amplifier U5 allows CW signal to drive loudspeaker directly.—H. Olson, CW Regenerator/Processor, *73 Magazine*, July 1976, p 80–82.

1-kHz TONE DECODER—Used as interface between amateur CW receiver and Motorola 6800 microcomputer to copy any code speed from 3 to 60 WPM while adjusting automatically to irregularities of hand-sent code. Translation program given in article requires about 600 bytes of memory. Algorithm can be converted to run on almost any 8-bit microprocessor. Tone decoder is 567 phase-locked loop tuned to center frequency of about 1 kHz, with bandwidth of about 100 Hz. Circuit will switch fast enough for most code speeds. PLL gives noise immunity. Optimum input level is about 200 mV. Output rests at +5 V, dropping near ground when tone of correct frequency is detected.—R. D. Grappel and J. Hemenway, Add This 6800 MORSER to Your Amateur Radio Station, *BYTE,* Oct. 1976, p 30–35.

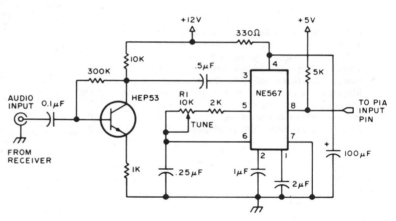

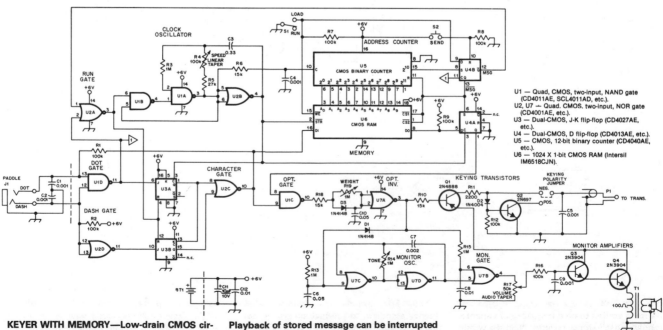

U1 — Quad, CMOS, two-input, NAND gate (CD4011AE, SCL4011AD, etc.).
U2, U7 — Quad. CMOS. two-input, NOR gate (CD4001AE, etc.).
U3 — Dual-CMOS, J-K flip-flop (CD4027AE, etc.).
U4 — Dual-CMOS, D flip-flop (CD4013AE, etc.).
U5 — CMOS, 12-bit binary counter (CD4040AE, etc.).
U6 — 1024 X 1-bit CMOS RAM (Intersil IM6518CJN).

KEYER WITH MEMORY—Low-drain CMOS circuit permits storage in RAM of message being keyed, for repeated later use by pushing button. Includes monitor, simple weight control, and both positive and negative keying outputs. Playback of stored message can be interrupted by closing either paddle contact. 1024-bit memory will hold two runs of alphabet, two sets of numbers, and several punctuation marks. Dot is stored as 1 followed by 0; dash is three 1s followed by 0. Free-running clock ensures that spaces will be recorded.—C. B. Opal, The Micro-TO Message Keyer, *QST,* Feb. 1978, p 11–14.

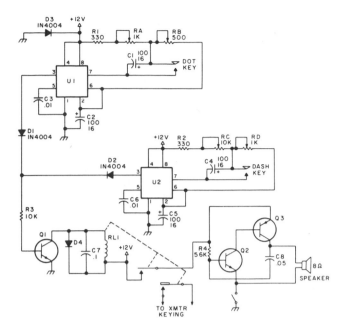

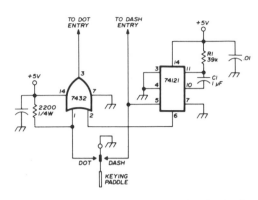

TWO-TIMER KEYER—Characters are self-completing and fully adjustable as to speed and length of character. When dot key is closed, NE555 timer U1 becomes astable MVBR with speed determined by RB and dot duration by RA. Identical timer U2 provides longer character lengths for dashes. All diodes are 1N4004 or equivalent 400 V PIV at 1 A. Q1 and Q2 are 2N2222, Q3 is 2N5964, and RL1 is 12-V reed relay. Capacitor values are in microfarads. All pots are linear. Power is not regulated.—A. Ring, Build the World's Simplest Keyer, *73 Magazine*, May 1977, p 46—47.

STOPPING PADDLE BOUNCE—Simple circuit prevents generation of erroneous dots by paddle contact bounce in keyers having dot memory. Uses 74121 mono MVBR and 7432 AND gate. Output of 74121 stays low if paddle is not in use or if dots or dashes are being sent. Release of dash paddle makes 74121 transmit high-level pulse to AND gate, long enough to block dot caused by bounce. Suitable only for keyers using +5 V.—B. Locher, Keyer Modification, *Ham Radio*, Aug. 1976, p 80.

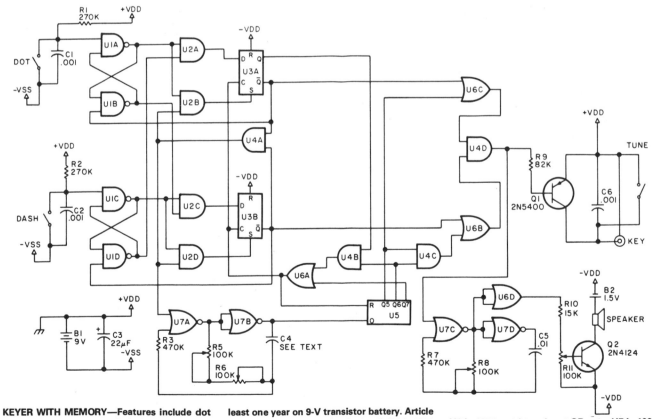

KEYER WITH MEMORY—Features include dot and dash memory, gated clock, low standby and key-down currents, built-in sidetone oscillator with loudspeaker, and keying circuit for grid-block keyed transmitter. Will operate at least one year on 9-V transistor battery. Article describes circuit operation in detail. U1 is 4011 quad two-input NAND gate. U2 and U4 are 4081 quad two-input AND gates. U3 is 4013 dual D flip-flop. U5 is 4024 seven-stage binary counter. U6 is 4071 quad two-input OR gate. U7 is 4001 quad two-input NOR gate.—E. A. Pfeiffer, MINI-MOS—the Best Keyer Yet?, *73 Magazine*, Aug. 1976, p 38—40 and 42—43.

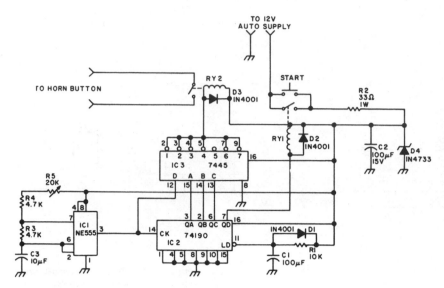

AUTOMATIC HI ON HORN—Pressing start button of circuit momentarily makes it send letters HI in Morse code on automobile horn, as friendly signal to another ham on road. Uses NE555 timer as oscillator, acting with counter IC2, decoder IC3, power-supply latch, and regulator. Space between four dots of H and two dots of I is achieved by not using pin 5 of decoder. RY1 should pull in at 5 V and 16 mA maximum, while RY2 should pull in at 5 V and 80 mA maximum and have contacts for switching 0.5-A inductive load of horn.—J. F. Reid, Sending HI, *73 Magazine*, May 1977, p 90.

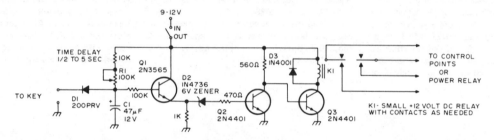

QUASI-BREAK-IN— Amateur station stays in receive mode until operator starts to send code. Tapping on key makes transmitter switch into transmit mode and stay there after last character is sent, for delay of several seconds (determined by R1) before transmitter is deenergized. Developed for use with cathode-keyed transmitters.—F. E. Hinkle, Jr., KOX for CW, *73 Magazine*, Feb. 1975, p 129–130.

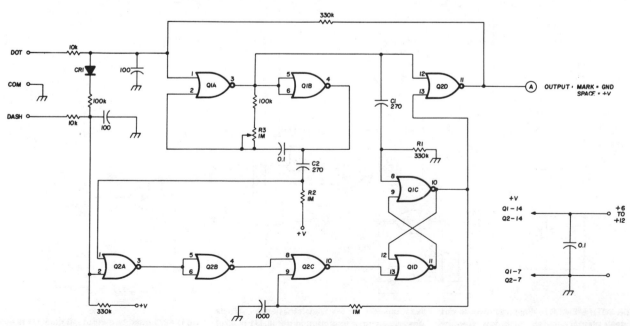

CMOS KEYER—Uses two CD4001AE quad two-input NOR gates. Q1A-Q1B form time-base MVBR, and Q1C-Q1D form dash flip-flop. Three of remaining gates synthesize three-input NOR gate for dash control. Q2D controls time-base MVBR and provides keyer output. Speed is adjustable from below 10 to over 70 WPM with R3.—C. J. Bader, Improved CW Transceiver for 40 and 80, *Ham Radio*, July 1977, p 18–22.

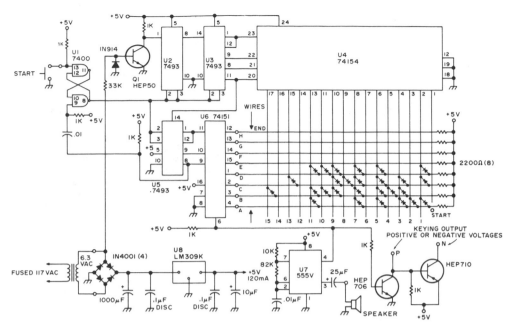

CW IDENTIFIER—Circuit automatically generates call letters for FCC-required code identification for FM repeaters and RTTY, when started by pushbutton or by pulse from other equipment. Audio output can be fed to loudspeaker as monitor or used to modulate FM repeater. Circuit shown is programmed for DE K4EEU by installing diodes at locations where tone is wanted on matrix. Article gives construction and programming details.—B. Kelley, A Super Cheapo CW IDer, *73 Magazine,* Dec. 1976, p 46–48.

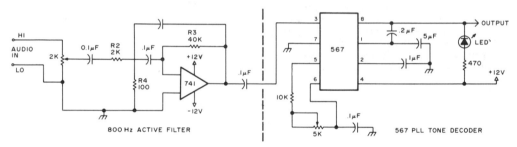

WEAK-SIGNAL DECODER—Combination of narrow-bandpass 800-Hz active filter and phase-locked loop of tone decoder permits copying very weak signals in Morse code. LED provides visual indication supplementing conventional output for headphones or loudspeaker.—Circuits, *73 Magazine,* July 1977, p 35.

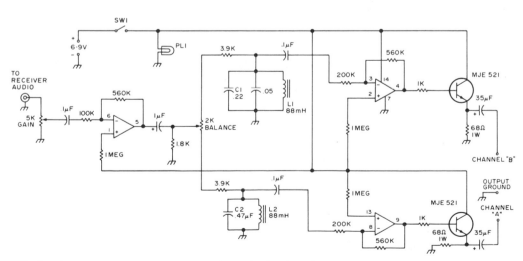

CW STEREO FILTER—Developed to enhance ability to read CW despite heavy contest traffic or other QRM. Two high-Q filters, one at each end of 400-Hz CW filter in receiver, create separate audio channels to give effect of stereo. Transistors at outputs of channels provide extra current gain for driving low-impedance stereo headphones. CW signal at 800 Hz then appears to come from left, 1200-Hz signal from right, and in-between frequencies at various azimuth angles. Illusion of direction makes it easier for operator to concentrate on desired signal in presence of others having slightly different frequencies. L1 and C1 form filter for 1200-Hz channel, while L2 and C2 form 800-Hz filter for other channel.—R. L. Anderson, Stereo—a New Type of CW Filter, *73 Magazine,* March 1976, p 48–50.

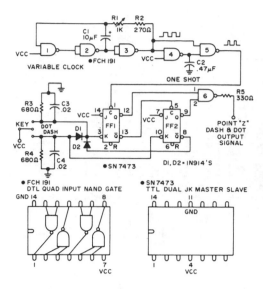

1:3 DOT-DASH KEYER—Gives accurate 1:3 dot-dash ratio at any desired keying speed, with self-completing characters. NAND gates 1, 2, and 3 of first FCH191 form variable-frequency square-wave oscillator, with C1, R1, and R2 determining frequency. With values shown, frequency is adjustable from about 150 to 1500 Hz, equivalent to code speed range of about 4 to 40 WPM. NAND gates 4 and 5 form mono MVBR. Flip-flops FF1 and FF2 are SN7473 TTL JK master-slave, acting with D1, D2, and NAND gate 6 to generate dots and dashes.—H. P. Fischer, Versatile IC Keyer, *73 Magazine,* Sept. 1973, p 69–71.

SIMPLE KEYER—Based on rapid charging of capacitors and controlled discharge through relay coil. When C_2 has discharged to relay release voltage, relay drops out and cycle starts over again as long as dot side of paddle is pressed. Dashes are similarly formed by C_1. R_4 adjusts speed from 10 to 40 WPM. K_1 is DPDT plate-current relay having 1K to 10K resistance.—J. J. Russo, An Inexpensive Electronic Keyer, *CQ,* Aug. 1971, p 58.

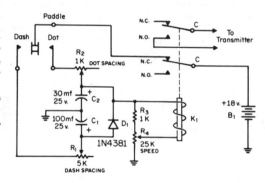

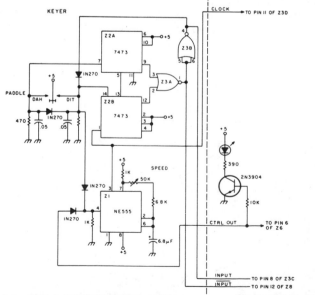

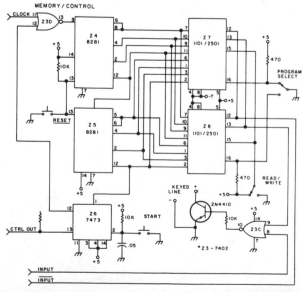

KEYER WITH MEMORY—Clock Z1 is NE555 timer giving keying speed range of 5 to 35 WPM. Flip-flops Z2A and Z2B count clock pulses to provide self-completing dits and dahs with spaces. Z4 and Z5 are 4-bit binary counters used for addressing static 256 × 1 bit RAM. To program keyer, switch to write, hit START button, and feed in message on keyer paddle. To send message back, switch to read and hit START again. To clear address counter if error is made, or for changing message, hit RESET switch and start over again.—D. W. Sewhuk, Contest Special Keyer, *73 Magazine,* Feb. 1977, p 38.

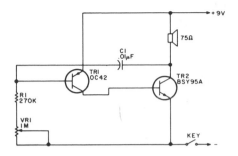

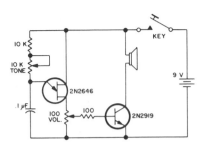

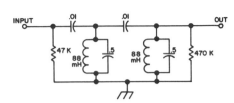

CODE PRACTICE—Simple AF oscillator drives loudspeaker for producing audio tone when key in negative supply lead is closed. Adjust VR1 for most pleasing tone.—Circuits, *73 Magazine*, July 1975, p 154.

PRACTICE OSCILLATOR—Simple design provides for adjustment of both volume and tone.—Circuits, *73 Magazine*, July 1974, p 81.

PASSIVE CW FILTER—Uses inexpensive 88-mH toroids to give very sharp 35-Hz bandwidth at 3 dB down. Filter has high insertion loss. Keyed waveshape has slow rise and fall, so CW signals have pronounced ringing that may be objectionable.—A. F. Stahler, An Experimental Comparison of CW Audio Filters, *73 Magazine*, July 1973, p 65–70.

CW SIDETONE—Audio sidetone oscillator serves for monitoring CW keying. Changes in transmitter operating frequency do not affect sidetone circuit. Hartley AF oscillator Q2 is turned on by diode rectifier and DC amplifier Q1. Antenna coupling shown is adequate from 160 through 10 meters. For VHF use on 6 and 2 meters, small tuned circuit and pickup antenna are usually required to get enough RF input for monitor. Audio pitch is adjusted by changing value of C1.—J. Fisk, Circuits and Techniques, *Ham Radio*, June 1976, p 48–52.

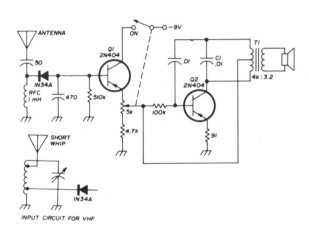

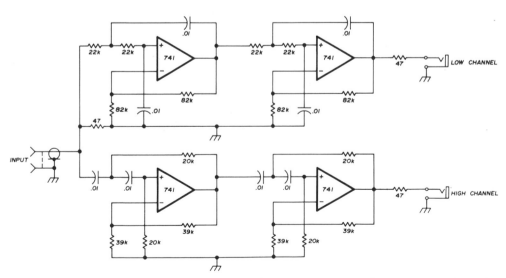

BINAURAL SYNTHESIZER FOR CW—Provides two channels for feeding stereo phones or two loudspeakers. When interference occurs a few hundred hertz from desired frequency, receiver is tuned so desired signal appears to be midway between loudspeakers, leaving interfering signals at right or left. Left or low channel has low-pass active filter and right or high channel has high-pass filter, with crossover at 750 Hz. Synthesizer is designed for low-impedance drive, as from loudspeaker output of receiver. Resistors in output channels prevent oscillation when 8-ohm phones or loudspeakers are directly connected to outputs. Opamps will drive 2000-ohm phones with ample volume and give moderate volume levels with 8-ohm loads.—D. E. Hildreth, Synthesizer for Binaural CW Reception, *Ham Radio*, Nov. 1975, p 46–48.

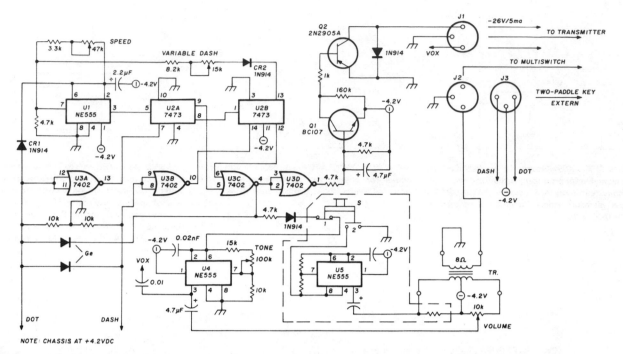

NOTE: CHASSIS AT +4.2VDC

3–50 WPM KEYER—Uses NE555V U1 as switchable dot generator providing accurate 1:1 ratio. SN7473 U2 forms variable dash circuit. NE555V sidetone generator U4 has tone range of about three octaves and easily drives 3-W 4-ohm loudspeaker through small transistor-radio transformer TR. NE555V U5 acts with U4 to provide two-tone oscillator for SSB tuning. Output section Q1-Q2 easily handles keying bias of −26 V at 5 mA.—H. Seeger, Micro-TO Keyer Mods, *Ham Radio,* July 1976, p 68–69.

CHAPTER 14
Comparator Circuits

Used to compare two values of voltage, frequency, phase, or digital inputs and provide logic output for driving variety of control circuits and indicating devices. See also Logic, Logic Probe, and Voltage-Level Detector chapters.

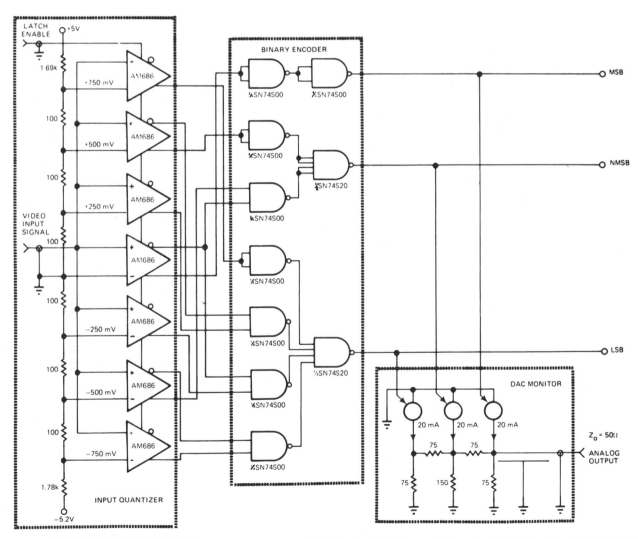

LATCH COMPARATORS FORM 3-BIT A/D CONVERTER—Seven Advanced Micro Devices AM686 comparators are arranged for direct parallel conversion of rapidly changing input signals, without prior sample-and-hold conditioning. Comparators feed Schottky TTL binary encoder logic for encoding to 3-bit offset binary. Quantization process is monitored by D/A converter. Article describes operation in detail and gives performance graphs which show freedom from output glitching at conversion speeds under 12 ns.—S. Dendinger, Try the Sampling Comparator in Your Next A/D Interface Design, *EDN Magazine*, Sept. 20, 1976, p 91—95.

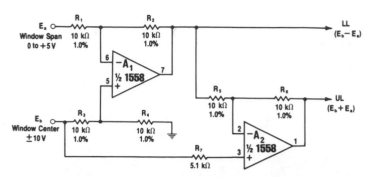

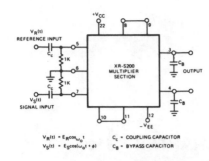

VARIABLE WINDOW—Single comparator can be programmed for wide variety of applications. One reference input voltage positions center of window, and other sets width of window. Sum or difference of reference voltages must not exceed ±10 V; if larger voltages must be handled, add voltage divider to scale them down into comparison range. A_1 is subtractor, generating voltage $E_b - E_a$ for use as lower limit voltage. Lower limit is added to $2E_b$ at A_2 to derive upper limit voltage $E_b + E_a$.—W. G. Jung, "IC Op-Amp Cookbook," Howard W. Sams, Indianapolis, IN, 1974, p 232–233.

PHASE COMPARATOR—High-level reference or carrier signal and low-level reference signal are applied to multiplier inputs of Exar XR-S200 PLL IC. If both inputs are same frequency, DC output is proportional to phase angle between inputs. For low-level inputs, conversion gain is proportional to input signal amplitude. For high-level inputs (V_S above 40 mVRMS), conversion gain is constant at about 2 V/rad.—"Phase-Locked Loop Data Book," Exar Integrated Systems, Sunnyvale, CA, 1978, p 9–16.

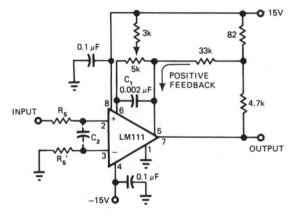

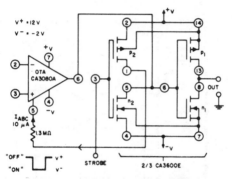

SUPPRESSING OSCILLATION—Use of positive feedback to pin 5 of comparator gives sharp and clean output transitions even with slow triangle-wave inputs, with no possibility of comparator bursting into oscillation near crossing point. Input resistors should not be wirewound. Circuit will handle triangle-wave inputs up to several hundred kilohertz.—P. Lefferts, Overcome Comparator Oscillation Through Use of Careful Design, *EDN Magazine*, May 20, 1978, p 123–124.

STROBED COMPARATOR—Combination of CA3080A opamp and two CMOS transistor pairs from CA3600E array gives programmable micropower comparator having quiescent power drain of about 10 µW. When comparator is strobed on, opamp becomes active and circuit draws 420 µW while responding to differential-input signal in about 8 µs. Common-mode input range is −1 V to +10.5 V. Voltage gain of comparator is typically 130 dB.—"Linear Integrated Circuits and MOS/FET's," RCA Solid State Division, Somerville, NJ, 1977, p 279.

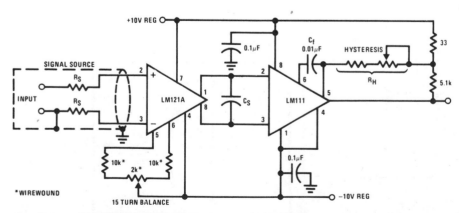

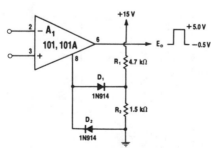

MICROVOLT COMPARATOR—Combination of National LM121A preamp and LM111 comparator serves for comparing DC signal levels that are only within microvolts of each other. With bias network shown, preamp has open-loop temperature-stable voltage gain close to 100. Separation of preamp from comparator chip minimizes effects of temperature variations. Circuit hysteresis is 5 µV, which under certain conditions can be trimmed to 1 µV.—"Linear Applications, Vol. 2," National Semiconductor, Santa Clara, CA, 1976, LB-32.

5-V CLAMPED COMPARATOR—R_1 and R_2 provide +3.8 V bias for D_1, clamping positive output of comparator opamp to +5 V. D_2 limits negative output swing to −0.5 V. Open-loop circuit means that output voltage will vary in proportion to load current.—W. G. Jung, "IC Op-Amp Cookbook," Howard W. Sams, Indianapolis, IN, 1974, p 226–228.

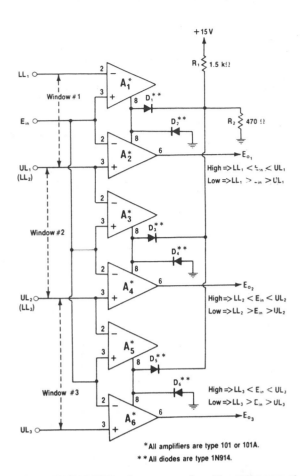

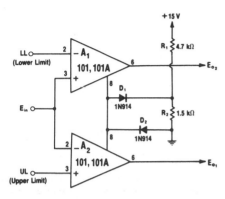

STAIRCASE WINDOW COMPARATOR—Cascading of 101-type window comparators for sequential operation indicates which of three windows input voltage is in. Input voltage is applied in parallel to all comparators. Output goes high only for comparator whose range includes voltage value of input. Lamp or other indicator can be added to each output line to give visual indication of voltage range.—W. G. Jung, "IC Op-Amp Cookbook," Howard W. Sams, Indianapolis, IN, 1974, p 233–234.

INTERNALLY GATED WINDOW COMPARATOR—Operation is based on fact that source and sink currents available at pin 8 of 101 opamp are unequal, with negative-going drive being larger. Voltage at pin 8 is low if either comparison input (A_1 or A_2) so dictates. Both A_1 and A_2 must have high outputs for pin 8 to be high. Outputs of A_1 and A_2 thus follow pin 8 since opamps have unity gain. D_1 and D_2 form clamp network. Either output of A_1 or A_2 can be used. Outputs go to +5 V only when input voltage is in window established by upper and lower voltage limits.—W. G. Jung, "IC Op-Amp Cookbook," Howard W. Sams, Indianapolis, IN, 1974, p 231–232.

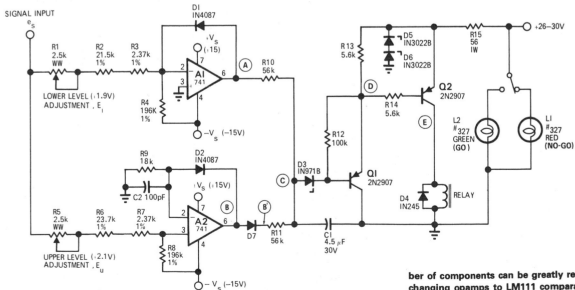

1.9–2.1 V WINDOW COMPARATOR—When positive input voltage is between levels set by R1 and R5, relay is actuated and green indicator lamp is turned on. Red lamp is on for voltages outside limits of window. Article gives design equations and traces operation of circuit. Number of components can be greatly reduced by changing opamps to LM111 comparators.—J. C. Nirschl, 'Window' Comparator Indicates System Status, *EDN/EEE Magazine*, June 15, 1971, p 49–50.

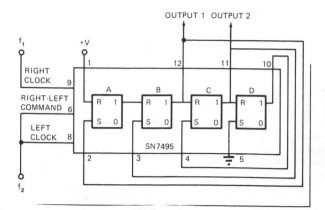

FREQUENCY/PHASE UP TO 25 MHz—Universal shift register such as 5495/7495 is connected to compare both frequency and phase of two carrier signals anywhere in range from DC to 25 MHz. When f_1 is greater than f_2, output is 1; when f_1 is less than f_2, output is 0. For $f_1 = f_2$, output is square wave whose duty cycle varies linearly with phase difference between f_1 and f_2. Comparisons are almost instantaneous, requiring at most two carrier cycles.—J. Breese, Single IC Compares Frequencies and Phase, *EDN Magazine,* Sept. 15, 1972, p 44.

VARIABLE BIPOLAR CLAMPING—Precision comparator provides independent regulation of both output voltage limits without connection to comparison inputs. A_2 and A_3 are complementary precision rectifiers having independent positive and negative reference voltages, with both rectifiers operating in closed loop through A_1. A_2 senses positive peak of E_o and maintains it equal to $+V_{clamp}$ by adjusting voltage applied to D_1. A_3 and D_3 perform similar function on negative peaks. Feedback network around output stage of A_1 regulates output voltage independently of inputs to A_1.—W. G. Jung, "IC Op-Amp Cookbook," Howard W. Sams, Indianapolis, IN, 1974, p 228–229.

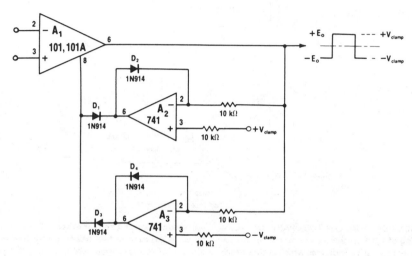

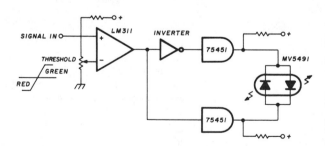

LEVEL-CROSSING DISPLAY—Uses Monsanto MV5491 dual red/green LED, with 220 ohms in upper lead to +5 V supply and 100 ohms in lower +5 V lead because red and green LEDs in parallel back-to-back have different voltage requirements. Circuit requires SN75451 driver ICs and one section of SN7404 hex inverter, with LM311 comparator. All operate from single +5 V source. Provides indicator change from red to green with input change of only a few millivolts.—K. Powell, Novel Indicator Circuit, *Ham Radio,* April 1977, p 60–63.

FREQUENCY COMPARATOR—Can be used with wide range of clock frequencies up to 5.3 MHz to provide output frequency that is equal to absolute difference between input frequencies f_1 and f_2. Article traces operation of circuit and gives design equations.—P. B. Morin, Frequency Comparator Provides Difference Frequency, *EEE Magazine,* April 1971, p 65–66.

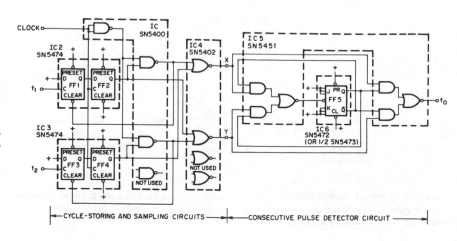

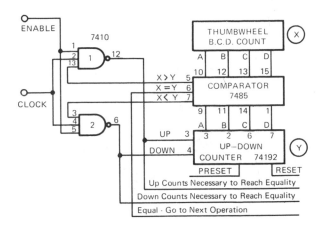

4-BIT BCD COMPARATOR—Provides less than, equal to, or greater than comparison between setting of BCD thumbwheel switch at X and BCD input digit at Y (Y is count preset into 74192 up/down counter). If equality does not exist, circuit will count up or down until it reaches equality, and thereby calculate difference between BCD values. Separate register can be used to store up or down counts required to reach equality.—R. A. Scher, Digital Comparator Is Self-Adjusting, *EDN Magazine,* Sept. 1, 1972, p 51.

INDEPENDENT SIGNALS—Single AD521 instrumentation amplifier compares two independent signal levels from sources having no common reference point. When one differential signal is applied to usual input of opamp and other to reference input, output is proportional to difference. Positive feedback provides small amount of hysteresis, to eliminate ambiguity and reduce noise susceptibility. Stable threshold of about 25 mV is derived from AD580 low-voltage reference circuit. Reference voltage is 2.5 V, but values used for R_S and R_G are in ratio of 1:100 so comparator output switches when normal input is about 1/100 of reference input. Output is negative when normal input is zero, and switches positive when input exceeds threshold. Output swings ±12 V as inputs go through critical ratio. R_3 and D_1 provide TTL-compatible second output.—A. P. Brokaw, You Can Compare Two Independent Signal Levels with Only One IC, *EDN Magazine,* April 5, 1975, p 107–108.

VOLTAGE COMPARATOR—Motorola MC1539 opamp provides excellent temperature characteristics and very high slewing rate for comparator applications. Zener connected to pin 5 limits positive-going waveform at output to about 2 V below zener voltage. Silicon diode connected to output limits negative excursion of output to give protection for logic circuit being driven. Parallel RC network in output provides impedance matching and minimizes output current overload problems.—E. Renschler, "The MC1539 Operational Amplifier and Its Applications," Motorola, Phoenix, AZ, 1974, AN-439, p 18.

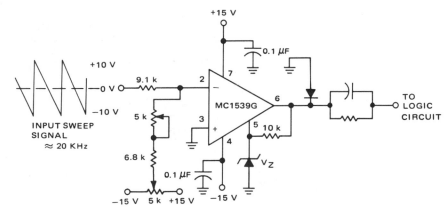

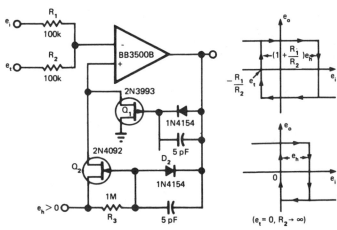

VOLTAGE-CONTROLLED HYSTERESIS—Precise, independent control of comparator trip point and hysteresis is achieved by switching hysteresis control signal e_h to comparator input with Q_1 and Q_2 when opamp changes state. Circuit avoids hysteresis feedback error while achieving inherent 0.01% trip-point accuracy of comparator. Control voltage e_t determines first trip point. When opamp output is negative, Q_2 is held off and Q_1 is on for connecting noninverting input to ground. Output switching occurs when input signal e_i drives input of inverting amplifier to zero.—J. Graeme, Comparator Has Precise, Voltage-Controlled Hysteresis, *EDN Magazine,* Aug. 20, 1975, p 78 and 80.

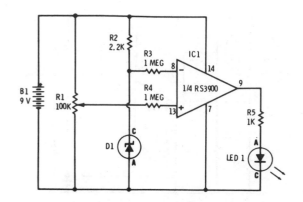

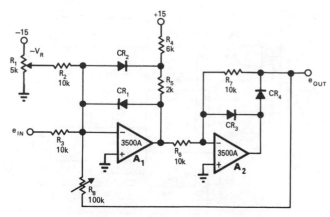

ZENER REFERENCE—One section of RS3900 quad opamp is connected as comparator using zener D1 for reference voltage. When voltage applied to pin 13 by R1 exceeds breakdown voltage of zener D1, comparator amplifies difference voltage to produce output voltage high enough to turn on LED. Can be used for classroom demonstration of comparator action. Zener breakdown should be under 9 V. LED can be Radio Shack 276-041.—F. M. Mims, "Semiconductor Projects, Vol. 2," Radio Shack, Fort Worth, TX, 1976, p 35–42.

INDEPENDENT HYSTERESIS ADJUSTMENT—Trip point and hysteresis of comparator opamp A_1 can be adjusted independently, with trip point being determined by setting of R_1 or programmed by DC voltage applied to R_2. Opamp A_2 provides polarity inversion and rectification of A_1 output. Hysteresis control R_8 is in feedback path from A_2 back to A_1. Amount of hysteresis is determined by ratio of R_3 to R_8. With values shown, circuit output levels are 0 and 5 V.—G. Tobey, Comparator with Noninteracting Adjustments, *EDN/EEE Magazine*, Oct. 1, 1971, p 43.

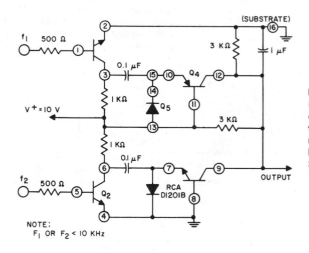

FREQUENCY COMPARATOR—Circuit using CA3096 transistor array plus one discrete diode develops DC output voltage that is proportional to difference between frequencies of input signals f_1 and f_2. Maximum input frequency is 10 kHz.—"Circuit Ideas for RCA Linear ICs," RCA Solid State Division, Somerville, NJ, 1977, p 17.

SLEW RATE—Circuit measures slew rate of input signal with Am685 comparator in circuit having delay-line length under 10 ns. When slew rate exceeds predetermined limit set by R_6, comparator changes state and latches, turning on LED. Pushing reset switch restores normal operation. Based on comparison of input signal with time-delayed counterpart. Derivative of input signal, equal to its instantaneous slew rate, is measured accurately for swings of 6 V P-P as found in most 50-ohm video signals. Action is fast enough to detect glitches.—R. C. Culter, Slew-Rate Limit Detector Is Simple, yet Versatile, *EDN Magazine*, Aug. 20, 1977, p 140–141.

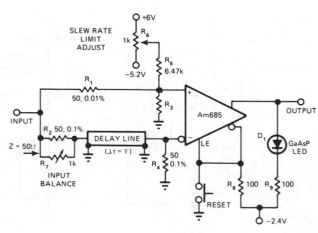

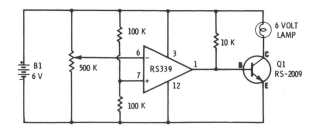

COMPARATOR DRIVES LAMP—Classroom demonstration circuit for comparator action uses transistor to amplify output of one section of RS339 quad comparator, to boost output current enough for driving 60-mA lamp. Lamp comes on when voltage at movable arm of 500K pot is greater than half of supply voltage.—F. M. Mims, "Integrated Circuit Projects, Vol. 6," Radio Shack, Fort Worth, TX, 1977, p 33–41.

STROBED MICROPOWER—Uses CA3080A variable opamp and CA3600E CMOS transistor array. Quiescent power drain from ± 12 V supply is only 10 μW, increasing to 420 μW when comparator is strobed on to make CA3080A active.—"Circuit Ideas for RCA Linear ICs," RCA Solid State Division, Somerville, NJ, 1977, p 16.

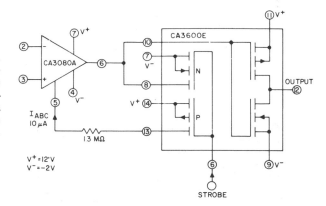

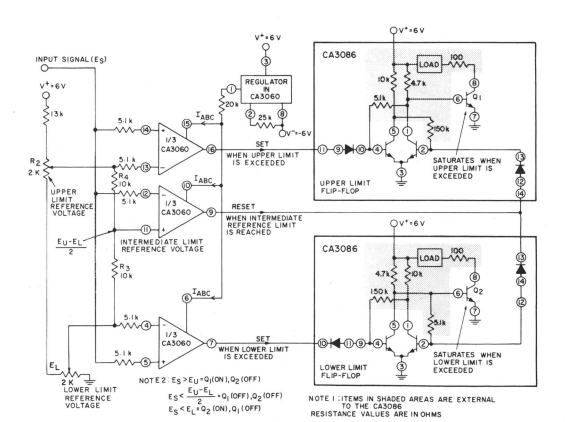

THREE-LEVEL COMPARATOR—All three sections of CA3060 three-opamp array are used with CA3086 transistor arrays to provide three adjustable limits for comparator. If upper or lower limit is exceeded, appropriate output is activated until input signal returns to preselected intermediate limit. Suitable for many types of industrial control applications.—"Circuit Ideas for RCA Linear ICs," RCA Solid State Division, Somerville, NJ, 1977, p 17.

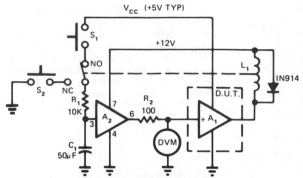

A_1 (D.U.T.): NON–INVERTING COMPARATOR WITH
10% HYSTERESIS UNDER TEST
A_2: VOLTAGE FOLLOWER, LM310
L_1: RELAY, COUCH 2X10B460A

MEASURING THRESHOLDS—Upper and lower thresholds of noninverting comparator under test (A_1) are read on DVM at end of capacitor charge and discharge cycles initiated by S_1 and S_2. With C_1 discharged, relay L_1 is energized. Closing S_1 allows C_1 to charge toward V_{CC}. When upper threshold is reached, relay drops out and meter is read. Closing S_2 starts discharge cycle which stops at lower threshold. Reverse relay connections when testing inverting comparator.—E. S. Papanicolaou, Comparator Is Part of Its Own Measuring System, *EDN Magazine,* Aug. 5, 1974, p 76.

COMPARATOR DRIVES LED—Simple classroom demonstrator of comparator action uses one section of RS339 quad comparator. Reference voltage applied to positive input of comparator is half of supply voltage. R1 serves as voltage divider applying variable voltage to inverting input. When voltage applied to pin 6 by R1 exceeds reference voltage on pin 7, comparator switches on and LED lights. R4 is chosen for use with Radio Shack 276-041 red LED.—F. M. Mims, "Integrated Circuit Projects, Vol. 6," Radio Shack, Fort Worth TX, 1977, p 33–41.

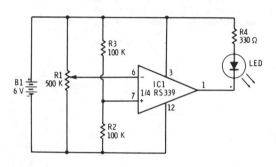

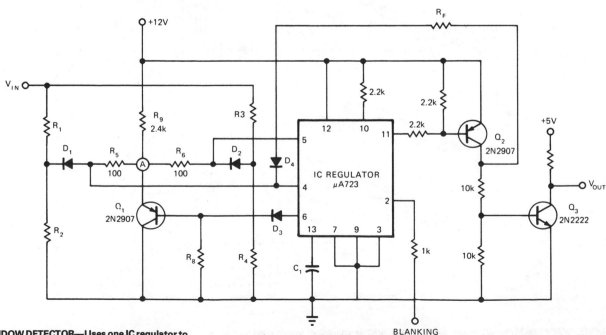

WINDOW DETECTOR—Uses one IC regulator to compare output voltages of two separate voltage dividers with fixed reference voltage. Resulting absolute error signal is amplified and converted to TTL-compatible logic signal. Voltage divider for lower limit of window detector is R_1-R_2 and for upper limit is R_3-R_4. Article covers circuit operation in detail.—N. Pritchard, Window Detector Uses One IC Regulator, *EDN Magazine,* May 20, 1973, p 81 and 83.

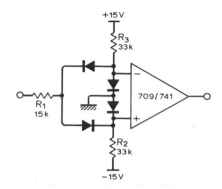

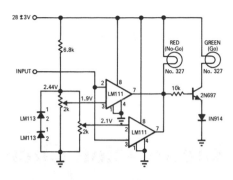

DUAL LIMITS—Opamp used without frequency compensation gives positive output only when input voltage exceeds 8.5 V in either polarity. Resistors in supply leads determine limit points. For inverted output, reverse inputs to opamp. Diodes are 1N914.—K. Pickard, Dual Limit Comparator Using Single Op-Amp, *Wireless World*, Dec. 1974, p 504.

VOLTAGE-WINDOW COMPARATOR—Use of LM111 opamps minimizes number of components required to turn on green indicator lamp when input voltage is between predetermined limits set by 2K pots. Similar circuit using 741 opamps requires total of 31 components. Improved circuit draws only 120 nA from voltage level being monitored, and operates within 0.3% threshold level stability using single unregulated supply varying ±3 V from 28 V.—D. Priebe, Comparators Compared, *EDN/EEE Magazine*, Oct. 1, 1971, p 61.

CHAPTER 15
Contact Bounce Suppression Circuits

Used to solve bounce problems of switch and relay contacts during closing or opening.

BOUNCELESS SQUARE OUTPUT—NE555 timer eliminates need for gates to suppress contact bounce. Timer can provide pulse at least 5 ms long (much longer if desired) and can remain on as long as trigger input (key pulse) is low (grounded). Timer triggers on negative-going edge of low-going pulse, such as key down to ground. Common negative is isolated from ground. V_{CC} can be 5 to 15 VDC. Timer output can be connected directly to exciter keying input for negative grid keying. Because of square-wave output on make or break (100 ns each), circuits must be added in exciter or between keying transistors to provide at least 5-ms rise and fall times for Morse or RTTY keying.—B. Conklin, Improving Transmitter Keying, *Ham Radio*, June 1976, p 44–47.

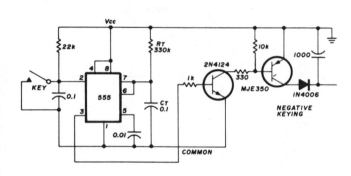

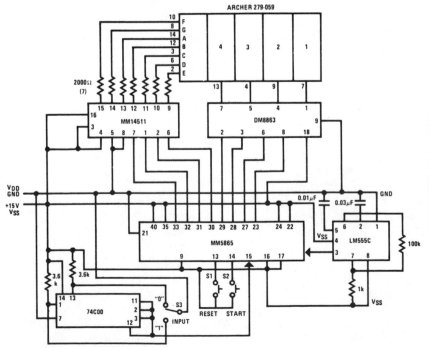

SWITCH-CLOSURE COUNTER—National MM5865 universal timer and counter chip is used with 74C00 debouncer and LM555C timer to drive digital display that counts closures of manual switch S3. Reset transition restores display to 0000. BCD segment outputs of MM5865 feed LED 4-digit display through MM14511 interface, while digit enable outputs go to display through DM8863 driver.—"MOS/LSI Data-book," National Semiconductor, Santa Clara, CA, 1977, p 2-23–2-32.

DEBOUNCER—Generates single pulse on switch closure, provided wiper of switch bounces only between contact and an open. Output A goes low when switch is pushed, and at same time output B goes high.—E. E. Hrlvnak, House Cleaning the Logical Way, *73 Magazine,* Aug. 1974, p 85–90.

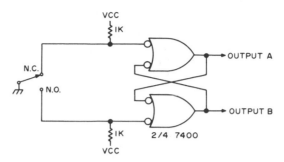

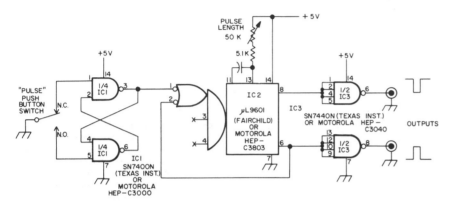

ONE PULSE PER PUSH—Circuit generates only one rectangular pulse for each actuation of pushbutton switch, even if contacts bounce. TTL gates IC1 are wired as RS flip-flop (latch) that triggers mono MVBR IC2 having fixed-duration positive and negative output pulses. Output drives are increased by TTL inverting buffer gates.—H. Olson, Further Adventures of the Bounceless Switch, *73 Magazine,* Feb. 1975, p 111–114.

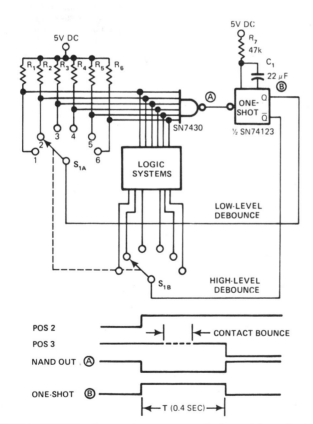

ROTARY SWITCH DEBOUNCE—Outputs from mono (one-shot) provide common returns for rotary switch. Multi-input NAND gate, tied to normally high signals from one deck of rotary switch, instantly detects opening of one contact and triggers mono. Mono then simulates open contact for interval determined by values used for R7 and C1; for values shown, delay is 400 ms.—E. S. Peltzman, Circuit Eliminates Rotary-Switch Bounce Problems, *EDN Magazine,* April 20, 1978, p 132.

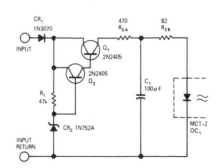

BOUNCELESS ISOLATOR—Integrating filter C_1-R_2 eliminates effects of contact bounce that may be superimposed on digital input signal feeding optoisolator. Photodiode in optoisolator drives Schmitt trigger that makes output to TTL circuits change state when LED is turned on by input signal.—C. E. Mitchell, Optical Coupler and Level Shifter, *EDN/EEE Magazine,* Feb.1, 1972, p 55.

DEBOUNCING WITH COUNTER—Circuit uses CMOS counter/decoder with any inexpensive 200-Hz or higher clock such as CMOS two-gate oscillator or 555 timer. Signal to be debounced is fed directly to reset input of counter, with no preconditioning. When contact is made by switch, counter unclears and starts counting up. Each bounce of contact resets counter, so it cycles between states 0 and 1 until contacts settle. Counter then delivers clean nonoverlapping pulses to remaining output lines, any of which may be used as conditioned output signal. When counter reaches state 7, it inhibits itself to prevent repeated pulsing of output lines. When switch is opened, cycling action is repeated during bounces, with output never going higher than state 1. After bouncing, counter is held in clear state ready for next closing.—L. T. Hauck, Solve Contact Bounce Problems Without a One-Shot, *EDN Magazine*, Sept. 5, 1975, p 80 and 82.

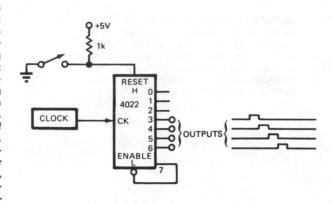

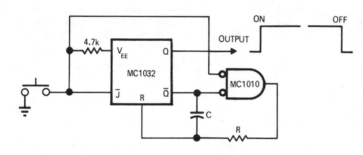

BOUNCELESS MAKE/BREAK—Circuit eliminates switch bounce problems during closing as well as opening. When switch is closed, Q output of flip-flop goes to logic 1 for delay period determined by RC time constant. Releasing switch operates NAND gate, making its output go to logic 1. This charges C through R until reset level is reached. Flip-flop then resets, changing Q output to logic 0. Values for R and C are chosen according to bounce duration of switch used. For typical 1-A SPST switch, 10,000 ohms and 0.47 μF were used.—L. F. Walsh and T. W. Hill, Make-and-Break Bounceless Switching, *EDN/EEE Magazine*, July 15, 1971, p 49.

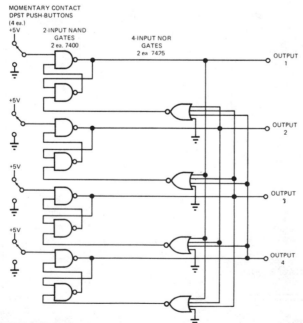

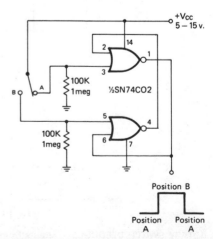

BOUNCE-FREE INTERLOCKING—Arrangement provides low-cost equivalent of mechanically interlocked switch assembly, while providing TTL compatibility and freedom from switch bounce. Momentary pressing of any pushbutton restores its associated RS flip-flop to normal and makes output of that channel high. Arrangement uses cross-coupled two-input NAND gates for each flip-flop, connected so each actuation produces an output and resets all other flip-flops. If two or more buttons are pushed simultaneously, all their channels will go high, but only last one released will stay on. Any number of channels may be added.—B. Brandstedt, Digital Interlocking Switch Is Inexpensive to Build, *EDN Magazine*, Dec. 15, 1972, p 42.

LATCHING GATES—SN74C02 quad two-input NOR gate forms latching circuit in which first noise pulse produced by switch latches circuit, making it immune to contact bounce.—I. Math, Bounceless Switch, *CQ*, July 1976, p 50.

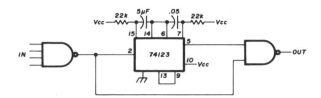

DELAYED START—Keyed output of RTTY terminal equipment or other keys and relays is delayed by 74123 dual mono for at least 5 ms while contact bounce settles down. Can be used for calculator keyboards, flip-flop testers, and other applications in which final clean pulse length is not highly important.—B. Conklin, Improving Transmitter Keying, *Ham Radio,* June 1976, p 44–47.

KEYBOARD BOUNCE ELIMINATOR—Dual 9602 mono MVBR is used with Harris HD-0165 keyboard encoder to generate delayed strobe pulse St', with delay set at about 10 ms by first mono. Pulse width is determined by second mono and should be set to meet system requirements. Circuit eliminates effects of arcing or switch bounce and provides proper encoding under two-key rollover conditions.—"Linear & Data Acquisition Products," Harris Semiconductor, Melbourne, FL, Vol. 1, 1977, p 6-4.

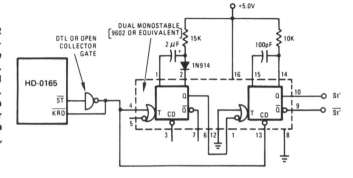

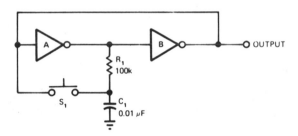

GATES FORM SWITCH—Each time pushbutton switch is closed momentarily, voltage on C_1 makes inverter A change state, with positive feedback from inverter B, to give alternate ON and OFF action. R_1 delays charging and discharging of C_1, making circuit essentially immune to contact bounce. Switch works equally well with either CMOS or TTL gates. Values of R_1 and C_1 are not critical.—T. Tyler, Inverters Provide "ON-OFF" from Momentary Switch, *EDN Magazine,* June 20, 1976, p 126.

CHAPTER 16
Converter Circuits—Analog-to-Digital

Includes circuits for converting DC, audio, and video analog inputs to linearly related binary, BCD, or Gray-code digital outputs. Some circuits have autoranging or some type of input compression, input multiplexing, and input buffering.

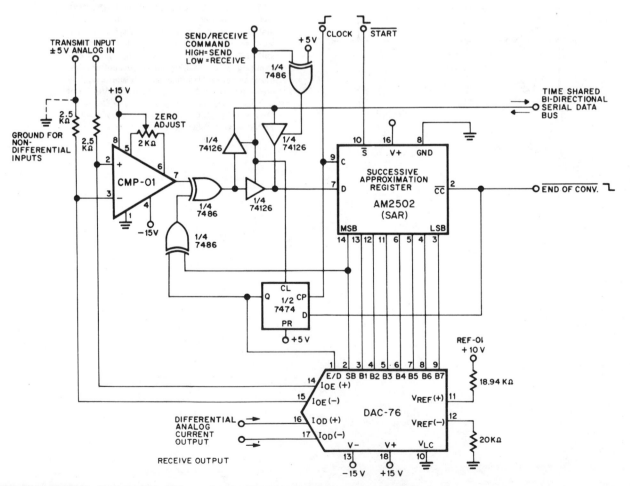

SERIAL DATA OUTPUT—Precision Monolithics ICs form transceiving converter suitable for use in control systems incorporating 8-bit microprocessors. Output conforms with Bell-System μ-255 logarithmic law for PCM transmission. Applications include servocontrols, stress and vibration analysis, digital recording, and speech synthesis. Start must be held low for one clock cycle to begin send or receive cycle. Conversion is completed in nine clock cycles, and output is available for one full clock cycle. Other half of system is identical.—"COMDAC Companding D/A Converter," Precision Monolithics, Santa Clara, CA, 1977, DAC-76, p 12.

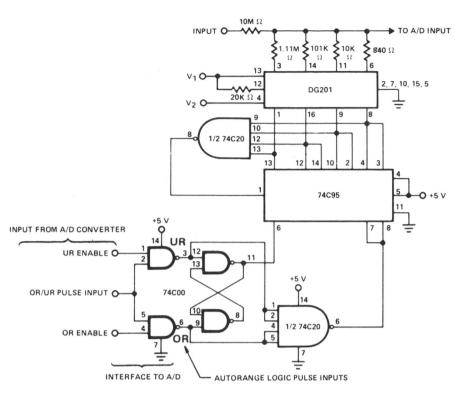

SELF-CONTROLLED AUTORANGING—DG201 quad analog switch inserts one of four attenuator resistors in input circuit of Siliconix LD130 or comparable A/D converter under control of autoranging pulse output derived from converter. Control logic includes 74C00 quad two-input NAND gate with two sections connected as flip-flop, 74C95 4-bit right-shift left-shift register, and 74C20 dual four-input NAND gate.—"Analog Switches and Their Applications," Siliconix, Santa Clara, CA, 1976, p 6-28—6-29.

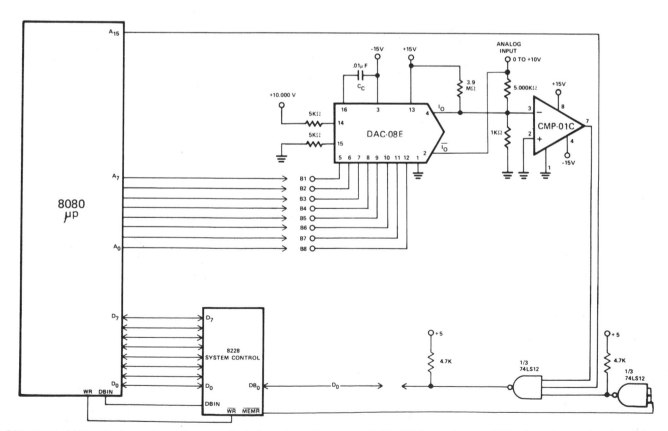

SOFTWARE CONTROL—Innovative software for Intel 8080A microprocessor eliminates need for peripheral isolation devices when using Precision Monolithics DAC-08E D/A converter and CMP-01C comparator for 8-bit A/D conversion. Technique can easily be expanded to 10-bit or 12-bit conversions and adapted to other microprocessors. Logic of microprocessor replaces conventional successive-approximation register. 8 lowest-order address bits control data bit input to DAC, using software given in article.—W. Ritmanich and W. Freeman, "Software Controlled Analog to Digital Conversion Using DAC-08 and the 8080A Microprocessor," Precision Monolithics, Santa Clara, CA, 1977, AN-22, p 3.

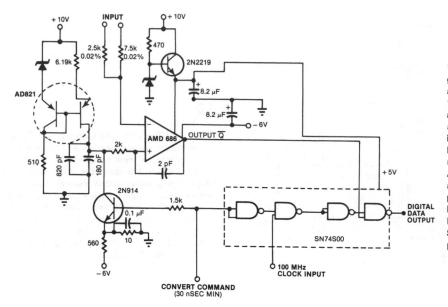

10-BIT ACCURACY—Single-slope A/D converter gives high-speed conversion of DC input voltage to digital data output. For 0–10 V input, 1024 pulses of 100-MHz clock appear at full scale and 512 at half scale. When command pulse is applied, 2N914 transistor resets 1000-pF capacitor (820 and 180 in parallel) to 0 V. Capacitor begins to charge linearly on falling edge of command pulse, to 2.5 V. 10-μs ramp is applied to AMD686 for comparison with unknown voltage. Output of opamp is pulse whose width is proportional to input voltage and can therefore be used to gate 100-MHz clock.—J. Williams, Low-Cost, Linear A/D Conversion Uses Single-Slope Techniques, *EDN Magazine*, Aug. 5, 1978, p 101–104.

VIDEO COMPRESSOR—Nonlinear function amplifier IC-2 compresses video input signals as required to compensate for inefficient quantization where there are too many levels for small signals and too few levels for large signals. Designed to feed 6-bit analog-to-digital converter, IC-1 attenuates input −20 dB and shifts level. Output of IC-2 is amplified by IC-3 to voltage range comparable to that of input signal. IC-4 acts as temperature compensator and output level shifter. R_7 nulls small output offsets.—J. B. Frost, Non-Linear Function Amplifier, *EEE Magazine*, March 1971, p 78.

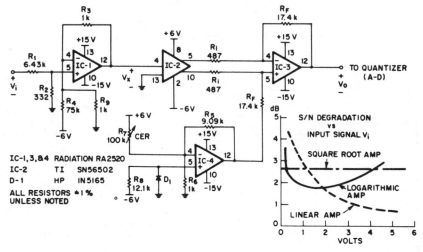

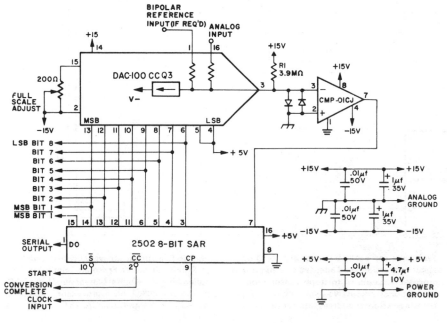

8-BIT SUCCESSIVE APPROXIMATION—Uses Precision Monolithics DAC-100 CCQ3 D/A converter and CMP-01CJ fast precision comparator in combination with Advanced Micro Devices AM2502PC or equivalent successive approximation register to compare analog input with series of trial conversions. Clamp diodes minimize settling time and prevent large inputs from damaging DAC output. Digital output is available in serial nonreturn-to-zero format at data output DO shortly after each positive-going clock transition.—D. Soderquist, "A Low Cost, Easy-to-Build Successive Approximation Analog-to-Digital Converter," Precision Monolithics, Santa Clara, CA, 1976, AN-11, p 3.

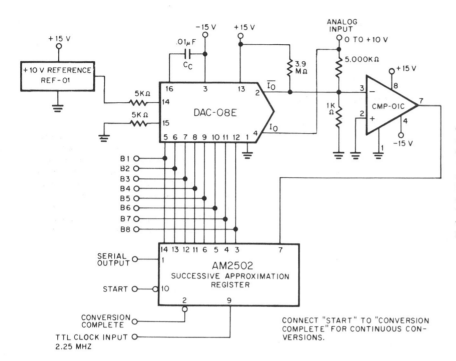

4-μs CONVERSION TIME—Provides conversion of analog input to 8-bit digital output by successive approximation, with conversion time of 4 μs. Advanced Micro Devices AM2502 successive-approximation register contains logic for Precision Monolithics DAC-08E and CMP-01C comparator.—D. Soderquist and J. Schoeff, "Low Cost, High Speed Analog-to-Digital Conversion with the DAC-08," Precision Monolithics, Santa Clara, CA, 1977, AN-16, p 3.

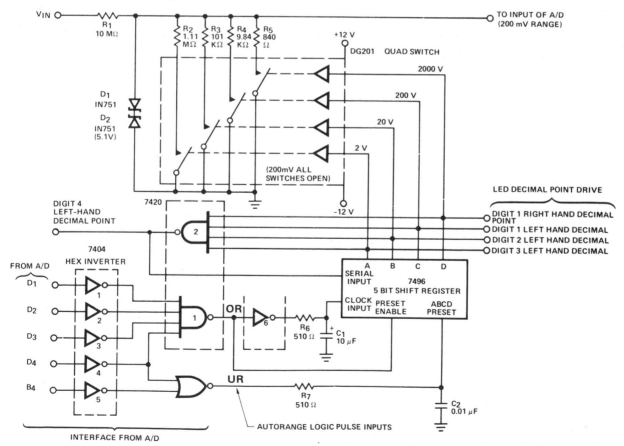

AUTORANGING—Digitally controlled attenuator uses DG201 quad analog switch as input ladder attenuator switches for A/D converter. Switches are controlled by digital logic that detects overrange and underrange information from A/D converter and closes appropriate attenuator path. Circuit is suitable for Siliconix LD110/111 or LD111/114 A/D converter.—"Analog Switches and Their Applications," Siliconix, Santa Clara, CA, 1976, p 6-28.

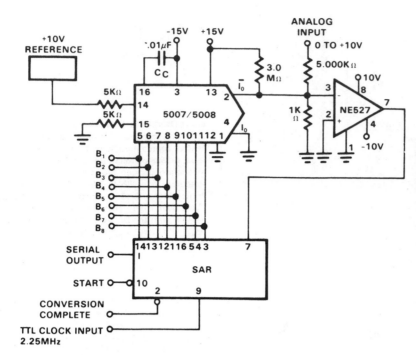

0–10 V ANALOG INPUT—Used to provide digital input to computer for processing and storage of analog signals. Requires only three ICs in addition to external +10 V reference and 2.25-MHz TTL clock. Successive-approximation register (SAR) can be Motorola MC1408 or equivalent. For continuous conversions, connect pins 10 and 2 of SAR.—"Signetics Analog Data Manual," Signetics, Sunnyvale, CA, 1977, p 677–685.

HIGH-IMPEDANCE BUFFER—Two sections of Motorola MC3403 quad opamp serve as voltage followers for differential inputs of third section connected as buffer for MC1505 A/D converter. Dual transistor Q1, connected as dual diode, provides 0.6-V offset at inputs of voltage followers, to obtain temperature tracking and predictable performance at low bias currents of opamp.—D. Aldridge and S. Kelley, "Input Buffer Circuits for the MC1505 Dual Ramp A-to-D Converter Subsystem," Motorola, Phoenix, AZ, 1976, EB-24A.

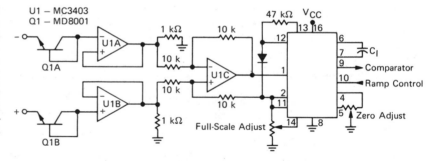

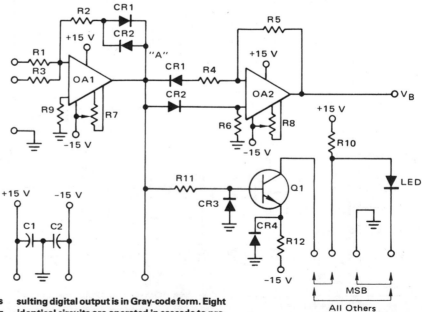

R1, 3	10 k 0.1%
R2, 4, 5, 6	20 k 0.1%
R7, 8	10 k Pot
R9	3.9 k
R10, 12	1.2 k
R11	10 k
OA1, 2	MC1456C
CR1	MSD6100
CR2	MSD6150
CR3, 4	1N914
Q1	MPS6415
C1, 2	0.1
LED	MLED 630

CYCLIC CONVERTER—Unknown voltage is successively compared to reference voltage for determining each digital bit. After determining bit, voltage difference between unknown and reference is operated on, then sent to successive stages to determine less significant bit. Resulting digital output is in Gray-code form. Eight identical circuits are operated in cascade to provide 8-bit A/D converter having accuracy within 1 LSB and full-scale range of 0-8 V. Circuit requires only two MC1456CG opamps per stage, with MPS6514 transistor as comparator. Switching diode CR1 is MSD6100, and CR2 is MSD6150. Other diodes are 1N914.—J. Barnes, "Analog-to-Digital Cyclic Converter," Motorola, Phoenix, AZ, 1974, AN-557, p 7.

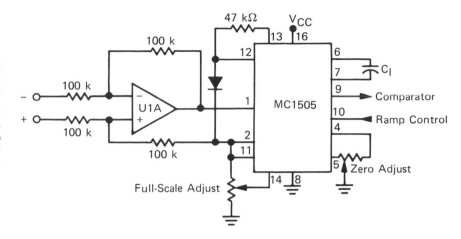

DIFFERENTIAL OPAMP AS BUFFER—Section of Motorola MC3403 quad opamp, operating from single supply, serves as low-cost unity-gain buffer for MC1505 dual-ramp A/D converter. Opamp is used as differential amplifier referenced to MC1505 reference voltage of 1.25 V.— D. Aldridge and S. Kelley, "Input Buffer Circuits for the MC1505 Dual Ramp A-to-D Converter Subsystem," Motorola, Phoenix, AZ, 1976, EB-24A.

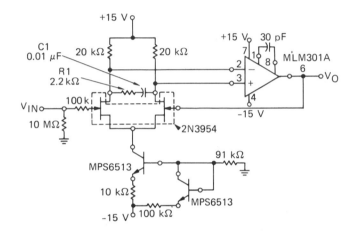

FET-INPUT BUFFER—Used ahead of Motorola MC1505 A/D converter to provide input impedance of 10 megohms. FETs are connected as differential amplifier having common source leads returned to constant-current generator built from bipolar transistor, with similar transistor providing temperature compensation. Temperature drift of amplifier is well under 1 mV from 0 to 50°C.—D. Aldridge and S. Kelley, "Input Buffer Circuits for the MC1505 Dual Ramp A-to-D Converter Subsystem," Motorola, Phoenix, AZ, 1976, EB-24A.

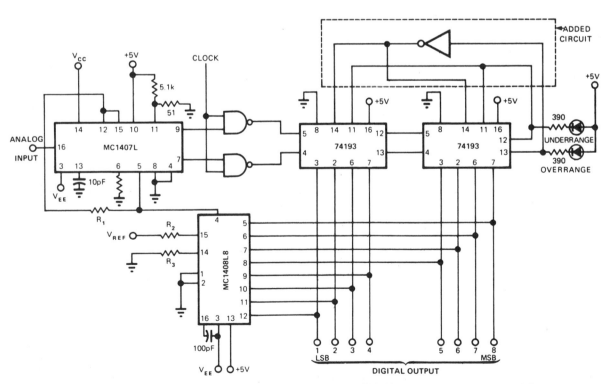

TRACKING A/D CONVERTER—Addition of one gate to tracking or servo-type A/D converter, as shown in dashed box, overcomes instability problems otherwise occurring when input voltages are less than zero or greater than full scale. With 8-bit converter shown, count of 11111111 when counting up makes carry output and load inputs go low, holding counter in this state so subsequent up clocks are ignored. When count is all 0s, borrow output goes low and clear input goes high, so counter is free to count up only.— A. Helfrick, Tracking A/D Converters Need Another Look, *EDN Magazine,* June 20, 1975, p 118 and 120.

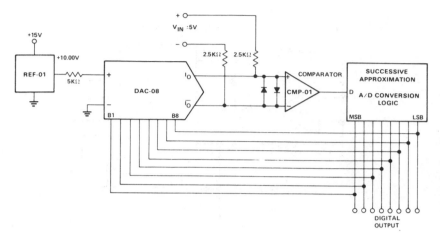

DIFFERENTIAL CONVERSION—Uses high current output capability of Precision Monolithic DAC-08 D/A converter and high common-mode voltage rejection of CMP-01 comparator to give differential-input ADC without input signal conditioning. Successive-approximation conversion logic is obtained with REF-01 +10 V reference and 2502-type successive-approximation register, driven by DAC and comparator. Analog input is converted in less than 2 μs. Differential input range is 5 V. Diodes are 1N4148.—J. Schoeff and D. Soderquist, "Differential and Multiplying Digital to Analog Converter Applications," Precision Monolithics, Santa Clara, CA, 1976, AN-19, p 5.

REPETITIVE-MODE OPERATION—Quicker conversion is obtained in Teledyne Philbrick 4109 or 4111 A/D converter by restarting converter within a few microseconds after status signal, using sure-start circuit shown. Reset pulse is fed to converter when status signal is held at low DC level. When status command is high, oscillator A-B is disabled. If reset pulse is not obeyed and status signal remains low, oscillator starts up until conversion does occur.—R. W. Jacobs, "Repetitive Mode Operation for Models 4109/4111 Integrating A/D Converters," Teledyne Philbrick, Dedham, MA, 1977, AN-28.

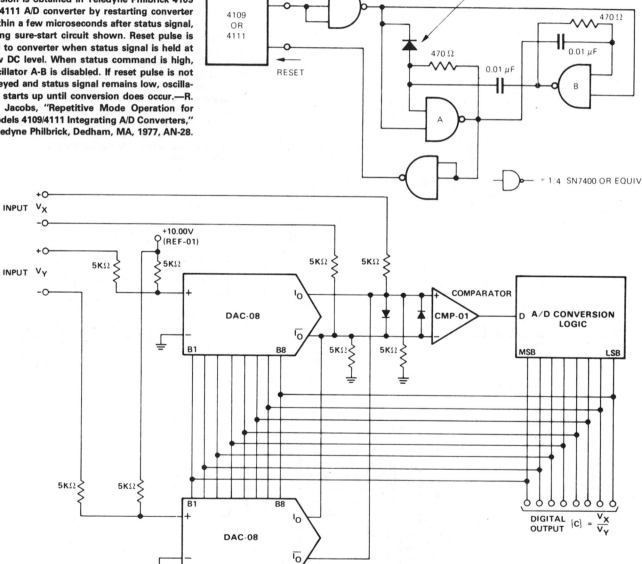

FOUR-QUADRANT RATIOMETRIC—Uses Precision Monolithics DAC-08 D/A converters and CMP-01 comparator to drive successive-approximation conversion logic using REF-01 +10 V reference and 2502-type successive-approximation register. Imputs V_X are connected conventionally, and inputs V_Y are connected in multiplying fashion. I_{REF} for both DACs is modulated between 1 and 3 mA. Resulting output currents are differentially transformed into voltages by 5K resistors at comparator inputs and compared with V_X differential input. When conversion process is complete (comparator inputs differentially nulled to less than ½ LSB), digital output corresponds to quotient V_X/V_Y. Diodes are 1N4148.—J. Schoeff and D. Soderquist, "Differential and Multiplying Digital to Analog Converter Applications," Precision Monolithics, Santa Clara, CA, 1976, AN-19, p 5.

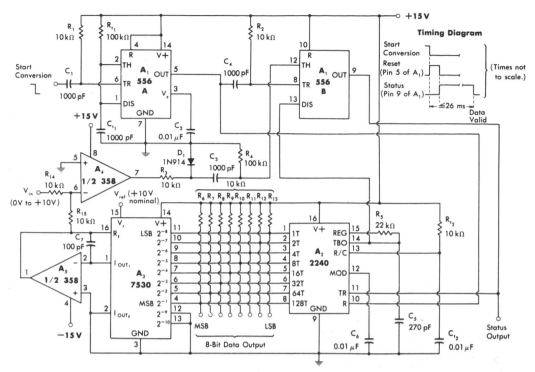

BINARY OUTPUT—Converts analog signal in range of 0–10 V to 8-bit binary word having all 0s for 0 V and all 1s for full-scale input of +9.960 V. Output is 15-V CMOS-compatible but can be adapted for TTL compatibility. Maximum conversion time is about 26 ms. A_2 and A_3 form negative-going staircase generator for which start-conversion signal is formed by one section of 556. Opamp A_4 compares negative output of 7530 with input voltage V_{in}. When 7530 output voltage equals input voltage, comparator output goes positive and resets control flip-flop to complete conversion.—W. G. Jung, "IC Timer Cookbook," Howard W. Sams, Indianapolis, IN, 1977, p 226–228.

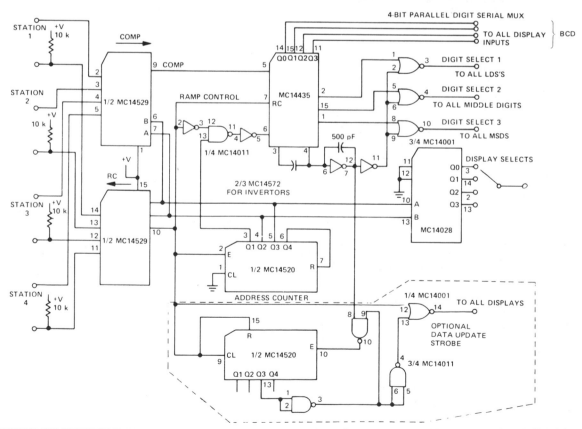

FOUR-CHANNEL INPUT MULTIPLEXING—Conversion process is divided between central station and remote locations having analog sensors. Each station transmits two noise-immune low-frequency digital signals under control of central multiplexer. System is much more economical than having separate A/D converter at each sensor. Can be extended to 32 channels. Multiplexing is performed under control of clock in Motorola MC14435, operating between 100 kHz and 1 MHz. At 500 kHz, each conversion takes about 15 ms.—S. Kelley, "Analog Data Acquisition Network for Digital Processing Using the MC1405-MC14435 A/D System," Motorola, Phoenix, AZ, 1975, EB-58.

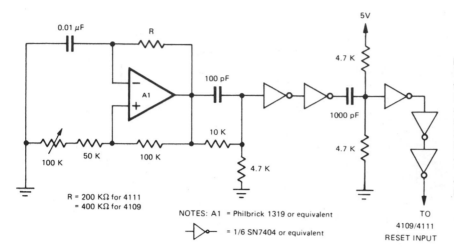

EXTERNAL TRIGGER—Generates pulse with 100-ns minimum width in range of 125–250 Hz for application to reset input of Teledyne Philbrick A/D converter in applications requiring unattended operation with continuous conversion. Adjust 100K pot to give 125 Hz for 4109 or 250 Hz for 4111. Successive stages of SN7404 inverter provide required sharpening of pulse. A1 is positive-starting MVBR.—R. W. Jacobs, "Repetitive Mode Operation for Models 4109/4111 Integrating A/D Converters," Teledyne Philbrick, Dedham, MA, 1977, AN-28.

R = 200 KΩ for 4111
  = 400 KΩ for 4109

NOTES: A1 = Philbrick 1319 or equivalent

▷○ = 1/6 SN7404 or equivalent

TO 4109/4111 RESET INPUT

MOSFET-INPUT BUFFER—Uses Motorola MC14007 dual complementary pair plus inverter, with two of MOSFETs connected as differential amplifier for buffering opamp and third serving as current source for differential amplifier. Arrangement gives high input impedance required in some applications of MC1505 A/D converter for which buffer was designed. 1-megohm pot controls gate voltage for current source. Temperature drift is well under 2 mV over range of 0–50°C. Pin 14 of MC14007 should be tied to +5 V.—D. Aldridge and S. Kelley, "Input Buffer Circuits for the MC1505 Dual Ramp A-to-D Converter Subsystem," Motorola, Phoenix, AZ, 1976, EB-24A.

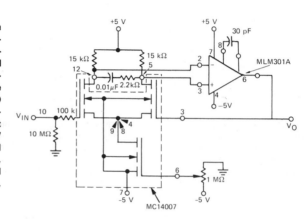

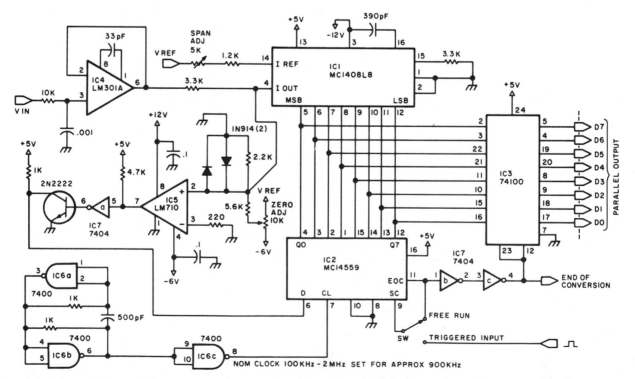

VOICE DIGITIZER—Uses 8-bit ADC capable of sampling AF input signal 100,000 times per second when using 900-kHz clock. 100-kHz clock gives 9000 samples per second, about minimum for human voice. Digital output is stored in computer memory for later conversion back to analog form for such applications as synthesis of speech from phonemes and providing voice answers to queries. Requires about 10,000 bytes in memory for 1 s of voice data. Pin 7 of IC4 is +12 V, and pin 4 is −6 V. For IC6 and IC7, pin 14 is +5 V and pin 7 is ground. 8080 assembler programs are given for input and output of memory.—S. Ciarcia, Talk to Me! Add a Voice to Your Computer for $35, *BYTE*, June 1978, p 142–151.

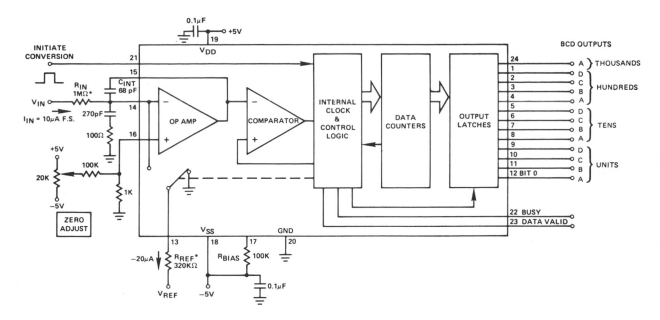

BCD OUTPUT—Latched nonmultiplexed parallel BCD outputs from Teledyne 8750 3½-digit CMOS analog-to-digital converter are suitable for liquid crystal and gas-discharge displays. 2-mA drain on ±5 V supply permits battery operation. Features include high linearity, noise immunity, and 3½-digit resolution within 0.025% error. Circuit is based on switching number of current pulses needed to bring analog current to zero at input of opamp, then determining digital equivalent by counting these pulses. Values shown are for full-scale voltage input of 10 V and voltage reference of −6.4 V.— CMOS A-D Converter Provides BCD Output, *Computer Design,* Nov. 1977, p 156 and 158.

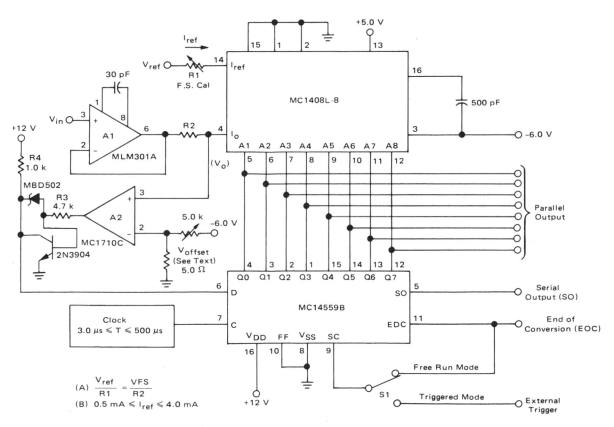

$$\text{(A)} \quad \frac{V_{ref}}{R1} = \frac{VFS}{R2}$$

$$\text{(B)} \quad 0.5 \text{ mA} \leq I_{ref} \leq 4.0 \text{ mA}$$

HIGH-SPEED SUCCESSIVE-APPROXIMATION—Total conversion time for 8-bit system is about 4.5 µs. Clock rate is up to 2 MHz. Serial output is used for transmission to one or more other locations.—T. Henry, "Successive Approximation A/D Conversion," Motorola, Phoenix, AZ, 1974, AN-716, p 5.

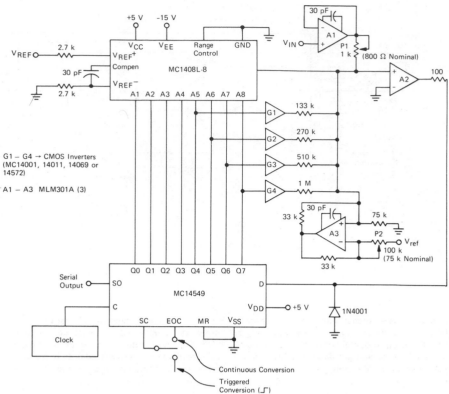

2-DIGIT BCD—Uses Motorola MC14549 successive-approximation register and MC1408L-8 D/A converter to give full-scale value of 0.99 V in 10-mV increments. Input is buffered by opamp A1 connected as voltage follower, with pot P1 set to give output current proportional to unknown input voltage. This current is compared to that required by total BCD A/D converter. Pins 1-4 and 12-15 of MC14549 provide required 2-digit parallel BCD output. Clock frequency can be 100 kHz.—D. Aldridge, "Successive Approximation BCD A/D Converter," Motorola, Phoenix, AZ, 1975, EB-51.

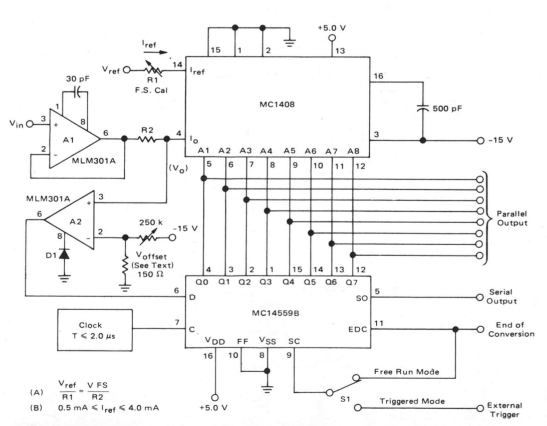

(A) $\dfrac{V_{ref}}{R1} = \dfrac{V\,FS}{R2}$

(B) $0.5\ \text{mA} \leq I_{ref} \leq 4.0\ \text{mA}$

8-BIT SUCCESSIVE-APPROXIMATION—Requires only four ICs. For each cycle, most significant bit is enabled first, with comparator giving output signifying that input signal is greater or less in amplitude than output of Motorola MC1408. If output is greater, bit is reset or turned off. Process is repeated for next most significant bit until all bits have been tried, completing conversion cycle. Conversion time is 18 μs, total propagation delay is about 1.5 μs, and overall operational figure is about 2 μs per bit for 8-bit system.—T. Henry, "Successive Approximation A/D Conversion," Motorola, Phoenix, AZ, 1974, AN-716, p 4.

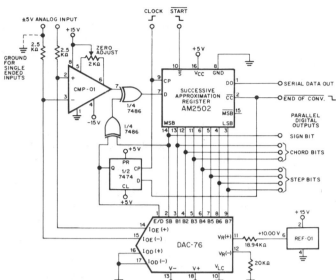

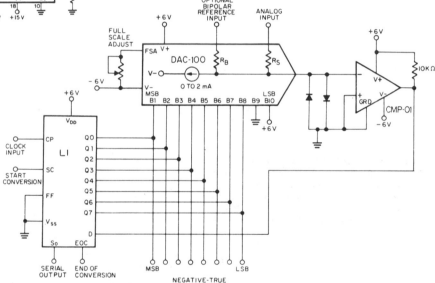

COMPRESSING A/D CONVERSION—Step size increases as output changes from zero scale to full scale, in contrast to conventional linear converter in which step size is constant percentage of full scale. Uses Precision Monolithic DAC-76 D/A converter in combination with CMP-01 comparator, any standard EXCLUSIVE-OR gate, and successive-approximation register for conversion logic. Encoding sequence begins with sign-bit comparison and decision. Bits are converted with successive-removal technique, starting with decision at code 011 1111 and turning off bits sequentially until all decisions have been made. Conversion is completed in nine clock cycles.—"COMDAC Companding D/A Converter," Precision Monolithics, Santa Clara, CA, 1977, DAC-76, p 12.

CMOS-COMPATIBLE SUCCESSIVE-APPROXI-MATION—Converts analog input to 8-bit digital output by using MC14559 CMOS successive-approximation register with Precision Monolithics DAC-100 D/A converter and CMP-01 comparator. Conversion sequence is initiated by applying positive pulse, with width greater than one clock cycle, to START CONVERSION input. Analog input is then compared successively to ½ scale, ¼ scale, and remaining binarily decreasing bit weights until it has been resolved within ½ LSB. END OF CONVERSION then changes to logic 1 and parallel answer is present in negative-true binary-coded format at register outputs.—D. Soderquist, "Interfacing Precision Monolithics Digital-to-Analog Converters with CMOS Logic," Precision Monolithics, Santa Clara, CA, 1975, AN-14, p 4.

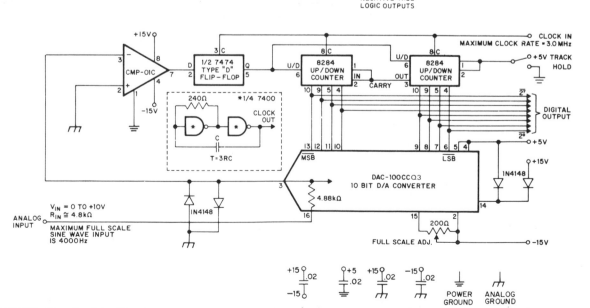

8-BIT TRACKING—Uses Precision Monolithics DAC-100 CCQ3 D/A converter and CMP-01CJ fast precision comparator to make digital data continuously available at output while tracking analog input. Diode clamps hold DAC output near zero despite input and turn-on transients. Unused least significant digital inputs of 10-bit DAC are turned off by connecting to +5 V as shown. Simple clock circuit shown in dashed box is stable over wide range of temperatures and supply voltages. D/A converter is used in feedback configuration to obtain A/D operation.—"A Low Cost, High-Performance Tracking A/D Converter," Precision Monolithics, Santa Clara, CA, 1977, AN-6, p 2.

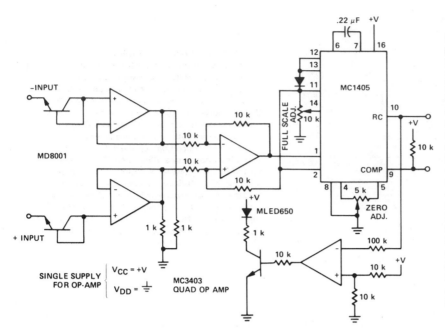

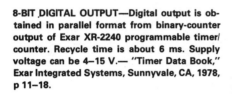

SINGLE SUPPLY FOR OP-AMP | $V_{CC} = +V$ | $V_{DD} = \perp$

REMOTE STATION—Multiplexing of large number of analog voltages from widely separated locations in large industrial control systems is simplified by transmitting two noise-immune low-frequency digital signals from each remote to central multiplexer driving display and microprocessor. Central station using MC14435 controls direction of integration in each remote-station MC1405 through ramp control output. At beginning of conversion, integrator of MC1405 integrates upward for 1000 counts of central-station clock. Integrator then ramps down while comparator remains high, with clock continuing until comparator threshold is again crossed. Counts during down ramp are latched by counter when comparator goes low, and circuits are reset for next conversion. Analog input voltage is thus transmitted to central-station MC14435 as two digital signals.—S. Kelley, "Analog Data Acquisition Network for Digital Processing Using the MC1405-MC14435 A/D System," Motorola, Phoenix, AZ, 1975, EB-58.

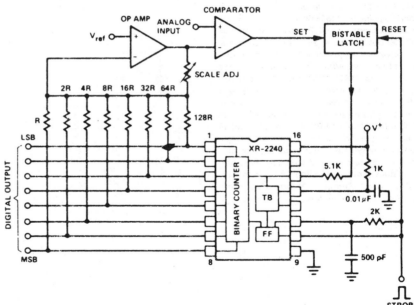

8-BIT DIGITAL OUTPUT—Digital output is obtained in parallel format from binary-counter output of Exar XR-2240 programmable timer/counter. Recycle time is about 6 ms. Supply voltage can be 4–15 V.— "Timer Data Book," Exar Integrated Systems, Sunnyvale, CA, 1978, p 11–18.

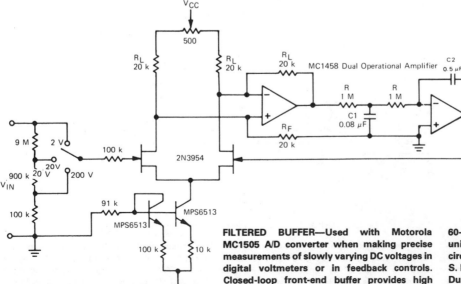

FILTERED BUFFER—Used with Motorola MC1505 A/D converter when making precise measurements of slowly varying DC voltages in digital voltmeters or in feedback controls. Closed-loop front-end buffer provides high input impedance and reduces stray noise and 60-Hz pickup. Two-pole filter is included in unity-feedback loop of buffer. Front-end scaling circuit is included with buffer.—D. Aldridge and S. Kelley, "Input Buffer Circuits for the MC1505 Dual Ramp A-to-D Converter Subsystem," Motorola, Phoenix, AZ, 1976, EB-24A.

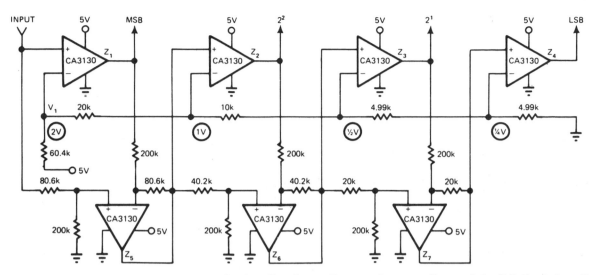

4-BIT CLOCKLESS—Simple and low-cost arrangement of seven CA3130 opamps gives conversion times fast enough for tracking sine-wave signals well up into audio range. Even with relatively slow 741 opamps, signals up to 300 Hz were easily tracked. Additional bits are easily cascaded.—B. P. Vandenberg, Tracking-Type A/D Requires No Clock Oscillator, *EDN Magazine,* Jan. 20, 1977, p 92 and 94.

CHAPTER 17
Converter Circuits—DC to DC

Use inverters typically operating from DC supplies in range of 2–15 V to generate AC voltage at frequency typically in range of 16–25 kHz, for step-up by voltage-doubling rectifier or transformer-rectifier combination to give desired new positive or negative DC supply voltage that can be as high as 10 kV.

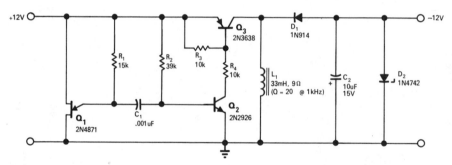

+12 V TO −12 V—Transformerless inverting DC-to-DC converter has above 55% efficiency and can withstand output shorts lasting up to several minutes. UJT Q_1 and base-emitter diode of transistor Q_2 form free-running MVBR whose 25-kHz output is amplified by Q_2 to drive switching-mode converter Q_3-L_1-D_1-C_2. Zener D_2 regulates output for variations in input voltage or output loads up to 40 mA.—G. Bank, Transformerless Converter Supplies Inverted Output, *EDN/EEE Magazine*, July 1, 1971, p 48.

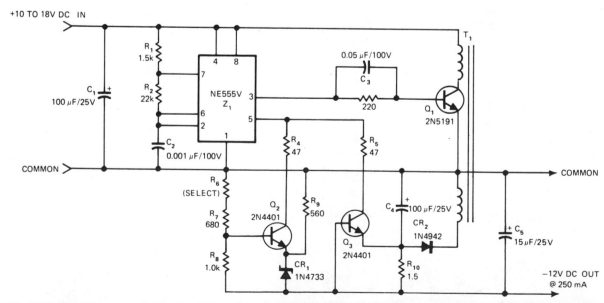

+12 V TO −12 V—Transforms unregulated +12 VDC to current-limited regulated −12 VDC. Front end of 555 is connected in astable configuration, with R_2 selected to give about 25 kHz at pin 3. Control of modulation input to pin 5 gives voltage regulation and current limiting. Circuit tolerates continuous operation under short-circuit conditions. With 10-V nominal output, line regulation is within ±0.05% for input and output voltage ranges of 0.3 to 10 V. Load regulation is 0.2% for loads from 10 μA to 10 mA when load impedance is 10 ohms.—R. Dow, Build a Short-Circuit-Proof +12V Inverter with One IC, *EDN Magazine*, Sept. 5, 1977, p 177–178.

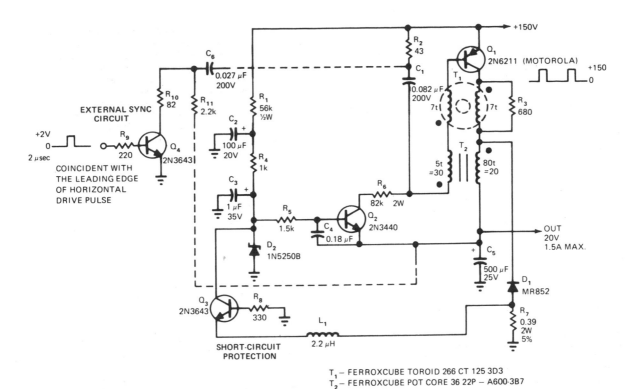

T_1 — FERROXCUBE TOROID 266 CT 125 3D3
T_2 — FERROXCUBE POT CORE 36 22P — A600·3B7

2 V TO 20 AND 150 V—Use of 7-turn toroidal transformer in self-excited ringing-choke blocking-oscillator circuit improves efficiency of converter circuit by providing fast switching time. Circuit is practical only when input and output voltages differ significantly. Blocking oscillator is formed by Q_1, T_2, C_1, R_2, and base-bias network R_6-Q_2. Q_4 makes possible external synchronization, permitting use in television systems for triggering regulator with leading edge of horizontal drive pulse. This ensures completion of cycle within blanking interval.—N. Tkacenko, Transformer Increases DC-DC Converter Efficiency to 80%, *EDN Magazine,* May 5, 1976, p 110 and 112.

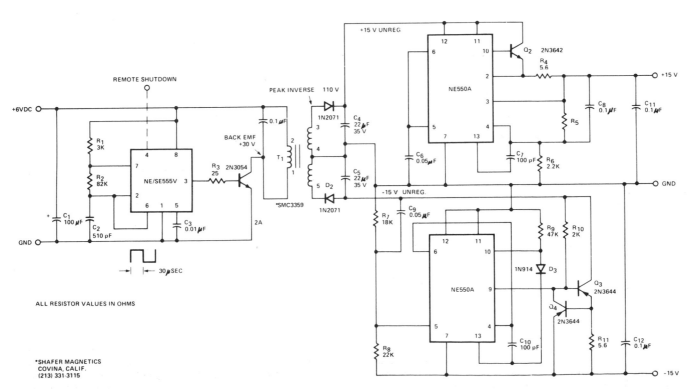

ALL RESISTOR VALUES IN OHMS

*SHAFER MAGNETICS
COVINA, CALIF.
(213) 331·3115

+6 V TO ±15 V—Combination of 555 timer and two NE550A precision adjustable regulators gives 0.1% line and load regulation. Timer operates as oscillator driving step-up transformer which feeds full-wave rectifier.—"Signetics Analog Data Manual," Signetics, Sunnyvale, CA, 1977, p 726–727.

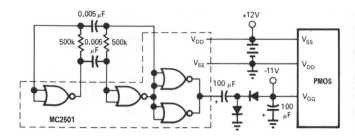

12 V TO −11 V WITH CMOS IC—Bipolar inverter and rectifier together provide −11 V from 12-V auto battery for operating high-threshold MOS logic of portable or automotive equipment. Diode types are not critical. Inverter draws only about 1 mA from 12-V battery on standby and supplies 2 mA from −11 V terminal.—B. Fette, Inexpensive Inverters Generate V_{GG} for Portable MOS Applications, *EDN/EEE Magazine*, Dec. 15, 1971, p 51.

BATTERY-LIFE EXTENDER—Conserves battery life by charging capacitor from 0 V at efficiencies over 80% and by allowing battery to be used to lower endpoint voltage. Will generate voltages above or below battery voltages. When used in capacitor-discharge ignition system, power conversion efficiency is so high that heatsink is unnecessary and only one power transistor is needed. Gives full output voltage even when car battery voltage is less than half nominal value, as during cold starting. Article describes operation of circuit in detail. Tr_1 may require series RC protection between collector and emitter.—R. M. Carter, Variable Voltage-Ratio Transistor Converter, *Wireless World*, Nov. 1975, p 519.

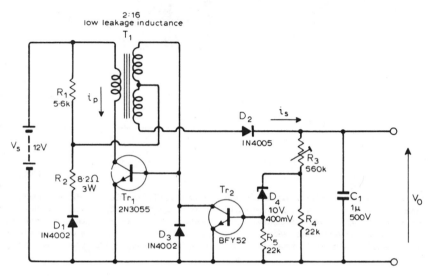

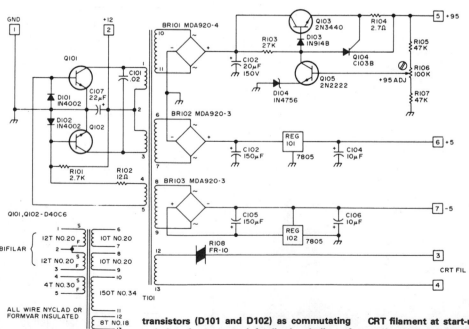

CRO LOW-VOLTAGE SUPPLY—Developed for use as one of supplies for portable CRO, operating from battery using sealed rechargeable cells supplying 12 V at 2–5 Ah. High-efficiency inverter uses two GE D40C6 or RCA 2N5294 transistors (D101 and D102) as commutating switches for untapped feedback winding of power transformer. R102 then determines drive, while R101 produces required unbalanced starting bias. Thermistor R108 in series with CRT filament has cold resistance of 10 ohms to counteract very low cold resistance of CRT filament at start-up and prevent inverter malfunction. Article gives instructions for winding T101, along with high-voltage supply circuit and all other circuits of CRO covering DC to 10 MHz.—G. E. Friton, Eyes for Your Shack, *73 Magazine*, Nov./Dec. 1975, p 74–76, 78–88, and 90–94.

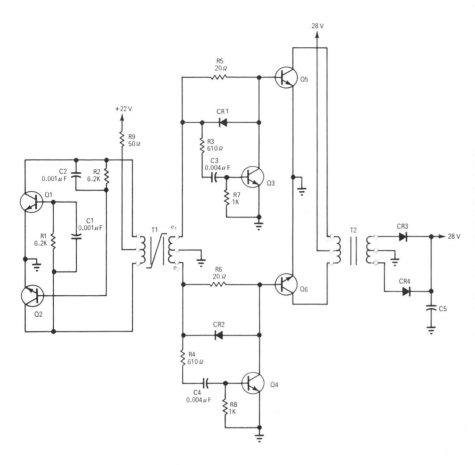

28-V HIGH-EFFICIENCY—Uses driven-type converter in which signal source is simple two-transistor oscillator Q1-Q2. Turn-on delay technique eliminates overlap current otherwise flowing in 2N1016 push-pull power transistors Q5 and Q6 when one is still on in storage state while other is driven on. Efficiency can approach 90%. Q3 and Q4 prevent off transistor from conducting until opposite device has turned off. Values for T2, CR3, and CR4 in output circuit are chosen to give desired DC output voltage.—R. F. Downs, Minimize Overlap to Maximize Efficiency in Saturated Push-Pull Circuits, *EDN/EEE Magazine*, Feb. 1, 1972, p 48–50.

12 V TO −11 V WITH TRANSISTORS—Bipolar inverter and rectifier together provide −11 V from 12-V auto battery for operating high-threshold MOS logic of portable or automotive equipment. Transistor and diode types are not critical. Multivibrator draws only 1.2 mA from battery on standby and supplies 12 mA from negative output terminal.—B. Fette, Inexpensive Inverters Generate V_{GG} for Portable MOS Applications, *EDN/EEE Magazine*, Dec. 15, 1971, p 51.

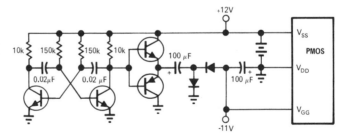

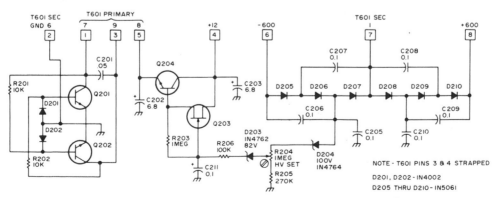

CRO HIGH-VOLTAGE SUPPLY—Controlled inverter operates from 12-V battery and feeds positive and negative triplers for producing ±600 V required for portable CRO. T601 high-voltage transformer has 22K, 5.2K, and 600-ohm windings, all center-tapped, often marked "Lionel" when available in surplus shops. Q201, Q202, and Q204 are 2N697 or 2N2219. Q203 is 2N4302 or 2N5457.—G. E. Friton, Eyes for Your Shack, *73 Magazine*, Nov./Dec. 1975, p 74–76, 78–88, and 90–94.

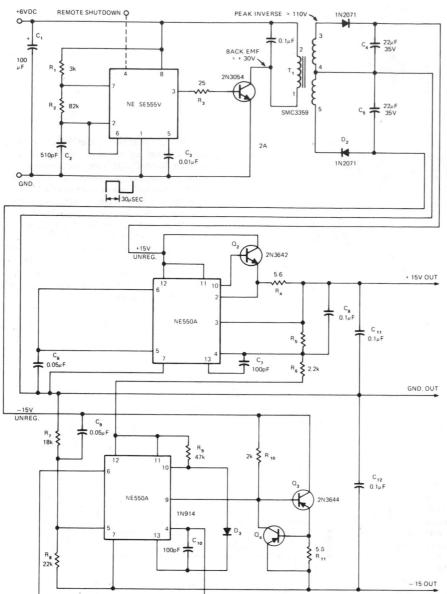

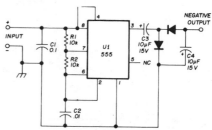

6 V TO ±15 V—Combination of 555 timer and two NE550 voltage regulators provides voltage multiplication along with regulation of independent DC outputs. Selected oscillator frequency of 17 kHz optimizes performance of transformer. Can be used to power opamps from either TTL supplies or 6-V batteries. Line and load regulation are 0.1%, while power efficiency at full load of 100 mA is better than 75%.—R. Solomon and R. Broadway, DC-to-DC Converter Uses IC Timer, *EDN Magazine*, Sept. 5, 1973, p 87, 89, and 91.

+12 V TO −8 V—Developed for use with mobile equipment when DC voltage is required with opposite polarity to that of auto battery. U1 is 555 timer operated as free-running square-wave oscillator. Frequency is determined by R1, R2, and C2; with values shown, it is about 6 kHz. C1 reduces 6-kHz signal radiated back through input lines. For 12-V input, typical outputs are −8.4 V at 10 mA, −7.9 V at 20 mA, and −5.7 V at 50 mA. All diodes are 1N914, 1N4148, or equivalent.—G. A. Graham, Low-Power DC-DC Converter, *Ham Radio*, March 1975, p 54–56.

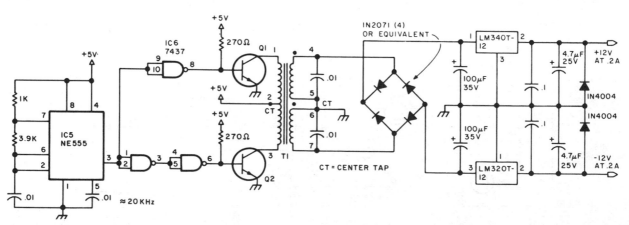

±12 V FROM +5 V—NE555 timer connected as 20-kHz oscillator drives pair of D44H4 transistors through 7437 quad two-input NAND buffer to produce full 200 mA of regulated output for each polarity. Circuit uses push-pull inverter technique to generate AC for driving transformer constructed by rewinding 88-mH toroid to have 40 turns No. 20 center-tapped for primary and 350 turns No. 26 center-tapped for secondary.—S. Ciarcia, Build a 5 W DC to DC Converter, *BYTE*, Oct. 1978, p 22, 24, 26, 28, and 30–31.

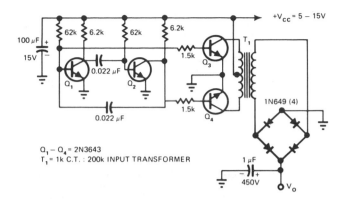

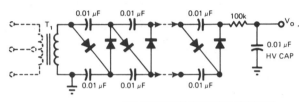

- ADD ADDITIONAL STAGES AS REQUIRED
- DIODES ARE 1N649 OR EQUIVALENT
- CAPACITORS ARE CERAMIC DISC 1 kV

5 V TO 400 V—Astable MVBR operating at 2.174 kHz for values shown drives push-pull transistor pair feeding primary of audio input transformer T₁. Secondary voltage is rectified by diode bridge to provide DC output voltage ranging from 100 to 400 V depending on load resistance and exact value of supply voltage V_{cc}. Bridge rectifier can be replaced by 40-stage multiplier as shown in lower diagram, to give 10-kVDC output.—A. M. Hudor, Jr., Power Converter Uses Low-Cost Audio Transformer, *EDN Magazine*, April 20, 1977, p 139.

280 V TO 600 V—Cascode push-pull transistor switch conversion circuit uses low-voltage transistors and provides automatic equalization of transistor storage time. Drive-signal input to cascode push-pull switch is symmetrical 50-kHz 15 V P-P square wave from 50-ohm source. Q1 and Q2 each see only half of DC source voltage because C1 and C2, in series across 280-V input, charge to 140 V each. Circuit is adaptable to wide range of output voltages and currents because identical units can be connected in series or parallel to obtain desired rating.—L. G. Wright and W. E. Milberger, HV Building Block Uses Series Transistor Switches, *EDN Magazine*, Feb. 15, 1971, p 39–40.

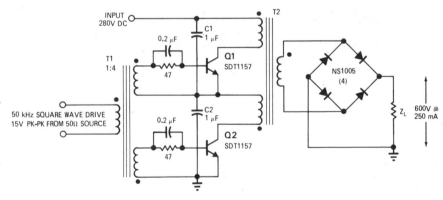

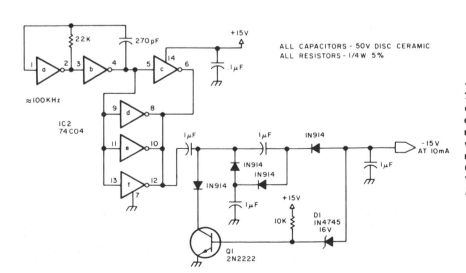

−15 V FROM +15 V—First two sections of 74C04 hex inverter form 100-kHz oscillator, with other sections connected to provide inversion of standard microprocessor source voltage as required for some interfaces and some D/A converters. Shunt regulator formed by D1 and Q1 maintains output voltage relatively constant. Changing zener D1 to 13 V makes output −12 V.—S. Ciarcia, Build a 5 W DC to DC Converter, *BYTE*, Oct. 1978, p 22, 24, 26, 28, and 30–31.

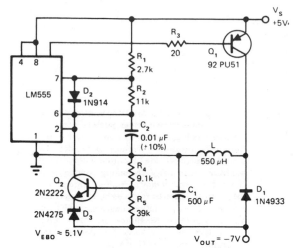

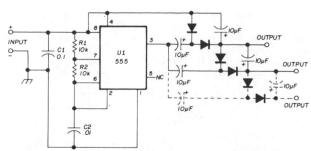

+5 V TO −7 V—Uses LM555 timer as variable-duty-cycle pulse generator controlling transistor switch Q_1 which in turn drives flyback circuit. Regulator Q_2-D_3-R_4-R_5 varies duty cycle according to load, and flyback circuit L-D_1-C_1 develops negative output voltage. When Q_1 is on, current flows through L to ground. When Q_1 turns off, polarity across L reverses, diode becomes forward-biased, and negative voltage appears across C_1 and load. When Q_1 turns on again, voltage across L reverses for start of new cycle. Circuit eliminates separate transformer supply for negative supply of microprocessor. Efficiency is about 60%, load regulation 1.3%, and supply rejection 30 dB. Article gives design equations.—P. Brown, Jr., Converter Generates Negative μP Bias Voltage from +5V, *EDN Magazine*, Aug. 5, 1977, p 42, 44, and 46.

DC MULTIPLIERS—Voltage output from positive voltage booster (+12 VDC to +20 VDC) is increased by using diode-capacitor voltage-doubler sections as shown. Diodes are 1N914, 1N4148, or equivalent. Doubling is achieved at expense of available current. Same technique may be used to increase output of DC/DC converter having negative output voltage.—G. A. Graham, Low-Power DC-DC Converter, *Ham Radio*, March 1975, p 54–56.

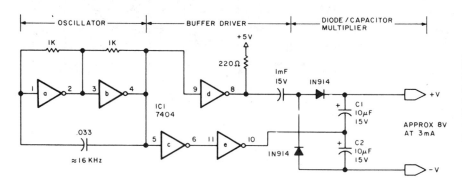

+8 V FROM +5 V—Oscillator operating at about 16 kHz steps up 5-V supply voltage of microprocessor to 8 V for driving special interface circuits. Sections c, d, and e of 7404 hex inverter form buffer and driver for voltage-doubling rectifier.—S. Ciarcia, Build a 5 W DC to DC Converter, *BYTE*, Oct. 1978, p 22, 24, 26, 28, and 30–31.

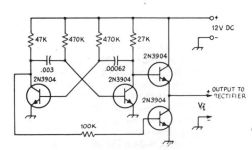

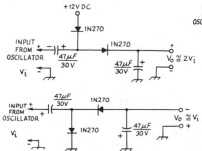

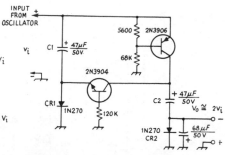

POLARITY REVERSER—Simple RC oscillator operating at about 1200 Hz can be used with choice of rectifier circuits to provide negative or positive voltages equal to or higher than DC supply, without use of transformer. Output transistors connect load alternately to positive supply and to ground for high operating efficiency. Two-diode voltage doubler with connection to 12-V supply gives positive output. Other diode rectifier circuit doubles oscillator output and gives negative supply. Negative doubler uses switching transistors. All three rectifier circuits provide common ground from supply to output.—J. M. Pike, Negative and High Voltages from a Positive Supply, *QST*, Jan. 1974, p 23–25.

±15 V FROM +5 V—Provides positive and negative higher voltages required by some interface devices used with microprocessors. NE555
timer is connected as 100-kHz oscillator that
switches transistor on and off, inducing current
in primary of T1. High-voltage spike reflected
back to collector of transistor by pulse transformer is routed through D1 to filter-regulator
for providing positive output.—S. Ciarcia, Build
a 5 W DC to DC Converter, *BYTE*, Oct. 1978, p
22, 24, 26, 28, and 30–31.

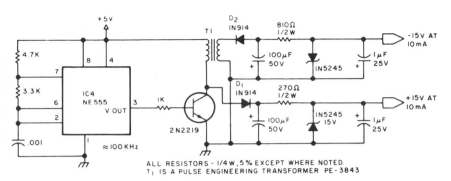

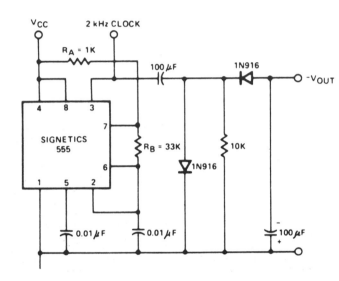

**TRANSFORMERLESS POSITIVE TO NEGA-
TIVE**—Used to derive negative supply voltage
from positive supply voltage, while at same
time generating 2-kHz clock signal. Negative
output voltage tracks DC input voltage linearly,
but magnitude is about 3 V lower. Circuit does
not provide regulation.—"Signetics Analog
Data Manual," Signetics, Sunnyvale, CA, 1977,
p 729.

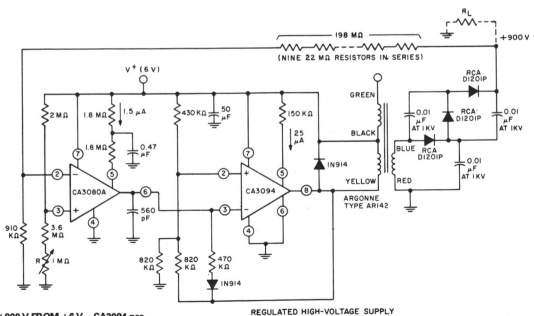

REGULATED HIGH-VOLTAGE SUPPLY

REGULATED +900 V FROM +6 V—CA3094 programmable opamp is connected as oscillator for
driving step-up transformer that develops suitable high voltage for rectification in diode network. Sample of +900 V regulated output is fed
to CA3080A variable opamp through 198-meg-
ohm resistor of voltage divider to control pulse
repetition rate of oscillator. Magnitude of regulated output is controlled by pot R. Regulation
is within 1% for loads of 5 to 26 μA. DC-to-DC
conversion efficiency is about 50%.—"Circuit
Ideas for RCA Linear ICs," RCA Solid State Division, Somerville, NJ, 1977, p 19.

0 TO −10 V FROM +12 V—Variable-output converter using NE555 timer delivers negative output voltage required by some interface devices and D/A converters used with microprocessors.—S. Ciarcia, Build a 5 W DC to DC Converter, *BYTE*, Oct. 1978, p 22, 24, 26, 28, and 30–31.

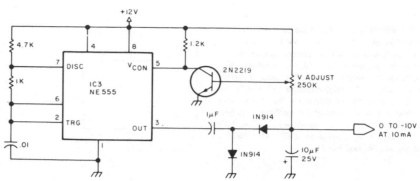

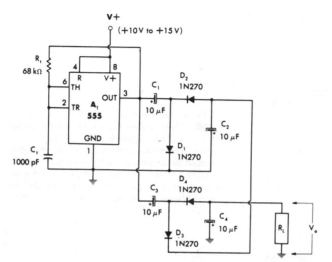

15 V TO −24 V—Voltage doubler is used in combination with 555 astable MVBR and two peak-to-peak detectors to give high negative voltages from positive voltage source. Load current capability is about 10 mA. Output drops to about −14.5 V when using supply of +10 V.—W. G. Jung, "IC Timer Cookbook," Howard W. Sams, Indianapolis, IN, 1977, p 197–201.

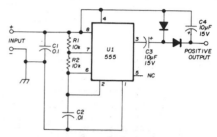

+12 V TO +20 V—Developed for use with mobile equipment when DC voltage higher than that of auto battery is needed. One application is trickle-charging 12-V nickel-cadmium batteries. U1 is 555 timer operated as free-running square-wave oscillator. Frequency is determined by R1, R2, and C2; with values shown, it is about 6 kHz. C1 reduces 6-kHz signal radiated back through input lines. If converter is used with high-frequency receiver, insert 100-μH RF chokes in power leads to suppress harmonics of 6 kHz. For 12-V input, typical outputs are 20.4 V at 10 mA, 19.9 V at 20 mA, and 17.7 V at 50 mA. All diodes are 1N914, 1N4148, or equivalent.—G. A. Graham, Low-Power DC-DC Converter; *Ham Radio*, March 1975, p 54–56.

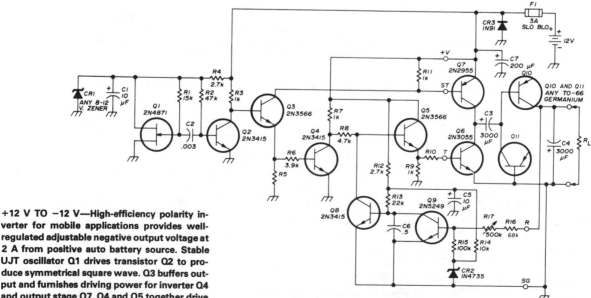

+12 V TO −12 V—High-efficiency polarity inverter for mobile applications provides well-regulated adjustable negative output voltage at 2 A from positive auto battery source. Stable UJT oscillator Q1 drives transistor Q2 to produce symmetrical square wave. Q3 buffers output and furnishes driving power for inverter Q4 and output stage Q7. Q4 and Q5 together drive Q6 into complete saturation. Q6 and Q7 form complementary-symmetry output operating in saturation mode, with only one transistor turned on at a time. As they are alternately switched on and off, square wave alternating between ground and nearly battery potential is applied to C3. Q10 and Q11 are connected as diodes for clamping square wave negatively. Output voltage is regulated by transistor feedback loop Q8 and Q9, with zener CR2 providing stable reference. R10 is 100 ohms for 2 A maximum; increasing its value improves efficiency but reduces maximum current.—J. R. Laughlin, Medium Current Polarity Inverter, *Ham Radio*, Nov. 1973, p 26–30.

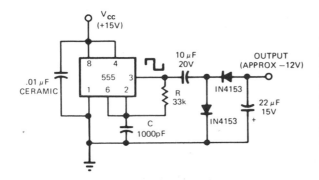

+15 V TO −12 V—Simple transformerless power converter uses 555 timer in self-triggered mode as square-wave generator, followed by voltage-doubling rectifier. Values shown for R and C give frequency of about 20 kHz, which permits good filtering with relatively small capacitors. Maximum load current is about 80 mA.—M. Strange, IC Timer Makes Transformerless Power Converter, *EDN Magazine*, Dec. 20, 1973, p 81.

±15 V FROM 12 V—Steps up output of 12-V battery to voltages required by PLL such as NE561. Uses 900-Hz sine-wave oscillator and LM380N AF amplifier to drive voice-coil side of standard 500-ohm to 3.2-ohm output transformer having bridge rectifier across center-tapped primary. With 10-mA loads, maximum ripple is 15 mV P-P. With receiver quiet, 900-Hz hum is audible, but is normally lost under background noise. Oscillator choke (about 700 mH) is 800 turns of No. 44 magnet wire in Ferroxcube 3C pot core.—R. Megirian, Build a Noise-Free Power Supply, *73 Magazine*, Dec. 1977, p 208–209.

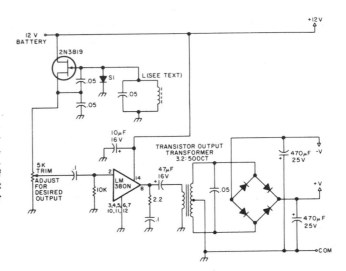

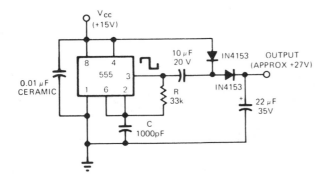

+15 V TO +27 V—Uses 555 timer in self-triggered mode as square-wave generator operating at about 20 kHz, followed by voltage-doubling rectifier. Provides approximate doubling of voltage without use of transformer. Maximum load current is about 80 mA.—M. Strange, IC Timer Makes Transformerless Power Converter, *EDN Magazine*, Dec. 20, 1973, p 81.

CHAPTER 18
Converter Circuits—Digital-to-Analog

Includes circuits for converting variety of digital inputs to linearly related analog output voltage or current, providing analog sum of two digital inputs, or converting stored digital speech back to analog form.

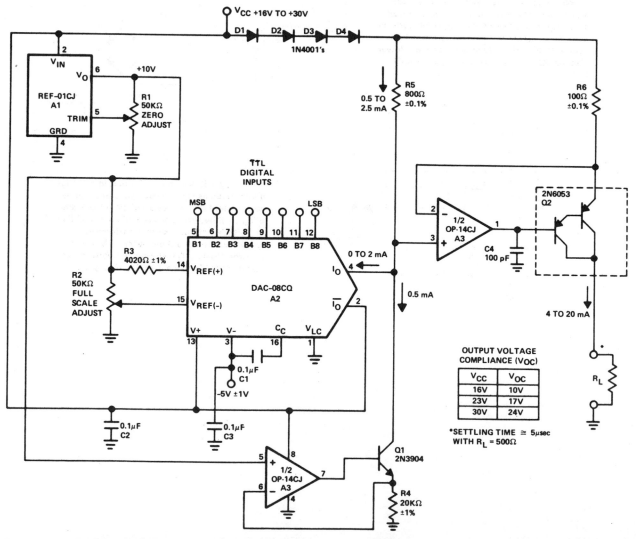

8-BIT BINARY TO PROCESS CURRENT—Uses only three Precision Monolithics ICs operating from −5 V and +23 V supplies to convert 8-bit binary digital input to process current in range of 4-20 mA. Fixed current of 0.5 mA is added to DAC output current varying between 0 and 2 mA, with resulting total current multiplied by factor of 8 to give up to 20 mA through 500-ohm load.—D. Soderquist, "3 IC 8 Bit Binary Digital to Process Current Converter with 4-20 mA Output," Precision Monolithics, Santa Clara, CA, 1977, AN-21.

HIGH-SPEED OUTPUT OPAMP—Precision Monolithics OP-17F opamp optimizes DAC-08E D/A converter for highest speed in converting DAC output current to output voltage up to 10 V under control of digital input. Settling time is 380 ns.—G. Erdi, "The OP-17, OP-16, OP-15 as Output Amplifiers for High Speed D/A Converters," Precision Monolithics, Santa Clara, CA, 1977, AN-24, p 2.

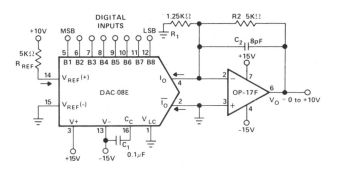

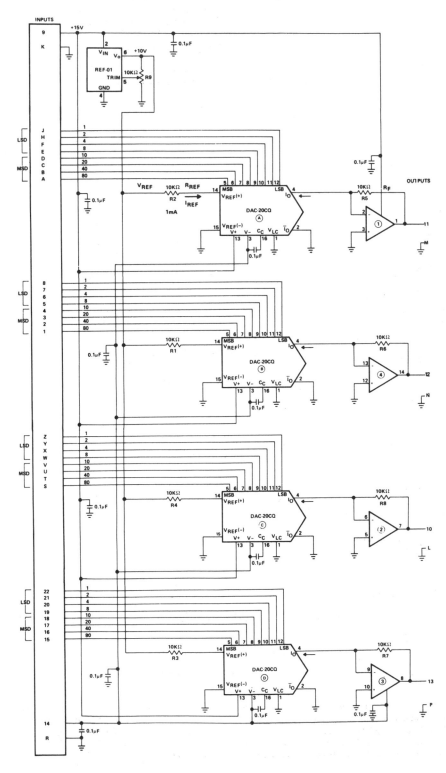

FOUR-CHANNEL BCD—Uses four Precision Monolithics DAC-20CQ 2-digit BCD D/A converter, OP-11FY precision quad opamp, and REF-01HJ +10 V voltage reference to convert 2-digit BCD input coding to proportional analog 0 to +10 V output for each of four channels. Same configuration will handle binary inputs, as covered in application note. For output range of 0 to +5 V, change voltage reference to REF-02.—D. Soderquist, "Low Cost Four Channel DAC Gives BCD or Binary Coding," Precision Monolithics, Santa Clara, CA, 1977, AN-26, p 3.

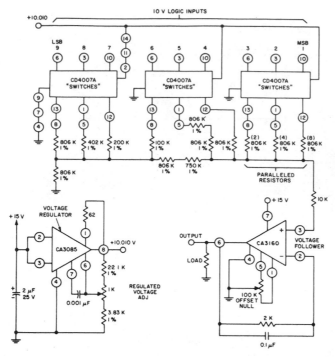

9-BIT USING DIGITAL SWITCHES—Combination of CD4007A multiple-switch CMOS ICs, ladder network of discrete metal-oxide film resistors, CA3160 voltage-follower opamp, and CA3085 voltage regulator gives digital-to-analog converter that is readily interfaced with 10-V logic levels of CMOS input. Required resistor accuracy, ranging from ±0.1% for bit 2 to ±1% for bits 6-9, is achieved by using series and parallel combinations of 806K resistors.—"Linear Integrated Circuits and MOS/FET's," RCA Solid State Division, Somerville, NJ, 1977, p 267–268.

6 BITS TO ANALOG—Uses Motorola MC1723G voltage regulator to provide reference voltage and opamp for MC1406L 6-bit D/A converter. Output current can be up to 150 mA. Full-scale output is about 10 V, but can be boosted as high as 32 V by increasing value of R₂ and increasing +15 V supply proportionately to maximum of 35 V.—D. Aldridge and K. Huehne, 6-Bit D/A Converter Uses Inexpensive Components, *EDN Magazine*, Dec. 15, 1972, p 40–41.

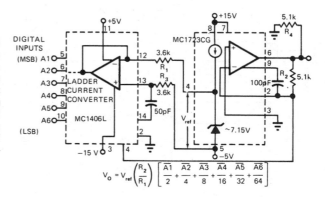

$$V_O = V_{ref}\left(\frac{R_2}{R_1}\right)\left[\frac{\overline{A1}}{2} + \frac{\overline{A2}}{4} + \frac{\overline{A3}}{8} + \frac{\overline{A4}}{16} + \frac{\overline{A5}}{32} + \frac{\overline{A6}}{64}\right]$$

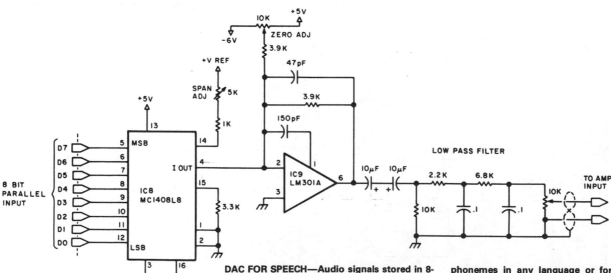

DAC FOR SPEECH—Audio signals stored in 8-channel digital form in computer are converted back into analog form for feed through low-pass filter to input of audio amplifier. Can be used for computer-controlled synthesis of speech from phonemes in any language or for providing voice replies to queries. Pin 7 of IC9 is +12 V, and pin 4 is −12 V.—S. Ciarcia, Talk to Me! Add a Voice to Your Computer for $35, *BYTE*, June 1978, p 142–151.

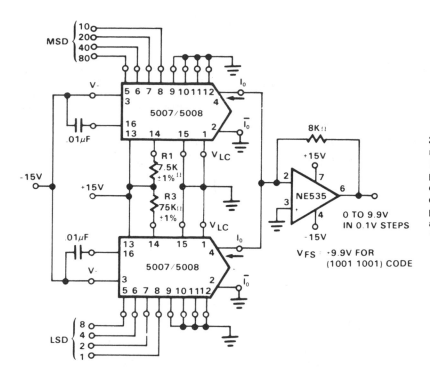

2-DIGIT BCD INPUT—Each Signetics 5007/5008 multiplying D/A converter serves one digit of input voltage to give output current that is product of digital input number and input reference current. Opamp combines currents and converts them to analog output voltage proportional to digital input value.—"Signetics Analog Data Manual," Signetics, Sunnyvale, CA, 1977, p 677–685.

ANALOG SUM OF DIGITAL NUMBERS—Two Precision Monolithics DAC-100 D/A converters and OP-01 opamp combine conversion with adding to give high-precision DC output voltage. 200-ohm pots are adjusted initially to give exactly desired output for input of all 0s.—"8 & 10 Bit Digital-to-Analog Converter," Precision Monolithics, Santa Clara, CA, 1977, DAC-100, p 5.

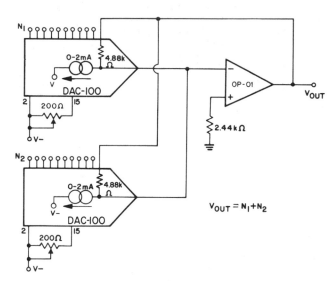

SIMPLE DAC—Transistors are either saturated or cut off by outputs of clock-controlled SN7490 BCD counter. Portions of emitter voltages of the four transistors are added in ratios 1:2:4:8 by 741 summing opamp to obtain analog output. Article tells how two such circuits can be combined for use in two-digit DVM.—D. James, Simple Digital to Analogue Converter, *Wireless World,* June 1974, p 197.

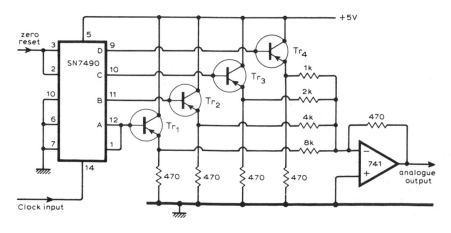

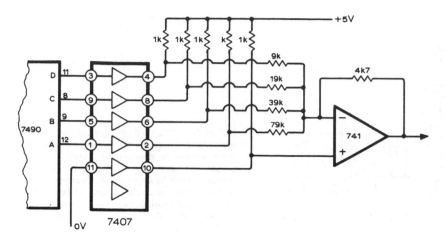

TEMPERATURE COMPENSATION—Use of 7407 hex buffer following SN7490 of D/A converter permits satisfactory performance over reasonably wide temperature range even when driving several TTL stages. Noninverting input of 741 opamp is connected to output of unused buffer at logic 0. Circuit is modification of D/A converter developed by D. James for use in simple two-digit DVM.—R. J. Chance, Improved Simple D. to A. Converter, *Wireless World,* Dec. 1974, p 503.

2-DIGIT BCD—Output current of Precision Monolithics DAC-100 D/A converter can be adjusted to exactly desired value with 200-ohm pot for each DAC; adjustment is made with input of all 0s. Circuit can be expanded to 3 digits by adding third DAC and adding 99 to current divider.—"8 & 10 Bit Digital-to-Analog Converter," Precision Monolithics, Santa Clara, CA, 1977, DAC-100, p 5.

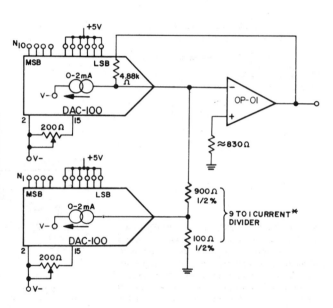

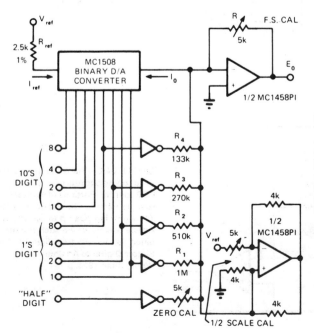

2½-DIGIT INPUT FOR 199 COUNT—Addition of ½-digit circuit to basic 2-digit BCD DAC increases count from 99 to 199. Circuit sequences to 99 while ½-digit section of MC14009 hex two-input NOR gate has low output, and goes through steps 100 to 199 while ½-digit output is high. Reference voltage is 5.0 V. Calibration procedure is given.—T. Henry, Binary D/A Converters Can Provide BCD-Coded Conversion, *EDN Magazine,* Aug. 5, 1973, p 70–73.

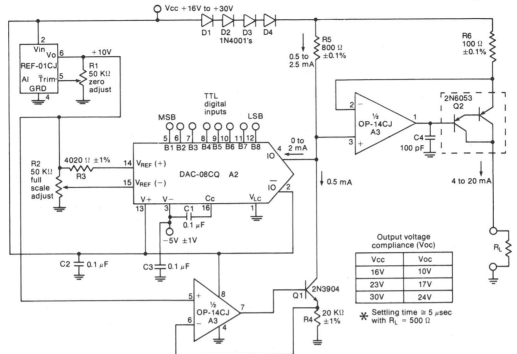

CURRENT CONVERTER—Converts 8-bit TTL digital inputs to process current in range of 4 to 20 mA, for microprocessor control of industrial operations. Fixed 0.5-mA current is added to DAC output current varying between 0 and 2.0 mA and multiplied by factor of 8 to produce final output current of 4–20 mA. To calibrate, connect ammeter between output and ground, then apply +23 V ± 7 V and −5 V ± 1 V to converter. Make digital inputs all 0s (less than +0.8 V). Adjust R1 until output current is 4.0 mA. Change digital inputs to all 1s (greater than +2.0 V), and adjust R2 until output current is 20 mA.—D. Soderquist, Build Your Own 4-20 mA Digital to Analog Converter, *Instruments & Control Systems,* March 1977, p 57–58.

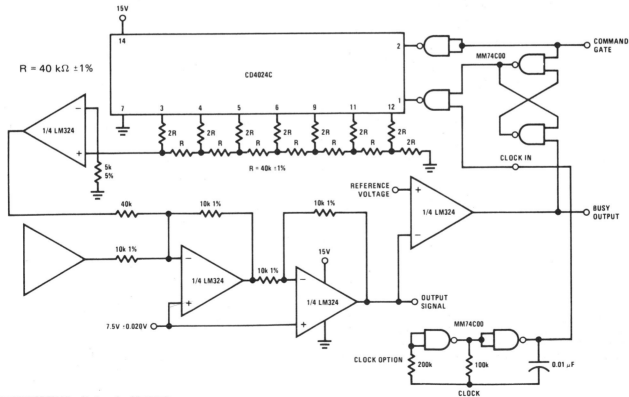

AUTOREFERENCE—National CD4024C converter is used with logic and summer elements to eliminate virtually all offset errors induced by time and temperature changes in process control system fed by transducer. Best suited for applications having short repeated duty cycles, each containing reference point. Examples include weighing scale in which transducer is load cell, pressure control systems, fuel pumps, and sphygmomanometers. Circuit eliminates warm-up errors.—"Pressure Transducer Handbook," National Semiconductor, Santa Clara, CA, 1977, p 7-4–7-8.

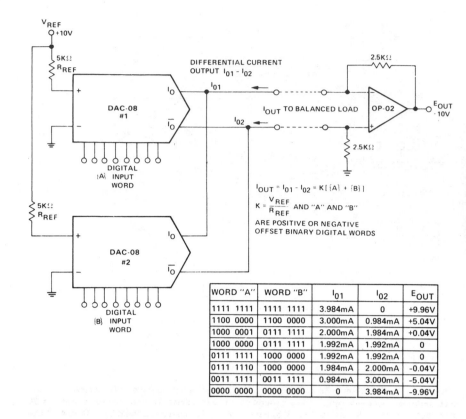

$$I_{OUT} = I_{01} - I_{02} = K[(A) + (B)]$$

$$K = \frac{V_{REF}}{R_{REF}} \text{ AND "A" AND "B"}$$

ARE POSITIVE OR NEGATIVE
OFFSET BINARY DIGITAL WORDS

FOUR-QUADRANT ALGEBRAIC—Two Precision Monolithics DAC-08 D/A converters perform fast algebraic summation of two digital input words and feed OP-02 opamp that provides direct analog output which is algebraic sum of words A and B in all four quadrants.—J. Schoeff and D. Soderquist, "Differential and Multiplying Digital to Analog Converter Applications," Precision Monolithics, Santa Clara, CA, 1976, AN-19, p 7.

WORD "A"	WORD "B"	I_{01}	I_{02}	E_{OUT}
1111 1111	1111 1111	3.984mA	0	+9.96V
1100 0000	1100 0000	3.000mA	0.984mA	+5.04V
1000 0001	0111 1111	2.000mA	1.984mA	+0.04V
1000 0000	0111 1111	1.992mA	1.992mA	0
0111 1111	1000 0000	1.992mA	1.992mA	0
0111 1110	1000 0000	1.984mA	2.000mA	-0.04V
0011 1111	0011 1111	0.984mA	3.000mA	-5.04V
0000 0000	0000 0000	0	3.984mA	-9.96V

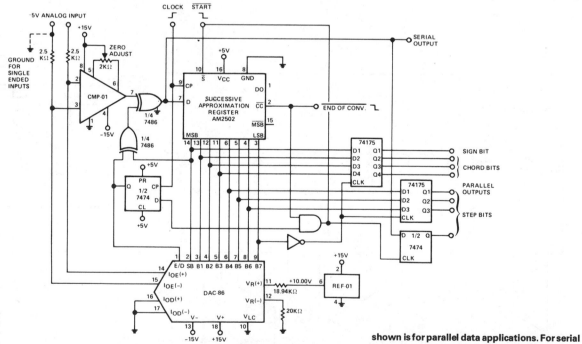

BELL-SYSTEM μ-255 COMPANDING LAW—Precision Monolithics DAC-86 is used in circuit that provides 15-segment linear approximation by using 3 bits to select one of eight binarily related chords, then using 4 bits to select one of sixteen linearly related steps within each chord. Sign bit determines signal polarity, and encode/decode select bit determines operation. Circuit shown is for parallel data applications. For serial data, omit inverter, two 74175 chips, and half of 7474. Power supplies should be well bypassed.—"COMDAC Companding D/A Converter," Precision Monolithics, Santa Clara, CA, 1977, DAC-86, p 6.

CHAPTER 19
Converter Circuits—General

Includes V/F, V/I, V/pulse width, V/time, F/V, 7-segment/BCD, BCD/7-segment, Gray/BCD, Gray/binary, binary/BCD, time/V, pulse height/time, I/V, and other converter circuits for changing one parameter linearly to another. See also other Converter chapters.

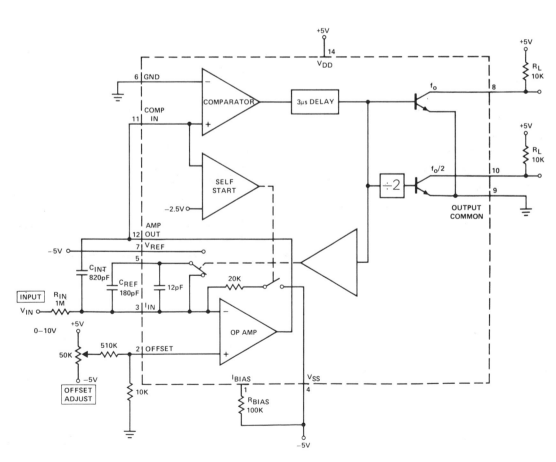

10 Hz TO 10 kHz V/F—External circuit shown for Teledyne 9400 voltage-to-frequency converter provides means for trimming zero location and full-scale frequency value of output. For 10-kHz full-scale value, set V_{IN} to 10 mV and trim with 50K offset adjust pot to get 10-Hz output, then set V_{IN} to 10.000 V and trim either R_{IN}, V_{REF}, or C_{REF} to obtain 10-kHz output.—M. O. Paiva, "Applications of the 9400 Voltage to Frequency Frequency to Voltage Converter," Teledyne Semiconductor, Mountain View, CA, 1978, AN-10, p 3–5.

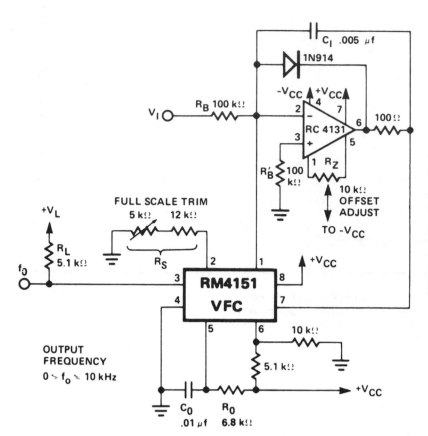

V/F CONVERTER WITH 0.05% LINEARITY—Raytheon RM4151 converter is used with integrator opamp to give highly linear conversion of inputs up to −10 VDC to proportional frequency of square-wave output. With maximum input of −10 V, adjust 5K full-scale trimpot for maximum output frequency of 10 kHz. Set offset adjust pot to give 10-Hz output for input of −10 mV. To operate from single positive supply, change opamp to RC3403A.—"Linear Integrated Circuit Data Book," Raytheon Semiconductor Division, Mountain View, CA, 1978, p 7-38.

0.1 Hz–100 kHz V/F—Uses NE556 timer in dual mode in combination with opamp and FET for linear voltage-to-frequency conversion with output range from 0.1 Hz to 100 kHz. Operating frequency is 0.91/2RC where R is resistance of FET.—K. Kraus, Linear V-F Converter, *Wireless World,* May 1977, p 80.

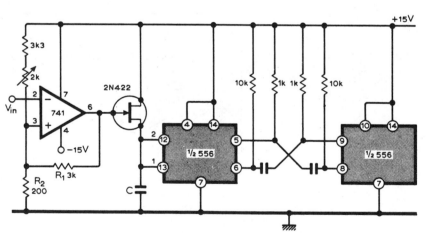

BCD	GRAY	X-3 GRAY
0000	0000	0010
0001	0001	0110
0010	0011	0111
0011	0010	0101
0100	0110	0100
0101	0111	1100
0110	0101	1101
0111	0100	1111
1000	1100	1110
1001	1101	1010

EXCESS-THREE GRAY CODE TO BCD—Developed for use with shaft encoder providing excess-three Gray-code output. Requires only two TTL ICs, connected as shown. To convert regular Gray code to BCD, omit SN7483 4-bit adder. Tabulation shows how circuit accomplishes conversion for both types of Gray codes.—D. M. Risch, Two ICs Convert Excess-Three Gray Code to BCD, *EDN Magazine,* Nov. 1, 1972, p 44.

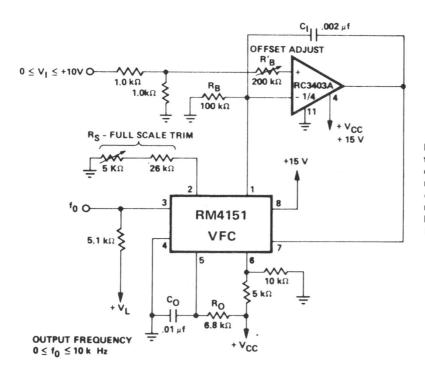

HIGH-PRECISION V/F CONVERTER—Active integrator using one section of RC3403A quad opamp improves linearity, frequency offset, and response time of Raytheon RM4151 converter operating from single supply. Opamp develops null voltage.—"Linear Integrated Circuit Data Book," Raytheon Semiconductor Division, Mountain View, CA, 1978, p 7-38.

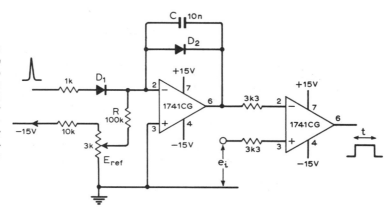

DC VOLTAGE TO TIME—Opamp connected as integrator feeds opamp comparator to produce output pulse whose width is proportional to magnitude of DC input voltage. Circuit shown is for positive inputs only; for both positive and negative inputs, article tells how to add another comparator. Circuit can then be used to generate start and stop pulses applied to digital timer of digital voltmeter.—G. B. Clayton, Experiments with Operational Amplifiers, *Wireless World*, Sept. 1973, p 447–448.

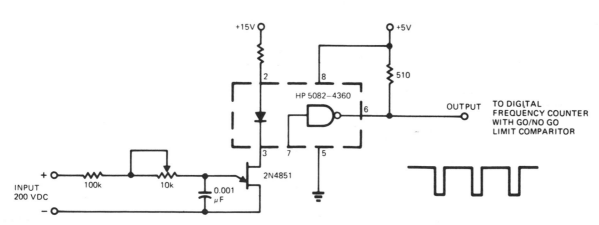

VOLTAGE-TO-FREQUENCY GO/NO-GO—Single UJT is used as V/F converter to provide completely isolated inputs and outputs for high-voltage go/no-go test monitor. When voltage exceeds predetermined limit, output to digital frequency counter exceeds corresponding frequency limit. Output can be fed directly into digital frequency-limit detector that provides go/no-go indication.—T. H. Li, VFC Used in Isolated GO/NO GO Voltage Monitor, *EDN Magazine*, July 5, 1974, p 75.

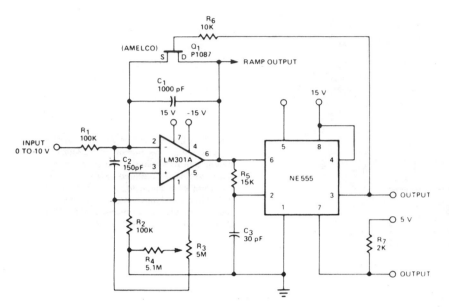

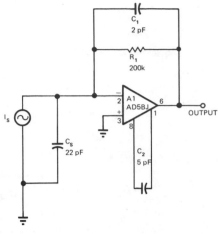

VOLTAGE TO FREQUENCY—Input voltage range of 0 to −10 VDC is converted by opamp and timer to proportional frequency with good linearity. Circuit is TTL-compatible. Accuracy is 0.2%.—"Signetics Analog Data Manual," Signetics, Sunnyvale, CA, 1977, p 727–729.

CURRENT TO VOLTAGE—Developed for use with current-output transducers such as silicon photocells. For widest frequency response, circuit values may need some adjusting for source current and capacitance. C_1, across feedback resistor of opamp, eliminates ringing around 500 kHz. If input coupling capacitor is added to reduce DC gain, circuit can be used with inductive source such as magnetic tape head.—R. S. Burwen, Current-to-Voltage Converter for Transducer Use, *EDN Magazine,* Dec. 15, 1972, p 40.

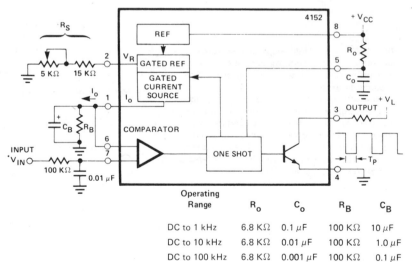

0.01–10 V to 1, 10, OR 100 kHz—Simple voltage-to-frequency converter uses Raytheon 4152 operating from single 15-V supply to convert analog input voltage to proportional frequency of square-wave output. Maximum output frequency depends on values used for resistors and capacitors, as given in table. Suitable for applications where input dynamic range is limited and does not go to zero.—"Linear Integrated Circuit Data Book," Raytheon Semiconductor Division, Mountain View, CA, 1978, p 7-45–7-46.

Operating Range	R_O	C_O	R_B	C_B
DC to 1 kHz	6.8 KΩ	0.1 μF	100 KΩ	10 μF
DC to 10 kHz	6.8 KΩ	0.01 μF	100 KΩ	1.0 μF
DC to 100 kHz	6.8 KΩ	0.001 μF	100 KΩ	0.1 μF

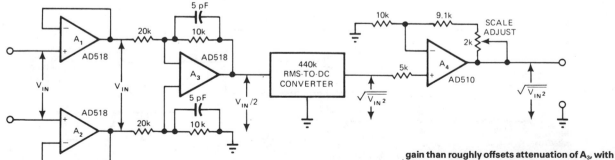

20-kHz SQUARE-WAVE TO DC—Provides accuracy within 0.1% for square-wave inputs of 3 to 7 VRMS in frequency range of 5 to 20 kHz when duty cycle is 50%. Opamps A_1 and A_2 are connected as differential input voltage-followers to provide high input impedance. A_3 converts to single-ended output as required by Model 440 converter IC. A_4 provides adjustable gain than roughly offsets attenuation of A_3, with 2K pot being adjusted to provide desired ratio of DC output voltage to RMS value of input.—J. Renken, Differential High-Z RMS-DC Converter Has 0.1% Accuracy, *EDN Magazine,* May 5, 1977, p 114 and 116.

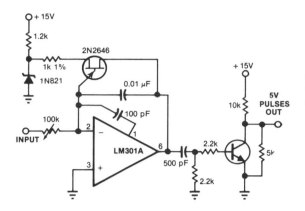

SINGLE-SLOPE V/F CONVERSION—UJT forms reference that determines reset point of LM301A integrator for converting analog input voltage to proportional frequency. Output of integrator ramps negative until UJT switches and drives output positive at high slew rate. Positive edge of integrator output is differentiated by RC network and level-shifted by NPN bipolar transistor to provide logic-compatible pulse.—J. Williams, Low-Cost, Linear A/D Conversion Uses Single-Slope Techniques, *EDN Magazine*, Aug. 5, 1978, p 101–104.

GRAY TO BINARY—Converts first 4 bits of Gray-code word to binary output. Uses two MC7496 shift registers and logic elements to transfer data serially from input register through MC1812 EXCLUSIVE-OR IC to output register. One requirement is that strobe on pin 8 of input register must complete its function before clock appears on pin 1 of register. When this and other timing conditions are satisfied, converter will work at speeds up to about 10 megabits per second.—J. Barnes, "Analog-to-Digital Cyclic Converter," Motorola, Phoenix, AZ, 1974, AN-557, p 9.

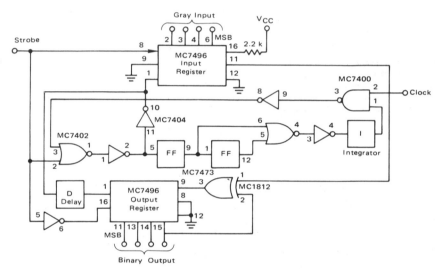

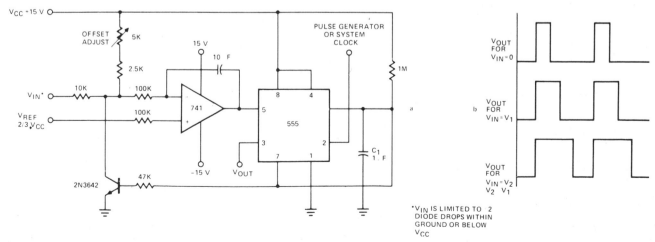

VOLTAGE TO PULSE WIDTH—Opamp and timer together convert input voltage level to width of output pulse with accuracy better than 1%. Output is at same frequency as input.— "Signetics Analog Data Manual," Signetics, Sunnyvale, CA, 1977, p 726–727.

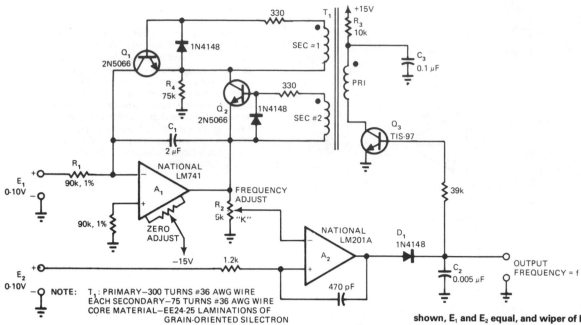

DIVIDING ANALOG VOLTAGES—Charge-and-dump V/F conversion technique is used to obtain quotient of two analog voltages digitally. Applications include measurement of total mass flow of gas under constant pressure. With Q_1 and Q_2 normally off, C_1 charges at rate proportional to input E_1, producing negative ramp at output of integrating opamp A_1. When R_2 wiper voltage exceeds E_2, output of A_2 goes high, turning on Q_3 and temporarily driving Q_1 and Q_2 into conduction. This process discharges C_1, resetting output of A_1 to zero. With values shown, E_1 and E_2 equal, and wiper of R_2 at midpoint, circuit output is 10,000 pulses per hour, suitable for driving electromechanical counter. Output frequency is proportional to average DC value of E_1, even when input changes rapidly.—H. L. Trietley, Voltage-to-Frequency Converter Performs Division, *EDN Magazine,* Jan. 5, 1978, p 79–80.

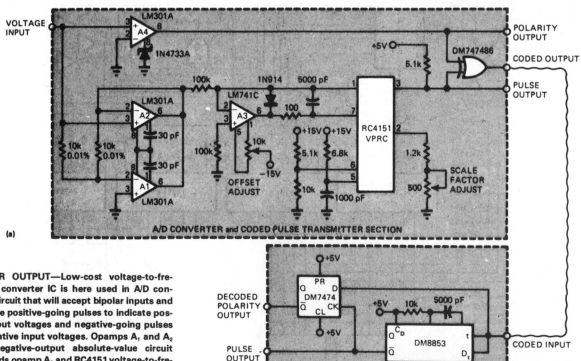

(a) A/D CONVERTER and CODED PULSE TRANSMITTER SECTION

(b) PULSE RECEIVER/DECODER SECTION

BIPOLAR OUTPUT—Low-cost voltage-to-frequency converter IC is here used in A/D converter circuit that will accept bipolar inputs and generate positive-going pulses to indicate positive input voltages and negative-going pulses for negative input voltages. Opamps A_1 and A_2 form negative-output absolute-value circuit that feeds opamp A_3 and RC4151 voltage-to-frequency converter, for generating 10-μs negative-going pulses with repetition-rate scale factor of 1 kHz/V. Full-scale input is 20 V P-P. Opamp A_4 is ground-referenced voltage comparator having zener clamp, providing TTL-compatible logic output for indicating input polarity. Pulse-receiver/decoder section takes digitally coded signals arriving from transmitter, for processing by data line of D flip-flop and two trigger inputs of dual-edge retriggerable mono MVBR. This generates 25-μs negative-going output pulse for each 10-μs input pulse from data line. Logic 1 on complementary output of flip-flop indicates that original input voltage to transmitter was positive.—E. J. DeWath, Low-Cost A/D Converter Transmits and Receives, *EDN Magazine,* Jan. 5, 1977, p 35.

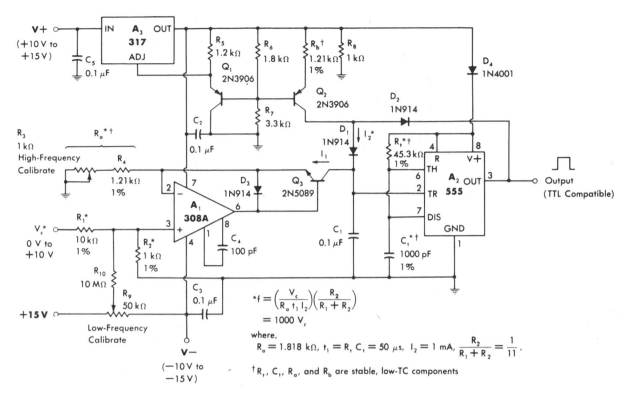

$$*f = \left(\frac{V_c}{R_a t_1 I_2}\right)\left(\frac{R_2}{R_1 + R_2}\right)$$
$$= 1000 V_c$$

where,

$R_a = 1.818$ kΩ, $t_1 = R_1 C_1 = 50$ μs, $I_2 = 1$ mA, $\dfrac{R_2}{R_1 + R_2} = \dfrac{1}{11}$.

$\dagger R_1$, C_1, R_a, and R_b are stable, low-TC components

POSITIVE-INPUT V/F—Input voltages from 0 to 10 V are divided by R_1 and R_2 for application to noninverting input of current source A_1. 555 timer A_2 provides functions of precision mono MVBR and level sensor. Regulator A_3 acts as gated current source and provides stabilized voltage output for 555 and 308A.—W. G. Jung, "IC Timer Cookbook," Howard W. Sams, Indianapolis, IN, 1977, p 184–192.

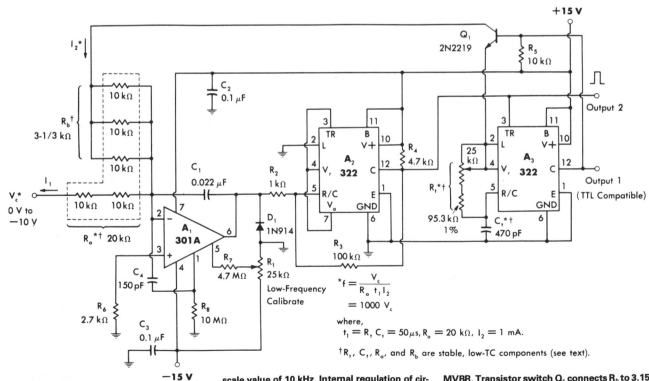

$$*f = \frac{V_c}{R_a t_1 I_2}$$
$$= 1000 V_c$$

where,

$t_1 = R_1 C_1 = 50 \mu$s, $R_a = 20$ kΩ, $I_2 = 1$ mA.

$\dagger R_1$, C_1, R_a, and R_b are stable, low-TC components (see text).

−10 V GIVES 10 kHz—Control voltage input in range of 0 to −10 V is converted linearly to frequency of digital output pulse train having full-scale value of 10 kHz. Internal regulation of circuit makes operation essentially independent of ±15 V supply level. A_1 is opamp integrator, A_2 is comparator, and A_3 is precision mono MVBR. Transistor switch Q_1 connects R_b to 3.15-V reference voltage during t_1 timing period of A_3.—W. G. Jung, "IC Timer Cookbook," Howard W. Sams, Indianapolis, IN, 1977, p 184–192.

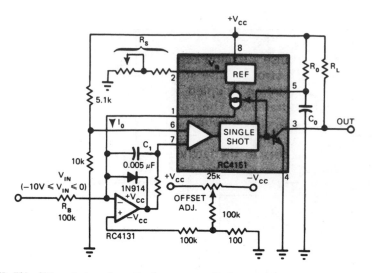

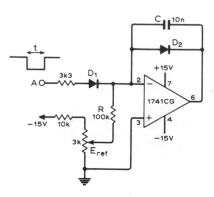

V/F AND F/V—Although based on Raytheon 4151 IC voltage-to-frequency converter, circuit is readily adapted to other modern V/F converters now costing under $10 each. With values shown, input of 0 to −10 VDC provides proportional frequency change from 0 to 10 kHz at output. Design equations are given. Article also covers F/V operation of same IC for demodulating FSK data.—T. Cate, IC V/F Converters Readily Handle Other Functions Such as F/V, A/D, *EDN Magazine*, Jan. 5, 1977, p 82–86.

TIME TO VOLTAGE—Time period of negative gating signal determines amplitude of linear output ramp generated by integrator opamp. Amplitude of ramp, proportional to input time, is observed on calibrated screen of oscilloscope.—G. B. Clayton, Experiments with Operational Amplifiers, *Wireless World*, Sept. 1973, p 447–448.

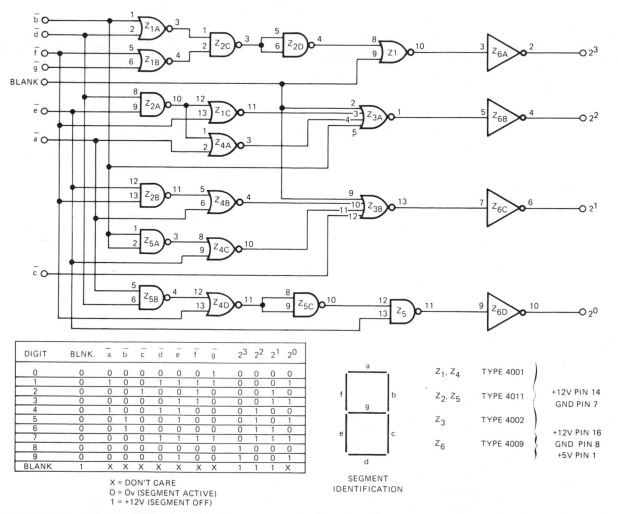

DIGIT	BLNK.	$\bar{a}$	$\bar{b}$	$\bar{c}$	$\bar{d}$	$\bar{e}$	$\bar{f}$	$\bar{g}$	2^3	2^2	2^1	2^0
0	0	0	0	0	0	0	0	1	0	0	0	0
1	0	1	0	0	1	1	1	1	0	0	0	1
2	0	0	0	1	0	0	1	0	0	0	1	0
3	0	0	0	0	0	1	1	0	0	0	1	1
4	0	1	0	0	1	1	0	0	0	1	0	0
5	0	0	1	0	0	1	0	0	0	1	0	1
6	0	0	1	0	0	0	0	0	0	1	1	0
7	0	0	0	0	1	1	1	1	0	1	1	1
8	0	0	0	0	0	0	0	0	1	0	0	0
9	0	0	0	0	1	0	0	1	0	0	1
BLANK	1	X	X	X	X	X	X	X	1	1	1	X

X = DON'T CARE
O = 0v (SEGMENT ACTIVE)
1 = +12V (SEGMENT OFF)

SEGMENT IDENTIFICATION

Z_1, Z_4	TYPE 4001	+12V PIN 14
Z_2, Z_5	TYPE 4011	GND PIN 7
Z_3	TYPE 4002	+12V PIN 16
Z_6	TYPE 4009	GND PIN 8 +5V PIN 1

7-SEGMENT TO BCD—Uses six CMOS packages to convert 7-segment display to corresponding four-line positive-logic BCD code for digits 0-9. Added feature is blank input which, when high, forces blank code (1110 or 1111) into readout, for use in suppressing leading zeros with some types of data storage. Use 4010 in place of 4009 for Z_6 when negative-logic BCD output is required.—R. Sturla, Real-Time 7-Segment to BCD Converter, *EDN Magazine*, June 20, 1973, p 89.

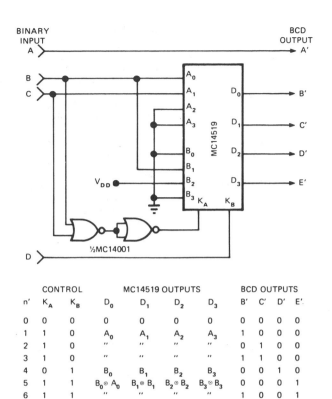

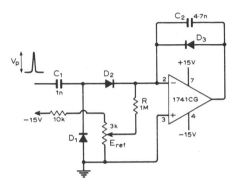

CONTROL		MC14519 OUTPUTS				BCD OUTPUTS			
n'	K_A	K_B	D_0	D_1	D_2	D_3	B'	C'	D'	E'
0	0	0	0	0	0	0	0	0	0	0
1	1	0	A_0	A_1	A_2	A_3	1	0	0	0
2	1	0	"	"	"	"	0	1	0	0
3	1	0	"	"	"	"	1	1	0	0
4	0	1	B_0	B_1	B_2	B_3	0	0	1	0
5	1	1	$B_0 \oplus A_0$	$B_1 \oplus B_1$	$B_2 \oplus B_2$	$B_3 \oplus B_3$	0	0	0	1
6	1	1	"	"	"	"	1	0	0	1
7	1	1	"	"	"	"	0	1	0	1

4-BIT BINARY TO 5-BIT BCD—Converts binary number within machine to BCD value from 0 to 15, for driving visual displays. Requires only quad two-channel data selector with EXCLUSIVE-NOR function, available in IC packages.

Article gives truth tables and traces operation step by step.—J. Barnes and J. Tonn, Binary-to-BCD Converter Implements Simple Algorithm, *EDN Magazine*, Jan. 5, 1975, p 56, 57, and 59.

PULSE HEIGHT TO TIME—Simple opamp circuit produces time interval proportional to height of positive input pulse. Opamp is connected as integrator whose output is held at about zero by negative feedback through D_3. Positive input pulse charges C_1 and C_2, amplifier output steps down, and D_3 is reverse-biased. Time for output to charge back up to zero, as observed on oscilloscope, is then directly proportional to input pulse height. Article gives design equations.—G. B. Clayton, Experiments with Operational Amplifiers, *Wireless World*, Sept. 1973, p 447–448.

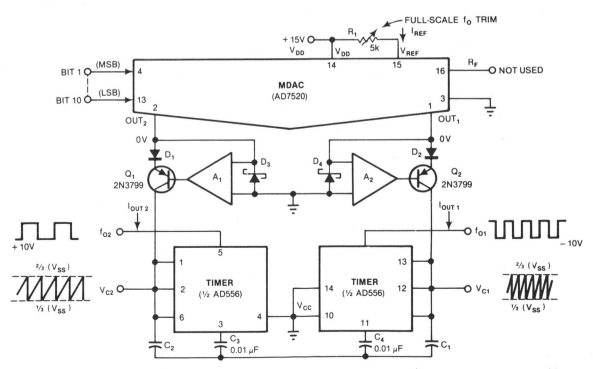

DIGITAL TO FREQUENCY—Combination of multiplying DAC and 556 dual timer provides complementary output frequencies under control of digital input. Opamp and diode types are not critical. Output frequency of each timer depends on supply voltages, capacitor values, and setting of R_1.—J. Wilson and J. Whitmore, MDAC's Open Up a New World of Digital-Control Applications, *EDN Magazine*, Sept. 20, 1978, p 97–105.

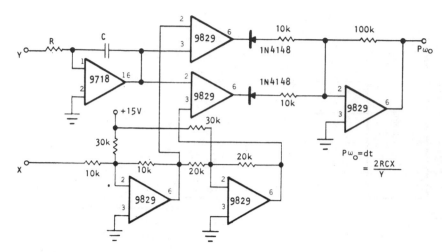

VOLTAGE TO PULSE DURATION—Optical Electronics 9829 opamps are used as fast comparators and 9718 FET opamp as fast integrator to give high precision at high speed for converting analog voltage to pulse duration for such applications as A/D conversion, delta code generation, motor speed control, and pulse-duration modulation. Output pulse durations can be as short as 1 μs. Conversion linearity is better than 0.1%. Minimum pulse duration is 100 ns, and maximum dynamic range is 40 dB. Reference voltages are determined by X input; if X is 3 V, reference voltages differ by 6 V. Two 9829 opamps present reference voltages to two comparator opamps. Fifth 9829 sums comparator outputs and gives positive output.—"Voltage to Pulse Width Converter," Optical Electronics, Tucson, AZ, Application Tip 10230.

VOLTAGE TO CURRENT—Circuit is capable of supplying constant alternating current up to 1 A to variable load. Actual value of load current is determined by input voltage, values of R_1-R_3, and value of R_5. Input of 250 mV gives 0.5 A through load (RMS values) with less than 0.5% total harmonic distortion. Applications include control of electromagnet current.—"Audio Handbook," National Semiconductor, Santa Clara, CA, 1977, p 4-21–4-28.

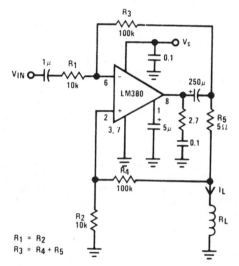

$$R_1 = R_2$$
$$R_3 = R_4 + R_5$$

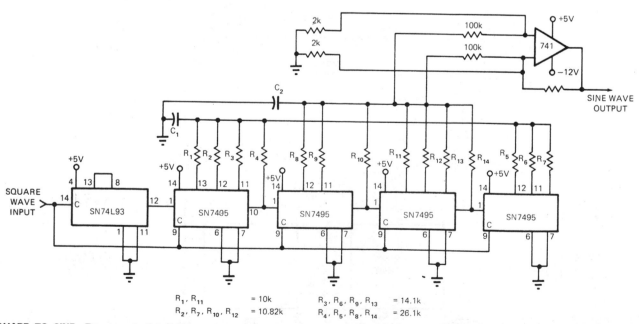

R_1, R_{11}	= 10k
R_2, R_7, R_{10}, R_{12}	= 10.82k
R_3, R_6, R_9, R_{13}	= 14.1k
R_4, R_5, R_8, R_{14}	= 26.1k

SQUARE TO SINE—Transversal digital filter suppresses harmonics present on input square wave, to give pure sine wave. Resistors weight data as it passes through 16-bit shift register, so sine wave is sampled at 16 times its frequency and theoretically has no harmonics below the 16th. Simple RC filter removes remaining harmonics. Input is clock whose repetition rate is 16 times desired frequency. SN74L93 4-bit ripple counter divides this down to provide square wave of desired frequency. Square wave is sampled 16 times per cycle and shifted down SN7495 16-bit shift register. C_1 and C_2 are selected to eliminate higher harmonics. Sine-wave output has harmonic distortion of less than −50 dB.—L. J. Mandell, Sine-Wave Synthesizer Has Low Harmonic Distortion, *EDN Magazine*, Aug. 15, 1972, p 52.

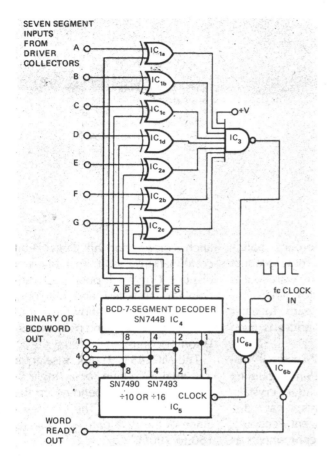

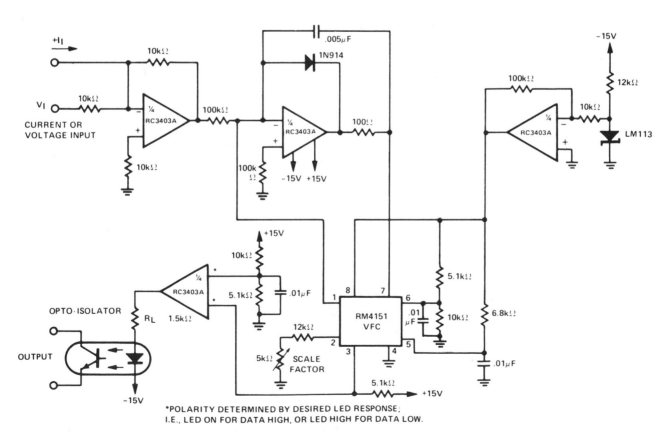

7-SEGMENT TO BCD—Arrangement uses SN7448 BCD to 7-segment decoder IC$_4$ as lookup table for inverse decoding technique. When desired 7-segment code is applied to input of decoder and does not match output code from IC$_4$, gate IC$_3$ output is logic 1. This allows pulses from clock to advance BCD counter IC$_5$ until its decoded state from IC$_4$ matches that of input code. With coincidence, output of IC$_3$ goes low, holding proper BCD code in IC$_5$ and indicating by means of IC$_{6b}$ that BCD information is ready. With 100-kHz clock, correct code is available for at least 90% of digit display time.—J. P. Cater, 7-Segment to BCD Decoder, *EDN Magazine,* Feb. 20, 1973, p 92–93.

*POLARITY DETERMINED BY DESIRED LED RESPONSE;
I.E., LED ON FOR DATA HIGH, OR LED HIGH FOR DATA LOW.

OPTICALLY COUPLED V/F—Input voltage range of 0–10 V is converted to proportional frequency at output of optoisolator with high linearity by RM4151 converter used in combination with RC3403A quad opamp that provides functions of inverter, integrator, regulator, and LED driver.—"Linear Integrated Circuit Data Book," Raytheon Semiconductor Division, Mountain View, CA, 1978, p 7-40–7-41.

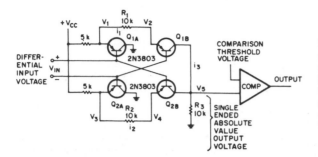

DIFFERENTIAL TO ABSOLUTE VALUE—Used in comparing differential level to threshold level with good common-mode rejection. Input impedance is maintained high to avoid overloading differential input. Output voltage remains positive when input polarity is reversed.—R. L. Wiker, Differential to Absolute Value Converter, *EEE Magazine*, Jan. 1971, p 65.

RMS TO DC—Single AD534 analog multiplier and two opamps compute RMS value of input signal as square root of sum of squares. Input is first squared at X_2 and Y_1, then time-averaged by integrator. Closing output loop back to X_1 and Y_2 completes square-rooting function. Crest factors up to 10 do not appreciably affect accuracy as long as input limits of multiplier are not exceeded. Accuracy is maintained up to 100 kHz. Article gives calibration procedure.—R. Frantz, Analog Multipliers—New IC Versions Manipulate Real-World Phenomena with Ease, *EDN Magazine*, Sept. 5, 1977, p 125–129.

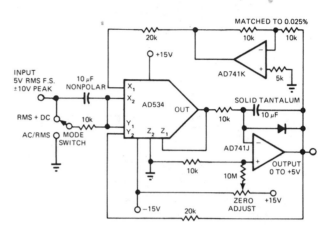

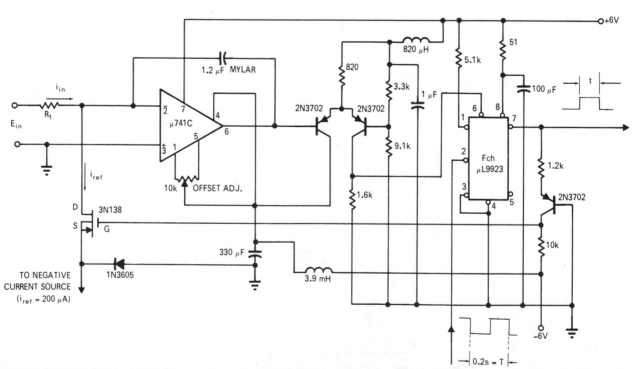

ANALOG TO PULSE WIDTH—Stripped-down version of dual-slope A/D converter integrates input current constantly but switches reference current into integrator each time clock pulse occurs. Accuracy of 0.1% makes circuit suitable for use in digital voltmeter. Reference current is switched out of integrator when output voltage reaches +4.5V. With values shown, using 100 kilohms for R_1, maximum input current is 80 μA and full-scale voltage is 8 V. Article includes timing diagram and design equations.—N. A. Robin, Analog-to-Pulse-Width Converter Yields 0.1% Accuracy, *EDN Magazine*, Nov. 1, 1970, p 42–43.

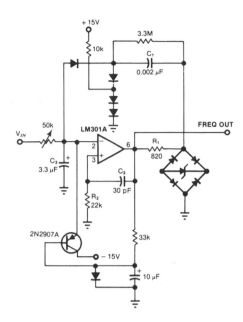

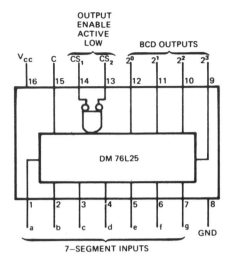

CHARGE-DISPENSING V/F CONVERSION— Output state of opamp switches C_1 between reference voltage provided by diode bridge and its inverting input. Network R_2-C_2 reinforces direction of opamp output change. Circuit can deliver 0–10 kHz output with 0.01% linearity for 0–10 V input. —J. Williams, Low-Cost, Linear A/D Conversion Uses Single-Slope Techniques, *EDN Magazine,* Aug. 5, 1978, p 101–104.

BCD FROM 7-SEGMENT DISPLAY— Single National DM76L25 read-only memory provides conversion from 7-segment outputs of MOS chip driving display to BCD inputs for data processing. Typical power dissipation is 75 mW. Access time is 70 ns when using 5-V supply. Article gives truth table for all standard and special characters of 7-segment display.—U. Priel, 7-Segment-to-BCD Converter: The Last Word?, *EDN Magazine,* Aug. 20, 1974, p 94–95.

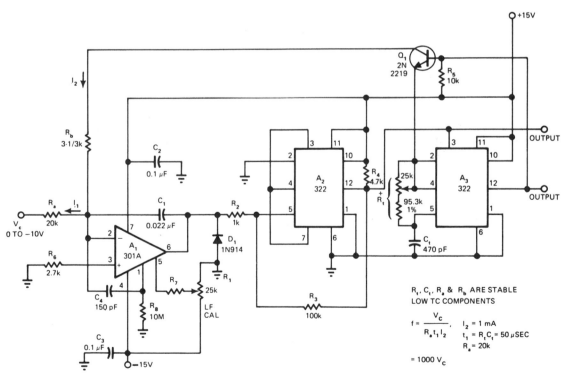

VOLTAGE-TO-FREQUENCY USING IC TIMERS— Two 322 IC timers and single 301A opamp provide all functions required for charge-balancing type of voltage-to-frequency converter, including integrator, level sensor or comparator, precision mono, and gated current source. Circuit accepts control voltage inputs of 0 to −10 V, corresponding to output pulse stream range of 0 to 10 kHz. Article describes operation in detail. R_4 should be 4.7 megohms. Output pulses of comparator A_2 trigger mono A_3, which generates pulse having duration t_1 that saturates Q_1, to force reference current I_2 into summing point of opamp integrator.—W. G. Jung, Take a Fresh Look at New IC Timer Applications, *EDN Magazine,* March 20, 1977, p 127–135.

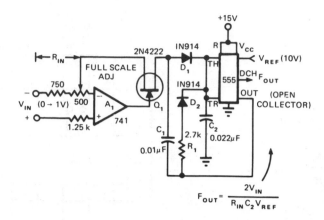

$$F_{OUT} = \frac{2V_{IN}}{R_{IN}C_2 V_{REF}}$$

V/F GIVES 10 to 10,000 Hz—Current proportional to input voltage is balanced via periodic charging of C_1 to precisely repeatable voltage by opamp A_1 and FET Q_1. With values shown, nominal scale factor is 10 kHz/V. Input of 0 to 1 V gives output of 10 to 10,000 Hz with better than 0.05% linearity. Article gives operating details and design equations.—W. S. Woodward, Simple 10 kHz V/F Features Differential Inputs, *EDN Magazine*, Oct. 20, 1974, p 86.

0–10 VDC TO 0–10 kHz—Single-supply voltage-to-frequency converter produces square-wave output at frequency varying linearly with input voltage. Linearity error is typically only 1%. For values shown, response time for step change of input from 0 to +10 V is 135 ms. Uses Raytheon 4151 converter. Supply can be 15 V.—"Linear Integrated Circuit Data Book," Raytheon Semiconductor Division, Mountain View, CA, 1978, p 7-38.

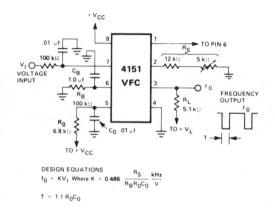

DESIGN EQUATIONS
$f_0 = KV_1$ Where $K = 0.486 \dfrac{R_S}{R_B R_0 C_0} \dfrac{kHz}{V}$

$T = 1.1 R_0 C_0$

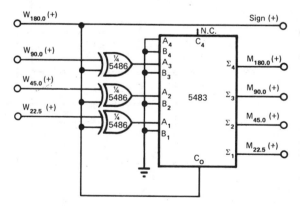

0–360° to 0–180°—Used for converting angular information in 360° wrap-around code to ±180° sign-plus-magnitude code. For values under 180°, converter outputs and inputs are identical. For larger input angles, output code is complement of input plus one. Used for interfacing shaft encoders and synchro-to-digital converters to digital display. Article gives truth table showing which lines are high and which are low at input and at output for angular increments of 22.5°.—J. N. Phillips, Convert Wrap-Around Code to Sign-Plus-Magnitude, *EDN Magazine*, Jan. 5, 1973, p 103.

CHAPTER 20
Converter Circuits—Radio

Various combinations of RF oscillator and mixer circuits convert wide range of incoming long-wave and shortwave signals to correct input or intermediate frequencies for broadcast or communication receivers not originally covering those bands.

120–150 MHz FOR TRANSISTOR RADIO—Values shown cover bands for aircraft radio, 2-m amateur radio band, and other services. Circuit is regenerative converter, with incoming signal tuned by L1-C2 and mixed in 2N2222 or equivalent transistor connected as oscillator with frequency controlled by L3 and C7. Difference frequency is adjusted to fall in standard broadcast band, for pickup by radio when converter is mounted close to ferrite loop. For local stations, antenna of converter can be 19-in length of wire. L2 and L4 are 100-μH chokes or about 20 in of fine wire wound on 100K resistor.—S. Kelly, Simple VHF Monitor, *73 Magazine,* July 1976, p 160.

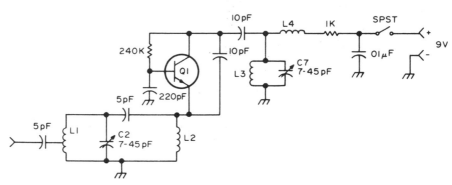

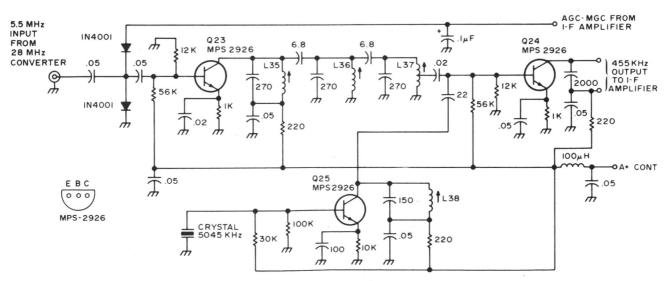

5.5 MHz TO 455 kHz—Developed for use as second converter in all-band double-conversion superheterodyne receiver for AM, narrow-band FM, CW, and SSB operation. IF amplifier Q23 is followed by triple-tuned filter feeding second mixer Q24, with Q25 as crystal oscillator. Supply is 13.6 V regulated. Article gives all circuits of receiver.—D. M. Eisenberg, Build This All-Band VHF Receiver, *73 Magazine,* Jan. 1975, p 105–112.

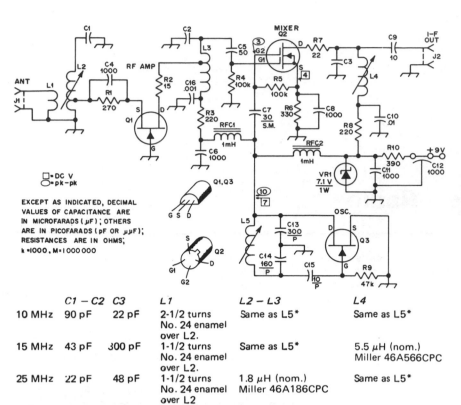

EXCEPT AS INDICATED, DECIMAL
VALUES OF CAPACITANCE ARE
IN MICROFARADS (μF); OTHERS
ARE IN PICOFARADS (pF OR $\mu\mu$F);
RESISTANCES ARE IN OHMS;
k =1000, M=1 000 000

□ = DC V
○ = pk-pk

	C1 – C2	C3	L1	L2 – L3	L4
10 MHz	90 pF	22 pF	2-1/2 turns No. 24 enamel over L2.	Same as L5*	Same as L5*
15 MHz	43 pF	300 pF	1-1/2 turns No. 24 enamel over L2.	Same as L5*	5.5 μH (nom.) Miller 46A566CPC
25 MHz	22 pF	48 pF	1-1/2 turns No. 24 enamel over L2	1.8 μH (nom.) Miller 46A186CPC	Same as L5*

*L5 — 2.42-2.96 μH, Miller 46A276CPC

WWV CONVERTER—Designed for use with amateur receiver for reception of NBS stations WWV or WWVH on 10, 15, or 25 MHz. Receiver is tuned to 4, 14, or 21 MHz to serve as IF amplifier, detector, and audio stages. Current drain of converter is 15 mA, low enough for operation from 9-V transistor-radio battery. Table gives tuned-circuit values for frequency desired. Restriction to single frequency eliminates band-switching. Q1 can be any common-gate JFET RF amplifier providing 8-dB gain. Mixer is 40673 MOSFET. Oscillator transistor is not critical. Oscillator output serves for all three WWV frequencies.—C. Watts, NBS—Ears for Your Ham-Band Receivers, QST, June 1976, p 25–26.

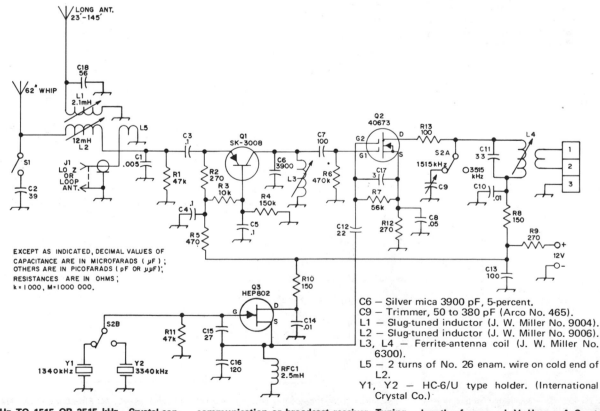

EXCEPT AS INDICATED, DECIMAL VALUES OF
CAPACITANCE ARE IN MICROFARADS (μF);
OTHERS ARE IN PICOFARADS (pF OR $\mu\mu$F);
RESISTANCES ARE IN OHMS;
k = 1000, M= 1000 000.

C6 — Silver mica 3900 pF, 5-percent.
C9 — Trimmer, 50 to 380 pF (Arco No. 465).
L1 — Slug-tuned inductor (J. W. Miller No. 9004).
L2 — Slug-tuned inductor (J. W. Miller No. 9006).
L3, L4 — Ferrite-antenna coil (J. W. Miller No. 6300).
L5 — 2 turns of No. 26 enam. wire on cold end of L2.
Y1, Y2 — HC-6/U type holder. (International Crystal Co.)

175 kHz TO 1515 OR 3515 kHz—Crystal-controlled VLF converter covering 1750-meter band gives choice of two outputs, selected by S2, for communication or broadcast receiver. Tuning range is 160 to 190 kHz. Connect general-coverage receiver to IF output terminals with length of coax.—J. V. Hagan, A Crystal-Controlled Converter and Simple Transmitter for 1750-Meter Operation, QST, Jan. 1974, p 19–22.

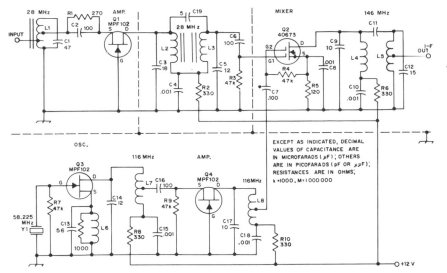

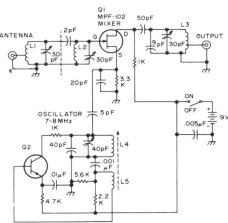

C11 — Two 1/2-inch pieces. No. 18 insulated hookup wire, twisted together 1/2 turn.

L1, L2, L3 — 18 turns No. 28 enam. wound on Amidon T-30-6 core. L1 tapped at 6 turns and 11 turns from ground end.

L4, L5 — 5 turns No. 20 enam., formed by using threads of 1/4-20 bolt as a guide. L5 is tapped 2 turns from the ground end.

L6 — 10 turns No. 24 enam. close wound on the body of a 1000-ohm 1/2-watt resistor.

L7, L8 — 5 turns No. 20 enam., formed the same as L4. Both are tapped 2 turns from the hot end.

Y1 — 58.225-MHz crystal. International Crystal third-overtone type in FM-1 (wire leads) or FM-2 (pins) holder.

28 MHz TO 144 MHz—Addition of small upconverter to 144-MHz (2-meter) SSB transceiver permits reception of 10-meter signals from Oscar satellite on single transceiver. Output of mixer, between 145.85 and 145.95 MHz, is fed to antenna terminal of receiver.—T. McMullen, An Up Converter for Oscar Reception, *QST*, March 1975, p 41–44.

152–165 MHz TO 146.94 MHz—Permits listening to public service band with any good 2-meter FM receiver or transceiver. Local oscillator has tuning range of 7–8 MHz and uses germanium PNP high-frequency transistor. No RF stage is needed for full quieting from stations 10 miles away when using ground-plane antenna. To avoid burning out converter, do not transmit while converter is connected to transceiver. Article gives coil-winding data.—H. Schoenbach, Public Service Band Converter, *73 Magazine*, Dec. 1974, p 78–79.

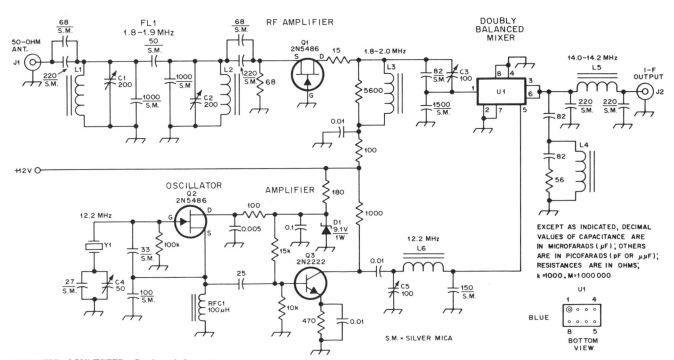

160-METER CONVERTER—Designed for use with receiver covering 20-meter band. Uses up-conversion techniques to get from 1.8 MHz of 160-meter band to 14-MHz tunable IF of receiver. Butterworth bandpass filter at input of converter covers 1.8–1.9 MHz. L1 and L2 each have 31 turns No. 22 enamel on T68-6 toroid core to give 5.1 μH. L3 is 50 μH, using 66 turns No. 18 enamel on T68-1 toroid. Other three coils each use T50-6 toroid core, with L4 having 7 turns No. 24 enamel, L5 11 turns No. 24 enamel, and L6 26 turns No. 28 enamel. U1 is SRA-1, CM-1, or ML-1 diode-quad double-balanced mixer module.—M. Arnold and D. DeMaw, Build This High-Performance Top-Band Converter, *QST*, Oct. 1978, p 22–24 and 38.

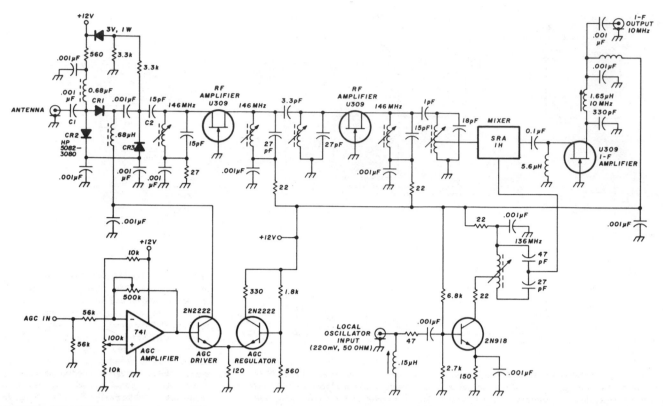

2-METER FOR 10-MHz IF—Designed for extreme linearity and selectivity while keeping noise figure below 5 dB. Circuit has +15 dBm intercept point and 16-dB power gain. Five tuned circuits at input frequency give overall bandwidth of 4 MHz, with image suppression of 60 dB for 10-MHz IF and 80 dB for 30-MHz IF. Converter uses grounded-gate FET circuit.—U. Rohde, High Dynamic Range Two-Meter Converter, *Ham Radio*, July 1977, p 55–57.

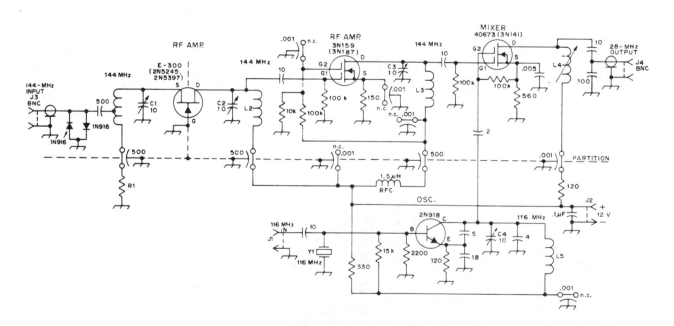

144 MHz TO 28 MHz—Brings 2-meter band to input range of ordinary amateur receiver. Crystal eliminates need for multiplier stages that can generate spurious responses. Signal can be injected from external source if crystal is removed.—Construction Hints for VHF Converters, *QST*, Sept. 1975, p 32–33 and 39.

C1-C4, incl. — 10-pF tubular ceramic trimmer (Centralab 829-10).
L1 — 6 turns No. 16, 3/8-inch dia, spaced wire dia. Tap at 2-1/2 turns from bypassed end, or for best noise figure.
L2 — 4-3/4 turns, like L1.
L3 — 4 turns No. 22, 1/4-inch dia, 5/16 inch long.
L4 — 2.7 to 4.2-μH slug-tuned coil (Miller 4307).
R1 — Adjust for 5 mA drain current, or lowest noise figure. Final value in original unit, 220 ohms.
Y1 — 116-MHz overtone crystal (International Crystal Mfg. Co.).

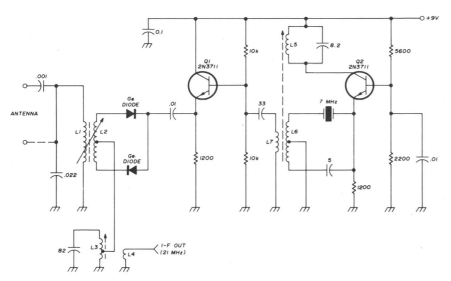

TUNABLE VLF—Gives tuning range of 10 kHz to 150 kHz without bandswitching, for WWV transmissions on 20 and 60 kHz and for operation on no-license amateur band around 1750 meters. Uses inductive tuning with toroidal ferrite core L1-L2 that is magnetically biased by pair of ½-inch diameter button-type permanent magnets. Rotating one of magnets with respect to other varies flux through toroid, changing its permeability and inductance. Toroid uses 100 turns of stranded wire to give inductance variation from 100 μH to 12 mH (120:1 range). Ferrite cores with higher permeability require fewer turns. Converter output on 15 meters feeds into communication receiver. Local oscillator uses FT-243 7-MHz crystal in third-overtone mode to give 21 MHz. Antenna is directly coupled or coupled through capacitor to improve matching to long antenna.—G. Ruehr, Tuned Very Low-Frequency Converter, *Ham Radio*, Nov. 1974, p 49–51.

L1,L2	magnetically tuned inductor (see text)
L3	10 turns no. 20 on ¼'' (6-mm) slug-tuned form, tapped 5 turns from cold end
L4	2 turns no. 20 around cold end of L3
L5	15 turns no. 20 on ¼'' (6-mm) slug-tuned form
L6	4 turns no. 20, center tapped, around cold end of L5
L7	2 turns no. 20 around cold end of L5

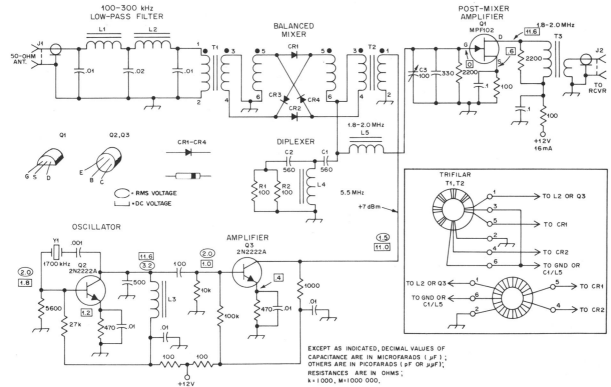

EXCEPT AS INDICATED, DECIMAL VALUES OF CAPACITANCE ARE IN MICROFARADS (μF); OTHERS ARE IN PICOFARADS (pF or μμF); RESISTANCES ARE IN OHMS; k = 1000, M=1000 000.

L1-L3, incl. — 40 turns no. 30 enam. wire wound on a T50-3 core.
L4 — 17 turns no. 28 enam. wire wound on a T50-2 core.

L5 — 70 turns no. 30 enam. wire wound on a T50-2 core.
T1, T2 — Broadband transformer. For conventional style winding: primary, 27 turns no. 30 enam. wire wound over secondary turns. Secondary, 54 turns no. 30 enam. wire wound on an FT-50-43 core. For trifilar winding: three individual windings of no. 30 enam. wire

on an FT-50-43 core. Connect as shown in inset drawing.
T3 — Broadband transformer. Primary, 50 turns no. 30 enam. wire on an FT-50-72 core. Secondary, 7 turns no. 28 enam. wire wound over primary turns.
Y1 — General-purpose crystal. 1700 kHz, 32-pF load capacitance.

100–200 kHz TO 1.8–2 MHz—High-performance low-frequency converter picks up experimental CW, SSB, RTTY, and beacon signals in 160–190 kHz band for conversion to tunable IF range of modern communication receiver. Double-balanced diode-ring mixer has conversion loss of 6–8 dB and will stand up against strong signals without causing overloading and cross-modulation. Use 1N914 matched diodes. Diplexer at mixer output is tuned to 3 times converter IF. Article covers construction and alignment.—D. DeMaw, A High-Performance Low-Frequency Converter, *QST*, June 1977, p 23–26.

PASSIVE LONGWAVE—Uses VFO of amateur-band transmitter to supply heterodyne for bringing in frequencies below 450-kHz broadcast band on amateur receiver, such as Omega navigation station on 10.2–13.6 kHz, NAA Teletype on 17.8 kHz, and GBR time signals on 16 kHz. L_1-C_1-C_2 100-kHz input wavetrap for Loran C can be omitted at far-inland locations. C_2 is two-gang broadcast variable capacitor with both sections in parallel. Mount L_2 at right angle to L_3. Converter should be well shielded. Operation involves tuning receiver to bottom of any amateur radio band, on frequency equal to difference between VFO and that to which receiver is tuned. If VFO is on 7 MHz and receiver is tuned to 7.085 MHz, combination will be set for 85-kHz station. In USA, unlicensed transmission on 160–190 kHz is permitted with 1-W input and 50-foot antenna including length of transmission line.—M. Muench, Longwave Simplified, *CQ*, March 1976, p 41–42.

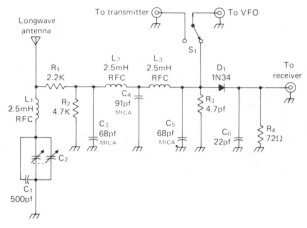

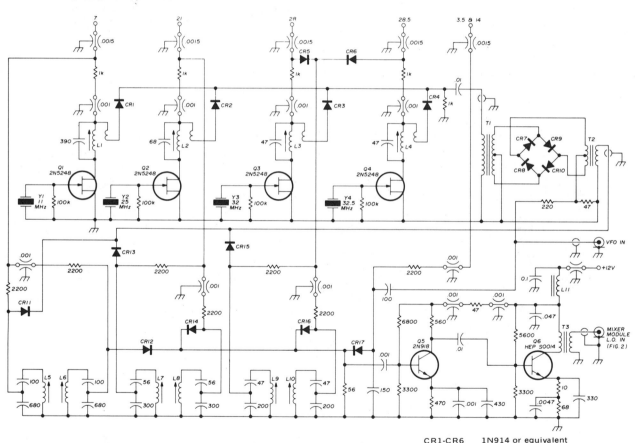

VFO CONVERTER—Used in solid-state five-band communication receiver. VFO input (5–5.5 MHz) goes directly to amplifiers Q5 and Q6 when bandswitch is on 3.5 or 14 MHz. When VFO signal is applied to balanced mixer CR7-CR10, product is at 9 MHz. Diodes should be carefully selected for equal voltage drops ±20 mV at various current values such as 0.75, 2, 10, and 20 mA. When bandswitch is on 7, 21, or 28 MHz, VFO signal is mixed with output of FET crystal oscillator and filtered before being applied to Q5 and Q6. FET oscillators Q1-Q4 are energized by +12 V from bandswitch, with diodes CR1-CR4 selecting output. Crystals are parallel-resonant with 32-pF load. Y2, Y3, and Y4 are third-overtone type.—P. Moroni, Solid-State Communications Receiver, *Ham Radio*, Oct. 1975, p 32–41.

CR1-CR6	1N914 or equivalent
CR7-CR10	Selected 1N270 diodes (see text)
CR11-CR17	1N914 or equivalent
L1-L4	0.6 μH (10 turns no. 22 (0.6mm) enamelled on 3/8" (9mm) diameter slug-tuned forms. Link is 3 turns no. 22 (0.6mm)
L5,L6	1.2 μH. 20 turns no. 28 (0.3mm) on 3/4" (9mm) diameter slug-tuned forms
L7-L10	0.6 μH (same as L1 - L4)
T1, T2	10 turns no. 32 (0.2mm), trifilar wound on Amidon T50-6 toroid core
T3	10 turns no. 32 (0.2mm), trifilar wound on Amidon T50-6 toroid core. Collector winding has two windings in series to give 2:1 ratio

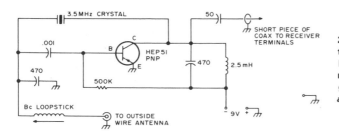

25–500 kHz TO 3.5–4 MHz—When receiver is tuned to 3.5 MHz and converter is peaked for loudest signal, combination is tuning 25-kHz range. With receiver tuned to 4 MHz, converter gives coverage at 500 kHz.—Circuits, *73 Magazine,* May 1977, p 19.

20 METERS TO 40 METERS—Used with 40-meter receiver for which circuit is also given. Converter output is in 40-meter band, for direct feed to input of receiver. L4 is 12 turns No. 26 enamel on Amidon FT37-61 toroid, L5 is 24 turns No. 26 enamel on Amidon T-50-6 toroid, and T3 uses Amidon T-50-6 toroid with 2 turns No. 26 enamel for primary and 21 turns for secondary.—D. DeMaw, The Mini-Miser's Dream Receiver, *QST,* Sept. 1976, p 20–23.

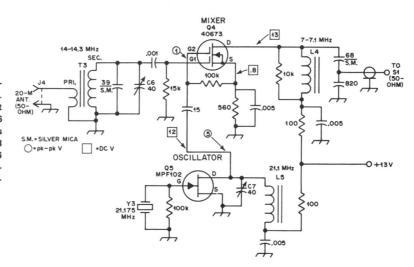

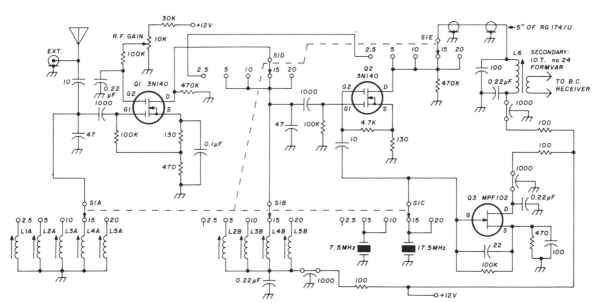

L1	61 - 122 μH (Cambion X2060-7)	L5	0.83 - 1.6 μH (Delevan 4000-10)
L2	10 - 18 μH (Cambion X2060-4)	L6	28 - 63 μH (Cambion X2060-6), 10 turn secondary
L3	2 - 3.7 μH (Cambion X2060-1)	Q1,Q2	3N140, MFE3006, HEP F2004, RCA 40673
L4	1.3 - 2.5 μH (Delevan 4000-12)	Q3	MPF102, HEP802, HEP F0015

WWV FET CONVERTER—Receives WWV on 2.5, 5, 10, 15, and 20 MHz, using modified transistor AM broadcast receiver operating straight-through for 2.5-MHz reception and serving as IF amplifier for converter when tuned to higher WWV and WWVH frequencies. Only two crystals are needed because each allows reception of two WWV frequencies; thus, 10 and 20 MHz are image frequencies when receiving 5 and 15 MHz. Loopstick antenna of radio is replaced with small slug-tuned coil L6 to use 2.5-MHz image frequency when radio is tuned to 1590 kHz. Converter uses dual-gate MOSFETs in RF stage Q1 and mixer Q2, with JFET Q3 as oscillator. Antenna is short piece of wire.—H. Olson, Five-Frequency Receiver for WWV, *Ham Radio,* July 1976, p 36–38.

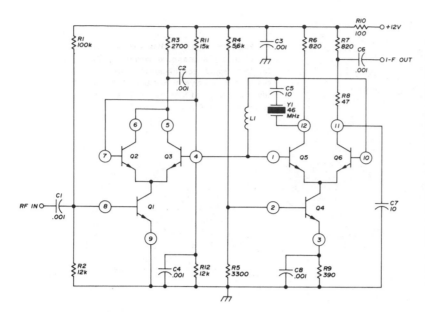

SINGLE-CRYSTAL FOR 46 TO 420 MHz—Covers all VHF amateur bands by using mixer-generated harmonics of 66-MHz crystal oscillator frequency for mixing action. IF can be tuned with any communication receiver. Fundamental is used directly, third harmonic of 138 MHz serves for 2 meters, fifth of 230 MHz for 220-MHz band, and ninth of 414 MHz for 420-MHz band. Q1-Q3 are broadband RF preamp. Y1 is plated overtone crystal oscillating at 46 MHz in series-resonant mode. Q5 and Q6 form differential-amplifier oscillator, and Q4 is mixer driver. No tuning is required in converter, but external tuning is required to prevent device from working on all bands at once. All transistors are part of RCA CA3049T IC, for which pin numbers are circled. L1 is 72 inches of No. 30 enamel, doubled and twisted 1 turn per inch and wound on 1-megohm ½-W resistor to form quarter-wave transmission line.—S. Smith, Four-Band VHF Receiving Converter, *Ham Radio*, Oct. 1976, p 64–66.

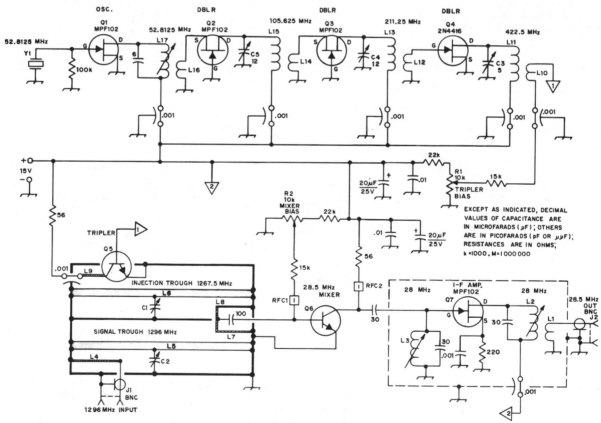

C1, C2 — Nut or copper disk on end of adjusting screw, with lock nut on top side of assembly. Makes variable capacitance to midpoint of half-wave trough-line inner conductor.
C3 — 0.5- to 5-pF glass trimmer.
C4, C5 — 12-pF ceramic trimmer.
L1 — 2 turns insulated hookup wire around L2.
L2, L3 — 13 turns No. 28 enam, 1/2 inch (13 mm) iron-slug form.
L4 — L-shaped coupling loop, 3/4 inch (19 mm) long, No. 18, adjacent to L5.
L5 L6 — 1/4-inch (6 mm) copper tubing, 4-1/4 inches (114 mm) long.
L7, L8 — U-shaped double coupling loop, No. 14, 1/2 inch (13 mm) wide, 1/2 inch long, centered in opening in partition P5.

L9 — Like L4, except adjacent to L6. Collector lead of Q5 is part of this loop.
L10, L12, L14 — 1 turn No. 14, 3/8 inch (10 mm) dia.
L11 — 1-1/4 turns No. 14, 3/8 inch dia.
L13 — 3 turns No. 14, 3/8 inch (10 mm) dia.
L15 — 5 turns No. 14, 3/8 inch (10 mm) dia.
L16 — 2 turns insulated hookup wire around L17.
L17 — 16 turns No. 28 enam, 1/4 inch (6 mm) iron-slug form.
Q5, Q6 — Uhf transistor. Use best available low-noise type for Q6.
R1, R2 — 10,000-ohm linear-taper control.
RFC1, RFC2 — Ferrite-bead choke.
Y1 — Third-overtone crystal, 52.8125 MHz, or to suit i-f range used.

1296 MHz TO 28.5 MHz—Uses UHF transistors in active mixer and in final stage of injection chain, for lower noise figure and useful conversion gain. Doubler and tripler stages are individually shielded.—L. Crutcher, An Active-Mixer Converter for 1296 MHz, *QST*, Aug. 1974, p 11–14.

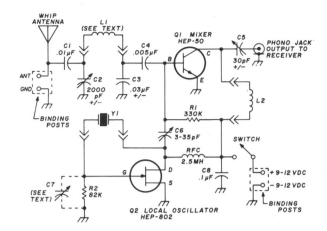

BELOW BROADCAST BAND—Simple solid-state converter can be used with any good communication receiver covering 3.5–4 MHz to bring in stations from 5–550 kHz (200 meters and up). Input coil L1 is changed from 0.28 H for 5–11 kHz to 120 μH for 250–550 kHz in eight steps, as given in article. C2 consists of two 3-gang variable capacitors with stators wired in parallel, gang-tuned with dial cords. Trimmer C7 is 1–12 pF, adjusted to give reliable starting of FT-243 3500-kHz crystal. L2 is 80–90 μH for 80 meters, and loopstick is for broadcast band.—K. Cornell, 200 Meters and Up Receiving Converter for Low Frequencies, *Ham Radio*, Nov. 1976, p 24–26.

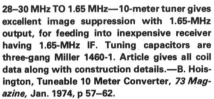

28–30 MHz TO 1.65 MHz—10-meter tuner gives excellent image suppression with 1.65-MHz output, for feeding into inexpensive receiver having 1.65-MHz IF. Tuning capacitors are three-gang Miller 1460-1. Article gives all coil data along with construction details.—B. Hoisington, Tuneable 10 Meter Converter, *73 Magazine*, Jan. 1974, p 57–62.

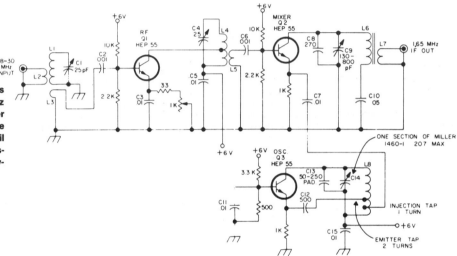

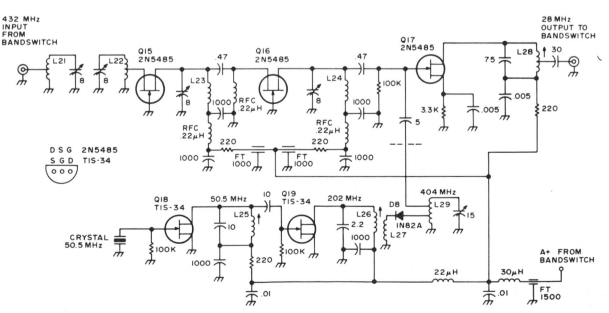

432 MHz TO 28 MHz—Contains bandpass filter, grounded-grid RF amplifier stages Q15-Q16, mixer Q17, and crystal oscillator Q18-Q19. De- veloped for use in all-band double-conversion superheterodyne receiver for AM, narrow-band FM, CW, and SSB operation. Supply is 13.6 V regulated. Article gives all circuits of receiver.— D. M. Eisenberg, Build This All-Band VHF Re- ceiver, *73 Magazine*, Jan. 1975, p 105–112.

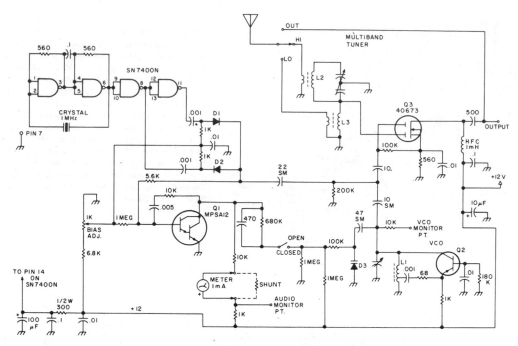

8–30 MHz AMATEUR BANDS—Will convert any frequency in tuning range to IF value between 3.5 and 4 MHz. Requires only three transistors and one IC. D1 and D2 can be germanium or silicon, such as 1N914. D3 is rectifier diode. Q2 can be almost any general-purpose high-frequency transistor. Two of gates in SN7400N TTL IC serve as crystal oscillator, and other two gates are buffers for detector diodes D1 and D2. Diodes are modulated by VCO Q2. Detector output is amplified and filtered by Q1 to produce control voltage for tuning D3. Output from mixer Q3 is untuned, with RFC as drain load element.—R. Megirian, High Frequency Utility Converter, *73 Magazine*, June 1977, p 50–53.

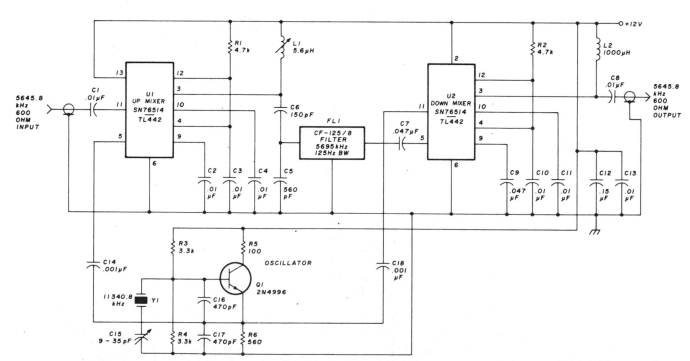

UP/DOWN—Circuit shown was developed for use in 5645.8-kHz IF amplifier of Drake R-4B amateur-band receiver, to utilize high-performance characteristics of Sherwood Engineering CF-125/8 CW crystal filter having bandwidth of only 125 Hz. Texas Instruments TL442 double-balanced mixers convert IF signal to 5695-kHz center frequency of filter and convert filter output back to IF value. Same crystal oscillator serves for both upconversion and downconversion. Gives true single-signal reception. Article covers procedures for interfacing any crystal filter with any receiver IF value.—H. Sartori, An Up/Down Filter Converter, *Ham Radio*, Dec. 1977, p 20–25.

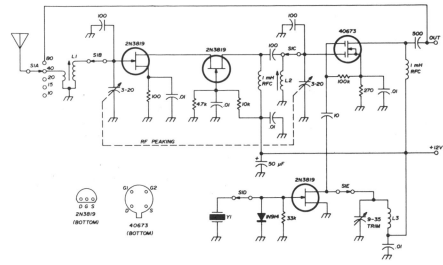

10–40 METERS TO 3.5–4 MHz—Five-band converter is designed for use with miniaturized communication receiver tuning from 3.5 to 4 MHz. Signals for 80-meter band are fed directly to receiver. Two-gang tuning capacitor used to peak converter front end is film-dielectric type taken from transistor FM radio.—R. Megirian, Design Ideas for Miniature Communications Receivers, Ham Radio, April 1976, p 18–25.

band	L1	L2	L3	Y1
40	20 turns no. 36 (0.13mm), 2 turn link	20 turns no. 36 (0.13mm)	4.7 μH	11 MHz
20	12 turns no. 28 (0.3mm), 1.5 turn link	12 turns no. 28 (0.3mm)	2.2 μH	18 MHz
15	7 turns no. 28 (0.3mm), 1 turn link	7 turns no. 28 (0.3mm)	1.5 μH	25 MHz
10	4 turns no. 28 (0.3mm), 1 turn link	4 turns no. 28 (0.3mm)	1.5 μH	25 MHz

WWV ON AC/DC RADIO—When fed into IF amplifier of ordinary broadcast-band radio, simple converter circuit gives choice of WWV on 10 or 15 MHz, for reception of time signals and radio propagation reports. C2 is 1.5–10 pF; C3 and C6 are 7–60 pF; C4 is 7–100 pF (all compression trimmers); and C5 is 1.8–8.7 pF miniature variable capacitor.—W. C. Powis, Notes on Converting the AC/DC for WWV, 73 Magazine, Oct. 1974, p 116.

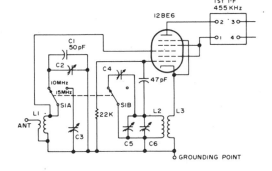

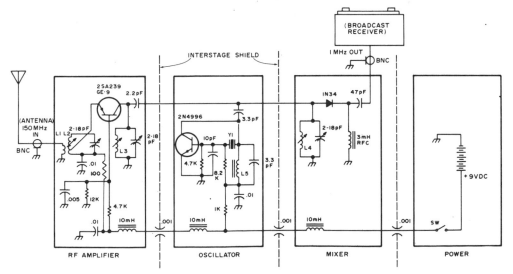

2 m TO BROADCAST BAND—Permits tuning to 2-m (146-MHz) amateur band with ordinary AM auto radio, for monitoring FM repeaters and other 2-m amateur stations. Article stresses importance of shielding, compartmentalization, and RF blocking along power lead to prevent bleed-through of broadcast stations. Separate 9-V battery gives long life if converter is turned off when not in use, because drain is only 25 mA. L2 is 4 turns No. 20 on 7-mm slug-tuned form, with 2-turn link L1 at low end and tap 1½ turns from low end. L3 and L4 are 3 turns No. 20 on 7-mm slug-tuned form. L5 is 20 turns No. 30 on 4-mm solid ferrite form. Y1 is 48.5-MHz third-overtone crystal. Converter gives good reception of both AM and FM stations on 2 m, with sharpness of receiver IF tuning determining ability of radio to slope-detect FM signals.—J. R. Johnson, New Improved Repeater Monitor, 73 Magazine, Dec. 1976, p 106–109.

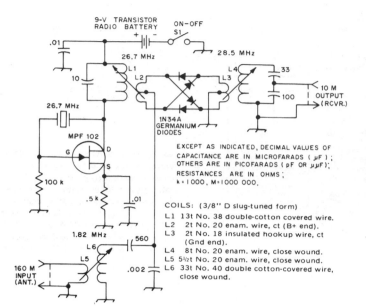

160 METERS TO 10 METERS—Simple converter adds 160-meter band capability to older CW or AM receiver. Passive mixer is adequate. High output frequency eliminates IF feedthrough and image signals. Crystal oscillates on third overtone and feeds directly into mixer.—A. Bloom, A Simple 160-Meter Converter, *QST*, Feb. 1975, p 46.

EXCEPT AS INDICATED, DECIMAL VALUES OF CAPACITANCE ARE IN MICROFARADS (μF); OTHERS ARE IN PICOFARADS (pF OR μμF); RESISTANCES ARE IN OHMS; k = 1 000, M = 1 000 000.

COILS: (3/8" D slug-tuned form)
L1 13t No. 38 double-cotton covered wire.
L2 2t No. 20 enam. wire, ct (B+ end).
L3 2t No. 18 insulated hookup wire, ct (Gnd end).
L4 8t No. 20 enam. wire, close wound.
L5 5½t No. 20 enam. wire, close wound.
L6 33t No. 40 double cotton-covered wire, close wound.

144 MHz TO 14 MHz—Oscillator uses 43.333-MHz overtone-cut crystal feeding class A tripler that injects 130-MHz signal into gate of MPF-102 mixer for combining with 144-MHz output of IGFET RF amplifier to give 14 MHz for amateur-band or general-coverage receivers. Article covers construction and alignment, including detailed coil-winding data.—C. Klinert, A Two Meter Converter, *73 Magazine*, Sept. 1973, p 65–67.

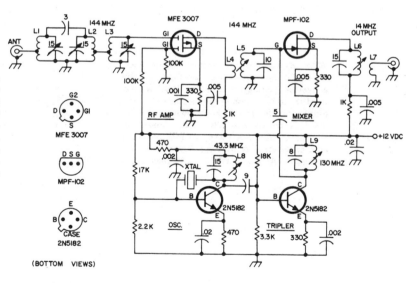

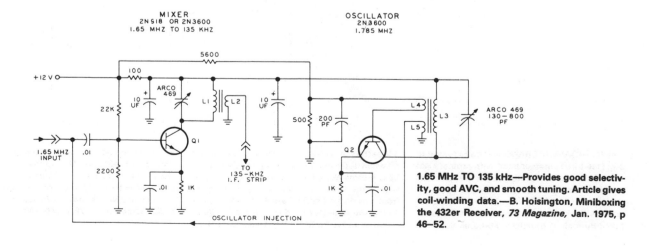

1.65 MHz TO 135 kHz—Provides good selectivity, good AVC, and smooth tuning. Article gives coil-winding data.—B. Hoisington, Miniboxing the 432er Receiver, *73 Magazine*, Jan. 1975, p 46–52.

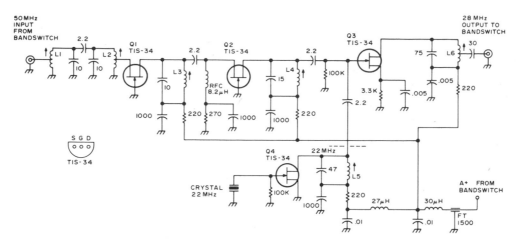

50 MHz TO 28 MHz—Contains bandpass filter, two grounded-grid RF amplifier stages Q1-Q2, mixer Q3, and crystal oscillator Q4. Developed for use in all-band double-conversion superheterodyne receiver for AM, narrow-band FM, CW, and SSB operation. Supply is 13.6 V regulated. Article gives all circuits of receiver.—D. M. Eisenberg, Build This All-Band VHF Receiver, *73 Magazine*, Jan. 1975, p 105–112.

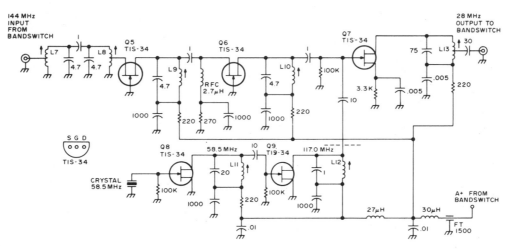

144 MHz TO 28 MHz—Contains bandpass filter, grounded-grid RF amplifier stages Q5-Q6, mixer Q7, and crystal oscillator Q8-Q9. Developed for use in all-band double-conversion superheterodyne receiver for AM, narrow-band FM, CW, and SSB operation. Supply is 13.6 V regulated. Article gives all circuits of receiver.—D. M. Eisenberg, Build This All-Band VHF Receiver, *73 Magazine*, Jan. 1975, p 105–112.

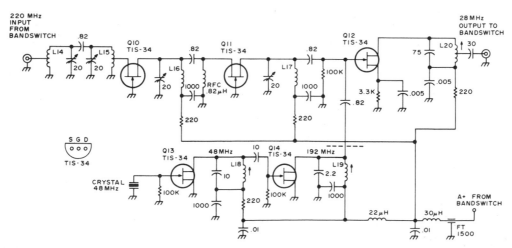

220 MHz TO 28 MHz—Contains bandpass filter, grounded-grid RF amplifier stages Q10-Q11, mixer Q12, and crystal oscillator Q13-Q14. Developed for use in all-band double-conversion superheterodyne receiver for AM, narrow-band FM, CW, and SSB operation. Supply is 13.6 V regulated. Article gives all circuits of receiver.—D. M. Eisenberg, Build This All-Band VHF Receiver, *73 Magazine*, Jan. 1975, p 105–112.

VLF CONVERTER—Uses low-pass filter instead of usual tuned circuit, so only associated receiver need be tuned. Measured threshold sensitivity is about 20 μV. Transistors used in dual-gate MOSFET mixer and FET oscillator are not critical. Crystal can be any frequency compatible with tuning range of receiver used. With 3.5-MHz crystal, 3.5 MHz on receiver dial corresponds to 0 kHz and 3.6 MHz to 100 kHz.—R. N. Coan, VLF Converter, *Ham Radio*, July 1976, p 69.

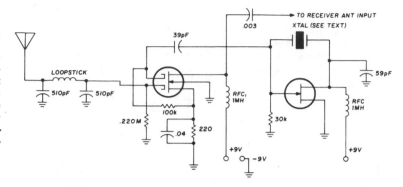

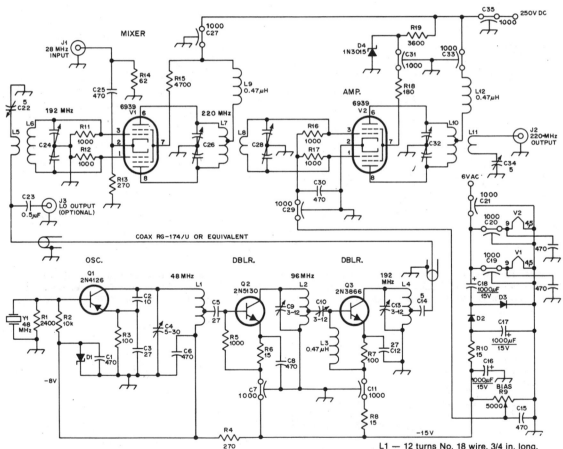

28 MHz TO 220 MHz FOR TRANSMIT—Permits use of 2-meter transceiver to transmit in 220-MHz band with minimum of 6-W power output for 1-W drive on 28 MHz. Local-oscillator output at 192 MHz can be used for receiving converter as well. Use 8-pF butterfly-type air variable (Johnson 160-028-001) for C24, C26, C28, and C32. D1 is GE ZD8.2 8-V 1-W zener.—F. J. Merry, A 220-MHz Transmit Converter, *QST*, Jan. 1978, p 16–20.

L1 — 12 turns No. 18 wire, 3/4 in. long, 1/4 in. diameter. Tap at 1-3/4 turns.
L2 — 6 turns No. 18 wire, 3/4 in. long, 1/4 in. diameter. Tap at 2 turns.
L4 — 3 turns No. 18 wire, 3/4 in. long, 1/4 in. diameter. Tap at 1 turn.
L3 — 0.47-μH rf choke (Miller).
L5 — 1 turn, 1/2-in. dia, 3/4-in. leads; No. 22 insulated wire.
L6 — 1-1/2 turns, 1/2-in. dia, 1-in. leads; No. 16 wire.
L7 — 1-1/2 turns, 5/8-in. dia, 3/16-in. leads; No. 16 wire.
L8 — 1/2 turn, 5/8-in. dia, 3/16-in. leads; No. 16 wire.
L10 — 1-1/2 turns, 5/8-in. dia, 1-1/4-in. leads; No. 16 wire.
L11 — 1 turn, 1/2-in. dia. 3/4-in. leads; No. 22 insulated wire.

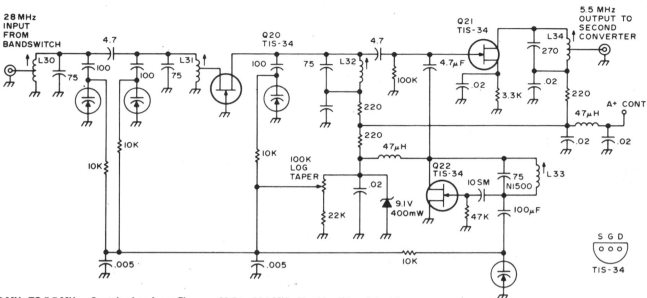

28 MHz TO 5.5 MHz—Contains bandpass filter, grounded-grid RF stage Q20, mixer Q21, and oscillator Q22, with all tuning accomplished by variable-capacitance diodes. Oscillator covers 22.5 to 24.5 MHz. Used in all-band double-conversion superheterodyne receiver for AM, narrow-band FM, CW, and SSB operation. Supply is 13.6 V regulated. Article gives all circuits of receiver.—D. M. Eisenberg, Build This All-Band VHF Receiver, *73 Magazine,* Jan. 1975, p 105–112.

CHAPTER 21
Counter Circuits

Includes circuits for counting events and pulses over various ranges from 0 to 1.2 GHz singly, by 4s, or by decades, along with counting-rate meter, up/down, multifunction, anticoincidence, PROM-controlled, free-running classroom-demonstration, and switch-closure counters driving multiplexed or continuous digital displays. See also Frequency Counter chapter.

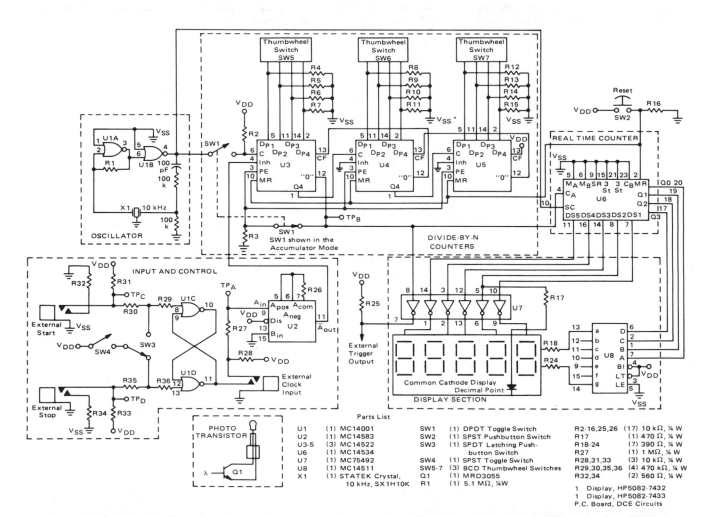

5-DIGIT PRESET COUNTER—Basis of circuit is Motorola CMOS real-time MC14534 five-decade counter containing five ripple-type decade counters whose outputs are time-multiplexed by internal scanner. Time-base oscillator provides 10-kHz crystal reference for clocking counters. Total current drain of system is 65 mA from 5-V supply. When used to control quantity of items placed in carton, each item interrupts light beam of photoelectric system to give count. External trigger output is connected to control mechanism that advances conveyor belt when box is full. Quantity of items desired per box is dialed on thumbwheel switches. Display is used to indicate number of boxes filled. Other applications include count and display of number of interruptions of light beam, measurement of conveyor speed, and measurement of log lengths in sawmill.—A. Mouton, "Five Digit Accumulator/Elapsed Time Indicator," Motorola, Phoenix, AZ, 1975, AN-743, p 3.

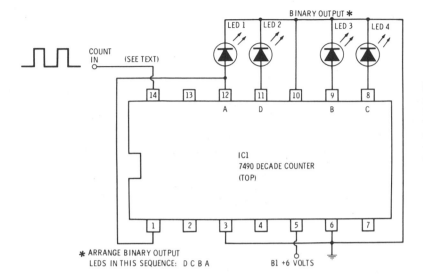

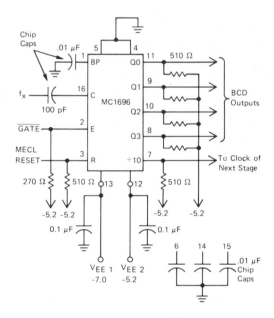

FOUR-LED BCD DISPLAY—Square-wave input pulses are counted by 7490 IC that drives LEDs indicating count in binary format up to 10 and then recycling. Can be used for classroom demonstrations of counters, flip-flop action, and binary counting. Pulses can be obtained from UJT clock circuit operating at audio rate.—F. M. Mims, "Computer Circuits for Experimenters," Radio Shack, Fort Worth, TX, 1974, p 85–93.

1.2-GHz DECADE COUNTER—Motorola MC1696 BCD-output counter provides direct counting of events at up to 1.2 GHz without prescaling. Connection shown is for AC coupling of input signals. Decoupling capacitors are used on power supplies and all unused pins. MC1696 provides division by 10, with output driving cascaded MC10138 biquinary counters and associated latches connected to drive five-decade display as covered in report.—J. Roy, "Event Counter and Storage Latches for High-Frequency, High-Resolution Counters," Motorola, Phoenix, AZ, 1975, EB-47.

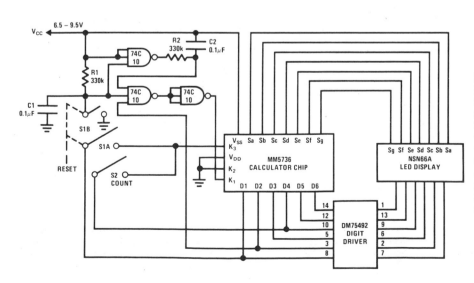

SELF-STARTING COUNTER—Addition of three logic elements eliminates need for separate starting switch when using National MM5736 calculator chip as counter driving LED display. When reset switch is returned to normal position after pushing it to clear calculator, additional parts serve to generate delayed pulse that gates digit output 2 into calculator and thus enters a 1. This action resets counter with single manual operation.—M. Watts, "Calculator Chip Makes a Counter," National Semiconductor, Santa Clara, CA, 1974, AN-112, p 4.

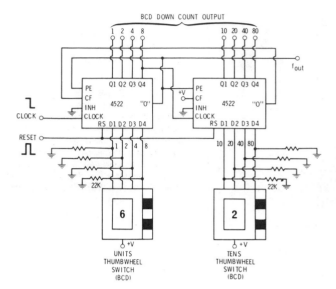

CASCADED DOWN COUNTER—4522 decimal divide-by-N counter is used with BCD thumbwheel switch for each decade. Output is in BCD format, going down from preset number in range of 0-99. Decoded 0 output of tens stage is connected to CF or carry-forward input of units stage. Only when both counters are in 0 state is 0 output provided. Preset number is then reloaded into counters.—D. Lancaster, "CMOS Cookbook," Howard W. Sams, Indianapolis, IN, 1977, p 311-312.

UP/DOWN COUNTER—Cascading of 4192 decade up/down counters and use of two clocks give fully synchronous system for adding or subtracting count. Both clocks are normally held high. Low on up clock advances count. Low on down clock subtracts 1 from count. Clocking takes place on trailing or positive edge of negative pulse. Parallel loading inputs are used to preset counter to any desired number.—D. Lancaster, "CMOS Cookbook," Howard W. Sams, Indianapolis, IN, 1977, p 309-310.

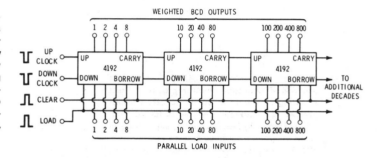

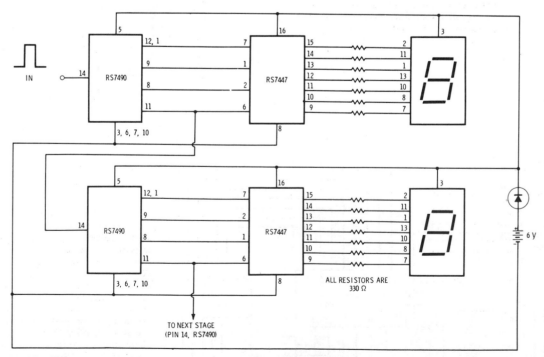

MULTIDIGIT DEMONSTRATION COUNTER—Simple interconnection of RS7490 decade counter, RS7447 decoder, and 7-segment digital display for each desired digit makes ideal counter for classroom demonstrations and Science Fair exhibits. With two additional stages added, display reaches 9999 before recycling. Use 1N914 diode in series with battery to protect against polarity reversal and reduce supply to 5 V for ICs.—F. M. Mims, "Integrated Circuit Projects, Vol. 6," Radio Shack, Fort Worth, TX, 1977, p 53-63.

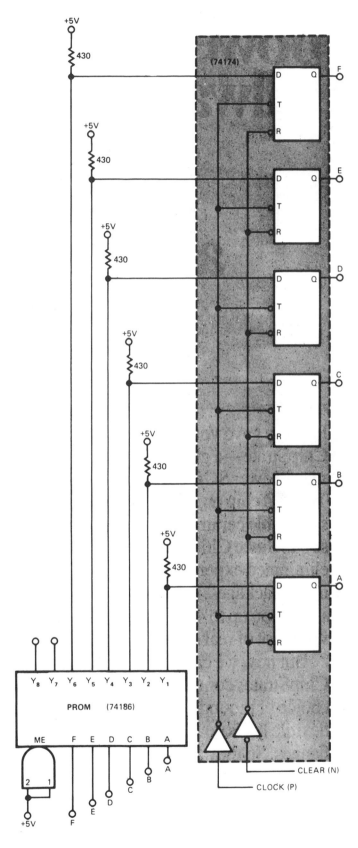

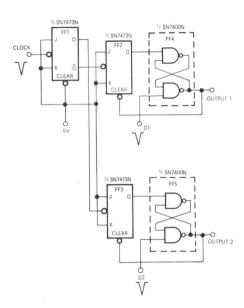

ANTICOINCIDENCE—Developed for use with bidirectional counter circuits to avoid counting errors when up and down pulses occur simultaneously. Operation is based on knowing maximum frequency of separate data pulses. Outputs 1 and 2 will be separated by at least one clock period even if inputs D1 and D2 occur simultaneously. Article gives operating details.—J. H. Burkhardt, Jr., Anti-Coincidence Circuit Prevents Loss of Data, *EDN/EEE Magazine,* Jan. 1, 1972, p 73.

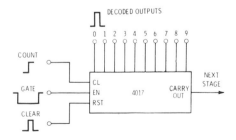

10-POINT STEPPER—4017 divide-by-10 counter routes input clock signal sequentially to each of ten output lines, with only selected output going high. Internal circuit of IC is self-clearing walking ring that is glitch-free, with minimum overlap between outputs. Counters can be cascaded to provide more steps.—D. Lancaster, "CMOS Cookbook," Howard W. Sams, Indianapolis, IN, 1977, p 309.

PROM-CONTROLLED COUNTER—Universal counter can be set to count in any desired sequence ranging from simple binary to pseudorandom, using only one programmable read-only memory chip and one 74174 edge-triggered flip-flop register chip. Version shown is 6-bit 64-state counter for which PROM is organized in 64 8-bit words. Pull-up resistors are required. Article covers application details, including expansion techniques. PROM outputs serve as data input to register chip, and register outputs provide PROM address inputs.—T. M. Farr, Jr., Read-Only Memory Controls Universal Counter, *EDN Magazine,* May 5, 1976, p 114.

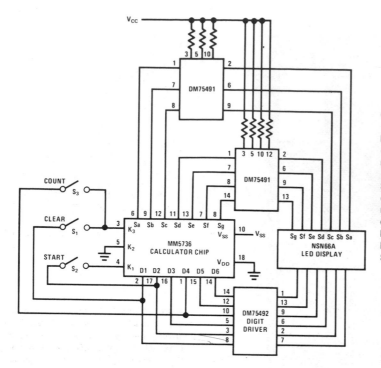

CALCULATOR AS COUNTER—National MM5736 calculator chip is used with two DM75491 segment drivers and DM75492 digit driver for LED 6-digit display. Switches provide manual control of counter. To reset, push S_1 to clear calculator, push S_2 to enter a 1, then push S_3 when new count is to be started. Current drive to LEDs is supplied by V_{CC} through current-limiting resistors, giving power saving because V_{CC} can be less than V_{SS}. Will drive large LED display.—M. Watts, "Calculator Chip Makes a Counter," National Semiconductor, Santa Clara, CA, 1974, AN-112, p 4.

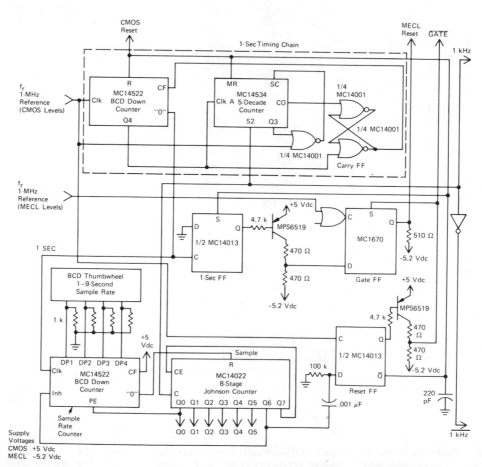

TIME BASE FOR 1.2-GHz COUNTER—Provides 1-s gate, latch strobing signals, 1-kHz signal for multiplexing displays, and digital sample rate control for high-frequency high-resolution counters. Timing chain divides 1-MHz external reference signal by 10^6 to give 1-Hz output. MC14534 five-decade counter generates 1-kHz multiplexing frequency with 20% duty cycle for blanking. Digital sample rate control is programmed on BCD thumbwheel switch in increments ranging from 1 to 9 s, using single MC14522 BCD down counter.—J. Roy, "A Time Base and Control Logic Subsystem for High-Frequency, High-Resolution Counters," Motorola, Phoenix, AZ, 1975, EB-48.

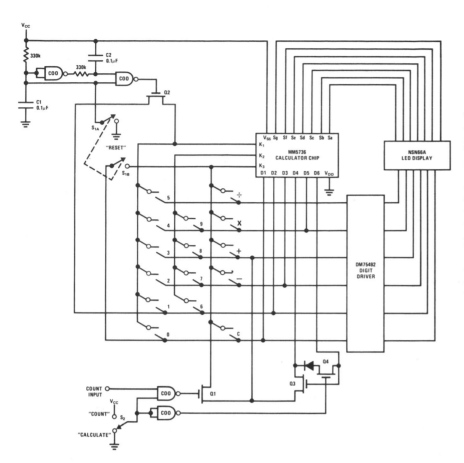

CALCULATOR/COUNTER—Normal arithmetic functions of National MM5736 calculator chip are preserved while providing counting capability, through use of MOS transistors Q1-Q4. When reset switch is pushed, pin D1 is connected to pin K_3 of calculator and calculator is cleared. C1 and C2 are discharged while S_1 is closed but are charged when it is released, generating negative-going delayed pulse that causes a 1 to be entered into calculator. Delay allows clear function to be debounced by calculator chip. When S_2 is in count mode, Q4 is turned on and D6 is tied to D4, for doubling maximum counting rate. Input pulse will now turn Q1 on, making calculator perform addition. Additional pulse adds 1 to sum. When S_2 is returned to calculate position, keyboard logic is returned to normal state. MM74C00 NAND gates can be replaced with MM74C02 NOR gates, and MOS transistors can be replaced with MM5616 CMOS switch.—M. Watts, "Calculator Chip Makes a Counter," National Semiconductor, Santa Clara, CA, 1974, AN-112, p 6.

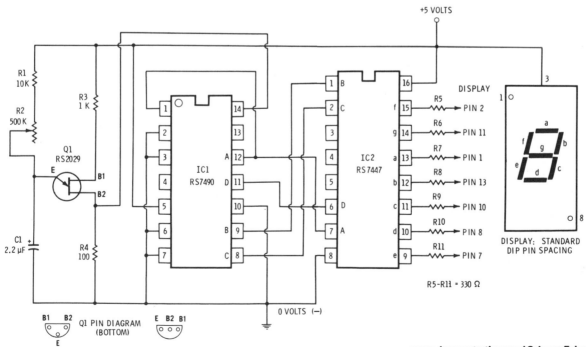

SELF-DRIVING COUNTER—UJT relaxation oscillator Q1 supplies series of pulses to input pin 14 of RS7490 decade counter at frequency determined by setting of R2 and value used for C1. Counter feeds corresponding BCD outputs to BCD input pins of RS7447 decoder for conversion into 7-segment decimal format for driving Radio Shack 276-052 LED display. Ideal for classroom demonstrations and Science Fair exhibits. 6-V battery with 1N914 diode in series can be used in place of 5-V supply.—F. M. Mims, "Integrated Circuit Projects, Vol. 2," Radio Shack, Fort Worth, TX, 1977, 2nd Ed., p 41—56.

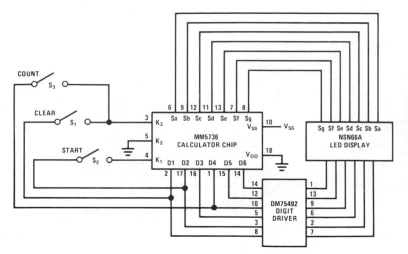

TWO-CHIP COUNTER—Combination of National MM5736 calculator chip and DM75492 digit driver for 6-digit LED display is suitable for applications where typical maximum counting rate can be about 100 Hz. Counter is reset manually by closing S_1 to clear calculator and closing S_2 to enter a 1. Operator now controls start of new count by pressing S_3, without need for gating count input.—M. Watts, "Calculator Chip Makes a Counter," National Semiconductor, Santa Clara, CA, 1974, AN-112, p 2.

2-BIT BINARY—Sections of RS7473 dual flip-flop are connected to form simple counter that counts to three in binary with LEDs. By adding more flip-flop stages, count can be extended to higher values. If OFF LED represents 0 and ON LED is 1, combinations 00, 01, 10, and 11 represent 0, 1, 2, and 3, respectively. Input is restricted to low audio frequency so LED changes can be readily observed during demonstrations.—F. M. Mims, "Integrated Circuit Projects, Vol. 6," Radio Shack, Fort Worth, TX, 1977, p 23–32.

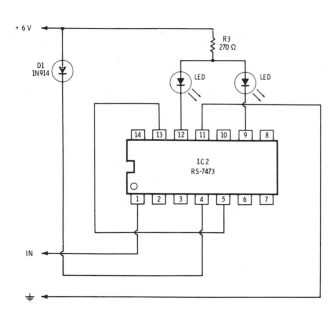

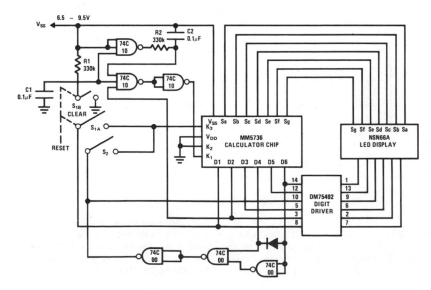

CALCULATOR COUNTS UP TO 300 Hz—Logic elements used with MM5736 calculator chip provide self-starting counting action in range from 80 to 300 Hz. Increase in counting rate is obtained by feeding digit output 6 back to digit output 4, to bypass some internal logic of calculator.—M. Watts, "Calculator Chip Makes a Counter," National Semiconductor, Santa Clara, CA, 1974, AN-112, p 4.

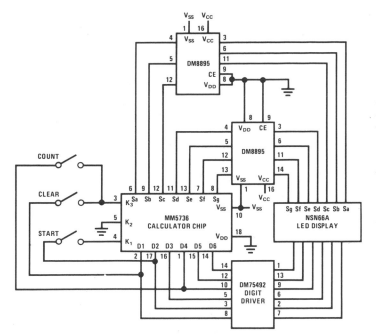

LARGE LED DISPLAY—National MM5736 calculator chip is used with DM8895 segment driver that can be mask-programmed to source several values of current in range from 5 to 17 mA per segment of LED display, permitting use of fairly large display. Display current comes from V_{CC} supply terminal of DM8895 rather than from calculator chip. Combination serves as 6-decade counter driving 6-digit display.—M. Watts, "Calculator Chip Makes a Counter," National Semiconductor, Santa Clara, CA, 1974, AN-112, p 3.

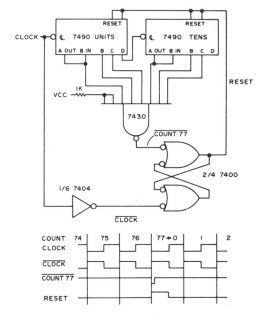

CLEAN RESET—Adding latch consisting of two 7400 NAND gates to reset circuit of divide-by-77 counter guarantees good reset. Reset pulse will always be half a clock period wide.—E. E. Hrivnak, House Cleaning the Logical Way, *73 Magazine,* Aug. 1974, p 85—90.

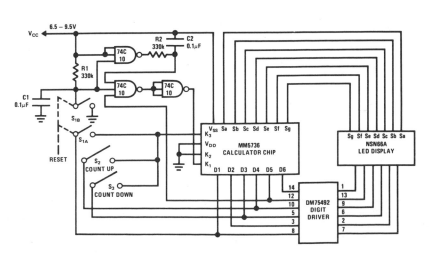

CALCULATOR COUNTS BY 4s—Connection shown for National MM5736 calculator chip counts either up or down by 4s, as might be required for keeping track of inventory in bin when parts are packaged in groups of 4. To count by numbers other than 1, desired number is entered into calculator during manual start operation. When S_2 is pushed, counter adds 4 to accumulated total. When S_3 is pushed, counter subtracts 4 from accumulated total. Logic elements provide self-starting action of counter.— M. Watts, "Calculator Chip Makes a Counter," National Semiconductor, Santa Clara, CA, 1974, AN-112, p 5.

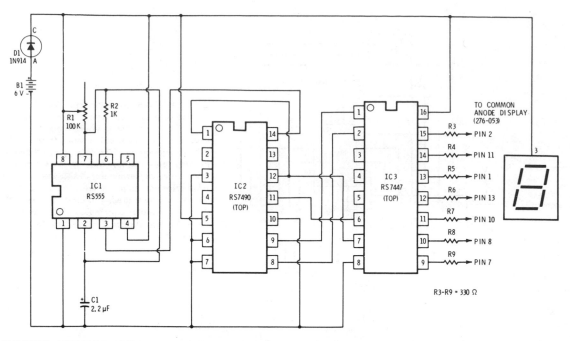

DIGITAL COUNTING DEMONSTRATOR—555 timer serves as clock for driving RS7490 decade counter feeding RS7447 BCD to 7-segment decoder that drives 7-segment digital display. R1 is adjusted to give clock frequency that makes display cycle slowly through digits 0-9 and repeat, for classroom demonstrations.—F. M. Mims, "Integrated Circuit Projects, Vol. 6," Radio Shack, Fort Worth, TX, 1977, p 53–63.

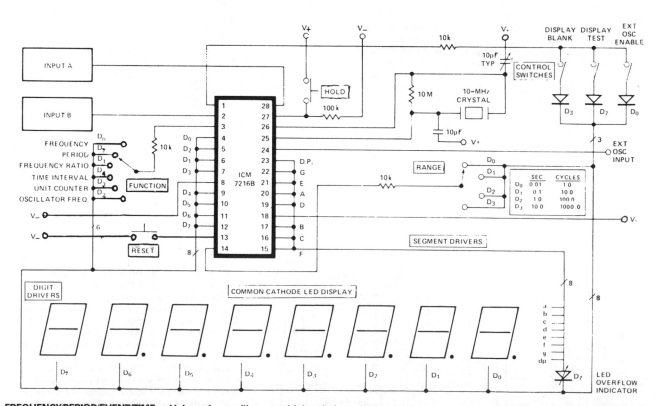

FREQUENCY/PERIOD/EVENT/TIME—Universal counter with 10-MHz maximum frequency provides multiple functions with minimum number of components. Range of time period measurements is 0.5 μs to 10 s. Includes 10-MHz crystal oscillator, multiplex timing with interdigit and leading-zero blanking, as well as overflow indication. Decimal position is selectable. Eight-digit multiplexed LED display outputs of IC can switch up to 250 mA per digit for handling large displays. Maximum supply voltage is 6 V.—Low Cost Universal Counter Performs Wide Range of Functions, *Computer Design*, Aug. 1978, p 168 and 170.

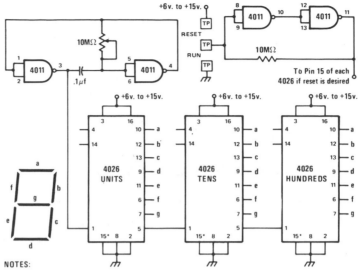

FREE-RUNNING COUNTER DISPLAY—Attention-getting circuit simply counts at predetermined rate while driving 3-digit display using 7-segment LEDs. Circuit uses two sections of 4011 CMOS quad NAND gate to generate pulses at rate controlled by 100-megohm pot. Pulses trigger 4026 counters connected as shown, with outputs a-g of each going to 7-segment LED display. When all three displays reach 9, next pulse resets all to 0 and count continues. Auxiliary circuit at upper right uses remaining sections of 4011 as flip-flop controlled by touch-plate switches; bridging gap between center and grounded plates with finger makes counter run. Bridging other gap resets counter to 0 and holds it there. If reset is not used, connect input pins 8, 9, 12, and 13 of unused gates to pin 14.—J. A. Sandler, 9 Easy to Build Projects under $9, *Modern Electronics*, July 1978, p 53–56.

NOTES:
+6v. to +15v. to Pin 14 of 4011 return to Pin 7 of 4011
No dropping resistors needed for most .3 to .8 inch LED displays.

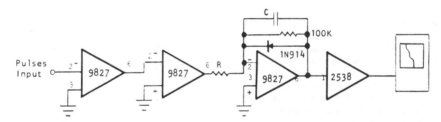

COUNTING-RATE METER—Uses three Optical Electronics 9827 opamps to amplify, square up, and integrate input pulses from event detector, to give integrated DC voltage that is function of counting rate. This voltage is compressed by 2538 DC logamp having 60-dB dynamic range for driving chart recorder. Values of R and C depend on counting rate. Well-regulated power supply is required because this determines amplitude of squared pulses that drive integrator. Applications include counting photons of photomultiplier or nuclear particles of solid-state detector. Logamp compresses output of integrator to eliminate need for scale changing while giving constant accuracy over wide dynamic range of counting rates.—"Logarithmic Counting Rate Readout," Optical Electronics, Tucson, AZ, Application Tip 10106.

CHAPTER **22**
Current Control Circuits

Includes fixed, adjustable, and voltage-controlled current sources, bilateral
sources, current limiters, current regulators, current sink, current-controlled
oscillator, power supply monitor, and electronic fuse.

5 mA WITH VARIABLE-SLOPE START-UP—
Low-cost NSL5022 LED typically has 1.6-V drop
at 2 mA, to produce constant 0.9 V across 390-
ohm emitter resistors in circuit shown. Use of
two current sources, each feeding other's LED
reference, eliminates all voltage defects except
for small voltage-dependent changes in transistor parameters. Adding 240K slope resistor cancels these changes, holding current constant
within 0.1% over supply voltage range of 5–20
V. Applications include use as voltage divider
with gain, Q multiplier for tuned circuits, and
bias compensation.—P. Lefferts, Variable Slope
Current Source Starts at 2.5 V, *EDN Magazine*,
Nov. 5, 1975, p 100.

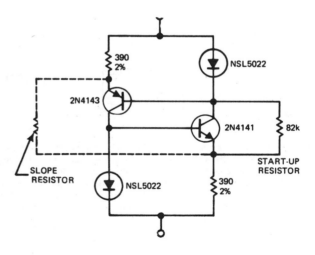

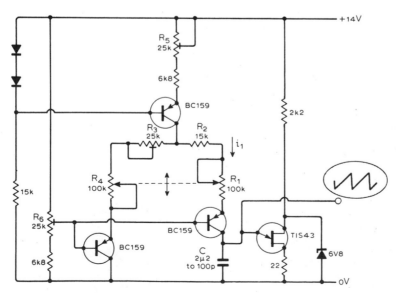

POT CONTROL—Circuit makes current through
linear pot a linear function of rotational angle of
pot. Article gives design equations. Current i_1
through R_1 is used to charge C, which is periodically discharged by UJT when trigger voltage
is reached. Frequency of output sawtooth is
proportional to i_1 and hence to angle of rotation
of pots. To set up, adjust pots to give maximum
sawtooth frequency and adjust preset R_5 for required maximum frequency. Set pots to other
extreme and reset R_3 for required minimum frequency.—A. Armit, Linear Current/Rotation
Control, *Wireless World*, Dec. 1975, p 576.

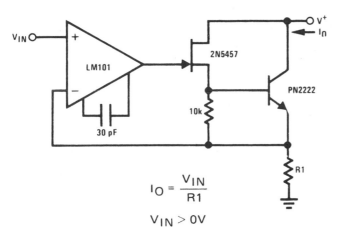

$$I_O = \frac{V_{IN}}{R1}$$

$$V_{IN} > 0V$$

PRECISION CURRENT SINK—R1 serves as current-sensing resistor providing negative feedback for opamp to enhance true current-sink nature of circuit. Both JFET and bipolar have inherently high output impedance as required for high-accuracy current sink.—"FET Databook," National Semiconductor, Santa Clara, CA, 1977, p 6-26—6-36.

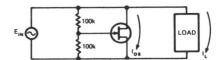

JFET CURRENT SINK—Simple circuit effectively raises load operating point of current-sensitive device by shunting current through JFET having nonlinear action. JFET type is not critical. Applications include improvement of thyristor noise performance by diverting current around load.—V. Gregory, FET Current Sinks Raise Operating Points, *EDN Magazine,* Feb. 20, 1974, p 81.

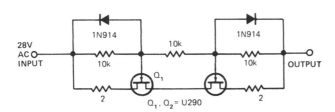

28-VAC CURRENT LIMITER—Dual JFETs in voltage-sharing arrangement protect output of 28-VAC power amplifier. Transistors should be matched for I_{DSS}. During positive half-cycle of input, Q_2 operates as current limiter and Q_1 as source follower. If Q_1 does not supply enough current, drain voltage of Q_2 drops and makes Q_1 turn on further. Conversely, if Q_1 supplies too much current, Q_2 drain voltage rises and tends to turn Q_1 off. On negative half-cycle, Q_1 becomes limiter and Q_2 is source follower.—J. P. Thompson, Current Limiter Protects Amplifier from Load Faults, *EDN Magazine,* June 5, 1978, p 148 and 150.

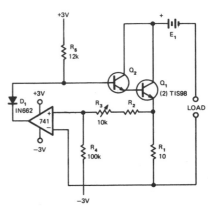

20-mA FLOATING SOURCE—Battery-operated circuit shown gives adequate stability for strain-gage bridge. Uses four alkaline penlight cells to provide ± 3 V for 741 opamp. E_1 is chosen to give adequate voltage for intended load at maximum load current. Temperature stability is 0.7 μA/°C from 0 to 50°C.—R. Tenny, Isolated Current Source, *EDN Magazine,* April 20, 1973, p 85.

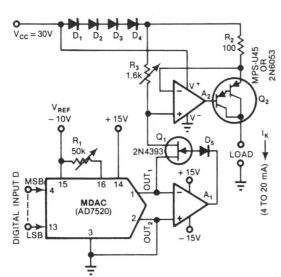

4–20 mA SOURCE—Digital input to multiplying digital-to-analog converter determines load current in range of 4 to 20 mA with 15.6-μA resolution. R_1 adjusts ratio of full-scale to zero-scale current at output 1 of MDAC, and R_3 sets circuit offset and span to give correct end-range currents for load. Maximum load compliance is 25 V. Opamp types are not critical.—J. Wilson and J. Whitmore, MDAC's Open Up a New World of Digital-Control Applications, *EDN Magazine,* Sept. 20, 1978, p 97–105.

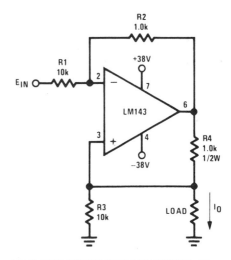

HIGH-COMPLIANCE CURRENT SOURCE—Noninverting input of LM143 high-voltage opamp senses current through R4 to establish output current that is proportional to input voltage. With ± 38 V supply, compliance of current source is ± 28 V.—"Linear Applications, Vol. 2," National Semiconductor, Santa Clara, CA, 1976, AN-127, p 3.

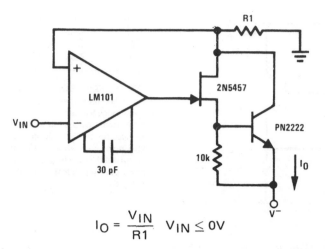

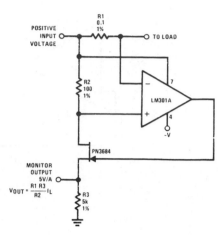

PRECISION CURRENT SOURCE—2N5457 JFET and PN2222 bipolar transistor serve as isolators between output and current-sensing resistor R1. LM101 opamp provides high loop gain to assure that circuit acts as current source.—"FET Databook," National Semiconductor, Santa Clara, CA, 1977, p 6-26–6-36.

POWER SUPPLY MONITOR—R1 senses output current of power supply. PN3684 JFET is used as buffer because source and drain currents are equal, so monitor output voltage accurately reflects current flow of power supply.—"FET Databook," National Semiconductor, Santa Clara, CA, 1977, p 6-26–6-36.

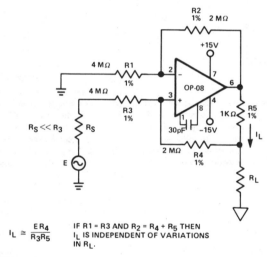

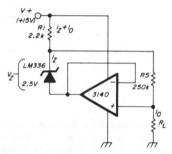

BILATERAL CURRENT SOURCE—Output current through load is constant within 2% of value related to input voltage and resistor values, regardless of variations in load from 10 to 2000 ohms. Circuit is built around Precision Monolithics OP-08 opamp.—"Precision Low Input Current Op Amp," Precision Monolithics, Santa Clara, CA, 1978, OP-08, p 7.

CURRENT REGULATOR—Combines zener with opamp in bootstrap configuration. Regulated output current I_o can be any value less than I_z but must be much greater than opamp bias current. Current in zener is set by R1 to provide minimum of 1 mA. Performance can be improved by using Motorola MC1403 or other 2.5-V three-terminal voltage reference in place of zener.—W. Jung, An IC Op Amp Update, *Ham Radio,* March 1978, p 62–69.

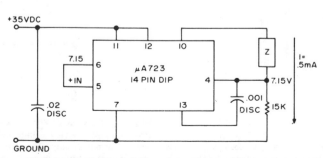

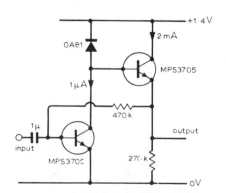

0.5 mA FOR 0–50 KILOHMS—Current source uses Fairchild μA723 voltage regulator operating from ordinary unregulated supply not over 40 VDC. Regulator has built-in 7.15-V reference. Output current is well within 1% of 0.5 mA for load impedances from 0–50K.—L. Nickel, Constant Current Sources, *73 Magazine,* March 1974, p 29.

CURRENT SOURCE AS TRANSISTOR LOAD—Reverse-biased germanium diode serves as voltage-independent current source for loading silicon transistors in linear amplifier having voltage gain of 50 and −3 dB bandwidth of 16–4000 Hz. In addition to low cost, circuit design permits reliable operation of reliable micropower circuits over wide temperature range at optimum current drain.—M. G. Baker, Low-Current Source, *Wireless World,* April 1976, p 61.

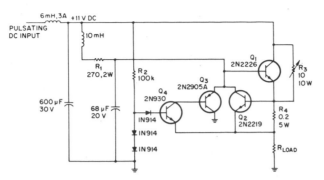

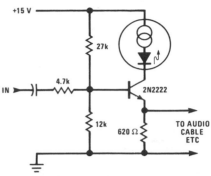

3-A LIMITER—Simple current limiter protects itself from overdissipation during shorted output, while handling capacitor or cold-filament loads that momentarily act like shorts. R_3 is adjusted so starting current is high enough to begin heating cold filament. As filament voltage increases to about 100 mV, Q_4 and Q_3 turn off, allowing load current to rise to 3-A limiting value.—L. G. Wright, Short-Protected Current Limiter Ignores Inrush Currents, *EEE Magazine*, Sept. 1970, p 89–90.

CONSTANT-CURRENT LED—National NSL4944 LED having built-in current control features can be used in simple circuit shown to provide current limiting and short-circuit protection for 15-V supply. Even with output shorted, LED draws only a little more than rated current.—"Linear Applications, Vol. 2," National Semiconductor, Santa Clara, CA, 1976, AN-153, p 3.

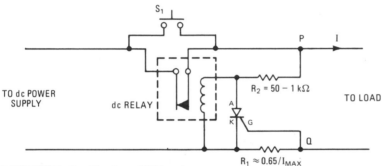

ELECTRONIC FUSE—Combination of SCR and line relay gives faster action than circuit breaker for protection against current overload. Closing S_1 momentarily energizes relay, completing current path from supply to load. Overload current increases voltage drop across R_1 to above 0.65 V, switching on SCR and thereby shorting relay coil to make it open. S_1 must be pressed again to reset relay. For adjustable dropout, gate of SCR can be connected to pot placed across R_1.—R. Quong, Resettable Electronic Fuse Consists of SCR and Relay, *Electronics*, Sept. 15, 1977, p 117.

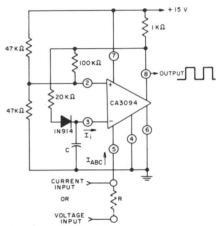

CURRENT-CONTROLLED OSCILLATOR—Makes use of proportional relationship between input current I_i and amplifier input bias current I_{ABC} of CA3094 programmable opamp. Linearity is within 1% over middle half of characteristic. Circuit can be used for voltage input if voltage is applied to pin 5 through appropriate dropping resistor R. Output is square wave.—"Circuit Ideas for RCA Linear ICs," RCA Solid State Division, Somerville, NJ, 1977, p 4.

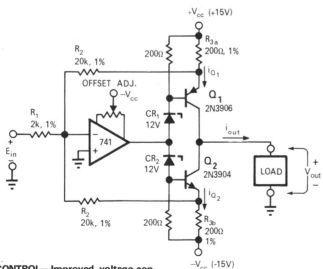

VOLTAGE CONTROL—Improved voltage-controlled current source uses complementary transistors in opamp feedback loop. Common-mode voltage at input to opamp is always near zero. Circuit was designed for use in integrator having ground-referenced integrating capacitor, to produce 1 mA/V. R_{3a} and R_{3b} sense current through Q_1 and Q_2, so voltage proportional to difference is fed back to input of opamp for comparison with input voltage. Zener voltages determine quiescent-current level. Frequency response is limited to 1 MHz by performance of specified opamp.—P. T. Skelly, Voltage-Controlled Current Source, *EDN/EEE Magazine*, Aug. 1, 1971, p 45–46.

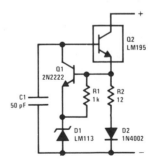

100-mA CURRENT REGULATOR—Two-terminal circuit using LM195 power transistor has low temperature coefficient and operates down to 3 V. 2N2222 controls voltage across current-sensing resistor R2 and diode D1. Voltage across sense network is base-emitter voltage of 2N2222 plus 1.2 V from LM113. R1 sets current through LM113 to 0.6 mA.—R. Dobkin, "Fast IC Power Transistor with Thermal Protection," National Semiconductor, Santa Clara, CA, 1974, AN-110, p 6.

CHAPTER 23
Data Transmission Circuits

Includes line driver, line receiver, modem, bit-rate generator, coder-decoder, FSK demodulator, signal conditioner, optoisolator, PDM telemetry, active bandpass filter, and other circuits used for transmitting digital data and digital speech over twisted-pair, coaxial, or balanced line.

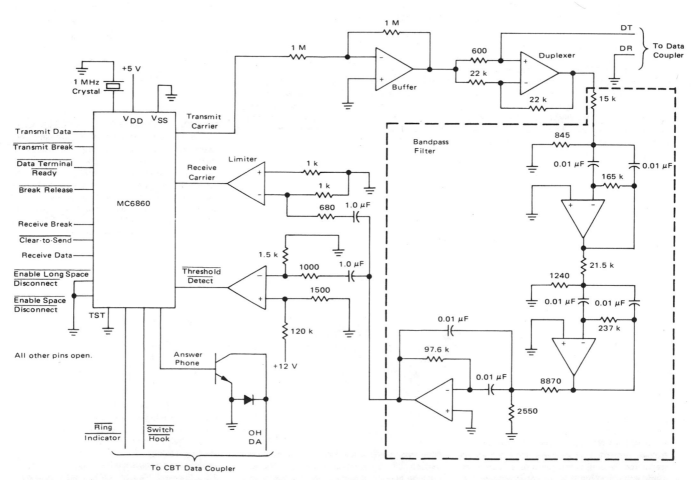

ANSWER MODEM—Transmits on upper channel (mark 2225 Hz and space 2025 Hz) and receives on lower channel (mark 1270 Hz and space 1070 Hz). Buffer and duplexer provide modem interface to transmission network. Bandpass filter allows only desired receive signals to be seen by limiter and demodulator. Motorola MC6860 modem IC contains modulator, demodulator, and supervisory control functions.—G. Nash, "Low-Speed Modem Fundamentals," Motorola, Phoenix, AZ, 1974, AN-731, p 6.

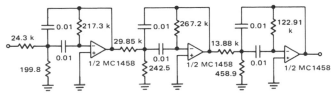

BANDPASS ORIGINATE FILTER—Provides gain of over 15 dB between 1975 and 2275 Hz, to accept 2025–2225 Hz signals of low-speed modem system using Motorola MC6860 IC.—J. M. DeLaune, "Low-Speed Modem System Design Using the MC6860," Motorola, Phoenix, AZ, 1975, AN-747, p 13.

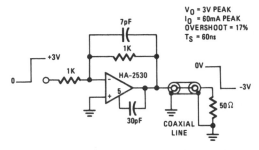

FAST-SETTLING COAX DRIVER—Suitable for use as radar pulse driver, video sync driver, or pulse-amplitude-modulation line driver. Uses Harris HA-2530/2535 wideband amplifier having high slew rate. Usable bandwidth is about 100 kHz when connected for noninverting operation as shown. Driver output is 60 mA into 60-ohm load. 5% settling time is 60 ns.—"Linear & Data Acquisition Products," Harris Semiconductor, Melbourne, FL, Vol. 1, 1977, p 7-54 (Application Note 516).

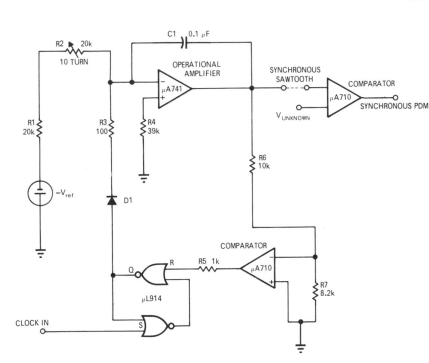

SYNCHRONOUS SAWTOOTH FOR PDM TELEMETRY—Circuit generates highly linear ramp that is reset to zero by each clock pulse. When ramp exceeds analog value of unknown input voltage, pulse is terminated. R1, R2, and C1 form integrating network around opamp. Varying R2 changes slope of ramp output.—J. Springer, Build a Sawtooth Generator with Three ICs, *EDN Magazine,* Nov. 15, 1970, p 49.

INTERFACES FOR 100-OHM LINE—Permits transferring data signals from SA900/901 diskette storage drive to location of MC6800 microprocessor up to maximum of 20 feet away through 100-ohm coax. Data line drivers used are capable of sinking 100-mA in logic true state with maximum voltage of 0.3 V with respect to logic ground. When line driver is in logic false state, driver transistor is cut off and voltage at output of driver is at least 3 V with respect to logic ground.—"Microprocessor Applications Manual" (Motorola Series in Solid-State Electronics), McGraw-Hill, New York, NY, 1975, p 5-211–5-212.

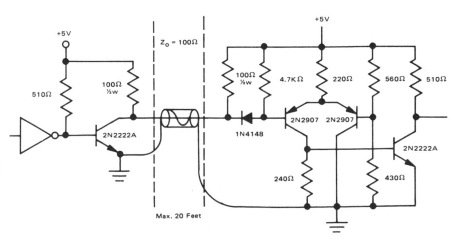

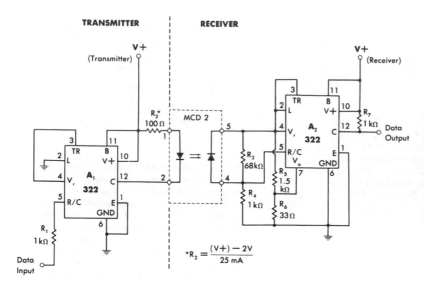

OPTICALLY COUPLED DATA LINK—322 comparator at transmitter end of link drives LED of MCD 2 optoisolator which accepts TTL input. Receiver is similar comparator having additional biasing to match photodiode output of optoisolator. Complete system is noninverting, with delay of about 2 μs. Receiver can have any supply within 4.5–40 V range of 322. Transmitter should be matched to its supply voltage by selecting R_2 according to equation shown.—W. G. Jung, "IC Timer Cookbook," Howard W. Sams, Indianapolis, IN, 1977, p 156–158.

$$*R_2 = \frac{(V+) - 2V}{25\ mA}$$

RECEIVE FILTER—Used as prefilter having controlled group-delay distortion, ahead of receiving modem in data transmission system. Values shown are for 950–1400 Hz answer filter. For 1900–2350 Hz originate filter, change critical values to those given in parentheses.—D. Lancaster, "TV Typewriter Cookbook," Howard W. Sams, Indianapolis, IN, 1976, p 180–182.

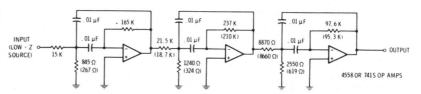

950-1400 Hz = NORMAL VALUES = ANSWER FILTER
1900-2350 Hz = (PARENTHETICAL VALUES) ORIGINATE FILTER

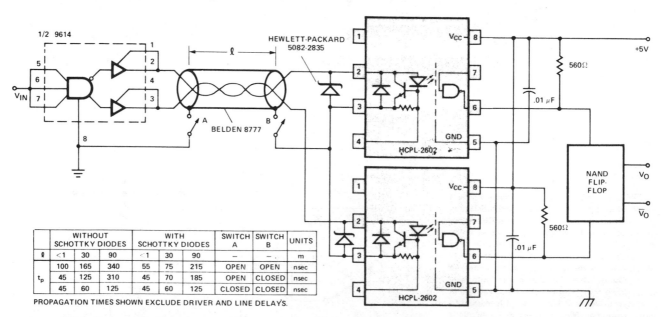

	WITHOUT SCHOTTKY DIODES			WITH SCHOTTKY DIODES			SWITCH A	SWITCH B	UNITS
ℓ	<1	30	90	<1	30	90	—	—	m
	100	165	340	55	75	215	OPEN	OPEN	nsec
t_p	45	125	310	45	70	185	OPEN	CLOSED	nsec
	45	60	125	45	60	125	CLOSED	CLOSED	nsec

PROPAGATION TIMES SHOWN EXCLUDE DRIVER AND LINE DELAYS.

POLARITY-REVERSING SPLIT-PHASE DRIVE—Half of 9614 polarity-reversing line driver feeds pair of Hewlett-Packard HCPL-2602 optically coupled line receivers through coax cable. Cable-grounding switches A and B change performance. Closing only switch B enhances common-mode rejection but reduces propagation delay slightly. Closing both switches optimizes data rate. Schottky diodes at receiver inputs improve data rate. NAND flip-flop at output greatly improves system noise rejection in split-phase termination of line.—"Optoelectronics Designer's Catalog 1977," Hewlett-Packard, Palo Alto, CA, 1977, p 158–159.

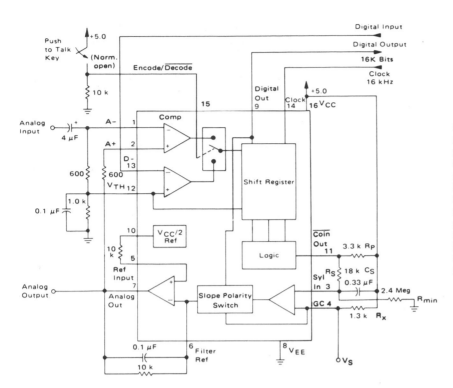

CVSD ENCODER FOR SECURE RADIO—Motorola MC3417 continuously variable slope delta modulator-demodulator IC is used as 16-kHz simplex voice coder-decoder for systems requiring digital communication of analog signals. Clock rate used depends on bandwidth required and can be 9.6 kHz or less for voice-only systems. Analog output uses single-pole integration network formed with 0.1 μF and 10K. Report covers circuit operation in detail for various applications.—"Continuously Variable Slope Delta Modulator/Demodulator," Motorola, Phoenix, AZ, 1978, DS 9488.

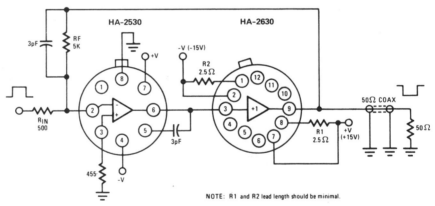

5-MHz COAX LINE DRIVER—Combination of Harris HA-2530 wideband inverting amplifier and HA-2630 unity-gain current amplifier provides 20-dB gain with extremely high slew rate and full power bandwidth even under heavy output loading conditions.—"Linear & Data Acquisition Products," Harris Semiconductor, Melbourne, FL, Vol. 1, 1977, p 2-47–2-50.

NOTE: R1 and R2 lead length should be minimal.

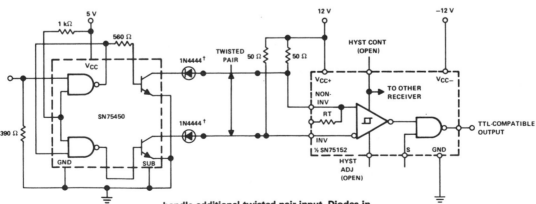

BALANCED-LINE TRANSMISSION—Transmits data at rates up to 0.5 MHz over twisted pair to Texas Instruments SN75152 dual-line receiver. Other section of receiver is identical and can handle additional twisted-pair input. Diodes in lines are required only for negative common-mode protection at driver outputs. System has high common-mode voltage capability. SN75450 is dual peripheral driver for high-current switching at high speeds.—"The Linear and Interface Circuits Data Book for Design Engineers," Texas Instruments, Dallas, TX, 1973, p 8-78.

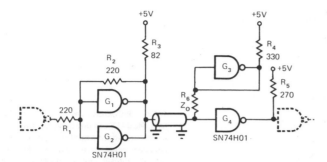

COAX DRIVER AND RECEIVER—Uses two TTL gates of SN74H01 package to form either driver or receiver for transmitting data over RG59 or RG174 coax at rates exceeding 10 megabits per second, with distance increasing from 400 meters at 10 Mb/s to over 1000 meters at 100 kb/s for RG59 and lesser distances for RG174. Can also be used for twisted-pair lines but at lower data rates. Bias gate G_3 exhibits low output impedance, for terminating channel load resistor R_6.—R. W. Stewart, Two TTL Gates Drive Very Long Coax Lines, *EDN Magazine,* Oct. 1, 1972, p 49.

103-COMPATIBLE MODEM—Motorola 4412 IC converts serial data, usually to and from universal asynchronous receiver-transmitter, into tones suitable for telephone communication. In originate mode, logic 0 is transmitted as 1070 Hz and logic 1 as 1270 Hz. In answer mode, logic 0 is transmitted as 2025 Hz and logic 1 as 2225 Hz. Modems are used in pairs, with receiver responding to tone group not being transmitted. Speed capability is up to 300-baud data rate. Output is 300 mVRMS into 100K load.—D. Lancaster, "CMOS Cookbook," Howard W. Sams, Indianapolis, IN, 1977, p 133.

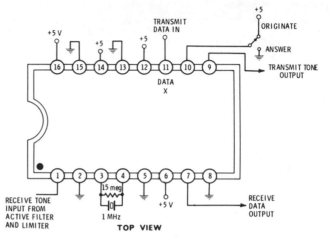

MODEM—Developed as part of TV terminal for microprocessor, to permit communication over telephone line with time-sharing computer system. Uses Motorola MC14412 modem chip for full-duplex FSK modulation having originate frequencies of 1270 Hz for mark and 1070 Hz for space, with answer frequencies of 2225 Hz for mark and 2025 Hz for space. AY-5-1012 UART serves as parallel interface to microprocessor. Article covers operation, construction, testing, connection to telephone lines, and use of modem.—R. Lange, Build the $35 Modem, *Kilobaud,* Nov. 1977, p 94–96.

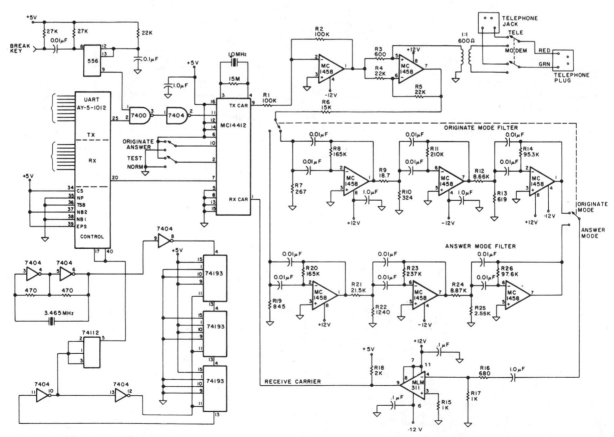

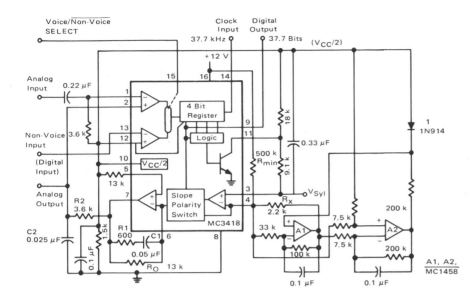

TELEPHONE-QUALITY CODER-DECODER—
Uses Motorola MC3418 continuously variable
slope delta modulator-demodulator IC to give
over 50 dB of dynamic range for 1-kHz test at
37.7K bit rate. At this rate, 40 voice channels can
be multiplexed on standard 1.544-megabit telephone carrier facility. IC includes active companding control and double integration for improved performance in encoding and decoding
digital speech. Opamp types are not critical.—
"Continuously Variable Slope Delta Modulator/
Demodulator," Motorola, Phoenix, AZ, 1978, DS
9488.

REZEROING AMPLIFIER—Used where input
signal has unknown and variable DC offset, as
in telemetry applications. Rezero command line
is enabled while ground reference signal is applied to input, making C1 charge to level proportional to DC offset of system. When rezero
line is deactivated, amplifier becomes conventional inverter, subtracting system offset and
giving true ground-referenced output. For 10-V
full-scale system requiring 0.1% (10-mV) accuracy, amplifier needs rezeroing reference every
100 ms.—"Linear Applications, Vol. 1," National Semiconductor, Santa Clara, CA, 1973,
AN-63, p 1–12.

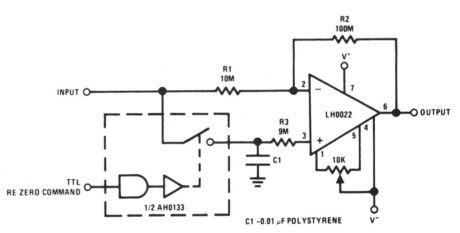

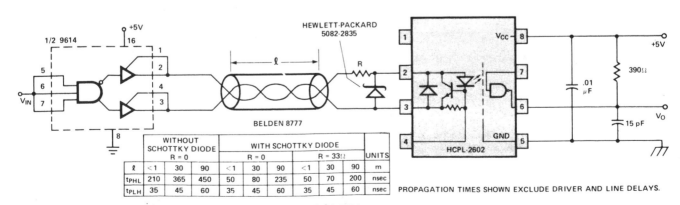

| | WITHOUT SCHOTTKY DIODE | | | WITH SCHOTTKY DIODE | | | | | | |
	R = 0			R = 0			R = 33Ω			UNITS
ℓ	<1	30	90	<1	30	90	<1	30	90	m
t_{PHL}	210	365	450	50	80	235	50	70	200	nsec
t_{PLH}	35	45	60	35	45	60	35	45	60	nsec

PROPAGATION TIMES SHOWN EXCLUDE DRIVER AND LINE DELAYS.

POLARITY-REVERSING DRIVE—Half of 9614
polarity-reversing line driver feeds Hewlett-
Packard HCPL-2602 optically coupled line receiver through shielded, twisted-pair, or coax
cable. Data rate is improved considerably by
using Schottky diode at input of receiver. Best
data rates are achieved when t_{PHL} (propagation
delay time to low output level) and t_{PLH} (propagation delay time to high output level) are clos-
est to being equal.—"Optoelectronics Designer's Catalog 1977," Hewlett-Packard, Palo
Alto, CA, 1977, p 158–159.

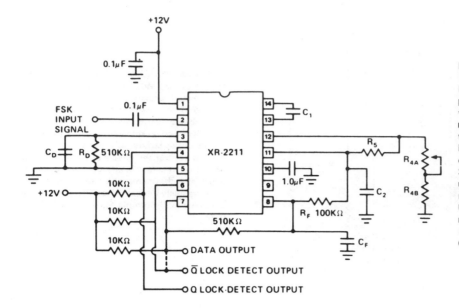

FSK DEMODULATOR WITH CARRIER DETECT— Exar XR-2211 FSK demodulator operating with PLL provides choice of outputs when carrier is present; pin 5 goes low and pin 6 goes high when carrier is detected. With pins 6 and 7 connected, output from these pins provides data when FSK is applied but is low when no carrier is present. Circuit performance is independent of input signal strength over range of 2 mV to 3 VRMS. Center frequency is $1/C_1 R_4$ Hz, with values in farads and ohms. Choose frequency to fall midway between mark and space frequencies. Used in transmitting digital data over telecommunication links.—"Phase-Locked Loop Data Book," Exar Integrated Systems, Sunnyvale, CA, 1978, p 57–61.

FSK DETECTOR—Exar XR-S200 PLL IC is connected as modem suitable for Bell 103 or 202 data sets operating at data transmission rates up to 1800 bauds. Input frequency shift corresponding to data bit reverses polarity of DC output voltage of multiplier. DC level is changed to binary output pulse by gain block connected as voltage comparator.—"Phase-Locked Loop Data Book," Exar Integrated Systems, Sunnyvale, CA, 1978, p 9–16.

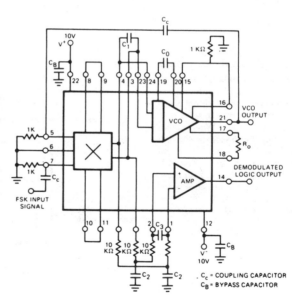

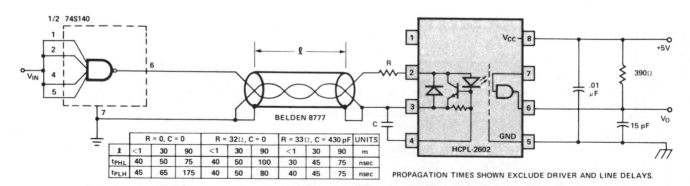

PROPAGATION TIMES SHOWN EXCLUDE DRIVER AND LINE DELAYS.

ℓ	R = 0, C = 0			R = 32Ω, C = 0			R = 33Ω, C = 430 pF			UNITS
ℓ	<1	30	90	<1	30	90	<1	30	90	m
t_{PHL}	40	50	75	40	50	100	30	45	75	nsec
t_{PLH}	45	65	175	40	50	80	40	45	75	nsec

POLARITY-NONREVERSING DRIVE—Hewlett-Packard HCPL-2602 optically coupled line receiver handles high data rates from shielded, twisted-pair, or coax cable fed by 74S140 line driver. Reflections due to active termination do not affect performance. Peaking capacitor C and series resistor R can be added to achieve highest possible data rate. C should be as large as possible without preventing regulator in line receiver from turning off during negative excursions of input signal. Highest data rates are achieved by equalizing t_{PHL} (propagation delay time to low output level) and t_{PLH} (propagation delay time to high output level).—"Optoelectronics Designer's Catalog 1977," Hewlett-Packard, Palo Alto, CA, 1977, p 158–159.

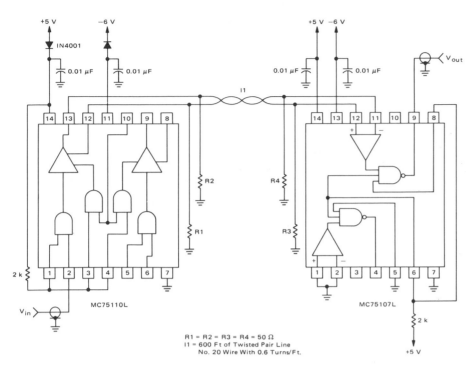

DIFFERENTIAL LINE DRIVER—Uses Motorola MC75110L line driver and MC75107L receiver with twisted-pair transmission line having attenuation of 1.6 dB per 100 feet at 10 MHz. Clock rate is 18.5 MHz. With push-pull driver shown, single pulse corresponds to transmission of 1 followed by series of 0s; one line is then at ground and the other at −300 mV. Arrangement is suitable for party-line or bus applications.— T. Hopkins, "Line Driver and Receiver Considerations," Motorola, Phoenix, AZ, 1978, AN-708A, p 11.

R1 = R2 = R3 = R4 = 50 Ω
I1 = 600 Ft of Twisted Pair Line
No. 20 Wire With 0.6 Turns/Ft.

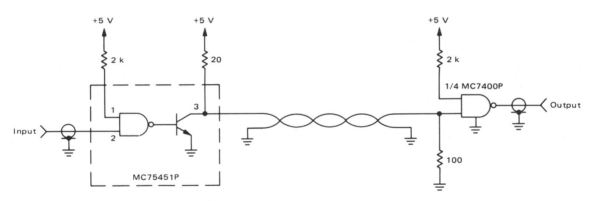

SINGLE-ENDED LINE DRIVER—Supplies 4.2-V input pulse to twisted-pair transmission line for point-to-point system. Requires only single +5 V supply.—T. Hopkins, "Line Driver and Receiver Considerations," Motorola, Phoenix, AZ, 1978, AN-708A, p 14.

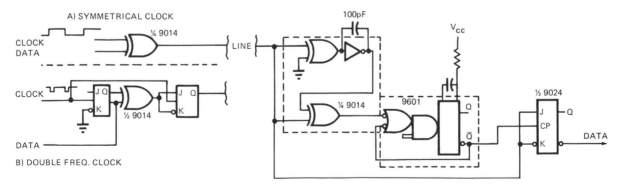

EXCLUSIVE-OR GATES—Use of retriggerable mono with EXCLUSIVE-OR gates simplifies design of both transmitter and receiver for handling binary phase-modulated digital data over single line. With 50% duty-cycle clock at transmitter, clock and data signals are applied to inputs of 9014 to generate output signal for line. At receiver, clock and data stream are regenerated by 9601 adjusted to 75% of data-bit time and connected in nonretriggerable mode. One EXCLUSIVE-OR gate and an EXCLUSIVE-NOR gate connected as inverting delay element will trigger 9601. System remains synchronized as long as pulse width of mono is between 50% and 100% of data-bit time.—P. Alfke, Exclusive-OR Gates Simplify Modem Designs, *EDN Magazine*, Sept. 15, 1972, p 43.

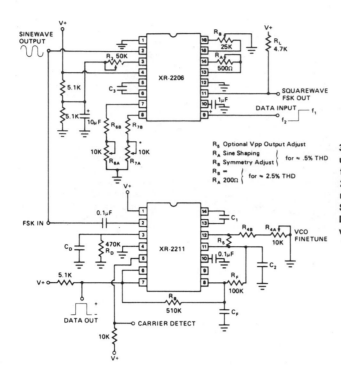

3-KILOBAUD FULL-DUPLEX FSK MODEM—Values shown are for 13-kHz bandwidth, 1070 Hz for mark and 1270 Hz for space, using Exar XR-2206 function generator and XR-2211 FSK demodulator. Report gives design procedure. Supply can be +12 V.—"Phase-Locked Loop Data Book," Exar Integrated Systems, Sunnyvale, CA, 1978, p 57–61.

DUAL-LINE RECEIVER FOR COAX—Single Texas Instruments SN75152 IC contains two receiver sections, each taking input from separate coax. Other receiver section (not shown) is identical and provides similar TTL output for its coax. Driver shown has OR capability for feeding single coax. Receiver has adjustable noise immunity and continuously adjustable hysteresis control (not shown).—"The Linear and Interface Circuits Data Book for Design Engineers," Texas Instruments, Dallas, TX, 1973, p 8-78.

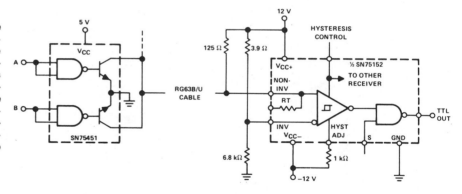

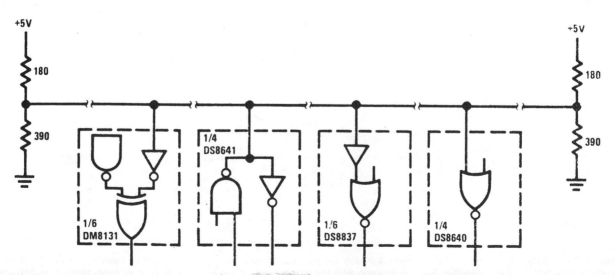

BUS TRANSCEIVER—Designed for use in bus-organized data transmission systems interconnected by terminated 120-ohm lines. Up to 27 driver/receiver pairs can be connected to common bus. One two-input NOR gate is included in National DS8641 quad unified bus transceiver package to disable all drivers in package simultaneously.—"Interface Integrated Circuits," National Semiconductor, Santa Clara, CA, 1975, p 3-17–3-18.

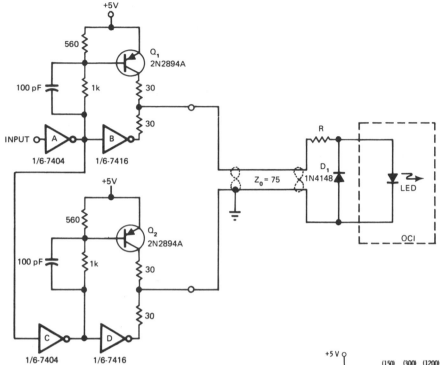

LINE DRIVER FOR LED—Single-ended input is converted to balanced differential drive for feeding 75-ohm transmission line terminated by LED serving as input for optically coupled line receiver. Logic 1 input is inverted to logic 0 by inverter A, turning on Q_1 and turning off output of gate B. At same time, output of inverter A is logic 1, which inhibits turn-on of Q_2 and makes output of inverter D go low. Thus, logic 1 input means that current is sourced into line and LED by Q_1, then sunk by output of D. Similarly, logic 0 input results in current being sourced into line by Q_2 and sunk by inverter B, making diode D_1 conduct and turn off LED of OCI receiver.—K. Erickson, Line Driver Is Compatible with OCI Line Receiver, *EDN Magazine*, Oct. 5, 1976, p 106.

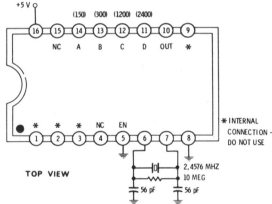

BIT-RATE GENERATOR—Fairchild 4702 IC synthesizes frequencies most often used in serial data communication, particularly with UARTS. With connections shown, output is 1760 Hz which is 16 × 110-baud rate of serial teletypes. Grounding only pin A generates 16 × 150 bauds. Grounding only pin B gives 16 × 300 bauds, grounding pin C gives 16 × 1200 bauds, and grounding pin D gives 16 × 2400 bauds. Will drive one regular TTL load at supply drain of 1 mA.—D. Lancaster, "CMOS Cookbook," Howard W. Sams, Indianapolis, IN, 1977, p 155.

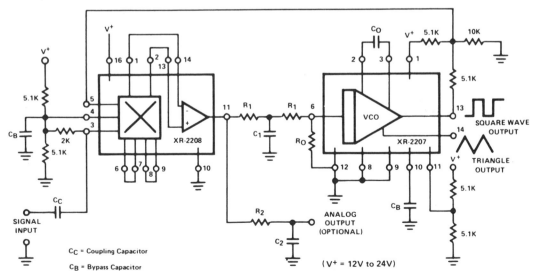

SINGLE-SUPPLY HIGH-PRECISION PLL—Combination of Exar XR-2207 VCO and XR-2208 operational multiplier is connected for operation from single 12–24 V supply for data communication and signal conditioning applications. Operating frequency range is 0.01 Hz to 100 kHz. Timing resistor R_0 should be in range of 5K to 100K, and R_1 should be greater than R_0. For 10-kHz center frequency, C_0 can be 0.01 μF and R_0 can be 10K. R_1 and C_1, which determine tracking range and low-pass filter characteristics, are 45K and 0.032 μF.—"Phase-Locked Loop Data Book," Exar Integrated Systems, Sunnyvale, CA, 1978, p 62–64.

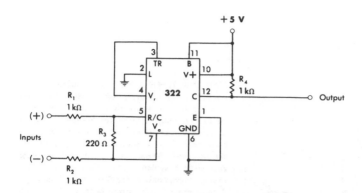

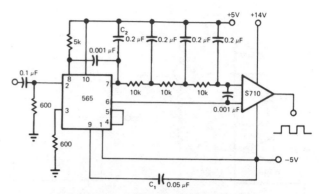

DIFFERENTIAL LINE RECEIVER—Responds to balanced-input drive signals fed to both comparator inputs of 322. Output is undisturbed even with up to 1 V of common-mode noise on input lines. TTL-compatible output is in phase with positive input. Overall delay is about 1 μs.—W. G. Jung, "IC Timer Cookbook," Howard W. Sams, Indianapolis, IN, 1977, p 153–155.

ANALOG PLL IN FSK DEMODULATOR—Developed for frequency-shift keying used in data transmission over wires, in which inputs vary carrier between two preset frequencies corresponding to low and high states of binary input signal. Circuit uses elaborate filter to separate modulated signal from carrier signal passed by PLL. 565 PLL provides reference for S710 comparator. Article gives design equations.—E. Murthi, Monolithic Phase-Locked Loops—Analogs Do All the Work of Digitals, and Much More, *EDN Magazine,* Sept. 5, 1977, p 59–64.

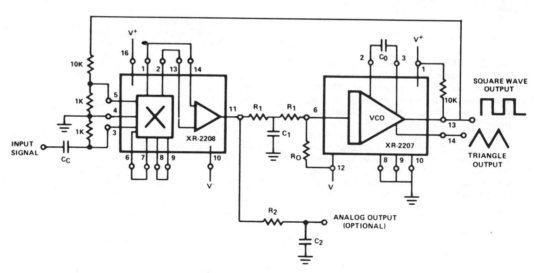

PLL FOR 0.01 Hz TO 100 kHz—Highly stable and precise phase-locked loop system using Exar XR-2207 VCO and XR-2208 operational multiplier is suitable for wide range of applications in data transmission and signal conditioning. Supply voltage range is ±6 V to ±13 V. For 10-kHz center frequency, R_O is 10K and C_O is 0.01 μF. R_1 and C_1, which determine tracking range and low-pass filter characteristics, are 45K and 0.032 μF.—"Phase-Locked Loop Data Book," Exar Integrated Systems, Sunnyvale, CA, 1978, p 62–64.

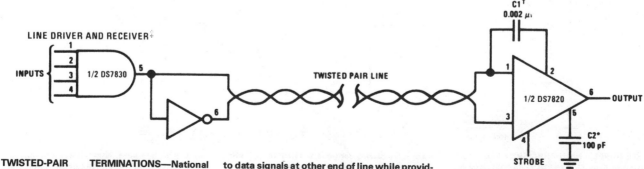

TWISTED-PAIR TERMINATIONS—National DS7830 line driver applies digital data to twisted-pair transmission line in high-noise environment, and DS7820 line receiver responds to data signals at other end of line while providing immunity to noise spikes. Exact value of C1 depends on line length. Supply voltage is 4.5 to 5 V for both receiver and driver. C2 is optional and controls response time.—"Interface Integrated Circuits," National Semiconductor, Santa Clara, CA, 1975, p 8-1–8-16.

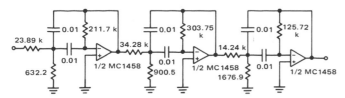

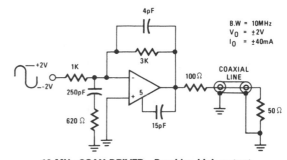

All capacitors are in μF.

BANDPASS ANSWER FILTER—Provides gain of 15 dB over bandwidth of 1020 to 1320 Hz for low-speed modem system using Motorola MC6860 IC. Attenuation is 35 dB at 2225 Hz, as required for answer-only modem system. Equations for values of filter components are given.—J. M. DeLaune, "Low-Speed Modem System Design Using the MC6860," Motorola, Phoenix, AZ, 1975, AN-747, p 10.

10-MHz COAX DRIVER—Provides high output current to coaxial line over bandwidth limited only by single-pole response of feedback components. Response is flat with no peaking and distortion is low. Uses Harris HA-2530/2535 wideband amplifier having high slew rate.— "Linear & Data Acquisition Products," Harris Semiconductor, Melbourne, FL, Vol. 1, 1977, p 7-54 (Application Note 516).

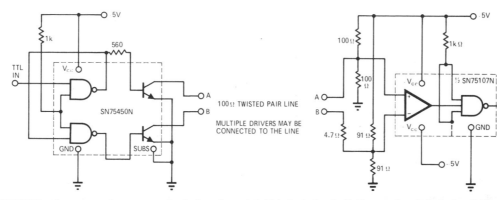

WIRED-OR TERMINALS—Arrangement permits connecting several IC line drivers in parallel for feeding single 100-ohm twisted-pair data line. With wired-OR transmitting capability, TTL output of receiver at right is logic 1 only if all paralleled drivers are transmitting logic 1. If any one or all of drivers transmit logic 0, output of receiver is logic 0.—D. Pippenger, Termination Is the Key to Wired-OR Capability, *EDN/EEE Magazine*, Dec. 15, 1971, p 17.

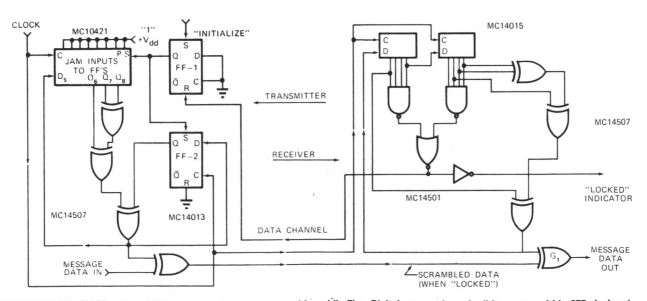

PSEUDORANDOM CMOS—Uses MC14021 8-bit shift register in conjunction with MC14507 EXCLUSIVE-OR gates to generate pseudorandom digital code. To develop code pattern, 1st, 6th, 7th, and 8th bits are sent through EXCLUSIVE-OR gates and fed back to shift-register input. Output can be used as random test signal or for protecting messages sent over public channels or stored in public files. Digital message is scrambled by mixing it with output of code generator in EXCLUSIVE-OR gate. Functionally identical 255-bit random generator is used at receiver to unscramble data. Decoding circuit must have access to sending clock and means for synchronizing so as to put both registers into all-1 state. Register in receiver goes through all its states within 255 clock pulses; when it reaches all-1 state, signal is fed back to sender for releasing FF-1 so scrambling can commence. Article traces operation in detail.— J. Halligan, Pseudo-Random Number Generator Uses CMOS Logic, *EDN Magazine*, Aug. 15, 1972, p 42–43.

SINGLE-SUPPLY LINE DRIVER—Motorola MC75451P driver and external components shown provide differential signal for twisted-pair transmission line from single +5 V supply. External gate provides required input phase reversal to gate G2 of IC. Each output of IC varies between 0.5 V and 3.6 V, so net differential voltage driven into line is about 6 V. Only receiver end of line is terminated in its characteristic impedance, since arrangement is intended only for point-to-point transmission.—T. Hopkins, "Line Driver and Receiver Considerations," Motorola, Phoenix, AZ, 1978, AN-708A, p 12.

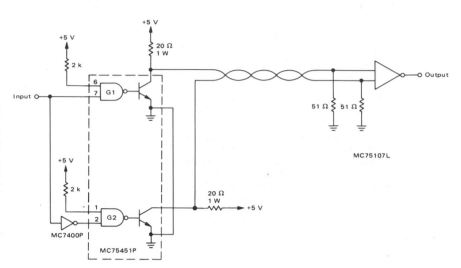

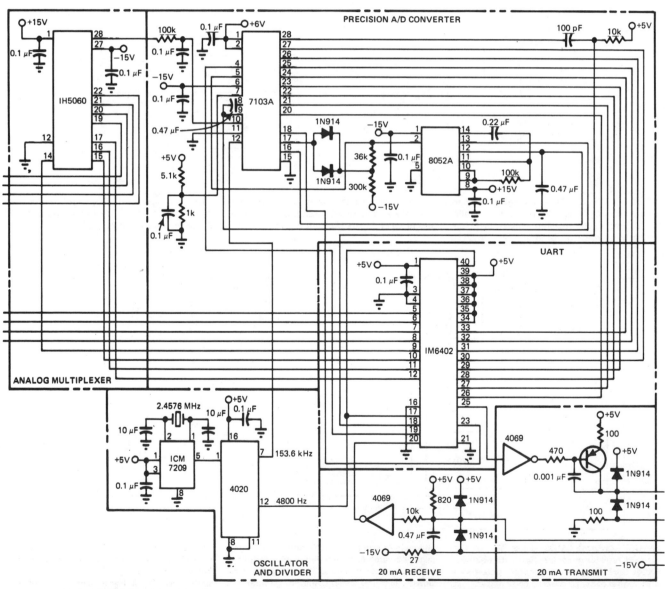

REMOTE DATA STATION—Circuit monitors DC voltages applied to pins 19 and 20 of IH5060 multiplexer and converts them to digital format for transmission as serial data to remote micro-processor. IM6100 remote host processor sends control signals to IM6402 UART to select individual multiplexer channels. Single 7209 oscillator provides clock signal for A/D converter and UART. Developed for use with Intercept Jr. microprocessor system.—S. Osgood, Remote Data Station Simplifies Data Gathering, *EDN Magazine,* Jan. 20, 1978, p 38 and 41.

CHAPTER 24

Digital Clock Circuits

Provide 12- or 24-hour time on LED, LCD, gas-discharge, or fluorescent digital displays for watches and clocks. Some also have calendar display and alarm-tone generator. Special circuits provide battery backup for AC power failure, multiplexing of display to reduce battery drain, stopwatch, and tide clock. Clock-pulse generators for logic and microprocessor applications are given in Clock Signal chapter.

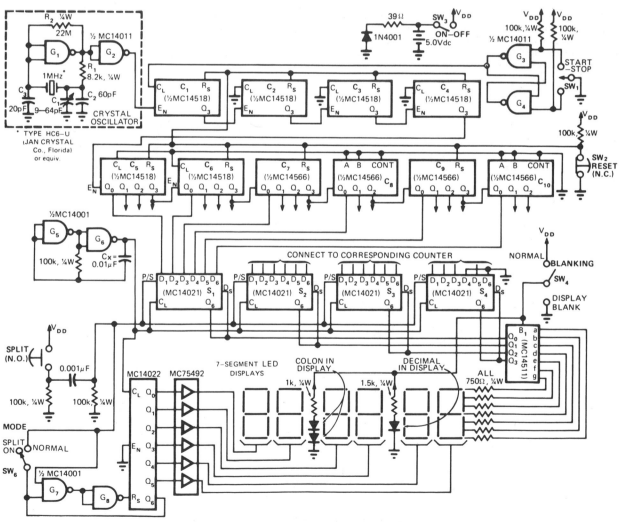

6-DIGIT STOPWATCH—Low-cost battery-powered electronic stopwatch with 6-digit LED display uses readily available complex-function CMOS ICs to minimize component count. Time range is up to 59 min and 59.99 s. Multiplexing by time-sharing counters through one display-driving decoder cuts battery drain because each digit is on for only one-sixth of time. Article traces operation of circuit step by step. Maximum error is only 0.001 s/h. Four rechargeable nicad batteries last 500 h per charge if displays are blanked when not being read, and about 6 h without blanking.—A. Mouton, Build Your Own Digital Stopwatch with Strobed LED Readout, *EDN Magazine,* April 5, 1974, p 55–57.

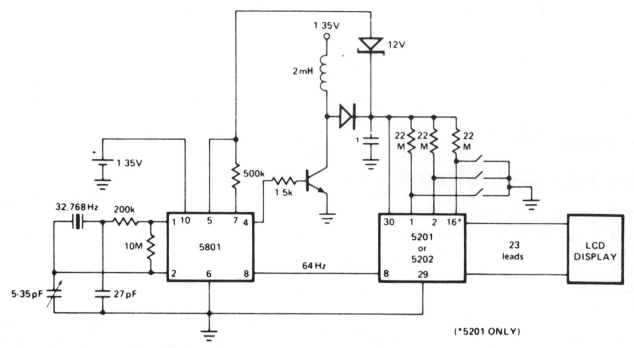

LCD WRISTWATCH—Inverter section of Intel 5801 oscillator/divider is used with 32,768-Hz crystal to produce time base. First divider in 5801 reduces this to 1024 Hz for driving upconverter transistor. Feedback from transistor through 12-V zener is used to regulate and control pulse width of 1024-Hz signal. Upconverter also provides 12–15 V required by LCD and 5201 decoder/driver IC. Output to each LCD segment and to common backplate is 32-Hz square wave. Separate drive flashes colon at 1-Hz rate.—M. S. Robbins, "Electronic Clocks and Watches," Howard W. Sams, Indianapolis, IN, 1975, p 128–130.

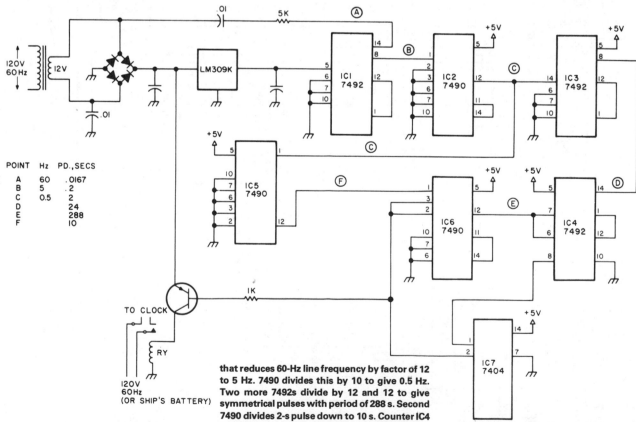

POINT	Hz	PD.,SECS
A	60	.0167
B	5	.2
C	0.5	2
D		24
E		288
F		10

TIDE CLOCK—Circuit shuts off electric clock of any type for 5 s out of every 144 s, to give loss of 50 min in 24 h as required for making high tides conform to clock readings. Regulated 5-V supply shown drives TTL 7492 frequency divider that reduces 60-Hz line frequency by factor of 12 to 5 Hz. 7490 divides this by 10 to give 0.5 Hz. Two more 7492s divide by 12 and 12 to give symmetrical pulses with period of 288 s. Second 7490 divides 2-s pulse down to 10 s. Counter IC4 inhibits 5-s counter by feeding low output into one gate of IC7 hex inverter. When IC4 counts up to 144 s, its output goes high and resets IC6 to low for start of 5-s low period of that counter. Article gives timing waveforms. Switching transistor is used to control relay that opens clock circuit. Set tide clock at 12:00 for high tide at location of use, and it will be 12:00 at high tide thereafter. Low tide will then be at 6:00.—J. F. Crowther, Time and Tide—Digitally, *73 Magazine*, Aug. 1978, p 156–157.

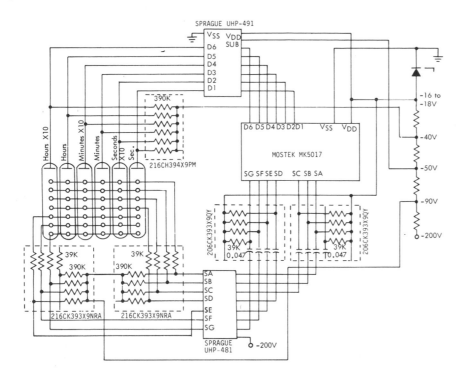

12-h WITH SECONDS—Combination of Mostek clock IC and Sprague high-voltage display drivers, acting through 206C and 216C single in-line resistor network, provides drive for conventional seven-element gas-discharge digital clock display showing hours, minutes, and seconds. Requires −200 V supply. Display can be Burroughs Panaplex, Cherry Plasma-Lux, or Beckman SP series.—"Integrated Circuits Data Book—1," Sprague, North Adams, MA, 1978, p 3-5.

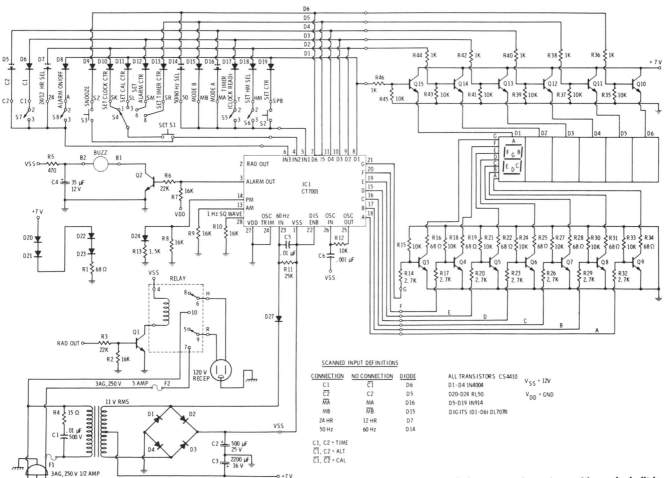

6-DIGIT WITH CALENDAR AND ALARMS—Circuit is built around Cal-Tex CT7001 IC that includes outputs for displaying day of month along with time on Litronix DL707 LED readouts. Transistor switch Q1 and relay form timer triggered by IC to control radio or other appliance drawing up to 5 A from AC line. Dual-voltage power supply provides 7 and 14 VDC. Includes snooze alarm along with regular built-in transistor-driven buzzer.—M. S. Robbins, "Electronic Clocks and Watches," Howard W. Sams, Indianapolis, IN, 1975, p 103–104 and 116–117.

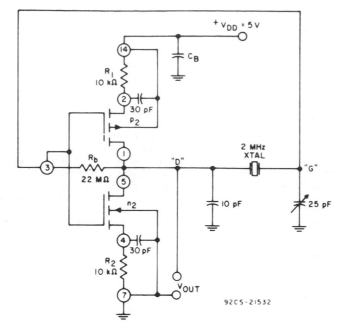

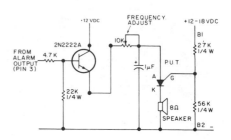

2-MHz CRYSTAL USING CMOS PAIR—One CMOS transistor pair from CA3600E array is connected with feedback pi network to give stable oscillator performance with 2-MHz crystal.

Low power drain makes circuit ideal for use in digital clocks and watches.—"Linear Integrated Circuits and MOS/FET's," RCA Solid State Division, Somerville, NJ, 1977, p 280.

ALARM FOR DIGITAL CLOCK—Uses transistor as driver to turn on programmable unijunction transistor (PUT) oscillator feeding 8-ohm loudspeaker. Pitch of tone can be adjusted with 10K pot. Input is from alarm pin of digital clock IC (pin 3 for Fairchild FCM7001 equivalent of Cal-Tex CT7001). PUT is Radio Shack 276-119 or equivalent.—W. J. Prudhomme, CT7001 Clockbuster, *73 Magazine*, Dec. 1976, p 52–54 and 56–58.

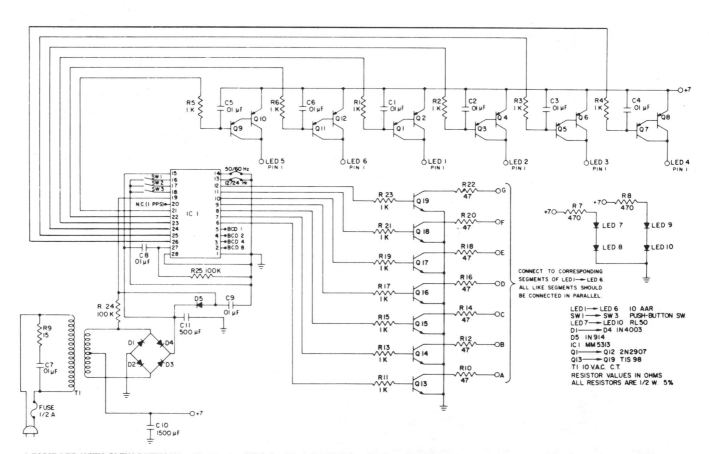

6-DIGIT LED WITH SLEW BUTTONS—National MM5313 PMOS digital clock IC drives display which includes four discrete LEDs mounted on readout panel to form colons between hours, minutes, and seconds. AC supply provides 14 VDC for IC and 7 VDC for displays. Hold pushbutton SW1 stops count to give precise seconds setting. Slow-slew button SW2 advances time at 1 min/s for precise setting, and fast-slew button SW3 advances time 1 h/s. Digit drivers Q1- Q12 are Darlington-connected pairs of PNP transistors. Segment drivers Q13-Q19 are single PNP transistors.—M. S. Robbins, "Electronic Clocks and Watches," Howard W. Sams, Indianapolis, IN, 1975, p 103 and 113.

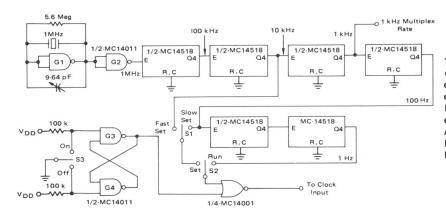

1-Hz REFERENCE—Output of 1-MHz crystal oscillator is stepped down to 1 Hz by CMOS decade divider chain using Motorola MC14518 dual decade counters. Circuit also generates 1-kHz multiplex rate for display used with 24-h industrial clock. Supply is +5 V.—D. Aldridge and A. Mouton, "Industrial Clock/Timer Featuring Back-Up Power Supply Operation," Motorola, Phoenix, AZ, 1974, AN-718A, p 5.

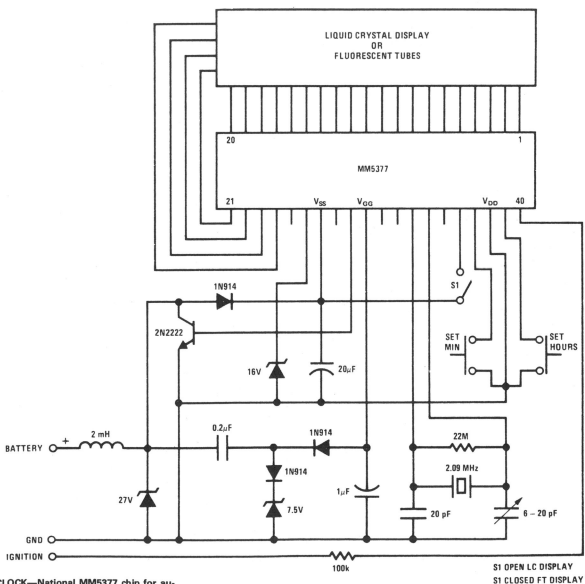

AUTO CLOCK—National MM5377 chip for automobile clock interfaces directly with 4-digit liquid-crystal or fluorescent-tube display. 12-h format includes leading-zero blanking and colon indication. Voltage-sensitive output drives energy-storage network serving as voltage doubler/regulator. Crystal oscillator is referenced time base.—"MOS/LSI Databook," National Semiconductor, Santa Clara, CA, 1977, p 1-33–1-37.

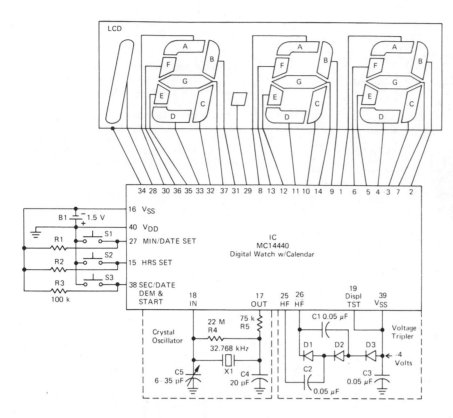

1.5-V LCD—Will operate over 1 year on single 1.5-V AAA battery with accuracy of ±1 min. Basic timekeeping functions are provided by Motorola MC14440 CMOS device that includes calendar. 32.768-kHz NT-cut quartz crystal and trimming capacitor provide reference frequency. Output of 1.5-V alkaline cell is increased to 4 V for display by voltage tripler using MBD101 Schottky diodes.—J. Roy and A. Mouton, "A Cordless, CMOS, Liquid-Crystal Display Clock," Motorola, Phoenix, AZ, 1977, EB-56.

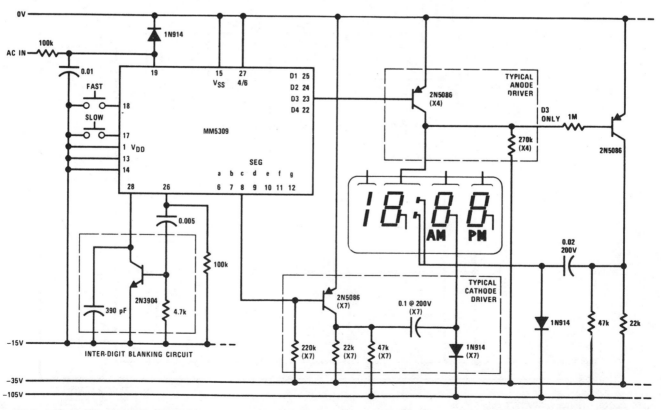

GAS-DISCHARGE DISPLAY—National MM5309 digital clock gives choice of 12- or 24-h display and 50- or 60-Hz operation for driving 4-digit gas-discharge display having colon and AM/PM indications. Separate cathode driver and separate anode driver are required for each digit.— "MOS/LSI Databook," National Semiconductor, Santa Clara, CA, 1977, p 1-2–1-8.

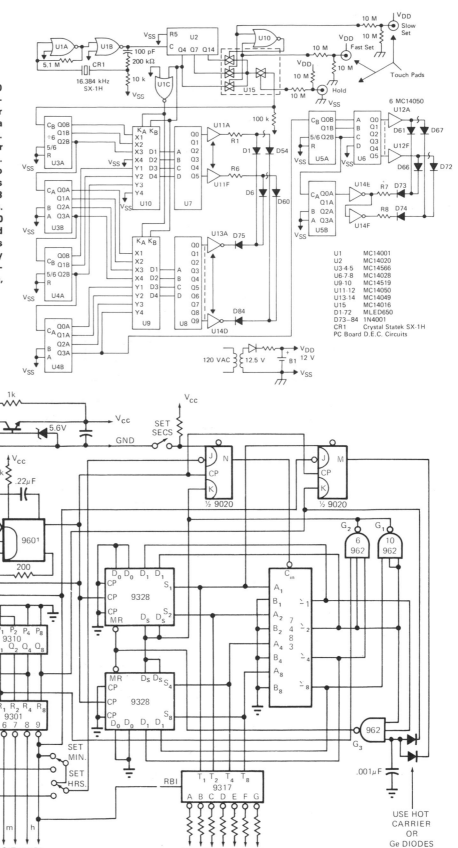

CIRCULAR LED ARRAY—Arrangement of 60 LEDs sequencing in outer ring to indicate seconds and minutes, combined with 12 in inner ring to indicate hours, is driven by Motorola MC14566 CMOS industrial time-base generator. Time reference is 16.384-kHz crystal oscillator consisting of two NOR gates and Statek crystal. Reference frequency is divided by 2^{14} in U2 to give 1-s pulse rate for driving accumulators U3A-U5B. Maximum error is 1 s per month. U3 counts seconds, U4 minutes, and U5 hours. Multiplexing is required because same set of 60 LEDs serves for minutes and seconds. Fast and slow touch pads eliminate need for switches when setting time. Single 12-V nicad battery provides backup for AC line failure.—A. Mouton, "The LED Circular Timepiece," Motorola, Phoenix, AZ, 1975, EB-41.

U1	MC14001
U2	MC14020
U3-4-5	MC14566
U6-7-8	MC14028
U9-10	MC14519
U11-12	MC14050
U13-14	MC14049
U15	MC14016
D1-72	MLED650
D73-84	1N4001
CR1	Crystal Statek SX-1H
	PC Board D.E.C. Circuits

MULTIPLEXED CLOCK DISPLAY—Multiplexed display suitable for LED readouts is provided by circuit using TTL counters to count 60-Hz line. When count reaches 10 o'clock, flip-flop M is set on every cycle. Gate G_3 then detects when time goes to 13 o'clock, and clears shift register. Carry flip-flop remains set, so 1 is loaded into hours digit to accomplish transition from 12:59:59 to 1:00:00. Seven-segment decoder driver looks at shift register output and drives segment lines of LED. Leading hours digit is blanked, using RBI input on 9317.—G. Smith, Novel Clock Circuit Provides Multiplexed Display, *EDN Magazine*, Sept. 1, 1972, p 50–51.

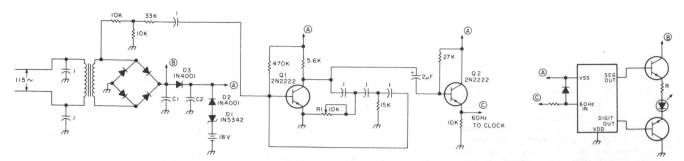

STANDBY SUPPLY—Phase-shift oscillator Q1 operates from AC line through bridge-rectifier power supply and provides line-synchronized 60-Hz power to standard digital clock through isolating emitter-follower Q2. During power outage, oscillator is switched automatically to battery by diode network and provides reasonably accurate signal for operating clock. Free-running oscillator is adjusted to be slightly low, such as 59.9 Hz. For reasonably long power outage, say 4 h, this 0.1-Hz error is equivalent to 0.167% error in time, so clock loses only 24 s during outage. C1 and C2 are 200 to 300 μF. Adjust R1 to give output just below 60 Hz on battery operation. To minimize battery drain, LEDs on digital clock are not energized during standby.—R. S. Isenson, Digital Clock Fail-Safe, *73 Magazine*, July 1977, p 168–169.

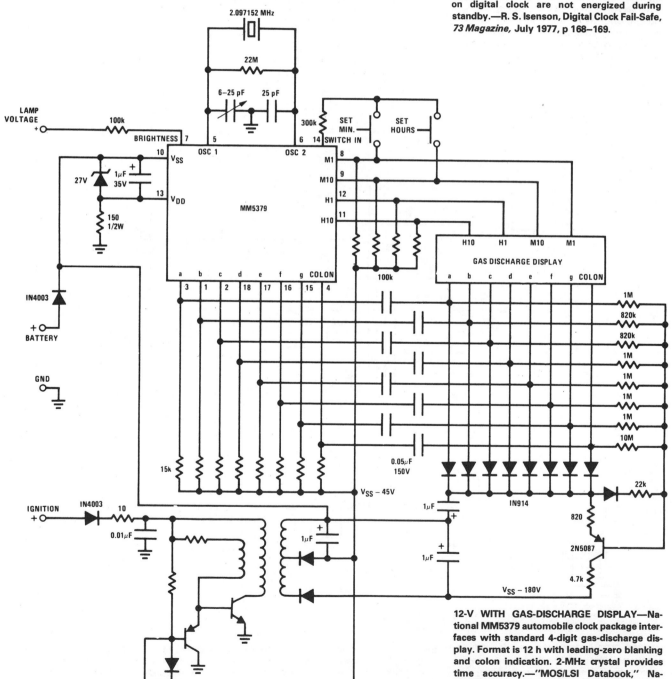

12-V WITH GAS-DISCHARGE DISPLAY—National MM5379 automobile clock package interfaces with standard 4-digit gas-discharge display. Format is 12 h with leading-zero blanking and colon indication. 2-MHz crystal provides time accuracy.—"MOS/LSI Databook," National Semiconductor, Santa Clara, CA, 1977, p 1-38–1-42.

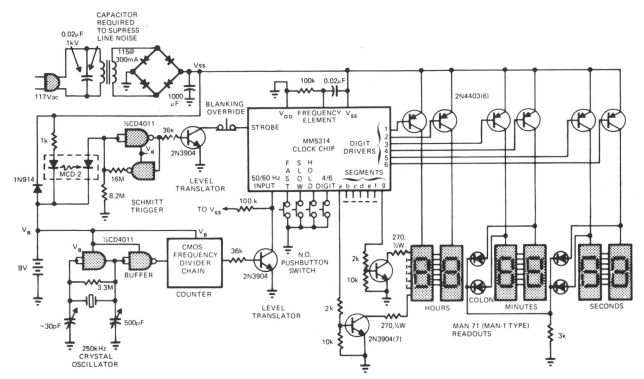

AC/DC CLOCK—When AC power fails, MCD-2 optoisolator senses voltage drop and makes Schmitt trigger force strobe input of clock chip to ground, blanking display and reducing current drain from 200 mA on AC to 12 mA on 9-V standby battery. Clock will run for days on 1000-mAh battery. Two LED pairs that form colons between time digits are operated from digit strobe lines and remain lit when display is blanked, but draw only 1 mA.—S. I. Green, Digital Clock Keeps Counting Even When AC Power Fails, *EDN Magazine*, Dec. 20, 1974, p 49–51.

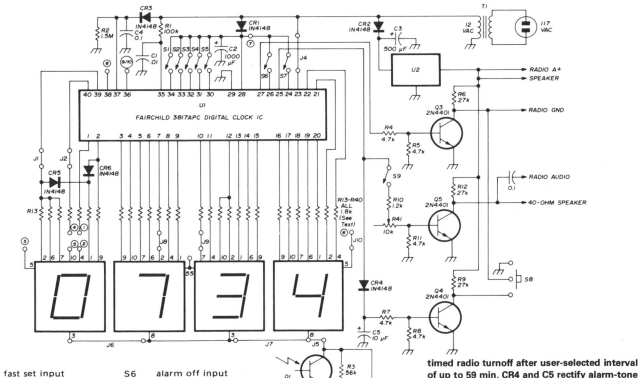

S1	fast set input	S6	alarm off input
S2	slow set input	S7	snooze input
S3	seconds display input	S8	alarm tone on/off
S4	alarm display input	S9	alarm output on/off
S5	sleep display input	R41	tone amplitude control

DIGITAL ALARM—Direct drive offered by Fairchild 3817 IC allows design of simple low-cost clock radio providing display drive, alarm, and sleep-to-music features in 12- or 24-h formats. Display is Fairchild FND500 LED. Either 50- or 60-Hz input may be used. U2 is 7800-series IC voltage regulator rated to meet requirements of radio used. Q3 provides active low output for timed radio turnoff after user-selected interval of up to 59 min. CR4 and C5 rectify alarm-tone output for amplification by Q4 to give active low output for timed radio turn-on when coincidence is detected by alarm comparators. Q5 provides alarm-tone output at level sufficient to drive 40-ohm loudspeaker with ample wake-up volume. If radio is used, omit loudspeaker. Article covers construction and adjustment.—D. R. Schmieskors, Jr., Low-Cost Digital Clock, *Ham Radio*, Feb. 1976, p 26–30.

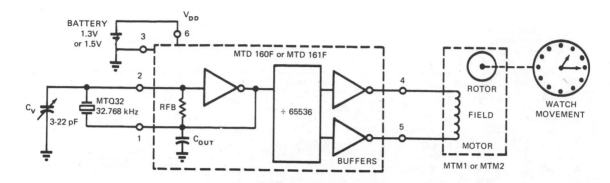

C_V = Trimmer capacitance
C_{OUT} = Integrated oscillator output capacitance
         $\approx$ 20 pF
R_{FB} = Integrated oscillator feedback resistance
         $\approx$ 40 M

QUARTZ-MOTOR WRISTWATCH—Uses one 32.768-kHz crystal at input of Motorola MTD 160F or 161F custom CMOS chip, with stepper motor at output of chip for driving conventional watch hands. Chip contains three-inverter oscillator, 16 counting flip-flops, and motor drive buffers.—B. Furlow, CMOS Gates in Linear Applications: The Results Are Surprisingly Good, *EDN Magazine,* March 5, 1973, p 42–48.

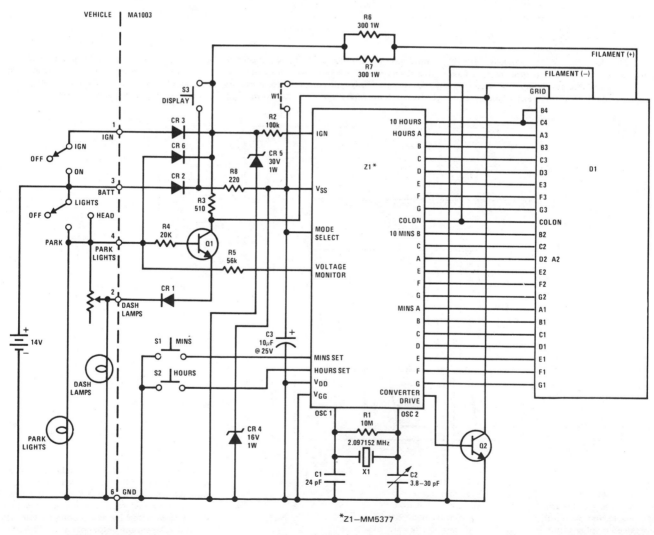

12-V AUTO CLOCK—National MA1003 automotive/instrument clock module combines MM5377 MOS LSI clock with 4-digit 0.3-inch green vacuum fluorescent display, 2.097-MHz crystal, and discrete components on single printed-circuit board to give complete digital clock. Brightness control logic blanks display when ignition is off, reduces brightness to 33% when parking or headlight lamps are on, and follows dash-lamp dimming control setting. Display has leading-zero blanking. For portable applications, display can be activated by closing display switch momentarily.—"MOS/LSI Databook," National Semiconductor, Santa Clara, CA, 1977, p 13-8—13-10.

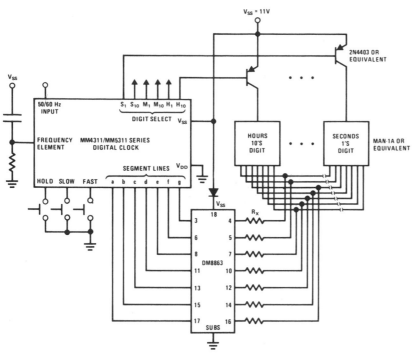

6-DIGIT DISPLAY—National DM8863 8-digit LED driver serves as segment driver for common-anode display of hours, minutes, and seconds, replacing total of 14 resistors and 7 transistors.—C. Carinalli, "Driving 7-Segment LED Displays with National Semiconductor Circuits," National Semiconductor, Santa Clara, CA, 1974, AN-99, p 11.

$R_x \approx 200$, VARIABLE DEPENDING ON DESIRED DISPLAY BRIGHTNESS.

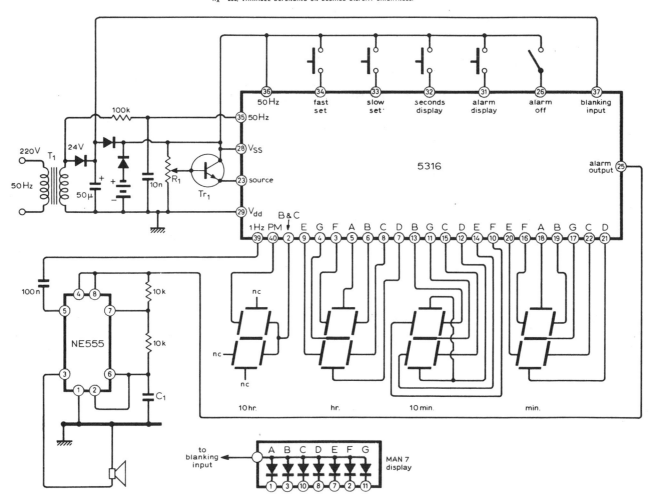

AC DIGITAL CLOCK WITH STANDBY BATTERY—Uses MM5316 alarm-clock IC, originally designed to drive LCD or fluorescent displays, but modified here for LED display. Diodes and batteries provide power if AC fails, with blanking of display to extend battery life. Accuracy is poor on batteries but batteries make resetting of time and alarm easier after AC interruption. Alarm uses 555 multivibrator to produce frequency-shift warble on output tone. Time is set by fast and slow buttons, and alarm is set with same buttons while depressing alarm-display button. Transistor type is not critical.—M. F. Smith, Digital Alarm Clock, *Wireless World*, Nov. 1976, p 62.

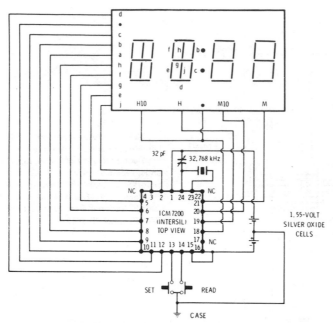

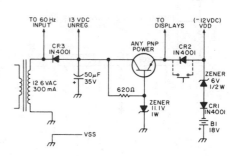

DIGITAL WRISTWATCH—Single Intersil ICM7200 IC drives multiplexed display giving choice of hours and minutes, seconds, and day/date. CMOS chip divides 32.768-kHz crystal output in long internal binary divider to produce basic 1-s clock rate. Further division gives other elements of display. Pressing read button once gives hours and minutes; pressing second time gives day and date; and pressing third time gives seconds.—D. Lancaster, "CMOS Cookbook," Howard W. Sams, Indianapolis, IN, 1977, p 377–378.

BATTERY BACKUP—During normal operation, all power for digital clock is provided by AC power supply. During power failure, clock continues operating from battery backup using two 9-V batteries in series. Battery drain is limited by diode CR2 that blocks power flow to displays. Optional switch may be installed across diode to short it for momentary viewing of display.—W. J. Prudhomme, CT7001 Clockbuster, *73 Magazine,* Dec. 1976, p 52–54 and 56–58.

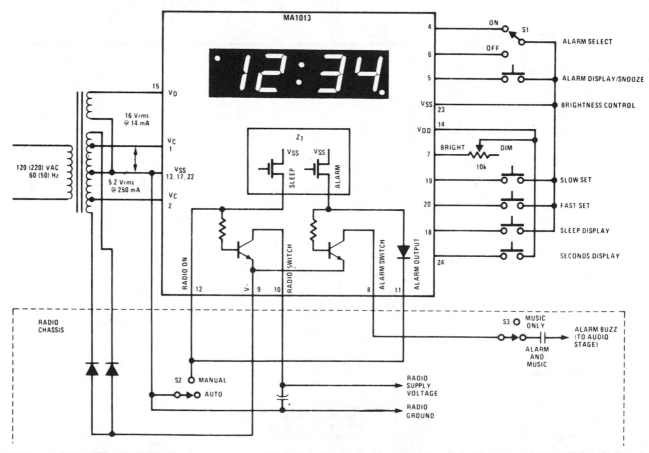

4-DIGIT 0.7-INCH LED DISPLAY—National MA1013 clock module contains MOS LSI clock IC, display, power supply, and associated discrete components on single printed-circuit board that is easily connected to radio. Operates from either 50-Hz or 60-Hz inputs, and gives either 12- or 24-h display format. Nonmultiplexed LED drive eliminates RF interference. Display is flashed at 1-Hz rate after power failure of any duration, to indicate need for resetting clock. Zero appearing in first digit is blanked. On 12-h version, dot in upper left corner is energized to indicate PM.—"MOS/LSI Databook," National Semiconductor, Santa Clara, CA, 1977, p 13-23–13-28.

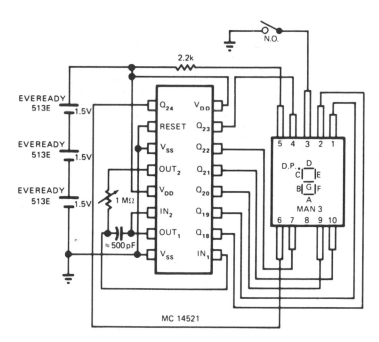

4-h DIGITAL WATCH—Single Motorola MC 14521 CMOS IC drives single-digit MAN 3 LED display in such a way that time range of 4 h is obtained with 1.875 min resolution. Can be built into old watch case at cost under $10 for parts. Oscillator frequency of 1.165 kHz can be tweaked to adjust clock, or crystal oscillator can be added for high accuracy. Analog/binary format of readout provides deciphering challenge to user, even though article gives diagram showing which segments of LED are lit for each time reading. Time intervals represented by each lit segment of display are: B = 2 h; C = 1 h; A = 30 min; F = 15 min; G = 7.5 min; E = 3.75 min; D = 1.875 min.—R. M. Steimle, Small CMOS Digital Watch Has Analog LED Output, *EDN Magazine*, Aug. 20, 1976, p 86.

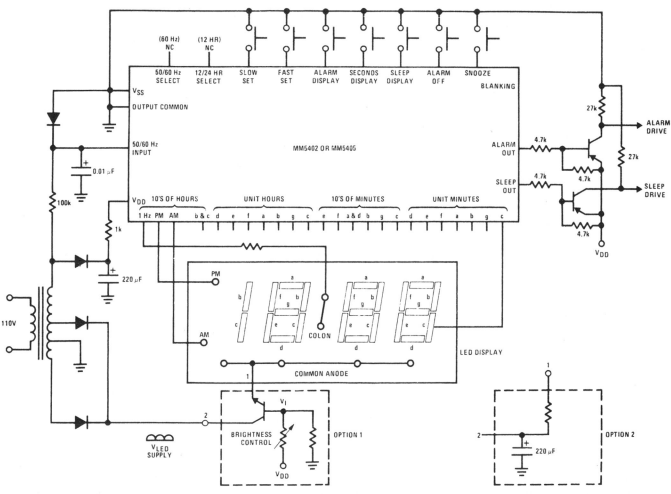

12-h ALARM—General-purpose digital clock with alarm uses National MM5402 or MM5405 MOS IC to drive 3½-digit LED display and provide drive for alarm. Brightness control is optional. Sleep output can be used to turn off radio after desired time interval of up to 59 min.—"MOS/LSI Databook," National Semiconductor, Santa Clara, CA, 1977, p 1-68–1-73.

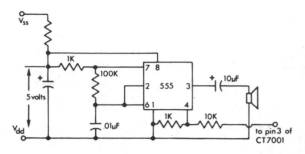

ALARM GENERATOR—Simple 555 timer generates alarm tone driving small loudspeaker, for use with Cal-Tex CT7001 and other similar digital clocks which do not have internal tone generator. Circuit requires +5 V, but supply can be higher value if suitable dropping resistor is used.—M. S. Robbins, "Electronic Clocks and Watches," Howard W. Sams, Indianapolis, IN, 1975, p 91.

0–9 s DIGITAL READOUT—Can be used for classroom demonstration of digital logic driving 7-segment LED or as attention-getting desk display. Time base Q1 feeds sequential timing pulses to 7490 decade counter. Pulses are counted in binary mode, and bit pattern corresponding to digits 0-9 is fed to 7447 binary-to-decimal decoder/driver connected to 7-segment readout. Calibrate with watch or with timing reference signals from WWV, adjusting R1 so display advances 1 digit per second.—F. M. Mims, "Electronic Circuitbook 5: LED Projects," Howard W. Sams, Indianapolis, IN, 1976, p 72–75.

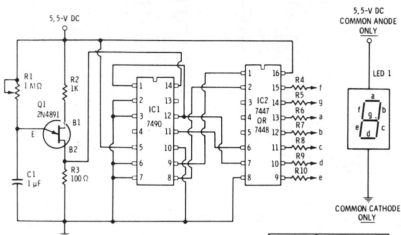

R4-R10	SEGMENT CURRENT
370 Ω	10 mA
185 Ω	20 mA

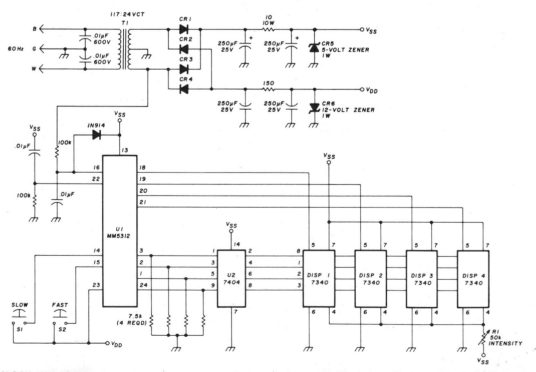

SIMPLE 24-h CLOCK—Use of 60-Hz power frequency as time base simplifies design while still giving long-term accuracy comparable to that of crystal time base. Four-digit display uses Hewlett-Packard 5082-7340 displays requiring only simple four-line BCD input. National MM5312N IC divides line frequency down to one pulse per minute and advances its internal storage register at same rate. Output of register is in binary form at pins 1, 2, 3, and 24, synchronized with digit-enable outputs at pins 18, 19, 20, and 21. Binary data is thus applied to all four displays in parallel, with enable lines controlling data feed. SN7404N inverter converts binary output data to TTL level required by displays. Power supply provides +5 V and −12 V for ICs and 60-Hz reference for clock check. CR5 is Radio Shack 276-561, CR6 is 276-563, and CR1-4 are 276-1146.—K. Powell, 24-Hour Clock with Digital Readout and Line-Frequency Time Base, *Ham Radio*, March 1977, p 44–48.

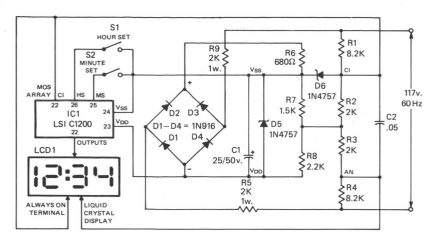

2-INCH LCD NUMERALS—Uses C1200 clock IC made by LSI Computer Systems, having time set, logic, division for seconds, minutes, and hours, 7-segment decoding, and display drivers and switches. Four-digit liquid crystal display panel (LCD) is MGC-50. S1 and S2 advance minutes or hours on display at 2-Hz rate for setting time. To use as elapsed-time indicator, close S1 and S2 simultaneously to generate reset pulse that sets timing change to zero. When both switches are released simultaneously, time count starts from zero.—R. F. Graf and G. J. Whalen, A Giant LCD Clock, *CQ*, Feb. 1978, p 18–23 and 76.

DIVIDE BY 5000 FOR CLOCK—Counter chain uses CD4017 that divides by integer from 2 to 10, selected by connecting appropriate output to reset. Extra gates recommended by RCA are not needed. Used in digital clock that changes automatically to battery operation when AC power fails. Clock operates on either 50 or 60 Hz.—S. I. Green, Digital Clock Keeps Counting Even When AC Power Fails, *EDN Magazine*, Dec. 20, 1974, p 49–51.

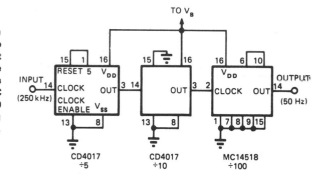

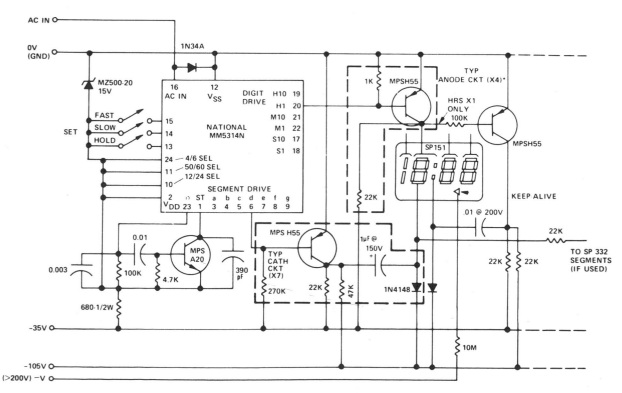

*(X4) FOR HRS, MINS, (X6) FOR HRS, MINS, SECS

4-DIGIT GAS DISPLAY—CMOS clock IC drives multidigit gas-discharge display. Simple circuit does not include alarm, flashing colon, and AM/PM features. Seven segment-driver circuits and four digit-driver circuits are required, although only one of each is shown. Additional drivers are needed if seconds display is desired. Required supply voltages can be obtained from transformer-type supply driving diode bridge; regulation is not needed.—M. S. Robbins, "Electronic Clocks and Watches," Howard W. Sams, Indianapolis, IN, 1975, p 68–71.

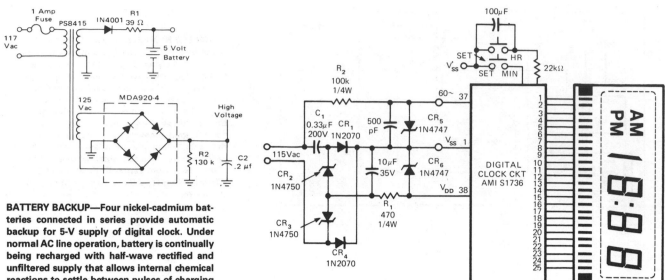

BATTERY BACKUP—Four nickel-cadmium batteries connected in series provide automatic backup for 5-V supply of digital clock. Under normal AC line operation, battery is continually being recharged with half-wave rectified and unfiltered supply that allows internal chemical reactions to settle between pulses of charging energy. R1 is chosen to make average charging current about 5% of battery rating.—D. Aldridge and A. Mouton, "Industrial Clock/Timer Featuring Back-Up Power Supply Operation," Motorola, Phoenix, AZ, 1974, AN-718A, p 7.

12- OR 24-h CLOCK—Single American Microsystems AMI S1736 clock chip drives liquid-crystal readout to give either 12-h display with AM/PM indicator or 24-h digital display by changing only three connections.—LSI in Consumer Applications, Round 2: Clocks on a Chip, *EDN Magazine*, May 5, 1973, p 22—23.

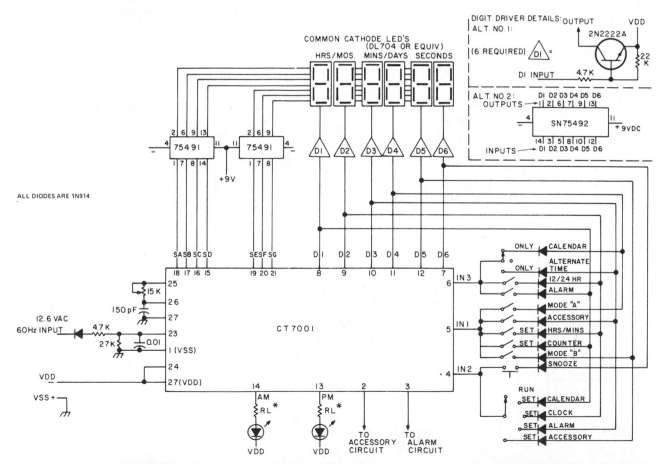

CALENDAR CLOCK—Uses Fairchild FCM7001 IC equivalent of Cal-Tex CT7001 clock chip (which is no longer available) to drive six 7-segment LEDs that can be switched to show 12- or 24-h time and 28/30/31 calendar, along with alarm features. Article gives construction details. Each SN75491 driver chip has pins 3, 5, 10, and 12 connected to pin 11 through 150-ohm resistor. RL is typically 2.7K, chosen to limit LED current to less than 5 mA.—W. J. Prudhomme, CT7001 Clockbuster, *73 Magazine*, Dec. 1976, p 52—54 and 56—58.

CHAPTER 25
Display Circuits

Drives and controls for LEDs used singly or in arrays, as well as liquid crystal,
gas-discharge, fluorescent, incandescent, bar-graph, and Nixie displays.
Includes controls for brightness, zero-suppression, strobing, and multiplexing.
For displays on cathode-ray screens, see Cathode-Ray chapter.

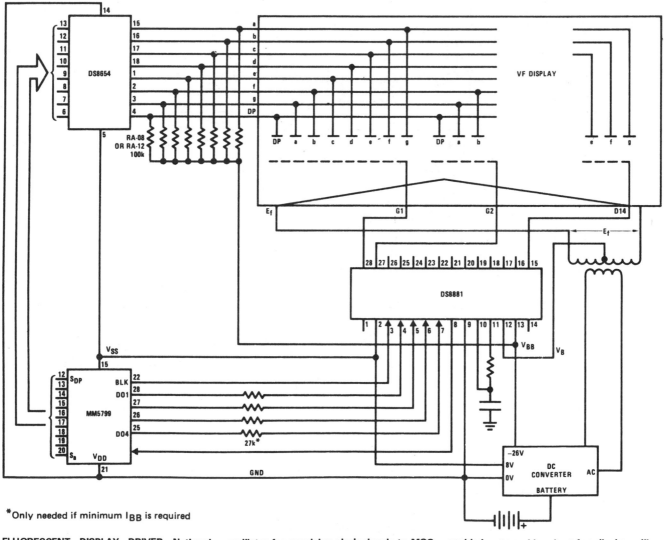

*Only needed if minimum I_{BB} is required

FLUORESCENT DISPLAY DRIVER—National DS8881 vacuum fluorescent display driver handles 16-digit grids. Decode inputs select 1 of 16 outputs to be pulled high. Driver also contains oscillator for supplying clock signals to MOS circuit, filament-bias zener, and 50K pulldown resistors for each grid. Outputs will source up to 7 mA. Supply is 9 V. Interdigit blanking with enable input provides ghost-free display.—"Interface Databook," National Semiconductor, Santa Clara, CA, 1978, p 5-57—5-60.

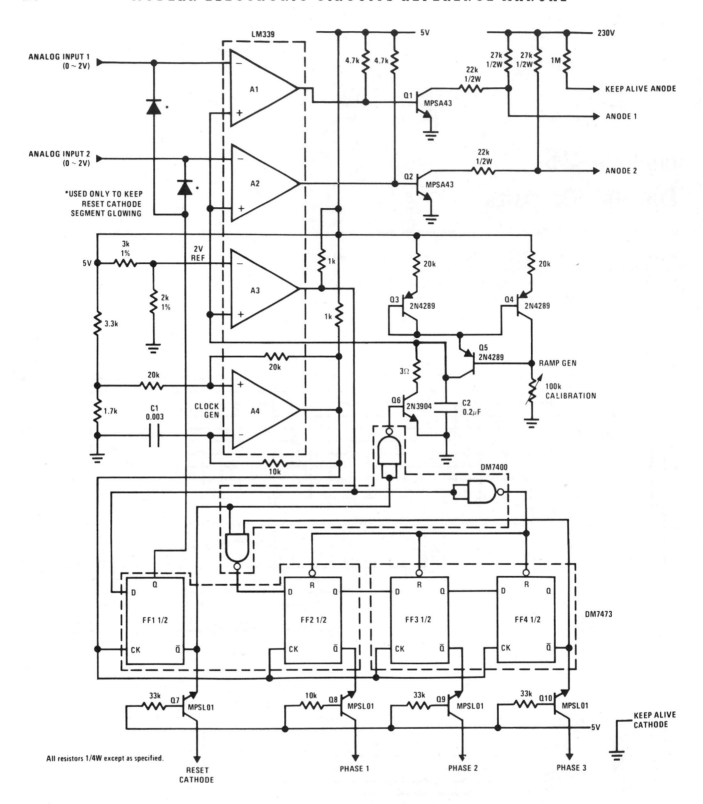

BAR-GRAPH DRIVE—Use of National LM339 quad comparator minimizes number of components needed to drive Burroughs 200-segment gas-discharge bar display. Every third electrode of display is tied together, so only three lines (phase 1, 2, and 3) control all segments. When phase lines are driven by consecutive pulses, glow of gas-discharge element is propagated continuously along array. Anode voltage is gated so number of glowing segments is proportional to analog input. Comparators A1 and A2 generate gated anode signals with durations proportional to inputs. A3 compares ramp signal to 2-V reference and generates end-of-scan signal when ramp exceeds reference. A4 generates clock having period of about 60 μs.—S. N. Kim, "Driving Burroughs' Bar Graph Display," National Semiconductor, Santa Clara, CA, 1975, DB-4.

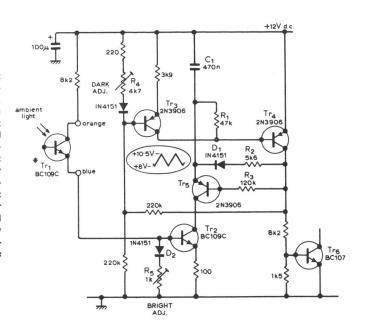

AUTOMATIC BRIGHTNESS CONTROL—Circuit adjusts brightness of LED digital display by altering mark-space ratio of LED supply voltage. Ambient-light input is sensed by BC109C transistor with top taken off. Normal display current of 20 mA is reduced to 2 mA when darkened room makes brightness unnecessary, conserving battery life. Tr_3 and Tr_4 operate as Schmitt trigger, with mark time of 1.5 ms determined by R_2 and space time controlled by charging current through Tr_5 and Tr_2 as affected by ambient light on Tr_1. Article gives complete circuits for driving 11-LED array. To add brightness control circuit, break ground connection of LED supply transistors and insert saturated transistor Tr_6.— G. Kalanit, Analogue to Digital Meter, *Wireless World*, July 1976, p 53–57.

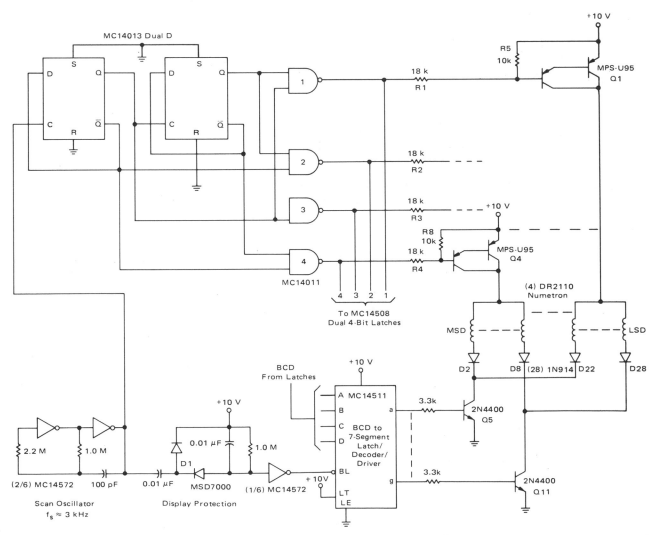

4-DIGIT INCANDESCENT—Circuit serves for interfacing CMOS logic to multiplexed 4-digit incandescent display. Scan decoder requires only two input NAND gates since blanking is not required. Incandescent display requires 4.5 V at 24 mA per segment when direct-driven; with multiplexing, instantaneous power must be 9 V at 48 mA to maintain same average power per segment. Display protection circuit monitors scan oscillator and blanks display if oscillator fails, to prevent high peak current from degrading display when applied continuously to 1 digit.—A. Pshaenich, "Interface Considerations for Numeric Display Systems," Motorola, Phoenix, AZ, 1975, AN-741, p 25.

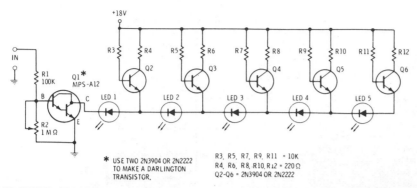

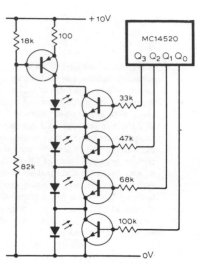

BAR-GRAPH READOUT—Transistors switch row of LEDs on in succession to give rising-bar display indicating input voltage. R2 can be adjusted from minimum range of 0.1 to 0.5 V in 0.1-V increments for five LEDs to maximum of 1.0 to 10.0 V in 1-V increments. Input resistance of circuit is above 100,000 ohms.—F. M. Mims, "Electronic Circuitbook 5: LED Projects," Howard W. Sams, Indianapolis, IN, 1976, p 86–88.

REDUCING LED POWER DRAIN—Arrangement of LEDs in groups of four with constant-current source greatly eliminates wastage of battery power. Circuit shows utilization of this technique to display 4-bit binary number from CMOS counter.—T. R. Owen, L.E.D. Display, *Wireless World,* June 1976, p 72.

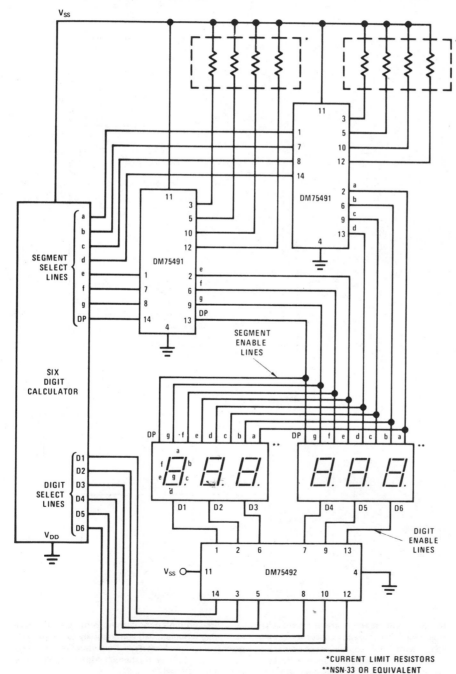

6-DIGIT DRIVE—Uses two National DM75491 four-segment drivers for multiplex-mode display of MOS calculator. Total of eight segment drivers provides drive for each one of seven segments plus logic control for decimal point.—C. Carinalli, "Driving 7-Segment LED Displays with National Semiconductor Circuits," National Semiconductor, Santa Clara, CA, 1974, AN-99, p 9.

*CURRENT LIMIT RESISTORS
**NSN-33 OR EQUIVALENT

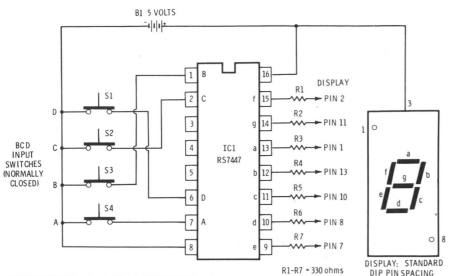

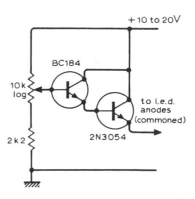

BCD DECODER—Radio Shack RS7447 BCD to 7-segment decoder converts settings of four BCD input switches to corresponding 0–9 digit on 7-segment common-anode LED display. Display is Radio Shack 276-053. Battery can be four AA cells in series, with 1N914 diode inserted in positive lead to reduce voltage to 5 V.—F. M. Mims, "Integrated Circuit Projects, Vol. 2," Radio Shack, Fort Worth, TX, 1977, 2nd Ed., p 27–40.

LED BRIGHTNESS CONTROL—Uses 10K logarithmic pot to vary brightness simultaneously for all LEDs in digital display.—S. F. Bywaters and J. E. West, Peak-Reading Audio Level Indicator, *Wireless World,* Aug. 1975, p 357–361.

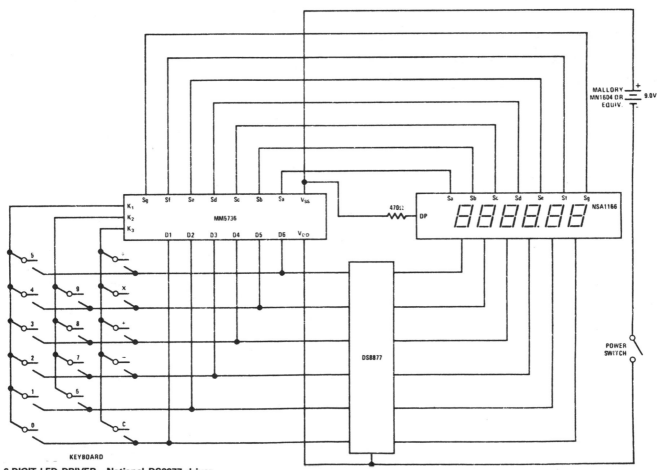

6-DIGIT LED DRIVER—National DS8877 driver is shown in configuration for use with 6-digit calculator. Digit current is in range of 5–50 mA. Driver requires no standby power and operates from either 4.5 V, 6 V, or 9 V.—"Interface Databook," National Semiconductor, Santa Clara, CA, 1978, p 5-52–5-53.

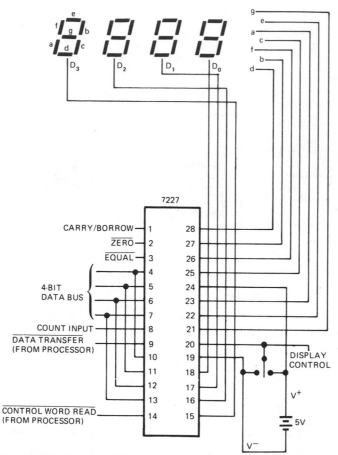

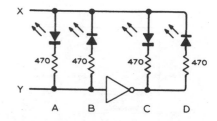

BINARY LINE STATES—Simple circuit using four LEDs and one inverter (which may be a transistor or spare gate) displays all four possible states on two binary lines. When levels of lines X and Y are the same, A and B will be off. Inverter then places C and D at different levels so one LED (C or D) will be on. Reverse situation occurs when X and Y are at different levels.—D. Straker, Binary State Indicator, *Wireless World*, Feb. 1977, p 44.

TIMER DRIVES LED DISPLAY—Intersil 7227 microprocessor-controlled timer provides direct drive for LED display under supervision of microprocessor. Tri-state 4-bit data bus serves to read in control word such as up/down, store, reset, or load, then deliver counter data, feed in settable register word, or preset counter to initial value.—B. O'Neil, IC Timers—the "Old Reliable" 555 Has Company, *EDN Magazine*, Sept. 5, 1977, p 89–93.

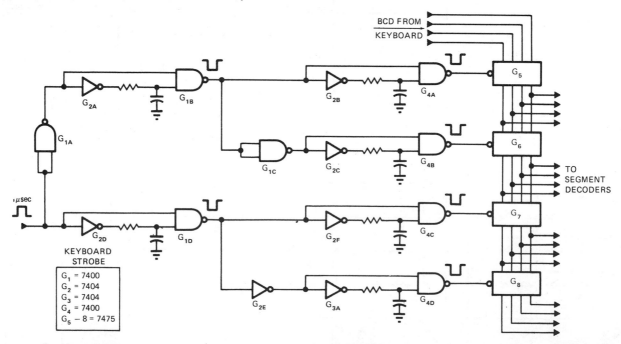

DIGIT SHIFTER FOR DISPLAY—Circuit takes BCD output from 10-key keyboard and shifts each number, as entered, from right to left on display panel. Internal clock is not used. Keyboard strobe is delayed 2 ms to allow time for keyboard switches to stop bouncing. BCD outputs from G_5-G_8 go directly to 7- or 10-segment decoder driver, such as SN7447 decoders driving RCA DR-2100 series low-voltage readouts. All resistors are 220 ohms. Capacitors for G_{1B} and G_{1D} are 1000 pF, and capacitors for other gates are 240 pF. Article traces circuit operation.—T. O'Toole, Transfer Parallel Information Without a Clock, *EDN Magazine*, Aug. 1, 1972, p 59.

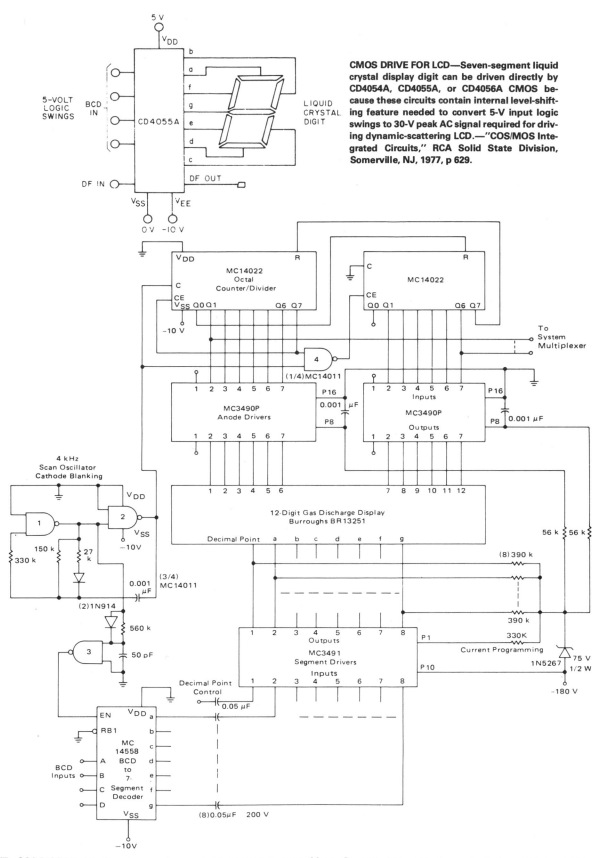

CMOS DRIVE FOR LCD—Seven-segment liquid crystal display digit can be driven directly by CD4054A, CD4055A, or CD4056A CMOS because these circuits contain internal level-shifting feature needed to convert 5-V input logic swings to 30-V peak AC signal required for driving dynamic-scattering LCD.—"COS/MOS Integrated Circuits," RCA Solid State Division, Somerville, NJ, 1977, p 629.

12-DIGIT GAS-DISCHARGE—Display anodes are referenced to ground and cathodes to −180 V because number of digits in display is greater than number of segment drivers. Positive-logic CMOS address circuits are powered by −10 V, with Motorola MC14558 decoder outputs coupled to MC3491 segment drivers. Scan circuit is directly coupled to MC3490P anode drivers. Digit scanning is derived from two cascaded MC14022 octal counter/dividers. Required 12 sequenced output pulses are achieved by resetting counters with Q7 output of second counter. Counter output also controls system multiplexer (not shown) to give synchronization of entire display system.—A. Pshaenich, "Interface Considerations for Numeric Display Systems," Motorola, Phoenix, AZ, 1975, AN-741, p 23.

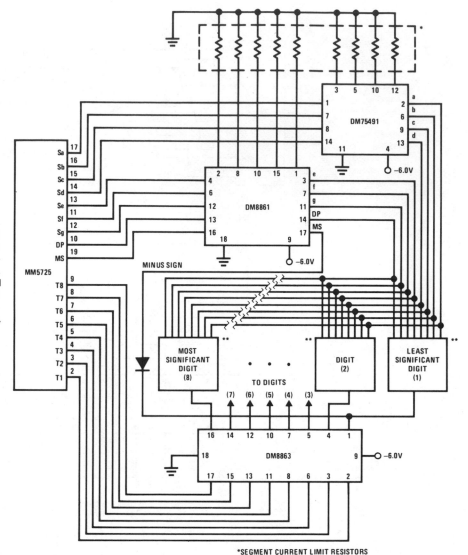

8-DIGIT LED DRIVE—National DM8863 8-digit LED driver is used in conjunction with DM75491 and DM8861 drivers for driving eight common-mode LED digits operating in multiplex mode. Circuit also provides logic control for decimal point.—C. Carinalli, "Driving 7-Segment LED Displays with National Semiconductor Circuits," National Semiconductor, Santa Clara, CA, 1974, AN-99, p 10.

*SEGMENT CURRENT LIMIT RESISTORS
**EITHER SINGLE DIGIT, MULTI-DIGIT, NSN 33, OR EQUIVALENT

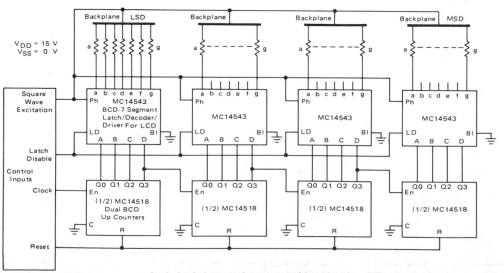

4-DIGIT DIRECT-DRIVE LCD—Each digit of liquid crystal display has separate counter, latch, decoder, and driver. Excitation signal also feeds LCD backplane. When segment is to be deenergized, backplane and segment drive signals have same phase and magnitude so there is no voltage across display. When segment is to be energized, signals are 180° out of phase so square-wave voltage is twice IC supply value. BCD inputs are generated from cascaded MC14518 dual BCD up counters.—A. Pshaenich, "Interface Considerations for Numeric Display Systems," Motorola, Phoenix, AZ, 1975, AN-741, p 5.

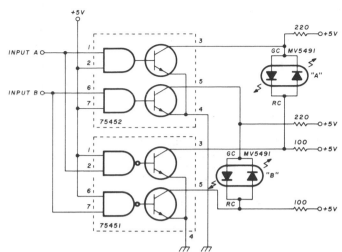

TWO-INPUT RED/GREEN LED—Uses Monsanto MV5491 having red and green LEDs in same housing, connected inversely in parallel so current in one direction gives green and reverse current gives red. Two different drivers are used, SN75452 noninverting and SN75451 inverting. Each LED pair shows one color for correct polarity at its driver input and other color for opposite polarity.—K. Powell, Novel Indicator Circuit, *Ham Radio*, April 1977, p 60–63.

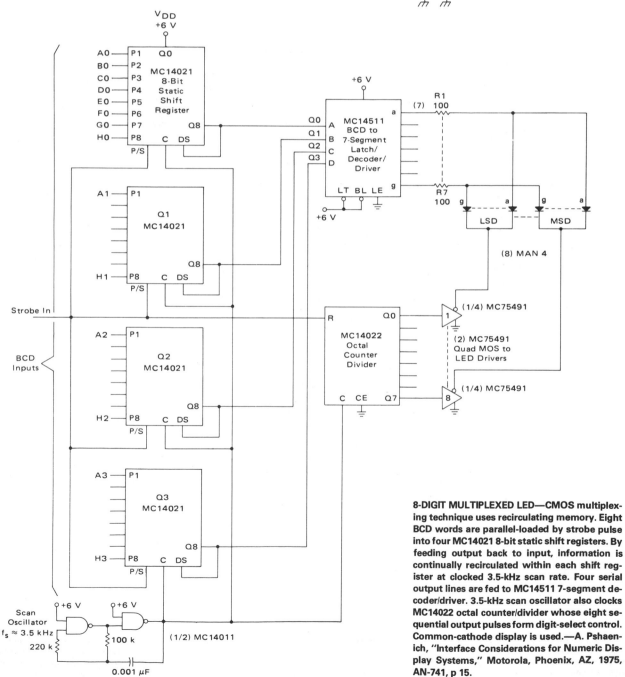

8-DIGIT MULTIPLEXED LED—CMOS multiplexing technique uses recirculating memory. Eight BCD words are parallel-loaded by strobe pulse into four MC14021 8-bit static shift registers. By feeding output back to input, information is continually recirculated within each shift register at clocked 3.5-kHz scan rate. Four serial output lines are fed to MC14511 7-segment decoder/driver. 3.5-kHz scan oscillator also clocks MC14022 octal counter/divider whose eight sequential output pulses form digit-select control. Common-cathode display is used.—A. Pshaenich, "Interface Considerations for Numeric Display Systems," Motorola, Phoenix, AZ, 1975, AN-741, p 15.

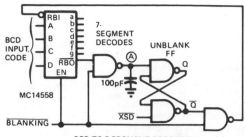

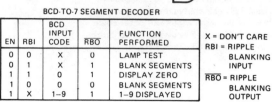

ZERO SUPPRESSION—Simple CMOS circuit using MC14011 quad two-input NAND gate and MC14558 IC provides zero suppression for multiplexed displays in which scanning is left to right for leading-zero suppression and right to left for trailing zeros. Article covers operation of circuit.—J. J. Roy, Eliminate Excess Zeros in Multiplexed Displays, *EDN Magazine,* Sept. 5, 1975, p 77.

BCD-TO-7 SEGMENT DECODER

EN	RBI	BCD INPUT CODE	RBO	FUNCTION PERFORMED
0	0	X	0	LAMP TEST
0	1	X	1	BLANK SEGMENTS
1	1	0	1	DISPLAY ZERO
1	0	0	0	BLANK SEGMENTS
1	X	1–9	1	1–9 DISPLAYED

X = DON'T CARE
RBI = RIPPLE BLANKING INPUT
$\overline{RBO}$ = RIPPLE BLANKING OUTPUT

TIMING FOR ZERO SUPPRESSION CIRCUIT
NEW BCD DATA TO 14558 INPUTS**

BLANKING (EN)
XSD***
RBI
$\overline{RBO}$
A NO CAPACITOR
A 100 pF CAPACITOR
UNBLANK FF, $\overline{Q}$
TIME WHEN DISPLAY BLANKED

**FIRST AND SECOND DIGITS ARE "0" AND THIRD DIGIT IS NONZERO IN EXAMPLE TIMING
***$\overline{XSD}$ = $\overline{MSD}$ FOR LEADING ZERO SUPPRESSION
XSD = LSD FOR TRAILING ZERO SUPPRESSION

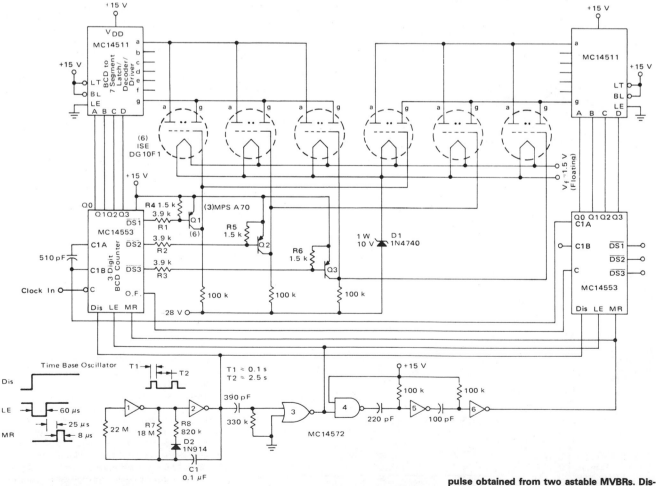

6-DIGIT FLUORESCENT TRIODES—Uses two sets of cascaded counters and decoders with series switching of positive voltage to anode with MC14511 ICs. Digit scanning is accomplished by turning on grid control transistors Q1-Q3 with negative-going digit select outputs of one MC14553. Timing for counters is derived from MC14572 logic elements, with disable pulse obtained from two astable MVBRs. Display digits can be packaged individually or in single envelope.—A. Pshaenich, "Interface Considerations for Numeric Display Systems," Motorola, Phoenix, AZ, 1975, AN-741, p 9.

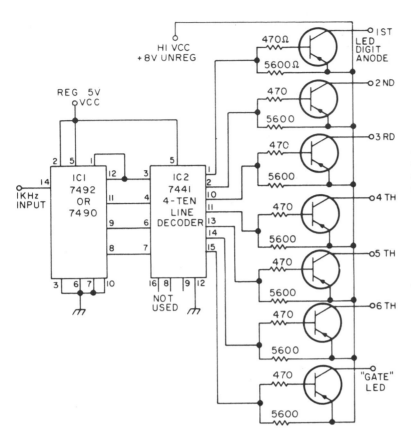

STROBING LED DISPLAY—Applies power in sequence to segments of display, so fast that eye cannot detect flicker, to reduce drain on power supply. Input of 1000 Hz can be taken from timing chain of circuit that is driving display. 7492 divide-by-12 counter gives scan frequency of 83.3 Hz for display. Binary output of 7492 is converted to 1-in-10 output by 7441 decoder for sequential drive of 2N3904 PNP pass transistor that grounds LED which is to be lighted.—W. K. McKellips, Strobing Displays Is Cool, *73 Magazine,* Nov./Dec. 1975, p 49–50.

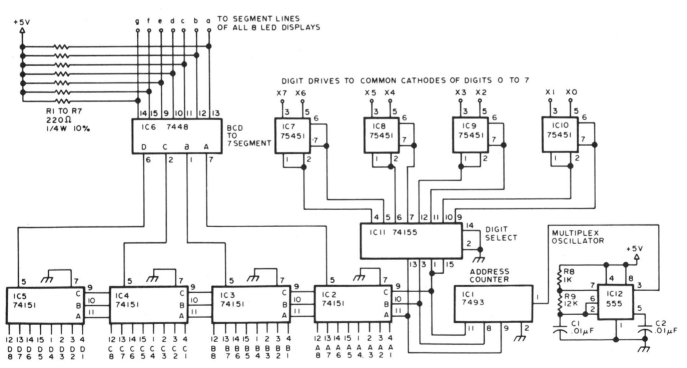

MULTIPLEXING EIGHT DIGITS—Uses only one 7-segment driver for eight digits of parallel BCD data on eight-LED display that can use MAN 4 or DL764 7-segment LEDs. Power is supplied to only one digit at a time but is switched at high enough rate so all digits appear to be on. Uses one eight-channel 74151 multiplexer for each of the 4 data input bits. Multiplexers and demultiplexer are addressed by 7493 counter that is incremented at about 4 kHz by 555 oscillator. IC11 is connected for three- to eight-line demultiplexing. IC7-IC10 are peripheral interface gates, each sinking up to 300 mA for its LED. 7448 decoder/driver converts BCD data to 7-segment code for driving segments of LEDs. For 74151 and 74155, pin 16 goes to +5 V and pin 8 to ground. Pin 8 of 75451 goes to +5 V and pin 4 to ground. Pin 5 of 7493 goes to +5 V and pin 10 to ground.—J. Hogenson, Multiplex Your Digital LED Displays, *BYTE,* March 1977, p 122–126 and 128.

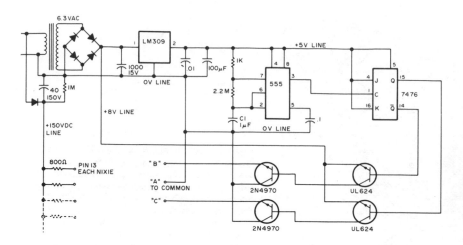

TWO MESSAGES WITH NIXIES—Circuit flashes two messages alternately on same Burroughs giant Nixie B7971 display. Lighted segments needed on individual Nixies to form desired wording are divided into three strings. Segments A are common to both sets of letters and numbers. Segments B are those required with A segments to form first message. Segments C are those required with A segments to form second message. Changeover from segments B to C is done with switching transistors controlled by 555 timer and 7476 or 7473 flip-flop. Decimal or other punctuation is formed with NE2 neon and 100K resistor wired in series between pin 13 of a Nixie and B or C. Article gives construction details.—J. Grimes, Put Your Name in Lights, *73 Magazine,* Nov. 1976, p 60–61.

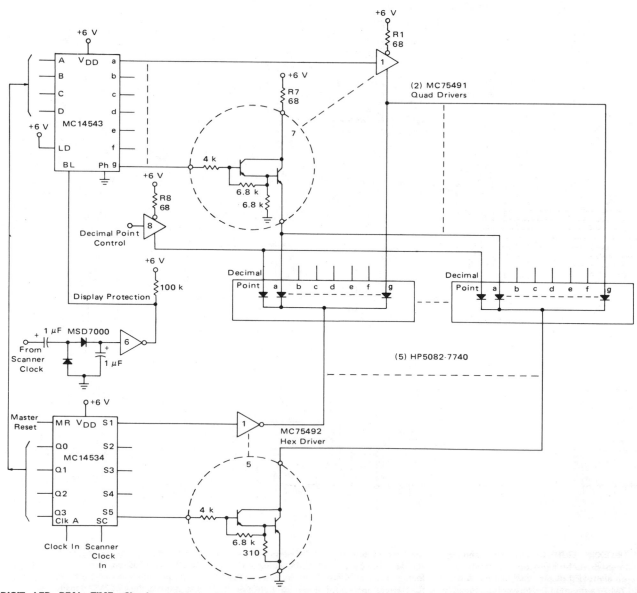

5-DIGIT LED REAL TIME—Circuit provides strobing of LEDs so peak current and light output are greater for same average current. Peak forward current for display is about 40 mA. All like anode segments of common-cathode displays are driven by emitter outputs of MC75491 quad drivers.—A. Pshaenich, "Interface Considerations for Numeric Display Systems," Motorola, Phoenix, AZ, 1975, AN-741, p 13.

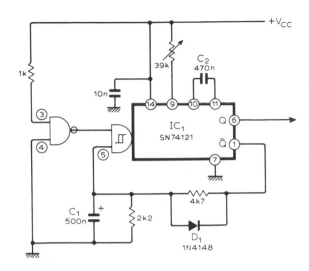

PWM BRIGHTNESS CONTROL—Single TTL IC combines functions of oscillator and modulator to provide intensity control of solid-state display by pulse-width modulation. Fan-out of 10 is available from Q output, suitable for displays such as Hewlett-Packard 7300 series, and smaller fan-out is available from other Q terminal.—C. Bartram, P.W.M. Oscillator to Vary Display-Intensity, *Wireless World,* March 1976, p 89.

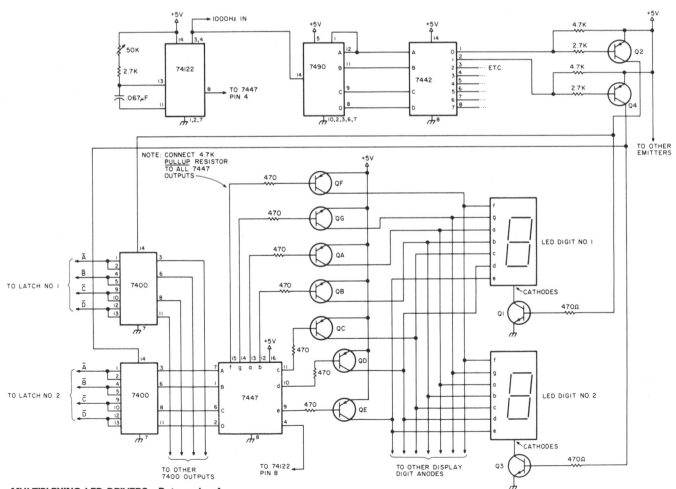

MULTIPLEXING LED DRIVERS—Duty cycle of each display digit can be varied from 10% (full on with strobing) to less than 1% (almost off). Circuit uses 7490 and 7442 as 1-of-10 multiplex driver to strobe cathodes of display digits through Q1, Q3, etc and turn on required 7400 multiplex gate through Q2, Q4, etc. Outputs of 7447 are polarity-inverted by QA, QB, etc, which can be Sylvania ECG 159 rated at 200 mA. Q1, Q3, etc can be ECG 123 or HEP S0002, while Q2, Q4, etc can be any silicon PNP transistor.—B. Hart, Current-Saver Counter Display, *73 Magazine,* June 1977, p 174–176.

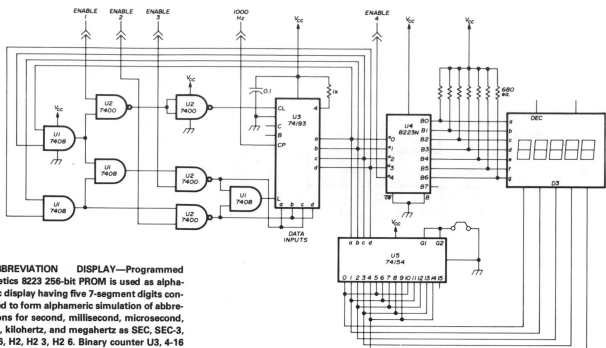

SI-ABBREVIATION DISPLAY—Programmed Signetics 8223 256-bit PROM is used as alphameric display having five 7-segment digits connected to form alphameric simulation of abbreviations for second, millisecond, microsecond, hertz, kilohertz, and megahertz as SEC, SEC-3, SEC-6, H2, H2 3, H2 6. Binary counter U3, 4-16 line decoder, and 5-digit parallel-connected Hewlett-Packard display form simple multiplexer that addresses memory U4 one word at a time. External 1000-Hz square-wave oscillator drives counter and sets scanning rate. Requires only one 5-V supply. Article gives truth table for memory and circuit for programmer required to set it up.—J. W. Springer, Function/Units Indicator Using LED Displays, *Ham Radio*, March 1977, p 58–63.

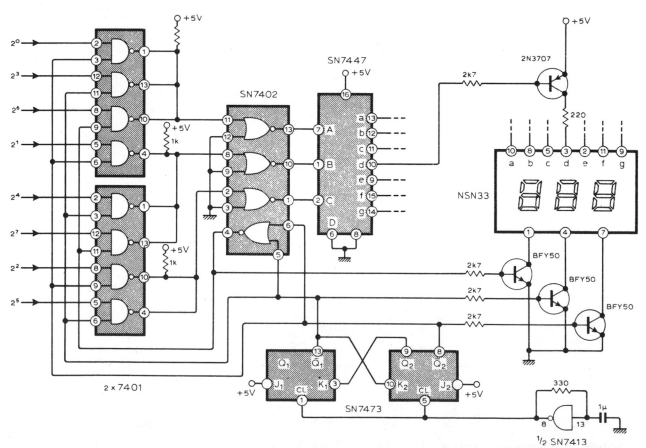

OCTAL DISPLAY—Circuit provides display of 8-bit data word in conventional octal form for convenience in experimentation with microprocessors and small computers. Three-state counter built around synchronously operated JK flip-flops provides digit and data selection. The two or three bits appropriate to each display are steered to 7-segment decoder by wired-AND gates and inverters. Oscillator multiplexing frequency is about 2 kHz.—R. D. Mount, Octal Display for Microprocessors, *Wireless World*, March 1977, p 41.

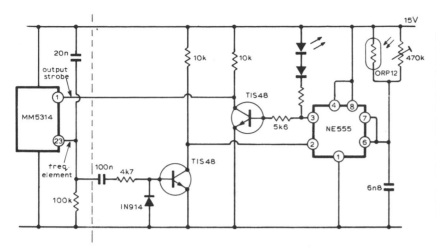

MULTIPLEXED BRIGHTNESS CONTROL—Developed for use with single-chip digital clocks in which several displays are multiplexed. Provides automatic brightness control by using variable duty cycle and switching it on and off in synchronism with display of time. Uses 555 timer in monostable mode, triggered by multiplex oscillator to determine off time of display. When ambient light is bright, resistance of ORP12 photocell is low and display is on most of time. Set 470K pot to give low light output without mistriggering under dark conditions. Timer can also drive decimal point directly and give matched brightness.—M. G. Martin, Automatic Display-Brightness Control, *Wireless World,* April 1976, p 61.

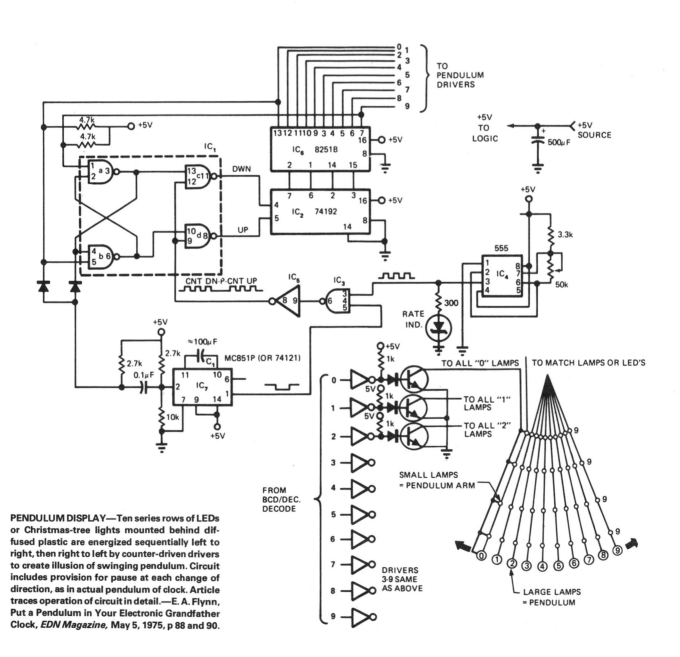

PENDULUM DISPLAY—Ten series rows of LEDs or Christmas-tree lights mounted behind diffused plastic are energized sequentially left to right, then right to left by counter-driven drivers to create illusion of swinging pendulum. Circuit includes provision for pause at each change of direction, as in actual pendulum of clock. Article traces operation of circuit in detail.—E. A. Flynn, Put a Pendulum in Your Electronic Grandfather Clock, *EDN Magazine,* May 5, 1975, p 88 and 90.

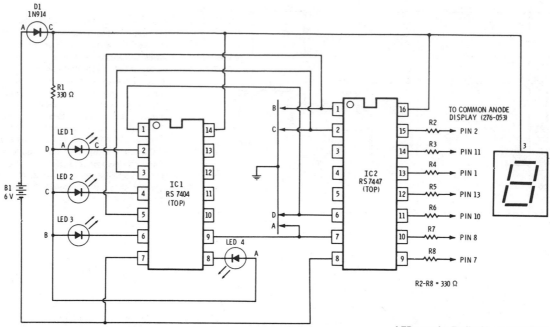

BCD DECODER—Drive for 7-segment common-anode digital display is provided by RS7447 BCD to 7-segment decoder. Binary input indicator using RS7404 hex inverter and four LEDs shows input in binary form for decimal digits 0–9. Developed for classroom demonstrations. Red LEDs can be Radio Shack 276-1805, and display is 276-053.—F. M. Mims, "Integrated Circuit Projects, Vol. 6," Radio Shack, Fort Worth, TX, 1977, p 42–52.

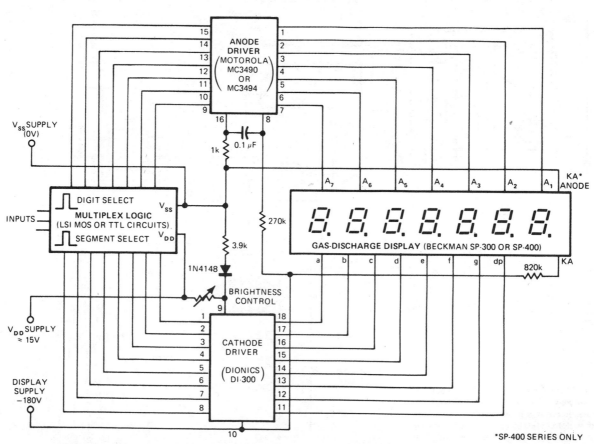

MULTIPLEXING 7 DIGITS—Uses Motorola MC3490 anode driver for active-high inputs (MC3494 for active-low inputs) to accept digit-select signals from multiplex logic source and drive display anodes directly. Constant cathode currents are maintained for gas-discharge display by Dionics D1-300 IC, to provide constant brightness without using supply-voltage regulation. For each digit added to display, equal number of anode drivers is required. Only one cathode driver is needed because all cathodes are bused together.—D. Sien, Multiplex Display Circuit Features Minimum Parts Count, *EDN Magazine*, May 5, 1977, p 112.

NANOSECOND PULSE DETECTOR—Used to provide visual indication of presence of a nonrepetitive digital pulse having microsecond or nanosecond width. Bistable IC transfers pulse information from its data input to the Q output on positive-going edge of clock pulse, to energize LED indicator.—P. V. Prior, Digital Pulse Detector, *Wireless World,* March 1976, p 90.

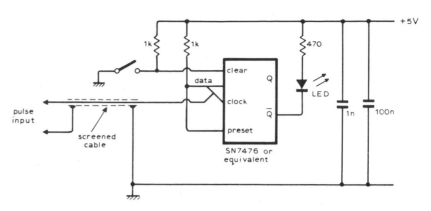

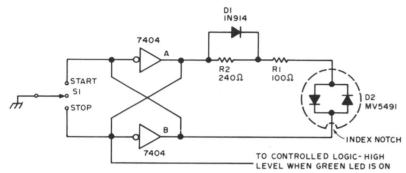

TWO-COLOR LED—Dual LED D2 shows green when normally off momentary switch S1 is moved to START and shows red when moved to STOP. Latching circuit using two 7404 TTL inverters serves as run and halt flip-flop and also debounces switch. Momentary contact at START toggles latch, biasing green LED. D1 shorts R2, leaving R1 to limit forward current to about 20 mA for green. R1 and R2 limit current for D1 and brighter red LED to about 10 mA for momentary contact at STOP.—E. W. Gray, LEDs Light Up Your Logic, *BYTE,* Feb. 1976, p 54–57.

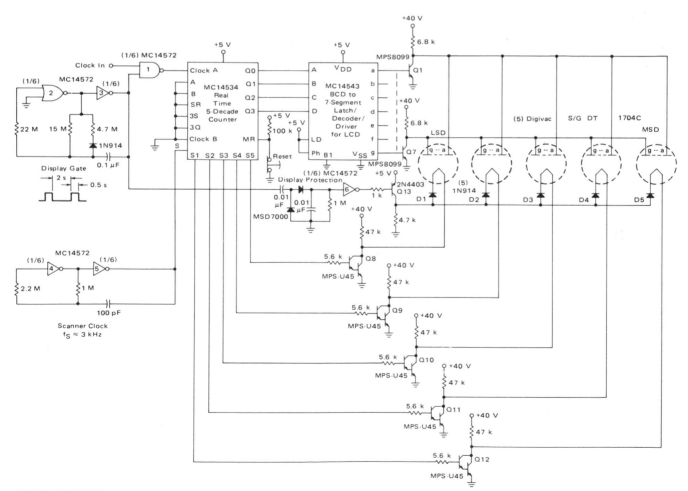

5-DIGIT FLUORESCENT DIODES—Real-time drive for five-decade counter requires only three ICs. MC14534 contains five-decade ripple counter with output time multiplexed by internal scanner. Scanning rate is controlled by inverters 4 and 5 of MC14572. Multiplexed BCD outputs go to MC14543 7-segment decoder whose outputs drive fluorescent diodes.—A. Pshaenich, "Interface Considerations for Numeric Display Systems," Motorola, Phoenix, AZ, 1975, AN-741, p 10.

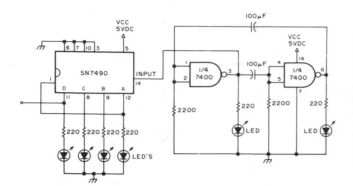

BINARY DEMO—Two sections of SN7400 quad gate form low-frequency MVBR serving as pulse source for SN7490 decade counter. Rate is low enough so blinking of LED status indicators in MVBR can be seen, as indication of pulse generation. Similarly, LEDs of counter blink to indicate counts of 8, 4, 2, and 1 from left to right, with combinations of lights coming on to display binary values 0 to 15 before recycling. Ideal for Science Fair exhibit.—A. MacLean, How Do You Use ICs?, *73 Magazine*, Dec. 1977, p 56–59.

PULSED LED—Circuit can generate peak currents above 1 A with pulse widths greater than 10 ms at repetition rates of 12 kHz with efficiency better than 90%, for 100-mA current drain from 2.5-V battery. Rise time of pulses is 0.2 ms. Can be used in low-light-level TV systems where high peak radiation gives better resolving power than constant illumination having same average power. Also useful for LED pilot lamps in battery-operated equipment and as low-power strobe for studying mechanical motions.—J. Dimitrios, Current-Pulse Generator for LED's, *EDN/EEE Magazine*, July 1, 1971, p 51.

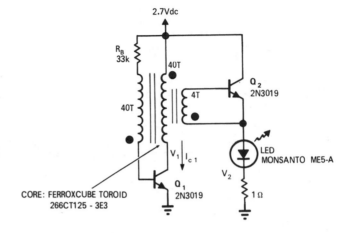

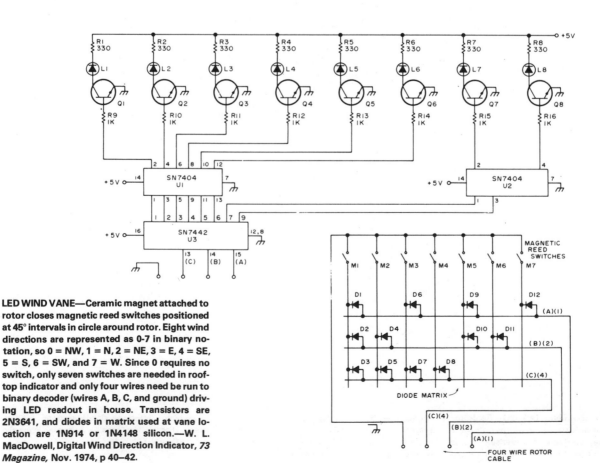

LED WIND VANE—Ceramic magnet attached to rotor closes magnetic reed switches positioned at 45° intervals in circle around rotor. Eight wind directions are represented as 0-7 in binary notation, so 0 = NW, 1 = N, 2 = NE, 3 = E, 4 = SE, 5 = S, 6 = SW, and 7 = W. Since 0 requires no switch, only seven switches are needed in rooftop indicator and only four wires need be run to binary decoder (wires A, B, C, and ground) driving LED readout in house. Transistors are 2N3641, and diodes in matrix used at vane location are 1N914 or 1N4148 silicon.—W. L. MacDowell, Digital Wind Direction Indicator, *73 Magazine*, Nov. 1974, p 40–42.

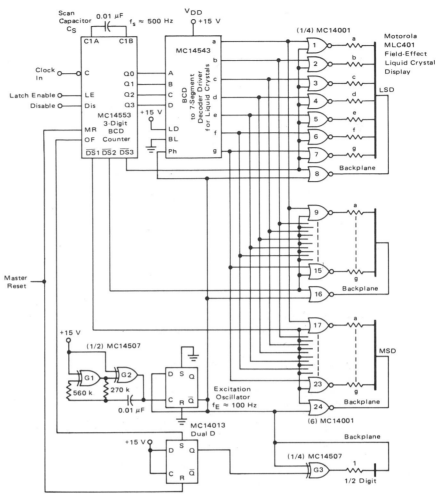

3½-DIGIT MULTIPLEXED LCD—Uses MLC401 field-effect liquid crystal display, which is more suitable for multiplexing than dynamic-scattering LCD. Counters, latches, multiplexer, and scan circuits are all in MC14553 3-digit BCD counter whose outputs feed MC14543 decoder and driver for display. Excitation frequency of 100 Hz is derived from square-wave oscillator G1-G2 having exactly 50% duty cycle. Scan frequency is about 500 Hz, giving display refresh rate of 170 Hz, which is well beyond detectable flicker rate.—A. Pshaenich, "Interface Considerations for Numeric Display Systems," Motorola, Phoenix, AZ, 1975, AN-741, p 6.

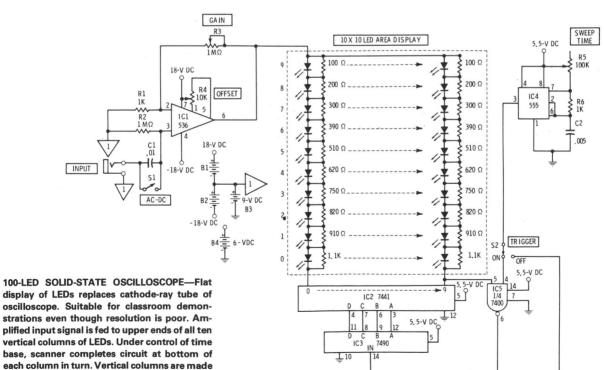

100-LED SOLID-STATE OSCILLOSCOPE—Flat display of LEDs replaces cathode-ray tube of oscilloscope. Suitable for classroom demonstrations even though resolution is poor. Amplified input signal is fed to upper ends of all ten vertical columns of LEDs. Under control of time base, scanner completes circuit at bottom of each column in turn. Vertical columns are made voltage-sensitive by resistors paralleling LEDs. Gate circuit using section of 7400 provides triggering when desired. Voltage-sensitive bargraph readouts formed by vertical columns are scanned by 7490 decade counter and 7441 decoder. Sweep rate of display is adjustable from 1 to 20 vertical columns per second with R5.—F. M. Mims, "Electronic Circuitbook 5: LED Projects," Howard W. Sams, Indianapolis, IN, 1976, p 92–96.

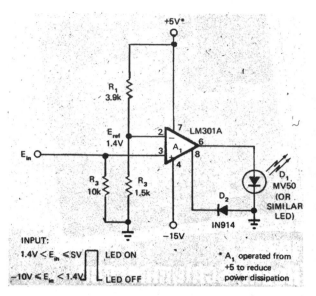

OPAMP DRIVES LED—LM301A is used as open-loop voltage comparator, with LED receiving total source current of about 20 mA from opamp. Input is TTL-compatible, with R_1-R_2 reference divider biasing opamp in center of TTL output transition region. Circuit realizes full open-loop speed of opamp since it is uncompensated and its internal voltage amplifier stages are kept out of saturation by clamping of D_2 and by inherent current-limiting action. Response times for toggling LED are in microsecond range.—W. G. Jung, Poor Man's LED Driver Is TTL Compatible, *EDN Magazine*, Feb. 5, 1973, p 86.

POSITIVE INPUT GIVES RED—Uses Monsanto MV5491 dual LED having red and green light-emitting diodes connected inversely in parallel, so current in one direction gives green light and reverse current gives red. Circuit uses single SN75452 IC driver and one section of SN7404 hex inverter. High or positive input gives red indication, while low input gives green. Current-limiting resistors R1 and R2 have different values because voltage and current specifications of parallel LEDs are different. Indicator appears to change color as input changes.—K. Powell, Novel Indicator Circuit, *Ham Radio*, April 1977, p 60–63.

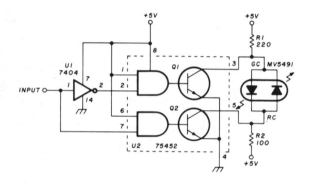

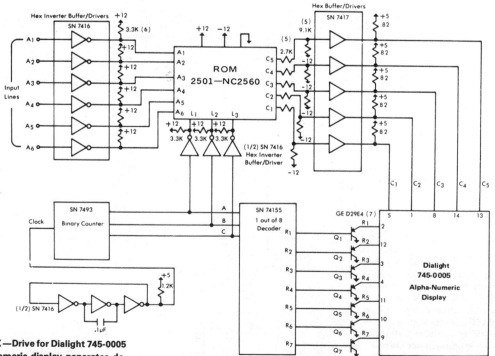

5 × 7 DOT MATRIX—Drive for Dialight 745-0005 64-character alphameric display generates desired character in response to pattern of 0s and 1s on input lines A1-A6. Timing of sequential scanning operation for seven horizontal rows of matrix is controlled by clock that drives binary counter having row-selecting outputs A, B, and C. Outputs C1-C5 of ROM correspond to vertical rows of dots enabled by 1-out-of-8 decoder.—"Readout Displays," Dialight, Brooklyn, NY, 1978, Catalog SG745, p 24–26.

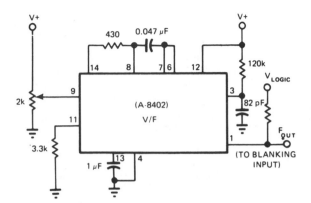

LED DIMMER—Intech/Function Modules A-8402 is used in voltage-to-frequency mode to provide controllable dimming of LED display by varying frequency of blanking input to display driver. Display is pulsed on and off rapidly; the higher the duty cycle, the brighter the display. At highest input voltage the converter is forced out of linear region, making its mono remain on continuously for brightest display.—P. Pinter and D. Timm, Voltage-to-Frequency Converters—IC Versions Perform Accurate Data Conversion (and Much More) at Low Cost, *EDN Magazine,* Sept. 5, 1977, p 153–157.

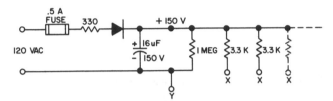

GIANT-NIXIE SIGN—Simple power supply drives any desired number of alphameric Nixie characters each 2½ inches high to form illuminated house numbers, "ON THE AIR" sign for amateur station, or "BAR IS CLOSED" sign for party room. Connect 3.3K resistor from +150 V to pin 13 of each Nixie, and connect to point Y (−150 V) each segment to be lighted. Sign can be changed at any time by resoldering connec-tions to segments. Diode must handle AC line voltage.—J. Grimes, Display Yourself in a Big Way, *73 Magazine,* Nov./Dec. 1975, p 186–188.

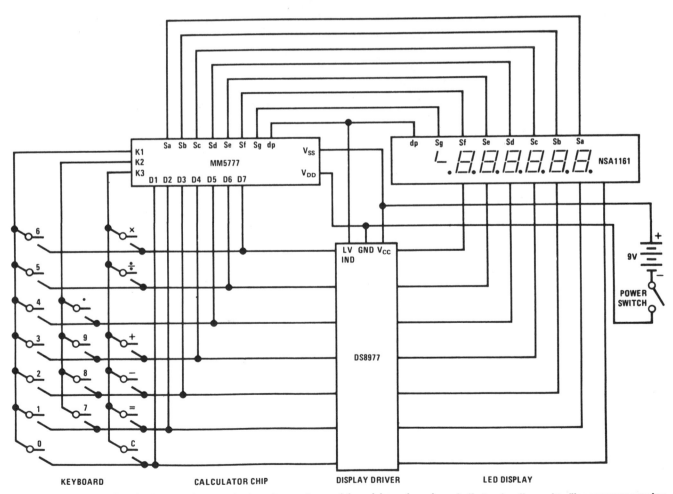

6-DIGIT FOUR-FUNCTION—National MM5777 calculator chip requires only keyboard, NSA1161 LED display, DS8977 digit driver, and 9-V battery to provide add, subtract, multiply, and divide functions. Calculator chip includes keyboard encoding and key debouncing circuits, along with all clock and timing generators. LED segments can be driven directly, without multiplexing. Seventh digit position is used for negative sign of 6-digit number and as error indicator. Leading and trailing zero suppression is included.—"MOS/LSI Databook," National Semiconductor, Santa Clara, CA, 1977, p 8-84–8-89.

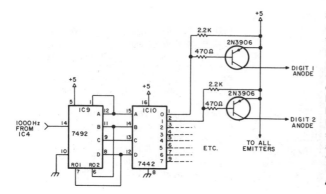

STROBING LED DISPLAY—Sequential strobing of individual LED displays, at rate fast enough to eliminate flicker (about 10% duty cycle), cuts power requirements of LEDs and eliminates need for power-wasting resistors in series with digit segments. Circuit uses 7492 binary counter connected to divide by 10, continuously clocked by 1000-Hz signal from external counter time base. Each of the 10 counter states is decoded by 7442 decoder for use in turning on PNP switch transistor connected in series with anode of each 7-segment LED digit. Digits are thus turned on for 10% of time at 100-Hz rate.—B. Hart, Current-Saver Counter Display, *73 Magazine,* June 1977, p 174—176.

POSITIVE INPUT GIVES GREEN—High or positive input to circuit gives green indication in Monsanto MV5491 dual red/green LED, and low input gives red. Circuit uses single SN75452 IC driver and one section of SN7404 hex inverter.—K. Powell, Novel Indicator Circuit, *Ham Radio,* April 1977, p 60—63.

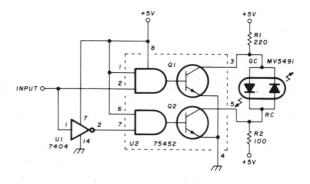

CHAPTER 26
Fiber-Optic Circuits

Includes LED modulators and photodiode or phototransistor receivers for single- or multiple-fiber data links handling audio, data, and teleprinter signals. Circuits are also given for infrared receivers and transmitters, high-voltage isolator links, laser-diode modulator, Manchester-code demodulator, and fiber light-transmission checker.

DATA COUPLER WITH ISOLATION—Length of fiber or polystyrene rod determines amount of voltage isolation provided between digital or analog signal input and Fairchild FPT 100 photodetector driving Optical Electronics 9720 opamp having 100-mA output for driving cables, relays, or loudspeakers. LED can be Monsanto MV50 handling up to 200 mA. Output of opamp is zero for no light. Pulse-duration modulation should be used for transmission of analog data.—"High Voltage Optically Isolated Data Coupler," Optical Electronics, Tucson, AZ, Application Tip 10266.

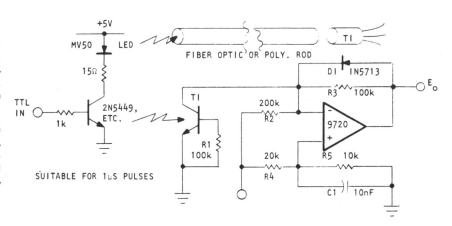

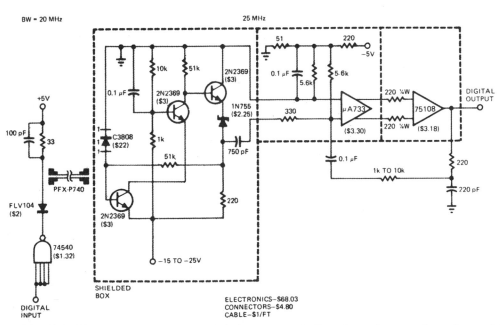

24-MEGABIT DATA LINK—High data-rate capability for square-wave pulses is achieved by increasing complexity of receiver feeding digital output to microprocessor from remote teleprinter. Preamp design compensates for noise over limited frequency range, giving uniform S/N ratio to about 20 MHz. With demonstration setup, visible-spectrum LED and photodetector shown performed acceptably with 40-foot cable.—O. E. Marvel and J. C. Freeborn, A Little Hands-On Experience Illuminates Fiber-Optic Links, *EDN Magazine*, Nov. 5, 1977, p 71–75.

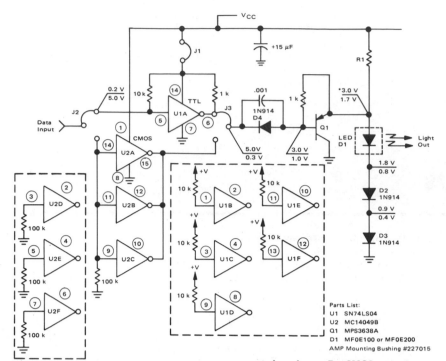

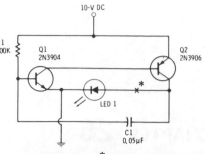

Parts List:
U1 SN74LS04
U2 MC14049B
Q1 MPS3638A
D1 MF0E100 or MF0E200
AMP Mounting Bushing #227015

HIGH-CURRENT INFRARED LED PULSER—Circuit operates as regenerative amplifier for delivering 10-μs pulses with amplitude of 1.1 A and repetition rate of 1.4 kHz to infrared LED. Suitable for infrared beacon in fiber-optic communication and optical radar applications. Drain is 100 mA from 10-V supply. Use gallium arsenide LED such as SSL-55C or TIL32 for high-output infrared transmitter. Q2 can be changed to germanium transistor such as 2N1305 to give peak current of 2 A at pulse width of 15 μs and 750-Hz repetition rate.—F. M. Mims, "Electronic Circuitbook 5: LED Projects," Howard W. Sams, Indianapolis, IN, 1976, p 33–35.

FIBER-OPTIC TRANSMITTER—Will handle NRZ data rates to 10 megabits or square waves to 5 MHz. Input is TTL- or CMOS-compatible depending on circuit selected. Transmitter draws only 150 mA from 5-V supply for TTL or from 5-15 V supply for CMOS. Choose R1 to give LED drive current for proper operation of system. For TTL operation, jumpers J1, J2, and J3 are connected as shown. For CMOS operation, remove J1 and transfer J2 and J3 to alternate positions for connecting to U2. Choice of LED depends on system length and desired data rate. Power supply can be HP6218A or equivalent. DC voltages shown are for TTL interface, with upper value for LED on at 50 mA and lower value for LED off.—"Basic Experimental Fiber Optic Systems," Motorola, Phoenix, AZ, 1978.

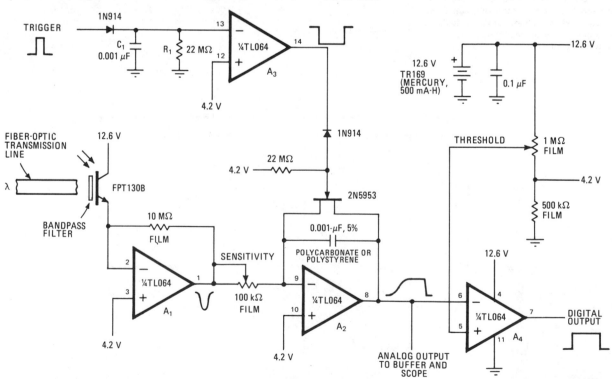

TL064: TEXAS INSTRUMENTS QUAD BI-FET AMPLIFIER OR EQUIVALENT

LIGHT TRANSMISSION CHECKER—Phototransistor and quad opamp serve as total-energy detector of pulsed-light signals propagated through fiber-optic cable of communication system. Can be used for checking and comparing condition of long fibers if light intensity of source is held constant. Will also detect changes in light intensity and changes in pulse width. Circuit gives linear response to light levels from 100 to 10,000 ergs/cm^2 if minimum pulse width is at least 10 μs. A$_2$ acts as RC integrator, giving voltage proportional to total light energy received. A$_4$ provides comparison to fixed level.—E. W. Rummel, Low-Level-Light Detector Checks Optical Cables Fast, *Electronics*, April 27, 1978, p 148 and 150.

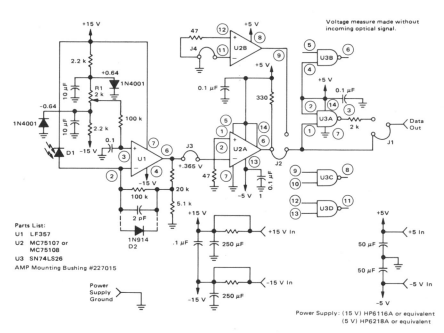

FIBER-OPTIC RECEIVER—Uses MF0D100 PIN photodiode as optical detector for handling megabit data rates. Minimum photocurrent required to drive LF357 opamp U1 is 250 nA. Voltage comparator U2 inverts output of U1 and provides standard TTL output. For CMOS output, quad two-input NAND gate U3 is wired into circuit, with jumper J1 connected from U3 output to output terminal of receiver. Adjust R1 to give accurate reproduction of 1-MHz square wave with 50% duty cycle at receiver output.—"Basic Experimental Fiber Optic Systems," Motorola, Phoenix, AZ, 1978.

LASER-DIODE SOURCE—With transistor switching circuit shown for RCA SG2007 laser diode, pulses as short as 10 ns are possible at repetition rates above 100 kHz. Used in optical communication system in which fiber bundle or single fiber is attached directly to laser pellet.—J. T. O'Brien, Laser Diodes Provide High Power for High-Speed Communications Systems, *Electronics*, Aug. 5, 1976, p 94—96.

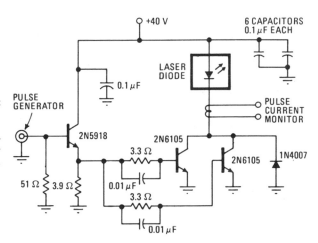

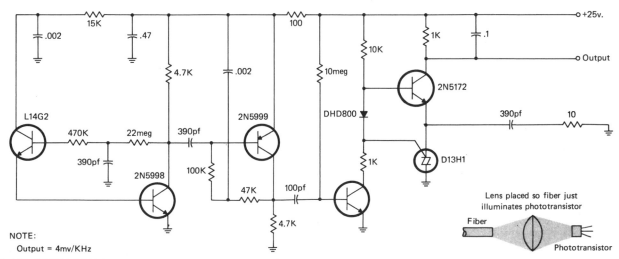

NOTE:
Output = 4mv/KHz

50-kHz FM OPTICAL RECEIVER—Designed for pulse-rate modulation system in which transmitter varies pulse rate of modulated light beam in optical-fiber cable above and below center frequency of 50 kHz. L14G2 GE phototransistor converts modulated optical light to RF signal for demodulation and reconstruction of original audio. Based on circuit in "General Electric Opto-Electronics Manual."—I. Math, Math's Notes, *CQ*, July 1977, p 67—68 and 90.

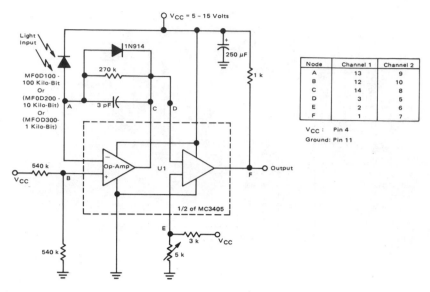

1/10/100-KILOBIT FIBER-OPTIC RECEIVER— Choice of input device determines operating speed of receiver. MC3405 contains two opamps and two comparators, permitting use as two-channel receiver. Table gives pin connections for each channel.—"Basic Experimental Fiber Optic Systems," Motorola, Phoenix, AZ, 1978.

Node	Channel 1	Channel 2
A	13	9
B	12	10
C	14	8
D	3	5
E	2	6
F	1	7

V_{CC}: Pin 4
Ground: Pin 11

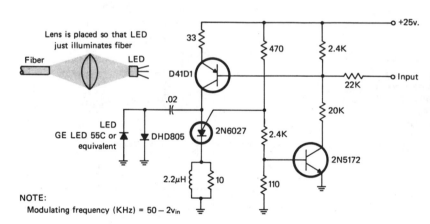

NOTE:
Modulating frequency (KHz) = $50 - 2v_{in}$

50-kHz FM OPTICAL TRANSMITTER— Uses pulse-rate modulation system with center frequency of 50 kHz. Audio fed into transmitter varies pulse rate, for driving LED coupled to optical fiber. Phototransistor at other end of fiber receives and demodulates light signal for reconstruction of audio.—I. Math, Math's Notes, *CQ*, July 1977, p 67–68 and 90.

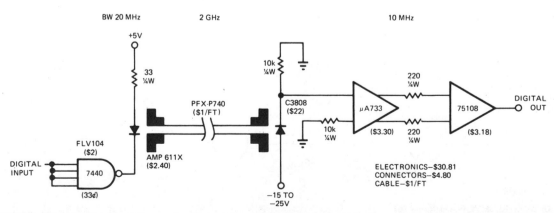

10-MEGABIT LINK— Transmitter and receiver for fiber-optic data link between teleprinter and microprocessor utilize wide bandwidth of cable for transmitting data at 10-megabit rate. Receiver input requires C3808 PIN photodiode.—O. E. Marvel and J. C. Freeborn, A Little Hands-On Experience Illuminates Fiber-Optic Links, *EDN Magazine*, Nov. 5, 1977, p 71–75.

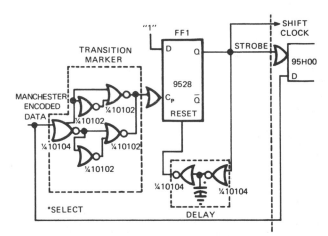

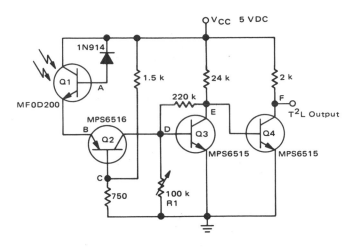

1-GHz MANCHESTER DECODER—Use of ECL flip-flop with toggle rates above 1 GHz makes decoding of bit rates approaching gigabit speeds feasible. Article gives step-by-step design procedure for 48-Mb telemetry application using PCM over single optical-fiber cable.—B. R. Jarrett, Could You Design a High-Speed Manchester-Code Demodulator?, *EDN Magazine*, Aug. 20, 1974, p 75—80.

20-KILOBIT FIBER-OPTIC RECEIVER—Phototransistor driving three-transistor amplifier provides TTL output for data rates up to 20 kilobits.—"Basic Experimental Fiber Optic Systems," Motorola, Phoenix, AZ, 1978.

IR DETECTOR—Photodiode transforms light-signal output of fiber-optic cable to electric signal. Spectral response of detector closely matches that of IR-emitting diode at other end of cable, for maximum system efficiency. Rise and fall times of detector can be less than 35 ns when properly biased and loaded by receiver circuit. Developed by Augat, Inc., Attleboro, MA, as part of fiber-optic evaluation kit for TTL applications.—Fiber-Optic Kit Allows Engineering Evaluation of Complete Interconnection System, *Computer Design*, Nov. 1977, p 27 and 30.

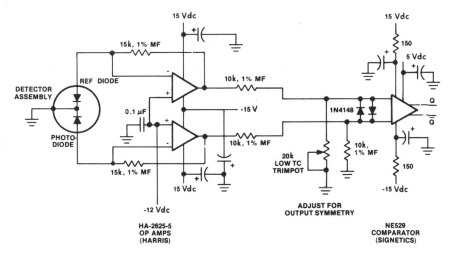

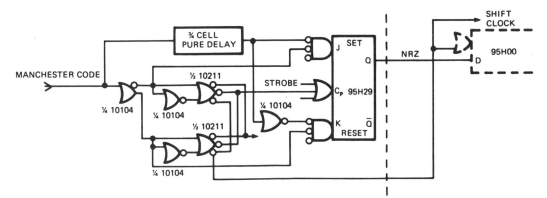

MANCHESTER-CODE DEMODULATOR—Digital approach using ECL provides maximum speed, is self-synchronizing for alternate bit-pairs, and has minimum complexity. Developed for optically coupled 25-channel PCM telemetry system used over single optical-fiber channel. Undesired transitions in input data are masked by creating strobe. Approach recognizes distinction between identical sequences that would give some output except for time-of-occurrence restriction. Article gives step-by-step design procedure, waveforms, and excitation table.—B. R. Jarrett, Could You Design a High-Speed Manchester-Code Demodulator?, *EDN Magazine*, Aug. 20, 1974, p 75—80.

1-MHz LED PULSE MODULATION—Circuit provides required low driving-point impedance for fast turn-on of gallium arsenide phosphide LED used as source for high-speed pulse modulation of fiber-optic or other light beam. Q_1 supplies DC level and modulation information to emitter-follower output stage Q_3. Output current is sensed and limited to about 30 mA by Q_2. Turn-on time for full brightness is 12 ns.—G. Schmidt, LED Modulator, *EDN/EEE Magazine*, June 15, 1971, p 57.

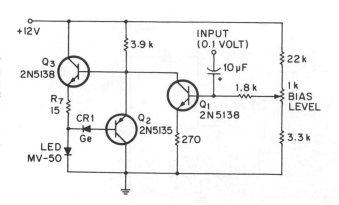

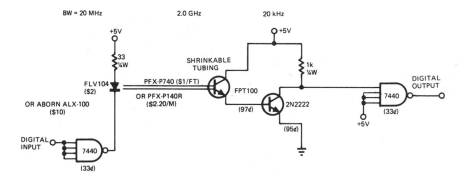

TTY LINK FOR MICROPROCESSOR—Demonstration circuit illustrates use of fiber-optic cable with low-cost components for relatively narrow-band application, to provide feed from remote teleprinter to microprocessor.—O. E. Marvel and J. C. Freeborn, A Little Hands-On Experience Illuminates Fiber-Optic Links, *EDN Magazine*, Nov. 5, 1977, p 71–75.

CHAPTER 27
Filter Circuits—Active

Includes low-pass, high-pass, bandpass, notch, state-variable (2 to 4 functions), tracking, and equalizing filters covering from 1 Hz to limits of audio spectrum, along with gyrator, Q multiplier and variable-Q circuits, crossover networks for loudspeakers, and RF circuits providing frequency emphasis.

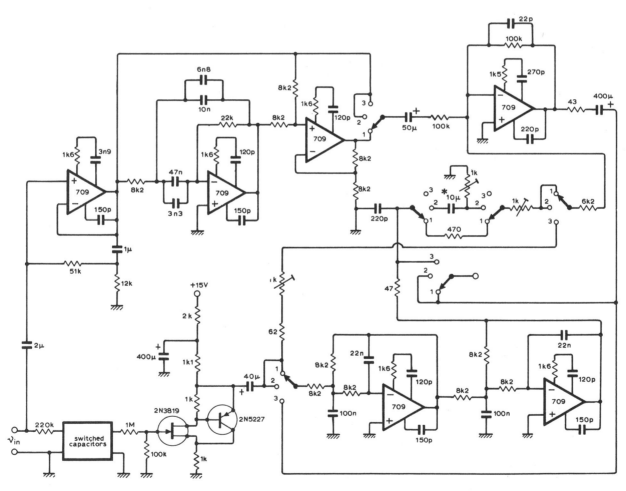

*non-polarized polycarbonate

TRACKING LINE-FREQUENCY FILTER—Improvements in commutating RC network filter extend dynamic range without sacrificing signal bandwidth, for reducing interference at fundamental of power-line frequency and harmonics up to fifth. Although values in circuit are for British 50-Hz mains frequency, circuit can readily be adapted for 60-Hz rejection. Operation involves commutating 16 capacitors electronically at 16 times line frequency. Article gives one method of doing this, by driving two 8-way multiplexers alternately. Each multiplexer has eight MOSFETs, each switched on in turn by consecutive input clock pulses. Circuit details, design equations, and performance graphs are given. Three-position switch gives choice of filter characteristics.—K. F. Knott and L. Unsworth, Mains Rejection Tracking Filter, *Wireless World,* Oct. 1974, p 375–379.

10-kHz VARIABLE-Q—Second-order state-variable filter having center frequency of 10 kHz uses all four sections of OP-11FY quad opamp. Center frequency can be tuned by varying 1.6K feedback resistors or by changing 0.01-μF feedback capacitors. Value of feedback resistor for opamp D determines Q of filter, for adjusting circuit bandwidth or damping. For higher-frequency operation, use high-speed opamps such as OP-15 or OP-16.—D. Van Dalsen, Need an Active Filter? Try These Design Aids, *EDN Magazine*, Nov. 5, 1978, p 105–110.

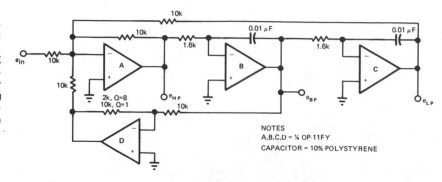

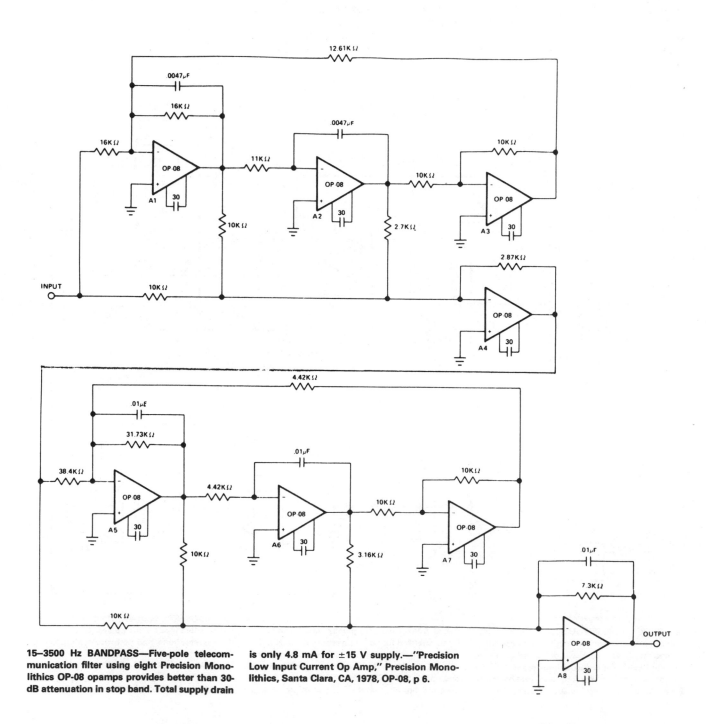

15–3500 Hz BANDPASS—Five-pole telecommunication filter using eight Precision Monolithics OP-08 opamps provides better than 30-dB attenuation in stop band. Total supply drain is only 4.8 mA for ±15 V supply.—"Precision Low Input Current Op Amp," Precision Monolithics, Santa Clara, CA, 1978, OP-08, p 6.

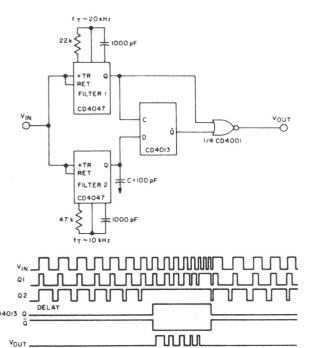

10–20 kHz BANDPASS—Two CD4047A low-pass filters, one connected for 10-kHz cutoff and other for 20-kHz cutoff, drive CD4013A flip-flop. If output of filter 2 is delayed by C, flip-flop clocks high only when input pulse frequency exceeds 10-kHz cutoff of filter 2. Waveforms show performance when input signal is swept through passband.—"CQS/MOS Integrated Circuits," RCA Solid State Division, Somerville, NJ, 1977, p 619.

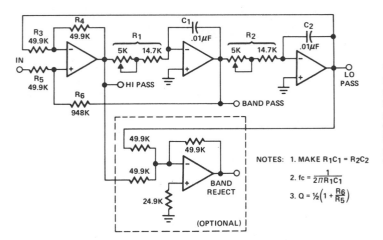

1-kHz STATE-VARIABLE WITH Q OF 10—Use of all four sections of Harris HA-4602/4605 quad opamp gives four types of 1-kHz second-order filtering simultaneously. Pot adjustments permit matching of various RC products allowing for noninteractive adjustment of Q and center frequency.—"Linear & Data Acquisition Products," Harris Semiconductor, Melbourne, FL, Vol. 1, 1977, p 2-84.

NOTES: 1. MAKE $R_1C_1 = R_2C_2$

2. $fc = \dfrac{1}{2\pi R_1 C_1}$

3. $Q = \frac{1}{2}\left(1 + \dfrac{R_6}{R_5}\right)$

5-kHz SERIES-SWITCHED BANDPASS—N-path filter having N of 4, Q of 500, and voltage gain of 2 uses DG509 four-channel CMOS differential multiplexer having necessary pairs of analog switches and decode logic. Dual D flip-flop generates 2-bit binary sequence from 20-kHz clock signal. Bandwidth is about 10 Hz for 3 dB down, centered on 5 kHz.—"Analog Switches and Their Applications," Siliconix, Santa Clara, CA, 1976, p 5-15–5-17.

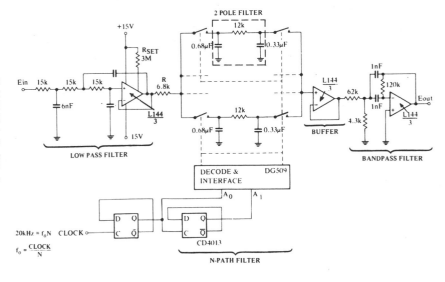

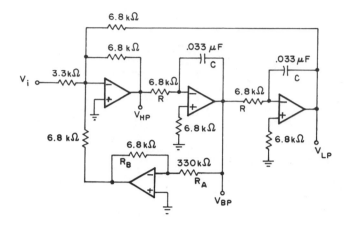

700-Hz STATE-VARIABLE—Provides voltage gain at center frequency of 100 (40 dB) and Q of 50. Used when simultaneous low-pass, high-pass, and bandpass output responses are required. Cutoff frequency of low-pass and high-pass responses is equal to center frequency of bandpass response. Opamps can be 741. Based on use of 5% resistors.—H. M. Berlin, "Design of Active Filters, with Experiments," Howard W. Sams, Indianapolis, IN, 1977, p 184–187.

1.5-kHz NOTCH—Unity-gain state-variable filter consists of low-pass and high-pass sections combined with two-input summing amplifier to give notch response for suppression of 1.5-kHz signals. Opamps can be 741.—H. M. Berlin, "Design of Active Filters, with Experiments," Howard W. Sams, Indianapolis, IN, 1977, p 186–189.

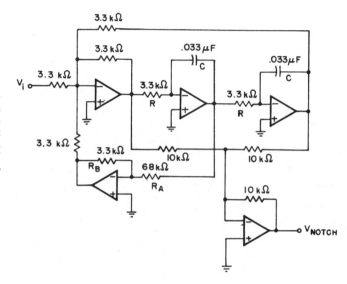

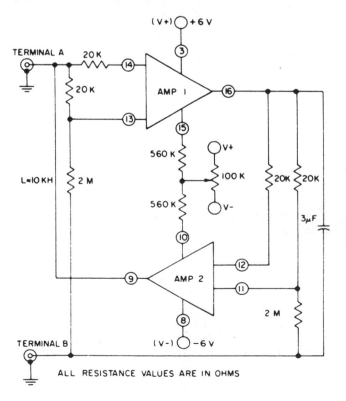

10-kH GYRATOR—Active filter circuit uses two sections of CA3060 three-opamp array as gyrator that makes 3-μF capacitor function as floating 10-kilohenry inductor across terminals A and B. Q of inductor is 13. 100K pot tunes inductor by changing gyration resistance.—"Linear Integrated Circuits and MOS/FET's," RCA Solid State Division, Somerville, NJ, 1977, p 152.

ALL RESISTANCE VALUES ARE IN OHMS

60-Hz NOTCH FILTER—Design is based on pass-band gain of 3 and Q of 6. Resistors can be 5%. Opamps can be 741. Notch response is obtained by subtracting output signal of bandpass filter from its input signal with R₆.—H. M. Berlin, "Design of Active Filters, with Experiments," Howard W. Sams, Indianapolis, IN, 1977, p 155.

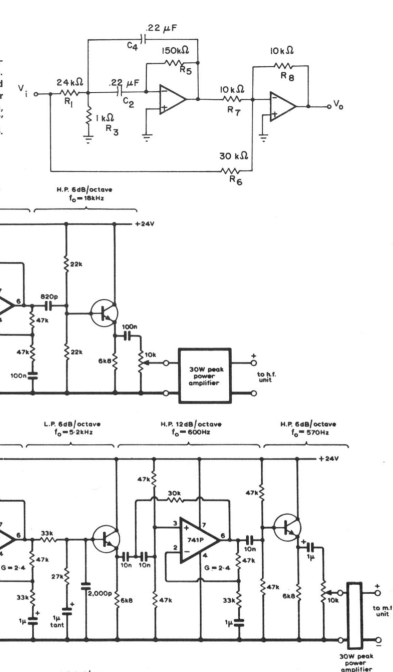

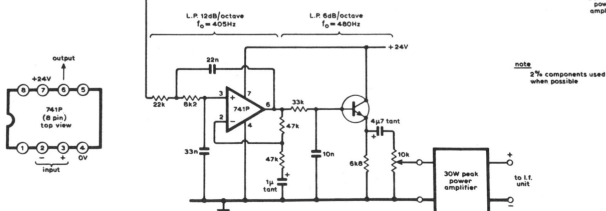

THREE-LOUDSPEAKER CROSSOVERS—Active filter network splits AF input into three frequency bands each feeding separate 30-W power amplifier. Design allows adjustment of any part of frequency characteristic to any de-sired level and gives choice of slopes in any part of frequency band. Article gives design equations and construction details. NPN transistors can be BC107 or 2N3904; PNP transistors can be BCY70, BCY71, BCY72, or 2N3906. Article also gives circuit of suitable 30-W amplifier.—D. C. Read, Active Filter Crossover Networks, *Wireless World*, Dec. 1973, p 574–576.

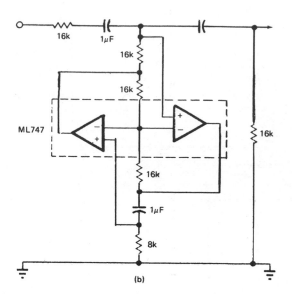

(b)

10-Hz HIGH-PASS—Equiterminated Butterworth high-pass ladder filter has corner frequency of 10 Hz and output impedance level of 16K. Opamps are matched pair in single ML747 package. Article covers design procedure based on use of generalized impedance converters and gives frequency response curve.—L. T. Burton and D. Treleaven, Active Filter Design Using Generalized Impedance Converters, *EDN Magazine*, Feb. 5, 1973, p 68–75.

10-kHz VOLTAGE-TUNED—High-Q circuit using Optical Electronics 9831 opamp has sharp resonance, as required for analysis of spectrum of incoming signal. Reverse-biased silicon junctions serve as voltage variable capacitors for sweeping center frequency over 3:1 range. Values shown for three resistors in twin-T network give center frequency of 10 kHz.—"Voltage Tuned High-Q Filter," Optical Electronics, Tucson, AZ, Application Tip 10207.

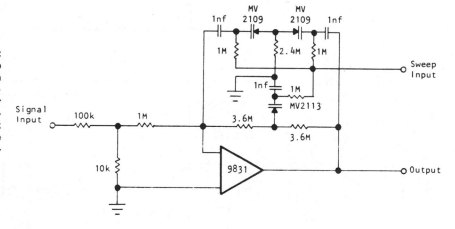

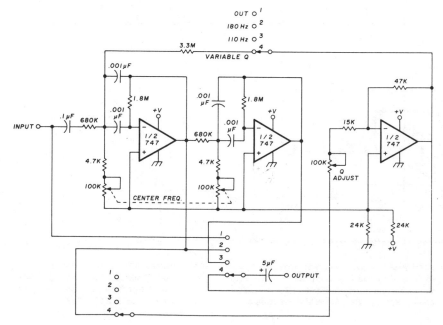

ACTIVE CW FILTER—Modifications made on MFJ Enterprises CWF-2 active audio filter permit maximum flexibility. Circuit provides fixed bandwidth of 180 or 110 Hz centered on 750 Hz, or optional variable bandwidth for which center frequency can be adjusted in range of 280 to 1590 Hz.—H. M. Berlin, Increased Flexibility for the MFJ Enterprises CW Filters, *Ham Radio*, Dec. 1976, p 58–60.

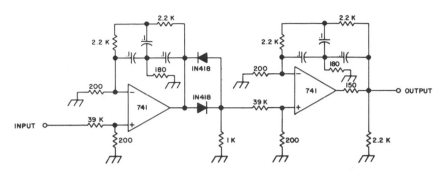

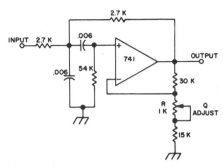

TWO-STAGE CW—Uses diode threshold detector between stages to prevent weak undesired signals from passing through until CW signal of desired frequency is present, so as to provide quiet tuning between signals. Bandwidth of filter is sharp (16 Hz), and keyed waveform is good. Gain is near unity, and frequency and Q are both fixed.—A. F. Stahler, An Experimental Comparison of CW Audio Filters, *73 Magazine,* July 1973, p 65–70.

VARIABLE Q FOR CW—Fixed-frequency active filter gives slowly rising and falling keyed waveform with good slope considering narrowness of bandwidth, which is 75 Hz at 3 dB down. Adjusting Q with 1K pot changes bandwidth.—A. F. Stahler, An Experimental Comparison of CW Audio Filters, *73 Magazine,* July 1973, p 65–70.

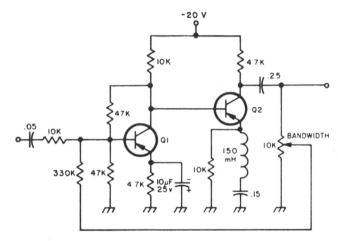

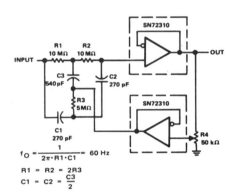

NARROW BANDPASS FOR SPEECH—Simple audio filter provides about 20-dB gain at bandwidth of 80 Hz. Bandwidth can be narrowed to limits of unintelligibility by adjusting 10K pot. Input is plugged into phone jack of receiver, and headphones are connected to output. Transistors are SK3004 or equivalent.—Circuits, *73 Magazine,* Jan. 1974, p 125.

60-Hz ADJUSTABLE-Q NOTCH—Connection shown for two SN72310 voltage-follower opamps provides attenuation of 60-Hz power-line frequency. Setting of R4 determines Q of filter.—"The Linear and Interface Circuits Data Book for Design Engineers," Texas Instruments, Dallas, TX, 1973, p 4-39.

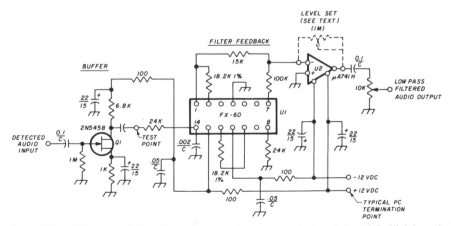

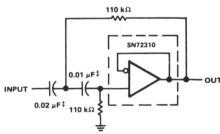

ACTIVE AF FOR SSB AND CW—Uses Kinetic Technology FX-60 IC (culled from FS-60, FS-65, and FS-61 production by manufacturer) as 2.5-kHz tunable detected-audio low-pass filter for SSB. Provides inexpensive hybrid active filter using multiloop negative feedback for low-pass transfer functions. External resistors tune filter frequency and give choice of Q. High-impedance buffer Q1 provides nominal gain while isolating filter from previous receiver stages. Opamp U2 boosts overall gain.—M. A. Chapman, Audio Filters for Improving SSB and CW Reception, *Ham Radio,* Nov. 1976, p 18–23.

100-Hz HIGH-PASS—Metallized polycarbonate capacitors are required for good temperature stability in high-pass active filter using voltage-follower opamp. Cutoff frequency is 100 Hz.— "The Linear and Interface Circuits Data Book for Design Engineers," Texas Instruments, Dallas, TX, 1973, p 4-39.

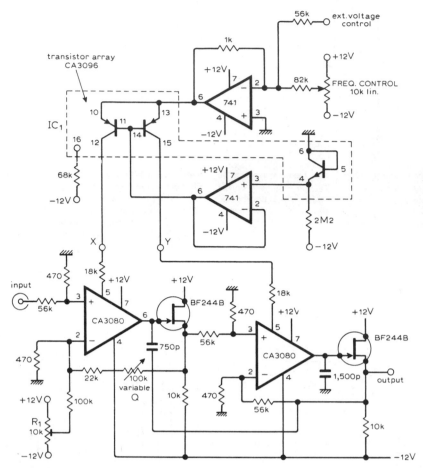

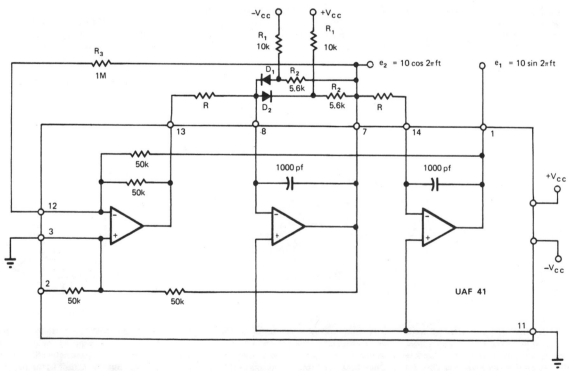

CONTROLLABLE 12 dB PER OCTAVE ROLL-OFF—Frequency at which rolloff starts can be set in range between 15 and 15,000 Hz by external voltage or current. If only manual control of frequency is required, short points X and Y and connect them to wiper of 10,000-ohm logarithmic potentiometer that is positioned between −12 V and ground.—T. Orr, Voltage/Current Controlled Filter, *Wireless World,* Nov. 1976, p 63.

QUADRATURE OSCILLATOR—Addition of diode limiter and positive-feedback resistor to UAF41 universal active filter gives precision quadrature oscillator.—Y. J. Wong, Design a Low Cost, Low-Distortion, Precision Sine-Wave Oscillator, *EDN Magazine,* Sept. 20, 1978, p 107–113.

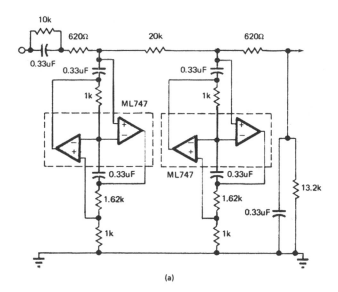

(a)

480-kHz LOW-PASS—Butterworth low-pass active filter uses pair of dual opamps with external resistors and capacitors to give corner frequency of 480 kHz and output impedance level of 1K. Article presents design procedure in detail and gives frequency response curve.—L. T. Burton and D. Treleaven, Active Filter Design Using Generalized Impedance Converters, *EDN Magazine*, Feb. 5, 1973, p 68–75.

300–3000 Hz WIDEBAND—Used in voice communication systems where signals below 300 Hz and above 3000 Hz must be rejected. Second-order Butterworth stopband responses are achieved by combining low-pass and high-pass sections of equal-component voltage-controlled voltage-source filters. Overall passband gain is 8 dB. Opamps can be 741.—H. M. Berlin, "Design of Active Filters, with Experiments," Howard W. Sams, Indianapolis, IN, 1977, p 148–151.

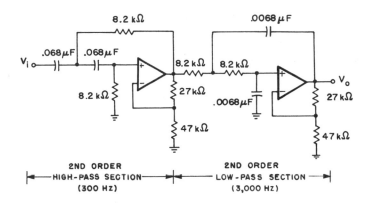

18 dB PER OCTAVE EMPHASIS—Circuit shown is result of design procedure given in article for active filter that provides frequency emphasis at rate of 18 dB per octave between 5 and 15 kHz. Emphasis does not exceed 40 dB at 20 kHz. Design equations include parameters for closed-loop gain of opamp. Scale factor is applied to input and feedback networks individually after design, to give reasonable component values.—B. Brandstedt, Tailor the Response of Your Active Filters, *EDN Magazine*, March 5, 1973, p 68–72.

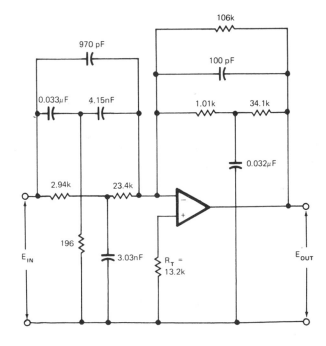

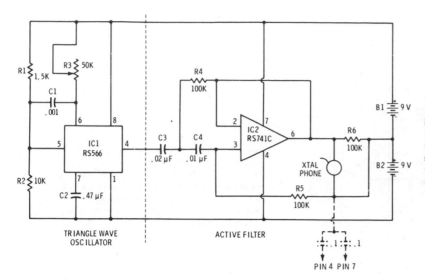

TRIANGLE WAVE OSCILLATOR   ACTIVE FILTER

PIN 4   PIN 7

150-Hz HIGH-PASS—Circuit includes variable high-frequency source supplying triangle-wave input to filter for demonstrating high-pass action. If long supply leads cause oscillation, connect 0.1-μF capacitors between ground and supply pins 4 and 7 as shown.—F. M. Mims, "Integrated Circuit Projects, Vol. 4," Radio Shack, Fort Worth, TX, 1977, 2nd Ed., p 87–94.

750-Hz SIXTH-ORDER BANDPASS—Provides passband gain of 6 (15.6 dB) and Q of 8.53 by cascading three identical second-order filter sections. Each section uses multiple feedback.—H. M. Berlin, "Design of Active Filters, with Experiments," Howard W. Sams, Indianapolis, IN, 1977, p 147–148.

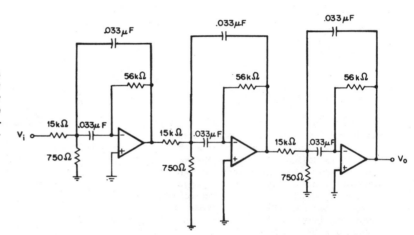

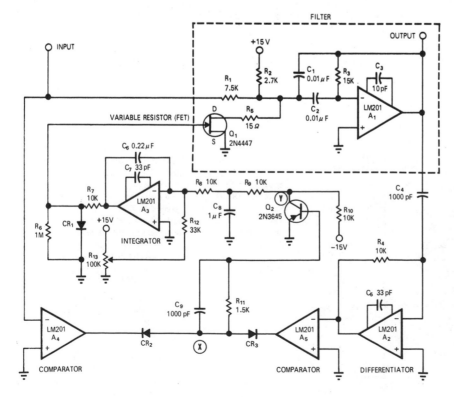

2–20 kHz SELF-TUNING BANDPASS—Center frequency of filter is automatically adjusted to track signal frequency, for optimum noise rejection when input frequency varies over wide range as it does with many types of vibrating transducers. Requires no reference frequency and no internal oscillator or synchronizing circuits. Frequency range can be extended in decade steps by capacitor switching. When filter is not tuned to input frequency, phase shift is not 180° and phase detector applies error signal to gate of FET to control its drain-source resistance. Phase detector A_4-A_5-CR_2-CR_3 and FET form part of negative-feedback loop around filter, so error in phase changes resistance of FET and thereby retunes filter. Article gives design equations, operational details, and waveforms at various points in circuit.—G. J. Deboo and R. C. Hedlund, Automatically Tuned Filter Uses IC Operational Amplifiers, *EDN/EEE Magazine*, Feb. 1, 1972, p 38–41.

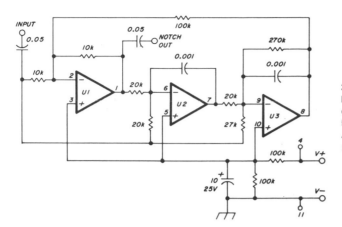

3-kHz NOTCH—Uses three sections of National LM324 quad opamp to provide fixed center frequency of 3 kHz for notch. Single supply can be 5–25 V.—P. A. Lovelock, Discrete Operational Amplifier Active Filters, *Ham Radio*, Feb. 1978, p 70–73.

STATE-VARIABLE DESIGN—Universal filter network using three opamps can provide low-pass, high-pass, or bandpass audio response for CW and SSB reception. Filter uses one summing block U1, two identical integrators, U2 and U3, and one damping network. Cutoff frequencies are same as center frequency for bandpass response. Article gives design equations and graph for choosing values to give optimum performance for type of response desired. For unity-gain second-order Butterworth filter with low-pass or high-pass cutoff of 700 Hz, R is 6800 ohms and C is 0.033 μF. Q must be fixed at 0.707 and R_A must equal $1.12 \times R_B$. Thus, if R_B is 2700 ohms, R_A should be 3000.—H. M. Berlin, The State-Variable Filter, *QST*, April 1978, p 14–16.

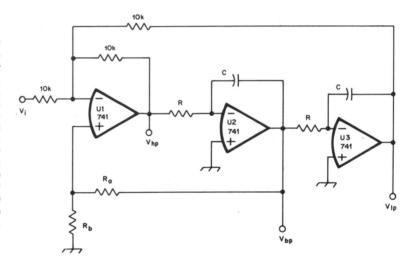

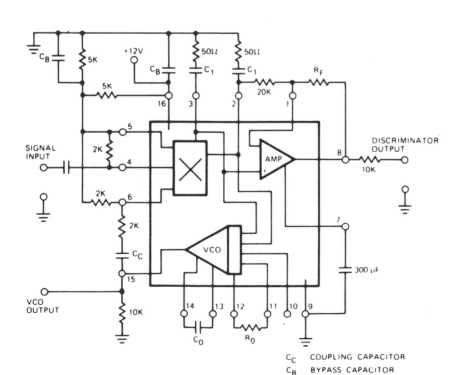

3:1 TRACKING FILTER—Connection shown for Exar XR-215 PLL IC tracks input signal over 3:1 frequency range centered on free-running frequency of VCO. Tracking range is maximum when pin 10 is open. R_0 is typically between 1K and 4K. C_1 is between 30 and 300 times C_0 where timing capacitor C_0 depends on center frequency. System can also be operated as linear discriminator or analog frequency meter covering same 3:1 change of input frequency. R_F can be 36K.—"Phase-Locked Loop Data Book," Exar Integrated Systems, Sunnyvale, CA, 1978, p 21–28.

C_C COUPLING CAPACITOR

C_B BYPASS CAPACITOR

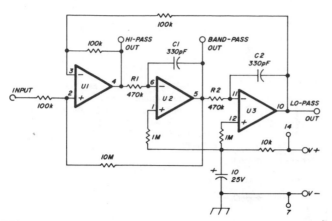

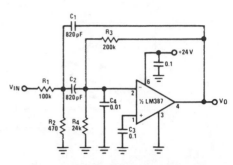

1-kHz THREE-FUNCTION—Three-function fixed-frequency active filter uses three sections of RCA CA3401E, Motorola MC3301P, or National LM3900N quad opamp. Circuit uses high-value series resistors for noninverting inputs to limit bias current to between 10 and 100 μA.—P. A. Lovelock, Discrete Operational Amplifier Active Filters, *Ham Radio*, Feb. 1978, p 70–73.

20-kHz BANDPASS—Provides bandwidth of 2000 Hz and midband gain of 1 for applications requiring narrow-bandwidth bandpass active filter. Design procedure is given.—"Audio Handbook," National Semiconductor, Santa Clara, CA, 1977, p 2-52–2-53.

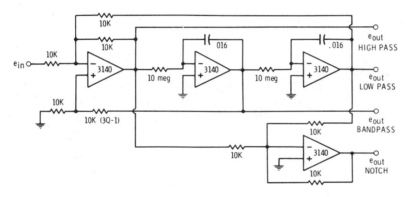

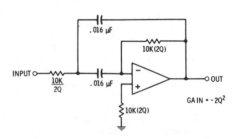

1-Hz STATE-VARIABLE FILTER—Universal filter has simultaneous low-pass, bandpass, high-pass, and notch outputs all with cutoff frequency of 1 Hz. To scale circuit up to 1-kHz cutoff, replace 10-megohm resistors with 10K.—D. Lancaster, "CMOS Cookbook," Howard W. Sams, Indianapolis, IN, 1977, p 343–344.

1-kHz MULTIPLE-FEEDBACK BANDPASS—Single 741 or equivalent opamp is suitable for applications where bandwidth is less than 100%. Gain is fixed at $-2Q^2$, where Q is reciprocal of damping d and ranges from less than 1 to above 100. Q is changed by varying ratio of input and feedback resistors while keeping their product constant. For Q of 3, feedback resistor should be 36 times value of input resistor.—D. Lancaster, "Active-Filter Cookbook," Howard W. Sams, Indianapolis, IN, 1975, p 150–154.

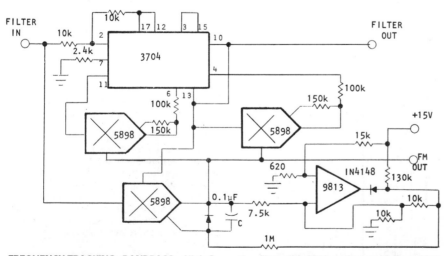

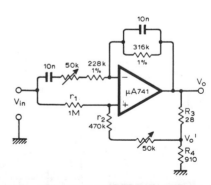

FREQUENCY-TRACKING BANDPASS—High-Q active bandpass filter automatically tracks input signal frequency over 10:1 range in presence of noise. When signal goes outside tracking range, circuit sweeps between low- and high-frequency limits until suitable signal reappears. Circuit is basically voltage-controlled bandpass filter using Optical Electronics 3704 active filter with 5898 analog multipliers. 9813 opamp connected as Schmitt trigger is main element of scanning circuit. Frequency range is 160 to 1600 Hz, and FM bandwidth of error-voltage output is 20 Hz.—"Frequency Tracking Active Filter," Optical Electronics, Tucson, AZ, Application Tip 10270.

50-Hz WIEN-BRIDGE NOTCH FILTER—Uses opamp in circuit having essentially zero output impedance, making additional buffer amplifier unnecessary. Article gives design theory and covers many other types of notch filters.—Y. Nezer, Active Notch Filters, *Wireless World*, July 1975, p 307–311.

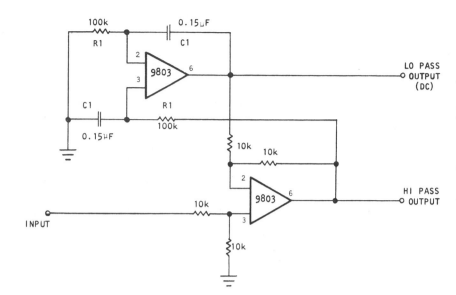

DC LEVEL SHIFTER FOR AF—Circuit using Optical Electronics 9803 opamps separates AF input signal into two outputs. Low-pass output contains DC to 10 Hz, and high-pass output has frequency content above 10 Hz to upper frequency limit approaching 10 MHz for opamp used. Dynamic output impedance of both outputs is less than 1 ohm. Both outputs have DC continuity. DC output of high-pass terminal is equal to offset voltage of integrator. DC output of low-pass terminal equals DC input plus offset voltages of both opamps.—"Automatic DC Level Shifter," Optical Electronics, Tucson, AZ, Application Tip 10226.

VOLTAGE-CONTROLLED BANDPASS—Two Optical Electronics 5898 four-quadrant analog multipliers and 3704 state-variable active filter permit use of voltage control for changing filter characteristics remotely without having noise pickup problems on control lines. Analog multipliers serve as variable-gain blocks that change current levels in resistors and in effect change resistor values. Circuit has linear relationship of frequency to control voltage, constant gain and Q with frequency, and good temperature stability.—"Voltage-Controlled Active Bandpass Filter," Optical Electronics, Tucson, AZ, Application Tip 10269.

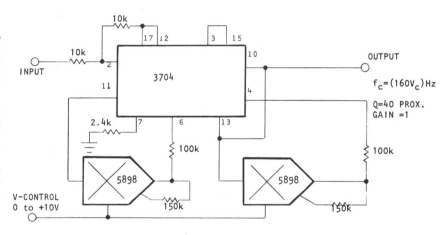

$$f_c = (160 V_c)\,Hz$$
$$Q = 40\ PROX.$$
$$GAIN = 1$$

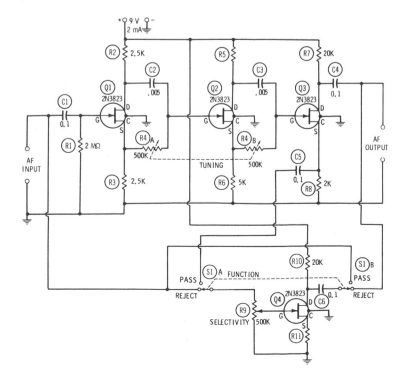

PASS/REJECT TUNABLE NOTCH—Full rotation of ganged tuning control R4 tunes circuit from 100 Hz to 10 kHz, with position of switch S1 determining whether circuit passes or rejects frequency to which it is tuned. Maximum selectivity, corresponding to maximum height of pass curve or depth of reject curve and minimum width of either curve, is obtained when R9 is set for maximum gain in FET Q4. If R9 is advanced far enough with switch set to pass, circuit will oscillate and give sine-wave output at tuned frequency.—R. P. Turner, "FET Circuits," Howard W. Sams, Indianapolis, IN, 1977, 2nd Ed., p 71–73.

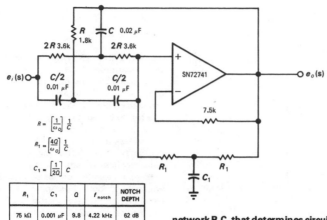

$$R = \left[\frac{1}{\omega_0}\right]\frac{1}{C}$$

$$R_1 = \left[\frac{4Q}{\omega_0}\right]\frac{1}{C}$$

$$C_1 = \left[\frac{1}{2Q}\right]C$$

R_1	C_1	Q	f_{notch}	NOTCH DEPTH
75 kΩ	0.001 μF	9.8	4.22 kHz	62 dB
150 kΩ	660 pF	18.4	4.22 kHz	62.7 dB
220 kΩ	360 pF	25.0	4.22 kHz	62 dB

4.22-kHz NOTCH—Circuit consists of positive unity-gain opamp, RC twin-T network, and T

network R_1C_1 that determines circuit Q. Variable Q feature is controlled by single passive RC network. Center frequency of notch filter is about 4.22 kHz. Table gives values of R_1 and C_1 for three different values of Q.—H. T. Russell, *Notch Filter Has Passive Q Control, EDN/EEE Magazine*, July 1, 1971, p 43 and 45.

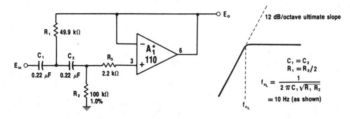

10-Hz HIGH-PASS UNITY-GAIN—Low cutoff frequency is 10 Hz in active filter using opamp as voltage-controlled voltage source. Alterna-

tive opamps can be 1556 and 8007.—W. G. Jung, "IC Op-Amp Cookbook," Howard W. Sams, Indianapolis, IN, 1974, p 331–333.

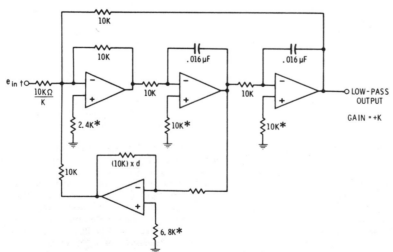

† must return to ground via low-impedance dc path.
*optional offset compensation, may be replaced with short in noncritical circuits.

1-kHz VARIABLE-GAIN STATE-VARIABLE—Damping signal is inverted with fourth opamp to make gain and damping as well as frequency independently adjustable. Damping is in range of 0–2, with critical value of 1.414 giving flattest response. For high pass, take output from first

opamp; for bandpass, take output from second opamp. Value of input resistor is 10K (10,000 ohms) when gain K is 1.—D. Lancaster, "Active-Filter Cookbook," Howard W. Sams, Indianapolis, IN, 1975, p 135–136.

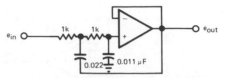

10-kHz LOW-PASS SALLEN-KEY—Article gives design equations from which values of components were obtained. Critical damping (Q = 0.71) is provided. Frequency can be tuned over range of two decades by changing resistor values simultaneously. Opamp can be one section of OP-11FY. For equivalent high-pass filter, transpose positions of resistors and capacitors. Gain is unity.—D. Van Dalsen, Need an Active Filter? Try These Design Aids, *EDN Magazine*, Nov. 5, 1978, p 105–110.

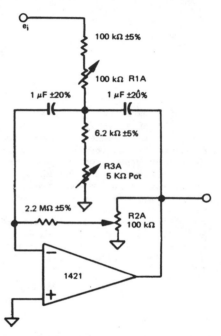

1 Hz WITH 0.1-Hz BANDWIDTH—Three pots provides easy trimming to precise values desired. Use R2A to trim bandwidth to exactly 0.100 Hz. Use R1A to trim gain to exactly 10.00. Finally, trim center frequency to exactly 1.000Hz. Adjustments are almost perfectly non-interacting if made in sequence given.—R. A. Pease, "Band-Pass Active Filter with Easy Trim for Center Frequency," Teledyne Philbrick, Dedham, MA, 1972, Applications Bulletin 4.

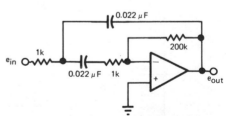

100-Hz BANDPASS SALLEN-KEY—Uses one section of OP-11FY quad opamp or equivalent in circuit having Q of 4.7 and providing closed-loop gain of 200 or 46 dB. Opamp selected should have open-loop gain of 5 to 10 times required gain at resonance. Adjust resistor values to tune center frequency.—D. Van Dalsen, Need an Active Filter? Try These Design Aids, *EDN Magazine*, Nov. 5, 1978, p 105–110.

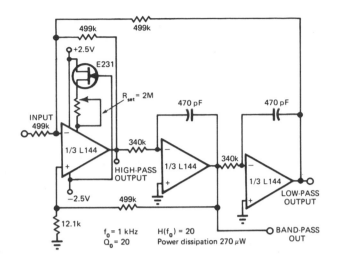

1-kHz STATE-VARIABLE—Low-power filter uses three opamps to provide simultaneous high-, low-, and bandpass outputs. Article presents complete design procedure for keeping current drain at minimum while providing desired gain-bandwidth product of 240 kHz.—L. Schaeffer, Op-Amp Active Filters—Simple to Design Once You Know the Game, *EDN Magazine*, April 20, 1976, p 79–84.

$f_0 = 1$ kHz
$Q_0 = 20$
$H(f_0) = 20$
Power dissipation 270 μW

1 Hz–500 kHz VOLTAGE-TUNED BANDPASS—Coupling FET opamps with analog multiplier gives simple two-pole bandpass filter that can be tuned by external voltage of 0–10 VDC to give center frequency anywhere in range from 1 Hz to 500 kHz with components shown. Article gives design equations.—T. Cate, Voltage Tune Your Bandpass Filters with Multipliers, *EDN Magazine*, March 1, 1971, p 45–47.

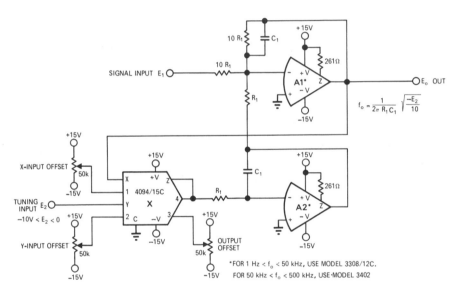

$$f_0 = \frac{1}{2\pi R_1 C_1}\sqrt{\frac{-E_2}{10}}$$

*FOR 1 Hz < f_0 < 50 kHz, USE MODEL 3308/12C.
FOR 50 kHz < f_0 < 500 kHz, USE MODEL 3402

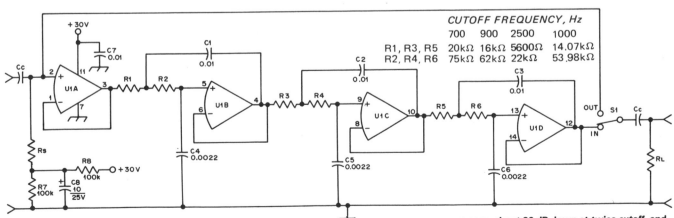

CUTOFF FREQUENCY, Hz			
	700	900	2500	1000
R1, R3, R5	20kΩ	16kΩ	5600Ω	14.07kΩ
R2, R4, R6	75kΩ	62kΩ	22kΩ	53.98kΩ

LOW-PASS AF—Can be used to attenuate undesired high-frequency audio response in superhet or direct-conversion receivers having inadequate IF selectivity, to improve CW or SSB reception. Resistor values determine cutoff frequency; 700 and 900 Hz are for CW and 2500 Hz for SSB. Insert filter at point having low audio level. Filter has input buffer, three cascaded active low-pass filter stages, and IN/OUT switch. Overall gain is unity. U1 is Fairchild μA4136, Raytheon RC4136, or equivalent quad opamp. Overall response is 1.5 dB down at cutoff frequency, about 36 dB down at twice cutoff, and about 60 dB down at three times cutoff. R7 and R8 provide pseudoground of half supply voltage, to eliminate need for negative supply. Will operate with supply from 6 to 36 V, drawing about 7 mA.—T. Berg, Active Low-Pass Filters for CW or SSB, *QST*, Aug. 1977, p 40–41.

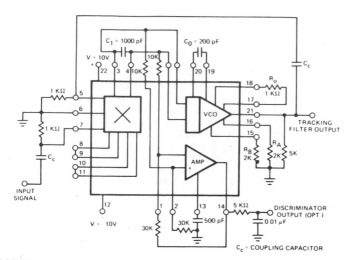

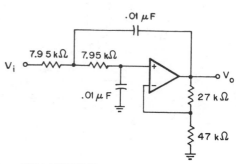

1-MHz TRACKING FILTER—Exar XR-S200 PLL IC is connected to function as frequency filter when phase-locked loop locks on input signal, to produce filtered version of input signal frequency at VCO output. Because circuit can track input over 3:1 range of frequencies around free-running frequency of VCO, it is known as tracking filter. Optional wideband discriminator output is also provided.—"Phase-Locked Loop Data Book," Exar Integrated Systems, Sunnyvale, CA, 1978, p 9–16.

2-kHz LOW-PASS—Voltage-controlled voltage-source filter uses equal-value input resistors and equal-value capacitors, simplifying selection of components. Equation for cutoff frequency then simplifies to $f = 1/6.28RC$ or $1/(6.28)(7950)(0.01)(10^{-6})$. Opamp can be 741.—H. M. Berlin, "Design of Active Filters, with Experiments," Howard W. Sams, Indianapolis, IN, 1977, p 85–86.

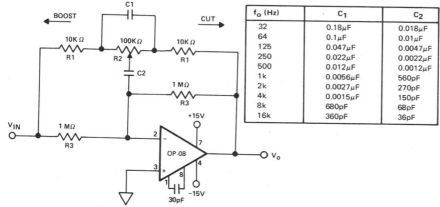

f_o (Hz)	C_1	C_2
32	0.18μF	0.018μF
64	0.1μF	0.01μF
125	0.047μF	0.0047μF
250	0.022μF	0.0022μF
500	0.012μF	0.0012μF
1k	0.0056μF	560pF
2k	0.0027μF	270pF
4k	0.0015μF	150pF
8k	680pF	68pF
16k	360pF	36pF

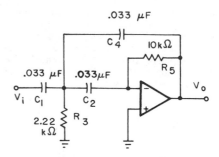

OCTAVE AUDIO EQUALIZER—R2 provides up to 12-dB boost or cut at center frequency determined by values of C1 and C2 as given in table. Uses Precision Monolithics OP-08 opamp. Low input bias current of opamp permits scaling resistors up by factor of 10, to reduce values of C1 and C2 at low-frequency end. Same circuit is used for all 10 sections of equalizer, which together draw only 6 mA maximum from supply.—"Precision Low Input Current Op Amp," Precision Monolithics, Santa Clara, CA, 1978, OP-08, p 7.

1-kHz HIGH-PASS UNITY-GAIN—Passband gain of 741 or equivalent opamp circuit is set by ratio of C_4 to C_i rather than by resistors. Values shown give unity gain for passband above 1-Hz cutoff. Circuit uses multiple feedback.—H. M. Berlin, "Design of Active Filters, with Experiments," Howard W. Sams, Indianapolis, IN, 1977, p 100–102.

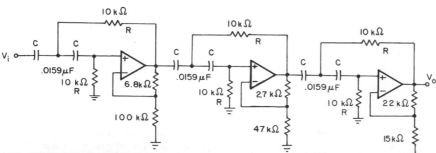

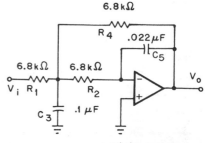

1-kHz SIXTH-ORDER HIGH-PASS—Formed by cascading three different second-order sections. Passband gain is 12.5 dB. Opamps can be 741 or equivalent. Used when high rejection is needed for signals just below passband, in application where such rejection justifies cost of extra filter sections.—H. M. Berlin, "Design of Active Filters, with Experiments," Howard W. Sams, Indianapolis, IN, 1977, p 122–125.

500-Hz LOW-PASS UNITY-GAIN—Multiple-feedback filter using 741 or equivalent opamp has unity gain in passband below 500-Hz cutoff. Resistors can be 5% tolerance.—H. M. Berlin, "Design of Active Filters, with Experiments," Howard W. Sams, Indianapolis, IN, 1977, p 99–100.

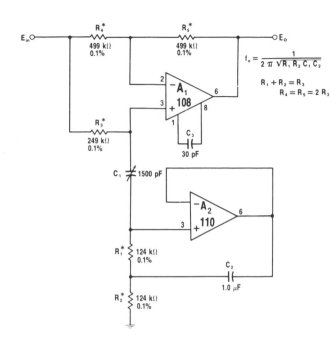

$$f_o = \frac{1}{2\pi \sqrt{R_1 R_2 C_1 C_2}}$$

$$R_1 + R_2 = R_3$$
$$R_4 = R_5 = 2R_3$$

AF NOTCH—Notch frequency is easily tuned at frequencies below 1 kHz with single capacitor C_1 or by replacing R_1 and R_2 with 249K pot. For higher frequencies, use 118 opamp for A_1 and 5K for R_3 while lowering other resistances in proportion to R_3. Indicated resistance tolerances are necessary for optimum notch depth.—W. G. Jung, "IC Op-Amp Cookbook," Howard W. Sams, Indianapolis, IN, 1974, p 340–341.

1.4-kHz TWIN-T BANDPASS—Combination of passive twin-T bandpass filter and 741 opamp gives simple audio filter for amplifying narrow frequency band (about 300 Hz wide) centered on 1.4 kHz. Filter can be tuned to other frequencies by replacing R1 and R2 with 10K pots. Frequency is equal to 1/6.28RC where R is value in ohms of R1 and R2 and C is capacitance in farads of C1 and C2. R3 is half of R1.—F. M. Mims, "Integrated Circuit Projects, Vol. 2," Radio Shack, Fort Worth, TX, 1977, 2nd Ed., p 71–80.

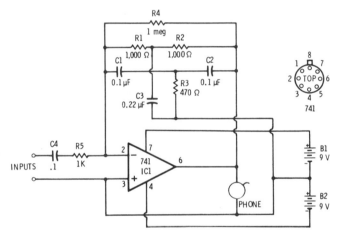

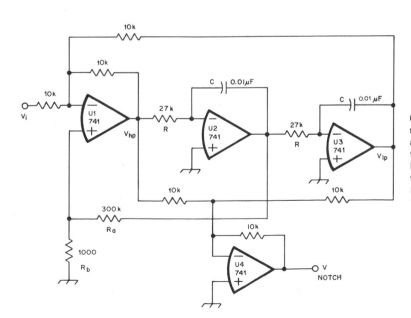

600-Hz NOTCH—With values obtained from design equations and graph in article, state-variable or universal filter provides Q of 100 with four opamps. Notch filter is achieved by adding low-pass and high-pass outputs equally, for feed to dual-input summing amplifier.—H. M. Berlin, The State-Variable Filter, *QST*, April 1978, p 14–16.

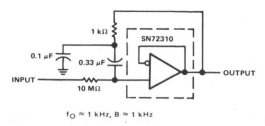

$f_O \approx 1$ kHz, $B \approx 1$ kHz

1-kHz BANDPASS—Simple circuit using voltage-follower opamp provides bandpass of 1 kHz centered on 1 kHz, to give output range of 500– 1500 Hz.—"The Linear and Interface Circuits Data Book for Design Engineers," Texas Instruments, Dallas, TX, 1973, p 4-39.

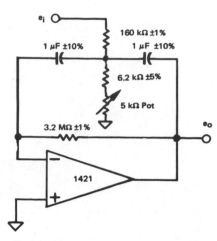

1-Hz BANDPASS—Single pot provides easy trimming to exact center frequency desired without change in bandwidth or gain. Q is 10. Design equations are given.—R. A. Pease, "Band-Pass Active Filter with Easy Trim for Center Frequency," Teledyne Philbrick, Dedham, MA, 1972, Applications Bulletin 4.

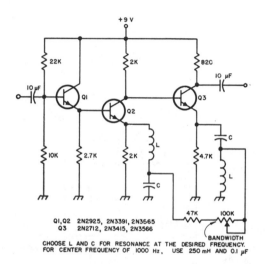

Q1,Q2 2N2925, 2N3391, 2N3565
Q3   2N2712, 2N3415, 2N3566

CHOOSE L AND C FOR RESONANCE AT THE DESIRED FREQUENCY. FOR CENTER FREQUENCY OF 1000 Hz, USE 250 mH AND 0.1 μF

1-kHz BANDPASS—Three-stage audio filter uses two series resonant circuits to give very narrow audio passband. Amount of feedback determines Q and bandwidth.—*Circuits, 73 Magazine,* March 1974, p 89.

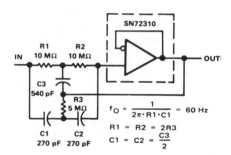

$$f_O = \frac{1}{2\pi \cdot R1 \cdot C1} = 60 \text{ Hz}$$

$$R1 = R2 = 2R3$$

$$C1 = C2 = \frac{C3}{2}$$

60-Hz HIGH-Q NOTCH—Input network for SN72310 voltage-follower opamp provides attenuation of 60-Hz power-line frequency. Use high-quality capacitors for maximum Q.—"The Linear and Interface Circuits Data Book for Design Engineers," Texas Instruments, Dallas, TX, 1973, p 4-39.

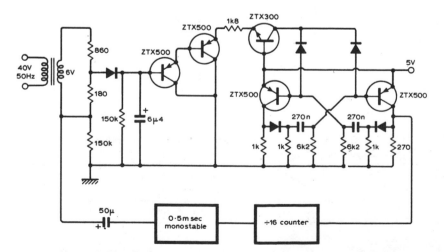

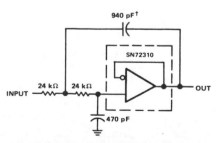

CLOCK FOR COMMUTATING RC FILTER—Circuit synchronizes multivibrator with line frequency to provide clock waveform required for switching capacitors electronically in n-path active filter for rejecting line frequency and harmonics up to fifth. Article gives complete circuit of active filter and describes operation. Clock serves to switch 16 MOSFETs on in turn for commutating 16 capacitors electronically at 16 times line frequency.—K. F. Knott and L. Unsworth, Mains Rejection Tracking Filter, *Wireless World,* Oct. 1974, p 375–379.

LOW-PASS WITH 10-kHz CUTOFF—Simple circuit uses only one Texas Instruments SN72310 voltage-follower opamp. For good temperature stability, use silvered mica capacitors.—"The Linear and Interface Circuits Data Book for Design Engineers," Texas Instruments, Dallas, TX, 1973, p 4-39.

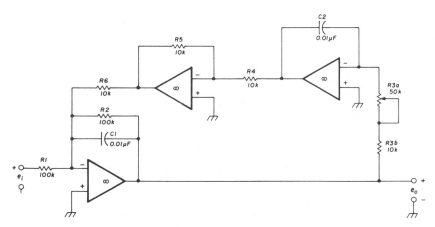

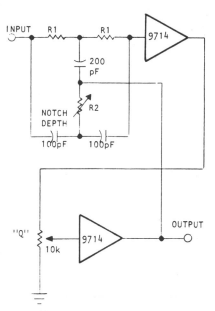

1-kHz CASCADED-OPAMP BANDPASS—Bandwidth of only 71 Hz is achieved by using four identical three-opamp filters in series, for increased selectivity in communication or RTTY receiver. Values for R1 and R2 are made variable for first filter stage, but pot is used for R3 in all four stages so they can be tuned to same center frequency. Batteries are used for supply, as filter draws only 17 mA.—F. M. Griffee, RC Active Filters Using Op Amps, *Ham Radio*, Oct. 1976, p 54–58.

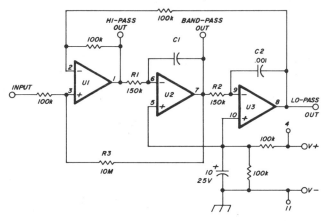

1-kHz THREE-FUNCTION—Uses National LM324 quad opamp, with appropriate biasing for single supply of +5 to +25 VDC. Values of R1 and R2 establish f_c at 1000 Hz, while R3 gives Q of 50. Values of R1 and R2 for other bandpass center and cutoff frequencies f_c can be calculated from $R = 15 \times 10^7/f_c$. Fourth opamp may be used as output amplifier or for summing high-pass and low-pass outputs. C1 is same as C2.—P. A. Lovelock, Discrete Operational Amplifier Active Filters, *Ham Radio*, Feb. 1978, p 70–73.

19-kHz NOTCH—Used in commercial FM transmitters to eliminate 19-kHz program material from stereo encoder. Uses Optical Electronics 9714 opamp in circuit that gives unity passband gain below center frequency, 0.7 gain above center frequency, and less than 0.001 gain at notch frequency. Provides adjustments for notch rejection level and Q. R1 is 84K, and R2 is 36K in series with 10K pot.—"Precision Notch Filter," Optical Electronics, Tucson, AZ, Application Tip 10255.

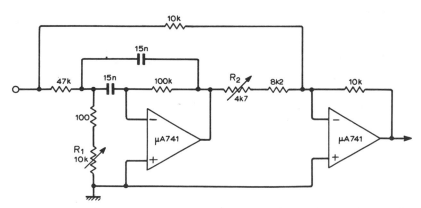

TUNABLE NOTCH—Opamp circuit requires only one pot (R_1) to vary notch frequency. R_2 is used to set noise rejection to maximum. With values shown, filter tunes from 170 Hz to 3 kHz, with 3-dB bandwidth of 230 Hz and notch rejection better than 40 dB over entire range. Circuit can be voltage-tuned by replacing R_1 with FET operated as voltage-variable resistor.—R. J. Harris, Simple Tunable Notch Filter, *Wireless World*, May 1973, p 253.

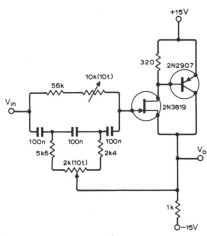

TUNABLE NOTCH FILTER—Simple pot-tuned active notch filter has tuning range of 200 Hz in audio band and 3-dB rejection bandwidth of 10 Hz, as required for tuning out whistle or power-line hum that is interfering with radio program. Article gives design theory for many other types of notch filters.—Y. Nezer, Active Notch Filters, *Wireless World*, July 1975, p 307–311.

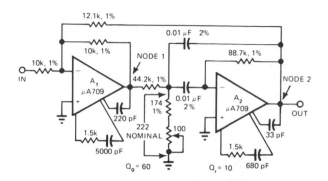

Q MULTIPLIER—Article gives design procedure and equations for utilizing Q multiplication to simplify circuit for active bandpass filter. With values shown, center frequency is 3.6 kHz and Q of 10 is multiplied by gain of 6 to give effective Q of 60.—A. B. Williams, Q-Multiplier Techniques Increase Filter Selectivity, *EDN Magazine,* Oct. 5, 1975, p 74 and 76.

1-kHz FIFTH-ORDER LOW-PASS—Uses single first-order section and two different second-order sections to give passband gain of 10.3 dB. Opamps can be 741 or equivalent.—H. M. Berlin, "Design of Active Filters, with Experiments," Howard W. Sams, Indianapolis, IN, 1977, p 119–122.

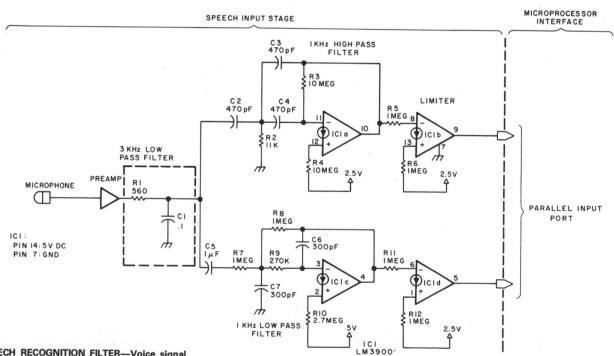

SPEECH RECOGNITION FILTER—Voice signal picked up by microphone is preamplified and sent through 3-kHz low-pass passive filter C1-R1 to 1-kHz high-pass active filter and 1-kHz low-pass active filter using sections of LM3900 quad opamp. Diode symbols on opamps indicate use of current mirrors for noninverting inputs. Outputs are sampled about 60 times per second to implement speech recognition algorithm of computer, which counts number of high-pass and low-pass zero crossings per second and compares results with series of word models in memory to determine most likely match.—J. R. Boddie, Speech Recognition for a Personal Computer System, *BYTE,* July 1977, p 64–68 and 70–71.

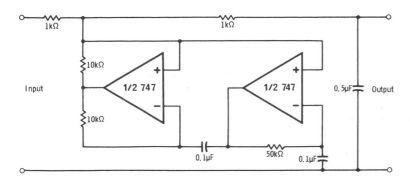

320-Hz LOW-PASS—Frequency-dependent negative-resistance circuit uses 747 dual opamp. Signal source used as input should have low resistance, and load should have high resistance. Voltage-follower stages can be used to isolate both input and output of filter.—R. Melen and H. Garland, "Understanding IC Operational Amplifiers," Howard W. Sams, Indianapolis, IN, 2nd Ed., 1978, p 104–105.

VOLTAGE-TUNED STATE-VARIABLE—Provides choice of high-pass, bandpass, and low-pass outputs, each with cutoff frequency variable between 1 and 6 kHz by varying control voltage between −10 V and +15 V. Output load resistor sets voltage gain between input and output. Gain-control input varies gain from maximum set by load resistor down to zero. Input signals must be limited to 100 mV because input circuit is differential amplifier operating without feedback.—D. Lancaster, "Active-Filter Cookbook," Howard W. Sams, Indianapolis, IN, 1975, p 203–205.

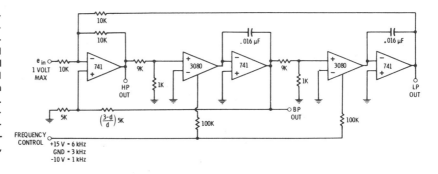

1-kHz THIRD-ORDER LOW-PASS—Circuit using 741 or equivalent opamp consists of unity-gain first-order section followed by equal-component voltage-controlled voltage-source second-order section.—H. M. Berlin, "Design of Active Filters, with Experiments," Howard W. Sams, Indianapolis, IN, 1977, p 113–114.

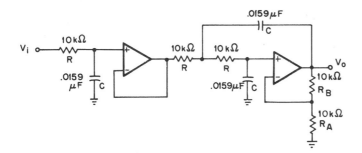

1-kHz BIQUAD BANDPASS—Three 741 opamps are connected to give two integrators and inverter. Overall gain is −Q, determined by value of input resistor used. Circuit is tuned by varying capacitors in steps. Absolute bandwidth remains constant as frequency changes. Chief applications are in telephone systems, where identical absolute-bandwidth channels are required.—D. Lancaster, "Active-Filter Cookbook," Howard W. Sams, Indianapolis, IN, 1975, p 159–164.

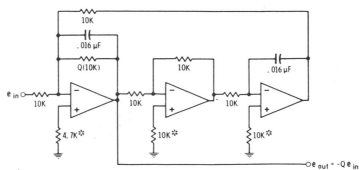

*offset resistors may be replaced with shorts in noncritical circuits.

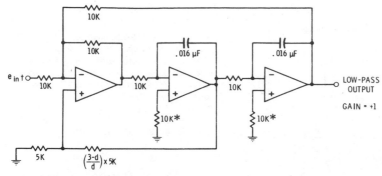

† must return to ground via low-impedance dc path.

* optional offset compensation, may be replaced with short in noncritical circuits.

1-kHz STATE-VARIABLE—Circuit using three 741 opamps offers low sensitivity, voltage-controlled tuning, and easy conversion to high pass or bandpass. For high pass, take output from first opamp. For bandpass, take output from second opamp. To increase frequency, change 10K resistors to identical higher values. 10:1 resistance change produces 10:1 frequency change. Damping d is adjustable; critical value of 1.414 gives maximum flatness of response without overshoot. Design equations are given.—D. Lancaster, "Active-Filter Cookbook," Howard W. Sams, Indianapolis, IN, 1975, p 129–135.

VARIABLE-BANDWIDTH AF—Audio filter using 1000-Hz Wien bridge provides bandwidths from 70 to 600 Hz. Transistors can be SK3004, GE-2, or HEP-254.—Circuits, *73 Magazine,* Jan. 1974, p 124.

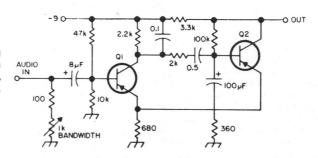

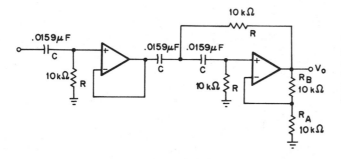

1-kHz THIRD-ORDER HIGH-PASS—Passband gain is 6 dB for Butterworth filter above 1-kHz cutoff. Damping factor is 1.000 for both sections, each using 741 or equivalent opamp.—H. M. Berlin, "Design of Active Filters, with Experiments," Howard W. Sams, Indianapolis, IN, 1977, p 115–116.

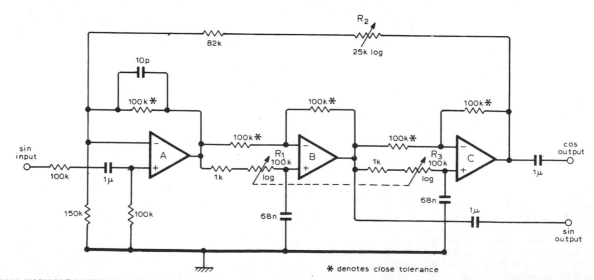

* denotes close tolerance

20–2000 Hz VARIABLE BANDPASS—High-Q active bandpass filter can be adjusted over wide frequency range (100:1) while maintaining Q essentially constant over 100. Two-phase output is available. Opamps can be 741 or equivalent. Cascaded all-pass networks B and C each have 0 to 180° phase variation and unity gain at all frequencies. These are driven by opamp A whose feedback signal is sum of input and output of all-pass networks. R_2 adjusts Q, and ganged log pots change center frequency.—J. M. Worley, Variable Band-Pass Filter, *Wireless World,* April 1977, p 61.

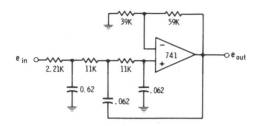

250-Hz THIRD-ORDER LOW-PASS—Values shown place cutoff at 250 Hz, with 1-dB dip in response curve. Input must be returned to ground with low-impedance DC path.—D. Lancaster, "Active-Filter Cookbook," Howard W. Sams, Indianapolis, IN, 1975, p 146.

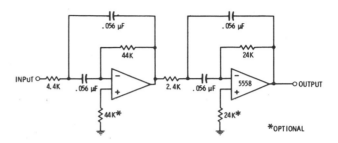

200–400 Hz PASSBAND—Design is based on use of 3.2 for value of Q, to hold passband dip at 1 dB for two-pole filter. Multiple feedback is used for each pole. First opamp can be 741 or equivalent. Center frequency is 283 Hz.—D. Lancaster, "Active-Filter Cookbook," Howard W. Sams, Indianapolis, IN, 1975, p 166.

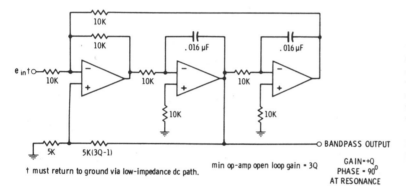

† must return to ground via low-impedance dc path.

min op-amp open loop gain = 3Q

GAIN = +Q
PHASE = 90°
AT RESONANCE

1-kHz STATE-VARIABLE BANDPASS—With three 741 opamps or equivalent, circuit gain is Q (reciprocal of damping). Frequency is changed by changing 10K coupling resistors between opamps while keeping their values equal. Increasing resistors 10 times increases frequency 10 times. High-pass output is obtained from first opamp and low pass from second opamp.—D. Lancaster, "Active-Filter Cookbook," Howard W. Sams, Indianapolis, IN, 1975, p 156–159.

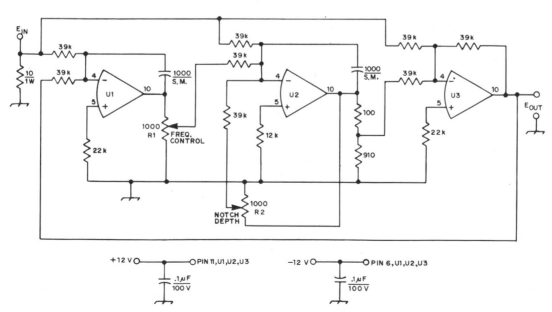

AF NOTCH—Center frequency of notch can be varied with single control R1; upper limit is about 4 kHz. Circuit Q and notch depth are constant over range. R2 is adjusted initially for best notch depth. All opamps are 741 14-pin DIP, such as Motorola MC1741L. U1 and U2 are integrators with DC gain of about 2500, and U3 is summing device. Notch depth is at least 50 dB. Input to filter is taken from loudspeaker or headphone jack of receiver. High-impedance headset may be connected directly across output, or buffer stage can be added to drive lower-impedance loudspeaker or headset. Use with AGC off.—A. Taflove, An Analog-Computer-Type Active Filter, *QST*, May 1975, p 26–27.

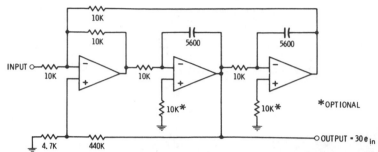

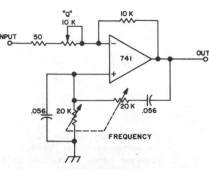

3-kHz STATE-VARIABLE BANDPASS—Design is based on value of 30 for Q, corresponding to 0.033 for damping factor d. Opamps can be 741.—D. Lancaster, "Active-Filter Cookbook," Howard W. Sams, Indianapolis, IN, 1975, p 166.

VARIABLE Q AND FREQUENCY—Bandwidth can be made extremely sharp (less than 9 Hz) or very broad (greater than 300 Hz). Adjusting Q to change bandwidth also changes gain of filter. Center frequency of filter is independently adjustable.—A. F. Stahler, An Experimental Comparison of CW Audio Filters, *73 Magazine*, July 1973, p 65–70.

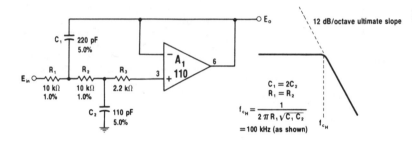

100-kHz LOW-PASS UNITY-GAIN—Opamp serves as active element in voltage-controlled voltage-source second-order filter. Other opamps having required high input resistance, low input current, and high speed are 1556 and 8007.—W. G. Jung, "IC Op-Amp Cookbook," Howard W. Sams, Indianapolis, IN, 1974, p 331–333.

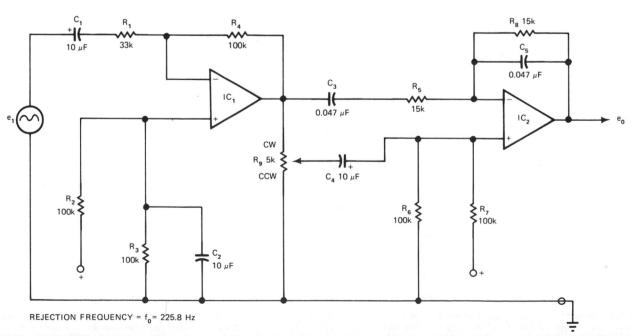

REJECTION FREQUENCY = f_0 = 225.8 Hz

225.8-Hz REJECTION—Provides extremely sharp adjustable-depth notch with only two low-gain opamps. Suitable for single-ended supplies. Article gives equation for transfer function.—R. Carter, Sharp Null Filter Utilizes Minimum Component Count, *EDN Magazine*, Sept. 20, 1976, p 110.

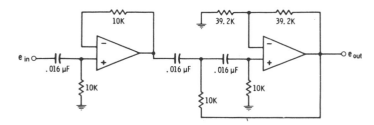

1-kHz HIGH-PASS FLATTEST-RESPONSE—Values are chosen for flattest possible response obtainable with third-order configuration of two 741 opamps. Tripling capacitance values cuts cutoff frequency by one-third and vice versa. Component tolerance can be 10%. Gain is 2.—D. Lancaster, "Active-Filter Cookbook," Howard W. Sams, Indianapolis, IN, 1975, p 186.

2.4-kHz LOW-PASS/HIGH-PASS—Three 741 opamps are connected to provide separate low-pass and high-pass outputs simultaneously for complex synthesis problem requiring state-variable filter. Gain is 1.—D. Lancaster, "Active-Filter Cookbook," Howard W. Sams, Indianapolis, IN, 1975, p 192–193.

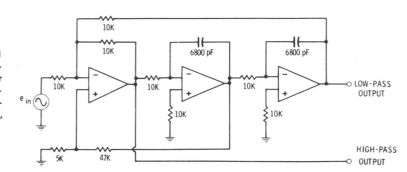

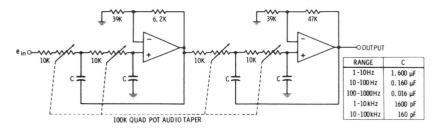

TUNABLE FOURTH-ORDER LOW-PASS—Use of four ganged pots permits varying cutoff frequency over 10:1 range. Table gives ranges obtainable with five different values for C. Opamps can be 741 or equivalent. Tracking of 5% for pots calls for expensive components, but ordinary snap-together pots may prove satisfactory if tuning range is restricted to 3:1 or less and more capacitor switching is used.—D. Lancaster, "Active-Filter Cookbook," Howard W. Sams, Indianapolis, IN, 1975, p 195–197.

RANGE	C
1-10 Hz	1.600 µF
10-100 Hz	0.160 µF
100-1000 Hz	0.016 µF
1-10 kHz	1600 pF
10-100 kHz	160 pF

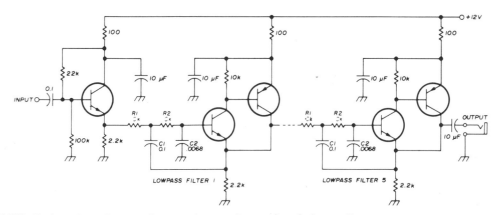

AF LOW-PASS FOR CW—Design using 10% tolerance components gives sufficiently wide bandwidth while maintaining steep skirt response for CW reception in direct-conversion communication receiver. Filter has five identical three-transistor sections, each peaked at cutoff frequency. Q of each section is about 1.9, which gives 6-dB bandwidth of about 200 Hz. With center frequency at 540 Hz, attenuation is 75 dB at 1200 Hz. Net gain of system is 28 dB at resonance. NPN transistors are 2N3565, 2N3904, or similar; PNP transistors are 2N3638, 2N3906, or similar.—W. Howard, Simple Active Filters for Direct-Conversion Receivers, *Ham Radio*, April 1974, p 12–15.

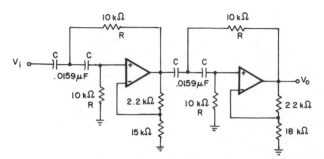

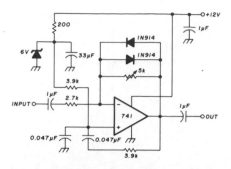

1-kHz FOURTH-ORDER HIGH-PASS—First section is second-order high-pass filter having gain of 1.2 dB, and second section has gain of 7 dB. Opamps are 741 or equivalent.—H. M. Berlin, "Design of Active Filters, with Experiments," Howard W. Sams, Indianapolis, IN, 1977, p 116–117.

800-Hz BANDPASS—Active filter has 800-Hz center frequency for optimum CW reception. Bandwidth is adjustable. Back-to-back diodes provide noise-limiting capability.—U. L. Rohde, IF Amplifier Design, *Ham Radio,* March 1977, p 10–21.

1.8–1.9 MHz BANDPASS—Butterworth bandpass filter, suitable for use with broadband preamp, helps reject out-of-band signals. Filter also protects preamp from signals across response range from broadcast band through VHF. C1 and C2 are mica trimmers. L1 and L2 have 30 turns No. 22 enamel on Amidon T68-6 toroid cores to give 5.1 µH.—D. DeMaw, Beat the Noise with a "Scoop Loop," *QST,* July 1977, p 30–34.

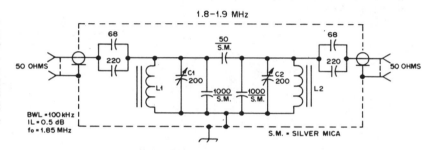

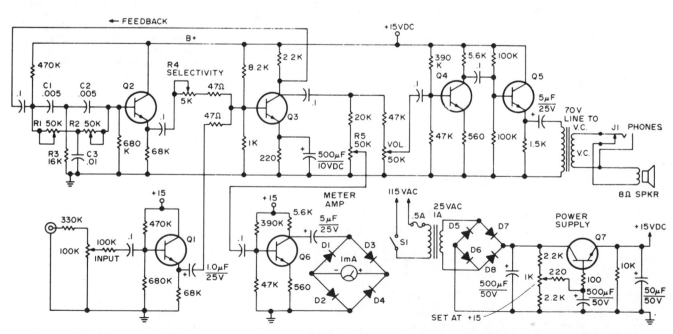

VARIABLE-Q AF—Consists of tuned amplifier having inverse feedback, connected so bandwidth at −6 dB is variable from 50 to 400 Hz for center frequency of 1 kHz. Improves selectivity of amateur receivers. Audio from receiver is applied to inverter Q3 through Q1. Part of inverted signal is fed back to twin-T network R1-R2-R3-C1-C2-C3 which has high impedance to ground except at its resonant frequency. Unattenuated signal goes through Q2 for adding to uninverted output at base of Q3. Degree of cancellation by two out-of-phase signals feeding Q3 is controlled by R4 to adjust selectivity. Filtered output is boosted by Q4 and Q5. Article covers construction, calibration, and operation. Q1-Q6 are GE-20, and Q7 is GE-14 or GE-28.—C. Townsend, A Variable Q Audio Filter, *73 Magazine,* Feb. 1974, p 54–56.

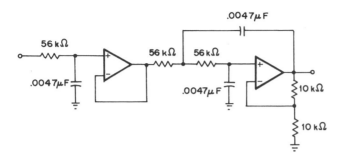

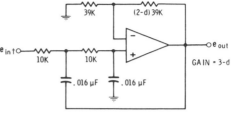

† must return to ground via low-impedance dc path.

600-Hz THIRD-ORDER LOW-PASS—Butterworth filter using 741 or equivalent opamp provides gain of 6 dB in passband below 600-Hz cutoff. All components can be 5% tolerance.—H. M. Berlin, "Design of Active Filters, with Experiments," Howard W. Sams, Indianapolis, IN, 1977, p 114–116.

SECOND-ORDER 1-kHz LOW-PASS—Circuit using 741 opamp has equal-value series input resistors and high-pass capacitors. Cutoff frequency can be increased by changing 10K resistors to higher values while keeping their values identical. 10:1 resistance change provides 10:1 frequency change. Damping d is adjustable; critical value of 1.414 gives maximum flatness of response without overshoot. Interchange 10K resistors and 0.016-μF capacitors to convert circuit to 1-kHz high-pass filter.—D. Lancaster, "Active-Filter Cookbook," Howard W. Sams, Indianapolis, IN, 1975, p 127–129.

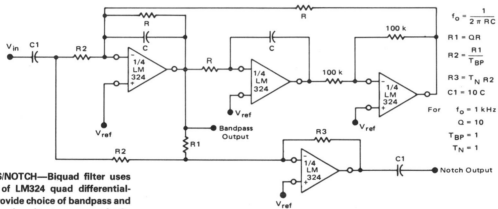

1-kHz BANDPASS/NOTCH—Biquad filter uses all four sections of LM324 quad differential-input opamp to provide choice of bandpass and notch outputs. Supply voltage range can be 3–32 V, with reference voltage equal to half of supply value used. For center frequency of 1 kHz, R is 160K, C is 0.001 μF, and R1-R3 are 1.6 megohms. Coupling capacitors C1 can be 10 times value used for C.—"Quad Low Power Operational Amplifiers," Motorola, Phoenix, AZ, 1978, DS 9339 R1.

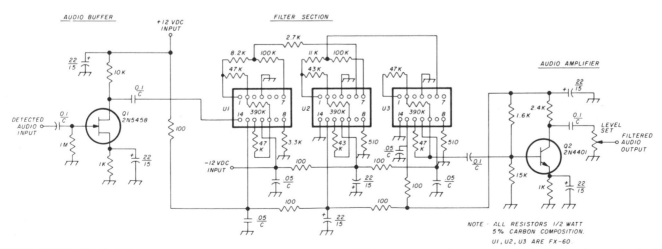

BANDPASS FOR CW—Sophisticated audio processing system for CW bandpass, using communication receiver, has actual bandpass center frequency between 900 and 950 Hz. Bandwidth is about 150 Hz. Minimum relative attenuation is above 20 dB. Uses three Kinetic Technology FX-60 ICs (culled from FS-60, FS-65, and FS-61 production by manufacturer).—M. A. Chapman, Audio Filters for Improving SSB and CW Reception, *Ham Radio,* Nov. 1976, p 18–23.

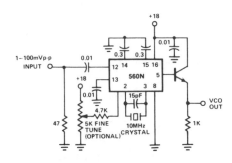

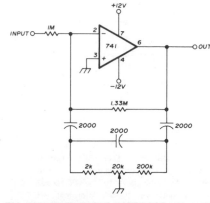

10-MHz TRACKING—Connection shown for 560N PLL tracking filter uses crystal to keep free-running frequency at desired value for signals near 10 MHz. Lock range varies with input amplitude, from about 0.3 kHz for 1 mV P-P input to about 3 kHz for 100 mV P-P.—"Signetics Analog Data Manual," Signetics, Sunnyvale, CA, 1977, p 850–851.

700–2000 Hz TUNABLE BANDPASS—Uses RC notch circuit as feedback element for active-filter opamp. With tuning pot set for center frequency of 1000 Hz, 3-dB bandwidth is 23 Hz and 10-dB bandwidth is 68 Hz. At 1000 Hz, voltage gain is 36 dB. High-frequency rolloff is good, being 43 dB down at 2000 Hz, so circuit converts 1000-Hz square wave into sine wave. Article gives design equations.—C. Hall, Tunable RC Notch Filter, *Ham Radio*, Sept. 1975, p 16–20.

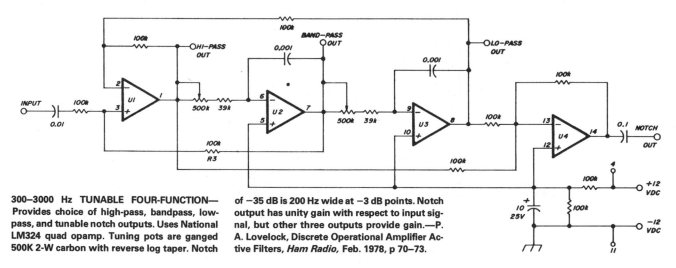

300–3000 Hz TUNABLE FOUR-FUNCTION—Provides choice of high-pass, bandpass, low-pass, and tunable notch outputs. Uses National LM324 quad opamp. Tuning pots are ganged 500K 2-W carbon with reverse log taper. Notch of −35 dB is 200 Hz wide at −3 dB points. Notch output has unity gain with respect to input signal, but other three outputs provide gain.—P. A. Lovelock, Discrete Operational Amplifier Active Filters, *Ham Radio*, Feb. 1978, p 70–73.

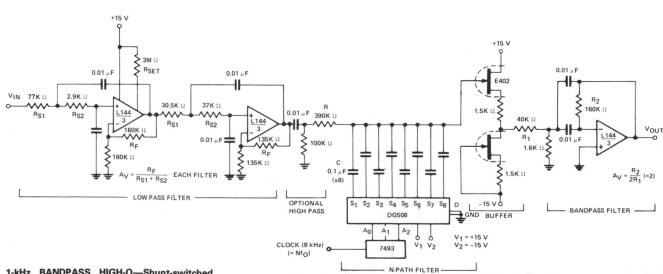

1-kHz BANDPASS HIGH-Q—Shunt-switched bandpass filter with Q of 1000 and voltage gain of about 7 uses DG508 CMOS multiplexer containing required analog switches, interface circuits, and decode logic for 8-path filter. Bandwidth for 3 dB down is 1 Hz centered on 1 kHz, with asymptotic slope of 6 dB per octave. Clock controls shunt-switched filter action.—"Analog Switches and Their Applications," Siliconix, Santa Clara, CA, 1976, p 5-12–5-14.

CHAPTER 28
Filter Circuits—Passive

Includes low-pass, high-pass, and bandpass AF filters for improving reception of voice, CW, SSB, and RTTY signals, along with higher-frequency circuits for suppressing broadcast-band interference in communication receivers and minimizing other types of interference.

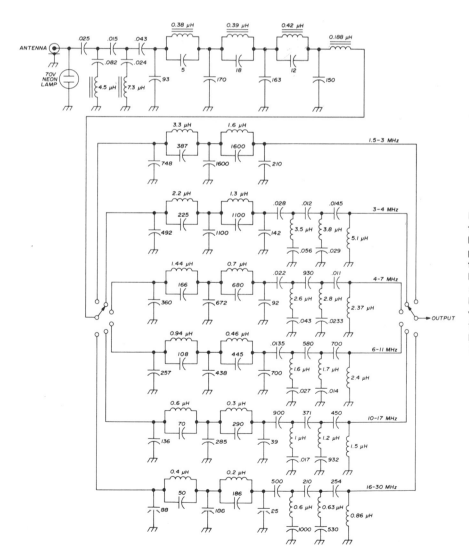

ELLIPTIC HIGH-PASS/LOW-PASS—Covers 1.45 to 32 MHz in six steps, for use at front end of high-frequency communication receiver to suppress unwanted broadcast signals. Low-pass filter section, acting with one of six following low-pass sections, gives over 90-dB image suppression. Special Bessel-Cauer elliptic filter having Chebyshev response in passband provides required 50-ohm impedance matching so filters can be cascaded.—U. L. Rohde, Optimum Design for High-Frequency Communications Receivers, *Ham Radio,* **Oct. 1976, p 10–25.**

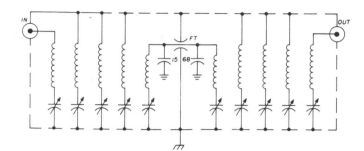

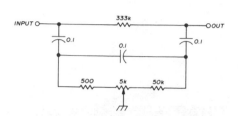

50.5-MHz BANDPASS—Provides 60% bandwidth with only 4-dB insertion loss. Each coil is about 2.2 μH, and trimmer capacitors are 1.5–7 pF. Sweep signal generator and 5-in CRO are essential for alignment.—P. H. Sellers, 50-MHz Bandpass Filter, *Ham Radio*, Aug. 1976, p 70–71.

60-Hz TUNABLE NOTCH—Can be used to minimize hum pickup from AC line. Circuit tunes from 40 to 120 Hz with single pot. Article gives design equations. With unmeasured ceramic disk capacitors and 5% resistors, notch depth at 60 Hz was 44.5 dB. By selecting capacitors with equal values and replacing 333K with 500K trimpot, careful adjustment increases notch depth to 57 dB.—C. Hall, Tunable RC Notch Filter, *Ham Radio*, Sept. 1975, p 16–20.

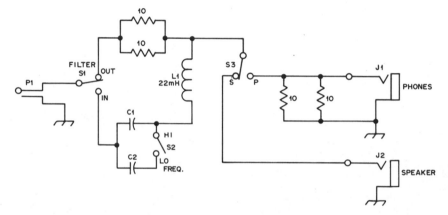

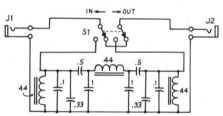

CW FILTER FOR INTERFERENCE—Audio bandpass filter, designed for connection between loudspeaker jack of receiver and external loudspeaker or phone, has half-power bandwidth of about 70 Hz but rolls off gradually without causing ringing. Series LC combination, connected in hot line to loudspeaker, looks like 5-ohm resistance at resonance, cutting signal amplitude about in half. At lower frequencies filter looks like large capacitive reactance and at higher frequencies it resembles large inductive reactance, both causing high attenuation. Filter thus discriminates against all except switch-selected resonant frequency, either 760 or 1070 Hz. Choose best frequency for particular receiving situation by trial.—F. Noble, A Passive CW Filter to Improve Selectivity, *QST*, Nov. 1977, p 34–35.

BANDPASS FOR CW—Provides bandwidth of about 400 Hz (3 dB down) centered on 875 Hz, for improving reception of CW signals with amateur receiver. Uses three 44-mH toroids.—D. C. Rife, Low-Loss Passive Bandpass CW Filters, *QST*, Sept. 1971, p 42–44.

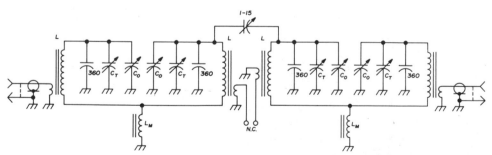

| L | 33 turns no. 20 enamelled on Amidon T-106-2 toroid cores (approximately 14 μH). Links are each 2 turns | C_o | 4-section air variable, 10-160 pF per section |
| L_m | 6 turns no. 20 enamelled on Amidon T-30-2 toroid core | C_T | 35-pF trimmer capacitors |

160-METER BANDPASS—Four-resonator filter is tunable from 1.8 to 2 MHz and has insertion loss of 5 dB. 3-dB bandwidth is 30 kHz, and 6-dB shape factor is 4.78. Stopband attenuation is over 120 dB. Key to high performance is use of high-Q toroid cores. Article covers theory, construction, and adjustment.—W. Hayward, Bandpass Filters for Receiver Preselectors, *Ham Radio*, Feb. 1975, p 18–27.

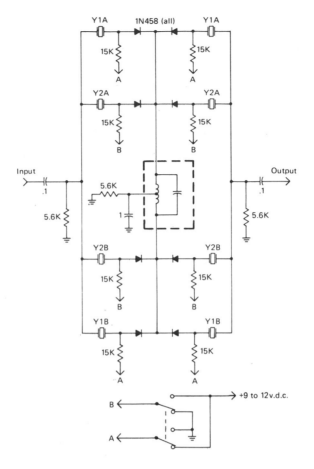

DIODE-SWITCHED FOUR-CRYSTAL IF FILTER— Application of 9–12 VDC to control points A or B gives choice of two different selectivities for IF amplifier in amateur communication receiver. For 500-Hz bandwidth at 455 kHz, frequencies of crystals in use should be 300 Hz apart for CW, 1.8 kHz apart for 2.7-kHz SSB bandwidth, and 1.25 kHz apart for 2.1-kHz SSB bandwidth. Article gives design graphs.—J. J. Schultz, Economical Diode-Switched Crystal Filters, *CQ*, July 1978, p 33–35 and 91.

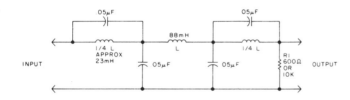

2125-Hz LOW-PASS—Used with AFSK keyer to convert 2125-Hz square wave to sine wave by removing third and fifth harmonics. All three coils are toroids, with its two windings in series for 88 mH and in parallel for 23 mH.—L. J. Fox, **Dodge That Hurricane!**, *73 Magazine*, Jan. 1978, p 62–69.

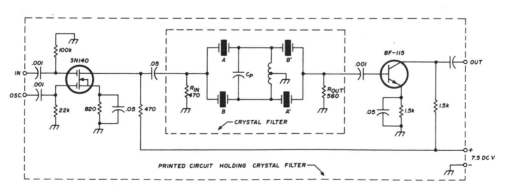

FOUR-CRYSTAL FILTER—Uses two matched sets of crystals, with each pair having maximum frequency difference of 25 Hz. Transistors serve as input and output isolating stages. Each matched pair, such as A-A', should be from same manufacturer and have same nominal parallel capacitance for circuit, same activity, and same resonant frequency within 25 Hz. Article gives detailed instructions for grinding crystal to increase resonant frequency when necessary for matching. Use frequency counter for checking frequency. Values given in circuit are for 5.645-MHz crystal filter with −6 dB bandpass of 1.82 kHz and insertion loss of about 5 dB. Crystals used are 5.644410 MHz and 5.644416 MHz for A and A', and 5.645627 MHz and 5.645641 MHz for B and B'. Coil has 7 + 7 turns No. 28 enamel bifilar wound on 10.7-MHz IF transformer having 2.4-mm slug diameter. C_p is 39 to 47 pF.—J. Perolo, Practical Considerations in Crystal-Filter Design, *Ham Radio*, Nov. 1976, p 34–38.

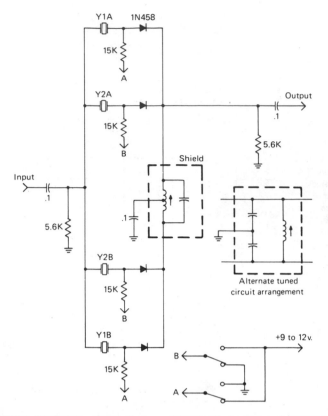

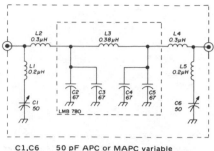

C1,C6   50 pF APC or MAPC variable

C2,C3   67 pF ±5%, 7500 working volts dc

C4,C5   (Centralab type 850S ceramic capacitor, 6 for $1.00 from John Meshna, P.O. Box 62, E. Lynn, Massachusetts 01904)

L1,L5   0.2 μH, 3 turns no. 16 or no. 14 enamelled, ½ inch (13mm) ID, spaced 1/8 inch (3mm) per turn

L2,L4   0.3 μH, 5½ turns no. 14 enamelled, ½ inch (13mm) ID, spaced 1/8 inch (3mm) per turn

L3   0.38 μH, 7 turns no. 14 enamelled, ½ inch (13mm) ID, spaced 1/8 inch (3mm) per turn

LOW-PASS WITH 42.5-MHz CUTOFF—Designed for insertion in antenna coax of amateur radio station up to 1 kW, to cure TVI problems. Provides 60-dB attenuation on channel 2. Filter uses m-derived terminating half-sections at each end, with two constant-K midsections. End sections are tuned either to channel 2 (55 MHz) or channel 3 (61 MHz). Article covers construction and tune-up.—N. Johnson, High-Frequency Lowpass Filter, *Ham Radio,* March 1975, p 24–27.

DIODE-SWITCHED CRYSTALS—1N458 diodes switch crystals in pairs to provide two different degrees of selectivity for 455-kHz IF filter. For 500-Hz bandwidth in amateur communication receiver, spacing between crystal frequencies should be 300 Hz, which is obtained with 455.150 kHz for Y1A and 454.850 kHz for Y1B. Provides adequate CW selectivity for transceiver having good SSB filter.—J. J. Schultz, Economical Diode-Switched Crystal Filters, *CQ,* July 1978, p 33–35 and 91.

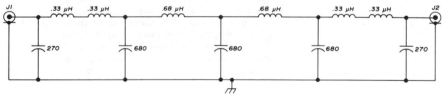

2.955-MHz HIGH-PASS—Used in offset frequency-measuring system for amateur-band signals. Nine-section Chebyshev high-pass filter with 1-dB passband ripple attenuates undesired 2.045–2.245 MHz image 16 dB while selecting desired 2.955–3.155 MHz signal. Filter has sharper cutoff characteristic, for given number of sections, than Butterworth or image parameter designs.—J. Walker, Accurate Frequency Measurement of Received Signals, *Ham Radio,* Oct. 1973, p 38–55.

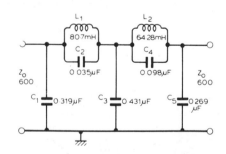

L1   253 turns 28 swg   35mm VINKOR LA1211

L1-C2   Tune to 3kHz

L2   226 turns 28 swg   35mm VINKOR LA1211

L2-C4   Tune to 2kHz

1-kHz FIFTH-ORDER LOW-PASS—Used with 1-kHz signal generator to remove unwanted harmonics, leaving pure sine wave as required for measuring distortion in modern audio amplifiers. Attenuation peaks are carefully positioned to coincide with second and third harmonics, giving 65-dB attenuation of these harmonics and at least 50-dB attenuation of higher harmonics.—J. A. Hardcastle, 1 kHz Source Cleaning Filter, *Wireless World,* Oct. 1978, p 59.

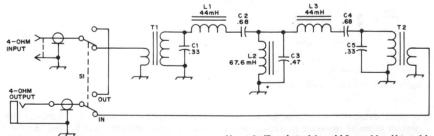

VOICE BANDPASS—Used between 8-ohm output of communication receiver and 8-ohm loudspeaker or low-impedance phones, to suppress Continuous Random Unwanted Disturbances on voice transmissions. Passband is 355 to 2530 Hz at 3-dB points. L1 and L3 are 44-mH toroids. L2 is 88-mH toroid with 94 turns removed. T1 and T2 are 88-mH toroids with 100 turns No. 28 enamel wound over original winding of each for primary.—R. M. Myers, The SSB Crud-O-Ject, *QST,* May 1974, p 23–25 and 56.

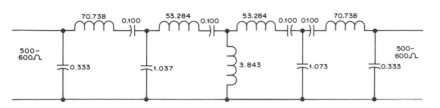

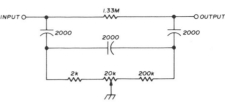

225-Hz BANDPASS RTTY—Used ahead of limiter in 170-Hz-shift RTTY receiving converter. Chebyshev mesh configuration with 0.1-dB ripple uses inductor to ground for sharpening lower skirt, with capacitive coupling for sharpening upper skirt, to give good symmetry for response curve. Capacitors should be high-Q types, well matched. Take turns off inductors as required to move passband higher if initially low in frequency. Insertion loss is 6.6 dB and 3-dB bandwidth is 225 Hz, which makes mark and space tones only 1.5 dB down.—A. J. Klappenberger, A High-Performance RTTY Band-Pass Filter, QST, Jan. 1978, p 33.

693–2079 Hz TUNABLE NOTCH—Requires only one tuning pot to cover entire frequency range. Developed for use in tunable narrow-band audio amplifier. Article gives design equations. Depth of notch is greater than 50 dB. Doubling capacitor values changes tuning range to 355–1028 Hz, while cutting values in half gives range of 1340–4110 Hz.—C. Hall, Tunable RC Notch Filter, Ham Radio, Sept. 1975, p 16–20.

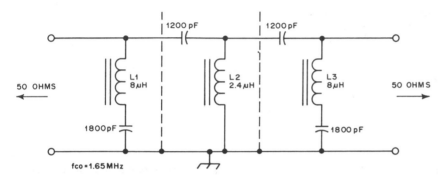

AM BROADCAST REJECTION—Seven-element m-derived high-pass filter provides 30-dB rejection at AM broadcast-band frequencies while passing signals in 160-meter band. Midsection m-derived branch of circuit was eliminated to simplify construction, but can be added and tuned to particular broadcast station that presents difficult interference problem. L1 and L3 are 40 turns No. 30 enamel wound on T50-2 powdered-iron toroid. L2 has 22 turns No. 30 on T50-2 core.—D. DeMaw, Low-Noise Receiving Antennas, QST, Dec. 1977, p 36–39.

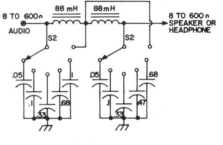

SWITCHABLE AF FILTER—Provides wide range of switch-selected capacitor values for varying cutoff frequencies, to permit use of filter for either phone or CW reception. On CW, circuit improves reception by eliminating higher frequencies that are largely interference.—J. J. Schultz, The Quiet Maker, 73 Magazine, March 1974, p 81–84.

L 35 turns no. 22 enamelled on Amidon T-68-2 toroid cores (7 µH). Input link is 4 turns, output link is 3 turns

C 210 pF, 1% silver mica

C_V 60-pF mica compression trimmers

80-METER BANDPASS—Four-resonator filter for use in 80-meter amateur band has 100-kHz bandwidth, 4.4-dB insertion loss, and 6–60 dB shape factor of 5.16. Filter was designed and aligned at 3.75 MHz; realignment at 3.6 and 3.9 MHz yielded similar results. Article covers theory, construction, and adjustment.—W. Hayward, Bandpass Filters for Receiver Preselectors, Ham Radio, Feb. 1975, p 18–27.

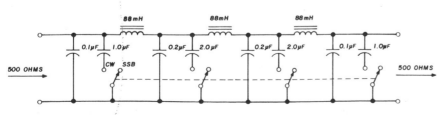

LOW-PASS PI-SECTION AF—Four-pole double-throw switch gives choice of 650-Hz cutoff for CW or 2000 Hz for SSB. Filter capacitors are matched. Response decreases continuously beyond cutoff frequency, with no loss of attenuation.—E. Noll, Circuits and Techniques, Ham Radio, April 1976, p 40–43.

CHAPTER 29
Fire Alarm Circuits

Sensors used may respond to gas, ionization, flame, or smoke associated with fire, for triggering circuits driving variety of alarm devices.

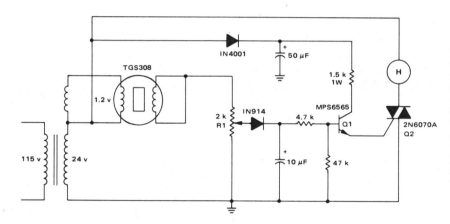

TRIAC GAS/SMOKE DETECTOR—Conductivity of Taguchi TGS308 gas sensor increases in presence of combustible gases, increasing load voltage across R1 from normal 3 VRMS to as much as 20 V. Rise in voltage trips comparator to turn on transistor Q1 that supplies trigger current to 2N6070A sensitive-gate triac. Resulting full-wave drive of Delta 16003168 24-VAC horn gives sound output of 90 dB at 10 feet. Horn stops automatically when gas clears sensor.—A. Pshaenich, "Solid State Gas/Smoke Detector Systems," Motorola, Phoenix, AZ, 1975, AN-735, p 4.

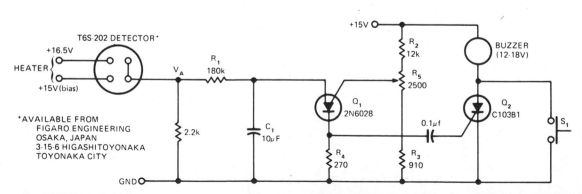

GAS/SMOKE SENSOR—Sensor is based on selective absorption of hydrocarbons by N-type metal-oxide surface. Heater in sensor burns off hydrocarbons when gas or smoke disappears, to make sensor reusable. Requires initial warm-up time of about 15 min in hydrocarbon-free environment. When gas or smoke is present, V_A quickly rises and triggers programmable UJT Q_1. Resulting voltage pulse across R_4 triggers Q_2 and thereby energizes buzzer. S_1 is reset switch. R_1 and C_1 give time delay that prevents triggering by small transients such as smoke from cigarette. R_5 adjusts alarm threshold. Use regulated supply.—S. J. Bepko, Gas/Smoke Detector Is Sensitive and Inexpensive, *EDN Magazine,* Sept. 20, 1973, p 83 and 85.

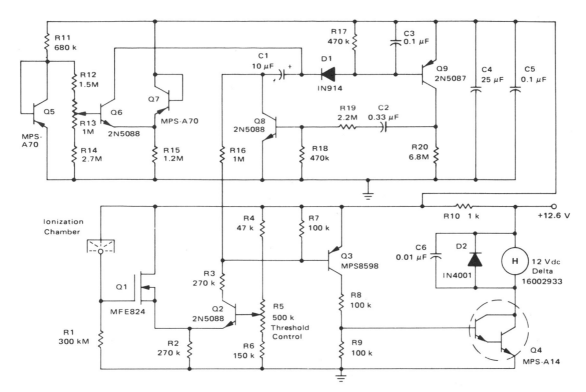

IONIZATION ALARM USING TRANSISTORS—
Use of continuous smoke alarm signal rather than beeping horn simplifies transistor circuits needed to trigger fire alarm and low-battery alarm. When high impedance of ionization chamber is lowered by smoke or gas, amplifier Q1-Q2-Q3 supplies 100-μA base current to Darlington Q4 for powering horn continuously as long as smoke content exceeds that set by threshold control R5. Low-battery circuit is tripped at voltage range between 9.8 and 11.2 V, as determined by R13, to energize MVBR Q8-Q9 for driving horn 0.7 s, with 50-s OFF intervals. Battery is chosen to last at least 1 year while furnishing standby current of about 70 μA.—A. Pshaenich, "Solid State Gas/Smoke Detector Systems," Motorola, Phoenix, AZ, 1975, AN-735, p 8.

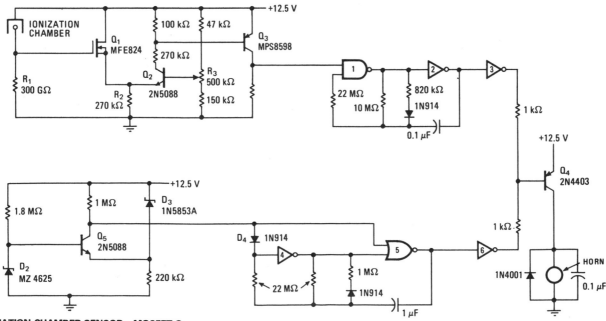

IONIZATION-CHAMBER SENSOR—MOSFET Q$_1$
with high input impedance monitors voltage level at divider formed by R$_1$ and ionization chamber, with output of Q$_1$ going to Q$_2$ which forms other half of differential amplifier. With smoke level of 2% or higher, Q$_3$ is turned on and applies logic 1 to one input of NAND gate 1 in asymmetrical astable MVBR. Capacitor in MVBR charges quickly and discharges slowly, making alarm horn sound during discharge via inverter 3 and driver transistor Q$_4$. Comparator circuit Q$_5$ drives second MVBR to energize horn through inverter 6 and same driver Q$_4$ when battery is low, but with distinctive 1-s toot every 23 s to conserve energy remaining in battery and differentiate from fire warning.—A. Pshaenich and R. Janikowski, Gas and Smoke Detector Uses Low-Leakage MOS Transistor, *Electronics*, Nov. 28, 1974, p 124–125.

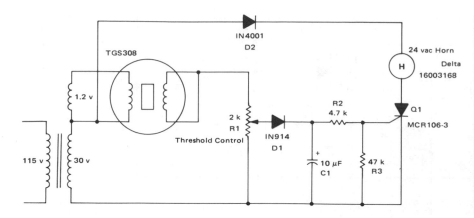

SCR GAS/SMOKE DETECTOR—Simple circuit uses Taguchi TGS308 gas sensor with SCR Q1 for half-wave control of 24-VAC alarm horn. Sensor is based on adsorptive and desorptive reaction of gases on tin oxide semiconductor surface encased in noble-metal heater that serves also as electrode. Combustible gases increase conductivity of sensor, thereby increasing load voltage enough to trip comparator and initiate alarm. Output voltage across R1 is normally about 3 VRMS. With gas or smoke, voltage can rise to 20 V. When gas or smoke has cleared sensor, SCR turns off at first zero crossing. Drawbacks are absence of time delay for preventing false alarm when power is turned on and reduced sound level of horn with half-wave operation.—A. Pshaenich, "Solid State Gas/ Smoke Detector Systems," Motorola, Phoenix, AZ, 1975, AN-735, p 3.

FLAME DETECTOR DRIVES TTL LOAD—Sensor is silicon Darlington phototransistor Q₁ having peak response near infrared bands. Filter is required to reduce interference from visible light sources. Circuit is sensitive enough to pick up hydrogen flames that emit no visible light. Article describes operation of circuit and gives design equations. Output can go directly to input port of microprocessor.—A. Ames, This Flame Detector Interfaces Directly to a μP, *EDN Magazine*, Oct. 20, 1976, p 122 and 124.

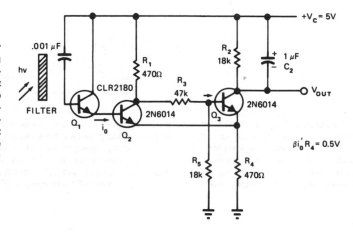

$$\beta i_0' R_4 = 0.5V$$

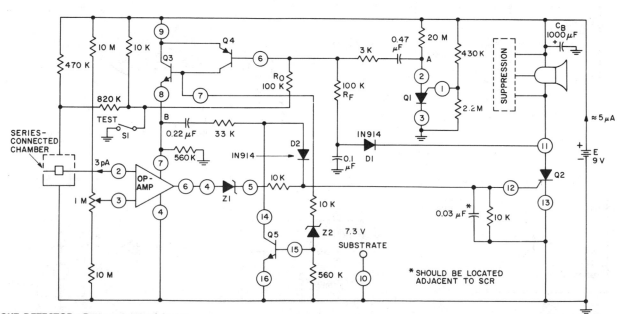

SMOKE DETECTOR—Battery-operated ionization-type smoke detector uses RCA CA3130 opamp as interface for ionization chamber that provides picoampere currents. With opamp in pulsed mode (on for 20 ms of 20-s period), IC draws only 0.6 μA average instead of 600 μA. Other active components and zener, all on RCA CA3097 array, provide low-battery monitor and horn-driver functions. When chamber detects smoke, combination of R_F and D1 provides sufficient base current to keep Q3 and Q4 on. Opamp is then powered continuously, and steering diode Z1 supplies continuous current to gate of Q2 for energizing horn. Battery drain is only 5 mA in monitoring mode.—G. J. Granieri, Bipolar-MOS and Bipolar IC's Building Blocks for Smoke-Detector Circuits, *IEEE Transactions on Consumer Electronics*, Nov. 1977, p 522–527.

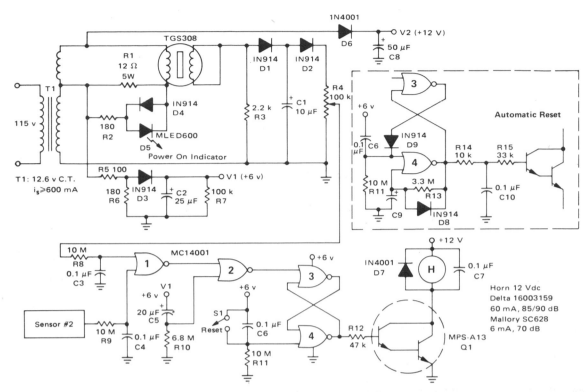

GAS/SMOKE DETECTOR WITH LATCH—CMOS latching logic provides 2-min time delay to prevent false alarm when power is first applied to fire alarm using Taguchi TGS308 gas sensor whose conductivity increases in presence of combustible gases. Normal voltage of 3 VRMS across R4 increases to about 20 V in presence of fire. Half of 12.6-V center-tapped transformer secondary is used for 6-V supply and full 12.6 V for DC horn supply. Latch is reset manually with S1 to turn off alarm after gas level drops. Optional circuit shown can be used for automatic reset.—A. Pshaenich, "Solid State Gas/Smoke Detector Systems," Motorola, Phoenix, AZ, 1975, AN-735, p 5.

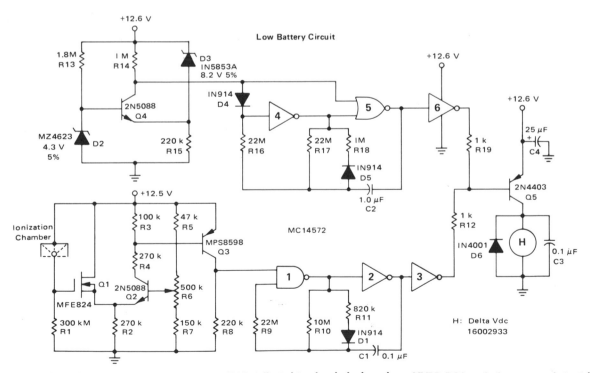

IONIZATION ALARM—Gates in Motorola MC14572 CMOS IC form two alarm oscillators, one energized in presence of smoke at ionization chamber and other for low battery. Standby currents of circuits are low enough to give at least 1 year of operation from 750-mAh battery. R6 is adjusted to give desired smoke detection sensitivity. Gates 1 and 2 form MVBR that drives horn at astable rate of 2.5 s on and 0.2 s off in presence of smoke. When battery is low, comparator Q4-D2-D3 trips (about 10.5 V) and energizes inverter 4 of low-battery astable MVBR. DC horn is then powered at astable rate of about 1 s every 23 s to give early warning of need to change battery.—A. Pshaenich, "Solid State Gas/Smoke Detector Systems," Motorola, Phoenix, AZ, 1975, AN-735, p 7.

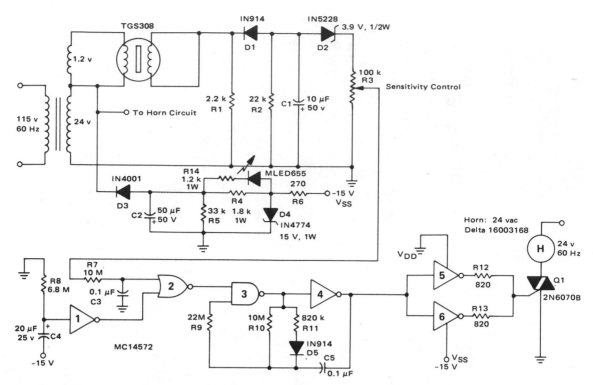

GAS/SMOKE DETECTOR WITH BEEPING HORN—Taguchi TGS308 gas sensor increases voltage across R3 when sensor conductivity is increased by combustible gases. After time delay provided to prevent power turn-on false alarms, CMOS astable MVBR using gates 3 and 4 is energized to fire triac and drive AC horn to give distinctive repetitive sound lasting about 2.5 s, with 0.2-s intervals between beeps. Triac gate drivers operate from −15 V supply derived from 24-V winding of power transformer.—A. Pshaenich, "Solid State Gas/Smoke Detector Systems," Motorola, Phoenix, AZ, 1975, AN-735, p 6.

CHAPTER 30
Flasher Circuits

Provide fixed or variable flash rates for LEDs, incandescent lamps, or fluorescent lamps used as indicators, alarms, warnings, and for such special effects as Christmas-light shimmer. See also Game and Lamp Control chapters.

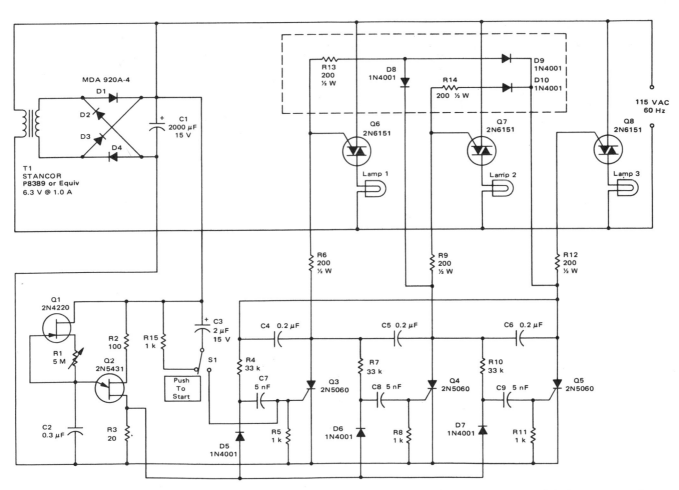

SEQUENTIAL AC FLASHER—Uses simple ring counter in which triac gates form part of counter load. Incandescent lamps come on in sequence, with only one lamp normally on at a time. Pulse rate for switching lamp can be adjusted from about 1 every 0.1 s to 1 every 8 s. Circuit enclosed in dashed rectangle can be added to keep previous lamps on when next lamp is turned on. Only three stages are shown, but any number of additional stages can be added.—"Circuit Applications for the Triac," Motorola, Phoenix, AZ, 1971, AN-466, p 11.

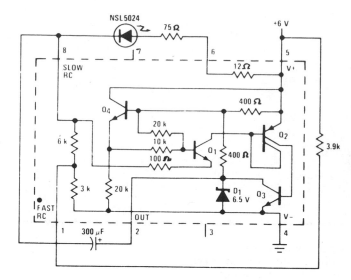

VARIABLE FLASHER FOR LED—Terminal connections of National LM3909 flasher IC give choice of three different flash rates for LED used as indicator in battery portable equipment. External resistors provide additional adjustments of flash rate. Appropriate connections to pins 1 and 8 make flash-controlling internal resistance 3K, 6K, or 9K. Flasher operates at any supply voltage above 2 V, with low duty cycle to give long battery life.—P. Lefferts, Power-Miser Flasher IC Has Many Novel Applications, *EDN Magazine*, March 20, 1976, p 59—66.

SHIMMER FOR CHRISTMAS LIGHTS—Circuit uses half of AC cycle to power lights conventionally. On other half-cycle, C charges and builds up voltage on gate of SCR. When firing point is reached, SCR conducts and allows remainder of this half-cycle to pass through light string. Result is flash that gives shimmer or strobe effect. C is 100-μF 50-V electrolytic, R1 is 2.7K, R2 is 22K, R3 is 3.3K, R4 is 100K pot, and R5 is 1K. Diodes are Motorola HEP R0053. SCR is GE C106B1 or Motorola HEP R1221 mounted on heatsink.—R. F. Graf and G. J. Whalen, Add Shimmer to Your Christmas Lights, *Popular Science*, Dec. 1973, p 124.

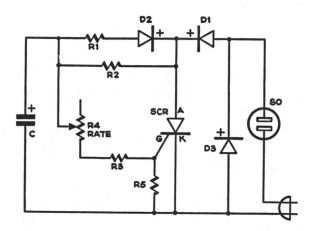

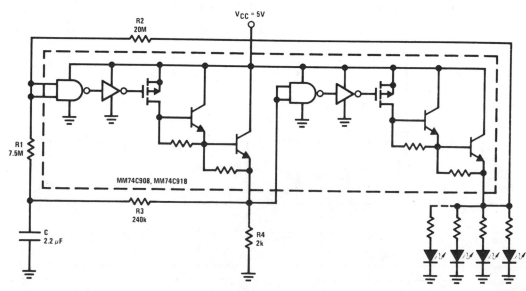

DRIVING LED ARRAY—National MM74C908/ MM74C918 dual CMOS driver has sections connected as Schmitt-trigger oscillator, with R1 and R2 used to generate hysteresis. R3 and C are inverting feedback timing elements, and R4 is pulldown load for first driver. Output current drive capability is greater than 250 mA, making circuit suitable for driving array of LEDs or lamps.—"CMOS Databook," National Semiconductor, Santa Clara, CA, 1977, p 5-38—5-49.

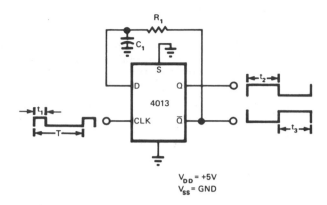

1-Hz LAMP BLINKER—Single CMOS flip-flop generates approximately constant low-frequency signal from variable high-frequency signal. RC network in feedback loop determines output frequency, which is independent of rate at which flip-flop is clocked if output frequency is lower than clock frequency. If clock frequency is lower, output transitions occur at half of clock frequency. Provides two outputs, approximately equal in duty cycle but opposite in phase. Circuit was developed to blink lamp at 1 Hz to indicate presence of active digital signal having variable duty cycle in range of 100 to 3000 Hz.—V. L. Schuck, Generate a Constant Frequency Cheaply, *EDN Magazine,* Aug. 20, 1975, p 80 and 82.

3-V STROBE—Flash rate of 1767 lamp can be adjusted from no flashes to continuously on, in circuit using National LM3909 flasher IC with external NPN power transistor rated at 1 A or higher. Can be used as variable-rate warning light, for advertising, or for special effects. With lamp in large reflector in dark room, flashes several times per second are almost fast enough to stop motion of dancer.—P. Lefferts, Power-Miser Flasher IC Has Many Novel Applications, *EDN Magazine,* March 20, 1976, p 59–66.

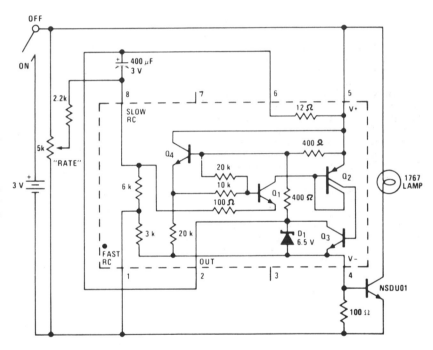

OUT-OF-PHASE DOUBLE FLASHER—Sections of National MM74C908/MM74C918 dual CMOS driver are connected as Schmitt-trigger oscillator, with LEDs at output of each section so LEDs will flash 180° out of phase. High output current capability makes circuit suitable for driving two LED arrays.—"CMOS Databook," National Semiconductor, Santa Clara, CA, 1977, p 5-38—5-49.

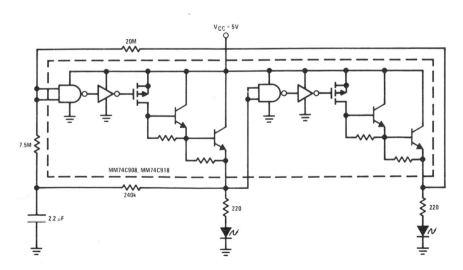

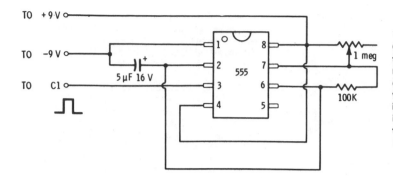

CLOCK DRIVE FOR FLIP-FLOP FLASHER—555 timer connected as astable MVBR generates series of timing pulses at rate determined by value of capacitor and setting of 1-megohm pot. Provides automatic string of input pulses for driving flip-flop of dual flasher. Pulse output goes to input capacitor C1 of flip-flop.—F. M. Mims, "Integrated Circuit Projects, Vol. 5," Radio Shack, Fort Worth, TX, 1977, 2nd Ed., p 30–37.

SEQUENTIAL SWITCHING OF LOADS—Ring counter using four-layer diodes D_N provides sequential switching of loads under control of input pulse-train signal. Indicator lamps are shown, but any load from 15 to 200 mA can be switched. After power is applied, reset switch must be pressed to establish current through L. When switch is released, this current flows through C_2 and breaks down D_2, allowing current to flow through first lamp I_1. Input pulse to transistor Q (normally held off by current through R_1) turns Q off and removes power from diode circuits, thus turning I_1 and D_2 off. At end of input pulse, Q comes on and restores power to diode circuits, but all loads will be turned off. Voltage on C_3 now adds to 6 V normally across D_4, making D_4 break down and turn on I_2. Next input pulse will break down D_6 in same manner. Output signals may be picked up as negative pulses at A or B or by current-sensing at C if required for controlling larger loads.—J. Bliss and D. Zinder, "4-Layer and Current-Limiter Diodes Reduce Circuit Cost and Complexity," Motorola, Phoenix, AZ, 1974, AN-221, p 5.

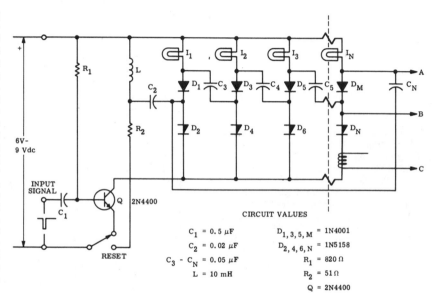

CIRCUIT VALUES

$C_1 = 0.5\ \mu F$ $D_{1,3,5,M} = 1N4001$

$C_2 = 0.02\ \mu F$ $D_{2,4,6,N} = 1N5158$

$C_3 - C_N = 0.05\ \mu F$ $R_1 = 820\ \Omega$

$L = 10\ mH$ $R_2 = 51\ \Omega$

$Q = 2N4400$

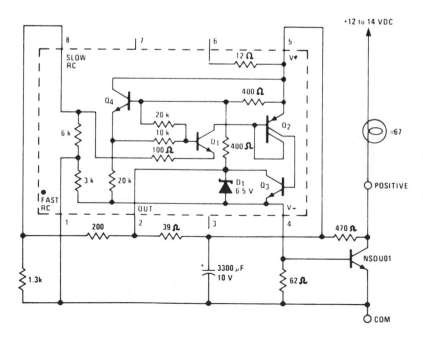

1-Hz AUTO FLASHER—Lamp drawing nominal 600 mA is flashed at 1 Hz by National LM3909 flasher IC operating from 12-V automotive battery. Use of 3300-μF capacitor makes flasher IC immune to supply spikes and provides means of limiting IC supply voltage to about 7 V.—P. Lefferts, Power-Miser Flasher IC Has Many Novel Applications, *EDN Magazine*, March 20, 1976, p 59–66.

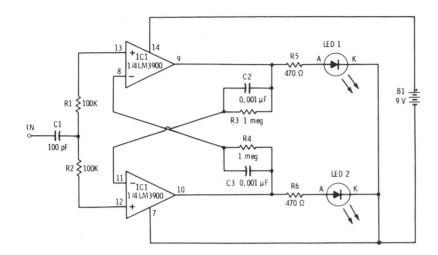

DEMONSTRATION FLIP-FLOP—Two sections of LM3900 quad opamp form bistable MVBR for flip-flop having two stable states. When input is grounded momentarily, output of one of opamps swings completely on and turns other opamp off. LED indicates which opamp is on at any particular time. Next grounding of input reverses conditions. Ideal for classroom demonstrations.—F. M. Mims, "Integrated Circuit Projects, Vol. 5," Radio Shack, Fort Worth, TX, 1977, 2nd Ed., p 30–37.

SCR FLASHES LED—UJT oscillator Q1 provides timing pulses for triggering SCR driving red Radio Shack 276-041 LED. Circuit draws only 2 mA from 9-V battery when producing 12 flashes per second. SCR is 6-A 50-V 276-1089.—F. M. Mims, "Semiconductor Projects, Vol. 2," Radio Shack, Fort Worth, TX, 1976, p 78–84.

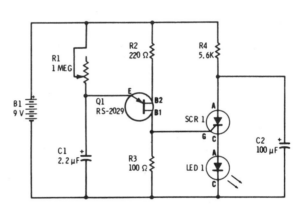

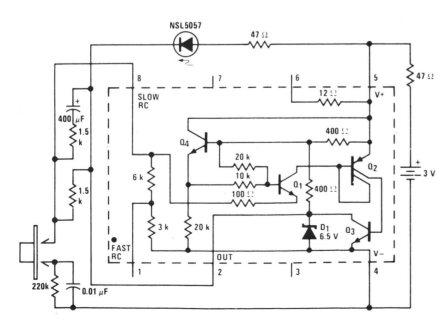

SINGLE-FLASH LED—Mono MVBR connection of National LM3909 IC produces 0.5-s flash with LED each time pushbutton makes momentary contact.—"Linear Applications, Vol. 2," National Semiconductor, Santa Clara, CA, 1976, AN-154, p 9.

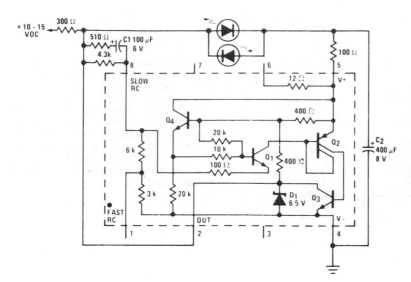

ALTERNATING RED/GREEN—National LM3909 IC is connected as relaxation oscillator for flashing red and green LEDs alternately. With 12-VDC supply, repetition rate is about 2.5 Hz. Green LED should have its anode or positive lead toward pin 5 as shown for lower LED, where shorter but higher-voltage pulse is available. LED types are not critical.—"Linear Applications, Vol. 2," National Semiconductor, Santa Clara, CA, 1976, AN-154, p 3.

ALARM-DRIVEN FLASHER—Simple two-transistor flasher circuit for annunciator system is activated by alarm. Operator acknowledges alarm condition by depressing S$_A$, which changes lamp from flashing to steady ON condition. 6-V incandescent lamp draws about 0.3 A through Q$_2$, but 1K load resistor for Q$_1$ limits current of this transistor to about 6 mA so smaller transistor can be used.—T. Stehney, Flasher Design Cuts Extra Components, *EDN Magazine*, Sept. 20, 1978, p 144.

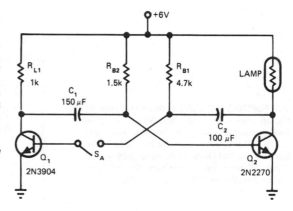

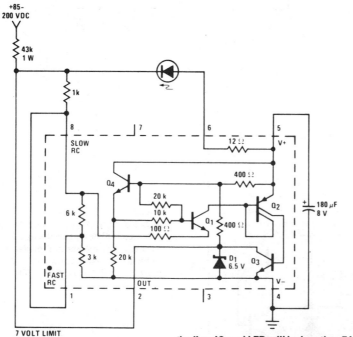

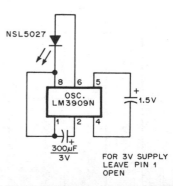

FLASHING LED IS REMOTE MONITOR—Circuit uses National LM3909 flasher IC to drive LED for monitoring remotely located high-voltage power supply. When 43K dropping resistor is located at power supply, all other voltages on the line, IC, and LED will be less than 7 V above ground, for safe remote monitoring. Use any LED drawing less than 150 mA.—P. Lefferts, Power-Miser Flasher IC Has Many Novel Applications, *EDN Magazine*, March 20, 1976, p 59–66.

1.5-V OR 3-V INDICATOR—Digi-Key LM3909N flasher/oscillator drives LED serving as ON/OFF indicator for battery-operated devices. At flash rate of 2 Hz, battery life almost equals shelf life.—C. Shaw, ON-OFF Indicator for Battery Device, *QST*, March 1978, p 41–42.

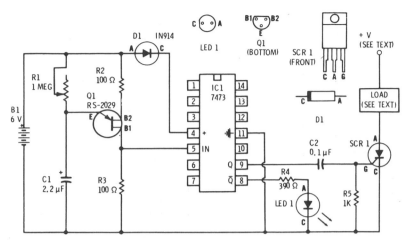

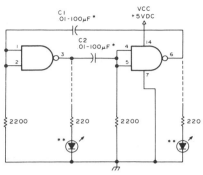

FLIP-FLOP DRIVES SCR—UJT relaxation oscillator Q1 serves as clock for driving section of 7473 dual flip-flop. One output of flip-flop flashes Radio Shack 276-041 red LED to indicate operating status. Other output alternately triggers SCR which can be 6-A 50-V 276-1089, for flashing lamp load. Load and SCR supply voltage depend on application but must be within SCR rating.—F. M. Mims, "Semiconductor Projects, Vol. 2," Radio Shack, Fort Worth, TX, 1976, p 62–70.

LED BLINKER—Two sections of SN7400 quad gate form MVBR operating at low enough frequency so LED status indicators come on and off slowly for visual observation of MVBR. LEDs are optional and do not affect operation of MVBR. Capacitors must be same value. Ideal for student demonstration in classroom or as Science Fair exhibit.—A. MacLean, How Do You Use ICs?, *73 Magazine*, Dec. 1977, p 56–59.

RED/GREEN LED FLASHER—One section of LM324 quad opamp is connected as square-wave generator giving about 1 flash per second for each LED. Series resistors for LEDs have different values because they have different forward voltage requirements. If LED 2 glows between flashes, increase value of R6 slightly. Too large a value for R6 reduces flash brilliance of LED 2. Supply can be 5 or 6 V.—F. M. Mims, "Semiconductor Projects, Vol. 1," Radio Shack, Fort Worth, TX, 1975, p 69–74.

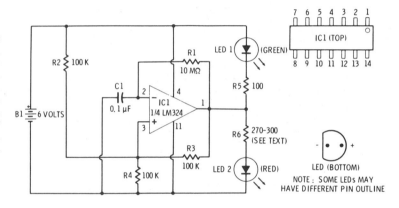

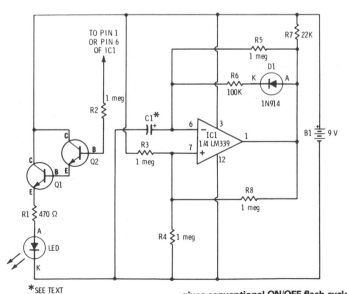

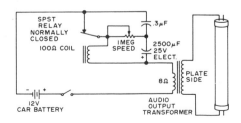

COMPARATOR LED FLASHER—One section of LM339 quad comparator drives two RS2016 NPN transistors having LED load, to give simple flasher for classroom demonstrations. Circuit can be duplicated with other three sections to give four flashers. Connecting R2 to pin 1 of IC gives conventional ON/OFF flash cycle in which LED turns on and off rapidly. Connecting R2 to pin 6 makes LED turn on rapidly and turn off very slowly. C1 controls flash interval; typical value is 0.01 μF.—F. M. Mims, "Integrated Circuit Projects, Vol. 5," Radio Shack, Fort Worth, TX, 1977, 2nd Ed., p 45–51.

12-V FLUORESCENT—Relay acts as mechanical DC/AC converter operating off 12-V car battery. Each time relay opens, inductive kick in relay coil is stepped up by output transformer to high enough voltage for ionizing 24-inch fluorescent tube, giving flash that can serve as emergency flasher when car breaks down.—Circuits, *73 Magazine*, June 1975, p 175.

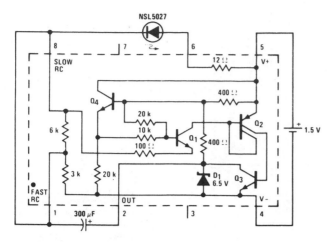

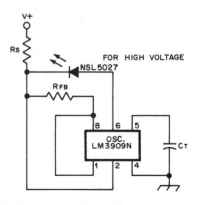

1.5-V LED FLASHER—National LM3909 IC operating from 1.5-V battery drives NSL5027 LED in such a way that current is drawn by LED only about 1% of time. External 300-μF capacitor sets flash rate at about 1 Hz.—"Linear Applications, Vol. 2," National Semiconductor, Santa Clara, CA, 1976, AN-154, p 2.

LED FLASHER—Requires only LM3909 IC and external capacitor operating from 1.25-V nicad or other penlight cell. Circuit can be duplicated for as many additional flashing LEDs as are desired for display. Optional charging circuit uses silicon solar cells and diode for daytime charging of battery automatically.—J. A. Sandler, 11 Projects under $11, *Modern Electronics*, June 1978, p 54–58.

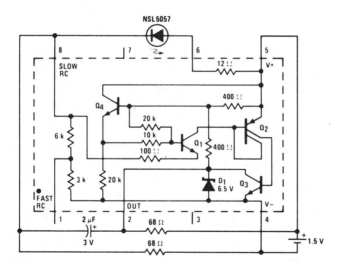

2-kHz FLASHER FOR LED—Single 1.5-V cell provides power for National LM3909 flasher IC that operates at high enough frequency to appear on continuously, for use as indicator in battery portable equipment. Duty cycle and frequency of current pulses to LED are increased by changing external resistors until average energy reaching LED provides sufficient light for application. At 2 kHz, no flicker is noticeable.—P. Lefferts, Power-Miser Flasher IC Has Many Novel Applications, *EDN Magazine*, March 20, 1976, p 59–66.

6-V OR 15-V INDICATOR—Uses Digi-Key LM3909N flasher/oscillator to drive LED at 2 Hz as ON/OFF indicator for battery-operated devices. For 6-V battery, C_T is 400 μF, R_S is 1000 ohms, and R_{FB} is 1500 ohms. For 15 V, corresponding values are 180, 3900, and 1000. Battery life is essentially same as shelf life.—C. Shaw, ON-OFF Indicator for Battery Device, *QST*, March 1978, p 41–42.

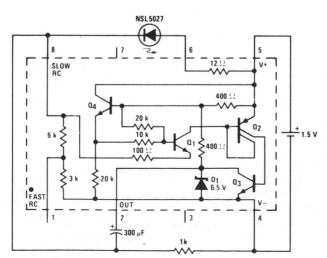

FAST 1.5-V BLINKER—Addition of 1K resistor between pins 4 and 8 of National LM3909 IC increases flash rate to about 3 times that obtainable when 300 μF is connected between pins 1 and 2. Modification of external connections gives choice of 3K, 6K, or 9K for internal RC resistors.—"Linear Applications, Vol. 2," National Semiconductor, Santa Clara, CA, 1976, AN-154, p 2.

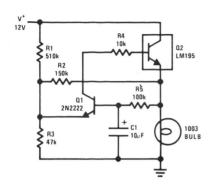

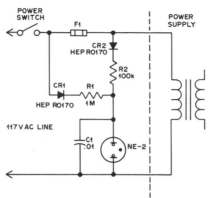

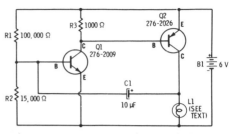

1-A LAMP FLASHER—National LM195 power transistor is turned on and off once per second for flashing 12-V lamp. Current limiting in LM195 prevents high peak currents during turn-on even though cold lamp can draw 8 times normal operating current. Current-limiting feature prolongs lamp life in flashing applications.—R. Dobkin, "Fast IC Power Transistor with Thermal Protection," National Semiconductor, Santa Clara, CA, 1974, AN-110, p 5.

BLOWN-FUSE BLINKER—Neon lamp NE-2 glows steadily when fuse is good and flashes when fuse opens. Flash rate, determined by R1 and C1, is about 10 flashes per second for values shown.—T. Lincoln, A "Smart" Blown-Fuse Indicator, QST, March 1977, p 48.

AUTO-BREAKDOWN FLASHER—Two-transistor amplifer with regenerative feedback sends 60-ms pulses of currents up to several amperes through low-voltage lamp to give high-brilliance flashes without destroying lamp. L1 can be PR-2 lamp (Radio Shack 272-1120).—F. M. Mims, "Transistor Projects, Vol. 1," Radio Shack, Fort Worth, TX, 1977, 2nd Ed., p 27–32.

CHAPTER 31

Frequency Counter Circuits

Used to indicate frequency value directly on digital display by counting number of cycles in period of exactly 1 second. Included are preamps, time bases, and prescalers for extending counting range to as high as 500 MHz.

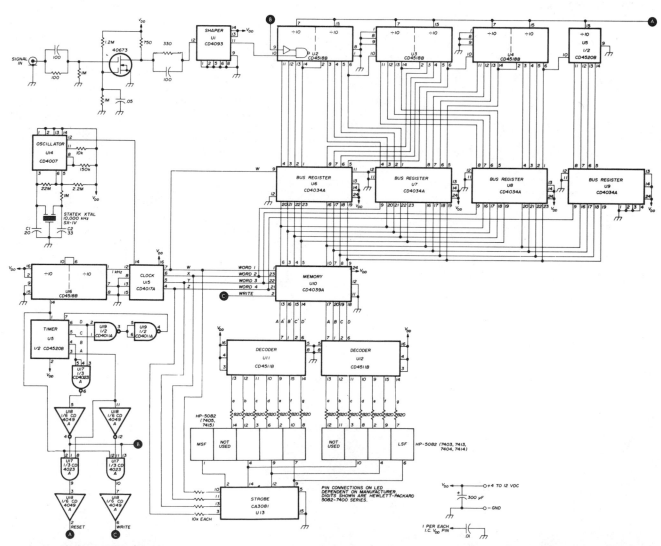

4-MHz COUNTER—Portable frequency counter using RCA CMOS logic draws only 300 mW (12 V at 25 mA) yet operates to well above 4 MHz. Supply voltage can be between 4 and 15 V, loosely regulated, without affecting accuracy. Display uses multiplexing with 10% duty cycle to minimize battery drain. One multiplexed output is for three least significant figures and other for four most significant figures. Article describes operation in detail. Applications include setting RTTY mark and space tones, FM repeater tones, signal-generator frequencies for TV alignment, tuning musical instruments, and serving as tachometer or speedometer in car.— R. M. Mendelson, Milliwatt Portable Counter, *Ham Radio,* Feb. 1977, p 22–25.

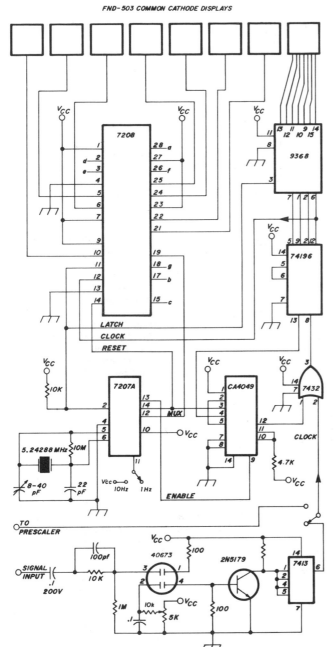

FND-503 COMMON CATHODE DISPLAYS

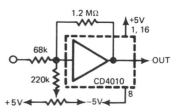

0–3 MHz PREAMP—Provides wide frequency response required for amplifying 100 mV P-P input signals to 5-V level for driving CMOS logic of frequency counter.—R. Tenny, Counter Pre-Amp Matches CMOS Logic Capability, *EDN Magazine,* Sept. 20, 1976, p 114 and 116.

50 MHz WITH 1-Hz RESOLUTION—Combination of CMOS and TTL devices reduces chip count for digital frequency counter that provides 1-Hz resolution from below 20 Hz to above 50 MHz. Use of 10:1 prescaler, also given in article, extends range to above 300 MHz with 10-Hz resolution. Uses Intersil 7208 CMOS seven-decade counter that includes multiplexer, decoder, drivers, and other controls for Fairchild FND-503 8-digit display. High-stability 5.24288-MHz crystal oscillator and frequency divider provide 1-s gate required for counting, outputs for synchronizing multiplexer, and short pulses for latching and resetting counters. Resolution can be decreased by factor of 10 by connecting pin 11 of 7207A to V_{CC}, which is regulated 5 V.—H. E. Harris, Simplifying the Digital Frequency Counter, *Ham Radio,* Feb. 1978, p 22–25.

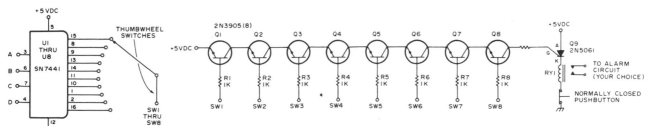

PRESET-FREQUENCY ALARM—When selected frequency occurs in Nixie-driving counter, alarm circuit triggers and locks until reset manually. Requires one SN7441 Nixie decoder/driver, one decimal-type thumbwheel switch, and one 2N3905 transistor for each digit of display in counter. Four connections are made to each counter stage to get BCD inputs A, B, C, and D for 7441. Connections can be made to 7475 quad latch in typical counter. Circuits are for 8-digit display. When display reaches digit to which switch is set, switch output is grounded. When all switch outputs are grounded, all transistors are turned on and SCR fires to actuate alarm relay. If latching is undesirable, use medium-power NPN transistor in place of SCR.—W. L. MacDowell, Frequency Detector for Your Counter, *73 Magazine,* Oct. 1976 p 50–51.

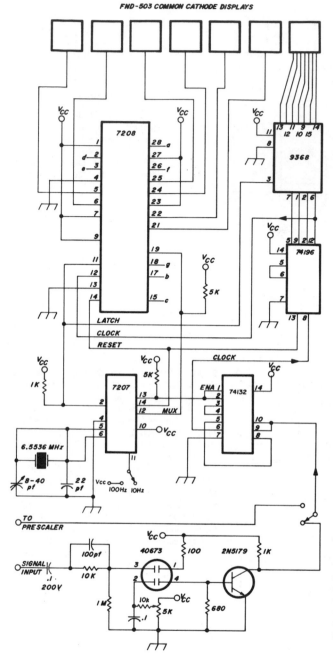

FND-503 COMMON CATHODE DISPLAYS

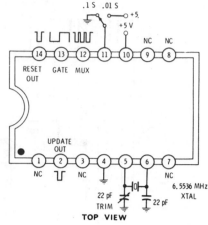

TOP VIEW

TIME BASE—Intersil 7207 IC generates clock and housekeeping pulses required for frequency counter. With pin 11 grounded, gate output is high for 0.1 s and low for 0.1 s. With pin 11 high, gate is high 0.01 s and low 0.01 s. 1.6-kHz square wave at pin 12 is useful for multiplexing displays. Update output is narrow negative-going pulse coincident with rising edge of gate output, for use in transferring count to display latches. Reset output is used to reset counter.—D. Lancaster, "CMOS Cookbook," Howard W. Sams, Indianapolis, IN, 1977, p 161.

30 MHz WITH 10-Hz RESOLUTION—Simplified counter design using low chip count provides multiplexing of seven digits in Fairchild display, for applications where 1-Hz resolution is not needed. 7207 oscillator/timer gives counting interval of 0.1 s, for updating display 5 times per second. Article also gives circuit of 10:1 prescaler that increases frequency limit to 300 MHz, though with 100-Hz resolution. Total counter current drain is 300 mA from regulated 5-V supply.—H. E. Harris, Simplifying the Digital Frequency Counter, *Ham Radio*, Feb. 1978, p 22–25.

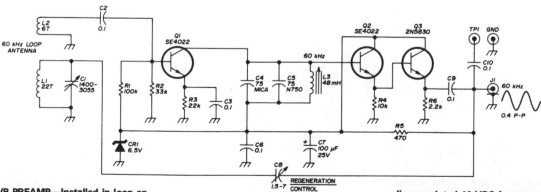

60-kHz WWVB PREAMP—Installed in loop antenna to boost strength of 60-kHz standard-frequency broadcasts from NBS station at Boulder, Colorado, enough to drive digital frequency counter for which circuit is also given in article. Although construction details apply to double-copper shielded 54-inch-diameter circular loop, preamp can also be used with simple unshielded wood-frame loop. Output coax supplies regulated 10 VDC for preamp. Article includes techniques for minimizing interference from nearby TV receivers.—H. Isenring, WWVB Signal Processor, *Ham Radio*, March 1976, p 28–34.

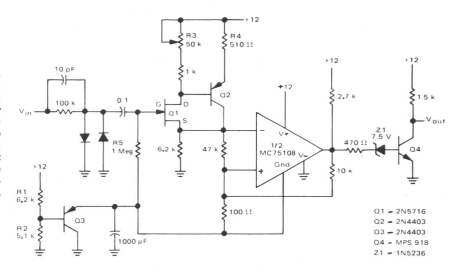

5-MHz FRONT END—Used ahead of 5-MHz frequency counter to make input signal swing from logic 0 of 0 V to logic 1 of about 10 V as required for accurate counting of frequency for input signal having any input waveform shape and level. Input of front end has high impedance to minimize effect on input waveform. FET transistor Q1 and bipolar buffer Q2 drive Schmitt trigger using half of Motorola MC75108 dual line receiver.—D. Aldridge, "Battery-Powered 5-MHz Frequency Counter," Motorola, Phoenix, AZ, 1974, AN-717, p 5.

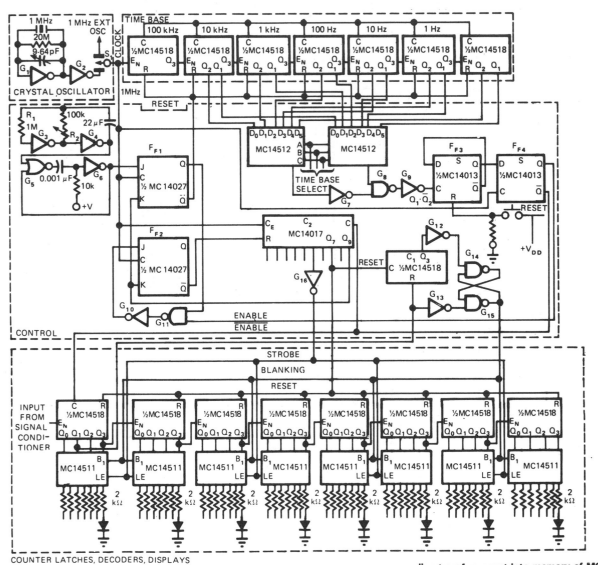

COUNTER LATCHES, DECODERS, DISPLAYS

12-V 5-MHz COUNTER—Portable counter is designed with low-power logic to minimize battery drain, yet provides good performance. Since most of milliwatt power drain is taken by digital readout, circuit blanks out LED display when there is no input signal. Time base divides 1-MHz crystal oscillator frequency down to desired enable time, up to 10 s, using 3½ MC14518 dual decade counters connected in ripple-through mode. Actual counting of input signal codes is also done with MC14518 counters. Latches and BCD to 7-segment decoders use MC14511s. Enable line turns first counter on and off for precise enable time period. Strobe line transfers count into memory of MC14511 latch decoder, and control line resets MC14518 decade counters for next count cycle. Displays are Monsanto MAN-4 LEDs. Article traces circuit operation in detail and gives timing diagram.—D. Aldridge, CMOS Counter Circuitry Slashes Battery Power Requirements, *EDN Magazine*, Oct. 20, 1974, p 65–71.

300-MHz PRESCALER—Uses Fairchild 95H90 IC to divide input signal frequency by factor of 10 up to 320 MHz. Full-wave diode limiter at input prevents damage to IC. R_A is chosen to bias IC at point of maximum sensitivity; typical value is 680 ohms. Transistor amplifier provides 2–3 V P-P output. Bias resistor R_B is set to make collector-base voltage 3 V; typical value is 620 ohms. Wind one lead of 4.7-μH choke around nail 4 times, then remove nail and slip ferrite bead over end of wire before connecting it to pin 1. Keep all leads as short as possible. Article covers construction and alignment in detail.—I. Math, Math's Notes, *CQ*, May 1975, p 42–44 and 64.

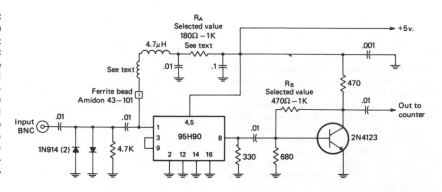

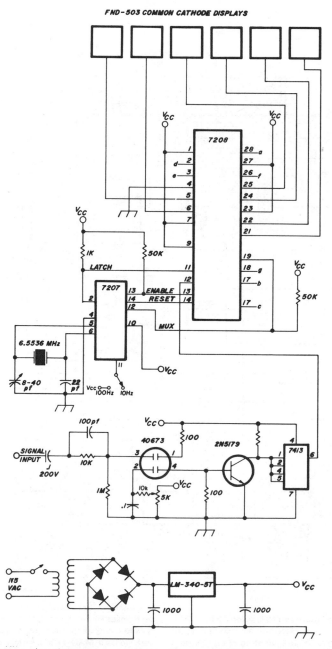

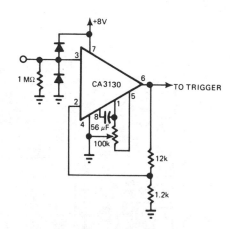

DIRECT-COUPLED PREAMP—Provides frequency response from 0 to 1 MHz at very low power levels, as required for driving CMOS logic of frequency counter. Diodes protect input from overload. Output impedance of frequency source should be kept below 50K to minimize noise pickup.—R. Tenny, Counter Pre-Amp Matches CMOS Logic Capability, *EDN Magazine*, Sept. 20, 1976, p 114 and 116.

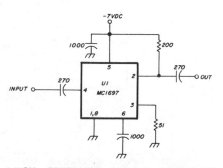

1.5-GHz PRESCALER—Motorola MC1697 IC provides division by 4 to extend operating range of 400-MHz counter above 1.5 GHz. Circuit will operate on input signals as low as 1 mW. Requires 60-mA power supply at −7 V. Article gives construction and test details.—J. Hinshaw, 1.5 GHz Divide-by-Four Prescaler, *Ham Radio*, Dec. 1978, p 84–86.

6 MHz WITH 10-Hz RESOLUTION—Intersil 7208 CMOS counter provides multiplexing of six digits in Fairchild display operating from 5-V regulated supply. Uses MOSFET 4673 input stage.—H. E. Harris, Simplifying the Digital Frequency Counter, *Ham Radio*, Feb. 1978, p 22–25.

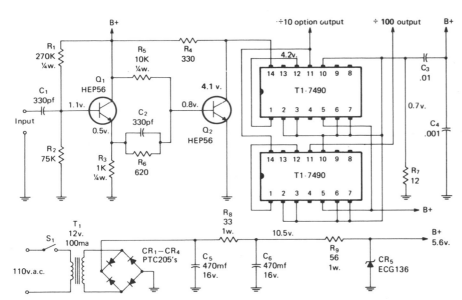

TWO-DECADE SCALER—Solid-state frequency scaler extends range of older frequency counters by factors of 10 and 100, or up to 10 MHz for 100-kHz counter. Emitter-follower Q_1 provides matching from high input impedance to low impedance for driving sensitive clipper Q_2 that operates class B and presents 4 V P-P square wave to decade counter. Input accepts 1 to 14 V P-P.—D. Peck, A Solid State Scaler for Frequency Counters, *CQ*, April 1974, p 24–27.

200-MHz BUFFER—Developed for use ahead of prescaler in 200-MHz autoranging frequency counter. Provides high input impedance to count-sensing device. Circuit includes Schmitt trigger action. Sensitivity is about 50 mV P-P.—T. Balph, "A 200 MHz Autoranging MECL—McMOS Frequency Counter," Motorola, Phoenix, AZ, 1975, AN-742, p 10.

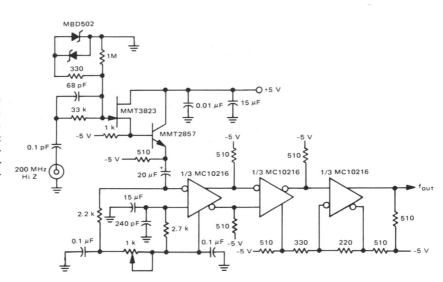

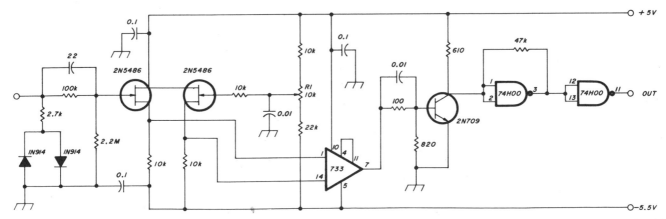

IMPROVED PREAMP—Replaces inefficient input circuit of inexpensive frequency counter, to ensure accurate counting from DC to over 60 MHz. Circuit brings input signal waveform to TTL level of 3.5 V P-P while providing required perfect square waves down to lowest-frequency input signal. Input circuit is balanced FET source-follower having extremely high input impedance. Back-to-back diodes provide overload protection. Input stage drives 733 differential video amplifier having 100-MHz bandwidth and gain of 400. 2N709 switching transistor squares preamp signal for TTL translator using two sections of 74H00 high-speed quad NAND gate. Circuit requires dual-polarity supply delivering at least 63 mA; regulation is optional.—G. Beltrami, High-Impedance Preamp and Pulse Shaper for Frequency Counters, *Ham Radio*, Feb. 1978, p 47–49.

25–250 MHz PRESCALER—Based on use of Fairchild 95H90 decade counter, with preamp Q1 and associated components selected for 25–250 MHz range. 1N914N diodes prevent overloading of input. Voltage regulator is LM340T-5. L1 is 8 turns No. 28 wound on body of 1000-ohm or larger ½-W resistor, with ends soldered to resistor leads. Quarter-wave whip antenna at input will pick up adequate signal from 1-W 146-MHz transceiver hand-held 6–10 feet away. Counter provides division by 10.—R. D. Shriner, Prescaler Updates the DVM/Frequency Counter, *QST*, Sept. 1978, p 22–24 and 37.

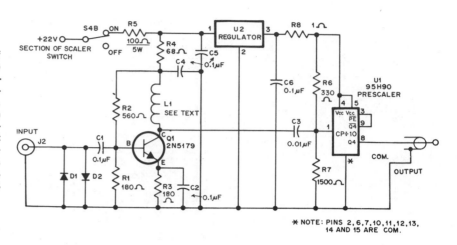

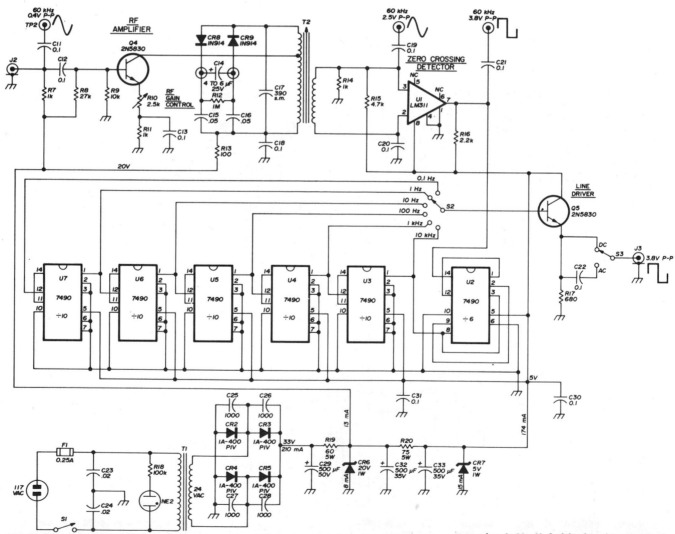

COUNTER DRIVE FOR WWVB—Uses LM311 as true zero-crossing detector for 60-kHz carrier of NBS standard-frequency station at Boulder, Colorado. Resulting square wave is fed to chain of 7490 dividers whose outputs are selected by S2 to serve as gate for frequency counter. T1 is 24-V 500-mA power transformer, and T2 is 40-kHz cup-core slug-tuned RF transformer as used in many TV remote controls. Primary inductance (7.5–46 mH) of T2 is tapped at 0.5 mH; secondary is 20 μH. Article also gives circuit of preamp that can be built into 60-kHz loop antenna to build up signal strength to 0.4 V as required for input to processor.—H. Isenring, WWVB Signal Processor, *Ham Radio*, March 1976, p 28–34.

10:1 PRESCALER FOR 500 MHz—Uses Fairchild 11C06 D flip-flop and Fairchild 95H91 divide-by-5 counter. Input sensitivity is less than 100 mV from 10 to 500 MHz. Back-to-back diodes protect 11C06 input from overloads. Output is fed to 50-MHz frequency counter. Use regulated supply.—W. C. Ryder, 500-MHz Decade Prescaler, *Ham Radio*, June 1975, p 32–33.

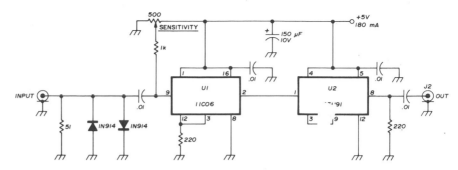

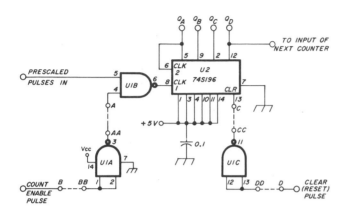

UPDATE TO 100 MHz—Simple counter stage can be added to input of existing frequency counter to extend direct counting range to 100 MHz, preparing it for use with 1-GHz prescaler. Use 74S196 presettable decade counter and 74S00 NAND-gate IC. If existing counter has positive reset pulse, connect C to CC and D to DD; if reset pulse is negative, connect C to D. If count enable pulse is negative, connect A to AA and B to BB; if positive, connect A to B. Power supply bypass capacitor should be shunted by 47-μF 10-V tantalum or electrolytic. Article covers modifications required in some counter input stages.—I. MacFarlane, How to Modify Your Frequency Counter for Direct Counting to 100 MHz, *Ham Radio*, Feb. 1978, p 26–29.

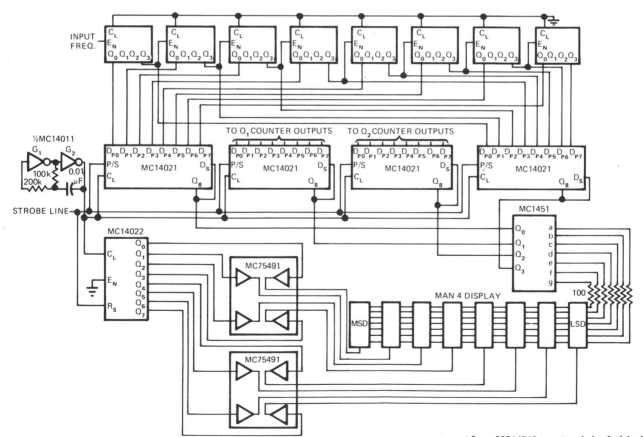

MULTIPLEXED DISPLAY—Used with battery-operated frequency counter to reduce battery drain by multiplexing single decoder driver between all of MAN-4 displays. Readout is integrated by eye over total time period, making display appear continuous. Display operates at peak of 20 mA but duty cycle is only 12.5%. Counter is operated from 6-V supply. Four MC14021 8-bit shift registers implement multiplexing and provide latches needed to store count from MC14518 counter chain. Article describes operation in detail and gives circuit of companion front end and 5-MHz counter.—D. Aldridge, CMOS Counter Circuitry Slashes Battery Power Requirements, *EDN Magazine*, Oct. 20, 1974, p 65–71.

500 MHz WITH 100-Hz RESOLUTION—Circuit provides separate 0–50 MHz preamp Q1-Q7 and 50–500 MHz prescaler for Intersil ICM7207A 7-digit CMOS frequency counter. 500-MHz prescaler uses Fairchild 11C90 that drives TTL directly, with 2N5179 transistor as preamp. L1 and output capacitance of 2N5179 form low-Q resonant circuit. U3 is 50-MHz prescaler for both preamps. Crystal frequency is 5.242880 MHz. 5-V regulators are MC7805, and 12-V regulator is MC7812. Article covers construction and adjustment.—J. H. Bordelon, Simple Front-Ends for a 500-MHz Frequency Counter, *Ham Radio*, Feb. 1978, p 30–33.

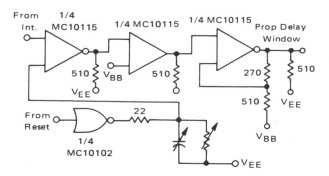

RAMP GENERATOR/COMPARATOR—Developed for frequency counter using standard ECL components. One input to comparator is from integrator stage, and other is from ramp generator driven by reset signal from UJT oscillator.—W. R. Blood, Jr., "Measure Frequency and Propagation Delay with High Speed MECL Circuits," Motorola, Phoenix, AZ, 1972, AN-586, p 3.

INPUT BUFFER FOR 100-MHz COUNTER—Can be used with 500-ohm probe for wide range of high-frequency input signal levels and waveforms, as part of frequency counter using standard emitter-coupled logic. Opamps used have 50-ohm input impedance. 450-ohm resistor in series with coax gives 10:1 attenuation factor (80 mV at amplifier input when measuring 800-mV ECL swing).—W. R. Blood, Jr., "Measure Frequency and Propagation Delay with High Speed MECL Circuits," Motorola, Phoenix, AZ, 1972, AN-586, p 3.

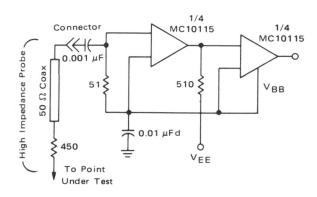

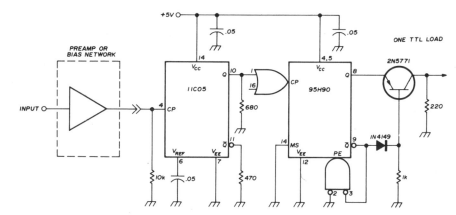

40:1 SCALER FOR 1200 MHz—Uses Fairchild 11C05 divide-by-4 counter and 95H90 decade divider. Unused CP input is tied to ground. Transistor translates ECL level to TTL for driving one unit load. Operates from single regulated power supply. Input may be AC or DC coupled so either input amplifier or simple bias network (also given in article) may be used. 10K resistor from pin 4 to ground eliminates noise triggering in middle frequency ranges.—D. Schmieskors, 1200-MHz Frequency Scalers, *Ham Radio*, Feb. 1975, p 38–40.

SCALER FOR CB—Low-cost prescaler for low-range frequency counter permits accurate monitoring of 450-MHz CB transceiver. Fairchild 11C90 decade counter gives division by 10 for counters covering up to 45 MHz. For lower-range counter, add 74196 TTL decade counter as shown to give total division by 100 for conversion to 4.5-MHz output. D1 and D2 should be fast-switching diodes such as 1N914 or 1N4148. Keep input signal under 1 V to avoid damaging 11C90. Will operate from 5-V supply or four D cells.—P. A. Stark, 500 MHz Scaler, *73 Magazine*, Oct. 1976, p 62–63.

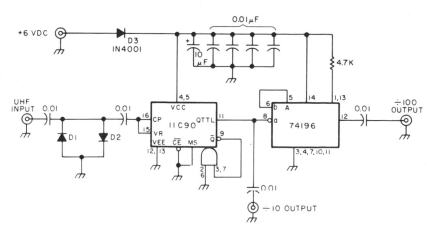

10:1 SCALER—Used to increase range of frequency counter. Sensitivity is 20 mV at 175 MHz, 40 mV at 220 MHz, and 90 mV at 250 MHz. Fairchild IC is used. Simple L-section filter at output rolls off frequencies above 30 MHz, allowing scaler to be used up to 250 MHz without erroneous counting of second or third harmonics of square-wave output of scaler if counter in use will respond to 60 MHz or more.—E. Guerri, Frequency Pre-Scaler, *Ham Radio,* Feb. 1973, p 57.

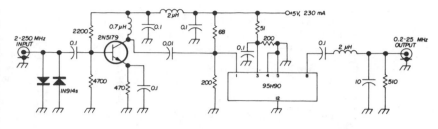

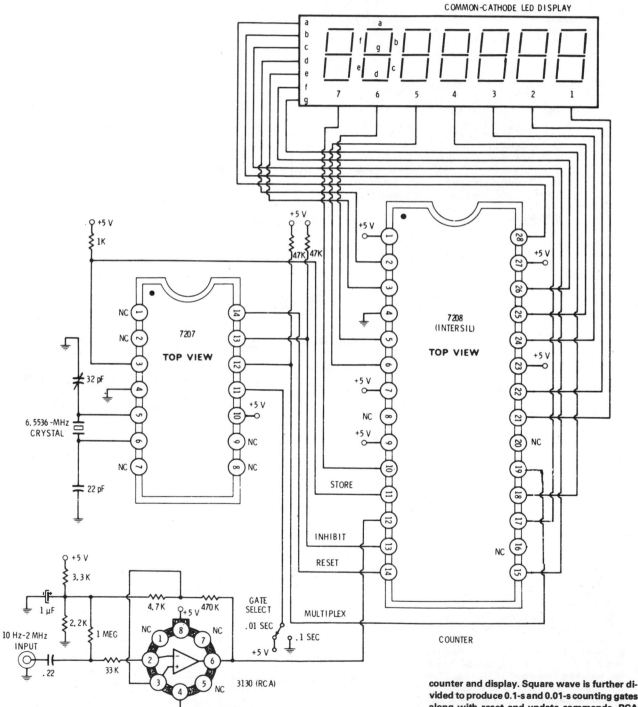

10 Hz TO 2 MHz—Seven-decade Intersil 7208 latched and multiplexed frequency counter with direct digit and display drive obtains timing waveform from 7207 IC which divides 6.5536-MHz crystal oscillator output by 2^{12} to produce 1600-Hz square wave for multiplexing counter and display. Square wave is further divided to produce 0.1-s and 0.01-s counting gates along with reset and update commands. RCA 3130 opamp is used for conditioning of input signal.—D. Lancaster, "CMOS Cookbook," Howard W. Sams, Indianapolis, IN, 1977, p 380–382.

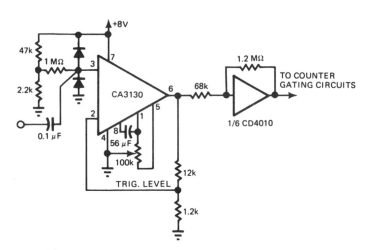

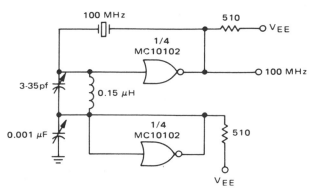

PREAMP FOR CMOS LOGIC—Combination of CA3130 and CD4010 ICs provides broad frequency response at very low power levels, as required for driving frequency counter. Diodes protect input from overvoltage. Amplifier offset control pot is used as trigger level control. Input sensitivity of amplifier/trigger combination is 50 mV P-P from 1 Hz to 1 MHz.—R. Tenny, Counter Pre-Amp Matches CMOS Logic Capability, *EDN Magazine,* Sept. 20, 1976, p 114 and 116.

100-MHz CRYSTAL OSCILLATOR—Developed for frequency counter that uses standard ECL components. Crystal can be changed to 10 MHz when measuring TTL performance.—W. R. Blood, Jr., "Measure Frequency and Propagation Delay with High Speed MECL Circuits," Motorola, Phoenix, AZ, 1972, AN-586, p 3.

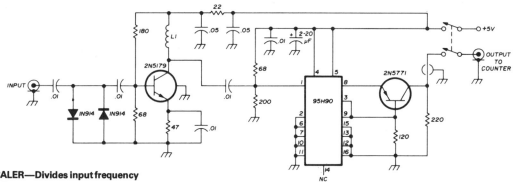

300-MHz PRESCALER—Divides input frequency by 10 for extending range of digital frequency counter up to prescaler limit of 300 MHz. Reading of counter must be multiplied by 10. Article also gives circuits of high-resolution counters using CMOS TTL devices.—H. E. Harris, Simplifying the Digital Frequency Counter, *Ham Radio,* Feb. 1978, p 22–25.

THREE TIME-BASE WINDOWS—Intersil 7207A crystal-controlled timer generates precision gate windows of 10 ms, 100 ms, and 1 s for use as time bases, calibration markers, or gate timers for frequency counters such as 7208.—B. O'Neil, IC Timers—the "Old Reliable" 555 Has Company, *EDN Magazine,* Sept. 5, 1977, p 89–93.

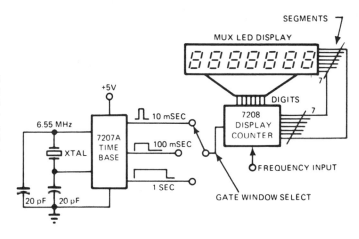

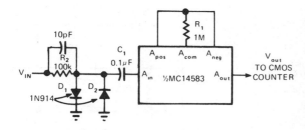

LOW-DRAIN 2-MHz FRONT END—Simple CMOS linear front end for 5-MHz battery-operated counter reduces power drain and makes it proportional to input frequency. With no input, drain is only a few microamperes. Half of MC14583 CMOS Schmitt trigger forms front end operating from single 6-V battery used in counter. Upper frequency limit is about 3 MHz, and input sensitivity is 400 mV.—D. Aldridge, CMOS Counter Circuitry Slashes Battery Power Requirements, *EDN Magazine,* Oct. 20, 1974, p 65–71.

CHAPTER 32
Frequency Divider Circuits

Provide division ratios in range from 2 to 29 for clock-signal generators, receivers, transmitters, and event counters. See also Clock Signal, Digital Clock, Frequency Multiplier, Frequency Synthesizer, and Logic chapters.

DIVIDE BY 7—Requires only two different types of chips. Input clock is alternately inverted and noninverted by gates operating in conjunction with 3 bits of storage using 852, to give square-wave output at one-seventh of clock frequency.—C. W. Hardy, Reader Responds to Odd Modulo Divider in July 1st EDN, *EDN Magazine*, Oct. 1, 1972, p 50.

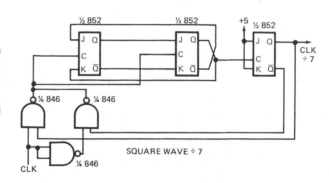

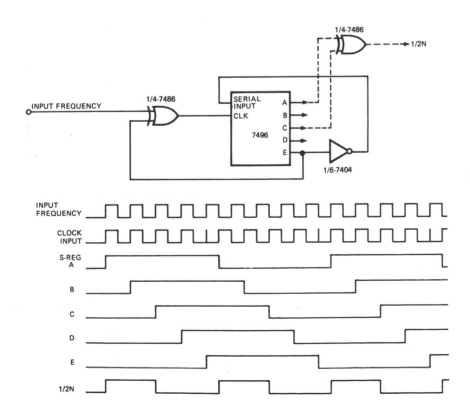

DIVIDE BY 9 WITH SHIFT REGISTER—Uses 7496 as 5-bit shift register, 7486 as EXCLUSIVE-OR gate, and 7404 as inverter to give division of square-wave input frequency by 9 while maintaining 50% duty cycle at output. Article covers connection changes needed for other division ratios. With 8-bit shift register, circuit will divide by as much as 15. Addition of 7486 EXCLUSIVE-OR gate across any outputs, as shown by dashed lines, makes effective output half that of basic TTL circuit.—J. N. Hobbs, Jr., Divide-by-N Uses Shift Register, *EDN Magazine*, Oct. 5, 1976, p 108.

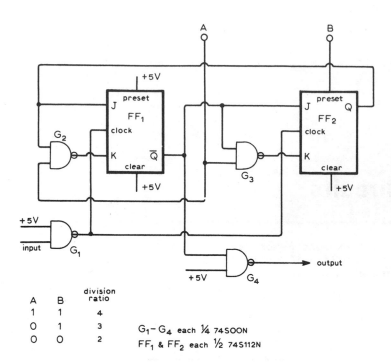

DIVIDING 40–60 MHz BY 2, 3, OR 4—Division ratio is controlled externally by making control terminals A and B high (1) or low (0), as given in table. Developed for use in receiver requiring local oscillator covering 10 to 30 MHz. Counter simplifies tuner design.—C. Attenborough, Fast Modulo-3 Counter, *Wireless World,* Aug. 1976, p 52.

A	B	division ratio
1	1	4
0	1	3
0	0	2

G_1–G_4 each ¼ 74S00N
FF_1 & FF_2 each ½ 74S112N

AF DIVISION BY 2 TO 11—Ratio of C2 to C1 determines division ratio, as given in table. When C2 charges to peak point firing voltage of Q2, it fires and discharges C2, so C1 charges to line voltage. Q2 then turns off. Next cycle begins with another positive pulse on base of Q1, discharging C1. Division range can be changed by utilizing programmable aspect of PUT Q2 and changing ratio of resistances.—R. J. Haver and B. C. Shiner, "Theory, Characteristics and Applications of the Programmable Unijunction Transistor," Motorola, Phoenix, AZ, 1974, AN-527, p 9.

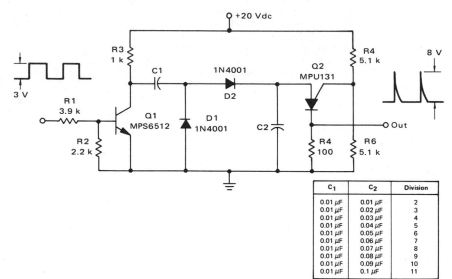

C_1	C_2	Division
0.01 μF	0.01 μF	2
0.01 μF	0.02 μF	3
0.01 μF	0.03 μF	4
0.01 μF	0.04 μF	5
0.01 μF	0.05 μF	6
0.01 μF	0.06 μF	7
0.01 μF	0.07 μF	8
0.01 μF	0.08 μF	9
0.01 μF	0.09 μF	10
0.01 μF	0.1 μF	11

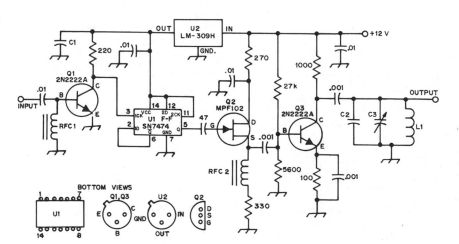

DIVIDER FOR 7 MHz—Used after 7-MHz VFO of 40-meter transmitter, to provide 3.5 MHz as required for operation in 80-meter band. Half of 7474 TTL D flip-flop U1 is connected in divide-by-2 configuration. U2 provides required well-regulated 5-V source. Q1 clips negative-going portion of 7-MHz sine wave to prevent damage to 7474. Square-wave 3.5-MHz output from U1 is applied to source-follower Q2 which drives class A amplifier output stage Q3. RFC1 and RFC2 are 10 μH. C1 is 6.8-μF 10-V tantalum. C2 is 100-pF mica. L1 is 41 turns No. 26 enamel spaced to fill entire T-80-2 core, to give 10 μH.— S. Creason, A VFO Frequency Divider, *QST,* Nov. 1976, p 23–24.

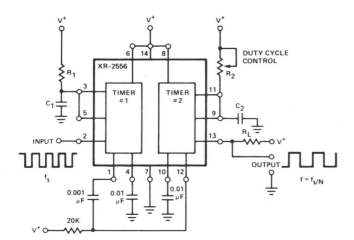

DIVIDER WITH PULSE SHAPER—Half of Exar XR-2556 dual timer divides input pulse frequency by 2 or 3, and other half shapes output pulse by controlling duty cycle over range that can be adjusted from 1% to 99% with R₂. Supply voltage can be 4.5–16 V.—"Timer Data Book," Exar Integrated Systems, Sunnyvale, CA, 1978, p 23–30.

DUAL-TIMER TONE STEPPER—One section of RS556 dual timer is connected as free-running astable MVBR for supplying pulses to trigger input of other section connected as mono MVBR driving loudspeaker. When both MVBRs are adjusted so one trigger pulse initiates each timing period and no trigger pulses occur during timing periods, output tone has frequency of free-running MVBR. With two trigger pulses per timing cycle, every other trigger pulse is ignored and tone is at half frequency. With three trigger pulses per cycle, output is one-third of frequency. Can be used for classroom demonstration of electronic music; settings of R1 and R3 can be adjusted to give tones resembling violin, bagpipes, or almost any other instrument.—F. M. Mims, "Integrated Circuit Projects, Vol. 6," Radio Shack, Fort Worth, TX, 1977, p 70–78.

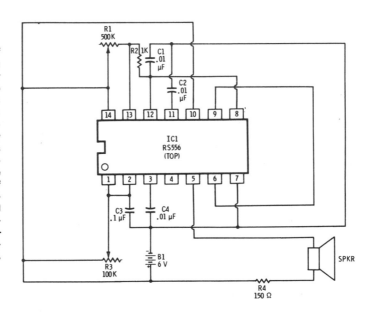

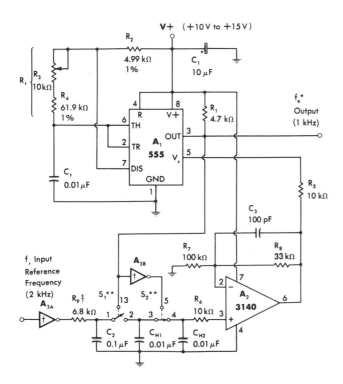

PLL DIVIDER—Simple phase-locked loop is suitable for generating integral submultiples M of input frequency. Values shown give M of 2. Square-wave input reference is limited in amplitude to supply voltage by first CMOS inverter A_{3A}. RC network R_9-C_2 integrates output to give 2 V P-P triangle across C_2 for sampling by sample-and-hold switch sections S_1 and S_2 of 4016 CMOS analog switch. Sampled error voltage of loop, stored on C_{H2}, is read out by FET amplifier A_2. Amplified error voltage is applied to A_1 through R_5 to induce changes in center frequency of A_1 as required to maintain locked condition.—W. G. Jung, "IC Timer Cookbook," Howard W. Sams, Indianapolis, IN, 1977, p 220–224.

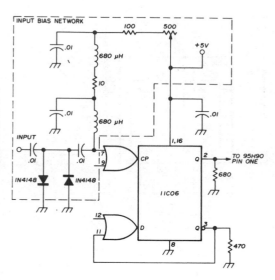

UHF PRESCALER—Uses Fairchild 11C06 700-MHz D flip-flop as divide-by-20 UHF prescaler with toggle rates in excess of 550 MHz from 0 to 75°C. Amplifier may be used in place of input bias network shown. Developed for use with 95H90 decade divider. Unused CP and D inputs are tied to ground.—D. Schmieskors, 1200-MHz Frequency Scalers, *Ham Radio*, Feb. 1975, p 38–40.

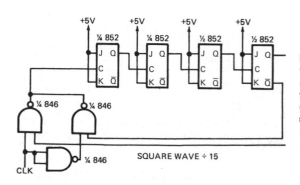

DIVIDE BY 15—Input clock is alternately inverted and noninverted by gates operating in conjunction with 4 bits of storage using 852 JK flip-flops, to give square-wave output at 1/15 of clock frequency.—C. W. Hardy, Reader Responds to Odd Modulo Divider in July 1st EDN, *EDN Magazine*, Oct. 1, 1972, p 50.

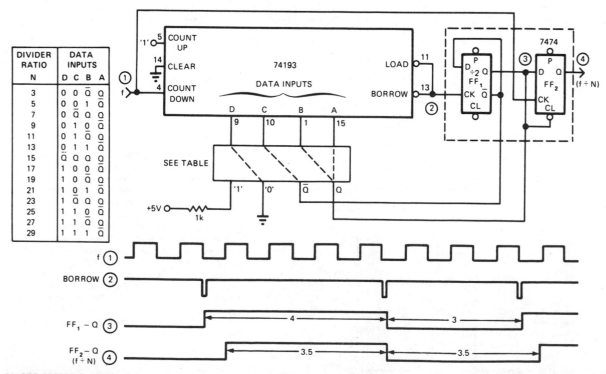

DIVIDER RATIO N	DATA INPUTS			
	D	C	B	A
3	0	0	$\bar{Q}$	Q
5	0	0	1	$\bar{Q}$
7	0	$\bar{Q}$	Q	$\bar{Q}$
9	0	1	0	$\bar{Q}$
11	0	1	$\bar{Q}$	Q
13	0	1	1	$\bar{Q}$
15	$\bar{Q}$	Q	Q	$\bar{Q}$
17	1	0	0	$\bar{Q}$
19	1	0	$\bar{Q}$	Q
21	1	0	1	$\bar{Q}$
23	1	$\bar{Q}$	Q	Q
25	1	1	0	$\bar{Q}$
27	1	1	$\bar{Q}$	Q
29	1	1	1	$\bar{Q}$

3 TO 29 ODD-MODULO—Basic divider using 74193 4-bit up/down counter and single 7474 dual D flip-flop provides any odd number of divider ratios from 3 to 29 by changing feedback connections as shown in table, all with symmetrical output waveforms. Based on writing any odd number N as $N = M + (M + 1)$, where M is integer. Circuit forces counter to divide alternately by M and M + 1. Connection shown is for divide-by-7.—V. R. Godbole, Simplify Design of Fixed Odd-Modulo Dividers, *EDN Magazine*, June 5, 1975, p 77–78.

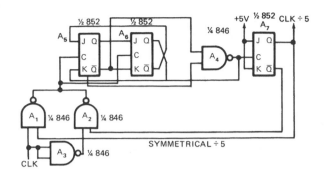

DIVIDE BY 5—Requires only two digital chip types. Input clock is alternately inverted and noninverted for clocking divide-by-3 counter, to give effect of dividing by 2½ which toggles A_3 to give symmetrical divide-by-5 output with 50% duty cycle for pulses. Article gives timing diagram and traces operation of circuit.—C. W. Hardy, Reader Responds to Odd Modulo Divider in July 1st EDN, *EDN Magazine*, Oct. 1, 1972, p 50.

SQUARE-WAVE DIVIDER—Divides input square wave by 1, 2, 5, or 10 depending on which switch is open. Signal at OUT-1 is inverted with respect to input, and OUT-2 is noninverted.—Circuits, *73 Magazine*, June 1977, p 49.

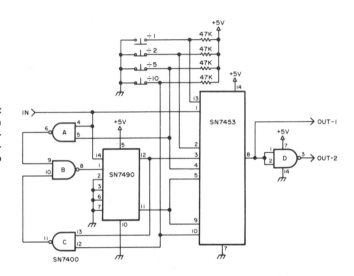

PROGRAMMABLE COUNTER—Input frequency can be divided by any number between 1 and 16 by pressing appropriate key on keyboard connected to National MM74C922 16-key encoder. Output frequency is symmetrical for odd and even divisors. Can be used for simple frequency synthesis or as keyboard-controlled CRO trigger. Operates over standard CMOS supply range of 3–15 V. Typical upper frequency limit is 1 MHz with 10-V supply. Circuit uses two MM74C74 dual D flip-flops and MM74C86 EXCLUSIVE-OR package.—"CMOS Databook," National Semiconductor, Santa Clara, CA, 1977, p 5-50–5-51.

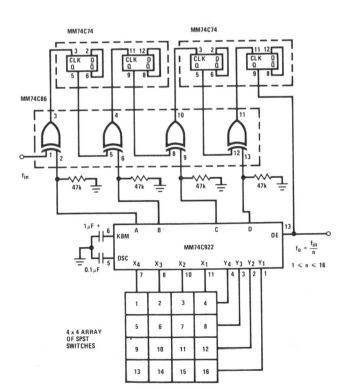

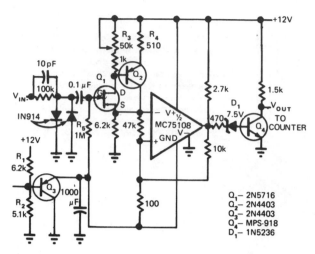

10-MHz FRONT END—Front-end design for battery-operated 5-MHz counter consists of FET and bipolar buffer followed by Schmitt trigger made from MC75108 dual line receiver. Circuit operates linearly up to 10 MHz with 25-mV input signal. Requires swings from logic 0 (0 V) to logic 1 (about 10 V), for which suitable counter circuit is given in article. Accepts any input waveform shape and level.—D. Aldridge, CMOS Counter Circuitry Slashes Battery Power Requirements, *EDN Magazine*, Oct. 20, 1974, p 65–71.

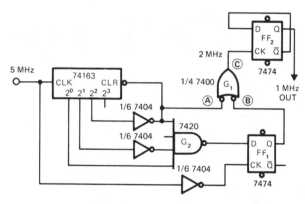

SYMMETRICAL DIVIDE-BY-5 CLOCK—Uses 74163 counter to generate two phases of 1-MHz clock pulse with 50% duty cycle from 5-MHz system reference. One phase is decode of binary 4 from counter, while other is decode of 1 clocked at midbit time. Both phases are recombined in gate G_1 to give 2-MHz clock that toggles FF_2 to generate desired 1-MHz output.—L. A. Mann, Divider Circuit Maintains Pulse Symmetry, *EDN Magazine*, July 1, 1972, p 54–55.

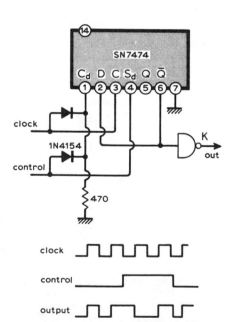

SWITCHED DIVIDER FOR BINARY COUNTER—Simple circuit provides method of switching division by two into or out of stream of clock pulses. Output is in phase with input and free of spikes. Switching requires only one D-type flip-flop and one inverter. When control is high, logic action gives normal connection for division by two, using D-type flip-flop; inverter then restores phase.—J. M. Firth, Control of a Binary Counter for Division by One or Two, *Wireless World*, Jan. 1975, p 12.

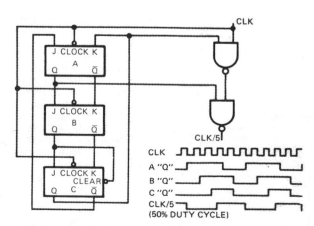

DIVIDE BY 5 WITH TWO GATES AND 3 BITS—Arrangement shown for dividing clock input frequency by 5 requires only two gates from 846 IC and 3 bits of 852 JK flip-flop storage to give square-wave output pulses having 50% duty cycle.—C. L. Maginniss, Another Reader Responds to Odd Modulo Divider, *EDN Magazine*, Oct. 15, 1972, p 57.

CHAPTER 33
Frequency Measuring Circuits

Includes direct-reading heterodyne frequency meters, synchroscopes, dip meters, tuning indicators, frequency-to-voltage converters, tachometers, and monitors showing when input frequency or pulse rate is above or below reference, for variety of frequencies in range from 1 Hz through power-line and audio values to 150 MHz. See also Frequency Counter and Frequency Multiplier chapters.

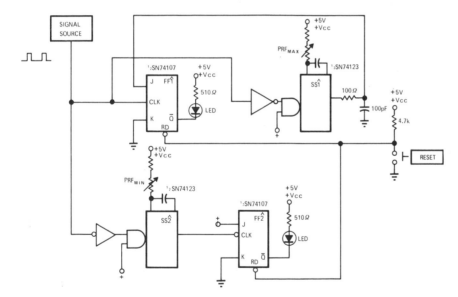

PRF MONITOR—Upper channel latches when pulse repetition frequency of train of pulses is higher than specified limits, turning on above-limit LED driven by JK flip-flop FF_1. Lower channel latches and turns on its LED when PRF is below second specified limit. Upper channel also detects single noise pulse, while lower channel detects single missing pulse. After off-limit indication, circuit must be reset.—L. Birkwood and D. Porat, PRF Monitor with Adjustable End Limits, *EDN/EEE Magazine,* Feb. 1, 1972, p 57–58.

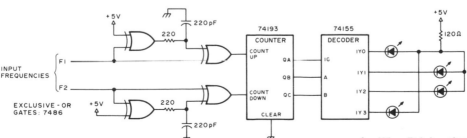

BEAT-FREQUENCY DISPLAY—Apparent rotation of dot on four-LED display gives indication of beat frequency between two tone oscillators. When F1 is greater than F2, dot rotates clockwise. When F1 is less than F2, dot rotates counterclockwise. When F1 equals F2, dot does not move.—*Circuits, 73 Magazine,* July 1977, p 35.

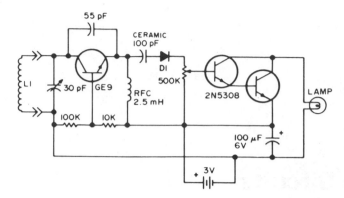

GRID-DIP METER—Uses ordinary No. 48 or 49 pilot lamp as resonance indicator. Will oscillate at frequencies up to 12 MHz. Wind L1 to cover desired frequency ranges.—Circuits, *73 Magazine,* April 1973, p 133.

LED SYNCHROSCOPE—Circuit uses four LEDs to indicate direction of phase error as correct setting is approached when tuning oscillator to standard frequency. Lamps form display that rotates once per cycle at reference frequency, with brightness of each lamp being modulated at frequency of oscillator being adjusted. Display thus appears to have frequency equal to difference between two signal frequencies, rotating in direction indicative of sense of frequency difference. Mount lamps on smallest possible circle. Diode and LED types are not critical.—R. H. Pearson, An L.E.D. Synchroscope, *Wireless World,* Sept. 1974, p 321.

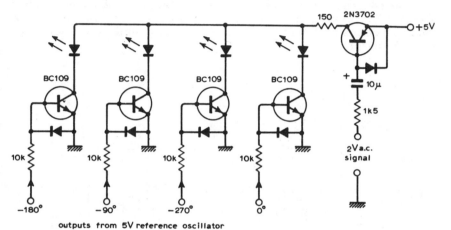

outputs from 5V reference oscillator

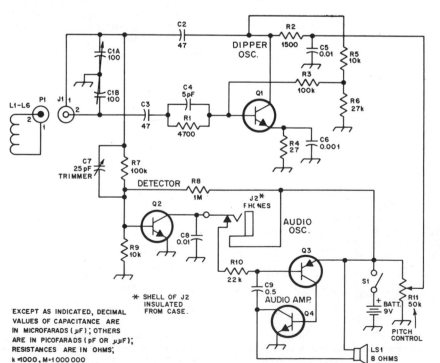

AUDIBLE DIPPER—Designed for use by blind radio amateurs, but tone indication has advantage of permitting anyone to concentrate on equipment while checking antenna, tracking parasitics, or neutralizing amplifier with dip meter. Plug-in coils L1-L6 are Heathkit parts 40-1689 through 40-1695. Q1, Q2, and Q4 are Radio Shack RS-2011 or equivalent, and Q3 is RS-2021 or equivalent. Pitch of tone heard from loudspeaker drops sharply when tuned circuit of dipper becomes loaded by external source.—W. E. Quay, An Auditory Dip Oscillator, *QST,* Sept. 1978, p 25—27.

EXCEPT AS INDICATED, DECIMAL VALUES OF CAPACITANCE ARE IN MICROFARADS (μF); OTHERS ARE IN PICOFARADS (pF OR $\mu\mu$F); RESISTANCES ARE IN OHMS; k =1000, M=1 000 000

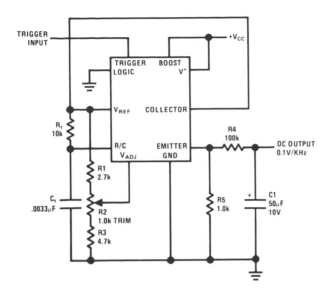

FREQUENCY-TO-VOLTAGE CONVERTER—National LM122 timer is used as tachometer by averaging output pulses with simple filter. Pulse width is adjusted with R2 to provide initial calibration at 10 kHz. Linearity is about 0.2% for output range of 0–1 V. Analog meter can be driven directly by connecting it in series with R5. Supply can range from 4.5 to 40 V.—C. Nelson, "Versatile Timer Operates from Microseconds to Hours," National Semiconductor, Santa Clara, CA, 1973, AN-97, p 10.

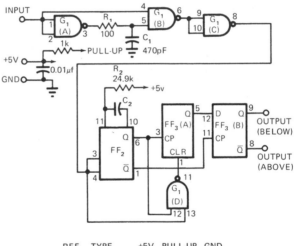

RATE DETECTOR—Only three ICs are used to sense pulse rate of input signal with high accuracy. For monitoring frequency, two such circuits can be used, with one set to upper frequency limit and other to lower limit. Output is high when input pulse rate is above set point and low for frequencies below set point. Frequency of set point is reciprocal of monostable delay time ($f_0 = 1/0.32R_2C_2$).—J. W. Poore, Three IC's Accurately Sense Pulse Rate, *EDN Magazine*, Aug. 15, 1972, p 53.

REF	TYPE	+5V	PULL-UP	GND
G_1	SN7400N	14		7
FF_2	SN54121N	14	5	7
FF_3	SN7474N	14	2, 4, 10, 13	7

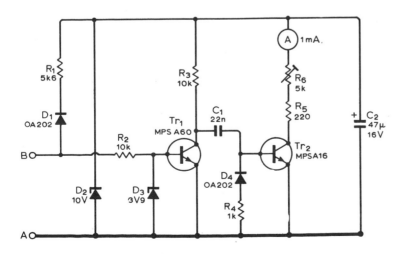

STANDBY-GENERATOR FREQUENCY METER—Developed for use with 10–60 VAC generator driven by lawn mower engine, as guide for adjusting speed manually to give correct power-line frequency. Output of alternator, connected to A and B, is converted to regulated 10 VDC by R_1, D_1, D_2, and C_2. Same input voltage is squared by Tr_1 and fed to Tr_2 through differentiating circuit. Current pulses developed in collector circuit of Tr_2 have constant width and varying repetition rate depending on input frequency. Inertia of meter movement provides integration required to give steady reading that changes only with input frequency. Meter scale is calibrated from 0 to 100 Hz, with R_6 adjusted to give correct reading when 10–60 VAC line voltage is applied to input. Power transformer must be used to boost output of alternator to correct AC line voltage.—J. M. Caunter, Low-Cost Emergency Power Generator, *Wireless World*, Feb. 1975, p 75–77.

TRANSISTORS FOR BC-221—Old BC-221 frequency meter can be modernized by replacing its three now-scarce tubes with four 2N3819 N-channel JFETs and changing supply to single 9-V battery. VT167 (6K8) mixer-oscillator is replaced by two JFETs with R2, R3, and C1 mounted inside octal plug. Resistance values may need some adjustment. Cut and insulate original leads to pins 2, 4, and 7, and connect top-cap clip of mixer tube to pin 4. VT116-B (6SJ7-Y) tube used for VFO is replaced by single 2N3819 connected as for Q1. Add R1 in parallel with plate load resistor; value depends on particular FET used, and can range from 1 to 6800 ohms (1500 is typical). VT116 (6SJ7) beat-frequency amplifier is replaced by 2N3819 mounted same as for Q1. Place 4K across 15K load resistor of VT116-B and replace 300-ohm cathode resistor with one giving 1-mA source current (typically 1K to 3.3K). Total current drain is about 3 mA. Try 3–10 pF capacitor between gate and drain of Q3 if circuit does not oscillate.—R. S. N. Rau, Solid-State BC-221 Frequency Meter, *QST*, Feb. 1977, p 35–36.

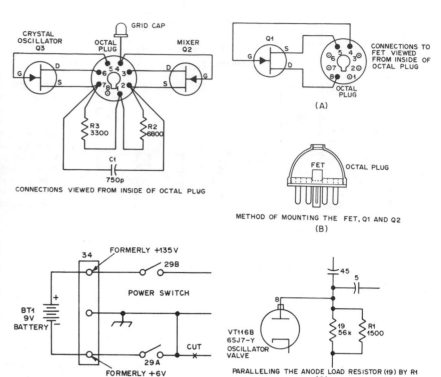

CONNECTIONS VIEWED FROM INSIDE OF OCTAL PLUG

CONNECTIONS TO FET VIEWED FROM INSIDE OF OCTAL PLUG

(A)

METHOD OF MOUNTING THE FET, Q1 AND Q2

(B)

PARALLELING THE ANODE LOAD RESISTOR (19) BY R1

(C)

Frequency range (MHz)	Coil diameter and turns				Wire size and coil const.
	1-1/2	1-3/8	1-1/4	1 (inches)	
	38	35	32	25 (mm)	
.08 — 0.2	700	750	800	1000	No. 30 enam., 5 pies.
0.205 — 0.6	220	240	256	310	No. 30 enam., 5 pies.
0.5 — 1.4	90	100	110	140	No. 30 enam., close wound.
0.95 — 3	40	45	49	64	No. 22 enam., close wound.
2.6 — 6	18	20	22	29	No. 22 S. C. enam., 1-3/4" long (44 mm)
5.5 — 15	7	8	9	12	No. 22 S. C. enam., 1-1/2" long (38 mm)
14 — 35	2 (11/16") (17 mm)	2 (3/4") (19 mm)	2 (1") (25 mm)	2-1/2 (1") (25 mm)	No. 22 S. C. enam., length indicated at bottom of column.

GATE DIPPER—Used to determine resonant frequency of tuned circuit, provide signal for receiver alignment, and make antenna measurements. Table gives winding data for plug-in coils L1. Parts values are not critical. T2 is transistor interstage audio transformer with 10,000-ohm primary and 2,000-ohm secondary in meter circuit. JFET Q1 is used in common-drain circuit followed by PNP bipolar transistor, with gate junction of JFET acting as rectifier. Dip meter M1 measures gate current. When tuned circuit of dipper is loaded by coupling it to external circuit, power is absorbed and meter reads dip occurring when L1-C1 is tuned to resonance with external circuit. R1 is regeneration control. Audio amplifier Q3-Q4 using 2N4125 or HEP52, optional, helps in listening to signals picked up by tuned circuit or enhances display on CRO. Can be used as field-strength meter if antenna is plugged into J1.—B. Clark, A Hybrid Gate-Dip Oscillator, *QST*, June 1974, p 33–37.

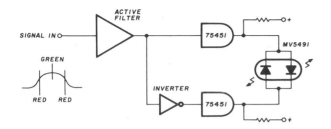

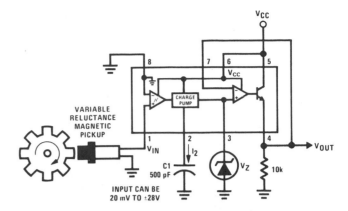

FREQUENCY INDICATOR—Circuit furnishes green indication at resonance and red for either side of resonance. Uses Monsanto MV5491 dual red/green LED, with 220 ohms in upper lead to +5 V supply and 100 ohms in lower +5 V lead because red and green LEDs in parallel back-to-back have different voltage requirements. Useful for SSTV, RTTY, or subaudio-tone indication for control purposes on FM. Circuit requires two driver ICs and one section of hex inverter IC, with any suitable active filter used to form level detector for signals at desired frequency.—K. Powell, Novel Indicator Circuit, *Ham Radio*, April 1977, p 60–63.

FREQUENCY-DOUBLING TACHOMETER—Connection shown for National LM2907 IC provides output pulse each time sine-wave input from magnetic pickup crosses zero, for use in digital control system. Width of each pulse is determined by size of C1 and supply voltage used. Circuit serves for doubling frequency presented to microprocessor control system.—"Linear Applications, Vol. 2," National Semiconductor, Santa Clara, CA, 1976, AN-162, p 12–13.

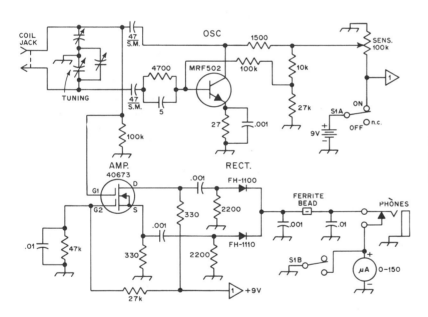

DIPPER—Circuit of Heath HD-1250 solid-state dip meter covers 1.6–250 MHz range with six plug-in coils.—The Heath HD-1250 Dip Meter, *QST*, Jan. 1976, p 38–39.

BICYCLE SPEED ALARM—Useful for long-distance bicycling, to indicate when rider drops below predetermined minimum speed. Speed sensor is reed switch attached to frame and tripped once per revolution by permanent magnet mounted on wheel. Rate at which switch closes determines level of DC voltage produced by circuit. When voltage drops below preset level determined by 100K pot, output transistor comes on and energizes relay controlling bicycle horn or other signaling device. Supply can by 9-V transistor battery. Transistor reading should be high enough to handle relay used.— J. Sandler, 9 Projects under $9, *Modern Electronics*, Sept. 1978, p 35–39.

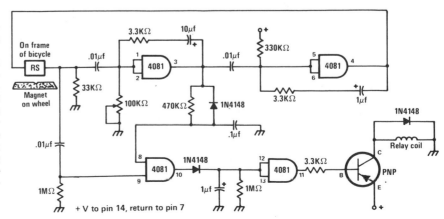

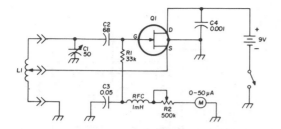

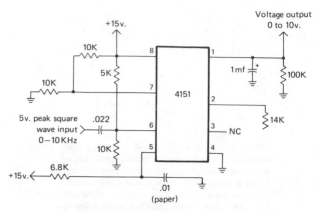

GATE DIPPER—Solid-state version of vacuum-tube grid-dip meter gives dip from 50 to about 20 μA on most bands in range of 1.8 to 150 MHz when dipper is held 1 inch away from resonant circuit under test. Uses Siliconix 2N5398 UHF JFET, but MPF107 (2N5486) can also be used. Coil tap position is more critical at higher frequencies; adjust tap for most pronounced dip. Article gives coil data for five frequency ranges.—C. G. Miller, Gate-Dip Meter, *Ham Radio*, June 1977, p 42–43.

0–10 kHz TO 0–10 VDC—Raytheon 4151VFC voltage-to-frequency converter is used in reverse as linear frequency-to-voltage converter. Applications include use in pairs as complete data transmission system, for remote monitoring of DC voltage such as output of SWR bridge located at junction of antenna with transmission line. DC voltage is changed into audio voltage at remote location, sent over lines, then changed back to DC at readout location. Line characteristics do not affect frequency of audio signal.—J. J. Schultz, A Voltage-to-Frequency Converter IC with Amateur Applications, *CQ*, Jan. 1977, p 39–41 and 75.

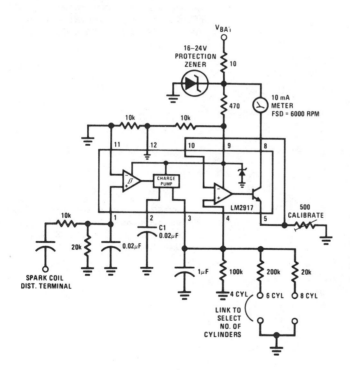

SPARK-COIL TACHOMETER—Input to National LM2917 tachometer IC is taken from spark-coil distributor terminal of gasoline engine. Frequency of input signal is converted to voltage for driving meter. Circuit is set up for number of cylinders on engine by adding link for appropriate timing resistor. Zener protects IC from transients found in auto battery circuit.—"Linear Applications, Vol. 2," National Semiconductor, Santa Clara, CA, 1976, AN-162, p 9–10.

HETERODYNE FREQUENCY METER—Circuit consists of 1–2 MHz oscillator Q1, untuned mixer X1, and AF beat-note amplifier Q2. C4 is calibrated to read directly in frequency from 1 to 2 MHz, using accurate unmodulated RF signal generator. After calibration, unknown RF signal input frequency is fed into meter for zero-beating with harmonics of calibrated oscillator. Magnetic headphones plugged into J1 make beat note audible. On second harmonic, dial of C4 covers 2–4 MHz; on twentieth harmonic, coverage is 20–40 MHz. L1 is 65 turns No. 28 enamel on 1-inch form, tapped 20 turns from ground. L2 is 10 turns No. 28 enamel close-wound around center of L1.—R. P. Turner, "FET Circuits," Howard W. Sams, Indianapolis, IN, 1977, 2nd Ed., p 144–146.

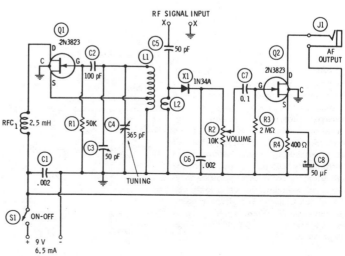

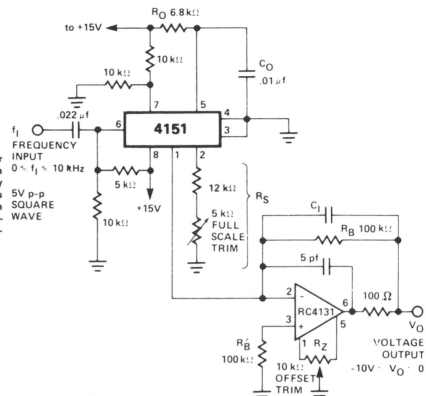

HIGH-PRECISION F/V—Use of integrator opamp with frequency-to-voltage connection of RM4151 converter gives increased accuracy and linearity for converting square-wave inputs of 0–10 kHz to proportional output voltage in range of −10 V to 0 V.—"Linear Integrated Circuit Data Book," Raytheon Semiconductor Division, Mountain View, CA, 1978, p 7-39.

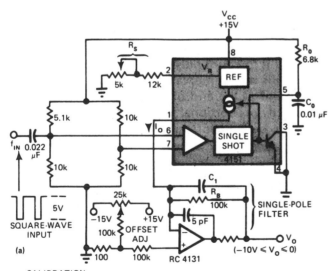

CALIBRATION:
1. SET f_{IN} TO 10 Hz. ADJUST OFFSET FOR −10 mV OUTPUT.
2. SET f_{IN} TO 10 kHz. ADJUST OFFSET FOR −10V OUTPUT.

F/V WITH 4151—Uses Raytheon 4151 as frequency-to-voltage converter for generating current pulses having precise amplitude and width. Average value of output pulse train is directly proportional to input frequency. Article gives design equations. Response time can be improved and ripple reduced by using second-order (double-pole) low-pass filter as shown in diagram (b). Ripple is less than 0.1 V P-P over range of 10 to 10,000 Hz when R_1 and R_2 are 100K and C_1 and C_2 are 0.1 μF.—T. Cate, IC V/F Converters Readily Handle Other Functions Such as F/V, A/D, *EDN Magazine*, Jan. 5, 1977, p 82–86.

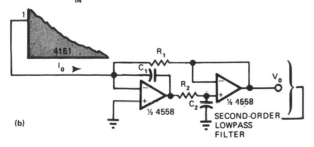

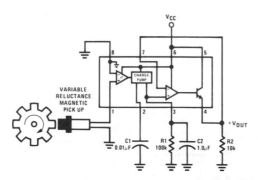

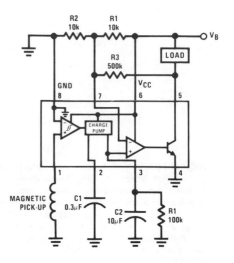

TACHOMETER USING MAGNETIC PICKUP— Signal frequency proportional to shaft speed being measured is fed into National LM2907 IC for conversion to output voltage that is proportional to input frequency. Output is zero at zero frequency. Quality of timing capacitor C1 determines accuracy of unit over temperature range. Use equivalent zener-regulated LM2917 IC if output voltage must be independent of variations in supply voltage.—"Linear Applications, Vol. 2," National Semiconductor, Santa Clara, CA, 1976, AN-162, p 3–4.

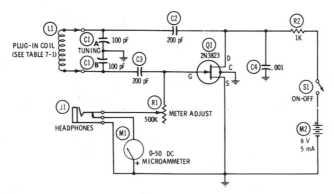

FET DIP OSCILLATOR—High input impedance of FET makes performance comparable to that of tube-type grid-dip oscillator. Six plug-in coils are wound on Millen 45004 1-inch 4-pin forms or equivalent. Use 150 turns No. 32 enamel for 1.1–2.5 MHz, 77 turns No. 28 for 2.5–5 MHz, 35 turns No. 22 for 5–11 MHz, 17 turns No. 22 spaced to 1 inch for 10–25 MHz, 8.5 turns No. 22 spaced to 1 inch for 20–45 MHz, and 4.5 turns No. 22 spaced to 1 inch for 40–95 MHz. Adjust R1 to set meter pointer to desired portion of scale before tuning for dip. R1 provides some control of volume when using headphones.—R. P. Turner, "FET Circuits," Howard W. Sams, Indianapolis, IN, 1977, 2nd Ed., p 134–136.

SPEED SWITCH—National LM2907 tachometer IC is used as switch to energize load when input frequency exceeds value corresponding to predetermined speed limit. Automotive applications include use as overspeed warning that activates audible and/or visual indicator when auto speed exceeds legal limit or other desired value. Another application is increasing intensity of auto or taxi horn above predetermined speed such as 45 mph. Input is variable-reluctance magnetic pickup positioned against teeth of gear wheel; in typical setup, pickup output is 16.6 Hz at 60 mph. Values shown for comparator-controlling components R1 and C1 (below IC) give switching operation at about 16.6 Hz at input. Report gives design procedure for other frequencies.—"Linear Applications, Vol. 2," National Semiconductor, Santa Clara, CA, 1976, AN-162, p 8.

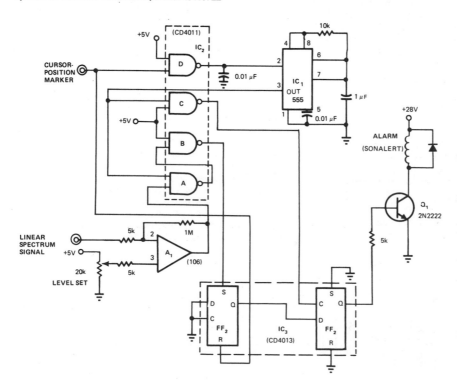

ALARM FOR SPECTRUM ANALYZER—Circuit drives audible alarm when frequency of interest appears in spectrum range. Display cursor can be preset to initiate narrow search band in which f_x is expected to appear. 100-μs pulse representing cursor position in display sweep triggers mono IC_1 so its output becomes window whose time-out is equivalent to band in which f_x is center. Comparator A_1 supplies high output when f_x appears. Simultaneous arrival of this signal and timer window at gate A sets output of left flip-flop high. At end of window period, right flip-flop also goes high and initiates alarm via Q_1. Loss of f_x stops alarm.—R. L. Messick, Alarm Simplifies Spectrum-Analyzer Measurements, *EDN Magazine*, June 5, 1978, p 152.

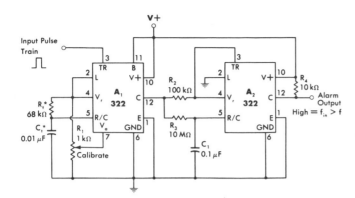

SPEED ALARM—Frequency detector using two IC timers provides alarm output when input frequency is greater than reference frequency, corresponding to overspeed. Calibrated mono MVBR A₁ produces fixed-width positive pulse across R₂, with average voltage of pulse varying linearly with input pulse train frequency. Comparator A₂ changes states when integrated output of R₃-C₁ on pin 5 goes above or below 2-V voltage threshold of A₂. With values shown, desired frequency is 1 kHz and circuit detects frequency variation of less than 1%. If low-frequency alarm is desired, connect logic input pin 2 of A₂ to reference voltage (pin 4) instead of to ground.—W. G. Jung, "IC Timer Cookbook," Howard W. Sams, Indianapolis, IN, 1977, p 228–230.

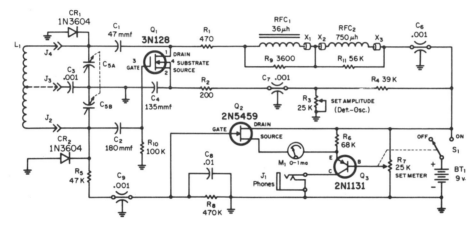

GRID-DIP OSCILLATOR—Millen 90652 solid-state grid-dip oscillator uses MOSFET operating in split-Colpitts circuit with resonating tank connected between drain and gate. Circuit is tuned by split-stator variable capacitor with rotor grounded, chosen to cover 1.7 to 300 MHz with seven plug-in coils. Oscillator also functions as Q multiplier that increases sensitivity. RF voltage across tuned circuit is indicated by meter whose reading dips for resonance with coupled test circuit. Full-wave rectifier CR₁-CR₂ provides DC voltage for meter and some over-load protection for MOSFET. Meter is suppressed-zero type, with readings only for upper portion of current range. J₃ is provided for use with low-frequency coils.—W. M. Scherer, CQ Reviews: The Millen Model 90652 Solid-State Dipper, CQ, Sept. 1971, p 63–64, 66, and 96.

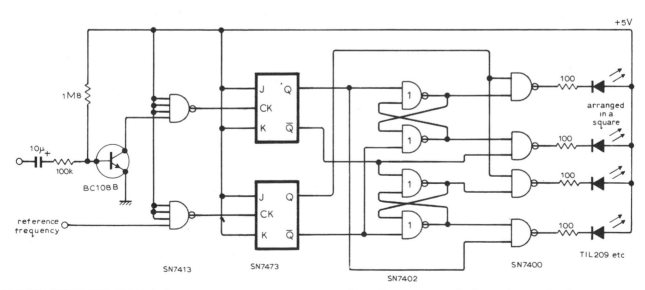

HI-LO LED FREQUENCY DISPLAY—Apparent rotation of flashing LEDs around square indicates whether input frequency is above or below reference frequency. Input and reference waveforms may have any shape, because Schmitt triggers reshape both to give rectangular pulses which flip-flops divide by two so outputs are square waves with mark-space ratios at half original frequency. Square waves are gated together to produce rectangular pulse train having mark-space ratio that depends on phase difference between the two square waves. Logic is arranged to drive LEDs at rotation rate of half the difference between input and reference frequencies. Correct positioning of LEDs in square is shown on small diagram. Reference frequency input should be via BC108B (not shown), same as for input frequency.—C. Clapp, Beat-Frequency Indicator, Wireless World, Nov. 1976, p 63.

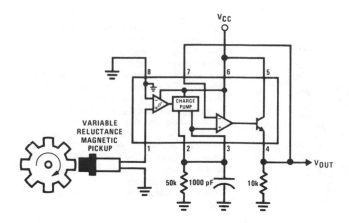

TACHOMETER WITH SQUARE-WAVE OUTPUT—National LM2907 tachometer IC provides square-wave output at same frequency as sine wave generated by magnetic pickup, for use as line driver in automatic control system.—"Linear Applications, Vol. 2," National Semiconductor, Santa Clara, CA, 1976, AN-162, p 12–13.

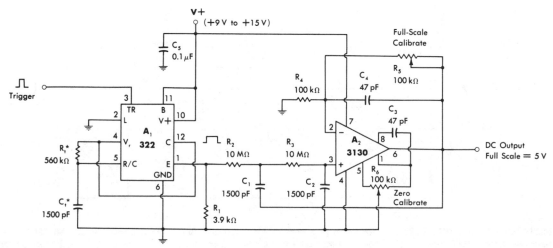

FREQUENCY METER—High-precision frequency-to-voltage converter can be used as frequency meter in laboratory. Input frequency range up to 1 kHz is converted to corresponding full-scale voltage value of +5 V. Two-pole filter removes ripple from positive output pulses across R_1 before signal is fed to 3130 opamp that provides gain and zero adjustments.—W. G. Jung, "IC Timer Cookbook," Howard W. Sams, Indianapolis, IN, 1977, p 192–196.

$$^*\text{Let } T = R_1 C_1 \cong \frac{0.95}{f_{full\,scale}}$$

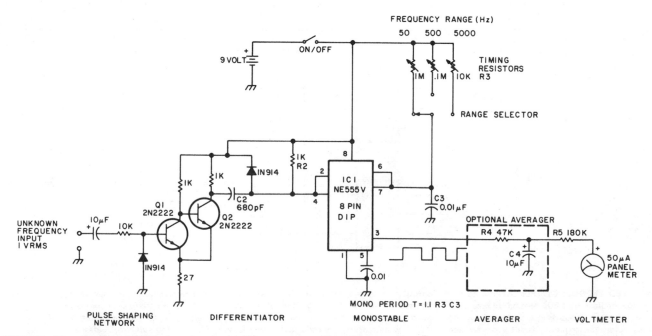

AF METER—Timer IC1 forms basis for linear frequency meter covering audio spectrum. Mono MVBR puts out fixed-width pulse when triggered by unknown input frequency. Article covers operation and calibration. Errata: pin 4 of 555 should be connected to pin 8 instead of to pin 2.—G. Hinkle, IC Audio Frequency Meter, *73 Magazine*, Holiday issue 1976, p 61.

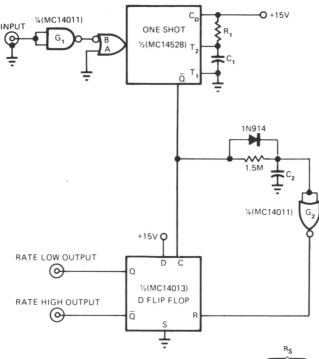

PULSE-RATE DETECTOR—Operates from 1 Hz to 2 MHz, providing one logic level when input rate crosses set point and opposite logic level when input rate falls below set point. Set-point rate is reciprocal of MVBR time, or $1/R_1C_1$. Two periods of input signal are sufficient for response to rate change. Value of C_2 is $R_1C_1/1.5 \times 10^6$.—J. M. Toth, Versatile Circuit Forms Accurate Pulse-Rate Detector, *EDN Magazine*, Aug. 20, 1977, p 142–143.

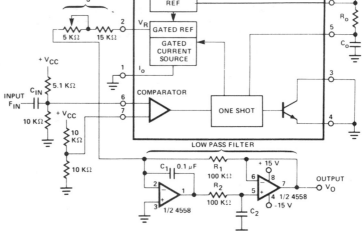

F/V WITH OUTPUT FILTER—Two-pole low-pass active filter improves dynamic range and response time of Raytheon 4152 frequency-to-voltage converter. Ripple in output is less than 0.02 V P-P above 100 Hz. Requires ±15 V supply. Maximum input frequency is 10 kHz when C_{IN} is 0.002 μF, R_0 is 6.8K, and C_0 is 0.01 μF.—"Linear Integrated Circuit Data Book," Raytheon Semiconductor Division, Mountain View, CA, 1978, p 7-48.

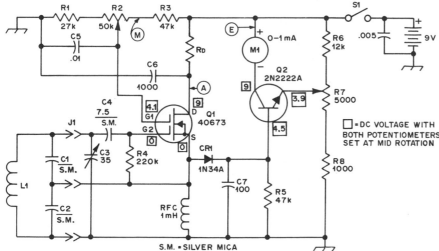

FREQ. RANGE	C1	C2	L1
MHz	pF	pF	TURNS
2.3-4	15	15	71-1/2
3.4-5.1	33	10	39-1/2
4.8-8	10	33	25-1/2
7.9-13	10	33	14-1/2
12.8-21.2	10	33	6-1/2
21-34	10	33	4-1/2
34-60	10	33	2-1/2
60-110	10	33	*
90-200	Not used	Not used	**

*denotes a 1-1/2-turn coil of No. 18 enam. wire wound on a 1/2-inch form spaced 1/8 inch between turns. It should be placed so that the coil is near the top of the coil form.
**denotes a hairpin loop made from flashing copper, 3/8-inch wide X 1-7/8-inch total length.
   All other coils are wound with No. 24 enam. wire.

MOSFET DIP METER—Output of grounded-drain Colpitts oscillator using RCA N-channel dual-gate MOSFET Q1 is detected by CR1 and amplified by Q2 for driving meter. Frequency of oscillation depends on C1, C2, C3, and L1, and reaches 250 MHz when L1 is reduced to hairpin. Table gives values of plug-in assembly L1-C1-C2 for nine frequency ranges. Circuit was designed for 12-V supply but works well with 9-V battery shown if drain resistor R_D is shorted. Battery drain is about 20 mA. All coils are wound on Millen 45004 forms.—F. Bruin, A Dual-Gate MOSFET Dip Meter, *QST*, Jan. 1977, p 16–17.

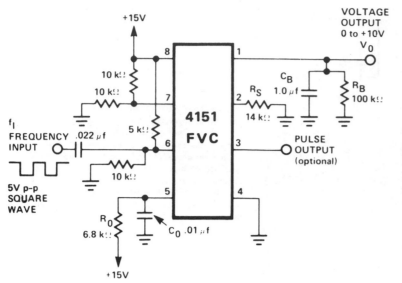

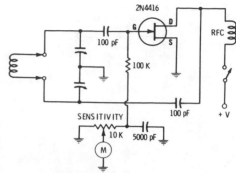

DESIGN EQUATIONS:

$$V_0 = f_1 K^{-1} \quad \text{Where } K = 0.486 \frac{R_S}{R_B R_0 C_0} \frac{Hz}{V}$$

$$T = 1.1 R_0 C_0$$

F/V CONVERTER—Single-supply circuit uses frequency-to-voltage connection of RM4151 converter to make output voltage vary between 0 and +10 V as frequency of 5 V P-P square-wave input varies between 0 and 10 kHz.—"Linear Integrated Circuit Data Book," Raytheon Semiconductor Division, Mountain View, CA, 1978, p 7-39.

GATE-DIP FET OSCILLATOR—Meter indicates gate current, which drops whenever resonant load is placed on tank circuit of oscillator by bringing plug-in input coil near frequency source being checked. By opening switch to power supply, circuit can be used as absorption wavemeter; when signal at resonant frequency of dip-meter tank circuit is picked up, gate-source circuit of FET operates as diode detector for producing increase in meter reading. Values of plug-in coil and tuning capacitor depend on frequency range of interest.—E. M. Noll, "FET Principles, Experiments, and Projects," Howard W. Sams, Indianapolis, IN, 2nd Ed., 1975, p 213–214.

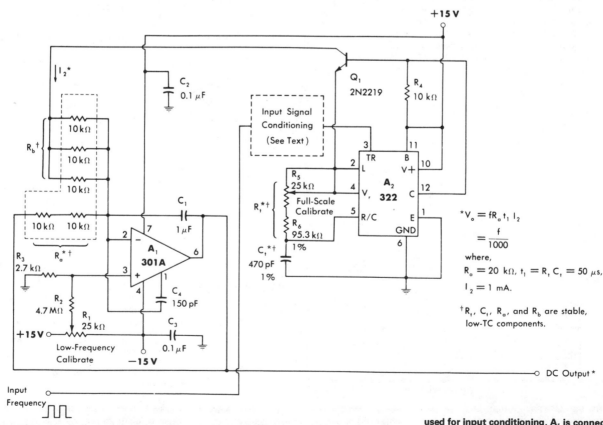

$$*V_o = f R_o t_1 I_2$$

$$= \frac{f}{1000}$$

where,

$$R_o = 20 \text{ k}\Omega, \ t_1 = R_t C_t = 50 \ \mu s,$$

$$I_2 = 1 \text{ mA}.$$

†R_t, C_t, R_o, and R_b are stable, low-TC components.

HIGH-PRECISION F/V—Components of V/F converter are reconnected to provide F/V function. Input frequency up to 10 kHz is fed to 322 mono MVBR A_2 either directly if pulsed or indirectly after conditioning. For low-frequency or slowly changing waveforms, zero-crossing detector is used for input conditioning. A_1 is connected as scaling amplifier and filter.—W. G. Jung, "IC Timer Cookbook," Howard W. Sams, Indianapolis, IN, 1977, p 192–196.

CHAPTER 34
Frequency Modulation Circuits

Covers FM circuits used in broadcast receivers and transmitters for monophonic or stereo transmissions, along with FM radio communication circuits and power-line FM carrier systems. Includes tuning indicators, stereo decoders, SCA demodulator, and FM deviation meter.

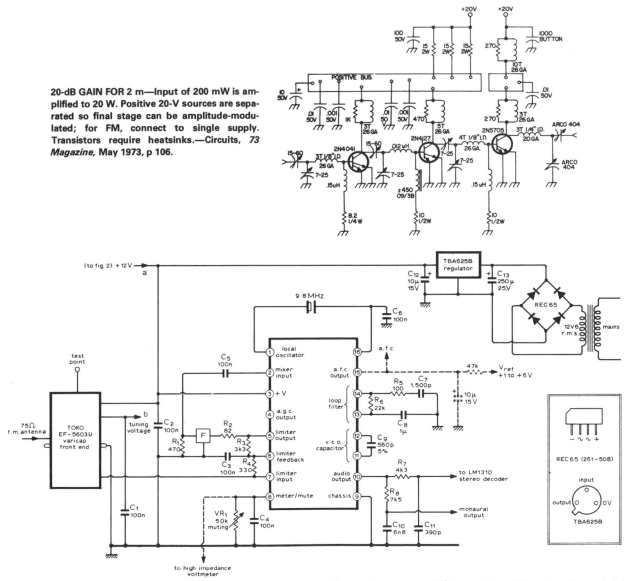

20-dB GAIN FOR 2 m—Input of 200 mW is amplified to 20 W. Positive 20-V sources are separated so final stage can be amplitude-modulated; for FM, connect to single supply. Transistors require heatsinks.—Circuits, *73 Magazine,* May 1973, p 106.

TUNER USES IC—Availability of Signetics NE563 IC having about 180 transistors greatly simplifies construction of high-quality FM tuner. IC includes circuits for converting IF output signal to lower frequency for driving phase-locked loop of demodulator. Use of varicap front end permits switched or continuous tuning with 100K Helipot or with switched preset 100K pots connected between +12 V and ground. Tuning controls can be remotely located. After 60 dB of amplification in NE563 IC, signal passes through ceramic filter F (Vernitron FM-4 or Toko CFS) before being fed back through C₅ to IC for mixing with crystal-controlled 9.8-MHz local oscillator. Article covers construction and operation of tuner in detail.— J. B. Dance, High-Quality F.M. Tuner, *Wireless World,* March 1975, p 111–113.

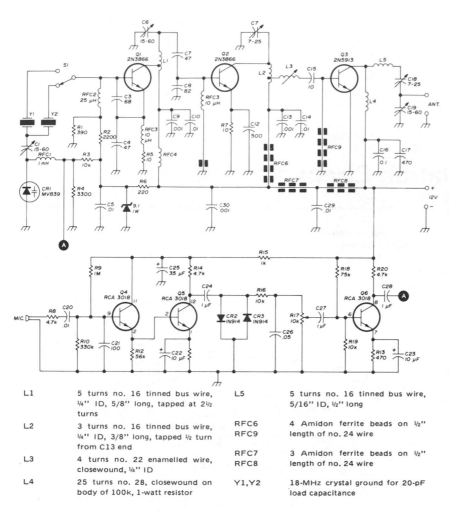

144-MHz FM TRANSMITTER—Low-power circuit was developed for use with double-conversion continuous-tuning FM receiver suitable for either fixed or mobile communication on 2-meter amateur band. Q4-Q6 are part of RCA CA3018 IC. Power output with 12-V supply is about 1.5 W. Two crystals are selected by slide switch; tuning can be compromised to use crystals whose 2-meter outputs are 1 MHz apart. Article also gives all circuits for receiver.—J. H. Ellison, Compact Package for Two-Meter FM, *Ham Radio,* Jan. 1974, p 36–44.

L1	5 turns no. 16 tinned bus wire, ¼" ID, 5/8" long, tapped at 2½ turns	L5	5 turns no. 16 tinned bus wire, 5/16" ID, ½" long
L2	3 turns no. 16 tinned bus wire, ¼" ID, 3/8" long, tapped ½ turn from C13 end	RFC6 RFC9	4 Amidon ferrite beads on ½" length of no. 24 wire
L3	4 turns no. 22 enamelled wire, closewound, ¼" ID	RFC7 RFC8	3 Amidon ferrite beads on ½" length of no. 24 wire
L4	25 turns no. 28, closewound on body of 100k, 1-watt resistor	Y1,Y2	18-MHz crystal ground for 20-pF load capacitance

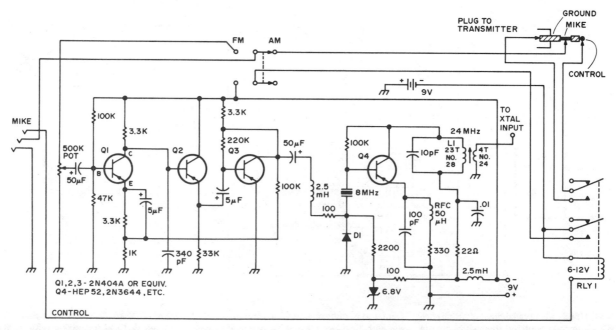

FM MODULATOR—Developed to permit FM operation on AM transceiver. Consists of microphone preamplifier, driver amplifier, and 8-MHz crystal oscillator providing 24-MHz output. Audio from modulator drives variable-capacitance diode D1 (which can be silicon switching diode) in oscillator circuit. Adapter feeds AM transmitter in which frequency multipliers increase deviation to about 8.5 kHz. To reduce deviation for narrow-band FM, adjust 500K pot in preamp.—R. Orozco, Jr., Put That AM Rig on FM, *73 Magazine,* April 1976, p 34–35.

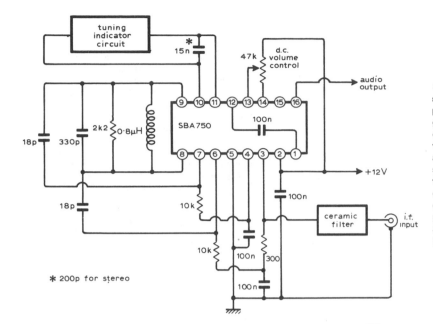

SINGLE-LED TUNING INDICATOR—Circuit shown, when driven by IF output of FM tuner, permits tuning for maximum brightness of single LED such as 5082-4403. Article gives choice of two tuning indicator circuit arrangements that can be used with the SBA750 limiting IF amplifier and detector IC. Recommended version of tuning indicator uses Plessey SL3046 five-transistor array with discrete resistors and capacitor to drive LED. Arrangement gives very clear, sensitive indication of correct tuning point to within a few millivolts.—J. A. Skingley, Sensitive F.M. Tuning Indicator, *Wireless World,* June 1974, p 173–174.

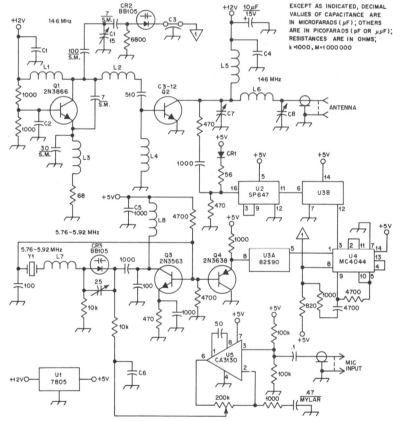

2-W 2-METER PHASE-LOCKED FM TRANSMIT-TER—Operating frequency of 144–148 MHz is generated directly, without using frequency-multiplier stages. Oscillator stability is achieved by phase-locking oscillator to crystal. Tuning range of 143–149 MHz corresponds to varactor control voltage of 1–4 V, which maintains proper loop gain across entire band. ECL decade divider U2 is Plessey SP647 driving Schottky-clamped divider U3B to give overall division of 50. Phase detector and loop amplifier functions are in U4. Pierce crystal oscillator Q3 feeds buffer Q4 interface with TTL levels. Microphone preamp U5 is slightly overdriven so speech waveform is clipped or limited. With phase-locked circuit, frequency stability is as good as that of crystal used in reference oscillator. Divide desired operating frequency by 25 to get crystal frequency.—A. D. Helfrick, A Phase-Locked 2-Meter FM Transmitter, *QST,* March 1977, p 37–39.

C1-C6, incl. — 1,000-pF ceramic feedthrough capacitors.
C7, C8 — 14- to 150-pF ceramic trimmer (Arco 424).
CR2, CR3 — BB105 or Motorola MV839 Varicap diode, 82 pF nominal capacitance, 73.8- to 90.2-pF total range.
L1, L3, L7 — 33-μH molded inductor (Miller 9230-56).
L2 — 1-1/2 turns no. 20 enameled wire, 1/4-inch diameter, 1/2-inch long.
L4 — 3 turns no. 28 enameled wire through ferrite bead.
L5 — 2.2-μH molded inductor (Miller 9230-28).
L6 — 1-1/2 turns no. 20 enameled wire, 3/8-inch diameter, 1-inch long.
L8 — 100-μH molded inductor (Miller 9230-68).

Q1 — RCA 2N3866 or Motorola HEP S3008 transistor.
Q2 — C3-12, manufactured by Communications Transistor Corp., a division of Varian. An RCA 2N5913 may be substituted.
Q3, Q4 — RCA transistor.
U1 — 5-volt, 1-ampere fixed positive regulator. An LM309K may be substituted.
U2 — Plessey Semiconductors integrated circuit.
U3 — Signetics 82S90 or National DM73LS196 integrated circuit.
U4 — Motorola MC4044 integrated circuit.
U5 — RCA CA3130 integrated circuit.
Y1 — Overtone Crystal, 5.76-5.92 MHz, International Crystal Mfg. Co. Type GP. Crystal frequency is discussed in the text.

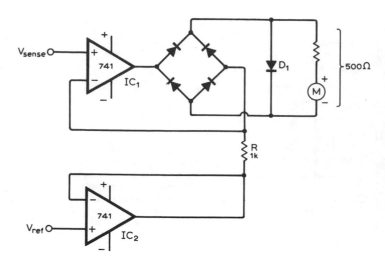

TUNING NULL INDICATOR—Uses standard left-zero meter as tuning indicator connected in basic opamp AC voltmeter configuration using IC_1, with reference buffered by opamp IC_2. DC output voltage of tuner is compared with non-zero reference voltage; as these voltages approach each other during tuning, meter pointer moves toward zero, and abruptly reverses direction as tuning null point is passed. Diode D_1 protects meter from overload. Use any low-leakage diodes for bridge.—A. S. Holden, Sensitive Null Indicator, *Wireless World*, Oct. 1974, p 381.

ADD-ON FM DETECTOR—Suitable for any communication receiver. Other IF values can be handled by changing values of L and C. Connecting C_2 to pin 10 instead of pin 9 may improve performance. Circuit is easy to construct and align; adjust slug-tuned coil for maximum recovered audio when receiving FM signal.—I. Math, Math's Notes, *CQ*, April 1975, p 37–38 and 62.

Final I.F.	C_1	C_2	C_3	R_1	(Miller 44000 Series or Similar) L_1
455 kHz	680 pF	3 pF	.005	22K	135-240 μHy
2 Mhz	300 pF	3 pF	.005	22K	16-30 μHy
4.5 Mhz	120 pF	3 pF	.003	22K	7-14 μHy
5.5 Mhz	100 pF	3 pF	.003	22K	5-8 μHy
10.7 Mhz	100 pF	4.7 pF	.01	3.9K	1.5-3 μHy

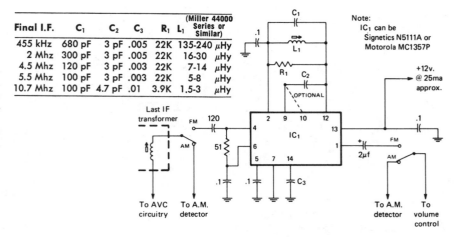

Note: IC_1 can be Signetics N5111A or Motorola MC1357P

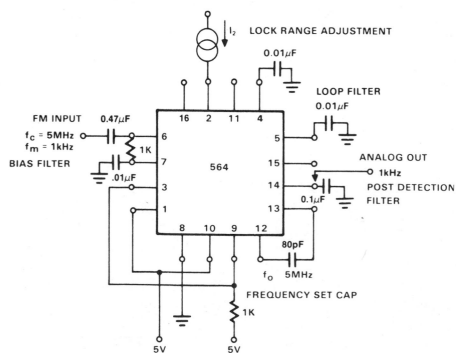

FM DEMODULATOR—Uses Signetics NE564 PLL having postdetection processor, operating from 5-V supply. Conversion gain is low so frequency deviation in input signal should be at least 1%.—"Signetics Analog Data Manual," Signetics, Sunnyvale, CA, 1977, p 828–830.

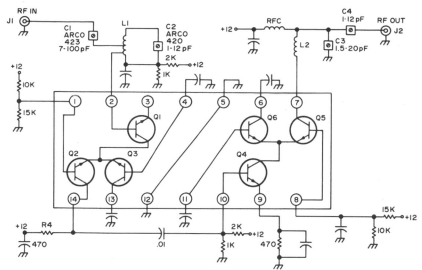

2-METER IC RF AMPLIFIER—High-gain double compound amplifier using RCA CA3102E has low noise, excellent stability, and only two tuned circuits. Ideal for 2-meter FM RF stage, but can be used from DC up to 500 MHz by changing tuned circuits. Article covers construction, with emphasis on proper shielding.— B. Hoisington, Two High Gain RF Stages in One IC for Two Meter FM, *73 Magazine*, May 1974, p 47–50 and 52.

FM DEMODULATOR—Uses RCA CA3046 IC IF amplifier connected as highly linear voltage-controlled oscillator, in phase-locked loop configuration capable of handling 10.7-MHz amplitude-limited FM input as FM demodulator. Output AF signal is about 20 mV for 75-kHz deviation. FET serves as synchronous-chopper type of phase-sensitive detector.—J. L. Linsley Hood, Linear Voltage Controlled Oscillator, *Wireless World*, Nov. 1973, p 567–569.

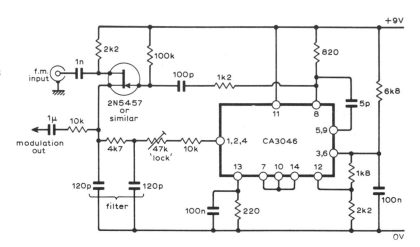

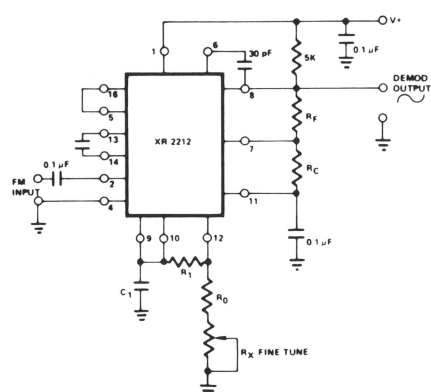

LINEAR FM DEMODULATOR—Exar XR-2212 precision PLL IC provides linear demodulation for both narrow-band and wideband FM signals. Article gives circuit design procedure. With +12 V supply voltage and 67-kHz carrier frequency having ±5 kHz frequency deviation, R_0 is 18K fixed resistor in series with 5K pot. C_0 (between pins 13 and 14) is 746 pF, R_1 is 89.3K, C_1 is 186 pF, R_F is 100K, and R_C is 80.6K. These values give ±4 V P output swing. All values except R_0 can be rounded off to nearest standard value.—"Phase-Locked Loop Data Book," Exar Integrated Systems, Sunnyvale, CA, 1978, p 35–40.

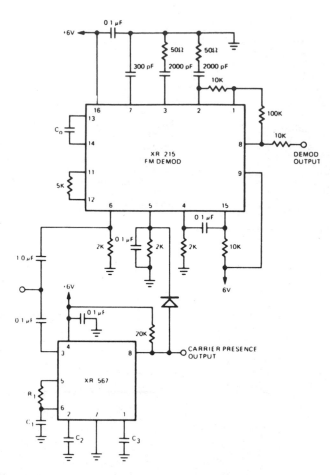

DEMODULATOR WITH CARRIER DETECT— Exar XR-567 PLL system is used with XR-215 FM demodulator to detect presence of carrier signal in narrow-band FM demodulation applications where bandwidth is less than 10% of carrier frequency. Output of XR-567 is used to turn off FM demodulator when no carrier is present, giving squelch action. Circuit will detect presence of carrier up to 500 kHz.—"Phase-Locked Loop Data Book," Exar Integrated Systems, Sunnyvale, CA, 1978, p 41–48.

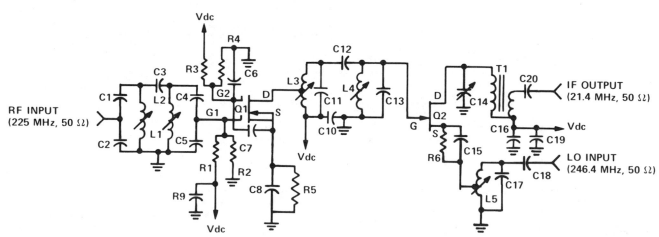

C1 — 8.2 pF	Ceramic disc capacitor	R1 — 300 kilohm, 1/4 W carbon resistor	
C2 — 43 pF	Dipped, silvered mica capacitor	R2 — 27 kilohm, 1/4 W carbon resistor	
C3 — 0.2 pF	Ceramic tubular capacitor	R3 — 62 kilohm, 1/4 W carbon resistor	
C4 — 12 pF	Ceramic disc capacitor	R4 — 56 kilohm, 1/4 W carbon resistor	
C5 — 15 pF	Dipped, silvered mica capacitor	R5 — 220 ohm, 1/4 W carbon resistor	
C6, C7, C8, C9, C10		R6 — 910 ohm, 1/4 W carbon resistor	
C16 — 220 pF	Ceramic disc capacitor	L1 — 2-1/2 turns no. 20 enameled wire on 9/32" plastic form with brass slug (70 nH)	
C11 — 10 pF	Dipped, silvered mica capacitor	L2, L4 — 2-1/6 turns no 20 enameled wire on 9/32" plastic form with brass slug (60 nH)	
C12 — 0.47 pF	Ceramic tubular capacitor	L3 — 1-5/6 turns no. 20 enameled wire on 9/32" plastic form with brass slug, tapped 1/2 turn from ground (55 nH)	
C13 — 3.6 pF	Ceramic tubular capacitor	L5 — 2-1/2 turns no. 20 enameled wire on 9/32" plastic form with brass slug, tapped 2/3 turn from ground (70 nH)	
C14 — 8–60 pF	Mica compression trimmer, Arco 404	Q1 — 3N201 MOSFET	
C15, C19 — 0.01 μF	Ceramic disc capacitor	Q2 — 2N5486 JFET	
C17, C18 — 3 pF	Dipped, silvered mica capacitor	T1 — Primary = 18 turns no. 24 enameled wire on T44-6 Micrometals toroid core	
C20 — 24 pF	Dipped, silvered mica capacitor	Secondary = 4 turns no. 24 enameled wire twisted around last 4 turns of ground end of primary	

225-MHz FRONT END—RF stage, mixer, and tuned circuits are designed for use in FM communication receiver having local oscillator input of 246.4 MHz, for IF of 21.4 MHz. Supply voltage is 12.5 V. Spurious-response rejection is 100 dB, image rejection is 97 dB, and noise figure is 12 dB.—J. Hatchett and B. Morgan, "Economical 225 MHz Receiver Front End Employs FETs," Motorola, Phoenix, AZ, 1978, EB-22.

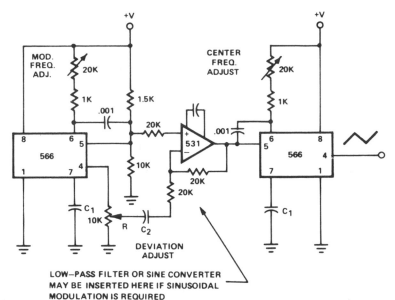

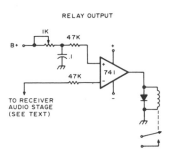

CARRIER-OPERATED RELAY—Relay is energized when carrier is present in FM receiver, to activate transmitter for repeater applications or turn on tape recorders, alarms, or other devices. Will work with either solid-state or tube-type receivers. Inverting (minus) input of opamp is connected directly to collector of audio preamp transistor or any other point having voltage change between signal and no signal. If voltage change is in wrong direction, reverse leads to opamp input. Use 1K pot to set reference voltage so relay trips reliably on incoming signal. Relay can be reed-type drawing less than 75 mA.—S. Uhrig, The 5 Minute COR, *73 Magazine,* Dec. 1976, p 152–153.

0.5 MHz WITH 100% DEVIATION—Carrier frequency under 0.5 MHz is generated by 566 function generator at right, for modulation by other 566 connected as triangle generator whose output is boosted by 531 opamp to give deviations up to ±100% of carrier frequency. Capacitors C₁ control frequency range of each function generator.—"Signetics Analog Data Manual," Signetics, Sunnyvale, CA, 1977, p 852–854.

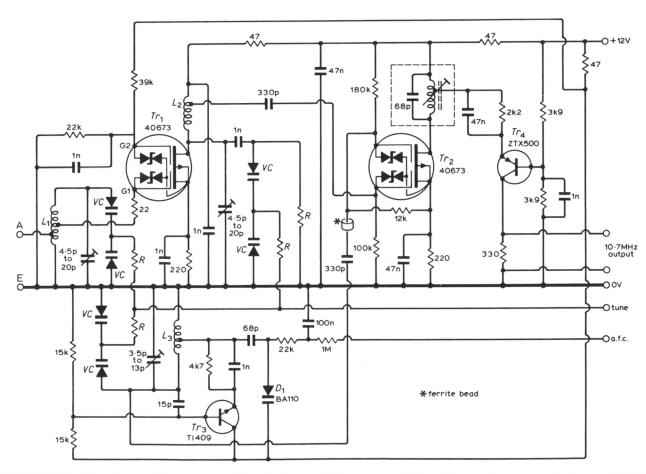

VARICAP TUNER—Uses silicon variable-capacitance diodes to provide voltage tuning over FM band of 87.5 to 108 MHz. Article covers construction and adjustment and gives circuit of stable noise-free regulated power supply that also provides required DC tuning voltage of 2 to 30 V. All six varicap diodes are Siemens BB103 of same color selection (all green or all blue). Resistors R can be any value between 100K and 1 megohm.—L. Nelson-Jones, F.M. Tuner Design—Two Years Later, *Wireless World*, June 1973, p 271–275.

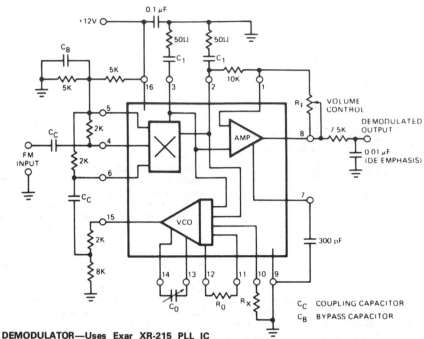

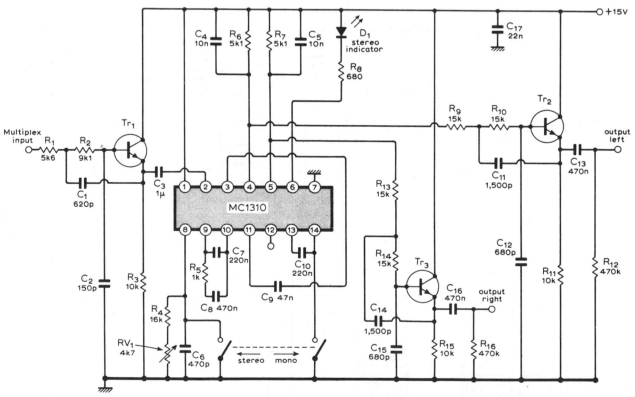

65–130 MHz DIODE RECEIVER—Tunable version of basic crystal detector is useful for FM broadcast work and for checking output of frequency multiplier. L2 is 5 turns center-tapped No. 12 copper wire air-wound to 5/8-inch diameter and 1-inch length. C1 is 30-pF miniature tuning capacitor.—B. Hoisington, Tuned Diode VHF Receivers, *73 Magazine*, Dec. 1974, p 81–84.

DEMODULATOR—Uses Exar XR-215 PLL IC connected for frequency-selective demodulation of FM signals. Value of C_0 depends on carrier frequency. C_1 determines selectivity; for 1–10 MHz, range of C_1 is 10–30 times C_0. For operation below 5 MHz, R_x can be opened; above 5 MHz, use about 750 ohms.—"Phase-Locked Loop Data Book," Exar Integrated Systems, Sunnyvale, CA, 1978, p 21–28.

STEREO DECODER—Improved circuit for FM tuner uses active filters to eliminate subcarrier harmonics as well as birdlike interference sounds (birdies) experienced under certain conditions. Stereo reception normally involves demodulation of stereo channel at 38 kHz by square-wave switching, a process that also demodulates signals around odd harmonics of 38 kHz. The first two of these, at 114 and 190 kHz, can produce audible signals from adjacent channels at 100 and 200 kHz away from wanted station. Resulting interference, centered on 14 kHz and 10 kHz, sounds like high-pitched twittering sounds of birds. Tr_1 serves as active filter for suppressing these sounds. This is followed by phase-locked loop type of IC decoder, operation of which is described in article that also gives complete circuit and construction details for entire FM tuner. All transistors are BC109 or equivalent. LED is 5082-4403. Tr_1 has roll-off response at 18 dB per octave above 53 kHz, while active filters Tr_2 and Tr_3 remove harmonics of 38 kHz from outputs.—J. A. Skingley and N. C. Thompson, Novel Stereo F.M. Tuner, *Wireless World*, Part 2—May 1974, p 124–129 (Part 1—April 1974, p 58–62).

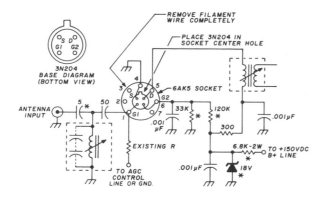

MOSFET RF STAGE—Changing 6AK5 tube to 3N204 dual-gate MOSFET improves sensitivity and lowers noise in older VHF FM communication receiver using tubes. Break off center grounding pin of tube socket and cut wires soldered to pin, then connect transistor circuit to tube socket as shown. Replace original resistor going to pin 6 with 120K and run 37K resistor from pin 6 to ground. Move antenna input lead to top of RF input coil, and remove 6-V filament wiring from socket. If tube filaments were in series, replace 6AK5 filament with 36-ohm 2-W resistor. Conversion increases sensitivity to 0.3 μV for 20-dB quieting.—H. Meyer, How to Improve Receiver Performance of Vacuum-Tube VHF-FM Equipment, *Ham Radio,* Oct. 1976, p 52–53.

TWIN-LED TUNING INDICATOR—Provides maximum sensitivity at correct tuning point and indicates direction of mistuning. Both lamps are in feedback loop of one opamp, connected to serve as highly sensitive null detector. When set is tuned correctly, output of this opamp is at midpoint of supply voltage and neither LED is lit. Circuit is used with RCA CA3089 IF chip in which AFC output is a current. Capacitor across first 741 opamp removes modulation components from this input.—M. G. Smart, F.M. Tuning Indicators, *Wireless World,* Dec. 1974, p 497.

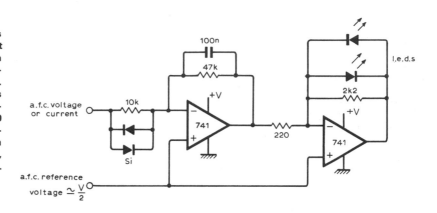

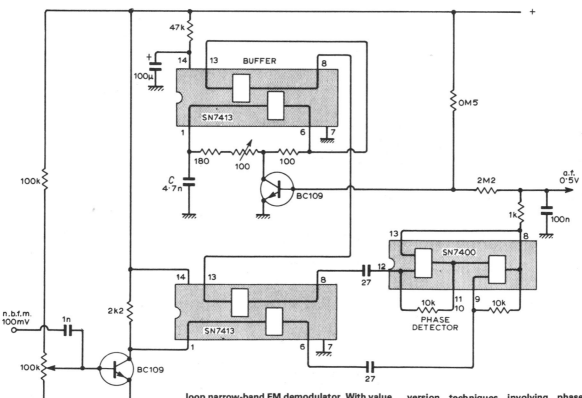

NARROW-BAND DEMODULATOR—Low-cost TTL ICs are connected to form phase-locked loop narrow-band FM demodulator. With value shown for C, circuit is suitable for IF value around 470 kHz. Article covers advantages of synchronous detection and various direct conversion techniques involving phase-locked loop.—P. Hawker, Synchronous Detection in Radio Reception, *Wireless World,* Nov. 1972, p 525–528.

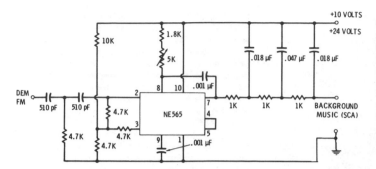

SCA DEMODULATOR—VCO of NE565 PLL is set at 67 kHz and is locked in by incoming 67-kHz subsidiary-carrier component used for transmitting uninterrupted commercial background music by FM broadcast stations. Circuit demodulates FM sidebands and applies them to audio input of commercial sound system through suitable filter. 5K pot is used to lock VCO exactly on frequency. Frequency response extends up to 7000 Hz.—E. M. Noll, "Linear IC Principles, Experiments, and Projects," Howard W. Sams, Indianapolis, IN, 1974, p 212–213.

STEREO DECODER—Single Sprague ULN-2122A IC is driven by composite signal derived at output of standard FM detector, to give original left- and right-channel audio signals for driving audio amplifiers of FM stereo receiver.—E. M. Noll, "Linear IC Principles, Experiments, and Projects," Howard W. Sams, Indianapolis, IN, 1974, p 263–266.

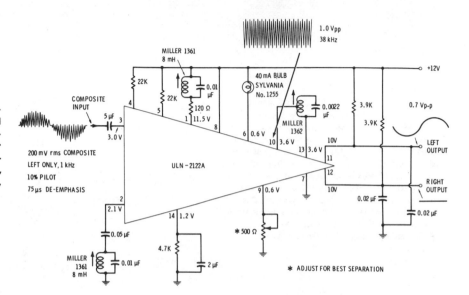

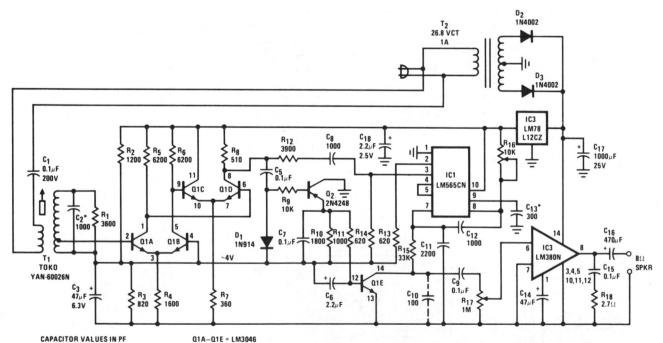

CAPACITOR VALUES IN PF
RESISTOR VALUES IN Ω
***SELECT FOR CARRIER FREQ.**

f_0	C_2	C_{13}
200 kHz	1000	300
100 kHz	3900	620

CARRIER-SYSTEM RECEIVER—Used to detect, amplify, limit, and demodulate FM carrier modulated with audio program, for feeding up to 2.5 W to remote loudspeaker. Can be plugged into any AC outlet on same side of distribution transformer. Carrier signal is taken from line by tuned transformer T_1. Output of two-stage limiter amplifier Q1A-Q1D is applied directly to mute peak detector D1-Q2-C7. Limiter output is reduced to 1 V P-P for driving National LM565CN PLL detector which operates as narrow-band tracking filter for input signal and provides low-distortion demodulated audio output. Mute circuit quiets receiver in absence of carrier.—J. Sherwin, N. Sevastopoulos, and T. Regan, "FM Remote Speaker System," National Semiconductor, Santa Clara, CA, 1975, AN-146.

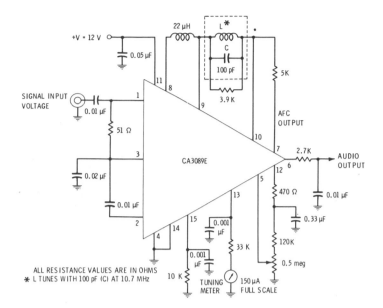

SINGLE-TUNED DETECTOR—RCA CA3089E IC serves as communication receiver subsystem providing three-stage FM IF amplifier/limiter channel, with signal-level detectors for each stage, and quadrature detector that can be used with single-tuned detector coil. Detector also supplies drive to AFC amplifier whose output can be used to hold local oscillator on correct frequency. Level-detector stages supply signal for tuning meter. Values shown are for 10.7-MHz IF.—E. M. Noll, "Linear IC Principles, Experiments, and Projects," Howard W. Sams, Indianapolis, IN, 1974, 347–349.

QUADRATURE DEMODULATOR—Quadrature coil associated with balanced-mixer demodulation system is connected to pin 6 of National LM373 IC, and output signal is taken from pin 7. Good output is obtained with only ±5 kHz deviation at either 455 kHz or 10.7 MHz. Can be operated as wideband or narrow-band circuit by choosing appropriate interstage and output LC and RC components.—E. M. Noll, "Linear IC Principles, Experiments, and Projects," Howard W. Sams, Indianapolis, IN, 1974, p 350–351.

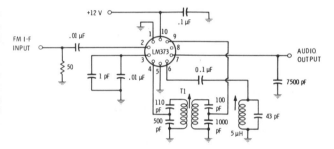

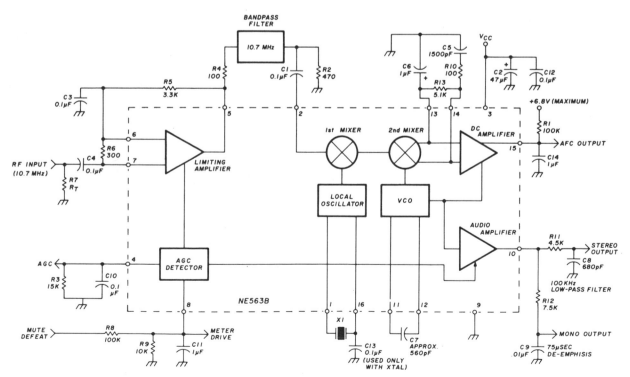

PLL IF AND DEMODULATOR—Signetics NE563B IC (in dashed lines) serves as complete IF amplifier and demodulator for FM broadcast receiver. Circuit uses downconversion from 10.7 MHz to 900 kHz, where phase detector operates. Ceramic bandpass filter provides IF selectivity at 10.7 MHz. X1 can be 9.8-MHz ceramic resonator, LC network, crystal, or capacitor.—H. Olson, FM Detectors, *Ham Radio*, June 1976, p 22–29.

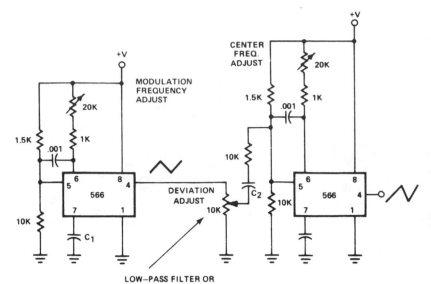

0.5 MHz WITH 20% DEVIATION—One 566 function generator serves for generating relatively low-frequency carrier (center frequency less than 0.5 MHz), and other 566 serves as modulator producing triangle output with frequency determined by C_1. Combination is suitable for deviations up to ±20% of carrier frequency.—"Signetics Analog Data Manual," Signetics, Sunnyvale, CA, 1977, p 852–853.

LOW–PASS FILTER OR SINE CONVERTER MAY BE INSERTED HERE IF SINUSOIDAL MODULATION IS REQUIRED

FM DETECTOR—Single IC can be added to any receiver not having FM detector. Moving C_2 from pin 9 to pin 10 gives higher audio output. Receivers having less than 5 kHz IF bandwidth can be broadened by stagger-tuning IF strip slightly to improve audio clarity. Adjust tuned circuit of detector for maximum recovered audio.—I. Math, Math's Notes, *CQ*, June 1972, p 49–51 and 80.

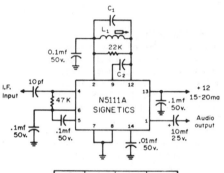

I.F.	C_1 (pf)	C_2 (pf)	L_1 (μH)
10.7 mHz	120	4.7	1.5 - 3
4.5 mHz	120	3.0	7 - 14
2 mHz	300	3.0	16 - 30
455 kHz	650	3.0	135 - 240

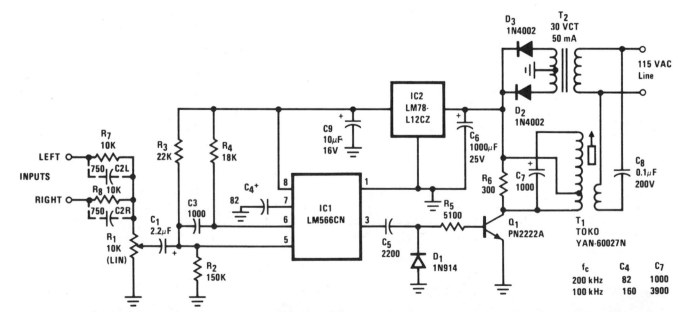

CARRIER-SYSTEM TRANSMITTER—Used to convert audio program material into FM format for coupling to standard power lines. Modulated FM signal can be detected at any other outlet on same side of distribution transformer, for demodulation and drive of loudspeaker. Input permits combining stereo signals for mono transmission to single remote loudspeaker. Uses National LM566CN VCO. Frequency response is 20–20,000 Hz, and total harmonic distortion is under 0.5% With 120/240 V power lines, system operates equally well with receiver on either side of line. Transmitter input can be taken from monitor or tape output jack of audio system.—J. Sherwin, N. Sevastopoulos, and T. Regan, "FM Remote Speaker System," National Semiconductor, Santa Clara, CA, 1975, AN-146.

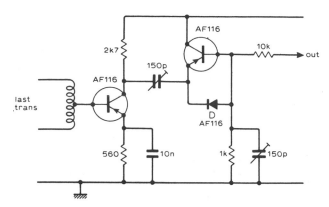

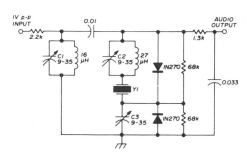

TRANSISTOR-PUMP DISCRIMINATOR—Used with 10.7-MHz IF strip of high-quality FM tuner built from discrete components. Circuit is placed between last IF stage and stereo decoder.—W. Anderson, F. M. Discriminator, *Wireless World*, April 1976, p 63.

CRYSTAL DISCRIMINATOR—Inexpensive third-overtone CB crystal used at 9-MHz fundamental serves as high-performance discriminator for VHF FM receiver. Adjust C3 for zero voltage with unmodulated carrier at or near center frequency. Adjust C1 and C2 with AF sine wave applied to FM signal generator, using CRO to check distortion of recovered sine wave. With 1 V P-P IF signal at 9 MHz and 5-kHz deviation, recovered audio will be about 1 V P-P at lower audio frequencies. Good limiter is required ahead of discriminator for AM rejection.—G. K. Shubert, Crystal Discriminator for VHF FM, *Ham Radio*, Oct. 1975, p 67—69.

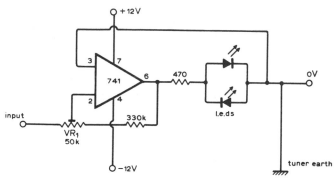

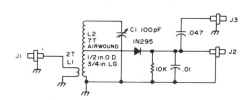

LED TUNING INDICATOR—One LED is mounted at each end of tuning scale. Tuning pointer is moved away from whichever LED is on, to dead spot at which both are off, to obtain correct tuning point. Advantages of lights-off tuning include minimum current drain and indication of even very slight mistuning by having one light come on even slightly. Adjust VR₁ to give wide enough dead spot so LEDs do not flicker on loud speech or music.—H. Hodgson, Simpler F.M. Tuning Indicator, *Wireless World*, Sept. 1975, p 413.

21—75 MHz DIODE RECEIVER—Covers 6-meter band and most 2-meter FM receiver oscillators near 45 MHz. Circuit is essentially that of crystal detector. Jack J3 gives AF output, and J2 gives DC output for meter.—B. Hoisington, Tuned Diode VHF Receivers, *73 Magazine*, Dec. 1974, p 81—84.

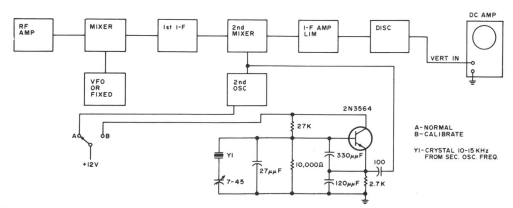

DEVIATION METER—Uses simple crystal oscillator combined with fixed or tunable FM receiver and CRO to show carrier shift on either side of center frequency. Vertical amplifier of CRO should be direct-coupled. To calibrate, tune oscillator either 10 or 15 kHz above or below second oscillator of receiver, and calibrate screen of CRO accordingly. One calibration oscillator is sufficient since transmitter usually deviates equally well both ways.—V. Epp, FM Deviation Meters, *73 Magazine*, March 1973, p 81—83.

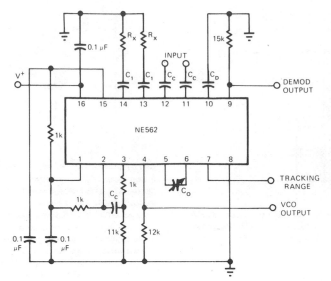

ANALOG PLL AS FM DEMODULATOR—Upper frequency limit of about 50 MHz for NE562 and other monolithic analog phase-locked loops complicates construction of FM telemetry receivers that directly demodulate standard 88–108 MHz FM broadcast signals. Circuit shown solves problem, with only small amount of signal preconditioning, by first converting RF carrier to 10.7 MHz with conventional superheterodyne front end, then applying signal to phase detector of PLL with VCO set to run free at 10.7 MHz. Input sensitivity is less than 30 μV, and audio output is greater than 100 mV.—E. Murthi, Monolithic Phase-Locked Loops—Analogs Do All the Work of Digitals, and Much More, *EDN Magazine,* Sept. 5, 1977, p 59–64.

OPAMP DRIVE FOR LED TUNING LAMPS—Opamp with 100K feedback resistor gives gain of 10 as optimum compromise for driving two LED tuning indicators in FM receiver.—R. D. Post, F.M. Tuning Indicator, *Wireless World,* May 1975, p 220.

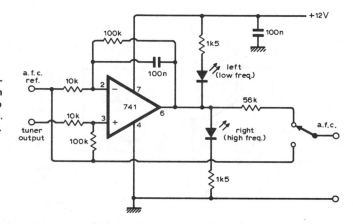

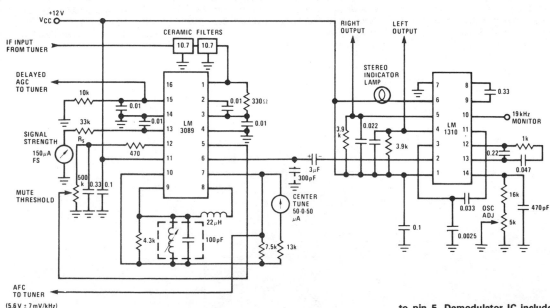

IF AND STEREO DEMODULATOR—National LM3089 IC and LM1310 PLL FM stereo demodulator provide all circuits required between FM tuner and inputs to power amplifiers of stereo receiver. Use of 10.7-MHz ceramic filters eliminates all but one IF alignment step. AFC output from pin 7 of IF strip drives center-tune meter. Wide bandwidth of detector and audio stage in IF strip is more than adequate for stereo receivers. Audio stage can be muted by input voltage to pin 5. Demodulator IC includes automatic stereo/monaural switching and 100-mA stereo indicator lamp driver. Optional 300-pF capacitor on pin 6 of LM3089 can be used to limit bandwidth.—"Audio Handbook," National Semiconductor, Santa Clara, CA, 1977, p 3-18–3-23.

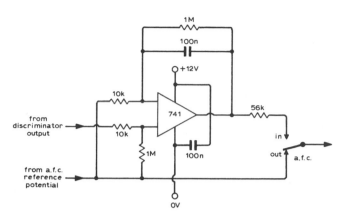

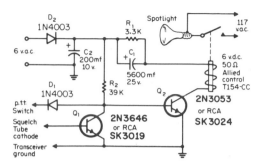

AFC AMPLIFIER—Simple DC amplifier can be added to AFC circuit of FM tuner to eliminate tuning errors over entire lock-in range.—J. S. Wilson, Improved A.F.C. for F.M. Tuners, *Wireless World,* July 1974, p 239.

CALL ALERT—Developed to trigger relay when signal arrives at squelch tube in GE Progress Line 2-meter FM receiver. Relay is held energized about 2 s, determined by C_1-R_1, then de-energized for at least 25 s. Used for flashing red spotlight in room that is too noisy for hearing bell or buzzer. Circuit is easily adapted for any other FM receiver having squelch stage. Control circuit responds to small change in voltage at cathode of squelch tube. With no carrier present, tube conducts and places positive voltage at face of Q_1, making it conduct and turn off Q_2. When carrier arrives, Q_1 restores bias to Q_2, turning on relay. Connection to push-to-talk switch keeps lamp from flashing during transmission.—L. Waggoner, The WA0QPM "Call Alert," *CQ,* May 1971, p 48–49.

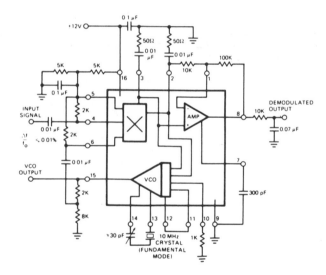

CRYSTAL FM DETECTOR—Exar XR-215 PLL IC is operated as crystal-controlled phase-locked loop by using crystal in place of conventional timing capacitor. Crystal should be operated in fundamental mode. Typical pull-in range is ±1 kHz at 10 MHz.—"Phase-Locked Loop Data Book," Exar Integrated Systems, Sunnyvale, CA, 1978, p 21–28.

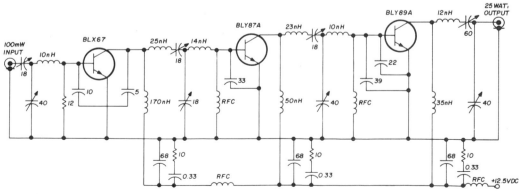

VHF POWER AMPLIFIER—Three-stage 25-W 225-MHz power amplifier module for FM applications uses three Amperex power transistors. Input and output are 50 ohms. With 100-mW input signal, output is 25 W. Four capacitive dividers serve for input, output, and interstage matching. Collectors are shunt-fed. Three decoupling networks prevent self-oscillation. Amplifier can withstand output mismatches as high as 50:1 without damage.—E. Noll, VHF/UHF Single-Frequency Conversion, *Ham Radio,* April 1975, p 62–67.

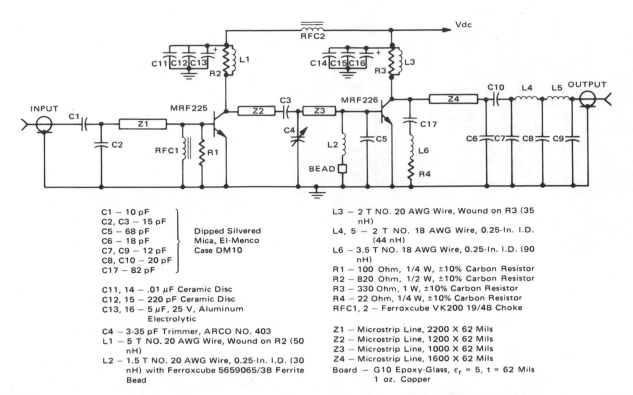

C1 — 10 pF
C2, C3 — 15 pF
C5 — 68 pF
C6 — 18 pF   } Dipped Silvered
C7, C9 — 12 pF   Mica, El-Menco
C8, C10 — 20 pF   Case DM10
C17 — 82 pF

C11, 14 — .01 μF Ceramic Disc
C12, 15 — 220 pF Ceramic Disc
C13, 16 — 5 μF, 25 V, Aluminum
　　　　　Electrolytic
C4 — 3-35 pF Trimmer, ARCO NO. 403
L1 — 5 T NO. 20 AWG Wire, Wound on R2 (50
　　　nH)
L2 — 1.5 T NO. 20 AWG Wire, 0.25-In. I.D. (30
　　　nH) with Ferroxcube 5659065/3B Ferrite
　　　Bead

L3 — 2 T NO. 20 AWG Wire, Wound on R3 (35
　　　nH)
L4, 5 — 2 T NO. 18 AWG Wire, 0.25-In. I.D.
　　　(44 nH)
L6 — 3.5 T NO. 18 AWG Wire, 0.25-In. I.D. (90
　　　nH)
R1 — 100 Ohm, 1/4 W, ±10% Carbon Resistor
R2 — 820 Ohm, 1/2 W, ±10% Carbon Resistor
R3 — 330 Ohm, 1 W, ±10% Carbon Resistor
R4 — 22 Ohm, 1/4 W, ±10% Carbon Resistor
RFC1, 2 — Ferroxcube VK200 19/4B Choke

Z1 — Microstrip Line, 2200 X 62 Mils
Z2 — Microstrip Line, 1200 X 62 Mils
Z3 — Microstrip Line, 1000 X 62 Mils
Z4 — Microstrip Line, 1600 X 62 Mils
Board — G10 Epoxy-Glass, ϵ_r = 5, t = 62 Mils
　　　1 oz. Copper

225-MHz 13-W AMPLIFIER—Suitable for use in FM transmitters for 220–225 MHz amateur radio band. Bandwidth is about 10 MHz for ±0.5 dB. Low-pass filter provides about 60-dB atten- uation of second harmonic. Microstrip match- ing network simplifies construction. Supply voltage is 12.5 V.—J. Hatchett and T. Sallet, "13-Watt Microstrip Amplifier for 220-225 MHz Operation," Motorola, Phoenix, AZ, 1975, AN-728, p 3.

CHAPTER 35
Frequency Multiplier Circuits

Emphasis is on frequency doublers, but includes circuits providing multiplying factors up to 10 for sine, square, and other waveforms in audio range and in RF systems extending well above 400 MHz.

SIMPLE DOUBLER—Performance is good up to about 10 kHz. R_2 is adjusted to set FET just at cutoff under no-signal conditions, to give operation in square-law region. With R_1 correctly adjusted, using scope or third harmonic distortion monitor to obtain minimal distortion, harmonic content of output can be made to approach that of sine-wave input. Article gives design equations.—R. Williams and J. Dunne, Frequency Doubler, *Wireless World,* Dec. 1975, p 575.

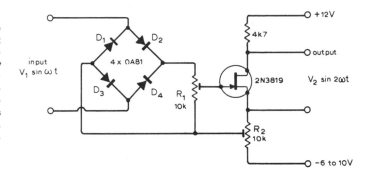

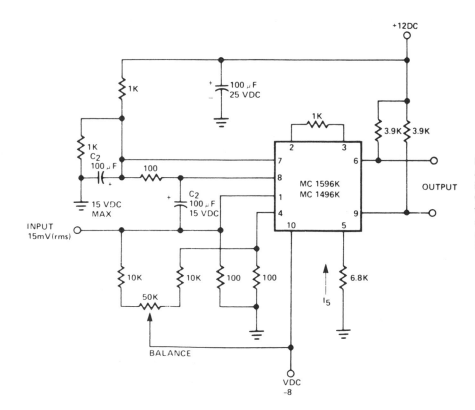

LOW-FREQUENCY DOUBLER—Signetics balanced modulator-demodulator transistor array is connected much like phase detector circuit. Output contains sum component which is twice frequency of input signal because same input signal frequency goes to both sections of balanced modulator.—Signetics Analog Data Manual," Signetics, Sunnyvale, CA, 1977, p 758.

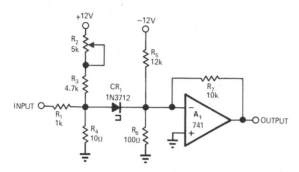

SQUARER—Simple tunnel-diode circuit doubles frequency efficiently without use of tuned circuits. Fundamental and other harmonics of input are at least 30 dB below level of frequency-doubled output. Circuit operates from DC to upper frequency limit of opamp used. Adjust R_2 so diode current is at peak of its bias current, to eliminate offset at amplifier output.—R. Kincaid, Squaring Circuit Makes Efficient Frequency Doubler, *EDN/EEE Magazine*, Aug. 15, 1971, p 45.

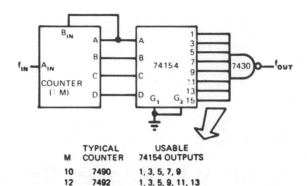

M	TYPICAL COUNTER	USABLE 74154 OUTPUTS
10	7490	1, 3, 5, 7, 9
12	7492	1, 3, 5, 9, 11, 13
16	7493	1, 3, 5, 7, 9, 11, 13, 15

MULTIPLES OF 2.5 MHz—Three TTL circuits provide integral frequency-multiplication ratios between 1 and 8. BCD outputs of counter having modulus M are fed to inputs of 74154 4-line to 16-line decoder. As outputs of counter change, at rate equal to input frequency divided by counter modulus M, each goes low at same rate. Output of NAND gate thus goes high once for each input to gate from decoder. If 7490 decade counter is used and input is 1 MHz, BCD outputs of 7490 limit usable outputs of 74154 to lines 1, 3, 5, 7, and 9. Since inputs to 74154 change at 100-kHz rate, output from gate will be n × 100 kHz. With input of 25 MHz, output is integral multiple of 2.5 MHz.—R. S. Stein, Three TTL IC's Provide Frequency Multiplication, *EDN Magazine*, April 5, 1975, p 117 and 119.

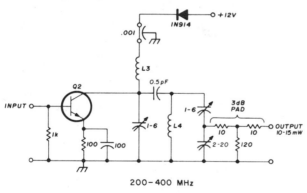

200–400 MHz

L3,L4   2 turns no. 22 (0.6mm), air core, 1/8'' (3mm) diameter, 1/4'' (6.5mm) long

Q1,Q2   Fairchild 2N5179 recommended but 2N2857, 2N918, FMT2060 or equivalent may be substituted

DOUBLING 200 MHz—Recommended for use with VHF/UHF converters having inputs of 180 to 220 MHz, with 5–10 mW output. Diode in series with power supply prevents damage if polarity is reversed.—J. Reisert, VHF/UHF Techniques, *Ham Radio*, March 1976, p 44–48.

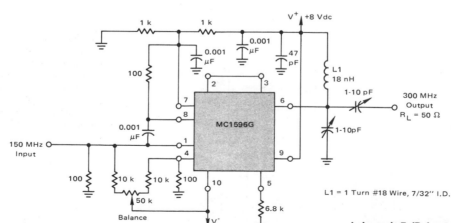

L1 = 1 Turn #18 Wire, 7/32'' I.D.

DOUBLING 150 MHz—Motorola MC1596G balanced modulator is connected for doubling at RF and UHF. With output filtering shown, all spurious outputs are at least 20 dB below desired 300-MHz output. Suppression of spurious outputs is poorer for higher input frequencies, being only 7 dB down for 400-MHz output, but performance is still superior to that of conventional transistor doubler.—R. Hejhall, "MC1596 Balanced Modulator," Motorola, Phoenix, AZ, 1975, AN-531, p 10.

24.5 MHz TO 147 MHz—Uses Q1 as tripler to 73.5 MHz for frequency-modulated input of 24.5 MHz, and Q2 as doubler whose output tank is tuned to 147 MHz by C6 and L3. Output is about 200 mW of RF. L1 is 20 turns No. 26 with center tap, on 0.5-cm form. L2 is 8 turns No. 22 on 0.8-cm form. RFC is 25 turns on 0.5-cm form. Article covers troubles likely to be encountered.—B. Hoisington, Frequency Multiplication the Easy Way, *73 Magazine*, Oct. 1973, p 69–71.

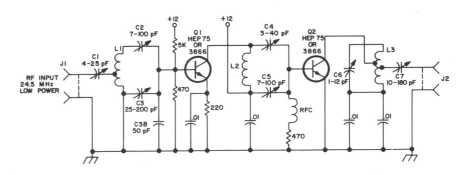

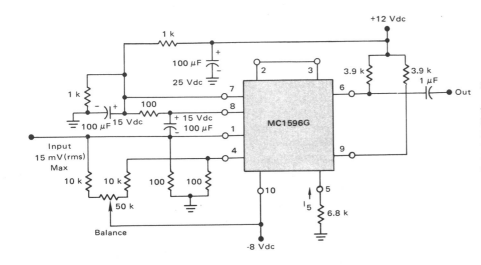

BROADBAND LOW-FREQUENCY DOUBLER—Motorola MC1596G balanced modulator functions as frequency doubler when same signal is injected into both input ports (pins 1 and 8). Doubling occurs in audio range and up to about 1 MHz.—R. Hejhall, "MC1596 Balanced Modulator," Motorola, Phoenix, AZ, 1975, AN-531, p 10.

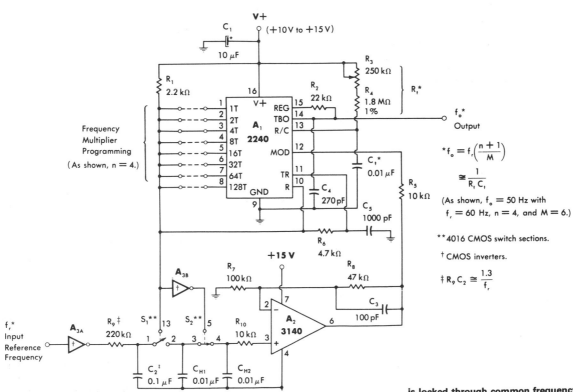

$$f_o = f_r\left(\frac{n+1}{M}\right)$$

$$\cong \frac{1}{R_1 C_1}$$

(As shown, $f_o = 50$ Hz with $f_r = 60$ Hz, $n = 4$, and $M = 6$.)

** 4016 CMOS switch sections.

† CMOS inverters.

‡ $R_9 C_2 \cong \dfrac{1.3}{f_r}$

PROGRAMMABLE PLL SYNTHESIZER/MULTIPLIER—Uses programmable timer/counter A₁ as VCO for generating frequencies both above and below that of square-wave reference. Phase-locked output frequency is not direct multiple of reference frequency. 2240 can lock on programmable multiple or on subharmonic reference. For values shown, phase-locked loop is locked through common frequency submultiple of 10 Hz, to give sampling rate of 10 Hz for reference input.—W. G. Jung, "IC Timer Cookbook," Howard W. Sams, Indianapolis, IN, 1977, p 220–224.

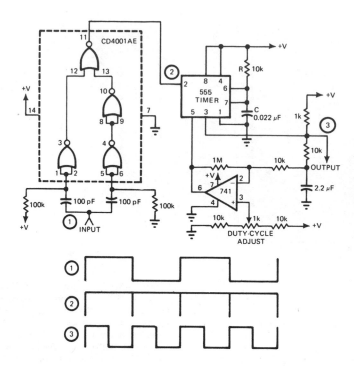

500–5000 Hz SQUARE-WAVE DOUBLER—Circuit shows virtually no deviation from 50% duty cycle over entire frequency range. Four NOR gates in CD4001AE IC form edge detector that presents negative pulse to 555 IC timer on both rising and falling edges of input square wave, to achieve frequency doubling. High-gain 741 opamp amplifies any difference between DC level at timer output and reference equal to half of supply voltage, to send correction voltage to pin 5 of timer for forcing output to 50% duty cycle.—L. P. Kahhan, Frequency Doubler Outputs Square Wave with 50% Duty Cycle, *EDN Magazine*, June 5, 1977, p 211–212.

PUSH-PUSH DOUBLER—Useful in VFO output circuits where oscillator operates at half output frequency of doubler. Circuit helps reduce oscillator instability during load changes while having about same efficiency as straight amplifier. Uses two sections of RCA CA3028A differential amplifier as doubler (U1C is not used). Values of tuned circuit depend on frequency, which can be up to 120 MHz.—D. DeMaw, Understanding Linear ICs, *QST*, Feb. 1977, p 19–23.

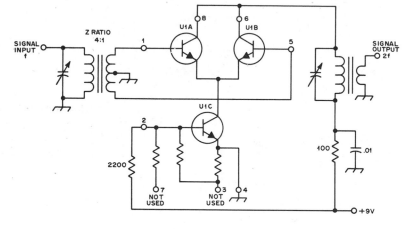

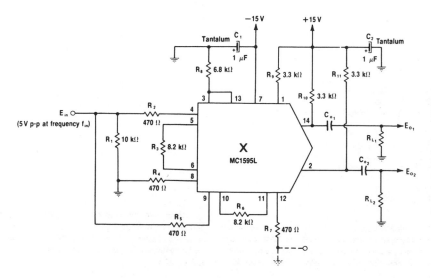

WIDEBAND DOUBLER—AC-coupled multiplier block is connected in squaring mode to provide second harmonic of input frequency with no tuned circuits. Circuit operates over wide bandwidths without adjustment. Output is low-distortion sine wave; total harmonic distortion is typically 1%. Output can be taken from pin 2 or 14, depending on phase desired. Circuit will work with R_7 grounded, but offset adjustment can be used to minimize distortion. Maximum operating frequency is several megahertz.—W. G. Jung, "IC Op-Amp Cookbook," Howard W. Sams, Indianapolis, IN, 1974, p 258–259.

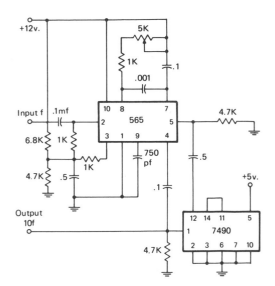

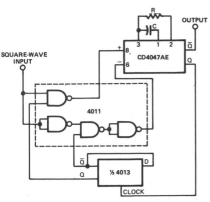

PLL MULTIPLIES BY 10—Used with frequency counter to measure very low frequencies. Two such circuits can be cascaded to give multiplication by 100. Requires +5 V and +12 V supplies.—H. S. Laidman, Upgrading Inexpensive Counters, *CQ*, Aug. 1975, p 16–22.

MULTIPLIER/DIVIDER—Choice of values for R and C determines multiplication or division factor acting on square-wave input frequency. Output of 4013 flip-flop sets gates of 4011 to steer input clock pulse of RCA CD4047AE mono MVBR to proper inputs. When rising edge of input triggers mono, Q output of mono goes high and switches flip-flop, preparing mono to accept falling-edge trigger. Since 4047 locks out inputs until it times out, mono triggers only on first falling edge occurring after its output goes low. Mono pulse length is about 2.5RC. With 60-Hz input clock, mono pulse length less than 8.33 ms allows triggering on every transition, to give 120-Hz output.—P. A. Lawless, One-Shot Forms Frequency Multiplier, *EDN Magazine*, Aug. 5, 1978, p 72.

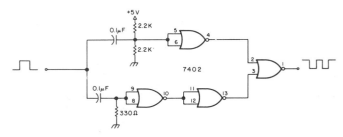

SQUARE-WAVE DOUBLER—Circuit locks onto both rise and fall of input square wave, to give identical square-wave output at doubled frequency. For high input frequencies, use smaller capacitance values.—Circuits, *73 Magazine*, April 1977, p 164.

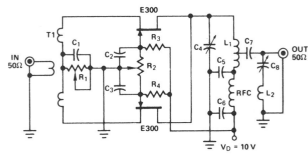

C_1, C_5, C_6 — 1500 pF	R_3, R_4 — 220KΩ, ¼W
C_2, C_3 — 1000 pF	L_1 — 4T #18 AWG, 5/16 DX 5/16 LG
C_4 — 8–35 pF	TAPPED 3/4T FROM COLD END
C_7 — 30 pF	L_2 — 2T #16 AWG 5/16 DX 3/16 LG
C_8 — 2.3–20 pF	RFC — 1.2 μHy
R_1 — 1KΩ	T1 — RELCOM BT-9
R_2 — 10KΩ	50Ω IN-400CT-400Ω OUT

FET DOUBLER—Siliconix E300 matched FETs are connected as common-gate amplifiers in balanced push-push circuit giving up to 100% efficiency as frequency multiplier in UHF range. Series-tuned output trap L_2C_8 increases rejection of third-order harmonics to greater than 70 dB. Positive bias of 0.5 V is applied to FET gates to permit inclusion of balance control R_2. Gain of doubler is about 1 dB.—"Analog Switches and Their Applications," Siliconix, Santa Clara, CA, 1976, p 7–52.

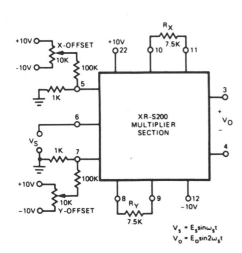

$$V_S = E_s \sin\omega_s t$$
$$V_O = E_O \sin 2\omega_s t$$

SINE-WAVE DOUBLER—Frequency of sinusoidal input signal V_S is doubled to give sine-wave output with total harmonic distortion less than 0.6%. With input of 4 V P-P at 10 kHz, output is 1 V P-P at 20 kHz. X and Y offset adjustments are nulled to minimize harmonic content of output.—"Phase-Locked Loop Data Book," Exar Integrated Systems, Sunnyvale, CA, 1978, p 9–16.

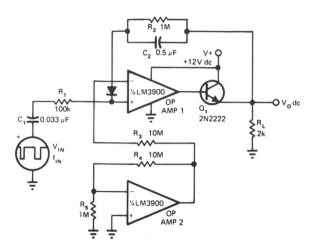

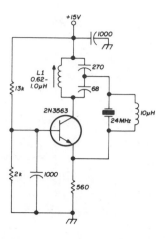

DOUBLER FOR TACHOMETER—Frequency of input from tachometer is doubled by charging and discharging of C_1 to reduce ripple in DC output voltage of tachometer circuit. Opamp 2 provides bias current for opamp 1, while Q_1 drives large load currents and provides DC level shift required for bringing output voltage to zero when input frequency is zero.—T. Frederiksen, Frequency-Doubling Tach Operates from a Single Supply, *EDN Magazine*, June 5, 1977, p 208.

73.333 MHz ON THIRD OVERTONE—Simple crystal oscillator circuit requires only one tripler for multiplying to 220-MHz amateur band. Mode suppression is provided by 10-μH coil which, with 4.5-pF capacitance of crystal holder, is series resonant at 24 MHz.—H. Olson, Frequency Synthesizer for 220 MHz, *Ham Radio*, Dec. 1974, p 8–14.

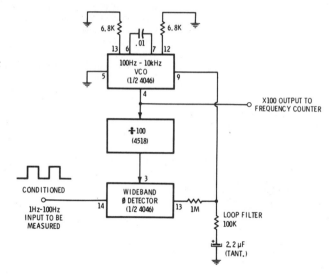

AF MULTIPLIER—Multiplies 1–100 Hz input signal by 100 to permit measuring frequency with ordinary counter. Half of 4046 PLL is connected as 100–10,000 Hz VCO whose output is divided by 100 in 4518 dual divide-by-10 counter for comparison with input signal in other half of PLL connected as wideband phase detector. Output of detector goes to loop filter and to VCO for locking VCO at 100 times input frequency.—D. Lancaster, "CMOS Cookbook," Howard W. Sams, Indianapolis, IN, 1977, p 364–366.

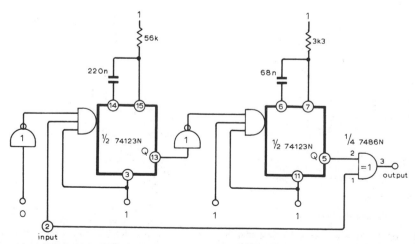

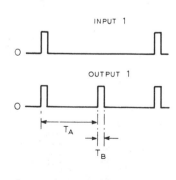

PULSE DOUBLER—Generates pulses at twice pulse input frequency. Input pulse at first monostable of 74123N makes it run for time T_A. Negative edge, terminating T_A, triggers second monostable which runs for time T_B. If T_A equals half of input period and T_B equals width of input pulse, desired result is achieved wherein additional pulse is generated between input pulses.

EXCLUSIVE-OR gate combines both pulses at output. Values shown for R and C will double frequency of 800-μs-wide input pulses having repetition rate of about 130 per second.—K. R. Brooks, Pulse Rate Doubler, *Wireless World*, April 1976, p 63.

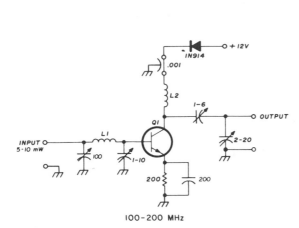

100–200 MHz

L1   12 turns no. 28 (0.3mm) on Amidon T25-12 toroid
     core
L2   7 turns no. 24 (0.5mm), air core, closewound on 0.1″
     (2.5mm) diameter

DOUBLING 100 MHz—Recommended for use with VHF/UHF converters having inputs of 90 to 120 MHz. Diode in series with power supply prevents damage if the polarity is reversed.—J. Reisert, VHF/UHF Techniques, *Ham Radio*, March 1976, p 44–48.

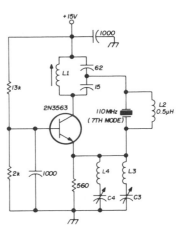

C3-   series resonant at approximately 5/7(110
L3   MHz), 1 μH and 0.6-10 pF
C4-   series resonant at approximately 3/7(110
L4   MHz), 2.2 μH and 0.6-10 pF
L2   0.5 μH (parallel resonant at 110 MHz
      with 4.5-pF holder capacitance)

110 MHz ON SEVENTH OVERTONE—Requires only one doubler for use in 220-MHz amateur band. Series-resonant traps are at frequencies of undesired lower modes.—H. Olson, Frequency Synthesizer for 220 MHz, *Ham Radio*, Dec. 1974, p 8–14.

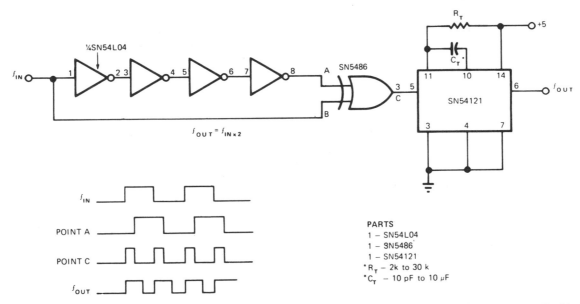

$f_{OUT} = f_{IN \times 2}$

PARTS
1 – SN54L04
1 – SN5486
1 – SN54121
*R$_T$ – 2k to 30 k
*C$_T$ – 10 pF to 10 μF

DOUBLER FOR 1 Hz TO 12 MHz—Simple arrangement of EXCLUSIVE-OR, mono, and hex inverter ICs provides extremely accurate frequency doubling in digital systems, along with waveform symmetry. Article gives design equation. Series inverters create about 120 ns of delay.—V. Rende, Frequency Doubler Operates from 1 Hz to 12 MHz, *EDN Magazine*, Aug. 20, 1976, p 85.

CHAPTER 36
Frequency Synthesizer Circuits

Covers methods of generating up to 2500 different discrete frequencies in audio and RF spectrums, generally by setting thumbwheel switches or by keyboard control, for use in test equipment, receivers, and transceivers.

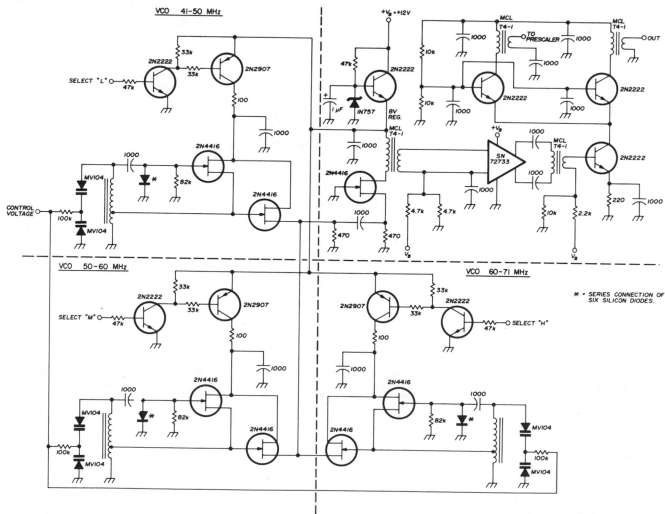

41–50, 50–60, AND 60–71 MHz—Three independent low-noise VCOs are used in 41–71 MHz frequency synthesizer. Control voltage is obtained from phase-comparator output of synthesizer. Outputs are chosen by selector switch. Texas Instruments SN72733 wideband amplifier is used for decoupling. Cascade arrangement provides two independent outputs at low impedance.—U. L. Rohde, Modern Design of Frequency Synthesizers, *Ham Radio*, July 1976, p 10–23.

1.1–1.6 MHz IN 10-Hz STEPS—Data input requirement is parallel BCD with 10-V CMOS levels and five digits. Reference input is 1-MHz sine or square with at least 1 V P-P. VCO covers 110–160 MHz in 1-kHz steps, operating in loop having 1-kHz reference. VCO signal is divided by 100 to give final output in 10-Hz steps. L1 is 6 turns No. 22 on 3-mm form, tapped at 2 turns. RFC is 6 turns No. 28 on F754-1-06 ferrite bead. T1 is Mini-Circuits Lab T16-1 broadband RF transformer.—R. C. Petit, Frequency Synthesized Local-Oscillator System for the High-Frequency Amateur Bands, *Ham Radio*, Oct. 1978, p 60–65.

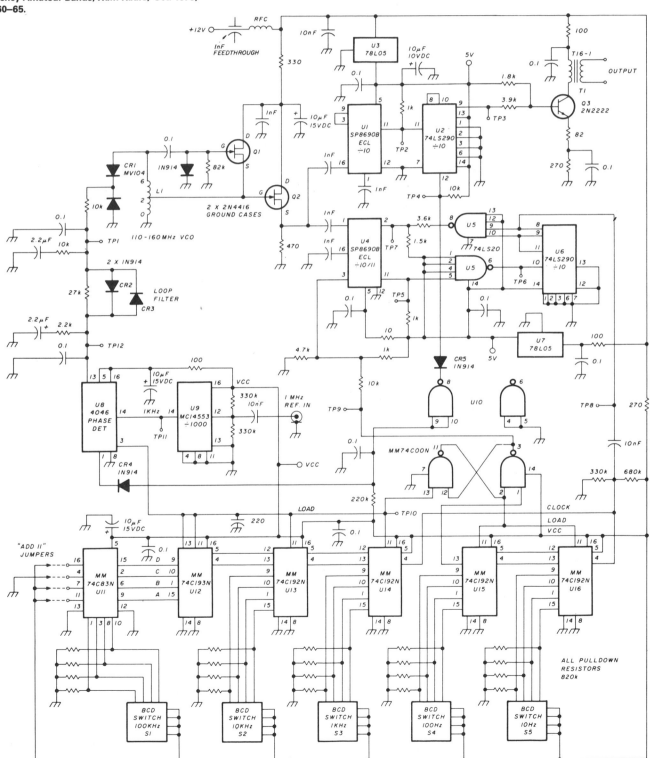

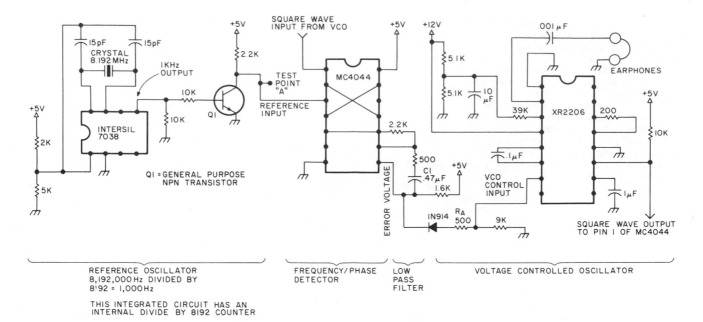

REFERENCE OSCILLATOR
8,192,000 Hz DIVIDED BY
8192 = 1,000 Hz

THIS INTEGRATED CIRCUIT HAS AN
INTERNAL DIVIDE BY 8192 COUNTER

FREQUENCY/PHASE DETECTOR

LOW PASS FILTER

VOLTAGE CONTROLLED OSCILLATOR

1–9 kHz PLL—Simple experimental phase-locked loop circuit synthesizes frequencies in audio range, for easy monitoring with headphones. Article gives theory of operation and setup procedure.—G. R. Allen, Synthesize Yourself!, *73 Magazine*, Oct. 1977, p 182–188.

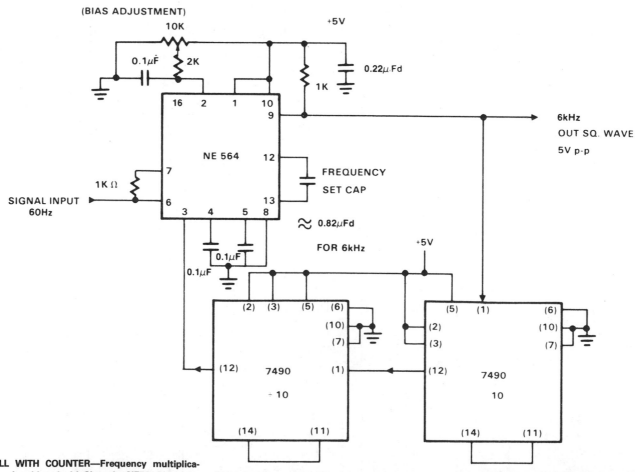

PLL WITH COUNTER—Frequency multiplication is achieved with Signetics NE564 PLL by inserting counter in loop between VCO and phase comparator. VCO is then running at multiple of input frequency determined by counter; with connections and values shown, multiplication factor for 60-Hz input signal is 100, giving 6-kHz square-wave output.—"Signetics Analog Data Manual," Signetics, Sunnyvale, CA, 1977, p 830–831.

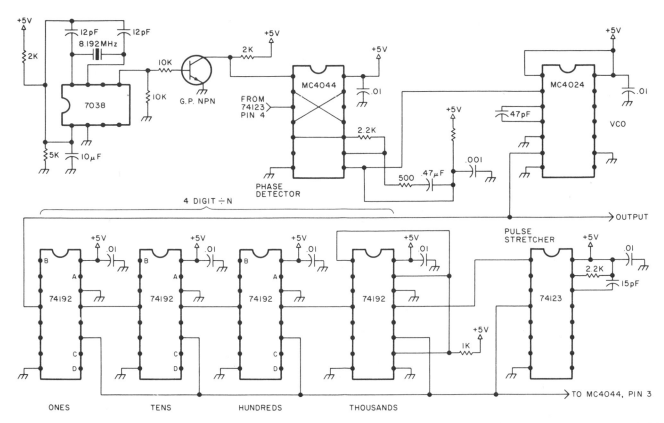

7.000–7.999 MHz PLL—Provides output in 1-kHz steps under digital programming, except that first digit is hard-wired to 7 and does not change. VCO is Motorola MC4024, which generates square-wave output. For sine output, use low-pass filter at output of VCO to eliminate all frequencies above 7.999 MHz, or use different VCO. 74123 mono lengthens reset pulse generated by divide-by-N circuit. Terminals A, B, C, and D of 74192s go to grounding switches that are set to give desired division ratio. Article gives theory of PLL synthesizers.—G. R. Allen, Synthesize Yourself!, *73 Magazine*, Oct. 1977, p 182–188.

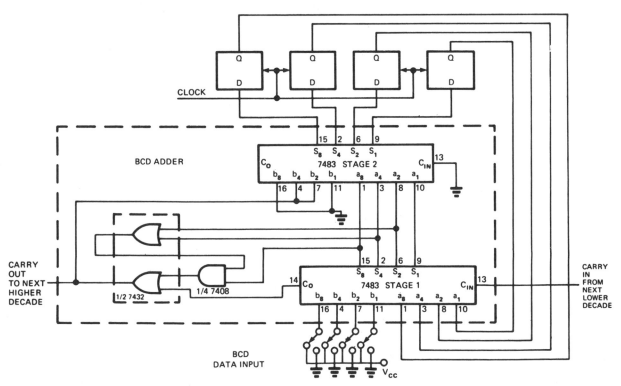

SWITCH-CONTROLLED ADDER—Direct BCD input from thumbwheel switch and use of standard crystal frequencies are primary advantages of accumulator stage of synthesizer, one decade of which is shown. BCD adder drives four D flip-flops whose outputs are fed back and added to switch states. Frequency range depends on number of decades used. Output pulse may be used directly for synchronization. If square wave is needed, clock frequency can be doubled and output of accumulator used to clock flip-flop.—D. W. Coulbourn, Set Frequency Synthesizer with Thumbwheel Switches, *EDN Magazine*, April 5, 1975, p 115 and 117.

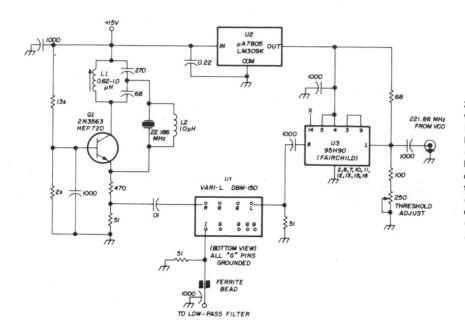

220-MHz PHASE-LOCKED—Fairchild 95H90 divide-by-10 counter U3 is used to divide 221.86-MHz VCO output frequency by 10. Resulting 22-MHz output of U3 is compared in phase with output of 22-MHz crystal-controlled oscillator by phase comparator U1, which is standard double-balanced mixer. Output of phase detector is passed through active low-pass filter for control of VCO. Article gives filter and VCO circuits. U2 is 5-V voltage regulator for 95H90.—H. Olson, Frequency Synthesizer for 220 MHz, *Ham Radio,* Dec. 1974, p 8–14.

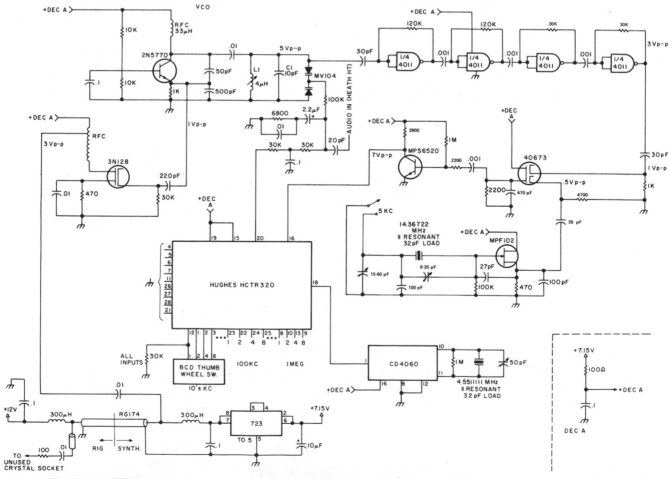

140–150 MHz IN 5-kHz STEPS—Developed for use with amateur 2-meter radios, to give direct choice of frequency by setting thumbwheel or lever switches. Phase-locked loop gives precise high-purity output. Input frequency to system is 4.5511111-MHz reference signal from CD4060 crystal oscillator. Digital edge-triggered Hughes HCTR320 phase comparator maintains inputs of both frequency and phase coherence at lock; lock range is thus capture range, making locking on harmonics impossible. Article describes operation of circuit in detail, and gives construction details as well as circuit for keyboard entry system. Power supply is 723 precision regulator giving 7.15 V.—M. I. Cohen, A Practical 2m Synthesizer, *73 Magazine,* Sept. 1977, p 146–151.

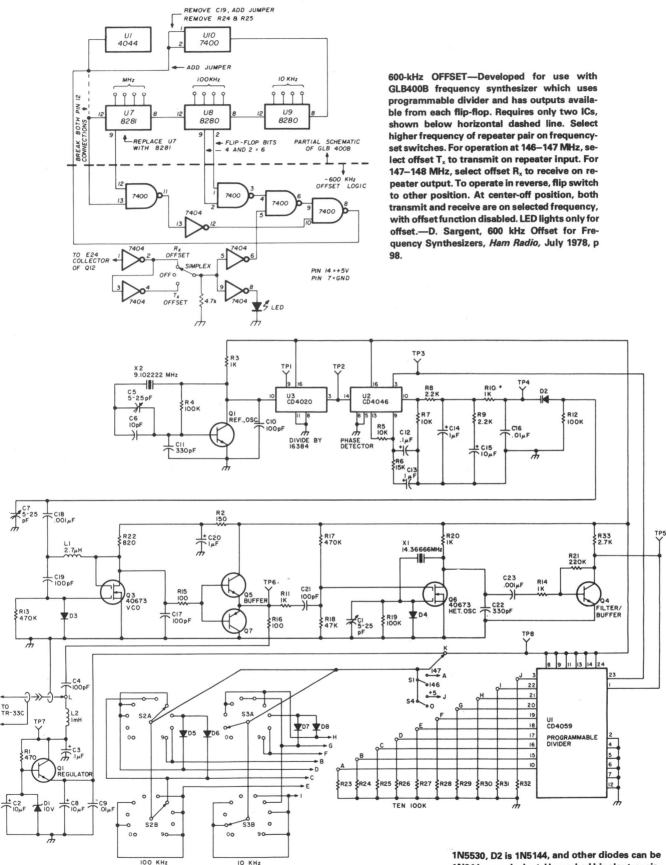

600-kHz OFFSET—Developed for use with GLB400B frequency synthesizer which uses programmable divider and has outputs available from each flip-flop. Requires only two ICs, shown below horizontal dashed line. Select higher frequency of repeater pair on frequency-set switches. For operation at 146–147 MHz, select offset T_x to transmit on repeater input. For 147–148 MHz, select offset R_x to receive on repeater output. To operate in reverse, flip switch to other position. At center-off position, both transmit and receive are on selected frequency, with offset function disabled. LED lights only for offset.—D. Sargent, 600 kHz Offset for Frequency Synthesizers, *Ham Radio*, July 1978, p 98.

146.000–147.995 kHz SYNTHESIZER—Designed for use with Drake TR-33C transceiver. Circuit has built-in offset providing choice of any 5-kHz-spaced channel in frequency range for transmit and receive frequencies. Only two crystals are required. Desired frequency can be entered with BCD thumbwheel switches. Tune-up and testing procedures are given. D1 is 1N5530, D2 is 1N5144, and other diodes can be 1N914 or equivalent. Unmarked bipolar transistors are fast-switching silicon types; NPNs can be 2N2222, and PNPs can be 2N4403.—J. Moell, Super Deluxing the TR-33, *73 Magazine*, April 1978, p 72–74.

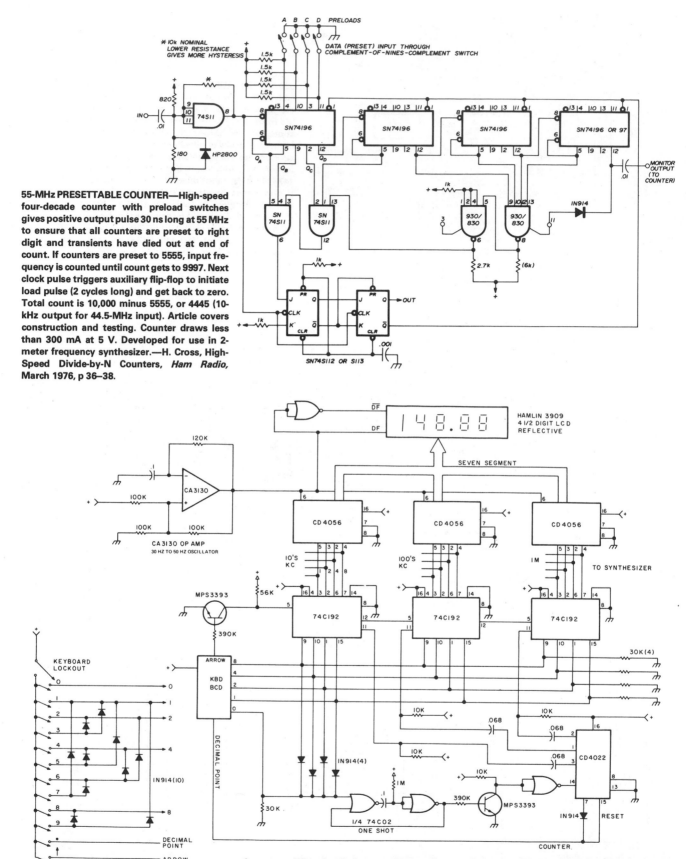

55-MHz PRESETTABLE COUNTER—High-speed four-decade counter with preload switches gives positive output pulse 30 ns long at 55 MHz to ensure that all counters are preset to right digit and transients have died out at end of count. If counters are preset to 5555, input frequency is counted until count gets to 9997. Next clock pulse triggers auxiliary flip-flop to initiate load pulse (2 cycles long) and get back to zero. Total count is 10,000 minus 5555, or 4445 (10-kHz output for 44.5-MHz input). Article covers construction and testing. Counter draws less than 300 mA at 5 V. Developed for use in 2-meter frequency synthesizer.—H. Cross, High-Speed Divide-by-N Counters, *Ham Radio*, March 1976, p 36–38.

KEYBOARD ENTRY WITH 4½-DIGIT DISPLAY— Developed to give keyboard entry of desired frequency for 2-meter frequency synthesizer, as alternative to thumbwheel-switch setting of frequency. When key is depressed, mono (one-shot) fires, causing CD4022 counter to increment. At same time, keyswitch places appropriate BCD data on input lines of 74C192 presettable decade counters. Output from counters goes to synthesizer input and to display decoder/driver. LED display may be used in place of LCD display if current drain is not important. Keyboard lockout switch prevents accidental change of frequency.—M. I. Cohen, A Practical 2m Synthesizer, *73 Magazine*, Sept. 1977, p 146–151.

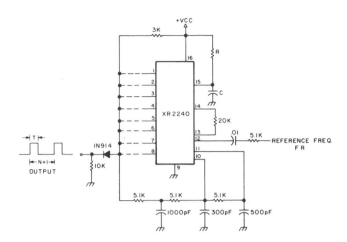

XR2240 SYNTHESIZER—Circuit uses XR2240 programmable timer/counter for simultaneous multiplication of input frequency FR by factor of M and division of input frequency by factor of N + 1, where M and N are integers selected by appropriate connections of binary pins 1-8 to common output bus. Output frequency is then FR(M)/(1 + N) where M is between 1 and 10 inclusive and N is between 1 and 255 inclusive. VCC is 4–15 V.—H. M. Berlin, IC Timer Review, *73 Magazine*, Jan. 1978, p 40–45.

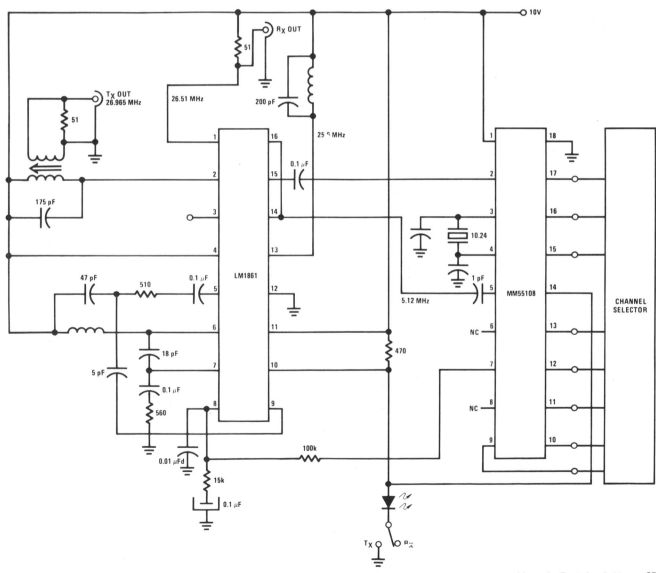

SINGLE-CONVERSION SYNTHESIZER—Used in single-conversion CB transceiver in which VCO operates at channel frequency during transmit and 455 kHz below channel frequency during receive. 5.120 MHz is quintupled to 25.600 MHz to mix and provide input to programmable divider.—L. Sample, A Linear CB Synthesizer, *IEEE Transactions on Consumer Electronics*, Aug. 1977, p 200–206.

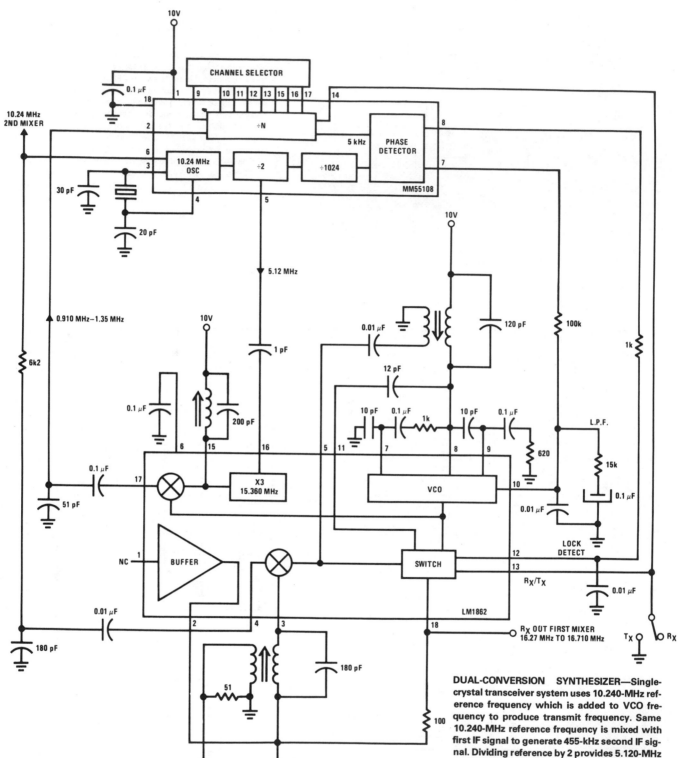

DUAL-CONVERSION SYNTHESIZER—Single-crystal transceiver system uses 10.240-MHz reference frequency which is added to VCO frequency to produce transmit frequency. Same 10.240-MHz reference frequency is mixed with first IF signal to generate 455-kHz second IF signal. Dividing reference by 2 provides 5.120-MHz signal that is lightly coupled to multiply-by-3 buffer whose output is tuned to third harmonic (15.360 MHz). This is mixed with VCO frequency to provide input signal for programmable divider. When VCO is operating as first mixer and local oscillator on CB channel 1 (16.270 MHz), difference frequency is 910 kHz. Programmable divider divides by 182 to give necessary 5-kHz input to phase detector. If VCO moves off frequency, divided input to phase detector moves away from 5 kHz and action of loop pulls VCO back on frequency.—L. Sample, A Linear CB Synthesizer, *IEEE Transactions on Consumer Electronics*, Aug. 1977, p 200–206.

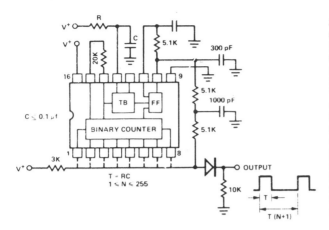

255-FREQUENCY SYNTHESIZER—Circuit as shown for programmable counter section of Exar XR-2240 programmable timer/counter provides square-wave outputs at 255 discrete frequencies from given internal time-base setting. Output is positive pulse train with pulse width T determined by values of R and C. Period is equal to (N + 1)T where N is programmed count in counter of IC. Counter output connections to output bus determine value of N; if pins 1, 3, and 4 are connected to bus, N is 1 + 4 + 8 or 13 and period is 14T. Supply voltage range is 4–15 V. If counter cannot be triggered when using supply above 7 V and less than 0.1 μF for C, connect 300 pF from pin 14 to ground.—"Timer Data Book," Exar Integrated Systems, Sunnyvale, CA, 1978, p 11–18.

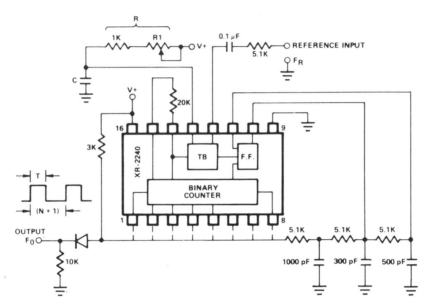

2500 FREQUENCIES WITH SYNCHRONIZATION—EXAR XR-2240 programmable timer/counter containing 8-bit programmable binary counter and stable time-base oscillator can generate over 2500 discrete frequencies from single input reference frequency. Circuit simultaneously multiplies input frequency by factor M and divides by N + 1, where M and N are adjustable integer values. Output frequency F_O is equal to input frequency F_R multiplied by M/(1 + N). M and N can be externally adjusted over broad range, with M between 1 and 10 and N between 1 and 255. Multiplication factor M is obtained by locking on harmonics of reference. Division factor N is determined by preprogrammed count in binary counter section, established by wiring appropriate pins 1-8 to output bus. Input reference is 3 V P-P pulse train with pulse duration ranging from 30% to 80% of time-base period T. R_1 determines value of M. C is in range of 0.005 to 0.1 μF, and R is between 1K and 1 megohm for maximum output frequency of about 200 kHz. With M = 5 and N = 2, 100-Hz clock synchronized to 60-Hz line frequency is obtained.—"Timer Data Book," Exar Integrated Systems, Sunnyvale, CA, 1978, p 31–32.

CHAPTER 37

Function Generator Circuits

Used for generating various combinations of sine, square, and triangle waveforms, usually with manual or external variations of frequency in AF or RF ranges by DC control voltage. Also includes circuits for generating cubic, quadratic, hyperbolic, trigonometric, ramp, and other mathematical waveforms, as well as circuits for converting one of these waveforms to one or more others. See also Multivibrator, Oscillator, Pulse Generator, Signal Generator, and Sweep chapters.

FSK SINE-SQUARE-TRIANGLE GENERATOR— Exar XR-2206 modulator-demodulator (modem) is connected as function generator providing high-purity sinusoidal output along with triangle and square outputs, for FSK applications. Circuit has excellent frequency stability along with TTL and CMOS compatibility. Total harmonic distortion in 3 V P-P sine output is about 2.5% untrimmed, but can be trimmed to 0.5%. High-level data input signal selects frequency of $1/R_6C_3$ Hz, while low-level input selects $1/R_7C_3$ Hz. For optimum stability, R_6 and R_7 should be in range of 10K to 100K. Adjust R_8 and R_9 for minimum distortion.—"Phase-Locked Loop Data Book," Exar Integrated Systems, Sunnyvale, CA, 1978, p 57–61.

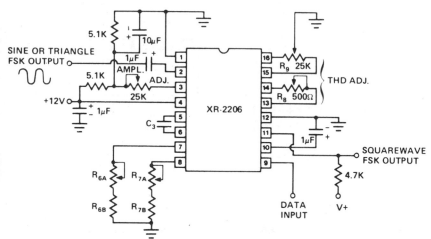

HYPERBOLIC A/X FUNCTION—Uses Precision Monolithics DAC-20EX D/A converter with OP-17G opamp to generate extended-range hyperbolic functions of the type A/X, where A is analog constant and X represents decimally expressed digital divisor. R5 provides simultaneous adjustment of scale factor and output amplifier offset voltage. Same circuit serves for −A/X function if DAC reference amplifier and output opamp terminals are reversed.—W. Ritmanich, B. Blair, and B. Debowey, "Digital-to-Analog Converter Generates Hyperbolic Functions," Precision Monolithics, Santa Clara, CA, 1977, AN-23, p 2.

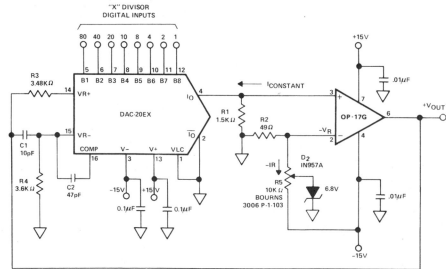

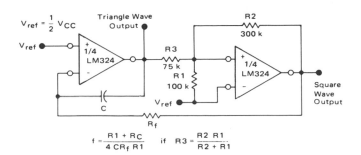

$$f = \frac{R1 + RC}{4\,CR_f\,R1} \quad \text{if} \quad R3 = \frac{R2\,R1}{R2 + R1}.$$

BASIC SQUARE-TRIANGLE—Requires only two sections of LM324 quad differential-input opamp to provide choice of triangle or square-wave outputs at frequency determined by values of components. Supply voltage range is 3–32 V.—"Quad Low Power Operational Amplifiers," Motorola, Phoenix, AZ, 1978, DS 9339 R1.

WAVEFORM GENERATOR—Two Optical Electronics 9008 integrators and 9813 comparator together generate choice of sine, square, and triangle waveforms suitable for system testing and display generation. Square wave is typically ±13.5 V with 20-μs transition time. Triangle wave is ±10 V with better than 0.1% triangle linearity. Comparator senses zero crossings of sine-wave output to produce square waves, thus completing feedback loop. Integrators are commanded at pin 8 for zero output, so triangle and sine outputs can be made to start from zero.—"Waveform Generator," Optical Electronics, Tucson, AZ, Application Tip 10257.

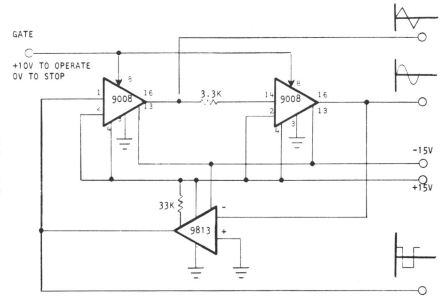

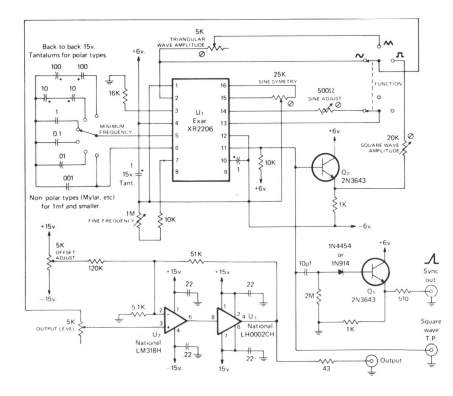

0.01 Hz TO 100 kHz—Variable DC offset permits adjustment of average value of sine, square, or triangle waveform to any arbitrary plus or minus value within voltage swing capability of opamp U_3. Buffer stage U_2 is inside feedback loop. Simple emitter-follower differentiator provides positive-going 1-V 0.5-μs output at sink terminal. Square-wave output is buffered by emitter-follower Q_2.—H. Olson, The Function Generator, *CQ*, July 1975, p 26–28 and 71–72.

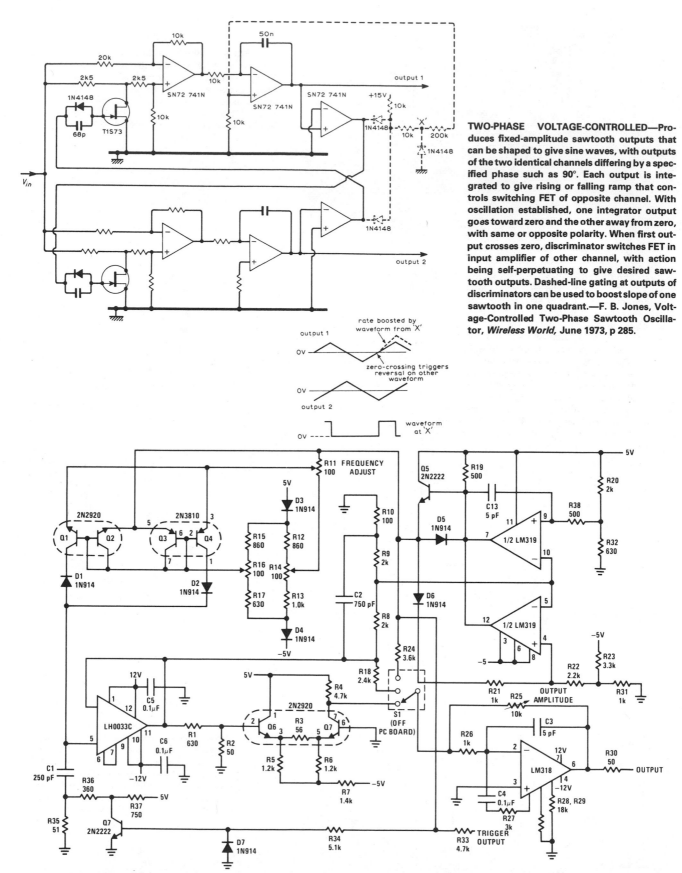

TWO-PHASE VOLTAGE-CONTROLLED—Produces fixed-amplitude sawtooth outputs that can be shaped to give sine waves, with outputs of the two identical channels differing by a specified phase such as 90°. Each output is integrated to give rising or falling ramp that controls switching FET of opposite channel. With oscillation established, one integrator output goes toward zero and the other away from zero, with same or opposite polarity. When first output crosses zero, discriminator switches FET in input amplifier of other channel, with action being self-perpetuating to give desired sawtooth outputs. Dashed-line gating at outputs of discriminators can be used to boost slope of one sawtooth in one quadrant.—F. B. Jones, Voltage-Controlled Two-Phase Sawtooth Oscillator, *Wireless World*, June 1973, p 285.

10 Hz TO 2 MHz—Triangle wave is generated by switching current-source transistors to charge and discharge timing capacitor. Precision dual comparator sets peak-to-peak amplitude. Sine converter requires close amplitude control to give low-distortion output from triangle input. Square-wave output is obtained at emitter of Q5, for driving current switches Q1-Q4 and LM318 output amplifier. Scaling permits adjusting all three waveforms to ±10 V. Waveforms are symmetrical up to 1 MHz, and output is usable to about 2 MHz.—R. C. Dobkin, "Wide Range Function Generator," National Semiconductor, Santa Clara, CA, 1974, AN-115.

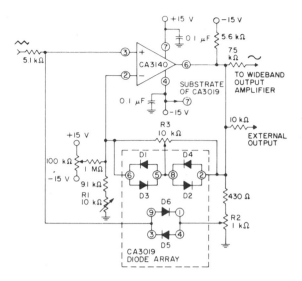

SINE-WAVE SHAPER—Uses CA3140 opamp as voltage follower, acting with diodes from CA3019 array to convert triangle output of function generator or other source to sine wave having total harmonic distortion typically less than 2%.—"Circuit Ideas for RCA Linear ICs," RCA Solid State Division, Somerville, NJ, 1977, p 5.

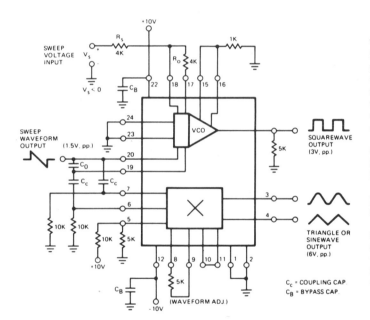

VOLTAGE-TUNED WITH 10:1 FREQUENCY RANGE—Exar XR-S200 PLL IC is connected to generate basic periodic square or sawtooth waveform. Multiplier section, used as linear differential amplifier, converts differential sawtooth input waveform to triangle wave. 5K pot connected between pins 8 and 9 rounds peaks of triangle to give low-distortion sine wave with less than 2% total harmonic distortion. Output frequency can be swept or frequency-modulated by applying proper analog control input. For linear frequency modulation with less than 10% deviation, modulation is applied between pins 23 and 24. For larger deviations, negative-going sweep voltage V_s is applied to pin 18 as shown. Digital control input pins 15 and 16 can be used for FSK applications; if this is not desired, pins are disabled by connecting to ground through current-limiting resistor.—"Phase-Locked Loop Data Book," Exar Integrated Systems, Sunnyvale, CA, 1978, p 9–16.

SINE/COSINE—Uses National SK0003 sine/cosine look-up table kit consisting of four MOS ROMs and three output adders. Combination implements equation $\sin \theta = \sin M \cos L + \cos M \sin L$. Worst-case error is 1 5/8 bits in least significant bit. Cosine is approximated with loss in resolution of ½ bit in 11-bit input or ¼ bit in 10-bit input.—"Memory Databook," National Semiconductor, Santa Clara, CA, 1977, p 6-98–6-99.

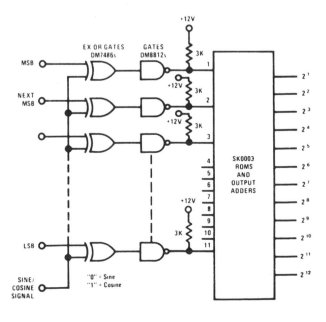

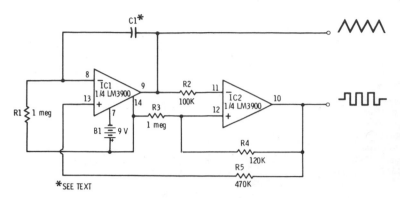

SQUARE-TRIANGLE AF—Two sections of LM3900 quad opamp are connected to generate dual-polarity triangle- and square-wave AF outputs while operating from single supply, by using current mirror circuit at noninverting input. Value used for C1 determines frequency and pulse width; frequency ranges from 0.5 Hz with 1 μF to 3800 Hz with 0.0001 μF and 21 kHz with C1 omitted. Pulse-width range is 35 μs without C1 to 1.6 s with 1 μF.—F. M. Mims, "Integrated Circuit Projects, Vol. 5," Radio Shack, Fort Worth, TX, 1977, 2nd Ed., p 57–63.

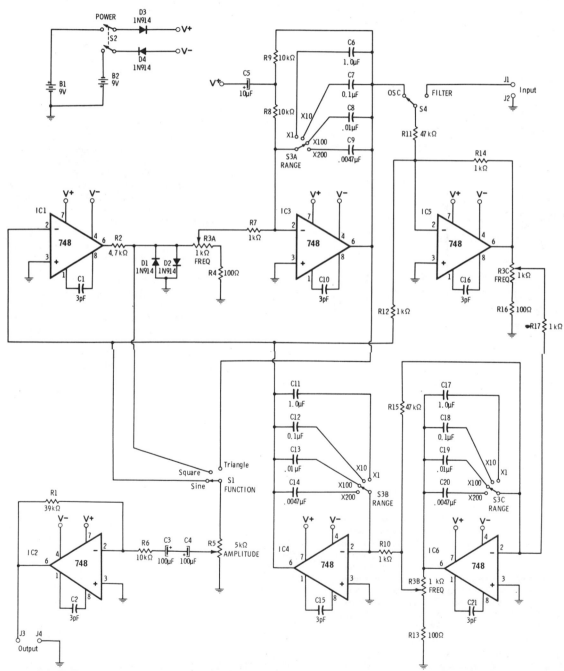

AF SINE-SQUARE-TRIANGLE—Can be tuned over entire audio spectrum in four ranges for generation of low-distortion waves for laboratory use. IC1 converts sine wave to square wave. IC3 acts as integrator converting square-wave output of IC1 to triangle wave. IC4-IC6 form state-variable filter for removing sine-wave component from triangle wave. IC2 is simple inverting amplifier for output.—R. Melen and H. Garland, "Understanding IC Operational Amplifiers," Howard W. Sams, Indianapolis, IN, 2nd Ed., 1978, p 130–134.

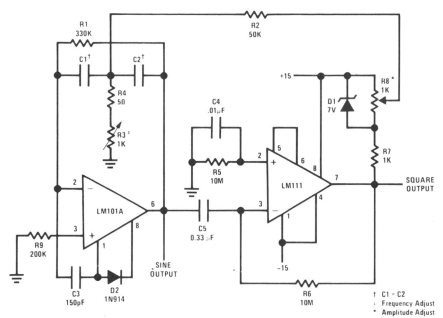

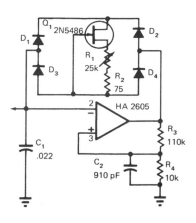

20–20,000 Hz SINE-SQUARE—Opamp is used as tuned circuit driven by square wave from voltage comparator. Frequency is controlled by R1-R3, C1, and C2, with R3 providing tuning. Comparator is fed with resulting sine wave to obtain square wave for feedback to input of tuned circuit, to cause oscillation. Zener stabilizes amplitude of square wave that is fed back.

R6 and C5 provide DC negative feedback around comparator to ensure starting. Values of C1 and C2 are equal, and range from 0.4 μF for 18–80 Hz to 0.002 μF for 4.4–20 kHz.—"Easily Tuned Sine Wave Oscillators," National Semiconductor, Santa Clara, CA, 1971, LB-16.

† C1 = C2
‡ Frequency Adjust
* Amplitude Adjust

$$F_0 = \frac{1}{2 \cdot C_1 \sqrt{R_3 R_1}}$$

0.5–25 kHz TRIANGLE—Diode bridge and FET form constant-current source charging C_1, to make voltage across C_1 change at linear rate as required for triangle output across C_1. Frequency can be adjusted from 500 Hz to above 20 kHz with constant output amplitude, by means of R_1. Short-term stability is better than 1 part in 10,000. Since the same R and C are used to generate both sections of the waveform, positive and negative slopes are identical. Diodes are HP 5082-2810.—G. R. Begault, Op Amp Makes Variable-Frequency Triangular Wave Generator, *EDN Magazine,* Sept. 15, 1972, p 42–43.

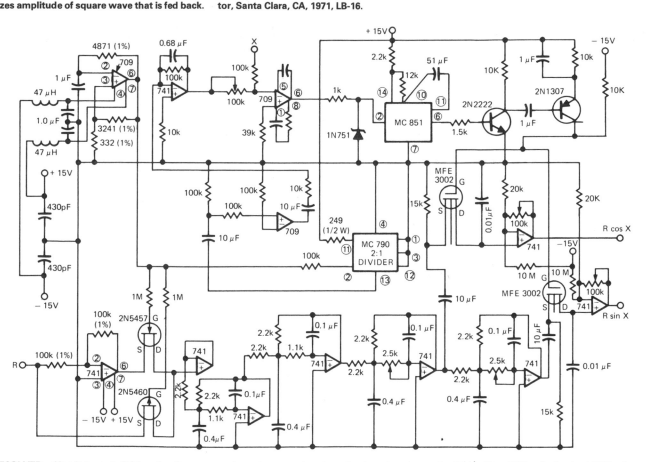

RESOLVER—Circuit accepts DC input voltages R and X and generates two DC output voltages R sin X and R cos X. Can be used in guidance computers to solve coordinate conversion problems (polar to rectangular) and in feedback systems to convert rectangular to polar coordinates. Sine wave is generated by chopping input signal R and filtering resulting square wave. Sine wave is then sampled at time controlled by X to generate R sin X. Cosine output is obtained by shifting first output 90° in phase. Circuit also generates proper sampling pulses and contains two sample-and-hold circuits on outputs.—W. H. Licata, Solid-State Resolver, *EDN Magazine,* July 20, 1973, p 82–83.

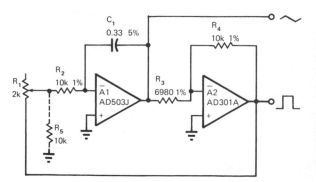

100-Hz SQUARE-TRIANGLE—Two-opamp oscillator delivers ±13 V square waves and ±10 V triangle waves simultaneously at 100 Hz for values shown. By scaling R_1, R_2, and C_1 wide range of frequencies can be covered down to 0.1 Hz (increase R_2 to 10 megohms for frequencies near lower limit). Square-wave rise time is about 1.5 μs and fall time 0.5 μs. Opamp A_1 operates as integrator and A_2 as Schmitt trigger.—R. S. Burwen, *Triangular and Square Wave Generator Has Wide Range,* *EDN Magazine,* Dec. 1, 1972, p 59.

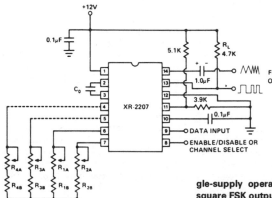

Logic Level		Active Timing Resistor	Output Frequency
Pin 8	Pin 9		
L	L	Pin 6	$\dfrac{1}{C_0 R_1}$
L	H	Pins 6 and 7	$\dfrac{1}{C_0 R_1} + \dfrac{1}{C_0 R_2}$
H	L	Pin 5	$\dfrac{1}{C_0 R_3}$
H	H	Pins 4 and 5	$\dfrac{1}{C_0 R_3} + \dfrac{1}{C_0 R_4}$

FSK SQUARE-TRIANGLE GENERATOR—Uses Exar XR-2207 FSK modulator connected for single-supply operation, to produce triangle or square FSK outputs for either single-channel or two-channel multiplex operation. Used in transmitting digital data over telecommunication links. Table gives equations for selecting timing resistors R_1-R_4; resistor values are in ohms, C_0 is in farads, and frequency is in hertz. For optimum stability, R_1 and R_3 should be in range of 10K to 100K. For two-channel multiplex, make connections shown by dotted lines.—"Phase-Locked Loop Data Book," Exar Integrated Systems, Sunnyvale, CA, 1978, p 57–61.

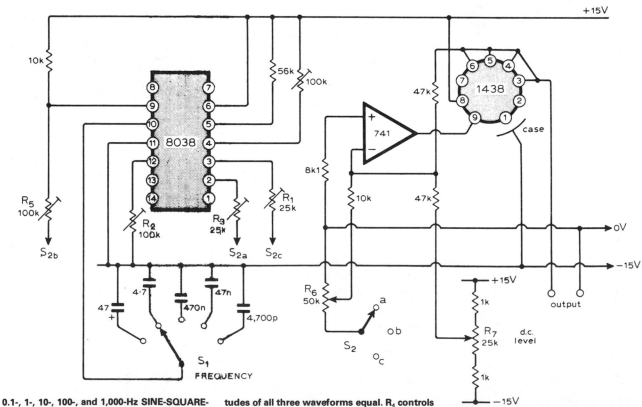

0.1-, 1-, 10-, 100-, and 1,000-Hz SINE-SQUARE-TRIANGLE—Provides choice of five spot frequencies switched in decades by S_1. Setting of S_2 determines shape of output waveform. Adjust R_1, R_3, and R_5 to make peak-to-peak amplitudes of all three waveforms equal. R_4 controls symmetry. R_2 is adjusted for minimum distortion of sine-wave output. Output may be set up to 100 mA by R_6, and is short-circuit-proof. DC level may be set anywhere between ±14 V by R_7. Motorola 1438 IC and 741 opamp boost output of 8038 IC sufficiently to drive most laboratory loads.—G. R. Wilson, *Low-Frequency Generator,* *Wireless World,* Feb. 1977, p 44.

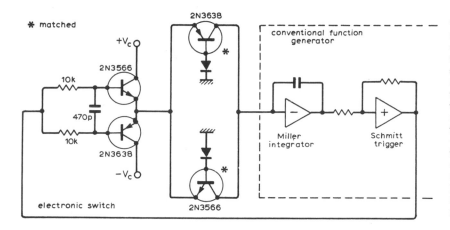

1000:1 FREQUENCY SWEEP—Permits varying output frequency of function generator over wide frequency range by using pot to vary control voltage V_c. Network consisting of two transistors and two diodes replaces usual charging resistor of Miller integrator in function generator, and has output current varying exponentially with input voltage. Electronic switch using pair of transistors is controlled by Schmitt trigger of function generator, which connects $+V_c$ and $-V_c$ alternately to charging circuit. If frequency pot is mechanically connected to strip-chart recorder, Bode plots of audio equipment can be made over entire audio range.—P. D. Hiscocks, Function Generator Mod. for Wide Sweep Range, *Wireless World,* Aug. 1973, p 374.

CURRENT-CONTROLLED SQUARE-TRIANGLE GENERATOR—CA3080 opamp is connected as current-controlled integrator of both polarities for use in current-controlled triangle oscillator. Frequency depends on values of C and opamp bias current and can be anywhere in audio range of 20 Hz to 20 kHz. Square-wave output is obtained by using LM301A opamp as Schmitt trigger.—S. Franco, Current-Controlled Triangular/Square-Wave Generator, *EDN Magazine,* Sept. 5, 1973, p 91.

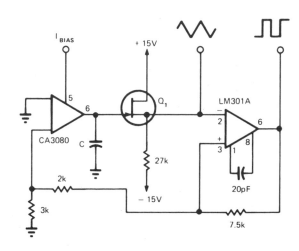

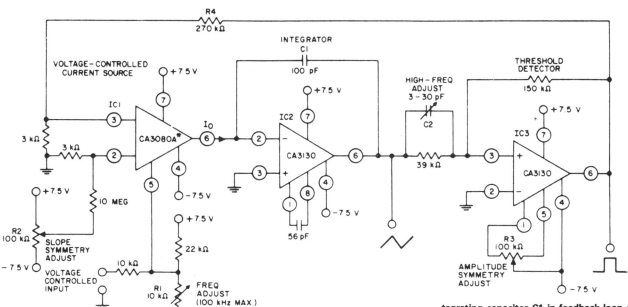

SINGLE CONTROL FOR 1,000,000:1 FREQUENCY RANGE—Uses two RCA CA3130 opamps and CA3080A operational transconductance amplifier to generate square and triangle outputs that can be swept over range of 0.1 Hz to 100 kHz with single 100K pot R1. Alternate voltage-control input is available for remote adjustment of sweep frequency. IC1 is operated as voltage-controlled current source whose output current is applied directly to integrating capacitor C1 in feedback loop of integrator IC2. R2 adjusts symmetry of triangle output. IC3 is used as controlled switch to set excursion limits of triangle output when square wave is desired.—"Linear Integrated Circuits and MOS/FET's," RCA Solid State Division, Somerville, NJ, 1977, p 236–244.

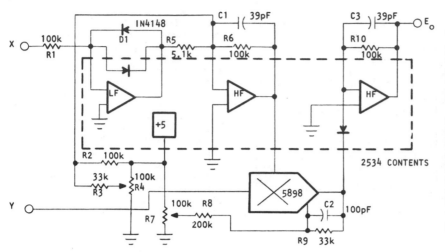

2534 CONTENTS

VOLTAGE-CONTROLLED NONLINEAR—Circuit produces function $E_0 = X^{Y/2}$, where X is input voltage in range of +10 mV to +10 V and Y is analog programming voltage in range of −0.4 V to −10 V. Uses Optical Electronics 2534 temperature-compensated log feedback elements, +5 V reference, two high-frequency opamps, and one low-frequency opamp. 2534 produces log conversion of input signal. 5898 multiplier serves to vary scale factor of log signal. With offsets used as shown, +10 V input will always produce +10 V output regardless of Y input. To set up, adjust R4 until output does not change with Y for +10 V input, then adjust R7 for +10 V output with +10 V input.—"Voltage-Controlled Non-Linear Function Generator," Optical Electronics, Tucson, AZ, Application Tip 10263.

0.5 Hz TO 1 MHz SINE-SQUARE-TRIANGLE—Uses Exar XR-2206 IC function generator in simple circuit that operates from dual supply ranging from ±6 V to ±12 V. With 1-μF capacitor for C, 2-megohm frequency control covers range of 0.5–1000 Hz. Range is 5–10,000 Hz with 0.1 μF, 50 Hz to 100 kHz with 0.01 μF, and 500 Hz to 1 MHz with 0.001 μF. Designed for experiments with active filters.—H. M. Berlin, "Design of Active Filters, with Experiments," Howard W. Sams, Indianapolis, IN, 1977, p 9–10.

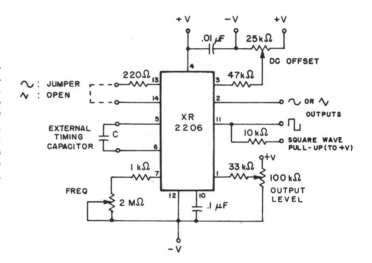

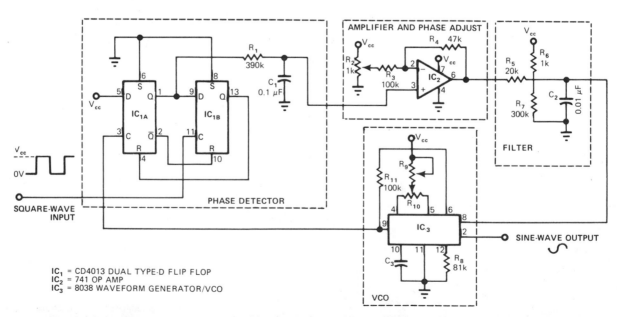

IC$_1$ = CD4013 DUAL TYPE-D FLIP FLOP
IC$_2$ = 741 OP AMP
IC$_3$ = 8038 WAVEFORM GENERATOR/VCO

SQUARE TO SINE WITH PLL—8038 waveform generator simultaneously generates synthesized sine wave and square wave. Square-wave output closes phase-locked loop through 741 opamp IC$_2$ and dual flip-flop IC$_1$, while sine-wave output functions as converted output. Center frequency is 0.15/R$_9$C$_3$. R$_{10}$ should be at least 10 times smaller than R$_9$. If center frequency is 400 Hz, capture range is half that or ±100 Hz. When input is applied, phase comparator generates voltage related to frequency and phase difference of input and free-running signals. IC$_2$ amplifies and offsets phase-difference signal. Sine output has less than 1% distortion, DC component of 0.5 V$_{CC}$, and minimum amplitude of 0.2 V$_{CC}$ P-P.—L. S. Kasevich, PLL Converts Square Wave into Sine Wave, *EDN Magazine*, June 20, 1978, p 128.

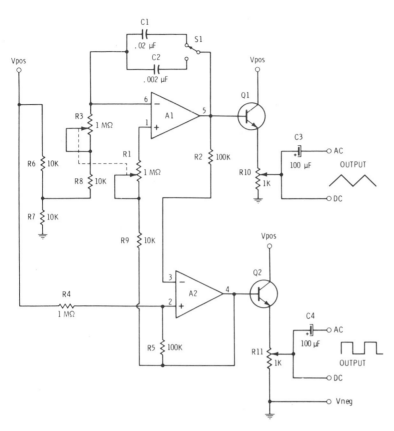

VARIABLE SQUARE-TRIANGLE—Dual pot R1-R3 varies frequency over range of 15–500 Hz when C1 is in circuit and 150–4800 Hz when C2 is in circuit. Each output has amplitude control. Opamps are Motorola MC3401P or National LM3900, and transistors are 2N2924 or equivalent NPN. Supply can be 12 VDC.—C. D. Rakes, "Integrated Circuit Projects," Howard W. Sams, Indianapolis, IN, 1975, p 19–20.

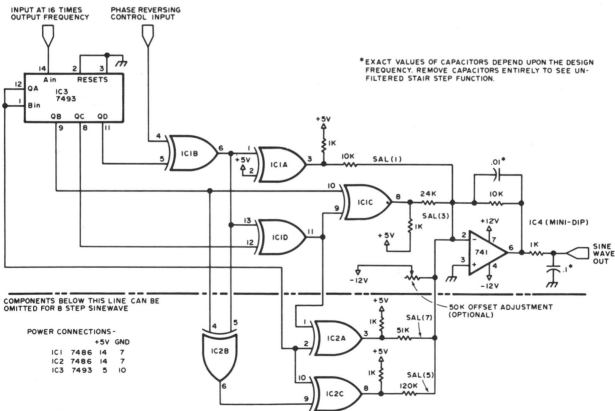

DIGITAL FOURIER—Sine-wave generator produces Walsh-function approximation of sine function. Frequency of sine wave is set by square-wave input to pin 14 of 7493. Filter components of opamp help smooth staircase waveform generated by summing Walsh-function components as weighted by resistors. Circuit is converter consisting of digital expander that expands input square wave into variety of digital waveforms and analog combiner that adds these waveforms to produce periodic analog output. Negative signs of Walsh harmonics are handled with digital inverter, and magnitudes are handled by choice of resistor value in summing junction. Signs and magnitudes are under microprocessor control. Net output is stairstep approximation to desired output, which can be smoothed by low-pass filter.—B. F. Jacoby, Walsh Functions: A Digital Fourier Series, *BYTE,* Sept. 1977, p 190–198.

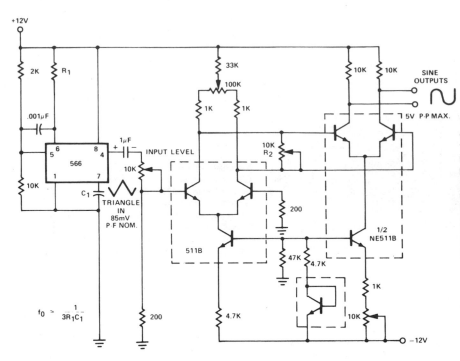

TRIANGLE-TO-SINE CONVERTER—Nonlinear emitter-base junction characteristic of 511B transistor array is used for shaping triangle output of 566 function generator to give sine output having less than 2% distortion. Amplitude of triangle is critical and must be carefully adjusted for minimum distortion of sine wave by varying values of R_1, R_2, and input level pot while monitoring output with Hewlett-Packard 333A distortion analyzer.—"Signetics Analog Data Manual," Signetics, Sunnyvale, CA, 1977, p 851–853.

$$f_0 \approx \frac{1}{3R_1C_1}$$

PULSE/SAWTOOTH GENERATOR—Pulse output is obtained from Exar XR-2006C function-generator IC when pin 9 is shorted to square-wave output at pin 11. Pulse duty cycle, along with rise and fall times of ramp from pin 2, is determined by values of R1 and R2. Both can be adjusted from 1 to 99% by proper selection of resistor values as given in formulas alongside diagram.—E. Noll, VHF/UHF Single-Frequency Conversion, *Ham Radio,* April 1975, p 62–67.

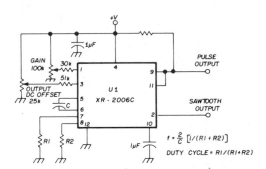

$$f = \frac{2}{C}\left[1/(R1 + R2)\right]$$

DUTY CYCLE = R1/(R1+R2)

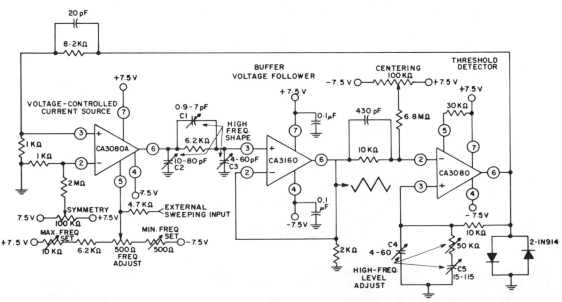

SINGLE FREQUENCY CONTROL—Adjustment range of over 1,000,000 to 1 for frequency is achieved by using CA3080A as programmable current source, CA3160 opamp as voltage follower, and CA3080 variable opamp as high-speed capacitor. Variable capacitors C1-C3 shape triangle waveform between 500 kHz and 1 MHz. C4 and C5 with 50K trimmer in series with C5 maintain constant amplitude within 10% up to 1 MHz.—"Circuit Ideas for RCA Linear ICs," RCA Solid State Division, Somerville, NJ, 1977, p 6.

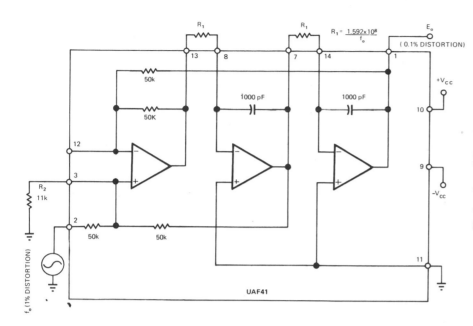

$$R_1 = \frac{1.592 \times 10^8}{f_o}$$

REDUCING DISTORTION—Use of UAF41 universal active filter at output of function generator reduces distortion of sine-wave output by eliminating some of harmonics. In typical application, two-pole active filter reduces 1% distortion down to 0.1%, using low-pass configuration. Article gives design equations. For 1-kHz cutoff, R_1 should be 159.2K.—Y. J. Wong, Design a Low Cost, Low-Distortion, Precision Sine-Wave Oscillator, *EDN Magazine,* Sept. 20, 1978, p 107–113.

1 MHz—Simple sinusoidal generator using Exar XR-2206C IC provides sine, triangle, or square outputs. For sine output, S1 is closed and R_A and R_B are adjusted for minimum distortion. Exact output frequency f is 1/RC where R is about 2 megohms from pin 7 to ground and C is connected between pins 5 and 6. FM output is obtained when modulating input is applied to either pin 7 or 8. For AM output, modulation is applied to pin 1.—E. Noll, VHF/UHF Single-Frequency Conversion, *Ham Radio,* April 1975, p 62–67.

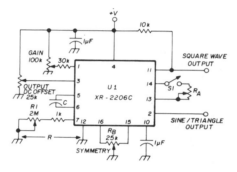

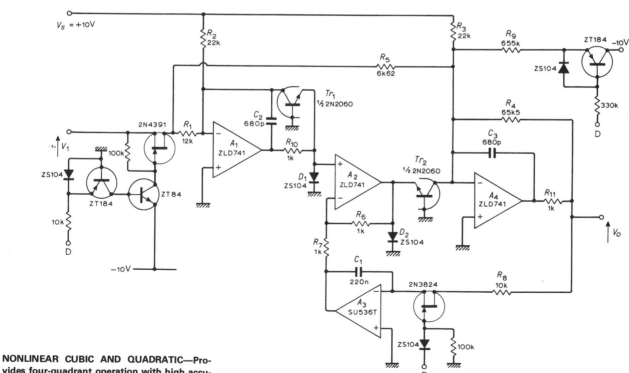

NONLINEAR CUBIC AND QUADRATIC—Provides four-quadrant operation with high accuracy over input amplitude range of several decades. Applications include analog computations for radar and ballistic problems, linearization of transducer characteristics, and teaching theory of quadratic equations. Article gives design equations and complete design procedure.—H. McPherson, Non-Linear Function Generator, *Wireless World,* Oct. 1972, p 485–487.

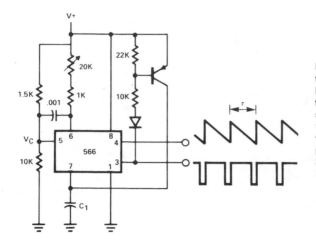

NEGATIVE RAMP—Connection shown for 566 function generator gives negative output ramp having period equal to 1/2f where f is normal free-running frequency of 566 as determined by supply voltage and RC values used. Ramp has very fast reset because PNP transistor charges timing capacitor C_1 rapidly at end of discharge period. Short output pulse is available at pin 3.—"Signetics Analog Data Manual," Signetics, Sunnyvale, CA, 1977, p 851.

THREE-WAVEFORM—Gives simultaneous sine, square, and triangle outputs with low distortion (1%), high linearity (0.1%), 0.05 Hz to 1 MHz frequency range, and duty cycle of 2% to 98%. Intersil 8038 waveform generator feeds buffer amplifier using 2N3709 transistor, switched to desired output waveform. Timing capacitors C1-C8, determining frequency decades of signal generator, start with 500 μF for 0.05 Hz to 0.5 Hz and decrease in submultiples of 10 to 500 pF for 50 kHz to 500 kHz. C8 is 250 pF for final range of 100 kHz to 1 MHz.—H. P. Fisher, Precision Waveform Generator, *73 Magazine,* Dec. 1973, p 41–43.

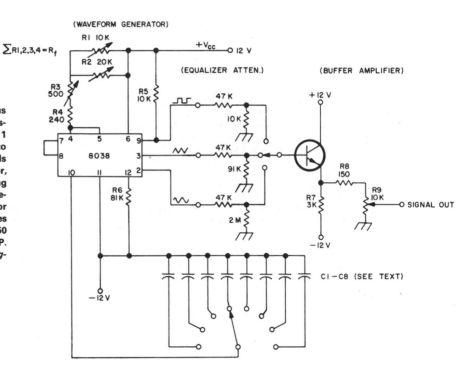

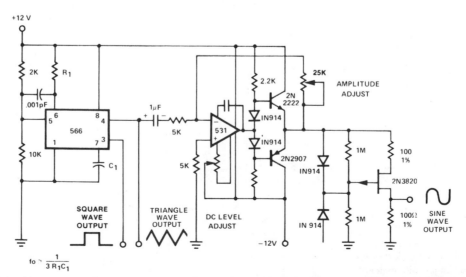

FET TRIANGLE TO SINE CONVERTER—Use of nonlinear transfer characteristic of P-channel junction FET to shape triangle output of 566 function generator gives sine wave having less than 2% distortion. Amplitude of triangle wave is critical and must be carefully adjusted for minimum distortion of sine output.—"Signetics Analog Data Manual," Signetics, Sunnyvale, CA, 1977, p 851–852.

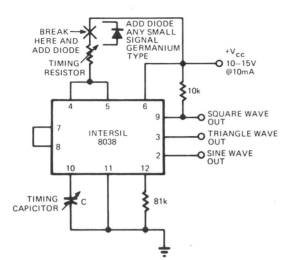

DIODE CANCELS SUPPLY CHANGES—Adding any small-signal germanium diode to Intersil 8038 sine-square-triangle function generator as shown will compensate for changes in supply voltage. When using diode, change from 10 to 15 V produces only 5-Hz change in output over frequency range of 100–10,000 Hz. Technique can be applied to other IC function generators, such as Signetics 565, as well as to 555 timers.—R. Liebman, Single Diode Compensates IC Oscillator, *EDN Magazine,* April 20, 1974, p 87.

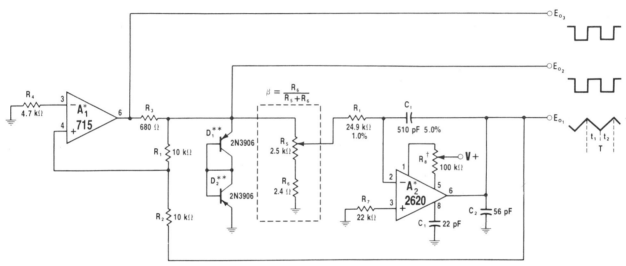

20–20,000 Hz SQUARE-TRIANGLE—R_t and C_t are chosen for upper frequency limit of 20 kHz, and oscillator is adjustable down to lower limit of 20 Hz with R_5. Circuit will operate up to 100 kHz if component values are suitably changed. A_2 should be offset-nulled by adjusting for best symmetry at lowest frequency. Total width of T of output waveform varies between 50 μs and 50 ms at frequency range covered.—W. G. Jung, "IC Op-Amp Cookbook," Howard W. Sams, Indianapolis, IN, 1974, p 381–383.

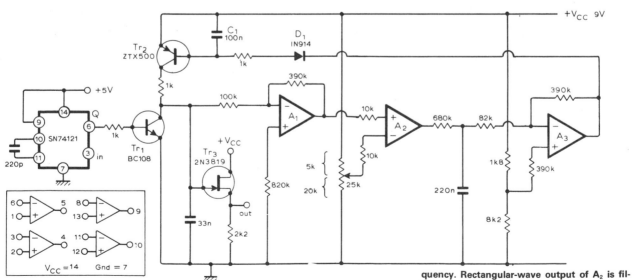

DRIVEN CONSTANT-AMPLITUDE SAW-TOOTH—Gives constant-amplitude output over input frequency range of 2–100 kHz. Input signal from SN74121 IC is 300-ns pulse that drives basic sawtooth generator Tr_1-Tr_2. Resulting sawtooth waveform is amplified by opamp A_1 of MC3401P four-opamp package and fed to A_2 which acts as comparator for amplitude-sensing. 25K threshold-setting pot is adjusted for maximum linearity of amplitude versus frequency. Rectangular-wave output of A_2 is filtered to give control voltage that is shifted in level by A_3 and D_1 to meet input voltage requirements of Tr_2. Desired sawtooth output appears at source of Tr_3.—J. N. Paine, Constant Amplitude Sawtooth Generator, *Wireless World,* Oct. 1975, p 473.

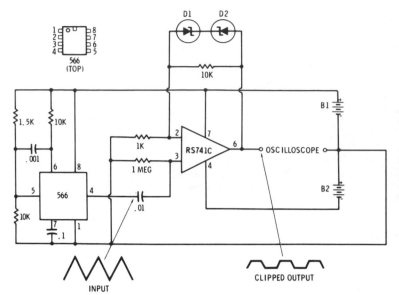

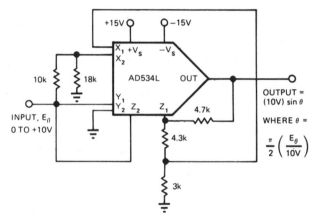

TRIANGLE-WAVE CLIPPER—Triangle-wave generator using 566 function generator is connected to RS741C opamp for clipping positive and negative peaks of triangle waves. Output as seen on CRO is modified square wave with sloping side. Clipping level depends on voltage ratings of zeners used. Supply voltages can range from 1.5 to 9 V each. Can be used for classroom demonstrations.—F. M. Mims, "Integrated Circuit Projects, Vol. 4," Radio Shack, Fort Worth, TX, 1977, 2nd Ed., p 37–44.

TRIANGLE TO SINE—Expression approximating sine function from 0 to 90° is generated by function fitting and duplicated by using AD534L analog multiplier and appropriate close-tolerance (0.1%) resistors. Accuracy of sine wave is within ±0.5% at all points. Linearly increasing voltage of triangle develops rising sinusoidal output. Conversely, linearly decreasing input generates mirror of rising sinusoids. Increasing triangle waveform, then bringing it back to zero again, completes full cycle of sine-wave output.—R. Frantz, Analog Multipliers—New IC Versions Manipulate Real-World Phenomena with Ease, *EDN Magazine*, Sept. 5, 1977, p 125–129.

$$\text{OUTPUT} = (10V)\sin\theta$$

$$\text{WHERE } \theta = \frac{\pi}{2}\left(\frac{E_\theta}{10V}\right)$$

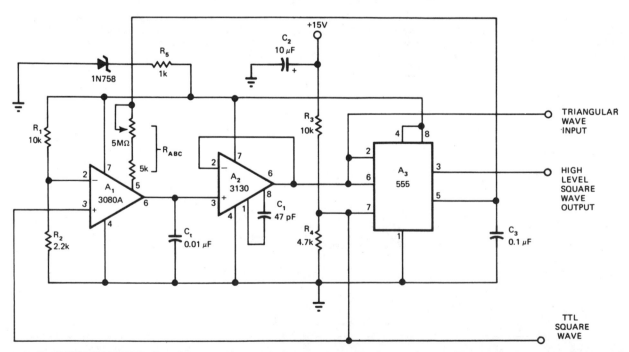

30–20,000 Hz SQUARE-TRIANGLE—Uses 555 timer as bilevel threshold detector, together with A_1 as bidirectional constant-current source and A_2 as buffer amplifier. If buffered triangle output is not needed, opamp A_2 can be omitted.

A_1 charges C_t between +5 and +10 V threshold points of 555 to give linear triangle output for buffering by A_2. Simultaneously, toggling of 555 between its high and low output generates square wave of about 13 V at pin 3, along with

5-V TTL output at pin 7 for additional square-wave output and for controlling state of A_1.—W. G. Jung, Build a Function Generator with a 555 Timer, *EDN Magazine*, Oct. 5, 1976, p 110.

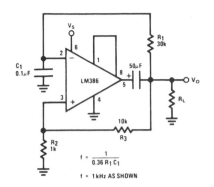

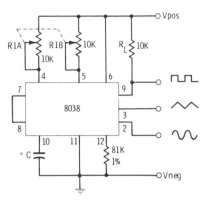

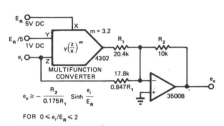

1 kHz SQUARE-TRIANGLE—National LM386 opamp operating from 9-V supply generates 1-kHz square wave at 0.5 W for driving 8-ohm loudspeaker or other load. Exact frequency is determined by values used for R_1 and C_1. Triangle output can be taken from pin 2. Pins 1 and 8 can be shorted because DC offset voltages are unimportant.—"Audio Handbook," National Semiconductor, Santa Clara, CA, 1977, p 4-30–4-33.

90–900 Hz SINE-SQUARE-TRIANGLE—Ganged 10K dual pot covers range when C is 0.25 μF for 8038CC waveform generator. Other values of C (from 0.005 μF to 2.2 μF) give different frequency ranges between 10 Hz and 50 kHz. Linearity of waveforms depends on tracking precision of dual pot.—C. D. Rakes, "Integrated Circuit Projects," Howard W. Sams, Indianapolis, IN, 1975, p 116–120.

HYPERBOLIC SINE—Burr-Brown 4302 multifunction converter and opamp generate hyperbolic sine transfer function with response matching ideal curve within 0.7%. Technique permits setting powers and roots at fractional as well as integer values. Converter shown is set for exponent of 3.2. Choice of amplifier gain and reference voltage scales response for given input and output signal levels. Article gives design equations.—J. Graeme, Sinh Generator Boasts 0.7% Error, *END Magazine*, Aug. 5, 1978, p 70 and 72.

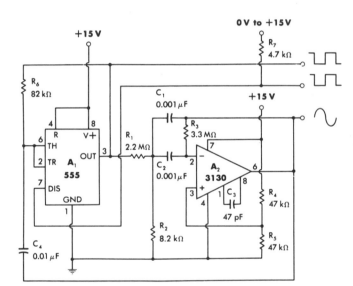

1-kHz SINE-SQUARE—555 timer starts as astable MVBR but then acts with A_2 to form multiple-feedback bandpass filter that removes harmonics from square wave to give sine-wave output with distortion less than 2%. Sine output is fed back to timer through C_4 which now operates as Schmitt trigger, shaping sine wave to give output square wave. Frequency is determined primarily by filter components C_1, C_2, R_1, and R_2. Output is about 9 V P-P.—W. G. Jung, "IC Timer Cookbook," Howard W. Sams, Indianapolis, IN, 1977, p 203–204.

SQUARE-TRIANGLE VCO—With DC control voltage of 5 mV to 5 V, circuit controls frequency of both square and triangle outputs with good linearity. Peak value of triangle output is precisely set at 2.44 V and 0 V by reference voltages at noninverting inputs of comparators. Comparator A2 drives load for low outputs, while comparator A1 drives load when output is high. Article tells how circuit works.—R. C. Dobkin, Comparators Can Do More Than Just Compare, *EDN Magazine*, Nov. 1, 1972, p 34–37.

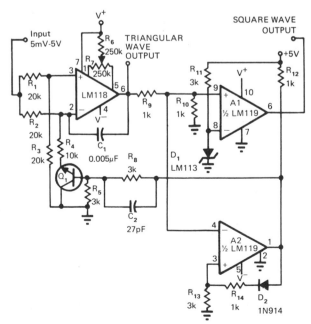

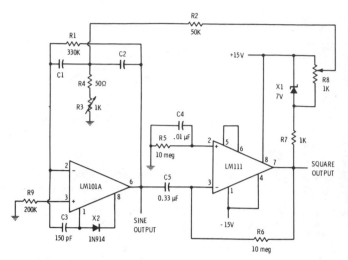

C1, C2	MIN FREQ	MAX FREQ
0.47 μF	18 Hz	80 Hz
0.1 μF	80 Hz	380 Hz
.022 μF	380 Hz	1.7 kHz
.0047 μF	1.7 kHz	8 kHz
.002 μF	4.4 kHz	20 kHz

18–20,000 Hz SINE-SQUARE—Circuit uses two opamps to obtain necessary positive feedback for sustaining oscillation. RC network of first stage acts as tuned circuit, permitting operation only at frequency determined by values of R and C. Pot R3 provides tuning over significant frequency range. Pot R8 controls amplitude of sine output. Zener X1 stabilizes amplitude of square output. Sine signal is applied to LM111 acting as limiter to provide desired square wave. Table gives values of C1 and C2 for five different frequency ranges.—E. M. Noll, "Linear IC Principles, Experiments, and Projects," Howard W. Sams, Indianapolis, IN, 1974, p 123–124.

POSITIVE RAMP—NPN transistor across timing capacitor C_1 of 566 function generator gives fast charging of capacitor at end of discharge period, for positive ramp having very fast reset. Period of ramp is equal to 1/2f where f is normal free-running frequency of 566 as determined by supply voltage and RC values used.—"Signetics Analog Data Manual," Signetics, Sunnyvale, CA, 1977, p 851.

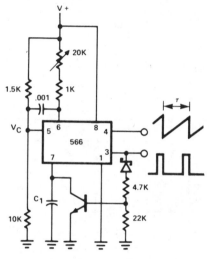

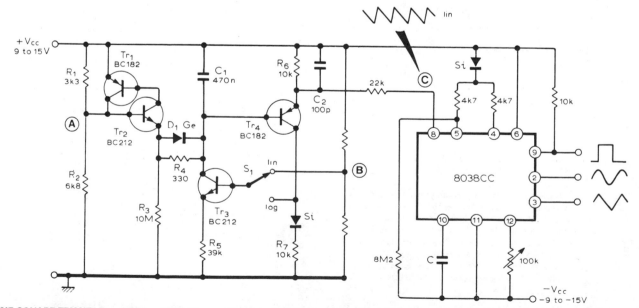

SINE-SQUARE-TRIANGLE WITH LIN/LOG SWEEP—Four-transistor circuit provides choice of linear or logarithmic sweeps for Intersil 8038 IC function generator. In linear mode, constant-current generator Tr_3 charges C_1 almost linearly, with Tr_1-Tr_2 resetting C_1 when its voltage reaches about one-third V_{cc} plus 0.9 V. In logarithmic mode, positive feedback provides exponential charging of C_1. Voltage at B must be set experimentally because it depends on V_{cc}. For overall frequency control, make R_5 variable. Point A has short positive pulse that can be used to reset capacitor C of IC, and to sync an oscilloscope.—S. Villone, Linear/Logarithmic Sweep Generator, *Wireless World*, Dec. 1976, p 42.

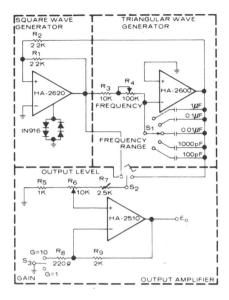

2.5 Hz TO 250 kHz SQUARE-TRIANGLE—Five switched frequency ranges each give continuous variation of frequency over one decade and adjustment of output amplitude from 0.2 to 20 V P-P. Slope of triangle is highly linear, and rise time of square wave is less than 100 ns. Square-wave generator is simple hysteresis circuit triggered by triangle generator. Output voltage is clamped to desired level by diodes connected to bandwidth control point. Output opamp is selected for high slew rate. S_3 gives choice of 1 or 10 for gain. Maximum output current should be limited to 20 mA.—"Linear & Data Acquisition Products," Harris Semiconductor, Melbourne, FL, Vol. 1, 1977, p 7–25 (Application Note 510).

1 Hz TO 100 kHz SQUARE-TRIANGLE—Wide-range function generator built around LM111 comparator provides two different output waveforms whose frequency can be varied over five decades by R_1, from 1 Hz to 100 kHz. Two transistor pairs are used to vary charging current of timing capacitor exponentially. Output current from transistor pairs is controlled by linear pot, so rotation of pot is proportional to log of output frequency. Sensistor R_2 provides temperature compensation for transistor pairs.—R. C. Dobkin, Comparators Can Do More Than Just Compare, *EDN Magazine,* Nov. 1, 1972, p 34–37.

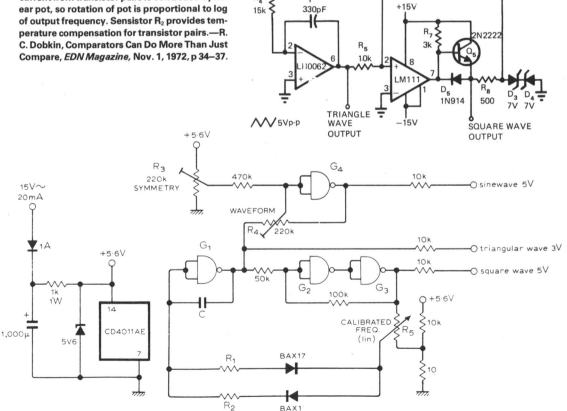

SINE-SQUARE-TRIANGLE AT 3–5 V—Uses CD4011 IC operating from 15-V AC line as at left. NAND gates of IC are connected as at right, with G_1 serving as integrator with variable delay time, G_2-G_3 as Schmitt trigger, and G_4 as triangle to sine-wave converter. Sine-wave approximation, depending on transfer function of G_4, is calibrated by R_3 and R_4. Values of R_1 and R_2 may be varied between 0.01 and 10 megohms and C between 100 pF and 2.2 µF to obtain desired sawtooth and pulse waveforms at desired frequency determined by setting of R_5.—J. W. Richter, Single I.C. Function Generator, *Wireless World,* Nov. 1976, p 61.

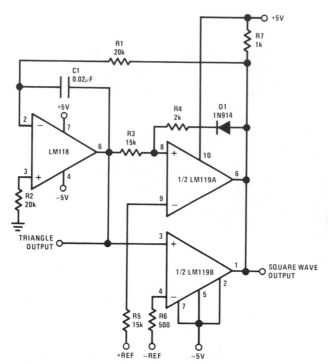

HIGH-PRECISION TRIANGLE—Opamp circuit provides easily controlled peak-to-peak amplitude of triangle wave suitable for use in sweep circuits and test equipment. Positive and negative peak amplitudes are controllable to accuracy of about ±0.01 V by DC input. Output frequency is likewise easily adjusted over range of two decades. Circuit consists of integrator and two comparators. One comparator sets positive peak, and other sets negative peak. Operating frequency depends on R1, C1, and reference voltages. Maximum difference in reference voltages is 5 V. Frequency limit is about 200 kHz.—R. C. Dobkin, "Precise Tri-Wave Generation," National Semiconductor, Santa Clara, CA, 1973, LB-23.

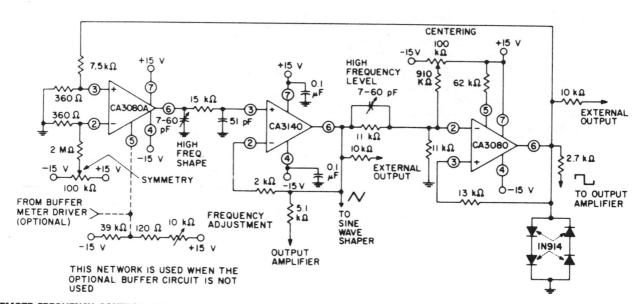

REMOTE FREQUENCY CONTROL—Frequency of square and triangle outputs can be adjusted over range of 1,000,000:1 with 10K pot or by varying DC voltage applied to pin 5 of CA3080A over wire line from remote location. CA3140 serves as noninverting readout amplifier for triangle wave developed across integrating capacitor network at output of CA3080A current source. Second CA3080 acts as high-hysteresis switch having trip level established by four diodes, to give desired square-wave output.— "Linear Integrated Circuits and MOS/FET's," RCA Solid State Division, Somerville, NJ, 1977, p 248–254.

CHAPTER 38
Game Circuits

Included are chip connections, VHF modulators, score generators, and sound effects for variety of TV games, along with electronic dice, roulette wheel, coin tosser, robot toy, model railroad switch, six-note chimes, and attention-getting LED displays.

RIFLE—Developed for use with General Instruments AY-3-8500-1 TV game chip to simulate target practice with rifle. Player aims at bright target spot moving randomly across TV screen. If gun is on target when trigger is pulled, phototransistor in barrel picks up light from target and generates pulse for producing sound effect of hit and incrementing player's score. PT-1 can be TIL64 or equivalent phototransistor. 4098 is dual mono, and 4011 is quad two-input NAND gate. Pulse outputs go to pins of game chip. Article gives all circuits but covers construction only in general terms.—S. Ciarcia, Hey, Look What My Daddy Built!, 73 Magazine, Oct. 1976, p 104–108.

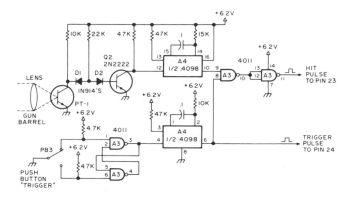

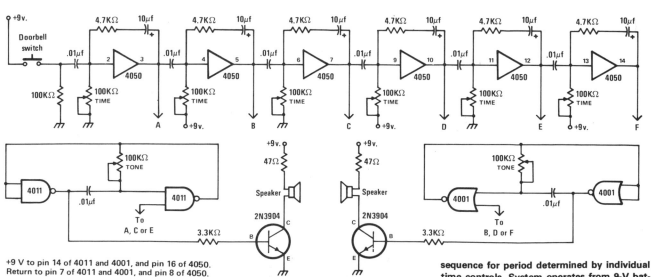

+9 V to pin 14 of 4011 and 4001, and pin 16 of 4050.
Return to pin 7 of 4011 and 4001, and pin 8 of 4050.

SIX-TONE CHIME—Separate AF oscillators, gated on by six-stage time-delay circuit, generate six different chime tones. Loudspeakers can be mounted so each tone comes from different location in house. When doorbell button is pushed, each tone generator is turned on in sequence for period determined by individual time controls. System operates from 9-V battery, with CMOS logic drawing very little standby current.—J. Sandler, 9 Projects under $9, Modern Electronics, Sept. 1978, p 35–39.

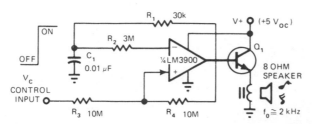

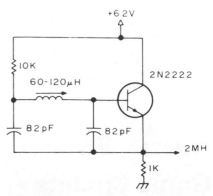

DIGITAL NOISEMAKER—Simple sound-effect generator for video games, electronic cash registers, and electronic toys uses one-fourth of LM3900 quad opamp chip as 2-kHz signal generator that can be turned on or off by input control voltage. Suitable for applications that do not require pure sine wave. Output transistor Q_1, needed with low-impedance voice coil, is not critical as to type. For smaller acoustic output, Q_1 can be replaced by 100-ohm resistor if 100-ohm voice coil is used, to avoid overloading IC.—T. Frederiksen, Build a Transformerless Tone Annunciator, *EDN Magazine,* April 5, 1977, p 141–142.

2-MHz MASTER CLOCK—Developed for use with General Instruments AY-3-8500-1 TV game chip, which contains dividers that deliver required 60-Hz vertical and 15.75-kHz horizontal sync signals for video signal going to TV set. Coil is Miller 9055 miniature slug-tuned. Article gives other circuits for game.—S. Ciarcia, Hey, Look What My Daddy Built!, *73 Magazine,* Oct. 1976, p 104–108.

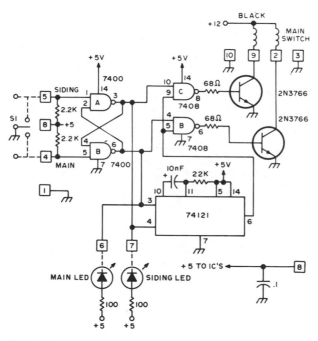

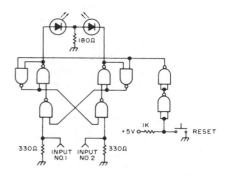

WHO'S FIRST?—One of LEDs comes on to indicate which of two people pushes button first after event such as stopping of music. Circuit requires two 7400 quad gates.—Circuits, *73 Magazine,* Nov. 1974, p 142.

[1], [3], [10]   GROUND

[2]   TO SWITCH COIL—MAIN   } SEPARATE + 12V SUPPLY 1
[9]   TO SWITCH COIL—SIDING   } AMP, (UNREGULATED) USED FOR SWITCHES

[4]   CONTROL—SHORT TO GROUND TO THROW SWITCH TO MAIN LINE

[5]   CONTROL—SHORT TO GROUND TO THROW SWITCH TO SIDING

[6]   LED TO +5 TO INDICATE SWITCH IN MAIN (THIS POINT LOW)

[7]   LED TO +5 TO INDICATE SWITCH IN SIDING

[8]   +5 VOLTS IN FOR ICs

POINTS 4 AND 5 CAN BE PARALLEL TO MANUAL MOMENTARY SWITCHES AND LOGIC SWITCHES—ANY PULSE (LOW) WILL WORK, HOLDING POWER ON ABOUT ½ SECOND, 74121 WITH RESISTOR AND CAP CONTROL TIME.

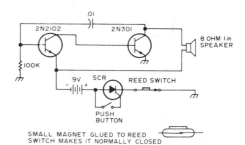

HOWLING BOX—Tone oscillator driving loudspeaker is sealed into wood or plastic box, with reed switch mounted on one face of box and pushbutton of other switch projecting out through hole in box. Place "DO NOT TOUCH" label on button. When button is pushed despite warning, SCR latches and applies power to AF oscillator. Only way to turn off howling is to hold large permanent magnet against location of reed switch, to oppose field of magnet glued on switch and make reed contacts open. If mercury switch is used in box in place of pushbutton, alarm goes off when box is picked up.—P. Walton, Now What Have I Done?, *73 Magazine,* May 1975, p 81.

MODEL RAILROAD SWITCHING—Control circuit is used to drive solenoid-operated track switches of typical HO train layout. Input can be pair of complementary TTL signals from 8008 or other computer or can be from manual switch S1. 74121 mono MVBR controls time that switch is energized in given direction. Output transistors are rated at 20 W, enough for driving solenoids taking 1 A at 12 V. Use protective diodes across coils of solenoids.—H. De Monstoy, Model Railroad Switch Control Circuit, *BYTE,* Oct. 1975, p 87.

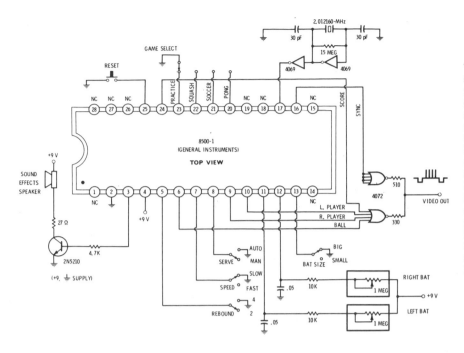

TV GAME CONTROLLER—Single General Instruments 8500 IC contains most of electronics needed for pong, hockey, squash, or practice games using screen of TV set. Desired game is selected by grounding one of pins 20-23. Connect ball, player, and score outputs to four-input OR circuit to generate composite video for combining with sync output. Final output can be fed directly to video amplifier of TV set or fed to suitable RF modulator. Sound output is fed to loudspeaker through transistor audio amplifier. No connection on pin 5 gives two rebound angles, while grounding gives four rebound angles. Open pin 7 gives fast speed, and grounding gives slow speed. Open pin 13 gives small bats, and grounding gives large bats.—D. Lancaster, "CMOS Cookbook," Howard W. Sams, Indianapolis, IN, 1977, p 166.

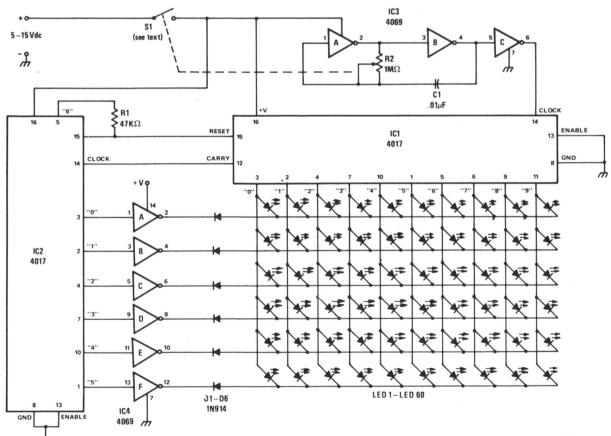

60-LED HYPNOTIC SPIRAL—LEDs are mounted on display board in spiral arrangement and wired in matrix connected to ICs so each LED is lighted in sequence as IC1 and IC2 carry out counting function. IC3 is square-wave oscillator with frequency determined by C1 and setting of R2. Output pulses are used to clock IC1 to advance count, with carry output of IC1 clocking IC2 every tenth count. At end of 60 counts, both ICs reset to zero for new sequence. Inherent current limiting of ICs makes dropping resistors unnecessary for LEDs.—F. Blechman, Digitrance, *Modern Electronics*, Dec. 1978, p 29–31.

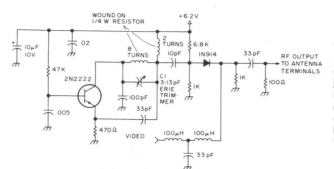

VHF MODULATOR—Developed as interface between General Instruments AY-3-8500-1 TV game chip and antenna terminal of TV set. Adjust C1 to frequency of unused channel to which receiver is set for playing games. Article gives all circuits but covers construction only in general terms.—S. Ciarcia, Hey, Look What My Daddy Built!, *73 Magazine*, Oct. 1976, p 104–108.

OSCILLATOR FOR CHANNELS 2–6—Transmitter serving as interface between video game and TV set can be tuned with L₁ to vacant channel in low TV band. Regular antenna should be disconnected when output of oscillator is fed to TV set via twin-line, to avoid broadcasting game signals. L₁ is 4 turns No. 18 spaced 3/8 inch on ¼-inch slug-tuned form.—B. Matteson, "King Pong" Game Offers Hockey and Tennis Alternatives to TV Re-Runs, *EDN Magazine*, Aug. 5, 1975, p 47–55.

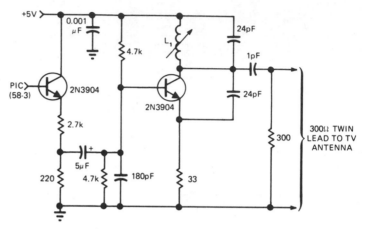

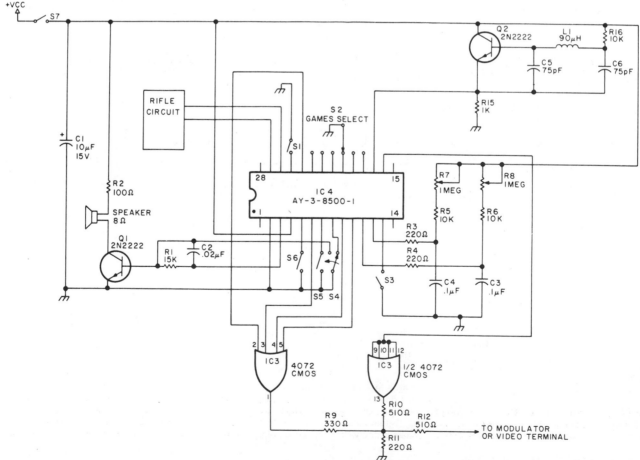

SIX-GAME VIDEO—General Instruments AY-3-8500-1 MOS chip gives choice of hockey, squash, tennis, two types of rifle shoot, and practice games, all with sound effects and automatic scoring on 0-15 display at top of TV screen. Can be used with standard TV receiver (using RF modulator circuit) or with video monitor. S4 grounds base of Q1 when in manual-serve mode, to eliminate steady boing when ball leaves playing field. R5-R8 position players on field. Article covers operation in detail and gives suitable rifle circuit. Supply is +6 V. 2-MHz clock is at upper right.—A. Dorman, Six Games on a Chip, *Kilobaud*, Jan. 1977, p 130, 132, 134, 136, and 138.

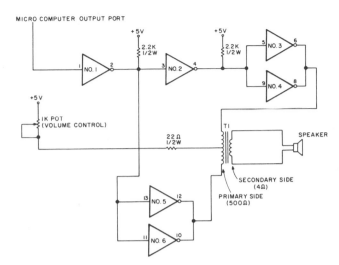

PLOP FOR GAMES—Section 1 of 7406 TTL hex inverter can be attached to output port and driven by program loop, to provide sound effects for computer games. When output port goes to logic 1 (greater than +2 V), action of inverter is such that paralleled inverters 3 and 4 go to 0 and draw current through primary of T1, making loudspeaker produce single plopping sound. When output port goes to 0, another plop is produced. If output port is switched between 0 and 1 fast enough, loudspeaker output will be tone at switching frequency.—D. Parks, Adding "Plop" to Your System, *Kilobaud*, May 1977, p 98.

HEADS/TAILS FLIPPER—Uses only half of 7400 quad NAND gate as gated clock driving half of 7473 JK flip-flop. With power switch closed, LEDs representing heads and tails flash on and off at clock frequency. Closing FLIP switch stops clock randomly, leaving one LED on to give equivalent of tossing coin for heads/tails call.— G. Young, JK Flip-Flops and Clocked Logic, *Kilobaud*, July 1977, p 66–70 and 72–73.

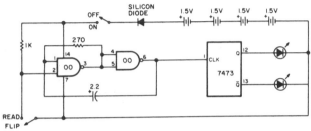

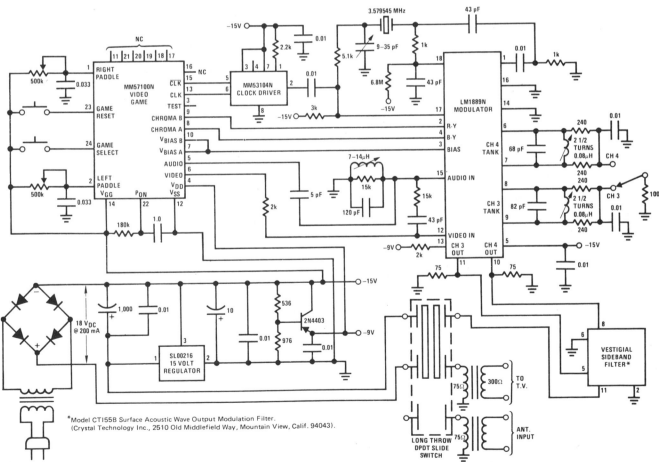

*Model CT155B Surface Acoustic Wave Output Modulation Filter. (Crystal Technology Inc., 2510 Old Middlefield Way, Mountain View, Calif. 94043).

HOCKEY/TENNIS/HANDBALL—Uses National MM57100 TV game chip to provide logic for generating backgrounds, paddles, ball, and digital scoring. All three games are in color and have sound. Circuit generates all necessary timing (sync, blanking, and burst) to interface with circuit of standard TV receiver. With addition of chroma, audio, and RF modulator, circuit will interface directly to antenna terminals of set.— "MOS/LSI Databook," National Semiconductor, Santa Clara, CA, 1977, p 4-37–4-47.

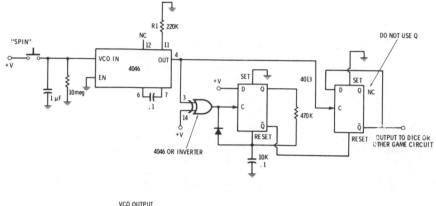

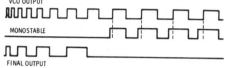

DICE OR ROULETTE RUNDOWN—4046 PLL connected as VCO is set at twice desired maximum rate for dice or roulette-wheel counters. Pressing spin button momentarily to start action charges 1-μF capacitor to supply voltage and jumps VCO to highest frequency. Output frequency then decreases rapidly as capacitor is discharged by 10-megohm resistor. Output is stopped by using retriggerable mono to drive other half of 4013 dual D flip-flop. When frequency drops below value at which mono times out, mono resets flip-flop and holds it to stop display.—D. Lancaster, "CMOS Cookbook," Howard W. Sams, Indianpolis, IN, 1977, p 252–254.

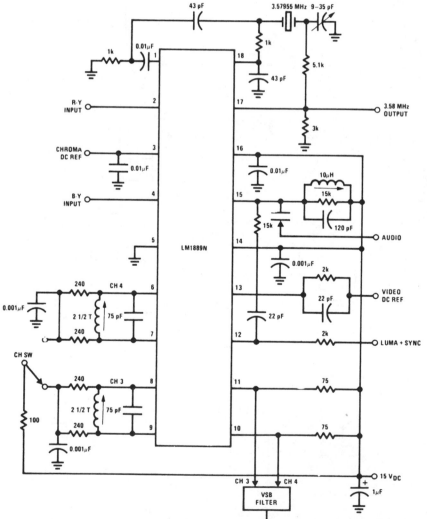

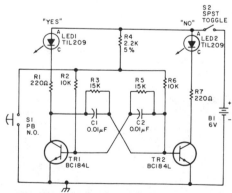

VIDEO MODULATOR—National LM1889N serves to interface audio, color difference, and luminance signals to antenna terminals of TV receiver. Circuit allows video information from video games, test equipment, videotape recorders, and similar sources to be displayed on black-and-white or color TV receivers. LM1889N consists of sound subcarrier oscillator, chroma subcarrier oscillator, quadrature chroma modulators, and RF oscillators and modulators for two low VHF channels.—"MOS/LSI Databook," National Semiconductor, Santa Clara, CA, 1977, p 4-48–4-49.

COIN FLIPPER—One of LEDs comes on when S1 is pressed, to simulate tossing of coin. LEDs can be labeled HEADS and TAILS if desired. Transistor types are not critical. For true random results, voltage between collectors of transistors should be 0 V with S2 closed and S1 open.—Circuits, 73 Magazine, June 1975, p 161.

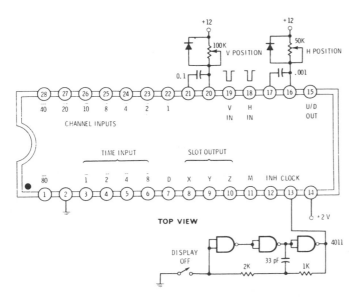

SCORE DISPLAY—National 5841 IC is used for display of video game scores on TV receiver, as well as for time and channel number displays. Properly conditioned H and V pulses must be applied to pins 18 and 19 to interface TV. Output video on pin 15 must be buffered and summed into existing video inside TV set. Display position is controlled by H and V pots. Horizontal display size depends on clock frequency. Grounding M input gives only channel number. Positive voltage at M gives both channel and time. Grounding D input provides 5-slot time display, while positive D input gives 8-slot time display. Channel inputs are applied continuously in negative-logic form, with time inputs multiplexed externally.—D. Lancaster, "CMOS Cookbook," Howard W. Sams, Indianapolis, IN, 1977, p 158.

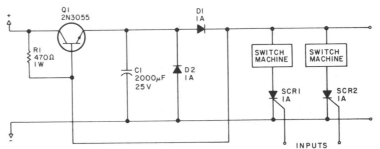

MODEL-TRAIN SWITCHING—Individual SCRs are triggered by logic-level signals independently to initiate discharge of large capacitor C1 through solenoid of model railroad track switch.—D. W. Zimmerli, Two Hobbies: Model Railroading and Computing, *Kilobaud,* Aug. 1978, p 62–68.

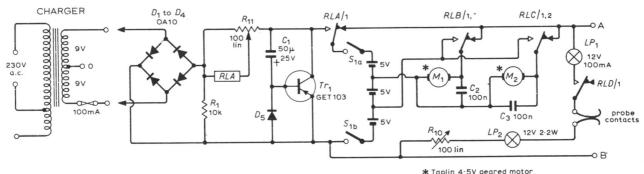

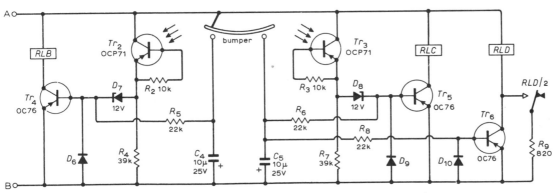

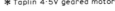

DUAL-MOTOR ROBOT—Battery-operated toy car roams around room, reversing whenever it hits wall or obstacle, and returns automatically to home base when batteries are in need of charge. Small geared motor, such as Meccano No. 11057 or 4.5-V Taplin, is used for each rear wheel so reversal of one motor provides steering. Single free-swiveling caster is at front of machine. With head-on collision, both contacts of bumper close to reverse both motors so machine backs away, turns, and proceeds in new direction. With glancing collision, motor on opposite side is reversed so machine sheers away. White tape on floor, leading to charger having female jacks, is sensed by two phototransistors used to control motors so machine follows tape until probes at opposite end from bumper enter jacks. Circuit permits search mode for recharging only when relay D senses low battery voltage and energizes lamps that illuminate white tape. Article gives operation and construction details.—M. F. Huber, Free Roving Machine, *Wireless World,* Dec. 1972, p 593–594.

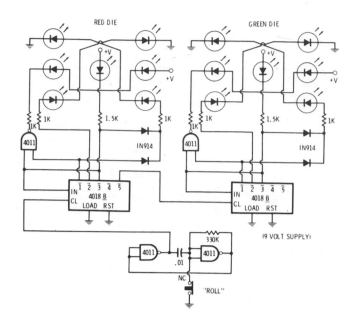

DICE SIMULATOR—Two 4018B synchronous counters are connected in modulo-6 walking-ring sequences for driving LEDs to produce familiar die patterns. Pressing roll button starts gated astable that cycles first die hundreds of times and second die dozens of times, for randomizing of result. When roll button is released, final state of each die is held.—D. Lancaster, "CMOS Cookbook," Howard W. Sams, Indianapolis, IN, 1977, p 324–325.

RANDOM-FLASHING NEONS—Neon glow lamps such as Radio Shack 272-1101 flash in unpredictable sequences at various rates that are determined by values of R and C used for each lamp, to give attention-getting display for classrooms and Science Fairs. Value of R1 can be as low as 2200 ohms for higher repetition rates, but battery drain increases. When circuit is energized, each neon receives full voltage and fires. Lamp capacitor begins charging, decreasing voltage across lamp until lamp goes out and cycle starts over. Use of different capacitor values makes lamps recycle at different rates. T1 is 6.3-VAC filament transformer used to step up oscillator voltage.—F. M. Mims, "Transistor Projects, Vol. 2," Radio Shack, Fort Worth, TX, 1974, p 43–52.

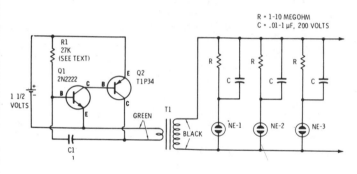

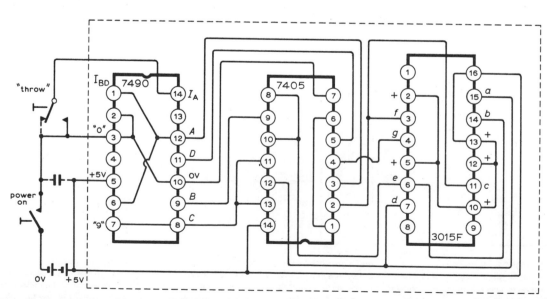

DICE—Simple low-cost arrangement of three ICs operating from 5-V battery (four nickel-cadmium or alkaline cells) provides bar display corresponding to spots on six sides of die. Uses SN7490N TTL decade counter with SN7405 hex inverter to drive Minitron 3015F seven-segment display. Article describes operation in detail and suggests variations for Arabic and binary displays.—G. J. Naaijer, Electronic Dice, *Wireless World*, Aug. 1973, p 401–403.

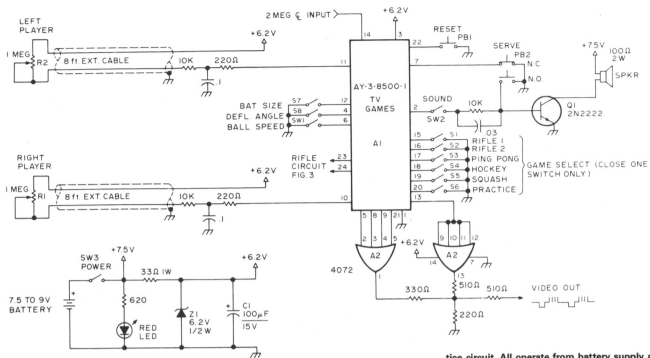

SIX-GAME CHIP—General Instruments AY-3-8500-1 TV game chip and associated circuits give choice of six different games. Article gives additional circuits required, including that for 2- MHz master clock whose output is divided in chip to get vertical and horizontal sync frequencies, VHF modulator used between game and antenna terminal of TV set, and rifle target practice circuit. All operate from battery supply at lower left. Article covers construction only in general terms.—S. Ciarcia, Hey, Look What My Daddy Built!, *73 Magazine,* Oct. 1976, p 104–108.

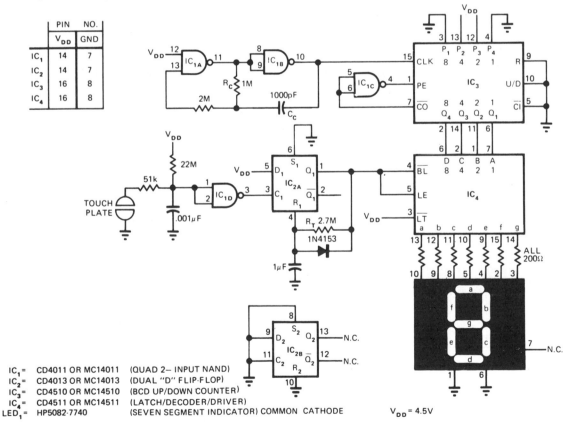

PIN NO.		
	V_{DD}	GND
IC_1	14	7
IC_2	14	7
IC_3	16	8
IC_4	16	8

IC_1 =	CD4011 OR MC14011	(QUAD 2-INPUT NAND)
IC_2 =	CD4013 OR MC14013	(DUAL "D" FLIP-FLOP)
IC_3 =	CD4510 OR MC14510	(BCD UP/DOWN COUNTER)
IC_4 =	CD4511 OR MC14511	(LATCH/DECODER/DRIVER)
LED_1 =	HP5082-7740	(SEVEN SEGMENT INDICATOR) COMMON CATHODE

V_{DD} = 4.5V

LED DIE—When positive bias on input of IC_{1D} NAND gate is pulled to ground by skin resistance of finger, D flip-flop IC_{2A} connected as mono is triggered. Pin 1 goes high for about 2 s, making IC_4 latch outputs of counter IC_3 and unblank LED display. Random time that finger is on touch plate determines randomness of number displayed. Number can be between 1 and 9, between 1 and 6 for die, or between 1 and 2 to represent heads or tails. Change BCD value of jam inputs of IC_3 to highest random number desired. Values shown are for 4.5-V supply and display current of 10 mA per segment. LED is blanked until plate is touched. Standby current drain of 10 μA on three AA alkaline cells is so low that ON/OFF switch is unnecessary.—C. Cullings, Electronic Die Uses Touchplate and 7-Segment LED Display, *EDN Magazine,* May 20, 1975, p 70 and 72.

CHAPTER 39
IF Amplifier Circuits

Gives circuits for most common IF values used in single-conversion and double-conversion superheterodyne receivers, including noise blanker, CW filter, Q multiplier, and T attenuator variations. See also Frequency Modulation, Receiver, Single-Sideband, Television, and Transceiver chapters.

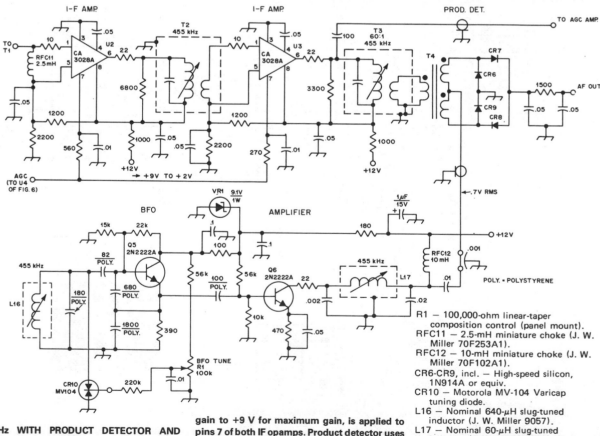

455-kHz WITH PRODUCT DETECTOR AND BFO—Used in 1.8–2 MHz communication receiver having wide dynamic range. Input comes from diode-switched bandpass filter giving choice of 400-Hz or 2.1-kHz bandwidths. Output for AGC amplifier is taken from primary of T3. AGC voltage, ranging from +2 V for minimum gain to +9 V for maximum gain, is applied to pins 7 of both IF opamps. Product detector uses quad 1N914A diodes. Varicap CR10 and R1 vary BFO output from 453 to 457 kHz. Two-part article gives all other circuits of receiver.—D. DeMaw, His Eminence—the Receiver, *QST*, Part 2—July 1976, p 14–17 (Part 1—June 1976, p 27–30).

R1 — 100,000-ohm linear-taper composition control (panel mount).
RFC11 — 2.5-mH miniature choke (J. W. Miller 70F253A1).
RFC12 — 10-mH miniature choke (J. W. Miller 70F102A1).
CR6-CR9, incl. — High-speed silicon, 1N914A or equiv.
CR10 — Motorola MV-104 Varicap tuning diode.
L16 — Nominal 640-μH slug-tuned inductor (J. W. Miller 9057).
L17 — Nominal 60-μH slug-tuned inductor (J. W. Miller 9054).
T2, T3 — 455-kHz i-f transformer. See text. (J. W. Miller 2067).
T4 — Trifilar broadband transformer. 15 trifilar turns of No. 26 enam. wire on Amidon T-50-61 toroid core.
U2, U3 — RCA IC.
VR1 — 9.1-V, 1-W Zener diode.

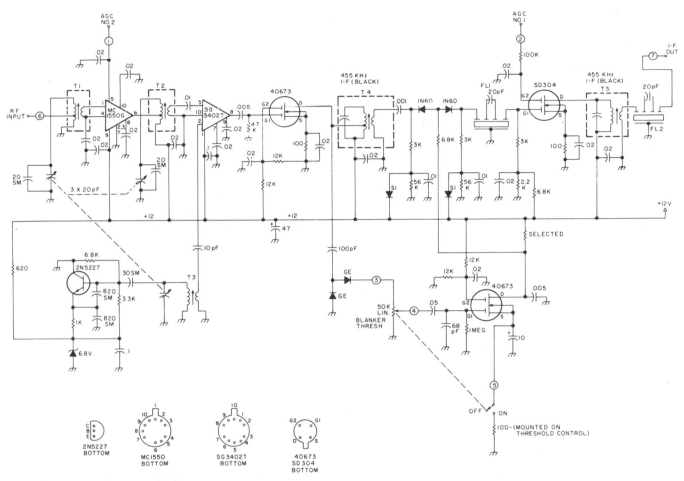

3.5–4 MHz TUNABLE IF WITH NOISE BLANK-ING—MC1550G is followed by Silicon General SG3402T mixer that provides good conversion gain with very light oscillator loading. Mixer output is fed to 40673 amplifier that builds up noise spikes for blanker consisting of 1N60 diode gate and 40673 pulse amplifier. Blanker is fed from envelope detector that controls gate feeding FL1 dual ceramic filter providing IF selectivity. IF stage following blanker uses SD304 dual-gate MOSFET, transformer-coupled to ceramic filter FL2 at IF output. Article gives construction details.—R. Megirian, The Minicom Receiver, *73 Magazine*, April 1977, p 136–149.

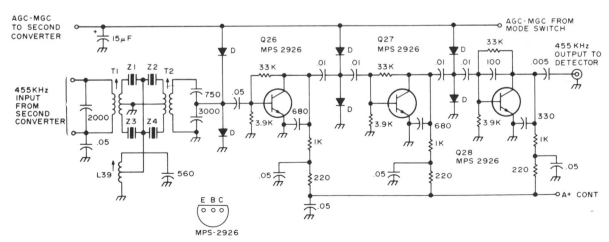

455-kHz SECOND IF—Used in all-band double-conversion superheterodyne receiver for AM, narrow-band FM, CW, and SSB operation. Input is fed through 455-kHz ceramic filter to high-gain amplifier using three MPS 2926 transistors all having automatic gain control and master gain control. Use of silicon rectifiers in interstage networks of IF amplifier gives economical wide-range AGC circuit. Supply is 13.6 V regulated. Article gives all circuits of receiver.—D. M. Eisenberg, Build This All-Band VHF Receiver, *73 Magazine*, Jan. 1975, p 105–112.

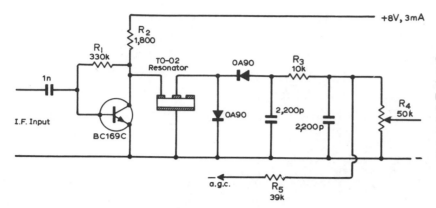

COUPLING TO HIGH-IMPEDANCE DETECTOR— Final IF stage of receiver uses piezoelectric overtone resonator connected backwards for coupling to high-impedance detector. Arrangement provides useful voltage step-up as well, about 2.5 times. Resonator can be Brush Clevite Transfilter.—G. W. Short, Reversed Operation of 'Transfilter,' *Wireless World*, Aug. 1971, p 386.

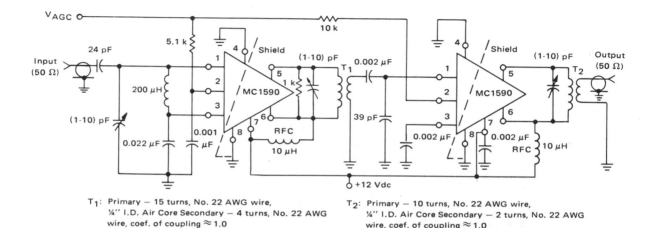

T$_1$: Primary — 15 turns, No. 22 AWG wire, ¼" I.D. Air Core Secondary — 4 turns, No. 22 AWG wire, coef. of coupling ≈ 1.0

T$_2$: Primary — 10 turns, No. 22 AWG wire, ¼" I.D. Air Core Secondary — 2 turns, No. 22 AWG wire, coef. of coupling ≈ 1.0

60 MHz WITH 80-dB POWER GAIN—Two-stage tuned IF amplifier achieves maximum gain and output signal swing capability by using differential-mode coupling for interstage and output network. Overall bandwidth is 1.5 MHz. Resistors in series with AGC pins 2 of opamp stages provide more efficient AGC action.—B. Trout, "A High Gain Integrated Circuit RF-IF Amplifier with Wide Range AGC," Motorola, Phoenix, AZ. 1975, AN-513, p 8.

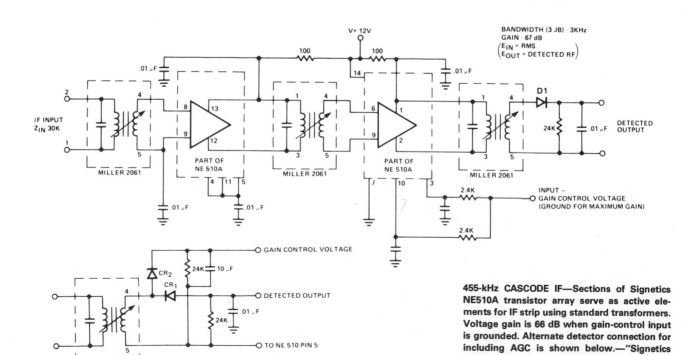

455-kHz CASCODE IF—Sections of Signetics NE510A transistor array serve as active elements for IF strip using standard transformers. Voltage gain is 66 dB when gain-control input is grounded. Alternate detector connection for including AGC is shown below.—"Signetics Analog Data Manual," Signetics, Sunnyvale, CA, 1977, p 746–747.

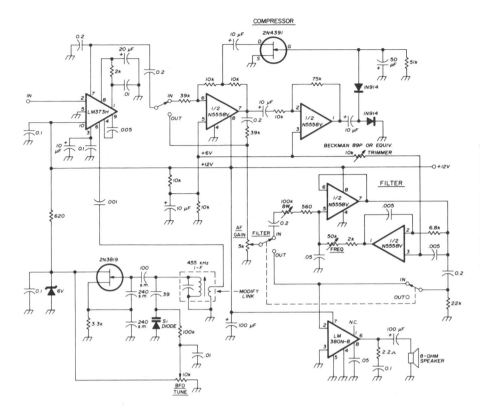

455-kHz IF WITH AF COMPRESSOR—Combination of IF amplifier, audio compressor, tunable audio filter, and audio output system operates from single supply. Compressor and filter each use N5558V dual opamps or equivalent units. Tuning range of filter is about 500 to 2000 Hz. IF input goes directly to pin 2 of LM373H. Use coupling capacitor to prevent shorting pin 2 to ground and damaging IC.—R. Megirian, Design Ideas for Miniature Communications Receivers, *Ham Radio*, April 1976, p 18–25.

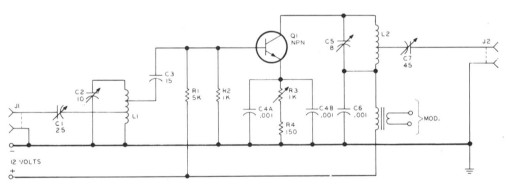

120–144 MHz—General-purpose amplifier can be used around 120 MHz as microwave IF strip and up to 144 MHz for RF. Transistor type is not critical.—B. Hoisington, DC Isolation, *73 Magazine*, Peterborough NH 03458, July 1974, p 55–62.

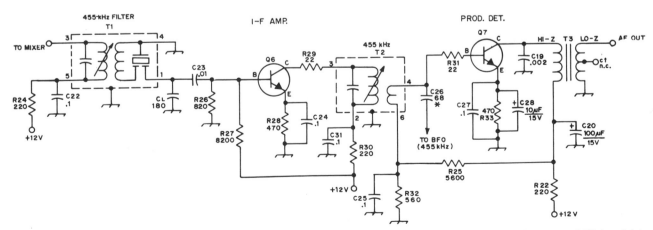

455 kHz WITH PRODUCT DETECTOR—Bipolar transistor Q6 provides about 20-dB gain at 455 kHz, which is adequate for handling wide range of signal amplitudes without changing audio gain setting in receiver without AGC. Product detector produces output in audio range when its inputs are 455-kHz IF signal and BFO signal near 455 kHz. Transistors can be 2N222, 2N3641, 2N4123, or equivalent. T1 is J. W. Miller 8814 455-kHz IF transformer/filter, T2 is miniature 455-kHz IF transformer, and T3 is miniature audio transformer with 10K primary and 2K center-tap secondary (CT not used).—D. DeMaw and L. McCoy, Learning to Work with Semiconductors, *QST*, Aug. 1974, p 26–30.

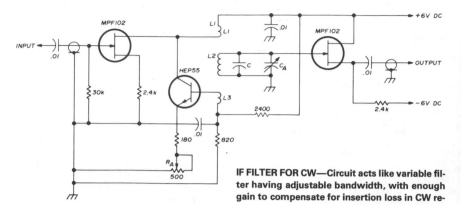

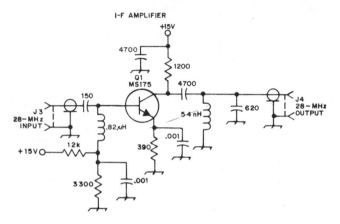

I-F AMPLIFIER

IF FILTER FOR CW—Circuit acts like variable filter having adjustable bandwidth, with enough gain to compensate for insertion loss in CW receiver. Used to isolate weak CW signals despite noise and interference, as required in low-power amateur work. FET input is directly coupled to collector of HEP55 serving as Q-multiplier regenerative amplifier. Transformer L1-L3 provides feedback. Filter is connected in series with input end of IF strip, following 1.2-kHz mechanical filter. When two CW signals are received, one can be eliminated by adjusting CA to recenter passband of filter. Current drain of 6 mA can be supplied by two 6-V batteries. For 455-kHz IF, C is 470 pF, CA is 7—45 pF, and L1-L2-L3 are 12-115-24 turns No. 32 enamel on Amidon T44-15 core.—S. M. Olberg, Vari-Q Filter, *Ham Radio*, Sept. 1973, p 62—65.

28-MHz LOW-NOISE—Developed for use with 2304-MHz balanced mixer. Provides required match between 50-ohm mixer output and input of 28-MHz IF amplifier in UHF receiver. Input and output connections are made with short lengths of RG-58/U coax. Noise figure is less than 1.5 dB.—L. May and B. Lowe, A Simple and Efficient Mixer for 2304 MHz, *QST*, April 1974, p 15—19 and 31.

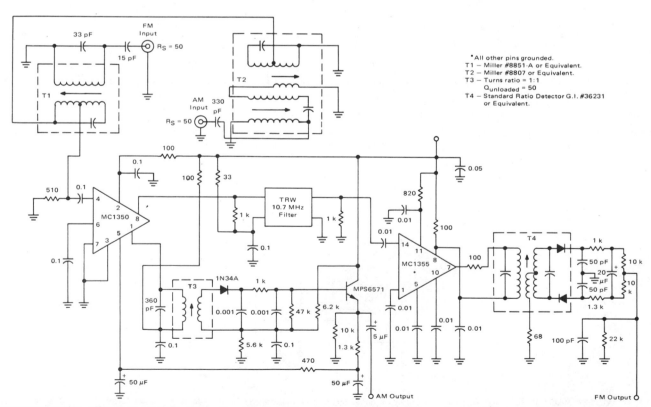

AM/FM WITH AGC—Operates from single +15 V supply. Standard 455-kHz IF is used for AM to feed 1N34A diode detector. One output of MC1350 is used for FM signal component and the other for AM component. External transistor is needed because MC1350 requires up to 0.2 mA of AGC drive and this is more than can be furnished by diode detector.—"Integrated Circuit IF Amplifiers for AM/FM and FM Radios," Motorola, Phoenix, AZ, 1975, AN-543A, p 10.

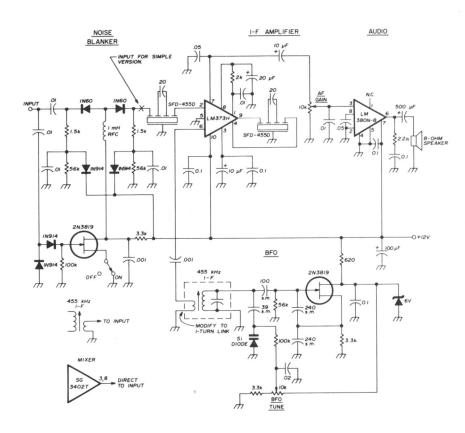

IF WITH NOISE BLANKER—Addition of BFO and noise blanker to 455-kHz IF amplifier gives setup for testing new tuners and front ends. Two methods of coupling into amplifier are shown. LM373H IC with two Murata SFD-455D ceramic filters fulfills requirements for IF amplifier, detector, and AGC functions.—R. Megirian, Design Ideas for Miniature Communications Receivers, *Ham Radio*, April 1976, p 18–25.

T ATTENUATOR—When inserted between stages of IF amplifier, circuit acts as three-section attenuator with dynamic range greater than 60 dB. Can be controlled by positive voltage from AVC system of receiver or manually with 100K pot. Use PIN diodes.—Super Circuits, *73 Magazine*, Aug. 1975, p 140.

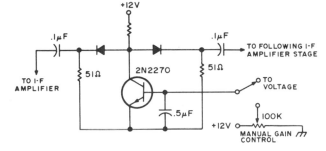

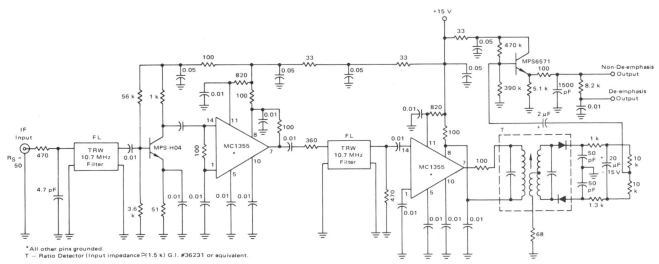

10.7-MHz IF for FM—Motorola dual MC1355 limiting gain blocks are used with two TRW five-pole linear phase filters and external ratio detector to give complete high-performance IF amplifier for FM receiver. MPS-H04 discrete transistor is used after first filter block to reduce noise figure for overall system. Input and output impedances are 235 ohms.—"Integrated Circuit IF Amplifiers for AM/FM and FM Radios," Motorola, Phoenix, AZ, 1975, AN-543A, p 4.

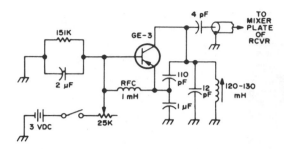

Q MULTIPLIER—Transistorized Q multiplier can be connected to plate of mixer in receiver having IF in range of 1400–1500 kHz. Iron-core coil should have high Q. Setting of pot depends on transistor used, which could also be HE-3 or 2N1742.—Q & A, *73 Magazine*, April 1977, p 165.

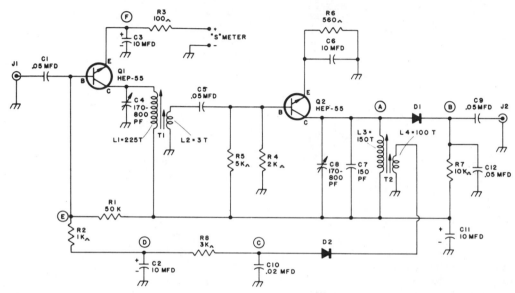

NOTE: ENCIRCLED LETTERS DESIGNATE TEST POINTS

135-kHz STRIP—Developed for use in all-band VHF/UHF/S-band receiver. Operates from 12-V supply, connected to positive terminal of C11 through 100-ohm resistor. Article covers design and construction.—B. Hoisington, Building a 135 kHz I-F Strip, *73 Magazine*, Sept. 1975, p 127–130 and 132.

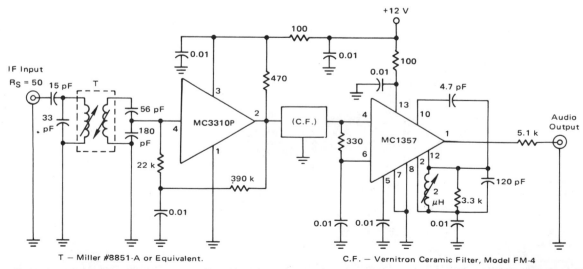

T — Miller #8851-A or Equivalent.

C.F. — Vernitron Ceramic Filter, Model FM-4

FM AUTO RADIO IF—Uses MC1357 quadrature detector after ceramic filter to give IF bandwidth required for good stereo reproduction. Sensitivity is 18 μV for 3% total harmonic distortion.—"Integrated Circuit IF Amplifiers for AM/FM and FM Radios," Motorola, Phoenix, AZ, 1975, AN-543A, p 6.

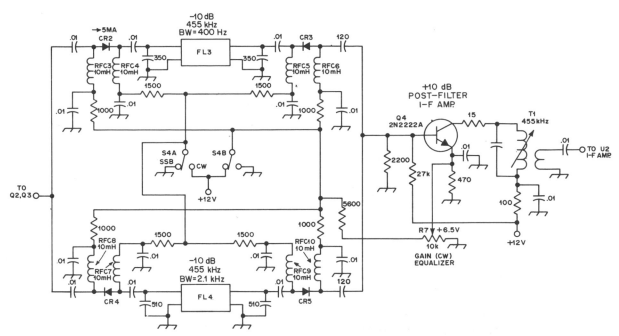

DIODE-SWITCHED IF FILTER—Used in 1.8–2 MHz communication receiver having wide dynamic range. 1N914 diodes select Collins mechanical filter F455FD-04 FL3 (400-Hz bandwidth) or F455FD-25 FL4 (2.5-kHz bandwidth). Reverse bias is applied to nonconducting diodes to lessen leakage through switching diodes. Filter is located between IF preamp and main IF strip of receiver. Two-part article gives all other circuits of receiver.—D. DeMaw, His Eminence—the Receiver, *QST*, Part 1—June 1976, p 27–30 (Part 2—July 1976, p 14–17).

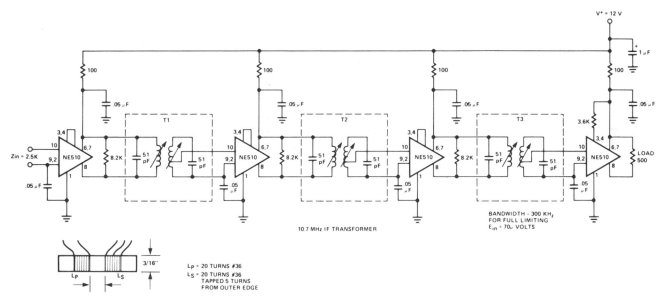

10.7-MHz LIMITING AMPLIFIER—Uses Signetics NE510 transistor arrays in common-collector common-base configuration as IF strip for commercial FM broadcast receiver. Bandwidth is 300 kHz, achieved by adjusting transformers for 600-kHz bandwidth and using dual cup-core transformers originally designed for tubes. Windings were changed to give critical coupling. Full limiting is provided by circuit with input voltage of 70 μVRMS.—"Signetics Analog Data Manual," Signetics, Sunnyvale, CA, 1977, p 747–748.

CHAPTER 40

Instrumentation Circuits

Includes DC, AF, and wideband RF amplifiers with such special features as automatic nulling and automatic calibration, for use with resistance-bridge, photocell, strain-gage, and other input transducers. Applications include measurement of ionization, radiation, small currents, liquid flow and level, light level, pH, power, torque, weight, and wind velocity. Metal detectors and proximity detectors are also covered. See also chapters covering measurement of Capacitance, Frequency, Resistance, and Temperature.

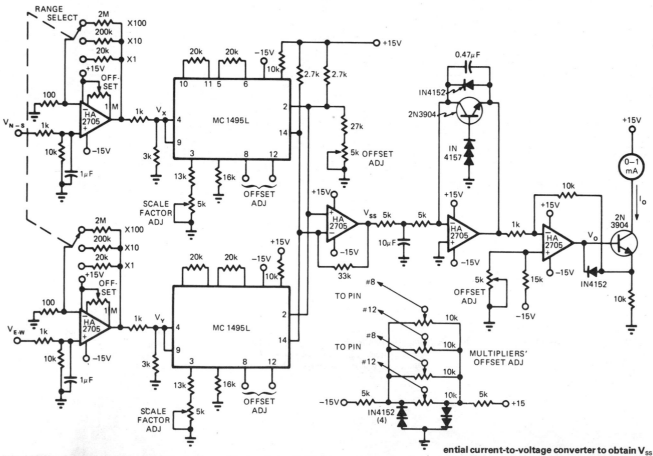

WIND SPEED—Developed to give magnitude of wind velocity over wide range of values when its two measured vectors are expressed as voltages. Output is in logarithmic form for easy adaptation to data processors. N-S and E-W vector voltages from strain-gage sensors are converted to normalized values V_X and V_Y which are squared by MC1495L four-quadrant transconductance multipliers. Output currents are then summed, and HA2705 opamp is used as differential current-to-voltage converter to obtain V_{SS} as sum-of-squares of V_X and V_Y. Range covered is 1–100 mph. Article covers operation of circuit in detail.—J. A. Connelly and M. B. Lundberg, Analog Multipliers Determine True Wind Speed, *EDN Magazine*, April 20, 1974, p 69–72.

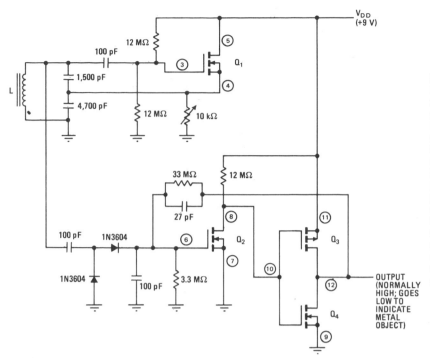

PROXIMITY DETECTOR—Output changes from high (9 V) to low (0 V) when conducting object moves within 1 cm from open end of 150-turn coil L (No. 34 enamel) mounted in half of Ferroxcube 1811-PL00-3B7 core set. Can be used as contactless limit switch or tachometer pickup. Q_1-Q_4 are CMOS MOSFETs in CD4007A package. Q_1 and pickup coil form 100-kHz oscillator. Diodes develop DC voltage proportional to peak-to-peak value of oscillator signal, for application to Schmitt trigger Q_2-Q_4. Conductive object near coil absorbs energy from magnetic field, lowering oscillator amplitude and turning Schmitt trigger off. 10K pot adjusts sensitivity. Circuit drives CMOS logic directly. For TTL drive, use buffer.—M. L. Fichtenbaum, Inductive Proximity Detector Uses Little Power, *Electronics,* Jan. 22, 1976, p 112.

0–15 MHz WITH 100-dB CMR—Differential inputs are applied to Optical Electronics 9715 opamp through 9714 voltage followers. Current booster using 9810 opamp raises load current to ±100 mA. Complete amplifier has very high differential and common-mode input impedance. Common-mode rejection can be trimmed to greater than 100 dB at 1 kHz for unity gain. Gain is determined by value of resistor RG connected between points A and B and is equal to (2R2/R1)(1 + 2R2/RG). Settling time is 500 ns. Accuracy is maintained from −55°C to +85°C.—"Instrumentation Amplifier," Optical Electronics, Tucson, AZ, Application Tip 10240.

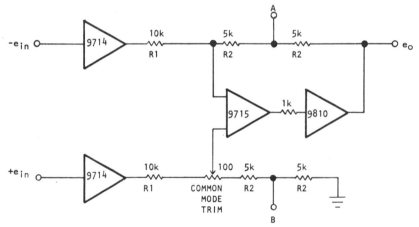

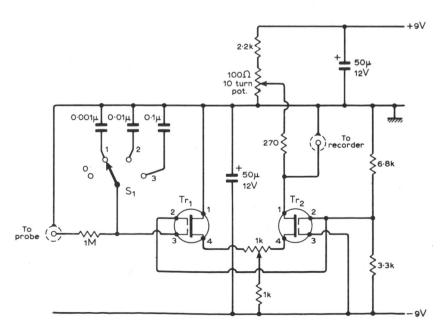

MOSFET DIFFERENTIAL AMPLIFIER—Developed to monitor chemical process of titration, by recording probe output voltages between 100 and 400 mV when internal impedance of probe is in gigohm range. Either 40673 or 3N187 dual-gate MOSFETs connected as differential amplifier are suitable for meeting high input resistance requirement. Transistor level drifts because of temperature are in opposition and tend to cancel each other. Overall power gain of amplifier is about 70 dB. Circuit is suitable for other electrometer applications as well.—D. R. Bowman, Automatic Titration Potentiometer, *Wireless World,* Aug. 1971, p 400–401.

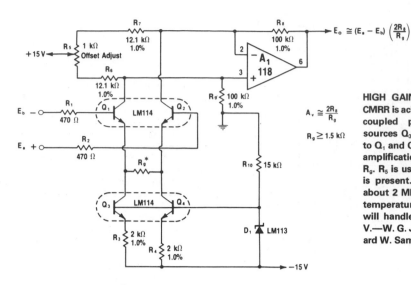

$$E_o \cong (E_a - E_b)\left(\frac{2R_8}{R_9}\right)$$

$$A_v \cong \frac{2R_8}{R_9}$$

$$R_9 \geq 1.5\ k\Omega$$

HIGH GAIN WITH WIDE BANDWIDTH—High CMRR is achieved by using Q_1 and Q_2 as emitter-coupled pair biased by constant-current sources Q_3 and Q_4. Differential signals applied to Q_1 and Q_2 appear across 100K resistor R_9 for amplification by factor inversely proportional to R_9. R_5 is used to null opamp A_1 when no input is present. Bandwidth, determined by A_1, is about 2 MHz. Gain is flat at about 40 dB over temperature range of $-55°C$ to $+125°C$. Circuit will handle common-mode inputs up to ±10 V.—W. G. Jung, "IC Op-Amp Cookbook," Howard W. Sams, Indianapolis, IN, 1974, p 243–245.

*R_9 varied to adjust input sensitivity and gain; can be greater than 200 kΩ if attenuation desired.

TORQUE WRENCH—Micro Networks MN2200 instrumentation amplifier is used with strain gage to create digital-readout torque wrench. Strain gages having nominal impedance of 120 ohms are bonded to torque-sensing member at 45° to longitudinal axis, so gages in opposite bridge arms are under simultaneous tension or compression for given direction of torque. Bridge power is taken from 5-V digital panel meter supply. Instrumentation amplifier will work with any voltage from ±5 to ±15 V. Variable gain-adjust resistor G (10-turn 50K pot) is set so DPM reads 200 ft-lb of torque at full scale in increments of 0.1 ft-lb.—R. Duris, Instrumentation Amplifiers—They're Great Problem Solvers When Correctly Applied, *EDN Magazine*, Sept. 5, 1977, p 133–135.

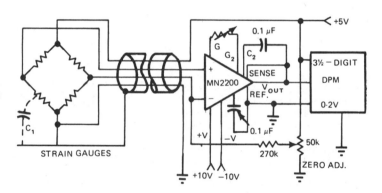

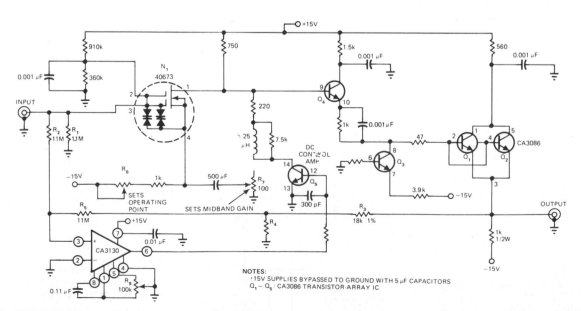

0–35 MHz WITH GAIN OF 10—Wideband amplifier handles inputs up to 100 mV P-P and drives 1-kilohm load, to meet requirements of oscilloscope preamps, instrumentation and pulse signal amplifiers, and video signal processors. High-frequency gain is provided by 40673 dual-gate MOSFET. Low-frequency gain with DC stabilization is provided by CA3130 CMOS opamp. Transistors Q_1-Q_5 are part of CA3086 transistor-array IC. Values of R_3 and R_4 in feedback path establish amplifier gain. R_6 sets operating point of N_1 for 10-mA drain current. Base resistor of Q_3 is 1 kilohm.—H. A. Wittlinger, CMOS Op Amp, MOSFET Implement Wideband Amplifier, *EDN Magazine*, June 20, 1977, p 114.

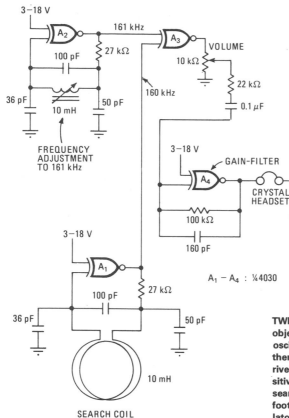

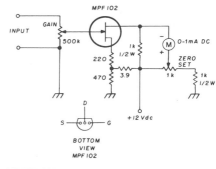

METER AMPLIFIER—Junction FET in simple DC amplifier circuit converts 0–1 mA DC milliammeter to 0–100 μA DC microammeter. Adjust zero-set control for zero meter current with no input, then apply input signal and adjust gain to desired value.—N. J. Foot, Electronic Meter Amplifier, *Ham Radio,* Dec. 1976, p 38–39.

TWIN-OSCILLATOR METAL DETECTOR—Metal object near search coil changes frequency of oscillator A_1 which is initially tuned to 160 kHz, thereby changing frequency of 1-kHz output derived by mixing with 161-kHz output of A_2. Sensitivity, determined largely by dimensions of search coil, is sufficient to detect coins about 1 foot away.—M. E. Anglin, C-MOS Twin Oscillator Forms Micropower Metal Detector, *Electronics,* Dec. 22, 1977, p 78.

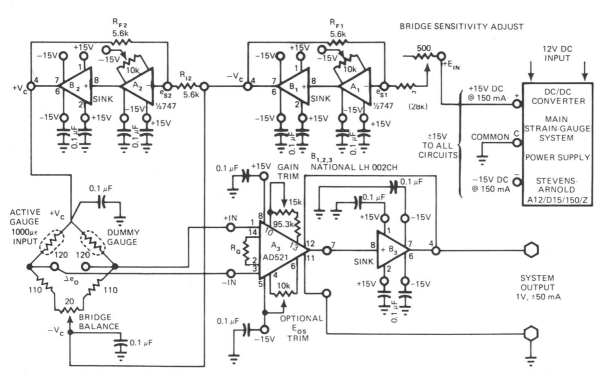

STRAIN-GAGE AMPLIFIER—Optimum performance is achieved in fully portable system by utilizing combination of 747 opamps for A_1 and A_2 with National LH002CH opamp for B_1-B_3 and special AD521K instrumentation amplifier for output stage. Bypass capacitors suppress undesirable high-frequency signals. Stevens-Arnold DC/DC converter operating from 12-V storage battery provides required regulated ±15 VDC for system while giving excellent power isolation.—D. Sheehan, Strain-Gauge Transducer System Uses Off-the-Shelf Components, *EDN Magazine,* Nov. 5, 1977, p 79–81.

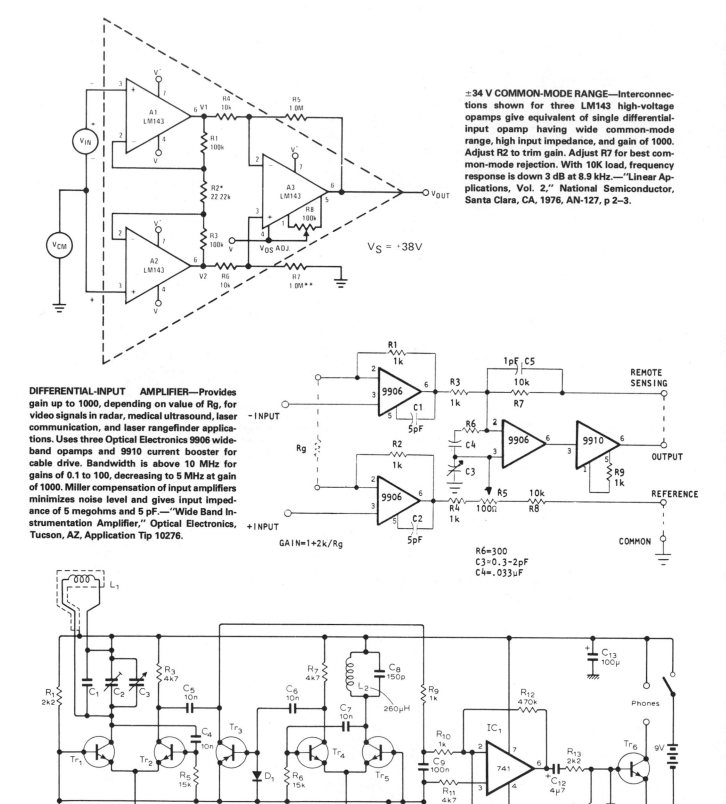

±34 V COMMON-MODE RANGE—Interconnections shown for three LM143 high-voltage opamps give equivalent of single differential-input opamp having wide common-mode range, high input impedance, and gain of 1000. Adjust R2 to trim gain. Adjust R7 for best common-mode rejection. With 10K load, frequency response is down 3 dB at 8.9 kHz.—"Linear Applications, Vol. 2," National Semiconductor, Santa Clara, CA, 1976, AN-127, p 2–3.

$V_S = +38V$

DIFFERENTIAL-INPUT AMPLIFIER—Provides gain up to 1000, depending on value of Rg, for video signals in radar, medical ultrasound, laser communication, and laser rangefinder applications. Uses three Optical Electronics 9906 wideband opamps and 9910 current booster for cable drive. Bandwidth is above 10 MHz for gains of 0.1 to 100, decreasing to 5 MHz at gain of 1000. Miller compensation of input amplifiers minimizes noise level and gives input impedance of 5 megohms and 5 pF.—"Wide Band Instrumentation Amplifier," Optical Electronics, Tucson, AZ, Application Tip 10276.

$GAIN = 1 + 2k/Rg$

$R6 = 300$
$C3 \approx 0.3 - 2pF$
$C4 = .033\mu F$

METAL DETECTOR—Will detect small coin up to about 5 inches underground and larger metal objects at much greater depths. Frequency of search oscillator Tr_1-Tr_2 depends on values used for three paralleled capacitors, search coil, and metal objects in vicinity of coil. Mixer Tr_3 feeds difference between search oscillator and reference oscillator Tr_4-Tr_5 to opamp and Tr_6 for driving phones or loudspeaker. Article gives construction and adjustment details, including dimensions for search coil. Reference oscillator is set to 625 kHz. C_1 is 560 pF, C_2 150 pF, and C_3 10 pF variable. C_2 is used for coarse tuning, and C_3 for fine adjustment to get beat note. Diodes are 1N4148. Tr_3 is BC308, BCY72, or equivalent, and other transistors are BF238, BC108, or equivalent.—D. E. Waddington, Metal Detector, *Wireless World*, April 1977, p 45–48.

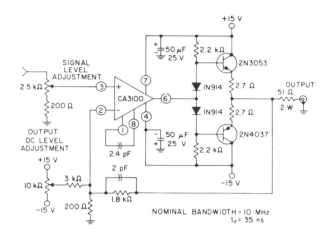

50-OHM LINE DRIVER—CA3100 bipolar MOS opamp operates as high-slew-rate wideband amplifier that provides 18 V P-P into open circuit or 9 V P-P into 50-ohm transmission line. Slew rate is 28 V/μs.—"Circuit Ideas for RCA Linear ICs," RCA Solid State Division, Somerville, NJ, 1977, p 13.

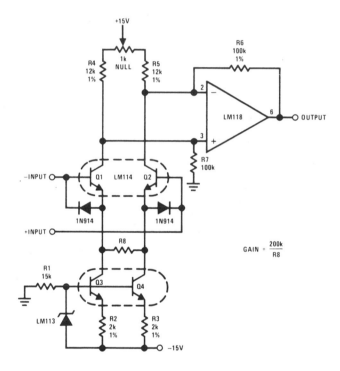

ADJUSTABLE-GAIN WIDEBAND AMPLIFIER— Single resistor R8 adjusts gain from less than 1 to over 1000, with gain value equal to 200,000 divided by value in ohms used for R8. Common-mode rejection ratio is about 100 dB, independent of gain. Q1-Q2 are operated open-loop as floating differential input stage. Current sources Q3 and Q4 set operating current of input transistors.—"Linear Applications, Vol. 2," National Semiconductor, Santa Clara, CA, 1976, LB-21.

$$\text{GAIN} = \frac{200k}{R8}$$

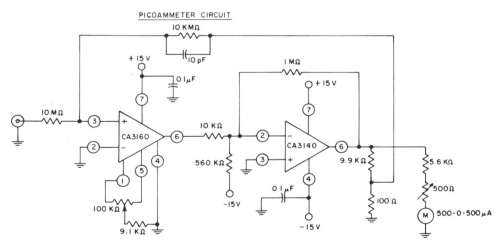

OPAMP PICOAMMETER—Current-to-voltage converter connection for CA3160 and CA3140 bipolar MOS opamps provides full-scale meter deflection for ±3 pA. CA3160 is operated in guarded mode to reduce leakage current. CA3140 provides gain of 100 for driving zero-center microammeter. With suitable switching, full-scale current ranges of 3 pA to 1 nA can be handled with single 10,000-megohm resistor in overall feedback path.—"Circuit Ideas for RCA Linear ICs," RCA Solid State Division, Somerville, NJ, 1977, p 14.

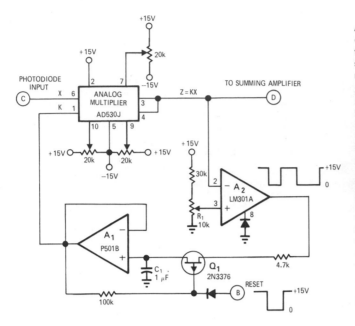

AUTOMATIC CALIBRATOR—Automatic scaling circuit permits frequent and fast recalibration for precision optical measurements, to compensate for variations in light intensity due to thermal cycling of lamp filament, dirty optics, and gain variations between photodetectors and between amplifiers. With reset pulse at point B, comparator A_2 compares output of multiplier to preset reference voltage on R_1. If A_2 input voltage is greater than reference applied to pin 3 by R_2, output switches to zero and remains there until C_1 has discharged enough to lower output of A_1 and output of multiplier below reference on pin 3. If input at pin 2 of A_2 is less than reference on pin 3, A_2 will switch to 15 V and output of multiplier will be adjusted upward until voltage on pin 2 of A_2 again exceeds that on pin 3. Output of A_2 is thus continually switching between 15 V and 0 V during reset or scaling. After reset pulse is removed, scale factor K is maintained constant by multiplier during measuring.—R. E. Keil, Automatic Scaling Circuit for Optical Measurements, *EDN/EEE Magazine,* Nov. 15, 1971, p 49–50.

HIGH GAIN FOR WEAK SIGNALS—National LM121 differential amplifier is operated open-loop as input stage for input signals up to ±10 mV. Input voltage is converted to differential output current for driving opamp acting as current-to-voltage converter with single-ended output. R4 is adjusted to set gain at 1000. Null pot R3 serves for offset adjustment.—"Linear Applications, Vol. 2," National Semiconductor, Santa Clara, CA, 1976, AN-79, p 7–8.

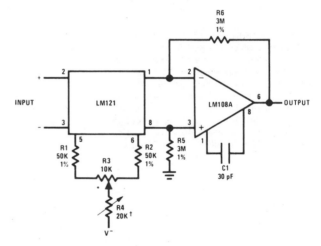

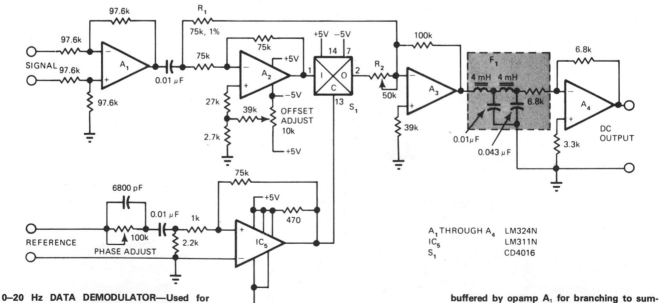

0–20 Hz DATA DEMODULATOR—Used for measuring and monitoring suppressed-carrier signal modulation from aircraft control systems. Provides data frequency response within 0.1 dB from DC to 20 Hz, with linearity better than 0.1%. In-phase reference voltage applied to comparator IC_5 controls gating of CD4016 MOS switch S_1. Suppressed-carrier signal is buffered by opamp A_1 for branching to summing junction of A_3. Article describes operation of circuit.—J. A. Tabb and M. L. Roginsky, Instrumentation Signal Demodulator Uses Low-Power IC's, *EDN Magazine,* Jan. 20, 1976, p 80.

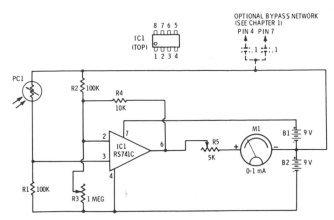

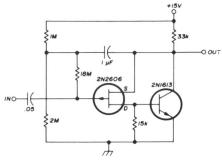

HIGH INPUT Z—Suitable for use as active probe for CRO, as electrometer, and for instrumentation applications. Combination of unipolar and bipolar transistors gives desirable amplifying features of each solid-state device.—I. M. Gottlieb, A New Look at Solid-State Amplifiers, *Ham Radio,* Feb. 1976, p 16—19.

PHOTOCELL BRIDGE—Radio Shack 276-116 cadmium sulfide photocell is connected in Wheatstone bridge circuit. When bridge is balanced, RS741C opamp connected to opposite corners of bridge receives no voltage and meter reads zero. Light on photocell unbalances bridge and gives meter deflection. Can be used as high-sensitivity light meter. Adjust R3 until meter reads zero with photocell covered while R5 is at maximum resistance, adjust R5 until needle moves away from zero, rezero with R3, and repeat procedure until meter can no longer be brought to zero. Sensitivity is now maximum, and uncovered photocell will detect flame from candle at 20 feet.—F. M. Mims, "Integrated Circuit Projects, Vol. 4," Radio Shack, Fort Worth, TX, 1977, 2nd Ed., p 29—35.

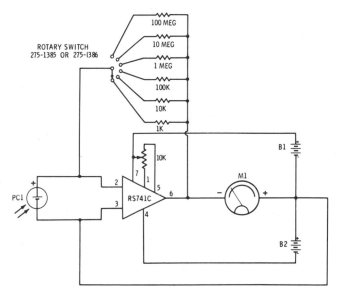

SIX-RANGE LIGHT METER—Switching of feedback resistors for opamp driven by Radio Shack 276-115 selenium solar cell gives multirange linear light meter. With 1000-megohm resistor for highest sensitivity, star Sirius will produce photocurrent of about 25 pA when solar cell is shielded from ambient light with length of cardboard tubing. Supplies are 9 V, and meter is 0—1 mA.—F. M. Mims, "Integrated Circuit Projects, Vol. 4," Radio Shack, Fort Worth, TX, 1977, 2nd Ed., p 45—53.

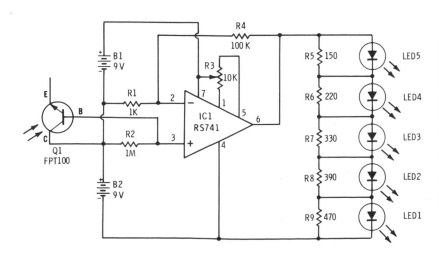

LIGHT METER WITH LED READOUT—Light on phototransistor Q1 (Radio Shack 276-130) produces voltage change across R2 for amplification by opamp whose output drives array of five LEDs forming bar graph voltage indicator. Adjust R3 initially for highest sensitivity by turning off room lights and rotating until LED 1 just stops glowing. Now, as light is gradually increased on sensor, LEDs come on one by one in upward sequence and stay on until all five are lit. Solar cells or selenium cells can be used in place of phototransistor.—F. M. Mims, "Optoelectronic Projects, Vol. 1," Radio Shack, Fort Worth, TX, 1977, 2nd Ed., p 85—93.

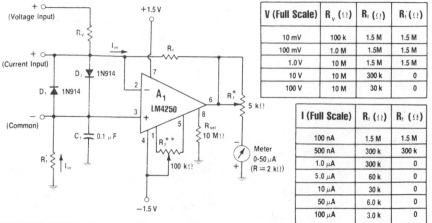

V (Full Scale)	R_v (Ω)	R_i (Ω)	R_i' (Ω)
10 mV	100 k	1.5 M	1.5 M
100 mV	1.0 M	1.5M	1.5 M
1.0 V	10 M	1.5 M	1.5 M
10 V	10 M	300 k	0
100 V	10 M	30 k	0

I (Full Scale)	R_i (Ω)	R_i' (Ω)
100 nA	1.5 M	1.5 M
500 nA	300 k	300 k
1.0 μA	300 k	0
5.0 μA	60 k	0
10 μA	30 k	0
50 μA	6.0 k	0
100 μA	3.0 k	0

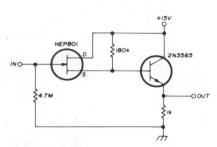

NANOAMMETER—Programmable amplifier operating from ±1.5 V supply such as D cells is used as current-to-voltage converter. Offset null of A_1 is used to minimize input offset voltage error. If programmed for low bias current, amplifier can convert currents as small as 100 nA with less than 1% error. Resistor values for variety of current and voltage ranges are given in tables. Adjust R_1 to calibrate meter, and adjust R_2 to null input offset voltage on lowest range. Not suitable for higher current ranges because power drain is excessive above 100 μA.—W. G. Jung, "IC Op-Amp Cookbook," Howard W. Sams, Indianapolis, IN, 1974, p 414–417.

FET-BIPOLAR DARLINGTON—Can be used as meter interface amplifier, impedance transformer, coax driver, or relay actuator. Combination of unipolar and bipolar transistors gives desirable amplifying features of each solid-state device.—I. M. Gottlieb, A New Look at Solid-State Amplifiers, *Ham Radio,* Feb. 1976, p 16–19.

DIGITAL pH METER—3130 CMOS opamp gives required high input impedance for pH probe at low cost. Output of probe, ranging from positive generated DC voltage for low pH to 0 V for pH 7 and negative voltages for high pH values, is amplified in circuit that provides gain adjustment to correct for temperature of solution being measured. For analog reading, output of opamp can be fed directly to center-scale milliammeter through 100K calibrating pot. For digital display giving reading of 7.00 for 0-V output, pH output is converted to calibrated current for summing with stable offset current equal to 700 counts. This is fed to current-to-frequency converter driving suitable digital display. Standard pH buffer solutions are used for calibration.—D. Lancaster, "CMOS Cookbook," Howard W. Sams, Indianapolis, IN, 1977, p 347–349.

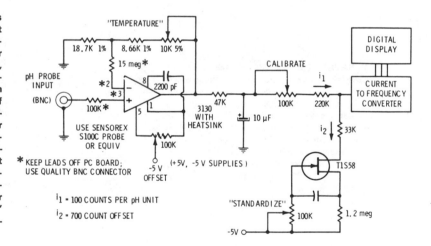

DIFFERENTIAL-INPUT VARIABLE-GAIN—Gain of A_3 is varied by modifying feedback returned to R_4. A_4 serves as active attenuator in feedback path, presenting constant zero-impedance source to R_4 as required for maintaining good balance and high CMRR. With values shown, gain can be varied from unity to 300.—W. G. Jung, "IC Op-Amp Cookbook," Howard W. Sams, Indianapolis, IN, 1974, p 238–239.

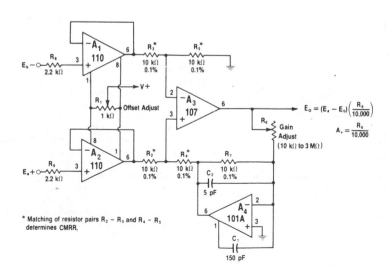

$$E_o = (E_a - E_b)\left(\frac{R_6}{10,000}\right)$$

$$A_v = \frac{R_6}{10,000}$$

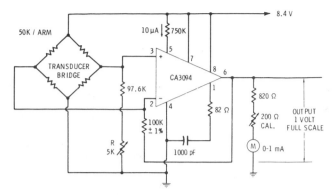

SENSOR-BRIDGE AMPLIFIER—RCA CA3094 combination power switch and amplifier can be used with variety of transducer bridges for instrumentation and other applications. Circuit delivers output of 1 V full-scale for driving meter. Pot R serves as centering or reference control. Can be used as thermometer if one leg of bridge is thermistor and meter scale is calibrated in degrees.—E. M. Noll, "Linear IC Principles, Experiments, and Projects," Howard W. Sams, Indianapolis, IN, 1974, p 311–313.

HIGH CMRR—Use of two Precision Monolithics OP-05 opamps feeding OP-01 opamp gives input impedance of about 100 gigohms and high common-mode rejection for instrumentation applications.—"Instrumentation Operational Amplifier," Precision Monolithics, Santa Clara, CA, 1977, OP-05, p 7.

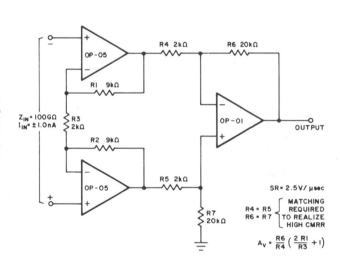

$$A_V = \frac{R6}{R4}\left(\frac{2\,R1}{R3}+1\right)$$

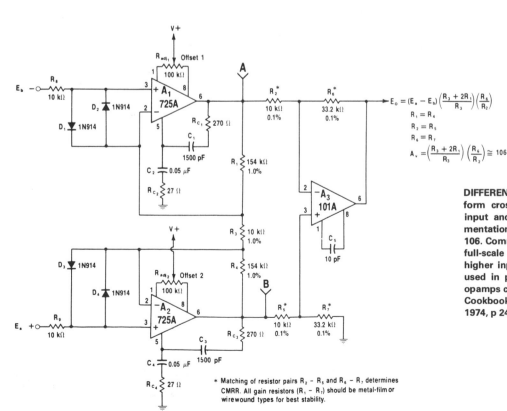

$$E_o = (E_a - E_b)\left(\frac{R_3 + 2R_1}{R_3}\right)\left(\frac{R_6}{R_2}\right)$$

$$R_1 = R_4$$
$$R_2 = R_5$$
$$R_6 = R_7$$

$$A_v = \left(\frac{R_3 + 2R_1}{R_3}\right)\left(\frac{R_6}{R_2}\right) \cong 106$$

DIFFERENTIAL PREAMP—Opamps A_1 and A_2 form cross-coupled preamp with differential input and differential output, driving instrumentation opamp A_3 to provide overall gain of 106. Common-mode input range is ±10 V, and full-scale differential input is ±100 mV. For higher input impedance, 108 opamps can be used in preamp. For higher speed, all three opamps can be 118.—W. G. Jung, "IC Op-Amp Cookbook," Howard W. Sams, Indianapolis, IN, 1974, p 241–243.

* Matching of resistor pairs $R_2 - R_5$ and $R_6 - R_7$ determines CMRR. All gain resistors ($R_1 - R_7$) should be metal-film or wirewound types for best stability.

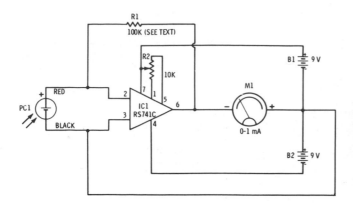

LINEAR LIGHT METER—Uses Radio Shack 276-115 selenium solar cell or equivalent photocell with high-gain RS741C opamp to drive meter. Sensitivity is sufficient to detect individual stars at night without magnifying lens if photocell is shielded from ambient light with length of cardboard tubing. Increasing value of R1 increases gain and sensitivity of circuit. R2 sets meter needle to zero when sensor is dark.—F. M. Mims, "Integrated Circuit Projects, Vol. 4," Radio Shack, Fort Worth, TX, 1977, 2nd Ed., p 45–53.

HODOSCOPE AMPLIFIER—Charge amplifier using Teledyne Philbrick 102601 opamp was developed for use with each Geiger counter of 132-counter array for ionization hodoscope used in tracing paths of cosmic rays. Charge-sensitive stage A_1 converts input charge pulse to voltage pulse significantly larger than noise of second stage. With 616-pF load capacitor, output is 12 V for input of 10 mV. Cost of charge amplifier is about $50.—H. C. Carpenter, Low Cost Charge Amplifier, *EDN Magazine,* May 20, 1973, p 83 and 85.

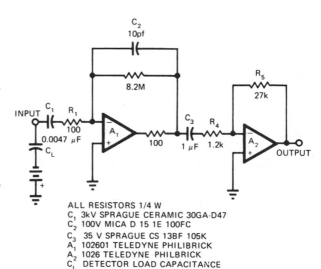

ALL RESISTORS 1/4 W
C_1 3kV SPRAGUE CERAMIC 30GA-D47
C_2 100V MICA D 15 1E 100FC
C_3 35 V SPRAGUE CS 13BF 105K
A_1 102601 TELEDYNE PHILIBRICK
A_2 1026 TELEDYNE PHILBRICK
C_L DETECTOR LOAD CAPACITANCE

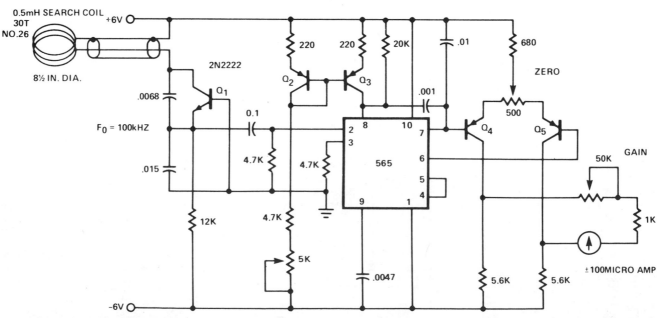

PLL DETECTOR FOR ALL METALS—Frequency change produced in Colpitts oscillator by metal object near tank coil is indicated by 565 PLL connected as frequency meter. Oscillator frequency increases when search coil is brought near non-ferrous metal object. Oscillator frequency decreases, as indicated by lower meter reading, when coil is brought near ferrous object.—"Signetics Analog Data Manual," Signetics, Sunnyvale, CA, 1977, p 856–858.

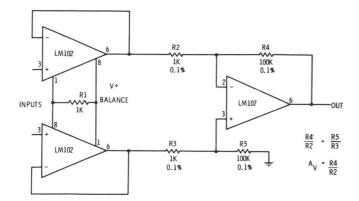

THREE-STAGE OPAMP—Responds to difference between two applied signals. Differential output voltage of LM102 pair is applied to balanced differential input of LM107 opamp. Output can be metered or used in any other desired manner. Voltage gain is equal to ratio R4/R2 and is 100 for values shown.—E. M. Noll, "Linear IC Principles, Experiments, and Projects," Howard W. Sams, Indianapolis, IN, 1974, p 126.

$$\frac{R4}{R2} = \frac{R5}{R3}$$

$$A_V = \frac{R4}{R2}$$

FEEDBACK OPAMP FOR BRIDGE—Uses CA3094 programmable opamp to convert differential input signal from resistor bridge to single-ended 1-V output signal. Circuit provides feedback for opamp. RC network between terminals 1 and 4 of opamp provides compensation to improve stability.—"Circuit Ideas for RCA Linear ICs," RCA Solid State Division, Somerville, NJ, 1977, p 13.

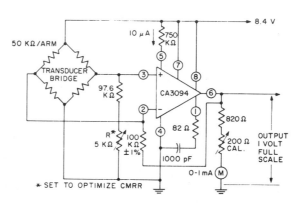

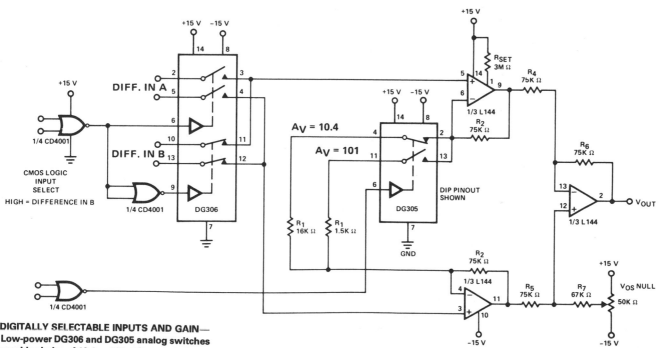

DIGITALLY SELECTABLE INPUTS AND GAIN—Low-power DG306 and DG305 analog switches provide choice of 10.4 or 101 gain and choice of two differential input channels for instrumentation applications. Highest gain is obtained when control logic is high.—"Analog Switches and Their Applications," Siliconix, Santa Clara, CA, 1976, p 7-91.

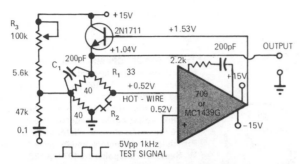

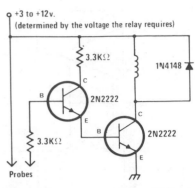

FLOW METER—Simple opamp circuit with one transistor gives reliable hot-wire anemometer for measuring flow of gases or liquids. R_2 is heated above ambient temperature in Wheatstone bridge including overheat resistor R_1 which is calibrated to be 30% larger than cold resistance of R_2. Bridge is fed from power transistor which is within feedback loop of opamp that senses bridge unbalance. Output of bridge is fed back to power transistor in correct phase for maintaining constant-temperature condition in which R_2 is approximately equal to R_1. Article covers construction of hot-wire probe made from Wollaston wire.—W. Bank, Build Your Own Constant-Temperature Hot-Wire Anemometer, *EDN Magazine*, Aug. 1, 1972, p 43.

NONLATCHING RELAY—When liquid rises above level determined by positions of probes, circuit is triggered and relay, buzzer, or other indicator is energized. Alarm stops when liquid drops below preset level again. Use any operating voltage from 3 to 12 V that will actuate load employed.—J. A. Sandler, 9 Easy to Build Projects under $9, *Modern Electronics*, July 1978, p 53–56.

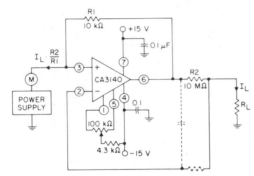

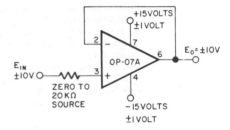

SMALL-CURRENT AMPLIFIER—CA3140 bipolar MOS opamp serves as high-gain current amplifier. Input current through load is increased by ratio of R2 to R1, which is 1000 for values shown, for reading by meter M. Dashed lines show method of decoupling circuit from effects of high output-lead capacitance.—"Circuit Ideas for RCA Linear ICs," RCA Solid State Division, Somerville, NJ, 1977, p 13.

LARGE-SIGNAL BUFFER—Unity-gain connection of Precision Monolithics OP-07A opamp provides high accuracy (0.005% worst case) over temperature range of −55°C to +125°C for buffer applications for ±10 V signals.—D. Soderquist and G. Erdi, "The OP-07 Ultra-Low Offset Voltage Op Amp—a Bipolar Op Amp That Challenges Choppers, Eliminates Nulling," Precision Monolithics, Santa Clara, CA, 1975, AN-13, p 8.

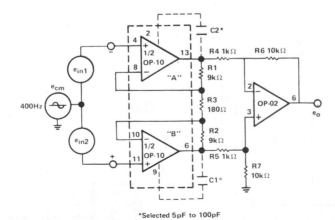

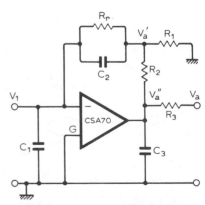

400-Hz AMPLIFIER WITH 95-dB CMRR—Precision Monolithics OP-10 dual opamp driving OP-02 opamp gives high common-mode rejection ratio. CMRR is optimized by selecting C1 and C2 in range of 5 to 100 pF for minimum output e_0 as viewed on CRO while feeding ±10 V signal at 400 Hz to common connection of inputs.—"Linear & Conversion I.C. Products," Precision Monolithics, Santa Clara, CA, 1977–1978, p 15-2.

PICOAMMETER—Highly stable circuit uses Valvo CSA70 chopper-stabilized opamp. Required high feedback resistance is provided by R_1-R_2 in feedback loop. Article gives design equations. R_r' and R_2 are 1 megohm, R_1 is 10 ohms, and all capacitors are 0.1 μF.—K. Kraus, High-Speed Picoammeter, *Wireless World*, May 1976, p 78.

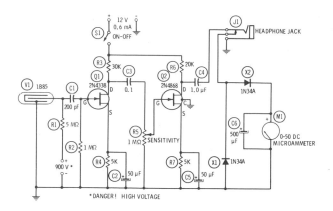

GEIGER COUNTER—Output signal of Victoreen 1B85 G-M tube biased at 900 VDC is proportional to beta-gamma particle count. Signal is amplified by high-gain AF amplifier Q1-Q2 for driving AC meter circuit. Closed-circuit jack is provided for alternate use of headphones. Count-rate range of instrument is determined by exposing G-M tube to different calibrated radioactive samples and marking meter scale for each. Bias for counter can be obtained from three 300-V photoflash batteries in series or equivalent supply capable of providing up to 10 mA.—R. P. Turner, "FET Circuits," Howard W. Sams, Indianapolis, IN, 1977, 2nd Ed., p 152–153.

DIFFERENTIAL TO SINGLE-ENDED—Conversion from differential input signal of thermocouple to single-ended output signal is achieved without feedback by using CA3094A programmable opamp. Output is ±4.7 V at 8.35 mA. Preamp gain is 180. For linear operation, differential input must be equal to or less than ±26 mV.—"Circuit Ideas for RCA Linear ICs," RCA Solid State Division, Somerville, NJ, 1977, p 13.

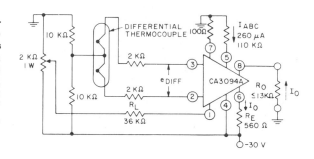

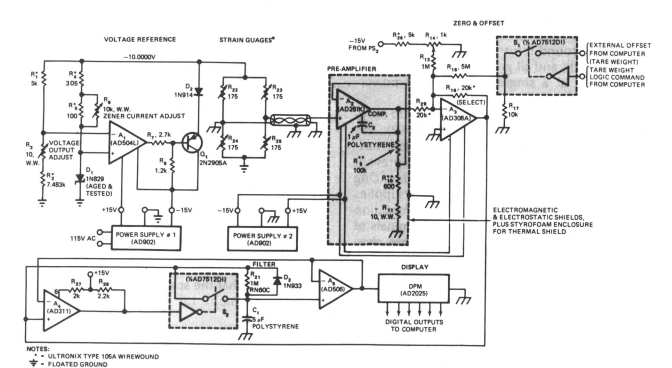

NOTES:
- * = ULTRONIX TYPE 105A WIREWOUND
- ⏚ = FLOATED GROUND
- ⏚ = INSTRUMENT (EDISON) GROUND
- ** = 100k, 0.005%, TYPE R-44, JULIE RESEARCH LABS
- *** = 600, 0.1%, R-44, JULIE RESEARCH LABS
      ALL OP AMPS AND POWER SUPPLIES = ANALOG DEVICES
- ▪ = SCHEMATIC FOR STRAIN GUAGES IS SIMPLIFIED

0.02% WEIGHING ACCURACY—Analog instrument covers up to 300 lb with resolution of 0.01 lb, for monitoring changes in body weight during clinical study. Bonded strain gages distributed symmetrically on platform of scale form bridge network R_1-R_2-R_4-R_5 serving as input for circuit that displays weight on digital panel meter and provides digital outputs to computer. Proper grounding is critical; all ground returns should go to single point at each power-supply common line. Article covers circuit operation in detail.—J. Williams, This 30-PPM Scale Proves That Analog Designs Aren't Dead Yet, *EDN Magazine,* Oct. 5, 1976, p 61–64.

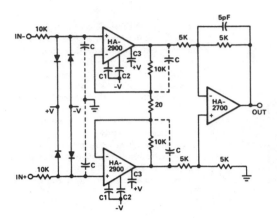

HIGH-IMPEDANCE DIFFERENTIAL-INPUT— Two Harris HA-2900 chopper-stabilized opamps feed HA-2700 high-performance opamp for instrumentation applications. Circuit provides excellent rejection of ±10 V common-mode input signals. Protection diodes prevent voltages at input terminals from exceeding either power supply. Supply can be ±15 V.—"Linear & Data Acquisition Products," Harris Semiconductor, Melbourne, FL, Vol. 1, 1977, p 7-70 (Application Note 518).

FERRITE-BEAD CURRENT TRANSFORMER— No. 27 ferrite bead (Ferronics 11-122-B) wound with 25 turns No. 30 enameled wire and shunted by 50-ohm ¼-W resistor, gives low-cost transformer that can be used in range of 3 kHz to 30 MHz. Current-conversion ratio is 1 V/A into 50-ohm coaxial-cable termination, with excellent linearity from milliamperes up into amperes. Wire carrying current to be measured is passed once through core, to serve as single-turn primary of transformer.—M. Salvati, Ferrite Bead Makes Cheap Current Transformer, *EDN Magazine,* March 20, 1974, p 85.

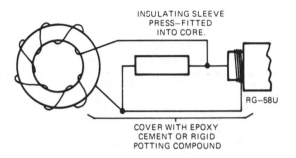

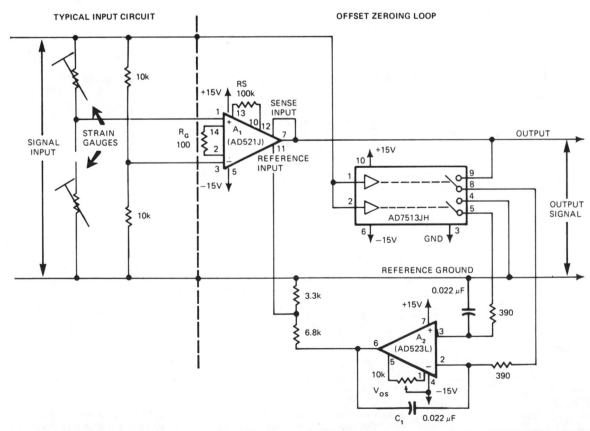

AUTOMATIC NULLING—Simple offset zeroing loop reduces effective input offset of instrumentation amplifier to less than a microvolt by using zero-and-hold that nulls output of instrumentation opamp A₁ to reference ground. Article describes operation in detail. Input signal source drives logic input, for nulling up to 4 V without using external nulling pot. Typical input circuit using strain gages is shown at left of dashed line.—M. Cerat, Zeroing Loop Reduces Instrumentation Amplifier Offsets, *EDN Magazine,* March 20, 1976, p 100 and 102.

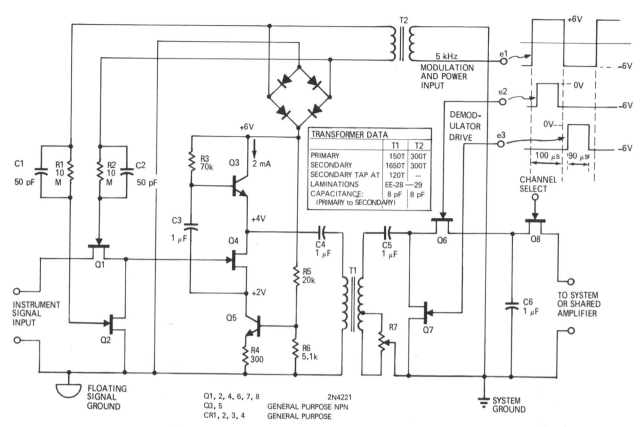

TRANSFORMER DATA

	T1	T2
PRIMARY	150T	300T
SECONDARY	1650T	300T
SECONDARY TAP AT	120T	—
LAMINATIONS	EE-28	—29
CAPACITANCE: (PRIMARY to SECONDARY)	8 pF	8 pF

Q1, 2, 4, 6, 7, 8   2N4221
Q3, 5   GENERAL PURPOSE NPN
CR1, 2, 3, 4   GENERAL PURPOSE

ANALOG PREAMP—Combination of 5-kHz FET-chopper amplifier Q1-Q2 with transformer isolation of signal and system grounds gives low-cost analog instrumentation amplifier that will process millivolt DC signals while rejecting hundreds of volts of common-mode DC. Signal accuracy is 0.1% for inputs between 50 and 500 mV. Input impedance is 4 megohms, drift is only 0.2 μV/°C, and DC common-mode rejection rate is better than 120 dB. Low-impedance output of followers Q3-Q4-Q5 is sent through T1 to synchronous FET demodulator Q6-Q7. R7 adjusts system scaling.—C. A. Walton, High-CMR, Low-Cost DC Instrumentation Preamp, *EDN Magazine*, Jan. 15, 1971, p 47—48.

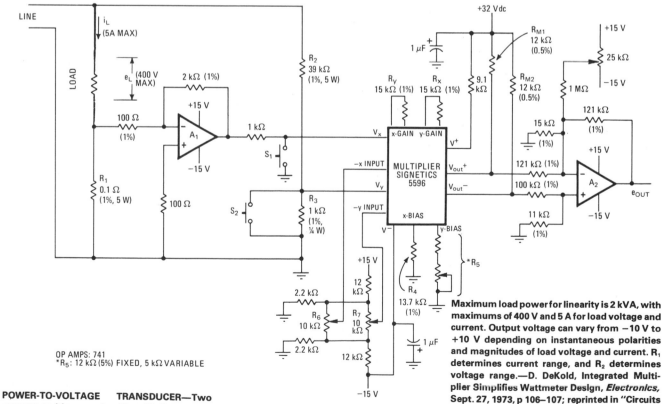

OP AMPS: 741
*R5: 12 kΩ (5%) FIXED, 5 kΩ VARIABLE

POWER-TO-VOLTAGE TRANSDUCER—Two opamps and inexpensive IC multiplier provide output voltage directly proportional to instantaneous power through load. Frequency response extends from DC to several kilohertz. Maximum load power for linearity is 2 kVA, with maximums of 400 V and 5 A for load voltage and current. Output voltage can vary from −10 V to +10 V depending on instantaneous polarities and magnitudes of load voltage and current. R1 determines current range, and R2 determines voltage range.—D. DeKold, Integrated Multiplier Simplifies Wattmeter Design, *Electronics*, Sept. 27, 1973, p 106—107; reprinted in "Circuits for Electronics Engineers," *Electronics*, 1977, p 175—176.

CHAPTER 41
Integrator Circuits

Provide output that is integral of input with respect to time. Special features include logic-reset, analog start/stop, fast dump, and fast recovery.

INTEGRATE AND DUMP—Transistor is used as switch, without power supply. Simple RC integrator will dump (discharge C) completely in about 1 μs when dump input is logic 1 (+5 V). Values of R and C determine time constant of integrator. Without power supply, circuit can only drive high-impedance load; for low-impedance load, add FET-input opamp such as Analog Devices 40J as buffer.—R. Riordan, Integrate and Dump Circuit Uses No Power Supply, *EDN Magazine,* Feb. 20, 1973, p 93.

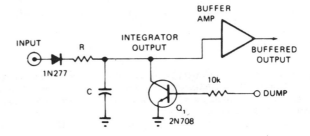

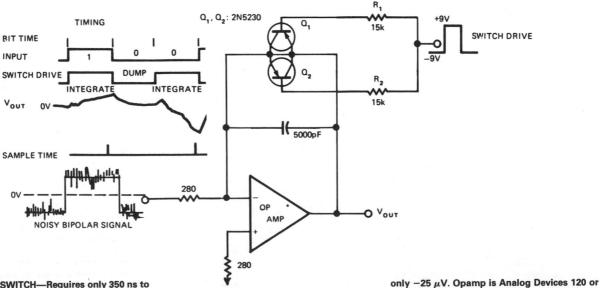

FAST DUMP SWITCH—Requires only 350 ns to dump 6-V output to level of 3 mV. Transistors are connected so one of them is biased in forward mode independently of output polarity. Both transistors turn on during dump interval. Transistor operating in forward mode determines initial discharge rate until it saturates, after which inverted-mode transistor continues to discharge capacitor. Offset voltage error is only −25 μV. Opamp is Analog Devices 120 or equivalent having unity-gain bandwidth above 100 MHz and slew rate above 200 V/μs.—F. Tarico, Fast Bipolar Dump Switch Has Low Offset, *EDN Magazine,* Nov. 5, 1974, p 66.

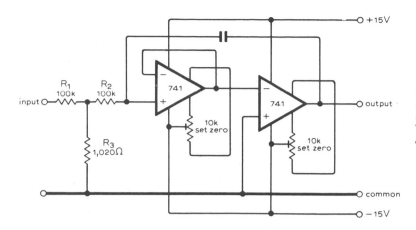

HIGH INPUT IMPEDANCE—Two-opamp connection shown gives input resistance of 200 megohms and drift time of 90 min for 0 to 10 V.—N. G. Boreham, Op-Amp Integrator, *Wireless World,* March 1977, p 42.

FAST RECOVERY—Two diodes and two zeners clamp output of integrator below saturation level of opamp, making recovery time approximately equal to slew rate. With values shown, integration time constant R_3C_1 is 35.6 ms and output is clamped at +23 V. Output linearity is ±1%, and threshold range of circuit is −3 V to −10 V.—K. S. Wong, Fast-Recovery Integrator with Adjustable Threshold, *EEE Magazine,* Aug. 1970, p 77.

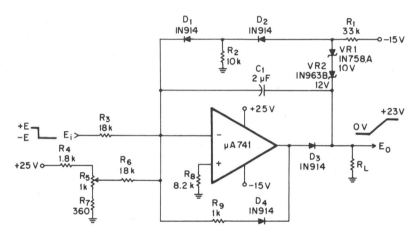

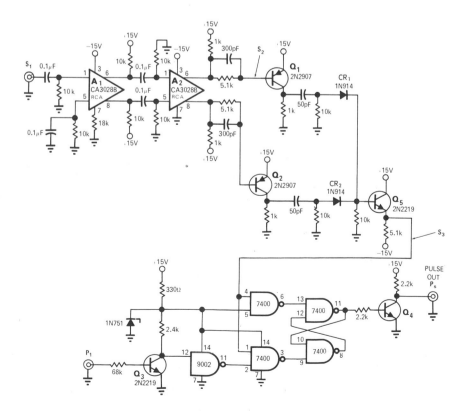

ELIMINATING RINGING—High-Q bandpass integrator reduces ringing significantly by amplifying AC input signal in two broadband differential amplifier stages, A_1 and A_2, before differentiation and selection by D_1 and D_2. Input gate pulse P_1 and differentiated pulses drive AND logic that generates output pulse P_s coincident with zero crossings of AC input. Leading edge of P_s will always occur at first zero crossing after P_1 initiates gating action. If output drives balanced diode bridge, gating pedestal and AC signal transient are eliminated; high-Q bandpass integrator then has fast settling time, permitting faster repetition rates.—R. J. Turner, Reduce Integrator Transients with Synchronized Gate Signals, *EDN/EEE Magazine,* Jan. 15, 1972, p 46–47.

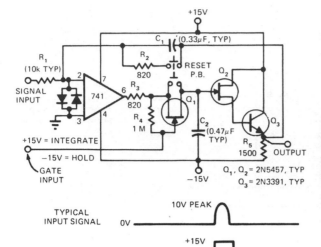

STORING INTEGRATOR OUTPUT—Modified sample-and-hold circuit with capacitive feedback combines integrate, sample, and hold functions, for use in temporarily storing output of integrator. Integrating amplifier is 741 opamp; for critical applications, FET opamp is preferable.—E. Crovella, Circuit Combines Integrate, Sample and Hold Functions, *EDN Magazine*, Oct. 20, 1974, p 90.

JFET WITH AC COUPLING—Connection shown gives very high voltage gain. Use of C1 as Miller integrator or capacitance multiplier allows simple circuit to handle very long time constant.—"FET Databook," National Semiconductor, Santa Clara, CA, 1977, p 6-26—6-36.

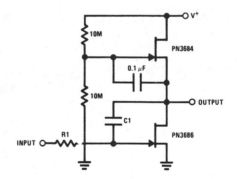

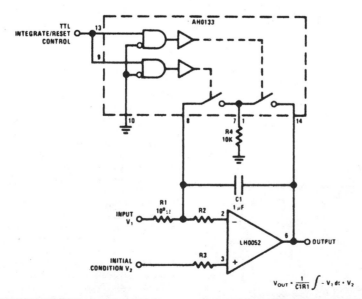

$$V_{OUT} = \frac{1}{C1R1}\int -V_1\,dt + V_2$$

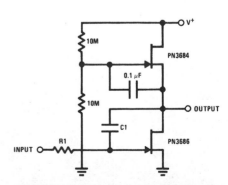

PRECISION INTEGRATOR—Low input bias current of National LH0052 opamp makes it suitable for applications requiring long time constants. R1 is selected so total leakage current at summing mode is sufficiently smaller than signal current to ensure required accuracy. R2 is included to protect input circuit during reset transient but can be omitted for low-speed applications. R3, used to balance resistance in inputs, should be equal to sum of R2 and 100-ohm resistance of reset switch.—"Linear Applications, Vol. 1," National Semiconductor, Santa Clara, CA, 1973, AN-63, p 1–12.

AC-COUPLED JFET—Use of C1 as Miller integrator or capacitance multiplier allows simple circuit to handle very long time constant while providing high voltage gain. Circuit also offers low distortion with low noise and high dynamic range.—"FET Databook," National Semiconductor, Santa Clara, CA, 1977, p 6-26.

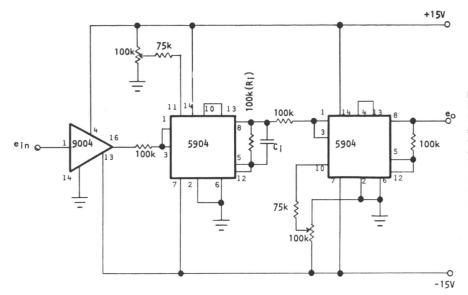

RMS CONVERTER—Converts analog voltage to RMS equivalent by squaring operation followed by integration and square-rooting. Bipolar input signal is first converted into linear absolute value with Optical Electronics 9004 absolute-value module, as required for processing by 5904 unipolar devices. Pots are used to establish 10-V full-scale level. R_iC_i is integration time constant.—"Simple RMS Converter," Optical Electronics, Tucson, AZ, Application Tip 10246.

ANALOG START/STOP AND RESET—One section of DG300 dual analog switch serves for discharging integrator capacitor C through R_L when resetting integrator, with start/stop switch section being held open by control logic. When both switches are open, output of integrator is held.—"Analog Switches and Their Applications," Siliconix, Santa Clara, CA, 1976, p 7-81.

$$V_o = 1/RC \int V_{IN}\, dt$$

$R = 50\,\Omega$
$50 < R_L < 100$ FOR 15 VOLT OUTPUT SWINGS

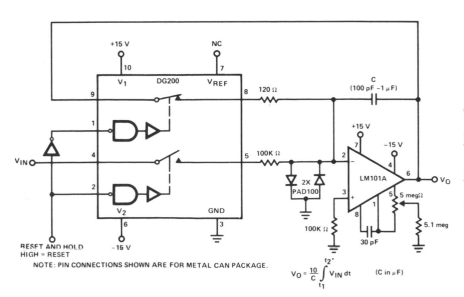

LOGIC-RESETTABLE INTEGRATOR—DG200 CMOS analog switch serves for discharging integrator capacitor C rapidly for high logic input pulse. Other section of switch disconnects integrator from analog input when logic goes high. When logic input is returned to low, integrator is triggered. Diodes prevent capacitor from charging to over 15 V.—"Analog Switches and Their Applications," Siliconix, Santa Clara, CA, 1976, p 7-68.

RESET AND HOLD
HIGH = RESET

NOTE: PIN CONNECTIONS SHOWN ARE FOR METAL CAN PACKAGE.

$$V_O = \frac{10}{C} \int_{t_1}^{t_2} V_{IN}\, dt \qquad (C \text{ in } \mu F)$$

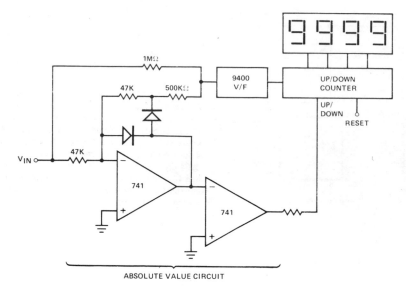

ABSOLUTE VALUE CIRCUIT

BIPOLAR INPUT FOR V/F CONVERTER—Absolute-value integrator circuit gives effect of generating negative frequencies when input signal is negative by making counter count up for positive voltage and count down for negative voltage. Diode types are not critical.—M. O. Paiva, "Applications of the 9400 Voltage to Frequency Frequency to Voltage Converter," Teledyne Semiconductor, Mountain View, CA, 1978, AN-10, p 3.

CHAPTER **42**

Intercom Circuits

Covers one-way and two-way basic intercom circuits, four-station two-way system, induction receiver for paging, and private telephone system. Audio amplifier circuits suitable for intercoms are also given.

AUDIO INDUCTION RECEIVER—Used to pick up audio signal being fed to low-impedance single-wire loop encircling room or other area to be covered. Pickup loop L1 is 100–500 turns wound around plastic case of receiver. Opamp sections are from Motorola MC3401P or National LM3900 quad opamp. Supply can be 9–15 V. Requires no FCC license. Can be used as private paging system if audio amplifier of transmitter has microphone input.—C. D. Rakes, "Integrated Circuit Projects," Howard W. Sams, Indianapolis, IN, 1975, p 23–25.

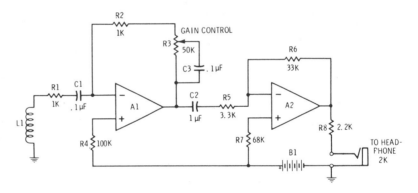

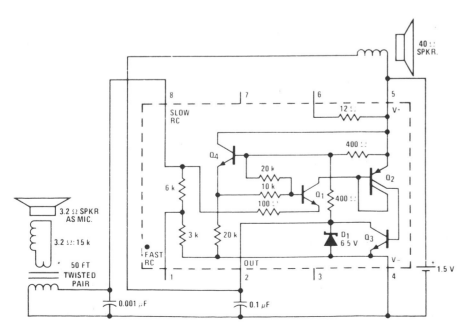

MICROPOWER ONE-WAY INTERCOM—National LM3909 IC operating from single 1.5-V cell serves as low-power one-way intercom suitable for listening-in on child's room and meeting other room-to-room communication needs. Battery drain is only about 15 mA. Person speaking directly into 3.2-ohm loudspeaker used as microphone delivers full 1.4 V P-P signal to 40-ohm loudspeaker at listening location.—"Linear Applications, Vol. 2," National Semiconductor, Santa Clara, CA, 1976, AN-154, p 9.

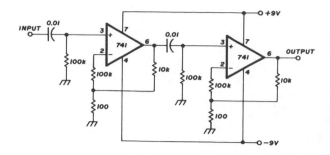

CASCADED 741 OPAMPS—Two opamps in series provide 80 dB of audio gain with bandwidth of about 300 to 6000 Hz. Gain of each opamp is set at 100. With three stages, bandwidth would be 5100 Hz. Output will drive loudspeaker at comfortable room level, if fed through 1-μF nonpolarized capacitor to output transformer having 500-ohm primary and 8-ohm secondary.—C. Hall, Circuit Design with the 741 Op Amp, *Ham Radio*, April 1976, p 26–29.

2 W WITH IC—Inexpensive audio amplifier using 14-pin DIP provides adequate power for small audio projects and audio troubleshooting. Pins 3, 4, 5, 10, 11, and 12 are soldered directly to foil side of printed-wiring board used for construction, to give effect of heatsink.—J. Schultz, An Audio Circuit Breadboarder's Delight, *CQ*, Jan. 1978, p 42 and 75.

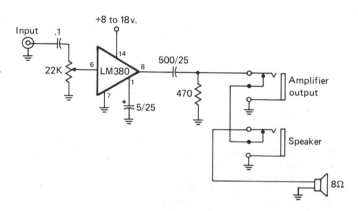

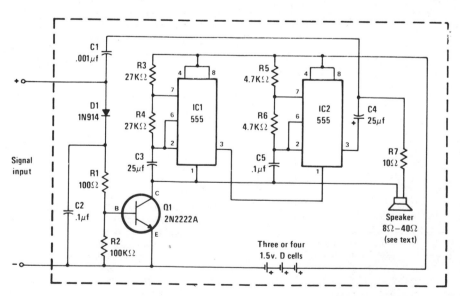

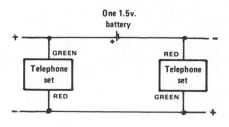

BEEPER—Private two-station telephone system for home requires only two wires between ordinary telephone sets, with 1.5-V battery in series with one line, but this voltage is not enough to actuate ringers in sets. Beeper in parallel with each set, with polarity as shown, serves same purpose as ringer. 555 timer IC1 turns on IC2 about once every 3 s, and IC2 then generates 1000-Hz beep for about 1 s as ringing signal. No switches are required, because telephone handsets provide automatic switching. When both telephones are hung up, 1.5-V battery splits equally between beepers and resulting 0.75 V is not enough to turn on Q1 in either set. When one telephone is picked up, beeper at other telephone receives close to 1.5 V and Q1 turns on IC1 to initiate beeping call. When other telephone is picked up, beeping automatically stops because 1.5 V is again divided between sets.—P. Stark, Private Telephone: Simple Two-Station Intercom, *Modern Electronics*, July 1978, p 32–34.

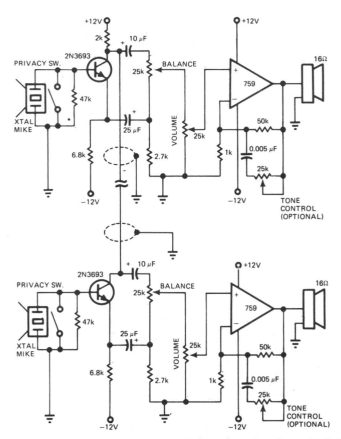

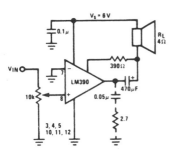

1 W AT 6 V—Battery-operated power amplifier using National LM390 IC provides ample power for loudspeaker despite operation from 6-V portable battery.—"Audio Handbook," National Semiconductor, Santa Clara, CA, 1977, p 4-41.

BIDIRECTIONAL INTERCOM—Uses 759 power opamps to provide 0.5 W for 16-ohm loudspeakers. Crystal microphones feed NPN transistors that provide both in-phase and 180° out-of-phase signals. Balance-adjusting circuits of amplifier cancel out the two signals, so only out-of-phase signal goes to receiving unit. Privacy switch across microphone eliminates audio feedback while listening. Article tells how to calculate heatsink requirements.—R. J. Apfel, Power Op Amps—Their Innovative Circuits and Packaging Provide Designers with More Options, *EDN Magazine,* Sept. 5, 1977, p 141–144.

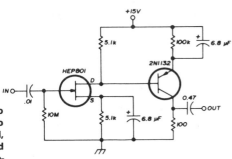

DIRECT-COUPLED AF—Combination of unipolar and bipolar transistors gives desirable amplifying features of each solid-state device. Can be used as speech amplifier and for other low-level audio applications.—I. M. Gottlieb, A New Look at Solid-State Amplifiers, *Ham Radio,* Feb. 1976, p 16–19.

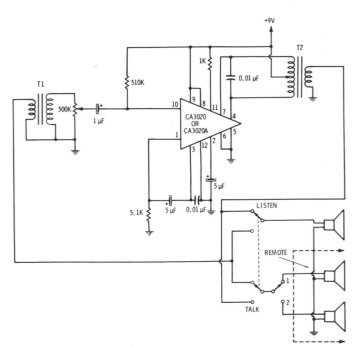

SINGLE IC WITH TRANSFORMERS—CA3020 differential amplifier uses AF input transformer T1 to match loudspeakers (used as microphone) to higher input resistance of IC. AF output transformer T2 similarly matches IC to loudspeakers operating conventionally.—E. M. Noll, "Linear IC Principles, Experiments, and Projects," Howard W. Sams, Indianapolis, IN, 1974, p 100–101.

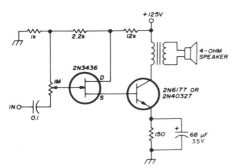

AF OUTPUT—Operates directly from 125-V rectified AC line voltage. Combination of unipolar and bipolar transistors gives desirable amplifying features of each solid-state device.—I. M. Gottlieb, A New Look at Solid-State Amplifiers, *Ham Radio,* Feb. 1976, p 16–19.

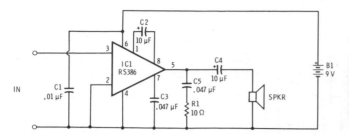

0.25-W AMPLIFIER—Single Radio Shack RS386 IC powered by 6–9 V from battery provides gain of about 200 with sufficient power to drive 8-ohm loudspeaker when speaking closely into small dynamic microphone of type used with portable tape recorders.—F. M. Mims, "Integrated Circuit Projects, Vol. 2," Radio Shack, Fort Worth, TX, 1977, 2nd Ed., p 87–95.

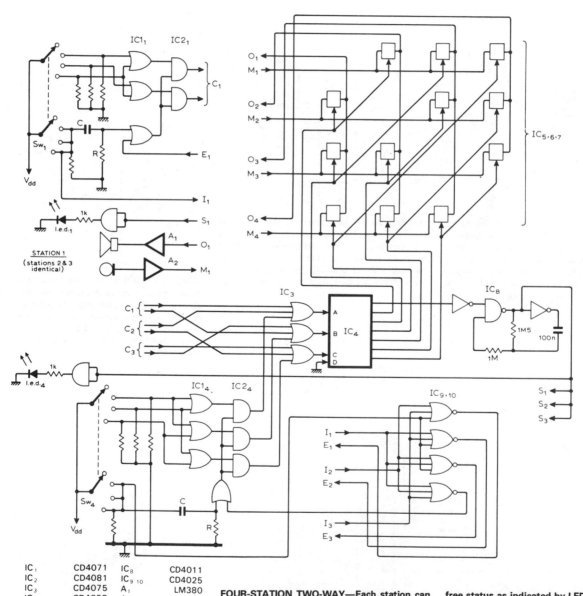

IC₁	CD4071	IC₈	CD4011
IC₂	CD4081	IC₉,₁₀	CD4025
IC₃	CD4075	A₁	LM380
IC₄	CD4028	A₂	741
IC₅,₆,₇	CD4016		

Station links	Code
1 to 2	001
1 to 3	010
1 to 4	011
2 to 3	100
2 to 4	101
3 to 4	110

FOUR-STATION TWO-WAY—Each station can communicate privately with any one of others. All four stations have identical inputs as at upper left, with fourth station having master circuit. Each two-station combination is assigned 3-bit code as given in table, for selection by switches Sw₁–Sw₄. All station codes are ORed and decoded by IC₄ to drive matrix of analog switches for coupling appropriate audio inputs and outputs. Code 000 is used for system-free status as indicated by LEDs 1 and 4 being on. LEDs flash for system-busy status. When code is selected, enable inputs of nonselected stations go low to prevent generation of further codes. System can be expanded to six stations by using 4-bit code and CD4514 decoder with larger matrix of analog switches.—B. Voynovich, Multiple Station Two-Way Intercom, *Wireless World,* March 1978, p 59.

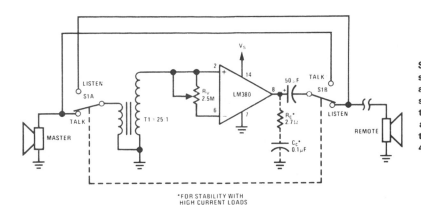

SINGLE OPAMP—When switch S1 is in talk position as shown, loudspeaker of master station acts as microphone, driving opamp through step-up transformer T1. Switch at remote station must then be in listen position. Supply voltage range is 8–20 V.—"Audio Handbook," National Semiconductor, Santa Clara, CA, 1977, p 4-21–4-28.

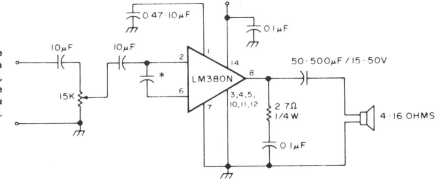

2-W LM380 POWER AMPLIFIER—Complete basic circuit for most audio or communication purposes uses minimum of external parts. C3, used to limit high frequencies, can be in range of 0.005 to 0.05 μF.—A. MacLean, How Do You Use ICs?, *73 Magazine*, June 1977, p 184–187.

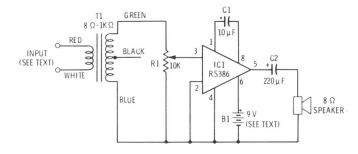

0.5-W AF IC—Simple audio power stage drives 8-ohm loudspeaker for producing greater volume with pocket radio or for intercom applications. Supply range is 4–12 V. For long life, 6-V lantern batteries are recommended. Transformer is Radio Shack 273-1380.—F. M. Mims, "Integrated Circuit Projects, Vol. 5," Radio Shack, Fort Worth, TX, 1977, 2nd Ed., p 38–44.

HIGH-GAIN INTERCOM—Internal bootstrapping in National LM388 audio power amplifier IC gives output power levels above 1 W at supply voltages in range of 6–12 V, with minimum parts count. AC gain is set at about 300 V/V, eliminating need for step-up transformer normally used in intercoms. Optional RC network suppresses spurious oscillations.—"Audio Handbook," National Semiconductor, Santa Clara, CA, 1977, p 4-37–4-41.

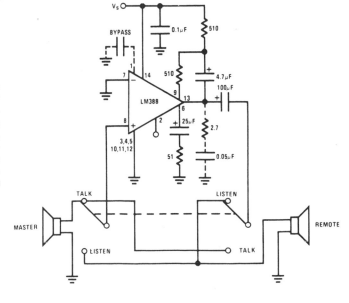

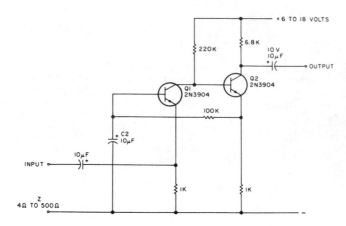

LOW-Z INPUT—Can be used with low-impedance source, such as 4- to 16-ohm loudspeaker or telephone earphone used as mike. If loudspeaker is put out in yard, sensitivity is sufficient to pick up sounds made by prowlers. Can be fed into input of any high-fidelity amplifier.— E. Dusina, Build a General Purpose Preamp, *73 Magazine,* Nov. 1977, p 98.

CHAPTER 43

Keyboard Circuits

Includes interface circuits required for converting keyboard operation to
ASCII, BCD, hexadecimal, teleprinter, Baudot, Morse code, and other formats
serving as inputs for microprocessors, PROMs, CW transmitters, hard-copy
printers, TV typewriters, and other code-driven applications. See also
Microprocessor and Telephone chapters.

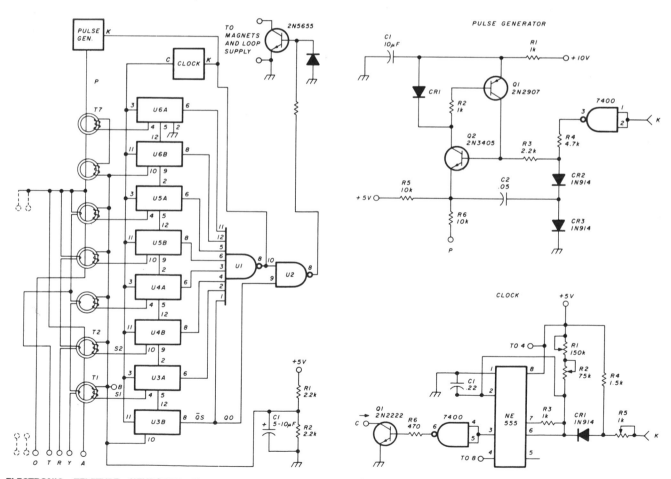

ELECTRONIC TELETYPE KEYBOARD—Uses eight 7474 shift register sections in combination with pulse generator that is discharged through appropriate toroid core to create correct mark-space coding for energizing magnet drivers of Teletype. Developed to permit communication by handicapped people. Simplified keyboard has one set of alphabetic characters and five numerics for BCD input. Outputs of shift registers drive 7430 NAND gate U2. If keyboard is to be used at 60 WPM, adjust R1 so 555 oscillates at 45 Hz. Toroids are Indiana General CF-102 having 10-turn primaries.—L. A. Stapp, Electronic Teleprinter Keyboard, *Ham Radio,* Aug. 1978, p 56—57.

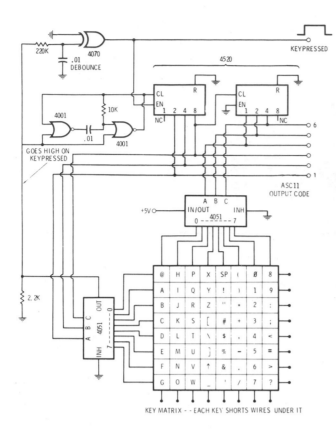

SCANNING ASCII ENCODER—Converts key action into composite parallel ASCII code. Circuit includes debouncing and two-key rollover. Two 4001 quad two-input NOR gate sections form 50-kHz clock that is gated. When clock is allowed to run, two cascaded 4520 binary counters are driven for continuous cycling through all their counts. Slower counter produces 1-of-8 decoded output for 4051 1-of-8 switch. Faster counter drives second switch that monitors sequential rows of characters. When key is pressed, output from +5 V through both selectors stops gated oscillator and holds count. Resulting ASCII output is then routed to external output logic for control and shift operations. When key is released, scanning resumes and continues until new key is pressed. If second key is pressed before first is released, nothing happens until first key is released. Scanning then resumes and stops at second key location, to give two-key rollover permitting faster typing with minimum error.—D. Lancaster, "CMOS Cookbook," Howard W. Sams, Indianapolis, IN, 1977, p 358–359.

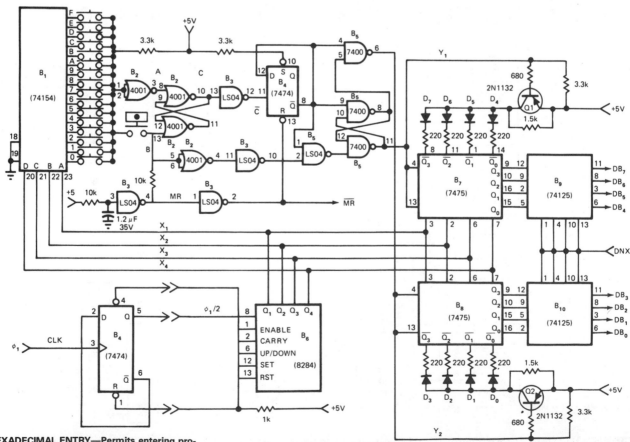

HEXADECIMAL ENTRY—Permits entering program into microprocessor in hexadecimal notation, in much less time than is required with binary notation. Binary switches from input port of μP are replaced with 16-key keyboard shown. To enter hexadecimal number 3B, turn on power; press 3 button, press decimal bar, then press B button; operate loading switch, then press decimal bar again to set keyboard for next entry. Article traces keyboard operation through circuit.—B. K. Erickson, Talk to Your μP with a Hex-Latching Keyboard, *EDN Magazine*, Nov. 20, 1976, p 319–320.

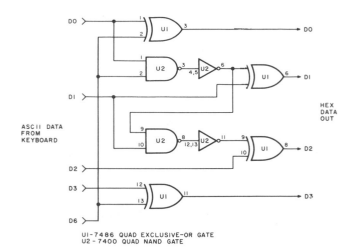

ASCII TO HEX CONVERTER—Simple two-chip circuit takes ASCII data from keyboard and converts characters 0–9 and A-F to 4-bit hexadecimal machine-language format, as required for loading operating system initially into 1802-based microprocessor system. Once loaded, further code conversion can be achieved with software.—E. Copes, One Keyboard: Hex and ASCII, *Kilobaud*, June 1978, p 57.

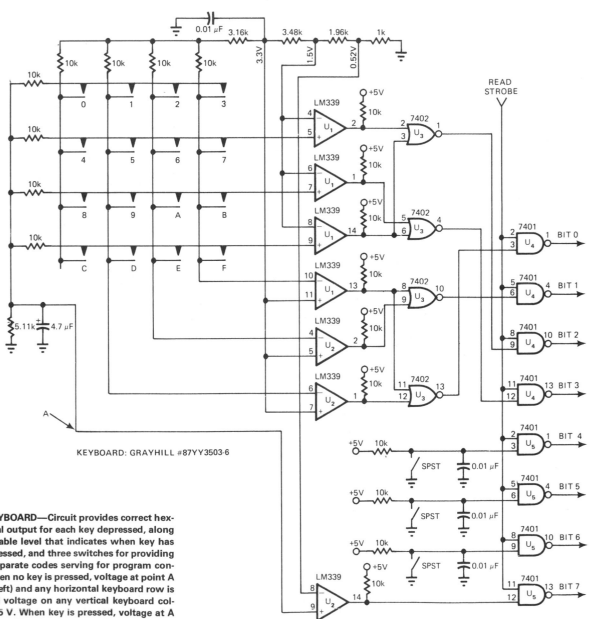

HEX KEYBOARD—Circuit provides correct hexadecimal output for each key depressed, along with enable level that indicates when key has been pressed, and three switches for providing eight separate codes serving for program control. When no key is pressed, voltage at point A (lower left) and any horizontal keyboard row is 0 V and voltage on any vertical keyboard column is 5 V. When key is pressed, voltage at A becomes 1 V and voltage at selected row and column becomes 3 V. Comparator outputs for U₁ and U₂ are decoded with NOR gates and strobed onto data lines with open-collector NAND gates.—J. F. Czebiniak, Simple Hex Keyboard Provides Program Control, *EDN Magazine*, Jan. 5, 1978, p 27–28.

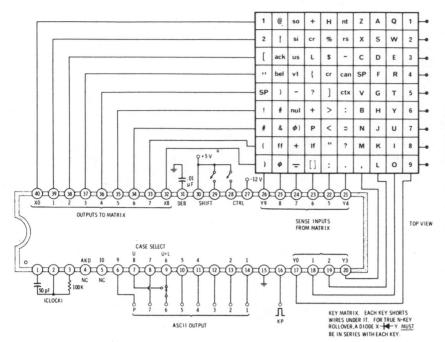

SCANNING KEYBOARD ENCODER—Uses RCA CA3600 IC for sampling 90 keys and providing parallel ASCII output with parity. Network on pins 1-3 sets clock at 50 kHz. Capacitor on pin 31 sets debounce time at about 8 ms. Uppercase only or both uppercase and lowercase can be selected by switching between pins 7 and 9. Provides two-key rollover; N-key rollover can be obtained by adding diodes to keys. Each output will drive one TTL load.—D. Lancaster, "TV Typewriter Cookbook," Howard W. Sams, Indianapolis, IN, 1976, p 38.

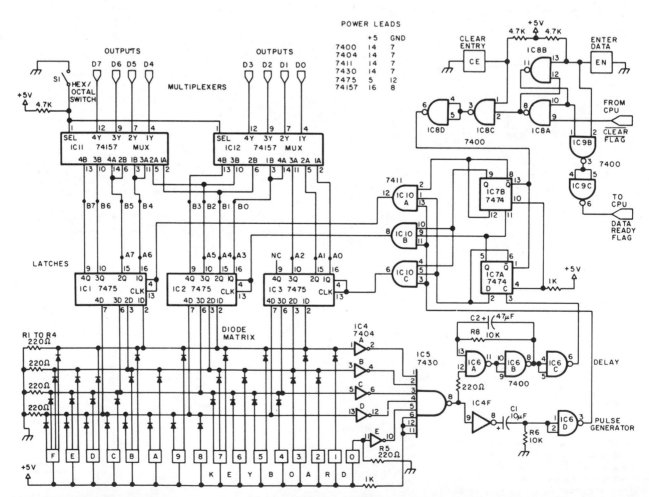

CALCULATOR-KEYBOARD INPUT—Uses diode matrix that encodes 16 hexadecimal input keys as 4-bit code for microprocessor. Register holds conversion results. Multiplexer gives switch-selected 3-digit octal or 2-digit hexadecimal interpretation to inputs. Control logic serves for keyboard debouncing, clearing, and entering data. Circuit eliminates need for entering programs with front-panel switches, by using keyboard to enter data in octal or hexadecimal form. Choice of form is achieved with 74157 multiplexers IC11 and IC12, set by S1. Article covers circuit operation in detail and gives 8008 full keyboard input program that defines memory address with first 2 bytes, then enters loop that loads memory byte by byte in ascending address sequence. RGS-008A interface logic is used to control interface for RGS-008A computer.—J. Hoegerl, Calculator Keyboard Input for the Microcomputer, *BYTE*, Feb. 1977, p 104–107.

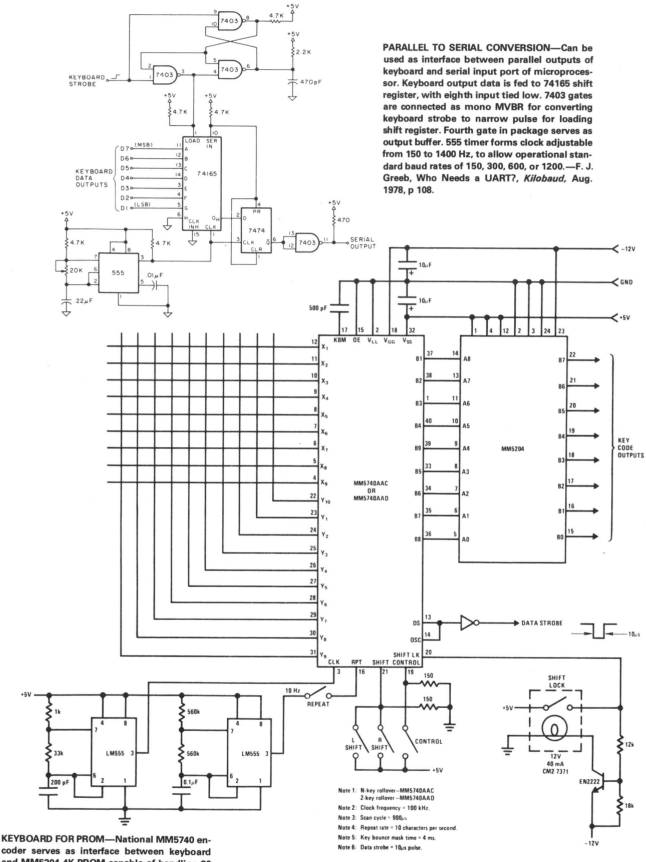

PARALLEL TO SERIAL CONVERSION—Can be used as interface between parallel outputs of keyboard and serial input port of microprocessor. Keyboard output data is fed to 74165 shift register, with eighth input tied low. 7403 gates are connected as mono MVBR for converting keyboard strobe to narrow pulse for loading shift register. Fourth gate in package serves as output buffer. 555 timer forms clock adjustable from 150 to 1400 Hz, to allow operational standard baud rates of 150, 300, 600, or 1200.—F. J. Greeb, Who Needs a UART?, *Kilobaud,* Aug. 1978, p 108.

Note 1: N-key rollover—MM5740AAC
         2-key rollover—MM5740AAD
Note 2: Clock frequency = 100 kHz.
Note 3: Scan cycle = 900μs.
Note 4: Repeat rate = 10 characters per second.
Note 5: Key bounce mask time = 4 ms.
Note 6: Data strobe = 10μs pulse.

KEYBOARD FOR PROM—National MM5740 encoder serves as interface between keyboard and MM5204 4K PROM capable of handling 90 four-mode keys. Encoder includes all logic needed for key validation, two-key or N-key rollover, bounce masking, mode selection, and strobe generation. Key code outputs can be defined by user. Bit-paired coding system of encoder has five common bits (B1-B4 and B9) and four variable bits (B5-B8) for each key. Each keyswitch is defined by one X drive line and one Y sense line of encoder. Combination gives total of 360 9-bit codes.—"Memory Applications Handbook," National Semiconductor, Santa Clara, CA, 1978, p 5-5—5-8.

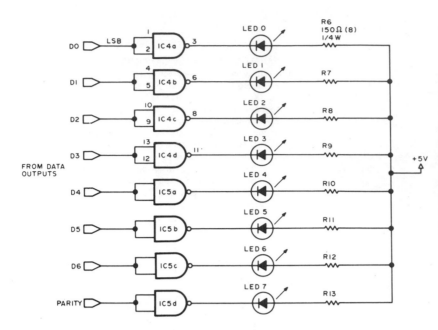

LED DISPLAY FOR ASCII CODE—Shows code for any ASCII keyboard character as aid in debugging keyboard matrix wiring. Requires only two 7400 quad NAND gates, eight resistors, and eight LEDs. High input to gate forces output low, grounding LED and lighting it. On standby with positive logic, all LEDs are lit; with negative logic, all are dark on standby. Arrange LEDs on keyboard so rightmost represents least significant bit (DO) and leftmost LED represents parity bit. Using positive logic, ASCII code for any depressed key will show as lit LED for 1 and dark LED for 0. Pin 14 of 7400 is +5 V, and pin 7 is ground.—B. Brehm, Using a Keyboard ROM, *BYTE*, May 1977, p 76–82.

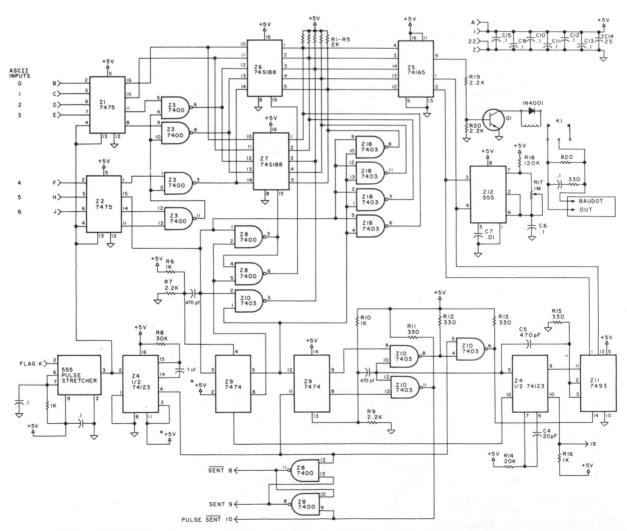

ASCII TO BAUDOT CONVERTER—Converts parallel ASCII output of keyboard to serial Baudot format for driving Baudot teleprinters. Conversion circuit can save several hundred bytes of computer or microprocessor storage, making this memory space available for other purposes. Article gives conversion table for all capital and lowercase alphabetic characters, numerics, and punctuation.—J. A. Lehman and R. Graham, ASCII to Baudot ... er ... Murray (the Hard Way), *Kilobaud*, June 1978, p 80–83.

TYPEWRITER KEYS CW TRANSMITTER—Allows operator to send perfect Morse code simply by typing messages on alphameric keyboard. Accuracy is determined only by typing skill of operator. One RC oscillator controls mark-space ratio of Morse characters and duration of character and word spaces, while clock oscillator is divided down and switched to give sending-speed choices of 6, 12, 24, and 48 words per minute. Each square on keyboard diagram is SPST switch, and each cross at intersection is silicon switching diode such as 1N914. Outputs of keyboard switches are converted into 15-bit code for feeding into 64-character first-in first-out memory using four MP3812B ICs for storage and for generating corresponding Morse characters. Article gives construction and adjustment details.—C. I. B. Trusson, Morse Keyboard and Memory, *Wireless World*, Jan. 1977, p 55—59.

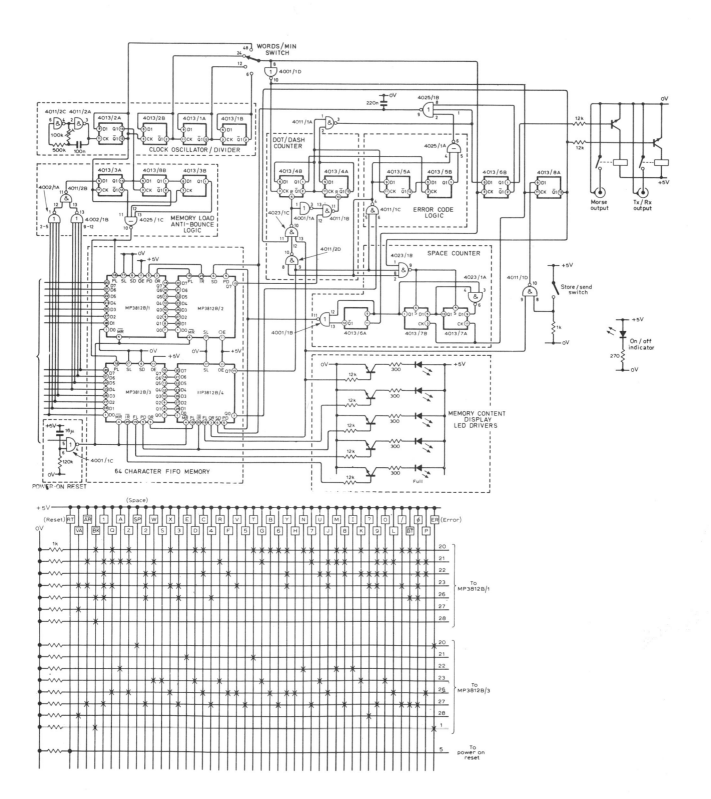

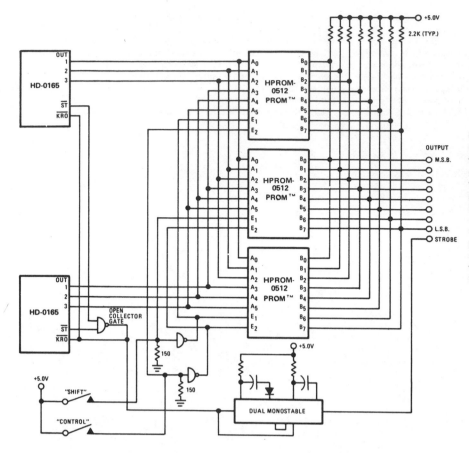

UNIVERSAL KEYBOARD ENCODER—Harris HD-0165 keyboard encoders are combined with three read-only memories wired in parallel to generate universal code that can be translated to desired output code by using suitable electric programming of ROMs. Key-to-encoder wiring is arbitrary as long as each key operation produces unique 6-bit output code from encoders. One ROM is programmed to contain all 64 output words for unshifted mode, second for shifted mode, and third ROM contains all words for control mode.—"Linear & Data Acquisition Products," Harris Semiconductor, Melbourne, FL, Vol. 1, 1977, p 7-2–7-8 (Application Note 204).

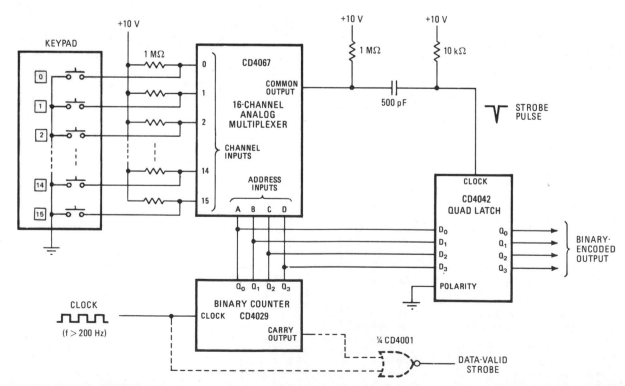

DATA-ENTRY KEYPAD ENCODER—Designed for use in point-of-sale and other data-entry applications requiring bounceless binary encoder. Keypad is scanned by CD4067 16-channel analog multiplexer. When key is depressed, appropriate channel is driven low. When counter addresses low input of multiplexer, common output goes low for one clock cycle. Differentiating network changes this to negative spike for strobing counter data into quad latch where it remains until another key is depressed. Clock should be above 200 Hz.—M. E. Keppel, Multiplexer Scans Keyboard for Reliable Binary Encoding, *Electronics,* March 17, 1977, p 99.

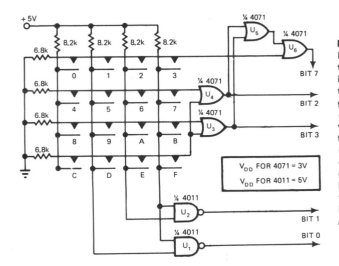

HEX ENCODER FOR KEYBOARD—Standard 16-button keyboard is used with CMOS packages to provide hex output bits. Pressing one button increases voltage on its horizontal line from 0 to 2.2 V. If button is in other than first row, OR gate U_3 or U_4 is activated and either bit 2 or 3 changes from LOW to HIGH. Simultaneously, voltage of vertical column of button decreases from 5 V to 2.2 V. If button is in other than first column, NAND gate U_1 or U_2 is activated and either bit 0 or 1 changes from LOW to HIGH. OR gates U_5 and U_6 provide button-pressed signal by changing bit 7 from LOW to HIGH upon activation of any button.—W. H. Hailey, CMOS Logic Implements Keyboard Encoder, *EDN Magazine*, Aug. 5, 1978, p 54 and 56.

SCANNING KEYBOARD WITH MEMORY—Standard keyboard is updated to 53 keys as required for character feed in microprocessor-based TV typewriter system. Features include two-key rollover, along with choices of letter case, strobe polarity, clear key, and repeat key. Space output is produced automatically during clear.—D. Lancaster, "TV Typewriter Cookbook," Howard W. Sams, Indianapolis, IN, 1976, p 150–151.

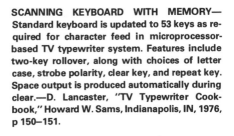

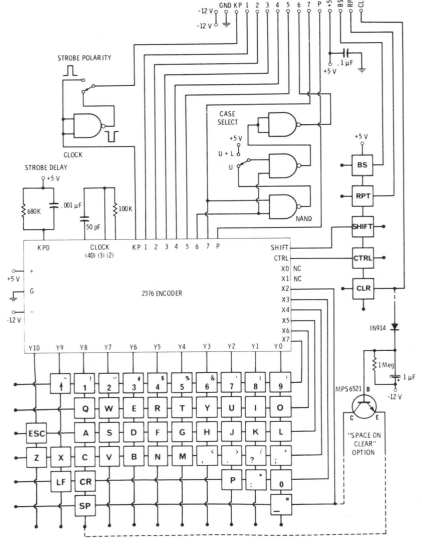

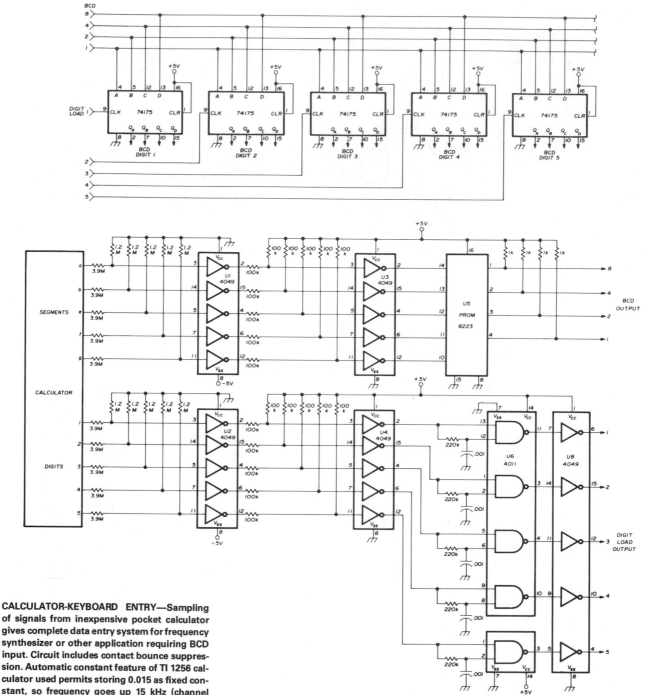

CALCULATOR-KEYBOARD ENTRY—Sampling of signals from inexpensive pocket calculator gives complete data entry system for frequency synthesizer or other application requiring BCD input. Circuit includes contact bounce suppression. Automatic constant feature of TI 1256 calculator used permits storing 0.015 as fixed constant, so frequency goes up 15 kHz (channel spacing on 2 meters) each time + key is pressed. Seven-segment signal normally used for calculator display is decoded by programming 74S188 (8223) PROM, using simplified code conversion table given in article. 74175 ICs provide required demultiplexing of segment data. Article gives construction details.—B. McNair and G. Williman, Digital Keyboard Entry System, *Ham Radio,* Sept. 1978, p 92–97.

CHAPTER 44
Lamp Control Circuits

Covers methods of triggering triacs and silicon controlled rectifiers for turning on, dimming, and otherwise regulating lamp loads in response to photoelectric, acoustic, logic, or manual control at input. Starting circuits for fluorescent lamps are also given.

AC CONTROL WITH TRIAC—Decoder outputs of microprocessor feed 7476 JK flip-flop that drives optocoupler which triggers triac for ON/OFF control of lamp or other AC load. LED and cadmium sulfide photocell are mounted in light shield. When light from LED is on photocell, cell resistance drops and allows control voltage of correct direction and amplitude to trigger gate of triac, turning it on. When light disappears, triac remains on until voltage falls near zero in AC cycle.—R. Wright, Utilize ASCII Control Codes!, *Kilobaud,* Oct. 1977, p 80–83.

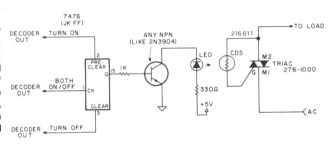

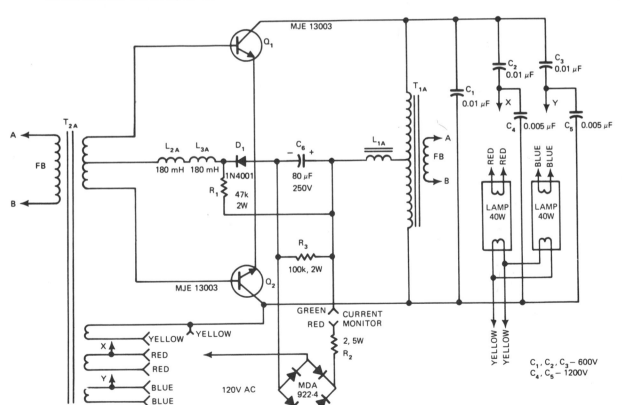

40-W RAPID-START BALLAST—AC line voltage is rectified by diode bridge and filtered by C_6-L_{1A}. Transistors Q_1-Q_2 with center-tapped tank coil T_{1A} and C_1-C_3 make up power stage of 20-kHz oscillator that develops 600 V P sine wave across T_{1A}. When fluorescent lamps ionize, current to each is limited to about 0.4 A. Lamps operate independently, so one stays on when other is removed. Feedback transformer T_{2A} supplies base drive for transistors and filament power for lamps. Article gives transformer and choke winding data.—R. J. Haver, The Verdict Is In: Solid-State Fluorescent Ballasts Are Here, *EDN Magazine,* Nov. 5, 1976, p 65–69.

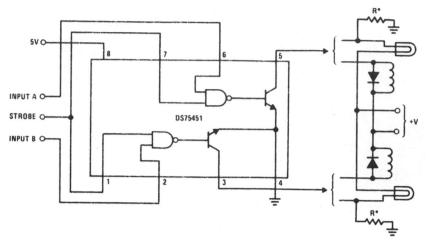

DUAL LAMP DRIVER—National DS75451 dual peripheral AND driver using positive logic provides up to 300 mA per section for driving incandescent lamps. Optional keep-alive resistors R maintain OFF-state lamp current at about 10% of rated value to reduce surge current. Lamp voltage depends on lamps used. Relays shown, with diodes across solenoids, can be used in place of lamps if desired.—"Interface Databook," National Semiconductor, Santa Clara, CA, 1978, p 3-20—3-30.

COMPLEMENTARY FADER—Control unit for stage lighting fades out one lamp while simultaneously increasing light output of another with accurate tracking. Gate of silicon controlled rectifier SCR₁ is driven by standard external phase control circuit. Interlock network connected to output of SCR₁ provides complementary signal for trigger of SCR₂. If lamps larger than 150 W are required, use larger value for C_1.—M. E. Anglin, Complementary Lighting Control Uses Few Parts, *Electronics*, Dec. 12, 1974, p 111; reprinted in "Circuits for Electronics Engineers," *Electronics*, 1977, p 78.

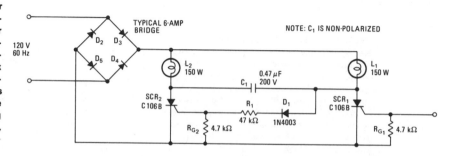

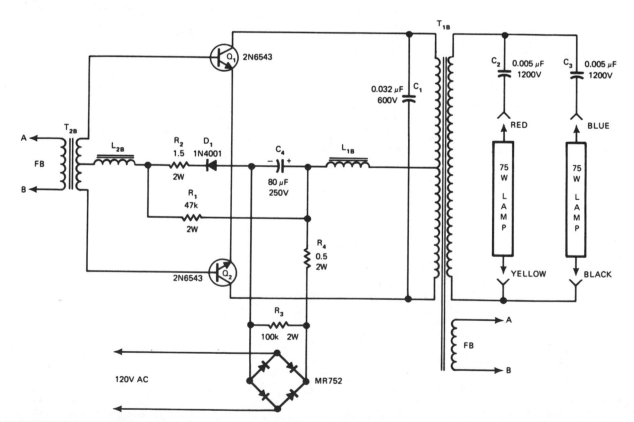

75-W INSTANT-START BALLAST—DC voltage for 20-kHz two-transistor oscillator is obtained from AC line. Secondary is added to center-tapped tank coil of T_{1B} to provide 1 kV P starting voltage required by 96-inch instant-start lamps. Article gives transformer and choke winding data along with circuit details and performance data. Lamps operate independently, so one stays on when other is removed.—R. J. Haver, The Verdict Is In: Solid-State Fluorescent Ballasts Are Here, *EDN Magazine*, Nov. 5, 1976, p 65—69.

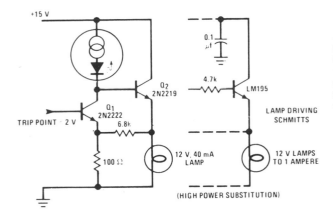

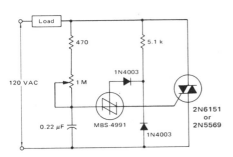

ACTIVE LOAD—National NSL4944 constant-current LED serves as current source for collector resistor of Schmitt trigger to provide up to 12-V output at 40 mA for lamp load. When lamp and Q_2 are off, most of LED current flows through 100-ohm resistor to determine circuit trip point of 2 V. When control signal saturates Q_1, Q_2 provides about 1 V for lamp to give some preheating and reduce starting current surge. When control is above trip point, Q_2 turns on and energizes lamp.—"Linear Applications, Vol. 2," National Semiconductor, Santa Clara, CA, 1976, AN-153, p 3.

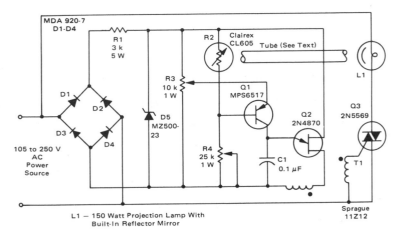

L1 — 150 Watt Projection Lamp With Built-In Reflector Mirror

PROJECTION-LAMP VOLTAGE REGULATOR—Circuit will regulate RMS output voltage across lamp to 100 V ± 2% for input voltages between 105 and 250 VAC. Light output of 150-W projection lamp is sensed indirectly for use as feedback to firing circuit Q1-Q2 that controls conduction angle of triac Q3. Light pipe, painted black, is used to pick up red glow from back of reflector inside lamp, which has relatively large mass and hence has relatively no 60-Hz modulation.—"Circuit Applications for the Triac," Motorola, Phoenix, AZ, 1971, AN-466, p 12.

800-W TRIAC DIMMER—Simple circuit uses Motorola MBS-4991 silicon bilateral switch to provide phase control of triac. 1-megohm pot varies conduction angle of triac from 0° to about 170°, to give better than 97% of full power to load at maximum setting. Conduction angle is the same for both half-cycles at any given setting of pot.—"Circuit Applications for the Triac," Motorola, Phoenix, AZ, 1971, AN-466, p 5.

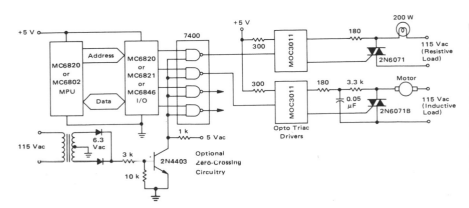

INTERFACE FOR AC LOAD CONTROL—Standard 7400 series gates provide input to Motorola MOC3011 optoisolators for control of triac handling resistive or inductive AC load. Gates are driven by MC6800-type peripheral interface adapters. If second input of two-input gate is tied to simple transistor timing circuit as shown, triac is energized only at zero crossings of AC line voltage. This extends life of incandescent lamps, reduces surge-current effect on triac, and reduces EMI generated by load switching.—P. O'Neil, "Applications of the MOC3011 Triac Driver," Motorola, Phoenix, AZ, 1978, AN-780, p 6.

HELMET-LAMP DIMMER—Provides lossless variation in brightness of incandescent lamp by using duty-cycle modulation. All three sections of 4025 triple three-input NOR gate turn lamp on and off rapidly at rate determined by setting of brightness control pot in astable MVBR circuit. Output transistor rating must be sufficient to handle lamp current.—D. Lancaster, "CMOS Cookbook," Howard W. Sams, Indianapolis, IN, 1977, p 231.

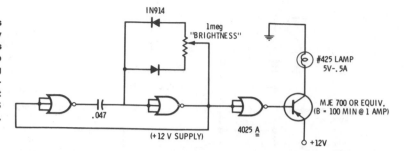

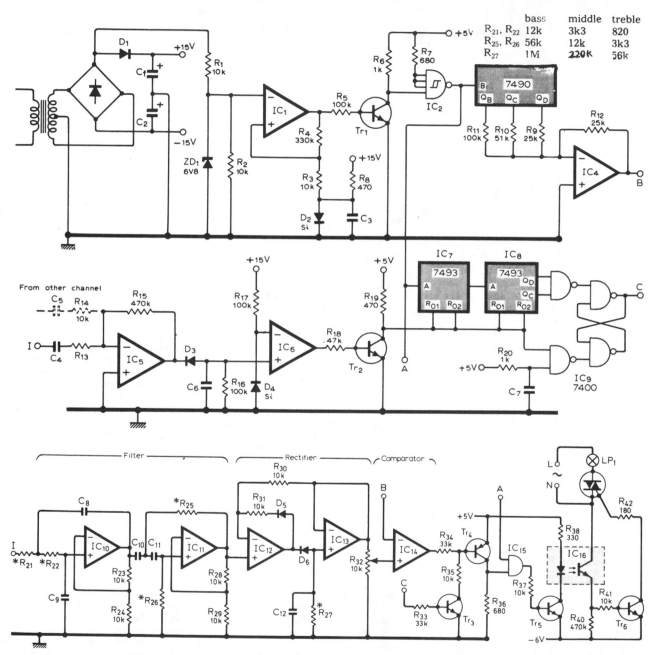

	bass	middle	treble
R_{21}, R_{22}	12k	3k3	820
R_{25}, R_{26}	56k	12k	3k3
R_{27}	1M	220k	56k

SOUND-CONTROLLED LAMP—Zero-voltage switching achieves interference-free proportional control of lamp intensity by sound source. Both inputs to AND gate IC_{15} must be high for triac to turn on. One input is from zero-crossing detector IC_1, Tr_1, and IC_2, which produces 100-Hz series of positive-going pulses. Other input is provided by filter/rectifier/comparator circuit. Inverting input of comparator

IC_{14} is fed by DAC IC_4 which produces stepped ramp waveform from outputs of 7490 counter IC_3. Counter is connected to count to 5 before resetting internally, giving five possible brightness levels for lamp. Opamps IC_5 and IC_6 detect when audio input falls below about 10 mV and then release IC_7-IC_9 from reset stage so the two 4-bit counters start counting 100-Hz waveform. Resetting occurs again when audio input next

passes 10-mV level. Lamp automatically turns on when music stops. All ICs are 741 or equivalent except as marked. Unmarked diodes are 1N4148, C_1 and C_2 are 100-nF polyester electrolytics, and all transistors are general-purpose types. Resistor values in table are for three-channel system, but more channels can be used if desired.—A. R. Ward, Sound-to Light Unit, *Wireless World*, July 1978, p 75.

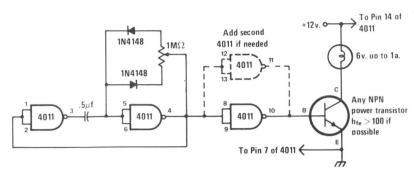

LANTERN-BATTERY EXTENDER—Life of lantern battery can be tripled without reducing light by chopping current while doubling voltage at 50% duty factor. 6-V lamp is connected across chopped 12-V supply built around 4011 CMOS quad NAND gate. First two gates form chopping oscillator, while third serves as interface to any NPN high-gain power transistor. If lamp draws more than 1 A, add fourth gate as shown in dashed lines. If gate is not used, tie its input leads to pin 14 of IC. Duty cycle is varied with 1-megohm pot; set at midrange before applying power, then adjust for normal lamp brilliance.—J. A. Sandler, 9 Easy to Build Projects under $9, *Modern Electronics,* July 1978, p 53–56.

FULL-WAVE CONTROL—Monsanto MCS6200 dual SCR optocoupler provides direct full-wave control of 15-W lamp or other AC device when driven by output logic voltages of microprocessor. LEDs are connected in series and photo-SCRs in reverse parallel to create equivalent of triac.—H. Olson, Controlling the Real World, *BYTE,* March 1978, p 174–177.

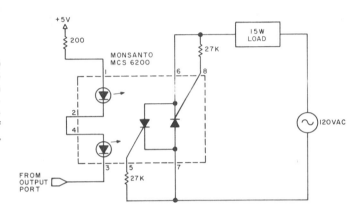

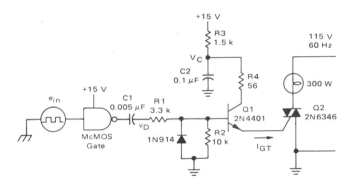

CMOS LOGIC CONTROL OF 300-W LAMP—Storage capacitor C2 in interface transistor circuit for typical CMOS gate charges to full +15 V supply voltage in time determined by R3 and C2, after which Q1 is fired by positive-going differentiated pulse derived from input square wave. C2 then dumps its charge through R4 and Q1, to fire triac Q2 and energize AC load. For maximum load power, triac should be fired early in conduction angle. With 1-kHz input square wave, output power is over 98% of maximum possible.—A. Pshaenich, "Interface Techniques Between Industrial Logic and Power Devices," Motorola, Phoenix, AZ, 1975, AN-712A, p 13.

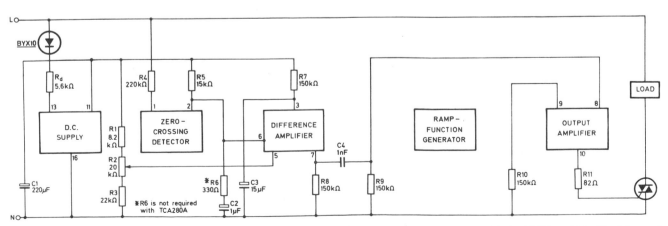

PHASE-CONTROLLED DIMMER—Mullard TCA280A trigger module is connected to compare amplitude of ramp waveform with controllable DC voltage in difference amplifier. At point of coincidence, trigger pulse is produced in output amplifier for triggering triac that controls lamp load. Choice of triac depends on load. Values shown for C4 and R9 give 100-μs pulse.—"TCA280A Trigger IC for Thyristors and Triacs," Mullard, London, 1975, Technical Note 19, TP1490, p 12.

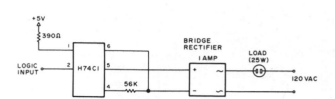

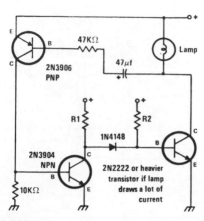

LOGIC CONTROLS 25-W LAMP—Ordinary 1-A bridge is used with H74C1 optoisolator to pass full current to 25-W lamp when logic input goes low (to ground, so full 5 V is applied to light source in optoisolator).—D. D. Mickle, Practical Computer Projects, *73 Magazine,* Jan. 1978, p 92–93.

LAMP SURGE SUPPRESSOR—Circuit limits turn-on current through cold filament, which is major cause of lamp failure but provides normal current when filament reaches operating temperature. Developed primarily for use with lamps in locations where replacement is extremely difficult. Values shown are primarily for low-voltage pilot lamps such as No. 44 and No. 47 but can be applied to any lamp within voltage and current ratings of transistors used.—J. A. Sandler, 11 Projects under $11, *Modern Electronics,* June 1978, p 54–58.

CHAPTER 45
Limiter Circuits

Covers clamps, clippers, and limiters used to keep signal peaks below predetermined limits for positive swings or for both positive and negative swings. See also Audio Control and Automatic Gain Control chapters.

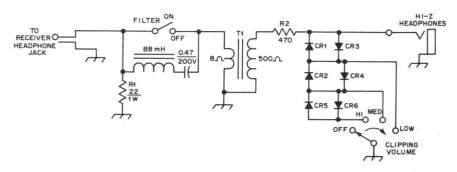

AUDIO CLIPPER/FILTER—Improves selectivity of communication receiver and prevents uncomfortably loud volume in 500-ohm headphones. 88-mH toroid and 0.47-μF capacitor form series resonant circuit at about 750 Hz with 6-dB bandwidth of 75 Hz. R2 and diodes form audio clipper whose level is determined by forward conduction voltage of diodes. With germanium diodes for CR1-CR4 and silicon for CR5-CR6, each successive switch position boosts volume 6 dB. For low-impedance headphones, omit T1 and use 8 ohms for R2.—A. R. Bloom, An Audio Clipper/Filter, *QST*, Aug. 1977, p 48.

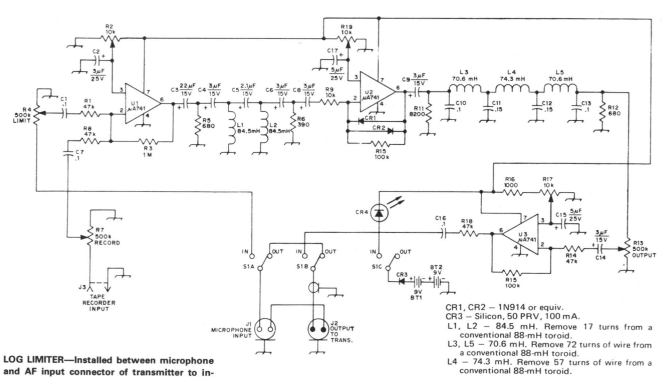

CR1, CR2 – 1N914 or equiv.
CR3 – Silicon, 50 PRV, 100 mA.
L1, L2 – 84.5 mH. Remove 17 turns from a conventional 88-mH toroid.
L3, L5 – 70.6 mH. Remove 72 turns of wire from a conventional 88-mH toroid.
L4 – 74.3 mH. Remove 57 turns of wire from a conventional 88-mH toroid.

LOG LIMITER—Installed between microphone and AF input connector of transmitter to increase average level of human voice and corresponding SSB transmitter output signal. Preamp U1 increases audio input level to overcome loss of following high-pass filter. Low-pass filter after logamp U2 eliminates all energy above about 2950 Hz. Last opamp U3 sets output level of processor for driving station transmitter properly. Overall amount of amplitude limiting is determined by setting of preamp gain control R4. High-pass filter eliminates 60-Hz hum developed in audio system or in accessory equipment and reduces harmonic distortion generated by deeply pitched voice or by hum picked up from tape recorder or phone patch.

Gain of opamp is determined by CR1-CR2, which give logarithmic response to audio input amplitude. Adjust R2, R17, and R19 for minimum distortion while applying 1-kHz sine wave to input.—R. Myers, A Quasi-Logarithmic Analog Amplitude Limiter with Frequency-Domain Processing, *QST*, Aug. 1974, p 22–25 and 40.

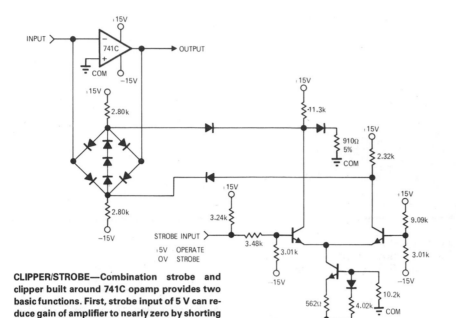

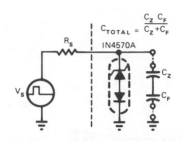

FAST CLAMP—Smaller capacitance of temperature-compensated reference diodes allows clamping of faster pulses than is possible with ordinary zeners. Diagram shows positive 6.4-V diode clamp at ground potential. For smaller zener voltages, connect diode to more negative potential. Device limits only positive peaks; for negative peaks, reverse connections to 1N4570A. To clamp both polarities, use two units in parallel, oppositely connected.—W. Walloch, Clamp Fast Pulses with One Component, *EDN Magazine*, April 5, 1974, p 76.

CLIPPER/STROBE—Combination strobe and clipper built around 741C opamp provides two basic functions. First, strobe input of 5 V can reduce gain of amplifier to nearly zero by shorting output to input. Second, clipping function holds all outputs below predetermined level. When used in integrator, circuit provides constant rate of discharge for integrating capacitor. Clipping is done by four-diode bridge having three extra diodes acting as zener. Strobe function is provided by four-diode bridge alone, with strobing input circuit having single-ended input and differential output. All transistors are 2N2219, and all diodes are 1N914.—L. Strahan, Op Amp Control Without Relays, *EDN/EEE Magazine*, Aug. 15, 1971, p 41–42.

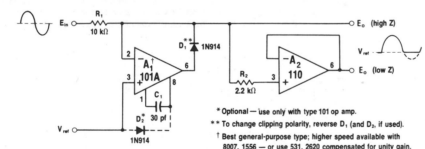

* Optional — use only with type 101 op amp.
** To change clipping polarity, reverse D₁ (and D₂, if used).
† Best general-purpose type; higher speed available with 8007, 1556 — or use 531, 2620 compensated for unity gain.

SHUNT CLIPPER—When input voltage is above reference voltage, D_1 is reverse-biased and input voltage passes through R_1 to output. When negative peak of E_{in} exceeds V_{ref}, opamp A_1 turns D_1 on to absorb input current from R_1, thereby clamping output at level of V_{ref}. If low output impedance is required, use 110 high-speed voltage follower as buffer amplifier.—W. G. Jung, "IC Op-Amp Cookbook," Howard W. Sams, Indianapolis, IN, 1974, p 189–190.

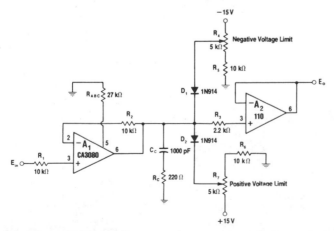

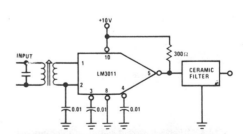

VOLTAGE LIMITER WITH BUFFER—CA3080 operational transconductance amplifier is used in combination with D_1 and D_2 that are normally reverse-biased at low levels, to provide abrupt or hard limiting. When output peaks from A_1 exceed voltage limits set by R_4 and R_7, diodes conduct and absorb A_1 output current, to limit its voltage swing. Buffer A_2 is required with this type of limiter to maintain loop gain and ensure constant limiting level for varying loads. Limiter action is absolute, in that there can be no further output voltage change when current limit is reached. Limiter can handle input/output levels of ±10 V or more when using ±15 V supply.—W. G. Jung, "IC Op-Amp Cookbook," Howard W. Sams, Indianapolis, IN, 1974, p 466–467.

CERAMIC-FILTER DRIVE—National LM3011 gain block provides three differential stages and current-source output suitable for driving 300-ohm ceramic filter in IF amplifier of FM receiver. Circuit provides 60 dB of power gain to matched load.—"Audio Handbook," National Semiconductor, Santa Clara, CA, 1977, p 3-11–3-12.

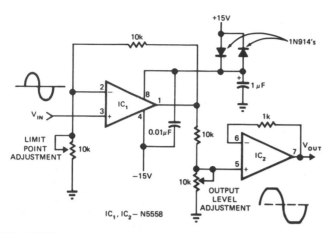

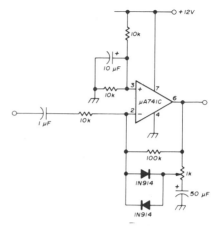

HARD-LIMITING OPAMP—Based on limited output swing of opamp. Single pot adjusts both positive and negative limit points. First opamp produces limiting. Second opamp serves as voltage follower for isolating attenuator and reducing output impedance. Matching of output between limit points is within about 100 mV at output level of 2 V P-P. Opamps specified are adequate for 3-kHz bandwidth.—E. E. Barnes, Ease Hard-Limiter Design with Op Amps, *EDN Magazine,* Aug. 5, 1975, p 76.

INVERTER/LIMITER—Developed for direct-conversion receivers that lack AGC, to provide limiting for CW reception. Below adjustable limiting threshold, amplifier is linear with voltage gain of 10. When output is high enough for silicon diodes to conduct, gain drops below unity. Amplifier should be preceded by several sections of filtering and followed by single-section low-pass filter to eliminate harmonic distortions generated in limiting process.—W. Hayward, Simple Active Filters for Direct-Conversion Receivers, *Ham Radio,* April 1974, p 12–15.

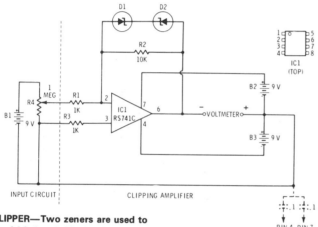

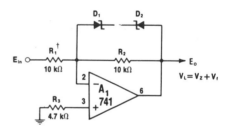

OPAMP AS CLIPPER—Two zeners are used to clip both sides of AC signal. Clipping level is determined by rating of zeners used, which can be 6, 9, 12, or 15 V depending on application. Ratio of R2 to R1 determines amplification. If long supply leads cause oscillation, connect 0.1-μF capacitors between ground and supply pins 4 and 7 as shown.—F. M. Mims, "Integrated Circuit Projects, Vol. 4," Radio Shack, Fort Worth, TX, 1977, 2nd Ed., p 37–44.

BIPOLAR ZENER—Circuit limits opamp output swing in either direction to sum of zener- and forward-breakdown voltages of D_1 and D_2. With matched zeners, positive and negative limiting levels are symmetrical. With 10-V zeners, limiting occurs at 10.6 V (10.0 + 0.6 V) to allow linear ±10 V swing without saturating A_1. R_2 is selected according to gain requirement. Diodes are matched pair chosen to provide desired limit voltage, such as 1N758 for ±10 V output swing.—W. G. Jung, "IC Op-Amp Cookbook," Howard W. Sams, Indianapolis, IN, 1974, p 201.

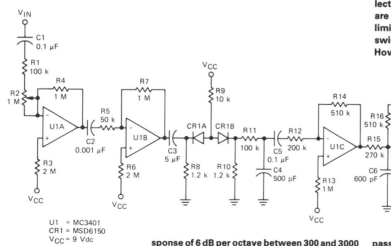

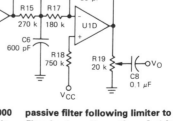

LIMITING PREAMP FOR FM—All four sections of Motorola MC3401 quad amplifier are used as interface between high-impedance microphone and FM modulator to provide preemphasis response of 6 dB per octave between 300 and 3000 Hz, with 6-dB per octave rolloff beyond. Includes amplitude limiter to prevent peak deviation of transmitter from exceeding allowed maximum. U1D forms active filter acting with passive filter following limiter to give low-pass filter having attenuation of at least 12 dB per octave above 3 kHz.—D. Aldridge, "An Economical FM Transmitter Voice Processor from a Single IC," Motorola, Phoenix, AZ, 1975, EB-57.

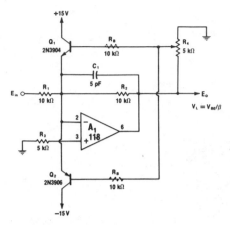

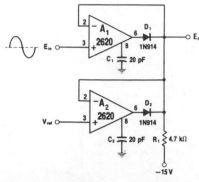

SYMMETRICAL BIPOLAR—Circuit uses transistor feedback, with current returned to opamp summing point through emitter-followers Q_1 and Q_2. When output E_0 rises to positive voltage high enough so tap on R_4 makes Q_1 turn on, output of A_1 is limited to value corresponding to setting of R_4. Q_2 performs similar limiting function for negative swings of input signal. R_4 sets limiting level for both polarities, for variable symmetrical limiting. Transistors and opamps shown are selected for high-speed operation. For single-polarity limiting, insert large resistance R_B in base lead of remaining transistor to prevent it from conducting heavily on output voltage swings.—W. G. Jung, "IC Op-Amp Cookbook," Howard W. Sams, Indianapolis, IN, 1974, p 203–204.

SELF-BUFFERED SERIES CLIPPER—Provides negative clipping of sine-wave input, using 2620 opamps with diodes as linear OR gate. When input voltage is greater than reference voltage, A_1-D_1 turns on and input voltage appears at output. When E_{in} falls below V_{ref}, A_2-D_2 provide V_{ref} as output signal. With 2620, circuit gives good performance to above 10 kHz. For positive clipping, reverse diode connections and return R_1 to +15 V. Circuit provides low output impedance.—W. G. Jung, "IC Op-Amp Cookbook," Howard W. Sams, Indianapolis, IN, 1974, p 187–189.

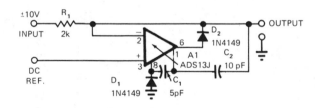

CLIPPING OPAMP—Simple circuit limits excursion of DC input voltage to level precisely equal to DC reference input voltage. When reference is 0 V, circuit can be used as half-wave rectifier up to 100 kHz. Feedback overcomes breakover characteristics of D_2, to provide sharp clipping at levels from millivolts to volts. At 10 kHz, with D_1 omitted, range of levels is 70 mVRMS to 7 VRMS. With D_1, circuit is useful at levels down to 0.3 VRMS at 100 kHz.—R. S. Burwen, Precision Clipper Operates from Millivolts to Volts, *EDN Magazine*, Dec. 1, 1972, p 57 and 59.

CHAPTER 46
Logarithmic Circuits

Combinations of logamps and opamps provide squaring, cubing, log, and antilog functions for analog signals, along with logarithm of ratio of two input values. Also includes logamp test circuit.

80-dBV INPUT RANGE—Logamp using Texas Instruments SN56502 or SN76502 in combination with three SN52741 opamps can handle input voltage range greater than 80 dB with respect to 1 V P-P. Inputs are limited by reducing supply voltages of input amplifiers to ±4 V. Gains of input amplifiers are adjusted to achieve smooth transitions.—"The Linear and Interface Circuits Data Book for Design Engineers," Texas Instruments, Dallas, TX, 1973, p 7-45.

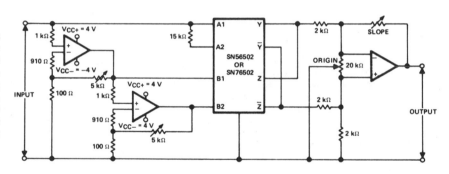

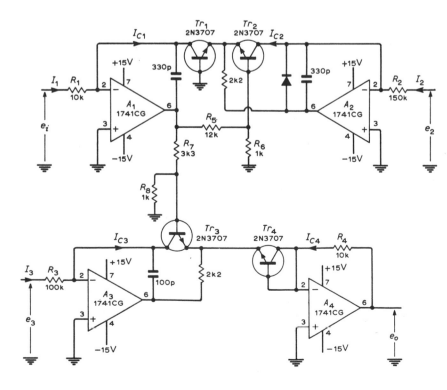

CUBE GENERATOR—Combination of temperature-compensated opamp log converter and antilog converter generates output signal e_o proportional to cube of input signal e_i. Article gives design equations.—G. B. Clayton, Experiments with Operational Amplifiers, *Wireless World*, Feb. 1973, p 91–93.

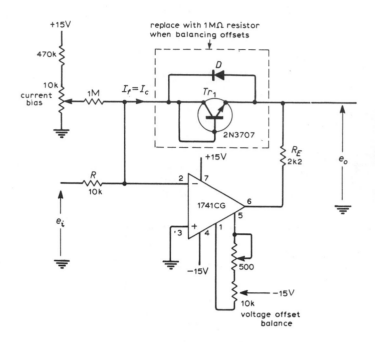

BASIC LOG CONVERTER—Suitable for positive input voltages. Negative feedback is applied to opamp through diode-connected transistor, with additional diode D protecting transistor from excessive inverse voltage caused by wrong input polarity. For negative inputs, reverse transistor and diode connections. Article gives design equations and application procedures.—G. B. Clayton, Experiments with Operational Amplifiers, *Wireless World,* Jan. 1973, p 33–35.

LOGAMP TESTER—Circuit sweeps logamp under test (LAUT) over its dynamic range while automatically canceling offset voltage at input, with output serving as indication of this offset voltage. Exponential decay voltage generated is accurate from 10 V to 100 μV. At time zero, pulse T2 is applied to Optical Electronics 9729 opamp connected as exponential generator, charging NPO feedback capacitor to 10 V. At end of pulse, opamp output voltage decays exponentially. Input pulse also generates ±10 V precision reference ramp in pair of 9813 opamps. Output of LAUT is compared with reference ramp; any difference is proportional to nonlinearity of logamp.—"Testing Logarithmic Amplifiers," Optical Electronics, Tucson, AZ, Tech Tip 10268.

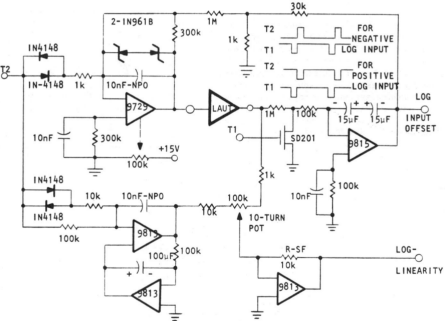

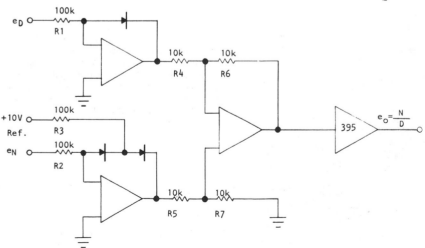

ANALOG DIVIDER—Optical Electronics 2457 logarithmic module containing two pair of bipolar log elements and three opamps drives single 395 opamp to provide analog division function with high accuracy and wide dynamic range. Input resistors set full-scale input voltage. R3 determines AC accuracy. Connect inputs together for drive with signal varying between 100 mV and 10 V (1% to 100% of full scale), and adjust R3 for straight-line output at 1 V. For negative inputs, reverse diode connections and make reference voltage negative.—"Accurate Logarithmic Analog Divider," Optical Electronics, Tucson, AZ, Application Tip 10213.

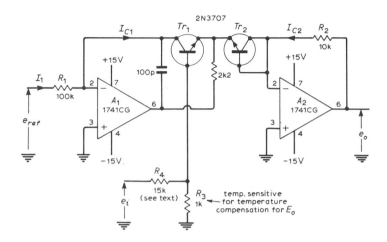

TEMPERATURE-COMPENSATED ANTILOG CONVERTER—Circuit provides antilog conversion with high degree of temperature compensation. Article gives design equations and application instructions. Circuit is for positive inputs; for negative inputs, use PNP transistors such as 2N4058 in place of those shown. Adjust value of R_4 to make output of opamp A_2 exactly 10 times input voltage e_i when input is −1 V.— G. B. Clayton, Experiments with Operational Amplifiers, *Wireless World,* Jan. 1973, p 33–35.

TEMPERATURE-COMPENSATED LOG CONVERTER—Two opamps and two logging transistors give output proportional to logarithm of input with high degree of temperature compensation. Article gives design equations and application instructions. Circuit is for positive inputs; for negative inputs, use PNP transistors such as 2N4058 in place of those shown.—G. B. Clayton, Experiments with Operational Amplifiers, *Wireless World,* Jan. 1973, p 33–35.

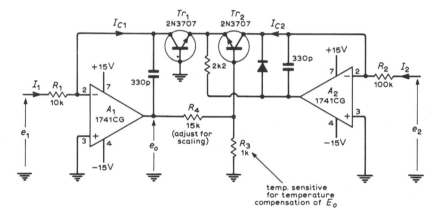

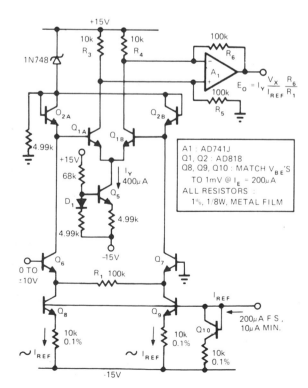

VARIABLE-TRANSCONDUCTANCE DIVIDER— Practical analog divider follows ideal division equation over typical 20:1 range of reference current and operates in two quadrants. Circuit is analyzed in terms of logarithmic behavior of its elements. Bandwidths up to 5 MHz can be achieved. Article gives design equations.—L. Counts and D. Sheingold, Analog Dividers: What Choice Do You Have?, *EDN Magazine,* May 5, 1974, p 55–61.

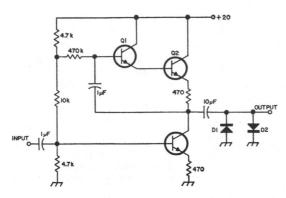

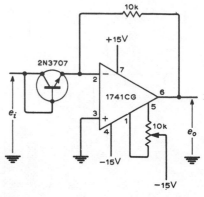

LOGAMP—Based on fact that back-to-back diodes driven by current generator give output that varies logarithmically with input signal. With values shown, relation is logarithmic over 60-dB range. Transistors are 2N2924, SK3019, GE-10, or HEP-54, and diodes are 1N914.—Circuits, *73 Magazine,* April 1974, p 34.

ANTILOG CONVERTER—Basic opamp circuit with diode-connected transistor as logging element performs antilog conversion for positive input signals. For negative inputs, reverse transistor connections.—G. B. Clayton, Experiments with Operational Amplifiers, *Wireless World,* Jan. 1973, p 33–35.

40-dB LOGAMP—Uses Optical Electronics 2457 logarithmic module containing two pairs of bipolar log elements and three opamps. Connection shown produces pure $\log_{10}$ function on positive inputs. Zero point is set by R2 or by reference e. Reference must be positive for positive logs and negative for negative logs. Trim 110K resistors for exactly 10-V output.—"Two-Decade Precision Logarithmic Amplifier," Optical Electronics, Tucson, AZ, Application Tip 10212.

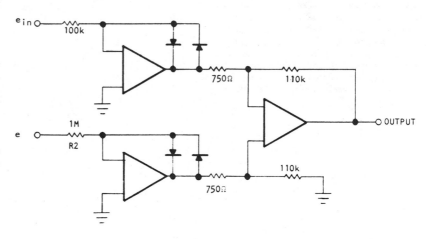

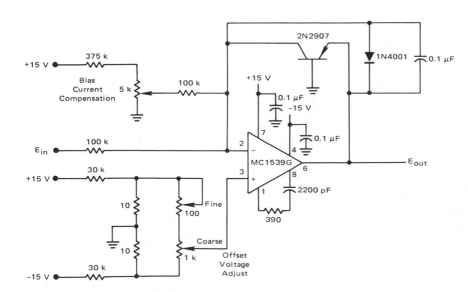

LOW-COST USING OPAMP—Motorola MC1539G opamp is connected with PNP transistor as logarithmic element. Circuit requires external compensation, has about 200-nA bias current, and accommodates wide range of input voltages when appropriate networks are used to compensate for errors. To adjust bias current initially, replace transistor with 500K resistor and adjust 5K pot for gain of 5 over input signal range.—K. Huehne, "Transistor Logarithmic Conversion Using an Integrated Operational Amplifier," Motorola, Phoenix, AZ, 1971, AN-261A, p 4.

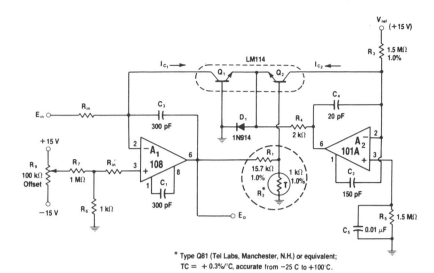

* Type Q81 (Tel Labs, Manchester, N.H.) or equivalent;
TC = + 0.3%/°C, accurate from −25 C to +100°C.

100-dB DYNAMIC RANGE—Circuit generates log ratio of currents I_{C1} and I_{C2} with accuracy within 3% from 10 nA to 1 mA (100-dB range) when I_{c2} is fixed at 10-μA reference value. Accuracy increases to 1% for current inputs between 40 nA and 400 μA (80 dB). A_2 supplies constant reference current to Q_2. Q_1 is operated as transdiode with Q_2 providing temperature compensation of offset voltage.—W. G. Jung, "IC Op-Amp Cookbook," Howard W. Sams, Indianapolis, IN, 1974, p 213–214.

ANTILOG GENERATOR—Basic log generator circuit is rearranged to perform inverse operation of antilog (exponential) generation. Exponential current generated by Q_2 is summed at current-to-voltage converter A_2. Q_1 voltage drive to Q_2 is such that collector current of Q_2 is exponentially related to voltage at base of Q_1. Temperature-compensating divider scales input sensitivity to 1 V per decade. Feed-forward connections at A_1 and A_2 optimize circuit for speed.—W. G. Jung, "IC Op-Amp Cookbook," Howard W. Sams, Indianapolis, IN, 1974, p 214–216.

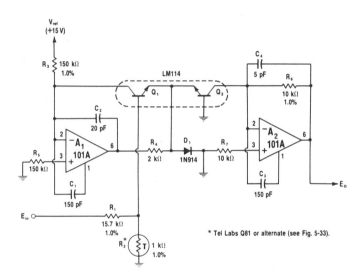

* Tel Labs Q81 or alternate (see Fig. 5-33).

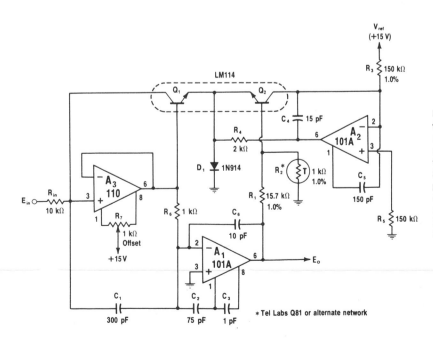

* Tel Labs Q81 or alternate network

FAST LOG GENERATOR—Circuit is designed for settling time of only 10 μs. Voltage-follower opamp A_3 buffers base current of Q_1 and input current to A_1, allowing Q_1 to operate as diode while demanding minimum input current to A_3. Scale factor is 1 V per decade. Zero-crossover point of output swing is controlled by R_3. Accuracy is 1% for logging range of 100 nA to 1 mA.—W. G. Jung, "IC Op-Amp Cookbook," Howard W. Sams, Indianapolis, IN, 1974, p 214–215.

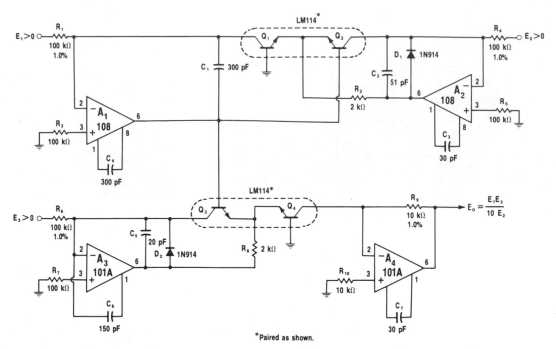

*Paired as shown.

MULTIPLIER/DIVIDER—Upper half of circuit is log converter in which output at A_1 is logarithmic ratio of E_1 and E_2. A_3 and Q_3 form second log converter for E_3 input. Log output of second converter is added to that of upper circuit, producing log (E_1E_3/E_2) at emitter of Q_4. Q_4 and A_4 take antilog to give final output equal to $E_1E_3/10E_2$. If only multiplication is desired, E_2 can be reference voltage; R_4 will then establish reference current.—W. G. Jung, "IC Op-Amp Cookbook," Howard W. Sams, Indianapolis, IN, 1974, p 216–217.

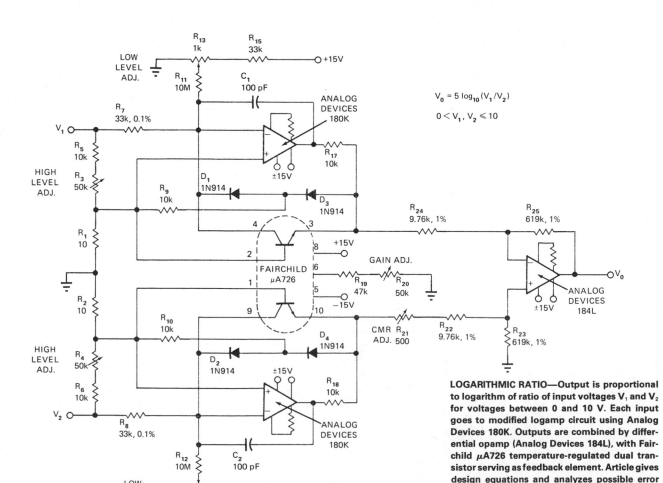

$$V_0 = 5 \log_{10}(V_1/V_2)$$

$$0 < V_1, V_2 \leqslant 10$$

LOGARITHMIC RATIO—Output is proportional to logarithm of ratio of input voltages V_1 and V_2 for voltages between 0 and 10 V. Each input goes to modified logamp circuit using Analog Devices 180K. Outputs are combined by differential opamp (Analog Devices 184L), with Fairchild μA726 temperature-regulated dual transistor serving as feedback element. Article gives design equations and analyzes possible error sources.—D. R. Morgan, Get the Most Out of Log Amplifiers by Understanding the Error Sources, *EDN Magazine,* Jan. 20, 1973, p 52–55.

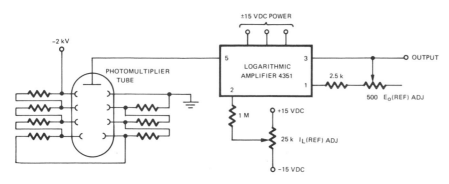

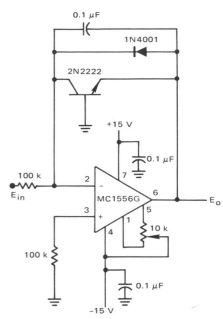

RECORDING PHOTOMULTIPLIER OUTPUT— Wide range of output data from photomultiplier is fed through Teledyne Philbrick 4351 logarithmic amplifier for compression of data to range of ±5 VDC for feed to tape recorder. Report covers calibration procedure for obtaining overall accuracy of ±2 dB.—"How to Specify Parameters of Nonlinear Circuits," Teledyne Philbrick, Dedham, MA, 1974, AN-15, p 4.

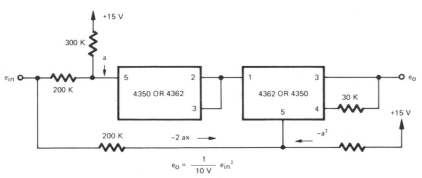

$$e_o = \frac{1}{10\ V}\ e_{in}^2$$

NOTE: TO OBTAIN NEGATIVE OUTPUT VOLTAGE USE TWO 4351'S AND THE –15 V SUPPLY.

TWO-QUADRANT SQUARING— Teledyne Philbrick log modules are used in simple two-quadrant squarer in which input is offset because modules accept only one polarity. Circuit shown provides positive output voltage.—"Applications for Models 4350/4351 & 4362/4363 Logarithmic Amplifiers," Teledyne Philbrick, Dedham, MA, 1974, AN-14.

LOGAMP WITH OFFSET ADJUSTMENT— Motorola MC1556G opamp provides accurate operation down to millivolt input levels without bias current compensation. Input offset is adjusted with 10K pot. 0.1-μF capacitor is required across feedback transistor to reduce AC gain. Diode protects transistor from polarity reversal of input voltage. Power supplies should be bypassed as close as possible to amplifier socket. Positive input voltage gives negative output voltage. IC includes output short-circuit protection.—K. Huehne, "Transistor Logarithmic Conversion Using an Integrated Operational Amplifier," Motorola, Phoenix, AZ, 1971, AN-261A, p 3.

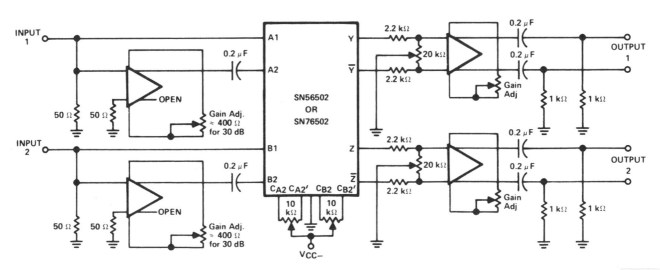

10-MHz DUAL-CHANNEL LOGAMP— Circuit using Texas Instruments SN56502 or SN76502 in combination with four SN52741 opamps will handle 50-dB input range per channel at all frequencies up to 10 MHz. Suitable for data compression, analog computation, radar and infrared detection systems, and weapons systems. Differential output voltage levels are generally less than 0.6 V. Output swing and slope of output response are adjusted by varying gain in each channel. Coordinate origin is adjusted with offset pots of output buffers.—"The Linear and Interface Circuits Data Book for Design Engineers," Texas Instruments, Dallas, TX, 1973, p 7-46.

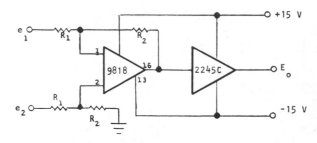

DIFFERENTIAL LOGARITHM—Optical Electronics 9818 opamp and 2245C logamp together give logarithm of differential input voltage or current. Combined transfer function is $E_0 = K \log (R_2/R_1)(e_2 - e_1)$. R_1 can be zero for differential input current circuit. For unity-gain preamp, R_1 and R_2 should be 10K. For 1-V full-scale input, use 10K for R_1 and 100K for R_2. For 100-mV full-scale input, use 10K for R_1 and 1 megohm for R_2.—"How to Obtain a Differential Logarithm," Optical Electronics, Tucson, AZ, Application Tip 10126.

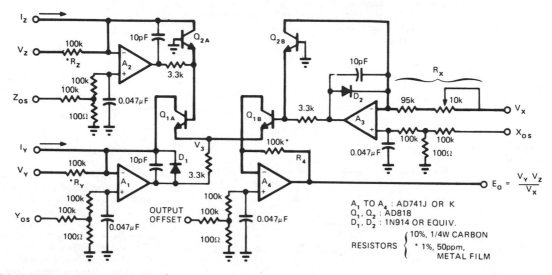

LOG-ANTILOG DIVIDER—Can be used in applications where both numerator and denominator are restricted to single polarity (to one quadrant). Input variables X, Y, and Z are applied to three independent transdiode log amplifiers (A_1-Q_{1A}, A_2-Q_{2A}, and A_3-Q_{2B}). Outputs of logamps are proportional to logarithms of input variables. R_4 in feedback circuit of A_4 converts collector current of Q_{1B} to output voltage proportional to $V_Z V_Y/V_X$. Circuit performs multiplication and division simultaneously and with equal accuracy. Overall nonlinearity can be as low as 0.05%. Article gives design equations.—L. Counts and D. Sheingold, Analog Dividers: What Choice Do You Have?, *EDN Magazine*, May 5, 1974, p 55–61.

CHAPTER 47

Logic Circuits

Includes interfaces for different types of logic, along with gates, Schmitt triggers, and other types of logic circuits that are responsive to sudden or gradual changes in input logic levels. Also covered are pulse sequence, pulse coincidence, and pulse-width detectors, along with pulse memories. See also Logic Probe, Memory, Microprocessor, and Operational Amplifier chapters.

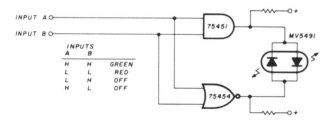

COINCIDENCE DETECTOR—If inputs A and B are both high, indication will be green. If A and B are both low, indication will be red. If inputs are out of phase, so one is high and the other low, indicator will be off. Suitable for monitoring complex logic circuits. Uses Monsanto MV5491 dual red/green LED, with 220 ohms in upper lead to +5 V supply and 100 ohms in lower +5 V lead because red and green LEDs in parallel back-to-back have different voltage requirements. Drivers are SN75451 and SN75454.—K. Powell, Novel Indicator Circuit, *Ham Radio*, April 1977, p 60–63.

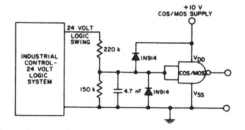

INTERFACE FOR INDUSTRIAL CONTROL—Simple resistive divider circuit provides interface between 24-V logic swing of industrial control system and CMOS logic operating from 10-V supply. Filter capacitor enhances excellent noise immunity of CMOS logic. Clamp diodes ensure that input signal voltage is between V_{DD} and V_{SS}.—"COS/MOS Integrated Circuits," RCA Solid State Division, Somerville, NJ, 1977, p 628.

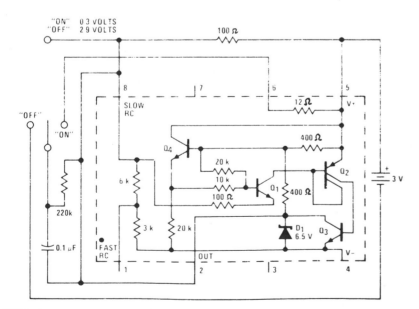

LATCH—National LM3909 IC operating from 3-V battery requires only momentary contact by switch to change logic level of output and hold that level.—"Linear Applications, Vol. 2," National Semiconductor, Santa Clara, CA, 1976, AN-154, p 9.

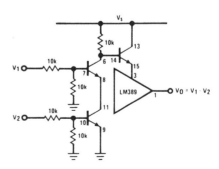

AND LOGIC FOR MUTING—Connection shown for National LM389 combination of three transistors with opamp gives standard AND circuit for controlling muting transistor in audio system. Shorting pin 12 of opamp to ground gives NAND logic.—"Audio Handbook," National Semiconductor, Santa Clara, CA, 1977, p 4-33–4-37.

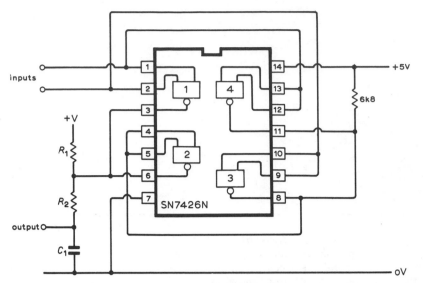

PHASE-SENSITIVE DETECTOR—Uses quadruple two-input NAND-gate IC with minimum of external components. DC output level is absolutely linear with phase difference at inputs, making circuit suitable for phase-locked loops and phase-shift keyed demodulation. Output is rectangular wave whose mark-space ratio is proportional to phase difference between input square waves. This output is applied to low-pass filter R_2-C_1 having values chosen to suit operating frequency and required output resistance. R_1 is chosen to give required output swing up to maximum of 15 V.—R. A. Harrold, Inexpensive P.S.D., *Wireless World,* Jan. 1973, p 32.

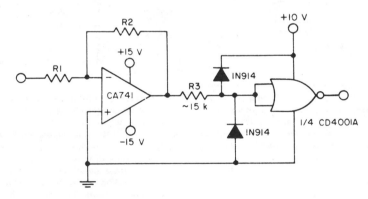

CMOS INTERFACE FOR OPAMP—Clamp diodes and single resistor provide interface between CMOS circuit and opamp operating between normal ±15 V supply rails. Diodes ensure that CMOS input voltage does not go outside permissible range.—"COS/MOS Integrated Circuits," RCA Solid State Division, Somerville, NJ, 1977, p 629.

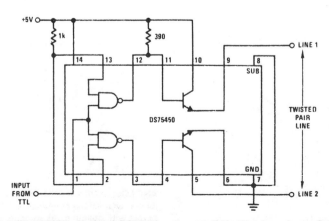

BALANCED LINE DRIVER—National DS75450 dual peripheral driver serves as interface between TTL and twisted-pair line. Output line 1 is terminated to ground through half of line impedance, and line 2 is terminated to +5 V through half of line impedance. Output current is 300 mA.—"Interface Databook," National Semiconductor, Santa Clara, CA, 1978, p 3-20–3-30.

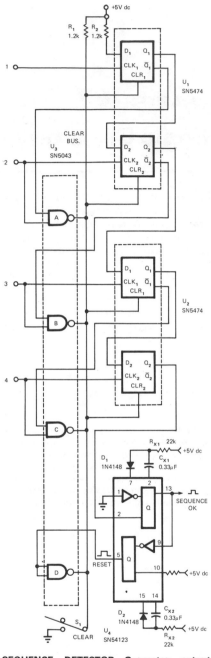

SEQUENCE DETECTOR—Generates output pulse if, and only if, sequence of input pulses is in prescribed order. Any other sequence inhibits output pulse and clears circuit at instant of first out-of-order pulse. Circuit also clears itself at end of correct sequence, by generating sequence-OK pulse. Developed for use in control systems, electronic combination locks, and any other applications requiring sequence of pulses. To detect more than 4 bits in sequence, NAND gates and flip-flops can be added. If TTL input pulses occur in correct T_1-T_2-T_3-T_4 order, 1 state at U_1D_1 will propagate down chain of D flip-flops until U_2Q_2 output is reached. Simultaneously, 0 is propagated in similar manner to hold clear bus at 1. Article traces circuit operation in detail.—M. J. Gallagher, Self-Clearing Digital Sequence Detector, *EDN Magazine,* April 5, 1973, p 88–89.

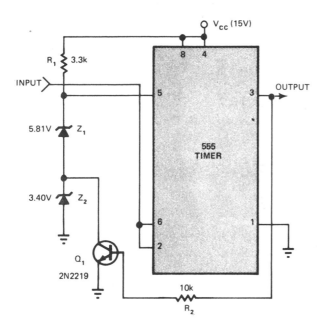

VARIABLE-HYSTERESIS SCHMITT—Uses standard 555 timer with only five additional components to give fully TTL/DTL-compatible Schmitt trigger that responds to slow input ramps as well as straight DC levels. When input is 0, output goes high and turns on Q_1. When input increases to 5.8 V, output goes low and Q_1 turns off. Decreasing output to 4.7 V (lower threshold point) makes output high again. For values shown, hysteresis is thus 1.1 V. Threshold points and hysteresis value can be adjusted as required by changing zeners used.—M. K. Lalitha and P. R. Chetty, Variable-Threshold Schmitt Trigger Uses 555 Timer, *EDN Magazine,* Sept. 20, 1976, p 112 and 114.

POLARITY REVERSER—Logic input controls operation of DG303 low-power analog switch providing polarity reversal for output of opamp. Low input logic gives noninverting operation because input signal then goes to pin 12 of opamp.—"Analog Switches and Their Applications," Siliconix, Santa Clara, CA, 1976, p 7-91.

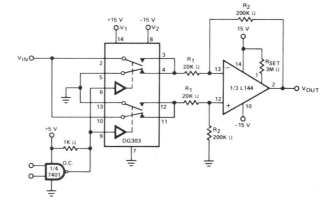

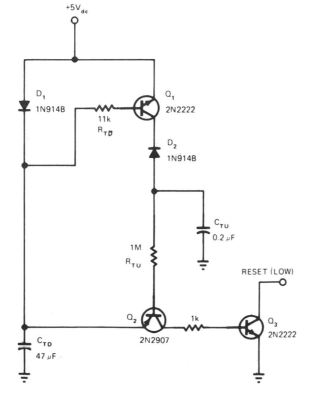

POWER-UP/POWER-DOWN RESET—Output of circuit goes high at end of time interval $R_{TU}C_{TU}$ required for charging of C_{TU} by applied 5-VDC input signal. When 5-V power fails, Q_1 drives Q_2 and Q_3 off, making output low for time interval determined by $R_{TD}C_{TD}$. When C_{TD} is discharged, output goes high again. Interruption of power thus produces negative-going pulse having variable width determined by values chosen for resistors and capacitors.—S. Rummel, Reset Circuit Detects Power Drop-Out, *EDN Magazine,* May 20, 1976, p 94.

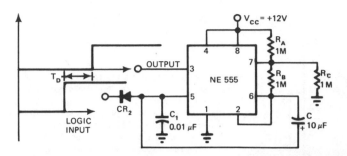

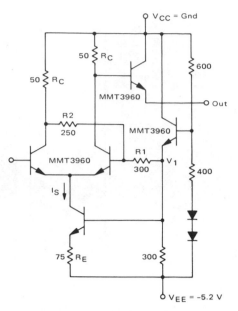

TIME-DELAY SWITCH—Oscillator connection shown for 555 timer provides delay for switching action controlled by logic input. When logic input is 0 V, timer output is low. When logic input goes high, output remains low for delay of $0.693R_BC$ s and then switches high. Output remains high as long as input is high.—K. D. Dighe, Rearranged Components Cut 555's Initial-Pulse Errors, *EDN Magazine*, Jan. 5, 1978, p 82 and 84.

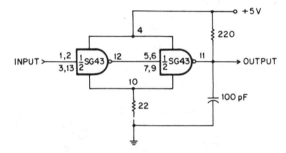

CURRENT-MODE SCHMITT TRIGGER—Motorola MMT3960 transistor array is connected as differential amplifier with feedback to produce Schmitt trigger having input hysteresis of 500 mV. Small rise and fall times make circuit suitable for driving flip-flop over wide range of frequencies.—B. Broeker, "Micro-T Packaged Transistors for High Speed Logic Systems," Motorola, Phoenix, AZ, 1974, AN-536, p 6.

TWO-GATE SCHMITT—TTL inverter gates are connected in series, with small feedback resistor in common ground lead, so gates are always in opposing logic states. Resulting constant voltage drop across 22-ohm resistor produces constant offset voltage that is corrected by 220-ohm resistor at output terminal. With values shown, positive-going threshold is 2.4 V and negative-going threshold is 2 V.—C. J. Ulrick, Schmitt Trigger Uses Two Logic Gates, *EEE Magazine*, Dec. 1970, p 54.

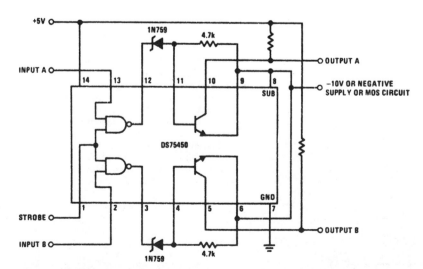

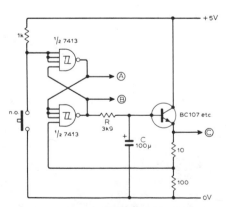

TTL-TO-MOS DRIVER—Uses National DS75450 dual peripheral driver having 300-mA output current capability to provide high-speed switching while providing compatibility between different logic types. Operates from single 5-V supply.—"Interface Databook," National Semiconductor, Santa Clara, CA, 1978, p 3-20—3-30.

PULSE-COMPLETING SCHMITT—Half of 7413 dual four-input NAND Schmitt trigger forms RS bistable which ensures that cycle is completed after switch is opened. Low-impedance exponential sawtooth output is produced at point C. Point A is high when oscillator is running and can be used as control signal.—T. P. Hopkins, Improved Schmitt Trigger Oscillator, *Wireless World*, Jan. 1978, p 58.

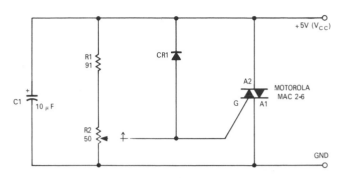

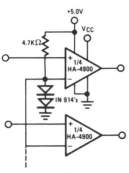

LOGIC PROTECTOR—Simple bidirectional triac crowbar can be set so positive voltages above 6 V and negative voltages greater than 1.5 V cannot reach digital logic. Article covers initial adjustment of R2.—D. L. Sporre, Bidirectional Crowbar Protects Logic, *EDN Magazine*, Dec. 15, 1970, p 37.

TTL-TO-CMOS TRANSLATOR—Two sections of Harris HA-4900/4905 precision quad comparator provide interface between TTL drive and CMOS output circuits. Supply is ±15 V.—"Linear & Data Acquisition Products," Harris Semiconductor, Melbourne, FL, Vol. 1, 1977, p 2-95.

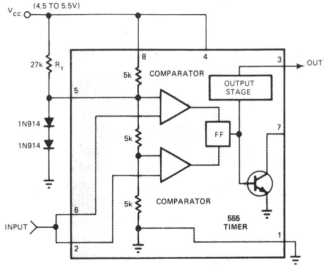

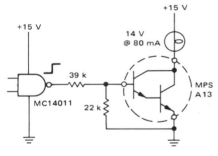

SCHMITT FROM 555 TIMER—Only three additional parts need be added to standard 555 timer to give fully TTL/DTL-compatible Schmitt trigger that responds to slow input ramps as well as straight DC levels. With output initially high, increasing input from 0 to about 1.35 V drives output low. Conversely, input can decrease to about 0.7 V before output goes high again.—M. K. Lalitha and P. R. Chetty, Variable-Threshold Schmitt Trigger Uses 555 Timer, *EDN Magazine*, Sept. 20, 1976, p 112 and 114.

HIGH-LEVEL ACTIVATION BY CMOS—High output of typical CMOS gate drives complementary MPS-A13 Darlington transistor having 80-mA lamp load.—A. Pshaenich, "Interface Techniques Between Industrial Logic and Power Devices," Motorola, Phoenix, AZ, 1975, AN-712A, p 11.

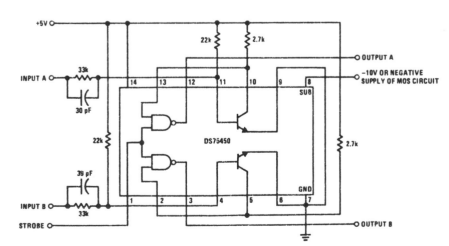

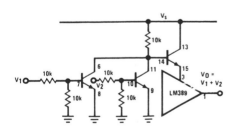

MOS-TO-TTL DRIVER—National DS75450 dual peripheral driver serves as interface between different logic types while providing high-speed switching and 300-mA output current per section. Requires only single 5-V supply.—"Interface Databook," National Semiconductor, Santa Clara, CA, 1978, p 3-20–3-30.

OR LOGIC FOR MUTING—Connection shown for National LM389 combination of three transistors with opamp gives standard OR circuit for controlling muting transistor in audio system. Shorting pin 12 of opamp to ground gives NOR logic.—"Audio Handbook," National Semiconductor, Santa Clara, CA, 1977, p 4-33–4-37.

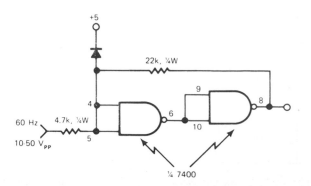

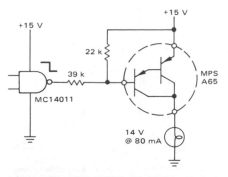

SCHMITT USING 7400 GATES—Uses two sections of 7400 quad NAND gate. Will accept input voltages of 10–50 V P-P with values shown, but can use line voltage directly if input resistor is 22K and feedback resistor is 220K. Diode in DC supply limits positive-going input to 5.7 V for protection of input circuit. Used as interface between 60-Hz line and frequency divider having TTL logic, when 1-s time base is required for timing applications.—W. A. Palm, Connect a 7400 Gate as a Schmitt Trigger, *EDN Magazine*, Aug. 20, 1976, p 84.

LOW-LEVEL ACTIVATION BY CMOS—Typical CMOS gate interfaces directly with small-signal Darlington transistor driving 80-mA lamp load.—A. Pshaenich, "Interface Techniques Between Industrial Logic and Power Devices," Motorola, Phoenix, AZ, 1975, AN-712A, p 11.

PULSE-WIDTH DETECTOR—Logic terminal of National LM122 timer is driven simultaneously with trigger input to give high-accuracy pulse-width detector. Output changes state only when trigger input stays high for longer than time period set by R_t and C_t; resulting output pulse width is then equal to input trigger width minus $R_t C_t$. C_L filters out narrow spikes that would occur at output due to interval delays during switching. Supply can range from 4.5 to 40 V.—C. Nelson, "Versatile Timer Operates from Microseconds to Hours," National Semiconductor, Santa Clara, CA, 1973, AN-97, p 9.

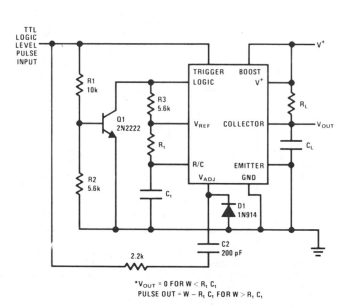

*$V_{OUT} = 0$ FOR $W < R_t C_t$
PULSE OUT $= W - R_t C_t$ FOR $W > R_t C_t$

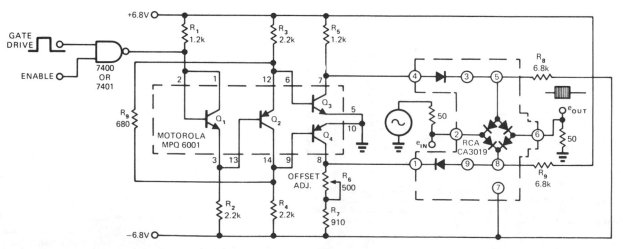

TTL INTERFACE—Motorola MPQ 6001 quad complementary-pair transistor IC serves as interface between TTL gate and diode bridge used as signal gate. Propagation delay from leading edge of gate drive pulse until bridge gate opens is 30 ns, with negligible delay between complementary outputs. Used in providing low-level burst of input signal when making response time measurement.—R. W. Hilsher, Universal Interface: TTL to Diode Array, *EDN Magazine*, March 5, 1975, p 74 and 76.

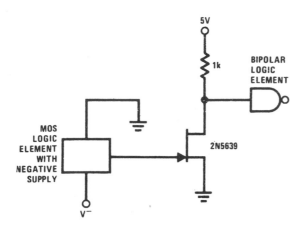

LEVEL AND POLARITY SHIFTER—Simple FET circuit provides for level shifting from MOS logic element having negative supply to TTL or other bipolar logic level operating from positive supply and ground. Transistor has fast switching time.—"FET Databook," National Semiconductor, Santa Clara, CA, 1977, p 6-26–6-36.

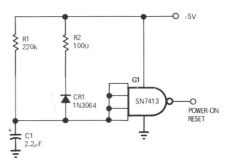

POWER-ON RESET—One Schmitt trigger and three discrete components ensure correct initial state of logic circuits when power is applied. During charge time of C1, output of gate G1 is high. When C1 reaches 1.5 V, gate output goes low and terminates power-on reset.—R. C. Snyder, Single-Voltage Circuit Generates "Power-On" Reset Pulse, *EDN/EEE Magazine,* Jan. 1, 1972, p 72.

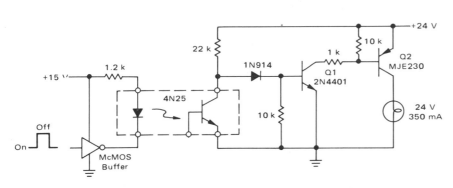

CMOS INTERFACE USING OPTOISOLATOR—Provides logic control of 350-mA lamp. High level on input of typical CMOS inverter energizes 4N25 optoisolator, to clamp Q1 off. This removes drive from Q2, deenergizing load. Logic 0 at input reverses conditions, turning on lamp. With values shown, 10 mA at optoisolator input controls completely isolated 350-mA load.—A. Pshaenich, "Interface Techniques Between Industrial Logic and Power Devices," Motorola, Phoenix, AZ, 1975, AN-712A, p 16.

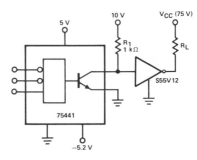

ECL INTERFACE FOR VMOS—S55V12 VMOS (identical to S55V01 except for higher breakdown/saturation voltage) is used to buffer ECL-compatible SN75441 peripheral driver. Combination is capable of handling up to 90 V at 2 A. SN75441 has open-collector output, so interface with VMOS requires only pull-up resistor R_1.—L. Shaeffer, VMOS Peripheral Drivers Solve High Power Load Interface Problems, *Computer Design,* Dec. 1977, p 90, 94, and 96–98.

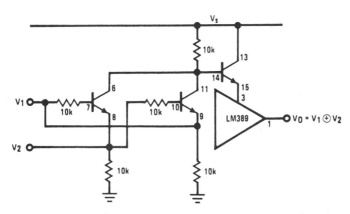

EXCLUSIVE-OR LOGIC FOR MUTING—Connection shown for National LM389 combination of three transistors with opamp gives standard EXCLUSIVE-OR circuit for controlling muting transistor in audio system. Shorting pin 12 of opamp to ground gives EXCLUSIVE-NOR logic.—"Audio Handbook," National Semiconductor, Santa Clara, CA, 1977, p 4-33–4-37.

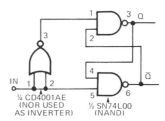

LOW-POWER SCHMITT—Uses two NAND gates from SN74L00 NAND package and one NOR gate (used as inverter) from CD4001AE CMOS package to make low-power Schmitt trigger. NAND gates are connected to form RS flip-flop. Q goes high when input voltage is greater than 2.1 V, and other output does not go low until input voltage is less than 1.2 V. Both polarities of output signal are available.—R. Cox, CMOS and LPTTL Gates Make Low-Power Schmitt Trigger, *EDN Magazine,* Oct. 1, 1972, p 48.

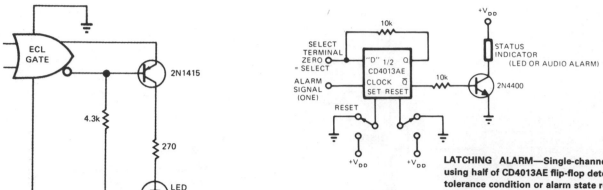

ECL INTERFACE FOR LED—PNP germanium transistor serves as interface for driving LED from emitter-coupled logic. Same interface can be used to drive 7-segment or other arrays that have common-cathode configuration, such as Fairchild FND10 or Monsanto MAN3. Optoisolator can be used in place of LED.—G. A. Altemose, One Transistor Provides ECL to LED Interface, *EDN Magazine,* Oct. 15, 1972, p 54.

LATCHING ALARM—Single-channel monitor using half of CD4013AE flip-flop detects out-of-tolerance condition or alarm state represented by logic 1 on clock input. Flip-flop then changes to alarm state that turns on transistor for energizing alarm device, and holds alarm condition until operator reacts by applying voltage to SET input. Logic 0 on D terminal activates flip-flop when monitoring is desired.—J. C. Nichols, CMOS "D" Flop Makes Latching "AND" Gate, *EDN Magazine,* April 20, 1974, p 89 and 91.

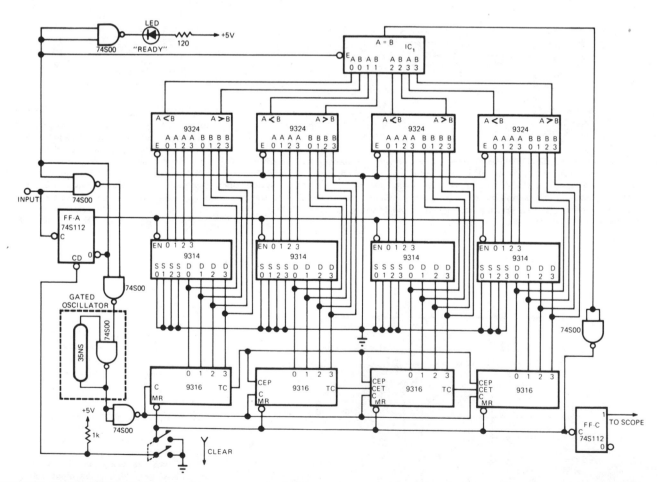

STORING SINGLE PULSE—Designed to take single-event positive-going TTL pulse, occurring only once when series of logic conditions is met, and recreating pulse accurately on CRO as square wave in which half of cycle represents original pulse width. FF-A and 9316 binary counters are initially cleared. Input pulse of interest gates delay-line oscillator (lower left) on for duration of pulse width. During this time, binary representation of pulse width is accumulated in 9316 counters, then stored in 9314 D-latches. At same time, final 9324 comparator IC₁ is enabled and oscillator is gated on again to reset 9316s and toggle FF-C. Square-wave output of FF-C then represents original pulse width within 35 ns (one clock).—N. L. White, Don't Miss That Single Event Pulse—Store It, *EDN Magazine,* Sept. 20, 1975, p 70 and 72.

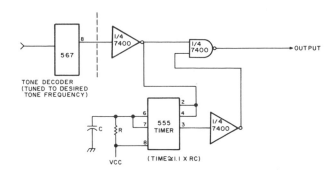

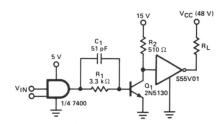

TONE DETECTOR—Output goes low only when input tone has been continuous at desired frequency for interval exceeding duration of pulse from 555 timer. Circuit can be used to reset an alarm system or to detect TTL level that exceeds predetermined time duration.—Circuits, *73 Magazine*, April 1977, p 164.

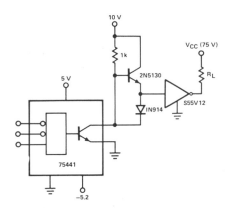

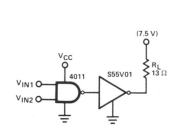

TOTEM-POLE ECL INTERFACE FOR VMOS—Transistor and diode in totem-pole configuration improve performance of SN75441 ECL-compatible peripheral driver for S55V12 VMOS.—L. Shaeffer, VMOS Peripheral Drivers Solve High Power Load Interface Problems, *Computer Design,* Dec. 1977, p 90, 94, and 96–98.

CMOS INTERFACE FOR VMOS—Simple 4011 CMOS gate connection provides required logic interface for S55V01 VMOS that is capable of handling up to 1 A. Switching time is about 25 ns but can be doubled by connecting four CMOS gates in parallel. V_{CC} can be either 10 or 15 V.—L. Shaeffer, VMOS Peripheral Drivers Solve High Power Load Interface Problems, *Computer Design,* Dec. 1977, p 90, 94, and 96–98.

AMPLIFIED TTL INTERFACE FOR VMOS—Bipolar voltage amplifier Q_1 translates TTL output swing of 0.4 to 2.4 V from 7400 interface to 15-V drive signal for S55V01 VMOS peripheral driver. Circuit can be driven by any low-level signal, including ECL, if comparator such as AM686 is used in place of SN7400 quad NAND gate. Switching times of circuit are less than 40 ns in both directions.—L. Shaeffer, VMOS Peripheral Drivers Solve High Power Load Interface Problems, *Computer Design,* Dec. 1977, p 90, 94, and 96–98.

CHAPTER **48**
Logic Probe Circuits

Provide LED and/or audible indication of logic-level status at terminal on which test probe is held for troubleshooting or other purposes. One circuit shows status at eight separate terminals on CRO display.

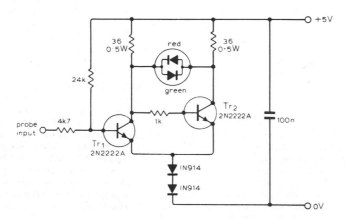

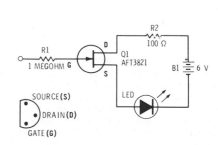

TTL-STATE PROBE—Uses voltage drop of LED in Schmitt trigger to indicate high, low, open-circuit, and pulse-train conditions at probe input. Indicator is Monsanto MV5491 dual red-green LED package. High input saturates Tr_1, cuts off Tr_2, and turns on red LED. Low input

cuts off Tr_1 and turns on green LED. For high impedance at input, both LEDs are off. Rectangular waves up to about 1 MHz turn on both LEDs, with relative brightness giving rough indication of mark-space ratio.—J. C. Flower, Logic Probe, *Wireless World,* Sept. 1976, p 72.

FET LOGIC PROBE—Field-effect transistor with very high input resistance makes LED glow when logic 1 is present at input, without loading circuit being monitored.—F. M. Mims, "Computer Circuits for Experimenters," Radio Shack, Fort Worth, TX, 1974, p 35–43.

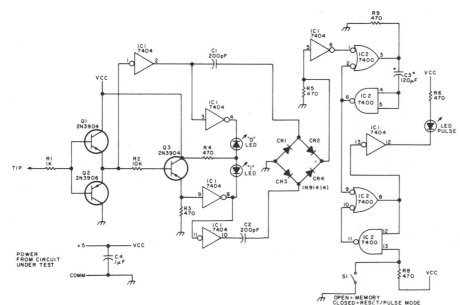

LOGIC PROBE—Provides almost as much information, when working with TTL or DTL digital circuits, as costly CRO or logic analyzer. Can be built into length of plastic tubing, with probe tip projecting at one end and two supply leads coming from other end. One LED flashes for high to low transition, and other for low to high transition. Flash is visible even with very narrow pulses, because probe circuit stretches pulse width. With S1 open (memory mode), pulse LED at right stays on until reset by operator, for capturing any stray pulse. If probe tip is held on open circuit or on chain of floating inputs, no LED will light.—C. W. Andreasen, Superprobe, *73 Magazine*, Holiday issue 1976, p 92–93.

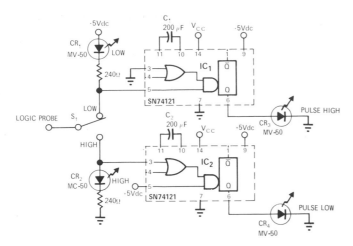

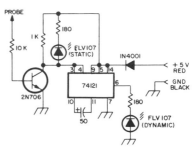

LOGIC-STATE PROBE—Useful for low-repetition-rate applications having single narrow pulses, as in computer interfaces and logic control systems. Pulses as narrow as 50 ns are stretched by mono so they provide clear LED indication. If steady state of circuit under test is low, CR$_1$ will light when S$_1$ is in low position. If steady state is high, CR$_2$ will light when switch is high. Other two lamps indicate presence of pulse superimposed on steady-state signal. If repetition frequency of pulse is high, pulse light appears on continuously.—J. W. Hamill, Low-Speed Logic Probe, *EDN/EEE Magazine*, Nov. 1, 1971, p 51.

RTL/TTL PROBE—Static LED indicates logic level at probe tip. Dynamic LED indicates presence of high or 1 pulse at probe tip even though momentary, because circuit stretches pulse to about 50 μs for easy visibility. Pulse stretcher requires at least 100-ns 4-V pulse. 1N4001 diode protects against reversed power leads. Probe requires +5 V from circuit under test. LEDs and transistor are not critical.—Circuits, *73 Magazine*, April 1974, p 31.

LOGIC PROBE—Pocket-size battery-powered probe using two inexpensive CMOS ICs has such low standby current drain that ON/OFF switch is unnecessary. Drain is appreciable only when LEDs are on during use. Mallory TR-133 4.2-V mercury battery gives 2-V threshold for TTL compatibility. LEDs come on to indicate logic 1. Pulse-stretching circuit ensures detecting positive or negative pulses as short as 250 ns.—J. Edrington, Battery-Powered Logic Probe Needs No ON/OFF Switch, *EDN Magazine*, May 20, 1975, p 72 and 74.

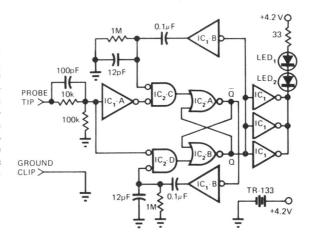

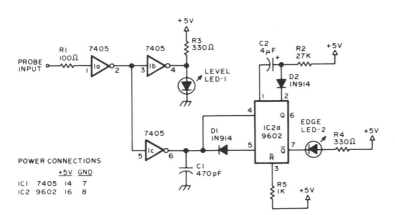

BUILT-IN LOGIC PROBE—Permanently wired LEDs indicate TTL levels and changes in levels. LEVEL LED-1 comes on for TTL 1 input and goes out for TTL 0 input. EDGE LED-2 comes on momentarily for input changes. Will detect levels, steps, single pulses, and pulse trains. Only half of each IC is used, so dual tester can be made if desired. ICs 1a and 1b form noninverting driver for 9602 mono which is triggered on both positive-going and negative-going edges by C1 and D1. Pulse train input having period less than width of flash pulse will keep EDGE indicator on.—K. W. Christner, The Built-In Logic Tester, *BYTE*, Jan. 1977, p 82–83.

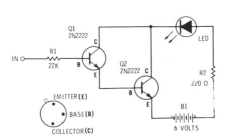

DARLINGTON LOGIC PROBE—Two-transistor Darlington connection provides very high input impedance that does not load logic circuit being monitored, while driving LED that glows when logic 1 is present at input.—F. M. Mims, "Computer Circuits for Experimenters," Radio Shack, Fort Worth, TX, 1974, p 35–43.

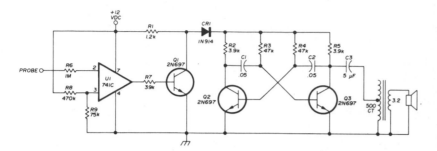

AUDIBLE LOGIC INDICATOR—Audio oscillator Q1-Q2-Q3 is isolated from TTL by opamp U1 wired as Schmitt trigger. Opamp acts as high-input-impedance inverter. Reference level is set at +1.6 V by R8-R9 for TTL-compatible logic (about midway between high and low logic levels). When probe input is below +1.6 V, opamp output of about 10.5 V saturates Q1 and disables Q2-Q3 to cut off tone. When probe voltage is above +1.6 V, U1 output is about 2 V which cuts off Q1 and allows Q2-Q3 to generate tone indicating high logic.—H. F. Batie, Versatile Audio Oscillator, *Ham Radio,* Jan. 1976, p 72–74.

RED/GREEN LEVEL DISPLAY—Level detector can be used with TTL, DTL, and RTL, as probe for troubleshooting. Indicator is Monsanto MV5491 dual red/green LED, with 220 ohms in upper lead to +5 V supply and 100 ohms in lower +5 V lead because red and green LEDs in parallel back-to-back have different voltage requirements. Will furnish green indication on high or plus signal and red indication on low or false signal. Supply voltage of +5 V can be taken from equipment under test. Circuit requires SN75451 driver ICs and two sections of SN7404 hex inverter.—K. Powell, Novel Indicator Circuit, *Ham Radio,* April 1977, p 60–63.

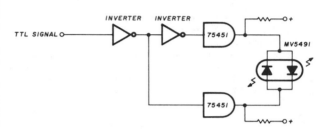

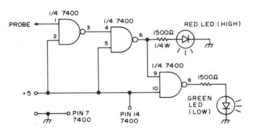

PROBE—Red LED comes on to indicate that test point is at high logic level, while green LED signifies low level. Circuit uses one 7400 quad dual-input NAND gate.—S. Uhrig, Check Logic with This Simple Probe, *73 Magazine,* Dec. 1974, p 76.

LED DISPLAY FOR TTL—Useful for observing TTL levels when CRO is not available. A1 and A2 are 74123 dual retriggerable mono MVBRs with clear, used to turn on LEDs when input transitions are detected. Even very short pulses are made visible because mono stretches pulse length. Table shows how LED indications are interpreted.—B. Voight, The TTL One Shot, *73 Magazine,* Feb. 1977, p 56–58.

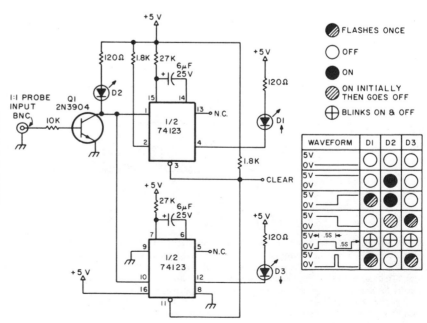

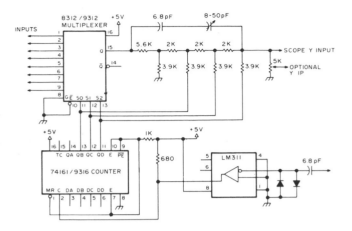

8-TRACE LOGIC DISPLAY—Adapter for standard oscilloscope shows time relationship between pulses at eight different locations in digital circuit, for troubleshooting and isolation of glitches. Almost any general-purpose CRO can be used, but triggered sweep improves usefulness. Multiplexer feeds each input in turn to Y input of CRO, under control of counter. Article covers operation and use of adapter.—R. A. Johnson, Eight Trace Scope Adapter, *73 Magazine,* Sept. 1976, p 108–110.

CMOS LOGIC PROBE—Designed for wiring into microprocessor to show status of an important terminal, as troubleshooting aid. LEVEL LED is on for TTL input of 1 and off for input of 0. EDGE LED lights momentarily when input changes from 1 to 0 or 0 to 1. Use of CMOS inverting buffer 3a at input prevents probe from affecting microprocessor.—F. A. Weissig, A CMOS Logic Probe, *BYTE,* Oct. 1977, p 11.

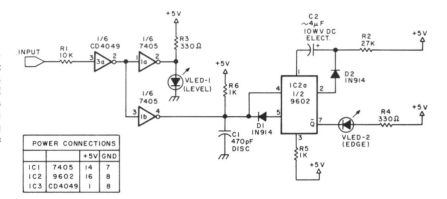

POWER CONNECTIONS		
	+5V	GND
IC1 7405	14	7
IC2 9602	16	8
IC3 CD4049	1	8

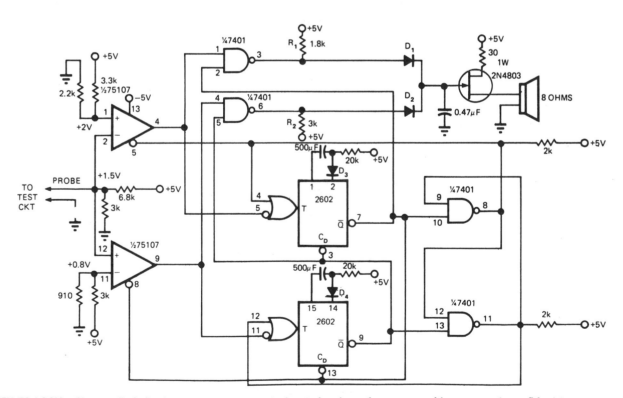

LISTEN TO LOGIC—Gives audio indication of TTL states when probe is held on IC pin. Input level above 2 V makes 2N4803 UJT oscillate at about 400 Hz and give tone from loudspeaker.

Logic 0 at probe input gives lower frequency. Logic pulse transitions trigger 2602 dual monos for 1 s, modulating tones at about 1-s rate. Monos also drive two-gate latch that prevents

either mono from firing two consecutive times.—I. Simon, Audio Output Eases Logic Level Checking, *EDN Magazine,* June 20, 1975, p 116 and 118.

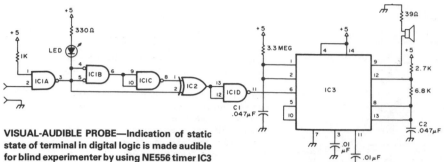

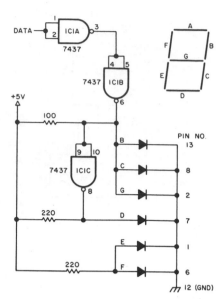

VISUAL-AUDIBLE PROBE—Indication of static state of terminal in digital logic is made audible for blind experimenter by using NE556 timer IC3 as oscillator controlled by IC1 (SN74132N) and IC2 (SN7486N). Any logic transition from 1 to 0 or 0 to 1, lasting at least 50 ns, is detected and indicated by audio beep. Visual indication is provided by LED that comes on when input is logic 1.—T. Lincoln, A Logic Probe You Can Hear, *73 Magazine,* Aug. 1976, p 106.

OUTPUT STATUS DISPLAY—Monitors state of single bit and shows H or L on 7-segment display depending on status of data input. Uses two 7437 inverters and one DL-704 common-cathode display. Diode symbols represent segments of display (segment A is not needed for H or L). Power connections to 7437 inverter are +5 V to pin 14 and ground to pin 7.—G. Tomalesky, Bit Status Display, *BYTE,* Dec. 1977, p 197.

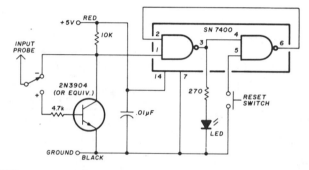

TTL PROBE—LED comes on, and stays on until circuit is reset, if input probe receives low-level (negative-going) pulse when polarity switch is set as shown. For other position of switch, LED comes on for high-level or +5 V pulse.—R. B. Shreve, Troubleshooting Logic Circuits, *Ham Radio,* Feb. 1977, p 56–59.

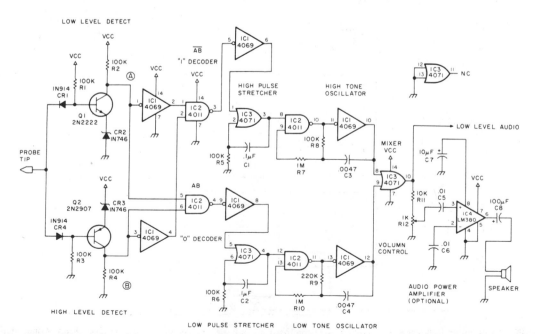

AUDIBLE CMOS PROBE—Eliminates need for watching meter while applying probe tip in turn to large number of closely spaced terminals during troubleshooting. Produces high tone for logic high, low tone for logic low, and no sound for open or floating string. Supply is 12 V. Article describes circuit operation in detail.—C. W. Andreasen, The Best Probe Yet?, *73 Magazine,* April 1978, p 134–135.

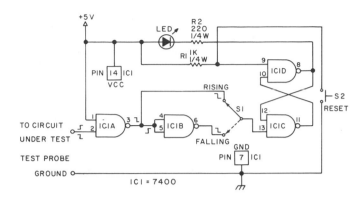

FAST TTL PROBE—RS flip-flop wired from NAND gate detects pulses as short as combined gate delays of NAND sections forming flip-flop (about 30 ns). Circuit changes state at start of pulse, with LED monitoring flip-flop output. After pulse has been detected, circuit must be reset with S2. Will work with either positive-or negative-going pulse, as selected by S1.—W. A. Walde, Build a TTL Pulse Catcher, *BYTE,* Feb. 1976, p 58 and 60.

AUDIBLE PROBE—Speeds troubleshooting by eliminating need to look at meter, and prevents possible damage to logic under test by minimizing possibility of shorts occurring if probe slips when looking at meter. Components used with A-8402 IC give 1-kHz output for 6-V input. Binary counter divides this signal to give output in low audio range and converts to square wave more suitable for driving loudspeaker. Probe can also serve as input for digital voltmeter. When probing TTL circuits, probe gives low frequency for 0 state, high frequency for 1, and middle frequency for open circuit. Output transistor type is not critical.—G. E. Row, Audible Logic Probe Doubles as DVM Input Section, *EDN Magazine,* Oct. 20, 1977, p 82.

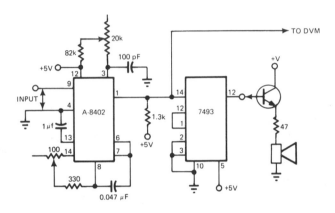

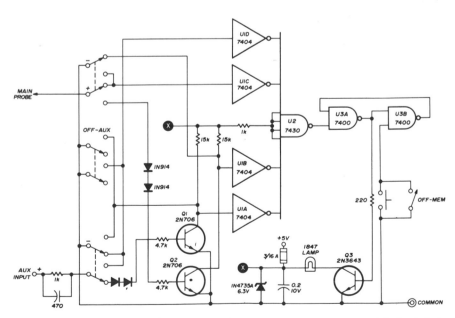

TEST PROBE—Checks binary levels and pulse coincidence. Indicator lamp, driven by switching transistor Q3, is bright enough to be seen in sunlight. Close OFF-MEM switch when using probe as binary level indicator, to eliminate need for pushing button continuously. At AUX position of main switch, two inputs are needed at same time. To check for coincidence, connect patch cord from AUX jack to second point being checked in logic circuit. All signal diodes are 1N914. Probe drain is about 160 mA; changing to LED would cut drain to 60 mA.—R. H. Fransen, Improved Logic Test Probe, *Ham Radio,* Dec. 1973, p 53–55.

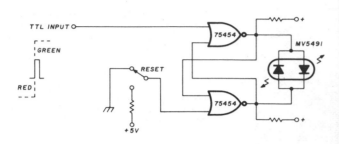

LATCHING PROBE—TTL pulse catcher can be used as logic probe for IC troubleshooting. Uses Monsanto MV5491 dual red/green LED, with 220 ohms in upper lead to +5 V supply and 100 ohms in lower +5 V lead because red and green LEDs in parallel back-to-back have different voltage requirements. SN75454 driver circuit is cross-coupled to form latch or memory element. Positive-going pulse sets latch and makes LED change color to indicate arrival of pulse. Latch must be reset manually with RESET switch. Also useful for locating intermittents such as glitches; probe can be left connected to circuit under test, to see if latch has been set by unwanted signal.—K. Powell, Novel Indicator Circuit, *Ham Radio*, April 1977, p 60–63.

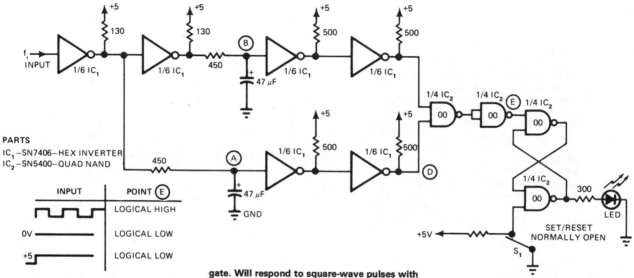

PARTS
IC₁—SN7406—HEX INVERTER
IC₂—SN5400—QUAD NAND

INPUT	POINT Ⓔ
⊓⊔⊓⊔	LOGICAL HIGH
0V ───	LOGICAL LOW
+5 ⌐─	LOGICAL LOW

PULSE FREQUENCY DETECTOR—Can be used as digital-logic probe or as frequency detector for test equipment. Requires only two ICs, SN7406 hex inverter and SN5400 quad NAND gate. Will respond to square-wave pulses with 50% duty cycles up to 3 MHz. When pulse appears at input, points A and B detect logic high level and make point E go high so latch sets and turns on LED. Without an input frequency, A and B will be complementary and E will go low, resetting latch and turning off LED.—V. Rende, Digital Frequency Detector Uses Only Two IC's, *EDN Magazine*, April 20, 1976, p 114.

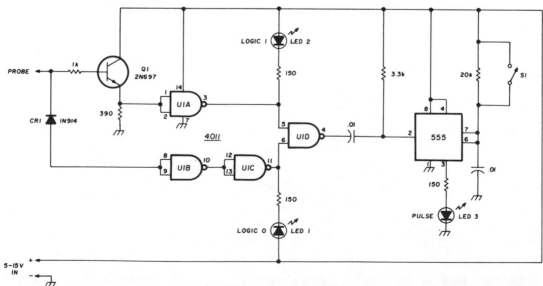

LOGIC PROBE—Designed to indicate logic states in TTL and CMOS circuits. Will substitute for high-speed triggered CRO in indicating presence of positive- and negative-going pulses. Input of logic 0 lights LED1 and LED3, with LEDs staying on only 200 ms. With logic 1, only LED2 lights. With S1 closed, LED1 and LED3 stay on if pulse is positive-going, while LED2 and LED3 stay on if pulse is negative-going.—H. M. Berlin, A TTL and CMOS Logic Probe, *Ham Radio*, March 1978, p 114.

CHAPTER 49
Medical Circuits

Includes circuits for telemetering and processing of heart, brain, muscle, and other bioelectric potentials, recording data from joggers, monitoring therapeutic radiation, synthesizing speech, and providing audible indications for blind persons of light level, voltage, logic status, bridge null, and other measurable parameters.

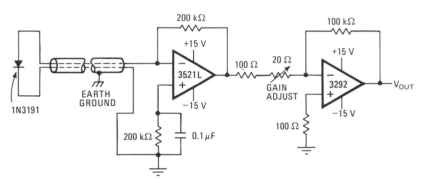

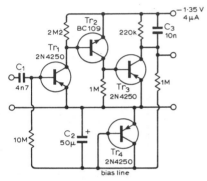

RADIATION MONITOR—1N3191 commercial diode serves as sensor in high-accuracy dosage-rate meter for gamma rays and high-energy X-rays used in radiotherapy. Diode is small enough for accurate mapping of radiation field. Output voltage varies linearly from 0.1 V to 10 V as dose rate increases from 10 to 1000 rads per minute. Low-drift FET-input 3521L opamp

amplifies detector current to usable level for 3292 chopper-stabilized opamp that provides additional gain while minimizing temperature errors.—P. Prazak and W. B. Scott, Radiation Monitor Has Linear Output, *Electronics*, March 20, 1975, p 117; reprinted in "Circuits for Electronics Engineers," *Electronics*, 1977, p 106.

IMPLANT AMPLIFIER—Designed for use in implanted transmitters monitoring brain and heart potentials. Requires only 4 μA at 1.35 V. Voltage gain is 2000, and equivalent input noise only 10 μV P-P with 10-megohm source impedance. Tr$_1$ is current-starved, but resulting limited bandwidth of about 5 kHz is acceptable for biological applications.—C. Horwitz, Micropower Low-Noise Amplifier, *Wireless World*, Dec. 1974, p 504.

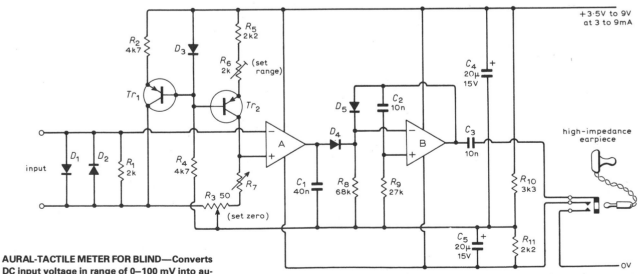

AURAL-TACTILE METER FOR BLIND—Converts DC input voltage in range of 0—100 mV into audible indication that is produced at instant when measured voltage exceeds reference voltage as set by decade switches of R$_7$. Blind person can then read Braille markings at switch

settings to get input voltage. Opamp B is connected as free-running MVBR that generates AF signal for earpiece. Use germanium transistors such as OC45 or OC71. Opamps are Motorola

1435. Use silicon diodes such as 1N914, BA100, or OA200. R$_7$ can alternatively be wirewound pot.—R. S. Maddever, Meter for Blind Students, *Wireless World*, Jan. 1973, p 36—37.

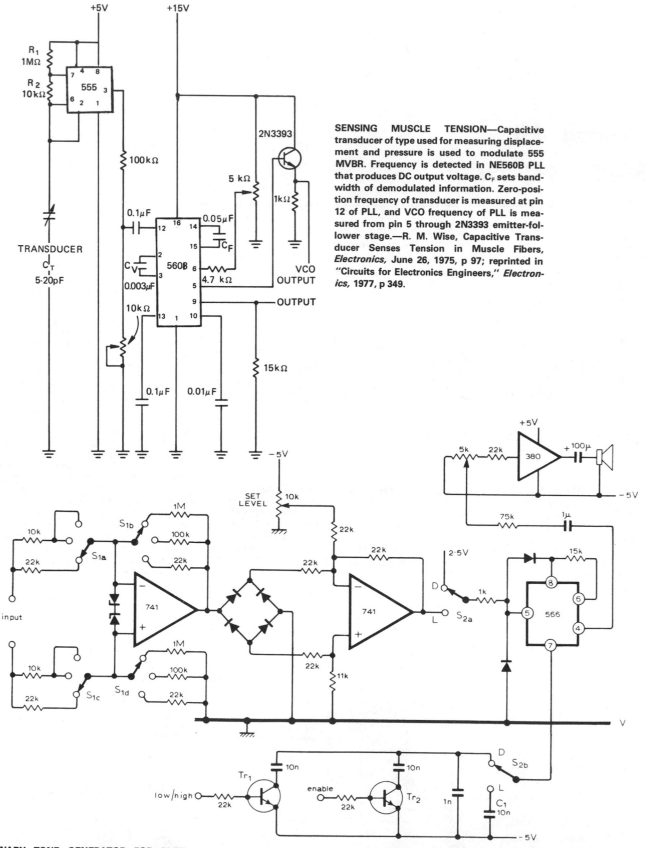

SENSING MUSCLE TENSION—Capacitive transducer of type used for measuring displacement and pressure is used to modulate 555 MVBR. Frequency is detected in NE560B PLL that produces DC output voltage. C_F sets bandwidth of demodulated information. Zero-position frequency of transducer is measured at pin 12 of PLL, and VCO frequency of PLL is measured from pin 5 through 2N3393 emitter-follower stage.—R. M. Wise, Capacitive Transducer Senses Tension in Muscle Fibers, *Electronics,* June 26, 1975, p 97; reprinted in "Circuits for Electronics Engineers," *Electronics,* 1977, p 349.

BINARY TONE GENERATOR FOR BLIND— When low/high input is voltage in binary form, as obtained from converter circuit (also given in article) fed by digital voltmeter, circuit produces low pitch for binary 0 and high pitch for binary 1 when S_2 is set at D for digital voltmeter mode. Recognition of binary digits in tone form can be learned by blind person much as learning of Morse code. Uses LM566 IC as tone-generating VCO that feeds loudspeaker through LM380 IC amplifier and 5K volume control. With S_2 at position L, circuit serves as audio null detector for bridge connected to input terminals; S_1 is used to increase sensitivity of 741 opamp as null is approached. Article covers operation of circuits in detail.—R. A. Hoare, An Audible Voltmeter and Bridge-Indicator, *Wireless World,* Sept. 1976, p 87–89.

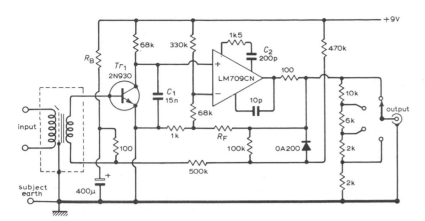

ELECTROMYOGRAM AMPLIFIER—Used to amplify voltages in range of several microvolts to several millivolts in frequency spectrum of 20 to 5000 Hz, as picked up with 13-mm thin silver disks placed on skin over muscle being studied. Article also covers electrocardiographic applications involving source impedances as high as 50 kilohms (as with one electrode on each wrist). Maximum output capability is 9 V P-P. Voltage gain is 1000. R_F is 800K pot, adjusted to give 12 dB per octave dropoff above turnover frequency.—R. E. George, Simple Amplifier for Muscle Voltages, *Wireless World*, Oct. 1972, p 495–496.

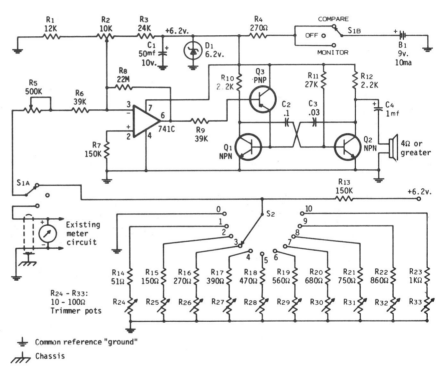

AUDIBLE VOLTMETER—Voltage-controlled audio oscillator produces 400-Hz tone for 0 V, with frequency of tone increasing with voltage over two-octave range to 1600 Hz for maximum or full-scale voltage. Ten-resistor voltage divider produces calibrated reference tones corresponding to main 0–10 divisions of meter scale for aural comparison. Simple square-wave audio oscillator Q_1-Q_2 is voltage-controlled by Q_3, which in turn is driven by opamp whose gain is set by R_5. Article covers adjustment of sensitivity pot R_5 and frequency pots R_{24}-R_{33} so VCO tracks voltage being measured and tones coincide at MONITOR and COMPARE positions of S_1 for each meter division.—H. F. Batie, An Audible Meter for the Blind Amateur, *CQ*, Dec. 1973, p 26–31.

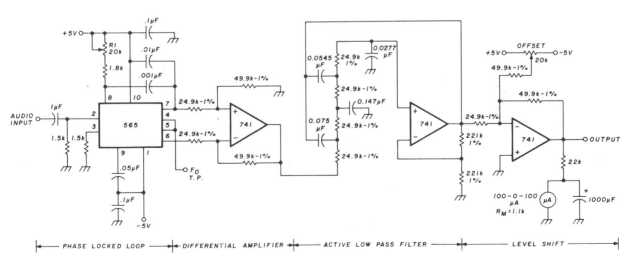

EKG FM DEMODULATOR—Developed as part of system using satellite for relaying electrocardiograms and other medical data having bandwidth of 0.6 to 50 Hz. Audio signal serving as source of FM is applied to voltage-controlled 1-kHz oscillator having ±40% deviation for full-scale input. Corresponding audio signal at receiving location is fed to input of 565 phase-locked loop. Error voltage of loop, at pin 7, contains data being sought as well as undesirable DC and AC components. DC component of error signal is removed by 741 differential amplifier following PLL. Following four-pole active RC low-pass filter eliminates high-frequency AC components and determines bandwidth of demodulator. Cutoff frequency is 100 Hz. Final 741 opamp scales and shifts output to reasonable value. Recorded output could not be distinguished from original EKG by doctors.—D. Nelson, Medical Data Relay via Oscar Satellite, *Ham Radio*, April 1977, p 67–73.

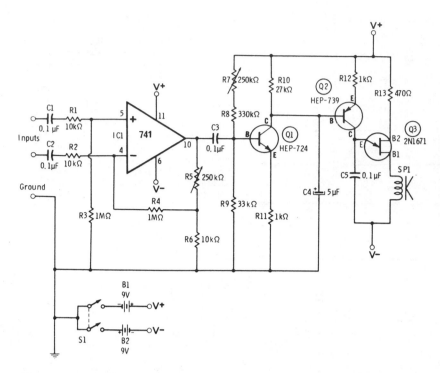

AUDIO EMG MONITOR—Used to measure very small voltages that appear on surface of skin over body muscle. Instead of recording voltage in form of electromyogram (EMG), opamp drives transistor circuit to produce audible note that varies in pitch as EMG signal varies in amplitude. Applications include use by stroke patient as aid to learning reuse of muscle group affected by stroke. Q1 rectifies and averages amplified EMG signal. Q2 controls charging current of C5 for varying frequency of UJT oscillator Q3.—R. Melen and H. Garland, "Understanding IC Operational Amplifiers," Howard W. Sams, Indianapolis, IN, 2nd Ed., 1978, p 125–127.

10-Hz LOW-PASS—Filter design for biomedical experiment has 10-Hz cutoff, tolerable transient and overshoot response, and at least 30-dB rejection of all frequencies above 15 Hz. All components should have 2% tolerance.—D. Lancaster, "Active-Filter Cookbook," Howard W. Sams, Indianapolis, IN, 1975, p 147.

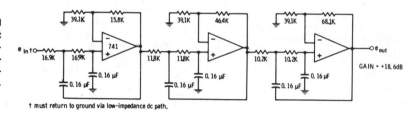

† must return to ground via low-impedance dc path.

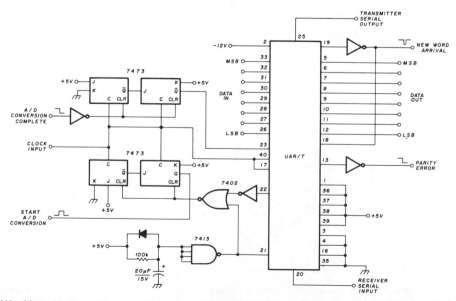

UART FOR EKG RELAY—After electrocardiogram is converted to digital form by commercial A/D converter, circuit shown takes 8-bit word output of converter for processing by universal asynchronous receiver-transmitter (UART) to give required serial asynchronous code for transmitter of satellite relay system, with start, stop, and parity bits added to data under control of 19.2-kHz external clock. This serial output is then used to control FSK oscillator that switches between two discrete audio frequencies to give signal required for transmission through satellite. Article covers operation of UART in detail.—D. Nelson, Medical Data Relay via Oscar Satellite, *Ham Radio,* April 1977, p 67–73.

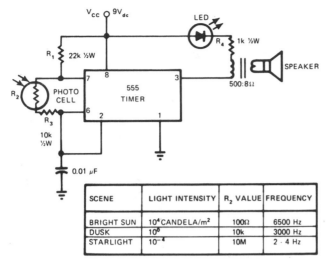

timer connected so frequency increases directly with intensity of light. Free-running frequency and duty cycle of timer operating in astable mode are controlled by two resistors and one capacitor. R_3 sets upper frequency limit at about 6.5 kHz, and dark resistance of photocell R_2 sets lower limit at about 1 Hz. Loudspeaker provides audio output, while LED flashes for visual indication when frequency goes below about 12 Hz. Applications include detection of lightning flashes, use as optical radar for blind, and use as sunrise alarm.—C. R. Graf, Build a Light Sensitive Audio Oscillator, *EDN Magazine,* Aug. 5, 1976, p 83.

SCENE	LIGHT INTENSITY	R_2 VALUE	FREQUENCY
BRIGHT SUN	10^4 CANDELA/m^2	100Ω	6500 Hz
DUSK	10^0	10k	3000 Hz
STARLIGHT	10^{-4}	10M	2 - 4 Hz

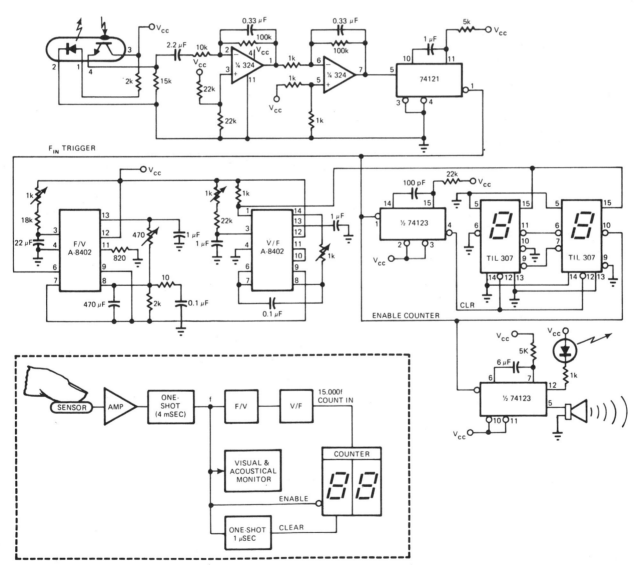

HEART-RATE MONITOR—Measures instantaneous frequency of such slow signals as heart beats (1 Hz) or 33-rpm motors (0.5 Hz) by measuring period T and inverting that quantity to obtain f. Operates from single 5-V supply for portable operation. Fast response time gives reading of heartbeat rate on digital display in two or three pulses. Optoisolator serving as sensor can be taped to almost any part of body because it responds to reflectivity changes caused by changing blood pressure. Accuracy is near 1%.—G. Timmermann, Heartbeat-Rate Monitor Captures VLF Signals, *EDN Magazine,* Oct. 20, 1977, p 79–80.

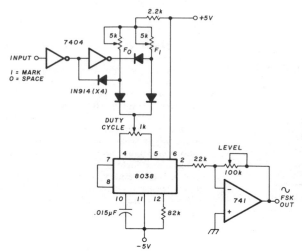

FSK OSCILLATOR FOR EKG RELAY—Used in satellite system for relaying electrocardiograms in digital form. Input consists of 8-bit words obtained in serial form from universal asynchronous receiver-transmitter. Uses 8038 function generator that is switched between two adjustable trimmer resistors giving independently adjustable discrete audio frequencies for mark and space. Output is phase-coherent even though switching does not necessarily take place at zero-crossing points of sine wave. Operation is much like that of FSK RTTY.—D. Nelson, Medical Data Relay via Oscar Satellite, *Ham Radio,* April 1977, p 67–73.

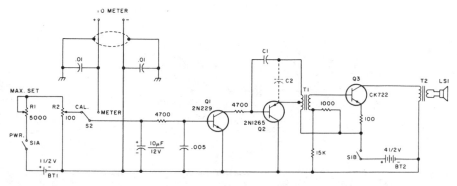

AUDIBLE METER READER—Analog meter terminals are connected to input of DC amplifier Q1 for feeding audio oscillator Q2 and output amplifier Q3. Frequency of oscillator is directly proportional to reading of meter. At calibrate position of S2, DC amplifier is fed by voltage divider R1-R2 and R2 is adjusted until tones heard are identical for both positions of S2. Developed for use by blind person. Knob of R2 sweeps over large scale having markings in Braille for reading of setting at which tones match. Alternatively, R2 can be preset to desired reading and equipment under test adjusted to give tone match. Article covers construction and calibration. C1 is chosen in range of 0.002 to 0.1 μF to give desired minimum frequency. C2, if required, is in same range. T1 is transistor driver transformer (10,000 to 2000 ohms), and T2 is transistor output transformer (500 to 3.2 ohms).—N. Rosenberg, Tune-Up Aids for the Blind, *73 Magazine,* Feb. 1978, p 64–67.

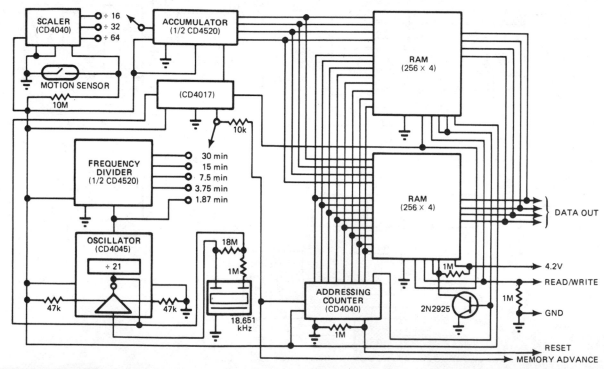

JOGGER DATA COLLECTION—Portable data acquisition system using microcomputer to drive digital cassette tape transport operates from 12-V rechargeable battery and fits in backpack having total weight of only 8 lb. Sample rate can be set between 20 and 100 Hz, with 2 min of continuous data being stored at fast rate. Recorded data is played into PDP-11 minicomputer later for analysis. Motion sensor shown can be replaced by other types of transducers for measuring desired physiological phenomena during jogging, walking, or running.—P. G. Schreier, Physiological Data Acquisition Presents Unusual Problems, Solutions, *EDN Magazine,* June 20, 1978, p 25–26, 28, and 30.

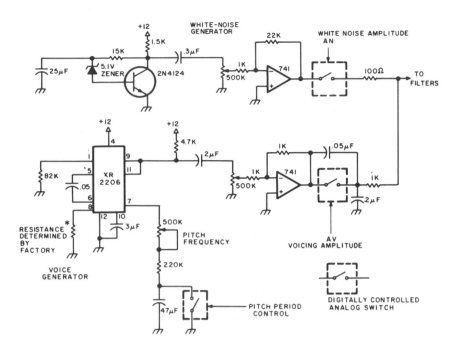

SPEECH SYNTHESIZER—Based on analog simulation of vocal tract. Rush of air through vocal passages is simulated by white-noise generator, while action of larynx is simulated in lower branch of circuit. Article covers problems involved in achieving transitions from phoneme to phoneme, along with automatic emphasis of leading or terminating consonants and intonation of rhythm associated with importance or placement of word in speech. ASCII symbols are given for 33 phonemes generated in Ai Cybernetic Systems model 1000 speech synthesizer, which uses circuit shown in combination with 10 active filters composed of 15 opamps, vocal excitation circuits, ASCII character decoders, and phoneme memories.—W. Atmar, The Time Has Come to Talk, *Byte,* Aug. 1976, p 26–30 and 32–33.

EKG TELEMETER—Developed for experimentation or educational demonstrations in which audience listens to electrocardiograph signal voltage as fed through LM4250 opamp for modulating NE566 connected as VCO driving small loudspeaker. Acoustic output can be picked up by microphone for telemetry purposes if desired. Connection to patient can be made with standard adhesive monitoring electrodes or with small metal disks held on wrists with rubber bands. Tone shifts frequency with each pulse beat.—M. I. Leavey, Inexpensive EKG Encoder, *73 Magazine,* Feb. 1978, p 20–23.

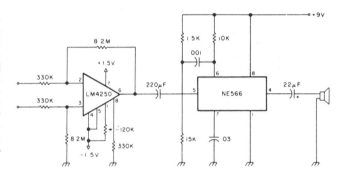

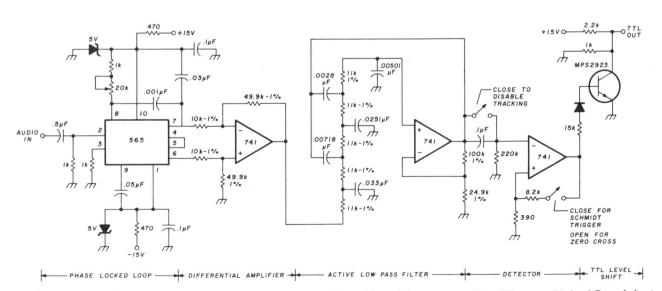

FSK DEMODULATOR FOR EKG RELAY—Used at receiving end of satellite system for relaying EKGs, to convert received audio FSK signal to TTL level-shifting output from which original EKG can be obtained. Phase-locked loop tracks input signal frequency and feeds appropriate error signal through differential amplifier to five-pole Butterworth low-pass filter having 1500-Hz cutoff. DC offset is removed by capacitor coupling, for use in zero-crossing detector or Schmitt-trigger detector. Signal is next converted into TTL-compatible level. Recorded output could not be distinguished from original EKG by doctors.—D. Nelson, Medical Data Relay via Oscar Satellite, *Ham Radio,* April 1977, p 67–73.

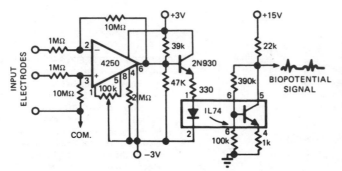

ISOLATED PREAMP—Optoisolator in electrocardiograph preamp circuit prevents circulating ground currents from shocking patients under test. Can be used with practically all other types of AC line-operated equipment in medical environments.—R. R. Ady, Let's Take an Illuminating Look at Latest Developments in LED's, *EDN Magazine*, Aug. 5, 1975, p 30–35.

AURAL SWR INDICATOR—Permits blind amateur radio operator to check standing waves on transmission line and adjust for best possible impedance match between source and load. Darkened areas are foil strips 6 × 70 mm, 1.5 mm apart, forming inductive trough that transfers RF energy from transmission line to simple aural monitor. Rectified RF energy changes bias on base of Q1, which makes tone increase in pitch with increasing voltage. Idling tone is about 500 Hz for values shown. Operates from three penlight batteries. Transmitter is peaked for maximum output on rising pitch, and matchbox antenna tuner is adjusted for minimum SWR on descending pitch. To lower audio tone, increase size of 82K resistor.—C. G. Bird, Aural SWR Indicator for the Visually Handicapped, *Ham Radio*, May 1976, p 53–53.

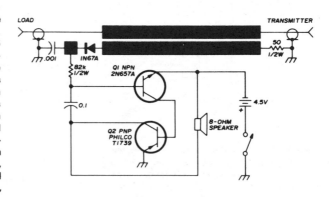

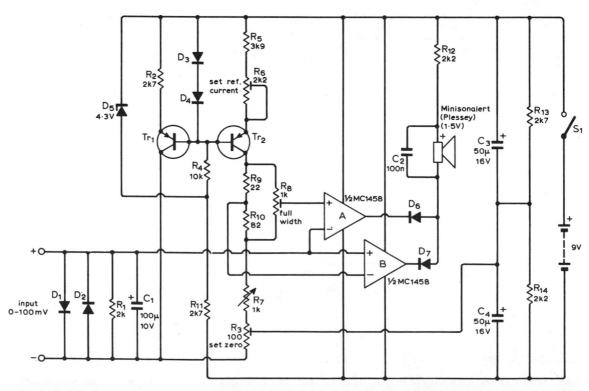

MULTIMETER FOR BLIND—Uses small electric horn to produce sound when DC voltage being measured is different from reference voltage value determined by setting of linear wirewound pot R7. Blind person adjusts R7 for null in sound, then reads Braille dots for that setting to get value of voltage being measured. Use PNP silicon transistors, such as BC177 or BC187. D5 is 4.3-V 400-mW zener, such as BZX79/C4V3. Other diodes are small-signal silicon, such as BA100 or 1N914A.—G. P. Roberts, Multimeters for Blind Students, *Wireless World*, April 1974, p 73–74.

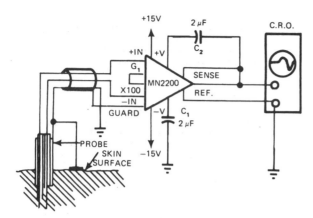

BIOELECTRIC VOLTMETER—Used to measure bioelectric phenomena involving both DC and waveform characteristics with amplitudes of about 10 mV. Since electrodes have impedance of 20,000 to 100,000 ohms, guard terminal must be used to drive input shield. Bias-current return comes from ground plate on skin. Fixed gain of 1000 gives absolute measure of input-voltage magnitude.—R. Duris, Instrumentation Amplifiers—They're Great Problem Solvers When Correctly Applied, *EDN Magazine*, Sept. 5, 1977, p 133–135.

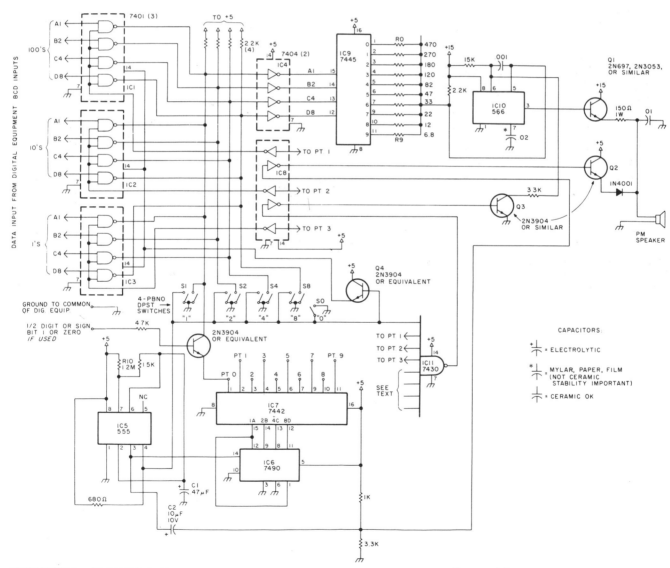

TONE OUTPUT FOR DIGITAL DISPLAY—Converts BCD Input from digital test gear to sequence of 10 different tones representing 0 to 9, for recognition of reading on digital display by blind radio operator or experimenter. Length of tone sequence equals number of digits displayed, plus sign indicator or half-digit if desired. Circuit shown is for 3½ digits. Article describes operation of circuit and gives construction details. Resistors R0-R9 (values in kilohms from 6.8K to 470K), determining frequencies of generated tones, are switched into VCO IC10 by IC9.—D. R. Pacholok, Digital to Audio Decoder, *73 Magazine*, Oct. 1977, p 178–180.

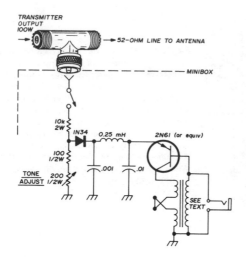

AUDIBLE TUNING FOR BLIND—Transmitter or exciter output is sampled at coax line and high-resistance voltage divider. Rectified voltage of divider, which varies during transmitter tuning, is fed to relaxation oscillator whose output varies in pitch with voltage; low voltage gives high-pitched tone, and high voltage gives low-pitched tone. Input divider draws about 1 W from 100-W transmitter; for higher power, such as 1 kW, change 10K to about 100K. Diode feeds about 2 V to emitter of transistor. Any audio-type PNP transistor can be used. For NPN device, reverse diode connections. Transformer is from 5-W transistor amplifier, with 22-ohm high-impedance winding. Other two windings, in series aiding, are 4 ohms each.—D. H. Atkins, Tuning Aid for the Sightless, *Ham Radio,* Sept. 1976, p 83.

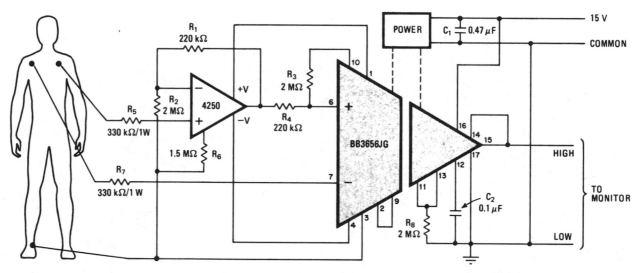

HEART MONITOR—Electrocardiograph amplifier uses Burr-Brown BB3656 isolation amplifier to protect electrocardiograph from inadvertent applications of defibrillation pulses while pa-tient is being monitored. Heart pulses are accurately amplified over frequency range of DC to 3 kHz. Resistors must be carbon-composition types.—B. Olschewski, Unique Transformer Design Shrinks Hybrid Isolation Amplifier's Size and Cost, *Electronics,* July 20, 1978, p 105–112.

CHAPTER 50
Memory Circuits

Includes circuits for read, write, refresh, one-time programming, and interfacing various types of microprocessor memories, LED display showing PROM status, wave synthesizer, punched-card converter, and addition of memory capacity to calculators. See also Microprocessor chapter.

LEVEL SHIFT WITH IC—Level shifter for dynamic MOS random-access memory uses SN7406 IC with two-transistor booster to convert TTL levels to MOS levels. Booster can be omitted for data input lines because they drive such low capacitive loads in typical arrays.—M. E. Hoff, Designing an LSI Memory System That Outperforms Cores—Economically, *Computer Hardware* (section of *EDN Magazine*), Jan. 15, 1971, p 6–15 (p 000110–001111).

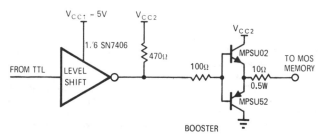

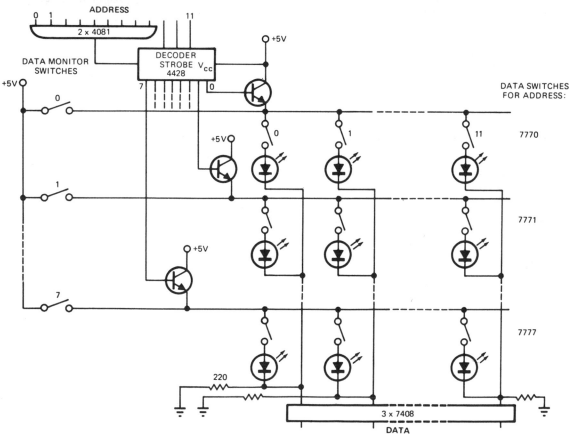

PROM WITH LED DISPLAY—Developed for use in debugging small microprocessor systems. Uses LEDs in place of diodes as OR gates of 8 × 12-bit diode matrix memory which displays memory-cell content when word is addressed during execution of a program. Monitor switches can be used for data display when program is inserted. System was built to debug Intersil IM6100 μP. Since voltage output of diode array is too small for direct input to MOS circuit, 7408 gates are used to boost high-level output. Article gives instructions for use of display.—K. S. Hojberg, Light-Emitting Memory Aids μP Debugging, *EDN Magazine*, May 5, 1977, p 107.

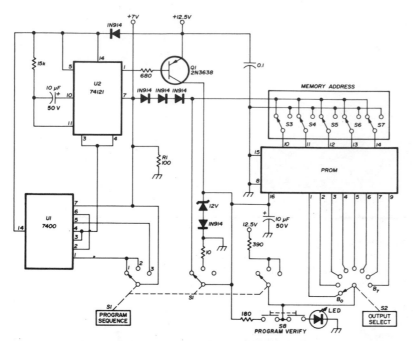

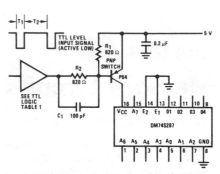

PROGRAMMER FOR SIGNETICS 8223— Bounceless switch U1 triggers mono MVBR U2, both operating at 7 V above ground. When Q1 is saturated by pulse from U2, it applies 250-ms 12.5-V programming pulse to V_{CC} terminal (pin 16) of memory chip and opens fuse at previously addressed bit to make it logic 1. Separate regulators are required for 7 V and 12.5 V. Used with alphameric display having five 7-segment

digits in circuit serving as function/units indicator for interval timer/counter, where it forms simulation of abbreviations for time and frequency units. Article gives step-by-step instructions for mistake-free operation of programmer.—J. W. Springer, Function/Units Indicator Using LED Displays, *Ham Radio,* March 1977, p 58–63.

PROM POWER-DOWN—Conserves power in applications when data is required from PROM for only small percentage of system cycle. Circuit turns PROM off automatically when not needed, with access time increased only by 80-ns delay of power-down circuit. PNP switch can be PN4313 or 2N3467 pass transistor. With two 74S04 sections in series at input, active low selection is obtained. 74S00 at input gives active high selection. When logic input to R_2 goes low, PNP switch is turned on and +5 V is applied to National DM74S287 256 × 4 PROM.—"Memory Applications Handbook," National Semiconductor, Santa Clara, CA, 1978, p 5-9–5-12.

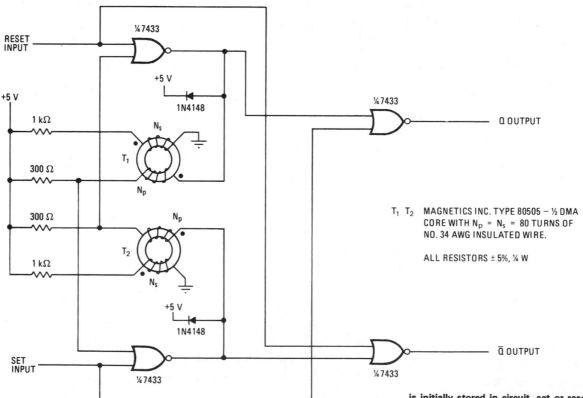

T$_1$ T$_2$ MAGNETICS INC. TYPE 80505 – ½ DMA CORE WITH N_p = N_s = 80 TURNS OF NO. 34 AWG INSULATED WIRE.

ALL RESISTORS ± 5%, ¼ W

LATCH MEMORY—Use of saturable transformers makes memory nonvolatile and immune to false command signals. Transformers

stay magnetically biased without voltage supply to provide reference state to which memory latch returns when power is reapplied. When bit

is initially stored in circuit, set or reset pulse must have minimum width of 35 μs and minimum separation of 65 μs.—G. E. Bloom, Saturable Core Transformers Harden Latch Memories, *Electronics,* March 31, 1977, p 104–105.

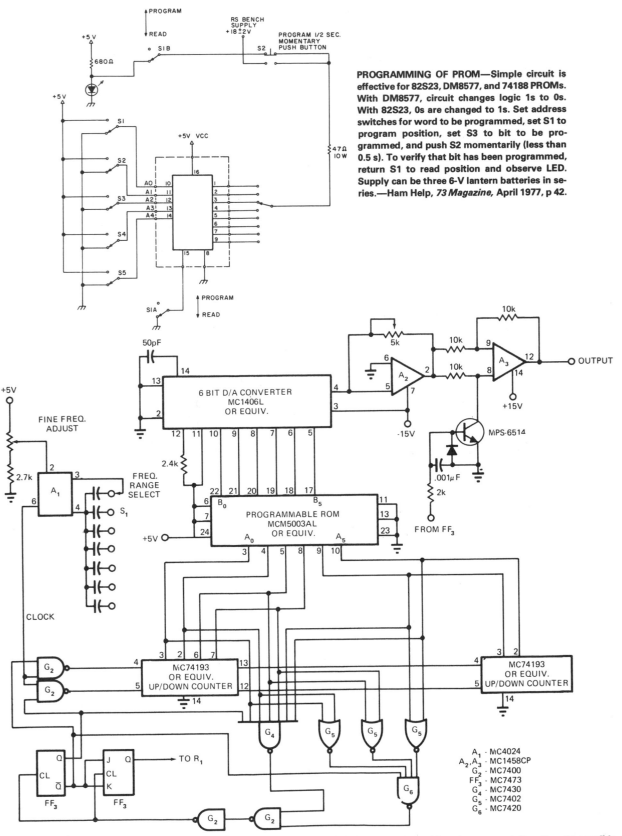

PROGRAMMING OF PROM—Simple circuit is effective for 82S23, DM8577, and 74188 PROMs. With DM8577, circuit changes logic 1s to 0s. With 82S23, 0s are changed to 1s. Set address switches for word to be programmed, set S1 to program position, set S3 to bit to be programmed, and push S2 momentarily (less than 0.5 s). To verify that bit has been programmed, return S1 to read position and observe LED. Supply can be three 6-V lantern batteries in series.—Ham Help, *73 Magazine*, April 1977, p 42.

A₁ - MC4024
A₂,A₃ - MC1458CP
G₂ - MC7400
FF₃ - MC7473
G₄ - MC7430
G₅ - MC7402
G₆ - MC7420

WAVE SYNTHESIZER—Virtually any symmetrical waveform can be generated by using only IC counters, a read-only memory, and a monolithic D/A converter. Only first 90° of waveform need be digitized; this information can be manipulated to generate other 270° and repeated as often as necessary. To digitize desired waveform, divide first 90° into 64 points, calculate sine or other function for each, multiply each result by 63 to normalize, round off, convert each to 6-bit binary equivalent, take complements, and use results for programming ROM. Article describes operation of circuit in detail. Use of MC1480 8-bit monolithic D/A converter gives better resolution than is possible with MC1406L 6-bit D/A converter because 8-bit words give 256 discrete output levels instead of 64.—K. Huehne, Programmable ROMs Offer a Digital Approach to Waveform Synthesis, *EDN Magazine*, Aug. 1, 1972. p 38–41.

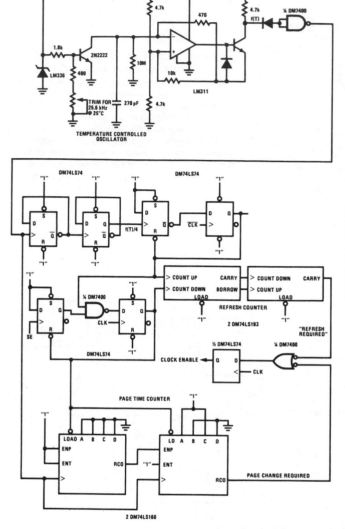

ADAPTIVE REFRESH—Circuit monitors system utilization of National MM2464 64-kilobit charge-coupled device (CCD). Refresh time and maximum page times are determined by two counters that obtain clock signals from temperature-controlled oscillator.—"Memory Applications Handbook," National Semiconductor, Santa Clara, CA, 1978, p 7-1–7-10.

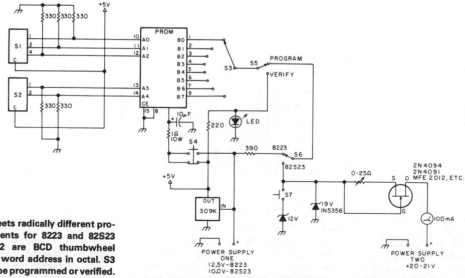

PROM BURNER—Meets radically different programming requirements for 8223 and 82S23 PROMs. S1 and S2 are BCD thumbwheel switches that select word address in octal. S3 selects output bit to be programmed or verified. S4 is pushed momentarily to burn out programmed bit, and S5 verifies programming. S6 is set to PROM type, and S7 puts 12-V zener across 19-V zener for current calibration. For 82S23, adjust power supplies to 10 and 21 V and adjust 25-ohm pot to give meter reading of 65 mA when S7 is closed.—W. J. Hosking, Finally! A Simple PROM Burner!, *73 Magazine*, Dec. 1977, p 186–187.

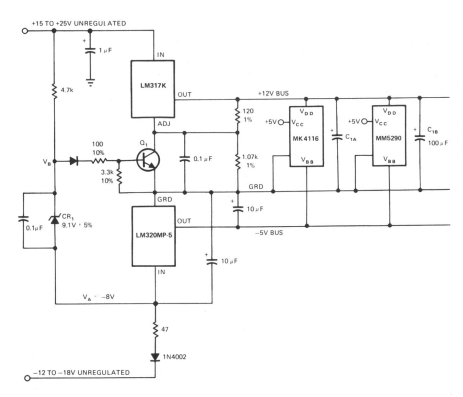

MEMORY-PROTECTING REGULATOR—Developed for MOS RAM in which accidental removal of −5 V bias supply would allow buildup of currents flowing between +12 V and ground to value sufficient to open up metal paths and destroy memory. Protection is achieved by feeding unregulated negative voltage to on-card local regulator using LM320MP-5 to provide −5 V regulated bias for all memory chips. LM317K is used as +12 V regulator for delivering up to 1.5 A. If −12 V gradually drops out of regulation, Q_1 turns on and pulls LM317K adjust pin to ground so output of this regulator drops to +1.3 V and logic circuits are undamaged. If −12 V shorts to ground, capacitors on −5 V line hold up for several hundred microseconds so Q_1 has time to turn off +12 V regulator. Use heatsinks for regulators.—R. Pease, Safe Supply Manages MOS Memories, *EDN Magazine,* Oct. 20, 1978, p 82 and 84.

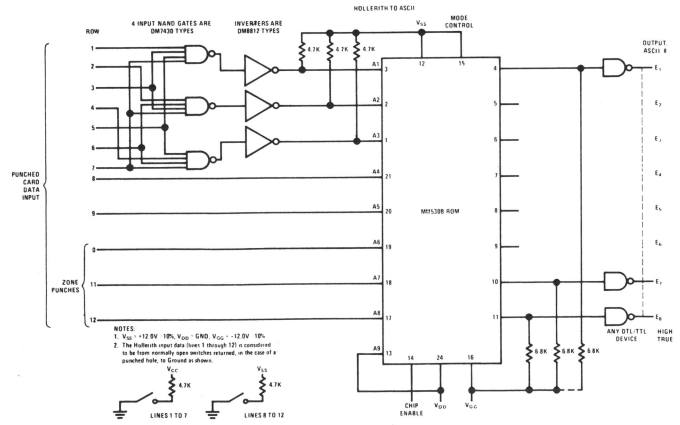

CARD CONVERTER—National MM530B ROM forms basis for conversion of 12-line Hollerith punched-card code to 8-line ASCII. All 12 inputs from cards are presented to programmable logic array (PLA). Invalid input produces all-high output state because it is not recognizable product term. First 7 Hollerith lines, which are ordinary decimally coded lines, are encoded to 3 binary lines with additional logic elements shown, before being presented into common 8-input ROM.—"Memory Databook," National Semiconductor, Santa Clara, CA, 1977, p 11-49–11-56.

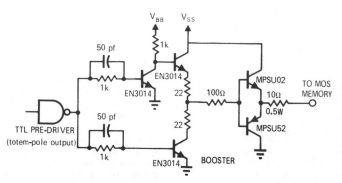

LEVEL SHIFTER FOR RAM—Uses predriver with three transistors, followed by two-transistor booster, to convert TTL levels to those required by dynamic MOS random-access memory. Booster can be omitted for data input lines because they drive such low capacitive loads in typical arrays.—M. E. Hoff, Designing an LSI Memory System That Outperforms Cores—Economically, *Computer Hardware* (section of *EDN Magazine*), Jan. 15, 1971, p 6–15 (p 000110–001111).

ADDING FOUR FUNCTIONS—Capability of 8-digit four-function Novus 850 calculator made by National Semiconductor (also marketed as Montgomery Ward P50) can be doubled by adding four SPST switches and connecting as shown. These provide additional functions of memory store, memory recall, percent, and constant. Switches can be put on front panel above display, at corners of battery. Cutler Hammer SA1BV20 SPST switches with SW53AA1 caps can be squeezed in.—D. Arnett, Add Memory, Constant and % to a 4-Function Calculator, *EDN Magazine,* Aug. 20, 1975, p 82.

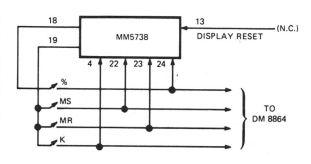

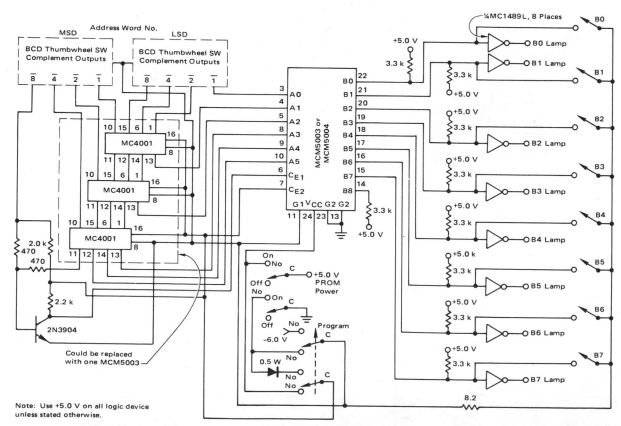

512-BIT PROM PROGRAMMER—Simple manual programmer requires minimum equipment for fusing memory links of Motorola MCM5003 or MCM5004 programmable read-only memory. One link is fused at a time. MC1489 quad line receivers show contents of output bits by driving 5-V 20-mA lamps. Address word number is selected with two BCD thumbwheel switches. Three MC4001 ROMs convert BCD code to that required at address inputs. Program/verify switch must simultaneously be set along with proper output switch for each used bit.—J. E. Prioste, "Programming the MCM5003/5004 Programmable Read Only Memory," Motorola, Phoenix, AZ, 1974, AN-550, p 4.

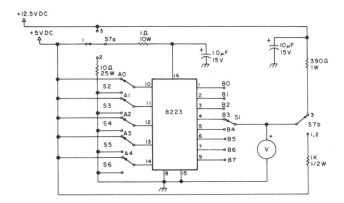

PROM PROGRAMMER—Can be used to blow links one by one, as required, on 8223 programmable read-only memory and on 82S23 Schottky version. Article covers construction, pretesting, and operation.—R. M. Stevenson, An 82S23 PROM Programmer!, *73 Magazine,* June 1977, p 82–83.

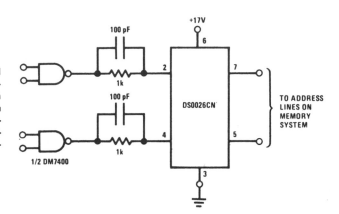

CLOCK DRIVER—Circuit uses National DS0026CN monolithic clock driver as direct-coupled driver for address or precharge lines on MM1103 RAM at frequencies above 1 MHz. Can be driven by standard TTL gates and flip-flops.—"Memory Databook," National Semiconductor, Santa Clara, CA, 1977, p 11-27–11-36.

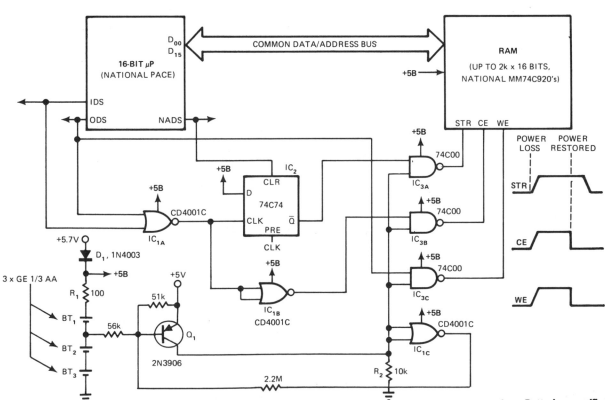

ON-BOARD SUPPLY PROTECTION FOR RAM—To prevent loss of memory data when RAM is removed from microprocessor control system during system design or maintenance, three miniature nickel-cadmium batteries are mounted directly on memory board. Data-protection circuit senses loss of 5.7-V main supply and disables STR, CE, and WE lines of RAM to prevent memory loss. Batteries specified provide protection for up to 40 days.—F. R. Quinlivan, On-Board Backup Supply Protects Volatile RAM Data, *EDN Magazine,* April 5, 1978, p 120 and 122.

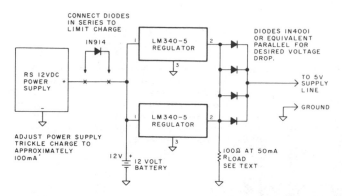

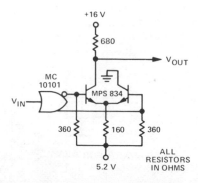

MEMORY SAVER—Standby battery takes over automatically during power failure to prevent memory loss. Use any rechargeable battery that can handle load. Connect one LM340-5 regulator in circuit for each 1.2 A of load current drawn by circuits to be protected. In normal operation, output diodes are biased off by slightly higher voltage from computer. During power failure, they are switched on by removal of bias. Current is then supplied by battery. Output should be loaded with resistor drawing about half of trickle-charge current, so battery has small continuous current flow and stays charged.—C. R. Carpenter, Protect Your Memory Against Power Failure, *Kilobaud,* March 1978, p 73.

LOW-CAPACITANCE DRIVER FOR RAM—Suitable for data input lines of memory system operating from 16-V supply with memory input logic swing of 16 V. Used with Motorola 1103 dynamic RAM for which transition times of data input signals should be 20 ns to give maximum memory speed. Suitable for maximum capacitances per line as high as 20 pF.—D. Brunner, "A MECL 10,000 Main Frame Memory System Employing Dynamic MOS RAMs," Motorola, Phoenix, AZ, 1975, AN-583, p 13.

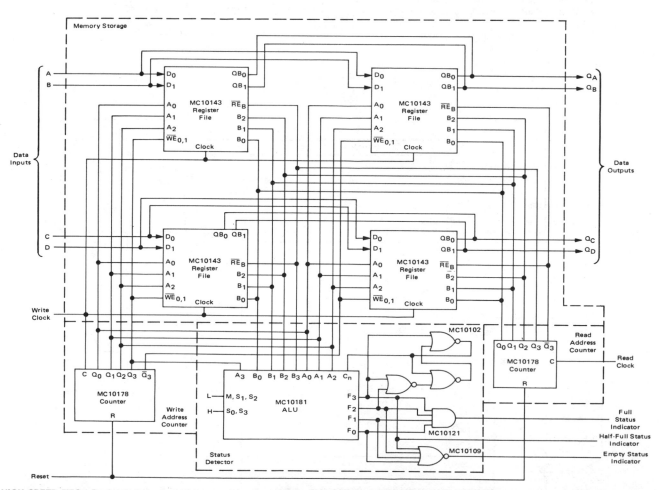

HIGH-SPEED FIFO—Design is based on Motorola MC10143 register file, with each IC holding 8 words by 2 bits. Circuit includes write and read enable inputs for cascading two register file packages to memory depth of 16 words. Full master-slave flip-flop operation allows simultaneous read and write. Reset is applied initially to drive both address counters to empty state. To enter data, write clock input is enabled with negative-going pulse. Write addressing is controlled by MC10178 binary counter. Used for stack registers of computing systems when register outputs are read sequentially in same order that data was entered (first-in first-out).— B. Blood, "A High Speed FIFO Memory Using the MECL MC10143 Register File," Motorola, Phoenix, AZ, 1974, AN-730, p 5.

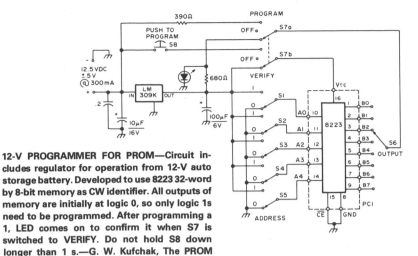

12-V PROGRAMMER FOR PROM—Circuit includes regulator for operation from 12-V auto storage battery. Developed to use 8223 32-word by 8-bit memory as CW identifier. All outputs of memory are initially at logic 0, so only logic 1s need to be programmed. After programming a 1, LED comes on to confirm it when S7 is switched to VERIFY. Do not hold S8 down longer than 1 s.—G. W. Kufchak, The PROM Zapper, *73 Magazine*, Sept. 1976, p 112.

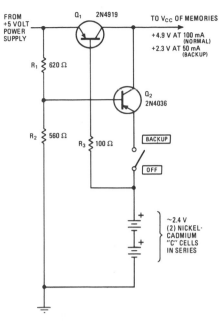

BATTERY BACKUP—Delivers 2.3 V to microprocessor memory automatically in event of supply failure, to prevent loss of data. On standby, batteries receive charge of about 20 mA through R_3 and Q_1. When power supply fails, Q_1 isolates it from load and Q_2 conducts to provide changeover to battery power. Standby switch (optional) permits defeating battery backup.—R. N. Bennett, 2.4-V Battery Backup Protects Microprocessor Memory, *Electronics*, Feb. 3, 1977, p 109; reprinted in "Circuits for Electronics Engineers," *Electronics*, 1977, p 304.

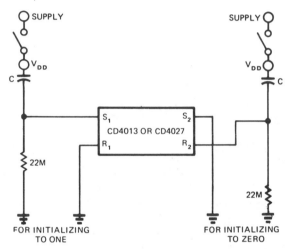

INITIALIZING CMOS STORAGE—Simple RC circuit initializes all storage elements (flip-flops, registers, and counters) of CMOS system to all 1s (switch at left) or all 0s (switch at right) when power supply is turned on. For most CMOS storage elements, 30 pF for C ensures setting or resetting when power is applied by closing switch. If power supply is turned on while supply line is directly connected, C should be 1000 to 1500 pF. Article also shows how to get set or reset function after initialization by using pair of CD4016A or CD4066A transmission gates.—O. Bismarck, A Simple Method for Initializing CMOS Storage Elements, *EDN Magazine*, Feb. 20, 1974, p 83.

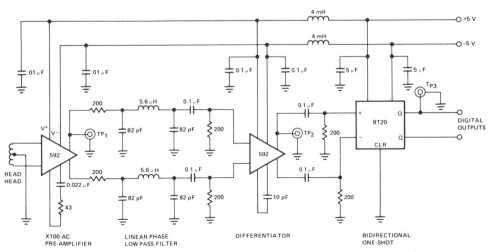

DISK-FILE DECODER—Provides preconditioning of readback data from disk or drum files by using NE592 video amplifier coupled to 8T20 bidirectional mono MVBR through low-pass filter and second 592 serving as low-noise differentiator/amplifier. Mono provides required output pulses at zero-crossing points of differentiator. Designed for reading 5-MHz phase-encoded data.—"Signetics Analog Data Manual," Signetics, Sunnyvale, CA, 1977, p 708–710.

PROGRAMMER FOR SCIENTIFIC CALCULATOR—National MM5766 dynamic key sequence programmer can be added to MM5758 scientific calculator chip to provide learn-mode programmability. Circuit memorizes any combination of key entries up to 102 characters while in load mode, then automatically plays back programmed sequence as often as desired in run mode. Halt key programs variable data entry points at which control is temporarily returned to operator in run mode.—"MOS/LSI Databook," National Semiconductor, Santa Clara, CA, 1977, p 8-76–8-79.

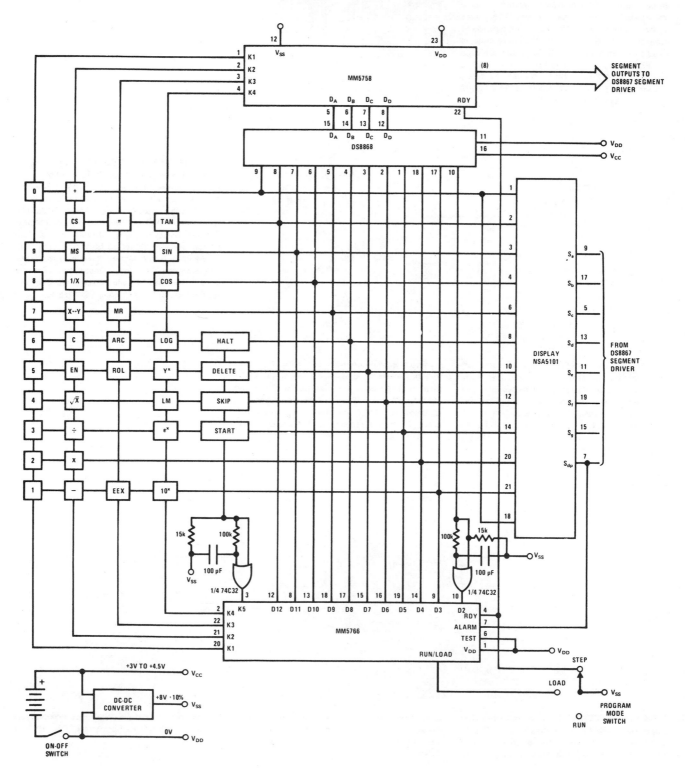

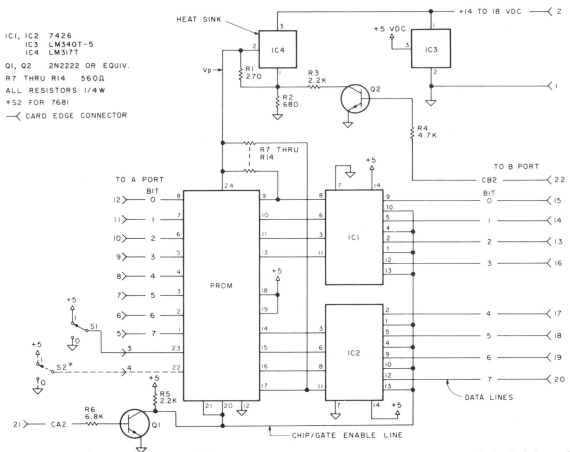

IC1, IC2 7426
IC3 LM340T-5
IC4 LM317T

Q1, Q2 2N2222 OR EQUIV.

R7 THRU R14 560Ω

ALL RESISTORS 1/4W

*S2 FOR 7681

—< CARD EDGE CONNECTOR

PROM PROGRAMMER—Developed specifically for programming Harris 7641-5 and 7681-5 PROMs, which come with logic 1 in all bit positions. Programmer is used to burn bit at each position requiring logic 0 by process equivalent to blowing fusible link on PROM chip. Process is not reversible, so one error ruins PROM. Circuit uses two regulator ICs (IC3 and IC4) and two TTL gates (IC1 and IC2). Article gives programming procedure in detail. Software for generating critical timing and control signals with microprocessor is also given.—T. Hayek, Simple and Low-Cost PROM Programmer, *Kilobaud,* July 1978, p 94–96 and 98–99.

CHAPTER 51

Microprocessor Circuits

Includes handshaking, time-logging, power-failure alarm, motor control, light pen, interface, and troubleshooting circuits for microcomputers, minicomputers, and computers as well as for microprocessors. See also Clock Signal, Game, Keyboard, and Programmable chapters.

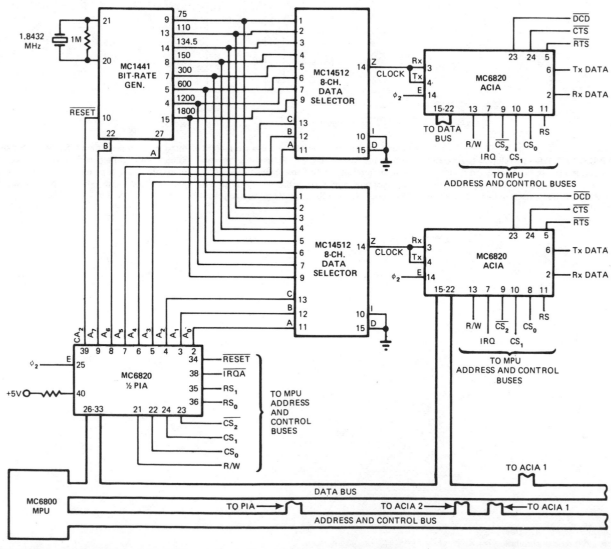

PROGRAMMED BIT RATES—Choice of 37 different bit rates in range from 75 to 1800 b/s, each multiplied by 1, 8, 16, or 64, can be obtained from Motorola MC1441 bit-rate generator, two 8-channel data selectors, half of peripheral interface adapter (PIA) and two asynchronous communication interface adapters (ACIA) by appropriate programming of Motorola MC6800 microprocessor used in data communication system. Article gives operating details and example of initialization program.—G. Nash, Microprocessor Software Programs Bit-Rate Generator, *EDN Magazine,* Aug. 20, 1977, p 134, 136, and 137.

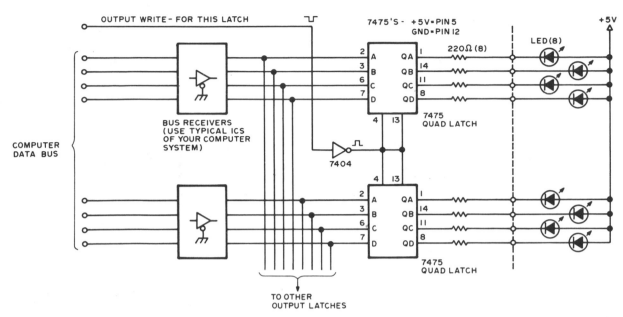

MOVING-LIGHT DISPLAY—Computer-controlled blinking of LEDs arranged in circle or other pattern gives illusion of motion. One section of quad 7475 latch is assigned to each of several output ports of microprocessor. Decoding logic of output port determines when latch is addressed for output. Decoded WRITE signal latches data presented at bus receivers. Interface can be extended to many registers in groups of eight, limited only by input/output addressing capability and power available for LEDs. Software is written to turn on LEDs in desired sequence and provide desired variable delays between blinks.—C. Helmers, There's More to Blinking Lights than Meets the Eye, *BYTE*, Jan. 1976, p 52–54.

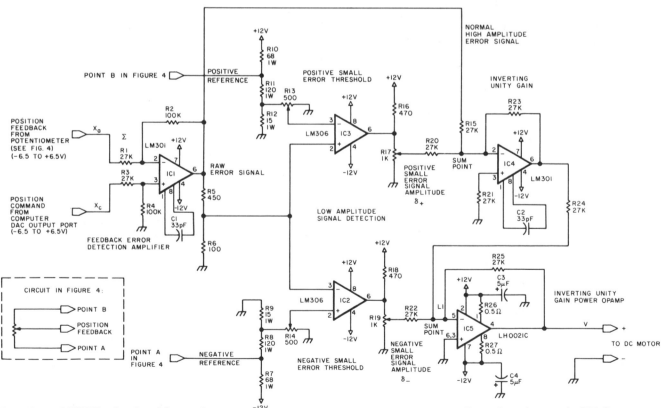

DC MOTOR CONTROL—Developed for use in realistic lunar lander simulation display. Throttle signal and altitude signal serve as inputs to microprocessor. Feedback position-measuring pot is geared to 12-VDC motor so full travel of pot shaft occurs while lunar module traverses full altitude range. Circuit provides minimum drive voltage required by DC motor for motion to occur. Output of difference amplifier IC1 goes to summing amplifier IC4 as one component of final motor voltage. Comparators IC2 and IC3 sense when difference voltage is larger than small positive voltage set by R13 or smaller than small negative voltage set by R14. Comparator output then becomes 12 V, and portion of this (about 2 V) drives motor into operating range. IC5 is high-power opamp delivering 1 A at 12 V.—L. Sweer, T. Dwyer, and M. Critchfield, Controlling Small DC Motors with Analog Signals, *BYTE*, Aug. 1977, p 18–20, 22, and 24.

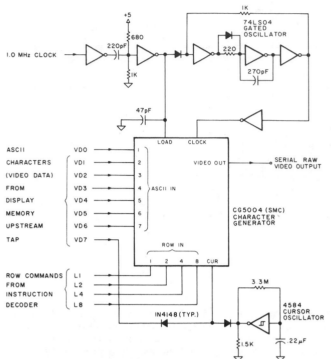

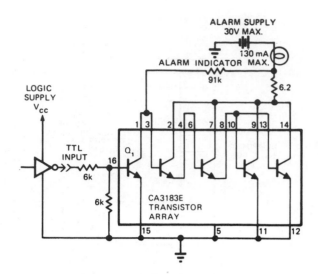

FAIL-SAFE ALARM—Circuit monitors status of microprocessor system and energizes lamp or other alarm indicator to indicate when CPU halts or power is lost. Input is TTL-compatible, and output can drive loads up to 130 mA at 30 V.—J. Elias, Alarm Driver Is Fail-Safe, *EDN Magazine,* May 20, 1975, p 76.

UPPERCASE/LOWERCASE DRIVE FOR TV— Standard Microsystems CG5004 alphameric data-to-video converter provides both uppercase and lowercase characters and all numerals in serial video form for display as 7 × 9 character matrix on TV screen under microprocessor control. IC requires only single +5 V supply. Winking underline cursor is produced automatically by cursor oscillator. Internal shift register is part of IC. Raw video requires predistorting for clarity and addition of sync pulses before it can be fed to TV set.—D. Lancaster, TVT Hardware Design, *Kilobaud,* Jan. 1978, p 64–68.

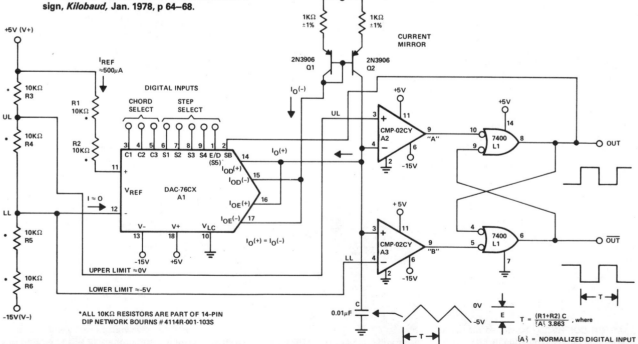

OSCILLATOR CONTROL—Digital inputs from microprocessor to Precision Monolithics DAC-76CX 8-bit companding D/A converter provide 8159:1 frequency range for AF oscillator, from 2.5 to 20,000 Hz. DAC functions as programmable current source that alternately charges and discharges capacitor between precisely controlled upper and lower limits. Since both limits are derived by dividing power supply voltages, frequency is independent of changes in supply voltage. Design equations are given.— D. Soderquist, "Exponential Digitally Controlled Oscillator Using DAC-76," Precision Monolithics, Santa Clara, CA, 1977, AN-20, p 1.

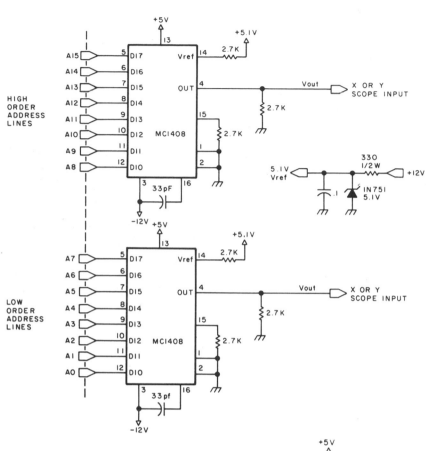

ADDRESS STATE ANALYZER—Dynamic fluctuations of 16-bit memory address bus are displayed on CRO for troubleshooting. Two MC1408 8-bit DACs drive inputs of CRO with analog equivalents of eight high-order and eight low-order address lines. Display serves as visual picture of computer in action, in which accessing of unexpected memory locations is instantly visible. Incoming address lines can be connected to MC1408s in any order. Article covers evaluation of scope patterns.—S. Ciarcia, A Penny Pinching Address State Analyzer, *BYTE*, Feb. 1978, p 6, 8, 10, and 12.

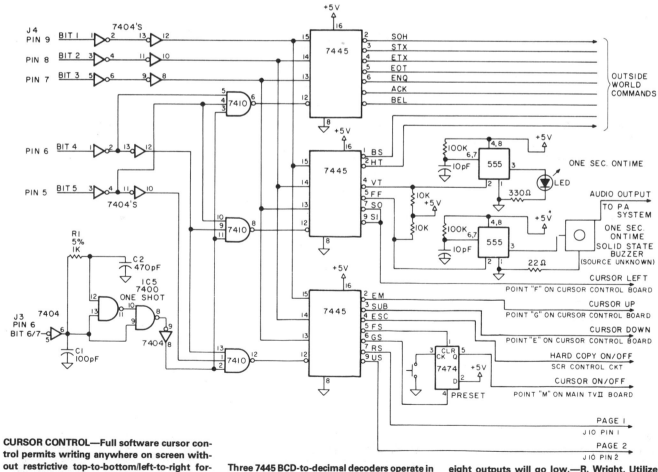

CURSOR CONTROL—Full software cursor control permits writing anywhere on screen without restrictive top-to-bottom/left-to-right format. System uses 18 of possible 32 ASCII control codes in TV II system having 8K BASIC. Three 7445 BCD-to-decimal decoders operate in 3-line-to-8-line mode wherein pin 12 becomes chip enable. When pin 12 goes low, one of the eight outputs will go low.—R. Wright, Utilize ASCII Control Codes!, *Kilobaud*, Oct. 1977, p 80–83.

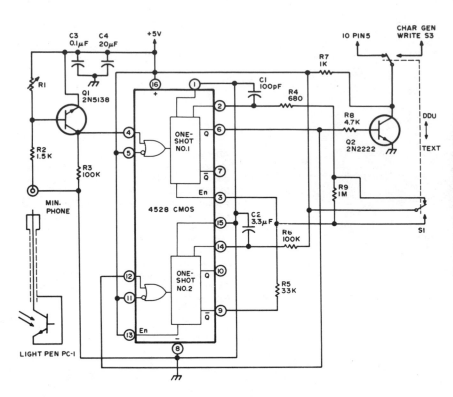

PICTURE-DRAWING LIGHT PEN—Circuit improves ability to draw pictures on display screen with light pen by using short data lockout period to avoid smearing. Value of R1 depends on light pen; use 1 megohm for Texas Instruments H-35. One-shot No. 1 produces constant-amplitude 200-ns pulse for storing 1 or 0 bit in 2102 memory of CRO graphics interface. One-shot No. 2 delays generation of another write command 0.25 s, giving operator time to withdraw or move pen before double dot is formed. R4 and C1 control length of write pulse, while R5 and C2 control wait time.—S. S. Loomis, Let There Be Light Pens, *BYTE,* Jan. 1976, p 26–30.

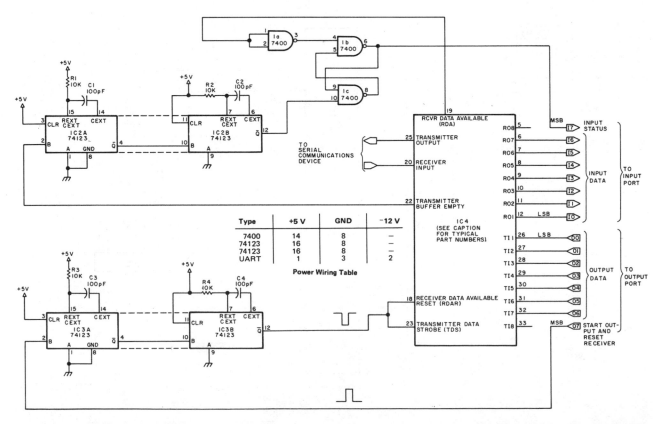

Type	+5 V	GND	−12 V
7400	14	8	—
74123	16	8	—
74123	16	8	—
UART	1	3	2

Power Wiring Table

HANDSHAKING—Circuit sets up operating connection between computer and UART (universal asynchronous receiver-transmitter). Eighth bits of I/O ports indicate when data has been successfully transmitted and system is ready to transmit more information. Can be adapted to any 8-bit computer. IC4 is standard UART such as Signetics 2536, General Instruments AY 5 1012, Texas Instruments TMS 6011, or American Microsystems S1883. Receiver of UART has seven data lines connected to input port. Article covers handshaking operation in detail and gives typical software routines for parallel I/O handshaking. Technique permits running UART at any desired clock speed, as long as all clocks in system are matched.—T. McGahee, Save Software: Use a UART for Serial IO, *BYTE,* Dec. 1977, p 164–166.

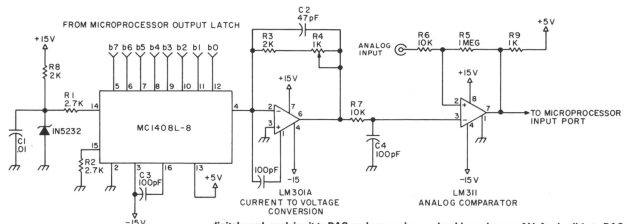

ANALOG/DIGITAL CONVERSION—Circuit can be used for either ramp or successive-approximation method of converting analog input to digital word, applying it to DAC, and comparing analog output of DAC to analog input to be converted. Results of comparison determine next digital word to be generated for DAC by microprocessor. LM301A changes 0–2 mA output of DAC into 0–5 V for LM311 comparator. To calibrate, apply all 0s to DAC; pin 6 of LM301A should now be near 0 V. Apply all 1s to DAC and adjust output of LM301A to 5.00 V with R4. Conversion routines implementing these functions are given for Motorola MC6800 and Intel 8008 microprocessors.—R. Frank, Microprocessor Based Analog/Digital Conversion, *BYTE*, May 1976, p 70–73.

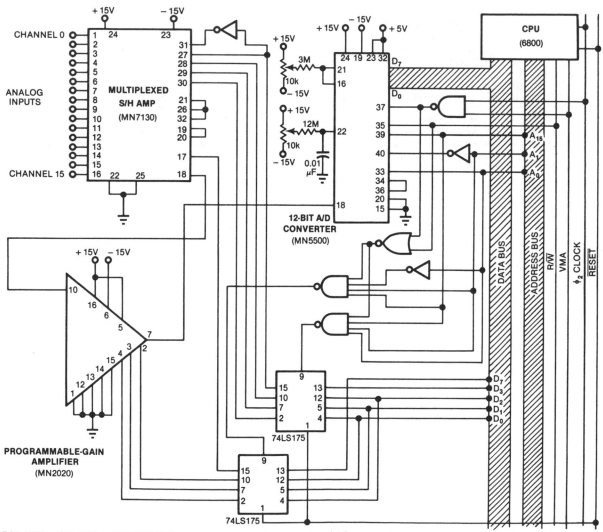

AUTORANGING FOR DATA ACQUISITION—Circuit accepts up to 16 channels of analog information and provides full-scale input ranges from ±78 mV to ±10 V when used with 6800 CPU. Resolution is 0.01%. MN7130 multiplexed sample-and-hold amplifier includes eight full-differential multiplexers for 16 channels and instrumentation opamp. MN2020 programmable-gain amplifier provides choice of eight gains ranging from 1 to 128 in binary progression. Autoranging operation of data acquisition system allows channel-sampling plans that depend on random events. With known input signal levels, required gain information can be stored in microprocessor memory for use in place of autoranging.—R. Duris and J. Munn, PGA's Give Your DAS Designs Autoranging, Wide Dynamic Range, *EDN Magazine*, Sept. 5, 1978, p 137–141.

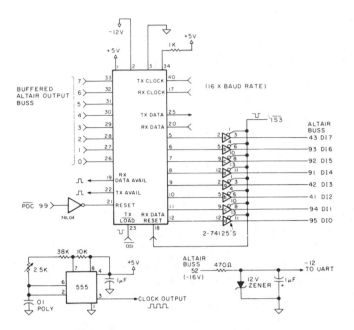

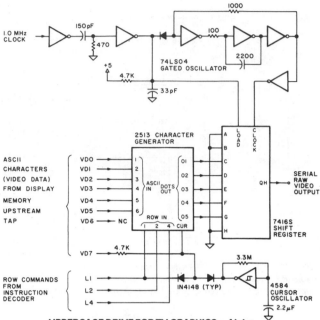

UART INTERFACE—Uses TMS-6011 UART to convert parallel data into serial data and back again for Altair 8800 microprocessor. UART mates directly to computer bus, because all outputs from UART are three-state buffers with separate enable lines provided for status bits and 8 bits of parallel output. Pin 22 is high when UART can accept another character for conversion. Pin 18 must be pulsed low to reset pin 19 so it can signal receipt of another character. Connections to pins 35–39 depend on I/O devices used, as covered in article.—W. T. Walters, Build a Universal I/O Board, Kilobaud, Oct. 1977, p 102–108.

UPPERCASE DRIVE FOR TV GRAPHICS—Alphameric data-to-video converter using 2513 character generator accepts ASCII words from microprocessor memory and three line commands from instruction decoder. Five dots are outputted simultaneously, corresponding to one row on 5 × 7 dot-matrix character. 7416S eight-input one-output shift register converts dots into serial output video. Input repeats to generate all seven dot rows in row of characters. Shift register is driven by high-frequency timing circuit that delivers LOAD pulse once each microsecond along with CLOCK output running continuously at desired dot rate. Optional cursor uses 4584 5-Hz oscillator for cursor winking rate. If ASCII input bit 8 is high, cursor input goes high and output is white line on leads 01 through 05. Right diode mades this line blink, while left diode allows winking cursors only during valid character times.—D. Lancaster, TVT Hardware Design. Kilobaud, Jan. 1978, p 64–68.

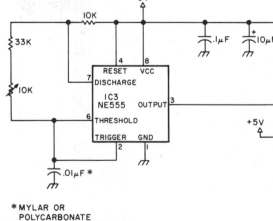

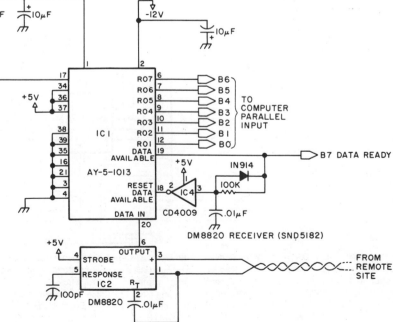

*MYLAR OR POLYCARBONATE

RECEIVER FOR REMOTE TERMINAL—National DM8820 receiver at computer location is connected by twisted-pair line to remote terminal. NE555 oscillator is set at 1760 Hz within 1% with aid of frequency counter, to match corresponding clock in remote terminal. Serial bits coming over line are converted to parallel bits for computer by AY-5-1013 UART. Article gives circuit for remote terminal and covers operation in detail.—S. Ciarcia, Come Upstairs and Be Respectable, BYTE, May 1977, p 50–54.

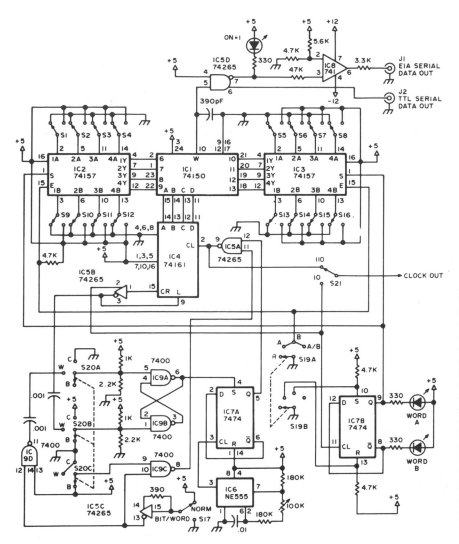

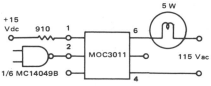

MALFUNCTION ALARM—Motorola MOC3011 optoisolator serves as interface between CMOS logic of microprocessor and 5-W 115-VAC lamp. Input logic is connected to energize infrared LED of optoisolator by providing up to 50 mA. Once triggered, indicator lamp remains on until current drops below holding value of about 100 μA.—P. O'Neil, "Applications of the MOC3011 Triac Driver," Motorola, Phoenix, AZ, 1978, AN-780, p 2.

SERIAL ASCII GENERATOR—Provides choice of two different words in standard serial ASCII asynchronous data format for troubleshooting and testing code converters and other computer peripherals. S19 gives choice of four data output patterns. R gives logic high for all 8 bits. A gives pattern determined by settings of S1-S8. B gives pattern determined by settings of S9-S16. A/B alternates words A and B. S20 gives choice of three different output modes. Mode B generates words 1 bit at a time. Mode W produces single word at rate of 110 bauds. Mode C produces continuous output of selected word pattern, for testing teleprinters and other output devices. Article covers construction and operation of circuit. IC power (+5 V) and ground pins are: 74150—24 and 12; 74157, 74161, and 74265—16 and 8; 555—8 and 1; 7474 and 7400—14 and 7.—R. J. Finger, Build a Serial ASCII Word Generator, *BYTE*, May 1976, p 50–53.

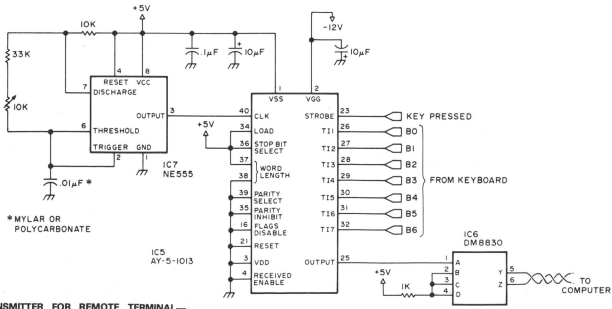

TRANSMITTER FOR REMOTE TERMINAL—Consists of AY-5-1013 UART attached to keyboard, with twisted-pair cable running to receiver unit at computer location. Coaxial extension cable for monitor is only other connection to computer system because terminal has own power supply. Transmission is in one direction only. NE555 oscillator is set at 1760 Hz ± 1% with aid of frequency counter, for 110-b/s serial rate. IC6 is 5-V National DM8830 differential line driver or equivalent. Pin 14 of IC6 goes to +5 V and pin 7 to ground.—S. Ciarcia, Come Upstairs and Be Respectable, *BYTE*, May 1977, p 50–54.

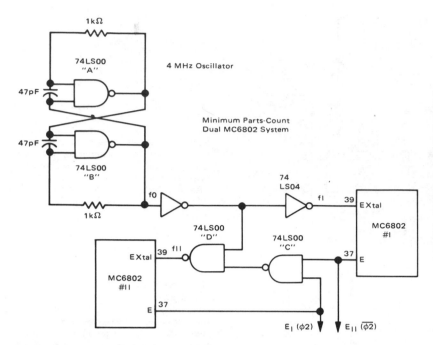

4-MHz NAND-GATE OSCILLATOR SYNCHRONIZES TWO MPUs—Low-cost NAND-gate sections A and B form low-cost oscillator for driving two Motorola MC6802 microprocessors. NAND gates C and D function as phase-locked loop, with D ensuring that phases of enable outputs are 180° apart. Small drifts in oscillator frequency do not affect synchronization. Circuit allows each MPU to operate during half-cycle that other MPU has disabled, to provide additional computing power of two microprocessors while maintaining system costs of one data bus.—J. Farrell, "Synchronizing Two Motorola MC6802s on One Bus," Motorola, Phoenix, AZ, 1978, AN-783.

LIGHT PEN WITH INTERFACE—Any high-quality photodiode mounted in discarded housing of marking pen serves as pickup for holding against screen of video display. If diode is mounted in plastic lens, flatten end of lens with emery cloth to give narrower angle of acceptance. Developed for use with VDM-1 display terminal. Use CRO to monitor output as pen is moved across screen. Dark area on screen gives 5-VDC level, and white area gives dips. Article covers use in program design, editing memory dumps, and arranging complex displays.—J. Webster and J. Young, Add a $3 Light Pen to Your Video Display, *BYTE*, Feb. 1978, p 52, 54, 56, and 58.

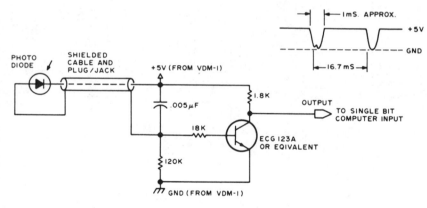

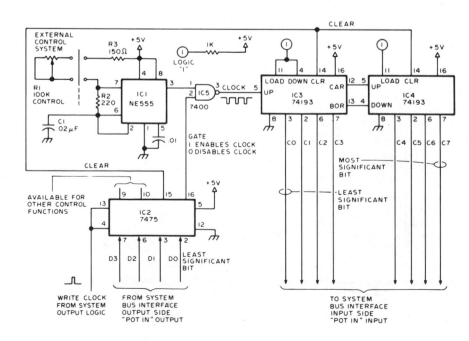

POT INTERFACE—Circuit converts resistance of pot setting to frequency with NE555 timer IC1. Frequency is measured under direct control of microprocessor program, using 8-bit counter with CPU clock as time base. Processor is programmed to clear counter, turn on counter, wait 2 ms, turn off counter, and read count. Result is number of cycles in 2-ms period, ranging from 1 to about 240. Relationship of frequency to control position is accurate enough for game control. R1, R2, R3, and C1 are chosen so frequency varies from about 0.75 to 122 kHz as R1 is varied from 100K to 0. Use audio-taper pot to improve conversion linearity. Article gives subroutines for Motorola 6800 and Intel 8080 microprocessors.—C. Helmers, Getting Inputs from Joysticks and Slide Pots, *BYTE*, Feb. 1976, p 86–88.

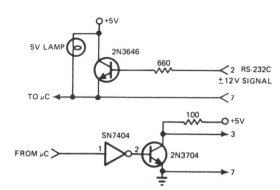

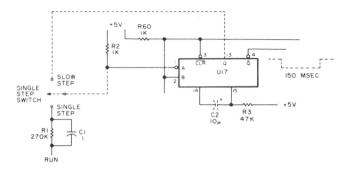

TERMINAL INTERFACE—Developed for use between computer terminal and microprocessor development board. Provides interface between teleprinter terminal using EIA RS-232C standard and input of microcomputer (upper diagram) and interface between computer and teleprinter using 20-mA current-loop standard for actuating printout. Logic 1 is −3 to −9 V or less, and logic 0 is +3 to +9 V or more.—P. Snigier, Constructing a Low-Cost Terminal Interface, *EDN Magazine*, June 5, 1977, p 205–206.

SLOW STEPPER—Addition of slow-stepping switch position to single-step circuit of microprocessor eliminates need for pushing single-step switch repeatedly while executing endless loop program and watching address and data lights to see where program or hardware fails during debugging run. Circuit uses 74123 mono MVBR to give 1.5-ms pulses on single step and 150-ms on slow step.—H. R. Bendrot, The Slow-Stepping Debugger, *Kilobaud*, April 1977, p 60.

LIGHT PEN—Photocell in tip of light pen senses when dot is written on screen at its location by becoming conductive and biasing Q1 so it feeds short pulse through C1 to base of Q2. If pulse is greater than 0.6 V, Q2 is driven into saturation and output of pen drops to 0.3 V. Output line goes to pin 5 of digital display unit (DDU), which writes 1 or 0 (dot or no dot) on screen at instant that electron beam of CRT terminal reaches position of pen. Sensitivity control can be adjusted so illuminated dot just ahead of pen can be used to create new dot in adjacent dark space. If screen is dark all around pen, footswitch-controlled auxiliary circuit can be used to override Z-axis control and flood screen with light momentarily by feeding logic 1 to Z input. This mode can be used for creating or correcting graphics.—S. S. Loomis, Let There Be Light Pens, *BYTE*, Jan. 1976, p 26–30.

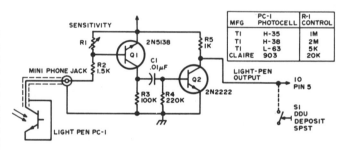

MFG	PC-I PHOTOCELL	R-I CONTROL
TI	H-35	IM
TI	H-38	2M
TI	L-63	5K
CLAIRE	903	20K

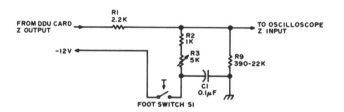

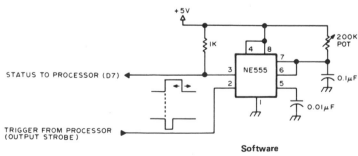

DIGITIZING POT POSITION—Converts position of pot arm into digital value, using NE555 timer and several bytes of program in 8008 or 8080 microprocessor having 2.5-μs clock. NE555 is triggered at pin 2 by OUT TRIGGER instruction. Program monitors output at pin 3 in loop that increments B register. When NE555 times out, program exits from subroutine and B register contains digital representation of pot position.—J. M. Schulein, Pot Position Digitizing Idea, *BYTE*, March 1976, p 79.

```
              Software
POTPOS:   MVI   B,0
          OUT   TRIGGER
CONT:     INR   B
          IN    STATUS
          ANA   A        ;Sets sign flag
          JM    CONT
          RET
```

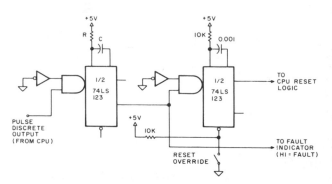

SOFTWARE MONITOR—Detects large percentage of random faults, such as overflow conditions. Based on fact that part of program passes through reentry on predictable repeat basis. For each reentry, pulse is fed from CPU to input of 74LS123 retriggerable mono MVBR. On each pass through program, programmer sets and then resets MVBR. Period of mono is made longer than longest normal time between programmed pulse outputs, to prevent false alarms. Any system fault that makes program repeat instructions endlessly or lose control will give false indication.—D. Brickner, Get a Watchdog to Monitor Those Real-Time Operations, *Kilobaud,* April 1978, p 118–119.

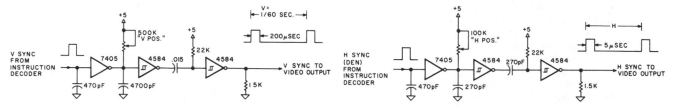

POSITION CONTROL FOR GRAPHICS—Six inverters serve for moving entire character display to any position on TV screen. Display is produced by microprocessor through alphameric data-to-video converter. Circuit requires continuous feed of H and V signals from instruction decoder of microprocessor.—D. Lancaster, TVT Hardware Design, *Kilobaud,* Jan. 1978, p 64–68.

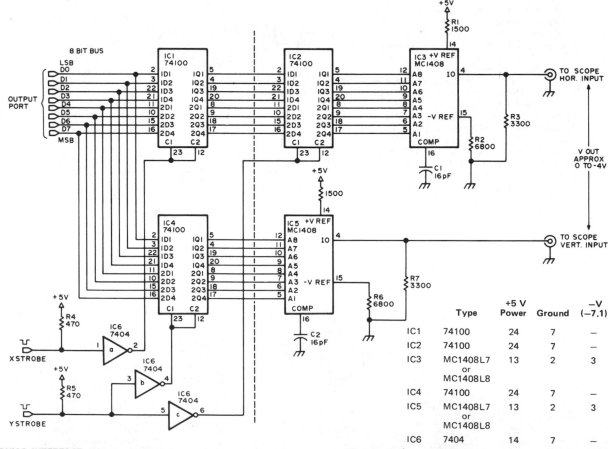

Type	+5 V Power	Ground	−V (−7.1)
IC1	74100	24	7	—
IC2	74100	24	7	—
IC3	MC1408L7 or MC1408L8	13	2	3
IC4	74100	24	7	—
IC5	MC1408L7 or MC1408L8	13	2	3
IC6	7404	14	7	—

GRAPHICS INTERFACE—Used between computer and ordinary CRO to create images with array of 512 × 512 dots stored in computer memory. Location of each dot is specified by two voltages, for application to V and H inputs of CRO. Computer provides voltage values by outputting two binary words to pair of DACs giving voltages proportional to numerical values of words. Dot pattern is repeated many times per second to give steady nonflickering image. Dot brightness can be increased by storing in several locations so it is refreshed more often than other points. Article gives listing for Intel 8080 graphics drive program. Program provides for interrupts once per scan to give keyboard-controlled drawing mode.—P. Nelson, Build the Beer Budget Graphics Interface, *BYTE,* Nov. 1976, p 26–29.

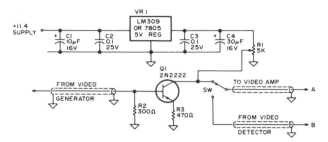

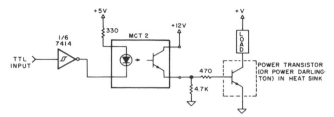

TV AS DISPLAY TERMINAL—Simple switch inserted in ordinary TV receiver serves for feeding video output of microprocessor directly into video amplifier of set, to give low-cost display terminal. TV must have transformer-type power supply. Excellent set for monitor use is 12-inch Hitachi model P-04, having Hitachi SX chassis. This set has very wide bandwidth, giving sharp display with line widths up to 80 characters.—G. Runyan, The Great TV to CRT Monitor Conversion, *Kilobaud,* July 1977, p 30–31.

DC CONTROL BY TTL I/O—Output of microprocessor drives LED of photocoupler through one section of 7414 hex Schmitt-trigger-input inverter. Phototransistor switches power transistor or power Darlington on and off for control of direct current through load. With transistor having current gain of 30 and 20-mA control current, load current can be 500 mA. With power Darlington having higher current gain, load can be several amperes. Since power device is either off or saturated, heatsink can be small.—M. Boyd, Interfacing Tips, *Kilobaud,* Feb. 1978, p 72–74.

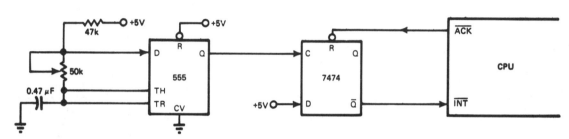

LOGGING EXACT TIME—Used in microcomputer applications requiring recording of exact time of each event by data-logging printer. Uses 555 timer and 7474 D flip-flop to produce interrupt request every 16.67 ms. Requires use of software routine that acknowledges interrupt, increments counter in known location to serve as time-of-day clock, and resumes interrupted program. Article diagrams software routine required.—Real-Time Software Keeps Program Segments on Schedule, *EDN Magazine,* Nov. 20, 1976, p 277–283.

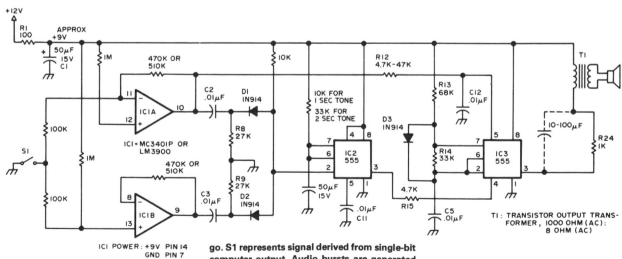

TWO-TONE ALARM—Provides audible backup for visual display in microprocessor-controlled system that requires human intervention under certain conditions, to alert operator who may be watching machinery rather than display. When input line goes high, device emits 1-s beep that means stop. When input goes low, 1-s lower-frequency boop sounds to indicate go. S1 represents signal derived from single-bit computer output. Audio bursts are generated by 555 timers. IC2 is wired as mono MVBR to determine tone duration, set by C4 and R11. Negative-going pulse on pin 2 triggers mono on. If microprocessor circuit creates pulse rather than level change, input should go to pin 2. Tone frequency is set by C5, R13, and R14 of IC3. Trigger uses IC1A as inverting opamp and IC1B as noninverting opamp. Trigger outputs are differentiated to give negative-going spike for each input level change. Spikes are ORed by D1 and D2 for trigger input of IC2. Different tones are achieved by using IC1A to change input voltage to pin 5 of tone generator IC3. Optional electrolytic capacitor across R24 will increase volume.—C. F. Douds, Audible Interrupts for Humans, *BYTE,* Feb. 1977, p 54 and 58.

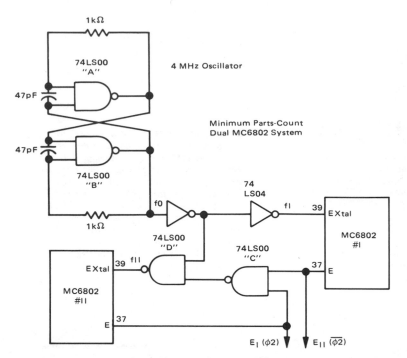

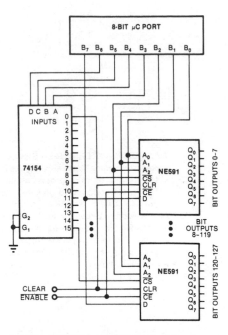

SYNCHRONIZING TWO MICROPROCES-SORS—4-MHz oscillator using 74LS00 quad two-input NAND gates serves in place of crystal source of individual MC6802 microprocessors to ensure synchronizing so one microprocessor operates during half-cycle when other microprocessor is disabled. Arrangement gives computing power of two MPUs with system cost of one data bus. No time is sacrificed since half-cycle used would normally be dead time on bus.—J. Farrell, "Synchronizing Two Motorola MC6802s on One Bus," Motorola, Phoenix, AZ, 1978, AN-783.

8 BITS CONTROL 128 BITS—Combination of 74154 demultiplexer and 16 NE591 latches permits control of 128 single-bit outputs with standard 8-bit parallel output of microprocessor. Latches are driven from three low-order bits of microprocessor. Next four higher bits drive 74154 which in turn drives chip-enable inputs of NE591s. Eighth bit of microprocessor drives data inputs of all latches and sets addressed bits appropriately.—R. D. Grappel, Control 128 Bits with an 8-Bit μC Port, *EDN Magazine,* Sept. 5, 1978, p 70 and 72.

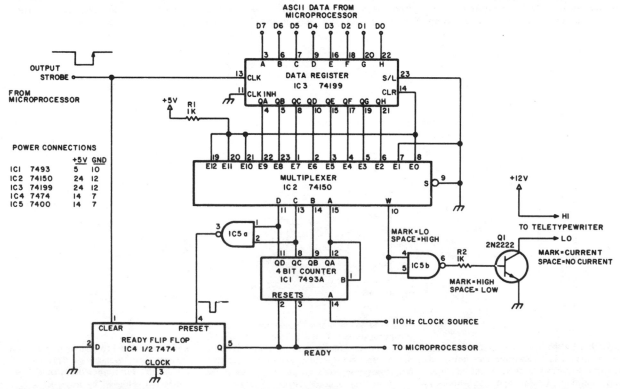

PARALLEL-TO-SERIAL CONVERTER—Multiplexer IC2 selects formatting and data bits according to state of IC1. IC3 is output latch. Teletypewriter current loop is driven by Q1 from output of multiplexer. 110-Hz clock gives transfer rate of 10 characters per second. Provides standard asynchronous format of 1 start bit, 8 data bits, and 2 stop bits for teletypewriter without using universal asynchronous receiver-transmitter.—G. C. Jewell, How to Drive a Teletype Without a UART, *BYTE,* Jan, 1977, p 32.

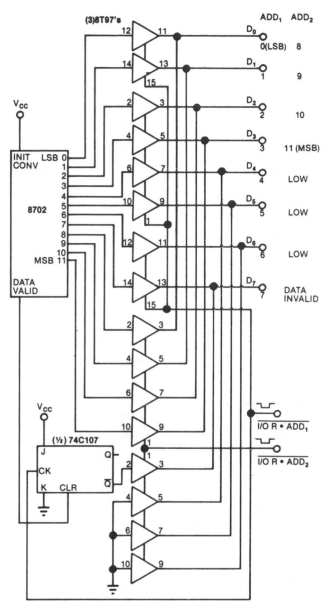

NOTE: UNUSED INPUTS OF 74C107
MUST BE TIED TO V_{CC} OR GND

12 BITS ON 8-BIT BUS—Arrangement shown speeds reading of 12 bits of data onto 8-bit microprocessor bus by simplifying checking procedure. Output of 74C107 flip-flop becomes DATA INVALID bit and is placed on bus during second read cycle. Simultaneously, circuit pulls 3 remaining bits (D_4-D_6) low through use of spare 8T97 gates. Article gives simplified 8080 subroutine required.—D. W. Taylor, Speed-Read 12 Bits onto an 8-Bit Bus, *EDN Magazine*, Sept. 5, 1978, p 70.

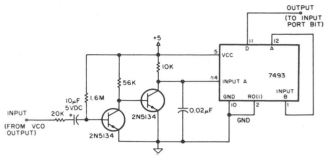

CALIBRATING MARK/SPACE VCO—Simple buffer/counter provides accurate calibration of FSK circuit used in cassette interface of 8080 microprocessor to generate mark and space frequencies. Audio FSK waveform is squared and divided down in 7493 4-bit counter. Resulting pulses are fed to input port of microprocessor for software pulse counting. Software sets VCO frequency, waits until pulse starts, counts each pulse occurrence, and displays resultant count. Each 7493 count is 29 Hz or 14.5 μs, so 371 counts correspond to 2975 Hz for space. Mark frequency of 2125 Hz gives pulses separated by 519 counts.—D. R. Bourdeau, Cassette Interface First Aid, *Kilobaud*, July 1977, p 49.

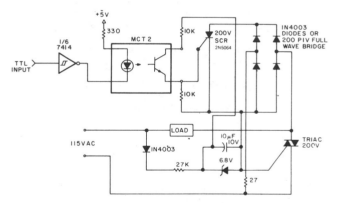

AC CONTROL BY TTL I/O—Output of microprocessor drives LED of photocoupler through one section of 7414 hex Schmitt-trigger-input inverter. Output of phototransistor controls SCR connected across full-wave diode bridge. Current flow through bridge and SCR turns on triac, allowing alternating current to flow through load. Choose triac rated for handling required load current.—M. Boyd, Interfacing Tips, *Kilobaud*, Feb. 1978, p 72–74.

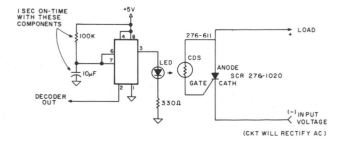

AC/DC CONTROL WITH SCR—Decoder output of microprocessor feeds 555 timer driving LED that is mounted in light shield with cadmium sulfide photocell. Combination serves as optocoupler for triggering gate of SCR to energize load for short time interval determined by value of resistor and capacitor used. Serves to control small DC motor such as is used to raise or lower transmitter power in small increments.—R. Wright, Utilize ASCII Control Codes!, *Kilobaud*, Oct. 1977, p 80–83.

CHAPTER 52
Modulator Circuits

Covers circuits that vary amplitude or some other characteristic of carrier signal or pulse train in accordance with information contained in modulating signal. Includes PCM, PDM, PPM, duty-cycle, and other types of pulse modulators, light-beam modulators, delta modulators, and various types of AM, SSB, and suppressed-carrier modulators. See also Frequency Modulation, Transceiver, and Transmitter chapters.

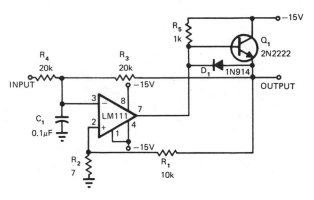

PULSE-RATIO MODULATOR—LM111 comparator serves with single transistor to provide pulse-train output whose average value is proportional to input voltage. Frequency of output is relatively constant but pulse width varies. Pulse-ratio accuracy is 0.1%. Circuit can be used to drive power stage of high-efficiency switching amplifier, or as pulse-width/pulse-height multiplier. Article tells how circuit works.—R. C. Dobkin, Comparators Can Do More than Just Compare, *EDN Magazine*, Nov. 1, 1972, p 34–37.

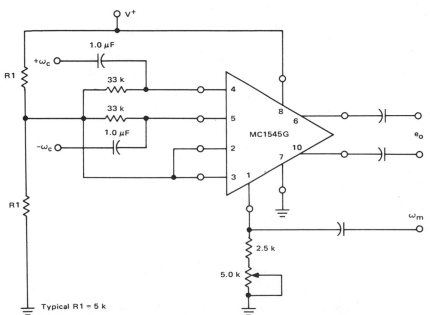

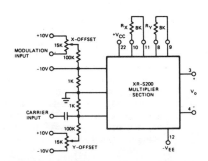

SINGLE-SUPPLY AM—Motorola MC1545 gated video amplifier is connected as amplitude modulator operating from single supply. Artificial ground is established for IC at half of supply voltage by 5K resistors R1, which should draw much more than bias current of 15 μA. All signals must be AC coupled to prevent application of excessive common-mode voltage to IC.— "Gated Video Amplifier Applications—the MC1545," Motorola, Phoenix, AZ, 1976, AN-491, p 15.

DOUBLE-SIDEBAND AM—Connection shown for multiplier section of Exar XR-S200 PLL IC gives double-sideband AM output. X-offset adjustment for modulation input sets carrier output level, and Y-offset adjustment of carrier input controls symmetry of output waveform. Modulation input can also be used as linear automatic gain control (AGC) for controlling amplification with respect to carrier input signals.—"Phase-Locked Loop Data Book," Exar Integrated Systems, Sunnyvale, CA, 1978, p 9–16.

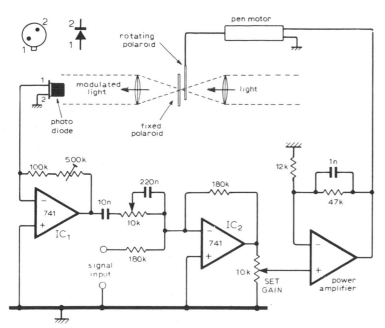

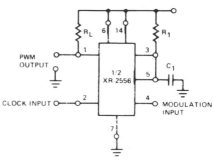

LIGHT-BEAM MODULATOR—Intensity of light beam is modulated by rotating Polaroid vane driven by small motor. Since amplitude is not constant with change in frequency between 10 and 100 Hz, compensation is provided by sampling modulated beam with silicon photodiode that is linearized by IC₁. Input and feedback sig-nals are mixed by summing amplifier IC₂ which drives noninverting power amplifier consisting of 741 opamp driving two OC28 power transis-tors connected in closed feedback loop having gain of 5. Power amplifier drives pen motor of modulator.—R. F. Cartwright, Constant Ampli-tude Light Modulator, *Wireless World*, Sept. 1976, p 73.

PULSE-DURATION MODULATOR USES TIMER—Half of Exar XR-2556 dual timer is con-nected to operate in monostable mode, for trig-gering with continuous pulse train. Output pulses are generated at same rate as input, with pulse duration determined by R₁ and C₁. Supply voltage is 4.5–16 V.—"Timer Data Book," Exar Integrated Systems, Sunnyvale, CA, 1978, p 23–30.

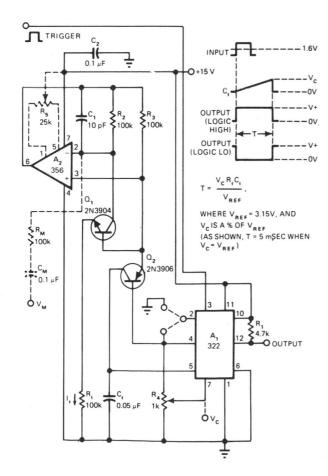

$$T = \frac{V_C R_t C_t}{V_{REF}}$$

WHERE V_{REF} = 3.15V, AND V_C IS A % OF V_{REF} (AS SHOWN, T = 5 mSEC WHEN $V_C = V_{REF}$)

VOLTAGE TO PULSE WIDTH—Constant-cur-rent source Q₂ produces linear timing ramp across C₁ in circuit of 322 IC timer A₁, for com-parison internally with 0–3.15 V applied to pin 7. Pulse is thus linearly variable function of con-trol voltage V_C over dynamic range of more than 100:1. Circuit is highly flexible, permitting use of many other operational modes as covered in article. When AC waveform is applied to V_M, circuit operates as linear pulse-width modula-tor.—W. G. Jung, Take a Fresh Look at New IC Timer Applications, *EDN Magazine*, March 20, 1977, p 127–135.

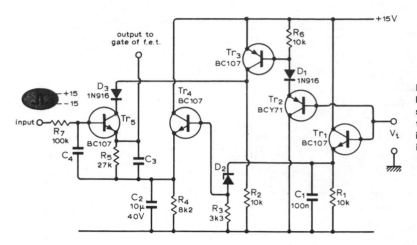

PULSE HEIGHT MODULATOR—Used ahead of FET gate to reduce spike feedthrough. Voltage swing on FET gate is limited to difference between V_i and pinchoff voltage of FET. Zener D_2 is matched to measured pinchoff voltage of FET in use.—M. D. Dabbs, Pulse Height Modulator, *Wireless World,* April 1975, p 176.

SUPPRESSED-CARRIER MODULATOR—Mullard TCA240 dual balanced modulator-demodulator provides suppression of carrier frequency at output, as required for SSB or DSB operation of transmitter. Bias resistor R7 is adjusted for minimum carrier output to correct imbalance. Can be used as conventional AM modulator if biasing of circuit sections is deliberately unbalanced.—"Applications of the TCA240," Mullard, London, 1975, Technical Note 18, TP1489.

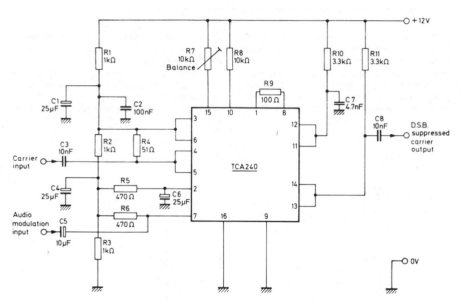

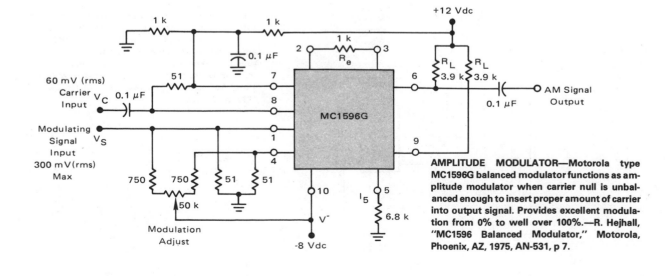

AMPLITUDE MODULATOR—Motorola type MC1596G balanced modulator functions as amplitude modulator when carrier null is unbalanced enough to insert proper amount of carrier into output signal. Provides excellent modulation from 0% to well over 100%.—R. Hejhall, "MC1596 Balanced Modulator," Motorola, Phoenix, AZ, 1975, AN-531, p 7.

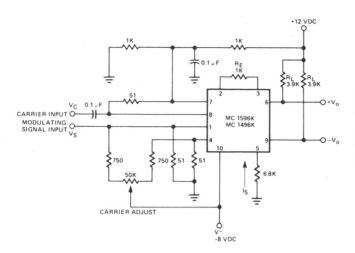

SINGLE-IC AM—Adjustable carrier offset is added to carrier differential pairs to provide carrier-frequency output that varies in amplitude with strength of modulation signal.—"Signetics Analog Data Manual," Signetics, Sunnyvale, CA, 1977, p 757.

DSB BALANCED MODULATOR—Provides excellent gain and carrier suppression by operating upper (carrier) differential amplifiers of Motorola MC1596G balanced modulator at saturated level and lower differential amplifier in linear mode. Recommended input levels are 60 mVRMS for carrier and 300 mVRMS maximum for modulating signal.—R. Hejhall, "MC1596 Balanced Modulator," Motorola, Phoenix, AZ, 1975, AN-531, p 3.

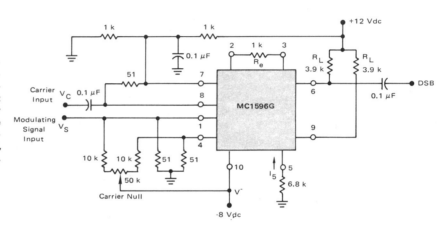

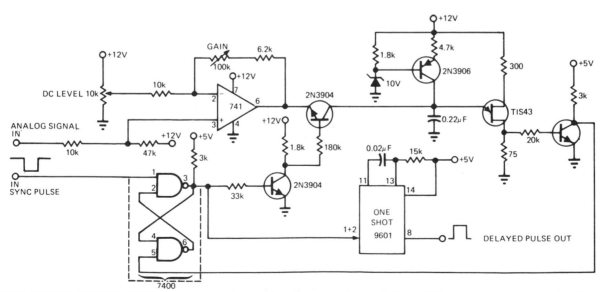

PPM WITH ANALOG CONTROL OF DELAY—Opamp, UJT, and two TTL packages generate pulse whose delay, following sync pulse, is controlled by amplitude of analog input signal at time of sync pulse. Opamp precharges timing capacitor to level depending on analog signal. Sync pulse disconnects opamp, after which timing capacitor charges up to UJT firing point. UJT output pulse then resets circuit, giving desired delayed output pulse through 9601 mono MVBR.—J. Taylor, Analog Signal Controls Pulse Delay, *EDN Magazine*, Feb. 5, 1974, p 96.

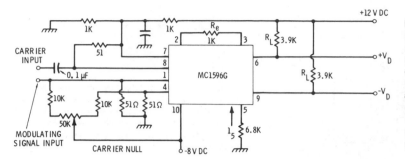

DOUBLE-SIDEBAND SUPPRESSED-CARRIER— Motorola MC1596G double-balanced modulator has carrier input between pins 8 and 7 and modulation between pins 1 and 4. Balancing carrier-null circuit, also connected between pins 1 and 4, contributes to excellent carrier rejection at output. For unbalanced output, ground one of push-pull output terminals. Requires two supplies.—E. M. Noll, "Linear IC Principles, Experiments, and Projects," Howard W. Sams, Indianapolis, IN, 1974, p 138–139.

DELTA MODULATOR—Uses LM111 comparator in basic pulse-ratio modulator circuit, with output pulse width and transition time fixed by external clock signal applied to gate of JFET switch Q_2. Average value of output is always proportional to input voltage.—R. C. Dobkin, Comparators Can Do More than Just Compare, *EDN Magazine*, Nov. 1, 1972, p 34–37.

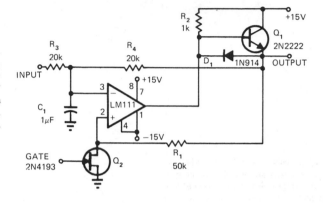

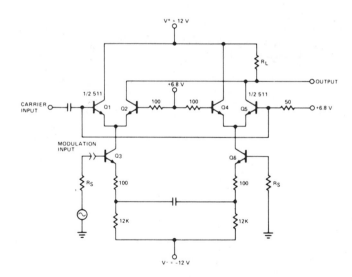

SUPPRESSED-CARRIER AM—Double-balanced modulator using Signetics 511 transistor array gives output consisting of sum and difference frequencies of carrier and modulation inputs along with related harmonics. Circuit is self-balancing, eliminating need for pots. Output includes small amounts of carrier and modulating signal. Capacitor between emitters of Q3 and Q6 is selected to have low reactance at lowest modulating frequency.—"Signetics Analog Data Manual," Signetics, Sunnyvale, CA, 1977, p 750–751.

BALANCED MODULATOR—High-performance balanced modulator for 80-meter SSB transceiver uses Motorola MC1496 IC. Adjust 50K pot for maximum carrier suppression of double-sideband output.—D. Hembling, Solid-State 80-Meter SSB Transceiver, *Ham Radio*, March 1973, p 6–17.

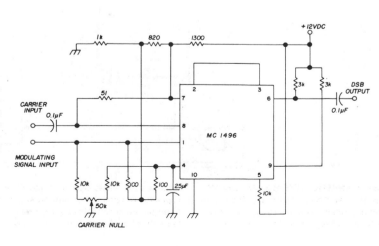

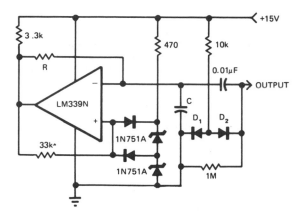

DUTY-CYCLE MODULATOR—Uses half of LM339N or LM3302N quad comparator. With no modulation signal, output is symmetrical square wave generated by one of comparators. Constant-amplitude triangle wave is generated at inverting input of second comparator, and is relatively independent of supply voltage and frequency changes. Modulating signal varies switching points to produce duty-cycle modulated wave for such applications as class D amplification for servo and audio systems.—H. F. Stearns, Voltage Comparator Makes a Duty-Cycle Modulator, *EDN Magazine,* June 5, 1975, p 76–77.

FET BALANCED MODULATOR FOR SSB—AF modulating signal is applied to gates of matched FETs in push-pull through T1 having accurately center-tapped secondary, and RF carrier is applied to sources in parallel through C3. Carrier is canceled in output circuit, leaving two sidebands. R3 is adjusted to correct for unbalance in circuit components.—R. P. Turner, "FET Circuits," Howard W. Sams, Indianapolis, IN, 1977, 2nd Ed., p 90–91.

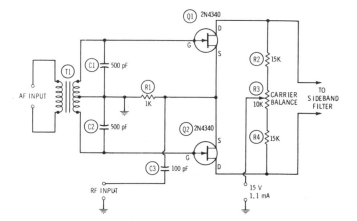

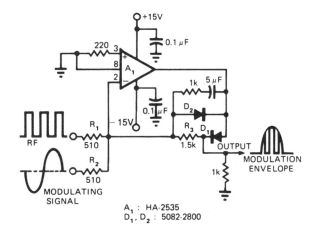

100% MODULATION OF DIGITAL SIGNALS—Developed to produce positive-going half-sine envelope modulated 100% by digital RF signal. Modulator uses loop gain of opamp to reduce diode drops very nearly to ideal zero level. D_2 prevents opamp output terminal from swinging more negative than diode drop of 0.3 V, which is not apparent at output. RF input amplitude must be sufficient to provide 100% modulation; this can be achieved by providing about 20% overdrive to give safety factor. Use hot-carrier diodes such as HP-2800 series.—D. L. Quick, Improve Amplitude Modulation of Fast Digital Signals, *EDN Magazine,* Sept. 20, 1975, p 68 and 70.

DOUBLE-SIDEBAND SUPPRESSED-CARRIER—Signetics MC1496 balanced modulator-demodulator transistor array provides carrier suppression while passing sum and difference frequencies. Gain is set by value used for emitter degeneration resistor connected between pins 2 and 3. Output filtering is used to remove unwanted harmonics.—"Signetics Analog Data Manual," Signetics, Sunnyvale, CA, 1977, p 756–757.

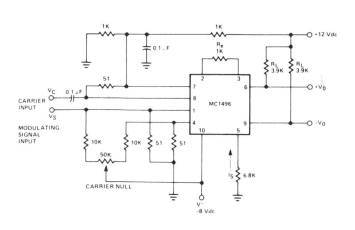

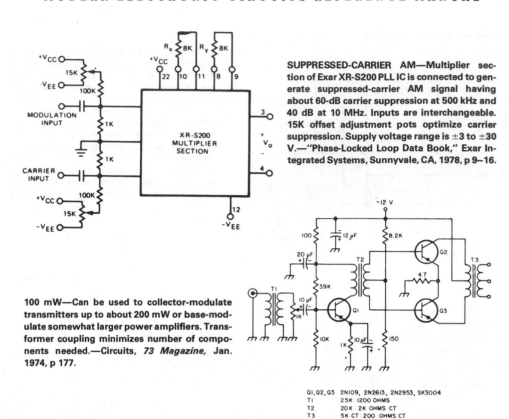

SUPPRESSED-CARRIER AM—Multiplier section of Exar XR-S200 PLL IC is connected to generate suppressed-carrier AM signal having about 60-dB carrier suppression at 500 kHz and 40 dB at 10 MHz. Inputs are interchangeable. 15K offset adjustment pots optimize carrier suppression. Supply voltage range is ±3 to ±30 V.—"Phase-Locked Loop Data Book," Exar Integrated Systems, Sunnyvale, CA, 1978, p 9–16.

100 mW—Can be used to collector-modulate transmitters up to about 200 mW or base-modulate somewhat larger power amplifiers. Transformer coupling minimizes number of components needed.—Circuits, *73 Magazine,* Jan. 1974, p 177.

Q1,Q2,Q3   2N109, 2N2613, 2N2953, SK3004
T1   25K : 1200 OHMS
T2   20K : 2K OHMS CT
T3   5K CT : 200 OHMS CT

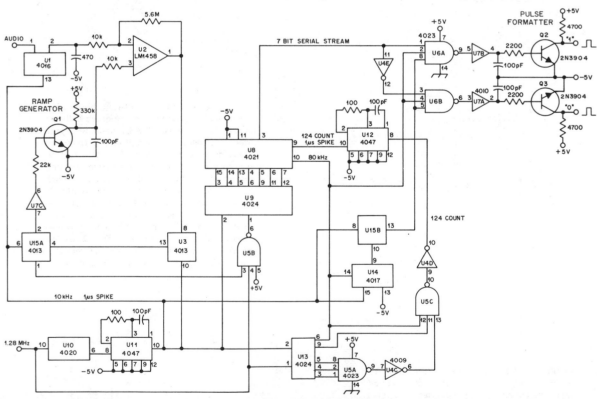

PCM FOR MICROWAVE TRANSMITTER—Modern pulse-code modulator can be used for experimentation above lowest legal frequency of 2.3 GHz, such as for satellite and moonbounce communication. Single voice channel is encoded by using CMOS technology having low power consumption, good noise immunity, and moderate cost. Audio is sampled 10,000 times per second for conversion to 7-pulse code plus synchronizing bit, for 123 levels of encoding frequencies up to 5 kHz. When 10-kHz sampling spike (derived from external 1.28-MHz oscillator by frequency divider U10 and mono U11) arrives at pin 13 of transmission gate U1, AF voltage of pin 1 appears at pin 2. 470-pF capacitor charges to this voltage and holds charge until next sample. This voltage is compared by U2 to linear ramp started by Q1 at sampling instant. When ramp voltage exceeds sampled voltage, U2 triggers, setting flip-flop U3 and resetting ramp generator to −5 V. At same time, binary counter U9 is stopped by reset flip-flop U15A, and binary equivalent of sample appears at pins 3, 4, 5, 6, 9, 11, and 12 of U9. Remainder of circuit converts bits to serial form for transmission. Article explains circuit operation in detail and gives corresponding decoder circuit for receiver.—V. Biancomano, A Prototype Pulse-Code Modulation System, *QST,* Jan. 1977, p 24–29.

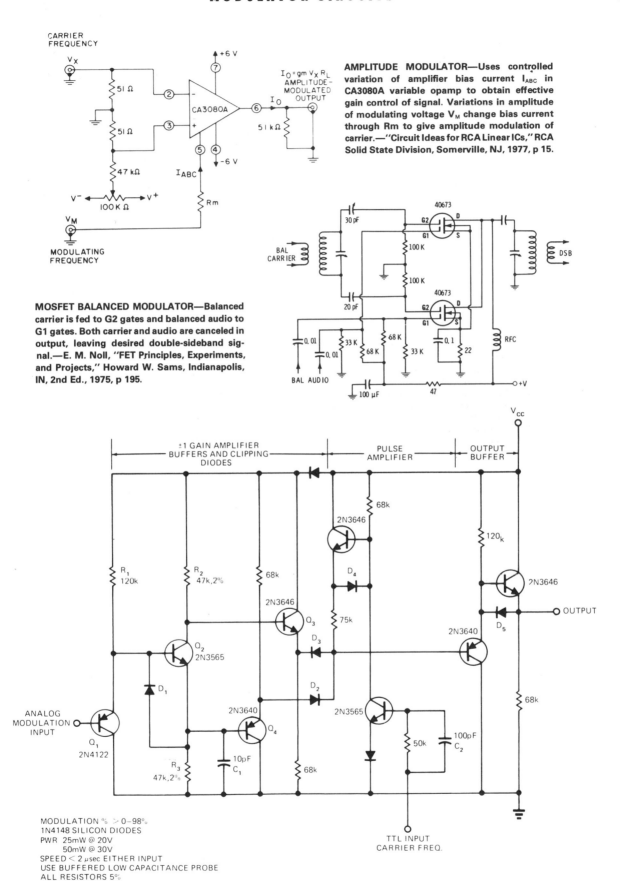

AMPLITUDE MODULATOR—Uses controlled variation of amplifier bias current I_{ABC} in CA3080A variable opamp to obtain effective gain control of signal. Variations in amplitude of modulating voltage V_M change bias current through Rm to give amplitude modulation of carrier.—"Circuit Ideas for RCA Linear ICs," RCA Solid State Division, Somerville, NJ, 1977, p 15.

MOSFET BALANCED MODULATOR—Balanced carrier is fed to G2 gates and balanced audio to G1 gates. Both carrier and audio are canceled in output, leaving desired double-sideband signal.—E. M. Noll, "FET Principles, Experiments, and Projects," Howard W. Sams, Indianapolis, IN, 2nd Ed., 1975, p 195.

MODULATION % > 0–98%
1N4148 SILICON DIODES
PWR 25mW @ 20V
50mW @ 30V
SPEED < 2 μsec EITHER INPUT
USE BUFFERED LOW CAPACITANCE PROBE
ALL RESISTORS 5%

SQUARE-WAVE MODULATOR—Analog modulation input goes through buffer Q_1-R_1 to amplifier Q_2-R_2-R_3. Outputs of Q_2 are buffered by separate emitter-followers Q_3 and Q_4 and fed to clipping diodes D_2 and D_3 acting on top and bottom of amplified carrier wave. Following pulse amplifier converts TTL signal to square wave. Two opposed emitter-followers and D_5 form buffer that can either source or sink current from load. Other three diodes are used to speed up circuit operation. Provides modulation range of 0–98%.—E. Burwen, Low-Power Square-Wave Modulator Is TTL Compatible, *EDN Magazine*, Nov. 20, 1973, p 91 and 93.

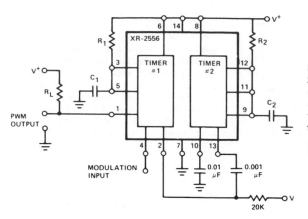

PDM WITH CLOCK—First section of Exar XR-2556 dual timer operates as pulse-duration modulator and second section as clock generator, eliminating need for external clock. Supply voltage is 4.5–16 V. Values of R and C determine frequency and pulse duration of output.—"Timer Data Book," Exar Integrated Systems, Sunnyvale, CA, 1978, p 23–30.

FOUR-QUADRANT MULTIPLIER—Provides amplitude modulation for applications where low power consumption is more important than accuracy. Uses CA3080A variable opamp in combination with transistors of CA3018A array and 2N4037 amplifier for bias current of opamp. Accuracy is within 7% full scale.—"Circuit Ideas for RCA Linear ICs," RCA Solid State Division, Somerville, NJ, 1977, p 15.

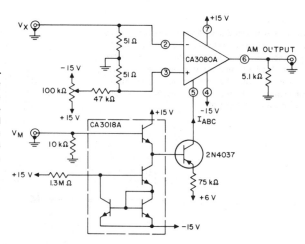

CHAPTER 53
Motor Control Circuits

Speed control circuits for various types and sizes of AC and DC motors, including three-phase motors. Some use tachometer feedback to maintain desired constant speed. Includes stepper motor drives, phase sequence detector, braking control, facsimile phase control, and revolution-counting control. Many respond to logic inputs. See also Antenna, Lamp Control, Power Control, Servo, and Temperature Control chapters.

SWITCHING-MODE CONTROLLER—Developed for driving 0.01-hp motor M at variable speeds with minimum battery drain. Circuit uses pulses with low duty cycle to set up continuous current in motor approximating almost 200 mA when average battery drain is 100 mA for output voltage of 3.5 V. Voltage comparator A_1 serves as oscillator and as duty-cycle element of controller. C_1 and R_1 provide positive feedback giving oscillation at about 20 kHz, with duty-cycle range of 10% to 70% controlled by feedback loop Q_1-R_1-C_3-R_3. D_2 is used in place of costly large capacitor for filtering.—J. C. Sinnett, Switching-Mode Controller Boosts DC Motor Efficiency, *Electronics*, May 25, 1978, p 132.

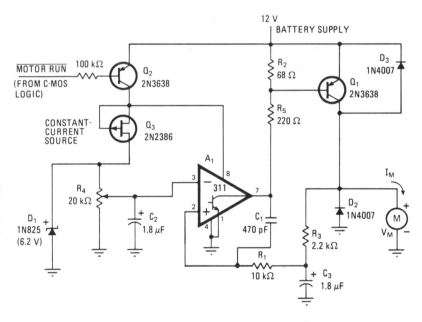

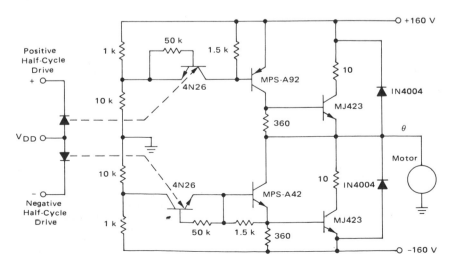

PWM SPEED CONTROL—Power stage using Motorola 4N26 optoisolators and push-pull transistors drives fractional-horsepower single-phase AC motor over speed range of 5% to 100% of base speed. Input drives are provided by pulse-width modulation inverter using stored program in ROM to generate sine-weighted pulse trains to provide variable-frequency drive.—T. Mazur, "A ROM-Digital Approach to PWM-Type Speed Control of AC Motors," Motorola, Phoenix, AZ, 1974, AN-733, p 12.

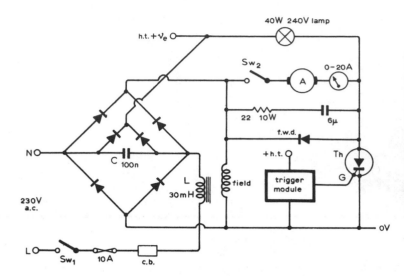

2-hp THYRISTOR CONTROL—Provides smooth variation in speed of shunt-wound DC motor from standstill to 90% of rated speed. Use thyristor rated 30 A at 600 V. Outer diodes of bridge are 35-A 600-PIV silicon power diodes, as also is thyristor diode, and inner diodes are 5-A 600-V silicon power diodes. Article gives complete circuit of trigger pulse generator used to control speed by varying duty cycle of thyristor. Larger motors can be controlled similarly by uprating thyristor and diodes. Controller will also handle other types of loads, including lamps and heaters.—F. Butler, Thyristor Control of Shunt-Wound D.C. Motors, *Wireless World,* Sept. 1974, p 325–328.

TAPE-LOOP SPEED CONTROL—Shunt rectifier-capacitor circuit was developed for speed control of permanent split-capacitor fractional-horsepower induction motor used in some motion-picture projectors. Light-dependent resistor LDR makes Q_2 conduct when light from lamp is not blocked by tape loop. Split capacitor C_1 for motor provides both run and speed-control functions without switching. Values are: C_2 0.01 μF; D 1N4004; Q_1 2N4987; Q_2 C106B; R_1 330K; R_2 100; R_3 10.—T. A. Gross, Control the Speed of Small Induction Motors, *EDN Magazine,* Aug. 20, 1977, p 141–142.

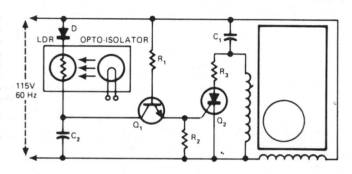

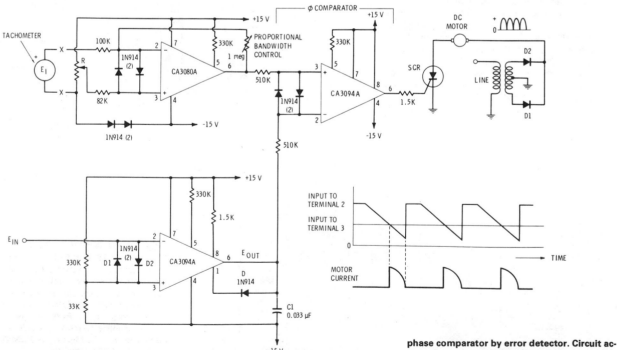

DC MOTOR SPEED CONTROLLER—Tachometer driven by motor produces output voltage proportional to speed for application to CA3080A voltage comparator after rectification and filtering. Output of CA3080A is applied to upper CA3094A phase comparator that is receiving reference voltage from another CA3094A connected as ramp generator. Output of phase comparator triggers SCR in motor circuit. Amount of motor current is set by time duration of positive signal at pin 6, which in turn is determined by DC voltage applied to pin 3 of phase comparator by error detector. Circuit action serves to maintain constant motor speed at value determined by position of pot R. Input to ramp generator is pulsating DC voltage used to control rapid charging of C1 and slower discharging to form ramp.—E. M. Noll, "Linear IC Principles, Experiments, and Projects," Howard W. Sams, Indianapolis, IN, 1974, p 321–323.

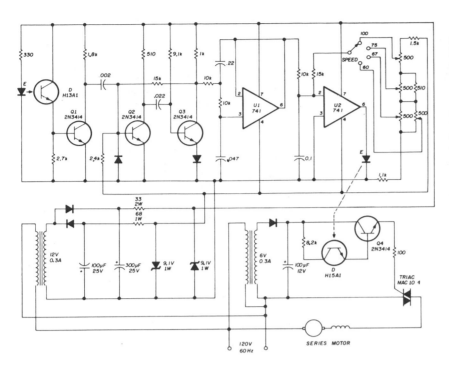

SERIES-MOTOR SPEED CONTROL—Adjustable-speed solid-state motor drive replaces governor in Kleinschmidt RTTY page printer, to give knob-controlled speed range of 60 to 100 WPM. Notched (33-slot) sheet-aluminum disk serving as pulse wheel is mounted on motor shaft and rotates in gap between LED and phototransistor of GE H13A1 optical coupler to form motor-speed sensor or tachometer. Pulses from tachometer, squared by Q1, trigger mono MVBR Q2-Q3 which converts signal to constant-amplitude constant-width pulses having repetition rate proportional to motor speed. Opamp U1 forms three-pole Butterworth active filter that develops required average DC voltage from pulse train. Output current of U1 is compared to reference current derived from speed control circuit, for switching U2 sharply on and off as speed varies above and below desired value. U2 in turn switches motor on and off through H15A1 optical coupler and Q4 in gate circuit of triac. Second coupler isolates control circuit from AC line.—K. H. Sueker, Electronic Speed Control for RTTY Machines, *Ham Radio,* Aug. 1974, p 50–54.

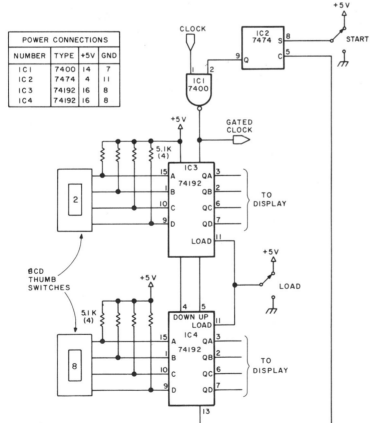

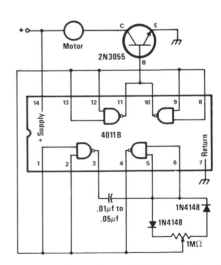

STEP COUNTER FOR STEPPER—Used to deliver selected number of pulses to stepper motor when start button is pushed, in microprocessor application where number of steps is more important than precise speed. Thumbwheel switch inputs can be I/O port lines of microprocessor. LOAD line transfers into counter the desired count as set up by switches. Article gives flowcharts and software routines for microprocessor to be used for controlling stepper motor.—R. E. Bober, Taking the First Step, *BYTE,* Feb. 1978, p 35–36, 38, 102, 104, 106, and 108–112.

SPEED CONTROL FOR 3-V MOTOR—Designed for use with hobby or toy motors running at about 10,000 rpm and powered by 3-V to 6-V batteries. Uses 4011 CMOS NAND gate with diodes and power transistor to provide variable duty cycle, so adjustment of 1-megohm speed control varies average voltage applied to motor without affecting peak voltage. Motor battery is connected between + terminal and ground of circuit.—J. A. Sandler, 11 Projects under $11, *Modern Electronics,* June 1978, p 54–58.

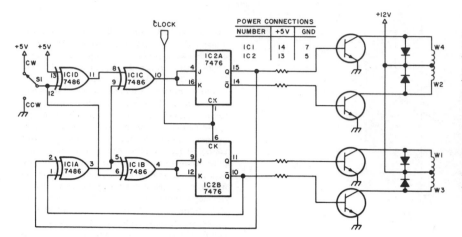

POWER CONNECTIONS		
NUMBER	+5V	GND
IC1 IC2A 7476	14	7
IC1 IC2	13	5

FOUR-PHASE STEPPER DRIVE—EXCLUSIVE-OR gates of 7486 provide steering, while 7476 flip-flops provide memory for generating drive patterns of bidirectional logic stepper motor that is controlled by microprocessor. Output transistors, diodes, and resistors are chosen to meet power requirements for each phase of motor. Speed is controlled by frequency of clock input. Use 555 for coarse control or crystal oscillator for accurate control. S1, which can be an I/O line of microprocessor, controls direction of rotation. Frequency can be obtained from digitally controlled oscillator whose setting is determined by DAC.—R. E. Bober, Taking the First Step, *BYTE*, Feb. 1978, p 35–36, 38, 102, 104, 106, and 108–112.

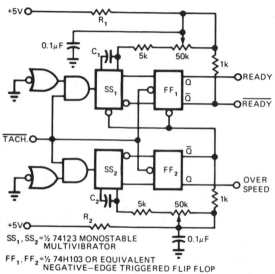

UNDER/OVERSPEED LOGIC—Provides signal (READY output high) only when tachometer pulses from motor are within specific upper and lower limits. Also provides overspeed output signal when upper limit has been exceeded. Single-action triggering eliminates instability at decision point. Article covers circuit operation in detail and gives timing diagram.—W. Bleher, Circuit Indicates Logic "Ready," *EDN Magazine*, March 5, 1974, p 72 and 74.

SS_1, SS_2 = ½ 74123 MONOSTABLE MULTIVIBRATOR

FF_1, FF_2 = ½ 74H103 OR EQUIVALENT NEGATIVE–EDGE TRIGGERED FLIP FLOP

C_1, C_2 = SELECTED FOR DESIRED TIMING RANGE

R_1 = SELECTED FOR UNDER/NOMINAL SPEED HYSTERESIS

R_2 = SELECTED FOR NOMINAL/OVER SPEED HYSTERESIS

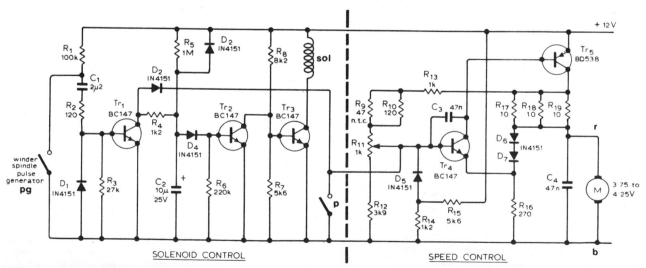

SOLENOID CONTROL   SPEED CONTROL

CASSETTE DRIVE CONTROLLER—Used in high-quality stereo cassette deck operating from AC line or battery. Combines current source for cassette-retaining solenoid with speed control for drive motor. As motor turns, associated motor-driven pulse-generating switch keeps Tr_1 conducting; this cuts off Tr_2 and allows current to flow through Tr_3 for energizing solenoid. When motor stops, pulse-generating switch also stops and Tr_1 stops conducting. After 3-s delay determined by C_2 and R_5, Tr_2 conducts and solenoid is deenergized, releasing cassette. In speed control circuit, Tr_5 acts as constant-current source for motor, using feedback from its collector to base of Tr_4. Back EMF developed by motor is applied to emitter of Tr_4, reducing its forward bias and reducing current in the base of Tr_5 so as to stabilize motor. Article gives all other circuits of cassette deck and describes operation in detail.—J. L. Linsley Hood, Low-Noise, Low-Cost Cassette Deck, *Wireless World*, Part 3—Aug. 1976, p 55–56 (Part 1—May 1976, p 36–40; Part 2—June 1976, p 62–66).

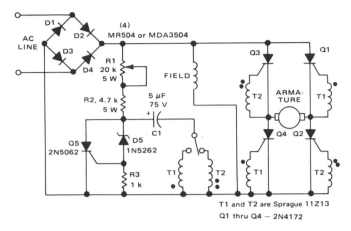

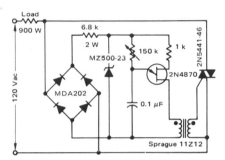

SHUNT-WOUND MOTOR—Switch provides direction control and R1 controls speed of fractional-horsepower shunt-wound DC motor. Field is placed across rectified supply, and armature windings are in four-SCR bridge circuit. Switch determines which diagonal pair of SCRs is turned on, to control direction of rotation. Triggering circuit consisting of Q5, D5, and C1 is controlled by R1, for changing conduction angle of triggered SCR path.—"Direction and Speed Control for Series, Universal and Shunt Motors," Motorola, Phoenix, AZ, 1976, AN-443.

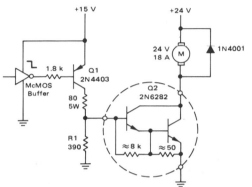

LOW-LEVEL CMOS CONTROL—Low-level output of CMOS buffer turns on DC motor through Q1 and 20-A Darlington power transistor Q2.—A. Pshaenich, "Interface Techniques Between Industrial Logic and Power Devices," Motorola, Phoenix, AZ, 1975, AN-712A, p 18.

900-W FULL-WAVE TRIGGER—Uses UJT for phase control of triac. Suitable for control of shaded-pole motors driving loads having low starting torque, such as fans and blowers.—D. A. Zinder, "Electronic Speed Control for Appliance Motors," Motorola, Phoenix, AZ, 1975, AN-482, p 4.

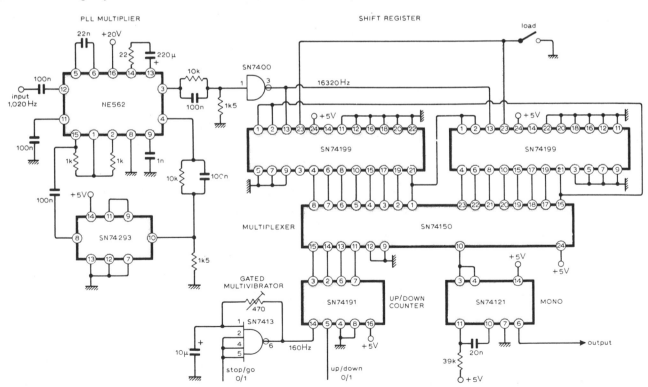

FACSIMILE PHASE CONTROL—Circuit provides accurate phasing of 51-pole-pair phonic/synchronous motor in facsimile transmitter, and can readily be adapted for similar applications. A 16-stage shift register loaded with 1 bit and connected as ring counter is clocked at 16 times required drive motor frequency. This gives pulse train with 1:15 mark-space ratio and repetition rate equal to drive frequency. Multiplexer used as single-pole 16-way switch can select output for any stage of shift register; each clockwise switch step gives 360/16 or 22.5° phase advance. Article describes circuit operation in detail.—P. E. Baylis and R. J. Brush, Synchronous-Motor Phase Control, *Wireless World,* April 1976, p 62.

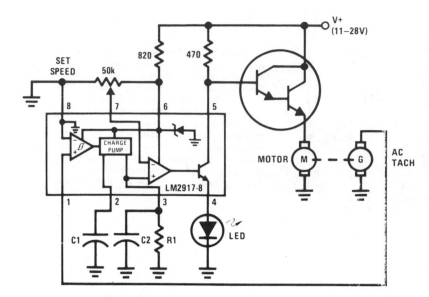

SHUNT-MODE SPEED CONTROL—AC tachometer on shaft of DC motor serves as input for National LM2917N-8 IC acting as shunt-mode regulator with LED indicator. Output of Darlington power transistor provides analog drive to motor. As motor speed approaches reference level set by values chosen for R1, C1, and C2, current to motor is proportionally reduced so motor comes gradually up to speed and is maintained there without operating in switching mode. Advantage of this arrangement is absence of electric noise normally generated during switching-mode operation.—"Linear Applications, Vol. 2," National Semiconductor, Santa Clara, CA, 1976, AN-162, p 10–11.

PAPER-TAPE FEED—High or 1 bit at output port of microprocessor turns on LED of optocoupler to energize solenoid of pinch-roller drive for paper tape of tape reader. Circuit will control reader from computer keyboard. Optoisolator is essential to keep grounds separate, since mechanical devices are electrically noisy and can generate garbage in computer. Article gives software for tape input routine on 8008 microprocessor.—D. Hogg, The Paper Taper Caper, *Kilobaud,* March 1977, p 34–40.

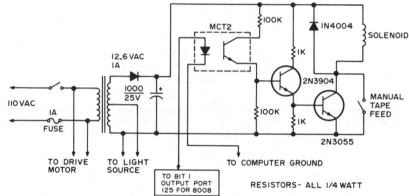

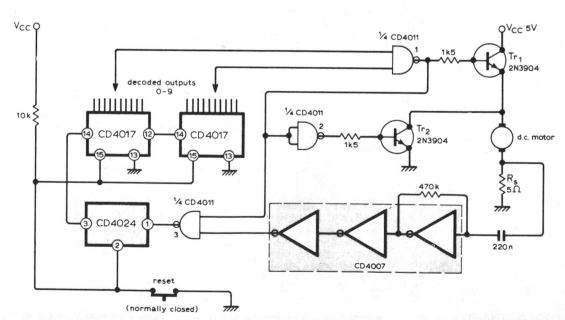

REVOLUTION-COUNTING CONTROL—When desired number of revolutions is reached by DC motor, as determined by preset counter, Tr₁ is turned off to interrupt path to 5-V motor supply, while TR₂ is turned on to brake motor rapidly. Voltage developed across 5-ohm resistor R_s in series with motor contains frequency component related to speed of rotation and number of armature coils. This signal is amplified by CD4007 CMOS inverter for feeding to counters through signal-squaring inverters. Counter outputs are decoded by gate 1. Motor slowdown by heavy loads does not affect accuracy of revolution-counting.—R. McGillivray, Motor Revolutions Control, *Wireless World,* Jan. 1977, p 76.

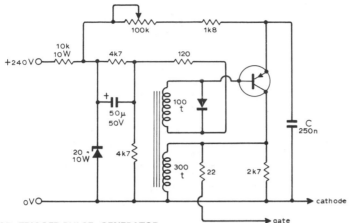

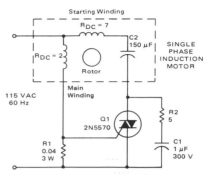

THYRISTOR TRIGGER-PULSE GENERATOR—
Used with thyristor speed control for 2-hp shunt-wound DC motor. Circuit provides train of pulses with variable delay with respect to zero-crossing instants of AC supply, for feeding to cathode and gate of thyristor to vary duty cycle. Use Mullard BFX29 silicon PNP transistor or equivalent, and any small-signal silicon diode. Output pulses are suitable for triggering all types of thyristors up to largest. Article also gives motor control circuit.—F. Butler, Thyristor Control of Shunt-Wound D.C. Motors, *Wireless World,* Sept. 1974, p 325–328.

TRIAC STARTING SWITCH FOR ½-hp MOTOR—
Triac replaces centrifugal switch normally used to control current through starting winding of single-phase induction motor. Value of R1 is chosen so triac turns on only when starting current exceeds 12 A. When motor approaches normal speed, running current drops to 8 A and triac blocks current through starting winding.—"Circuit Applications for the Triac," Motorola, Phoenix, AZ, 1971, AN-466, p 8.

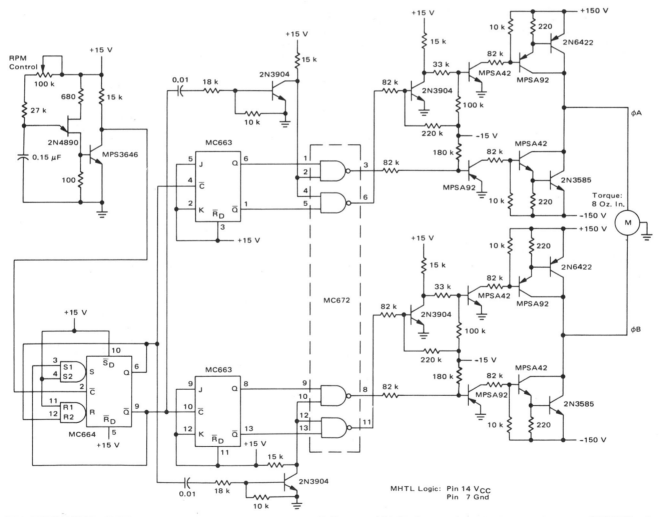

INDUCTION-MOTOR SPEED—
Uses variable-frequency UJT oscillator at upper left to toggle MC664 RS flip-flop which in turn clocks MC663 JK flip-flops. Quadrature-phased JK outputs are combined with fixed-width pulses in MC672 to provide zero-voltage steps of drive signals for phase A and phase B. Outputs of RS flip-flops are differentiated and positive-going transitions amplified by pair of 2N3904 transistors, with pulse width of about 500 μs. NAND-gate outputs are then translated by small-signal amplifiers to levels suitable for driving final transistors having complementary NPN/PNP pairs. Circuit will provide speed range of 300 to 1700 rpm for permanent-split capacitor motor.—T. Mazur, "Variable Speed Control System for Induction Motors," Motorola, Phoenix, AZ, 1974, AN-575A, p 6.

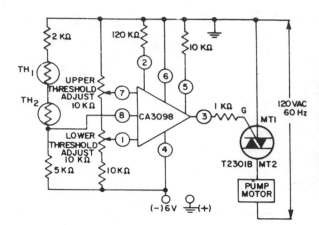

WATER-LEVEL CONTROL—Two thermistors operating in self-heating mode are mounted on sides of water tank. Thermistors change resistance when water level rises so liquid rather than air conducts heat away. Threshold adjustment pots are set so RCA CA3098 programmable Schmitt trigger turns on pump motor when water level rises above thermistor mounted near upper edge of tank, to remove water from tank and prevent overflow. Motor stays on to pump water out of tank until water level drops below location of lower thermistor inside tank.—"Linear Integrated Circuit and MOS/FET's," RCA Solid State Division, Somerville, NJ, 1977, p 218–221.

OPAMP SPEED CONTROL—Provides fine speed control of DC motor by using 0.25-W 6-V motor as tachogenerator giving about 4 V at 13,000 rpm. Opamp (RCA 3047A or equivalent) provides switching action for transistor in series with controlled motor, up to within a few volts of supply voltage. Choose transistor to meet motor current requirement.—N. G. Boreham, D.C. Motor Controller, *Wireless World*, Aug. 1971, p 386.

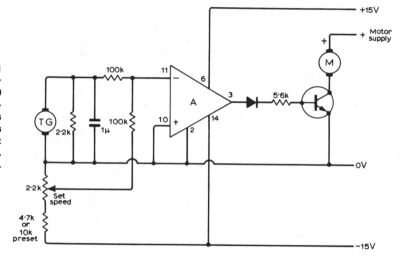

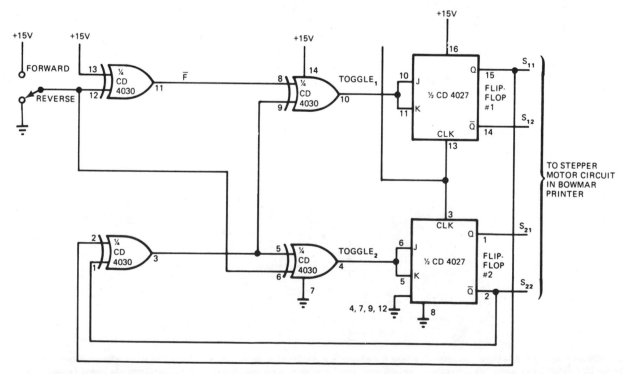

STEPPER MOTOR DRIVE—Two CMOS packages provide the four feed signals required for controlling forward/reverse drive of stepper motor for carriage drive and paper advance of Bowmar Model TP 3100 thermal printer. Outputs of flip-flops are above 10 V, enough to drive stepper motor directly. Each clock pulse to JK flip-flop advances carriage one step in direction commanded.—R. Bober, Stepper Drive Circuit Simplifies Printer Control, *EDN Magazine*, April 5, 1976, p 114.

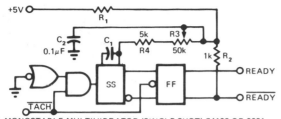

UP-TO-SPEED LOGIC—Simple speed-sensing circuit fed by tachometer pulses makes READY output high when rotating device reaches desired minimum or threshold speed. Single-action triggering eliminates instability at decision point. Circuit also provides hysteresis, for separating pull-in and drop-out points any desired amount as determined by ratio of R_1 to R_2 in timing network. Article covers circuit operation and gives timing diagram.—W. Bleher, Circuit Indicates Logic "Ready," *EDN Magazine,* March 5, 1974, p 72 and 74.

SS=MONOSTABLE MULTIVIBRATOR (SINGLE SHOT) 74122 OR 9601 OR ½ 74123 OR 9602
FF=NEGATIVE EDGE–TRIGGERED FLIP FLOP, 74H103 OR EQUIVALENT
R_1=0 TO 20% OF R_2, SELECTED FOR REQUIRED HYSTERESIS
C_1=SELECTED FOR REQUIRED TIMING RANGE.

CONTINUOUS-DUTY BRAKE—High or 1 bit at output port of microprocessor energizes brake solenoid of paper-tape reader through optocoupler and amplifier. When tape is to be stopped, brake solenoid is energized and tape is squeezed between top of solenoid and flat iron brake shoe that is attracted by solenoid.—D. Hogg, The Paper Taper Caper, *Kilobaud,* March 1977, p 34–40.

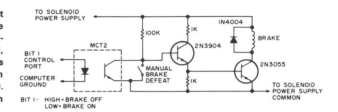

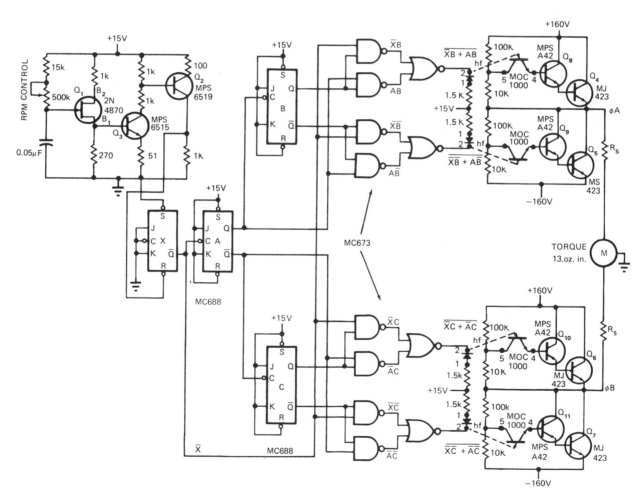

SPEED CONTROL FOR INDUCTION MOTOR—Uses UJT oscillator Q_1 to generate frequency in range from 40 to 1200 Hz for feeding to divide-by-4 configuration that gives motor source frequency range of 10 to 300 Hz. With induction motor having two pairs of poles, this gives theoretical speed range of 300 to 9000 rpm with essentially constant torque. Speed varies linearly with frequency. Circuit uses pair of flip-flops (MC673) operated in time-quadrature to perform same function as phase-shifting capacitor so motor receives two drive signals 90° apart. Article covers operation of circuit in detail. Optoisolators are used to provide bipolar drive signals from unipolar control signals. Each output drive circuit is normally off and is turned on only when its LED is on. If logic power fails, drives are disabled and motor is turned off as fail-safe feature.—T. Mazur, Unique Semiconductor Mix Controls Induction Motor Speed, *EDN Magazine,* Nov. 1, 1972, p 28–31.

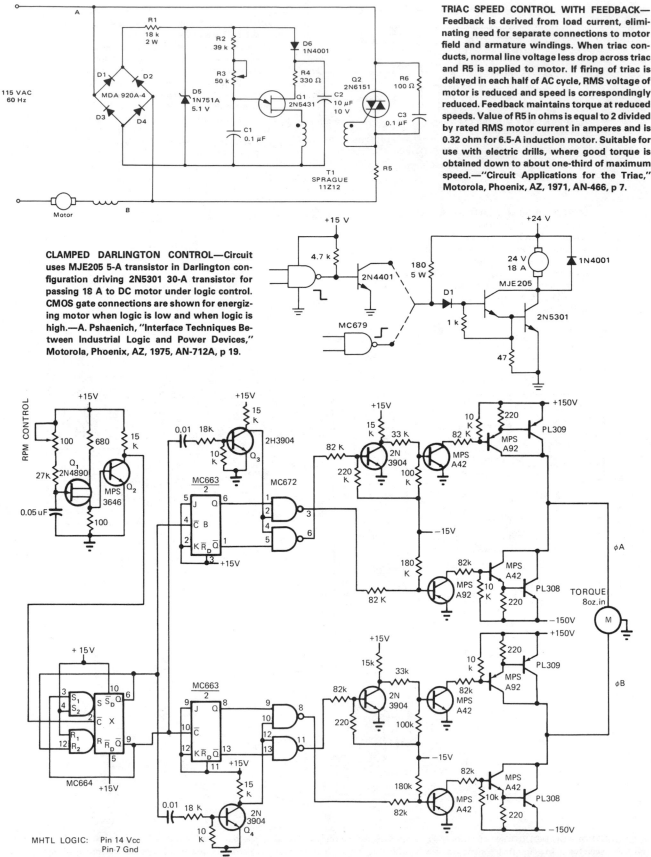

TRIAC SPEED CONTROL WITH FEEDBACK— Feedback is derived from load current, eliminating need for separate connections to motor field and armature windings. When triac conducts, normal line voltage less drop across triac and R5 is applied to motor. If firing of triac is delayed in each half of AC cycle, RMS voltage of motor is reduced and speed is correspondingly reduced. Feedback maintains torque at reduced speeds. Value of R5 in ohms is equal to 2 divided by rated RMS motor current in amperes and is 0.32 ohm for 6.5-A induction motor. Suitable for use with electric drills, where good torque is obtained down to about one-third of maximum speed.—"Circuit Applications for the Triac," Motorola, Phoenix, AZ, 1971, AN-466, p 7.

CLAMPED DARLINGTON CONTROL— Circuit uses MJE205 5-A transistor in Darlington configuration driving 2N5301 30-A transistor for passing 18 A to DC motor under logic control. CMOS gate connections are shown for energizing motor when logic is low and when logic is high.—A. Pshaenich, "Interface Techniques Between Industrial Logic and Power Devices," Motorola, Phoenix, AZ, 1975, AN-712A, p 19.

MHTL LOGIC: Pin 14 Vcc
              Pin 7 Gnd

FREQUENCY CONTROLS SPEED— Circuit generates variable frequency between 10 and 300 Hz at constant voltage for changing speed of induction motor between theoretical limits of 300 and 9000 rpm without affecting maximum torque. Direct coupling between control and drive circuits is used; if motor noise affects control logic circuits, optoisolators should be used between control and drive sections. Article tells how circuit works and gives similar circuit using optical coupling.—T. Mazur, Unique Semiconductor Mix Controls Induction Motor Speed, *EDN Magazine*, Nov. 1, 1972, p 28–31.

2-hp THREE-PHASE INDUCTION—Speed is controlled by applying continuously variable DC voltage to VCO of control circuit for 750-VDC 7-A bridge inverter driving three sets of six Delco DTS-709 duolithic Darlingtons. Bridge inverter circuit for other two phases is identical to that shown for phase AA'. VCO output is converted to three-phase frequency varying from 5 Hz at 50 VDC to 60 Hz at 600 VDC for driving output Darlingtons. Optoisolators are used for base drive of three switching elements connected to high-voltage side of inverter.—"A 7A, 750 VDC Inverter for a 2 hp, 3 Phase, 480 VAC Induction Motor," Delco, Kokomo, IN, 1977, Application Note 60.

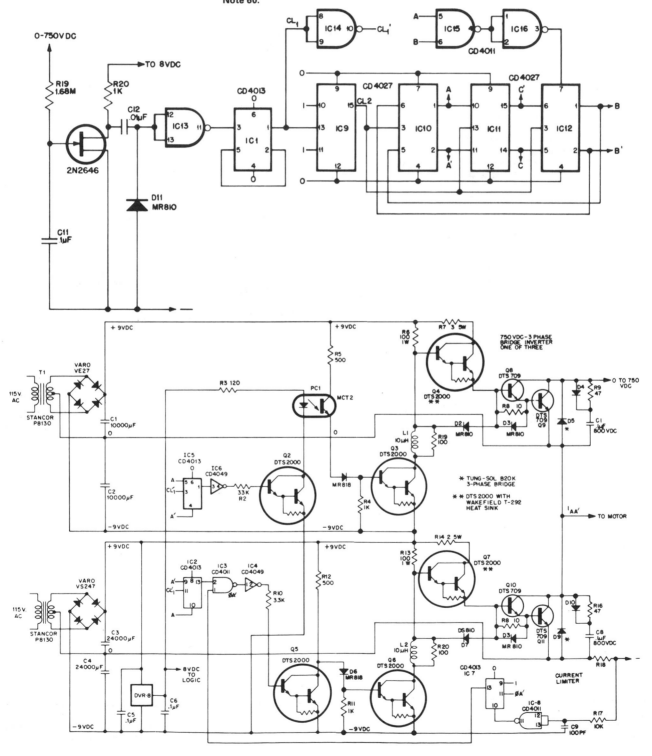

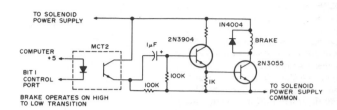

PULSED BRAKE—Transition from high (1) to low (0) at control port of microprocessor energizes brake solenoid of paper-tape reader in pulses lasting several microseconds, with time determined by size of capacitor used. Energizing of solenoid squeezes tape between top of solenoid and flat iron brake shoe that is attracted by solenoid. Unmarked resistor is 1K.— D. Hogg, *The Paper Taper Caper, Kilobaud,* March 1977, p 34–40.

SERIES-WOUND MOTOR—Provides both direction and speed control for fractional-horsepower series-wound or universal DC motors as long as motor current requirements are within SCR ratings. Q1-Q4, connected in bridge, are triggered in diagonal pairs. S1 determines which pair is turned on, to provide direction control. Pulse circuit is used to drive SCRs through T1 or T2. When C1 charges to breakdown voltage of zener D5, zener passes current to gate of SCR Q5 and turns it on. This discharges C1 through T1 or T2 to create desired triggering pulse. Q5 stays on for duration of half-cycle. R1 controls motor speed by changing time required to charge C1, thereby changing conduction angle of Q1-Q4 or Q2-Q3.—"Direction and Speed Control for Series, Universal and Shunt Motors," Motorola, Phoenix, AZ, 1976, AN-443.

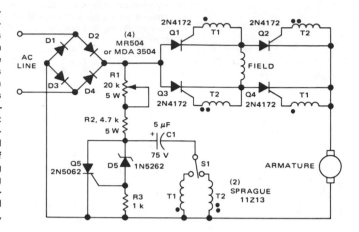

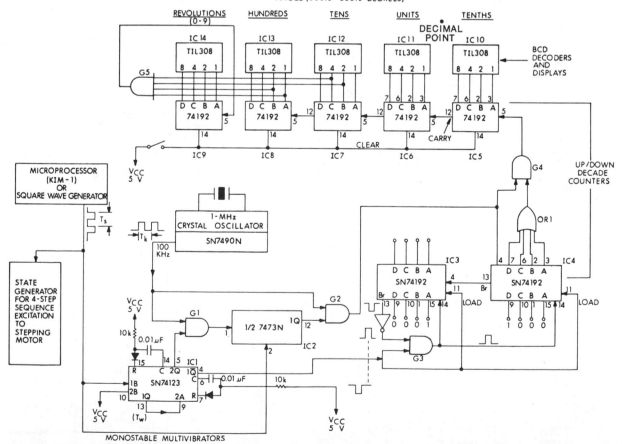

MOTOR STEP-ANGLE DISPLAY—Digital display circuit tracks stepper-motor shaft movements. Up/down decade counters read out four BCD digits as travel angle (000.0 to 360.0) in degrees and number of completed revolutions (0 to 9). Stepper under study is driven by state generator that produces high-current square-wave pulses under control of clock used for display, which can be external square-wave generator or clock output of microprocessor such as KIM-1. Power source for digital display is 5 V at 1.2 A. Applications include monitoring movements of incremental plotters, precision film camera drives, numerical control machines, and precision start-stop motions of fuel control rods in nuclear reactor.—H. Lo, *Digital Display of Stepper Motor Rotation, Computer Design,* April 1978, p 147–148 and 150–151.

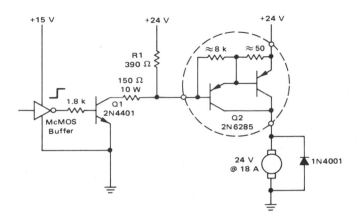

HIGH-LEVEL CMOS CONTROL—When output of CMOS buffer goes high, Q1 turns on and sinks 150-mA base current of power Darlington Q2, to activate motor load. Used in logic-controlled industrial applications.—A. Pshaenich, "Interface Techniques Between Industrial Logic and Power Devices," Motorola, Phoenix, AZ, 1975, AN-712A, p 18.

TELEFAX PHASING—Simple coincidence circuit provides reliable synchronization of Telefax machine in which 2500-Hz signal is generated by photoelectric scanning of paper placed on revolving drum. Circuit uses 7402 quad two-input NOR gate. If alternate connection enclosed in dashed line is not used, connect pin 8 to ground at pin 7. Q1 is S0014 silicon or equivalent. If relay contacts will handle motor voltage and current, they can be connected directly across points of test switch on machine, with switch left open for phasing circuit to work.— W. C. Smith, A Logic Circuit for Phasing the Telefax, *QST*, Nov. 1978, p 33–34.

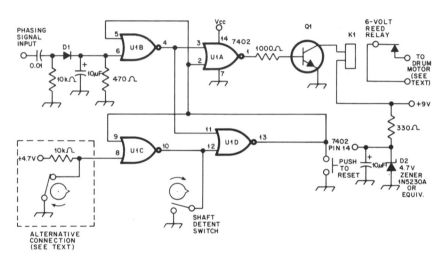

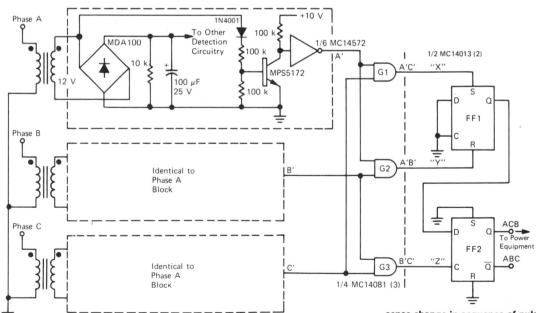

PHASE-REVERSAL DETECTOR—Used in three-phase applications in which direction of rotation of phases is critical, as in three-phase motors where reversal of two phases can provide disastrous reversal of motor. Line voltages are stepped down and isolated by control-type transformers. Each phase is half-wave rectified and shaped by 1N4001 diode and MPS5172 transistor, with additional shaping by MC14572 inverter. Shaped outputs of all three phases are combined in AND gates G1-G3 to give pulse outputs sequentially. D flip-flops are connected to sense change in sequence of pulses caused by reversal of one or more input phases. Flip-flop output can be used to trip relay or other protective device for removing air conditioner or other equipment from line before it is damaged.—T. Malarkey, "A Simple Line Phase-Reversal Detection Circuit," Motorola, Phoenix, AZ, 1975, EB-54.

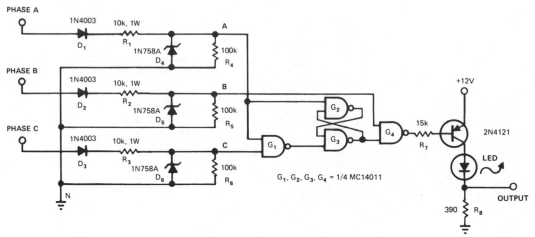

PHASE SEQUENCE DETECTOR—Circuit detects incorrect phase sequence of motor driving pump, compressor, conveyor, or other equipment that can be damaged by reverse rotation. Circuit also protects motor from phase loss that could cause rapid temperature rise and heat damage. LED is on when phasing is correct. For phase loss or incorrect sequence, output goes low and LED is dark. Diodes and zeners change sine waves for all phases to rectangular logic-level pulses that feed gates. When phases are correct, output of G_4 is train of rectangular pulses about 2.5 ns wide. Output is zero for incorrect sequences. Since leading edge of output pulse coincides with positive zero crossing of phase B, output pulses can be used to trigger SCR connected across phase B and driving relay-coil load. SCR then energizes relay only when sequence is correct.—H. Normet, Detector Protects 3-Phase-Powered Equipment, *EDN Magazine*, Aug. 5, 1978, p 78 and 80.

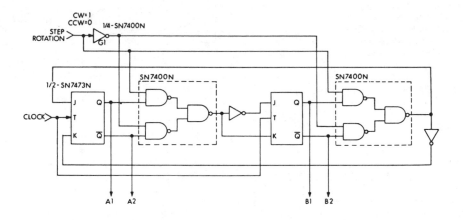

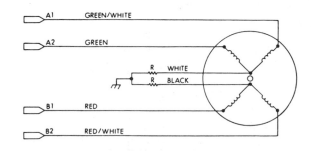

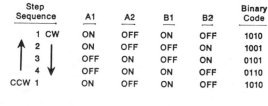

Step Sequence		A1	A2	B1	B2	Binary Code
1	CW	ON	OFF	ON	OFF	1010
2		ON	OFF	OFF	ON	1001
3		OFF	ON	OFF	ON	0101
4		OFF	ON	ON	OFF	0110
CCW 1		ON	OFF	ON	OFF	1010

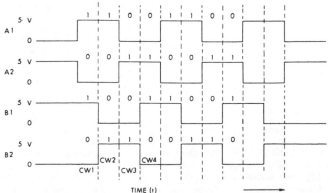

STATE GENERATOR FOR STEPPER—Generates high-current square-wave pulses and provides correct switching sequence for exciting stepper motor when digital display is required to show instantaneous step angle and total revolutions traveled by shaft of stepper motor. If microprocessor is used, speed and direction of motor rotation can be controlled by programming period and level of output pulses. Clock signals trigger SN7473N JK flip-flop that changes ON/OFF states of four outputs as shown in table. Clock signal is obtained from external square-wave generator or from microprocessor such as KIM-1. Article also gives digital display circuit driven by same clock.—H. Lo, Digital Display of Stepper Motor Rotation, *Computer Design*, April 1978, p 147–148 and 150–151.

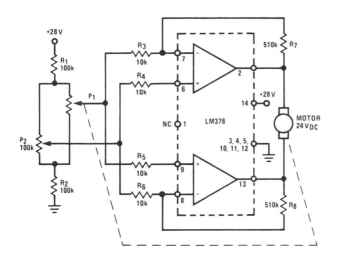

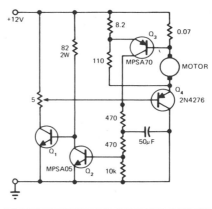

MOTOR VOLTAGE	TRANSISTOR VOLTAGE	Q₃ EMITTER VOLTS	LIMIT CURRENT
0	12	0	10A
6	6	0.42	16A
11.5	0.5	0.20	21.4A

24-VDC PROPORTIONAL SPEED CONTROL—National LM378 amplifier IC is basis for low-cost proportional speed controller capable of furnishing 700 mA continuously for such applications as antenna rotors and motor-controlled valves. Proportional control results from error signal developed across Wheatstone bridge R_1-R_2-P_1-P_2. P_1 is mechanically coupled to motor shaft as continuously variable feedback sensor. As motor turns, P_1 tracks movement and error signal becomes smaller and smaller; system stops when error voltage reaches 0 V.—"Audio Handbook," National Semiconductor, Santa Clara, CA, 1977, p 4-8–4-20.

STALLED-MOTOR PROTECTION—Modification of basic speed control circuit for small DC permanent-magnet motors provides maximum current limit under normal conditions and reduced current limit under stall conditions, to limit dissipation of series transistor Q_4 to safe value. When motor stalls, motor voltage falls, reducing voltage and motor current required to turn on Q_3 and thereby limiting stalled-motor current.—D. Zinder, Current Limit and Foldback for Small Motor Control, *EDN Magazine,* May 5, 1974, p 77 and 79.

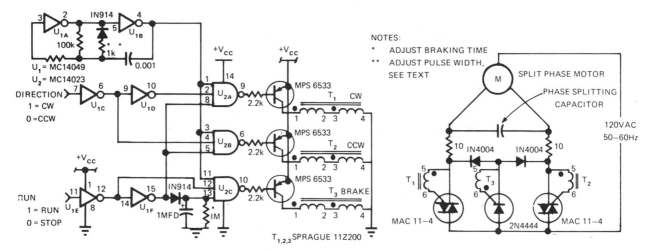

SPLIT-PHASE CONTROL WITH BRAKING—Use of CMOS logic to gate direction-controlling triacs and turn on SCR for braking provides low-cost switchless control of split-phase motor used in place of brush-type DC motor. Applications include control of ball valves and other throttling functions in process control. With shaft-position encoders, circuit generates feedback information. Overshoot and other stability problems are easily controlled by strong braking function. CMOS logic provides complete noise immunity. Oscillator pulse width is adjusted with 1K resistor in series with 1N914, and brake duration is controlled by 1-megohm resistor at input of U_{2C}. With values shown, brake is applied for about 1 s. Circuit works reliably on supply voltages of 5 to 15 V.—V. C. Gregory, Split-Phase Motor Control Accomplished with CMOS, *EDN Magazine,* Oct. 5, 1974, p 65–67.

CHAPTER 54
Multiplexer Circuits

Includes circuits for multiplexing of data or communication channels, analog switches, filters, displays, and sensor channels under control of logic signals. See also Cathode-Ray, Data Transmission, Display, Instrumentation, and Logic chapters.

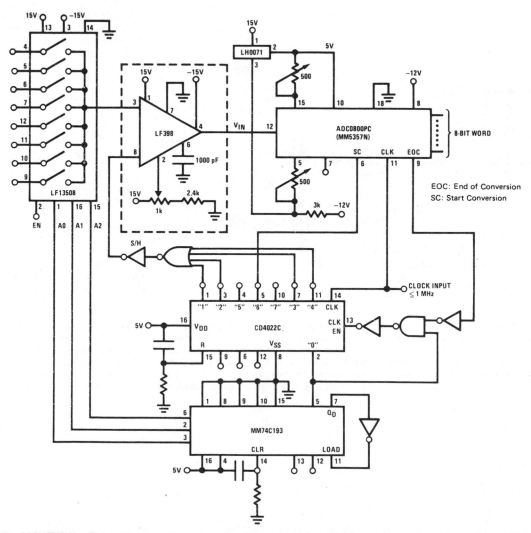

EIGHT-CHANNEL SEQUENTIAL—Eight different analog inputs are sampled by National LF13508 multiplexer and converted into digital words for further processing. Maximum throughput rate of system is 2800 samples per second per channel. Output will settle to ±0.05 mV in 1 μs after hold command.—"FET Databook," National Semiconductor, Santa Clara, CA, 1977, p 5-77–5-78.

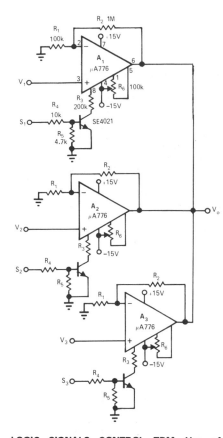

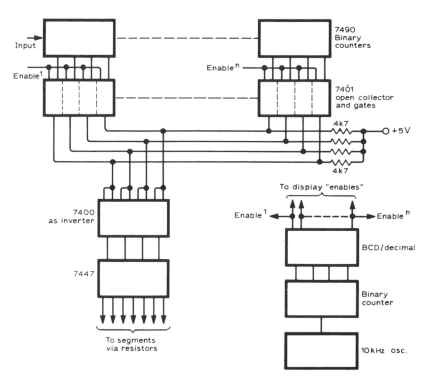

LOGIC SIGNALS CONTROL TDM—Use of μA776 opamps eliminates need for FET switches in time-division multiplexing and signal conditioning. Multiplex inputs S_1-S_2-S_3 can be driven directly by TTL or DTL devices. Each amplifier is controlled by either supplying or denying its 70-μA master-bias input, thus turning it on or off. All stages have identical component values.—M. K. Vander Kooi, Multiplexing Without FETs, *EDN/EEE Magazine,* Aug. 1, 1971, p 47.

MULTIPLEXING OF DISPLAYS—Three or more displays can be multiplexed simply by gating 7490 counter output via 7401, then using OR connection for outputs. 7401s are switched on in rotation by positive enable signal which also switches displays on in turn. If using common-cathode displays, segment and display enable signals must be inverted. With individual display units, segments must be paralleled.—G. A. Bobker, Simplified Multiplexing, *Wireless World,* Feb. 1978, p 59.

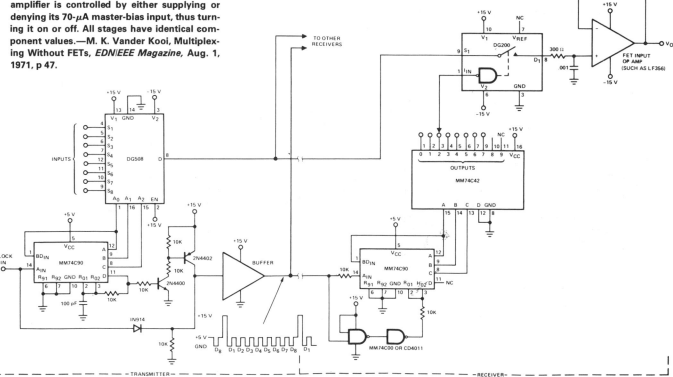

EIGHT-CHANNEL TELEMETER—Multiplex transmission system permits monitoring eight different inputs at remote location. 5-V pulse train is sent down separate channel to receiving location to perform timing and synchronizing functions for outputs of MM74C42 BCD-to-decimal decoder. 15-V reset pulse is superimposed on 5-V clock. Other receivers can be added if monitoring is desired at more than one location.—"Analog Switches and Their Applications," Siliconix, Santa Clara, CA, 1976, p 7-71–7-72.

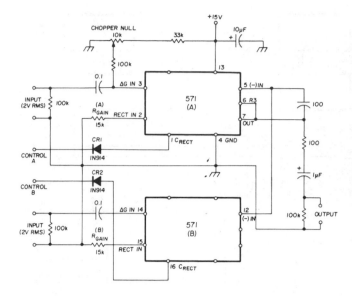

TWO-INPUT FSK MULTIPLEXER—Uses Signetics NE571 or NE570 analog compandors. Gain of each channel is unity, as determined by R_{GAIN} value for channel. When complementary control signals are provided, FSK generator switches between the two signal inputs. Outputs, when on, are summed by opamp in IC. Each channel is gated off by low control logic input. For FSK or alternate-channel use, CONTROL A and CONTROL B signals should be complementary. Control signal suppression is optimized with chopper null pot. Suppression is better than 60 dB after trimming. Circuit can also be used as summing switch, with both signals on at any given instant.—W. G. Jung, Gain Control IC for Audio Signal Processing, *Ham Radio,* July 1977, p 47–53.

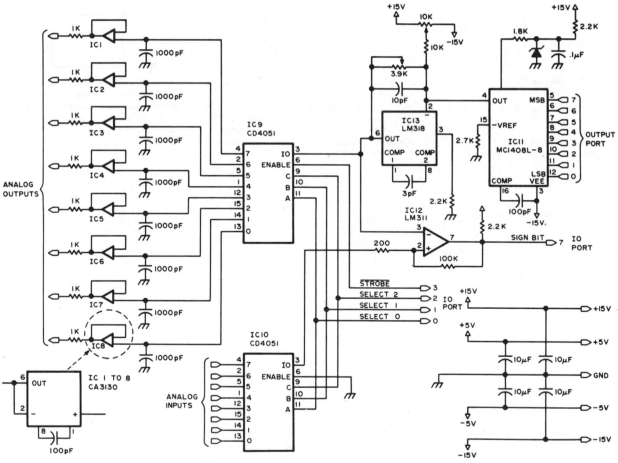

Number	Type	+5 V	GND	−15 V	+15 V	−5 V
1 to 8	CA3130	7				4
9	CD4051	16	8			7
10	CD4051	16	8			7
11	MCI408L-8	13	2			
12	LM311		1	4	8	
13	LM318			4	8	

MULTIPLEXED A/D-D/A CONVERTER INTERFACE—Time-multiplexed interface minimizes hardware required for applications of personal computer system. Useful in interactive games, equipment testing, and electronic music. Optimized for 0.1-100 Hz signals. Bypass each power pin with 0.01 μF to suppress stray spikes caused by power surges. Use of LM318 opamp minimizes response time of MC1408L-8 DAC.—D. R. Kraul, Designing Multichannel Analog Interfaces, *BYTE,* June 1977, p 18–23.

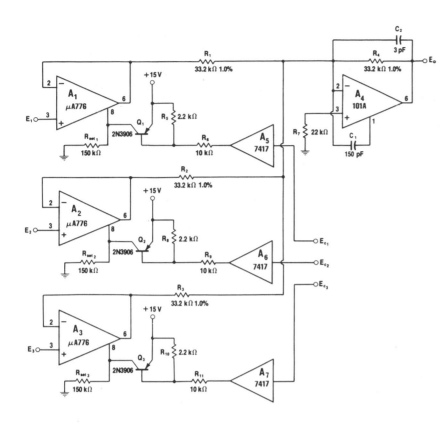

ANALOG SUMMING—ON/OFF programming of μA776 opamps allows any or all inputs to be on at given time. Switched outputs are combined in summing inverter A_4. Voltage followers A_1-A_3 are programmed from on to off by Q_1-Q_3 and A_5-A_7. Noise gain of stage A_4 is minimized when input channel is switched off. If sign inversion by A_4 is undesirable, add inverter stage following A_4. Any number of additional channels can be added. Programming pulses are applied to inputs of A_5, A_6, and A_7.—W. G. Jung, "IC Op-Amp Cookbook," Howard W. Sams, Indianapolis, IN, 1974, p 419–421.

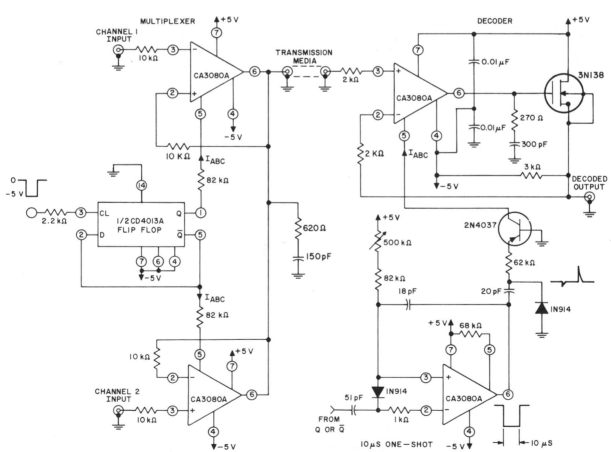

DATA MULTIPLEXER/DECODER—CD4013A flip-flop switches channel opamps alternately to transmission line under control of −5 V clock pulses for multiplexing of inputs to line. Any number of input channels can be added by extending circuit. At receiving end, one CA3080A variable opamp is used as mono MVBR to provide 10-μs delay for input signal to settle before being sampled by sample-and-hold decoder. Either output of flip-flop can be used to trigger MVBR.—"Circuit Ideas for RCA Linear ICs," RCA Solid State Division, Somerville, NJ, 1977, p 15.

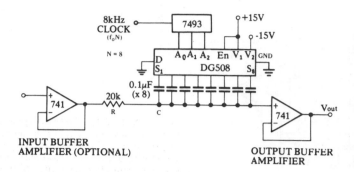

1-kHz COMB FILTER—DG508 eight-channel CMOS multiplexer is used in comb filter having fundamental frequency of 1 kHz. Sampling action provides response at each harmonic multiple except at 8 and 16 kHz (no response at Nf_0 or $2Nf_0$). Used in selective filtering of periodic signals from background of nonperiodic noise interference. 7493 TTL binary counter provides necessary 3-bit binary count sequence from 8-kHz clock. Q is 50.—"Analog Switches and Their Applications," Siliconix, Santa Clara, CA, 1976, p 5-17–5-18.

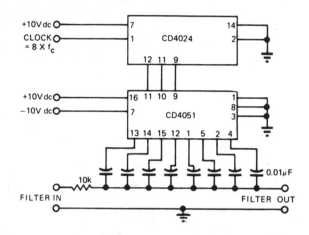

COMMUTATING BANDPASS FILTER—CD4051 analog multiplexer serves for commutation and switching of eight low-pass filter sections. Multiplexer is driven by CD4024 binary counter that is clocked at 8 times desired 100-kHz center frequency. Can be tuned by varying commutating frequency.—J. Tracy, CMOS Offers New Approach to Commutating Filters, *EDN Magazine*, Feb. 5, 1974, p 94–95.

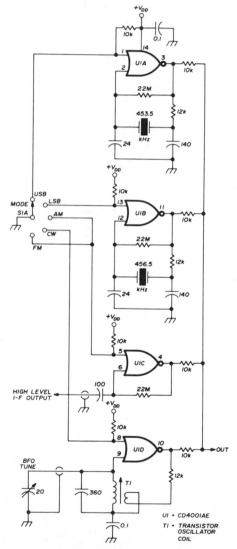

BFO MULTIPLEXER—Signal 455-kHz multimode detection system using RCA CD4001AE quad NOR gate functions as upper-sideband or lower-sideband crystal oscillator, tunable BFO for CW, or limiter of IF signal for FM or synchronous AM reception. Desired oscillator or limiter is gated on by grounding its digital control line with S1A. Multimode reception occurs when multiplexed output of oscillators and limiter is applied to product detector.—J. Regula, BFO Multiplexer for a Multimode Detector, *Ham Radio*, Oct. 1975, p 52–55.

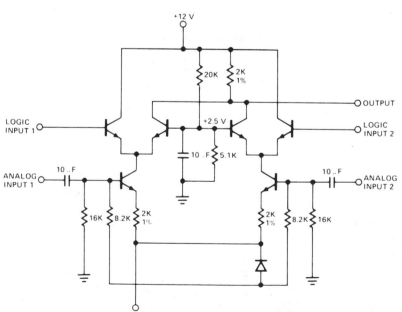

ANALOG SWITCH—Circuit using Signetics 511 transistor array provides digital selection of either of two analog signals. When logic input at left is zero, signal at analog input 1 goes to output and other analog input signal is rejected. Similarly, when logic 0 is applied to logic input 2, analog input 2 goes to output. Eight-channel analog multiplex switch can be formed by combining four 511 analog switches with Signetics 8250 binary-to-octal decoder. Analog signals up to 200 kHz are switched without amplitude degradation.—"Signetics Analog Data Manual," Signetics, Sunnyvale, CA, 1977, p 753–754.

64-CHANNEL TWO-LEVEL—Two-level multiplexing system increases effective switching speeds when transmitting 64 analog signals on single transmission line. Four clock phases are generated with DM7473 2-bit counter that toggles on high-to-low clock edge. DM7400 NAND gates decode flip-flop outputs into required four clock phases. As clock phase goes from low to high state, DG181 analog switch fed by it turns off and corresponding DM7493 4-bit binary counter is triggered to next address state for sampling of that input channel at output. Reset is used to set system for starting on first channel when power is applied.—"Analog Switches and Their Applications," Siliconix, Santa Clara, CA, 1976, p 7-11–7-13.

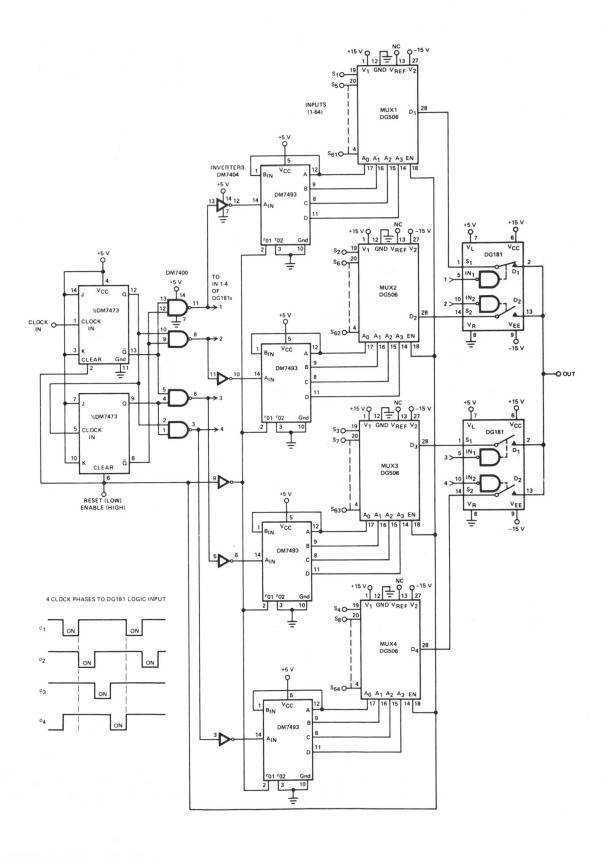

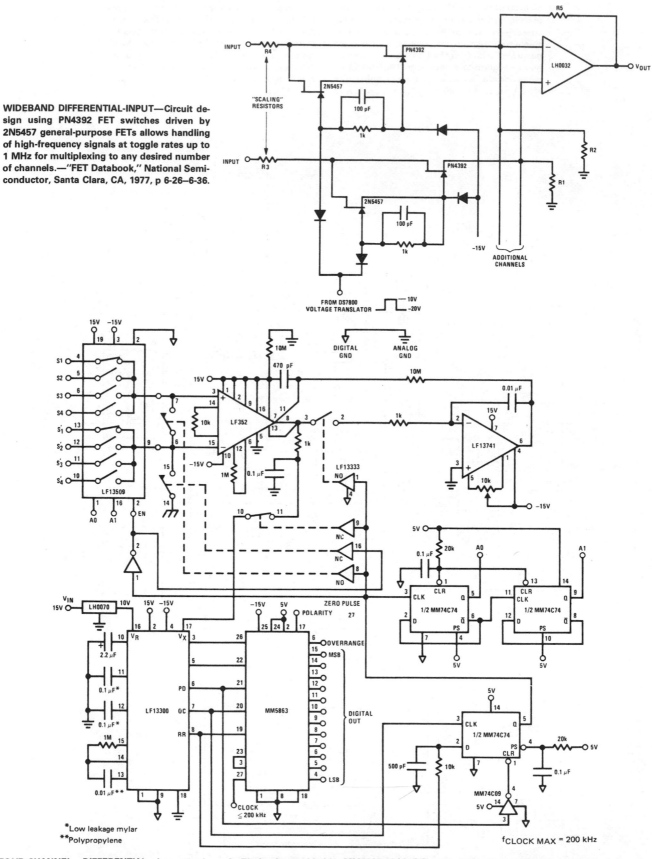

WIDEBAND DIFFERENTIAL-INPUT—Circuit design using PN4392 FET switches driven by 2N5457 general-purpose FETs allows handling of high-frequency signals at toggle rates up to 1 MHz for multiplexing to any desired number of channels.—"FET Databook," National Semiconductor, Santa Clara, CA, 1977, p 6-26—6-36.

FOUR-CHANNEL DIFFERENTIAL—Low-speed high-accuracy data acquisition unit acquires analog input signal differentially with LF13509 IC and preconditions it through LF352 instrumentation amplifier having automatic zeroing circuit. Timing is provided by MM5863 12-bit A/D converter and lower MM74C74 flip-flop. Upper two flip-flops form 2-bit up counter for channel select. Instrumentation amplifier is zeroed at power-up and after each conversion. Maximum clock frequency depends on required accuracy and minimum zeroing time of instrumentation amplifier.—"MOS/LSI Databook," National Semiconductor, Santa Clara, CA, 1977, p 5-2—5-22.

64-CHANNEL TWO-LEVEL HIGH-SPEED—Four DG506 16-channel multiplexers serve for first multiplexer level, and two DG304 high-speed dual analog switches serve in second level for switching DG506 outputs to single output of multiplexer. As one multiplexer is being sampled at output, other multiplexers are being switched to next address line; this shortens overall system transition time from 1.5 μs to 0.25 μs. Two-level system also lowers output node capacitance and output leakage. CMOS digital logic controls entire system.—"Analog Switches and Their Applications," Siliconix, Santa Clara, CA, 1976, p 7-82–7-84.

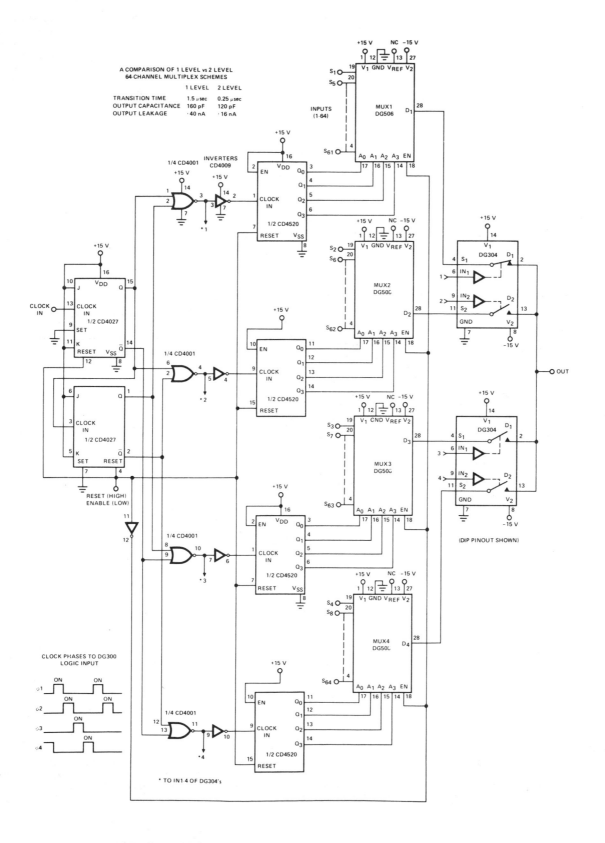

A COMPARISON OF 1 LEVEL vs 2 LEVEL
64-CHANNEL MULTIPLEX SCHEMES

	1 LEVEL	2 LEVEL
TRANSITION TIME	1.5 μsec	0.25 μsec
OUTPUT CAPACITANCE	160 pF	120 pF
OUTPUT LEAKAGE	40 nA	16 nA

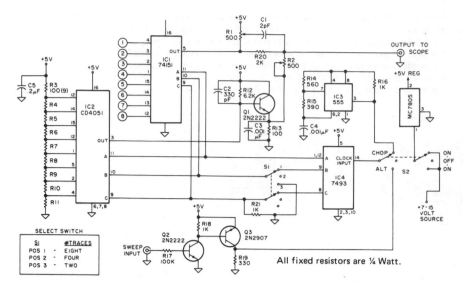

8-CHANNEL CRO MULTIPLEXER—Simple adapter converts any single-trace oscilloscope into professional 8-channel model by multiplexing eight input signals into one output for vertical amplifier. Discrete voltages are picked off resistor divider chain sequentially and added to digital input signal so trace is shifted fast enough to produce eight individual traces. Addressing for 74151 digital multiplexer requires 3 bits of binary code to address all eight channels in sampling sequence. As each channel is sampled, its logic level (1 or 0) appears at pin 5 of 74151 for feed to vertical input of CRO. Developed for troubleshooting in digital circuits.— W. J. Prudhomme, Build an Eight Channel Multiplexer for Your Scope, *Kilobaud,* April 1977, p 29–32.

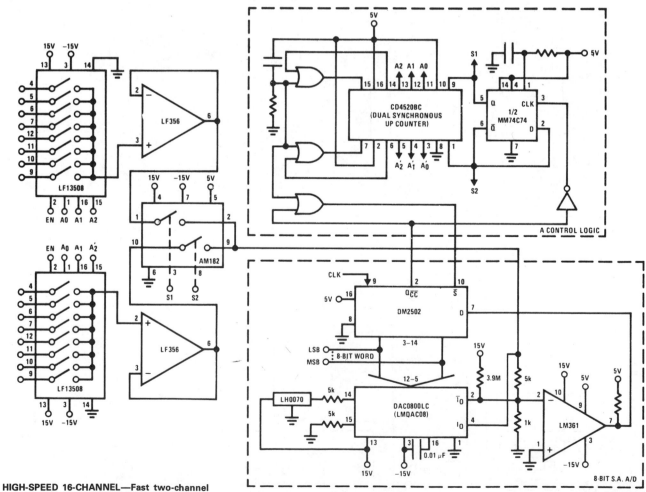

HIGH-SPEED 16-CHANNEL—Fast two-channel multiplexer using National AM182 dual analog switch provides second-level multiplexing by accepting outputs of each LF13508 eight-channel multiplexer and feeding these outputs sequentially into 8-bit successive-approximation A/D converter. Technique makes throughput rate of system independent of analog switch speed. With maximum clock frequency of 4.5 MHz, throughput rate is 31,250 samples per second per channel.—"FET Databook," National Semiconductor, Santa Clara, CA, 1977, p 5-79–5-80.

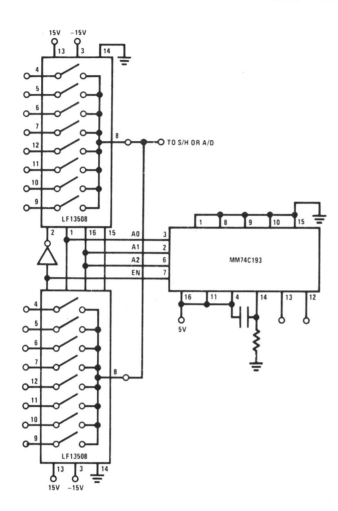

16-CHANNEL SIMPLIFIED SEQUENTIAL MULTIPLEXING—Two National LF13508 eight-channel multiplexers are connected so enable pins are used to disconnect one multiplexer while the other is sampling. Any number of eight-channel multiplexers can be connected in this way if speed is not prime system requirement.—"FET Databook," National Semiconductor, Santa Clara, CA, 1977, p 5-79—5-81.

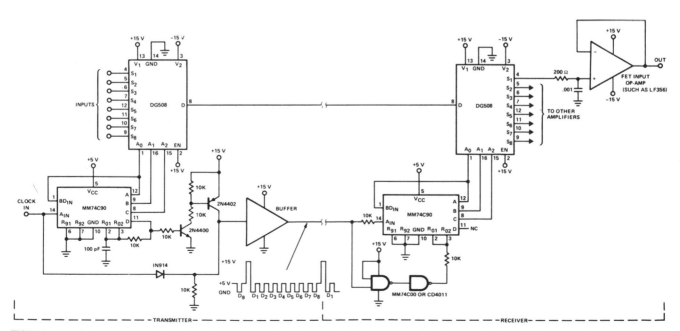

EIGHT-CHANNEL MUX/DEMUX—Provides for monitoring of all eight channels continuously at remote location instead of scanning channels at receiver. Each output of DG508 eight-channel analog multiplexer in receiver feeds its own opamp for driving readout. Similar DG508 at transmitter feeds inputs over single-wire line under control of MM74C90 presettable decade counter which also feeds pulse train over separate channel to receiver for timing and synchronization. 15-V reset pulse superimposed on 5-V clock pulses keeps channels synchronized.—"Analog Switches and Their Applications," Siliconix, Santa Clara, CA, 1976, p 7-71—7-74.

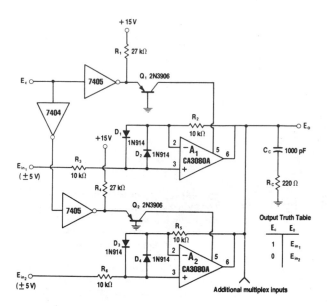

TWO-CHANNEL LOGIC-CONTROLLED—Two CA3080A operational transconductance amplifiers are programmed on and off alternately by control logic, for multiplexing two inputs each up to ±5 V into single output. When control line E_c is high or 1, input 1 appears at output; for logic 0, input 2 appears at output. Both channels operate at unity gain. Control logic is TTL-compatible.—W. G. Jung, "IC Op-Amp Cookbook," Howard W. Sams, Indianapolis, IN, 1974, p 458–462.

Output Truth Table

E_c	E_o
1	E_{in_1}
0	E_{in_2}

FOUR-CHANNEL ANALOG MULTIPLEXER—Buffered input and output provide very high input impedance and low output impedance for analog signal selection or time-division multiplexing with Harris HA-2400/HA-2405 IC that combines functions of analog switch and high-performance opamp. Bandwidth is about 8 MHz. Signal is fed back from output to digitally selected input channel to place that channel in voltage-follower configuration with noninverting unity gain.—"Linear & Data Acquisition Products," Harris Semiconductor, Melbourne, FL, Vol. 1, 1977, p 7-36–7-38 (Application Note 514).

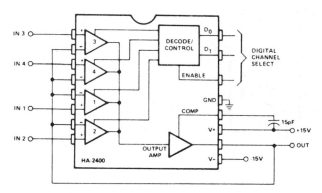

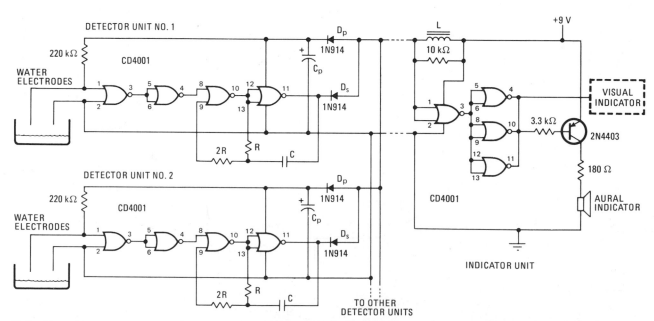

WATER-LEAK ISOLATOR—Master indicator monitors large number of two-electrode water-leak detectors simultaneously, with each detector indicating presence of water by producing unique tone signal. Only two wires are needed between detectors and master indicator unit. Each detector location has quad NOR gate. First gate is water sensor, second is inverter, and last two form astable MVBR. When sensing electrodes are dry, resistance between them is above 500K and MVBR is disabled. When water drops resistance below 100K, MVBR oscillates at audio frequency determined by its RC time constant, equal to 1/1.4RC where R is in ohms and C is in farads. C_P value in farads should be 1/1000 of lowest frequency used by detectors. L in henrys is about 1/600 of lowest frequency.—F. E. Hinkle, Multiplexed Detectors Isolate Water Leaks, *Electronics*, Dec. 11, 1975, p 116–117.

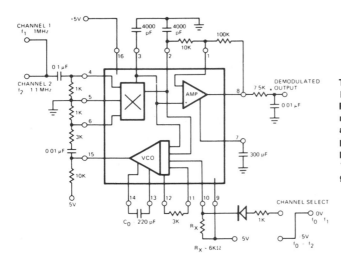

TIME-MULTIPLEXING TWO FM CHANNELS— Digital programming capability of Exar XR-215 PLL IC makes possible time-multiplexing demodulator between two FM channels, at 1.0 and 1.1 MHz. Channel-select logic signal is applied to pin 10, and both input channels are applied simultaneously to PLL input pin 4.— "Phase-Locked Loop Data Book," Exar Integrated Systems, Sunnyvale, CA, 1978, p 21–28.

THREE-CHANNEL FOR DATA— Each input channel uses CA3060 variable opamp as high-impedance voltage follower driving output MOSFET serving as buffer and power amplifier. Cascade arrangement of opamps with MOSFET provides open-loop voltage gain in excess of 100 dB.—"Circuit Ideas for RCA Linear ICs," RCA Solid State Division, Somerville, NJ, 1977, p 16.

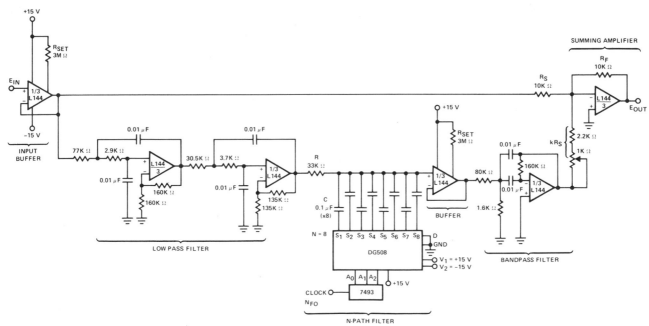

1-kHz N-PATH NOTCH FILTER— Combination of DG508 eight-channel CMOS multiplexer with low-pass and bandpass active filters provides 1-kHz notch filter having Q of 1330 and 3-dB bandwidth of 0.75 Hz at 1 kHz. Low-pass filter introduces 180° phase shift at 1 kHz. Amplifier sums original signal in phase-shifted bandpass output from N-path filter, canceling 1-kHz components in original signal to produce desired notch characteristic.—"Analog Switches and Their Applications," Siliconix, Santa Clara, CA, 1976, p 5-18—5-20.

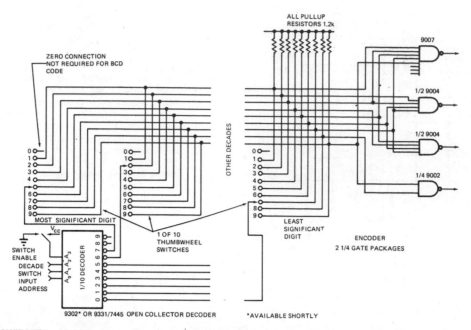

MULTIPLEXING BCD SWITCHES—Multiplexing technique reduces number of interconnections between thumbwheel switches and counters, displays, or industrial control equipment being programmed remotely. Ten decades of BCD switches require only 10 interconnections, as compared to 50 without multiplexing. All 10 outputs of low-cost single-pole decade switches are paralleled, with wiper arm connections being brought out separately. Parallel outputs are fed into simple encoder using four NAND gates to generate 4-bit BCD output code. Wiper of each switch is addressed from active low open-collector decoder. In operation, 3-bit input address determines which decade switch is addressed, and switch position then determines which encoder NAND gates are activated.—E. Breeze, Putting the "Thumb" on Thumbwheel Switch Multiplexing, *EDN Magazine*, Aug. 1, 1972, p 56.

CHAPTER 55
Multiplier Circuits

Opamps and analog-multiplier ICs are used separately or together to provide a variety of operations involving multiplication, including analog product, square, square root, root-sum-square, difference of squares, trigonometric approximations, and vector sum. See also Frequency Divider, Frequency Multiplier, and Logarithmic chapters.

$A^2 - B^2$—Transfer function $V_{OUT} = (A^2 - B^2)/10$ is easily generated by AD534 analog multiplier. Differential inputs on Z terminals permit addition of feedback attenuator to decrease scale factor (or increase signal gain) from 40 to nominal value of 10. Feedback attenuation increases output offset proportionally; to make offset adjustment, connect 4.7-megohm resistor between Z_1 and wiper of 50K pot connected across power supplies.—R. Frantz, Analog Multipliers—New IC Versions Manipulate Real-World Phenomena with Ease, *EDN Magazine*, Sept. 5, 1977, p 125–129.

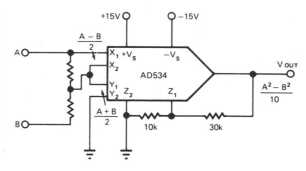

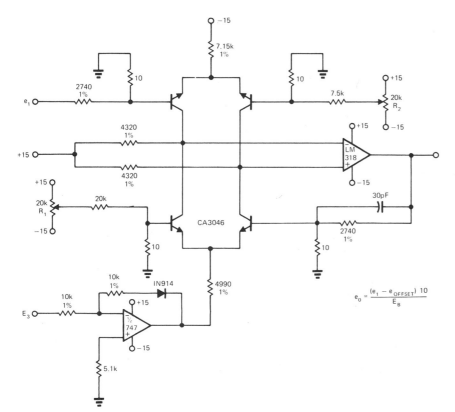

XY ÷ Z CIRCUIT—Developed for controlling amplitude of video signal from photodiode ε ¹ᵛ over 2 to 1 range. Circuit provides 3-dB bandwidth of 3 MHz, using CA3046 transistor array combined with LM318 opamp connected as analog divider, with half of 747 opamp serving as current source. Article covers initial adjustments of circuit. Basic circuit may be used as full XY ÷ Z multiply/divide element with differential bipolar signals on one numerator input.—A. R. Kopp, An Analog Multiplier/Divider Circuit, *EDN Magazine*, May 5, 1973, p 74–75.

$$e_0 = \frac{(e_1 - e_{OFFSET})\,10}{E_B}$$

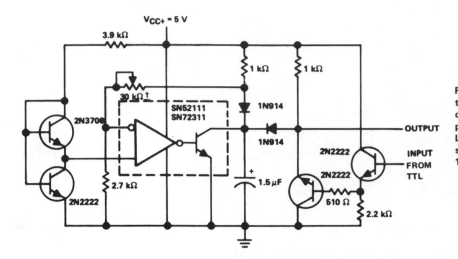

PRECISION SQUARER—Circuit using differential comparator IC accepts input from TTL and delivers square of input signal voltage to output. Adjust 30K pot to set clamp level.—"The Linear and Interface Circuits Data Book for Design Engineers," Texas Instruments, Dallas, TX, 1973, p 6-16.

LINEARIZING X INPUT—Adding resistors as shown to IC transconductance multiplier gives major improvement in X-input linearization. Article gives adjustment procedure. SMILE and FROWN terminal notes refer to X feedthrough pattern observed on CRO during setup, telling which position requires addition of R_{SEL} resistor.—L. Counts, Reduce Multiplier Errors by up to an Order of Magnitude, *EDN Magazine*, March 20, 1974, p 65–68.

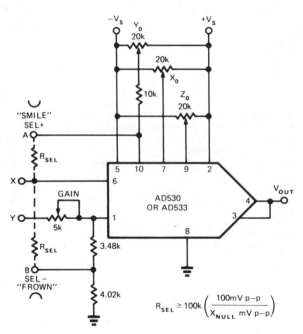

$$R_{SEL} \cong 100k \left(\frac{100mV \ p{-}p}{X_{NULL} \ mV \ p{-}p} \right)$$

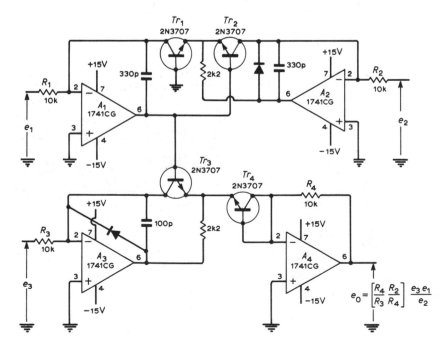

$$e_O = \left[\frac{R_4}{R_3} \frac{R_2}{R_4} \right] \frac{e_3 e_1}{e_2}$$

MULTIPLIER/DIVIDER—Combination opamp-transistor circuit may be used for either multiplication or division. All signals must be of same polarity (positive). For multiplication, use inputs e_1 and e_3. For division, use e_1 and e_2, with e_3 being adjusted to give desired scaling factor. Log output at base of Tr_2 is connected directly to antilog circuit at base of Tr_3. Article gives design equations. Circuit shown was developed to measure current gain of PNP transistor over range of operating currents.—G. B. Clayton, Experiments with Operational Amplifiers, *Wireless World,* Feb. 1973, p 91–93.

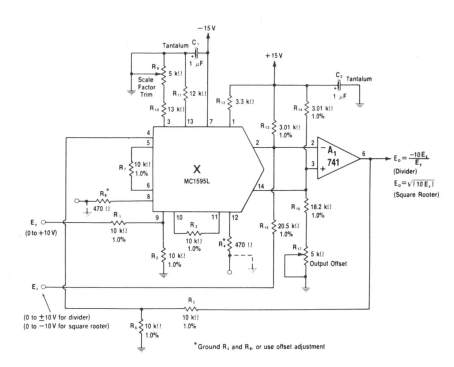

DIVIDER/SQUARE-ROOTER—Modification of multiplier circuit gives divider in which negative output voltage is equal to 10 times ratio E_x/E_y. To use as square-rooter for E_x, connect pins 4 and 9 together and omit R_1 and R_2 at E_y input. Circuit uses multiplier block driving 741 current-to-voltage converter.—W. G. Jung, "IC Op-Amp Cookbook," Howard W. Sams, Indianapolis, IN, 1974, p 257–258.

APPROXIMATING SINES—Analog Devices 433 multiplier/divider IC approximates sine of angle to less than 0.25% in just two terms (one quadrant). Arrangement requires only single opamp. Article gives analysis of theoretical errors and shows error curve.—D. H. Sheingold, Approximate Analog Functions with a Low-Cost Multiplier/Divider, *EDN Magazine,* Feb. 5, 1973, p 50–52.

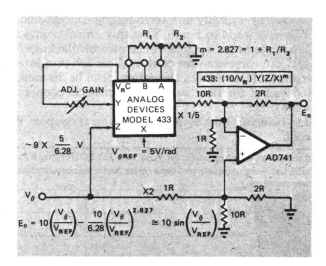

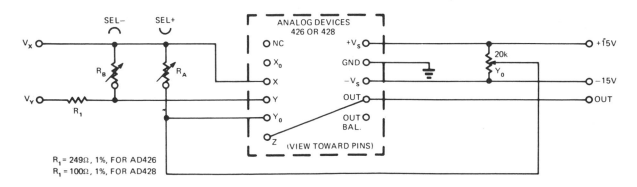

TYPICAL RANGE OF $R_A + R_B$: 25kΩ TO 200kΩ, 1%, METAL FILM RESISTOR

LINEARIZING MODULAR MULTIPLIER—Adding three external resistors to conventional transconductance multiplier reduces feedthrough, decreases average nonlinearity, and cuts overall error in half. Article gives adjustment procedure.—L. Counts, Reduce Multiplier Errors by up to an Order of Magnitude, *EDN Magazine,* March 20, 1974, p 65–68.

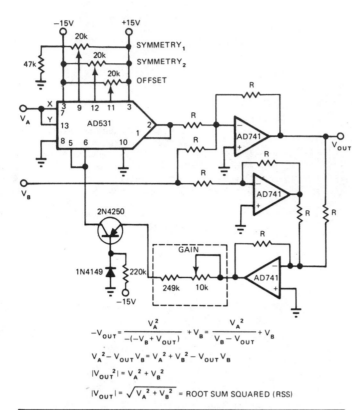

ROOT-SUM-SQUARED—Vector summation circuit uses AD531 variable-gain analog multiplier. Starting with trimpots centered and input V_B grounded, apply specified DC voltages to input V_A and adjust trimpots for output specified in table.—R. Frantz, Analog Multipliers—New IC Versions Manipulate Real-World Phenomena with Ease, *EDN Magazine*, Sept. 5, 1977, p 125–129.

$$-V_{OUT} = \frac{V_A^2}{-(-V_B + V_{OUT})} + V_B = \frac{V_A^2}{V_B - V_{OUT}} + V_B$$

$$V_A^2 - V_{OUT} V_B = V_A^2 + V_B^2 - V_{OUT} V_B$$

$$|V_{OUT}^2| = V_A^2 + V_B^2$$

$$|V_{OUT}| = \sqrt{V_A^2 + V_B^2} = \text{ROOT SUM SQUARED (RSS)}$$

ADJUSTMENT PROCEDURE			
STEP	V_A	ADJUST	V_{OUT}
1	+10.0V, −10.0V	SYMMETRY$_1$	EQUAL OUTPUTS FOR ±IN
2	+1.0V, −1.0V	SYMMETRY$_2$	EQUAL OUTPUTS FOR ±IN
3	+1.0V	OFFSET	+1.0V
4	+10.0V	GAIN	+10.0V

$-10\,V \le V_x \le +10\,V$
$-10\,V \le V_y \le +10\,V$

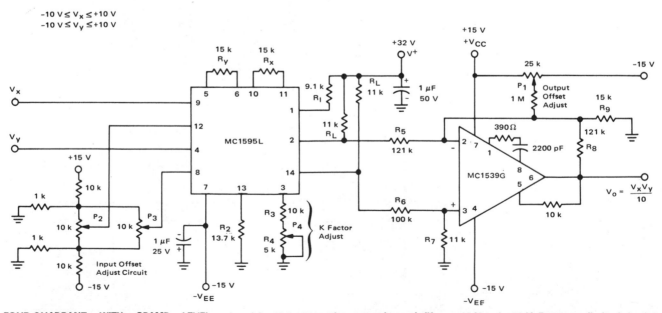

FOUR-QUADRANT WITH OPAMP LEVEL SHIFT—Connections shown for Motorola MC1595L linear four-quadrant multiplier are used in applications requiring level shift to ground reference. Common-mode voltage is re- duced by 10:1 attenuation networks, and dif- ferential output voltage is fed to opamp having closed-loop gain of 10. Resulting output is still $V_x V_y$/10, which appears single-ended above ground reference. Each input can be between −10 V and +10 V. Frequency limit of circuit is about 50 kHz for signal swings approaching ±10 V.—E. Renschler, "Analysis and Basic Opera- tion of the MC1595," Motorola, Phoenix, AZ, 1975, AN-489, p 9.

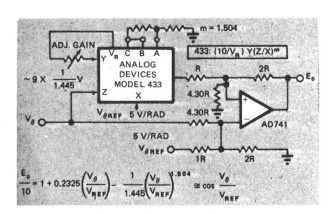

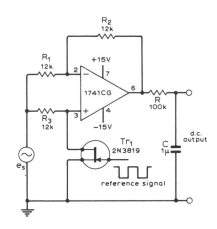

APPROXIMATING COSINES—Analog Devices 433 multiplier/divider IC approximates cosine of angle to better than 1%, by computing nonintegral exponents. Only one opamp is needed. Approximation uses arbitrary exponent as one term of cosine θ plus a linear term and a constant term, as described in article.—D. H. Sheingold, Approximate Analog Functions with a Low-Cost Multiplier/Divider, *EDN Magazine,* Feb. 5, 1973, p 50—52.

PHASE-SENSITIVE DETECTOR—Circuit using single opamp produces DC output proportional to both amplitude of AC input signal and cosine of its phase angle relative to reference signal. Can be used as synchronous rectifier in chopper-type DC amplifier or for accurate measurement of small AC signals obscured by noise. Article gives design equations.—G. B. Clayton, Experiments with Operational Amplifiers, *Wireless World,* July 1973, p 355—356.

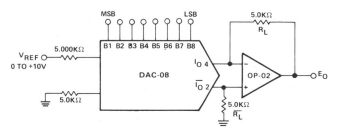

TWO-QUADRANT—Bipolar digital multiplier has output polarity controlled by offset-binary-coded digital input word. Precision Monolithics DAC-08 D/A converter drives OP-02 opamp. Output is symmetrical about ground.—J. Schoeff and D. Soderquist, "Differential and Multiplying Digital to Analog Converter Applications," Precision Monolithics, Santa Clara, CA, 1976, AN-19, p 2.

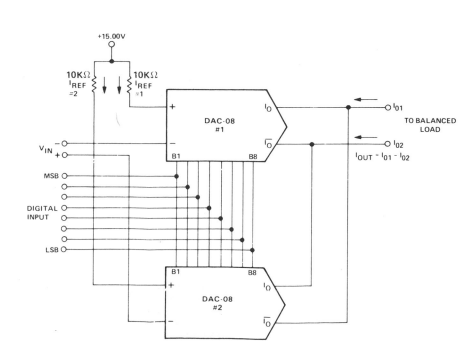

FOUR-QUADRANT MULTIPLYING DAC—Combination of two Precision Monolithics DAC-08 D/A converters accepts differential input voltage and produces differential current output. Output opamp is not normally required. Output analog polarity is controlled by analog input reference or by offset-binary digital input word. Common-mode current present at output must be accommodated by balanced load. Differential input range is 10 V.—J. Schoeff and D. Soderquist, "Differential and Multiplying Digital to Analog Converter Applications," Precision Monolithics, Santa Clara, CA, 1976, AN-19, p 3.

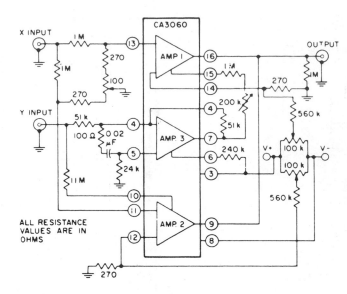

FOUR-QUADRANT WITHOUT LEVEL SHIFT— CA3060 three-opamp array provides four-quadrant multiplication without level shift between input and output. Circuit includes adjustments associated with differential input and adjustment for equalizing gains of amplifiers 1 and 2. Amplifier 3 is connected as unity-gain inverter.—"Linear Integrated Circuits and MOS/FET's," RCA Solid State Division, Somerville, NJ, 1977, p 153.

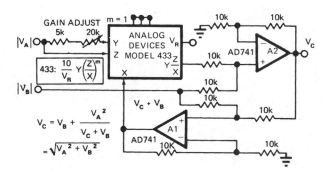

$$V_C = V_B + \frac{V_A{}^2}{V_C + V_B}$$
$$= \sqrt{V_A{}^2 + V_B{}^2}$$

APPROXIMATING VECTOR SUMS—Combination of two opamps and Analog Devices 433 multiplier/divider IC provides output voltage equal to vector sum of two input voltages, by computing square root of sum of squares.—D. H. Sheingold, Approximate Analog Functions with a Low-Cost Multiplier/Divider, *EDN Magazine,* Feb. 5, 1973, p 50–52.

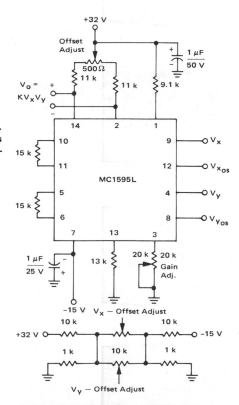

FOUR-QUADRANT—Motorola MC1595L linear four-quadrant multiplier takes two different input voltages, each between −10 V and +10 V, and gives output equal to one-tenth of their product. Circuit can be operated in either AC or DC mode. Design and setup procedures are given.—E. Renschler, "Analysis and Basic Operation of the MC1595," Motorola, Phoenix, AZ, 1975, AN-489, p 8.

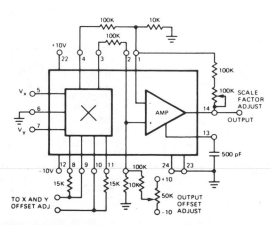

ANALOG MULTIPLIER—Multiplier and amplifier sections of Exar XR-S200 PLL IC are combined to perform analog multiplication without need for DC level shifting between input and output. Single-ended output is at ground level.—"Phase-Locked Loop Data Book," Exar Integrated Systems, Sunnyvale, CA, 1978, p 9–16.

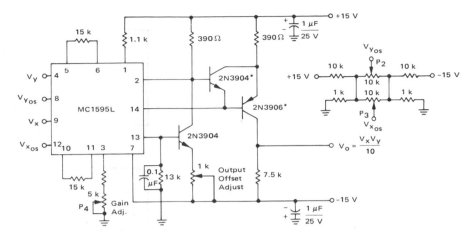

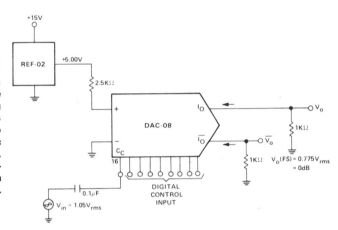

OUTPUT LEVEL SHIFTER—Transistors connected to Motorola MC1595L linear four-quadrant multiplier perform level shifting for applications requiring output having ground reference. Temperature sensitivity of circuit is minimized by using complementary transistors in same package, such as MD6100, in place of upper two transistors. If high output impedance and low current drive are drawbacks, opamp can be connected as source-follower output stage.—E. Renschler, "Analysis and Basic Operation of the MC1595," Motorola, Phoenix, AZ, 1975, AN-489, p 10.

AC-COUPLED MULTIPLICATION—Combination of Precision Monolithics REF-02 voltage reference and DAC-08 D/A converter uses compensation capacitor terminal C_C as input. With full-scale input code, output V_O is flat to above 200 kHz and 3 dB down at 1 MHz, for multiplying applications far beyond audio range. Circuit has high input impedance, as often required to avoid loading high source impedance. Dynamic range is greater than 40 dB.—J. Schoeff and D. Soderquist, "Differential and Multiplying Digital to Analog Converter Applications," Precision Monolithics, Santa Clara, CA, 1976, AN-19, p 4.

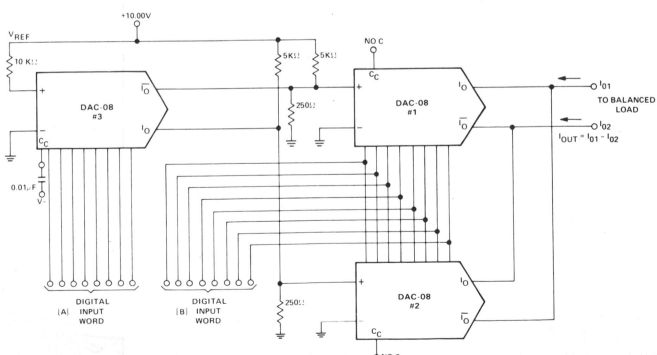

FOUR-QUADRANT 8-BIT—Requires only three Precision Monolithics DAC-08 D/A converters to provide high-speed multiplication of two 8-bit digital words and give analog output.—J. Schoeff and D. Soderquist, "Differential and Multiplying Digital to Analog Converter Applications," Precision Monolithics, Santa Clara, CA, 1976, AN-19, p 7.

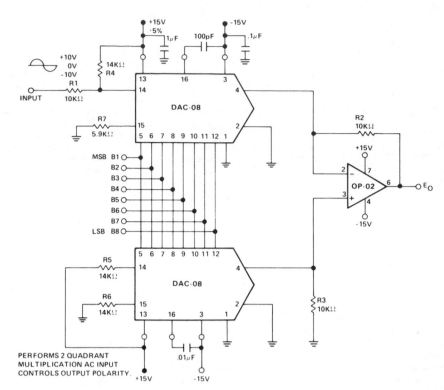

BIPOLAR ANALOG TWO-QUADRANT—Bipolar reference voltage for upper Precision Monolithics DAC-08 D/A converter modulates reference current by ±1.0 mA around quiescent current of 1.1 mA. Lower DAC-08 has same 1.1-mA reference current and effectively subtracts out quiescent 1.1 mA of upper reference current at all input codes since voltage across R3 varies between −10 V and 0 V. Output voltage E_0 is thus product of digital input word and bipolar analog reference voltage.—J. Schoeff and D. Soderquist, "Differential and Multiplying Digital to Analog Converter Applications," Precision Monolithics, Santa Clara, CA, 1976, AN-19, p 3.

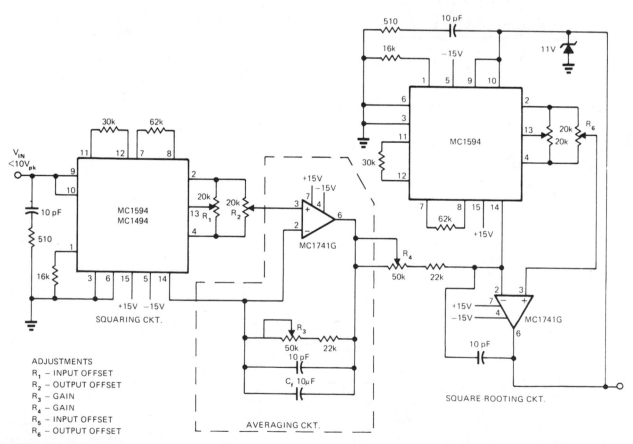

SQUARING FOR RMS—Combination of two MC1594 multipliers and two opamps gives RMS detector for squaring instantaneous input values, averaging over time interval, then taking square root to give RMS value of input waveform. First multiplier, used to square input waveform, delivers output current to first opamp for conversion to voltage and for averaging by means of capacitor in feedback path. Second opamp is used with second multiplier as feedback element for taking square root. Technique eliminates thermal response time drawback of most other RMS measuring circuits. Input voltage range for circuit is 2 to 10 V P-P; for other ranges, input scaling can be used. Since direct coupling is used, output voltage includes DC components of input. Maximum input frequency is about 600 kHz, and accuracy is about 1%.—K. Huehne and D. Aldridge, True RMS Measurements Using IC Multipliers, *EDN Magazine,* March 20, 1973, p 85–86.

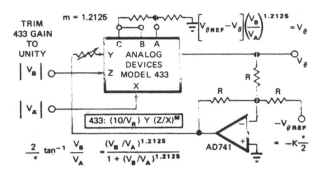

APPROXIMATING ARC TANGENTS—Analog Devices 433 multiplier/divider IC approximates arc tangent to 0.75%. Article presents mathematical basis for approximation used.—D. H. Sheingold, Approximate Analog Functions with a Low-Cost Multiplier/Divider, *EDN Magazine*, Feb. 5, 1973, p 50–52.

FOUR-QUADRANT MULTIPLIER/SQUARER—Basic 1595 multiplier block and 741 current-to-voltage converter convert input voltages E_x and E_y to output equal to one-tenth of their product when connected as shown for multiplier use. To operate as squarer of E_x, connect pins 4 and 9 together and omit R_5 and R_6 at E_y input. Output is then one-tenth of square of E_x.—W. G. Jung, "IC Op-Amp Cookbook," Howard W. Sams, Indianapolis, IN, 1974, p 255–257.

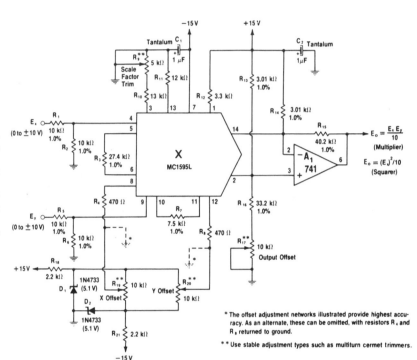

* The offset adjustment networks illustrated provide highest accuracy. As an alternate, these can be omitted, with resistors R_4 and R_8 returned to ground.

** Use stable adjustment types such as multiturn cermet trimmers.

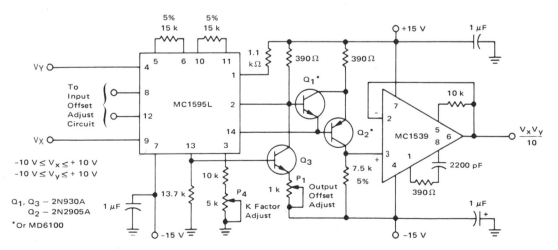

LEVEL SHIFTER WITH HIGH CURRENT DRIVE—Motorola MC1539 opamp is used as source-follower output stage for three-transistor level shifter of MC1595L linear four-quadrant multiplier, to improve current drive capabilities. Output voltage is in range of ±10 V. Input offset adjusting circuits are 10K pots in series with 10K resistors between ±15 V, with 1K to ground from each side of paralleled pots.—E. Renschler, "Analysis and Basic Operation of the MC1595," Motorola, Phoenix, AZ, 1975, AN-489, p 11.

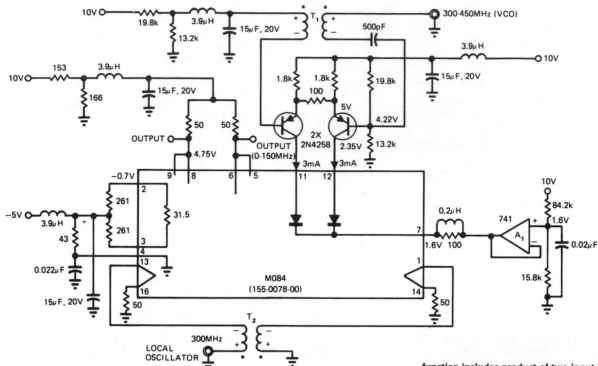

BROADBAND MIXER—Uses Tektronix MO84 multiplier as broadband mixer having linear output within −3 dB limit from 2 MHz to above 150 MHz. Current-gain cell in multiplier takes advantage of logarithmic relationship between current and voltage in a semiconductor. Output function includes product of two input signals, a 300-450 MHz swept VCO signal and a 300-MHz local oscillator signal.—M. Jaffe, Build a Low-Cost Wideband Mixer with a Monolithic Multiplier, *EDN Magazine,* May 20, 1975, p 63–64.

CHAPTER 56
Multivibrator Circuits

Includes circuits in which one section is cut off when the other conducts. In an astable or free-running MVBR, frequency of spontaneous transition is determined by time and/or external control voltage. In a monostable MVBR, external trigger signal forces circuit into unstable state, with circuit constants determining time for return to stable state. In a bistable MVBR or flip-flop, external trigger is required for each transition. Use chiefly for generating square-wave pulses and signals.

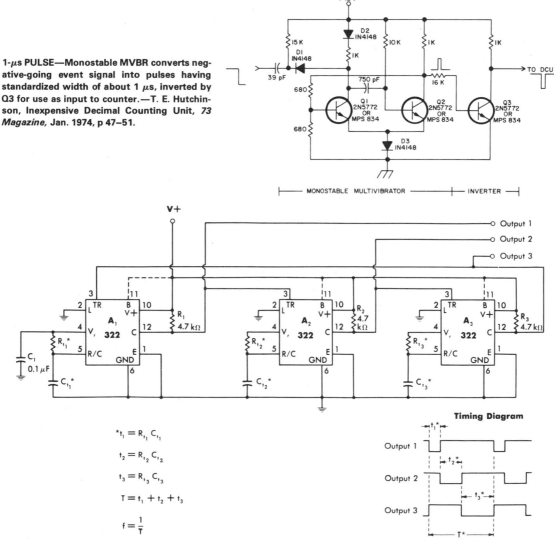

1-μs PULSE—Monostable MVBR converts negative-going event signal into pulses having standardized width of about 1 μs, inverted by Q3 for use as input to counter.—T. E. Hutchinson, Inexpensive Decimal Counting Unit, *73 Magazine*, Jan. 1974, p 47–51.

$$*t_1 = R_{t_1} C_{t_1}$$

$$t_2 = R_{t_2} C_{t_2}$$

$$t_3 = R_{t_3} C_{t_3}$$

$$T = t_1 + t_2 + t_3$$

$$f = \frac{1}{T}$$

CHAINED ASTABLES—Cross-connected 322 mono MVBRs, operating as astables, are interconnected so individual timing periods are generated in sequence as shown. Total period T is sum of three individual periods, after which cycle repeats itself. Chain may be extended further if desired. Useful when prescribed sequence of timing events is required. Equations give values of R and C for timing periods ranging from 10 μs to several minutes.—W. G. Jung, "IC Timer Cookbook," Howard W. Sams, Indianapolis, IN, 1977, p 125–128.

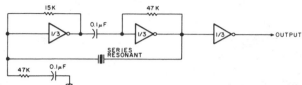

CRYSTAL MONO USING INVERTERS—Uses all three sections of CD4049 triple inverter, with series-resonant crystal connection. Supply can be in range of 3 to 15 V. Serves as compact low-power portable RF oscillator having low battery drain.—W. J. Prudhomme, CMOS Oscillators, *73 Magazine*, July 1977, p 60–63.

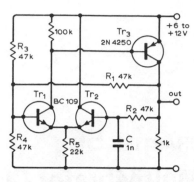

20-kHz ASTABLE—Single-capacitor circuit is reliable over wide range of temperatures, voltages, and transistor gains. Frequency varies only by 0.05% for supply voltage changes between 6 and 12 V. Timing can be changed with R_1, R_2, and C. Duty cycle depends on ratio of R_3 to R_4, and is 50% for values shown.—C. Horwitz, Tolerant Astable Circuits, *Wireless World*, Feb. 1975, p 93.

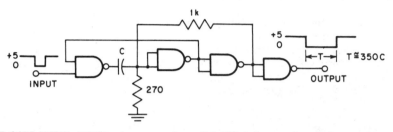

FOUR-GATE MONO—NAND-gate mono using Texas Instruments SN7400 package provides cleaner, more stable output. Feedback resistor eliminates tendency to oscillate. Output pulse width T is equal to 1.3 RC; when R is 270 ohms, T is 350 C. Input pulse widths over 30 ns can initiate output. C can be 100 pF to 100 μF.—J. E. McAlister, Single NAND Package Improves One-Shot, *EEE Magazine*, Aug. 1970, p 78.

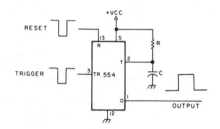

554 MONO—Uses one section of 554 quad monostable timer, connected to give output pulse for negative-going trigger pulse. Width of output pulse in seconds is equal to RC. Trigger must be narrower than output pulse. VCC is 4.5–16 V at 3–10 mA.—H. M. Berlin, IC Timer Review, *73 Magazine*, Jan. 1978, p 40–45.

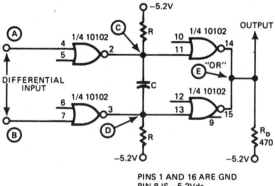

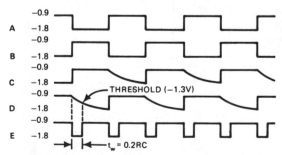

PINS 1 AND 16 ARE GND
PIN 8 IS −5.2Vdc
BYPASS PIN 8 TO GND WITH 0.1 μF

BIDIRECTIONAL MONO—Requires only one IC, three resistors, and one capacitor. Will trigger on both positive- and negative-going transitions, as required in critical timing applications involving pulses narrower than 50 ns. Capacitor alternately discharges through one pulldown resistor to threshold, then the other. Output gates are tied together to form common output. Width of pulse is defined by values of components.—W. A. Palm, Bidirectional ECL One-Shot Uses a Single IC, *EDN Magazine*, Jan. 5, 1977, p 41–42.

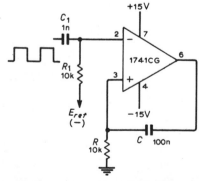

VOLTAGE-CONTROLLED MONO—Timing period of opamp operating as monostable multivibrator is controlled by magnitude of DC reference voltage. With square-wave input shown, differentiating action by C_1-R_1 gives positive pulses that cause mono to make transitions. Article gives design equation and typical waveforms.—G. B. Clayton, Experiments with Operational Amplifiers, *Wireless World*, May 1973, p 241–242.

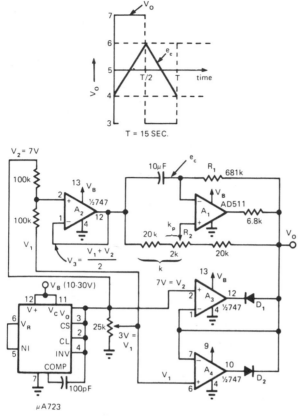

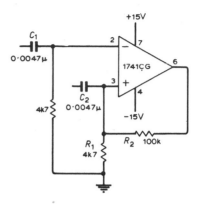

TRIGGERED BISTABLE—Positive feedback applied to opamp through R_2 and R_1 causes amplifier output to remain at either its positive or negative saturation limit. Triggering pulses for changing state of output may be applied to either input terminal, through C_1 or C_2; pulse polarity required to produce transition depends on state of circuit, which should be verified experimentally.—G. B. Clayton, Experiments with Operational Amplifiers, *Wireless World,* May 1973, p 241–242.

15-s ASTABLE—Precision opamp/diode clamp circuit simulating zener reduces cost of astable multivibrator having long time constant and good temperature stability. Circuit operates from single-ended supply, with μA723 providing 7-V reference for clamp amplifiers A_3 and A_4.

A_2 provides oscillator with reference voltage V_3 halfway between V_1 and V_2. R_2 allows frequency of oscillator to be adjusted about ±6%. Article gives design equations.—L. Drake, Long Time-Constant Oscillator Uses Precision Clamps, *EDN Magazine,* Dec. 20, 1974, p 51–52.

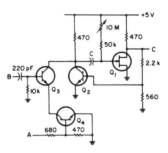

VERSATILE MONO—Uses standard digital IC voltage levels as inputs, and can be enabled or inhibited at any time without causing output pulse. Input gate Q_3-Q_4 is enabled with logic 1 at point A and inhibited with logic 0. Logic 1 at B starts timing cycle. Q_1 is 2N3819 JFET, and all other transistors are 2N3704.—R. Tenny, Versatile One-Shot, *EEE Magazine,* Sept. 1970, p 89.

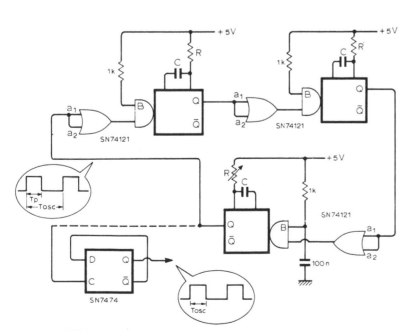

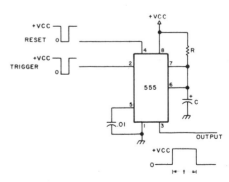

0.01 Hz to 7 MHz—Generates square waves suitable for clock signals in sequential digital circuits, with values of RC pairs determining period in range from about 150 ns to 120 s. To obtain equal mark-space ratio, set oscillator to half the required period and add bistable SN7474 divider as shown by dashed line.—P. J. Best, Monostable Ring Oscillator, *Wireless World,* March 1976, p 89.

POSITIVE-OUTPUT MONO—Timer is triggered by negative-going pulse to give positive output pulse whose width t in seconds is 1.1RC. VCC is 4.5–16 V at 3–10 mA.—H. M. Berlin, IC Timer Review, *73 Magazine,* Jan. 1978, p 40–45.

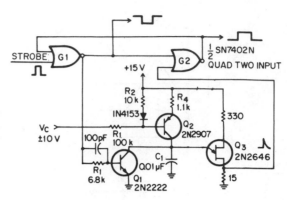

PWM MONO—Circuit provides pulse-width modulation with high duty cycles and complementary output. Strobe input to gate G_1 drives output of gate to binary 0, turning Q_1 off and letting voltage across C_1 build up until UJT Q_3 fires, discharging C_1. Output of UJT drives output of G_2 to binary 0. Article gives timing diagrams.—G. Lewis, Simple One Shot Has Complementary Outputs, *EEE Magazine*, Oct. 1970, p 78–79.

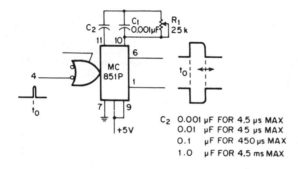

C_2 0.001 µF FOR 4.5 µs MAX
       0.01 µF FOR 45 µs MAX
       0.1   µF FOR 450 µs MAX
       1.0   µF FOR 4.5 ms MAX

VARIABLE PULSE WIDTH—R_1 and C_2 together provide wide range of pulse widths from Motorola MC851P mono. Rise and fall times of complementary output pulses are better than 100 µs. With only four switched capacitors in combination with R_1, pulse widths can be varied between maximum of 4.5 ms and minimum well under 4.5 µs.—C. W. Stoops, Wide-Range Variable Pulse-Width Monostable, *EEE Magazine*, Dec. 1970, p 56.

LOW-POWER TTL MONO—Simple monostable circuit using DM74L03 draws only 800-µA standby current yet delivers pulses up to 1 s wide. Uses RC time control and regenerative feedback, with values of C_2 and C_3 determining frequency. Pulse width increases from 0.1 s to 0.55 s as C_2 and C_3 are increased from 10 µF to 60 µF.—C. Gilbert and C. Davis, LPTTL One-Shot Yields Wide, Clean Pulses, *EDN/EEE Magazine*, May 15, 1971, p 47–48.

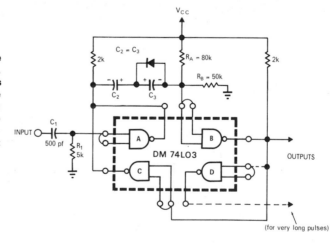

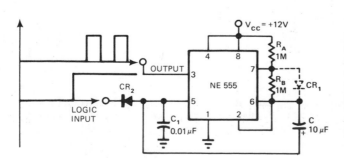

LOW OUTPUT FOR POWER-ON—Logic signal controls both turn-on and turnoff of 555 timer used as oscillator. When input signal at cathode of CR_2 goes low, oscillator remains off and output at pin 3 is low. When input goes high, oscillator starts with its first state low so there are no initial pulse errors.—K. D. Dighe, Rearranged Components Cut 555's Initial-Pulse Errors, *EDN Magazine*, Jan. 5, 1978, p 82 and 84.

DUAL-EDGE TRIGGERING—Although 9602 multivibrator IC can be triggered normally either on leading or falling edge of square wave, but not on both, addition of two resistors and one capacitor provides double-edge triggering. When input goes low, negative-going pulse through C_1 triggers 9602 and makes it deliver one output pulse. When input goes high again, high-going pulse is delivered directly to pin 12 of 9602, triggering it again so it produces another pulse.—J. P. Yang, Circuit Triggers One-Shot on Both Edges of Square Wave, *EDN Magazine*, Nov. 15, 1972, p 49.

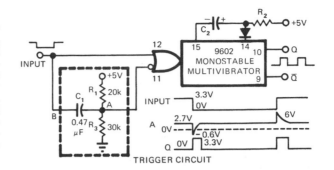

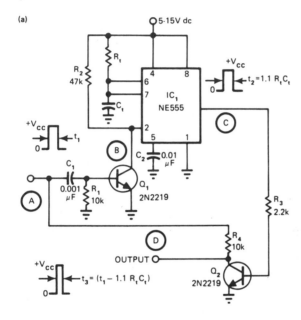

PULSE-WIDTH DETECTOR—Connections as shown for 555 timer give output only if trigger pulse width is greater than time constant ($t_2 = 1.1R_tC_t$) of mono MVBR circuit. Q_1 is normally off. Pin 2 of 555 is then high. At start of trigger pulse, output at point C is low. Positive trigger drives Q_1 on for time determined by R_1C_1, feeding negative-going pulse to trigger pin 2. Timer then acts as normal mono, driving Q_2 on for time t_2. If input pulse is still high at end of t_2, it appears at output D since Q_2 is now off. Output pulse width is thus equal to input trigger width less $1.1R_tC_t$. For greater accuracy, insert delay between point A and R_4 equal to inherent propagation delay of timer.—S. Sarpangal, Build a Pulse-Width Detector with a 555 Timer, *EDN Magazine*, Oct. 5, 1977, p 93 and 96.

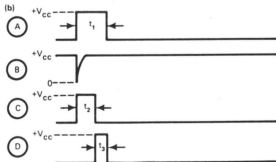

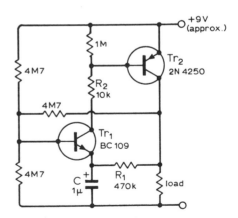

2-Hz ASTABLE PULSER—Single-capacitor circuit operates at very low duty cycles, in range of 10% to 1%. Battery drain is low because off current is about 1 μA for 50-mA on current. R_2 and C determine on time, while R_1 and C set off time. Circuit pulses about twice per second, which is suitable for animal temperature and heart-rate studies. Can be used with implanted transmitters operating from single mercury button cell for more than one year with suitable resistor values.—C. Horwitz, Tolerant Astable Circuits, *Wireless World*, Feb. 1975, p 93.

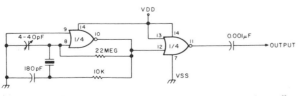

CRYSTAL WITH NOR GATES—Uses two sections of CD4001 quad NOR gate to give mono multivibrator operating in frequency range from 10 kHz up to top limit of about 10 MHz, with exact frequency depending on values used for R and C.—W. J. Prudhomme, CMOS Oscillators, *73 Magazine*, July 1977, p 60–63.

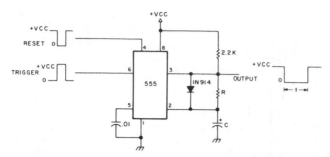

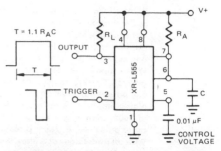

NEGATIVE-OUTPUT MONO—Timer is wired to give negative output pulse for positive-going input trigger pulse. Width of output pulse in seconds is 1.1RC. Input pulse must be narrower than desired output pulse width. When reset pin is momentarily grounded, output returns to stable state. VCC is 4.5–16 V at 3–10 mA.—H. M. Berlin, IC Timer Review, *73 Magazine*, Jan. 1978, p 40–45.

MICROPOWER MONO—Uses Exar XR-L555 having typical power dissipation of only 900 μW at 5 V, serving as direct replacement for 555 timer in micropower circuits. Time delay is controlled by one external resistor and one capacitor (R_A and C) which determine output pulse duration. Can be triggered or reset on falling waveform. Output will drive TTL circuits or source up to 50 mA.—"Timer Data Book," Exar Integrated Systems, Sunnyvale, CA, 1978, p 7–8.

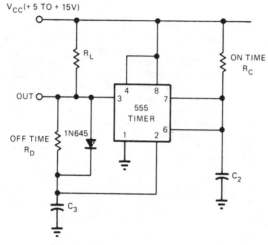

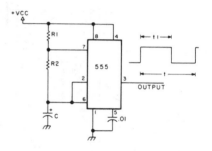

ASTABLE OSCILLATOR—Circuit for astable operation of 555 timer provides completely independent ON and OFF times. Time constant for one mode is 1.1 $R_C C_2$ and for other mode is 1.1 $R_C C_3$. Free-running period is sum of these time constants.—J. P. Carter, Astable Operation of IC Timers Can Be Improved, *EDN Magazine*, June 20, 1973, p 83.

555 ASTABLE—Produces repetitive rectangular output at frequency equal to $1.443/(R_1 + 2R_2)C$ hertz. Duty cycle is determined by values of R_1 and R_2; R_2 must be much larger than R_1 to obtain nearly a 50% duty cycle. Normal range for duty cycle is 51 to 99%. VCC is 4.5–16 V at 3–10 mA.—H. M. Berlin, IC Timer Review, *73 Magazine*, Jan. 1978, p 40–45.

LOW-POWER MONO—555 timer provides low-drain monostable operation suitable for interfacing with CMOS 4011B NAND gates. Standby drain is less than 50 μA. When mono is on, current drawn is 4.5 mA for pulse duration of T = 1.1RC.—"Signetics Analog Data Manual," Signetics, Sunnyvale, CA, 1977, p 733.

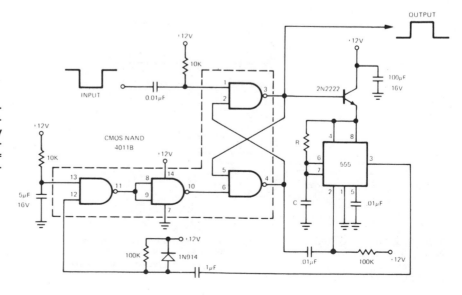

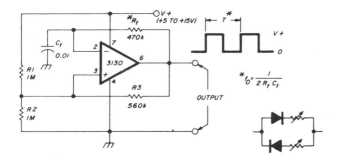

1 Hz TO 1 MHz—Opamp-based astable MVBR generates square waves over extremely wide range, with suitable changes in circuit values. RCA 3130 opamp has CMOS output stage for driving either 5-V TTL or 10–15 V CMOS logic stages directly. Values are for 100 Hz. R_t and C_t can be readily scaled for different ranges. To control symmetry, replace R_t with two resistors in series with reverse-connected diodes as at lower right.—W. Jung, An IC Op Amp Update, *Ham Radio*, March 1978, p 62–69.

$$*_t = \frac{1}{2R_f C_f}$$

PROGRAMMABLE ASTABLE—4016 CMOS analog switch selects 1.5-megohm timing resistor R_{t1} when control input line is high, to give negative-going 100-Hz output pulses. When input is low, CMOS switch S_2 is on, selecting 1.2-megohm timing resistor R_{t2} to give 120-Hz output.—W. G. Jung, "IC Timer Cookbook," Howard W. Sams, Indianapolis, IN, 1977, p 136–137.

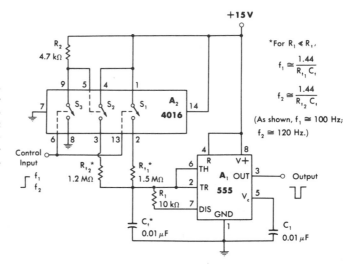

*For $R_1 \ll R_{t}$,

$$f_1 \cong \frac{1.44}{R_{t_1} C_t}$$

$$f_2 \cong \frac{1.44}{R_{t_2} C_t}$$

(As shown, $f_1 \cong 100$ Hz; $f_2 \cong 120$ Hz.)

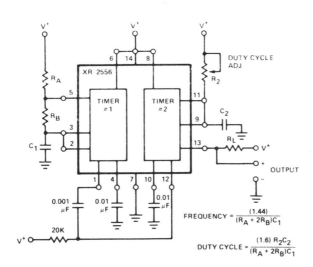

$$FREQUENCY = \frac{(1.44)}{(R_A + 2R_B)C_1}$$

$$DUTY \ CYCLE = \frac{(1.6) R_2 C_2}{(R_A + 2R_B)C_1}$$

VARIABLE DUTY CYCLE—First section of Exar XR-2556 dual timer operates as astable MVBR whose frequency is equal to $1.44/(R_A + 2R_B)C_1$, with output used to trigger timer 2 connected in monostable mode. Time delay T_2 of timer 2 is made less than period of timer 1 waveform, so both timers have same frequency. Duty cycle is determined by timing cycle of timer 2, adjustable from 1% to 99% with R_2. Supply voltage is 4.5–16 V.—"Timer Data Book," Exar Integrated Systems, Sunnyvale, CA, 1978, p 23–30.

INVERTED MONO—Connection shown for 555 timer accepts positive trigger pulses and delivers negative output pulses. Duty cycles above 99% are possible without jitter. Heavy loads can be driven from pin 7 without loss of accuracy, but excessive loading of pin 3 can affect timing accuracy. Width of output pulse is 1 ms for values of R_t and C_t shown. Trigger must be held below two-thirds of supply voltage for standby and raised above two-thirds of supply momentarily (not longer than pulse width) for triggering.—W. G. Jung, "IC Timer Cookbook," Howard W. Sams, Indianapolis, IN, 1977, p 89.

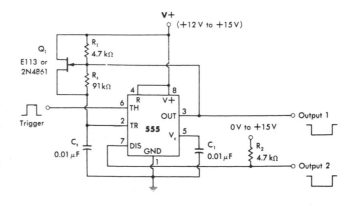

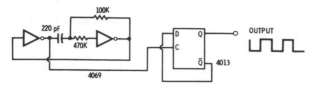

DIVIDING FOR SYMMETRY—4013 dual D flip-flop is used as binary divider at output of astable MVBR to give 50/50 symmetry for output frequency half that of MVBR.—D. Lancaster, "CMOS Cookbook," Howard W. Sams, Indianapolis, IN, 1977, p 232–234.

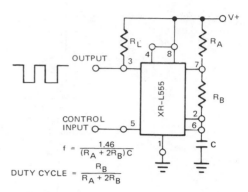

$$f = \frac{1.46}{(R_A + 2R_B)C}$$

$$\text{DUTY CYCLE} = \frac{R_B}{R_A + 2R_B}$$

MICROPOWER CLOCK—Free-running frequency and duty cycle are controlled by R_A, R_B, and C in astable MVBR connection of Exar XR-L555 micropower equivalent of 555 timer. With 5-V supply, power dissipation is only 900 μW.—"Timer Data Book," Exar Integrated Systems, Sunnyvale, CA, 1978, p 7–8.

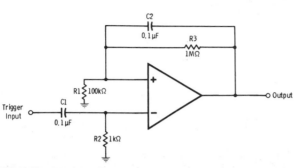

AC-COUPLED FLIP-FLOP—When leading edge of 2-V positive trigger pulse is applied to negative input of 741 or equivalent opamp, this input becomes more positive than positive input and opamp swings into negative saturation. This condition is held by positive feedback until trailing edge of next trigger pulse makes opamp swing back into positive saturation. C2 prevents trailing edge of first pulse from driving opamp back into positive saturation. Value shown for C2 should be increased if pulses are longer than 50 ms.—R. Melen and H. Garland, "Understanding IC Operational Amplifiers," Howard W. Sams, Indianapolis, IN, 2nd Ed., 1978, p 118–119.

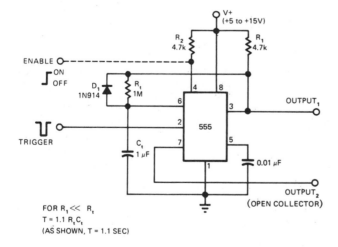

FOR $R_1 \ll R_t$
$T = 1.1 R_t C_t$
(AS SHOWN, T = 1.1 SEC)

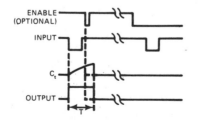

BASIC 555 MONO—Circuit variation shown for original IC timer prototype has same timing equation and input trigger requirements for standard connection but provides two outputs rather than one. This is achieved by using pin 7 of IC as an open-collector output that can be referred to any supply voltage between 0 and +15 V regardless of voltage used for timer. Article describes operation in detail and points out possible drawbacks, including possibility of timing error for high duty cycle operation because C_1 takes longer to discharge than in conventional monostable MVBR.—W. G. Jung, Take a Fresh Look at New IC Timer Applications, *EDN Magazine*, March 20, 1977, p 127–135.

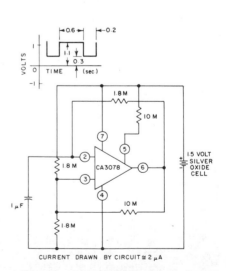

TIMING-PULSE GENERATOR—Astable MVBR uses CA3078 micropower opamp to develop timing pulses for driving other low-power circuits. Current drain is only about 2μA from 1.5-VDC supply.—"Circuit Ideas for RCA Linear ICs," RCA Solid State Division, Somerville, NJ, 1977, p 4.

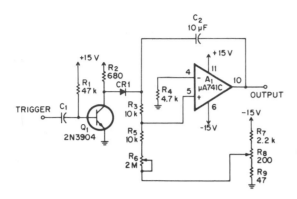

RETRIGGERABLE MONO—Circuit provides pulse widths up to 60 s, has short reset time, and can be retriggered during timing cycle. Pulse width is determined by C_2, R_3, R_5, and R_6. If trigger pulse arrives while output of A_1 is high, C_2 discharges to its original triggered state for initiating completely new timing cycle.—D. Pantic, Retriggerable Monostable, *EDN/EEE Magazine*, May 15, 1971, p 50.

TRIGGERED MVBR—Each trigger input produces fixed number of pulses, between 2 and 30 depending on setting of 1-megohm frequency control. Monostable feeds gated astable, both realized with single CD4001 IC. Use dashed circuit with pushbutton for manual operation in place of trigger pulse.—K. Padmanabhan, N-Stable Multivibrator, *Wireless World*, April 1977, p 61.

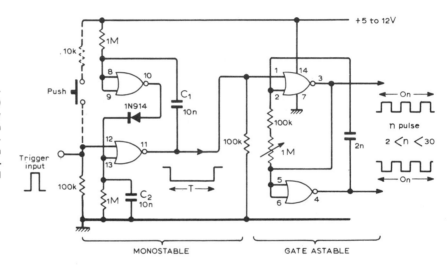

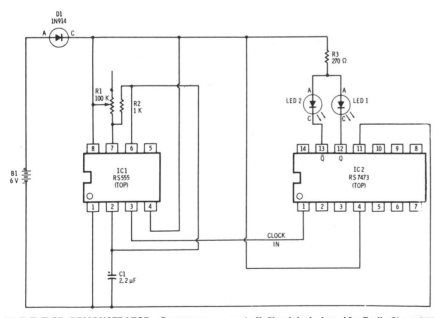

IC FLIP-FLOP DEMONSTRATOR—Demonstration circuit using RS7473 dual flip-flop incorporates 555 clock circuit providing sequential train of input pulses at AF rate to JK master-slave flip-flop section for toggling LED loads of flip-flop back and forth between ON and OFF states. R1 controls rate at which LEDs flash on and off. Circuit is designed for Radio Shack 276-041 or equivalent red LEDs. If clock is omitted, state of flip-flop is changed by grounding clock input pin 1 momentarily.—F. M. Mims, "Integrated Circuit Projects, Vol. 6," Radio Shack, Fort Worth, TX, 1977, p 23–32.

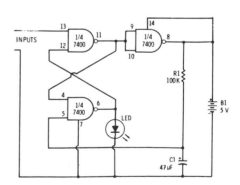

MONO WITH NORMALLY ON LED—Connection shown for three gates of 7400 quad NAND gate inverts operating mode, so LED is normally on. Trigger pulse at input extinguishes LED for time determined by R1 and C1 while making gates change states. Gates revert to original states after delay also determined by values of R1 and C1.—F. M. Mims, "Integrated Circuit Projects, Vol. 2," Radio Shack, Fort Worth, TX, 1977, 2nd Ed., p 19–26.

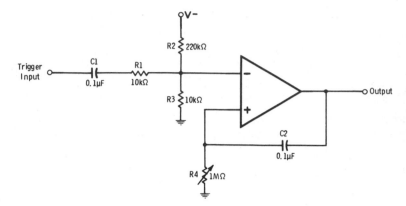

BASIC MONO—Opamp is normally in positive saturation because of negative voltage provided by voltage divider R2-R3. When 2-V positive trigger pulse is applied to input, output of opamp swings into negative saturation but automatically returns to positive saturation after time interval determined by values used for C2 and R4. With R4 set at maximum resistance, this time is about 1 s. Increase size of C2 for longer time periods. Opamp can be 741 or equivalent.—R. Melen and H. Garland, "Understanding IC Operational Amplifiers," Howard W. Sams, Indianapolis, IN, 2nd Ed., 1978, p 120–121.

ASTABLE USING CMOS TRANSISTOR PAIR—One transistor pair from CA3600E array is used with CA3080 operational transconductance amplifier to give precise timing and threshold for square waves. Quiescent power consumption is typically 6 mW with values shown.—"Linear Integrated Circuits and MOS/FET's," RCA Solid State Division, Somerville, NJ, 1977, p 279.

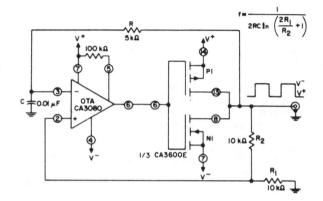

PERFECT SQUARE WAVES FOR LAB—Two sections of 4049 hex inverting buffer are connected as 10:1 variable-frequency astable MVBR to feed chain of four divide-by-10 counters using 4518 dual counters. Frequency division provides perfect symmetry for square-wave output.—D. Lancaster, "CMOS Cookbook," Howard W. Sams, Indianapolis, IN, 1977, p 232–234.

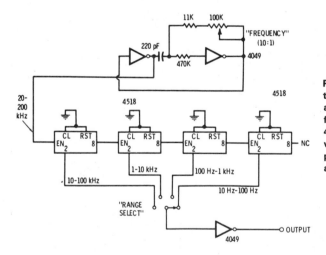

17-Hz SQUARE WAVES—Uses three comparators from MC3302P IC to generate three symmetrical square waves 120° apart. Inverting outputs of comparators gives waves 60° apart. Operating voltage can be anywhere from 4 to 12 V.—L. J. Bell, Three Coupled Astables, *Wireless World,* Feb. 1977, p 44.

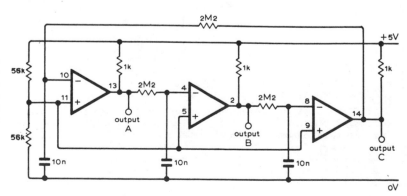

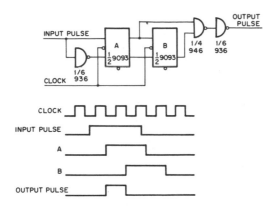

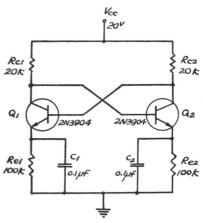

CLOCK-SYNCHRONIZED—Circuit generates pulse that is two clock pulses wide, in synchronism with clock, from random input pulse whose width is more than 5 times that of clock pulse. Flip-flops A and B are connected as shift register. When clock pulse falls, input of flip-flop A goes to 1 and sets it. B follows state of A with delay of one clock pulse. Output pulse can occur only once during a particular input strobe.—F. E. Nesbitt, Synchronized One Shot, *EDN/IEEE Magazine*, May 15, 1971, p 50.

DIRECT-COUPLED ASTABLE—Collectors and bases of both emitter-biased transistors are directly coupled to each other. Switching action takes place by means of capacitor in each emitter circuit. Triangle waves are generated at emitters. Neither transistor can remain permanently cut off. Instead, circuit has two quasistates, with switching action achieved by charging and discharging capacitor between these states. Single 0.1-μF capacitor can be used between emitters in place of C_1 and C_2.—S. Chang, Two New Direct-Coupled Astable Multivibrators, *Proceedings of the IEEE*, March 1973, p 390–391.

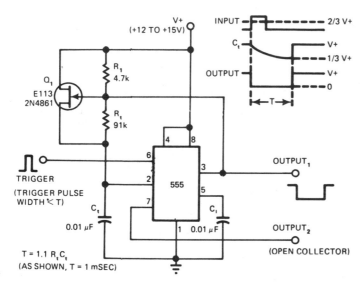

NEGATIVE-GOING DUAL-OUTPUT 555—Circuit triggers on positive-going pulses and delivers negative-going output timing pulses. C_1 charges when JFET switch Q_1 is held on by high output state of timer. When output of timer goes low, C_t discharges to ground through R_t. Timing accuracy is good, and duty cycles above 99% are possible without jitter.—W. G. Jung, Take a Fresh Look at New IC Timer Applications, *EDN Magazine*, March 20, 1977, p 127–135.

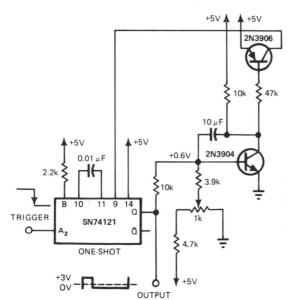

DUTY-CYCLE CONTROL—Feedback loop through two transistors automatically adjusts timing of MVBR to hold duty cycle constant over wide range of triggering rates. 2N3904 acts as integrator with time constant much longer than pulsing period. If duty cycle increases or decreases, current into integrator becomes positive or negative and DC voltage at its collector slowly decreases or increases. This collector voltage drives 2N3906 operating as current generator for adjusting automatically to give chosen duty cycle as selected by 1K pot. Range is 17% to above 50%.—J. L. Engle, Regulate Duty Cycle Automatically, *EDN Magazine*, Nov. 5, 1978, p 122.

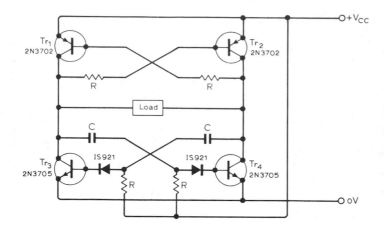

LOW BATTERY DRAIN—Combination of astable and bistable MVBRs, connected so diagonally opposite transistors switch on and off together, minimizes current drain in battery-powered signal generator. Period of square wave is approximately equal to 1.4CR, and peak load current can be up to 70 mA with 24-V battery supply. Circuit will tolerate wide range of values for CR.—J. C. Hopkins, Efficient Square-Wave Oscillator, *Wireless World,* June 1977, p 58.

PRECISION MONO—Negative-going pulse triggers mono, making output go LOW for duration of eight clock pulses at frequency determined by values of R and C in 4011 clock generator IC_1. R_1 is greater than $2R_T$, and clock frequency is $1/13.8R_TC$. Width of output pulse depends on number of stages in shift register IC_2 and clock frequency.—B. Bong, Two CMOS IC's Yield Precision One-Shot, *EDN Magazine,* Aug. 5, 1978, p 82.

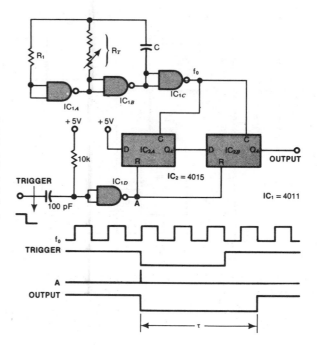

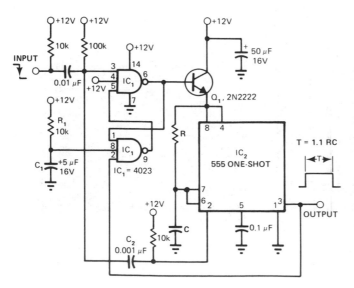

LOW STANDBY POWER—Basic 555 timer circuit is combined with control logic to keep drain from 12-V supply down to 1 μA during standby. Drain increases to 6 mA when input signal makes output pulse go high. Circuit can be interfaced with CMOS logic. Negative-going input pulse triggers SR flip-flop, which in turn saturates Q_1 and applies power to 555. Simultaneously, C_2 feeds trigger to trigger input pin 2 of 555, to make output pulse go high. At end of time delay determined by values of R and C, timer output goes low and transition resets flip-flop for standby operation.—K. J. Imhof, 555 One-Shot Circuit Features Low-Power Standby Mode, *EDN Magazine,* April 20, 1978, p 134.

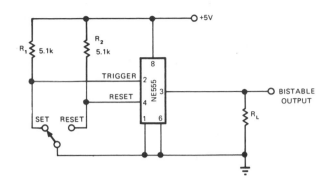

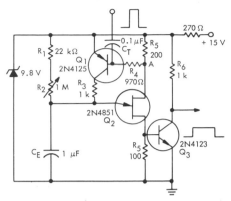

TRUTH TABLE

TRIGGER (PIN 2)	RESET (PIN 4)	OUT
↓	HIGH	↑
↑	HIGH	HIGH
HIGH	↓	↓
HIGH	↑	LOW

555 TIMER AS FLIP-FLOP—Eliminating RC timing network of 555 timer and tying threshold low makes output states depend on trigger and reset inputs. These are pulled high through R_1 and R_2, then pulled low either with switch or TTL level of 0 on reset input pin 4. Output then stays low until reset goes high and trigger goes low; this bistable action prevents contact bounce from switching output erroneously. Circuit will source or sink 200 mA.—R. L. Gephart, Mini-DIP Bistable Flip-Flop Sinks or Sources 200 mA, *EDN Magazine*, Oct. 5, 1974, p 76 and 78.

UJT MONO—UJT Q_2 is normally on, with its emitter saturation current supplied by transistor Q_1 which is also on. Application of positive trigger pulse to base of Q_1 turns both off to start timing cycle. C_E starts charging each time from saturation voltage of UJT. When capacitor voltage becomes high enough to fire UJT, Q_1 turns on and supplies emitter current required to keep UJT on. Output transistor Q_3 delivers pulse having duration related to value used for C_E.— "Unijunction Transistor Timers and Oscillators," Motorola, Phoenix, AZ, 1974, AN-294, p 5.

MOSFET ASTABLE—RCA 40841 dual-gate N-channel depletion-type MOSFETs alternate between high and low conduction states in between dormant periods when C_1 is charging or discharging through R_1. Circuit switches state when voltage level at gate of Q_1 makes gain high enough for regeneration to occur. D_1 reduces voltage across R_4 to give TTL drive capability.—D. R. Armstrong, Wide-Frequency Astable Multivibrator Uses One R-C Network, *EDN Magazine*, Aug. 5, 1977, p 54.

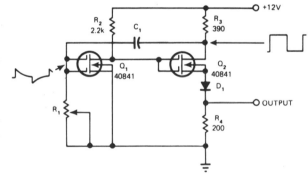

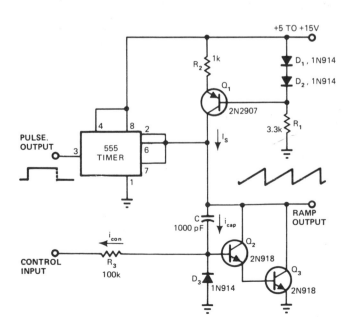

VOLTAGE-CONTROLLED MONO—Circuit gives choice of linear sawtooth and constant pulse-width outputs over frequency range from DC to 50 kHz. Output frequency and pulse repetition rate vary linearly with control current. Applications include audio synthesizers, variable time bases, and current-to-frequency converters.— S. Wetenkamp, Minor Changes Turn VCO into Voltage-Controlled One-Shot, *EDN Magazine*, March 5, 1978, p 67–69.

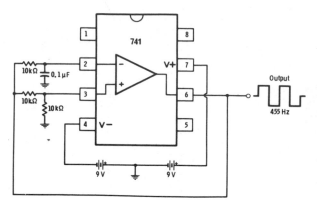

455-Hz ASTABLE—Frequency of square-wave output depends on values used for external capacitor and resistors. Very low frequencies can be obtained by using large values for both. High-frequency performance is limited by slew rate of opamp.—R. Melen and H. Garland, "Understanding IC Operational Amplifiers," Howard W. Sams, Indianapolis, IN, 2nd Ed., 1978, p 119–120.

GATED ASTABLE—With values shown, circuit produces positive output pulses at about 1 kHz when gated on by positive pulse at pin 4. Supply voltage for 555 timer can be 15 V.—W. G. Jung, "IC Timer Cookbook," Howard W. Sams, Indianapolis, IN, 1977, p 135–136.

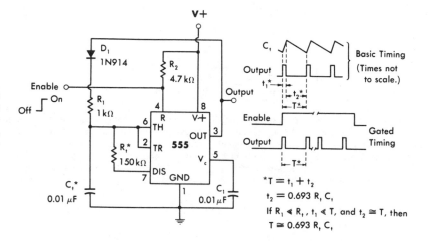

PULSE-WIDTH CONTROL—DC input voltage controls width of rectangular output pulse of opamp operating as free-running multivibrator, by injecting additional current into phase-inverting input of opamp. This current serves to increase one timing period and decrease the other. Circuit also provides similarly controllable sawtooth output (at pin 2). Output circuit uses diode bridge and zener for symmetrically clamping output voltage limits of amplifier when this feature is required.—G. B. Clayton, Experiments with Operational Amplifiers, *Wireless World*, May 1973, p 241–242.

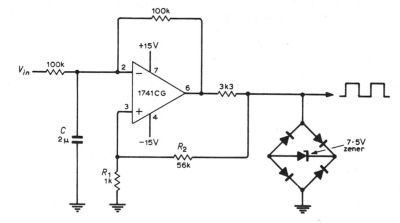

SEQUENTIAL TIMING GIVES DELAYED MONO MVBR—Output of first timer section of Exar XR-2556 dual timer is capacitively coupled to trigger pin of second timer section. When input trigger is applied, output 1 goes high for duration $T_1 = 1.1R_1C_1$, then goes low and triggers timer 2 through C_C. Output at pin 13 then goes high for duration $T_2 = 1.1R_2C_2$ to give performance of delayed mono MVBR. Supply voltage is 4.5–16 V. Choose R_L to keep timer output below 200 mA for supply voltage used.—"Timer Data Book," Exar Integrated Systems, Sunnyvale, CA, 1978, p 23–30.

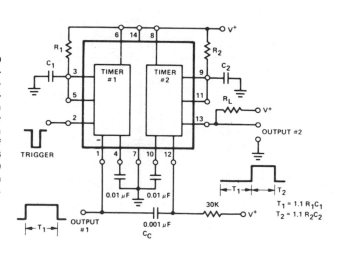

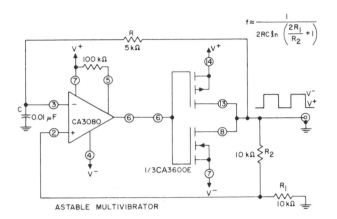

$$f \approx \frac{1}{2RC \ln \left(\frac{2R_1}{R_2} + 1\right)}$$

ASTABLE MULTIVIBRATOR

ASTABLE SQUARE-WAVE—CA3080 variable opamp drives one inverter/amplifier section of CA3600E inverter array. Quiescent power drain is typically 6 mW. Supply voltage range is ±3 to ±15 V.—"Circuit Ideas for RCA Linear ICs," RCA Solid State Division, Somerville, NJ, 1977, p 5.

FREE-RUNNING—Positive feedback is applied to noninverting input terminal of opamp through voltage divider R_1-R_2, to make amplifier switch regeneratively and repetitively between saturated states. Charging time of C controls duration of each state, to give desired free-running multivibrator providing rectangular (pin 6), trapezoidal (pin 3), and sawtooth (pin 2) symmetrical waveforms. Article gives design equations and waveforms. For nonsymmetrical waveforms, use alternative circuit in place of R; here, diodes switch two different timing resistors into circuit alternately.—G. B. Clayton, Experiments with Operational Amplifiers, *Wireless World*, May 1973, p 241–242.

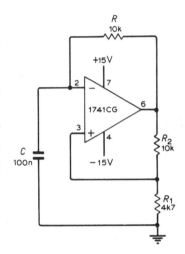

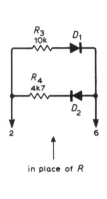

in place of R

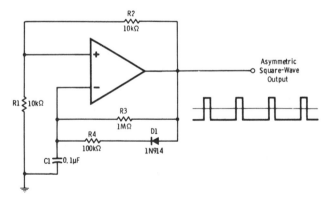

ASYMMETRICAL SQUARE WAVES—Addition of D1 and R4 to astable MVBR connection of 741 or equivalent opamp results in two different charging rates for C1, depending on whether opamp is in positive or negative saturation. Positive and negative peaks of output pulse then have different widths.—R. Melen and H. Garland, "Understanding IC Operational Amplifiers," Howard W. Sams, Indianapolis, IN, 2nd Ed., 1978, p 122–123.

SQUARE-WAVE BURSTS—When pushbutton switch is closed, 555 timer generates square-wave tone bursts for duration depending on how long voltage on pin 4 exceeds threshold value. R_1, R_2, and C_1 control astable action of timer.—"Signetics Analog Data Manual," Signetics, Sunnyvale, CA, 1977, p 726.

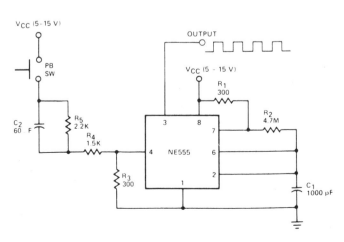

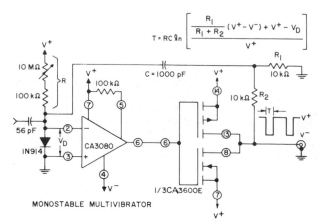

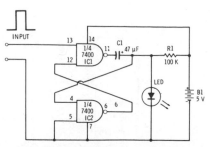

MONOSTABLE MULTIVIBRATOR

MONO SQUARE-WAVE—Stable characteristics of differential amplifier in CA3080 variable opamp assure precise timing and threshold for output waveform. Opamp drives one inverter/ amplifier section of CA3600E inverter array. Supply voltage range is ±3 to ±15 V.—"Circuit Ideas for RCA Linear ICs," RCA Solid State Division, Somerville, NJ, 1977, p 5.

BASIC MONO DRIVES LED—Two sections of 7400 quad NAND gate are connected as monostable MVBR having one stable state and one unstable state. Incoming pulse at pin 13 changes state of gate IC1. Since output of this gate goes to input of other gate, that also changes state. After interval determined by values of C1 and R1, gates automatically return to original states. LED flashes when input pulse is applied, for duration also determined by R1 or C1. If C1 is increased to 470 μF, LED will stay on for over 1 s before fading out gradually as capacitor discharges.—F. M. Mims, "Integrated Circuit Projects, Vol. 2," Radio Shack, Fort Worth, TX, 1977, 2nd Ed., p 19–26.

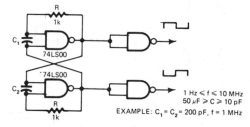

1Hz TO 10 MHz—Simple circuit operating from single 5-V supply provides TTL output levels with reliable starting over wide frequency range. When capacitors are equal, period of oscillation is equal to $5 \times 10^3 C$ s. By changing ratio of C_1 to C_2, duty cycle can be made as low as 20%.—D. B. Arnett, One-Chip TTL Oscillator Requires One 5V Supply, *EDN Magazine*, Jan. 5, 1978, p 96.

LOGIC CONTROL—External circuit modification shown for 555 timer makes initial pulse more nearly equal to subsequent pulses and makes circuit deliver low output when power is applied. Optional diode ensures 50% duty cycle.—K. D. Dighe, Rearranged Components Cut 555's Initial-Pulse Errors, *EDN Magazine*, Jan. 5, 1978, p 82 and 84.

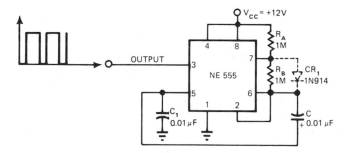

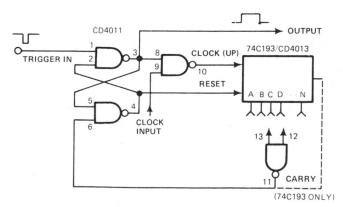

PULSE-TRIGGERED MONO—Combination of two ICs generates pulse having width precisely determined by external clock frequency and countdown factor N of 74C193 binary counter; time t in seconds is equal to N divided by frequency in hertz. With 74C193, 11 values of N are possible (1–6, 8–10, 12, and 16). With CD4013 14-stage counter, values of N can range from 1 to 24,576. Input pulse must be shorter than output pulse. Value of N depends on which two counter outputs are connected to two-input NAND gate.—R. L. Anderson, Digital One-Shot Produces Long, Accurate Pulses, *EDN Magazine*, March 5, 1978, p 127.

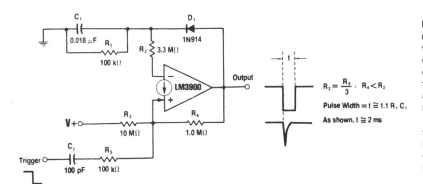

LM3900 AS MONO—R_4 holds output high normally, so C_1 is charged almost to V+ level through D_1. Negative input trigger forces output of current-differencing amplifier low, and C_1 discharges through R_1. When decreasing current through R_2 approaches current in R_3 (when voltage across C_1 is about one-third of V+), output switches to high and returns circuit to standby state. Pulse width, equal to $1.1R_1C_1$, can be programmed easily by using pot or some form of manual or electronic switching for R_1.—W. G. Jung, "IC Op-Amp Cookbook," Howard W. Sams, Indianapolis, IN, 1974, p 510–512.

EDGE-TRIGGERED MONO—Output pulse width is precisely determined by external clock frequency and countdown factor N of 74C193 binary counter (t = N/f). Pulse is generated on rising edge of input, with output remaining high until count N is reached by binary counter. Counter resets and output returns instantly to zero if input pulse goes to zero before count of N. Value of N depends on which two counter outputs are connected to two-input NAND gate. With 74C193, 11 values of N between 1 and 16 are possible; with CD4013, values of N can range from 1 to 24,576.—R. L. Anderson, Digital One-Shot Produces Long, Accurate Pulses, *EDN Magazine,* March 5, 1978, p 127.

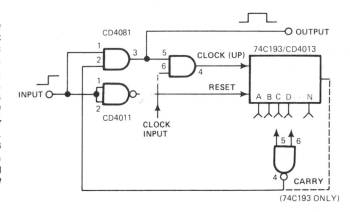

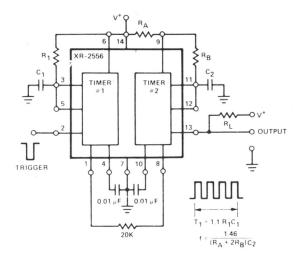

KEYED FREE-RUNNING MVBR—One section of Exar XR-2556 dual timer is operated in free-running mode, with other section used to provide ON/OFF keying. Frequency of oscillator is set by R_A, R_B, and C_2. Timer 1 operates as mono MVBR with output connected to reset pin 8 of timer 2. Trigger drives pin 1 of timer 1 high, keying timer 2 on and producing tone-burst output for duration set by R_1 and C_1. Supply voltage is 4.5–16 V.—"Timer Data Book," Exar Integrated Systems, Sunnyvale, CA, 1978, p 23–30.

Music Circuits

Includes organ, piano, trombone, bell, theremin, bird-call, and other sound and music synthesizer circuits, along with circuits giving warble, fuzz, three-part harmony, reverberation, tremolo, attack, decay, rhythm, and other musical effects. Joystick control for music, active filters, contact-pickup preamp, metronomes, and tuning aids are also given.

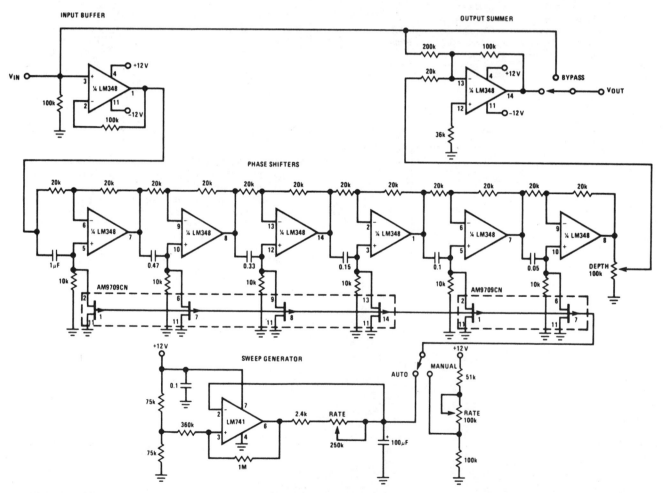

SIMULATION OF FLANGING—Sound-effect circuit sometimes called phase shifter simulates playing of two tape recorders having same material while varying speed of one by pressing on flange of tape reel. Resulting time delay causes some signals to be summed out of phase and canceled. Effect is that of rotating loudspeaker or of Doppler characteristic. Uses two LM348 quad opamps, two AM9709CN quad JFET devices, and one LM741 opamp. Phase-shift stages are spaced one octave apart from 160 to 3200 Hz in center of audio spectrum, with each stage providing 90° shift at its frequency. JFETs control phase shifters. Gate voltage of JFETs is adjusted from 5 V to 8 V either manually with foot-operated rheostat or automatically by LM741 triangle-wave generator whose rate is adjustable from 0.05 Hz to 5 Hz.—"Audio Handbook," National Semiconductor, Santa Clara, CA, 1977, p 5-10–5-11.

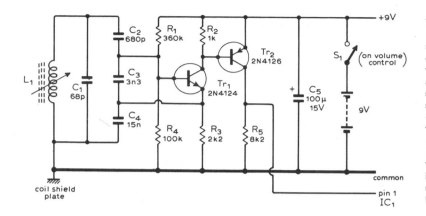

TUNING FOR EQUAL TEMPERAMENT—Instrument described enables anyone to tune such instruments as organ, piano, and harpsichord in equal temperament with accuracy approaching that of professional tuner. Only requirement is ability to hear beats between two tones sounded together. Master oscillator circuit shown generates 250.830 kHz for feeding to first of five ICs connected as programmable divider that provides 12 notes of an octave as 12 equal semitones differing from each other by factor of 1.0594. Article gives suitable power amplifier to fit along with divider connections and detailed instructions for construction, calibration, and use.—W. S. Pike, Digital Tuning Aid, *Wireless World,* July 1974, p 224–227.

TREMOLO CONTROL—National LM324 opamp connected as phase-shift oscillator operates at variable rate between 5 and 10 Hz set by speed pot. Portion of oscillator output is taken from depth pot and used to modulate ON resistance of two 1N914 diodes operating as voltage-controlled attenuators. Input should be kept below 0.6 V P-P to avoid undesirable clipping. Used for producing special musical effects.—"Audio Handbook," National Semiconductor, Santa Clara, CA, 1977, p 5-11–5-12.

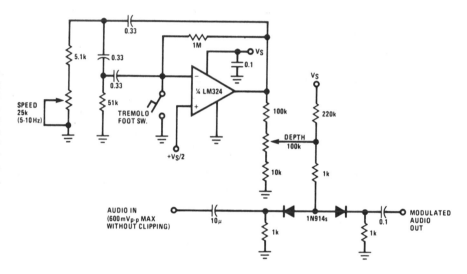

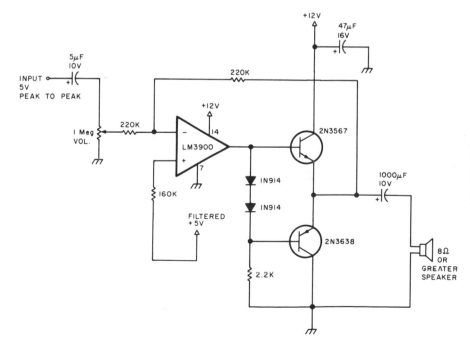

AUDIO FOR COMPUTER MUSIC—Wideband low-power audio amplifier was developed for use with DAC and low-pass active filter to create music with microprocessor.—H. Chamberlin, A Sampling of Techniques for Computer Performance of Music, *BYTE,* Sept. 1977, p 62–66, 68–70, 72, 74, 76–80, and 82–83.

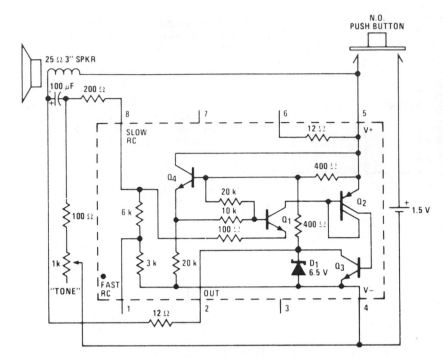

TROMBONE CIRCUIT—Unique arrangement for driving 25-ohm loudspeaker with National LM3909 IC operating from 1.5-V cell permits generation of slide tones resembling those of trombone. Operation is based on use of voltage generated by resonant motion of loudspeaker voice coil as major positive feedback for IC. Loudspeaker is mounted in roughly cubical box having volume of about 64 in³, with one end of box arranged to slide in and out like piston. Positioning of piston and operation of pushbutton permit playing reasonable semblance of simple tune. IC, loudspeaker, and battery are mounted on piston, with 2½-in length of ⁵⁄₁₆-in tubing provided to bleed air in and out as piston is moved, without affecting resonant frequency. Frequency of oscillator becomes equal to resonant frequency of enclosure.—"Linear Applications, Vol. 2," National Semiconductor, Santa Clara, CA, 1976, AN-154, p 6.

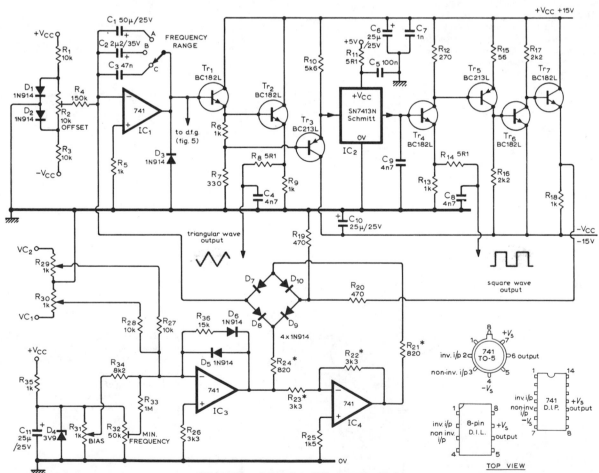

VCO SOUND SYNTHESIZER—Developed for use in instrument capable of duplicating variety of sounds ranging from bird distress calls and engine noises to spoken words and wide variety of musical instruments. Three-part article gives all circuits and describes their operation in detail. Heart of oscillator is triangle and square-wave generator built around IC Schmitt trigger. Ramp rate and operating frequency are varied by changing drive voltage or gain of integrator. Similar VCO in synthesizer also produces sine, pulse, and ramp waveforms.—T. Orr and D. W. Thomas, Electronic Sound Synthesizer, *Wireless World*, Part 1—Aug. 1973, p 366–372 (Part 2—Sept. 1973, p 429–434; Part 3—Oct. 1973, p 485–490).

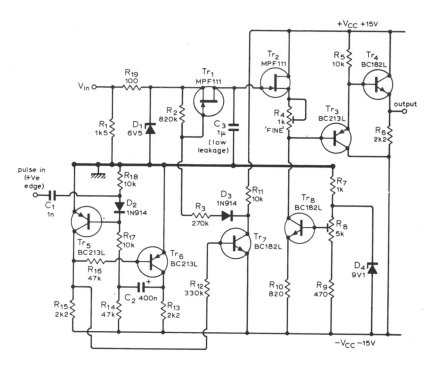

ANALOG MEMORY—Used in synthesizer for generating wide variety of musical and other sounds, to provide constant control signal for sounds requiring long fadeout. Positive input pulse initiates sampling of analog signal for pre-set time, with signal being held for unspecified period. Input voltage range is from about −0.5 V to +6.5 V, being deliberately limited by D_1. Three-part article describes operation in detail and gives all other circuits used in synthesizer.—T. Orr and D. W. Thomas, Electronic Sound Synthesizer, *Wireless World,* Part 3—Oct. 1973, p 485–490 (Part 1—Aug. 1973, p 366–372; Part 2—Sept. 1973, p 429–434).

TREMOLO AMPLIFIER—Provides amplitude modulation at subaudio rate (usually between 5 and 15 Hz) of audio-frequency input signal. Uses National LM389 array having three transistors along with power amplifier. Transistors form differential pair having active current-source tail to give output proportional to product of two input signals. Gain control pot is adjusted for desired tremolo depth. Interstage RC network forms 160-Hz high-pass filter, requiring that tremolo frequency be less than 160 Hz.—"Audio Handbook," National Semiconductor, Santa Clara, CA, 1977, p 4-33–4-37.

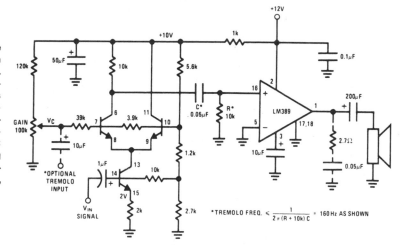

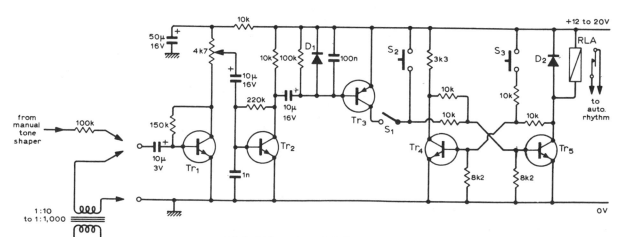

AUTOMATIC REMOTE RHYTHM CONTROL—When added to electronic organ, circuit is activated by audio signal from lower manual or pedal, to initiate start of rhythm accompaniment. High-impedance input connection through 100K is made to toneshaper output, and transformer connection is used with electromechanical Hammond organ. Transistor and diode types are not critical. If S_1 is closed, current passes through to Tr_5 and triggers bistable that pulls in relay. S_2 and S_3 are used for manual start and stop of rhythm.—K. B. Sorensen, Touch Start of Automatic Rhythm Device, *Wireless World,* Oct. 1974, p 381.

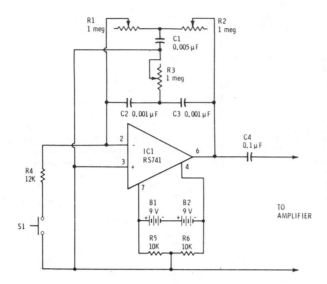

MUSICAL BELLS—Opamp connected as active filter simulates attack followed by gradual decay as produced when bell or tuning fork is struck. Filter portion of circuit uses twin-T network adjusted so active filter breaks into oscillation when slight external disturbance is introduced by closing S1 momentarily. Circuit feeds external audio amplifier and loudspeaker for converting ringing frequency into audible sound. Set R3 just below oscillation point. R1 and R2 can be adjusted to give sounds of other musical instruments, such as drums, bamboo, and triangles.—F. M. Mims, "Electronic Music Projects, Vol. 1," Radio Shack, Fort Worth, TX, 1977, 2nd Ed., p 71–80.

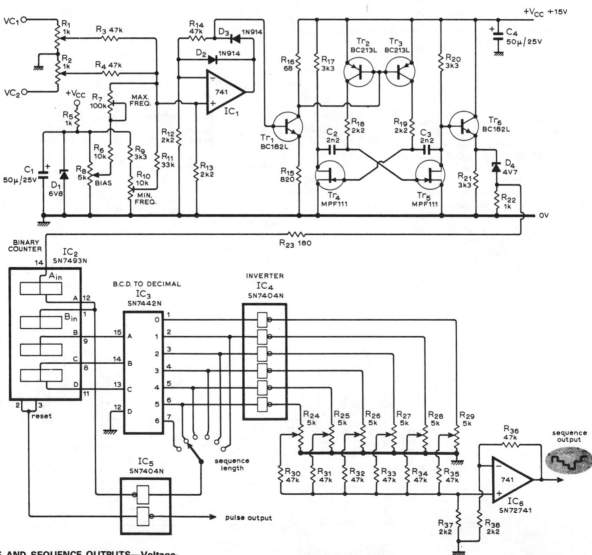

PULSE AND SEQUENCE OUTPUTS—Voltage-controlled oscillator produces sequence of steps, with amplitude of each step individually controllable up to maximum of six steps. Circuit also generates series of pulses having 1:1 mark-space ratio, each coincident with leading edge of a step. Pair of summing inputs controls os-cillator, with exponential frequency-voltage relationship extending in one range from subsonic frequencies to over 20 kHz. Used in sound synthesizer described in three-part article that gives all circuits and operating details. Applications include synthesizing sounds ranging from bird distress calls and engine noises to spoken words and wide variety of musical instruments.—T. Orr and D. W. Thomas, Electronic Sound Synthesizer, *Wireless World,* Part 2—Sept. 1973, p 429–434 (Part 1—Aug. 1973, p 366–372; Part 3—Oct. 1973, p 485–490).

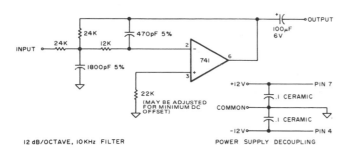

12 dB/OCTAVE, IOKHZ FILTER

POWER SUPPLY DECOUPLING

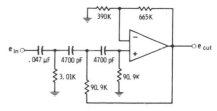

10-kHz LOW-PASS FILTER—Suitable for use at both input and output of A/D-D/A converter in digital audio system for synthesizing speech or music. Serves for smoothing steps of output waveform and suppressing background noise on output when small signals are being processed with 8-bit linear encoding.—T. Scott, Digital Audio, *Kilobaud,* May 1977, p 82–86.

327-Hz HIGH-PASS—Developed to make third harmonic of 130.81 Hz (C3 note) minimum of 30 dB stronger than fundamental, to give sawtooth output for use in electronic music system. Design uses third-order filter with 3-dB dips in response. Opamp can be 741.—D. Lancaster, "Active-Filter Cookbook," Howard W. Sams, Indianapolis, IN, 1975, p 192.

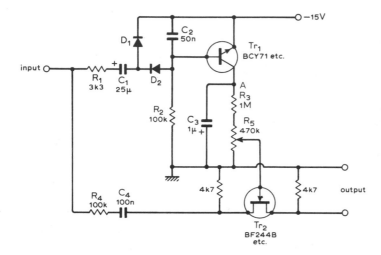

PIANO MUSIC FROM ORGAN—Simple add-on circuit for electronic organ attenuates output of oscillator exponentially to zero in manner suitable for mimicking waveform of piano. Circuit is self-triggering, so exponential decay starts only when output of multivibrator is applied; this eliminates need for extra contacts on keyboard. With no input, Tr_1 is on and point A is at supply voltage. Input signal turns Tr_1 off, discharging C_3 through R_3 and R_5. Voltage across C_3 controls gate of FET, with R_5 being adjusted so FET just switches off when C_3 is fully discharged. Tr_1 then conducts and C_3 charges rapidly, to permit fast piano playing.—C. J. Outlaw, Electronic Organ to Piano, *Wireless World,* Feb. 1975, p 94.

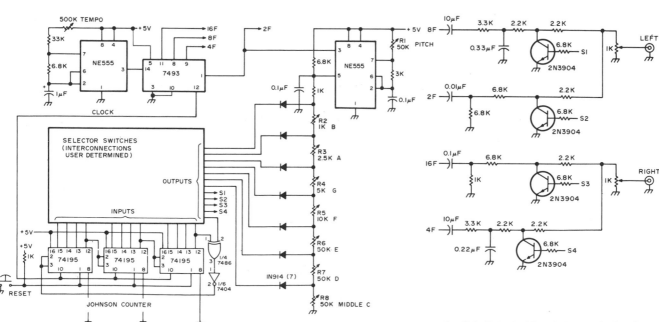

RANDOM MUSIC—Uses Johnson counter as special shift register producing almost random bit patterns of 18 to 3255 12-bit words under control of clock operating in range of about 1–10 Hz. Oscillator (upper right) uses NE555 as voltage-controlled square-wave generator playing one of eight musical notes (C, D, E, F, G, A, B, or C), depending on state of seven note-selector lines coming from selector switches. Oscillator is divided down in frequency by three-stage ripple counter to provide four octaves of range. R1-R8 serve for tuning each note to pitch. Outputs 8F and 2F are paired to drive left input of stereo amplifier, while outputs 16F and 4F are similarly paired for right channel. Article covers construction, tune-up, and creation of pleasing musical sequences.—D. A. Wallace, The Sound of Random Numbers, *73 Magazine,* Feb. 1976, p 60–64.

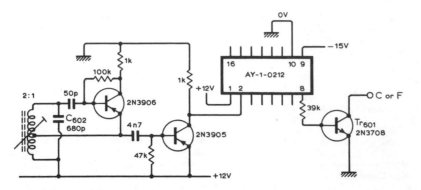

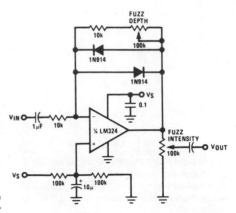

PIANO TONE GENERATOR—RF oscillator combined with General Instrument AY-1-0212 IC master tone generator replaces 12 conventional RC oscillators otherwise required in electronic piano. Frequencies generated are within 0.1% of equal-temperament scale, so piano will work well without being tuned. Three-part article gives all circuits and construction details for simple portable touch-sensitive electronic piano.—G. Cowie, Electronic Piano Design, *Wireless World*, Part 3, May 1974, p 143–145.

FUZZ CIRCUIT—Two diodes in feedback path of LM324 opamp create musical-instrument effect known as fuzz by limiting output voltage swing to ±0.7 V. Resultant square wave contains chiefly odd harmonics, resembling sounds of clarinet. Fuzz depth pot controls level at which clipping begins, and fuzz intensity pot controls output level.—"Audio Handbook," National Semiconductor, Santa Clara, CA, 1977, p 5-11.

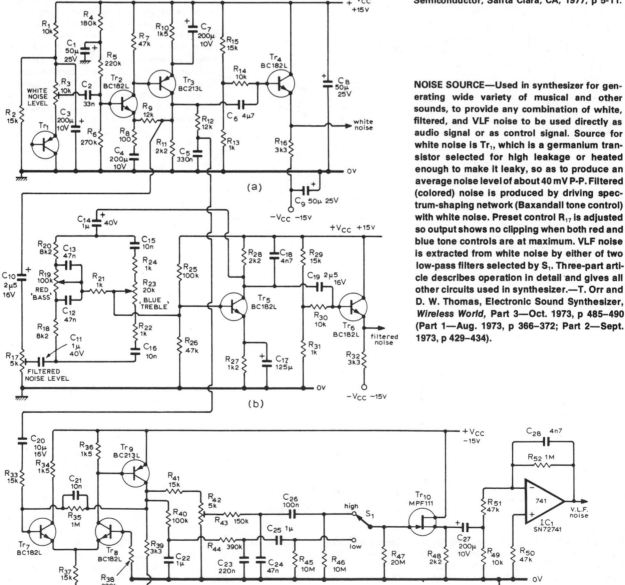

NOISE SOURCE—Used in synthesizer for generating wide variety of musical and other sounds, to provide any combination of white, filtered, and VLF noise to be used directly as audio signal or as control signal. Source for white noise is Tr_1, which is a germanium transistor selected for high leakage or heated enough to make it leaky, so as to produce an average noise level of about 40 mV P-P. Filtered (colored) noise is produced by driving spectrum-shaping network (Baxandall tone control) with white noise. Preset control R_{17} is adjusted so output shows no clipping when both red and blue tone controls are at maximum. VLF noise is extracted from white noise by either of two low-pass filters selected by S_1. Three-part article describes operation in detail and gives all other circuits used in synthesizer.—T. Orr and D. W. Thomas, Electronic Sound Synthesizer, *Wireless World*, Part 3—Oct. 1973, p 485–490 (Part 1—Aug. 1973, p 366–372; Part 2—Sept. 1973, p 429–434).

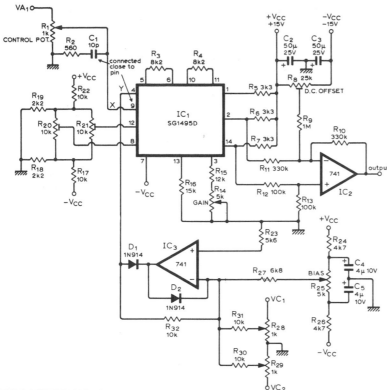

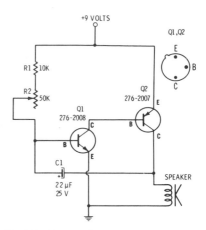

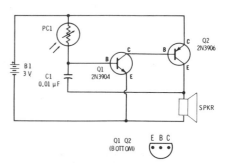

VOLTAGE-CONTROLLED AMPLIFIER—Gain is linearly controlled by sum of input control voltages and a bias voltage, to provide amplitude modulation as required for synthesizer used to generate wide variety of sounds. Heart of circuit is linear four-quadrant multiplier IC. Output is taken between two load resistors, with differential amplifier IC_2 removing common-mode signal. Article describes operation in detail and gives all other circuits of synthesizer, along with procedure for aligning preset controls R_8, R_{14}, R_{20}, and R_{21}.—T. Orr and D. W. Thomas, Electronic Sound Synthesizer, *Wireless World,* Part 2—Sept. 1973, p 429–434 (Part 1—Aug. 1973, p 366–372; Part 3—Oct. 1973, p 485–490).

CLICKING METRONOME—Basic lamp-flashing circuit is used to produce sharp click in loudspeaker each time Q2 is turned on by RC oscillator Q1. R2 adjusts repetition rate over range of 20–280 beats per minute. Changing value of C1 varies tone of clicks.—F. M. Mims, "Transistor Projects, Vol. 1," Radio Shack, Fort Worth, TX, 1977, 2nd Ed., p 33–39.

LIGHT-SENSITIVE THEREMIN—Tone of loudspeaker increases and decreases in frequency as flashlight is moved in vicinity of photocell in darkened room. Use Radio Shack 276-116 cadmium sulfide photocell. Cell resistance decreases with light, increasing frequency of audio oscillator. Continuously changing frequency resembles that produced by hand-controlled theremin.—F. M. Mims, "Electronic Music Projects, Vol. 1," Radio Shack, Fort Worth, TX, 1977, 2nd Ed., p 91–95.

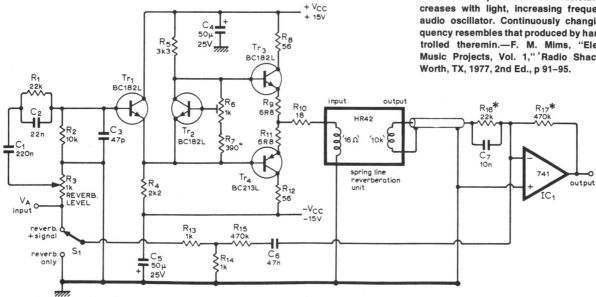

REVERBERATION—Used in sound synthesizer developed for generating wide variety of musical and other sounds. Four-transistor driver feeds spring-type reverberation unit at up to about 4 kHz, with switch giving choice of reverberation only or reverberation combined with input signal at V_A. Amount of reverberation can be controlled manually with R_3 or automatically with voltage-controlled amplifier or voltage-controlled filter of synthesizer. Three-part article gives all circuits and describes operation in detail.—T. Orr and D. W. Thomas, Electronic Sound Synthesizer, *Wireless World,* Part 2—Sept. 1973, p 429–434 (Part 1—Aug. 1973, p 366–372; Part 3—Oct. 1973, p 485–490).

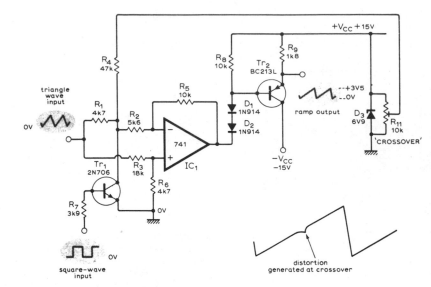

RAMP FUNCTION—Circuit combines triangle and square-wave inputs from VCO in differential amplifier having switched gain, to generate ramp function for use with variety of other waveforms in sound synthesizer designed for duplicating wide variety of sounds. Three-part article gives all circuits and operating details.—T. Orr and D. W. Thomas, Electronic Sound Synthesizer, *Wireless World*, Part 1—Aug. 1973, p 366–372 (Part 2—Sept. 1973, p 429–434; Part 3—Oct. 1973, p 485–490).

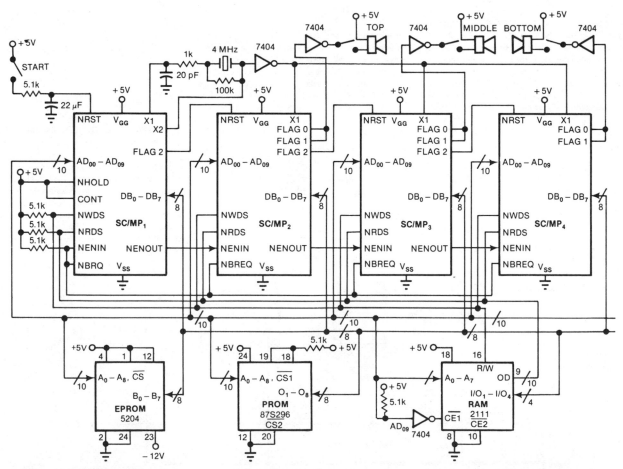

THREE-PART HARMONY—Four SC/MP microprocessors, one serving as conductor and three as instrumentalists, generate multiple parts for harmony feeding common loudspeaker system. Microprocessors have paralleled address and data buses, with 4K RAM connecting to lowest 4 bits of data bus. Each microprocessor is supplied with list of notes by note number and note lengths as part of software. At end of each basic note length, SC/MP₁ checks each other processor to see if it is time to proceed to next note in list. If it is, next note is played by other processors until signaled by conductor via memory. Article gives software listing.—T. Doone, Quartet of SC/MP's Plays Music for Trios, *EDN Magazine,* Sept. 20, 1978, p 57–58 and 60.

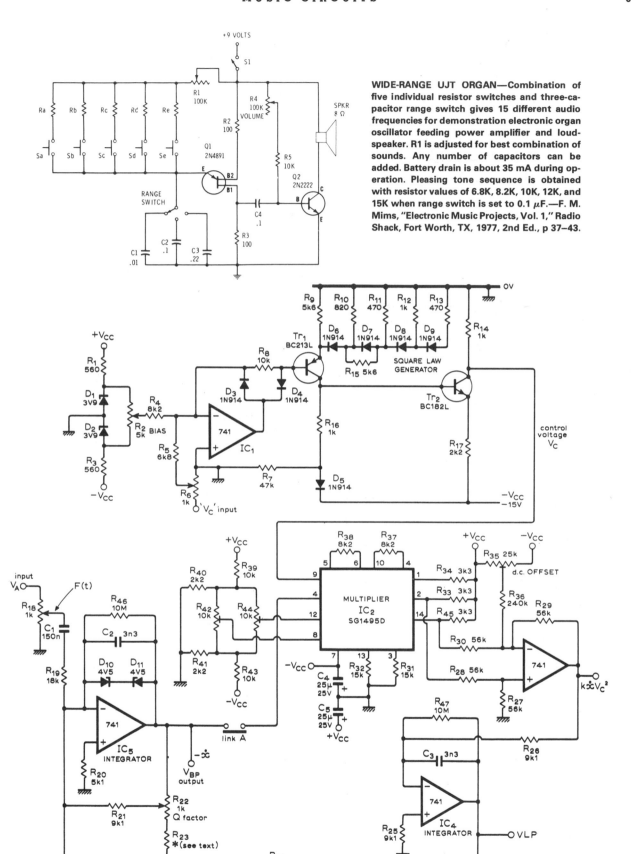

WIDE-RANGE UJT ORGAN—Combination of five individual resistor switches and three-capacitor range switch gives 15 different audio frequencies for demonstration electronic organ oscillator feeding power amplifier and loudspeaker. R1 is adjusted for best combination of sounds. Any number of capacitors can be added. Battery drain is about 35 mA during operation. Pleasing tone sequence is obtained with resistor values of 6.8K, 8.2K, 10K, 12K, and 15K when range switch is set to 0.1 μF.—F. M. Mims, "Electronic Music Projects, Vol. 1," Radio Shack, Fort Worth, TX, 1977, 2nd Ed., p 37—43.

VOLTAGE-CONTROLLED FILTER—Used in elaborate sound synthesizer developed for generating wide variety of sounds. Serves as bandpass filter for which resonant frequency is linearly proportional to sum of input control voltages and a bias voltage. Can also be used as notch filter or as spectrum analyzer. Three-part article describes operation in detail and gives all other circuits of synthesizer. Supply voltages are 15 V, with polarity as indicated.—T. Orr and D. W. Thomas, Electronic Sound Synthesizer, *Wireless World,* Part 2—Sept. 1973, p 429—434 (Part 1—Aug. 1973, p 366—372; Part 3—Oct. 1973, p 485—490).

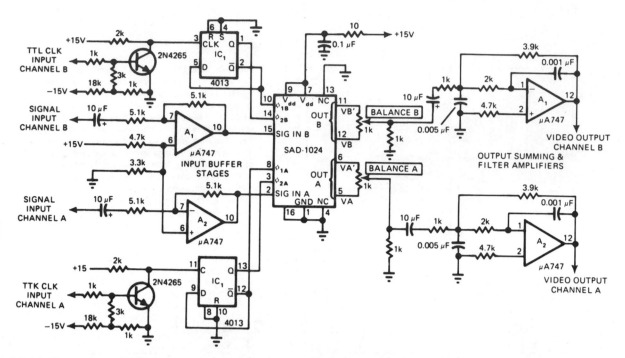

CCD DELAY FOR SPECIAL EFFECTS—Basic bucket-brigade device incorporated in Reticon Corp. SAD-1024 charge-coupled-device delay line can synthesize such interesting audio-system delay effects as reverberation enhancement, chorus, and vibrato generation. Other applications include speech compression and voice scrambling. Evaluation circuit shown was developed by manufacturer. Input clock frequency is 200 kHz, and signal input is 5-kHz sine wave. Article describes operation of evaluation circuit in detail and presents variety of practical applications.—R. R. Buss, CCD's Improve Audio System Performance and Generate Effects, *EDN Magazine,* Jan. 5, 1977, p 55–61.

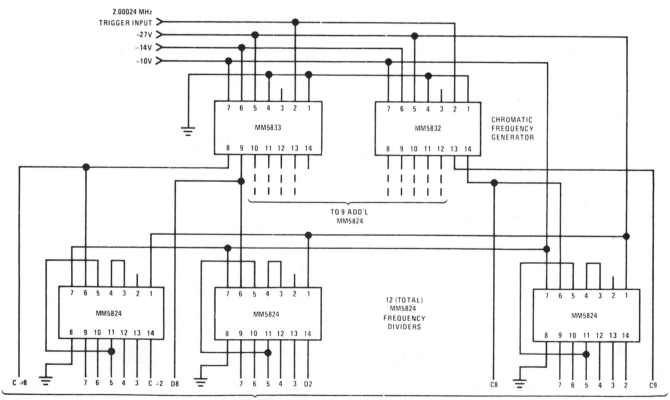

ORGAN TONE GENERATOR—National MM5832 and MM5833 chromatic frequency generators are used with 12 MM5824 frequency dividers to generate 85 musical frequencies fully spanning equal-tempered octave. Can also be used as celeste or chorus tone generator and as electronic music synthesizer. Square-wave input for organ is 2.00024 MHz but can be as low as 7 kHz for other applications.—"MOS/LSI Databook," National Semiconductor, Santa Clara, CA, 1977, p 3-11–3-13.

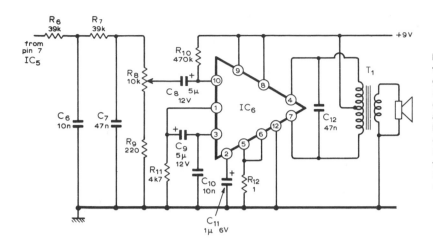

PIANO-TUNING AMPLIFIER—Used with battery-powered digital tuning aid that provides 12 equal semitones of octave, between 261.6625 and 493.8833 Hz, for equal-temperament tuning of such keyboard instruments as organ, piano, and harpsichord. IC used is part of RCA amplifier kit KC-4003 which includes T_1 and other discrete components. Article gives circuits for oscillator and programmable divider, along with instructions for construction, calibration, and use.—W. S. Pike, Digital Tuning Aid, *Wireless World,* July 1974, p 224–227.

WARBLER—One 555 timer is connected as low-frequency square-wave generator that modulates second timer producing higher-frequency tone, to give warbling tone that can be varied with R1 and R4 to simulate siren or songs of certain birds. Will operate over supply range of 4.5–18 V. Use 8-ohm miniature loudspeaker, with optional volume control R7 in series.—F. M. Mims, "Electronic Music Projects, Vol. 1," Radio Shack, Fort Worth, TX, 1977, 2nd Ed., p 29–35.

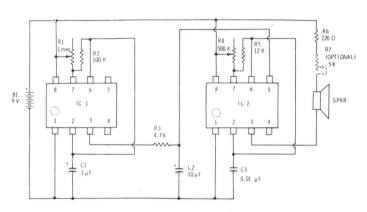

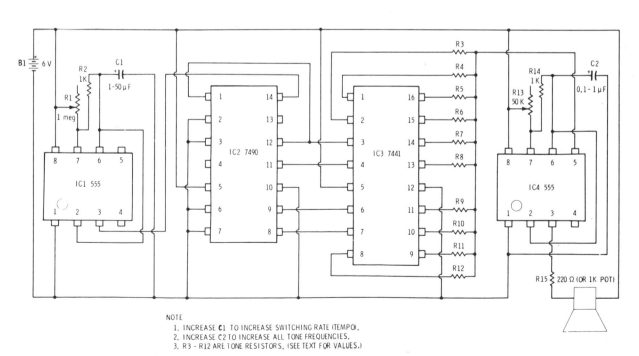

NOTE
1. INCREASE C1 TO INCREASE SWITCHING RATE (TEMPO).
2. INCREASE C2 TO INCREASE ALL TONE FREQUENCIES.
3. R3 - R12 ARE TONE RESISTORS, (SEE TEXT FOR VALUES.)

MUSIC SYNTHESIZER—555 timer is used as clock to set beat, adjustable with R1. Beat pulses drive flip-flop chain in 7490, which provides running total in binary format to 7441 1-of-10 decoder for conversion to decimal output. Each of ten outputs feeds through tone-controlling resistor to modulation input terminal of another 555 used as voltage-controlled AF oscillator feeding loudspeaker. Values used for resistors determine frequencies of ten notes that are generated in sequence repeatedly. R3-R12 can be 1000-ohm pots, so ten-note tune being played can be easily changed.—F. M. Mims, "Electronic Music Projects, Vol. 1," Radio Shack, Fort Worth, TX, 1977, 2nd Ed., p 61–70.

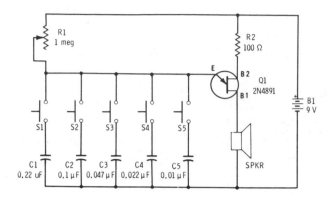

BASIC UJT ORGAN—Switches give choice of five audio frequencies for simple transistor oscillator driving loudspeaker. Adjust R1 for best combination of sounds. Capacitor values can be changed to give other frequencies. Ideal for classroom demonstrations.—F. M. Mims, "Electronic Music Projects, Vol. 1," Radio Shack, Fort Worth, TX, 1977, 2nd Ed., p 37–43.

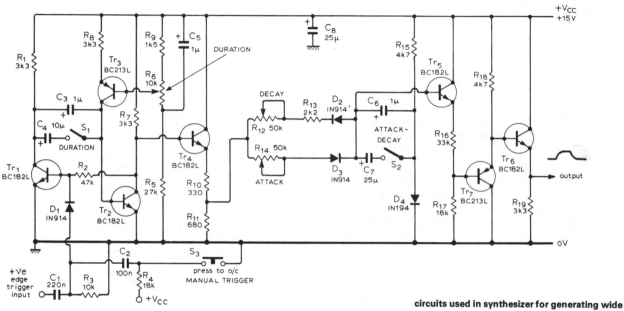

ATTACK/DECAY—Waveform generator produces approximate rectangular waveform having exponential rise (attack) and exponential decay, initiated either by manual trigger or electronic signal derived from other circuits of sound synthesizer. All characteristics of waveform are arbitrarily variable. Three-part article describes circuit operation and gives all other circuits used in synthesizer for generating wide variety of musical and other sounds.—T. Orr and D. W. Thomas, Electronic Sound Synthesizer, *Wireless World,* Part 3—Oct. 1973, p 485–490 (Part 1—Aug. 1973, p 366–372; Part 2—Sept. 1973, p 429–434).

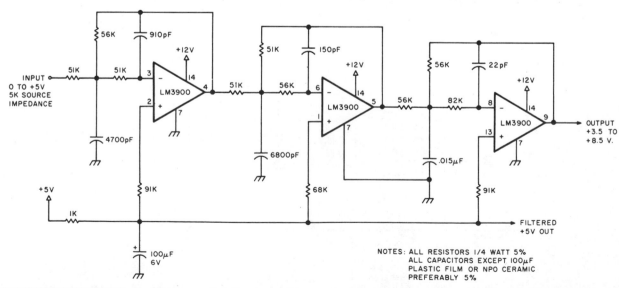

LOW-PASS WITH 3-kHz CUTOFF—Used in computer music system to suppress high-fidelity distortion resulting from steps in sine-wave output of DAC. Article covers complete computer synthesis of music by microprocessor and gives frequency table, program for generating four simultaneous musical voices, and song table for encoding "The Star Spangled Banner" in four-part harmony, using 5 bytes per musical event.—H. Chamberlin, A Sampling of Techniques for Computer Performance of Music, *BYTE,* Sept. 1977, p 62–66, 68–70, 72, 74, 76–80, and 82–83.

ATTACK-DECAY GENERATOR—Designed for polytonic electronic music system handling more than one note at a time. Each note to be controlled is sent through voltage-controlled amplifier (VCA) whose gain is set by charge on capacitor. Attack is changed by varying charging rate. Discharge rate sets decay of individual note. To avoid having separate adjustment pot for each VCA, duty-cycle modulation is used to change charging current through resistors. Attack pulses are generated by upper three inverters forming variable-symmetry astable MVBR. Decay pulses are generated by lower three inverters connected as half-mono MVBR. Additional half-monos can be added as needed

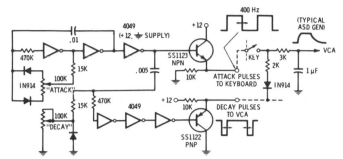

for percussion, snubbing, and other two-step decay effects.—D. Lancaster, "CMOS Cook-book," Howard W. Sams, Indianapolis, IN, 1977, p 231–232.

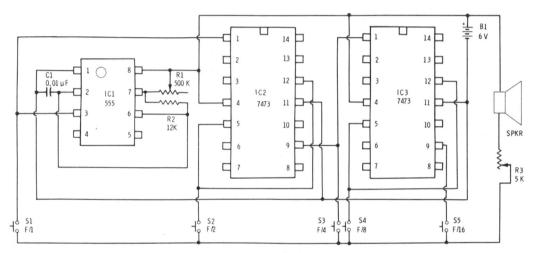

FOUR-OCTAVE ORGAN—Two 7473 dual flip-flops provide four frequency dividers for 555 timer connected as master tone generator. S1 gives fundamental frequency, and each succeeding switch gives tones precisely one octave lower. Four organ applications, pushbutton switches are added to timer circuit for switching frequency-controlling capacitors or resistors to give desired variety of notes.—F. M. Mims, "Electronic Music Projects, Vol. 1," Radio Shack, Fort Worth, TX, 1977, 2nd Ed., p 45–53.

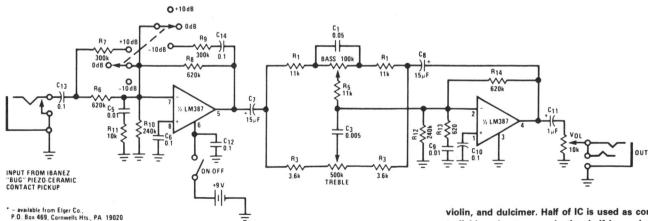

* – available from Elger Co.,
P.O. Box 469, Cornwells Hts., PA 19020

PREAMP FOR ACOUSTIC PICKUP—National LM387 dual opamp provides switchable gain choice of ±10 dB along with bass/treble tone control and volume control. Used with flat-response piezoceramic contact pickup for acoustic stringed musical instruments such as guitar, violin, and dulcimer. Half of IC is used as controllable gain stage, and other half is used as active two-band tone-control block.—"Audio Handbook," National Semiconductor, Santa Clara, CA, 1977, p 5-12.

AUDIBLE/VISIBLE METRONOME—Produces uniformly spaced beats in synchronism with flashes of LED, at rate that can be adjusted with R1 from one beat every few seconds to ten or more beats per second. Use red Radio Shack 276-041 or similar LED. Add switch in series with battery to avoid disturbing setting of R1. R3 serves as volume control. Add 5–10 μF capacitor across loudspeaker to mellow beat sound if desired.—F. M. Mims, "Electronic Music Projects. Vol. 1," Radio Shack, Fort Worth, TX, 1977, 2nd Ed., p 55–59.

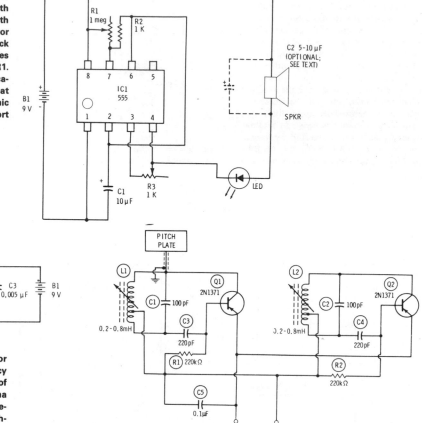

HAND-WAVING THEREMIN—Single-transistor RF oscillator is tuned to generate frequency about 455 kHz above oscillator frequency of transistor radio. With aluminum-foil antenna away from nearby objects and radio tuned between stations, R1 is adjusted until high-pitched tone is heard from radio. Now, as hand is brought toward and away from foil antenna, wailing sounds are produced. With practice, musician can produce recognizable melodies by vibrating hand. Primary controls of frequency are C1 (10–365 pF broadcast radio tuning capacitor) and adjustable antenna coil L1 (Radio Shack 270-1430). Radio can be up to 15 feet away from theremin. Rotate radio for maximum pickup from L1.—F. M. Mims, "Electronic Music Projects, Vol. 1," Radio Shack, Fort Worth, TX, 1977, 2nd Ed., p 81–89.

THEREMIN—Two transistor oscillator stages generate separate low-power RF signal in broadcast band, for pickup by AM broadcast receiver. Movement of hand toward or away from metal pitch plate varies frequency of Q1, making audio output of receiver vary correspondingly as beat frequency changes. Both circuits are Hartley oscillators, using Miller 9012 or equivalent slug-tuned coils. To adjust initially, place next to radio and set tuning slug of L1 about two-thirds out of its winding. Set slug of L2 about one-third out of its winding. Tune radio until either oscillator signal is heard. Signal can be identified by whistle if on top of broadcast station or by quieting of background noise if between stations. Adjust slugs so whistle is heard at desired location of quieting signal. Pitch of whistle should change now as hand is brought near pitch plate.—J. P. Shields, "How to Build Proximity Detectors & Metal Locators," Howard W. Sams, Indianapolis, IN, 2nd Ed., 1972, p 154–156.

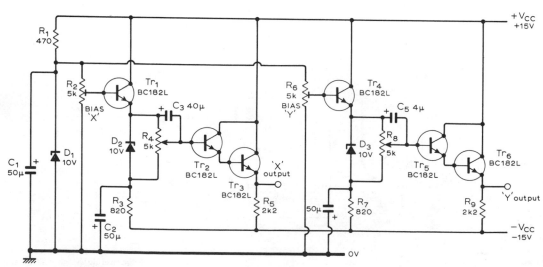

JOYSTICK CONTROL—Mechanically controlled voltage source generates two independent control voltages, proportional to stick position, to serve as one of controls for elaborate sound synthesizer used for generating wide variety of musical and other sounds. Three-part article describes circuit operation and gives all other circuits used in synthesizer.—T. Orr and D. W. Thomas, Electronic Sound Synthesizer, *Wireless World,* Part 3—Oct. 1973, p 485–490 (Part 1—Aug. 1973, p 366–372; Part 2—Sept. 1973, p 429–434).

CHAPTER 58
Noise Circuits

Includes many types of noise limiters, blankers, and filters for audio, IF, RF, and digital applications, along with suppression of noise from arcing contacts and motors. Circuits for white-noise and pink-noise test-signal generators are also given.

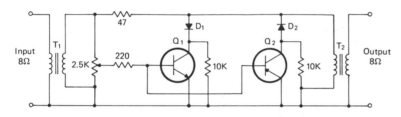

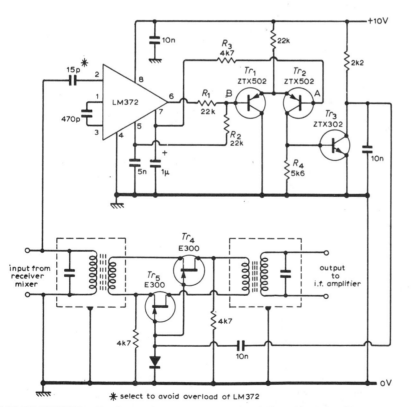

AF-POWERED CLIPPER—Designed for use just ahead of 8-ohm loudspeaker in receiver covering lower amateur phone bands (75 and 40 meters). Reduces hissing noise caused by short-wave diathermy, electric motors, and fluorescent lighting, as well as impulse noise generated by auto ignition system or atmospheric interference. T_1 and T_2 are transistor radio output transformers with 500:4 or 600:8 ohm impedance. Q_1 is 2N2222 NPN transistor. Q_2 is 2N2907 PNP transistor. D_1 and D_2 are 1N270.—C. Laster, An Audio Powered Noise Clipper, *CQ*, May 1976, p 26–27.

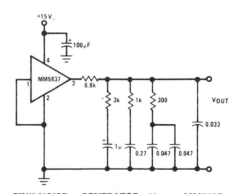

PINK-NOISE GENERATOR—Uses MM5837 broadband white-noise generator with −3 dB per octave filter from 10 Hz to 40 kHz to give pink-noise output having flat spectral distribution over entire audio band from 20 Hz to 20 kHz. Output is about 1 V P-P of pink noise riding on 8.5-VDC level. Used as controlled source of noise for adjusting octave equalizer to optimum settings for specific listening area.—"Audio Handbook," National Semiconductor, Santa Clara, CA, 1977, p 2-53–2-59.

AM NOISE SILENCER—Circuit samples mixer output (IF input) of AM receiver and, when noise pulse is detected, interrupts IF input signal for duration of noise pulse. Uses National LM372 IC having AGC loop with range of about 69 dB, for accommodating wide range of input levels. Article describes operation of circuit in detail. For frequencies above 2 MHz, use LM373 in place of LM372.—T. A. Tong, Noise Silencer for A.M. Receivers, *Wireless World*, Oct. 1972, p 483–484.

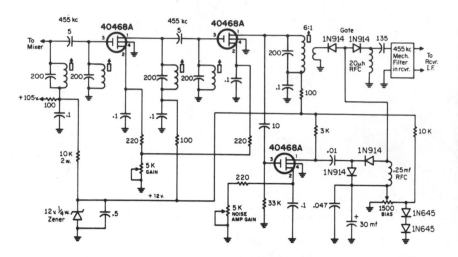

IF NOISE BLANKER—Used ahead of 455-kHz IF strip of communication receiver to provide about 40-dB attenuation of ignition and other noise pulses that can interfere with reception in 2- and 6-meter amateur bands. Two paths for noise pulses, one AC and the other DC, must be balanced for good operation. Resistor and capacitor values in noise rectifier are chosen to select sharp noise pulses in preference to signals. DC noise pulses are amplified by pulse amplifier and converted to AC noise pulses. Settings of pots are optimized for best noise blanking. Circuit requires 12-V supply, which can be obtained from receiver with appropriate dropping resistors and zener as shown for +105 V, or from separate source.—F. C. Jones, Experimental I.F. Noise Blankers, *CQ*, March 1971, p 81–83.

EXCESS-NOISE SOURCE—Develops about 18 dB of excess noise in region of 50–300 MHz for optimizing converter or receiver for best noise figure. Can also be used for noise optimizing of TV receivers and for peaking UHF TV front ends. Q1 and Q2 form cross-coupled 700-MHz MVBR. C1 is greater than C2 to favor conduction of Q2. When Q2 is on, Q3 turns on and makes current flow through broadband noise diode CR1. Diode is forward-biased because available gating voltage does not generate enough noise in reverse-bias mode. If noise output is too great, insert 2000-ohm attenuator as shown.—T. E. Hartson, A Gated Noise Source, *QST*, Jan. 1977, p 22–23.

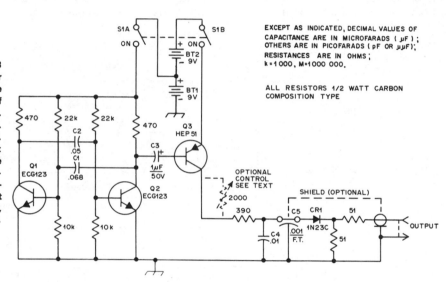

EXCEPT AS INDICATED, DECIMAL VALUES OF CAPACITANCE ARE IN MICROFARADS (μF); OTHERS ARE IN PICOFARADS (pF OR μμF); RESISTANCES ARE IN OHMS; k = 1 000, M = 1000 000.

ALL RESISTORS 1/2 WATT CARBON COMPOSITION TYPE

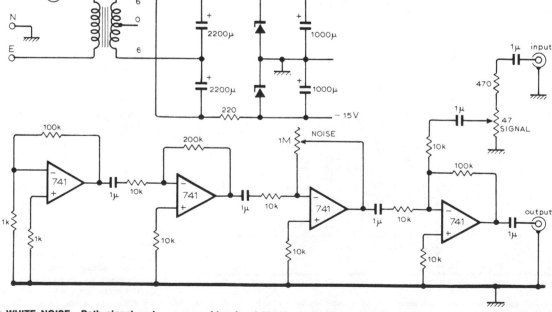

50–5000 Hz WHITE NOISE—Both signal and noise levels are continuously and independently variable from zero to maximum in simple noise generator developed to demonstrate recovery of low-level 500-Hz signal from noise. Circuit gives maximum noise output into 1500-ohm load; for lower load impedances, reduce noise level to prevent oscillation. Opamps require ±15 V supply, which can be simple voltage doubler without regulation.—J. E. Morris, Simple Noise Generator, *Wireless World*, April 1977, p 62.

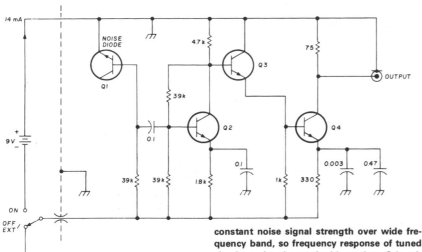

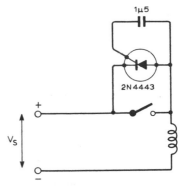

NOISE GENERATOR—Used with communication receiver to measure Q of tuned circuit without disconnecting circuit. Noise diode gives constant noise signal strength over wide frequency band, so frequency response of tuned circuit can be observed with receiver. Q is calculated after using S-meter of receiver to find −6 dB bandwidth. All transistors are 2N2368; select noisiest for Q1. Article covers test setup for measuring Q.—R. C. Marshall, Q Measurement and More, *Ham Radio*, Jan. 1977, p 49–51.

THYRISTOR SPARK QUENCHER—Suppresses arcing at contacts when switching large inductive loads. Use SCR capable of operating at well over twice supply voltage and passing full load current during switch-off. SCR shown will handle up to 500 V at up to 80 A provided current pulses are under 8 ms.—E. Potter, Switch Spark Quench for Inductive Loads, *Wireless World*, Dec. 1973, p 605.

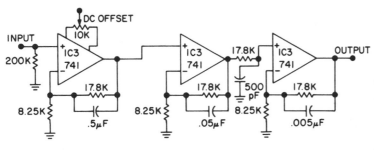

PINK-NOISE FILTER—Used in acoustics for measuring transducer characteristics, absorption-reflection and transmission coefficients of materials, and room parameters such as reverberation time. Offsets falloff of detected noise signal at low frequencies by using filter shown to convert random-noise source from constant energy per hertz (white-noise frequency spectrum) to constant energy per octave (pink-noise response). Filter covers audio range from 10 Hz to 20 kHz, providing −20 dB per decade transmission characteristics with three 741 opamp stages. Frequency characteristic is independent of source and load impedances. Supply voltages can be from ±6 to ±18 V.—R. Mauro, Simple Pink Noise Filter, *Audio*, March 1977, p 36 and 38.

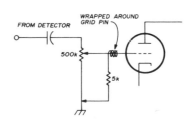

INPUT NOISE TEST CIRCUIT—Used for measuring noise immunity of emitter-coupled logic to transient signal on input line. Supply voltages used permit terminating logic outputs to ground through 50-ohm CRO probe. Accurate bias provided by power supply is used to set input logic levels. 450-ohm resistor is used in series with 50-ohm input of CRO to isolate input; this gives 10:1 amplitude attenuation while still providing accurate picture of input noise.—B. Blood, "AC Noise Immunity of MECL 10,000 Integrated Circuits," Motorola, Phoenix, AZ, 1972, AN-592.

CURE FOR NOISY CONTROL—Connecting 5K resistor between grid of first AF stage and ground as shown substantially reduces noise generated by worn volume-control pot in older tube-type communication receiver. Modification can be made from top of chassis by wrapping piece of wire around grid pin of audio tube, bringing wire up alongside tube and out through top of shield, then soldering 5K resistor between wire and chassis.—J. Schroeder, Temporary Fix for Noisy Volume Controls, *Ham Radio*, Aug. 1974, p 62.

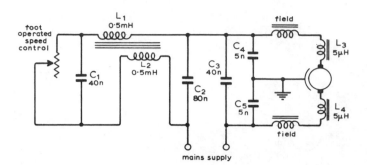

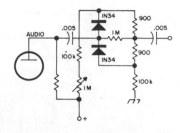

SEWING-MACHINE SUPPRESSION—Circuit is used to suppress clicks from speed control as well as interference produced by motor itself in sound and television broadcast bands.—A. S.

McLachlan, J. H. Ainley, and R. J. Harry, Radio Interference—a Review, *Wireless World,* June 1974, p 191—195.

AF NOISE LIMITER—Trough limiter eliminates background noise that is normally passed by conventional limiters, to permit use of higher volume level without annoying static when monitoring single radio channel continuously.—Circuits, 73 *Magazine,* Dec. 1973, p 120.

ZENER GENERATOR—Uses National LM389 array having three transistors along with opamp. Application of reverse voltage to emitter of one grounded-base transistor breaks it down in avalanche mode to give action of zener diode. Reverse voltage characteristic, typically 7.1 V, is used as noise source for amplification by second transistor and power opamp. Third transistor (not shown) can be used to gate noise generator if desired.—"Audio Handbook," National Semiconductor, Santa Clara, CA, 1977, p 4-33–4-37.

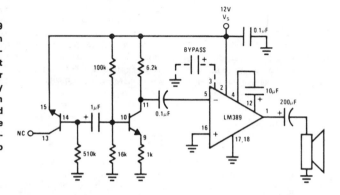

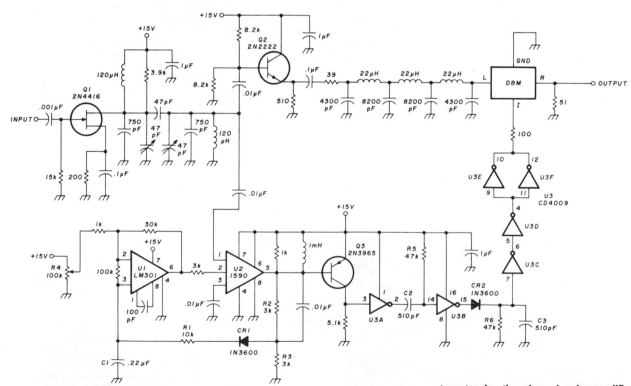

NOISE BLANKER—Minimizes effects of short-duration high-amplitude low-repetition-rate noise such as auto ignition noise, power-line arcing, and make-or-break switching. Developed for use in Collins ARR-41 receiver, where it is inserted between plate of second mixer and 500-kHz first IF amplifier. Q1 and its double-tuned drain circuit form low-gain bandpass amplifier that removes remaining local oscillator signal and sets bandwidth at about 50 kHz. Signal is then split into two channels. Q2 in main channel drives 50-ohm low-pass delay network with 700-kHz cutoff, feeding double-balanced mixer DBM operated as current-controlled attenuator. In other channel, noise amplifier U2 drives pulse detector Q3 and AGC detector CR1. Opamp U1 amplifies AGC and controls gain of U2. R4 is threshold adjustment. Gates U3B and U3C form mono used with gates of U3 to develop proper phase and current amplitude for operating blanking gate.—W. Stewart, Noise Blanker Design, *Ham Radio,* Nov. 1977, p 26—29.

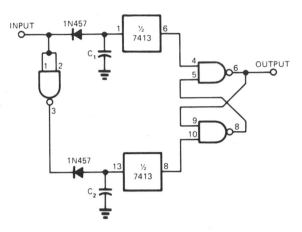

LOW-PASS DIGITAL FILTER—Used to retrieve pulse train data from noisy signal line. Filtering is achieved with SN7400 quad two-input NAND gate, SN7413 dual four-input Schmitt trigger, two diodes, and two capacitors. One gate of SN7400 is used as inverter driving pulse delay operating on negative-going transition of input signal. Other Schmitt trigger, diode, and capacitor provide delay on positive-going transition. Any additional pulses occurring during delay-circuit time-out resets delay time without affecting output.—T. H. Haydon, Low-Pass Digital Filter, *EDN Magazine,* Nov. 20, 1973, p 85.

$$C_1 = C_2 = \frac{480}{F + 200} \times 10^{-6}$$
$$F = \text{CUTOFF FREQUENCY}$$

UNUSED

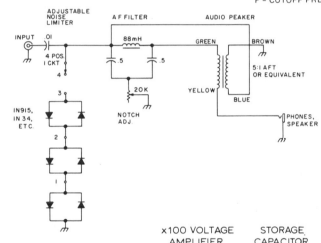

NOISE CONTROL—Circuit is plugged into headphone jack of amateur receiver. Four-position rotary switch selects desired combination of noise-limiting diodes for handling progressively more severe noise pulses. Adjustable AF T-notch filter limits passband over sufficient range for both phone and CW. Inductor is common 88-mH toroid. Audio peaker circuit overcomes insertion losses of filters.—S. T. Rappold, Noise Rejector, *73 Magazine,* Sept. 1977, p 116.

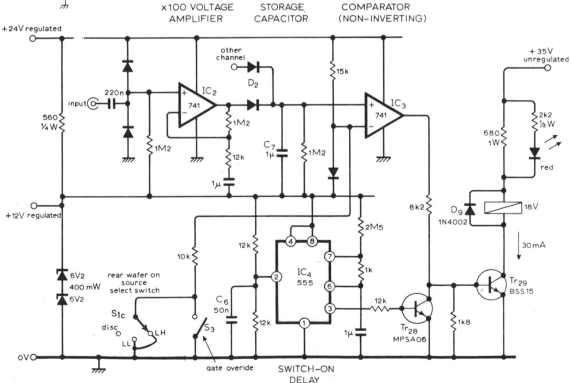

NOISE GATE FOR AF PREAMP—Used in high-performance phono preamplifier to mute output when there is no signal at phono input. Opamps each provide gains of about 100. Circuit controls muting reed relay serving both stereo channels of preamp. Delay switch-on using 555 IC overrides noise-gate opamps. Unmarked diodes are 1N914 or equivalent, and red LED is TIL209 or equivalent. Article covers circuit operation in detail and gives all other circuits of preamp.—D. Self, Advanced Preamplifier Design, *Wireless World,* Nov. 1976, p 41–46.

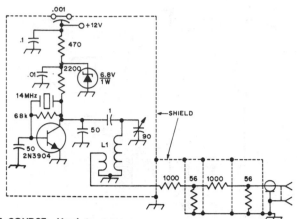

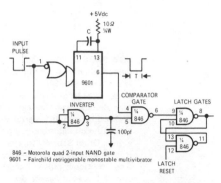

LOW-LEVEL RF SOURCE—Used to measure noise floor of receiver under test. RF source is simple, well-shielded crystal-controlled oscillator that is decoupled from battery supply. After attenuator resistors are adjusted to provide about S7 signal in receiver, oscillator housing is sealed with solder. Once calibrated, RF source is comparable to commercial signal generator, as leakage is quite low. Output is −112 dBm at 14 MHz. L1 has 24 turns enamel on Amidon T50-6 toroid, with 1 turn for output link.—W. Hayward, Defining and Measuring Receiver Dynamic Range, *QST*, July 1975, p 15–21 and 43.

NOISE-RESISTANT LATCH—False triggering of latch gates by noise spikes is prevented by generating pulse T whose width is equal to minimum width of desired input pulse. Values used for RC combination set T. If R is chosen as 10 kilohms, C should be T/3.424, where C is in picofarads and T is in nanoseconds.—S. R. Martin, Latching Circuit Provides Noise Immunity, *EDN/EEE Magazine*, Feb. 1, 1972, p 56.

AF NOISE LIMITER—Operation is similar to that of delay line. Voltage developed across voltage divider at output of 1N34 germanium diode is instantaneous, while DC voltage at output of circuit is delayed. If no pulses are present and 0.1-μF capacitor is not at ground, 1N914 silicon diode will have floating voltage. High positive pulses charge capacitor, and silicon diode shorts audio voltage. Negative pulses disable germanium diode directly. Circuit thus acts as noise blanker in both directions. Used in European communication receivers. Transistor type is not critical.—U. L. Rohde, IF Amplifier Design, *Ham Radio*, March 1977, p 10–21.

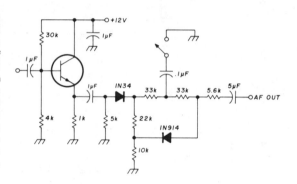

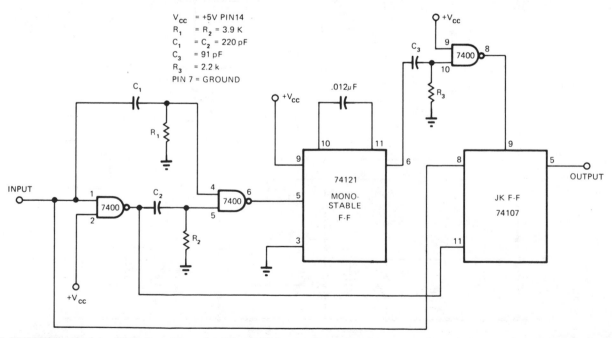

SPIKE REJECTION—Used to eliminate noise that may be present on signal line. Based on sampling input line at fixed time after each detected transition. If transition was due to noise spike, spike will no longer be present and true signal level will be sampled. If transition was caused by desired legitimate signal, sampled waveform represents true signal delayed by pulse width of mono MVBR. Mono pulse width is about 12 μs. Article gives circuit waveforms and describes operation in detail.—A. S. Bozorth, Pulse Verification Yields Good Noise Immunity, *EDN Magazine*, Nov. 5, 1973, p 75 and 77.

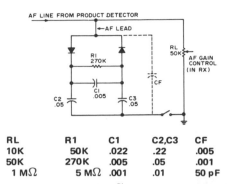

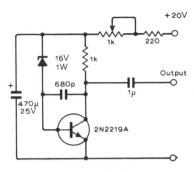

RL	R1	C1	C2,C3	CF
10K	50K	.022	.22	.005
50K	270K	.005	.05	.001
1 MΩ	5 MΩ	.001	.01	50 pF

AF NOISE LIMITER—Audio signals rectified by diodes develop bias across R1 and C1 such that diodes are back-biased. Diodes thus conduct and clip only when noise signal peaks exceed bias level. Component values depend on impedance of audio circuit; table gives values for three common load resistors. Diodes are fast-switching silicon such as 1N916. To minimize residual clipping distortion, use value for CF that gives 3-dB rolloff at about 2.5 kHz.—P. Lovelock, The Audio Bishop, *73 Magazine,* Sept. 1974, p 75–76.

EMERGENCY NOISE GENERATOR—Simple circuit generates noise in audio range at wideband level adjustable with 1K pot from 0 to over 1 V. If 680-pF capacitor is omitted, noise output goes up to 30 MHz with wideband level more than 5 V.—D. Di Mario, Simple Noise Generator, *Wireless World,* May 1978, p 70.

SSB CW NOISE LIMITER—Simple limiter is easy to install in receiver having good product detector. In place of dual-diode tube, semiconductor diodes having high front-to-back ratio may be used.—Novice Q & A, *73 Magazine,* Holiday issue 1976, p 20.

CHAPTER 59
Operational Amplifier Circuits

Versatility of modern opamps is illustrated by variety of amplification, control, signal processing, and other general-purpose functions involving frequencies ranging from DC to many megahertz. More specific applications will be found in practically all other chapters.

VARIABLE DEAD-BAND RESPONSE—Diode bridge in feedback loop of opamp provides controlled amount of dead-band response. As value of R_2 is increased from 0 ohms, voltage developed across R_2 serves to raise dead-band level at which bridge opens and circuit amplifies with normal gain of R_3/R_1. Below dead-band level, bridge is blocked and circuit gain is equal to parallel combination of R_2 and R_3 divided by R_1. Use matched diodes such as CA3019 for peaks below ±7 V; for higher peaks, use 1N914s.—W. G. Jung, "IC Op-Amp Cookbook," Howard W. Sams, Indianapolis, IN, 1974, p 207.

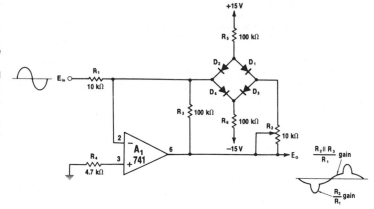

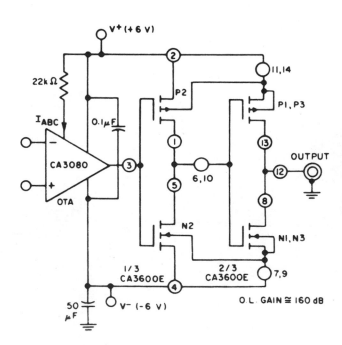

TWO-STAGE POSTAMPLIFIER—Connections shown for CA3600E CMOS transistor-pair array give total open-loop gain of about 160 dB for system. Open-loop slew rate is about 65 V/μs.—"Linear Integrated Circuits and MOS/FET's," RCA Solid State Division, Somerville, NJ, 1977, p 278–279.

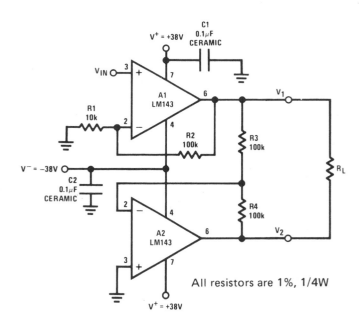

130 V P-P DRIVE—Two LM143 high-voltage opamps operating from 38-V supply can provide up to 138 V P-P unclipped into 10K floating load when connected as shown to give noninverting voltage amplifier followed by unity-gain inverter. Power supplies should be bypassed to ground with 0.1-μF capacitors.—"Linear Applications, Vol. 2," National Semiconductor, Santa Clara, CA, 1976, AN-127, p 1–3.

All resistors are 1%, 1/4W

BUFFERED OPAMP—NPD8301 dual FET is ideal low-offset low-drift buffer for LM101A opamp. Matched sections of FET track well over entire bias range, for improved common-mode rejection.—"FET Databook," National Semiconductor, Santa Clara, CA, 1977, p 6-26–6-36.

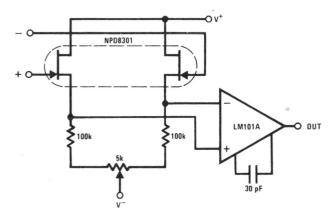

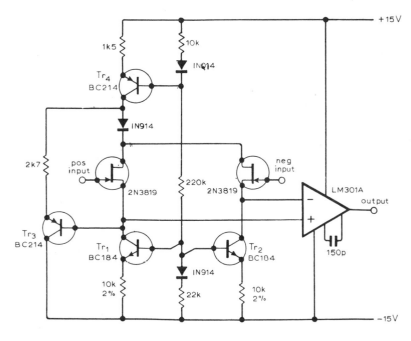

LOW-COST FET-INPUT—Uses two 2N3819 FETs as differential source-follower operated at constant source current of 200 μA provided by Tr_1 and Tr_2. Input performance is comparable to that of more expensive commercial units. Match FETs to reduce thermal drift. Trim input offset voltage to zero by adding resistor in appropriate FET source. Input impedance of circuit is greater than 10^{13} ohms.—J. Setton, F.E.T.-Input Operational Amplifier, *Wireless World*, Nov. 1976, p 61.

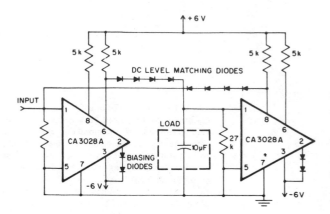

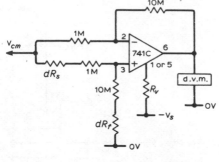

12-μH GYRATOR—Two RCA opamps in gyrator loaded with 10-μF capacitor give effective 12-μH inductor that remains constant in value over range from 10 Hz to almost 1 MHz. Q varies from 1 at 10 Hz to maximum of 500 at 10 kHz. Article gives design equations.—A. C. Caggiano, Simple Gyrator for L from C, *EEE Magazine,* Aug. 1970, p 78.

OPTIMIZING CMR—Article covers procedures for optimizing common-mode rejection when opamp is used to drive digital voltmeter. Value of R_v is determined by using resistance box connected between negative supply and pin 1 or 5 while other pin is shorted to negative supply, choosing pin which gives voltage swing in right direction on meter, then adjusting resistance box for zero output. Resistance box is similarly used at dR_s and dR_f locations.—R. J. Isaacs, Optimizing Op-Amps, *Wireless World,* April 1973, p 185–186.

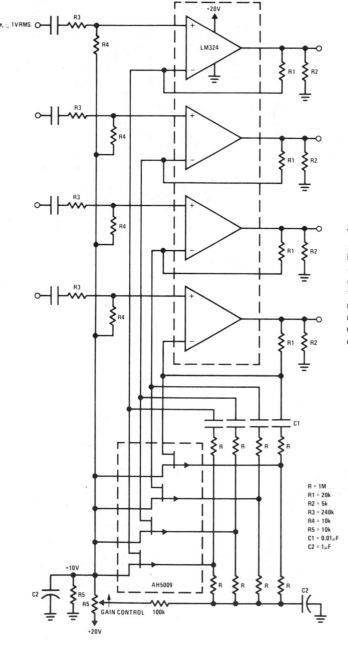

TRACKING QUAD GAIN CONTROL—Basic circuit for each channel uses section of National LM324 quad opamp with section of AH5009 quad FET in feedback path. Each channel is AC coupled and has 40-dB range (gain range of 1 to 100). Bandwidth is minimum of 10 kHz, and S/N ratio is better than 70 dB with 4.3-VRMS maximum output.—J. Sherwin, "A Linear Multiple Gain-Controlled Amplifier," National Semiconductor, Santa Clara, CA, 1975, AN-129, p 6.

R = 1M
R1 = 20k
R2 = 5k
R3 = 240k
R4 = 10k
R5 = 10k
C1 = 0.01μF
C2 = 1μF

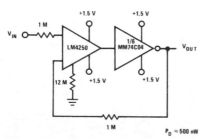

POSTAMPLIFIER FOR OPAMP—High input impedance of National MM74C04 inverter makes it ideal for isolating load from output of LM4250 micropower opamp operating from single dry cell.—"Linear Applications, Vol. 2," National Semiconductor, Santa Clara, CA, 1976, AN-88, p 2.

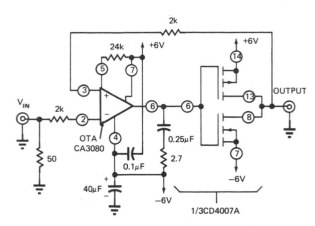

CMOS DRIVER FOR OPAMP—CMOS inverter pair (one-third of CD4007A) is used in closed-loop mode as unity-gain voltage follower for CA3080 opamp. Slew rate is 1 V/μs. Output current capability of 6 mA can be increased by par- alleling two other sections of CMOS.—B. Furlow, CMOS Gates in Linear Applications: The Results Are Surprisingly Good, *EDN Magazine,* March 5, 1973, p 42–48.

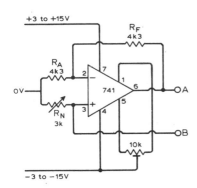

NEGATIVE R—Negative-resistance connection of 741 opamp is suitable for both AC and DC applications. Requires floating power supply because 0-V terminal floats with respect to both output terminals. For DC use, adjust 10K pot to cancel offset voltage of amplifier. Value of negative resistance is varied with R_N or by adjusting ratio of R_F to R_A. Can be used to make LC circuits operate at subaudio frequencies.—D. A. Miller, Negative Resistor, *Wireless World,* June 1974, p 197.

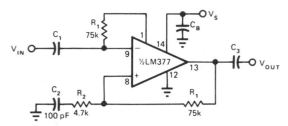

UNITY-GAIN AF CURRENT AMPLIFIER—External components are used with National LM377/378/379 family of opamps to provide stability at unity gain. Article gives design equations. At frequencies above audio band, gain rises with frequency, to well above 10 at 340 kHz for values shown.—D. Bohn, AC Unity-Gain Power Buffers Amplify Current, *EDN Magazine,* May 5, 1977, p 113–114.

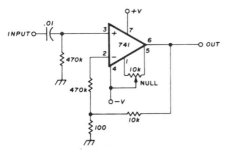

741 OPAMP—Power supply and null pot connections for TO-5 metal-can package and 8-lead DIP package are shown. Maximum rated power supply voltages are ±18 V, but lower voltages may be used. 9-V transistor battery is often used for each supply, but higher voltages will permit larger output signal swing. Pin 3 is inverting input, and pin 4 is noninverting input. With values shown, both input terminals see about same resistance, and output offset can be nulled to zero. Gain of circuit is about 100.—C. Hall, Circuit Design with the 741 Op Amp, *Ham Radio,* April 1976, p 26–29.

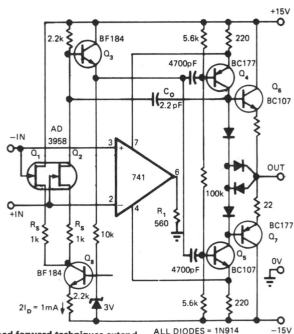

FASTER 741—Feed-forward techniques extend dynamic response of differential opamp to give unity-gain bandwidth of 18 MHz, slew rate over 200 V/μs, and DC gain above 10^7 V/V, while preserving latchup-free operation and wide input voltage range. Composite amplifier uses fast symmetrical four-transistor output stage that is symmetrically driven by DC-coupled 741 and by AC-coupled feed-forward amplifier. Performance depends on use of nonstandard pin connections for 741, as shown. Developed for processing fast analog data in frequency domain.—J. Dostal, 741 + Feedforward = Fast-Differential Op Amp, *EDN Magazine,* Aug. 20, 1974, p 90.

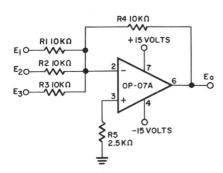

SUMMING AMPLIFIER—Provides output equal to sum of all input voltages, with high precision. Use of Precision Monolithics OP-07A opamp makes circuit adjustment-free.—"Ultra-Low Offset Voltage Op Amp," Precision Monolithics, Santa Clara, CA, 1977, OP-07, p 7.

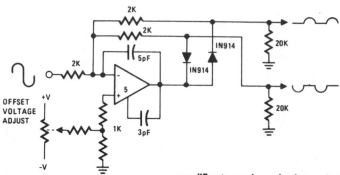

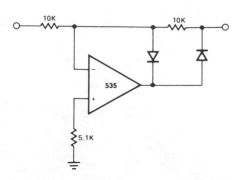

SIGNAL SEPARATOR—Circuit shown for Harris HA-2530 opamp separates input voltage into its positive and negative components. Diodes steer components to separate outputs. Applications include feeding outputs into differential amplifier to produce absolute-value circuit for multiplying or averaging functions. For bandwidth of 1 MHz, dynamic range is 100 mV to 10 V peak.—"Linear & Data Acquisition Products," Harris Semiconductor, Melbourne, FL, Vol. 1, 1977, p 7-54–7-55 (Application Note 516).

HALF-WAVE RECTIFIER—Provides accurate half-wave rectification of incoming signal. Gain is 0 for positive signals and −1 for negative signals. Diode types are not critical. Polarity can be inverted by reversing both diodes. With opamp shown, circuit will function up to 10 kHz with less than 5% distortion.—"Signetics Analog Data Manual," Signetics, Sunnyvale, CA, 1977, p 641–643.

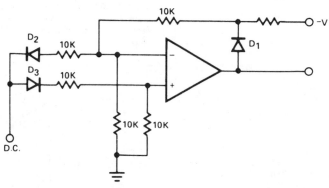

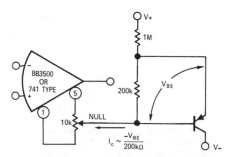

ABSOLUTE-VALUE AMPLIFIER—Generates positive output voltage for either polarity of DC input. Opamp and diode types are not critical. Accuracy is highest for input voltages greater than 1 V. Opamp is noninverting on positive signals and inverting on negative signals.—"Signetics Analog Data Manual," Signetics, Sunnyvale, CA, 1977, p 641–643.

TEMPERATURE-COMPENSATED OFFSET CONTROL—Drift effects of offset adjustment are removed by deriving correction current from emitter-base voltage of PNP signal transistor to develop appropriate temperature compensation. Correction current is divided with conventional control pot used for adjusting offset voltage. Article gives design equations.—J. Graeme, Offset Null Techniques Increase Op Amp Drift, *EDN Magazine*, April 1, 1971, p 47–48.

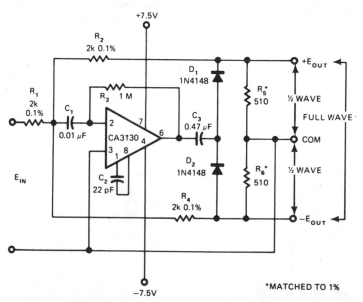

*MATCHED TO 1%

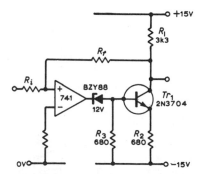

RECTIFIER WITHOUT DC OFFSET—Avoids drawback of large nonlinearity at low signal levels, by isolating AC of opamp from DC output. Circuit has wide bandwidth, as required for rectifying 20-kHz input signal with high precision. Output coupling capacitor C_3 is low-leakage Mylar; for low-frequency operation, it can be replaced with two back-to-back low-leakage tantalums. D_1 and D_2 should be matched for forward voltage at peak load current. Use Hewlett-Packard 5082-2810 hot-carrier diodes instead to improve operation at millivolt signal levels or at higher frequencies.—D. Belanger, Single Op Amp Full-Wave Rectifier Has No DC Offset, *EDN Magazine*, April 5, 1977, p 144 and 146.

FASTER SLEWING—Single transistor stage at output of opamp increases slewing rate by factor equal to gain of transistor stage. Choose R_1 to meet output impedance requirements and current rating of supply. R_2 is then made equal to R_1 divided by desired gain of transistor stage. Collector of Tr_1 should be at 0 V when output of opamp is 0 V, assuming feedback loop is not closed by R_f. Article gives design equations.—L. Short, Faster Slewing Rate with 741 Op-Amp, *Wireless World*, Jan. 1973, p 31.

SIGNAL CONDITIONER—FET-buffered opamp circuit will operate from source impedances up to 100 megohms while providing voltage gain of 5. Offset adjustment is provided for initial calibration of circuit. Developed for use with high-impedance sensors such as pH electrodes.—"Industrial Control Engineering Bulletin," Motorola, Phoenix, AZ, 1973, EB-4.

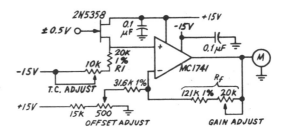

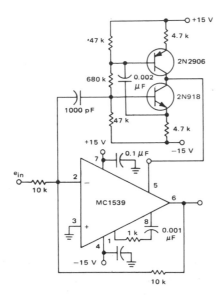

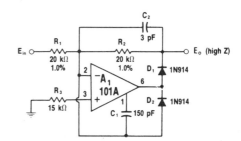

HIGH-SPEED HALF-WAVE RECTIFIER—Produces inverted half-wave replica of input signal with low error at frequencies up to 100 kHz. C_1 provides feed-forward compensation. For negative-going output, reverse connections to diodes.—W. G. Jung, "IC Op-Amp Cookbook," Howard W. Sams, Indianapolis, IN, 1974, p 191–192.

UNITY-GAIN FEED-FORWARD—Provides 10 V P-P output signal at 2 MHz when gain of feed-forward amplifier is increased to give closed-loop gain of 10. Provides fast response to step-function input, with slow settling. High-frequency circuit takes over completely when input frequency is too high for input stage to respond.—E. Renschler, "The MC1539 Operational Amplifier and Its Applications," Motorola, Phoenix, AZ, 1974, AN-439, p 20.

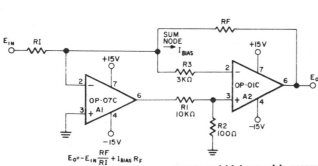

$$E_0 = -E_{IN}\frac{R_F}{RI} + I_{BIAS} R_F$$

SUMMING AMPLIFIER—Combination of Precision Monolithics OP-07C and OP-01C opamps gives 18 V/μs slew rate. Can be used as current-output summing amplifier for D/A converter because it requires no zero scale offset adjustments and high speed is preserved.—D. Soderquist and G. Erdi, "The OP-07 Ultra-Low Offset Voltage Op Amp—a Bipolar Op Amp That Challenges Choppers, Eliminates Nulling," Precision Monolithics, Santa Clara, CA, 1975, AN-13, p 9.

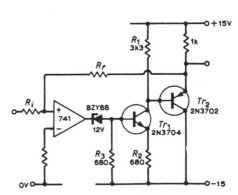

FAST SLEWING AND LOW IMPEDANCE—With values shown, Tr_1 increases slewing rate of opamp by factor of 5, and Tr_2 connected as emitter-follower reduces output impedance to meet requirements of following circuit. Feedback is taken from emitter of Tr_1 to noninverting input of opamp.—L. Short, Faster Slewing Rate with 741 Op-Amp, Wireless World, Jan. 1973, p 31.

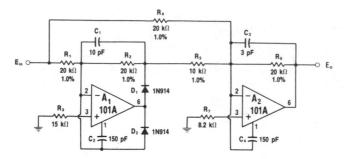

100-kHz FULL-WAVE RECTIFIER—Feed-forward connection of opamps gives high-speed full-wave rectification of signals up to 100 kHz for measurement and analysis.—W. G. Jung, "IC Op-Amp Cookbook," Howard W. Sams, Indianapolis, IN, 1974, p 193–194.

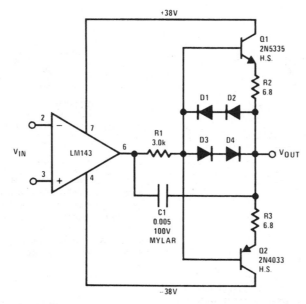

100-mA CURRENT BOOSTER—Provides short-circuit protection along with current boosting for LM143 high-voltage opamp. Diodes are 1N914. Use Thermalloy 2230-5 or equivalent heatsinks with transistors. Output is ±33 V P-P into 400-ohm load.—"Linear Applications, Vol. 2," National Semiconductor, Santa Clara, CA, 1976, AN-127, p 4.

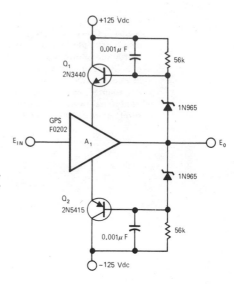

BOOSTING VOLTAGE RATING—Bootstrapping technique permits operation of low-voltage unity-gain opamp from high-voltage DC supply for handling large input signal voltage swings, while retaining gain and voltage stability of opamp. Allowable input-voltage range depends entirely on transistor rating. With 1000-V transistors, circuit can handle input signals of ±475 V. Input capability for values shown is ±100 V P-P for DC to 10 kHz. Output capability is 5 mA P at ±100 V. Input impedance is 10 teraohms.—S. A. Jensen, High-Voltage Source Follower, *EDN/EEE Magazine*, Feb. 1, 1972, p 58.

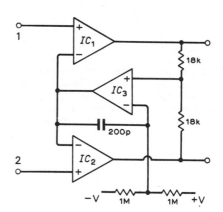

DIFFERENTIAL I/O—Arrangement shown for three 741 opamps gives amplifier having differential output as well as differential input. Circuit is designed primarily to drive meter with signal of either polarity when center-tap power supply is not available. Article covers operation and adjustment of circuit.—A. D. Monstall, Differential Input and Output with Op-Amps, *Wireless World*, Jan. 1973, p 31.

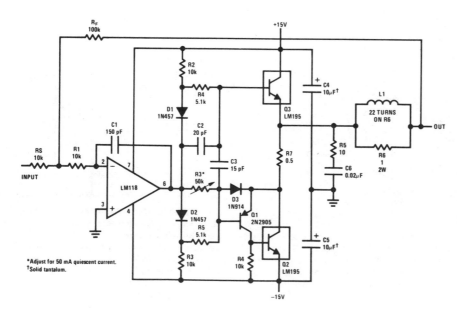

*Adjust for 50 mA quiescent current.
†Solid tantalum.

POWER OPAMP—Transistor Q1 and power transistor IC Q2 form equivalent of power PNP transistor for use with NPN LM195 power transistor IC serving as output stage for opamp. Circuit is stable for almost any load. Bandwidth can be increased to 150 kHz with full output response by decreasing C1 to 15 pF if there is no capacitive load to cause oscillation.—"Linear Applications, Vol. 2," National Semiconductor, Santa Clara, CA, 1976, AN-110, p 5–6.

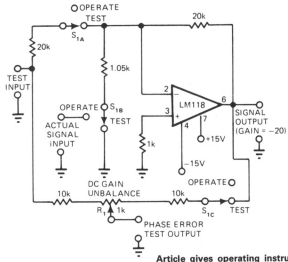

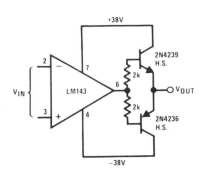

PHASE-ERROR TESTER—Circuit reveals significant phase errors at relatively low frequencies, even for high-speed opamps. Technique applies to most opamps and almost any signal gain. Article gives operating instructions based on observation of null with XY CRO connected to phase-error test output.—R. A. Pease, Technique Trims Op-Amp Amplifiers for Low Phase Shift, *EDN Magazine*, Aug. 20, 1977, p 138.

POWER BOOSTER—Simple two-transistor power stage increases power output of LM143 high-voltage opamp. Intended for loads less than 2K. Drawbacks are noticeable crossover distortion and lack of short-circuit protection. Transistors should be used with Thermalloy 2230-5 or equivalent heatsinks.—"Linear Applications, Vol. 2," National Semiconductor, Santa Clara, CA, 1976, AN-127, p 3.

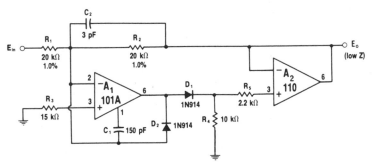

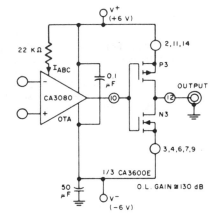

100-kHz BUFFERED RECTIFIER—High-speed 110 voltage follower is used within feedback loop of A_1 to maintain low output impedance for precision half-wave rectifier. When input signal is positive, D_1 and R_4 rectify signal and A_2 follows this signal. On opposite alternations, D_1 is off and feedback loop of A_2 is closed through D_2 so output terminal is maintained at low impedance. For opposite output polarity, reverse diode connections.—W. G. Jung, "IC Op-Amp Cookbook," Howard W. Sams, Indianapolis, IN, 1974, p 192.

POSTAMPLIFIER—CMOS transistor pair from CA3600E transistor array provides additional 30-dB gain above 100-dB gain of CA3080 opamp to give total of 130 dB. Current output is about 10 mA. Remaining transistor pairs of array can be paralleled pair shown to give greater output.—"Linear Integrated Circuits and MOS/FET's," RCA Solid State Division, Somerville, NJ, 1977, p 278–279.

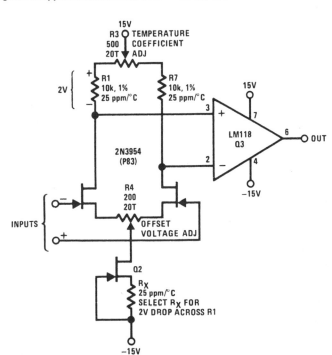

LOW TEMPERATURE COEFFICIENT—Use of National 2N3954 dual FET as input device for opamp gives fast response to thermal transients, making it possible to adjust R3 and R4 so temperature coefficient is less than 5 μV/°C from −25°C to +85°C. Common-mode rejection ratio is typically greater than 100 dB for input voltage swings of 5 V. Drain current level is set by Q2 which is 2N5457 FET.—"FET Databook," National Semiconductor, Santa Clara, CA, 1977, p 6-4–6-7.

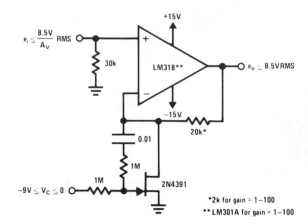

1–1000 GAIN RANGE—Control voltage of 0 to −9 V changes gain of amplifier over complete range while providing maximum output level of 8.5 VRMS and bandwidth of over 20 kHz at maximum gain. If gain range of 100 is sufficient, amplifier can be changed to LM301A; 20K resistor is then changed to 2K.—"Linear Applications, Vol. 2," National Semiconductor, Santa Clara, CA, 1976, AN-129, p 5.

JFET INPUT—U401 dual JFET acting as preamp for standard bipolar opamp uses CR033 N-channel JFET as 330-μA current source. R4 is used to null initial offset. R3 is adjusted for minimum drift.—"Analog Switches and Their Applications," Siliconix, Santa Clara, CA, 1976, p 7-51.

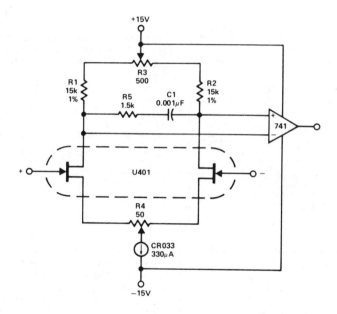

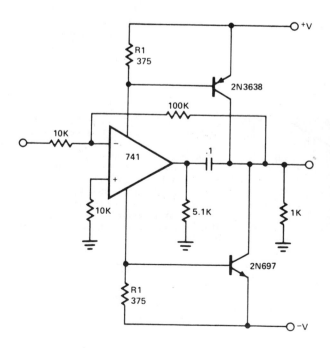

POWER BOOSTER—Opamp power booster is used after conventional opamp when greater power-handling capability is required. 741 opamp circuit shown will drive moderate loads. Other opamps may be substituted in power stage if value of R1 is appropriately changed.—"Signetics Analog Data Manual," Signetics, Sunnyvale, CA, 1977, p 640–642.

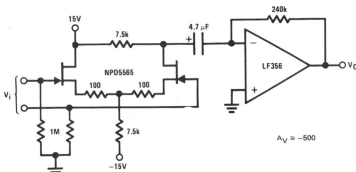

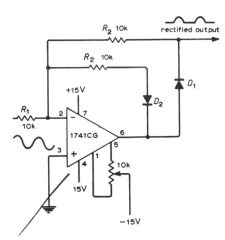

DIFFERENTIAL JFET INPUT—Differential connection of National NPD5565 dual JFET is used when balanced inputs and low distortion are main requirements for AC amplifier. Combination with LF356 opamp shown gives gain of about 500. Noise is somewhat higher than with single-ended JFET.—"FET Databook," National Semiconductor, Santa Clara, CA, 1977, p 6-17–6-19.

PRECISE RECTIFICATION—Use of opamp in combination with silicon diode overcomes nonlinearity of diode at forward voltages under about 0.5 V. Offset-voltage pot is adjusted for symmetrical output waveform for small input voltages. D_1 is connected in opamp feedback path so initial forward voltage drop required to make diode conduct is supplied by amplifier output. Second feedback path through D_2 prevents output saturation on input half-cycles for which D_1 is reverse-biased.—G. B. Clayton, Experiments with Operational Amplifiers, *Wireless World*, June 1973, p 275–276.

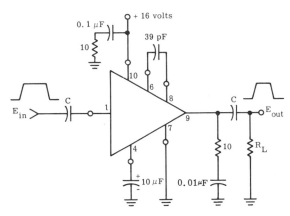

3-W PULSE AMPLIFIER—Motorola MC1554 power amplifier provides voltage gain of 18 for peak pulse power output up to 3 W. Maximum peak output current rating of 500 mA for IC should not be exceeded during peak of output pulse.—"The MC1554 One-Watt Monolithic Integrated Circuit Power Amplifier," Motorola, Phoenix, AZ, 1972, AN-401, p 3.

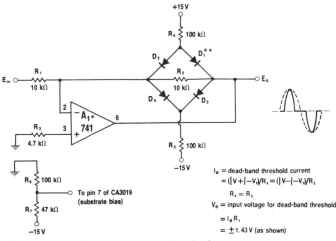

I_B = dead-band threshold current
$$= (|V+|-V_b)/R_4 = (|V-|-V_b)/R_5$$
$$R_4 = R_5$$
V_B = input voltage for dead-band threshold
$$= I_B R_1$$
$$= \pm 1.43 \text{ V (as shown)}$$

* Or other op amp compensated for unity gain.

** D_1 – D_4 are matched monolithic diodes, such as the CA3019.
For peak voltage higher than ±7 V, use 1N914s.

DEAD-BAND RESPONSE—With bridge in feedback loop of opamp, low-level input signals give essentially 100% feedback around A_1 so there is very little output voltage. When input current through R_1 rises above allowable current limit of circuit, bridge opens and output voltage jumps to new level determined by R_2. Input is then amplified by ratio of R_2/R_1 in normal linear manner. Circuit thus has dead-band property for low levels. Value of R_1 sets threshold level.—W. G. Jung, "IC Op-Amp Cookbook," Howard W. Sams, Indianapolis, IN, 1974, p 206–207.

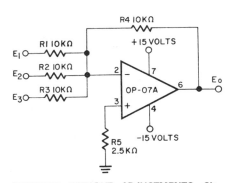

SUMMING WITHOUT ADJUSTMENTS—Single-stage opamp for analog computation provides high-precision output that is function of multiple input variables. Circuit drift is less than 2 μV per month, eliminating need for periodic calibration while ensuring long-term accuracy. Opamp is Precision Monolithics OP-07A.—D. Soderquist and G. Erdi, "The OP-07 Ultra-Low Offset Voltage Op Amp—a Bipolar Op Amp That Challenges Choppers, Eliminates Nulling," Precision Monolithics, Santa Clara, CA, 1975, AN-13, p 11.

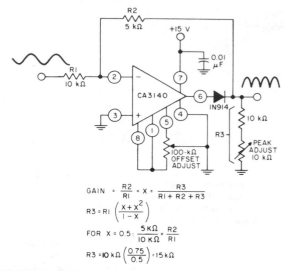

$$\text{GAIN} = \frac{R2}{R1} = X = \frac{R3}{R1 + R2 + R3}$$

$$R3 = R1\left(\frac{X + X^2}{1 - X}\right)$$

$$\text{FOR } X = 0.5: \frac{5K\Omega}{10K\Omega} = \frac{R2}{R1}$$

$$R3 = 10\,k\Omega\left(\frac{0.75}{0.5}\right) = 15\,k\Omega$$

ABSOLUTE-VALUE RECTIFIER—Use of CA3140 bipolar MOS opamp in inverting gain configuration gives symmetrical full-wave output when equality of design equations is satisfied. Bandwidth for −3 dB is 290 kHz, and average DC output is 3.2 V for 20 V P-P input.—"Circuit Ideas for RCA Linear ICs," RCA Solid State Division, Somerville, NJ, 1977, p 18.

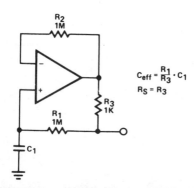

SIGN CHANGER—When switch S_1 grounds pin 3 of opamp, circuit becomes inverter providing 180° phase shift. When S_1 is at position A, input voltage acts on both inputs of A_1 and no current flows through R_1 and R_2; output voltage is then equal to input voltage. Switch permits remote programming of phase reversal. For higher input impedance, 1556 opamp can be used.— W. G. Jung, "IC Op-Amp Cookbook," Howard W. Sams, Indianapolis, IN, 1974, p 208–209.

SINGLE-ENDED JFET—Basic JFET amplifier is virtually free from popcorn noise problems of bipolar transistors and bipolar-input opamps. Combining JFET transconductance amplifier with current-to-voltage opamp adds high voltage gain and simplifies circuit applications. Gain-limiting 7.5K FET drain resistor is bypassed and removed from gain equation. Parameter variation problems are minimized by biasing FET source through 15.1K resistance to negative supply. Gain variations are minimized by leaving 100 ohms of this resistance unbypassed.—J. Maxwell, FET Amplifiers—Take Another Look at These Devices, *EDN Magazine*, Sept. 5, 1977, p 161–163.

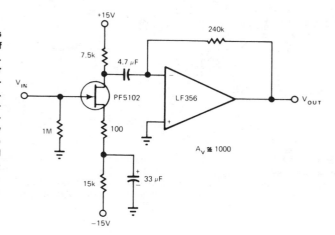

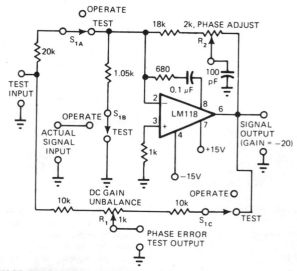

MINIMIZING PHASE ERROR—Phase compensation circuit trimmed by R_2 keeps phase error of LM118 opamp well below 1° from DC to 200 kHz. In-phase error due to gain peaking is also low. Feed-forward network connected to pin 8 improves stability, making feedback capacitor unnecessary. Step response has about 30% overshoot, and sine response has about +1 dB of peaking before going 3 dB down at about 2 MHz.—R. A. Pease, Technique Trims Op-Amp Amplifiers for Low Phase Shift, *EDN Magazine*, Aug. 20, 1977, p 138.

$$C_{eff} = \frac{R_1}{R_3} \cdot C_1$$

$$R_S = R_3$$

CAPACITANCE MULTIPLIER—Resistance ratio determines factor by which value of C_1 is multiplied when used in simple opamp circuit shown. With values shown, ratio is 1000 and 10-μF capacitor provides effective capacitance of 10,000 μF. Q of circuit is limited by effective series resistance, so R_1 should be as large as practical. Opamp type is not critical.—"Signetics Analog Data Manual," Signetics, Sunnyvale, CA, 1977, p 640–641.

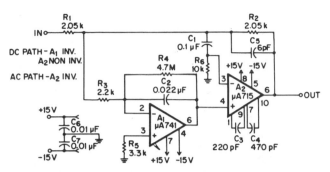

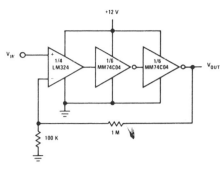

FEED-FORWARD OPAMP—DC input characteristics are determined by A_1, which is bypassed at high frequencies, while AC-coupled A_2 determines dynamic performance. Resulting composite amplifier combines such desired properties as low input current and drift, large bandwidth and slew rate, and fast settling time. Compensation network C_3-C_4-C_5 is chosen first to give desired bandwidth. Composite rolloff of 6 dB per octave is then obtained by narrowbanding A_1 with R_4 and C_2, so gain-bandwidth product is equal to ratio between unity-gain crossover frequency of A_2 and open-loop gain.—Fairchild Linear IC Contest Winners, *EEE Magazine,* Jan. 1971, p 48–49.

SINGLE-SUPPLY POSTAMPLIFIER—Use of two sections of MM74C04 as postamplifier for LM324 single-supply amplifier gives open-loop gain of about 160 dB. Additional CMOS inverter sections can be paralleled for increased power to drive higher current loads; each MM74C04 section is rated for 5-mA load.—"Linear Application, Vol. 2," National Semiconductor, Santa Clara, CA, 1976, AN-88, p 2.

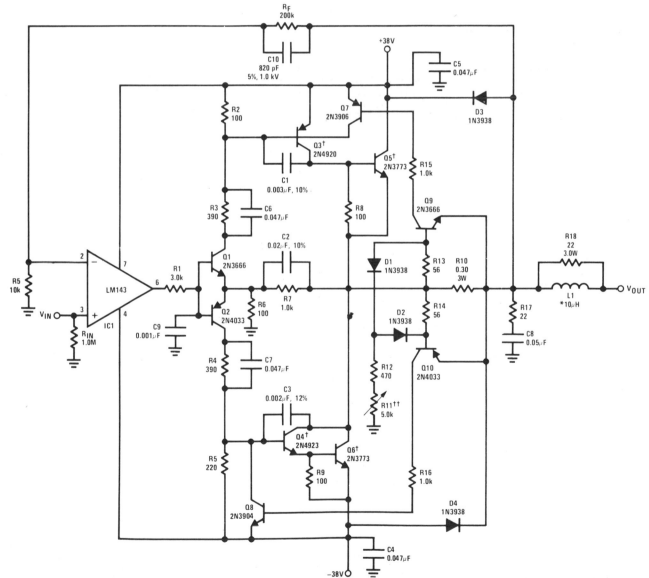

8-A CURRENT BOOSTER—High-compliance power stage for LM143 high-voltage opamp provides very high peak drive currents along with output voltage swings to within 4 V of ±38 V supply under full load. Maximum output current depends on setting of current-adjusting pot R11 and on output voltage. Limit ranges from 14 A when R11 is 0 down to about 4 A for 5K. Maximum power output is 144 WRMS, for which frequency response is 3 dB down at 10 kHz. Voltage gain is 21. Q3-Q6 should be on common Thermalloy 6006B or equivalent heatsink.—"Linear Applications, Vol. 2," National Semiconductor, Santa Clara, CA, 1976, AN-127, p 5–6.

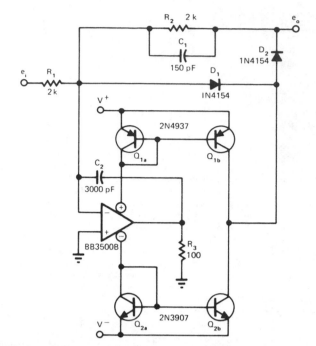

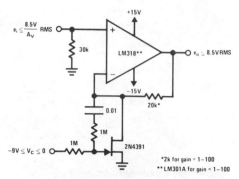

60-kHz PRECISION RECTIFIER—Usable full-power response of typical opamp is boosted to 60 kHz while giving 300-kHz small-signal bandwidth. Circuit uses transistors to provide speed-boosting gain during transition from one precision rectifier diode to the other in feedback loop of opamp. Added stage is driven from power-supply current drains of opamp. Article traces operation of circuit in detail.—J. Graeme, Boost Precision Rectifier BW above That of Op Amp Used, *EDN Magazine*, July 5, 1974, p 67–69.

GAIN-CONTROLLED AMPLIFIER—Control voltage in range of 0 to −9 V provides gain range of 1 to 1000 for National LM318 opamp using FET in feedback path. Bandwidth is better than 20 kHz at maximum gain. Applications include remote or multichannel gain control, volume expansion, and volume compression/limiting.—J. Sherwin, "A Linear Multiple Gain-Controlled Amplifier," National Semiconductor, Santa Clara, CA, 1975, AN-129, p 5.

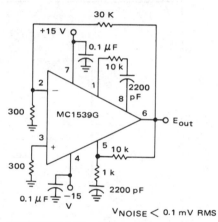

$V_{NOISE} < 0.1$ mV RMS

LOW-NOISE 5-kHz—Values shown are for operation of Motorola MC1539G opamp in closed-loop mode with noninverting gain of 100 and source impedance of about 300 ohms. Circuit bandwidth is about 5 kHz.—E. Renschler, "The MC1539 Operational Amplifier and Its Applications," Motorola, Phoenix, AZ, 1974, AN-439, p 19.

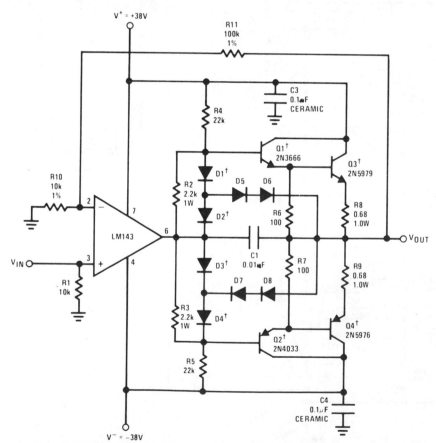

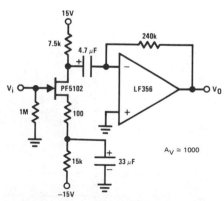

$A_V \cong 1000$

1-A CURRENT BOOSTER—Used with LM143 high-voltage opamp to increase output current while providing short-circuit protection and low crossover distortion. With 40-ohm load, output voltage can swing to + 29.6 V and −28 V. All four transistors should be on Thermalloy 6006B or equivalent common heatsink. All diodes are 1N3193.—"Linear Applications, Vol. 2," National Semiconductor, Santa Clara, CA, 1976, AN-127, p 4–5.

FET DRIVE—National PF5102 JFET is combined with LF356 opamp to give low noise and high gain, for use as wide-bandwidth AC amplifier. Typical gain for combination shown is about 1000. Any other opamp can be used as long as it meets slew rate and bandwidth requirements.—"FET Databook," National Semiconductor, Santa Clara, CA, 1977, p 6-17–6-19.

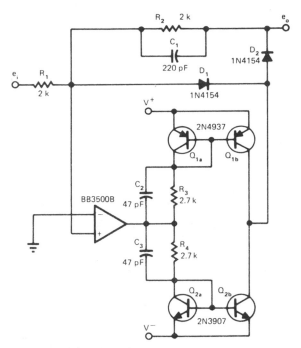

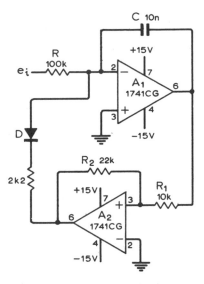

PRECISION RECTIFIER WITH GAIN—Gain is selectively added during open-loop switching transition of precision rectifier diodes D_1 and D_2 in feedback loop of opamp, to boost speed while maintaining feedback stability following switching. Q_1 and Q_2 add gain of about 250 up to 30 kHz during switching, because D_1 and D_2 are then off and do not shunt output of added stage. Following transition, one of diodes conducts heavily, shunting high output impedance of stage and dropping its gain below unity. Article covers circuit operation in detail.—J. Graeme, Boost Precision Rectifier BW above That of Op Amp Used, *EDN Magazine*, July 5, 1974, p 67–69.

VOLTAGE/FREQUENCY CONVERTER—Uses opamp A_1 as integrator and A_2 as regenerative comparator with hysteresis, to generate sequence of pulses with repetition frequency proportional to DC input voltage. Article gives design equations and typical waveforms. Input voltage range is 10 mV to 20 V for linear operation.—G. B. Clayton, Experiments with Operational Amplifiers, *Wireless World*, Dec. 1973, p 582.

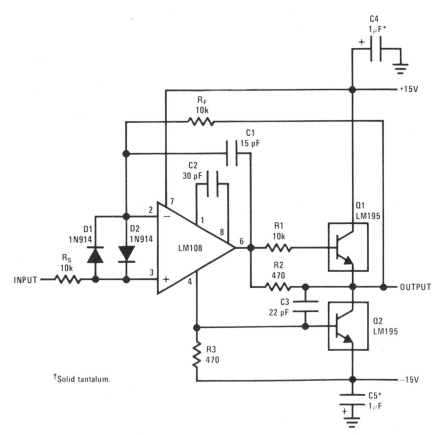

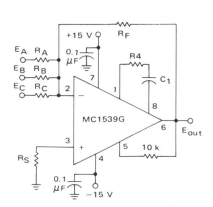

300-Hz VOLTAGE FOLLOWER—Simple LM195 power output stage provides 1-A output for voltage-follower connection of LM108 opamp.—R. Dobkin, "Fast IC Power Transistor with Thermal Protection," National Semiconductor, Santa Clara, CA, 1974, AN-110, p 6.

SUMMING OPAMP—Motorola MC1539 serves as closed-loop summing amplifier having very small loop-gain error because of high open-loop gain. R_S should equal parallel combination of R_A, R_B, R_C, and R_F.—E. Renschler, "The MC1539 Operational Amplifier and Its Applications," Motorola, Phoenix, AZ, 1974, AN-439, p 18.

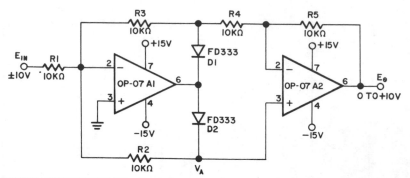

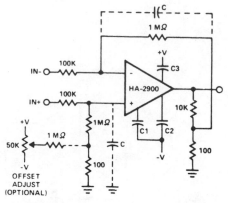

PRECISION ABSOLUTE VALUE—Circuit using two Precision Monolithics OP-07 opamps provides precise full-wave rectification by inverting negative-polarity input voltages and operating as unity-gain buffer for positive-polarity inputs. Applications include positive-peak detectors, single-quadrant multipliers, and magnitude-only measuring systems. For positive inputs, circuit simply operates as two unity-gain am-plifier stages. Negative input turns D1 off and D2 on, changing resistor currents precisely enough to give overall circuit gain of −1. Design equations are given.—D. Soderquist and G. Erdi, "The OP-07 Ultra-Low Offset Voltage Op Amp—a Bipolar Op Amp That Challenges Choppers, Eliminates Nulling," Precision Monolithics, Santa Clara, CA, 1975, AN-13, p 10.

1000 GAIN AT 2 kHz—Uses Harris HA-2900 chopper-stabilized opamp. Either input terminal may be grounded, giving choice of inverting or noninverting operation, or inputs may be driven differentially. Symmetrical input networks eliminate chopper noise, limiting total input noise to about 30 μVRMS when C is 0. Noise can be further reduced, at expense of bandwidth, by adding optional capacitors C as shown. Without these capacitors, bandwidth is 2 kHz.—"Linear & Data Acquisition Products," Harris Semiconductor, Melbourne, FL, Vol. 1, 1977, p 7-69 (Application Note 518).

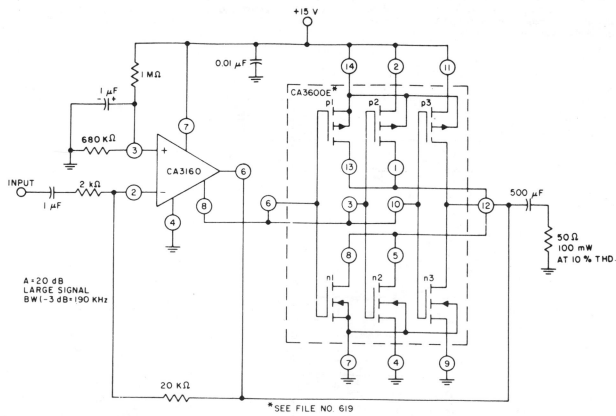

NOTE:
TRANSISTORS p1, p2, p3 AND n1, n2, n3 ARE PARALLEL-CONNECTED WITH Q8 AND Q12, RESPECTIVELY, OF THE CA3160

POWER BOOSTER—CA3600E CMOS transistor array provides parallel-connected transistors for power-boosting capability with CA3160 opamp. Feedback is used to establish closed-loop gain of 20 dB. Typical large-signal bandwidth (−3 dB) is 190 kHz.—"Linear Integrated Circuits and MOS/FET's," RCA Solid State Division, Somerville, NJ, 1977, p 271–273.

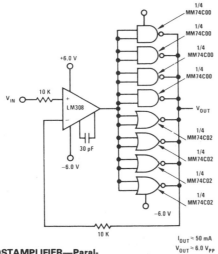

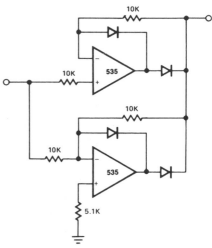

COMPLEMENTARY POSTAMPLIFIER—Paralleled NAND gates provide buffering for LM308 opamp while increasing current drive to about 50 mA for 6 V P-P output. MM74C00 NAND gates supply about 10 mA each from positive supply while MM74C02 gates supply same amount from negative supply.—"Linear Applications, Vol. 2," National Semiconductor, Santa Clara, CA, 1976, AN-88, p 2–3.

FULL-WAVE RECTIFIER—Circuit provides accurate full-wave rectification of input signal, with distortion below 5% up to 10 kHz. Reversal of all diode polarities reverses polarity of output. Output impedance is low for both input polarities, and errors are small at all signal levels.—"Signetics Analog Data Manual," Signetics, Sunnyvale, CA, 1977, p 641–643.

GUARDED FULL-DIFFERENTIAL—Extremely high input impedance is achieved by intercepting leakage currents with guard conductor placed in leakage path and operated at same voltage as inputs. A2 serves as guard drive amplifier, with R5 and R6 developing proper voltage for guard at their junction. R7 balances detector R5 plus R6 without degrading closed-loop common-mode rejection.—"Linear Applications, Vol. 1," National Semiconductor, Santa Clara, CA, 1973, AN-63, p 1–12.

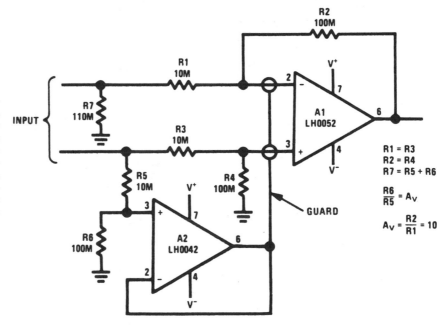

CHAPTER 60

Optoelectronic Circuits

Basic voltage-isolating applications for optoisolators. Includes bar-code reader circuits. Other chapters may include optoisolators in circuits having specific applications.

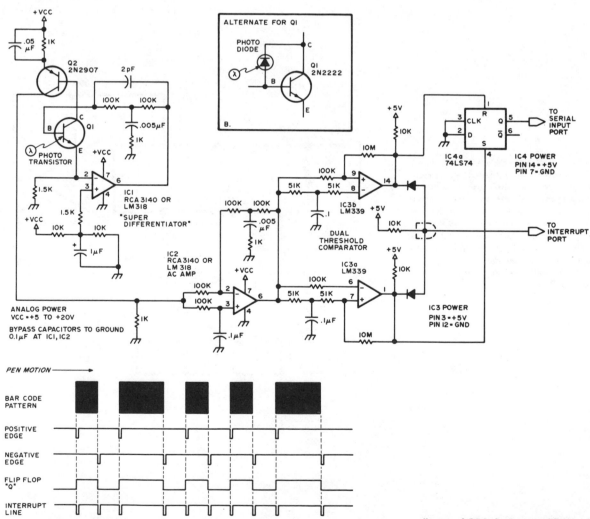

BAR-CODE READER—Edge-sensitive circuit outputs short pulses at each black-to-white or white-to-black transition. Timing diagram shows outputs corresponding to bar-code pattern indicated. Direct-current level at base of Q1 is held constant by DC servo action despite changes in temperature, ambient light, or background of pattern. Alternate sensor uses photodiode and 2N2222 transistor for increased bandwidth. Amplified differentiated signal from collector of Q2 is further amplified by IC2 and fed to dual threshold comparator. Output of comparator is short pulse for each transition, suitable for feed to microprocessor.—F. L. Merkowitz, Signal Processing for Optical Bar Code Scanning, *BYTE,* Dec. 1976, p 77–78 and 80–84.

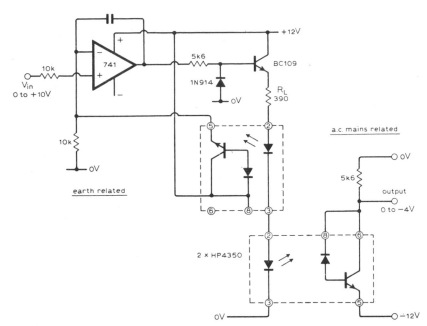

DC/DC OPTOISOLATOR—Designed to provide input isolation for thyristor converters. Linearity is within 2%. Loop gain of opamp makes diode turn-on voltage insignificant.—R. J. Haney, Linear D.C./D.C. Opto Isolator, *Wireless World,* June 1976, p 72.

FAST OPTICALLY ISOLATED SWITCH—Uses almost any standard optoisolator. Less than 20 μA is needed from photodiode D2 to turn LM195 power transistor fully on. Returning cathode of D2 to separate positive supply rather than to collector of Q1 eliminates collector-base capacitance of diode and increases switching speed to 500 ns for 40-V 1-A load.—R. Dobkin, "Fast IC Power Transistor with Thermal Protection," National Semiconductor, Santa Clara, CA, 1974, AN-110, p 5.

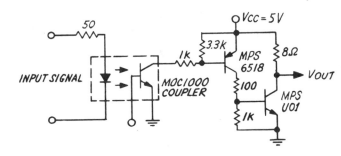

PULSE AMPLIFIER—Motorola MOC1000 optoisolator permits coupling digital logic to system having different supply voltages or unequal grounds while providing essentially complete isolation. Circuit provides transfer characteristics needed in instrumentation applications and has sufficient drive for handling low input impedances.—"Industrial Control Engineering Bulletin," Motorola, Phoenix, AZ, 1973, EB-4.

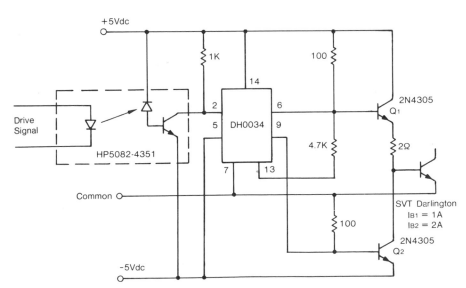

OPTICAL DRIVE FOR SWITCHING TRANSISTOR—Base driver circuit for TRW SVT6062 power Darlington switching transistor uses separate isolated bias supplies for each transistor to provide performance characteristics of driver transformer at lower cost. Bias supplies can use small 60-Hz transformers with bridge rectifiers and light filtering. Control isolation is provided by high-speed optical coupler that can be controlled directly from logic. DH0034 IC amplifies coupler output and provides level shifting as required for driving transistors Q_1 and Q_2.—D. Roark, "Base Drive Considerations in High Power Switching Transistors," TRW Power Semiconductors, Lawndale, CA, 1975, Application Note No. 120, p 8.

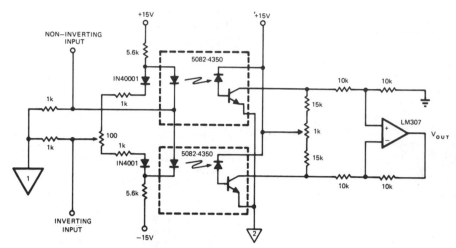

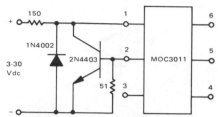

OPTOISOLATOR INPUT PROTECTION—Combination of diode and transistor limits input current to LED of Motorola MOC3011 optoisolator to safe maximum of less than 15 mA for input voltage range of 3–30 VDC. Circuit also protects LED from accidental reversal of polarity.—P. O'Neil, "Applications of the MOC3011 Triac Driver," Motorola, Phoenix, AZ, 1978, AN-780, p 4.

DC ISOLATOR WITH HARMONIC SUPPRESSION—Two isolators operating like push-pull amplifier minimize harmonic generation. When input signal is applied, upward change of incremental gain in one isolator is balanced by downward change in other to give harmonic cancellation. Circuit gain is about unity. Bandwidth is 2 MHz for signals below 2 V P-P. Input signals of either polarity may be applied at either inverting or noninverting input.—H. Sorensen, Opto-Isolator Developments Are Making Your Design Chores Simpler, *EDN Magazine,* Dec. 20, 1973, p 36–44.

SET-RESET LATCH—Provides almost complete isolation between each input and the output, as well as between inputs. Applying 2-V pulse at 14 mA momentarily to SET terminals allows up to 150 mA to flow between output terminals. This current flows until about 2 V at 15 mA is applied to RESET terminals or until load voltage is reduced enough to drop load current below 1 mA.—R. N. Dotson, Set-Reset Latch Uses Optical Couplers, *EDN Magazine,* Jan. 5, 1973, p 107.

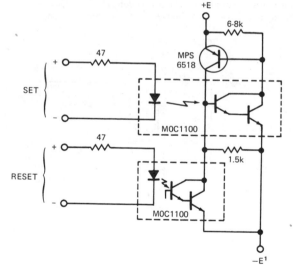

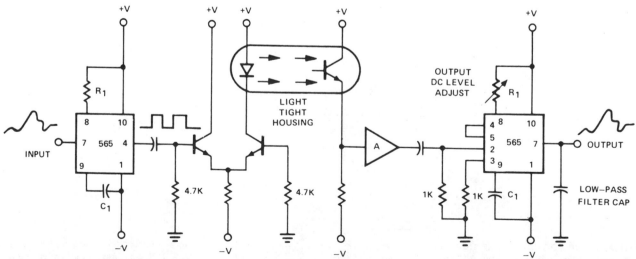

ANALOG ISOLATOR—Circuit is basically FM transmission system in which light is used as transmission medium. Transmitter uses 565 PLL as VCO for flashing LED of optoisolator at rate proportional to input voltage. Phototransistor drives amplifier having sufficient gain to apply 200 mV P-P signal to input of receiving 565 acting as FM detector for re-creating input to transmitter. Supply can be ±6 V to ±12 V.— "Signetics Analog Data Manual," Signetics, Sunnyvale, CA, 1977, p 846–847.

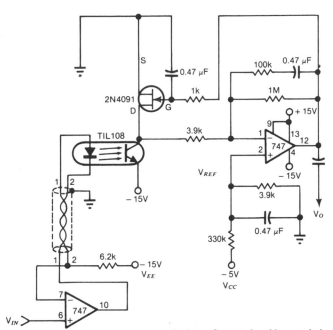

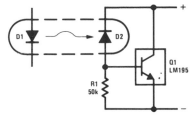

OPTOISOLATOR DRIVES 1-A POWER TRANSISTOR IC—Practically any standard optoisolator provides sufficient output to meet input current requirement of power transistor IC capable of handling 1 A. With no drive, R1 absorbs base current of Q1, holding it off. When power is applied to LED D1, less than 20 μA from photodiode D2 is sufficient to turn LM195 fully on. Supply can be up to 42 V.—"Linear Applications, Vol. 2," National Semiconductor, Santa Clara, CA, 1976, AN-110, p 5.

ISOLATION WITH GAIN COMPENSATION—Provides total harmonic distortion under 1% while automatically adjusting for temperature-produced or other DC gain variations in optoisolator. Output signal is sampled and fed back to FET to maintain constant AC gain. Design equations are given.—A. Billings, Optocoupler Provides Analog Isolation, *EDN Magazine,* Nov. 5, 1978, p 121–122.

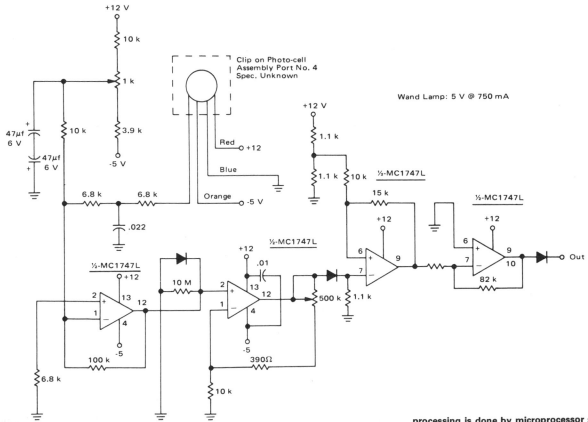

UPC WAND-SIGNAL CONDITIONER—Used in recovering analog output signal of photocell assembly for reading bars of universal product code. All four sections of two MC1747 dual opamps are used to amplify and condition photocell output so conditioned output of circuit provides TTL level 1 while wand is scanning black and 0 while scanning white. Additional processing is done by microprocessor such as MC6800.—"Microprocessor Applications Manual" (Motorola Series in Solid-State Electronics), McGraw-Hill, New York, NY, 1975, p 5-16–5-17.

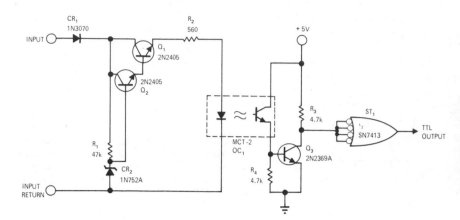

DIGITAL ISOLATION—Optical coupling provides complete electrical isolation between two digital circuits. Input signal as low as +4 V can make output change state, yet circuit safely handles input peaks up to +100 V without breakdown. Q_1 and Q_2 form current regulator that limits loop current through input of optoisolator to 7 mA. Zener CR_2 provides reference voltage that defines current through R_2. Schmitt trigger ST_1 in output eliminates oscillations that could otherwise occur when slow-rise-time signal is applied to fast TTL circuits. Output changes state when input signal lights LED in optoisolator.—C. E. Mitchell, Optical Coupler and Level Shifter, *EDN/EEE Magazine*, Feb. 1, 1972, p 55.

ANALOG ISOLATOR—Uses Mullard CNY44 optoisolator to transmit analog signals between units of equipment having unequal ground potentials. Circuit has 3-dB rolloff point at 6 Hz and 80 kHz. Total harmonic distortion at 8 V P-P output is less than 1.5% between 100 Hz and 20 kHz. Output transistor TR2 is not critical.—"Photocouplers," Mullard, London, 1974, Technical Information 4, TP1477, p 12.

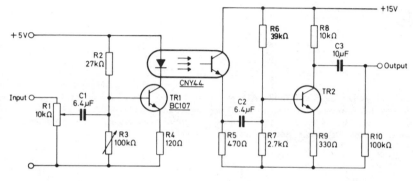

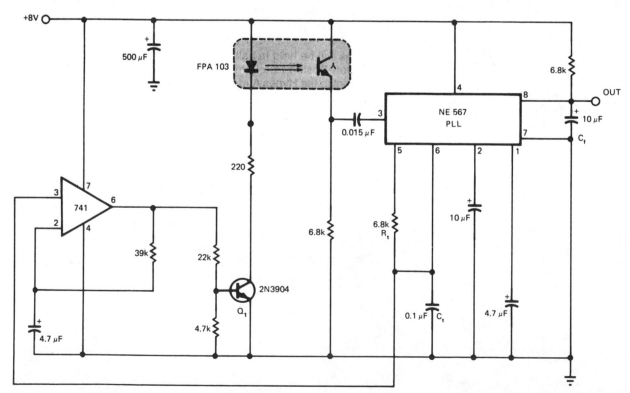

MODULATED OPTOISOLATOR—Circuit provides modulation of Fairchild FPA 103 optoisolator at about 1400 Hz and demodulation of signal from detector, to make optoisolator insensitive to strong fluorescent light without compromising performance. R_t and C_t set VCO of NE567 phase-locked loop IC at about 1400 Hz, and 741 opamp converts triangle wave at pin 6 of PLL to square wave with 50% duty cycle for driving LED of optoisolator through Q_1. Also useful with visible light systems.—R. Oliver, Improve Photo Sensors with a Phase-Locked Loop IC, *EDN Magazine*, April 5, 1976, p 112.

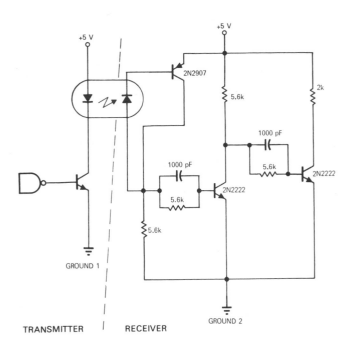

GROUND ISOLATION—Optoisolator such as HP4320 provides ground isolation up to 200 V between systems used in spacecraft. Arrangement is effective over bandwidth of DC to 1 MHz for both DTL- and TTL-driven circuits.—W. C. Milo, Simple Scheme Isolates System Grounds Optically, *EDN Magazine*, Sept. 15, 1970, p 64.

400-VDC SWITCH—Optically isolated photo-SCR serves for switching high-voltage DC. Turn-off of SCR occurs when Q_3 in MCA2 photo-Darlington shunts load current through gate, bypassing gate-cathode junction within SCR. Circuit can be operated by pulsing appropriate LEDs to turn SCR on or off. Without input signal, inverter maintains current through LED of MCA2 to keep SCR clamped off.—G. C. Riddle, Opto-Isolators Switch High-Voltage DC Current, *EDN Magazine*, Feb. 5, 1975, p 54.

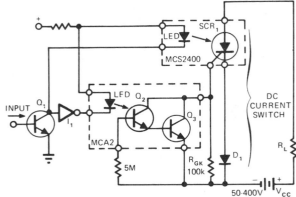

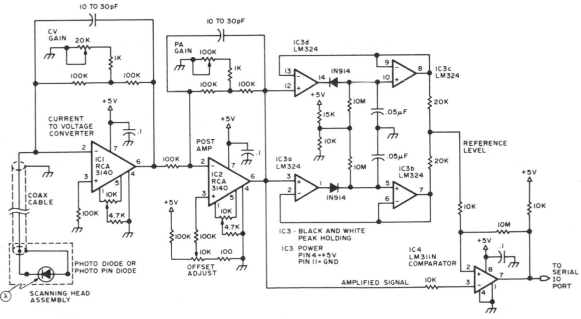

BAR-CODE SIGNAL CONDITIONER—Processes low-level signal from photodiode of bar-code scanner by converting its current output to voltage in IC1 for further amplification in IC2. Amplified signal is routed to peak holding circuits that set reference level and to comparator that outputs 0 or 1 based on reference level established. Peak values of white level and black level are held long enough to read through coded bar pattern. Difference between peak values is divided by 2 and fed to one input of comparator, while amplified signal level goes to inverting input. If signal level is greater than reference level, comparator output is 0. If signal level is less than reference level (black bar), output is 1.—F. L. Merkowitz, Signal Processing for Optical Bar Code Scanning, *BYTE*, Dec. 1976, p 77–78 and 80–84.

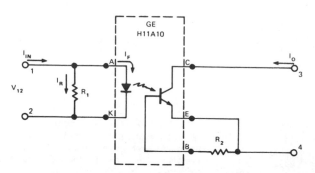

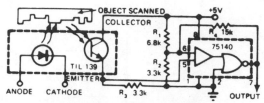

ISOLATED THRESHOLD SWITCH—Standard photocoupler programmed with 150-ohm resistor R_1 provides threshold switching function for separating high-level noise from switching-signal pulses as short as 10 μs. Current-transfer ratio of phototransistor coupler is made practically zero at some arbitrary input current, and changed rapidly back to 10% or more at slightly higher level. Programming range for threshold value extends from 60 mA for 10 ohms at R_1, to 3 mA for 400 ohms. Use of 2.7-megohm resistor R_2 across base-emitter terminals of coupler reduces low-current gain of phototransistor. Noise currents up to 5 mA on sensing line are rejected while operating currents as low as 10 mA are accepted.—J. Cook, Photocoupler Makes an Isolated Threshold Switch, *EDN Magazine*, Oct. 5, 1974, p 72, 74, and 76.

OPTOISOLATOR AS SCANNER—Consists essentially of Texas Instruments TIL 139 source/sensor assembly and common 75140 line receiver. Applications include response to reflected or interrupted light. With 5-V supply, output is at standard TTL levels. To make sensitivity adjustable, insert 500-ohm pot between R_1 and R_2. To invert output polarity, connect pin 7 of 75140 to pin 3 and take output from pin 1.—W. Grenlund, Low-Cost Photo Scanner Yields High Performance, *EDN Magazine*, Nov. 20, 1976, p 320.

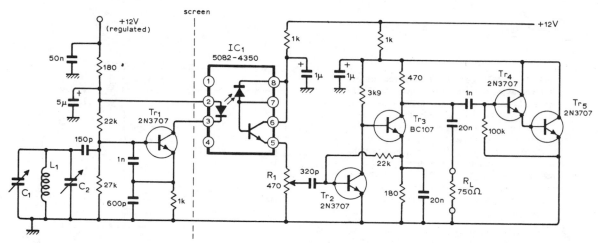

1.5–5.7 MHz OPTICALLY ISOLATED VFO—Isolation gives long-term frequency stability despite changes in ambient temperature, and eliminates effect of fluctuating load on frequency. Oscillator is emitter-coupled Colpitts using low-noise 2N3707 transistor. Article also gives circuit for output amplifier and automatic limiting control, along with alternative versions using ICs in place of transistors. Designed for use in amateur radio equipment.—A. K. Langford, Optically Coupled V.F.O., *Wireless World*, Nov. 1974, p 455–457.

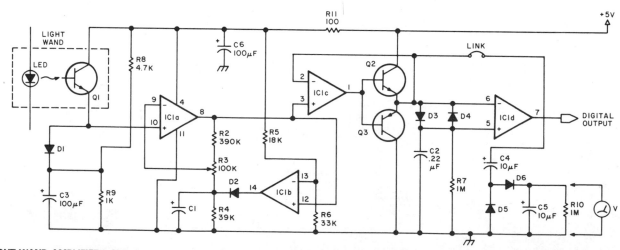

LIGHT-WAND AMPLIFIER—Signal processor is independent of most variables involved in reading printed bar data. Amplifier uses exponential forward conduction properties of silicon diode D1 to transform output of wand to logarithmically varying voltage having peak-to-peak value proportional to ratio of white and black photo-currents and independent of absolute photocurrent. White-level output of amplifier IC1a is clamped at fixed level by comparator IC1b and peak detector D2-C1. Amplified and clamped signal is converted to binary digital output required by microprocessor. Article traces operation of circuit step by step. IC1 is National LM324 quad opamp. All diodes are 1N4148 silicon or equivalent. Q2 is MPS6513 or equivalent, and Q3 is MPS6517 or equivalent. Output is TTL-compatible.—R. C. Moseley, A Low Cost Light Wand Amplifier, *BYTE*, May 1978, p 92 and 94–95.

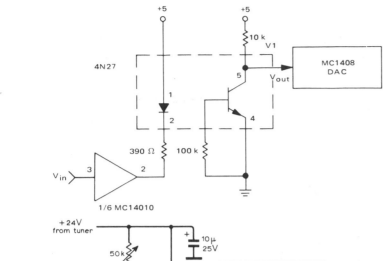

1500-V ISOLATION FOR DAC—Motorola 4N27 optoisolator provides required isolation between DAC of programmable power supply and remotely located CMOS MC14010 noninverting buffer.—D. Aldridge and N. Wellenstein, "Designing Digitally-Controlled Power Supplies," Motorola, Phoenix, AZ, 1975, AN-703, p 9.

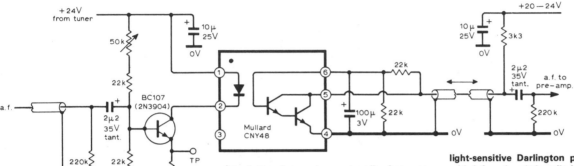

HUM-BLOCKING OPTOISOLATOR—Optoelectronic isolator for audio feed in TV set prevents circulation of ground currents at line frequency, for protection of low-level signal runs from hum interference. Used in tuner providing quality sound and video outputs, circuits for which are given in four-part article. Optoisolator uses light-sensitive Darlington pair in conjunction with infrared-emitting diode. Diode current is adjusted with 50K variable resistor to give best compromise between noise and distortion.—D. C. Read, Television Tuner Design, *Wireless World,* Jan. 1976, p 51—57.

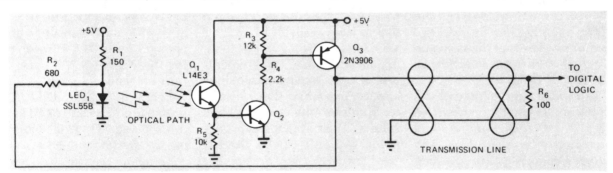

BUILT-IN HYSTERESIS—Will operate at all speeds in range from 20 kHz down to zero while still having suitable rise times for driving digital logic. When optical path is blocked, all three transistors are off and output is low. As light on Q_1 increases, Q_2 and Q_3 begin turning on; rising collector of Q_3 adds more current through R_2 to LED, giving Q_1 more light and driving Q_3 into saturation. When light dims, Q_1 begins to turn off and extra current is cut off, driving Q_3 off. With this hysteresis action, there is no constant light level at which circuit will oscillate.—D. C. Hoffman, Optical Sensor Has Built-In Hysteresis, *EDN Magazine,* June 5, 1973, p 91.

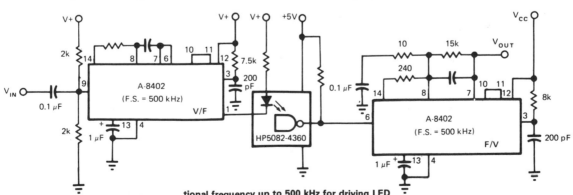

30-kHz BANDWIDTH—Isolation amplifer circuit uses Intech/Function Modules A-8402 voltage-to-frequency converter having linearity of ±0.05% to convert input voltage to proportional frequency up to 500 kHz for driving LED of optoisolator. Similar IC converts output of optoisolator back to proportional DC voltage. Supply for converters is nominally 12 V, but can be 5 to 18 V.—P. Pinter and D. Timm, Voltage-to-Frequency Converters—IC Versions Perform Accurate Data Conversion (and Much More) at Low Cost, *EDN Magazine,* Sept. 5, 1977, p 153–157.

CHAPTER 61
Oscillator Circuits—AF

Includes variety of Wien-bridge, phase-shift, voltage-controlled, and multivibrator types of oscillators producing output at audio and ultrasonic frequencies. Other audio oscillators can be found in Code, Frequency Synthesizer, Function Generator, Pulse Generator, Signal Generator, Staircase Generator, Sweep, and Test chapters.

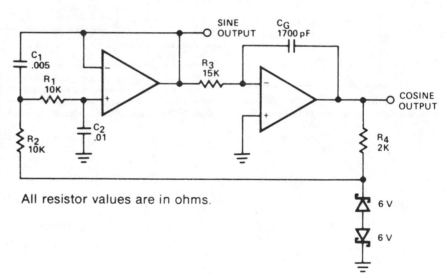

2-kHz TWO-PHASE—Dual opamp circuit uses two-pole Butterworth bandpass filter followed by phase-shifting single-pole stage that is fed back through zener voltage limiter. Circuit provides simultaneous sine and cosine outputs. Distortion is about 1.5% for sine output and about 3% for cosine. Component values shown are for 741 opamp. For higher frequencies, use 531 opamps to reduce distortion due to slew limiting.—"Signetics Analog Data Manual," Signetics, Sunnyvale, CA, 1977, p 642–644.

All resistor values are in ohms.

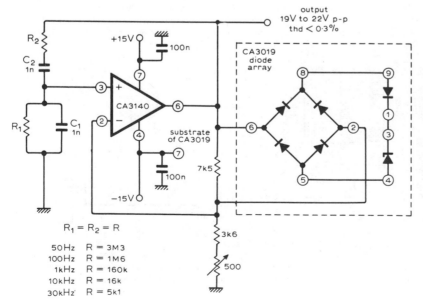

$R_1 = R_2 = R$

50Hz	R = 3M3
100Hz	R = 1M6
1kHz	R = 160k
10kHz	R = 16k
30kHz	R = 5k1

BASIC MOS OSCILLATOR—Output is 50 Hz when R_1 and R_2 are 3.3 megohms, increasing to 30 kHz as resistor values are reduced to 5100 ohms. Circuit has no inherent lower frequency limit; with 22-megohm resistors and 1-μF capacitors for C_1 and C_2, sine-wave output is 0.007 Hz. Article gives basic equations for circuit. Features include high input impedance, fast slew rate, and high output voltage capability. Combination of bridge rectifier with monolithic zener diodes in regulating system provides practically zero temperature coefficient.—M. Bailey, Op-Amp Wien Bridge Oscillator, *Wireless World*, Jan. 1977, p 77.

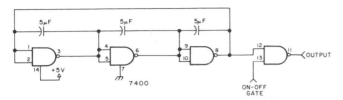

1000 Hz WITH ONE CHIP—Quad NAND gate gives sawtooth output waveform at 800 to 1000

Hz for driving other TTL circuits.—Circuits, *73 Magazine*, June 1977, p 49.

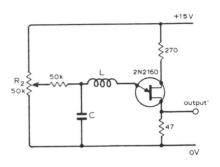

1—50 kHz SINE-WAVE—Uses unijunction transistor as negative resistance in simple RLC circuit. Maximum output with good waveform is about 200 mV. Exact frequency depends on values used for L and C.—R. P. Hart, Simple Sine-Wave Oscillator, *Wireless World*, July 1976, p 34.

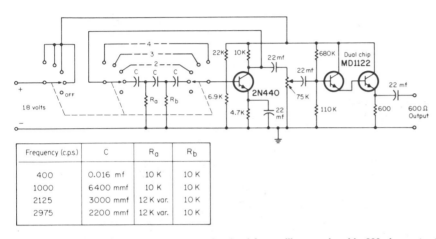

Frequency (c.p.s.)	C	R_a	R_b
400	0.016 mf	10 K	10 K
1000	6400 mmf	10 K	10 K
2125	3000 mmf	12 K var.	10 K
2975	2200 mmf	12 K var.	10 K

TEST TONES—Provides preset frequencies of 400, 1000, 2125, and 2975 Hz. Circuit consists of RC phase-shift oscillator driving Darlington emitter-follower that provides high-impedance

load for oscillator and stable 600-ohm output impedance.—S. Kelly, A Simple Audio Test Oscillator, *CQ*, Oct. 1970, p 50 and 90.

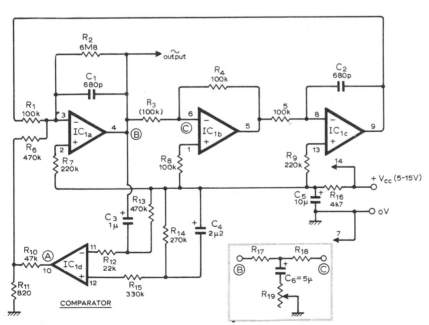

2.34-kHz SINE-WAVE—Uses low-cost LM3900N quad differential amplifier IC in low-distortion oscillator for which third harmonic distortion is typically 0.5%. Peak-to-peak amplitude of sine-wave output is typically 25% of source voltage V_{cc}. Frequency can be changed

by altering single component, R_3, or by inserting between points B and C an RC network and pot connected as shown in inset. Article gives design equations for frequency and Q.—T. J. Rossiter, Sine Oscillator Uses C.D.A., *Wireless World*, April 1975, p 176.

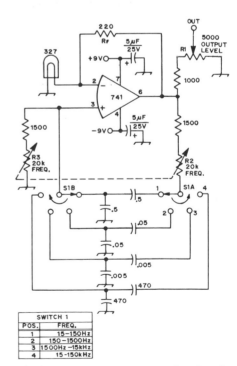

SWITCH 1	
POS.	FREQ.
1	15–150Hz
2	150–1500Hz
3	1500Hz –15kHz
4	15–150kHz

15 Hz TO 150 kHz IN FOUR RANGES—Switch gives choice of ranges, with R2 and R3 varying frequency in each. Circuit draws only 4 mA from two 9-V batteries and provides moderate output at 4–5 V. Connections shown are for TO-5 case of 741.—T. Schultz, Audio Oscillator, *QST*, Nov. 1974, p 43.

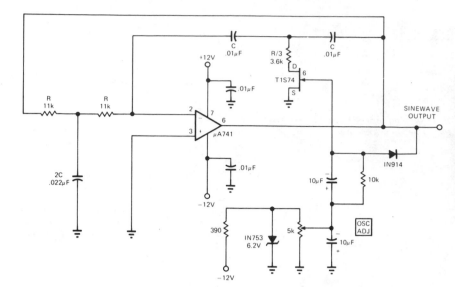

STABILIZED SINE-WAVE—Peak detector is used with FET operated in voltage variable-resistance mode, in combination with standard double-integration circuit having regenerative feedback, to give 1.46-kHz sine-wave output into 500-ohm load at 10 V P-P. Will operate at power supply voltages of 8 to 18 V without appreciable variation in output amplitude or frequency. Output varies less than 1.5% in frequency and 6% in amplitude over temperature range of 10 to 65°C. Circuit can be modified for other frequencies.—F. Macli, FET Stabilizes Sine-Wave Oscillator, *EDN Magazine*, June 5, 1973, p 87.

1-kHz LOW-DISTORTION—Total harmonic distortion is only 0.01% in amplitude-stabilized oscillator delivering 7 VRMS. Opamp A_1 has closed-loop gain of 3. Regenerative feedback through bandpass filter C_1-C_2-R_1-R_5 determines frequency of oscillation. Output is stabilized by multiplier whose control voltage is derived from integrator A_2.—R. Burwen, Ultra Low Distortion Oscillator, *EDN/EEE Magazine*, June 1, 1971, p 45.

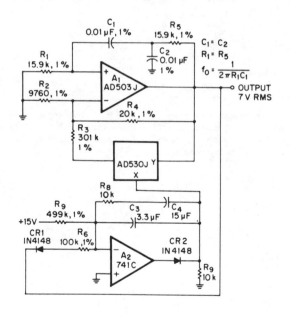

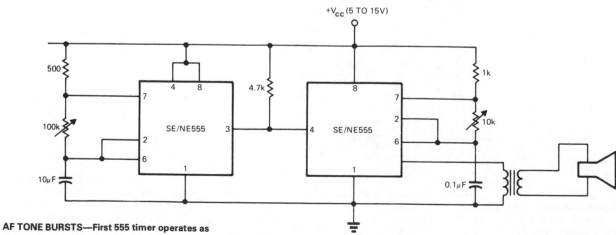

AF TONE BURSTS—First 555 timer operates as slow astable multivibrator whose output is used to gate second timer operating as AF oscillator. Arrangement provides repeatable tone-burst generation.—E. R. Hnatek, Put the IC Timer to Work in a Myriad of Ways, *EDN Magazine*, March 5, 1973, p 54–58.

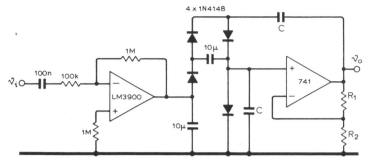

CURRENT-CONTROLLED WIEN—Small variations in input voltage to National LM3900 current-mode amplifier change frequency of four-diode current-controlled Wien bridge over range from 10 to 50,000 Hz, with frequency being proportional to control current. Value of C is 700 pF. Ratio of R_2 to sum of R_1 and R_2 should be greater than 3 to give voltage gain needed.—K. Kraus, Oscillator with Current-Controlled Frequency, *Wireless World*, Aug. 1974, p 272.

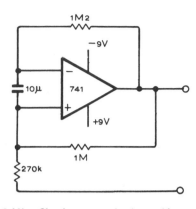

3.8 kHz—Simple opamp circuit provides convenient sine-wave AF signal.—J. S. Lucas, Unusual Sinewave Generator, *Wireless World*, May 1977, p 81.

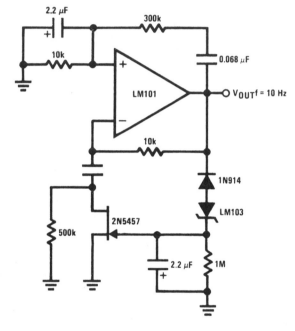

Peak output voltage
$$V_p \cong V_z + 1V$$

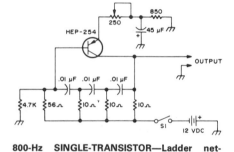

800-Hz SINGLE-TRANSISTOR—Ladder network determines frequency. For higher frequencies, decrease values of capacitors in network. Circuit also works with OC-2, SK-3004, and AT30H transistors.—Circuits, *73 Magazine*, May 1977, p 31.

10-Hz WIEN-BRIDGE—JFET serves as voltage-variable resistor in feedback loop of opamp, as required for producing low-distortion constant-amplitude sine wave. LM103 zener provides voltage reference for peak amplitude of sine wave; this voltage is rectified and fed to gate of JFET to vary its channel resistance and loop gain of opamp.—"FET Databook," National Semiconductor, Santa Clara, CA, 1977, p 6-26–6-36.

DOT GENERATOR—Can be used by amateur radio operator to "talk" himself onto frequency while listening on downlink passband of Oscar satellite, without causing interference to other stations using satellite. Generates audio dots at rate of 12 per second. Frequency of free-running MVBR Q1-Q2 is determined by values of C1, C2, R1, and R2. Emitter-follower Q3 drives 1500-Hz audio oscillator Q4. C1 and C2 are 1-μF 16-V electrolytics. Q1-Q4 are 2N2222 or equivalent NPN general-purpose transistors.—M. Righini and G. Emiliani, Audio Dot Generator Eases OSCAR SSB Spotting, *QST*, Nov. 1977, p 45.

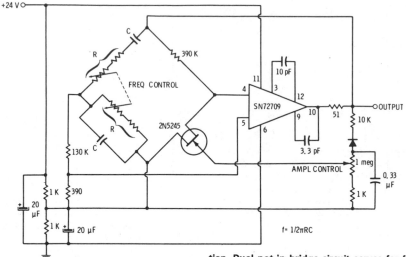

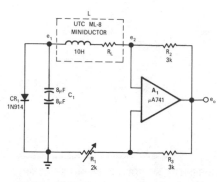

WIEN-BRIDGE AF/RF—Single JFET in basic Wien-bridge circuit drives Texas Instruments linear opamp serving as output stage. Feedback path from output of IC to base of JFET stabilizes output and provides temperature compensation. Dual pot in bridge circuit serves for frequency control. Circuit performs well as either AF or RF oscillator depending on values used for R and C.—E. M. Noll, "FET Principles, Experiments, and Projects," Howard W. Sams, Indianapolis, IN, 2nd Ed., 1975, p 213–214.

25-Hz SINE-WAVE—Output voltage is 8 V P-P at about 25 Hz for values shown, with total harmonic distortion less than 0.5%. Circuit will operate from 15 Hz to 100 kHz by using other values. Set regeneration control R_1 at minimum value needed to sustain oscillation.—J. C. Freeborn, Simple Sinewave Oscillator, *EDN/EEE Magazine*, Sept. 1, 1971, p 44.

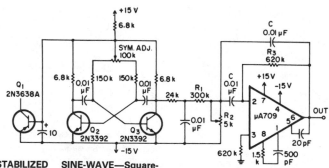

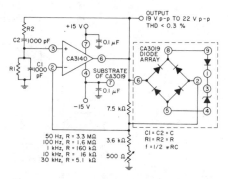

350-Hz STABILIZED SINE-WAVE—Square-wave oscillator Q_2-Q_3 stabilized by Q_1, followed by passive filter and active filter using μA709, produces amplitude-stabilized sine wave at 350 Hz, for which third harmonic is 39 dB down and other harmonics are insignificant.—E. Neugroschel and A. Paterson, Amplitude-Stabilized Audio Oscillator, *EEE Magazine*, April 1971, p 65.

50–30,000 Hz WIEN-BRIDGE—Wide-range audio oscillator utilizes high input impedance, high slew rate, and high voltage characteristics of CA3140 opamp in combination with CA3019 diode array. R1 and R2 are same value, chosen for frequency desired as given in table.—"Circuit Ideas for RCA Linear ICs," RCA Solid State Division, Somerville, NJ, 1977, p 4.

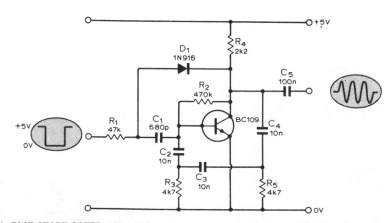

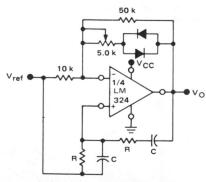

1-kHz FAST-START GATED—Circuit is conventional phase-shift oscillator in which frequency is determined by C_2, C_3, C_4, R_5, R_3, and input impedance of transistor. When input is +5 V, almost 100% negative feedback blocks oscillator. When input drops to 0 V, D_1 is reverse-biased and negative feedback is removed. At same time, edge of input pulse is applied to transistor base to kick off oscillator on its first half-cycle, which is always in phase with falling edge of input signal.—G. F. Butcher, Gated Oscillator with Rapid Start, *Wireless World*, Aug. 1974, p 272.

1-kHz WIEN-BRIDGE—Simple circuit uses only one section of LM324 quad opamp having true differential inputs. Supply voltage range is 3–32 V. Reference voltage is half of supply voltage. Values of R and C determine frequency according to equation f = 1/6.28RC. For 16K and 0.01 μF, frequency is 1 kHz. Diode types are not critical.—"Quad Low Power Operational Amplifiers," Motorola, Phoenix, AZ, 1978, DS 9339 R1.

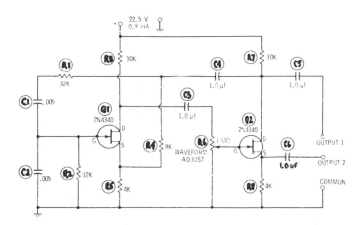

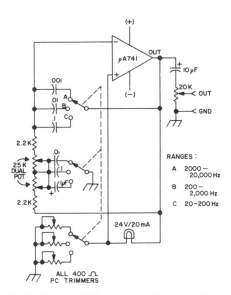

100-Hz WIEN-BRIDGE—Simple RC-tuned oscillator uses only two resistors (R1 and R2) and two capacitors (C1 and C2) to set frequency. Feedback path covers both FET stages. Set R6 for best sine-wave output. For other audio frequencies, change value of R in ohms and C in farads in equation f = 1/6.28RC where frequency is in hertz, R = R1 = R2, and C = C1 = C2.—R. P. Turner, "FET Circuits," Howard W. Sams, Indianapolis, IN, 1977, 2nd Ed., p 48–50.

20–20,000 Hz—Wide-range audio oscillator covers AF spectrum in three switch-selected ranges, with harmonic distortion as low as 0.15%, for quick checks of audio equipment. Drain is only 6 mA from two 9-V batteries. Circuit is Wien-bridge oscillator using 741 opamp. Article covers construction and calibration, including optional connection for operation from single 9-V battery with AF output reduced to 2 V.—J. J. Schultz, Wide Range IC Audio Oscillator, *73 Magazine,* Jan. 1974, p 25–28.

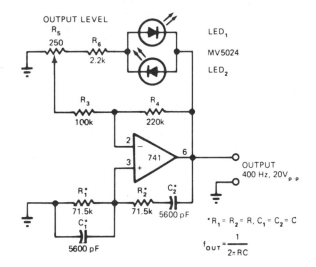

400-Hz LED-OPAMP SINE-WAVE—Uses LEDs as nonlinear-resistance diodes in Wien-bridge configuration with opamp operating from 15-V supply. Circuit will operate over wide range of other frequencies if values of R and C are changed. R_5 adjusts output amplitude from 10 to 20 V P-P. Total harmonic distortion is 1%.—W. G. Jung, LED's Do Dual Duty in Sine-Wave Oscillator, *EDN Magazine,* Aug. 20, 1976, p 84–85.

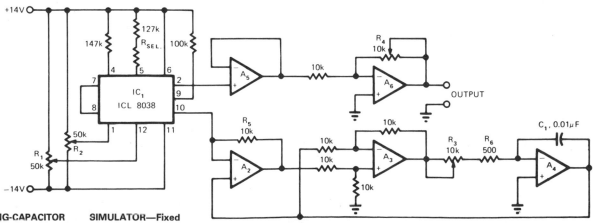

TUNING-CAPACITOR SIMULATOR—Fixed biasing network is used with Intersil 8038 variable-frequency sine-wave oscillator. Frequency is varied between 175 and 3500 Hz by circuit components forming capacitor simulator. Adjusting R_3 varies equivalent capacitor value from 500 pF to 0.01 μF. Distortion is less than 1% over frequency range. Buffer opamp A_5 provides high load impedance to IC_1 and low source impedance to variable-gain opamp A_6. All opamps are 741.—R. Gunderson, Variable-Frequency Oscillator Features Low Distortion, *EDN Magazine,* Aug. 5, 1974, p 76 and 78.

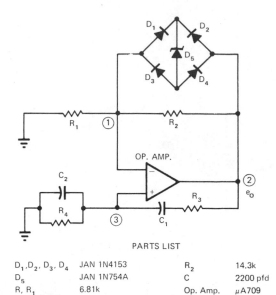

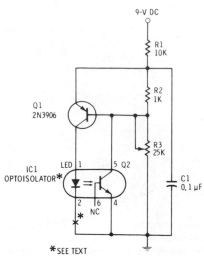

PARTS LIST

D_1, D_2, D_3, D_4	JAN 1N4153	R_2	14.3k
D_5	JAN 1N754A	C	2200 pfd
R, R_1	6.81k	Op. Amp.	μA709

ZENER CONTROLS BRIDGE—Amplitude of 10.5-kHz Wien-bridge oscillator output is maintained symmetrical above ground by using single zener with diode bridge. As output e_0 approaches soft knee threshold of conduction for zener, its impedance decreases and shunts R_2. This violates oscillator requirement that $R_2 =$ $2R_1$, so output begins decreasing sinusoidally. As swing decreases, gain increases until e_0 reaches negative threshold. Signal then reverses and again starts going positive.—W. B. Crittenden and E. J. Owings, Jr., Zener-Diode Controls Wien-Bridge Oscillator, *EDN Magazine*, Aug. 1, 1972, p 57–58.

NEGATIVE-RESISTANCE LED OSCILLATOR—Covers frequency range of about 3.2–8 kHz with values shown. Will drive loudspeaker inserted at point X. For lower frequency (range of 120–1800 Hz) and louder sound, change C1 to 1 μF. Negative-resistance portion of circuit includes Q1, Q2, LED, R2, and R3. Optoisolator can be MCT-2 or equivalent.—F. M. Mims, "Electronic Circuitbook 5: LED Projects," Howard W. Sams, Indianapolis, IN, 1976, p 26–29.

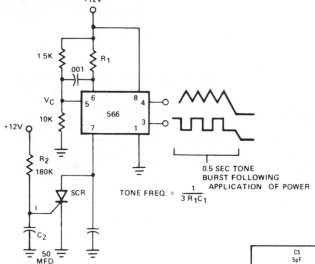

TONE FREQ. $\approx \dfrac{1}{3\,R_1 C_1}$

0.5 SEC TONE BURST FOLLOWING APPLICATION OF POWER

0.5-s TONE BURSTS—Simple 566 function generator circuit supplies audio tone for 0.5 s after power is applied, for use as communication-network alert signal. SCR is gated on when C_2 charges up to its gate voltage, which takes 0.5 s, to shunt timing capacitor between pin 7 and ground and thereby stop tone. If SCR is replaced by NPN transistor, tone can be switched on and off manually at transistor base terminal.—"Signetics Analog Data Manual," Signetics, Sunnyvale, CA, 1977, p 852–853.

20–20,000 Hz LOW-DISTORTION—Opamp at right is driven by square-wave output of comparator at left, with feedback between opamps providing oscillation. Frequency range covered by tuning control R3 is determined by equal-value capacitors C1 and C2, which range from 0.4 μF for 18–80 Hz to 0.002 μF for 4.4–20 kHz. Distortion ranges from 0.2% to 0.4% when 20% clipping of sine wave is provided by zeners. Both positive and negative supplies should be bypassed with 0.1-μF disk ceramic capacitors.—"Easily Tuned Sine Wave Oscillators," National Semiconductor, Santa Clara, CA, 1971, LB-16.

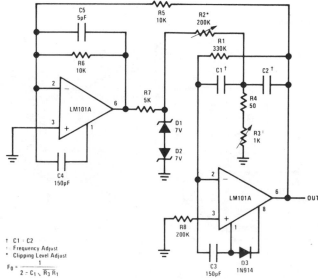

† C1 = C2
‡ Frequency Adjust
* Clipping Level Adjust

$$F_0 = \frac{1}{2 \cdot C_1 \sqrt{R_3 R_1}}$$

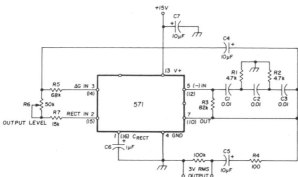

PHASE-SHIFT SINE-WAVE—Uses NE571 analog compandor as phase-shift oscillator, with internal inverting amplifier serving to sustain oscillation. Cl, C2, and C3 are timing capacitors, while R1 and R2 serve for phase-shift network. Suitable for use only as spot-frequency AF oscillator, with frequency being varied by changing values of Cl, C2, and C3. Total harmonic distortion is only 0.01% at 3-V output.—W.G. Jung, Gain Control IC for Audio Signal Processing, *Ham Radio*, July 1977, p 47–53.

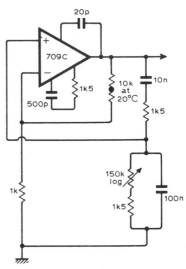

SINGLE-POT WIEN—Can be tuned from 340 to 3400 Hz with single 150K logarithmic pot. Output is constant over tuning range. Opamp can also be 741. Components in the two arms of the Wien bridge have large ratio to each other, so attenuation of network is only slightly affected by change in one of resistors.—P. C. Healy, Wien Oscillator with Single Component Frequency Control, *Wireless World*, Aug. 1974, p 272.

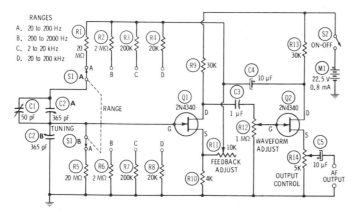

20 Hz TO 200 kHz—Variable-frequency RC-tuned oscillator uses FETs with Wien-bridge frequency-determining network. Identical resistors accurate to at least 1% are switched in pairs to change range. Dual 365-pF variable capacitor C2 is used for tuning in each range. Can be calibrated against standard audio frequency with CRO set up for Lissajous figures, or calibrated with high-precision AF meter connected to AF output terminals.—R. P. Turner, "FET Circuits," Howard W. Sams, Indianapolis, IN, 1977, 2nd Ed., p 132–134.

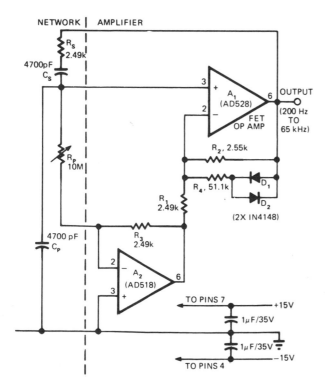

200–65,000 Hz WIEN—Adding single opamp to Wien-bridge oscillator gives wide-range oscillator having single-control tuning. R_4, D_1, and D_2 together stabilize output amplitude by providing controlled nonlinearity that reduces gain at high signal levels. AD528 opamp A_1 is FET-input complement to AD518 A_2 and has bandwidth required for wide output frequency range. R_P sweeps output from 200 Hz to 65 kHz. Since oscillation frequency is inversely proportional to square root of R_P, frequency changes rapidly near low-resistance end of pot. Use of pot with audio or log taper makes tuning more linear.—P. Brokaw, FET Op Amp Adds New Twist to an Old Circuit, *EDN Magazine*, June 5, 1974, p 75–77.

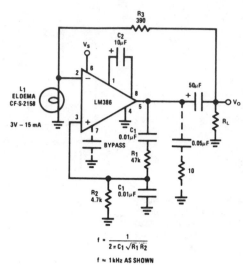

$$f = \frac{1}{2\pi C_1 \sqrt{R_1 R_2}}$$

f ≈ 1kHz AS SHOWN

1-kHz WIEN-BRIDGE—Closed-loop gain of 10, fixed by ratio of R_1 to R_2, is sufficient to avoid spurious oscillations. Frequency is easily changed by using different values for capacitors C_1. R_3 and lamp L_1 provide amplitude-stabilizing negative feedback. Supply can be 9 V.—"Audio Handbook," National Semiconductor, Santa Clara, CA, 1977, p 4-30—4-33.

WIEN SINE-WAVE—Uses NE571 analog compandor in oscillator circuit based on Wien network formed by R1-Cl and R2-C2, placed around output amplifier of section A to make it bandpass amplifier. Section B serves as inverting amplifier with nominal gain of 2. Total harmonic distortion is below 0.1%. Operating frequency is about 1.6 kHz for values shown, but can be varied from 10 Hz to 10 kHz. Frequency is $1/2\pi RC$ for R = R1 = R2 and C = Cl = C2. R should be kept between 10K and 1 megohm and C between 1000 pF and 1 μF. Useful as fixed-frequency oscillator but can be tuned if matched dual pot is used for R1-R2.—W. G. Jung, Gain Control IC for Audio Signal Processing, *Ham Radio*, July 1977, p 47–53.

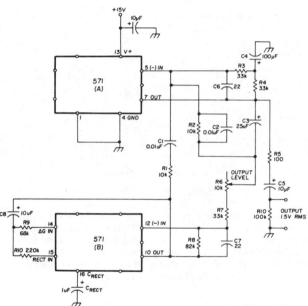

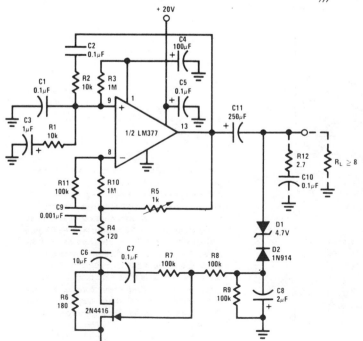

WIEN-BRIDGE 2-W—Uses half of LM377 IC connected as oscillator, with FET amplitude stabilization in negative feedback path. Total harmonic distortion is under 1% up to 10 kHz. With values shown, maximum output is 5.3 VRMS at 60 Hz. R12 and C10 are added if necessary to prevent high-frequency instability.—"Audio Handbook," National Semiconductor, Santa Clara, CA, 1977, p 4-8—4-20.

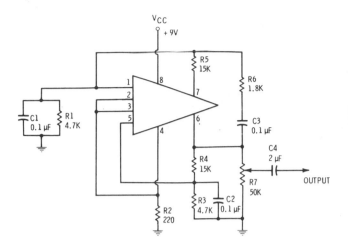

1–2 kHz TONE GENERATOR—Simple feedback circuit converts HEP 580 IC to emitter-coupled MVBR producing reasonably sinusoidal output somewhere between 1 and 2 kHz. Supply is 9-V battery.—E. M. Noll, "Linear IC Principles, Experiments, and Projects," Howard W. Sams, Indianapolis, IN, 1974, p 64–65.

136.5-Hz TONE—Uses 2.235-MHz crystal with counter chip to produce 136.5-Hz subaudible tone for amateur transmitter. Low-frequency square-wave output of IC is put through multi-stage low-pass filter to develop sine wave. Tone should be introduced into transmitter just after audio processing (after deviation control). Alternate filter is also shown; use whichever gives best performance. For other tone frequency, use crystal that is 16,384 times frequency desired.—E. Gellender and M. Marcel, P/L Tone Generator, QST, Aug. 1976, p 43.

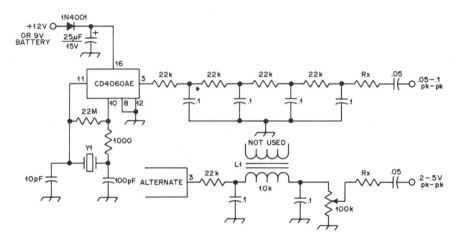

1850-Hz DIGITAL IC—Two sections of RS7404 hex inverter are connected as astable MVBR operating at frequency determined by values used for C1 and C2. Output drives loudspeaker as shown to produce audible tones or can be connected to flash LEDs. D1 reduces battery voltage to 5 V required by IC. Developed for classroom demonstrations. Circuit produces nearly square waves with amplitude of about 3 V and pulse width of about 100 μs if used as square-wave generator.—F. M. Mims, "Integrated Circuit Projects, Vol. 6," Radio Shack, Fort Worth, TX, 1977, p 64–69.

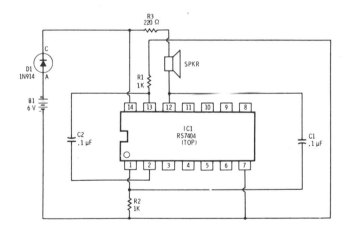

10-kHz SINE-COSINE—Combination of SN72310 voltage-follower opamp and SN 72301A high-performance opamp gives two outputs differing in phase by 90°. Supply is ±18 V.—"The Linear and Interface Circuits Data Book for Design Engineers," Texas Instruments, Dallas, TX, 1973, p 4-40.

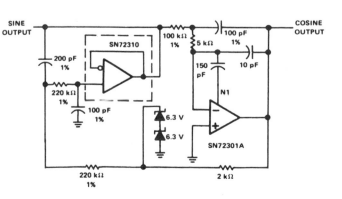

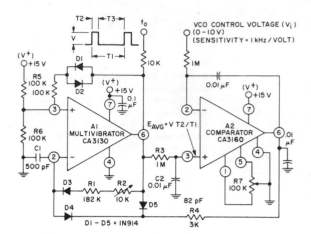

1 kHz/V FOR VCO—Voltage-controlled oscillator uses CA3130 opamp as MVBR and CA3160 opamp as comparator. Tracking error is about 0.02%, and temperature coefficient is 0.01% per degree C.—"Circuit Ideas for RCA Linear ICs," RCA Solid State Division, Somerville, NJ, 1977, p 4.

SINE-WAVE WIEN—Uses CA3140 opamp and diode array to generate low-distortion sine waves. Table gives values recommended for R and C to obtain frequencies from 50 Hz to 30 kHz. Use of zener diode clamp for amplitude control gives fast AGC.—W. Jung, An IC Op Amp Update, *Ham Radio,* March 1978, p 62–69.

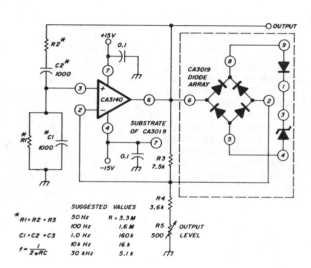

SUGGESTED VALUES

*R1 = R2 = R3	50 Hz	R = 3.3 M
	100 Hz	1.6 M
C1 = C2 = C3	1.0 Hz	160 k
	10 k Hz	16 k
$f = \dfrac{1}{2\pi RC}$	30 kHz	5.1 k

CHAPTER 62
Oscillator Circuits—RF

Includes fixed and tunable Clapp, Colpitts, crystal, LC, RC, Pierce, relaxation, and wobbulator oscillators having sine or square outputs in range from AF spectrum to 200 MHz. Some can be changed in frequency by digital control or diode switching of crystals.

SECONDARY STANDARD FOR 100 AND 10 kHz—Combination of 100-kHz crystal oscillator and 10-kHz MVBR provides 100-kHz harmonics far up into high-frequency spectrum, with each 100-kHz interval subdivided by harmonics of MVBR using two FETs. Oscillator is tuned to crystal frequency with Miller 42A223CBI or equivalent slug-tuned coil L1. C1 adjusts crystal frequency over narrow range for standardizing against WWV transmissions. Synchronizing 100-kHz voltage is injected into MVBR through R5.—R. P. Turner, "FET Circuits," Howard W. Sams, Indianapolis, IN, 1977, 2nd Ed., p 127–129.

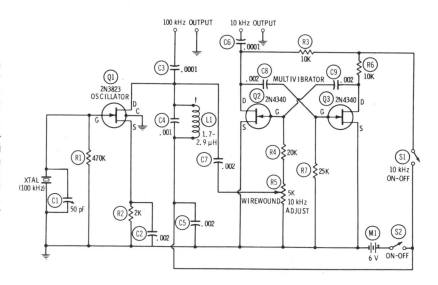

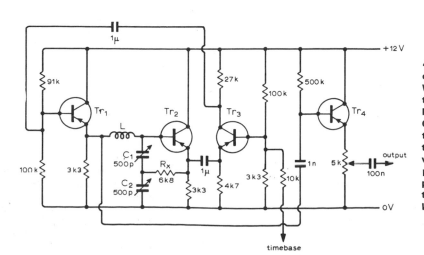

450–500 kHz WOBBULATOR—Center frequency of sweep is adjusted with C_1 and C_2. With appropriate coil, operation can be extended up to 10.7 MHz. Transistors can be BC107, BF115, BF194, or other equivalent. Choose value of R_x to give best waveform with transistor types used. Feedback for VCO is taken via Tr_3 without phase change. If control voltage for base of Tr_3 is derived from ramp output of oscilloscope time base, wobbulator output will follow variations in sweep voltage of time base.—E. C. Lay, Wobbulator, *Wireless World,* May 1975, p 226.

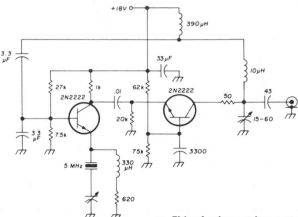

5-MHz LOW-NOISE CRYSTAL—Extremely low-noise series-mode crystal oscillator is designed for use in high-quality communication receiv-ers. Either fundamental or overtone crystals can be used.—U. L. Rohde, Effects of Noise in Receiving Systems, *Ham Radio,* Nov. 1977, p 34–41.

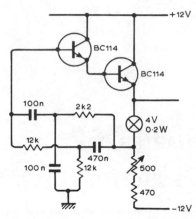

RC CONTROL—Chief advantage is absence of attenuation at zero phase shift in passive RC network used to define frequency of oscillation. Output is 20 V P-P. Pilot lamp stabilizes loop gain to unity, eliminating need for thermistor.—W. R. Jackson, Oscillator Uses Passive Voltage-Gain Network, *Wireless World,* April 1975, p 175.

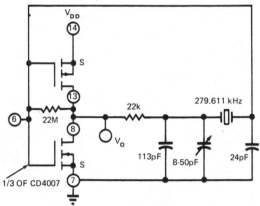

279.611-kHz CRYSTAL—DT-cut quartz crystal operating in CMOS inverter pair circuit serves as efficient timing circuit. Supply voltage can be from 5 to 15 V. With TA5987 low-voltage equiv-alent of 4007, supply can be 2.5 to 5 V. Stability is 4.3 PPM, not including temperature varia-tions.—B. Furlow, CMOS Gates in Linear Appli-cations: The Results Are Surprisingly Good, *EDN Magazine,* March 5, 1973, p 42–48.

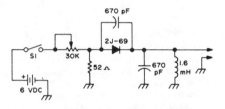

100-kHz SINE—Tunnel-diode sine-wave oscil-lator uses single GE 2J-69. Frequency is stable provided there are no drastic temperature changes, but for long-term accuracy and stabil-ity a crystal oscillator is recommended.—Cir-cuits, *73 Magazine,* May 1977, p 31.

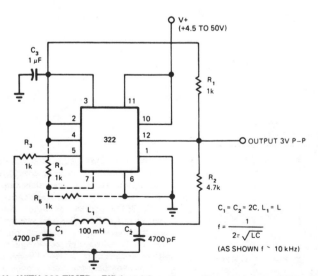

$C_1 = C_2 = 2C, L_1 = L$

$$f = \frac{1}{2\pi\sqrt{LC}}$$

(AS SHOWN f ~ 10 kHz)

UP TO 100 kHz WITH 322 TIMER—Efficient LC oscillator uses IC timer as inverting comparator, with pi-network LC tank as resonant circuit. Output square wave is regulated to 3 V in am-plitude, independently of supply voltage; upper supply limit should be 40 V instead of value shown. Sine-wave output of oscillator may also be used externally by adding single-supply opamp as buffer. Values shown give 10 kHz, but upper limit is 100 kHz.—W. G. Jung, Take a Fresh Look at New IC Timer Applications, *EDN Magazine,* March 20, 1977, p 127–135.

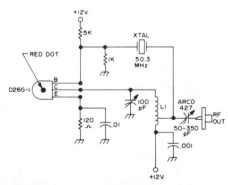

L1-25 TURNS 28 AWG TAPPED 4 TURNS FROM COLD END, WOUND ON 1/8 in. DIAMETER FORM, APPROXIMATELY 1/2 in. LG.

50-MHz CRYSTAL—Uses microtransistor as os-cillator handling 100-mW input power and giv-ing 40–50% efficiency. Article covers construc-tion with microcomponents and gives other microtransistor circuits for low-power amateur radio use and possible bugging applications.—B. Hoisington, Introduction to "Microtransis-tors," *73 Magazine,* Oct. 1974, p 24–30.

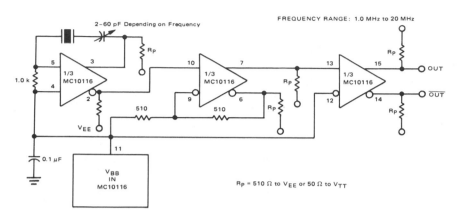

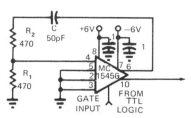

1–20 MHz FUNDAMENTAL CRYSTAL—Oscillator requires no resonant tank circuit for frequencies below 20 MHz. Use of noninverting output makes oscillator section of Motorola MC10116 IC function simply as amplifier. Second section is connected as Schmitt trigger to improve signal waveform. Third section is buffer providing complementary outputs.—B. Blood, "IC Crystal Controlled Oscillators," Motorola, Phoenix, AZ, 1977, AN-417B, p 4.

GATED 5-MHz RELAXATION—Output always starts in same phase with respect to gating signal. Frequency-selective network R_1-R_2-C provides positive feedback around MC 1545G gate-controlled wideband amplifier.—F. Macli, IC Op Amp Makes Gated Oscillator, *EDN Magazine,* Sept. 1, 1972, p 52.

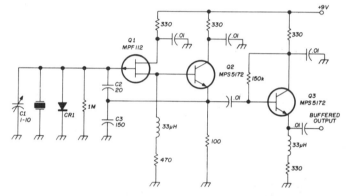

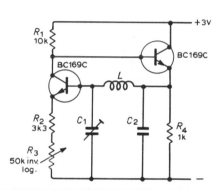

10–20 MHz CRYSTAL—Modification of basic Colpitts crystal oscillator has excellent load capacitance correlation and temperature stability. Crystal will oscillate very close to its series resonant point. Component values are optimized for 10–20 MHz. Emitter-follower Q2 provides power gain for feedback energy and gives high crystal activity without changing phase angle of signal. Output buffer Q3 prevents loading of oscillator. Q1 is low-cost Motorola JFET, but practically any other JFET will work. CR1 is 1N914 or 1N4148.—D. L. Stoner, High-Stability Crystal Oscillator, *Ham Radio,* Oct. 1974, p 36–39.

GENERAL-PURPOSE UP TO 10 MHz—Variation of Colpitts oscillator uses negative feedback at all frequencies at which LC network does not provide phase inversion and voltage step-up. Choose values for coil and capacitors to give frequency desired. R_3 serves as regeneration control and for changing waveform of output.—G. W. Short, Good-Tempered LC Oscillator, *Wireless World,* Feb. 1973, p 84.

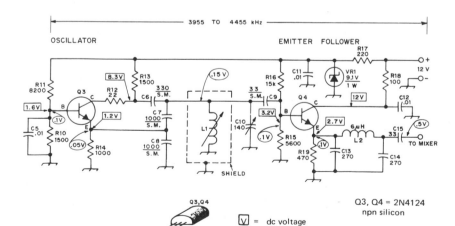

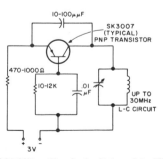

3.955–4.455 MHz VFO—Basic Colpitts LC oscillator designed for 80-meter receiver with 455-kHz IF uses zener in supply line to minimize frequency drift. Emitter-follower buffer contributes to stability by isolating oscillator from mixer. Low-pass filter C13-L2-C14 attenuates harmonic currents developed in Q3 and Q4. L1 is Miller 4503 1.7–2.7 μH variable inductor. L2 is 48 turns No. 30 enamel closewound on ¼-inch wood dowel or polystyrene rod. Main tuning capacitor C10 can be 365-pF unit with six of rear rotor plates removed.—D. DeMaw and L. McCoy, Learning to Work with Semiconductors, *QST,* June 1974, p 18–22 and 72.

UP TO 30 MHz—Simple single-transistor RF oscillator is easily assembled from noncritical parts. Tuning capacitor and coil determine frequency.—Circuits, *73 Magazine,* July 1977, p 35.

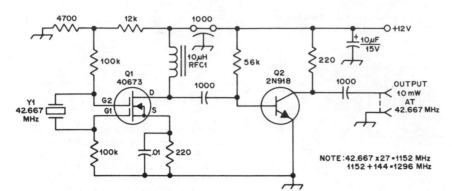

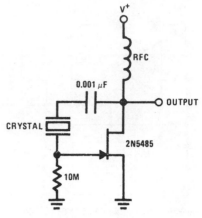

42.667-MHz MOSFET—Unusual crystal-controlled oscillator, similar to modified Pierce oscillator that uses crystal between grids 1 and 2 of tetrode tube, can be used as local oscillator in VHF and UHF converters. No trimming or tuning is required to get overtone frequency. If fundamental of crystal is desired, increase RFC1 to 100 μH or replace it with 1K resistor. Stability is excellent. Circuit works well with supply as low as 4 V.—G. Tomassetti, Dual-Gate MOSFET Offers an Unusual Crystal-Controlled Oscillator Concept, *QST*, June 1976, p 39.

JFET PIERCE CRYSTAL—Basic JFET oscillator circuit permits use of wide frequency range of crystals. High Q is maintained because JFET gate does not load crystal, thereby ensuring good frequency stability.—"FET Databook," National Semiconductor, Santa Clara, CA, 1977, p 6-26–6-36.

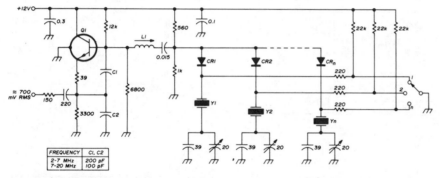

SWITCHED CRYSTALS—High stability is combined with multichannel selection by diode switching of crystals in range of 2–20 MHz, used in series-resonant mode. L1 is about 30 μH at 2 MHz and 1 μH at 20 MHz. Q1 is 2N708, HEP50, BC108, or similar NPN RF type. Diodes are switching types such as BAY67.—U. Rohde, Stable Crystal Oscillators, *Ham Radio*, June 1975, p 34–37.

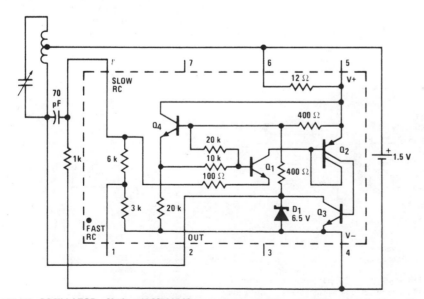

800-kHz OSCILLATOR—National LM3909 IC operating from single 1.5-V cell is used with standard AM radio ferrite antenna coil having tap 40% of turns from one end, with standard 365-pF tuning capacitor across coil. Developed for demonstrating versatility of this low-voltage IC.—"Linear Applications, Vol. 2," National Semiconductor, Santa Clara, CA, 1976, AN-154, p 8.

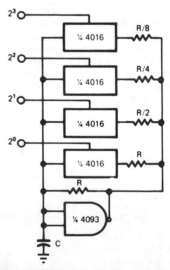

DIGITAL CONTROL TO 100 kHz—Schmitt trigger function of CD4093B IC gives oscillator operation over four decades of frequency without changing C. Basic frequency value is equal to k/RC, with k equal to 1.3 up to about 5 kHz and decreasing gradually to 1.0 at 100 kHz. Use of CD4016 quad transmission gate permits remote switching in of additional resistors to provide direct digital control of frequency. Arrangement shown gives choice of five unrelated frequencies, but binary selection of binary-weighted resistors will give choice of 16 unrelated frequencies.—R. Tenny, CMOS Oscillator Features Digital Frequency Control, *EDN Magazine*, June 5, 1976, p 114 and 116.

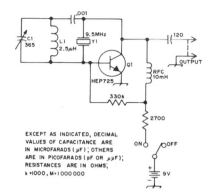

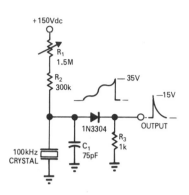

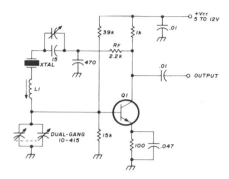

INCREASING CRYSTAL FREQUENCY—Adding parallel resonant circuit across crystal, tuned slightly above crystal frequency, makes oscillator frequency increase. Some plated crystals will work better than others in this circuit; third-overtone types operating on their fundamental generally give best results. Article covers theory of operation.—L. Lisle, The Tunable Crystal Oscillator, *QST*, Oct. 1973, p 30–32.

100-kHz CRYSTAL-DIODE RELAXATION—Crystal-controlled relaxation oscillator uses 1N3304 four-layer diode as active element. R_1 adjusts RC time constant so oscillator locks at fundamental frequency of crystal or at half this frequency.—R. D. Clement and R. L. Starliper, Crystal-Controlled Relaxation Oscillator, *EDN/EEE Magazine*, Oct. 15, 1971, p 62 and 64.

VARIABLE CRYSTAL—Maximum frequency shift is almost 10 kHz at 5 MHz. Use crystal made especially for variable operation. Frequency stability is good even at extremes of shift. Use 5–20 μH for L1 with crystals from 6–15 MHz, and 20–50 μH for 3–6 MHz. Q1 is 2N3563, 2N3564, 2N5770, BC107, BC547, BF115, BF180, SE1010, or equivalent.—R. Harrison, Survey of Crystal Oscillators, *Ham Radio*, March 1976, p 10–22.

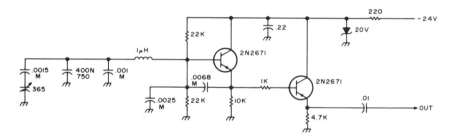

5 MHz ± 250 kHz—Simple and stable circuit using PNP transistors has tuning range of about 250 kHz in any segment of 5–9 MHz range, depending on how oscillator coil is set. Wind coil on ceramic form or use air-wound coil. Capacitors marked M should be mica for stability. Tuning capacitor is 365 pF, from AM radio. 400/N750 temperature-compensating capacitor can be replaced by 400-pF mica unless VFO is used in mobile application.—An Accessory VFO—the Easy Way, *73 Magazine*, Aug. 1975, p 103 and 106–108.

C2 — Double-bearing variable capacitor, 50 pF.
C3 — Miniature 30-pF air variable.
CR1 — High-speed switching diode, silicon type 1N914A.
L18 — 17- to 41-μH slug-tuned inductor, Q_u of 175 (J. W. Miller 43A335CBI in Miller S-74 shield can).
L19 — 10- to 18.7-μH slug-tuned pc-board inductor (J. W. Miller 23A155RPC).
RFC13, RFC14 — Miniature 1-mH rf choke (J. W. Miller 70F103AI).
VR2 — 8.6-V, 1-W Zener diode.

2.255–2.455 kHz LOCAL OSCILLATOR—Used in 1.8–2 MHz communication receiver having wide dynamic range. Oscillator has good stability, with circuit noise at least 90 dB below fundamental output. Amplifier Q14 provides required +7 dBm for injection into balanced mixer of receiver. Two-part article gives all other circuits of receiver.—D. DeMaw, His Eminence—the Receiver, *QST*, Part 1—June 1976, p 27–30 (Part 2—July 1976, p 14–17).

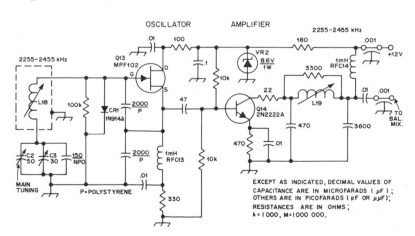

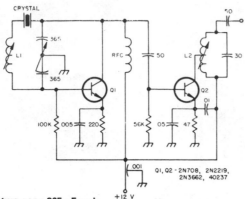

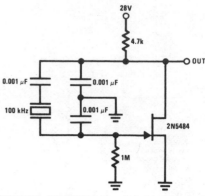

8 MHz ± 5 kHz—Tuning two-gang 365-pF variable capacitor through its range provides frequency change up to 5 kHz in output of 8-MHz crystal oscillator. L1 is 16–24 μH Miller 4507, and L2 is 40 turns No. 36 tapped at 13 turns, on ¼-inch slug-tuned form.—Circuits, *73 Magazine,* Jan. 1974, p 128.

CRYSTAL COLPITTS—Circuit is ideal for low-frequency crystal oscillators because JFET circuit loading does not vary with temperature. Output frequency is determined by threshold used.—"FET Databook," National Semiconductor, Santa Clara, CA, 1977, p 6-26–6-36.

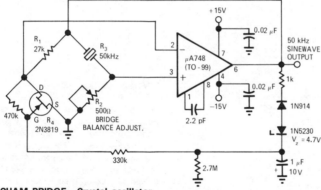

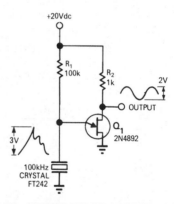

50-kHz MEACHAM BRIDGE—Crystal oscillator using Meacham bridge requires no transformers for producing low-distortion sine-wave output. Quartz crystal should be cut for operation in series-resonant mode. With minor modifications, same circuit can be used for 100- and 200-kHz crystals. By adding single-transistor stage, oscillator can be used as clock generator for TTL circuits.—K. J. Peter, Stable Low-Distortion Bridge Oscillator, *EDN/EEE Magazine,* Nov. 15, 1971, p 50–51.

100-kHz CRYSTAL-FET RELAXATION—Adding crystal in frequency-determining circuit improves frequency stability of UJT relaxation oscillator. With charging capacitor replaced by 100-kHz quartz crystal, measured output frequency was 99.925 kHz.—R. D. Clement and R. L. Starliper, Crystal-Controlled Relaxation Oscillator, *EDN/EEE Magazine,* Oct. 15, 1971, p 62 and 64.

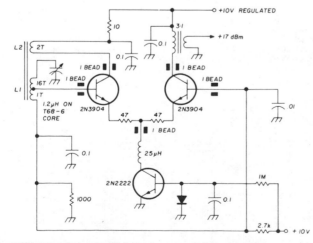

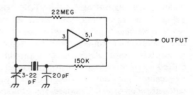

LOW-NOISE 5-MHz—Very low-noise high-Q LC oscillator operating at 5 MHz is designed for use in high-performance communication receivers. Oscillator uses two stages, one operating in class A and the other operating as limiter that also serves as feedback path.—U. L. Rohde, Effects of Noise in Receiving Systems, *Ham Radio,* Nov. 1977, p 34–41.

CRYSTAL WITH CMOS INVERTER—Simple mono multivibrator circuit using MC14007 or CD4007 operates in frequency range from 10 kHz up to top limit of about 10 MHz, with exact frequency depending on values used for R and C. Pin 7 of IC is VSS and pin 14 is VDD. Pins 5 and 1 must be connected together for proper operation.—W. J. Prudhomme, CMOS Oscillators, *73 Magazine,* July 1977, p 60–63.

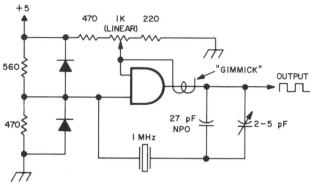

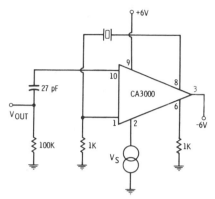

1 MHz WITH ONE GATE—Crystal oscillator uses only one section of SN7408 TTL quad AND gate. Use series-resonant crystal having 30-pF series capacitance. Adjust 1K pot for reliable start-up and symmetrical square-wave output. Diodes are 1N34A or 1N914. Gimmick is 1 or 2 turns of insulated wire wrapped around output lead.—Clyde E. Wade, Jr., An Even Simpler Clock Oscillator, *73 Magazine*, Nov./Dec. 1975, p 164.

MODULATED CRYSTAL—CA3000 differential amplifier is operated as efficient crystal-controlled oscillator. Output frequency depends on crystal. If desired, RF output can be modulated with low-frequency tone applied between pin 2 and ground.—E. M. Noll, "Linear IC Principles, Experiments, and Projects," Howard W. Sams, Indianapolis, IN, 1974, p 91.

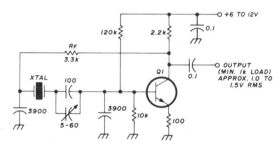

50–500 kHz CRYSTAL—Parallel-mode low-frequency oscillator makes excellent BFO for 455 kHz. If oscillator will not start, reduce value of feedback resistor R_F. Increasing R_F reduces harmonic output, but oscillator may then take up to 20 s to reach full output. For crystals with specified load capacitance of 30 or 50 pF, remove 100-pF capacitor C1 in series with crystal. Q1 is 2N2920, 2N2979, 2N3565, 2N3646, 2N5770, BC107, or BC547.—R. Harrison, Survey of Crystal Oscillators, *Ham Radio*, March 1976, p 10–22.

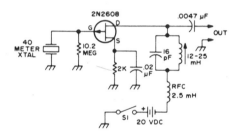

7 MHz—Uses single Siliconix 2N2608 FET. Keep leads short. Coil can be air-wound or permeability-tuned. If tuning capacitor is variable, coil value can be fixed. RF output level depends on circuit voltages and on activity of crystal used.—Q & A, *73 Magazine*, April 1977, p 165.

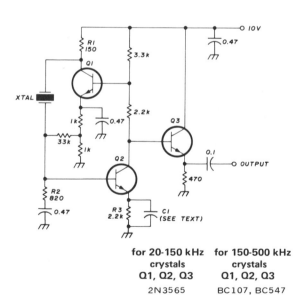

for 20-150 kHz crystals Q1, Q2, Q3	for 150-500 kHz crystals Q1, Q2, Q3
2N3565	BC107, BC547
2N2920	2N3565
2N2979	2N5770
	2N2222

20–500 kHz CRYSTAL—Series-mode oscillator requires no tuned circuit, gives choice of sine or square output, and has good frequency and mode stability. Works nicely with troublesome FT241 crystals. If any crystal fails to start reliably, increase R1 to 270 ohms and R2 to 3.3K. For square-wave operation, C1 is 1-μF nonelectrolytic. Omit C1 for sine-wave operation; harmonic output is then quite low, with second harmonic typically −30 dB. Output is about 1.5-VRMS sine wave or 4-V square wave.—R. Harrison, Survey of Crystal Oscillators, *Ham Radio*, March 1976, p 10–22.

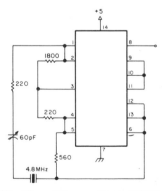

4.8 MHz—Uses all four sections of 7400 quad dual-input NAND gate to give 4.8 MHz output at pin 8, as harmonic-rich square wave. Can cause severe television interference during testing. Article gives five other crystal oscillator circuits using same IC.—A. MacLean, How Do You Use ICs?, *73 Magazine*, Oct. 1976, p 38–41.

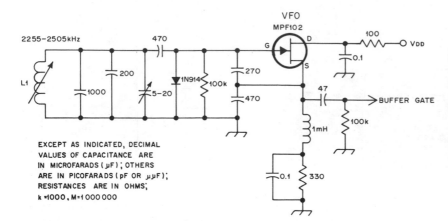

PRECISION VFO—Permeability-tuned oscilla-tor provides stability and linearity at low cost for receivers with 160-meter tunable IF stages. L1 has 28 turns No. 36 enamel closewound on J. W. Miller form 64A022-2. Article covers con-struction of tuning dial, incuding contouring of L1 core to give good dial linearity. Frequency coverage is 2.255–2.505 MHz. Direct-reading dial is accurate within 1.5 kHz over entire 250-kHz tuning range.—W. A. Gregoire, Jr., A Permeability-Tuned Variable-Frequency Oscil-lator, *QST*, March 1978, p 26–28.

7 MHz ± 50 kHz—Requires no tuning capaci-tors. Collector-to-base junctions of two 2N3053 transistors perform function of varactor diodes to provide tuning over range of about 50 kHz centered on 7 MHz. Capacitors marked M should be mica.—An Accessory VFO—the Easy Way, *73 Magazine*, Aug. 1975, p 103 and 106–108.

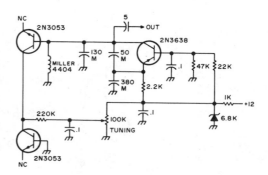

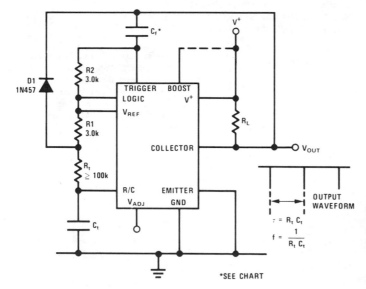

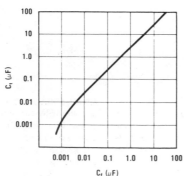

TIMER AS OSCILLATOR—Output of National LM122 timer is fed back to trigger input through capacitor to give self-starting oscillator. Fre-quency is $1/R_tC_t$. Output is narrow negative pulse having duration of about $2R2C_f$. Conser-vative value for C_f for optimum frequency sta-bility can be chosen from graph based on size of timing capacitor C_t.—C. Nelson, "Versatile Timer Operates from Microseconds to Hours," National Semiconductor, Santa Clara, CA, 1973, AN-97, p 10.

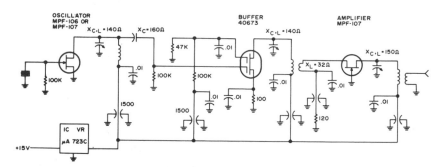

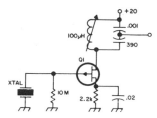

STABLE CRYSTAL—Stability is good enough for microwave transmitter frequency control. Will operate with fundamental or overtone crystals from 1.6 to 160 MHz, with coils and capacitors being chosen for frequency in use.—Circuits, *73 Magazine,* May 1973, p 105.

1-MHz FET PIERCE—Field-effect transistor serves in place of vacuum triode in Pierce oscillator. Circuit values are for 1 MHz, but tuned circuit can be adjusted to other desired frequency. Q1 can be 2N4360 or TIM12.—Circuits, *73 Magazine,* March 1974, p 89.

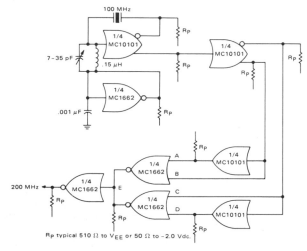

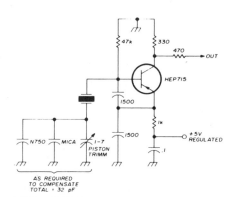

200 MHz WITH OSCILLATOR/DOUBLER—One section of Motorola MC10101 is connected as 100-MHz crystal oscillator having crystal in series with feedback loop. LC tank circuit tunes 100-MHz harmonic of crystal and can be used to adjust circuit to exact frequency. Second section of IC serves as buffer and gives complementary 100-MHz signals for frequency doubler having two MC10101 gates as phase shifters and two MC1662 NOR gates. Outputs of MC1662s are wired-OR connected to give 200-MHz signal. One of remaining MC1662 gates is used as bias generator for oscillator.—B. Blood, "IC Crystal Controlled Oscillators," Motorola, Phoenix, AZ, 1977, AN-417B, p 5.

4-MHz CRYSTAL—High-stability crystal oscillator uses two 1500-pF capacitors to swamp out internal impedance changes that might cause frequency drift. For best stability when used as frequency standard, choose high-accuracy 4-MHz crystal.—B. Kelley, Universal Frequency Standard, *Ham Radio,* Feb. 1974, p 40–47.

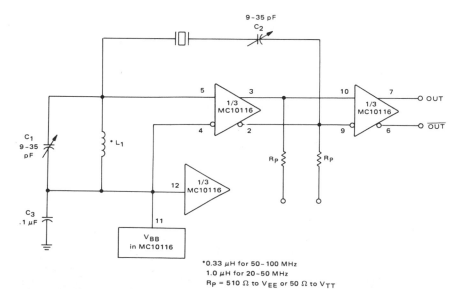

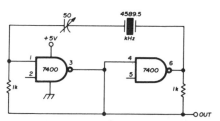

20–100 MHz OVERTONE CRYSTAL—Adjustable tank circuit C_1L_1 ensures operation at desired crystal overtone. Reference voltage for differential amplifier is supplied internally by Motorola 10116 IC and is nominally −1.3 V.—B. Blood, "IC Crystal Controlled Oscillators," Motorola, Phoenix, AZ, 1977, AN-417B, p 3.

TTL 4.59-MHz CRYSTAL—Uses FT243 crystal hand-ground to 4.5895 MHz, with 50-pF series capacitor allowing frequency to be trimmed to exactly 4.59 MHz for use in AFSK generator.—J. Nugues, AFSK Generator, *Ham Radio,* July 1976, p 69.

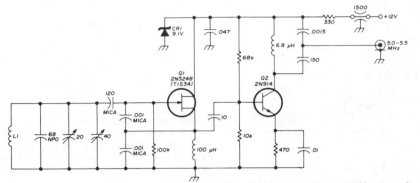

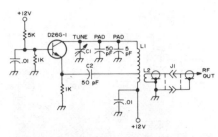

5–5.5 MHz VFO—Used in solid-state five-band communication receiver. Temperature compensation is provided by 20-pF trimmer that sets band center. L1 is 34 turns No. 24 on Amidon T50-6 toroid core.—P. Moroni, Solid-State Communications Receiver, *Ham Radio*, Oct. 1975, p 32–41.

51–55 MHz—Tunable local oscillator is padded to tune over range required for use with 1.65-MHz IF in 6-meter receiver, using Johnson type U 14-plate tuning capacitor. Can also serve as test transmitter putting out up to 20 mW. L1 is 9 turns No. 26 tapped 1 turn from low end, and L2 is 1 or 2 turns. Article covers construction in 1¼ × 1¼ × ½ inch box.—B. Hoisington, A Real Hot Front End for Six, *73 Magazine*, Nov. 1974, p 88–90 and 92–94.

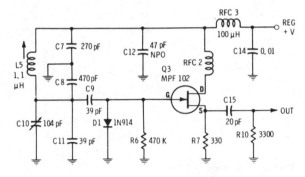

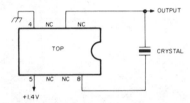

10-MHz VFO—Values shown for high-stability variable-frequency oscillator give operation in 10-MHz range. Stable supply voltage is essential. Use silver mica capacitors in gate circuit for maximum stability.—E. M. Noll, "FET Principles, Experiments, and Projects," Howard W. Sams, Indianapolis, IN , 2nd Ed., 1975, p 193–194.

465-kHz FOR IF TUNE-UP—Simple crystal oscillator using National LM3909N is adjusted to exactly desired frequency with capacitor in series with pin 8. Drain from AA cell is less than 0.5 mA at 1.2 V. Use 465-kHz crystal and couple oscillator to receiver input with 100-pF capacitor. With 100-kHz crystal, circuit will generate strong harmonics beyond 30 MHz; to zero-beat with WWV, use about 10 pF in series with crystal.—I. Queen, Simple Crystal Oscillator, *Ham Radio*, Nov. 1977, p 98.

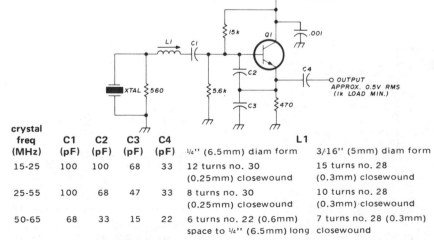

crystal freq (MHz)	C1 (pF)	C2 (pF)	C3 (pF)	C4 (pF)	L1 ¼" (6.5mm) diam form	3/16" (5mm) diam form
15-25	100	100	68	33	12 turns no. 30 (0.25mm) closewound	15 turns no. 28 (0.3mm) closewound
25-55	100	68	47	33	8 turns no. 30 (0.25mm) closewound	10 turns no. 28 (0.3mm) closewound
50-65	68	33	15	22	6 turns no. 22 (0.6mm) space to ¼" (6.5mm) long	7 turns no. 28 (0.3mm) closewound

15–65 MHz IMPEDANCE-INVERTING—Uses third-overtone crystals. L1 trims crystal frequency. Resistor across crystal prevents oscillation at undesired modes. Starting is reliable and stability is good. Q1 is 2N3563, 2N3564, 2N5770, BF180, BF200, or SE1010.—R. Harrison, Survey of Crystal Oscillators, *Ham Radio*, March 1976, p 10–22.

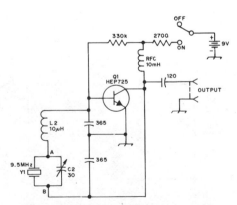

9.5-MHz TUNABLE CRYSTAL—Clapp oscillator with inductance in series with crystal can be tuned with C2 as much as 100 kHz below rated frequency of crystal. Based on making crystal act as capacitive reactance below its series-resonant frequency. Circuit can be adapted to other amateur bands by keeping reactances of various components approximately the same.—L. Lisle, The Tunable Crystal Oscillator, *QST*, Oct. 1973, p 30–32.

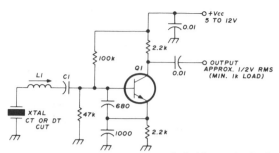

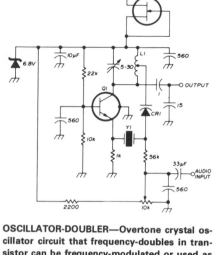

150–500 kHz CRYSTAL—Circuit is series-mode if C1 is 0.01 μF. Parallel-mode crystals can be used if C1 is equal to specified load capacitance (30, 50, or 100 pF) for crystal. Harmonic output is usually better than −30 dB. Circuit is particularly good for crystals prone to oscillate un-desirably at twice fundamental frequency. L1 is 800–2000 μH for 150–300 kHz, and 360–1000 μH for 300–500 kHz. Adjusting slug in L1 pulls crystal frequency. Q1 is 2N3563, 2N3564, 2N3693, BC107, BC547, or SE1010.—R. Harrison, Survey of Crystal Oscillators, *Ham Radio*, March 1976, p 10–22.

OSCILLATOR-DOUBLER—Overtone crystal oscillator circuit that frequency-doubles in transistor can be frequency-modulated or used as stable voltage-controlled crystal oscillator. Tuning range with 70-MHz third-overtone crystal is typically 30 kHz at crystal frequency or 60 kHz at output. L1 is resonant with C1 at desired output frequency. Tap for varactor CR1 (Motorola BB105B or BB142) is at one-fourth total number of turns. Q1 is 2N918, BF115, HEP709, or equivalent.—U. Rohde, Stable Crystal Oscillators, *Ham Radio*, June 1975, p 34–37.

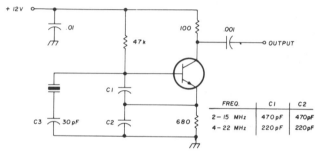

FREQ.	C1	C2
2 – 15 MHz	470 pF	470 pF
4 – 22 MHz	220 pF	220 pF

2–22 MHz FUNDAMENTAL-MODE—International Crystal OF-1 oscillator for fundamental-mode crystal has no LC tuned circuits and requires no inductors. With 28.3-MHz third-over-tone crystal, output is at fundamental of crystal or about 9.43 MHz.—C. Hall, Overtone Crystal Oscillators Without Inductors, *Ham Radio*, April 1978, p 50–51.

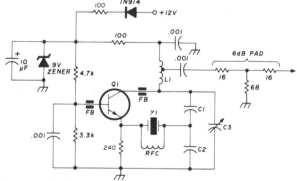

C1	10 pF mica
C2	20 to 60 pF mica. Use as high value as possible (until circuit just oscillates reliably when C3 is tuned through resonance)
C3	20 pF piston or miniature trimmer
L1	8 turns no. 24 (0.5mm) on Amidon T37-12 toroid core, tapped 3 turns from cold end
Q1	Fairchild 2N5179 recommended but 2N2857, 2N3563, 2N918 or equivalent may be substituted
RFC	0.39 μH. Resonates with crystal holder capacitance (4 to 6 pF typical) for parallel resonance at crystal frequency
Y1	90 to 125 MHz, 5th or 7th overtone, series-resonant, HC-18/U crystal. Cut leads as short as possible (¼″ or 6mm maximum)

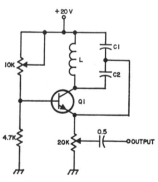

Q1 - 2N2925, 2N3392

FREQUENCY	C1	C2	L
50 kHz	3500 pf	1500 pf	10 mH
80 kHz	2200 pf	910 pf	6.2 mH
100 kHz	1800 pf	750 pf	4.7 mH
200 kHz	910 pf	390 pf	2.2 mH
455 kHz	390 pf	160 pf	1 mH
1000 kHz	180 pf	75 pf	0.47 mH

90–125 MHz CRYSTAL—Recmmended for VHF/UHF converters. Output is 5 to 15 mW. Crystal should be high-quality fifth- or seventh-overtone type. Ferrite bead FB prevents undesired oscillation above 500 MHz. For best stability, allow crystal to operate at its natural series-resonant frequency and use regulated power supply.—J. Reisert, VHF/UHF Techniques, *Ham Radio*, March 1976, p 44–48.

50–1000 kHz—Simple single-transistor circuit provides extremely stable beat-frequency oscillator for which frequency can be changed by using tank-circuit components listed in table.—Circuits, *73 Magazine*, Feb. 1974, p 101.

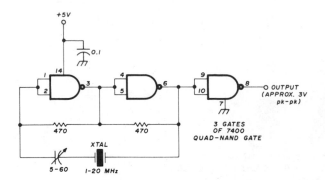

NAND-GATE TTL CRYSTAL—Overcomes problems of poor starting performance and has upper frequency limit of 20 MHz. Suitable for applications requiring high-output aperiodic oscillator. Excellent as frequency marker.—R. Harrison, Survey of Crystal Oscillators, *Ham Radio*, March 1976, p 10–22.

2–20 MHz VXO—Variable-frequency crystal oscillator plus buffer, using Signetics N7404A hex inverter or equivalent, covers 2–20 MHz. Only three inverters are used, two forming oscillator and one as output buffer. V_{cc} is +5 V. Crystals can operate at fundamental, third, or fifth overtone. Frequency-limiting capacitor C_P can be 15 pF. Only higher-frequency crystals can be moved useful amounts without creating instability problems. Article gives design equations and tables showing frequencies obtained with various crystals for various values of frequency controls C_V (0–100 pF) and L_V (0–17 μH).—B. King, Hex Inverter VXO Circuit, *Ham Radio*, April 1975, p 50–55.

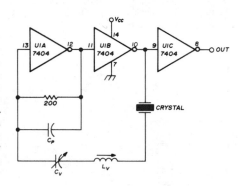

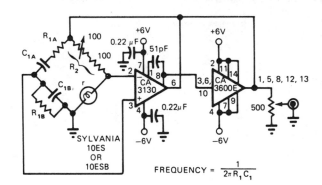

$$FREQUENCY = \frac{1}{2\pi R_1 C_1}$$

CAPACITIVELY TUNED WIEN—Output of amplifier is connected to apex of Wien bridge. Positive feedback is taken from junction of C_{1A} and C_{1B} for noninverting input of first opamp, while negative feedback is taken from other junction of bridge for inverting input. Oscillation is sustained when $R_2 = 2r$. Nonlinearity of lamp r provides stabilization of oscillator. Frequency depends on values used for bridge components.—H. D. Olson, Wien-Bridge Oscillator Is Capacitively Tuned, *EDN Magazine*, Aug. 5, 1975, p 74.

65–110 MHz OVERTONE—Uses fifth- or seventh-overtone crystals. RF choke formed by L2 is wound on low-value resistor to suppress lower-frequency resonances of crystal. Buffer is recommended. Circuit is slightly frequency-sensitive to supply voltage variations, so use well-regulated supply. Q1 is 2N3563, 2N3564, 2N5770, BF180, BF200, or SE1010.—R. Harrison, Survey of Crystal Oscillators, *Ham Radio*, March 1976, p 10–22.

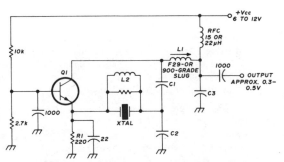

L1	65-85 MHz:	7 turns no. 22 (0.6mm) or no. 24 (0.5mm) enamelled, closewound on 3/16" (5mm) diameter form		
	85-110 MHz:	4 turns no. 22 (0.6mm) or no. 24 (0.5mm) enamelled, on 3/16" (5mm) diameter form, turns spaced one wire diameter		
L2		10 turns no. 34 (0.2mm) closewound on low-value ¼-watt resistor		
C1	65-85 MHz:	15 pF	85-110 MHz:	10 pF
C2	65-85 MHz:	150 pF	85-110 MHz:	100 pF
C3	65-85 MHz:	100 pF	85-110 MHz:	68 pF

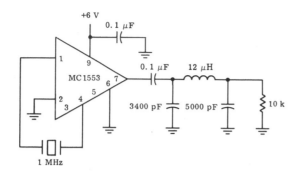

1-MHz SERIES-MODE CRYSTAL—Motorola MC1553 video amplifier provides wide bandwidth and output swing capability needed for high-frequency master clock or local oscillator in many system designs. Positive feedback is injected through crystal to input pin 1. Output is taken from pin 7 which is buffered internally from oscillator by gain and emitter-follower stages. Brute-force pi filter at output extracts desired fundamental frequency.—"A Wide Band Monolithic Video Amplifier," Motorola, Phoenix, AZ, 1973, AN-404, p 9.

9-MHz LINEAR VCO—U1A and U1C of RCA CA3046 transistor array form emitter-coupled oscillator. Portion of U1A current is diverted through U1B and L1, producing magnetic flux that reduces effective inductance of resonating coil L2. Output frequency is varied in direct proportion to voltage applied at A. L1 is 23 turns on ¾-inch Teflon form 2 inches long, with 4 turns wound between windings for L2. VR1 is 1N3828 6.2-V zener. Circuit must be well grounded and shielded to avoid hum pickup by input, which could modulate output.—D. G. Stephenson, A Second Look at Linear Tuning, *QST*, March 1977, p 40–41.

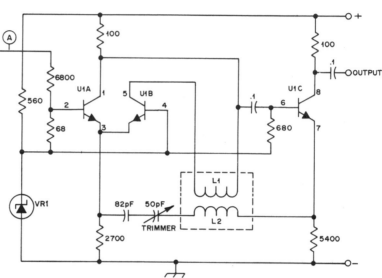

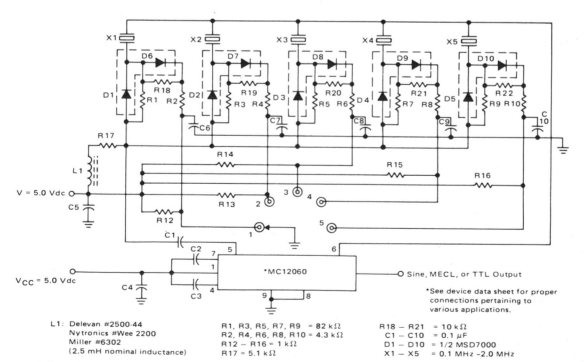

L1: Delevan #2500-44
Nytronics #Wee 2200
Miller #6302
(2.5 mH nominal inductance)

R1, R3, R5, R7, R9 = 82 kΩ
R2, R4, R6, R8, R10 = 4.3 kΩ
R12 − R16 = 1 kΩ
R17 = 5.1 kΩ

R18 − R21 = 10 kΩ
C1 − C10 = 0.1 μF
D1 − D10 = 1/2 MSD7000
X1 − X5 = 0.1 MHz −2.0 MHz

CRYSTAL-SWITCHING DIODES—Circuit for Motorola MC12060 crystal oscillator uses diodes as RF switches giving choice of five different crystal frequencies. Forward bias is applied to diode associated with desired crystal and reverse bias to diodes for other four crystals. Diode switching eliminates need to run high-frequency signals through mechanical switch, permits control of switching from remote location, and is readily adapted to electronic scanning. Requires only single 5-V supply. Frequency pulling is minimized.—J. Hatchett and R. Janikowski, "Crystal Switching Methods for MC12060/MC12061 Oscillators," Motorola, Phoenix, AZ, 1975, AN-756.

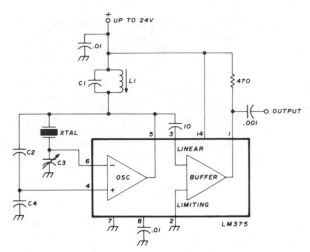

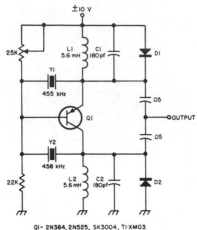

IC CRYSTAL—Uses LM375 IC with crystals from 3 to 20 MHz in parallel mode. Will oscillate with only 4-V supply, but output voltage increases with supply voltage. L1-C1 is resonant at crystal frequency. Adjust L1 only for maximum output, not for trimming frequency. If C3 is 3–30 pF, it can be used to adjust frequency of crystal.—R.

crystal freq (MHz)	C2/C3 (pF)	C4 (pF)
3-10	22	180
10-20	10	82

Harrison, Survey of Crystal Oscillators, *Ham Radio,* March 1976, p 10–22.

DUAL-FREQUENCY CRYSTAL—Uses two different crystals, with frequency being changed by reversing supply voltage. Transistor then inverts itself and gain reduces to about 2, which is adequate for oscillator operation. Provides two frequencies from single stage with minimum of switching.—Circuits, *73 Magazine,* Feb. 1974, p 101.

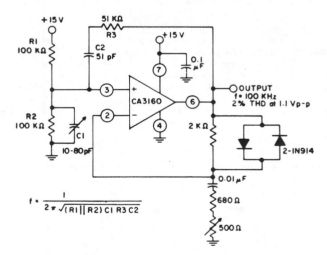

$$f = \frac{1}{2\pi\sqrt{(R1\|R2)\,C1\,R3\,C2}}$$

100-kHz WIEN-BRIDGE—CA3160 opamp in bridge circuit operates from single 15-V supply. Parallel-connected diodes form gain-setting network that stabilizes output voltage at about 1.1 V. 500-ohm pot is adjusted so oscillator always starts and oscillation is maintained.— "Linear Integrated Circuits and MOS/FET's," RCA Solid State Division, Somerville, NJ, 1977, p 271–272.

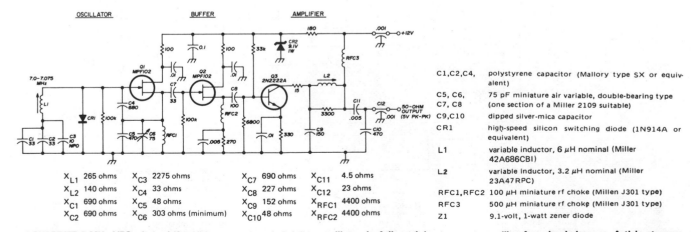

X_{L1} 265 ohms	X_{C3} 2275 ohms	X_{C7} 690 ohms	X_{C11} 4.5 ohms
X_{L2} 140 ohms	X_{C4} 33 ohms	X_{C8} 227 ohms	X_{C12} 23 ohms
X_{C1} 690 ohms	X_{C5} 48 ohms	X_{C9} 152 ohms	X_{RFC1} 4400 ohms
X_{C2} 690 ohms	X_{C6} 303 ohms (minimum)	X_{C10} 48 ohms	X_{RFC2} 4400 ohms

C1,C2,C4,	polystyrene capacitor (Mallory type SX or equivalent)
C5, C6,	75 pF miniature air variable, double-bearing type (one section of a Miller 2109 suitable)
C7, C8	
C9,C10	dipped silver-mica capacitor
CR1	high-speed silicon switching diode (1N914A or equivalent)
L1	variable inductor, 6 μH nominal (Miller 42A686CBI)
L2	variable inductor, 3.2 μH nominal (Miller 23A47RPC)
RFC1,RFC2	100 μH miniature rf choke (Millen J301 type)
RFC3	500 μH miniature rf choke (Millen J301 type)
Z1	9.1-volt, 1-watt zener diode

LOW-DRIFT 7-MHz VFO—Low-drift solid-state design for 40-meter band has maximum change of only 25 Hz from cold start to full warm-up at 25°C. After stabilization, maximum hunting is 5 Hz. Drift is minimized by paralleling two or more capacitors in critical parts of circuit. Series-tuned Colpitts oscillator is followed by two buffer stages, with second providing enough amplification for practical amateur work while further improving isolation of oscillator. Low-impedance output network minimizes oscillator pulling from load changes. Article stresses importance of choosing and using components that minimize drift.—D. DeMaw, VFO Design Techniques for Improved Stability, *Ham Radio,* June 1976, p 10–17.

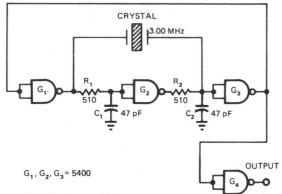

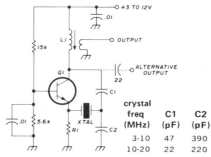

SURE-STARTING CRYSTAL—Loop-within-a-loop oscillator design ensures reliable starting without use of critical components. Frequency depends on crystal, which can be anywhere in range from 1 to 20 MHz. IC can be 54L00 for 1–2 MHz, standard 5400 for 2–6 MHz, and 54H00 or 54S00 for 6–20 MHz. Temperature stability is adequate for crystal clocks and other digital-system applications.—J. E. Buchanan, Crystal-Oscillator Design Eliminates Start-Up Problems, *EDN Magazine*, Feb. 20, 1978, p 110.

3–20 MHz CRYSTAL—Circuit is series-mode oscillator, but parallel-mode crystals can be used if trimmer in series with crystal is replaced by short-circuit. Adjust feedback by varying ratio of C1 to C2. Use grid-dip oscillator to resonate L1 with C1 when crystal is shorted; then remove short and tune slug of L1 to pull crystal exactly to frequency. R1 should be between 100 and 1000 ohms. The lower its value, the lower the crystal power dissipation and the better is stability. Q1 is 2N918, 2N3564, 2N5770, BF200, SE1001, or equivalent.—R. Harrison, Survey of Crystal Oscillators, *Ham Radio*, March 1976, p 10–22.

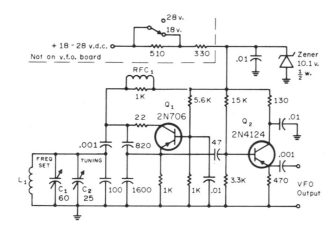

160-METER VFO—Standard Colpitts oscillator Q₁ with emitter-follower Q₂ gives dependability and adequate isolation from later stages. Zener regulation provides stability even with weak battery. Output is about 0.7 VRMS. Low-level parasitic oscillation may occur about 150 kHz below operating frequency but is suppressed by tuned stages following VFO. L₁ is 52 turns No. 28 enamel on Amidon T-50-2 toroid. RFC₁ is 850 μH.—A. Weiss, Design Notes on a Moderate Power Solid State Transmitter for 1.8 MHz, *CQ*, Nov. 1972, p 18–22, 24, 98, 100, and 102.

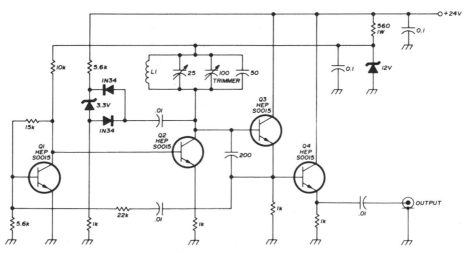

STABLE 3.5–3.8 MHz VFO—Oscillator Q1-Q2, emitter-follower output Q3, and buffer Q4 provide 5 V P-P into 200-ohm load, with good isolation between oscillator and load. Total drift is less than 10 Hz from turn-on, and less than 330 Hz as supply voltage varies between 15 and 30 V. Amplitude stability is within 1 dB over tuning range. Oscillator amplitude is stabilized by two 1N34 diodes and 3.3-V zener. L1 is 25 turns No. 18 closewound on 1.5-in form.—J. Fisk, Circuits and Techniques, *Ham Radio*, June 1976, p 48–52.

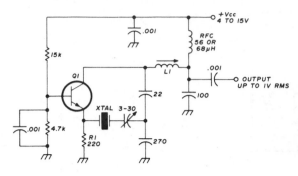

15–65 MHz THIRD-OVERTONE—Crystal starting is reliable, and power dissipation of crystal is well below allowable maximum. Q1 is 2N918, 2N3564, 2N5770, BF180, or BF200. L1 resonates at crystal frequency with 22 pF (1 μH for 15–30 MHz or 0.5 μH for 30–65 MHz). Stability is as good as that of fundamental-frequency oscillator. Set L1 roughly to frequency with no supply voltage by shorting crystal and dipping L1 with grid-dip oscillator. Now apply power and tune L1 close to marked crystal frequency while monitoring output frequency. Remove short and trim to frequency with 3–30 pF trimmer.— R. Harrison, Survey of Crystal Oscillators, *Ham Radio,* March 1976, p 10–22.

18–60 MHz THIRD-OVERTONE—International Crystal OF-1 oscillator for third-overtone crystals requires no inductors. Crystal operates near series resonance, making capacitor unnecessary in series with crystal. With 28.3-MHz third-overtone crystal, circuit delivers 28.3 MHz when C1 is 100 pF and C2 is 18 pF. Using larger values given in table produces oscillation at fundamental of 9.43 MHz.—C. Hall, Overtone Crystal Oscillators Without Inductors, *Ham Radio,* April 1978, p 50–51.

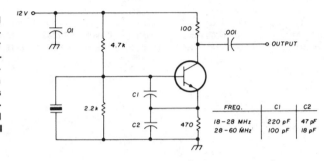

FREQ.	C1	C2
18–28 MHz	220 pF	47 pF
28–60 MHz	100 pF	18 pF

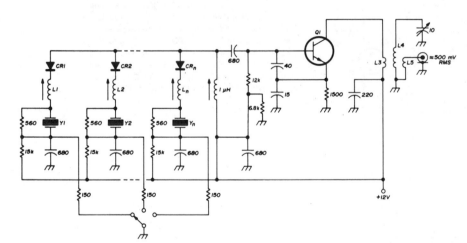

SWITCHED OVERTONE CRYSTALS—Uses third-overtone crystals between 20 and 80 MHz, with diode switching and with frequency doubling in transistors. L1-Ln are series resonant with 10 pF at each crystal frequency. L4 is resonant with 10 pF at desired output frequency. L3 and L5 have one-third as many turns as L4. Q1 is 2N918, BF115, HEP709, or equivalent. Diodes are switching types such as BAY67.—U. Rohde, Stable Crystal Oscillators, *Ham Radio,* June 1975, p 34–37.

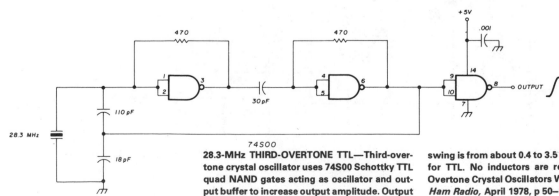

28.3-MHz THIRD-OVERTONE TTL—Third-overtone crystal oscillator uses 74S00 Schottky TTL quad NAND gates acting as oscillator and output buffer to increase output amplitude. Output swing is from about 0.4 to 3.5 V P-P, as required for TTL. No inductors are required.—C. Hall, Overtone Crystal Oscillators Without Inductors, *Ham Radio,* April 1978, p 50–51.

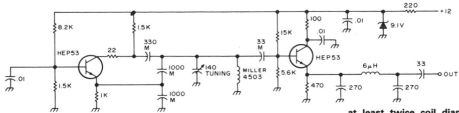

4–4.6 MHz TUNABLE—Emitter-follower buffer stage for isolation has low-pass filter for reducing harmonic output and giving better sine-wave output. Oscillator coil should be in shield at least twice coil diameter.—*An Accessory VFO—the Easy Way, 73 Magazine,* Aug. 1975, p 103 and 106–108.

CHAPTER 63

Phase Control Circuits

Includes circuits for measuring, shifting, comparing, and digitally controlling phase of signal. Many use phase-locked loops. See also Lamp Control, Motor Control, Power Control, and Temperature Control chapters.

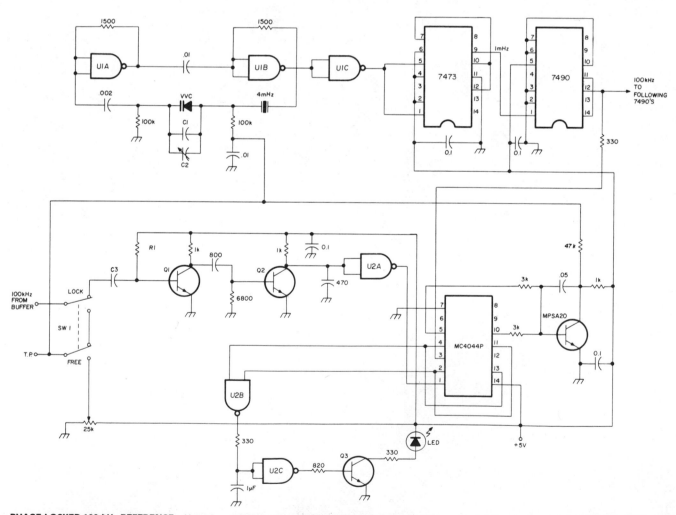

PHASE-LOCKED 100-kHz REFERENCE—Uses 4-MHz crystal in oscillator, with voltage-variable capacitor VVC in parallel with fixed and variable capacitors for setting frequency precisely. Varicap or silicon diode can also be used for VVC. Control voltage for VVC is developed by Motorola MC4044P phase-frequency detector and associated MPSA20 amplifier and filter. 7473 and 7490 ICs divide 4-MHz signal by 4 and then by 10 to give 100 kHz. Main output can be further divided with additional 7490s, down to 60 Hz for driving electric clock if desired. Adjust C3 and R1 for symmetrical square wave at pin 1 of MC4044P, with clean leading and trailing edges. Typical values are 68 pF for C3 and 300K for R1, but values will depend on transistors used.

Transistor types are not critical. Gates U2B and U2C with Q3 form lock indicator circuit that turns on LED when 4-MHz oscillator is phase-locked to output of external high-stability 100-kHz frequency standard. U1 and U2 are SN7400.—C. A. Harvey, How to Improve the Accuracy of Your Frequency Counter, *Ham Radio,* Oct. 1977, p 26—28.

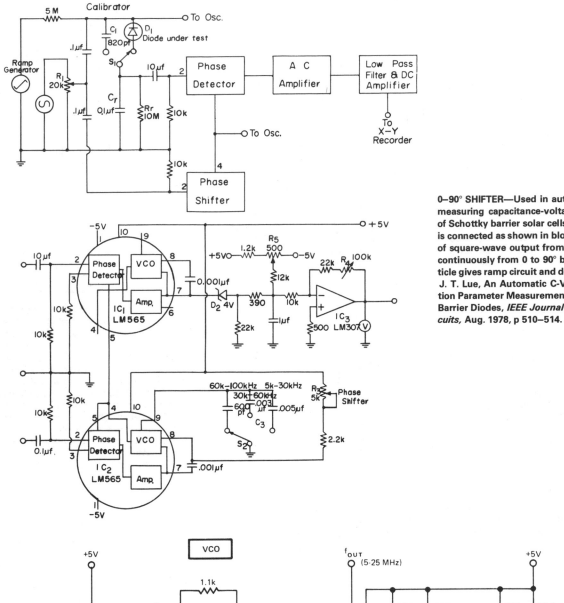

0–90° SHIFTER—Used in automatic plotter for measuring capacitance-voltage characteristics of Schottky barrier solar cells. Diode under test is connected as shown in block diagram. Phase of square-wave output from IC_2 can be shifted continuously from 0 to 90° by adjusting R_3. Article gives ramp circuit and design equations.— J. T. Lue, An Automatic C-V Plotter and Junction Parameter Measurements of MIS Schottky Barrier Diodes, *IEEE Journal of Solid-State Circuits,* Aug. 1978, p 510–514.

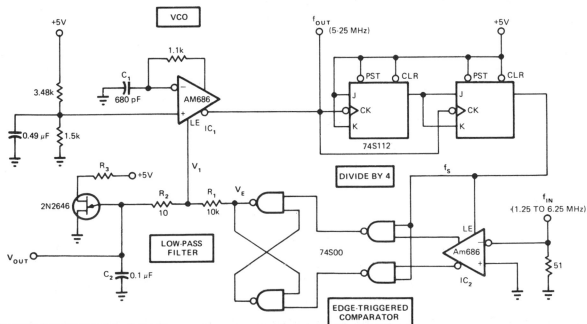

WIDE CAPTURE RANGE FOR PLL—Fast wideband phase-locked loop uses one Am686 latching comparator as voltage-controlled oscillator, while other is coupled with TTL latch to produce edge-triggered comparator. VCO and comparator combined with low-pass filter R_1-R_2-C_2 form PLL. When locking fails, UJT causes V_{OUT} to scan, repetitively sweeping all frequencies in VCO range until lock is restored. Capture and locking ranges are both equal at ±60% for 5-MHz input.—M. C. Hahn, PLL's Capture Range Equals Its Locking Range, *EDN Magazine,* Sept. 20, 1977, p 117 and 119–120.

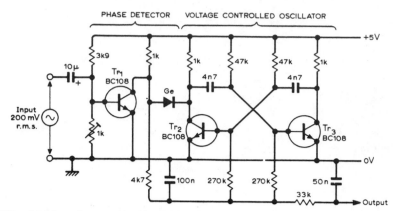

PHASE DETECTOR   VOLTAGE CONTROLLED OSCILLATOR

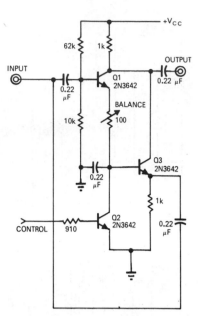

AF PLL—Addition of components to conventional two-transistor MVBR gives simple phase-locked loop. Tr_1 and diode form logic gate that conducts during alternate half-cycles of input and VCO waveforms respectively. Output of this phase detector, when filtered, is most negative when waveforms are in phase, and most positive when they are out of phase. Once phase lock has been established, it is maintained by VCO over range of 100 to 3000 Hz.—J. B. Cole, Simple Phase-Locked Loop, *Wireless World*, June 1977, p 56.

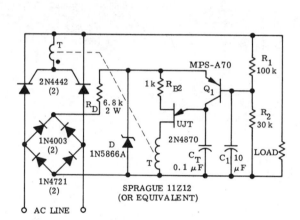

FULL-WAVE FEEDBACK—Used when average load voltage is desired feedback variable for full-wave phase control of load power. Circuit requires use of pulse transformer T.—D. A. Zinder, "Unijunction Trigger Circuits for Gated Thyristors," Motorola, Phoenix, AZ, 1974, AN-413, p 4.

VOLTAGE-CONTROLLED PHASE SHIFTER—Circuit shifts carrier 180° by sensing polarity of modulating voltage. Operating range is 5 kHz to 10 MHz. Circuit can also be used to convert unipolar pulses to alternate bipolar pulses or vice versa when synchronized square wave is supplied to control input. With 0 V at base of Q2, Q3 will amplify RF voltage applied to input, without phase shift. To actuate switch and provide 180° phase shift, positive voltage is applied to base of Q2 so it saturates and cuts off, allowing Q1 to conduct. Output then appears across load with phase reversed.—A. H. Hargrove, Simple Circuits Control Phase-Shift, *EDN Magazine*, Jan. 1, 1971, p 39.

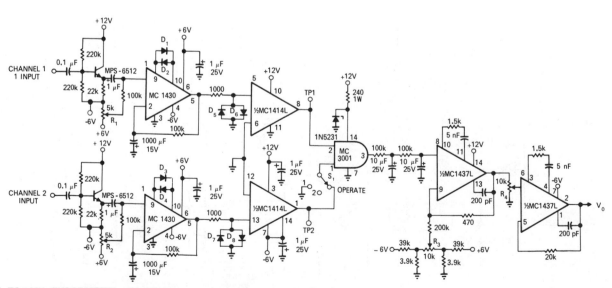

100 Hz TO 1 MHz PHASE METER—Provides better than 2% accuracy over most of frequency range, as required for making Bode plots. Based on squaring two sine waves and comparing amount of overlap to total period of an input wave. This gives directly the amount of phase difference between input wave trains, up to 180°. Instead of measuring periods, overlap is integrated over total period to give average of ON to OFF times that can be read as phase difference on voltmeter. Article gives performance specifications and describes circuit operation in detail.—D. Kesner, IC Phase Meter Beats High Costs, *EDN/EEE Magazine*, Oct. 15, 1971, p 49–52.

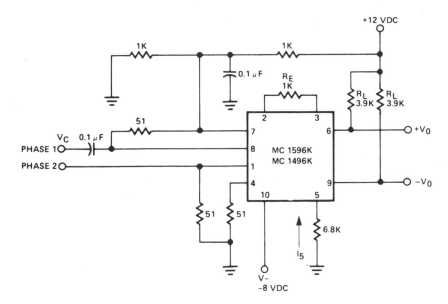

PHASE COMPARATOR—Signetics balanced modulator-demodulator transistor array is connected as phase detector in which output contains term related to cosine of phase angle. Equal-frequency input signals are multiplied together by IC to produce sum and difference frequencies. Difference component becomes DC, while undesired sum component is filtered out. DC component is related to phase angle, with cosine becoming 0 at 90° and having maximum positive or negative value at 0° and 180° respectively. Balanced modulator provides excellent conversion linearity along with conversion gain.—"Signetics Analog Data Manual," Signetics, Sunnyvale, CA, 1977, p 757–758.

0–360° PHASE SHIFTER—Each J202 JFET stage provides up to 180° phase shift under control of 1-megohm pot. Ganged pots give full range of control. JFETs specified are ideal for circuit because they do not load phase-shift networks.—"FET Databook," National Semiconductor, Santa Clara, CA, 1977, p 6-26–6-36.

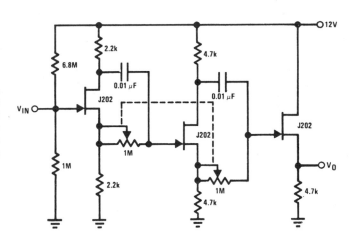

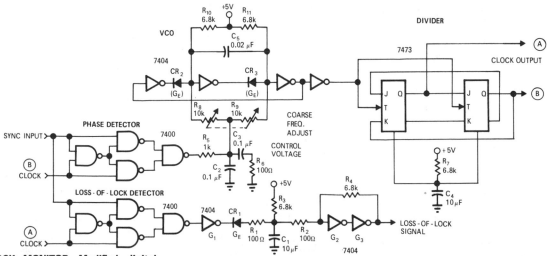

LOSS-OF-LOCK MONITOR—Modified digital phase-locked loop includes second phase detector that provides alarm signal when loop gets out of lock. Output may also be used to disable other circuits. Voltage-controlled multivibrator VCO, operating at 4 times desired clock frequency, drives two-stage switch-tailed ring counter that provides two-phase internal clock signals A and B for detectors. Article describes operation of circuit and gives timing diagrams.—C. A. Herbst, Digital Phase-Locked Loop with Loss-of-Lock Monitor, *EDN/EEE Magazine*, Oct. 15, 1971, p 64–65.

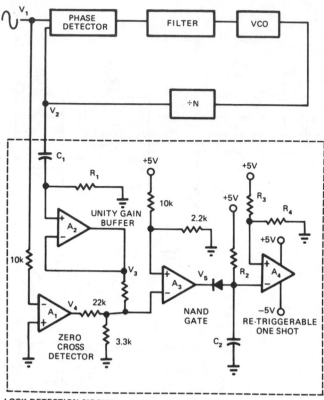

PLL LOST-LOCK INDICATOR—Developed for use with phase-locked loop to indicate both acquisition and loss of lock. Based on concept that lock exists as long as static-phase error is less than 90°. Uses quad opamp package such as RC4136, with A_4 feeding retriggerable one-shot; output of one-shot is low when lock exists. When lock is lost, output of one-shot immediately goes high and remains high until lock is reacquired plus time duration of one-shot. Developed for use in systems where certain processes must be interrupted immediately upon loss of lock.—J. C. Hanisko, PLL Lock Indicator Uses a Single IC, *EDN Magazine,* Oct. 5, 1976, p 104.

LOCK DETECTION CIRCUIT

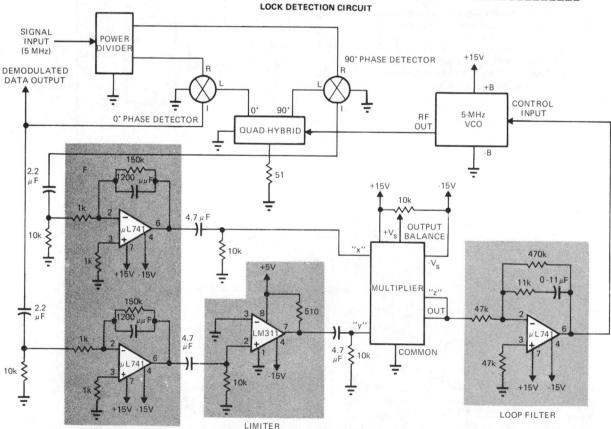

1200-Hz LOW PASS FILTER

LIMITER

LOOP FILTER

DPSK ON DSBSC—Differential phase-shift keyed double-sideband suppressed-carrier signal is demodulated by reinsertion of missing carrier, using synchronous or coherent detection. Receiver input signal is multiplexed by locally generated carrier, accurately controlled in frequency and phase. This is followed by low-pass filtering. Demodulation at output is by frequency/phase controlled loop that automatically locks local oscillator in frequency and phase to received vestige of carrier. This extracts phase information from modulated signal. Power divider is RF Associates H22, quad hybrid is Merrimac Research QHT-2, 0° and 90° phase detectors are Relcom M6A, and multiplier is Analog Devices 4281. Article covers theory and operation of circuit in detail.—R. Hennick, Demodulate DPSK Signals Coherently Using a Costas Phase-Lock Loop, *EDN Magazine,* July 1, 1972, p 44–47.

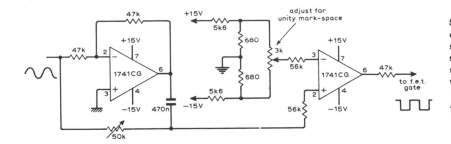

SHIFTING AND SQUARING—Circuit uses two opamps to derive phase-shifted reference square wave and DC output signal of phase-sensitive detector from same sine-wave signal source. Article gives theory of operation and waveforms for various operating conditions.—G. B. Clayton, Experiments with Operational Amplifiers, *Wireless World,* July 1973, p 355–356.

VOLTAGE FEEDBACK—Used when quantity to be sensed is isolated varying DC voltage e_s such as output of tachometer. Operating point is determined by setting of R_c. Output of voltage feedback circuit goes to thyristor in series with load.—D. A. Zinder, "Unijunction Trigger Circuits for Gated Thyristors," Motorola, Phoenix, AZ, 1974, AN-413, p 4.

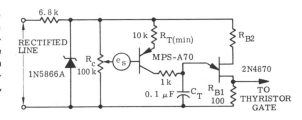

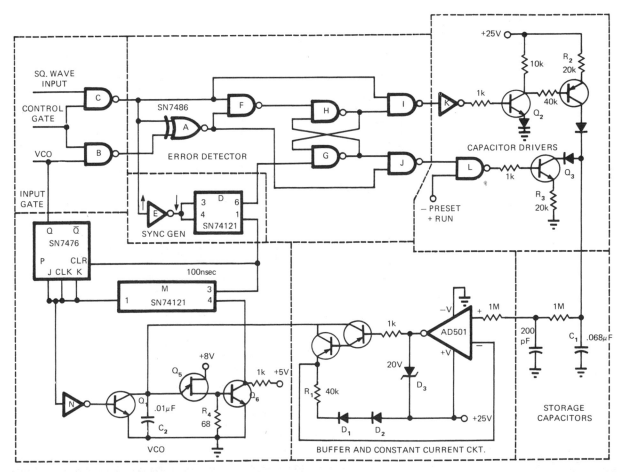

FASTER PHASE LOCK—Circuit was developed to reduce the normally long acquisition time of phase-locked loops when measuring frequency of short signal bursts. Synchronization of VCO to input phase allows correction pulses to be developed in correct polarity only, to give lockup time less than 10 cycles of input when using idling frequency of 12 kHz for VCO. Input signals are compared to those of VCO at EXCLUSIVE-OR gate A. Gating of error pulses by gate F and flip-flop G-H allows I or J to drive current pulses of correct polarity into C_1. Voltage correction on C_1, controlled by values of R_2 and R_3, is proportional to width of error pulses. Article covers circuit operation in detail.—R. Bohlken, A Synchronized Phase Locked Loop, *EDN Magazine,* March 20, 1973, p 84–85.

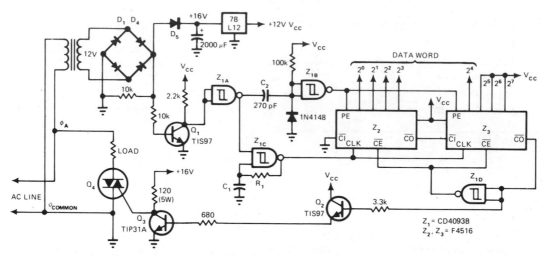

DIGITAL CONTROL OF PHASE ANGLE—Circuit transforms 5-bit digital control word into phase angle over full range. Resolution is proportional to length of control word. Developed for stage lighting control. Z_{1C} serves as clock oscillator that is periodically synchronized with zero crossings of AC line; R_1 and C_1 set clock frequency, which for 5-bit control word must be 64 times line frequency or 3.84 kHz. Load requirements determine choice of triac for Q_4. Required signal for generating triac drive is produced by D_1-D_4 and Q_1.—R. Tenny, Circuit Provides Digital Phase Control of AC Loads, *EDN Magazine*, Oct. 5, 1977, p 99–101.

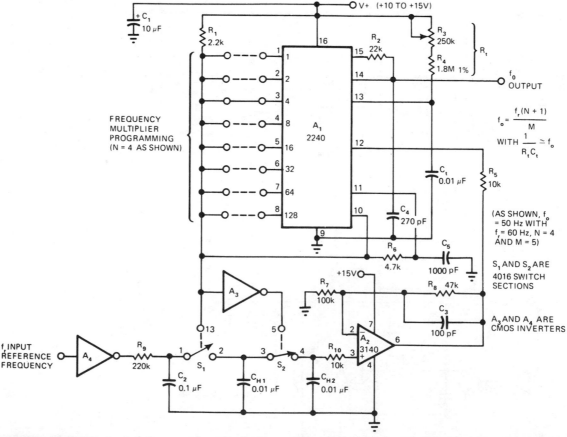

$$f_o = \frac{f_r(N+1)}{M}$$

$$\text{WITH } \frac{1}{R_t C_t} \simeq f_o$$

(AS SHOWN, f_o = 50 Hz WITH f_r = 60 Hz, N = 4 AND M = 5)

S_1 AND S_2 ARE 4016 SWITCH SECTIONS

A_3 AND A_4 ARE CMOS INVERTERS

PLL WITH IC TIMER—Uses 2240 programmable timer/counter as combination voltage-controlled oscillator and frequency divider, with CMOS analog switches serving as sample-and-hold phase detectors. Incoming reference frequency is amplified and limited by CMOS inverter, then integrated into reference triangle waveform by R_9-C_2. Triangle is sampled by S_1 and S_2 which with C_{H1} and C_{H2} form cascaded sample-and-hold network that holds only last instantaneous voltage on C_{H1} as error voltage. This error is amplified by FET-input 3140 opamp A_2 for driving pin 12 of 2240 timer as correction voltage, to establish lock. Reference and output frequencies need not have direct harmonic relationship; with circuit values shown, output is 50 Hz for reference input of 60 Hz. Output frequency can go as high as 100 kHz by using programmability of divider chain.—W. G. Jung, Take a Fresh Look at New IC Timer Applications, *EDN Magazine*, March 20, 1977, p 127–135.

CHAPTER 64
Phonograph Circuits

Includes RIAA-equalized preamps for all types of mono and stereo phono pickups, along with power amplifiers, tone controls, rumble and scratch filters, and test circuits. See also Audio Amplifier and Audio Control chapters.

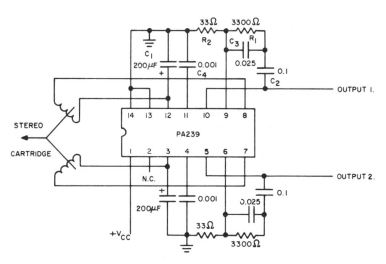

MAGNETIC-CARTRIDGE PREAMP—Uses Signetics PA239 dual low-noise amplifier designed specifically for low-level low-noise applications. Stereo channel separation at 1 kHz is typically 90 dB, and total harmonic distortion without feedback is 0.5%. Circuit matches amplifier response with RIAA recording characteristic. Supply voltage can be between 9 and 15 V at 22 mA. Article gives design equations.—A. G. Ogilvie, Construct a Magnetic-Cartridge Preamp, *Audio*, June 1974, p 40 and 42.

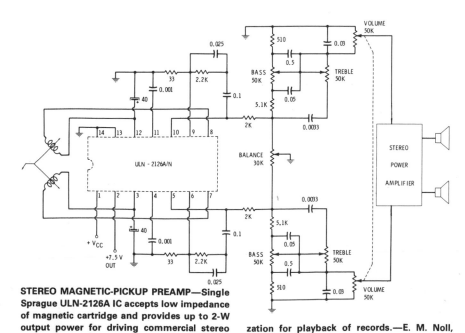

STEREO MAGNETIC-PICKUP PREAMP—Single Sprague ULN-2126A IC accepts low impedance of magnetic cartridge and provides up to 2-W output power for driving commercial stereo power amplifier. Circuit includes balance control and all tone controls along with ganged volume control. Values shown give proper equalization for playback of records.—E. M. Noll, "Linear IC Principles, Experiments, and Projects," Howard W. Sams, Indianapolis, IN, 1974, p 237 and 242.

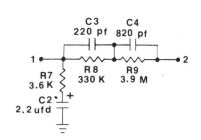

NEW RIAA NETWORK—Values of R7 and C2 have been changed as shown in standard network for phonograph playback equalization. Tantalum electrolytic rated at least 20 V is recommended for C2. Network can also be used as inverse RIAA equalizer for testing preamps, with signal applied to terminal 2 and output to preamp taken from terminal 1. New standard extends playback equalization to 20,000 Hz and specifies that equalization be 3 dB down from previous standard at 20 Hz, with rolloff at 6 dB per octave below 20 Hz.—W. M. Leach, New RIAA Feedback Network, *Audio*, March 1978, p 103.

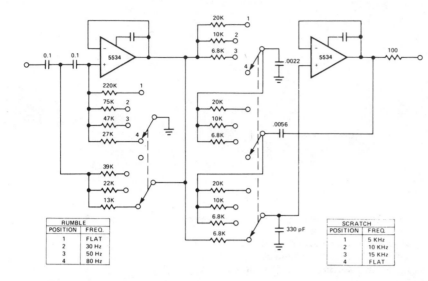

RUMBLE	
POSITION	FREQ.
1	FLAT
2	30 Hz
3	50 Hz
4	80 Hz

SCRATCH	
POSITION	FREQ.
1	5 KHz
2	10 KHz
3	15 KHz
4	FLAT

RUMBLE/SCRATCH FILTER—Used after pre-amp in high-quality audio system to improve reproduction of phonograph records. Two-pole Butterworth design has switchable breakpoints providing any desired degree of filtering.—"Signetics Analog Data Manual," Signetics, Sunnyvale, CA, 1977, p 638–639.

SCRATCH FILTER—Provides passband gain of 1 and corner frequency of 10 kHz for rolling off excess high-frequency noise appearing as hiss, ticks, and pops from worn records. Design procedure is given.—"Audio Handbook," National Semiconductor, Santa Clara, CA, 1977, p 2-49–2-52.

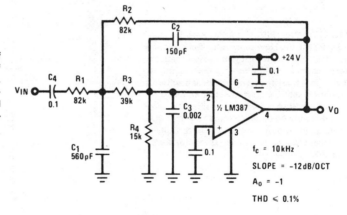

$f_c = 10 \text{ kHz}$

$\text{SLOPE} = -12 \text{dB/OCT}$

$A_0 = -1$

$\text{THD} \leqslant 0.1\%$

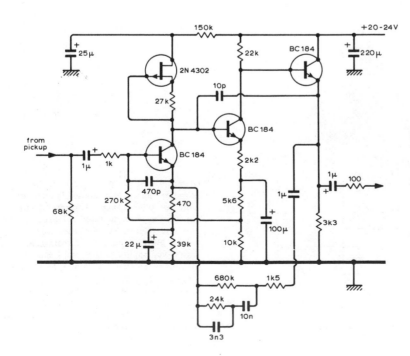

RIAA PREAMP—Low-noise circuit (below −70 dB referred to 5-mV input from pickup) has high overload capability and low distortion (below 0.05% intermodulation at 2 VRMS output). Arrangement of first stage gives improved transient reponse over usual feedback pair. Second stage provides gain of 10.—S. F. Bywaters, RIAA-Equalized Pre-Amplifier, *Wireless World*, Dec. 1974, p 503.

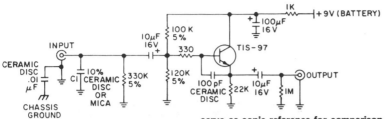

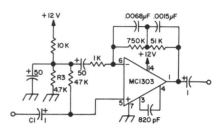

INPUT BUFFER FOR PREAMP—Used between cartridge and preamp of each stereo channel to make comparison testing of phonograph preamps more nearly independent of cartridge and cable capacitances. Buffer terminates cartridge in 47K in parallel with C1. Buffer can then serve as sonic reference for comparison with preamps for which input impedance is unknown. Article tells how to determine correct value of C1 for cartridge used and covers preamp test procedures in detail.—T. Holman, New Tests for Preamplifiers, *Audio*, Feb. 1977, p 58, 60, 62, and 64.

MAGNETIC-CARTRIDGE PREAMP—Uses dual opamp for stereo, other half of which is connected exactly the same but with connections to pin numbers changed to those in parentheses: 6 (5), 5 (8), 3 (11), 4 (10), and 1 (13).—Circuits, *73 Magazine*, Sept. 1973, p 143.

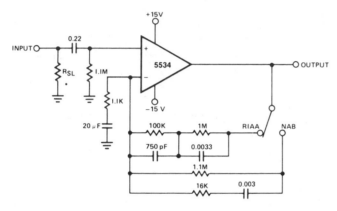

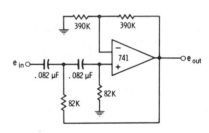

EQUALIZED PREAMP—Low-frequency boost is provided by inductance of magnetic cartridge, acting with RC network to approximate theoretical RIAA or NAB compensation as determined by position of compensation switch. Input resistor is selected to provide specified loading for cartridge. Output noise is about 0.8 mVRMS with input shorted.—"Signetics Analog Data Manual," Signetics, Sunnyvale, CA, 1977, p 638–639.

20-Hz HIGH-PASS RUMBLE FILTER—Second-order rumble filter for phonograph amplifier has 1-dB peak and 20-Hz cutoff frequency. Design uses large resistance values to permit use of smaller and lower-cost capacitors.—D. Lancaster, "Active-Filter Cookbook," Howard W. Sams, Indianapolis, IN, 1975, p 191–192.

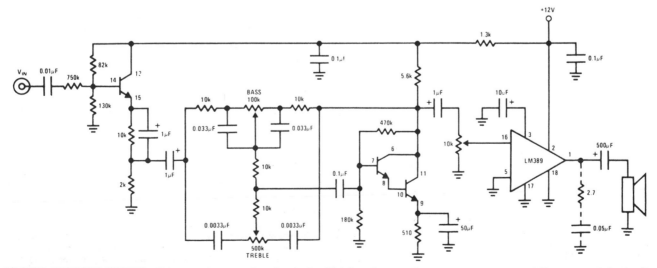

CERAMIC-CARTRIDGE SYSTEM—Circuit using National LM389 opamp having three transistors on same chip provides required high input impedance for ceramic cartridge because input transistor is wired as high-impedance emitter-follower. Remaining transistors form high-gain Darlington pair used as active element in low-distortion Baxandall tone-control circuit.—"Audio Handbook," National Semiconductor, Santa Clara, CA, 1977, p 4-33–4-37.

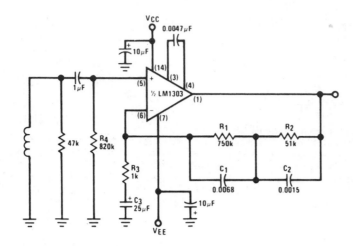

SPLIT-SUPPLY PHONO PREAMP—Low-noise circuit using LM1303 provides RIAA response and operates over supply voltage range of ±4.5 to ±15 V. 0-dB reference gain (1 kHz) is about 34 dB. Input is from magnetic cartridge.—"Audio Handbook," National Semiconductor, Santa Clara, CA, 1977, p 2-25–2-31.

SCRATCH/RUMBLE FILTER—Single active filter provides two widely differing turnover frequencies, as required in audio amplifier used with phonograph. For values shown, insertion loss of filter is −6 dB at 37 Hz and at 23 kHz. Components may be switched to provide different turnover frequencies, but complete removal of filter requires considerably more complicated switching.—P. I. Day, Combined Rumble and Scratch Filter, *Wireless World*, Dec. 1973, p 606.

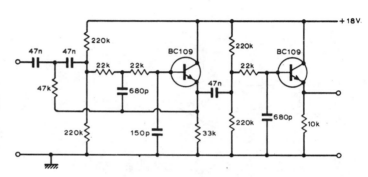

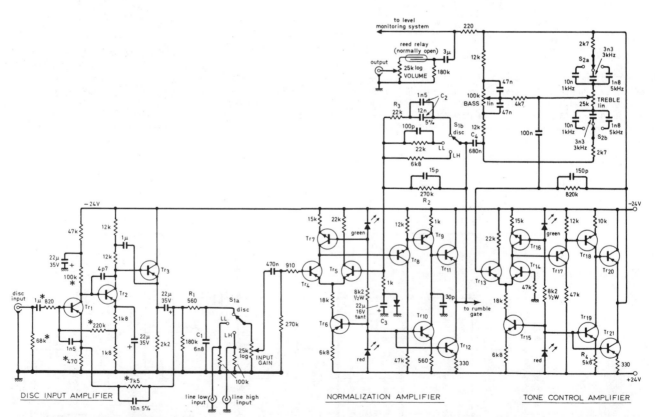

NO-COMPROMISE PHONO PREAMP—Distortion figure is below 0.002 percent, overload margin is about 47 dB, and S/N ratio is 71 dB for phono amplifier. This feeds normalization amplifier whose output is set at 0 dBm by setting input gain control. Feedback components R₂, R₃, and C₂ provide RIAA bass boost. Tone-control circuit is based on Baxandall system but has bass control turnover frequency which decreases as control approaches flat position. This allows small amount of boost at low end of audio spectrum to correct for transducer shortcomings. Article describes circuit operation in detail and gives additional circuits used for tape output, level detection, noise gate, and power supply. Transistors Tr₁-Tr₆ and Tr₁₃-Tr₁₅ are BCY71; Tr₇-Tr₉ and Tr₁₆-Tr₁₈ are MPS A06; Tr₁₀-Tr₁₂ and Tr₁₉-Tr₂₁ are MPS A56; Tr₉ is BFX85 or equivalent. Circuit is duplicated for other stereo channel.—D. Self, Advanced Preamplifier Design, *Wireless World*, Nov. 1976, p 41–46.

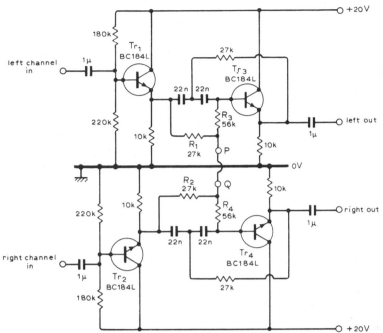

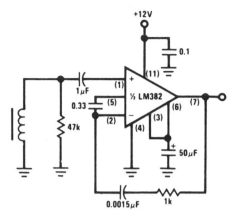

12-V PHONO PREAMP—Low-noise circuit has midband 0-dB reference gain of 46 dB. Designed for RIAA response. Internal resistor matrix of IC minimizes parts count. Input is from magnetic cartridge.—"Audio Handbook," National Semiconductor, Santa Clara, CA, 1977, p 2-25–2-31.

RUMBLE FILTER—Used when rumble from cheaper turntable or record extends above 100 Hz, causing disconcerting out-of-phase loudspeaker signals. Circuit is based on fact that human ear is not sensitive to directional information below about 400 Hz, making it permissible to remove stereo (L − R) signal at low frequencies and thus remove stereo rumble without losing stereo separation. Emitter-followers feed high-pass filters having 200-Hz breakpoint frequencies and Butterworth characteristics. Attenuation of filter is 12 dB at 100 Hz. Filter circuit can be disabled by placing switch between points P and Q.—M. L. Oldfield, Stereo Rumble Filter, *Wireless World*, Oct. 1975, p 474.

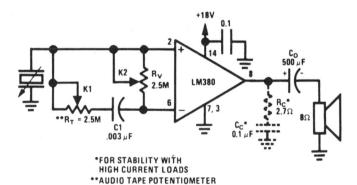

*FOR STABILITY WITH HIGH CURRENT LOADS
**AUDIO TAPE POTENTIOMETER (10% OF R_T AT 50% ROTATION)

COMMON-MODE VOLUME AND TONE CONTROL—Eliminates attenuation of signal by conventional voltage-divider type of volume control and gives maximum input impedance. Used with transducers having high source impedance, but will also serve with low-impedance transducers.—"Audio Handbook," National Semiconductor, Santa Clara, CA, 1977, p 4-21–4-28.

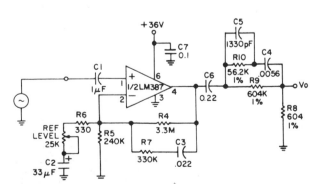

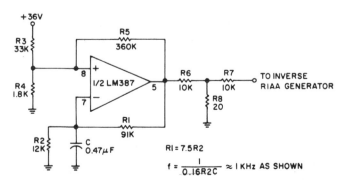

INVERSE RIAA RESPONSE GENERATOR— Used in design, construction, and testing of phonograph preamp. Provides opposite of playback characteristic. Passive filter is added to output of National LM387, used as flat-response adjustable-gain block. Gain range is 24 to 60 dB, set in accordance with 0-dB reference gain (1 kHz) of preamp under test. Input is from 1-kHz square-wave generator, which can be built with other half of LM387 connected as also shown.— D. Bohn, Inverse RIAA/Square Wave Generator, *Audio*, Feb. 1977, p 65–66.

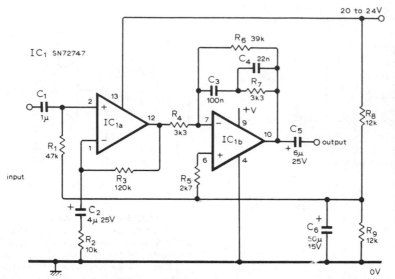

MAGNETIC-PICKUP PREAMP—Circuit uses type 747 dual opamp, but individual 741 opamps may be used instead. Input signal is first amplified flat, after which equalization acts on both signal and noise to give improved S/N ratio. Adjust first opamp for gain of 13. Series feedback is used to minimize noise since impedance of magnetic pickup is low compared to opamp input impedance. Second opamp has frequency-dependent series feedback for RIAA compensation. Gain here is unity at 1 kHz. Output is about 70 mV for modern pickup having output of about 5 mV.—B. S. Wolfenden, Magnetic Pick-Up Preamplifier, *Wireless World*, Sept. 1976, p 81–82.

RUMBLE FILTER—Used to roll off low-frequency noise associated with worn turntable and tape transport mechanisms. Gain is 1. Design procedure is given. For values shown, corner frequency is 50 Hz and slope is −12 dB per octave.—"Audio Handbook," National Semiconductor, Santa Clara, CA, 1977, p 2-49–2-52.

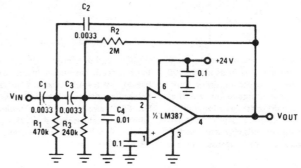

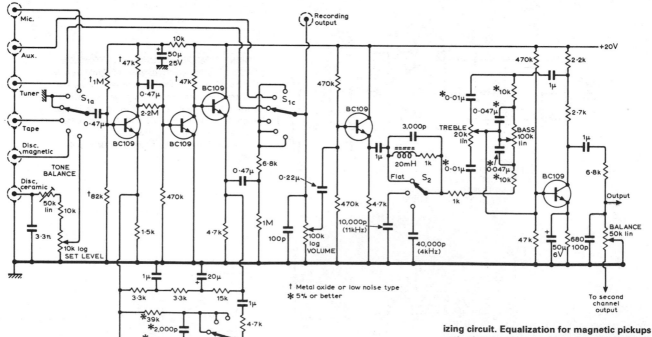

PREAMP WITH EQUALIZATION—Based on 1966 high-performance Bailey preamp design with improved filter and tone control circuits and additional complete ceramic-pickup equal-izing circuit. Equalization for magnetic pickups and other types of inputs is automatically selected by three-deck input selector switch. To avoid overloading input stage, adjust set level control to give comfortable listening level for given input when main volume control is at about half its maximum rotation. Article also gives lower-cost version for ceramic-pickup equalization and changes required in this for operation from negative supply.—B. J. Burrows, Ceramic Pickup Equalization, *Wireless World*, Aug. 1971, p 379–382.

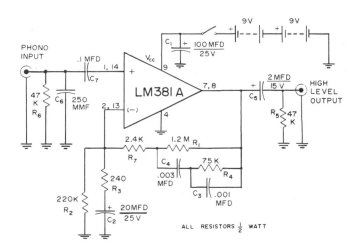

LOW-NOISE PREAMP—Provides dynamic range of about 80 dB for phonograph playback system, even when using highest-quality cartridge having low output. Source voltage is reduced to 18 V for National LM381A, which still provides ample signal for 2-V high-level input of stereo channel. Cross-channel isolation is better than 60 dB from 20 to 20,000 Hz.—J. P. Holm, A Quiet Phonograph Preamplifier, *Audio,* Oct. 1972, p 34—35.

CERAMIC-CARTRIDGE AMPLIFIER—Single National LM380 forms simple amplifier with tone and volume controls for driving 8-ohm loudspeaker at outputs above 3 W. Supply voltage range is 12–22 V, with higher voltage giving higher power. Tone control changes high-frequency rolloff.—"Audio Handbook," National Semiconductor, Santa Clara, CA, 1977, p 4-21–4-28.

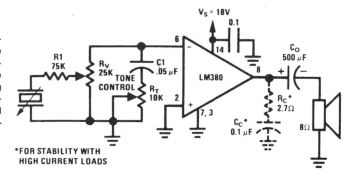

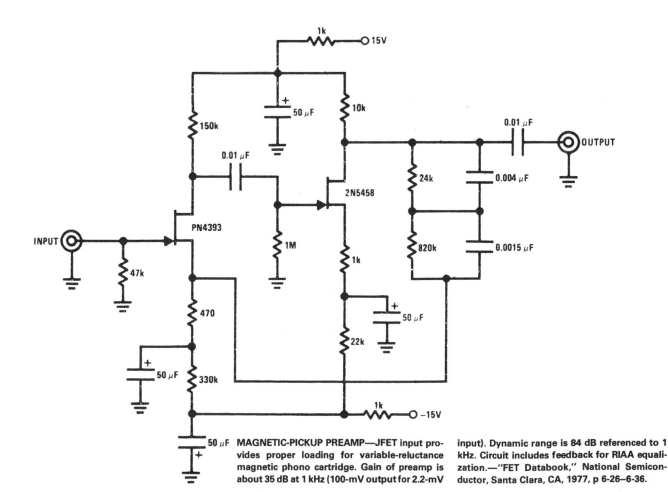

MAGNETIC-PICKUP PREAMP—JFET input provides proper loading for variable-reluctance magnetic phono cartridge. Gain of preamp is about 35 dB at 1 kHz (100-mV output for 2.2-mV input). Dynamic range is 84 dB referenced to 1 kHz. Circuit includes feedback for RIAA equalization.—"FET Databook," National Semiconductor, Santa Clara, CA, 1977, p 6-26–6-36.

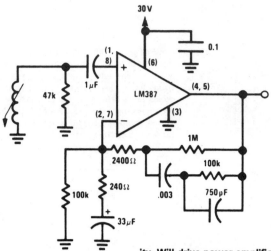

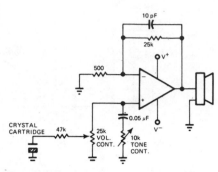

RIAA PHONO PREAMP—Design procedure is given for operation from 30-V supply, using magnetic cartridge having 0.5 mV/cm/s sensitivity. Will drive power amplifier having 5 VRMS input overload limit.—"Audio Handbook," National Semiconductor, Santa Clara, CA, 1977, p 2-25–2-31.

5-W POWER OPAMP—Low-cost phono amplifier using only single 591 power opamp provides 5 W into 8-ohm load with only 0.2% total harmonic distortion. With crystal cartridge, circuit has fixed gain of 50.—R. J. Apfel, Power Op Amps—Their Innovative Circuits and Packaging Provide Designers with More Options, *EDN Magazine,* Sept. 5, 1977, p 141–144.

CHAPTER 65

Photoelectric Circuits

Covers circuits involving change in light on photocell or other light-sensitive device, including punched-tape reader, transmission of voice or data signals on light beam, and solar-power oscillator. See also Burglar Alarm, Fiber-Optic, Instrumentation, Lamp Control, and Optoelectronic chapters.

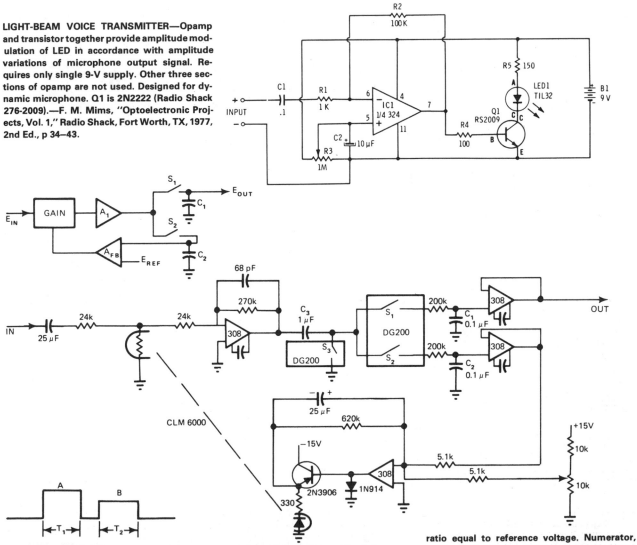

LIGHT-BEAM VOICE TRANSMITTER—Opamp and transistor together provide amplitude modulation of LED in accordance with amplitude variations of microphone output signal. Requires only single 9-V supply. Other three sections of opamp are not used. Designed for dynamic microphone. Q1 is 2N2222 (Radio Shack 276-2009).—F. M. Mims, "Optoelectronic Projects, Vol. 1," Radio Shack, Fort Worth, TX, 1977, 2nd Ed., p 34–43.

RATIO OF TWO UNKNOWNS—Developed for use when two signals are time-shared on same input line, such as exists when two LEDs alternately illuminate single photocell. Measures ratio of amplitudes of unknowns with accuracy better than 1%. During time period T_1, input is sampled through S_2 and stored on C_2 for comparison with reference voltage. Result is applied through switchable amplifier network A_{FB} to gain control element which is LED-photoresistor coupled pair (CLM 6000). This closed loop adjusts signal gain to make denominator of ratio equal to reference voltage. Numerator, corresponding to time T_2, is multiplied by same gain so numerator output is proportional to desired ratio B/A of unknowns. Article describes circuit operation in detail.—R. E. Bober, Here's a Low-Cost Way to Measure Ratios, *EDN Magazine*, March 5, 1976, p 108, 110, and 112.

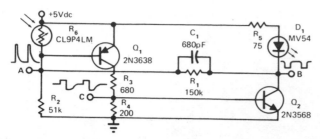

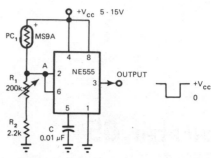

5-kHz PHOTOCELL OSCILLATOR—Provides 5-V pulses at about 5 kHz only if photocell is illuminated by its companion LED. Repetition rate varies with illumination, so interruption or attenuation of light produces easily detected frequency change that can be used as control signal. Applications include fail-safe interruption monitor and illumination transducer. Oscillation stops if beam is completely interrupted or if strong ambient light falls on photocell.—H. L. Hardy, FM Pulsed Photocell Is Foolproof, *EDN Magazine*, March 5, 1975, p 72.

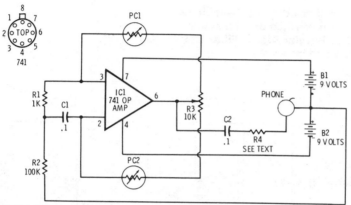

AUDIBLE LIGHT SENSOR—741 opamp is connected as audio oscillator with Radio Shack 276-677 photocells in feedback circuits. When light strikes PC1, its resistance decreases and frequency of audio tone in headphone decreases correspondingly. When light strikes PC2, which is connected to noninverting input of 741, increase in illumination serves to increase frequency. Choose R4 to reduce volume to desired level. R3 is balancing control for photocells.—F. M. Mims, "Integrated Circuit Projects, Vol. 2," Radio Shack, Fort Worth, TX, 1977, 2nd Ed., p 81–86.

PUNCHED-TAPE READER—Connection of 555 timer as Schmitt trigger produces output pulses with sharp rise and fall times that are independent of tape speed. Output is compatible with TTL or CMOS circuits. When scanning light beam hits hole in punched card or tape, resistance of light-sensitive resistor drops sharply and voltage at pins 2 and 6 rises above 0.67 V_{CC}. Voltage at output pin 3 then drops sharply from V_{CC} to 0 V. When PC_1 goes dark, circuit switches rapidly back to original state. Reverse PC_1 and R_1-R_2 for positive edge-triggered logic.—S. Sarpangal, 555 Timer Implements Tape Reader, *EDN Magazine*, Jan. 5, 1978, p 86 and 90.

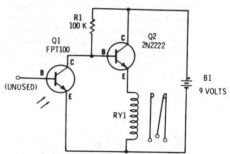

PHOTOTRANSISTOR RELAY—With phototransistor Q1 dark, R1 biases Q2 into conduction and miniature SPDT relay (Radio Shack 275-004) is energized. When light falls on Q1, Q2 is turned off and relay drops out. Battery drain is about 5 mA in darkness, dropping almost to 0 mA with light.—F. M. Mims, "Transistor Projects, Vol. 3," Radio Shack, Fort Worth, TX, 1975, p 69–74.

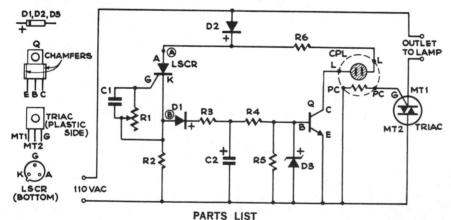

PARTS LIST

C1—0.1mfd capacitor
C2—10mfd @150V capacitor
R1—1-meg. carb. potentiometer
R2—82,000-ohm, ½w resistor
R3—390-ohm, 1w resistor
R4—2.2-megohm, ½w resistor

R5—560,000-ohm, ½w resistor
R6—22,000-ohm, ½w resistor
D1,D2—Diode (Motorola HEP 156)
D3—Zener diode, 6.2V (Motorola HEP 103 or equiv.)
Q—High-voltage transistor

(Motorola HEP S3022)
LSCR—Light-op. SCR, 200V (Radio Shack 276-1081)
Triac—Mot. HEP R1725
CPL—Light coupler Sigma 301T1-120A1 (SW Tech. Prod., 219 W. Rhapsody, San Antonio, Tex.)

GARAGE-LIGHT CONTROL—When mounted on far wall in garage, controller picks up headlight beams as car is driven in at night and turns on one or more garage lights long enough (3 min) for driver to get out of car and reach exit. Controller then flickers lights as warning and begins dimming them out. With parts specified, will handle up to 800 W of lamps. Adjust sensitivity control R1 so light in optocoupler CPL comes on when headlights strike light-operated SCR. Controller must be kept out of direct sunlight. For manual control, connect pushbutton switch between points A and B. To increase time delay, increase value of C2. With 20 µF, time will be doubled.—C. R. Lewart, Automatic Garage Light Control, *Popular Science*, July 1973, p 110.

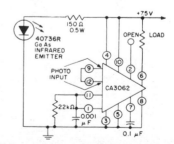

ON/OFF CONTROL—RCA CA3062 combination photodetector and power amplifier provides ON/OFF output in response to light signal. Output transistors in IC should be either saturated or blocked to avoid heat rise in silicon chip. Complementary outputs give choice of load normally on or normally off when light from infrared emitter falls on photo input of IC. Interruption of light path then produces opposite load condition.—"Linear Integrated Circuits and MOS/FET's," RCA Solid State Division, Somerville, NJ, 1977, p 156.

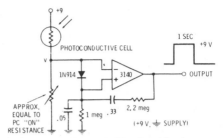

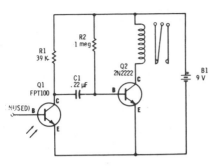

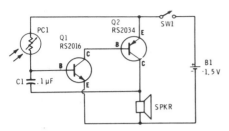

LIGHT-CHANGE DETECTOR—Combination amplifier and detector using 3140 opamp responds only to sudden changes in light on photocell while ignoring slow changes in ambient light. When beam is suddenly broken, opamp output swings positive and stays positive for delay time set by recharging of 0.05-μF capacitor on positive input. Delay locks out spurious signals until photocell resets itself to normal illumination. Values shown give time-out delay of about 1 s, with clean conditioned rectangular output pulse.—D. Lancaster, "CMOS Cookbook," Howard W. Sams, Indianapolis, IN, 1977, p 346–347.

LIGHT-CHANGE SENSOR DRIVES RELAY—Capacitive coupling between phototransistor and bipolar transistor makes circuit respond only to interruptions or rapid changes in light while ignoring normal gradual changes in ambient light as caused by clouds or at sunrise. Relay pulls in when flash of light occurs and drops out when light is removed. Use Radio Shack 275-004 miniature relay.—F. M. Mims, "Transistor Projects, Vol. 3," Radio Shack, Fort Worth, TX, 1975, p 69–74.

AUDIBLE LIGHT METER—Low light on cadmium sulfide photocell (Radio Shack 276-116) produces series of clicks in miniature 8-ohm loudspeaker. As light increases, clicks merge into audio tone that increases in frequency as light intensity increases. Can be used for classroom demonstrations or as sunrise alarm clock. Circuit is quiet in total darkness.—F. M. Mims, "Optoelectronic Projects, Vol. 1," Radio Shack, Fort Worth, TX, 1977, 2nd Ed., p 61–66.

MODULATED-LIGHT RECEIVER—Two FET stages amplify chopped or smoothly modulated output signal of silicon solar cell. With 1000-Hz modulation of 5-lm/ft² light beam, circuit will produce 1 VRMS at output when R4 is set for maximum gain. Can be used for light-beam communication and for alarm systems.—R. P. Turner, "FET Circuits," Howard W. Sams, Indianapolis, IN, 1977, 2nd Ed., p 113–114.

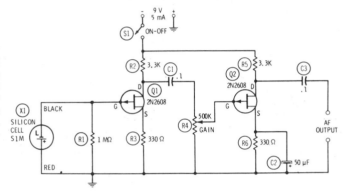

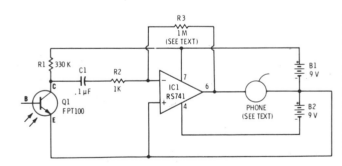

LIGHT-BEAM RECEIVER—Converts amplitude-modulated light beam back to audio signal for driving transistor radio earphone having resistance of 500–1000 ohms. Miniature 8-ohm loudspeaker can be used by adding output transformer such as Radio Shack 273-1380. Gain of opamp is controlled by R3, which can be trimmer resistor or pot. Designed for use with transmitter providing amplitude modulation of LED, for short-range voice communication.—F. M. Mims, "Optoelectronic Projects, Vol. 1," Radio Shack, Fort Worth, TX, 1977, 2nd Ed., p 44–54.

END-OF-TAPE DETECTOR—Self-compensating sensor automatically compares short-term light variations produced by beginning and end markers on digital magnetic recording tape against long-term variations of ambient light, to improve reliability of sensing marker when there are reflections from blank tape. Low-pass filter R3-C1, having time constant about 5 times expected 10-ms incoming pulse width, stores long-term light level without reacting to short signal pulse. Low-pass filter R4-C2, having 1/20 time constant of incoming pulse width, reduces spurious noise without deteriorating incoming pulses.—C. A. Herbst, Optical Tape-Marker Detector, *EEE Magazine*, March 1971, p 79.

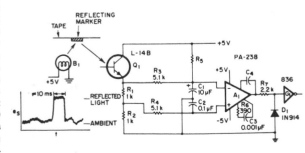

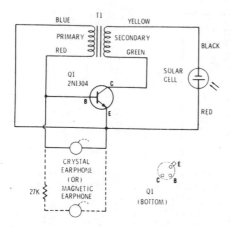

SOLAR-POWER OSCILLATOR—Supply voltage for single-transistor audio oscillator is generated by Radio Shack 276-115 selenium solar cell that produces about 0.35 V in bright sunlight. With cell 3 feet away from 75-W incandescent lamp, oscillator frequency is about 2400 Hz. Frequency drops as light increases. Transformer is 273-1378.—F. M. Mims, "Transistor Projects, Vol. 2," Radio Shack, Fort Worth, TX, 1974, p 53–58.

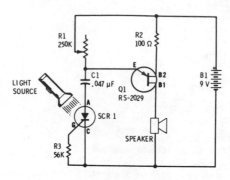

LASCR-CONTROLLED OSCILLATOR—UJT relaxation oscillator having loudspeaker load produces single click each time flash of light falls on light-activated SCR. Setting of R1 determines whether circuit produces series of pulses or tone burst during time light is on. Oscillator frequency increases with light intensity.—F. M. Mims, "Semiconductor Projects, Vol. 2," Radio Shack, Fort Worth, TX, 1976, p 71–77.

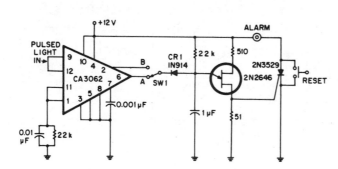

MISSING-PULSE ALARM—Developed for sensing missing light pulses or detecting absence of object on moving conveyor belt. CA3062 combination light sensor and amplifier detects light pulses synchronized to 60-Hz line. With SW1 at A, each pulse resets 20-ms timing network of 2N2646 UJT at 16.7-ms intervals, preventing UJT from firing. If light beam is interrupted by object, UJT is allowed to fire and trigger 2N3529 SCR that turns on alarm. With SW1 at B, circuit detects interruptions in steady light beam and sounds alarm only when interruption does not occur.—J. F. Kingsbury, Double Duty Photo Alarm, *EDN/EEE Magazine*, May 15, 1971, p 51.

CHAPTER 66
Photography Circuits

Includes adjustable or programmable timers for enlargers and printers, photoflash, slave flash, strobe, and controlled-sequence flash circuits, exposure meters, and gray-scale control for CRT. See also Instrumentation, Lamp Control, and Timer chapters.

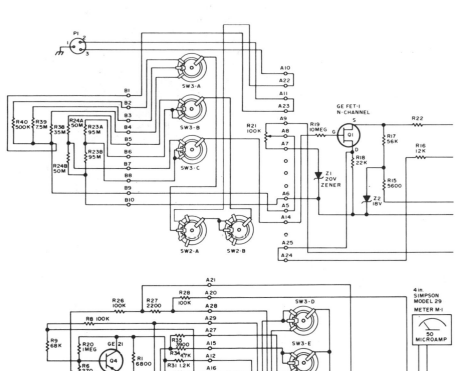

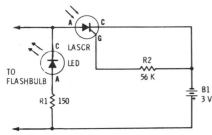

SLAVE FLASH—Remote flashtube having no connection with camera is fired by light-activated SCR (LASCR) when triggered by main flash of camera. Used to provide illumination at greater depth than main flash range, to soften sharp shadows, and to provide backlighting for flash photographs. LED is indicator showing that circuit has been triggered, reminding photographer that new flash lamp should be inserted.—F. M. Mims, "Transistor Projects, Vol. 1," Radio Shack, Fort Worth, TX, 1977, 2nd Ed., p 79–85.

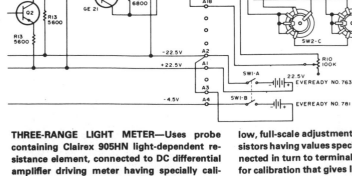

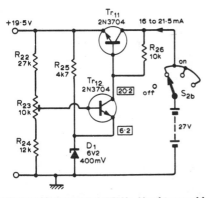

THREE-RANGE LIGHT METER—Uses probe containing Clairex 905HN light-dependent resistance element, connected to DC differential amplifier driving meter having specially calibrated scale. Article gives calibration procedure. Switching circuit provides constant check on voltage of 22.5-V battery. If 4.5-V battery is low, full-scale adjustment cannot be made. Resistors having values specified in article are connected in turn to terminals of photocell jack P1 for calibration that gives linear scale reading.—J. L. Mills, Jr., Light Right?—Do-It-Yourself Photo Exposure Meter, *73 Magazine*, Sept. 1978, p 204–206 and 208–211.

19.5 V FROM 27-V BATTERY—Used to provide precise voltage levels required by portable trigger unit designed to fire up to five different flash units at equal intervals that may range from 11 ms to 11 s. Article gives all circuits.—R. Lewis, Multi-Flash Trigger Unit, *Wireless World*, Nov. 1973, p 529–532.

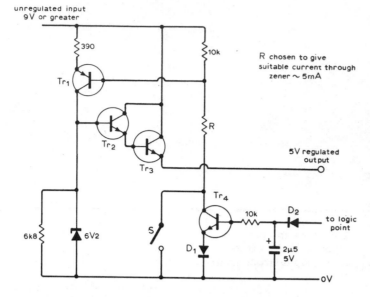

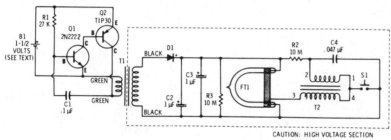

TIMER SWITCH-OFF FOR ENLARGER—Circuit shows power supply designed to operate digital exposure timer using TTL. Since timer logic is needed only when enlarger lamp is on, power supply circuit will be turned off automatically when timer goes low and turns off enlarger lamp at end of exposure. Switch S is closed when timer cycle is activated, setting timer output at 5 V and turning on lamp. Transistor and diode types are not critical. Since D_2 is connected to logic point of timer, power supply remains on when S is released, until completion of timer cycle.—E. R. Rumbo, Automatic Switch-Off Power Supply, *Wireless World*, Feb. 1976, p 77.

XENON STROBE—Two-transistor oscillator generates pulses at about 500 Hz for step-up by 300-mA filament transformer T1 (Radio Shack 273-1384) to charge storage capacitors C2 and C3, which are 250-V electrolytics. Simultaneously, C4 is charged through R2. After allowing sufficient time for capacitors to charge, S1 is pressed to discharge C4 through 272-1146 flashtube trigger transformer T2, which steps up voltage pulse to about 4000 V for ionizing gas in 272-1145 xenon flashtube FT1. C2 and C3 now discharge through ionized gas to produce brilliant flash of white light lasting only a few microseconds, as required for photography of objects moving at high speed. Circuit may require two cells in series for reliable operation.—F. M. Mims, "Transistor Projects, Vol. 3," Radio Shack, Fort Worth, TX, 1975, p 49—60.

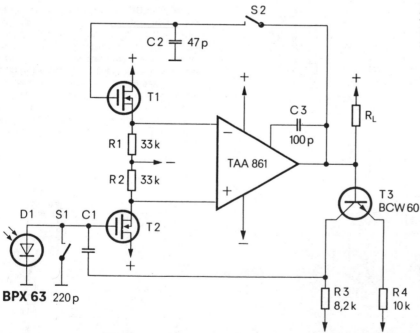

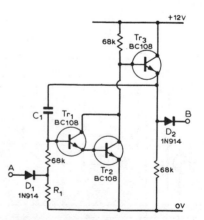

LOW-LEVEL EXPOSURE METER—Uses Siemens BPX 63 photodiode having sensitivity of 10 nA per lux in circuit which ensures that aperture setting is affected only by useful light and not by noise signals. When used at low light levels, circuit recovers quickly from temporary light bursts. Switches S1 and S2 are closed when camera shutter is not open; opamp output is then connected to its inverting input through FET T1. At commencement of exposure, S1 and S2 open to give amplification of over 3000. Integrating capacitor C1 is then charged by photocurrent, making output voltage vary linearly with time. Base-emitter junction of T3 begins to conduct at output voltage of 1 V. Exposure is completed when C1 provides feedback via T3 so no current flows through load resistor R_L. Supply is ±3 V.—"Photodiode BPX 63—All It Needs Is Starlight," Siemens, Iselin, NJ.

2-min RAMP—Used in multiple timer for development of photographic paper, in which six independent timers are started in sequence as each sheet of exposed paper is placed in developer. C_1 is 1 μF and R_1 is 11 megohms for 2-min timer having accuracy within 5 s. Article gives all other circuits required and suggests modifications to meet other needs. Output B drives meter and trigger circuit for audible alarm. Timer is started by input switch connected to A.—R. G. Wicker, Photographic Development Timer, *Wireless World*, April 1974, p 87—90.

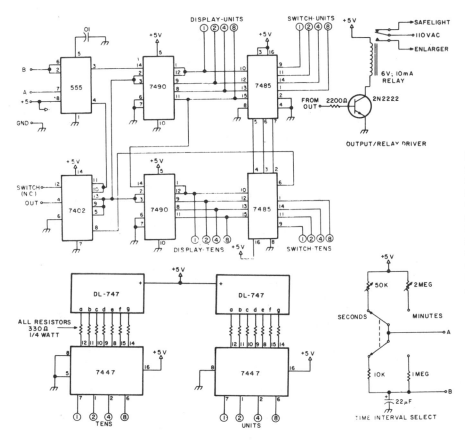

BCD THUMBWHEEL-SET 99-min TIMER—Provides timing in seconds to 99 s, and timing in minutes to 99 min, with 2-digit LED indicator showing elapsed time. Desired interval is set with BCD thumbwheel switches. LED readout counts up to preset time, then resets automatically to zero. Switch giving choice of seconds or minutes has center-off position that stops count temporarily for burning in portion of negative. Article gives construction details.—M. I. Leavey, Build a Unique Timer, *73 Magazine*, Aug. 1977, p 66–71.

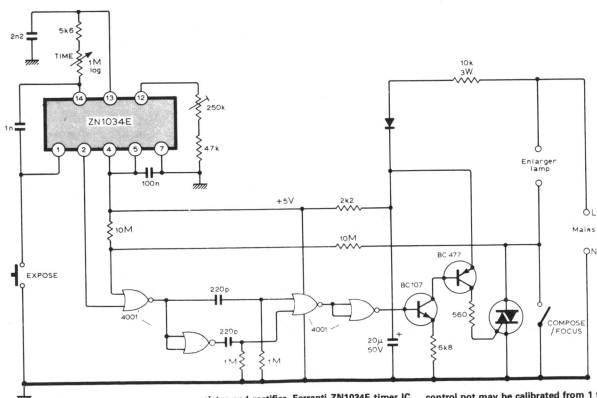

ENLARGER TIMER—Requires no transformer-type power supply because circuit operates from 1 mA taken from AC line through 10K resistor and rectifier. Ferranti ZN1034E timer IC generates delay and supplies 5 V for 4001 CMOS gates. Triac is triggered with 100-μs 60-mA pulses at zero-crossing point. Logarithmic time-control pot may be calibrated from 1 to 120 s. Choose triac to handle current drawn by enlarger lamp used.—M. J. Mayo, Transformerless Enlarger Timer, *Wireless World*, May 1978, p 68.

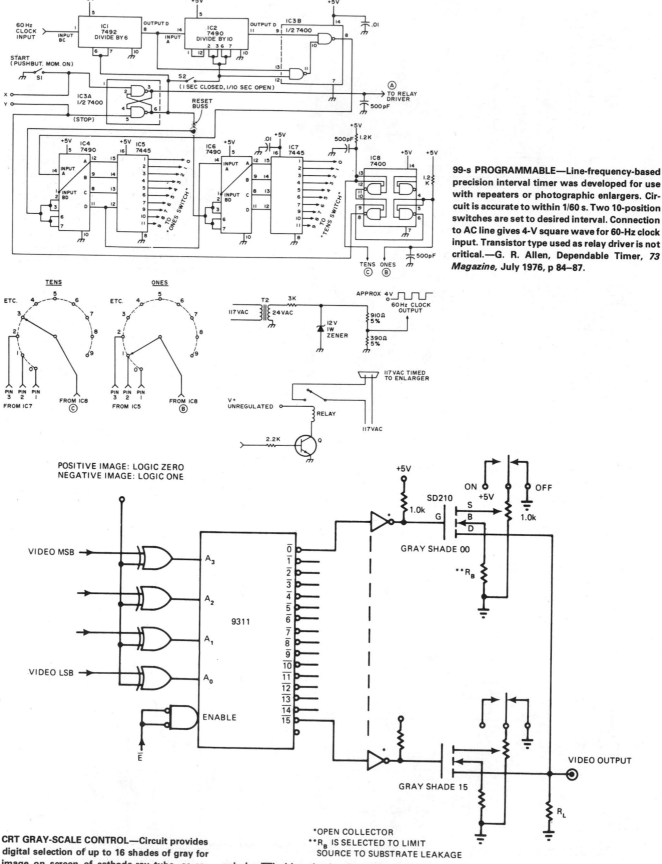

99-s PROGRAMMABLE—Line-frequency-based precision interval timer was developed for use with repeaters or photographic enlargers. Circuit is accurate to within 1/60 s. Two 10-position switches are set to desired interval. Connection to AC line gives 4-V square wave for 60-Hz clock input. Transistor type used as relay driver is not critical.—G. R. Allen, Dependable Timer, *73 Magazine*, July 1976, p 84–87.

CRT GRAY-SCALE CONTROL—Circuit provides digital selection of up to 16 shades of gray for image on screen of cathode-ray tube, as required for different imaging requirements or different photographic films. DMOS FETs provide fast switching times so data rate is limited only by TTL drive circuits. Four bits of digital data stored in 9311 memory are used for selecting desired scale. Output of circuit is used to control beam intensity. Circuit also permits complete video inversion for negative images.—K. R. Peterman, Fast CRT Intensity Selector Adjusts the Gray Scale, *EDN Magazine*, March 20, 1976, p 98 and 100.

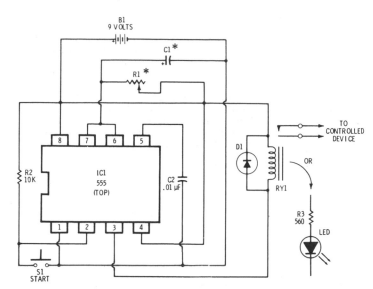

BASIC 555 TIMER—Closing switch S1 momentarily applies activating signal to trigger input pin 2 of timer, to start charging of C1. When C1 charges to two-thirds of supply voltage, timer discharges it to complete timing cycle. Duration of charging interval can be varied from several microseconds to over 5 min by changing values of R1 and C1. With 1K for R1, capacitor values of 0.01 to 100 μF give time range of 10 μs to 100 ms. With 100 megohms and 1 μF, time increases to 10 s. Once timer starts, closing S1 again has no effect. Timing cycle can be interrupted only by applying reset pulse to pin 4 or opening power supply. Circuit will drive LED directly or can be used with miniature relay (Radio Shack 275-004) to control larger loads. Can be used as darkroom timer if LED is kept several feet away from photographic paper. Diode is 1N914.—F. M. Mims, "Integrated Circuit Projects, Vol. 2," Radio Shack, Fort Worth, TX, 1977, 2nd Ed., p 57–65.

MULTIFLASH SWITCH—When ramp output of flash trigger circuit (given in article) is applied to input at A, flash at output of switch circuit is tripped when ramp voltage reaches level determined by setting of R_{12}. Similar voltage-operated switches are required for other flashes. Used for taking sequence photographs such as springboard diver in flight. Settings of R_{12} for different switches are chosen for equal times between flashes, with intervals from 11 ms to 11 s. Article gives all circuits and setup procedure. Regulated 19.5-V supply is required—R. Lewis, Multi-Flash Trigger Unit, *Wireless World,* Nov. 1973, p 529–532.

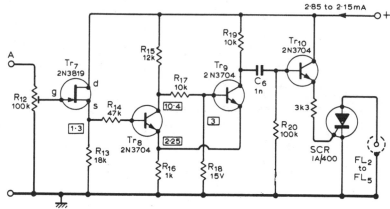

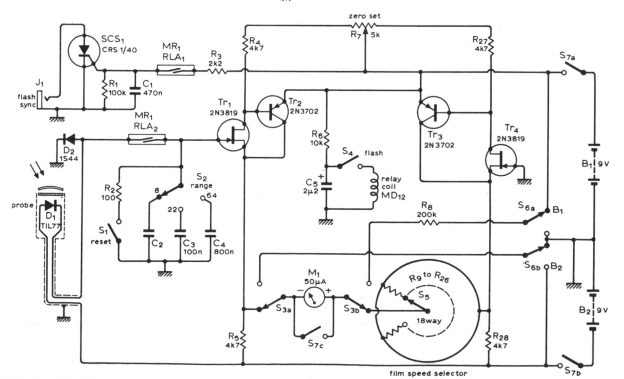

f-NUMBER FLASHMETER—Used to measure light produced at subject position by electronic flashlamps prior to actual taking of picture. Meter is calibrated to read correct f-number setting of lens aperture. Three ranges are provided, from f/2 to f/64, while film speed selector covers films from ASA 12 to 650. Texas Instruments TIL77 photodiode is used as sensing element in probe. Article covers construction, operation, and calibration of meter in detail. Table in article gives values for 18 resistors (one for each film speed) selected by S_5. Examples are 20K for ASA 64 and 51K for ASA 25.—R. Lewis, Photographic Flashmeter, *Wireless World,* Aug. 1974, p 273–278.

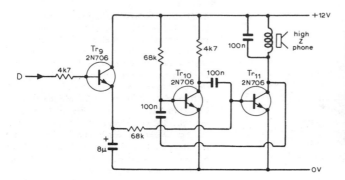

AUDIBLE ALARM FOR TIMER—Used with 2-min timer for developing photographic paper, to produce short warning bleep indicating end of developing time. Input D is taken from output of Schmitt trigger that changes state when 2-min ramp generator times out. Tr_9 and C_4 together lengthen short reset pulse so MVBR Tr_{10}-Tr_{11} oscillates long enough for signal to be heard.—R. G. Wicker, Photographic Development Timer, *Wireless World,* April 1974, p 87–90.

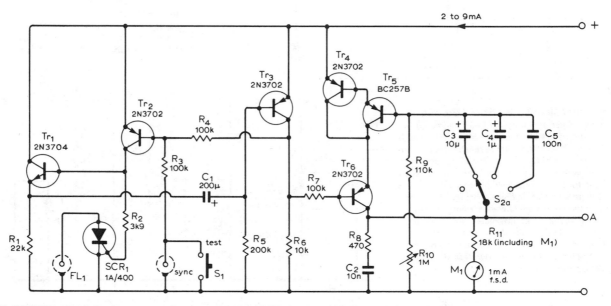

FLASH TRIGGER—Used in instrument designed to trigger up to five individual flash units at equal increments of time that can range from 11 ms to 11 s, as required for such assignments as taking sequence photographs of springboard diver in flight. Transistors Tr_1, Tr_2, and Tr_3 form monostable MVBR that is switched to unstable state by negative pulse applied to base of Tr_2 by SCR_1 when camera shutter contacts FL_1 are closed. Timing circuit Tr_4-Tr_5-Tr_6 provides ramp output at A for feeding voltage-operated switches set to trip at different points of ramp waveform as required for triggering flashes in sequence. Article gives all circuits and setup procedure. Regulated 19.5-V supply is required.—R. Lewis, Multi-Flash Trigger Unit, *Wireless World,* Nov. 1973, p 529–532.

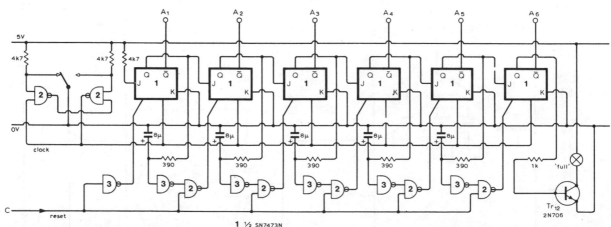

MULTIPLE TIMER FOR PRINTS—Six independent 2-min timers, each using half of SN7473N IC, are set in sequence by unique input switch as sheets of exposed paper are inserted in developer at about 20-s intervals. When capacity of six prints is reached, Tr_{12} turns on light to tell operator that no more prints should be inserted until control logic activates alarm signifying 2-

1 ½ SN7473N
2 ¼ SN7400N
3 ⅙ SN7404N

min time for first sheet inserted. Audible bleep is repeated as each subsequent sheet reaches its 2-min development time. Article gives all circuits and explains operation in detail. Two-input NAND gates (each ¼ of SN7400N) and inverters (each ⅙ of SN7404N) are used to steer reset pulses. Similar two-input NAND gates are used to form fully compatible input pulses from input switch control, each having correct level, rise time, and fall time, without contact bounce that might cause spurious starting of several timers simultaneously.—R. G. Wicker, Photographic Development Timer, *Wireless World,* April 1974, p 87–90.

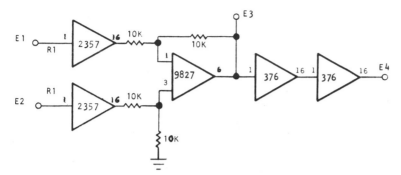

DENSITY AND EXPOSURE—Circuit converts transmission parameter of spectrophotometer to more useful density parameter, which in turn can be converted to exposure parameter. Optical Electronics 2357 opamps at input provide 90-dB dynamic range for DC to 1 kHz or 40-dB range for DC to 100 kHz, operating basically as current amplifiers. 9827 is used as wideband opamp in unity-gain subtracter configuration. Additional 376 opamps are used only for converting to exposure parameter. Use 1000 ohms for R1 with 10-V full-scale inputs.—"Conversion of Transmission to Density and Density to Exposure," Optical Electronics, Tucson, AZ, Application Tip 10133.

100-Ws PHOTOFLASH—Uses AC supply and large storage capacitors to give intense flash lasting only about 250 ms, as required for stop-motion photography of fast-moving objects such as bullets. For battery-powered operation, T1 can be replaced by solid-state chopper circuit. Contacts can be in camera or in external control device.—W. E. Hood, Lightning in a Bottle, *73 Magazine*, Sept. 1974, p 109–112.

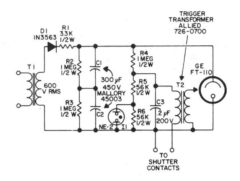

CHAPTER 67
Power Control Circuits

Included are general-purpose circuits capable of handling many types of resistive or inductive loads, as contrasted to specialized circuits given in Lamp Control, Motor Control, Servo, and Temperature Control chapters. Although most circuits are solid-state relays, conventional relay controls are also shown. Inputs can respond to logic levels, pulses, or sensing transducers. Output devices are chiefly SCRs or triacs. Many circuits have zero-crossing action for suppressing RFI, as well as optoisolators at input or output.

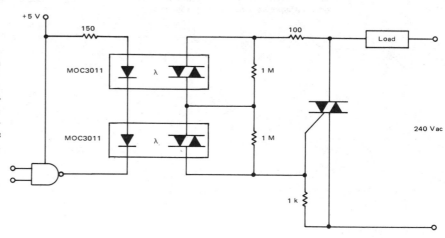

DRIVING 240-VAC TRIAC—Two Motorola MOC3011 optoisolators are used in series as interface between logic and triac controlling 240-VAC load. 1-megohm resistors across optoisolators equalize voltage drops across them. Choice of triac depends on load to be handled.— P. O'Neil, "Applications of the MOC3011 Triac Driver," Motorola, Phoenix, AZ, 1978, AN-780, p 5.

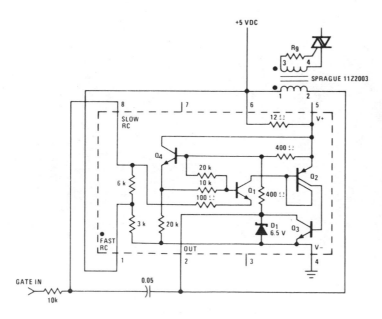

TRIAC TRIGGER—National LM3909 IC is connected as pulse-transformer driver operating from standard 5-V logic supply. IC is biased off when logic input is high. With low logic input, IC provides 10-μs pulses for transformer at about 7 kHz. Trigger is not synchronized to zero crossings but will trigger within 8 V of zero for resistive load and 115-VAC line. Triggering occurs at about 1 V, but trigger level can be changed by using other input resistors or bias dividers.—"Linear Applications, Vol. 2," National Semiconductor, Santa Clara, CA, 1976, AN-154, p 7.

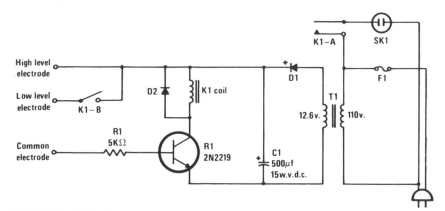

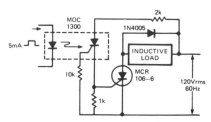

SUMP-PUMP CONTROL—Impurities in water provide conductivity for completing circuit of transistor when water reaches level of sensing electrode, energizing relay that starts pump motor. Extra set of contacts on relay keeps motor running until water drops to predeter-mined lower level. Diodes are 1N4001 or equivalent, rated 1 A. Fuse should be chosen to pass normal motor current. Use 12-V double-pole relay. T1 is 300-mA filament transformer.—J. H. Gilder, Automatic Turn-On, *Modern Electronics,* Dec. 1978, p 78.

HALF-WAVE CONTROL—Simple AC relay operates during positive alternations of AC source, with optoisolator providing complete isolation between control circuit and SCR handling inductive load. When input LED is energized by control pulse, photo-SCR of optoisolator conducts and provides gate current for turning on power SCR. 1N4005 diode protects SCR from back EMF transients of inductive load.—T. Mazur, Solid-State Relays Offer New Solutions to Many Old Problems, *EDN Magazine,* Nov. 20, 1973, p 26–32.

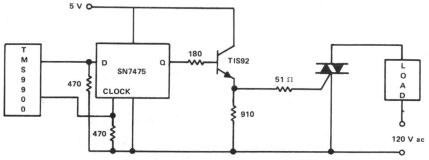

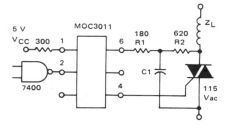

LOGIC-TRIGGERED TRIAC—Pulsed output from microprocessor controls gate drive of triac through SN7475 clock and transistor. Pulse from one output port of microprocessor is applied to D input of clock simultaneously with pulse from communications register unit (CRU) going to clock input, to raise Q output of clock to logic 1. Output remains high until another pulse from CRU returns it to zero, thus giving latching action. High output turns on transistor and supplies about 100-mA gate drive to TIC263 25-A triac.—"Thyristor Gating for μP Applications," Texas Instruments, Dallas, TX, 1977, CA-191, p 4.

LOGIC DRIVE FOR INDUCTIVE LOAD—When output of NAND gate goes high and furnishes 10 mA to LED of Motorola MOC3011 optically coupled triac driver, output of optoisolator provides necessary trigger for triac controlling inductive load. C1 is 0.22 μF for load power factor of 0.75 and 0.33 μF for 0.5 power factor. Omit C1 for resistive load. R1, R2, and C1 serve as snubber that limits rate of rise in voltage applied to triac.—P. O'Neil, "Applications of the MOC3011 Triac Driver," Motorola, Phoenix, AZ, 1978, AN-780, p 2.

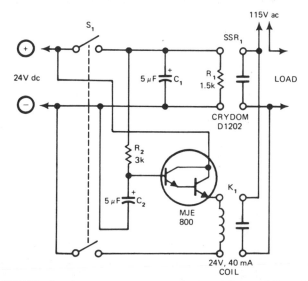

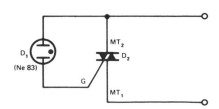

PERFECT AC SWITCH—Developed for use in computerized equipment to prevent generation of severe noise spikes if contact closure can occur at any point in AC cycle. Closing S_1 gates solid-state relay SSR_1, which noiselessly switches load at next zero crossing. During this time, C_2 charges through R_2. After time T = $3R_2C_2$, MJE800 Darlington is turned on, pulling in relay K_1 to follow up SSR_1 with hard contacts. When S_1 is later opened, K_1 drops out immediately but C_1 discharges through gate of SSR_1 to hold it on for about T = $6R_1C_1$. Load is then switched off at next zero crossing after this delay.—E. Woodward, This Circuit Switches AC Loads the Clean Way, *EDN Magazine,* Nov. 20, 1975, p 160 and 162.

VOLTAGE-SENSITIVE SWITCH—RCA 40527 triac is triggered by small neon. After breakdown occurs bidirectionally at 88 V, triac takes over as short-circuit. D_1 can be any other voltage breakdown device, such as diac or zener, and thyristor can be used in place of triac to give unilateral switching. Applications include use as power crowbar, with breakdown level set by artificial resistance-controlled zener.—L. A. Rosenthal, Breakdown and Power Devices Form Unusual Power Switch, *EDN Magazine,* July 5, 1974, p 74–75.

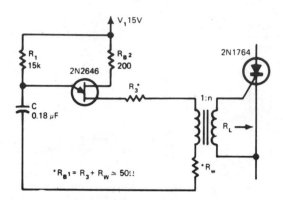

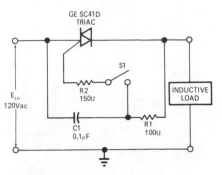

400-Hz TRIGGER FOR SCR—Simple UJT oscillator combined with pulse transformer provides pulses required for firing 2N1764 SCR. Article gives design data for pulse transformer, along with design equations.—W. Dull, A. Kusko, and T. Knutrud, Pulse and Trigger Transformers—Performance Dictates Their Specs, *EDN Magazine,* Aug. 20, 1976, p 57–62.

TRIAC FOR INDUCTIVE LOADS—Simple triac gating circuit applies AC power to inductive load when low-power switch S1 is closed. R1 and C1 provide dv/dt suppression.—C. A. Farel and D. M. Fickle, Triac Gating Circuit, *EDN/IEEE Magazine,* Jan. 1, 1972, p 72–73.

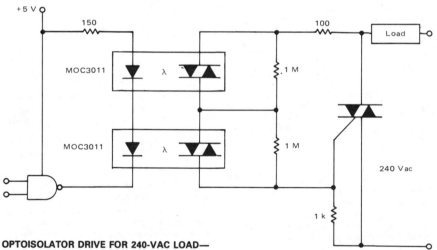

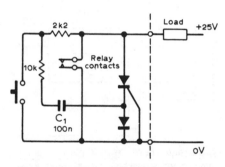

OPTOISOLATOR DRIVE FOR 240-VAC LOAD—Two Motorola MOC3011 optically coupled triac drivers are used in series to overcome voltage limitation of single coupler when triggering triac connected to control 240-VAC load. Two 1-megohm resistors equalize voltage drops across couplers.—P. O'Neil, "Applications of the MOC3011 Triac Driver," Motorola, Phoenix, AZ, 1978, AN-780, p 5.

THYRISTOR SWITCH—When circuit of conventional P-gate thyristor is grounded by switch, negative-going pulse is applied to thyristor cathode, which reverse-biases the diode. When thyristor conducts, diode is forward-biased and has only about 0.7-V drop. Use low-voltage diode, rated for full load current. Opening of relay contacts makes circuit switch off.—R. V. Hartopp, Grounded Gate Thyristor, *Wireless World,* Feb. 1977, p 45.

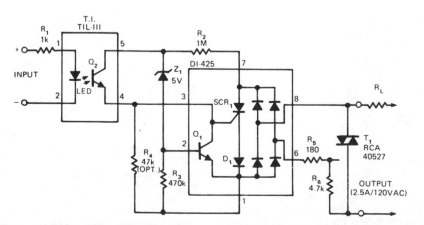

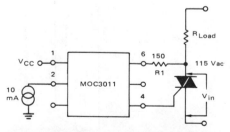

TRIAC CONTROL WITH OPTOISOLATOR—Dionics DI425 switchable bridge circuit controls 120-VAC line in optically isolated zero-crossing solid-state relay that can be used as trigger for power triac. Small AC devices, drawing under 5 W, can be switched directly in either random or zero-crossing mode.—High-Voltage Monolithic Technology Produces 200V AC Switching Circuit, *EDN Magazine,* April 5, 1975, p 121.

TRIAC DRIVE—Motorola MOC3011 optoisolator serves as interface between 10-mA input circuit and gate of triac controlling AC load. Choice of triac depends on load being handled. Optoisolator detector chip responds to infrared LED; once triggered on, optoisolator stays on until input current drops below holding value of about 100 μA.—P. O'Neil, "Applications of the MOC3011 Triac Driver," Motorola, Phoenix, AZ, 1978, AN-780, p 2.

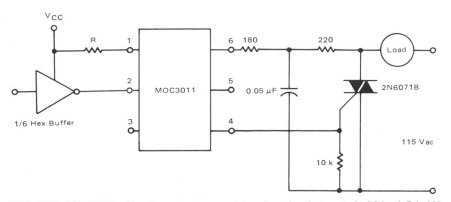

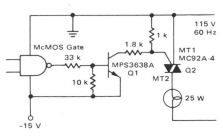

MOS DRIVE FOR TRIAC—Circuit uses one section of MC75492 hex buffer to boost 0.5-mA output of CMOS logic gate to 10 mA required for LED at input of Motorola MOC3011 optically coupled triac driver. When MOS input goes high, optoisolator provides output voltage for triggering triac that controls AC load. R is 220 ohms for 5-V supply and 600 ohms for 10-V supply. For 15 V, use MC14049B buffer and 910 ohms for R.—P. O'Neil, "Applications of the MOC3011 Triac Driver," Motorola, Phoenix, AZ, 1978, AN-780, p 4.

ACTIVE-HIGH TRIAC INTERFACE—Typical CMOS logic gate operating from negative supply triggers triac on negative gate current of 8 mA for control of 25-W AC load. High supply lines for both logic gate and interface transistor are grounded.—A. Pshaenich, "Interface Techniques Between Industrial Logic and Power Devices," Motorola, Phoenix, AZ, 1975, AN-712A, p 12.

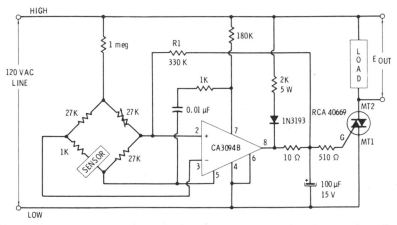

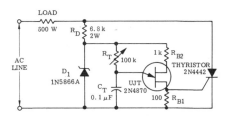

BRIDGE-TRIGGERED TRIAC—Developed for use with AC sensor in one leg of bridge. CA3094 is shut down on negative half-cycles of line. When bridge is unbalanced so as to make pin 2 more positive than pin 3, IC is off at instant that AC line swings positive; pin 8 then goes high and drives triac into conduction. Triac conduction is maintained on next negative half-cycle by energy stored in 100-μF capacitor. Bridge unbalance in opposite direction does not trigger triac.—E. M. Noll, "Linear IC Principles, Experiments, and Projects," Howard W. Sams, Indianapolis, IN, 1974, p 313–314.

600-W HALF-WAVE—UJT serves as trigger for thyristor in circuit that provides power control for load only on positive half-cycles. Thyristor acts also as rectifier, providing variable power determined by setting of R_T during positive half-cycle and no power to load during negative half-cycle.—D. A. Zinder, "Unijunction Trigger Circuits for Gated Thyristors," Motorola, Phoenix, AZ, 1974, AN-413, p 3.

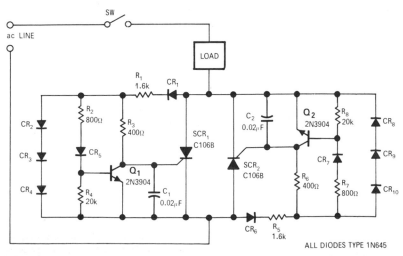

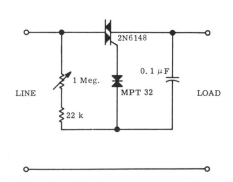

ALL DIODES TYPE 1N645

LINE-POWERED SWITCH—When AC line switch is closed, power is not applied to load until after line voltage next goes through zero. Identical circuits control each half of AC cycle. Transistor turn-on at 1.4 V prevents SCR from triggering until 0.013 ms (less than one-third electrical degree) after next zero-crossing point.—A. S. Roberts and O. W. Craig, Efficient and Simple Zero-Crossing Switch, *EDN/IEEE Magazine*, Aug. 15, 1971, p 46–47.

FULL-WAVE POWER CONTROL—Bidirectional three-layer trigger for triac allows triggering on both half-cycles at point determined by setting of 1-megohm pot. Triac rating determines size of load that can be handled.—"SCR Power Control Fundamentals," Motorola, Phoenix, AZ, 1971, AN-240, p 6.

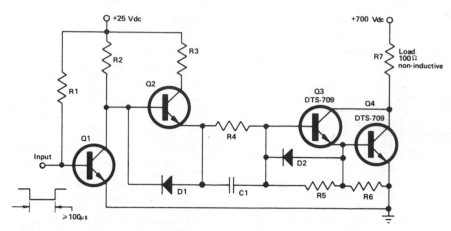

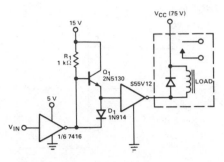

SWITCHING 4500 W AT UP TO 10 kHz—Darlington connection of Delco DTS-709 transistors will switch 7 A at 700 V with 1-μs switching time. Suitable for motor speed control, switching regulator, and inverter applications. Can be operated directly from 440-VAC line. Q1 and Q2

are 2N6100. Diodes are 1N4001. C1 is 4 μF at 15 V. R1 is 510 ohms, R2 is 100, R3 is 12, R4 is 10, R5 is 1K, R6 is 47, and R7 is 100.—"Low Cost 'Duolithic Darlington' Switches 4500 Watts at up to 10 kHz," Delco, Kokomo, IN, 1973, Application Note 54, p 2.

FAST-SWITCHING TTL INTERFACE FOR VMOS—Totem-pole TTL interface drive for S55V01 VMOS gives appreciably faster switching times (less than 30 ns). To achieve fast turn-on time without unduly small pull-up resistor, which dissipates considerable power when switch is in OFF state, emitter-follower Q₁ drives high peak currents into capacitive VMOS input.—L. Shaeffer, VMOS Peripheral Drivers Solve High Power Load Interface Problems, *Computer Design,* Dec. 1977, p 90, 94, and 96–98.

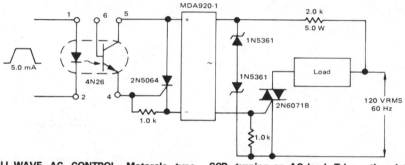

FULL-WAVE AC CONTROL—Motorola type MDA920-1 bridge rectifier provides full-wave rectification of AC line voltage for 2N5064 SCR placed across DC output of bridge. When positive logic pulse from CMOS circuit energizes optoisolator, SCR conducts and completes path for triac gate trigger current through bridge and

SCR, turning on AC load. Triac rating determines size of load. Drawback of circuit is generation of EMI if logic signal occurs at other than zero crossings of AC line.—A. Pshaenich, "Interface Techniques Between Industrial Logic and Power Devices," Motorola, Phoenix, AZ, 1975, AN-712A, p 17.

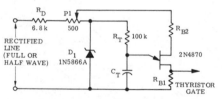

LINE-VOLTAGE COMPENSATION—Can be used with either half-wave or full-wave phase control circuit to make load voltage independent of changes in AC line voltage. P1 is adjusted to provide reasonably constant output over desired range of line voltage. As line voltage increases, P1 wiper voltage increases. This has effect of charging C_T to higher voltage so more time is taken to trigger UJT. Additional delay reduces thyristor conduction angle and thereby maintains desired average voltage.—D. A. Zinder, "Unijunction Trigger Circuits for Gated Thyristors," Motorola, Phoenix, AZ, 1974, AN-413, p 4.

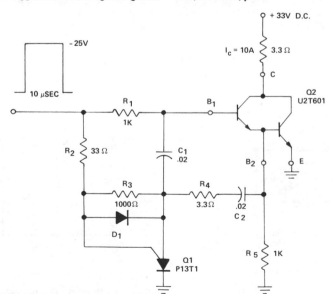

125-ns POWER SWITCH—Developed for repetitive pulse applications in which rise, fall, and storage times of pulse must be kept at absolute minimum. Circuit provides very high gain of Unitrode U2T601 Darlington and switching speeds up to 5 times greater than conventional

techniques. Load power up to 10 A is typically applied within 125 ns. Applications include drive for laser diode and for radar circuits.—"Designer's Guide to Power Darlingtons as Switching Devices," Unitrode, Watertown, MA, 1975, U-70, p 19.

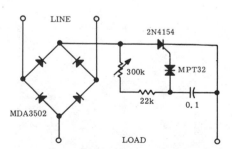

600-W TRIGGERED SCR—2N4154 SCR is operated from DC output of bridge rectifier and triggered by MPT32 at setting determined by position of 300K pot. Circuit provides full-wave DC control of lamp and other loads up to 600 W, using relaxation oscillator operating from DC source.—"SCR Power Control Fundamentals," Motorola, Phoenix, AZ, 1971, AN-240, p 6.

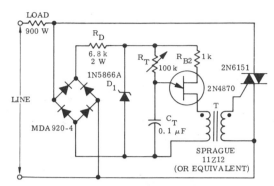

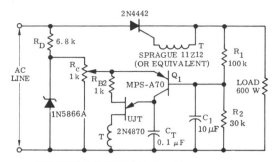

900-W FULL-WAVE—Combination of bridge rectifier, pulse transformer, and triac allows 100K pot R_T to control power to resistive load on both positive and negative half-cycles. Triac is triggered through transformer by 2N4870 UJT.—D. A. Zinder, "Unijunction Trigger Circuits for Gated Thyristors," Motorola, Phoenix, AZ, 1974, AN-413, p 3.

HALF-WAVE FEEDBACK—Provides phase control of power to resistive load in applications where average load voltage is desired feedback variable. RC network R_1-R_2-C_1 averages load voltage so it can be compared with set point on R_c by Q_1.—D. A. Zinder, "Unijunction Trigger Circuits for Gated Thyristors," Motorola, Phoenix, AZ, 1974, AN-413, p 4.

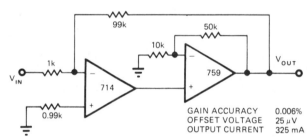

GAIN ACCURACY	0.006%
OFFSET VOLTAGE	25 μV
OUTPUT CURRENT	325 mA

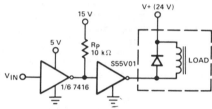

POWER OPAMP FOR CONTROL—Precision 714 opamp drives 759 power opamp to give ultra-precision power amplifier system. High current capability (up to 500-mA peak output current) makes circuit suitable for such control applications as driving motors, relays, solenoids, or transmission lines. Article tells how to calculate heatsink requirements for opamp.—R. J. Apfel, Power Op Amps—Their Innovative Circuits and Packaging Provide Designers with More Options, *EDN Magazine*, Sept. 5, 1977, p 141–144.

TTL INTERFACE FOR VMOS—By using open-collector 7416 TTL interface with its output pulled up to 15 V, S55V01 VMOS will switch up to 2 A easily. Inductive load such as relay requires diode across relay coil.—L. Shaeffer, VMOS Peripheral Drivers Solve High Power Load Interface Problems, *Computer Design*, Dec. 1977, p 90, 94, and 96–98.

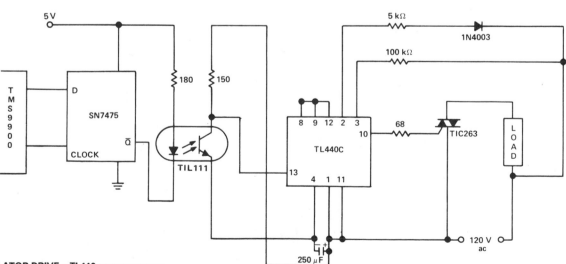

OPTOISOLATOR DRIVE—TL440 zero-crossover switch uses pulsed gate drive. Switch supplies phase-locked driving pulses during zero-voltage crossover period, to minimize electromagnetic interference generated during turn-on of triac. Works well with resistive loads, but inductive loads can create phase shift that affects firing time of triac.—"Thyristor Gating for μP Applications," Texas Instruments, Dallas, TX, 1977, CA-191, p 6.

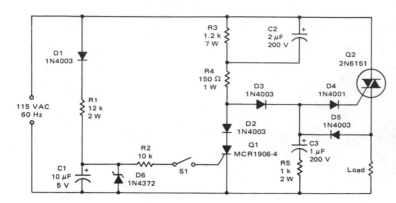

ZERO-POINT SWITCH—Used to control resistive loads. With S1 open, triac Q2 is turned on very close to zero on initial part of positive half-cycle because of large current flow into C2. Once Q2 is on, C3 charges through D5. When line voltage goes through zero and starts negative, C3 is still discharging into gate of Q2 to turn it on near zero of negative half-cycle. Load current thus flows for most of both half-cycles. When S1 is closed, Q1 is turned on and shunts gate current away from Q2 during positive half-cycles. Q2 cannot turn on during negative half-cycle because C3 cannot charge, which makes load current zero.—"Circuit Applications for the Triac," Motorola, Phoenix, AZ, 1971, AN-466, p 12.

OUTPUT CONTROL FOR CLOCK COMPARATOR—Circuit triggers 10-A triac when Q output of comparator-driven flip-flop is logic 1. LED in optoisolator is then energized, activating phototransistor pair for driving gate circuit of triac through diode bridge. Trigger voltage of triac is positive for first quadrant and negative for third quadrant, to give maximum sensitivity of triac control.—D. Aldridge and A. Mouton, "Industrial Clock/Timer Featuring Back-Up Power Supply Operation," Motorola, Phoenix, AZ, 1974, AN-718A, p 7.

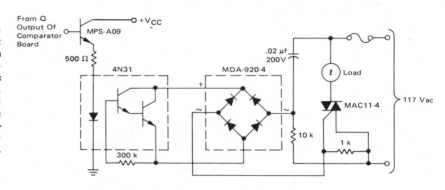

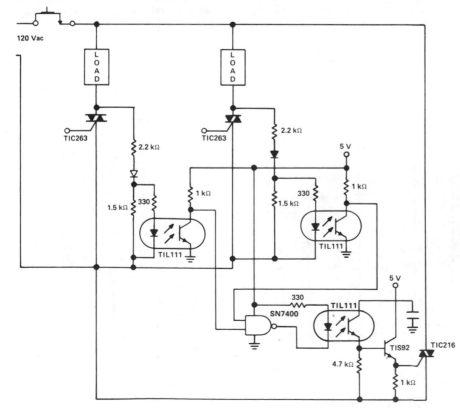

PROTECTION AGAINST SIMULTANEOUS OPERATION OF TRIACS—Optoisolators provide cross-connection between solid-state triac relay circuits to eliminate possibility that two or more triacs come on at same time due to circuit malfunction or component failure. Circuit shuts system down when this occurs.—"Thyristor Gating for μP Applications," Texas Instruments, Dallas, TX, 1977, CA-191, p 5–9.

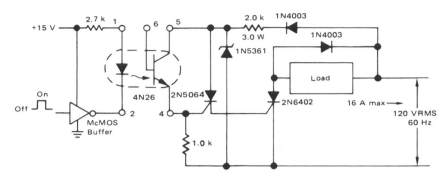

HALF-WAVE AC CONTROL—Motorola 4N26 optoisolator serves as interface for static series switch in gate circuit of 2N5064 SCR. When logic input goes high, optoisolator is energized and first SCR is triggered on. Resulting current turns on power SCR for passing load current on that positive half-cycle of AC line voltage. When logic goes low, load current stops at next zero crossing of AC source. 5 mA of isolated DC control current thus controls up to 16 A for half-wave load.—A. Pshaenich, "Interface Techniques Between Industrial Logic and Power Devices," Motorola, Phoenix, AZ, 1975, AN-712A, p 16.

LOGIC CONTROLS 15-A LOAD—Load is energized when logic input drops to 0. Can be used to drive solenoid or electromagnet from 48-VDC supply, for stopping paper tape in high-speed tape reader. If relay is to be activated by high or 1 level, add inverter at input as shown.—D. D. Mickle, Practical Computer Projects, *73 Magazine,* Jan. 1978, p 92–93.

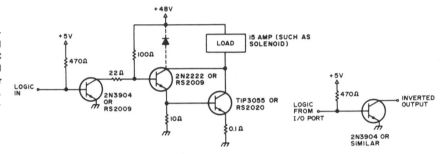

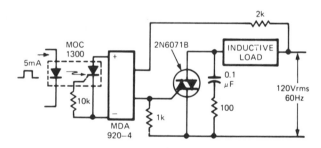

FULL-WAVE CONTROL—Uses triac to provide current for inductive load during both positive and negative alternations of AC source, with optoisolator providing complete isolation for logic control circuit. MDA920-4 diode bridge provides pulsating DC voltage for photo-SCR of optoisolator so gate current is supplied to triac for both halves of AC cycle.—T. Mazur, Solid-State Relays Offer New Solutions to Many Old Problems, *EDN Magazine,* Nov. 20, 1973, p 26–32.

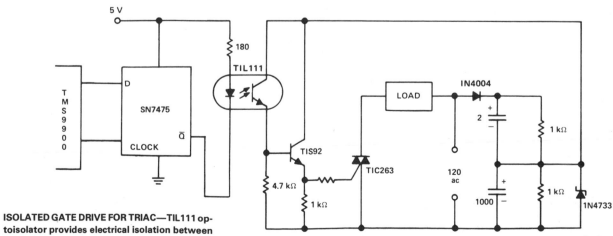

ISOLATED GATE DRIVE FOR TRIAC—TIL111 optoisolator provides electrical isolation between control logic and gate drive for triac at low cost, with faster switching than is possible with re-lays. Transistor provides direct current drive for gate of triac.—"Thyristor Gating for μP Applications," Texas Instruments, Dallas, TX, 1977, CA-191, p 4–5.

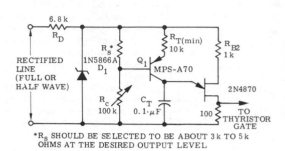

*R_S SHOULD BE SELECTED TO BE ABOUT 3 k TO 5 k OHMS AT THE DESIRED OUTPUT LEVEL

FEEDBACK CONTROL—Replacement of manual phase control pot with sensor and transistor provides automatic control of load power in response to stimuli such as heat, light, pressure, or magnetic fields. Output of feedback control circuit goes to thyristor in series with load. R_c establishes desired operating point for sensing resistor R_s. As R_s increases in resistance, more current flows into C_T and makes 2N4870 UJT trigger at smaller phase angle so more power is applied to load. For opposite effect, interchange R_s and R_c.—D. A. Zinder, "Unijunction Trigger Circuits for Gated Thyristors," Motorola, Phoenix, AZ, 1974, AN-413, p 4.

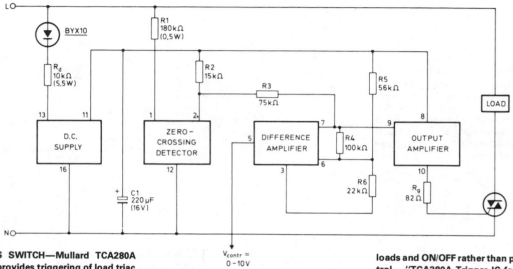

SYNCHRONOUS SWITCH—Mullard TCA280A trigger module provides triggering of load triac at zero crossings of AC line voltage. Values shown are for triac requiring gate current of 100 mA; for other triacs, values of R_d, R_g, and C_1 may need to be changed. Designed for resistive loads and ON/OFF rather than proportional control.—"TCA280A Trigger IC for Thyristors and Triacs," Mullard, London, 1975, Technical Note 19, TP1490, p 10.

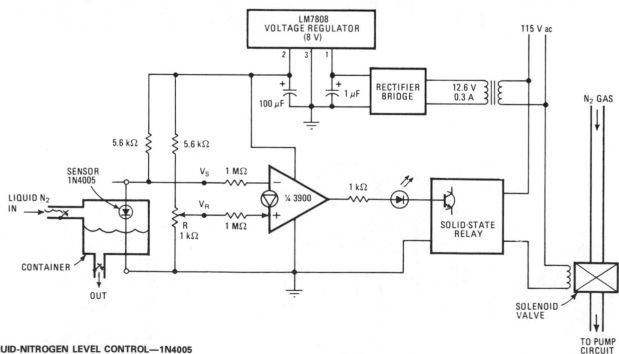

LIQUID-NITROGEN LEVEL CONTROL—1N4005 diode serves as sensor. Junction voltage of diode increases from 0.7 V at room temperature to 1.05 V when in liquid nitrogen. Voltage change is used to activate amplifier that controls solenoid valve through solid-state relay. 1K pot R adjusts circuit sensitivity; this can be set so pump starts refilling of container when liquid is as much as 2 inches below diode, to eliminate frequency recycling. LED indicates valve status.—V. J. H. Chiu, Diode Sensor and Norton Amp Control Liquid-Nitrogen Level, *Electronics*, Feb. 2, 1978, p 117.

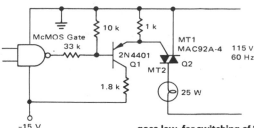

ACTIVE-LOW TRIAC INTERFACE—With connection shown for interface transistor Q1, typical CMOS gate triggers triac when gate output goes low, for switching of 25-W lamp load.—A. Pshaenich, "Interface Techniques Between Industrial Logic and Power Devices," Motorola, Phoenix, AZ, 1975, AN-712A, p 12.

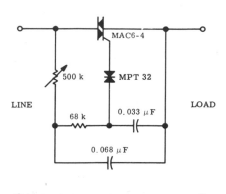

FULL-RANGE CONTROL—Triggered triac is used with double phase-shift network to obtain reliable triggering at conduction angles as low as 5°, as required for control of incandescent lamps and some motors. Triac rating determines size of load.—"SCR Power Control Fundamentals," Motorola, Phoenix, AZ, 1971, AN-240, p 6.

PUT CONTROLS SCR—Programmable unijunction transistor Q2 provides phase control for both halves of AC line voltage by triggering SCR connected across bridge. Relaxation oscillator formed by Q2 varies conduction interval of Q1 from 1 to 7.8 ms or from 21.6° to 168.5°, to give control over 97% of power available to load.—R. J. Haver and B. C. Shiner, "Theory, Characteristics and Applications of the Programmable Unijunction Transistor," Motorola, Phoenix, AZ, 1974, AN-527, p 10.

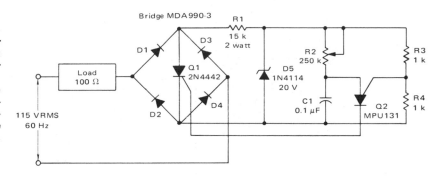

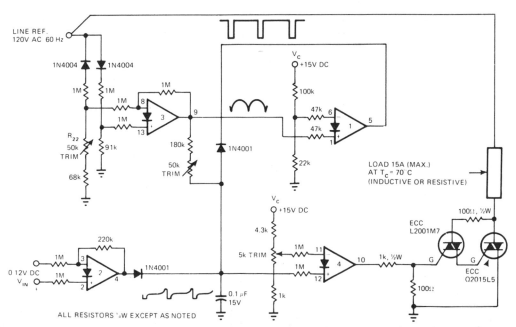

GROUND-REFERENCED RAMP-AND-PEDESTAL CONTROL—Need for transformer is eliminated by applying alternate half-cycles to inverting and noninverting inputs of section 3 of LM3900 quad opamp, so full-wave-rectified waveform is referenced to ground. Comparator opamp 1 discharges timing capacitor at zero line voltage and synchronizes circuit with line frequency. Buffer opamp 2 scales input and provides linear pedestal for capacitor. Opamp 4 is comparator serving as output driver whose output is high when capacitor is charged to level selected by high-end trimming pot. Output is sufficient for optoisolators and logic triacs.—J. C. Johnson, Ramp-And-Pedestal Phase Control Uses Quad Op Amp, *EDN Magazine,* June 5, 1977, p 208 and 211.

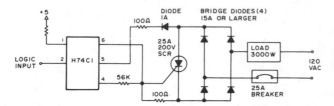

LOGIC DRIVES SCR—Uses light-activated SCR in H74C1 optoisolator to trigger larger SCR for controlling loads up to 3000 W through bridge diodes. When logic input goes low (to ground), load is energized. Limit for inductive loads is 8 A or about 1000 W if using 25-A SCR.—D. D. Mickle, Practical Computer Projects, *73 Magazine*, Jan. 1978, p 92–93.

LOGIC DRIVES TRIAC—H74C1 optoisolator combined with saturation characteristic of ordinary filament transformer serves to trigger full-wave triac on or off under control of logic input, for energizing AC loads up to rating of triac. Logic 0 (ground level) turns load on, and logic 1 turns it off.—D. D. Mickle, Practical Computer Projects, *73 Magazine*, Jan. 1978, p 92–93.

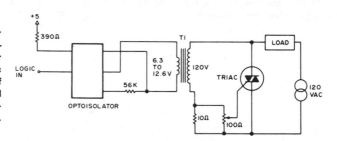

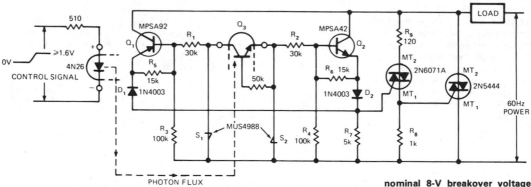

ZERO-CROSSING CONTROL—When control signal calls for power, optoisolator energizes circuit that provides load turn-on at zero-voltage time of AC waveform. If phototransistor Q_3 of optoisolator is illuminated after S_1 drops to 1-V conduction voltage of MUS4988, triacs will not be turned on. Circuit thus provides relay-enabling voltage window, lower limit of which is point at which all components involved in turning on triacs are forward-biased. Upper limit is nominal 8-V breakover voltage of unilateral switch S_1. S_2 performs similar function on negative voltage alternations. Load-controlling triac is rated 40 A.—T. Mazur, Solid-State Relays Offer New Solutions to Many Old Problems, *EDN Magazine*, Nov. 20, 1973, p 26–32.

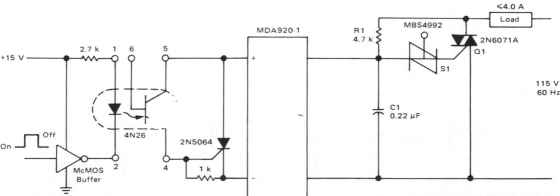

4-A FULL-WAVE CONTROL—When logic input to CMOS buffer goes high, load is off. Low input logic deenergizes optoisolator; clamp formed by bridge rectifier and SCR is then removed from C1, allowing it to charge through R1. When voltage across C1 reaches triggering voltage of S1 (about 8 V), MBS4992 silicon bidirectional switch fires, allowing C1 to dump charge into gate of triac. Triac and load are then turned on. R1 and C1 are chosen to give time constant small enough to fire triac early in its conduction angle (near zero crossing), to maximize load power while minimizing EMI.—A. Pshaenich, "Interface Techniques Between Industrial Logic and Power Devices," Motorola, Phoenix, AZ, 1975, AN-712A, p 17.

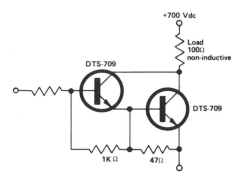

SWITCHING 4500 W BELOW 5 kHz—Simple Darlington connection of Delco DTS-709 transistors serves for switching of up to 700 V at 7 A for low-speed motor control, regulator, and inverter applications.—"Low Cost 'Duolithic Darlington' Switches 4500 Watts at up to 10 kHz," Delco, Kokomo, IN, 1973, Application Note 54, p 3.

RELAY DRIVE—Logic input of 0 turns first transistor off, allowing base of next transistor to go high so it turns on and energizes relay for achieving desired control function.—D. D. Mickle, Practical Computer Projects, 73 Magazine, Jan. 1978, p 92–93.

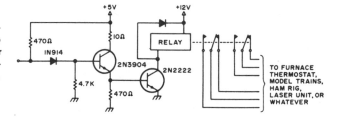

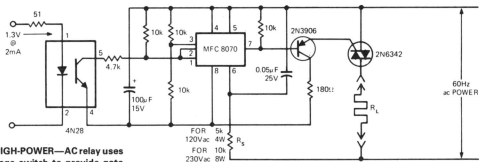

ZERO-CROSSING HIGH-POWER—AC relay uses MFC8070 zero-voltage switch to provide gate current pulses for triac under control of differential input voltages derived from output of 4N28 optoisolator. When LED is off, voltage at pins 1 and 2 of switch is positive with respect to pin 3, inhibiting switch so no current pulses go to triac gate. When LED is energized by logic input, gate current pulses are generated at zero-voltage excursions of AC power source. Transistor ensures adequate gate drive at low temperatures. For normally-on configuration, interchange input connections 3 and 1-2 of switch. Triac can be selected to handle resistive loads from 4 to 40 A.—T. Mazur, Solid-State Relays Offer New Solutions to Many Old Problems, EDN Magazine, Nov. 20, 1973, p 26–32.

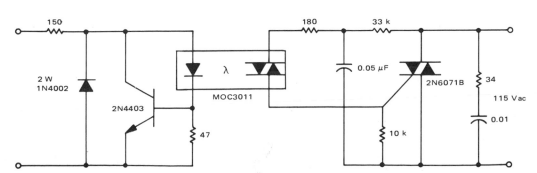

OPTOISOLATOR AS SOLID-STATE RELAY—Circuit provides input protection of LED from overvoltage and reverse polarity, along with snubber network for handling inductive AC loads. Triac should be chosen to handle load. Safe input voltage range is 3–30 VDC.—P. O'Neil, "Applications of the MOC3011 Triac Driver," Motorola, Phoenix, AZ, 1978, AN-780, p 6.

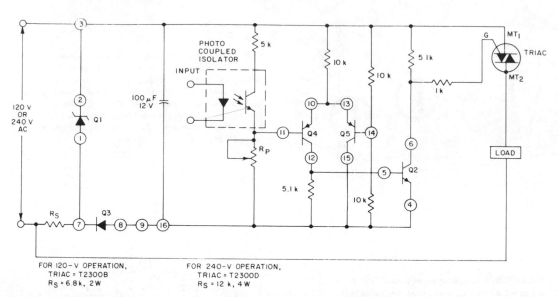

FOR 120-V OPERATION,
  TRIAC = T2300B
  R_S = 6.8 k, 2W

FOR 240-V OPERATION,
  TRIAC = T2300D
  R_S = 12 k, 4 W

ALARM CONTROL—Any input suitable for driving LED of optoisolator triggers triac for energizing load such as alarm gong. Transistors Q4 and Q5 of CA3096 array serve as comparator. Diode-connected transistor Q3, zener-connected transistor Q1, and 100-μF capacitor develop DC supply voltages from AC line.—"Circuit Ideas for RCA Linear ICs," RCA Solid State Division, Somerville, NJ, 1977, p 9.

CHAPTER **68**

Power Supply Circuits

Includes unregulated circuits for changing AC input voltage to variety of DC voltages ranging from 1.5 V to 3 kV. Also includes inverter circuits containing oscillator operating from DC supply and providing AC voltage at 60 Hz or 400 Hz, along with RMS AC regulator. See also Converter—DC to DC, Regulated Power Supply, Regulator, and Switching Regulator chapters.

12-V TRANSFORMERLESS PREREGULATOR—AC line voltage is converted to regulated 12 VDC by varying firing angle of 10-A SCR. Circuit provides reliable operation for AC line voltages between 50 and 140 V. Key element in triggering of SCR is programmable unijunction transistor that provides variable and accurate control of firing time. Developed for use in power supply that uses digital techniques of sample-and-hold switching to achieve high degree of isolation between power line and load without using transformer.—J. A. Dickerson, Transformerless Power Supply Achieves Line-to-Load Isolation, *EDN Magazine,* May 5, 1976, p 92–96.

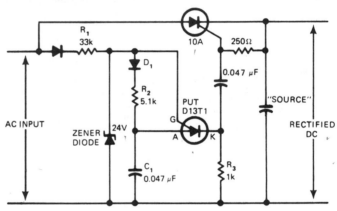

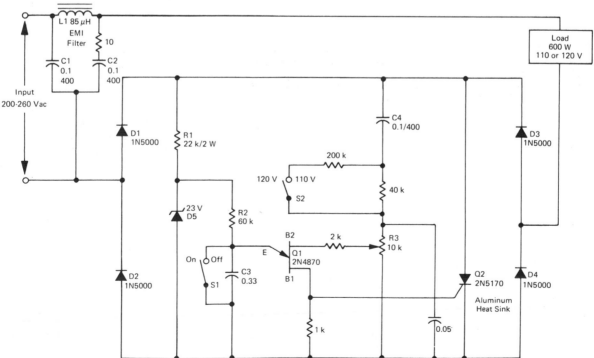

110/120 VAC ± 2.5 V AT 600 W—Simple open-loop voltage compensator for small conduction angles operates from 200–260 VAC input and provides true RMS output voltage for sensitive equipment such as photographic enlargers, oven heaters, projection lights, and certain types of AC motors. Full-wave bridge D1-D4 and SCR Q2 provide full-wave control, with UJT Q1 serving as trigger. Triggering frequency is determined by charge and discharge of C3 through R2. As input voltage increases, required trigger voltage also increases, retarding firing point of SCR to compensate for change in input.—D. Perkins, "True RMS Voltage Regulators," Motorola, Phoenix, AZ, 1975, AN-509, p 3.

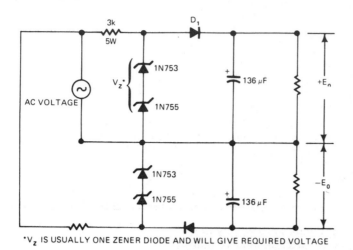

*V_z IS USUALLY ONE ZENER DIODE AND WILL GIVE REQUIRED VOLTAGE

TRANSFORMERLESS ±12 V AT 15 mA—Developed to provide bias voltage for six 741 opamps. Circuit connects directly across 120-V 60-Hz AC line. Article gives design procedure to meet performance requirements. For values shown, ripple is 1.1 V. Diode types are not critical.—C. Venditti, Build this Transformerless Low-Voltage Supply, *EDN Magazine,* Feb. 5, 1977, p 102.

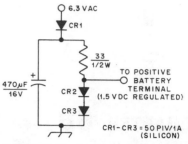

1.5 V FOR VTVM—Simple rectifier circuit replaces battery in vacuum-tube voltmeter. Provides good regulation and eliminates need for frequent battery replacement. Remove battery before using supply. AC source can be 6.3-V secondary of filament transformer or terminals of 6.3-V pilot lamp in any AC equipment.—P. Alexander, Battery Replacement Circuit for VTVM, *QST,* Jan. 1976, p 42–43.

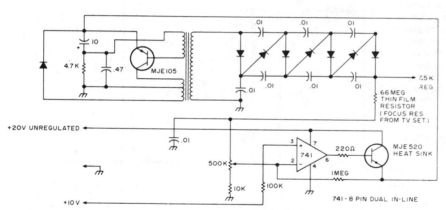

7.5-kV REGULATED SUPPLY—Power transformer is special design, but commercial unit delivering 5 to 10 kV can be used. Inverter circuit uses MJE105 transistor driving primaries of transformer. 741 opamp and transistor provide regulation for 7.5-kV output used in slow-scan TV monitor. Diodes are 1 kV, such as 1N4007. Article gives circuit of complete monitor, including low-voltage supply.—L. Pryor, Homebrew This SSTV Monitor, *73 Magazine,* June 1975, p 22–24, 26–28, and 30.

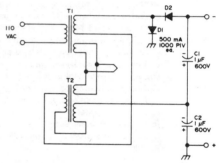

1000 V FOR CRT—Unique connection of two TV booster transformers having 125-V secondaries gives high-voltage supply for small monitor scope. T1 is connected conventionally, with its 6.3-V winding going to heater of CRT. 6.3-V winding of T2, also connected to CRT, serves as primary for second transformer. Remaining windings of T2 and high-voltage secondary of T1 are connected in series aiding to give about 367 VAC for doubling by D1-D2 and C1-C2. Since CRT drain is low, filter charges to very nearly peak voltage of 1027 VDC.—W. P. Turner, Cheap Power Supply for a CRT, *73 Magazine,* March 1974, p 53.

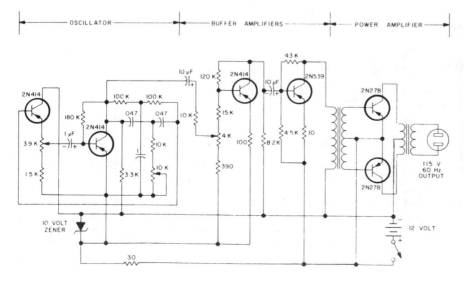

100-W SINE AT 60 Hz—Consists essentially of 60-Hz sine-wave oscillator with 10K frequency-control pot, two buffer stages, and push-pull power amplifier. Circuit eliminates noise problems of square-wave inverters when operating 115-V radio receiver or cassette player in car.—G. C. Ford, Power Inverter with Sine Wave Output, *73 Magazine,* May 1973, p 29–32.

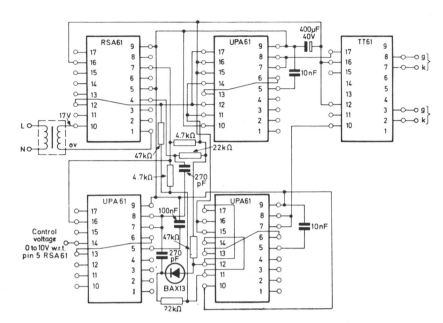

12 V TO 6 V—Permits operation of older 6-V VHF FM mobile equipment from 12-V storage battery. With transistor mounted on suitable heatsink, maximum output is 15 A. If positive and negative lines are isolated from chassis, converter may be used with either negative or positive ground.—E. Noll, Circuits and Techniques, *Ham Radio,* April 1976, p 40–43.

PARALLEL INVERTER DRIVE—Uses Mullard modules for converting DC power to AC at high power levels for such applications as driving induction motors at higher speeds than are obtainable with line frequency. DC control voltage of 0–10 V varies output frequency up to 400 Hz. UPA61 modules provide functions of level detector, pulse generator, ramp generator, capacitor discharge circuit, and bistable MVBR for parallel inverter system. RSA61 and TT61 are trigger modules, with RSA61 also providing power supplies for other modules.—"Universal Circuit Modules for Thyristor Trigger Systems (61 Series)," Mullard, London, 1978, Technical Information 66, TP1660, p 19.

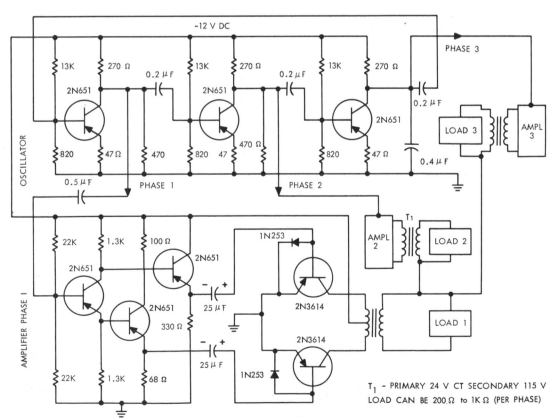

12 VDC TO 115 VAC AT 400 Hz—Provides three-phase output at 20 W by using RC coupling to oscillator in such a way that 120° phase difference exists at collectors of 2N651 transistors of oscillator. Emitter-follower amplifier driving push-pull power output transistors is shown only for phase 1; other two phases use similar amplifiers. Power transistors are operated in saturated switching mode.—R. J. Haver, "The ABC's of DC to AC Inverters," Motorola, Phoenix, AZ, 1976, AN-222, p 15.

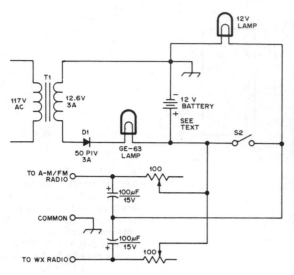

12-V EMERGENCY POWER—Trickle-charge circuit and 12-V motorcycle battery provide reliable emergency power for battery-operated weather radio, portable AM/FM receiver, or hand-held transceiver for many hours. 100K pots drop voltage to 9 V for each receiver. Lamp can be auto dome light. GE-63 pilot lamp in charging circuit acts as current limiter and charge indicator.—J. Rice, Simple Emergency Power, *QST*, March 1978, p 42.

130 AND 270 V FOR CRT—High-voltage power supply provides 270 V required for deflection plates of 2AP1-A CRT used as RTTY tuning indicator, as well as 130 V for high-voltage amplifier. Large capacitor keeps ripple voltage low.—R. R. Parry, RTTY CRT Tuning Indicator, *73 Magazine*, Sept. 1977, p 118–120.

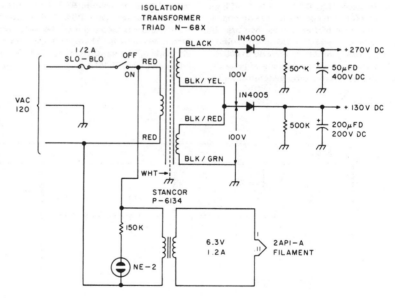

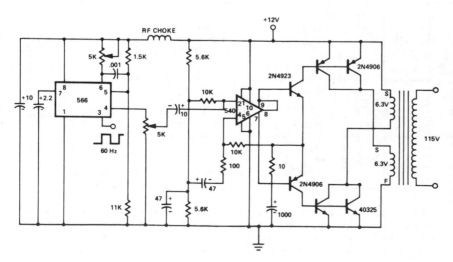

12 VDC TO 115 VAC AT 100 W—566 function generator provides triangle output at 60 Hz with frequency stability better than ±0.02%/°C. 540 power driver feeds six-transistor power output stage. Transformer load attenuates third harmonic, giving output very close to pure 60-Hz sine wave. 566 also provides square-wave output for other purposes.—"Signetics Analog Data Manual," Signetics, Sunnyvale, CA, 1977, p 853–854.

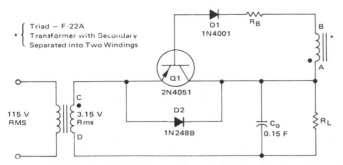

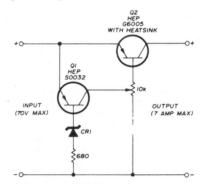

HALF-WAVE SYNCHRONOUS RECTIFIER— Transistor Q1 is synchronously biased on by AC input voltage to give efficient low-voltage regulation. When points A and C are positive with respect to points B and D, base-emitter junction of Q1 is forward-biased and collector current flows through load R_L. On negative alternations, Q1 is reverse-biased and transistor is blocked.— B. C. Shiner, "Improving the Efficiency of Low Voltage, High-Current Rectification," Motorola, Phoenix, AZ, 1973, AN-517, p 3.

230 VAC FROM 115 VAC— Connect 6.3-V filament transformers back-to-back as shown to get 230 V when step-up transformer is not available. 115-V windings must be phased properly in series; if wrong, output voltage will be zero. Output power rating at 230 V is somewhat less than twice the power (E × I) rating of smallest filament transformer. If 6.3-V 10-A transformers are used, power rating would be about 100 W (less than 2 × 6.3 × 10).—A. E. McGee, Jr., Cheap and Easy 230 Volt AC Power Supply, *73 Magazine,* Aug. 1974, p 64.

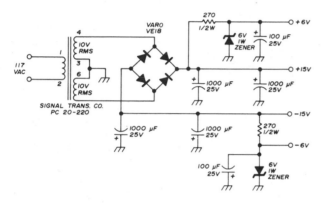

±6 V AND ±15 V— Suitable for use when frequency or some other critical parameter of load is not dependent on voltage. Developed for use in CMOS IC function generator.—R. Megirian, Inegrated-Circuit Function Generator, *Ham Radio,* June 1974, p 22–29.

TRANSIENT ELIMINATOR— Used between DC power supply and load to eliminate supply transients that might damage semiconductor devices. Zener rating should be about 10% higher than supply voltage so Q1 is normally turned off. Q2 is normally conducting. When voltage spike is present on input line, zener conducts and turns Q1 on. Q1 then places positive bias on 10K pot to turn off Q2 and protect load during transient.—J. Fisk, Circuits and Techniques, *Ham Radio,* June 1976, p 48–52.

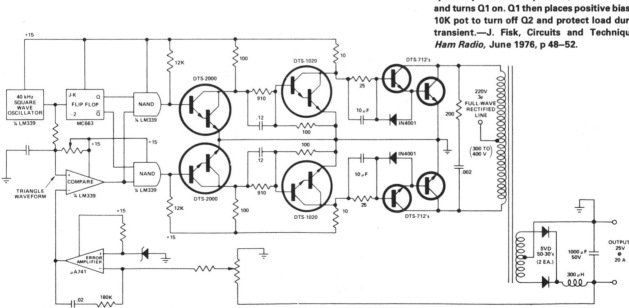

500 W AT 20 kHz— Uses four Delco DTS-712 transistors in push-pull Darlington configuration, with pulse-width modulation on push-pull inverter providing regulation. Can be operated from 220-VAC three-phase full-wave rectified line. Efficiency is up to 80%. Square-wave output of 40-kHz primary oscillator drives JK flip-flop that generates complementary square waves and divides frequency by 2 with necessary symmetry. NAND gates establish primary ON/OFF periods of power stage. Portion of output signal is compared to reference voltage, and error signal is fed to NAND gates to give regulation better than 0.1% for load range of 200–500 W or line range of 300–400 V.—"A 20 kHz, 500 W Regulating Converter Using DTS-712 Transistors," Delco, Kokomo, IN, 1974, Application Note 55.

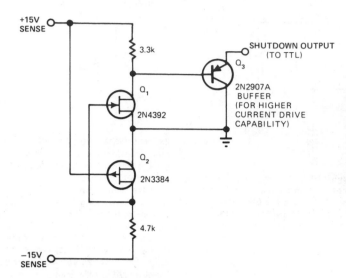

SHUTDOWN PROTECTION—Used with digital logic to prevent generation of false logic signals when power supply is turned on or turned off. FETs sense +15 V and −15 V supplies and conduct when either supply drops below pinch-off voltage, activating shutdown output. With values shown, shutdown output is disabled when supplies exceed about 4 V, to provide normal operation.—E. Burwen, Power-Supply Monitor Suppresses False Output Signals, *EDN Magazine*, Nov. 5, 1977, p 110 and 112.

117 VAC FROM 24–60 VDC—Will operate from either 24- or 32-V storage battery or from 60-VDC source. Circuit shown is set up for 60-V operation. For 24/32 V, remove F1 and F4 and insert F2 and F3, then switch S3 to 32 V. T1 is 117-V 20-A Variac with bifilar primary winding added; use 38 bifilar turns of No. 8 for 24 V and 48 turns for 32 V. Commutating capacitor C1 consists of ten 120-μF 400-VDC oil-filled capacitors (do not use electrolytics). SCRs Q1 and Q2, rated 100 A at 800 PIV (Poly Paks 92CU1167), are switched by Cornell-Dubilier 98600 60-Hz 12-V vibrator. VR1, which limits voltage across vibrator coil, consists of two 6.8-V 10-W zeners. With 60-V operation, use 1-ohm 200-W resistor in series with inverter while starting, but short it out while inverter is running. Inverter output voltage varies from 150 VAC no-load to 110 VAC with 1650-W resistive load. CR1 is 400-PIV 200-A silicon. CR2 and CR3 are 800-PIV 250-A silicon.—R. Dunaja, A High-Power SCR Inverter, *QST*, June 1974, p 36–37.

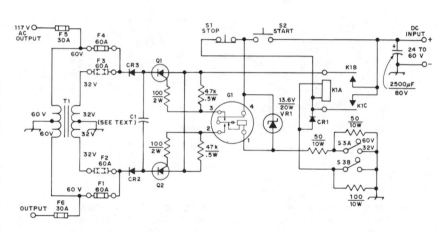

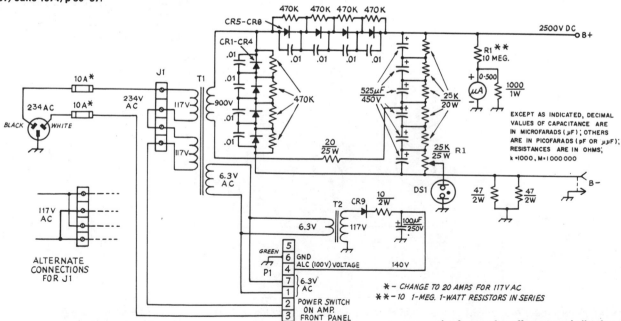

EXCEPT AS INDICATED, DECIMAL VALUES OF CAPACITANCE ARE IN MICROFARADS (μF); OTHERS ARE IN PICOFARADS (pF OR μμF); RESISTANCES ARE IN OHMS; k =1000, M=1 000 000

* – CHANGE TO 20 AMPS FOR 117 V AC
** – 10 1-MEG. 1-WATT RESISTORS IN SERIES

2500 V AT 500 mA—Meets power requirements for 2-kW linear amplifier using pair of 8873 conduction-cooled triodes for SSB transmitter service. Power transformer is Hammond 101165. Diodes are 1000 PIV at 2.5 A, such as Motorola HEP170. T2 is Stancor P-8190 rated 6.3 V at 1.2 A. DS1 is 117-V neon pilot lamp. Set tap on R1 5000 ohms from B−lead. Make adjustments only after turning off power and allowing time for capacitors to discharge; output voltages are dangerous.—R. M. Myers and G. Wilson, 8873s in a Two-Kilowatt Amplifier, *QST*, Oct. 1973, p 14–19.

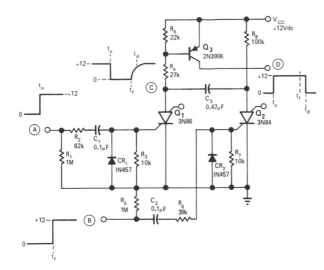

LOW STANDBY DRAIN—Positive 12-V pulse at input A triggers SCS Q_1 and turns on transistor switch Q_3. Positive pulse at input B gates SCS Q_2 on and turns off Q_3. Current drain is essentially zero (typically 3 μA). Circuit was designed to supply up to 7 mA of switched current from 12-VDC supply.—D. B. Heckman, Bistable Switch with Zero Standby Drain, *EDN/EEE Magazine*, Oct. 1, 1971, p 42.

200 W AT 25 kHz—Two Delco DTS-403 high-voltage silicon transistors are connected as push-pull oscillator operating on 150-VDC bias. Efficiency is 78% at full load. Diodes serve alternately as steering and clamp diodes.—"25 kHz High Efficiency 200 Watt Inverter," Delco, Kokomo, IN, 1971, Application Note 47.

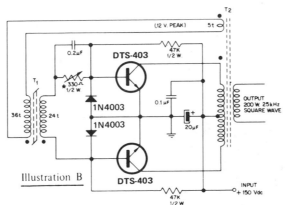

Illustration B

T_1 Pri: 36 t#30 AWG
        Sec: 24 t#25 AWG
        Core: Ferroxcube
             266 T 125-3E2A
             Ferrite toroid

T_2 Pri: 126 t tapped (@) 63t,
             40 strands #38 AWG litz
     *Sec: 2.38 V/t
Feedback: 5t #25 AWG
       Core: Ferroxcube (ferrite toroid)
               528T500-3C5

★ Adjust value of resistor for maximum efficiency at full load.

* To be determined by individual requirements.

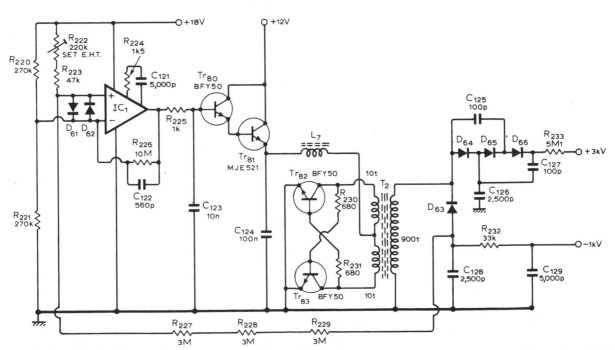

3 kV FOR CRO—Circuit also provides 1-kV negative supply at 2 mA, as required for cathode-ray tube of oscilloscope. Positive supply furnishes 50 μA at 3 kV. Design uses transistor inverter operating at about 20 kHz to simplify filtering. Tr_{82} and Tr_{83} form current-switched class D oscillator producing sine waves at high efficiency. Current multiplication is provided by Tr_{80} and Tr_{81} for 709 IC opamp.—C. M. Little, A 50 MHz Oscilloscope, *Wireless World*, July 1975, p 319–322.

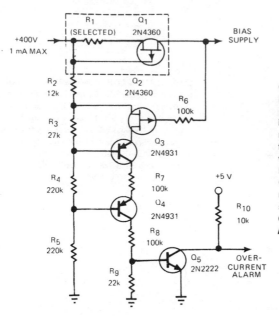

OVERCURRENT PROTECTION FOR 400-V SUPPLY—R_1 and Q_1 form current detector for bias supply. At normal current levels, voltage drop is very small and Q_2 is reverse-biased. When current reaches 400 μA, voltage drop across R_1 forces gate of Q_1 to near pinchoff. Combined voltage drop across R_1 and Q_1 then becomes large so Q_2 is forced almost to full conduction. Q_3 and Q_4 then turn on Q_5, to provide overcurrent-alarm signal for activating logic circuit that shuts off power supply.—J. P. Thompson, Overcurrent Alarm Protects HV Supply, *EDN Magazine*, Nov. 20, 1978, p 321–322.

FULL-WAVE SYNCHRONOUS RECTIFIER—Transistors are biased on alternately by AC input voltage, to supply load current on alternate half-cycles. Silicon diodes D1 and D2 protect transistors from charging current of capacitive load when circuit is turned on. Capacitor discharge problems are minimized by use of diodes D3 and D4 in base circuits of transistors.—B. C. Shiner, "Improving the Efficiency of Low Voltage, High-Current Rectification," Motorola, Phoenix, AZ, 1973, AN-517, p 4.

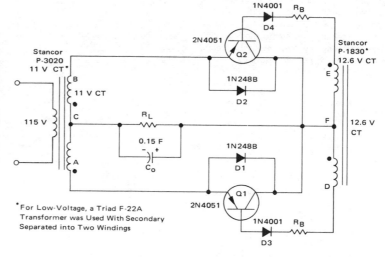

*For Low-Voltage, a Triad F-22A Transformer was Used With Secondary Separated into Two Windings

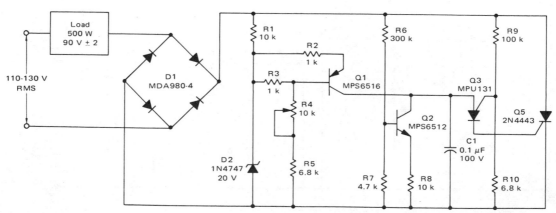

90 VRMS AT 500 W—Open-loop RMS voltage regulator acts with full-wave bridge to provide good AC voltage regulation for AC load over line voltage range of 110–130 VAC. As input voltage increases, voltage across R10 increases and serves to increase firing point of PUT Q3. This delays firing of SCR Q5 to hold output voltage fairly constant as input voltage increases. Delay network of Q1 prevents circuit from latching up at beginning of each charging cycle for C1.—R. J. Haver and B. C. Shiner, "Theory, Characteristics and Applications of the Programmable Unijunction Transistor," Motorola, Phoenix, AZ, 1974, AN-527, p 11.

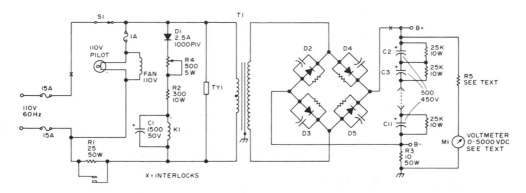

3-kV SUPPLY—Circuit uses full-wave bridge rectifier D2-D5, with each diode stack constructed from two 1000-PIV 2.5-A diodes in series. Each diode pair is shunted by 470K 1-W resistor and 0.01-μF 1000-V disk capacitor. C2-C11 are 500 μF at 450 VDC. Capacitor combination thus gives equivalent of 50 μF for filter, rated 4500 V. When using 500-μA movement for output voltmeter, R5 should be ten 1-megohm resistors in series. Thyrector TY1 is GE 6R520SP4B4. T1 has 2200-V secondary rated 500 mA. K1 is 24-V relay. Article covers construction and stresses safety precautions.—E. H. Hartz, 3000 VDC Supply, *73 Magazine*, July 1974, p 69–72.

CHAPTER 69

Programmable Circuits

Circuits having 1 to 11 digital control inputs provide switch, logic, or computer-programmed choice of values for variables such as attenuation, division ratio, filter center frequency, math function, pulse width, or output frequency. See also Microprocessor chapter.

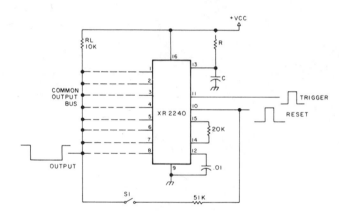

TIMER/COUNTER—Basic circuit using XR2240 programmable timer/counter acts as programmable mono when S1 is closed, with output pulse width being a multiple in binary of RC seconds. With 8-bit binary counter, time delays range from 1 RC to 255 RC seconds. As an example, if only pin 6 (dividing input frequency by 32) is connected to common output bus, duration of output pulse will be 32 RC seconds. Similarly, with pins 1, 2, 5, and 7 connected to bus, delay is 83 RC seconds. With S1 open for astable operation, output frequency is 1/t hertz where t is multiple of RC from 1 to 255. VCC is 4–15 V.—H. M. Berlin, IC Timer Review, *73 Magazine,* Jan. 1978, p 40–45.

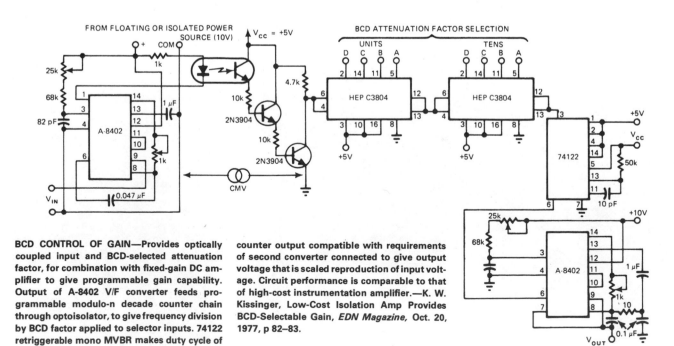

BCD CONTROL OF GAIN—Provides optically coupled input and BCD-selected attenuation factor, for combination with fixed-gain DC amplifier to give programmable gain capability. Output of A-8402 V/F converter feeds programmable modulo-n decade counter chain through optoisolator, to give frequency division by BCD factor applied to selector inputs. 74122 retriggerable mono MVBR makes duty cycle of counter output compatible with requirements of second converter connected to give output voltage that is scaled reproduction of input voltage. Circuit performance is comparable to that of high-cost instrumentation amplifier.—K. W. Kissinger, Low-Cost Isolation Amp Provides BCD-Selectable Gain, *EDN Magazine,* Oct. 20, 1977, p 82–83.

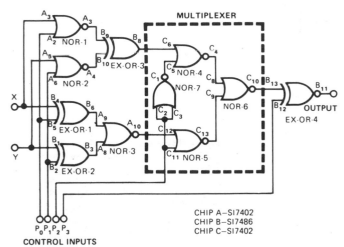

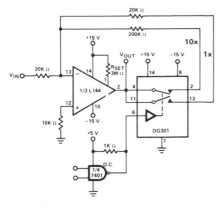

CHIP A—SI7402
CHIP B—SI7486
CHIP C—SI7402

CONTROL INPUTS

DIGITALLY SELECTABLE GAIN—TTL controls operation of DG301 low-power analog switch at output of inverting opamp. Low logic gives gain of 1, and high logic gives gain of 10.—"*Analog Switches and Their Applications,*" Siliconix, Santa Clara, CA, 1976, p 7-90.

P_3	P_2	P_1	P_0	OUTPUT
\multicolumn{4}{c}{PROGRAMMING INPUT}	
0	0	0	0	$\overline{X+Y}$
0	0	0	1	$X \cdot \overline{Y}$
0	0	1	0	$\overline{X} \cdot Y$
0	0	1	1	$X \cdot Y$
0	1	0	0	X Y
0	1	0	1	$\overline{Y}$
0	1	1	0	$\overline{X}$
0	1	1	1	1
1	0	0	0	$X+Y$
1	0	0	1	$\overline{X}+Y$
1	0	1	0	$X+\overline{Y}$
1	0	1	1	$\overline{X \cdot Y}$
1	1	0	0	$\overline{X} \quad \overline{Y}$
1	1	0	1	Y
1	1	1	0	X
1	1	1	1	0

16 FUNCTIONS OF X AND Y—With only three IC chips, circuit provides choice of any one of 16 possible functions of two Boolean variables. Table shows output states for all programming combinations of control inputs P.—S. Murugesan, Programmable Logic Circuit Has Versatile Outputs, *EDN Magazine,* Feb. 5, 1975, p 57.

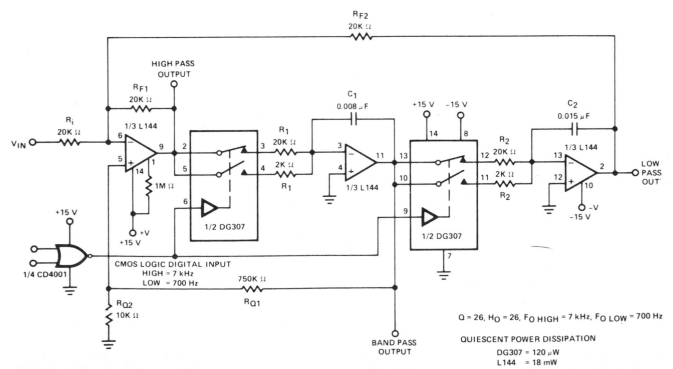

Q = 26, H_O = 26, F_O HIGH = 7 kHz, F_O LOW = 700 Hz

QUIESCENT POWER DISSIPATION
DG307 = 120 µW
L144 = 18 mW

PROGRAMMABLE-FREQUENCY STATE-VARIABLE—Provides choice of low-pass, high-pass, and bandpass outputs with logic-selectable center frequency of 700 or 7000 Hz. Logic input controls DG307 low-power dual analog switch for changing values of frequency-determining resistors R_1 and R_2.—"*Analog Switches and Their Applications,*" Siliconix, Santa Clara, CA, 1976, p 7-86.

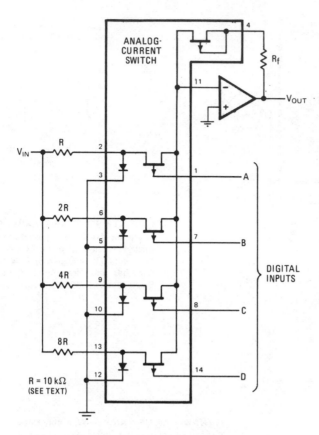

$$V_{OUT} = -V_{IN} \frac{R_f}{R} \left[\overline{A}(2^0) + \overline{B}(2^{-1}) + \overline{C}(2^{-2}) + \overline{D}(2^{-3}) \right]$$

GAIN-PROGRAMMABLE AMPLIFIER—National AH5010 4-bit current-mode analog switch for TTL input is used with general-purpose opamp such as LM118 to give multiplying D/A converter at low cost. For CMOS control logic, use AM97C10 switch. Use of 10K for gain-programming resistor R gives compromise between switch resistance and switch leakage. Use 0.2% tolerance resistors for R and 2R, 0.5% for 4R, and 5% for highest resistance, with 0.1% tolerance for feedback resistor R_f which is also 10K, to give overall accuracy within 0.2%.—J. Maxwell, Analog Current Switch Makes Gain-Programmable Amplifier, *Electronics*, Feb. 17, 1977, p 99 and 101.

FOUR-STATE ATTENUATOR—HA-2400 four-channel programmable amplfiier is used as non-inverting four-state attenuator controlled by logic inputs 0 and 1 to D_0 and D_1. Output voltage for each logic combination is given in truth table. Values shown provide gains of 1, ½, ¼, and ⅛.—W. G. Jung, "IC Op-Amp Cookbook," Howard W. Sams, Indianapolis, IN, 1974, p 429–431.

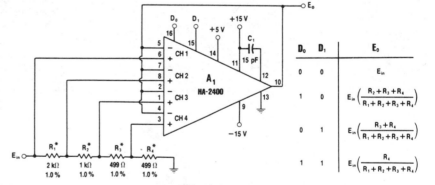

D_0	D_1	E_o
0	0	E_{in}
1	0	$E_{in}\left(\dfrac{R_2+R_3+R_4}{R_1+R_2+R_3+R_4}\right)$
0	1	$E_{in}\left(\dfrac{R_3+R_4}{R_1+R_2+R_3+R_4}\right)$
1	1	$E_{in}\left(\dfrac{R_4}{R_1+R_2+R_3+R_4}\right)$

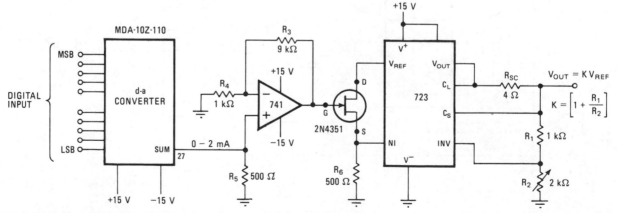

CONTROLLING REGULATOR OUTPUT—Digital control of D/A converter determines output voltage of regulator, with FET serving as voltage-variable resistor. Applications include generation of sequence of voltages for testing components or equipment. Analog Devices MDA-10Z-110 converter generates 0–2 mA output with resolution determined by 10-bit digital input. 741 opamp transforms current to 0–6 V output for varying output of 723 regulator over range of 7–37 V at 150 mA maximum.—C. Viswanath, D-A Converter Controls Programmable Power Source, *Electronics*, July 21, 1977, p 125.

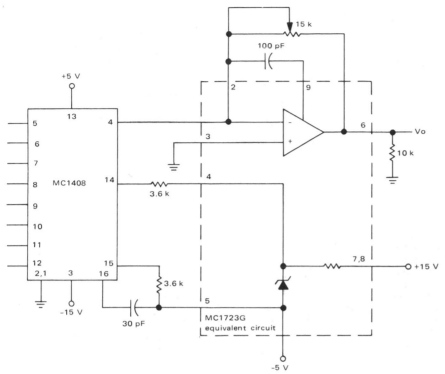

0–12 V PROGRAMMABLE—Combination of Motorola MC1408 DAC and MC1723 regulator gives digitally programmable voltages in 0.1-V increments at currents in excess of 100 mA. Can be used as programmable lab power supply, computer-controlled supply for automatic test equipment, or in industrial control systems. Requires ±5 V and ±15 V supplies. Voltage range can be increased to 25.5 V if positive supply is increased to 28.5 V or higher.—D. Aldridge and N. Wellenstein, "Designing Digitally-Controlled Power Supplies," Motorola, Phoenix, AZ, 1975, AN-703, p 3.

PROGRAMMABLE ASTABLE—Square-wave output frequency of Exar XR-2240 programmable timer/counter is made digitally programmable by use of 4051 CMOS multiplexer having three-line channel-select control ABC. Lines select one of eight possible switching paths by binary combination. When all three inputs are zero, highest of eight basic output frequencies of 2240 is obtained, as shown in truth table. Circuit will yield outputs with periods of 1, 2, 4, 8, 16, 32, 64, and 128 s.—W. G. Jung, "IC Timer Cookbook," Howard W. Sams, Indianapolis, IN, 1977, p 123–125.

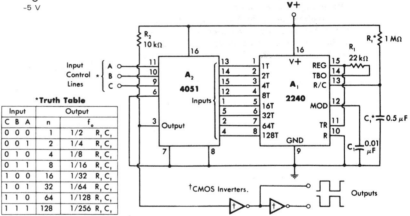

*Truth Table

Input			Output	
C	B	A	n	f_o
0	0	0	1	1/2 $R_1 C_1$
0	0	1	2	1/4 $R_1 C_1$
0	1	0	4	1/8 $R_1 C_1$
0	1	1	8	1/16 $R_1 C_1$
1	0	0	16	1/32 $R_1 C_1$
1	0	1	32	1/64 $R_1 C_1$
1	1	0	64	1/128 $R_1 C_1$
1	1	1	128	1/256 $R_1 C_1$

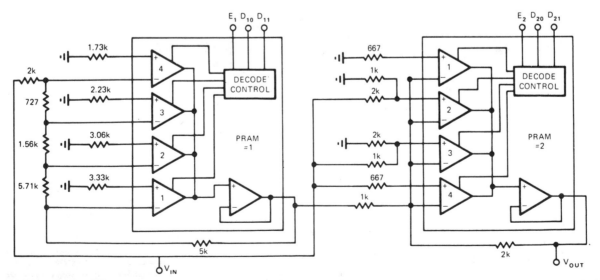

PROGRAMMABLE-GAIN OPAMP—Cascading two HA-2400 digitally programmed amplifiers, each combining functions of analog switches and high-performance opamps on single IC chip, gives 16 different programmable gains in unit steps. Article gives truth table showing total gain obtained for 16 combinations of 0 and 1 on control lines D_{10}, D_{11}, D_{20}, and D_{21}. Enable lines are normally at 1, and E_2 is made 0 only when total gain must be zero. Applications include digital AGC and digital control of servosystems and level detectors.—J. A. Connelly, N. C. Currie, and D. S. Bonnet, Op Amp Has 16-Step Digital Gain Control, *EDN Magazine*, May 5, 1974, p 75 and 77.

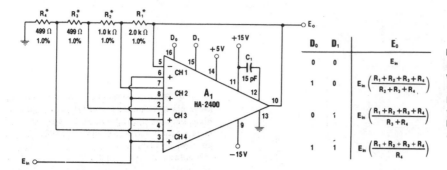

D_0	D_1	E_0
0	0	E_{in}
1	0	$E_{in}\left(\dfrac{R_1+R_2+R_3+R_4}{R_2+R_3+R_4}\right)$
0	1	$E_{in}\left(\dfrac{R_1+R_2+R_3+R_4}{R_3+R_4}\right)$
1	1	$E_{in}\left(\dfrac{R_1+R_2+R_3+R_4}{R_4}\right)$

FOUR-STATE AMPLIFIER—HA-2400 four-channel programmable amplifier is used with tapped voltage divider in feedback loop to give gains of 1, 2, 4, and 8 controlled by logic inputs to D_0 and D_1 as shown in truth table. Amplifier is noninverting.—W. G. Jung, "IC Op-Amp Cookbook," Howard W. Sams, Indianapolis, IN, 1974, p 431–432.

PROGRAMMED 0–25.5 V—Uses D/A converter and 2.5-V zener to form digitally programmed voltage reference. Binary-coded TTL information selects voltage ranges of 0–2.55 V or 0–25.5 V. Can be used as lab source having 10 mV per step in low range and 100 mV per step in high range. Output is adjusted with 8-bit input control. R3 determines basic voltage range, being set at 1280 ohms for low range and 12.8K for high range. Although 741 or other general-purpose opamp is adequate for low range, higher-voltage single-supply opamp such as 759 is better for high range and for both ranges, because it gives higher output current on low range.—W. Jung, An IC Op Amp Update, *Ham Radio*, March 1978, p 62–69.

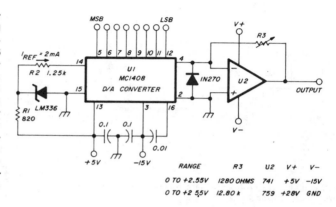

RANGE	R3	U2	V+	V−
0 TO +2.55V	1280 OHMS	741	+5V	−15V
0 TO +2.55V	12.80 k	759	+28V	GND

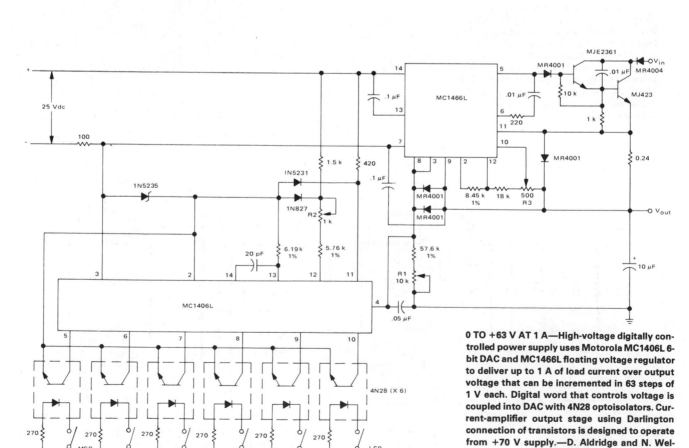

0 TO +63 V AT 1 A—High-voltage digitally controlled power supply uses Motorola MC1406L 6-bit DAC and MC1466L floating voltage regulator to deliver up to 1 A of load current over output voltage that can be incremented in 63 steps of 1 V each. Digital word that controls voltage is coupled into DAC with 4N28 optoisolators. Current-amplifier output stage using Darlington connection of transistors is designed to operate from +70 V supply.—D. Aldridge and N. Wellenstein, "Designing Digitally-Controlled Power Supplies," Motorola, Phoenix, AZ, 1975, AN-703, p 6.

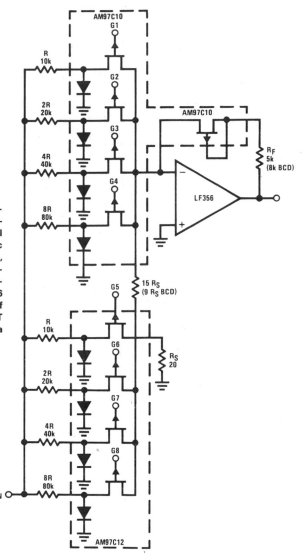

GAIN-PROGRAMMED AMPLIFIER—8-bit multiplying D/A converter using cascaded 4-bit sections provides logic-controlled gain for signal preconditioning, level control, and dynamic range expansion. Logic 0 turns JFET switch on, and logic 1 turns switch off. Series FET in feedback path of opamp compensates for ON resistance of JFET switch. Circuit has gain of 0.9996 (binary) with 5K feedback resistor and gain of 0.99 (BDC) with 8K feedback resistor.—"FET Databook," National Semiconductor, Santa Clara, CA, 1977, p 6-47–6-49.

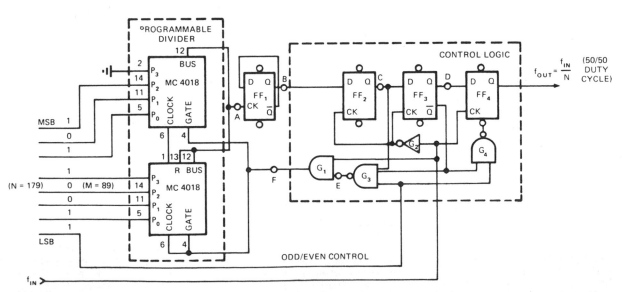

PROGRAMMED DIVIDE-BY-179—Produces symmetrical output waveforms even if divider ratios are large, variable, and even or odd. Circuit is set up for output of N = 179, for which M = 89 is programmed into divider and odd/even control of logic is a 0. Control logic can be simplified, depending on particular requirements; thus, if perfect symmetry is not essential, G_4 and FF$_4$ can be eliminated. Article tells how to program for any other value of N.—V. R. Godbole, Programmable Divider Maintains Output Symmetry, *EDN Magazine*, July 5, 1974, p 72–74.

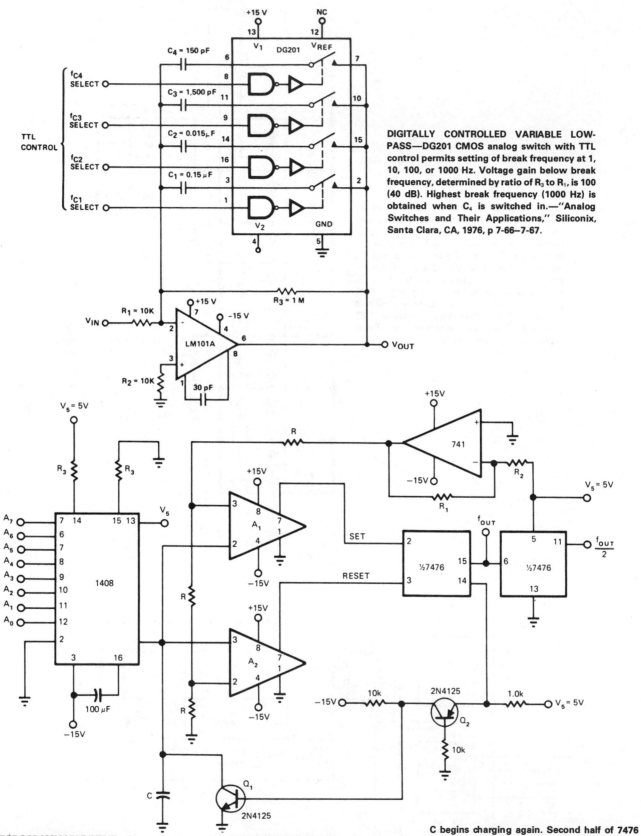

DIGITALLY CONTROLLED VARIABLE LOW-PASS—DG201 CMOS analog switch with TTL control permits setting of break frequency at 1, 10, 100, or 1000 Hz. Voltage gain below break frequency, determined by ratio of R_3 to R_1, is 100 (40 dB). Highest break frequency (1000 Hz) is obtained when C_4 is switched in.—"Analog Switches and Their Applications," Siliconix, Santa Clara, CA, 1976, p 7-66—7-67.

8-BIT PROGRAMMABLE INPUT—Serves as digitally programmable frequency source covering 10 to 2550 Hz for systems under microprocessor control. Frequency change is essentially instantaneous, assuring immediately valid data. Uses MC1408L8 8-bit monolithic D/A converter to supply constant current for charging C in neg-ative direction. When capacitor voltage exceeds lower negative threshold voltage at pin 3 of LM311 high-impedance comparator A_1, comparator changes state and sets 7476 flip-flop. This turns on Q_1 through level-shifter Q_2, discharging C until it exceeds higher threshold voltage at pin 2 of A_2. Flip-flop then resets and C begins charging again. Second half of 7476 serves as divider. To cover above frequency range, values should be: R 3.9K; R_1 27K; R_2 10K; R_3 2.2K; C 0.1 μF. Circuit gives 8-bit accuracy. Design equations are presented in article.—A. Helfrick, Eight-Bit Frequency Source Suited for μP Control, *EDN Magazine*, Sept. 20, 1976, p 116 and 118.

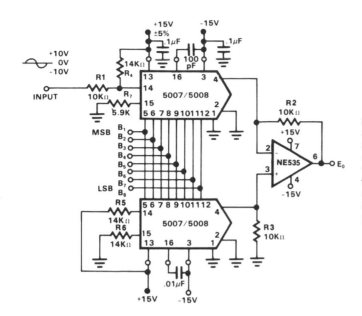

PROGRAMMABLE GAIN—Combination of two Signetics 5007/5008 multiplying D/A converters and NE535 opamp gives digital control of attenuation and gain in audio system. AC input controls polarity of output, giving output of ± 10 V for inputs from DC to 10 kHz and ± 5 V if response goes up to 20 kHz.—"Signetics Analog Data Manual," Signetics, Sunnyvale, CA, 1977, p 677–685.

GAIN-PROGRAMMABLE—Gain of noninverting opamp can be programmed with standard digital logic levels. With input at 0 V, Q_1 is turned on but is held out of saturation by Schottky diode D_2. Resulting opamp gain is $(R_1 + R_2)/R_2$. If R_2 is 1.13 kilohms and R_1 10 kilohms, gain is 10. When input is 5 V, Q_1 is off and gate of Q_2 is driven to -15 V. Gain now becomes $(R_1 + R_2 + R_3)/(R_2 + R_3)$. If R_3 is 8.87 kilohms, gain is 2.—K. Karash, Gain-Programmable Amplifier, *EEE Magazine*, Sept. 1970, p 89.

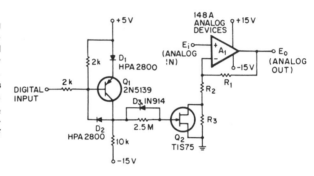

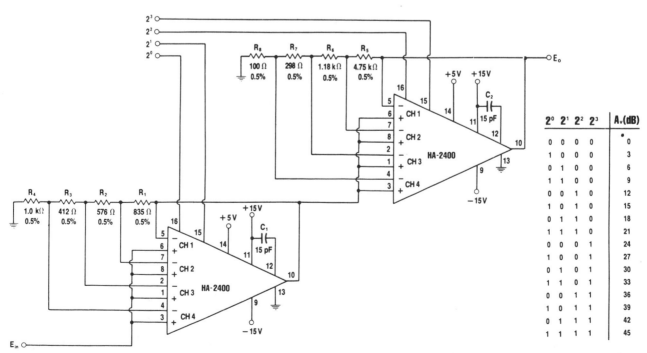

2^0	2^1	2^2	2^3	A_v(dB)
0	0	0	0	0
1	0	0	0	3
0	1	0	0	6
1	1	0	0	9
0	0	1	0	12
1	0	1	0	15
0	1	1	0	18
1	1	1	0	21
0	0	0	1	24
1	0	0	1	27
0	1	0	1	30
1	1	0	1	33
0	0	1	1	36
1	0	1	1	39
0	1	1	1	42
1	1	1	1	45

MULTISTAGE PROGRAMMABLE AMPLIFIER—Cascading of two HA-2400 four-channel programmable amplifiers gives choice of 16 different values of gain, ranging from 0 to 45 dB, by applying logic pulses to control inputs for pins 15 and 16 in accordance with truth table shown.—W. G. Jung, "IC Op-Amp Cookbook," Howard W. Sams, Indianapolis, IN, 1974, p 433–435.

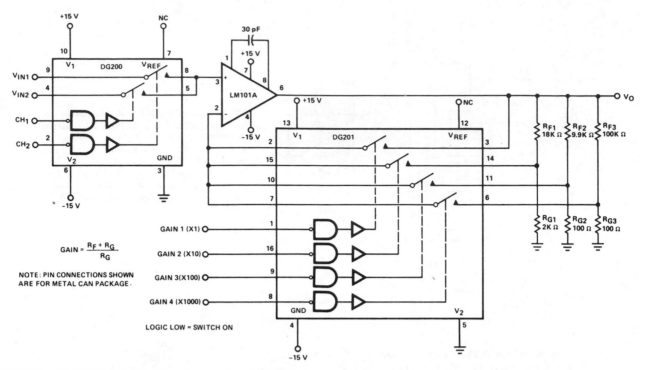

GAIN = $\dfrac{R_F + R_G}{R_G}$

NOTE: PIN CONNECTIONS SHOWN ARE FOR METAL CAN PACKAGE.

LOGIC LOW = SWITCH ON

DIGITALLY PROGRAMMED INPUTS AND GAINS—DG200 CMOS analog switch gives programmable choice of two inputs to opamp, and DG201 switch gives choice of four different gain values (1, 10, 100, or 1000) for opamp. Full opamp output range of ±12 V is provided even for unity-gain position of switch.—"Analog Switches and Their Applications," Siliconix, Santa Clara, CA, 1976, p 7-67.

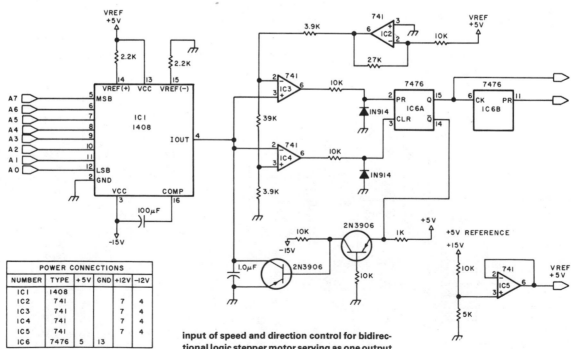

POWER CONNECTIONS					
NUMBER	TYPE	+5V	GND	+12V	-12V
IC1	1408				
IC2	741			7	4
IC3	741			7	4
IC4	741			7	4
IC5	741			7	4
IC6	7476	5	13		

OSCILLATOR CONTROL—Digitally controlled oscillator generates frequency proportional to integer output to 1408 DAC, for use as clock input of speed and direction control for bidirectional logic stepper motor serving as one output of microprocessor. Used to provide speed control for stepper. Number of bits determines number of speed selections available under computer control.—R. E. Bober, Taking the First Step, *BYTE*, Feb. 1978, p 35–36, 38, 102, 104, 106, and 108–112.

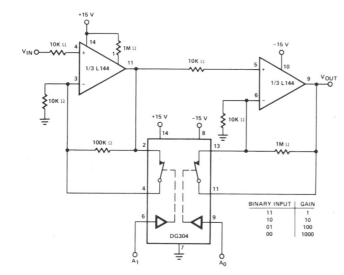

BINARY CONTROL OF GAIN—Gain of amplifier increases by decades from 1 to 1000 as binary input to A_1 and A_0 of DG304 low-power analog switch decreases from 1,1 to 0,0. Power dissipation of switch is less than 0.1 mW.—"Analog Switches and Their Applications," Siliconix, Santa Clara, CA, 1976, p 7-86.

BINARY INPUT	GAIN
11	1
10	10
01	100
00	1000

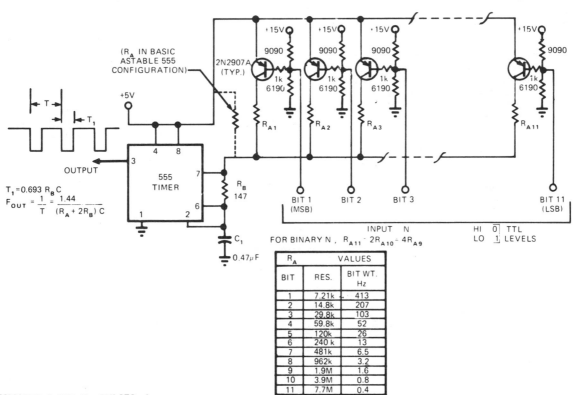

$$T_1 = 0.693\, R_B C$$
$$F_{OUT} = \frac{1}{T} = \frac{1.44}{(R_A + 2R_B)\,C}$$

FOR BINARY N, $R_{A11} = 2R_{A10} = 4R_{A9}$

	HI	0	TTL
	LO	1	LEVELS

R_A	VALUES	
BIT	RES.	BIT WT. Hz
1	7.21k	413
2	14.8k	207
3	29.8k	103
4	59.8k	52
5	120k	26
6	240k	13
7	481k	6.5
8	962k	3.2
9	1.9M	1.6
10	3.9M	0.8
11	7.7M	0.4

PROGRAMMABLE 0–825 Hz PULSES—Inexpensive pulse generator is programmable in 0.4-Hz steps from 0 to 825 Hz, and can be modified to extend range to 200 kHz. Circuit uses 555 connected in astable mode, with timing resistor R_A replaced by 11 sets of timing resistors and switching transistors. Inputs and outputs are TTL-compatible. When bit input is high (0), its associated transistor is turned off. When bit input is low (1), transistor is on, allowing C_1 to charge. When more than one input is low, charging is through parallel combination of resistors. Width of output pulse T_1 is constant over frequency range.—E. G. Laughlin, Inexpensive Pulse Generator Is Logic Programmable, *EDN Magazine*, Aug. 20, 1974, p 92.

CHAPTER 70
Protection Circuits

Provide protection of equipment and components from overvoltage or overcurrent conditions, ground fault, loose ground, contact arcing, and inductive transients. Also included are digital-coded or tone-coded controls for doors, auto ignition switches, and equipment ON/OFF switches, along with fail-safe interlocks and power-outage indicators. See also Burglar Alarm, Fire Alarm, Power Control, Power Supply, Regulated Power Supply, Regulator, and Siren chapters.

CROWBAR—When output of regulator for microprocessor power supply exceeds maximum safe voltage as determined by zener Q1, SCR Q2 is triggered on and conducts heavily, blowing fuse rapidly to protect equipment. Fuse rating is 125% of nominal load. Choose SCR to meet voltage and current requirements. Choose zener for desired trip voltage. Each germanium diode in series with Q1 will add 0.3 V to trip voltage, and silicon diodes add 0.6 V. To calibrate, place 1K resistor temporarily in series with Q2 and measure drop across it to see if SCR fires and produces surge on meter at desired V_{CC}.— J. Starr, Want to Buy a Little Insurance?, *Kilobaud,* Oct. 1978, p 89.

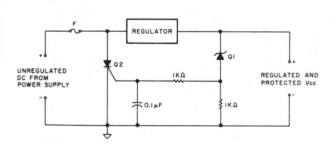

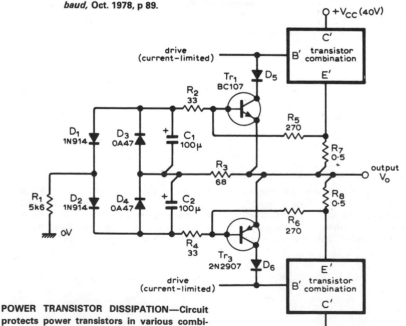

POWER TRANSISTOR DISSIPATION—Circuit protects power transistors in various combinations without limiting capabilities of AF amplifier when driving reactive loudspeaker load. With continuous signal drive into normal load, R_1 draws current from C_1 through D_1, in opposition to R_5. This gives drops of about 0.12 V across C_1 and C_2, allowing full drive. With short-circuited load, however, capacitor drops increase to about 0.55 V, thereby limiting average current in each output transistor to about 1.1 A. Diodes D_5 and D_6 are not critical, and simply prevent current flow from base to collector of transistors.—M. G. Hall, Amplifier Output Protection, *Wireless World,* Jan. 1977, p 78.

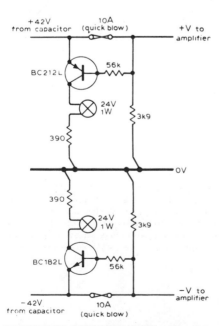

BLOWN-FUSE INDICATOR—Used with quick-blow fuses in high-power audio amplifier using split power supply. When fuse blows, transistor shunting it is turned on and passes current to corresponding indicator lamp. Maximum current in blown-fuse condition is less than 1 mA.—I. Flindell, Amplifier Blown-Fuse Indicator, *Wireless World,* Sept. 1976, p 73.

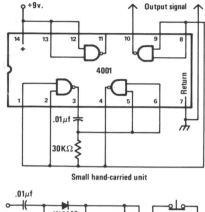

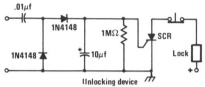

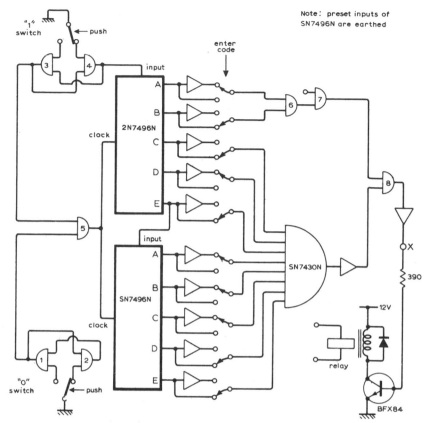

Note: preset inputs of
SN7496N are earthed

3-kHz TONE LOCK—Electric door lock opens only when signal voltage of about 3 kHz is applied to two exposed terminals by holding compact single-IC AF oscillator against terminals. Will not respond to DC or 60-Hz AC. Pocket oscillator operates from 9-V transistor radio battery, with current drawn only when output prongs are held against lock terminals. SCR can be any type capable of handling current drawn by electric lock.—J. A. Sandler, 11 Projects under $11, *Modern Electronics*, June 1978, p 54–58.

10-DIGIT CODED SWITCH—Uses seven Texas Instruments positive-logic chips. NAND gates 1-4 and 5-8 are from two SN7400N packages. Two SN7404N packages each provide six of inverting opamps shown. Desired code is set up as combination of 0s and 1s by presetting ten 2-position switches. To open lock, switches at input for 0 and 1 must be pushed in sequence of code. Arrangement gives 1024 possible combinations but provides much greater protection unless intruder knows that 10 digits are required. Article describes operation of circuit. One requirement of the 2N7496N shift registers is that information be present at serial input before clocking pulse occurs.—K. E. Potter, Ten-Digit Code-Operated Switch or Combination Lock, *Wireless World,* May 1974, p 123.

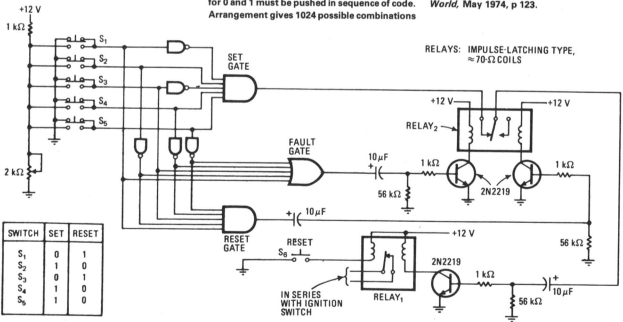

SWITCH	SET	RESET
S_1	0	1
S_2	1	0
S_3	0	1
S_4	1	0
S_5	1	0

ELECTRONIC LOCK—Correct combination of switches S_1-S_5 must be actuated to energize relay in series with ignition switch of auto or any other type of electric lock. If wrong combination is used, lock cannot be opened until resetting combination is entered. When car ignition is turned off, ignition relay should be reset (contact opened) by pressing S_6. With connections shown, switches S_2, S_4, and S_5 must be depressed simultaneously to open (set) lock. If error is made, output of fault gate goes to logic 1 and contacts of relay 2 will open. After error, S_1 and S_3 must be depressed simultaneously to reset lock before opening combination can be used again. Switches can be connected for any other desired combinations.—L. F. Caso, Electronic Combination Lock Offers Double Protection, *Electronics*, June 27, 1974, p 110; reprinted in "Circuits for Electronics Engineers," *Electronics*, 1977, p 346.

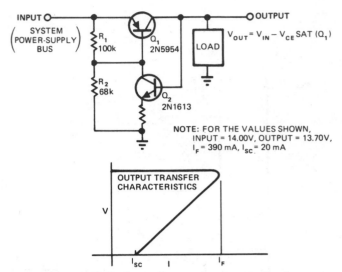

NOTE: FOR THE VALUES SHOWN,
INPUT = 14.00V, OUTPUT = 13.70V,
I_F = 390 mA, I_{sc} = 20 mA

$$V_{OUT} = V_{IN} - V_{CE} \, SAT \, (Q_1)$$

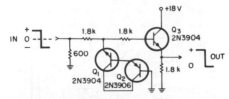

600-Hz CLAMP—Polar clamp was developed to provide overvoltage input protection for ±6 VDC teleprinter signals at 10 mA. Circuit will withstand input transients up to 120 VDC at 20 mA. When input exceeds emitter-base breakdown voltage of Q_1, Q_2 becomes forward-biased for clamping of input. With excessive negative input, Q_1 is forward-biased and emitter-base path in Q_2 completes clamping action.—R. R. Breazzano, A Polar Clamp, *EDN/EEE Magazine*, June 15, 1971, p 59.

FOLDBACK CURRENT LIMITER—Provides overload and short-circuit protection for load while isolating malfunctioning circuit from other loads on common supply bus. In normal operation, Q_1 is saturated. When load attempts to draw more than this saturation value, base current of Q_1 cannot maintain saturation so voltage across unmarked resistor drops and current through Q_1 drops correspondingly. When load is shorted, Q_2 goes off and short-circuit current folds back to safe lower value. Choose value of unmarked resistor to ensure saturation of Q_1 at load current.—S. T. Venkataramanan, Simple Circuit Isolates Defective Loads, *EDN Magazine*, Jan. 20, 1978, p 114.

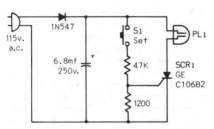

AC LINE MONITOR—Detects AC line failures of any duration and turns off neon lamp PL1 to indicate that clocks require resetting. Circuit is plugged into AC outlet, and S1 is pushed to trigger SCR on and send current through lamp.—J. R. Nelson, Some Ideas for Monitoring A.C. Power Lines, *CQ*, July 1973, p 56.

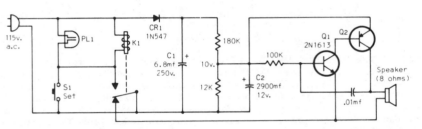

AUDIBLE LINE MONITOR—Audio oscillator coupled to simple relay circuit gives alerting tone when power fails even momentarily. C2 determines duration of tone. With 2900 μF, tone lasts about 1 s, as warning that clocks will need resetting. Q2 is any PNP audio power transistor, K1 is 115-V SPDT relay, and PL1 is neon lamp.—J. R. Nelson, Some Ideas for Monitoring A.C. Power Lines, *CQ*, July 1973, p 56.

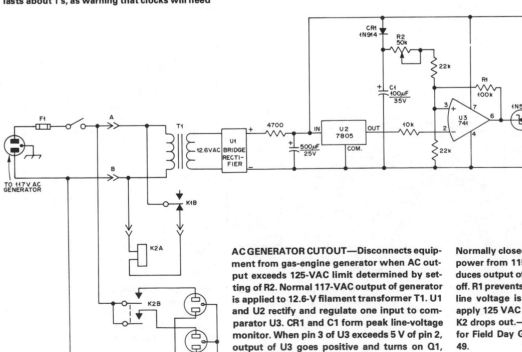

AC GENERATOR CUTOUT—Disconnects equipment from gas-engine generator when AC output exceeds 125-VAC limit determined by setting of R2. Normal 117-VAC output of generator is applied to 12.6-V filament transformer T1. U1 and U2 rectify and regulate one input to comparator U3. CR1 and C1 form peak line-voltage monitor. When pin 3 of U3 exceeds 5 V of pin 2, output of U3 goes positive and turns on Q1, which applies power to small 12-VDC relay K1. Normally closed contacts of K1 open, removing power from 115-VAC relay K2. 1N523 zener reduces output of U3 enough so Q1 can be turned off. R1 prevents relays from chattering when AC line voltage is close to threshold. To adjust, apply 125 VAC between A and B, and set R2 so K2 drops out.—P. Hansen, Overvoltage Cutout for Field Day Generators, *QST*, March 1977, p 49.

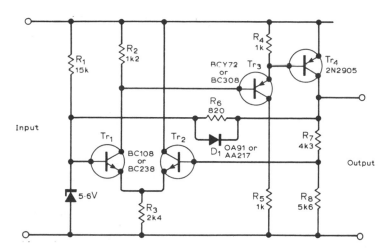

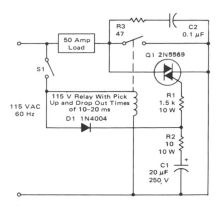

REGULATOR OVERLOAD—When output is shorted, germanium diode D_1 turns on and draws current through R_1, removing reference voltage across zener. Tr_1 is then held off and turns Tr_3 and Tr_4 off to block load current. When short is removed, circuit recovers automatically.—D. E. Waddington, Germanium Diode for Regulator Protection, *Wireless World,* March 1977, p 42.

TRIAC SUPPRESSES RELAY ARCING—Circuit prevents arcing at contacts of relay for loads up to 50 A, by turning on as soon as it is fired by gate current; this occurs after S1 is closed but before relay contacts close. Once contacts are closed, load current passes through them rather than through triac. When S1 is opened, triac limits maximum voltage across relay contacts to about 1 V. Circuit permits use of smaller relay since it does not have to interrupt full load current.—"Circuit Applications for the Triac," Motorola, Phoenix, AZ, 1971, AN-466, p 8.

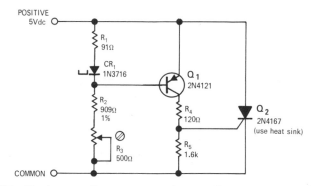

5-V CROWBAR—Simple overvoltage protection circuit for 5-V 1-A logic supply can be adjusted to trigger at 10% overvoltage or 5.5 V. Tunnel diode CR_1 senses level. At 5.5 V, diode switches slightly past its valley point, and voltage across diode biases Q_1 into saturation. Q_1 then supplies gate current to SCR Q_2, which fires and continues conducting until power supply is disconnected. Power supply must include current-limiting circuit and fuse. R_3 adjusts trip point.—L. Strahan, Logic-Supply Crowbar, *EDN/EEE Magazine,* Nov. 15, 1971, p 51.

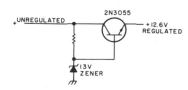

TRANSCEIVER-SAVER—Simple circuit has no effect on normal operation of CB transceiver or other solid-state equipment in auto but provides overvoltage protection if voltage regulator in auto fails. Use heatsink with transistor if transmit current is above 2 A. Choose resistance value to give output of 12.6 V during normal operation.—Circuits, *73 Magazine,* March 1977, p 152.

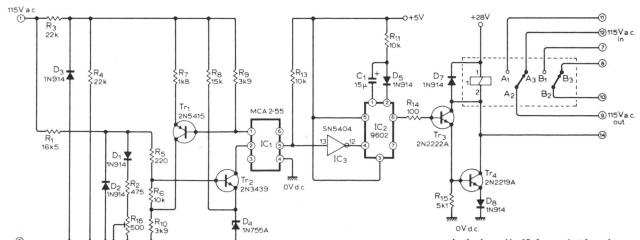

AC OVERVOLTAGE—Used to protect delicate equipment from sustained high AC line voltage, by disconnecting supply when it exceeds preset level selected by R_{16}. When base-emitter voltage of Tr_2 exceeds 7.5 V, optocoupler switches Tr_1 on to provide fast switching action. Output pulse is shaped by IC_3 for use in triggering mono IC_2. When line falls below preset level, mono reverts to stable state and switches on AC supply again.—F. E. George, A.C. Line Sensor, *Wireless World,* March 1977, p 42.

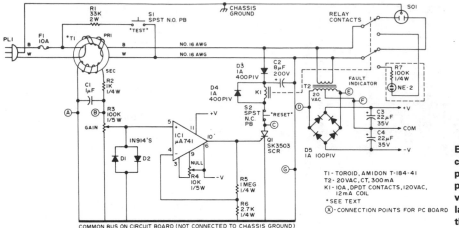

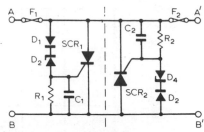

GROUND-FAULT INTERRUPTER—Compares current in ungrounded side of power line with current in neutral conductor. If currents are not equal, ground fault exists because portion of line current is taking an unintended return path through leaky electric appliance or human body. Voltage induced in toroid by unbalanced current is amplified for energizing relay K1 to break circuit. Toroid uses Amidon T-184-41 core, with 600 turns No. 30 for secondary, and 12 turns No. 16 solid twisted-pair for primary. Circuit operates on fault current of 4 mA, well below danger limit for children.—W. J. Prudhomme, The Unzapper, *73 Magazine*, Nov./Dec. 1975, p 151–156.

EQUIPMENT INTERFACE PROTECTION—When circuit shown is used to transfer signal from one piece of equipment to another, desired signal passes with very little degradation. Component values can be chosen to make thyristor SCR_1 latch at any desired voltage between A and B that is greater than 0 V, giving desired protective isolation. On other side of circuit, SCR_2 will latch and blow F_2 when voltage exceeds limiting value set by diode D_2 and zener D_4. Zeners are 10-V CV7144, diodes are CV9637 small-signal silicon, resistors are 10K, capacitors are 0.047 μF, and thyristors are 2N4147.—S. G. Pinto and A. P. Bell, Thyristor Protection Circuit, *Wireless World*, Oct. 1975, p 473.

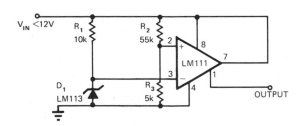

12-V OVERVOLTAGE LIMITER—Single LM111 comparator is basis for simple overvoltage protection of circuits drawing less than 50 mA. Fraction of input supply is compared to 1.2-V reference. When input exceeds reference level, power is removed from output.—R. C. Dobkin, Comparators Can Do More than Just Compare, *EDN Magazine*, Nov. 1, 1972, p 34–37.

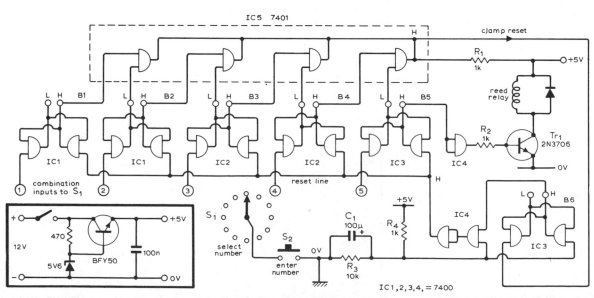

CODED LOCK—Five-digit combination lock uses five low-cost ICs operating from 5-V supply that can be derived from 12-V auto battery as shown in inset. Six set/reset bistable circuits are formed by cross-coupling pairs of dual-input NAND gates, so 0-V input is needed to change state of each. Five of bistables serve for combination, and sixth prevents operation by number in incorrect sequence. After S_1 is set to one number of code, S_2 is pushed to enter that number, with process being repeated for other four numbers of combination. Final correct number sets B5 and turns on Tr_1, to operate relay that can be used to open door.—S. Lamb, Simple Code-Operated Switch or Combination Lock, *Wireless World*, June 1974, p 196.

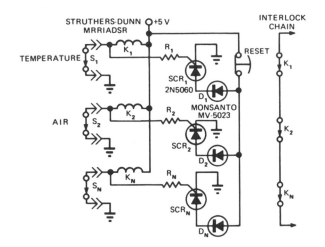

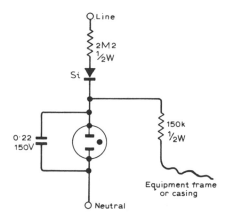

FAIL-SAFE INTERLOCK—Developed for protecting people and equipment at 40-kW RF accelerator station. All interlock switches (air flow, water flow, water pressure, temperature, etc) are normally closed, grounding one side of each relay coil. All relays are normally pulled in, to provide complete interlock chain. Failure of any component, including power supply, breaks chain and places system in safe mode.— T. W. Hardek, Interlock Protection Circuit Is Simple and Fail-Safe, *EDN Magazine,* May 20, 1975, p 74.

LOOSE-GROUND FLASHER—Uses ordinary neon lamp in series with silicon diode, with lamp normally dark. Gives warning by flashing if ground wire is accidentally or purposely disconnected from chassis of oscilloscope or other test instrument.—R. H. Troughton, Earth Warning Indicator, *Wireless World,* April 1977, p 62.

CURRENT SENSOR—Load current is sensed across base-emitter junction of output transistor. R_1 controls OFF time and R_2 controls ON time. Capacitor should be electrolytic rated above 16 V.—M. Faulkner, Two Terminal Circuit Breaker, *Wireless World,* March 1977, p 41.

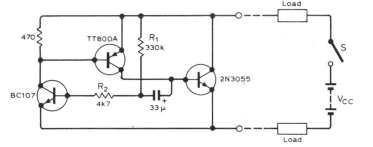

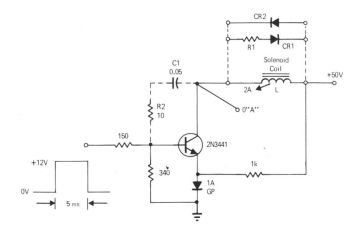

BYPASSING SOLENOID TRANSIENTS—Feedback from collector to base of power transistor through C1 and R2 protects device from destructive transients generated when inductive load such as solenoid is turned off. Alternative use of diode CR2 or CR1-R1 across coil would limit voltage transient but would increase solenoid release time.—D. Thomas, Feedback Protects High-Speed Solenoid Driver, *EDN Magazine,* Jan. 1, 1971, p 40.

CHAPTER 71
Pulse Generator Circuits

Generate square waves with fixed or variable width and duty cycle, at audio and radio frequencies up to 100 MHz. Includes tone-burst generators, strobe, pulse delay, PCM decoder, and single-pulse generators. See also Frequency Divider, Frequency Multiplier, Frequency Synthesizer, Function Generator, Multivibrator, Oscillator, and Signal Generator chapters.

PULSE-STRETCHING MONO—Section of CMOS MM74C04 inverter accepts positive input pulse by going low and discharging C. Capacitor is rapidly discharged, driving input of MM74C14 Schmitt trigger low. Output of Schmitt then goes positive for interval T_O which is equal to input pulse duration plus interval T that depends on values used for R, C, and supply voltage.—"CMOS Databook," National Semiconductor, Santa Clara, CA, 1977, p 5-30–5-35.

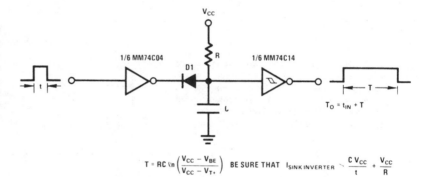

$$T = RC \ln\left(\frac{V_{CC} - V_{BE}}{V_{CC} - V_{T_+}}\right) \quad \text{BE SURE THAT} \quad I_{SINK\,INVERTER} > \frac{C\,V_{CC}}{t} + \frac{V_{CC}}{R}$$

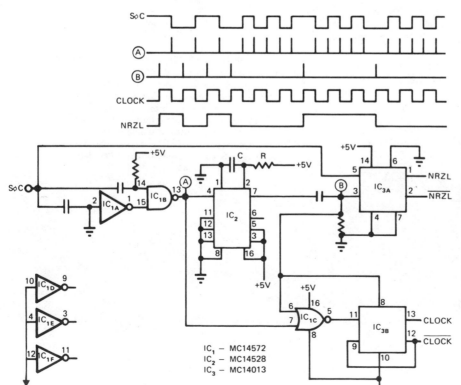

PCM DECODER—Three CMOS ICs provide decoding of Manchester (split-phase) PCM signals by generating missing mark which should occur at each change of level data to recover original clock frequency. Retriggerable mono MVBR times out at slightly longer than half of original clock frequency. Signal levels are TTL-compatible. Values of C and R depend on system frequency. Other resistors are 15K, and other capacitors are 470 pF.—M. A. Lear, M. L. Roginsky, and J. A. Tabb, PCM Signal Processor Draws Little Power, *EDN Magazine,* April 20, 1975, p 70.

IC_1 — MC14572
IC_2 — MC14528
IC_3 — MC14013

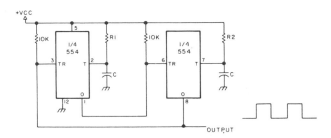

554 ASTABLE—Two sections of 554 quad monostable timer are used. Output frequency is $1/(R_1 + R_2)C$ hertz, and output duty cycle is $100R_2/(R_1 + R_2)$. When R_1 is equal to R_2, symmetrical square wave is obtained. VCC is 4.5–16 V at 3–10 mA.—H. M. Berlin, IC Timer Review, *73 Magazine,* Jan. 1978, p 40–45.

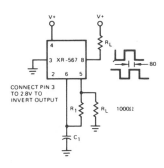

QUADRATURE OUTPUTS—Exar XR-567 tone decoder is connected as precision oscillator providing separate square-wave outputs that are very nearly in quadrature phase. Typical phase shift between outputs is 80°. Supply voltage range is 5–9 V.—"Phase-Locked Loop Data Book," Exar Integrated Systems, Sunnyvale, CA, 1978, p 41–48.

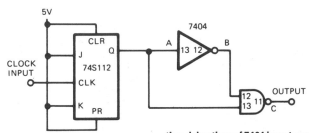

100 MHz—Developed for measuring impulse response of surface acoustic wave devices, for which pulse width had to be under 10 ns for frequency spectrum of about 100 MHz. Propagation delay time of 7404 inverter establishes output pulse width.—R. J. Lang, W. A. Porter, and B. Smilowitz, Simple Circuit Generates Nanosecond Pulses, *EDN Magazine,* Sept. 5, 1975, p 77–78.

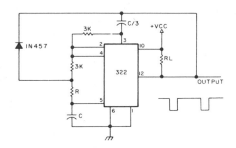

LM322 ASTABLE—National LM322 timer generates narrow negative pulse whose width is approximately 2RC seconds. VCC is 4.5–20 V. Will drive loads up to 5 mA.—H. M. Berlin, IC Timer Review, *73 Magazine,* Jan. 1978, p 40–45.

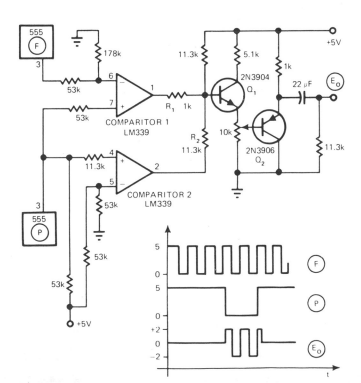

BIPOLAR PULSE TRAINS—Output of Signetics 555 timer F, consisting of unipolar waveform varying from ground to +5 V, is converted to bipolar pulse train having duration equal to that of output pulse from lower 555 timer. While P is high, comparator 2 is on, forcing R_2 to ground and placing base of Q_1 at 2.5 V (because comparator 1 is off, forcing R_1 high). Comparator 2 goes off when timer P goes low, and action of comparator 1 is turned on and off by timer F to produce bipolar pulse train at E_0.—G. L. Assard, Derive Bipolar Pulses from a Unipolar Source, *EDN Magazine,* April 5, 1977, p 144.

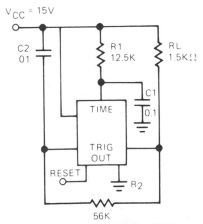

400 Hz—One section of Signetics NE558 quad timer is used as nonprecision audio oscillator providing square-wave output of about 400 Hz with values and supply voltage shown. Output frequency is affected by changes in supply voltage.—"Signetics Analog Data Manual," Signetics, Sunnyvale, CA, 1977, p 738.

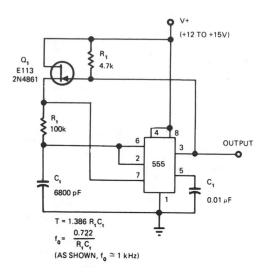

$$T = 1.386\,R_t C_t$$
$$f_0 = \frac{0.722}{R_t C_t}$$
(AS SHOWN, $f_0 \simeq 1$ kHz)

50% DUTY CYCLE WITH 555—Provides pure square-wave output without sacrificing allowable range of timing resistance. Q_1 replaces conventional timing resistor going to V+. Pull-up resistor R_1 is required to switch Q_1 fully on when it is driven by output of timer.—W. G. Jung, Take a Fresh Look at New IC Timer Applications, *EDN Magazine*, March 20, 1977, p 127–135.

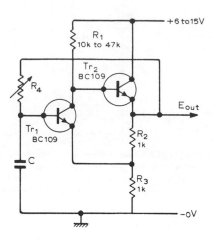

VARIABLE-FREQUENCY UP TO 0.5 MHz—Frequency is determined by choice of values for C and frequency-control potentiometer R_4. Square-wave output has almost equal mark-space ratio over wide frequency range. Regenerative action is rapid, reducing transition times. When circuit is switched on, C is uncharged and Tr_2 is on. C charges until Tr_1 begins to conduct, cutting off Tr_2 and discharging C through R_4 until Tr_1 cuts off and cycle repeats.— J. L. Linsley Hood, Square-Wave Generator with Single Frequency-Adjustment Resistor, *Wireless World*, July 1976, p 36.

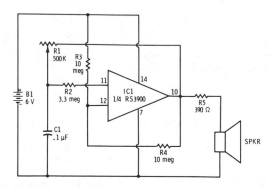

30–4000 Hz WITH OPAMP—Frequency is determined by pot R1 in feedback path. Square-wave output pulse amplitude is about 5 V. Circuit will generate almost perfect sine waves if 0.1-μF capacitor is connected between pin 12 and ground; R1 must be properly adjusted to give output of about 220 Hz.—F. M. Mims, "Integrated Circuit Projects, Vol. 6," Radio Shack, Fort Worth, TX, 1977, p 89–95.

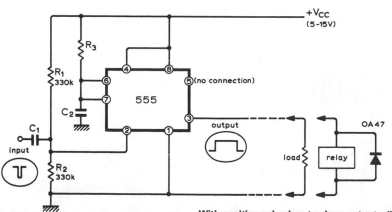

CONTROLLED-DURATION PULSES—Economical Signetics IC provides output pulse currents up to 200 mA at duration ranging from microseconds to many minutes depending on values used for R_3 and C_2. Input pulses may have duration under a microsecond, negative-going. With positive-going input pulses, output will be delayed until trailing edge occurs. Diode is required across output relay coil to suppress transients that might damage IC and cause automatic retriggering.—J. B. Dance, Simple Pulse Shaper or Relay Driver, *Wireless World*, Dec. 1973, p 605–606.

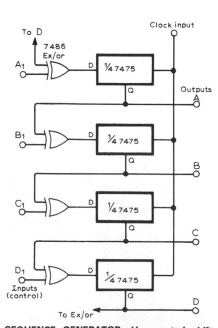

SEQUENCE GENERATOR—Uses gated shift register assembled from 7475 D-type latch, along with four EXCLUSIVE-OR gates. Clock pulse should be narrow to avoid race-around effects.—P. D. Maddison, Sequence Generator, *Wireless World*, Dec. 1977, p 80.

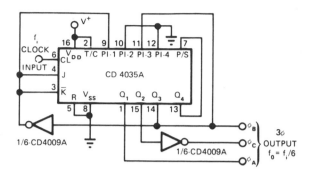

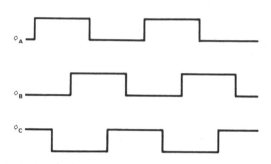

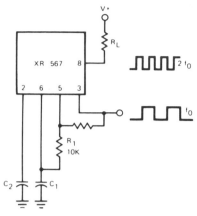

FREQUENCY-DOUBLED OUTPUT—Current-controlled oscillator section of Exar XR-567 tone decoder is connected to double frequency of square-wave output by feeding portion of output at pin 5 back to input at pin 3 through resistor. Quadrature detector of IC then functions as frequency doubler to give twice output frequency at pin 8. Supply voltage range is 5–9 V.—"Phase-Locked Loop Data Book," Exar Integrated Systems, Sunnyvale, CA, 1978, p 41–48.

THREE-PHASE PULSE GENERATOR—Requires only CMOS 4-bit shift register and two CMOS inverters. Register is connected to operate as divide-by-6 Johnson counter giving glitch-free outputs. Circuit is driven by square-wave clock signal having frequency 6 times that of desired output frequency.—C. Rutschow, Simple CMOS Circuit Generates 3-Phase Signals, *EDN Magazine,* June 20, 1976, p 128.

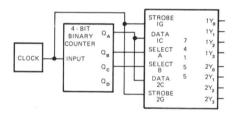

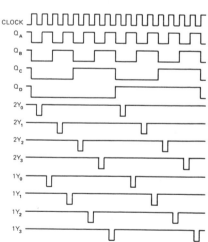

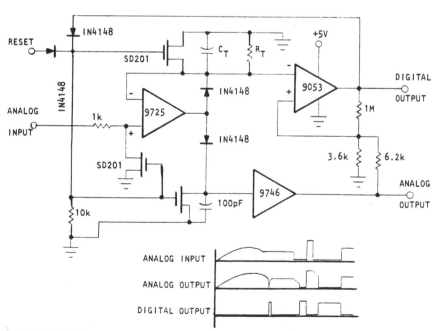

PULSE STRETCHER—Circuit also serves as analog one-shot memory and as peak sense-and-hold with automatic reset. Digital output is logic 0 until C_T and R_T have decayed by one time constant, when it goes to logic 1. Pulse duration is $R_C T_C$. Analog output amplitude is equal to input amplitude, and duration is same as digital pulse. Optical Electronics 9053 comparator automatically resets circuit after timing interval unless reset is performed manually. Analog output can be used as pulse stretcher with known and controllable pulse duration.—"Analog 'One-Shot'—Pulse Stretcher," Optical Electronics, Tucson, AZ, Application Tip 10292.

UNAMBIGUOUS STROBING—Combination of 74155 two-line to four-line decoder/demultiplexer with any conventional 4-bit binary counter provides family of strobe pulses staggered in such a way that pulse-edge ambiguity is impossible. Clock pulses at input serve to strobe 74155 as well as drive counter. Q_A of counter acts as data input, while Q_B and Q_C act as select lines. Action is such that edges of various 2Y pulses do not coincide with each other, with edges of 1Y pulses, or with edges of Q_B, Q_C, or Q_D pulses. Result is hazard-free strobing.—D. McLaughlin and C. Fanstini, End Edge Ambiguities with Two ICs, *EDN Magazine,* April 5, 1973, p 88.

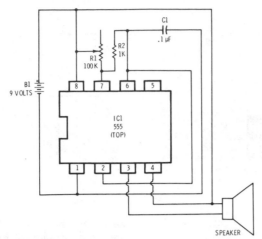

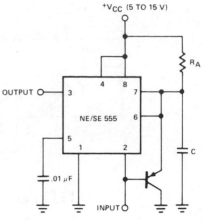

555 AS TONE GENERATOR—Connection of 555 timer as astable MVBR starts next timing cycle automatically, generating sequential square-wave output pulses in audio-frequency range with sufficient power to drive miniature 8-ohm loudspeaker. R1 controls frequency of tone.—F. M. Mims, "Integrated Circuit Projects, Vol. 2," Radio Shack, Fort Worth, TX, 1977, 2nd Ed., p 66–70.

MISSING-PULSE DETECTOR—Timing cycle of 555 timer is continuously reset by input pulse train. Change in input frequency or missing pulse allows completion of timing cycle, producing change in output level. Component values should be chosen so time delay is slightly longer than normal time between pulses.— "Signetics Analog Data Manual," Signetics, Sunnyvale, CA, 1977, p 723.

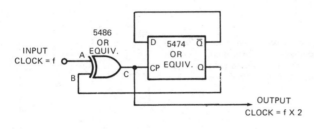

WAVEFORM-EDGE PULSER—Circuit generates square-wave output pulse for each edge of square-wave input. EXCLUSIVE-OR gate is used as programmable inverter that returns point C to quiescent low state following each transfer of data through 5474 IC. When used for frequency-doubling, input waveform should be symmetrical because output is proportional to propagation delay of flip-flop plus delay of 5486 EXCLUSIVE-OR gate.—D. Giboney, Double-Edge Pulser Uses Few Parts, *EDN Magazine*, Dec. 15, 1972, p 41.

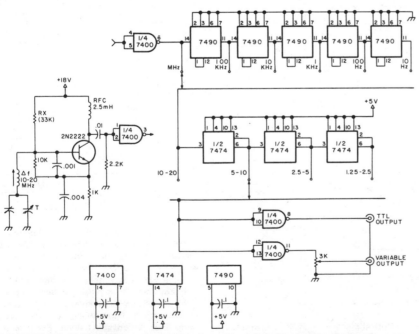

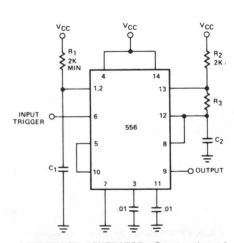

SUBAUDIO TO 20 MHz—Square-wave signal source covers wide frequency range in fully tunable decade steps, as TTL signal source for experimentation with counters, microprocessors, and other logic circuits. Uses tunable 2N2222 transistor oscillator operating at 10–20 MHz, with switchable decade dividers for range selection and switchable binary dividers for band selection. Article covers construction and calibration.—A. G. Evans, Digital Signal Source, *73 Magazine*, Dec. 1977, p 150–151.

TONE-BURST GENERATOR—One section of 556 dual timer is connected as mono MVBR and other section as oscillator. Pulse established by mono turns on oscillator, allowing generation of AF tone burst.—"Signetics Analog Data Manual," Signetics, Sunnyvale, CA, 1977, p 723–724.

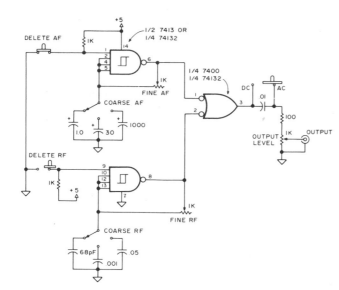

AF/RF SQUARE-WAVE—Use of feedback resistor between input and output of each gate produces oscillation in each Schmitt-trigger oscillator, one operating at audio frequencies and one operating at radio frequencies. Both AF and RF can be fed into NAND gate to give modulated RF, or outputs can be used separately as clocks for microprocessor.—B. Grater and G. Young, Build a Pulse Generator, *Kilobaud,* June 1977, p 49.

AF SQUARE WAVES—With value shown for C1, frequency of output square wave is 530 Hz. For 5300 Hz, use 0.001 μF; for 53 Hz, use 0.1 μF. Circuit will drive ordinary crystal earphone or crystal microphone used as earphone.—F. M. Mims, "Integrated Circuit Projects, Vol. 5," Radio Shack, Fort Worth, TX, 1977, 2nd Ed., p 52–56.

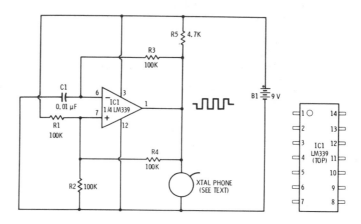

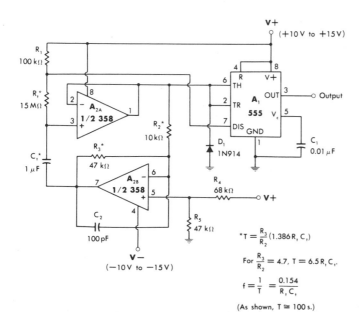

$$^*T = \frac{R_3}{R_2}(1.386\,R_1\,C_1)$$

For $\frac{R_3}{R_2} = 4.7$, $T = 6.5\,R_1\,C_1$.

$$f = \frac{1}{T} = \frac{0.154}{R_1\,C_1}$$

(As shown, T ≅ 100 s.)

EXTENDED-RANGE ASTABLE—Square-wave output is extended in frequency by combining buffer A_{2A} with opamp A_{2B} functioning as capacitance multiplier for 555 timer connected as astable MVBR. Value of 1-μF timing capacitor C_t is increased in effective value by ratio of gain of A_{2B} stage, equal to R_3/R_2. Output frequency thus corresponds to that of 4.7-μF capacitor. Negative supply should be equal and opposite to positive supply.—W. G. Jung, "IC Timer Cookbook," Howard W. Sams, Indianapolis, IN, 1977, p 118–121.

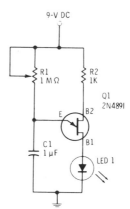

UJT/LED PULSER—Rise time of output pulse is about 200 ns and width is about 25 μs when using 1 μF for C1. Reducing value of C1 reduces pulse width. C1 charges through R1 until voltage across C1 is high enough to bias UJT into conduction. C1 then discharges through UJT and LED and cycle repeats. LED can be any common type.—F. M. Mims, "Electronic Circuitbook 5: LED Projects," Howard W. Sams, Indianapolis, IN, 1976, p 30–32.

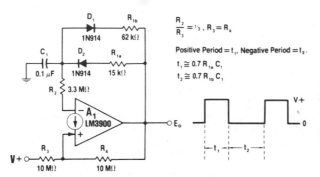

$\dfrac{R_2}{R_3} = {}^1\!/_3 \, , \, R_3 = R_4$

Positive Period $= t_1$, Negative Period $= t_2$.

$t_1 \cong 0.7 \, R_{1a} \, C_1$

$t_2 \cong 0.7 \, R_{1b} \, C_1$

ASYMMETRICAL PULSE GENERATOR— Charge and discharge paths of timing capacitor C_1 in LM3900 IC connected as astable oscillator are individually controlled by D_1 and D_2. Value of R_{1a} controls charge rate of C_1 and period t_1, while R_{1b} controls discharge rate and period t_2. Resistors can be pots for providing variable pulse width and repetition rate. For constant frequency with variable duty cycle, R_1 can be single pot with ends going to D_1 and D_2 and tap going to output. For values shown, t_1 is 1 ms and t_2 is 4 ms.—W. G. Jung, "IC Op-Amp Cookbook," Howard W. Sams, Indianapolis, IN, 1974, p 505.

1-MHz SQUARE-WAVE FOR TDR—Fast-rise-time 1-MHz pulse generator serves with wide-band CRO and T connector for time-domain reflectometry (TDR) setup used to pinpoint exact location of fault in transmission line. Will also locate multiple faults along line, measure SWR, and measure characteristic impedance of cable. With 1-MHz square-wave source having 500-ns duration for positive portions of wave, cables up to 150 feet long can be tested. R1 should equal characteristic impedance of line being tested. U1 is Signetics N7400A or equivalent quad NAND/NOR gate. Article gives instructions for use.—W. Jochem, An Inexpensive Time-Domain Reflectometer, *QST*, March 1973, p 19–21.

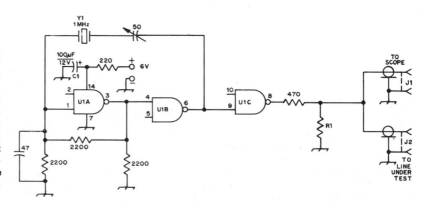

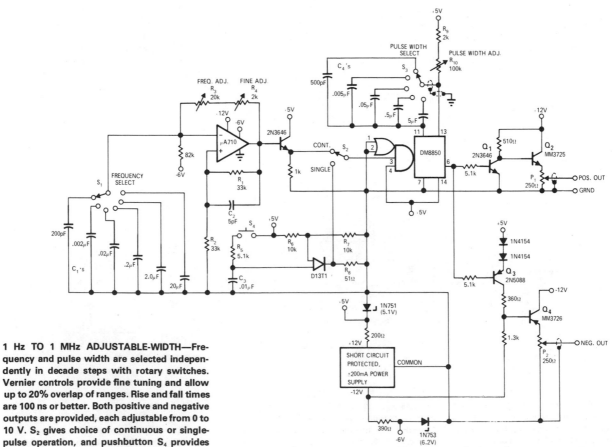

1 Hz TO 1 MHz ADJUSTABLE-WIDTH—Frequency and pulse width are selected independently in decade steps with rotary switches. Vernier controls provide fine tuning and allow up to 20% overlap of ranges. Rise and fall times are 100 ns or better. Both positive and negative outputs are provided, each adjustable from 0 to 10 V. S_2 gives choice of continuous or single-pulse operation, and pushbutton S_4 provides single-pulse outputs. μA710 comparator connected as astable MVBR provides trigger inputs for DM8850 retriggerable mono. Article gives circuit details and design equations.—C. Brogado, Versatile Inexpensive Pulse Generator, *EDN/EEE Magazine*, Oct. 1, 1971, p 37–38.

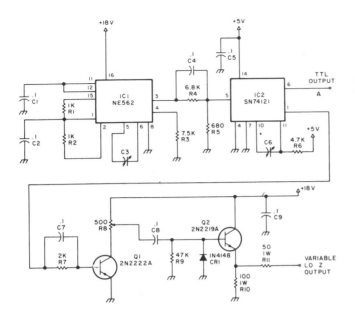

900 kHz TO 10 MHz—Pulse width is variable from about 50 ns to over 500 ms by adjusting only two components. Uses VCO portion of Signetics NE562 as pulse generator and 74121 mono MVBR to adjust pulse width. Variable capacitors C3 and C6 are broadcast-band type. VCO will operate to 30 MHz, limiting factor being stray capacitance and minimum of tuning capacitor. Low-frequency limit of VCO is about 1 Hz, obtained when C3 is 300 µF.—A. Plavcan, Pulses Galore!, *73 Magazine*, Jan. 1978, p 194–195.

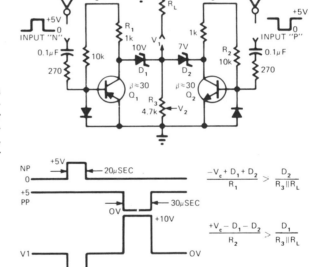

HIGH-SPEED PULSES—TTL circuit provides dual-polarity microsecond pulses. Pulse amplitude is adjusted by changing zeners D_1, D_2, or R_3. Design overcomes slew-rate problems associated with most opamps.—L. Johnson, Dual-Polarity Pulses from TTL Logic, *EDN Magazine*, April 20, 1974, p 91.

$$\frac{-V_c + D_1 + D_2}{R_1} > \frac{D_2}{R_3 \| R_L}$$

$$\frac{+V_c - D_1 - D_2}{R_2} > \frac{D_1}{R_3 \| R_L}$$

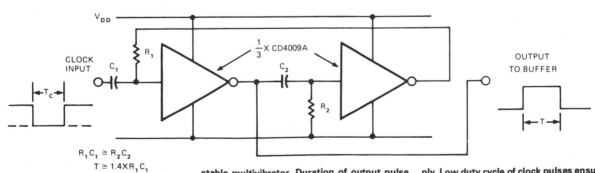

$$R_1 C_1 \cong R_2 C_2$$
$$T \cong 1.4 \times R_1 C_1$$
$$T_c \geq T$$

MONO PULSE-SHRINKER—Duty cycle of clock pulse is shortened by two CMOS inverters used to form negative-transition triggered monostable multivibrator. Duration of output pulse T is about $1.4 R_1 C_1$. Output pulse occurs each time input clock goes from high to low. Used with foldback current limiting for short-circuit protection in clock-driven regulated power supply. Low duty cycle of clock pulses ensures positive full-load starting of supply.—J. L. Bohan, Clocking Scheme Improves Power Supply Short-Circuit Protection, *EDN Magazine*, March 5, 1974, p 49–52.

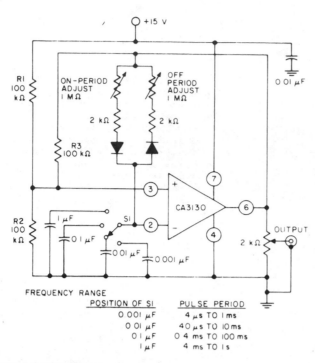

INDEPENDENT ON AND OFF PERIODS—High input resistance of CA3130 opamp permits use of high RC ratios in timing circuits, to give pulse period range of 4 μs to 1 s with switch-selected capacitors.—"Circuit Ideas for RCA Linear ICs," RCA Solid State Division, Somerville, NJ, 1977, p 5.

FREQUENCY RANGE

POSITION OF SI	PULSE PERIOD
0.001 μF	4 μs TO 1 ms
0.01 μF	40 μs TO 10 ms
0.1 μF	0.4 ms TO 100 ms
1 μF	4 ms TO 1 s

ADJUSTABLE SQUARE WAVES—Q_1 and Q_2 form flip-flop, with UJT Q_3 connected as time delay. When power is applied, one flip-flop transistor conducts and C_1 charges through one pot and diode. When C_1 reaches firing voltage of UJT, it conducts and resulting output pulse triggers flip-flop. Sequence of events now repeats, with C_1 charging through other diode. By proper selection of C_1 and pot values, circuit becomes square-wave generator with each pot controlling duration of one half-cycle. With one pot replaced by fixed resistor, circuit becomes pulse generator with other pot controlling pulse-repetition rate. If equal-value fixed resistors replace pots and R_1 is changed to pot, circuit becomes symmetrical square-wave generator with pot controlling frequency.—I. Math, Math's Notes, *CQ,* April 1974, p 64–65 and 91–92.

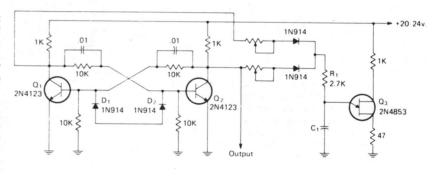

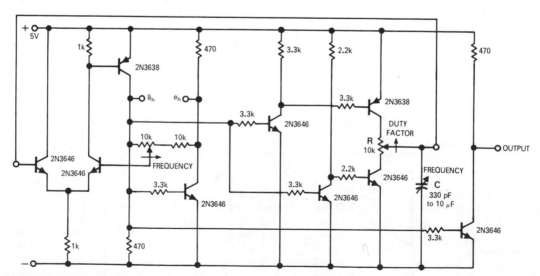

HYSTERESIS-AND-DELAY OSCILLATOR—Separate noninteracting frequency and duty-factor controls permit construction of simple telemetry oscillators having inherently linear transfer function. Absolute synchronization of independent and dependent variables is obtainable with relatively simple pulse-generating circuits. Synchronization cannot be lost. Average value of threshold voltage is maintained constant. Adjustment of hysteresis gap width moves threshold voltage limits symmetrically about average value. Resistance portion of RC delay is switched from positive to negative voltage symmetrically also. Article covers circuit operation in detail.—W. H. Swain, True Digital Synchronizer Employs Hysteresis-and-Delay Element, *EDN Magazine,* Jan. 1, 1971, p 33–35.

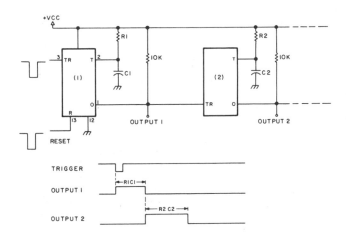

SEQUENTIAL PULSES—Any number of sections of 554 quad monostable timer can be cascaded as shown to give sequential series of output pulses of widths determined by values of R and C. No coupling capacitors are required because timer is edge-triggered. Negative reset pulse simultaneously resets all sections. Varying control voltage (in range of 4.5–16 V) affects period of all timer sections simultaneously.—H. M. Berlin, IC Timer Review, *73 Magazine*, Jan. 1978, p 40–45.

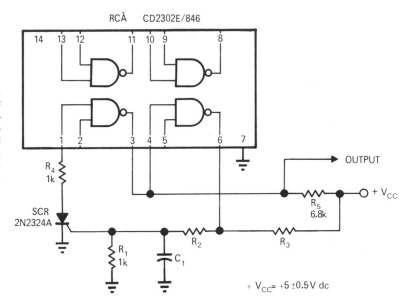

AF RECTANGULAR-WAVE—Frequency can be adjusted over wide AF range, with ON and OFF times of rectangular output signal independently varied between 35 and 60% on by choice of values for C_1 (0.05 to 40 μF), R_2 (1K or 2K), and R_3 (7.3K to 27K). Minimum value of R_3 is 6K.—D. E. Manners, Adjustable Rectangular-Wave Oscillator Interfaces with IC Logic, *EDN/IEEE Magazine*, Sept. 15, 1971, p 46.

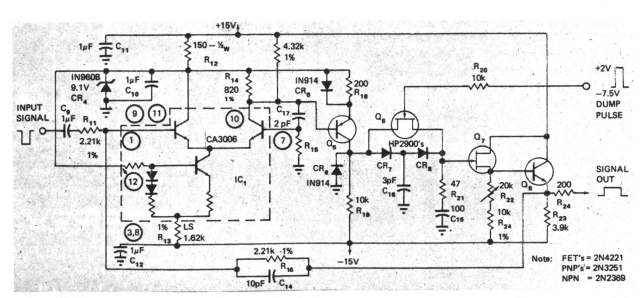

PULSE WIDENER—Peak detection diodes CR_7 and CR_8 in feedback loop of unity-gain CA3006 differential opamp form peak holder that maintains amplitude of narrow video pulses while stretching output pulses as much as 6000 times (from 50 ns to as much as 300 μs). Gain of circuit is unity. Article describes timing and control circuits required in conjunction with peak holder to achieve predictable termination times for stretched pulses. These external circuits include μA710 used as threshold limiter, 9602 dual monostable used as delay-pulse and dump-pulse timing generators, and discrete transistor stage serving as dump-pulse output stage.—B. Pearl, Peak Holder Stretches Narrow Video Pulses, *EDN Magazine*, Feb. 5, 1973, p 46–47.

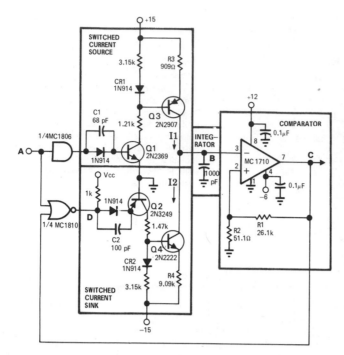

11X PULSE STRETCHER—Provides negative output pulse width equal to positive input pulse multiplied by 1 + R4/R3, which is 11 for values shown. Output pulses are TTL- or DTL-compatible. Minimum output pulse width is 70 ns, and maximum is 1/11 of pulse repetition rate. Circuit consists of switched current source, switched current sink, integrating capacitor, and comparator. Q1 and Q2 act as switches for current sources Q3 and Q4, while C1 and C2 reduce turn-on and turnoff times of switches. CR1 and CR2 provide temperature compensation for Q3 and Q4. AND gate compensates for propagation delay in NOR gate, to ensure that current sink is switched on by trailing edge of input pulse. Add inverter if output must be same polarity as input.—F. Tarico, Linear Circuit Multiplies Pulse Width, *EDN/IEEE Magazine*, Dec. 1, 1971, p 45–46.

COMPLETING LAST CYCLE—Developed for applications requiring that gated oscillator must always complete its timing cycle. Circuit uses only two NAND gates and two diodes, none of which are critical as to type. With no input at A, oscillator output B is low. When A is driven high, D goes low initially and drives output B high. If input at A is removed, regenerative feedback is applied from B through diode D_2 to C until normal timing cycle is finished. Then, with B low, D becomes high and keeps output B low.—L. P. Kahhan, Gated Oscillator Completes Last Cycle, *EDN Magazine*, Jan. 5, 1977, p 43.

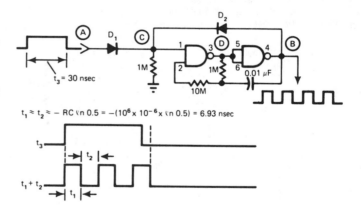

$$t_1 \approx t_2 \approx - RC \ln 0.5 = -(10^6 \times 10^{-6} \times \ln 0.5) = 6.93 \text{ nsec}$$

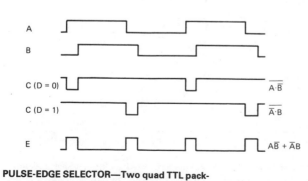

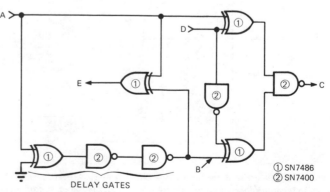

PULSE-EDGE SELECTOR—Two quad TTL packages form simple circuit that generates output pulse at C as function of either leading or trailing edge of input pulse at A, depending on logic level at terminal D. Additional output at E supplies pulses coinciding with both leading and trailing edges of input, independently of logic level at D. Maximum input frequency is 10 MHz, and edge pulses are about 35 ns wide. IC_1 is quad two-input EXCLUSIVE-OR gate, and IC_2 is quad two-input NAND gate.—C. F. Reeves, A Programmable Pulse-Edge Selector, *EDN Magazine*, April 20, 1973, p 85 and 87.

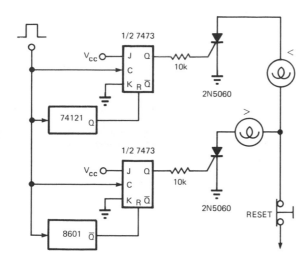

PULSE-WIDTH MONITOR—Circuit turns on upper pilot lamp when pulse width is less than predetermined minimum value, because upper 7473 JK flip-flop is clocked when pulse falls to ground before 74121 mono recovers, triggering upper SCR on. Similarly, 8601 mono is set to coincide with specified maximum pulse width; if pulse falls to ground after this mono recovers, its JK flip-flop is clocked and lower (greater than) lamp is turned on. Fault indication is held until reset button is pushed.—J. Kish, Jr., Three ICs Monitor Pulse Width, *EDN Magazine*, March 20, 1973, p 86.

60 Hz WITH 50% DUTY CYCLE—Adding single resistor R_2 to standard oscillator connection of 555 timer permits operation with 50% duty cycle independently of frequency as determined by value of C_1. For 60-Hz output, V_{CC} is 10 V, C_1 is 1 μF, R_1 is 10K, and R_2 is 75K.—R. Hofheimer, One Extra Resistor Gives 555 Timer 50% Duty Cycle, *EDN Magazine*, March 5, 1974, p 74–75.

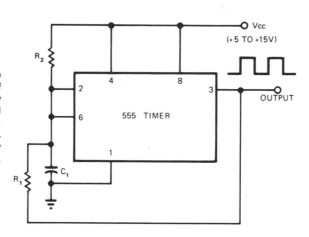

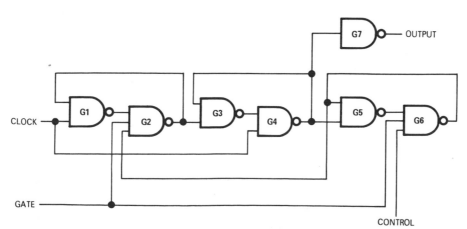

GATED PULSE TRAIN—When control is logic 0, circuit transmits train of complete clock pulses to output, beginning with first clock pulse that starts to rise after application of gate signal and ending with last clock pulse that starts before gate signal falls. When control is logic 1, circuit transmits one complete clock pulse after logic 1 gate signal rises. To send another single pulse, gate signal must be removed and reapplied. Gates are Fairchild LPDTμL9047 triple three-input NAND and 9046 quad two-input NAND; other compatible DTL or TTL NAND gates can also be used.—J. V. Sastry, Gated Clock Generates Pulse Train or Single Pulse, *EDN/IEEE Magazine*, July 1, 1971, p 50.

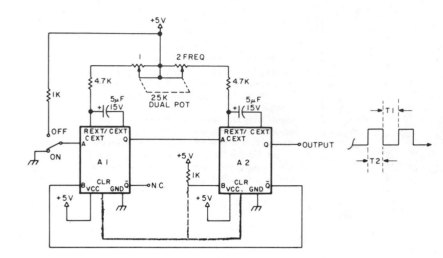

SQUARE-WAVE GENERATOR—Uses two 74122 retriggerable mono MVBRs with clear. Two single pots may be used in place of dual 25K pot if up and down times of output must be independently adjustable.—B. Voight, The TTL One Shot, *73 Magazine*, Feb. 1977, p 56–58.

120 kHz TO 4 MHz—Square-wave output of about 3.5 V can be obtained with SN7400 quad NAND gate, quartz crystal of desired frequency, and single resistor. One of unused gates may be used to gate generator output. Insertion of crystal in socket shocks crystal into oscillation at its resonant frequency, for generating square-wave output over most of frequency range. Waveform approaches clipped sine wave near 4 MHz. Output is suitable for triggering SN7490 decade counters reliably, with normal fanout.—E. G. Olson, 2 Gates Make Quartz Oscillator, *EDN Magazine*, May 5, 1973, p 74.

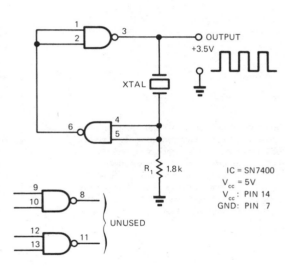

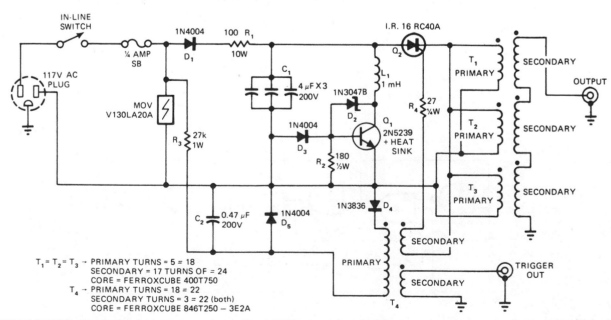

KILOVOLT PULSES—Simple circuit generates 1.5-kV pulses at fixed rate equal to line frequency. Used to drive small piezoelectric transducers for sound velocity measurements. Absence of power transformer minimizes cost, size, and weight. During half of AC cycle, C_1 charges. During other half, C_1 discharges through Q_2 into primaries T_1, T_2, and T_3 to provide output pulse. R_3, C_2, D_5, D_4, and T_4 provide trigger pulse for turning on Q_2. Shunt regulator formed by D_2, D_3, R_2, Q_1, and L_1 clamps voltage across C_1 at 130 V to ensure constant amplitude of output pulses.—S. Anderson, Portable Generator Produces Kilovolt Pulses, *EDN Magazine*, Oct. 20, 1977, p 102.

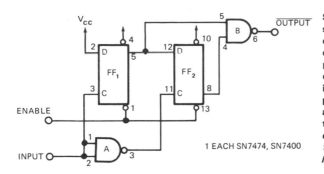

SINGLE-PULSE SELECTOR—Circuit is used to select any desired single pulse from wavetrain continuously applied to input terminal. When enable pulse (not exceeding width of input pulse) is applied, flip-flop FF₁ clocks on leading edge of next input pulse and FF₂ clocks on trailing edge. Output pulse thus has same width as pulses in input wavetrain. Edge-triggering characteristics of D flip-flops prevent operation if they are enabled during input pulse; in this case, next input pulse is delivered as output.—S. J. Cormack, Pulse Catcher Uses Two ICs, *EDN Magazine*, Jan. 5, 1973, p 109.

PULSE STRETCHER WITH ISOLATION—Motorola MOC1000 optoisolator provides safe interfacing with digital logic while stretching input pulse. Circuit uses phototransistor of optoisolator as one of transistors in mono MVBR. With input pulse width of 3 μs, output pulse width is about 1.2 ms.—"Industrial Control Engineering Bulletin," Motorola, Phoenix, AZ, 1973, EB-4.

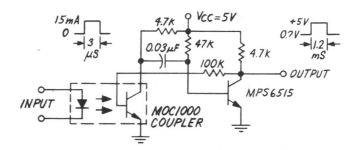

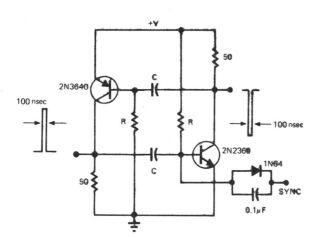

SYNCHRONIZATION TO 10 MHz—Free-running pulse generator circuit uses diode to inhibit operation until sync signal is applied. Circuit then pulses until sync signal returns to original state. Complementary outputs having pulse widths of 100 ns swing essentially from ground to power supply voltage that can be anywhere in range from 0.65 V to 15 V. Values used for R and C determine frequency. For oscillation, R must be in range of 1 kilohm to 1 megohm. For 5-V supply, frequency is 1.2/RC.—B. Shaw, Oscillator Provides Fast, Low Duty-Cycle Pulses, *EDN Magazine*, March 20, 1975, p 73.

VARIABLE-WIDTH TO 12.85 MHz—Single IC circuit uses two monostables to form pulse generator that covers over eight decades (0.054 Hz to 12.85 MHz) with only eight capacitors. Similarly, only eight capacitors cover pulse width range of over eight decades (60 ns to 18 s). Voltage control of frequency and pulse width can be obtained by connecting R₂ and R₄ to individual 1.5–4.5 V control voltage lines instead of to V_cc. Frequency will then vary almost linearly with control voltage, while pulse width will vary almost inversely with control voltage. Capacitor values range from 1 pF to over 100 μF.—M. J. Shah, Wide-Range Pulse Generator Uses Single IC, *EDN Magazine*, Jan. 5, 1973, p 107 and 109.

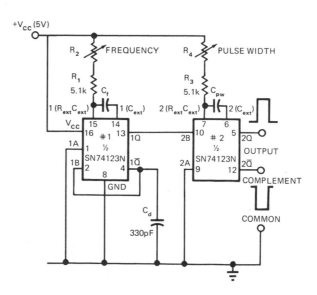

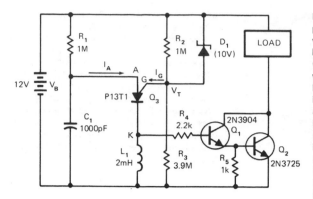

MEMORY REFRESHER—Delivers high-energy pulses in extremely low duty-cycle mode. Developed to provide memory-refresh pulses for MOS memory system from standby battery. Circuit is basically programmable UJT oscillator with divider R_2-R_3 setting threshold or trigger level. To complete one cycle and start next, UJT is turned off by zener. Current through L_1 drops, and resulting negative transient at base of Darlington cuts it off. This generates high-power pulse with very sharp rise and fall times. Charging sequence then starts again to give sustained oscillation. Article gives design equations.—J. P. Stein, PUT Delivers Ultra-Low-Power, High-Energy Pulses, *EDN Magazine*, Sept. 1, 1972, p 51–52.

DELAYED PULSE—Both time and duration of output pulse are programmable by selection of RC networks. Original and delayed clock pulses can both be used in gating circuits. With values shown, output pulse is 1 ms wide and is delayed 15 μs from trailing edge of clock pulse.—D. T. Anderson, Operational Amplifier Makes a Simple Delayed Pulse Generator, *EDN Magazine*, July 1, 1972, p 55.

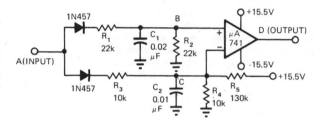

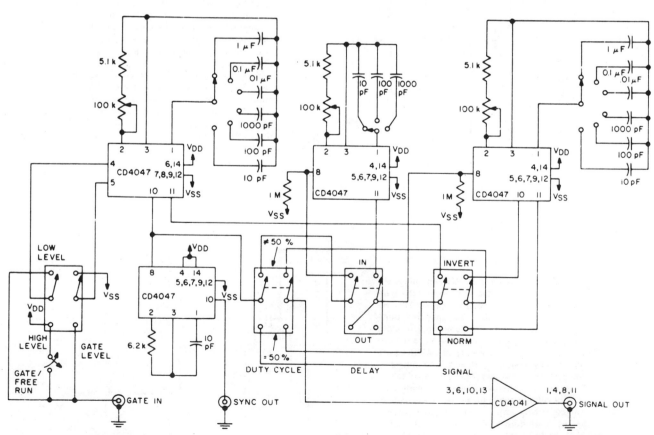

2 Hz TO 1 MHz—General-purpose laboratory pulse generator uses four CD4047A MVBRs to provide six overlapping ranges of frequencies along with pulse-width control, delayed sync, and gating from high-level or low-level input.

When delay switch is IN, MVBR at upper center produces variable output delay from 1.5 μs to 250 ms with respect to sync pulse. When delay pulse is OUT, this mono MVBR is bypassed and inherent delay is then about 400 ns. Signal out-put is buffered with CD4041A for driving any required load. Supply voltage is not critical and can be taken from device under test.—"COS/MOS Integrated Circuits," RCA Solid State Division, Somerville, NJ, 1977, p 619–620.

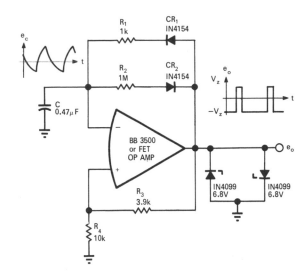

0.1–99.9% DUTY CYCLE—Single opamp circuit provides precise duty-cycle control of pulse train over wide dynamic range by choice of values for R_1 and R_2. Opamp forms gain element of astable MVBR, with pulse and space time intervals determined by feedback elements. Values shown give duty-cycle ratio of 0.001 and period of 1 s.—J. Graeme, Pulse Generator Offers Wide Range of Duty Cycles, *EDN/EEE Magazine*, Sept. 1, 1971, p 42–43.

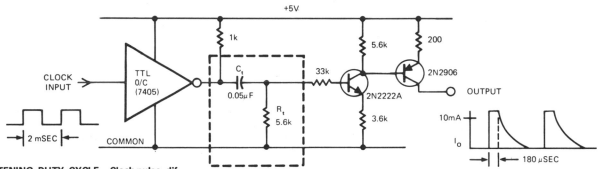

SHORTENING DUTY CYCLE—Clock-pulse differentiator/buffer shortens duty cycle of 500-Hz clock signal having positive period of 1000 μs. Open-collector 7405 inverter acts as buffer. C_t and R_t are primary differentiating components.

Saturation of transistors provides some stretching of output and gives output pulse width of about 180 μs. Developed for driving regulated power supply having clock-controlled

short-circuit protection.—J. L. Bohan, Clocking Scheme Improves Power Supply Short-Circuit Protection, *EDN Magazine*, March 5, 1974, p 49–52.

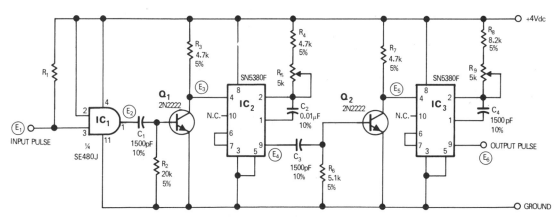

VARIABLE WIDTH AND DELAY—Produces variable-width blanking pulse at selectable delay time after triggering by input pulse. With values

shown, output pulse range is 8 to 12 μs, and delay range is independently adjustable from 26 to 36 μs with R_5.—D. E. Norris, Variable Delay

Blanking-Pulse Generator, *EDN/EEE Magazine*, Dec. 1, 1971, p 49.

CHAPTER 72

Receiver Circuits

Individual stages and complete circuits for entertainment, amateur radio, commercial communication, and other types of AM and FM receivers, including ultrasonic, radiotelescope, and satellite communication receivers. See also Amplifier, Antenna, Audio Amplifier, Code, Frequency Modulation, IF Amplifier, Single-Sideband, Squelch, Television, and Transceiver chapters.

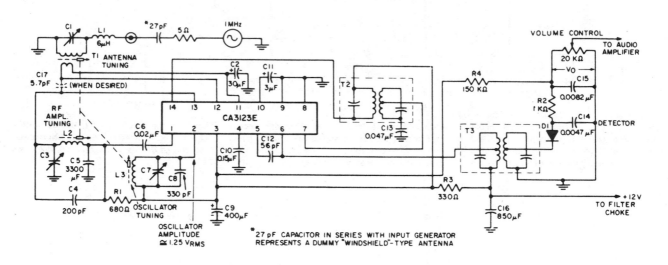

Transformer	Symbol	Frequency	Inductance μh ($\approx$)	Capacitance pF ($\approx$)	Q ($\approx$)	Total Turns To Tap Turns Ratio	Coupling
First IF: Primary	T_2	262 kHz	2840	130	60	none	critical $\approx 0.017 \approx 1/Q$
Secondary			2840	130	60	or 30:1 31:1	
Second IF: Primary	T_3	262 kHz	2840	130	60	8.5:1	—
Secondary			2840	130	60	8.5:1	critical $\approx 0.017 \approx 1/Q$
Antenna: Primary	T_1	1 MHz	195	$(C_1)-130$	65		
Secondary		Adjusted to an impedance of 75 Ω with primary resonant at 1 MHz. Coupling should be as tight as practical. Wire should be wound around end of coil away from tuning core.					
Coils	L_1	7.9 MHz	6		50		
	L_2	1 MHz	55		50		
	L_3	1.262 MHz	41		40		

AM SUPERHET SUBSYSTEM—RCA CA3123E provides all active elements needed up to audio volume control. Table gives values of components for tuned circuits. Operates from single 12-V supply, making subsystem particularly suitable for auto radios. IF value is 262 kHz. 1-MHz signal generator shown in input circuit is used only for initial tuning.—"Linear Integrated Circuits and MOS/FET's," RCA Solid State Division, Somerville, NJ, 1977, p 361–362.

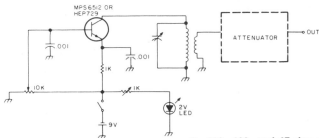

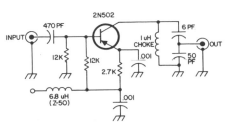

RECEIVER-CHECKING VFO—Simple variable-frequency oscillator is combined with attenuator network to generate signal of about 1 μV for checking performance of amateur radio receiver quickly. Attenuator is series arrangement of 47-, 100-, 100-, and 47-ohm resistors, with 1-ohm resistors going from each of the three junctions to ground. LC combinations are chosen for amateur band desired. Circuit will work down to at least 2 meters.—Is It the Band or My Receiver?, *73 Magazine*, Oct. 1976, p 132–133.

6-METER PREAMP—Simple transistor circuit requires no tuning, draws less than 50 mW from 9-V supply, and increases sensitivity of low-priced receiver without complicated impedance matching.—E. R. Davisson, Simple Six Pre-Amp, *73 Magazine*, Oct. 1974, p 111–112.

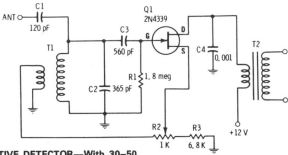

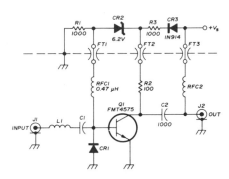

FET REGENERATIVE DETECTOR—With 30–50 foot antenna wire, circuit gives sufficient volume for driving headphones connected to secondary of Lafayette AR-104 or equivalent audio driver transformer T2, for reception of broadcast stations when tuned over AM broadcast band with C2. Feedback control R2 is backed off slightly from point of oscillation, for maximum sensitivity in removing modulation from incoming carrier. When used for CW reception, circuit is left in oscillation and audible difference frequency is produced in output corresponding to marks and spaces. T1 is Miller 2004 or equivalent antenna transformer.—E. M. Noll, "FET Principles, Experiments, and Projects," Howard W. Sams, Indianapolis, IN, 2nd Ed., 1975, p 235–237.

C1 100-180 pF miniature dipped mica or ceramic

C2 1000 pF miniature ceramic disc

CR1 hot-carrier diode (Hewlett-Packard 5082-2810)

CR2 6.2 volt zener diode (1N4735)

CR3 silicon diode (1N914)

FT1-FT3 feedthrough capacitors, 470-1000 pF

J1,J2 SMA-type coaxial connectors (see text)

L1 4 turns no. 24 on 0.1" (2.5mm) diameter, spaced wire diameter (approximately 30 nH)

Q1 Fairchild FMT 4575 low-noise transistor (see text)

R2 100 ohms, ¼ watt (see text)

RFC1 0.47 μH miniature rf choke (Nytronics SWD=0.47)

RFC2 0.2-0.47 μH miniature rf choke or Ohmite Z-460 (value not critical)

432-MHz LOW-NOISE PREAMP—Uses Fairchild FMT4575 transistor having 1.25-dB noise figure, equaling performance of best paramps at 432 MHz. Input matching circuit is low-loss low-Q L matching section L1-CR1. Value of blocking capacitor C1 is not critical, but should be low-loss high-Q type. Hot-carrier diode CR1 in matching section adds about 0.75 pF to circuit, and serves also as low-loss limiter that protects transistor from excessive RF. Zener-diode biasing permits direct grounding of emitter, is insensitive to transistor current gain, provides some DC protection to transistor, and requires no adjustments.—J. H. Reisert, Jr., Ultra Low-Noise UHF Preamplifier, *Ham Radio*, March 1975, p 8–19.

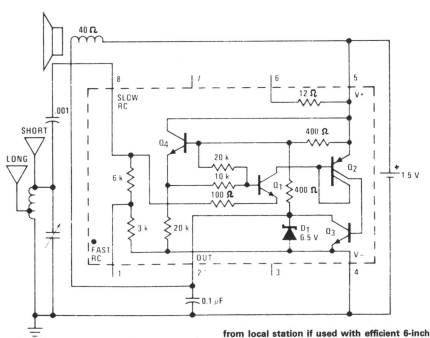

SINGLE-IC RADIO—National LM3909 IC is connected as detector-amplifier driving loudspeaker, with extremely low power gain giving continuous operation for 1 month from D cell. Tuning capability is comparable to that of simple crystal set. Provides acceptable volume from local station if used with efficient 6-inch 40-ohm loudspeaker. Coil is standard AM radio ferrite loopstick having tap 40% of turns from one end. Short antenna can be 10–20 feet, and long antenna can be 30–100 feet.—"Linear Applications, Vol. 2," National Semiconductor, Santa Clara, CA, 1976, AN-154, p 8–9.

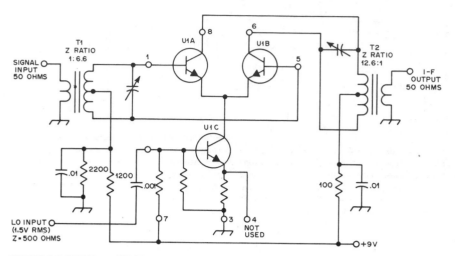

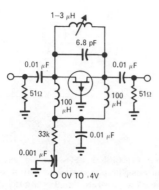

SINGLE-BALANCED MIXER—Uses RCA CA3028A differential amplifier U1 to provide conversion gain of about 30 dB for signal inputs up to 120 MHz. Values of tuned circuits depend on frequency used. Unmarked resistors are on IC.—D. DeMaw, Understanding Linear ICs, *QST,* Feb. 1977, p 19–23.

50-dB ATTENUATION CONTROL—Low-cost FET in pi network serves as resistive attenuator providing up to 50-dB attenuation of 30-MHz level-controlled signal source by varying DC control voltage over range of 0 to 4 V. Tank circuit across source-drain leads of FET keeps phase shift under 2° over entire attenuation range. FET type is not critical.—E. E. Baldwin, Voltage-Controlled Attenuator Has Minimum Phase Shift, *EDN/EEE Magazine,* Nov. 15, 1971, p 40.

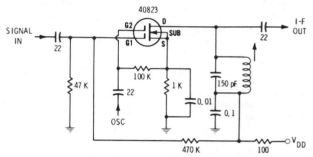

DUAL-GATE MOSFET MIXER—RF and oscillator signals are applied to gates G1 and G2 for mixing in MOSFET. Choice of sum or difference frequency is determined by values used in tank circuit and by tuned circuits of IF amplifier.—E. M. Noll, "FET Principles, Experiments, and Projects," Howard W. Sams, Indianapolis, IN, 2nd Ed., 1975, p 141–142.

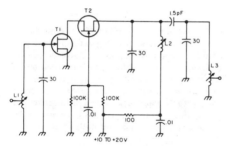

10-METER PREAMP—Simple preamp can also be used on 2 meters with appropriate change of coils. Needs no neutralization. Developed for use with receiver capable of receiving satellite transmissions on 29.45 to 29.55 MHz. Transistors can be MPF 102, MPF 106, or 2N4416. All coils are 1.2 μH having 7 turns No. 26 enamel on 3/16-inch slug-tuned form. L1 and L3 have tap at 3 turns.—G. L. Tater, CQ OSCAR 7, *73 Magazine,* Feb. 1975, p 54–56 and 58–60.

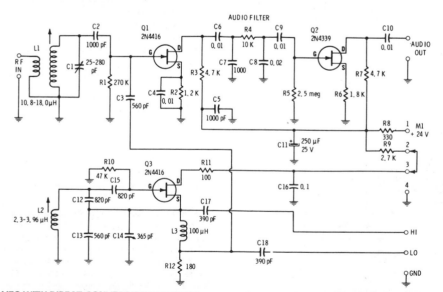

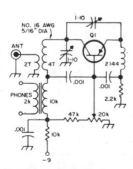

VFO WITH DIRECT-CONVERSION DETECTOR—Used in demodulating SSB and CW signals. Incoming signal is heterodyned with VFO output to give direct conversion to audio. Audio filter R4-C7-C8 allows only AF components to be transferred to audio amplifier Q2. Oscillator can be operated alone for other purposes by removing jumper between output terminals 2 and 3, then applying +24 V to terminal 3 and negative supply to terminal 4. Load on low-level output of oscillator has negligible effect on frequency. Load on high output may change frequency, but this can be corrected by retuning oscillator if load is constant. Values shown are for 80-meter band.—E. M. Noll, "FET Principles, Experiments, and Projects," Howard W. Sams, Indianapolis, IN, 2nd Ed., 1975, p 165–173.

SUPERREGENERATIVE—Simple single-transistor superregenerative receiver is adequate for copying many local signals in 2-meter amateur band. With components shown, tuning range is about 90 to 150 MHz. Transistor can be GE-9 or HEP-2.—Circuits, *73 Magazine,* Feb. 1974, p 100.

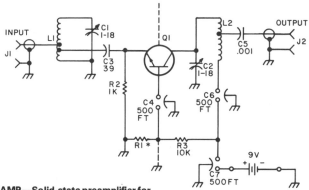

144-MHz PREAMP—Solid-state preamplifier for 2-m band equals performance of best tube designs. Value of R1 is chosen for optimum gain versus noise figure. Two 500-pF feedthrough capacitors (FT) serve as convenient terminals for connections. Q1 is 2N2708, 2N4936, or equivalent. Article covers construction and tune-up.—C. Sondgeroth, Really Soup Up Your 2m Receiver, *73 Magazine*, Feb. 1976, p 40–42.

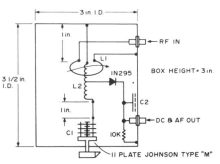

100–200 MHz DIODE RECEIVER—Hybrid tuned-diode version of basic crystal detector uses line cavity cut from sheet copper and soldered into box to give high Q up to 200 MHz. Useful for checking local oscillators around 135 MHz and transmitters around 147 MHz, along with 2-meter transmitters and transceivers. L2 is 4 turns air-wound ½ inch diameter and ½ inch long, with L1 as 1 turn adjustable around it. C2 is 1 × 2 inch brass plate with 0.005-inch teflon sheet with nylon bolts.—B. Hoisington, Tuned Diode VHF Receivers, *73 Magazine*, Dec. 1974, p 81–84.

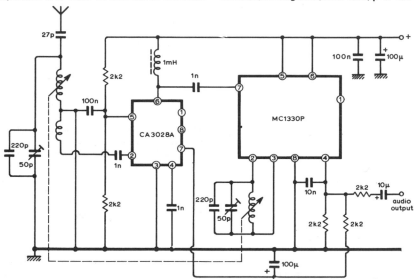

AM HOMODYNE—Circuit uses ICs to provide product demodulator that overcomes interstation tuning whistles or heterodynes by deriving local oscillator source from incoming signal carrier. This AM carrier is amplified before modulation is stripped off, and used in homodyne system in which converter output (IF signal) is the same as audio signal. Motorola MC1330P IC, originally developed as color TV video detector, is here connected as synchronous demodulator. RF amplifier IC is RCA CA3028A operating in cascode mode, with permeability tuning of input and RF choke and input of MC1330P together forming output load. Article covers circuit operation in detail and gives alignment procedures. Although developed for broadcast band, basic circuit can be adapted for communication and FM receivers as well.—J. W. Herbert, A Homodyne Receiver, *Wireless World*, Sept. 1973, p 416–419.

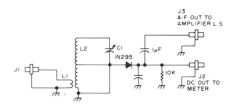

2–12 MHz DIODE RECEIVER—Basic crystal detector circuit can be tuned over 4-MHz range and can also serve as AM detector for 10.7-MHz IF strip. L2 is 64 turns on ½-inch diameter air-wound, center-tapped, with 2 turns around it for L1. C1 is about 365 pF. Unmarked C can be 0.01 μF.—B. Hoisington, Tuned Diode VHF Receivers, *73 Magazine*, Dec. 1974, p 81–84.

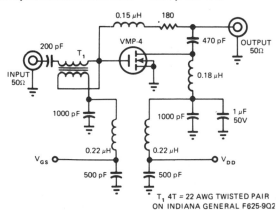

T₁ 4T = 22 AWG TWISTED PAIR ON INDIANA GENERAL F625-9Q2

40–275 MHz BROADBAND VMOS—Response is flat within 1 dB over entire frequency range for 12-dB output. Circuit requires no initial adjustments and cannot be damaged by mismatched loads. Designed primarily for communication applications. Uses Siliconix VMP-4 vertical MOS power transistor.—E. Oxner, Will VMOS Power Transistors Replace Bipolars in HF Systems?, *EDN Magazine*, June 20, 1977, p 71–75.

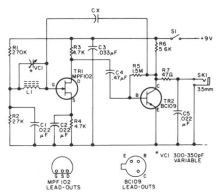

TWO-TRANSISTOR RECEIVER—Suitable for reception of local AM broadcast stations. L1 is standard ferrite-rod antenna or suitable winding on 5/16-inch ferrite rod 3½ inches long, for use with broadcast-band tuning capacitor. Choose value of CX that gives maximum sensitivity.—Circuits, *73 Magazine*, July 1975, p 170.

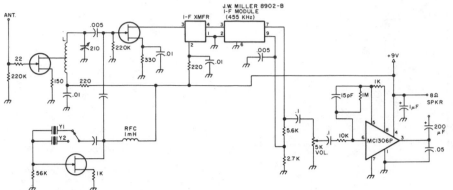

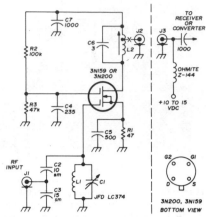

WWV RECEIVER—Solid-state components simplify construction of good HF utility-type receiver suitable for monitoring WWV and other station transmissions or for checking specific frequencies in HF bands. Receiver is single-conversion superheterodyne with FET front end, crystal-controlled. No bandswitching is required over 6–15 MHz range. All transistors are MPF 102 or HEP 802. IF transformer is part of Miller IF module. L is 26 turns No. 26 on ¼-inch form, tapped at 13 turns. Y1 is 9.545 MHz for 10-MHz WWV, and Y2 is 14.545 MHz for 15-MHz WWV.—Build a Useful HF Receiver, *73 Magazine*, Dec. 1977, p 216–217.

C1   part of JFD LC374 tank circuit (see text)

C2   10-pF silver-mica with 3N159, 8-pF silver mica with 3N200

C3   15-pF silver-mica with 3N159, 12-pF silver mica with 3N200

C4   235-pF mica button

C5   500-pF mica button

L1   JFD LC374 tank circuit (contains C1)

L2   6 turns no. 22 enamelled on a 5-mm (0.2") diameter slug-tuned form, tap at 1 turn

144-MHz PREAMP—High-performance circuit for 2-meter receiver provides 15-dB gain and low noise figure. Low-loss low-noise input circuit uses tapped-capacitor coupling. L1 and C1 are JFD LC374 tank circuit with C1 serving mainly to support L1 and one transistor lead. C1 also serves for adjusting resonant frequency at input. Dual-gate MOSFET requires no neutralization. Power is fed into preamp through RF output connector. Power feed arrangement shown at right can be located inside unit with which preamp is used.—R. E. Guentzler, A Good Two-Meter Preamplifier, *Ham Radio*, June 1974, p 36–38.

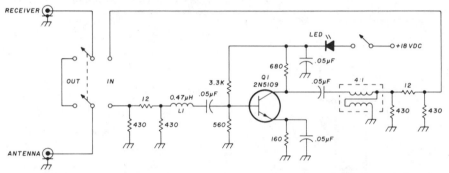

PREAMP COVERS 1–50 MHz—Can be used to improve performance of old communication receivers as well as modern equipment. L1 is several ferrite beads over short loop of about No. 20 insulated wire. Transformer is made by winding about 10 bifilar turns of about No. 20 wire on 0.5-inch toroid core.—E. Pacyna, Wideband Preamp, *Ham Radio*, Oct. 1976, p 60–61.

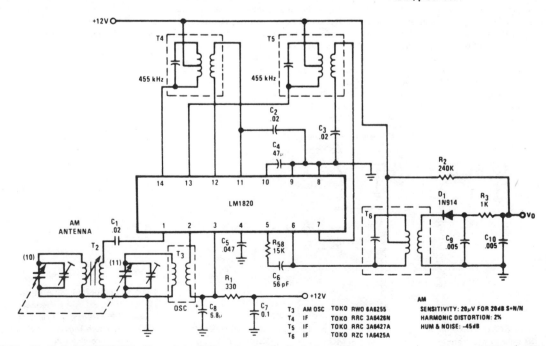

AM MIXER/IF IC—Single National LM1820 chip provides all active stages for oscillator, mixer, IF amplifier, and AGC detector of superheterodyne AM broadcast radio. Omission of RF stage reduces cost at some sacrifice in sensitivity and stability, along with more noise, but careful layout can minimize stability problems. Total gain is 88 dB.—"Audio Handbook," National Semiconductor, Santa Clara, CA, 1977, p 3-4–3-8.

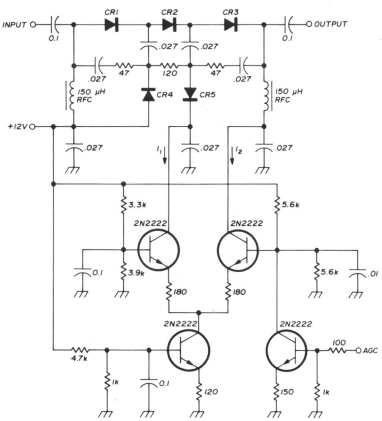

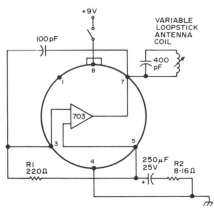

455-kHz BFO—Simple basic oscillator produces signal that can be mixed with signals in all-band radio to give beat frequency for CW or SSB reception. By itself, circuit can be used as low-power (QRP) phone or CW transmitter or as signal source for other purposes.—R. L. Price, 99¢ IC BFO, *73 Magazine*, Jan. 1976, p 201.

INPUT ATTENUATOR—Low-distortion automatic input attenuator for modern communication receiver is activated at input signal levels above 100 μV. For range of 1–30 MHz, use PIN diodes such as HP5082-3081. Intermodulation distortion products of attenuator are about 85 dB down for two 1-V signals.—U. L. Rohde, Optimum Design for High-Frequency Communications Receivers, *Ham Radio*, Oct. 1976, p 10–25.

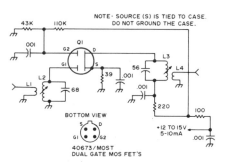

29.45-MHz PREAMP—Reduces noise figure of average communication receiver 2.5 dB and adds up to 20 dB of gain, as required for reception of 29.45-MHz satellite beacon signals. Q1 is RCA 40673 or almost any other dual-gate MOSFET, with shield partition across the device. L2 is 10 turns No. 24E spaced on ¼-inch slug-tuned core, with 2 turns over cold end for L1. L3 is 10 turns No. 24E closewound on ¼-inch slug-tuned core, with 2 turns over cold end for L4.—J. D. Colson, An Oscar Preamp That Works Wonders, *73 Magazine*, July 1975, p 31–32.

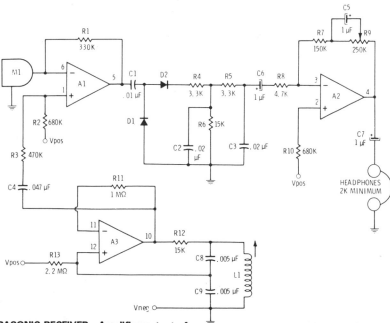

ULTRASONIC RECEIVER—Amplifies output of 40-kHz ultrasonic transducer M1 by mixing in opamp A1 with signal of local oscillator A3 to produce AF signal for further amplification by A2 which drives headphones. Opamp sections are from Motorola MC3401P quad opamp. Diodes are 1N914 or equivalent. L1 is Miller 6315 4–30 mH. Supply can be 9–12 V. Transmitter can be wide-range audio amplifier capable of handling 38–42 kHz, driving similar ultrasonic transducer.—C. D. Rakes, "Integrated Circuit Projects," Howard W. Sams, Indianapolis, IN, 1975, p 26–29.

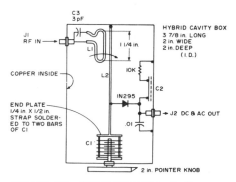

160–500 MHz DIODE RECEIVER—Cavity version of basic crystal detector was developed for use chiefly in 220-MHz and 450-MHz amateur bands. C1 is 25-pF tuning capacitor, and C2 is 1 × 2 inch brass plate insulated from sheet-copper cavity by 0.005-inch Teflon sheet or mica. L1 is 3-inch length of 1-inch copper strap.—B. Hoisington, Tuned Diode VHF Receivers, *73 Magazine*, Dec. 1974, p 81–84.

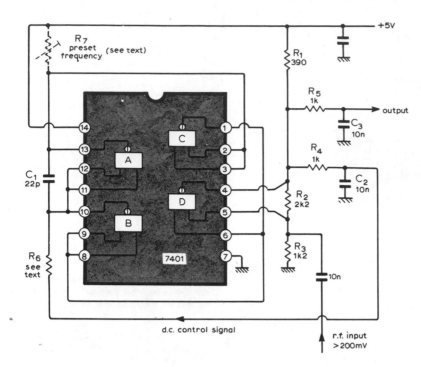

SIMPLE PLL DEMODULATOR—Requires only single IC to provide synchronous detection. Performance is satisfactory for most requirements of most amateur radio experimenters. Gates A, B, and C in IC form relaxation-type VCO whose output frequency is determined by C_1 and positive current sources supplying pins 10 and 13 of IC. When pin 6 is high, gate D is biased by R_2 and R_3 to operate as linear amplifier for input signal. In operation, pin 6 is made alternately high and low by oscillator output, so D acts as amplifying phase detector. Output goes through low-pass filter R_4-C_2 to VCO, completing phase-locked loop. Separate filter R_5-C_3 provides AF output. When C_1 is 22 pF, circuit operates at about 10 MHz. With 270 ohms for R_6, lock is maintained over range of 2 MHz; with 10 kilohms, locking range is 300 kHz. R_7 is optional, for fine adjustment of frequency.—R. King, Phase-Locked Loop Demodulator, *Wireless World*, July 1973, p 337.

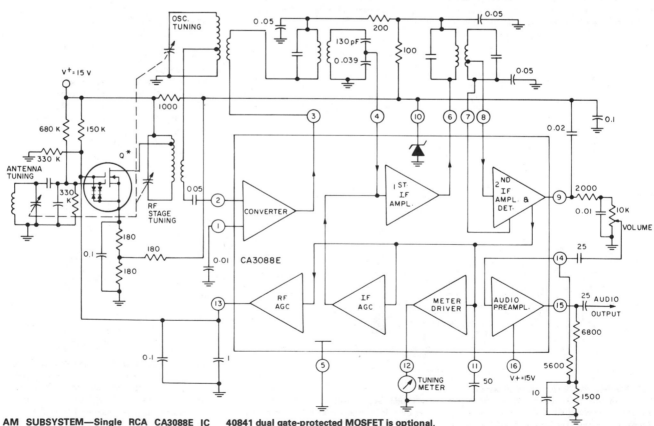

AM SUBSYSTEM—Single RCA CA3088E IC serves as AM converter, IF amplifier, detector, and preamp for AM broadcast or communication receiver. RF amplifier stage using RCA 40841 dual gate-protected MOSFET is optional. IC also provides internal AGC for first IF stage and delayed AGC for optional RF stage. Internal buffer stage can be used to drive tuning meter.—"Linear Integrated Circuits and MOS/FET's," RCA Solid State Division, Somerville, NJ, 1977, p 348–349.

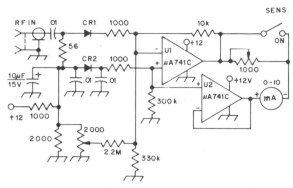

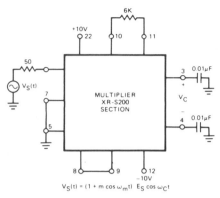

RF METER—Simple square-law detector can detect and measure signals as low as −26 dBm, at microwatt levels. CR1 is biased with about 20 μA by opamp U1 serving as low-impedance DC source. CR2 provides temperature compensa-tion, and U2 serves as low-impedance reference for 10-mA meter. Diodes can be hot-carrier types or 1N914s.—W. Hayward, Defining and Measuring Receiver Dynamic Range, *QST*, July 1975, p 15–21 and 43.

SYNCHRONOUS AM DETECTOR—Input signal is applied to multiplier section of Exar XR-S200 PLL IC with pins 5 and 7 grounded. Detector gain and demodulated output linearity are then de-termined by resistor connected between pins 10 and 11, in range of 1K to 10K for carrier am-plitudes of 100 mV P-P or greater. Multiplier out-put can be low-pass filtered to obtain de-modulated output. For typical 30% modulated input with 10-MHz carrier and 1-kHz modula-tion, output is clean 1-kHz sine wave.—"Phase-Locked Loop Data Book," Exar Integrated Sys-tems, Sunnyvale, CA, 1978, p 9–16.

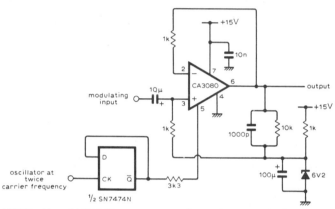

BALANCED MIXER—Uses CA3080 IC transcon-ductance amplifier as precise low-frequency single balanced mixer with inherent carrier bal-ance and accurately defined conversion gain. Binary divider IC halves oscillator frequency, giving carrier waveform having highly accurate unity mark-space ratio. Divided carrier is used to switch amplifier on as unity-gain voltage fol-lower. Conversion loss is 4 dB.—R. J. Harris, Single Balanced Mixer, *Wireless World*, May 1976, p 79.

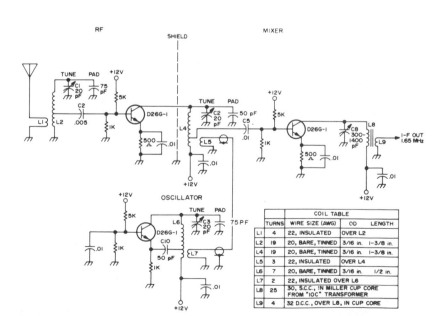

COIL TABLE			
TURNS	WIRE SIZE (AWG)	OD	LENGTH
L1	4	22, INSULATED	OVER L2
L2	19	20, BARE, TINNED	3/16 in.	1-3/8 in.
L4	19	20, BARE, TINNED	3/16 in.	1-3/8 in.
L5	3	22, INSULATED	OVER L4
L6	7	20, BARE, TINNED	3/16 in.	1/2 in.
L7	2	22, INSULATED OVER L6
L8	25	30, S.C.C., IN MILLER CUP CORE FROM "IOC" TRANSFORMER
L9	4	32 D.C.C., OVER L8, IN CUP CORE

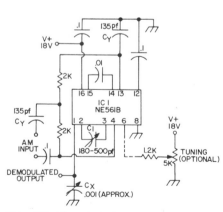

6-METER FRONT END—Developed for use as converter with any communication receiver having 1.65-MHz IF. Article covers construction and tune-up. Use of GE microtransistors per-mits miniaturization.—B. Hoisington, A Real Hot Front End for Six, *73 Magazine*, Nov. 1974, p 88–90 and 92–94.

PLL AM—Phase-locked loop of Signetics NE561B is locked to AM signal carrier fre-quency, and output of VCO in IC is used as local oscillator signal for product detector. Tuned RF stage will generally be required, along with good antenna and ground. Simple one-transis-tor audio amplifier will suffice for driving loud-speaker. Circuit can be adapted for other fre-quencies outside of broadcast band, from 1 Hz to 15 MHz, by changing values of C_Y and C_1.—E. Kanter, PLL IC Applications for Hams, *73 Mag-azine*, Sept. 1973, p 47–49.

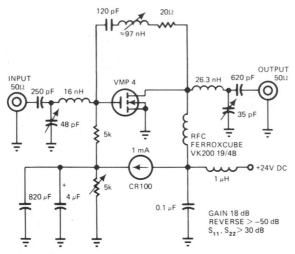

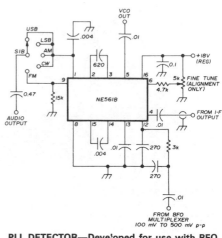

200-MHz NEUTRALIZED—Provides 18.2-dB gain and −50 dB reverse isolation for communication applications. Noise figure is low. Uses Siliconix VMP-4 vertical MOS power transis-tor.—E. Oxner, Will VMOS Power Transistors Replace Bipolars in HF Systems?, *EDN Magazine,* June 20, 1977, p 71–75.

PLL DETECTOR—Developed for use with BFO multiplexer in 455-kHz multimode detection system using NE561 phase-locked loop IC. Circuit provides required 90° phase-shift network in series with output of BFO multiplexer, to compensate for lockup of NE561 in quadrature with signal at input of phase detector during AM reception. IF input level to NE561 should be below about 100 mVRMS for minimum distortion. Audio output level will then be at least half that for narrow-band FM, about same for SSB and CW, and about double for AM if both sidebands are passed by IF filters. FM audio output level is proportional to percent deviation and cannot be increased by increasing signal level. Two 0.004-μF capacitors limit audio bandwidth to about 4 kHz. VCO output of NE561 is 0.6 V P-P square wave at AM carrier or BFO frequency.—J. Regula, BFO Multiplexer for a Multimode Detector, *Ham Radio,* Oct. 1975, p 52–55.

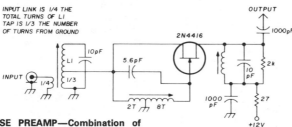

2-m LOW-NOISE PREAMP—Combination of grounded-gate and grounded-source connections uses bridge arrangement for neutralizing feedback capacitance between gate and drain. Input impedance is transformed in parallel between gate and ground to provide necessary wideband characteristic. Noise figure is between 1 and 2 dB, with gain of about 15 dB. Circuit is unconditionally stable, and combines optimum matching for best noise, lowest input SWR, and high power gain.—U. Rohde, High Dynamic Range Two-Meter Converter, *Ham Radio,* July 1977, p 55–57.

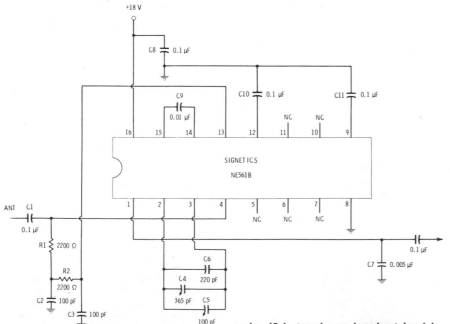

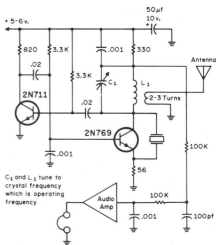

PLL AS AM DEMODULATOR—Single phase-locked loop IC provides audio output signal when connected to suitable antenna for broadcast band. Demodulation is achieved without use of input tuned circuits because control oscillator of PLL is locked to frequency of incoming carrier. IC is tuned over broadcast band by changing frequency of internal VCO with external variable capacitor C4. By changing capacitor limits, circuit can be used to cover long-wave and shortwave bands.—E. M. Noll, "Linear IC Principles, Experiments, and Projects," Howard W. Sams, Indianapolis, IN, 1974, p 303–305.

CRYSTAL-CONTROLLED SUPERREGENERA-TIVE—Two high-frequency transistors connected as 20-kHz MVBR provide switching action at same rate for RF oscillations generated in crystal feedback path. Received AM signal induced in tank circuit of C_1 will modulate exact switching point of circuit at rate directly proportional to modulation component of received signal. Choose L_1, C_1, and crystal for frequency desired. If at 10 MHz, standard WWV time broadcasts can be picked up.—I. Math, Math's Notes, *CQ,* Sept. 1972, p 36–37.

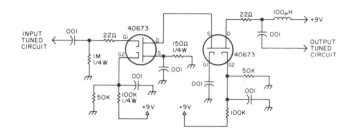

PREAMP BOOSTS GAIN 20 dB—Two RCA MOS-FETs in cascode provide extra 20 dB of gain when used ahead of older Radio Shack AX-190 shortwave receiver. Input and output tuned circuits, gang-tuned, are part of receiver preselector. Article covers construction and tune-up.— P. J. Dujmich, Improve the AX-190 Receiver, *73 Magazine,* Jan. 1978, p 106–107.

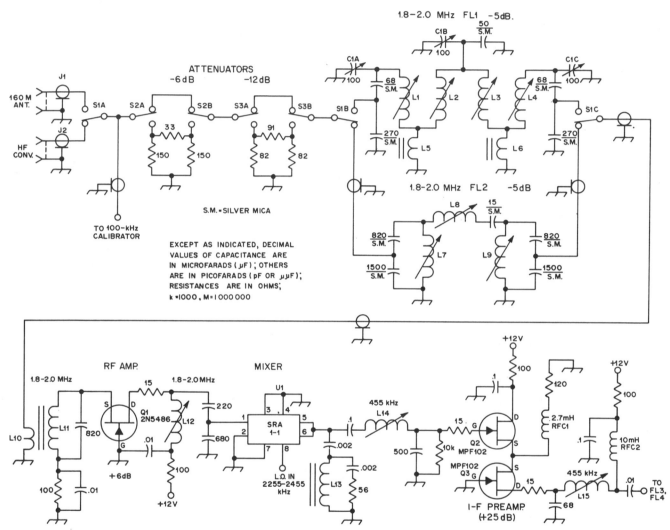

C1 — Three-section variable, 100 pF per section. Model used here obtained as surplus.
J1 — SO-239.
J2 — Phono jack.
L1, L4 — 38 to 68 μH, Q_u of 175 at 1.8 MHz, slug-tuned (J. W. Miller 43A685CBI in Miller S-74 shield can).
L2, L3 — .95 to 187 μH, Q_u of 175 at 1.8 MHz, slug tuned (J. W. Miller 43A154CBI in S-74 shield can).
L5, L6 — 1.45-μH toroid inductor, Q_u of 250 at 1.8 MHz.
15 turns No. 26 enam. wire on Amidon T-50-2 toroid.

L7, L9 — 13-μH slug-tuned inductor (J. W. Miller 9052).
L8 — 380-μH slug-tuned inductor (J. W. Miller 9057).
L10 — 16 turns No. 30 enam. wire over L11 winding.
L11 — 45 turns No. 30 enam. wire on Amidon T-50-2 toroid, 8.5 μH.
L12 — 42-μH slug-tuned inductor, Q_u of 50 at 1.8 MHz. (J. W. Miller 9054).
L13 — 8.7-μH toroidal inductor. 12 turns No. 26 enam. wire on Amidon FT-37-61 ferrite core.
L14 — 120- to 280-μH, slug-tuned inductor

(J. W. Miller 9056).
L15 — 1.3- to 3.0-mH, slug-tuned inductor (J. W. Miller 9059).
Q1, Q2, Q3 — Motorola JFET.
RFC1 — 2.7-mH miniature choke (J. W. Miller 70F273AI).
RFC2 — 10-mH miniature choke (J. W. Miller 70F102AI).
S1 — Three-pole, two-position phenolic wafer switch.
S2, S3 — Two-pole, double-throw miniature toggle.
U1 — Mini-Circuits Labs. SRA-1-1 doubly balanced diode mixer (2913 Quentin Rd., Brooklyn, NY 11229).

1.8–2 MHz FRONT END—Includes enough attenuation for comfortable listening even when nearby high-power amateur station comes on air. Used with downconverter to cover 80 meters through 10 meters. Fixed-tuned 1.8–2 MHz bandpass filter FL2 eliminates need for repeating three-pole tracking filter FL1 when tuning in band. RF amplifier Q1 compensates for filter loss by giving maximum of 6-dB gain. Double-balanced diode-ring mixer U1 handles high signal levels and has good port-to-port signal isolation. High-pass diplexer network at output of

IC mixer U1 improves noise performance without degrading 455-kHz IF. Output goes to IF filters. Two-part article gives all other circuits of receiver.—D. DeMaw, His Eminence—the Receiver, *QST,* Part 1—June 1976, p 27–30 (Part 2—July 1976, p 14–17).

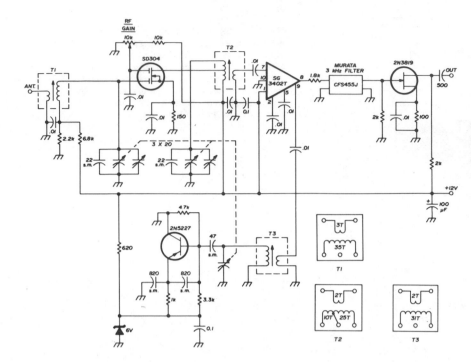

80-METER TUNER—RF stage uses dual-gate N-channel enhancement-mode Signetics SD304 operating with positive bias. With 0–6 V applied to gate 2, AGC range is about 40 dB, but circuit shown uses manual RF gain control. Extra stage of IF overcomes insertion loss of 3-kHz ceramic ladder filter. SG3402T IC is used in mixer; remove pin 6. Transformers T1, T2, and T3 are wound on standard ⅜-in IF forms.—R. Megirian, Design Ideas for Miniature Communications Receivers, *Ham Radio,* April 1976, p 18–25.

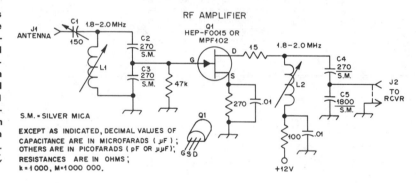

20-dB PREAMP FOR 160 METERS—Provides badly needed extra gain when using Beverage or other inefficient low-noise receiving antennas. Gate of common-source JFET is tapped down on tuned circuit by capacitive divider C3-C4 to prevent self-oscillation. Mica compression trimmer C1 provides match to antenna. L1 and L2 are J. W. Miller 43-series slug-tuned coils; L1 has tuning range of 36–57 μH, and L2 has 24–40 μH range. For 160-meter band, L1 and L2 can be peaked at 1827 kHz to provide maximum gain in 1825–1830 kHz DX window.—D. DeMaw, Build This "Quickie" Preamp, *QST,* April 1977, p 43–44.

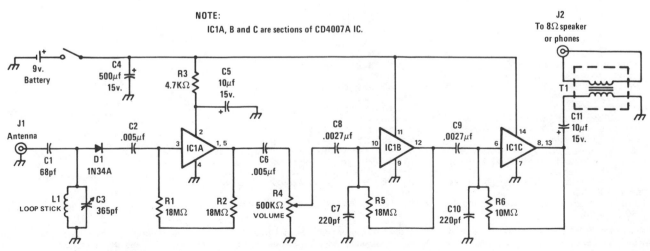

ALL-CMOS RECEIVER—Uses CD4007A IC, having complementary pair of opamps and inverter, to provide all circuits for AM broadcast radio capable of driving headphones or 8-ohm loudspeaker. Selectivity is provided by single tuned circuit and can be improved by optimizing value of C1 to adjust antenna loading. Tune with C3, adjusting L1 if necessary to get stations at low end of band.—C. Green, Easy-to-Build CMOS Radio Receiver, *Modern Electronics,* Sept. 1978, p 40–41, 46, and 59.

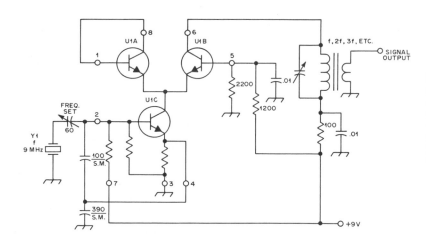

9-MHz CRYSTAL WITH MULTIPLIER—Uses two sections of RCA CA3028A differential amplifier as Colpitts oscillator U1C feeding U1B which can be either amplifier or multiplier depending on values used for output tuned circuit. U1A is not used. Unmarked resistors are on IC.—D. DeMaw, Understanding Linear ICs, *QST,* Feb. 1977, p 19—23.

TTL DIGITAL MIXER—Uses two of 7400 TTL gates as crystal oscillator and other two gates as input buffers to 7474 D flip-flop serving as mixer. RF input signal must be lower than crystal frequency, and IF signal must be less than half crystal frequency. With 8-MHz crystal and 6.75-MHz RF signal, IF is 1.25 MHz. Common TTL 7474 can be used up to 25 MHz, 74H74 to 43 MHz, and 74S74 Schottky version to 100 MHz; Motorola MC12000 is good to 250 MHz.—G. H. Schrick, Introduction to the Digital Mixer, *Ham Radio,* Dec. 1973, p 42—43.

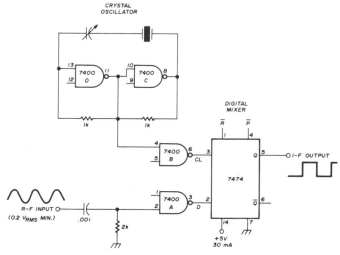

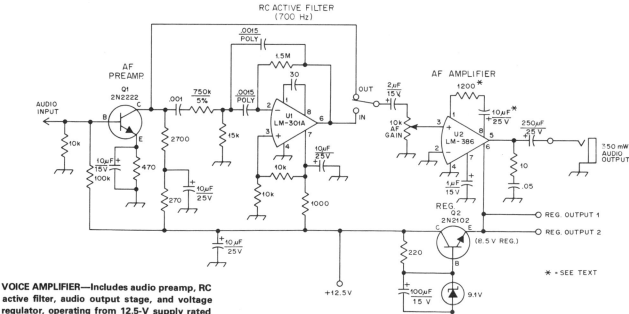

VOICE AMPLIFIER—Includes audio preamp, RC active filter, audio output stage, and voltage regulator, operating from 12.5-V supply rated 300 mA or more. Developed as low-distortion audio amplifier for communication receiver. Two taps for regulated supply provide regulated 8.5 V at 250 mA for other circuits. With filter out, changing input frequency from 300 to 3000 Hz has little effect on output. Switching in audio filter should attenuate all frequencies not in 700-Hz passband of filter. Gain is adjustable over wide range. Output will drive small loudspeaker of 4—16 ohms or headphones of 4—2000 ohms. Can also be used as test bench audio amplifier, intercom, or with code-practice oscillator.—J. Rusgrove, A General-Purpose Audio Amplifier, *QST,* Nov. 1976, p 32—34.

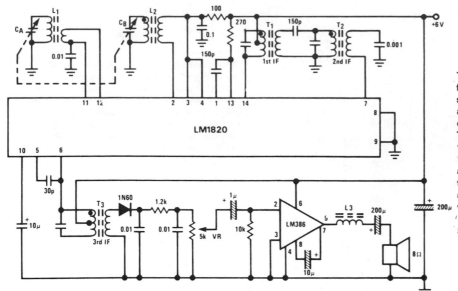

TWO-IC AM RADIO—National LM1820 IC serves for RF, oscillator, mixer, and IF stages of AM superheterodyne radio while LM386 IC is audio amplifier driving loudspeaker. Double-tuned circuit at output of mixer provides selectivity. Total gain from base of input stage to diode detector is 95 dB. C_A is 140 pF, C_B is 60 pF, L_1 has 110 and 5 turns for broadcast band, L_2 has 98 and 12 turns for oscillator, T_1 has 140 turns center tap and 2 turns, T_2 has 142 turns and 7 turns, and T_3 has 142 turns center tap and 71 turns. IF value is 455 kHz. L_3 has 3 turns on ferrite bead.— '''Audio Handbook,'' National Semiconductor, Santa Clara, CA, 1977, p 3-4–3-8.

2-METER SINGLE-VMOS—Provides 5-W PEP output at 146 MHz, with noise figure of only 2.35 dB. Developed for amateur radio applications. Uses Siliconix VN65AJ transistor.—E. Oxner, Will VMOS Power Transistors Replace Bipolars in HF Systems?, *EDN Magazine,* June 20, 1977, p 71–75.

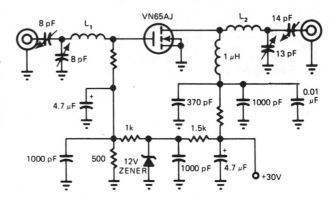

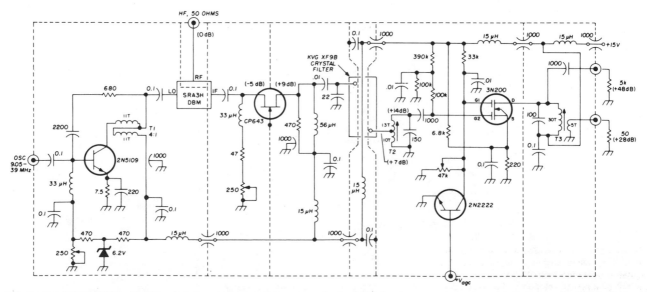

DOUBLE-BALANCED MIXER—Uses grounded-gate CP643 preamp having high dynamic range, 2N5109 oscillator injection amplifier, and 3N200 IF amplifier in combination with Minilabs SRA3H double-balanced mixer. Third-order intercept point is +30 dBm. Oscillator requirement is −1 to +2 dBm (200 to 280 mV across 50 ohms). AGC range is greater than 50 dB. Levels shown in parentheses are for 0 dBm (224 mV) at input and zero AGC voltage.—U. L. Rohde, High Dynamic Range Receiver Input Stages, *Ham Radio,* Oct. 1975, p 26–31.

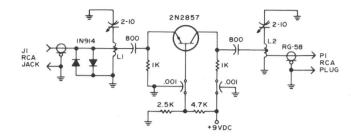

450-MHz PREAMP—Provides up to 10-dB extra gain for older tube-type 450-MHz receivers. Gives significant improvement in receiver sensitivity and quieting. Use of trough-line inductors simplifies construction. 1N914 diodes in parallel at input jack protect transistor from burnout by nearby transmitter. L1 and L2 are made from ¼-inch diameter copper tubing, 8.6 cm long. Article covers construction and operation.—C. Klinert, Easy Preamp for 450 MHz, *73 Magazine,* May 1973, p 33 and 36–38.

9-MHz TUNED OUTPUT—Motorola MC1596G balanced modulator connected as double-balanced mixer has 3-dB bandwidth of 450 kHz at output. Local oscillator signal LO is injected at upper input port and modulated signal of about 15 VRMS maximum at lower input port. Conversion gain is 13 dB for 30-MHz input and 39-MHz LO.—R. Hejhall, "MC1596 Balanced Modulator," Motorola, Phoenix, AZ, 1975, AN-531, p 7.

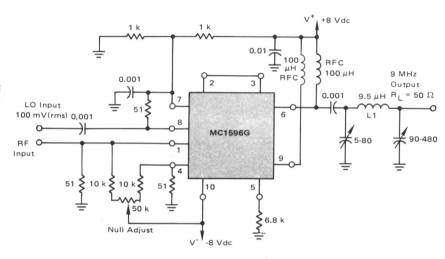

PRECISION FULL-WAVE DIODE DETECTOR—Uses opamp to reduce input voltage at which transfer curve of diode detector becomes nonlinear by factor equal to open-loop gain of opamp. Chief drawback is that delay for positive input signals, which are inverted and amplified 2 times, is twice that for composite signals. Because of delay difference, signals do not subtract in phase and high-frequency performance suffers. Values shown are for test purposes, with low-pass active filter having 2-kHz cutoff.—H. Olson, Diode Detectors, *Ham Radio,* Jan. 1976, p 28–34.

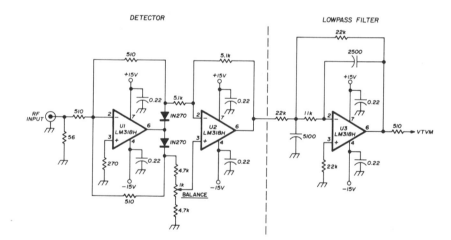

WWV REGENERATIVE—Tunes from 4.7 to 15.5 MHz, covering three WWV frequencies, 20- and 40-meter amateur bands, and several foreign broadcast bands. Draws only 1.5 mA from single D cell when using headphones with 2000-ohm or higher impedance. Performs well with AM, CW, or SSB. When oscillating, detector provides own BFO signal. L1 is 3.8 μH, with emitter tap 1 turn from ground. Use clip for adjusting antenna tap. Tuning requires two hands.—C. Hall, Simple Regenerative WWV Receiver, *Ham Radio,* April 1973, p 42–45.

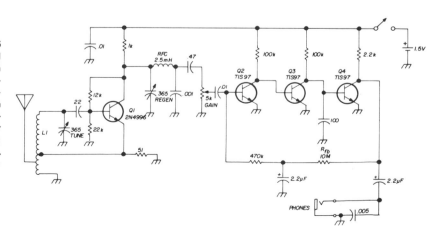

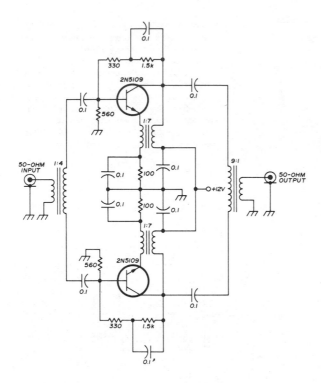

LOW-NOISE RF INPUT—Low-noise version of transistorized push-pull RF stage uses emitter feedback through transformer to give extremely high input and output impedances. Noise figure is below 2 dB. Developed for use in high-quality communication receiver.—U. L. Rohde, Optimum Design for High-Frequency Communications Receivers, *Ham Radio,* Oct. 1976, p 10–25.

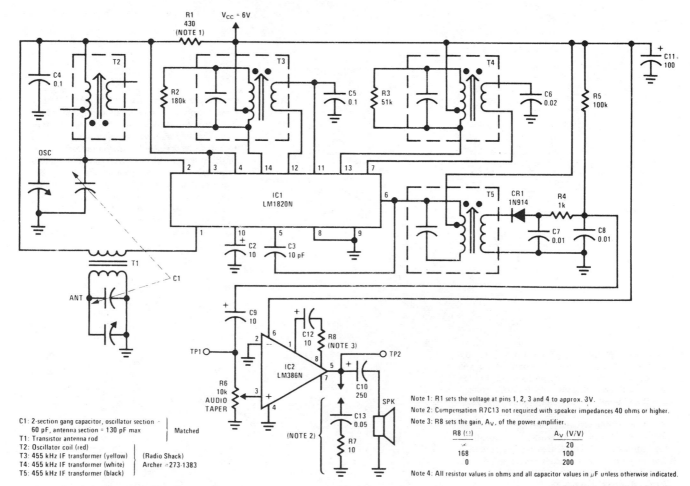

Note 1: R1 sets the voltage at pins 1, 2, 3 and 4 to approx. 3V.
Note 2: Compensation R7C13 not required with speaker impedances 40 ohms or higher.
Note 3: R8 sets the gain, A_V, of the power amplifier.

R8 (Ω)	A_V (V/V)
∞	20
168	100
0	200

Note 4: All resistor values in ohms and all capacitor values in μF unless otherwise indicated.

C1: 2-section gang capacitor, oscillator section = 60 pF, antenna section = 130 pF max   } Matched
T1: Transistor antenna rod
T2: Oscillator coil (red)
T3: 455 kHz IF transformer (yellow)   (Radio Shack)
T4: 455 kHz IF transformer (white)   Archer =273-1383
T5: 455 kHz IF transformer (black)

AM RADIO—National LM1820N IC provides all sections of superheterodyne broadcast-band radio up to second detector, with diode and power opamp forming rest of receiver. Output is ¼ W into 8-ohm loudspeaker when operating from 6-V supply. Total current drain is about 10 mA, making battery operation feasible.—E. S. Papanicolaou and H. H. Mortensen, "Low-Cost AM-Radio System Using LM1820 and LM386," National Semiconductor, Santa Clara, CA, 1975, LB-29.

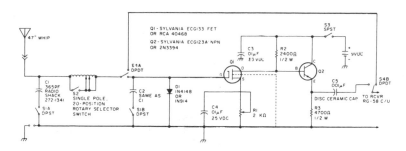

ALL-BAND PREAMP WITH WHIP—Combination of two-stage preamp and 47-inch telescoping antenna gives overall gain of over 30 dB from 160 to 10 meters, for use with communication and SWL receivers when frequent travel precludes erection of fixed antennas. Use type F, BNC, or SO-239 antenna connector. Tuning coil has 20 taps on 150 turns of No. 28 enamel wire wound on ½-inch dowel, with taps at 3, 7, 12, 18, and 25 turns and then about every 10 or 11 turns. Keep leads of Q1 shorted during handling and soldering, to avoid damage by static charges.—K. T. Thurber, Jr., Build A Vacation Special, *73 Magazine*, Aug. 1977, p 62–63.

14–30 MHz PRESELECTOR—Simple self-powered preselector using FET improves overall noise figure of shortwave receiver along with sensitivity in 14–30 MHz portion of HF band. Also helps reduce cross-modulation from strong out-of-band shortwave broadcast stations. C1 and C2 are 50–500 pF Miller 160B. L1 is 10 turns No. 22 on T50-10 Micrometals core with 1-turn link. L2 is 10 turns No. 22 with center tap and 2-turn link on T50-10 core. Q1 is MPF102, HEP-802, or HEP-F0015. D1 and D2 are 1N4002 or HEP-R0051.—H. Olson, The S38 Is Not Dead!, *73 Magazine*, Nov. 1976, p 88–89.

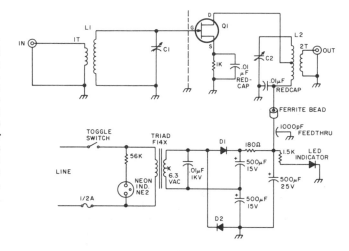

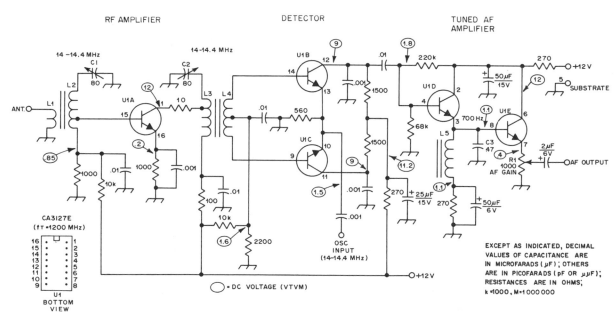

RF AMPLIFIER   DETECTOR   TUNED AF AMPLIFIER

EXCEPT AS INDICATED, DECIMAL VALUES OF CAPACITANCE ARE IN MICROFARADS (µF); OTHERS ARE IN PICOFARADS (pF OR µµF); RESISTANCES ARE IN OHMS; k =1000, M=1 000 000

C1, C2 — 8- to 60-pF mica or ceramic trimmer (Arco 404 or JFD DV11PS60Q suitable).
C3 — 0.47-µF Mylar capacitor.
L1 — Two-turn link of No. 24 enam. wire over L2.
L2 — 25 turns No. 24 enam. wire on T50-6 powdered-iron toroid core. Tap 4 turns up from low-Z end. (See *QST* ads for toroid suppliers, Amidon, G. R.

Whitehouse and Palomar Eng.) Mount L1/L2 on opposite side of pc board from L3/L4. L2 = 2.5 µH.
L3 — 25 turns No. 24 enam. wire on T50-6 toroid core. Tap 10 turns from C2 end. L3 = 2.5 µH.
L4 — 6 turns No. 24 enam. wire, center tapped. Wind over L3.
L5 — Pot-core inductor, 110 mH. Wind 172

turns No. 28 enam. wire on bobbin. Core kit is Amidon PC-2213-77.
R1 — 1000-ohm linear-taper composition control, panel-mounted.
U1 — RCA CA3127E npn transistor-array IC.

20-METER DIRECT-CONVERSION CW/SSB—Simple direct-conversion or synchrodyne receiver uses RCA CA3127E five-transistor array. Product detector follows 14-MHz RF stage. Low drain makes receiver ideal for battery operation, but circuit has no AGC. AF output will drive headphones adequately for strong 20-meter signals, but not loudspeaker. Local-oscillator energy at 14–14.4 MHz for product detector at 1.5–2 VRMS must be furnished by external BFO.—D. DeMaw, Understanding Linear ICs, *QST*, Jan. 1977, p 11–15.

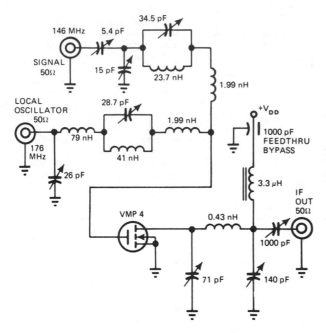

2-METER VMOS ADDITIVE MIXER—Single-ended circuit for amateur band can deliver 0.5 W of power to IF amplifier while providing conversion gain of 18 dB and compression level of 10 dBm. Noise figure is 5.2 dB. Traps in both signal and noise feeds to Siliconix VMP-4 power transistor prevent radiation of unwanted signals.—E. Oxner, Will VMOS Power Transistors Replace Bipolars in HF Systems?, *EDN Magazine,* June 20, 1977, p 71–75.

PRODUCT DETECTOR—Excellent isolation is provided by dual-gate MOSFET. Used for demodulating SSB or CW signals. Input resonant circuit is tuned to IF value. High-frequency components of signal are filtered out by drain output circuit. RC low-pass filter passes voice frequencies to succeeding audio amplifier.—E. Noll, MOSFET Circuits, *Ham Radio,* Feb. 1975, p 50–57.

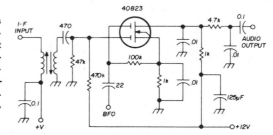

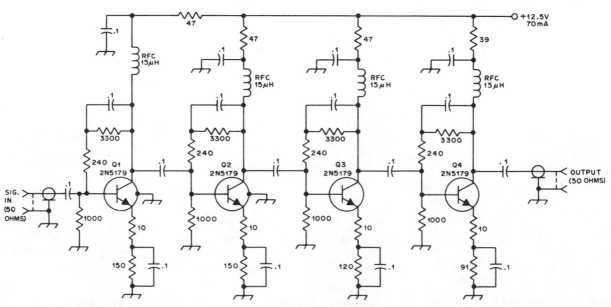

160-METER PREAMP—Broadband 40-dB preamp has response range extending from broadcast band through VHF. Upper 3-dB point of amplifier is at 65 MHz. Heavy feedback stabilizes gain and provides 50-ohm characteristic.—D. DeMaw, Beat the Noise with a "Scoop Loop," *QST,* July 1977, p 30–34.

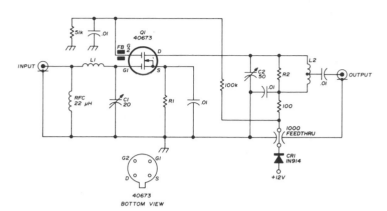

C1 20 pF trimmer (JFD DVJ300 or equivalent ceramic trimmer)

C2 50 pF trimmer (JFD DVJ305 or equivalent ceramic trimmer)

FB Ferrite bead (56-590/65/3B or equivalent)

L1 25 turns no. 24 (0.5mm) on Amidon T50-10 toroid core

L2 22 turns no. 24 (0.5mm) on Amidon T50-10 toroid core, tapped 7 turns from cold end

R1 150 ohms typical (see text)

R2 2000 ohms typical (see text)

28–30 MHz SATELLITE PREAMP—Low-noise design provides up to 25-dB gain and typical noise figure of 1 dB, using dual-gate MOSFET in cascode circuit. Adjust C1 and C2 for maximum output. Developed for use at input of communication receiver. Drain from 12-V power supply should be 3 to 7 mA; if too low or too high, adjust value of R1.—J. Reisert, Jr., Low Noise Figure 28–30 MHz Preamplifier for Satellite Reception, *Ham Radio*, Oct. 1975, p 48–51.

VHF REGENERATIVE—Covers 2-meter amateur band, 152–174 MHz public-service channels including 162.5-MHz weather service, and number of other services. Performance is good enough for use as emergency communication receiver. Supply should be six D cells in series; cheaper 9-V transistor radio batteries may have too much impedance and cause motorboating.—S. Kelly, A Solid State V.H.F. Regenerative Receiver, *CQ*, March 1970, p 63–64.

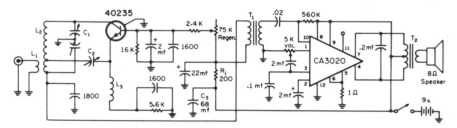

C₁—20 mmf split stator tuning capacitor.
C₂—9-35 mmf ceramic trimmer capacitor.
L₁—1t #10e., 1/4'' diam. coupled to the cold end of L₂.
L₂—6t #10e., 1/4'' diam. tapped at 3/4 turns.

L₃—14t #24e., wound on a 1 meg 1 watt resistor.
T₁—2K to 10K, Olson T-230 or equiv.
T₂—250 ohms center tapped to 8 ohms.

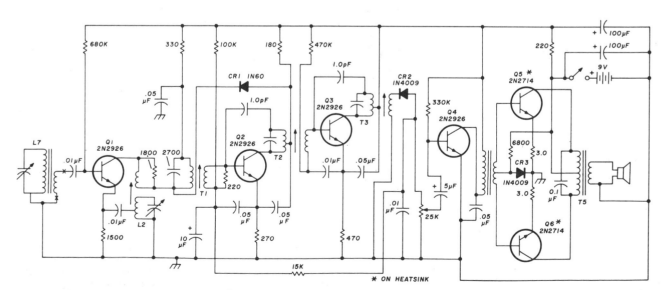

SIX-TRANSISTOR AM—Typical older Magnavox radio uses PNP germanium transistors. L7 is loopstick antenna. Article tells how to add FET converter to radio for use as standard-frequency receiver.—H. Olson, Five-Frequency Receiver for WWV, *Ham Radio*, July 1976, p 36–38.

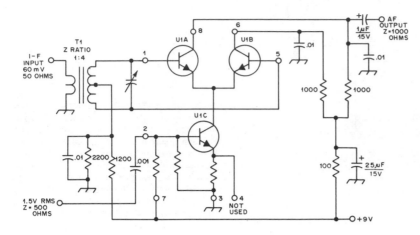

BALANCED PRODUCT DETECTOR—Uses RCA CA3028A differential amplifier U1 to provide conversion gain of about 18 dB for commonly used IF values. Values of tuned circuits depend on frequency used. Unmarked resistors are on IC.—D. DeMaw, *Understanding Linear ICs, QST,* Feb. 1977, p 19–23.

144-MHz PREAMP—Low-noise 2-meter preamp has 18-dB gain and typical noise figure of 1.7 dB. L1 and L2 are each 3½ turns No. 18 wound ½ in long on ⅜-in form and tapped 1 turn from cold end. MOSFET Q1 is MEM554C, 3N159, 3N140, or 3N141. Avoid static charges until MOSFET is connected. Developed for use with tube-type 2-meter converter.—E. Noll, Circuits and Techniques, *Ham Radio,* April 1976, p 40–43.

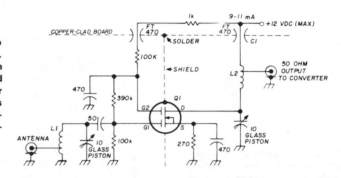

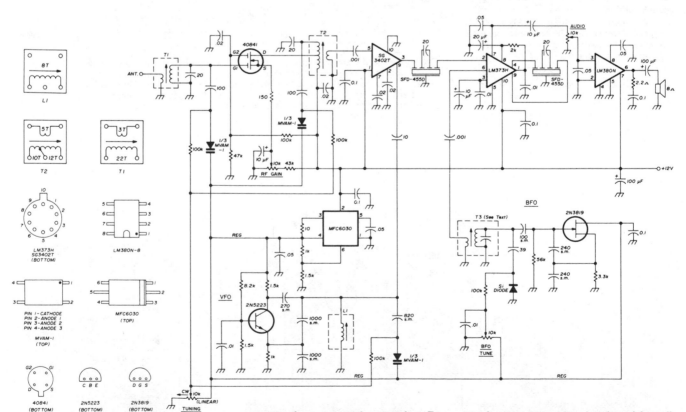

3.5–4 MHz WITH TUNING DIODES—Miniaturized communication receiver was developed for use as tunable IF fed by external converter for all-band coverage up to 30 MHz. Motorola MVAM-1 triple tuning diode serves in place of customary three-gang tuning capacitor. Two Murata SFD-455D ceramic filters provide IF selectivity. MFC6030 voltage regulator provides around 7 VDC with regulation required for diode tuning. Regulator also supplies VFO and BFO. Standard 3/8-in diameter 455-kHz transistor IF transformers were stripped and used for coil forms. T3 is 455-kHz IF with secondary changed to 1 turn. Remove pins 4, 6, and 8 from Silicon General SG3402T mixer.—R. Megirian, Design Ideas for Miniature Communications Receivers, *Ham Radio,* April 1976, p 18–25.

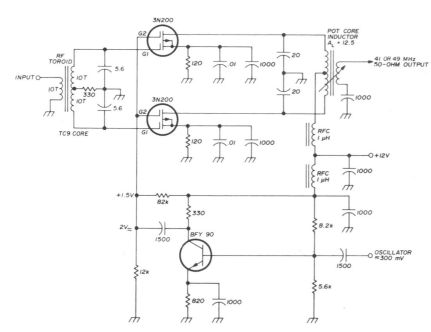

FET MIXER—Double-balanced mixer developed for use in high-quality high-fidelity communication receiver has high input impedance (about 1000 ohms). Two-tone 176-mV signal produces third-order intermodulation distortion 68 dB down.—U. L. Rohde, Optimum Design for High-Frequency Communications Receivers, *Ham Radio,* Oct. 1976, p 10–25.

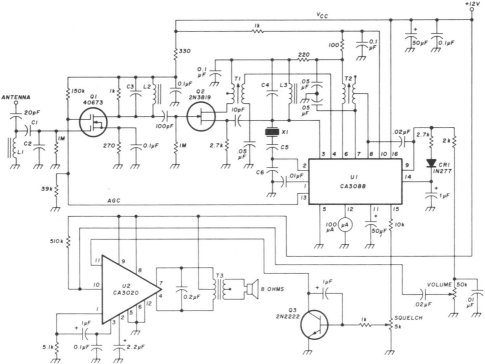

station	freq (MHz)	crystal freq (MHz)	C1 value (pF)	C2 value (pF)	C3,C4 values (pF)	C5 value (pF)	C6 value (pF)	turns	L1&L2 AWG	(mm)	L3 AWG	(mm)	coil cores
WWV WWVH	2.50	2.955	300	820	220	30	150	66	32	(0.2)	32	(0.2)	T37-2
WWV WWVH	5.00	5.455	120	680	100	30	150	49	32	(0.2)	32	(0.2)	T37-2
WWV WWVH	10.00	10.455	56	330	47	30	150	40	32	(0.2)	32	(0.2)	T25-2
WWV WWVH	15.00	15.455	33	330	30	30	150	37	30	(0.25)	30	(0.25)	T25-6
WWV WWVH	20.00	20.455	30	330	27	short	10	29	32	(0.2)	32	(0.2)	T25-6
WWV WWVH	25.00	25.455	24	300	22	short	10	26	32	(0.2)	32	(0.2)	T25-6
CHU	3.33	3.785	300	820	220	30	150	50	30	(0.25)	30	(0.25)	T37-2
CHU	7.34	7.795	68	350	56	30	150	44	32	(0.2)	32	(0.2)	T37-2
CHU	14.67	15.125	33	330	30	30	150	36	32	(0.2)	32	(0.2)	T25-6

10-MHz FIXED FOR WWV—Fixed-frequency receiver has high sensitivity, portability, low power consumption, and low cost. Number of parts is minimized by using RCA CA3088 IC for converter, IF, detector, audio preamp, AGC, and tuning-meter output, along with RCA CA3020 as audio amplifier. Table gives crystal frequencies and tuned-circuit values for all nine frequencies on which frequency calibration data, propagation forecasts, geophysical alerts, time signals, and storm warnings are broadcast by American and Canadian governments. Core type numbers are for Amidon Associates cores. IF transformers come as Radio Shack set 273-1383; use only T1 (gray core) and T2 (white core). Specify load capacitance as 32 pF when ordering crystals. Use overtone crystals for 20 and 25 MHz with C5 replaced by short and C6 reduced to 10 pF.—A. M. Hudor, Jr., Fixed-Frequency Receiver for WWV, *Ham Radio,* Feb. 1977, p 28–33.

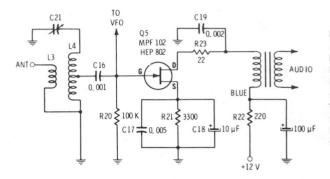

DIRECT-CONVERSION PRODUCT DETECTOR— Antenna is matched to high-impedance gate input of JFET with resonant input transformer. Demodulating carrier is applied to same gate. RC filter and audio transformer in output circuit of JFET recover demodulating audio while filtering out RF signals and undesired mixing components.—E. M. Noll, "FET Principles, Experiments, and Projects," Howard W. Sams, Indianapolis, IN, 2nd Ed., 1975, p 155.

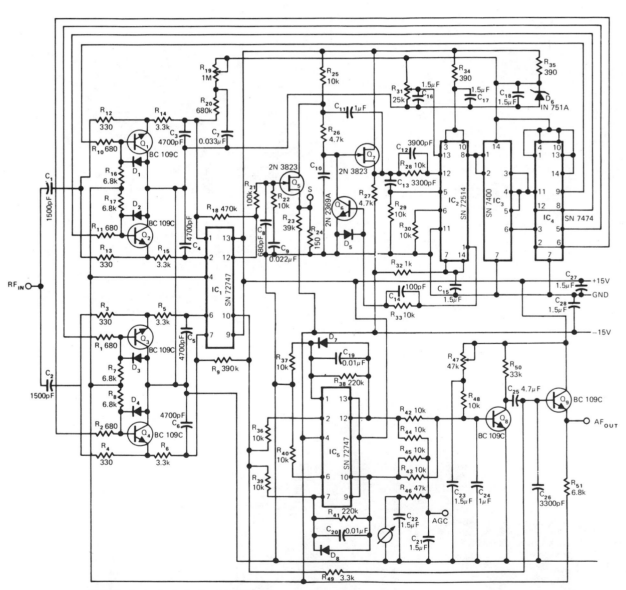

NOTE: D_1, D_2, D_3, D_4, D_5, D_7, D_8 = High-speed Ge types

PLL IN AM RECEIVER—Phase-locked loops provide required stability for synchronous detection to improve reception quality of commercial double-sideband AM transmissions. Signal input and output of VCO are multiplied in phase-sensitive detector or multiplier that produces voltage proportional to phase difference between input and VCO signals. After filtering and amplifying, this voltage is used to control frequency of VCO to make it synchronize with incoming signal. Features include absence of image responses since IF is 0 Hz, almost complete immunity to selective fading, and conversion of RF to audio at very low signal levels so overall receiver gain is achieved mainly in audio amplifier. Article traces development and operation of receiver in detail.—T. Mollinga, Solve Phase Stability Problem in AM Receivers with PLL Techniques, *EDN Magazine,* Feb. 20, 1975, p 51–56.

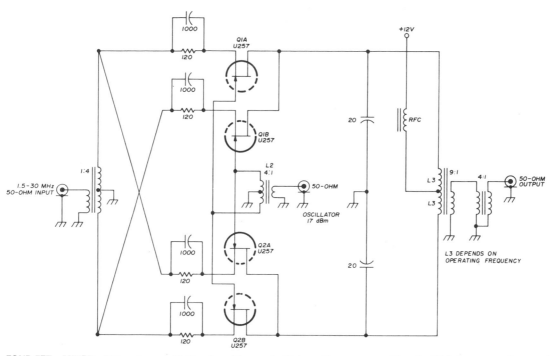

BALANCED FOUR-FET MIXER—Uses two matched FET pairs to bring third-order intermodulation distortion suppression down to 71 dB. Developed for use in high-quality communication receiver.—U. L. Rohde, Optimum Design for High-Frequency Communications Receivers, *Ham Radio,* Oct. 1976, p 10–25.

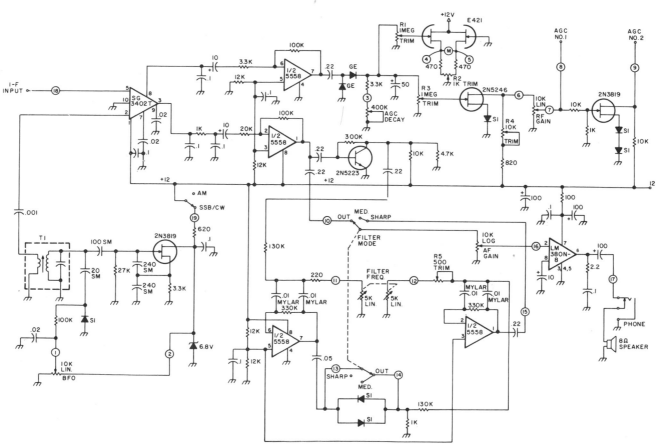

AF FOR AM/SSB/CW—Uses SG3402T as detector, with BFO disabled for AM. Pin 3 of detector output is main audio source, feeding preamp using half of dual opamp whose output goes to AF gain control except when CW filter is in use. Filter has two identical 400–1600 Hz active bandpass sections joined by threshold detector. LM380N-8 AF power amplifier is rated at 600-mW output. Audio from pin 8 of detector is amplified about 30 times in second half of dual opamp before rectification for use as AGC voltage. Circuit includes S-meter fed by AGC section. Article gives construction details of complete receiver.—R. Megirian, The Minicom Receiver, *73 Magazine,* April 1977, p 136–149.

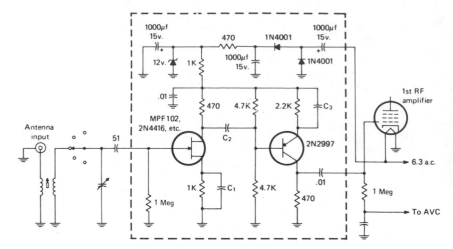

RF PREAMP—Boosts sensitivity of older tube-type communication receiver when added ahead of first RF tube. Has low noise figure. Values of C_1, C_2, and C_3 are varied to suit receiver being used. Using 0.01 μF for these gives 20-dB gain from 0.5 to 30 MHz; if this overloads receiver on lower frequency ranges, try smaller values.—I. Math, Math's Notes, *CQ*, April 1975, p 37—38 and 62.

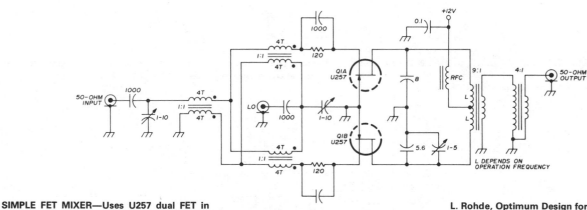

SIMPLE FET MIXER—Uses U257 dual FET in double-balanced circuit having 50-ohm input impedance, for high-quality communication receiver. Gives excellent third-order intermodulation distortion suppression (68 dB down).—U. L. Rohde, Optimum Design for High-Frequency Communications Receivers, *Ham Radio*, Oct. 1976, p 10—25.

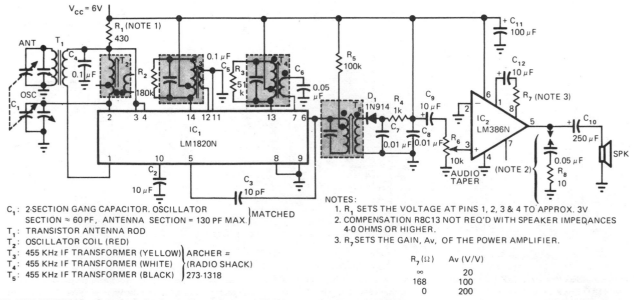

C_1: 2-SECTION GANG CAPACITOR. OSCILLATOR SECTION ≈ 60 PF. ANTENNA SECTION = 130 PF MAX.}MATCHED
T_1: TRANSISTOR ANTENNA ROD
T_2: OSCILLATOR COIL (RED)
T_3: 455 KHz IF TRANSFORMER (YELLOW)}ARCHER =
T_4: 455 KHz IF TRANSFORMER (WHITE)}(RADIO SHACK)
T_5: 455 KHz IF TRANSFORMER (BLACK)}273-1318

NOTES:
1. R_1 SETS THE VOLTAGE AT PINS 1, 2, 3 & 4 TO APPROX. 3V
2. COMPENSATION R8C13 NOT REQ'D WITH SPEAKER IMPEDANCES 4-0 OHMS OR HIGHER.
3. R_7 SETS THE GAIN, A_v, OF THE POWER AMPLIFIER.

$R_7 (\Omega)$	A_v (V/V)
∞	20
168	100
0	200

TWO-CHIP AM RADIO—Current drain of only 10 mA makes operation from 6-V battery feasible. National LM1820N IC serves for oscillator/mixer, two IF stages, and AGC, and LM386N AF chip provides power output of 0.25 W into 8-ohm loudspeaker. D_1 is diode detector.—E. S. Papanicolaou and H. H. Mortensen, Low Cost AM Radio Uses Only Two IC's, *EDN Magazine*, Jan. 20, 1976, p 82 and 84.

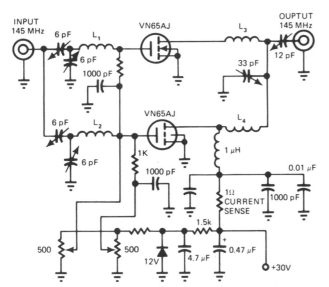

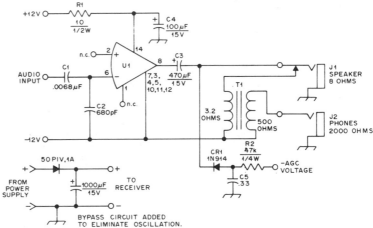

2-METER DUAL-VMOS—Provides 10-W PEP output at 146 MHz for amateur applications. Noise figure is only 2.35 dB, and two-tone IMD is −30 dBC.—E. Oxner, Will VMOS Power Transistors Replace Bipolars in HF Systems?, *EDN Magazine,* June 20, 1977, p 71–75.

AUDIO BOOST—LM380 power amplifier operating on 12 V is well suited for communication receiver having only limited audio gain. Circuit provides excellent headphone volume and enough loudspeaker output for small room. If signal at full secondary winding of product detector AF coupling transformer in receiver overloads U1, take signal from center tap of transformer winding that feeds volume control in amplifier.—H. L. Ley, Jr., More Audio for *QST* Course Receiver, *QST,* Oct. 1977, p 45.

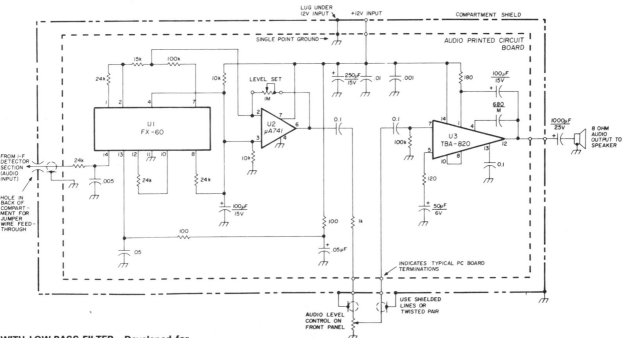

2-W WITH LOW-PASS FILTER—Developed for use in dual-conversion amateur receiver. Detected audio is passed through active low-pass filter-opamp arrangement U1-U2 and further amplified by 2-W audio amplifier U3. Simple voltage-divider circuit on pin 3 of U2 establishes artificial ground for U1 and U2. Low-pass rolloff starts at 2500 Hz, with about 20-dB attenuation of higher audio frequencies. IF heterodyne hiss is greatly attenuated and overall S/N ratio of receiver enhanced. Level-set 1-megohm pot between pins 2 and 7 of U2 establishes output gain for U1 and U2 together at about 0.8.—M. A. Chapman, High-Performance 20-Meter Receiver with Digital Frequency Readout, *Ham Radio,* Oct. 1977, p 48–61.

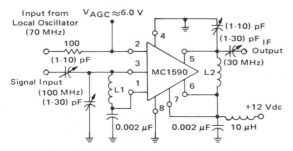

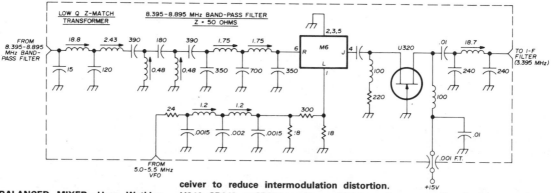

L1 = 5 Turns, #16 AWG, 1/4" ID,
5/8" Long
L2 = 16 Turns, #20 AWG Wire on a Toroid
Core, (T 44-6 Micro Metal or Equiv)

100-MHz MIXER—With local oscillator frequency of 70 MHz, opamp provides difference frequency of 30 MHz at high conversion gain. Isolation between oscillator and signal source is excellent.—B. Trout, "A High Gain Integrated Circuit RF-IF Amplifier with Wide Range AGC," Motorola, Phoenix, AZ, 1975, AN-513, p 9.

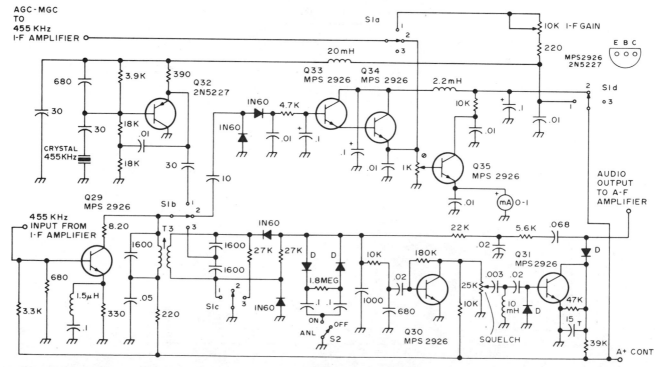

DOUBLE-BALANCED MIXER—Uses Watkins-Johnson M6 low-level hot-carrier-diode double-balanced mixer as replacement for FET second mixer in amateur-band dual-conversion receiver to reduce intermodulation distortion. U310, CP643, or CP651 can be used in place of U320 high-transconductance JFET. Pi-network output circuit couples 2000-ohm output of JFET stage to 2000-ohm IF filter. All inductance values are in microhenrys.—A. J. Burwasser, Reducing Intermodulation Distortion in High-Frequency Receivers, *Ham Radio*, March 1977, p 26–30.

LIMITER-DETECTOR—Used in all-band double-conversion superheterodyne receiver for AM, narrow-band FM, CW, and SSB operation. Q29 acts as limiter on FM but on AM is 455-kHz amplifier whose RF output is coupled to 1N60 AGC rectifier pair connected as voltage doubler that provides bias for AGC amplifier Q33-Q34. Output of Q29 is coupled to detector by T3. Although detector is actually phase discriminator, mode switch connects circuit as half-wave rectifier for AM and CW/SSB. On CW/SSB (S1 on 1), AGC rectifier is disconnected and AGC diodes receive bias from manual gain-control pot. BFO is then energized and connected to T3. Output of diode detector feeds squelch and audio stages. Supply is 13.6 V regulated. Article gives all circuits of receiver.—D. M. Eisenberg, Build This All-Band VHF Receiver, *73 Magazine*, Jan. 1975, p 105–112.

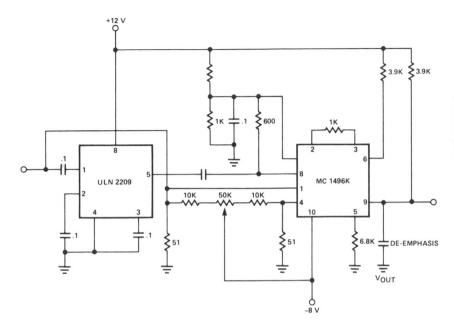

AM DEMODULATOR—Signetics ULN2209 IC provides 55-dB gain for input signal and symmetrical limiting above 400 μV. Limited carrier is then applied to MC1496K balanced modulator-demodulator transistor array for demodulation. Output filtering is required to remove high-frequency sum components of carrier from audio signal. Output amplitude is maximum when phase difference between signal and carrier inputs is 0°.—"Signetics Analog Data Manual," Signetics, Sunnyvale, CA, 1977, p 757–758.

432-MHz PREAMP—Developed as part of 400-MHz radiotelescope, for low-noise operation. Gain is 12 dB, and noise figure is below 2 dB. Circuit is basic common-emitter amplifier with tuned input and output. Neither neutralization nor shielding are needed. Supply can be 9-V transistor radio battery. Current drain is 3 mA. L1 and L2 are each 1 turn No. 16 wire ⅜ inch in diameter, with L2 center-tapped.—S. A. Maas, An Inexpensive Low Noise Preamplifier for 432 MHz, *QST*, Jan. 1975, p 21–22.

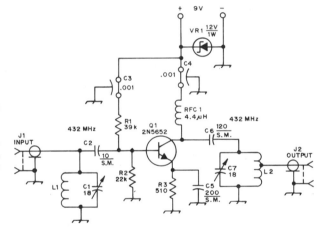

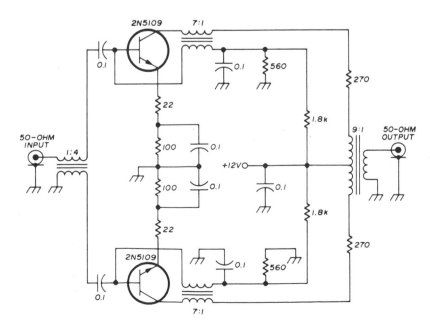

RF INPUT STAGE—Push-pull transistorized RF stage for communication receiver uses voltage and current feedback to minimize intermodulation distortion. Transformers serve to stabilize impedance. Circuit is basically constant-current device. Second-order intermodulation distortion products are suppressed almost 40 dB more than with single-transistor stage. To apply AGC, replace the two 270-ohm resistors with single PIN-diode shunt regulator.—U. L. Rohde, Optimum Design for High-Frequency Communications Receivers, *Ham Radio*, Oct. 1976, p 10–25.

PUSH-PULL RF—Uses VHF power transistors to obtain wide dynamic range. Transformers are trifilar wound on Indiana General F625-9-TC9 toroid cores. Circuit has extremely low VSWR at both input and output, along with low noise figure. Second-order intermodulation products can be suppressed nearly 40 dB over single stage. Either RCA 2N5109 or Amperex BFR95 transistors can be used. Gain is about 11 dB. Current feedback is used through unbypassed 6.8-ohm emitter resistor, voltage feedback through unbypassed 330-ohm base-to-collector resistor, and transformer feedback through third winding on wideband transformer to stabilize input and output impedances.—U. L. Rohde, High Dynamic Range Receiver Input Stages, *Ham Radio,* Oct. 1975, p 26–31.

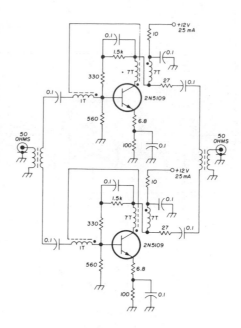

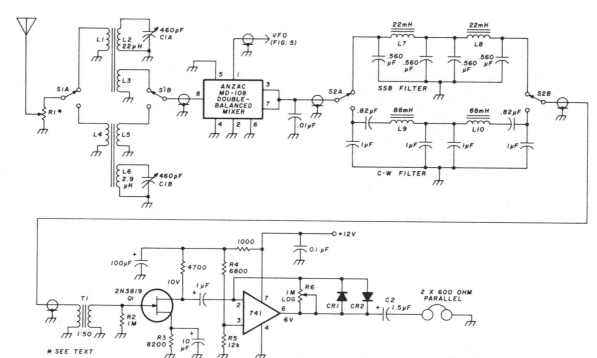

DIRECT CONVERSION—Simple direct-conversion amateur receiver uses VFO and mixer to produce AF signal directly, with no IF amplifier or second detector. For SSB reception, VFO is tuned to frequency of suppressed carrier. For CW, VFO is detuned enough to give note of desired pitch. Not suitable for AM or FM reception. Separate input tuned circuits are used for 15–40 meters and for 80–160 meters. Use ferrite or powdered iron toroid cores for coils, with turns determined experimentally. L7 and L8 are 88-mH toroids with series-connected windings. R1 is used to attenuate strong signals. Article gives circuit for VFO and buffer amplifier. Separate VFO is used for each band (160, 80, 40, 20, and 15 meters). Construction details are given, along with advantages and drawbacks of direct conversion.—D. Rollema, Direct-Conversion Receiver, *Ham Radio,* Nov. 1977, p 44–55.

CHAPTER 73
Regulated Power Supply Circuits

Various combinations of line-powered rectifiers and voltage regulators provide highly regulated fixed and variable positive and negative outputs ranging from 0 to ±35 V at maximum currents from 24 mA to 24 A. Dual-output supplies may have tracking. See also Power Supply, Regulator, and Switching Regulator chapters.

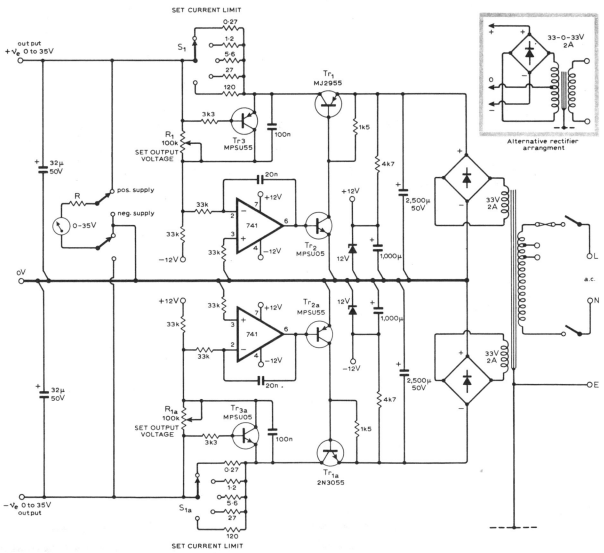

0 to ±35 V—Twin stabilized DC supply uses ganged pots R_1 and R_{1a} to set both positive and negative regulated outputs at any desired value up to 35 V. Input supplies from bridge rectifiers also provide ±12 V lines for 741 opamps. Load regulation is within 2 mV from no load to full 2-A maximum output. Output hum, noise, and ripple are together only 150 μV and independent of load.—J. L. Linsley Hood, Twin Voltage Stabilized Power Supply, *Wireless World,* Jan. 1975, p 43–45.

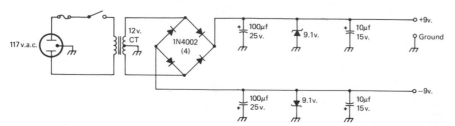

±9 V—Developed for use with demodulator of teleprinter. Regulation is provided by zeners.—I. Schwartz, An RTTY Primer, *CQ*, Feb. 1978, p 31–36.

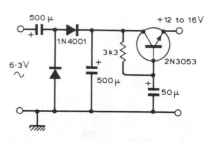

12–16 V FROM 6.3 VAC—Designed for use with transistor or IC amplifier being fed by tube-type preamp having 6.3-V power transformer winding for filament supply.—K. D. James, Balanced Output Amplifier, *Wireless World*, Dec. 1975, p 576.

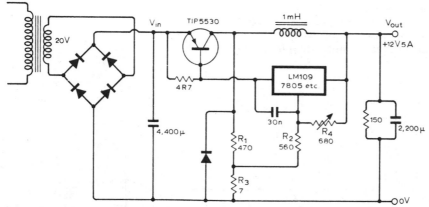

ADJUSTABLE SWITCHED REGULATOR—Circuit shows method of using LM109, 7805, or other IC voltage regulator to provide output voltage that is higher than rated output of IC. Voltage pedestal is developed across R_2 and R_3 for adding to normal regulated output of IC. R_4 adjusts amount of added voltage. Divider R_1-R_2 provides positive feedback into pedestal circuit of regulator, to allow switching of IC and transistor.—V. R. Krause, Adjustable Voltage-Switching Regulator, *Wireless World*, May 1976, p 80.

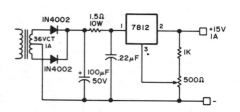

15 V AT 1 A—Developed for operating CRO from AC line. Can also be used for recharging batteries of portable CRO if pot is set to correct charging voltage for cells being used. Use good heatsink with 7812 regulator.—G. E. Friton, Eyes for Your Shack, *73 Magazine*, Jan. 1976, p 66–69.

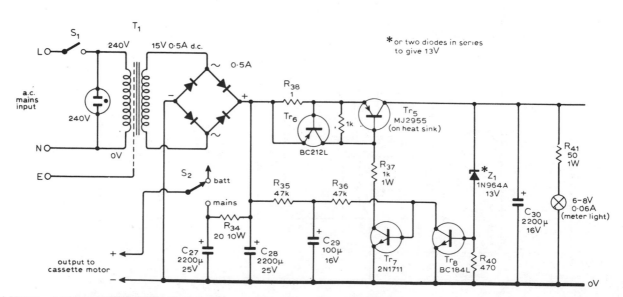

14 V AT 250 mA FOR CASSETTE DECK—Used in high-quality stereo cassette deck operating from AC line or battery. For U.S. applications, use 120-V power transformer. Power for cassette motor is taken directly from power-supply filter capacitors through 20-ohm 10-W resistor, with negative return line connecting directly to filter capacitors instead of chassis, to eliminate noise originating from pulsating current of cassette-drive motor-control circuit. Article gives all other circuits of cassette deck and describes operation in detail.—J. L. Linsley Hood, Low-Noise, Low-Cost Cassette Deck, *Wireless World*, Part 2—June 1976, p 62–66 (Part 1—May 1976, p 36–40; Part 3—Aug. 1976, p 55–56).

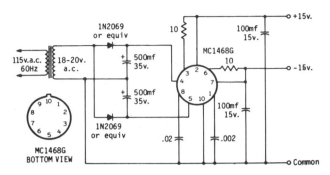

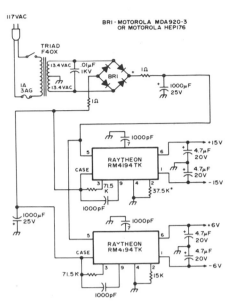

±15 V—Provides positive and negative supply voltages required by some opamps. Supply is short-circuit-proof and protects itself against overloads.—I. Math, Math's Notes, *CQ*, Jan. 1974, p 68–69.

±6 AND ±15 V—Developed for use with function generator. Mount regulators on heatsinks insulated from chassis by mica wafers. Article covers construction and adjustment to give exactly desired outputs.—H. Olson, Build This Amazing Function Generator, *73 Magazine*, Aug. 1975, p 121–124.

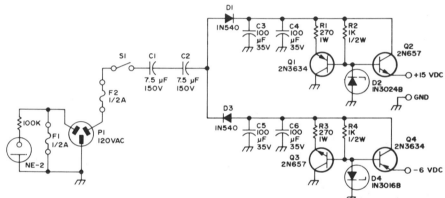

+15 V AND −6 V TRANSFORMERLESS—Transistorized regulator provides good voltage regulation with low ripple. Second ground prong is connected through fuse to grounded center conductor of AC line to guard against faulty AC wiring. If wiring is reversed, fuse will disable power supply and neon fault indicator will come on. At currents up to 55 mA, −6 V output had 0.1-V ripple and +15 V output had 0.05-V ripple.—D. Kochen, Transformerless Power Supplies, *73 Magazine*, Sept. 1971, p 14–17.

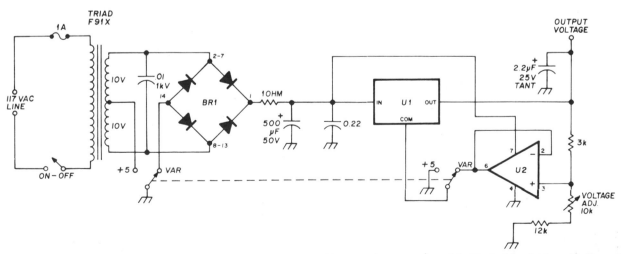

+5 V AT 200 mA OR 7–20 V AT 100 mA—Uses National LM741 opamp as noninverting follower to sample output of voltage divider and drive common terminal of National LM340-05 three-terminal voltage regulator. Heatsink tab of regulator U1 must be connected to floating heatsink. BR1 is Adva bridge.—H. Olson, Second-Generation IC Voltage Regulators, *Ham Radio*, March 1977, p 31–37.

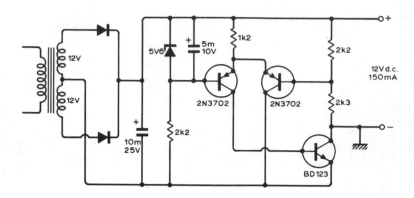

12-V LOW-RIPPLE—Three-transistor feedback circuit gives low-cost voltage stabilizer in which ripple is low and regulated output is very little less than unstabilized input voltage.—R. H. Pearson, Novel 5-Watt Class A Amplifier Uses Three-Transistor Feedback Circuit, *Wireless World,* March 1974, p 18.

+5 V WITH UNREGULATED +15 V—Developed for use with audio decoder that converts BCD output of digital display to audio tones that can be recognized by blind radio operator or experimenter.—D. R. Pacholok, Digital to Audio Decoder, *73 Magazine,* Oct. 1977, p 178–180.

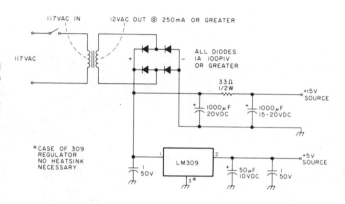

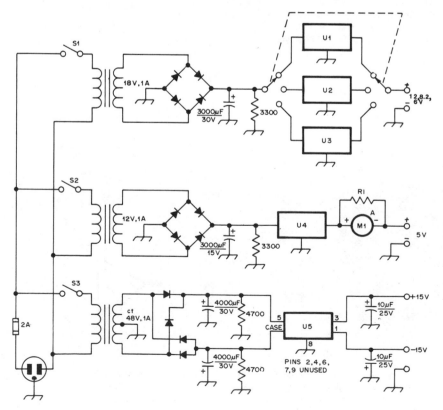

POPULAR-VOLTAGES SUPPLY—Provides most common fixed voltages required for transistor and IC projects. Eliminates cost and nuisance of replacing batteries. Provides ±15 V at 100 mA, +5 V at 1 A, and choice of +6.0, +8.2, and +12.0 V at 1 A. Separate grounds (not chassis grounds) permit connecting supplies in series to get combination voltages. Rectifier diodes are 100 PIV at 1 A. Use meter and shunt to give full scale at 1 A; for 150-mA meter, use 0.08 ohm for R1 (six 0.5-ohm resistors in parallel). U1 is LM340K-12, U2 is LM340K-8, U3 is LM340K-6, U4 is LM340K-5, and U5 is 4195.—C. J. Appel, A Combination Fixed-Voltage Supply, *QST,* Nov. 1977, p 36–37.

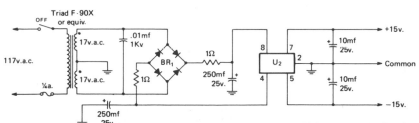

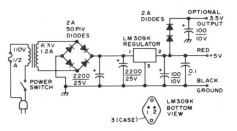

±15 V—Developed for use in two-tone AF generator for testing SSB equipment. BR₁ is Motorola HEP176 or MDA-920-2. U₂ is Raytheon 4195DN.—H. Olson, A One-Chip, Two Tone Generator, *CQ*, April 1974, p 48–49.

5 V AT 1 A—Can handle over 30 TTL ICs in frequency counter if LM309K regulator is mounted directly on aluminum heatsink. Case of regulator is grounded, so mica insulation is not needed. Provides excellent regulation with practically no output ripple and is short-circuit-proof. Circuit also shuts itself off if temperature gets too high.—P. A. Stark, A Simple 5 V Power Supply for Digital Experiments, *73 Magazine*, Oct. 1974, p 43–44.

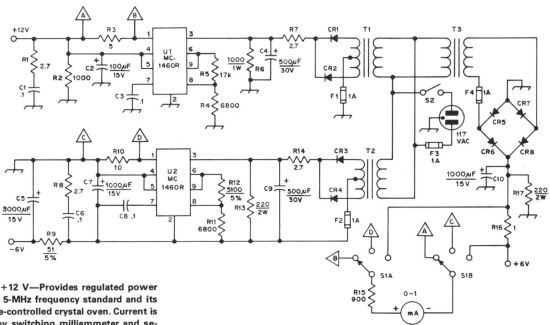

±6 V AND +12 V—Provides regulated power required by 5-MHz frequency standard and its temperature-controlled crystal oven. Current is measured by switching milliammeter and series resistor across current-limiting resistors R3 and R10, which also serve as meter shunts. DC input voltage at terminal 3 of MC1460R regulator should be at least 3 V greater than output voltage but should not exceed 20-V rating. Adjust R5 and R12 as required to get correct output voltages. Diodes are 200 PIV at 0.5 A. T1 and T2 have 117-V primary and 25.2-V secondary at 0.3

A. T3 has 117-V primary and 6.3-V secondary at 1 A.—R. Silberstein, An Experimental Frequency Standard Using ICs, *QST*, Sept. 1974, p 14–21 and 167.

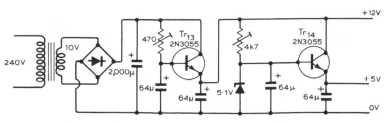

12 AND 5 V—Used with multiple photographic development timer to provide 12 V at about 5 mA and 5 V at about 300 mA for logic, control, and audible alarm circuits.—R. G. Wicker, Photographic Development Timer, *Wireless World*, April 1974, p 87–90.

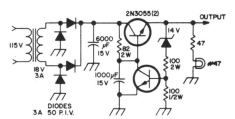

12 V FOR TRANSCEIVER—Permits operation of 144-MHz transceiver from AC line at base station. If 18-V transformer at 3 A is not available, use 12-V 3-A unit and add 26 turns No. 20 teflon-coated wire to secondary in proper phase. Choose transformer having enough room for these extra turns.—W. W. Pinner, Midland 2 Meter Base or Portable, *73 Magazine*, Aug. 1974, p 61–63.

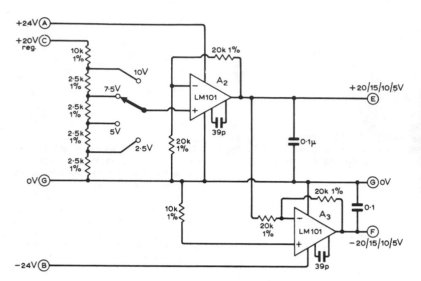

5/10/15/20 V SWITCH—Reference voltage selected by switch is applied to noninverting input of opamp having voltage gain of 2, to provide both positive and negative regulated voltages at desired value. Any standard opamp, such as µA709 or µA741, can be used in place of National LM101.—T. D. Towers, Elements of Linear Microcircuits, *Wireless World,* July 1971, p 342–346.

±4 V TO ±25 V—Arrangement permits varying both positive and negative regulated output voltages simultaneously with single control, with maximum load current of 400 mA for both regulators. Positive supply controls negative slave regulator to provide tracking within 0.05 V at full output. Developed for use in lab to observe effect of varying supply voltages on circuits under development.—J. A. Agnew, Dual Power Supply Delivers Tracking Voltages, *EDN Magazine,* Oct. 15, 1970, p 51.

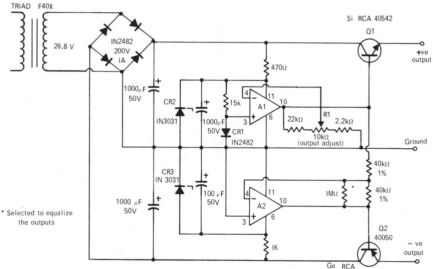

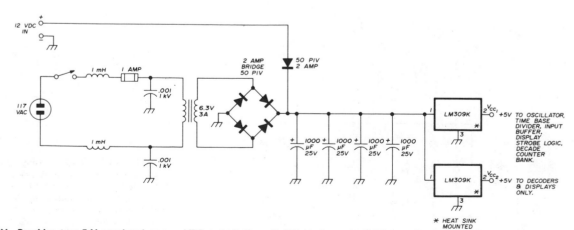

DUAL 5-V—Provides two 5-V regulated supplies for frequency counter, operating either from 9-VDC outputs of AC supply or from 12-VDC auto battery. Splitting of supply divides current demand so regulators operate well below maximum ratings and provide decoupling between sections of load.—J. Pollock, Six Digit 50-MHz Frequency Counter, *Ham Radio,* Jan. 1976, p 18–22.

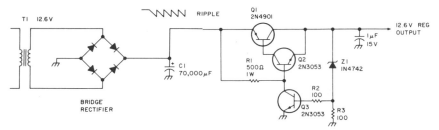

12.6 V AT 3 A—Article gives step-by-step procedure for designing simplest possible regulated supply to meet specific requirements in general service. Power transformer rated 12.6 V at 3 A delivers about 18 V P to bridge rectifier rated 50 V at 5 A. Value of C1 is chosen to keep voltage to regulator above 15-V limit at which circuit would drop out of regulation.—C. W. Andreasen, Practical P. S. Design, *73 Magazine*, June 1977, p 84–85.

+5 V AND ±6 OR ±12 V—Three power supplies for experimental use are achieved with only one transformer. LM309K regulates 5-V supply. Other two supplies are regulated by 6.2-V zeners in conventional regulator; shorting out one zener in each with gang switch reduces output to 6 V.—Design a Circuit Designer!, *73 Magazine*, Oct. 1977, p 152–153.

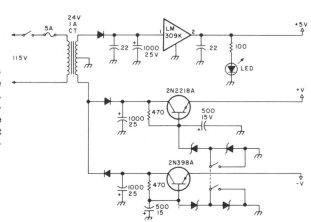

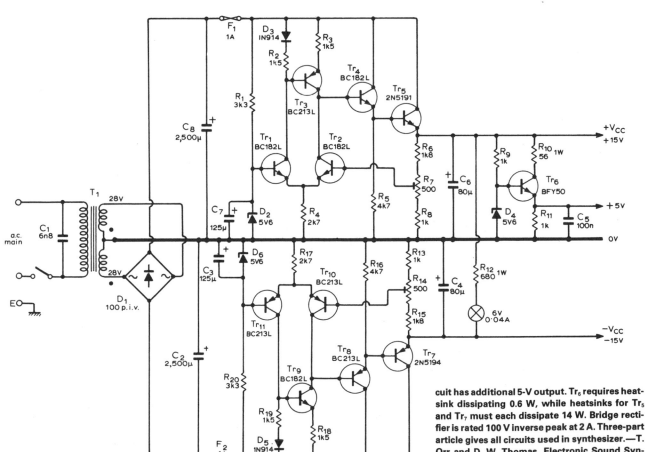

±15 V FOR SOUND SYNTHESIZER—Provides highly stabilized voltages required by elaborate sound synthesizer developed for generating wide variety of musical and other sounds. Circuit has additional 5-V output. Tr6 requires heatsink dissipating 0.6 W, while heatsinks for Tr5 and Tr7 must each dissipate 14 W. Bridge rectifier is rated 100 V inverse peak at 2 A. Three-part article gives all circuits used in synthesizer.—T. Orr and D. W. Thomas, Electronic Sound Synthesizer, *Wireless World*, Part 3—Oct. 1973, p 485–490 (Part 1—Aug. 1973, p 366–372; Part 2—Sept. 1973, p 429–434).

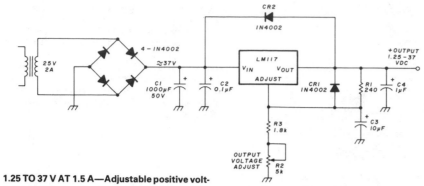

1.25 TO 37 V AT 1.5 A—Adjustable positive voltage regulator used with simple bridge rectifier and capacitor-input filter delivers wide range of regulated voltages, all with current and thermal overload protection. Load regulation is about 0.3%.—H. Berlin, A Simple Adjustable IC Power Supply, *Ham Radio,* Jan. 1978, p 95.

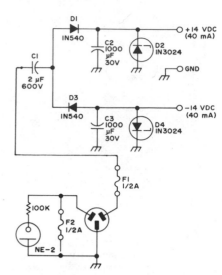

±14 V TRANSFORMERLESS—Simple low-current regulated supply requires no power transformer. Output current can be increased by using better filtering. Second ground prong is connected through fuse to grounded center conductor of AC line to guard against faulty AC wiring. If wiring is reversed, fuse will disable power supply and neon fault indicator will come on.—D. Kochen, Transformerless Power Supplies, *73 Magazine,* Sept. 1971, p 14—17.

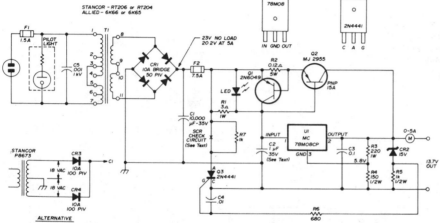

13.7 V AT 5 A—Output is constant within 0.7 V for AC line range of 98 to 128 VAC, and regulation is within tenths of a volt from 0 to 5 A. Design includes short-circuit, overcurrent, and overvoltage protection. Uses series-pass transistor to increase current-carrying capability of regulator. Transistors are mounted on but insulated from heatsink. C2 is essential to prevent oscillation under certain conditions. Use gallium arsenide phosphide LED. Article tells how to determine exact trip point of SCR crowbar.—B. Meyer, Low-Cost All-Mode-Protected Power Supply, *Ham Radio,* Oct. 1977, p 74—77.

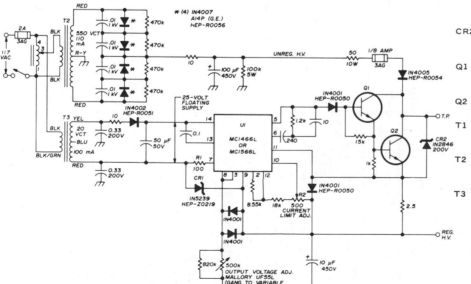

CR2	200-volt, 50-watt zener diode (heatsink to chassis)
Q1	Motorola HEP244 or MJE340 (heatsink to chassis)
Q2	Motorola HEP707 or MJ413 (heatsink to chassis)
T1	0-132 volt, 2.25 A (0.3 VA) variable auto-transformer (Superior Electric 10B)
T2	550 volts center-tapped, 110 mA (Triad R112A or R12A, filament windings not used)
T3	20 volts center-tapped, 100 mA (Triad F90X)

50—300 V VARIABLE AT 100 mA—Solid-state version of regulated high-voltage supply for tube circuit has adjustable current-limiting, instant turn-on, and long component life. Small variable autotransformer in primary circuit of high-voltage transformer is mechanically ganged to DC voltage-control pot connected to pin 8 of U1 to keep input-to-output voltage difference nearly constant. Differential voltage across Q1 never exceeds 100 V so power dissipation of Q1 is only 5 W maximum. Regulator circuit is designed around Motorola MC1466L or MC1566L floating regulator powered by 25-V supply having no common connection to ground. Use 600-V rating for 0.33 μF from T3 to ground.—H. Olson, Regulated, Variable Solid-State High-Voltage Power Supply, *Ham Radio,* Jan. 1975, p 40—44.

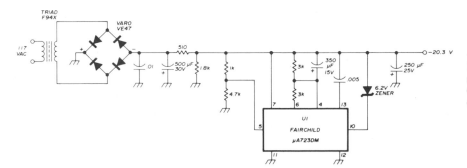

−20 V FOR VARACTORS—Precision low-ripple bias supply for varactor tuning applications can provide up to 20-mA output current.—M. A. Chapman, Multiple Band Master Frequency Oscillator, *Ham Radio*, Nov. 1975, p 50–55.

±15 V AT 10 A—Uses two Motorola positive voltage regulators, each having separate 18–24 VRMS secondary winding on power transformer T₁. Current-limiting resistor R_{SC} is in range of 0.66 to 0.066 ohm. Use copper wire about 50% longer than calculated length and shorten step by step until required pass current is obtained; thus, start with 25 ft of No. 16, 15 ft of No. 18, 10 ft of No. 20, or 6 ft of No. 22.—G. L. Tater, The MPC1000—Super Regulator, *Ham Radio*, Sept. 1976, p 52–54.

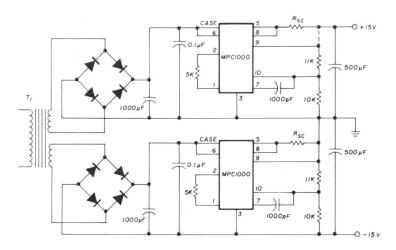

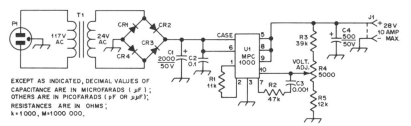

EXCEPT AS INDICATED, DECIMAL VALUES OF CAPACITANCE ARE IN MICROFARADS (μF); OTHERS ARE IN PICOFARADS (pF OR μμF); RESISTANCES ARE IN OHMS; k = 1000, M = 1000 000.

28 V AT 10 A—Developed for 60-W UHF linear amplifier. C1 and C4 are computer-grade electrolytics. CR1-CR4 are Motorola 1N3209 100-PIV 10-A silicon diodes. U1 is Motorola MPC100 or equivalent voltage regulator mounted on heatsink. T1 is Stancor P-8619 or equivalent 24-V 8-A transformer.—J. Buscemi, A 60-Watt Solid-State UHF Linear Amplifier, *QST*, July 1977, p 42–45.

12 V AT 10 A FOR HOUSE—Power supply is more than adequate for handling 12-V FM transceiver and even small amplifier. Series combination of three 6.3-V 10-A filament transformers drives 12-A 50-PIV bridge rectifier supplying 18 VDC to National LM305 regulator and pass transistors. Output voltage is at least 4 V less. Circuit provides foldback-current limiting for protection against load shorts. 500-ohm pot varies output from 11.2 to 14.1 V. Q1 is 2N2905, Q2 is 2N3445, and Q3 is 2N3772. T1-T3 are 6.3 V at 10 A (Essex Stancor P-6464 or equivalent).—C. Carroll, That's a Big 12 Volts, *QST*, Aug. 1976, p 26–27.

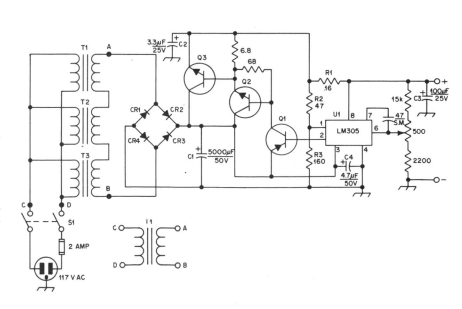

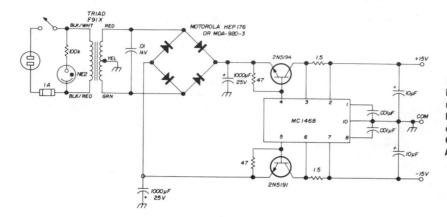

±15 V TRACKING—Uses Motorola dual-polarity regulator to provide balanced positive and negative voltages, with series-pass transistor handling major part of output current. Developed for use with audio signal generator.—H. Olson, Integrated-Circuit Audio Oscillator, *Ham Radio,* Feb. 1973, p 50–54.

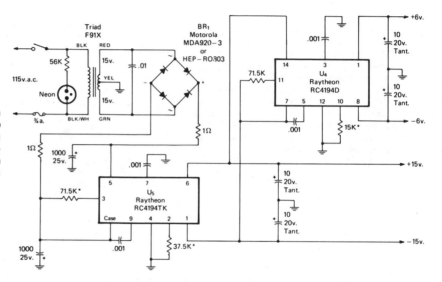

±6 V AND ±15 V—Developed for use with wide-range function generator requiring these voltages for transistors and ICs. Voltage-setting 15K and 37.5K resistors are adjusted to give desired output voltages.—H. Olson, The Function Generator, *CQ,* July 1975, p 26–28 and 71–72.

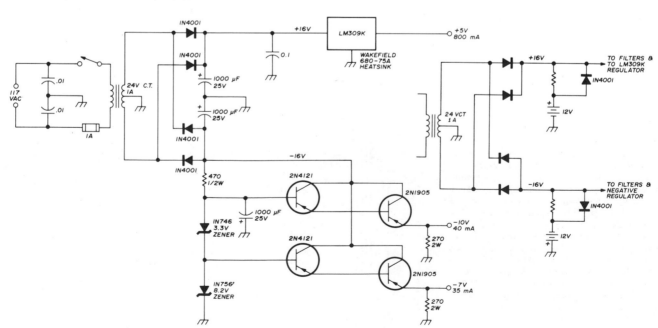

+5, −7, AND −10 V—Developed to meet power requirements of RTTY message generator having TTL and Numitrons requiring +5 V and MOS RAM requiring negative voltages. Diagram shows how to add 12-V storage batteries to prevent loss of programming if AC power fails momentarily.—B. Kelley, Random Access Memory RTTY Message Generator, *Ham Radio,* Jan. 1975, p 8–15.

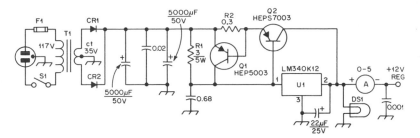

12 V AT 5 A—Uses National LM340K-12 mounted on external heatsink, with series-pass transistor Q2 boosting current rating to 5 A. Provides complete protection from load shorts; output drops suddenly to nearly zero when current exceeds 5 A. R2 is several feet of No. 22 enamel wound on phenolic form to make 0.3-ohm 60-W resistor. CR1 and CR2 are HEP R0103 or equivalent. Transformer is rated 18 V at 8 A.— C. R. Watts, A Crowbar-Proof 12-V Power Supply, *QST,* Aug. 1977, p 36–37.

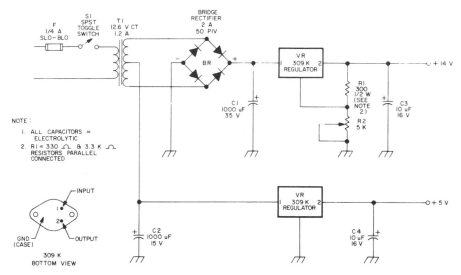

5 AND 14 V AT 1 A—Regulated dual-voltage power supply serves for experimenting with TTL, CMOS, and linear IC projects. Higher-voltage regulator must be insulated from heatsink.—A. Lorona, Dual Voltage Power Supply, *73 Magazine,* Holiday issue 1976, p 146–147.

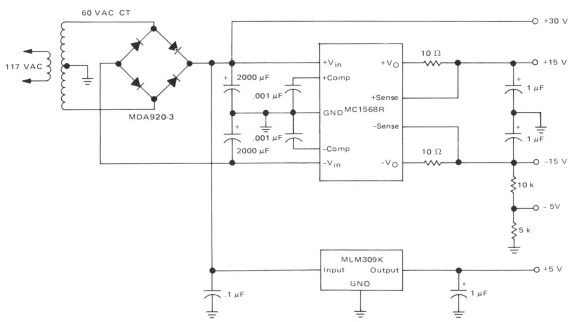

±5, ±15, AND +30 V—Provides all voltages needed for digitally controlled power supply that has voltage range from 0 to 25.5 V in 0.1-V increments. Highest positive voltage of +30 V is well above maximum output voltage that can be programmed.—D. Aldridge and N. Wellenstein, "Designing Digitally-Controlled Power Supplies," Motorola, Phoenix, AZ, 1975, AN-703, p 4.

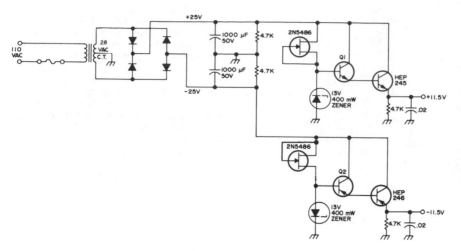

±11.5 V—Developed for experimentation with IC audio amplifiers, delivering up to 5 W, where good regulation is required to prevent oscillation caused by feedback through power supply to input stage. Q1 and Q2 are inexpensive silicon transistors, serving also as low-cost fuses because they burn out first when power supply is overloaded. Use heatsinks with silicone grease for output transistors.—D. J. Kenney, Integrated Circuit Audio Amplifiers, *73 Magazine*, Feb. 1974, p 25–30.

±0 TO 15 V AT 200 mA AND 3.8 TO 5 V AT 2 A— Developed for use with slow-scan television. Design provides equal positive and negative voltages that track each other with one manual control, adjustable from 0 to 15 V for opamps. Current limiting is provided for both positive and negative outputs. Low digital voltage can be adjusted to 3.8 V for RTL or 5 V for TTL, without current limiting. Use transformer having 35-V secondary, center-tapped, rated at 3 A.—D. Miller and R. Taggart, Popular SSTV Circuits, *73 Magazine,* March 1973, p 55–60, 62, and 64–67.

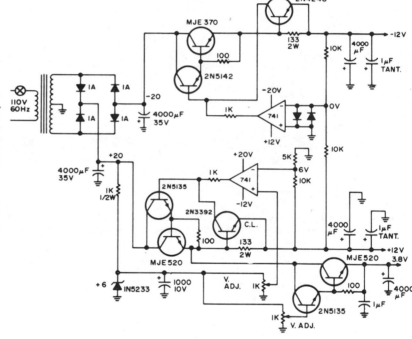

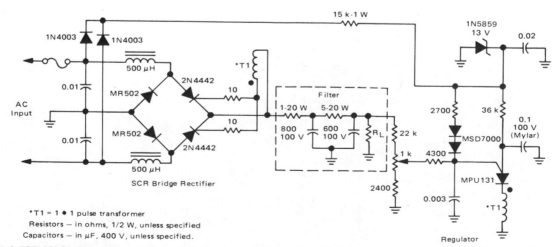

*T1 = 1 • 1 pulse transformer
Resistors — in ohms, 1/2 W, unless specified
Capacitors — in μF, 400 V, unless specified.

80 V AT 1.5 A FOR COLOR TV—Holds output voltage across R_L within 2% over line-voltage range of 105 to 140 V. Designed for use in 19-inch color TV receiver having 700-V flyback horizontal system. Bridge rectifier has two 2N4442 SCRs that control amount of output voltage by using variable duty cycle. Regulator uses MPU131 programmable UJT, which also serves for gating SCRs. 1K pot provides control of PUT gate voltage, which in turn determines output voltage across R_L.—R. J. Valentine, "A Low-Cost 80 V–1.5 A Color TV Power Supply," Motorola, Phoenix, AZ, 1974, AN-725, p 2.

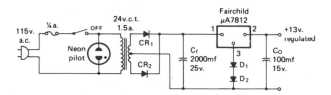

+13 V AT 1 A—Regulated output voltage can be varied upward about 0.6 V per diode by placing silicon diodes between pin 3 and ground. Two diodes boost output of regulator from 11.9 V to about 13 V. Insulate regulator from heatsink with mica washers. CR_1 and CR_2 are 50-PIV 3-A diodes. Motorola equivalent of regulator is MC7812.—A. M. Clarke, Simple, Superregulated, 12 Volt Supply, *CQ*, April 1974, p 61–62.

12 V AT 1 A—Simple supply furnishes up to 1 A with excellent regulation. Bottom of chassis can be used for heatsink. Connect additional 0.1-μF capacitor between pin 1 of VR_1 and ground. T_1 is Radio Shack 273-1505 with 12.6-V CT 1.2-A secondary, and F_1 is 0.5-A fuse.—A. Pike, Radio Shack Power Supply, *CQ*, Sept. 1977, p 66.

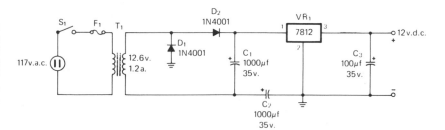

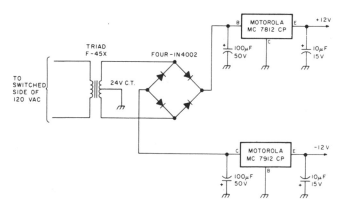

±12 V—Simple circuit provides power required for two 741 opamps used in CRT tuning indicator circuit for RTTY receiver.—R. R. Parry, RTTY CRT Tuning Indicator, *73 Magazine*, Sept. 1977, p 118–120.

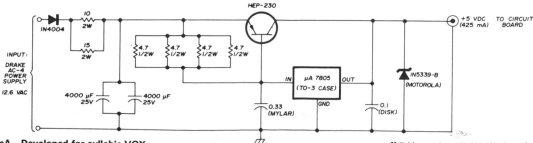

+5 V AT 425 mA—Developed for syllabic VOX system used with Drake T-4XB and R-4B transmitter and receiver. Input is taken from 12.6-VAC transformer winding of power supply for Drake, so equipment power switch also turns off 5-V supply.—R. W. Hitchcock, Syllabic VOX System for Drake Equipment, *Ham Radio,* Aug. 1976, p 24–29.

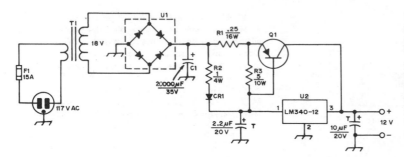

12 V AT 10 A—Permits AC operation of 12-V FM transceiver. Article tells how to rewind 12-V TV power transformer rated above 120 W with No. 12 enamel to get required 18-V secondary. If original winding has 2 turns per volt, new secondary will need 36 turns. Q1 is HEP233, HEP237, or similar transistor rated 10 A or higher, with heatsink. U1 is 25-A 100-PIV bridge rectifier, and U2 is National LM340K-12 regulator. CR1 can be any rectifier rated at least 3 A at 35 V.—L. McCoy, The Ugly Duckling, *QST*, Nov. 1976, p 29–31.

5 V AT 1 A—Simple lab supply provides voltage required for digital ICs. Rectifier is 6-A 50-PIV bridge. Power transformer has 12.6-V secondary rated 1 A, such as filament transformer.—G. McClellan, Give That Professional Look to Your Home Brew Equipment, *73 Magazine*, Feb. 1977, p 28–31.

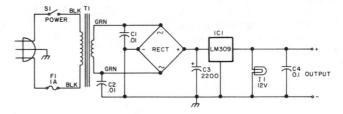

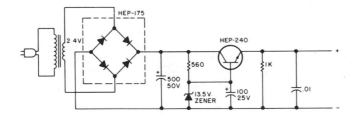

12 V FOR TRANSCEIVER—Output voltage varies only 0.2 V between transmit and receive. Transistor can be mounted directly on side of metal minibox for heatsinking. Transformer secondary is 24 V at 5 A.—Circuits, *73 Magazine*, March 1977, p 152.

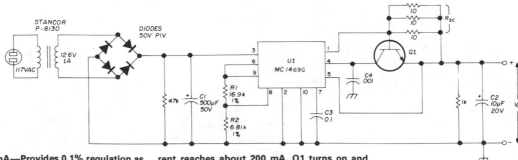

+12 V AT 50 mA—Provides 0.1% regulation as required for PLL RTTY tuning unit and other critical applications. R_{SC} and Q1 provide short-circuit protection for regulator. When output current reaches about 200 mA, Q1 turns on and limits regulator output. U1 can be Motorola MC1469G or HEP C6049G, and Q1 is any general-purpose NPN silicon transistor.—E. Lawrence, Precision Voltage Supply for Phase-Locked Terminal Unit, *Ham Radio*, July 1974, p 60–61.

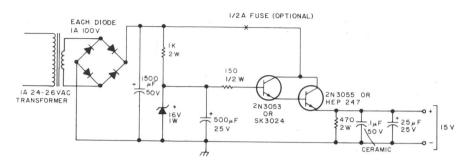

15 V AT 600 mA—Developed for 2-meter FM transceiver used as repeater. Output voltage is well filtered. Regulator allows voltage to drop only 0.1 V when repeater goes from standby to transmit. Use heatsink on 2N3055 series-regulator transistor.—H. Cone, The Minirepeater, *73 Magazine*, June 1975, p 55–57, 60–62, and 64–65.

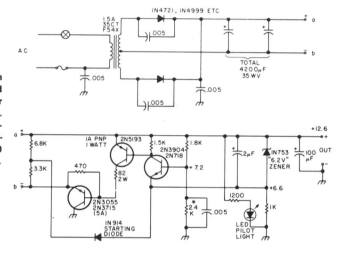

12.6 V AT 3 A—Will handle typical 15-W 2-m transceiver. Short-circuit protection is provided by 82-ohm resistor. Adjust value of resistor marked 2.4K to give desired output voltage. Transformer secondary is nominally 35 V center-tapped at 1.5 A. Output capacitor can be tantalum-slug electrolytic with any value above 10 μF.—H. H. Cross, The Chintzy 12, *73 Magazine*, Feb. 1977, p 40–41.

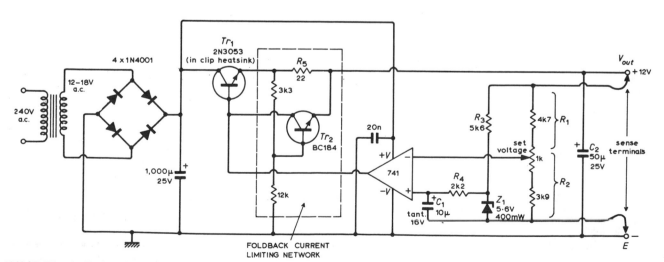

12 V AT 150 mA—Designed for use with audio preamps, FM tuners, and stereo decoders for which minimum ripple, minimum noise, good regulation, and good temperature stability are important. Uses 5.6-V reference zener that is fed from output but is inside feedback loop. Unregulated input can be up to 36 V.—M. L. Oldfield, Regulated Power Supplies, *Wireless World*, Nov. 1972, p 520–521.

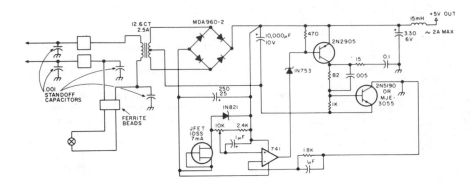

5 V AT 2 A—Developed as supply for receiver frequency counter having LED display. FET having I_{DSS} of about 7.5 mA serves in place of 1200-ohm resistor as current regulator. Power transformer is Triad F-26X with secondary rated 12.6 V center-tapped at 2.5 A.—H. H. Cross, The Chintzy 12, *73 Magazine,* Feb. 1977, p 40–41.

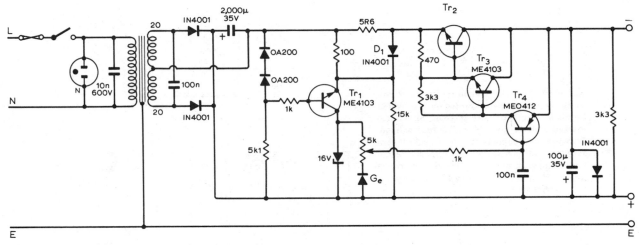

0–15 V BENCH SUPPLY—Provides up to 175 mA with ripple less than 1 mV. Choose Tr_2 to handle load current. Current limiting is provided by 5.6-ohm (5R6) resistor and D_1; when resistor drop exceeds about 1.2 V, current source Tr_1 produces less current and output voltage is reduced.—J. A. Roberts, Bench Power Supply, *Wireless World,* May 1973, p 253.

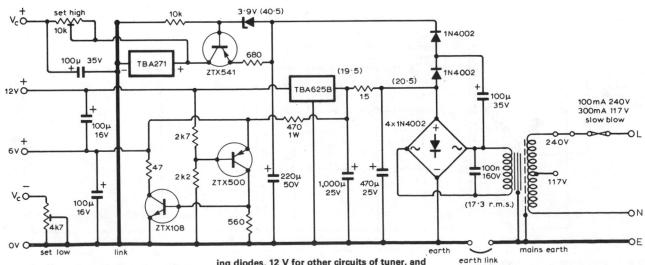

6, 12, AND 30 V FOR FM TUNER—Provides regulated 30 V for voltage-controlled varicap tuning diodes, 12 V for other circuits of tuner, and optional 6 V for stereo decoder. Uses SGS IC regulators.—L. Nelson-Jones, F.M. Tuner Design—Two Years Later, *Wireless World,* June 1973, p 271–275.

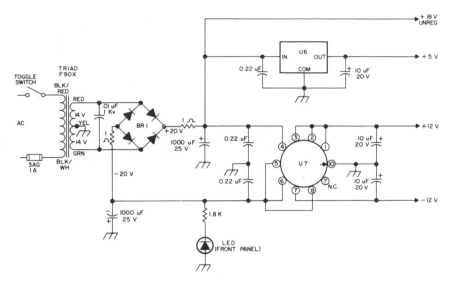

5 V AND ±12 V—Also provides 18 V unregulated for use with code regenerator driving automatic Morse-code printer. BR1 is Motorola MDA920-3 or HEP-R0802 bridge. LED is HP5082-4882 or HEP-P2000. U6 is LM341-5, MC7805, or HEP-C6110P. U7 is LM326H with TO5 finned clip-on heatsink.—H. Olson, CW Regenerator/Processor, *73 Magazine,* July 1976, p 80–82.

13 V AT 2 A WITH NPN TRANSISTORS—Q1 is reference voltage source and Q2 is series-pass regulator for basic supply suitable for running mobile FM transceiver or other 12-V portable equipment in home. Transformer secondary is 16–19 V, or can be 6-V and 12-V filament transformers in series. R1 protects rectifier diodes from surge current generated when supply is turned on. Article tells how to adapt circuit for other output voltages.—R. B. Joerger, Power Supply, *73 Magazine,* Holiday issue 1976, p 40–41.

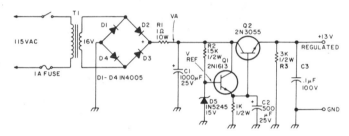

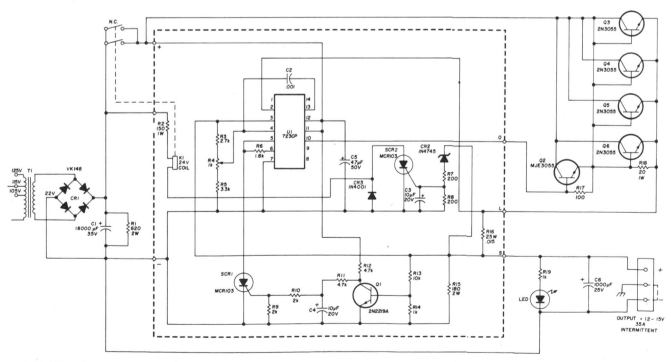

12–15 V AT 500 W—Developed to permit operation of high-power mobile solid-state amateur transmitter in home. Current sensing is done with 15-milliohm resistor R16. Short-circuit cutoff is provided by regulator along with current limiting through R16. Output voltage begins dropping as load exceeds 35 A. When voltage drops below 8 V, Q1 turns off and SCR1 turns on, cutting output power. Power supply must be turned off to unlatch SCR1. For overvoltage shutdown, CR2 starts conducting above 16 V, turning on SCR2 and activating relay K1 to cut off main DC supply. Article gives construction details. T1 has 22-V secondary.—C. C. Lo, 500-Watt Regulated Power Supply, *Ham Radio,* Dec. 1977, p 30–32.

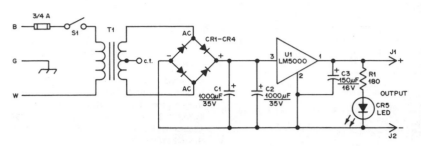

5 V AT 3 A—National LM5000 voltage regulator having built-in overload protection is basis of small bench supply for TTL work. Filament transformer rated 12.6 V at 3 A feeds full-wave bridge rectifier rated 200 PIV at 6 A, such as Radio Shack 276-1172. U1 requires heatsink insulated from chassis. Output filter C3 should be mounted directly on regulator terminals to minimize circuit oscillation. Output should read within 100 mV of 5 V. Radio Shack 276-047 LED serves as output indicator. Use 0.22-μF bypass between pins 2 and 3 of U1.—K. Powell, The 5 × 3 Power Supply, *QST*, May 1977, p 25–26.

13 V AT 2 A WITH PNP TRANSISTORS—Reference voltage source Q1 is 2N301, while series-pass regulator Q2 is 2N1523. D5 is 1N5245 15-V zener. Secondary of T1 is 16–19 V, or can be 6-V and 12-V filament transformers in series. Article tells how to adapt circuit for other output voltages.—R. B. Joerger, Power Supply, *73 Magazine*, Holiday issue 1976, p 40–41.

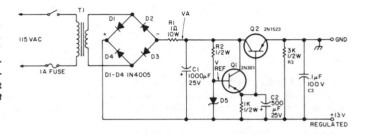

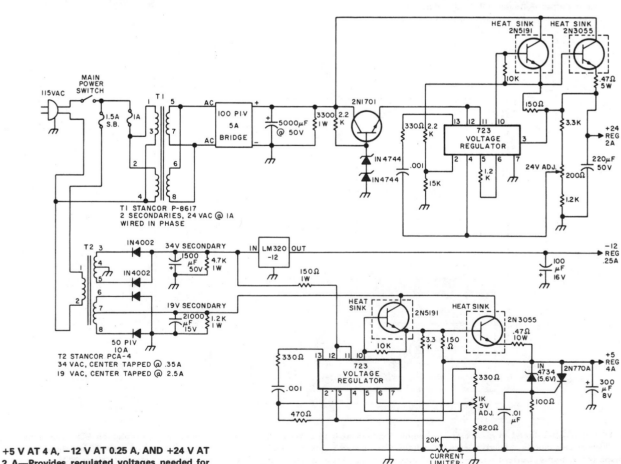

+5 V AT 4 A, −12 V AT 0.25 A, AND +24 V AT 2 A—Provides regulated voltages needed for Sykes 7158 floppy disk and its interface controller, used in Southwest Technical Products MP-68 computer system. Circuit provides adjustable current limiting and overvoltage protection on 5-V supply. Output voltage adjustments are provided for 5-V and 24-V supplies.— P. Hughes, Interfacing the Sykes OEM Floppy Disk Kit to a Personal Computer, *BYTE*, March 1978, p 178–185.

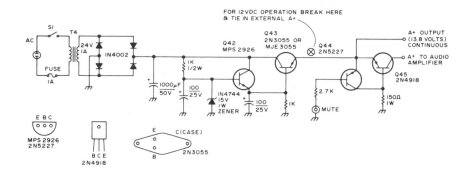

13.6 V AT 1 A—Used in all-band double-conversion superheterodyne receiver for AM, narrow-band FM, CW, and SSB operation. Simple transformer-rectifier-filter circuit is followed by zener-referenced Darlington pair. When transmitter of amateur station is on air, muting is accomplished by grounding base of Q44 through 2.7K resistor, which turns off Q45 and kills A+ to audio amplifier.—D. M. Eisenberg, Build This All-Band VHF Receiver, *73 Magazine*, Jan. 1975, p 105–112.

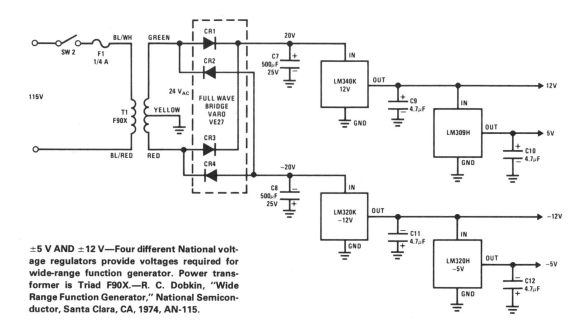

±5 V AND ±12 V—Four different National voltage regulators provide voltages required for wide-range function generator. Power transformer is Triad F90X.—R. C. Dobkin, "Wide Range Function Generator," National Semiconductor, Santa Clara, CA, 1974, AN-115.

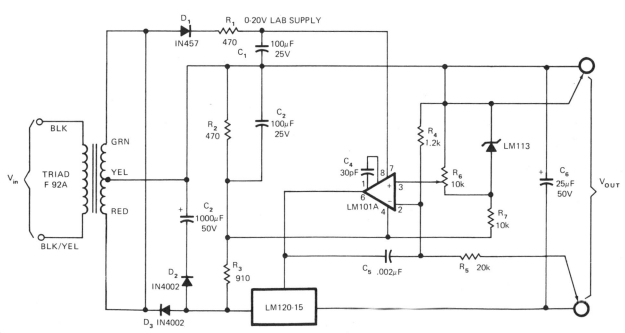

0–20 V AT 1 A—Variable-output regulated supply for lab use maintains output voltage within 2 mV of desired value for outputs up to 1 A. Arrangement uses National LM120 negative regulator as pass element, LM101A opamp as error amplifier, and LM113 zener as reference. Circuit provides complete protection against load shorts. LM120 requires adequate heatsink for continuous operation.—C. T. Nelson, Power Distribution and Regulation Can Be Simple, Cheap and Rugged, *EDN Magazine*, Feb. 20, 1973, p 52–58.

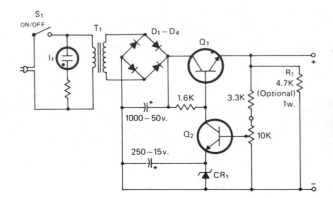

6–30 V AT 500 mA—Zener used for CR_1 should be rated 1 V less than desired minimum voltage, at 300 mW. R_1 improves regulation at low current levels. Current-limiting value is about 1 A. Diodes are 50-PIV 1-A silicon. I_1 is 117-V neon lamp. Q_1 is any 15-W NPN power transistor. Q_2 is 2N697 or equivalent. T_1 is power transformer with 24-V secondary at 0.5 A.—J. Huffman, The Li'l Zapper—a Versatile Low Voltage Supply, *CQ,* Nov. 1977, p 44.

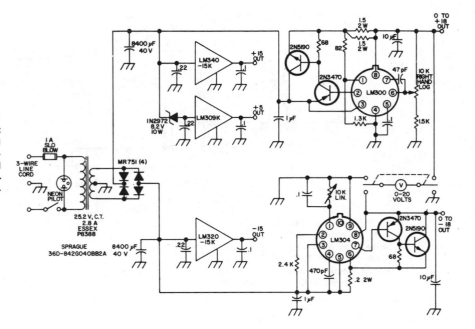

UNIVERSAL SUPPLY—Provides three different fixed voltages and two variable, each regulated and each current-limited at 1.5 A for use on experimenter's bench. Use heatsinks for fixed voltage regulators and for output transistors.— N. Calvin, Universal Power Supply, *73 Magazine,* Aug. 1974, p 65–66.

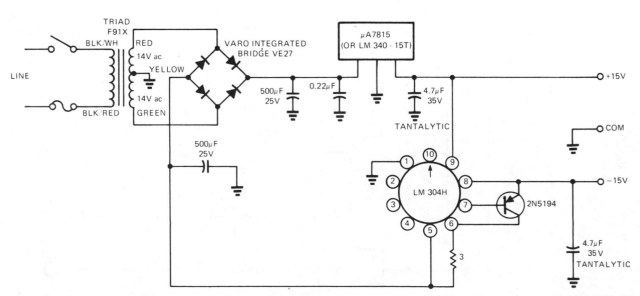

100-mA TRACKING—Circuit uses +15 V from μA7815 positive fixed-output regulator as external reference for LM304 negative regulator operating with outboard current-carrying PNP transistor. Arrangement requires only one center-tapped transformer winding yet gives required tracking of voltages. Output can be boosted to 200 mA by using larger bridge rectifier section.—H. Olson, Simple ±15V Regulated Supply Provides Tracking, *EDN Magazine,* March 20, 1973, p 87.

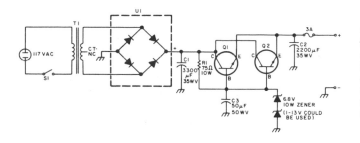

12 V AT 2.8 A—Simple supply was developed for use with 2-meter FM transceiver when operating in home. Power transistors are Radio Shack 276-592 rated 40 W. T1 is 12.6 V at 3 A, and U1 is 276-1171 rated 100 V at 6 A. Article covers construction.—M. L. Lovell, 12 Inexpensive Volts for Your Base Station, *73 Magazine,* Sept. 1976, p 60–62.

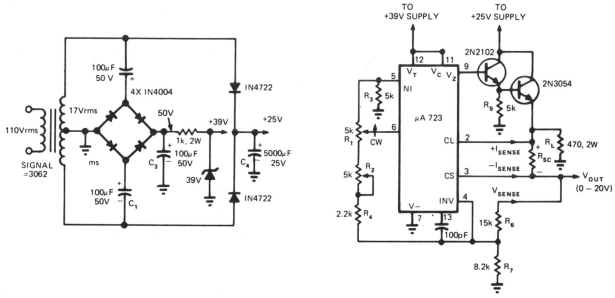

0–20 V CURRENT-LIMITING—Novel full-wave voltage doubler formed by diode bridge and C₁-C₂-C₃ provides 39 V required by μA723 regulator whose output is continuously variable with R₁.

Initially, R₂ is adjusted for minimum output voltage when R₁ is maximum counterclockwise, to balance bridge R₁-R₂-R₃-R₄ when output voltage is zero. Value used for R_{SC} determines short-cir-

cuit current. Raw DC supply provides separate 25 V for pass transistors.—L. Drake, Variable Voltage Power Supply Uses Minimum Components, *EDN Magazine,* Aug. 5, 1974, p 80 and 82.

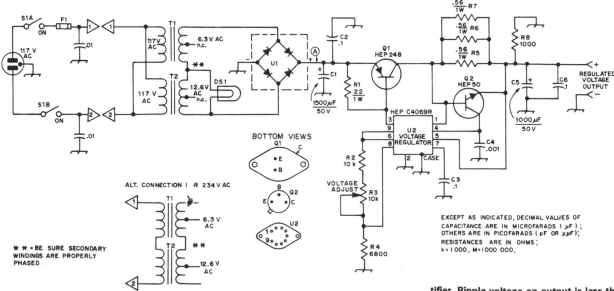

12 V AT 2 A—Will operate 10-W 220-MHz portable FM transceiver from AC line. Output volt-

age is adjustable from 9 to 13 V. DC voltage at point A is about 30 V. U1 is 50-V 10-A bridge rec-

tifier. Ripple voltage on output is less than 30 mV P-P.—E. Kalin, A No-Junkbox Regulated Power Supply, *QST,* Jan. 1975, p 30–33.

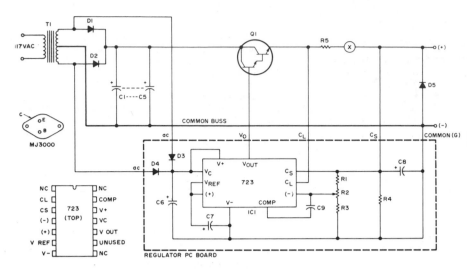

12 V AT 5 A—Uses MJ3000 Darlington power device as pass element providing gain of 1000 at 5 A. Output is set to current-limit at 6.5 A. Fuse at X is desirable. Values: R1 is 1.8K; R2 is 2.5K trimpot; R3 is 2.7K; R4 is 1.5K; R5 is 0.1 ohm at 5 W; C1-C5 are 4000 μF each at 20 V; C6 is 250 μF at 25 V; C7-C8 are 1.2 μF at 35 V; C9 is 220 pF; D1-D2 are MR1120 or equivalent rated 6 A; D3-D4 are 1N4607 or equivalent; D5 is 1N4002 or equivalent; and T1 is 24–28 V CT secondary at 4 A. Article gives design procedure for increasing regulated output to as much as 100 A.—C. Anderton, A Hefty 12 Volt Supply, *73 Magazine*, May 1975, p 85–87.

5 V FROM AC OR DC—Developed for use with secondary frequency standard to permit checking frequency of amateur radio transmitter at station or in field. Any battery capable of delivering 250 mA at 9–15 V is suitable.—T. Shankland, Build a Super Standard, *73 Magazine*, Oct. 1976, p 66–69.

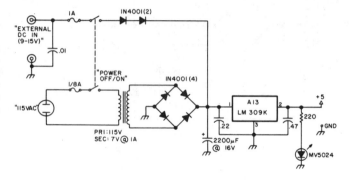

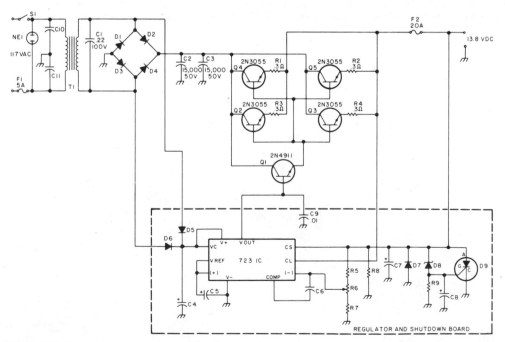

+13.8 VDC AT 18 A—Developed for use with amateur radio transceiver. Transformer secondary is rated 25 VAC at 12 A. When output voltage exceeds 15 VDC, zener D8 (1N965A or equivalent) conducts and fires 2N4441 SCR to crowbar supply and protect transceiver. Parts values are: R5 1.8K, R6 2.5K, R7 2.7K, R8 1.5K, R9 1K, C4 250 μF, C5-C6 1.2 μF, C7 220 pF, C8 100 μF, C9-C11 0.01 μF, D1-D4 1N3492 or equivalent with 100 PIV at 18 A, D5-D6 1N4607 or equivalent, and D7 1N4002 or equivalent.—T. Lawrence, Build a Brute Power Supply, *73 Magazine*, Aug. 1977, p 78–79.

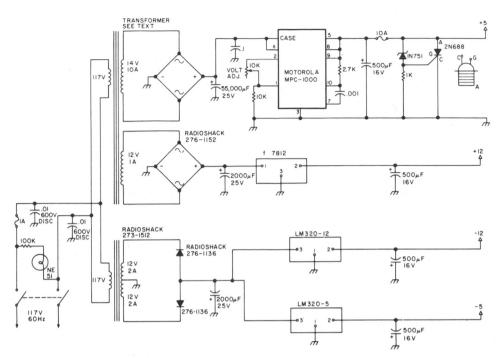

±5 AND ±12 V FOR COMPUTER—Provides all voltages required for 8080-4BD microcomputer system marketed by The Digital Group (Denver, CO). Transformer for positive supplies is 6.3-V 20-A unit with secondary replaced by two new windings giving required voltage and current. Crowbar circuit using 2N688 SCR protects ICs in memory and CPU. Use of at least 50,000 μF in filter of 5-V supply prevents noise problems in computer. MPC-1000 5-V 10-A regulator should be mounted on large heatsink at rear of computer housing in open air.—L. I. Hutton, A Ham's Computer, *73 Magazine*, Dec. 1976, p 78–79 and 82–83.

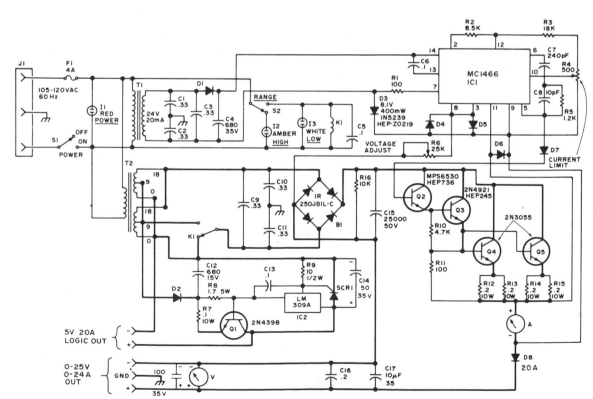

5 V AT 20 A AND 0–25 V AT 0–24 A—Developed as lab supply for experimenting with high-current TTL circuits. Motorola MC1466 monitors voltage and current requirements continuously, providing output proportional to parameters called for by front-panel controls of supply. D2 and D8 are 50-PIV 20-A diodes, and all other diodes except D3 are 1N4002 or equivalent. Article gives construction details.—J. W. Crawford, The Smart Power Supply, *73 Magazine*, March 1976, p 96–98 and 100–101.

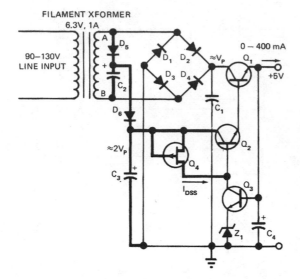

5 V WITH DOUBLER—Doubling permits use of inexpensive 6.3-V filament transformer without risking loss of regulation when line voltage drops below about 105 V. With values shown, output varied only 6 mV for line voltage range of 95 to 135 V. Doubler circuit consists of C_2, C_3, D_1, D_3, D_5, and D_6.—A. Paterson, Voltage Doubler Prevents Supply from Losing Regulation, *EDN Magazine,* Nov. 1, 1972, p 46.

COMPONENT VALUES

D_1 THRU D_4	:	1N4001
D_5, D_6	:	1N914
C_1	:	8000μF, 15V
C_2	:	50μF, 25V
C_3	:	250μF, 25V
C_4	:	100μF, 10V
Q_1	:	MJE 521 (HEAT SINK)
Q_2, Q_3	:	2N3392
Q_4	:	CONSTANT CURRENT DIODE, (e.g. 2N5033, 2-3 mA)
Z_1	:	REF. DIODE, 4.3 V @ 2-3 mA (e.g. LVA43A)

12–14 V AT 3 A—Basic circuit for operating mobile equipment off AC line uses IC voltage regulator in conjunction with series-pass transistor.—Circuits, *73 Magazine,* Holiday issue 1976, p 170.

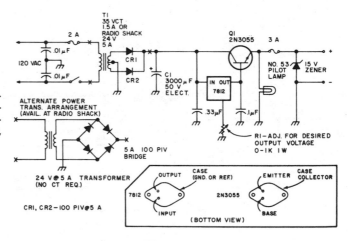

CHAPTER 74
Regulator Circuits

Used at outputs of unregulated power supplies to provide highly regulated fixed and variable positive and negative output voltages ranging from 0 to ±65 V for solid-state applications and up to 1000 V at 100 W for other purposes. Maximum current ratings range from 5 mA to 20 A. Some regulators have overvoltage crowbar or foldback current limiting. Dual-output regulators may have tracking. Current regulators are included. See also Regulated Power Supply and Switching Regulator chapters.

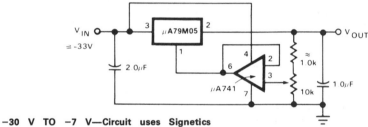

−30 V TO −7 V—Circuit uses Signetics µA79M05 adjustable voltage regulator in combination with 741 opamp to give wide negative output voltage range. Regulator includes thermal overload protection and internal short-circuit protection. Input voltage should be at least 3 V more negative than maximum output voltage desired.—"Signetics Analog Data Manual," Signetics, Sunnyvale, CA, 1977, p 670.

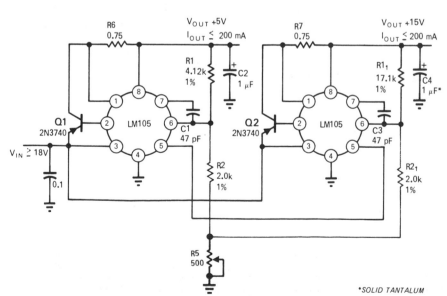

*SOLID TANTALUM

5 AND 15 V SINGLE CONTROL—Single potentiometer serves for adjusting two regulators simultaneously. Accuracy depends on output voltage differences of regulators; error decreases when output voltages are closer. Article gives design equation and covers other possible sources of error.—R. C. Dobkin, One Adjustment Controls Many Regulators, *EDN Magazine,* Nov. 1, 1970, p 33–35.

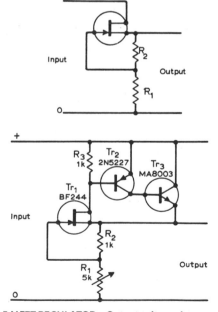

5-V FET REGULATOR—Output voltage changes less than 0.1 V for load current change from 0 to 60 mA. Output voltage changes caused by change in load resistance affect gate-source voltage of FET Tr_1 via R_1 and R_2, causing compensating change in drain current. Additional transistors serve to reduce output resistance and increase output current without affecting stabilization ratio of about 1000.—C. R. Masson, F.E.T. Voltage Regulator, *Wireless World,* Aug. 1971, p 386.

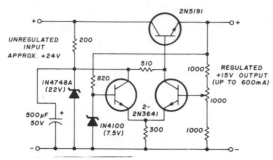

+15 V WITH DIFFERENTIAL AMPLIFIER—Series regulator uses differential amplifier as control circuit in which one side is referenced to zener and other to fraction of output voltage. Second zener provides coarse regulated voltage to differential pair.—H. Olson, Power-Supply Servicing, *Ham Radio*, Nov. 1976, p 44–50.

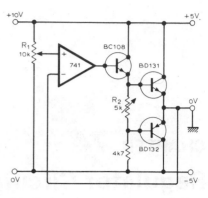

REGULATED DIVIDER FOR ±5 V—Used at output of adjustable regulated power supply providing up to 15 V, to give lower positive and negative voltages that remain steady despite changes in load current. To get +5 V and −5 V from +10 V, set R_1 at midposition and adjust R_2 for 20 mA through output transistors. Uses 741 opamp.—C. H. Banthorpe, Voltage Divider, *Wireless World*, Dec. 1976, p 41.

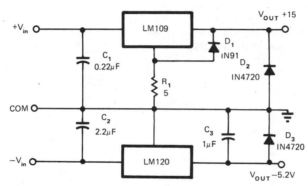

DUAL −5.2 V AND +15 V—Output voltages are equal to preset values of regulator ICs in basic arrangement shown. R_1 and D_1 ensure startup of LM109 when common load exists across supplies. D_1 should be germanium or Schottky having forward voltage drop of 0.4 V or less at 50 mA. D_2 and D_3 protect against polarity reversal of output during overloads.—C. T. Nelson, Power Distribution and Regulation Can Be Simple, Cheap and Rugged, *EDN Magazine*, Feb. 20, 1973, p 52–58.

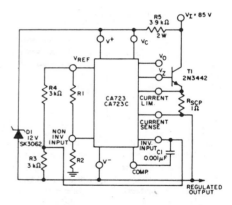

+50 V FLOATING—RCA CA723 regulator operating from 85-V supply delivers 50 V with line regulation of 15 mV for 20-V supply change and load regulation of 20 mV for 50-mA load current change.—"Linear Integrated Circuits and MOS/FET's," RCA Solid State Division, Somerville, NJ, 1977, p 61.

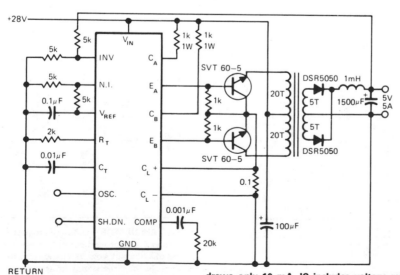

5 V AT 5 A WITH IC SWITCHER—Uses Silicon General SG1524 IC as pulse-width-modulated regulator for which operating frequency remains constant, with ON time of each pulse adjusted to maintain desired output voltage. Operating range extends above 100 kHz but device draws only 10 mA. IC includes voltage reference, oscillator, comparator, error amplifier, current limiter, pulse-steering flip-flop, and automatic shutdown for overload.—P. Franson, Today's Monolithic Switching Circuits Greatly Simplify Power-Supply Designs, *EDN Magazine*, March 20, 1977, p 47–48, 51, and 53.

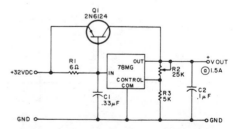

5–30 V AT 1.5 A—External series-pass transistor boosts 500-mA rated output of 78MG or 79MG regulator to 1.5 A for use as adjustable power supply in lab. Circuit has no short-circuit protection for safe-area limiting for external pass transistor, but article shows how to add protective transistor for this purpose.—J. Trulove, A New Breed of Voltage Regulators, *73 Magazine*, March 1977, p 62–64.

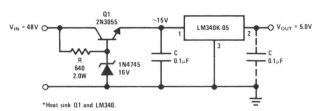

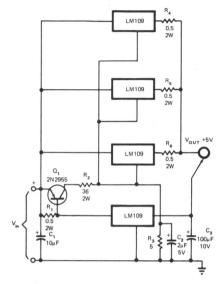

*Heat sink Q1 and LM340.

5 V FROM 48 V—Combination of zener and resistor R gives equivalent of power zener as solution to regulator protection problem when input voltage is much higher than rated maximum of regulator. Maximum load is 1 A. With optional capacitor, circuit noise is only 700 μV P-P.—"Linear Applications, Vol. 2," National Semiconductor, Santa Clara, CA, 1976, AN-103, p 10.

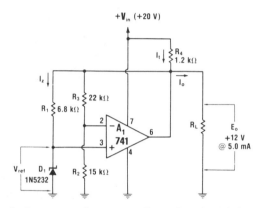

12-V SHUNT AT 5 mA—Low-power shunt regulator uses opamp to absorb excess load current. Value of R_1 is chosen to step up reference voltage of 5.6-V zener to +12 V at 5 mA. Design procedure for other output voltages is given. Output impedance is 0.01 ohm at 100 Hz, giving 120-Hz ripple-frequency filtering comparable to that of 100,000-μF capacitor.—W. G. Jung, "IC Op-Amp Cookbook," Howard W. Sams, Indianapolis, IN, 1974, p 166–168.

PARALLELING REGULATORS—Current-sharing problem is overcome without sacrificing ripple rejection or load regulation, by using bottom regulator as control device that supplies most of load current until current through this regulator reaches about 1.3 A. At this point Q_1 turns on and raises output voltage of other regulators to supply additional load current demands. Circuit shown will supply up to 6 A for minimum input voltage of 8 V. For optimum regulation, minimum load current should be 1 A.—C. T. Nelson, Power Distribution and Regulation Can Be Simple, Cheap and Rugged, *EDN Magazine*, Feb. 20, 1973, p 52–58.

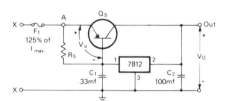

12 V AT 20 A—Regulator conducts and regulates until current demand is such that IR drop across R_S is sufficient to overcome base-emitter junction potential of switch transistor Q_S, which is two 2N174 germanium transistors in parallel. Use 2 ohms for R_S. Q_S is then turned on, with current/voltage regulation to its base controlled by regulator. Input voltage of 7812 regulator should be 2 V more than desired output voltage. Article gives three different rectifier circuits suitable for use with regulator.—A. M. Clarke, Regulated 200 Watt-12 Volt D.C. Power Supply, *CQ*, Oct. 1975, p 28–30 and 78–79.

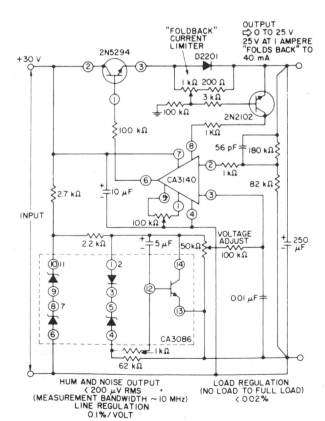

HUM AND NOISE OUTPUT
< 200 μV RMS
(MEASUREMENT BANDWIDTH ~ 10 MHz)
LINE REGULATION
0.1%/VOLT

LOAD REGULATION
(NO LOAD TO FULL LOAD)
< 0.02%

0–25 V WITH FOLDBACK CURRENT LIMITING—When D2201 diode senses load current of 1 A at maximum regulated output of 25 V, 2N2102 current-sensing transistor provides foldback of output current to 40 mA. Arrangement permits use of 2N5294 transistor as series-pass element, using only small heatsink. High-impedance reference-voltage divider across 30-V supply serves CA3140 connected as noninverting power opamp.—"Linear Integrated Circuits and MOS/FET's," RCA Solid State Division, Somerville, NJ, 1977, p 248–257.

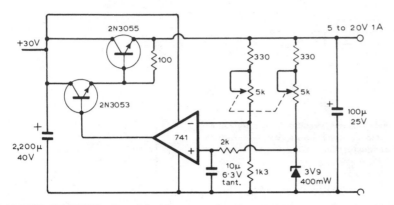

5–20 V ZENER-STABILIZED—Use of dual linear pot simplifies problem of feeding reference zener diode from variable-voltage supply.—L. J. Baughan, Variable Power Supply with Zener Stabilization, *Wireless World,* Nov. 1975, p 520.

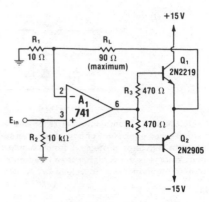

100-mA CURRENT REGULATOR—741 opamp is connected as noninverting voltage-controlled current source feeding transistors that boost output and provide bidirectional current capability in load R_L. If single-polarity current flow is sufficient, omit opposite-polarity transistor.— W. G. Jung, "IC Op-Amp Cookbook," Howard W. Sams, Indianapolis, IN, 1974, p 173.

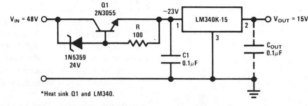

*Heat sink Q1 and LM340.

+15 V FROM HIGH INPUT VOLTAGE—Zener is used in series with resistor R to level-shift input voltage higher than rated maximum of LM340K-15 regulator. Typical load regulation is 40 mV for 0–1 A pulsed load, and line regulation is 2 mV for 1-V change in input voltage for no load. With optional output capacitor, circuit noise is only 700 μV P-P.—"Linear Applications, Vol. 2," National Semiconductor, Santa Clara, CA, 1976, AN-103, p 9–10.

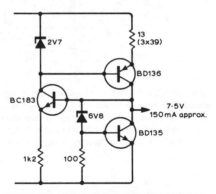

HUM-FREE CASSETTE RECORDER SUPPLY— Designed for tape recorder feeding into AF amplifier, to permit operation of recorder from power supply of amplifier without having hum due to positive feedback through shared ground connection. Circuit provides up to 150 mA at 7.5 V from supply ranging from 12 to 24 V. Transistors are connected as constant-current source in series with constant-voltage sink. Use three 39-ohm resistors in parallel as 13-ohm resistor.—G. Hibbert, Avoiding Power Supply Hum, *Wireless World,* Oct. 1973, p 515.

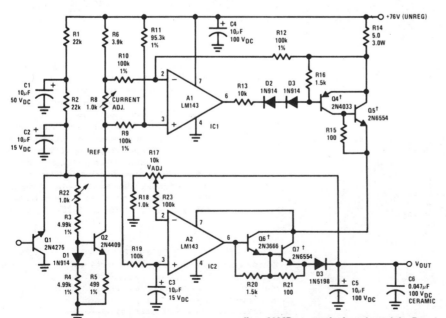

7.1–65 V AT 0–1 A—Provides continuously variable output voltage and adjustable output current range. Q1 is connected as zener to give 6.5-V reference voltage. Darlington current boosters Q4-Q7 should be on common Thermalloy 6006B or equivalent heatsink. Developed for use with pulsed loads. For input voltage range of 46–76 V, regulation is within 286 mV for 500-mA DC output.—"Linear Applications, Vol. 2," National Semiconductor, Santa Clara, CA, 1976, AN-127, p 8–10.

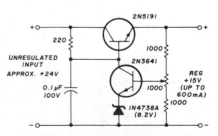

+15 V WITH FEEDBACK—Fraction of output voltage is fed back to base of 2N3641 regulator transistor. Difference between this voltage and zener diode voltage is amplified to control base of 2N5191 series transistor.—H. Olson, Power-Supply Servicing, *Ham Radio,* Nov. 1976, p 44–50.

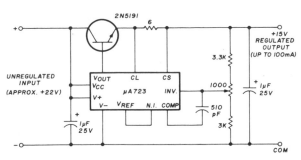

+15 V WITH μA723—Series power transistor and Fairchild IC voltage regulator provide up to 100 mA. Article covers troubleshooting and repair of all types of regulators.—H. Olson, Power-Supply Servicing, *Ham Radio,* Nov. 1976, p 44–50.

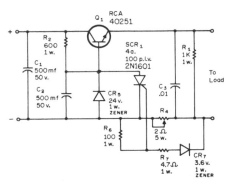

OVERLOAD PROTECTION—When critical current is exceeded, SCR_1 conducts and reduces base-ground voltage of Q_1, cutting it off. Load current then drops to very low value, and Q_1 is protected. Operation is restored by turning off current supply to power transformer after clearing short-circuit condition.—R. Phelps, Jr., Protective Circuits for Transistor Power Supplies, *CQ,* March 1973, p 44–48 and 92.

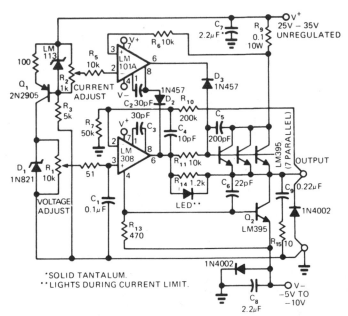

*SOLID TANTALUM.
**LIGHTS DURING CURRENT LIMIT.

25 V AT 10 A FOR LAB—Circuit uses no large output capacitors yet has good response as constant-voltage or constant-current source. LM395 units (7 in parallel) act as current-limited thermally limited high-gain power transistor. Mount all on same heatsink for good current sharing, since 300 W will be dissipated under worst-case conditions. Only two control opamps are needed, one for voltage control and one for current control.—R. C. Dobkin, General-Purpose Power Supply Furnishes 10A and 25V, *EDN Magazine,* March 5, 1975, p 70.

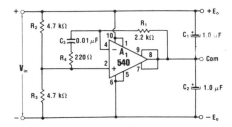

CONVERTING TO DUAL SUPPLY—With equal values for R_2 and R_3, input of 30 V is converted to ±15 V at output. If desired, R_2 and R_3 can be scaled for unequal voltage drops. Circuit uses 540 power IC having 100-mA rating for each output, for handling load imbalances up to 100 mA.—W. G. Jung, "IC Op-Amp Cookbook," Howard W. Sams, Indianapolis, IN, 1974, p 170–171.

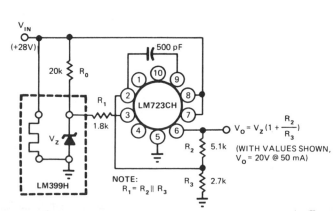

$$V_O = V_Z \left(1 + \frac{R_2}{R_3}\right)$$

(WITH VALUES SHOWN, V_O = 20V @ 50 mA)

NOTE: $R_1 = R_2 \| R_3$

10 PPM/°C—Connections shown convert LM723CH regulator into precision power reference having excellent long-term stability and temperature stability. LM399H replaces internal reference of LM723 with low-noise 6.9 V to give desired performance over temperature range from +15 to +65°C.—B. Welling, High-Stability Power Supply Uses 723 Regulator, *EDN Magazine,* Jan. 20, 1978, p 114 and 116.

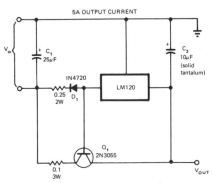

5 A AT −5 TO −15 V—Use of 2N3055 pass transistor boosts current output of LM120 regulator IC. Minimum differential between input and output voltages is typically 2.5 V, so supply voltage must be 2.5 V higher than preset output voltage of regulator chosen from National LM120 series.—C. T. Nelson, Power Distribution and Regulation Can Be Simple, Cheap and Rugged, *EDN Magazine,* Feb. 20, 1973, p 52–58.

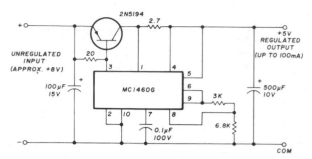

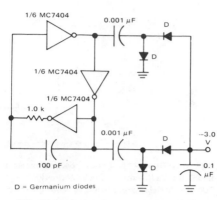

+5 V WITH MC1460G—Series power transistor and Motorola IC voltage regulator provide up to 100 mA. IC shown has been replaced by MC1469. Equivalents made by other manufacturers can also be used.—H. Olson, Power-Supply Servicing, *Ham Radio,* Nov. 1976, p 44–50.

−3 V—Circuit using three sections of Motorola MC7404 operates from +5 V supply and generates −3 V at up to 100 μA, as one of supply voltages required by Motorola MCM6570 8192-bit character generator using 7 × 9 matrix.—"A CRT Display System Using NMOS Memories," Motorola, Phoenix, AZ, 1975, AN-706A, p 5.

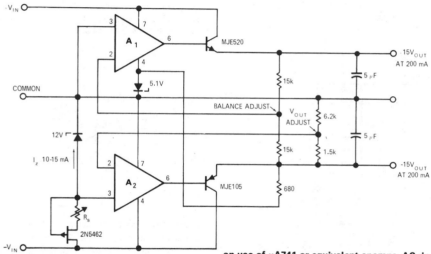

±15 V AT 200 mA—Two-opamp regulator gives dual-polarity tracking outputs that can be balanced to within millivolts of each other or can be offset as required. Negative voltage is regulated, and positive output tracks negative. Article gives step-by-step design procedure based on use of μA741 or equivalent opamps. AC ripple is less than 2 mV P-P. Conventional full-wave bridge rectifier with capacitor-input filter can be used to provide required unregulated 36 VDC for inputs.—C. Brogado, IC Op Amps Simplify Regulator Design, *EDN/EEE Magazine,* Jan. 15, 1972, p 30–34.

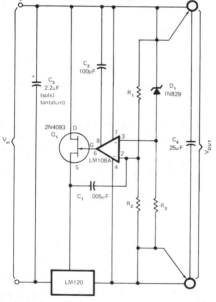

1 A WITH 0.005% VOLTAGE ACCURACY—Use of National LM120 negative regulator with LM108A low-drift opamp and 1N829 precision reference diode gives extremely tight regulation, very low temperature drift, and full overload protection. Bridge arrangement sets output voltage and holds reference diode current constant. FET is required because 4-mA maximum ground current of regulator exceeds output current rating of opamp. R_1 and R_2 should track to 1 PPM or less. R_3 is chosen to set reference current at 7.5 mA. For output of 8 to 14 V, use LM120-5.0; for 15–17 V, use LM120-12.—C. T. Nelson, Power Distribution and Regulation Can Be Simple, Cheap and Rugged, *EDN Magazine,* Feb. 20, 1973, p 52–58.

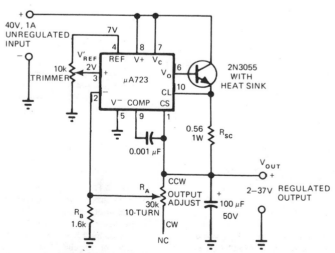

2–37 V—Simple circuit gives fine linear control with 10-turn pot over wide voltage range by first using 10K trimmer pot to divide 7-V reference down to 2 V.—G. Dressel, Regulator Circuit Provides Linear 2-37 V Adjustment Range, *EDN Magazine,* March 5, 1978, p 122.

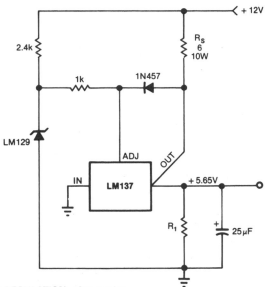

POSITIVE SHUNT REGULATION—Connection shown for LM137 negative series regulator provides high-reliability positive shunt regulation for applications having high-voltage spike on raw DC supply. Output is 5.65 V.—P. Lefferts, Series Regulators Provide Shunt Regulation, *EDN Magazine*, Sept. 5, 1978, p 158 and 160.

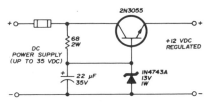

+12 V AT 2 A—Developed for unregulated 12-VDC supplies used by some amateurs with low-power VHF FM equipment, where no-load voltage may be 18 V or more. During transmit, voltage drops to about 12 V, but on receive may exceed voltage ratings of small-signal transistors in transceiver. Use heatsink with transistor, and use 2-A fuse to protect transistor from shorted load.—J. Fisk, Circuits and Techniques, *Ham Radio*, June 1976, p 48–52.

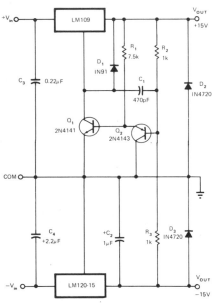

±15 V WITH TRACKING—In arrangement shown for National regulator ICs, positive output voltage tracks negative voltage to better than 1%. Ripple rejection is 80 dB for both outputs. Load regulation is 30 mV at 1 A for negative output and less than 10 mV for positive output. Circuit works well for output in range of ±6 to ±15 V. C₁ provides stability.—C. T. Nelson, Power Distribution and Regulation Can Be Simple, Cheap and Rugged, *EDN Magazine*, Feb. 20, 1973, p 52–58.

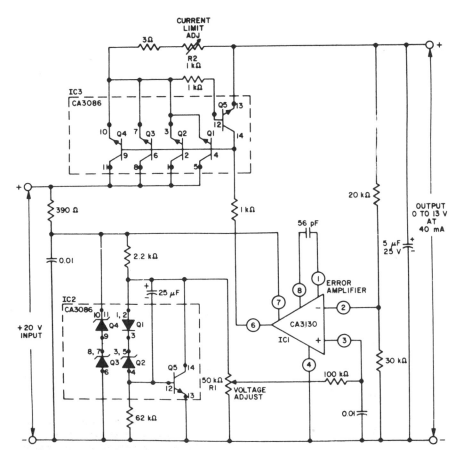

0–13 V AT 40 mA—Combination of RCA CA3130 opamp and two CA3086 NPN transistor arrays provides better than 0.01% regulation from no load to full load and input regulation of 0.02%/V. Hum and noise output is less than 25 μV up to 100 kHz.—"Linear Integrated Circuits and MOS/FET's," RCA Solid State Division, Somerville, NJ, 1977, p 236–243.

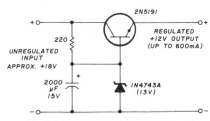

+12 V SERIES EMITTER-FOLLOWER—Base-emitter voltage is more or less constant without use of feedback because base is held at constant voltage by zener diode. Ripple at base is reduced by RC filter.—H. Olson, Power-Supply Servicing, *Ham Radio*, Nov. 1976, p 44–50.

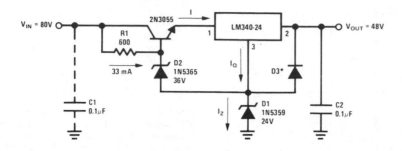

*Germanium signal diode

48 V FROM 80 V—Level-shifting transistor-zener combination R1-D2 is used with zener D1 to keep voltage across LM340-24 regulator below maximum rated value. Addition of zeners has drawback of increasing output noise to about 2 mV P-P. Load regulation is 60 mV for pulsed load change from 5 mA to 1 A. Line regulation is 0.01%/V of input voltage change for 500-mA load.—"Linear Applications, Vol. 2," National Semiconductor, Santa Clara, CA, 1976, AN-103, p 10–11.

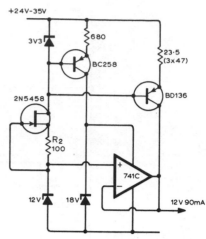

HUM-FREE TUNER SUPPLY—Permits operation of high-quality FM tuner from amplifier supply without having hum due to positive feedback through shared ground connection. Circuit provides up to 90 mA at 12 V from any supply ranging from 24 to 34 V. Low output impedance eliminates all likely sources of feedback and suppresses ripple. Circuit requires careful initial adjustment to limit current sunk by 741C opamp to less than 15 mA; coarse adjustment is made by varying number of 47-ohm resistors in parallel serving as BD136 emitter resistor, and fine adjustment by changing R_2.—G. Hibbert, Avoiding Power Supply Hum, *Wireless World*, Oct. 1973, p 515.

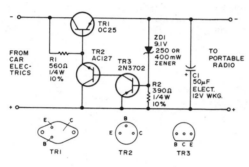

9 V FROM 12 V—Developed for economical operation of 9-V portable radio from 12-V storage battery of car.—Circuits, *73 Magazine*, March 1975, p 136.

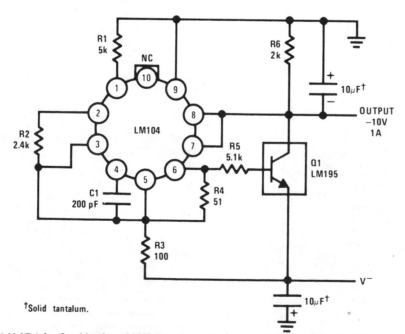

†Solid tantalum.

−10 V AT 1 A—Combination of LM195 power transistor IC and standard LM104 regulator gives negative output voltage with full overload protection and better than 2-mV load regulation. Input voltage must be only 2 V greater than output voltage.—"Linear Applications, Vol. 2," National Semiconductor, Santa Clara, CA, 1976, AN-110, p 4–5.

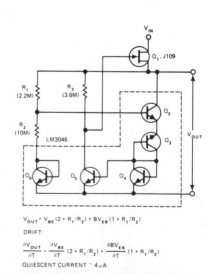

$$V_{OUT} = V_{BE}(2 + R_1/R_2) + BV_{EB}(1 + R_1/R_2)$$

DRIFT:

$$\frac{\partial V_{OUT}}{\partial T} = \frac{\partial V_{BE}}{\partial T}(2 + R_1/R_2) + \frac{\partial BV_{EB}}{\partial T}(1 + R_1/R_2)$$

QUIESCENT CURRENT ∼ 4 μA

JFET SERIES-PASS—Use of JFET as series-pass element for LM3046 voltage regulator IC minimizes battery drain in microprocessor system applications. Pass element needs no preregulation because drive comes from regulated output. Gate source is isolated from line by drain and thus provides excellent line regulation.—J. Maxwell, Voltage Regulator Bridges Gap Between IC's and Zeners, *EDN Magazine*, Sept. 5, 1977, p 178–179.

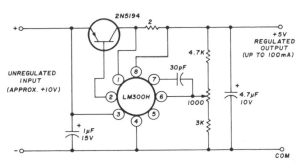

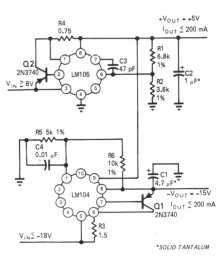

+5 V WITH LM300H—Series power transistor and National IC voltage regulator provide up to 100 mA. Improved version of regulator, LM305H, may be substituted.—H. Olson, Power-Supply Servicing, *Ham Radio*, Nov. 1976, p 44–50.

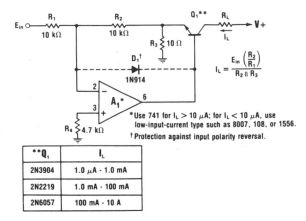

$$I_L = \frac{E_{in}\left(\frac{R_2}{R_1}\right)}{R_2 \parallel R_3}$$

*Use 741 for $I_L > 10\ \mu A$; for $I_L < 10\ \mu A$, use low-input-current type such as 8007, 108, or 1556.

†Protection against input polarity reversal.

Q₁	I_L
2N3904	1.0 μA - 1.0 mA
2N2219	1.0 mA - 100 mA
2N6057	100 mA - 10 A

NEGATIVE-INPUT CURRENT REGULATOR—Opamp is used as inverter starting current-boosting transistor to provide positive supply voltage. Load current range depends on transistor used. R_3 forces Q_1 to conduct much heavier current than feedback current, as required for high load current. Current gain depends on ratio of R_2 to R_3.—W. G. Jung, "IC Op-Amp Cookbook," Howard W. Sams, Indianapolis, IN, 1974, p 176–177.

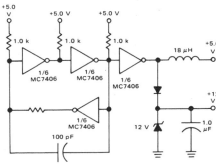

−15 V TRACKING +5 V—LM104 negative regulator is used with inverting gain to give negative output voltage that is greater than positive reference voltage. Noninverting input is tied to divider R5-R6 between negative output and ground. Positive reference determines line regulation and temperature drift, with negative output tracking.—R. C. Dobkin, One Adjustment Controls Many Regulators, *EDN Magazine*, Nov. 1, 1970, p 33–35.

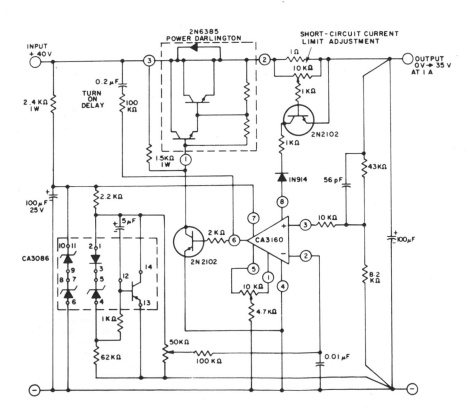

+5 AND +12 V AT 6 mA—Circuit using four sections of Motorola MC7406 provides +12 V supply required by MCM6570 8192-bit character generator using 7 × 9 matrix, along with conventional +5 V.—"A CRT Display System Using NMOS Memories," Motorola, Phoenix, AZ, 1975, AN-706A, p 5.

0.1–35 V AT 1 A—CA3160 serves as error amplifier in continuously adjustable regulator that functions down to vicinity of 0 V. RC network between base of 2N2102 output drive transistor and Input source prevents turn-on overshoot. Input regulation is better than 0.01%/V, and regulation from no load to full load is better than 0.005%. Hum and noise output is less than 250 μVRMS.—"Linear Integrated Circuits and MOS/FET's," RCA Solid State Division, Somerville, NJ, 1977, p 267–269.

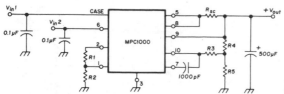

PARAMETER VALUES FOR BEST RESULTS		
	$2V < V_O < 7V$	$7V < V_O < 35V$
R1	$\approx \frac{R2(7-V_O)}{V_O}$	$\approx \left[\frac{R4\,R5}{R4+R5}\right]$
R2	$10k < R1+R2 < 100k$	∞
R3	$\approx \left[\frac{R1\,R2}{R1+R2}\right]$	—
R4	—	$\approx \frac{R5(V_O-7)}{7}$
R5	∞	$10k < R5 < 100k$
R_{sc}	$\approx \frac{0.66}{i_{sc}}$ @ $T_J = 25°C$	

+2 TO +35 V AT 10 A—Provides fixed output voltage at value determined by choice of resistance values, computed as given in table. Heatsink should have very low thermal resistance. For similar range of negative voltages, Motorola MPC900 regulator can be used, with circuit modified slightly as set forth in article.—H. Olson, Second-Generation IC Voltage Regulators, *Ham Radio*, March 1977, p 31–37.

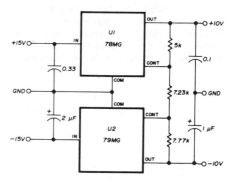

+0.5 TO 1 V BIAS—Motorola MC1723G regulator, 2N5991 current-boost transistor, and base-emitter junction of 2N5190 transistor CR1 serve as adjustable bias voltage source for 300-W solid-state power amplifier. R3 sets current limiting at about 0.65 A. Measured output-voltage variations are about ±6 mV for load changes of 0 to 600 mA.—H. O. Granberg, One KW—Solid-State Style, *QST*, April 1976, p 11–14.

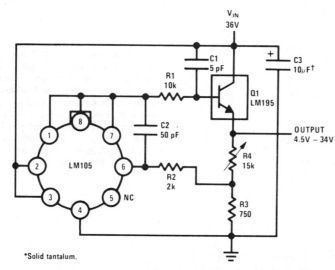

*Solid tantalum.

4.5–34 V AT 1 A—Combination of LM195 power transistor IC and standard LM105 regulator gives better than 2-mV load regulation with overload protection. Differential between input and output voltages is only 2 V.—"Linear Applications, Vol. 2," National Semiconductor, Santa Clara, CA, 1976, AN-110, p 4.

±10 V TRACKING—Fairchild 78MG and 79MG positive and negative voltage-regulator ICs provide up to 500-mA output, with protection against short-circuits and thermal overloads.—D. Schmieskors, Adjustable Voltage-Regulator ICs, *Ham Radio*, Aug. 1975, p 36–38.

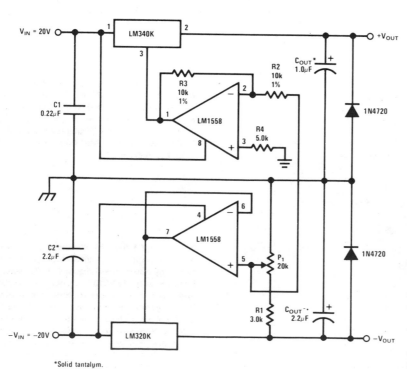

*Solid tantalum.

±5 TO ±18 V WITH TRACKING—Ground pin of LM340K-15 positive regulator is lifted by LM1558 inverter, while ground pin of negative LM320K-15 is lifted by LM1558 voltage follower. Positive regulator is made to track negative regulator within about 50 mV over entire output range. At ±15 V, typical load regulation is between 40 and 80 mV for 0–1 A pulsed load.—"Linear Applications, Vol. 2," National Semiconductor, Santa Clara, CA, 1976, AN-103, p 8–9.

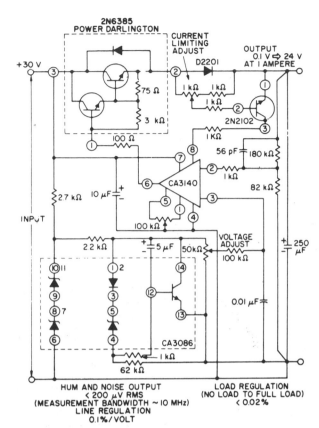

0.1–24 V AT 1 A—High-impedance reference-voltage divider across 30-V supply serves CA3140 connected as noninverting power opamp with gain of 3.2. 8-V reference input gives maximum output voltage of about 25 V. D2201 high-speed diode serves as current sensor for 2N2102 current-limit sensing amplifier. Current-limiting point can be adjusted over range of 10 mA to 1 A with single 1K pot. Power Darlington serves as series-pass element.— "Linear Integrated Circuits and MOS/FET's," RCA Solid State Division, Somerville, NJ, 1977, p 248–256.

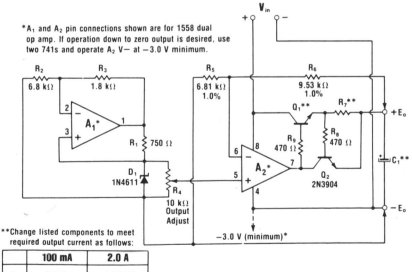

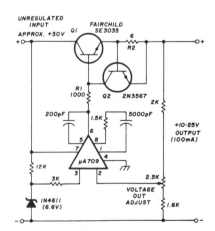

*A_1 and A_2 pin connections shown are for 1558 dual op amp. If operation down to zero output is desired, use two 741s and operate A_2 V– at –3.0 V minimum.

**Change listed components to meet required output current as follows:

	100 mA	2.0 A
Q_1	2N3766	2N6057†
R_7	4.7 Ω	0.25 Ω
C_1	100 μF	1000 μF

†Heat sink required.

0–15 V AT 2 A—Basic zener-opamp regulator output of 6.6 V is scaled up to maximum of 15 V, adjusted with R_4, by adding buffer opamp A_2 and current-boosting transistors. Q_2 provides short-circuit protection by sensing load current through R_7. Large output capacitor C_1 maintains low output impedance at high frequencies where gain of A_2 falls off.—W. G. Jung, "IC Op-Amp Cookbook," Howard W. Sams, Indianapolis, IN, 1974, p 158–159.

+10 TO +25 V AT 100 mA—Series regulator uses opamp as differential amplifier and extra transistor Q2 as current limiter. When 100 mA is drawn, 0.6 V is developed across R2 to make Q2 conduct, pulling Q1 base in negative direction. This action prevents excessive current from being passed by Q1.—H. Olson, Power-Supply Servicing, *Ham Radio*, Nov. 1976, p 44–50.

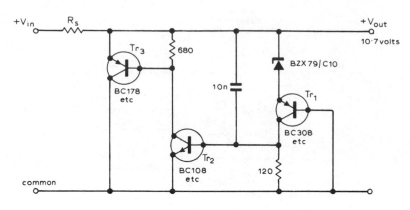

RIPPLE-PREAMP SUPPLY—Shunt regulator removes virtually all AC line ripple without using large capacitor, making it ideal for audio applications where freedom from ripple is more important than precise supply voltage level. Circuit cannot be damaged by short-circuits. Tr_3 may be power transistor or Darlington.—P. S. Bright, Ripple Eliminator, *Wireless World*, April 1977, p 62.

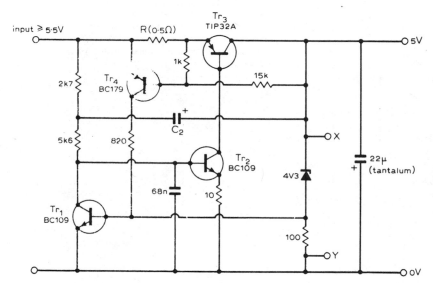

LOW COST WITH DISCRETE ELEMENTS—Performance is comparable to that of combined discrete and monolithic circuits, with load regulation of 0.01%, line regulation of 0.05%, ripple rejection of 0.1%, and output ripple and noise of 1 mV. Output is 1 A at 5 V. Foldback short-circuit protection is provided by Tr_4, with maximum current determined by value of R. C_2, which can be 100 μF, gives extra ripple rejection by introducing more AC feedback into loop. TIP32A is plastic series transistor, and is not critical; many other types will work equally well.—K. W. Mitchell, High Performance Voltage Regulator, *Wireless World*, May 1976, p 83–84.

TRIMMED DUAL SUPPLY

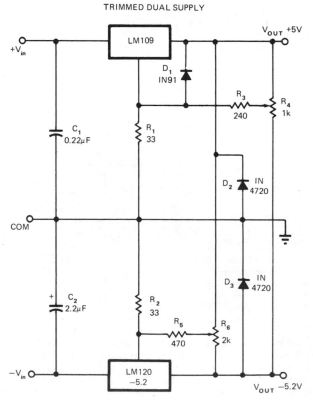

DUAL OUTPUTS WITH TRIMMING—Trimming pots connected across outputs provide positive or negative currents for producing small trimming voltages across 33-ohm ground-leg resistors of National regulators. Same components can be used for higher output voltages, but resistance values of pots should be increased if power dissipation becomes problem.—C. T. Nelson, Power Distribution and Regulation Can Be Simple, Cheap and Rugged, *EDN Magazine*, Feb. 20, 1973, p 52–58.

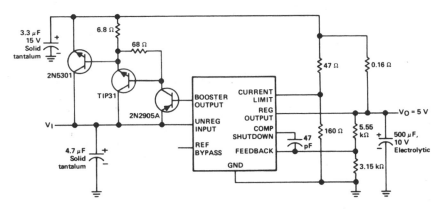

5 V AT 10 A WITH CURRENT LIMITING—Combination of three transistors and SN52105 or SN72305 regulator provides foldback current limiting for overload protection. Input voltage can be up to 40 V greater than 5-V output. Load regulation is about 0.1%, and input regulation is 0.1%/V. Regulators are interchangeable with LM105 and LM305 respectively.—"The Linear and Interface Circuits Data Book for Design Engineers," Texas Instruments, Dallas, TX, 1973, p 5-9.

0 TO ±6.6 V TRACKING AT 5 mA—Master-slave regulator combination is used to make second regulator provide mirror image of first while output of first is varied over full range from 0 to zener limit with R_4. Accuracy of tracking depends on match between R_5 and R_6, which should be 1% film or wirewound.—W. G. Jung, "IC Op-Amp Cookbook," Howard W. Sams, Indianapolis, IN, 1974, p 160–162.

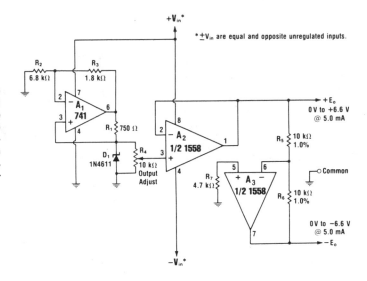

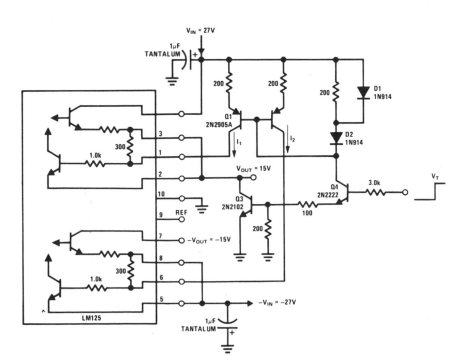

ELECTRONIC SHUTDOWN—Both sections of National LM125 dual tracking regulator are shut down by TTL-compatible control signal V_T which shorts internal reference voltage of regulator to ground. Q3 acts only as current sink.—T. Smathers and N. Sevastopoulos, "LM125/LM126/LM127 Precision Dual Tracking Regulators," National Semiconductor, Santa Clara, CA, 1974, AN-82, p 15.

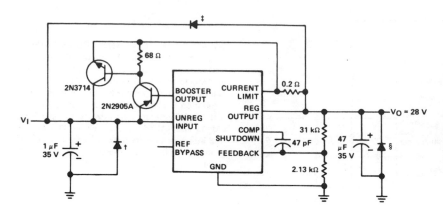

28 V AT 1 A—Circuit uses SN52105 or SN72305 regulator with three protective diodes. Feedback diode at top protects against shorted input and inductive loads on unregulated supply. Input diode protects against input voltage reversal. Output diode protects against output voltage reversal. Maximum input voltage is 50 V.—"The Linear and Interface Circuits Data Book for Design Engineers," Texas Instruments, Dallas, TX, 1973, p 5-9.

15 V AT 5 A WITH PROTECTION—External boost transistor is used with National LM340T-15 regulator to boost output current capability to 5 A without affecting such features as short-circuit current limiting and thermal shutdown. Short-circuit current is held to 5.5 A. Heatsink for Q1 should have at least 4 times capacity of heatsink for IC.—"Linear Applications, Vol. 2," National Semiconductor, Santa Clara, CA, 1976, AN-103, p 3–4.

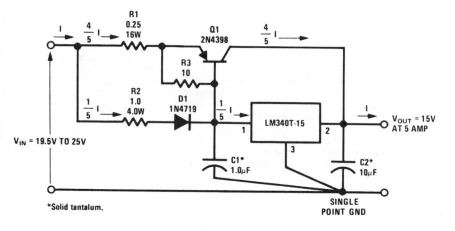

±15 V AT 10 A WITH FOLDBACK CURRENT LIMITING—Combination of Darlington pass transistors and current limiting is used with National LM125 dual tracking regulator to give high output currents with protection from short-circuits.—T. Smathers and N. Sevastopoulos, "LM125/LM126/LM127 Precision Dual Tracking Regulators," National Semiconductor, Santa Clara, CA, 1974, AN-82, p 11.

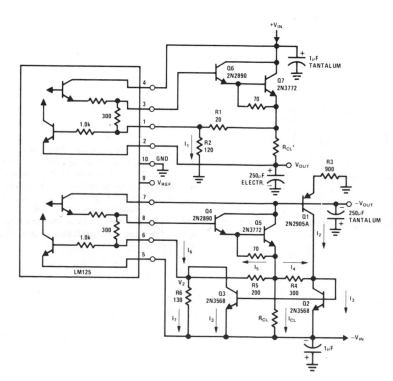

2–35 V VARIABLE—Wide voltage range is achieved by using μA723 regulator IC in simple feedback arrangement requiring only single pot to vary output voltage continuously and linearly from 2 to 35 V. Resistors R_3 and R_4 divide output voltage by 5, so inverting input of regulator sees one-fifth of output voltage. R_1 is connected between 7-V reference of IC and ground to present any intermediate voltage to noninverting input. IC acts to keep these two voltages equal. Maximum input voltage limit is 40 V; if possibility of higher voltages exists in lab applications, protect IC with 40-V zener across it.—J. Gangi, *Continuously Variable Voltage Regulator, EDN Magazine*, Feb. 20, 1973, p 91.

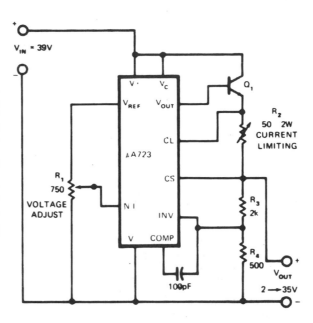

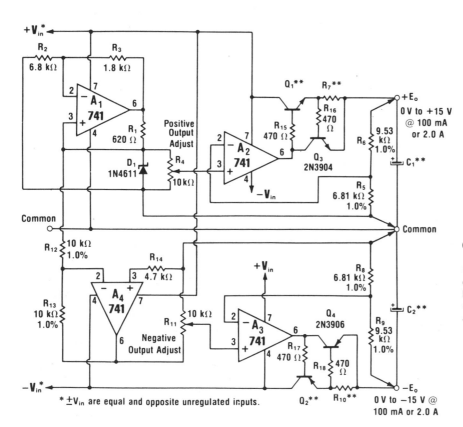

0 TO ±15 V INDEPENDENTLY VARIABLE—Common zener reference serves for both regulators. Buffer A_3 uses negative reference voltage developed from 6.6-V positive voltage across D_1 by inverter A_4. Both regulators provide 100 mA or 2 A depending on transistors used.—W. G. Jung, "IC Op-Amp Cookbook," Howard W. Sams, Indianapolis, IN, 1974, p 162–164.

* ±V_{in} are equal and opposite unregulated inputs.

**Change listed components to meet required output current as follows:

	100 mA	**2.0 A**
Q_1	2N3766	2N6057†
Q_2	2N3740	2N6050†
R_7, R_{10}	4.7 Ω	0.25 Ω
C_1, C_2	100 μF	1000 μF

† Heat sink required.

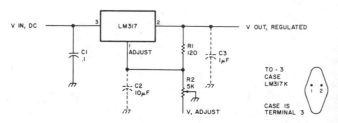

1.2–37 V AT 1.5 A—Uses National LM317 adjustable three-terminal positive voltage regulator. Output voltage is determined by ratio of R1 and R2. Output can be adjusted from 37 V down to 1.2 V with R2. If DC input is 40 V, regulation is about 0.1% at all settings when going from no load to full load. Regulator includes overload and thermal protection. If current limit is exceeded, regulator shuts down. C2 and C3 are optional; C2 improves ripple rejection, and C3 prevents instability when load capacitance is between 500 and 5000 pF.—Adjustable Bench Supply, *73 Magazine,* Dec. 1977, p 192–193.

+5 V AT 3 A—Uses Motorola MPC1000 positive voltage regulator to provide high current required for large TTL project. Current-limiting resistor R_{SC} is in range of 0.66 to 0.066 ohm. Use copper wire about 50% longer than calculated length and shorten step by step until required pass current is obtained; thus, start with 25 ft of No. 16, 15 ft of No. 18, 10 ft of No. 20, or 6 ft of No. 22.—G. L. Tater, The MPC1000—Super Regulator, *Ham Radio,* Sept. 1976, p 52–54.

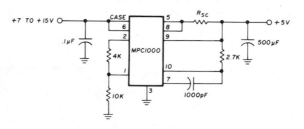

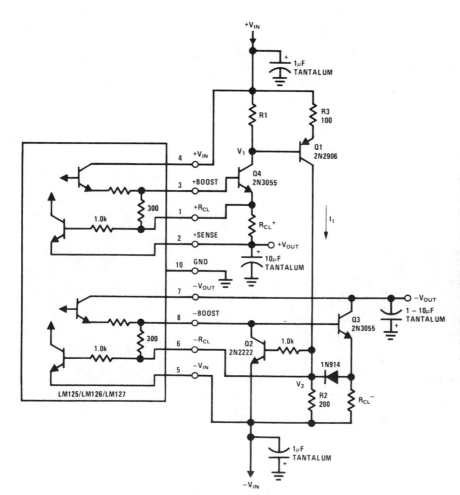

SIMULTANEOUS CURRENT LIMITING—Limiting action of circuit depends on output current of positive regulator but acts simultaneously on both positive and negative outputs of National LM125 dual tracking regulator. Positive output current produces voltage drop across R1 that makes Q1 conduct. When increase in current makes voltage drop across R2 equal negative current limit sense voltage, negative regulator will current-limit. Positive regulator closely follows negative output down to level of about 700 mV. Q2 turns off negative pass transistor during simultaneous current limiting. Output voltages are ±15 V.—"Linear Applications, Vol. 2," National Semiconductor, Santa Clara, CA, 1976, AN-82, p 12–13.

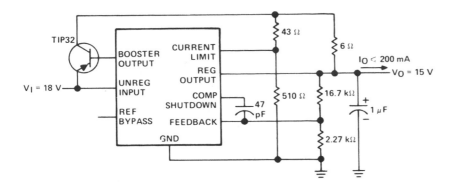

15 V AT 200 mA—Linear regulator using Texas Instruments SN52105, SN72305, or SN72376 is connected for foldback current limiting. Regulators are interchangeable with LM105, LM305, and LM376 respectively. Load regulation is 0.1%, and input regulation is 0.1%/V.—"The Linear and Interface Circuits Data Book for Design Engineers," Texas Instruments, Dallas, TX, 1973, p 5-9.

5 V AT 1 A—Use of Darlington at output boosts power rating of standard opamp voltage regulator circuit. Article gives step-by-step design procedure. With μA741 opamp, circuit gives good regulation along with short-circuit protection. AC ripple is less than 2 mV P-P. Required input of 30 V is obtained from conventional full-wave bridge rectifier with capacitor-input filter.—C. Brogado, IC Op Amps Simplify Regulator Design, *EDN/EEE Magazine*, Jan. 15, 1972, p 30–34.

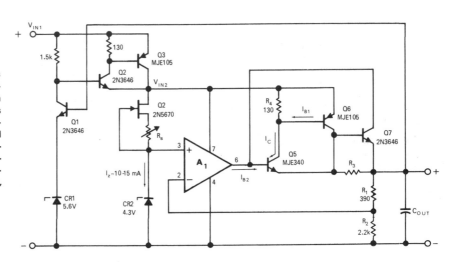

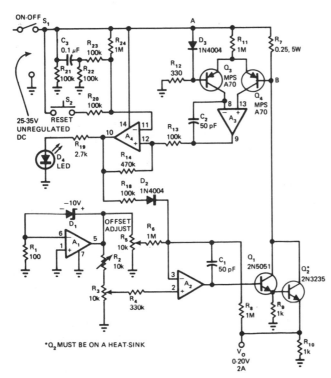

0–20 V AT 2 A—R_3 provides control of output voltage for regulator built around LM3900 quad Norton opamp. Output is well regulated against both line and load variations and is free of ripple. Opamp sections A_3 and A_4 provide overcurrent sensing and shutdown functions; after output fault is cleared, S_2 is closed momentarily to restore output power. Article describes circuit operation and initial setup in detail.—J. C. Hanisko and W. Wiseman, Variable Supply Built Around Quad Amp Outputs 2A, *EDN Magazine*, June 20, 1976, p 128 and 130.

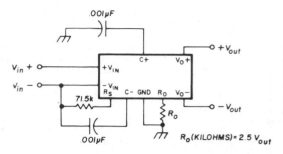

VARIABLE DUAL-POLARITY—External resistor R_0 determines values of positive and negative regulated output voltages provided by Silicon General SG3501 dual regulator.—H. Olson, Second-Generation IC Voltage Regulators, *Ham Radio,* March 1977, p 31–37.

$R_0 (KILOHMS) = 2.5 \ V_{out}$

5 V AT 200 mA—Article gives step-by-step design procedure for developing special opamp regulator when commercial unit meeting desired specifications is not available. Opamp is μA741. Circuit gives good regulation along with short-circuit protection, with less than 2 mV P-P AC ripple. Required input of 20 V is obtained from conventional full-wave bridge rectifier with capacitor-input filter.—C. Brogado, IC Op Amps Simplify Regulator Design, *EDN/EEE Magazine,* Jan. 15, 1972, p 30–34.

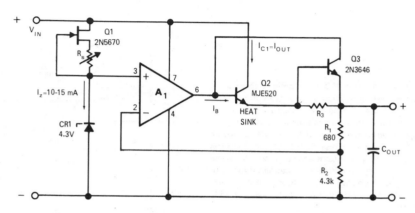

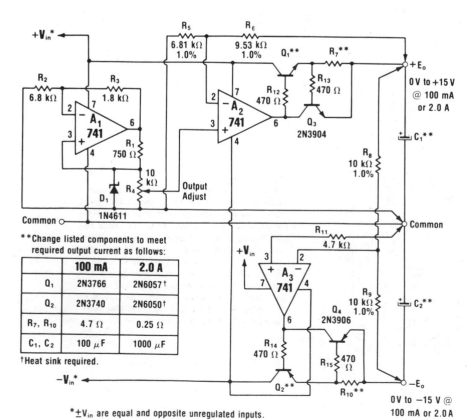

0 TO ±15 V TRACKING AT 100 mA OR 2 A—Basic tracking regulator is combined with transistors to extend output to voltages higher than zener reference and provide higher output currents. Choice of transistors for Q_1 and Q_2 determines maximum load current.—W. G. Jung, "IC Op-Amp Cookbook," Howard W. Sams, Indianapolis, IN, 1974, p 161–163.

**Change listed components to meet required output current as follows:

	100 mA	2.0 A
Q_1	2N3766	2N6057†
Q_2	2N3740	2N6050†
R_7, R_{10}	4.7 Ω	0.25 Ω
C_1, C_2	100 μF	1000 μF

†Heat sink required.

*±V_{in} are equal and opposite unregulated inputs.

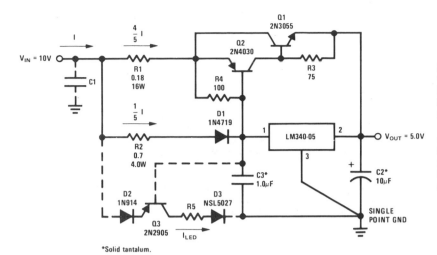

5 V AT 5 A FOR TTL—Typical load regulation is 1.8% from no load to full load. Q1 and Q2 serve in place of single higher-cost power PNP boost transistor. Dotted lines show how to add over-load indicator using National NSL5027 LED and R2 as overload sensor. When load current exceeds 5 A, Q3 turns on and D3 lights. Circuit includes thermal shutdown and short-circuit protection.—"Linear Applications, Vol. 2," National Semiconductor, Santa Clara, CA, 1976, AN-103, p 5.

NEGATIVE SHUNT REGULATION—Connection shown for LM117 positive series regulator provides spike-suppressing negative shunt regulation of −5 V output. With capacitor shown, regulator will withstand 75-V spikes on raw DC supply. For larger spikes, increase capacitor value.—P. Lefferts, Series Regulators Provide Shunt Regulation, *EDN Magazine*, Sept. 5, 1978, p 158 and 160.

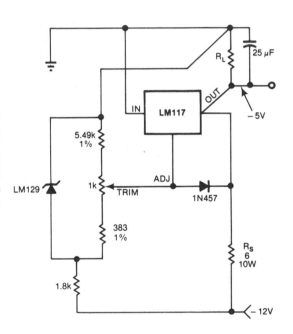

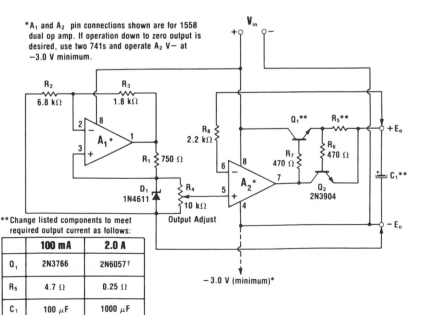

0–6.6 V AT 2 A—High-power circuit is suitable for low-voltage logic devices that require high current at supply voltages between 3 and 6 V. Maximum output of 2 A is obtained with 2N6057 Darlington pair for Q_1. Single 2N3766 can be used if load is only 100 mA. Q_2 provides short-circuit protection for Q_1. Since supply does not have to be adjusted down to 0 V, negative supply for A_2 can go to common negative of circuit. Optional connection to −3 V is used only when voltage range must go down to 0 V.—W. G. Jung, "IC Op-Amp Cookbook," Howard W. Sams, Indianapolis, IN, 1974, p 157–158.

**Change listed components to meet required output current as follows:

	100 mA	**2.0 A**
Q_1	2N3766	2N6057†
R_5	4.7 Ω	0.25 Ω
C_1	100 μF	1000 μF

†Heat sink required.

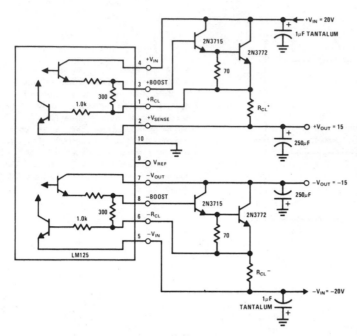

±15 V AT 7 A—External Darlington stages boost output currents of LM125 dual tracking regulator and increase minimum input/output voltage differential to 4.5 V. Maximum output current is limited by power dissipation of 2N3772. Typical load regulation is 40 mV from no load to full load.—T. Smathers and N. Sevastopoulos, "LM125/LM126/LM127 Precision Dual Tracking Regulators," National Semiconductor, Santa Clara, CA, 1974, AN-82, p 6.

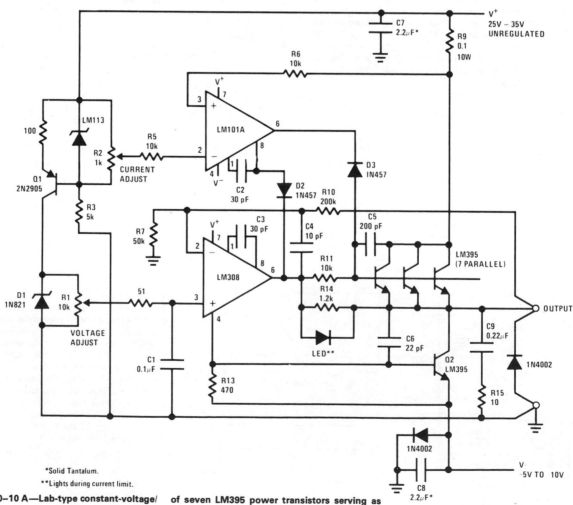

*Solid Tantalum.

**Lights during current limit.

0–25 V AT 0–10 A—Lab-type constant-voltage/constant-current power supply using standard ICs achieves high current output by paralleling of seven LM395 power transistors serving as pass element. Current limiting is provided on LM395 chip for complete overload protection.—"Linear Applications, Vol. 2," National Semiconductor, Santa Clara, CA, 1976, LB-28.

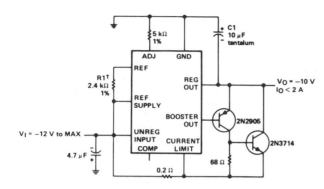

−10 V AT 2 A—Negative-voltage regulator using SN52104 or SN72304 accepts input voltage of −12 V to −40 V and uses only single external resistor to provide regulated output of −10 V with typical load regulation of 1 mV and input regulation of 0.06%. ICs are interchangeable with LM104 and LM304 respectively.— "The Linear and Interface Circuits Data Book for Design Engineers," Texas Instruments, Dallas, TX, 1973, p 5-5.

±15 V TRACKING—Single NE/SE5554 dual tracking regulator is used with pass transistors to give higher output current than 200-mA limit for each section of regulator, with close-tolerance tracking.—"Signetics Analog Data Manual," Signetics, Sunnyvale, CA, 1977, p 672–673.

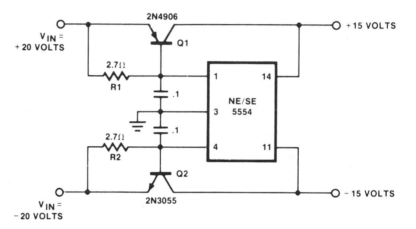

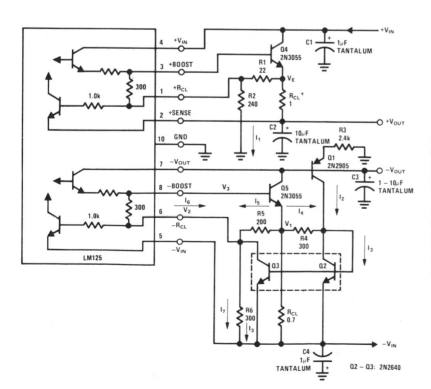

FOLDBACK CURRENT LIMITING—Reduces short-circuit output current of National LM125 dual tracking regulator sections to fraction of full-load output current, avoiding need for larger heatsink. Programmable current source is used to give constant voltage drop across R5 for negative regulator. Simple resistor divider serves same purpose for positive regulator. Design examples are given.—T. Smathers and N. Sevastopoulos, "LM125/LM126/LM127 Precision Dual Tracking Regulators," National Semiconductor, Santa Clara, CA, 1974, AN-82, p 7.

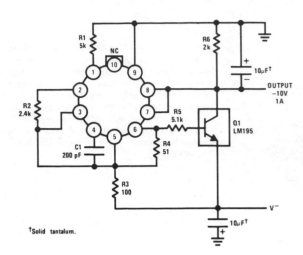

−10 V AT 1 A—National LM195 power transistor, used with LM105 regulator, provides full overload protection. Load regulation is better than 2 mV. Circuit requires only 2-V differential between input and output voltages.—R. Dobkin, "Fast IC Power Transistor with Thermal Protection," National Semiconductor, Santa Clara, CA, 1974, AN-110, p 5.

†Solid tantalum.

15 V AT 1 A WITH LOGIC SHUTDOWN—Arrangement shown provides practical method of shutting down LM340T-15 or similar regulator under control of TTL or DTL gate. Pass transistor Q1 operates as saturated transistor when logic input is high (2.4 V minimum for TTL) and Q2 is turned on. When logic input is low (below 0.4 V for TTL), Q2 and Q1 are off and regulator is in effect shut down.—"Linear Applications, Vol. 2," National Semiconductor, Santa Clara, CA, 1976, AN-103, p 11.

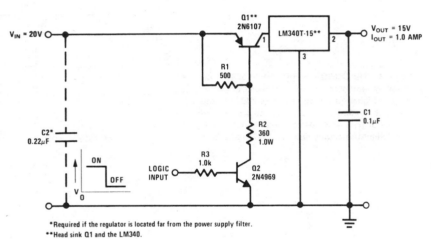

*Required if the regulator is located far from the power supply filter.
**Head sink Q1 and the LM340.

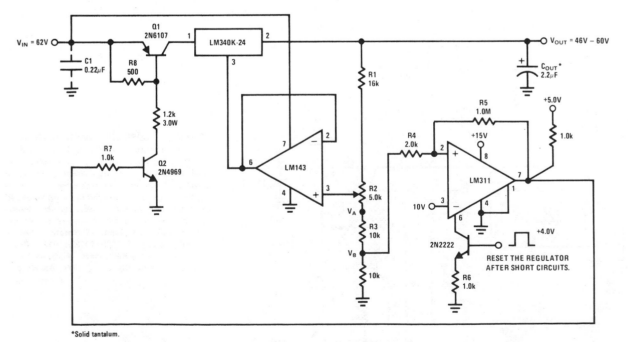

*Solid tantalum.

46–60 V FROM 62 V—Variable-output high-voltage regulator includes short-circuit and overvoltage protection. When LM340K-24 regulator has been shut down by shorted load, LM311 must be activated by applying 4-V strobe pulse to 2N2222 transistor to make Q1 close again and start regulator.—"Linear Applications, Vol. 2," National Semiconductor, Santa Clara, CA, 1976, AN-103, p 11–12.

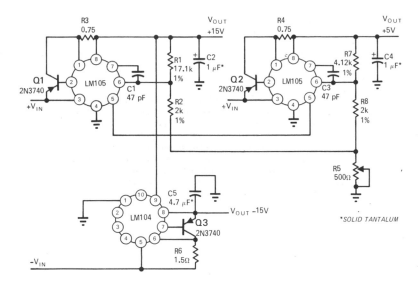

+15, +5, AND −15 V—Single potentiometer R5 serves for adjusting all three regulated output voltages simultaneously. Accuracy of adjustment is within 2%.—R. C. Dobkin, One Adjustment Controls Many Regulators, *EDN Magazine,* Nov. 1, 1970, p 33–35.

*SOLID TANTALUM

0–15 V—Addition of 307 or 301A opamp and three inexpensive components to standard three-terminal voltage regulator provides programming capability from maximum terminal voltage down to zero. With adequate heatsink, output current can be up to 1 A. Opamp A_2 provides floating reference voltage to normally grounded common terminal of A_1, with pot allowing ground to be positioned anywhere along voltage drop of 15 V across pot. Unregulated negative supply is not critical, and drain is 10 mA.—W. G. Jung, Three Components Program Regulator from Maximum to Zero, *EDN Magazine,* May 20, 1977, p 126 and 128.

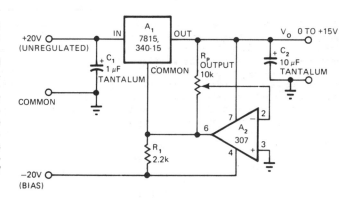

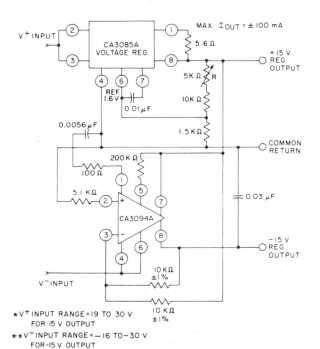

±15 V TRACKING AT 100 mA—Provides line and load regulation of 0.075% by using CA3094A programmable opamp and CA3085A series voltage regulator. V+ input range is 19 to 30 V for 15-V output, while V− input range is −16 to −30 V for −15 V output.—"Circuit Ideas for RCA Linear ICs," RCA Solid State Division, Somerville, NJ, 1977, p 18.

*V⁺ INPUT RANGE=19 TO 30 V
 FOR 15 V OUTPUT

**V⁻ INPUT RANGE=−16 TO−30 V
 FOR−15 V OUTPUT

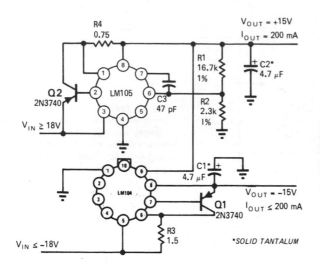

±15 V TRACKING—Arrangement uses LM104 negative regulator to track positive regulator, with both regulators adjusted simultaneously by changing R1. Inverting opamp can be added to provide negative output voltage while using positive voltage as reference.—R. C. Dobkin, One Adjustment Controls Many Regulators, *EDN Magazine*, Nov. 1, 1970, p 33–35.

BOOSTING OUTPUT CURRENT—External NPN pass transistor is added to each section of LM125 precision dual tracking regulator to increase maximum output current by factor equal to beta of transistor. To prevent overheating and destruction of pass transistors and resultant damage to regulator, series resistor R_{CL} is used to sense load current. When voltage drop across R_{CL} equals current-limit sense voltage in range of about 0.3 to 0.8 V (related to junction temperature), regulator will current-limit. Maximum load current is about 1 A for 25°C junction and 0.6 ohm for R_{CL}. LM125 provides ±15 V, LM126 provides ±12 V, and LM127 provides +5 V and −12 V.—T. Smathers and N. Sevastopoulos, "LM125/LM126/LM127 Precision Dual Tracking Regulators," National Semiconductor, Santa Clara, CA, 1974, AN-82, p 5.

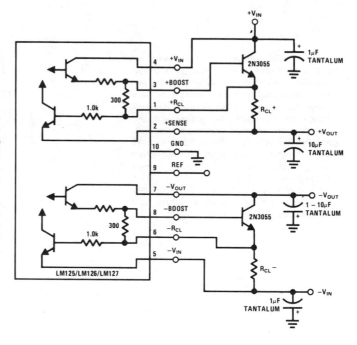

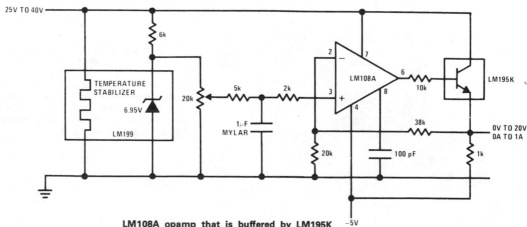

0–20 V HIGH-PRECISION—National LM199 temperature-stabilized 6.95-V reference feeds LM108A opamp that is buffered by LM195K power transistor IC which provides full overload protection.—"Linear Applications, Vol. 2," National Semiconductor, Santa Clara, CA, 1976, AN-161, p 6.

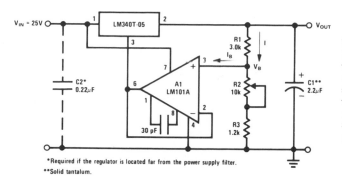

7–23 V AT 1.2–2 A—Ground terminal of LM340T-05 regulator is raised by amount equal to voltage applied to noninverting (+) input of opamp, to give output voltage set by R2 in resistive divider. Short-circuit protection and thermal shutdown are provided over full output range.—"Linear Applications, Vol. 2," National Semiconductor, Santa Clara, CA, 1976, AN-103, p 6–7.

*Required if the regulator is located far from the power supply filter.
**Solid tantalum.

0 TO −6.6 V AT 5 mA—Voltage follower A_2 buffers output that can be adjusted over full range from 0 V to zener limit with R_4. Positive supply of A_2 must go to voltage slightly more positive than +3 V common if linear output operation is required over full range.—W. G. Jung, "IC Op-Amp Cookbook," Howard W. Sams, Indianapolis, IN, 1974, p 159–160.

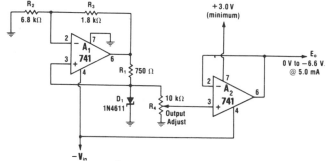

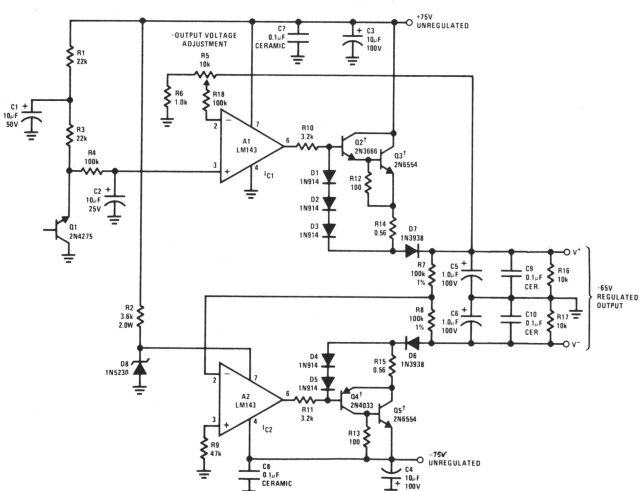

±65 V TRACKING AT 1 A—Circuit uses two LM143 high-voltage opamps in combination with zener reference and discrete power-transistor pass elements. Q1 is transistor used as stable 6.5-V zener voltage reference. Opamp A1 amplifies reference voltage from 1 to about 10 times for application through R10 to Darlington-connected transistors Q2 and Q3. Feedback resistor R5 is made variable so positive output voltage can be varied from 6.5 V to about 65 V. This output is applied to unity-gain inverting power opamp A2 to generate negative output voltage. Q2-Q5 should be on common Thermalloy 6006B or equivalent heatsink. Supply includes short-circuit protection. Maximum shorted load current is about 1.25 A.—"Linear Applications, Vol. 2," National Semiconductor, Santa Clara, CA, 1976, AN-127, p 8–9.

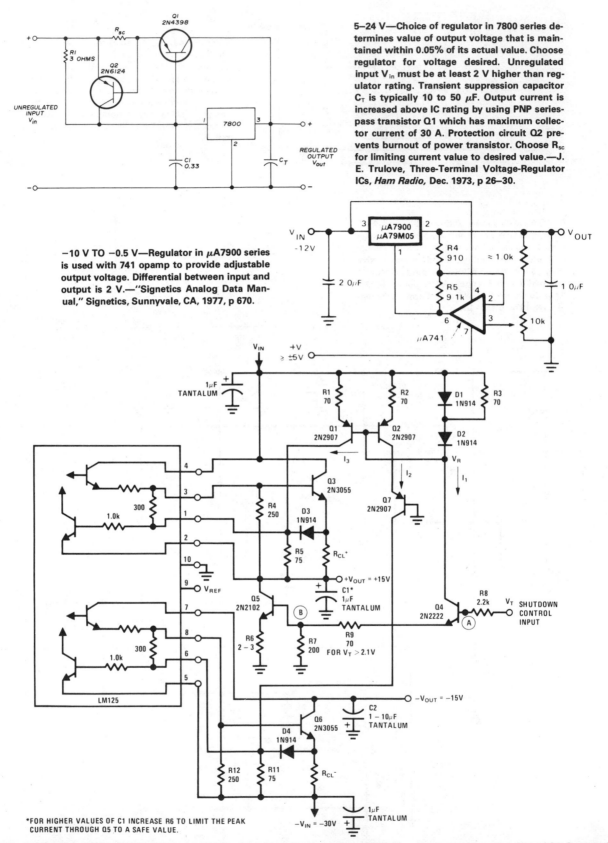

5–24 V—Choice of regulator in 7800 series determines value of output voltage that is maintained within 0.05% of its actual value. Choose regulator for voltage desired. Unregulated input V_{in} must be at least 2 V higher than regulator rating. Transient suppression capacitor C_T is typically 10 to 50 μF. Output current is increased above IC rating by using PNP series-pass transistor Q1 which has maximum collector current of 30 A. Protection circuit Q2 prevents burnout of power transistor. Choose R_{sc} for limiting current value to desired value.—J. E. Trulove, Three-Terminal Voltage-Regulator ICs, *Ham Radio,* Dec. 1973, p 26–30.

–10 V TO –0.5 V—Regulator in μA7900 series is used with 741 opamp to provide adjustable output voltage. Differential between input and output is 2 V.—"Signetics Analog Data Manual," Signetics, Sunnyvale, CA, 1977, p 670.

*FOR HIGHER VALUES OF C1 INCREASE R6 TO LIMIT THE PEAK CURRENT THROUGH Q5 TO A SAFE VALUE.

CURRENT BOOSTING WITH ELECTRONIC SHUTDOWN—Circuit provides complete shutdown for both sections of National LM125 dual tracking regulator without affecting unregulated inputs that may be powering additional equipment. Shutdown control signal is TTL-compatible, but regulator may be shut down at any desired level by adjusting values of R8 and R9. Control signal is used to short internal reference voltage of regulator to ground, thereby forcing positive and negative outputs to about +700 mV and +300 mV respectively. When shutdown signal is applied, Q4 draws current through R3 and D2, establishing voltage V_R that starts current sources Q1 and Q2. Currents I_1, i_2, and I_3 are then equal so both sides of regulator are shut down simultaneously.—T. Smathers and N. Sevastopoulos, "LM125/LM126/LM127 Precision Dual Tracking Regulators," National Semiconductor, Santa Clara, CA, 1974, AN-82, p 14.

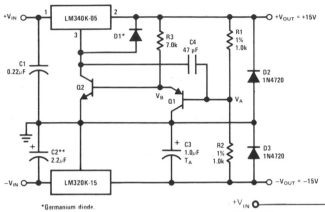

±15 V TRACKING AT 1 A—Positive regulator tracks negative. If Q1 and Q2 are perfectly matched, tracking action is unchanged over full operating temperature range, with tracking better than 100 mV. Regulation from no load to full load is 10 mV for positive side and 45 mV for negative side.—"Linear Applications, Vol. 2," National Semiconductor, Santa Clara, CA, 1976, AN-103, p 8–9.

*Germanium diode.
**Solid tantalum.

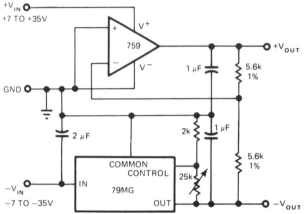

DUAL TRACKING—Uses one 759 power opamp for positive output, connected to track with 79MG negative voltage regulator having adjustable output. Common-mode range of 79MG includes ground, permitting operation from single supply. Circuit can also be built with two power opamps, one inverting and the other noninverting.—R. J. Apfel, Power Op Amps—Their Innovative Circuits and Packaging Provide Designers with More Options, *EDN Magazine*, Sept. 5, 1977, p 141–144.

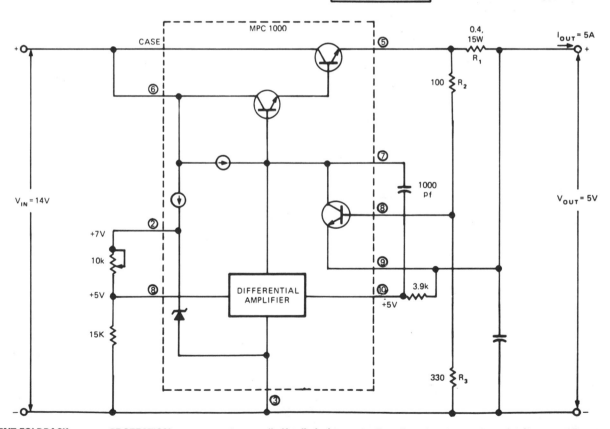

CURRENT-FOLDBACK PROTECTION—MPC1000 hybrid regulator provides regulated output of 5 V at 5 A from 14-V input. Values of components are based on foldback current of 6 A and short-circuit current of 2 A; this ensures that dissipation of regulator on short-circuit is less than dissipation at rated load. Short-circuit current is controlled by diode drop across R_1 and foldback current by drop across R_2. Article gives design equations and procedure for obtaining other output voltages. Circuit also serves to limit starting surges into capacitive load, and reduces heatsink size and transistor ratings. Returning R_3 to pin 2 of MPC1000 instead of pin 3 gives lower short-circuit current, improves efficiency, and reduces heat generation. Foldback protection is not suitable for variable-output supplies because foldback current is proportional to output voltage.—R. L. Haver, Use Current Foldback to Protect Your Voltage Regulator, *EDN Magazine,* Aug. 20, 1974, p 69–72.

28 V AT 7 A—Uses Motorola MPC1000 positive voltage regulator to provide regulated voltage for aircraft radio equipment being used at ground station. Current-limiting resistor R_{SC} is in range of 0.66 to 0.066 ohm. Use copper wire about 50% longer than calculated length and shorten step by step until required pass current is obtained; thus, start with 25 ft of No. 16, 15 ft of No. 18, 10 ft of No. 20, or 6 ft of No. 22. Input

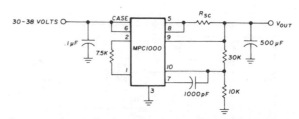

voltage is obtained from 30-V transformer and bridge rectifier.—G. L. Tater, The MPC1000— Super Regulator, *Ham Radio,* Sept. 1976, p 52–54.

4.5–34 V VARIABLE AT 1 A—National LM195 power transistor is used with LM105 regulator to give fully adjustable range of output voltages with overload protection and only 2-V input-to-output voltage differential. Load regulation is better than 2 mV.—R. Dobkin, "Fast IC Power Transistor with Thermal Protection," National Semiconductor, Santa Clara, CA, 1974, AN-110, p 4.

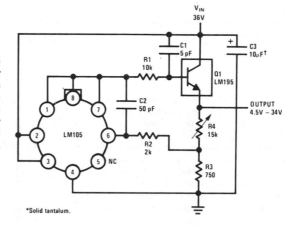

*Solid tantalum.

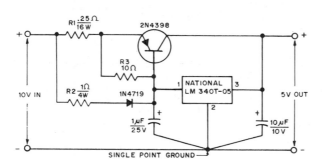

5 V AT 5 A—Current-sharing design provides short-circuit protection, safe-operating-area protection, and thermal shutdown. Typical load regulation is 1.4%.—W. R. Calbo, A High-Current, Low-Voltage Regulator for TTL Circuits, *QST,* Sept. 1975, p 44.

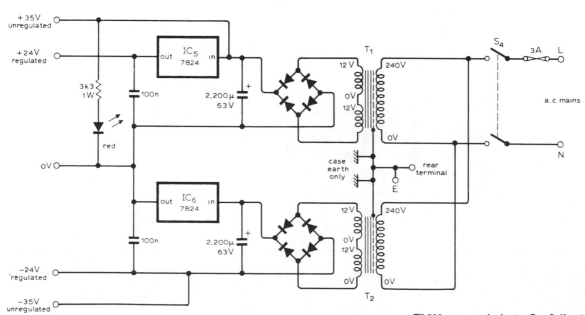

±24 V REGULATED AND ±35 V UNREGU-LATED—Developed for use with high-performance stereo preamp. Each IC regulator requires about 7 cm² of heatsink area. Red LED is TIL209 or equivalent.—D. Self, Advanced Preamplifier Design, *Wireless World,* Nov. 1976, p 41–46.

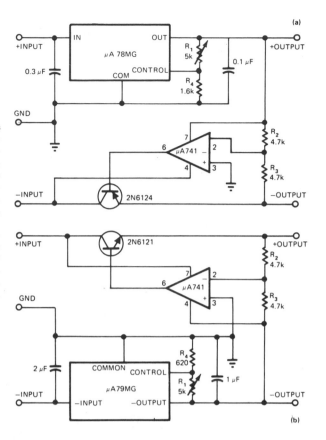

SLAVED DUAL TRACKING REGULATOR—Uses Fairchild μA78MG adjustable four-terminal regulator with opamp and power transistor for delivering output currents up to 0.5 A per side, with output voltages adjustable from ±5 V to ±20 V for component values shown. Positive side functions independently of negative side, but negative output is slave of positive output. To slave positive side, use μA79MG and 2N6121 NPN transistor as at (b). Opamp functions as inverting amplifier driving power transistor serving as series-pass element for opposite side of regulator, with R_1 adjusting both output voltages simultaneously.—A. Adamian, Dual Adjustable Tracking Regulator Delivers 0.5A/Side, *EDN Magazine*, Jan. 5, 1977, p 42.

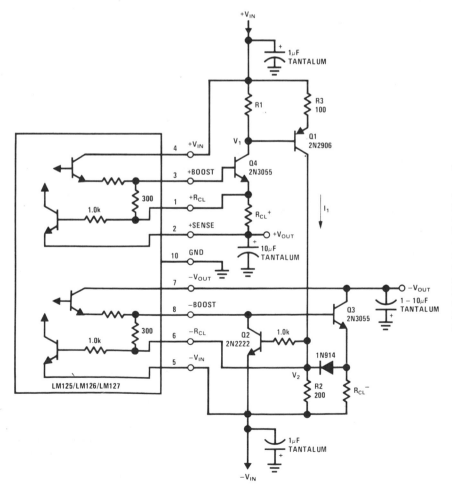

TRACKED CURRENT LIMITING—Simultaneous limiting scheme for both sections of National dual tracking regulator depends on output current of positive regulator. Voltage drop produced across R1 by positive regulator brings Q1 into conduction, with positive load current I_1 increasing until voltage drop across R2 equals negative current-limit sense voltage. Negative regulator will then current-limit, and positive side will closely follow negative output down to level of about 700 mV.—T. Smathers and N. Sevastopoulos, "LM125/LM126/LM127 Precision Dual Tracking Regulators," National Semiconductor, Santa Clara, CA, 1974, AN-82, p 13.

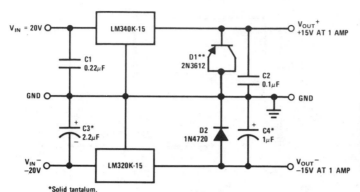

±15 V SYMMETRICAL AT 1 A—Connection shown gives same line and load regulation characteristics as for individual regulators. D1 ensures start-up of LM340K-15 under worst-case conditions of common load and 1-A load current over full temperature range.—"Linear Applications, Vol. 2," National Semiconductor, Santa Clara, CA, 1976, AN-103, p 8.

*Solid tantalum.

**Germanium diode (using a PNP germanium transistor with the collector shorted to the emitter).

Note: C1 and C2 required if regulators are located far from power supply filter.

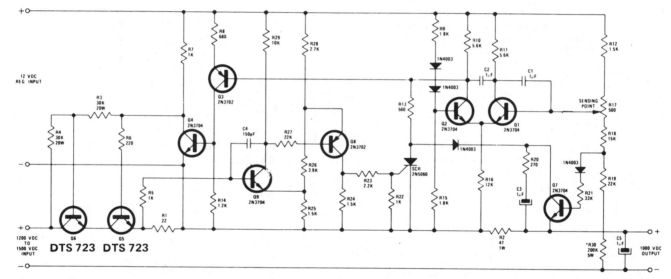

1000 V AT 100 W—Two Delco DTS-723 transistors in series function as pass element of regulator in which differential amplifier Q1-Q2 senses output voltage and compares it with reference voltage at base of Q2. Difference signal is amplified by Q3-Q4 for feed to Q5. 12-V regulated supply is referenced to high side of output voltage through R2. R1 is chosen so regulator shuts down when load current reaches 120 mA and triggers Schmitt trigger Q8-Q9 which fires SCR. When overload is removed, circuit returns to normal operation. Input voltage range of 1200–1500 V gives 0.1% regulation at full load.—"1000-Volt Linear DC Regulator," Delco, Kokomo, IN, 1974, Application Note 45.

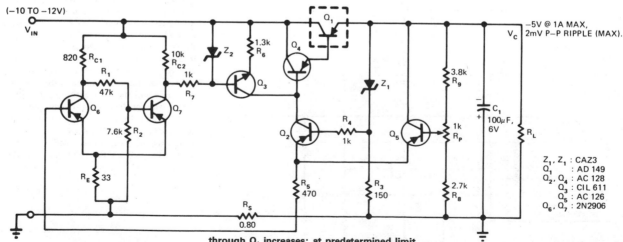

−5 V WITH PROTECTION—Switching-type short-circuit protection network uses R_7 connected to Schmitt trigger Q_6-Q_7. Ground is provided by Q_7 which is normally conducting. If output of regulator is short-circuited, current through Q_1 increases; at predetermined limit, Q_6 conducts and cuts off Q_7, breaking ground connection of R_7 and thus cutting off Q_3. Power transistor Q_1 is also cut off, and output current begins to decrease. When load current drops below another predetermined level, Q_6 again goes off and Q_7 turns on to begin another ON/OFF cycle, with switching process continuing until short-circuit is removed.—H. S. Raina and R. K. Misra, Novel Circuit Provides Short Circuit Protection, *EDN Magazine*, June 5, 1974, p 84.

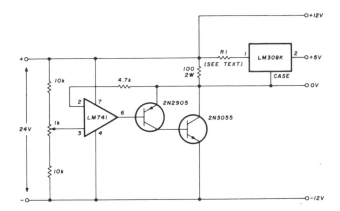

VOLTAGE ADAPTER—Bench power supply provides ±12 V and +5 V from single regulated 24-V source, for use with many ICs. Both 12-V supplies can be adjusted in same direction by varying 24-V source or in opposite directions by adjusting 1K pot. R1 is used to decrease power dissipated in LM309K voltage regulator and is normally 2.2 ohms.—J. A. Piat, Voltage Adapter for MSI/LSI Circuits, *Ham Radio*, March 1978, p 115.

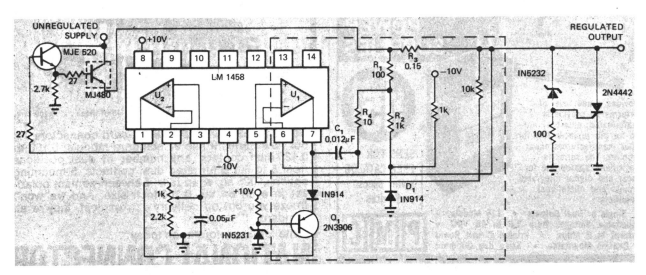

OVERVOLTAGE CROWBAR—Components within dashed lines protect regulator IC from overcurrent condition frequently encountered when zener-SCR crowbar is used across output.

Regulated output is 5 V with IC shown. Article gives operating details of circuit and equation for shutdown time, which is about 1 s.—S. J.

Pirkle, Circuit Protects Power Supply Regulator from Overcurrent, *EDN Magazine*, Feb. 5, 1973, p 89.

CHAPTER 75

Remote Control Circuits

Wired, wireless, light-beam, and other techniques are given for controlling transmitters, transceivers, receivers, motors, and other switched devices from a distance, including use of tone coders and decoders. See also Data Transmission, Instrumentation, Optoelectronic, and Repeater chapters.

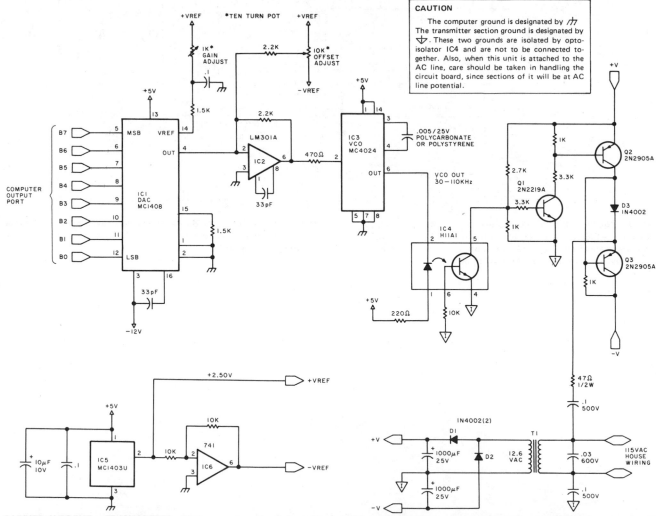

CARRIER-CURRENT TRANSMITTER—Modulates existing house wiring with high-frequency signals that can be detected by special receivers plugged into any AC outlet, for control of appliances by home computer. Applications include turning house lights on and off during owner's absence on elaborate time schedule programmed into computer. IC1 converts 8-bit data word from computer to proportional analog output current. This is converted to voltage by IC2 for control of VCO IC3 that gives frequency proportional to voltage. With values shown, range is about 30 to 110 kHz, with 256 discrete increments of frequency. Thus, input code 00000000 gives 30 kHz, 00000001 gives 30.3 kHz, and 01000000 (decimal 64) gives 49.2 kHz. Signal is applied to house wiring by 0.5-W power amplifier Q1-Q3, using optical coupling through IC4 to prevent computer circuit from interacting with house wiring. Supply voltage ±V is 11 to 13 V. T1 is 12.6-VAC 300-mA filament trans-former. IC5 is 2 to 2.5 V reference chip such as MC1403U. System uses one frequency to turn receiver on and frequency 4 kHz above or below in 8-kHz band to turn it off, for maximum of ten control channels in system. Article covers calibration of transmitter.—S. Ciarcia, Tune in and Turn on, Part 1: A Computerized Wireless AC Control System, *BYTE,* April 1978, p 114–116, 118, 120, and 122–125 (Part 2—May 1978, p 97–100 and 102).

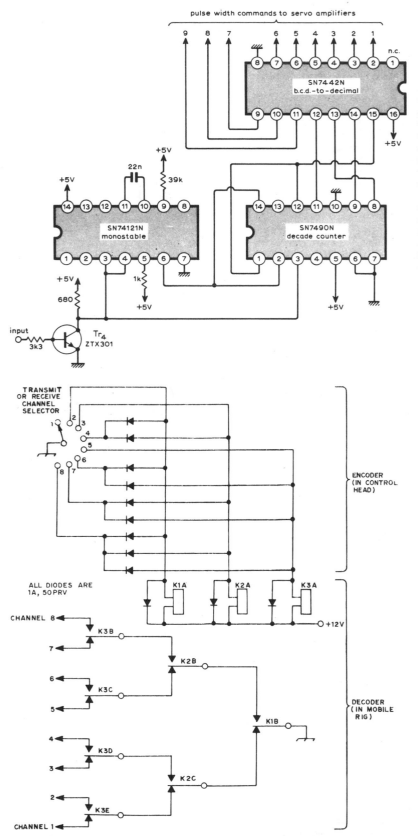

pulse width commands to servo amplifiers

NINE-CHANNEL DECODER—Circuit accepts serial information arriving over data link as series of nine varying-width pulses followed by fixed-width sync pulse, and after detection passes the nine individual commands to their respective servoamplifiers. Use of TTL ICs gives low component count for remote control system. Detection of sync pulse is done by comparing length of inverted input pulses with output of 0.6-ms monostable reference. All command pulses exceed 0.6 ms, so only 0.5-ms sync pulse clears counter to prepare for next channel-1 command pulse. Article gives operating details of system and circuits for coder and servoamplifier.—M. F. Bessant, Multi-Channel Proportional Remote Control, *Wireless World,* Oct. 1973, p 479–482.

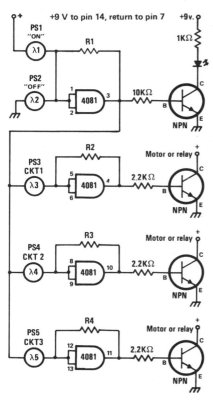

LIGHT BEAM FOR CONTROL OF MOVING TOY—Battery-powered CMOS logic is switched on and off by aiming flashlight beam at photocell, for turning small motors of model train or other powered toy on and off. Transistors can be 2N2222A for most small motors, but larger motors will require power transistors. Use high-intensity flashlight, with shield over lens to restrict beam width, so only one of five photocells is illuminated at a time. LED shows ON/OFF status of circuit. Values of R1-R4 are chosen so each gate flips logic state only when associated photocell is illuminated.—J. Sandler, 9 Projects under $9, *Modern Electronics,* Sept. 1978, p 35–39.

8 CHOICES WITH 3 WIRES—Provides remotely selected choice of eight functions, such as channels in mobile FM station, with only three wires running from control head to controlled equipment that can be in front of car. System involves converting 8-position switch selection in control head to 3-bit binary form for three control wires going to three-relay arrangement for decoding back to 8-position format. Relays are two-pole and four-pole double-throw 12-V units.—G. D. Rose, Independent 8-Channel Frequency Selection with Only Three Wires, *QST,* Aug. 1974, p 36–40.

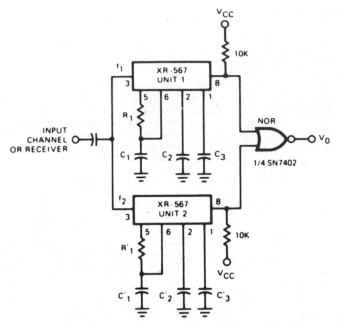

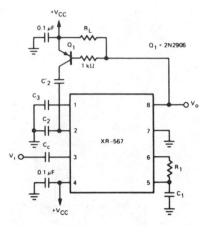

DUAL TONE DECODER—Used in communication systems where control or other information is transmitted as two simultaneous but separate tones. Circuit uses two Exar XR-567 PLL units in parallel, with resistor and capacitor values of each PLL decoder selected to provide desired center frequencies and bandwidth requirements. Supply voltage is 5–9 V.—"Phase-Locked Loop Data Book," Exar Integrated Systems, Sunnyvale, CA, 1978, p 41–48.

DUAL TIME-CONSTANT TONE DECODER—Exar XR-567 PLL system is connected as decoder having narrow bandwidth and fast response time. Circuit has two low-pass loop filter capacitors, C_2 and C'_2. With no input, pin 8 is high, Q_1 is off, and C'_2 is out of circuit. Filter then has only C_2, which is kept small for minimum response time. When in-band input tone signal is detected, pin 8 goes low, Q_1 turns on, and C'_2 is in parallel with C_2 to give narrow bandwidth. Supply voltage can be 5–9 V.—"Phase-Locked Loop Data Book," Exar Integrated Systems, Sunnyvale, CA, 1978, p 41–48.

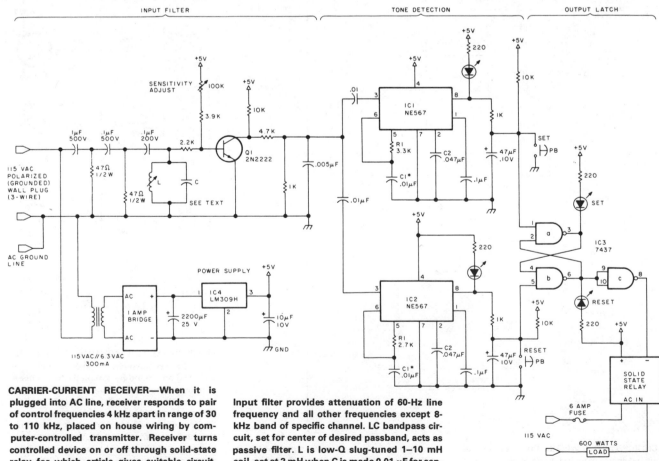

CARRIER-CURRENT RECEIVER—When it is plugged into AC line, receiver responds to pair of control frequencies 4 kHz apart in range of 30 to 110 kHz, placed on house wiring by computer-controlled transmitter. Receiver turns controlled device on or off through solid-state relay for which article gives suitable circuit. Tuned bandpass filter amplifies only that pair of frequencies assigned to its receiver, attenuating all other frequency pairs used in system. Amplified signal is sent to tone decoders IC1 and IC2, one responding to each frequency. Input filter provides attenuation of 60-Hz line frequency and all other frequencies except 8-kHz band of specific channel. LC bandpass circuit, set for center of desired passband, acts as passive filter. L is low-Q slug-tuned 1–10 mH coil, set at 2 mH when C is made 0.01 μF for center frequency of 35 kHz. Article covers operation in detail and gives procedure for determining values of R1, C1, and C2 for each detector. Solid-state output relay can be Sigma 226 RE1-5A1, rated 6 A.—S. Ciarcia, Tune in and Turn on, Part 2: An AC Wireless Remote Control System, *BYTE*, May 1978, p 97–100 and 102 (Part 1—April 1978, p 114–116, 118, 120, and 122–125).

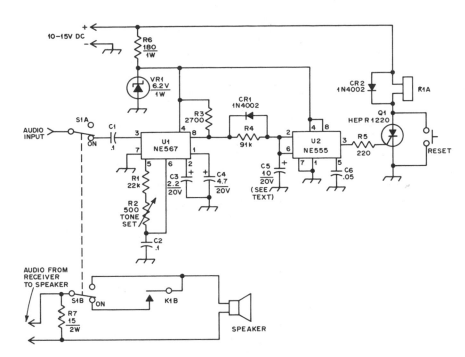

TIMED-TONE DECODER—Uses NE567 PLL and 555 timer to activate muted monitoring receiver until alerting audio tone of correct frequency and duration is received. Can be applied to almost any receiver for weather emergency alert warnings, paging calls, and similar services without having to listen continously to other traffic on channel. If received tone is within bandwidth of tone decoder, output of U1 goes nearly to zero and C5 starts to discharge through R4. When voltage at pins 2 and 6 of U2 reaches one-third of supply voltage, output of U2 goes high and triggers SCR Q1, energizing 12-V relay K1. Values shown for C4 and C5 give 1-s delay, which means triggering tone must be on at least 1 s. Once SCR is triggered, it holds relay on even after tone ceases. Pushbutton switch shorts SCR and releases relay when reset is desired. Zener provides regulated 6.2 V required for decoder. Values shown for R1, R2, and C2 give response to 450-Hz tone. Avoid use of Touch-Tone frequency, to prevent accidental triggering by those using Touch-Tone system.—J. S. Paquette, A Time-Delayed Tone Decoder, *QST*, Feb. 1977, p 16–17.

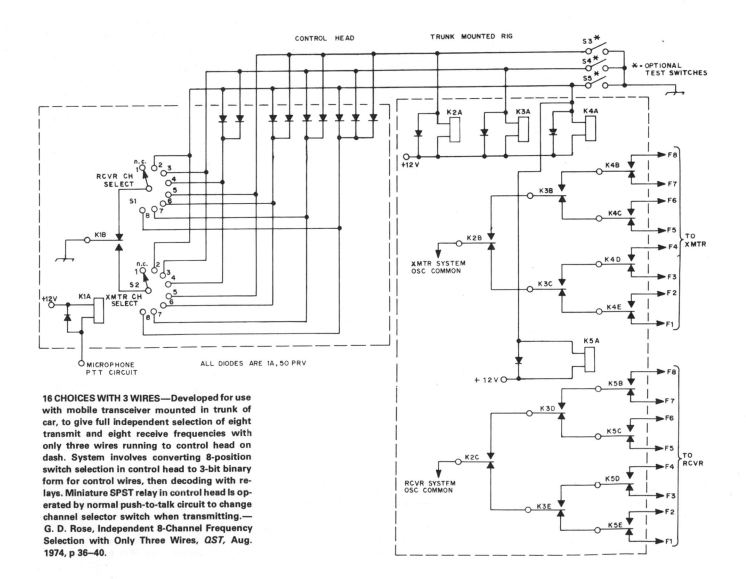

16 CHOICES WITH 3 WIRES—Developed for use with mobile transceiver mounted in trunk of car, to give full independent selection of eight transmit and eight receive frequencies with only three wires running to control head on dash. System involves converting 8-position switch selection in control head to 3-bit binary form for control wires, then decoding with relays. Miniature SPST relay in control head is operated by normal push-to-talk circuit to change channel selector switch when transmitting.—G. D. Rose, Independent 8-Channel Frequency Selection with Only Three Wires, *QST*, Aug. 1974, p 36–40.

REMOTE SWITCHING—Uses four flip-flops, each having one 4-input and one 2-input CMOS NAND gate. Momentarily grounding any input drives corresponding output high and all other outputs low. Unless power is interrupted, additional pulses on same input have no effect; circuit remains stable until some other input is momentarily grounded. Outputs can be used to drive other logic devices directly or through buffer if current required exceeds 10 mA. Can be used for remote frequency control of VHF transceiver and for other applications requiring remote selection of mutually exclusive functions.—P. Shreve, Remote-Switching Circuit, *Ham Radio*, March 1978, p 114.

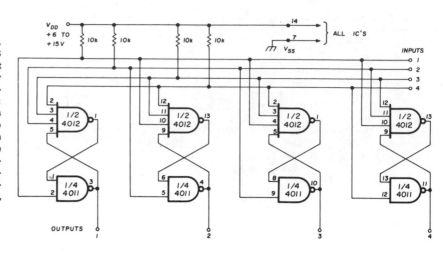

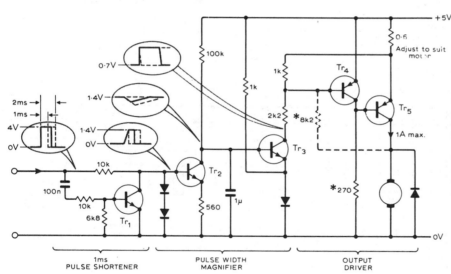

RADIO CONTROL FOR MOTOR—Proportional control system produces control pulses every 20 ms, with length of each adjustable between 1 and 2 ms. Circuit removes first 1 ms of pulse and expands remainder to produce 0–20 ms pulses for driving motor. Pulsing of motor gives smoother control than resistors, particularly at very low speeds. Transistor types are not critical. Tr_5 can be OC28. Optional dashed connection of 8.2K resistor provides foldback current/voltage protection.—M. Weston, Variable-Speed Radio Control Motor, *Wireless World*, Feb. 1978, p 59.

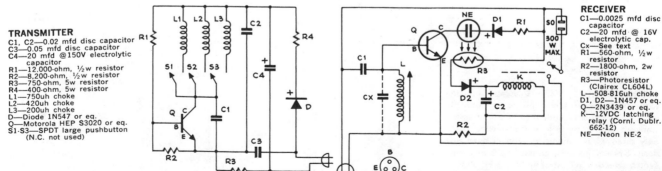

TRANSMITTER
C1, C2—0.02 mfd disc capacitor
C3—0.05 mfd disc capacitor
C4—20 mfd @150V electrolytic capacitor
R1—12,000-ohm, ½w resistor
R2—8,200-ohm, ½w resistor
R3—750-ohm, 5w resistor
R4—400-ohm, 5w resistor
L1—750uh choke
L2—420uh choke
L3—200uh choke
D—Diode 1N547 or eq.
Q—Motorola HEP S3020 or eq.
S1-S3—SPDT large pushbutton (N.C. not used)

RECEIVER
C1—0.0025 mfd disc capacitor
C2—20 mfd @ 16V electrolytic cap.
Cx—See text
R1—560-ohm, ½w resistor
R2—1800-ohm, 2w resistor
R3—Photoresistor (Clairex CL604L)
L—508-816uh choke
D1, D2—1N457 or eq.
Q—2N3439 or eq.
K—12VDC latching relay (Cornl. Dublr. 662-12)
NE—Neon NE-2

WIRELESS CONTROL—Choice of three lamps or appliances anywhere in house and garage, or even in neighboring home if on the same power transformer, can be turned on or off individually with three-channel transmitter that plugs into any wall outlet. Transmitter injects one of three tones (depending on button pushed) into house wiring. Receivers at locations of controlled devices are each tuned to one of carrier tones. Correct tone for receiver energizes neon lamp, and resulting light is picked up by photoresistor that energizes latching relay K for turning on controlled device. Relay is released by sending same tone again. Values of CX can be 0.005, 0.01, and 0.02 μF. Adjust slug of L2 for each receiver so neon comes on when assigned tone for that receiver arrives.—W. J. Hawkins, Three-Channel Wireless Switch—Use It Anywhere, *Popular Science*, Sept. 1973, p 98–99 and 121.

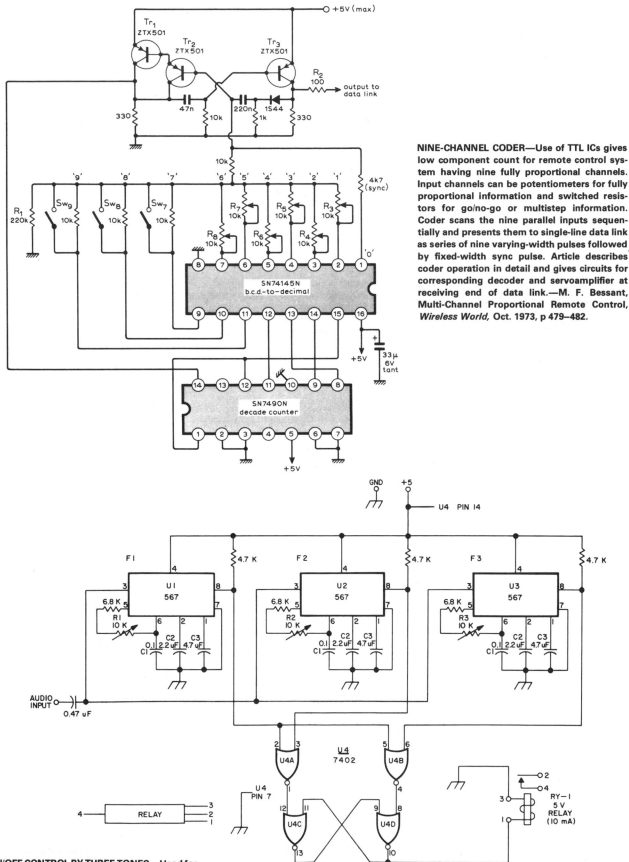

NINE-CHANNEL CODER—Use of TTL ICs gives low component count for remote control system having nine fully proportional channels. Input channels can be potentiometers for fully proportional information and switched resistors for go/no-go or multistep information. Coder scans the nine parallel inputs sequentially and presents them to single-line data link as series of nine varying-width pulses followed by fixed-width sync pulse. Article describes coder operation in detail and gives circuits for corresponding decoder and servoamplifier at receiving end of data link.—M. F. Bessant, Multi-Channel Proportional Remote Control, *Wireless World*, Oct. 1973, p 479–482.

ON/OFF CONTROL BY THREE TONES—Used for decoding two Touch-Tone digits to give operation or release of relay by remote control over wire line. Three 567 tone decoders and 7402 quad gate are adjusted to recognize tones corresponding to any two keys in given row or column on Touch-Tone keyboard. As example, * key generates 941 and 1209 Hz, and circuit can be adjusted so these two frequencies energize relay. Similarly, pushing of # key generates 941 and 1477 Hz that can be used for deenergizing relay.—W. J. Hosking, Simple New TT Decoder, *73 Magazine*, April 1976, p 52–53.

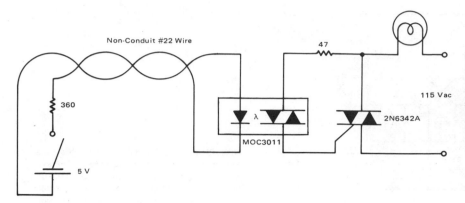

LAMP CONTROL WITHOUT CONDUIT—Motorola MOC3011 optoisolator permits control of large lamp, motor, pool pump, and other AC loads from remote location with low-voltage signal wiring while meeting building codes. Choice of triac depends on load being handled.—P. O'Neil, "Applications of the MOC3011 Triac Driver," Motorola, Phoenix, AZ, 1978, AN-780, p 5.

WIRELESS REMOTE TUNING—Frequency-to-voltage converter for transceiver responds to AF output of control receiver and feeds corresponding DC voltage to varactor tuning diode in VFO of transceiver, for remote wireless tuning. In most cases only a few volts of DC variation across varactor are sufficient, so variable audio oscillator at remote-control location need have range of only a few kilohertz.—J. Schultz, H.F. Operating—Remote Control Style, *CQ*, March 1978, p 22–23 and 90.

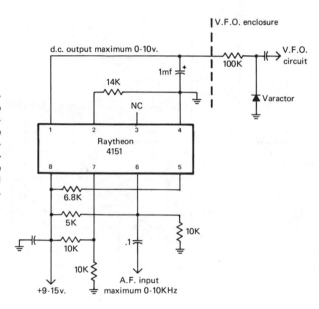

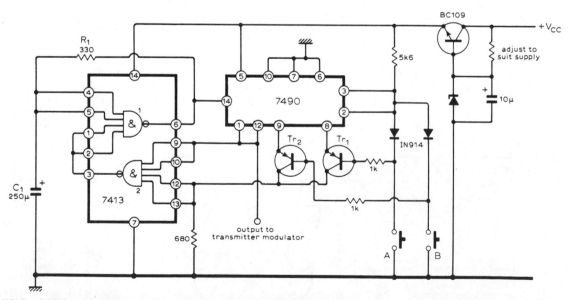

BLIP-AND-HOLD RADIO CONTROL—Coder uses two ICs to generate sequence of pulses suitable for actuators of radio control system. During standby, oscillator formed by NAND gate 1 operates at 0.5 Hz as determined by C_1 and R_1 and all four outputs of 7490 IC are zero. When switch A is closed, 7490 is clocked by negative edge of oscillator waveform and Tr_1 becomes forward-biased. Output of NAND gate 2 then drops to zero, stopping oscillator and holding outputs of 7490. When switch A is opened, outputs of 7490 again drop to zero. Many different blip-and-hold combinations can be obtained by suitable arrangement of switches and gates.—G. D. Southern, Sequence Generator for Radio Control, *Wireless World*, Jan. 1976, p 60.

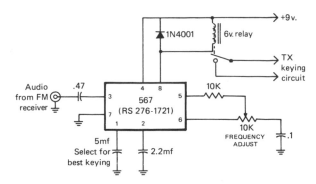

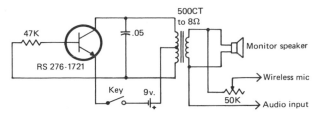

PLL TONE DECODER—Used in simple wireless FM remote control set up for keying transmitter. Keyed 500-Hz tone output of FM receiver at transmitter site acts through 567 PLL to operate 6-V relay whose contacts are in keying circuit of transmitter.—J. Schultz, H.F. Operating—Remote Control Style, *CQ,* March 1978, p 22–23 and 90.

500-Hz CONTROL TONE—Developed for use as wireless FM remote control for keying transmitter at another location by sending keyed audio tone over radio link, acoustic link fed by loudspeaker, or audio line. Frequency is about 500 Hz.—J. Schultz, H.F. Operating—Remote Control Style, *CQ,* March 1978, p 22–23 and 90.

TWO-TONE CONTROL—Used to perform simple ON/OFF auxiliary function via repeater input. Two 567 decoders energize relay for input tone of 1800 Hz, with latching, and release it for 1950 Hz. Diodes are 1N4001. Relay can be 12 or 24 V. Q1 is 2N3905, 2N3906, MPS6521, or 2N2222.— W. Hosking, A Single Tone Can Do It, *73 Magazine,* Nov. 1977, p 184–185.

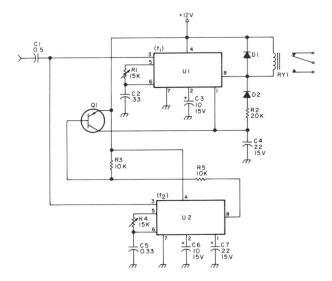

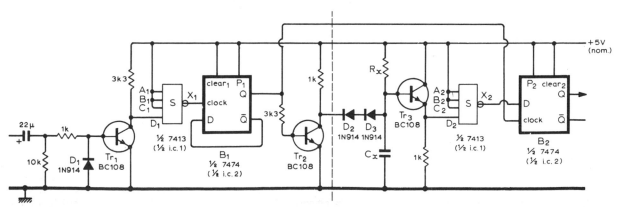

TONE DECODER—Replaces resonant reeds commonly used in multichannel radio-controlled models to detect modulation frequency being transmitted. Use of IC logic has advantage that range of audio frequencies can exceed an octave, whereas reeds cannot because they respond to second harmonic. Decoder has dig-ital high-pass characteristic that is passed through inverter to give digital low-pass char-acteristic. Values of R_x and C_x determine critical frequency; for 900 Hz, use 150,000 ohms and 0.015 μF. To obtain *n* nonoverlapping bandpass characteristics, *n* − 1 basic elements with dif-ferent critical frequencies are required; com-ponents to left of dashed line may be common to all these elements. Article covers multichan-nel systems in detail, along with use of time-division multiplexing.—C. Attenborough, Radio Control Tone Decoder, *Wireless World,* Dec. 1973, p 593–594.

CHAPTER 76
Repeater Circuits

Audible and subaudible tone generators and decoders provide access to desired FM repeater and give autopatch for telephone connections. Also included are time-out timers, phone-ring-counting control, microprocessor control, carrier-operated control, VOX, and lightning detector for remote site.

TIME-OUT WARNING—Provides either visual or audible warning after preset interval up to 3 min, when pin 1 of first NE555 is grounded by push-to-talk switch on microphone of mobile transmitter. When preset limit is reached, pin 3 provides ground for second NE555 that makes it pulse at 2-Hz rate to warn that time limit for use of repeater is being approached.—F. Sharp, Time-Out Warning, *QST*, Oct. 1976, p 67.

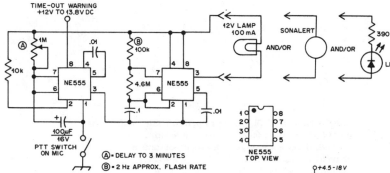

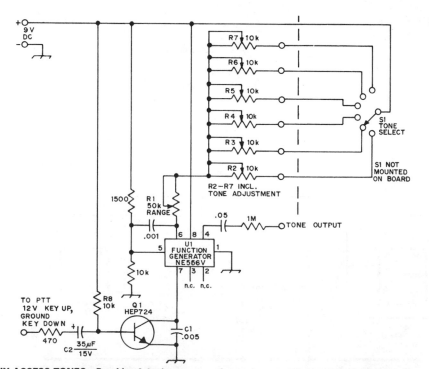

SIX ACCESS TONES—Provides 0.4-s bursts at choice of six audio frequencies, for access to up to six different repeaters. Value of C1 in transistor circuit determines duration of burst. AF oscillator uses Signetics NE566V phase-locked loop, with tone frequencies determined by C1 and R1 plus R2 through R7. To adjust initially, remove Q1 from circuit so oscillator runs con-tinuously, connect frequency counter to junction of 0.05-μF capacitor and 1-megohm resistor, set all pots at minimum, adjust R1 for 2500 Hz, set selector switch to position 1, and adjust corresponding control for desired frequency. Repeat for other pots.—G. M. Dickson, A Tone-Burst Generator for Repeater Access, *QST*, April 1974, p 30–31.

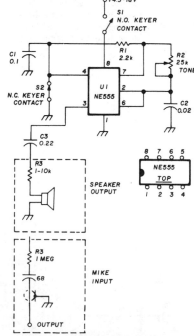

TIMER IS AUTOPATCH KEYER—Simple keyer oscillator using NE555 timer was designed for autopatch in repeater having decoder bandpass from 2980 to 3080 Hz. Adjust R2 to 3042 Hz. Output options for loudspeaker and microphone are shown. Adjust R3 for required input/output level; use variable resistor if desired. Normally closed keyer contacts can also be connected between pin 7 and ground. Supply can be 9-V transistor radio battery.—E. Noll, Circuits and Techniques, *Ham Radio*, April 1976, p 40–43.

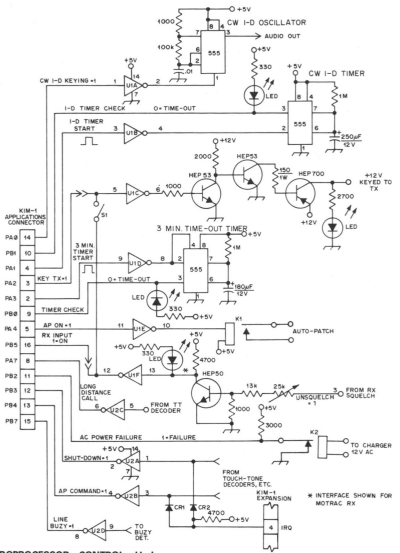

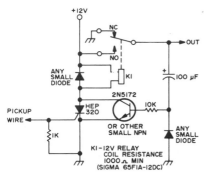

LIGHTNING DETECTOR—Uses 20-foot wire strung around repeater house to pick up pulses induced by lightning. Keep wire well away from antenna and transmitter. Pulse at SCR gate turns it on and energizes relay that activates signaling device at desired location. Circuit automatically resets itself after capacitor discharges through 10K resistor.—P. A. Stark, Simple Lightning Detector, *73 Magazine,* April 1973, p 85.

MICROPROCESSOR CONTROL—Under program control, interface for MOS Technology KIM-1 microcomputer turns on transmitter of repeater when signal arrives at receiver; provides 3-min timer, CW ID timer, and tail or delay at end of transmission; and generates Morse code CW ID. Article gives flow charts and programming for basic functions, along with complete autopatch control routine. CR1 and CR2 are small-signal silicon diodes such as 1N914. U1 and U2 are 7404 TTL hex inverters.—C. M. Robbins, The Microprocessor and Repeater Control, *QST,* Jan. 1977, p 30–34.

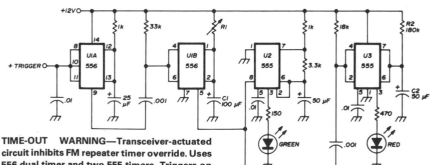

TIME-OUT WARNING—Transceiver-actuated circuit inhibits FM repeater timer override. Uses 556 dual timer and two 555 timers. Triggers on positive step-input voltage from transceiver when PTT switch is pushed. If negative triggering is preferred, omit U1A. Values of R1 and C1 provide delay that is 10 s less than repeater timer. With 60-s repeater, delay should be 50 s. Flip-flop U2 flashes green LED 80 times per minute during this delay. After 50 s, U2 is disabled and U3 flashes red LED for 10 s as indication that transmission must stop to avoid timing out repeater. At end of 10 s, red LED goes out and cycle is completed. If transmission time is less than that of repeater timer, indicator is recycled when PTT switch is pressed again. R1-C1 determine green flash time, and R2-C2 determine red flash time.—H. M. Berlin, Time-Out Warning Indicator for FM Repeater Users, *Ham Radio,* June 1976, p 62–63.

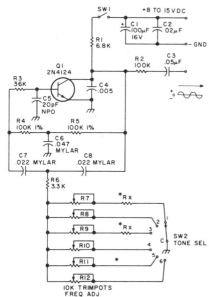

SUBAUDIO TONES—Six-channel subaudible encoder uses twin-T oscillator covering 93 to 170 Hz. Tones are adjustable with 20-turn 10K trimpots. Used with 2-meter amateur transmitters to access and maintain signal through repeater having subaudible tone decoder. When transmitted signal opens up receiver of repeater, subaudible tone on incoming audio closes relay and permits transmitter to key up and repeat signal. Choice of tones permits use of different repeaters in given area. For 93–107 Hz, use 12K for R_x; for 98–116 Hz, use 8.2K; and for 114–170 Hz, use jumper. Article gives construction details.—W. G. Moneysmith, Subaudible Tone Encoder, *73 Magazine,* Oct. 1977, p 52–53.

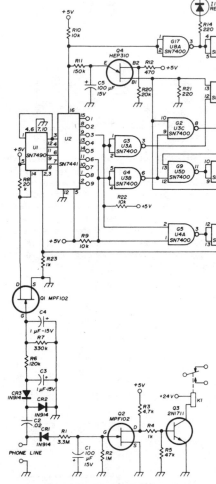

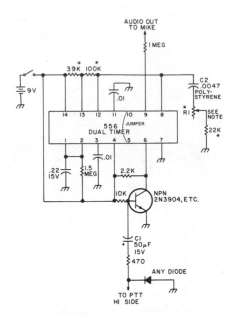

PHONE-RING REMOTE CONTROL—Repeater or other unattended equipment can be turned on or off with ordinary telephone. Phone at remote station is called and allowed to ring three times.

Caller then hangs up, waits 20 s, redials number and lets it ring three times again, then hangs up. Circuit then performs desired control function. Any combination of rings can be used as long as total is less than nine. Decoder U2 is programmed by moving two jumper wires to various outputs of U2. Relay K2 is chosen to give desired momentary, latching, or stepping function. Relay K1 is used for validating phone line. If remote station keying voltage is taken through contacts of K1, interruption of phone line prevents activation of transmitter. C1 stores voltage during brief interruptions such as when phone is ringing. Article gives detailed explanation of ring-counting circuit. LEDs I1-I4 indicate status of control sequence and aid in troubleshooting. K1 and K2 are sensitive DPDT relays with 8000-ohm coils. R11 is selected for desired time setting.—R. C. Heptig, Automatic Telephone Controller for Your Repeater, *Ham Radio,* Nov. 1974, p 44–48.

1800-Hz TONE BURST—Developed to provide access to repeater requiring accurate tone frequency. Half of 556 dual timer serves as mono MVBR having ON time of about 400 ms. Other half is free-running oscillator that is disabled when mono goes low. Transistor starts tone burst when push-to-talk switch is closed. For frequency stability, resistors with asterisk should be cermet or wirewound. R1 is 15- or 20-turn trimmer pot having low temperature coefficient and giving about 30 Hz change per turn.—L. Meyer, One IC Tone Burster, *73 Magazine,* April 1976, p 55.

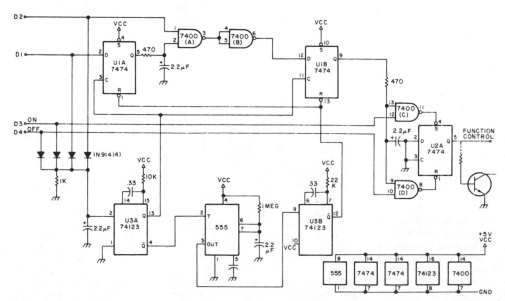

FUNCTION DECODER—Simple controller for single repeater uses 3-digit control code (523) to turn repeater on and 524 to turn it off. Digits must be in correct sequence. Digit decoder (not shown) uses TTL-compatible inputs which go high when digit is decoded. Four 1N914 diodes form OR gate that triggers U3A to create clock pulse with each digit received. Other output of U3A triggers 555 timer U5, set for delay of about 8 s. At end of delay, timer triggers U3B to reset all logic except for output stage. Regulated VCC of 5 V is obtained from 7805 regulator connected to 12 V.—W. J. Hosking, Simple Sequential Decoder, *73 Magazine,* Jan. 1978, p 166–167.

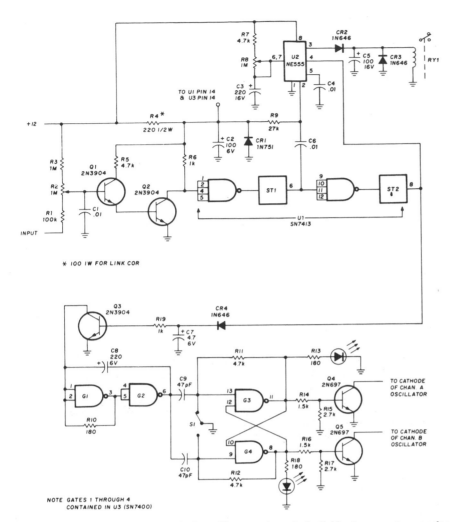

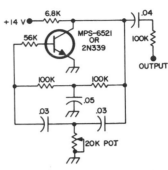

100 Hz—Simple and stable subaudible tone generator serves for access to FM repeaters. Frequency is adjusted to that of repeater with 20K pot. Will operate from car battery.—Circuits, *73 Magazine*, Sept. 1973, p 143.

COR FOR LINKING—Carrier-operated relay will operate repeater, serve as guard receiver for repeater input channels, and provide loudspeaker muting when no station is being received. Simple search-lock feature following CR4 controls two channels, for linking two repeaters or using two-channel drive receiver. Q1 and Q2 are connected as Darlington amplifier for negative-going control signals, as found in vacuum-tube receivers. Dual Schmitt trigger U1 provides positive ON/OFF action. Time-out is controlled by setting of R8 and value of C3. To monitor repeater or simplex channel for call without listening to other conversations, set timer for about 5 s. When call comes in, first few words will be at normal volume so call can be identified. At time-out, volume will drop to low level. If call is for you, disable COR for normal listening. Search-lock uses single SN7400, with gates G1 and G2 connected as oscillator and gates G3 and G4 as dual D flip-flop. Q3 acts as lock to stop oscillator. With no input carrier, Q3 is off and oscillator makes Q4 and Q5 switch between channels A and B alternately. If signal arrives on one channel, oscillator stops on it and relay closes, bringing repeater transmitter on.—R. C. Heptig and R. D. Shriner, Carrier-Operated Relay for Repeater Linking, *Ham Radio*, July 1976, p 57–59.

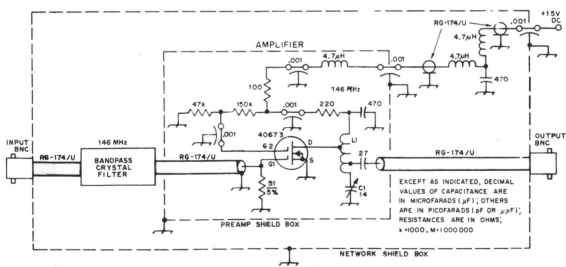

146-MHz RECEIVER PREAMP—Uses 146-MHz bandpass crystal filter to suppress front-end intermodulation-distortion products (IMD) in VHF repeater circuits while providing gain of 6–8 dB to overcome filter insertion loss. Also helps front-end overload problems from strong adjacent-channel amateur signals. Filter response is down 40 dB at ±38 kHz and down over 50 dB beyond 60 kHz from filter center frequency. Filter used is Piezo Technology TM-4133VBP. Input and output for filter should be exactly 50 ohms.—J. M. Hood, Monolithic Crystal Filter Application in Amateur VHF Repeaters, *QST*, July 1975, p 27–29 and 48.

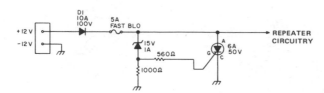

CROWBAR—Developed to protect portable repeaters from reverse or excessive voltage when operating on emergency power supply. Zener voltage rating determines maximum voltage that can reach repeater. Diode prevents damage by incorrect polarity. Use fast-blowing fuse.—Circuits, *73 Magazine*, April 1977, p 164.

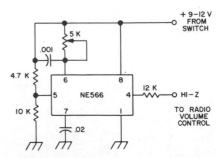

TONE BURST FOR MIKE—Uses NE566 PLL function generator to provide tone entry into 2-meter repeaters. Output is fed into volume control of any AM transistor radio, and loudspeaker is held in front of transceiver microphone when tone is desired.—F. J. Derfler, An Acoustically Coupled Digital Keyed Squeaker for Tone Burst Entry, *73 Magazine*, Aug. 1973, p 27–30.

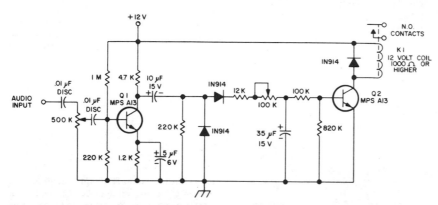

DELAYED VOX—Minimizes unintentional triggering of repeater by using timer requiring 3 s of continuous carrier and audio on input receiver of repeater to turn on transmitter. Repeater is then controlled by its own carrier-operated relay until 20-s lapse in transmission occurs. System then requires another 3-s carrier to restore normal operation. Q1 is common-emitter amplifier with gain adjusted by 500K pot. Audio peaks from Q1 feed full-wave diode detector through 10-μF electrolytic. Series resistance in path through 35-μF electrolytic controls its charge-up time and thus controls delay of relay pickup. Adjust 100K pot to give desired turn-on delay in range from fraction of second to about 4 s. Resistance divider shown provides relay dropout delay of about 20 s, which prevents VOX circuit from dropping out repeater during lapses in speech.—J. Everhart, A Delayed VOX for Repeaters, *73 Magazine*, April 1974, p 17 and 19–20.

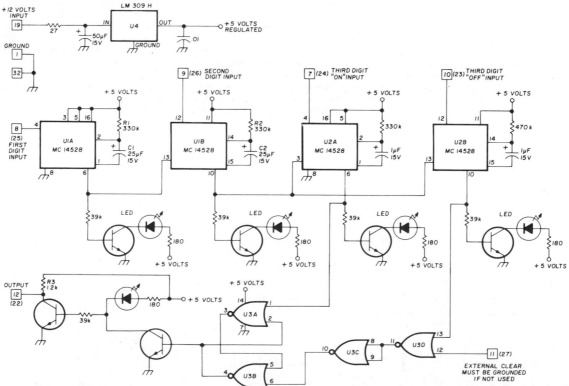

CONTROL-FUNCTION DECODER—Circuit detects predetermined 3-digit sequence of Touch-Tone signals and sets flip-flop to provide required output for activating desired function. Another 3-digit sequence, differing in only third digit, resets flip-flop for turning off controlled device. Circuit uses two dual mono MVBRs. Output of U1A goes to reset terminal of U1B. Output of U1B goes to reset terminals of monos U2A and U2B. Mono outputs drive RS flip-flop to provide output required for desired control function, such as control of autopath, switching repeater mode from carrier to tone-access, and switching to remote receiver. Q1-Q6 can be almost any NPN silicon transistors, such as 2N3904. U3 can be MC14001 or CD4001. Numbers in boxes and parentheses refer to edge connector pins.—T. E. Doyle, Control Function Decoder, *Ham Radio*, March 1977, p 66–67.

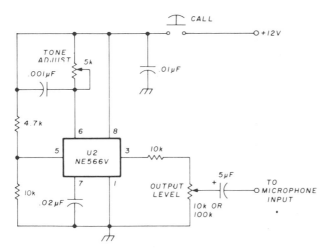

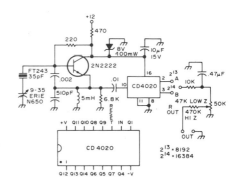

CONTROL-TONE GENERATOR—Uses NE566V PLL as tone generator, directly connected to microphone input of transmitter for activating loudspeaker in receiver being called. Operation is similar to that of paging units using selective call tones. Output level of PLL is adjusted so tone is same amplitude as voice.—K. Wyatt, Private Call System for VHF FM, *Ham Radio*, Sept. 1977, p 62–64.

SUBAUDIBLE TONE ENCODER—Simple crystal-controlled oscillator drives CD4020 CMOS divider to give output frequency below normal voice range, for providing tone access to repeater. With 1.120-MHz crystal and division ratio of 8192, output is 136.7 Hz. Output circuit is RC filter that converts square wave to triangle, with pot setting level. Article tells how to choose crystals for other output frequencies and division ratios.—C. Haines, Jr., Go Tone for Ten, *73 Magazine*, Dec. 1976, p 22–23.

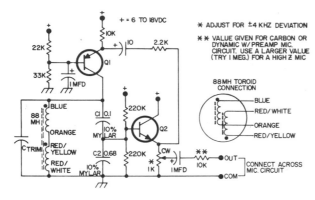

1800-Hz COMMAND OSCILLATOR—Connects across microphone leads of FM transceiver, to produce control (command) tone for entry to repeater or for other purposes. Switch may be placed in supply lead if desired. Q1 can be 2N404 or one of transistors in 2N1303, 2N2904, 2N3638, 2N6516, or 2N6533 series. Q2 can be 2N1308, 2N2712, 2N3565, 2N3569, 2N6513, or equivalent. C_{TRIM} is 0.0062 μF.—Circuits, *73 Magazine*, April 1973, p 132.

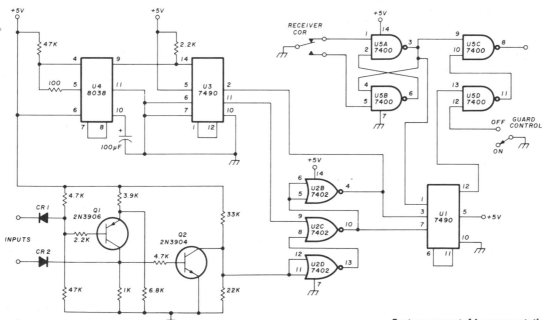

ACCESS CONTROL FOR OPEN REPEATER—Permits repeater to run open, for user access without access tone, when there is no outside interference on input frequency. Carrier-operated relay (COR) pulse is shaped by 7400 gates U5A and U5B, for keying transmitter through U5C as long as output of U5D is high. (Turning guard control switch off latches output of U5D high, letting repeater run open.) When control switch is on, repeater can be accessed only by use of guarded input applied to Q1, such as 2000-Hz tone burst, 1336-Hz Touch-Tone signal, or 110.9-Hz private line. Repeater then remains open for 5 s after duration of each transmission. System prevents fringe-area station from blocking repeater access for local users. When receiver squelch is operated 3 times in succession by signal not having one of access tones, input is automatically guarded for 15 min by timer U3-U4 unless accepted access tone arrives.—R. B. Shreve, Troubleshooting Logic Circuits, *Ham Radio*, Feb. 1977, p 56–59.

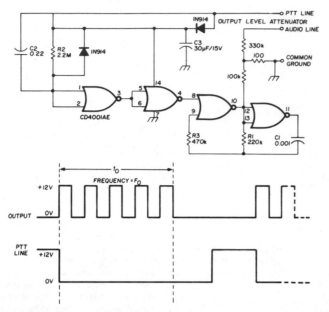

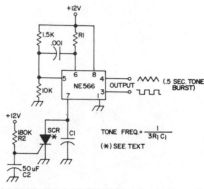

ACCESS TONE-BURST—Generates 1-s tone burst at specified audio frequency such as 1800 Hz each time transmitter is energized, to provide access to a desired repeater. Uses CMOS CD4001AE quad NOR gate, which is small enough so entire circuit can be fitted in microphone housing. Gates are connected as astable MVBR, with ON time of tone burst determined by R2C2 time constant, triggered by push-to-talk switch (PTT). Large capacitor C3 powers MVBR during burst, with capacitor normally charged to 12 V through PTT circuit during receive. Supply voltage is taken through PTT relay in transmitter.—G. Hinkle, Tone-Burst Generator for Repeater Accessing, *Ham Radio*, Sept. 1977, p 68–69.

ACCESS-TONE GENERATOR—Produces audio burst for 0.5 s at frequency determined by R1 and C1, for use with FM transceiver as subaudible tone generator to access repeaters. Provides buffered outputs of square waves as well as triangle waves. Uses NE566 function generator as voltage-controlled oscillator having excellent stability and linearity. R1 should be between 2K and 20K. SCR should be type that triggers at 70 μA.—E. Kanter, PLL IC Applications for Hams, *73 Magazine*, Sept. 1973, p 47–49.

TIME-OUT TIMER—Prevents unnecessary repeater timeouts by generating warning tone in loudspeaker of mobile transceiver when transmitter is on too long. Normal push-to-talk line is broken and connected to points A and D. Pushing talk button starts timer. Time in seconds is 1.1 R3C3, with R3 in megohms and C3 in microfarads. Thus, 1 megohm and 27 μF give 30 s, while 2 megohms and 81 μF give 180 s.—J. A. Kvochick, Keeping the Wind Down, *73 Magazine*, Feb. 1977, p 50.

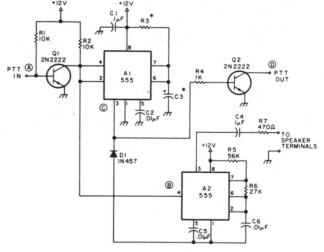

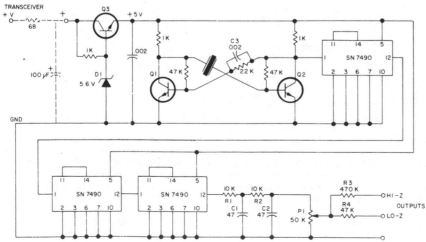

65–240 Hz—Output of 65–240 kHz crystal oscillator is divided by 1000 in three SN7490s to give 65–240 Hz tone required for access to amateur FM repeater. Choose crystal to give exactly desired frequency. Q1 and Q2 are MPS 6513, and Q3 is 2N1613. Supply can be 9 to 15 VDC from transceiver or from 9-V battery.—P. H. Wiese, Rock Solid Subaudible Tone Generator, *73 Magazine*, April 1974, p 79–81.

CHAPTER 77
Resistance Measuring Circuits

Includes ohmmeter circuits of various types for measuring resistances from under 0.001 ohm to 500 megohms, along with continuity checkers providing audible or LED indications, potentiometer tester, RLC bridges, and AC ohmmeter for nonpolarized measurements. See also Instrumentation and Test chapters.

AUDIBLE OHMMETER—Circuit is built around 555 timer. Resistance detection range is from 0 ohms to about 10 megohms. At high end of range, output is series of clicks from loudspeaker, and at low end is high-pitched tone. Intermediate resistance values produce different tones. Current through unknown resistance is only a few microamperes, so semiconductor junctions can be checked without damage. R2 sets frequency for 0 ohms. Can also be used as code practice oscillator if R_x terminals are shorted and ground lead of pin 1 is keyed.—J. Schultz, An Ohmmeter Potpourri, *CQ*, June 1978, p 32–33.

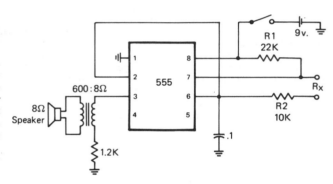

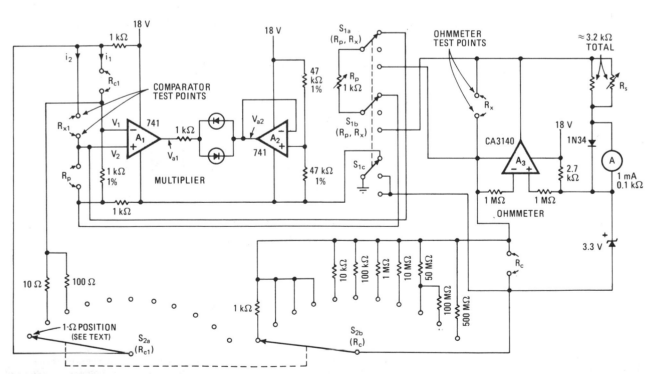

500-MEGOHM LINEAR-SCALE OHMMETER—Resistance multiplier using 741 opamp reduces current drain when measuring low resistance values (below 10 ohms). LEDs indicate voltage difference between outputs of A_1 and A_2, as guide for minimizing difference during measurement. R_s sets meter for full-scale deflection when measuring resistor value equal to standard R_c.—E. H. Armanino, Extending the Range of the Linear-Scale Ohmmeter, *Electronics*, Dec. 22, 1977, p 93–94.

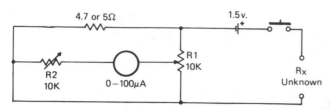

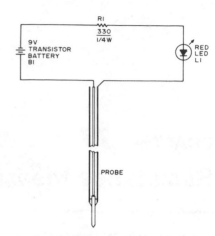

BRIDGE FOR 0.1–5 OHMS—Set meter reading to zero with multiturn pot R1 when test leads are shorted. Simple one-point calibration is made by using known low resistance for about midrange value and setting meter to conve-nient scale marking by adjusting R2. Calibration curve is then made for meter scale by using other normal resistances.—J. Schultz, An Ohmmeter Potpourri, *CQ*, June 1978, p 32–33.

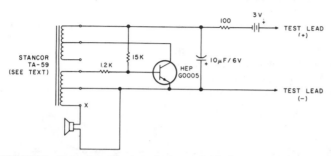

MOISTURE TESTER—Simple probe tells when plants need watering. Amount of moisture and minerals in soil together determine current available for LED. Almost full brilliance indicates adequate moisture, and no illumination means plant needs water badly. Probe can be No. 14 wire filed to point and centered with epoxy in ³/₁₆-inch copper tubing, or simply two stiff probe wires about 1 inch apart.—W. L. MacDowell, The Violet Tester, *73 Magazine*, May 1975, p 52–53.

AUDIBLE OHMMETER—Volume and/or pitch of one-transistor audio oscillator varies with resis-tance across test terminals, to give audible in-dication of continuity and relative resistance without looking at meter. Can also be used as transistor and diode tester, signal injector, code practice oscillator, and CW monitor. Oscillator uses transistor-type transformer having center-tapped windings, connected as shown to give three-winding transformer. Operates from two 1.5-V cells in series. Will respond to resistance values from short to about 100K. Volume in-creases and pitch rises as resistance is in-creased. For very low resistance, tone resem-bles croaking of frog.—Build the El Sapo Tester, *73 Magazine*, Dec. 1977, p 184–185.

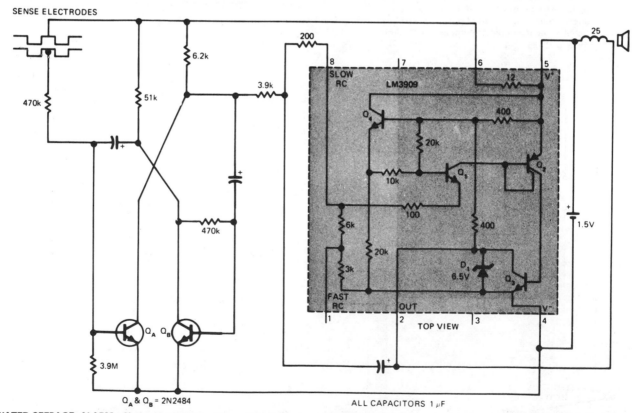

WATER-SEEPAGE ALARM—National LM3909 flasher IC operating from 1.5-V cell provides safe water-seepage alarm for potentially damp floors because there is no connection to power line. When sensing electrodes pass about 0.25 μA through moisture, pair of 2N2484 transistors (Q_A and Q_B) become astable MVBR operating at rate that starts at 1 Hz and increases with leak-age between electrodes. Pulse waveform ap-plied to pin 8 of IC varies timing current of flasher, resulting in distinctively modulated tone output for loudspeaker of alarm. Sensors can be two strips of stainless steel on insulators or zigzag path in copper of printed-wiring board. Place damp finger across gap to test alarm.—P. Lefferts, Power-Miser Flasher IC Has Many Novel Applications, *EDN Magazine*, March 20, 1976, p 59–66.

LINEAR-SCALE OHMMETER—Unknown resistance value is arithmetic product of standard resistor value and current reading in milliamperes. With 1K standard resistor, deflections from 0 to 1 A correspond to resistance readings from 0 to 1K. Requires no calibration and no zero adjustment. Can be made multirange by switching in different standard resistors.—J. Schultz, An Ohmmeter Potpourri, *CQ*, June 1978, p 32–33.

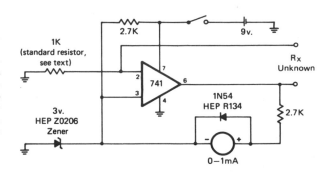

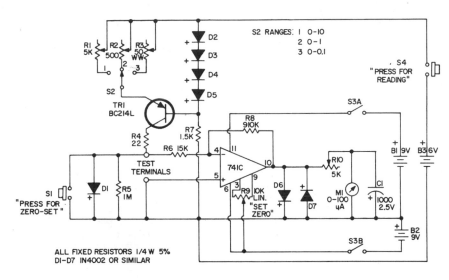

THREE-RANGE MILLIOHMMETER—Solid-state design measures resistance values accurately down to less than 0.001 ohm. Full-scale values for ranges are 10, 1, and 0.1 ohms.—Circuits, *73 Magazine*, July 1974, p 80.

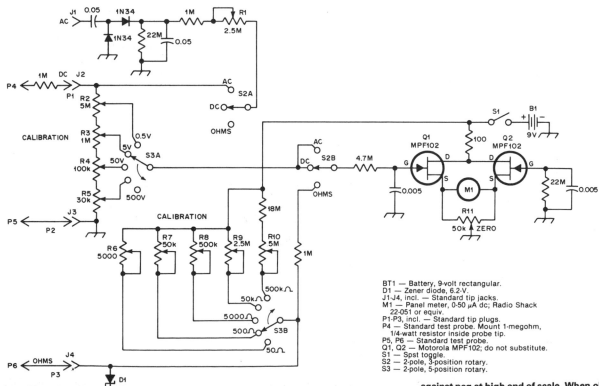

BT1 — Battery, 9-volt rectangular.
D1 — Zener diode, 6.2-V.
J1-J4, incl. — Standard tip jacks.
M1 — Panel meter, 0-50 μA dc; Radio Shack 22-051 or equiv.
P1-P3, incl. — Standard tip plugs.
P4 — Standard test probe. Mount 1-megohm, 1/4-watt resistor inside probe tip.
P5, P6 — Standard test probe.
Q1, Q2 — Motorola MPF102; do not substitute.
S1 — Spst toggle.
S2 — 2-pole, 3-position rotary.
S3 — 2-pole, 5-position rotary.

FET VOM—Use two FETs in balanced circuit. Meter reads zero when circuit is balanced with R11. Values being measured produce imbalance linearly proportional to output voltage of bridge. Resistance measurements use linear ohms-readout system, with single meter scale serving for all resistance and voltage measurements. In ohms position, meter will rest gently against peg at high end of scale. When ohmmeter leads are shorted, zero pot is adjusted so meter reads 0 ohms. Article covers construction and calibration.—J. Rusgrove, An FET Volt-Ohmmeter with Linear Ohms Readout, *QST*, March 1978, p 16–18.

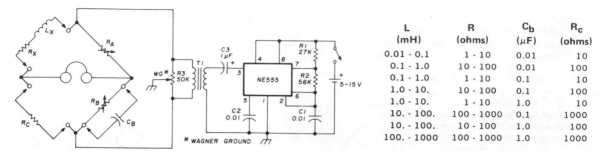

L (mH)	R (ohms)	C_b (μF)	R_c (ohms)
0.01 - 0.1	1 - 10	0.01	10
0.1 - 1.0	10 - 100	0.01	100
0.1 - 1.0	1 - 10	0.1	10
1.0 - 10.	10 - 100	0.1	100
1.0 - 10.	1 - 10	1.0	10
10. - 100.	100 - 1000	0.1	1000
10. - 100.	10 - 100	1.0	100
100. - 1000	100 - 1000	1.0	1000

RLC BRIDGE—Maxwell bridge uses only one reactive element for measuring resistance, inductance, and capacitance. Wagner ground balances stray internal capacitances to ground to obtain perfect null. Measurement ranges are shown in table. Over fixed range, R_A can be calibrated to read inductance values directly. R_A and R_B can be calibrated initially over their variable ranges by using standard resistors. Measurements are not affected by frequency of driving source. Circuit is set up as shown for measuring inductance. If standard inductance is used in place of L_x, unknown capacitor at C_B can be measured. Signetics NE555 is connected as astable oscillator running at about 1000 Hz with values shown for R and C, drawing 6.5 mA from 9-V battery. Article covers construction and calibration and gives balance equations for all measurements.—J. H. Ellison, Universal L, C, R Bridge, *Ham Radio*, April 1976, p 54–55.

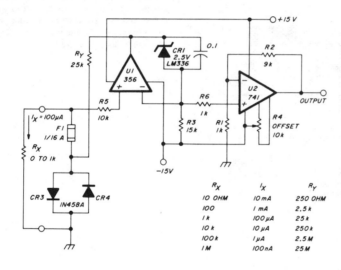

R_X	I_X	R_Y
10 OHM	10 mA	250 OHM
100	1 mA	2.5 k
1 k	100 μA	25 k
10 k	10 μA	250 k
100 k	1 μA	2.5 M
1 M	100 nA	25 M

LOW-VOLTAGE OHMMETER—Combines stable constant-current source U1-CR1-R_Y with DC amplifier U2 having gain of 10 to keep applied voltage down to 0.1 V. Output is linearly proportional to unknown resistance. Resistances well below 1 ohm can be measured accurately. U2 scales 0–100 mV unknown voltage to 0–1 V at output, so 1K resistor under test can be read as 1.000K on DVM scale. U2 should be offset-nulled to eliminate zero error, for best low-scale accuracy, by shorting input and adjusting R4 for 0.000 V out of U2. Full-scale calibration involves trimming individual range values of R_Y for correct output, while using reference value for R_X. Fuse and clamp diodes protect range resistors, and R5 protects opamp.—W. Jung, An IC Op Amp Update, *Ham Radio*, March 1978, p 62–69.

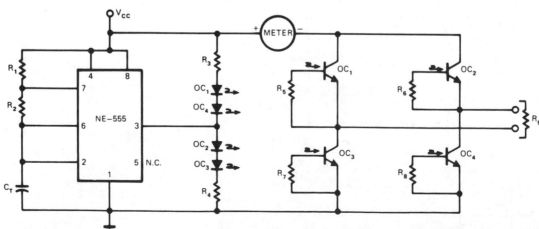

AC OHMMETER—Optoisolator circuit operating from single battery develops alternating current for measuring resistance of soils and construction materials without errors due to polarization and earth-current effects. 555 IC timer controls output at frequency determined by R_1, R_2, and C_T. R_1 is made very much less than R_2 but should not be below about 1 K. Frequency value is 1.44/(R_1 + 2R_2)C_T. Output switching matrix is controlled by timer so OC_1 and OC_4 are on for one half-cycle and OC_2 and OC_3 are on for other half. Output current is independent of frequency and duty cycle up to 150 Hz. With Monsanto MCT-2 optoisolators, R_3 and R_4 are 330 ohms and R_5-R_8 are each 22K.—D. J. Beckwitt, AC Ohmmeter Provides Novel Use for Opto-Isolators, *EDN Magazine*, July 5, 1974, p 70.

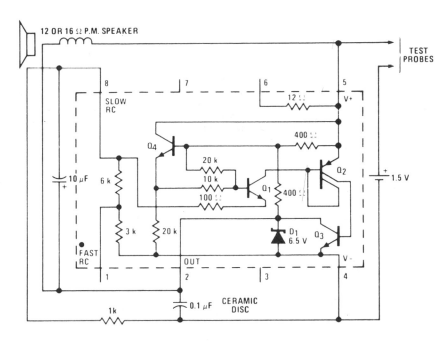

CONTINUITY CHECKER—National LM3909 IC operating from 1.5-V cell provides enough audio power to drive loudspeaker when probes are shorted by resistance up to about 100 ohms. By probing two points in rapid succession, small differences in resistance can be detected by noticeable differences in tone; this feature is useful for identifying windings of transformers.—"Linear Applications, Vol. 2," National Semiconductor, Santa Clara, CA, 1976, AN-154, p 4—5.

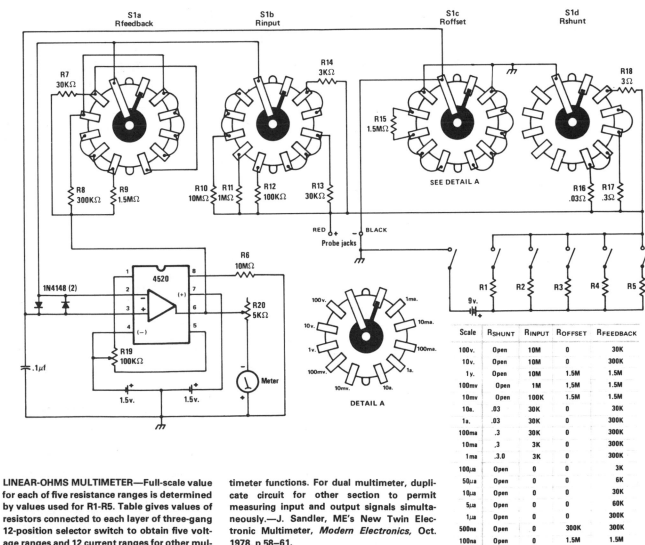

Scale	R_{SHUNT}	R_{INPUT}	R_{OFFSET}	$R_{FEEDBACK}$
100v.	Open	10M	0	30K
10v.	Open	10M	0	300K
1v.	Open	10M	1.5M	1.5M
100mv.	Open	1M	1.5M	1.5M
10mv.	Open	100K	1.5M	1.5M
10a.	.03	30K	0	30K
1a.	.03	30K	0	300K
100ma.	.3	30K	0	300K
10ma.	.3	3K	0	300K
1ma.	.3.0	3K	0	300K
100μa	Open	0	0	3K
50μa	Open	0	0	6K
10μa	Open	0	0	30K
5μa	Open	0	0	60K
1μa	Open	0	0	300K
500na	Open	0	300K	300K
100na	Open	0	1.5M	1.5M

LINEAR-OHMS MULTIMETER—Full-scale value for each of five resistance ranges is determined by values used for R1-R5. Table gives values of resistors connected to each layer of three-gang 12-position selector switch to obtain five voltage ranges and 12 current ranges for other multimeter functions. For dual multimeter, duplicate circuit for other section to permit measuring input and output signals simultaneously.—J. Sandler, ME's New Twin Electronic Multimeter, *Modern Electronics*, Oct. 1978, p 58—61.

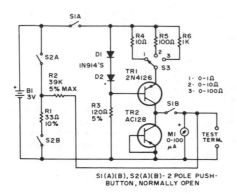

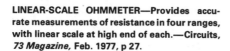

DOWN TO 0.05 OHM—Switch S3 gives choice of three ranges in linear ohmmeter circuit developed for measuring very small values of resistance. Values of R4, R5, and R6 require adjustment during initial calibration.—Circuits, *73 Magazine*, June 1975, p 161.

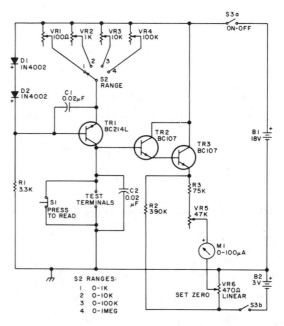

LINEAR-SCALE OHMMETER—Provides accurate measurements of resistance in four ranges, with linear scale at high end of each.—Circuits, *73 Magazine*, Feb. 1977, p 27.

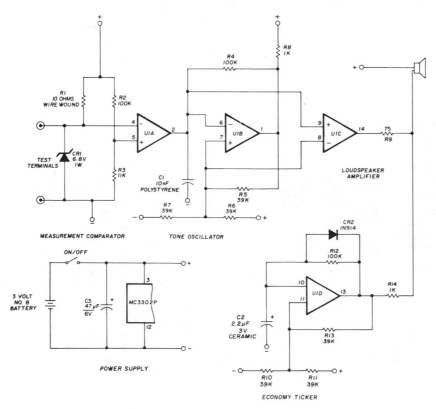

CIRCUIT TRACER—Continuity tester delivers continuous audio tone when its test terminals are connected by resistance less than about 1 ohm. Circuit under test does not receive more than 3 V or 300 mA, depending on resistance between terminals. Tester ticks softly when switched on but is open-circuited, as reminder that battery drain is then about 0.8 mA. Tester uses sections of Motorola low-power MC3302P quad comparator as measurement comparator, 600-Hz tone oscillator, AF amplifier, and ticker. Pins 3 and 12 of IC are used in power supply.—R. C. Marshall, Continuity Bleeper for Circuit Tracing, *Ham Radio*, July 1977, p 67–69.

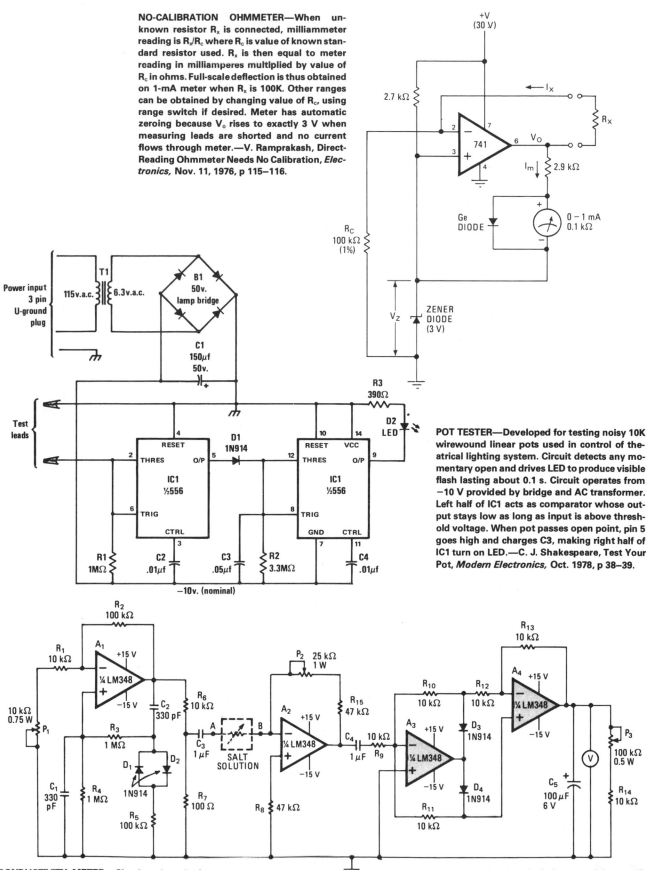

NO-CALIBRATION OHMMETER—When unknown resistor R_x is connected, milliammeter reading is R_x/R_c where R_c is value of known standard resistor used. R_x is then equal to meter reading in milliamperes multiplied by value of R_c in ohms. Full-scale deflection is thus obtained on 1-mA meter when R_x is 100K. Other ranges can be obtained by changing value of R_c, using range switch if desired. Meter has automatic zeroing because V_o rises to exactly 3 V when measuring leads are shorted and no current flows through meter.—V. Ramprakash, Direct-Reading Ohmmeter Needs No Calibration, *Electronics,* Nov. 11, 1976, p 115–116.

POT TESTER—Developed for testing noisy 10K wirewound linear pots used in control of theatrical lighting system. Circuit detects any momentary open and drives LED to produce visible flash lasting about 0.1 s. Circuit operates from −10 V provided by bridge and AC transformer. Left half of IC1 acts as comparator whose output stays low as long as input is above threshold voltage. When pot passes open point, pin 5 goes high and charges C3, making right half of IC1 turn on LED.—C. J. Shakespeare, Test Your Pot, *Modern Electronics,* Oct. 1978, p 38–39.

CONDUCTIVITY METER—Circuit using single quad opamp measures relative change in concentration of salt solution by monitoring its conductance. Use of alternating current through solution eliminates errors caused by electrolysis effect. Wien-bridge oscillator having R_4C_1 and R_2R_3 as arms of bridge generates 1-kHz signal for driving amplifier A_2 through solution. P_1 controls oscillator amplitude, and P_2 adjusts gain of A_2. A_3-A_4 form precision rectifier giving output voltage equal to absolute value of input voltage.—M. Ahmon, One-Chip Conductivity Meter Monitors Salt Concentration, *Electronics,* Sept. 15, 1978, p 132–133.

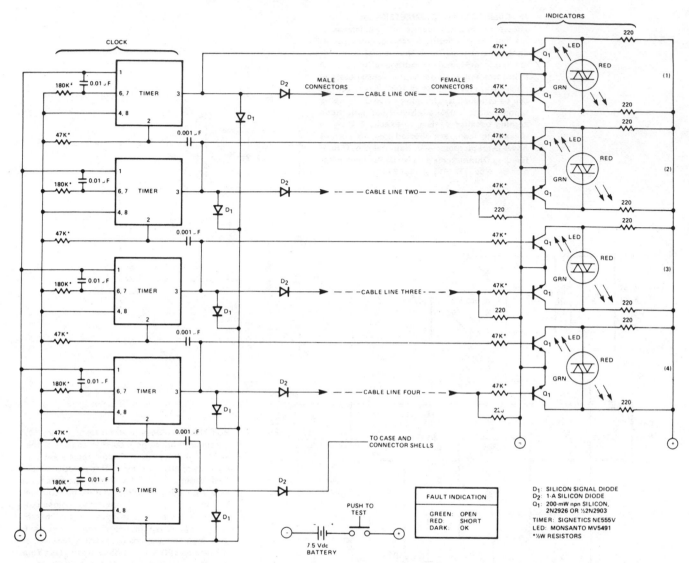

CABLE TESTER—Five Signetics NE555V timers check all lines of four-conductor cable for opens and for short-circuit conditions. Differential transistor pair at one end of each cable line re- mains balanced as long as clock pulses at opposite ends of line are identical. Clock pulse at timer end of one line turns on green LED to indicate open in line. Clock pulse only at transistor end of line turns on red LED to indicate that line is shorted. With good cable line, neither LED is on.—"Signetics Analog Data Manual," Signetics, Sunnyvale, CA, 1977, p 730.

CHAPTER **78**
Sampling Circuits

Methods of sampling and holding analog signals, including digital selection of
sampling level and long-term storage of signals in digital form as substitute
for storage CRO.

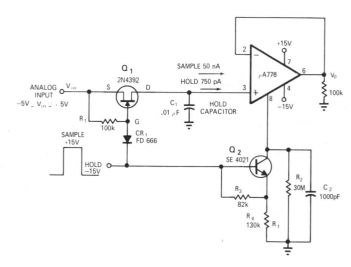

TWO SLEW RATES—Cost of sample-and-hold
circuits is reduced by using high slew rate only
during sample period. Programmable μA776
opamp permits switching from high rate requiring 50-nA input bias current to holding amplifier mode requiring only 750-pA input bias
current. Output level is held constant within 1%
for about 2 s, making circuit ideal for digital
readouts.—M. K. Vander Kooi, Low Cost Sample-and-Hold Circuit, *EDN/EEE Magazine,* Nov.
1, 1971, p 46.

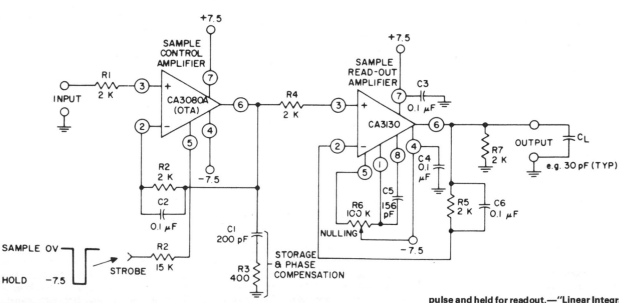

SAMPLE READOUT AMPLIFIER—RCA CA3080A
operational transconductance amplifier feeds
CA3130 to give amplification of sampled signal.
Input voltage is sampled for duration of strobe
pulse and held for readout.—"Linear Integrated
Circuits and MOS/FET's," RCA Solid State Division, Somerville, NJ, 1977, p 165–170.

881

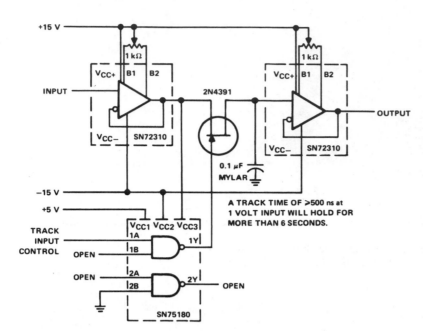

TRACK AND HOLD—When track input control is 0.8 V or less, gate in SN75180 holds 2N4391 transistor source-drain path closed so input signal goes to output unchanged. When control voltage is increased to 2 V, gate opens path through transistor, so signal voltage stored at that instant in 0.1-μF capacitor is held at output. Track time is greater than 500 ns for 1-V input, giving hold time over 6 s. Circuit uses two SN72310 wideband voltage-follower op-amps.—"The Linear and Interface Circuits Data Book for Design Engineers," Texas Instruments, Dallas, TX, 1973, p 4-41.

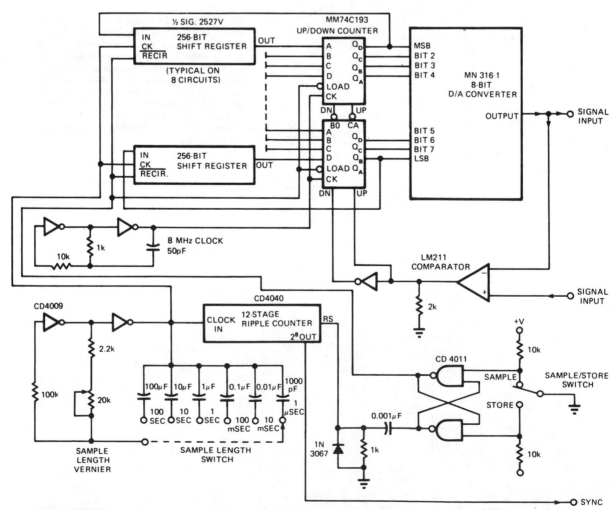

256 8-BIT SAMPLE/STORE—Low-cost substitute for storage oscilloscope can be used to study analog variables in speech synthesis, transient signal analysis, and destructive testing of components. Circuit is basically a tracking A/D converter whose digital output is fed into shift register holding 256 8-bit words. Separate clock for shift register is continuously adjustable from about 250 kHz down to about 4 s per cycle, with output going to 12-stage ripple counter. At 250 kHz, shift register stores input signal for 1 ms. Article gives details of circuit operation.—K. P. Roby, Transient Signal Analyzer Has Multiple Uses, *EDN Magazine*, Oct. 20, 1974, p 46–48.

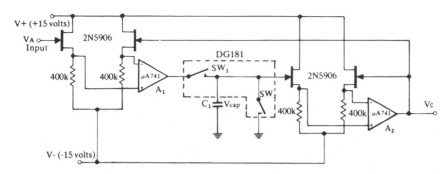

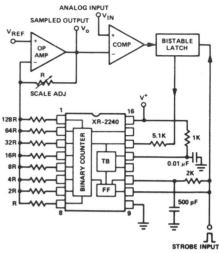

NONINVERTING SAMPLE AND HOLD— Matched pair of FETs gives high input resistance for analog input signal greater than 10^{12} ohms, while output resistance of FET pair is under 12K. Opamp A_1 acts as buffer and allows C_1 to charge rapidly. Use of DG181 analog switches limits leakage current flowing into or out of C_1, while SW_2 provides fast resetting of capacitor voltage to zero. Similar FET pair and opamp provide output voltage proportional to sampled value.—"Analog Switches and Their Applications," Siliconix, Santa Clara, CA, 1976, p 4-7–4-8.

DIGITAL SAMPLE AND HOLD—When strobe input is applied, RC low-pass network between reset and trigger inputs of Exar XR-2240 programmable timer/counter resets and then triggers timer, sets output of bistable latch to high state, and activates counter. Circuit generates staircase voltage at opamp output. When staircase level reaches that of analog input to be sampled, comparator changes state, activates bistable latch, and stops count. Opamp output voltage level then corresponds to sampled analog input. Sample is held until next strobe signal. Minimum recycle time is about 6 ms. Supply voltage can be 4–15 V.—"Timer Data Book," Exar Integrated Systems, Sunnyvale, CA, 1978, p 11–18.

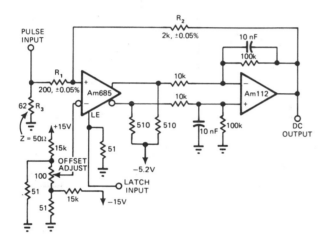

ANALOG SAMPLE-AND-HOLD—Uses AM685 comparator for continuous sampling of analog voltage at summing node formed by R_1 and R_2. Complementary logic outputs of comparator drive differential indicator formed by AM112 opamp. When error voltage at summing node is positive, comparator latches in high state and causes output opamp to integrate toward more negative voltage. When error voltage at latch time (determined by pulse input) is negative, integrator voltage ramps to more positive value. Circuit soon reaches equilibrium, at which output voltage is equal to -10 times value of sampled waveform. Article gives performance waveforms for sampling video pulses.—S. Dendinger, High-Speed Analog Sampler Uses Only Two IC's, *EDN Magazine*, May 20, 1977, p 128 and 130.

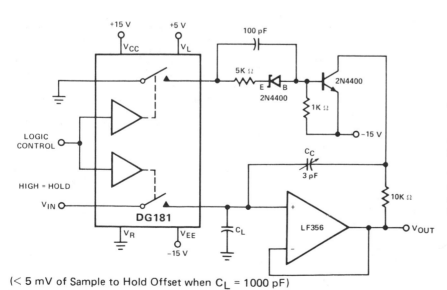

(< 5 mV of Sample to Hold Offset when C_L = 1000 pF)

NEUTRALIZATION OVER ±7.5 V RANGE— Switching transients are attenuated in sample-and-hold circuit using DG181 FET analog switch by adding neutralization derived from complementary signal coupled through upper switch of DG181. Charge transferred from second switch is then opposed to that from main channel. Circuit is controlled by input logic signal. With compensation, change in transferred charge is less than 5 picocoulombs for input signal range from -7.5 V to $+7.5$ V.—"Analog Switches and Their Applications," Siliconix, Santa Clara, CA, 1976, p 7-61.

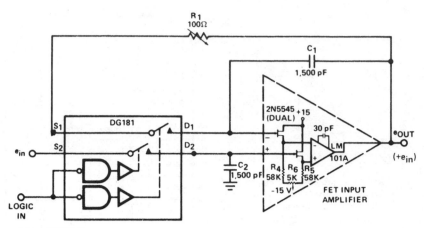

NONINVERTING SAMPLE AND HOLD—DG181 JFET analog switch provides best combination of settling speed and inherent charge transfer accuracy for 2N5545 FET-input opamp for high-speed sample-and-hold applications.—"Analog Switches and Their Applications," Siliconix, Santa Clara, CA, 1976, p 7-60.

TIMED SLOPE-SAMPLING—Circuit measures rate of signal change for slowly varying signals (changing less than 1 V/min) by using sample-and-hold circuit to store instantaneous sample. After compatible time interval, sample is compared with new input current value; difference is then the desired slope. One limitation is that offset errors added to stored signal near zero voltage can cause large errors in the derivative. Circuit is highly sensitive to noise spikes during sampling.—R. E. Bober, This Derivative Circuit Handles Slowly Varying Signals, *EDN Magazine,* Jan. 20, 1976, p 82.

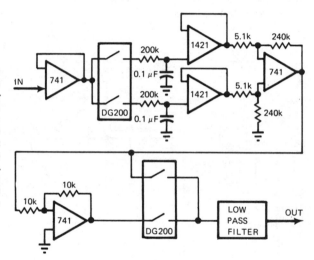

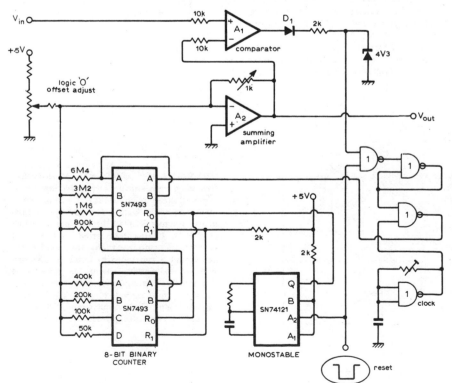

256-LEVEL HOLD—Uses digital approximation to hold sampled analog voltage for long periods. Cascaded SN7493 ICs form 8-bit binary counter providing 256 discrete voltage levels from opamp A_2, while input voltage provides varying reference to opamp A_1 serving as comparator. Apply 0 at reset input to clear counter for period determined by monostable IC. Counter now feeds staircase waveform to A_1 through A_2 until staircase reaches V_{in}, when counter goes high and disables counter clock. Count is then held and sampled voltage appears at output.—N. Macdonald, Digital Sample and Hold, *Wireless World,* May 1976, p 78.

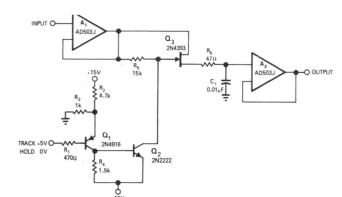

TRACK-AND-HOLD UP TO 4 kHz—Developed for tracking ±10 V AF input signal when control input is +5 V. When control drops to 0 V, series FET Q_3 opens and input voltage at that time is stored on C_1 for transfer to output through high-impedance opamp A_2.—R. S. Burwen, Track-and-Hold Amplifier, *EDN/EEE Magazine,* Sept. 1, 1971, p 43.

SAMPLE AND HOLD—CA3140 bipolar MOS opamp serves as readout amplifier for storage capacitor C1 which is charged by CA3080A variable opamp serving as input buffer and low-feedthrough transmission switch. CA3140 also provides offset nulling.—"Circuit Ideas for RCA Linear ICs," RCA Solid State Division, Somerville, NJ, 1977, p 17.

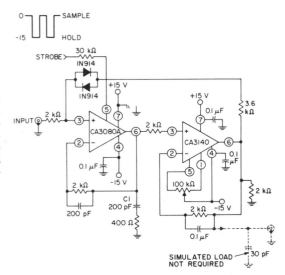

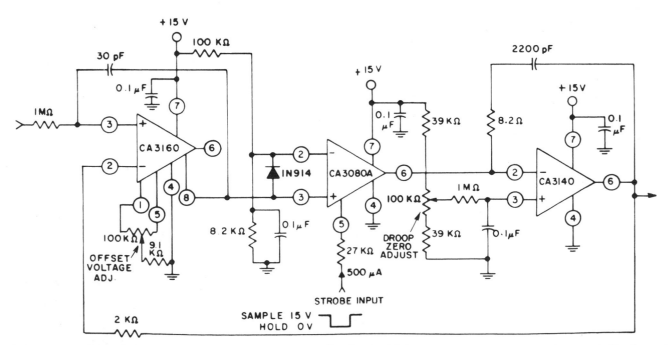

15-V SINGLE-SUPPLY—CA3160 opamp provides high input impedance and input voltage range of 0–10 V. CA3080A functions as strobed current source for CA3140 output integrator and storage capacitor. Pulse droop during hold interval can be reduced to zero by adjusting 100K pot.—"Linear Integrated Circuits and MOS/FET's," RCA Solid State Division, Somerville, NJ, 1977, p 271–272.

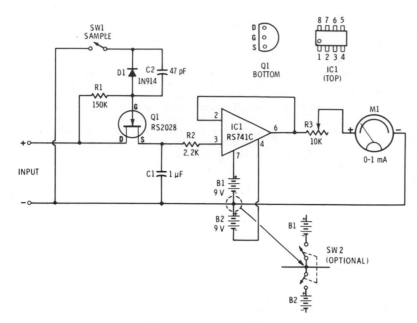

FET-OPAMP SAMPLE AND HOLD—Meter indicates output signal when input is present and stores input in C1 when sample switch is open. Opamp is connected as unity-gain voltage follower. Charge on C1 will be drained within a few minutes by opamp shown. Charge can be held longer by changing to FET-input opamp or by opening both battery circuits with alternative DPDT switch SW2. Developed for classroom demonstrations.—F. M. Mims, "Integrated Circuit Projects, Vol. 4," Radio Shack, Fort Worth, TX, 1977, 2nd Ed., p 61—69.

LOW-DRIFT SAMPLE AND HOLD—JFETs provide complete buffering to sample-and-hold capacitor C1. During sample, Q1 is turned on to provide charging path. During hold, Q1 and Q2 are turned off so discharge paths through transistors for C1 are each less than 100 pA. Q2 also serves as buffer for opamp so feedback and output current are supplied only from opamp source.—"FET Databook," National Semiconductor, Santa Clara, CA, 1977, p 6-26—6-36.

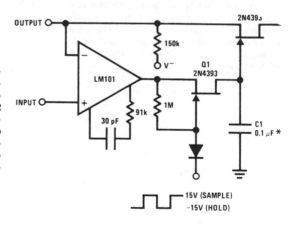

*Polycarbonate dielectric capacitor

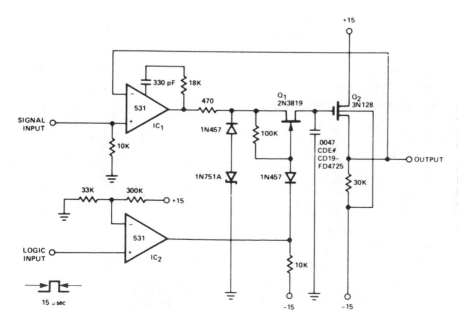

FAST SAMPLE-AND-HOLD—Strobe pulse developed from logic input of 531 opamp IC$_2$ turns on JFET Q$_1$ to complete feedback loop to IC$_1$, Q$_1$, and Q$_2$. C$_1$ charges to voltage equal to that of input signal plus gate-to-source offset voltage of Q$_2$. At end of strobe time, feedback loop is broken and C$_1$ holds voltage until time of next strobe pulse. Decay in output voltage between samplings is 1 mV/s.—"Signetics Analog Data Manual," Signetics, Sunnyvale, CA, 1977, p 643—644.

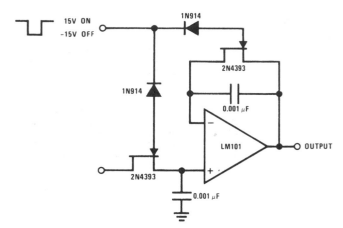

JFET SAMPLE AND HOLD—Logic voltage is applied simultaneously to sample-and-hold JFETs. By matching input impedance and feedback resistance and capacitance, errors due to ON resistance of JFETs are minimized.—"FET Databook," National Semiconductor, Santa Clara, CA, 1977, p 6-26–6-36.

FEEDBACK REDUCES DRIFT—FET Q_2 serves as buffer for hold capacitor C_1, minimizing droop error. Switching transistor Q_1 is placed in feedback loop. A_1 serves as input signal buffer and as driver for Q_1 and C_1. When sample line is raised to +15 V by control pulse, D_1 is reverse-biased and Q_1 is turned on. C_1 now charges until output terminal reaches equilibrium with input (so A_1 is tracking input). When sample pulse goes low, feedback loop of A_1 is opened; output of A_1 stays at voltage last sampled since C_1 retains its charge and Q_2 buffers this voltage while presenting it to output. Optional voltage follower can be used if more output current is needed to feed low-impedance load.—W. G. Jung, "IC Op-Amp Cookbook," Howard W. Sams, Indianapolis, IN, 1974, p 198–200.

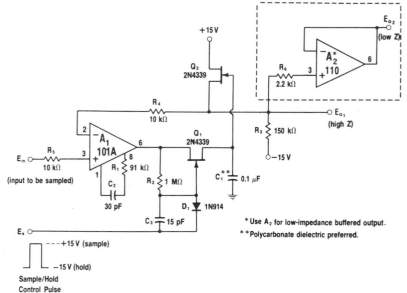

* Use A_2 for low-impedance buffered output.

**Polycarbonate dielectric preferred.

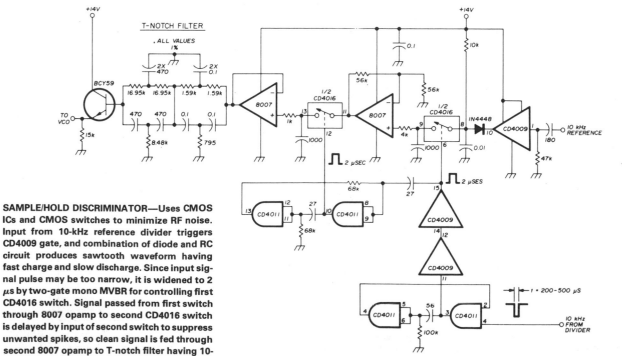

SAMPLE/HOLD DISCRIMINATOR—Uses CMOS ICs and CMOS switches to minimize RF noise. Input from 10-kHz reference divider triggers CD4009 gate, and combination of diode and RC circuit produces sawtooth waveform having fast charge and slow discharge. Since input signal pulse may be too narrow, it is widened to 2 μs by two-gate mono MVBR for controlling first CD4016 switch. Signal passed from first switch through 8007 opamp to second CD4016 switch is delayed by input of second switch to suppress unwanted spikes, so clean signal is fed through second 8007 opamp to T-notch filter having 10-kHz reference frequency for one leg and 20 kHz for other leg. Notch depth can be 60 dB. Filter drives VCO of frequency synthesizer through BCY59 emitter-follower transistor.—U. L. Rohde, Modern Design of Frequency Synthesizers, *Ham Radio*, July 1976, p 10–23.

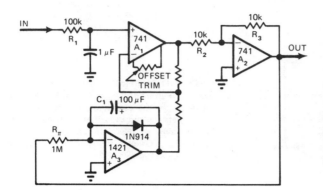

DERIVATIVE SLOPE-SAMPLING—Analog derivative circuit forces voltage across C_1 to follow slowly changing input signal. Current required to keep capacitor voltage equal to signal voltage is proportional to rate of change of signal voltage. RCA 3033 FET-input opamp will work equally as well as Teledyne Philbrick 1421 shown. Careful selection of values for R_π and C_1 will set rate limit that will reject spikes.—R. E. Bober, This Derivative Circuit Handles Slowly Varying Signals, *EDN Magazine*, Jan. 20, 1976, p 82.

INVERTING SAMPLE AND HOLD—Total offset error can be adjusted to much less than 1 mV in 2N5545 FET-input opamp by using compensation circuit R_3-C_2-C_3 with DG181 JFET analog switch. Switch operation occurs consistently at constant voltage, reducing aperture time jitter. Designed for high-speed sample-and-hold requirements.—"Analog Switches and Their Applications," Siliconix, Santa Clara, CA, 1976, p 7-59.

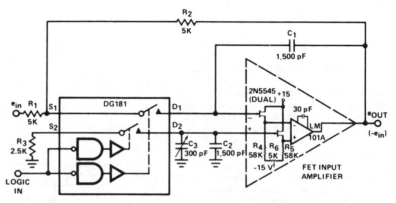

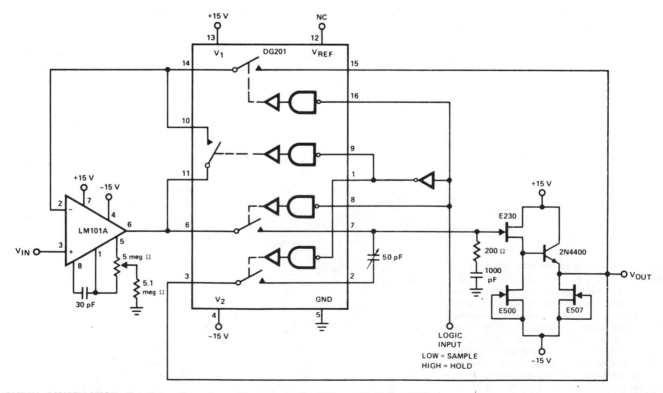

GLITCH CANCELLATION—Fourth section of DG201 quad CMOS analog switch provides cancellation of coupled charges (glitches), to keep sample-and-hold offset below 5 mV over analog voltage range of −10 V to +10 V. Acquisition time is 25 μs for opamp shown but can be improved by using faster-slewing opamp. Aperture time is typically 1 μs.—"Analog Switches and Their Applications," Siliconix, Santa Clara, CA, 1976, p 7-68.

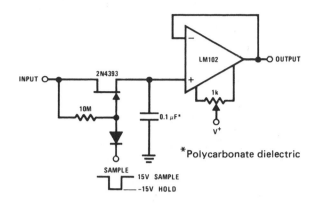

SAMPLE AND HOLD WITH OFFSET ADJUST-MENT—Use of 2N4393 JFET at input of opamp gives simple high-performance circuit having low leakage. Offset is easily adjusted with 1K pot.—"FET Databook," National Semiconductor, Santa Clara, CA, 1977, p 6-26—6-36.

CHAPTER **79**
Servo Circuits

Includes logic-controlled preamps and power amplifiers for driving two-phase, stepper, and other types of 60-Hz and 400-Hz servomotors in either direction for correct time at correct speed for bringing servo shaft exactly to desired new position. See also Motor Control chapter.

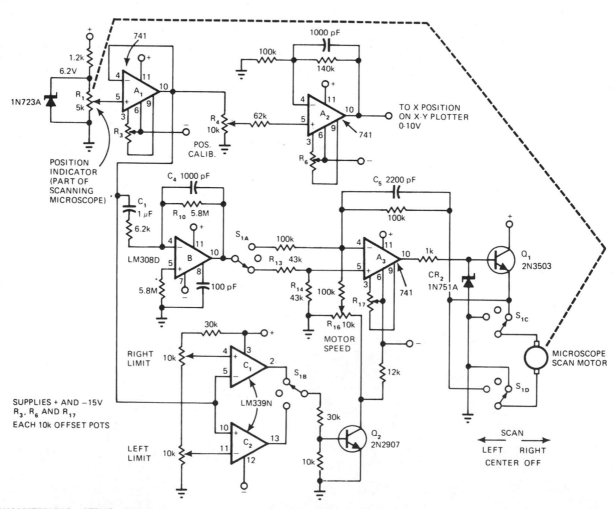

TACHOMETERLESS SERVO—Developed to provide speed control for motor enclosed in such a way that tachometer cannot be used for feedback. Position pot R_1 and differentiator B substitute for tachometer in controlling rate of scanning-microscope eyepiece used for measuring CRT line width. Buffer A_1 feeds X input of XY plotter through opamp A_2, and also feeds differentiator B and limit-detector voltage comparator C_1-C_2. S_1 switches A_3 between inverting and noninverting operation each time scanning direction changes, to keep feedback negative.— H. F. Stearns, Differentiator and Position Pot Sub for Tachometer, *EDN Magazine*, Aug. 5, 1977, p 50–52.

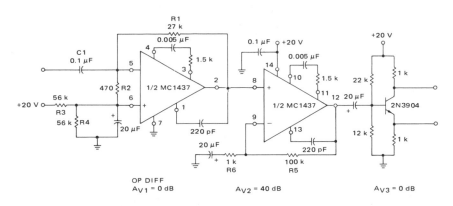

DUAL-OPAMP PREAMP—First section of Motorola MC1437 dual opamp is connected as operational differentiator driving direct-coupled noninverting opamp. Single-ended output is converted to push-pull by following phase-splitting amplifier for driving power amplifier of 115-V 60-Hz servomotor.—A. Pshaenich, "Servo Motor Drive Amplifiers," Motorola, Phoenix, AZ, 1972, AN-590.

SERVO DRIVE—Combination of Fairchild μA795 multiplier and μA741 opamp generates AC error signal for driving two-phase servomotor. Phase-shifted signal from R_1-C_1 is applied to input pin 4 of multiplier, DC signal input is applied to pin 9, and servo position signal goes to pin 12. Multiplier takes difference between signals on 9 and 12, multiplies this by signal on pin 4, and feeds resulting sine wave from pin 14 to opamp for amplification and transfer to servo driver. When servomotor action makes voltages on 9 and 12 equal, system is nulled.—Fairchild Linear IC Contest Winners, *EEE Magazine*, Jan. 1971, p 48–49.

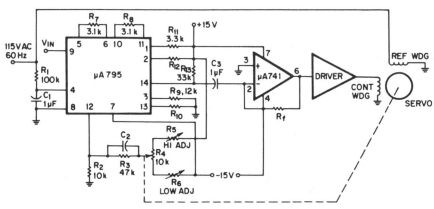

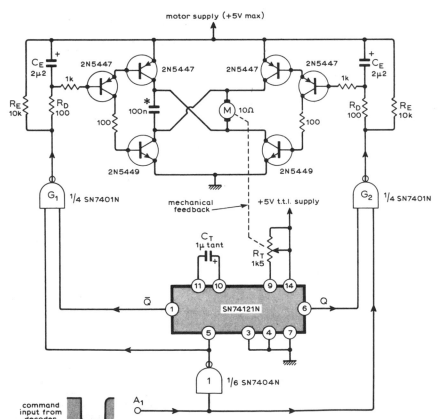

*= motor commutating capacitor

TTL SERVO CONTROL—Used in nine-channel remote control system having nine identical servos fed by decoder at receiving end of data link. Variable-width pulse command from decoder is fed into TTL IC pulse-width comparator that feeds bridge-type motor drive. Command pulse controls both direction and duration of motor rotation. Article describes operation in detail and gives associated coder and decoder circuits.—M. F. Bessant, Multi-Channel Proportional Remote Control, *Wireless World*, Oct. 1973, p 479–482.

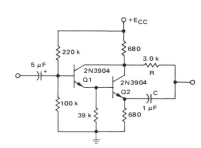

DARLINGTON PHASE SHIFTER—Basic 90° push-pull RC phase shifter using discrete transistors is connected as phase-splitting amplifier. Used at output of follow-up pot in servoamplifier driving 115-V 60-Hz servomotor. Supply is 28 V. Motorola MPSA13 Darlington IC can be used if 39K resistor is omitted.—A. Pshaenich, "Servo Motor Drive Amplifiers," Motorola, Phoenix, AZ, 1972, AN-590.

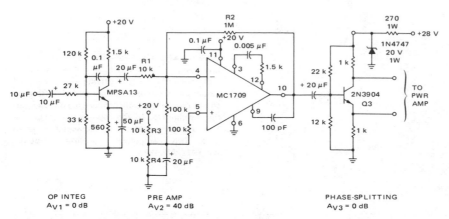

OP INTEG
A_{V1} = 0 dB

PRE AMP
A_{V2} = 40 dB

PHASE-SPLITTING
A_{V3} = 0 dB

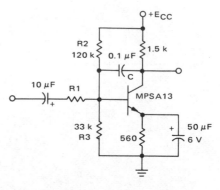

OPERATIONAL-INTEGRATOR PHASE SHIFT-ER—Motorola MPSA13 Darlington IC provides 90° phase shift required in servoamplifier for 115-V 60-Hz servomotor. Two cascaded 2N3904 discrete Darlingtons can be used in place of IC.—A. Pshaenich, "Servo Motor Drive Amplifiers," Motorola, Phoenix, AZ, 1972, AN-590.

PHASE-SPLITTING PREAMP—Uses Motorola MPSA13 operational integrator to provide 90° phase shift for MC1709 inverting opamp, with single-ended output complemented by phase-splitting amplifier to provide push-pull drive for power amplifier of 115-V 60-Hz servomotor. Voltage gain is about 40 dB.—A. Pshaenich, "Servo Motor Drive Amplifiers," Motorola, Phoenix, AZ, 1972, AN-590.

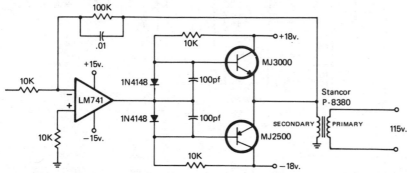

20 W AT 60 Hz—Adding high-current complementary transistors to opamp gives servoamplifier with 115-V output. Opamp drives low impedance of 10-V filament transformer connected in reverse to boost output to 115 V for driving servo. Use heatsink for transistors. Bringing opamp feedback resistor to actual output point makes nonlinearities and crossover point between transistors insignificant by placing them in feedback loop.—I. Math, Math's Notes, *CQ*, Jan. 1978, p 53–54 and 70.

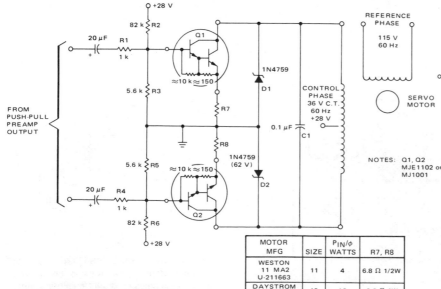

MOTOR MFG	SIZE	P_{IN}/ϕ WATTS	R7, R8
WESTON 11 MA2 U-211663	11	4	6.8 Ω 1/2W
DAYSTROM U-207263	18	10	3.3 Ω 1W

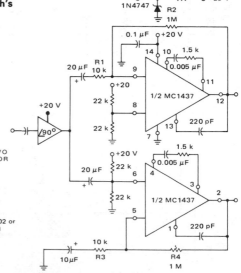

PARALLEL-OPAMP PREAMP—Provides differential output required for driving power amplifier of 115-V 60-Hz servomotor. One opamp section is connected inverting and the other noninverting to give required complementary outputs. Voltage gain is 40 dB, operating from single 20-V zener-regulated supply. High DC feedback gives excellent DC stability. Bandwidth is about 6 kHz. Input is driven by 90° phase shifter.—A. Pshaenich, "Servo Motor Drive Amplifiers," Motorola, Phoenix, AZ, 1972, AN-590.

28-V PUSH-PULL POWER AMPLIFIER—Power Darlingtons are used in common-emitter configuration to give high current gain for driving control phase of 60-Hz servo while providing high input impedance for preamp. No transformers are required. Darlingtons require heat-sinks. Suitable for driving size 11 servo at 4 W and size 18 at 10 W if emitter resistors R7 and R8 are changed as in table.—A. Pshaenich, "Servo Motor Drive Amplifiers," Motorola, Phoenix, AZ, 1972, AN-590.

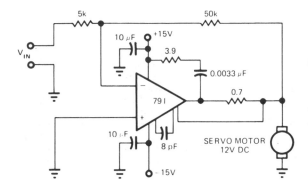

12-VDC DRIVE—Circuit uses 791 power opamp in inverting configuration with gain of 10 for driving size 8 12-VDC servomotor in either direction. Article tells how to calculate heatsink requirements for opamp.—R. J. Apfel, Power Op Amps—Their Innovative Circuits and Packaging Provide Designers with More Options, *EDN Magazine*, Sept. 5, 1977, p 141–144.

DIFFERENTIAL INPUT AND OUTPUT—Preamplifier for servosystem uses 90° operational integrator to drive MC1420 opamp having differential input and differential output connected in inverting configuration. With values shown, voltage gain is about 38 dB. Bandwidth is about 4 kHz, giving stability when using 510-pF compensating capacitors. Zener provides 12 V required for opamp operation from single supply.—A. Pshaenich, "Servo Motor Drive Amplifiers," Motorola, Phoenix, AZ, 1972, AN-590.

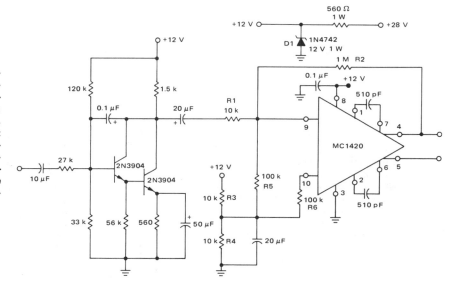

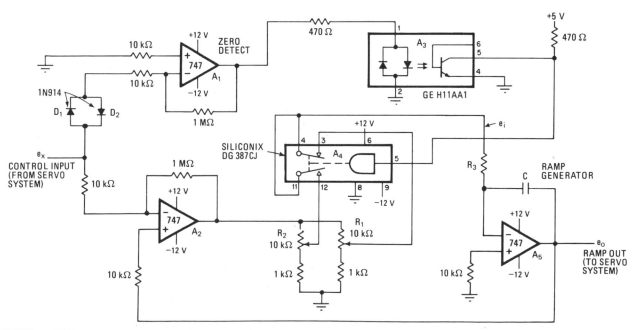

UP/DOWN RAMP CONTROL—Siliconix DG387CJ solid-state relay A_4 provides switching from up ramp to down ramp for decelerating servo when it zeroes in on correct new position. Slopes are determined by settings of R_1 and R_2. Arrangement ensures optimum servo system response at low cost. A_1 detects that input is other than 0 V and energizes optoisolator A_3 for switching A_4. Resulting positive-going ramp from A_5 moves system load toward desired position, making feedback voltage of servo reduce control-input voltage. When this drops to within 0.7 V of ground, A_1 goes low and A_3 turns off. A_4 now initiates down-ramp waveform to decelerate system to stop. For ramp rate of 20 V/s, C can be 0.33 μF and R_3 1.8 megohms.—R. E. Kelly, Up-Down Ramp Quickens Servo System Response, *Electronics*, July 20, 1978, p 121 and 123.

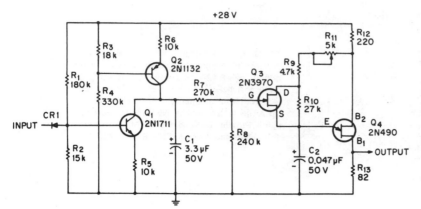

STEP-SERVO CONTROL—Variable UJT oscillator generates train of pulses under control of digital input logic levels, at 1000 pulses per second for logic 1 or 4400 pulses per second for logic 0, with smooth transitions between rates when logic changes, for driving stepping servomotor. Q_1 and Q_2 are constant-current sources. JFET Q_3 acts as voltage-controlled variable resistor in parallel with R_{10}, controlling pulse rate of UJT oscillator Q_4.—C. R. Forbes, Step-Servo Motor Slew Generator, *EEE Magazine,* Oct. 1970, p 76–77.

TWO-PHASE SERVO DRIVE—Both sections of National LM377 power amplifier are connected to provide up to 3 W per phase for driving small 60-Hz two-phase servomotor. Power is sufficient for phonograph turntable drive. Lamp is used in simple amplitude stabilization loop. Motor windings are 8 ohms, tuned to 60 Hz with shunt capacitors.—"Audio Handbook," National Semiconductor, Santa Clara, CA, 1977, p 4-8–4-20.

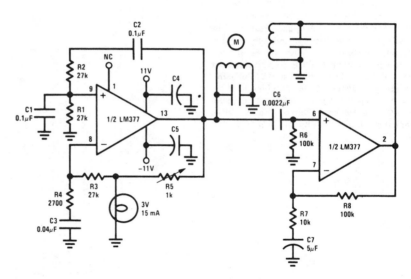

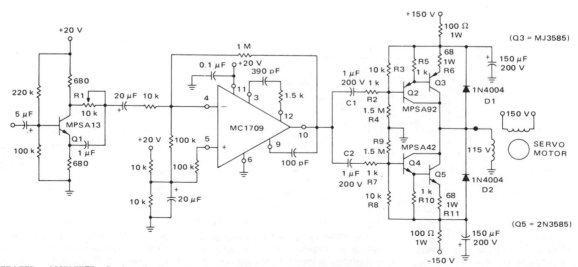

LINE-OPERATED AMPLIFIER—Push-pull RC phase shifter, single-ended preamp, and push-pull class B power amplifier all obtain supply voltages from AC supply that can either use power transformer or operate directly from line with diode rectifiers. Power output is enough to drive size 18 servomotor at 10 W. Larger servomotors can be used if reduced supply voltages can be tolerated. Suitable power supply circuits are given.—A. Pshaenich, "Servo Motor Drive Amplifiers," Motorola, Phoenix, AZ, 1972, AN-590.

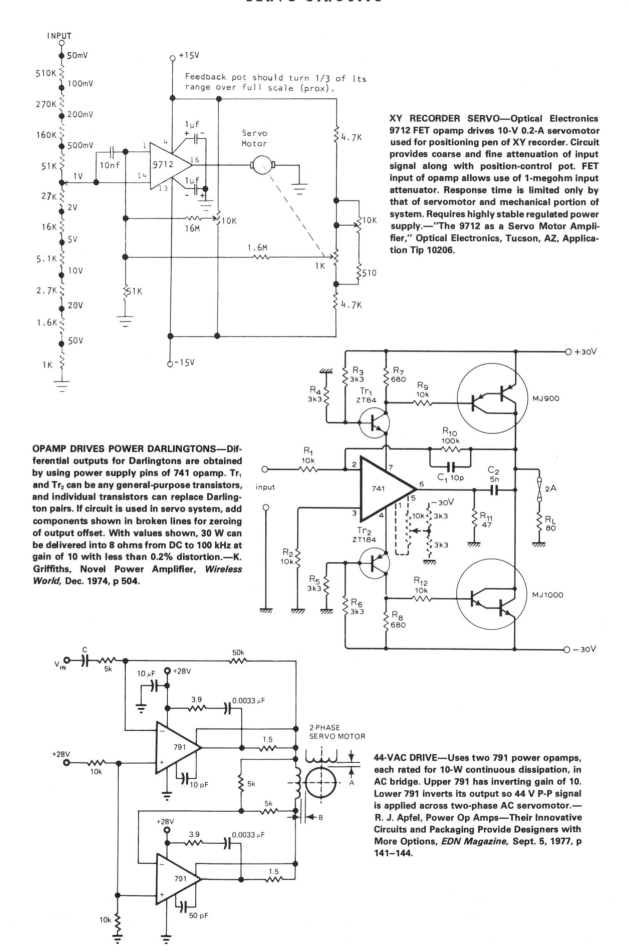

XY RECORDER SERVO—Optical Electronics 9712 FET opamp drives 10-V 0.2-A servomotor used for positioning pen of XY recorder. Circuit provides coarse and fine attenuation of input signal along with position-control pot. FET input of opamp allows use of 1-megohm input attenuator. Response time is limited only by that of servomotor and mechanical portion of system. Requires highly stable regulated power supply.—"The 9712 as a Servo Motor Amplifier," Optical Electronics, Tucson, AZ, Application Tip 10206.

OPAMP DRIVES POWER DARLINGTONS—Differential outputs for Darlingtons are obtained by using power supply pins of 741 opamp. Tr_1 and Tr_2 can be any general-purpose transistors, and individual transistors can replace Darlington pairs. If circuit is used in servo system, add components shown in broken lines for zeroing of output offset. With values shown, 30 W can be delivered into 8 ohms from DC to 100 kHz at gain of 10 with less than 0.2% distortion.—K. Griffiths, Novel Power Amplifier, *Wireless World,* Dec. 1974, p 504.

44-VAC DRIVE—Uses two 791 power opamps, each rated for 10-W continuous dissipation, in AC bridge. Upper 791 has inverting gain of 10. Lower 791 inverts its output so 44 V P-P signal is applied across two-phase AC servomotor.—R. J. Apfel, Power Op Amps—Their Innovative Circuits and Packaging Provide Designers with More Options, *EDN Magazine,* Sept. 5, 1977, p 141–144.

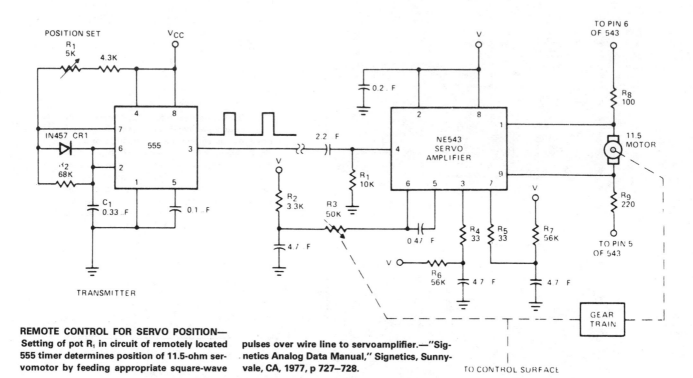

REMOTE CONTROL FOR SERVO POSITION— Setting of pot R_1 in circuit of remotely located 555 timer determines position of 11.5-ohm servomotor by feeding appropriate square-wave pulses over wire line to servoamplifier.—"Signetics Analog Data Manual," Signetics, Sunnyvale, CA, 1977, p 727–728.

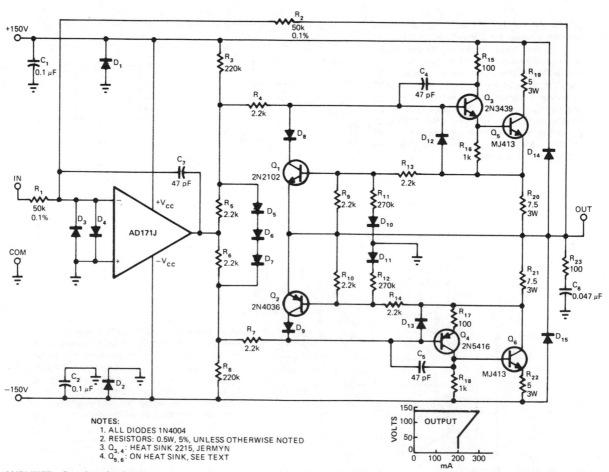

NOTES:
1. ALL DIODES 1N4004
2. RESISTORS: 0.5W, 5%, UNLESS OTHERWISE NOTED
3. $Q_{3,4}$: HEAT SINK 2215, JERMYN
4. $Q_{5,6}$: ON HEAT SINK, SEE TEXT

400-Hz AMPLIFIER—Developed to increase output power of digital-to-synchro converter systems while providing stable and accurate output and overall gain even with reactive loads. Includes overload protection. Delivers 95 VRMS at 400 Hz continuously into 500-ohm load. Power bandwidth is about 20 kHz. Foldback current limiting drops short-circuit current to 200 mA when load exceeds 300 mA.—F. H. Cattermolen and J. A. Pieterse, Digital/Synchro Amplifier Features Overload Protection, *EDN Magazine*, Nov. 5, 1977, p 107–108.

CHAPTER 80
Signal Generator Circuits

Includes fixed-frequency and tunable sine-wave and square-wave oscillator circuits operating in various portions of spectrum from 1 Hz to 1296 MHz, all of which can be accurately calibrated. Used in adjusting, testing, and troubleshooting tuned circuits. Includes band-edge marker generators and FM signal generators. See also Frequency Synthesizer, Function Generator, Staircase Generator, and Sweep chapters.

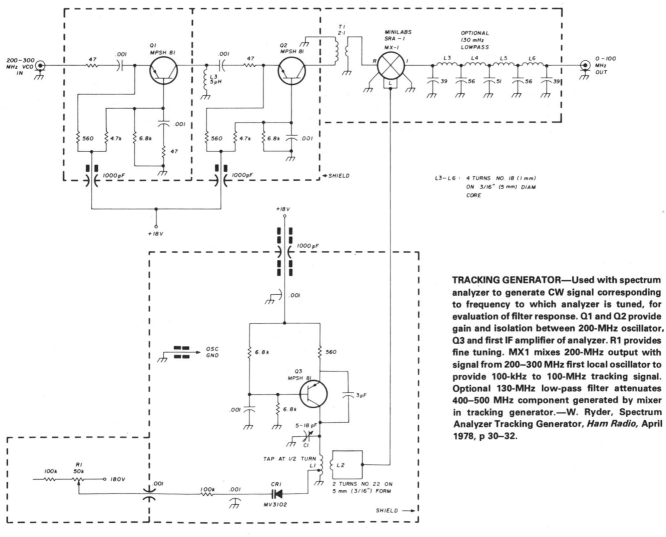

TRACKING GENERATOR—Used with spectrum analyzer to generate CW signal corresponding to frequency to which analyzer is tuned, for evaluation of filter response. Q1 and Q2 provide gain and isolation between 200-MHz oscillator, Q3 and first IF amplifier of analyzer. R1 provides fine tuning. MX1 mixes 200-MHz output with signal from 200–300 MHz first local oscillator to provide 100-kHz to 100-MHz tracking signal. Optional 130-MHz low-pass filter attenuates 400–500 MHz component generated by mixer in tracking generator.—W. Ryder, Spectrum Analyzer Tracking Generator, *Ham Radio,* April 1978, p 30–32.

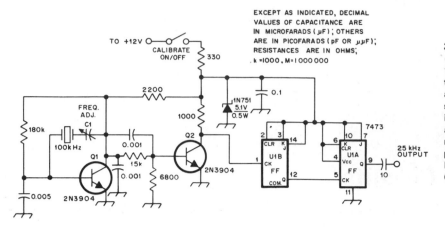

EXCEPT AS INDICATED, DECIMAL VALUES OF CAPACITANCE ARE IN MICROFARADS (μF); OTHERS ARE IN PICOFARADS (pF OR μμF); RESISTANCES ARE IN OHMS; k =1000, M =1 000 000

25-kHz CALIBRATOR—Addition of one 7473 dual JK flip-flop IC to circuit of Radio Shack 28-140 100-kHz calibrator kit provides conversion to 25 kHz for checking frequency settings of amateur receivers. Amplitude of output is 5 V P-P square wave with rich harmonic content. Originally designed for use with HW-8 Heathkit amateur receiver. Output of calibrator is coupled to receive side of antenna relay through 10-pF capacitor. Article covers initial calibration.—D. Karpiej, A 25-kHz Calibrator for the HW-8, *QST*, Oct. 1978, p 20–21.

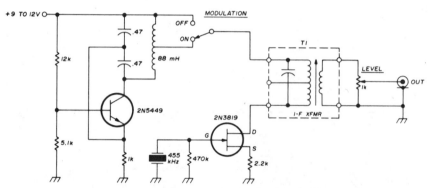

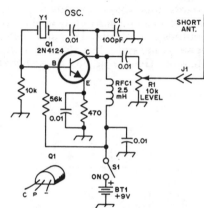

455-kHz FOR IF ALIGNMENT—Simple crystal-controlled signal generator serves for aligning IF strips. Amplitude modulator uses Colpitts 1-kHz oscillator circuit, with surplus 88-mH toroid in tank circuit; tie two adjacent leads together to provide center tap. T1 is 455-kHz IF transformer from AM transistor radio, used to tune drain circuit and obtain low output impedance. Current drain is about 7 mA with 12-V supply and 5 mA with 9-V battery.—C. Hall, 455-kHz I-F Alignment Signal Generator, *Ham Radio*, Feb. 1974, p 50–52.

UNIVERSAL TEST OSCILLATOR—Crystal is in feedback path of Pierce oscillator, between base and collector of Q1, with 2.5-mH RF choke in place of tuned collector circuit. Oscillator works from 400 kHz to 20 MHz, depending on crystal. Designed for fundamental crystals; third overtone types will oscillate but at fundamental. Value of C1 is for 1 MHz and higher; increase to 330 pF for lower frequencies. Antenna can be 20-inch wire; increasing length increases signal radiation. Can be used as signal source for receiver alignment (with either radiated or probe-coupled signal), as marker generator, or in combination with station receiver as code-practice oscillator.—D. DeMaw, Build a UTO-1, *QST*, Oct. 1977, p 19–21.

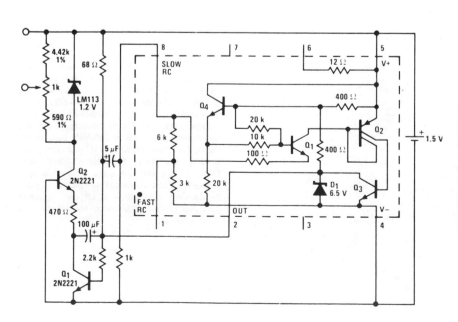

LAB GENERATOR/CALIBRATOR—Portable design operating from single flashlight D cell uses National LM3909 flasher IC to produce clean rectangular wave that can be adjusted to exactly 1 V. Pulse width is 1.5 ms and OFF interval between pulses is 5.5 ms. Useful for calibrating oscilloscopes and adjusting their probes. Article describes operation of circuit in detail. Current drain is low enough to give 500 h of operation.—P. Lefferts, Power-Miser Flasher IC Has Many Novel Applications, *EDN Magazine*, March 20, 1976, p 59–66.

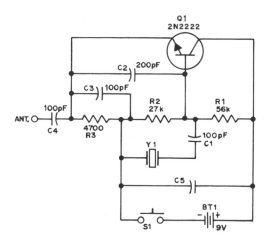

BAND-EDGE MARKER—Series-tuned Colpitts crystal oscillator feeding 10-inch insulated-wire antenna provides sufficient signal radiation for pickup by nearby communication receiver. Used to provide band-edge marker for calibrating receiver tuning dial so receiver meets FCC rules for checking transmitter frequency when using VFO rather than crystal control for Novice transmitter in amateur bands. Crystal can be either for 40- or 80-meter band. Although band-edge frequency is convenient for warning when transmitter is going off frequency, calibration can be done with any frequency in or near band of interest. C5 is 0.25 μF.—K. Negoro, A Band-Edge Marker Generator, *QST*, April 1973, p 16–17.

SINE/COSINE OSCILLATOR—Two oscillators in cascade with positive feedback generate two sine waves in quadrature (differing in phase by 90°). Limiting network D_1-D_2-R_5 is used around A_2 to prevent oscillator from stabilizing at saturation limit of A_2. R_5 is used to set output at any level above zener limits of D_1-D_2. Frequency is 1 kHz for values shown.—W. G. Jung, "IC Op-Amp Cookbook," Howard W. Sams, Indianapolis, IN, 1974, p 371–372.

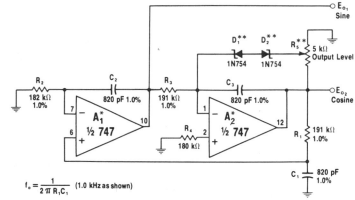

$$f_o = \frac{1}{2\pi R_1 C_1} \quad (1.0 \text{ kHz as shown})$$

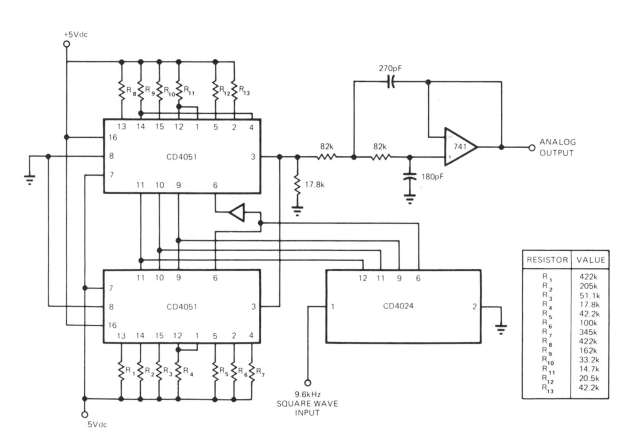

RESISTOR	VALUE
R_1	422k
R_2	205k
R_3	51.1k
R_4	17.8k
R_5	42.2k
R_6	100k
R_7	345k
R_8	422k
R_9	162k
R_{10}	33.2k
R_{11}	14.7k
R_{12}	20.5k
R_{13}	42.2k

WAVEFORM SYNTHESIZER—Values of weighting resistors connected to inputs of multiplexer chips determine waveform of analog output. CD4024 binary counter sequences multiplexers through all states at 16 times fundamental frequency of desired waveform. Active filter using 741 opamp removes components of sampling frequency. For near-approximation to sine wave, weighting resistors range from about 15K to 425K.—J. R. Tracy, CMOS Circuits Generate Arbitrary Periodic Waveforms, *EDN Magazine,* Aug. 20, 1973, p 86–87.

30-kHz MARKERS FOR 2-METER FM—Crystal is placed in loop of standard TTL MVBR. Circuit is modified so 32-pF parallel-mode unit will work into effective load of 32 pF. Series 220-pF capacitor raises crystal frequency enough to permit accurate frequency adjustment by trimmer. Oscillator output is fed to two decade dividers; output of second decade IC$_3$ is 30-kHz square wave with 20% duty factor, coinciding with standard 2-meter FM channels. Regulated supply is 5 VDC at 110 mA.—G. E. Zook, Channel Marker Generator, *CQ*, April 1972, p 41–42.

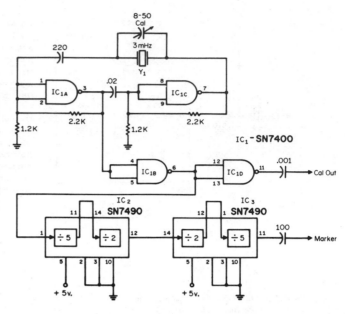

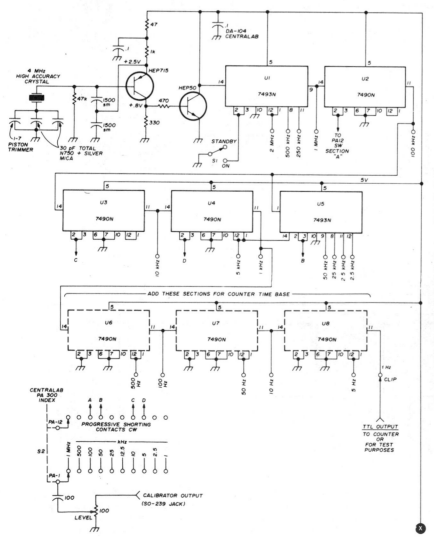

FREQUENCY STANDARD—Uses high-performance TTL ICs operating from regulated 5-V supply furnishing 260 mA, connected to point X. Provides choice of 18 precision frequencies if all ICs are used, or 8 marker frequencies if only upper three ICs are used. Adjustable level control permits matching output of frequency calibrator to incoming signals such as from WWV, or turning full on for strong, clear markers. HEP715 oscillator transistor is coupled to TTL by HEP50 transistor. 7493 binary dividers U1 and U5 divide by factors of 2, with 7490 decade dividers making up remainder of logic. Reset pins 2 and 3 control operation of logic, either with S1 or with progressively shorted contacts of rotary switch S2. Crystal should be ordered for 0.0005% tolerance, F-700 or SP7-P holder, 32-pF load, and 4 MHz at room temperature.—B. Kelley, Universal Frequency Standard, *Ham Radio*, Feb. 1974, p 40–47.

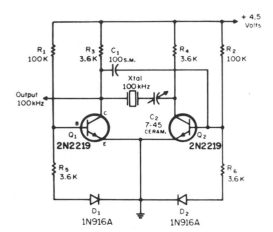

100-kHz CRYSTAL—Drives JK or RS flip-flops to provide markers at 10-kHz or 20-kHz intervals for calibrating transmitter, receiver, or transceiver. D_1 and D_2, used to stabilize output, can be eliminated if desired; R_5 and R_6 are then grounded directly. Transistor and diode types are not critical.—G. F. Moynahan, An Improved Crystal Calibrator Using Solid-State Techniques, *CQ*, May 1972, p 18 and 20.

15 Hz TO 40 kHz IN FOUR RANGES—Tunable wide-range Wien-bridge audio oscillator is switched to cover 15–200 Hz, 150–2000 Hz, 1.5–20 kHz, and 3–40 kHz. U1 is high-input-impedance opamp in bridge circuit using thermistor as nonlinear feedback element. C1 is adjustable in series branch of bridge to compensate for capacitance (about 10 pF) of ungrounded common terminal of dual tuning capacitor. Use ±15 V dual regulated supply.—H. Olson, Integrated-Circuit Audio Oscillator, *Ham Radio,* Feb. 1973, p 50–54.

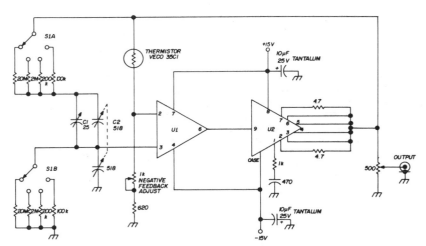

C1	4.5 - 25 pF trimmer capacitor
C2	Dual 518-pF tuning capacitor (Jackson Brothers 5084/2/518HO)

U1	Fairchild µA740C, Signetics NE536T, National NH0042C or Intersill ICL8007C
U2	Motorola MC1438R or MC1538R

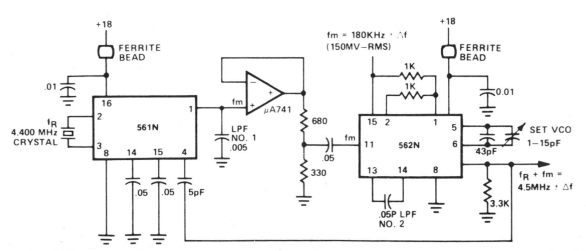

PRECISION 4.5-MHz FM FOR TV IF—Translation loop made from Signetics 561N and 562N PLLs produces 4.5-MHz signal with deviation of ±25 kHz, using 4.400-MHz crystal to control reference frequency. Modulation frequency is 400 Hz.—"Signetics Analog Data Manual," Signetics, Sunnyvale, CA, 1977, p 843–845.

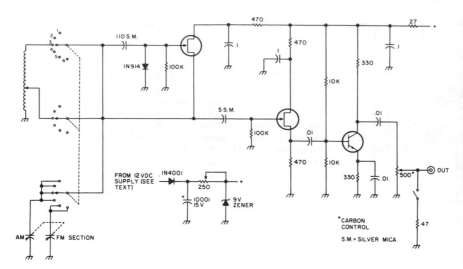

SIX-RANGE RF—Covers from 100 kHz to about 75 MHz in six bands, for checking low-frequency IF circuits on up to VHF circuits (using harmonics up to 220 MHz). Omission of frequency read-out dial scale simplifies design without affecting usefulness for troubleshooting. FETs can be HEP-802 or MPF102, and output transistor is 2N3866 or 2N706. First FET is Hartley oscillator lightly coupled through 5-pF capacitor to FET source-follower isolation stage. Last stage boosts signal level up to about 1 V on most bands. Tuning capacitor is broadcast-band type having 300-pF AM section and 25-pF FM section. Set of six coils can be purchased as Conar CO-69 through CO-74 from National Radio Institute or can be wound as suggested in article to give high-end band limits of 0.57, 1.4, 4.5, 17, 39, and 75–80 MHz. For portable use, 12-V battery pack or 9-V transistor radio battery can be used to give constant 9-V supply.—Brew Up a Signal Generator, *73 Magazine*, Jan. 1978, p 50–52.

30 Hz TO 100 kHz—Wien-bridge sine-wave oscillator using two RCA CA3140 opamps covers frequency range with less than 0.5% total harmonic distortion. Adjust 10K pot for best output waveform. Maximum output into 600-ohm load is about 1 VRMS. Opamps are direct replacements for 741 but have higher input impedance and better slew rate.—C. Hall, New Op Amp Challenges the 741, *Ham Radio*, Jan. 1978, p 76–78.

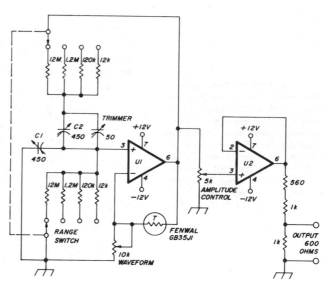

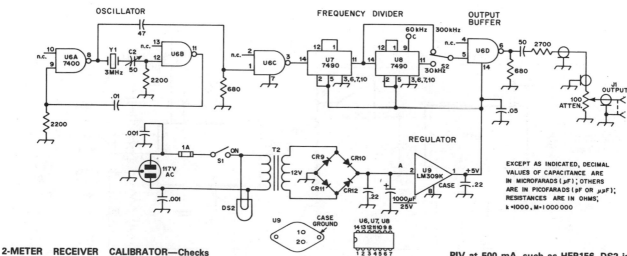

2-METER RECEIVER CALIBRATOR—Checks both frequency and sensitivity of amateur FM receiver. Starts with 3-MHz crystal and provides markers every 30 kHz or every 300 kHz. Energy levels are so low that only simple 100-ohm attenuator is needed. Article covers construction and calibration. CR9-CR12 are silicon rated 200 PIV at 500 mA, such as HEP156. DS2 is 117-V neon. T2 is 12 V at 0.3 A.—H. Lukoff, A 2-Meter Frequency and Sensitivity Calibrator, *QST*, Feb. 1976, p 34–36.

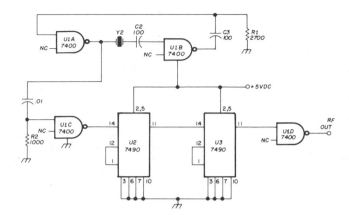

ALIGNMENT OSCILLATOR—With 500-kHz crystal, output can be used as 5-kHz markers in sweep alignment procedure. If SN7490P decade dividers are omitted and 455-kHz crystal is chosen, TTL circuit can be used to supply low IF value used by some receivers. Circuit will oscillate up to several megahertz.—J. Carr, VHF FM Receiver Alignment Techniques, *Ham Radio*, Aug. 1975, p 14–22.

FM SIGNAL GENERATOR OR WOBBULATOR—With sine-wave input, RCA CA3046 transistor array connected as VCO can be used as low-distortion FM signal generator. With sawtooth input, same arrangement serves as wobbulator. Increasing size of timing capacitor reduces operating frequency, permitting use down to audio frequencies as voltage-controlled oscillator in electronic organ.—J. L. Linsley Hood, Linear Voltage Controlled Oscillator, *Wireless World*, Nov. 1973, p 567–569.

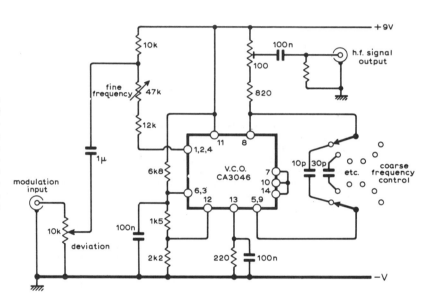

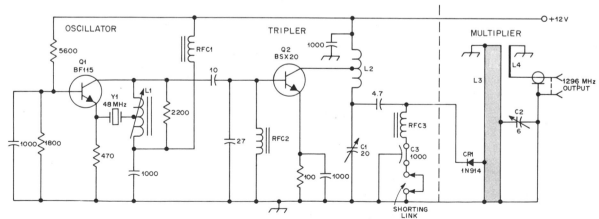

C1, C2 — Glass piston trimmer or other high-*Q* variable.
CR1 — 1N914 silicon high-speed switching diode.
L1 — 12 turns no. 28 enam. wire on 1/4-

inch diameter slug-tuned form. Tap 1 turn from cold end.
L2 — 4 turns no. 18 copper wire, 1/4-inch ID by 7/16-inch long, center tapped.
L3, L4 — See Fig.

RFC1-RFC3, incl. — 2-1/2 turns no. 28 enam. wire on ferrite bead with 950 permeability (Amidon miniature beads suitable).
Y1 — Third-overtone crystal, 48 MHz.

1296 MHz—Can be used as signal source for receiver adjustment and antenna testing, or as minibeacon on 1296 MHz. 48-MHz oscillator and clipper feed 144 MHz to 1N914 diode which multiplies frequency by 9. Half-wavelength stripline tank L3-C2 rejects other harmonics. Shorting link below RFC3 is removed for measuring diode current.—A 1296-MHz Signal Source, *QST*, March 1977, p 26.

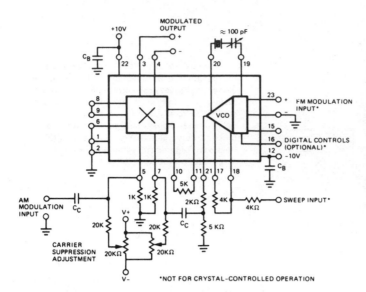

AM, FM, AND SWEEP—Oscillator and multiplier sections of Exar XR-S200 PLL IC are connected as general-purpose voltage-tuned AM/FM radio-frequency signal generator. Can also serve as high-stability carrier or reference generator if crystal at desired frequency is connected between pins 19 and 20 as shown. Multiplier section introduces amplitude modulation on carrier signal generated by VCO. Balanced multiplier allows suppressed-carrier or double-sideband modulation. Typical carrier suppression is above 40 dB for frequencies up to 10 MHz. With timing capacitor used in place of crystal, oscillator section can provide highly linear FM or frequency sweep. Digital control terminals of oscillator can be used for frequency-shift keying.—"Phase-Locked Loop Data Book," Exar Integrated Systems, Sunnyvale, CA, 1978, p 9–16.

1-kHz SQUARE WAVE—Useful for signal-tracing from audio frequencies to several megahertz because 1000-Hz square-wave output of 555 timer is rich in harmonics. Use 5-V supply. Developed for checking audio, IF, and RF stages of amateur receiver operating on 160- to 40-m bands.—J. J. Carr, How to Become a Troubleshooting Wizard, *73 Magazine*, Jan. 1976, p 138–143.

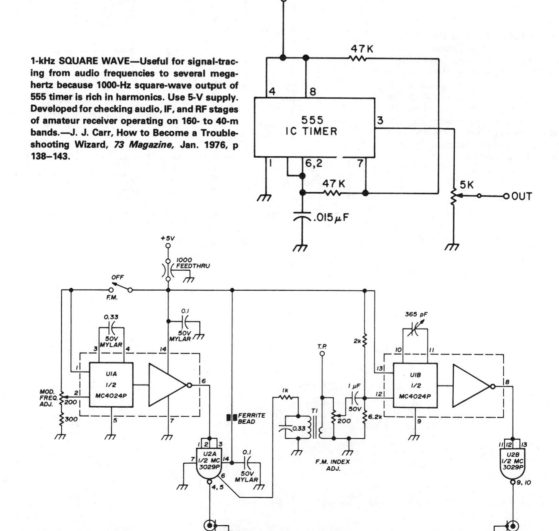

600 kHz TO 12 MHz—Uses Motorola MC4024P or HEP3805P dual voltage-controlled MVBR or VCO. One half is used to produce rectangular RF output and other half to generate rectangular 1-kHz modulation frequency. RF output frequency is proportional to 1/C, with 365-pF variable capacitor providing tuning over 20:1 range from 600 kHz to 12 MHz. Use large dial for calibration. Half of MC3029P line-driver NAND gate follows each of MVBRs in MC4024P to provide isolation and to drive 50-ohm lines with either output. Output voltage is well over 1 V P-P. T1 is 88-mH toroid with 30 turns No. 26 enamel wound over it as secondary. Use regulated supply.—H. Olson, Wide Range RF Signal Generator, *Ham Radio*, Dec. 1973, p 18–21.

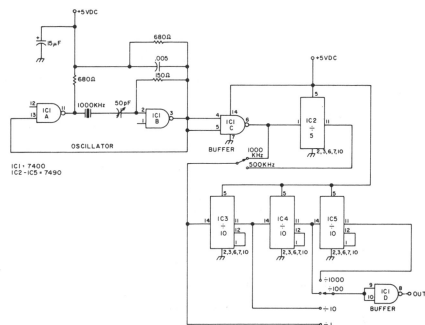

TTL CRYSTAL CALIBRATOR—Easily assembled from low-cost TTL digital ICs, for use as troubleshooting signal generator. Almost any frequency can be obtained by correct choice of oscillator crystal and/or division ratio. If zeroed against frequency standard such as WWV, circuit gives accurate frequency check.—J. J. Carr, How to Become a Troubleshooting Wizard, *73 Magazine,* Jan. 1976, p 138–143.

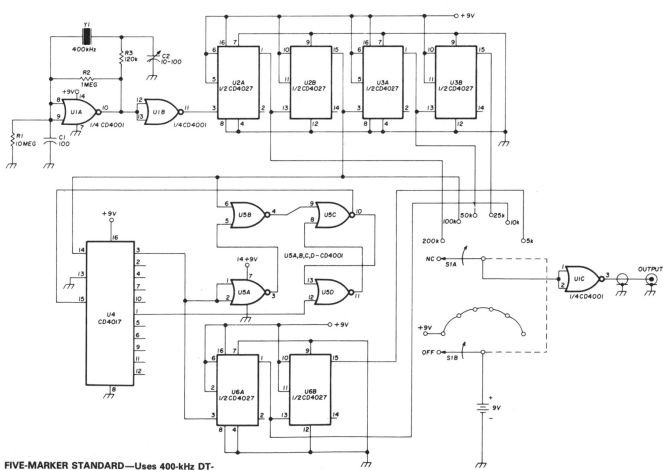

FIVE-MARKER STANDARD—Uses 400-kHz DT-cut crystal in NOR-gate oscillator U1A and divider chain U2-U6 to provide calibration markers at 200, 100, 50, 25, 10, and 5 kHz. U2A-U3B are wired as D flip-flops for dividing by 2. U4 is divide-by-N counter, with latch arrangement of U5 used to reset selected divide-by-5 logic. Output of 100 kHz is divided by 5 and then by 2 to give symmetrical 10-kHz output for division by 2 to provide 5 kHz. CMOS CD4000-series logic elements reduce power consumption from 9-V battery to 2.8 mA but allow sufficient switching speed and harmonic energy for good response throughout HF bands.—F. M. Griffee, Frequency-Marker Standard Using CMOS Logic, *Ham Radio,* Aug. 1977, p 44–45.

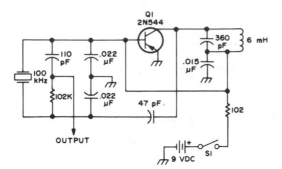

100-kHz CALIBRATOR—Simple crystal-controlled single-transistor oscillator can be used to calibrate amateur radio transceiver. Output should be connected to antenna input side of receiver, not to antenna terminal normally used.—Novice Q & A, *73 Magazine*, March 1977, p 187.

RECEIVER CHECKER—Single-transistor 28-MHz crystal oscillator and carefully designed attenuator network serve to generate signal of about 1 μV for checking performance of amateur radio receiver. Can be used on any band down to 6 meters by appropriate choice of crystal and LC circuit components. Coil is CTC LS5 form having 15 turns No. 22 enamel and 2-turn link.—Is It the Band or My Receiver?, *73 Magazine*, Oct. 1976, p 132–133.

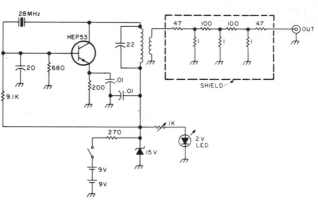

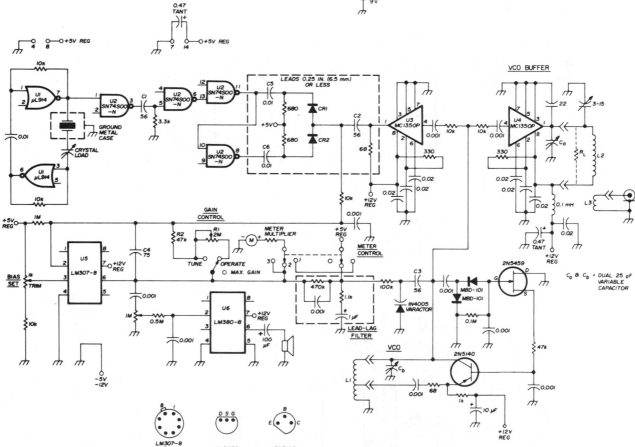

6–36 MHz HARMONIC GENERATOR—Phase-locked loop is used with short-duration pulses from 1-MHz crystal reference oscillator to produce highly accurate harmonics. SN74S00N Schottky U2 changes reference waveform to harmonic-rich 100-ns pulses for feed to 1N914 phase-detector diodes CR1 and CR2. Buffer U3 delivers output of 2N5140 VCO to diodes for phase comparison, and phase-frequency output is fed to opamp U5 that locks VCO more tightly to reference-oscillator output by increasing its control of varactor. VCO frequency will then be harmonic of reference oscillator. Meter used to monitor control voltage to varactor can have full-scale value of 100 μA to 1 mA, with its multiplier resistor adjusted to read 5 V at midscale. When opamp is capturing VCO, meter needle will flop from side to side but will return to midscale after lock is achieved. Article covers construction, tuning, and operation.—K. W. Robbins and J. R. True, Crystal-Controlled Harmonic Generator, *Ham Radio*, Nov. 1977, p 66–69.

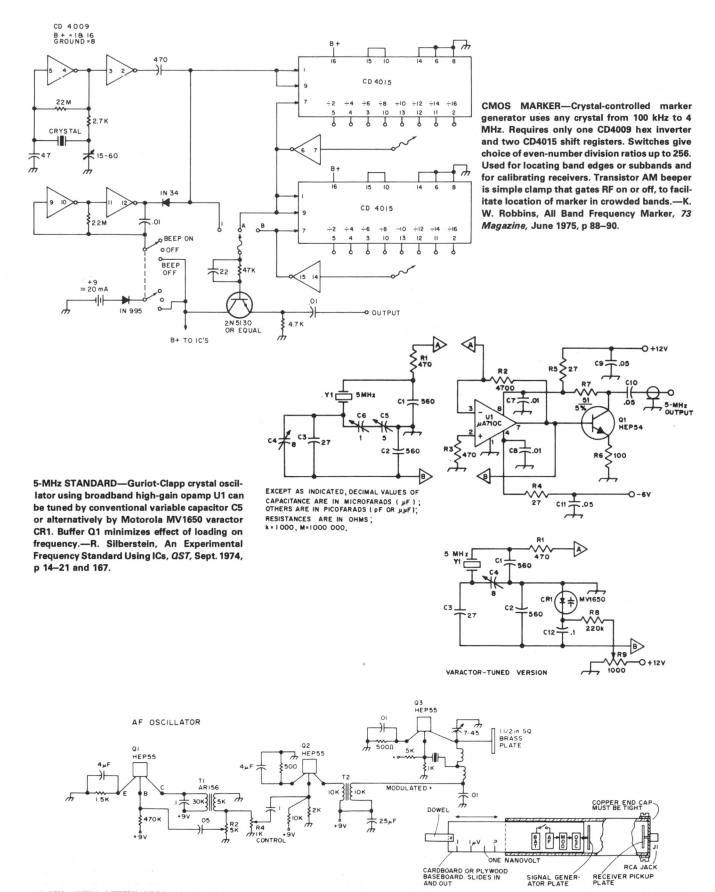

CMOS MARKER—Crystal-controlled marker generator uses any crystal from 100 kHz to 4 MHz. Requires only one CD4009 hex inverter and two CD4015 shift registers. Switches give choice of even-number division ratios up to 256. Used for locating band edges or subbands and for calibrating receivers. Transistor AM beeper is simple clamp that gates RF on or off, to facilitate location of marker in crowded bands.—K. W. Robbins, All Band Frequency Marker, *73 Magazine*, June 1975, p 88–90.

EXCEPT AS INDICATED, DECIMAL VALUES OF CAPACITANCE ARE IN MICROFARADS (μF); OTHERS ARE IN PICOFARADS (pF OR μμF); RESISTANCES ARE IN OHMS; k = 1000, M = 1000 000.

5-MHz STANDARD—Guriot-Clapp crystal oscillator using broadband high-gain opamp U1 can be tuned by conventional variable capacitor C5 or alternatively by Motorola MV1650 varactor CR1. Buffer Q1 minimizes effect of loading on frequency.—R. Silberstein, An Experimental Frequency Standard Using ICs, *QST*, Sept. 1974, p 14–21 and 167.

VARACTOR-TUNED VERSION

50 MHz WITH ATTENUATOR—Positioning of miniaturized signal generator in 4½ × 2⅛ × 24 inch waveguide provides stable variable-strength signal that can be dropped gradually down to zero as generator is moved away from receiver pickup plate. Slide can be calibrated for measuring sensitivity of 6-meter receiver in tenths of a microvolt. Circuit consists of 50-MHz crystal oscillator, AF oscillator, and simple class A modulator.—B. Hoisington, Low-Cost Infinite Attenuator for Amateur Use, *73 Magazine*, Sept. 1974, p 107–108.

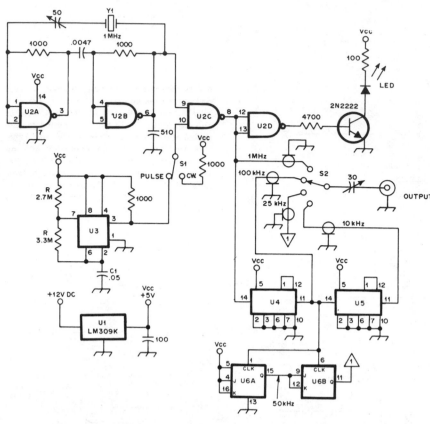

PULSED MARKER—Crystal calibrator circuit provides pulsed output for easy spotting, eliminating need for turning marker on and off repeatedly to identify it in crowded band. With values shown, switching rate is about 2 Hz. When marker is found, placing S1 in CW position keeps it on for zero-beating calibrator output more accurately. Reducing value of C1 increases switching speed of U3, thus increasing pulse rate. LED indicates either pulsed or CW output. ICs are conventional types used in crystal calibrators.—R. G. Brunner, Crystal Calibrator Has Pulsed Output, *QST*, Nov. 1977, p 45.

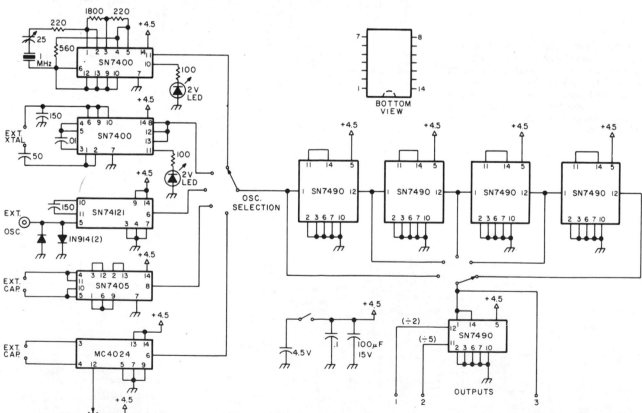

UNIVERSAL SIGNAL GENERATOR—Collection of IC oscillators and dividers generates square waves from HF down to subaudible AF, along with markers up into VHF. Selectable oscillator section feeds fixed string of four divide-by-10 stages. Extra SN7490 divider can be switched in at various points along string to add divide-by-5 and divide-by-2 functions. LED in 1-MHz crystal stage indicates that circuit is oscillating. Second stage can be used with any external crystal up into low VHF range. Third stage accepts and conditions external sine or square input for feeding to divider chain. External capacitor for fourth stage tunes square-wave generator from several hertz to several megahertz. Optional fifth stage is VCO for entire HF range up to 25 MHz.—J. Schultz, Updated Universal Frequency Generator, *73 Magazine*, Nov. 1976, p 50–52.

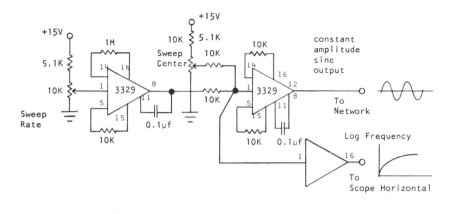

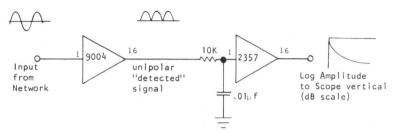

NETWORK TESTER—Sweep generator produces time-varying constant-amplitude frequency signal for network under test, with output of network converted to logarithmic DC voltage for display of amplitude or gain on decibel scale of CRO or recorder. Sawtooth voltage for sweep, generated by first Optical Electronics 3329 IC, drives another 3329 that converts sawtooth into logarithmic signal for log-frequency output. Detector uses 9004 absolute-value module as linear detector, with 2357 logamp converting output to decibel scale. Resulting display is Bode plot of frequency response.—"A Simple Sweep Generator," Optical Electronics, Tucson, AZ, Application Tip 10201.

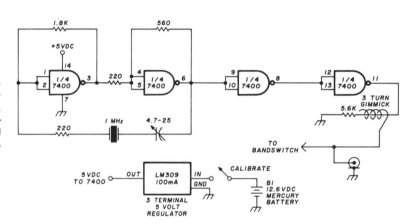

1-MHz CRYSTAL CALIBRATOR—Battery-powered crystal oscillator for checking frequency calibration of communication receiver uses TTL. Regulator is in TO-5 can, which is 100-mA version of LM309.—J. J. Carr, Resurrecting the Old War Horse: New Hope for the Old Receiver, *Ham Radio,* Dec. 1976, p 52–55.

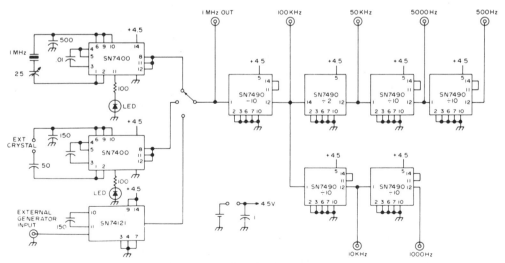

AF/RF FROM 1 MHz—String of SN7490 decade counters divides output of 1-MHz crystal oscillator by 10 or 2 to give choice of six fixed frequencies between 1000 Hz and 100 kHz along with undivided 1 MHz. One gate of SN7400 crystal oscillator drives LED to indicate that crystal is oscillating. When using external sine-wave source, input is squared by SN74121 MVBR for driving frequency divider chain. Circuit is easily modified to give other divider ratios. Applications include use as marker generator for receiver calibration up into VHF range or as signal generator for precise AF or RF square-wave signal at desired frequencies.—J. J. Schultz, Poor Man's Universal Frequency Generator, *73 Magazine,* July 1974, p 33 and 35–36.

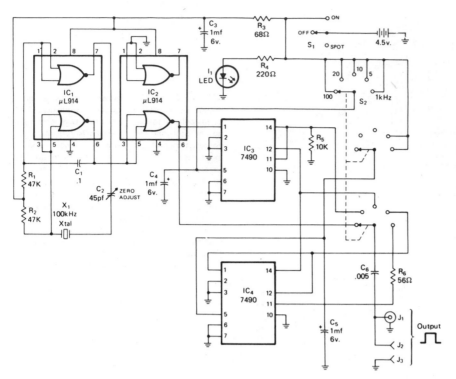

SECONDARY STANDARD—Provides switch-selected square-wave outputs of 100, 20, 10, 5, and 1 kHz, calibrated with receiver tuned to WWV frequency. 100-kHz clock signal is generated by crystal oscillator IC_1. Half of IC_2 is buffer between oscillator and first decade counter IC_3, used to generate 10- and 20-kHz outputs. Second decade counter IC_4 gives 1 and 5 kHz.—E. R. Spadoni, A Versatile Secondary Frequency Standard, *CQ*, Sept. 1975, p 31–32.

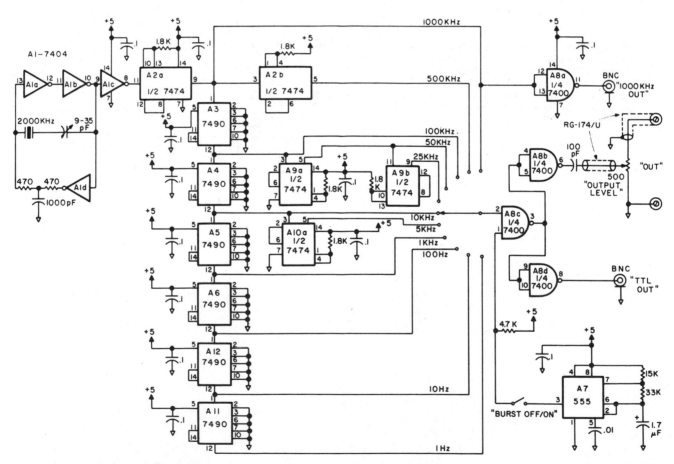

1 Hz TO 1 MHz—Low-cost secondary frequency standard generates marker signals of 1000, 500, 100, 50, 25, 10, 5, and 1 kHz and 100, 10, and 1 Hz, with harmonic markers usable well beyond 30 MHz. Two TTL output levels are available as clocks or signal injectors for checking TTL. Short-term accuracy is about 1 part in 10^6. Unit is easily aligned to WWV with shortwave receiver. Frequency-burst mode turns output on and off 10 times per second, for identification of markers in crowded band of receiver. Article covers construction and operation in detail, and gives circuit for suitable regulated power supply having standby battery.—T. Shankland, Build a Super Standard, *73 Magazine*, Oct. 1976, p 66–69.

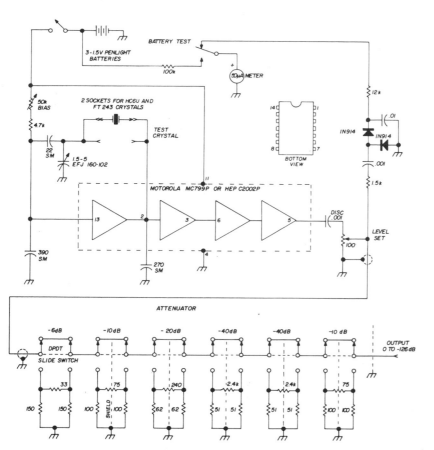

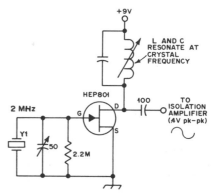

1–10 MHz CRYSTAL—Stable crystal test oscillator takes any crystal in frequency range with no tuning adjustments. Uses single Motorola MC799P IC in circuit that provides 32-pF crystal loading. Trimmer may be used to adjust crystal for exact frequency if desired. Circuit is not critical. Bias pot compensates for battery voltage changes. Output attenuator uses standard resistor values to provide up to 126 dB of output signal control. Meter serves as battery tester and gives instant indication of crystal activity when circuit is used for testing crystals.—A. A. Kelley, Crystal Test Oscillator and Signal Generator, *Ham Radio,* March 1973, p 46–47.

2-MHz STANDARD WITH DIVIDERS—Can be used for calibration of frequency meters, frequency counters, and amateur receivers. Two crystal oscillators (2 MHz and 100 kHz) feed two 7490 decade counters through isolation amplifier. Arrangement gives frequency division by 2, 4, 10, 20, and 100 for each oscillator, with all frequencies rich in harmonics and usable through 144 MHz. Counter reset gates at pins 2, 3, 6, and 7 and ground at pin 10 must be connected to common terminal for all modes of operation. Supply should not exceed 5.5 V.—J. M. Janicke, A Wide-Range Crystal-Controlled Frequency Standard, *QST,* July 1976, p 27–28.

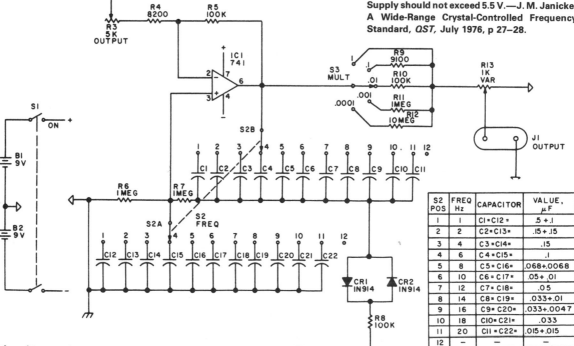

S2 POS	FREQ Hz	CAPACITOR	VALUE, μF
1	1	C1 = C12 =	.5 + .1
2	2	C2 = C13 =	.15 + .15
3	4	C3 = C14 =	.15
4	6	C4 = C15 =	.1
5	8	C5 = C16 =	.068 + .0068
6	10	C6 = C17 =	.05 + .01
7	12	C7 = C18 =	.05
8	14	C8 = C19 =	.033 + .01
9	16	C9 = C20 =	.033 + .0047
10	18	C10 = C21 =	.033
11	20	C11 = C22 =	.015 + .015
12	—	—	—

1–20 Hz SINE—Designed to complement usual lab sine-wave generator that goes down to only 20 Hz, by providing discrete switch-selected output frequencies of 1 Hz and 2–20 Hz in 2-Hz steps. Output attenuator uses pot and switch to set output at any value within range of five decades. Circuit uses 741 opamp in Wien-bridge oscillator having four-element RC network in positive feedback path of amplifier (R6 in parallel with capacitor selected by S2A, and R7 in series with capacitor of S2B), so oscillation occurs at frequency where phase shift occurs. Article gives construction details. Offset adjustment R2 may need touching up as batteries run down.—D. Hileman and L. Hileman, V-V-V LF Generator, *73 Magazine,* Holiday issue 1976, p 97–99.

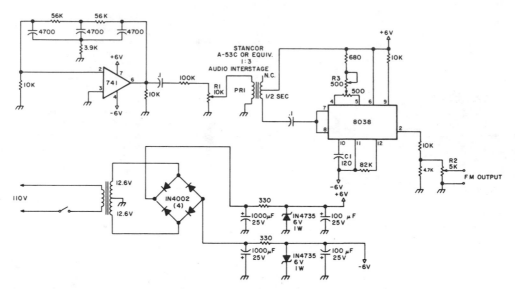

455-kHz FREQUENCY-MODULATED—Can be used to align IF amplifier and quadrature detector of FM receivers. Unit is stable and provides ample deviation for amateur receivers. Uses 8038 function generator and 741 opamp connected as audio oscillator to provide about 1000-Hz modulating voltage. Includes deviation control R1, output level control R2, and carrier frequency control R3. Adjust 500-ohm pot between pins 4 and 5 of 8038 for clean sine-wave output on CRO. Adjust R3 to give 455 kHz as measured by meter or frequency counter. To check audio oscillator, connect AC voltmeter or CRO across R1, which should have clean 1000-Hz sine wave of several hundred millivolts. Transformer in power supply can be two separate 12.6-V units with primaries in parallel and secondaries in series.—J. C. Chapel, Build This FM Signal Generator, *73 Magazine,* Jan. 1978, p 154–155.

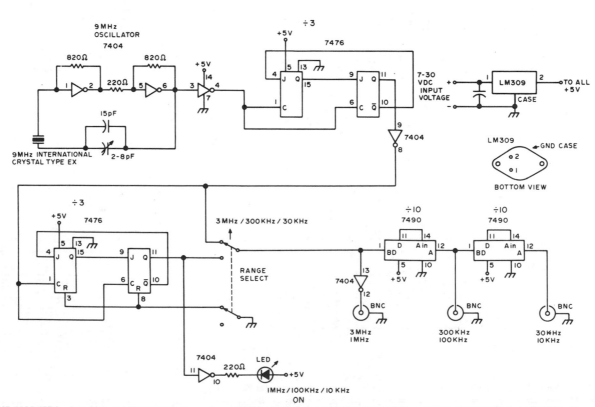

HF/VHF MARKERS—Provides markers needed for most amateur radio bands, including 30 and 300 kHz for VHF FM operation and 10 and 100 kHz for 2-m FM operation. When LED is on, outputs are 1 MHz, 100 kHz, and 10 kHz. When LED is off, outputs are 3 MHz, 300 kHz, and 30 kHz. Uses 7404 TTL hex inverter as crystal oscillator, with 2–8 pF trimmer for zeroing crystal with WWV. All 7476 TTL dual JK flip-flops are connected to divide by 3.—F. E. Hinkle, Inexpensive HF-VHF Frequency Standard, *73 Magazine,* April 1976, p 62–63.

CHAPTER 81
Single-Sideband Circuits

Includes audio clippers, shapers, and other circuits for improving speech readability, along with product detectors, sideband selectors, double-balanced mixers, direct-conversion receivers, and SSB test equipment.

PRODUCT DETECTOR—Developed for use in SSB receiver having 9-MHz IF amplifier. Values in parentheses are for receiver having 455-kHz IF amplifier. Diode types are not critical.—Circuits, *73 Magazine,* May 1973, p 105.

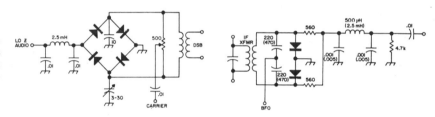

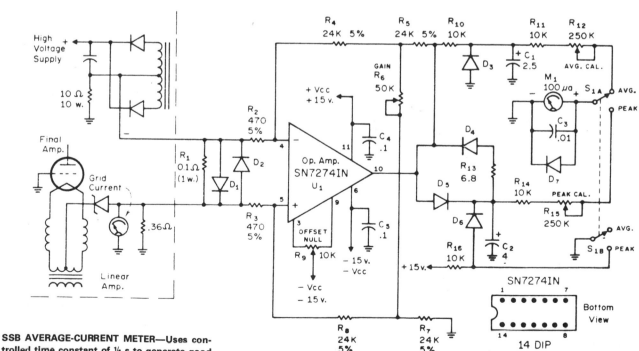

SSB AVERAGE-CURRENT METER—Uses controlled time constant of ¼ s to generate good approximation of average current in voice-modulated SSB signal. Circuit also gives peak current reading, with peaks measured and held for short time. Opamp isolates current from plate supply and converts it into signal that can be run through shaping network to get averages and peaks. D₁-D₇ are 1N3064 or equivalent. D₁ and D₂ eliminate spikes that might damage opamp. R₆ allows small adjustments of opamp gain, which is normally set at 100.—R. Sans, Make Your Meter Readings Count, *CQ,* Dec. 1972, p 28–29.

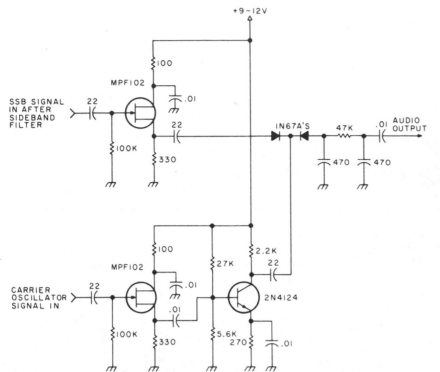

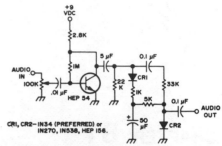

SPEECH PROCESSOR—Preamplification combined with clipping or compression gives higher average level and increased intelligibility of SSB communication. Degree of compression is controlled with 100K pot that adjusts input. Output of transistor feeds passive diode compressor. Amount of compression will vary with diode type, and experimentation is suggested. Article covers construction and adjustment of circuit.—B. Barrington, Simple Audio Preamp, *73 Magazine*, Feb. 1974, p 69–70.

SSB MONITOR—Requires two connections to SSB transmitter, at output of carrier oscillator and at output of sideband filter. FET isolation stages for each connection feed 2N4124 product detector that gives audio signal for monitoring directly with headphones or for feeding AF amplifier driving loudspeaker.—Clean Up Your Act, *73 Magazine*, Jan. 1978, p 136–137.

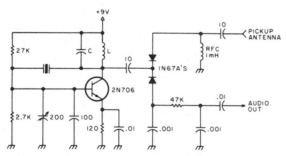

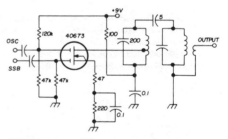

SIDEBAND MIXER—Used as transmitting mixer in SSB transceiver made by Sideband Associates for radiomarine communication in 2–23 MHz range. Low-frequency sideband signal and high-frequency oscillator signal are mixed to produce higher sum frequency at output. Double-tuned resonant circuit provides adequate output bandwidth and excellent skirt rejection of undesired frequency components.—E. Noll, MOSFET Circuits, *Ham Radio*, Feb. 1975, p 50–57.

OFF-AIR MONITOR—Single-frequency modulation monitor for SSB transmitter combines crystal oscillator with product detector, for checking audio at one point within band. If tuning and loading of SSB transmitter are same over rest of band, audio quality also remains constant. Use fundamental-mode crystal for desired band. LC circuit should be resonated to band being used. Oscillator signal is spotted with transceiver in receive mode, after which transceiver can be monitored during transmissions.—Clean Up Your Act, *73 Magazine*, Jan. 1978, p 136–137.

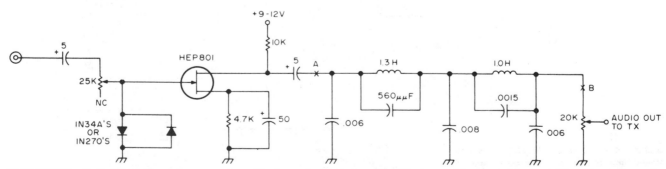

SOFT CLIPPER—Used after audio compressor to improve effective signal strength of SSB transmitter. Soft clipping is achieved by driving diode pair through resistance. Clipper is followed by low-noise FET voltage amplifier having broadband flat frequency response. Output filter sharply attenuates signals above about 3 kHz.—J. J. Schultz, Adding dBs to the Audio Compressor, *73 Magazine*, May 1974, p 21–23 and 25.

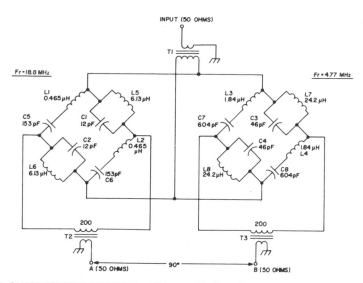

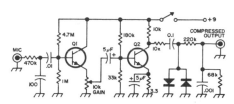

PREAMP WITH CLIPPING—Increases speech intelligibility, particularly with SSB amateur stations. Q1 and Q2 are HEP-54. Diodes are 1N456 or HEP-158.—Circuits, *73 Magazine*, Feb. 1974, p 100.

3–30 MHz QUADRATURE PHASE SHIFT—Wideband passive AF phase-shift network makes direct-conversion SSB generation possible. Bridge networks each provide 45° phase shift, to give differential phase shift of 90° over entire frequency range with maximum phase error of about 1°. Overall loss of network is about 6 dB. T1, T2, and T3 are wound on Neosid 1050-1-F14 of Indiana General F684-1 balun core. Twist together three 7-inch lengths of No. 26 enamel and wind 3 turns through the two holes. Connect two wires in series for 200-ohm windings. Article gives data for winding all other coils.—R. Harrison, A Review of SSB Phasing Techniques, *Ham Radio*, Jan. 1978, p 52–62.

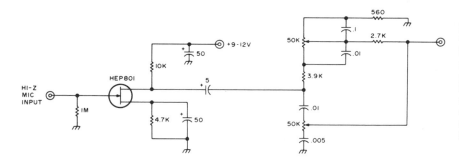

PRECOMPRESSION SHAPER—Improves effective signal strength of SSB transmitter by shaping AF frequency response ahead of audio compressor. Low-noise FET preamp providing initial gain for high-impedance microphone is followed by low- and high-rolloff circuit giving 15-dB boost or rolloff to frequencies centered at about 1 kHz.—J. J. Schultz, Adding dBs to the Audio Compressor, *73 Magazine*, May 1974, p 21–23 and 25.

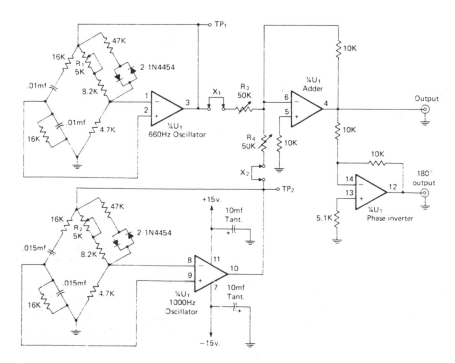

660 AND 1000 Hz FOR SSB TESTING—Produces two audio frequencies with all harmonics and crossproducts down 40 dB or more, as required for accurate testing of amateur SSB equipment. Two sections of Raytheon 4136D quad opamp serve as Wien-bridge audio oscillators, one at 1000 Hz and one at 660 Hz. Silicon signal diodes in each bridge act as nonlinear stabilization elements. Third section of opamp adds sine waves, and fourth section is simple inverter with gain of 1 for push-pull or balanced output. With all four pots at midvalue, adjust R_1 for 12 V P-P at TP_1 and adjust R_2 for 12 V P-P at TP_2. Open X_2 and adjust R_3 to give 12 V P-P from either output terminal to ground, then close X_2 and repeat for X_1 and R_4. Output should now be 660 and 1000 Hz added linearly as required, with no crossproducts. Use regulated ±15 V supply.—H. Olson, A One-Chip, Two Tone Generator, *CQ*, April 1974, p 48–49.

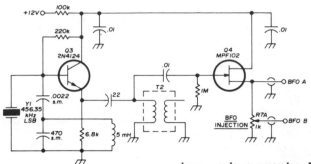

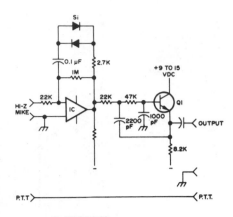

CRYSTAL BFO—Bipolar crystal oscillator is coupled to FET source-follower by miniature 455-kHz IF transformer T2. RF output is adjusted with R7A so BFO injection voltage can be set for maximum carrier suppression. BFO is 456.35 kHz for lower sideband operation or 453.75 kHz for upper sideband in SSB transceiver.—W. J. Weiser, Integrated Circuit SSB Transceiver for 80 Meters, *Ham Radio*, April 1976, p 48–52.

SPIKE CLIPPER—Improves efficiency of low-power SSB amateur transmitter by removing from voice waveform the spikes that cause overmodulation or give low average modulation level. When used as in-line microphone amplifier, circuit gives up to 20-dB equivalent gain at receiving location.—H. E. Weber, Increase Your SSB Efficiency, *73 Magazine*, Dec. 1973, p 71.

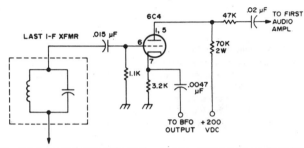

SSB DETECTOR—Can be switched in and out of most tube-type AM receivers for use in place of regular detector stage. Requires stable BFO.—Novice Q & A, *73 Magazine*, March 1977, p 187.

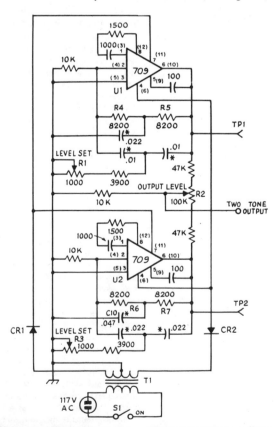

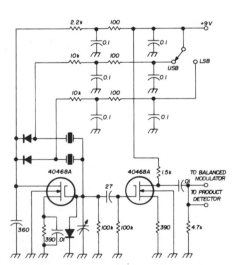

TWO-TONE BURSTS—Uses 709 or 741 opamps to generate 1850-Hz and 855-Hz tones simultaneously, pulsed at 60-Hz line rate for duty cycle slightly below 50%. Used to measure linearity of high-frequency amplifiers in low-power stages of SSB transmitter. Each opamp is powered by half-wave rectified AC, with opposite voltage polarities provided by 12.6-V CT 50-mA filament transformer. Tone frequency of each opamp is determined by symmetrical twin-T network. R1 sets level of 1850-Hz tone, and R3 controls 855-Hz tone. R2 mixes and balances tones. Diodes are 50-PIV 500-mA silicon.—B. Buus, A Technique for Burst Two-Tone Testing of Linear Amplifiers, *QST*, Aug. 1971, p 17–21.

CARRIER OSCILLATOR—Two MOSFETs serve with diode switching arrangement for selecting either upper or lower sideband. Circuit between gates couples oscillator to isolating output stage. Upper/lower sideband switch applies +9 V to anode of switching diode that closes feedback circuit for crystal to be activated. Output can be used as injection voltage in demodulation of incoming signal by product detector of transceiver or as basic carrier applied to balanced modulator for transmitting mode. Used in SSB transceiver made by Sideband Associates for radiomarine communication in 2–23 MHz range.—E. Noll, MOSFET Circuits, *Ham Radio*, Feb. 1975, p 50–57.

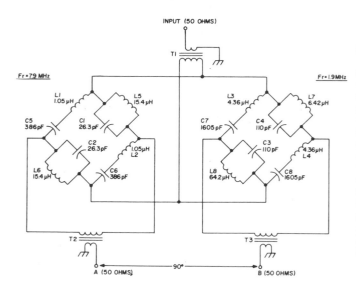

90° PHASE SHIFT WITH BRIDGES—Wideband passive AF phase-shift network makes direct-conversion SSB generation possible. Developed for use with circularly polarized antenna system. Bridge networks each provide 45° phase shift between 1 and 15 MHz, to give differential phase shift of 90° over that frequency range with phase error less than 1°. Amplitude difference between outputs is less than 0.5 dB over range. T1, T2, and T3 are wound on Neosid 1050-1-F14 of Indiana General F684-1 balun core. Twist together three 7-inch lengths of No. 26 enamel and wind 3 turns through the two holes. Connect two wires in series for windings going to bridges. Article gives data for winding all other coils.—R. Harrison, A Review of SSB Phasing Techniques, *Ham Radio,* Jan. 1978, p 52—62.

SSB/CW DEMODULATOR—LM373 communication IC uses balanced mixer as product detector, with reinserted carrier reapplied to pin 6. CW or SSB output is taken from pin 7. If desired, RF gain control can be inserted in AGC feedback path.—E. M. Noll, "Linear IC Principles, Experiments, and Projects," Howard W. Sams, Indianapolis, IN, 1974, p 350–351.

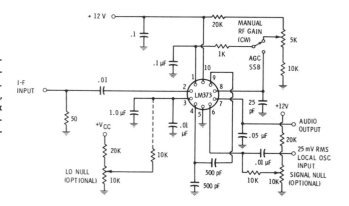

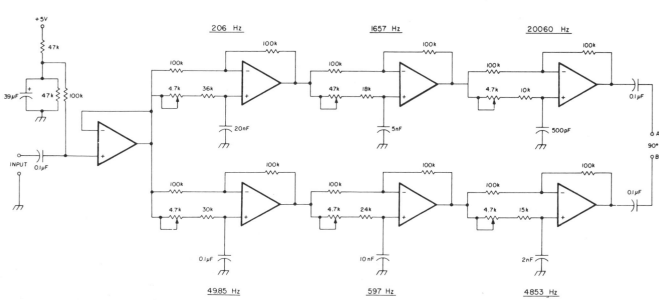

WIDEBAND ACTIVE PHASE SHIFTER—Active audio phase-shift network uses two LM324 quad opamps to provide equal-amplitude outputs differing in phase by 90° ± 2° from 100 Hz to 10 kHz. Each stage is adjusted with 4.7K trimpot to give 90° phase shift at frequency shown on diagram. Align with audio oscillator and CRO or phase meter. Operates from single 5-V supply. Overall gain of entire circuit is unity.—R. Harrison, A Review of SSB Phasing Techniques, *Ham Radio,* Jan. 1978, p 52—62.

MOSFET PRODUCT DETECTOR—SSB IF signal is applied to one gate of MOSFET and demodulating carrier to other gate. Linear demodulation is obtained without distortion components. RC filter connected into drain circuit removes IF and carrier components, leaving demodulated audio as output.—E. M. Noll, "FET Principles, Experiments, and Projects," Howard W. Sams, Indianapolis, IN, 2nd Ed., 1975, p 154.

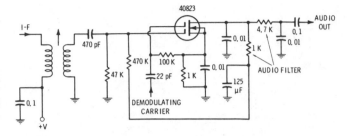

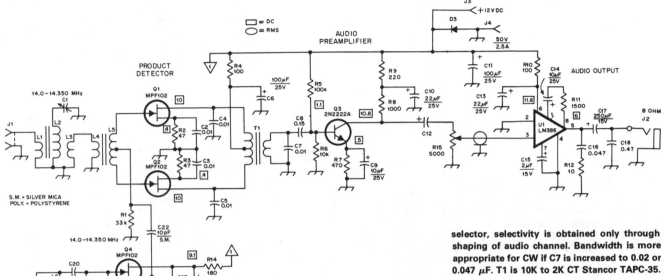

20-METER DIRECT-CONVERSION—Well-designed circuit provides pleasing polarity and depth of sound, with SSB signals seeming to stand out against nearly noiseless background. Covers entire 20-meter band. Use of balanced-product detector improves stability to reject strong broadcast-band AM signals. BFO energy from Q4 is injected through center tap of broadband toroidal transformer L4-L5. Except for pre-selector, selectivity is obtained only through shaping of audio channel. Bandwidth is more appropriate for CW if C7 is increased to 0.02 or 0.047 μF. T1 is 10K to 2K CT Stancor TAPC-35. L2 is 40 turns No. 30 enamel on T-37-6 core, with 2 turns No. 28 on it for L1 and 4 turns for L3. L5 is 16 turns No. 28 on FT-37-63 core, with center tap, with 4 turns No. 28 on it for L4. L6 is 19 turns No. 28 on T-37-6 core, tapped 7 turns above ground.—J. Rusgrove, A 20-Meter High-Performance Direct-Conversion Receiver, *QST*, April 1978, p 11–13.

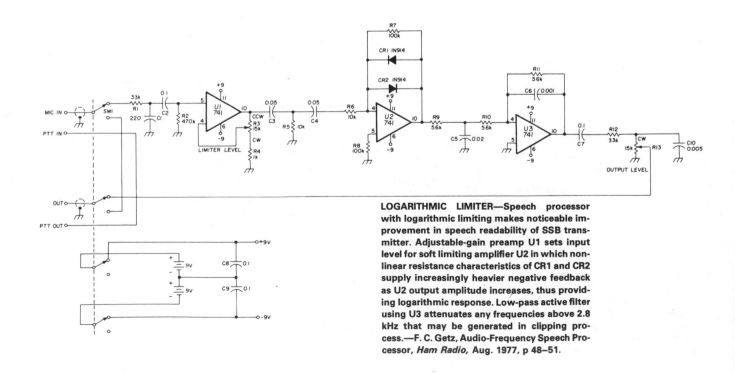

LOGARITHMIC LIMITER—Speech processor with logarithmic limiting makes noticeable improvement in speech readability of SSB transmitter. Adjustable-gain preamp U1 sets input level for soft limiting amplifier U2 in which nonlinear resistance characteristics of CR1 and CR2 supply increasingly heavier negative feedback as U2 output amplitude increases, thus providing logarithmic response. Low-pass active filter using U3 attenuates any frequencies above 2.8 kHz that may be generated in clipping process.—F. C. Getz, Audio-Frequency Speech Processor, *Ham Radio*, Aug. 1977, p 48–51.

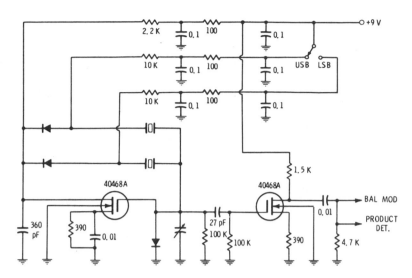

SIDEBAND SELECTOR—MOSFET circuit uses diode switching of crystals in carrier oscillator to select either upper or lower sideband. Second transistor serves as common-source isolating amplifier for driving modulator. Switch applies +9 V to anode of diode that closes feedback circuit for crystal to be activated. Developed for use in SSB transceiver; if operating in 9-MHz range, crystals can be 8.9985 MHz and 9.0015 MHz.—E. M. Noll, "FET Principles, Experiments, and Projects," Howard W. Sams, Indianapolis, IN, 2nd Ed., 1975, p 191–192.

CLIPPER—Combined speech amplifier and logarithmic clipper for use with SSB transmitters reduces speech bandwidth to about 500–3000 Hz, with very little distortion. Power can be obtained from separate battery or from transmitter itself.—Circuits, *73 Magazine,* March 1977, p 152.

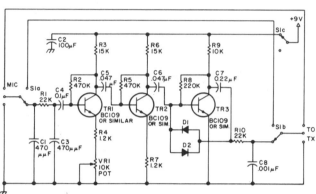

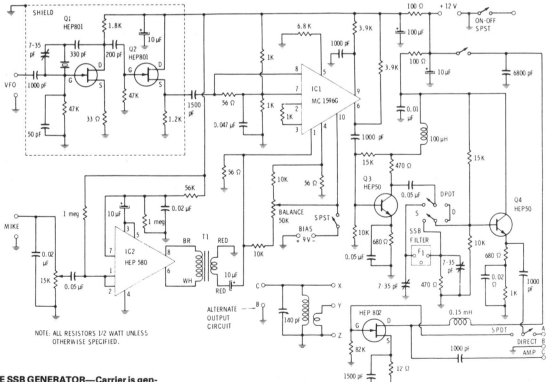

FILTER-TYPE SSB GENERATOR—Carrier is generated by FET crystal oscillator Q1 followed by buffer Q2. Modulating wave is applied to HEP 580 connected as two-stage audio amplifier feeding double-balanced modulator IC1 through transformer T1. Double-sideband signal from pin 6 of IC1 is applied to amplifier Q3-Q4 for straight-through amplification when double-sideband output is desired. For 9-MHz single-sideband operation, sideband filter is switched in between transistors.—E. M. Noll, "Linear IC Principles, Experiments, and Projects," Howard W. Sams, Indianapolis, IN, 1974, p 353–356.

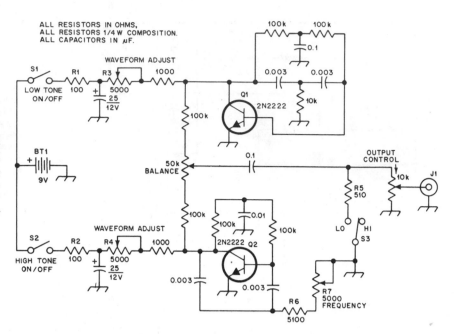

ALL RESISTORS IN OHMS,
ALL RESISTORS 1/4 W COMPOSITION.
ALL CAPACITORS IN μF.

TWO-TONE TESTER—Twin-T transistor oscillators generate two distinct sine-wave AF signals for use in adjusting SSB transmitters. Q1 is fixed at about 1000 Hz, and Q2 is adjustable between 1000 and 1300 Hz, giving frequency difference of 0–300 Hz for use with scopes having 60-Hz horizontal sweep rate that permits display of one to five cycles of RF envelope pattern. Switches permit use of either oscillator separately.—F. Brown, The Two-Tone Tester, *QST,* Nov. 1978, p 22–24.

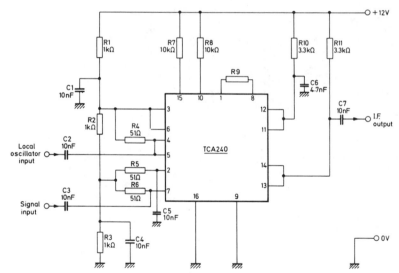

UNTUNED DOUBLE-BALANCED MIXER—Wideband mixer for high-frequency SSB circuits operates from single 12-V supply. Gain is controlled by R9 and increases as value of R9 is decreased. Output at pins 13 and 14 is product of local oscillator and signal input frequencies and contains desired IF value for receiver. Oscillator signal level should be kept below 20 mV to avoid undesired harmonics produced by limiting. Circuit uses Mullard TCA240 dual balanced modulator-demodulator.—"Applications of the TCA240," Mullard, London, 1975, Technical Note 18, TP1489.

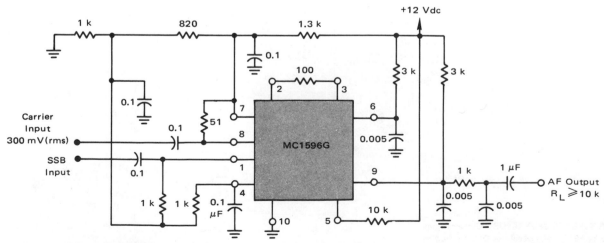

SINGLE-SUPPLY PRODUCT DETECTOR—Motorola MC1596G balanced modulator requires no carrier null adjustment because all frequencies except desired demodulated audio are in RF spectrum and easily filtered out. Circuit performs well with carrier input levels of 100–500 mVRMS. Provides good product detector performance from very low frequencies up to 100 MHz.—R. Hejhall, "MC1596 Balanced Modulator," Motorola, Phoenix, AZ, 1975, AN-531, p 7.

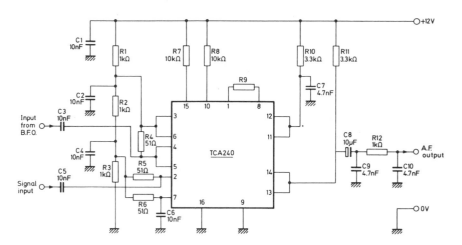

10-MHz PRODUCT DETECTOR—IF signal of SSB receiver is mixed with signal from beat-frequency oscillator in Mullard TCA240 dual balanced modulator-demodulator to give desired audio output signal. Simple low-pass filter R12-C9-C10 removes unwanted output signal.— "Applications of the TCA240," Mullard, London, 1975, Technical Note 18, TP1489.

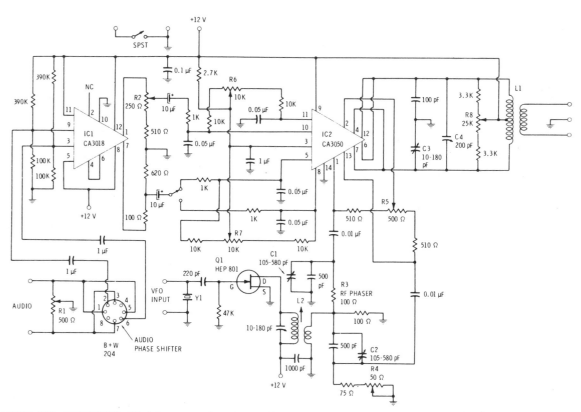

PHASING-TYPE SSB GENERATOR—Modulating wave is first applied to input audio phase shifter for generating audio components that are equal in magnitude but differ 90° in phase. After amplification in CA3018, these audio components are applied to CA3050 double-balanced modulator. SPST and SPDT switch settings determine which sideband will appear at output of modulator. Carrier is generated by FET crystal oscillator for application through RF phase-shift network to pins 1 and 13 of modulator. Both carrier and modulating frequencies are suppressed in balanced output circuit of modulator, leaving only desired sideband. Resonant output transformer provides low-impedance feed to succeeding linear amplifier. Designed for 160-meter band.—E. M. Noll, "Linear IC Principles, Experiments, and Projects," Howard W. Sams, Indianapolis, IN, 1974, p 356–357.

CHAPTER 82
Siren Circuits

Includes variety of circuits for simulating sounds of police and other emergency sirens. Battery-operated versions can be used in toys or as part of burglar or fire alarm system. Some have adjustments for frequency, whooping rate, and duration of rising and falling tones.

10-W AUTO ALARM SIREN—Generates force field of high-intensity sound inside car, painful enough to discourage thief from entering car after tripping alarm switch by opening door. Circuit produces square-wave output that sweeps up and down in frequency. Modulation is provided by triangle waveform generated by R1, D1, and C1. If sweep-frequency siren is prohibited, remove C1 to produce legal two-tone sound. Use efficient horn loudspeaker capable of handling up to 10 W. D2 is silicon rectifier rated 1 A at 50 PIV. Other diodes are general-purpose silicon.—A. T. Roderick III, New Protection for Your Car, *73 Magazine,* March 1978, p 76–77.

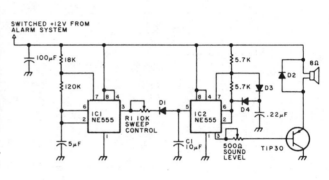

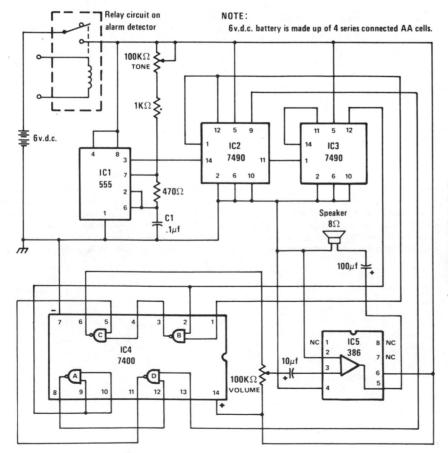

NOTE:
6 v.d.c. battery is made up of 4 series connected AA cells.

Speaker
8Ω

LOW-NOTE SIREN—Produces up/down blooping sounds characteristic of European police cars and now being used on some US emergency vehicles. Can be connected to burglar or theft alarm system for protection purposes, or used as portable sound box operated by momentary pushbutton switch. Includes volume control and tone control that varies both pitch and rate.—D. Heiserman, Whizbox, *Modern Electronics,* June 1978, p 67.

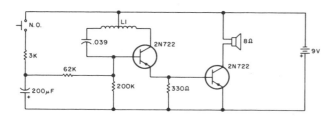

SIREN—Creates sounds resembling those of police-car siren in which air is forced through slots in motor-driven disk. L1 is half of audio transformer, using winding having 10K center tap.—Circuits, *73 Magazine*, April 1977, p 164.

FIRE SIREN USES FLASHER—Low-drain circuit operating from 1.5-V cell uses National LM3909 flasher IC to simulate fire-alarm siren. Pressing button produces rapidly rising wail, with tone coasting down in frequency after button is released. Sound from loudspeaker resembles that of motor-driven siren. Volume is adequate for child's pedal car.—P. Lefferts, Power-Miser Flasher IC Has Many Novel Applications, *EDN Magazine*, March 20, 1976, p 59—66.

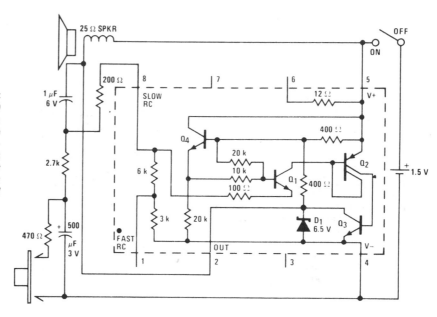

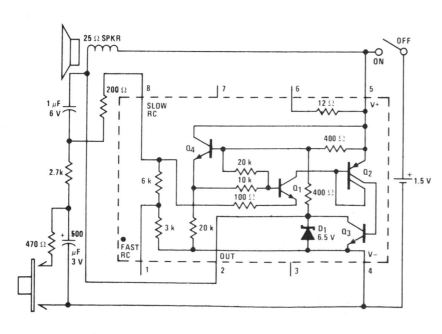

FIRE SIREN—Pressing button produces rapidly rising wail, and releasing button gives slower lowering of frequency resembling sounds of typical siren on fire engine. Circuit uses National LM3909 IC operating from 1.5-V cell for driving 25-ohm loudspeaker. 1-μF capacitor and 200-ohm resistor determine width of loudspeaker pulse, while 2.7K resistor and 500-μF capacitor determine repetition rate of pulses.—"Linear Applications, Vol. 2," National Semiconductor, Santa Clara, CA, 1976, AN-154, p 6—7.

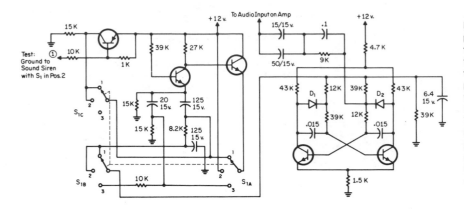

POLICE SIREN—Circuit used in Dietz siren-light police-car system gives distinctive tones. Position 1 of S_1 produces slow continuous rise and fall. Position 3 produces fast rising and falling tone. Position 2 rises slowly to full pitch when point 1 is grounded, then decays at same rate when point 1 is ungrounded. Position 3 gives most noticeable tone for break-in alarm on car. Terminal 1 goes to normally open door, hood, and other switches that complete circuit to ground when opened by intruder. Audio transistors and diode are general replacement types.—J. W. Crawford, The Ultimate Auto Alarm—Model II, *CQ,* Aug. 1971, p 54–57 and 96.

VARIABLE FREQUENCY AND RATE—Uses National LM380 opamp as astable oscillator with frequency determined by R_2 and C_2. Base of Q_1 is driven by output of LM3900 opamp connected as second astable oscillator, to turn output of LM380 on and off at rate fixed by R_1 and C_1. Transistor type is not critical. Circuit is ideal for experimenters.—"Audio Handbook," National Semiconductor, Santa Clara, CA, 1977, p 4-21–4-28.

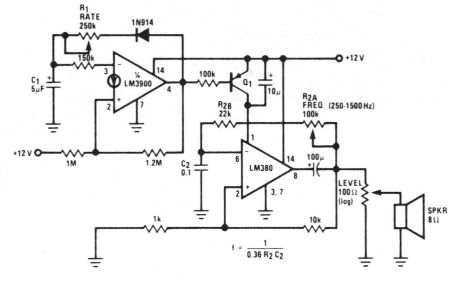

$$f = \frac{1}{0.36\,R_2\,C_2}$$

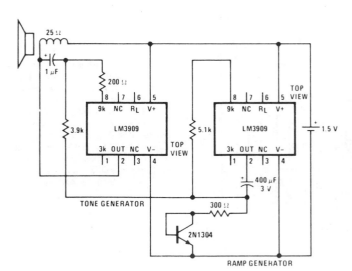

WHOOPER—Two National LM3909 ICs and single transistor generate rapidly modulated tone resembling that used on some police cars, ambulances, and airport emergency vehicles. Rapidly rising and falling modulating voltage is generated by IC having 400-μF capacitor. Diode-connected transistor forces this IC ramp generator to have longer ON periods than OFF periods, raising average tone of tone generator and making modulations seem more even.—"Linear Applications, Vol. 2," National Semiconductor, Santa Clara, CA, 1976, AN-154, p 7.

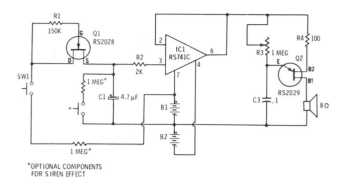

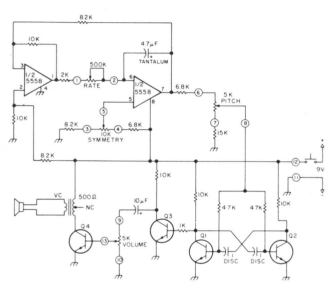

VARIABLE TONE USING VCO—Tone generator uses UJT and opamp in voltage-controlled oscillator. Frequency of audio output is determined by setting of R3. For two-tone siren effects, optional switches and resistors can be used. To speed up siren effect, use smaller value for C1.—F. M. Mims, "Integrated Circuit Projects, Vol. 4," Radio Shack, Fort Worth, TX, 1977, 2nd Ed., p 61—69.

LOUD BIKE SIREN—Uses 5558 dual opamp and four general-purpose NPN transistors to generate triangle wave that can be distorted by 10K symmetry control to give either fast or slow rise for sawtooth applied as base bias to astable MVBR Q1-Q2. Drain is reasonably low with 9-V radio battery. Repetition rate can be varied from long wail to rapid warble, and volume changed from soft to annoying. Article gives construction details, and recommends use of removable mounting on bike to avoid theft.—R. Megirian, Simple Electronic Siren, *73 Magazine*, Oct. 1977, p 176—177.

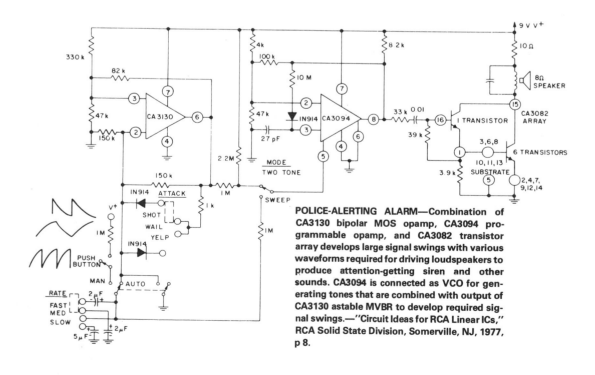

POLICE-ALERTING ALARM—Combination of CA3130 bipolar MOS opamp, CA3094 programmable opamp, and CA3082 transistor array develops large signal swings with various waveforms required for driving loudspeakers to produce attention-getting siren and other sounds. CA3094 is connected as VCO for generating tones that are combined with output of CA3130 astable MVBR to develop required signal swings.—"Circuit Ideas for RCA Linear ICs," RCA Solid State Division, Somerville, NJ, 1977, p 8.

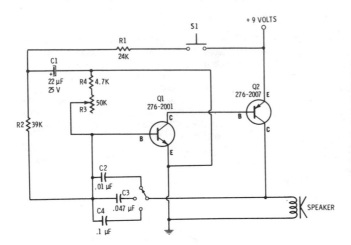

ADJUSTABLE SIREN—Tone is made adjustable by using multiposition switch to change capacitors in oscillator circuit. Speed (rate of change in frequency) of siren is adjusted with R3. 4700-ohm resistor in series with R3 keeps siren operational when R3 is rotated to minimum-resistance position. Siren is operated by pressing switch to produce rising wail, then releasing switch until wail drops down to cutoff.—F. M. Mims, "Transistor Projects, Vol. 1," Radio Shack, Fort Worth, TX, 1977, 2nd Ed., p 58–63.

POLICE SIREN USES FLASHER—Low-drain circuit operating from 1.5-V cell uses National LM3909 flasher ICs to simulate "whooper" sounds of electronic sirens used on some city police cars and ambulances. Two flashers are required for generating required rapidly rising and falling modulating voltage. Transistor is connected as diode to force ramp generator of IC to have longer ON periods than OFF periods, raising average tone and making modulation seem more even.—P. Lefferts, Power-Miser Flasher IC Has Many Novel Applications, *EDN Magazine,* March 20, 1976, p 59–66.

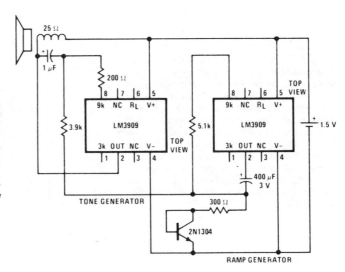

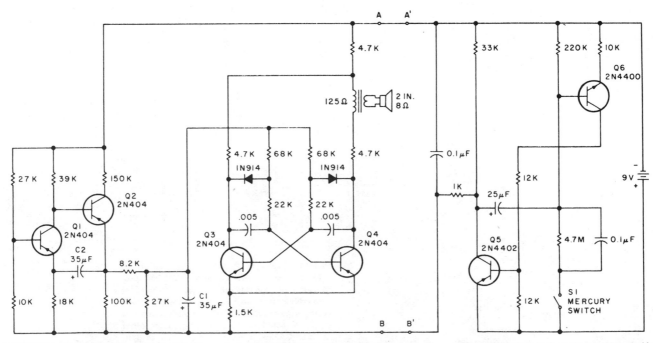

PORTABLE TOY SIREN—Can be assembled in small box as toy for small child. If mercury switch is used for S1, siren comes on automatically when box is picked up. MVBR Q1-Q2 controls rate at which siren wails, while Q3 and Q4 form AF MVBR that produces actual siren sound with frequency varied by triangle waveform on C1. MVBR Q5-Q6 is mono that conducts for preset time period when S1 is closed, for applying power to siren. Values shown give 12 s of operation before siren is shut off. When carried by child, siren is jostled enough so it keeps recycling.—J. H. Everhart, Super Siren, *73 Magazine,* Feb. 1978, p 96–97.

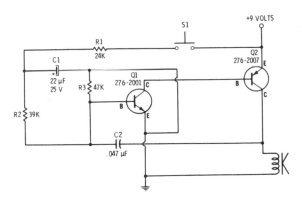

MANUALLY CONTROLLED SIREN—When switch is pressed, output tone of loudspeaker builds from low to high frequency. Releasing switch brings high frequency slowly back to low point and then cutoff. Siren sounds can be varied manually by pushing and releasing switch at different points in cycle. C2 controls pitch, and R3 determines speed at which pitch changes.—F. M. Mims, "Transistor Projects, Vol. 1," Radio Shack, Fort Worth, TX, 1977, 2nd Ed., p 58—63.

SIREN WITH MUTING—National LM389 array having three transistors and power opamp on same chip uses opamp as square-wave oscillator whose frequency is adjusted with R2B. One transistor is used in muting circuit to gate power amplifier on and off, while other two transistors form cross-coupled MVBR that controls rate of square-wave oscillator.—"Audio Handbook," National Semiconductor, Santa Clara, CA, 1977, p 4-33—4-37.

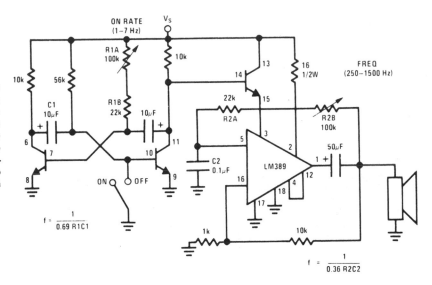

$$f = \frac{1}{0.69\,R1C1}$$

$$f = \frac{1}{0.36\,R2C2}$$

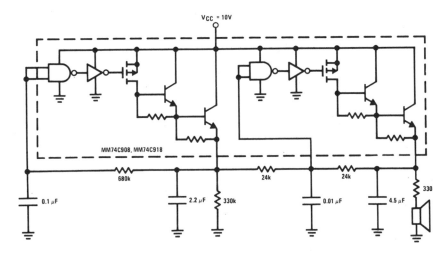

10-V SIREN CHIP—One section of National MM74C908/MM74C918 dual CMOS driver is used as audio VCO and other section as voltage ramp generator that varies frequency of VCO. Combination gives siren effect at low cost, with output current up to 250 mA for driving loudspeaker.—"CMOS Databook," National Semiconductor, Santa Clara, CA, 1977, p 5-38—5-49.

CHAPTER **83**
Squelch Circuits

Used to suppress background noise in transmitters and receivers during intervals between sentences and words, when tuning between stations or when carrier is absent. Also included are decoder circuits that unblock squelch of amateur receiver only when special tone is transmitted by desired station.

AF SQUELCH AMPLIFIER—Holds audio channel of receiver silent until receiver input signal reaches predetermined amplitude. DC control voltage can be derived from IF amplifier by rectification or from second detector of receiver. FET is biased to cutoff by DC gate voltage applied to threshold terminals. DC control voltage bucks this bias and activates amplifier whenever it exceeds predetermined threshold in range of 0–6 V. If receiver gives opposite polarity for DC voltages, use P-channel FET such as 2N2608, reverse C2, and change R4 and R5 as required.—R. P. Turner, "FET Circuits," Howard W. Sams, Indianapolis, IN, 1977, 2nd Ed., p 73–74.

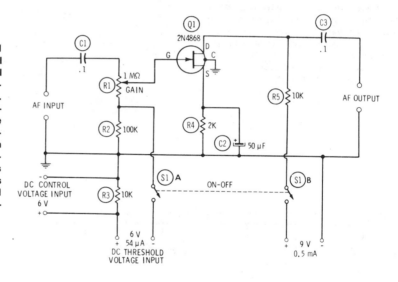

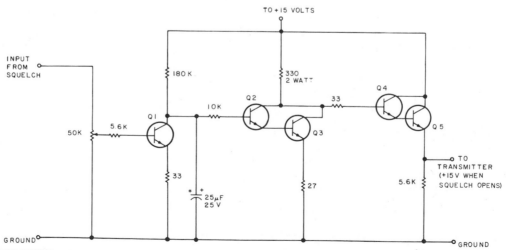

CARRIER-OPERATED SWITCH—Turns on transmitter of 2-meter FM transceiver (used as repeater) when squelch of receiver is broken by signal. Transmitter remains on about 1 s after received signal disappears; for longer delay, use larger electrolytic on collector of Q1. Q1-Q4 can be 2N3904, 2N3565, 2N2222, or other good NPN switching transistor. Q5 is 2N3054 or equivalent, capable of handling 25 V at 1 A.—H. Cone, The Minirepeater, *73 Magazine,* June 1975, p 55–57, 60–62, and 64–65.

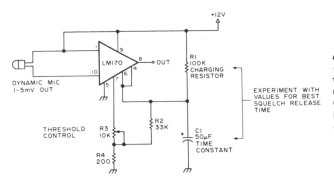

AUDIO SQUELCH—Used to suppress background noise during intervals between sentences and words when operating amateur radio station on VOX and using compressor. Circuit attenuates audio path below preset input level determined by setting of R3.—Circuits, *73 Magazine,* May 1977, p 19.

DIGITAL CTCSS OSCILLATOR—Uses two gates of CMOS quad NAND gate as 3.2-kHz oscillator, one gate as buffer, and one as amplifier serving in active bandpass filter. Requires only one precision capacitor, and uses ordinary carbon resistors in frequency-determining network. C_1 must be polystyrene, polycarbonate, Teflon, or silver mica. IC_2 divides oscillator frequency by binary multiple. Output is fed back to gate of IC_1 for converting square wave into sine wave by filtering out high-frequency harmonics. Provides continuous-tone-coded subaudible squelch (CTCSS) for amateur repeater system to protect input from interference on commonly shared channels. Voltage regulator can be replaced by zener. Use base-collector junction of 2N3638 or equivalent transistor as varactor in parallel with transmitter crystal of true FM transmitter, to modulate output frequency of crystal oscillator for CTCSS encoding.—D. Dauben, Miniature Solid State Tone Encoders to Replace Reeds, *CQ,* Dec. 1975, p 42–45 and 76.

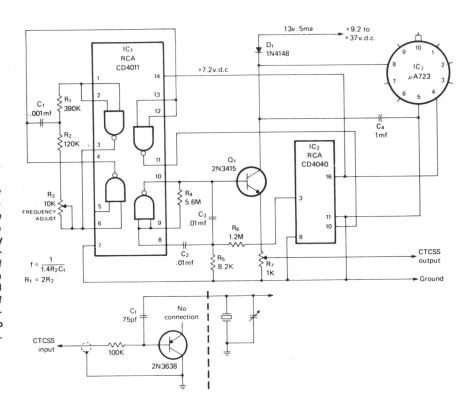

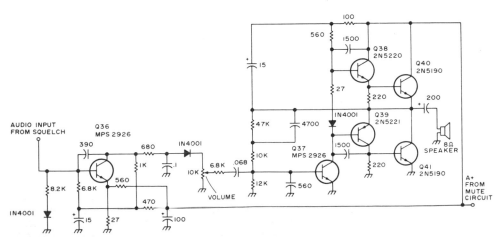

3-W CLASS AB—Used in all-band double-conversion superheterodyne receiver for AM, narrow-band FM, CW, and SSB operation. When only noise is present, first audio transistor Q36 is biased out of conduction by squelch and mutes loudspeaker. Supply is 13.6 V regulated. Article gives all circuits of receiver.—D. M. Eisenberg, Build This All-Band VHF Receiver, *73 Magazine,* Jan. 1975, p 105–112.

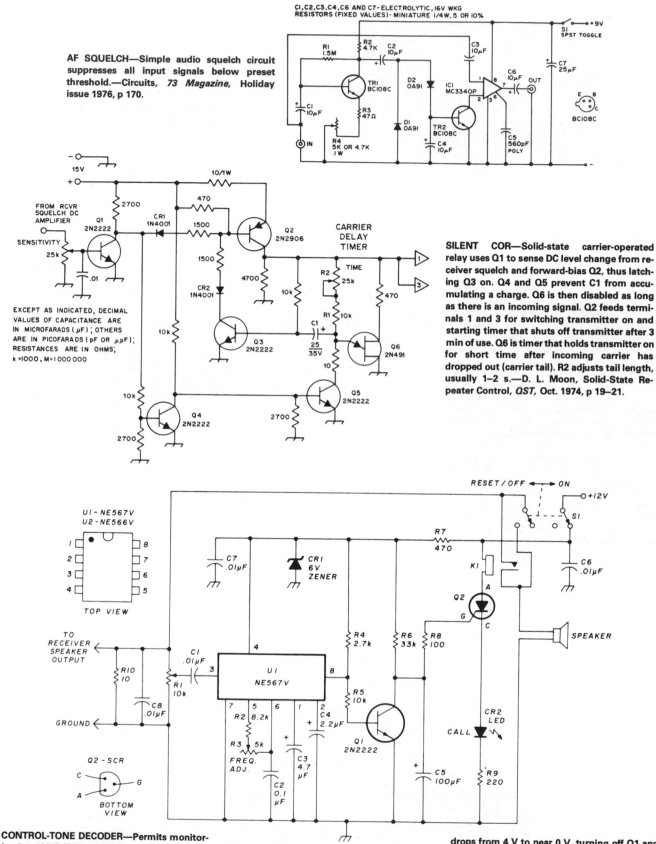

AF SQUELCH—Simple audio squelch circuit suppresses all input signals below preset threshold.—Circuits, *73 Magazine*, Holiday issue 1976, p 170.

SILENT COR—Solid-state carrier-operated relay uses Q1 to sense DC level change from receiver squelch and forward-bias Q2, thus latching Q3 on. Q4 and Q5 prevent C1 from accumulating a charge. Q6 is then disabled as long as there is an incoming signal. Q2 feeds terminals 1 and 3 for switching transmitter on and starting timer that shuts off transmitter after 3 min of use. Q6 is timer that holds transmitter on for short time after incoming carrier has dropped out (carrier tail). R2 adjusts tail length, usually 1–2 s.—D. L. Moon, Solid-State Repeater Control, *QST*, Oct. 1974, p 19–21.

CONTROL-TONE DECODER—Permits monitoring local VHF FM repeater for calls from friends without having to listen to chatter of others or to repeater noise. Operation is similar to that of Motorola paging units in which special tone is transmitted to disable squelch of receiver being called. Each friend has tone encoder for his transmitter, set at correct frequency for connecting loudspeaker so desired call can be heard. Red LED comes on to confirm that loudspeaker is connected. Audio from receiver loudspeaker is fed into pin 3 of NE567V PLL U1. When correct tone frequency is received, pin 8 drops from 4 V to near 0 V, turning off Q1 and turning on Q2. Q2 closes relay K1, to connect loudspeaker, and holds it on until RESET switch is operated. Q2 is Radio Shack 276-1059 or other small SCR. CR1 is 1N4735, and CR2 is red LED.—K. Wyatt, Private Call System for VHF FM, *Ham Radio*, Sept. 1977, p 62–64.

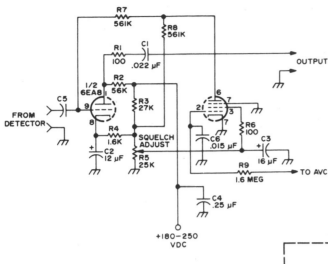

ONE-TUBE SQUELCH—Designed for insertion between second detector and AF volume control in tube-type AM receiver having AVC and well-filtered DC supply, to suppress noise when there is no input signal, such as when tuning between stations.—Circuits, *73 Magazine*, May 1977, p 31.

SQUELCH ADAPTER—Designed for use with any solid-state receiver having discrete transistor audio stages. Uses RCA CA3018 IC containing four NPN transistors, connected here to give noise amplifier that drives DC bias control stage acting on switching transistor for AF stages of receiver (one transistor in IC is unused). Designed for circuits having positive supply-voltage ground; for negative-ground circuits, reverse polarity of C6 and connect C4 to negative ground. R1 should be about 5 times resistance of volume control in receiver. B+1 should not exceed +12 VDC and can be as low as 6 V. B+2 is 3.5 V. Squelch is used chiefly when monitoring police bands on radio.—P. A. Lovelock, The Postage Stamp Squelcher, *73 Magazine*, May 1975, p 103–105.

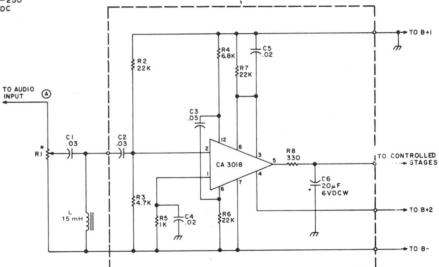

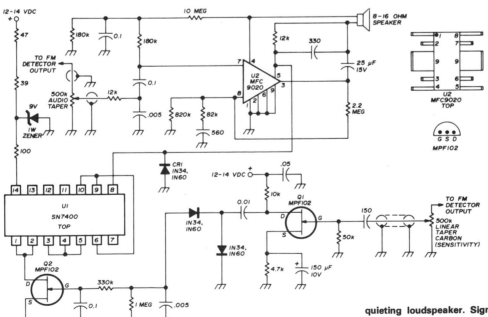

SQUELCH—Simple system with sharply defined threshold can be added to any FM receiver. Circuit includes conventional IC audio amplifier. Audio is taken from FM detector output by shielded audio line and filtered by U2 to drive loudspeaker. Similar arrangement (below) connects FM detector output to 500K squelch sensitivity control, for amplification by Q1 and rectification. Q2 is turned off at threshold level determined by sensitivity control. Q2 then begins logic toggling action through U1. Low on pin 8 of U1 clamps off portion of U2, quieting loudspeaker. Signal carrier reverses process, passing audio to loudspeaker. No-signal noise output voltage from FM detector should be at least 0.75 VAC. Circuit eliminates no-signal noise while allowing weakest desired signals to pass.—R. C. Harris, Versatile Squelch-Audio Amplifier for FM Receivers, *Ham Radio*, Sept. 1974, p 68–69.

100-Hz CTCSS OSCILLATOR—Stable Wienbridge oscillator provides continuous-tone-coded subaudible squelch (CTCSS) for amateur FM repeater system to protect input from interference on commonly shared channels and add security to input frequency. Tone can be heard in background but does not become irritating. Use film resistors for R_1 and R_2. C_1 and C_2 should be polystyrene, polycarbonate, Teflon, or silver mica. Select R_5 to give 8–10 V P-P sine wave when operating from 12-V supply. R_1 and R_2 may be varied slightly to adjust frequency.—D. Dauben, *Miniature Solid State Tone Encoders to Replace Reeds, CQ,* Dec. 1975, p 42–45 and 76.

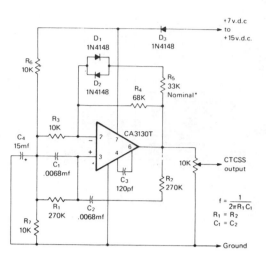

$$f = \frac{1}{2\pi R_1 C_1}$$
$$R_1 = R_2$$
$$C_1 = C_2$$

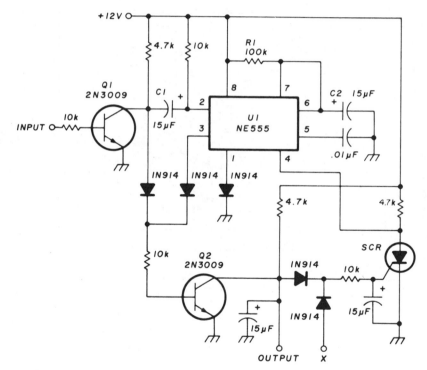

NOISE SUPPRESSOR—Eliminates repeater squelch tails from receiver having its own squelch, while allowing normal communications to pass through. Circuit goes between receiver squelch gate and point at which squelch acts on audio amplifier, to provide about 3-s delay before turning amplifier on. If received signal disappears before end of delay, radio remains silent and circuit resets itself. If received signal lasts longer than 3 s, as when repeater is interrogated, receiver operation is normal. Designed for receivers using low voltage level to squelch audio amplifier. NE555 timer is wired as mono MVBR that is triggered through inverter Q1 each time receiver squelch is tripped, provided SCR is off. SCR type is not critical. Input is taken from squelch gate in receiver.—R. K. Morrow, Jr., *Repeater Kerchunk Eliminator for Mobile Rigs, Ham Radio,* Oct. 1977, p 70–71.

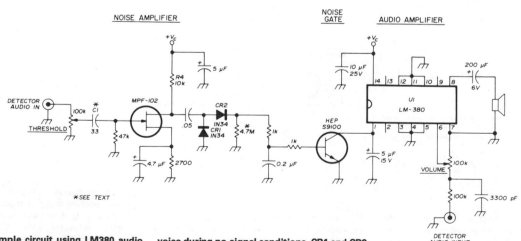

SQUELCH—Simple circuit using LM380 audio amplifier IC gives excellent performance. First transistor amplifies random noise which is greater in frequency than normal spectrum of voice during no-signal conditions. CR1 and CR2 rectify noise. Second transistor conducts and clamps U1 off when there is no signal. Increasing value of C1 increases gain of noise amplifier, but small value of C1 makes circuit less susceptible to heavy noise peaks.—R. Harris, *Another Squelch Circuit, Ham Radio,* Oct. 1976, p 78.

CHAPTER 84
Staircase Generator Circuits

Generate output voltage that increases or decreases in number of equal or unequal steps in range from 7 to 256 steps. Applications include curve tracers for semiconductor devices, video testers, production of gray scale for satellite weather pictures, and feed for one axis of XY recorder or storage CRO.

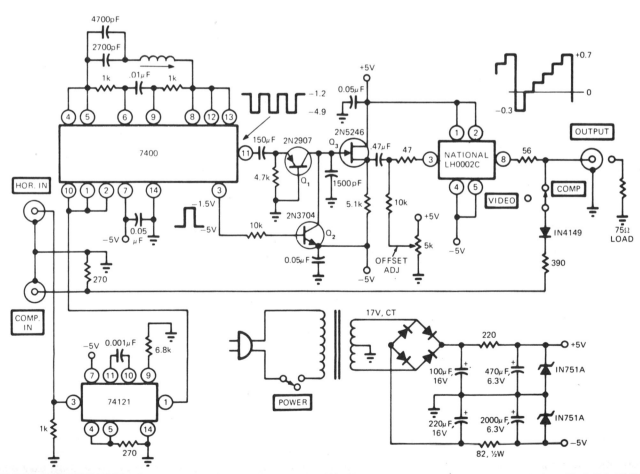

SIX-STEP COMPOSITE VIDEO—Circuit accepts negatively referenced output signals of TV sync generator and delivers 1 V P-P six-step composite video signal to 75-ohm load. 74121 mono changes wide horizontal blanking pulse to correct width for triggering oscillator IC. National LH002C current driver provides low-impedance drive capability for video signal. Used in testing TV sets and VTR decks.—M. J. Salvati, VFO Adds Versatility to TV Sync Generator, *EDN Magazine*, May 20, 1974, p 70 and 72.

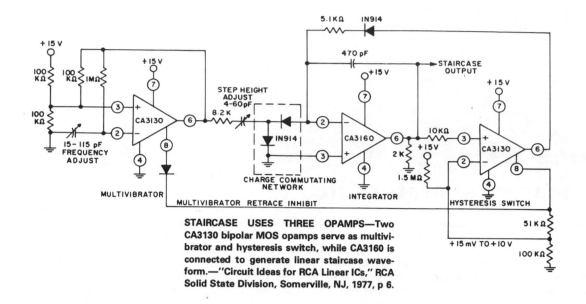

STAIRCASE USES THREE OPAMPS—Two CA3130 bipolar MOS opamps serve as multivibrator and hysteresis switch, while CA3160 is connected to generate linear staircase waveform.—"Circuit Ideas for RCA Linear ICs," RCA Solid State Division, Somerville, NJ, 1977, p 6.

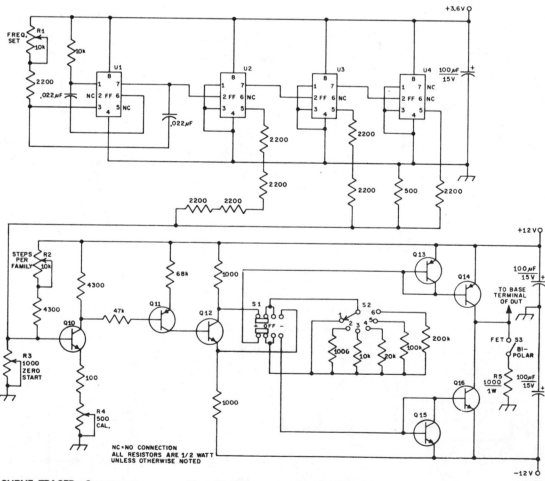

STEPS FOR CURVE TRACER—Square waves with 1-V amplitude and 1/60-s period, 2 V at 1/30-s, and 4 V at 11–15 s are generated by µL914 MVBR and µL923 flip-flops U2-U4, for combining in simple ladder network to give staircase waveform. Flip-flops count down MVBR output. Complementary-amplifier stage Q10-Q11 drives phase splitter Q12. Output of phase splitter goes through S1 to appropriate current source, Q13-Q14 or Q15-Q16, for supplying base terminal of device under test (DUT). Use 2N3904 for Q10, Q12, Q15, and Q16. Use 2N3906 for Q11, Q13, and Q14.—R. P. Ulrich, A Semiconductor Curve Tracer for the Amateur, QST, Aug. 1971, p 24–28.

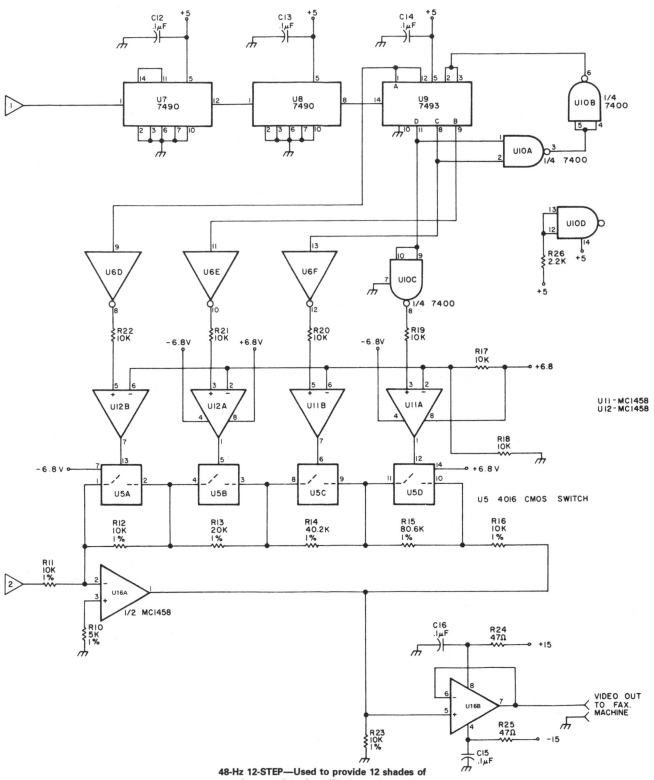

48-Hz 12-STEP—Used to provide 12 shades of gray for reception of satellite facsimile weather pictures. Input 1 is 2400-Hz square wave obtained from separate reference oscillator. Input 2 is 2400-Hz sine wave having 1 V P-P maximum voltage. U9 is 4-bit binary counter having special reset provided by U10A and U10B at count 12 to give desired 12 states. Outputs of U9 are used to adjust gain of U16A in 12 steps. Article describes operation in detail and gives circuits for reference oscillator and power supply.—R. Cawthon, Toward a More Perfect Weather Picture, *73 Magazine*, April 1978, p 116–118.

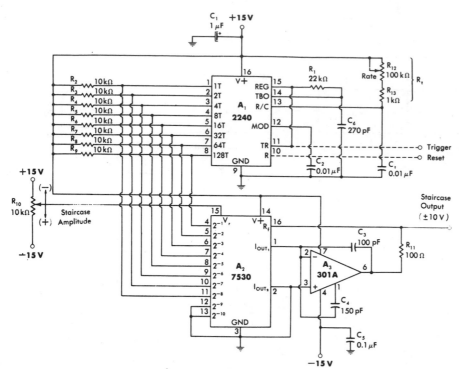

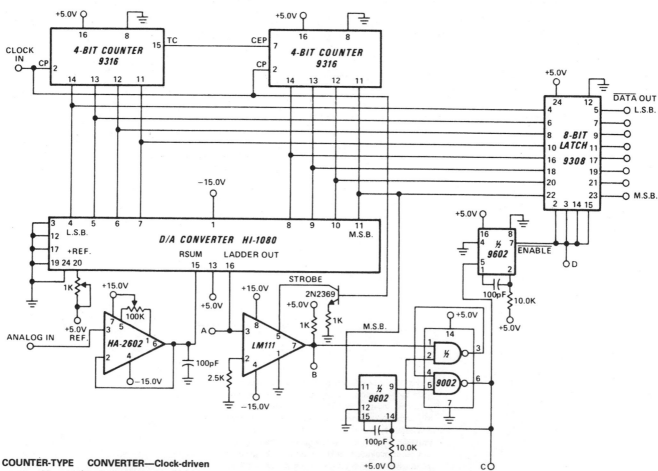

4–400 Hz BIPOLAR—Single 2240 serves as time-base generator with R_t determining frequency in range of 4–400 Hz. Digital output is converted to analog form by 7530 10-bit CMOS multiplying D/A converter. Reference voltage can be varied up to ±10 V to give variable-amplitude bipolar staircase output in same amplitude range, with 255 staircase steps corresponding to 8-bit count of 2240. Opamp A_3 serves as current-to-voltage converter for changing ±1 mA output current of 7530 to ±10 V swing for staircase.—W. G. Jung, "IC Timer Cookbook," Howard W. Sams, Indianapolis, IN, 1977, p 224–226.

COUNTER-TYPE CONVERTER—Clock-driven counter drives Harris HI-1080 D/A converter producing staircase voltage ramp. When converter output voltage equals analog input voltage as determined by HA-2602 comparator, comparator changes state. At that instant, state of counter represents 8-bit digital equivalent of analog input. Data output from latch is complement of digital value. Input range is 0–10 V. Other input ranges, positive or negative, are obtained by changing opamp gain or polarity or by adjusting 1K reference pot. Accuracy is maintained within ½ LSB at clock rates up to 330 kHz.—"Linear & Data Acquisition Products," Harris Semiconductor, Melbourne, FL, Vol. 1, 1977, p 7-33–7-35 (Application Note 512).

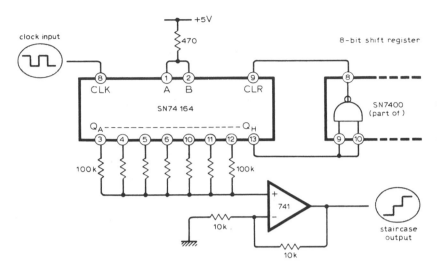

SEVEN STEPS—Circuit shown generates seven identical steps before waveform is repeated. Number of steps can be increased by cascading two or more SN74164 shift registers, or reduced by taking clear pulse from earlier Q output of register.—P. Cochrane, Simple Staircase Generator, *Wireless World,* April 1976, p 63.

NEGATIVE TRIGGERING—Standard transistor pump circuit is driven by differentiating and squaring circuit designed so each staircase block is triggered by negative edge rather than by pulse. Circuit was developed for FET curve tracer, and can be used in other applications where only resetting edge of normal sawtooth is available as trigger. R_{28} changes number of steps produced before staircase is reset. Tr_6 is Texas Instruments 43, IC opamp is SN72741P, transistors are BC182L or equivalent, and diodes are 1S44.—L. G. Cuthbert, An F.E.T. Curve Tracer, *Wireless World,* April 1974, p 101–103.

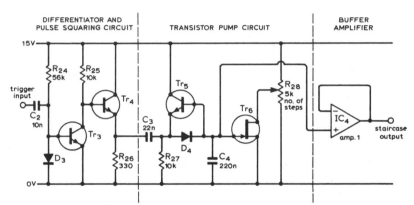

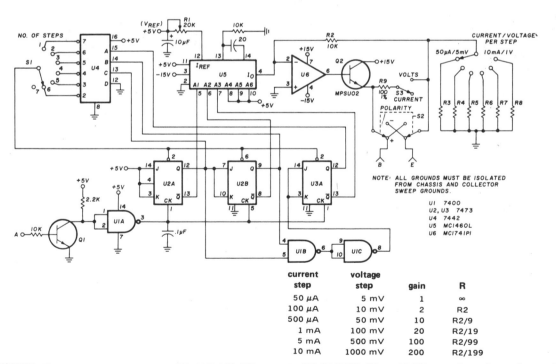

current step	voltage step	gain	R
50 μA	5 mV	1	∞
100 μA	10 mV	2	R2
500 μA	50 mV	10	R2/9
1 mA	100 mV	20	R2/19
5 mA	500 mV	100	R2/99
10 mA	1000 mV	200	R2/199

STEP GENERATOR—Base-step generator produces series of voltage or current steps synchronized with beginning of each collector voltage sweep, for application to base or gate of three-terminal semiconductor device while sweep voltage is applied to collector of curve tracer that displays current-voltage characteristics on CRO. Circuit is built around Motorola MC1406L 6-bit D/A converter U5. U1 (7400), U2, and U3 (both 7473) form synchronous divide-by-8 counter whose outputs are applied to A1-A3 inputs of U5. U6 (MC1741P1) and Q2 form current amplifier. Q1 is general-purpose NPN transistor having DC current gain of about 30. Point A goes to output of full-wave rectifier using two 50-PIV 1-A diodes connected across 26.4-V secondary of transformer. Table gives values for R3 through R8 as ratio of R2 for various gains and steps. Thus, for 500-mV steps (gain of 100), R7 is about 101 ohms. Accuracy depends on values of R2-R9 used. Never apply voltage steps to base of bipolar transistor.—H. Wurzburg, Integrated Circuit Base-Step Generator, *Ham Radio,* July 1976, p 44–46.

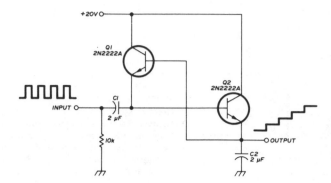

SQUARE TO STEP—Simple staircase generator circuit converts square wave into staircase voltage output. Each step approximates level of input pulse. First pulse charges C2 to amplitude of input pulse. After pulse, voltage across C2 acts through Q1 to charge C1 to same voltage. Next pulse adds to voltage across C2, doubling its charge. Each subsequent pulse steps up height of staircase until it reaches level of supply voltage.—J. Fisk, Circuits and Techniques, *Ham Radio*, June 1976, p 48–52.

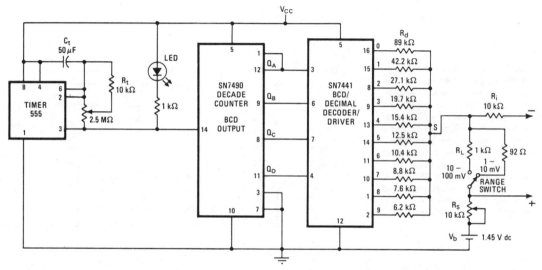

STEP-FUNCTION GENERATOR—Successively lower resistances at decoder outputs create stairstep function for testing various types of instruments. Steps are equally spaced and of equal height, covering range of 5–12 or 50–120 mV depending on setting of range switch and R_s. Spacing between steps ranges from 1.6 s to 6 min, so total time for complete 10-step staircase is 16 s to 60 min depending on setting of 2.5-megohm timer pot. Reference voltage V_b can be 1.45-V mercury cell. Resistor values shown for R_d provide fixed 10% increments in stairstep.—M. M. Lacefield, Simple Step-Function Generator Aids in Testing Instruments, *Electronics*, Dec. 26, 1974, p 103 and 105.

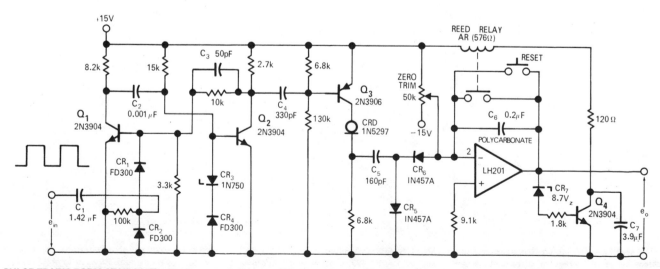

PULSE TRAINS FORM STAIRCASE—Circuit accepts pulse trains from pulse-generating position transducer and produces staircase waveform as analog input for horizontal axis of XY recorder or storage scope. Current-regulating diode serves as constant-current source for staircase generator that produces analog output proportional to number of input pulses. Mono switches on Q_3 for constant time duration with every pulse, to ensure that C_5 gets same amount of charge regardless of pulse rate. Relay resets integrator to zero when output voltage reaches almost 8.7 V, to prevent data from being lost if opamp saturates before data run is completed.—R. G. Warsinski, Staircase Generator Uses Current-Regulating Diode, *EDN/IEEE Magazine*, Aug. 1, 1971, p 46.

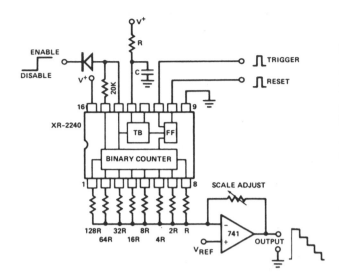

256 NEGATIVE STEPS—Interconnection of Exar XR-2240 programmable timer/counter with 741 opamp and precision resistor ladder forms staircase generator. Reset pulse drives output low. When trigger is applied, output goes high and circuit generates negative-going staircase having 256 equal steps. Duration of each step is equal to time-base period as determined by values used for R and C. Staircase is stopped by applying disable signal (less than 1.4 V) to pin 14 through steering diode. Supply voltage range is 4–15 V. If counter cannot be triggered when using supply above 7 V and less than 0.1 μF for C, connect 300 pF from pin 14 to ground.—"Timer Data Book," Exar Integrated Systems, Sunnyvale, CA, 1978, p 11–18.

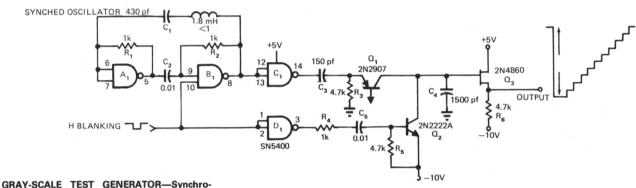

GRAY-SCALE TEST GENERATOR—Synchronized LC oscillator drives staircase generator, both of which are reset by horizontal-blanking input signal. Developed for testing video equipment. Synchronized oscillator uses two TTL gates of SN5400, biased at their linear range by negative feedback resistors R_1 and R_2. Oscillator always starts in same condition. Circuit generates staircase by integrating train of equally spaced pulses. Article covers theory and gives design equations.—E. E. Morris, Simple Stair-Step Generator Uses 1 IC and 3 Transistors, *EDN Magazine*, Oct. 1, 1972, p 48–49.

CHAPTER 85
Stereo Circuits

Includes amplifier and signal-processing circuits developed specifically for stereo FM, tape recorder, and phonograph systems. Many can be used singly in monophonic systems. Includes circuits for FM noise suppression, reverberation, rear-channel ambience, and loudspeaker phasing.

ACTIVE TONE CONTROLS—Provides ±20 dB gain with 3-dB corners at 30 and 10,000 Hz. Use of LM349 quad opamp means only one IC is needed for both stereo channels. Buffer at input gives high input impedance (100K) for source. Total harmonic distortion is typically 0.05% across audio band. Input-to-output gain is at least 5.—"Audio Handbook," National Semiconductor, Santa Clara, CA, 1977, p 2-40–2-49.

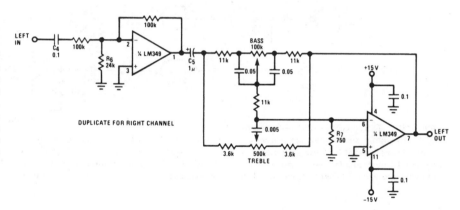

DUPLICATE FOR RIGHT CHANNEL

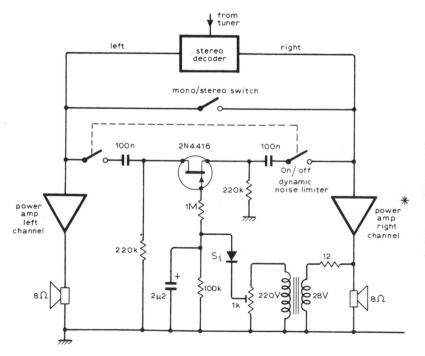

FM NOISE SUPPRESSOR—Circuit acts as noise limiter to help produce pseudostereo sound having reduced noise, to offset noise signal heard during weak passages during stereo reception of FM stations. FET short-circuits both audio channels when audio signal strength drops sufficiently to make noise objectionable. If this voltage is insufficient to drive FET, amplifier or transformer must be used.—J. W. Richter, Stereo Dynamic Noise Limiter, *Wireless World,* Oct. 1975, p 474.

(✻ includes; volume balance and tone controls)

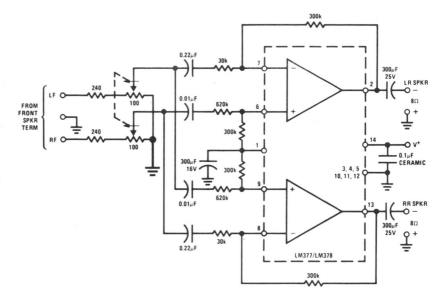

REAR-CHANNEL AMBIENCE—Can be added to existing left front and right front loudspeakers of stereo system to extract difference signal for combining with some direct signal (R or L) to add fullness for concert-hall realism during reproduction of recorded music. Very little power is required for pair of rear loudspeakers, and this can be furnished by National LM377/LM378 dual-amplifier IC operating from about 24-V supply.—"Audio Handbook," National Semiconductor, Santa Clara, CA, 1977, p 4-8—4-20.

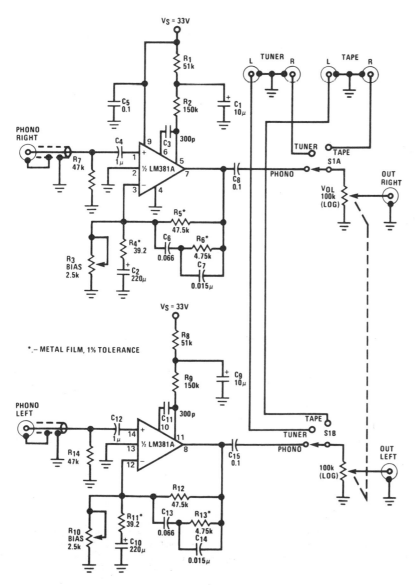

ULTRALOW-NOISE PREAMP—Complete preamp has inputs for magnetic-cartridge pickup, tuner, and tape, along with ganged volume control and ganged selector switch for both channels. Tone controls are easily added. RIAA frequency response is within ±0.6 dB of standard values. 0-dB reference gain at 1 kHz is 41.6 dB, producing 1.5-VRMS output from 12.5-mVRMS input. Signal-to-noise ratio is better than −85 dB referenced to 10-mV input level.—"Audio Handbook," National Semiconductor, Santa Clara, CA, 1977, p 2-25—2-31.

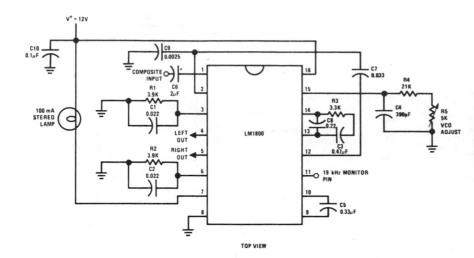

STEREO FM DEMODULATOR—Single National LM1800 IC converts composite AF input signal to left and right signals for audio power amplifiers. LED with series resistor can be used in place of 100-mA lamp.—"Audio Handbook," National Semiconductor, Santa Clara, CA, 1977, p 3-23–3-27.

53-dB PREAMP—RCA CA3052 quad AC amplifier serves for both channels of complete stereo preamp. Circuit is duplicated for other channel. Total harmonic distortion at 1-kHz reference and 1-V output is less than 0.3%. Gain at 1 kHz is 47 dB, with 11.5-dB boost at 100 Hz and 10 kHz. Cut at 100 Hz is 10 dB and at 10 kHz is 9 dB. Operates from single-ended supply. Inputs can be from tape recorders and magnetic-cartridge phonographs.—"Linear Integrated Circuits and MOS/FET's," RCA Solid State Division, Somerville, NJ, 1977, p 327–330.

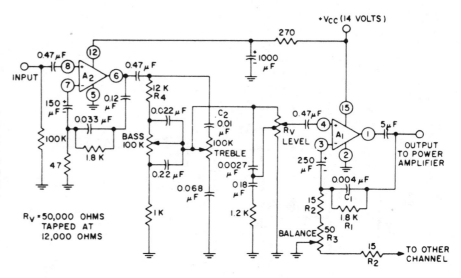

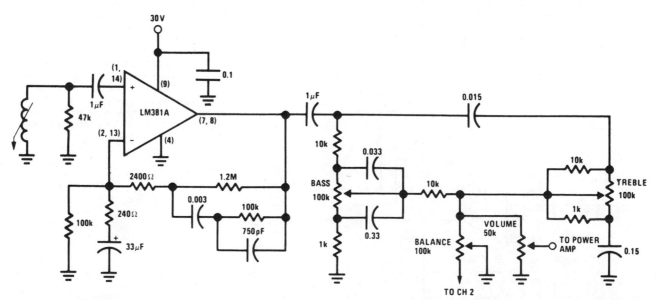

PREAMP WITH TONE CONTROLS—Use of LM381A selected low-noise preamp with passive bass and treble tone controls as phono or tape preamp gives superior noise performance while eliminating need for transistor to offset signal loss in passive controls. Circuit provides 20-dB boost and cut at 50 Hz and 10 kHz relative to midband gain. Design equations are given. Use log pots for tone controls. Other stereo channel is identical. Controls are ganged.—"Audio Handbook," National Semiconductor, Santa Clara, CA, 1977, p 2-40–2-49.

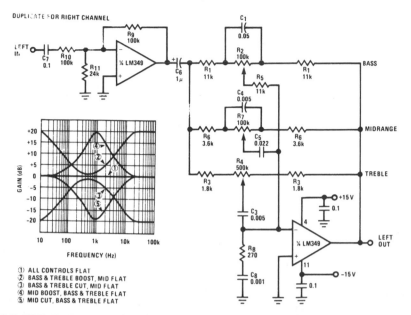

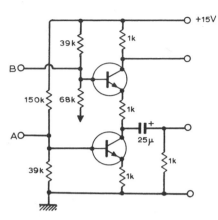

ACTIVE MIDRANGE TONE CONTROL—Addition of midrange tone control to active bass and treble control gives greater control flexibility. Center frequency of midrange control is determined by C_4 and C_5 and is 1 kHz for values shown. C_5 should have 5 times value of C_4.—"Audio Handbook," National Semiconductor, Santa Clara, CA, 1977, p 2-40–2-49.

SUM AND DIFFERENCE—Simple circuit using two BC109 or equivalent transistors is effective for summing and differencing two signals, as required in stereo and quadraphonic sound applications. For resistor values shown, upper output is $-\frac{1}{2}(A + B)$ and lower output is $-\frac{1}{2}(A - B)$. Will handle input signals up to 1.4 V. Bottom of 68K resistor should go to ground.—B. J. Shelley, Active Sum and Difference Circuit, *Wireless World,* July 1974, p 239.

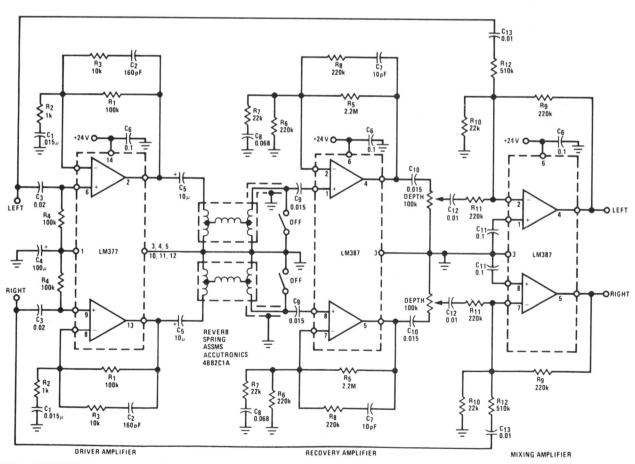

STEREO REVERBERATION—Uses National LM377 dual power amplifier as driver for springs acting as mechanical delay lines. Used to enhance performance of stereo music system by adding artificial reverberation to simulate reflection and re-reflection of sound off walls, ceiling, and floor of listening environment. Amplifier has frequency response of 100–5000 Hz, with rolloff below 100 Hz to suppress booming. Recovery amplifier uses LM387 low-noise dual preamp, and another LM387 provides mixing of delayed signal with original in inverting summing configuration. Output is about half of original signal added to all of delayed signal.—"Audio Handbook," National Semiconductor, Santa Clara, CA, 1977, p 5-7–5-10.

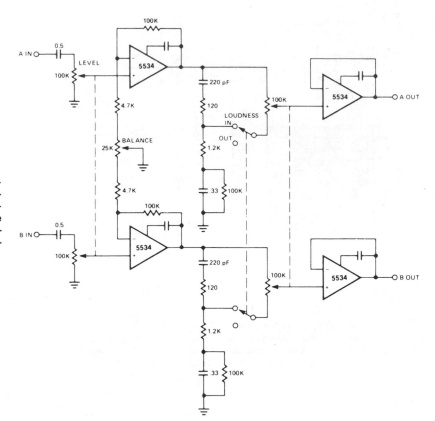

BALANCE AND LOUDNESS CONTROL—Provides bass boost at low listening levels to compensate for nonlinearity of human hearing system. Balance control permits equalizing volume from left and right loudspeakers at particular listening location.—"Signetics Analog Data Manual," Signetics, Sunnyvale, CA, 1977, p 640.

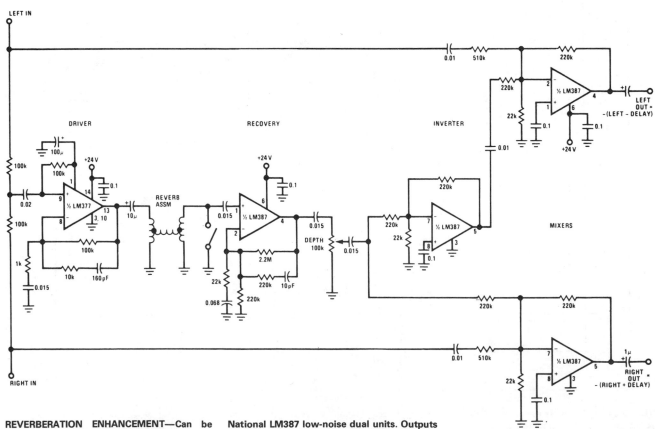

REVERBERATION ENHANCEMENT—Can be used to synthesize stereo effect from monaural source or can be added to existing stereo system. Requires only one spring assembly, which can be Accutronics 4BB2C1A. All opamps are National LM387 low-noise dual units. Outputs are inverted scaled sums of original and delayed signals; left output is left signal minus delay, while right output is right signal plus delay. With mono source, both inputs are tied together and outputs become input minus delay and input plus delay.—"Audio Handbook," National Semiconductor, Santa Clara, CA, 1977, p 5-7–5-10.

ACTIVE TONE CONTROLS USING FEEDBACK— Variation of Baxandall negative-feedback tone control circuit reduces number of capacitors required. Developed for stereo systems. R_4 and R_5 provide negative input bias for opamp, while C_0 prevents DC voltages from being fed back to tone control circuit. For other supply voltages, R_4 is only resistor changed; design procedure is given.—"Audio Handbook," National Semiconductor, Santa Clara, CA, 1977, p 2-40–2-49.

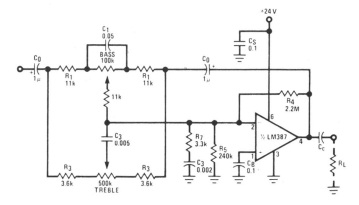

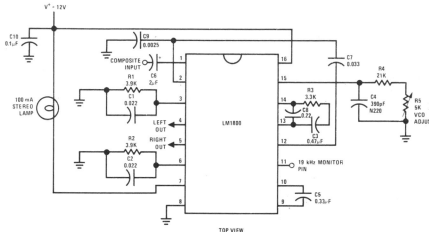

FM DEMODULATOR— National LM1800 PLL IC accepts composite IF output and converts it to separate audio signals for left and right channels. C8 has effect of shunting phase jitter to minimize channel separation problems. If free-running frequency of VCO is set at precisely 19 kHz with R5, separation remains constant over wide range of composite input levels, signal frequencies, temperature changes, and drift in component values.—"Linear Applications, Vol. 2," National Semiconductor, Santa Clara, CA, 1976, AN-81, p 7–8.

LOUDSPEAKER PHASING— Used to determine correct phasing of loudspeakers, microphones, amplifiers, and audio lines in complex stereo systems. Transmitter input feeds sawtooth waveform into stereo input jack of one channel, and receiver unit having microphone input and zero-center meter output is held in front of each loudspeaker in turn for same channel. Components are correctly phased when meter deflects in same direction for all loudspeakers. Procedure is then repeated for other channel. Sawtooth waveform is generated by Analog Devices AD537JD voltage-to-frequency converter. Microphone can be that used with portable cassette recorder. 741 opamp IC_1 with gain of 200 feeds dual peak detector D1-D2. Filtered DC signals are detected ramp and detected spike, with spike overriding ramp. Resulting DC level is amplified by 741 opamp having gain of 10, for driving meter. Microphones to be phased are plugged into J1 and connections noted for giving correct meter deflection. J2 is used for phasing amplifiers, lines, and other audio components. Article covers calibration and use.—C. Kitchin, Build an Audio Phase Detector, *Audio*, Jan. 1978, p 54 and 56–57.

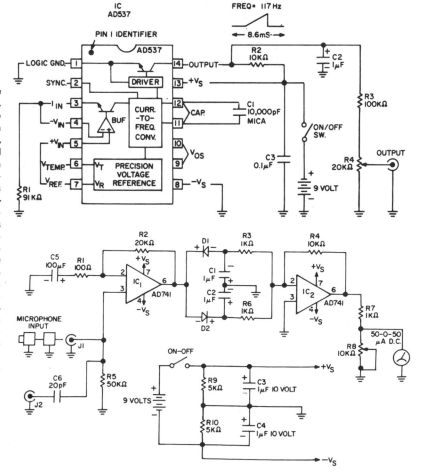

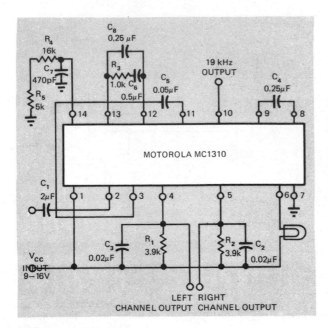

PLL DECODER—Motorola MC1310 phase-locked loop stereo decoder requires only one adjustment, by 5K pot R_5. With pin 2 open, adjust R_5 until reading of 19.00 kHz is obtained with frequency counter at pin 10. Alternatively, tune to stereo broadcast and adjust R_5 to center of lock-in range of stereo pilot lamp. Circuit gives 40-dB separation and about 0.3% total harmonic distortion.—B. Korth, Phase-Locked Loop Stereo Decoder Is Aligned Easily, *EDN Magazine*, Jan. 20, 1973, p 95.

PLL STEREO FM DEMODULATOR—National LM1800 IC uses phase-locked loop techniques to regenerate 38-kHz subcarrier. Automatic stereo/monaural switching is included. Supply voltage range is 10–18 V.—"LM1800 Phase Locked Loop FM Stereo Demodulator," National Semiconductor, Santa Clara, CA, 1974.

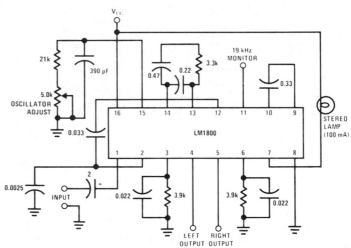

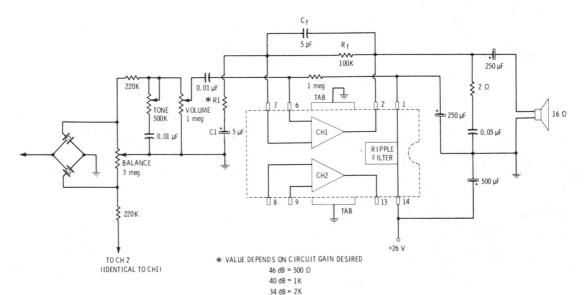

LOW-COST STEREO PHONOGRAPH—Uses single Sprague ULN-2277 IC containing two audio amplifiers each capable of driving loudspeaker directly, for input from high-impedance stereo cartridge. Connections are identical for other channel. Power output per channel is 2 W. Tone and volume controls are ganged with those for other channel, but balance control shown serves both channels.—E. M. Noll, "Linear IC Principles, Experiments, and Projects," Howard W. Sams, Indianapolis, IN, 1974, p 237–239.

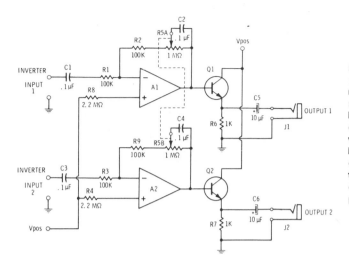

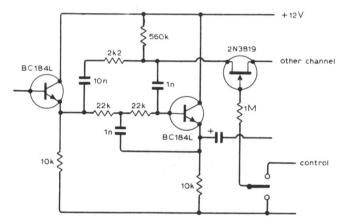

HEADPHONE AMPLIFIER—Designed to drive medium- to high-impedance headphones. Add matching transformers having 1000-ohm primaries if using low-impedance headphones. Dual 1-megohm pot controls gain in stereo channels over range of 1 to 100. Use 9–15 V well-filtered supply rated at least 20 mA. Use Motorola MC3401P or National LM3900 quad opamp and 2N2924 or equivalent NPN transistors.—C. D. Rakes, "Integrated Circuit Projects," Howard W. Sams, Indianapolis, IN, 1975, p 21–24.

FM HISS LIMITER—Uses low-pass filter to remove noise sometimes heard with weak passages during stereo reception of FM stations. FET driven by output of amplifier or tuner is used to switch low-pass filter into operation rather than switching over to mono. Based on fact that the hiss is an antiphase effect that can be removed with little detriment to overall signal.—G. Hibbert, Stereo Noise Limiter Improvement, *Wireless World,* March 1976, p 62.

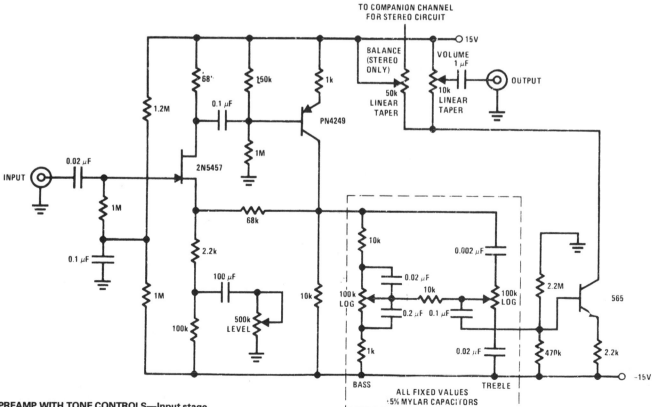

PREAMP WITH TONE CONTROLS—Input stage is JFET having high input impedance and low noise. Circuit parameters are not critical, yet harmonic distortion level is less than 0.05% and S/N ratio is over 85 dB. Tone controls allow 18 dB of cut and boost. Input of 100 mV gives 1-V output at maximum level. Identical preamp is used for other stereo channel.—"FET Databook," National Semiconductor, Santa Clara, CA, 1977, p 6-26–6-36.

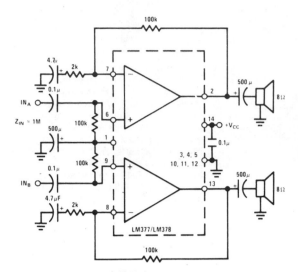

NONINVERTING POWER AMPLIFIER—Single National LM377/LM378 provides gain of 50 and 3 W per channel for driving loudspeakers. Supply is 24 V. High input impedance permits use of high-impedance tone and volume controls. Heatsink is required.—"Audio Handbook," National Semiconductor, Santa Clara, CA, 1977, p 4-8–4-20.

INVERTING POWER AMPLIFIER—Single National LM377 IC provides 2 W per channel with 18-V supply for driving loudspeakers when fed by stereo demodulator of FM receiver. Similar LM378 chip gives 3 W per channel with 24-V supply, and LM379 gives 4 W per channel with 28-V supply. Gain is 50 for all. Heatsink is required.—"Audio Handbook," National Semiconductor, Santa Clara, CA, 1977, p 4-8–4-20.

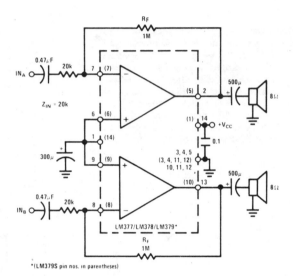

CHAPTER 86
Sweep Circuits

Includes circuits for generating linear, nonlinear, logarithmic, exponential, negative-starting, variable start/stop, bidirectional, and other types of ramps or sweeps at frequencies ranging from 0.2 Hz to 10.7 MHz for CRO and other applications. See also Cathode-Ray and Signal Generator chapters.

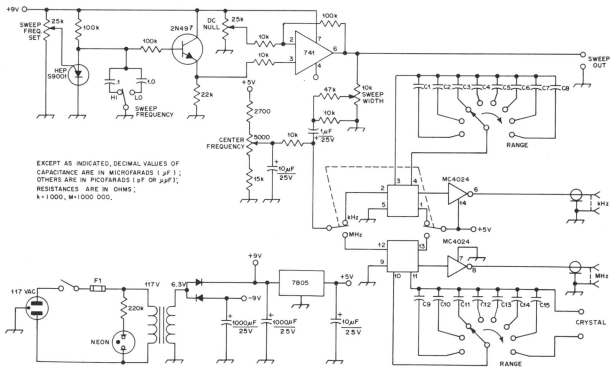

15-RANGE SWEEP—Serves for adjusting tuned circuits, aligning IF strips, and checking filter characteristics. Uses Motorola MC4024 IC containing two separate voltage-controlled MVBRs with output buffer for each. Frequency of oscillation is determined primarily by single switch-selected external capacitor that gives choice of 15 frequency ranges, with frequency within each range varied by applying DC control voltage to one pin of IC. For 3.5–5 V of control voltage, linearity is good. Output waveform is 4 V P-P at lower frequencies, becoming triangular at higher frequency ranges. HEP S9001 programmable UJT generates sweep signal. Switch gives choice of 100 Hz or 1 kHz sweep. Buffer and 741 opamp then give exponential sweep from about −1 to +1 V after DC level is set to 0 by 25K pot, for horizontal input of CRO. Sweep signal is also fed to MVBRs through controls giving independent width and center frequency adjustments. Diodes are 1-A 50-PIV silicon. T1 is 6.3 V at 1 A.—W. C. Smith, An Inexpensive Sweep-Frequency Generator, *QST*, Oct. 1976, p 17–19.

CAPACITANCE	FREQUENCY RANGE
C1 0.4 μF	0.5 – 1 kHz
C2 0.2 μF	1 – 2 kHz
C3 0.1 μF	2 – 4 kHz
C4 .05 μF	4 – 8 kHz
C5 .025 μF	8 – 16 kHz
C6 .0125 μF	16 – 32 kHz
C7 .0062 μF	32 – 64 kHz
C8 .0033 μF	64 – 130 kHz
C9 .00125 μF	0.15 – 0.3 MHz
C10 620 pF	0.3 – 0.6 MHz
C11 300 pF	0.6 – 1.2 MHz
C12 150 pF	1.2 – 2.4 MHz
C13 75 pF	2.4 – 5 MHz
C14 33 pF	5 – 10 MHz
C15 15 pF	10 – 20 MHz

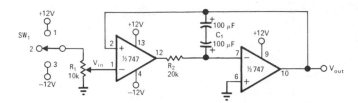

BIDIRECTIONAL RAMP—Originally used to vary reference voltage for DC servomotor to control acceleration and deceleration. R_2 and C_1 determine ramp rate, while R_1 controls ramp amplitude. With values shown, output takes 1 s to ramp from 0 to 10 V.—R. W. Currell, Linear Bidirectional Ramp Generator, *EDN/EEE Magazine*, Nov. 1, 1971, p 50–51.

LINEAR RAMP—Free-running ramp generator has excellent linearity and repetition rate independent of supply voltage. C_2 is charged at constant current through Q_1 and is discharged by Q_2. R_2 provides sync pulse during retrace. Repetition rate of ramp is controlled by R_3, from about 100 to 4000 Hz. Output voltage is 10 V P-P, and sync pulse amplitude is 5 V P-P.—J. J. Nagle, Voltage Independent Ramp Generator, *CQ*, Sept. 1972, p 61 and 98.

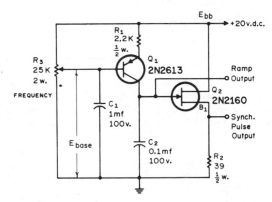

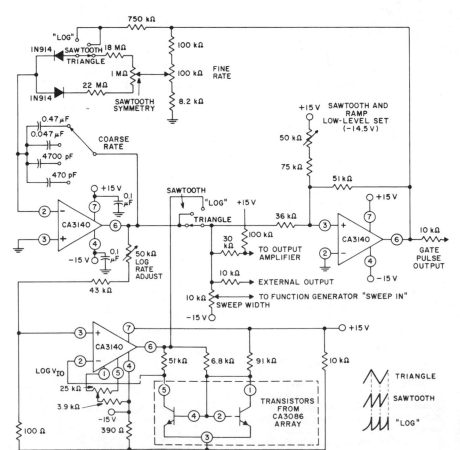

SWEEPING RAMP—Uses three CA3140 bipolar MOS opamps. One serves as integrator, another as hysteresis switch determining start and stop of sweep, and third as logarithmic shaping network for log function. Circuit generates rates and slopes as well as sawtooth, triangle, and logarithmic sweeps.—"Circuit Ideas for RCA Linear ICs," RCA Solid State Division, Somerville, NJ, 1977, p 7.

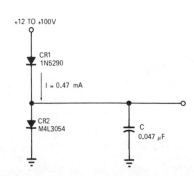

SIMPLEST SWEEP GENERATOR—Requires only constant-current generator CR1, Schottky diode CR2, and capacitor. Provides excellent linearity (0.07%) and stability over wide range of supply voltages and temperatures. Sweep rates as high as 100 kHz can be obtained by changing value of C. Article gives design equations.—D. R. Morgan, Sweep Generator Boasts Only Three Parts, *EDN Magazine*, Sept. 15, 1970, p 57.

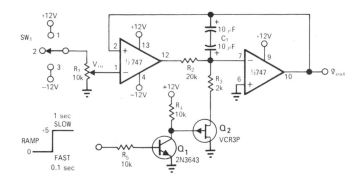

LOGIC-CONTROLLED RAMP RATE—Bidirectional linear ramp generator uses FET switch to slow ramp. With FET off, fast ramp has duration of 100 ms. With FET on, slow ramp is 1 s.—R. W. Currell, Linear Bidirectional Ramp Generator, *EDN/EEE Magazine,* Nov. 1, 1971, p 50–51.

0.2–20,000 Hz VOLTAGE-CONTROLLED RAMP—With values shown, frequency of ramp can be varied over range of about 20 kHz by changing DC input voltage. Lowest frequency is set by R_1. Adjust R_2 to make average output 0 V, and set desired output level with R_3. Uses 555 IC timer as astable MVBR, with charge current being supplied by transistor. Voltage/frequency relationship is logarithmic, making oscillator suitable for use in sound synthesizers.—J. L. Brice, Voltage-Controlled Ramp Generator, *Wireless World,* June 1976, p 72.

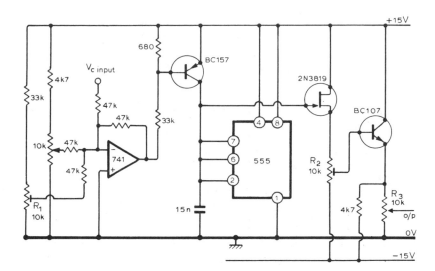

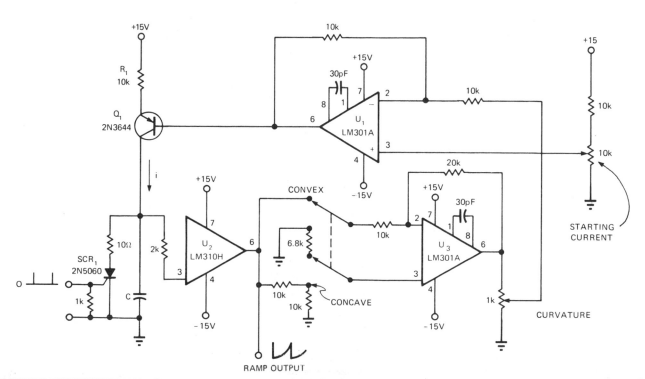

ADJUSTABLE NONLINEAR RAMP—Circuit provides predistortion of sweep with concavity or convexity as required to compensate for nonlinearity of circuit being driven. Q_1 operates as constant-current source that provides current proportional to voltage difference between ±15 V supply and base voltage of Q_1. Ramp output is linear when wiper of curvature pot is set to minimum position (ground). Period of ramp is same as that of trigger impulses that gate SCR on. Circuit uses DC coupling, avoiding need for large coupling capacitors. With 0.22 μF for C, period is 6 ms.—H. Olson, Ramp Generator Has Adjustable Nonlinearity, *EDN Magazine,* May 20, 1973, p 85 and 87.

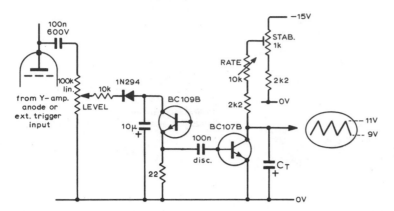

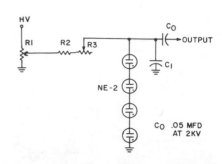

FAST-FLYBACK 2-V SAWTOOTH—Simple modern time-base circuit using transistors can be added to old oscilloscopes that have only a synchronized free-running sweep. Size of C_T determines sweep rate. When timing capacitor C_T charges to breakdown voltage of reverse-biased BC107B, capacitor is quickly discharged until voltage drops about 2 V and transistor assumes its high-resistance state again for start of next sweep.—K. Padmanabhan, Timebase Circuit, *Wireless World,* June 1974, p 196.

30-Hz SAWTOOTH—Uses neons as relaxation oscillator for producing sawtooth wave required for monitor scope of SSB transmitter. Rate at which C1 charges depends on its value and those of R2 and R3. When C1 charges to breakdown voltage of neon string, around 70 V per tube, neons fire and C1 discharges through them. C1 then starts charging again, to give sawtooth output. Voltage source should be about 500 VDC, and R1 at least 500K. For greater sweep width, increase number of neons in series.—D. Schmarder, A Simple Sweep Generator for Monitor Scopes, *73 Magazine,* Feb. 1974, p 32.

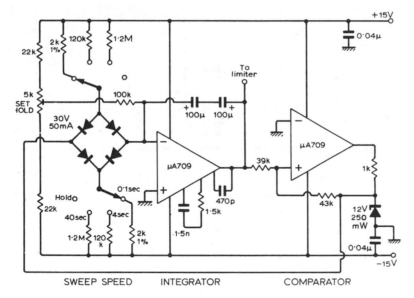

SWEEP SPEED   INTEGRATOR   COMPARATOR

10 Hz–100 kHz RAMP—Uses Miller integrator switched at selected rate by IC comparator in feedback loop. Ramp circuit was developed to drive FETs serving as voltage-dependent resistances in Wien-bridge oscillator of AF sweep generator. Article gives all circuits and construction details. Sweep linearity is better than 15% for all four ranges, covering 10 Hz to 100 kHz. For greatest accuracy, use 40-s sweep time; 4-s sweep is for long-persistence CRT, and 0.1-s sweep can be used only on upper three ranges.—F. H. Trist, Audio Sweep Generator, *Wireless World,* July 1971, p 335–338.

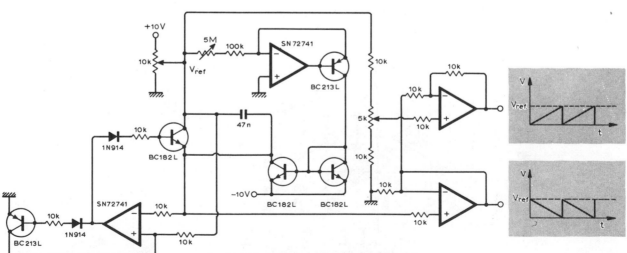

COMPLEMENTARY RAMPS—Opamp circuit provides independent controls over amplitude and frequency, as required for amplitude-modulating auditory signals to either ear for creating impression of left to right scan. High output impedance of ramps makes use of voltage follower output stage essential. Another opamp is used for inversion and level-shifting of complementary ramp, so both ramps are available from very low impedance sources. Article describes circuit operation in detail. For values shown, ramp output is variable from 400 mV to 8 V, and ramp time from 50 ms to 2 s.—L. J. Retallack, Complementary Ramp Generator with Independent Amplitude/Slope Control, *Wireless World,* Feb. 1975, p 94.

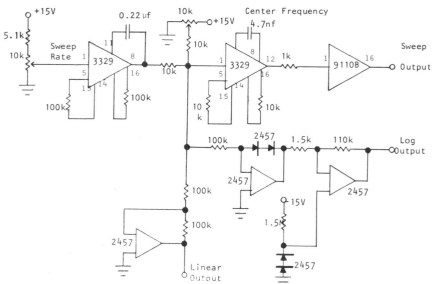

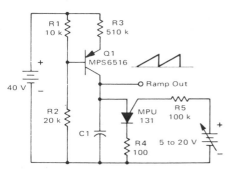

VOLTAGE-CONTROLLED RAMP—Current source Q1 and C1 together set duration time of ramp. As positive DC voltage at gate of MPU131 is increased, peak point firing voltage of PUT is changed and duration time is increased. With 0.01 μF, control voltage change from 5 V to 20 V increases duration time linearly from 2 ms to 7.2 ms.—R. J. Haver and B. C. Shiner, "Theory, Characteristics and Applications of the Programmable Unijunction Transistor," Motorola, Phoenix, AZ, 1974, AN-527, p 8.

LIN/LOG SWEEP—Sweep rate and center frequency are adjustable in versatile sweep generator providing sweep-frequency output and both linear and logarithmic voltages representing output frequency. Optical Electronics 3329 voltage-to-frequency transducer generates sweep voltage at frequency determined by external resistors, capacitor, and input voltage from 10K pot. Sweep waveform is triangular; for sawtooth waveform, change one of 100K timing resistors to 10K. Second 3329 delivers frequency-varying sine wave to 9110B buffer. 10K pot determines center frequency. Stable log f output is obtained with 2457 log module. Frequency and sweep rates can be range-switched from DC to 100 kHz by changing timing capacitor.—"Improved Sweep Generator," Optical Electronics, Tucson, AZ, Application Tip 10209.

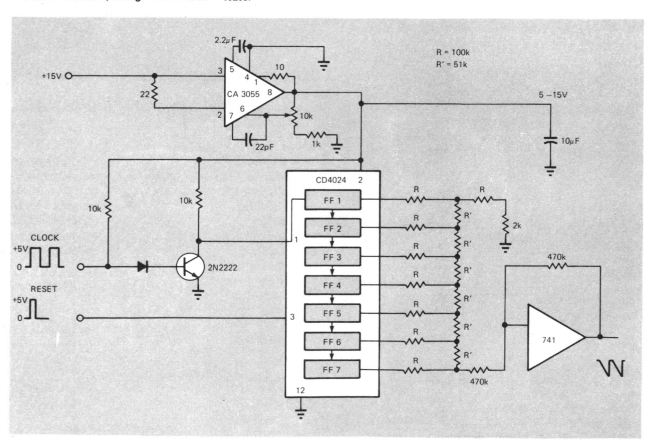

DIGITAL RAMP—Digital-to-analog technique using single CD4024A CMOS shift register eliminates temperature and linearity problems normally encountered when using RC circuit to drive VCO of digital ramp. Ramp is generated from 50-kHz clock and stopped by applying reset pulse to counters. Use of stable but variable supply for IC permits adjustment of ramp output amplitude. Ramp itself consists of large number of small steps; if these steps are too large, second CD4024A can be added and clock frequency increased. If response of 741 opamp is not adequate for very steep ramps, use opamp having higher slew rate.—K. Bower, CMOS Linear-Ramp Generator Has Amplitude Control, *EDN Magazine*, June 20, 1973, p 87.

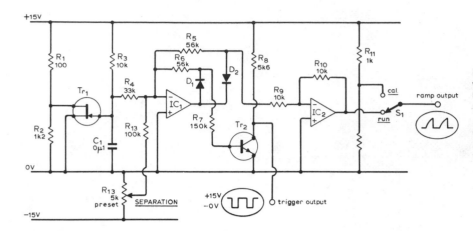

EXPONENTIAL RAMP—Used in curve tracer in which ramp does not need to be linear. Input UJT Tr₁ is Texas Instruments 43 or equivalent, ICs are SN72558P dual opamp or individual SN72741P opamps, and diodes are 1S44 or equivalent. Article gives other circuits and calibration procedure.—L. G. Cuthbert, An F.E.T. Curve Tracer, *Wireless World*, April 1974, p 101–103.

10.7-MHz SWEEP GENERATOR—Can be used with CRO for studying response of IF amplifier or filter. Greater dynamic range is obtained by using with spectrum analyzer.—*Circuits, 73 Magazine*, Holiday issue 1976, p 170.

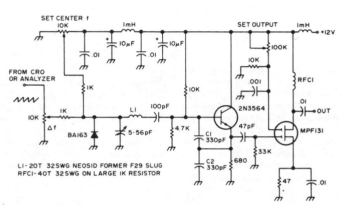

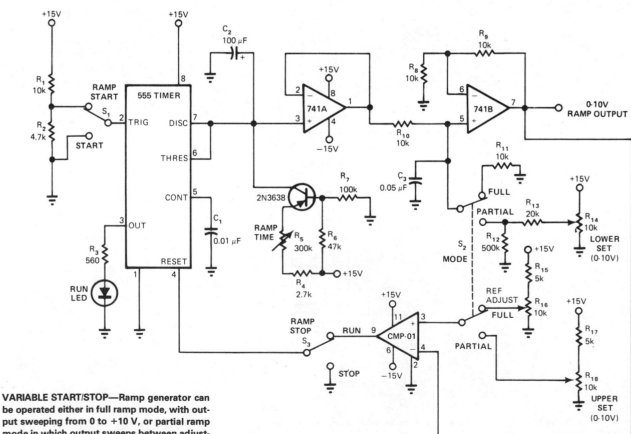

VARIABLE START/STOP—Ramp generator can be operated either in full ramp mode, with output sweeping from 0 to +10 V, or partial ramp mode in which output sweeps between adjustable starting point and adjustable stopping point. R₅ selects time period in both modes. Ramp is reset automatically when output reaches preset voltage limit. Values shown for C₂ and R₅ give 100-s charge time, but changing R₅ to 1 megohm increases charge period to 7 min. S₃ stops ramp and resets circuit at any point in ramp cycle.—D. Dantuono, Ramp Generator Features Variable Start/Stop Points, *EDN Magazine*, April 20, 1978, p 130 and 132.

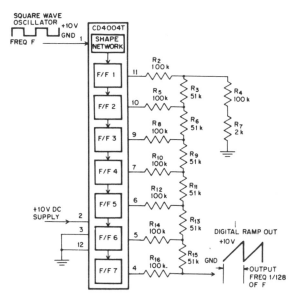

DIGITAL RAMP—RCA CD4004T IC, internally connected as ripple counter, provides flip-flop outputs corresponding to number of binary bits loaded into single input 1. Frequency range of counter is DC to 2.5 MHz, making it ideal for low-frequency operation. With R-2R ladder connected to flip-flop outputs, input square wave gives digitally stepped ramp at ladder output, with ramp frequency equal to $\frac{1}{128}$ of input frequency.—W. E. Peterson, Digital Ramp Generator, EEE Magazine, Jan. 1971, p 64–65.

RAMP FROM −10 V—Based on use of integrating opamp to generate triangle wave from square wave. Circuit goes one step further by converting triangle to ramp function having predetermined negative starting level of −10 V. When square-wave input signal changes in polarity from positive to negative, output of circuit begins to go positive as ramp function and C_4 charges with output voltage. When input changes from negative to positive and output begins to go negative, Q_1 conducts and drives noninverting input of opamp negative. Since square-wave input is positive at this time, opamp output is forced to go negative at its slew rate. Output then remains negative until square-wave input switches in negative direction for repeating cycle.—L. Wing, Op Amp and One Transistor Produce Ramp Function, EDN Magazine, Nov. 15, 1972, p 49.

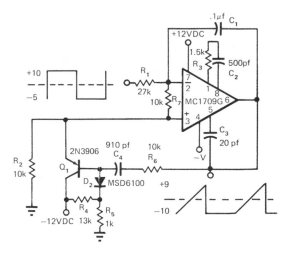

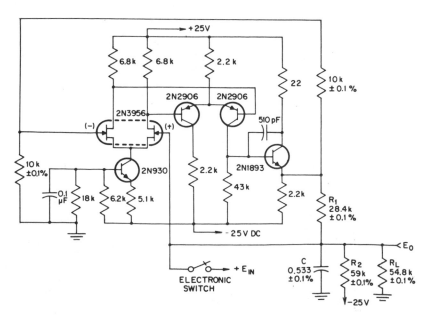

ULTRALINEAR SWEEP—Designed to generate horizontal and vertical sweep signals for military radar system. Linearity was so good that deviations could not be measured. With E_{in} positive as shown, circuit gives negative-going sweep. Reverse input polarity to get positive-going sweep. Circuit is immune to short-circuits.—R. C. Scheerer, Designing Linear Sweep Generators, EDN/EEE Magazine, July 1, 1971, p 39–42.

CHAPTER **87**
Switching Circuits

Includes circuits using switching opamps, switching transistors, analog-switch ICs, and other devices under control of logic or other input signals to provide solid-state SPST, SPDT, or DPDT switching functions for RF and AF signals.

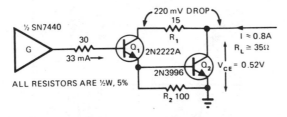

FAST ON AND OFF—Driver Q_1 is low-power device with fast switching time, while power transistor Q_2 handles power dissipation and amperes of current being switched. Used in TTL circuits requiring fast solid-state switches having known and repeatable switching times. Current reaches maximum in about 50 ns.—C. Venditti, Fast Power Switch Self-Corrects for Degradation, *EDN Magazine,* Jan. 20, 1975, p 59–60.

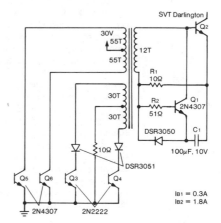

TRANSFORMER DRIVE FOR SWITCHING TRANSISTOR—Transformer provides isolated base drive for high-speed high-power TRW SVT6062 power Darlington Q_2. When 12-V secondary switches positive, C_1 charges rapidly, after which base drive current is maintained at level determined by base-emitter voltage of Q_2 and value of R_1. During turnoff, transformer secondary goes to zero due to shorting of transformer primary by Q_3 and Q_4. Base of Q_1 is then forward-biased by capacitor and turned on, discharging C_1 through base-emitter path of Q_2.—D. Roark, "Base Drive Considerations in High Power Switching Transistors," TRW Power Semiconductors, Lawndale, CA, 1975, Application Note No. 120, p 6.

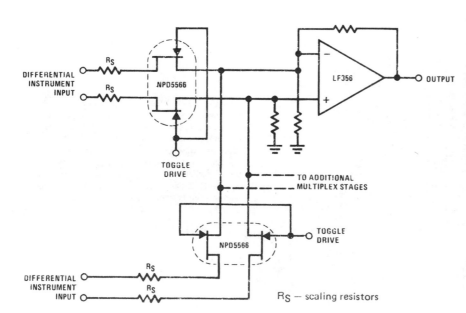

DIFFERENTIAL ANALOG SWITCH—NPD5566 dual JFETs provide high accuracy for differential multiplexer because JFET sections track at better than ±1% over wide temperature range. Close tracking reduces errors due to common-mode signals. Values of resistors depend on application and on type of opamp used.—"FET Databook," National Semiconductor, Santa Clara, CA, 1977, p 6-26–6-36.

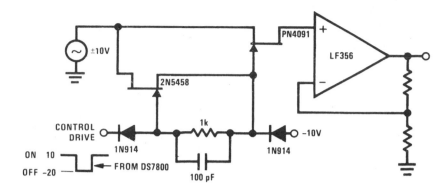

ANALOG WITH HIGH TOGGLE RATE—Simple commutator circuit provides low-impedance gate drive to PN4091 FET analog switch for both ON and OFF drive conditions. For high-frequency signal handling, circuit also approaches ideal gate drive conditions by providing low AC impedance for OFF drive and high AC impedance for ON drive.—"FET Databook," National Semiconductor, Santa Clara, CA, 1977, p 6-26– 6-36.

DPDT FET—With ON resistance of several ohms and OFF resistance of thousands of megohms, drain-source channel of field-effect transistor makes ideal low-frequency switch. Transistor capacitances are detrimental to high-frequency signal isolation and limit response times.— "Low Frequency Applications of Field-Effect Transistors," Motorola, Phoenix, AZ, 1976, AN-511A, p 5.

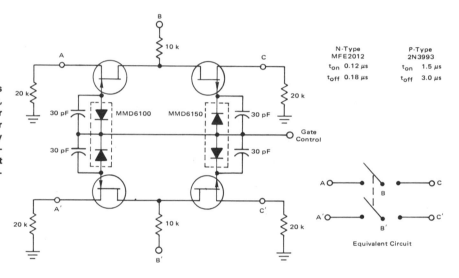

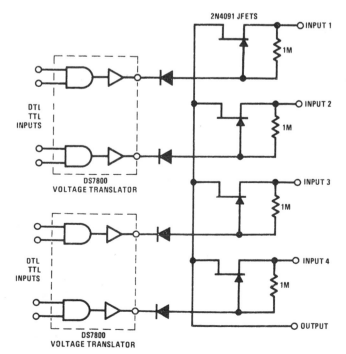

FOUR-CHANNEL COMMUTATOR—2N4091 JFETs give ON resistance of less than 30 ohms for each channel along with low OFF current leakage. DS7800 voltage translators provide gate drives of 10 V to −20 V for JFETs while giving DTL/TTL compatibility.—"FET Databook," National Semiconductor, Santa Clara, CA, 1977, p 6-26–6-36.

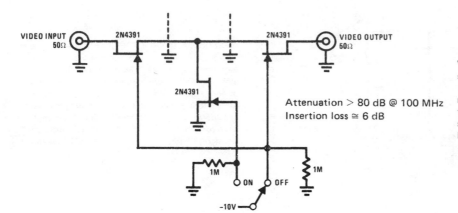

Attenuation > 80 dB @ 100 MHz
Insertion loss ≅ 6 dB

VIDEO SWITCH—2N4391 FETs provide ON resistance of only 30 ohms and OFF impedance less than 0.2 pF, to give performance comparable to that of ideal high-frequency switch. Attenuation is greater than 80 dB at 100 MHz. Insertion loss is about 6 dB.—"FET Databook," National Semiconductor, Santa Clara, CA, 1977, p 6-26—6-36.

FERRITE-CORE SWITCHING—Darlington transistor driven by low-power current stage serves for saturating groups of ferrite cores in phased arrays for radar systems or in read/write core switching for memory systems. Circuit provides fast rise and fall times.—"Designer's Guide to Power Darlingtons as Switching Devices," Unitrode, Watertown, MA, 1975, U-70, p 4.

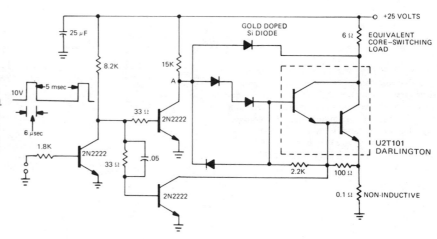

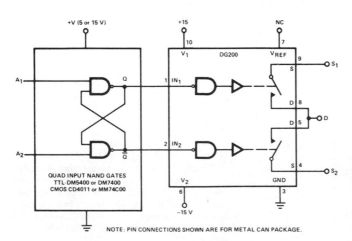

NOTE: PIN CONNECTIONS SHOWN ARE FOR METAL CAN PACKAGE.

LATCHING SPDT—DG200 CMOS analog switch is driven through pair of NAND gates connected for logic inputs. With inputs normally low, both switches are held in predetermined states. When either input receives high command pulse, switches assume states given in truth table. Both switches are off when both inputs are held high; after release of high commands, last input to go low determines states of switches.—"Analog Switches and Their Applications," Siliconix, Santa Clara, CA, 1976, p 7-69.

TRUTH TABLE

COMMAND		STATE OF SWITCHES AFTER COMMAND	
A_2	A_1	S_2	S_1
0	0 (normal)	same	same
0	1	OFF	ON
1	0	ON	OFF
1	1	INDETERMINATE	

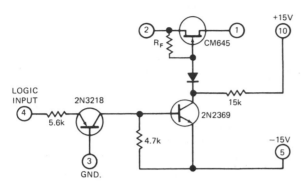

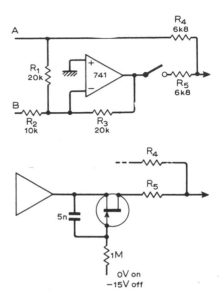

FET ANALOG SWITCH—Simple level-shifting driver provides analog switching. Input of logic 1 makes emitter and collector current flow in input PNP transistor, for shifting from logic to −15 V. This current makes NPN transistor turn on so its collector is −15 V, diode is forward-biased, and FET gate is about −14.3 V. At logic 0, both transistors are off and driver output is at +15 V. Diode is now reverse-biased, turning FET on to provide desired switching action between outputs 1 and 2.—J. Cohen, Solid-State Signal Switching: It's Getting Better All the Time, *EDN Magazine*, Nov. 15, 1972, p 22–28.

SWITCHING OPAMP—Circuit provides changeover function when only single pair of contacts is available. With switch open, input A goes to output. With switch closed, input B goes to output and signal of input A is inverted by opamp so as to cancel direct signal A. Gain is unity for both output signals. Switch can be replaced by FET as in lower diagram; here, capacitor prevents FET from cutting off during positive half-cycle above about 100 Hz. In multichangeover applications, opamp could be section of programmable opamp.—M. J. Sells, Electronic Changeover Switching, *Wireless World*, Dec. 1974, p 503.

ANALOG SWITCH PROTECTION—Current-limiting resistors are used in series with sections of DG300 dual analog switch to limit contact currents to 30 mA continuous or 100 mA pulsed for less than 1 ms. Values of limiting resistors depend on supply voltage used and are therefore determined by experimentation. Technique is suitable for applications in which DG300 serves for charging and discharging capacitor.—"Analog Switches and Their Applications," Siliconix, Santa Clara, CA, 1976, p 7-81.

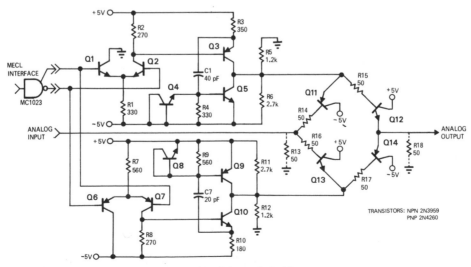

DIAMOND-BRIDGE ANALOG SWITCH—Analog signals up to 3 V P-P are switched in less than 3 ns to meet requirements of multiplexer and sample-and-hold portions of 100 Mb/s PCM telemetry encoder. Symmetrical drive circuits turn four-transistor diamond bridge on and off at 20-MHz clock frequency. Transient-coupled pullback transistors Q5 and Q9 speed turnoff. Typical rise time is 1.5 ns and fall time is 2 ns for 1-VDC analog input.—W. A. Vincent, Diamond Bridge Improves Analog Switching, *EDN Magazine*, Feb. 15, 1971, p 41–42.

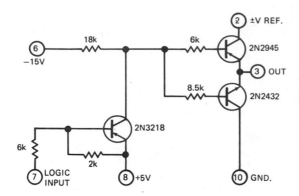

AC SPDT—Complementary NPN and PNP transistors provide single-pole double-throw switching action for AC signals, as required in some digital-to-analog converter applications. Circuit switches between ground and signals of up to ±5 V. Output transistor pair will toggle with unipolar drive.—J. Cohen, Solid-State Signal Switching: It's Getting Better All the Time, *EDN Magazine*, Nov. 15, 1972, p 22–28.

SWITCHING SINGLE-SUPPLY OPAMPS—DG301 low-power analog switch serves as interface between TTL control input and DC-coupled opamp pair. Logic level determines which opamp is connected to single output.—"Analog Switches and Their Applications," Siliconix, Santa Clara, CA, 1976, p 7-88—7-89.

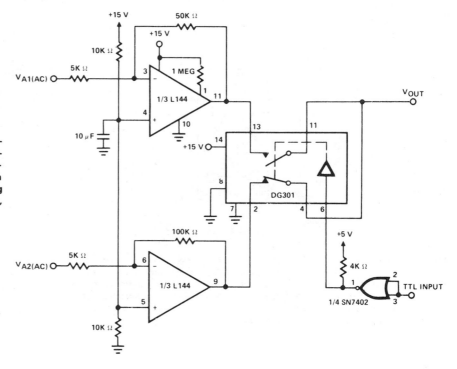

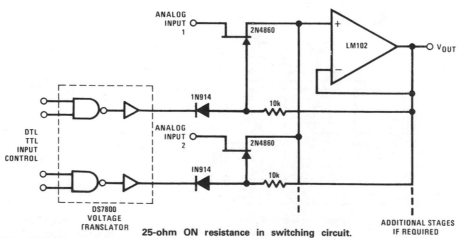

LOGIC-CONTROLLED ANALOG SWITCH—2N4860 JFETs were chosen for low leakage and 25-ohm ON resistance in switching circuit. LM102 opamp serves as voltage buffer. DS7800 IC provides switch drive under control of DTL or TTL levels.—"FET Databook," National Semiconductor, Santa Clara, CA, 1977, p 6-26—6-36.

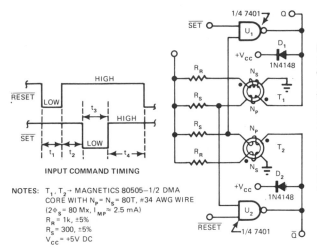

INPUT COMMAND TIMING

NOTES: T_1, $T_2 \rightarrow$ MAGNETICS 80505–1/2 DMA
CORE WITH $N_P = N_S = 80T$, #34 AWG WIRE
$(2\Phi_S = 80$ Mx, $I_{MP} \approx 2.5$ mA)
$R_R = 1k$, ±5%
$R_S = 300$, ±5%
$V_{CC} = +5V$ DC

NONVOLATILE LATCH—Design shown for latch gives immunity to interfering noise pulses on command line and prevents loss of essential data bit during unexpected power-line interruption. Subminiature saturable transformers in positive feedback paths between two gates of latch prevent instantaneous change of state. Transformers remain magnetically biased in positive or negative saturation even without circuit power, to provide pretransient-state reference to which latch must return when power is reapplied.—G. E. Bloom, Add Nonvolatility to Your Next Latch Design, *EDN Magazine*, Jan. 5, 1978, p 80 and 82.

CAPACITOR-COUPLED DRIVE FOR SWITCHING TRANSISTOR—TRW SVT6062 power Darlington switching transistor is used in grounded-emitter connection in which negative bias required for turnoff is created by charging C_1 during ON interval. Zener limits charge on C_1 and provides path for base drive current to Darlington. Diodes give faster response to input signal by preventing Q_3 from saturating. Grounding base of Q_1 makes Darlington conduct, whereas high input level to Q_1 initiates turnoff. Circuit will operate at pulse widths down to 5 μs.—D. Roark, "Base Drive Considerations in High Power Switching Transistors," TRW Power Semiconductors, Lawndale, CA, 1975, Application Note No. 120, p 7.

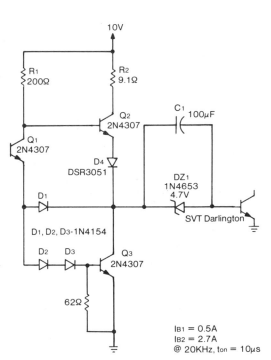

$I_{B1} = 0.5A$
$I_{B2} = 2.7A$
@ 20KHz, $t_{on} = 10\mu s$

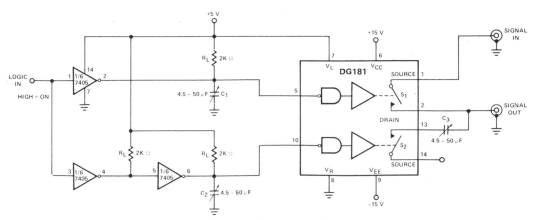

SYNCHRONIZED SWITCHING—Transients occurring during switching between two amplifier channels are attenuated by synchronizing turn-on of one switch in DG181 JFET analog switch with turnoff of other switch. Switching action is controlled by logic input.—"Analog Switches and Their Applications," Siliconix, Santa Clara, CA, 1976, p 7-61.

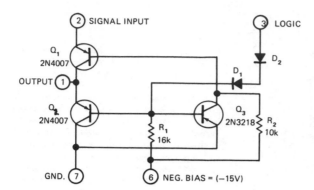

NEGATIVE SPDT—Developed for driving R/2R ladder network in D/A converters. Signal input or reference voltage range is 0 to −12 V. With logic 0, R_1 forward-biases Q_2 and Q_3; Q_2 then switches output to ground, and Q_3 clamps base of Q_1 to ground to keep it off. With logic 1, D_1 and D_2 conduct and make Q_1 switch on.—J. Cohen, Solid-State Signal Switching: It's Getting Better All the Time, *EDN Magazine,* Nov. 15, 1972, p 22–28.

FET DPDT—Uses FETs as switching elements for transferring VFO and carrier oscillator signals between first and second mixers of SSB transceiver. Transmitter key line is at +13 V on receive and 0 V on transmit, and "T" line has opposite voltages. On receive, Q_{304} and Q_{305} are pinched off by about +10 V, while Q_{303} and Q_{306} are conducting with only about +0.7 V on their gates. VFO signal then flows to first mixer and carrier oscillator to second mixer. On transmit, conditions are opposite.—J. Schultz, CQ Reviews: The Atlas 210 and 215 SSB Transceivers, *CQ,* May 1975, p 22–27 and 65.

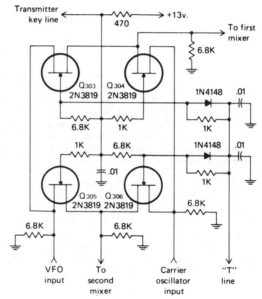

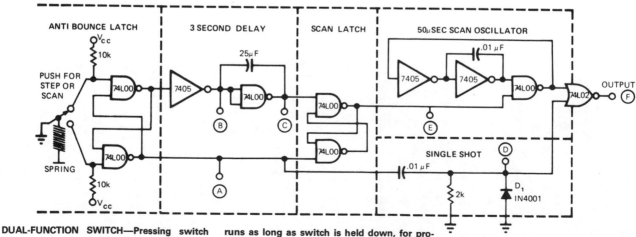

DUAL-FUNCTION SWITCH—Pressing switch for less than 3 s produces single output pulse about 5 μs wide. When switch is pressed longer than 3 s, single pulse is generated as before, and scan oscillator is turned on 3 s later. Oscillator runs as long as switch is held down, for producing repetitive stepping motion. Applications include positioning of test probes on single semiconductor chip or on wafer of several hundred chips, or indexing device either step by step or automatically through desired number of steps.—J. McDowell, Single Switch Controls Two Functions, *EDN Magazine,* April 5, 1974, p 78 and 80.

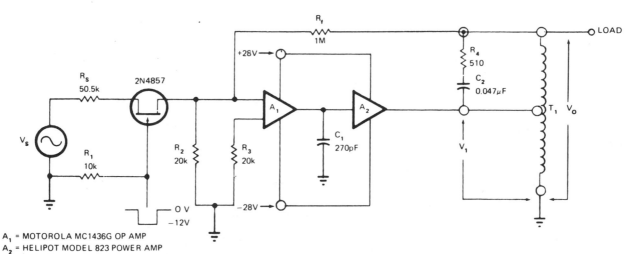

A_1 = MOTOROLA MC1436G OP AMP
A_2 = HELIPOT MODEL 823 POWER AMP
T_1 = MICROTRAN MT-11

80-VRMS ANALOG SWITCH—Developed for switching high-level analog signals with speed, accuracy, and reliability, for such applications as digital-to-synchro converters. Feedback network stabilizes output against changes in circuit parameters. For AC signal inputs between ±10 V, −12 VDC on gate of FET blocks input channel and R_2 grounds inverting input of opamp A_1 to prevent noise pickup and minimize voltage offset. Grounding gate of FET turns on input channel; input signal is then amplified by A_1 and fed to unity-gain power opamp A_2.—D. J. Musto, Analog Switch and IC Amp Controls 80V RMS, *EDN Magazine*, Feb. 20, 1973, p 91–92.

CHAPTER 88

Switching Regulator Circuits

Covers regulators in which DC input voltage is converted to pulse-width-modulated frequency in range of 9–100 kHz, with duty cycle or frequency being varied automatically to maintain essentially constant output voltage at desired value. Circuit may use discrete components or switching-regulator IC.

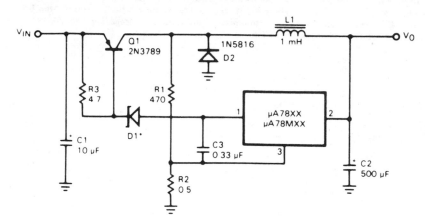

5–24 V SWITCHING—Choice of regulator in μA7800 series determines fixed output voltage. Devices are available for rated outputs of 5, 6, 8, 12, 15, 18, and 24 V, positive or negative, with output current ratings of 100 mA, 500 mA, or 1 A. If input voltage is greater than maximum input rating of regulator used, add voltage-dropping zener D1 to bring voltage between pins 1 and 3 down to acceptable level.—"Signetics Analog Data Manual," Signetics, Sunnyvale, CA, 1977, p 668.

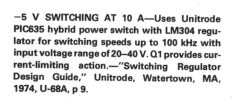

−5 V SWITCHING AT 10 A—Uses Unitrode PIC635 hybrid power switch with LM304 regulator for switching speeds up to 100 kHz with input voltage range of 20–40 V. Q1 provides current-limiting action.—"Switching Regulator Design Guide," Unitrode, Watertown, MA, 1974, U-68A, p 9.

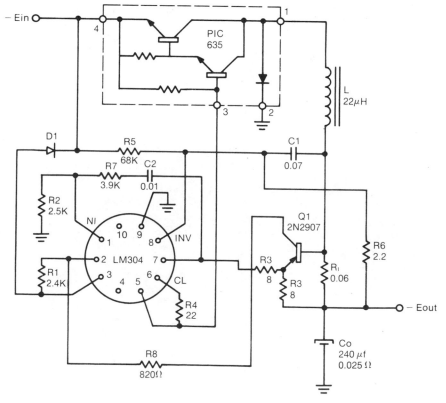

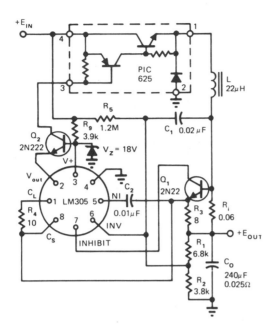

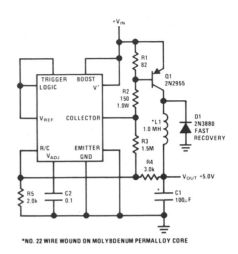

5-V FIXED OFF-TIME SWITCHING—Uses LM305 regulator and Unitrode hybrid power switch in PIC600 series. Operates in fixed OFF-time mode. Output ripple of 100 mV P-P is independent of input voltage range of 20 to 40 V for output of +5 V ± 1%. Switching speed is nominally 50 kHz but can go up to 100 kHz. Article covers theory of operation in detail.—L. Dixon and R. Patel, Designers' Guide to: Switching Regulators, *EDN Magazine,* Oct. 20, 1974, p 53—59.

5 V AT 1 A—National LM122 timer is connected as switching regulator by using internal reference and comparator to drive switching transistor Q1. Minimum input voltage is 5.5 V. Line and load regulation are less than 0.5%, and output ripple at switching frequency is only 30 mV. Output voltage can be adjusted between 1 V and 30 V by using appropriate values for R2-R5.—C. Nelson, "Versatile Timer Operates from Microseconds to Hours," National Semiconductor, Santa Clara, CA, 1973, AN-97, p 9.

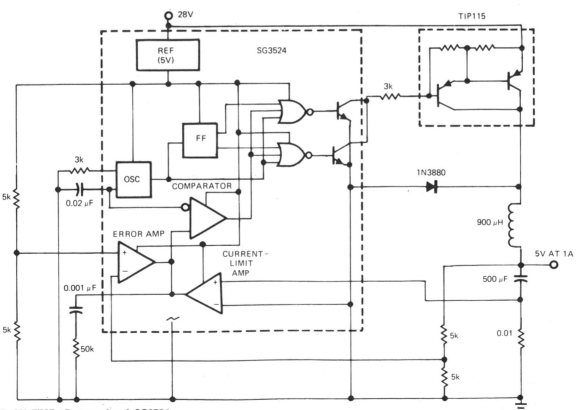

VARIABLE ON TIME—Duty cycle of SG3524 switching regulator is varied by modulating ON time while maintaining constant switching frequency, using pulse-duration-modulation control circuit.—J. Spencer, Monolithic Switching Regulators—They Fit Today's Power-Supply Needs, *EDN Magazine,* Sept. 5, 1977, p 117—121.

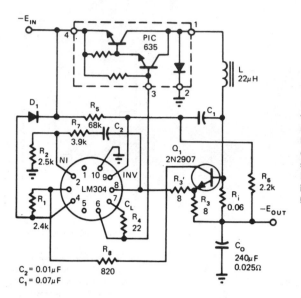

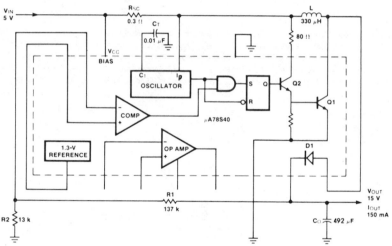

−10 V SWITCHING—Uses LM304 regulator and Unitrode hybrid power switch in PIC600 series to provide output of 10 A. R_1 and R_2 determine reference voltage. Current limiting is achieved by reducing reference voltage to ground instead of turning off base drive to power output switch. Article covers operating theory.—L. Dixon and R. Patel, Designers' Guide to: Switching Regulators, *EDN Magazine*, Oct. 20, 1974, p 53–59.

STEPPING 5 V UP TO 15 V—Fairchild μA78S40 switching regulator transforms 5 V to 15 V at efficiency of 80% for 150-mA load. Average input current is only 550 mA. Article gives design equations.—R. J. Apfel and D. B. Jones, Universal Switching Regulator Diversifies Power Subsystem Applications, *Computer Design*, March 1978, p 103–112.

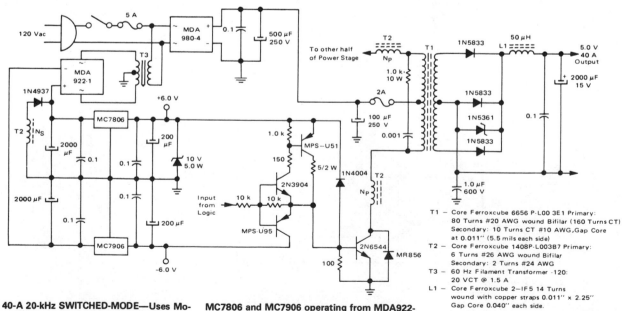

5-V 40-A 20-kHz SWITCHED-MODE—Uses Motorola 2N6544 power transistors operating with 3-A collector current (other half of power stage is identical). Bridge rectifier and capacitive filter connected directly to AC line form 150-VDC supply for inverter operating at 20 kHz. Regulators MC7806 and MC7906 operating from MDA922-1 bridge rectifier of 15-W filament transformer T3 provide ±6 V for logic circuits that provide pulse-width modulation for inverter. When logic signal is high, MPS-U51 saturates and supplies 1 A to base of 2N6544 inverter power transistor. When logic is low, MPS-U95 Darlington holds inverter transistor off.—R. J. Haver, "Switched Mode Power Supplies—Highlighting a 5-V, 40-A Inverter Design," Motorola, Phoenix, AZ, 1977, AN-737A, p 10.

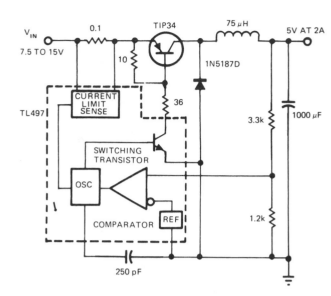

VARIABLE SWITCHING FREQUENCY—TL497 switching regulator operates at maximum frequency under maximum load conditions. For smaller loads, duty cycle is varied automatically by maintaining fixed ON time and varying switching frequency. Circuit optimizes efficiency at about 75% by reducing switching losses as load decreases.—J. Spencer, Monolithic Switching Regulators—They Fit Today's Power-Supply Needs, *EDN Magazine*, Sept. 5, 1977, p 117–121.

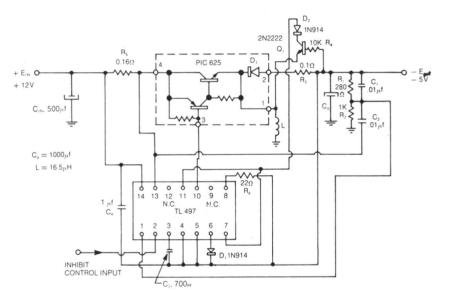

−5 V FLYBACK SWITCHING—Uses Unitrode PIC625 regulator operating at 25 kHz and TL497 control circuit operating in current-limiting mode to give line and load regulation of 0.2% for input voltage of 12 V ±25%. Efficiency is 75%. Short-circuit current is automatically limited to 3 A.—"Flyback and Boost Switching Regulator Design Guide," Unitrode, Watertown, MA, 1978, U-76, p 5.

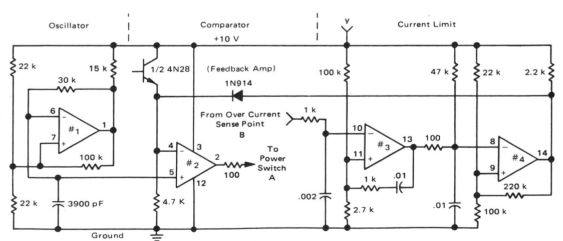

CONTROL FOR SWITCHING REGULATOR—Uses all four sections of Motorola MC3302 quad comparator. First section is connected as 20-kHz oscillator that supplies sawtooth output sweeping between voltage limits set by 100K positive feedback resistor and 15-V supply. Section 2 compares sawtooth output to feedback signal, to produce variable-duty-cycle output pulse for power switch of switching regulator. Sections 3 and 4 initiate current-limiting action; section 3 senses overcurrent and triggers section 4 connected as mono MVBR. Limiting occurs at about 4 A. When load short is removed, regulator resets automatically. Point A goes to push-pull drive for power switch of regulator, and point B goes to current-sensing resistor in output circuit of regulator. Point y goes to 10-V supply.—R. J. Haver, "A New Approach to Switching Regulators," Motorola, Phoenix, AZ, 1975, AN-719, p 7.

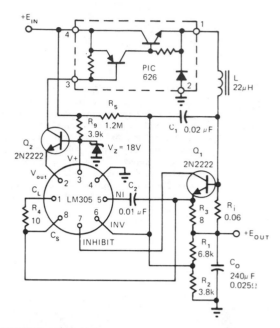

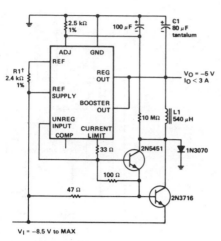

HIGH-VOLTAGE POSITIVE SWITCHING—Uses 18-V zener in series with 3.9K resistor to provide power for LM305 IC regulator. Q_2 provides base drive for PIC626 hybrid power switch and isolates output of LM305 from switch.—L. Dixon and R. Patel, Designers' Guide to: Switching Regulators, *EDN Magazine,* Oct. 20, 1974, p 53–59.

−5 V AT 3 A SWITCHING—Negative-voltage regulator using SN52104 or SN72304 accepts input voltage range of −8.5 V to −40 V and provides regulated output of −5 V with typical load regulation of 1 mV and input regulation of 0.06%. ICs are interchangeable with LM104 and LM304 respectively. L1 is 60 turns No. 20 on Arnold Engineering A930157-2 molybdenum permalloy core or equivalent.—"The Linear and Interface Circuits Data Book for Design Engineers," Texas Instruments, Dallas, TX, 1973, p 5-5.

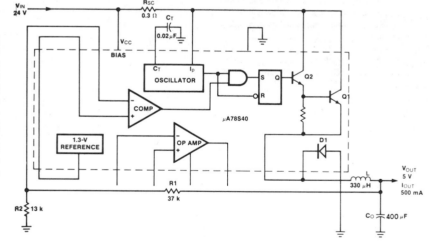

REDUCING 24 V TO 5 V—Uses Fairchild μA78S40 switching regulator having variety of internal functions that can provide differing voltage step-up, step-down, and inverter modes by appropriately connecting external components. Connections shown provide step-down from 24 V to 5 V at 500 mA with 83% efficiency. Applications include running TTL from 24-V battery. Output ripple is less than 25 V. Article gives design equations.—R. J. Apfel and D. B. Jones, Universal Switching Regulator Diversifies Power Subsystem Applications, *Computer Design,* March 1978, p 103–112.

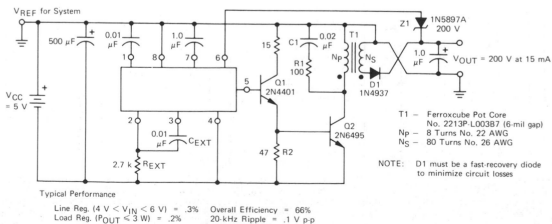

Typical Performance

Line Reg. (4 V < V_{IN} < 6 V) = .3%	Overall Efficiency = 66%
Load Reg. (P_{OUT} ≤ 3 W) = .2%	20-kHz Ripple = .1 V p-p

5 V TO 200 V WITH SWITCHING REGULATOR—Converts standard logic supply voltage to high voltage required by gas-discharge displays, using Motorola MC3380 astable MVBR as control element in switching regulator. Will drive up to 15 digits. Operating frequency is about 20 kHz.—H. Wurzburg, "Control Your Switching Regulator with the MC3380 Astable Multivibrator," Motorola, Phoenix, AZ, 1975, EB-52.

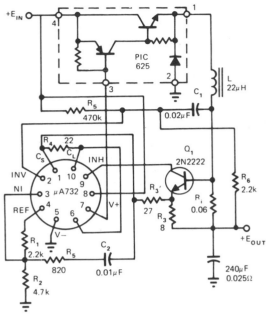

+10 V SWITCHING—Positive switching regulator circuit uses μA732 with Unitrode PIC625 hybrid power switch and single transistor, operating in fixed OFF-time mode. Article covers regulator theory of operation in detail.—L. Dixon and R. Patel, Designers' Guide to: Switching Regulators, *EDN Magazine*, Oct. 20, 1974, p 53–59.

BATTERY REGULATOR—Uses LM376N positive voltage regulator in switching mode to compensate for voltage changes of battery supply during discharge cycle, without adjusting series rheostat. Load regulation is 0.3% for unregulated input of 9 to 30 V, with R1 and R2 setting output voltage anywhere between 5 and 27 V. Maximum output current is 25 mA. Switching frequency of regulator is 33 kHz.—E. R. Hnatek and L. Goldstein, Switching Regulator Designed for Portable Eqiupment, *EDN/EEE Magazine*, Sept. 15, 1971, p 39–41.

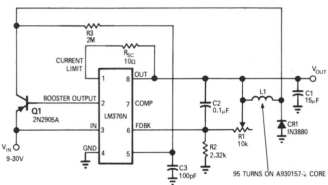

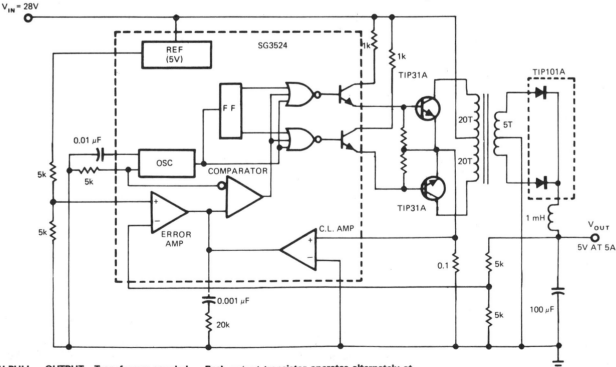

PUSH-PULL OUTPUT—Transformer-coupled push-pull output for SG3524 fixed-frequency pulse-duration-modulated switching regulator gives output flexibility, allowing for multiple outputs and wide range of output voltages. Each output transistor operates alternately at half of switching frequency. Switching regulator applies voltage alternately to opposite ends of transformer primary, making transformer perform as if it had AC input. TIP101A rectifier then provides desired 5-VDC output at 5A.—J. Spencer, Monolithic Switching Regulators—They Fit Today's Power-Supply Needs, *EDN Magazine*, Sept. 5, 1977, p 117–121.

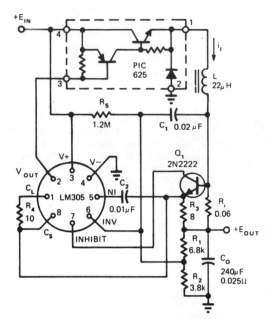

5-V SWITCHING—Fixed OFF-time mode of operation is used in switching regulator design to provide 5-V output that is constant within 100 mV P-P for input range of 20–40 V, for loads ranging from 10 A maximum to 2 A minimum. Switching frequency can be in range of 1–50 kHz. Operation above 20 kHz eliminates possibility of audio noise but with some drop in efficiency. Values shown are for 50 kHz. Article gives design equations and design procedure.—L. Dixon and R. Patel, Designers' Guide to: Switching Regulators, Part 2, *EDN Magazine*, Nov. 5, 1974, p 37–40.

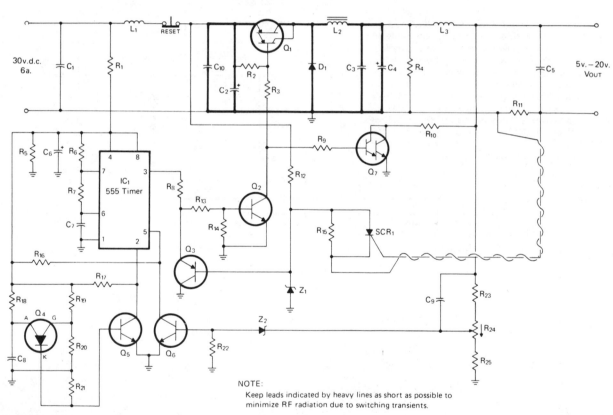

NOTE:
Keep leads indicated by heavy lines as short as possible to minimize RF radiation due to switching transients.

C_1, C_3, C_5, C_{10}—1.0 μF, Polycarb	R_8, R_{13}, R_{23}—1.2K, ½ W
C_2, C_6—100 μF, 50V	R_9—15K, ½ W
C_4—1000 μF, 50V	R_{10}—20 Ω, 10W
C_7—0.0082 μF	R_{11}—0.075 Ω, 6 watts
C_8—390pF	R_{12}—1.5K, 1W
C_9—0.002 μF	R_{14}—330 Ω, ½ W
D_1—1N3890	R_{15}, R_{19}—680 Ω, ½ W
L_1, L_3—10 μhy, 10 amps	R_{16}—22K, ½ W
L_2—180 μhy	R_{17}—4.7K, ½ W
Q_1—D45E2 (General Electric)—	R_{18}—120K, ½ W
Q_2, Q_5—D33D25	R_{20}—1K, ½ W
Q_3—D29E25	R_{21}—100 Ω, ½ W
Q_4—2N6027	R_{22}—18K, ½ W
Q_6—D32S4	R_{24}—1K, 1W Pot.
Q_7—D40K2—Use Thermalloy	R_{25}—390 Ω, ½ W
6063B heatsink	SCR-1—C103B
R_1, R_3, R_4, R_5—1.2K, ½ W	Z_1—1N5233B
R_2, R_7—110 Ω, ½ W	Z_2—1N5226B
R_6—4.7K, ½ W	IC-1—555 Timer

150-W SWITCH-MODE—Unregulated DC voltage is applied to power Darlington Q_1 serving as switch that chops voltage so rectangular waveform is applied to RLC output filter. Average voltage to filter depends on duty cycle of switch. 555 timer operates in mono MVBR mode as pulse generator and pulse-duration modulator. R_{24} applies varying voltage to pin 5 to modulate pulse duration linearly with respect to applied voltage. Actions of Q_1, Q_2, and Q_6 maintain constant 3.6 V at arm of control pot.

Q_4 and Q_5 provide 20-kHz clock pulse, above audible range. Overcurrent protection of transistors is provided by R_{11}, SCR, and Q_3. Adjust R_{11} so SCR turns on and shuts down circuit when current through R_{11} reaches 8 A. Circuit must be reset manually after overload. Q_7 and R_{10} load circuit to prevent oscillation at low output voltage and light load.—R. J. Walker, A 150 Watt Switch-Mode Regulator, *CQ*, March 1977, p 40–43 and 74–75.

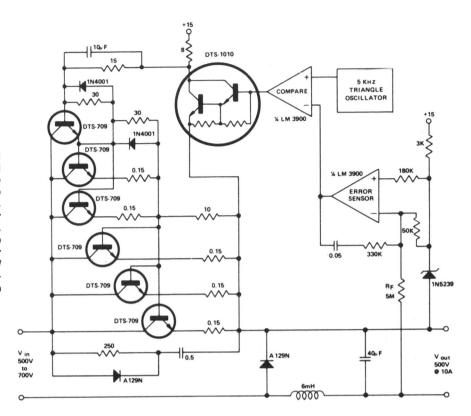

5-kW SWITCHING—Six Delco DTS-709 transistors are connected in progressive Darlington configuration to provide stable and efficient switching at high voltages. Can be operated from 480-V three-phase full-wave rectified line to minimize filter cost. Control circuit uses one LM3900 IC operating from isolated 15-V supply, along with 5-kHz triangle oscillator and error sensor feeding into comparator. In power stage, one DTS-709 drives two DTS-709s which drive three DTS-709s. Efficiency is better than 90% for all loads above 500 W.—"Economical 5 kW Switching Regulator Using DTS-709 Transistors," Delco, Kokomo, IN, 1974, Application Note 56.

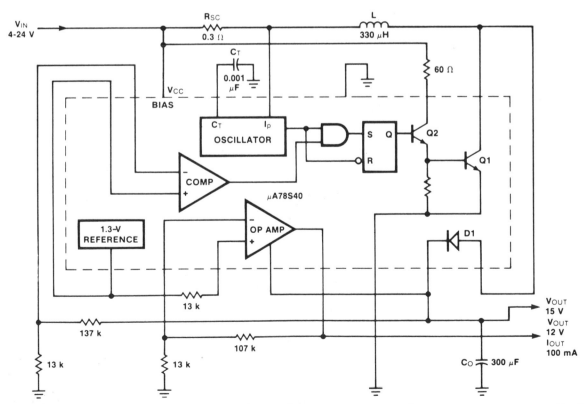

+12 V AND +15 V FROM 4–24 V—Connections shown for Fairchild μA78S40 switching regulator give universal regulator providing either step-up or step-down, for loads up to 100 mA.

Efficiency is about 50% for input extremes of 4 and 24 V, increasing to maximum of 75% for other input voltages. Output ripple is essentially eliminated at 12-V output.—R. J. Apfel

and D. B. Jones, Universal Switching Regulator Diversifies Power Subsystem Applications, *Computer Design*, March 1978, p 103–112.

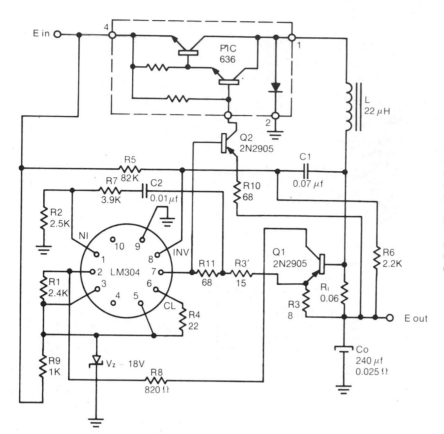

HIGH-VOLTAGE NEGATIVE-SWITCHING—Designed for operation from supply voltages above maximum of −40 V for LM304 regulator. Output is −5 V at up to 10 A. Q2 provides voltage isolation between regulator and Unitrode PIC636 hybrid power switch. R9 limits current through zener under steady-state and start-up conditions.—"Switching Regulator Design Guide," Unitrode, Watertown, MA, 1974, U-68A, p 9.

6 V FOR CALCULATOR—Can be mounted in housing of calculator or small transistor radio, for operation from AC line. D_1 and D_2 produce 15 VDC across filter capacitor C_2 as supply for inverter Tr_1 operating at 13 kHz. Transformer is wound with No. 37 wire on small core such as Phillips P14/8 337 pot core. Primary windings are bifilar. Use grounded shield to reduce radiated switching noise.—M. Faulkner, Miniature Switch Mode Power Supply, *Wireless World,* Oct. 1977, p 65.

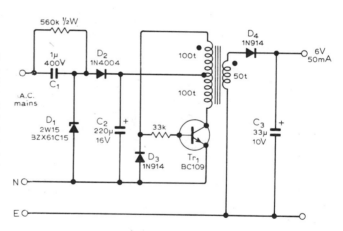

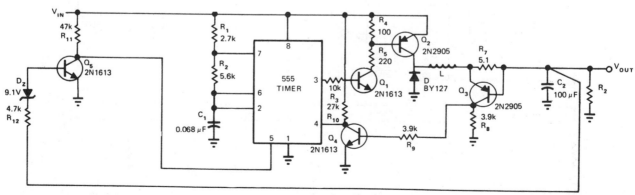

10-V SWITCHING AT 100 mA—Use of 555 timer as pulse-width-modulated regulator gives line regulation of 0.5% and load regulation of 1%.   Circuit includes current foldback. With 15-V input, output is 10 V.—P. R. K. Chetty, Put a 555 Timer in Your Next Switching Regulator Design, *EDN Magazine,* Jan. 5, 1976, p 72.

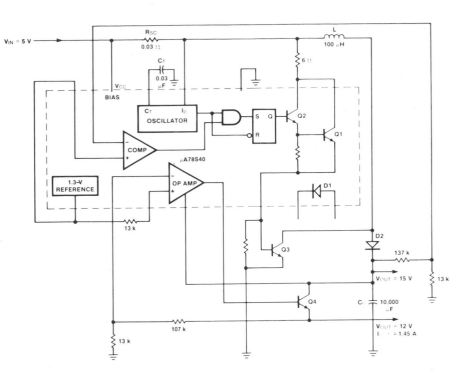

+12 V AND +15 V FROM 5 V—Uses Fairchild μA78S40 switching regulator having variety of internal functions that can provide differing voltage step-up, step-down, and inverter modes by appropriately connecting external components. External NPN transistor Q3 boosts step-up regulator, and NPN transistor Q4 increases series-pass regulator output well above 1 A. Total of 1.5 A is available from two outputs. Transistor and diode types are not critical. Efficiency is 80% for 15-V output and 64% for 12-V output.—R. J. Apfel and D. B. Jones, Universal Switching Regulator Diversifies Power Subsystem Applications, *Computer Design*, March 1978, p 103–112.

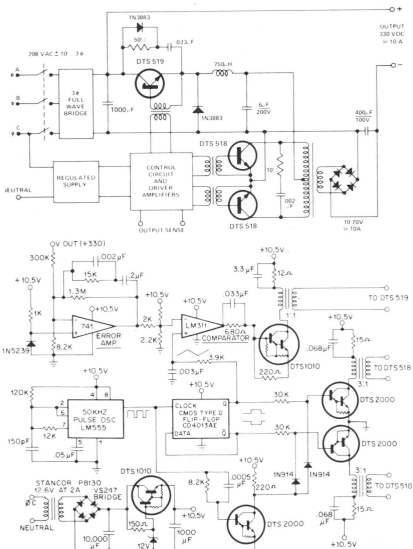

3.3-kW SWITCHING—Delco DTS-518 and DTS-519 power transistors in high-efficiency stacked supply are operated at 25-kHz switching rate to provide 330 VDC at 10 A. Control circuit operates at primary 50-kHz pulse frequency, with negative-going pulses having 2-μs duration. Flip-flop converts this to 25-kHz complementary square-wave signal driving Darlington DTS-2000s. Transformer cores are Magnetics EE No. 42510 each having 15-turn primary and 5-turn secondary for driving DTS-518s. Error amplifier compares portion of total output voltage to zener reference for control of DTS-519 power transistor switching at 25 kHz. Efficiency is 95% at full load.—"3.3kW High Efficiency Switch Mode Regulator," Delco, Kokomo, IN, 1977, Application Note 59.

4.5–30 V SWITCHING AT 6 A—LM105 positive regulator serves as amplifier-reference for LM195 power transistor IC in switching regulator. Duty cycle of switching action adjusts automatically to give constant output. Q2 consists of four LM195s in parallel since each is rated at only about 2 A. R8 serves as output voltage control.—"Linear Applications, Vol. 2," National Semiconductor, Santa Clara, CA, 1976, AN-110, p 4.

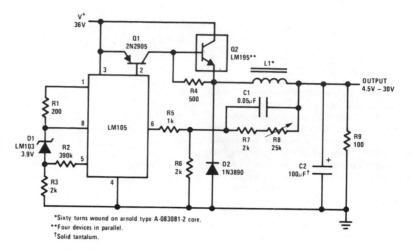

*Sixty turns wound on arnold type A-083081-2 core.
**Four devices in parallel.
†Solid tantalum.

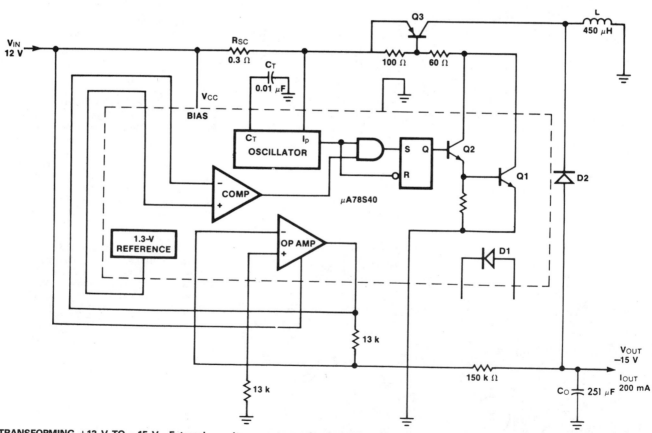

TRANSFORMING +12 V TO −15 V—External PNP transistor Q3 and catch diode D2 (types not critical) are used with Fairchild μA78S40 switching regulator so no pin of IC substrate has voltage more negative than substrate, which is grounded. Efficiency is 84% with 200-mA load. Output voltage ripple is 50 mV but can be reduced by increasing value of C₀.—R. J. Apfel and D. B. Jones, Universal Switching Regulator Diversifies Power Subsystem Applications, *Computer Design,* March 1978, p 103–112.

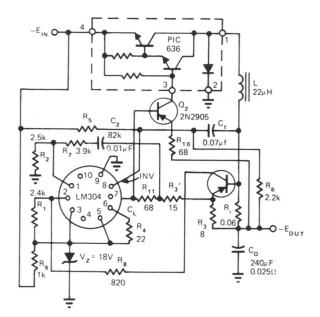

HIGH-VOLTAGE NEGATIVE SWITCHING—Uses zener to reduce supply voltage to acceptable level for LM304 IC regulator. Base drive and voltage isolation are provided by Q_2, R_{10}, and R_{11} for PIC636 hybrid power switch. Circuit operates in fixed OFF-time mode.—L. Dixon and R. Patel, Designers' Guide to: Switching Regulators, *EDN Magazine*, Oct. 20, 1974, p 53–59.

−12 V AT 300 mA FROM −48 V—Uses Fairchild μA78S40 switching regulator having variety of internal functions that can provide differing voltage step-up, step-down, and inverter modes by appropriately connecting external components. Efficiency is 86%, and output ripple is 300 mV. Extra opamp on chip is used to derive required reference voltage of −2.6 V from internal 1.3-V reference.—R. J. Apfel and D. B. Jones, Universal Switching Regulator Diversifies Power Subsystem Applications, *Computer Design*, March 1978, p 103–112.

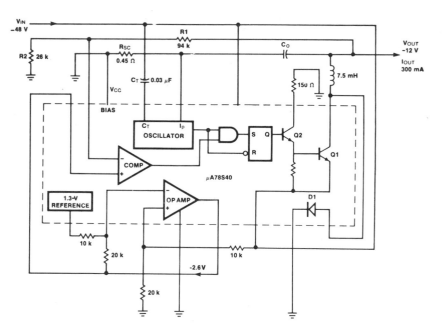

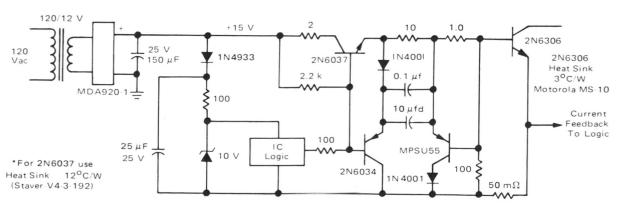

POWER SWITCH FOR SWITCHING REGULATOR—Circuit operating from 12-V step-down transformer includes push-pull driver providing interface between logic drive signal and 2N6306 high-voltage power transistor. Switching is provided at 3 A and 20 kHz, with artificial negative bias supply created from single positive supply to improve fall time. Current limiting is added to base current to limit overdrive and reduce storage time. Power switch is turned off by forcing IC to logic low. Used in 24-V 3-A switching-mode power supply operating from AC line.—R. J. Haver, "A New Approach to Switching Regulators," Motorola, Phoenix, AZ, 1975, AN-719, p 5.

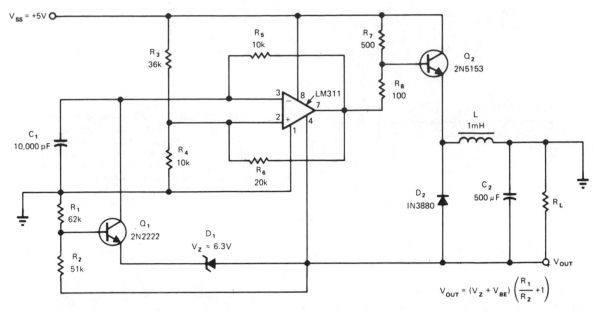

$$V_{OUT} = (V_Z + V_{BE})\left(\frac{R_1}{R_2} + 1\right)$$

+5 V TO −15 V—Use of switching regulator for voltage conversion permits generation of higher output voltage along with polarity reversal. LM311 operates as free-running MVBR with low duty cycle. Frequency is determined by C_1 and R_5 and duty cycle by divider R_3-R_4. Extra loop function performed by Q_1 and zener operating in conjunction with resistor network modifies oscillator duty cycle until desired output level is obtained. Nominal frequency is 6 kHz, duty cycle is 20% for −15 V output, and maximum load current is 200 mA. Design equations are given.—H. Mortensen, IC Comparator Converts +5 to −15V DC, *EDN Magazine*, Dec. 20, 1973, p 78–79.

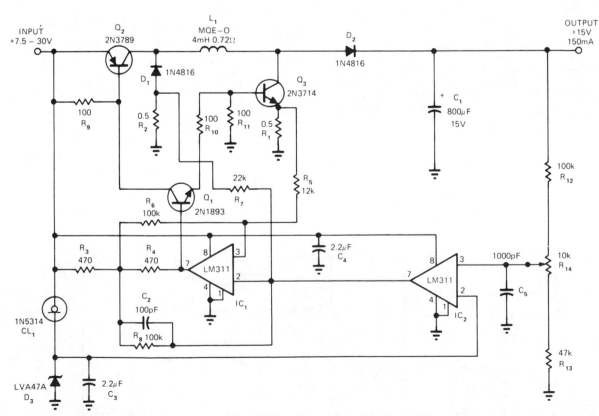

15 V FROM 7.5–30 V—Switching regulator operation is independent of input voltage level. When power is applied, Q_1 conducts and turns on Q_2 and Q_3. When linear rising current of Q_1 exceeds upper threshold as sensed by R_1, IC_1 switches to low output state and turns off all three transistors. Voltage across L_1 reverses, and current flows into C_1 through D_1 and D_2. When this current as sensed by R_2 falls below lower threshold, IC_1 switches back to its high output state. This oscillating action continues until output voltage as sensed by IC_2 rises above desired level, when IC_2 switches to its low output state and holds IC_1 low until output drops back below preset level to complete one cycle of oscillation.—A. Delagrange, Voltage Regulator Can Have Same Input and Output Level, *EDN Magazine*, Aug. 5, 1973, p 87 and 89.

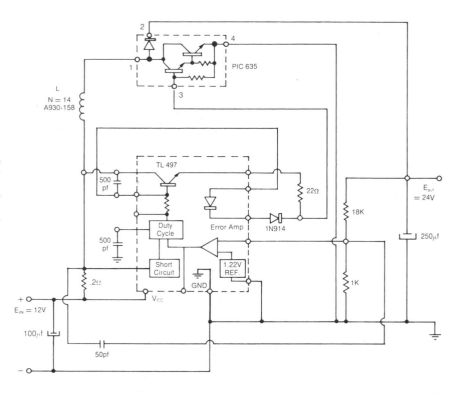

+24 V FROM +12 V AT 2 A—Combination of PIC635 boost switching regulator and TL497 control circuit accepts DC input voltage and provides regulated output voltage that must be greater than input voltage. When transistor switch is turned on, input voltage is applied across L. When transistor is turned off, energy stored in L is transferred through diode to load where it adds to energy transferred directly from input to output during diode conduction time. Output voltage is regulated by controlling duty cycle.—"Flyback and Boost Switching Regulator Design Guide," Unitrode, Watertown, MA, 1978, U-76, p 9.

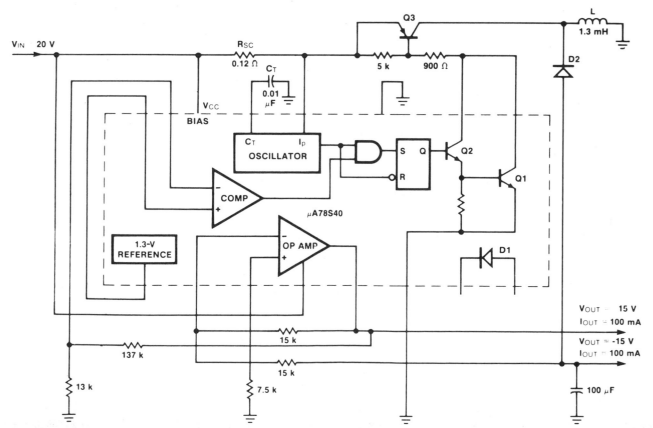

±15 V TRACKING—Dual-tracking connection for Fairchild μA78S40 switching regulator operates from single 20-V input. Efficiency is 75% for +15 V and 85% for −15 V, both at 100 mA. Output ripple is 30 mV.—R. J. Apfel and D. B. Jones, Universal Switching Regulator Diversi- fies Power Subsystem Applications, *Computer Design*, March 1978, p 103–112.

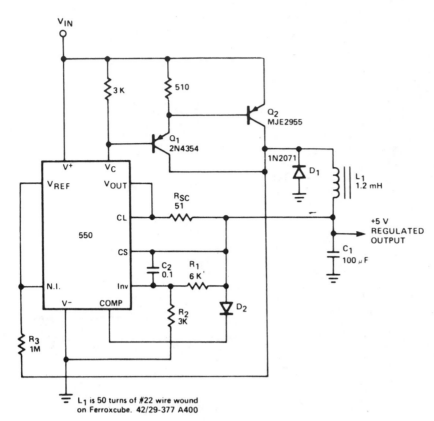

5-V SWITCHING—Darlington pair Q_1-Q_2 serves as switch for regulator using Signetics 550 as threshold detector. Design equations are given. Exact frequency of self-oscillating switching regulator depends primarily on parasitic components. If frequency is important, as in applications requiring EMI suppression, regulator may be locked to external square-wave drive signal fed to reference terminal.—"Signetics Analog Data Manual," Signetics, Sunnyvale, CA, 1977, p 661–662.

L_1 is 50 turns of #22 wire wound on Ferroxcube. 42/29-377 A400

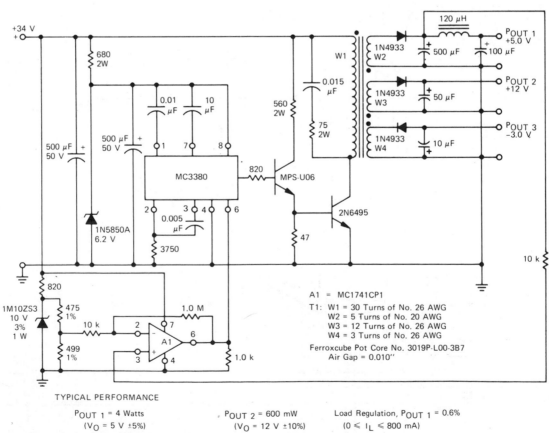

TYPICAL PERFORMANCE

$P_{OUT\ 1}$ = 4 Watts
(V_O = 5 V ±5%)
5-V Ripple Component = 50 mV
(120 Hz + 20 kHz)

$P_{OUT\ 2}$ = 600 mW
(V_O = 12 V ±10%)
$P_{OUT\ 3}$ = 3 mW
(V_O = 3 V ±10%)

Load Regulation, $P_{OUT\ 1}$ = 0.6%
(0 ≤ I_L ≤ 800 mA)

A1 = MC1741CP1
T1: W1 = 30 Turns of No. 26 AWG
W2 = 5 Turns of No. 20 AWG
W3 = 12 Turns of No. 26 AWG
W4 = 3 Turns of No. 26 AWG
Ferroxcube Pot Core No. 3019P-L00-3B7
Air Gap = 0.010"

MULTIPLE-OUTPUT SWITCHING REGULATOR—Additional outputs are obtained from switching regulator by adding secondary windings to power transformer. Motorola MC3380 astable MVBR serves as control element. Feedback is achieved by amplifying output error with opamp A1 and applying this voltage to pin 6. Report covers design of transformer and power circuit.—H. Wurzburg, "Control Your Switching Regulator with the MC3380 Astable Multivibrator," Motorola, Phoenix, AZ, 1975, EB-52.

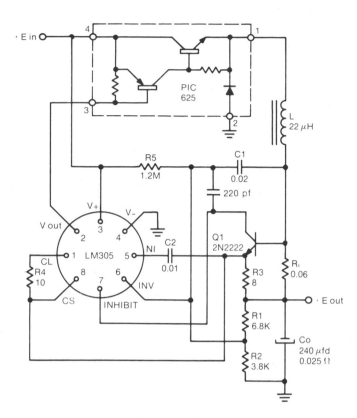

+5 V SWITCHING AT 10 A—Unitrode PIC625 hybrid power switch provides switching action for LM305 regulator at switching speeds up to 100 kHz for input voltage range of 20–40 V. Circuit operates in fixed OFF-time mode that makes output ripple independent of input voltage. Q1 provides current-limiting action.— "Switching Regulator Design Guide," Unitrode, Watertown, MA, 1974, U-68A, p 7.

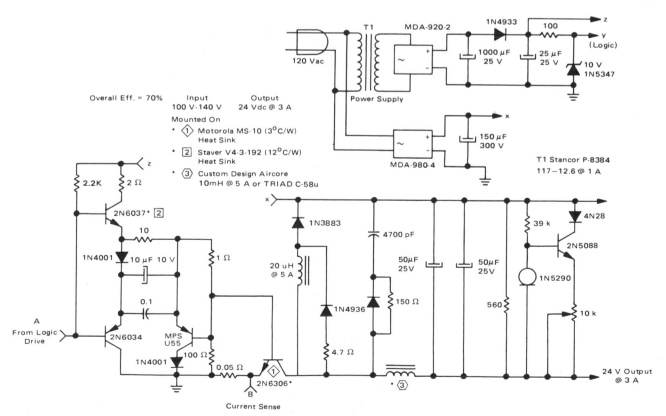

24-V 3-A SWITCHING-MODE—Circuit operates at 20 kHz from AC line with 70% efficiency. Control portion uses quad comparator and optoisolator and provides short-circuit protection.

Logic drive uses push-pull transistors to switch 2N6306 power transistor at 20-kHz rate. Load regulation is 0.8% over output range of 1.5 to 3 A with 120-VAC input. Line regulation is 3% at

3 A for input range of 100 to 140 VAC.—R. J. Haver, "A New Approach to Switching Regulators," Motorola, Phoenix, AZ, 1975, AN-719, p 11.

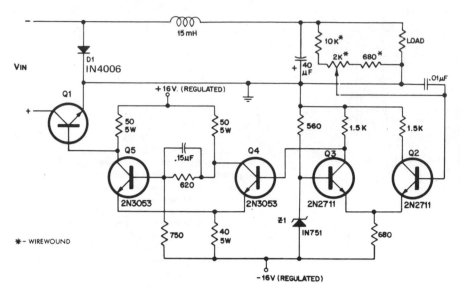

250 V AT 3 A—Single high-voltage silicon power transistor Q1 serves as series element in switching regulator, with regulation obtained by pulse-width modulation. Delco DTS-431 provides output of 250 V for maximum input of 325 V; other Delco transistors in same series give different combinations of output voltage and current in range of 300–750 W maximum output power. Efficiency is 92% at full load. Differential amplifier Q2-Q3 senses output voltage of regulator and feeds Schmitt trigger Q4-Q5 for turning series transistor Q1 on and off. Resulting square wave of voltage is smoothed by LC filter between Q1 and load.—"Pulse Width Modulated Switching Regulator," Delco, Kokomo, IN, 1972, Application Note 39, p 3.

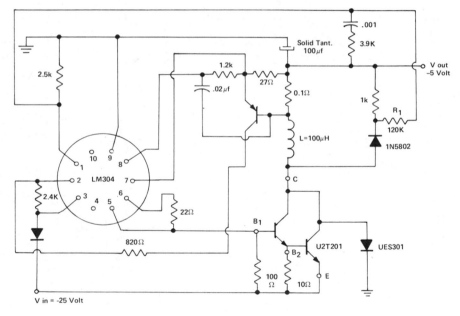

−5 V SWITCHING—Unitrode U2T201 Darlington serves as switching element for LM304 step-down switching regulator operating from input of −25 V. Operating frequency can be about 25 kHz. Darlington will handle peak currents up to 10 A.—"Designer's Guide to Power Darlingtons as Switching Devices," Unitrode, Watertown, MA, 1975, U-70, p 10.

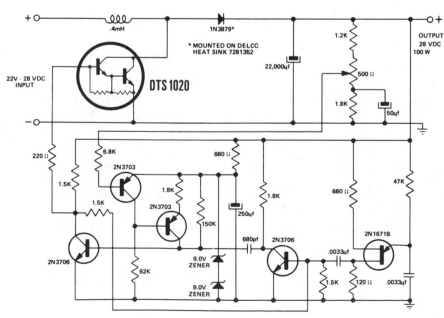

28 V AT 100 W—Circuit using Delco DTS-1020 Darlington silicon power transistor operates over input range of 22–28 V. Switching rate is 9 kHz. Efficiency is about 85% at full load. Output voltage is sensed to control pulse width of mono MVBR which is triggered at 9 kHz by oscillator.—"28 Volt Darlington Switching Regulator," Delco, Kokomo, IN, 1971, Application Note 49, p 4.

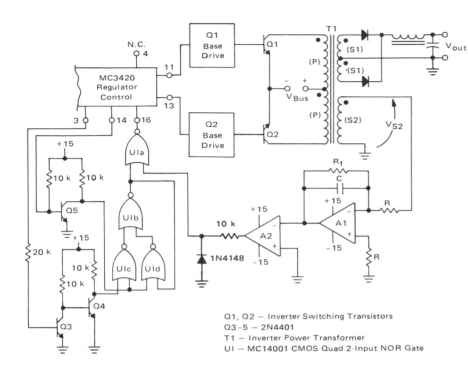

SYMMETRY CORRECTION—Low-cost external correction circuit for MC3420 switching-mode regulator ensures balanced operation of power transformer in push-pull inverter configuration. Circuit senses voltage impressed on primary of T1 through sensing secondary S2, for integration by opamp A1 so voltage on C represents volt-second product applied to T1. During conduction period of Q1, voltage on C ramps up to some positive value and output of A2 is low. Conduction period for output 2 then begins, Q2 turns on, and C ramps down to 0 V. A2 output then goes high, inhibiting output 2 and Q2. Times for C2 to charge and discharge are equal so conduction periods are equal.—H. Wurzburg, "A Symmetry Correcting Circuit for Use with the MC3420," Motorola, Phoenix, AZ, 1977, EB-66.

Q1, Q2 — Inverter Switching Transistors
Q3-5 — 2N4401
T1 — Inverter Power Transformer
UI — MC14001 CMOS Quad 2-Input NOR Gate

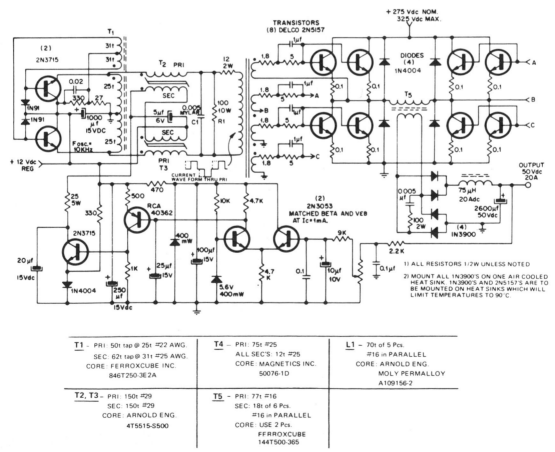

T1 —	PRI: 50t tap @ 25t #22 AWG.	T4 —	PRI: 75t #25	L1 —	70t of 5 Pcs.
	SEC: 62t tap @ 31t #25 AWG.		ALL SEC'S: 12t #25		#16 in PARALLEL
	CORE: FERROXCUBE INC.		CORE: MAGNETICS INC.		CORE: ARNOLD ENG.
	846T250-3E2A		50076-1D		MOLY PERMALLOY
					A109156-2
T2, T3 —	PRI: 150t #29	T5 —	PRI: 77t #16		
	SEC: 150t #29		SEC: 18t of 6 Pcs.		
	CORE: ARNOLD ENG.		#16 in PARALLEL		
	4T5515-S500		CORE: USE 2 Pcs.		
			FERROXCUBE		
			144T500-365		

50 V AT 1 kW—Switching regulator operating at 10 kHz uses pulse-width modulation to give 87% efficiency at full load. Input voltage is 275 VDC. Inverter output drives combination of eight Delco 2N5157 power transistors connected in paralleled pairs in each leg of bridge circuit. Clamp diodes in each bridge leg prevent reverse conduction through collector-base diodes of transistors. Regulator consists of differential amplifier and two-stage DC amplifier controlling direct current through windings of magnetic amplifier.—"One Kilowatt Regulated Power Converter with the 2N5157 Silicon Power Transistor," Delco, Kokomo, IN, 1972, Application Note 44, p 3.

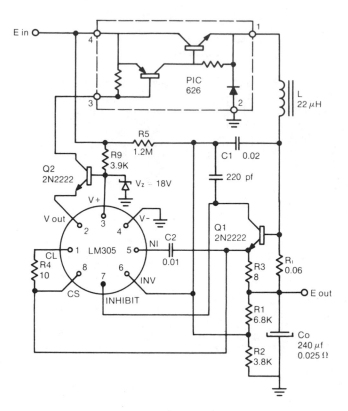

HIGH-VOLTAGE POSITIVE-SWITCHING—Designed for operation from supply voltages above 40-V maximum rating of LM305 regulator. Output is +5 V at up to 10 A. Circuit uses fraction of input voltage as determined by R9 and zener, with Q2 providing voltage isolation between regulator and Unitrode PIC626 hybrid power switch.—"Switching Regulator Design Guide," Unitrode, Watertown, MA, 1974, U-68A, p 9.

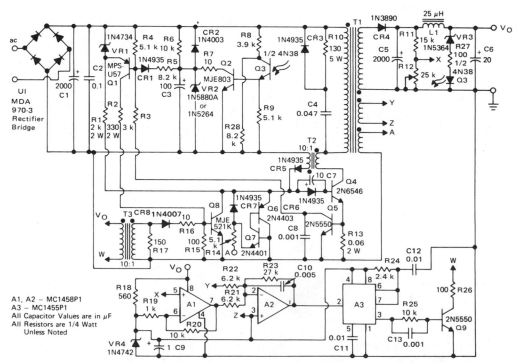

24 V AT 3 A FOR CATV—Switching regulator design meets requirements for cable television systems where small size, low weight, and high efficiency are prime considerations. Circuit operates above 18 kHz either from 40–60 V 60-Hz square-wave source (CATV power line from ferroresonant transformer) or from DC standby source. Control circuit consists of dual opamp and linear IC timer used to vary ON time of 2N6546 power transistor. At start-up, Q4 is saturated and full input voltage is applied to primary of power transformer T1. Current then ramps up linearly until Q4 is switched off by opamps A1 and A2 and timer A3. Power transistor is operated between saturation and OFF state at above 18 kHz, with ON time varied while OFF time is fixed, to maintain constant output voltage as sensed by A1.—J. Nappe and N. Wellenstein, "An 80-Watt Switching Regulator for CATV and Industrial Applications," Motorola, Phoenix, AZ, 1975, AN-752, p 5.

CHAPTER **89**

Tape Recorder Circuits

Includes interface circuits for recording and playback of instrumentation and microprocessor data signals, Morse code, and RTTY signals on inexpensive cassette deck, along with NAB-equalized preamps, erase/bias oscillator, AVC, dynamic range expansion, and VOX circuits for all types of mono and stereo tape recorders. Interface for keying CW transmitter with taped message is also given.

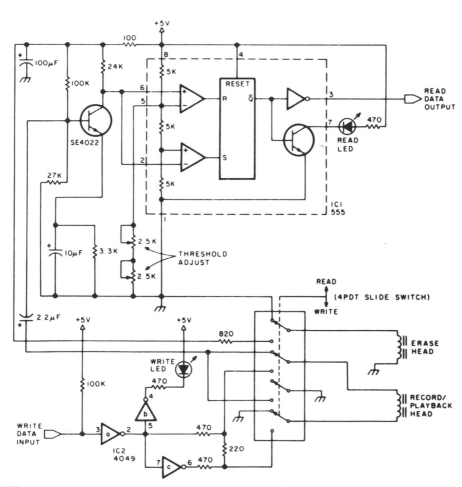

DIGITAL RECORDING WITH CASSETTES—Circuit shows modifications required for standard cassette recorder to bring read level up to about 1 V. Recorder works well over range of 100 to 1200 b/s. During write process, direct current is passed through record head to saturate tape, with polarity depending on direction of current. During read cycle, voltage is induced in head winding only when transition between oppositely polarized zones moves past head. 555 timer is used as combination level detector and flip-flop to recover serial data.—R. W. Burhans, A Simpler Digital Cassette Tape Interface, *BYTE*, Oct. 1978, p 142–143.

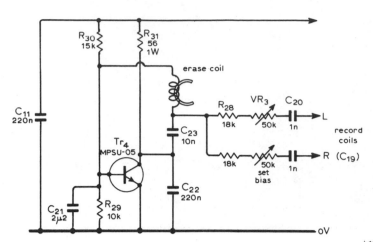

ERASE/BIAS OSCILLATOR—Used in high-quality stereo cassette deck operating from AC line or battery. Provides up to 33 VRMS at 50-kHz erase frequency, as required for completely erasing existing recording on tape when recording over it. Supply voltage should be in range of 12–14 V. Article gives all other circuits of cassette deck and describes operation in detail.— J. L. Linsley Hood, Low-Noise, Low-Cost Cassette Deck, *Wireless World*, Part 1—May 1976, p 36–40 (Part 2—June 1976, p 62–66; Part 3— Aug. 1976, p 55–56).

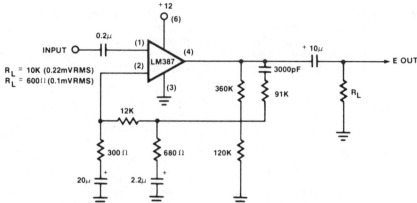

CASSETTE PREAMP—Provides gain of 81 dB and 0.22 mVRMS for 10K load. Gain drops to about 78 dB and output is 0.1 mVRMS for 600-ohm load. Gain values are for 100 Hz, with gain dropping above and below this value.—"Signetics Analog Data Manual," Signetics, Sunnyvale, CA, 1977, p 782.

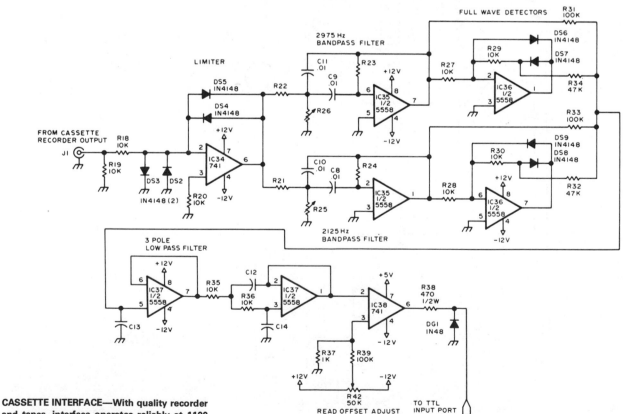

CASSETTE INTERFACE—With quality recorder and tapes, interface operates reliably at 1100 bauds, for loading 24K microprocessor system in 246 s. Cassette output is amplified and clipped by limiting amplifier IC34. Bandpass filters followed by full-wave detectors respond to 2125-Hz mark and 2975-Hz space frequencies and feed their outputs to summing junction at pin 5 of three-pole active low-pass filter IC37. 2975-Hz tones are rectified to positive voltage and 2125-Hz tones to negative voltage, with amplitudes varying from maximum at exact frequencies to sum voltage of 0 V at midfrequency of 2550 Hz. Output opamp IC38 delivers correct TTL level for reading by single-bit input port.— R. Suding, Why Wait? Build a Fast Cassette Interface, *BYTE*, July 1976, p 46–53.

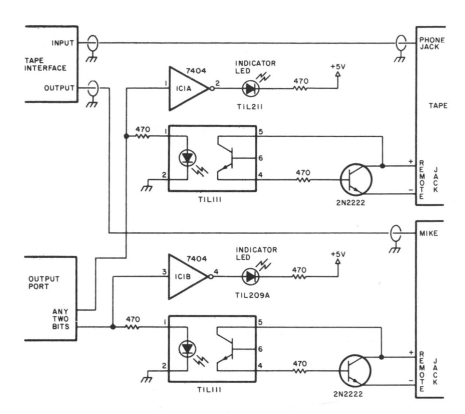

CASSETTE FILE UPDATE—Interface circuit controls two tape decks for updating mailing lists or other sequential files stored on magnetic tape in audio cassettes. Two cassette tape recorders are required, one for input (reading files) and one for output (writing files). Microprocessor tape input and output circuits are connected to appropriate tape unit as shown. Only one cassette operates at any given time. Optocouplers prevent polarity or voltage problems between tape motor and microprocessor. Tape functions are under software control. Software delay of about 1 s allows tape motor to come up to speed before recording starts. Records are in numerical or other logical sequence, so updating requires only one pass. On update, old cassette file is read into microprocessor for deletion, change, or addition of data, and corrected data is written on new cassette. Article covers use for maintaining Christmas card and other mailing lists, payroll records sequenced by Social Security number, and other sequential lists.—W. D. Smith, Fundamentals of Sequential File Processing, *BYTE*, Oct. 1977, p 114–116, 118, 120, 122, 124, and 126–127.

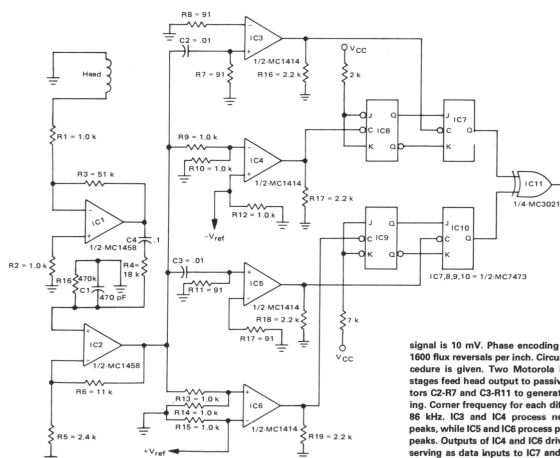

CASSETTE DATA READOUT—Uses separate circuits having threshold provisions for both positive and negative peaks, for reading data stored on cassette tape at 15 in/s. Head output signal is 10 mV. Phase encoding is used with 1600 flux reversals per inch. Circuit design procedure is given. Two Motorola MC1458 gain stages feed head output to passive differentiators C2-R7 and C3-R11 to generate zero crossing. Corner frequency for each differentiator is 86 kHz. IC3 and IC4 process negative-going peaks, while IC5 and IC6 process positive-going peaks. Outputs of IC4 and IC6 drive T flip-flops serving as data inputs to IC7 and IC10.—"The Recovery of Recorded Digital Information in Drum, Disk and Tape Systems," Motorola, Phoenix, AZ, 1974, AN-711, p 9.

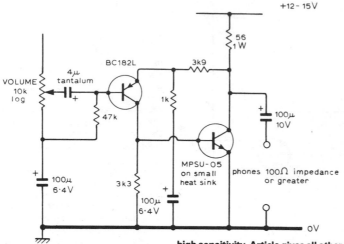

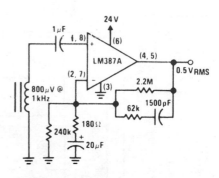

NAB PLAYBACK PREAMP—Provides standard NAB equalization for tape player requiring 0.5 VRMS from head having sensitivity of 800 μV at 1 kHz, with operating speed of 3¾ in/s. Design procedure is given. Voltage gain at 1 kHz is 56 dB.—"Audio Handbook," National Semiconductor, Santa Clara, CA, 1977, p 2-31–2-37.

HEADPHONE AMPLIFIER—Used in high-quality stereo cassette deck operating from AC line or battery. Provides gain of 5 in class A, for use with low-sensitivity headphones or low-impedance headphones down to 100 ohms. Replay amplifier output alone is adequate for headphones having 2000-ohm load impedance or high sensitivity. Article gives all other circuits of cassette deck and describes operation in detail. Input to volume control is taken from output of opamp in replay amplifier, nominally about +5 V.—J. L. Linsley Hood, Low-Noise, Low-Cost Cassette Deck, *Wireless World*, Part 2—June 1976, p 62–66 (Part 1—May 1976, p 36–40; Part 3—Aug. 1976, p 55–56).

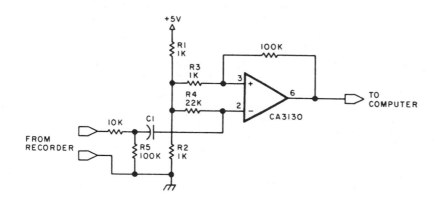

CASSETTE INTERFACE—Used between recorder and computer for loading data stored in tape cassette. Single divider network R1-R2 drives both opamp inputs and provides stabilized sensitivity. R3 isolates inputs.—B. E. Rehm, The TDL System Monitor Board, *BYTE*, April 1978, p 10, 12–14, and 16.

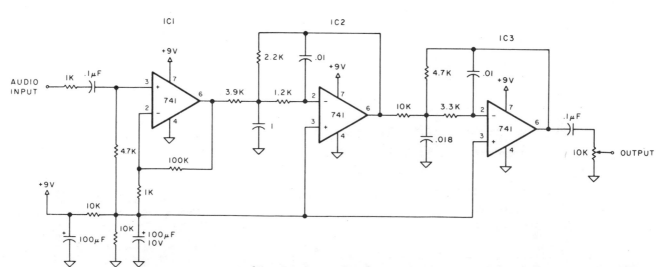

COPYING CASSETTE PROGRAMS—Controller serves for making duplicate copies of microprocessor programs recorded on magnetic tape, for insurance against accidental damage to master cassette during use. Used between audio output of cassette player and audio input of tape recorder. Opamp IC1 with gain of 100 overloads so output is constant-amplitude square wave regardless of input level from tape being copied. If program uses audio tones for digital data, eight cycles of 2400 Hz represents digital 1 and four cycles of 1200 Hz represents digital 0. Additional opamps act as four-pole Butterworth filter rejecting signals above 3000 Hz. 10K pot is adjusted so output level matches requirements of recorder.—P. A. Stark, Copying Computer Cassettes, *Kilobaud*, Aug. 1978, p 94–96.

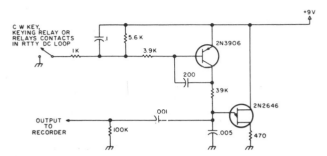

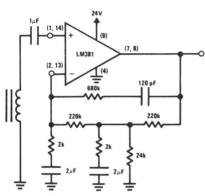

CW AND RTTY ON CASSETTES—Circuit provides conditioning of routine CW calls or RTTY test messages, as required for recording on endless-loop cassette. Keyed signal is filtered to remove contact bounce, then used to turn on 2N3906 which gates 2N2646 sawtooth oscillator operating at about 5 kHz when using 0.005-μF gate capacitor; for lower frequency, increase capacitor to 0.01 μF.—Cassette-Aided CW and RTTY, *73 Magazine*, Sept. 1977, p 122–123.

FAST TURN-ON PLAYBACK PREAMP—Turn-on for gain and supply voltage is only 0.1 s, as compared to 5 s normally required in preamp providing NAB tape playback response.—"Audio Handbook," National Semiconductor, Santa Clara, CA, 1977, p 2-31–2-37.

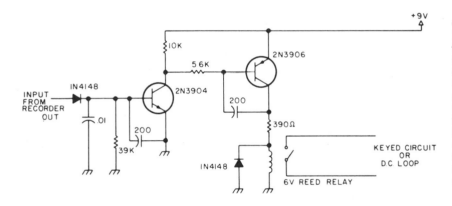

CASSETTE PLAYBACK OF CW AND RTTY—Playback-signal conditioning circuit is used between tape recorder and transmitter when routine CW calls or RTTY test messages are recorded on endless-loop cassette recorder. Recorded tone is rectified by 1N4148 and applied to RC timing circuit. Decay voltage developed across network when tone is removed turns on 2N3904 and 2N3906 stages. Output of 2N3906 drives reed relay in transmitter keying circuit. If resistor is used in place of relay, drop across it during key-down period can be used to drive electronic keyer.—Cassette-Aided CW and RTTY, *73 Magazine*, Sept. 1977, p 122–123.

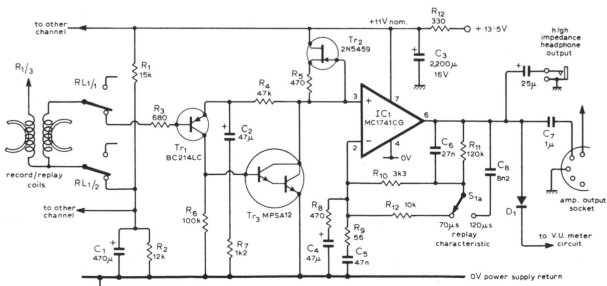

REPLAY AMPLIFIER—Used in high-quality stereo cassette deck operating from AC line or battery. Amplifier design is optimized for minimum noise voltage by using PNP silicon input transistor operated with lowest possible collector current (10 μA for Texas Instruments transistor specified). Motorola IC in second stage, similar to 741 but having 8-pin metal-can encapsulation, provides equalization required for replay. Output of amplifier is about 0.4 VRMS.

Article gives all other circuits of cassette deck and describes operation in detail.—J. L. Linsley Hood, Low-Noise, Low-Cost Cassette Deck, *Wireless World*, Part 1—May 1976, p 36–40 (Part 2—June 1976, p 62–66; Part 3—Aug. 1976, p 55–56).

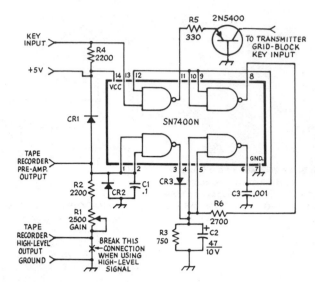

KEYING FROM TAPE—Simple envelope detector and wave-shaping circuit uses quad NAND gate for instant replay of recorded CW transmissions through transmitter. Diodes can be 1N270 or any other small-signal switching or general-purpose types. R3, C2, and CR3 provide envelope detection of amplified and clipped audio input from tape recorder.—A. H. Kilpatrick, Keying a Transmitter with a Tape Recorder, *QST,* Jan. 1974, p 45.

DIGITAL CASSETTE HEAD DRIVE—Provides saturation recording as required for digital data. Back-to-back zeners provide bipolar limiting at ±10 V. TTL-level inputs are applied to write data input, inverted by 7404, and fed to inverting input of opamp. Noninverting opamp input is referenced to +1.4 V so output will switch polarities when TTL level of input changes.—I. Rampil and J. Breimeir, The Digital Cassette Subsystem: Digital Recording Background and Head Interface Electronics, *BYTE,* Feb. 1977, p 24–31.

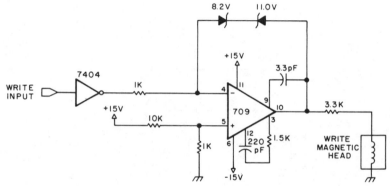

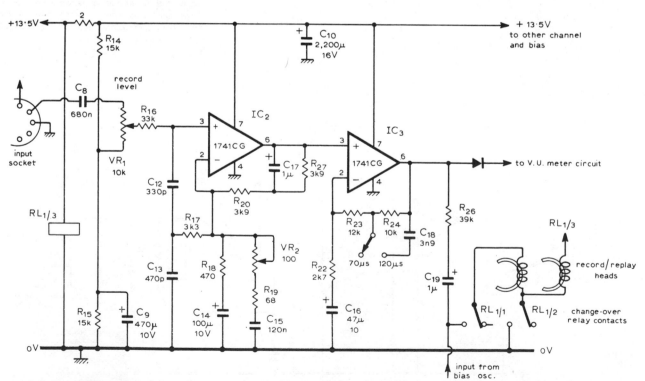

RECORDING AMPLIFIER—Used in high-quality stereo cassette deck operating from AC line or battery. Uses active RC circuit R_{16}-R_{17}-C_{12}-C_{13}-R_{19}-VR_2-C_{15} to provide required high-frequency recording characteristic for use with Garrard CT4 recording head; component values may have to be changed for other heads. C_{18} (3.9 nF) is switched in to change from basic 70-μs recording characteristic to 120 μs. C_{17} and R_{27} provide new cassette-standard bass preemphasis at 3,180 μs. Recording level is chosen as 0 VU at 660 Hz. Output feeds VU meter through silicon diode. Article gives all other circuits of cassette deck and describes operation in detail.—J. L. Linsley Hood, Low-Noise, Low-Cost Cassette Deck, *Wireless World,* Part 1—May 1976, p 36–40 (Part 2—June 1976, p 62–66; Part 3—Aug. 1976, p 55–56).

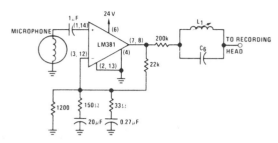

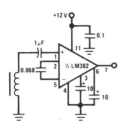

RECORDING AMPLIFIER—Designed for use with microphone having 10-mV peak output and recording head requiring 30-μA AC drive current. Output swing is 6 VRMS. High-frequency cutoff is 16 kHz, with circuit designed for slope of 6 dB per octave between 4 kHz and 16 kHz to compensate for falling frequency response of recording head starting at 4 kHz.— "Audio Handbook," National Semiconductor, Santa Clara, CA, 1977, p 2-31–2-37.

PLAYBACK PREAMP—Circuit is optimized for automotive use at supply of 10–15 V. Wideband 0-dB reference gain is 46 dB. NAB equalization is included. Tape speeds can be 1⅞ or 3¾ in/s.— "Audio Handbook," National Semiconductor, Santa Clara, CA, 1977, p 2-31–2-37.

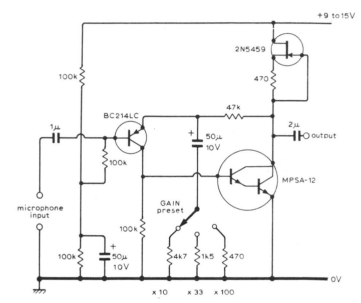

MICROPHONE PREAMP—Used in high-quality stereo cassette deck operating from AC line or battery. Provides three preset gain positions (10, 33, and 100) to meet amplification requirements of practically all types of microphones used with tape recorders. Recording input of cassette deck provides only enough gain for recording from audio amplifier or radio tuner delivering 50–100 mV at fairly low impedance, hence is not suitable for microphone input. Article gives all other circuits of cassette deck and describes operation in detail.—J. L. Linsley Hood, Low-Noise, Low-Cost Cassette Deck, *Wireless World*, Part 2—June 1976, p 62–66 (Part 1—May 1976, p 36–40; Part 3—Aug. 1976, p 55–56).

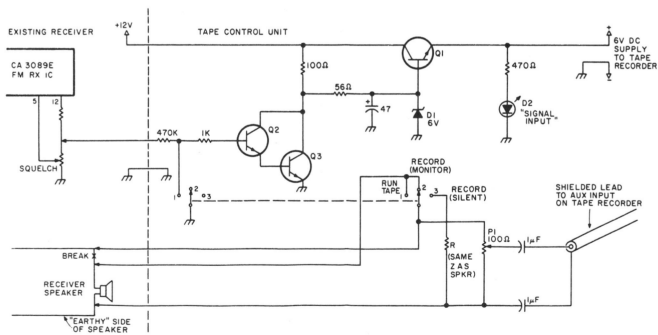

MESSAGE-CONTROLLED RECORDER—Circuit turns on tape recorder whenever input signal is present in receiver, and turns off recorder when signal goes off. Applications include monitoring local FM repeater for daily usage to obtain call signs of users, or unattended recording of messages left by other amateur stations. Uses cheap cassette tape recorder with autostop, operating at 6 V obtained from 12-V receiver supply by series regulator Q1 and zener D1. Connection to mute or squelch circuit of receiver is shown for set having CA3089E in IF tail end. Darlington pair Q2-Q3 effectively removes base supply for Q1 to turn recorder off. LED comes on when recorder is on. Q1 is NPN power transistor, while Q2 and Q3 are small-signal NPN transistors.—F. Johnson, Automatic Taping Unit, *73 Magazine*, May 1977, p 98–99.

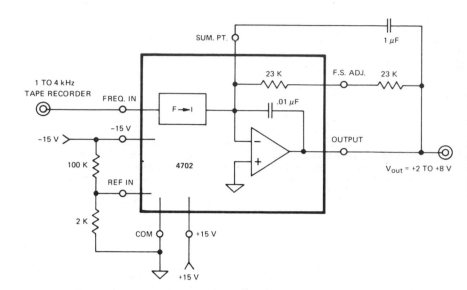

PLAYBACK OF PULSE TRAINS—Teledyne Philbrick 4702 frequency-to-voltage converter circuit provides ripple filter required for converting recorded square waves in frequency range of 0.5 to 5 kHz to desired analog output in range of 2 to 8 VDC. Report covers problems of recording and playing back pulse trains.—"V-F's, F-V's, and Audio Tape Recorders," Teledyne Philbrick, Dedham, MA, 1974, AN-11.

STEREO TAPE PLAYBACK—Single Sprague ULN-2126A IC provides preamplification for two channels along with 2-W output power for driving stereo power amplifier. Values shown give equalization required for tape playback. Single ganged tone control serves for both treble and bass adjustment.—E. M. Noll, "Linear IC Principles, Experiments, and Projects," Howard W. Sams, Indianapolis, IN, 1974, p 237 and 243.

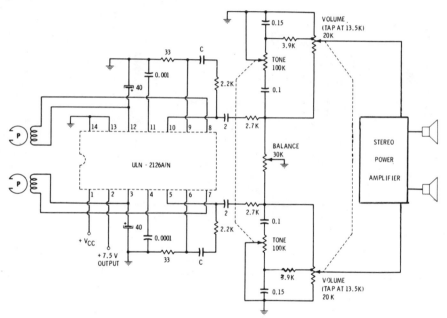

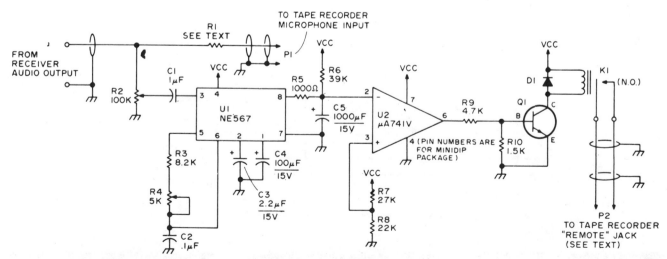

UNATTENDED RECORDER—Uses 567 tone decoder in circuit designed to respond to 1-kHz tone, to turn on recorder for taking message when receiver of amateur station is unattended. R6, C5, and 741 opamp U2 form timer that turns on RS-267-2016 transistor and pulls in relay to turn on tape recorder for recording about 30-s message. Relay then drops out. Use well-regulated 5-V supply. All transmitters using this service must have 1-kHz audio encoders for producing required control frequency. Article gives construction and adjustment details.—R. Perlman, The F.M. "Auto-Start," *73 Magazine,* April 1974, p 21 and 23–24.

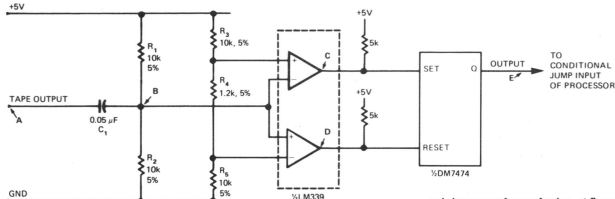

INTERFACE FOR AUDIO CASSETTES—Permits use of ordinary home cassette recorder to provide high-speed loading of assembler and source program into microprocessor. Data is recorded by using variation of phase encoding, which provides self-clocking and is independent of tape speed variation. Effective I/O rate is about 500 b/s or 5 times that of low-speed paper-tape punch or reader. Article covers phase-encoding procedure, gives flowchart, and shows waveforms of pulses at five points in circuit. Parity bits provide error correction and detection, using Hamming code.—S. Kim, An Inexpensive Audio Cassette Recorder Interface for μP's, *EDN Magazine*, March 5, 1976, p 83–86.

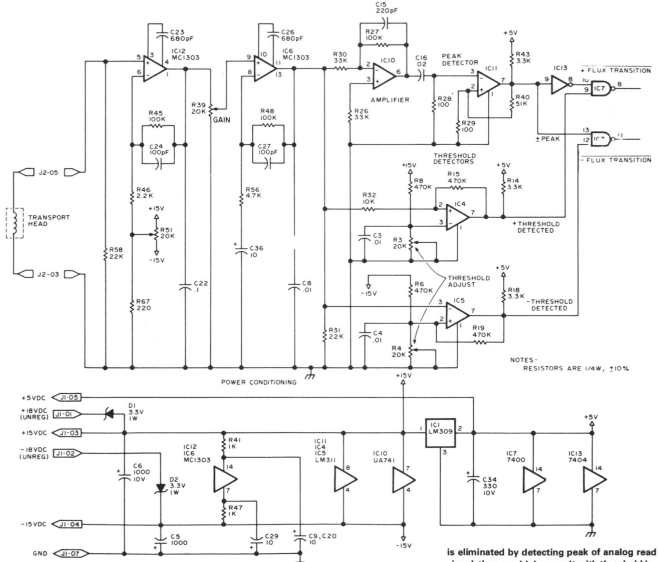

DIGITAL CASSETTE READ AMPLIFIER—Signal from magnetic head of digital tape cassette is amplified by two-stage MC1303 amplifier providing analog output of about 4 V P-P to μA741 opamp IC10 of LM311 peak detector IC11. Signal also goes to LM311 positive and negative threshold detectors IC4 and IC5, which give logic-level output. When input signal is below preset reference level, output of positive threshold detector is low; above reference, output is high. Negative threshold detector operates similarly for negative pulses. Time jitter in outputs is eliminated by detecting peak of analog read signal, then combining result with threshold information in peak detector. Circuit is used in Phi-Deck cassette system made by Economy Company, Oklahoma City.—I. Rampil and J. Breimeir, The Digital Cassette Subsystem: Digital Recording Background and Head Interface Electronics, *BYTE*, Feb. 1977, p 24–31.

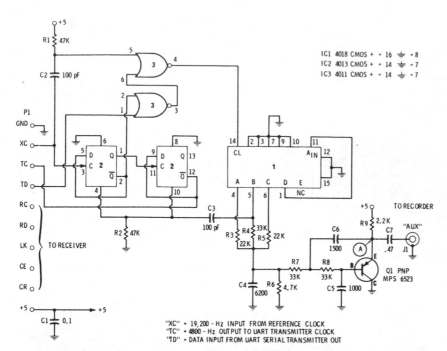

IC1 4018 CMOS + = 16 ⏚ = 8
IC2 4013 CMOS + = 14 ⏚ = 7
IC3 4011 CMOS + = 14 ⏚ = 7

"XC" = 19,200-Hz INPUT FROM REFERENCE CLOCK
"TC" = 4800-Hz OUTPUT TO UART TRANSMITTER CLOCK
"TD" = DATA INPUT FROM UART SERIAL TRANSMITTER OUT

FOUR-SPEED PLAYBACK PREAMP—Provides 0-dB reference gain of 34 dB. Supply can be ±4.5 to ±15 V. Values shown are for NAB equalization and 1⅞ or 3¾ in/s; for 7½ and 15 in/s, change values as indicated. Design equations are given.—"Audio Handbook," National Semiconductor, Santa Clara, CA, 1977, p 2-31–2-37.

300-BAUD BIT BOFFER TRANSMITTER—Permits recording of serial data on ordinary low-cost cassette tape recorders for bulk storage of data to be used later in microprocessor. Requires 19,200-Hz reference input to terminal XC from external clock. Feedback from sine-wave synthesizer IC1 to divide-by-4 counter IC2 automatically synchronizes system so sine waves automatically switch just before zero crossing each time serial data changes from 1 to 0 or back again. Output consists of 16 half sine waves at 2400 Hz for mark or digital 1 and 8 half sine waves at 1200 Hz for space or digital 0, for feed to input of cassette recorder. Circuit eliminates errors commonly encountered when attempting to record square waves on tape with low-cost recorder.—D. Lancaster, "TV Typewriter Cookbook," Howard W. Sams, Indianapolis, IN, 1976, p 167–171.

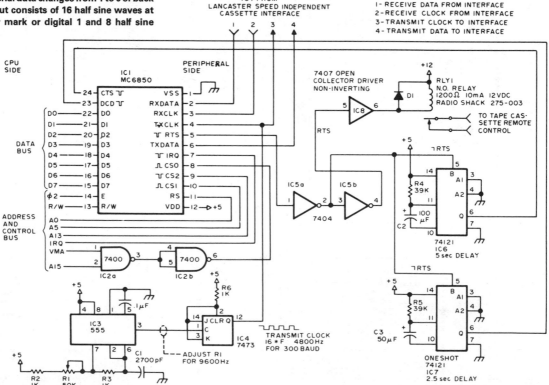

CASSETTE INTERFACE WITH ACIA—Permits use of audio pickup for mass storage in Motorola 6800 microcomputer system. Uses Motorola MC6850 asynchronous communication interface adapter (ACIA), which is specialized version of UART. All control, status, and data transfers in ACIA are made over single 8-bit bidirectional bus. Request-to-send line (RTS) controls tape recorder motor. When RTS is set high, input to IC8 is high and relay coil is not energized. IC6 gives 5-s delay following motor turn-on so long leader will be recorded at mark frequency. IC7 gives delay so reading starts 2.5 s before first data byte. Article covers circuit operation in detail and gives operating subroutines.—J. Hemenway, The Compleat Tape Cassette Interface, *BYTE*, March 1976, p 10–16.

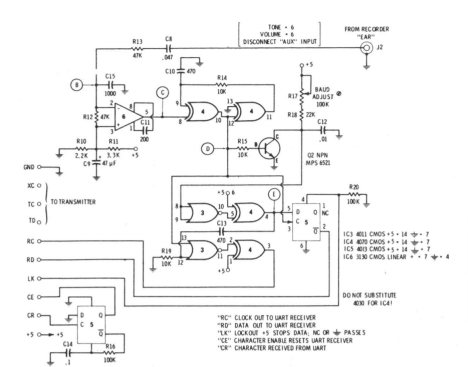

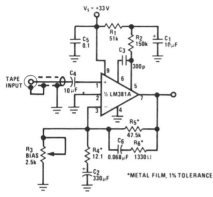

ULTRALOW-NOISE PLAYBACK PREAMP—Provides optimum noise performance at popular tape speeds of 1⅞ and 3¾ in/s. Reference gain for 0 dB is 41 dB, giving output level of 200 mV from head output of 1 mV at 1 kHz. Single-ended biasing and use of metal-film resistors reduce noise.—"Audio Handbook," National Semiconductor, Santa Clara, CA, 1977, p 2-31–2-37.

300-BAUD BIT BOFFER RECEIVER—Used with ordinary cassette recorder to convert half sine waves of recorded serial data to corresponding digital 1s and 0s. Output of recorder passes through filter and limiter IC6 to give square wave at point C whose zero crossings correspond to recorded sine wave. Leading and trailing edges of square wave are converted to narrow positive pulses by EXCLUSIVE-OR gate IC4 to give stream of pulses at D, one for each zero crossing. Transistor circuit forms retriggerable mono that is adjusted so point E goes positive three-fourths of way through low-frequency (1200 Hz) half-cycle. Point E then has stream of eight pulses for 0 and no pulses for 1. Final flip-flop provides recovery of data as 1s and 0s. Leading edge of waveform at D is shortened and combined with clock pulses to provide composite UART clock output. Boffer system eliminates errors commonly encountered when attempting to record square waves on tape with low-cost recorder.—D. Lancaster, "TV Typewriter Cookbook," Howard W. Sams, Indianapolis, IN, 1976, p 167–171.

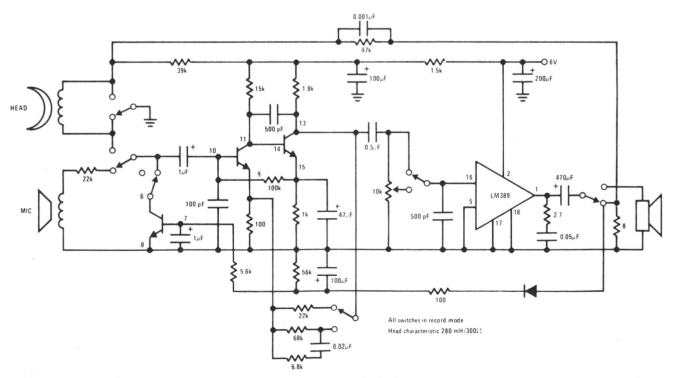

CASSETTE RECORD/PLAYBACK—National LM389 power amplifier chip includes three NPN transistors, to provide all circuits needed for complete recording and playback of cassette tapes. Two of internal transistors act as signal amplifiers while third is used for automatic level control when recording. Diode is also on chip.—"Audio Handbook," National Semiconductor, Santa Clara, CA, 1977, p 4-33–4-37.

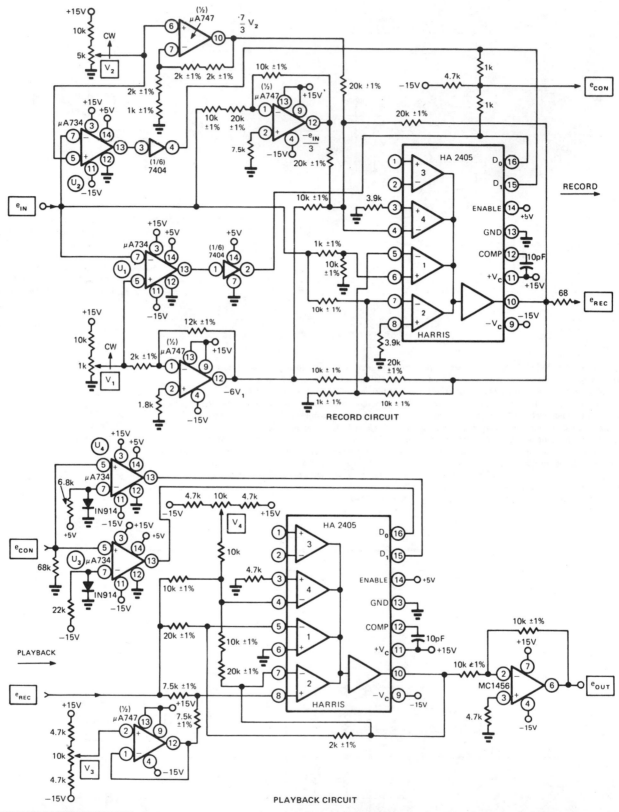

RECORD CIRCUIT

PLAYBACK CIRCUIT

AUTOMATIC RANGE EXPANSION—Instrumentation tape recorder technique folds recorded signal over and reuses same VCO range three times, at three different gains, for increasing dynamic recording range to over 10,000. Two comparators select one of three amplifier gains according to level of input signal and record selected gain on separate control track. During playback, control track signal e_{CON} is used to select corresponding inverse gain for unfolding recorded signal. Level of input signal e_{IN}, in range of 0–10 V, is sensed by comparators whose preset thresholds are determined by pots V_1 and V_2. If input is less than V_1, both comparator outputs are low and section 1 of HA2405 four-channel opamp is selected for recording at 10 times input. If input is greater than V_1 and less than V_2, section 2 having gain of −2 is selected so direction of e_{REC} is reversed. If e_{IN} is greater than V_2, both comparators are high and section 4 is selected for gain of +1/3, so e_{REC} again reverses to cross VCO range for third time. Outputs of comparators are summed to form three-level signal for recording on control track.—J. R. White, Comparator Technique Expands Tape Recorder's Range, *EDN Magazine*, April 5, 1975, p 111, 113, and 115.

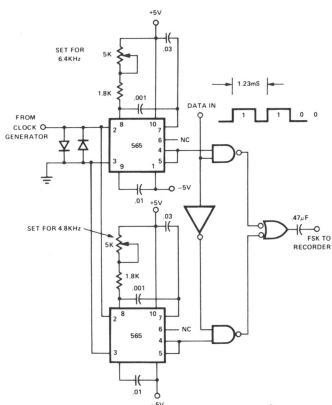

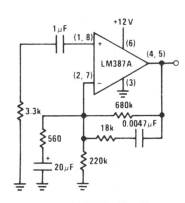

12-V PLAYBACK PREAMP—Provides standard NAB equalization. Gain is decreased gradually from 60 dB at 20 Hz to 32 dB at 20 kHz in accordance with NAB playback curve. Playback head is represented by 3.3K resistor.—"Audio Handbook," National Semiconductor, Santa Clara, CA, 1977, p 2-31–2-37.

FSK GENERATOR FOR CASSETTE DATA—Uses two 565 PLL ICs, locked to 800-Hz system clock but oscillating at 6.4 kHz and 4.8 kHz, to provide FSK signals for recording digital data on ordinary cassette tape. Harmonic suppression of square-wave output is taken care of automatically by high-frequency rolloff characteristic of tape recorder. Incoming data determines which oscillator feeds its signal to recorder.—"Signetics Analog Data Manual," Signetics, Sunnyvale, CA, 1977, p 859–860.

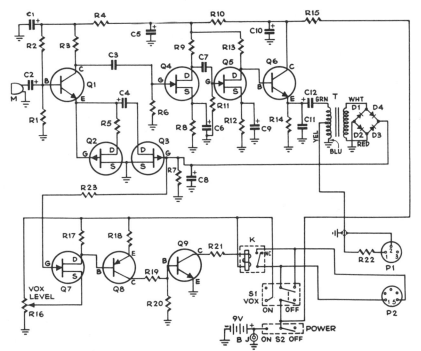

C1, 4, 8, 10, 12—50mfd electrolytic capacitor
C2—2.2mfd electrolytic capacitor
C3, 7, 11—0.1mfd ceramic capacitor
C5—220mfd electrolytic capacitor
C6, 9—100mfd electrolytic capacitor
R1—120,000-ohm, ¼ w resistor
R2—56,000-ohm, ¼ w resistor
R3, 9, 13—10,000-ohm, ¼ w resistor
R4, 10, 14—1000-ohm, ¼ w resistor
R5—15,000-ohm, ¼ w resistor
R6, 11—1 megohm, ¼ w resistor
R7—120,000-ohm, ¼ w resistor
R8, 12—2700-ohm, ¼ w resistor
R15—56-ohm, ¼ w resistor
R16—50,000-ohm, miniature potentiometer
R17—6800-ohm, ¼ w resistor
R18—1500-ohm, ¼ w resistor
R19—150-ohm, ¼ w resistor
R20—2200-ohm, ¼ w resistor
R21—330-ohm, ¼ w resistor
R22, 23—560,000-ohm, ¼ w resistor
D1, 2, 3, 4—diodes 1N266 or equiv.
Q1-Q6—NPN transistor Motorola HEP 50
Q2, 3—P-channel FET 2N3820
Q4, 5, 7—N-channel FET Motorola HEP 801
Q8—PNP transistor Motorola HEP 52
Q9—NPN transistor Motorola HEP 53
T—output transformer 1K-200K Radio Shack 273-1376
K—miniature relay Radio Shack 275-004

AVC AND VOX—Voice-operated ON/OFF switch uses microphone to sense normal background sound. Anything above background threshold preset by R16 energizes relay K for turning on recorder. Circuit provides about 2-s delay after subject stops talking, before releasing relay. Automatic volume control circuit keeps recorded signal essentially constant despite movements of loudspeaker toward or away from microphone.—G. Beard, Automatic Volume and VOX for Your Tape Recorder, *Popular Science*, Oct. 1973, p 134 and 136.

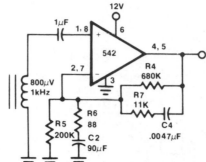

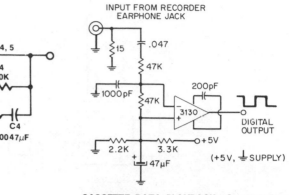

NAB TAPE PREAMP—One section of Signetics NE542 dual noise amplifier is used to provide 100-mV output level at 1 kHz following NAB equalization curve for tape speed of 7½ in/s.— "Signetics Analog Data Manual," Signetics, Sunnyvale, CA, 1977, p 780.

CASSETTE DATA PLAYBACK—Converts low-level digital signals from cassette recorder into CMOS-compatible 5-V square waves. Both inputs of 3130 opamp are biased to +2 V for use as open-loop comparator. RC input filter minimizes hum and bias interference.—D. Lancaster, "CMOS Cookbook," Howard W. Sams, Indianapolis, IN, 1977, p 345.

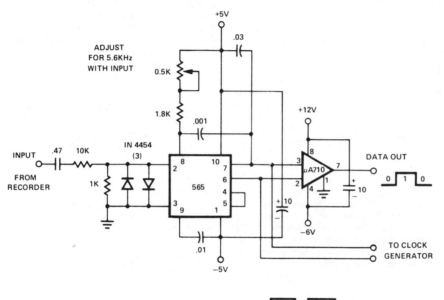

FSK DETECTOR FOR CASSETTE-RECORDED DATA—Connection shown for 565 PLL provides data output of 1 for 6.4 kHz and 0 for 4.8 kHz from ordinary cassette tape recorder having frequency response to 7 kHz. Report gives circuit of suitable recorder using return-to-zero FSK. System also requires 800-Hz clock generator for synchronizing to data. Up to seven 0s can occur in succession without making clock go out of sync. Odd parity is used.—"Signetics Analog Data Manual," Signetics, Sunnyvale, CA, 1977, p 857–859.

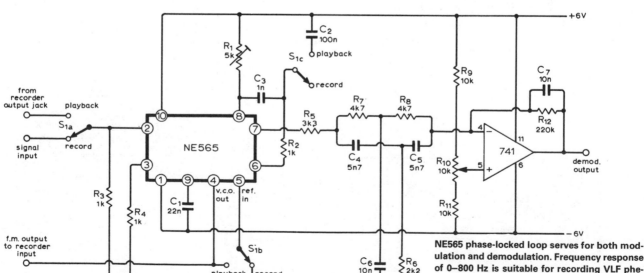

FM FOR INSTRUMENTATION—Frequency modulator-demodulator circuit using single IC and opamp converts ordinary low-cost tape recorder into instrumentation recorder. Signetics NE565 phase-locked loop serves for both modulation and demodulation. Frequency response of 0–800 Hz is suitable for recording VLF phenomena at tape speed of 9.1 cm/s (3.5 in/s). Carrier frequency is in midband, at 3 kHz. Article covers circuit operation in detail and gives design equations.—B. D. Jordan. Simple F.M. Modulator/Demodulator for a Magnetic Tape Recorder, *Wireless World*, March 1974, p 29–30.

CHAPTER 90

Telephone Circuits

Includes coders and decoders for standard Touch-Tone pairs of frequencies and for single-tone remote ON/OFF control, along with repeater autopatch circuits, Touch-Tone to dial converter, phone-call counter, ring detector, ring simulator, and busy-signal generator. See also Repeater chapter.

DIAL-TONE GENERATOR—Simultaneous pairs of Touch-Tone frequencies used by telephone company are generated by adjustment-free circuit using Motorola Touch-Tone dialer with external 1-MHz crystal. Internal circuits of IC select proper division rates and convert outputs to synthesized sine waves of correct frequencies. Grounding one of row inputs by pressing key gives lower-frequency tone, while grounding one of column inputs gives higher-frequency tone. Special Touch-Tone keyboard provides this grounding action automatically when single key is pressed.—D. Lancaster, "CMOS Cookbook," Howard W. Sams, Indianapolis, IN, 1977, p 239–240.

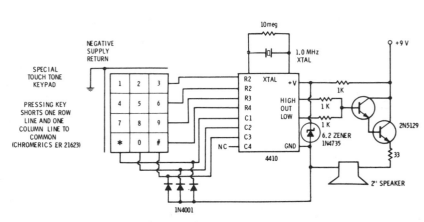

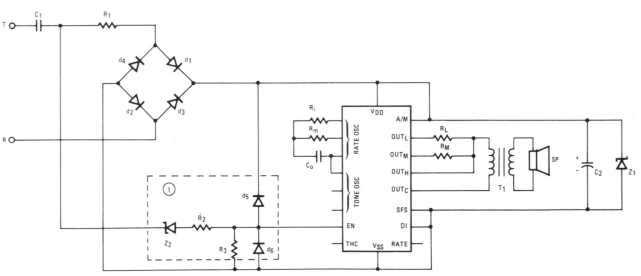

TYPICAL VALUES	C_1	1μF/200V	R_m	750kΩ	T_1	2000Ω/8Ω TRANSFORMER	Z_2	27V ZENER
	R_1	2kΩ	C_o	330pF	SP	8Ω SPEAKER	R_2	150kΩ
	d_1-d_4	1N 4004	R_L	18kΩ	C_2	47μF/25V	R_3	300kΩ
	R_i	200kΩ	R_M	3.3kΩ	Z_1	12V ZENER 1N4742	d_5, d_6	1N914

BELL SIMULATOR—Uses AMI S2561 CMOS IC to simulate effects of telephone bell by producing tone signal that shifts between two predetermined frequencies at about 16 Hz. In applications where dial pulse rejection is not necessary, network inside dashed lines can be omitted and pins EN and DI connected directly to V_{DD}, which is typically 10 V. Power is derived from telephone lines by diode-bridge supply. Values shown give tone frequencies of 512 and 640 Hz. Power output to 8-ohm loudspeaker is at least 50 mW, fed through 200:8 ohm transformer.—"Tone Ringer," American Microsystems, Santa Clara, CA, 1977, S2561, p 7.

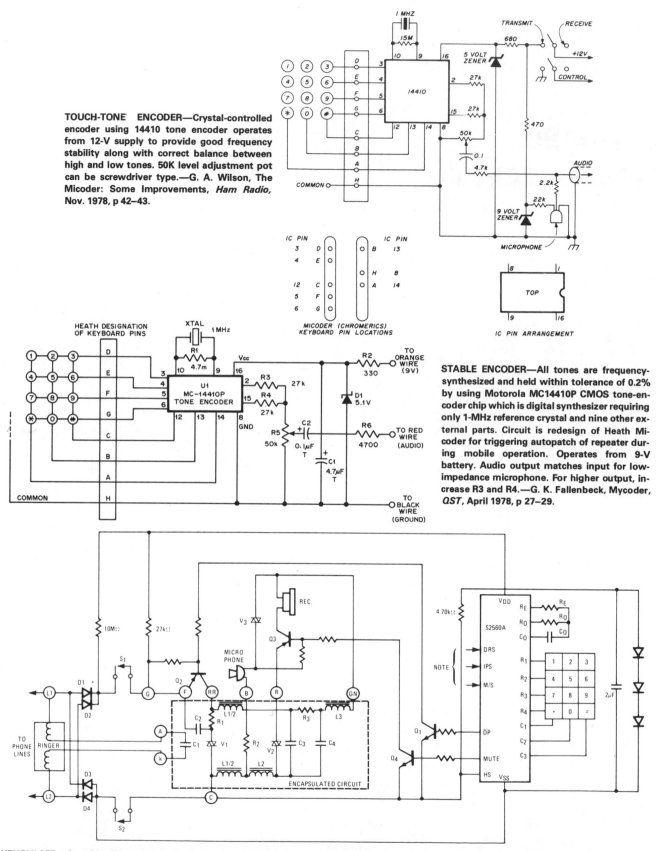

TOUCH-TONE ENCODER—Crystal-controlled encoder using 14410 tone encoder operates from 12-V supply to provide good frequency stability along with correct balance between high and low tones. 50K level adjustment pot can be screwdriver type.—G. A. Wilson, The Micoder: Some Improvements, *Ham Radio*, Nov. 1978, p 42–43.

STABLE ENCODER—All tones are frequency-synthesized and held within tolerance of 0.2% by using Motorola MC14410P CMOS tone-encoder chip which is digital synthesizer requiring only 1-MHz reference crystal and nine other external parts. Circuit is redesign of Heath Micoder for triggering autopatch of repeater during mobile operation. Operates from 9-V battery. Audio output matches input for low-impedance microphone. For higher output, increase R3 and R4.—G. K. Fallenbeck, Mycoder, *QST*, April 1978, p 27–29.

KEY PULSER—American Microsystems S2560A CMOS IC pulser converts pushbutton inputs to series of pulses suitable for telephone dialing, as replacement for mechanical telephone dial. Circuit shows typical connection to dial telephone set using 500-type encapsulated circuit. Dialing rate can be varied by changing dial rate oscillator frequency. IC includes 20-digit memory that makes last dialed number available for redialing until new number is entered. Entered digits are stored sequentially in internal memory, with dial pulsing starting as soon as first digit is entered. Arrangement permits entering digits much faster than output rate. Last number is redialed by going off hook and pressing # key.—"Key Pulser," American Microsystems, Santa Clara, CA, 1977, S2560A/S2560B, p 8.

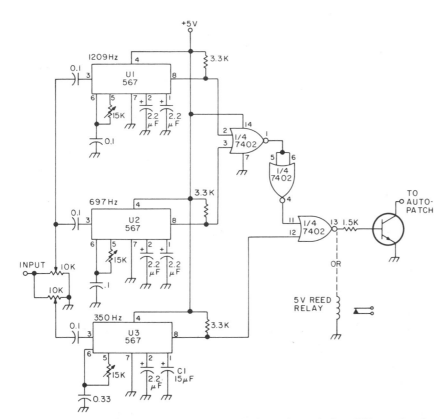

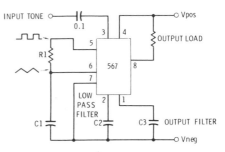

PLL SINGLE-TONE DECODER—Can be used for Touch-Tone decoding as well as for telephone-line and wireless control applications using single audio frequency. Operating center frequency depends on R1 and C1. R1 should be between 2K and 20K. C1 in microfarads is computed from $f = 1/R1C1$, where R is in megohms and f is in hertz. C2 is low-pass filter in range of 1–22 μF; the larger its value, the narrower its bandwidth. C3 is not critical and can be about twice C2.—C. D. Rakes, "Integrated Circuit Projects," Howard W. Sams, Indianapolis, IN, 1975, p 68–73.

TOLL-CALL KILLER—Prevents unauthorized direct long-distance dialing through repeater autopatch from areas where "1" must be dialed ahead of desired out-of-town phone number. Based on simultaneous detection of 350-Hz component of dial tone and 1209- and 697-Hz tones assigned to "1" in Touch-Tone system.

Circuit requires only three 567 tone decoders, 7402 quad gate, and either transistor or relay for controlling autopatch. Article covers installation and operation.—W. J. Hosking, Long Distance Call Eliminator, *73 Magazine*, April 1976, p 44–45.

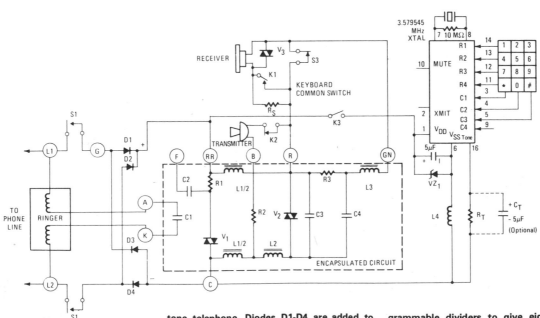

DUAL-TONE SIGNALING—American Microsystems S2559 digital tone generator IC at upper right interfaces directly with encapsulated 500-type telephone set to give pushbutton dual-tone telephone. Diodes D1-D4 are added to telephone set to ensure that polarity of direct voltage across device is unchanged even if connections to phone terminals are reversed. Generator IC requires external crystal feeding programmable dividers to give eight standard audio frequencies with high accuracy for combining in pairs as required for dual-tone signaling.—"Digital Tone Generator," American Microsystems, Santa Clara, CA, 1977, S2559, p 11.

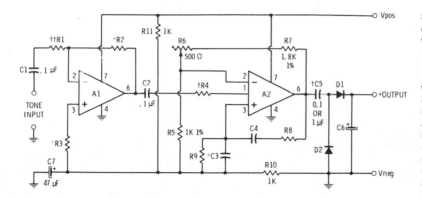

SINGLE-TONE DECODER—Used at receiving end of leased telephone line in which single tone frequency serves for alarm and other purposes. A1 is 741 opamp connected as inverting amplifier, with R1 and R3 chosen to match input impedance and R2 chosen to give gain required for available input signal level. For 10K input impedance, R1 and R3 are 10K and R2 in kilohms is 10 times required gain (500K for gain of 50). Actual tone decoding is performed by A2, which is also 741; here C3, C4, R8, and R9 are frequency-determining components and R6 is gain control. R4 is chosen to give desired bandwidth; use 470K for 5–10%, 1 megohm for 3–5%, and 2.2 megohms for 1–3%. R8 is same as R2, and R9 equals R3. Diodes are 1N914.—C. D. Rakes, "Integrated Circuit Projects," Howard W. Sams, Indianapolis, IN, 1975, p 60–66.

PHONE-CALL COUNTER—Circuit actuates solenoid that depresses R/S counting key of SR-56 calculator for each interrogation event consisting of sequence of pulse bursts each corresponding to ring of phone. Bursts are separated by 4-s pauses, so circuit includes time delay that prevents actuation of solenoid until line has remained quiescent for more than 5 s after burst. Article includes program that is inserted in calculator to total number of times R/S key is depressed. Applications include counting number of telephone calls received while away.—M. Bram, Hardware + Program Makes SR-56 Event Counter, *EDN Magazine,* Aug. 5, 1978, p 84 and 86.

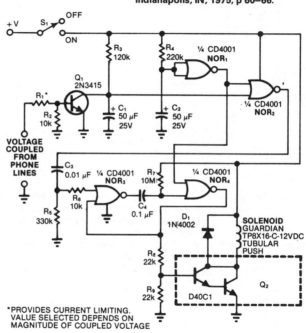

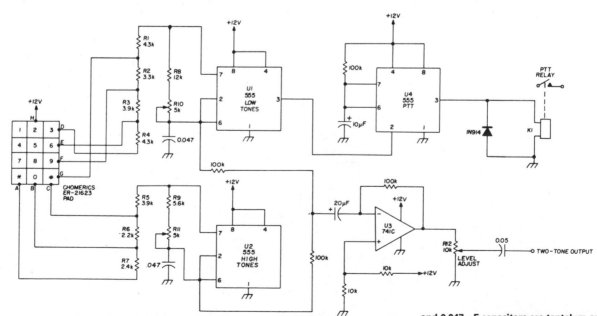

TOUCH-TONE ENCODER—Uses 555 timers to generate Touch-Tone frequencies in pairs using two of seven possible frequencies, under control of standard 12-button pad. Adjust R10 so low-group oscillator reads 941 Hz at pin 3 of U1 when * key is pressed. Frequencies of 852, 770, and 697 Hz will then be correct within 2% when 7, 4, and 1 are pressed, if 1% resistors are used and 0.047-μF capacitors are tantalum or Mylar. Automatic push-to-talk control uses U4 connected as 1-s mono MVBR driving relay K1.—H. M. Berlin, Homebrew Touch-Tone Encoder, *Ham Radio,* Aug. 1977, p 41–43.

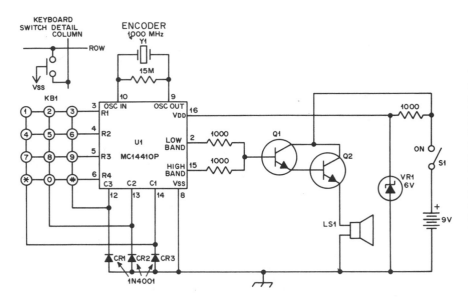

TOUCH-TONE DRIVE FOR LOUDSPEAKER—Encoder is held in front of microphone to access and use autopatch of repeater. Acoustic coupling eliminates need for opening new transceiver to make wire connections, which would void guarantee. Uses Motorola MC14410P IC with KB1 keyboard (Polypaks 92CU3149). Q1 and Q2 are 2N3643 or equivalent. Y1 is 1.000 MHz crystal (Mariann Labs ML18P or Sherold Crystal HC-6). Transistors Q1 and Q2 boost output enough to drive 8-ohm loudspeaker. Total current drain is 35 mA idling and 100 mA with full drive.—C. Gorski, A Low-Cost Touch-Tone Encoder, *QST*, Oct. 1976, p 36–37.

AUTOPATCH RELEASE—Control circuit automatically releases telephone autopatch at receiver when called party hangs up, by generating disconnect signal for patch control logic. Action is based on reversal of polarity of phone line when called party answers, and return of polarity to preanswered condition when called party hangs up. Article describes circuit operation and use.—T. R. Yocom, Automatic Autopatch Release, *73 Magazine*, April 1977, p 52.

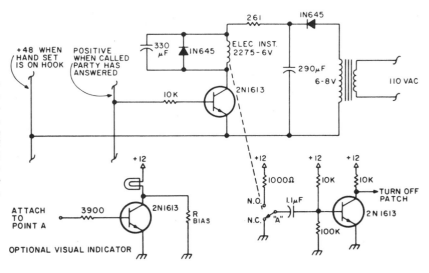

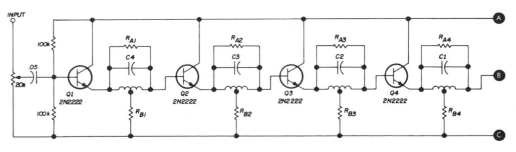

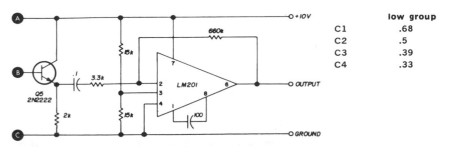

	low group	high group
C1	.68	.22
C2	.5	.18
C3	.39	.15
C4	.33	.1

TOUCH-TONE BAND-REJECT FILTER—Cascaded notch filters with active limiter at output provide 20-dB attenuation of either low (697–941 Hz) or high (1209–1633 Hz) groups of tones, as aid to decoding for repeater control functions. All coils are 88-mH toroid. R_A is between 5600 and 22,000 ohms, and R_B is 1000 to 3000 ohms. Article gives tuning procedure for selecting resistor values and adjusting toroids so each stage rejects different tone in its band.—B. Bretz, Multi-Function FM Repeater Decoder, *Ham Radio*, Jan. 1973, p 24–32.

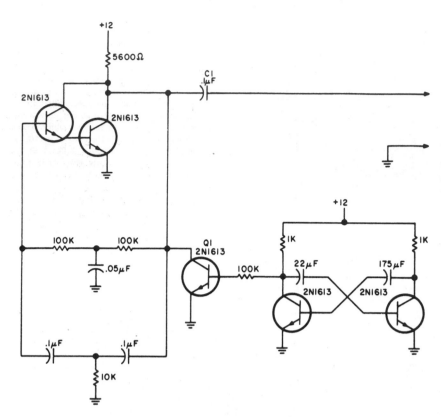

BUSY-SIGNAL GENERATOR—Conventional Bell System busy signal is provided by turning twin-T oscillator at left on and off with low-frequency asymmetrical square wave generated by transistor pair at right. Q1 acts as switch for turning oscillator on and off. Developed for use at repeater in home when autopatch connects to family telephone, to inhibit use of autopatch by mobile station when phone is in use. Article also covers connections to phone line and to repeater.—T. Yocom, An Autopatch Busy Signal, *73 Magazine,* Holiday issue 1976, p 148 and 150.

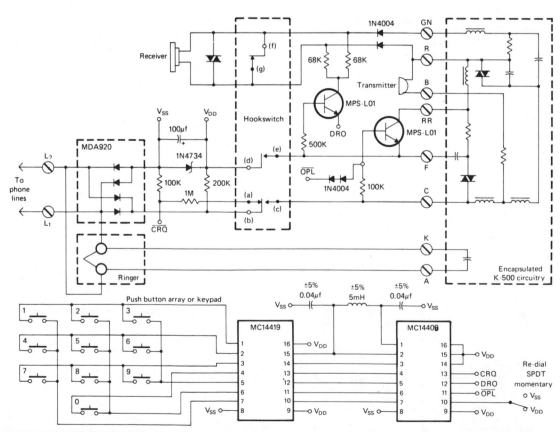

PUSHBUTTON-TO-DIAL CONVERTER—Combination of Motorola MC14419 keypad-to-binary converter and MC14408 BCD-to-dial telephone pulse converter is used with 10-switch pushbutton array to provide correct chain of pulses for dialing number on conventional dial-telephone system. Eleventh SPDT button is used for redial feature; if line called is busy, one press of redial button dials number over again. Number is stored for repeated use until new number is dialed. Check local telephone company regulations before making connections to telephone lines.—I. Math, A Push-Button to Dial Telephone Converter, *CQ,* Sept. 1976, p 36–37.

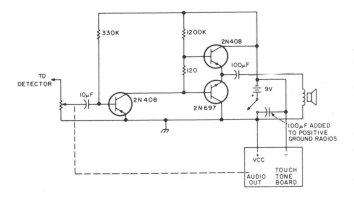

TOUCH-TONE ENCODER—Consists of SME Touch-Tone generator and keyboard made by Data Signal (Albany, GA) mounted on any small transistor radio. Only audio section is used, with output tones from loudspeaker being fed acoustically to microphone of FM amateur station. Article gives construction details.—D. Ingram, The Shirt Pocket Touch-Tone, *73 Magazine*, Nov. 1976, p 58–59.

45-kHz LOW-PASS STATE-VARIABLE FILTER—Used in precision telephone-network active equalizer. Damping value is 0.082, which requires 1% components. For high pass, take output from first opamp; for bandpass, take output from second opamp.—D. Lancaster, "Active-Filter Cookbook," Howard W. Sams, Indianapolis, IN, 1975, p 147.

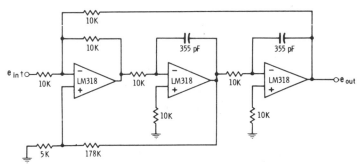

† must return to ground via low-impedance dc path.

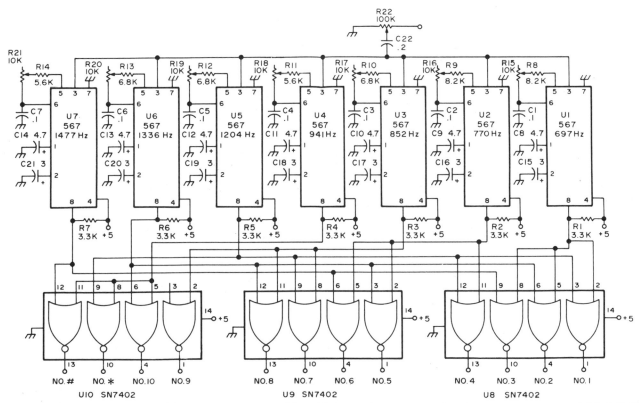

TOUCH-TONE DECODER—Uses seven National LM567 phase-locked loop decoders (U1–U7) having high noise rejection, immunity to false signals, and stable center frequency. Each 567 activates proper gate of SN7402, making output of gate go to high or 1 state for driving NPN transistor that can turn on LED labeled with corresponding Touch-Tone number. Alternatively, gate outputs can drive 12 relays, with relay contacts going to LEDs and/or to keyboard switches of ordinary calculator used as digital display. Article tells how to adjust 10K pot for each 567 for detection of desired frequency.—W. MacDowell, Touch-Tone Decoder, *73 Magazine*, June 1976, p 26–27.

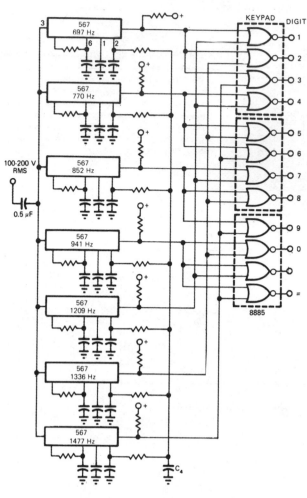

PLL TOUCH-TONE DECODER—Seven 567 PLLs sense presence of selected tones from common 100–200 VRMS input line, while 8885 NOR gates perform necessary decoding logic to generate decimal outputs. Circuit takes advantage of good frequency selectivity provided by lock-and-capture ranges of PLLs, as required for discriminating against many tones.—E. Murthi, Monolithic Phase-Locked Loops—Analogs Do All the Work of Digitals, and Much More, *EDN Magazine,* Sept. 5, 1977, p 59–64.

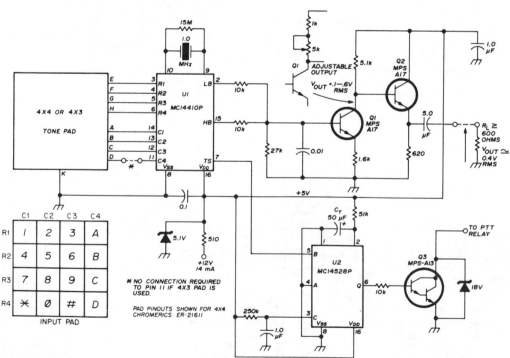

TONE ENCODER—Motorola MC14410 CMOS IC is basis of accurate low-power Touch-Tone encoder system providing full 2-of-8 encoding from basic 1-MHz crystal oscillator. Can be used with 2-of-7 or 2-of-8 keypad switch matrix such as Chromerics ER-21623 or ER-21611. Q1-Q2 form tone-amplifier/emitter-follower line driver. U2 is push-to-talk mono 1-s timer. Supply can be any voltage from 5 to 12 V if zener is used to supply 5 V to ICs. Article covers circuit operation in detail and gives tone-encoder frequency table.—J. DeLaune, Digital Touch-Tone Encoder for VHF FM, *Ham Radio,* April 1975, p 28–31.

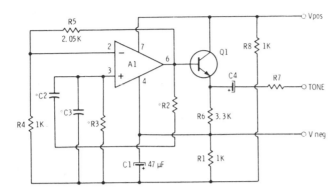

SINGLE-TONE SIGNALING—Wien-bridge oscillator using 741 opamp drives 2N2924 or equivalent NPN transistor to generate stable audio tone for signaling over telephone lines. Tuning capacitor (C2 and C3 are equal) and resistor (R2 and R3 are equal) values range from 0.1 μF and 15.9K for 100 Hz to 0.005 μF and 6.3K for 5000 Hz. For other frequencies, use f = 0.159/R2C2. With 12-V supply, tone output is about 7 V P-P. Select R7 to match impedance of driven circuit.—C. D. Rakes, "Integrated Circuit Projects," Howard W. Sams, Indianapolis, IN, 1975, p 55–60.

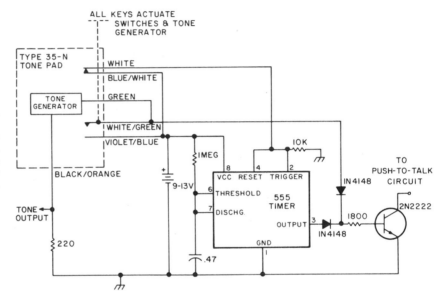

MOBILE AUTOPATCH—Circuit operates push-to-talk of mobile station automatically when any button on Touch-Tone pad is pushed for dialing telephone number after making autopatch, eliminating need for engaging microphone before dialing. Circuit remains active for about 2 s after Touch-Tone button is released.—Circuits, *73 Magazine,* May 1977, p 19.

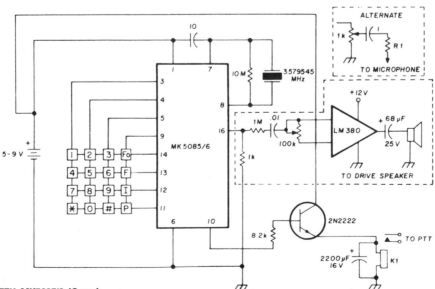

TOUCH-TONE IC—MOSTEK MK5085/6 IC and keyboard together form inexpensive Touch-Tone generator producing tones within 0.75% of required values. Uses 3.579545-MHz TV color-burst crystal. Pin 15 is grounded to provide dual tones only. Pin 10 provides output when keyboard entry has been made, for keying push-to-talk (PTT). Loudspeaker can be eliminated if output is fed directly into microphone input of transmitter. Choice of IC depends on type of keyboard used.—T. Ahrens, Integrated-Circuit Tone Generator, *Ham Radio,* Feb. 1977, p 70.

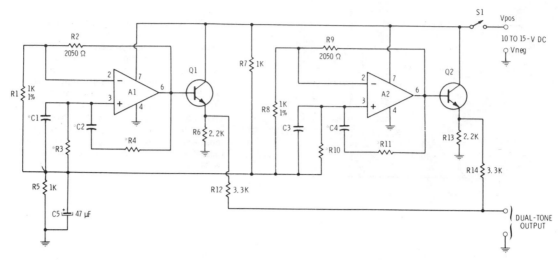

TWO-TONE ENCODER—741 opamps and 2N2924 transistors are connected as single-tone encoders producing different audio frequencies, with outputs connected together. For 2000-Hz tone, use 0.01 μF for C1 and C2 and use 8K for R3 and R4. Formula for frequency of each encoder is f = 0.159/RC where f is in hertz, R in megohms, and C in microfarads; R = R3 = R4 and C = C1 = C2. Frequencies can be chosen for Touch-Tone signaling.—C. D. Rakes, "Integrated Circuit Projects," Howard W. Sams, Indianapolis, IN, 1975, p 95—97.

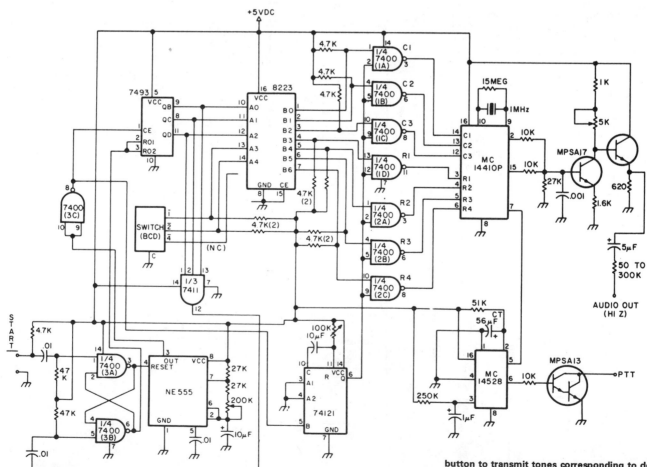

FOUR-NUMBER CALLER—Motorola MC14410 CMOS Touch-Tone generator chip forms basis for automatic dialer using BCD thumbwheel switch to choose telephone number desired. Numbers are stored in 256-bit PROM by conventional programming. Article shows how autopatch access and disconnect switches are added. To make telephone call from car through repeater, select number desired, push access button and, when dial tone is heard, push start button to transmit tones corresponding to desired number. Article covers circuit operation, programming, and coding, and gives additional circuit using 512-bit PROM to provide eight telephone numbers.—W. J. Hosking, Drive More Safely with a Mobile Dialer, *73 Magazine,* Feb. 1977, p 102—104.

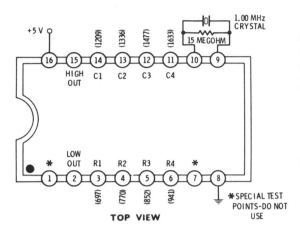

TOUCH-TONE DIALER—Single Motorola 4410 chip requires only two external components and 2-of-8 keyboard to generate two sine waves simultaneously for Touch-Tone dialing and telephone modem communication. Each key on keyboard grounds one of C inputs and one of R inputs. As example, when 6 key is pressed, R2 and C3 are grounded to give 770-Hz sine wave on pin 2 and 1477-Hz sine wave on pin 15. Designed for driving 1K load. Output voltage is about 600 mV P-P for low output and 800 mV P-P for high output.—D. Lancaster, "CMOS Cookbook," Howard W. Sams, Indianapolis, IN, 1977, p 132.

RING DETECTOR—Optoisolator using neon lamp and light-dependent resistor serves as interface between telephone line and line-operated remote bell. Neon fires reliably from nominal 100-VAC ring signal, while capacitor C_1 provides isolation required to prevent latch-up by sustaining voltages within range of phone-line quiet battery. If optional protective varistor R_{IC1} is added, rating of capacitor can be reduced to 400 V. Triac Q_1 in series with primary of line transformer provides synchronization to 20-Hz ringing frequency of phone system.—W. D. Kraengel, Jr., Ring Detector Optically Interfaces Phone, *EDN Magazine,* Aug. 5, 1978, p 80 and 82.

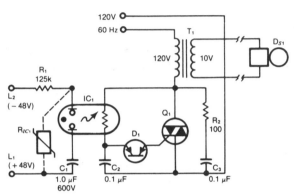

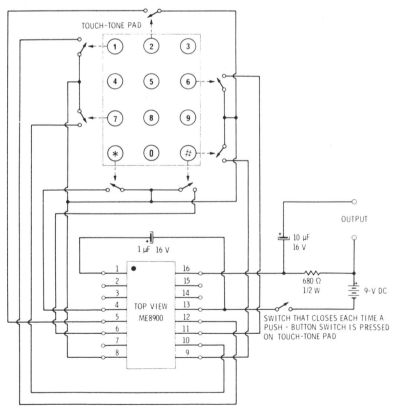

SINGLE-IC TOUCH-TONE ENCODER—Uses ME8900 tone generator made by Microsystems International, Ottawa, Canada, to generate pairs of audio frequencies for telephone signaling. Standard Touch-Tone pad provides column and row switch closures for pins 4, 5, 6, and 8 going to low-frequency parallel-T oscillators of IC and for pins 9, 10, 11, and 12 of high-frequency oscillator.—C. D. Rakes, "Integrated Circuit Projects," Howard W. Sams, Indianapolis, IN, 1975, p 100–101.

CHAPTER 91
Teleprinter Circuits

Includes tone generators and demodulators for FSK and AFSK used in wire and video Teletype systems having 170-Hz, 850-Hz, or other frequency shifts, as well as "QUICK BROWN FOX" and other test-character generators, RAM storage for up to 128 RTTY characters, autostart control, RTTY active filters and motor control, microprocessor and UART interfaces, and clock-signal generator for variety of keying speeds.

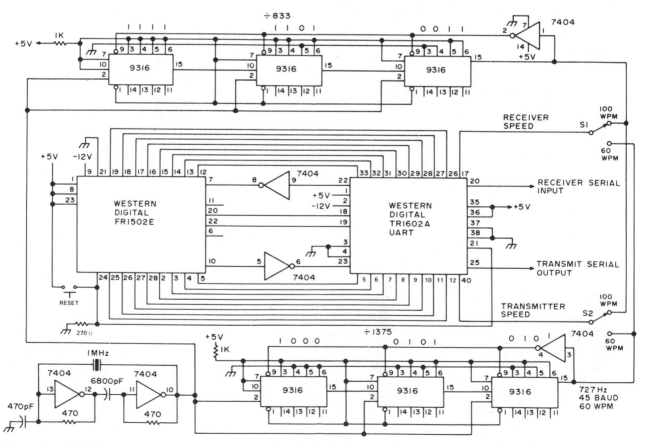

RTTY SPEED CONVERTER—Digital speed converter for amateur RTTY permits transmitting either above or below input speed from keyboard or tape. Uses FR1502E 40-character 9-bit FIFO storage chip, TR1602A universal asynchronous receiver-transmitter, and six Fairchild 9316 programmable dividers. Values shown give choice of 60 or 100 WPM for receiving and for transmitting, derived by dividing down from same 1-MHz clock. Input and output are TTL-compatible.—A. Sperduti, The 60 WPM Conversion, *73 Magazine*, April 1977, p 158–159.

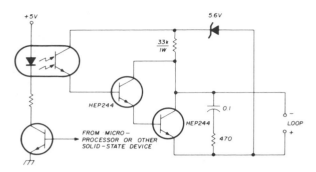

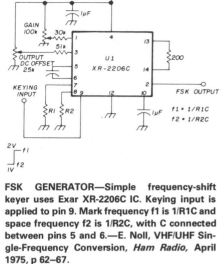

SAFE SWITCHING OF SOLENOIDS—Optoisolator provides protective interface between teleprinter and 8080A or other microprocessor when switching inductive loads of teleprinter.

RC filter across Darlington pair speeds release time of print magnets.—T. C. McDermott, Switching Inductive Loads with Solid-State Devices, *Ham Radio*, June 1978, p 99–100.

FSK GENERATOR—Simple frequency-shift keyer uses Exar XR-2206C IC. Keying input is applied to pin 9. Mark frequency f1 is 1/R1C and space frequency f2 is 1/R2C, with C connected between pins 5 and 6.—E. Noll, VHF/UHF Single-Frequency Conversion, *Ham Radio*, April 1975, p 62–67.

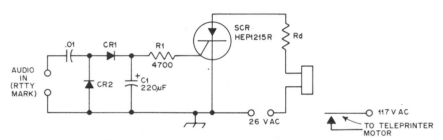

SCR CONTROLS RTTY MOTOR—Single SCR can replace several transistor or tube stages in RTTY, VOX, COR, and other relay control circuits. Threshold triggering effect of SCR means triggering is automatically suppressed on low-level noise and similar interference. Used in RTTY autostart and motor delay sections of de-

modulator. Pickup time is 1 s and dropout time is 3 s, determined by values of R1 and C1 and by SCR. Circuit keys only on 2125-Hz mark tone. Diodes are 50-PIV silicon. Rd is appropriate dropping resistor for relay, if needed.—D. Weeden, SCR Relay Control for RTTY, VOX, and COR, *QST*, July 1976, p 42.

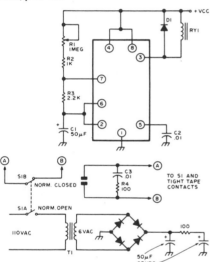

TAPE AS RTTY BUFFER—Developed to give constant-speed amateur radioteletype transmissions despite erratic keyboarding speeds. Uses NE555 timer chip as free-running MVBR whose speed can be varied by R1 down to about 1 character every 15 s or up to full machine speed. Can be used with any automatic send-receive machine. Keep enough slack in punched paper tape to permit backspacing and correcting errors before they are sent. With 5-V supply shown, 6-V SPST DC relay can be used.—B. Gulledge, Mechanical RTTY Buffer, *73 Magazine*, Oct. 1976, p 74.

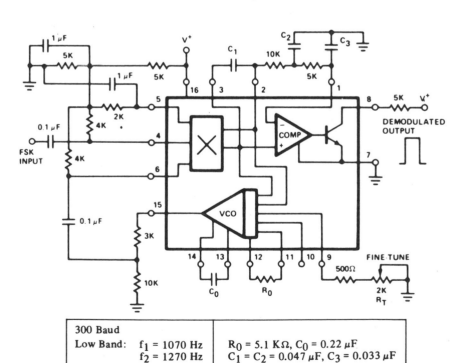

300 Baud		
Low Band:	$f_1 = 1070$ Hz	$R_0 = 5.1\ K\Omega, C_0 = 0.22\ \mu F$
	$f_2 = 1270$ Hz	$C_1 = C_2 = 0.047\ \mu F, C_3 = 0.033\ \mu F$
High Band:	$f_1 = 2025$ Hz	$R_0 = 8.2\ K\Omega, C_0 = 0.1\ \mu F$
	$f_2 = 2225$ Hz	$C_1 = C_2 = C_3 = 0.033\ \mu F$
1200 Baud		$R_0 = 2\ K\Omega, C_0 = 0.14\ \mu F$
	$f_1 = 1200$ Hz	$C_1 = 0.033\ \mu F, C_3 = 0.02\ \mu F$
	$f_2 = 2200$ Hz	$C_2 = 0.01\ \mu F$

FSK DEMODULATOR—Exar XR-210 FSK modulator-demodulator is connected as PLL system by providing AC coupling between VCO output pin 15 and pin 6. When input frequency is shifted, corresponding to data bit, polarity of DC voltage across phase detector output pins 2 and 3 is reversed. Voltage comparator and logic driver sections convert this DC level shift to binary pulse. C_1 serves as PLL filter. C_2 and C_3 are postdetection filters. Timing capacitor C_0 and fine-tune adjustment are used to set VCO midway between mark and space frequencies of input signal. Table gives typical values for 300-baud (103-type) and 1200-baud (202-type) modem applications. Supply can be 5–26 V.— "Phase-Locked Loop Data Book," Exar Integrated Systems, Sunnyvale, CA, 1978, p 17–20.

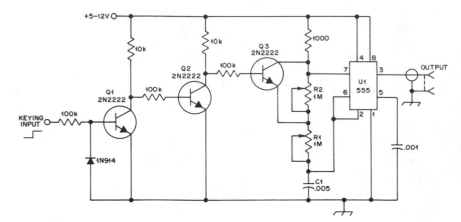

AFSK SHIFTS UP TO 20 kHz—Wide-range generator can be tuned from 50 Hz to 20 kHz, for shifting between two frequencies as much as 20 kHz apart. U1 is 555 timer connected as astable MVBR. When Q3 is biased off, charge/discharge currents for C1 flow chiefly through R1 and R2 to determine lower frequency of oscillation. When Q3 is on, R2 is effectively shorted and frequency is increased. Q1 acts as buffer and inverter so higher voltage at input gives higher tone. Keying occurs when input voltage exceeds 1 V with 5-V supply. If 10K supply resistor for Q1 is reduced to 1000 ohms, keying voltage increases to 3V.—T. M. Whittaker, Wide-Range AFSK Generator, *QST*, May 1977, p 48.

TTY RESETS CPU—Circuit uses break key on TTY as reset button for microprocessor. Retriggerable mono IC_1 monitors data input line from TTY, which goes low for spacing condition. During normal data input, constant spacing pulses in data retrigger mono, keeping IC_2 reset. When break key is depressed, input data goes to steady space and IC_1 times out. IC_2 then transfers high on its data input to its output to produce reset signal for CPU. Values of R_2 and C_1 give 150-ms period for mono, suitable for baud rates of 110 and higher.—C. Sondgeroth, Reset Your CPU from Your TTY's Break Key, *EDN Magazine*, May 5, 1978, p 39.

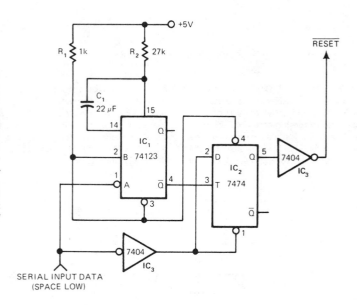

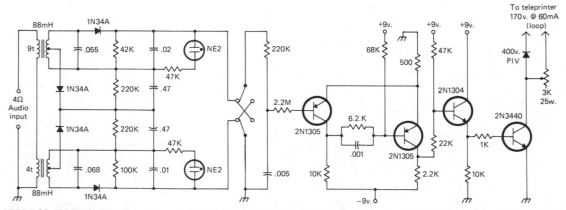

DEMODULATOR FOR 170-Hz SHIFT—Converts RTTY audio tones of 2125 and 2295 Hz to DC pulses required for driving selector magnets of teleprinter. Coupling links are added to standard 88-mH toroids as indicated.—I. Schwartz, An RTTY Primer, *CQ*, Feb. 1978, p 31–36.

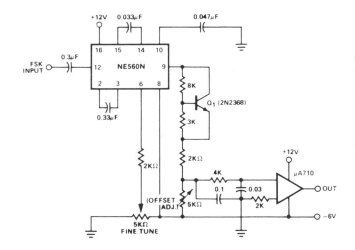

PLL FSK DEMODULATOR—Signetics NE560N phase-locked loop is used as receiving converter for demodulating carrier shifted between two preset frequencies, one corresponding to 0 and other to 1 of binary data signal. PLL provides shifting DC voltage to initiate 1 or 0 (mark or space) code elements. Circuit locks on and tracks output frequency of receiver. Input at pin 12 should be from 30 mV to 2 V P-P square or sine wave. Output of about 60 mVDC at pin 9 is amplified, conditioned, and fed to μA710 comparator to provide proper output voltages for interfacing with printer.—"Signetics Analog Data Manual," Signetics, Sunnyvale, CA, 1977, p 844–845.

FSK DECODER—Simple circuit for Signetics NE565 PLL locks to input frequency and tracks it between two values used, to produce corresponding DC shift at output. Values shown are for 1070-Hz and 1270-Hz FSK signals. Three-stage RC ladder filter removes sum frequency component. Band edge of filter is chosen to be about halfway between maximum keying rate (150 Hz) and twice input frequency (about 2200 Hz). Output is made logic-compatible by connecting voltage comparator to pin 6.—"Signetics Analog Data Manual," Signetics, Sunnyvale, CA, 1977, p 845.

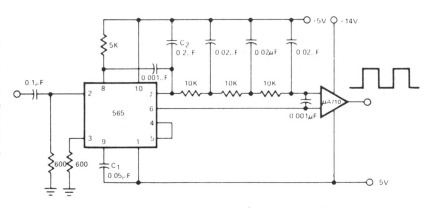

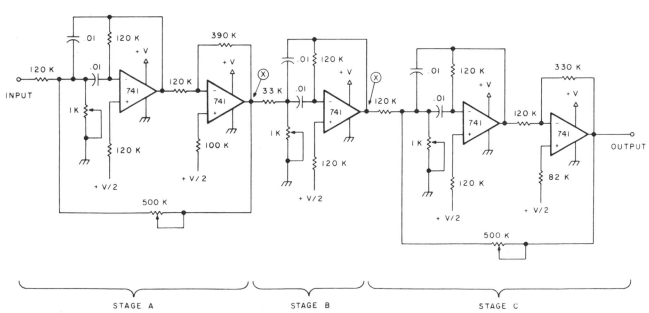

BANDPASS FOR 170-Hz RTTY SHIFT—Three-stage active Butterworth bandpass input filter is used in radioteletype demodulator to separate RTTY tones from each other and from noise. Filter is centered on 2200 Hz, and has bandpass of about 260 Hz to allow reception of some of audio sidebands produced by keying and allow for small drift. Five pots serve for trimming center frequency of each stage and Q of end stages. Article gives step-by-step design and alignment procedures. Use two 10K resistors between V and ground to get V/2 for bias when operating from single supply.—P. A. Stark, Design an Active RTTY Filter, 73 Magazine, Sept. 1977, p 38–43.

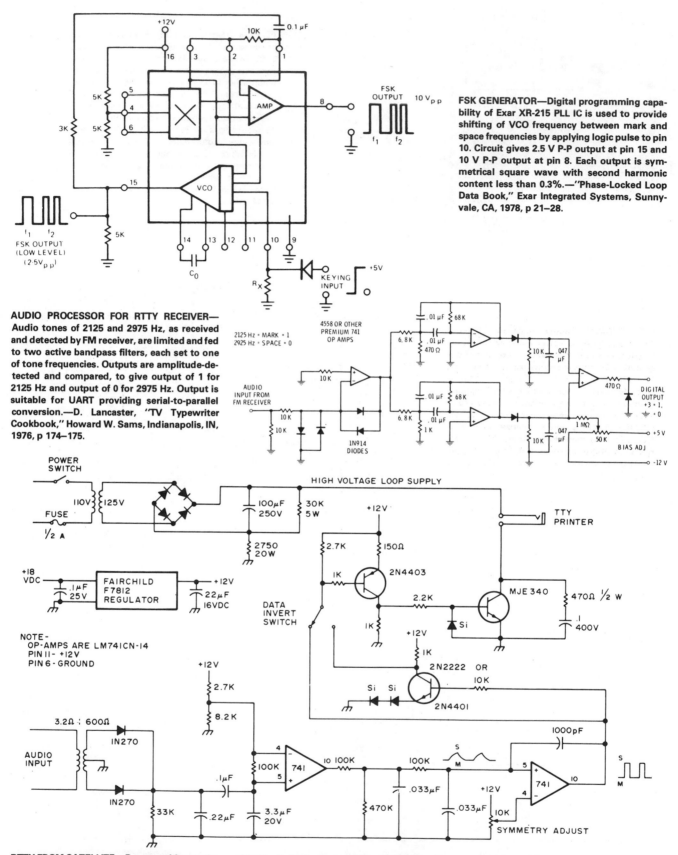

FSK GENERATOR—Digital programming capability of Exar XR-215 PLL IC is used to provide shifting of VCO frequency between mark and space frequencies by applying logic pulse to pin 10. Circuit gives 2.5 V P-P output at pin 15 and 10 V P-P output at pin 8. Each output is symmetrical square wave with second harmonic content less than 0.3%.—"Phase-Locked Loop Data Book," Exar Integrated Systems, Sunnyvale, CA, 1978, p 21–28.

AUDIO PROCESSOR FOR RTTY RECEIVER—Audio tones of 2125 and 2975 Hz, as received and detected by FM receiver, are limited and fed to two active bandpass filters, each set to one of tone frequencies. Outputs are amplitude-detected and compared, to give output of 1 for 2125 Hz and output of 0 for 2975 Hz. Output is suitable for UART providing serial-to-parallel conversion.—D. Lancaster, "TV Typewriter Cookbook," Howard W. Sams, Indianapolis, IN, 1976, p 174–175.

RTTY FROM SATELLITE—Developed for receiving RTTY transmitted from satellite as space-only keying. Receiver can be operated in CW or narrow-filter mode, to increase signal-to-noise ratio. Any receiver having CW filter with 400-Hz bandwidth can be used. Tune for audio output of 1 kHz. Audio is converted to varying DC voltage by envelope detector and amplified by 741 opamp that drives additional filter having high-level output for space and low-level output for mark. Slow rise and fall times of varying voltage are converted to ON/OFF keying signals by 741 used as comparator, for feeding to two-stage driver and high-voltage loop keying circuit of conventional design.—K. O. Learner and W. A. Kotras, Oscar RTTY Converter, *73 Magazine*, July 1975, p 53–54.

UART INTERFACE—Permits use of universal asynchronous receiver-transmitter with model 33 Teletype so keyboard can send and printer can receive at same time. Transmitter interface provides 20-mA current for mark or 1 and open circuit for space or 0. Receiver senses closed contact for mark or 1 and open contact for space or 0. Extra inverters are added to make codes correspond so 1 from UART is read as 1 by Teletype. Designed for 110-baud rate.—D. Lancaster, "TV Typewriter Cookbook," Howard W. Sams, Indianapolis, IN, 1976, p 162–164.

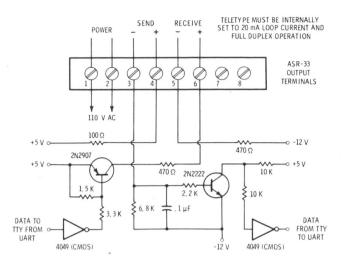

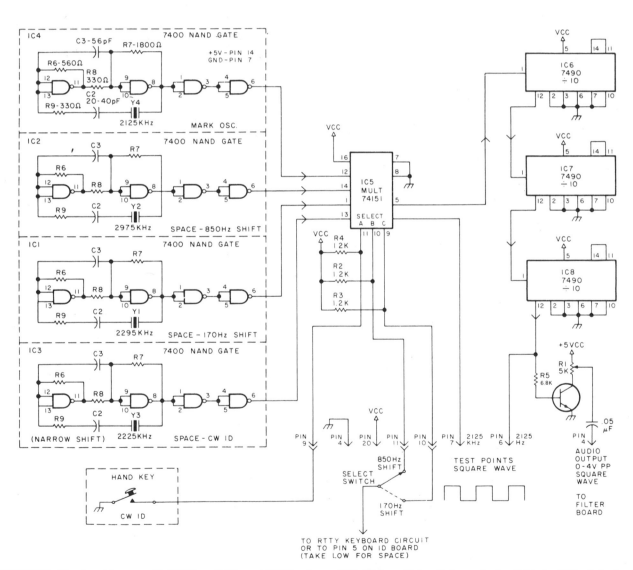

AFSK KEYER—Developed for use with 49-MHz FM transmitter to rebroadcast hurricane bulletins and other weather warnings to amateur RTTY stations. Crystals assure high precision in generating RTTY tones for 850-Hz shift, with extra crystals for 170-Hz shift and for narrow-shift CW identification. Frequency tolerance is ±1 Hz and requires no calibration. Circuit uses 7400 quad NAND-gate crystal oscillator, which works with almost any HC-6/U crystal. Frequency can be adjusted by changing value of C2. Outputs feed 74151 multiplexer. When all SELECT inputs are high (2.8–5 V), mark oscillator frequency appears at multiplexer output. When input B is low (0–0.8 V), multiplexer output changes to space frequency of 2975 kHz. Multiplexer feeds divide-by-1000 chain feeding 2125-Hz square wave to buffer transistor. Article gives circuit of low-pass filter that removes harmonics from output to give pure sine wave for modulating transmitter.—L. J. Fox, Dodge That Hurricane!, *73 Magazine*, Jan. 1978, p 62–69.

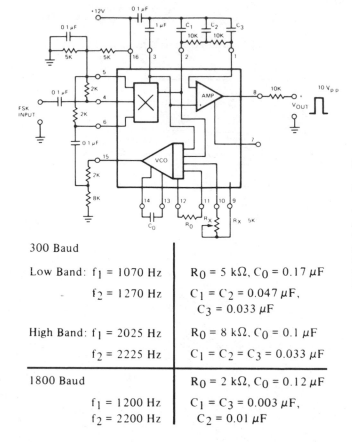

300 Baud	
Low Band: $f_1 = 1070$ Hz	$R_0 = 5$ kΩ, $C_0 = 0.17$ μF
$f_2 = 1270$ Hz	$C_1 = C_2 = 0.047$ μF, $C_3 = 0.033$ μF
High Band: $f_1 = 2025$ Hz	$R_0 = 8$ kΩ, $C_0 = 0.1$ μF
$f_2 = 2225$ Hz	$C_1 = C_2 = C_3 = 0.033$ μF
1800 Baud	
$f_1 = 1200$ Hz	$R_0 = 2$ kΩ, $C_0 = 0.12$ μF
$f_2 = 2200$ Hz	$C_1 = C_3 = 0.003$ μF, $C_2 = 0.01$ μF

300- AND 1800-BAUD FSK DEMODULATOR— Uses Exar XR-215 PLL IC having frequency range of 0.5 Hz to 35 MHz. When input frequency is shifted by data bit, DC voltage between pins 2 and 3 reverses polarity. Opamp section is connected as comparator for converting DC level shift to binary output pulse. C_1 serves as PLL filter. C_2 and C_3 are postdetection filters. Table gives typical values of components for two transmission speeds.—"Phase-Locked Loop Data Book," Exar Integrated Systems, Sunnyvale, CA, 1978, p 21–28.

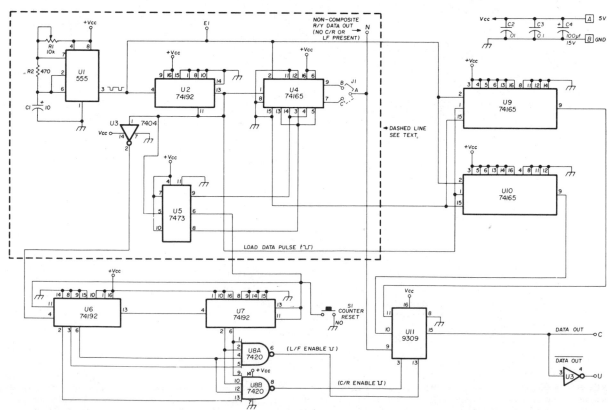

R-Y TEST GENERATOR—Uses ICs rather than PROM for automatic generation of sequence of 64 alternating Rs and Ys, plus Baudot codes for carriage return and line feed, as used for testing radioteletype equipment. Circuit in dashed lines generates codes for printing R and Y continuously without regard for line length. Jumper J1 gives choice of normal or inverted output data for keying transmitter with either mark-high or space-high signal. Operates at slightly less than 60 WPM. Adjust R1 so clock pulse generator U1 runs at 45.45 Hz. Article covers construction and operation.—J. Loughmiller, RTTY Test Generator, *Ham Radio*, Jan. 1978, p 64–66.

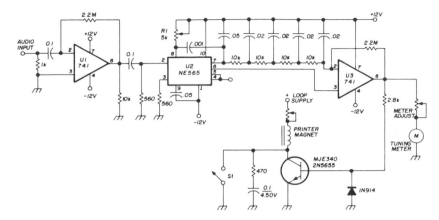

PLL RTTY TERMINAL—Uses 741 opamp as limiter, followed by NE565 phase-locked loop, another opamp U3 operating as voltage comparator or slicer, and keying transistor. Terminal requires no filters because incoming signal locks onto VCO whose frequency is placed between those of mark and space tones. As these tones alternate, output of PLL is made to produce plus and minus voltages by connecting voltage comparator to output of NE565. Resulting plus voltage corresponds to mark tone and minus voltage to space tone for use in keying loop circuit of teleprinter. R1 is only adjustment required; article covers adjustment for receiver in SSB mode and in CW mode.—N. Stinnette, Phase-Locked Loop RTTY Terminal Unit, *Ham Radio*, Feb. 1975, p 36–37.

FSK WITH SLOPE AND VOLTAGE DETECTION—Motorola MC1545G gated video amplifier is used with slope and differential voltage comparators to provide switching of output alternately between input signal f1 at 2975 Hz and f2 at 2125 Hz. With gate level on pin 1 of MC1545G high (greater than 1.5 V), signal applied between pins 4 and 5 is passed and signal between pins 2 and 3 is suppressed. With gate low (less than 0.5 V), situation is reversed. To avoid generation of spurious frequencies and noise, gate control voltage is allowed to change only when rate of change of f1 and f2 have same sign and values of f1 and f2 themselves have same sign and equal magnitude within several millivolts. Data rate is about 170 Hz.—"Gated Video Amplifier Applications—the MC1545," Motorola, Phoenix, AZ, 1976, AN-491, p 12.

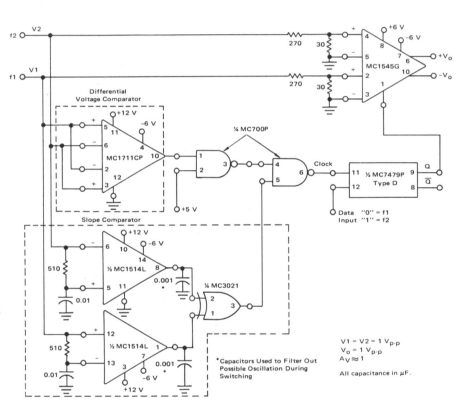

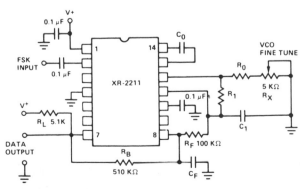

FSK BAND	COMPONENT VALUES	
300 Baud $f_1 = 1070$ Hz $f_2 = 1270$ Hz	$C_0 = 0.039\ \mu F$ $C_1 = 0.01\ \mu F$ $R_1 = 100\ K\Omega$	$C_F = 0.005\ \mu F$ $R_0 = 18\ K\Omega$
300 Baud $f_1 = 2025$ Hz $f_2 = 2225$ Hz	$C_0 = 0.022\ \mu F$ $C_1 = 0.0047\ \mu F$ $R_1 = 200\ K\Omega$	$C_F = 0.005\ \mu F$ $R_0 = 18\ K\Omega$
1200 Baud $f_1 = 1200$ Hz $f_2 = 2200$ Hz	$C_0 = 0.027\ \mu F$ $C_1 = 0.01\ \mu F$ $R_1 = 30\ K\Omega$	$C_F = 0.0022\ \mu F$ $R_0 = 18\ K\Omega$

FSK DECODER—R_0 and C_0 set PLL center frequency for Exar XR-2211 FSK demodulator/tone decoder. R_1 sets system bandwidth. C_1 sets loop filter time constant and loop damping factor. C_F and R_F form postdetection filter for FSK data output. Table gives values for most commonly used FSK bands.—"Phase-Locked Loop Data Book," Exar Integrated Systems, Sunnyvale, CA, 1978, p 29–34.

KEYER FOR AFSK—Uses one center-tapped 88-mH toroid tuned to desired RTTY space frequency by suitable value of C1 (0.0628 μF for 2295 Hz). When relay is closed, suitable value for C2 (0.0156 μF for 2125 Hz) is paralleled with C1 to give desired space frequency. Output is perfect sine wave. Plug output into audio input jack of transmitter. Plug relay coil directly into 150-V 60-mA loop of teleprinter. When loop current is turned on, relay closes and AFSK is on mark. Space frequency occurs when relay is opened by teleprinter keyboard.—J. B. Dillon, Audio-Frequency Shift Keyer, *Ham Radio,* Sept. 1976, p 45.

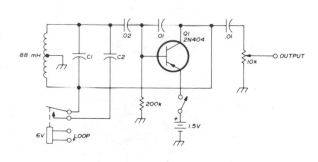

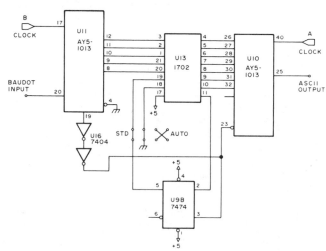

BAUDOT TO ASCII—Used with Baudot RTTY codes on amateur bands, to drive Teletype machine requiring ASCII code. Separate 555 timers are used as clocks running at 727 Hz and at 1760 Hz. Provides only one-way conversion for receiving capability, but article gives companion circuit for two-way code conversion as required for transmission with ASCII Teletype. On U13, pins 12, 13, 15, 22, and 23 all go to +5 V along with pin 17; only pin 14 is grounded.—J. G. Mills, Baudot to ASCII Converter, *73 Magazine,* Sept. 1977, p 80—85.

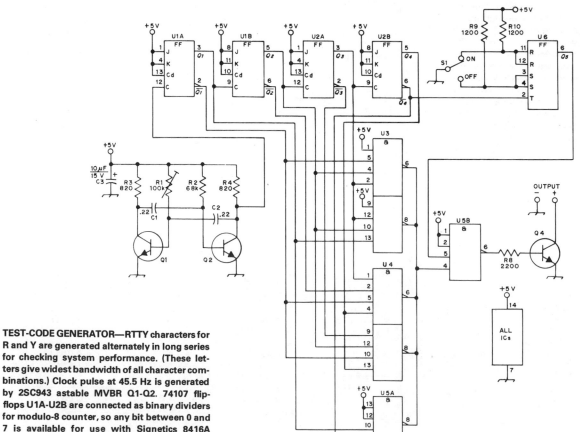

TEST-CODE GENERATOR—RTTY characters for R and Y are generated alternately in long series for checking system performance. (These letters give widest bandwidth of all character combinations.) Clock pulse at 45.5 Hz is generated by 2SC943 astable MVBR Q1-Q2. 74107 flip-flops U1A-U2B are connected as binary dividers for modulo-8 counter, so any bit between 0 and 7 is available for use with Signetics 8416A gates U3-U5 to feed desired character to 2SC372 output transistor. Automatic start/stop circuit using Fairchild 9945 clocked flip-flop U6 ensures that sequence always starts with R and ends with Y. Q4 conducts on mark and is cut off on space, for feeding frequency-shift keyer.—K.

Sekine, A Simple RY Code Generator for TTY, *QST,* Dec. 1974, p 20—24.

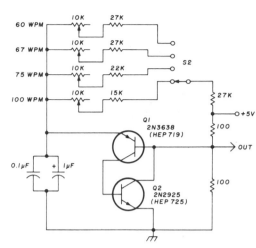

CLOCK FOR TEST MESSAGES—Generates negative-going pulse train at choice of four baud rates, to control RTTY test-message generator at four different speeds.—K. Ebneter and J. Romelfanger, RTTY Test-Message Generator, *Ham Radio*, Nov. 1976, p 30–32.

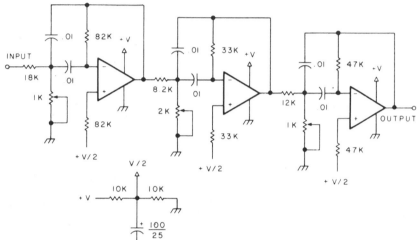

BANDPASS FOR 850-Hz RTTY SHIFT—Three-stage active Butterworth input filter passes 2125- and 2975-Hz tones plus modulation sidebands and allowance for drift in RTTY receiver. Inset shows how 741 opamps are biased when used with single power supply. Article gives step-by-step design and alignment procedures.—P. A. Stark, Design an Active RTTY Filter, *73 Magazine*, Sept. 1977, p 38–43.

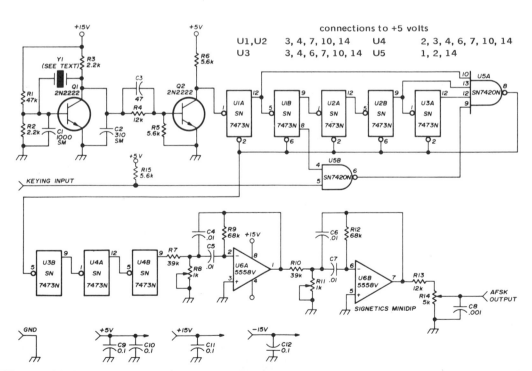

GENERATOR FOR 170-Hz SHIFT—Provides precise 2125- and 2295-Hz audio tones without requiring counter to establish correct frequency. Used for adjusting AFSK oscillator. Crystal can be 459.259 kHz (channel 48), which with appropriate divider chains gives output frequencies accurate within 2 Hz while preserving 170-Hz relative shift within 0.1 Hz. For even greater accuracy, order crystal that has been adjusted to exactly 459.000 kHz. When input is grounded, divide ratio is 25 to give 2295 Hz. When input is high, divide ratio is 27 to give 2125 Hz. Pin 11 of U1-U4 and pin 7 of U5 are grounded.—H. Nurse, Crystal Controlled AFSK Generator, *Ham Radio*, Dec. 1973, p 14–17.

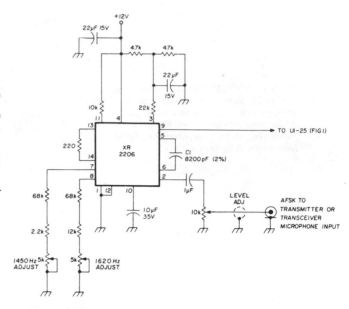

AFSK GENERATOR—Uses phase-continuous frequency shift to prevent out-of-band transients while generating radio frequencies of 1450 and 1620 Hz. Second harmonic is outside passband of modern SSB equipment. Frequency of sine wave is determined by C1 and total resistance connected to pin 7 or 8 of Exar XR2206.—E. Kirchner, Serial Converter for 8-Level Teleprinters, *Ham Radio,* Aug. 1977, p 67–73.

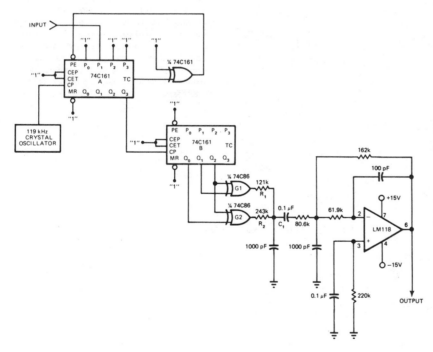

FSK FOR NRZ INPUT—Crystal-controlled frequency-shift keyer accepts nonreturn-to-zero digital input and generates 5-V P-P FSK output signal having less than 3% total harmonic distortion, at standard 2.125- and 2.975-kHz radioteletype frequencies. When input is low, counter A divides by seven; for high or logic 1 input, counter divides by five. Counter B divides by eight to produce required output frequencies. EXCLUSIVE-OR gates G_1 and G_2 generate first approximation to desired sine-wave output, for filtering by three-pole active Butterworth low-pass filter having 4.75-kHz cutoff.—K. Erickson, Frequency-Shift Keyer Features Rock-Steady Operation, *EDN Magazine,* Jan. 5, 1977, p 44.

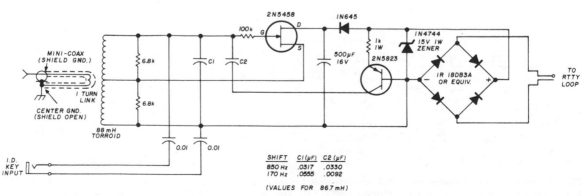

SHIFT	C1 (µF)	C2 (µF)
850 Hz	.0317	.0330
170 Hz	.0555	.0092

(VALUES FOR 86.7 mH)

170- OR 850-Hz SHIFT—Simple AF RTTY keyer uses 2N5823 silicon PNP transistor switch instead of optical coupler. Short piece of coax serves as 1-turn output link. Outer shield is grounded only at coaxial connector so braid acts as Faraday shield, eliminating capacitive signal and noise pickup from circuit.—E. Noll, Circuits and Techniques, *Ham Radio,* April 1976, p 40–43.

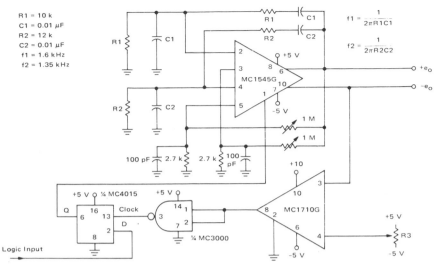

R1 = 10 k
C1 = 0.01 μF
R2 = 12 k
C2 = 0.01 μF
f1 = 1.6 kHz
f2 = 1.35 kHz

$$f_1 = \frac{1}{2\pi R_1 C_1}$$

$$f_2 = \frac{1}{2\pi R_2 C_2}$$

SELF-GENERATING FSK—Dual oscillators in Motorola MC1545G gated video amplifier are used with external frequency-determining components R1C1 and R2C2 to give 1.6 kHz for f1 and 1.35 kHz for f2. Logic switching network compares output to reference and updates gate input with each cycle of output. Circuit gives smaller switching transients than are possible with separate oscillators because one oscillator is driven at frequency of the other while first oscillator is off. R3 sets transition to any level desired.—"Gated Video Amplifier Applications—the MC1545," Motorola, Phoenix, AZ, 1976, AN-491, p 13.

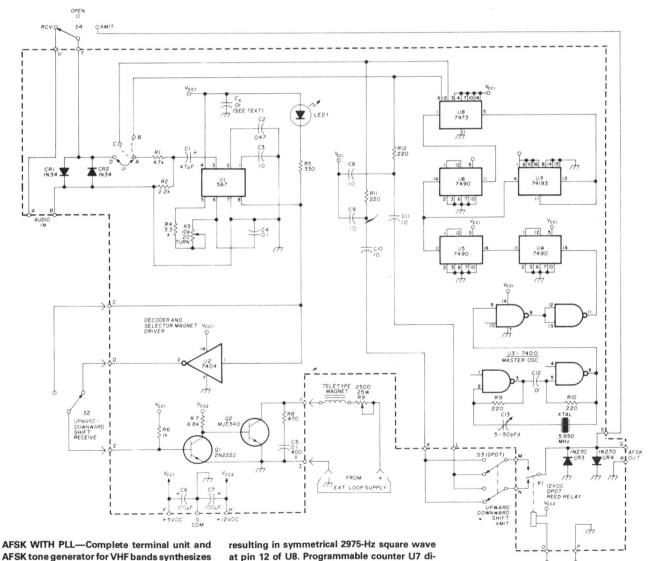

AFSK WITH PLL—Complete terminal unit and AFSK tone generator for VHF bands synthesizes tones digitally for 170-Hz narrow shift and 850-Hz standard shift found in VHF amateur RTTY bands. Additional feature is PLL circuit that follows drifting signal and copies signals from which mark or space information is missing. Precision AFSK generator consists of 5.95-MHz crystal oscillator U3, divide-by-100 ICs U4 and U5, divide-by-10 IC U6, and divide-by-2 IC U8, resulting in symmetrical 2975-Hz square wave at pin 12 of U8. Programmable counter U7 divides 59.5 kHz down to 4250 Hz. Other half of U8 divides this by 2, to give symmetrical 2125-Hz square wave at pin 9 of U8. Square waves are converted to trapezoids, with tops smoothed by CR3 and CR4, to give 1 V P-P quasi-sine waves at output. Demodulator U1 compares incoming frequencies to its internal current-controlled oscillator and generates digital signals when they are identical. Internal oscillator is locked to incoming signal if within detection bandwidth of about 220 Hz for 2125 Hz with 170-Hz shift. At 2975 Hz, detection bandwidth is about ±135 Hz.—J. Loughmiller, Digiratt—RTTY AFSK Generator and Demodulator, *Ham Radio*, Sept. 1977, p 26–28.

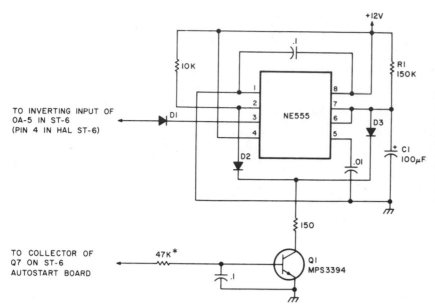

STOP FOR AUTOSTART—Uses 555 timer IC to make ST-6 autostart turn off motor of teleprinter if copying commercial station that does not drop its carrier when no text is being transmitted. Values shown for C1 and R1 give time of 15 s which, added to 25 s of ST-6 delay, gives about 40 s to turn off in presence of steady mark tone (carrier only). Useful when copying weather and press reports.—R. Bourgeois, Stop That Autostart, *73 Magazine,* May 1977, p 47.

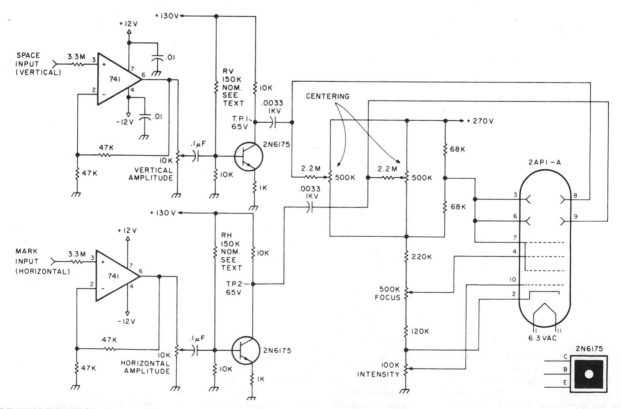

CRT TUNING INDICATOR—Crossed-ellipse display on CRT aids in tuning receiver to RTTY signal. Display shows at a glance if station is narrow or wide, or if other station is transmitting upside-down signals (mark and space frequencies reversed). Tuned for maximum amplitude of major axis of each ellipse; if transmitting station is wide or narrow, tune for equal amplitudes even though they are not maximized. Try different values of RV until T.P.1 voltage is 65 V, to center signal so it can swing equal amounts on either side. Adjust RH similarly.—R. R. Parry, RTTY CRT Tuning Indicator, *73 Magazine,* Sept. 1977, p 118–120.

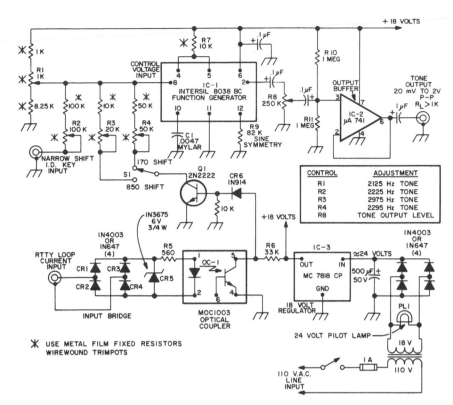

AFSK TONES—Generates tones needed for either 170- or 850-Hz frequency shift in automatic frequency-shift keying of RTTY equipment. Independent adjustments are provided for each tone. Sine-wave output has constant amplitude, with excellent tone frequency stability. Circuit permits plug-in operation in any RTTY loop, independent of loop polarity or grounding. Article covers construction and adjustment.—J. C. Roos, Universal AFSK Generator, *73 Magazine,* July 1974, p 37–40, 42, and 44–46.

CONTROL	ADJUSTMENT
R1	2125 Hz TONE
R2	2225 Hz TONE
R3	2975 Hz TONE
R4	2295 Hz TONE
R8	TONE OUTPUT LEVEL

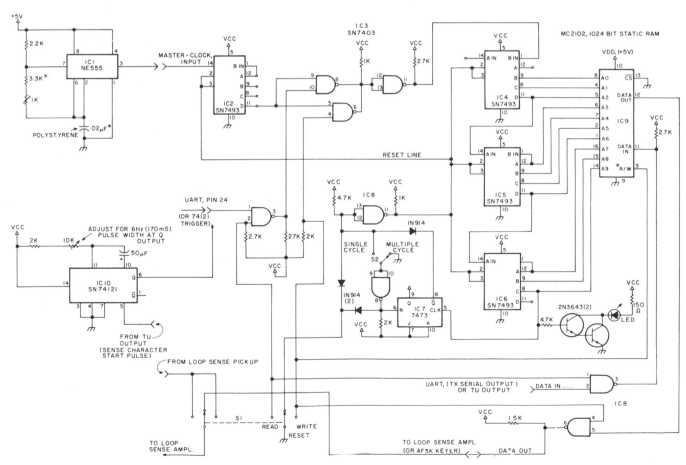

RAM FOR RTTY—Erasable MC2102 1024-bit RAM stores two Teletype lines (128 characters) of Baudot code for readout at machine speed. Can also serve in place of tape loop for frequently used code messages such as CQ calls. Values shown with IC1 timer are for 728-Hz master clock (16 × 45.45 bauds). Stored message is volatile, disappearing when power is turned off. For permanent storage, use ROM in place of RAM.—H. P. Fischer, RTTY Scratchpad Memory, *73 Magazine,* June 1977, p 54–55.

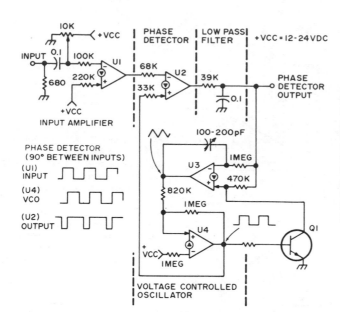

PLL FOR RTTY—Complete phase-locked loop uses all four sections of LM3900 quad linear opamp. Variable capacitor is set to give center frequency of 2.2 kHz for VCO. Once in lock, loop will maintain lock over range of 1.55 to 2.9 kHz, to cover tones normally used in RTTY. Additional keying circuit for TTY selector magnets and more filtering of output completes setup for driving printer. Q1 is 2N706, with 33K resistor in base circuit.—C. Sondgeroth, More PLL Magic, *73 Magazine*, Aug. 1976, p 56–59.

FSK DEMODULATOR—Uses IC originally developed for stereo multiplex decoders, containing phase-locked loop suitable for demodulating teleprinter FSK signals. Circuit shown requires only small input signal for phase lock, gives visual indication with lamp when phase lock has occurred, and requires only pair of 2N3055 drive transistors between outputs A-B and teleprinter receiving solenoids. Article also gives this output circuit and setup procedure.— K. S. Beddoe, Teleprinter Terminal Unit Uses Phase-Locked Loop, *Wireless World*, Dec. 1973, p 605.

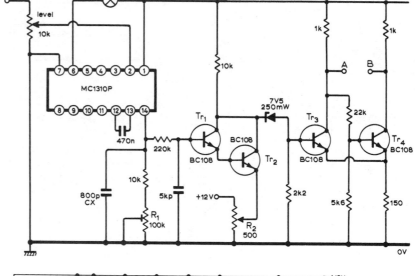

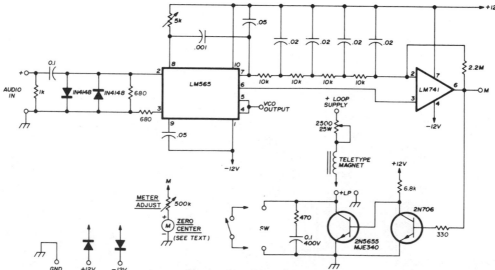

UPWARD-SHIFT RTTY DEMODULATOR—PLL demodulator serves for copying AFSK/FSK upward-shift RTTY signals. 2N706 switches 2N5655 on and off, reversing polarity of voltage from LM741 comparator on mark/hold as needed for smooth upward-shift copy. Works equally well on wide-shift or narrow-shift signals. Zero-center tuning meter will show full-scale minus (left) reading on mark/hold signal. Meter may not be needed on AFSK.—N. Stinnette, Update of the Phase-Locked Loop RTTY Demodulator, *Ham Radio*, Aug. 1976, p 16–17.

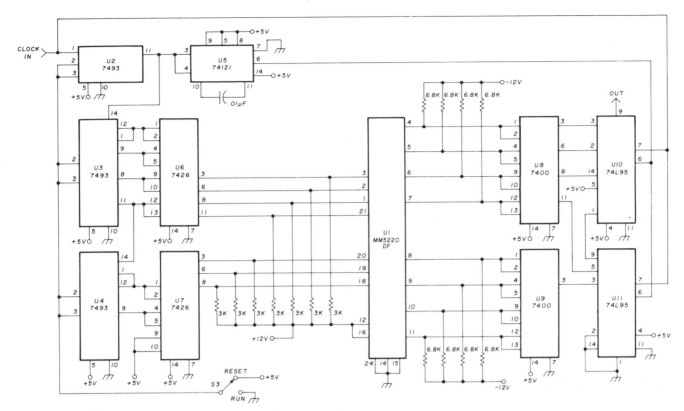

QUICK BROWN FOX GENERATOR—TTL ICs and National MM5220DF preprogrammed MOS read-only memory chip together generate standard RTTY test message: THE QUICK BROWN FOX JUMPS OVER THE LAZY DOG 1234567890 DE. Requires external clock providing sharp negative-going pulse train at frequency corresponding to RTTY speeds desired. Requires two supplies, for +5 V and ±12 V.—K. Ebneter and J. Romelfanger, RTTY Test-Message Generator, *Ham Radio,* Nov. 1976, p 30–32.

CHAPTER 92

Television Circuits

Covers circuits for black-and-white, color, industrial, and slow-scan amateur TV receivers, including infrared remote control, microprocessor interface, and test equipment. See also Cathode-Ray, Game, and Microprocessor chapters.

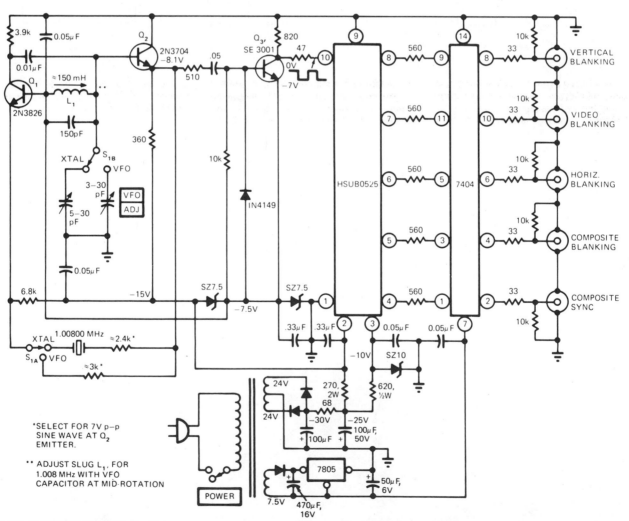

CRYSTAL/VFO SYNC GENERATOR—With crystal, can be used as system sync generator or as drive for staircase generators, custom pattern generators, and special TV test equipment. When VFO-controlled, circuit provides variable-frequency sync needed for determining pull-in range of sweep oscillators in TV sets and VTR decks. Clock-pulse section uses Q_1 and Q_2 as 1.00800-MHz Butler oscillator. On VFO operation, oscillator frequency can be varied ±3.5% from mean. Sine-wave output of Q_2 is converted to square wave by Q_3 for application to clock input of Hughes HSUB0525 sync generator. Each output of sync generator feeds one of inverters in 7404, providing 3.3 V P-P signal into 75-ohm load for each output.—M. J. Salvati, VFO Adds Versatility to TV Sync Generator, *EDN Magazine,* May 20, 1974, p 70 and 72.

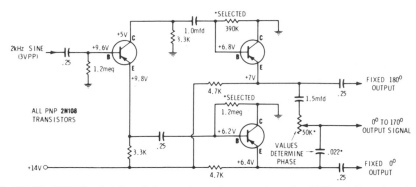

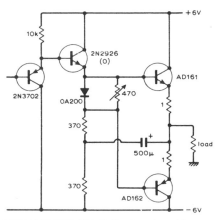

AF PHASE SHIFTER—Developed for testing chroma demodulator in color TV receiver. Audio oscillator is used as source of sine waves. First stage is phase inverter, followed by two emitter-followers. Resulting output signals of opposite phase are combined through a small capacitor (0.022 μF, selected for frequency used) from one channel and 50K variable resistor from other channel. 90° phases are judged by position of sine waves on screen of CRO; 90° is halfway between 0° and 180°.—C. Babcoke, Waveforms Explain Chroma Demodulators, *Electronic Servicing,* Sept. 1972, p 22–23, 26–28, and 30.

SSTV DEFLECTION DRIVE—Developed for use in line time-base amplifier of 4-Hz slow-scan TV system. Emitter-follower in driver stage is used with first transistor to match output impedance of UJT sawtooth oscillator. Will drive deflection coils of 17-inch CRT (coil resistance about 5 ohms).—M. Hadley, Deflection Coil Driver for Slow-Scan Television, *Wireless World,* March 1974, p. 18.

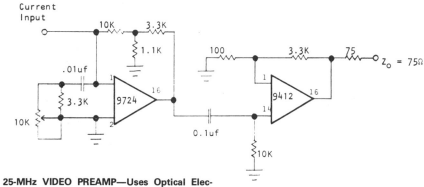

25-MHz VIDEO PREAMP—Uses Optical Electronics 9724 opamp for input stage and 9412 opamp for output stage. Current input can be from vidicon or image orthicon camera tube. Input compensation can be adjusted to provide aperture correction. Feedback network for input opamp minimizes effects of stray capacitance. Values shown give 1-V output for 1-μA input.—"A 25 MHz Video Preamplifier—Line Driver," Optical Electronics, Tucson, AZ, Application Tip 10195.

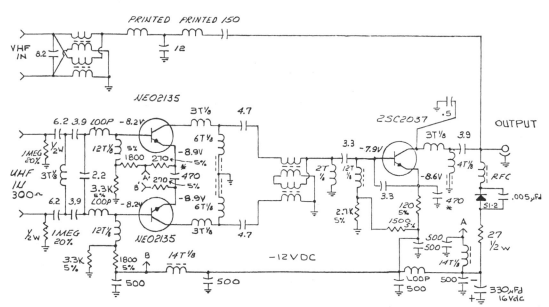

LOW-NOISE UHF PREAMP—Special push-pull input stage using low-noise UHF transistors gives average preamp noise figure as low as 2.2 dB. Can be used with 300-ohm line of broadband UHF antenna without usual balun transformer or differential input stage. Balun is used after amplifier to transform push-pull output to input of single-ended second stage without degrading noise figure. Developed for use with new deep-fringe-area UHF TV antenna having three flat in-line director elements, for over-the-air reception of UHF TV programs in areas previously having no watchable pictures.—J. E. Kluge, Advanced Antenna Design and an Ultra-low-Noise Preamplifier Extend UHF Viewing Area, *IEEE Transactions on Broadcasting,* March 1977, p 17–22.

CHROMA PROCESSOR—Combination of RCA CA3121E chroma amplifier/demodulator and CA3070 chroma signal processor provides automatic chroma control and color killer sensing along with other functions required for high-level B − Y, R − Y, and G − Y color difference signals having low impedances for driving high-level R, G, and B output amplifiers.—"Linear Integrated Circuits and MOS/FET's," RCA Solid State Division, Somerville, NJ, 1977, p 359–360.

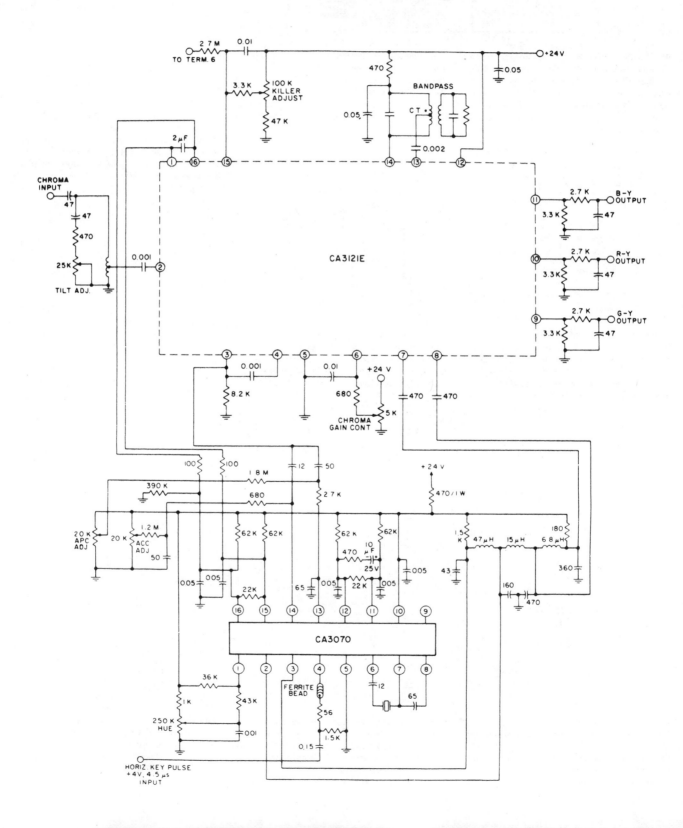

TV TURNOFF WITH WARNING BEEP—Timer providing turnoff delays up to 1 h gives warning beep about 30 s before turnoff to permit resetting if desired for watching remainder of particular program. Can be mounted inside TV set or in small chairside box connected to set by cable. Photocell can be substituted for ON/OFF switch to permit remote control with flashlight while leaving entire circuit in TV set. Can be operated from 9-V battery if this voltage is not available in receiver. Momentary closing of switch turns on TV and initiates timing cycle. With 3.6-megohm pot for R, maximum delay is 1 h. For 30-min delay, use 1.3 megohms. Setting of R determines exact delay.—J. Sandler, 9 Projects under $9, *Modern Electronics*, Sept. 1978, p 35–39.

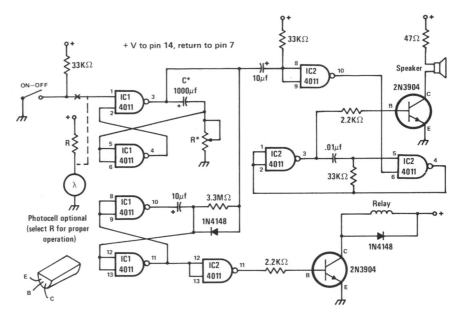

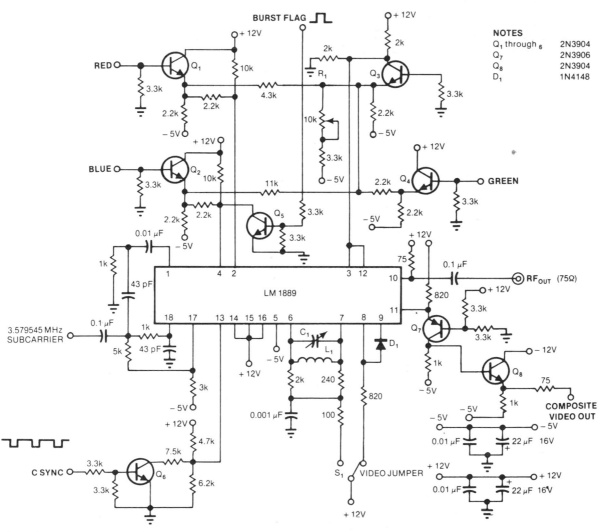

COMPOSITE COLOR SIGNAL GENERATOR—Single LM1889 encoder chip produces standard composite color video signal from separate sync, burst flag, 3.579545-MHz subcarrier, and 0–4 V red, green, and blue inputs. Subcarrier should be 1–5 V P-P. Modulated RF output can go to cable input of TV set through 75-ohm cable. Applications include TV mixing effects and video games.—L. Trottier and B. Matic, Signal Encoder Generates Composite Color, *EDN Magazine*, Aug. 20, 1978, p 148 and 150.

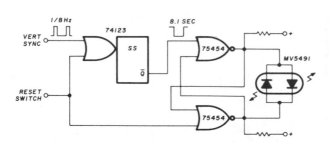

SSTV VERTICAL SYNC INDICATOR—Retriggerable mono MVBR using SN74123 is triggered by 0.125-Hz vertical sync pulse of amateur television system. Absence of pulse allows mono to time out and change state of LED from green to red. Vertical reset switch must then be used to restart vertical sweep and reset LED. Uses Monsanto MV5491 dual red/green LED, with 220 ohms in upper lead to +5 V supply and 100 ohms in lower +5 V lead because red and green LEDs in parallel back-to-back have different voltage requirements. Drivers for LEDs are SN75454.—K. Powell, Novel Indicator Circuit, *Ham Radio*, April 1977, p 60–63.

LINEAR AMPLIFIER—Motorola MHW-710 power module boosts 1-W output of VHF Engineering TX-432B crystal-controlled solid-state exciter to about 10 W for simple amateur TV transmitter. Interconnections are made with short lengths of RG-174 coax. Input jack J1 is connected to exciter by 50 ft of RG-174, and length is gradually reduced until proper drive level is obtained for linear operation. RFC uses 8 turns No. 22 enamel on 1-megohm 1-W resistor. To cover 400–440 MHz, use MHW-710-1; for 440–470 MHz, use MHW-710-2.—R. E. Taggart, Interested in Television?, *73 Magazine*, Oct. 1977, p 164–174.

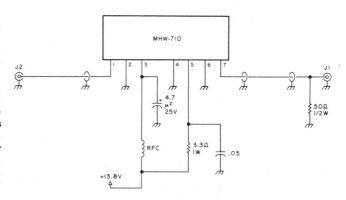

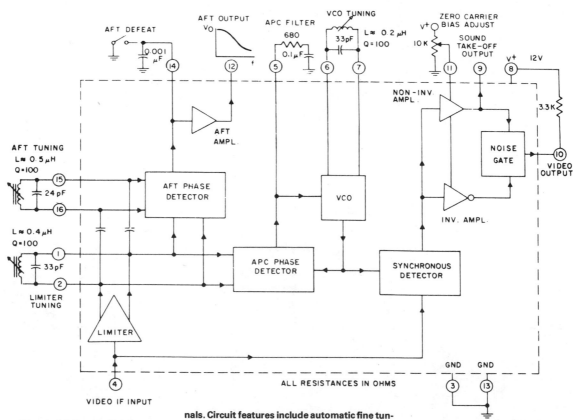

SYNCHRONOUS VIDEO DETECTOR—RCA CA3136E serves as video IF PLL synchronous detector for color TV receivers. Phase-locked oscillator demodulates 45.75-MHz video IF signals. Circuit features include automatic fine tuning voltage (AFT) for DC control of tuner, adjustment of zero-carrier DC level at video output terminal, amplifier arrangement for inverting noise impulses toward black level, and separate noninverting output terminal for sound IF. Requires single 12-V supply.—"Linear Integrated Circuits and MOS/FET's," RCA Solid State Division, Somerville, NJ, 1977, p 374.

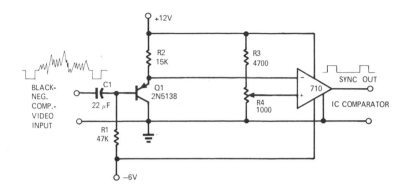

COMPARATOR SEPARATES SYNC PULSES—
By setting DC reference input of 710 comparator
at 0.15 VDC, only horizontal sync pulses are extracted from composite black-negative video
signal to appear at comparator output. Setting
reference level at 0.35 VDC gives only blanking
pulses at output.—R. G. Groom, IC Comparator
Separates Sync Pulses, *EDN Magazine,* Sept.
15, 1970, p 53–54.

VIDEO MODULATOR—Developed for use between solid-state TV camera and VHF Engineering TX-432B crystal-controlled solid-state
exciter to give simple amateur TV transmitter.
For tube-type cameras, add 1N914 or other
small-signal silicon diodes in series with input
until modulator provides proper video swing.
Article gives construction details.—R. E. Taggart, Interested in Television?, *73 Magazine,*
Oct. 1977, p 164–174.

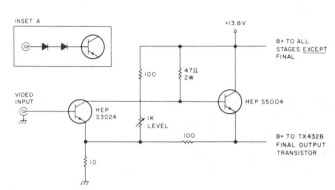

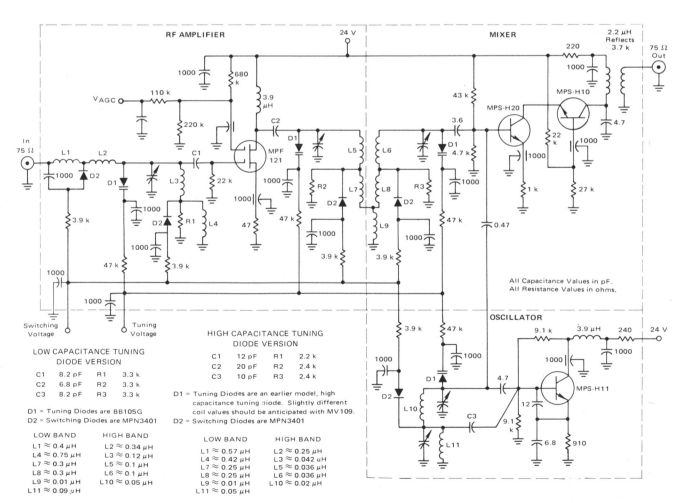

LOW CAPACITANCE TUNING DIODE VERSION

C1	8.2 pF	R1	3.3 k
C2	6.8 pF	R2	3.3 k
C3	8.2 pF	R3	3.3 k

D1 = Tuning Diodes are BB105G
D2 = Switching Diodes are MPN3401

LOW BAND	HIGH BAND
L1 ≈ 0.4 µH	L2 ≈ 0.34 µH
L4 ≈ 0.75 µH	L3 ≈ 0.12 µH
L7 ≈ 0.3 µH	L5 ≈ 0.1 µH
L8 ≈ 0.3 µH	L6 ≈ 0.1 µH
L9 ≈ 0.01 µH	L10 ≈ 0.05 µH
L11 ≈ 0.09 µH	

HIGH CAPACITANCE TUNING DIODE VERSION

C1	12 pF	R1	2.2 k
C2	20 pF	R2	2.4 k
C3	10 pF	R3	2.4 k

D1 = Tuning Diodes are an earlier model, high
capacitance tuning diode. Slightly different
coil values should be anticipated with MV109.
D2 = Switching Diodes are MPN3401

LOW BAND	HIGH BAND
L1 ≈ 0.57 µH	L2 ≈ 0.25 µH
L4 ≈ 0.42 µH	L3 ≈ 0.042 uH
L7 ≈ 0.25 µH	L5 ≈ 0.036 µH
L8 ≈ 0.25 µH	L6 ≈ 0.036 µH
L9 ≈ 0.01 µH	L10 ≈ 0.02 µH
L11 ≈ 0.05 µH	

VHF VARACTOR TUNER—DC bias voltages are
used in place of mechanical switches for channel selection. Values for tuned circuits depend
on varactor diode used. With high-capacitance
varactor diode, tuning voltage ranges from 4.3
V for channel 2 to 23 V for channel 13. Corresponding voltages for low-capacitance varactor
are 2.2 V and 20.4 V. Tuner noise figure is in
range of 4–5 dB.—J. Hopkins, "Printed Circuit
VHF TV Tuners Using Tuning Diodes," Motorola, Phoenix, AZ, 1972, AN-544A, p 4.

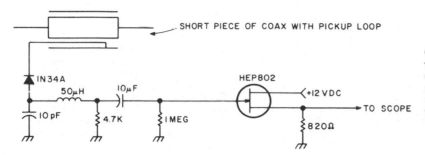

→ SHORT PIECE OF COAX WITH PICKUP LOOP

CRO AS TV MONITOR—Permits monitoring transmitted amateur television signals with oscilloscope, for such applications as checking sync levels and sync-pulse shape. Outgoing signal can be monitored while adjusting modulator.—Circuits, *73 Magazine,* March 1977, p 152.

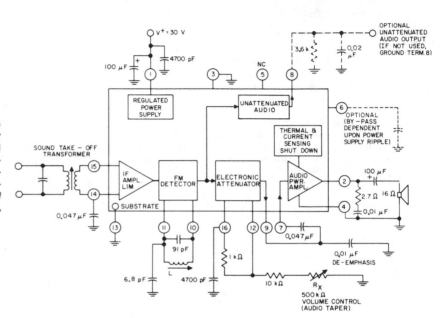

SOUND SUBSYSTEM—RCA CA3134 combines sound IF and audio output stages in single IC for use in TV receivers. Input is taken from sound IF output of receiver. Provides electronic volume control with improved taper. Alternate circuit shown provides unattenuated audio output.—"Linear Integrated Circuits and MOS/FET's," RCA Solid State Division, Somerville, NJ, 1977, p 368.

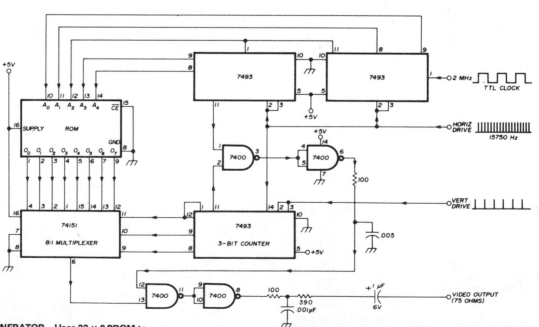

ATV CALL GENERATOR—Uses 32 × 8 PROM to generate up to six characters of amateur call. Squares in matrix are numbered 1–32 horizontally and 1–8 vertically starting from upper left, and black squares forming call letters are programmed as 1s in PROM. Pin connections shown for PROM are valid for AMI 27508/27509, 82S23/82S123, MM5330/MM5331, HPROM 8256, and IM5600/5610. Two 7493 binary counters address all 32 words in ROM, with clock rate (2–3 MHz) determining length of characters on screen. 74151 multiplexer advances to next ROM output once per scan line, under control of 7493 3-bit counter clocked by horizontal drive pulses from sync generator of ATV transmitter. Positive-going horizontal drive pulses reset 5-bit word counters, while positive-going vertical drive pulses reset 3-bit line counters, to make characters appear in same position on screen for all fields.—J. Pulice, Amateur Television Callsign Generator, *Ham Radio,* Feb. 1977, p 34–35.

SYNC SEPARATOR—Input video having negative synchronizing pulses is applied to Q_1 through 3.58-MHz notch filter L_t-C_t to remove color subcarrier components. Circuit is set up to conduct only on negative peaks, when Q_1, Q_2, and CR_1 are all on, so feedback is 100% at this time. Negative peaks of output then follow input exactly. C_2 acts as memory for negative peaks, storing their level between sync pulses.—W. Jung, An Operational Approach to Sync Separation, *EDN/EEE Magazine*, July 15, 1971, p 48–49.

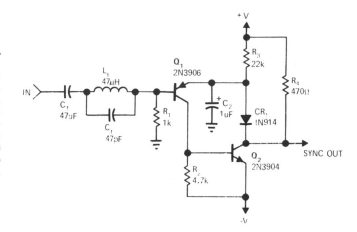

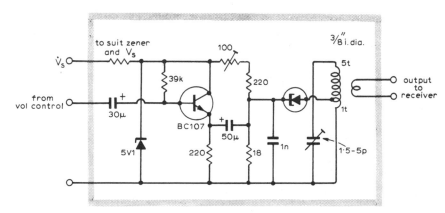

IMPROVING TV SOUND—AF signal from volume control of TV receiver is converted to FM signal by using BC107 transistor to frequency-modulate tunnel-diode oscillator operating within FM broadcast band. Oscillator output is fed through air-core transformer and coaxial line to FM receiver of high-fidelity sound system. Arrangement eliminates most of distortion introduced in power amplifier and loudspeaker of average TV set. Use shielding to keep unwanted FM radiation at minimum.—A. J. Smith, Improving Television Sound, *Wireless World*, Aug. 1973, p 373.

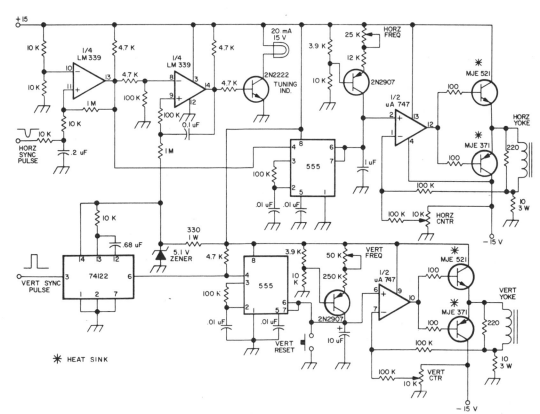

SWEEP FOR SSTV MONITOR—Uses two 555 timers, one as oscillator and other as linear sawtooth generator. Adjust R4 so oscillator period is slightly longer than interval between sync pulses. When sync is lost, oscillator runs very close to correct frequency and locks in again instantly on first good sync pulse. Circuit also has pulse stretcher, along with lamp driver that operates from horizontal sync pulses for use as tuning indicator.—R. L. Anderson, 555 Timer Sweep Circuit for SSTV, *73 Magazine*, May 1976, p 134–136.

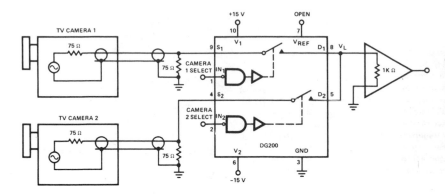

GENERAL-PURPOSE VIDEO SWITCH—When used for switching TV cameras, DG200 CMOS analog switch provides 45-dB isolation at 10 MHz between on and off cameras. Insertion loss of switch is 0.5 dB. For greater isolation, use additional analog switch in each camera line.—"Analog Switches and Their Applications," Siliconix, Santa Clara, CA, 1976, p 7-70.

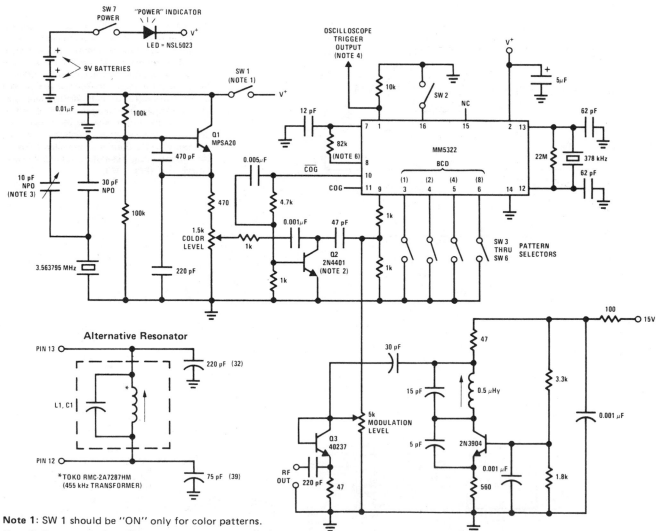

Note 1: SW 1 should be "ON" only for color patterns.

Note 2: Do not substitute Q2.

Note 3: Variable cap may be used to trim color crystal to exact frequency.

Note 4: SW 2 and 10k resistor on pins 16 and 1 are needed only if scope trigger pulse is desired.

Note 5: SW 2 selects "H" or "V" trigger output pulses.

Note 6: A 27k resistor in series with a 100k trimpot may be used in place of 82k resistor for variable vertical line width.

Note 7: Modulation level adjusted for best patterns as viewed on TV screen.

COLOR BAR GENERATOR—National MM5322 chip forms complete dot-bar and color hue generation system. Chip divides internal crystal-controlled oscillator frequency to provide timing, sync, and video information required for aligning color TV receivers. Composite video output serves for complete black-and-white dot-bar operation to give variety of screen patterns. Separate output is provided for precise gating of 3.56-MHz color bursts.—"MOS/LSI Databook," National Semiconductor, Santa Clara, CA, 1977, p 4-18–4-22.

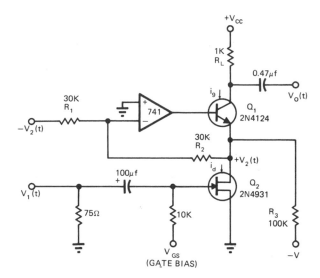

FET LINEAR MODULATOR—Circuit developed for closed-circuit industrial color television system uses linear portion of operating characteristic for 2N4931 FET to provide linear response at modulation frequencies from 1 MHz down to near zero. Article gives design equations.—G. R. Shapiro, Analog Multipliers Offer Solutions to Video Modulation Problems, *EDN Magazine,* Sept. 1, 1972, p 40–41.

IR TRANSMITTER FOR TV SOUND—Mono audio output of TV receiver is fed to infrared modulator using Intersil 8038 IC and transistor, to provide pulse-frequency-modulated infrared output that can be picked up by compact receiver built into headphones. S/N ratio is 58 dB in daylight in average living room having light walls and ceiling, but drops to 40 dB when receiver faces away from transmitter. Used in German TV receivers displayed at 1975 Berlin Exhibition.—International Radio and Television Exhibition, *Wireless World,* Nov. 1975, p 521–524 and 539.

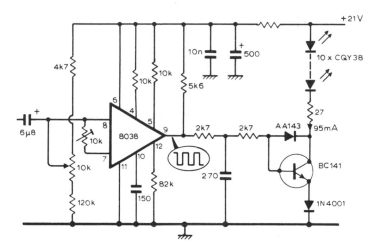

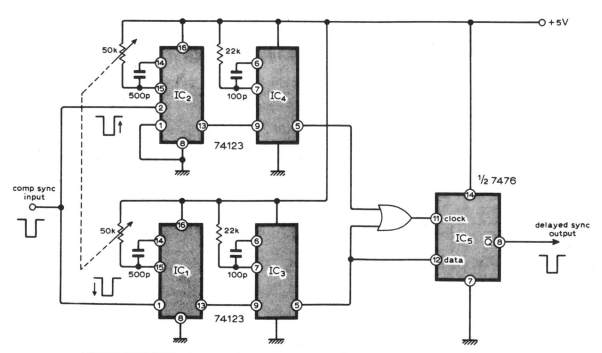

VARIABLE DELAY UP TO 7 μs—Used in television broadcasting when longer delay is required than can be achieved with passive elements for composite signal.—C. M. Wong, Sync-Pulse Delay, *Wireless World,* Feb. 1977, p 46.

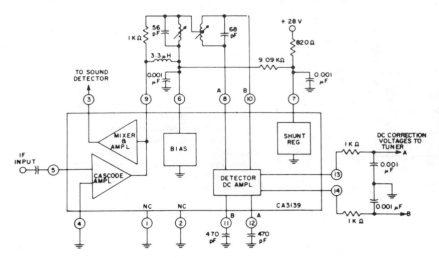

AFT SUBSYSTEM—RCA CA3139 automatic fine tuning IC combined with intercarrier mixer/amplifier for color and monochrome receivers provides AFT voltage for tuner correction and amplified 4.5-MHz intercarrier sound signal for external FM sound detector of receiver. Input is taken from output of IF amplifier in receiver.—"Linear Integrated Circuits and MOS/FET's," RCA Solid State Division, Somerville, NJ, 1977, p 381.

RGB OUTPUT—Motorola MDS21 high-voltage silicon transistors serve as output stages for red, green, and blue channels of color TV receiver, to provide video amplitude requirements for color picture tube. Transistors can be driven directly by most types of chroma demodulators.—"NPN Silicon Annular High Voltage Amplifier Transistors," Motorola, Phoenix, AZ, 1978, DS 3364.

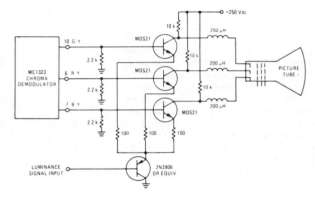

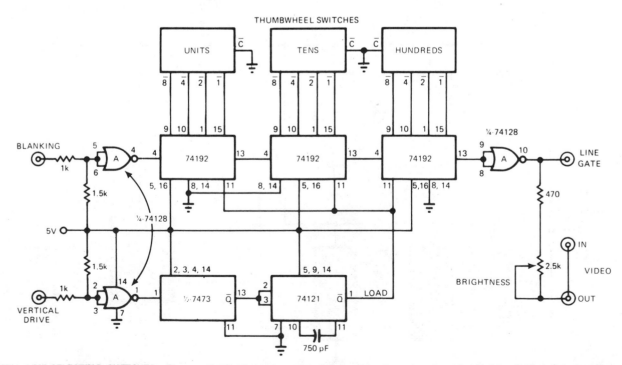

DIGITAL LINE-SELECTING SWITCHES—Three thumbwheel switches connected in binary mode control three 74192 counters, for selection of any desired line up to 999 in television field. Line-gate pulse injects into looped-through video for brightening selected line to make it visible on display. Circuit can also be used to determine exact number of active lines in each television field. Article describes operation in detail.—H. F. Stearns, Build a Thumbwheel-Switched Television Line Selector, *EDN Magazine,* June 20, 1976, p 124.

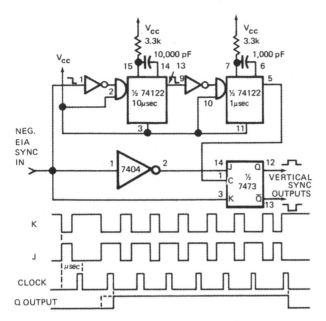

VERTICAL SYNC SEPARATOR—Arrangement uses controlled clocking sequence of JK flip-flop to detect presence of vertical sync interval in standard EIA television composite sync waveform. Complementary sync waveforms J and K are fed to J and K inputs of 7473 flip-flop, which is clocked by 1-μs pulse that is delayed slightly longer than 10-μs horizontal sync interval. First clock pulse after 11-μs interval changes flip-flop output Q to 1, where it stays for six clock periods before reverting to 0 state after vertical sync interval has passed.—W. G. Jung, Vertical Sync Separator Has No Integrating Network, *EDN Magazine*, Oct. 15, 1972, p 57.

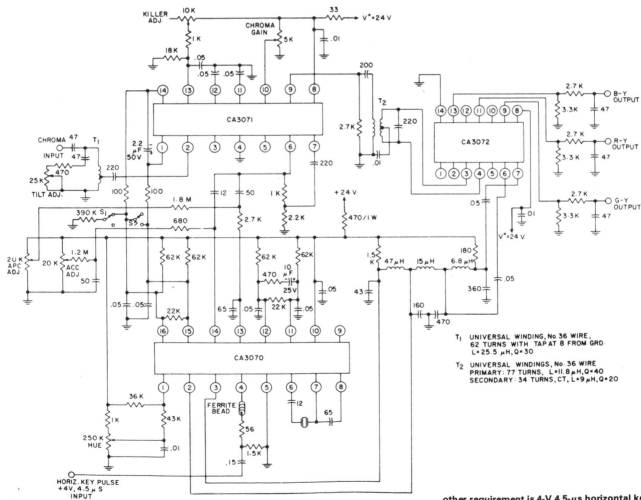

CHROMA SYSTEM—Uses RCA CA3070 as subcarrier regenerator, CA3071 as chroma amplifier, and CA3072 as chroma demodulator. Input can be taken from either first or second video stage of color TV receiver. Outputs from system are color difference signals for driving high-level amplifiers. Operates from single 24-V supply that should be maintained within 3 V. Only other requirement is 4-V 4.5-μs horizontal keying pulse centered on color burst. Crystal oscillator generates 3.579545 MHz.—"Linear Integrated Circuits and MOS/FET's," RCA Solid State Division, Somerville, NJ, 1977, p 345–346.

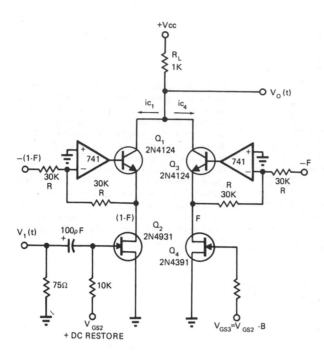

MODULATOR/MULTIPLIER—Balanced direct-coupled FET modulator/multiplier was developed for closed-circuit industrial color television system. Opamps handle modulation frequencies up to 1 MHz while providing linear response down to near zero modulation.—G. R. Shapiro, Analog Multipliers Offer Solutions to Video Modulation Problems, *EDN Magazine*, Sept. 1, 1972, p 40–41.

MICROPROCESSOR-SSTV INTERFACE—Digital-to-analog-to-frequency converter for slow-scan television permits direct generation of simple graphic and alphameric characters by microprocessor, without use of camera. U1 is CD4051 CMOS analog multiplexer, and U2A is one section of 74LS124 TTL dual voltage-controlled oscillator. Picture format uses 64 different lines, repeated once to give total of 128 lines, with maximum of 64 different picture elements per horizontal line each having one of four shades of gray. Separate pot is provided for setting each of five different levels so VCO oscillates at proper frequencies: sync—1200 Hz; black—1500 Hz; dark—1767 Hz; light—2033 Hz; white—2300 Hz. Article covers operation in detail and gives flow diagrams for microprocessor subroutines required.—B. Sanderson, SSTV Pictures from Your Microcomputer, *QST*, Oct. 1978, p 25–29.

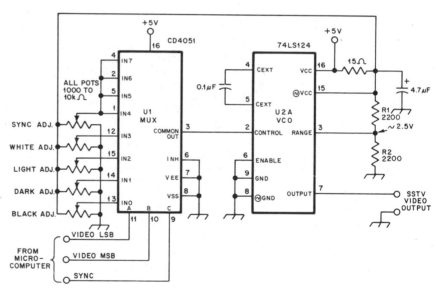

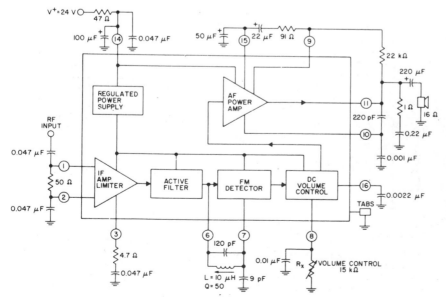

SOUND IF SUBSYSTEM—Single RCA CA1190GQ IC combines sound IF, FM detector, and complete audio amplifier for driving 8-, 16-, or 32-ohm loudspeaker in TV receiver. Nominal power output is 3 W. Electronic volume control on chip provides improved taper with single 15K wirewound control.—"Linear Integrated Circuits and MOS/FET's," RCA Solid-State Division, Somerville, NJ, 1977, p 301.

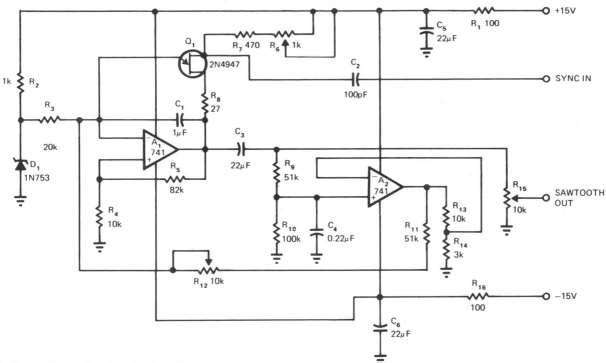

S-SHAPED RAMP—Developed for 60-Hz vertical deflection in high-resolution video display requiring highly linear ramp summed with second integral of ramp to give S shaping of deflection so sweep is linear on flat screen. Opamp A_1 is connected as integrator that takes integral of constant voltage across zener D_1. Period of integration is limited by UJT Q_1 that resets integrating capacitor C_1 when negative-going sync signal is applied to base 2 of Q_1. Sawtooth linearity can be trimmed by adjusting ratio of R_4 to R_5. Sync range is wide enough so external vertical hold can be eliminated. Amount of shaping can be adjusted with pot.—L. G. Smeins, "S"-Shaped Sawtooth Oscillator, *EDN Magazine,* Feb. 20, 1974, p 83 and 85.

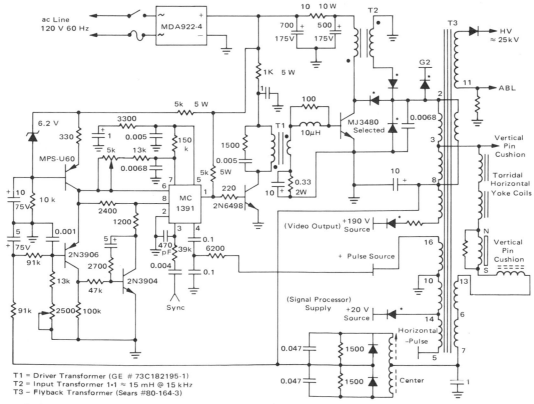

T1 = Driver Transformer (GE # 73C182195-1)
T2 = Input Transformer 1·1 ≈ 15 mH @ 15 kHz
T3 = Flyback Transformer (Sears #80-164-3)

*MR918 selected to 1400V

HORIZONTAL SYSTEM FOR 19-INCH COLOR—Self-regulating scan system includes short-circuit protection. Provides excellent high-voltage regulation at 25 kV. Vertical yoke current is also stabilized since it is powered from auxiliary flyback winding. System consumes 30% less power than more conventional circuit using SCR half-wave regulated supply.—R. J. Valentine, "A Self-Regulating Horizontal Scan System," Motorola, Phoenix, AZ, 1975, AN-750, p 7.

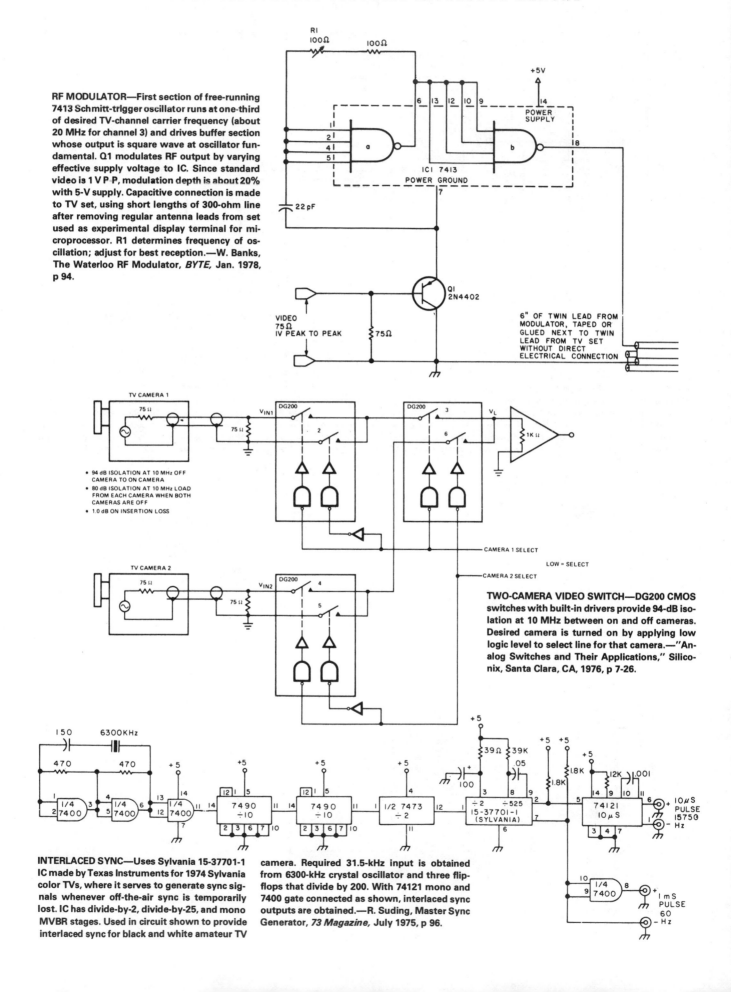

RF MODULATOR—First section of free-running 7413 Schmitt-trigger oscillator runs at one-third of desired TV-channel carrier frequency (about 20 MHz for channel 3) and drives buffer section whose output is square wave at oscillator fundamental. Q1 modulates RF output by varying effective supply voltage to IC. Since standard video is 1 V P-P, modulation depth is about 20% with 5-V supply. Capacitive connection is made to TV set, using short lengths of 300-ohm line after removing regular antenna leads from set used as experimental display terminal for microprocessor. R1 determines frequency of oscillation; adjust for best reception.—W. Banks, The Waterloo RF Modulator, *BYTE,* Jan. 1978, p 94.

TWO-CAMERA VIDEO SWITCH—DG200 CMOS switches with built-in drivers provide 94-dB isolation at 10 MHz between on and off cameras. Desired camera is turned on by applying low logic level to select line for that camera.—"Analog Switches and Their Applications," Siliconix, Santa Clara, CA, 1976, p 7-26.

INTERLACED SYNC—Uses Sylvania 15-37701-1 IC made by Texas Instruments for 1974 Sylvania color TVs, where it serves to generate sync signals whenever off-the-air sync is temporarily lost. IC has divide-by-2, divide-by-25, and mono MVBR stages. Used in circuit shown to provide interlaced sync for black and white amateur TV camera. Required 31.5-kHz input is obtained from 6300-kHz crystal oscillator and three flip-flops that divide by 200. With 74121 mono and 7400 gate connected as shown, interlaced sync outputs are obtained.—R. Suding, Master Sync Generator, *73 Magazine,* July 1975, p 96.

CHAPTER 93
Temperature Control Circuits

Variety of circuits maintain temperature at desired preset value to within as little as 0.0000033°C. Special features include overshoot compensation, anticipation control, differential control, and proportioning control. See also Power Control and Temperature Measuring chapters.

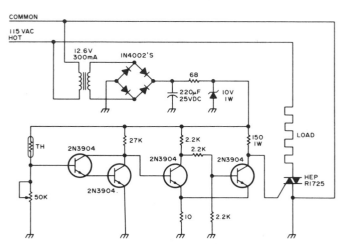

COOKER CONTROL—Uses Model K600A thermistor (Allied Electronics) placed in slow cooker to maintain ideal cooking temperature. Pot can be adjusted to provide triggering for ON/OFF control of heating element for cooker. Use polarized power plug for proper operation.—Circuits, *73 Magazine*, May 1977, p 19.

PROPORTIONING CONTROLLER—National LM122 timer is used as proportioning temperature controller with optical isolation and synchronized zero-crossing. R2 is used to set temperature to be controlled by thermistor R1. SCR used for Q2 is chosen to handle required load. D3 is rated at 200 V. R12, R13, and D2 implement synchronized zero-crossing feature.—C. Nelson, "Versatile Timer Operates from Microseconds to Hours," National Semiconductor, Santa Clara, CA, 1973, AN-97, p 8.

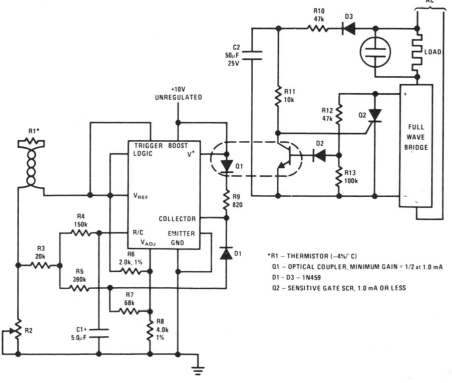

*R1 — THERMISTOR (−4%/°C)
Q1 — OPTICAL COUPLER, MINIMUM GAIN = 1/2 at 1.0 mA
D1 - D3 — 1N459
Q2 — SENSITIVE GATE SCR, 1.0 mA OR LESS

OVEN CONTROL—Simple circuit using RCA CA3059 zero-crossing switch regulates ON and OFF intervals of low-current SCR that controls solenoid in electric or gas oven. Sensor resistor has negative temperature coefficient. R_ν is set for desired control temperature.—E. M. Noll, "Linear IC Principles, Experiments, and Projects," Howard W. Sams, Indianapolis, IN, 1974, p 323.

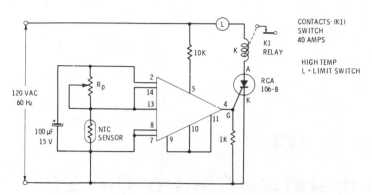

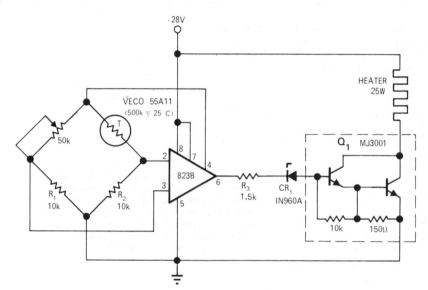

75–250°C OVEN CONTROL—Provides proportional temperature control of small oven to within 1°C over temperature range. Uses 823B voltage regulator operating from same 28-V source as oven. Temperature-setting pot should be 10-turn wirewound. Power transistor Q_1 operates either saturated or almost cut off, so no heatsink is required.—R. L. Wilbur, Proportional Oven-Temperature Controller, *EDN/EEE Magazine*, Sept. 15, 1971, p 45.

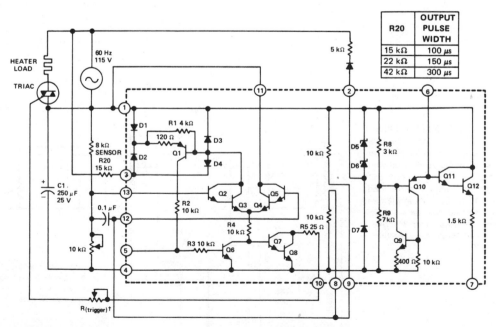

R20	OUTPUT PULSE WIDTH
15 kΩ	100 µs
22 kΩ	150 µs
42 kΩ	300 µs

ON/OFF HEATER CONTROL—Uses Texas Instruments SN72440 zero-voltage switch to trigger triac that turns AC heater on and off in accordance with demands of 8K resistance-type temperature sensor. One output pulse per zero crossing of AC line voltage is either inhibited or permitted by action of differential amplifier and resistance bridge circuit in IC. Width of output pulse at pin 10 is controlled by trigger pot R20 as given in table and should be varied to suit triggering characteristics of triac used.—"The Linear and Interface Circuits Data Book for Design Engineers," Texas Instruments, Dallas, TX, 1973, p 7-37.

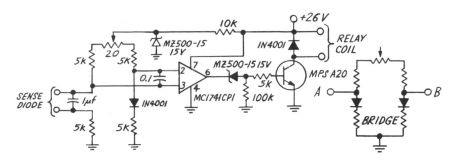

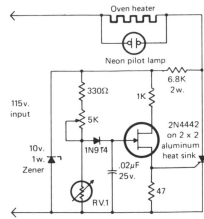

SILICON-DIODE SENSOR—Ordinary silicon diode having temperature coefficient of about −2 mV/°C over wide temperature range serves for sensing temperature differentials up to ±10°F with resolution of about 0.3°F. Two diodes connected in resistor bridge provide voltage proportional to temperature difference at terminals A and B. Pot supplies variable offset current corresponding to presettable temperature offset range. Low output voltage of bridge is amplified by opamp such as Motorola MC1741 which gives output swing of 30 V for input change of 0.3 mV. Buffer transistor is added for handling load such as motor control relay.—"Industrial Control Engineering Bulletin," Motorola, Phoenix, AZ, 1973, EB-4.

PROPORTIONAL CONTROL FOR OVEN—Thermistor RV1 and resistor-pot combination form voltage divider across 10-V zener, with output applied to UJT. Voltage across capacitor is ramp during positive half-cycles of AC line, slope of which is function of temperature and setting of 5K pot. When ramp reaches firing voltage of UJT, it turns on SCR and applies power to load. Negative half of AC input turns off SCR and cycle repeats. When oven temperature is low, SCR fires early in cycle to give more heat. When preset temperature is reached, SCR fires very late in cycle to compensate for heat lost by oven. Designed for 100°F environmental test chamber.—I. Math, Math's Notes, *CQ*, Sept. 1978, p 63 and 82–83.

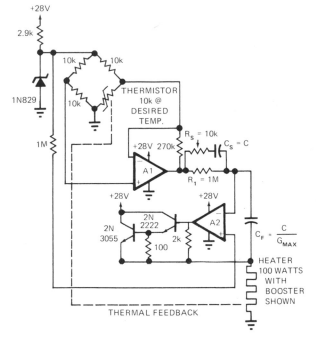

0.001°C ACCURACY—Simple design using thermistor bridge and two opamps controls temperature with high precision and has wide dynamic response as required for fast-changing ambient conditions. Circuit will not oscillate about desired temperature. Article covers design and operation of circuit in detail.—L. Accardi, Universal Temperature Controller, *EDN Magazine*, Dec. 1, 1972, p 53 and 55.

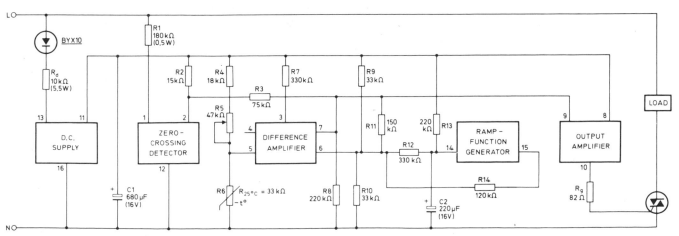

TIME-PROPORTIONAL CONTROL—Provides synchronous ON/OFF switching of resistive load under control of temperature-sensitive bridge formed by R4, R5, and negative temperature coefficient thermistor R6 in one bridge branch, with R9 and R10 in other branch. All required function elements are included in Mullard TCA280A trigger module. Values shown are for triac requiring gate current of 100 mA; for other triacs, values of R_d, R_g, and C_1 may need to be changed. Proportional band can be adjusted by changing value of R12. Triac triggering coincides with zero crossings of AC line voltage. Repetition time of internally generated sawtooth is about 30 s and can be adjusted by changing C2.—"TCA280A Trigger IC for Thyristors and Triacs," Mullard, London, 1975, Technical Note 19, TP1490, p 10.

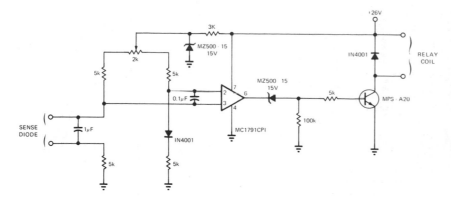

DIFFERENTIAL TO 10°F—Simple circuit senses difference between temperatures of two objects, as required for such control applications as turning on fans, turning off heaters, or operating mixing valves. Inexpensive 1N4001 silicon diode is used as sensor; with two such diodes in resistance bridge as shown, voltage proportional to temperature difference between reference and sensor diodes is applied to pins 4 and 7 of opamp. High-gain opamp is required because bridge output is only about 2 mV/°C of temperature differential. If output requires more than about 10 mA, buffer transistor is needed.—J. Barnes, Differential-Temperature Sensor is Very Inexpensive, *EDN Magazine*, April 5, 1973, p 90–91.

CRYSTAL OVEN CONTROL—Unbalance voltage produced in thermistor bridge when temperature drops below set point is sensed by differential opamp that feeds buffer Q1 and power amplifier Q2. Power dissipated in Q2 and its load R11 heats oven. Thermistor R4 has nominal resistance of 3600 ohms at 50°C (GE 1D53 or National Lead 1D053). Voltage divider R1-R2 reduces U1 input to safe level and makes thermistor operate at low current, minimizing self-heating effects. All arms of thermistor bridge except R7 (vernier temperature adjustment) are in oven.—R. Silberstein, An Experimental Frequency Standard Using ICs, *QST*, Sept. 1974, p 14–21 and 167.

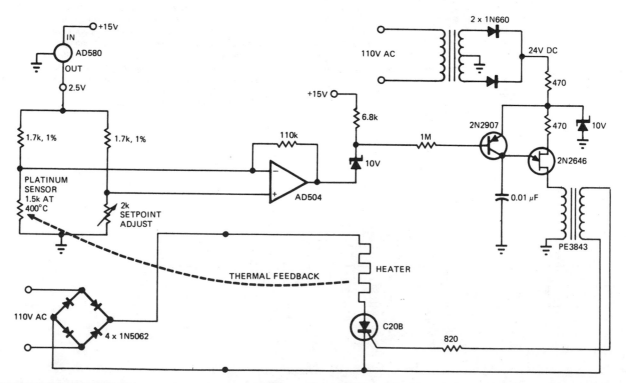

PHASE-FIRED SCR CONTROL—Can provide linear thermal control to 0.001°C at high power with good efficiency. Band-gap voltage reference of AD580 IC temperature transducer furnishes power to bridge circuit, while platinum sensor provides sensing function. AD504 opamp amplifies bridge output for biasing 2N2907 transistor which in turn controls 60-Hz synchronized UJT oscillator that drives gate of SCR through isolation transformer. Biasing action makes SCR fire at different points on AC waveform as required for precise control of oven heater. Possible drawback is RF noise generated because SCR chops in middle of waveform.—J. Williams, Designer's Guide to: Temperature Control, *EDN Magazine*, June 20, 1977, p 87–95.

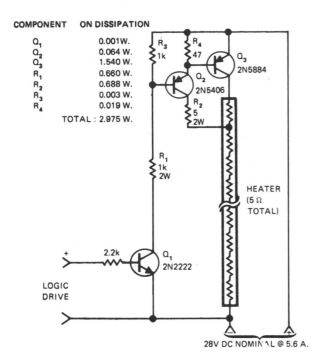

COMPONENT	ON DISSIPATION
Q_1	0.001 W.
Q_2	0.064 W.
Q_3	1.540 W.
R_1	0.660 W.
R_2	0.688 W.
R_3	0.003 W.
R_4	0.019 W.
TOTAL :	2.975 W.

LOW-DISSIPATION SWITCH—Logic-controlled power switch for 150-W instrument heater uses tap on heating element to force switch Q_3 and driver Q_2 into saturation and keep dissipation low. When input goes positive, Q_1 turns on and drives Q_2 and Q_3 on. Collector current of Q_2 and base drive of Q_3 are determined by R_2. Voltage drop across R_2 is proportional to supply voltage so drive for Q_3 is at optimum level over wide voltage range.—M. Strange, Increase Electronic Power Switch Efficiency, *EDN Magazine,* Aug. 20, 1975, p 78.

THERMOCOUPLE WITH ZERO-VOLTAGE SWITCH—Differential input connection of RCA CA3080A operational transconductance amplifier is used with thermocouple to drive CA3079 zero-voltage switch serving as trigger for triac handling AC load. Choose triac to match load being controlled. Supply voltage for opamps is not critical.—"Linear Integrated Circuits and MOS/FET's," RCA Solid State Division, Somerville, NJ, 1977, p 165–170.

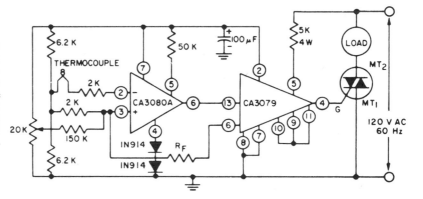

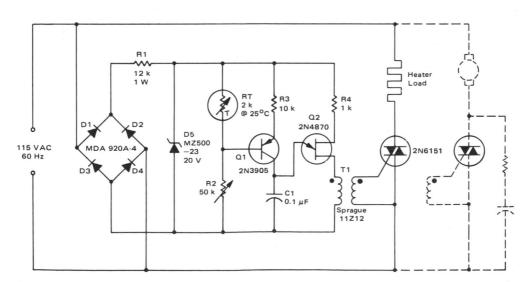

ROOM-HEATER CONTROL—Use of phase control for triac provides gradual reduction in heater load current as desired temperature is approached, eliminating large overshoots. R2 is adjusted so Q1 is off at desired temperature, turning Q2 off and preventing firing of triac. If temperature decreases, resistance of sensor RT increases and transistors initiate firing of triac. If RT continues to increase, C1 charges faster and triac is triggered earlier in each half-cycle, delivering more power to load. Dashed lines indicate alternate connections for controlling motor with constant load such as blower motor. For cooling applications, interchange RT and R2.—"Circuit Applications for the Triac," Motorola, Phoenix, AZ, 1971, AN-466, p 9.

75°C CRYSTAL OVEN—Proportional temperature controller using National LM3911 IC holds crystal oven temperature constant within 0.1°C of 75°C, to improve stability of oscillator used in frequency synthesizers and digital counters. Duty cycle of square-wave output of IC (ratio of OFF to ON time) varies with temperature of sensor in IC and with voltage at inverting input terminal. Duty cycle change makes average heater current change as required to bring temperature back to desired value. Square-wave frequency is determined by R4 and C1. 4N30 optocoupler drives power transistor having oven heater in collector circuit. During ON intervals of square wave, power transistor is driven to saturation, and during OFF intervals is cut off.—F. Schmidt, Precision Temperature Control for Crystal Ovens, *Ham Radio*, Feb. 1978, p 34–37.

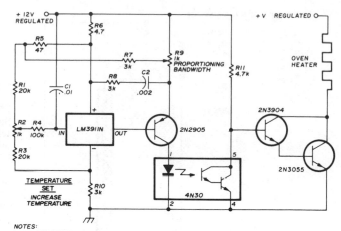

NOTES:
1. ALL RESISTORS 1/2 WATT COMPOSITION, EXCEPT R1 AND R3, WHICH SHOULD BE METAL FILM OR WIREWOUND.
2. VALUES OF R1, R2 AND R3 ARE FOR OVEN TEMPERATURE OF 75C.
3. +V SHOULD BE 5-6 VOLTS FOR 6.3 VOLT SURPLUS OVEN; 9-12 VOLTS FOR HOME BREW OVEN.

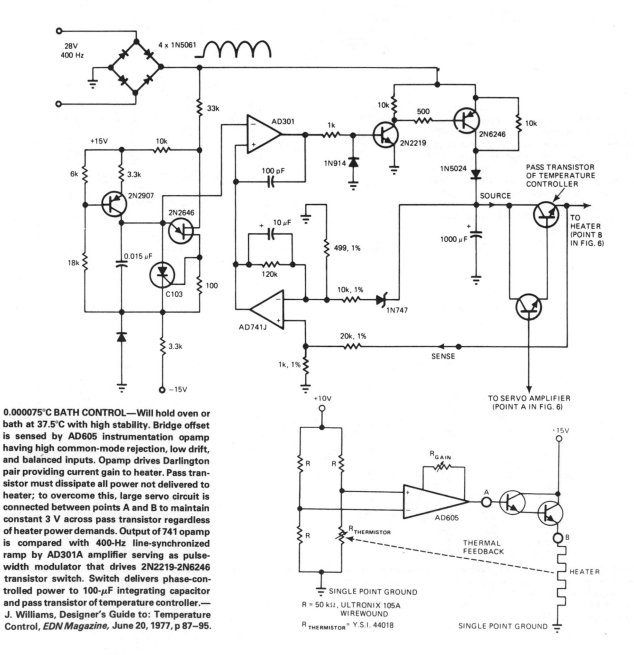

0.000075°C BATH CONTROL—Will hold oven or bath at 37.5°C with high stability. Bridge offset is sensed by AD605 instrumentation opamp having high common-mode rejection, low drift, and balanced inputs. Opamp drives Darlington pair providing current gain to heater. Pass transistor must dissipate all power not delivered to heater; to overcome this, large servo circuit is connected between points A and B to maintain constant 3 V across pass transistor regardless of heater power demands. Output of 741 opamp is compared with 400-Hz line-synchronized ramp by AD301A amplifier serving as pulse-width modulator that drives 2N2219-2N6246 transistor switch. Switch delivers phase-controlled power to 100-μF integrating capacitor and pass transistor of temperature controller.—J. Williams, Designer's Guide to: Temperature Control, *EDN Magazine*, June 20, 1977, p 87–95.

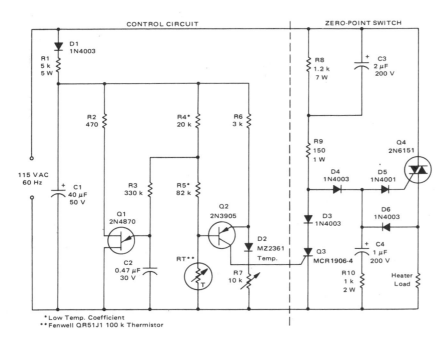

*Low Temp. Coefficient
**Fenwell QR51J1 100 k Thermistor

EMI-FREE PHASE CONTROL OF HEATER— Modulated triac zero-point switch eliminates electromagnetic interference generated by phase control while providing proportional ON/ OFF switching for accurate temperature regulation of heater load. Circuit at right of dashed line is basic zero-point switch that turns triac on almost immediately after each zero crossing between half-cycles. R7 is set so bridge in control circuit is balanced at desired temperature. When temperature overshoots, thermistor RT decreases in resistance and Q2 turns on to provide gate drive for SCR Q3. Q3 then turns on and shunts gate signal away from triac Q4, to remove power from load. When temperature drops, Q2 and Q3 turn off and full-wave power is applied to load. Modulation is achieved by superimposing sawtooth voltage from Q1 on one arm of bridge through R3, with sawtooth period equal to 12 cycles of line frequency. From 1 to 12 of these cycles can be applied to load for modulating power in 8% steps from 0% to 100% duty cycle.—"Circuit Applications for the Triac," Motorola, Phoenix, AZ, 1971, AN-466, p 13.

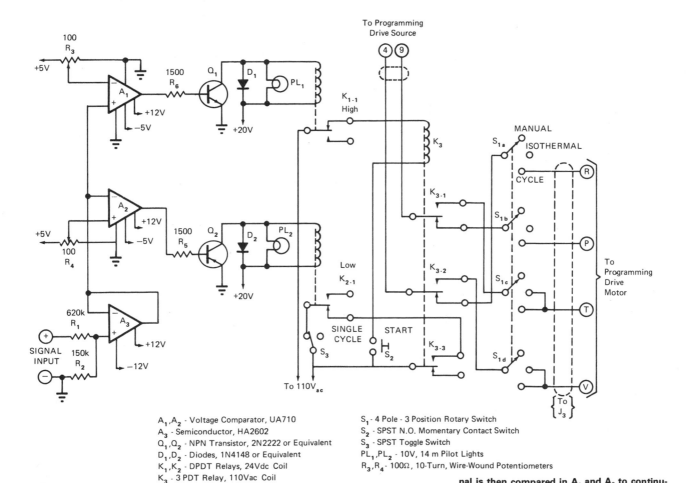

A_1, A_2 - Voltage Comparator, UA710
A_3 - Semiconductor, HA2602
Q_1, Q_2 - NPN Transistor, 2N2222 or Equivalent
D_1, D_2 - Diodes, 1N4148 or Equivalent
K_1, K_2 - DPDT Relays, 24Vdc Coil
K_3 - 3 PDT Relay, 110Vac Coil

S_1 - 4 Pole - 3 Position Rotary Switch
S_2 - SPST N.O. Momentary Contact Switch
S_3 - SPST Toggle Switch
PL_1, PL_2 - 10V, 14 m Pilot Lights
R_3, R_4 - 100Ω, 10-Turn, Wire-Wound Potentiometers

THERMAL CYCLER—Circuit allows operator to preselect upper and lower temperature limits for controller used in determining effect of continuous thermal cycling on properties of materials. Switching arrangement gives choice of modes ranging from manual to fully automatic continuous cycling. Operation of programming drive motor is controlled by contacts of relay K_3.

When relay is energized, motor runs in forward direction to increase temperature; when deenergized, motor is reversed. Condition of K_3 depends on which of limit relays K_1 or K_2 was most recently energized. Control circuit samples output of temperature programmer; this DC input signal is reduced to 5 V maximum by R_1-R_2 and amplified by voltage follower A_3. Sig-

nal is then compared in A_1 and A_2 to continuously variable reference voltage from 0 to 5 V preselected by 10-turn pots R_3 and R_4. Q_1 is cut off when input is below reference. When input exceeds reference, Q_1 goes on and energizes upper-limit relay K_1. Article gives initial setup procedure.—W.J. Dobbin, Variable Limit Switch Permits Hands-Off Equipment Cycling, *EDN Magazine,* Jan. 20, 1973, p 66–67.

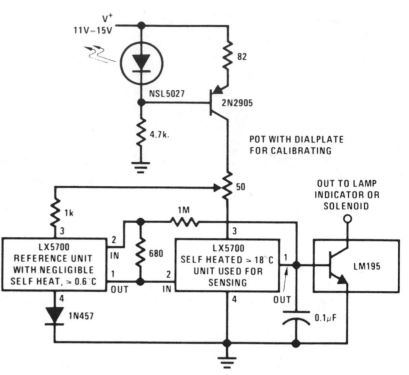

TEMPERATURE-DIFFERENCE DETECTOR—Pair of National LX5700 temperature transducers delivers output voltage proportional to temperature difference between transducers, as required for sensing temperature gradient in chemical processes, detecting failure of cooling fan, detecting movement of cooling oil, and monitoring other heat-absorbing phenomena. With sensing transducer in hot condition (out of liquid or in still air for 2 min), adjust 50-ohm pot to setting that just turns power output off. Next, with transducer in cool condition (in liquid or in moving air for 30 s), find setting that just turns output on. These settings overlap, but final setting between them will provide stable operation.—P. Lefferts, "A New Interfacing Concept; the Monolithic Temperature Transducer," National Semiconductor, Santa Clara, CA, 1975, AN-132, p 7.

Output "OFF" if sensing unit becomes hot, i.e., out of liquid or airstream.

Reference unit is 1 inch from the sensing unit in airstreams, and below the sensor in liquid sensing systems.

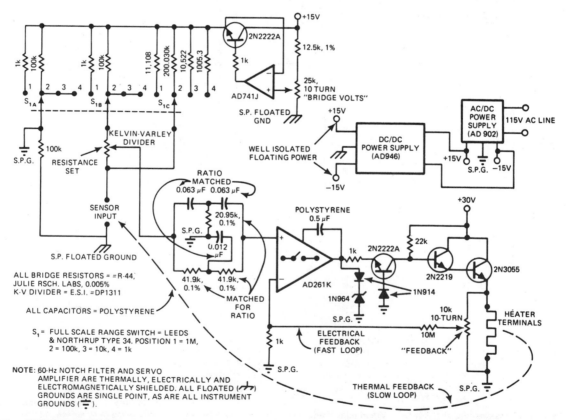

0.0000033°C CHOPPER-STABILIZED OVEN CONTROL—Uses chopper-stabilization techniques to provide ultimate in temperature control for laboratory oven. Multiranging bridge accommodates sensors from 10 ohms to 1 megohm, with Kelvin-Varley divider being used to dial sensor resistance control point directly to five digits. Use of floating power supply for bridge allows single-ended noninverting chopper-stabilized AD741J amplifier to take differential measurement and eliminates common-mode voltage error. Passive 60-Hz notch filter eliminates pickup noise at input of AD261K amplifier which in turn feeds 2N2222A transistor driving Darlington pair that provides up to 30 V across heater of oven. Article also gives circuit of 30-V regulated supply required for output transistors.—J. Williams, Designer's Guide to: Temperature Control, *EDN Magazine*, June 20, 1977, p 87–95.

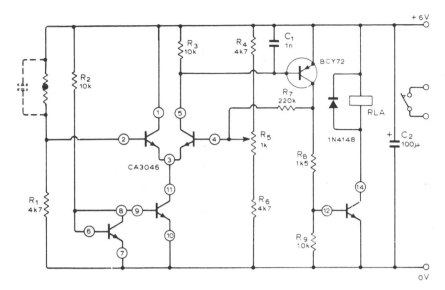

THERMISTOR BRIDGE—Bridge is formed by thermistor with R_1, R_4, R_5, and R_6. Unbalance is sensed by CA3046 IC having two matched pairs of transistors, with additional output transistor in IC. Positive feedback through R_7 prevents chatter as switching point is approached. R_5 sets switching temperature precisely. Relay comes on when temperature drops below predetermined point; for opposite function, reverse positions of thermistor and R_1. Value of R_1 is chosen to give approximately the desired control point.—D. E. Waddington, Thermistor Controlled Thermostat, *Wireless World*, July 1976, p 36.

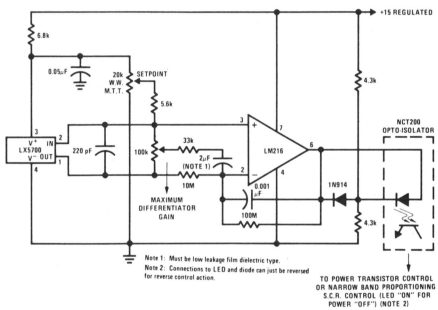

ANTICIPATING CONTROLLER—Circuit adds selected amount of phase leading signal to normal amplified output of National LX5700 temperature sensor to compensate at least partially for sensing lags. DC gain of LM216 opamp is set at 10 by 10-megohm and 100-megohm resistors to give opamp output of 1 V/°C. Output of opamp energizes optoisolator that feeds conventional temperature control system.—P. Lefferts, "A New Interfacing Concept; the Monolithic Temperature Transducer," National Semiconductor, Santa Clara, CA, 1975, AN-132, p 7.

Note 1: Must be low leakage film dielectric type.
Note 2: Connections to LED and diode can just be reversed for reverse control action.

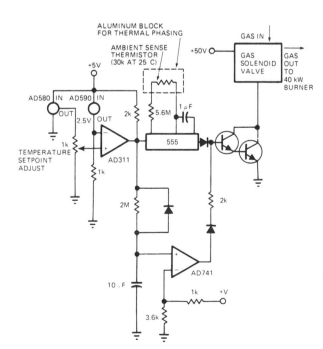

OVERSHOOT COMPENSATION—Used to control temperature of industrial gas-fired heater having very high thermal capacity. When AD311 opamp comparator trips at set-point temperature, 555 mono makes transistors turn on gas solenoid and light burner. When mono times out, burner goes off regardless of opamp output condition. Time constant of 555 compensates for lags in system by turning off heater before AD590 sensor reaches cutoff value. Thermistor across 555 mono compensates for changes in ambient temperature. During start-up, AD741 opamp and associated circuit effectively bypasses mono, and also turns on heater if mono fails to fire for any reason.—J. Williams, Designer's Guide to: Temperature Control, *EDN Magazine*, June 20, 1977, p 87–95.

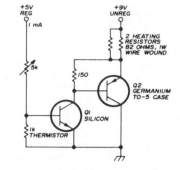

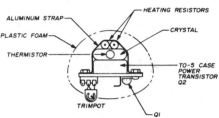

CRYSTAL OVEN—All components for proportional temperature control circuit are mounted on crystal, so total power of 2 W maximum serves for maintaining crystal temperature. Thermistor is about 1K at room temperature. Transistor types are not critical but should have low leakage currents. Thermistor current of about 1 mA should be much more than 0.1-mA base current of Q1. If Q2 is silicon, increase 150-ohm resistor to 680 ohms.—P. H. Mathieson, Simple Crystal Oven, *Ham Radio,* April 1976, p 66.

0.01°C CONTROL WITH OPAMP COMPARATOR—Uses platinum sensor in bridge configuration, with opamp connected across bridge differentially. When cold, sensor resistance is less than 500 ohms so opamp saturates to give positive output that turns on power transistor and heater. As oven warms, sensor resistance increases, bridge balance shifts, and heater is cut off.—J. Williams, Designer's Guide to: Temperature Control, *EDN Magazine,* June 20, 1977, p 87–95.

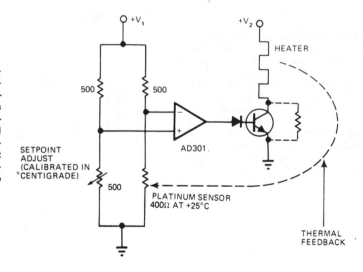

CHAPTER 94
Temperature Measuring Circuits

Convert temperature to frequency, voltage, or other parameter for driving meter or digital display that gives temperature value with desired accuracy. Includes wind-chill meter, air-velocity meter, position sensor, thermocouple multiplexer, integrator for soldering-energy pulses, and differential drive for strip-chart recorder.

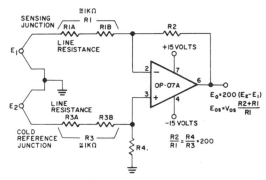

$$E_0 = 200(E_2 - E_1)$$
$$E_{os} = V_{os} \frac{R2+RI}{RI}$$
$$\frac{R2}{RI} = \frac{R4}{R3} = 200$$

THERMOCOUPLE AMPLIFIER—Precision Monolithics OP-07A opamp has high common-mode rejection ratio and long-term accuracy required for use with thermocouples having full-scale outputs under 50 mV, frequently located in high-noise industrial environments. CMRR is 100 dB over full ±13 V range when ratios R2/R1 and R4/R3 are matched within 0.01%. Circuit is useful in many other applications where small differential signals from low-impedance sources must be accurately amplified in presence of large common-mode voltages.—D. Soderquist and G. Erdi, "The OP-07 Ultra-Low Offset Voltage Op Amp—a Bipolar Op Amp That Challenges Choppers, Eliminates Nulling," Precision Monolithics, Santa Clara, CA, 1975, AN-13, p 11.

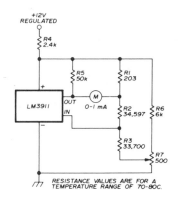

70–80°C THERMOMETER—Uses National LM3911 IC having built-in temperature sensor. If no thermometer is available for calibration, set pot R7 to its midpoint. Article gives equations for calculating resistance values for other temperature and meter ranges. Applications include monitoring of temperature in crystal oven. If permanently connected meter is not required, terminals can be provided for checking temperature with multimeter.—F. Schmidt, Precision Temperature Control for Crystal Ovens, *Ham Radio*, Feb. 1978, p 34–37.

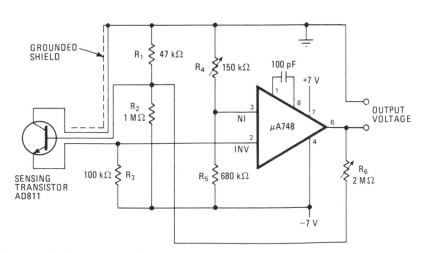

TRANSISTOR SENSOR—Use of bipolar supply for opamp makes electronic thermometer circuit fully linear even at low temperatures. Accuracy is within 0.05°C. Zero point is set by R₄ and gain by R₆.—C. J. Koch, Diode or Transistor Makes Fully Linear Thermometer, *Electronics*, May 13, 1976, p 110–112.

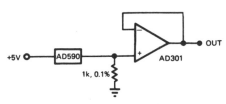

−125 to +200°C WITH 1° ACCURACY—Use of factory-trimmed AD590 IC temperature sensor gives wide temperature range with minimum number of parts. Other temperature scales can be obtained by offsetting AD301 buffer opamp.—J. Williams, Designer's Guide to: Temperature Measurement, *EDN Magazine*, May 20, 1977, p 71–77.

THERMISTOR THERMOMETER—Thermistor for desired temperature range is one leg of Wheatstone bridge driving microammeter through transistor to provide direct indication of temperature. Can also be used for control purposes if suitable amplifier and relay are used in place of meter. Thermistor cable can be ordinary parallel or twisted wires. To calibrate, immerse thermistor in water at various temperatures and measure water temperature with conventional high-accuracy thermometer. Calibration graph can then be prepared as guide for marking meter scale.—F. M. Mims, "Transistor Projects, Vol. 1," Radio Shack, Fort Worth, TX, 1977, 2nd Ed., p 86–93.

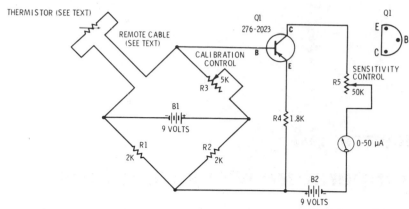

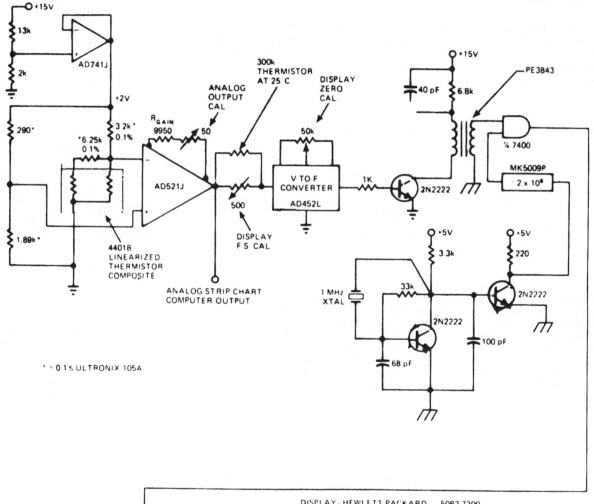

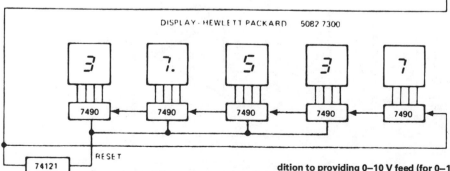

DISPLAY · HEWLETT PACKARD 5082 7300

5-DIGIT THERMOMETER—Temperature-to-frequency converter drives digital display providing 0.001°C resolution with 0.15°C absolute accuracy. Linearized thermistor network biases inverting input of AD521J instrumentation amplifier, while noninverting input is driven from same reference. Output can be fed directly to analog strip-chart recorder or computer, in addition to providing 0–10 V feed (for 0–100°C) to voltage-to-frequency circuit that drives display. Readout is updated at 2-s intervals.—J. Williams, Designer's Guide to: Temperature Measurement, *EDN Magazine*, May 20, 1977, p 71–77.

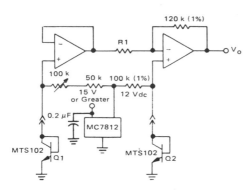

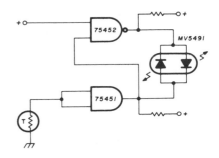

FOUR-THERMOCOUPLE MULTIPLEXING—Low power dissipation in DG306 analog switches means lower offset voltages added to thermocouple voltages by silicon in contact with aluminum in switches. Thermocouples are switched differentially to instrumentation amplifier driving meter, in order to cancel thermal offsets due to switch.—"Analog Switches and Their Applications," Siliconix, Santa Clara, CA, 1976, p 7-87.

RED/GREEN LED MONITOR—Set points are adjusted by trimming resistor shunted across thermistor, to give one color when desired temperature has been reached and other color when temperature is low. Uses Monsanto MV5491 dual red/green LED, with 220 ohms in upper lead to +5 V supply and 100 ohms in lower +5 V lead because red and green LEDs in parallel back-to-back have different voltage requirements. LED drivers are SN75452 and SN75451.—K. Powell, Novel Indicator Circuit, *Ham Radio,* April 1977, p 60–63.

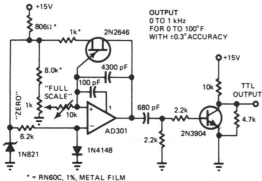

0–100°F GIVES 0–1 kHz OUTPUT—Circuit provides direct temperature-to-frequency conversion at low cost for applications where digital output is desired. Temperature sensor is 1N4148 diode having −2.2 mV/°C temperature shift, controlling AD301 opamp in relaxation oscillator circuit. Compensated 1N821 zener stabilizes against supply changes. Output network using 680 pF and 2.2K differentiates 400-ns reset edge of negative-going output ramp of opamp and drives single-transistor inverter to provide TTL output. Accuracy is within 0.3°F.—J. Williams, Designer's Guide to: Temperature Measurement, *EDN Magazine,* May 20, 1977, p 71–77.

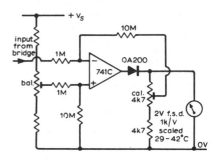

ZERO SUPPRESSION—Opamp is used in inverting configuration at output of temperature-sensing bridge, so noninverting input of opamp can be used for suppressing meter zero when temperature range for application is 29 to 42°C. Calibration control is set for gain of about 17.2 to make meter direct-reading. Article gives operation details and methods of improving temperature stability of circuit.—R. J. Isaacs, Optimizing Op-Amps, *Wireless World,* April 1973, p 185–186.

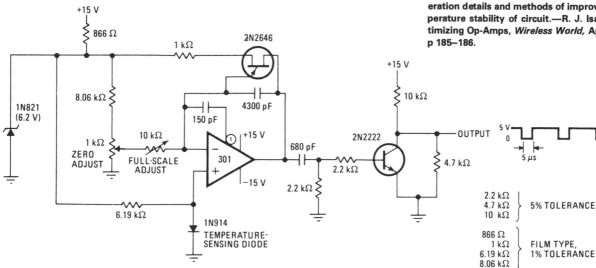

2.2 kΩ	
4.7 kΩ	5% TOLERANCE
10 kΩ	

866 Ω	
1 kΩ	FILM TYPE,
6.19 kΩ	1% TOLERANCE
8.06 kΩ	

TEMPERATURE-TO-FREQUENCY CONVERTER—Frequency of relaxation oscillator varies linearly with temperature-dependent voltage across 1N914 diode sensor, with range of 0–1000 Hz for 0–100°C. Frequency meter at output shows temperature directly with accuracy of ±0.3°C. Opamp is used as integrator, with 1N821 temperature-compensated diode providing voltage reference that determines firing point of UJT. Circuit functions as voltage-to-frequency converter. Calibrate at 100°C and 0°C, repeating until adjustments cease to interact. Output frequency is then 10 times Celsius temperature.—J. Williams and T. Durgavich, Direct-Reading Converter Yields Temperature, *Electronics,* April 3, 1975, p 101 and 103; reprinted in "Circuits for Electronics Engineers," *Electronics,* 1977, p 366.

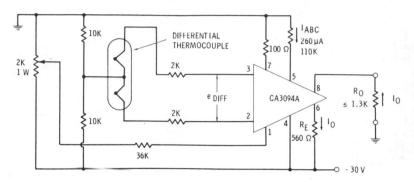

IC FOR DIFFERENTIAL THERMOCOUPLE— Amount of heat sensed by differential thermocouple is proportional to voltage between pins 2 and 3 of CA3094A programmable power switch/amplifier. Input swing of ±26 mV gives single-ended output current range of ±8.35 mA.—E. M. Noll, "Linear IC Principles, Experiments, and Projects," Howard W. Sams, Indianapolis, IN, 1974, p 314.

TEMPERATURE TRANSDUCER INTERFACE— Output of National LX5600 temperature-sensing transducer is inverted, level-shifted, and given extra voltage gain of 4 to give required output of 0 to +5 V for telemetry system or instrumentation recorder. Q1 furnishes constant current to thermometer, and Q2 provides inverting function. Resulting output signal is reinverted by LM201A opamp connected through zero-adjust divider to pin 3 which provides voltage reference.—P. Lefferts, "A New Interfacing Concept; the Monolithic Temperature Transducer," National Semiconductor, Santa Clara, CA, 1975, AN-132, p 3.

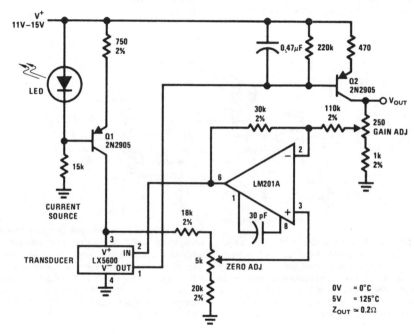

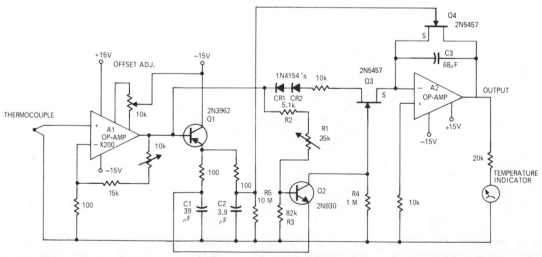

HEAT-ENERGY INTEGRATOR— Pulses of heat energy applied to solder preforms by tips of pulsed soldering machine are metered by integrate/hold-to-indicate circuit using thermocouple as input sensor. Temperature derived from area under time/temperature curve is indicated momentarily on output meter, as guide for operator when size of solder preform is changed. Article describes operation of circuit in detail and gives timing diagram.—C. Brogado, Heat-Energy Pulse Measured and Displayed, *EDN Magazine*, Sept. 15, 1970, p 61–62.

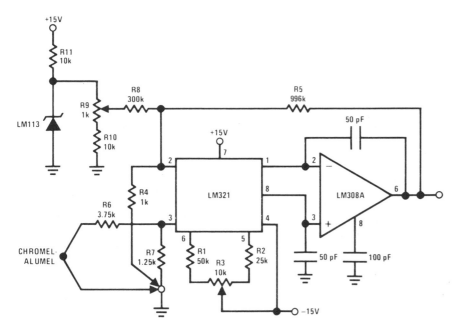

THERMOCOUPLE AMPLIFIER—Combination of LM321 preamp and LM308A opamp forms precision low-drift amplifier that includes compensation for ambient temperature variations. LM113 zener provides temperature-stable reference for offsetting output to read thermocouple temperature directly in degrees C. R4, R6, and R7 should be wirewound.—R. C. Dobkin, "Versatile IC Preamp Makes Thermocouple Amplifier with Cold Junction Compensation," National Semiconductor, Santa Clara, CA, 1973, LB-24.

TEMPERATURE-TO-FREQUENCY CONVERTER—Temperature sensor on chip of AD537 voltage-to-frequency converter IC minimizes number of external parts needed. Output frequency changes 10 Hz for each degree (kelvin or Celsius) change in temperature.—J. Williams, Designer's Guide to: Temperature Measurement, *EDN Magazine,* May 20, 1977, p 71–77.

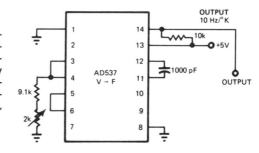

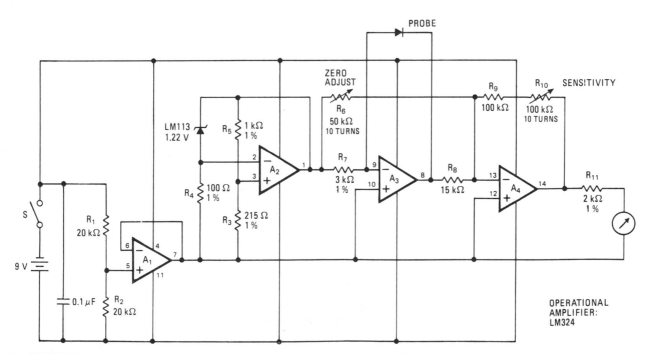

0.1°C PRECISION—Temperature sensor is LM113 diode in probe, with sections A_1 and A_2 of LM324 quad opamp maintaining constant current to diode to ensure that voltage changes across diode are direct result of temperature. 4.5-V output of A_1 is reference point for other opamps. Changes in output voltage of diode are reflected in output of A_4 through buffer A_3. Calibration involves adjusting R_6 for zero output voltage at low end of temperature range, then adjusting R_{10} for full-scale or other convenient reading at desired upper temperature limit. Use 1-mA meter movement.—Y. Nezer, Accurate Thermometer Uses Single Quad Op Amp, *Electronics,* May 26, 1977, p 126.

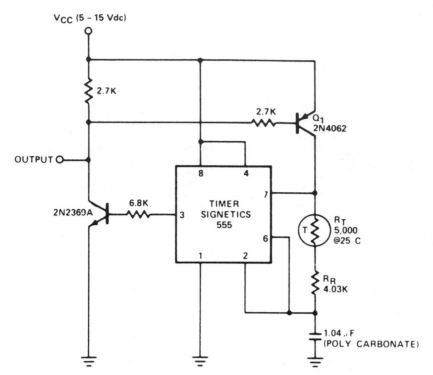

THERMISTOR-CONTROLLED TIMER—Thermistor and two transistors in charging network of 555 timer give output frequency that varies with temperature over 78°F range with accuracy of ±1 Hz.—"Signetics Analog Data Manual," Signetics, Sunnyvale, CA, 1977, p 731.

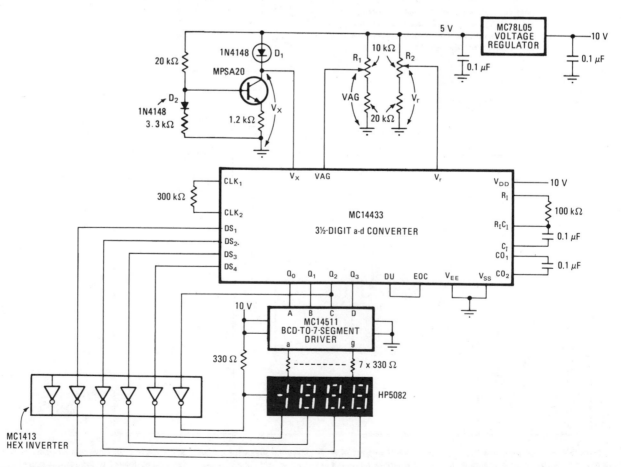

DIGITAL THERMOMETER—Diode D_2 serves as sensor for driving A/D converter directly, eliminating temperature-drift errors normally associated with amplifiers. Can be calibrated over temperature range of −199° to 199° in either Fahrenheit or Celsius scales. Accuracy is about 1°.—H. Wurzburg and M. Hadley, Digital Thermometer Circumvents Drift, *Electronics,* Jan. 5, 1978, p 176–177.

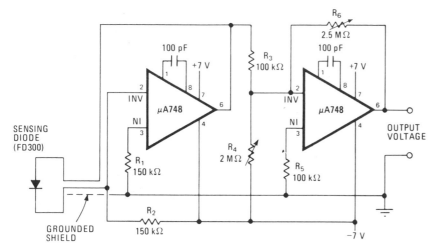

FULLY LINEAR DIODE SENSOR—First opamp acts as constant-current source for temperature-sensing diode, making voltage drop across diode depend only on temperature. Second opamp offsets diode voltage to whatever temperature range is desired and provides gain that is adjustable with R_6. R_4 is used to set output at zero for selected temperature such as for 0°C. Circuit can then be adjusted to give 1 V at 50°C.—C. J. Koch, Diode or Transistor Makes Fully Linear Thermometer, *Electronics*, May 13, 1976, p 110–112.

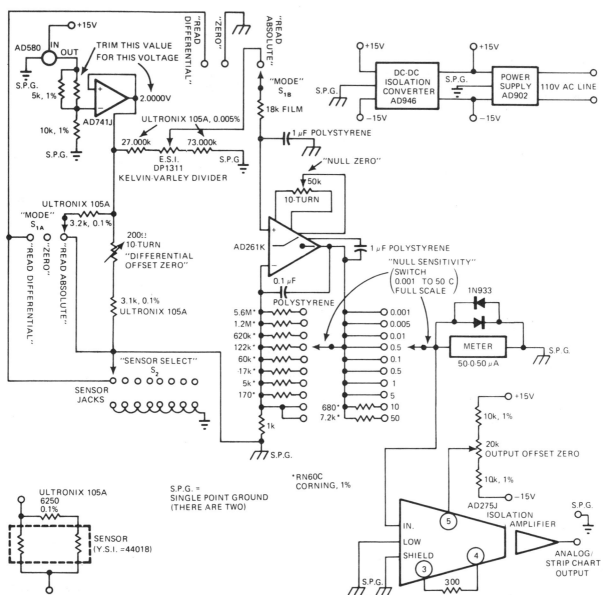

DIFFERENTIAL THERMOMETER—Temperature is directly dialed out on five-decade Kelvin-Varley voltage divider, and differences between dialed temperature and that of YSI 44018 sensor are read directly on meter. Full-scale sensitivity of meter is varied from 0.001 to 50°C by adjusting gain of AD261K chopper-stabilized null detector which drives both meter and AD275J isolation amplifier used to drive strip-chart recorder. Circuit can also be used to measure temperature difference between two sensors with 100-microdegree accuracy. Article describes other measuring modes as well, including techniques for measuring 200-nanodegree temperature shifts.—J. Williams, Designer's Guide to: Temperature Measurement, *EDN Magazine*, May 20, 1977, p 71–77.

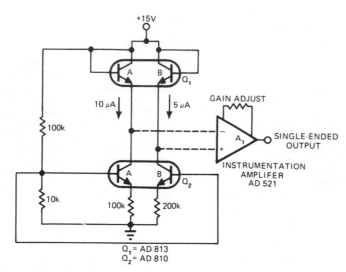

Q_1 = AD 813
Q_2 = AD 810

TRANSISTOR SENSOR—Current-ratio differential-pair temperature sensor uses dual transistor Q_1. Difference between base-emitter voltages of Q_{1A} and Q_{1B} varies linearly with temperature, when dual transistor Q_2 provides 10μA through Q_{1A} and 5μA through Q_{1B}. Instrumentation opamp provides single-ended output with better than 1°C accuracy over 300°C temperature range. Analog Devices AD 590 IC version of differential pair will operate over wire line thousands of feet away from instrumentation opamp, for remote sensing.—J. Williams, Designer's Guide to: Temperature Sensing, *EDN Magazine,* May 5, 1977, p 77–84.

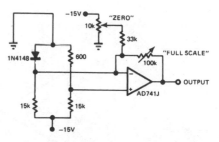

0–100°C WITH 1° ACCURACY—Low-cost diode serves as temperature sensor. To calibrate, place diode in 0°C environment and adjust zero pot for 0-V output, then place diode in 100°C environment and adjust full-scale pot for 10-V output. Repeat procedure until interaction between adjustments ceases.—J. Williams, Designer's Guide to: Temperature Measurement, *EDN Magazine,* May 20, 1977, p 71–77.

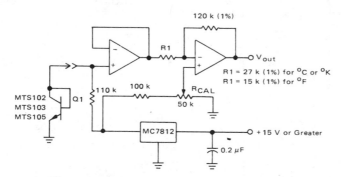

R1 = 27 k (1%) for °C or °K
R1 = 15 k (1%) for °F

ABSOLUTE-TEMPERATURE SENSING—Silicon temperature sensor (MTS102, MTS103, or MTS105) provides precise temperature-sensing accuracy over range of −40°C to +150°C. Sensor is essentially a transistor with base and collector leads connected together externally; base-emitter voltage drop then decreases linearly with temperature over operating range. Voltage change is amplified by two opamps in series, operating from regulated output of MC7812 regulator. Opamp types are not critical. With Q1 at known temperature, adjust 50K pot to give output voltage equal to TEMP × 10 mV. Output voltage is then 10 mV per degree in desired temperature scale.—"Silicon Temperature Sensors," Motorola, Phoenix, AZ, 1978, DS 2536.

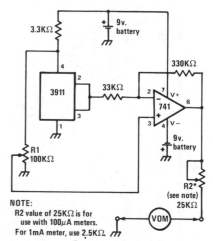

NOTE:
R2 value of 25KΩ is for use with 100μA meters.
For 1mA meter, use 2.5KΩ

THERMOMETER—Sensor is 3911 IC whose output is 10 mV/K (kelvin temperature scale). At 0°C, output is 2.73 V. Output swing is amplified by 741 opamp to 0.1 V/°C for driving volt-ohm-milliammeter or sensitive milliammeter. R2 adjusts scaling factor, for readout in °C or °F as desired.—J. Sandler, 9 Projects under $9, *Modern Electronics,* Sept. 1978, p 35–39.

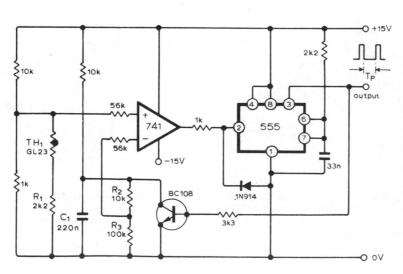

TEMPERATURE TO PULSE WIDTH—Temperature-dependent current through thermistor TH_1 develops voltage across R_1 that is compared with fraction of increasing voltage across C_1 by 741 opamp. When output of opamp goes negative, it triggers 555 IC connected as mono MVBR, to turn transistor on for about 100 μs and discharge C_1. Circuit is based on similarity between resistance-temperature curve of thermistor and inverse function of voltage across capacitor charging through resistor. For values shown, circuit gives 650-μs pulse width at 0°C, increasing 20 μs per degree with accuracy of ±1.2°C up to 60°C. If IC output is used to gate clock oscillator, number of oscillator output pulses will be directly proportional to temperature.—T. P. Y. Sander, Temperature to Pulse-Length Converter, *Wireless World,* Jan. 1977, p 76.

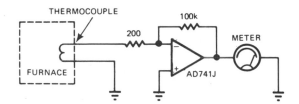

HOT-COLD METER—Output of unreferenced thermocouple drives meter through opamp that provides required gain, for monitoring temperature inside furnace when exact temperature value is not required. Meter is simply calibrated in terms of hot and cold.—J. Williams, Designer's Guide to: Temperature Measurement, *EDN Magazine,* May 20, 1977, p 71–77.

MATCHED-TRANSISTOR SENSOR—Precision Monolithics MAT-01H matched-transistor pair Q1 senses temperature over range of −55°C to +125°C with inherent linearity and long-term stability. Matched transistors Q2 (MAT-01GH) are current sources for sensing transistors. Transistor combination provides differential voltage output that is directly proportional to absolute temperature. Amplifier using OP-10CY changes this voltage difference to single-ended signal that can be used for measurement or control. Circuit will drive 10-V full-scale digital panel meter to give digital thermometer.—J. Simmons and D. Soderquist, "Temperature Measurement Method Based on Matched Transistor Pair Requires No Reference," Precision Monolithics, Santa Clara, CA, 1975, AN-12, p 4.

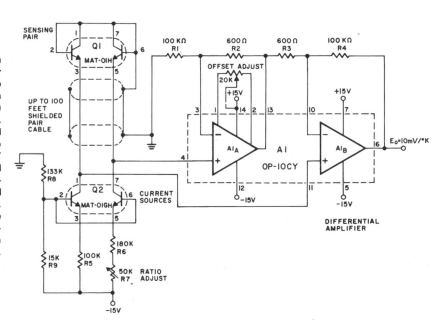

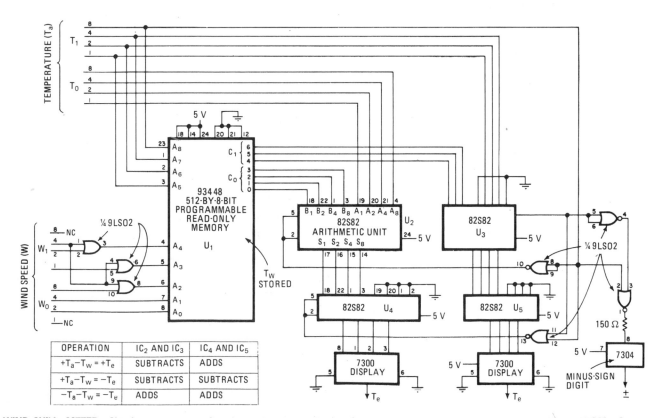

OPERATION	IC_2 AND IC_3	IC_4 AND IC_5
$+T_a - T_w = +T_e$	SUBTRACTS	ADDS
$+T_a - T_w = -T_e$	SUBTRACTS	SUBTRACTS
$-T_a - T_w = -T_e$	ADDS	ADDS

WIND-CHILL METER—Circuit measures and displays wind-chill equivalent temperature by combining air temperature and wind speed data. PROM is programmed to act in combination with arithmetic-logic units to generate output values corresponding to those of wind-chill temperature chart adopted by National Weather Service. Article gives listing of PROM contents.—V. R. Clark, PROM Converts Weather Data for Wind-Chill Index Display, *Electronics,* Jan. 5, 1978, p 158–159.

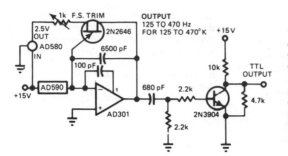

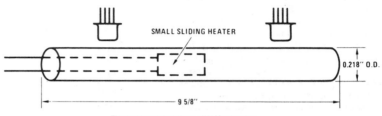

125–470 K GIVES 125–470 Hz—Use of AD590 current-ratioed differential-pair IC temperature transducer gives low parts count for temperature-to-frequency converter. Sensor controls AD301 opamp in relaxation oscillator, with negative-going output ramp being differentiated for driving single-transistor inverter giving TTL output.—J. Williams, Designer's Guide to: Temperature Measurement, *EDN Magazine*, May 20, 1977, p 71–77.

POSITION SENSOR—Position of small heating element sliding inside thin-wall brass tube is sensed by National LX5700 temperature transducer mounted outside of tube. With heater at center, transducers at both ends reach same temperature. With heater at one end of pipe, that transducer is about 50°C above ambient and other is near ambient. As heater moves toward one end, one thermometer becomes more sensitive and the other less. Circuit regulates heater power to keep position "gain" constant. Digital voltmeter gives average position. Applications include measuring average truck spring deflection while moving on rough road.—P. Lefferts, "A New Interfacing Concept; the Monolithic Temperature Transducer," National Semiconductor, Santa Clara, CA, 1975, AN-132, p 9.

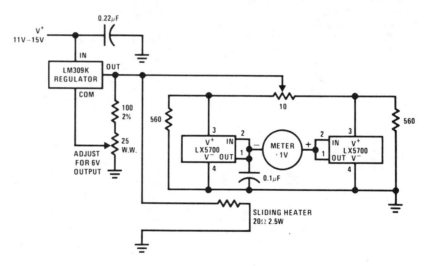

Thin wall brass tube with an LX5700 soldered on 7/8'' from each end. Mount horizontally in still air.

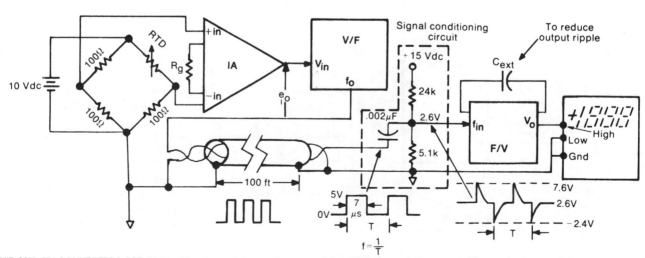

V/F AND F/V CONVERTERS FOR DPM—Signal transmitted as variations in frequency in 0–10 kHz range is converted back to voltage for driving digital panel meter to give indication of temperature value sensed by 100-ohm resistive thermal device (RTD) in bridge. V/F and F/V converters can be almost any commercial models designed for 0–10 V and 0–10 kHz. Instrument amplifier can be Datel AM201 or equivalent.—E. L. Murphy, Sending Transducer Signals over 100 Feet?, *Instruments & Control Systems*, June 1976, p 35–39.

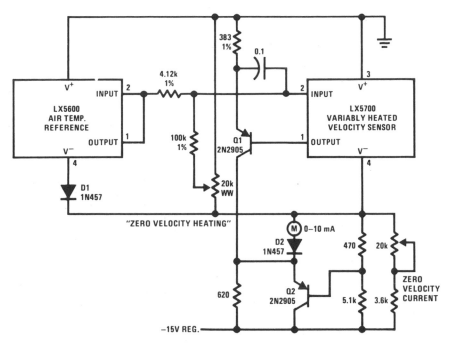

AIR VELOCITY METER—Uses National LX5600 air temperature reference, connected in unity-gain mode, in combination with LX5700 self-heated velocity sensor to convert wind velocity or airspeed to differential between heated and unheated transducers. As wind velocity rises, heating current required to hold velocity sensor predetermined number of degrees above ambient is measured. Calibration curve is drawn to show correlation between current and airspeed.—P. Lefferts, "A New Interfacing Concept; the Monolithic Temperature Transducer," National Semiconductor, Santa Clara, CA, 1975, AN-132, p 8.

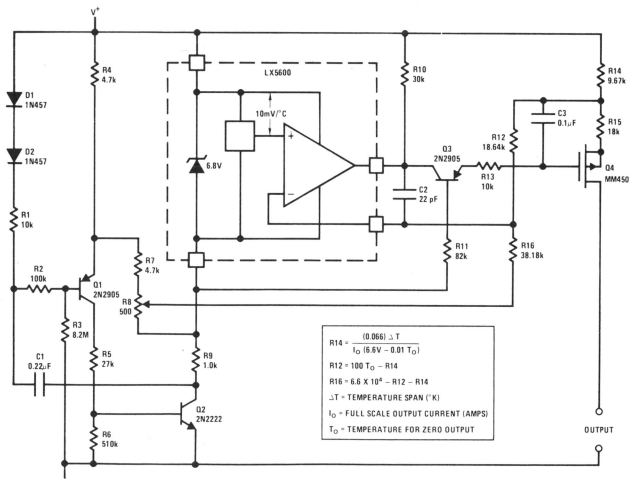

$$R14 = \frac{(0.066)\, \Delta T}{I_O\, (6.6V - 0.01\, T_O)}$$

$$R12 = 100\, T_O - R14$$

$$R16 = 6.6 \times 10^4 - R12 - R14$$

ΔT = TEMPERATURE SPAN (°K)

I_O = FULL SCALE OUTPUT CURRENT (AMPS)

T_O = TEMPERATURE FOR ZERO OUTPUT

MICROPOWER THERMOMETER—Low power consumption makes circuit attractive for battery-operated equipment. Uses National LX5600 temperature transducer covering −55°C to +125°C, whose output is directly proportional to absolute temperature at 10 mV/K. Both zero and scale factor are independently select-able. Thermometer is pulsed at low duty cycle to reduce power consumption, with sample and hold used to obtain continuous output between pulses. Supply range is 8–12 V; 8.4-V mercury battery will give over 1 year of operational life. Output can be used to drive meter for direct readout. MVBR Q1-Q2 drives LX5600 through R9. C1 and R3 control OFF time, and C1, R1, R4, and R7 control ON time. Q3 is sample transistor. Output is 0–50 μA for 50–100°F temperature change. Formulas in box give values for other ranges.—R. C. Dobkin, "Micropower Thermometer," National Semiconductor, Santa Clara, CA, 1974, LB-27.

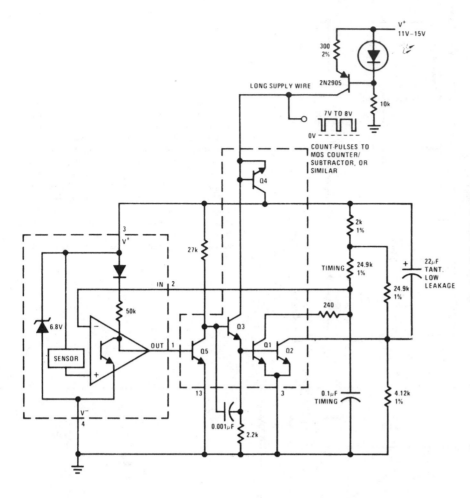

TEMPERATURE-TO-FREQUENCY CONVERTER—Transistors Q1-Q5 in National LM3046 transistor array form oscillator and ramp that together convert varying output voltages of LX5600 temperature transducer to proportional changes in frequency of square-wave pulses for feed to pulse counter.—P. Lefferts, "A New Interfacing Concept; the Monolithic Temperature Transducer," National Semiconductor, Santa Clara, CA, 1975, AN-132, p 4.

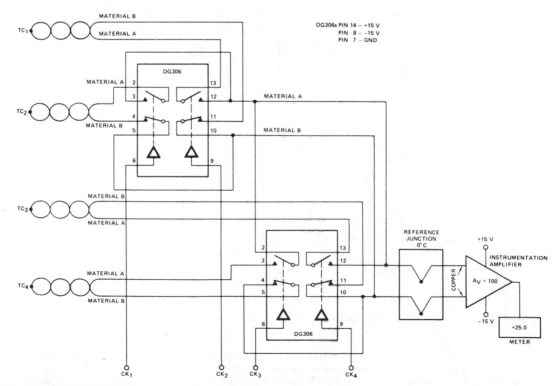

DIFFERENTIAL TEMPERATURE SENSOR—Responds to difference in temperatures of MTS102 silicon high-precision temperature sensors having range of −40°C to +150°C. With both sensors at same temperature, 100K pot is adjusted so output voltage is 0.000 V. Opamp types are not critical. R1 is 27K for measurements in Celsius or kelvin and 15K for Fahrenheit measurements.—"Silicon Temperature Sensors," Motorola, Phoenix, AZ, 1978, DS 2536.

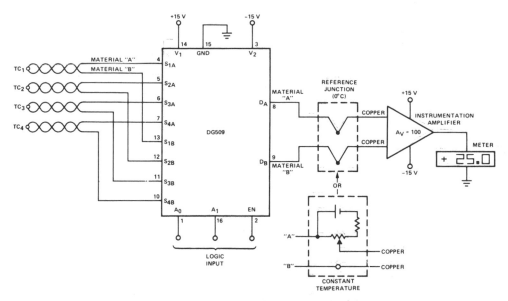

THERMOCOUPLE MULTIPLEXER—Under control of logic input, DG509 four-channel differential analog multiplexer connects selected one of four thermocouples to instrumentation amplifier driving digital or other readout. To decouple sensors from instrumentation amplifier, reference junction at 0°C can be used as shown. Alternatively, bucking voltage can be set at room temperature, but this arrangement will be sensitive to changes in ambient temperature.— "Analog Switches and Their Applications," Siliconix, Santa Clara, CA, 1976, p 7-77–7-78.

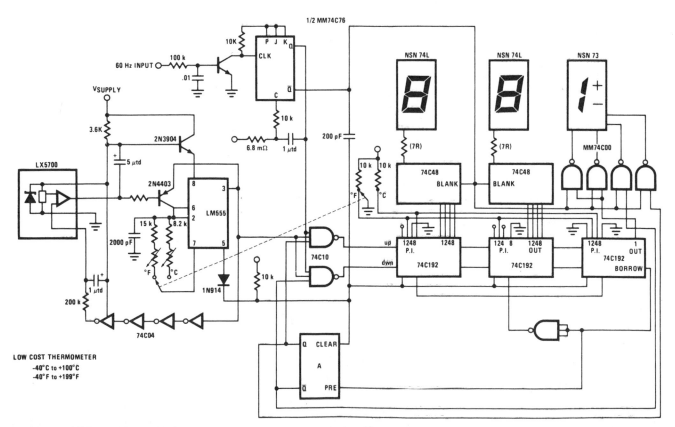

LOW COST THERMOMETER
−40°C to +100°C
−40°F to +199°F

FAHRENHEIT/CENTIGRADE LED THERMOMETER—National LX5700 temperature transducer provides input for code conversion circuit driving 3-digit LED display indicating temperature range from −40°C to +100°C or −40°F to +199°F under control of ganged switch.—"Linear Applications, Vol. 2," National Semiconductor, Santa Clara, CA, 1976, LB-30.

BRIDGE-TYPE SENSOR—CA3094 programmable opamp is connected as level-triggered MVBR at output of bridge, driving triac for temperature monitor or control applications. Sensor can be any temperature-dependent device. Load can be lamp, horn, or bell. For control applications, load is appropriate temperature-controlling device connected in feedback relationship to sensor.—"Circuit Ideas for RCA Linear ICs," RCA Solid State Division, Somerville, NJ, 1977, p 10.

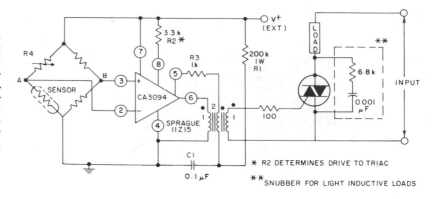

* R2 DETERMINES DRIVE TO TRIAC

** SNUBBER FOR LIGHT INDUCTIVE LOADS

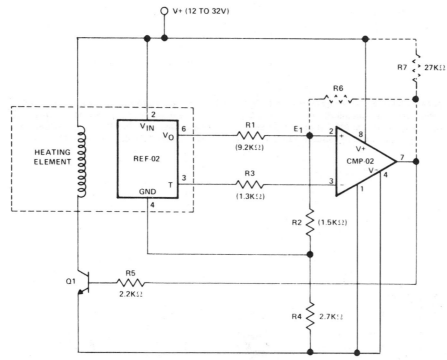

VOLTAGE-REFERENCE THERMOMETER—Precision Monolithics comparator CMP-02 turns on heating-element driver Q1 when temperature drops below set point determined by ratio of R1 to R2, as sensed by +5 V voltage reference REF-02 serving as thermometer. Circuit also provides adjustable hysteresis, determined by R6 and R7, if this feature is desired. Values in parentheses are for 60°C set point. REF-02 should be thermally connected to substance being heated. Design equations are given.—"Linear & Conversion I.C. Products," Precision Monolithics, Santa Clara, CA, 1977–1978, p 15-4.

SERVOED SHIELD FOR PROBE—Used when only part of temperature sensor can touch surface being measured. LM195H power transistor is main power amplifier and at same time serves as 23-W heater that is used to make copper shield track actual temperature of surface to be measured. Uses National LX5700 sensors. Diode in series with ground leg of one sensor permits adjusting pin 3 of that sensor over range of 40–80 mV to make it track with servo thermometer. Digital voltmeter is used to read temperature directly in degrees C.—P. Lefferts, "A New Interfacing Concept; the Monolithic Temperature Transducer," National Semiconductor, Santa Clara, CA, 1975, AN-132, p 6.

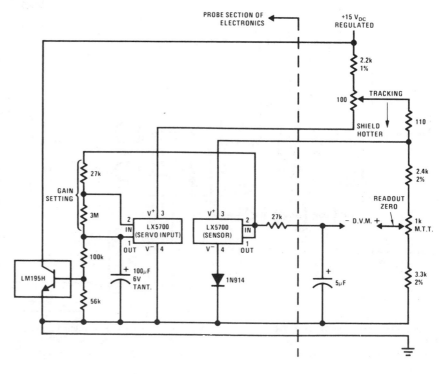

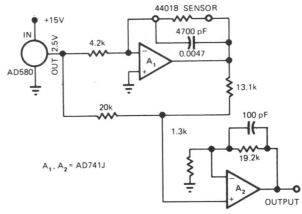

0–100°C WITH 0.15° ACCURACY—Low-cost YSI 44018 temperature sensor in feedback loop of 741J opamp gives accuracy approaching that of platinum sensors. Opamp is driven by AD580 band-gap reference. Voltage output of A_1 feeds similar opamp that provides zeroing and sets desired output gain.—J. Williams, Designer's Guide to: Temperature Measurement, *EDN Magazine,* May 20, 1977, p 71–77.

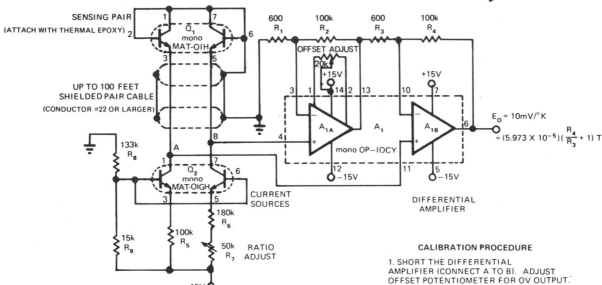

$$E_O = 10\text{mV}/°\text{K}$$
$$= (5.973 \times 10^{-5})(\frac{R_4}{R_3} + 1) T$$

R_1, R_2, R_3, R_4: 0.01%, GENERAL RESISTANCE ECONISTOR
R_5, R_6 : 0.1%, GENERAL RESISTANCE ECONISTOR
R_7, R_{10} : 10%, BOURNS TYPE 3006P
R_8, R_9 : 1%, TYPE RN55C

CALIBRATION PROCEDURE

1. SHORT THE DIFFERENTIAL AMPLIFIER (CONNECT A TO B). ADJUST OFFSET POTENTIOMETER FOR 0V OUTPUT. REMOVE THE INPUT SHORT

2. WITH SENSING PAIR AT KNOWN TEMPERATURE (e.g. ROOM TEMP.), ADJUST RATIO POTENTIOMETER FOR CORRECT OUTPUT READING.

±1 K ACCURACY FOR −55 to +125°C—Matched transistor pairs and opamps give high-accuracy temperature-measuring system that is easy to calibrate, has long-term stability, and can operate with sensor transistor pair up to 100 feet from rest of circuit. Common-mode rejection at amplifier input is greater than 100 dB. Output voltage is +2.18 V at −55°C (218 K), increasing to +3.98 V at +125°C (398 K).—J. Simmons and D. Soderquist, Temperature Measurement Method Requires No Reference, *EDN Magazine,* Aug. 5, 1974, p 78 and 80.

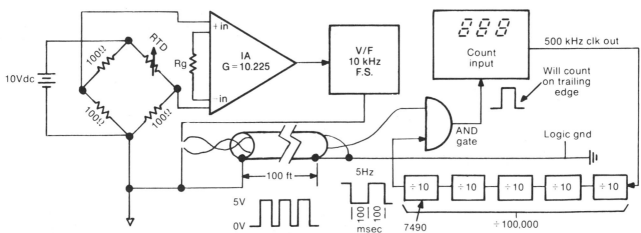

V/F CONVERTER FOR TRANSDUCER—Output of low-level transducer such as temperature bridge can be transmitted reliably over long wires (100 feet or more) in serial form if changes in 100-ohm resistive thermal device (RTD) are converted to corresponding changes in frequency with almost any commercially available V/F converter. Typical converter has 0–10 V full-scale analog input and 0–10 kHz output. If 5 V is applied to input, output pulse train will have rate of 5 kHz ± 0.5 Hz, which can be counted for 1 s or less and displayed on digital readout to show analog value. Instrument amplifier can be Datel AM201 or equivalent.—E. L. Murphy, Sending Transducer Signals over 100 Feet?, *Instruments & Control Systems,* June 1976, p 35–39.

CHAPTER 95
Test Circuits

Includes variety of circuits for checking diodes, transistors, opamps, ICs, coils, crystals, filters, and power supplies, along with curve tracers, signal injectors, signal tracers, power peak meter, printed-circuit ammeter, and pseudorandom digital generator.

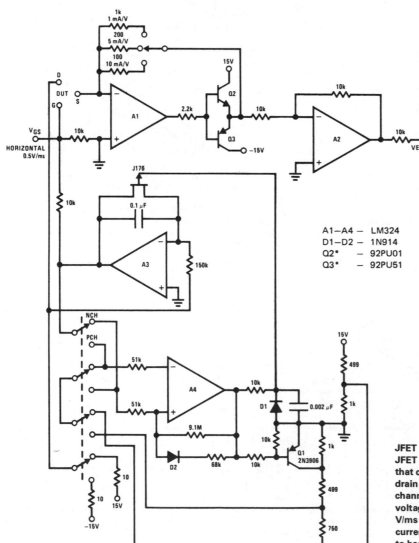

A1–A4 — LM324
D1–D2 — 1N914
Q2* — 92PU01
Q3* — 92PU51

*1W NPN, PNP

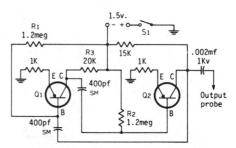

AF SIGNAL INJECTOR—Can be built into penlight housing, using single penlight cell or 1.5-V mercury cell for power. Output probe for feeding signal to audio circuit under test is about 1-inch length of stiff wire, pointed. For more output, run ground lead to equipment under test. Q_1 can be HEP253, 2N519, 2N741A, 2N2929, or equivalent. Q_2 can be HEP3, 2N1280, 2N2273, SK3005, or equivalent. Adjust R_1 and R_2 for good output, and adjust R_3 as required for good tone. To use, touch probe to input of any receiver or high-fidelity audio circuit. Tone should be heard from loudspeaker if circuit is good between probe and loudspeaker.—C. J. Schauers, Transistorized Signal Tracer, *CQ*, Sept. 1973, p 12 and 14.

JFET CURVE TRACER—Quad opamp and J176 JFET switch form basis of simple curve tracer that can be used with any CRO. Circuit displays drain current versus gate voltage for both P-channel and N-channel JFETs at constant drain voltage. Sweep time is 10 ms. Sweep rate is 0.5 V/ms with maximum gate voltage of ±5 V. Drain current is fed to vertical input and gate voltage to horizontal input.—"FET Databook," National Semiconductor, Santa Clara, CA, 1977, p 6-50–6-51.

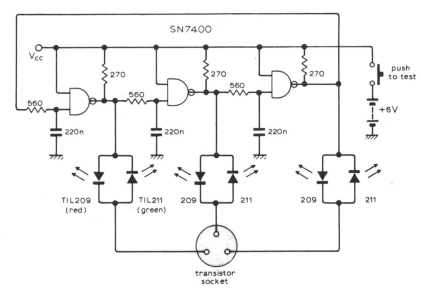

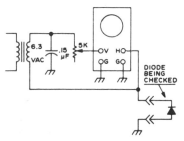

TRANSISTOR/DIODE TESTER—Checks for polarity, shorts, and opens in one measurement, using six LEDs as indicators. Circuit derives three-phase waveform from 2-kHz ring-of-three oscillator for application to device under test through LEDs. Oscillator waveform serves to make each pair of device terminals forward, reverse, and unbiased in turn for one-third of a cycle. Current flowing into device turns on red LED, and current flowing out turns on green LED, to indicate polarity and position of base lead.—N. E. Thomas, Semiconductor Tester, *Wireless World,* March 1977, p 43.

CHECKING DIODES WITH CRO—Simple oscilloscope setup checks and matches diodes. Sort diodes according to type, set pot to give desired trace size for good diode, then note relative sizes of traces obtained for unknown diodes. Reject diodes showing fuzz or ripple on oscilloscope trace.—Novice Q & A, *73 Magazine,* March 1977, p 187.

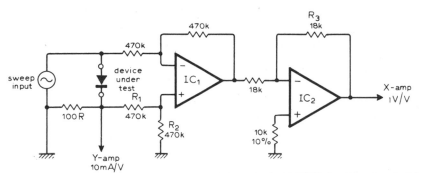

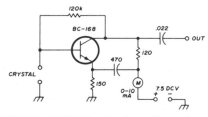

DIODE CURVE-TRACER—Circuit is designed to produce voltage-current characteristic curve of diode or other two-terminal device on oscilloscope. Sweep input can be any low-voltage AC source, such as 20-V Variac. Three-terminal devices may be traced if suitable external bias is provided. Opamps are 741.—S. Cahill, Diode Curve Tracer for Oscilloscope, *Wireless World,* Feb. 1976, p 76.

CRYSTAL CHECKER—Simple oscillator circuit checks crystal activity and resonant frequency, as required when choosing matched crystals for filters. For frequency check, signal from oscillator is injected into frequency counter. Values shown are for crystals around 5.5 MHz. For matching purposes, higher accuracy is obtained by reading harmonics of oscillator.—J. Perolo, Practical Considerations in Crystal-Filter Design, *Ham Radio,* Nov. 1976, p 34–38.

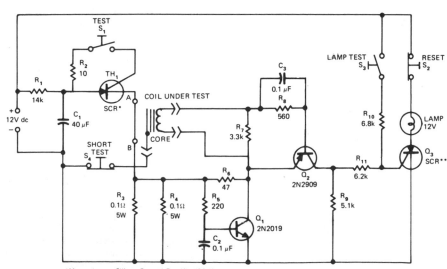

SEALED-COIL TESTER—Permits rapid nondestructive testing of hermetically sealed coils for shorted or open turns, coil-to-core shorts, and reversed polarity of connections. Can be used for simultaneous testing of all coils in recording heads for up to 18 tracks. Circuit develops test pulse having predetermined polarity, amplitude, and duration. Article gives details of circuit operation and test procedures. With multiple-coil units, lamp and detector circuit must be provided for each coil.—D. L. Uhls, Novel Method Nondestructively Tests Sealed Coils, *EDN Magazine,* March 20, 1976, p 102 and 104.

*Motorola-type Silicon Control Rectifier (SCR) or equivalent.
**Any SCR that can handle lamp-current requirements.

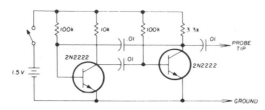

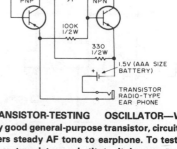

SIGNAL INJECTOR—Circuit is basically 1-kHz MVBR having high harmonic output through 50 MHz. Used with signal tracer for troubleshooting audio equipment. Practically any NPN transistors can be used. Article gives troubleshooting techniques for all types of equipment.—M. James, Basic Troubleshooting, *Ham Radio,* Jan. 1976, p 54–57.

TRANSISTOR-TESTING OSCILLATOR—With any good general-purpose transistor, circuit delivers steady AF tone to earphone. To test another transistor, substitute it in appropriate socket. No tone means it is bad. Low tone or chirp indicates questionable condition. If type (PNP or NPN) is unknown, try in both sockets. If leads of unknown cannot be identified, try all three possible positions in socket.—Circuits, *73 Magazine,* July 1977, p 35.

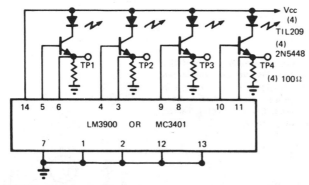

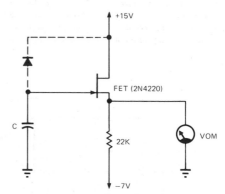

MATCHING OPAMPS—Simple circuit checks condition of quad Norton opamps (National LM3900 or Motorola MC3401). Can also be used to match or select devices for similar DC characteristics in critical applications. Amplifier under test is plugged into socket connected as shown. Good unit makes all four LEDs glow with about same brightness, and all four test-point voltages will agree within about 2 mV. If one section of amplifier is damaged, associated LED will glow very brightly or not at all. Wide variations between test-point voltages indicate partial damage. For critical applications, select amplifiers by matching average test-point voltages.—R. Tenny, Check Norton Amplifiers Quickly, *EDN Magazine,* March 5, 1974, p 72.

DIODE AND FET LEAKAGE—FET under test is connected with 22K resistor as source follower, with capacitor C across input from gate to ground. Leakage of FET charges capacitor at rate directly proportional to leakage and inversely proportional to capacitance. With 0.01 μF for C, each volt of change across C indicates stored charge of 10^{-8} coulomb. This can be interpreted as current in amperes if time for voltage on capacitor to rise 1 V is measured with stopwatch or timer while watching voltmeter. To test diode, connect as shown by dashed line and use good FET in circuit as shown. Article gives design equations; if voltage across C rises 1 V in 38.7 s, leakage current is 0.258 nA.—D. Dilatush, Leakage Testing of Diode and JFETs, *EDN Magazine,* May 5, 1973, p 72–73.

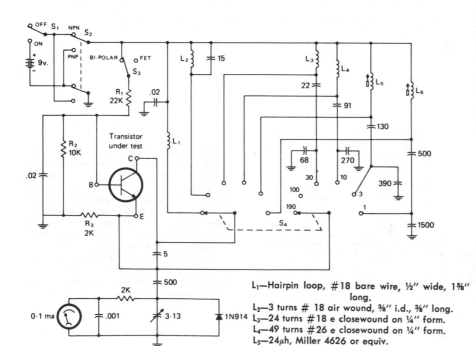

L₁—Hairpin loop, #18 bare wire, ½" wide, 1⅝" long.
L₂—3 turns # 18 air wound, ⅜" i.d., ⅜" long.
L₃—24 turns #18 e closewound on ¼" form.
L₄—49 turns #26 e closewound on ¼" form.
L₅—24μh, Miller 4626 or equiv.
L₆—62 μh, Miller 4630 or equiv.

RF TRANSISTOR TESTER—Tells if unknown bipolar or FET transistor is AF, RF, or VHF and whether it is NPN or PNP. Transistor to be tested is placed in frequency-switchable oscillator circuit, and amplitude of oscillation is noted on meter. Highest oscillation frequency corresponds to highest amplification frequency. Six switch positions cover frequency range of 1 to 190 MHz.—F. Brown, An R.F. Transistor Testor, *CQ,* April 1975, p 35–36 and 66.

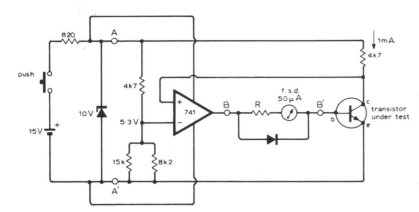

SINGLE-OPAMP TRANSISTOR TESTER—Meter scale is calibrated to read transistor gain directly for NPN devices. Addition of switch for reversing supply and meter polarities permits testing PNP devices as well. When reference voltage of 741 opamp is 5.3 V, circuit passes sufficient base current to make collector current 1 mA. Gain of transistor is then 1 mA divided by base current in microamperes; thus, 50-μA point on meter scale is marked for gain of 20 (1,000 divided by 50). Gain is 400 at 2.5 μA.—A. Rigby, Direct-Reading Transistor Tester, *Wireless World,* Aug. 1976, p 52.

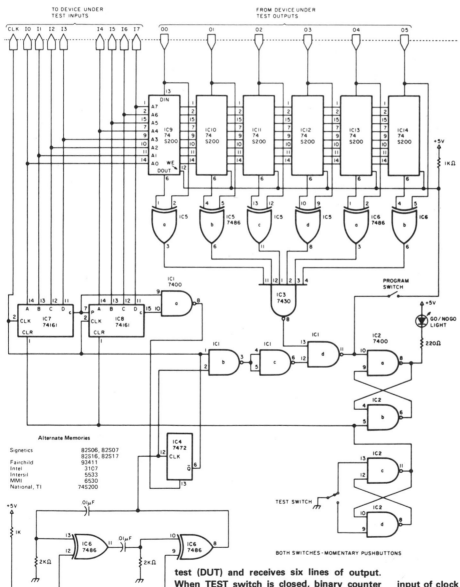

PROGRAMMABLE IC TESTER—Provides automatic, instantaneous, and exhaustive tests of most small-scale integration gates, inverters, flip-flops, etc, and medium-scale integration counters, latches, shift registers, etc. Circuit sends eight lines of input data to device under test (DUT) and receives six lines of output. When TEST switch is closed, binary counter driving DUT input lines is cleared and flip-flop driving GO/NO-GO light is set. Upon release of switch, counter increments through all 256 input conditions. Between counts, data on output lines is compared with data stored in memories IC9-IC14. If mismatch exists, GO/NO-GO flip-flop is cleared at terminal count, CLEAR input of clock oscillator flip-flop is driven low, and further counts are inhibited until TEST button is pushed again. If GO/NO-GO light stays on, component passes test. To program, hold PROGRAM button down while testing known good device. Article gives examples of various applications.—M. Thorson, A. Programmable IC Tester, *BYTE,* June 1978, p 28, 30, 32, and 35.

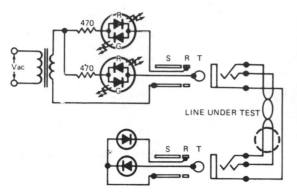

AF LINE TESTER—Gives complete check of shielded twisted-pair cable in one operation, indicating short-circuits between conductors and providing positive continuity check of each conductor. Tester using polarity-sensitive bicolor LEDs is connected to one end of cable under test, and two-diode plug is patched in at other end. If cable is good, only green LEDs come on. If a conductor in cable is open, one or both green LEDs will be off. One or both red LEDs will light for short between any combination of conductors or if cable is wired incorrectly. Signal diode types are not critical.—W. L. Mahood, Tester for Balanced Audio Lines, *EDN Magazine,* April 5, 1974, p 80 and 82.

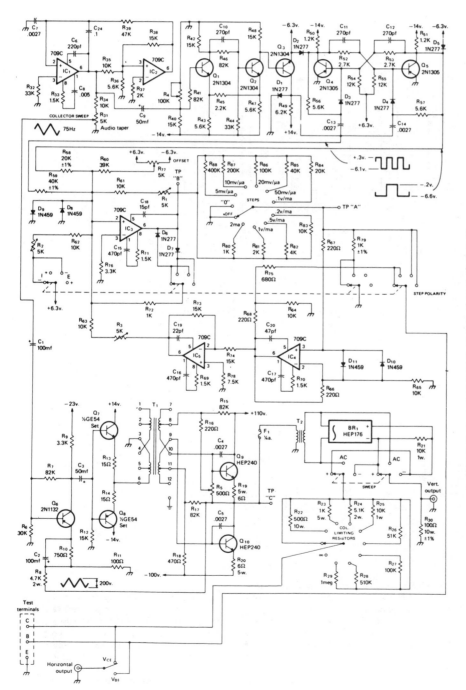

HIGH-ACCURACY CURVE TRACER—Can be used with any calibrated CRO, for matching and testing transistors or diodes by comparing performance curves. All opamps are 709C. Triangle wave generated by IC_1-IC_2 is fed to Schmitt trigger Q_1-Q_2, which generates square wave having transitions at zero voltage crossings of input triangle. Q_3 clamps square wave to 6.3 V P-P. Flip-flop Q_4-Q_5 generates same square wave but at half the frequency of triangle wave. Combining square waves gives three-step staircase voltage having steps precisely in phase with zero signal crossings of triangle wave. T_1 is UTC A-20 audio transformer, and T_2 is Stancor P-6411 I5-W 1:1 isolation transformer. Article covers construction, alignment, and use, and gives circuit of suitable regulated supply operating from ±110 and ±6.3 V available in AN/USM-140C military version of Hewlett-Packard 170 CRO.—A. J. Klappenberger, An Accurate Solid State Component Curve Tracer, *CQ,* July 1974, p 20–24 and 82.

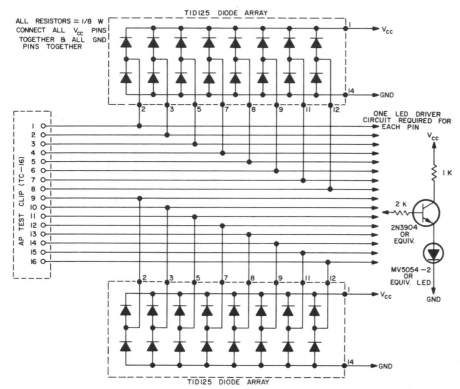

ALL RESISTORS = 1/8 W
CONNECT ALL V_CC PINS
TOGETHER & ALL GND
PINS TOGETHER

TID125 DIODE ARRAY

ONE LED DRIVER
CIRCUIT REQUIRED FOR
EACH PIN

AP TEST CLIP (TC-16)

TID125 DIODE ARRAY

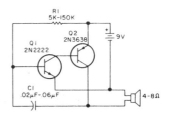

2N3904 OR EQUIV.

MV5054-2 OR EQUIV. LED

IC TEST CLIP—Provides in-circuit testing for all types of 16-pin ICs. LED array indicates logic status of each IC pin. Circuit uses Texas Instruments TID125 diode arrays on test clip to determine pin with highest voltage (V_CC) and pin with lowest voltage (GND). These pins are then used to supply power to LEDs. No batteries are needed. Position of clip on IC is unimportant. On 14-pin ICs, disregard LEDs for two unused pins. Circuit can be expanded for 24- or 40-pin ICs, although adding LEDs makes clip more difficult to use.—J. Errico and R. Baker, Powerless IC Test Clip, *BYTE*, Dec. 1975, p 26–27.

TRANSISTOR PIN-FINDER—Simple audio oscillator is assembled as shown and values of R1 and C1 adjusted for desired tone. General-purpose transistor to be tested is then substituted in circuit (NPN for Q1 or PNP for Q2) and rotated in socket until oscillator works again; pins then correspond to those of the good transistor. If oscillator will not work in any of three possible positions, transistor under test is bad.—Circuits, *73 Magazine*, July 1977, p 34.

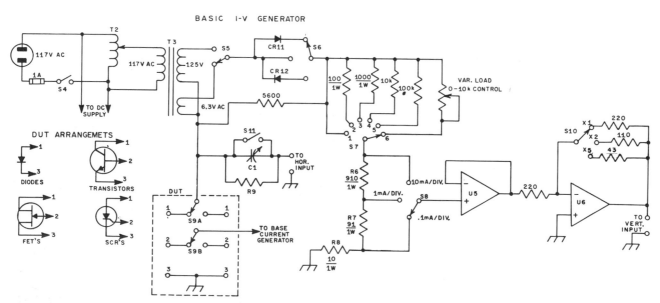

BASIC I-V GENERATOR

DUT ARRANGEMETS

DIODES
TRANSISTORS
FET'S
SCR'S

TRANSISTOR CURVE TRACER—When fed with staircase waveform of base-current generator, circuit generates series of current-voltage (I-V) curves as function of base current, for transistors and other three-terminal semiconductor devices. Cathode follower U5 and inverting controlled-gain amplifier U6 can be eliminated if correct sense of current indication is not essential. S11 switches multiplier R9 in and out; R9 is 18 megohms (about 9 times input resistance of CRO). C1 is 7–13 pF mica trimmer. Diodes are 1N4822. U5 and U6 are Fairchild 741. T2 is Knight 54A3800 or equivalent variable autotransformer rated 1 A. T3 is Knight 54A1410 or equivalent power transformer with 125-V 15-mA and 6.3-V 0.6-A secondaries.—R. P. Ulrich, A Semiconductor Curve Tracer for the Amateur, *QST*, Aug. 1971, p 24–28.

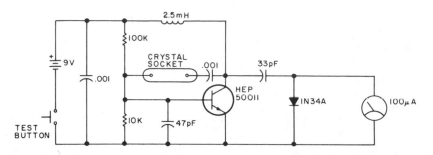

POCKET-SIZE CRYSTAL CHECKER—Provides quick check of condition when shopping for used or surplus crystals. Meter gives steady in- dication at about half scale when test button is pushed, if crystal is oscillating properly.—Circuits, *73 Magazine,* April 1977, p 164.

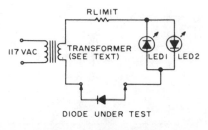

DIODE CHECKER—Requires only one resistor, two LEDs, and any small power transformer providing 3 to 25 VAC. If diode under test is open, neither LED lights. If diode is shorted, LED1 lights on one half-cycle and LED2 on other half-cycle, so both appear lit continuously. If diode is good, LED1 will light if anode of diode is toward transformer, and LED2 will light for other polarity of diode. Choose resistor to limit current through LEDs to about 10 mA.—M. D. Kitchens, Ultra Simple Diode Checker, *73 Magazine,* Oct. 1977, p 44—46.

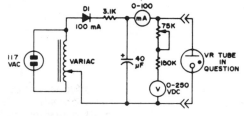

VR TUBE CHECKER—Increase output voltage of Variac gradually until VR tube fires, then read milliammeter and voltmeter. Good tubes will fire at their rated voltage and current values.— Circuits, *73 Magazine,* May 1977, p 31.

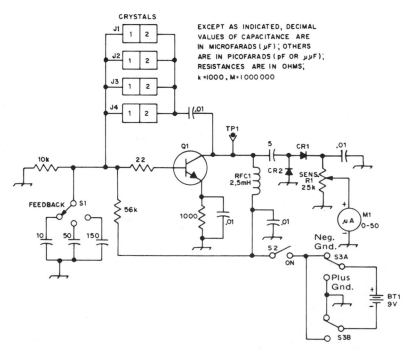

PORTABLE CRYSTAL TESTER—Pierce oscillator using 2N4124, MPS3563, or HEP53 NPN transistor gives indication of crystal activity on M1, from upper HF range down to at least 455 kHz. Increase feedback capacitance with S1 for lower frequency. Choose sockets J1-J4 for types of crystals to be tested. With known good crystal, circuit can also be used for checking bipolar transistors, with S3 providing correct polarity. Diodes are 1N34A germanium or equivalent.— D. DeMaw and C. Greene, A Pair of Handy Testers, *QST,* May 1973, p 24—27.

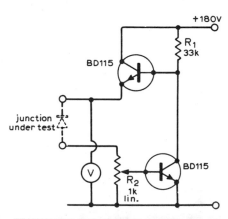

TRANSISTOR BREAKDOWN TESTER—Simple circuit measures breakdown voltages of most types of small-signal and power transistors, reverse breakdown voltages of small power diodes, and zener diode voltages. Two small 90-V batteries in series provide power. R_1 biases upper transistor into conduction. When voltage is applied to diode or transistor junction under test, junction breaks down and current flows through R_2. This makes lower transistor conduct, thereby dropping base voltage of upper transistor. R_2 may be used to set breakdown current over wide range. Voltmeter reads breakdown voltage of junction, since drop across R_2 is negligibly small.—J. W. Brown, Simple Breakdown Voltage Meter, *Wireless World,* July 1973, p 337.

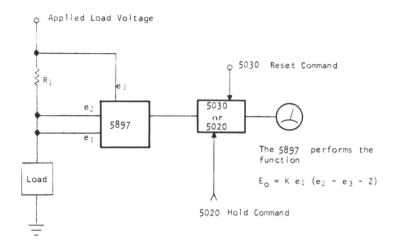

POWER PEAK METER—Optical Electronics 5897 four-quadrant multiplier generates product of load voltage and load current, while 5030 peak sense-and-hold module holds peak power for display on panel meter or other readout. Applications include measuring peak power applied to transistor, motor, lamp, or squib. If power peak at particular moment is required, such as that of transistor failure or squib detonation, 5020 sample-and-hold module is used in addition to or in place of 5030. Hold command can be obtained from flip-flop connected for triggering by abrupt change in power level.— "A Peak Reading/Sampled Reading Power Meter," Optical Electronics, Tucson, AZ, Application Tip 10083.

The 5897 performs the function

$$E_O = K\, e_1\, (e_2 - e_3 - Z)$$

TTL TESTER—Used to check quality, identify internal sections, and identify terminals of unmarked TTL ICs. Operates from 5-VDC source, which should have current-limited output for fuse protection against shorts. Used with ordinary CRO, for which horizontal and vertical jacks are shown on diagram. Article tells how eight different oscilloscope displays are interpreted, and gives procedure for identifying terminals of chip one by one as test probe is held on pins.—S. S. Smith, Jr., A TTL Tester, *73 Magazine*, Oct. 1976, p 110–111.

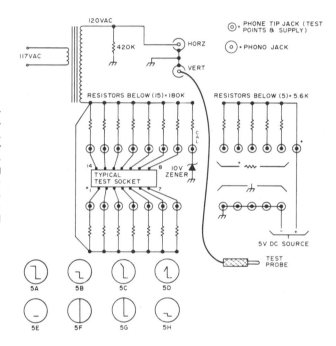

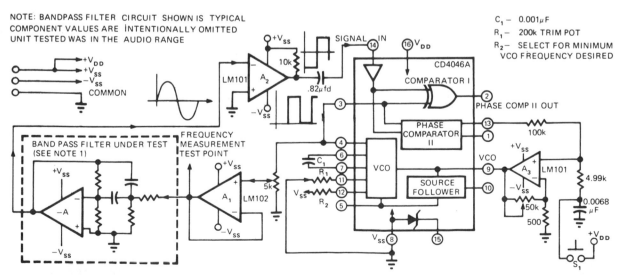

NOTE: BANDPASS FILTER CIRCUIT SHOWN IS TYPICAL
COMPONENT VALUES ARE INTENTIONALLY OMITTED
UNIT TESTED WAS IN THE AUDIO RANGE

C_1 — 0.001 µF
R_1 — 200k TRIM POT
R_2 — SELECT FOR MINIMUM VCO FREQUENCY DESIRED

BANDPASS-FILTER TESTER—Measures center frequency of active bandpass filter by measuring phase angle as function of frequency. Output of VCO excites bandpass filter under test. Output of filter serves as input for PLL comparator. When VCO and filter signals are in phase, PLL locks at center frequency of filter, corresponding to 0° phase shift. Accuracy is 1% for measurements in AF range.—M. P. Prongue, Phase-Lock Loops Test Bandpass Filters, *EDN Magazine*, June 20, 1974, p 76 and 78.

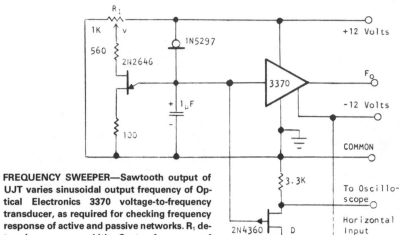

FREQUENCY SWEEPER—Sawtooth output of UJT varies sinusoidal output frequency of Optical Electronics 3370 voltage-to-frequency transducer, as required for checking frequency response of active and passive networks. R_1 determines sweep width. Center frequency of sweep output can be changed by adding voltage to sawtooth or adjusting 3370. Sweep speed can be increased by reducing value of 1-μF tantalum capacitor.—"Sweep Generator Using a Voltage-to-Frequency Transducer," Optical Electronics, Tucson, AZ, Application Tip 10059.

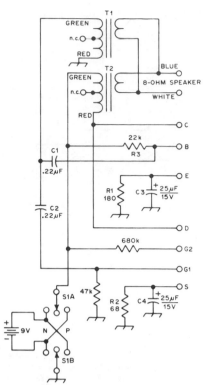

TRANSISTOR TESTER—Will test conventional bipolar transistors, JFETs, MOSFETs, Darlingtons and UJTs. Audible note between 1000 and 5000 Hz from connected loudspeaker indicates that device is functioning as amplifier and gives relative indication of gain and noise figure. Most devices can be tested in-circuit. Among similar JFETs or MOSFETs, those producing lowest tone pitch have lowest noise figure. Among similar devices of any type, those producing loudest tone have highest gain. Tester feeds back audio signal through two transformers to create sustained oscillation when amplifying device is connected to proper terminals. S1 applies positive or negative voltage through audio output transformer T2 to device under test. C3 and C4 must be nonpolarized electrolytics because R1 and R2 may produce either positive or negative voltage depending on device being tested. T1 and T2 have 1200-ohm primary and 8-ohm secondary (Calectro DI-724). Note above 10,000 Hz means device has some gain but does not meet specifications or is connected incorrectly. For MOSFETs, source and substrate are both connected to source terminal.—W. E. Anderson, A Universal Transistor Tester, *QST*, Dec. 1975, p 26–28.

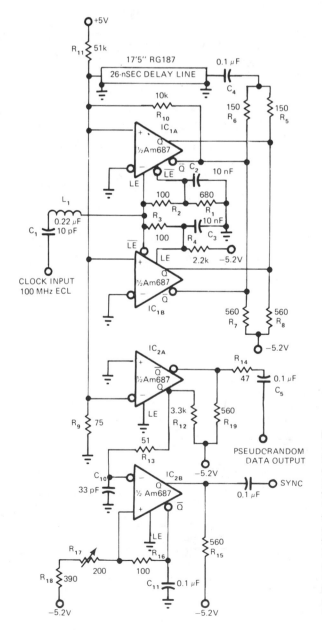

PSEUDORANDOM DIGITAL GENERATOR—Uses two Am687 dual sampling comparators to implement 200-Mb/s pseudorandom digital-sequence generator for checking and measuring performance of high-speed digital communication equipment. Produces sequence length of 127 bits, but delay-line length can be changed to give any desired other sequence length. Circuit accepts from 195 to 203 Mb/s, which is more than adequate for systems using crystal clock having ±0.1% frequency variation. Article describes operation and gives timing diagram.—G. L. Meyer, Sampling Comparators—They Sub for High-Speed Logic and Produce Power, Cost and Space Savings, *EDN Magazine*, Sept. 5, 1977, p 71–74.

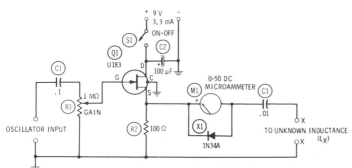

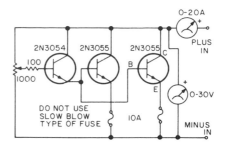

INDUCTANCE METER—When used with variable-frequency audio oscillator, FET circuit checks any inductance between 60 μH and 60,000 H by resonance method. With unknown coil connected to terminals XX, external oscillator is tuned for peak deflection of M1. Inductance is then calculated from $L = 1/(395 \times 10^{-9} f^2)$ where L is in henrys and f is in hertz. If desired, calibration graph can be prepared to eliminate calculations. High input impedance of FET minimizes oscillator loading. Adjust R1 for full-scale deflection of meter at resonance, to give maximum sensitivity.—R. P. Turner, "FET Circuits," Howard W. Sams, Indianapolis, IN, 1977, 2nd Ed., p 138–140.

POWER-SUPPLY TESTER—Serves as high-current solid-state resistor load for testing power supplies before use, to determine voltage and current under load. Darlington configuration of transistors reduces power-dissipating requirements of pot. Use large heatsink for 2N3055s because they must dissipate almost 200 W when power supply is delivering 15 A at 15 V.—E. Fruitman, The Smoke Tester, *73 Magazine*, Nov. 1976, p 159.

CRYSTAL TESTER—JFET Pierce oscillator will test any crystal from 50 kHz through 25-MHz upper frequency limit of fundamental-mode crystals without tuning, and drive counter for measuring crystal frequency. Will test overtone VHF crystals on their fundamental frequency. T1 is small output transformer from tube-type radio, having about 33:1 turns ratio, or 6.3-V filament transformer if 1N645 rectifiers are used in place of 50-μF filter capacitors to give full-wave voltage doubler providing required 9 V. RFC1 is 2.5 mH, and RFC2 is 150-mH miniature toroid.—F. Brown, A Universal Crystal Oscillator, *QST*, Feb. 1978, p 15–16.

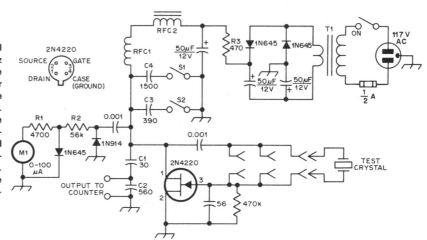

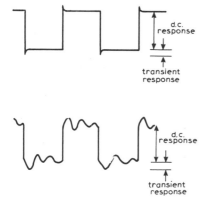

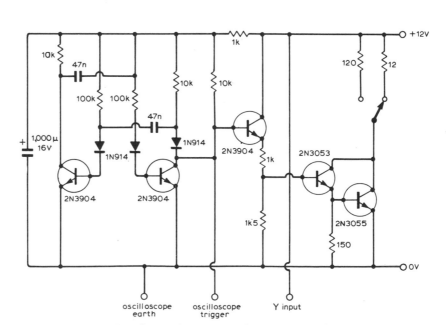

TRANSIENT RESPONSE OF REGULATED SUPPLIES—Developed for testing transient response of 12-V regulated power supply at loading of either 100 mA or 1 A, depending on switch position. Load resistors can be changed for other voltages and currents. Transients generated by supply may be observed on AC-coupled oscilloscope. Good transient response will show only small leading-edge peaks, as in upper waveform. Any tendency of power supply toward instability degrades waveform much more, as in lower diagram. Circuit consists of multivibrator using series diodes in base circuits to protect transistors from excessive voltage swings in switching cycle. Square-wave output is used for oscilloscope trigger and fed to other three transistors that provide load for power supply under test.—H. Macdonald, Transient Response Testing, *Wireless World*, July 1973, p 338.

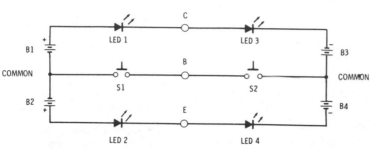

B1, B2, B3, B4 = 3 VOLTS

TRANSISTOR AND DIODE TESTER—Four pairs of AA penlight cells provide power for testing transistors and diodes quickly for opens and shorts. Circuit also distinguishes between PNP and NPN transistors and shows diode polarity. Leads of diode are inserted into base and collector jacks (B and C), and switches are pressed in succession. If LED 1 glows, diode is good and its anode lead is in collector jack. If LED 3 glows when S2 is pressed, anode of diode is in base jack. If both LEDs glow, diode is shorted. If neither LED glows, diode is open. With transistor in tester, unit is PNP if LED 1 and LED 2 glow. If LED 3 and LED 4 glow, unit is NPN. If one or no LEDs glow, transistor is open. If three or more LEDs glow, transistor has shorted junction. Any LEDs can be used.—F.M. Mims, "Transistor Projects, Vol. 3," Radio Shack, Fort Worth, TX, 1975, p 87–93.

FET TESTER—Can be used for measuring JFET pinchoff voltage, matching FETs of same generic type, and measuring bias range of FET. Opamps sense source current of FET under test. First 741 is buffer, while second is preset to 1 V and its output used to drive device under test (DUT) until source current is 100 μA. Polarity of VP is opposite that of VDD.—Circuits, 73 Magazine, June 1977, p 49.

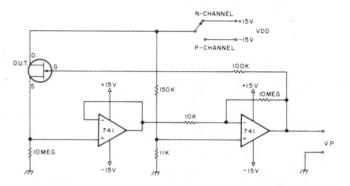

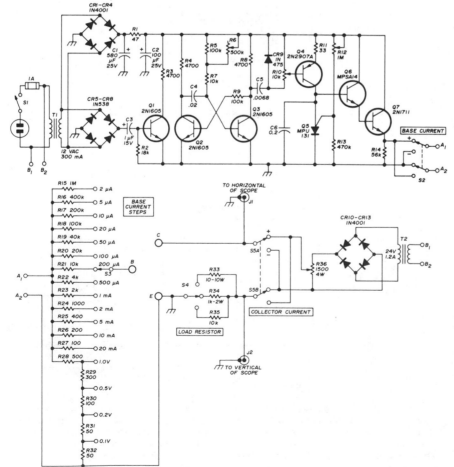

CURVE TRACER—Displays on CRO a family of six collector characteristics for transistors and voltage/current characteristics of diodes. Circuit varies base voltage in discrete steps while sweeping collector voltage from zero to maximum at each step. Collector voltage is 120-Hz rectified sine wave from bridge rectifier CR10-CR13, varied by R36. S5 selects proper polarity. Base voltage steps are synchronized to 120-Hz collector voltage by adjusting R6 in mono MVBR Q2-Q3 driven by Q1. R10 adjusts voltage between steps. Programmable UJT Q5 resets staircase generator to zero. R12 adjusts gate voltage. Staircase voltage is coupled to bias resistors R15-R32 for test transistor by Q6 and Q7, with bias polarity and value selected by S2 and S3. Connect device under test to points B, C, and E. Connect points A_1 and A_2 together. R15-R35 should have 5% accuracy.—D. Wright, Transistor Curve Tracer, Ham Radio, July 1973, p 52–55.

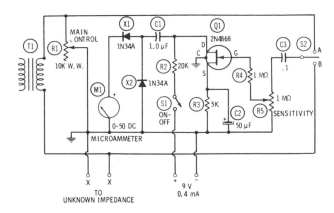

IMPEDANCE METER—External AF voltage is applied to unknown impedance through input transformer T1 and calibrated variable resistor R1. Voltage drops across impedance and R1 are checked separately with electronic voltmeter while R1 is varied. When drops are equal, unknown impedance is equal to setting of R1, read directly from its calibrated dial. Voltmeter uses FET followed by two-diode rectifier X1-X2 and microammeter M1. With S2 in position A, voltmeter reads drop across R1; in position B, voltmeter reads drop across unknown impedance. Adjust R5 so meter deflection is near full scale to increase comparison accuracy. Common test frequencies used are 400, 500, and 1000 Hz.—R. P. Turner, "FET Circuits," Howard W. Sams, Indianapolis, IN, 1977, 2nd Ed., p 143–144.

DIODE TESTER—Developed to demonstrate how 0.002-μF shunt capacitor increases efficiency of diode detector. Can also be used to compare performance of different diodes. Input terminates AM signal generator. IC with associated capacitors and resistors forms low-pass filter having cutoff at 2 kHz. R1 is detector load. Resistor at output of U1 prevents oscillation of opamp when using coax feed to VTVM. For higher input signal levels, shunt capacitor increases AF output up to 10 dB.—H. Olson, Diode Detectors, *Ham Radio,* Jan. 1976, p 28–34.

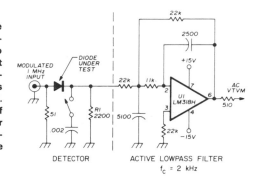

DETECTOR   ACTIVE LOWPASS FILTER
f_C = 2 kHz

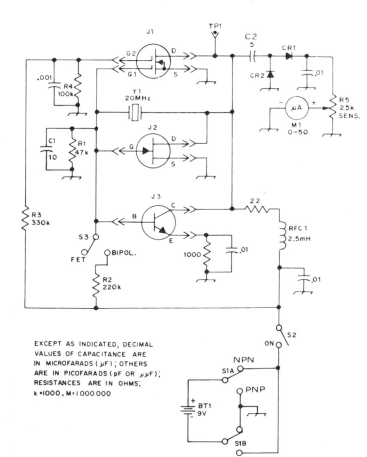

EXCEPT AS INDICATED, DECIMAL VALUES OF CAPACITANCE ARE IN MICROFARADS (μF); OTHERS ARE IN PICOFARADS (pF OR $\mu\mu$F); RESISTANCES ARE IN OHMS; k =1000, M=1 000 000

PORTABLE TRANSISTOR TESTER—Go/no-go tester shows relative condition of NPN and PNP transistors, junction FETs, and dual-gate MOSFETs. Not suitable for checking audio or high-power RF transistors. Crystal for upper range of HF spectrum is permanently wired; any HF crystal cut for fundamental-mode operation can be used. Rectified RF from oscillator is monitored on M1. S1 selects negative ground for testing N-channel FETs and NPN bipolars and provides positive ground for P-channel and PNP devices. If device is open, shorted, or extremely leaky, circuit will not oscillate and meter will not deflect. The higher the meter reading, the higher the gain of transistor at operating frequency. When testing MOSFETs that are not gate-protected, keep transistor leads shorted until device is in socket and replace short before removing device. Diodes are 1N34A or equivalent.—D. DeMaw and C. Greene, A Pair of Handy Testers, *QST,* May 1973, p 24–27.

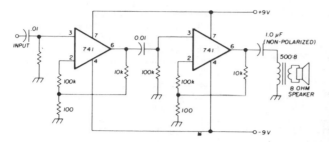

AF SIGNAL TRACER—High-gain audio amplifier with loudspeaker output serves for troubleshooting AF stages. With normal input to amplifier under test, presence of AF signal is tested in each stage in turn. If AF signal appears at input of stage but not at output, that stage has a defect. Signal source may also be signal injector having broadband output from audio through VHF.—M. James, Basic Troubleshooting, *Ham Radio*, Jan. 1976, p 54–57.

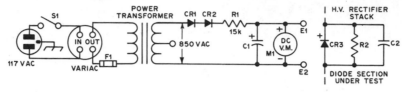

POWER DIODE TESTER—Provides reverse-voltage test of individual sections of rectifier stack at 1000 VDC. With test leads E1 and E2 clipped across diode section under test, Variac setting is increased from 0 until C1 is charged to 1000 V as indicated by voltmeter. If diode or capacitor in section under test is defective, meter will read low because of extra voltage loss across R1. Initial setup is made with good diode section. Open S1 before changing diode because voltage is lethal.—R. K. Dye, Testing "Dye-Odes," *QST*, Feb. 1976, p 44.

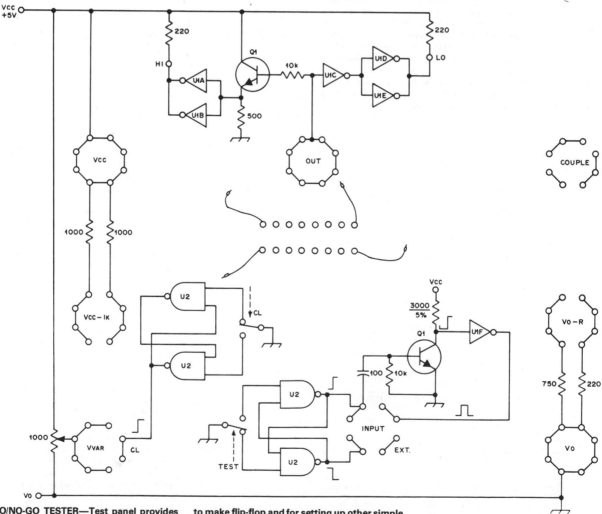

TTL GO/NO-GO TESTER—Test panel provides fast static test of surplus 7400 series TTL ICs. Each contact of 16-pin DIP socket has lead that can be plugged into array of seven other sockets carrying various supply voltages, loads, etc. Switches provide pulses of input current for toggling or clocking, counting, and resetting. Leads also serve for cross-coupling gates in IC to make flip-flop and for setting up other simple circuits. Input voltage control allows plotting of transfer functions and study of circuit operation under different signal-level conditions. External test meter can be connected when necessary. HI and LO indicators are LEDs that show level of terminal connected to output socket. Transistors are general-purpose NPN such as 2N2926. U1 is SN7404 hex inverter, and U2 is SN7400 quad two-input NAND gate. Tester is not suitable for complementary MOS devices requiring protection from static charges. Article gives detailed instructions for testing each type of TTL device.—J. S. Worthington, A Simple TTL Test Panel, *QST*, Dec. 1976, p 25–27.

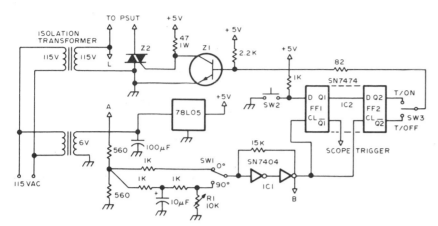

POWER-SUPPLY TESTER—Circuit switches power supply on or off either at peak or zero of line voltage while supply is connected to CRO. Spikes on trace then indicate supply defect. Developed for testing unknown power supplies before being placed in use on home computer systems, to ensure safe and reliable computer operation. Article gives detailed instructions for using circuit to check for transients, DC regulation, heat rise, and output impedance.—R. Tenny, Power Supply Testing, *73 Magazine*, July 1976, p 112–114 and 116–117.

DYNAMIC LOAD—R1 sets load current drawn from power supply under test to desired value, after R2 is chosen to give about 1-V output at maximum load current to be drawn from power supply under test (PSUT). Modulation input to C1 is obtained from external pulse generator, and serves to make load current increase and decrease over small range.—R. Tenny, Power Supply Testing, *73 Magazine*, July 1976, p 112–114 and 116–117.

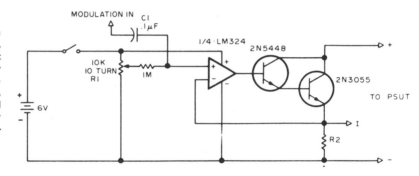

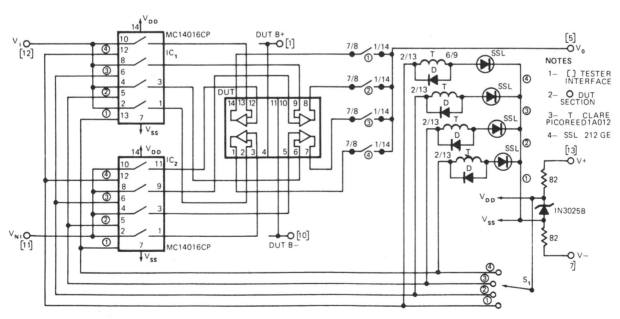

QUAD OPAMP TESTER—Interface circuit shown permits use of single-opamp tester for testing quads without major modification of tester's original function. Interface operates on power available from tester, which can be Teledyne/Philbrick 5102 or current Tektronix or ESI testers. CMOS input transmission gates IC$_1$ and IC$_2$ supply input signals to sections of device under test (DUT). Gate control signals are supplied by section selector S$_1$; switch can be automated by using two-line BCD selection. Same switch also activates reed relay that connects respective amplifier section to load and output monitoring circuits of tester. LEDs indicate section under test. Interface can be used with LM124, CA124, and MC3503 series of quad opamps.—A. C. Svoboda, Use a Single Op-Amp Tester for Quads, *EDN Magazine*, March 5, 1975, p 76.

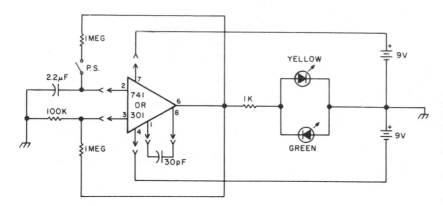

OPAMP TESTER—Developed for quick testing of 741 and 301 opamps. For good opamp, LEDs flash alternately with 1-s period. No flashing and no illumination indicate opamp output fault. If one of LEDs glows continuously, one of inputs is faulty. Asymmetrical blinking indicates leakage in opamp.—Circuits, *73 Magazine,* March 1977, p 152.

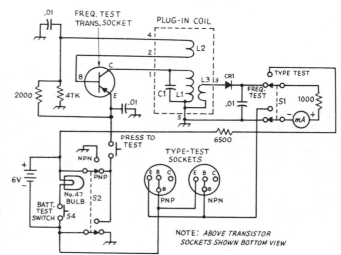

TRANSISTOR TESTER—Determines maximum frequency at which unknown transistor will still maintain reasonable current gain, and tells whether transistor is NPN or PNP. Based on fact that emitter-base junction of transistor is equivalent to crystal diode, conducting in one direction only. Direction depends on transistor type. Setting of S2 which gives meter reading identifies transistor type. Frequency-limit circuit is self-excited oscillator in which frequency depends on plug-in coil. If meter reads when 60-MHz coil is used, transistor is capable of handling 60 MHz; if no reading, change to lower-frequency coils one by one until reading is obtained. To check battery, press S4; lamp should have full brilliance. CR1 is 1N34A.—H. Hanson, How High Will It Go?, *QST,* April 1974, p 32–33 and 39.

Frequency (MHz)	L1 (Turns)	L2 (Turns)	L3 (Turns)	C1 (PF)
60	3	3	3	25
31	7	6	4	25
12	12	7	6	80
3	22	10	9	270
1	34	20	8	1000

Note: Above coils close-wound, 3/4-inch (19 mm) diameter.

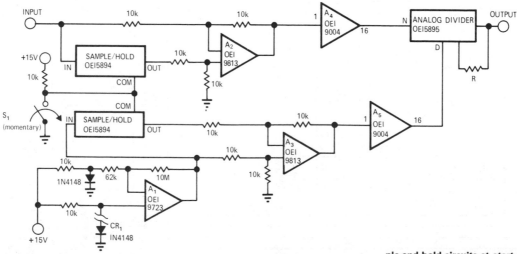

TEMPERATURE-COEFFICIENT COMPUTER—Circuit automatically measures and calculates temperature coefficients of analog circuits or devices. Silicon diode CR_1 is used as temperature probe having forward drop of about 2 mV/°C. R adjusts output scale factor. FET-input opamp A_1 converts forward voltage drop of temperature probe into high-level analog voltage that varies 325 mV/°C from −10 V at +55°C to 10 V at −5°C. Output of A_1 is applied to sample-and-hold circuit, while analog voltage from device under test is applied to second sample-and-hold. Momentary closing of S_1 causes voltage and temperature data to be stored in sample-and-hold circuits at start of test to bias A_2 and A_3. Outputs of these opamps can be positive or negative, but are made positive by unity-gain absolute-value opamps A_4 and A_5. From these outputs, analog divider calculates temperature coefficient.—R. C. Gerdes, Temperature-Coefficient Measuring Circuit, *EDN/EEE Magazine,* Feb. 1, 1972, p 54.

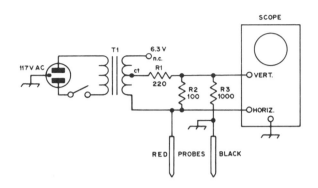

IN-CIRCUIT TESTER—Eliminates need for removing components one by one for testing. Voltages and currents used are low enough for almost any transistorized circuit. Will test for shorts and opens. Shows forward-reverse ratios on junction transistors and diodes. Lissajous figures and other combination displays on CRO facilitate analysis of circuits having reactive components, transistors, and ICs. Will detect high-resistance solder joint and check continuity of switches, fuses, lamps, and printed wiring. Displays form hands of clock or ovals. Vertical line indicates short, horizontal means open, slant indicates resistance, vertical oval is inductance, horizontal oval is capacitance, diode and highest-merit transistor show 3 o'clock, fair transistor shows 4 o'clock, and poor transistor shows 5 o'clock. For other patterns, compare with those obtained with known good components.—D. L. Ludlow, The Octopus, QST, Jan. 1975, p 40–42.

AF SIGNAL TRACER—Motorola MC1306P complementary power amplifier delivers ½ W into loudspeaker for 3-mVRMS input to preamp, for troubleshooting all types of audio equipment. Zero-signal current drain is only 4 mA with 9-V supply. For RF tracing of modulated AM or SSB signal, use demodulator probe at input.—W. M. Scherer, High-Gain Signal Tracer, CQ, July 1972, p 12 and 14.

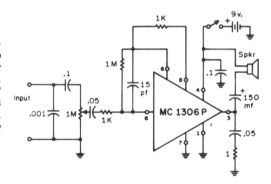

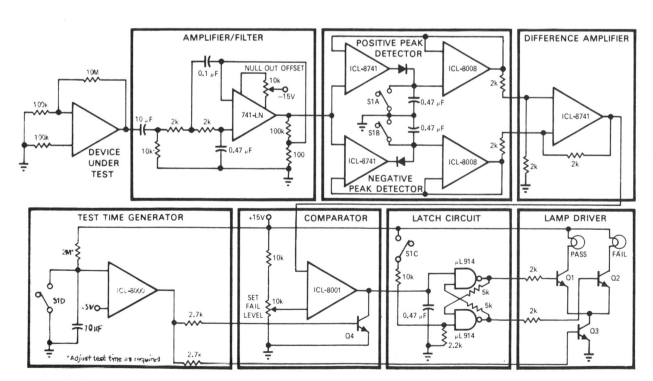

POPCORN NOISE TESTER—Developed at Intersil to test opamps for erratic low-frequency jumps between two or more stable states. After start-up switch S is closed, pass/fail lamps are inhibited by Q3. At end of preselected test time (typically 5 s), one of lamps will come on. If output from difference amplifier exceeds preset fail level at inverting input of comparator during test period, fail lamp is turned on by Q2. Q4 prevents triggering of latch by spurious signals after end of test time.—T. P. Rigoli, IC Op Amps, EDN Magazine, May 1, 1971, p 23–33.

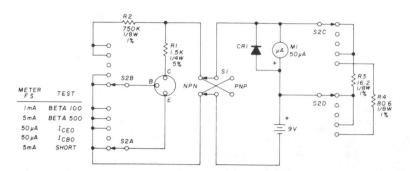

TRANSISTOR TESTER—Useful for troubleshooting and for checking small-signal transistors having no markings. Set four-pole five-position rotary switch S2 to SHORT (lowest position) before inserting transistor, then flip S1 back and forth. If meter shows any reading at all, reject transistor without further tests. If meter stays at zero, set S2 to I_{CBO} (collector-base current, emitter open). Discard transistor if reading is high for either position of S1; modern transistors pass only nanoamperes, but older types may give noticeable reading, particularly if germanium. Repeat test for I_{CEO} (collector-emitter current, base open), which should be greater than I_{CBO} by factor approximating current gain (beta) of device. Modern silicon transistors may give no reading here. For final beta test, older types show 100 or less and modern transistors like 2N3391 have beta readings between 300 and 400. CR1 is 1N4603.—D. Cheney, Shirt Pocket Transistor Tester, *Ham Radio*, July 1976, p 40–42.

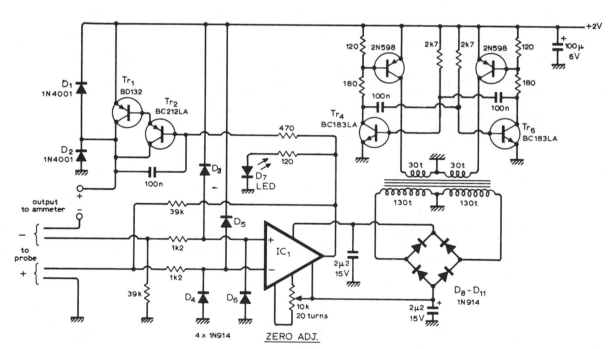

AMMETER FOR PRINTED-CIRCUIT WIRING—Permits measurement of current in single conductor on board without cutting it. Article gives design of probe having four projecting wires that are pressed on conductor being measured, and describes operation of circuit in detail. Opamp can be 741, but higher-cost 725C will improve performance. When all four wires of probe make contact, voltage drop appears at input of differential amplifier. Outer wires of probe carry current of opposite polarity passing through ammeter; because there is negative feedback loop in conductor, opamp input voltage will return to zero when outgoing current is equal to that of unknown current passing through printed-circuit conductor.—F. Andrews, P.C.B. Ammeter, *Wireless World*, July 1976, p 34.

CHAPTER 96

Timer Circuits

Includes circuits that give elapsed time between two events, produce desired switching action after predetermined adjustable delay, or perform switching actions at preset times. Time-of-day circuits are given in Digital Clock chapter. See also Burglar Alarm and Photography chapters.

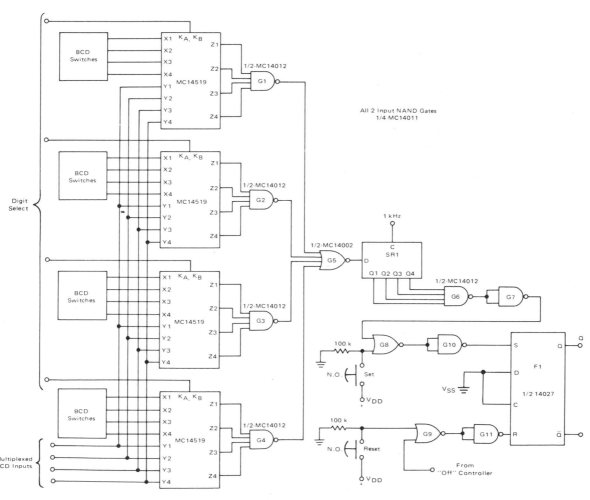

TIME COMPARATOR—Digital circuit compares time that has been preset on one set of BCD input switches to multiplexed BCD output of basic 24-h industrial clock. When time of day corresponds to preset time, output circuit of comparator turns controlled device on for preset period of time. Second set of comparators can be used to turn device off when time of day equals preset time. Only hour and minute digits are compared, using Motorola MC14519 4-bit AND/OR select ICs. Q output of F1 can be used to control load power through output control circuit driving triac, with optoisolator providing required isolation from AC line. Supply is +5 V.—D. Aldridge and A. Mouton, "Industrial Clock/Timer Featuring Back-Up Power Supply Operation," Motorola, Phoenix, AZ, 1974, AN-718A, p 6.

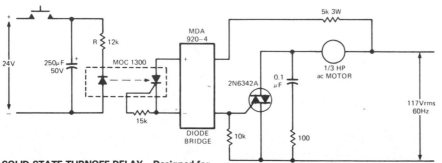

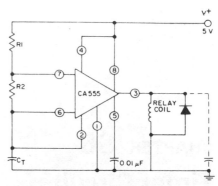

SOLID-STATE TURNOFF DELAY—Designed for applications where machine must remain energized for certain period of time after some other operation has stopped, as when pump motor must run long enough to clear pipes in chemical plant. When switch is closed, LED of optoisolator is forward-biased and turned on, making photo-SCR conduct and provide gate current via MDA920-4 diode bridge for triac. Values shown keep solid-state relay circuit on for about 5 s after pushbutton is released. Resistor and capacitor values can be changed to obtain different delay.—T. Mazur, Solid-State Relays Offer New Solutions to Many Old Problems, *EDN Magazine,* Nov. 20, 1973, p 26–32.

REPEAT-CYCLE—RCA CA555 is connected for astable operation in which total period is sum of individual perids t_1 and t_2, where $t_1 = 0.693(R_1 + R_2)C_T$ and $t_2 = 0.693R_2C_T$. With 5-V supply, output voltage has rectangular pulses with interval t_1 separated by interval t_2. With optional capacitor connected, voltage across capacitor is sawtooth that rises for interval t_1 and decays for interval t_2.—"Linear Integrated Circuits and MOS/FET's," RCA Solid State Division, Somerville, NJ, 1977, p 56.

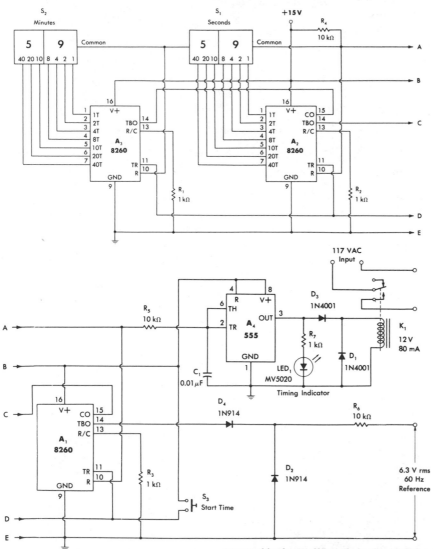

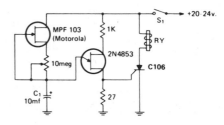

10-min DELAY—MPF103 JFET makes charging current of C_1 constant regardless of degree of charge, to give longer charging time and longer time delay. After delay determined by setting of 10-megohm pot, UJT conducts and discharges C_1 through 27-ohm resistor, triggering SCR and energizing relay. Open S_1 to reset circuit.—I. Math, Math's Notes, *CQ,* April 1974, p 64–65 and 91–92.

APPLIANCE TIMER—Controls intervals up to 1 h in 1-s increments as programmed by thumbwheel switches S_1 and S_2. Circuit is basically two-stage programmed counter driven by 1-s clock derived from 60-Hz power line. A_1 is connected as divide-by-60 counter triggered by 60-Hz signal developed across D_2. 1-Hz output from A_1 triggers A_2 which in turn triggers A_3, all programmable timers. When S_3 is closed, R_4 bus goes low to start timing cycle. Relay driver A_4 holds relay K_1 closed for application of AC power to device being controlled and energizing of LED to indicate active timing cycle. Applications include uses as kitchen and darkroom timers.—W. G. Jung, "IC Timer Cookbook," Howard W. Sams, Indianapolis, IN, 1977, p 214–218.

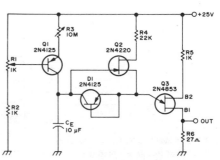

10-h FET—Long duration timer gives adjustable delays up to 10 hours before turning Q3 on to give output voltage.—Circuits, *73 Magazine,* Feb. 1974, p 101.

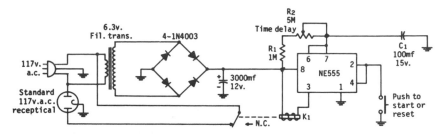

RADIO TURNOFF—Time-delay control R_2 can be set to turn off radio, TV, or other appliance at any desired interval between about 3 and 60 min after start button is pushed. Ideal for those who fall asleep to music. K_1 is 12-V relay drawing 200 mA or less.—P. Walton, An Electronic Timer for Less than $5.00, *CQ*, Aug. 1973, p 42 and 82.

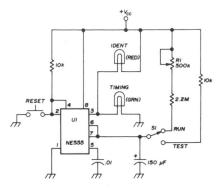

10-min ID TIMER—Red pilot lamp comes on at end of timing period, adjustable with R1 from 7 to 11 min, as reminder for amateur radio operator to make station identification required by FCC every 10 minutes. Green lamp indicates that timer is on and timing. Lamps should draw no more than 100 mA, to avoid overloading NE555 timer. Any 9–12 VDC supply can be used.—D. Backys, Identification Timer, *Ham Radio*, Nov. 1974, p 60–61.

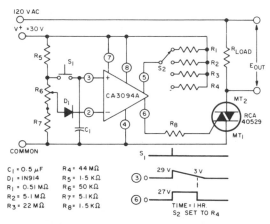

$C_1 = 0.5\,\mu F$	$R_4 = 44\,M\Omega$
$D_1 = 1N914$	$R_5 = 1.5\,K\Omega$
$R_1 = 0.51\,M\Omega$	$R_6 = 50\,K\Omega$
$R_2 = 5.1\,M\Omega$	$R_7 = 5.1\,K\Omega$
$R_3 = 22\,M\Omega$	$R_8 = 1.5\,K\Omega$

PRESETTABLE ANALOG TIMER—Switch S_2 gives choice of four delay intervals between closing of S_1 and triggering of triac by RCA CA3094A programmable power switch and amplifier. Pot R_6 is required for initial time set.— "Linear Integrated Circuits and MOS/FET's," RCA Solid State Division, Somerville, NJ, 1977, p 192–196.

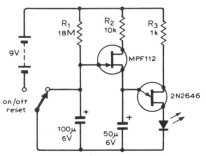

FLASHING-LED EGG TIMER—UJT oscillator controlled by FET timer makes LED flash after time delay determined by value of R_1. When 100-μF capacitor charges to about 1 V after switch is turned on, MPF112 FET switches on and forms part of charging circuit for 2N2646 UJT oscillator which then pulses LED at about 200 mA peak. Although developed as inexpensive egg timer, circuit has many other applications.—J. Jeffrey, Simple Flashing-L.E.D. Timer, *Wireless World*, Oct. 1974, p 381.

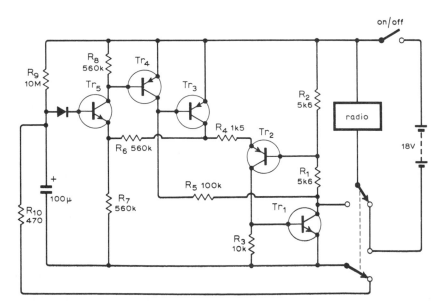

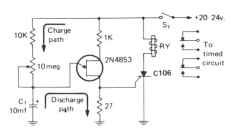

TRANSISTOR RADIO TURNOFF—Circuit switches radio off after delay of about 30 min, if ganged switch is set to other position and radio switch itself is left on. Current drain of timer circuit is negligible in both positions. Article describes timing action of transistors in detail. Tr_1 should have current gain above 25 with collector currents of 10 to 100 mA; 2N3706 can be used here and for Tr_5. Other transistors should have current gain above 50, as in 2N3702. For operation from 4.5 to 9 V, omit diode and cut values of R_4-R_7 in half.—S. Lamb, Delayed Switch Off for Transistor Radios, *Wireless World*, Aug. 1973, p 373.

1.5-min DELAY—When S_1 is closed, C_1 begins charging. After delay determined by setting of 10-megohm pot, UJT conducts and makes C_1 discharge through 27-ohm resistor, triggering SCR C106 and energizing relay. Open S_1 to reset circuit.—I. Math, Math's Notes, *CQ*, April 1974, p 64–65 and 91–92.

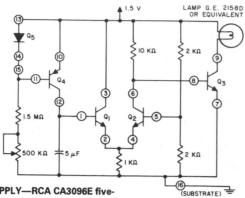

10 s USING 1.5-V SUPPLY—RCA CA3096E five-transistor array drives indicator lamp that comes on at end of timing interval. Q_5 is one of PNP transistors in array connected as diode.—

"Linear Integrated Circuits and MOS/FET's," RCA Solid State Division, Somerville, NJ, 1977, p 205–210.

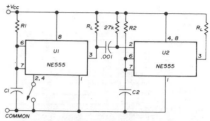

CASCADED TIMER—Two NE555 timers give sequential timing intervals for two separate loads. Time for U1 is set by R1 and C1, and for U2 by R2 and C2. Grounding pin 2 momentarily with switch starts timing. Once started, it cannot be retriggered. With pin 2 connected to reset input 4, both functions are obtained with one push of switch. If reset function is not wanted, connect 4 to 8. When pin 3 of U1 goes low at end of timing interval, negative pulse generated by 0.001-μF capacitor and 27K resistor goes to pin 2 of U2 to trigger second timer. With 15-V supply, each timer can handle 200-mA load.—H. Vordenbaum, Automatic Reset Timer, *Ham Radio*, Oct. 1974, p 50–51.

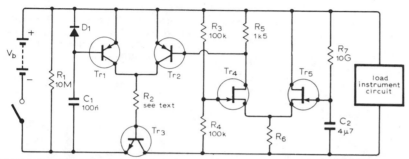

BATTERY SWITCH-OFF—Can be added to any battery-operated device to eliminate unnecessary running down of costly batteries when someone forgets to turn switch off manually. Circuit shown gives operating time of about 10 hours, permitting use of equipment for full working day without interruption. Normal operation can be restored after shut-off by turning manual switch off and then on again. When circuit has switched off, only battery drain is current through 10-megohm resistor R_1 and leakage through transistors. Time of switch-off can be changed by altering C_2 or R_7. Tr_1 and Tr_2 are 2N4061, BC478, or similar PNP silicon; Tr_3 is 2N3053, BC142, or similar medium-power NPN silicon; Tr_4 and Tr_5 are 2N3819; D_1 is any small silicon diode; R_6 is 4.7K for 9-V or 12-V battery, and 10K for 15-V to 27-V battery. R_2 is chosen to suit working current and battery voltage; suitable value is 15 V_b/I_{out}.—D. T. Smith, Automatic Battery Switch-Off Circuit, *Wireless World*, April 1976, p 76.

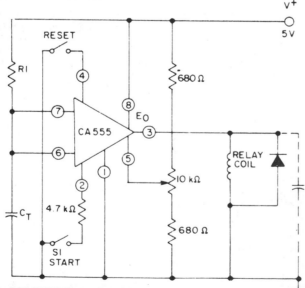

RESET MODE—RCA CA555 timer is connected so C_T is initially held in discharged state by transistor in IC. When start switch is closed, internal flip-flop clears short across C_T, driving output voltage high and energizing relay. Voltage across capacitor then increases exponentially with time constant R_1C_T. When capacitor voltage equals two-thirds of V+, flip-flop resets and discharges capacitor rapidly, driving output low and releasing relay. Timing interval is relatively independent of supply voltage variations. Applying negative pulse simultaneously to reset pin 4 and trigger pin 2 by closing both switches during timing cycle causes timing cycle to restart. Momentary closing of reset switch only serves to discharge C_T without restarting timer.—"Linear Integrated Circuits and MOS/FET's," RCA Solid State Division, Somerville, NJ, 1977, p 56.

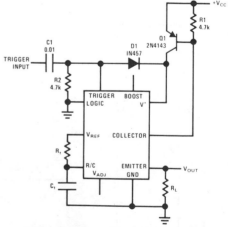

MINIMUM-DRAIN TIMER—National LM122 timer is connected to reduce supply drain to zero between timing cycles. External PNP transistor Q1 serves as latch between V+ terminal of timer and power supply. Between timing periods, Q1 is off and no current is drawn. Arrival of 5-V or larger trigger pulse turns Q1 on for duration of timing period set by R_t and C_t, which can range from microseconds to hours.—C. Nelson, "Versatile Timer Operates from Microseconds to Hours," National Semiconductor, Santa Clara, CA, 1973, AN-97, p 10.

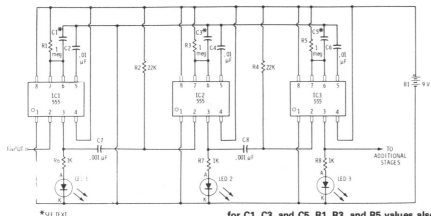

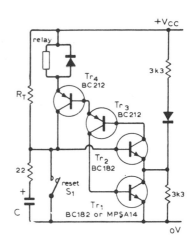

SEQUENCED TIMERS—Cascading of three 555 timers, each driving LED, gives sequenced flashes with individually adjustable durations. Times of timers are determined by values used for C1, C3, and C5. R1, R3, and R5 values also affect time delays; use pots if 1 μF is used for all three capacitors.—F. M. Mims, "Integrated Circuit Projects, Vol. 5," Radio Shack, Fort Worth, TX, 1977, 2nd Ed., p 64–75.

STABLE FOUR-TRANSISTOR TIMER—Circuit has good immunity to impulse noise because normal state of all transistors is on. This eliminates spurious timing cycles that sometimes occur in IC timers. At switch-on, C begins charging until its voltage makes Tr_2 start conducting; this in turn makes other three transistors switch on. Regeneration action then discharges C to about 0.6 V. Timer is started either by applying V_{CC} or opening S_1. Timing period depends on value of V_{CC}.—J. L. Linsley Hood, One-Shot Timer Circuit, *Wireless World*, Nov. 1975, p 520.

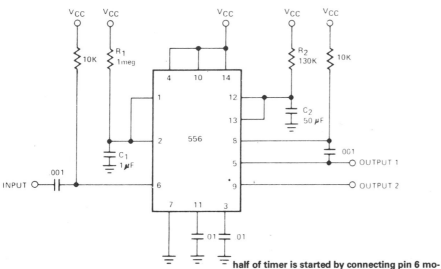

SEQUENTIAL TIMER—Output of first half of 556 dual timer is fed to input of second half through 0.001-μF coupling capacitor to give total delay equal to sum of individual timer delays. First half of timer is started by connecting pin 6 momentarily to ground. After interval determined by $1.1R_1C_1$, second timer starts its delay determined by $1.1R_2C_2$.—"Signetics Analog Data Manual," Signetics, Sunnyvale, CA, 1977, p 724.

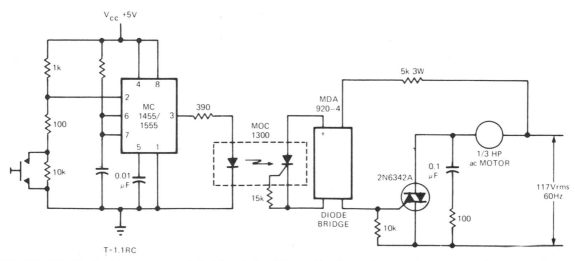

LONG TURNOFF DELAY—Combination of timer, optoisolator, and bridge-triggered triac keeps AC motor or other device energized for up to 1 h after control switch is depressed momentarily. Closing of switch drops voltage at pin 2 of timer below ⅓ V_{CC}, making timer output go high and thus turn LED on. At same time, capacitor at pin 7 begins charging. Output remains high until capacitor reaches ⅔ V_{CC}, when output is reset to low state and motor thereby turned off.—T. Mazur, Solid-State Relays Offer New Solutions to Many Old Problems, *EDN Magazine*, Nov. 20, 1973, p 26–32.

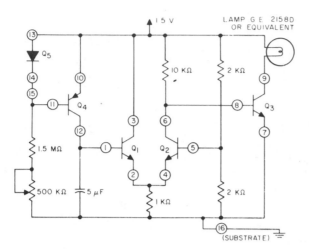

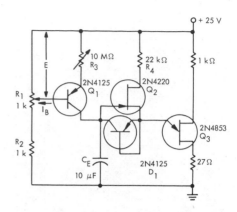

10-s TIMER—CA3096 transistor array provides all needed active devices. 5-μF capacitor charges through Q_4 until it turns on bistable switch Q_1-Q_2, which then triggers Q_3 to deliver current to lamp load to indicate end of timing interval.—"Circuit Ideas for RCA Linear ICs," RCA Solid State Division, Somerville, NJ, 1977, p 8.

10-h DELAYS—Long duration is achieved by separating peak current of timer from charging current. Q_1 and R_1-R_2-R_3 form constant-current source whose charging current can be adjusted to as low as several nanoamperes. Q_2 acts as source follower to supply current flowing in emitter lead prior to firing of UJT Q_3. Diode-connected transistor Q_1 provides low-impedance discharge path for timing capacitor C_E. Delay time varies linearly with setting of R_3.—"Unijunction Transistor Timers and Oscillators," Motorola, Phoenix, AZ, 1974, AN-294, p 5.

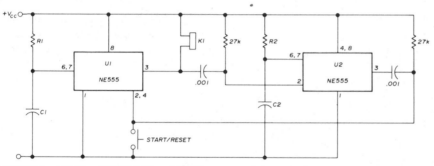

CASCADE WITH AUTOMATIC RETRIGGERING—Two timers, each controlling own load and having own time intervals (determined by R1C1 and R2C2), recycle automatically when start switch is closed momentarily. If desired, second timer can be set to control ON time of first timer. With 15-V supply, each timer can handle 200-mA load.—H. Vordenbaum, Automatic Reset Timer, *Ham Radio,* Oct. 1974, p 50–51.

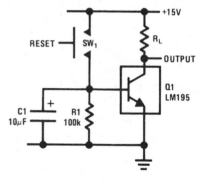

TIME DELAY USES POWER TRANSISTOR IC—Load is energized when switch is closed. C1 charges until voltage across R1 drops below 0.8 V, opening LM195 and deenergizing load. Long time delays can be obtained with small capacitor values since high resistance can be used.—"Linear Applications, Vol. 2," National Semiconductor, Santa Clara, CA, 1976, AN-110, p 4.

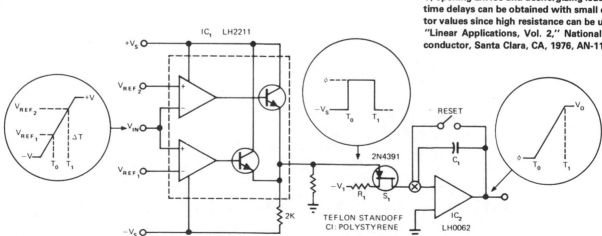

TIME-TO-VOLTAGE CONVERTER—Dual comparator, solid-state switch, and fast FET opamp provide flexibility, range, and accuracy required for using converter in computer-controlled test system. IC_2 operates as integrator and sample-and-hold circuit. Reference voltages of IC_1, V_{REF1}, and V_{REF2} allow time measurements of signal having either positive or negative voltage levels or both. Floating output stages of comparators provide voltage translation for FET switch S_1. With S_1 closed, IC_2 integrates for time period during which input signal is below reference voltages. With S_1 open, IC_2 holds final voltage value. Measurement range is from 1 μs to several hours.—C. Wojslaw, Wide Range Time Measurements Simplified, *EDN Magazine,* Feb. 5, 1974, p 95–96.

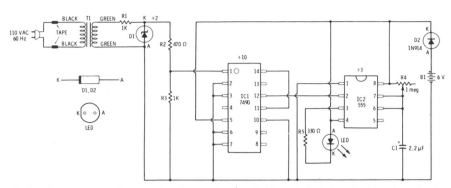

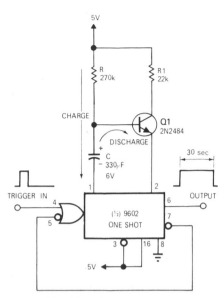

1-s HIGH-PRECISION—Accuracy is comparable to that of 60-Hz power-line frequency. After Radio Shack 276-561 6-V zener divides power-line frequency by 2, 7490 decade counter provides division by 10 to give 3 Hz. 555 timer connected as mono MVBR is then adjusted with R4

to divide by 3. LED then flashes at precise 1-s intervals, staying on about 100 ms. T1 is 6.3-V filament transformer.—F. M. Mims, "Integrated Circuit Projects, Vol. 5," Radio Shack, Fort Worth, TX, 1977, 2nd Ed., p 21–29.

30-s NONRETRIGGERABLE—With values shown, mono IC gives delay period of 30 s after triggering by input pulse and ignores other input pulses during timing period. Delay time in seconds is RC/3, where R is in ohms and C in farads. Delay can be reduced to as little as 10 ms by reducing R to 30 kilohms and C to 1 μF. Maximum delay is about 1 min, with 560 kilohms for R.—F. R. Shirley, Thirty-Second Timer Uses IC One-Shot, *EDN/EEE Magazine*, Jan. 1, 1972, p 73–74.

BATTERY-SAVER—Turns off battery-powered IC VTVM automatically about 3.5 min after turning on with S1, to prolong battery life even though user forgets to turn off instrument. Circuit can be retriggered at any time in timing cycle by switching S1 off and then on again. Q1 is programmable UJT that drives latch using two gates of CD4001AE quad NOR gate U1. With pin 3 high and pin 4 low, timing circuit and load are both turned off; battery drain by U1 is then only 0.001 μA. Values of R1 and C1 determine time interval. T1 is air-core pulse transformer having 600 turns (No. 36 to 40) enamel for primary and same number wound over primary for secondary.—R. Hardesty, Turn-Off Timer for Portable Equipment, *Ham Radio*, Sept. 1976, p 42–44.

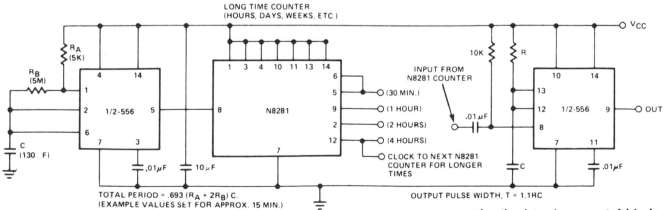

4-h SEQUENTIAL TIMER—Use of N8281 divider network between sections of 556 dual timer gives extremely long time delays without costly large low-leakage capacitors. First section op-

erates as oscillator having period of 1/f. Oscillator output is applied to divide-by-*N* network to give output with period of N/f for triggering second half of timer. Connection of divider to

second section determines amount of delay introduced by divider. Cascading of additional dividers increases maximum delay to days or even weeks.—"Signetics Analog Data Manual," Signetics, Sunnyvale, CA, 1977, p 724.

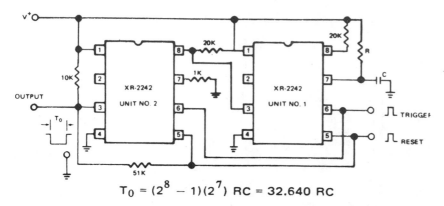

$$T_0 = (2^8 - 1)(2^7) \, RC = 32.640 \, RC$$

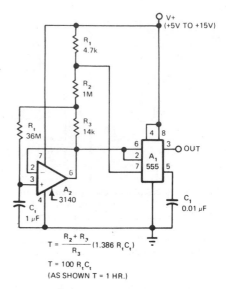

$$T = \frac{R_2 + R_3}{R_3}(1.386 \, R_1 C_1)$$

$$T = 100 \, R_1 C_1$$

(AS SHOWN T = 1 HR.)

1-YEAR TIMER—Cascaded operation of Exar XR-2242 long-range timers provides ultralong time delays, up to 1 year. Cascading of counter sections provides 32,640 clock cycles before output pin 3 of unit 2 changes state. Common pull-up resistor makes counter section of unit 2 trigger each time output of unit 1 makes posi- tive-going transition. Cascading additional timer with unit 2 extends time delay to 1.065 × 10⁹ RC. With RC values giving 0.1 s, time delay becomes 3.4 years.—"Timer Data Book," Exar Integrated Systems, Sunnyvale, CA, 1978, p 19–22.

MICROSECONDS TO HOURS—Timing range of 555 is increased 100 times by using 3140 FET-input opamp in circuit that effectively multiplies values of timing components and buffers timing network. Pin 7 of 555 switches applied voltage of timing network between V+ and ground. Ratio selected for R_2-R_3 can be varied over wide range to change multiplying ratio, provided square-wave voltage across R_3 is at least 50 mV. Output is essentially pure square wave. Supply voltage is not critical.—W. G. Jung, Take a Fresh Look at New IC Timer Applications, *EDN Magazine*, March 20, 1977, p 127–135.

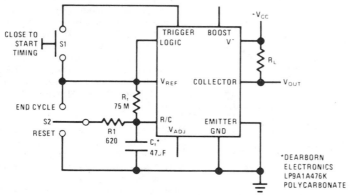

1 h WITH END-CYCLE SWITCH—National LM122 timer is connected with manual controls for start, reset, and intermediate termination of 1-h timing cycle started by closing S1. Once timing starts, S1 has no further effect. Moving S2 up ends cycle prematurely with appropriate change in output state. Moving S2 down resets timing capacitor to 0 V without changing output; releasing S2 starts new timing cycle.—C. Nelson, "Versatile Timer Operates from Microseconds to Hours," National Semiconductor, Santa Clara, CA, 1973, AN-97, p 9.

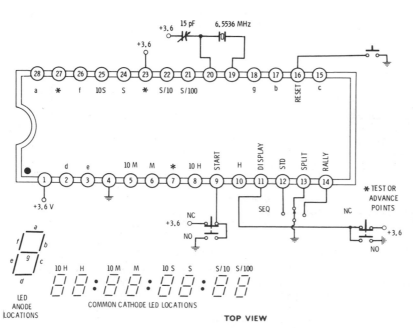

STOPWATCH—Intersil 7045 IC provides capability for driving digital display of time in hundredths of a second up through hours in four different operating modes selected by lower switches: sequential, standard, split, or rally. Grounding pin 9 momentarily with start switch initiates timing action. Repeated pressings of switch activate operating modes as selected. Grounding reset pin 16 clears stopwatch. IC can be connected directly to LED display, without drivers or resistors.—D. Lancaster, "CMOS Cookbook," Howard W. Sams, Indianapolis, IN, 1977, p 159.

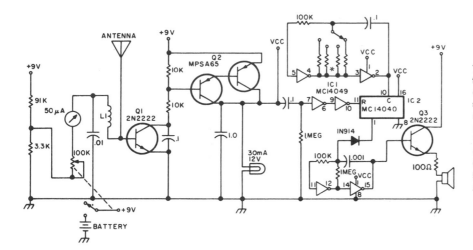

TALK TIMER—Switch gives choice of intervals from 0.5 to 5 min after going on air, before time-out alarm tone sounds. Circuit includes field-strength meter and on-the-air light. Releasing mike button for about 1 s resets timer, which consists of MC14049 hex inverter and MC14040 12-bit ripple counter. L1 is 3 turns No. 20 wire on 5/16-in form. Timing resistor values are 47K for 0.5 min, 100K for 1, 220K for 2, 390K for 3, and 510K for 5 min.—B. Fette, An FM Gadget, *73 Magazine,* April 1977, p 154–155.

90-s TALKING-LIMIT WARNING—Developed for AM radio transceivers making use of repeaters, to limit length of individual transmission so as to avoid being timed out at repeater. Uses NE-555 connected as timer, with C1 and R1 chosen to set timing at about 90 s. Point A is connected to terminal of TR switch that goes from neutral or ground on receive to 12 V on transmit. Timing cycle begins on transmit; when IC1 times out, it activates IC2 connected as 1000-Hz astable oscillator driving transceiver connected to B. Tone sounds until microphone button is released to reset timer.—S. Kraman, Try the Mini-Timer, *73 Magazine,* June 1977, p 48.

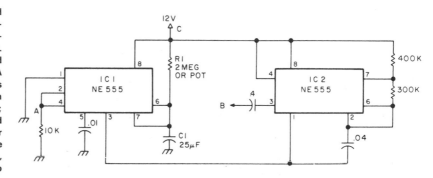

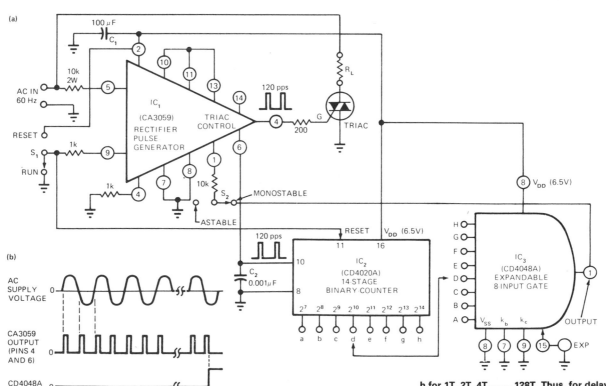

0.5333–136 s WITH LINE-FREQUENCY ACCURACY—Provides time delays in selected increments of 0.5333 s with accuracy essentially that of AC line frequency. IC_1 develops 120 pulses per second having 100-μs width at pins 4 and 6 for each zero crossing of line. First six stages of IC_2 determine basic timing period 1T. These stages produce pulse train with periodicity of 0.5333 s at input of seventh counter stage. Binary-ordered output signals are available at outputs a-h for 1T, 2T, 4T, . . . , 128T. Thus, for delay of 1 min (about 112T), use 64T + 32T, + 16T with AND-gate programming interconnections e-E, f-F, and g-G. Tie all unused AND-gate inputs to V_{DD} bus.—A. C. N. Sheng, Line-Operated Timer Couples High Accuracies with Long Time Delays, *EDN Magazine,* Jan. 5, 1976, p 37–40.

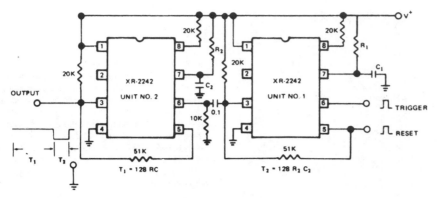

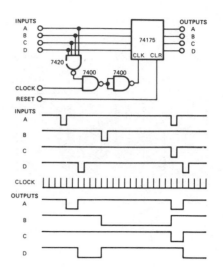

SEQUENTIAL TIMER—Second Exar XR-2242 long-range timer is triggered when first timer completes its cycle, length of which is equal to $128R_1C_1$. Output of second timer thus stays high for $T_1 = 128R_1C_1$ after trigger is applied, then goes low for duration $T_2 = 128R_2C_2$ corresponding to timing interval of unit 2. Circuit then reverts to rest state.—"Timer Data Book," Exar Integrated Systems, Sunnyvale, CA, 1978, p 19–22.

EVENT REGISTER—Inputs from sequential timer provide pulsed output and operate devices which stay on until next timed event. When an input goes low, output of four-input positive-NAND 7420 goes high, enabling the clock for 74175 4-bit D-type register. Since event-timer clock signal is used, outputs of register are coincident with clock. All outputs remain until one or more inputs goes low.—J. Glaab, Time Events with a Pulse Output Controller, *EDN Magazine*, Jan. 5, 1977, p 43.

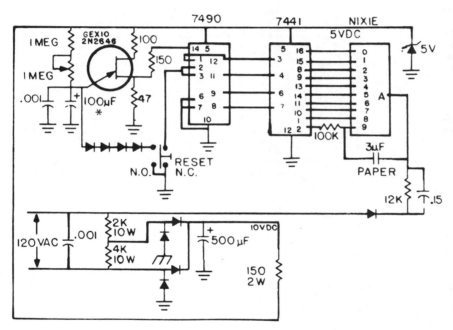

10 min WITH BLINKING—Station-identification timer uses single Nixie display to indicate elapsed time in minutes. After 9 min, numeral 9 blinks for 60 s before resetting to zero as visual reminder that amateur radio station identification should be made. Transistor can be GE X10 or 2N2646. Diodes are 1N4001 or equivalent. Numeral 9 or Nixie is connected as relaxation oscillator, flash rate of which can be changed by changing value of 100K resistor connected to pin 9.—W. Pinner, ID Timer, *73 Magazine*, Aug. 1974, p 95–96.

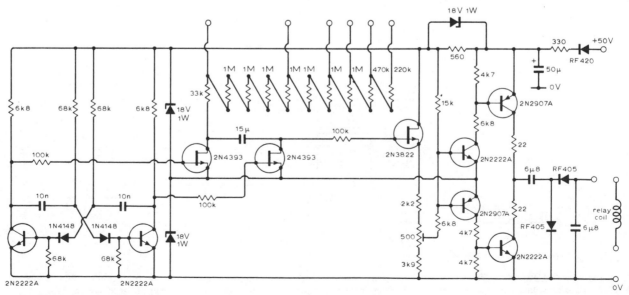

FAIL-SAFE TIMER FOR TRAINS—Provides delays up to 4 min, adjustable in 2-s steps, with accuracy better than 5%. Patented circuit was developed by ML Engineering to provide appropriate automatic braking or other action if engineer on train fails to respond to signal within period of delay.—W. E. Anderton, Computers, Communication and High Speed Railways, *Wireless World*, Aug. 1975, p 348–353.

ON OR OFF CONTROL—Circuit shows two ways of connecting 555 timer IC, for switching load on or switching load off at end of timing interval determined by setting of 5-megohm pot (1 to 60 s) and initiated by manual start switch. If desired, both loads can be connected to circuit for simultaneous switching.—E. R. Hnatek, Put the IC Timer to Work in a Myriad of Ways, *EDN Magazine,* March 5, 1973, p 54–58.

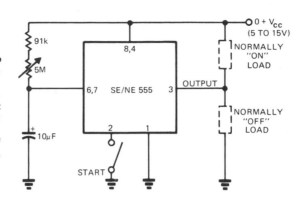

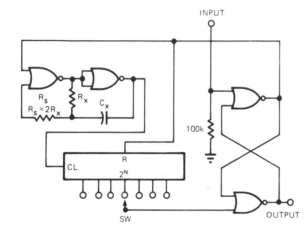

MICROSECONDS TO HOURS—Simple CMOS circuit serves as time-delay switch and general-purpose timer. Gated oscillator and latch are obtained from CD4001 quad two-input NOR gate, and 14-stage counter uses CD4020. T_{ON} is function of oscillator frequency as determined by $R_x C_x$ and proper 2^N output from counter. Pulse applied to latch input enables oscillator and counter. Latch output remains high until counter resets latch at end of count selected by switch. Further decoding is required for count or time variations finer than power of 2.—J. Chin, Low-Power Counter Is Programmable over Wide Range, *EDN Magazine,* March 20, 1974, p 83.

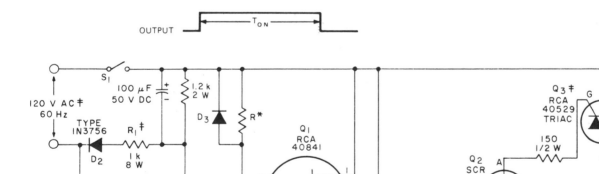

0–5 min DELAY—Value of resistor R controls duration of time delay provided by RCA 40841 dual-gate FET in SCR trigger circuit, with SCR in turn serving to trigger triac for handling high-current resistive or inductive AC loads. Maximum delay of 5 min is obtained when R is 60 megohms (IRC type CGH or equivalent resistor). Timing is accurate within 10% over temperature range of −25°C to +60°C. D_3 should be rated 60 V. Use any SCR capable of handling triac trigger current, rated 60 V.—"Linear Integrated Circuits and MOS/FET's," RCA Solid State Division, Somerville, NJ, 1977, p 435–437.

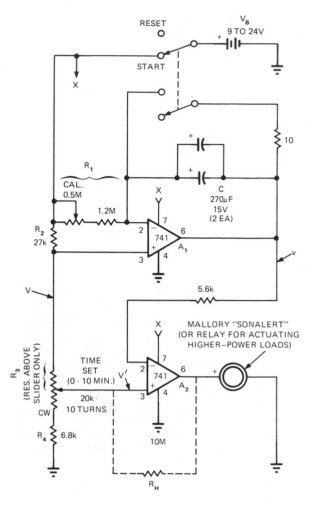

0–10 min WITH 1-s ACCURACY—After calibration, accuracy is independent of battery voltage because source voltage affects charging voltage of C and threshold of comparator A_2 equally. Time delay t for timer is CR_1R_3/R_2.—M. Strange, Simple Electronic Timer is Compact and Accurate, *EDN Magazine,* April 20, 1973, p 89 and 91.

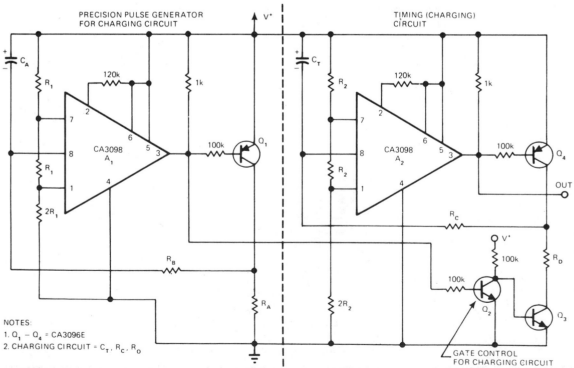

NOTES:
1. $Q_1 - Q_4$ = CA3096E
2. CHARGING CIRCUIT = C_T, R_C, R_O

LONG INTERVALS WITH SMALL C—Use of two CA3098 dual-input precision level detectors eliminates need for expensive high-capacitance low-leakage timing capacitors when delay intervals of several hours are required. For 4-h timer, C_T is only 16 μF if R_C is 22 megohms and R_D is 100 kilohms. Article traces circuit operation and gives design equations.—G. J. Granieri, Precision Level Detector IC Simplifies Control Circuit Design, *EDN Magazine,* Oct. 5, 1975, p 69–72.

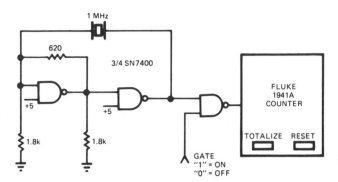

PROGRAM TIMER—Measures time between two points in microprocessor program while program is running. Gated 1-MHz crystal oscillator feeds Fluke 1941A counter used in totalize mode. Gate input is connected to unused bit of output port on microprocessor system. Instructions are then inserted in program under test to gate counter on at beginning of desired step and turn it off at end. Display then shows number of microseconds required by microprocessor to execute instructions.—M. M. Dodd, Benchmark Timer Eliminates Need to Total Individual Execution Times, *EDN Magazine*, Oct. 20, 1975, p 91–92.

1-h HOUSEHOLD TIMER—NE555 timer circuit turns off television set or other device at any desired time up to about 1 h after start switch is closed. Use IRC MR312C relay having coil resistance of 212 ohms, or other 12-V relay drawing less than 200 mA. With values shown, R2 gives time delay range of 3 to 58 min. For other ranges, change values of R2 and C1. Clockwise from top, pins on NE555 are 8, 3, 4, 2, 1, 7, and 6.—P. C. Walton, Build This $5 Timer, *73 Magazine,* Jan. 1976, p 129.

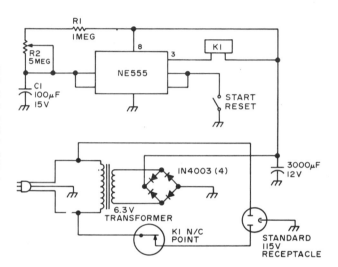

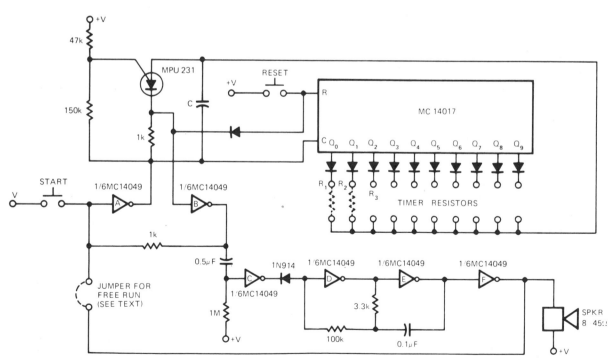

Diode needed only in FREE RUN MODE

10-INTERVAL TIMER—Ten independently predetermined time intervals run in sequence, with audible beep at end of each interval. Timer holds after each interval until start button is pressed to initiate next interval. If jumper is inserted, succeeding intervals start automatically. Values of R_1-R_9 and C determine time intervals. If same time is used for more than one step, diode outputs of those steps may be tied together to use same resistor. Supply can be in range of 5 to 18 V. Current drain is less than 100 μA, but increases to 40 mA during beep. Reset button can be depressed at any time, to restart timing at first interval.—T. Henry, Ten Step Sequential Interval Timer, *EDN Magazine,* March 20, 1974, p 78 and 80.

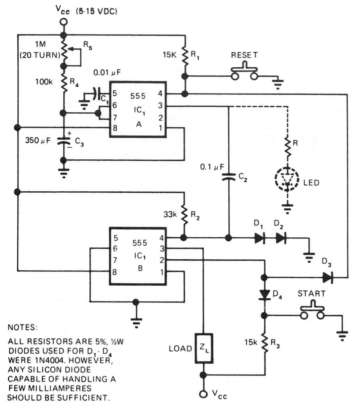

2–5 min STARTING DELAY—Energizing of load is delayed up to 5 min after start button is pushed, as required in some CMOS circuits and digital control systems. Uses pair of 555 timers, with A operating in straightforward timing mode and B connected as set-reset flip-flop. Pushing reset button initializes system, placing A in low state and making pin 3 of B high, leaving load unenergized. When start button is pushed, A goes high and begins timing out. After delay interval, output of B goes low, energizing load until system is reset. LED can be added to indicate that timing is in progress.—J. C. Nichols, Versatile Delay-on-Energize Timer Uses Two 555's, *EDN Magazine*, Oct. 5, 1975, p 76 and 78.

NOTES:

ALL RESISTORS ARE 5%, ½W DIODES USED FOR $D_1 \cdot D_4$ WERE 1N4004. HOWEVER, ANY SILICON DIODE CAPABLE OF HANDLING A FEW MILLIAMPERES SHOULD BE SUFFICIENT.

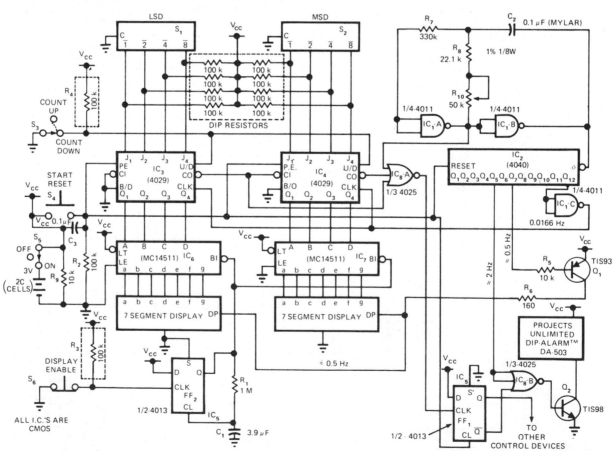

THUMBWHEEL-SET TO 99 min—Developed for timing events and for limiting event to predetermined interval that can be set up with 2-digit BCD-encoded thumbwheel switches. Digital display shows time remaining, as guide for speakers. Audible alarm indicates end of time interval. Flashing decimal point indicates counter is working. Designed for operation from two C cells. To conserve power, display is normally blanked; pressing display-enable switch turns on display for about 4 s. Article describes operation of circuit in detail.—R. A. Fairman, CMOS Lowers Timer Power Consumption, *EDN Magazine*, Oct. 5, 1975, p 78 and 80.

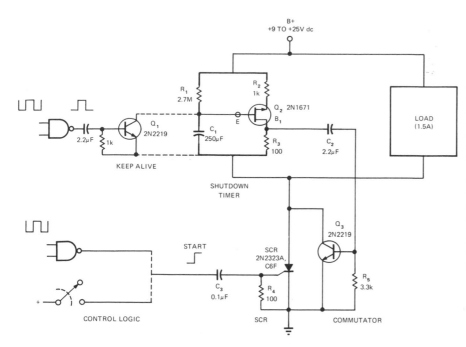

BATTERY-SAVING TIMER—Developed for use with alarms, remote controls, or unattended electronic equipment, to turn off battery automatically at predetermined interval after circuit is actuated by control logic or switch. Turn-on applies positive pulse that triggers SCR on, grounding load and UJT timer Q_2. After delay interval determined by values of C_1 and R_1, Q_2 fires and discharges C_1, producing pulse across R_3 that turns on Q_3. This in turn shunts SCR and commutates it off. Circuit is thus turned off, after which only very small leakage currents through reverse SCR junction will be drain on battery.—D. Weigand, Battery Saver Has Automatic Turn-Off, *EDN Magazine*, April 20, 1973, p 91.

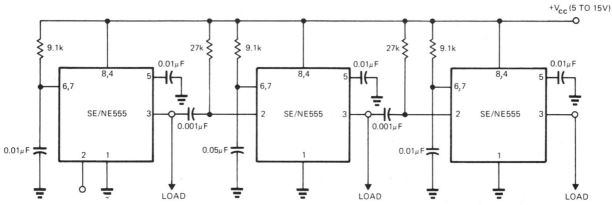

TEST SEQUENCING—Uses 555 timers connected sequentially. With values shown, first timer runs for 10 ms after starting with pulse at terminal 2 or by grounding 2. At end of timing cycle, second circuit runs for 50 ms before triggering third circuit having 10-ms delay. Each timer controls its own load, as required for sequencing of automatic tester.—E. R. Hnatek, Put the IC Timer to Work in a Myriad of Ways, *EDN Magazine*, March 5, 1973, p 54–58.

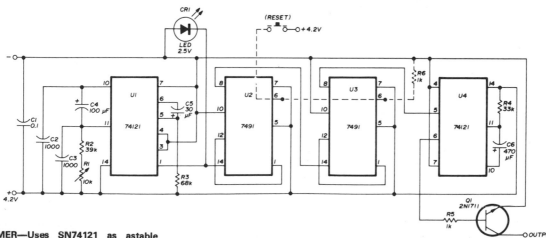

10-min TIMER—Uses SN74121 as astable MVBR generating pulses at 4-s intervals. U2 and U3 divide pulse train by 144 to give period of 576 s. U4 is then turned on, producing positive output pulse lasting 20 s that turns on Q1 for driving keyer, sidetone oscillator, lamp, or other signaling device as reminder for amateur radio operator to make 10-min station identification. R1 adjusts timing.—H. Seeger, Ten-Minute Timer, *Ham Radio*, Nov. 1976, p 66.

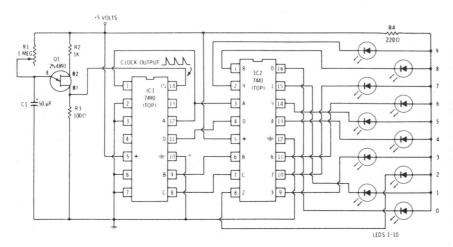

10 s TO 10 min—Array of ten LEDs serves for measuring time intervals up to 10 min, for timing phone calls, photographic exposures, and cooking. Pulse output rate of UJT oscillator Q1 is determined by value of C1 and setting of R1. Pulses are counted by 7490 which gives total count in binary form. 7490 recycles after each 10 counts. 7441 converts binary signals from 7490 to decimal outputs driving LED indicators. Each LED glows in sequence as count advances from 0 through 9 and repeats. For 10-min timer, adjust R1 until first LED stays on for exactly 1 min. For 10-s timer, adjust R1 for blink rate of 1 s per LED.—F. M. Mims, "Optoelectronic Projects, Vol. 1," Radio Shack, Fort Worth, TX, 1977, 2nd Ed., p 67–78.

UJT-SCR TIMER—Use of large capacitance for C1 in simple UJT relaxation oscillator provides time delay action for triggering SCR controlling relay. R1 provides convenient adjustment of delay. SCR can be 6-A 50-V Radio Shack 276-1089. Relay is 275-004.—F. M. Mims, "Semiconductor Projects, Vol. 2," Radio Shack, Fort Worth, TX, 1976, p 50–61.

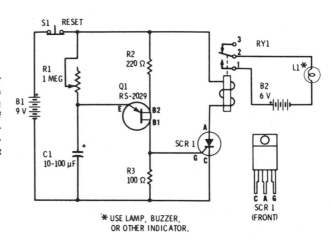

FAIL-SAFE LIGHT TIMER—National MM5309 clock IC is used as timer in circuit that maintains timing with adequate accuracy during periods of power-line failure and returns automatically to 60-Hz line as soon as power is restored. Applications include control of lights in unoccupied home. Timing action turns on lights for 4-h period every 24 h. When power is applied, internal multiplex circuit strobes each digit until digit with connected diode is accessed. This digit stops multiplex scanning, and BCD outputs present data from selected digit as control waveform whose edges determine timer data.—"MOS/LSI Databook," National Semiconductor, Santa Clara, CA, 1977, p 1-74–1-77.

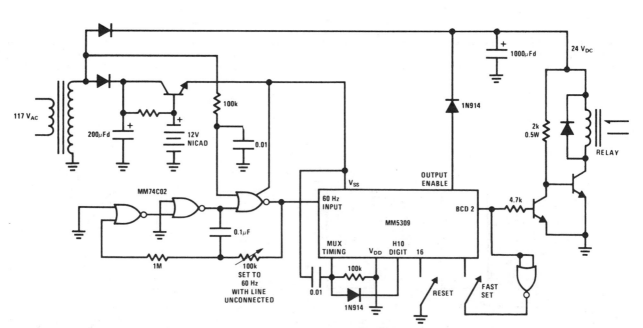

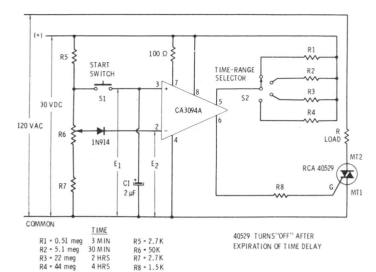

4-h CONTROL FOR TRIAC—Time at which triac is triggered by IC after momentarily pressing start switch is determined by resistor values used. When switch is released, charging capacitor C1 begins its long discharge interval. When voltage E_1 becomes less than E_2, pin 2 draws current and serves to reverse polarity of output at pin 6 for triggering triac. Diode limits maximum differential input voltage.—E. M. Noll, "Linear IC Principles, Experiments, and Projects," Howard W. Sams, Indianpolis, IN, 1974, p 316–317.

40529 TURNS "OFF" AFTER
EXPIRATION OF TIME DELAY

TIME

R1 = 0.51 meg	3 MIN	R5 = 2.7 K	
R2 = 5.1 meg	30 MIN	R6 = 50K	
R3 = 22 meg	2 HRS	R7 = 2.7 K	
R4 = 44 meg	4 HRS	R8 = 1.5 K	

3 min TO 4 h—Presettable analog timer achieves long time intervals by discharging C_1 into input terminal 3 of CA3094 programmable opamp, which provides sufficient output current for driving thyristors and other control devices.—"Circuit Ideas for RCA Linear ICs," RCA Solid State Division, Somerville, NJ, 1977, p 8.

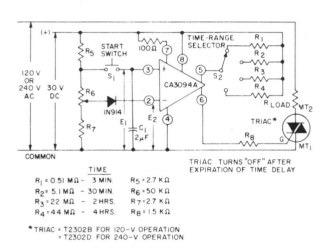

TRIAC TURNS "OFF" AFTER
EXPIRATION OF TIME DELAY

TIME

R_1 = 0.51 MΩ	3 MIN.	R_5 = 2.7 KΩ	
R_2 = 5.1 MΩ	30 MIN.	R_6 = 50 KΩ	
R_3 = 22 MΩ	2 HRS.	R_7 = 2.7 KΩ	
R_4 = 44 MΩ	4 HRS.	R_8 = 1.5 KΩ	

* TRIAC = T2302B FOR 120-V OPERATION
= T2302D FOR 240-V OPERATION

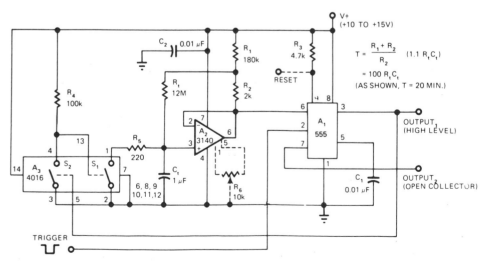

$$T = \frac{R_1 + R_2}{R_2} (1.1 R_t C_t)$$

$$= 100 R_t C_t$$

(AS SHOWN, T = 20 MIN.)

LONG-DELAY 555 MONO—FET-input 3140 opamp is used to multiply effective values of timing components R_t and C_t, eliminating need for high-value precision resistor and large low-leakage capacitor. Combination performs as standard 555 mono except that timing equation is $T = 100R_tC_t$ (for condition wherein division resistors R_1 and R_2 are chosen for 91-to-1 operation). Circuit has uncommitted open-collector output from pin 7 of 555, which can be referred to any voltage from 0 to +15 V. Pin 4 is between pins 6 and 8 on 555.—W. G. Jung, Take a Fresh Look at New IC Timer Applications, *EDN Magazine*, March 20, 1977, p 127–135.

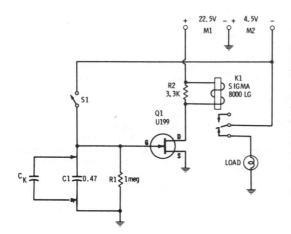

FET TIMER WITH RELAY—With values shown, circuit gives delay of several seconds. Increasing C1 by shunting with 20-μF capacitor delays energizing of relay to over 1 min. C1 is charged to −4.5 V when switch S1 is closed, biasing gate to cutoff and deenergizing relay. When S1 is open, capacitor begins discharging at rate determined by RC time constant of circuit. When voltage across capacitor drops to point at which Q1 conducts, relay is energized and power is applied to load.—E. M. Noll, "FET Principles, Experiments, and Projects," Howard W. Sams, Indianapolis, IN, 2nd Ed., 1975, p 215–216.

OVER 1 min—Circuit provides delays well over 1 min even with low operating voltages of ICs. When start pulse is applied to RS flip-flop A_1-A_2, Q_2 turns off and allows R_T to provide charging current for timing capacitor C_T. When voltage across C_T gets high enough, Q_1 turns on and resets flip-flop, terminating delay period. A_3 provides buffered complementary output.—R. W. Hilsher, Long-Delay Timer, *EEE Magazine*, Aug. 1970, p 79.

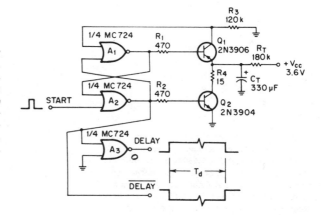

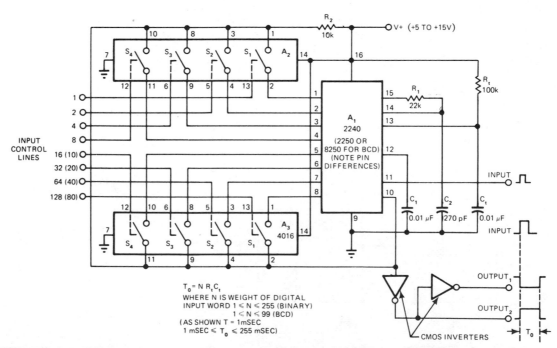

$$T_0 = N R_1 C_1$$
WHERE N IS WEIGHT OF DIGITAL
INPUT WORD $1 \leqslant N \leqslant 255$ (BINARY)
$1 \leqslant N \leqslant 99$ (BCD)
(AS SHOWN T = 1mSEC
1 mSEC $\leqslant T_0 \leqslant 255$ mSEC)

REMOTE DIGITAL PROGRAMMING OF TIMER—Either binary or BCD logic can be used for selecting delay interval of monostable timer A_1, with delays being integral multiples of shortest time. Timing is programmed by pair of 4016 CMOS analog switches, A_2 and A_3. Given timing tap is activated when corresponding digital input control line is high and deactivated when control is low. Programmable timing range is 1 to 255 ms for 2240, and 1 to 99 ms for 2250 or 8250 timer. Basic interval can be changed to suit other applications. CMOS output buffer stage ensures valid output logic levels. Although circuit will operate over supply range indicated, operation is optimum for supply of 10 to 15 V.—W. G. Jung, Take a Fresh Look at New IC Timer Applications, *EDN Magazine*, March 20, 1977, p 127–135.

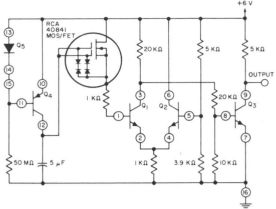

1 min WITH TRANSISTOR ARRAY—Circuit uses RCA CA3096AE five-transistor array in combination with dual-gate MOSFET to provide timing action that maintains accuracy within 7% for supply voltage variations of $\pm 10\%$. Q_5 is one of PNP transistors connected as diode.—"Linear Integrated Circuits and MOS/FET's," RCA Solid State Division, Somerville, NJ, 1977, p 205–210.

TWO INDEPENDENT DELAYS—Each timer section of Exar XR-2556 dual timer operates independently in mono MVBR mode to provide delays shown above output waveforms. Supply voltage range is 4.5 to 16 V.—"Timer Data Book," Exar Integrated Systems, Sunnyvale, CA, 1978, p 23–30.

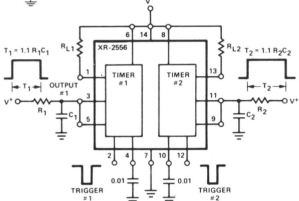

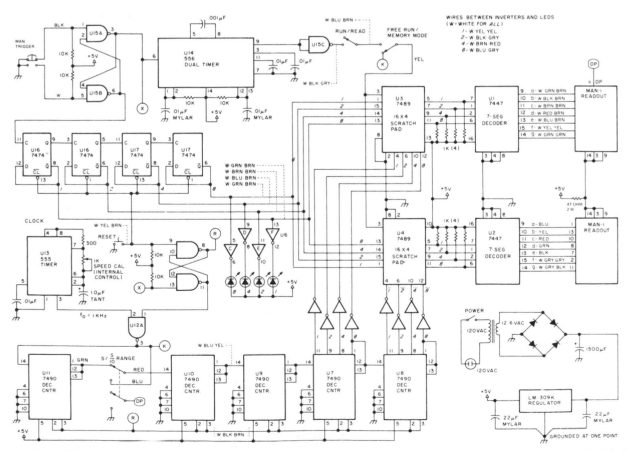

TIMER WITH MEMORY—Time elapsed since initial triggering at start of event is shown on 2-digit MAN-1 display in seconds or tenths of seconds and written into memory. Up to 16 event times can be stored for later readout. Free-running mode counts off seconds or tenths of seconds on display. Article covers construction and operation. Circuit was Science Fair winner. Gate and opamp types are not critical.—M. Jose, Event Timer with Memory, *73 Magazine*, June 1977, p 72–74.

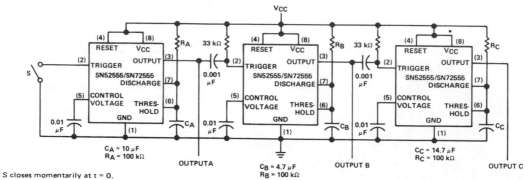

S closes momentarily at t = 0.

THREE-STEP SEQUENCE TIMER—Provides three different outputs at predetermined time intervals for initializing conditions during start-up or for activating test signals in sequence. Uses three Texas Instruments SN52555 or SN72555 timers which are interchangeable with other 555 timers. Values of R and C at output of each timer determine delays (T = 1.1RC). With values shown below timers, output A is 5 V for interval of 1.1 s after switch is closed. At end of this interval, output B goes to 5 V for 0.5 s, after which output C goes to 5 V for 1.5 s to complete sequence. Supply can be 5–15 V.— "The Linear and Interface Circuits Data Book for Design Engineers," Texas Instruments, Dallas, TX, 1973, p 7-53–7-61.

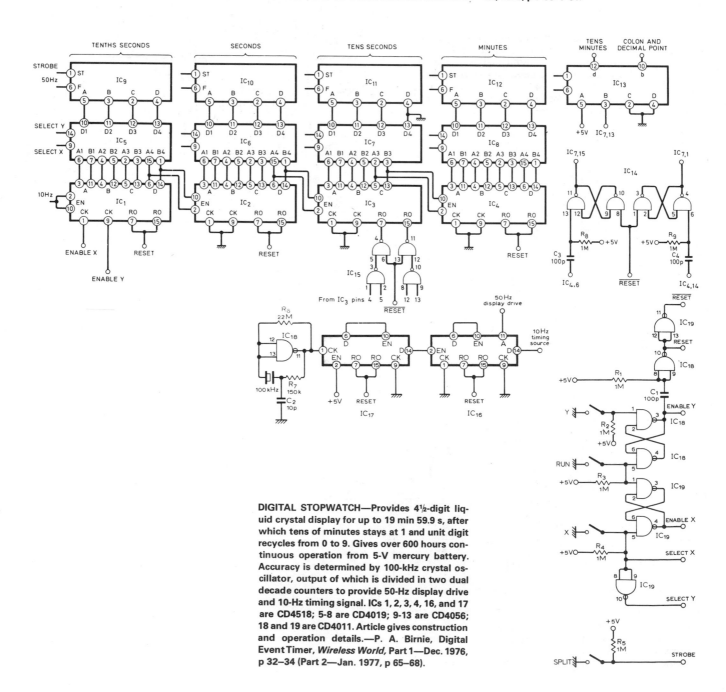

DIGITAL STOPWATCH—Provides 4½-digit liquid crystal display for up to 19 min 59.9 s, after which tens of minutes stays at 1 and unit digit recycles from 0 to 9. Gives over 600 hours continuous operation from 5-V mercury battery. Accuracy is determined by 100-kHz crystal oscillator, output of which is divided in two dual decade counters to provide 50-Hz display drive and 10-Hz timing signal. ICs 1, 2, 3, 4, 16, and 17 are CD4518; 5-8 are CD4019; 9-13 are CD4056; 18 and 19 are CD4011. Article gives construction and operation details.—P. A. Birnie, Digital Event Timer, *Wireless World*, Part 1—Dec. 1976, p 32–34 (Part 2—Jan. 1977, p 65–68).

CHAPTER 97
Touch-Switch Circuits

Includes circuits activated by skin resistance between two touch plates, by small AC voltage picked up by body and applied to single touch plate, or by changing of capacitance. Many circuits include debouncing.

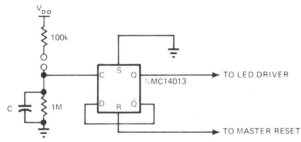

TOGGLING TOUCH SWITCH—Uses half of Motorola MC14013 as flip-flop that changes state each time contacts are bridged by resistance of finger. For status display, LED driven by 2N3903 transistor can be added. Possible drawback is bouncing if finger is carelessly applied.—V. Gregory, CMOS Touch Switches—Convenient, Less $ and Sexy, *EDN Magazine*, May 5, 1976, p 112.

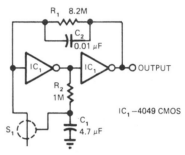

DEBOUNCE FOR TOUCH SWITCH—Two CMOS inverters respond to high-impedance path between electrodes of touch switch to provide finger-touch sensitivity and positive switching action with minimum components. Large time constant of R_1C_1 requires wait of about 4 s before attempting to retrigger circuit. C_2 prevents oscillation from 60-Hz pickup when electrodes are touched.—H. Manell, CMOS Inverters Implement Finger-Touch ON-OFF, *EDN Magazine*, Jan. 5, 1978, p 90.

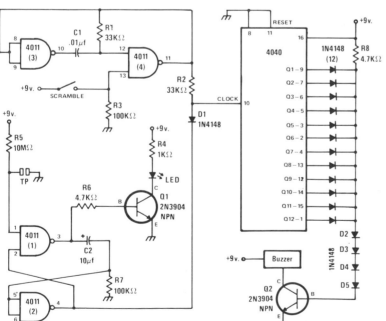

ADJUSTABLE-ODDS LOTTERY—Odds for lottery are set by diode matrix connected to output of 4040 counter. Range is from 1:2 to 1:1024. To use, close scramble switch for a second or two to make free-running oscillator drive counter at high rate, then let contestant hit touch plate to trigger flip-flop that advances counter one step.

If counter was frozen at next to last connected diode, next diode is biased off and Q2 energizes buzzer. For 50-50 odds, connect only diode at output Q1. For odds of 1 in 10, connect diodes only to outputs Q2 and Q4 for binary 1010 or 10. With all diodes connected as shown, odds are 1:1024.—J. Sandler, Play 'Random Chance,' *Modern Electronics*, Oct. 1978, p 42–43 and 88.

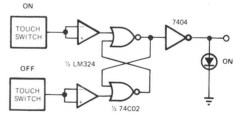

TOGGLE SWITCH—Touching one 0.5-inch-square copper-clad pattern on printed-circuit board turns switch circuit on by giving high output. Touching other plate turns switch off. LED between output and ground shows status of switch. For proper switching, circuit must connect to line-operated DC power supply.—R. D. Wood, Replace Bulky Mechanical Switches with Touch Controls, *EDN Magazine*, April 20, 1978, p 132–133.

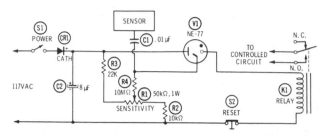

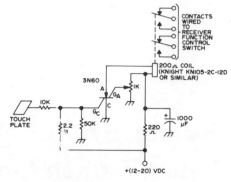

TOUCH SWITCH—Uses NE-77 neon lamp, which is similar to NE-2 but has third electrode for triggering. When person touches metal sensor plate of switch, AC voltage picked up by body is applied to trigger electrode of neon, making it fire and energize 5000-ohm relay K1 (Potter & Brumfield RS5D or equivalent). Relay remains energized until S2 is opened to reset circuit. Adjust R1 so voltage applied to center electrode of V1 is just below trigger point.—J. P. Shields, "How to Build Proximity Detectors & Metal Locators," Howard W. Sams, Indianapolis, IN, 2nd Ed., 1972, p 52–55.

TOUCH SWITCH—Performs function of switch by means of relay contacts when SCR is triggered by placing finger on touch plate. Values shown keep relay energized for 5–10 s after touch. Developed as replacement for switch-type controls on amateur radio receiver. Once SCR has fired, it conducts until charge on 1000-μF capacitor decreases enough to drop SCR current below minimum for conduction.—J. J. Schultz, Rapid Receiver Control Switching, *73 Magazine*, Dec. 1973, p 67–69.

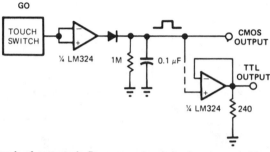

DATA ENTRY—Touch of operator's finger on input button produces CMOS output. Addition of one opamp section gives TTL output. Touch button can be 0.5-inch-square copper-clad pattern on printed-circuit board or machine-screw head having comparable area. For proper switching, circuit must connect to line-operated DC power supply.—R. D. Wood, Replace Bulky Mechanical Switches with Touch Controls, *EDN Magazine*, April 20, 1978, p 132–133.

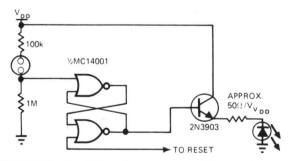

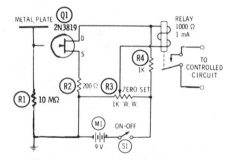

LATCHING TOUCH SWITCH—Uses LED as status display that substitutes for tactile feel of ordinary pushbutton switch. In reset state, LED is off. When touch contacts are bridged by resistance of finger, flip-flop changes state and LED comes on while output changes to high state.— V. Gregory, CMOS Touch Switches—Convenient, Less $ and Sexy, *EDN Magazine*, May 5, 1976, p 112.

TOUCH-PLATE RELAY—When 2-inch diameter disk of sheet metal or foil is firmly touched with finger, stray noise picked up by body and coupled into 10-megohm gate circuit of FET is sufficient to boost drain current to about 1.7 mA and close relay. Delayed dropout can be obtained by placing capacitor in parallel with relay coil; delay is about 0.8 s per 1000 μF of parallel capacitance.—R. P. Turner, "FET Circuits," Howard W. Sams, Indianapolis, IN, 1977, 2nd Ed., p 104–105.

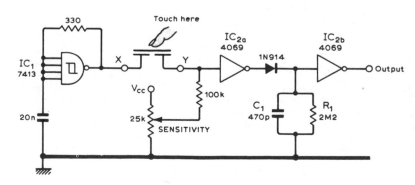

LOGIC SWITCH—Can be used with battery-powered circuits because CMOS touch switch does not require body pickup of AC line hum for switching action. Schmitt trigger IC$_1$ forms 100-kHz oscillator. IC$_{2a}$ amplifies oscillator output and charges C$_1$ through diode. When sensor is touched, oscillator output is severely attenuated, making C$_1$ discharge and thereby changing output state of level detector IC$_{2b}$. Sensor is 1-inch-square of double-sided printed-circuit board with lower side divided into two equal sections.—N. Sunderland, C.M.O.S. Touch Switch, *Wireless World*, May 1978, p 69.

PROXIMITY SWITCH—Hand brought near sensor plate induces 60-Hz power-line hum in section of quad two-input NOR gate. Hum is squared by gate and used to trip section of 4013 connected as retriggerable mono MVBR. Output of mono is clean from instant of first proximity until several milliseconds after moving hand away. Sensitivity depends on size of metal plate and on number of permissible false alarms from other noise sources nearby.—D. Lancaster, "CMOS Cookbook," Howard W. Sams, Indianapolis, IN, 1977, p 278–282.

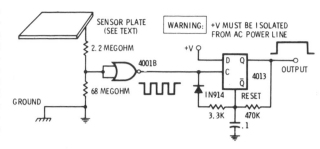

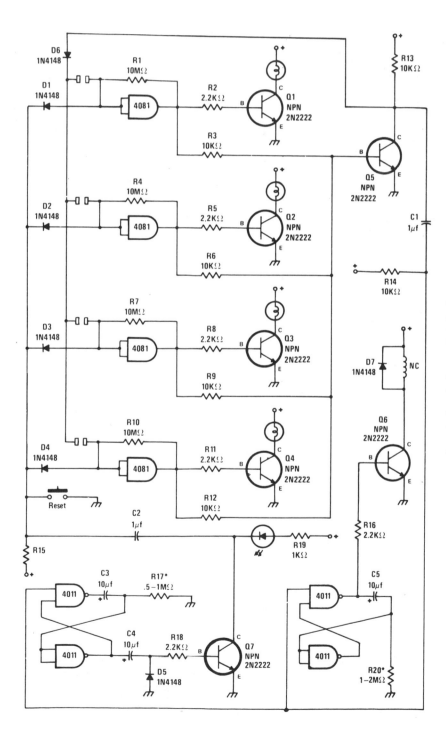

TV GAME CONTROL—Developed for use in game in which first person to recognize musical tune places finger on touch plate to energize his lamp. Action stops cassette player and locks out touch plates of other players. After 5-s delay, lockout is disabled so different player can have try at correct answer if first is wrong. After additional 5-s delay, relay is deenergized and music resumes. Additional reset switch is provided to reactivate all touch plates independently of delay. Supply is 12 V, and lamps are 12 V.—J. Sandler, Name That Tune, *Modern Electronics*, Dec. 1978, p 66 and 69–70.

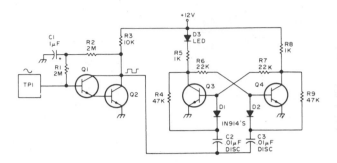

COIN TOSSER—Finger on touch plate TR1 feeds stray 60-Hz sine wave (picked up by body capacitance) to high-impedance Darlington pair Q1-Q2 for squaring. Output drives flip-flop Q3-Q4. LED conducts only when Q3 is on. Removal of finger leaves LED either on or off with random probability. Players can try in turn to match, play odd-man-buys, or play odd-man-out. For faster results, circuit can be duplicated so each player has own touch plate. Practically any transistor types can be used, as circuit is not critical.—J. H. Everhart, The Coffee Flipper, *73 Magazine,* Nov. 1976, p 162–163.

TOUCH SWITCH USES TIMER—Free-running or mono capabilities of Signetics 555 timer can be controlled through choice of trigger and reset inputs. Characteristics of output pulses can be adjusted over timing periods ranging from microseconds to hours. With 5-V supply, output is TTL-compatible and current drain of only 3 mA permits battery operation. Circuit is easily triggered by voltage differential between floating (ungrounded) human body and timer itself. Touch plate can be any conducting material, with virtually no size limitation. Once triggered by momentary touch, device cannot be retriggered until it has timed out. Duration of output pulse depends on RC time constant and on control voltage. Applications include switchless keyboards, burglar alarms, and bounce-free switches.—J. C. Heater, Monolithic Timer Makes Convenient Touch Switch, *EDN Magazine,* Dec. 1, 1972, p 55.

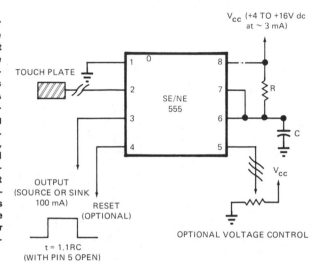

ANALOG SIGNAL CONTROL—CMOS logic gives bounceless operation of CD4016 analog switches by sensing of ambient signals at fingertip of operator. Connections for quadruple touch-switch array are shown below. Touch plates can be metal squares or disks up to 2 cm wide. If used in remote locations where power lines or other electromagnetic-field sources are not present, it may be necessary to provide grounded second contact at each sensor so slight conduction between contacts will assure triggering.—M. W. Hauser, C-MOS Touch-Switch Array Controls Analog Signals, *Electronics,* March 7, 1974, p 113–114; reprinted in "Circuits for Electronics Engineers," *Electronics,* 1977, p 357–358.

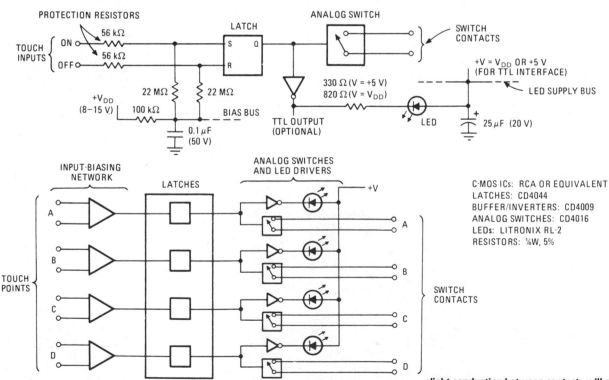

C-MOS ICs: RCA OR EQUIVALENT
LATCHES: CD4044
BUFFER/INVERTERS: CD4009
ANALOG SWITCHES: CD4016
LEDs: LITRONIX RL-2
RESISTORS: ¼W, 5%

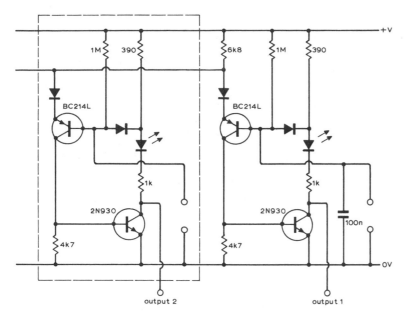

TOUCH BUTTONS—Based on detecting skin resistance between two contacts built into each touch button. Contact going to 0 V would normally be metal front panel of control. Any number of sections like that in dashed lines can be cascaded to handle more buttons. A particular button always comes on when power is applied, and is canceled by next button touched. LED identifies button currently activated; use any LED rated at 20 mA. Supply can be 20 to 30 V. Outputs may be used to drive FET analog switches, varactor tuning diodes, or relays.—P. G. Hinch, Self-Cancelling Touch Button Control, *Wireless World*, Oct. 1974, p 380.

TOUCH-CONTROLLED RELAY—Basic circuit uses Signetics 555 timer to make LED flash each time input is touched with finger. Replace LED with flip-flop (any type) and three sections of 8T90 hex power inverter to drive DC relay. Silicon diode suppresses voltage spikes generated when magnetic field of relay collapses. Use only as many paralleled sections of inverter as are required to operate relay. Input can be brass or copper plate at least 2 inches square.—G. Young, Voltage, Current, and Power Supplies, *Kilobaud*, Nov. 1977, p 76–78 and 80–82.

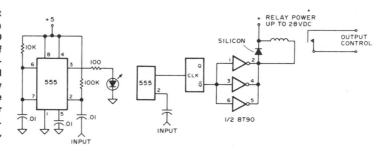

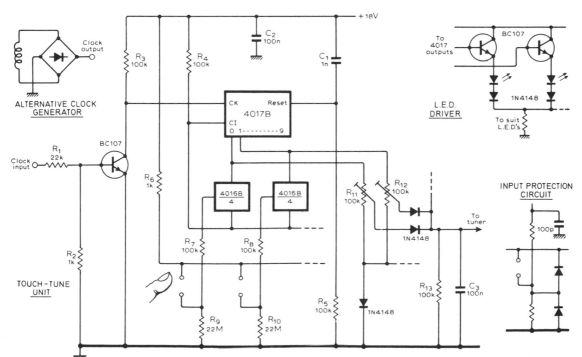

FM TOUCH-TUNE—Up to 10 channels can be tuned by turning on appropriate section of 4016 CMOS digital IC by finger contact that drives clock inhibit line low. 4017B then counts clock pulses until desired output goes high. C_1 and R_5 ensure that channel 0 is selected when circuit is turned on. Clock frequency is not critical and can range from 100 Hz to 19 kHz. For 120-Hz clock, wind several extra turns around power transformer of receiver and feed this voltage to bridge rectifier.—L. Crampin and R. van der Molen, Touch-Tune for F.M. Receivers, *Wireless World*, Jan. 1978, p 60.

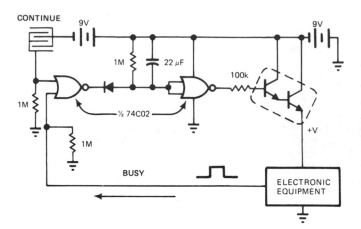

AUTOMATIC TURNOFF—Circuit removes bias from power Darlington about 15 s after both CONTINUE and BUSY signals go low, to conserve battery life in portable electronic equipment. Interleaved copper patterns on printed-circuit board form touch switch that must be reactivated every 15 s or kept closed by finger contact while equipment is being used.—R. D. Wood, Replace Bulky Mechanical Switches with Touch Controls, *EDN Magazine,* April 20, 1978, p 132–133.

CONTACTLESS KEYBOARD—Touching one of 16 metal pads at inputs of 74150 multiplexer produces corresponding 4-bit BCD output from 74194 shift register. During scanning of multiplexer inputs by counter, output is produced only when finger of operator is on corresponding fingertip-size touch pad. Requires 10-kHz pulse from external source to strobe multiplexer and serve as clock for counter. Duration of clock pulse must be more than 20 ns so untouched pads charge up to threshold voltage but not long enough to let touched pad charge.—D. Cockerell, TTL IC Serves as Touch Keyboard, *Electronics,* Feb. 20, 1975, p 108–109.

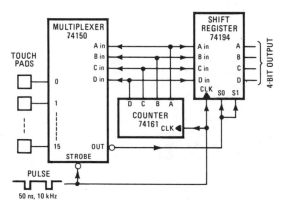

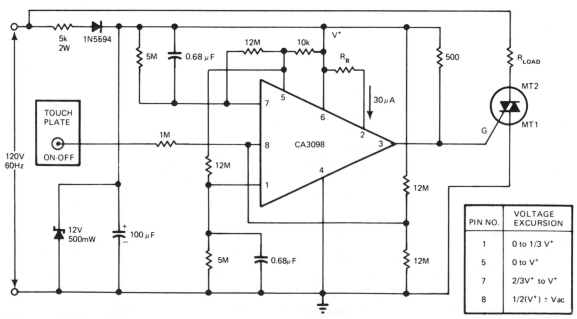

PIN NO.	VOLTAGE EXCURSION
1	0 to 1/3 V^+
5	0 to V^+
7	2/3V^+ to V^+
8	1/2(V^+) ± Vac

TOUCH SWITCH—Small AC signal momentarily introduced by finger contact on touch plate causes voltage at pin 8 of CA3098 dual-input precision level detector to be greater than high reference voltage. This toggles memory flip-flop in IC, making voltage high at pin 5. Voltage at pin 7 then increases exponentially to V+ in about 10 s. This 10-s delay is maximum that button can be touched; longer touch makes system oscillate between ON and OFF states until finger is removed. Shorter touch energizes load, placing pin 7 at V+. Next touch of plate turns circuit off.—G. J. Granieri, Precision Level Detector IC Simplifies Control Circuit Design, *EDN Magazine,* Oct. 5, 1975, p 69–72.

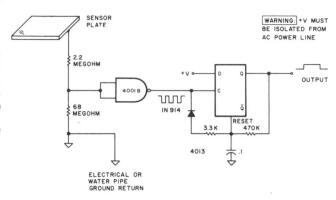

PROXIMITY SWITCH—Based on coupling of human body to 60-Hz power line. Hand held close to sensor plate induces hum into 4001B gate. This is squared and used to trip 4013 retriggerable mono MVBR. Output is clean from instant of first proximity until several milliseconds after release. Sensitivity depends on size of plate.—D. Lancaster, Clocked Logic, *Kilobaud*, May 1977, p 24–30.

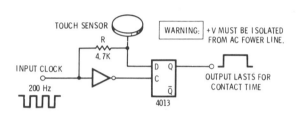

TOUCH SWITCH FOR MUSIC KEYBOARD—Touching metal sensor plate adds about 300 pF of capacitance between plate and ground, changing RC delay network that slows down clock waveform reaching D input of 4013 dual D flip-flop, making flip-flop output high for duration of contact. Circuit is repeated for each key in electronic music system.—D. Lancaster, "CMOS Cookbook," Howard W. Sams, Indianapolis, IN, 1977, p 278–282.

DEBOUNCING TOUCH SWITCH—Foolproof debouncing for touch switch using toggling flip-flop (half of Motorola MC14013) is provided by two gates connected as monostable pulse stretcher. Time constant of pulse stretcher is selected to match needs of application. For status display, LED driven by 2N3903 transistor can be connected to Q terminal of flip-flop.—V. Gregory, CMOS Touch Switches—Convenient, Less $ and Sexy, *EDN Magazine*, May 5, 1976, p 112.

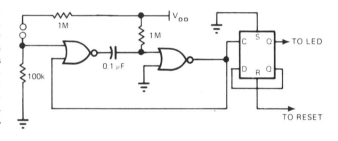

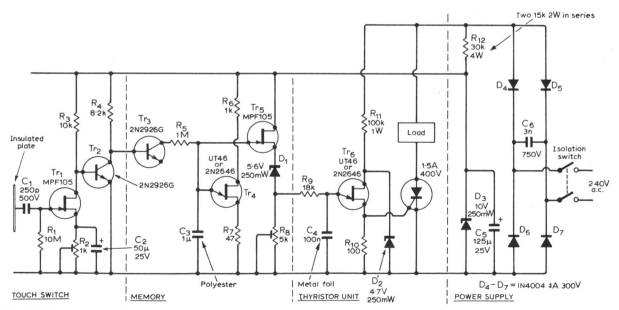

TOUCH SWITCH—Finger on insulated metal plate applies small AC voltage (picked up by body) to FET Tr_1 for amplification, to produce line-frequency square wave across R_4 for application to memory section Tr_3-Tr_5. Charging of C_3 through Tr_3 and R_5 produces DC output voltage across R_8 that is fed to UJT Tr_6 for triggering thyristor. C_4 is discharged at about 10-ms intervals by Tr_6 which operates from rectified AC line. For high voltage across R_8, such as 4 V, thyristor is triggered early in AC cycle and maximum power is supplied to load. Diodes D_4-D_7 ensure that control is provided over both positive and negative half-cycles of line. The longer a finger is held on touch switch, the greater is the voltage across R_8 and the more current there is through load. Removing finger turns off load, which can be lamps or other electric equipment.—R. Kreuzer, Touch-Switch Controller, *Wireless World*, Aug. 1971, p 389.

CHAPTER 98

Transceiver Circuits

Used in combined transmitters and receivers for amateur, CB, and other two-way communication applications. Includes voice-actuated TR switches, scanners, varactor tuners, and remote tuning systems. See also Antenna, Squelch, Receiver, and Transmitter chapters.

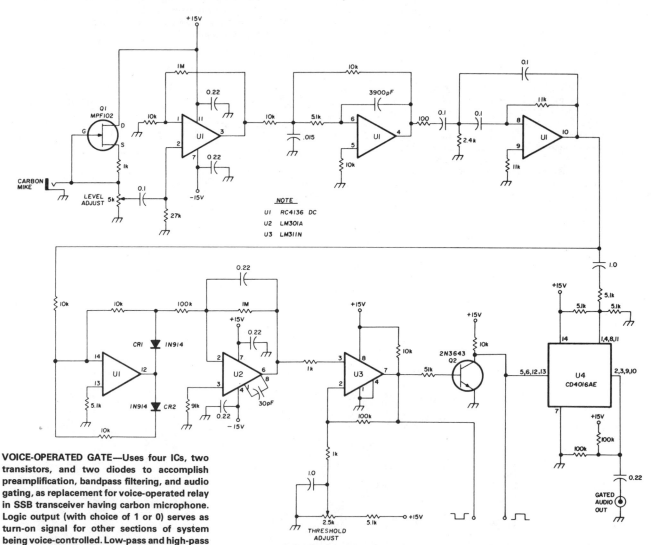

VOICE-OPERATED GATE—Uses four ICs, two transistors, and two diodes to accomplish preamplification, bandpass filtering, and audio gating, as replacement for voice-operated relay in SSB transceiver having carbon microphone. Logic output (with choice of 1 or 0) serves as turn-on signal for other sections of system being voice-controlled. Low-pass and high-pass active filter pair provides equivalent of 300–3000 Hz bandpass filter with 40 dB rolloff per decade at each edge, to discriminate against ambient noise. Q1 is FET constant-current source for carbon microphone, feeding preamp U1 that provides voltage gain of about 100 and output of about 3 VRMS. Next two sections of quad opamp U1 are active filters, and last section of U1 is active diode detector in which opamp linearizes detector CR1-CR2. Rectified audio is averaged by U2 to give smoothed long positive pulse with duration of audio burst. Schmitt trigger U3 sharpens pulse and makes it compatible with CMOS logic. Inverter Q2 turns on analog gate U4 when audio signal is present.—H. Olson, Voice-Operated Gate to Replace Voice-Operated Relays for Carbon Microphones, *Ham Radio,* Dec. 1977, p 35–37.

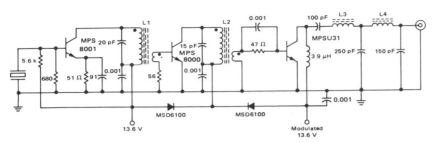

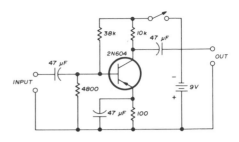

3.5-W TRANSMITTER—Class D circuit using economical plastic transistors operates from 12-V auto battery. 100% amplitude modulation requires about 2.5-W audio input. Modulator uses MSD6100 dual diode in modulated power supply system, with one diode in series with modulated supply voltage to MPS8000 driver to prevent driver from being down-modulated. Other diode maintains drive to final while final is being down-modulated, to make 100% modulation easy to obtain. All coils are wound on ¼-inch coil forms with No. 22 wire, with Carbonyl "J" ¼ × ⅜ inch cores in each. 2-turn secondaries are wound over bottom of primaries. L1 is 12 turns, L2 is 18, L3 is 7, and L4 is 5.—G. Young, "A Class D Citizen's Band Transmitter Using Low-Cost Plastic Transistors," Motorola, Phoenix, AZ, 1975, AN-596.

MATCHING LOW-Z MIKE—Single-transistor microphone impedance-matching circuit for low-impedance microphone feeds high-impedance input of amateur SSB transceiver. Circuit also boosts gain enough to meet transmitter input requirements.—C. Drumeller, Active Microphone Impedance Match, *Ham Radio*, Sept. 1973, p 67–68.

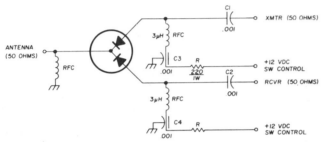

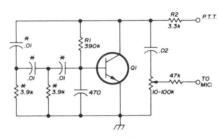

DIODE TR SWITCH—Microwave Associates MA8334 solid-state TR switch replaces conventional relays for switching antenna back and forth between transmitter and receiver. Handles up to 50 W CW at 144 MHz, and can be used at other frequencies up to 1000 MHz by proper choice of circuit constants. Measured insertion loss is 0.25 dB, and SWR is 1.23:1 when operated at 50 ohms.—T. Reddeck, Solid-State VHF-UHF Transmit/Receive Switch, *Ham Radio*, Feb. 1978, p 54.

END-OF-TRANSMISSION BEEPER—Release of push-to-talk (PTT) switch at end of radio conversation activates time delay for antenna changeover relay, keeping transmitter on air long enough to transmit 800-Hz tone burst indicating transmission termination. Tone is generated in simple one-transistor phase-shift oscillator powered by voltage present between PTT terminal and ground in receive mode of transceiver, which may be any voltage between 6 and 30 VDC. Transistor can be any small-signal NPN silicon with gain of at least 300 at 1 mA, such as 2N930. Parts marked with asterisk must be matched within 5%.—E. Hornbostel, Automatic Beeper for Station Control, *Ham Radio*, Sept. 1976, p 38–39.

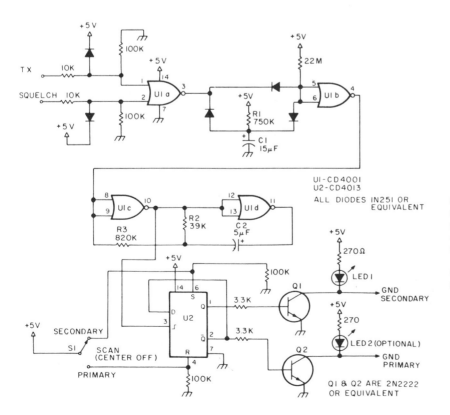

SCANNING ADAPTER—Developed for use with 2-meter transceiver having frequency synthesizer to provide automatic scanning that is disabled when transmitting and when receiver squelch is opened by transmission on one of channels being scanned. Will hang on to channel about 5 s after scanning is disabled, to allow starting of other side of communication. Scan rate is about 250 ms per channel. U1A generates 0 output when squelch is open, producing output of 1 for U1B that disables oscillator U1C-U1D. Oscillator drives D flip-flop that turns on Q1 and Q2 alternately. When Q1 is on, LED1 is lit to indicate that secondary channel is enabled. Article covers method of increasing number of scanned channels.—B. McNair, Add-a-Scanner, *73 Magazine*, Nov. 1978, p 116–119.

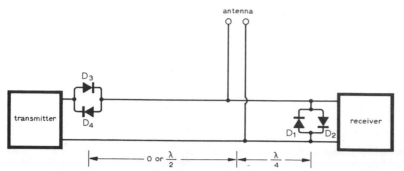

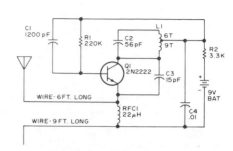

FOUR-DIODE TR SWITCH—Circuit requires only two pairs of high-frequency switching diodes having current ratings appropriate for transmitter power. With diode pairs spaced as shown, impedance at T junction looking toward transmitter is infinite during reception because there is open circuit half a wavelength away created by nonconducting D_3 and D_4. Line is matched in receiver direction so all incoming power from antenna goes into receiver. When transmitter is on, D_3 and D_4 conduct and power flows toward antenna, while D_1 and D_2 also conduct and place short-circuit across receiver input.—A. Lieber, Passive Solid-State Antenna Switch, *Wireless World*, Jan. 1975, p 12.

TUNE-UP AID—Superregenerative receiver circuit is modified to bring quenching frequency down into audio range, thereby giving many closely spaced carriers in region of 27 MHz. RF level across frequency range is essentially constant. Signal simplifies tune-up of front ends of CB units. Antenna is 6 feet of wire connected to emitter side of RFC, with 9 feet of wire on battery side of RFC as counterpoise. Combination, with circuit in center, can be hung vertically in tree if means can be provided for turning it off or removing battery when not in use. Drain is about 0.5 mA from 9-V battery.—E. A. Lawrence, Citizens Band Alignment Aid, *73 Magazine*, April 1973, p 87–88.

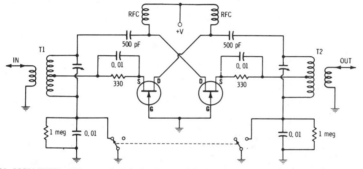

BILATERAL AMPLIFIER—When switch is in position shown, signal entering through T1 is amplified by first transistor and fed from its drain terminal to output through resonant transformer T2. Second transistor is not operational because it now has cutoff bias between gate and source. When switch position is reversed, incoming signal is applied to transistor at right through T2 and removed from left side of circuit to give changeover in signal direction. First transistor is inactive now.—E. M. Noll, "FET Principles, Experiments, and Projects," Howard W. Sams, Indianapolis, IN, 2nd Ed., 1975, p 198–199.

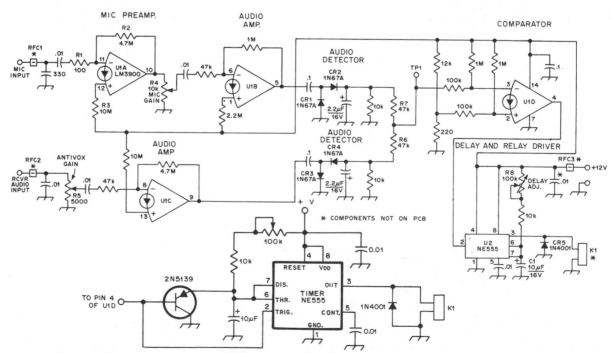

VOX FOR SSB—Uses LM3900 quad Norton opamp and NE555 timer operating from single supply. U1A and U1B amplify microphone signal. U1C amplifies audio sample obtained from station loudspeaker. Outputs of both amplifiers are converted to varying DC voltages by rectifiers in detector stages. Rectifier outputs are summed resistively by R6 and R7 for application to inverting input of voltage comparator U1D. Positive microphone signal drives comparator output low and triggers NE555, which in turn energizes 12-V relay K1 after delay set at about 10 ms by R8 to avoid losing first syllable. Same delay applies to relay dropout, to hold relay closed between words. If K1 drops out for fraction of a second at end of timing cycle even though audio is present, add 2N5139 transistor to NE555 input as shown.—D. A. Blakeslee, A VOX for a Very Small Box, *QST*, March 1976, p 24–26.

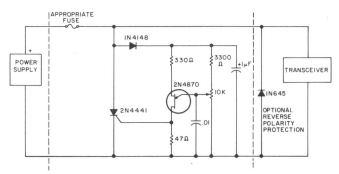

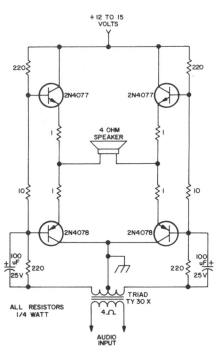

UJT-TRIGGERED CROWBAR—Circuit is used between transceiver and regulated 12-V power supply to protect transceiver from overvoltage or reverse polarity. UJT permits precise setting of overvoltage level at which 2N4441 SCR crow-bar operates. Fuse is blown within microseconds of overvoltage. Crowbar can be built into transceiver.—*Circuits, 73 Magazine,* July 1977, p 35.

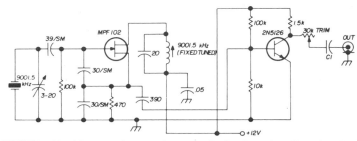

9-MHz CRYSTAL—Used in transmitter section of 80-meter 10-W SSB transceiver. Value of C1 is 50–330 pF, chosen for desired output range. Carrier level can be adjusted with slug-tuned coil or with 30K trimpot.—D. Hembling, Solid-State 80-Meter SSB Transceiver, *Ham Radio,* March 1973, p 6–17.

15-W ADD-ON—Increases usual 1-W audio output of transceiver to up to 15 W. Loudspeaker can be 8-ohm unit, but output will be somewhat reduced.—P. Bunnell, More Fun with the IC-230, *73 Magazine,* May 1975, p 45–46 and 48.

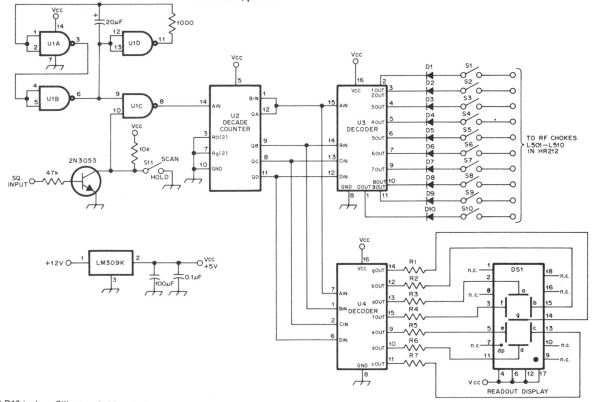

D1-D10 incl. — Silicon switching diodes, type 1N4154.
DS1 — Seven-segment readout display, Litronix type DL-747 or Radio Shack no. 276-056.

R1-R7 incl. — 220 ohm.
S1-S10 incl. — Spst switch.
U1 — Quad 2-input positive NAND gate, TTL type 7400.

U2 — Decade counter, TTL type 7490.
U3 — BCD-to-decimal decoder, TTL type 7445.
U4 — BCD-to-seven segment decoder, TTL type 7446.

10-CHANNEL SCANNER—Designed for Regency HR-212 2-meter transceiver but can be adapted for other transceivers. Features include automatic stop, start, and large LED 7-segment readout. Diodes D1-D10 prevent transceiver voltages from reaching scanner circuit. Squelch voltage input of 2N3053 is taken from transceiver. S1-S10 are used to switch out channels not monitored. Wires going to chokes in HR-212 should be connected to choke leads going to channel switch.—A. Little, 10-Channel Scanner for the Regency HR-212, *QST,* Feb. 1978, p 37.

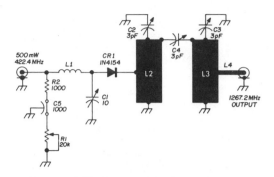

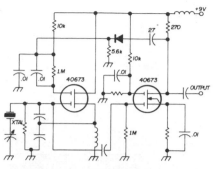

C1	1-10 pF concentric piston trimmer
C2,C3 C4	0.3-3 pF concentric piston trimmer
CR1	1N4154 high-speed switching diode
L1	2 turns no. 20, 0.1'' (2.5 mm) diameter, 0.25'' (6 mm) long
L2	micro-stripline, 0.3'' (7.5 mm) wide, 0.865'' (22 mm) long, grounded at bottom, tapped 0.20'' (5 mm) from ground end
L3	Same as L2 but tapped 0.25'' (6 mm) from ground end
L4	50-ohm micro-stripline, 0.1'' (2.5 mm) wide, any length
R1	20k, 10-turn trimpot

2–23 MHz UNTUNED OSCILLATOR—Two dual-gate MOSFETs operate in untuned Colpitts crystal oscillator. Used in SSB transceiver made by Sideband Associates for radiomarine communication in 2–23 MHz range. Oscillator feeds isolating amplifier. Small capacitor can be used for netting individual crystal to precise assigned frequency.—E. Noll, MOSFET Circuits, *Ham Radio,* Feb. 1975, p 50–57.

TRIPLING TO 1267.2 MHz—Diode tripler and filter combination is designed for double-clad glass epoxy printed-circuit board to simplify construction. Developed for use in 1296-MHz SSB transceiver for amateur 23-cm band. RF energy from 422.4-MHz power amplifier is applied to GE 1N4154 high-speed switching diode through L network. Harmonic comb at output of diode passes only desired frequency to output terminal going to mixer of transceiver.—H. P. Shuch, Easy-to-Build SSB Transceiver for 1296 MHz, *Ham Radio,* Sept. 1974, p 8–23.

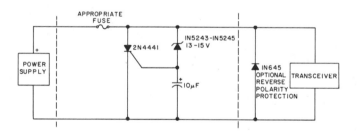

12-V CROWBAR—Uses overvoltage-sensing zener to trigger SCR and place it across power-supply line, blowing fuse within microseconds to protect transceiver. Optional diode provides protection from accidental reversal of supply polarity.—Circuits, *73 Magazine,* July 1977, p 35.

4-CHANNEL VHF FM SCANNER—Discrete components permit simpler readout, with any number of channels and any desired scan rate. Operates directly from 10–15 V power supply without regulation, and has low current drain. Can be used with either positive or negative logic from squelch circuit, so scanning can be stopped with either positive or ground signal. Any voltage from a few volts up will stop scanning. Q3 and Q4 form astable MVBR operating at about 10 pulses per second. Q2 turns on to stop MVBR when base of Q2 is high. Inverter Q1 provides proper polarity of signal to operate Q2. Q6-Q13 form 4-stage ring counter. Pulsing by Q5 serves to pass high output from stage to stage in endless ring pattern. When squelch stops pulsing action, counter stops stepping and output of one counter stage stays high, providing 5-V output for enabling corresponding oscillator and driving LED for that channel. Article gives connections to oscillator for almost any FM transceiver or receiver, along with modifications for changing scan speed and number of channels.—J. Vogt, Improved Channel Scanner for VHF FM, *Ham Radio,* Nov. 1974, p 26–31.

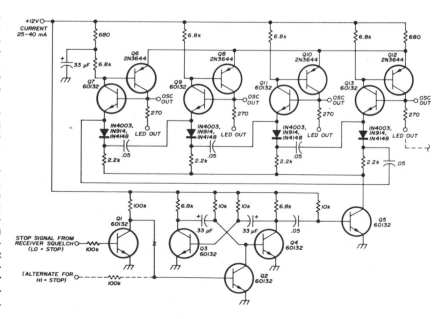

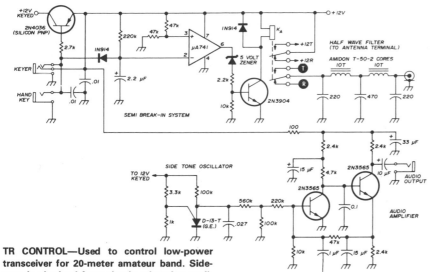

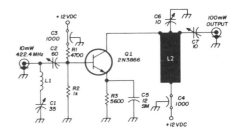

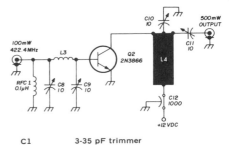

TR CONTROL—Used to control low-power transceiver for 20-meter amateur band. Side-tone is obtained from simple relaxation oscillator using GE D-13-T programmable UJT. Saw-tooth output is attenuated and applied to input of two-transistor audio amplifier. Transmitter keying is done with series switch using 2N4036 silicon transistor. Antenna relay applies +12 V appropriately to transceiver stages used during transmit (+12T) and receive (+12R), and provides switching of antenna between transmitter power chain and receiver RF amplifier. Use DPDT relay with 800-ohm 12-V coil. Transmit-receive logic uses μA741C opamp as differential

comparator. When key is closed, 2.2-μF capacitor is discharged, making opamp output switch to high state and saturate 2N3904 relay driver, pulling in relay for transmit operation. When key is released, capacitor begins to charge; at 6-V point (about 0.5 s with 220K timing resistor), opamp changes state again and relay opens for receiving.—W. Hayward, Low-Power Single-Band CW Transceiver, Ham Radio, Nov. 1974, p 8–17.

5 MHz ± 500 kHz—Developed for use as separate VFO control for transmitting and receiving frequencies in amateur transceiver. Capacitors marked M should be mica. Fine-tuning control covers ±20 kHz range. Will operate almost anywhere in HF range with appropriate change in coil. Output is 4 V P-P.—An Accessory VFO - the Easy Way, 73 Magazine, Aug. 1975, p 103 and 106–108.

C1	3-35 pF trimmer
C2	8-60 pF trimmer
C6-C11	10-pF concentric piston trimmers
L1	2 turns no. 18, wound on 1/4" (6 mm) mandrel, 1/8" (3 mm) long
L2,L4	brass strip, 0.5" (12.5 mm) wide, 1.5" (38 mm) long, mounted 1/8" (3 mm) above ground plane
L3	2 turns 1/8" (3 mm) wide brass strip, 0.1" (2.5 mm) diameter, 0.5" (12.5 mm) long

422.4-MHz POWER AMPLIFIER—Used in local oscillator chain of 1296-MHz SSB transceiver to boost 10-mW output of chain to 500 mW as required for driving final diode-type tripler stage. Sections are connected together with miniature 50-ohm coax.—H. P. Shuch, Easy-to-Build SSB Transceiver for 1296 MHz, Ham Radio, Sept. 1974, p 8–23.

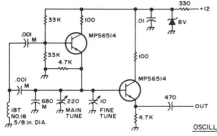

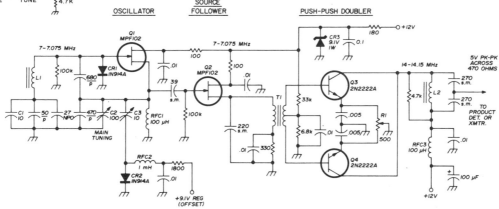

C1,C3	subminiature PC-board-mount air trimmers (Johnson 189-507-5 or T-9-5)
C2	100 pF miniature air variable (large gang of Miller 2109 suitable)
L1	7.6 μH toroidal inductor. 37 turns no. 24 (0.5mm) enamelled wire on Amidon T-68-2 toroidal core (see text)
L2	1 μH toroidal inductor. 14 turns no. 22 (0.6mm) enamelled wire on Amidon T-50-6 toroidal core

R1	500-ohm PC-board-mount control
T1	toroidal transformer. Primary, 2 μH. Use 23 turns no. 24 (0.5mm) enamelled wire on Amidon T-50-6 toroidal core. Secondary is 20 turns no. 24 (0.5mm) enamelled wire (center tapped) over primary winding. Observe same rotation sense when winding
Z1	9.1 volt, 1 watt, zener-diode regulator

14-MHz VFO USING DOUBLER—Developed for use with 20-meter low-power (QRP) transceiver. Push-push doubler avoids instability problems of 14-MHz oscillator and minimizes chirp during CW transmit periods. Uses low-

drift series-tuned Colpitts oscillator operating at 7 MHz, with source-follower buffer separating it from doubler. Adjust dynamic balance control R1 of doubler for best output waveform purity. Capacitors marked P are polystyrene,

and SM are silver mica. Article stresses importance of choosing and using components that minimize drift.—D. DeMaw, VFO Design Techniques for Improved Stability, Ham Radio, June 1976, p 10–17.

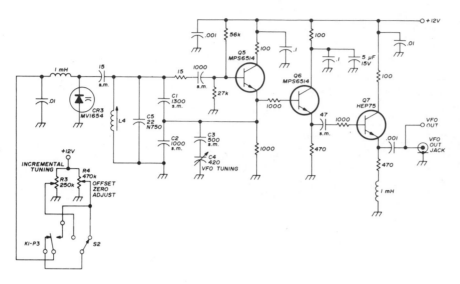

6.545–6.845 MHz VFO—Covers 40-meter amateur band of transceiver for SSB and CW with stable incremental tuning circuit using Motorola MV1654 varactor diode CR3. Tuner permits up to 10-kHz offset above or below VFO frequency. Varactor control voltage is set by offset tuning control R3. R4 compensates for differences in varactors and adjusts VFO for zero offset. Output buffering is provided by Q6 and Q7, with Q7 also serving as power amplifier for balanced mixer used in companion exciter of transmitter. S2 activates receiver offset. Relay K1 automatically turns off offset when receiver is in transmit or standby mode. Offset feature is needed only if there is frequency difference between transmitted and received signals. L4 has 5 turns No. 22 on ½-inch slug-tuned ceramic form.—W. J. Weiser, Simple SSB Transmitter and Receiver for 40 Meters, *Ham Radio,* March 1974, p 6–20.

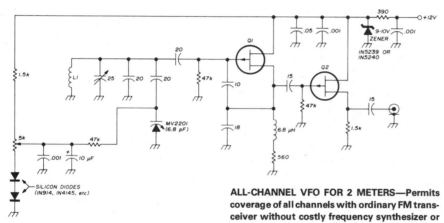

ALL-CHANNEL VFO FOR 2 METERS—Permits coverage of all channels with ordinary FM transceiver without costly frequency synthesizer or individual channel crystals. Can also be used for tuning input of repeater when made necessary

by malicious or accidental interference. Replaces first crystal oscillator in FM transceiver and tunes over required range, generally one-third of first injection frequency. Operates at about 45 MHz for most transceivers, but frequency can easily be changed. Tuning pot (10-turn 5K with digital dial is best) can be placed remotely. VFO uses FET oscillator tuned by tuning diode (also known as Varicap, varactor, or variable-capacitance diode). Source-follower output stage buffers oscillator. Supply is regulated by zener. Q1 and Q2 can be MPF102 FETs, but 2N5668 or 2N5669 are better. L1 is 4 turns No. 16 ½ inch in diameter. Article covers construction, testing, and installation in transceiver.—P. Franson, Simple Tunable Receiver Modification for VHF FM, *Ham Radio,* Oct. 1974, p 40–43.

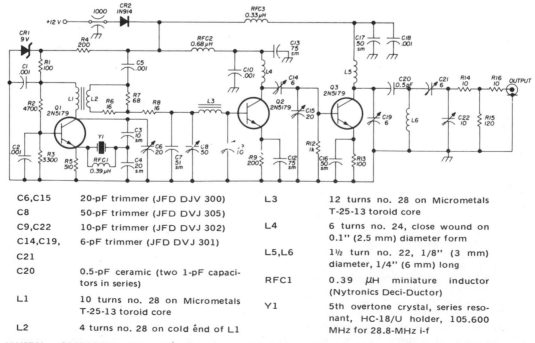

C6,C15	20-pF trimmer (JFD DJV 300)	L3	12 turns no. 28 on Micrometals T-25-13 toroid core
C8	50-pF trimmer (JFD DVJ 305)		
C9,C22	10-pF trimmer (JFD DVJ 302)	L4	6 turns no. 24, close wound on 0.1″ (2.5 mm) diameter form
C14,C19,	6-pF trimmer (JFD DVJ 301)		
C21		L5,L6	1½ turn no. 22, 1/8″ (3 mm) diameter, 1/4″ (6 mm) long
C20	0.5-pF ceramic (two 1-pF capacitors in series)	RFC1	0.39 μH miniature inductor (Nytronics Deci-Ductor)
L1	10 turns no. 28 on Micrometals T-25-13 toroid core	Y1	5th overtone crystal, series resonant, HC-18/U holder, 105.600 MHz for 28.8-MHz i-f
L2	4 turns no. 28 on cold end of L1		

422.4-MHz CRYSTAL OSCILLATOR—Uses 105.6-MHz crystal oscillator followed by frequency-doubling stages to give desired output

for driving external diode-type tripler for which circuit is also given in article. Developed for use in 1296-MHz SSB transceiver for 23-cm amateur

band.—H. P. Shuch, Easy-to-Build SSB Transceiver for 1296 MHz, *Ham Radio,* Sept. 1974, p 8–23.

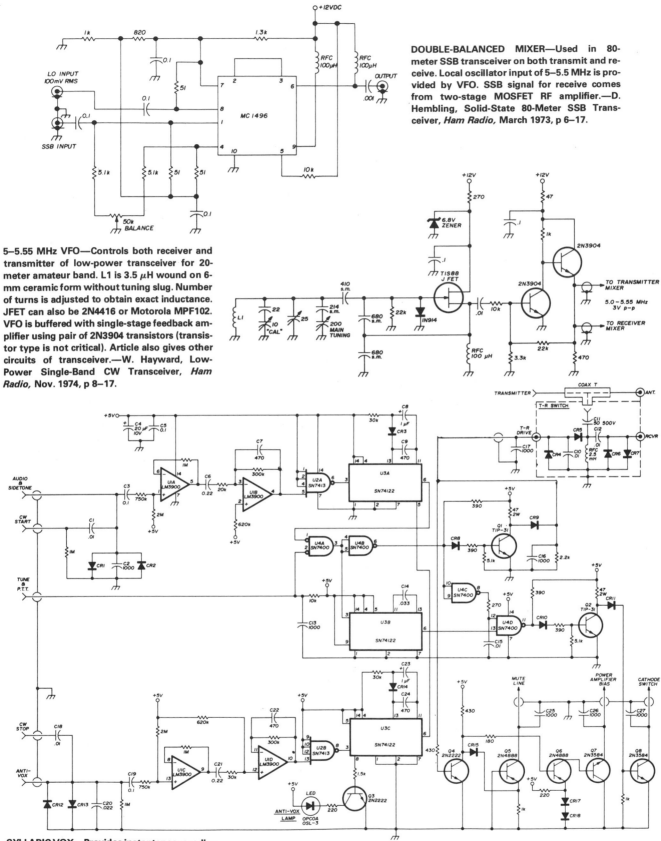

DOUBLE-BALANCED MIXER—Used in 80-meter SSB transceiver on both transmit and receive. Local oscillator input of 5–5.5 MHz is provided by VFO. SSB signal for receive comes from two-stage MOSFET RF amplifier.—D. Hembling, Solid-State 80-Meter SSB Transceiver, *Ham Radio,* March 1973, p 6–17.

5–5.55 MHz VFO—Controls both receiver and transmitter of low-power transceiver for 20-meter amateur band. L1 is 3.5 μH wound on 6-mm ceramic form without tuning slug. Number of turns is adjusted to obtain exact inductance. JFET can also be 2N4416 or Motorola MPF102. VFO is buffered with single-stage feedback amplifier using pair of 2N3904 transistors (transistor type is not critical). Article also gives other circuits of transceiver.—W. Hayward, Low-Power Single-Band CW Transceiver, *Ham Radio,* Nov. 1974, p 8–17.

SYLLABIC VOX—Provides instantaneous radiotelephone speech communication without conventional VOX relays. Switching transistors eliminate contact-bounce problems of relays. Designed for use with Drake T-4XB and R-4B transmitter and receiver. Also gives true break-in CW keying. Since only words and syllables of words go on air, there are no VOX RF transients and no extraneous local noise. Operating bias for final amplifier is applied only during transmission; at other times, final amplifier tubes are completely cut off. Diodes CR4-CR7, CR9, and CR11 are 1N4004 or equivalent. All other diodes are 1N914 or equivalent. RF choke is 2.5 mH rated 100 mA. Article describes operation in detail and gives installation and setup procedures.—R. W. Hitchcock, Syllabic VOX System for Drake Equipment, *Ham Radio,* Aug. 1976, p 24–29.

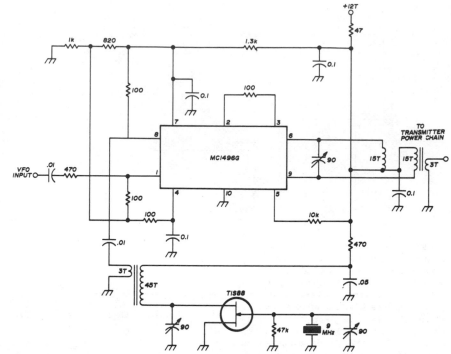

TRANSMIT MIXER—Used in low-power transceiver for 20-meter amateur band. VFO input, tunable from 5 to 5.55 MHz, is combined with 9-MHz output of crystal oscillator in Motorola MC1496G double-balanced modulator to give 14-MHz output for transmitter power chain. Supply voltage (+12 V) is applied only when transmitting. Transformers are wound on Amidon T-50-6 toroids or equivalent. Balance is maintained in mixer with center-tapped tuned circuit in output, made by putting 15-turn bifilar winding on toroid core and tuning series combination. Article gives all other circuits of transceiver.—W. Hayward, Low-Power Single-Band CW Transceiver, *Ham Radio,* Nov. 1974, p 8–17.

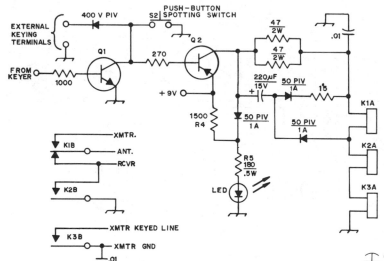

TR SWITCH—Consists of driver transistors Q1 and Q2 and reed relays K1-K3. K1 switches antenna from receiver to transmitter, K2 grounds receiver input, and K3 keys transmitter. Coil of K1 has 400 turns No. 32 enamel, while K2 and K3 each have 120 turns. Coils are wound directly on glass of reed relays and covered with epoxy cement. LED normally glows dimly, with brightness increasing when character is keyed.—J. H. Fox, An Integrated Keyer/TR Switch, *QST,* Jan. 1975, p 15–20.

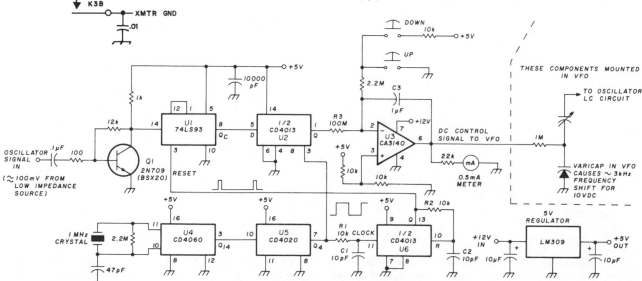

WARM-UP DRIFT COMPENSATOR—Warm-up drift of transceiver VFO is automatically corrected by using binary counter U1 to count oscillator frequency for interval of about 3.81 Hz determined by 1-MHz reference crystal oscillator and divider chain U4-U5. Switches labeled UP and DOWN are used to bring output of integrator R3-C3 into range manually after circuit switch-on.—K. Spaargaren, Drift-Correction Circuit for Free-Running Oscillators, *Ham Radio,* Dec. 1977, p 45–47.

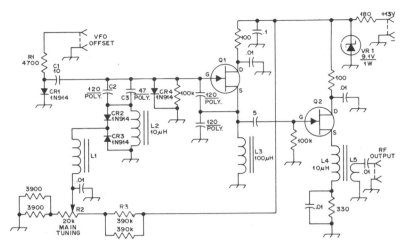

CR1-CR4 — Small-signal silicon diodes, 1N914 or equiv.
J1-J3 — Phono jack.
L1 — Modified rf choke (Radio Shack 273 102). Remove original turns and replace with approximately 100 turns no. 28 enam. wire.
L2, L4 — Rf choke (Radio Shack 273-101).

L3 — Rf choke (Radio Shack 273-102).
L5 — 1 turn no. 28 enam. wire over L4.
Q1, Q2 — JFET (Radio Shack RS 2036 or equiv.).
R2 — Slide potentiometer, 20 kΩ (from Radio Shack 271-1601).
VR1 — Zener diode, 9.1 V, 1 W.

7–7.1 MHz VFO—JFET Q1 serves as oscillator, with frequency determined by C2, L2, CR2, and CR3; diodes operate in reverse-bias regions as voltage-variable capacitors. Amount of reverse bias applied by R2 determines capacitance and frequency. VFO operates on both transmit and receive; on transmit, no voltage is applied to VFO offset circuit R1-C1-CR1 so it has little effect on oscillator. On receive, +12 V applied to R1 makes CR1 conduct and places C1 across frequency-determining network to shift VFO about 100 kHz away from operating frequency so receiver will not be blocked. Q2 is buffer between oscillator and transmitter. VR1 provides regulated 9.1 V for oscillator and buffer. (Project was named after chopped beef can in which it was mounted.)—J. Rusgrove, The CB Slider, *QST*, March 1977, p 15–17.

500-kHz SCAN ON 2 METERS—Circuit added to 2-meter transceiver sweeps 500-kHz segment of 2-meter band at 2-s intervals. When incoming signal is strong enough to trip receiver squelch, sweep stops and receiver locks on station. R4 and C3 determine scan rate. Adjust R1 for best lock-on. When signal is sensed, squelch is greater than 9 V on R3, driving output of U2 low, turning on LED, and removing charging voltage from R4. When signal disappears, output of U2 goes high and scanning continues.—W. Sward, Add Frequency Scan to a Receiver for $10, *QST*, March 1977, p 48.

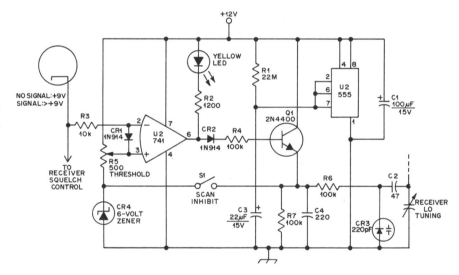

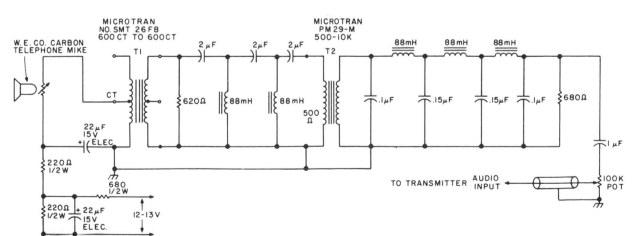

CARBON MIKE FOR MOBILE SERVICE—Filter enhances desirable characteristics of ordinary carbon mike taken from telephone, for use with mobile transceiver. Required excitation voltage of 3.5 V for mike is reduced from 12 V of car battery by resistor network having hash filter to keep alternator whine out of audio system. Output of 0.25 mV from mike-filter combination is reduced by 100K pot to value needed for transmitter.—S. Olberg, The Carbon Marvel, *73 Magazine*, April 1977, p 120.

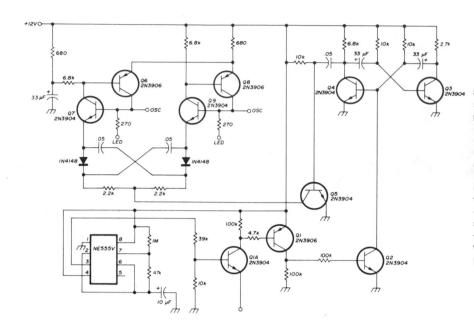

SCANNER WITH SEARCHBACK—Combines scanning between two repeater channels with periodic searchback, to prevent scanner from locking on one of channels during long periods of use. NE555 timer is added to squelch recognition circuit to provide automatic control of scanner so both frequencies are checked at least every 15 s. If scanner is extended to monitor four channels, none are unguarded for more than 1 min. Article shows how receiver section of transceiver is modified for diode switching by scanner of oscillator crystals for individual channels.—P. Shreve, Two-Channel Scanner for Repeater Monitoring, *Ham Radio,* Oct. 1976, p 48–51.

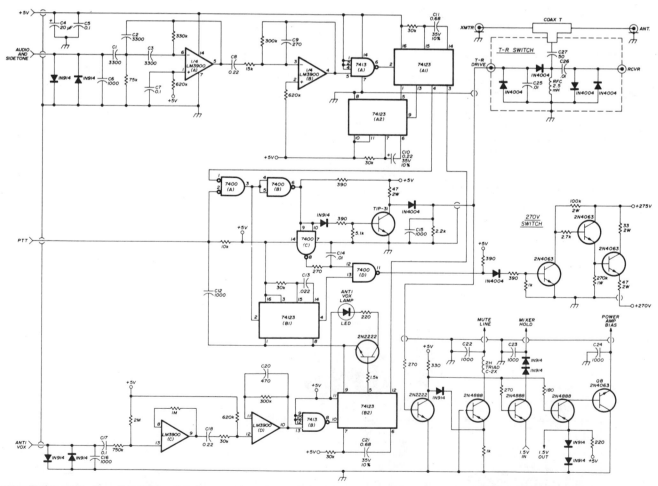

HIGH-SPEED VOX—Use of switching transistors for rapid, silent voice-controlled switching of transmitter-receiver functions improves on-the-air effectiveness of SSB station. Conversation is essentially the same as when using telephone. Each set of contacts that would open or close single circuit in relay-type VOX is replaced by switching transistor. TR switch is diode-biased antenna gate in which actual switching takes place 200 μs before RF appears, being accomplished by forward-biasing diode with DC voltage. Input to LM3900 is through high-pass filter. Operation of similar solid-state VOX circuit is described in detail in earlier article by same author (see author index).—H. R. Hildreth, Syllabic VOX System for the Collins S-Line, *Ham Radio,* Oct. 1977, p 29–33.

3–3.5 kHz VARACTOR TUNER—Variable oscillator for 80-meter SSB transceiver is tuned by 1N594 diode. MPF102 FET serves as source-follower buffer. With values shown, full excursion of R8 tunes oscillator from 3.045 to 3.545 MHz. Use well-regulated 12-V source. R9 allows synchronization of receive and transmit frequencies. K1 is 4PDT relay used for switching supply voltage and antenna from transmit to receive. L8 is 40 turns No. 32 on ¼-in slug-tuned form.— W. J. Weiser, Integrated Circuit SSB Transceiver for 80 Meters, *Ham Radio*, April 1976, p 48–52.

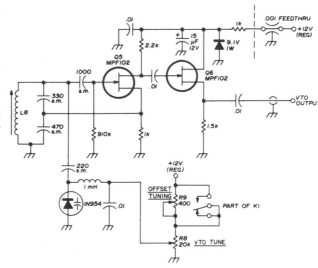

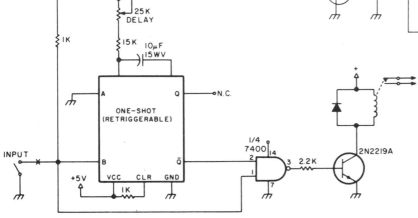

TR SWITCH—Simple transmit-receive switch uses 74122 retriggerable mono MVBR with clear and one NAND gate. When input of gate goes low (during key down or speech), transistor conducts and closes relay. Relay stays closed for period determined by delay pot, to maintain transmit condition between dots or dashes or between other pauses. Switch may be removed and any other TTL-compatible input applied at point X.—B. Voight, The TTL One Shot, *73 Magazine*, Feb. 1977, p 56–58.

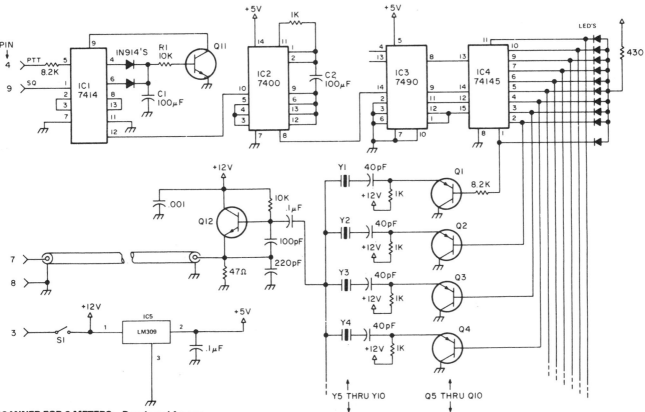

SCANNER FOR 2 METERS—Developed for use with frequency synthesizer to scan transmit and receive frequencies of four receivers plus six other channels in 2-meter amateur band. Pin connections at left go to socket provided on Icom IC-230 synthesizer for connecting external VFO operating between 11.255 and 12.255 MHz. Q1-Q10 can be 2N3638 or equivalent; Q11 is any NPN silicon such as 2N2102; and Q12 is 2N2102. LEDs operate from +12 V. In operation, scanner stops on active channel, and resumes scanning 5 s after channel goes off air. Article covers circuit operation, gives construction details, and tells how to calculate crystal frequency for each channel desired.—C. A. Kollar, Two Meter Scanner, *73 Magazine*, June 1977, p 46–48.

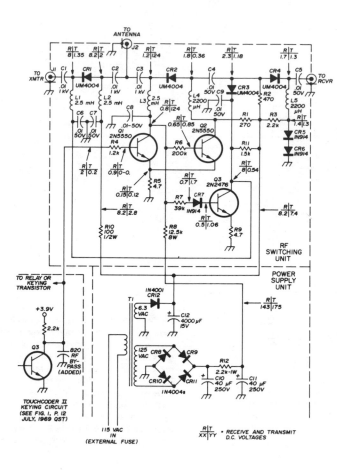

PIN-DIODE TR SWITCH—Solid-state TR switch operates at very high keying speeds and handles up to 100 W while transferring antenna between receiver and transmitter in accordance with transmitter keying demands. Uses Unitrode UM4004 PIN diodes to provide about 0.2-dB insertion loss when forward-biased and about 30-dB isolation when reverse-biased. CR1 is forward-biased for about 45 mA DC and CR2 is reverse-biased by 124 V during transmit, for minimum loss in CR1 and maximum isolation in CR2. Circuit is designed to operate from collector of keying-circuit transistor in Touchcoder II (in dotted lines at lower left), but any source providing required T (transmit) and R (receive) DC voltages shown on diagram will key circuit. Article covers construction in detail.—J. K. Boomer, PIN Diode Transmit/Receive Switch for 80-10 Meters, *Ham Radio*, May 1976, p 10—15.

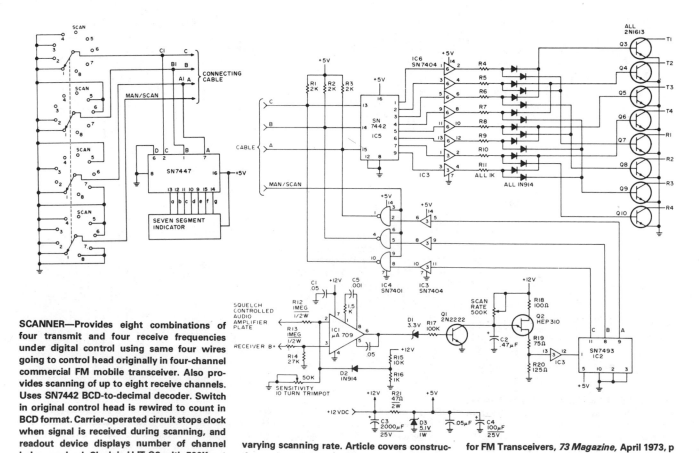

SCANNER—Provides eight combinations of four transmit and four receive frequencies under digital control using same four wires going to control head originally in four-channel commercial FM mobile transceiver. Also provides scanning of up to eight receive channels. Uses SN7442 BCD-to-decimal decoder. Switch in original control head is rewired to count in BCD format. Carrier-operated circuit stops clock when signal is received during scanning, and readout device displays number of channel being received. Clock is UJT Q2 with 500K pot varying scanning rate. Article covers construction and testing.—C. Durst, Scanning Adapter for FM Transceivers, *73 Magazine*, April 1973, p 73—78.

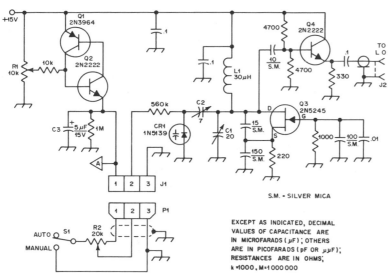

S.M. = SILVER MICA

EXCEPT AS INDICATED, DECIMAL
VALUES OF CAPACITANCE ARE
IN MICROFARADS (μF) ; OTHERS
ARE IN PICOFARADS (pF OR μμF) ;
RESISTANCES ARE IN OHMS ;
k =1000 , M=1 000 000

REMOTE TUNING—Simple sweep generator and 5–5.5 MHz VCO provide remote manual or automatic electronic tuning of 6-meter transceiver, for quick check of possible band openings a few kilohertz from frequency to which receiver is normally tuned. To adjust, set S1 on MANUAL and turn R2 fully counterclockwise. If signal at lowest received frequency is applied to antenna jack, signal can be centered within IF passband of receiver by adjusting C1. Next, turn R2 fully clockwise, apply signal at highest frequency to be received, and center signal within passband again by adjusting C2. With S1 in AUTO position, R1 determines highest frequency tuned. If sweep rate is too low, reduce value of C3. Point A is used to drive CRO through FET buffer stage, for displaying signals present within sweep range as pips on screen.—J. R. Bingham, Sweep 6 Meters and Really Clean Up! *QST,* April 1977, p 27–28.

76.25 AND 81.6 MHz—MC10102 ECL quad NOR gate provides convenient switching between two crystal oscillators, as required for change from receive to transmit in transceiver. Output level of about 0.8 V P-P can easily drive 50-ohm load and is fully buffered from oscillator sections. Gate A provides bias for oscillator gates B and C. Use 270 ohms for R_1-R_3. Crystals are fifth overtone; Y_B is 81.6 MHz with 97 nH for L_B and 39 pF for C_B, and Y_C is 76.25 MHz with 104 nH for L_C and 39 pF for C_C.—G. Griesmyer, Clocked CMOS One-Shot Has No RC Time Constant, *EDN Magazine,* May 20, 1978, p 164 and 172.

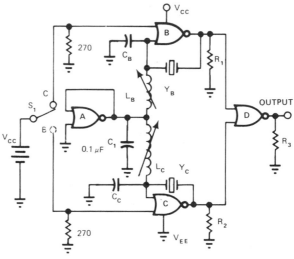

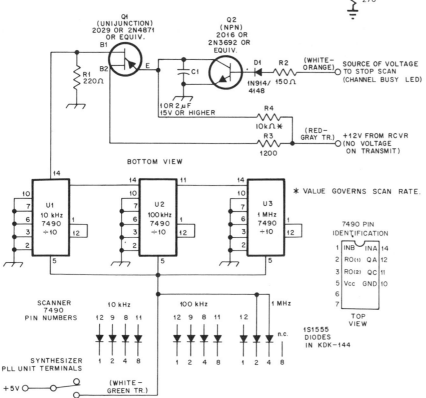

* VALUE GOVERNS SCAN RATE.

7490 PIN
IDENTIFICATION

1	INB	INA	14
2	RO(1)	QA	12
3	RO(2)	QC	11
5	Vcc	GND	10
6			
7			

TOP
VIEW

2-METER SCANNER—Designed for use with KDK-144 amateur 2-meter transceiver to provide automatic scanning between 146 and 147.990 MHz. When transceiver is switched to priority position, +5 VDC is applied to 7490 decade counters U1, U2, and U3 to activate scanner. Scanning stops when signal strong enough to open squelch turns on Darlington-connected transistors Q1 and Q2, shorting UJT timing capacitor C1 which is 1–2 μF.—R. W. Shoemaker, Jr., A Scanner for KDK, *QST,* Oct. 1978, p 36–37.

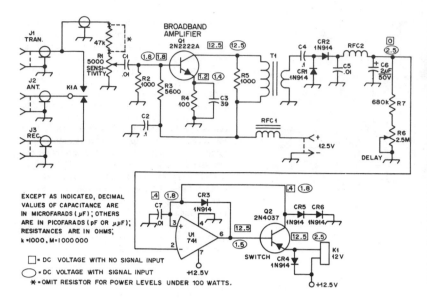

EXCEPT AS INDICATED, DECIMAL
VALUES OF CAPACITANCE ARE
IN MICROFARADS (µF); OTHERS
ARE IN PICOFARADS (pF OR µµF);
RESISTANCES ARE IN OHMS;
k = 1000, M = 1000 000

☐ = DC VOLTAGE WITH NO SIGNAL INPUT
◯ = DC VOLTAGE WITH SIGNAL INPUT
✱ = OMIT RESISTOR FOR POWER LEVELS UNDER 100 WATTS.

RF-SENSING TR SWITCH—System detects presence of RF at output of transmitter and changes antenna connection from receiver to transmitter instantly and automatically. No modifications are required for transmitter or receiver. RF voltage divider R1 permits use with transmitters of any power. Broadband amplifier Q1 feeds voltage doubler CR1-CR2 through broadband toroidal step-down transformer T1. CR6, R7, and R6 form adjustable timing network that governs hold-in time of relay K1. Inverting amplifier U1 turns on Q2 for energizing K1 and switching antenna to transmitter when RF is sensed. When no rectified RF reaches U1, Q2 is cut off and antenna is changed over to receiver. RFC1 and RFC2 have 42 turns No. 28 enamel on Amidon FT-50-43 core. T1 uses Amidon FT-50-43 core, with 25 turns No. 28 enamel for primary and 5 turns No. 28 over this for secondary.—D. DeMaw and J. Rusgrove, An RF-Sensed Antenna Changeover Relay, *QST*, Aug. 1976, p 21–23.

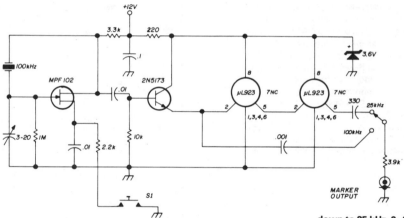

MARKER GENERATOR—Crystal-controlled frequency marker for 80-meter SSB transceiver provides front-panel control of either 25- or 100-kHz markers. S1 activates circuit by completing source circuit of FET momentarily. Two µL923 JK flip-flops divide 100-kHz crystal frequency down to 25 kHz. 3–20 pF trimmer is adjusted to zero-beat crystal against receiver tuned to WWV.—D. Hembling, Solid-State 80-Meter SSB Transceiver, *Ham Radio*, March 1973, p 6–17.

CHAPTER **99**

Transmitter Circuits

Covers circuits specifically developed for use in AM, FM, and CW communication transmitters for amateur, aircraft, marine, satellite relay, long-wave, and other applications. Power ratings range from fractions of watt for QRP low-power CW transmitters up to 2 kW for moonbounce transmitter. Circuits for measuring RF output power are included.

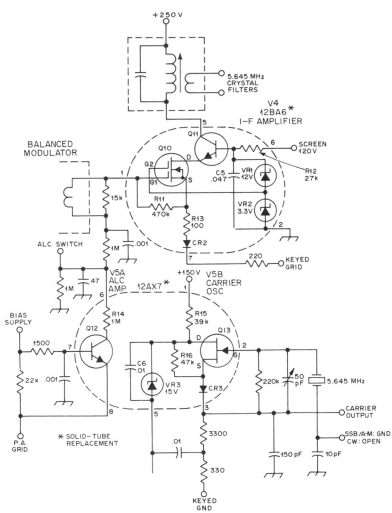

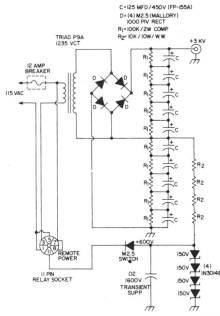

3 kV AT 2 kW—Developed as supply for 1-kW linear amplifier used in 2-meter moonbounce communication. Article covers construction, with emphasis on insulation requirements, and gives circuit of 1-kW amplifier using Eimac 5-500A pentode.—R. W. Campbell, *Kilowatt Linear Amplifier for 2 Meters, 73 Magazine,* Dec. 1973, p 29–35.

TRANSISTORS FOR OSCILLATOR AND IF TUBES—Article covers replacement of tubes in Drake T-4XB transmitter with solid-state equivalent circuits mounted in 7-pin and 9-pin miniature plugs. V4 uses dual-cascode MOSFET, with CR2 and zeners providing high-voltage protection from keyed grid. Carrier oscillator V5B uses single low-voltage high-μ JFET with zener voltage regulator in original grounded "plate" oscillator circuit. High keyed-ground voltages are isolated by CR3 and R16. Automatic level control amplifier V5A is single high-voltage transistor. R14 synthesizes 12AX7 plate resistance, to maintain same audio time response.—H. J. Sartori, *Solid-Tubes—a New Life for Old Designs, QST,* April 1977, p 45–50.

CR2, CR3 — General-purpose silicon diode, 300 PIV, 1N645 or equivalent.
Q10 — N-channel dual-gate MOSFET, 25 V$_{(BR)}$, 3N206 or equiv.
Q11 — Npn transistor, 300 V$_{(BR)}$, Texas Inst., TIS131.
Q12 — Npn transistor, 300 V$_{(BR)}$, Texas

Inst., A5T5058.
Q13 — N-channel JFET 30 V$_{(BR)}$, 2N5246 or equiv.
VR1 — Zener diode, 12 V, 400 mW, 1N759.
VR2 — Zener diode, 3.3 V, 1 W, 1N746.
VR3 — Zener diode, 15 V, 400 mW, 1N965.

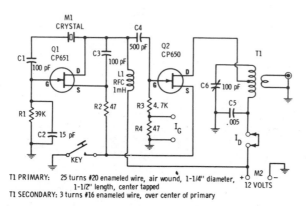

T1 PRIMARY: 25 turns #20 enameled wire, air wound, 1-1/4" diameter, 1-1/2" length, center tapped
T1 SECONDARY: 3 turns #16 enameled wire, over center of primary

5-W FET TRANSMITTER—Values shown give operation in 40-meter amateur band. Drain of power FET Q2 is connected to tap on primary of resonant tank circuit.—E. M. Noll, "FET Princi-ples, Experiments, and Projects," Howard W. Sams, Indianapolis, IN, 2nd Ed., 1975, p 188–189.

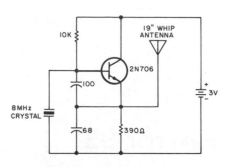

144-MHz LOW-POWER—Used as weak-signal source for tuning circuits of 2-m receiver or preamp when no stations are on air.—C. Sond-geroth, Really Soup Up Your 2m Receiver, *73 Magazine*, Feb. 1976, p 40–42.

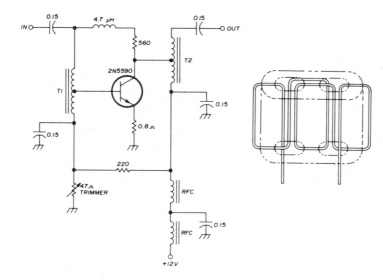

4-W LINEAR—Produces power output of 4 W across high-frequency RF range from 300 kHz to 30 MHz, for output of low-power QRP transmit-ter or as driver for final amplifier of higher-power transmitter. Gain is only 3 dB down at frequency limits and is still useful at 6 meters. Amplifier output may be shorted or left open in-definitely even with full drive. Stability and wide frequency response are achieved by add-ing considerable negative feedback to other-wise conventional broadband amplifier. T1 and T2 are wound on two-hole balun cores as found in TV sets, such as Phillips 4322-020-31520. Two lengths of No. 22 enamel are twisted about 3 times per inch and then wound through core as shown. One end of one wire is connected to op-posite end of other wire to serve as center tap for transformer. Transformers are responsible for wide frequency response of amplifier.—J. A. Koehler, Four-Watt Wideband Linear Amplifier, *Ham Radio*, Jan. 1976, p 42–44.

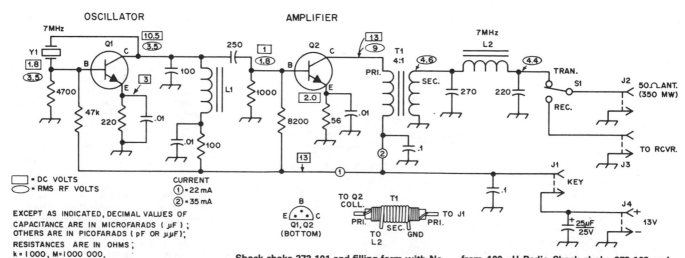

• = DC VOLTS
◯ = RMS RF VOLTS

CURRENT
① = 22 mA
② = 35 mA

EXCEPT AS INDICATED, DECIMAL VALUES OF CAPACITANCE ARE IN MICROFARADS (μF); OTHERS ARE IN PICOFARADS (pF OR μμF); RESISTANCES ARE IN OHMS; k = 1000, M = 1000 000.

250 mW FOR 40-METER CW—Two-transistor circuit is easily assembled on circular printed-circuit board small enough to fit into tunafish can, for low-power (QRP) operation. Simple Pierce crystal oscillator Q1 feeds class C ampli-fier Q2. L1 is made by unwinding 10-μH Radio Shack choke 273-101 and filling form with No. 28 or 30 enamel to give 24 μH. Similar choke is unwound so only 11 turns remain (1.36 μH), with turns spaced one wire thickness apart for L2. Adjust spacing of turns for maximum output during final tune-up with transmitter operating into 50-ohm load. For T1, remove all but 50 turns from 100-μH Radio Shack choke 273-102 and wind 25 turns No. 22 or 24 enamel over these. Supply can be nine Penlite, C, or D cells in series or 12-V or 13-V regulated DC supply. Q1 and Q2 are 2N2222A or equivalent. Y1 is 7-MHz funda-mental crystal.—D. DeMaw, Build a Tuna-Tin 2, *QST*, May 1976, p 14–16.

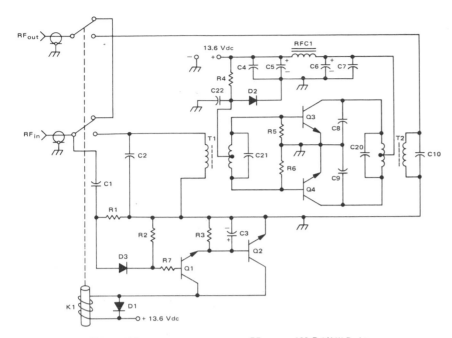

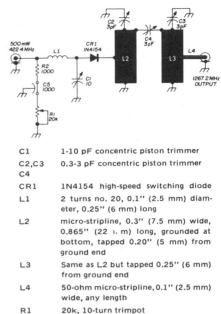

C1	1-10 pF concentric piston trimmer
C2,C3	0.3-3 pF concentric piston trimmer
C4	
CR1	1N4154 high-speed switching diode
L1	2 turns no. 20, 0.1'' (2.5 mm) diameter, 0.25'' (6 mm) long
L2	micro-stripline, 0.3'' (7.5 mm) wide, 0.865'' (22 ι. m) long, grounded at bottom, tapped 0.20'' (5 mm) from ground end
L3	Same as L2 but tapped 0.25'' (6 mm) from ground end
L4	50-ohm micro-stripline, 0.1'' (2.5 mm) wide, any length
R1	20k, 10-turn trimpot

C1	= 33 pF Dipped Mica		R7	= 100 Ω 1/4 W Resistor
C2	= 18 pF Dipped Mica		RFC1	= 9 Ferroxcube Beads on #18 AWG Wire
C3	= 10 μF 35 Vdc for AM operation, 100 μF 35 Vdc for SSB operation.		D1	= 1N4001
			D2	= 1N4997
C4	= .1 μF Erie		D3	= 1N914
C5	= 10 μF 35 Vdc Electrolytic		Q1, Q2	= 2N4401
C6	= 1 μF Tantalum		Q3, 4	= MRF454
C7	= .001 μF Erie Disc		T1,.T2	= 16:1 Transformers
C8, 9	= 330 pF Dipped Mica		C20	= 910 pF Dipped Mica
R1	= 100 kΩ 1/4 W Resistor		C21	= 1100 pF Dipped Mica
R2, 3	= 10 kΩ 1/4 W Resistor		C10	= 24 pF Dipped Mica
R4	= 33 Ω 5 W Wire Wound Resistor		C22	= 500 μF 3 Vdc Electrolytic
R5, 6	= 10 Ω 1/2 W Resistor		K1	= Potter & Brumfield KT11A 12 Vdc Relay or Equivalent

2–30 MHz 140-W LINEAR—Uses two Motorola MRF454 transistors Q3-Q4 in circuit providing relatively flat gain over frequency band, as required for power amplifier of amateur SSB transmitter. Bias diode D2 is mounted on heat-sink of Q3-Q4 for temperature tracking. Circuit includes carrier-operated relay driven by Q1 and Q2.—T. Bishop, "140W (PEP) Amateur Radio Linear Amplifier 2-30MHz," Motorola, Phoenix, AZ, 1976, EB-63.

100 W AT 432 MHz—Two-transistor 100-W PEP solid-state linear amplifier can be used for SSB activity in satellite relay service or for linear, CW, or FM service. Circuit uses Motorola MRF306 28-V 60-W 225–400 MHz power transistors in narrow-band parallel amplifier operating in class AB linear mode. Drive level is about 10 W PEP. Article covers construction and tune-up.—R. K. Olsen, 100-Watt Solid-State Power Amplifier for 432 MHz, *Ham Radio*, Sept. 1975, p 36–43.

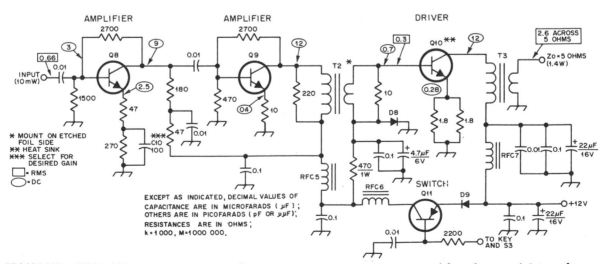

1.4-W BROADBAND LINEAR FOR 7 AND 14 MHz—Requires about 10-mW driving power. Frequency response is essentially flat over 7–14 MHz frequency range. Diodes are 1N4003. Q8 is 2N2222A, Q9 is 2N3866, Q10 is 2N2270, and Q11 is 2N4037. RF chokes use 18 turns No. 28 enamel on FT-37-43 ferrite toroid core. Primary of T2 is 30 turns of No. 28 enamel on FT-50-43 ferrite toroid core, with 4 turns No. 28 wound over cold end for secondary. T3 has 16 turns of No. 28 enamel for primary and 4 turns for secondary looped through BLN-43-302 ferrite core. Article gives test procedure.—D. DeMaw, Transmitter Design—Emphasis on Anatomy, *QST*, July 1978, p 23–25.

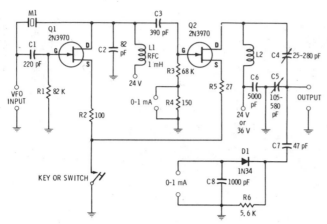

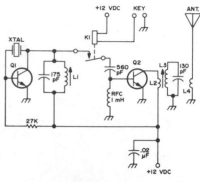

¼-W CW TRANSMITTER—Low-power two-stage FET transmitter for 80-meter amateur band uses Pierce crystal oscillator that requires no output resonant circuit. DC milliammeter can be connected across 150-ohm resistor in gate circuit of second transistor to indicate strength of oscillator output signal. Resonant circuit of RF amplifier Q2 uses toroid L2 (56 turns No. 24 enamel on ¹³/₁₆-in Permacor 57-1541 core) and two series-connected trimmers. Milliammeter is connected across C8 when level of RF output voltage is to be measured.—E. M. Noll, "FET Principles, Experiments, and Projects," Howard W. Sams, Indianapolis, IN, 2nd Ed., 1975, p 204–207.

2-METER QRP—Can supply up to 1 W of RF output on CW for portable or low-power (QRP) amateur radio operation. Provides chirpless keying with negligible backwave. Operates from 12-V car battery or lantern batteries. Oscillator uses 7-MHz fundamental crystal and 40080 transistor Q1, with 40081 in final stage. L1 is 20 turns No. 28 on ¼-inch slug-tuned form. L3 is 28 turns No. 28 on ¼-inch slug-tuned form, with 5 turns No. 24 wound on it for L2 and the same for L4. Article covers construction and operation.—C. Klinert, Simple QRP Transmitter, *73 Magazine*, Aug. 1973, p 65–67.

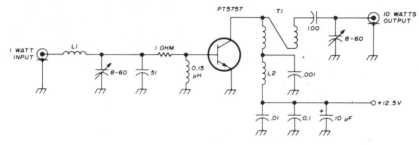

10 W ON 2 METERS—Single TRW PT5757 transistor provides 10-W output when operating from 12.5-V auto battery. L1 is 4 turns No. 20 enamel and L2 is 10 turns No. 20, both with ³/₃₂-inch inner diameter. T1 is 4:1 transmission-line transformer made from 3-inch length of twisted-pair No. 20 enamel.—J. Fisk, Two-Meter Power Amplifier, *Ham Radio*, Jan. 1974, p 67.

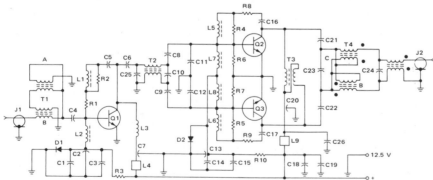

C1, C14, C18 — 0.1 µF ceramic.
C2, C7, C13, C20 — 0.001 µF feed through.
C3 — 100 µF/3V.
C4, C6 — 0.033 µF mylar.
C5 — 0.0047 µF mylar.
C8, C9 — 0.015 and 0.033 µF mylars in parallel.
C10 — 470 pF mica.
C11, C12 — 560 pF mica.
C15 — 1000 µF/3 V.
C16, C17 — 0.015 µF mylar
C19 — 10 pF 15 V
C21, C22 — two 0.068 µF mylars in parallel.
C23 — 330 pF mica
C24 — 39 pF mica
C25 — 680 pF mica
C26 — .01 µF ceramic

R1, R6, R7 — 10 Ω, 1/2 W carbon.
R2 — 51 Ω, 1/2 W carbon
R3 — 240 Ω, 1 wire W
R4, R5 — 18 Ω, 1 W carbon
R8, R9 — 27 Ω, 2 W carbon
R10 — 33 Ω, 6 W wire W

L1 — 0.22 µh molded choke
L2, L7, L8 — 10 µh molded choke
L5, L6 — 0.15 µh
L3 — 25 t, #26 wire, wound on a 100 Ω, 2 W resistor. (1.0 µh)
L4, L9 — 3 ferrite beads each.

T1 — 2 twisted pairs of #26 wire, 8 twists per inch. A = 4 turns, B = 8 turns. Core - Stackpole 57-9322-11, Indiana General F627-8Q1 or equivalent

T2 — 2 twisted pairs of #24 wire, 8 twists per inch, 6 turns. (Core as above.)

T3 — 2 twisted pairs of #20 wire, 6 twists per inch, 4 turns. (Core as above.)

T4 — A and B = 2 twisted pairs of #24 wire, 8 twists per inch. 5 turns each. C = 1 twisted pair of #24 wire, 8 turns. Core - Stackpole 57-9074-11, Indiana General F 624-19Q1 or equivalent.

Q1 — 2N6367

Q2, Q3 — 2N6368

D1 — 1N4001
D2 — 1N4997   J1, J2 — BNC connectors

80-W LINEAR FOR MOBILE SSB—Designed for operation from 12.5-V supply, using driver stage to provide total power gain of about 30 dB for 3–30 MHz band. Negative collector-to-base feedback provides gain compensation in both driver and output stages. Low circuit impedances make layout and construction more critical than with higher-voltage circuits.—H. Granberg, "Broadband Linear Power Amplifiers Using Push-Pull Transistors," Motorola, Phoenix, AZ, 1974, AN-593, p 7.

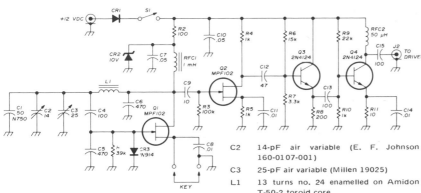

C1   50-pF, N750 temperature coefficient
      ceramic

C2   14-pF air variable (E. F. Johnson
      160-0107-001)

C3   25-pF air variable (Millen 19025)

L1   13 turns no. 24 enamelled on Amidon
      T-50-2 toroid core

RFC1 1-mH rf choke (Millen J300-1000)

RFC2 50-μH rf choke (Millen 00-50)

20-METER VFO—Tunes from 14.0 to 14.2 MHz, using stable Vackar design. Protective diode CR1 can be any silicon rectifier. Clamping diode CR3 improves stability by preventing conduction in gate of JFET oscillator Q1.—C. E. Galbreath, Low-Power Solid-State VFO Transmitter for 20 Meters, *Ham Radio*, Nov. 1973, p 6–11.

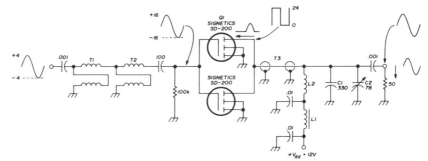

300-mW 25-MHz AMPLIFIER—Uses quarter-wavelength transmission line T3 in place of parallel-tuned traps to pass even-harmonic current freely while blocking odd harmonics. Circuit can be adapted to 300-mW walkie-talkie for 6 or 10 meters. At 25 MHz, efficiency is 73%.—F. H. Raab, High-Efficiency RF Power Amplifiers, *Ham Radio*, Oct. 1974, p 8–29.

C2   78-pF variable (E.F. Johnson 158-4)

L1   2.2 μH rf choke (Delevan 1025-28)

L2   106 nH (4 turns no. 26 wire on Perma-
      core 57-2656 or Micrometals T30-6 core)

T1,T2 11 turns no. 26 twisted pair on Perma-
       core 57-2656 or Micrometals T30-6 core

T3   piece of 125-ohm coaxial cable (RG-
      63B/U), 112.2'' (2.85 meters) long

D1 — 36-V, 1-W Zener diode.
J1-J4, incl. — Single-hole mount phono jack.
L1 — 100-μH choke (Radio Shack 273-102).
L2-L4, incl. — 10-μH choke (Radio Shack 273-101).

L5 — 12-μH inductor (Radio Shack 273-101 with 4 turns no. 26 enam. wire added).
L6 — 8.9-μH inductor (Radio Shack 273-101 with 3 turns removed).
S1 — Miniature spdt toggle or slide switch.

T1 — Broadband transformer (Radio Shack 273-101 for primary, with 5-turn secondary of no. 26 enam. wire over C6 end of primary).
Y1 — 80-meter fundamental type of crystal (crystal socket optional).

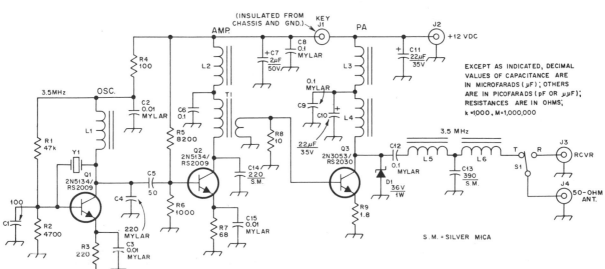

EXCEPT AS INDICATED, DECIMAL VALUES OF CAPACITANCE ARE IN MICROFARADS (μF); OTHERS ARE IN PICOFARADS (pF OR μμF); RESISTANCES ARE IN OHMS; k =1000, M=1,000,000

S.M. = SILVER MICA

80-METER CW FOR QRP—Low-power transmitter can be mounted on small can for operation from separate 12-V supply. Zener D1 protects Q3 by clamping on RF voltage peaks in excess of 36 V. Output tank of Q3 gives satisfactory operation from 3.5 to 3.75 MHz without tuning.—D. DeMaw, Build This "Sardine Sender," *QST*, Oct. 1978, p 15–17 and 38.

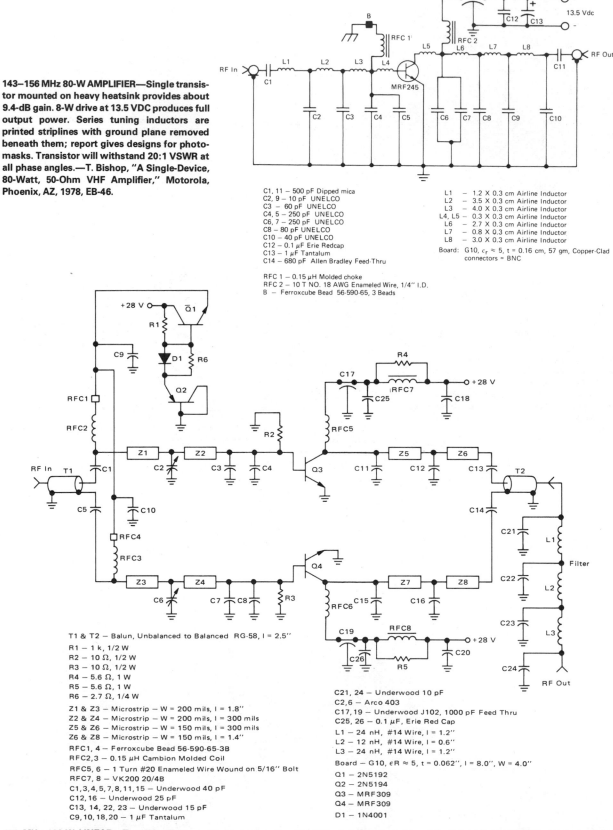

143–156 MHz 80-W AMPLIFIER—Single transistor mounted on heavy heatsink provides about 9.4-dB gain. 8-W drive at 13.5 VDC produces full output power. Series tuning inductors are printed striplines with ground plane removed beneath them; report gives designs for photomasks. Transistor will withstand 20:1 VSWR at all phase angles.—T. Bishop, "A Single-Device, 80-Watt, 50-Ohm VHF Amplifier," Motorola, Phoenix, AZ, 1978, EB-46.

C1, 11 — 500 pF Dipped mica
C2, 9 — 10 pF UNELCO
C3 — 60 pF UNELCO
C4, 5 — 250 pF UNELCO
C6, 7 — 250 pF UNELCO
C8 — 80 pF UNELCO
C10 — 40 pF UNELCO
C12 — 0.1 µF Erie Redcap
C13 — 1 µF Tantalum
C14 — 680 pF Allen Bradley Feed-Thru

RFC 1 — 0.15 µH Molded choke
RFC 2 — 10 T NO. 18 AWG Enameled Wire, 1/4" I.D.
B — Ferroxcube Bead 56-590-65, 3 Beads

L1 — 1.2 X 0.3 cm Airline Inductor
L2 — 3.5 X 0.3 cm Airline Inductor
L3 — 4.0 X 0.3 cm Airline Inductor
L4, L5 — 0.3 X 0.3 cm Airline Inductor
L6 — 2.7 X 0.3 cm Airline Inductor
L7 — 0.8 X 0.3 cm Airline Inductor
L8 — 3.0 X 0.3 cm Airline Inductor

Board: G10, $\epsilon_r \approx 5$, t = 0.16 cm, 57 gm, Copper-Clad
connectors = BNC

T1 & T2 — Balun, Unbalanced to Balanced RG-58, l = 2.5"

R1 — 1 k, 1/2 W
R2 — 10 Ω, 1/2 W
R3 — 10 Ω, 1/2 W
R4 — 5.6 Ω, 1 W
R5 — 5.6 Ω, 1 W
R6 — 2.7 Ω, 1/4 W

Z1 & Z3 — Microstrip — W = 200 mils, l = 1.8"
Z2 & Z4 — Microstrip — W = 200 mils, l = 300 mils
Z5 & Z6 — Microstrip — W = 150 mils, l = 300 mils
Z6 & Z8 — Microstrip — W = 150 mils, l = 1.4"

RFC1, 4 — Ferroxcube Bead 56-590-65-3B
RFC2,3 — 0.15 µH Cambion Molded Coil
RFC5, 6 — 1 Turn #20 Enameled Wire Wound on 5/16" Bolt
RFC7, 8 — VK200 20/4B
C1,3,4,5,7,8,11,15 — Underwood 40 pF
C12,16 — Underwood 25 pF
C13, 14, 22, 23 — Underwood 15 pF
C9,10,18,20 — 1 µF Tantalum

C21, 24 — Underwood 10 pF
C2,6 — Arco 403
C17,19 — Underwood J102, 1000 pF Feed Thru
C25, 26 — 0.1 µF, Erie Red Cap
L1 — 24 nH, #14 Wire, l = 1.2"
L2 — 12 nH, #14 Wire, l = 0.6"
L3 — 24 nH, #14 Wire, l = 1.2"
Board — G10, $\epsilon R \approx 5$, t = 0.062", l = 8.0", W = 4.0"
Q1 — 2N5192
Q2 — 2N5194
Q3 — MRF309
Q4 — MRF309
D1 — 1N4001

420–450 MHz 100-W LINEAR—Two Motorola MRF309 transistors in push-pull require only 16-W drive to deliver 100 W for transmitter applications. Circuit provides 8 dB of power gain at efficiency greater than 40% when operating from 28-V supply. Harmonic suppression inherent in push-pull operation is enhanced by seven-element low-pass filter at output. Q1 and Q2 are bias resistors and must be insulated from heatsink with mica washers. T1 and T2 are transformers constructed from RG58 coax. Use 3-inch lengths and prepare ¼ inch at each end to give total transformer length of 2½ inches.— H. Swanson and B. Tekniepe, "A 100-Watt PEP 420-450 MHz Push-Pull Linear Amplifier," Motorola, Phoenix, AZ, 1978, EB-67.

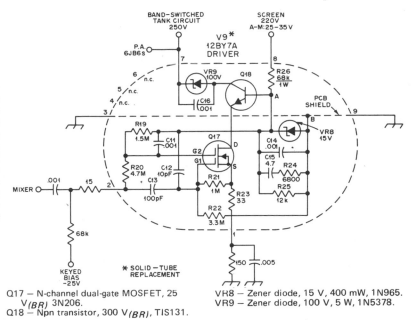

TRANSISTORS FOR DRIVER TUBES—Solid-state replacement for 12BY7A power amplifier runs much cooler than tube. Shield is required between input and output circuits. Gate 2 is biased very high and gate 1 is close to source voltage, to permit maximum signal range without changing parameters. Bypassed zener VR9 prevents 100-V collector signal swing from exceeding transistor breakdown voltage. Solid-state replacements for other tube types in Drake T-4XB transmitter are also given in article.—H. J. Sartori, Solid-Tubes—a New Life for Old Designs, *QST,* April 1977, p 45–50.

Q17 — N-channel dual-gate MOSFET, 25 V$_{(BR)}$ 3N206.
Q18 — Npn transistor, 300 V$_{(BR)}$, TIS131.

VR8 — Zener diode, 15 V, 400 mW, 1N965.
VR9 — Zener diode, 100 V, 5 W, 1N5378.

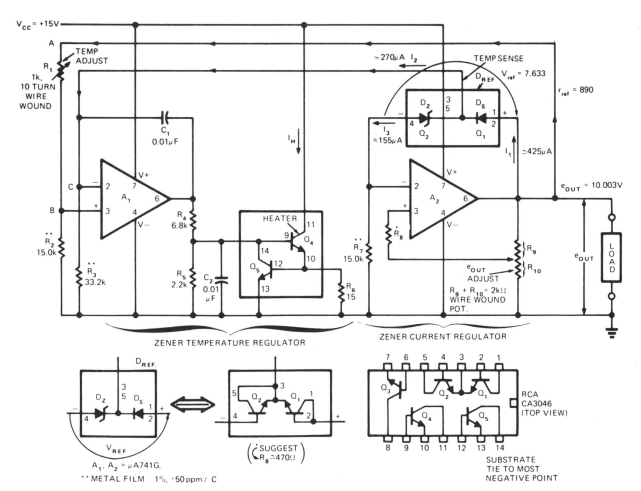

ZENER TEMPERATURE REGULATOR

ZENER CURRENT REGULATOR

A_1, A_2 = μA741G.
** METAL FILM 1%, ·50 ppm/ C

RCA CA3046 (TOP VIEW)

SUBSTRATE TIE TO MOST NEGATIVE POINT

10-V TEMPERATURE-STABILIZED—Self-regulation of substrate temperature of CA3046 five-transistor chip allows 10-V reference output voltage to rise only 0.5 mV when temperature increases from 27 to 62°C. Circuit requires only single 15-V supply. Zener-connected transistor Q_2 and diode-connected transistor Q_1 together provide temperature compensation by sensing voltage across Q_1 (D_S) and comparing it with temperature-reference voltage produced across R_1 by μA741G opamp A_1. Opamp drives Q_5 to control current through Q_4 which serves as chip heater. Opamp A_2 (μA741G) and associated components (including Q_1-Q_2 in feedback path) act as self-regulating (zener-current) voltage reference.—M. J. Shah, A Self-Regulating Temperature-Stabilized Reference, *EDN Magazine,* May 20, 1974, p 74 and 76.

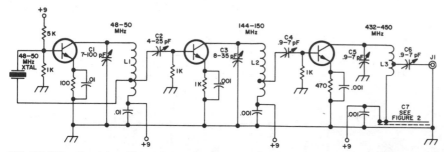

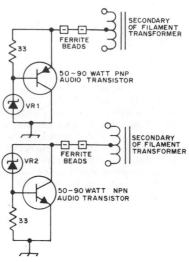

432–450 MHz—Crystal oscillator stage operating at 48–50 MHz puts out enough power to drive pair of triplers. All transistors are HEP-75. L1 is 20 turns No. 24 on 5-mm form, tapped 8 turns from cold end. L2 is 5 turns No. 20 air-wound to 8-mm diameter. L3 is 3 turns No. 20 air-wound to 5-mm diameter, with center tap. Article covers construction, adjustment, and uses.—B. Hoisington, Getting Started on 450 MHz, *73 Magazine,* Nov. 1973, p 21–24.

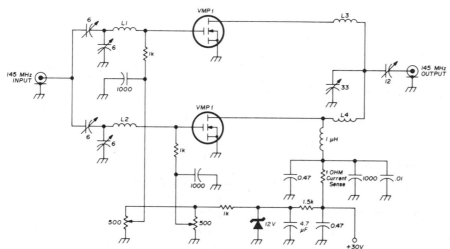

10 W ON 2 METERS—Linear power amplifier for 2-meter transverter delivers 10 W PEP using two Siliconix VMP1 Mospower FETs. L1 and L2 are 8 turns, and L3 and L4 are 5 turns, all close-wound with No. 20 enamel on 3-mm form. Transistor requires heatsink, insulated from chassis

AMPLIFIED ZENER—Combination of 1-W zener and 50–90 W audio transistor replaces 50-W zener in developing bias for high-power tube-type linear amplifier. Voltage rating of zener should be about 0.3 V less than desired bias voltage if using germanium transistor and about 0.7 V less for silicon transistor. Connections are shown for PNP and NPN transistors. Use chassis as heatsink for transistor, with mica insulating washer for NPN. Ferrite beads discourage parasitic oscillations.—An Alternative to High-Wattage Zener Diodes, *QST,* June 1975, p 45.

with 0.062-in beryllium oxide insulators. Efficiency is about 40%. —L. Leighton, Two-Meter Transverter Using Power FETs, *Ham Radio,* Sept. 1976, p 10–15.

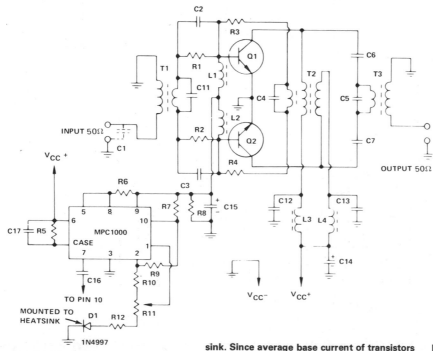

C1 - 100 pF
C2, C3 - 5600 pF
C4, C5 - 680 pF
C6, C7 - .10 μF
C11 - 470 pF
C12, C13 - .33 μF
C14 - 10 μF – 50 V electrolytic
C15 - 500 μF - 3 V electrolytic
C16 - 1000 pF
C17 - .1 μF
R1, R2 - 2 X 3.3Ω , 1/2-W in parallel
R3, R4 - 2 X 3.9Ω , 1/2-W in parallel
R5 - 47Ω, 5 W
R6 - 1.0Ω , 1/2 W
R7 - 1.0 k, 1/4 W
R8 - 100Ω, 1/4 W
R9 - 18 k, 1/4 W
R10 - 8.2 k, 1/4 W
R11 - 1.0 k Trimpot
R12 - 180Ω , 1/4 W
L1, L2 - Ferroxcube
VK200 20/4B
L3, L4 — 6 ferrite beads
each, Ferroxcube
56590 65/3B

2–30 MHz 300-W LINEAR—Motorola MRF422 high-power transistors connected in push-pull provide 300 W of PEP or CW output power across band. Uses MPC1000 regulator rated for 10 A and dissipation of 100 W with proper heat-sink. Since average base current of transistors is less than 500 mA, however, regulator can be used without heatsink. T1 and T3 have 9:1 impedance ratio, obtained with ⅛-inch copper-braid secondary through which 3 turns of No. 22 are wound for primary on Stackpole dual balun ferrite core 57-1845-24B. T2 has 5 turns of two twisted pairs No. 22 wound on Stackpole 57-9322 toroid.—H. Granberg, "Get 300 Watts PEP Linear Across 2 to 30 MHz from This Push-Pull Amplifier," Motorola, Phoenix, AZ, 1978, EB-27.

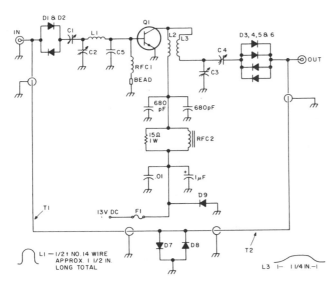

2-METER POWER AMPLIFIER—Provides 10-dB gain with full 30-W output at 160 MHz and about 0.5-dB more gain at 150 MHz, using Motorola MRF238 transistor. C1-C4 are Arco 463, 464, or 424. RFC1 is 10 turns No. 20 on 270-ohm ½-W resistor. C5 is three 90-pF silver mica in parallel. RFC2 is 6 to 8 turns No. 18 on toroid. L1 is ½ turn No. 14 1½ inch long. L2 is 4 turns No. 14 spaced on ¼ inch diameter. L3 is 1¼-inch curve of No. 14. D1-D8 are 1N4148. T1 and T2 are one quarter-wavelength of RG-174 or similar 50-ohm coax. D9 is 2-A silicon rectifier.—D. J. Lynch, Build a 2m Power Amp, *73 Magazine*, Nov. 1977, p 96—97.

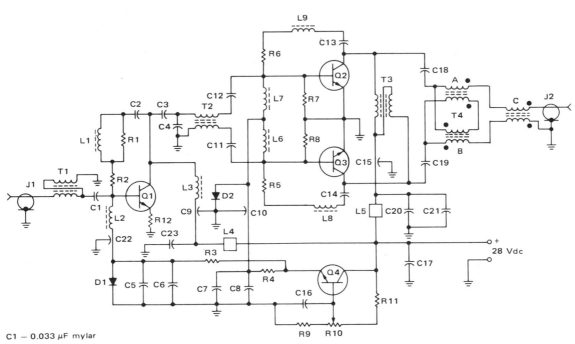

C1 — 0.033 μF mylar

C2, C3 — 0.01 μF mylar

C4 — 620 pF dipped mica

C5, C7, C16 — 0.1 μF ceramic

C6 — 100 μF/15 V electrolytic

C8 — 500 μF/6 V electrolytic

C9, C10, C15, C22 — 1000 pF feed through

C11, C12 — 0.01 μF

C13, C14 — 0.015 μF mylar

C17 — 10 μF/35 V electrolytic

C18, C19, C21 — Two 0.068 μF mylars in parallel

C20 — 0.1 μF disc ceramic

C23 — 0.1 μF disc ceramic

R1 — 220 Ω, 1/4 W carbon

R2 — 47 Ω, 1/2 W carbon

R3 — 820 Ω, 1 W wire W

R4 — 35 Ω, 5 W wire W

R5, R6 — Two 150 Ω, 1/2 W carbon in parallel

R7, R8 — 10 Ω, 1/2 W carbon

R9, R11 — 1 k, 1/2 W carbon

R10 — 1 k, 1/2 W potentiometer

R12 — 0.85 Ω (6 5.1 Ω or 4 3.3 Ω 1/4 W resistors in parallel, divided equally between both emitter leads)

T1 — 4:1 Transformer, 6 turns, 2 twisted pairs of #26 AWG enameled wire (8 twists per inch)

T2 — 1:1 Balun, 6 turns, 2 twisted pairs of #24 AWG enameled wire (6 twists per inch)

T3 — Collector choke, 4 turns, 2 twisted pairs of #22 AWG enameled wire (6 twists per inch)

T4 — 1:4 Transformer Balun, A&B — 5 turns, 2 twisted pairs of #24, C — 8 turns, 1 twisted pair of #24 AWG enameled wire (All windings 6 twists per inch). (T4 — Indiana General F624-19Q1, — All others are Indiana General F627-8Q1 ferrite toroids or equivalent.)

PARTS LIST

L1 — .33 μH, molded choke	Q1 — 2N6370
L2, L6, L7 — 10 μH, molded choke	Q2, Q3 — 2N5942
L3 — 1.8 μH (Ohmite 2-144)	Q4 — 2N5190
L4, L5 — 3 ferrite beads each	
L8, L9 — .22 μH, molded choke	D1 — 1N4001
	D2 — 1N4997
	J1, J2 — BNC connectors

160-W LINEAR SSB—Designed for operation at fixed land location, using 28-VDC supply. Circuit covers 3–30 MHz band, using driver stage to provide total power gain of about 30 dB. If heat-sinks are used, cooling fans are not normally required because average power for speech operation is about 15 dB below peak levels.—H. Granberg, "Broadband Linear Power Amplifiers Using Push-Pull Transistors," Motorola, Phoenix, AZ, 1974, AN-593, p 3.

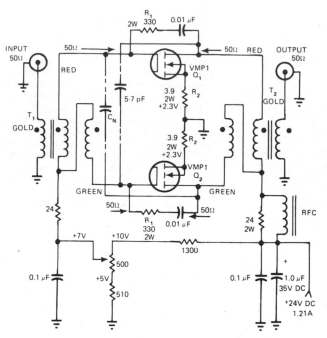

VMOS 8-W BROADBAND—Linear power amplifier provides 15-dB gain over entire range of 2 to 100 MHz. Negative feedback stabilizes gain and gives 50-ohm resistive input and output impedances. VSWR is 2:1 or less over frequency range. Can be used as low-power amplifier or driver for amateur radio transmitters, for boosting power level of standard signal generators, and as CB amplifier (with reduced supply voltage). Q_1 and Q_2 (Siliconix VMP1) combine with feedback resistors R_1 and R_2 to form separate broadband amplifiers, each delivering up to 5 W with 15-dB gain.—G. D. Frey, VMOS Power Amplifiers—This Broadband Circuit Outputs 8W with a 15 dB Gain, *EDN Magazine*, Sept. 5, 1977, p 83–85.

NOTES: T_1:   FERROXCUBE 0.375 IN. O.D. 3E2A FERRITE TOROID.
WIND SEVEN TURNS TRIFILAR #30, UNWIND ONE
TURN FROM EACH END OF RED AND GREEN WINDINGS.
CROSS CONNECT "FINISH" RED TO "START" GREEN
WIRE FOR CENTER TAP. USE CENTER "GOLD" WINDING FOR
50Ω UNBALANCED PORT.

           T_2:   STACKPOLE 57-9130 SLEEVE BALUN CORE CERAMAG
GRADE 11. SIMILAR CONSTRUCTION TO T_1 (COUNT TURNS
AS ONE PASS THROUGH BOTH HOLES = ONE TURN).

    RFC: STACKPOLE 57-9130 5T #30 WIRE (L ≈ 7.0 μH)

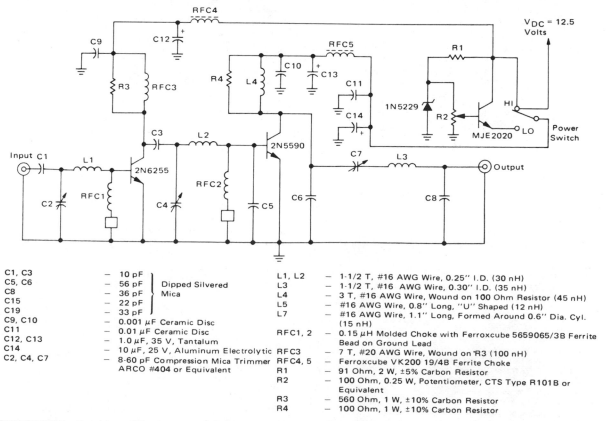

C1, C3	— 10 pF		
C5, C6	— 56 pF	Dipped Silvered	
C8	— 36 pF	Mica	
C15	— 22 pF		
C19	— 33 pF		
C9, C10	— 0.001 μF Ceramic Disc		
C11	— 0.01 μF Ceramic Disc		
C12, C13	— 1.0 μF, 35 V, Tantalum		
C14	— 10 μF, 25 V, Aluminum Electrolytic		
C2, C4, C7	— 8-60 pF Compression Mica Trimmer ARCO #404 or Equivalent		

L1, L2	— 1-1/2 T, #16 AWG Wire, 0.25" I.D. (30 nH)
L3	— 1-1/2 T, #16 AWG Wire, 0.30" I.D. (35 nH)
L4	— 3 T, #16 AWG Wire, Wound on 100 Ohm Resistor (45 nH)
L5	— #16 AWG Wire, 0.8" Long, "U" Shaped (12 nH)
L7	— #16 AWG Wire, 1.1" Long, Formed Around 0.6" Dia. Cyl. (15 nH)
RFC1, 2	— 0.15 μH Molded Choke with Ferroxcube 5659065/3B Ferrite Bead on Ground Lead
RFC3	— 7 T, #20 AWG Wire, Wound on R3 (100 nH)
RFC4, 5	— Ferroxcube VK200 19/4B Ferrite Choke
R1	— 91 Ohm, 2 W, ±5% Carbon Resistor
R2	— 100 Ohm, 0.25 W, Potentiometer, CTS Type R101B or Equivalent
R3	— 560 Ohm, 1 W, ±10% Carbon Resistor
R4	— 100 Ohm, 1 W, ±10% Carbon Resistor

10-W MARINE-BAND—Power amplifier operating in class C from 12.5-VDC supply is designed for 152–162 MHz VHF marine band. Switch permits reducing power output to 1 W or less. Tuning range of 144–175 MHz makes amplifier suitable for other applications such as amateur 2-meter and land-mobile radio. Power input is 180 mW, power gain is 17.4 dB, and efficiency is 44.5%.—J. Hatchett, "25-Watt and 10-Watt Marine Band Transmitters," Motorola, Phoenix, AZ, 1978, AN-595, p 4.

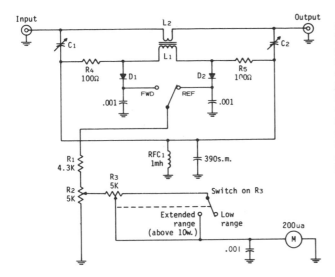

RF WATTMETER—Calibration is accurate on all HF bands because circuit is not frequency-sensitive. Sensitivity depends on meter movement, number of turns in primary coil, and resistive voltage divider. With values shown, pots can be adjusted for full-scale values from 1–14 W. C_1 and C_2 are 3–20 pF. Diodes are 1N34A, 1N60, or equivalent. L_1 is 46 turns No. 28 on Amidon T-50-2 toroid, with 2 turns No. 22 between ends of L_1 for L_2. To adjust, connect resistive dummy load to one coax receptacle and RF power source to other, with R_2 at maximum resistance. Place upper switch in position providing highest meter reading, and make that the FWD position. Switch to other position, which becomes REF, and adjust C_1 for null reading. Reverse RF source and load, leaving switch at FWD, and adjust C_2 for null. Wattmeter can now be calibrated.—A. Weiss, QRP Low-Low Power Operating, *CQ*, Jan. 1974, p 42–44 and 80.

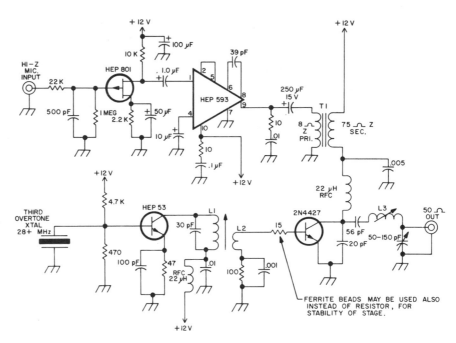

20-MHz PHONE—Colpitts oscillator using HEP 53 provides excellent stability with third-overtone crystal. Power amplifier stage uses 2N4427 in class C common-emitter stage modulated through collector circuit, to develop about 1.25-W output at 28 MHz. HEP 801 FET microphone amplifier provides high-impedance input for crystal microphone and drives HEP 593 IC to give about 1-W AF output. Article covers construction and tune-up.—B. Johnston, Little Bill, *73 Magazine*, July 1974, p 63–64 and 66–67.

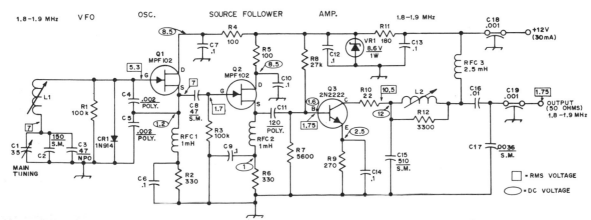

1.8–1.9 MHz VFO—Series-tuned Clapp oscillator using high-impedance JFET Q1 has good frequency stability. Diode stabilizes bias. Air variable C1 provides frequency spread of exactly 100 kHz. L1 is 25–58 μH slug-tuned (Miller 43A475CBI). L2 is 10–18.7 μH slug-tuned (Miller 23A155RPC).—D. DeMaw, More Basics on Solid-State Transmitter Design, *QST*, Nov. 1974, p 22–26 and 34.

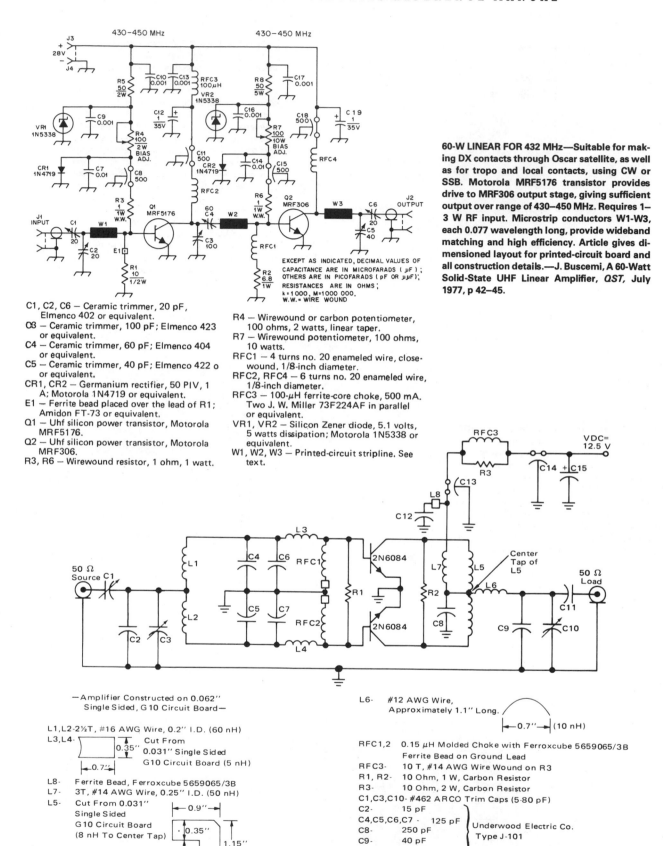

C1, C2, C6 — Ceramic trimmer, 20 pF, Elmenco 402 or equivalent.

C3 — Ceramic trimmer, 100 pF; Elmenco 423 or equivalent.

C4 — Ceramic trimmer, 60 pF; Elmenco 404 or equivalent.

C5 — Ceramic trimmer, 40 pF; Elmenco 422 o or equivalent.

CR1, CR2 — Germanium rectifier, 50 PIV, 1 A; Motorola 1N4719 or equivalent.

E1 — Ferrite bead placed over the lead of R1; Amidon FT-73 or equivalent.

Q1 — Uhf silicon power transistor, Motorola MRF5176.

Q2 — Uhf silicon power transistor, Motorola MRF306.

R3, R6 — Wirewound resistor, 1 ohm, 1 watt.

R4 — Wirewound or carbon potentiometer, 100 ohms, 2 watts, linear taper.

R7 — Wirewound potentiometer, 100 ohms, 10 watts.

RFC1 — 4 turns no. 20 enameled wire, close-wound, 1/8-inch diameter.

RFC2, RFC4 — 6 turns no. 20 enameled wire, 1/8-inch diameter.

RFC3 — 100-µH ferrite-core choke, 500 mA. Two J. W. Miller 73F224AF in parallel or equivalent.

VR1, VR2 — Silicon Zener diode, 5.1 volts, 5 watts dissipation; Motorola 1N5338 or equivalent.

W1, W2, W3 — Printed-circuit stripline. See text.

60-W LINEAR FOR 432 MHz—Suitable for making DX contacts through Oscar satellite, as well as for tropo and local contacts, using CW or SSB. Motorola MRF5176 transistor provides drive to MRF306 output stage, giving sufficient output over range of 430–450 MHz. Requires 1–3 W RF input. Microstrip conductors W1-W3, each 0.077 wavelength long, provide wideband matching and high efficiency. Article gives dimensioned layout for printed-circuit board and all construction details.—J. Buscemi, A 60-Watt Solid-State UHF Linear Amplifier, *QST*, July 1977, p 42–45.

—Amplifier Constructed on 0.062" Single Sided, G 10 Circuit Board—

L1,L2-2½T, #16 AWG Wire, 0.2" I.D. (60 nH)

L3,L4- Cut From 0.031" Single Sided G10 Circuit Board (5 nH) / 0.35" / 0.7"

L8- Ferrite Bead, Ferroxcube 5659065/3B

L7- 3T, #14 AWG Wire, 0.25" I.D. (50 nH)

L5- Cut From 0.031" Single Sided G10 Circuit Board (8 nH To Center Tap) / 0.9" / 0.35" / 0.45" / 1.15" / Center Tap

L6- #12 AWG Wire, Approximately 1.1" Long. / 0.7" / (10 nH)

RFC1,2 0.15 µH Molded Choke with Ferroxcube 5659065/3B Ferrite Bead on Ground Lead

RFC3   10 T, #14 AWG Wire Wound on R3

R1, R2- 10 Ohm, 1 W, Carbon Resistor

R3-   10 Ohm, 2 W, Carbon Resistor

C1,C3,C10- #462 ARCO Trim Caps (5-80 pF)

C2   15 pF

C4,C5,C6,C7 -   125 pF

C8   250 pF   } Underwood Electric Co.

C9   40 pF   } Type J-101

C11-   30 pF

C12-   0.1 µF, 75 V, Ceramic Disc

C13,C14- 680 pF, Allen Bradley Type FA5C

C15-   5.0 µF Tantalum

144–175 MHz 80-W SINGLE-STAGE FM MOBILE—Provides rated output into 50-ohm load. Can withstand open and shorted loads for all load phase angles without transistor damage. Uses Motorola 2N6084 land-mobile transistors optimized for 12.5-V FM operation. Transistors are used in parallel with single-ended input and output, isolated from each other by signal-splitting coils.—J. Hatchett, "VHF Power Amplifiers Using Paralleled Output Transistors," Motorola, Phoenix, AZ, 1972, AN-585, p 2.

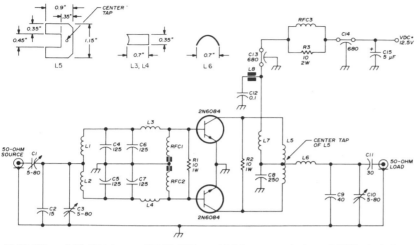

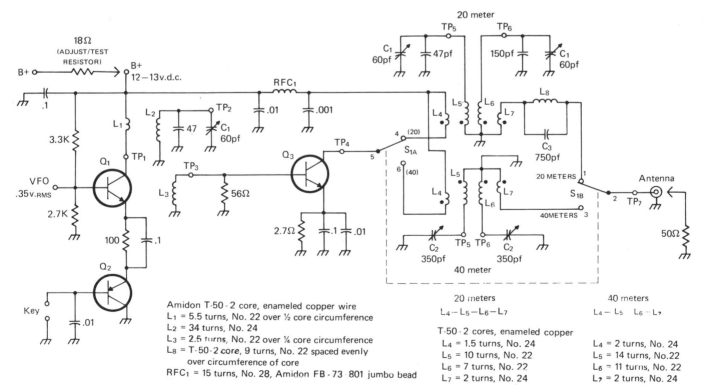

C1,C3,C10	5–80 pF trimmers (ARCO 462)		L3,L4	cut from 0.031" single-sided G10 circuit board (5 nH)
C2	15 pF metal clad (Underwood Electric type J-101*)		L5	cut from 0.031" single-sided G10 circuit board (8 nH to center tap)
C4,C5,C6,C7	125 pF metal clad (Underwood Electric type J-101)		L6	number-12 wire, approximately 1.1" long (10 nH)
C8	250 pF metal clad (Underwood Electric type J-101)		L7	3 turns number 14, 0.25" ID (50 nH)
C9	40 pF metal clad (Underwood Electric type J-101)		L8	ferrite bead (Ferroxcube 5659065/3B)
C11	30 pF metal clad (Underwood Electric type J-101)		RFC1,RFC2	0.15 μH molded choke with Ferroxcube 5659065/3B ferrite bead on ground lead
C12	0.1 μF, 75 V ceramic disc			
C13,C14	680 pF feedthrough (Allen Bradley type FA5C)		RFC3	10 turns number-14 wire wound around R3
C15	5.0 μF, 25v, aluminum electrolytic			
L1,L2	2½ turns number-16, 0.2" ID (60 nH)			

80 W ON 2 METERS—Single-stage design using two 2N6084 transistors combined with simple LC components can be tuned from 144 to 175 MHz. Typical input is 20 W for 80-W output at 144 MHz. Article shows how to add 2N6083 driver stage that reduces input drive requirement to 2.5 W. Power gain at 144 MHz is 6 dB. Article covers construction and adjustment.— J. Hatchett, A Solid 80 Watts for Two Meters, *Ham Radio,* Dec. 1973, p 6–12.

Amidon T-50-2 core, enameled copper wire
L₁ = 5.5 turns, No. 22 over ½ core circumference
L₂ = 34 turns, No. 24
L₃ = 2.5 turns, No. 22 over ¼ core circumference
L₈ = T-50-2 core, 9 turns, No. 22 spaced evenly over circumference of core
RFC₁ = 15 turns, No. 28, Amidon FB-73-801 jumbo bead

	20 meters L₄–L₅–L₆–L₇	40 meters L₄–L₅ L₆–L₇
	T-50-2 cores, enameled copper	
L₄	= 1.5 turns, No. 24	L₄ = 2 turns, No. 24
L₅	= 10 turns, No. 22	L₅ = 14 turns, No.22
L₆	= 7 turns, No. 22	L₆ = 11 turns, No. 22
L₇	= 2 turns, No. 24	L₇ = 2 turns, No. 24

1-W EXCITER FOR 7 AND 14 MHz—Developed for use in simple solid-state VFO transmitter covering 40 and 20 meters for low-power operation. Adequate drive can be provided by any 7-MHz VFO that develops 0.45 VRMS across 1000-ohm load. Circuit consists of class A buffer/amplifier Q₁ and keying switch Q₂. Q₃ is class C amplifier on 7 MHz (40 m) and frequency doubler on 14 MHz (20 m). Q₁ is MPS6514. Q₂ is 2N3906 or equivalent. Q₃ is MPS-U31. All transformer cores are Amidon T-50-2.—A. Weiss, QRP, *CQ,* Nov. 1977, p 54–58 and 88.

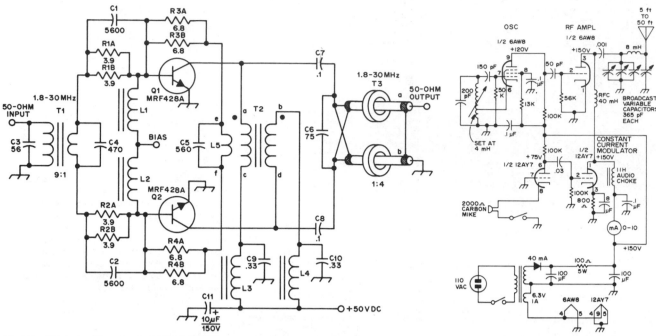

L1, L2 — Rf choke (Ferroxcube VK200-19/4B or equiv.)

L3, L4 — Rf choke (Ferroxcube 56-590-65/3B or equiv.)

T1 — Broadband 9:1 transformer on ferrite core (Stackpole 57-1845-24B or Fair-Rite Prod. 2973000201, or equiv. See text).

T2 — 7 bifilar turns of No. 20 enam wire on

Stackpole 57-9322 or Indiana General F627-8Q1 toroid core.

T3 — 14 turns Microdot 260-4118-000 25-ohm submin. coax cable (or equiv.) wound on each of two toroid cores. Cores are Stackpole 57-9074 or Indiana General F624-19Q1, or equiv.

300-W LINEAR SOLID-STATE—Class A circuit using two MRF428A transistors is emitter-ballasted to ensure even current-sharing. Requires separate 0.5–1 V regulated bias voltage source, circuit for which is given in article along with design procedure for amplifier. Second part of article (May 1976, p 28–30) tells how to combine four 300-W amplifiers to get 1-kW output for 1.8–30 MHz.—H. O. Granberg, One KW—Solid-State Style, QST, April 1976, p 11–14.

500 mW ON 180 kHz—Meets FCC requirement for amateur radio operation in 160–190 kHz band with 1-W maximum plate input power and antenna up to 50 feet long including lead-in. Working range is about 1 mile. Uses electron-coupled oscillator to minimize frequency shift during modulation. RF amplifier is self-biased. 40-mH choke in amplifier plate circuit is nearly self-resonant to 180 kHz.—C. Landahl, QRP on 180 kHz, 73 Magazine, May 1973, p 93–95.

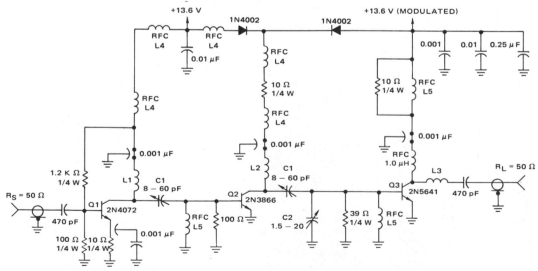

L1 = 6T #26 wire wound on toroid (micro-metals T30-13) with 3/32'' spacing

L2 = 2T #26 wire wound on toroid (see L1) with 1/8'' spacing

L3 = 2T #26 wire wound on toroid (see L1) with 5/16'' spacing

L4 = RF bead (one hole), 1/8''

L5 = Ferrite Choke (Ferroxcube VK-200)

C1 = 8-60 pF (Arco 404)

C2 = 1.5-20 pF (Arco 402)

2.5-W AIRCRAFT AM TRANSMITTER—Operates from 13.6-V supply, covers frequency range of 118–136 MHz without tuning, and has 50-ohm input and output terminations. Only three transistor stages are required. Diodes limit downward modulation to Q2. Upward modulation is 95%. Supply drain is 345 mA.—"A 13-W Broadband AM Aircraft Transmitter," Motorola, Phoenix, AZ, 1974, AN-507, p 5.

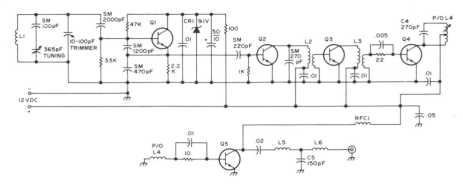

7 W FOR QRP—Operates at 7-W peak power for cutting through interference when operating on low power on any CW band from 80 to 10 meters with typical inefficient antenna systems of portable operation. Q1 is 2N709 VFO feeding 2N697 amplifier Q3 through 2N697 buffer Q2. Q4 is GE63 driver for Motorola HEP53001 final amplifier Q5. Keying can be introduced at Q2, Q3, or Q4. Article gives coil-winding data for all bands and covers construction and operation in detail.—J. Huffman, The Mini-Mite Allband QRP Rig, *73 Magazine,* July 1976, p 30–32 and 34–35.

MEASURING PEAK POWER—Addition of amplifier and rectifier circuits to Heath HM-102 or other similar RF wattmeter permits measurement of transmitter peak power output. DPDT toggle switch S2 is added to wattmeter to give choice of measurement desired. Circuit uses LM1458 or equivalent dual opamp. Current passing through 1N914 diode charges 0.1-μF capacitor at pin 3 of U1A, delaying meter return to zero long enough for reading of peak. When pointer just starts moving downward, next spoken word kicks it back up to peak value. To calibrate, set 10K pot so peak reading (S2 at PEP) is equal to normal reading (S2 at NORMAL) while using CW output of transmitter as test signal. CR1 is 2-A 50-PIV bridge rectifier. T1 has 12.6-V center-tap 100-mA secondary.—G. D. Rice, PEP Wattmeter—a la Heath, *QST,* Dec. 1976, p 30–31.

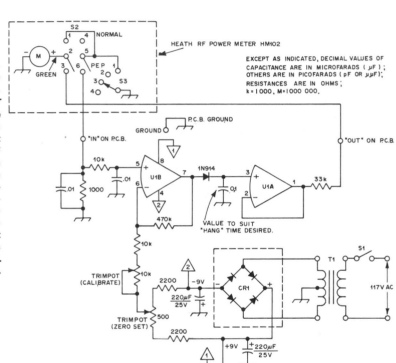

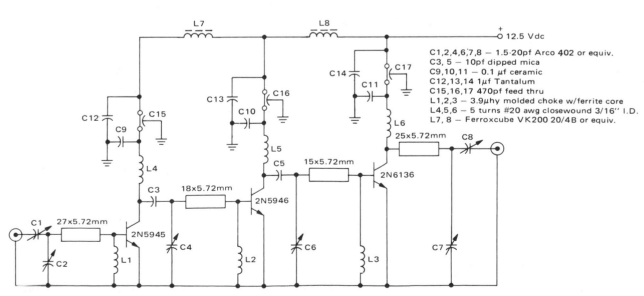

450–470 MHz AT 25 W—Power amplifier for land-mobile 12.5-V transmitter is constructed on double-sided microstrip substrate. Power gain at 470 MHz is 19.5 dB, and overall efficiency is 47%.—G. Young, "UHF Microstrip Amplifiers Utilizing G-10 Epoxy-Glass Laminate," Motorola, Phoenix, AZ, 1976, AN-578, p 4.

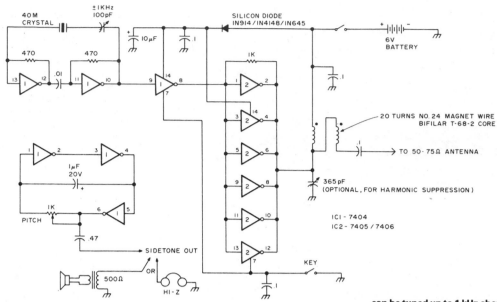

40-METER CW—Delivers about 250 mW of RF output, operating from 6-V battery. Sidetone can be monitored with high-impedance headphones or small loudspeaker. Carrier frequency can be tuned up to 1 kHz above and below nominal frequency of 40-meter crystal.—Circuits, *73 Magazine,* July 1977, p 34.

NOTES:
1. All resistors in Ohms
2. All capacitors in pf unless noted otherwise
3. All fixed value capacitors from 10 to 125 pF are Underwood Type J-101.
4. All trimmer capacitors are ARCO compression mica or equivalent.
5. Constructed on 0.062", single sided, G10, circuit board

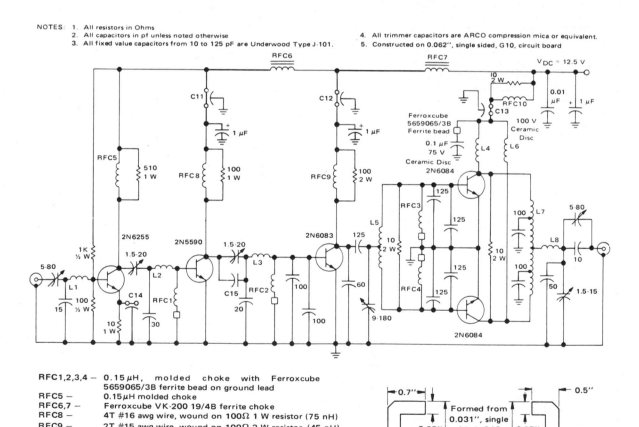

RFC1,2,3,4 — 0.15 µH, molded choke with Ferroxcube 5659065/3B ferrite bead on ground lead
RFC5 — 0.15 µH molded choke
RFC6,7 — Ferroxcube VK-200 19/4B ferrite choke
RFC8 — 4T #16 awg wire, wound on 100 Ω 1 W resistor (75 nH)
RFC9 — 2T #15 awg wire, wound on 100 Ω 2 W resistor (45 nH)
RFC10 — 10T #14 awg wire wound on 10 Ω 2 W resistor
L1,2,3 — 1T #18 awg, ¼" dia, ¾" L (25 nH)
L4,6 — 2T #15 awg wire, ¼" dia, ½" L (30 nH)
L5,7 — See outline diagram.
L8 — #12 awg wire approximately 1" Long (9 nH)
C11,12,13 — 680 pF, Allen Bradley Type FA5C
C14 — 470 pF, Allen Bradley Type SS5D
C15 — 5 pF, Dipped Silvered Mica

Outline Diagrams for Coils L5 and L7

175-MHz 80-W MOBILE FM—Uses Motorola transistors optimized for 12.5-V FM operation. All stages are class C. Signal-splitting techniques in input and output matching networks minimize problems of power output stage of unequal load sharing and of matching to extremely low impedance levels. Overall gain is 26 dB, and efficiency is 49.5%.—J. Hatchett, "Design Techniques for an 80 Watt, 175 MHz Transmitter for 12.5 Volt Operation," Motorola, Phoenix, AZ, 1972, AN-577, p 2.

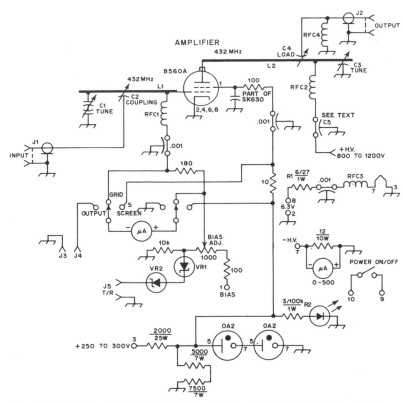

100-W LINEAR FOR 432 MHz—Medium-powered amplifier using 8560A conduction-cooled tetrode provides extra power needed for use of Oscar satellite in mode B on 432.15 MHz, with drive of only 7 W. Half-wave grid is fabricated from double-sided printed-circuit board. Capacitive probe to grid line serves for input coupling. Half-wave plate line is capacitively tuned by movable vanes. Article covers construction in detail.—T. McMullen, A Tramplifier for 432 MHz, *QST*, Jan. 1976, p 11–15.

L1 — 1-3/4 X 4-inch double-sided pc board, spaced 7/8-inch from chassis.
L2 — 3-1/2 X 6-1/4-inch double-sided pc board or aluminum strip. Length from tip of line to tube center is 7-1/8 inches.
C1 — 1.8- to 5.1-pF air variable, E. F. Johnson 160-0205-001. Mount on phenolic bracket.
C2 — 1/2-inch dia disk on center conductor of coaxial extension.

C3,C4 — Spring-brass flapper type tuning capacitors.
C5 — 2-1/2 X 4-inch pc board, single-sided, with .01-inch thick Teflon sheet for insulation to chassis. Copper-foil side mounted toward the chassis wall.
CR1 — 1/4-inch dia LED.
R1 — 27 ohm, 1-W resistor, 6 in parallel.
R2 — 100-kΩ 1-W resistor 3 in parallel.

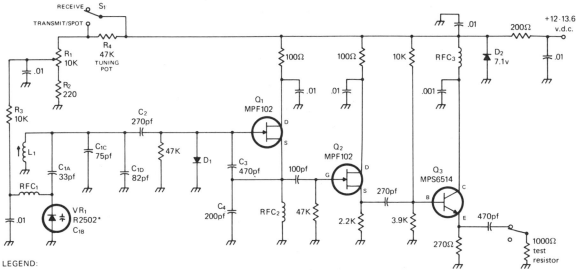

LEGEND:

   L1 = 17t No.24,.281" x .750" slug tuned form (J.W. Miller 23A014−3, green dot)
   VR1 = Motorola R2502/MV2105, 15pf varactor diode or HEP R2502
   RFC1 = 25t No. 28, Amidon FB-43-801 "jumbo bead" (or 100μh, Miller 4632-E)
   RFC2 = 14t No. 28, Amidon FB-43-801
   RFC3 = 22t No.28, Amidon FB-43-801
   D1 = 1N456, 1N914 or similar switching diode
   D2 = 6.8/9.1v. Zener, 1N4736/1N4739 or 1N757

VFO FOR 7 AND 14 MHz—Drift rate of Seiler variable-frequency oscillator can be less than 100 Hz if reasonable care is taken in board design and parts selection. Oscillator Q_1 is followed by buffer stages Q_2 and Q_3. Tuning control R_1 varies DC voltage applied to varactor tuning diode. Developed for use with 1-W exciter as solid-state transmitter for low-power (QRP) operation in 40-meter and 20-meter bands. VFO runs continuously to enhance stability; if oscillator signal leaks through receiver and interferes with desired incoming signal, switching S1 to RECEIVE puts R_4 in circuit to move oscillator from operating frequency. Offset resistor R_4 can also serve for improving bandspread on 20 meters.—A. Weiss, QRP, *CQ*, Dec. 1977, p 88–92 and 112.

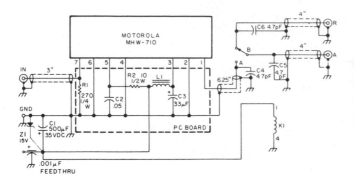

10 W AT 450 MHz—Uses Motorola MHW-710 sealed power module drawing 2.7 A on 13.8 VDC. Developed for use with fast-scan amateur TV transmitter having audio on video carrier and TR switching. Relay K1 is Archer (Radio Shack) 275-206. L1 is Ferroxcube VK200-20/4B.—B. J. Brown, Super Simple 450 MHz Rig, *73 Magazine*, Aug. 1976, p 72–75.

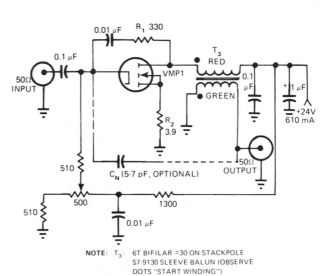

VMOS 5-W BROADBAND—Single-transistor broadband linear amplifier uses Siliconix VMP1 to provide 15-dB gain over entire frequency range of 2 to 100 MHz.—G. D. Frey, VMOS Power Amplifiers—This Broadband Circuit Outputs 8W with a 15 dB Gain, *EDN Magazine*, Sept. 5, 1977, p 83–85.

NOTE: T_3: 6T BIFILAR #30 ON STACKPOLE S7-9130 SLEEVE BALUN (OBSERVE DOTS "START WINDING")

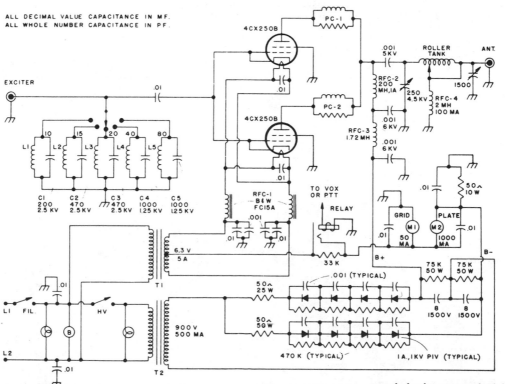

1200-W PEP POWER AMPLIFIER—Ceramic tetrodes are operated as low-mu triodes, with control grids tied to cathodes. Amplifier takes about 200-W drive. Tuned-cathode input circuit presents better load to exciter. Bias is developed through 33K cathode resistor that is shorted out by relay during operation. In standby mode, plate current is virtually zero. Article covers construction and operation, with emphasis on proper cooling of tubes. Separate tuned circuits are required for each amateur band. L1 and L2 are 0.15 μH, L3 and L4 are 0.31 μH, and L5 is 1.3 μH. PC-1 and PC-2 are 3 turns No. 16 enamel wound on 50-ohm 2-W carbon resistor.—S. W. Hochman, The Ample Amplifier, *73 Magazine*, March 1973, p 50–54.

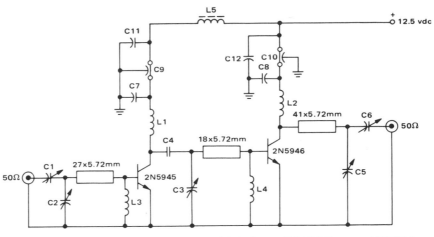

450–470 MHz AT 10 W—Power amplifier for land-mobile 12.5-V transmitter is constructed on double-sided microstrip substrate. Power gain at 470 MHz is 14.5 dB, and overall efficiency is 55%.—G. Young, "UHF Microstrip Amplifiers Utilizing G-10 Epoxy-Glass Laminate," Motorola, Phoenix, AZ, 1976, AN-578, p 3.

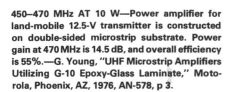

C1,2,3,5,6 — 1.5-20 pF, Arco 402 or equiv.
C4 — 10 pf dipped mica
C7, 8 — 0.1 μF ceramic
C9, 10 — 470 pf Feed thru
C11, 12 — 1 μf Tantalum

L1, 2 — 5 turns #20 AWG Closewound 3/16″ I.D.
L3, 4 — 3.9 μhy molded choke w/ferrite core
L5 — Ferroxcube V K200 20/4B or equiv.
Board is 1/16″ thick epoxy-glass
"G-10" Dielectric with 1oz copper on both sides

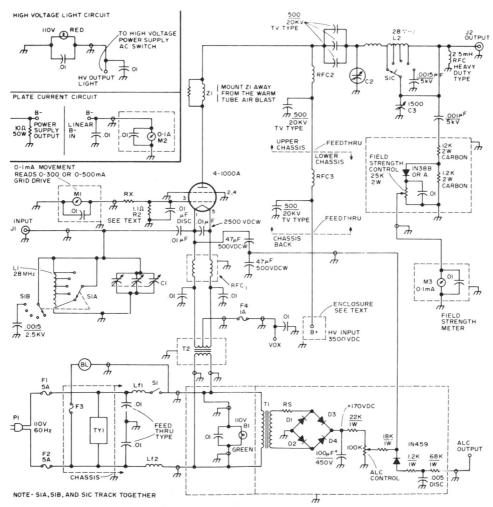

1-kW GROUNDED GRID—Class B linear amplifier for amateur transmitter can be switched to any band from 80 through 10 meters. Do not exceed 200-mA grid drive, and do not apply full excitation without plate voltage. Tube requires blower for air cooling. Article covers construction and adjustment.—E. Hartz, 4-1000 A Grounded Grid Linear, *73 Magazine,* July 1974, p 17, 19–20, 22–24, and 26.

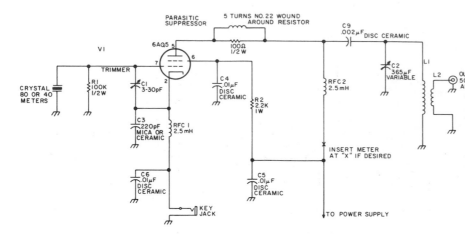

5 W ON 80 OR 40 m—Single 6AQ5 operating from 200-V supply can cover all states and some DX with CW on 40 meters as low-power amateur radio station. Antenna can be simple dipole 8 feet high. L1 is 15 turns No. 22 enamel on 1¼-inch plastic form, with 3 turns of insulated wire wound around cold end of L1 for L2. Article covers construction and operation and gives suitable 200-V voltage-doubling power supply circuit.—S. Dunn, QRP Fun on 40 and 80, *73 Magazine*, Oct. 1976, p 44–46.

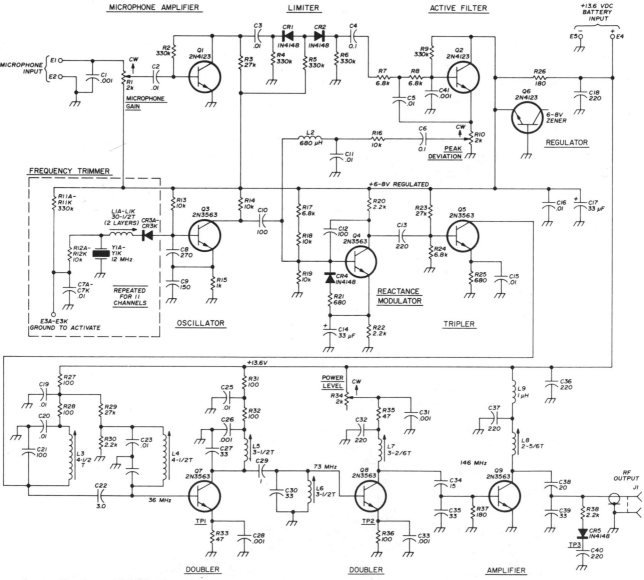

2-METER FM EXCITER—Includes both deviation and microphone gain controls. Low-pass filter following limiter eliminates raspy voice signal. Input takes either carbon or transistor-amplified dynamic microphones. Phase modulation used is suitable for multichannel operation and frequency synthesizers. Oscillator uses 12-MHz series-resonant crystals. Voltage regulator for oscillator, modulator, and audio stages minimizes effects of line-voltage variation and noise. Output power of 150–200 mW is enough to drive new TRW and Motorola RF power modules. Article gives construction and alignment details.—J. Vogt, High-Performance Two-Meter FM Exciter, *Ham Radio*, Aug. 1976, p 10–15.

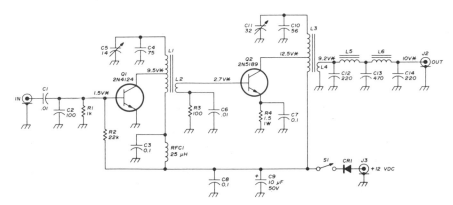

2 W FOR 20-METER CW—Motorola 2N4124 driver operates as class B amplifier. With no signal, collector current is near zero, minimizing current drain when key is up. Tank circuit of RCA 2N5189 final is similar to that of driver. Double-pi network in output assures good harmonic attenuation. RMS values of RF voltages are marked with asterisks. Protective diode CR1 is any silicon rectifier. For portable use, supply can be lantern battery.—C. E. Galbreath, Low-Power Solid-State VFO Transmitter for 20 Meters, *Ham Radio*, Nov. 1973, p 6–11.

C5	14-pF air variable (E. F. Johnson 160-0107-001)
C11	32-pF air variable (E. F. Johnson 160-0130-001)
L1	16 turns no. 24 enamelled on Amidon T-50-2 toroid core, tapped 6 turns from B+ end
L2	2 turns small insulated wire wound over B+ end of L1
L3	16 turns no. 20 enamelled on Amidon T-68-2 toroid core, tapped 3 turns from B+ end
L4	3 turns small insulated wire wound over B+ end of L3
L5	11 turns no. 20 enamelled on Amidon T-50-2 core
L6	
RFC1	25-μH rf choke (Millen J300-25)

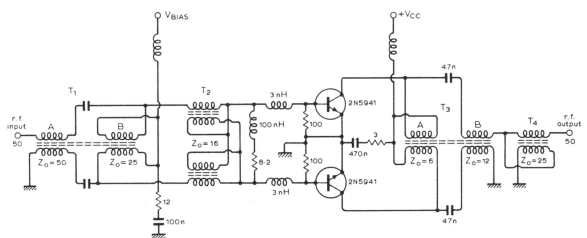

50-W PUSH-PULL—Single RF power amplifier stage uses broadband transmission-line transformers, operates between 50-ohm source and load impedances, and produces 50 W peak envelope power from 28-V supply over band of 2–30 MHz. Article gives design equations for toroid transformers.—W. P. O'Reilly, Transmitter Power Amplifier Design, *Wireless World*, Sept. 1975, p 417–422.

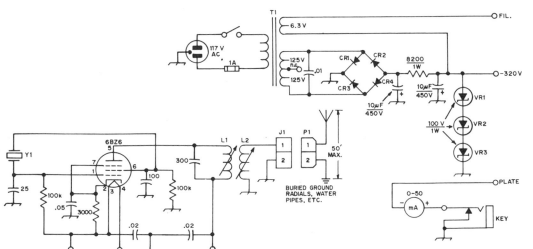

CR1-CR4, incl.	— 400 PRV, 1-A Silicon diode.
L1	— Slug-tuned inductor, 0.5 to 3 mH (J. W. Miller No. 9003).
L2	— Slug-tuned inductor, 2 to 8 mH (J. W. Miller No. 9004).
T1	— Power transformer, 250 V ct and 6.3 V ac (Stancor PS-8416 or equiv.).
VR1, VR2, VR3	— Zener diode, 100 volt.
Y1	— 160 to 190 kHz.

1 W ON 175 kHz—Simple one-tube circuit with zener-regulated power supply provides amateur CW operation in 30-kHz segment of longwave (VLF) spectrum. Adjust L1 and L2 to resonance with crystal used, then adjust coupling between them until meter between plate and ground reads correct current for legal limit of 1-W power input to final stage. Antenna is vertical mast insulated from ground, with transmitter directly at its base.—J. V. Hagan, A Crystal-Controlled Converter and Simple Transmitter for 1750-Meter Operation, *QST*, Jan. 1974, p 19–22.

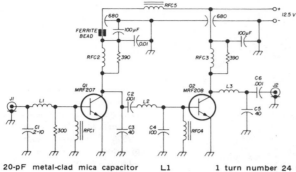

C1	20-pF metal-clad mica capacitor (El Menco MCM 01/002/-CA200DO)
C2,C6	0.001-μF metal-clad mica capacitor (El Menco MCM 01/002/-CA103DO)
C3,C5	40-pF metal-clad mica capacitor (El Menco MCM 01/002/-CA400DO)
C6	100-pF metal-clad mica capacitor (El Menco MCM 01/002/-CA102DO)

L1 L2	1 turn number 24 wire, 1/4" ID copper strap, 0.032" thick, 0.25" wide x 0.75" long
L3	0.8" lead of capacitor C6 (0.001-μF disc)
RFC1,RFC4	low-Q rf choke (Ferroxcube VK200-20/4B)
RFC5	
RFC2	2 turns no. 24 wound around 390-ohm, 1/4-watt resistor
RFC3	2 turns no. 20 wound around 390-ohm, 1-watt resistor

10 W FOR 220 MHz—Class C RF power amplifier for VHF FM transmitter has stable gain of 20 dB for operating bandwidth of 40 MHz. Article gives design procedure using Smith chart and covers construction and tune-up.—J. DuBois, 220-MHz RF Power Amplifier for VHF FM, *Ham Radio*, Sept. 1973, p 6–8.

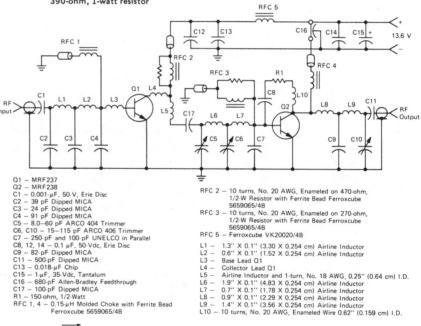

140–180 MHz AT 30 W—Two-transistor amplifier provides gain of over 20 dB for VHF marine, amateur, and commercial transmitters. Trimmers are tuned for peak output at center frequency in 10-MHz range of interest. Will operate into 30:1 mismatch without damage.—H. Burger and T. Bishop, "Two VHF Highband Gain Blocks Form 20-dB, 30-Watt Amplifier Chain," Motorola, Phoenix, AZ, 1975, EB-53.

Q1 — MRF237
Q2 — MRF238
C1 — 0.001-μF, 50-V, Erie Disc
C2 — 39 pF Dipped MICA
C3 — 24 pF Dipped MICA
C4 — 91 pF Dipped MICA
C5 — 8.0–60 pF ARCO 404 Trimmer
C6, C10 — 15–115 pF ARCO 406 Trimmer
C7 — 250-pF and 100-pF UNELCO in Parallel
C8, 12, 14 — 0.1 μF, 50-Vdc, Erie Disc
C9 — 82-pF Dipped MICA
C11 — 500-pF Dipped MICA
C13 — 0.018-μF Chip
C15 — 1-μF, 35-Vdc, Tantalum
C16 — 680-pF Allen-Bradley Feedthrough
C17 — 100-pF Dipped MICA
R1 — 150-ohm, 1/2-Watt
RFC 1, 4 — 0.15-μH Molded Choke with Ferrite Bead Ferroxcube 5659065/4B

RFC 2 — 10 turns, No. 20 AWG, Enameled on 470-ohm, 1/2-W Resistor with Ferrite Bead Ferroxcube 5659065/4B
RFC 3 — 10 turns, No. 20 AWG, Enameled on 270-ohm, 1/2-W Resistor with Ferrite Bead Ferroxcube 5659065/4B
RFC 5 — Ferroxcube VK20020/4B
L1 — 1.3" X 0.1" (3.30 X 0.254 cm) Airline Inductor
L2 — 0.6" X 0.1" (1.52 X 0.254 cm) Airline Inductor
L3 — Base Lead Q1
L4 — Collector Lead Q1
L5 — Airline Inductor and 1-turn, No. 18 AWG, 0.25" (0.64 cm) I.D.
L6 — 1.9" X 0.1" (4.83 X 0.254 cm) Airline Inductor
L7 — 0.7" X 0.1" (1.78 X 0.254 cm) Airline Inductor
L8 — 0.9" X 0.1" (2.29 X 0.254 cm) Airline Inductor
L9 — 1.4" X 0.1" (3.56 X 0.254 cm) Airline Inductor
L10 — 10 turns, No. 20 AWG, Enameled Wire 0.62" (0.159 cm) I.D.

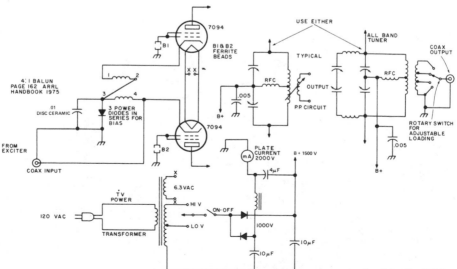

400-W PUSH-PULL—Grounded-grid linear push-pull power amplifier requires no neutralizing, uses balun for push-pull excitation, and can feed either one-band or all-band tuner for amateur radio bands. Balun is 8 turns of 72-ohm twin-line wound on Amidon 2-inch toroid core to give 4:1 ratio. Article covers construction and adjustment.—B. Baird, Build This Inexpensive 400 Watt Amplifier, *73 Magazine*, Holiday issue 1976, p 22–23.

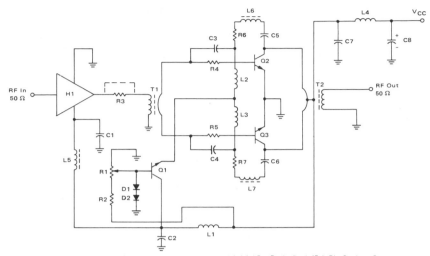

1.6–30 MHz 20-W HIGH-GAIN DRIVER—Broadband amplifier operating from 12-V supply uses Motorola MRF433 power transistors for class AB operation and MHW591 as predriver. For class A operation, power transistors should be MRF426. Q2 does not require heatsink because its peak dissipation is under 1 W. Power gain is 55 dB well beyond four-octave band of amplifier, and input VSWR is under 1.2.—H. O. Granberg, "Low-Distortion 1.6 to 30 MHz SSB Driver Designs," Motorola, Phoenix, AZ, 1977, AN-779, p 7.

L1, L4, L5 — Ferrite Beads (Fair-Rite Products Corp. #2643000101 or Ferroxcube #56 590 65/3B or equivalent)
L2, L3 — 10 μH Molded Choke
L6, L7 — 0.1 μH Molded Choke
Q1 — MJE240
Q2, Q3 — MRF433
H1 — MHW591
T1, T2 — 4:1 and 1:4 Impedance Transformers, respectively. (See discussion on transformers.) Ferrite Beads are Fair-Rite #2643006301 or equivalent)

R1 — 1 Ohm Trimpot
R2 — 1 k Ohm, 1/4 W
R3 — Optional
R4, R5 — 5. 6 Ohms, 1/4 W
R6, R7 — 47 Ohms, 1/4 W
C1, C2, C5, C6, C7 — 0.01 μF Chip
C3, C4 — 1800 pF Chip
C8 — 10 μF/35 V Electrolytic

1–2 MHz—Simple low-power AM transmitter uses low-impedance output transformer in reverse to drive 2N107 oscillator stage for short-range voice transmissions.—Circuits, *73 Magazine*, June 1977, p 49.

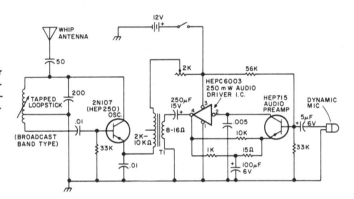

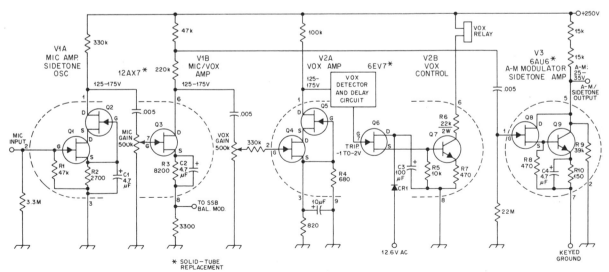

CR1 — General-purpose silicon diode, 300 PIV, 1N645 or equiv.
Q1, Q4 — N-channel JFET, 30 V$_{(BR)}$, 2N5246 or equiv.

Q2, Q3, Q5, Q8 — N-channel JFET, 300 V$_{(BR)}$, Texas Inst., A5T6449.
Q6 — N-channel JFET, 30 V$_{(BR)}$, 2N5950 or

equiv.
Q7, Q9 — Npn transistor, 300 V$_{(BR)}$, Texas Inst., A5T5058.

TRANSISTORS FOR AF TUBES—Article covers replacement of tubes in Drake T-4XB transmitter with solid-state equivalent circuits mounted in 7-pin and 9-pin miniature plugs. Numbers identify original tube pins. V1A and V2A use dual-cascode JFETs, while voice-operated transmitter relay control V2B and AM modulator V3 use high-voltage Darlington. Q9 collector voltage is set at 150 V during standby by adjusting R9. Circuit includes first and second audio stages. Voltages indicate proper operating points. Source resistors may require adjustment.—H. J. Sartori, Solid-Tubes—a New Life for Old Designs, *QST*, April 1977, p 45–50.

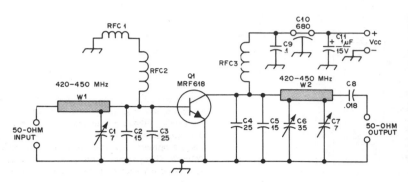

15-W POWER AMPLIFIER FOR 440 MHz—Power gain of 10 dB increases effective range of amateur transmitter. Narrow-band amplifier using Motorola MRF618 internally matched 12.5-V controlled-Q transistor can be tuned from 430 to 450 MHz. Multiple L sections using 50-ohm microstrip line and mica compression variable capacitors provide input-match and collector-load transformations. Article gives printed-circuit board layout for U-shaped 0.112-inch-wide stripline inductors W1 and W2. RFC1 is ferrite bead, RFC2 is 8 turns No. 22 enamel closewound on $\frac{1}{8}$-inch form, and RFC3 is 4 turns No. 22 enamel closewound on $\frac{1}{4}$-inch form.—R. Olsen, Build This Solid-State PA for 440 MHz, *QST*, Feb. 1977, p 37–38.

PTT LATCH—Eliminates need for holding down microphone switch continuously while transmitting. U1 is inverting hex buffer, and U2 is dual CMOS D flip-flop. When S1 is depressed, input to U1A goes low and its output goes high, making U2A and U1B together turn on Q2 and energize relay K1. Q3-Q5 serve as optional silent power switch for use when relay noise is objectionable.—B. Lambing, DC Latch Circuit, *Ham Radio*, Aug. 1975, p 42–44.

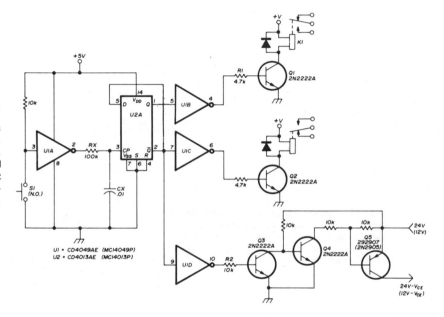

U1 = CD4049AE (MCl4049P)
U2 = CD4013AE (MCl4013P)

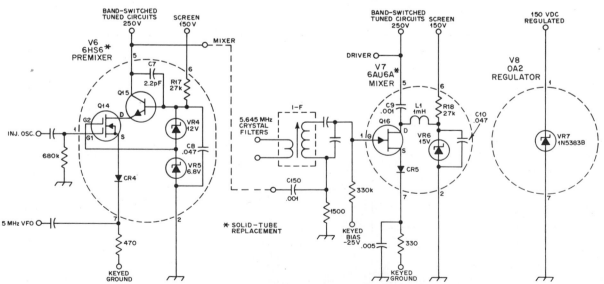

CR4-CR5 — General-purpose silicon diode, 300 PIV (1N645).
Q14 — N-channel dual-gate MOSFET, 25 V$_{(BR)}$ 3N206.

Q15 — Npn transistor 300 V$_{(BR)}$ TIS131.
Q16 — N-channel JFET 30 V$_{(BR)}$ 2N5950.
VR4 — Zener diode, 12 V, 400 mW, 1N759.
VR5 — Zener diode, 6.8 V, 400 mW, 1N754.

VR6 — Zener diode, 15 V, 400 mW, 1N965.
VR7 — Zener diode, 150 V, 5 W, Motorola 1N5383B.

TRANSISTORS FOR MIXER AND VR TUBES—Premixer V6 in Drake T-4XB transmitter is replaced by dual-cascode MOSFETs, with CR4 protecting MOSFET from keyed-ground circuit.

High-level mixer V7 operates over large dynamic range of signals coming from IF stage. Output of mixer is low level, about 4 VRMS maximum. Decoupling capacitor C9 and choke

L1 isolate JFET from high-voltage tube circuit.—H. J. Sartori, Solid-Tubes—a New Life for Old Designs, *QST*, April 1977, p 45–50.

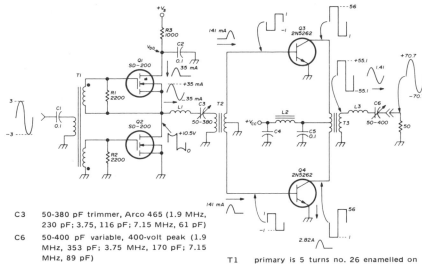

C3   50-380 pF trimmer, Arco 465 (1.9 MHz, 230 pF; 3.75, 116 pF; 7.15 MHz, 61 pF)

C6   50-400 pF variable, 400-volt peak (1.9 MHz, 353 pF; 3.75 MHz, 170 pF; 7.15 MHz, 89 pF)

L1   160 meters   63 turns no. 30 on Permacore 57-1753 core (30.5 μH)
     80 meters   44 turns no. 26 on Permacore 57-1753 core (15.4 μH)
     40 meters   36 turns no. 26 on Permacore 57-1677 core (8.1 μH)

L2   16 turns no. 26 on Permacore 56-3596 ferrite core (1.0 μH)

L3   160 meters   52 turns no. 26 on Permacore 57-1753 core (20.9 μH)
     80 meters   42 turns no. 26 on Permacore 57-1677 core (10.6 μH)
     40 meters   30 turns no. 26 on Permacore 57-1677 core (5.6 μH)

T1   primary is 5 turns no. 26 enamelled on Ferroxcube 226T125-3E2A ferrite core. Secondary windings are each 25 turns no. 26 on same core

T2   primary is 8 turns no. 20 enamelled wire wound through 6 Ceramic Magnetics CN-20 cores (two parallel stacks). Secondary is 4 turns no. 20, center-tapped, through same cores

T3   primary is 4 turns no. 20 enamelled wire, center-tapped to C5, wound through 12 Ceramic Magnetics CN-20 cores (two parallel stacks). Secondary is 4 turns no. 20 through same cores

35-W CLASS D ON 40, 80, OR 160 m—Can be used on any of three bands by changing values as set forth in parts table. Article gives circuit design procedure in detail. Power gain is about 27 dB. Almost any type of RF amplifier providing about 100 mW can be used as driver. V_S is 25 V or less, and V_{CC} is 28 V.—F. H. Raab, High-Efficiency RF Power Amplifiers, *Ham Radio,* Oct. 1974, p 8–29.

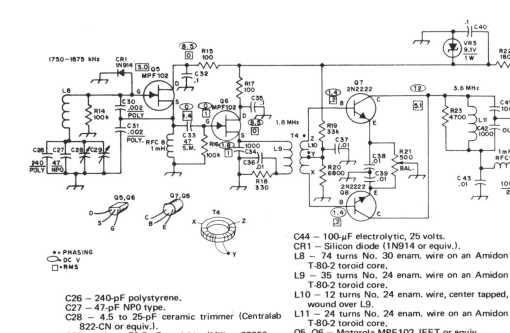

C44 — 100-μF electrolytic, 25 volts.
CR1 — Silicon diode (1N914 or equiv.).
L8 — 74 turns No. 30 enam. wire on an Amidon T-80-2 toroid core.
L9 — 35 turns No. 24 enam. wire on an Amidon T-80-2 toroid core.
L10 — 12 turns No. 24 enam. wire, center tapped, wound over L9.
L11 — 24 turns No. 24 enam. wire on an Amidon T-80-2 toroid core.
Q5, Q6 — Motorola MPF102 JFET or equiv.
Q7, Q8 — 2N2222 transistor.
R21 — 500-ohm control (Radio Shack 271-226 or equiv.).
RFC8, RFC9 — 1-mH rf choke (Millen J300-1000 or equiv.).
VR3 — Zener diode, 9.1 volt, 1 watt.

C26 — 240-pF polystyrene.
C27 — 47-pF NP0 type.
C28 — 4.5 to 25-pF ceramic trimmer (Centralab 822-CN or equiv.).
C29 — 4 to 53.5-pF variable (Millen 22050 or equiv.).
C30, C31 — .002-μF polystyrene.
C32, C35, C40 — 0.1 μF.
C33 — 47-pF silver mica.
C34, C41, C42 — .001 μF silver mica.
C36, C37, C38, C39, C43 — .01 μF.

80-METER VFO—Used in place of crystal-controlled oscillator in low-power (QRP) amateur transmitter. Tuning range is 1750–1875 kHz in 160-meter band. Colpitts oscillator uses JFET Q5 with series-tuned tank for good stability. Q6 provides isolation between oscillator and push- push class C doubler amplifier stage. Doubling gives desired 80-meter output.—D. DeMaw and J. Rusgrove, Learning to Work with Semiconductors, *QST,* Oct. 1975, p 38–42.

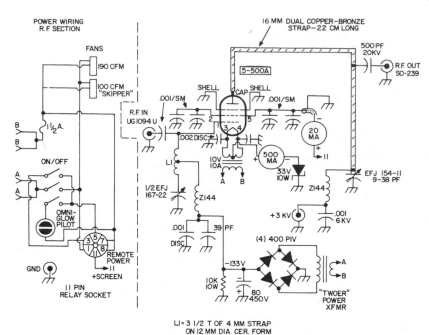

1 kW ON 2 METERS—Developed for moonbounce communication. Article covers construction, with emphasis on insulation and cooling, and gives circuit of 3-kV power supply required.—R. W. Campbell, Kilowatt Linear Amplifier for 2 Meters, *73 Magazine*, Dec. 1973, p 29–35.

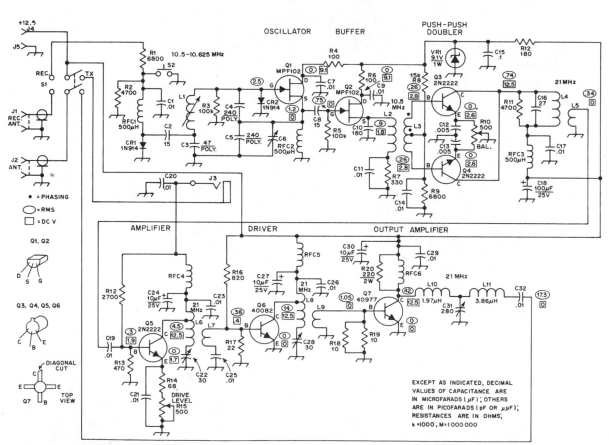

L1 — 6.05- to 12.5-μH adjustable coil (Miller 42A105CBI or equiv.).
L2 — 17 turns No. 28 enam. wire on Amidon T-50-6 core.
L3 — 10 turns No. 28 enam. wire, center tapped, wound over L2.
L4 — 17 turns No. 28 enam. wire on an Amidon T-50-6 core.

L5 — 5 turns No. 28 enam. wire wound over L4.
L6 — 30 turns No. 28 enam. wire on an Amidon T-50-6 core. Tap 10 turns above C23 end.
L7 — 4 turns No. 28 enam. wire wound over L6.
L8 — 30 turns No. 28 enam. wire on an

Amidon T-50-6 core. Tap 7 turns above C26 end.
L9 — 3 turns No. 28 enam. wire wound over L8.
L10 — 20 turns No. 22 enam. wire on an Amidon T-68-6 core.
L11 — 29 turns No. 22 enam. wire on an Amidon T-68-6 core.

21–21.25 MHz with VFO—Developed for low-power CW work in 15-meter amateur band. Colpitts oscillator Q1 runs continuously at 10.5–10.625 MHz during transmit and receive, for good frequency stability, so VFO frequency

must be shifted away from operating frequency during receive periods. Supply is 12 V at 1.3 A. C6 is 4–53.5 pF. RFC4 is 16 turns No. 28 enamel, RFC5 is 11 turns No. 22, and RFC6 is 6 turns No.

22, each on Amidon FT-50-61 core. Article covers construction and tune-up.—J. Rusgrove, A 15-Meter Goober Whistle, *QST*, Jan. 1976, p 16–19.

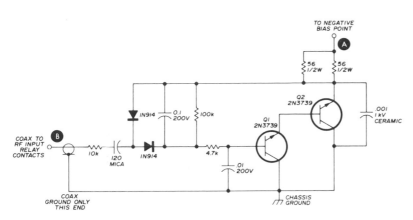

BIAS SWITCH—Automatic electronic bias switching improves efficiency of negatively biased linear class B RF power amplifier such as Heath SB-200 because no power is dissipated under no-signal conditions. Transistors are chosen to withstand maximum negative voltages switched, about −150 VDC. Capacitor across collector-base junction of Q1 can be adjusted to reduce turn-on time of switch. With no RF drive from transmitter, amplifier is biased to cutoff and plate current is zero. Switch will operate at RF threshold of about 2 V and apply class B bias voltage to amplifier. As RF drive is increased, plate current increases. With transmitter in SSB mode, plate current is zero with no speech. For speech, plate current increases with RF driving voltage.—F. E. Hinkle, Electronic Bias Switch for Negatively Biased Amplifiers, *Ham Radio*, Nov. 1976, p 27–29.

2–30 MHz SSB DRIVER—Two-stage complementary-symmetry amplifier combines single-ended impedance matching with high-gain push-pull design to provide up to 25 W PEP for driver applications. Provides good harmonic rejection and low intermodulation distortion. Supply voltage range is 22–30 V. Low-impedance windings of T1 and T2 use 1 turn of copper braid, with 2 turns No. 22 for primary of T1 and 4 turns No. 22 for secondary of T2.—H. Granberg, "A Complementary Symmetry Amplifier for 2-30 MHz SSB Driver Applications," Motorola, Phoenix, AZ, 1975, EB-32.

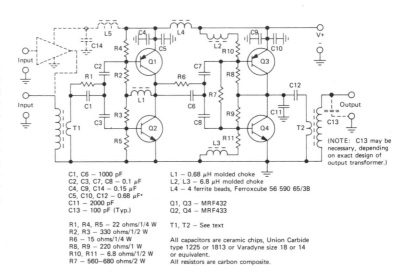

C1, C6 — 1000 pF
C2, C3, C7, C8 — 0.1 μF
C4, C9, C14 — 0.15 μF
C5, C10, C12 — 0.68 μF*
C11 — 2000 pF
C13 — 100 pF (Typ.)

L1 — 0.68 μH molded choke
L2, L3 — 6.8 μH molded choke
L4 — 4 ferrite beads, Ferroxcube 56 590 65/3B

Q1, Q3 — MRF432
Q2, Q4 — MRF433

R1, R4, R5 — 22 ohms/1/4 W
R2, R3 — 330 ohms/1/2 W
R6 — 15 ohms/1/4 W
R8, R9 — 220 ohms/1 W
R10, R11 — 6.8 ohms/1/2 W
R7 — 560–680 ohms/2 W

T1, T2 — See text

All capacitors are ceramic chips, Union Carbide type 1225 or 1813 or Varadyne size 18 or 14 or equivalent.
All resistors are carbon composite.

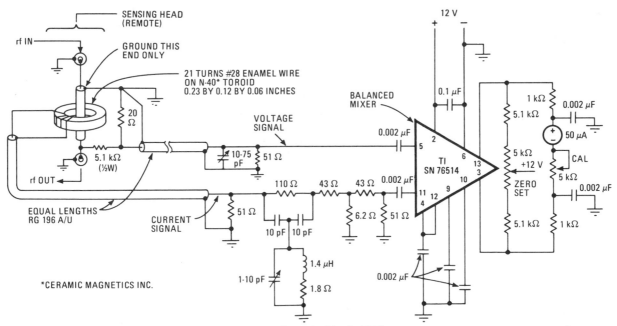

*CERAMIC MAGNETICS INC.

50-W RF—Direct-reading RF wattmeter developed for use at 27.12 MHz is accurate to within 1% of full scale. Circuit can be adapted for other frequencies up to about 100 MHz. Does not require subtraction of two readings to find power transferred to mismatched loads. RF line current and voltage are sensed by current transformer and voltage divider that can be remotely located. Meter is driven by IC balanced mixer functioning as four-quadrant analog multiplier. Average product of voltage and current appears as DC reading on microammeter.—F. C. Gabriel, Compact RF Wattmeter Measures up to 50 Watts, *Electronics*, Nov. 8, 1973, p 122.

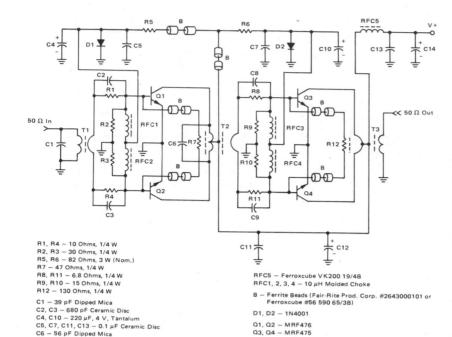

R1, R4 — 10 Ohms, 1/4 W
R2, R3 — 30 Ohms, 1/4 W
R5, R6 — 82 Ohms, 3 W (Nom.)
R7 — 47 Ohms, 1/4 W
R8, R11 — 6.8 Ohms, 1/4 W
R9, R10 — 15 Ohms, 1/4 W
R12 — 130 Ohms, 1/4 W

C1 — 39 pF Dipped Mica
C2, C3 — 680 pF Ceramic Disc
C4, C10 — 220 μF, 4 V, Tantalum
C5, C7, C11, C13 — 0.1 μF Ceramic Disc
C6 — 56 pF Dipped Mica
C8, C9 — 1200 pF Ceramic Disc
C12, C14 — 10 μF, 25 V Tantalum

RFC5 — Ferroxcube V K200 19/4B
RFC1, 2, 3, 4 — 10 μH Molded Choke
B — Ferrite Beads (Fair-Rite Prod. Corp. #2643000101 or
    Ferroxcube #56 590 65/3B)
D1, D2 — 1N4001
Q1, Q2 — MRF476
Q3, Q4 — MRF475
T1, T2 — 4:1 Impedance Transformer
T3 — 1:4 Impedance Transformer

1.6–30 MHz 20-W LINEAR DRIVER—Broadband amplifier using inexpensive plastic RF power transistors provides total power gain of about 25 dB for driving SSB transmitter power amplifiers to levels up to several hundred watts. Supply is 13.6 V. Circuit is stable even with load mismatches of 10:1.—H. O. Granberg, "Low-Distortion 1.6 to 30 MHz SSB Driver Designs," Motorola, Phoenix, AZ, 1977, AN-779, p 3.

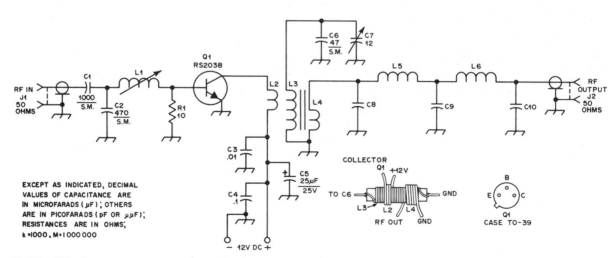

EXCEPT AS INDICATED, DECIMAL
VALUES OF CAPACITANCE ARE
IN MICROFARADS (μF); OTHERS
ARE IN PICOFARADS (pF OR μμF);
RESISTANCES ARE IN OHMS;
k =1000 , M =1 000 000

C8, C10 — 500 pF.
C9 — 1000 pF.
L1 — 11 turns No. 30 enamel wire spaced to occupy an entire Miller 4500-4 coil form.

L2 — 3 turns No. 22 insulated wire wound over L3.
L3 — Radio Shack choke (273-101).
L4 — 5 turns No. 22 insulated wire wound over ground end of L3.

L5, L6 — 18 turns No. 18 or 20 enamel wire wound on a 5/16-inch diameter plastic form. Space turns so that the length of each coil is 1-1/4 inches.
Q1 — Radio Shack transistor (RS2038).

40-METER 3.5-W AMPLIFIER—Designed for use with low-power (QRP) transmitter when band conditions are poor. Requires about 350-mW input. Half-wave filter at output keeps harmonics low. Use heatsink for Q1.—T. Mula, Codzila 1, *QST*, Feb. 1977, p 14–15.

CHAPTER 100

Voltage-Controlled Oscillator Circuits

Includes two-phase and quadrature oscillators, start-up control, reactance switching, remote fine tuning, and other methods of using DC control voltage to vary oscillator frequency over various portions of range from 5 Hz to 150 MHz. See also Frequency Synthesizer, Function Generator, Oscillator, Pulse Generator, Servo, and Sweep chapters.

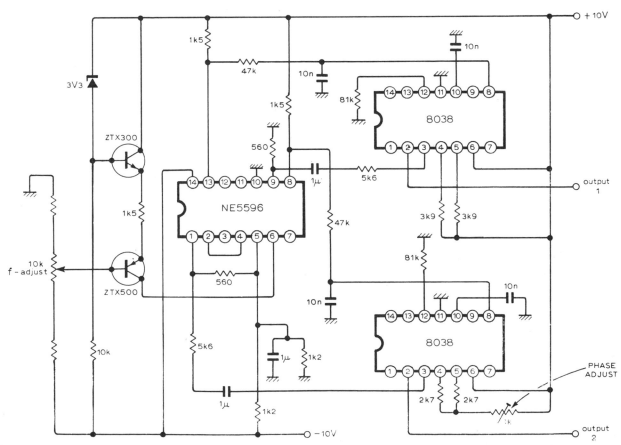

LOCKED 90° OUTPUTS—Delivers two-phase (sine and cosine) outputs locked together. Frequency can be varied over wide range by altering bias current with 10K pot that produces common-mode output voltage in NE5596 mul-tiplier IC driving 8038 ICs serving as VCOs. Triangular outputs of oscillators are fed to multiplier inputs for phase control. Lower-cost 566 VCO can be used if sinusoidal outputs are not needed. Phase error over tuning range is nom-inally zero, whereas with conventional phase-locked loop circuitry the capture range may be exceeded or phase error can be large.—J. M. Worley, Two-Phase V.C.O., *Wireless World*, Dec. 1976, p 41.

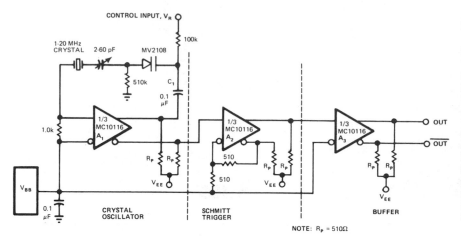

CONTROL INPUT, V_R

1-20 MHz CRYSTAL　2-60 pF　MV2108

100k

510k　C_1　0.1 μF

1.0k

1/3 MC10116 A_1

1/3 MC10116 A_2

1/3 MC10116 A_3

OUT

$\overline{OUT}$

R_P　R_P　510　R_P　R_P　R_P　R_P

V_{EE}　510　V_{EE}　V_{EE}

V_{BB}

0.1 μF

CRYSTAL OSCILLATOR　SCHMITT TRIGGER　BUFFER

NOTE: $R_P = 510\Omega$

REMOTE FINE TUNING—Addition of voltage-variable capacitance diode to crystal feedback path provides capacitance range of 50 to 12 pF with tuning voltage range of 0–30 V. Diode supplements 2–60 pF trimmer capacitor that adjusts oscillator frequency with respect to control-voltage input. Inverting input of A_1 connects to reference voltage V_{BB}, which is available on pin of MC10116 and is center voltage of output signal swing of amplifier. A_2 is connected as Schmitt trigger to give high-speed rise and fall times. Frequency deviation on either side of center is function of crystal frequency and ranges from ±50 to ±300 PPM for crystals between 1 and 20 MHz.—B. Blood, Fine-Tune This Oscillator with Voltage, *EDN Magazine*, Aug. 5, 1978, p 74.

100:1 FREQUENCY RANGE—Circuit provides good stability and excellent linearity over wide operating range. For values shown and +15 V supply, circuit transfer function is about 1 kHz/V over 100:1 frequency range, with linearity error less than 0.5%. Although circuit does not have sine-wave output, triangle output is easily converted to sine wave by filtering to remove harmonics. Article traces circuit operation.—G. Bank, A Wideband, Linear VCO, *EDN/EEE Magazine*, July 15, 1971, p 49–50.

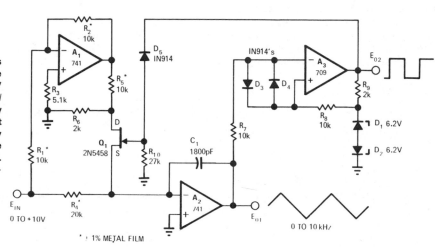

R_2^* 10k

A_1 741

R_3 5.1k

R_6 2k

R_1^* 10k

Q_1 2N5458

D_5 IN914

R_5^* 10k

D

S

R_{10} 27k

C_1 1800pF

R_7 10k

A_2 741

E_{O1}

IN914's

A_3 709

E_{02}

D_3　D_4

R_9 2k

R_8 10k

D_1 6.2V

D_2 6.2V

E_{IN} 0 TO +10V

R_4^* 20k

* 1% METAL FILM

0 TO 10 kHz

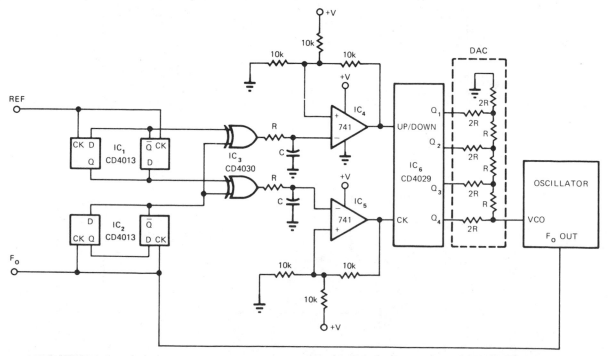

+V

10k

10k　10k

+V

REF

CK D　IC$_1$　$\overline{Q}$ CK

CD4013

Q　D

IC$_3$ CD4030

R

C

R

C

IC$_4$ 741

UP/DOWN

DAC

Q_1　2R

2R　R

Q_2　2R

R

Q_3　2R

R

Q_4　2R

OSCILLATOR

VCO

F_o OUT

D　IC$_2$　$\overline{Q}$

CK Q CD4013 D CK

F_o

IC$_5$ 741

+V

IC$_6$ CD4029

CK

10k　10k

10k

+V

FAST SYNCHRONIZING—Combination of phase splitter and DAC provides accurate synchronization of high-stability VCO with external reference frequency. IC$_1$ divides reference frequency by 4 and provides two signals 90° apart, while IC$_2$ divides VCO frequency similarly by 4. Phase relationship between outputs of IC$_3$ depends on whether VCO is higher or lower than reference, while frequency of IC$_3$ outputs depends on difference between oscillator and reference frequencies. Schmitt triggers IC$_4$ and IC$_5$ supply clock and up/down control to counter IC$_6$. If VCO frequency is low, IC$_6$ counts up at rate proportional to frequency difference and delivers increasing control voltage to VCO as required for increasing oscillator frequency.—H. W. Cooper, Oscillator Synchronizer Is Fast Acting, *EDN Magazine*, July 20, 1973, p 83–84.

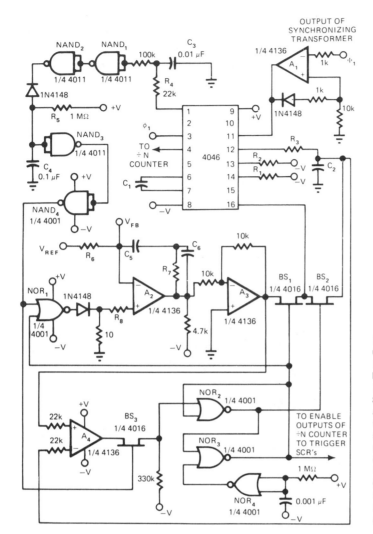

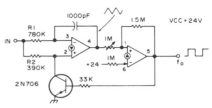

LINEAR VCO—Two sections of LM3900 quad linear opamp provide linear response for inputs of 2–12 VDC. Circuit can be adjusted with 1-megohm pot so 4-V input produces 400-Hz square wave at output, 5 V gives 500 Hz, etc. First opamp is connected as integrator and second as Schmitt trigger. When Schmitt output is high, transistor is turned on and diverts current away from noninverting input so integrator output ramps down toward ground.—C. Sondgeroth, More PLL Magic, *73 Magazine*, Aug. 1976, p 56–59.

START-UP CONTROL—Simple, smooth start-up circuit for phase-locked oscillator maintains synchronism with AC line despite presence of large transients, and maintains phase-angle limits as required for controlling firing angle of SCR. Article describes how two separate loops are used in circuit to achieve required locking with line. Values of R_6, R_7, R_8, C_5, and C_6 are chosen to meet system response time. Other unmarked values depend on operating factors; for 60-Hz line and 6X frequency multiplication by VCO, typical values are R_1 39K, R_2 27K, R_3 47K, C_1 0.1 μF, and C_2 0.22 μF.—J. C. Hanisko, Five IC's Make Ainsworth Oscillator with Start-Up Control, *EDN Magazine*, March 5, 1977, p 113 and 115.

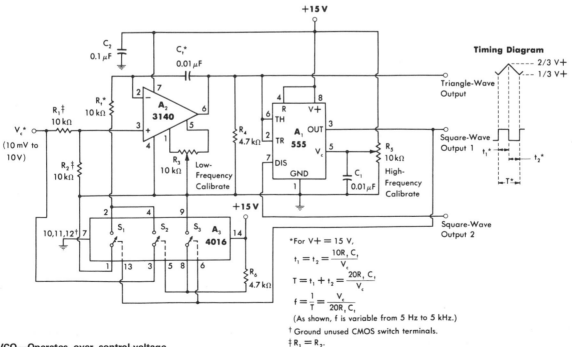

LINEAR VCO—Operates over control-voltage range of +10 mV to +10 V to provide either square or triangle outputs from 5 Hz to 5 kHz. Can be used for instrumentation or electronic music applications.—W. G. Jung, "IC Timer Cookbook," Howard W. Sams, Indianapolis, IN, 1977, p 174–179.

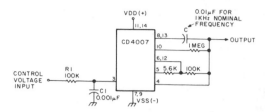

1-kHz VCO—Changes in control voltage input are used to vary nominal 1-kHz output of CD4007 CMOS voltage-controlled oscillator proportionately. Values of R and C can be changed to obtain other nominal frequencies.—W. J. Prudhomme, CMOS Oscillators, *73 Magazine*, July 1977, p 60–63.

0–10 kHz WITH 0–10 V CONTROL—CA3130 MVBR generates pulses of constant amplitude V and width T2. Average output voltage is applied to noninverting input of comparator through integrating network R3-C2. Comparator output signal from pin 6 is fed through R4 and D4 to inverting terminal 2 of A1 for adjusting MVBR interval T3 so E_{AVG} is equal to control voltage.—"Linear Integrated Circuits and MOS/FET's," RCA Solid State Division, Somerville, NJ, 1977, p 269.

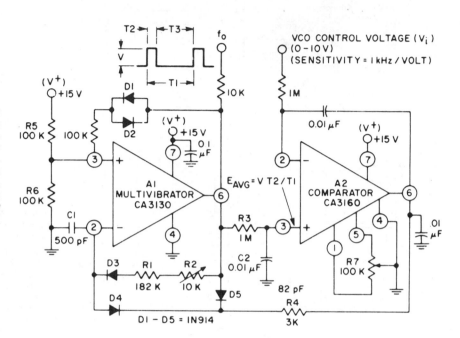

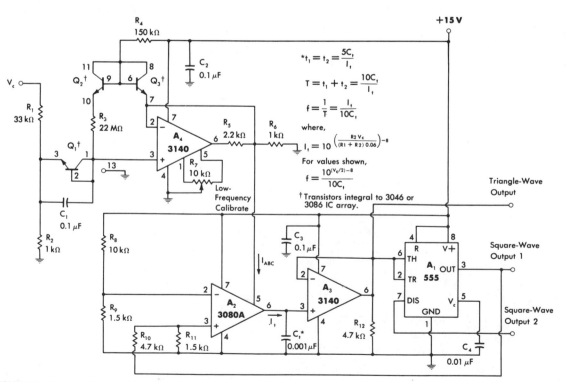

EXPONENTIAL VCO—Can be driven with linear time base of voltage and used with logarithmic frequency display, as in frequency-response tests. Useful range of circuit is four decades. Values shown give timing-current range of 10 nA to 100 μA, yielding frequency range of 1 Hz to 10 kHz. Input voltage range of 60 mV per decade is obtained from voltage divider R_1-R_2 to allow higher and more practical value for actual input voltage to circuit.—W. G. Jung, "IC Timer Cookbook," Howard W. Sams, Indianapolis, IN, 1977, p 174–179.

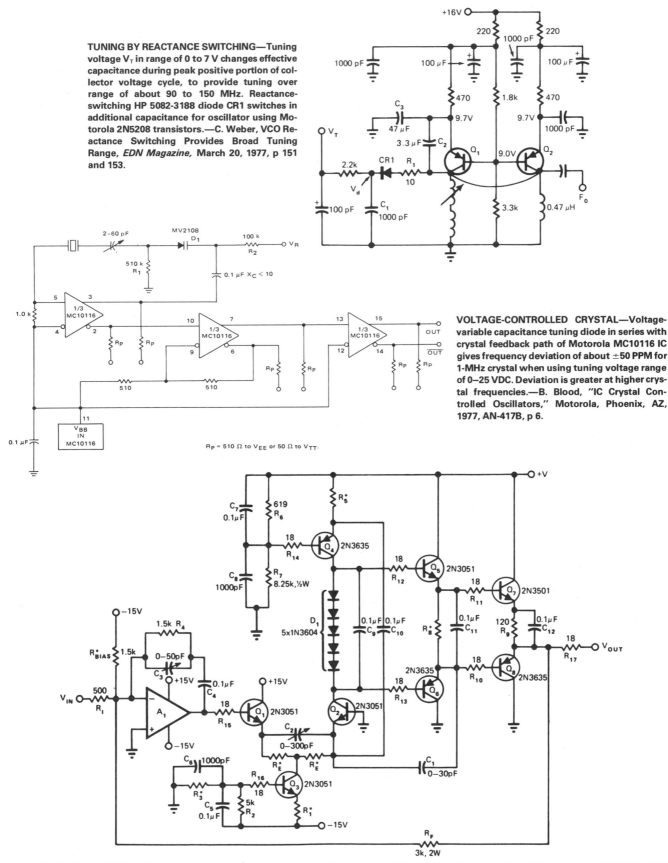

TUNING BY REACTANCE SWITCHING—Tuning voltage V_T in range of 0 to 7 V changes effective capacitance during peak positive portion of collector voltage cycle, to provide tuning over range of about 90 to 150 MHz. Reactance-switching HP 5082-3188 diode CR1 switches in additional capacitance for oscillator using Motorola 2N5208 transistors.—C. Weber, VCO Reactance Switching Provides Broad Tuning Range, *EDN Magazine,* March 20, 1977, p 151 and 153.

VOLTAGE-CONTROLLED CRYSTAL—Voltage-variable capacitance tuning diode in series with crystal feedback path of Motorola MC10116 IC gives frequency deviation of about ±50 PPM for 1-MHz crystal when using tuning voltage range of 0–25 VDC. Deviation is greater at higher crystal frequencies.—B. Blood, "IC Crystal Controlled Oscillators," Motorola, Phoenix, AZ, 1977, AN-417B, p 6.

$R_P = 510\ \Omega$ to V_{EE} or $50\ \Omega$ to V_{TT}.

FAST-SLEW VCO DRIVER—High-performance circuit slews at 4000 V/μs when operating from 80-V supply and provides output levels up to +30 VDC. Circuit handles large-signal modulation rates up to 20 MHz for 60-V varactors and small-signal bandwidths up to 86 MHz. Input opamp can be M. S. Kennedy Model 770 or other fast-input unit having −6 dB per octave rolloff. Operation in transimpedance configuration means associated buffer amplifier can have high gain. R_E is 250 ohms, R_1 is 100, R_3 is 4.3K, R_5 is 170, and R_8 is 90.—H. Bunin, Low Cost VCO Driver Amplifiers Really Perform If Designed Right, *EDN Magazine,* Oct. 5, 1974, p 51–55.

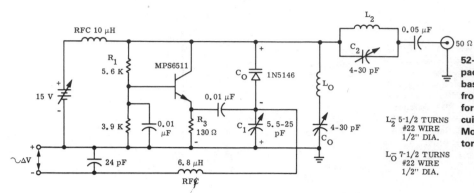

52-MHz WITH VVC FM—Voltage-variable capacitor C_O provides ±75 kHz modulation of basic 52-MHz transistor oscillator operating from 15-V supply. Modulation linearity is good for voltage inputs up to ±200 mV, making circuit suitable for commercial FM use.—"FM Modulation Capabilities of Epicap VVC's," Motorola, Phoenix, AZ, 1973, AN-210, p 2.

L_2^- 5-1/2 TURNS #22 WIRE 1/2" DIA.

L_O^- 7-1/2 TURNS #22 WIRE 1/2" DIA.

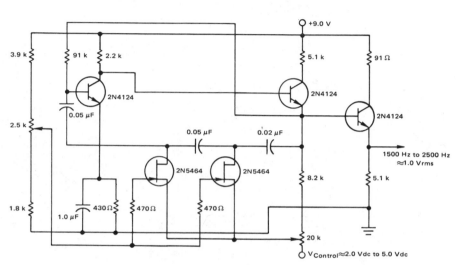

1.5–2.5 kHz SINE-WAVE—Three-section phase-shift oscillator is linear over its frequency range and has good sine waveform. Phase-shifting network is included in feedback loop of amplifier to give voltage-controlled oscillator action.—"Low Frequency Applications of Field-Effect Transistors," Motorola, Phoenix, AZ, 1976, AN-511A, p 8.

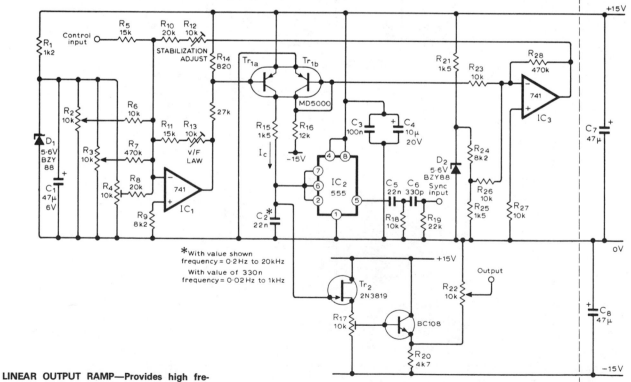

LINEAR OUTPUT RAMP—Provides high frequency stability as required for use in sound synthesizers. Can be synchronized to another oscillator. Uses 555 timer in astable mode, with Tr_{1a} supplying constant current to C_2. R_{12} and R_{13} should be multiturn pots. Synchronizing square-wave signal having 5–10 V peak can be fed in at R_{19} for differentiation, and resulting spikes used to control threshold voltage of 555. R_4 sets minimum frequency. R_{22} sets average output level to 0 V. R_2 and R_3 serve as coarse and fine frequency controls. Tr_1 can also be BFX11 or BFX36.—T. W. Stride, Voltage Controlled Oscillator, *Wireless World*, Oct. 1977, p 66.

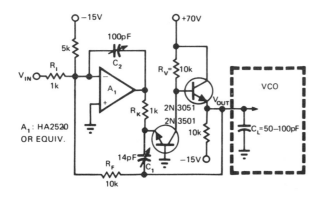

SIMPLE VCO DRIVER—Provides full output of 60 V P-P up to 1 MHz. Slew rate is 200 V/μs, and small-signal bandwidth is 5 MHz. Uses fast-input opamp, voltage buffer, and simple compensation technique. C_2 is trimmed for stability, while C_1 is adjusted to increase slew rate and bandwidth.—H. Bunin, Low Cost VCO Driver Amplifiers Really Perform If Designed Right, *EDN Magazine,* Oct. 5, 1974, p 51–55.

DOUBLING CONTROL RANGE—Circuit doubles frequency-deviation ratio of given VCO. Control voltage of MC1658 VCO, with range of 0 to −2 V, is attenuated and then applied to AM685 opamp comparator. When control voltage reaches an extreme and crosses over amplifier's reference voltage, detector switches to opposite state. Circuit output is thus either that of VCO or VCO divided by 2. Article describes operation in detail.—E. Kane, Expander Doubles VCO Frequency Deviation, *EDN Magazine,* Jan. 20, 1977, p 94.

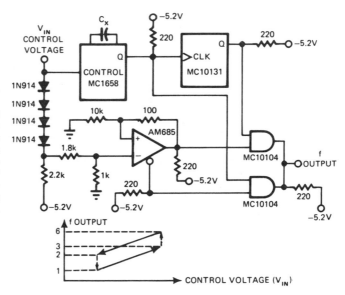

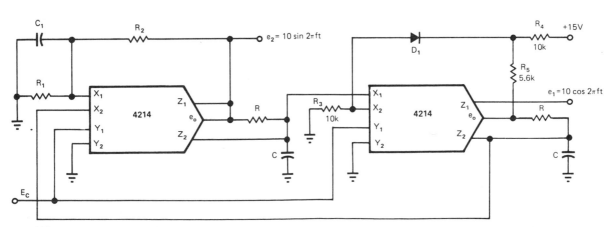

QUADRATURE OSCILLATOR USES MULTI-PLIERS—4214 differential multipliers eliminate need for opamps in quadrature oscillator in which frequency is controlled by external DC voltage. R_3, R_4, R_5, and D_1 form diode limiter,

$$f = \frac{E_c}{20\pi RC} \; ; \; 0.1V < E_c \leqslant 10V$$

while R_1, R_2, and C_1 provide positive feedback to sustain oscillation. R_1 should be about equal to R, R_2 about 20R, and C_1 about 10C. R_2 can be

readjusted for best compromise between distortion and speed of amplitude buildup.—Y. J. Wong, Design a Low Cost, Low-Distortion, Precision Sine-Wave Oscillator, *EDN Magazine,* Sept. 20, 1978, p 107–113.

CHAPTER 101
Voltage-Level Detector Circuits

Includes undervoltage, overvoltage, voltage-window, peak, trough, zero-crossing, and pulse-period-window detectors. See also Battery-Charging, Instrumentation, Power Control, Switching, and Voltage Measuring chapters.

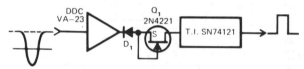

FAIL-SAFE TTL INTERFACE— Diode and FET protect SN74121 high-speed level detector from excessive opamp output voltage. If input of opamp goes too far negative, positive-going output will cause breakdown of TTL input. Protective interface makes circuit fail-safe without loss of operating speed. D_1 should be high-speed germanium diode with breakdown voltage above highest positive output of amplifier (usually about 15 V).—K. I. Wolfe, A Safer Analog-to-Digital Interface, *EDN Magazine,* March 5, 1974, p 74.

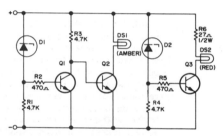

12-V MONITOR—Voltage-limit sensor gives visual indication that voltage in auto or boat electric system is satisfactory for operation of critical electronic equipment. Combination of zener diodes D1 and D2 acting with base-emitter voltage drops of Q1 and Q3 makes any voltage less than 13.5 V turn on amber No. 330 pilot lamp (14 V at 80 mA), while voltage above 15.2 V turns on red pilot lamp of same type. Transistors are Motorola MPS 3704. D1 is 1N5243B 13-V zener, and D2 is 1N5245B 15-V zener.—M. J. Moss, Voltage Limit Sensor, *73 Magazine,* May 1973, p 53–54.

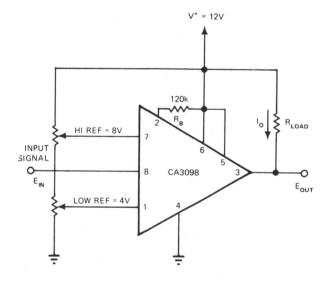

SEQUENCE	INPUT SIGNAL	OUTPUT VOLTAGE (PIN 3)
1	$4.0 \geq E_{IN} > 0$	0
2	$8.0 > E_{IN} > 4$	0
3	$E_{IN} > 8$	+12
2	$8.0 > E_{IN} > 4$	+12
1	$4.0 \geq E_{IN} > 0$	0

4–8 V WINDOW—CA3098 dual-input precision level detector tells if data input signal is above or below preset levels of 4 and 8 V. Table gives output states for various input levels. Output current can be up to 150 mA.—G. J. Granieri, Precision Level Detector IC Simplifies Control Circuit Design, *EDN Magazine,* Oct. 5, 1975, p 69–72.

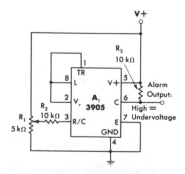

UNDERVOLTAGE ALARM—3905 timer output goes high when power supply drops below predetermined voltage level. Timer is connected as inverting comparator that compares fraction of supply voltage (as set by R_1) with fixed voltage-comparison threshold of 2 V for timer. Output can be used to drive suitable alarm indicator.— W. G. Jung, "IC Timer Cookbook," Howard W. Sams, Indianapolis, IN, 1977, p 230–231.

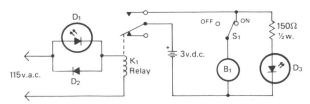

POWER-FAILURE ALARM—Buzzer sounds and red LED D₃ comes on when AC power fails, as reminder that clocks will need resetting. Green LED D₁ indicates that alarm is plugged in. D₂ is Radio Shack 276-1103 or equivalent silicon diode. B₁ is 1.5–6 VDC Radio Shack 273-004 or equivalent buzzer, and K₁ is Radio Shack 275-211 or equivalent 117-VAC SPDT relay.—C. R. Graf, The Powerlarm, *CQ*, Feb. 1977, p 47 and 73.

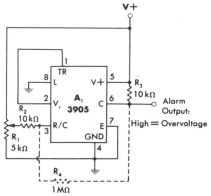

OVERVOLTAGE ALARM—Connection shown for 3905 timer makes output go high for energizing suitable alarm when supply voltage rises above predetermined level. Timer is connected as noninverting comparator that compares its fixed voltage-comparison threshold of 2 V with fraction of supply voltage determined by setting of R₁. Optional resistor R₄ can be added if some hysteresis is desirable to prevent tripping of alarm by momentary fluctuations of supply.—W. G. Jung, "IC Timer Cookbook," Howard W. Sams, Indianapolis, IN, 1977, p 230–231.

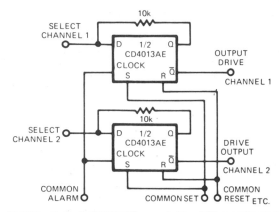

MULTICHANNEL ALARM—Half of CD4013AE flip-flop serves as latching AND gate in each channel being monitored for overvoltage, overtemperature, or any other out-of-tolerance condition that can be represented by logic 1 level applied to terminal that connects to clock inputs of all flip-flops. Any number of additional channels can be paralleled to common terminals. Each channel has own transistor driver and either LED or audio alarm. Alarm condition is held until operator resets system by applying voltage to common set terminal. Article shows how to obtain additional flexibility by adding NAND and AND gates to each select input and to common alarm input.—J. C. Nichols, CMOS "D" Flop Makes Latching "AND" Gate, *EDN Magazine*, April 20, 1974, p 89 and 91.

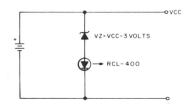

LED VOLTAGE MONITOR—Uses Litronix RCL-400 current-controlled LED having built-in voltage-sensing IC that turns on LED at 3 V and turns it off at 2 V. Use suitable zener or string of forward-biased silicon diodes to make VZ equal to 3 V less that VCC. Thus, for 4.5-V battery, put two silicon diodes in series with LED to make VZ 1.5 V across them.—S. W. Hawkinson, A Battery Voltage Monitor, *73 Magazine*, July 1977, p 52.

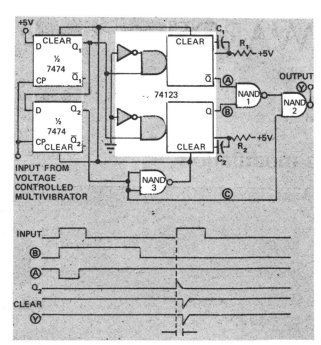

VCO SENSES VOLTAGE LIMITS—Used to indicate when pulse voltage goes outside preset limits for pulse period. Output pulse rate of voltage-controlled MVBR is monitored to implement double-ended limit detector consisting of 2-bit shift register, two monos, inverter, and two NAND gates. Circuit compares period of input pulses to preset maximum and minimum limits. Output at Y goes low whenever input pulse rate is outside limits, which are determined by R₁C₁ and R₂C₂ time constants.—B. Brandstedt, Double-Ended Limit Detector Senses Voltage with VCO, *EDN Magazine*, Nov. 15, 1972, p 47–48.

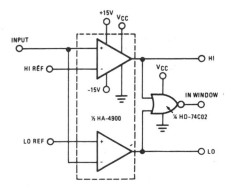

OUT-OF-LIMIT VOLTAGE SENSOR—High switching speed, low offset current, and low offset voltage of Harris HA-4900/4905 precision quad comparator make circuit well suited for industrial process control applications requiring fast, accurate decision-making based on voltage levels. Outputs can be used to drive alarm indicator or initiate corrective action.—"Linear & Data Acquisition Products," Harris Semiconductor, Melbourne, FL, Vol. 1, 1977, p 2-96.

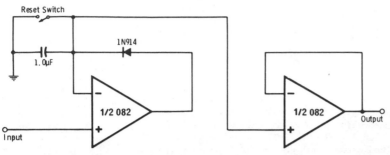

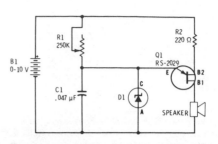

POSITIVE-PEAK DETECTOR—Circuit responds to and remembers peak positive excursions of input signal over period of time with first half of 082 dual opamp. Other half of opamp serves as voltage follower for isolating peak detector from output. Memory time is typically several minutes, depending on rate at which capacitor discharges due to its own leakage current, diode leakage current, opamp bias currents, and slight loading effect of voltage follower. Closing reset switch momentarily discharges capacitor in readiness for storing new peak value.—R. Melen and H. Garland, "Understanding IC Operational Amplifiers," Howard W. Sams, Indianapolis, IN, 2nd Ed., 1978, p 96–97.

LOW-VOLTAGE ALARM—UJT relaxation oscillator produces audio tone from loudspeaker when battery voltage drops below breakdown voltage of zener. For 9-V battery, zener can be 6-V unit (Radio Shack 276-561). When input voltage drops below zener breakdown, zener stops conducting and C1 begins charging, as required for oscillation. When battery is replaced, zener breaks down and prevents C1 from charging.—F. M. Mims, "Semiconductor Projects, Vol. 2," Radio Shack, Fort Worth, TX, 1976, p 43–49.

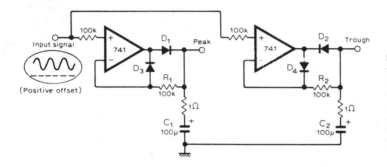

PEAK AND TROUGH DETECTOR—Uses only two opamps to detect peak and valley voltages of nonsymmetrical waveform. During valley period, D_2 conducts and discharges C_2 rapidly to lowest value of signal amplitude. C_2 charges only slightly through D_4 and R_2 during positive peaks, thus retaining minimum voltage.—C. Spain, Precision Peak and Trough Detector, *Wireless World*, Oct. 1977, p 65.

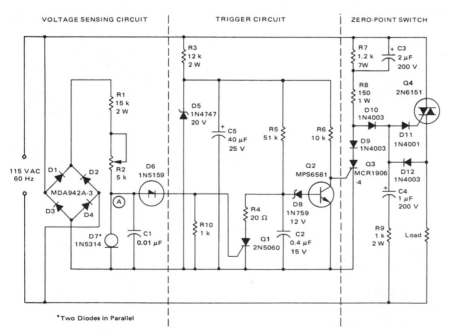

*Two Diodes in Parallel

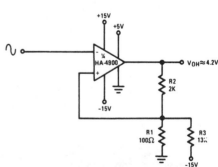

ZERO-POINT WITH OVERVOLTAGE PROTECTION—Used to protect voltage-sensitive load from excessive line voltage. Switch section operates conventionally to turn on triac almost immediately after each zero crossing between half-cycles. For normal line voltages, SCR Q3 is off. When overvoltage condition is sensed during any half-cycle, SCR Q1 is turned on, discharging C2 and turning Q2 off. This allows Q3 to turn on and divert triac gate drive, removing power from load. As long as overvoltage condition exists, Q1 is turned on each half-cycle and C2 is unable to charge enough to turn Q2 on. When overvoltage condition ceases, C2 charges to voltage set by D8 in about 20 ms, saturating Q2 so Q3 turns off and Q4 turns on. R2 can be set to allow line voltage variations from almost 0 to 11 V.—"Circuit Applications for the Triac," Motorola, Phoenix, AZ, 1971, AN-466, p 14.

ZERO DETECTOR WITH HYSTERESIS—Circuit using one section of Harris HA-4900/4905 precision quad comparator as Schmitt trigger has 100-mV hysteresis. Suitable for applications requiring fast transition times at output even though input signal approaches zero crossing slowly. Hysteresis loop also reduces false triggering by input noise. Output jumps to 4.2 V at instant when input reaches −100 mV after dropping to 0 V. Output drops from 4.2 V to 0 V when input passes through 0 V in positive direction and reaches +100 mV.—"Linear & Data Acquisition Products," Harris Semiconductor, Melbourne, FL, Vol. 1, 1977, p 2-96.

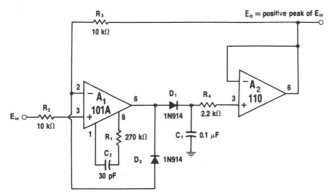

BUFFERED PEAK DETECTOR—Discharge current through C_1 is minimized by using low-input-current voltage follower A_2. R_3 allows A_1 to be clamped in OFF state by D_2 to give faster recovery. Circuit operates much like ideal diode, but with C_1 storing DC voltage equal to peak input voltage value. When input signal crosses zero, A_1 drives D_1 on and circuit output follows rising signal slope. When input signal reaches peak and reverses, C_1 is left charged. Reverse diode connections to detect negative peaks. Value of R_1 should be increased to 2.7 megohms if C_1 is increased to 1 μF to improve stability.— W. G. Jung, "IC Op-Amp Cookbook," Howard W. Sams, Indianapolis, IN, 1974, p 196–197.

LOGAMP ZERO-CROSSING DETECTOR—Feedback current for A_1 creates logarithmic output voltage due to diodes D_1 and D_2. A_1 is connected in feedforward mode to optimize speed and minimize phase error at high frequencies. Output voltage is nominally $\pm V_f$, where V_f is forward voltage drop of either diode. Dynamic range of circuit is about 70 dB. If higher or constant output voltages are required, add optional connection of saturated switch that delivers 0–5 V output.—W. G. Jung, "IC Op-Amp Cookbook," Howard W. Sams, Indianapolis, IN, 1974, p 229–230.

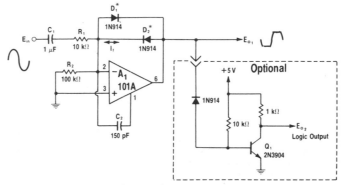

*For greater dynamic range, use a matched, monolithic transistor pair connected as diodes — e.g., CA3018.

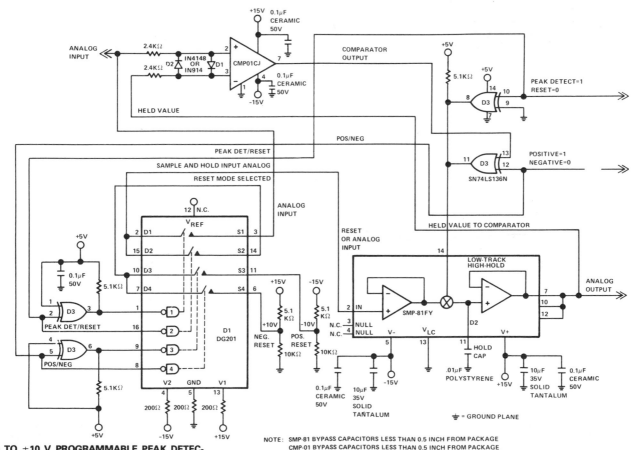

0 TO ±10 V PROGRAMMABLE PEAK DETECTOR—Principal components are Precision Monolithics CMP-01CJ voltage comparator, SMP-81FY sample-and-hold amplifier, SN74LS136N open-collector EXCLUSIVE-OR gate package, and DG201 quad analog switch. DC accuracy is within 5 mV at zero scale and within 10 mV at full scale. Resistors and diodes provide input overvoltage protection for comparator. Comparator continuously examines difference between analog input voltage and voltage peak held by sample-and-hold amplifier. If input exceeds held value, new input is held.—D. Soderquist, "Polarity Programmable Peak Detector," Precision Monolithics, Santa Clara, CA, 1978, AN-27.

Timing Diagram

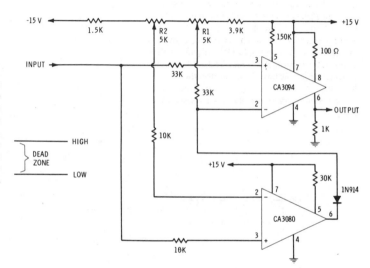

WINDOW DETECTOR—Connections shown for 322 comparators give high output only when input voltage is between thresholds set by R_2 and R_3 (within voltage window). Output of circuit goes low whenever input is below threshold 1 or above threshold 2.—W. G. Jung, "IC Timer Cookbook," Howard W. Sams, Indianapolis, IN, 1977, p 153.

DUAL-LIMIT DETECTOR—Provides 12-V output when applied DC input signal exceeds reference high limit established by setting of R1 or falls below reference low limit established by setting of R2. When input drops below low limit, CA3080 changes CA3094 to high-output condition. Output is low in voltage window between limits (dead zone).—E. M. Noll, "Linear IC Principles, Experiments, and Projects," Howard W. Sams, Indianapolis, IN, 1974, p 317–318.

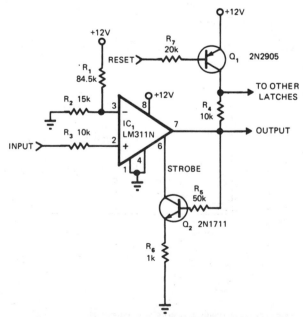

VOLTAGE-LEVEL LATCH—Circuit uses comparator to latch after input reaches predetermined threshold level. Output of IC_1 then goes high and enables input of strobe Q_2 to prevent output from going low. High level on reset input will turn off Q_1, removing supply voltage from open collector output of IC_1 and removing latch condition. Comparator will operate on supplies ranging from single 5-V level to dual ±15 V.—M. W. Bair, IC Comparator Doubles as a Latch, *EDN Magazine*, April 20, 1975, p 72.

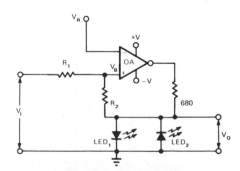

LEVEL INDICATOR—Visual indication of voltage level is achieved with two TIL203 LEDs, three resistors, and any opamp that can provide 15-mA output current. If input voltage momentarily or permanently exceeds most positive reference level, LED_1 is switched on. If voltage falls below negative or least positive reference level, LED_1 goes off and LED_2 comes on. Article gives design equations for determining values of R_1 and R_2. For levels of +2 V to turn LED_1 on and −1.2 V to turn LED_2 on, both R_1 and R_2 are 10K.—E. J. Richter, Op Amp Makes Visual Level Indicator, *EDN Magazine*, May 5, 1974, p 73.

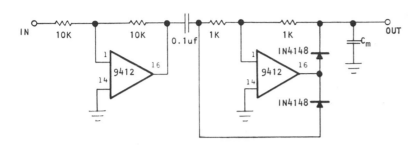

5-V PEAKS UP TO 2 MHz—Peak-to-peak detector using Optical Electronics 9412 opamps gives DC output voltage equal to peak-to-peak amplitude of sine-wave input voltage. Opamp charges memory capacitor C_m during negative half of input cycle and performs DC clamp (restoration) on positive half. Circuit has high input impedance. With 0.1-μF memory capacitor, 10-V pulse is acquired in 10 μs. For 5-V sine-wave input, maximum frequency is 0.8 MHz, but 0.01-μF memory capacitor boosts frequency capability to 2 MHz.—"A Wideband Peak-to-Peak Detector," Optical Electronics, Tucson, AZ, Application Tip 10176.

LED INDICATES SIGNAL LEVEL—Circuit is adjusted so opamp turns on LED at desired signal level as set by R1. Opamp is operated without feedback resistor to have maximum gain, so small input signal produces very large output signal. Values shown for R2 and R3 give turn-on voltage of 0.9 V for LED.—F. M. Mims, "Integrated Circuit Projects, Vol. 4," Radio Shack, Fort Worth, TX, 1977, 2nd Ed., p 70–75.

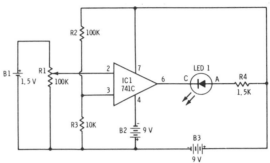

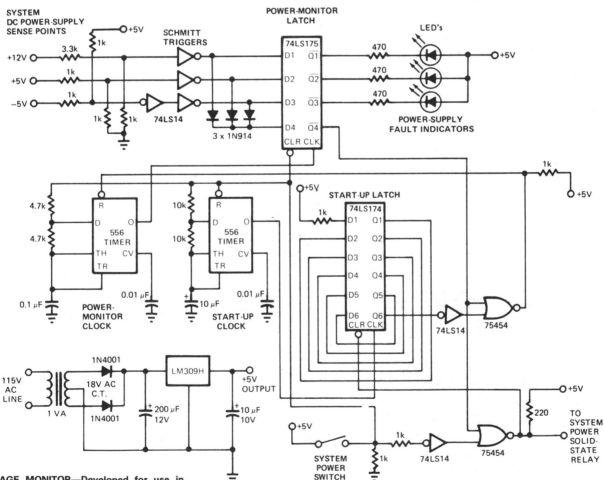

VOLTAGE MONITOR—Developed for use in systems having multiple DC bias voltages, to prevent damage when one supply voltage goes down while others remain normal. Control circuit includes its own independent AC/DC supply that ensures protection even when equipment containing RAMs and MOS devices is turned off. Failure of AC supply for monitor shuts down entire system. Can be applied to any number of supplies by adding resistive dividers, Schmitt triggers, diodes, and latches as required. Closing system power switch activates solid-state relay for applying AC line voltage to main power supplies. Half of 556 dual timer and 74LS174 hex D latch inhibit voltage monitor until all supplies have stabilized, about 500 ms later. Other half of 556 then clocks 74LS175 power-monitor latch. System then operates normally as long as all D inputs to monitor latch stay at logic 0. If one supply fails, logic 1 appears at its latch input and next clock pulse initiates shutdown of system. LED identifies supply that has failed.—J. E. Draut, Voltage Monitor Protects Against Power-Supply Failures, *EDN Magazine*, Nov. 20, 1977, p 239–240.

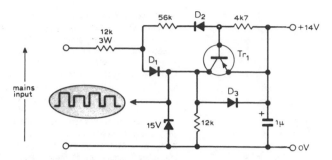

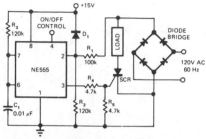

AC LINE ZERO-CROSSING DETECTOR—Positive-going half-cycles forward-bias D_1, allowing capacitor to charge through D_3 to 14 V. Negative half-cycles forward-bias D_2 to turn on Tr_1 and discharge capacitor. Output is about 1 V less on negative half-cycles. Transistor and diode types are not critical, except that D_1 must withstand full reverse voltage of AC line.—R. J. Torrens, Zero Crossing Detector, *Wireless World*, Jan. 1977, p 78.

555 TRIGGER—Low-cost 555 timer provides ON/OFF and proportional-control switching of AC loads without generating RFI or voltage spikes. Timer is used in monostable mode, re-triggering every half-cycle when voltage at pin 2 falls below about 1.67 V. R_3 and C_1 fix pulse width at about 1 ms, long enough to ensure firing SCR in next half-cycle yet short enough to turn SCR off at next zero-crossing without timing-cycle pulse. Pin 4 serves as ON/OFF control input. Varying duty cycle of square wave here gives proportional control for heating and other uses.—M. E. Anglin, Low Cost Zero-Cross Thyristor Trigger Uses a 555 IC, *EDN Magazine*, Sept. 5, 1977, p 180–181.

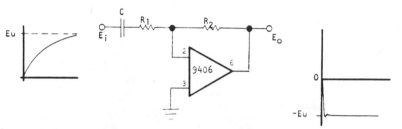

INSTANT ULTIMATE VALUE—Circuit instantly computes ultimate value of logarithmically increasing input signal E_i by performing augmented differentiation that gives step function equal to ultimate value E_u. Uses Optical Electronics 9406 opamp. Circuit values are computed from $E_u = E_o = -R_2E_i/R_1 - R_2Cde_i/dt$.—"Derivative Circuit Indicates Ultimate Value Instantly," Optical Electronics, Tucson, AZ, Application Tip 10179.

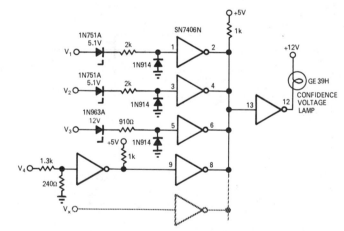

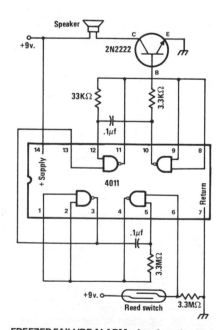

INPUT	NOMINAL VOLTAGE	CONFIDENCE–VOLTAGE LAMP:	
		TURNS OFF AT:	TURNS ON AT:
V_1	–5V	–4.8V	–4.9V
V_2	–5V	–4.3V	–4.4V
V_3	–12V	–11.3V	–11.5V
V_4	+8V	+6.6V	+6.7V

LOW-VOLTAGE ALARM—Simple indicator circuit uses hex inverter IC to monitor several different input voltages. Technique is flexible and easily modified for different voltage values (either positive or negative) and additional inputs. When negative input (V_1, V_2, or V_3) falls below breakdown voltage of its zener, logic 0 appears at inverter output (at wired-OR connection). Because lamp-driving inverter has logic 0 at its input, lamp goes out as no-go signal. When positive input V_4 falls below predetermined value, logic 0 again causes no-go indication.—R. J. Buonocore, Under-Voltage Sensing Circuit, *EDN/EEE Magazine*, Dec. 1, 1971, p 48–49.

FREEZER FAILURE ALARM—Loudspeaker is energized by 4011 audio oscillator and 2N2222 transistor operating from 9-V battery when ice melts and allows permanent magnet to drop on reed switch and close it. Magnet is bonded to wall inside of freezer with mixture of antifreeze and water.—J. A. Sandler, 11 Projects under $11, *Modern Electronics*, June 1978, p 54–58.

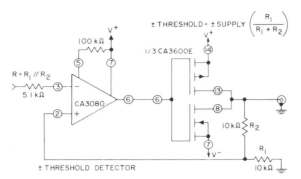

$$\pm \text{THRESHOLD} = \pm \text{SUPPLY} \left(\frac{R_1}{R_1 + R_2} \right)$$

± THRESHOLD DETECTOR

±1.5 TO ±7.5 V THRESHOLD—Precise timing and accurate threshold levels are assured by stable characteristics of input differential amplifier in CA 3080 variable opamp used to drive one of inverter/amplifier transistors in CA3600E array. For values shown, threshold voltage for given polarity is half of supply voltage used, in range of 3 to 15 V.—"Circuit Ideas for RCA Linear ICs," RCA Solid State Division, Somerville, NJ, 1977, p 16.

WINDOW DETECTOR—Unique voltage-range sensing circuit provides positive indications of high, low or acceptable input levels for voltage and includes adjustments for both threshold and hysteresis levels. R_1 and R_2 adjust upper and lower thresholds, while R_3 and R_4 adjust upper and lower hysteresis levels. If acceptable input range is 4.5 V to 5.5 V, output of opamp A_1 goes negative when e_{in} is greater than 5.5 V. This saturates Q_1 and Q_2, making Q_2 output go from 5 V to 0. TTL then indicates that input has exceeded 5.5 V. Upper hysteresis keeps A_1 output negative until input has dropped to setting of R_3, which might be 5.3 V. Similarly, when input drops below 4.5 V, output of A_2 goes positive and saturates Q_3.—I. Krell, Analog Monitor Has Threshold and Hysteresis Controls, *EDN Magazine*, Aug. 1, 1972, p 58.

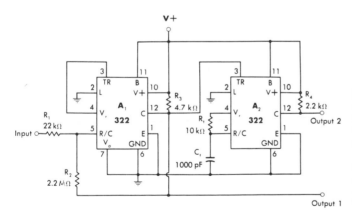

ZERO-CROSSING DETECTOR—Output 1 of 322 comparator A_1 is high when input signal is above zero and low when input is below zero. Output of comparator A_1 is thus square wave in phase with zero crossings of input. When R_1 is 22K, input can be up to ±10 V amplitude. A_2 is mono MVBR connected to fire when output 1 of A_1 goes high (at zero crossings). Resulting negative-going narrow pulses at output 2 are useful for time marks.—W. G. Jung, "IC Timer Cookbook," Howard W. Sams, Indianapolis, IN, 1977, p 152.

Timing Diagram

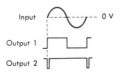

CHAPTER 102
Voltage Measuring Circuits

Gives voltmeter, multimeter, and electrometer circuits for measuring absolute, peak, RMS, or other values of AC, DC, and RF voltages. Indicators include meters, digital displays, loudspeakers, bar-graph displays, and frequency counter. Also includes automatic polarity circuits and voltage-null detectors. See also Audio Measuring, Logic Probe, Multiplier, Test, and Voltage-Level Detector chapters.

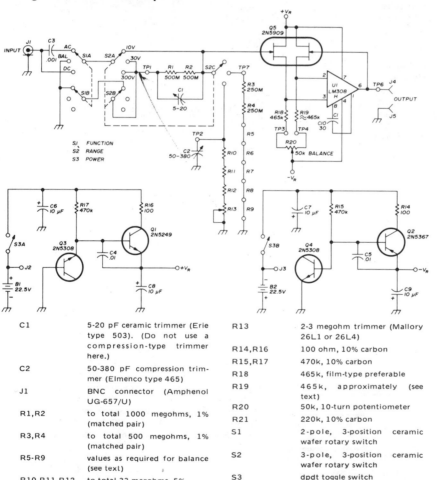

C1	5-20 pF ceramic trimmer (Erie type 503). (Do not use a compression-type trimmer here.)	R13	2-3 megohm trimmer (Mallory 26L1 or 26L4)
C2	50-380 pF compression trimmer (Elmenco type 465)	R14,R16	100 ohm, 10% carbon
		R15,R17	470k, 10% carbon
J1	BNC connector (Amphenol UG-657/U)	R18	465k, film-type preferable
		R19	465k, approximately (see text)
R1,R2	to total 1000 megohms, 1% (matched pair)	R20	50k, 10-turn potentiometer
		R21	220k, 10% carbon
R3,R4	to total 500 megohms, 1% (matched pair)	S1	2-pole, 3-position ceramic wafer rotary switch
R5-R9	values as required for balance (see text)	S2	3-pole, 3-position ceramic wafer rotary switch
R10,R11,R12	to total 33 megohms, 5%	S3	dpdt toggle switch

1-TERAOHM INPUT—High-accuracy meter-interface amplifier for AC and DC voltage measurements has input resistance of 1,000,000 megohms. Amplifier eliminates voltmeter errors due to loading by using special 2N5909 dual FET with exceptionally low gate leakage current. FET and opamp are connected as voltage follower with gain of 1. Accuracy on 0–10 V range is 0.1% or better. For higher voltage ranges, accuracy depends on that of resistive voltage divider used. Three ranges provided have full-scale values of 10, 30, and 300 V. AC RMS inputs are limited to 70% of DC ranges. Two voltage regulators are used with battery supply to permit use of batteries exceeding 18-V voltage rating of opamp, so battery voltages can drop considerably before replacement is required. Article covers construction and adjustment.—J. R. Laughlin, High-Impedance Meter Interface, *Ham Radio*, Jan. 1974, p 20–25.

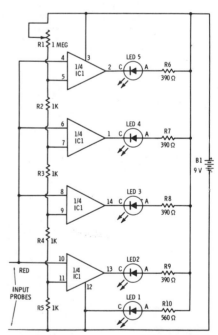

BAR GRAPH READOUT VOLTMETER—Sections of RS339 quad comparator each drive LED to give indications of four different input voltage levels, while LED 1 is connected to ground for use as zero indicator. Resistors shown are for Radio Shack 276-041 red LEDs; change R6-R9 to 270 ohms and R10 to 470 ohms for green LEDs. Pot R1 is used to calibrate voltage divider R2-R5. With R1 set at low resistance, comparators turn on at intervals of 1 V or more. With high resistance for R1, comparators turn on at fractional-volt intervals.—F. M. Mims, "Integrated Circuit Projects, Vol. 4," Radio Shack, Fort Worth, TX, 1977, 2nd Ed., p 76–85.

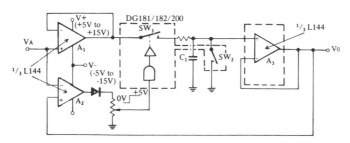

POSITIVE-PEAK DETECTOR—Combination of Siliconix triple L144 opamp and DG181 analog switch eliminates errors of conventional diode circuits. Third opamp acts as comparator providing logic drive for operating SW₁. Action of circuit is such that most positive analog input is stored. SW₂ serves as reset switch.—"Analog Switches and Their Applications," Siliconix, Santa Clara, CA, 1976, p 4-9.

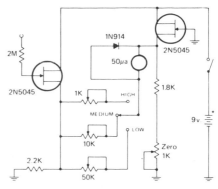

ELECTROMETER—Can be used for picoampere leakage measurements and nonloading voltage measurements. Bridge circuit has three pots for three ranges of sensitivity. Adjust for 0.5, 1.5, and 5 V full scale with appropriate input voltages. With 1000-megohm resistance between point 5 and probe tip, picoammeter gives full-scale deflection on 500 pA. For nonloading voltmeter, apply unknown voltage across same 1000-megohm resistor; now 0.5 V will give full-scale reading.—I. Math, Math's Notes, *CQ*, Oct. 1974, p 26–27.

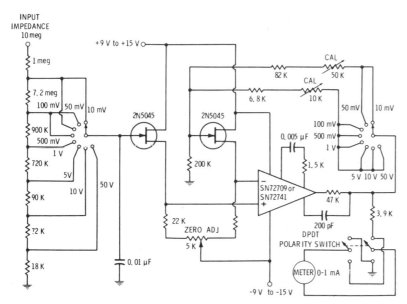

FET MILLIVOLTMETER—Eight-range meter uses pair of JFETs in bridge arrangement driving meter through opamp. FETs should be reasonably well matched, even though their operation can be balanced with 5K zero-adjust pot.—E. M. Noll, "FET Principles, Experiments, and Projects," Howard W. Sams, Indianapolis, IN, 2nd Ed., 1975, p 212–213.

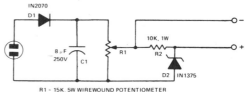

EXPANDED-RANGE AC VM—Line voltage is applied to D1, and resulting DC is filtered by C1. R1 delivers equivalent of RMS voltage to 100-V zener D2 through R2. Voltage is developed across R2 only when voltage applied by R1 exceeds 100 V, for reading with 1000-ohm-per-volt meter. To calibrate, measure AC line voltage with accurate AC voltmeter, then adjust R1 so meter across R2 reads 100 V less than this value.—W. P. Turner, Expanded Range Line Voltage Meter, *73 Magazine*, March 1974, p 54.

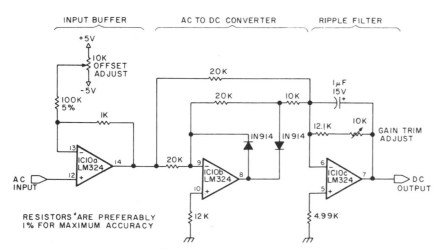

AC/DC CONVERTER—Used for measuring AC voltage with digital DC voltmeter. Resulting signal is equal to average RMS value of applied input signal. When 1-V peak 60-Hz is applied to converter, output should be +0.707 VDC. Connect pin 4 of LM324 to +5 V and pin 11 to −5 V.—S. Ciarcia, Add More Zing to the Cocktail, *BYTE*, Jan. 1978, p 37–39, 44, 46, 48, 50–52, and 54.

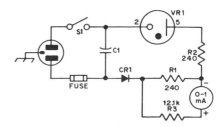

LINE-VOLTAGE MONITOR—0C3 (VR-105) voltage-regulator tube provides voltage offset that permits greater sensitivity in voltage range of interest. Meter scale covers 20-V range centered on about 115 VAC. Accuracy is much better than with AC range of ordinary multimeter. CR1 is 500-PIV 1-A silicon diode.—N. Johnson, An AC Line Monitor, *QST*, Jan. 1976, p 27.

4½-DIGIT VOLTOHMMETER—National type MM5330 IC provides logic circuits for implementing low-cost 4½-digit voltohmmeter. Display interface consists of TTL 7-segment decoder driver and four 2N4403 transistors. Operation is based on counting of up to 80,000 clock pulses. Circuit provides sign digit, either plus or minus, and numeral 1 for 10,000 to give full display of ±19,999 with decimal point.—"MOS/LSI Databook," National Semiconductor, Santa Clara, CA, 1977, p 5-23–5-29.

ZENER PROTECTS METER—Simple overvoltage protection circuit makes 10-V zener conduct when voltage E1 being measured goes over 20-V full-scale limit of voltmeter using milliammeter movement with multiplier resistor R2. This turns on Q1, drawing current through LED to give visual indication of overvoltage, while providing protective shunt path around meter.—H. Olson, Sensitive Meters Saved, *73 Magazine*, Oct. 1977, p 153.

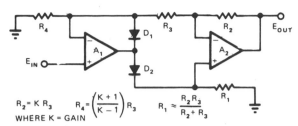

$$R_2 = KR_3 \qquad R_4 = \left(\frac{K+1}{K-1}\right)R_3 \qquad R_1 \approx \frac{R_2 R_3}{R_2 + R_3}$$
WHERE K = GAIN

ABSOLUTE VALUES—Positive output signal level is proportional to absolute value of input signal level, regardless of input polarity. Circuit combines simplicity with high input impedance, low output impedance, and greater than unity gain. Opamps A_1 and A_2 should have good CMRR and low offset and drift. D_1 and D_2 can be

1N914. For gain of 2.5, R_1 and R_3 are 1000 ohms, R_2 is 2500, and R_4 is 2333. For unity gain, R_4 is infinity and can be omitted. R_2 and R_3 are equal-value precision resistors. Value of R_1 is not critical at any gain.—R. Hofheimer, A Simple Absolute-Value Amplifier, *EDN Magazine*, June 20, 1974, p 78 and 80.

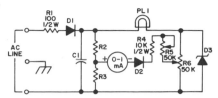

LINE-VOLTAGE MONITOR—AC line voltage is rectified by D1 and filtered by C1. R2 and R3 form voltage divider that holds one meter terminal at half of rectified line voltage. DC is also applied to low-voltage calibration pot R6 through 3-W 117-V lamp or equivalent resistor PL1 which limits zener current. Any increase in line voltage increases voltage at R2-R3 junction while voltage at slider of R6 remains constant, so bridge unbalances and meter reads upscale. Zener is 70 to 100 V at 10 W. R2 and R3 are equal and are from 8.2K to 15K. C1 is 50 to 100 μF at 200 V, and diodes are power silicon with PIV above 200 and 100-mA rating.—W. P. Turner, Expanded Range Line Voltage Monitor, *73 Magazine*, Jan. 1974, p 39.

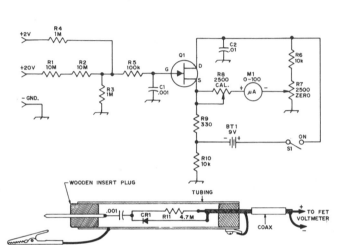

20-VDC FET VOLTMETER—Has high input impedance as required for accurate measurements in solid-state circuit. Uses Motorola MPF102, HEP802, or equivalent N-channel JFET. If meter cannot be zeroed, change R7 to 10,000 ohms for greater zeroing range. 2-V range gives extra flexibility. Half-wave RF probe using 1N914 or equivalent high-speed switching diode responds to peak RF voltage being measured. R11 reduces peak value to RMS value. Connect probe to known 10-VRMS source, then adjust R11 so meter reads 10 V.—D. DeMaw and L. McCoy, Learning to Work with Semiconductors, *QST*, April 1974, p 20–25 and 41.

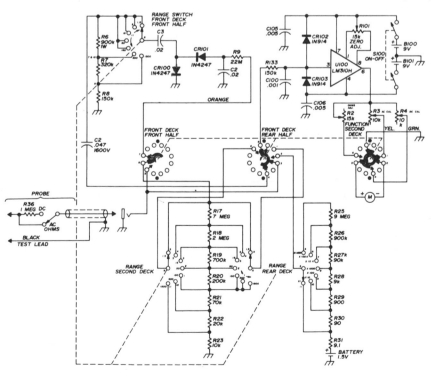

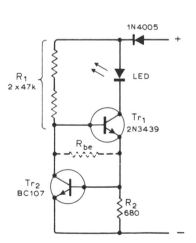

IC MODERNIZES VTVM—LM310H voltage-follower IC converts VTVM into battery-operated IC voltmeter. Input impedance and scale accuracy are unchanged. Conversion shown is for Heathkit IM-11 VTVM but will apply to most other VTVMs. Semiconductor diodes CR100 and CR101 replace original 6AL5 detector and

LM310H high-impedance unity-gain voltage follower replaces original 12AU7. C105 and C106 bypass battery supply and should be connected directly to U100. CR102 and CR103 provide overvoltage protection.—M. Kaufman, How to Convert Your VTVM to an IC Voltmeter, *Ham Radio*, Dec. 1974, p 42–44.

WIDE-RANGE VOLTAGE PROBE—Indicates presence of AC or DC voltages from 3 to 350 V with no range switching, using LED as indicator. Transistors serve essentially as constant-current supply for LED Voltage capability can be increased to 450 V by adding suitable base-emitter resistor R_{be}; typical value is 60 ohms, which somewhat impairs low-voltage operation.—G. Jones, Voltage Probe, *Wireless World*, Aug. 1976, p 52.

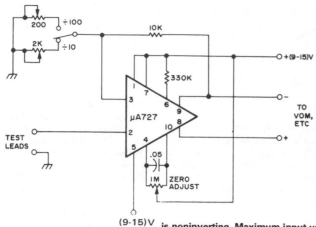

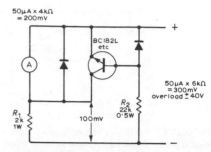

CALIBRATED-GAIN AMPLIFIER—Used to extend low range of VOM or CRO and provide very high input impedance (above 1000 megohms). Gain of amplifier from DC up to several hundred kilohertz is determined by ratio of 10K feedback resistor to 200- or 2000- ohm preset pot. Action is noninverting. Maximum input voltage range is ±10 V. Dual battery supply must be used. To calibrate, short input test leads and adjust 1-megohm pot for zero on VOM or other indicating instrument.—J. J. Schultz, Versatile Test Equipment Range Extender, *73 Magazine*, Nov. 1973, p 59–62.

METER OVERLOAD—When used with basic meter movement, circuit ensures that overloads merely make pointer run off scale in either direction in controlled manner and press gently against stop pin instead of winding around pin. Values shown will suit most meter movements, but article gives complete design procedure. If voltage being measured is between 350 and 700 mV, use two diodes in series in each position. For voltages between 700 mV and 1 V, use three diodes in series. Diode types are not critical.—C. Shenton, Meter Protection Circuit, *Wireless World*, Oct. 1972, p 475.

PEAK-TO-PEAK VOLTMETER—Two Optical Electronics opamps and two peak sense-and-hold modules form positive-peak-sense memory that adds negative peak amplitude to entire signal so as to bias negative peak to zero. Input must be restricted to ±5 V full scale (10 V P-P). Output is 0 V for −5 V input and +10 V for +5 V input. Unsymmetrical signals such as +2 V to −8 V still give +10 V output.—"Using Peak Sense Memories as Peak-to-Peak Detectors," Optical Electronics, Tucson, AZ, Application Tip 10275.

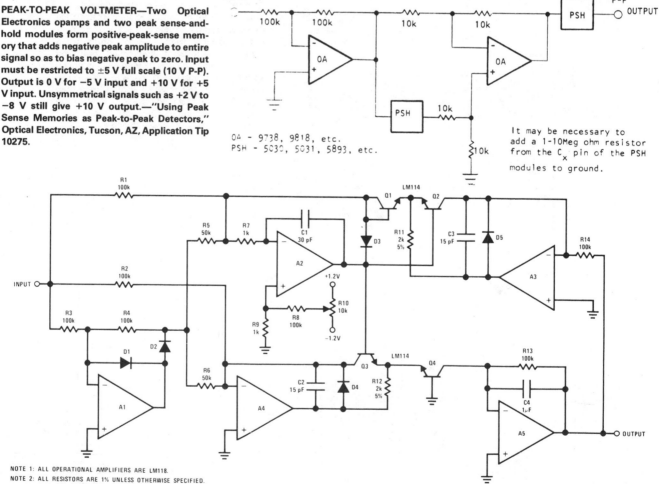

NOTE 1: ALL OPERATIONAL AMPLIFIERS ARE LM118.
NOTE 2: ALL RESISTORS ARE 1% UNLESS OTHERWISE SPECIFIED.
NOTE 3: ALL DIODES ARE 1N914.
NOTE 4: SUPPLY VOLTAGE ·15V.

TRUE RMS DETECTOR—Circuit using National LM118 opamps provides DC output equal to RMS value of sine, triangle, square, or other input waveform with 2% accuracy for 20 V P-P inputs from 50 Hz to 100 kHz. Circuit is usable up to about 500 kHz but with lower accuracy. Direct coupling of input provides true RMS equivalent of combined DC and AC signals. Absolute-value amplifier A1 provides positive input current to A2 and A4 independent of signal polarity. Amplifiers A2-A5 and transistors Q1-Q4 form log multiplier/divider. To calibrate, 10-VDC input signal is applied and R10 is adjusted for 10-VDC output. Transistors should be matched and mounted on common heatsink if possible.—R. C. Dobkin, "True RMS Detector," National Semiconductor, Santa Clara, CA, 1973, LB-25.

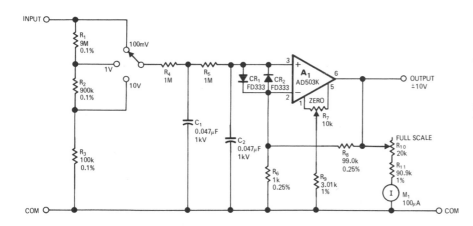

DC VOLTMETER—Opamp connected for closed-loop gain of 100 is used with attenuator network and 100-μA. microammeter to serve as general-purpose multirange laboratory voltmeter. Additional full-scale output of ±10 V is provided for driving chart recorder. Low-pass filter R_4-R_5-C_1-C_2 prevents amplifier from overloading on large AC input signals while allowing circuit to read DC component. Filter acting with diodes protects amplifier from input overloads up to 1000 V.—R. S. Burwen, Simple DC Voltmeter Uses Single Op Amp, *EDN/EEE Magazine,* Dec. 15, 1971, p 57.

IC MILLIVOLTMETER—Provides switched ranges of 5 mV to 500 V. Use 2% resistors. Pushbutton connection to positive supply gives internal calibration check on 5-V and 50-V ranges. Adjust meter initially with R_m to read 1.4 V on 5-V range. Back-to-back signal diodes provide overload protection. Power drain is so low that battery life is essentially shelf life.—D. A. Bundey, Where Is Your Simplified, Sensitive, Millivoltmeter?, *73 Magazine,* Sept. 1975, p 49–50.

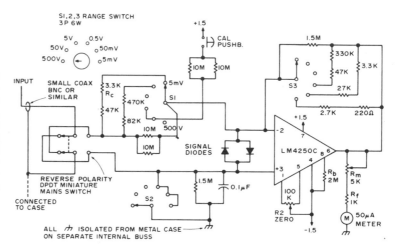

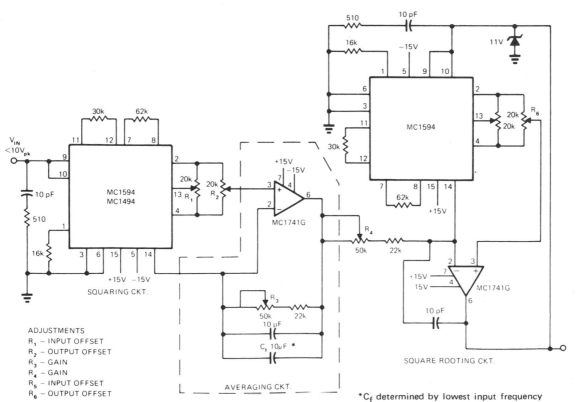

ADJUSTMENTS
R_1 – INPUT OFFSET
R_2 – OUTPUT OFFSET
R_3 – GAIN
R_4 – GAIN
R_5 – INPUT OFFSET
R_6 – OUTPUT OFFSET

*C_f determined by lowest input frequency

2–10 V P-P TRUE RMS TO 600 kHz—Input waveform is squared by first Motorola MC1594 multiplier, and current output is converted to voltage by opamp for driving second multiplier which has capacitor in feedback path to perform averaging function. Second opamp is used with second multiplier as feedback element to produce square-root configuration required for giving true RMS value. Accuracy is within 1% over input voltage range.—K. Huehne and D. Aldridge, "Multiplier/Op Amp Circuit Detects True RMS," Motorola, Phoenix, AZ, 1974, EB-20.

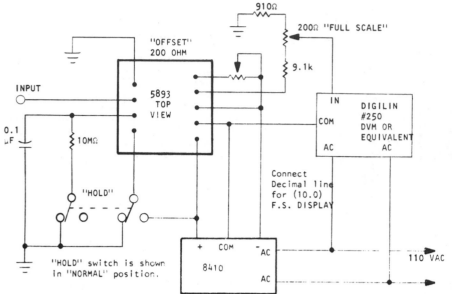

AC PEAKS—Optical Electronics 5893 peak sense-and-hold analog memory module senses input peaks of AC waveforms and produces smooth DC output voltage for driving 2½-digit digital voltmeter. Output of 5893 is divided by about 10 to give 1-V full-scale output for meter. External capacitor used with parallel resistor provides time constant required for steady display on DVM. Response time is 4 s for input change from 10 V to 0 and less than 1 s for rise from 0 to 10 V. Useful bandwidth is 20 Hz to 2 MHz. HOLD switch is operated when peak reading on meter is to be held several minutes.—"Digital Peak Reading AC Voltmeter," Optical Electronics, Tucson, AZ, Application Tip 10259.

DIFFERENTIAL INPUTS GIVE GROUND-REFERENCED OUTPUT—Circuit consists of differential-input controlled-current rectifying source A_1-A_2 and level-shifting voltage-to-current converter A_3. Feedback current of appropriate polarity is conducted to output opamp, while other feedback current is absorbed. Possible drawbacks are switching offset and bandwidth limitations common to precision rectifiers. Article gives design equations and theory of operation.—J. Graeme, Measure Differential AC Signals Easily with Precision Rectifiers, *EDN Magazine,* Jan. 20, 1975, p 45–48.

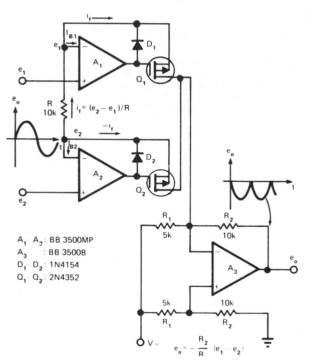

A_1, A_2: BB 3500MP
A_3 : BB 3500B
D_1, D_2: 1N4154
Q_1, Q_2: 2N4352

$$e_o = -\frac{R_2}{R}\,|e_1 - e_2|$$

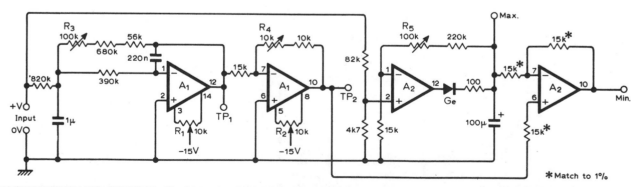

WAVEFORM PEAKS AND TROUGHS—Used in data-logging systems to measure. limits of waveform superimposed on DC level. Requires two LM747CN (dual 741) ICs. Measurements are made with conventional DC voltmeter. Input signal is fed to precision peak detector A_1. Same signal goes through active low-pass filter and inversion amplifier whose output at TP_2 is the mean value. Differential amplifier A_2 subtracts maximum from mean to give minimum value of input. For setup, short input and adjust R_1 for 0 V at TP_1, adjust R_2 for 0 V at TP_2, then apply +5 V to input and adjust R_3 for +5 V at TP_1, adjust R_4 for −5 V at TP_2, and adjust R_5 for +5 V at Max. output terminal.—K. R. Brooks, Peak and Trough Detector, *Wireless World,* Feb. 1977, p 45.

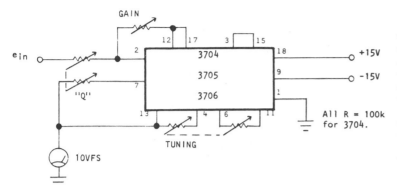

TUNED VOLTMETER—Optical Electronics active filter (3704 up to 5 kHz, 3705 up to 50 kHz, or 3706 up to 500 kHz) provides proper scale factor, impedance buffering, and isolation for measuring AC voltages at specific frequency. Circuit provides 100K input impedance and up to 10-mA drive for 10-V meter. IC provides independent gain (scale factor or sensitivity), tuning, and Q (selectivity) adjustments.—"Tuned Voltmeter," Optical Electronics, Tucson, AZ, Application Tip 10248.

HIGH-IMPEDANCE DIFFERENTIAL INPUTS—High impedance for both inputs of differential precision rectifier is provided by two opamps that produce current output for conversion to voltage by instrumentation amplifier A_2. Diode bridge in feedback path of opamp A_2 provides rectification with precise control for determining voltage drop across R. Design permits accurate measurement of differential AC inputs from millivolts to volts with AC voltmeter.—J. Graeme, Measure Differential AC Signals Easily with Precision Rectifiers, *EDN Magazine*, Jan. 20, 1975, p 45–48.

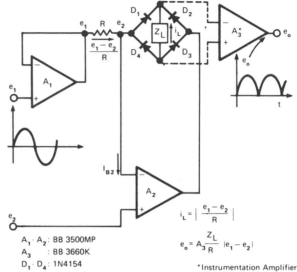

$A_1 \cdot A_2$: BB 3500MP
A_3 : BB 3660K
$D_1 \cdot D_4$: 1N4154

$$i_L = \left| \frac{e_1 - e_2}{R} \right|$$

$$e_o = A_3 \frac{Z_L}{R} |e_1 - e_2|$$

*Instrumentation Amplifier

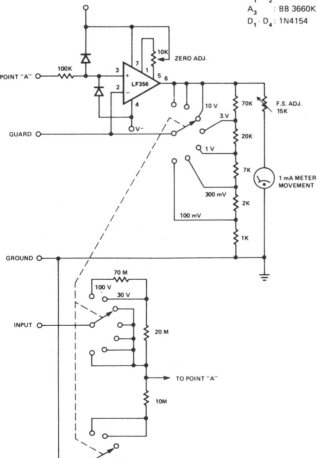

OPAMP DC VOLTMETER—Uses LF356 opamp in noninverting connection to give high input impedance along with diode protection against input overvoltage. On 100-V range, input impedance is 100 megohms.—"Signetics Analog Data Manual," Signetics, Sunnyvale, CA, 1977, p 640–641.

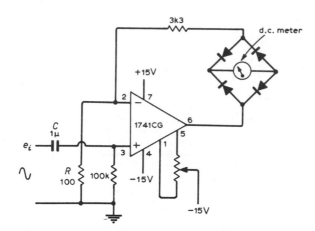

AC MILLIVOLTMETER—Combination of diode bridge and opamp forms basis for precise measurement of AC input voltages so small that they would be in nonlinear range of diodes alone. Article discusses linearity problems and gives output waveforms.—G. B. Clayton, Experiments with Operational Amplifiers, *Wireless World,* June 1973, p 275–276.

PEAK-READING RF PROBE FOR DC METER—Converts RF peaks of 1 mV to about 4 V, at any frequency up to over 100 MHz, to proportional DC voltage that can be fed to any multirange DC meter. Uses single CA3046 IC connected as two symmetrical Darlington pairs. Circuit must be mounted in small shielded housing with short probe tip and no IC socket. Temperature stability is excellent. Requires no DC offset adjustments.—Peak Reading R. F. Probe, *Wireless World,* Dec. 1976, p 42.

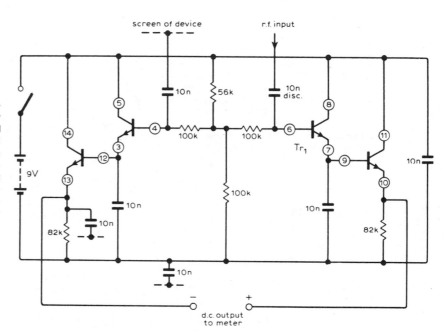

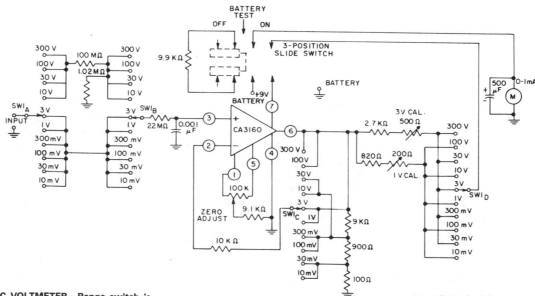

10-RANGE DC VOLTMETER—Range switch is ganged between input and output circuits of CA3160 bipolar MOS opamp to permit selection of proper output voltage for feedback to terminal 2 of opamp through 10K current-limiting resistor. Circuit operates from single 8.4-V mercury battery and draws about 500 μA plus meter current; at input for full-scale reading, total supply current drain is about 1.5 mA.— Circuit Ideas for RCA Linear ICs," RCA Solid State Division, Somerville, NJ, 1977, p 14.

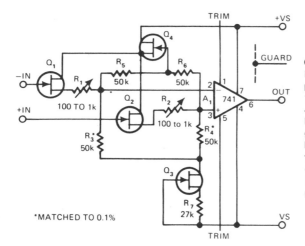

OPAMP ELECTROMETER—Use of FET input keeps input bias current down to 20 femtoamperes, with common-mode input resistance of 10^{15} ohms. Uses Analog Devices AD 832 dual JFET Q_1-Q_2 in source-follower connection, with low-cost general-purpose AD 3958 dual FET generating operating current and providing bootstrapping for Q_1-Q_2. Article covers guarding techniques used to minimize leakage currents.—J. Dostal, "Electrometer" Boasts Low Bias Current, *EDN Magazine,* Jan. 20, 1977, p 90 and 92.

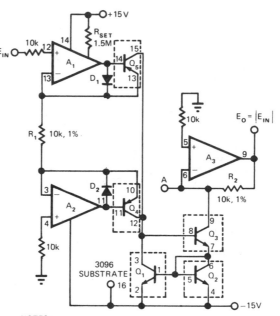

ABSOLUTE VALUES—Opamps A_1 and A_2 act with PNP transistors Q_4 and Q_5 to form operational rectifier having current-mode output. Current-to-voltage converter A_3 uses R_2 as scale factor. Input voltage range is determined by common-mode range of opamp and breakdown ratings of components. Circuit shown handles ± 10 V signal.—S. Smith, Full-Wave Rectifier Needs Only Two Precision Resistors, *EDN Magazine,* Jan. 5, 1975, p 56.

NOTES:
$D_{1,2}$ = 1N914 OR EQUIVALENT
$A_{1,2,3}$ = SILICONIX L144
$Q_{1,2,3,4,5}$ = RCA CA3096AE
ALL RESISTORS EXCEPT R_1 AND R_2 MAY BE $\pm 10\%$ TYPES

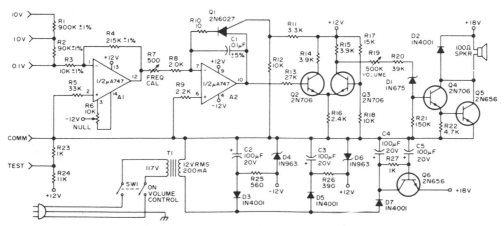

AUDIBLE VM—Voltage-controlled audio oscillator A2 serves for rough measurements of up to 10 VDC, allowing user to keep eyes on test probe during troubleshooting. Voltmeter circuit has input impedance of 100,000 ohms per volt. Separate input jacks provide full-scale ranges of 0.1, 1, and 10 V, with full-scale voltage for each producing 1000-Hz tone. Voltage less that full-scale produces proportionately lower frequency. Article describes circuit operation in detail.—S. Johnson, An Audible Voltmeter, *73 Magazine,* Aug. 1974, p 55 and 57–59.

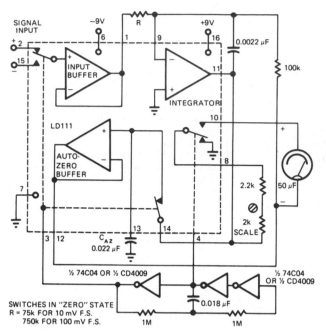

DVM IC DRIVES METER—Uses Siliconix LD111 IC analog processor section of digital voltmeter IC pair to combine desirable features of digital voltmeter with signal-averaging advantages of ordinary meter. Input range covered is 10 mV to 3 V, with resistive divider being required for larger input voltage. Differential inputs each have 1-gigohm input impedance. Circuit requires only two 9-V batteries. Article describes operation in detail.—B. Harvey, Digital Voltmeter IC Drives Analog Meter, *EDN Magazine*, June 20, 1977, p 113.

DVM POLARITY INDICATOR—Designed for use in low-cost digital voltmeters. With input polarity as indicated, arrows on connections indicate direction of current flow. When NPN Q_1 is on, LED 2 will be on; with PNP Q_2 on, LED 1 will be on. Darlingtons Q_1-Q_2 draw very little current, so choice of type is not critical. AC supply is usually available in lab but can be replaced by internal clock of DVM driving small transformer to give required floating AC source. A, B, C, and D identify nodes of bridge.—R. A. Snyder, Polarity Indicator Minimizes Parts Count, *EDN Magazine*, Feb. 20, 1977, p 121.

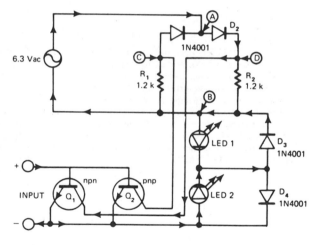

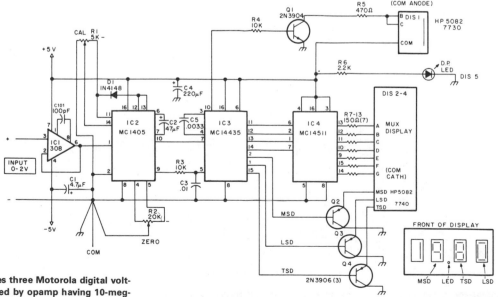

0–2 V DVM—Uses three Motorola digital voltmeter ICs preceded by opamp having 10-megohm input impedance, driving Hewlett-Packard HP5082 multiplexed digital display, with LED serving as decimal point. Input leads must be reversed to read negative voltages. Article gives construction and calibration details. Errata: move C3 upper connection to pin 9 of IC2, and transpose connections to pins 1 and 2 of IC3.—G. McClellan, DVMs Get Simpler and Simpler, *73 Magazine*, Feb. 1977, p 60–63.

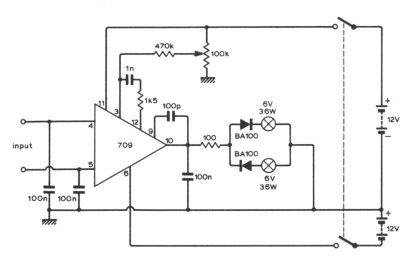

NULL INDICATOR—Opamp is driven open-loop so change of only 1 mV in input voltage makes output switch polarity. This is indicated by one of lamps. Both lamps go out to indicate null. If LEDs are used in place of lamps, diodes are not needed; adjust series resistance as required for full brilliance of LEDs.—B. P. Cowan, Miniature Null Indicator, *Wireless World*, June 1973, p 284.

PRECISION SIGNAL RECTIFIER—High input impedance at one of differential inputs of precision rectifier is achieved with opamp A_1 whose output is switched between inputs of instrumentation amplifier A_2 by diodes. This switching reverses polarity of gain provided by A_2 when signal polarity changes, so output signal is always positive.—J. Graeme, Measure Differential AC Signals Easily with Precision Rectifiers, *EDN Magazine*, Jan. 20, 1975, p 45–48.

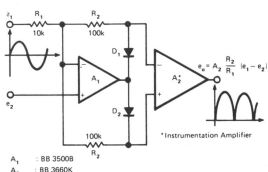

$e_o = A_2 \dfrac{R_2}{R_1} |e_1 - e_2|$

*Instrumentation Amplifier

A_1 : BB 3500B
A_2 : BB 3660K
D_1, D_2 : 1N4154

EIGHT-CHANNEL COMPUTERIZED 3½-DIGIT VM—Displays up to eight different DC voltages on CRT terminal of microprocessor under keyboard control, using BASIC commands and BASIC routine given in article. Uses Motorola MC14433 modified dual-lamp integrating analog-to-digital converter. Unknown voltage is applied to integrator having defined integration time constant for predetermined time limit, to give output voltage proportional to unknown voltage. Computer program substitutes −2.000 V reference from IC2, and circuit keeps track of time for integrator output to move back toward zero. Changing reference to 0.200 V makes same 1999 count represent 199.9 mV full scale. IC1 performs about 25 conversions per second. IC3 and IC4 are output buffers. IC5 is 7474 used as set-reset flip-flop. IC6 is eight-input CMOS multiplexer input.—S. Ciarcia, Try an 8 Channel DVM Cocktail!, *BYTE*, Dec. 1977, p 76, 78, 80, 92, 94, 96, and 98–103.

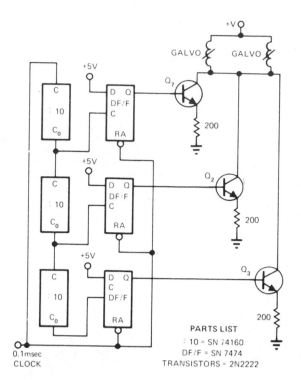

STRIP-CHART TIMING MARKS—Drives two galvanometers for generating three decades of timing marks in identical patterns on edges of chart. 10-ms marks are twice as long as 1-ms marks, and 100-ms marks are 3 times length of 1-ms marks. By placing ruler across equivalent marks on edges, exact time for any point on recorded pattern is easily and accurately determined.—S. Rummel, TTL Circuit Aids Evaluation of Oscillograph Data, *EDN Magazine,* Dec. 5, 1973, p 86.

PARTS LIST

10 = SN 74160
DF/F = SN 7474
TRANSISTORS = 2N2222

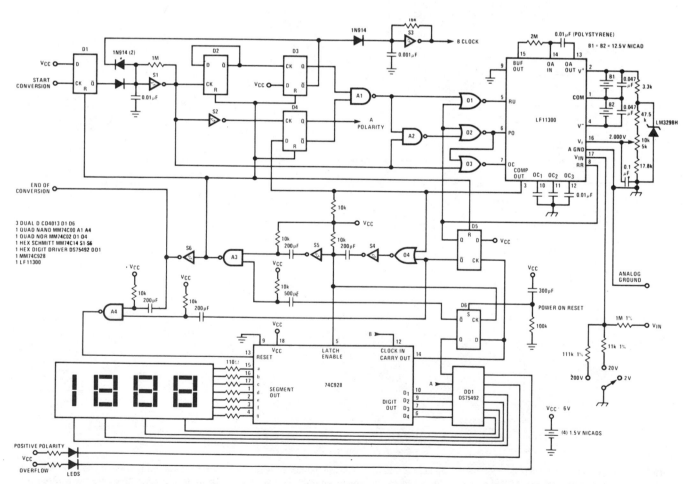

3½-DIGIT DVM—Combination of National LF11300 dual-slope analog building block and MM74C928 CMOS 3½-decade counter with 7-segment outputs gives automatic-zeroing automatic-polarity 3½-digit digital voltmeter. Counter drives LED display with multiplexed 7-segment information under control of internal free-running oscillator. Interface circuits provide nonoverlapping control signals to LF11300 for polarity determination and offset correction for every conversion cycle. Analog circuit draws 1.5 mA from each 12.5-V battery. Digital circuit draws about 40 mA from 6-V supply.—"CMOS Databook," National Semiconductor, Santa Clara, CA, 1977, p 5-36—5-37.

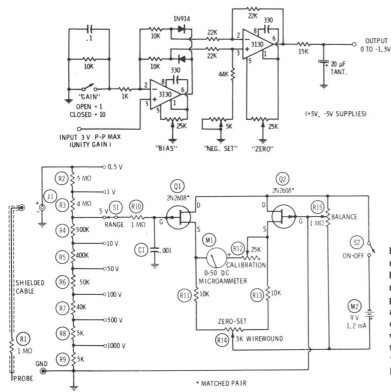

PRECISION RECTIFIER—Used in digital volt-meters to convert AC waveform to full-wave-rectified DC equivalent. First 3130 opamp is used as polarity separator, with negative-going signals appearing across upper 10K resistor and positive-going signals across lower 10K resistor. Output of opamp exceeds these voltage drops by exactly diode voltage drop. Second opamp stage recombines positive and negative peaks. 5K trimming pot is adjusted so both peaks are equal height. Output of second opamp is negative-going full-wave replica of input signal. After filtering, output is average DC value in range from 0 to −1.5 V for 0–3 V P-P input.—D. Lancaster, "CMOS Cookbook," Howard W. Sams, Indianapolis, IN, 1977, p 345–346.

BALANCED-FET DC VOLTMETER—Factory-matched FETs are connected in resistance bridge that is balanced by R14 to make meter read zero for 0–V input voltage. Voltage divider provides eight ranges, using 1% resistors for accuracy. Some must be made up by using two or more resistors in series. Balanced circuit has very low temperature drift, reducing number of times rebalancing is needed.—R. P. Turner, "FET Circuits," Howard W. Sams, Indianapolis, IN, 1977, 2nd Ed., p 119–122.

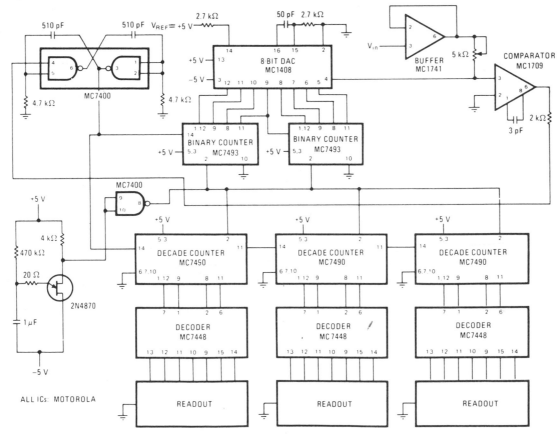

2⅔-DIGIT VOLTMETER—Closed-loop system designed around Motorola MC1408 8-bit D/A converter uses clocked binary counter feeding converter to produce staircase ramp function. Output of converter is compared to unknown input signal, and clock pulse is terminated when levels being compared are equal. Clock pulses are generated at 330 kHz by two cross-coupled NAND gates in MC7400. UJT oscillator resets both sets of counters so unknown voltage is resampled every 0.5 s. MC7448 BCD to 7-segment decoders convert outputs of BCD counters to format for LED displays. With values shown, meter can measure up to 2.55 V in 10-mV steps. Different full-scale values can be obtained by using input voltage dividers or by replacing unity-gain input buffer with suitable fixed-gain buffer.—D. Aldridge, "DAC Key to Inexpensive 2⅔ Digit Voltmeter," Motorola, Phoenix, AZ, 1975, EB-21.

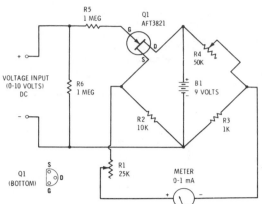

FET VOLTMETER—With FET in one leg of Wheatstone bridge, meter has input impedance of over 1 megohm. With no input voltage, adjust R4 so meter reads zero. With 9-V battery, R1 can be adjusted for full-scale meter reading of 8 V. With 12-V battery, meter range is 0–10 V.—F. M. Mims, "Transistor Projects, Vol. 2," Radio Shack, Fort Worth, TX, 1974, p 59–66.

METER AMPLIFIER—Meter in feedback path of opamp is connected in bridge circuit for measuring both AC and DC voltages. Input voltage is equal to meter current in amperes multiplied by 3 times value of R1 in ohms; with 3.3K and 0.1 mA, input is 0.99 V. Multiplying milliampere reading of meter by 10,000 thus gives input voltage. If long supply leads cause oscillation, connect 0.1-μF capacitors between ground and supply pins 4 and 7 as shown.—F. M. Mims, "Integrated Circuit Projects, Vol. 4," Radio Shack, Fort Worth, TX, 1977, 2nd Ed., p 54–60.

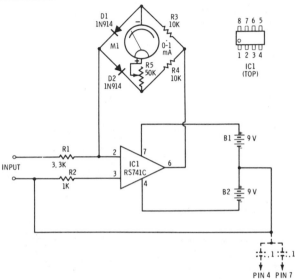

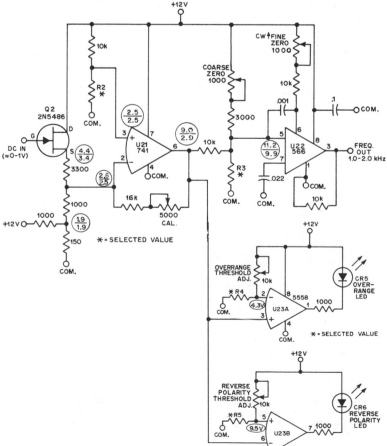

V/F CONVERTER—Can be used with any frequency counter. Only last three digits of display are read for voltage. VCO U22 runs at 1000 Hz when input is grounded and R3 is 56K. Counter is preset to 9000. For 0 V, count starts from 9000 and goes up to 10,000 on display, except that 1 at left overflows so reading is 0 V. If input is +1 V, U22 goes up to 2000 Hz, appearing as 1000 on display. Voltage divider ahead of input is needed to divide full-scale voltages of 10, 100, and 1000 V down to basic 0–1 V range. Range switch is wired to place decimal in appropriate position. Use 2.7K for R2. DC voltages are in circles; upper value is for input probe of electronic voltmeter on +12 V, and lower value for input probe grounded. Terminal A goes to overrange and reverse polarity indicators using 5558 dual opamp U23 and Archer (Radio Shack) 276-041 or equivalent LEDs. R4 and R5 depend on input-signal excursion range and exact value of supply; start with 2700 ohms for R4 and 18K for R5.—J. Hall and C. Watts, Learning to Work with Integrated Circuits, QST, June 1976, p 20–24; revised circuit in June 1977, p 20–21.

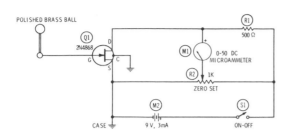

ELECTROSCOPE—Gate of FET floats, being connected only to smooth metal rod tipped with polished brass ball. Rod is insulated from housing with polystyrene washer in large hole. Static drain current of FET is balanced out of meter with R2 when ball is clear of operator's body or other object. Meter deflection then is proportional to intensity of charge on body brought near ball and on separation. Electroscope will respond to vigorously stroked paper or just-used comb.—R. P. Turner, "FET Circuits," Howard W. Sams, Indianapolis, IN, 1977, 2nd Ed., p 153–154.

HIGH INPUT IMPEDANCE—All transistors are on RCA CA 3095 transistor array. Q1-Q4 are connected to form bridge, with voltage to be measured applied to base 9. Circuit balance and calibration are achieved by varying DC voltage applied to base of Q2. Q7 and Q8 serve as constant-current source for cascode differential amplifier connection of Q1-Q4. Differential output of bridge is applied to differential input of CA3748 opamp driving meter. Switch gives choice of three voltage ranges.—E. M. Noll, "Linear IC Principles, Experiments, and Projects," Howard W. Sams, Indianapolis, IN, 1974, p 327.

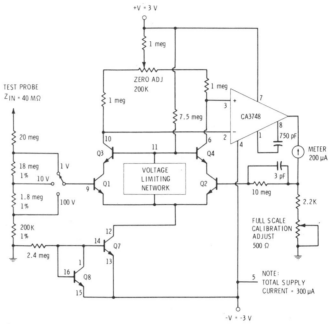

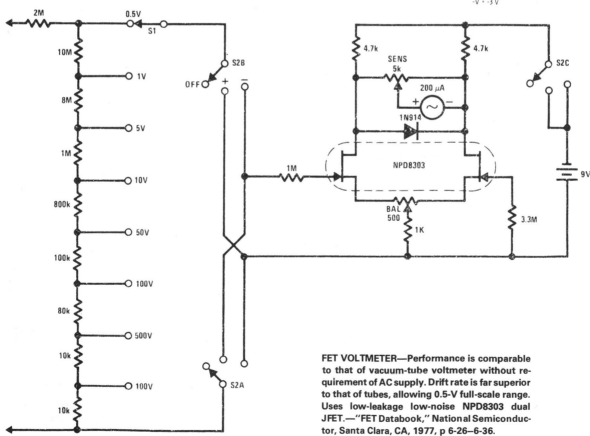

FET VOLTMETER—Performance is comparable to that of vacuum-tube voltmeter without requirement of AC supply. Drift rate is far superior to that of tubes, allowing 0.5-V full-scale range. Uses low-leakage low-noise NPD8303 dual JFET.—"FET Databook," National Semiconductor, Santa Clara, CA, 1977, p 6-26–6-36.

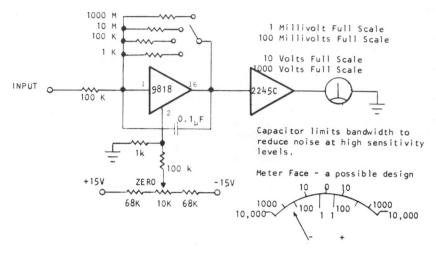

1 Millivolt Full Scale
100 Millivolts Full Scale

10 Volts Full Scale
1000 Volts Full Scale

Capacitor limits bandwidth to reduce noise at high sensitivity levels.

Meter Face - a possible design

NULL VOLTMETER—Logarithmic voltmeter using Optical Electronics 9818 opamp and 2245C four-decade bipolar logarithmic function can serve as output indicator of Wheatstone bridge, as solid-state galvanometer, or as indicator for differential voltmeter or comparison bridge. Meter scale values are relative; basic sensitivity of circuit corresponds to 1 on scale, representing 100 nV. With this sensitivity, 1 mV gives full-scale reading. Other three positions of range switch give 100 mV, 10 V, and 1000 V for full scale, when using 10–0–10 V meter. Inherent limiting of opamp protects circuit from overvoltage damage.— "A Logarithmic Null-Voltmeter Design," Optical Electronics, Tucson, AZ, Application Tip 10084.

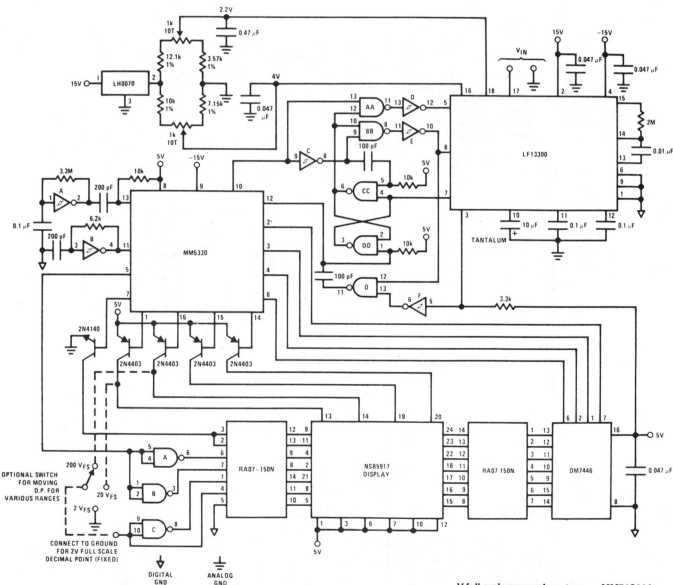

4½-DIGIT METER—National MM5330 BCD building block is used with LF13300 analog section of A/D converter to provide ±19,999 counts on NSB5917 display. Circuit contains counters, latches, and multiplexing system for four full digits of display with one decoder/driver, along with sign bit that is valid during overrange and 10,000-count numeral 1. LF13300 has automatic zeroing of all offset voltages. Operation is based on code conversion of number of counts made by MM5330 before comparator crossing is detected. Switch gives choice of 2-, 20-, and 200-V full-scale ranges. Inverters are MM74C14 hex Schmitt triggers. Two-letter NAND gates are MM74C00 CMOS quad NAND gates. One-letter NAND gates (A, B, etc) are DM7400 TTL quad NAND gates.—"MOS/LSI Databook," National Semiconductor, Santa Clara, CA, 1977, p 5-2–5-22.

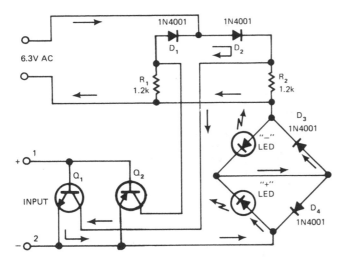

POLARITY INDICATOR—Q_1 and Q_2 at input should be Darlington transistors to minimize loading on input signal. With no input, current flows through R_1D_1 and R_2D_2 networks on alternate half-cycles. Positive signal at input 1 turns on Q_1, shunting sum of current through D_3 and lighting positive-indicating LED in diode bridge. Similarly, negative voltage on 1 turns on other LED. Supply requirement of 6.3 VAC can usually be obtained from digital multimeter with which indicator is used.—R. A. Snyder, Simple Polarity Indicator Suits DMM's or DPM's, *EDN Magazine*, Nov. 5, 1977, p 110.

DIFFERENTIAL AC SIGNALS—Precision full-wave rectification of differential voltage is achieved by transforming to current for rectification and reconversion to output voltage. One opamp serves as voltage-to-current converter and the other as rectifying current-to-current converter. Circuit is suitable for applications in which lower input impedance and lower frequency response are acceptable. Article gives design equations.—J. Graeme, Measure Differential AC Signals Easily with Precision Rectifiers, *EDN Magazine*, Jan. 20, 1975, p 45–48.

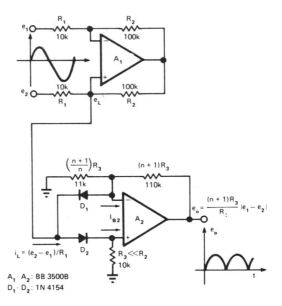

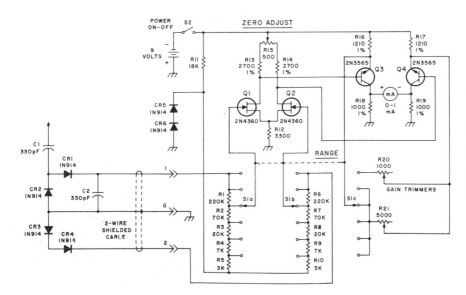

0.4–200 MHz VM—Battery-operated RF voltmeter has full-scale ranges from 0.03 V to 10 V and flat frequency response from 40 kHz to over 200 MHz. Circuit uses voltage-doubling rectifier-type probe CR1-CR2 followed by high-gain DC amplifier driving milliammeter. Article covers construction and calibration in detail.—J. M. Lomasney, Sensitive RF Voltmeter, *73 Magazine*, Dec. 1973, p 53–62.

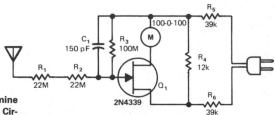

AC LINE POLARITY METER—Used to determine correctness of ground wiring in receptacle. Circuit compares voltage waveform on line conductors with AC potential of 10–40 V at 60 Hz picked up by antenna which can be human body. Circuit is synchronous demodulator that conducts on alternate half-cycles depending on whether gate voltage of JFET is positive with respect to source or drain. Zero-center DC milliammeter serves as readout. If plug is inserted into receptacle having balanced power line, milliammeter stays at center to indicate lack of ground. With properly grounded receptacle, meter swings full scale in either direction.—T. Gross, Indicator Shows Correct Wiring Polarity, *EDN Magazine,* Oct. 20, 1978, p 150 and 152.

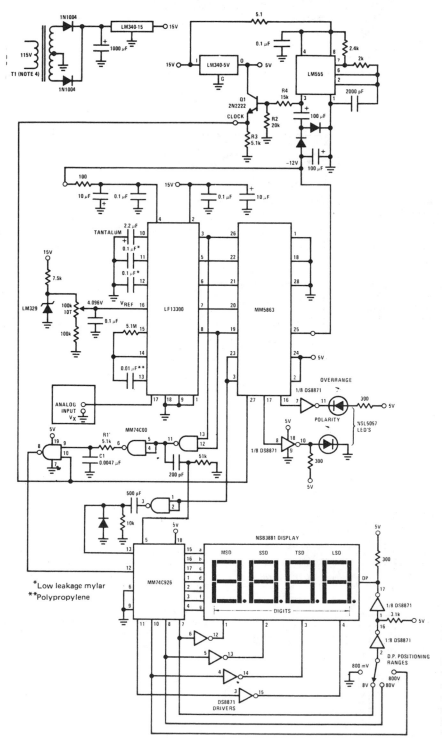

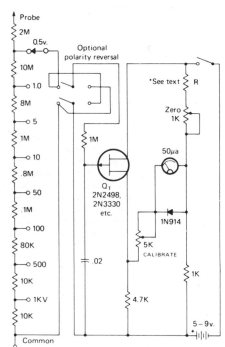

FET VOLTMETER—Voltage divider provides 22-megohm resistance for FET version of VTVM. JFET Q_1 is used as source follower. Meter is connected in bridge that is balanced with 1K zero-adjust pot. With proper selection of R, pot can also be used to set zero point of circuit to half scale. Accuracy depends primarily on divider chain. Total current drain rarely exceeds 1 mA, giving long life for almost any type of battery.—I. Math, Math's Notes, *CQ,* Oct. 1974, p 26–27.

3¾-DIGIT METER—National MM5863 12-bit binary building block is used with LF13300 analog section of A/D converter to provide ±8191 counts on NSB3881 display. MM74C926 CMOS counter is connected to count clock pulses during ramp reference cycle of LF13300. Counts are latched into display when comparator output trips and goes low. Operates from single 15-V supply with aid of DC/DC converter. LM555 serves as clock and generates required negative supply voltages. All diodes are 1N914.—"MOS/LSI Databook," National Semiconductor, Santa Clara, CA, 1977, p 5-2–5-22.

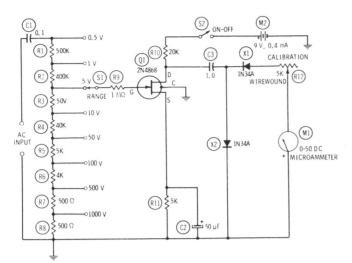

FET AC VOLTMETER—Covers 0–1000 VRMS in eight ranges. Frequency response referred to 1 kHz is down 3.5 dB at 50 Hz and down 2 dB at 50 kHz. Meter deflection is proportional to average value of AC signal voltage, but meter can be calibrated to read RMS voltages on sine-wave basis. Use 1% resistors for voltage divider. Useful for audio and ultrasonic measurements and tests.—R. P. Turner, "FET Circuits," Howard W. Sams, Indianapolis, IN, 1977, 2nd Ed., p 122–124.

AUTOMATIC POLARITY SWITCHING—Can be added to almost any high-impedance voltmeter to give automatic reversal of polarity as required during measurements. Additional contacts on relay can be used to switch polarity indicators. FET input prevents meter shunting. Feedback is used in opamp comparator to speed switching action.—H. Wedemeyer, Auto Polarity Switching for Voltmeters, *Wireless World,* Oct. 1974, p 380.

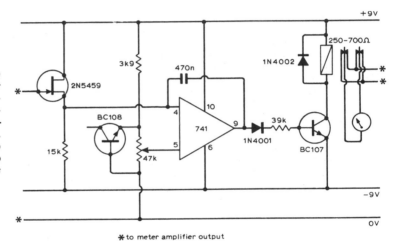

✳ to meter amplifier output

CHAPTER 103
Voltage Reference Circuits

Provides variety of fixed and variable positive and negative voltages up to 30 V for use in regulators and other circuits requiring highly stable reference voltage. Some of circuits can be used as exact replacements for standard cells. See also Regulated Power Supply, Regulator, and Voltage-Level Detector chapters.

VARIABLE REFERENCE—With 759 power opamp used as variable-output voltage regulator, output voltage can be varied over full range from zener maximum down to zero by varying voltage from zener. With 791 opamp, voltage can be adjusted down to 2 V. Since output voltage can be less than zener rating, simple bootstrapping cannot be used. Alternate biasing techniques are then required to improve line regulation. Arrangement is capable of supplying several hundred milliamperes while using only low-drift (5 PPM/°C) zener.—R. J. Apfel, Power Op Amps—Their Innovative Circuits and Packaging Provide Designers with More Options, *EDN Magazine*, Sept. 5, 1977, p 141–144.

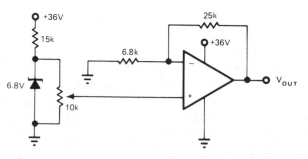

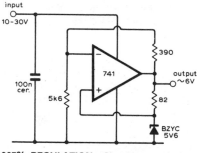

0.005% REGULATION—Simple opamp circuit changes less than 1 mV at output for input range of 10–30 V. Circuit is easily modified to give other output voltages, either positive or negative.—M. Walne, High Performance Reference, *Wireless World*, May 1974, p 123.

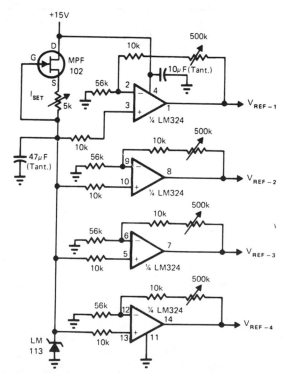

1.5–12 V FOUR-OUTPUT—Single LM113 1.22-V stable reference is driven by 1-mA FET constant-current source to provide highly stable low-voltage standard driving four adjustable-gain opamps. Gain of each is set to give desired output reference voltage in range from 1.5 to 12 V.

Use cermet trimmers and metal-oxide fixed resistors in opamp feedback circuits to achieve stabilities of several millivolts over 0 to 70°C range.—H. Olson, Two IC's and FET Provide Quad Stable Reference, *EDN Magazine*, Jan. 20, 1974, p 82.

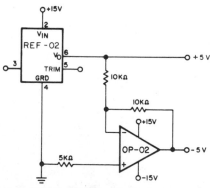

±5 V USING OPAMP—Precision Monolithics REF-02 voltage reference provides +5 V directly, while OP-02 inverting opamp provides −5 V.—"+5 V Precision Voltage Reference/ Thermometer," Precision Monolithics, Santa Clara, CA, 1978, REF-02, p 6.

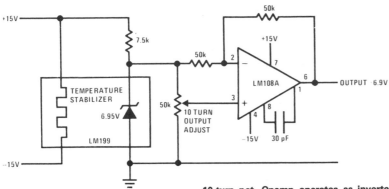

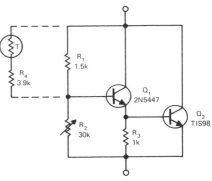

−6.9 V TO +6.9 V VARIABLE—National LM199 temperature-stabilized 6.95-V reference is converted to continuously variable bipolar output by LM108A opamp. Use precision wirewound 10-turn pot. Opamp operates as inverter for negative outputs but is noninverting for positive outputs.—"Linear Applications, Vol. 2," National Semiconductor, Santa Clara, CA, 1976, AN-161, p 6.

ADJUSTABLE REFERENCE—Two-transistor equivalent of zener is combined with Gulton 35TF1 thermistor to give voltage stability within 0.5% over 0–50°C range, with output voltage adjustable from 3.5 to 15.5 V with R_2. Dynamic impedance is only 1 ohm. Developed for regulator service in battery-powered MOS instruments.—R. Tenny, Compensated Adjustable Zener, *EDN Magazine,* May 5, 1973, p 72.

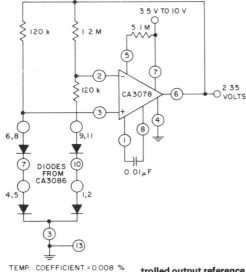

TEMP. COEFFICIENT = 0.008 %

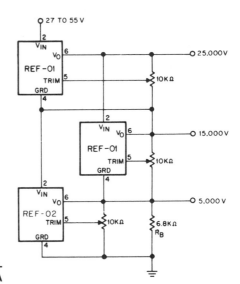

BAND-GAP PRECISION REFERENCE—Uses diodes from CA3086 array and CA3078 micropower opamp to develop 2.35-V precisely controlled output reference that is almost independent of temperature.—"Circuit Ideas for RCA Linear ICs," RCA Solid State Division, Somerville, NJ, 1977, p 18.

+5, +15, AND +25 V—Stacking of Precision Monolithics REF-02 5-V reference with two REF-01 10-V references gives outputs increasing in steps of 10.000 V from 5.000 V. Any number of additional references can be stacked in same way up to line-voltage limit of 130 V for references, provided total load current does not exceed about 21 mA. Input change from 27 to 55 V produces output change less than noise voltage of devices in circuit shown.—"+5 V Precision Voltage Reference/Thermometer," Precision Monolithics, Santa Clara, CA, 1978, REF-02, p 7.

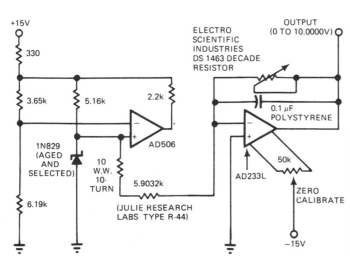

0–10.0000 V IN 100-μV STEPS—Constant current from AD506 opamp drives zener, with 5.16K resistor providing optimum current through zener for temperature-drift cancellation. Chopper-stabilized opamp scales output of zener over full range. Offset-voltage pot serves for zero calibration. Reference voltage is stable to about 11 PPM per year.—J. Williams, Don't Bypass the Voltage Reference That Best Suits Your Needs, *EDN Magazine,* Oct. 5, 1977, p 53–57.

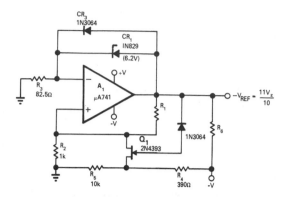

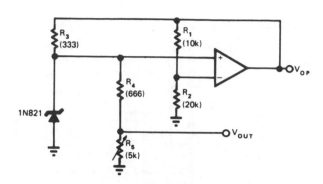

SELF-STABILIZING ZENER—Current through 7.5-mA zener CR_1 is independent of supply voltage, which may be as low as 10 V. Negative reference is 1.1 times zener rating, or −6.8 V for zener shown. Article describes circuit operation and gives design equations.—L. Accardi, Super-Stable Reference-Voltage Source, *EDN/EEE Magazine*, Oct. 1, 1971, p 41–42.

5 V AT 7.5 mA—Circuit uses single pot with standard opamp to adjust output voltage and simultaneously set current of 6.2-V zener at optimum value for temperature stability. With some opamps, emitter-follower may be needed at opamp output to supply necessary zener current. Technique eliminates need for separate zener current adjustment or permits use of lower-cost zener.—K. Hanna, Single Control Adjusts Voltage Reference, *EDN Magazine*, June 5, 1976, p 117.

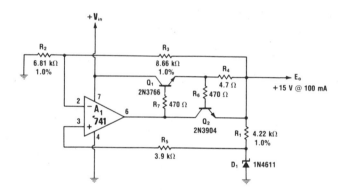

+15 V AT 100 mA—Boost transistor Q_1 is added inside feedback loop to amplify output current of A_1 to 100 mA at scaled-up reference of +15 V from 6.6-V value of zener. R_4 and Q_2 provide protection against load shorts.—W. G. Jung, "IC Op-Amp Cookbook," Howard W. Sams, Indianapolis, IN, 1974, p 152–155.

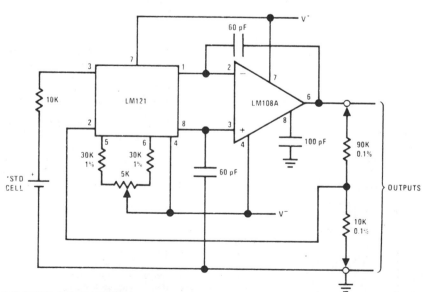

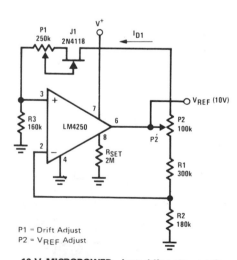

10 V USING STANDARD CELL—Low drift and low input current of National LM121 differential amplifier provide buffering for standard cell with high accuracy. Typical long-term drift for LM121 operating at constant temperature is less than 2 "V per 1000 h. Circuit should be shielded from air currents. When power is not applied, disconnect standard cell to prevent it from discharging through internal protection diodes.—"Linear Applications, Vol. 2," National Semiconductor, Santa Clara, CA, 1976, AN-79, p 8.

10-V MICROPOWER—Low-drift voltage reference has standby current less than 100 μA, using LM4250 opamp to convert zero-temperature-coefficient current to desired reference voltage output. Adjust P1 for low output temperature coefficient, and adjust P2 for exact reference desired.—"Linear Applications, Vol. 2," National Semiconductor, Santa Clara, CA, 1976, LB-34.

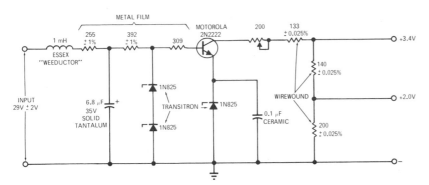

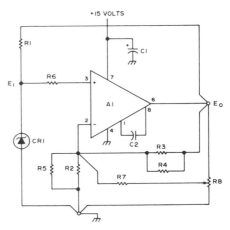

3.4-V RADIATION-HARDENED—Two-stage regulator is combined with special filter design that helps keep output voltage free of gamma-produced transients and RFI. Voltage divider using wirewound resistors provides 2-V output. Gamma radiation of 1,000,000 rads caused only 0.3% change in output voltages. Article covers procedure for designing and testing circuits that are to be operated in high-radiation environment.—A. J. Sofia, Designing a Radiation-Stable Voltage Reference, *EDN Magazine,* Sept. 15, 1970, p 39–41.

+10.000 V—Uses LM301A opamp with 1N825 6.2-V zener reference diode (not zener regulator diode) to maintain stable DC voltage under severe combinations of temperature, shock, and vibration. Gain resistors R2 and R3 should have same 0.01%/°C temperature coefficients as reference diode. R1 is RN55 511-ohm metal-film resistor, R2 is RN55E 6.04K metal-film resistor, and R3 is RN55E 3.57K metal-film resistor. R4 and R5 are gain trim resistors. R6 should equal parallel combination of R2 and R5. R8 is 10K cermet pot. with R7 (optional) 100K to 1 megohm. Article tells how to trim circuit for desired output and how to calculate values of resistors and temperature coefficients for other output voltages.—D. W. Ishmael, Precision +10.000 V DC Voltage Reference Standard, *73 Magazine,* Sept. 1975, p 124–126.

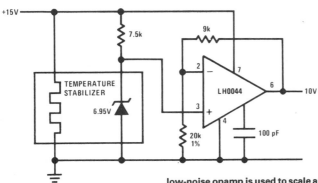

BUFFERED 10 V—Reference voltage developed by National LM199 temperature-stabilized IC is 6.95 V with very low temperature drift and excellent long-term stability. LH0044 precision low-noise opamp is used to scale and buffer reference to give required output of 10 V. Regulation of 15-V supply need be only about 1%.—"Linear Applications, Vol. 2," National Semiconductor, Santa Clara, CA, 1976, AN-161, p 5.

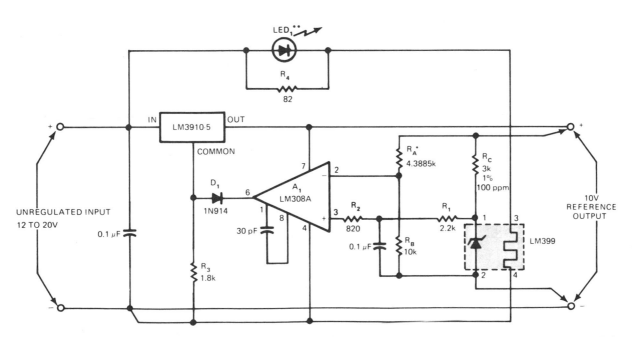

10-V HIGH-PRECISION—Use of LM399 thermally stabilized subsurface zener in state-of-the-art reference circuit keeps temperature error well under 2 PPM/°C over temperature range of 0 to 70°C. Article gives design equation and covers procedures for optimizing stability and minimizing power-supply rejection-ratio errors.—W. G. Jung, Precision Reference Source Features Minimum Errors, *EDN Magazine,* Aug. 5, 1976, p 80 and 82.

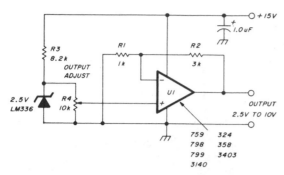

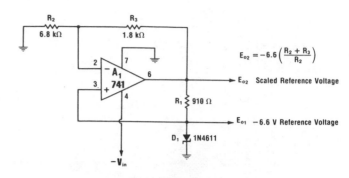

$$E_{o2} = -6.6 \left(\frac{R_2 + R_3}{R_2} \right)$$

E_{o2} Scaled Reference Voltage

E_{o1} −6.6 V Reference Voltage

VARIABLE 2.5–10 V—General-purpose opamp and zener, operating from single 15 V supply, serve as stable buffered voltage reference source that is readily adapted to wide range of output voltages and currents. R4 applies some fraction of zener's 2.5 V to opamp, which amplifies it by factor of 4 to give 2.5 to 10 V output. Output current rating depends on opamp and is about 10 mA for general-purpose types. 759 will handle up to 350 mA, and other devices can be buffered with NPN emitter-follower stage. For greater output range, use higher supply voltage and adjust R2 accordingly. R3 should be chosen to maintain about 1 mA in zener.—W. Jung, An IC Op Amp Update, *Ham Radio*, March 1978, p 62–69.

−6.6 V WITH 741 OPAMP—Reference output of −6.6 V, determined by breakdown voltage of zener, is scaled to more negative level at output of A_1. If 1558 dual opamp is used in place of 741, other section can be connected to zener as buffer that raises output current to 5 mA and lowers output impedance.—W. G. Jung, "IC Op-Amp Cookbook," Howard W. Sams, Indianapolis, IN, 1974, p 151.

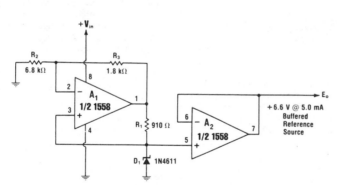

+6.6 V AT 5 mA—Half of 1558 dual opamp is used as buffer for basic opamp-zener voltage reference to raise output current and lower output impedance.—W. G. Jung, "IC Op-Amp Cookbook," Howard W. Sams, Indianapolis, IN, 1974, p 150–151.

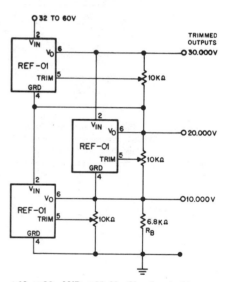

+10, +20, AND +30 V—Simple stacking arrangement of Precision Monolithics REF-01 voltage references gives near-perfect line regulation for inputs of 32 to 60 V, with output changes less than noise voltage of devices for input extremes. Any number of units can be stacked to obtain additional output voltages.— "+10 V Precision Voltage Reference," Precision Monolithics, Santa Clara, CA, 1977, REF-01, p 7.

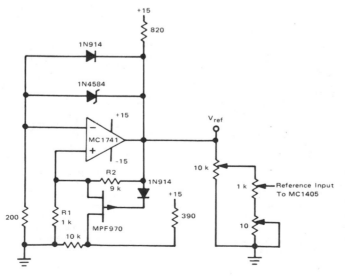

1-V HIGH-PRECISION—Drift is less than 1 mV over 20°C temperature range, and voltage divider reduces this to ±0.1 mV for 1.00-V reference required in 4½-digit meter. All three pots should be wirewound. Current of reference zener is regulated by opamp gain and zener voltage.—S. Kelley, "Applications of MC1405/MC14435 in Digital Meters," Motorola, Phoenix, AZ, 1975, AN-748, p 19.

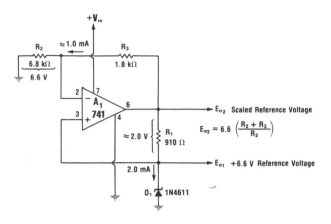

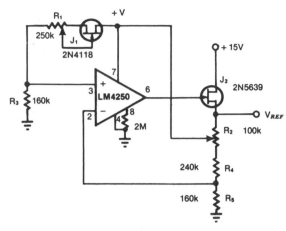

+6.6 V WITH 741 OPAMP—Uses combination of negative and positive feedback to maintain constant current of 2 mA in D_1, independent of variations in ambient temperature and unregulated input. Reference voltage of 6.6 V, determined by breakdown voltage of 1N4611 zener, is scaled up to more positive level at output of A_1. Scaled output has low impedance and can supply appreciable current without affecting reference voltage accuracy. Supply must be single-ended for reliable starting.—W. G. Jung, "IC Op-Amp Cookbook," Howard W. Sams, Indianapolis, IN, 1974, p 141–143.

LOW-DRIFT MICROPOWER—Uses JFET biased slightly below pinchoff in combination with micropower opamp to convert zero-temperature-coefficient drain current to correspondingly stable reference voltage. Additional JFET J_2 makes operation independent of value of unregulated input. Output impedance is low. For higher output impedance, reference voltage can be taken from wiper of R_2, but buffering could then be required. R_1 is adjusted to compensate for temperature coefficient arising from opamp supply current.—N. Sevastopoulos and J. Moyer, Micropower Reference Stays Stable, EDN Magazine, Sept. 5, 1978, p 158.

6.5-V REFERENCE—Reference amplifier uses mirror characteristic of noninverting input of current-mode opamp to determine zener current. Resulting voltage drop across zener provides, through R_2, current reference for other opamps or compensated voltage reference.—R. W. Fergus, Use Current-Mode Op Amps in Reference Circuits, EDN Magazine, June 20, 1974, p 80 and 83.

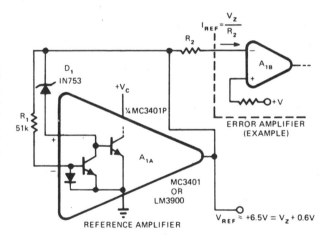

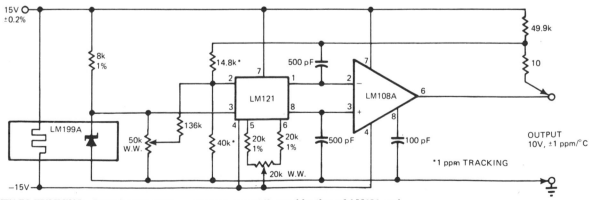

10 V WITH TC TRIMMING—Provides temperature-compensation trimming to give lowest possible reference-voltage drift for A/D converter. Reference zener is LM199A having 0.5 PPM/°C drift that is independent of operating current. Low-drift combination of LM121 and LM108A has drift predictably proportional to offset voltage, permitting use of potentiometers for trimming to better than 1 PPM/°C. Article gives details of trimming procedure to be used during temperature runs.—R. C. Dobkin, Don't Forget Reference Stability When Designing A-to-D Converters, EDN Magazine, June 20, 1977, p 105–108.

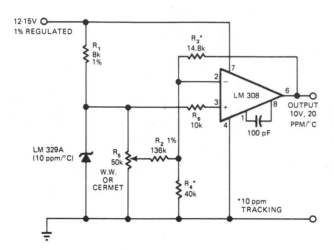

10 V WITH MODERATE DRIFT—Suitable for A/D converter applications in which output voltage can drift as much as 20 PPM/°C. Temperature-drift error is divided equally between zener and amplifier, permitting use of moderately low-drift components.—R. C. Dobkin, Don't Forget Reference Stability When Designing A-to-D Converters, *EDN Magazine,* June 20, 1977, p 105–108.

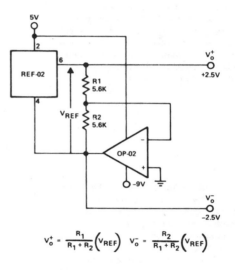

±2.5 V—Precision Monolithics REF-02 voltage reference and OP-02 inverting opamp provide desired references when used with supply voltages shown.—"+5 V Precision Voltage Reference/Thermometer," Precision Monolithics, Santa Clara, CA, 1978, REF-02, p 6.

$$V_o^+ = \frac{R_1}{R_1 + R_2}\left(V_{REF}\right) \qquad V_o^- = \frac{R_2}{R_1 + R_2}\left(V_{REF}\right)$$

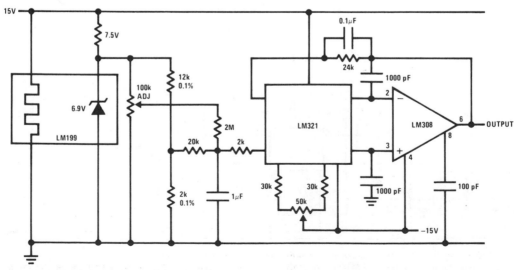

1.01-V STANDARD-CELL REPLACEMENT—National LM199 temperature-stabilized 6.95-V reference is applied to LM3308 opamp through LM321 preamp to give standard-cell replacement that can be adjusted to output of exactly 1.01 V. Null offset of opamp before adjusting for proper output voltage.—"Linear Applications, Vol. 2," National Semiconductor, Santa Clara, CA, 1976, AN-161, p 5.

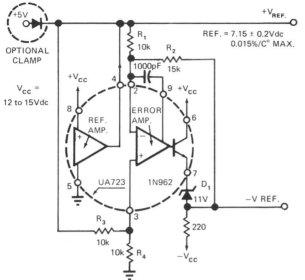

REF. = 7.15 ± 0.2Vdc
0.015%/C° MAX.

±7 V REFERENCE—Developed for analog applications requiring dual-polarity references. Both voltages are generated from single μA723 IC voltage regulator chip. Chief requirement is keeping inputs and outputs within dynamic range of amplifier, which is +2 to +9 V; this is done by shifting output voltage level upward with zener D_1 and shifting input error voltage with divider R_1-R_2. Changing ratio of R_3 to R_4 changes negative reference value.—D. Weigand, Dual 7V Reference Developed from a Single μA723, *EDN Magazine,* Nov. 1, 1972, p 47.

+15 V HIGH-PRECISION—Uses 725 opamp having low offset drift and high common-mode rejection in combination with low-drift version of 1N4611 zener to give highly stable operation at output currents up to 100 mA. Close-tolerance low-temperature-coefficient film or wirewound resistors are required for R_1, R_2, and R_3. Remote sensing at load corrects for wiring voltage drops.—W. G. Jung, "IC Op-Amp Cookbook," Howard W. Sams, Indianapolis, IN, 1974, p 152-155.

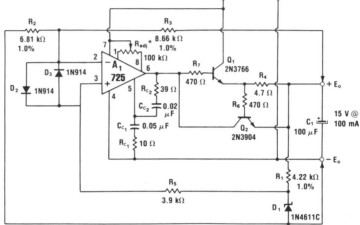

*Adjust for minimum input offset of A_1.

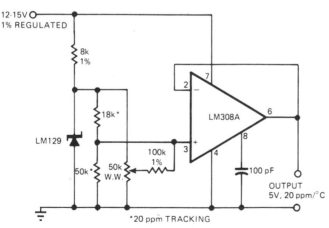

5 V FROM 15 V—Used as reference for A/D converter when reference voltage required is below that of zener, permitting simplified circuit design. Zener drift contributes proportionally to output temperature drift, while opamp offset drift contributes at greater rate. Opamp is unnecessary if high output impedance can be tolerated.—R. C. Dobkin, Don't Forget Reference Stability When Designing A-to-D Converters, *EDN Magazine,* June 20, 1977, p 105–108.

10.000 V WITH ZENER—Circuit provides stable current biasing of zener and adjustable output voltage by bootstrapping excitation current off output voltage. Commercial version of circuit (Analog Devices AD2700) achieves temperature drift of only 3 PPM/°C.—J. Williams, Don't Bypass the Voltage Reference That Best Suits Your Needs, *EDN Magazine,* Oct. 5, 1977, p 53–57.

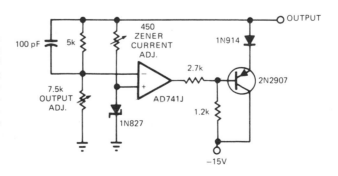

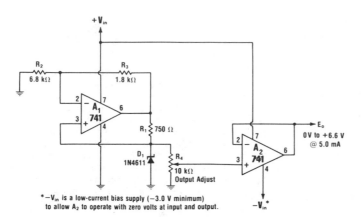

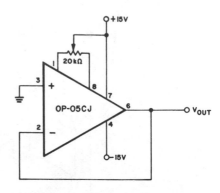

0–6.6 V AT 5 mA—R_4 at input of voltage-follower opamp permits varying reference voltage from 0 up to limit of zener. A_2 compensates for loading of zener by R_4. Dual opamp cannot be used because negative supply terminal of A_2 must be slightly more negative than −3 V common to permit linear output operation down to 0 V. If additional load current is required, NPN booster transistor can be used with A_2.—W. G. Jung, "IC Op-Amp Cookbook," Howard W. Sams, Indianapolis, IN, 1974, p 155–157.

OPAMP AS MILLIVOLT REFERENCE—High-performance bipolar-input Precision Monolithics OP-05CJ instrumentation opamp is connected as unity-gain buffer. Output is adjusted to desired reference voltage with pot connected to offset nulling terminals. Reference range is −3.5 mV to +3.5 mV, and long-term drift of zenerless source is less than 3.5 μV per month.—D. Soderquist, "Simple Precision Millivolt Reference Uses No Zeners," Precision Monolithics, Santa Clara, CA, 1975, AN-10.

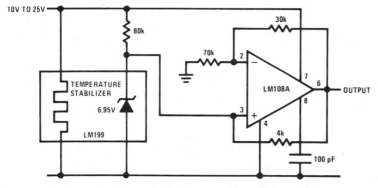

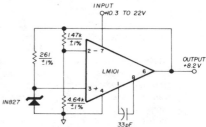

+8.2 V—Regulation is 0.01 mV/V for inputs of 10.3–22 V, and temperature stability is ±0.05% from −55°C to 125°C. Short-circuit protection for reference is provided internally. Uses National LM101, LM301A, or equivalent opamp.—D. W. Nelson, Introduction to Operational Amplifiers, *Ham Radio,* March 1978, p 48–60.

WIDE INPUT VOLTAGE RANGE—Power for LM199 temperature-stabilized voltage reference is obtained from output of LM108A buffer opamp, to permit use with 10–25 V supply. 80K resistor is used in unregulated input to make circuit start properly when power is applied.—"Linear Applications, Vol. 2," National Semiconductor, Santa Clara, CA, 1976, AN-161, p 6.

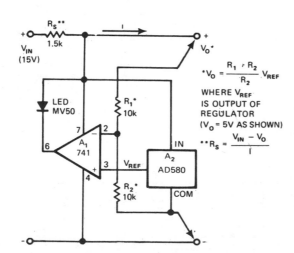

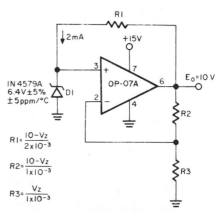

UPWARD SCALING—Output of AD580 three-terminal voltage regulator is multiplied in shunt regulator loop using opamp A_1. Resulting reference voltage V_{REF} is scaled upward by reciprocal of feedback divider R_1/R_2. Resulting reference is immune to input variations. Reference output range is from 5 V up to voltage limit of opamp used. If desired, reference value can be programmed via R_1. LED lights to show proper operation of circuit.—W. G. Jung, Programmable Voltage Reference Is Stable, yet Simple, *EDN Magazine,* Nov. 5, 1975, p 98 and 100.

+10 V WITH BOOTSTRAPPED OPAMP—High-stability Precision Monolithics OP-07A bipolar-input opamp with ultralow offset voltage provides precise 10 V virtually independent of changes in supply voltage, ambient temperature, and output loading. Choose value of R1 to maintain zener current at exactly 2 mA, using 5 PPM/°C resistor. All resistor values are determined from exact zener voltage V_z, as given by equations alongside circuit.—D. Soderquist and G. Erdi, "The OP-07 Ultra-Low Offset Voltage Op Amp—a Bipolar Op Amp That Challenges Choppers, Eliminates Nulling," Precision Monolithics, Santa Clara, CA, 1975, AN-13, p 8.

Name Index

Subject Index

1237